D1376876

HOW ANIMALS COMMUNICATE

How Animals Communicate

Edited by
THOMAS A. SEBEOK

INDIANA UNIVERSITY PRESS
Bloomington & London

Published in Canada by Fitzhenry & Whiteside Limited, Don Mills, Ontario
Manufactured in the United States of America

Library of Congress Cataloging in Publication Data
Main entry under title:
How animals communicate.
Bibliography
Includes indexes.
1. Animal communication. I. Sebeok, Thomas Albert, 1920-
QL775.H65 1977 591.5'9 76-48862
ISBN 0-253-32855-1 1 2 3 4 5 81 80 79 78 77

FOR HEINI HEDIGER

CONTENTS

PREFACE

This book is meant to replace both *Animal Communication: Techniques of Study and Results of Research,* published by Indiana University Press in 1968, and its companion volume, *Approaches to Animal Communication,* published by Mouton in 1969. All but one of the chapters are wholly new, and the organization of the contents has been substantially altered: a section on theoretical issues was added, while the one on interdisciplinary implications and applications was removed; and the section on communicative mechanisms was almost doubled. Perhaps most significantly—since this, in some manner, reflects the immense accretion of knowledge in this field—the section on communication in selected groups was expanded from nine chapters to twenty-five; for instance, whereas all the alloprimates could still be covered in one overview eight years ago, in this book the subject is barely covered in five separate chapters.

In addition to contributing chapter 9 to this book, Jack P. Hailman wrote a full-sized monograph on visual communication, which, in view of the importance of the topic and the excellence of his conspectus, is being published simultaneously by Indiana University Press, under the title *Optical Signals: Animal Communication and Light.*

In designing this volume, I received valuable counsel from Edward O. Wilson, whose personal review of many aspects of animal communication, in chapters 8–10 of his *Sociobiology,* is one of the more remarkable features of that monumental creation. I wish to thank May Lee, my editorial assistant for this project, for her devoted collaboration and for the two indexes with which this book concludes.

I have chosen to dedicate *How Animals Communicate* to Heini Hediger, who served as the director of the Zurich Zoo from 1954 until his recent retirement and was director of the Basel Zoo for ten years before that. Happily, he now continues to teach animal psychology and biology at Zurich University. Professor Hediger's work—embodied in numberless books and articles, both scientific and popular—is incomparably sensitive and subtle; his immense knowledge of the basic principles of animal communication, and particularly of the rules for two-way traffic between man and the multitude of speechless creatures, has been a model of scientific imagination applied to subjects of great human import for generations of his readers during forty-three productive years. His writings have been an inspiration to me from the beginning of my excursion into zoosemiotics, the more so since Hediger was the first student of animal behavior, in modern times, to appreciate the intimate relations of ethology and semiotics, and how the findings of each discipline can and must enrich the results of the other.

Bloomington, Indiana
November 1, 1976

THOMAS A. SEBEOK

ACKNOWLEDGMENTS

Barlow: "Much of what is written here was gestated over a number of years of research that was supported by the National Science Foundation. The labor of reading intensely was done mostly while a guest of the Zoology Department, Oxford University, which was made possible by the kind hospitality of Professors J. W. S. Pringle, F. R. S., and N. Tinbergen, F. R. S. I benefitted greatly, there, from discussions with J. M. Cullen, M. Dawkins, R. Dawkins, D. Morris, and M. Woolridge. For sending their unpublished articles, I am grateful to D. Franck, A. Manning, and W. M. Schleidt. The manuscript was born in Berkeley and helped through its difficult days by a critical reading from the seminar group in Animal Behavior and from S. E. Glickman and G. O. Barlow, to all of whom I say thanks."

Eisenberg and Golani: "The authors gratefully acknowledge the support of the National Zoological Park-Smithsonian Institution in the course of these studies. Golani was supported during the preparation of this manuscript by a Smithsonian postdoctoral award. Mr. L. Davies prepared the figures. Thanks are due to Dr. D. Kleiman and Mr. L. Davies for a critical reading of the manuscript."

Ewing: "I am grateful to Dr. L. S. Ewing and Dr. D. S. Saunders for their constructive criticisms of the manuscript, and to Dr. H. C. Bennet-Clark, Mr. M. Dow, and Mr. F. von Schilcher for helpful discussions on various aspects of dipteran communication. Mr. von Schilcher kindly allowed me to use his records of *D. melanogaster* courtship song and wing flicking, illustrated in Fig. 1. My thanks are also due to Dr. J. H. Mook of the Institute of Ecological Research, Arnhem, who provided Fig. 2."

Fine, Winn, and Olla: "The Office of Naval Research (Contract No. N00014-68-A-0215-0003) supported our bioacoustic research in the past. We thank Julie Fischer for her editorial assistance."

Geldard: "The experimental work reported from the Princeton Cutaneous Communication Laboratory was supported by a grant from National Institutes of Health, U. S. Department of Health, Education and Welfare. The preparation of this chapter was partially supported by NIH grant NS-04755."

Hailman: "A first draft of my chapter was helpfully criticized by Timothy Johnston and Dr. Gordon Stephenson, the final draft was checked by Judith Donmoyer and Edward Burtt, and Dr. Robert Jaeger patiently suffered through both versions. All illustrations are by Cheryle Hughes. Many others too numerous to name helpfully contributed in one way or another, and my wife, Liz, checked details at every stage from bibliographic search through final manuscript."

Hölldobler: "I would like to thank E. O. Wilson for reading the manuscript and my wife, Turid, for the illustrations. My own research, reported in this article, has been supported by grants from the Deutsche Forschungsgemeinschaft and National Science Foundation."

Hopkins: "This work has profited immeasurably from the association with Dr. Peter Marler and his colleagues at The Rockefeller University and with Dr. Theodore H. Bullock at the University of California, San Diego. The research was supported by NIH Training Grant #GM1789 from The National Institute of General Medical Sciences and from the Research Grant Program at the University of Minnesota Graduate School. I am deeply indebted to Eric Knudsen for permission to use the data illustrated in Fig. 2, to Michael Martinson for help with the analysis, and to Carol Gobar for preparation of the manuscript."

Klopfer: "I acknowledge the helpful critiques of Anne Clark, Lee McGeorge, Jay Russell, and Billy Seay, and the assistance of Lee Rosenson and Catherine Dewey. My work has been supported by grants from the National Institute of Mental Health, the National Institute of Child Health and Human Development, and a Research Scientist Award."

Lloyd: "I gratefully acknowledge the Alpha Helix Expeditions Program of the Scripps Institution of Oceanography (supported by the NSF), the National Science Foundation, the National Academy of Sciences, the Society of the Sigma Xi, and the National Institutes of Health for financial support of the research that forms the basis of much of this paper; Dr. Thomas J. Walker of the University of Florida for his helpful discussions and reading of the manuscript; the intellectual contributions of teachers and colleagues at Fredonia, N. Y., Ann Arbor, Michigan, Ithaca, N. Y., and Gainesville, Fla.; and H. H. Seliger and D. Otte for making unpublished manuscripts available to me."

Marler: " 'The Evolution of Communication' is a modified version of an article published in *Nonverbal Communication,* L. Krames, P. Pliner, and T. Alloway, eds. (New York: Plenum Press, 1974). Research was supported by grants from N.I.M.H. (MH14651) and N.S.F. (BN575-19431)."

Oppenheimer: "This paper was written while I was supported by U.S. Public Health Service Research Grant No. 5 RO AI 100048-13 from the National Institutes of Health to The Johns Hopkins University International Center for Medical Research. I wish to thank my colleagues for their help in sending reprints and preprints, and the Primate Information Center at University of Washington Regional Primate Research Center for its assistance. I am particularly grateful to Charles Southwick and John Eisenberg for their encouragement, to the latter for his help with *Alouatta, Ateles,* and *Lagothrix,* and to Dwain Parrack for his assistance in the preparation of the final draft in my absence. Audrey Baker graciously typed the manuscript, and Mahmooda Ghani Ahmed and Elizabeth C. Oppenheimer kindly helped with the proofing."

Poduschka: "These investigations were supported in part by a grant from the Austrian Council for Scientific Research, No. 1797. I am deeply appreciative of the help given to me by Friedrich Schaller (Zoological Institute I, University of Vienna) and John F. Eisenberg (Resident Scientist, National Zoological Park, Washington, D.C.) in providing suggestions and practical help during the course of the studies. Further, I must thank John F. Eisenberg and Devra G. Kleiman (National Zoological Park, Washington, D.C.) for reading the manuscript. I am also very much indebted to Gerald T. Williams, who agreed to undertake the translation of this report, and to his wife, Helga Williams, who helped in many ways. But I am especially indebted to my wife, Christl Poduschka, whose quiet, tireless, and unselfish help in caring for our numerous insectivores makes work and investigation possible."

Robertson: "Original work mentioned in this chapter was supported by NIH grant #HD-04722, NSF grant #GB-30784 A#1 and its writing was supported by a grant-in-aid from the Alfred P. Sloan Foundation."

Silberglied: "I wish to express my gratitude to numerous individuals for helpful criticism of an earlier draft, especially J. M. Burns, R. T. Cardé, F. M. Carpenter, B. A. Drummond III, T. Eisner, A. B. Forsyth, L. E. Gilbert, M. H. Robinson, K. D. Roeder, W. L. Roelofs, Hon. M. Rothschild, O. R. Taylor, D. S. Wilson, and E. O. Wilson. I am also grateful to G. Bergström, L. E. Gilbert, W. A. Haber, L. B. Hendry, R. C. Lederhouse, W. L. Roelofs, O. R. Taylor, and D. Windsor for access to unpublished data, and to B. Hölldobler, L. P. Lounibos, R. Robbins, and A. E. Treat for stimulating discussion and/or correspondence."

Smith: "Support during the period of writing came from the National Science Foundation grant GB 36772."

Walther: "I am grateful to the following institutions and foundations: Texas A & M University, Tanzania National Parks, Zoological Garden of Frankfurt a. M., Zoological Garden of Zurich, Zoological Garden of Naples, Fritz Thyssen Foundation, Caesar Kleberg Foundation, Frankfurt Zoological Society, Gertrud Rüegg Foundation, Smithsonian Institution, Deutsche Forschungsgemeinschaft, and the Research Council of the University of Missouri. I also wish to thank those persons who gave me definite stimulation and/or support in my research: H. Hediger, B. Grzimek, R. Faust, H. Dathe, and J. G. Teer."

Weygoldt: "The work was supported by grants from the Deutsche Forschungsgemeinschaft. I am grateful to P. D. Gabbutt, Manchester, and H.-O. von Hagen, Marburg, for proofreading the manuscript, and to H.-O. von Hagen and J. S. Rovner, Athens, Ohio, for information and suggestions."

Winn and **Schneider:** "Burney J. Le Boeuf, Ronald Schusterman, and John Terhune kindly criticized the manuscript. However, the authors must be held responsible for all statements. Mention of grey seal work by Schneider and the time spent on the manuscript were supported by Office of Naval Research Contract No. N00014-76-C-0226. The editorial work of Julie Fischer, Lois Winn, and Howard Winn's graduate students is kindly appreciated."

BIOGRAPHICAL SKETCHES

George W. Barlow (1929–) was born in Long Beach, California, and received his A.B., M.A., and Ph.D. from the University of California, Los Angeles, concentrating on comparative physiology and ichthyology. Thereafter he spent two postdoctoral years with Konrad Lorenz, learning about ethology. He taught for six years at the University of Illinois before moving on to the University of California, Berkeley, where he is now Professor. He has served as consultant to the National Science Foundation, is currently on the executive subcommittee of the International Ethological Committee, and is a Fellow of the Animal Behavior Society and of the American Association for the Advancement of Science. In 1976 he was chairperson of the Ecology Division of the American Society of Zoologists. His chief research interests center on the mechanisms of social behavior and the organization of social systems, as reflected in two lines of research, the ethology of cichlid fishes and community behavior of coral-reef fishes.

Gordon M. Burghardt (1941–) majored in biopsychology at the University of Chicago for both his undergraduate and graduate training and received his doctorate in 1966. After teaching biology for a year and a half at his alma mater, he moved to the Department of Psychology at the University of Tennessee, Knoxville, where he is now a professor. His research deals primarily with the behavior of reptiles, especially snakes, and bears. The problems that interest him include ontogeny, chemoreception, feeding, social behavior, behavioral evolution, and theoretical issues in ethology. He is author of numerous articles and book chapters and is currently (1976–77) at The Rockefeller University on a John Simon Guggenheim Foundation Fellowship.

Rene-Guy Busnel (1914–), a native of France, received his degree at the Faculty of Science in Paris. As a research worker at the National Center for Scientific Research from 1935 to 1948, he worked in the field of comparative physiology on the biochemical physiology of invertebrates and fish. In 1949 he organized and became director of the Acoustic Physiology Laboratory at the National Institute of Agronomic Research, investigating acoustic behavior throughout the animal kingdom, primarily in amphibians, insects, birds, and land and marine mammals. He has organized numerous international symposia on animal acoustic behavior (insects, 1953; mice, 1958; birds, 1962; biological sonar, 1966). In 1972 he was visiting professor at City College, C.U.N.Y. He is the author of numerous publications and the editor of *Animal Acoustic Behavior,* a volume which sums up much of the Institute's work.

David K. Caldwell (1928–), a native of Louisville, Kentucky, took degrees from Washington and Lee University and the universities of Michigan and Florida. Before completing his doctoral work at Florida, he engaged in special marine zoological study at the University of Miami. He has served as a Fishery Research Biologist with the then U.S. Fish and Wildlife Service, Curator of Marine Biology, Fishes, and Marine Mammals at the Natural History Museum of Los Angeles County and Curator and Director of Research at Marineland of Florida. Since 1970 he has been an associate professor with the Communication Sciences Laboratory of the University of Florida, and Head of that laboratory's Division of Biocommunication and Marine Mammal Research, located at the Cornelius Vanderbilt Whitney Marine Research Laboratory. His scientific papers deal with the biology and systematics of marine vertebrates, and now principally with cetacean biology, distribution, and intraspecific communication.

Melba C. Caldwell, was born in Augusta, Georgia. She received degrees from the universities of Georgia and California (Los Angeles) and has done special work in marine zoology at the Duke Marine Laboratory. She has been a Fishery Research Biologist with the U.S. Fish and Wildlife Service, a Research Associate with the Natural History Museum of Los Angeles County, a Staff Research Associate with the Allan Hancock Foundation of the University of Southern California, and Assistant Curator and Associate Director of Research at Marineland of Florida. Presently she is a Research Instructor with the Division of Biocommunication and Marine Mammal Research of the University of Florida Communication Sciences Laboratory, located at the University's Cornelius Vanderbilt Whitney Marine Research Laboratory. Her principal scientific papers deal with the biology of marine vertebrates, and especially with the biology, distribution, and intraspecific communication of cetaceans.

James A. Cohen (1952–) was born in New York City and earned a B.A. with special honors in Animal Psychology at Washington University, St. Louis, where he concentrated in Mammalian Ethology under the direction of M. W. Fox. In 1975 he went to India to study the behavior and ecology of the Asiatic Wild Dog. He later received an M.S. in zoology at the State University College of Environmental Science and Forestry in Syracuse, New York, where he studied displacement activities in the Norway rat. He is currently pursuing a Ph.D. in behavior and ecology at the Johns Hopkins University and is a research associate at the Institute for the Study of Animal Problems, a division of the Humane Society of the United States, in Washington, D.C.

John F. Eisenberg (1935–) did his undergraduate work in zoology and philosophy at Washington State University. He completed his M.A. and Ph.D. in zoology under Peter Marler at the University of California, Berkeley. He taught ethology for two years at the University of British Columbia before taking a faculty position at the University of Maryland, where he is now an adjunct professor. He has served as a staff scientist and director of research at the Smithsonian Institution's National Park since 1965. His research efforts have been devoted to the analysis of mammalian behavior patterns, especially the formation and maintenance of social structures. He has conducted field research in Panama, Madagascar, and Ceylon. His publications include: *The Behavior of Heteromyid Rodents, The Social Organizations of Mammals,* and (with E. Gould) *The Tenrecs, A Study in Mammalian Behavior and Evolution.*

Arthur W. Ewing (1935–) studied at the University of Edinburgh, Scotland, and has carried out postdoctoral research at the University of Oregon in Eugene. At present he is a lecturer in the Department of Zoology, University of Edinburgh, where he works on the neurophysiological and genetic bases of behavior in insects. He is also interested in the evolution of behavior patterns in insects and fish.

Michael L. Fine (1946–) received his B.S. degree in zoology from the University of Maryland and an M.A. degree in marine science from the Virginia Institute of Marine Science of the College of William and Mary. He received a Ph.D. in oceanography from the University of Rhode Island, where he worked on seasonal and geographical variation of toadfish sounds and on the evoking of sounds by brain stimulation. He has worked for the Naval Oceanographic Office on deep scattering layers and has studied faunal variation in pelagic *Sargassum* communities, and is currently working on neurophysiology of toadfish hearing at Cornell University.

Roger S. Fouts (1943–) was born in Sacramento, California, and earned his B.A. degree at California State College at Long Beach and his Ph.D. at the University of Nevada at Reno. He held a joint appointment as a research associate at the Institute for Primate Studies in Norman, Oklahoma, and the Psychology Department of the University of Oklahoma in Norman, Oklahoma, from 1970 to 1973. In 1973 he was appointed Assistant Professor of Psychology at the University of Oklahoma and is presently an associate professor. His major research interests are in examining the acquisition and usage of American Sign Language for the Deaf in nonhuman primates with special emphasis on the chimpanzee. He has been involved in this research since 1967.

Michael W. Fox (1937–), born in England, graduated from the Royal Veterinary College, London. He later obtained his Ph.D. from London University after working on brain and behavior development in the dog at Jackson Laboratory, Bar Harbor, Maine, and the Galesburg State Research Hospital, Galesburg, Illinois. He was awarded a D.Sc. in Ethology by London University for his studies of wild and domestic canids while at Washington University, St. Louis, where he was Associate Professor of Psychology. He now directs the Institute for the Study of Animal Problems, a division of the Humane Society of the United States.

A. Gautier (1940–) was born in France. She studied biology in Brittany, at the University of Rennes, where she became interested in ethology under the influence of Prof. Richard. A research assistant at the Centre National de la Recherche Scientifique since 1964, she has done a great deal of fieldwork in Gabon on the ecological and ethological study of the *Cercopithecus* species, principally of the talapoin, which was the subject of her doctorate. She is presently directing both a study in the natural environment in Gabon and a study in captivity at the Paimpont Biological Station on the problem of polyspecific associations in the *Cercopithecus* spp. This research is endeavoring to discover the specificity of the ecological niches and social behavior of species that associate with each other and the function of these associations.

J.-P. Gautier (1939–) was born in France. He pursued a degree in chemistry and physiology at Rennes, then a third degree in ethology with Prof. Richard. He was assistant on the Science Faculty at Rennes and then research attaché at the Centre National de la Recherche Scientifique. He began his work in Gabon in 1965, and was mainly interested in acoustic communication in the *Cercopithecus* spp., especially in close species living in association. The ontogenesis of the acoustic emissions of these woodland monkeys continues to be his principal research goal, and he has studied the same animals regularly for ten years in the Paimpont Biological Station compound.

Frank A. Geldard (1904–) was born in Worcester, Mass., and received all his formal education from schools in that city. His A.B., M.A., and Ph.D. degrees were all taken at Clark University, with graduate work in the fields of psychology and physiology. He also has an honorary D.Sc. from Washington and Lee University. He served 34 years on the faculty of the University of Virginia, ultimately as dean of the Graduate School of Arts and Sciences. His research has mainly been in the field of sensory psychophysiology, first in vision and subsequently in cutaneous sensitivity. His text, *The Human Senses,* first published in 1953, was revised and expanded in 1972. In 1962 he left Virginia to take the Stuart Professorship of Psychology at Princeton University, where a new research institute, The Cutaneous Communication Laboratory, was established. Now professor emeritus but continuing his research as Senior Research Psychologist, Dr. Geldard is actively concerned with problems of communication through cutaneous channels.

Ilan Golani (1938–), a native of Israel and a student of H. Mendelssohn, did his doctoral work in zoology at Tel-Aviv University on the nonmetric analysis of behavioral interaction sequences in captive jackals. After several years of fieldwork on the naturalistic behavior of jackals, he published a book on the topic with the Eshkol-Washman Movement Notation Society in Israel. He has done postdoctoral work at the Smithsonian Institution in Washington, D.C., and in the Department of Psychology at Dalhousie University, Halifax, Nova Scotia. He is on the faculty of the Department of Zoology at Tel-Aviv University.

Donald R. Griffin (1915–) was born at Southampton, New York, and earned his B.S., M.A., and Ph.D. degrees at Harvard University. He has held teaching appointments at Cornell University, Harvard University, where he served as chairman of the Department of Biology, and The Rockefeller University, where he is at present professor. He is a member of the American Academy of Arts and Sciences, the National Academy of Sciences, and the American Philosophical Society. His principal papers deal with bird navigation and the migrations, acoustic orientation, and orientation of bats by echolocation based on ultrasonic orientation sounds. Among his books are *Listening in the Dark,* for which he was awarded the Daniel Giraud Elliott Medal of the National Academy of Sciences, and *Bird Migration,* for which he received the 1965 Phi Beta Kappa Award in Science.

Jack P. Hailman (1936–) received degrees from Harvard and Duke universities and conducted postdoctoral research at the Universität Tübingen and the Rutgers Institute of Animal Behavior. He taught at the University of Maryland and is currently Professor of Zoology at the University of Wisconsin-Madison. His research interests include behavioral development, animal communication, and sensory processes, especially color vision. He has written or edited a number of monographs and books, including three with Peter H. Klopfer, and is currently American editor of the journal *Animal Behaviour.*

Bert Hölldobler (1936–) was born in Erling-Andechs (West Germany). He earned his Ph.D. at the University of Würzburg and his Habilitation degree (Dozent) at the University of Frankfurt. He has held a research associate fellowship at Harvard University, became Professor of Zoology at the University of Frankfurt in 1971, and in January 1973 joined the Harvard faculty as Professor of Biology. He is a member of the American Academy of Arts and Sciences and of the Deutsche Akademie der Naturforscher (Leopoldina). His major research interests are the analysis of physiological and ecological aspects of insect behavior. His principal scientific papers deal with the mechanisms of intra- and interspecific communication in ant societies, and orientation and territoriality in ants.

Carl D. Hopkins (1944–) is originally from Rochester, New York. He took his undergraduate training in physics at Bowdoin College and then switched to biology for his graduate training at The Rockefeller University. Under the influence of P. R. Marler and D. R. Griffin, he received his doctorate with a specialization in animal behavior. After a year's postdoctoral training in T. H. Bullock's laboratory in the Department of Neurosciences at the University of California, San Diego, he was appointed Assistant Professor of Ecology and Behavioral Biology at the University of Minnesota, Minneapolis, where he teaches animal behavior. His current research centers on the evolution of animal communication systems, from signal production to signal reception, with an emphasis on electrical communication.

A. Ross Kiester (1944–) was an undergraduate at the University of California at Berkeley and a graduate student at The Rockefeller University and Harvard University. This article was prepared while he was a Junior Fellow in the Society of Fellows at Harvard University. Currently he is an Assistant Professor in the Department of Biology at the University of Chicago.

Devra G. Kleiman (1942–) received a B.S. in biopsychology from the University of Chicago (1964). Her interest in ethology developed from undergraduate research on wolf behavior, which led to a year's study of canid behavior at the Zoological Society of London. She then joined the Wellcome Institute of Comparative Physiology at the London Zoo and entered the University of London, where she received her doctorate in zoology in 1969. This was followed by a two-year postdoctoral fellowship at the Institute of Animal Behavior, Rutgers. She taught animal behavior at Rutgers and the University of Maryland. Since 1972, she has held the position of Reproduction Zoologist at the National Zoological Park. She is also Adjunct Associate Professor, Department of Psychology, at George Washington University. Her major research interests are mammalian social and reproductive behavior.

Hans Klingel (1932–) studied at Carthage College, Illinois, and at the universities of Freiburg and Mainz, Germany. He received his doctorate in zoology at the University of Mainz, and is now professor at the University of Braunschweig, Germany. He did fieldwork on arachnids and centipedes in southeast Asia and on equids and other large mammals in Africa. His major research interests are social organization and social behavior of mammals in relation to environment, and behavior of land arthropods.

Peter H. Klopfer (1930–) is professor of zoology at Duke University, and Associate Director of the Primate Facility. He received his doctorate at Yale, under G. E. Hutchinson, and then worked with W. H. Thorpe at Cambridge. His interests include ecological problems relating to community structure and faunal diversity, and psychological problems of maternal-filial attachments and early learning. He is also an addicted cross-country runner and track competitor (Masters Division). Among his major publications are: *Behavioral Aspects of Ecology* and *An Introduction to Animal Behavior: Ethology's First Century* (with J. P. Hailman), and *On Behavior: Instinct Is a Cheshire Cat.*

Philip Lieberman (1934–) was born in New York. He received a B.S. and M.S. in Electrical Engineering from Massachusetts Institute of Technology in 1958, where he also received a Ph.D. in Linguistics in 1966, with a dissertation on the physiology, acoustics, and grammatical function of intonation in English. He was a member of the research staff at Air Force Cambridge Research Laboratories and Haskins Laboratories. From 1967 to 1974 he taught at the University of Connecticut and is now Professor of Linguistics at Brown University.

James E. Lloyd (1933–) was born in Oneida, New York, and after spending four years in the United States Navy as an aviation electronics technician and operator, he earned a B.S. at the State University of New York at Fredonia, an M.A. at the University of Michigan, and a Ph.D. at Cornell University. Following postdoctoral work in systematic and evolutionary biology at the University of Michigan, he joined the faculty of the University of Florida, where he is now Professor of Entomology and Nematology. His major research interests are in the systematics and behavior of luminescent animals, especially Lampyrid beetles (fireflies).

Peter Marler (1928–), a native of England, worked in plant ecology at the University of London before turning his attention to animal behavior. Under the influence of R. A. Hinde and W. H. Thorpe at the University of Cambridge, he studied problems of social communication in birds. In 1957 he joined the Department of Zoology at the University of California, Berkeley. In 1966 he moved to The Rockefeller University, where he is presently Director of the Field Research Center in Ecology and Ethology. He is a member of the American Academy of Arts and Sciences and the National Academy of Sciences. His writings deal mainly with the physical analysis of sounds of birds and primates, the role of genetic and environmental factors in their development, and general problems of animal communication. Together with W. J. Hamilton III, he has published a general textbook, *Mechanisms of Animal Behavior.*

Martin H. Moynihan (1928–) studied at Princeton and Oxford universities. After research appointments at Cornell and Harvard universities, he has served as Resident Naturalist, Director, and Senior Scientist at the Smithsonian Tropical Research Institute. His interests are the evolution, behavior, and ecology of birds, monkeys, and cephalopods.

Bori L. Olla (1937–) received his formal training at Fairleigh Dickinson University, the University of Hawaii, the University of Maryland and Seton Hall College of Medicine. For the past eleven years he has been at the Sandy Hook Laboratory in New Jersey, part of the National Marine Fisheries Service of the U.S. Department of Commerce, where he is Chief of Behavioral Investigations. His research has centered primarily on the comparative aspects of the behavior of marine fishes in both the field and the laboratory, with special emphasis on biorhythms, feeding and social behavior, and the effect of environmental stress on selected behavior patterns.

John R. Oppenheimer (1941–) was born in New York City and earned his B.A. at the University of Connecticut, and his M.S. and Ph.D. at the University of Illinois. His master's research was on the parental behavior of the mouthbreeding cichlid, *Tilapia melanotheron,* and his doctoral and postdoctoral research were on the behavior and ecology of the white-faced monkey, *Cebus capucinus,* on Barro Colorado Island. He has also observed the three other species in the genus *Cebus* in Colombia and Venezuela. Since 1968 he has been on the staff of the Johns Hopkins University and has spent two years in India in charge of the J. H. U. Field Station in Singur, West Bengal. Most of his time there was devoted to research on the daily and annual activity patterns of langurs, and on the ecology of dung beetles. His major interests are the organization of social behavior and the interaction between ecology and behavior.

Daniel Otte (1939–) was born in Natal, South Africa, and attended the University of Michigan, where he received his doctorate in zoology in 1968 under R. D. Alexander. After a year as research associate of R. D. Alexander doing systematic and biogeographical research on Australian crickets, he served on the zoology faculty at the University of Texas at Austin until 1975. He then became a research scientist in the Division of Systematic and Evolutionary Biology at The Academy of Natural Sciences of Philadelphia, where he is presently Associate Curator and Chairman of Entomology. His major research interests are comparative ethology, population ecology, biogeography, and systematics of Orthoptera.

Walter Poduschka (1922–) was born and lives in Vienna, Austria. He studied physical and cultural anthropology at the University of Vienna and earned his doctorate in 1955. Shortly afterwards he changed his course to zoology, turning his interest to mammalian behavior. From 1962 he specialized in insectivores, studying their behavior and their sensory and functional physiology with emphasis on comparative research. In 1971 he founded a free-lance insectivore research station in connection with the First Zoological Institute at the University of Vienna. Fieldwork has taken him to the Near and Middle East, to North Africa, to the Greater Antilles, and to Mexico.

Cheryl H. Pruitt (1946–) majored in psychology and zoology at Centre College and completed doctoral work in psychology at the University of Tennessee. Her dissertation was on the social behavior of young black bears. She is currently on the faculty of Northern Kentucky State College. Her research interests presently focus on the communicative aspect of social behavior, and particularly on its development in carnivores.

Randall L. Rigby (1945–) was graduated from Oklahoma State University in 1968, majoring in psychology. A career U.S. Army officer, Captain Rigby is presently serving as an instructor in the Department of Military Psychology and Leadership at the U.S. Military Academy, West Point, New York. Interested primarily in animal learning, he is currently undergoing graduate training in psychology at the University of Oklahoma.

Anthony Robertson (1941–) studied physiology at the University of Cambridge, where he received his B.A. in Natural Sciences in 1963. He worked at the National Physical Laboratory, Teddington, England, on pattern recognition and the physiology of visual neurones until 1968, then came to the University of Chicago. At present he is in the Department of Biophysics and Theoretical Biology, working on the control of development in the cellular slime molds.

Arcadio F. Rodaniche (1937–) studied at New York University and California State University at San Francisco. He has been Biological Technician and Biologist at the Smithsonian Tropical Research Institute. His work is concerned with the ecology and behavior of cephalopods and other marine organisms.

Jack Schneider (1946–) received his B.A. degree from Webster College, Missouri, and his M.S. from the University of Rhode Island. He is presently Curator of Exhibits at Mystic Marinelife Aquarium, Mystic, Connecticut; and is studying pinniped biology with an emphasis on the behavior and sounds of grey seals.

Kate Scow received her B.S. degree in biology from Antioch College in 1973. She is presently a biological technician at Northwestern University, Evanston, Illinois.

Thomas A. Sebeok (1920–) is a native of Budapest who has lived in the United States since 1937. After studying literary criticism, anthropology, and linguistics at The University of Chicago, he earned his doctorate at Princeton University. He has been a member of the Indiana University faculty in linguistics since 1943 and serves as Chairman of the university's Research Center for Language and Semiotic Studies. He is a Professor of Anthropology as well. In 1964, he was Director of the Linguistic Institute of the Linguistic Society of America. He was also designated Linguistic Society of America Professor for 1975. In 1968 he became Editor-in-Chief of *Semiotica.* He was a founding member of the Animal Behavior Society and has served as a member of its Executive Committee.

Harry H. Shorey (1931–) received his B.S. degree from the University of Massachusetts and his M.S. and Ph.D. degrees from Cornell University. He joined the faculty of the University of California, Riverside in 1959, where he is now Professor of Entomology and Professor of Biology. His major research interests concern animal communication by pheromones and the behavioral control of pest insects.

Robert E. Silberglied (1946–) studied entomology as an undergraduate at Cornell University. Under the influence of Thomas Eisner, he became interested in general problems of arthropod behavior and evolution. He received his doctorate at Harvard University, where he is now Assistant Professor of Biology and Assistant Curator in Entomology at the Museum of Comparative Zoology. He is also on the staff of the Smithsonian Tropical Research Institute in Panama. He is interested in all aspects of arthropod natural history. His research has centered on the role of ultraviolet reflection patterns in butterfly communication and in pollination systems.

W. John Smith (1934–) is a native of Canada and did his undergraduate work in biology at Carleton University, Ottawa. He earned his doctorate at Harvard University (1961) and spent the next three years associated with that university in a postdoctoral capacity. Much of this time was spent in South America, studying the evolutionary radiation of the bird family

Tyrannidae. In 1965 he joined the faculty of the University of Pennsylvania in the Department of Biology, and has subsequently also been appointed to the Department of Psychology. He is a research associate of the Smithsonian Tropical Research Institute, a consultant of the Penrose Research Laboratory of the Zoological Society of Philadelphia, and a research associate in ornithology of the Academy of Natural Sciences of Philadelphia. The comparative study of animal communication and other aspects of interactional behavior is now his major research interest.

Richard Tenaza (1939–) received his B.A. degree in biology from San Francisco State University and his Ph.D. (1974) in zoology from the University of California, Davis. In 1971 and 1972 he conducted research on the Mentawai Islands, west of Sumatra, evaluating relationships between resources and social organization in two primitive societies and studying vocalizations of gibbons and langurs. He has participated in studies of Adelie penguin social behavior in Antarctica, ecology and behavior of microtine rodents in Alaska and Bank's Island, behavior of slow lorises and gibbons in Thailand, and behavior of cormorants in the Farallon Islands. His principal interest is in the adaptive nature of animal and human social organizations. He is Assistant Professor of Biology, University of the Pacific, Stockton, California.

Fritz R. Walther (1921–) was born in Germany. He has lived in the United States since 1967 and is currently a professor at Texas A & M University. He studied medicine, pedagogy, philosophy, psychology, and zoology, mainly at the Johann Wolfgang Goethe University, Frankfurt/Main, where he received his doctorate in zoology in 1963. He has held positions at the Opel Zoo in Germany, the Zürich Zoo in Switzerland, and the Serengeti National Park, Tanzania. He has conducted research programs in Wyoming, Texas, Israel, South Africa, and East Africa. His primary interests are the ethology of game animals, especially horned ungulates, and problems of intraspecific aggression, reproductive behavior, and expressive behavior. His papers on the behavior of various artiodactyl species have been published in German, English, and African journals, and he has written two books on this subject in German. He also is a coeditor and author of several chapters in Grzimek's *Animal Life Encyclopedia.*

Christen Wemmer (1943–) received his undergraduate and masters degrees at San Francisco State College and his doctoral degree at the University of Maryland. He is presently Curator-in-Charge of the National Zoological Park's Conservation and Research Center, Front Royal, Virgina. His research interests are comparative ethology and communication of mammals.

Peter Weygoldt (1933–) was born in Wilhelmshaven, Germany. He studied at the Albert-Ludwigs-University in Freiburg and at the Christian-Albrechts-University in Kiel, where he received his doctorate. After five years as assistant professor at the Free University of Berlin he spent two years in the United States, first as a research associate at the Duke University Marine Laboratory and later as visiting professor at the Friday Harbor Laboratories. In 1967 he joined the staff of the Biology Department of the Albert-Ludwigs-University in Freiburg, where he is Professor of Zoology. His major research interests include embryology and biology of crustaceans and behavior, morphology, and evolution of arachnids.

Howard E. Winn (1926–) earned his B.A. degree at Bowdoin College and his M.S. and Ph.D. degrees at the University of Michigan. After ten years of teaching at the University of Maryland, in 1965 he joined the faculty of the University of Rhode Island as Professor of Zoology. Since 1967 he has held the position of Professor of Oceanography and Zoology there. His research interests center on bioacoustics and bioorientation with special reference to the toadfish, eel, and humpback whale.

Part I
Some Theoretical Issues

Chapter 1

THE PHYLOGENY OF LANGUAGE

Philip Lieberman

Since Charles Darwin's *On the Origin of Species* (1859), the study of hominid evolution has been based on two lines of complementary research: the comparative study of living species that we believe manifest certain aspects of the behavior of earlier hominids; and the inferences that we can make from the artifacts that have been found in association with extinct fossil hominids, as well as the direct examination of these fossils. The anatomy of extinct fossils is relevant to the study of their behavior because it is evident that certain aspects of behavior are predicated on the presence of particular anatomical specializations. Upright bipedal locomotion, for example, is not possible without certain specialized anatomical features that are present in modern *Homo sapiens.*

The study of the behavior of living nonhuman species is essential to our understanding of the function of "human" anatomical specializations. Electromyographic studies of chimpanzees and gorillas, for example, show that these animals, who lack the specialized pelvic and limbic anatomy of *Homo sapiens,* can stand upright only for short periods of time and with the expenditure of great muscular effort. In contrast, humans can stand upright or walk at a moderate pace with very little muscular effort (Basmajian and Tuttle, in press). Comparative studies of living nonhuman primates can serve as "experiments" that enable us to assess the function of particular anatomical specializations in *Homo sapiens.* All animals are, in effect, living "experimental preparations." Appropriate experiments can relate particular aspects of the morphology of different species to the behavior of these species.

We can project the relevance of the anatomical specializations that relate to upright posture to the study of human evolution when we note that the skeletal remains of particular extinct hominids also appear to be functionally adapted for upright posture and bipedal locomotion. The fossil remains of *Australopithecus africanus,* for example, show that these early hominids, who lived from one to four million years ago, also possessed the behavioral attribute of bipedal locomotion and upright posture (Campbell, 1966; Pilbeam, 1972). We can thus infer that these anatomical specializations were retained by the Darwinian process of natural selection because upright posture and bipedal locomotion were selective advantages to these early hominids.

The anatomical specializations that are necessary conditions for upright posture have a biologic "cost." The change in the pelvic area vis-à-vis animals who lack upright posture makes childbirth more difficult. The advantages of upright posture and bipedal locomotion must have outweighed the increased mortality in childbirth for these specializations to have been retained and elaborated over a period of millions of years. The presence of stone artifacts like the rounded stones that have been found in association with Australopithecine fossils (Leakey, 1971) thus take on a new significance when we consider the probable presence of upright posture and bipedal locomotion in Australopithecine culture. The stone "balls" may have served as projectiles in hunting small animals. The selective advantage of upright posture would have been enhanced if Australopithecines used projectiles. The use of projectiles in itself would not appear to be very significant if we again did not refer to the results of comparative studies on living nonhuman hominoids. Although living apes do, in fact, hurl stones and branches, they usually cannot hit particular targets. Instead they ineffectually hurl things about in some general direction (Beck, in press; van Lawick-Goodall, 1973). All humans can learn to hit targets regularly with projectiles. Small children learn to do so without any special instruction. It is a human attribute.

The ability to hit things with projectiles is not very interesting until we compare human behavior with that of living related species. The human "quality" of being able to hit targets must involve the presence of certain innately determined neural mechanisms and pathways that enable us to acquire this ability readily. The stone projectiles that have been found in association with Australopithecine fossil remains indicate that these early hominids probably had this ability. It would have made upright posture an asset by freeing the hands for throwing. The entire behavioral and physiologic complex—upright posture, stone throwing, hunting, pelvic and limbic anatomy, neural mechanisms coordinating the motions of the arms and hands with vision, etc.—thus can be viewed as an interdependent evolutionary pattern. It started with hominids like *Ramapithecus* (Pilbeam, 1972), who may have lacked upright posture and who might have hurled projectiles in some general direction, but it ultimately resulted in the evolution of hominids like the Australopithecines. The initial conditions for this evolutionary process can be seen in the physiology and behavior of present-day apes, who have the arms, hands, and acute vision that are necessary though not sufficient conditions for accurately throwing projectiles. There is, as Darwin (1859) claimed, a continuity and gradualness in the process of evolution.

Human language is at present a unique quality of *Homo sapiens.* However, like the unique human quality of stone throwing, it too must be the result of a gradual evolutionary process. We can profitably apply the same comparative techniques toward an understanding of its evolution. In fact, there already are a number of studies that bear either directly or indirectly on the evolution of language in general and of human language in particular. The claim is often made that human language has absolutely nothing in common with the communications of animals (Lenneberg, 1967). Human language is supposedly disjoint with the communications of animals. It is supposedly "referential," while animal communications are "emotive." In fact, there are no studies that demonstrate that the communications of animals are simply emotive. At best, there are tenuous arguments that claim that the neural embodiment of human language involves cortical-to-cortical pathways, which are present only in the human brain, whereas the communicative signals of animals supposedly involve cortical-subcortical pathways. Cortical-cortical pathways are supposedly involved in referential thought, while cortical-subcortical pathways are supposedly involved in the expression of emotions. These claims cannot be supported by the findings of modern neuroanatomy. The "referential" activity of the human brain probably involves many cortical-subcortical pathways. Nor is there any substantive basis for assigning the expression of emotion exclusively to cortical-subcortical pathways.

The supposed uniqueness of human language seems to me to be an echo of the traditional Cartesian view. Many ethological, behavioral, and linguistic studies of the communications of animals that are otherwise faultless are limited by their unquestioning acceptance of this Cartesian premise. The tendency is to draw negative conclusions concerning the linguistic ability of nonhuman animals, even though these conclusions are not supported by the data. Green (1973), for example, in an acoustic analysis of some of the vocalizations of nonhuman primates, uses spectrographic data that show that a number of these vocalizations form a "graded" series along what appear to be the acoustical dimensions of amplitude and fundamental frequency. Green concludes that the vocalizations are "graded" rather than "discrete." Nonhuman primate vocalizations, supposedly, are thus not "linguistic" signals, like the human sounds transcribed by the symbols /b/, /p/, /a/, etc.

However, the method that Green uses, if applied to human speech, would demonstrate that the discrete human phonetic elements /b/ and /p/ were also graded, nonlinguistic signals. The acoustic basis of the distinction between the English sounds /b/ and /p/ rests in the delay between the sound generated when the speaker's lips are opened and the start of phonatory vocal cord activity. If phonation starts within 20 msec after the speaker's lips open, the sound is perceived as a /b/. If the delay exceeds 20 msec, the sound is perceived as a /p/ (Lisker and Abramson, 1964). Human speakers, particularly young human speakers (Preston and Port, 1972), produce many /b/ sounds with phonation delays ranging over the 0–20 msec interval that defines these sounds. They also produce many /p/ sounds in which phonation delays vary from 20 to over 130 msec. If these sounds were examined using the same criteria that have been applied to the analysis of chimpanzee vocalizations, the discrete, categorical responses of human listeners to these sounds would not be apparent.

The basis of the discrete, categorical nature of these speech sounds appears to be the presence of an innately determined auditory mechanism—a neural "feature detector," which is even manifested in the behavior of four-week-old human infants (Eimas and Corbitt, 1973). The sounds /b/ and /p/, in other words, are discrete signals because human perception makes them discrete. Acoustic analysis, in itself, cannot reveal the presence of the neural mechanism that is the basis of the discrete nature of these speech sounds. We cannot assume that the primate vocalizations discussed by Green or in other accounts of primate communication are not discrete signaling units until we perform the appropriate perceptual experiments.

We really know very little about the communications and the possible "language" of various animals. Linguists have been rather anthropocentric when they attempt to limit the term "language" to human language. It would make as much sense to limit the term "swimming" to human swimming. The details of the way a human swims are different from the way a dog swims, but the end result is similar. Both animals move through water. The communication system of a dog is probably simpler than that of a human, but there may be common elements, and we can perhaps gain some insights into the nature of human language by studying the simpler, canid system.

We might want to restrict the term "language" to a special class of communication systems, but we really cannot limit the term to the language of present-day *Homo sapiens* without making the classificative function of the term unprofitably restrictive. The phonetic abilities of some fossil hominids like the classic Neanderthals, who inhabited parts of Europe and Asia about 70,000 years ago, would have precluded their speaking any of the languages of modern *Homo sapiens.* However, the stone artifacts that have been found in association with these fossils and the evidence of their cultural tradition make it evident that some form of language must have been a feature of Neanderthal society (Lieberman, 1975). The language of these advanced hominids would not have been a human language if we restrict the term "human" to modern *Homo sapiens.* However, there is no reason to believe that Neanderthal hominids could not have transmitted new, previously unanticipated information among themselves. Thus, the operational definition of language (Lieberman, 1973) is a communication system that permits the transmission of new information. This definition obviously would not fit the limited codes that simple animals like frogs appear to use. However, it would admit many possible languages that might differ substantially from the language of present-day *Homo sapiens.*

Defining language in terms of the properties of human language is fruitless because we do not know what they really are. Even if we knew the complete inventory of properties that characterize human language we probably would not want to limit the term "language" to communication systems that had all of these properties. For example, it would be unreasonable to state that a language that had all of the attributes of human languages except relative clauses really was not a language. The operational definition of language is functional rather than taxonomic. It is a productive definition insofar as it encourages questions about what animals can do with their communication systems and the relation of these particular systems to human language and to the intermediate levels of language that probably were associated with early hominids.

Neural Feature Detectors

As I noted earlier, the perception of human speech appears to involve the presence of neural mechanisms that are sensitive to particular acoustic events. In recent years a number of electrophysiological and behavioral studies have demonstrated that various animals have auditory detectors that are "tuned" to signals that are of interest to them. Wollberg and Newman (1972), for example, recorded the electrical activity of single cells in the auditory cortex of awake squirrel monkeys *(Saimiri sciureus)* during the presentation of recorded monkey vocalizations and other acoustic signals. They presented eleven calls, representing the major classes of this species' vocal repertoire, as well as a variety of acoustic signals designed to explore the total auditory range of these animals. Extracellular unit discharges were recorded from 213 neurons in the superior temporal gyrus of the monkeys. More than 80 percent of the neurons responded to the tape-recorded vocalizations. Some cells responded to many of the calls that had complex acoustic properties. Other cells, however, responded to only a few calls. One cell responded with a high probability only to one specific signal, the "isolation peep" call of the monkey.

The experimental techniques that are necessary in these electrophysiological studies demand great care and great patience. Microelectrodes that can isolate the electrical signal from a single neuron must be prepared and accurately positioned, the electrical signals must be amplified and recorded, and, most important, the experimenters must present the animal with a set of acoustic signals that explore the range of sounds it would encounter in its natural state. Demonstrating the presence of neural mechanisms matched to the constraints of the sound-producing systems of particular animals is therefore a difficult undertaking. The sound-producing possibilities and behavioral responses of most "higher" animals make comprehensive statements on the relationship between perception and production difficult. We can only explore part of the total system of signaling and behavior. "Simpler" animals, however, are useful in this respect since we can see the whole pattern of the animal's behavior.

The behavioral experiments of Capranica (1965) and the electrophysiological experiments of Frishkopf and Goldstein (1963), for example, demonstrate that the auditory system of the bullfrog *(Rana catesbeiana)* has single units that are matched to the formant frequencies of the species-specific mating call. Bullfrogs are among the simplest living animals that produce sound by means of a laryngeal source and a supralaryngeal vocal tract. The latter consists of a mouth, a pharynx, and a vocal sac that opens into the floor of the mouth in the male. Vocalizations are produced in the same manner as in primates. The vocal cords of the larynx open and close rapidly, emitting puffs of air into the supralaryngeal vocal tract, which acts as an acoustic filter. Frogs can make a number of different sounds (Bogert, 1960), including mating calls, release calls, territorial calls that serve as warnings to intruding frogs, rain calls, distress calls, and warning calls. The different calls have distinct acoustic properties.

The mating call of the bullfrog consists of a series of croaks. The duration of each croak varies from 0.6 to 1.5 sec, and the interval between each croak varies from 0.5 to 1.0 sec. The fundamental frequency of the bullfrog croak is about 100 Hz. The formant frequencies of the croak are about 200 Hz and 1,400 Hz. Capranica generated synthetic frog croaks by means of a fixed, POVO speech synthesizer (Stevens et al., 1955) that was designed to produce human vowels but that serves equally well for the synthesis of bullfrog croaks. In a behavioral experiment, Capranica showed that bullfrogs responded to synthesized croaks so long as the croaks had energy concentrations at either or both of the formant frequencies. The presence of acoustic energy at other frequencies inhibited the bullfrogs' responses (which consisted of joining in a croak chorus).

Frishkopf and Goldstein (1963), in their electrophysiologic study of the bullfrog's auditory system, found two types of auditory units. They found cells in units in the eighth cranial nerve of the anesthetized bullfrog that had maximum sensitivity to frequencies between 1,000 and 2,000 Hz. They found other units that had maximum sensitivity for frequencies between 200 Hz and 700 Hz. The units that responded to the lower frequency range, however, were inhibited by appropriate acoustic signals. Maximum response occurred when the two units responded to pulse trains at rates of 50 and 100 pulses per sec, with energy concentrations at or near the formant frequencies of bullfrog mating calls. Adding acoustic energy between the two formant frequencies, at 500 Hz, inhibited the responses of the low-frequency single units.

The electrophysiologic, behavioral, and acoustic data all complement each other. Bullfrogs have auditory mechanisms that are structured specifically to respond to the bullfrog mating call. They do not simply respond to any sort of acoustic signal as though it were a mating call, but respond to particular calls that can be made only by male bullfrogs; and they have neural mechanisms structured in terms of the species-specific constraints of the bullfrog sound-producing mechanism.

Plasticity and the Evolution of Human Speech

Frogs are rather simple animals but they have nonetheless evolved different species-specific calls. Of the thirty-four species whose mating calls failed to elicit responses from *Rana catesbeina,* some were closely related, others more distantly related. It is obvious that natural selection has produced changes in the mating calls of Anuran species. The neural mechanisms for the perception of frog calls are at the periphery of the auditory system. They apparently are not very plastic, since Capranica was not able to modify the bullfrogs' responses over the course of an eighteen-month interval. Despite this lack of plasticity, frogs have evolved different calls in the course of their evolutionary development.

Primates have more flexible and plastic neural mechanisms for the perception of their vocalizations. Recent electrophysiological data (Miller et al., 1972) show that primates like rhesus monkey *(Macaca mulatta)* will develop neural detectors that identify signals important to the animal. Receptors in the auditory cortex responsive to a 200 Hz sine wave were discovered after the animals were trained by the classic methods of conditioning to respond behaviorally to this acoustic signal. These neural detectors could not be found in the auditory cortex of untrained animals. The auditory system of these primates thus appears to be plastic. Receptive neural devices can be formed to respond to acoustic signals that the animal finds useful.These results are in accord with behavioral experiments involving human subjects in which "categorical" responses to arbitrary acoustic signals can be produced by means of operant conditioning techniques (Lane, 1965). They are also in accord with the results of classic conditioning experiments like those reported by Pavlov. The dogs learned to identify and respond decisively to the sound of a bell, which is an unnatural sound for a dog. The dog obviously had to learn to identify the bell.

The first hominid "languages" probably evolved from communication systems that resembled those of present-day apes. The social interactions of chimpanzees are marked by exchanges of facial and bodily gestures as well as vocalizations (van Lawick-Goodall, 1973). The recent successful efforts establishing "linguistic" communications between humans and chimpanzees by means of either visual symbols or sign language (Gardner and Gardner, 1969; Fouts, 1973; Premack, 1972) show that apes have the cognitive foundations for analytic thought. They also use tools, make tools, and engage in cooperative behavior (for example, hunting). All these activities have been identified as factors that may have placed a selective advantage on the evolution of enhanced linguistic ability (Washburn, 1968; Hill, 1972).

It is obviously impossible to determine directly what types of feature detectors may have existed in the brains of extinct hominids. We can, however, get some insights into the general evolutionary process that developed the phonetic level of hominid language by taking note of the evolution of the speech-producing anatomy. The "match" that exists between the constraints of speech production and speech perception in modern humans (Lieberman, 1970) as well as comparative data on other living species make this procedure reasonable. An extinct hominid could obviously not make use of a phonetic contrast that could not be produced by the species. The reader may still wonder precisely what insights we may gain on the general question of the evolution of language even if we can determine some of the phonetic constraints on the languages of earlier, now extinct hominids. The answer is that certain sounds that occur in the languages of present-day humans have important functional attributes. The presence or absence of these sounds can tell us something about the general level of linguistic ability in an extinct hominid species. I shall return to this topic, which involves the physiology of speech, after discussing some of the data that are available at present.

The Phonetic Ability of Neanderthal Hominids

Neanderthal hominids like those represented by the "classic" La Chapelle-aux-Saints and La Ferrasie fossils lived until comparatively recent times. They form a class of fossils that differ significantly, using quantitative statistical methods (Howells, 1970, in press), from other fossil hominid populations and from modern *Homo sapiens.* Neanderthal hominids were not primitive in the sense that they lacked culture. They produced complex stone tools and had a cultural tradition that has left traces of burial rituals and care for the infirm and aged. The data that form the basis of our inferences regarding Neanderthal culture consist of stone and bone tools, traces of fire sites, burial sites, and skeletal material that have survived between 40,000 and perhaps 100,000 years.

However, nothing remains of the soft tissue of the supralaryngeal vocal tract or the larynx. How can we then make any inferences about phonetic ability? Fortunately we can reconstruct the supralaryngeal vocal tract that is typical of these extinct fossils, making use of the methods first proposed by Darwin concerning the "affinities of extinct species to each other, and to living forms" (1859:329) and his observations with regard to embryology (1859:439–49). The basis for the reconstruction of the supralaryngeal vocal tract of Neanderthal hominids (Lieberman and Crelin, 1971) is the similarity between the skulls of the fossils and of newborn *Homo sapiens,* i.e., newborn modern man. At first this might seem implausible. How can an adult fossil skull that has massive supraorbital brow ridges, a huge massive mandible, and a generally prognathous aspect be compared with that of a newborn human? The answer is that certain aspects of the skeletal morphology of newborn human and Neanderthal skulls are similar, even though other aspects are not. The claim is not that newborn humans are little Neanderthalers, but that the two share certain skeletal features. Vlcek (1970), in his comparative study of the development of skeletal morphology in Neanderthal infants and children, independently arrived at similar conclusions.

There are a great many Neanderthal fossil skulls including those of infants and children, two to fourteen years old at the time of death. Vlcek was therefore able to study the ontogeny of Neanderthal skull development in relation to that of modern man. He concludes:

> Certain primitive traits that are present in the skeletons of Neanderthal forms occur again in different periods of the foetal life of contemporary man with different degrees of intensity. Thus we can observe the development and the presence of many morphological characteristics typical of the Neanderthal skeleton in the skeleton of contemporary man in the course of his ontogenetic development. [1970:150]

In Fig. 1 sketches of the skulls of adult and newborn *Homo sapiens* and the La Chapelle-aux-Saints fossil are presented. The La Chapelle-

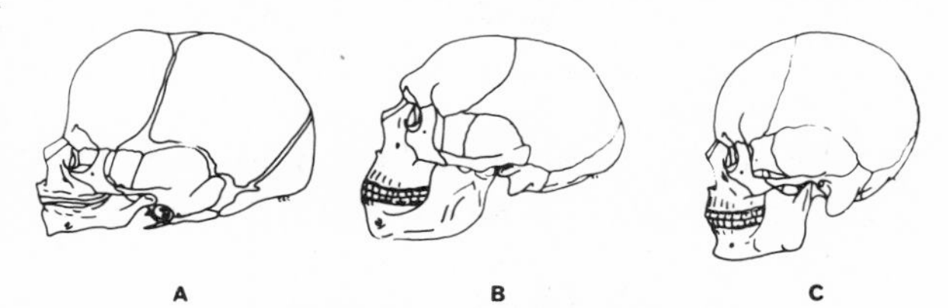

Fig. 1. A. Newborn human skull.
B. Neanderthal skull.
C. Adult human skull.

aux-Saints fossil is probably somewhere between 45,000 and 100,000 years old. The exact dating is not important since Neanderthalian fossils, for example, all those discussed in connection with Vlcek's study, persisted throughout this period. (The skulls in Fig. 1 have all been drawn to the same approximate size.) Note the basic similarity between the newborn human skull and the Neanderthal skull. The newborn human and the young Neanderthal skulls of Vlcek's study are very similar. The older skulls in Vlcek's study and the adult La Chapelle skull are even more similar in important features to the newborn human skull than the newborn human is to the adult human skull. The newborn human and Neanderthal skulls are relatively more elongated from front to back and relatively more flattened from top to bottom than that of adult *Homo sapiens.* The squamous part of the temporal bone is similar in newborn human and all Neanderthal skulls. A long list of similar anatomical features could be presented, but we are really only concerned with skeletal features that are directly relevant to the reconstruction of the supralaryngeal vocal tract of Neanderthal man.

We have to think in terms of the functional anatomy of the vocal tract. If we were to ignore the functional aspects of skeletal morphology we could be led astray, for example, by the fact that the mastoid process is absent in newborn humans and relatively small in the La Chapelle fossil, adding to their similarity to the skull of the adult *Homo sapiens* in Fig. 1. The mastoid process, however, plays no role in the reconstruction of the supralaryngeal vocal tract.

Most of the unsuccessful attempts at deducing the presence or absence of speech from skeletal structures were based on comparative studies that did not properly assess the functional roles of particular features. Vallois (1961) reviews many of these attempts, which were hampered by the absence of both a quantitative acoustic theory of speech production and suitable anatomical comparisons with living primates that lack the physical basis for human speech. The absence of prominent genial tubercles in certain fossil mandibles, for example, was taken as an absolute sign of the absence of speech, but genial tubercles are sometimes absent in normal adult humans who speak normally. They play a part in attaching the geniohyoid muscle to the mandible, but they are not in themselves crucial features. Indeed the notion of looking for crucial, isolated morphological features is not particularly useful. It is necessary to explore the complete relationship of the skeletal structure of the skull and mandible to the supralaryngeal vocal tract.

Fig. 2 shows lateral views of the skull, vertebral column, and larynx of newborn and adult *Homo sapiens* and the reconstructed La Chapelle-aux-Saints fossil. The Neanderthal skull is placed on top of an erect cervical vertebral column instead of on one sloping forward, as depicted by Boule (1911–13). This is in agreement with

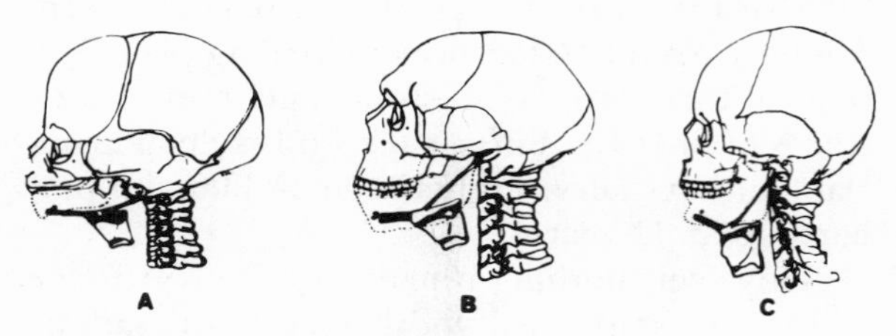

Fig. 2. A. Newborn human skull.
B. Reconstructed La Chapelle-aux-Saints fossil.
C. Adult human skull.

Straus and Cave (1957), who determined that the La Chapelle-aux-Saints fossil had suffered from arthritis, but this condition could not have affected his supralaryngeal vocal tract. Severe arthritis at advanced ages has virtually no effect on speech in modern man. (The La Chapelle-aux-Saints fossil was probably about forty years old at the time of his death.)

Since the second, third, and fourth cervical vertebrae were missing, they were reconstructed to conform with those of adult *Homo sapiens.* In addition, the spinous processes of the lower cervical vertebrae shown for the adult human in Fig. 2 are curved slightly upwards. They come from a normal vertebral column and were purposely chosen to show that the La Chapelle-aux-Saints vertebrae were not necessarily pongid in form, as

Boule (1911–13) claimed. Crelin's reconstruction (Lieberman and Crelin, 1971) is, in fact, purposely weighted toward making the La Chapelle-aux-Saints fossil more like modern man than like an ape. In all cases of doubt, the La Chapelle-aux-Saints supralaryngeal vocal tract reconstruction was modeled on that of the modern human vocal tract. Thus any conclusions that we will draw concerning limits on Neanderthal phonetic ability are conservative.

Note that the geniohyoid muscle in adult *Homo sapiens* runs down and back from the hyoid symphysis of the mandible. This is necessarily the case because the hyoid bone is positioned well below the mandible in adult *Homo sapiens.* The two anterior portions of the digastric muscle, which are not shown in Fig. 2, also run down and back from the mandible for the same reason. When the facets into which these muscles are inserted at the symphysis of the mandible are examined, it is evident that the facets are likewise inclined to minimize the shear forces for these muscles. Shear forces always pose a greater problem than do tensile forces in all mechanical systems since the shear strength of most materials is substantially smaller than the tensile strength. A stick of blackboard chalk, for example, has great tensile strength. It cannot be pulled apart easily if you pull on it lengthwise. However, it has an exceedingly small shear strength, and you can snap it apart with two fingers. The human chin appears to be a consequence of the inclination of the facets of the muscles that run down and back to the hyoid. The outward inclination of the chin in some human populations reflects the inclination of the inferior (inside) plane of the mandible at the symphysis. Muscles are essentially "glued" to their facets. In this light, tubercles and fossae may be simply regarded as adaptions that increase the strength of the muscle-to-bone bond by increasing the "glued" surface area. Their presence or absence is not very critical (DuBrul and Reed, 1960) since the inclination and form of the digastric and geniohyoid facets is the primary element in increasing the functional strength of the muscle-to-bone bond by minimizing shear forces. As Bernard Campbell (1966:2) succinctly notes, "Muscles leave marks where they are attached to bones, and from such marks we assess the form and size of the muscles."

You can easily feel the inclination of the inferior surface of the symphysis of your mandible. Whereas the chin is more prominent in some adult humans than in others, the inferior surface of the mandibular symphysis is always arranged to accommodate muscles that run down and back to a low hyoid position. As DuBrul (1958:42) correctly notes, the human mandible is unique, "The whole lower border of the jaw has swung from a pose leaning inward to one flaring outward." An examination of the collection of skulls at the Musée de l'Homme in Paris indicated that this is true regardless of race and sex for all normal adult humans. When the corresponding features are examined in newborn *Homo sapiens,* it is evident that the nearly horizontal inclination of the facets of the geniohyoid and digastric muscles is a concomitant feature of the high position of the hyoid bone (Negus, 1929; Crelin, 1969; Wind, 1970). These muscles are nearly horizontal with respect to the symphysis of the mandible in newborn *Homo sapiens,* and the facets therefore are nearly horizontal to minimize shear forces. Newborn *Homo sapiens* thus lacks a chin because the inferior surface of the symphysis of the mandible is not inclined to accommodate muscles that run down and back. When the mandible of the La Chapelle-aux-Saints fossil is examined, it is evident that the facets of these muscles resemble those of newborn *Homo sapiens.* The inclination of the styloid process away from the vertical plane is also a concomitant and coordinated aspect of the skeletal complex that supports a high hyoid position in newborn *Homo sapiens* and in the La Chapelle-aux-Saints fossil.

Enough of the base of the La Chapelle-aux-Saints styloid process remains to determine its original approximate size and location.

The skeletal features that support the muscles of the supralaryngeal vocal tract and mandible are all similar in the Neanderthal fossil and newborn *Homo sapiens.* When the bases of the skulls of newborn and adult *Homo sapiens* and the La Chapelle-aux-Saints fossil are examined, it is again apparent that the newborn *Homo sapiens* and the fossil forms have many common features that differ from adult *Homo sapiens.* These differences are all consistent with the morphology of the supralaryngeal airways of newborn *Homo sapiens,* in which the pharynx and the pharyngeal constrictor muscles lie behind the opening of the larynx. In Fig. 3 casts of the supralaryngeal vocal tracts of newborn *Homo sapiens,* the Neanderthal reconstruction, and adult *Homo sapiens* are shown. The details of the reconstruction as well as the general motivating constraints are discussed in Lieberman (1975) and in less detail in Lieberman and Crelin (1971) and Lieberman (1973).

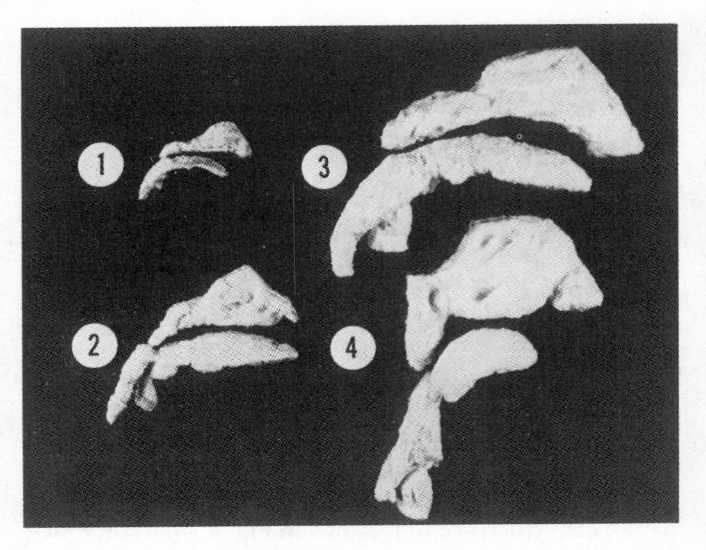

Fig. 3. Casts of the nasal, oral, pharyngeal, and laryngeal cavities of (1) newborn *Homo sapiens,* (2) adult chimpanzee, (3) La Chapelle-aux-Saints reconstruction, and (4) adult *Homo sapiens.* (After Lieberman et al., 1972.)

What are the phonetic consequences with respect to human speech of the reconstructed Neanderthal supralaryngeal vocal tract? Understanding the anatomical basis of human speech requires that we briefly review the source-filter theory of speech production (Chiba and Kajiyama, 1958; Fant, 1960). Human speech is the result of a source or sources of acoustic energy filtered by the supralaryngeal vocal tract. For voiced sounds, that is, sounds like the English vowels, the source of energy is the periodic puffs of air that pass through the larynx as the vocal cords rapidly open and shut. The rate at which the vocal cords open and close determines the fundamental frequency of phonation. Acoustic energy is present at the fundamental frequency and at higher harmonics. The fundamental frequency of phonation can vary about 80 Hz for adult males to about 500 Hz for children and some adult females. Significant acoustic energy is present in the harmonics of fundamental frequency to at least 3,000 Hz. The fundamental frequency of phonation is, within wide limits, under the control of the speaker, who can produce controlled variations by changing either the pulmonary air pressure or the tension of the laryngeal muscles (Lieberman, 1967). Linguistically significant information can be transmitted by means of these variations in fundamental frequency, as, for example, in Chinese, where they are used to differentiate words.

The main source of phonetic differentiation in human language, however, arises from the dynamic properties of the supralaryngeal vocal tract, which acts as an acoustic filter. The length and shape of the supralaryngeal vocal tract determines the frequencies at which maximum energy will be transmitted from the laryngeal source to the air adjacent to the speaker's lips. They are known as formant frequencies. A speaker can vary the formant frequencies by changing the length and shape of his supralaryngeal vocal tract. He can, for example, drastically alter the shape of the airway formed by the posterior margin of his tongue body in his pharynx. He can raise or lower the upper boundary of his tongue in his oral cavity. He can raise or lower his larynx and retract or extend his lips. He can open or close his nasal cavity to the rest of the supralaryngeal vocal tract by lowering or raising his velum. The speaker can, in short, continually vary the formant frequencies generated by his supralaryngeal vocal tract. The acoustic properties that differentiate the vowels [a] and [i], for example, are determined solely by the differences in shape and length that the supralaryngeal vocal tract assumes when these vowels are articulated. The situation is analogous to a pipe organ, where the length and type (open or closed end) of pipe determine the musical quality of each note. The damped resonances of the human supralaryngeal vocal tract are, in effect, the formant frequencies. The length and shape (more precisely the cross-sectional area as a function of distance from the laryngeal source) determine the formant frequencies.

The situation is similar for unvoiced sounds. Here the vocal cords do not open and close at a rapid rate to release quasiperiodic puffs of air, but the source of acoustic energy is the turbulence generated by air rushing through a constriction in the vocal tract. The vocal tract still acts as an acoustic filter but the acoustic source may not be at the level of the larynx; for example, in the sound [s] the source is the turbulence generated near the speaker's teeth.

Computer Modeling

The supralaryngeal vocal tract's filtering properties are completely specified by its shape and size, i.e., its cross-sectional area function. We can therefore determine the constraints that a particular vocal tract will impose on the phonetic repertoire independently of the possible limitations of such things as muscular ability or the properties of the larynx. We could, for example, make models of possible vocal tract shapes by pounding and forming brass tubes with cutters and brazing torches. We could record the actual formant frequencies that corresponded to particular shapes by exciting the tubes with an artificial larynx or a reed. We could thus determine the constraints that the supralaryngeal vocal tract of a Neanderthal fossil placed on the phonetic repertoire independently of the possible further limitations imposed by the extinct hominid's control or lack of control. The only difficulty that would obtain would be in making sure that we had explored the full range of vocal-tract shapes. The acoustic properties of the brass models would closely approximate the filtering properties of the vocal tract shapes they represented. This modeling technique was actually once the principal means of phonetic analysis. The technology of the late eighteenth and early nineteenth centuries was adequate for the fabrication of brass tubes with complex shapes and for making mechanical models. The speech synthesizers devised by Kratzenstein (1780) and von Kempelin (1791) (whose famous talking machine was one of the wonders of the time) generated acoustic signals by exciting tubes by means of mechanical reeds. The method employed by Lieberman and Crelin (1971) and Lieberman et al. (1972) simply makes use of the technology of the third quarter of the twentieth century.

We could, if we wished, continue to use mechanical models to assess the constraints of the supralaryngeal vocal tract on an animal's phonetic repertoire. We could determine the range of possible supralaryngeal vocal tract shapes by dissecting living animals that had similar vocal tracts, making casts of the air passages, and taking note of the musculature, soft tissue, and effects of the contraction of particular muscles.

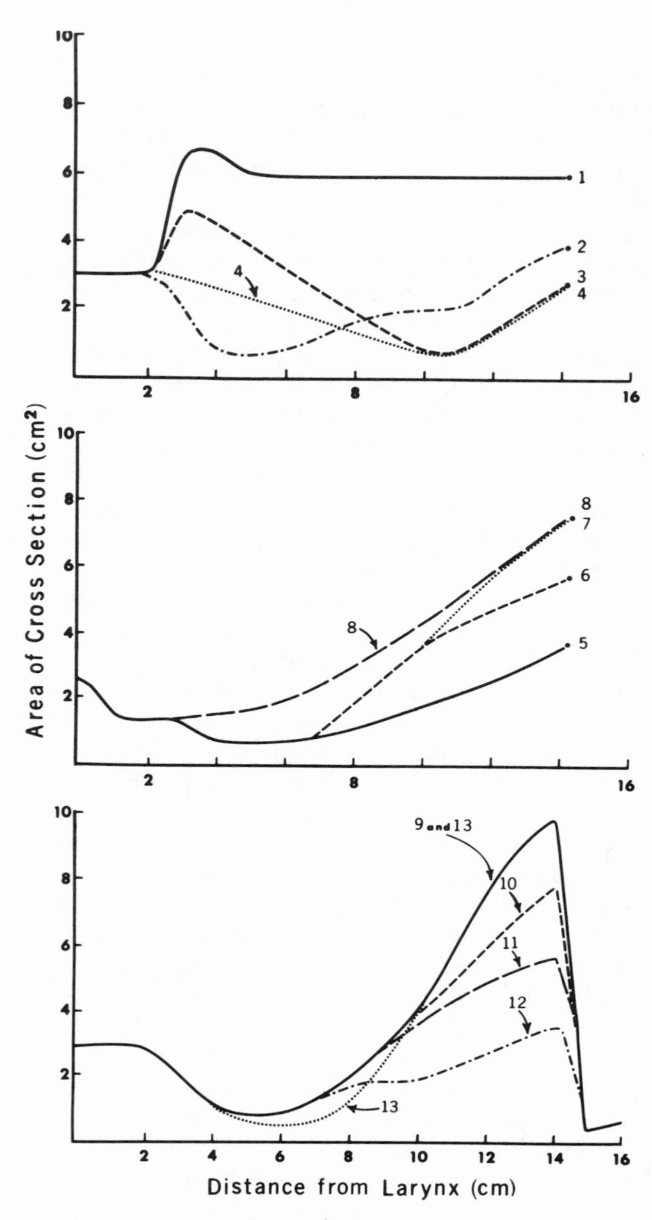

Fig. 4. A. Area functions of the supralaryngeal vocal tract of Neanderthal reconstruction modeled on computer. The area function from 0 to 2 cm is derived from Fant (1960) and represents the distance from the vocal folds to the opening of the larynx into the pharynx. Curve 1 is the unperturbed tract. Curves 2, 3, and 4 represent functions directed toward a "best match" to the human vowel /i/. Curves 5–8 are functions directed toward a best match to /a/, while curves 9–13 are directed toward /u/. (After Lieberman and Crelin, 1971.)

We could enhance our knowledge by making cineradiographs of the animal during episodes of phonation, respiration, and swallowing. It would then be possible, though somewhat tedious, to make models of possible supralaryngeal vocal-tract configurations. The models could even be made of plastic materials that approximated the acoustic properties of flesh. If they were excited by means of a rapid, quasiperiodic series of puffs of air (i.e., an artificial larynx), we would be able to hear the actual sounds that a particular vocal tract configuration produced. If we systematically made models that covered the range of possible vocal-tract configurations we could determine the constraints imposed by supralaryngeal vocal-tract morphology on phonetic repertoire independently of the further possible constraints of the extinct hominids' muscular or neural control, dialect, or habits. We would, of course, be restricted to continuant sounds, i.e., those that were not transient or interrupted, since we could not rapidly change the shape of our vocal tract model. We could, however, generalize our results to consonant-vowel syllables, like the sounds [bI] and [dæ], since we could model the articulatory configurations that occur at specified intervals of time when these sounds are produced.

In Fig. 4, area functions that could be generated by a Neanderthal vocal tract are plotted. These area functions were entered into a computer program (Henke, 1966) that essentially represents the supralaryngeal vocal tract by a series of contiguous cylindrical sections, each of fixed area. Each section can be described by a characteristic impedance and a complex propagation constant, both of which are known quanti-

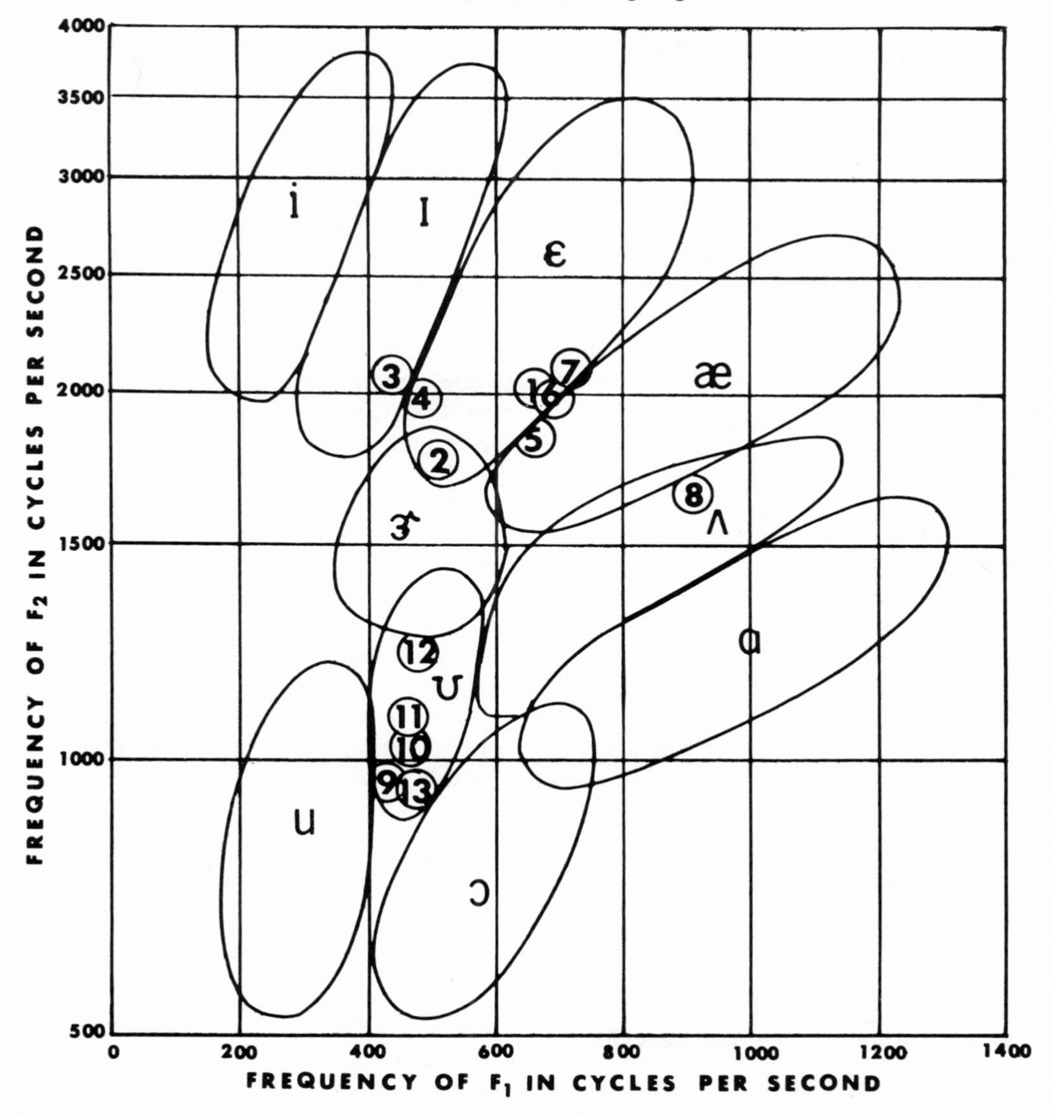

Fig. 4. B. Formant frequencies calculated by computer program for Neanderthal reconstruction.

ties for cylindrical tubes (Beranek, 1954). Junctions between sections satisfy the constraints of continuity of air pressure and conservation of air flow. In other words, the air pressure must be a continuous function, and air particles can neither disappear nor be created at section boundaries. The computer program calculated the three lowest formant frequencies for any area function specified. This arrangement made it possible to enter many area functions in a comparatively short time.

A number of area functions were sketched

into the computer, using its light pen and oscilloscope input system. The area functions plotted in Fig. 4 were directed toward producing the "best" Neanderthal approximations to the human vowels [i], [u], and [a]. The frequency scales and labeled loops are taken from the Peterson and Barney (1952) study of vowel production by adult men, adult women, and older children speaking American English. It is apparent that the reconstructed Neanderthal vocal tract cannot produce the vowels [i], [u], or [a]. Consonants like the dental and bilabial [d], [t], [s], [b], and [p] and other vowels would be phonetic possibilities for the Neanderthal vocal tract, but velar consonants like [g] and [k] would not (Lieberman and Crelin, 1971).

Reconstruction and modeling of the supralaryngeal vocal tract of *Australopitechs africanus* show similar phonetic restrictions (Lieberman, 1973, 1975). In contrast, the reconstructed vocal tract of the Es Skhul V fossil is essentially modern in character and would not restrict the phonetic ability of this hominid. These reconstructions are all the work of Edmund S. Crelin. His results are in accord with the independent univariate and multivariate analyses of Howells (1970, and in press), which demonstrate that Es Skhul V falls within the same class as modern humans, whereas certain measurements of the La Chapelle-aux-Saints fossil vary four to five standard deviations from those of modern skulls.

The Physiology of Human Speech

Since the time of Johannes Müller (1848), who initiated the modern study of the physiology of speech, it has been apparent that some sounds have a more central status than others. The vowels [i], [u], and [a] appear to occur in all human languages. Troubetzkoy (1969) notes that a language may have other vowels but that it always has one or more of these. Recent functional, i.e., physiologic, analyses of human speech show that these vowels really are more useful speech signals than other vowels. Stevens (1972), for example, shows that these vowels are acoustically stable. A speaker can make comparatively large articulatory errors in the production of these sounds without changing their acoustic character. The constricted part of the supralaryngeal vocal tract can, for example, vary over a two-centimeter range without perceptibly changing these vowels' formant frequencies. The formant frequency patterns that define these vowels are, moreover, maximally distinct from all other vowels. The formant frequencies are centered for [a], maximally high for [i], and maximally low for [u]. In effect, these three vowels are the "best" possible vowel sounds for vocal communication.

Natural selection would act to retain mutations that allowed these signals to be produced only if enhanced vocal communication were an advantage. The specialized anatomy that allows hominids like modern *Homo sapiens* and Es Skhul V (who is functionally modern) to produce these sounds is less suited for breathing, swallowing, and chewing than the vocal tract anatomy of Neanderthal and Australopithecine hominids and nonhuman primates. The human supralaryngeal airway, in which the pharynx is part of the direct path from the lungs, allows the tongue body to be shifted up, down, and back to form the abrupt discontinuities in the supralaryngeal area function that are necessary to produce vowels like [a], [i], and [u]. Other primates, who lack this anatomical complex, essentially have a "single" tube vocal tract, formed simply by the oral cavity, and the larynx and pharynx open independently into the oral cavity. This arrangement offers less resistance to air flow and allows respiration to go on when the oral cavity is full of fluid; the epiglottis can seal the oral cavity from the nasal-laryngeal pathway. In modern *Homo sapiens* the pharynx serves both as part of the air pathway and as part of the pathway for the ingestion of food. The adult human epiglottis cannot seal the oral cavity, and food lodged in the pharynx can block the air flow to the lungs. The only function for which the human supralaryngeal vocal tract is better adapted is speech production (Lieberman and Crelin, 1971; Lieberman et al., 1972; Negus, 1929; Kirchner, 1970; Lieberman, 1973, 1975).

Speech communication must have existed in late hominid forms like Neanderthal man. The mutations that yield enhanced phonetic ability in modern *Homo sapiens* would not have been retained unless vocal communication was already an established phonetic mode of language. However, there is an additional physiologic factor, which is related to the encoded nature of human speech, that would have resulted in strong selectional pressures for the retention of the mutations that allowed the vowels [a], [i], and [u] to be produced.

Speech Encoding and Decoding

Modern human speech communication achieves a high rate of speed by a process of speech encoding and a complementary process of speech decoding. Phonetic distinctions that differentiate meaningful words, e.g., the sounds symbolized by [b], [æ], and [t] in the word *bat,* are transmitted, identified, and stored at a rate of 20–30 segments per sec. It is obvious that human listeners cannot transmit and identify these sound distinctions as separate entities. The fastest rate at which sounds can be identified is about 7–9 segments per sec (Liberman, 1970). Sounds transmitted at a rate of 20 per sec merge into an undifferentiable "tone." That is why high-fidelity amplifiers and loudspeakers generally have an advertised lower frequency limit of 20 Hz. The human auditory system simply cannot temporally resolve auditory events that occur at a rate of 20 per sec. (The human visual system, incidentally, cannot work any faster either. A motion picture projector presents individual still frames at rates in excess of 16 frames per sec.) The linguist's traditional conception of phonetic elements as a set of "beads on a string" clearly is not correct at the acoustic level. How, then, is speech transmitted and perceived?

The answer to this question comes from work that was originally directed at making a reading machine for the blind. The machine was to identify alphabetic characters in printed texts and convert them into sounds that a blind person could listen to. It was not too difficult to devise a print-reading device, although that was not really necessary if the machine's use was to be initially restricted to the "reading" of new books and publications. At some stage in the preparation of a publication a machine with a keyboard is used. The talking machine could be connected to the keyboard so that it produced a different sound, or combination of sounds, for each typewriter or linotype key. The sequence of sounds could then be tape-recorded, and blind people could then listen to the recordings after the tapes were perhaps slowed down and edited to eliminate pauses and errors. A number of different systems were developed, but all of them were useless because the tapes had to be slowed down to rates about one-tenth that of normal human speech. The blind "readers" would forget what a sentence was about before they heard its end. It did not matter what sorts of sounds were connected to the typewriter keys. They all were equally bad. The basic rate of transmission and the inherent difficulty of these systems were about the same as listening to the traditional dots and dashes of the telegrapher's Morse Code. The systems would work, but they were very, very slow, and the listeners had to expend most of their attention simply keeping track of the message.

The obvious solution to this problem seemed to rest in making machines that would "glue" the phonetic elements of speech together to make words. There seemed to be no inherent problem if the linguists' traditional beads on a string were isolated, collected, and then appropriately strung together. The medium of tape recording seemed to be the solution. Carefully pronounced test words could be recorded, and the phonetic elements of these words could then be isolated by cutting up the magnetic tape (preferably by segmenting the tape with the electronic equivalent of a pair of scissors). The speaker, for example, would record a list of words that included *pet, bat, cat, hat,* etc. The experimenters would then theoretically be able to isolate the sounds [p], [b], [h], [k], [e], [æ], which would then be stored in a machine that could put them together in different patterns to form new words, for example, *get* and *pat.* The list of possible permutations would, of course, increase as the vocabulary of isolated stored phonetic elements increased. Systems of this sort were implemented at great expense and with enormous efforts (Peterson et al., 1958). They surprisingly produced speech that was scarcely intelligible. Despite many attempts to improve the technique by changing the methods used in isolating the phonetic elements, the system proved to be completely useless.

Though these studies failed to produce a useful reading machine, they demonstrated that phonetic elements could not be regarded as beads on a string. It was, in fact, impossible to isolate a consonant like [b] or [t] without also hearing the vowels that either preceded or followed it. It is in fact impossible to produce a stop consonant like [p] or [b] without pronouncing a vowel. The smallest segment of speech that can be pronounced is the syllable. If you try to say the sound [b] you will discover that it is impossible. You can say [bi], [bu], [bU], [ba], [bI], [bæ], [bIt], [bId], etc., but you cannot produce an isolated [b]. The results of the past twenty years of research on the perception of speech by humans demonstrate that individual sounds like [b], [I], and [d] are encoded, that is, "squashed together" into a single unit when we produce the syllable-sized unit [bIt] (the phonetic transcription of the English word *bit*). A human speaker in producing this syllable starts with his supralaryngeal vocal tract, i.e., his tongue, lips, velum, etc., in the positions characteristic of [b]. However, he does not maintain this articulatory configuration but instead moves his articulators toward the positions that would be attained if he were instructed to maintain an isolated, steady [I]. He never reaches these positions, however, because he starts toward the articulatory configuration characteristic of [t] before he ever reaches the "steady state" (isolated and sustained) vowel [I]. The articulatory gestures that would be characteristic of each isolated sound are never attained. Instead, the articulatory gestures are melded together into a composite, characteristic of the syllable.

The sound pattern that results from this encoding process is an indivisible composite. Just as there is no way of separating with absolute certainty the [b] articulatory gestures from the [I] gestures (you cannot tell exactly when the [b] ends and the [I] begins), there is no way of separating the acoustic cues that are generated by these articulatory maneuvers. The isolated sounds have a psychological status as motor control or "programming" instructions for the speech-production apparatus. The sound pattern that results is a composite, and the acoustic cues for the initial and final consonants are largely transmitted as modulations imposed on the vowel. The process is, in effect, a time-compressing system. The acoustic cues that characterize the initial and final consonants are transmitted in the time slot that would have been necessary to transmit a single, isolated [I] vowel.

The human brain decodes, that is, "unscrambles" the acoustic signal in terms of the articulatory maneuvers that were put together to generate the syllable. The individual consonants

[b] and [t], though they have no independent acoustic status, are perceived as discrete entities. The process of human speech perception inherently requires knowledge of the acoustic consequences of the possible range of human supralaryngeal vocal tract speech articulation and the size of the supralaryngeal vocal tract that produced the speech signal that is being decoded. A number of independent studies (Ladefoged and Broadbent, 1957; Rand, 1971; Nearey, 1975) have demonstrated that a human listener will interpret an identical acoustic stimulus as a different speech sound, e.g., the same acoustic signal may be "heard" as a token of the vowel [I], [æ], or [e]. The listener will perceive the sound as an [I] if he thinks the vocal tract he is listening to is large. If he thinks the vocal tract is small, he might perceive the same acoustic signal as a token of the vowel [æ].

Listeners can arrive at an estimate of the size of the vocal tract in several ways. They can listen to a stretch of speech and take note of the average range of formant frequencies. Larger and longer vocal tracts will tend to produce lower formant frequencies. A listener, however, can estimate the size of a vocal tract almost instantly if he knows what sound the speaker intended to make. The vowels [i], [u], and [a] have special acoustic properties that make them especially suited for this "vocal tract size calibrating" function (Lieberman, 1973, 1975). The formant frequency patterns that define vowels like [i] are "determinate," and listeners (or computer programs) can make use of these vowels to calibrate vowel perception (Nearey, 1975; Gerstman, 1967).

The absence of vowels like [i], [u], and [a] in the phonetic repertoire of Australopithecine and Neanderthal hominids is thus significant. Sounds that are inherently optimum signals for vocal communication and that facilitate fully encoded, rapid speech are absent. However, this deficiency cannot be taken as an indication either of the absence of vocal communication or of the total absence of encoding in the speech of these earlier hominids. The vocal-tract anatomy of present-day apes, for example, though it presents limits on the total phonetic repertoires of these animals, can produce many of the sound contrasts that convey meaningful information in human speech, i.e., the "phonetic features" that occur in human languages (Lieberman, 1973, 1975). Our knowledge of the vocal communication of living apes is rudimentary. We know virtually nothing about the perceptual factors that may structure their vocal communication, nor do we really have sufficient data that relate particular behavioral situations with the total vocal and gestural communicative output of living apes. The speech-producing anatomy of apes can be viewed as a factor that inherently sets an upper limit on the phonetic repertoires of these animals. It, however, would allow the production of at least the following phonetic features.

I shall start by discussing phonetic features that involve the laryngeal source. As Negus (1929) observed, as we ascend the phylogenetic scale in terrestrial animals, there is a continual elaboration of the larynx, which reflects, in part, adaptions for phonation. Studies like that of Kelemen (1948), which have attempted to show that chimpanzees cannot talk because of laryngeal deficiencies, are not correct. Kelemen shows that the chimpanzee's larynx is different from the larynx of a normal adult human male and will not produce the same range of fundamental frequencies; and, moreover, the spectrum of its glottal source will be different from that of a normal adult human male. The chimpanzee's voice thus would sound "harsh"—to a human listener! However, human listeners do not really count with regard to chimpanzees! Chimpanzees and other hominoids and New and Old World monkeys probably could produce the following phonetic features by making use of laryngeal and subglottal articulatory maneuvers.

VOICED VERSUS UNVOICED

The supralaryngeal vocal tract could be excited either by the quasi-periodic excitation of the larynx or by means of noise generated by air turbulence. Air turbulence will occur whenever the flow of air exceeds a critical value at any point in the vocal tract. During phonation the vocal cords are adducted, i.e., moved together and closed or nearly closed, and the flow of air through the larynx is relatively low. In humans, turbulent noise generally does not occur during the production of voiced vowels, i.e., vowels produced with normal phonation. In the production of a sound like [s], the vocal cords are in a more open position. The resulting air flow is much higher (Klatt et al., 1968), and noise is generated at the "dental" constriction orifice. It is clear that nonhuman primates can produce sounds that are either voiced or unvoiced (Lieberman, 1975).

HIGH FUNDAMENTAL VERSUS NORMAL FUNDAMENTAL FREQUENCY

Several studies (Van den Berg, 1960; Atkinson, 1973) have shown that the human larynx can be adjusted so that phonation occurs in the falsetto register, in which the fundamental frequency of phonation is higher than in the normal register. The mode of operation of the larynx is actually somewhat different in these two registers (Van den Berg, 1960; Lieberman, 1967). The spectrum of the glottal source also changes in falsetto, and comparatively little energy occurs at higher frequency harmonics of the fundamental. The larynges of nonhuman primates inherently should be capable of producing this distinction (Negus, 1929; Wind, 1970).

LOW FUNDAMENTAL VERSUS NORMAL FUNDAMENTAL FREQUENCY

The human larynx likewise may be adjusted to phonate at a low fundamental frequency. This lower register, termed "fry," produces very low fundamental frequencies (Hollien et al., 1966), which are very irregular (Lieberman, 1963).

DYNAMIC FUNDAMENTAL FREQUENCY VARIATIONS

Virtually all human languages make use of dynamic variations in the temporal pattern of fundamental frequency (Lieberman, 1967). In languages like Chinese, dynamic tone patterns, i.e., rapid changes in fundamental frequency, differentiate words. The spectrograms of chimpanzee and gorilla vocalizations often show fundamental frequency variations that could serve as the basis of phonetic features based on dynamic fundamental frequency variations (Lieberman, 1968). Vocalizations could be differentiated by means of rising or falling patterns or combinations of rising and falling contours with high or low fundamental frequencies.

STRIDENT LARYNGEAL OUTPUT

The high fundamental frequency cries mixed with breathy, i.e., noise excitation that can be observed in the spectrograms and oscillograms of the vocalizations of nonhuman primates and newborn humans, constitute a phonetic feature (Lieberman, 1975). Speakers of American English sometimes make use of this phonetic feature to convey emotional qualities. It does not have a strictly "linguistic" role in American English, since it is not used to convey different words, but that is not a crucial objection to our noting the possible use of this sound contrast as a phonetic feature. Many sound contrasts that serve as phonetic features in other languages are not used in English.

PHONATION ONSET

The anatomical basis of the phonetic feature of Phonation Onset rests in the independent nature of the laryngeal source and the supralaryngeal vocal tract. All primates thus can, in principle, make use of this sound contrast, which differentiates sounds like [b] and [p], as a phonetic feature.

STOP

Phonation Onset obviously involves articulatory maneuvers in the supralaryngeal vocal tract, which must be occluded to produce sounds like the English "stops" [p] and [b]. All primates inherently can do so to produce the phonetic feature Stop.

CONSONANTAL "PLACE OF ARTICULATION"

The point at which the supralaryngeal vocal tract can be occluded can vary. All primates can close their vocal tracts by moving their lips together. Thus a bilabial point of articulation is a possibility for all primates. A dental point of occlusion or constriction is effected in adult *Homo sapiens* by moving the tongue up toward the hard palate, and cineradiographic studies of the swallowing movements of newborn *Homo sapiens* (Truby, Bosma, and Lind, 1965) indicate that a dental point of articulation is a possibility for all primates. In all primates the supralaryngeal vocal tract can also be occluded at the level of the glottis, i.e., at the level of the vocal cords. This follows from one of the surviving, basic, vegetative functions of the larynx, which can close to protect the lungs from the intrusion of foreign material. A glottal point of articulation is thus a possibility for all primates.

A chimpanzee, therefore, has the speech-producing anatomy that would, *with the proper muscular controls,* be sufficient to allow the production of the English sounds [b], [p], [t], [d], the glottal stop [?], as well as prevoiced dental and bilabial stops like those that occur in Spanish. Glottal stops normally are not used to differentiate words in English, though they occur in many English dialects; they are used more extensively in many other languages, e.g., Danish. It is important to note that the phonetic feature of consonantal point of articulation is a multivalued feature and that we are simply discussing the upper bounds set by the gross anatomy of primates. An animal would have to possess the neural and muscular control necessary to position the tongue against the palate during speech at a precise moment if the dental point of articulation were to be realized.

CONTINUANT VERSUS INTERRUPTED

Sounds may be differentiated either by being prolonged without interruptions or by being interrupted. This phonetic feature can be effected either by direct modulation control of the laryngeal muscles to start and stop phonation or by occluding the supralaryngeal vocal tract.

FORMANT FREQUENCY RAISING

All primates can shorten the length of their supralaryngeal vocal tracts. They can shorten it at its "front" end by flaring and/or pulling their lips back, and adult *Homo sapiens* can shorten it at its "back" end by pulling the larynx upward as much as 20 mm during the course of a single word (Perkell, 1969). The mobility of the larynx is comparatively restricted in newborn *Homo sapiens* (Truby, Bosma, and Lind, 1965; Negus, 1929) and in nonhuman primates (Negus, 1949). The reduction in laryngeal mobility follows both from the position of the larynx with respect to the supralaryngeal vocal tract and from the fact that the hyoid bone is very close to the thyroid cartilage (Negus, 1929). The reduction in laryngeal mobility in forms other than adult humans can be observed in radiographic pictures of both speech and swallowing (Negus, 1929; Truby, Bosma, and Lind, 1965). During swallowing, for example, the larynx moves upward and forward in adult humans, whereas it only moves forward in newborn humans.

The acoustic consequences of shortening the supralaryngeal vocal tract—irrespective of the articulatory maneuvers that effect the shortening—is a rising formant frequency pattern.

FORMANT FREQUENCY LOWERING

All primates can also lengthen their supralaryngeal vocal tracts by protruding their lips or by moving their larynges downward or backward. Adult *Homo spaiens* again has more freedom in this regard since the human larynx has greater mobility. Closing the lips to produce a smaller orifice at the mouth has the same acoustic effect as increasing the length of the supralaryngeal vocal tract (Stevens and House, 1955; Fant, 1960). All these articulatory maneuvers generate a falling formant frequency pattern. In human speech formant transitions are the normal case. They may be rarer in the acoustic signals of nonhuman primates.

ORAL VERSUS NONORAL

Nonhuman primates can produce cries in which their oral cavities are closed by the epiglottis while the nose remains open.

AIR SAC VARIATIONS

Some nonhuman primates, e.g., howling monkey, have large air sacs above their vocal cords that can act as variable acoustic filters as their volume changes. The vocalizations of primates with these air sacs have not yet been subjected to quantitative acoustic analysis. Their calls, however, appear to be differentiated by modulations introduced by the air sacs.

The Uniqueness of Speech Encoding

Although the speech of modern *Homo sapiens* is fully encoded, the vocal communications of any animal that can produce formant transitions could be partially encoded. Since even early fossil hominids like *Australopithecus africanus* had supralaryngeal vocal tracts that were equivalent to those of modern apes, partial speech encoding may have existed at a very early period of hominid evolution. It is even possible that the communications of living nonhuman primates are partially encoded. The acoustic basis of speech encoding rests in the fact that the pattern of formant frequency variation of the supralaryngeal vocal tract must inherently involve transitions. The shape of the supralaryngeal vocal tract cannot change instantaneously. If a speaker utters a syllable that starts with the consonant [b] and ends with the vowel [æ], his vocal tract must first produce the shape necessary for [b] and then gradually move toward the [æ] shape. Formant transitions thus have to occur in the [æ] segment that reflects the initial [b] configuration. The transitions would be quite different if the initial consonant were a [d].

The nonhuman supralaryngeal vocal tract can, in fact, produce consonants like [b] and [d]. Simple encoding could be established using only bilabial and dental consonant contrasts. The formant transitions would either all be rising in frequency in the base of [dæ] or falling in frequency for [bæ]. It probably would be quite difficult, if not impossible, to sort the various intermediate vowels contrasts that are possible with the nonhuman vocal tract, but a simple encoding system could be built up using rising and falling formant transitions imposed on a general, unspecified vowel [V]. The resulting language would have only one vowel (a claim that has sometimes been made for the supposed ancestral language of *Homo sapiens:* Kuipers, 1960). The process of speech encoding and decoding and the elaboration of the vowel repertoire could build on vocal-tract normalization schemes that made use of sounds like [s] and could provide a listener or a digital computer program with information about the size of the speaker's vocal tract. Vocal-tract normalizing information could also be derived by listening to a fairly long stretch of speech and then computing the average formant frequency range. The process would be slower than simply hearing a token of [i] or [u], but it would be possible. There might have been a gradual path toward more and more encoding for all hominid populations as social structure and technology became more complex. If this were true, the preadaption of the bent pharyngeal-oral supralaryngeal vocal tract in some hominid populations would have provided an enormous selective advantage.

Conclusion

The differences between human speech and human language and the communication systems of other animals may not be qualitative. It is difficult to think of any aspect of human behavior that is really unique. Although language is seemingly a unique aspect of human behavior, qualitatively different from the communication system of any other living animal, the difference may be only a quantitative phenomenon. Qualitative behavioral differences can be the result of quantitative structural differences. Both an electronic desk calculator and a large general-purpose digital computer may be constructed using similar circuits and memory devices. However, the distinctions between the problems that can be solved using one device or the other will be qualitative as well as quantitative.

The differences between human and animal communication are more obvious because the intermediate stages of hominid evolution are no longer alive. It is possible that there are some qualitative differences, insofar as no other living species can presently make use of encoded acoustic signals. However, we have not examined the possible "speech" system of any other living animal sufficiently to demonstrate the absence of encoding. Quantitative acoustic analysis is still in its infancy, and we have still to develop a physiologic, i.e., functional, theory that explains the nature of human speech and human language. The study of the communications of species other than modern *Homo sapiens* is just as important for the insights we may gain into the nature of human language and its evolution as for our understanding of specific systems of animal communication.

References

Atkinson, J. R., 1973. Aspects of intonation in speech: implications from an experimental study of fundamental frequency. Ph.D. diss., University of Connecticut.

Basmajian, B. J., and Tuttle, R., in press. In: *Proceedings of the IXth International Congress of Anthropological and Ethnological Science,* Chicago. The Hague: Mouton.

Beck, B. B., in press. Primate tool behavior. In: *Proceedings of the IXth International Congress of Anthropological and Ethnological Science,* Chicago. The Hague: Mouton.

Beranek, L. L., 1954. *Acoustics.* New York: McGraw-Hill.

Bogert, C. M., 1960. The influence of sound on the behavior of amphibians and reptiles. In: *Animal Sounds and Communication,* W. E. Lanyon and W. N. Tavolga, eds. Arlington, Va.: American Institute of Biological Sciences.

Boule, M., 1911–1913. L'homme fosille de la Chapelle-aux-Saints. *Ann. Paleontol.,* 6:109; 7:21, 85; 8:1.

Campbell, B., 1966. *Human Evolution: An Introduction to Man's Adaptions.* Chicago: Aldine.

Capranica, R. R., 1965. *The Evoked Vocal Response of the Bullfrog.* Cambridge: M.I.T. Press.

Chiba, T., and Kajiyama, M., 1958. *The Vowel: Its Nature and Structure.* Tokyo: Phonetic Society of Japan.

Crelin, E. S., 1969. *Anatomy of the Newborn: An Atlas.* Philadelphia: Lea and Febiger.

Darwin, C., 1859. *On the Origin of Species,* facsimile ed. New York: Atheneum.

DuBrul, E. L., 1958. *Evolution of the Speech Apparatus.* Springfield, Ill.: Charles C. Thomas.

DuBrul, E. L., and Reed, C. A., 1960. Skeletal evidence of speech? *Amer. J. Phys. Anthropol.,* 18:153–56.

Eimas, P. D., and Corbitt, J. D., 1973. Selective adaption of linguistic feature detectors. *Cog. Psychol.,* 4:99–109.

Fant, G., 1960. *Acoustic Theory of Speech Production.* The Hague: Mouton.
Ferrein, C. J., 1741. *Mem. Acad. Paris,* 409–32 (Nov. 15).
Fouts, R. S., 1973. Acquisition and testing of gestural signs in four young chimpanzees. *Science,* 180:978–80.
Frishkopf, L. S., and Goldstein, M. H., Jr., 1963. Responses to acoustic stimuli from single units in the eighth nerve of the bullfrog. *J. Acoust. Soc. Amer.,* 35:1219–28.
Gardner, R. A., and Gardner, B. T., 1969. Teaching sign language to a chimpanzee. *Science,* 165:664–72.
Gerstman, L., 1967. Classification of self-normalized vowels. In: *Proceedings of IEEE Conference on Speech Communication and Processing.* New York: IEEE, pp. 97–100.
Green, S., 1973. Physiological control of vocalizations in the Japanese monkey: inferences from a field study. *J. Acoust. Soc. Amer.,* 53:310 (abstract).
Henke, W. L., 1966. Dynamic articulatory model of speech production using computer simulation. Ph.D. diss., Massachusetts Institute of Technology, Appendix B.
Hill, J. H., 1972. On the evolutionary foundations of language. *Amer. Anthropologist,* 74:308–17.
Hollien, H.; Moore, P.; Wendahl, R. W.; and Michel, J. F.; 1966. On the nature of vocal fry, *J. Speech Hearing Res.,* 9:245–47.
Howells, W. W., 1970. Mount Carmel man: morphological relationships. In: *Proceedings of the VIIIth International Congress of Anthropological and Ethnological Science,* Tokyo. Tokyo: Science Council of Japan. Vol. 1: *Anthropology,* pp. 269–72.
Howells, W. W., in press. Neanderthal man: facts and figures. In: *Proceedings of the IXth International Congress of Anthropological and Ethnological Science,* Chicago. The Hague: Mouton.
Kelemen, G., 1948. The anatomical basis of phonation in the chimpanzee. *J. Morphol.,* 82:229–56.
Kempelen, W. R. von, 1791. *Mechanismum der menschlichen Sprache nebst der Beschreibung seiner sprechenden Maschine.* J. B. Degen.
Kirchner, J. A., 1970. *Pressman and Kelemen's Physiology of the Larynx,* rev. ed. Washington, D.C.: American Academy of Ophthalmology and Otolaryngology.
Klatt, D. H.; Stevens, K. N.; and Mead, J.; 1968. Studies of articulatory activity and airflow during speech. *Ann. N.Y. Acad. Sci.,* 155:42–54.
Kratzenstein, C. G., 1780. Sur la naissance de la formation des voyelles. *J. Phys. Chim. Hist. Nat. Arts,* 21 (1782):358–81 (translated from *Acta Acad. Petrograd,* 1780).
Kuipers, A. H., 1960. *Phoneme and Morpheme in Kabardian.* The Hague: Mouton.
Ladefoged, P., and Broadbent, D. E., 1957. Information conveyed by vowels. *J. Acoust. Soc. Amer.,* 29:98–104.
Lane, H., 1965. Motor theory of speech perception: a critical review. *Psychol. Rev.,* 72:175–309.
Leakey, M. D., 1971. *Olduvai Gorge,* Vol. III. Cambridge: Cambridge University Press.
Lenneberg, E. H., 1967. *Biological Foundations of Language.* New York: Wiley.
Liberman, A. M., 1970. The grammars of speech and language. *Cog. Psychol.,* 1:301–23.
Lieberman, P., 1963. Some acoustic measures of the periodicity of normal and pathologic laryngea. *J. Acoust. Soc. Amer.,* 35:334–53.
Lieberman, P., 1967. *Intonation, Perception, and Language.* Cambridge: M.I.T. Press.
Lieberman, P., 1968. Primate vocalizations and human linguistic ability. *J. Acoust. Soc. Amer.,* 44:1574–84.
Lieberman, P., 1970. Towards a unified phonetic theory. *Linguist. Inquiry,* 1:307–22.
Lieberman, P., 1973. On the evolution of human language: a unified view. *Cognition,* 2:59–94.
Lieberman, P., 1975. *On the Origin of Language.* New York: Macmillan.
Lieberman, P., and Crelin, E. S., 1971. On the speech of Neanderthal man. *Linguist. Inquiry,* 2:203–22.
Lieberman, P.; Crelin, E. S.; and Klatt, D. H.; 1972. Phonetic ability and related anatomy of the newborn, adult human, Neanderthal man, and the chimpanzee. *Amer. Anthropologist,* 74:287–307.
Lisker, L., and Abramson, A. S., 1964. A cross-language study of voicing in initial stops: acoustical measurements. *Word,* 20:384–422.
Miller, J. M.; Sutton, D.; Pfingst, B.; Ryan, A.; and Beaton, R.; 1972. Single cell activity in the auditory cortex of rhesus monkeys: behavioral dependency. *Science,* 177:449–51.
Müller, J., 1848. *The Physiology of the Senses, Voice and Muscular Motion with the Mental Faculties,* W. Baly, trans. London: Walton and Maberly.
Nearey, T., 1975. Phonetic features for vowels. Ph.D. diss., University of Connecticut.
Negus, V. E., 1929. *The Mechanism of the Larynx.* New York: Heinemann.

Perkell, J. S., 1969. *Physiology of Speech Production: Results and Implications of a Quantitative Cineradiographic Study.* Cambridge: M.I.T. Press.

Peterson, G. E., and Barney, H. L., 1952. Control methods used in a study of the vowels. *J. Acoust. Soc. Amer.,* 24:175–84.

Peterson, G. E.; Wang, W. S.-Y.; and Sivertsen, E.; 1958. Segmentation techniques in speech synthesis. *J. Acoust. Soc. Amer.,* 30:739–42.

Pilbeam, D., 1972. *The Ascent of Man: An Introduction to Human Evolution.* New York: Macmillan.

Premack, D., 1972. Language in chimpanzee? *Science,* 172:808–22.

Preston, M., and Port, D., 1972. Early apical stop production. A voice onset time analysis. *Haskins Laboratories Status Reports* (New Haven, Ct.), SR 29/30:125–49.

Rand, T. C., 1971. Vocal tract normalization in the perception of stop consonants. *Haskins Laboratories Status Reports* (New Haven, Ct.), SR 25/26:141–46.

Stevens, K. N., 1972. Quantal nature of speech. In: *Human Communication: A Unified View,* E. E. David and P. B. Denes, eds. New York: McGraw-Hill.

Stevens, K. N.; Bastide, R. P.; and Smith, C. P.; 1955. Electrical synthesizer of continuous speech. *J. Acoust. Soc. Amer.,* 27:207.

Stevens, K. N., and House, A. S., 1955. Development of a quantitative description of vowel articulation. *J. Acoust. Soc. Amer.,* 27:484–93.

Straus, W. L., Jr., and Cave, A. J. E., 1957. Pathology and posture of Neanderthal man. *Quart. Rev. Biol.,* 32:348–63.

Troubetzkoy, N. S., 1969. *Principles of Phonology,* C. Baltaxe, trans. Berkeley: University of California Press.

Truby, H. M.; Bosma, J. F.; and Lind, J.; 1965. *Newborn Infant Cry.* Stockholm: Almqvist and Wiksell.

Vallois, H. V., 1961. The evidence of skeletons. In: *Social Life of Early Man,* S. L. Washburn, ed. Chicago: Aldine.

Van den Berg, J. W., 1960. Vocal ligaments versus registers. *Curr. Probl. Phoniat. Logoped.,* 1:19–34.

van Lawick-Goodall, J., 1973. Cultural elements in a chimpanzee community. In: *Symposia of the Fourth International Congress of Primatology,* vol. 1. White Plains, N.Y.: Karger.

Vlcek, E., 1970. Etude comparative onto-phylogenetique de l'enfant du Pech-de-l'Aze par rapport a d'autres enfants neandertaliens. In: *L'enfant du Pech-de-l'Aze,* D. Feremback et al., eds. Paris: Masson, pp. 149–86.

Washburn, S. L., 1968. *The Study of Human Evolution.* Portland: Oregon State System of Higher Education.

Wind, J., 1970. *On the Phylogeny and Ontogeny of the Human Larynx.* Groningen: Wolters-Noordhoff.

Wollberg, Z., and Newman, J. D., 1972. Auditory cortex of squirrel monkey: response patterns of single cells to species-specific vocalizations. *Science,* 175:212–14.

Chapter 2

EXPANDING HORIZONS IN ANIMAL COMMUNICATION BEHAVIOR

Donald R. Griffin

During the past thirty years we have been repeatedly surprised by discoveries about animal behavior, especially in the area of orientation and communication. The behavioristic tradition of Jacques Loeb (1964), John B. Watson (1929), and C. Lloyd Morgan (1894) had not prepared us for the possibilities that grew in this period from hesitant speculations to well-accepted facts. (For concise reviews of the historical background for current ideas about animal and human behavior see Klein, 1970, and Klopfer and Hailman, 1967.) Outstanding examples of eye-opening discoveries about orientation behavior are: (1) The orientation of birds by means of the sun or stars with at least approximate compensation for the apparent motions of these heavenly bodies across the sky (Matthews, 1968; Emlen, 1975). (2) The ability of bats and two species of cave-dwelling birds to orient rapid and agile flight through darkness by echolocation, including the ability of insectivorous bats to intercept flying insects by echolocation (Griffin, 1958; Griffin, Webster, and Michael, 1960; Simmons, Howell, and Suga, 1975). (3) The evolution in several different groups of insects of specialized auditory organs that warn them of approaching bats in time for successful evasive maneuvers (Roeder, 1970). (4) The orientation of certain electric fishes by sensing with their lateral line receptors changes in the electric fields produced by their own electric organs or by bioelectric potentials from other animals (Lissmann, 1958; Lissmann and Machin, 1958; Kalmijn, 1971; Bullock, 1973). (5) The ability of honeybees and many other invertebrate animals to orient themselves by the polarization patterns of the blue sky (von Frisch, 1950, 1967).

Comparable surprises have come from research on the communication behavior of animals. These were reviewed in *Animal Communication* (1968), and other chapters of the present volume bring the whole subject up to date. Of all these new discoveries about animal communication the two most impressive and significant are the "dance language" of honeybees (von Frisch, 1923, 1946, 1967, 1972, 1974), discussed along with the communication behavior of other insects in the chapter by Hölldobler, and the recent experiments on gestural communication with chimpanzees, reviewed in the chapter by Fouts. This improved understanding of communication among animals has profound implications that call into question some of our basic assumptions about the relationship between men and animals. This chapter will discuss some of these implications and advance tentative suggestions for their future analysis.

The waggle dances of honeybees are coded gestures used most commonly to signal the location of a food source, but they are also used under special conditions to convey information about the location of other things required by the mutually interdependent colony of bees. The dances specify not only quantitatively the distance and direction but also qualitatively the desirability of what a scout bee has located. The degree of desirability may reflect the sugar concentration of the nectar, or, as discussed below, other desiderata. It is not adequately appreciated that these communicative dances are not always exhibited in rigid fashion. When food is plentiful the returning foragers often do not dance at all. Odors are always important in recruiting the worker bees to newly available food sources: the odors of nectar and pollen and other odors from the area where food was gathered that are carried back by the dancing bee, as well as scents secreted by special glands that are often used to mark food sources. These odors always help to direct recruits to the exact location of new food sources, and often they appear to be sufficient. Sounds or vibrations are also involved, at least in conveying the arousal level of the dancing bee. Independent searching by individual foragers also seems to be adequate under many conditions when food is plentiful. Thus the dance communication system is called into play primarily when a colony of bees is in great need of food or something else.

It is important to recognize, as many readers of von Frisch's early writings did not, that the system is not rigidly linked to one kind of material needed by the colony. The same signaling system is used for such different things as nectar, pollen, water, and resinous exudations from plants (propolis) that are used to repair portions of the honeycomb. But most significant of all, scout bees from a swarm that has moved outside its former cavity use the same coded gestures to convey to other bees the locations of cavities suitable to serve as future locations for the entire colony (Lindauer, 1955). As in other situations where the dances are employed, the response of potential recruits varies with the vigor and intensity of the dance. Scout bees that have found a relatively poor cavity dance with reduced intensity and sometimes also pay attention to other, more vigorous dances. They may be strongly enough influenced to fly out to the cavity whose location has been indicated by the more intense dance of another scout, and after visiting this cavity may dance in a pattern and with an intensity indicating both its location and its quality. Thus the same bee can alternate within a short period of time between acting as a transmitter and a receiver of information, using the same gestural code.

The discovery of such complex communication behavior would have been surprising enough among mammals, but to find it in an insect was truly revolutionary. It is one thing for one of our closest phylogenetic relatives to share some of our basic patterns of communication behavior and be able to learn others, but bees are as distantly related to mammals as any animal; our closest common ancestors lived at least 500 million years ago. The use of symbolic communication by insects thus implies that flexible communication behavior is not restricted to any one phylogenetic group of animals. This discovery was such a surprising one that, as von Frisch (1972) so succinctly put the matter, "No good scientist should believe such a thing on first hearing." Many were skeptical at first, but convinced themselves by repeating von Frisch's basic observations that the properties of the waggle dances were indeed closely correlated with the location of a food source.

Some years later Wenner (1971) and his colleagues (Wells and Wenner, 1973) raised a more serious question—whether the evidence presented by von Frisch and his colleagues is sufficient to demonstrate the communication of information about distance and direction. They contend that site-specific odors account for the results of von Frisch's experiments and that bees simply convey to one another these odors, for which the recruits then search. They concede that the dance patterns are correlated with the distance and direction of the food source but interpret this as a sort of accidental epiphenomenon. This of course leaves them in the embarrassing position of having no explanation at all to offer for this remarkable coincidence. These critics also underemphasize the fact that many years previously von Frisch had described extensive experiments showing that odors are of great importance in recruiting bees to new food sources. Indeed, Wenner set up a sort of straw man by implying that von Frisch claimed that bees *always* dance or that they locate food *only* by information conveyed through the dances. Wilson (1971) is clearly correct that "the basic thrust of the criticism was wrong," and Michener (1974) concurs, even though he had previously been swayed by Wenner's criticisms. Yet the publication of Wenner's views has had the constructive effect of stimulating new and improved experiments that have established even more solidly that von Frisch was correct in his original interpretations of the dances. Gould (1974, 1975, 1976) has recently developed a new experimental technique by which he can cause dancing bees to signal a different location from the actual feeding place from which they have just returned. The result is that most recruits fly to the place indicated by the dances rather than the actual feeding place from which the dancer has carried back odors.

The intensity of the skepticism expressed by Wenner and his colleagues, and probably felt by many others, is related to the fact that the dance communication system of honeybees edges embarrassingly close to human language in its symbolism and flexibility. Many writers have objected to von Frisch's use of the term language for what can be more conservatively described as communication behavior. But the important point is not one of semantic detail but rather the basic philosophical question of what distinguishes human language from the communication behavior of animals. The prevailing view expressed by many linguists, philosophers, and even biologists is that human language is *the* primary, qualitative difference in kind that distinguishes human beings from animals (Anshen, 1957; Dobzhansky, 1967; Simpson, 1964; Thass-Thienemann, 1968). Indeed many philosophers base their fundamental definitions of humanity on very definite assertions (of a negative sort) about the communication behavior of animals (Adler, 1967).

Perhaps the most thoughtful of these philosophical discussions has been provided by Bennett (1964), who argues with charming erudition that bees are not rational, although he concedes that their dance communication has many attributes of language. In discussing what imaginary bees would have to do in order to satisfy his criteria of rationality, Bennett requires convincing evidence that they know what they are doing. Since his discussion seems to be based on the evidence presented by von Frisch up to 1950, he does not consider the implications of Lindauer's discovery that swarming bees exchange information about potential locations for a new colony. Bennett and others place much weight on the lack of any evidence that the dance communication system is used by one bee to deny a statement made by another. But the exchanges of information between scout bees returning to the swarm from more or less suitable cavities could be interpreted in terms of assertions and denials. Von Frisch and Lindauer have thus provided data that are critically important for the arguments of such philosophers as Bennett, but not sufficiently complete or detailed to answer the questions they have raised. This means, for one thing, that whatever students of animal communication have learned, or can learn in the future, about communication behavior in other species is directly relevant to major fundamental questions about linguistics and philosophy.

There is, of course, no question about the enormous difference in complexity, subtlety, and versatility that separates human speech, to say nothing of written language, from anything known or even speculatively suggested in the communication behavior of other species. But it has seemed of fundamental importance to many scholars and philosophers to insist on a "radical difference in kind" in the terminology, for example, of Adler (1967), rather than a quantitative difference in degree of elaboration. Hockett and Altmann (1968), Thorpe (1972), and others have struggled to formulate objective criteria by which human language can be qualitatively distinguished from animal communication. But this task has become increasingly difficult as more and more is learned about the communication behavior of honeybees, chimpanzees, and other animals.

These considerations force us to face squarely the question of behavioral and even mental continuity between animals and men. Hardly anyone doubts the historical fact of biological evolution. Continuity is a fundamental assumption underlying the use of animals as surrogates for studies of anatomy, physiology, biochemistry, and many types of behavior—investigations that are expected to throw some light on comparable phenomena in our own species. For example, it would be folly to draw any conclusions about human learning from studies of rats or pigeons if one believed that human learning was radically different in kind from that found in other animals. But when questions of language and communication behavior are under consideration, even hard-nosed behaviorists often seem to believe in a lack of continuity or perhaps a radical difference in kind that separates human language from the communication behavior of other species.

Close examination of most assertions that human language differs radically in kind from animal communication shows that this belief is based largely if not entirely on the assumption that human beings use language with conscious intent to convey information while animals are mechanical systems responding to stimuli but lacking any awareness of what they are doing. Similar assumptions, not limited to communication behavior, are thoughtfully stated by Dobzhansky (1969) and Eccles (1969). Recognition that this is a key assumption raises the question: On what do we base our opinions about the mental experiences or consciousness of other animals? The conventional position is either that animals have no mental experiences or, more cautiously, that we have no way of gathering information about the mental experiences of animals if these exist, and hence that the entire question lies outside the proper concerns of science. In either case the evidence is essentially negative, and logically it supports only an agnostic position. Lacking convincing evidence pro or con, we simply do not know what mental experiences animals may have.

While subjective experiences are vivid to each of us individually, all the deepest reflection and most thoughtful eloquence of philosophers have not sufficed to provide methods by which such experiences can be studied except through introspection and verbal reporting. This has heretofore made it appear out of the question to discover anything about mental experiences that animals might have, since it has been taken for granted that a language or communication system adequate for any sort of introspective report was impossible. Furthermore the behavioristic climate of opinion that banished even consideration of human mental experiences from psychology, together with faith in Occam's razor and Morgan's canon, has been so influential since the 1920s that behavioral scientists are now highly uncomfortable at the very thought of mental states or subjective qualities in animals. When such notions intrude into our scientific discourse, as they do remarkably often, we feel sheepish. And when we find ourselves, in spite of these inhibitions, using such words as fear, pain, pleasure, or the like we attempt to shield our reductionist egos behind a respectability blanket of quotation marks.

Yet as Fouts (1973, 1975) has pointed out, our gut feeling of species superiority has suffered a series of intellectual setbacks beginning with the Copernican and Darwinian revolutions, the second having a far more basic impact than the first. Moreover, when evidence is presented that some other species has achieved a previously proposed criterion for distinguishing human language, the list of such criteria is lengthened to exclude the threatening interloper. There is an obvious danger here that our attempts to buttress our anthropomorphic feelings of superiority will lead us into indefensibly circular reasoning.

If we examine the background of our current viewpoint, it is clear that it stems from a critical reaction roughly seventy years ago to an earlier tendency to ascribe human feelings to a wide variety of animals on the basis of anecdotal reports. This behavioristic reaction has served our science well, but so much has now been learned about animal behavior, and especially about orientation and communication, that it is perhaps time to reexamine this basic viewpoint and inquire whether what was once healthy and disciplined restraint may not now be limiting our perspectives and causing us to overlook important opportunities for further advances. There has been a tendency for what was originally an essentially agnostic position about mental experiences in animals to drift into a *de facto* denial that mental states or consciousness exist outside our own species. Here it seems appropriate to follow the example of Holloway (1974) and quote Daniel Yankelovich ("Smith," 1972), "The first step is to measure whatever can easily be measured. This is okay as far as it goes. The second step is to disregard that which can't be measured or give it an arbitrary quantitative value. This is artificial and misleading. The third step is to presume that what can't be measured easily isn't very important. This is blindness. The fourth step is to say that what can't be easily measured really doesn't exist. This is suicide."

Suppose that the discoveries of von Frisch about honeybee communication or those of the Gardners and Fouts about gestural communication with chimpanzees had been known in, say, 1910? Would Loeb, Watson, and others have taken such an adamant stand against considering animal behavior in terms more complex than reflexes, tropisms, and the like? The history of von Frisch's studies of the honeybee dances may also be considered in this light. In the early 1920s he noticed that under certain circumstances round dances occurred when he offered sugar solution in dishes, while waggle dances were exhibited by bees carrying pollen. Only twenty years later did he discover the far more significant correlation of dance pattern with distance and direction. Might not the dance communication of bees have been discovered in the 1920s if complex communication among insects had not been so utterly unthinkable?

The lack of any conceivable means of obtaining introspective reports from animals has always seemed to place hopelessly beyond our reach any fruitful study of the existence or nature of conscious intention in other species. But recent discoveries of versatile communication behavior in animals have changed the situation by opening up the possibility of investigating the fundamental question of mental continuity between animals and men. For it is now thinkable that communication with animals might eventually be developed to the point that it could be used to obtain from them simple but nevertheless significant introspective reports, if, contrary to prevailing opinion, they have mental experiences and are capable of conscious intentions (Griffin, 1976a, 1976b).

Indeed the two-way communication between investigators and chimpanzees has already progressed a small but significant step in this direction. Questions have been asked by Gardner and Gardner (1971) and others about short-term desires such as wanting to go out for a walk or wanting a particular type of food. And questions of this sort have been clearly answered by the chimpanzees. It may be objected that all such behavior can also be described in terms of stimuli and responses, but the same can be said of all human communication. The reason we tend to reject a strictly stimulus-response explanation of our own behavior is that we are convinced of our own conscious intentions. Most of us accept the existence of mental experience and conscious intention in our fellow human beings even though our only source of information about them comes through the imperfect channel of introspective verbal reporting. A similar channel is now potentially available to us for the same basic purpose, at least with chimpanzees. It is even possible that suitable methods can in time be developed to achieve two-way communication with other species. Models, mirror images, and tape recordings can elicit elements of communication behavior. But more versatile models will be needed to carry out significant two-way exchange of messages sufficient to ask questions and obtain introspective answers.

To many readers these suggestions will seem farfetched or even outrageous. One objection that can be anticipated with confidence is the exaggeration of the mere suggestion that animals may have any conscious awareness and intentions at all, into a sort of straw-man assertion that they can equal the richness and versatility of human thinking. But the question I am raising is a much more limited one: Do animals have any sort of mental awareness of probable future events, and do they make conscious choices with the intent to produce certain results? The traditional view is that animals have no such mental experiences or that if they do we can never hope to gather significant and meaningful information about them. I am simply suggesting that the recent developments discussed above open the way to asking animals direct questions about their mental experiences and intentions. If successful, this would bring the whole question within the reach of experimental testing. While new and difficult techniques will have to be perfected, it seems reasonable to hope for significant success if the problem can be attacked with experimental ingenuity comparable, for example, to that devoted in recent years to studies of learning. Once such experiments begin to yield consistent results we can begin to inquire how extensive and elaborate the mental experiences and intentions of animals may be. At this stage there might even begin to emerge, for the first time, a true science of comparative psychology.

References

(In many cases reprints or reviews are cited, but in all such cases these include references to the original sources.)

Adler, M. J., 1967. *The Difference of Man and the Difference It Makes.* New York: Holt, Rinehart and Winston.

Anshen, R. N., 1957. Language as idea. In: *Language: An Enquiry into Its Meaning and Function,* R. N. Anshen, ed. New York: Harper, p.3.

Bennett, J., 1964. *Rationality, an Essay towards an Analysis.* London: Routledge and Kegan Paul.

Bullock, T. H., 1973. Seeing the world through a new sense: electroreception in fish. *Amer. Sci.,* 61:316–25.

Dobzhansky, T., 1967. *The Biology of Ultimate Concern.* New York: New American Library.

Dobzhansky, T., 1969. The pattern of human evolution. In: *The Uniqueness of Man,* J. D. Roslansky, ed. Amsterdam: North Holland.

Eccles, J. C., 1969. The experiencing self. In: *The Uniqueness of Man,* J. D. Roslansky, ed. Amsterdam: North Holland.

Emlen, S. T., 1975. Migration: orientation and navigation. In: *Avian Biology,* vol. 5, D. S. Farner and J. R. King, eds. New York: Academic Press.

Fouts, R. S., 1973. Capacities for language in the great apes. *Proc. IXth International Congress of Anthropological and Ethnological Sciences.* The Hague: Mouton.

Fouts, R. S., 1975. Communication with chimpanzees. In: *Hominisation und Verhalten,* E. Eibl-Eibesfeldt and G. Kurth, eds. Stuttgart: Gustav Fischer Verlag.

Frisch, K. von, 1923. Über die "Sprache" der Bienen. *Zool. Jahrbücher, Abt. f. allgemeine Zool. und Physiol.,* 40:1–186.

Frisch, K. von, 1946. Die Tänze der Bienen. *Österreichische Zool. Zeitschr.,* 1:1–48.

Frisch, K. von, 1950. Die Sonne als Kompass im Leben der Bienen. *Experientia* (Basel), 6:210–21.

Frisch, K. von, 1967. *The Dance Language and Orientation of Bees,* L. Chadwick, trans. Cambridge: Harvard University Press.

Frisch, K. von, 1972. *Bees, Their Vision, Chemical Senses, and Language,* 2d ed. Ithaca: Cornell University Press.

Frisch, K. von, 1974. Decoding the language of the bee. *Science,* 185:663–68.

Gardner, B. T., and Gardner, R. A., 1971. Two-way communication with an infant chimpanzee. In: *Behavior of Non-human Primates,* A. M. Schrier and F. Stollnitz, eds. New York: Academic Press.

Gould, J. L., 1974. Honey bee communication: misdirection of recruits by foragers with covered ocelli. *Nature,* 252:300–301.

Gould, J. L., 1975. Honey bee communication: the dance-language controversy. *Science,* 189:685–93.

Gould, J. L., 1976. The dance-language controversy. *Q. Rev. Biol.* 51:211–244.

Griffin, D. R., 1958. *Listening in the Dark.* New Haven: Yale University Press. Reprint ed., New York: Dover Publications, 1974.

Griffin, D. R., 1976a. *The Question of Animal Awareness, Evolutionary Continuity of Mental Experience.* New York: Rockefeller University Press.

Griffin, D. R., 1976b. A possible window on the minds of animals. *Amer. Sci.*

Griffin, D. R.; Webster, F. A.; and Michael, C.; 1960. The echolocation of flying insects by bats. *Animal Behaviour,* 8:141–54.

Hockett, C. F., and Altmann, S. A., 1968. A note on design features. In: *Animal Communication,* T. A. Sebeok, ed. Bloomington: Indiana University Press.

Holloway, R. L., 1974. Review of Jerison, H. J., *Evolution of the Brain and Intelligence. Science,* 184:677–79.

Kalmijn, A. J., 1971. The electric sense of sharks and rays. *J. Exptl. Biol.,* 55:371–83.

Klein, D. B., 1970. *A History of Experimental Psychology.* New York: Basic Books.

Klopfer, P. H., and Hailman, J. P., 1967. *An Introduction to Animal Behavior.* Englewood Cliffs, N.J.: Prentice-Hall.

Lindauer, M., 1955. Schwarmbienen auf Wohnungssuche. *Z. f. vergl. Physiol.,* 37:263–324.

Lissmann, H. W., 1958. On the function and evolution of electric organs in fish. *J. Exptl. Biol.,* 35:156–91.

Lissmann, H. W., and Machin, K. E., 1958. The mechanism of object location in *Gymnarchus niloticus* and similar fish. *J. Exptl. Biol.,* 35:451–86.

Loeb, J., 1964. *The Mechanistic Conception of Life.* Reprint with preface by Fleming. Cambridge: Harvard University Press.

Matthews, G. V. T., 1968. *Bird Navigation.* (2d ed.) London: Cambridge University Press.

Michener, C. D., 1974. *The Social Behavior of the Bees.* Cambridge: Harvard University Press.

Morgan, C. L., 1894. *An Introduction to Comparative Psychology.* London: Morgan.

Roeder, K., 1970. Episodes in insect brains. *Amer. Sci.* 58:378–89.

Simmons, J. A.; Howell, D. J.; and Suga, N.; 1975. The information content of bat sonar echoes. *Amer. Sci.,* 63:204–15.

Simpson, G. G., 1964. *Biology and Man.* New York: Harcourt, Brace and World. Chap. 8.

"Smith, Adam," 1972. *Supermoney.* New York: Popular Library.

Thass-Thienemann, T., 1968. *Symbolic Behavior.* New York: Washington Square Press.

Thorpe, W. H., 1972. The comparison of vocal communication in animals and man. In: *Non-verbal Communication,* R. A. Hinde, ed. London: Cambridge University Press.

Watson, J. B., 1929. *Psychology from the Standpoint of a Behaviorist.* Philadelphia: Lippincott.

Wells, P. H., and Wenner, A. M., 1973. Do honey bees have a language? *Nature* (London), 241:171–75.

Wenner, A. M., 1971. *The Bee Language Controversy.* Boulder, Col.: Educational Programs Improvement Corp.

Wilson, E. O., 1971. *The Insect Societies.* Cambridge: Harvard University Press.

Chapter 3

CELLULAR COMMUNICATION

Anthony Robertson

Introduction

While I suspect that intercellular communication by propagated signals may be common in developing systems, there is no hard evidence except for the cellular slime molds, whose phylogenetic position is ambiguous. Indeed, it is difficult to avoid the feeling that the slime molds represent a cul-de-sac in evolution. Nonetheless, their development is intriguing in its own right, and there is good precedent for the usefulness of understanding thoroughly a process in a simple organism, even though it appears exceptional. There is fortunately no doubt that slime molds are Eucaryotes and that the majority of species possess a multicellular stage in their developmental cycle!

With this apology I shall devote the bulk of my article to the development of *Dictyostelium discoideum* and in particular to an analysis of intercellular communication during its development. At the end I shall try to put this analysis into the wider context of intercellular communication in developing embryos.

D. discoideum was discovered by Raper (1935), and his description of its morphogenesis and of experiments he performed is a classic paper (Raper, 1940). In the intervening years the popularity of *D. discoideum* has grown dramatically. While it was possible for Bonner (1967) to refer to all published work on *D. discoideum* in his book, that would now be beyond the scope of any one author. Early work was concentrated on descriptions and investigations of morphology and differentiation at or above the cellular level. More recent publications show a heavy bias toward biochemistry and the molecular biology of differentiation. Extensive reviews have been published by Shaffer (1962), Bonner (1967), Ashworth (1971), Robertson and Cohen (1972), Newell (1971), Sussman and Sussman (1969), Wright (1973), and Olive (1975), among others.

The single most provocative discovery was the probable identification of cyclic adenosine monophosphate (c-AMP) as the "Acrasin" or chemotactic attractant for aggregation in *D. discoideum* (Konijn et al., 1968).

Life Cycle of *D. discoideum*

D. discoideum amoebae hatch from elliptical spores about 6μ long, which germinate in moist places in the soil and on decaying vegetation (Cotter and Raper, 1966, 1968a, 1968b). The amoebae feed on bacteria and other microorganisms, which they find by chemotaxis (Potts, 1902;

Bonner, et al., 1970). Recently strains that will grow axenically have been produced. As long as food is available they divide about every four hours, somewhat less frequently in axenic media. When the food is exhausted the amoebae enter a period of differentiation called interphase (Bonner, 1963), which lasts for nine hours (Gerisch, 1968; Cohen and Robertson, 1972). During interphase the amoebae develop the competences that are required for aggregation, including the ability to respond to an extracellular signal, which is probably cyclic adenosine monophosphate (c-AMP) by chemotaxis (Konijn et al., 1968); the ability to relay a pulsatile signal of c-AMP when stimulated by a suprathreshold extracellular c-AMP concentration (Robertson, Drage, and Cohen, 1972); the ability to form EDTA stable intercellular contacts (DeHaan, 1959); and the ability, for a small proportion of the cells, to release periodic pulses of c-AMP autonomously (Cohen and Robertson, 1972). Development of each of these competences must involve changes in many cellular properties, in particular in the cell membrane. Such changes are being investigated by many workers (see Newell, 1971, for a review).

At the end of interphase some amoebae begin to release pulses of c-AMP autonomously. Their neighbors are competent to respond to autonomous signals by chemotaxis and by signal relaying. Each signal can travel only a limited distance because its amplitude is reduced by both diffusion and the action of an extracellular phosphodiesterase (PDE) (Chang, 1968). Just as there is a threshold concentration for signal relaying, there is also a critical density of amoebae below which a signal cannot be propagated (Cohen and Robertson, 1971a). This density corresponds to a distance of approximately 70μ between amoeba centers. After receiving a signal, an amoeba becomes refractory to further stimulation (Shaffer, 1957; Gerisch, 1968; Cohen and Robertson, 1971a). The refractory period is a function of age, decreasing from about nine or ten minutes at the beginning of aggregation to approximately two minutes within four hours. The refractory period ensures that signals propagate unisensally from a center and that a territorial boundary will form between neighboring centers as colliding signals annihilate each other. These properties of the system lead to the outward propagation of periodic waves of c-AMP concentration from centers, marked by inward waves of cell movement (Gerisch, 1968; Cohen and Robertson, 1971a, 1971b; Robertson, 1974).

Because each cell relays the signal, amoebae outside a center tend to move toward their nearest central neighbors, forming streams radiating from the center (Shaffer, 1957; Cohen and Robertson, 1972; Nanjundiah, 1973). An early aggregate therefore consists of a central mass of cells with randomly branching streams leading away from it. During the later stages of aggregation the central cell mass develops a nipple-shaped tip (Raper, 1940; Robertson, 1972). Cells within the tip remain there during the rest of morphogenesis (Takeuchi, 1969; Farnsworth, 1973). In late aggregation, cells in the aggregate secrete a mucopolysaccaride slime, which is liquid at the highest, most central region of the aggregate, but hardens as it flows downward (Shaffer, 1965). As cells are still entering the aggregate within streams, and as the slime is liquid only at the center, pressure in the center can be relieved only by the erection of a cylindrical column, bearing the tip and covered in slime. It grows until it becomes unstable and falls over to form the slug, or migrating pseudoplasmodium. Cells within the slug are joined by antero-posterior contacts and by lateral membrane interdigitations. They migrate as a unit within the slime sheath, which remains stationary with respect to the substrate and collapses behind the slug. Raper showed that the tip controls slug migration (Raper, 1940) and that extra tips grafted onto the side of a slug will take over cells posterior to the graft site. He concluded that the tips acted as though they were organizers, controlling cell movement and defining the developmental axis of the slug.

The distance that the slug migrates depends on environmental conditions (Raper, 1940; Newell, Telser, and Sussman, 1969). At the end of migration the cell mass rounds up and rotates so that the tip again moves to the top. At this stage the cells in the anterior third of the slug have become "pre-stalk" cells, showing histological and biochemical changes, while those in the posterior two thirds are "pre-spore" cells, again with a characteristic histology (Bonner, Chiquoine, and Kolderie, 1955; Gregg, 1965; Gregg and Badman, 1970; Hohl and Hamamoto, 1969; Maeda and Takeuchi, 1969) that has been developing since late aggregation. Their axial organization is retained on rotation of the cell mass. The tip is therefore on top of a mass of pre-stalk cells, which is on top of the pre-spore cells. Formation of the fruiting body ensues. Raper and Fennell (1952) have described this process in great detail. A tube of cellulose fibrils is formed with its top at the base of the tip. It extends downward until it makes contact with a group of cells at the base of the cell mass which will form the base-plate of the mature fruiting body.

At this stage the outer cells in the tip are organized radially, like a columnar epithelium, and it has been suggested that the stalk cellulose is secreted from their central faces (Bonner, 1967). Pre-stalk cells, which move inside the cellulose tube, vacuolate and increase in volume in addition to producing cellulose walls. This differentiation may be triggered by enclosure in the cellulose tube (Farnsworth, 1973). Cells are continually added to the top of the stalk, and the cellulose fibrils are continually laid down at the top of the cellulose cylinder. The stalk, therefore, elongates until the stock of pre-stalk cells is exhausted. We have noticed that the tip of the fruiting body shows periodic jerks, while elongation of the stalk itself by vacuolation of cells within it is continuous (Robertson, 1972). The period of tip jerks is between six and seven minutes, and its distribution is similar to that of signal periods from autonomous cells in early aggregation (Durston, 1974). We have therefore speculated that cell movement to the top of the stalk is periodic and under the control of the same signaling system that is responsible for cell movement during aggregation and during slug migration which also shows a periodic component although it is less well defined (Robertson and Cohen, 1972).

Tip Function

This description of the life cycle of *D. discoideum* shows that there is good reason to assume that the tip retains a special role throughout. Rubin and Robertson (Robertson et al., 1972; Rubin and Robertson, 1975) therefore repeated Raper's grafting experiments and further made grafts of tips from all developmental stages into all other stages. They also assayed tip function by grafting tips into fields of amoebae at different stages of interphase. Their experiments showed that tips from all stages have qualitatively similar properties. The signal a tip supplies is apparently not a function of its developmental age, but the response it evokes is characteristic of the recipient structure. Tip function remains constant until all tip cells have been used up in the process of fruiting body formation. The presence of a tip in a pseudoplasmodium inhibits further tip formation. Removal of a tip leads to the determination of a new tip within forty four minutes and its visible emergence within about an hour and a half, confirming Farnsworth's observation for the conus (Farnsworth, 1973) and Bonner's film (Bonner, 1959). Other pseudoplasmodial portions cannot replicate tip function until, as in the case of slug midportions, tip function reappears when, after a significant delay, the midportion regulates to produce its own tip. The signal from a tip corresponds in all respects to a continuous C-AMP signal.

The tip controls morphogenetic movement at all stages in the *D. discoideum* life cycle from late aggregation onward. The tip defines the direction of the developmental axis and controls a field of cells. It is able to do this by steadily secreting an attractant, presumably C-AMP, at concentrations above the threshold for signal relaying (Rubin and Robertson, unpublished results; Durston, 1974). This feature of its signal, fundamental to the tip's role as an organizer, allows it to "take over" and dominate centers of any type whose signals are initiated by autonomous cells or pulsing microelectrodes. Finally, a new tip or organizer can be produced by a regulative process when a tip is removed. A further implied, but not explicitly demonstrated, role of the tip is the control of differentiation. Regulation following tip renewal involves regulation of the proportions of pre-stalk and prespore cells (Raper, 1940; Bonner, 1967; Robertson, 1972). It is possible that the tip supplies positional information (Wolpert, 1969) by way of the gradient of signal, probably C-AMP, that it produces.

Thus the tip, like the autonomous cell in early aggregation, is the center of communication in the developing organism.

This rather formal description of the life cycle emphasizes the importance of the early stages of development during interphase. It is during this period that the amoebae undergo a sequence of differentiations, expressed finally as the ability to communicate by intercellular signaling, and leading inevitably to aggregation and later morphogenesis. We have therefore examined interphase in great detail in order to gain a suitable background for biochemical and genetic studies of differentiation.

Interphase

First, we wished to measure the rate of emergence of each competence: chemotactic sensitivity to C-AMP, the ability to relay a C-AMP signal, and the ability to release C-AMP signals autonomously. The last two have been measured exactly and the first roughly.

At the beginning of interphase amoebae are relatively insensitive to C-AMP, but within four hours all cells can respond chemotactically to a C-AMP signal provided it exceeds a threshold concentration of about 10^{-9} M. This is only a rough estimate, as all the amoebae begin to secrete an extracellular phosphodiesterase at the beginning of interphase and we have not yet taken the enzyme activity into account in our measurements. Its effect is to give artificially high threshold measurements, so 10^{-9} M may be taken as an upper limit. We made the measurements by observing the area of attraction toward a microelectrode that was releasing pulses of C-AMP into fields of amoebae going through interphase on an agar surface. The relaying threshold was measured by the same technique; it is approximately 10^{-8} M, which is significantly greater than the chemotactic threshold.

The technique was also used to measure the fraction of cells, $X_2(t)_9$ in a population capable of relaying a C-AMP signal. This measurement is more complicated. As cells in a given population differentiate, more and more become capable of relaying. However, a relayed signal cannot be propagated until the density of competent cells exceeds a critical value because the signal has a finite range. Thus the time at which relaying is first observed is a function of initial cell density. Measuring this time for a range of densities gives the crude rate of emergence of the relaying competence. However, critical density is itself a function of initial cell density and time, through the density and time dependence of PDE activity as PDE is continuously secreted by the amoebae.

Thus a correction to the raw data must be made to take account of the reduction of signal by increasing PDE activities. This has been done elegantly by Gingle (1976), who mixed populations of wild-type amoebae with cells of D1, a nonrelaying mutant. D1 has the same PDE activity of the wild-type, but never contributes to the population of relaying-competent cells. Using such mixtures, Gingle obtained not only a corrected $X_2(t)$ curve but also the time and density dependence of PDE secretion. $X_2(t)$ itself is only time dependent; there is no evidence for its enhancement by cellular interaction.

Very different results were obtained for $X_3(t)$, the proportion of cells, as a function of time, capable of signaling autonomously. For one thing, as was well known, only a few cells ever show this competence (Sussman and Ennis, 1959; Konijn and Raper, 1961). The technique for measuring $X_3(t)$ was to observe the initiation time for signaling in small populations of amoebae. $X_3(t)$ begins to increase from zero between seven and eight hours from the beginning of interphase, and saturates at different levels, which depend on both cell density and cell number, implying a cellular interaction for the development of this competence, at least for relatively long times. The most important results are that no more than one percent of cells ever become autonomous and that no cells, as far as we can judge, become autonomous unless they are already capable of relaying signals. In normal populations, aggregation begins at about nine hours because the first autonomous cells release signals into a field that is capable of relaying. It is only by the artificial confinement of small populations that we can follow the differentiation of autonomy beyond the tenth or eleventh hours, when cellular interactions become apparent. These interactions thus may represent an adaptation allowing efficient aggregation in sparse populations of cells that have been starved for a long time, but they are not important under normal circumstances.

To repeat, interphase is a period of differentiation during which the four competences essential for normal aggregation emerge (Cohen and Robertson, 1972):

(1) The ability to respond to a suprathreshold extracellular concentration of C-AMP by chemotaxis toward a C-AMP source, (2) the ability to relay a pulsatile signal of C-AMP when stimulated by a suprathreshold extracellular concentration of C-AMP higher than that for chemotaxis, (3) the ability to release periodic pulses of C-AMP autonomously, and (4) the ability to form EDTA stable polar intercellular contacts. The first three competences provide the basis for morphogenetic movement and its control during aggregation. Knowledge of the rates at which these competences appear during interphase is essential. Without it, biochemical and genetic studies of differentiation for aggregation are severely hampered (Cohen and Robertson, 1972), as are mathematical analyses of wave propagation and aggregation. This analysis now allows a much more detailed understanding of cellular behavior during the aggregation process, in particular wave propagation and the behavior of centers.

Wave Propagation

THE SIGNALING RANGE

The cellular event basic to the control of morphogenetic movement during aggregation and to the communication between cells in *D. discoideum* is the release of a burst of C-AMP molecules into the aqueous film surrounding each amoebae on, e.g., agar. The C-AMP molecules diffuse into the agar and, concurrently, become converted into linear AMP by phosphodiesterase (PDE), both bound to the plasma membrane and released into the agar. We have worked out the resulting somewhat complicated mathematical problem of determining the C-AMP concentration $C(r, t)$ at a distance r on the agar surface from the point of release and a time t after release. The result is qualitatively the same as found earlier, assuming a homogenous background of PDE in the agar (Cohen and Robertson, 1971a). The important points to note at present are that $C(r, t)$ has a maximum value $C_m(r)$ and that $C_m(r)$ is a monotonically decreasing function of distance, as sketched in Fig. 1 of Cohen and Robertson (1971a). Thus at distances larger than R, the signaling range (Cohen and Robertson, 1972), the value of $C(r, t)$ never rises above C^*, the threshold for the stimulation of signal relaying.

$$C_m(R) = C^*$$

Because $C(r, t)$ depends on amoeba density N and on time into interphase through the action of the PDE, so does R. Determination of R vs. N can lead in principle to determination of the PDE activity both in membrane-bound and in free form.

HIGH-DENSITY FIELDS

The significance of R is that an amoeba cannot stimulate another amoeba at a greater separation to relay its signal. R imposes a scale of distance on the mean amoeba separation. In the high-density case, when there are many amoebae within the range of a single signal, on average, i.e., when

$$\pi R^2 N >> 1, \quad (1)$$

the field of amoebae may be regarded as a continuous medium for the propagation of signaling waves.

After ten hours, when $X_2(t)$ has reached unity, the field is a sensitive medium (Robertson and Cohen, 1972; Robertson, 1972; Durston, 1973), in a condition to propagate waves in response to any suprathreshold signal of external origin. Such sensitive media are well understood (Wiener and Rosenblueth, 1946; Krinskii, 1968). Heart tissue, certain solutions of chemical reactants (Zaikin and Zhabotinsky, 1970; Winfree, 1972), multivibrator networks, and models of neural networks are known to be sensitive media. Wave propagation in such media is well understood (Krinskii, 1968; Goodwin and Cohen, 1969; Winfree, 1972). It follows the eikonal equation, but in addition possesses a Huygens construction without a superposition principle. Because of the existence of the refractory period, the boundary conditions are absorbing for propagation into a boundary. The eikonal equation admits the propagation of kinks in wave fronts. However, there are no shadows, as occur in geometric optics, because of the Huygens construction, and beams spread out to fill the medium.

A point source produces circular wave fronts propagating at a velocity, v, dependent on cell density and culture age. Spiral propagation occurs (Winfree, 1972, 1973) with the inner end of the most stable spiral describing a circle of perimeter vT_R, with v the propagation speed and T_R the refractory period.

What further remains to be understood is the propagation speed v. We have constructed a theory of v that relates it to the parameters of the signaling system: the number of C-AMP molecules released, n; the diffusion coefficient of C-AMP in agar or water, D; the threshold concentration for signal relaying, C^*; the delay time between signaling and receiving a suprathreshold signal, T_D; and the free and membrane-bound PDE activity. We have measured D, and T_D can be obtained by other measurements (see below). The density dependence of v can therefore in principle give values for n/C^* and the PDE activities.

LOW-DENSITY FIELDS

For low-density fields condition (1) does not apply. The field is patchy, density fluctuations are important, and the continuum treatment can no longer be used. The range of a signal is enhanced by cooperative effects within a wave front. That is, when several amoebae in a wave front signal together, the concentration produced near one of the signaling amoebae is larger than that produced by an individual amoeba alone. As the density is reduced, however, the randomness of the amoeba distribution breaks up the regular wave fronts so that cooperative effects are reduced. They are also reduced by the increasing separation of amoebae. Thus, as shown directly by R. P. Futrelle (unpublished) in computer simulations, simultaneous signaling by several amoebae does not have any significant quantitative effect on wave propagation at low densities. Under those circumstances, wave propagation can be considered a percolation process on a random medium (Shante and Kirkpatrick, 1971; Cohen and Robertson, 1972). Long-range propagation of signals cannot be sustained by the field unless the amoeba density exceeds a critical density N^* given by

$$\pi R(N^*)^2 N^* = 4.5. \quad (2)$$

For N less than N^*, signal propagation is restricted within isolated clusters of amoebae, the mean size of which falls rapidly with decreasing density. The criterion (2) for N^* derives from percolation theory (Shante and Kirkpatrick, 1971) and is only approximately applicable to our problem. However, the form of equation (2) should remain correct, the only change being in the numerical value of the right-hand side, which should decrease. Futrelle's simulations show that the decrease should be small, so I shall ignore it here.

The current best estimate of N^* from the literature (Konijn and Raper, 1961) and from our own measurements is about $2.5 \times 10^4 \mathrm{cm}^{-2}$, which yields a value of 70μ for R (Gingle and Robertson, 1976).

For densities just above N^*, wave propagation is restricted to narrow channels within the field. Indeed, some portions of these channels degenerate to strings of individual cells each one signaling only the next cell along the string as propagation proceeds. It is possible to identify signaling cells visually because they round up during the delay period between signaling and being signaled (Drage and Robertson, unpublished). The delay time, T_D, and the signaling range, $R(N)$, can thus be measured directly. Preliminary values of 15 sec and 70μ are consistent with earlier results.

Propagation along strings occurs whenever the number of sensitive cells just exceeds the critical density N^*. This happens during interphase when $X_2(t)N$ becomes greater than N^*. Identification of the onset of the capability of long-range wave propagation during interphase thus permits determination of $X_2(t)$, $R(N)$, and T_D. Similarly, the refractory period, T_R, has a fairly broad distribution of values at the time aggregation normally starts. Thus if a pulse is followed by another at an interval within the distribution of T_R such that the density of cells is again just above N^*, propagation along strings occurs. This permits the determination of the probability distribution of T_R, $R(N)$, and T_D. Such experiments are under way.

Pacemakers

In the preceding section, I have sketched the properties of a field of amoebae as a continuum or discrete sensitive medium. Without a source of cyclic AMP to initiate wave propagation, however, wave propagation simply does not occur in such a medium. This is observed in small-drop experiments (Raman, Hashimoto, Cohen, and Robertson, 1976). For example, a certain percentage of drops containing a small enough number of amoebae does not aggregate even though the density is well above N^*. Sources of c-AMP can be external, e.g., a microelectrode

(Robertson, Drage, and Cohen, 1972), an autonomous cell, or a more complex entity consisting of a group of cells. We call entities of this sort, which elicit a periodic response from the field, pacemakers. Durston (1973, 1974) has made a detailed study of natural pacemakers in *D. discoideum.*

He finds two geometries for wave propagation away from pacemakers, concentric and spiral. The latter have been observed to be single or double. Spiral to concentric switches, correlated with the emergence of the tip, have been observed.

Histograms were constructed for the intervals between successive waves for all waves observed, for various geometries, and for various stages of the life cycle. Time courses of intervals were determined for individual centers.

Durston's observations and analyses taken together with other observations and theoretical analyses suggested the following conclusions:

Autonomous cells contain an autonomous oscillator having a fairly sharply defined five-minute period, which is independent of developmental age. This oscillator is linked to the C-AMP release mechanism but is independent of it. Differentiation into autonomy could therefore take place in two ways. First, the oscillator could be present in all cells, e.g., a metabolic oscillator such as the glycolytic oscillator in yeast, and only the link need be constructed. Second, both the oscillator and the link must be constructed.

The refractory period decreases from about nine minutes at the onset of wave propagation to about two minutes after several hours. Initially, the refractory period is broadly distributed; the width of its distribution decreases markedly with time.

This rather broad initial probability distribution of values of refractory period means that the field is then spatially heterogeneous with regard to refractivity. Consequently, spirals are initiated in an otherwise homogeneous field at local maxima of refractivity by waves propagating from autonomous cells.

Finally, tips are steady sources of C-AMP.

Communication Systems in the Cellular Slime Molds

I have described a communication system that controls aggregation in the cellular slime mold *D. discoideum.* A small fraction of the cells are competent to release pulses of an acrasin (most likely C-AMP) autonomously with a period of about five minutes. Essentially all the cells are competent to relay a suprathreshold C-AMP signal. After releasing a signal itself a cell is refractory to further C-AMP signals for a period of about nine minutes, decreasing to two minutes, depending on developmental history. Cells respond chemotactically to a C-AMP signal above another threshold lower than the relaying threshold by a movement step toward the signal source. The step duration, about 100 seconds, substantially exceeds the duration of the signal itself. We have visual evidence from time-lapse films that the movement step is preceded by an approximately fifteen-second interval during which the cell tends to become hemispherical and all movement ceases, even the weak random movements common between successive movement steps. The agreement between the duration of fifteen-second period of stationarity and rounding with that of the delay period between signaling and being signaled (Cohen and Robertson, 1971a) as well as its preceding the movement step suggest that the two are coincident. This, in turn, implies a complementarity between morphogenetic movement and secretion in *D. discoideum* cells (Robertson, 1974).

Such a complementarity exists in all species of cellular slime molds for which the information is available. Before discussing the point further, however, reference should be made to Table 1, in which a classification of some of the cellular slime molds according to the known features of their aggregation control system, i.e., their intercellular communication system is presented (Cohen and Robertson, 1972). The various species are classified according to two main characteristics: whether they possess founder cells, i.e., source cells specialized and morphologically distinct from the remaining cells in the field, which act as centers of aggregation; and whether almost every cell in the field is a local source of acrasin (aggregative signal). The cells can be steady local sources, as in *D. lacteum,* or periodically relaying, as in *D. discoideum,* and this is indicated in a third column.

What is remarkable is that the founder cells are hemispherical in shape and do not move (Bonner, 1967: plate 2), precisely as for *D. discoideum* cells during the stage tentatively identified as the period between signaling and being signaled. It thus appears that in all cases where there is information, cells have similar morphologies and remain stationary while signaling. This suggests that founder cells in *P. violaceum* and *P. pallidum* are steady sources of acrasin and that the observed periodicity in their signal propagation derives from the refractory period of the field. Experiments to test this possibility are under way.

It can be seen from Table 1 that the diversity of known communication systems present in the cellular slime molds derives from only three binary choices: founder cells or no founder cells, local signal sources in the field or no local signal sources, steady local sources or pulsatile local relaying. These observations have considerable relevance to evolutionary and genetic questions.

Table 1
Communication systems of some cellular slime molds.

Species	Founder cell	Local signal production	Periodic relaying
D. minutum	+	–	–
D. Lacteum	–	+	–
P. violaceum	+	+	+
P. pallidum	+	+	+
D. rosarium	?	+	+
D. purpureum	–	+	+
D. mucoroides	–	+	+
D. discoideum	–	+	+

Discussion

What has this very detailed analysis of a developmental control system in one rather obscure organism to tell us about the control of development in general and intercellular communication in particular? It is intriguing to realize how many phenomena that are considered typical in multicellular development imply the existence of control systems that depend on intercellular communication. For example, regulation is a feature of all Metazoan embryos; by regulation, we mean the production of form independent of linear dimension, as is seen when normal tadpoles develop from the separated blastomeres of an amphibian embryo. I have reviewed many similar examples; see Robertson and Cohen (1972), Wolpert (1969) for a summary of the various theories that have been adduced to account for regulative processes. Here it is enough to say that both reliable development and regulation itself imply that cellular behavior in embryos may be subject to control by intercellular communication. While it is known that there are anatomical bases for such communication, nothing is known about how information is passed from cell to cell or what molecular species are involved.

In the slime molds we have a beautiful system in which the role of intercellular communication is clear and in which the molecule used is known. Further, the control of pseudoplasmodial morphogenesis depends on the tip, which is suspiciously like an "organizer." If the analogy between slime mold tips and Metazoan organizers can be sustained, we have a unique opportunity to investigate organizer action at every level of function, from the multicellular down, ultimately, to the control of gene expression. The most promising feature of the slime mold control system is that it can be subjected to a classical genetical analysis because mutations that disrupt the control system are not necessarily lethal. It is hard to imagine that analogous mutations would be either easy to obtain and study or relatively benign in, say, a chick embryo. Thus, if it is ever possible to analyze a control system in a vertebrate it is unlikely that classical genetics will be useful. Therefore, the more we can learn about the function of developmental control systems in simple organisms and the patterns of cell behavior to be expected from different kinds of intercellular communication, the easier it will be to classify cell behavior in Metazoan embryos and to guess the kinds of intercellular interaction on which they depend. The subject has become more exciting with our recent work which has shown evidence for an almost identical signaling system controlling cell movement in the early chick embryo (Robertson and Gingle, unpublished results).

References

Ashworth, J. M., 1971. Cell development in the cellular slime mould, *Dictyostelium discoideum. Symp. Soc. Exptl. Biol.,* 25:27–49.

Bonner, J. T. 1959. Fruiting in the cellular slime molds. New York: Precision Films, Inc.

Bonner, J. T., 1963. Epigenetic development in the cellular slime molds. *Symp. Soc. Exptl. Biol.,* 17:341–58.

Bonner, J. T., 1967. *The Cellular Slime Molds.* Princeton: Princeton University Press. 205pp.

Bonner, J. T.; Chiquoine, A. D.; and Kolderie, M. Q.; 1955. A histochemical study of differentiation in the cellular slime molds. *J. Exptl. Zool.,* 130:133–58.

Bonner, J. T.; Hall, E. M.; Sachsenmaier, W.; and Walker, B. K., 1970. Evidence for a second chemotactic system in the cellular slime mold *Dictyostelium discoideum. J. Bacteriol.,* 120:682–87.

Chang, Y. Y., 1968. Cyclic 3'5'-adenosine monophosphate phosphodiesterase produced by the slime mold *Dictyostelium discoideum. Science,* 160:57–59.

Cohen, M. H.; and Robertson, A.; 1971a. Wave propagation in the early stages of aggregation of cellular slime molds. *J. Theoret. Biol.,* 31:101–18.

Cohen, M. H., and Robertson, A., 1971b. Chemotaxis and the early stages of aggregation in cellular slime molds. *J. Theoret. Biol.,* 31:119–30.

Cohen, M. H., and Robertson, A., 1972. Differentiation for aggregation in the cellular slime molds. In: *Cell Differentiation,* R. Harris, P. Allin, and D. Viza, eds. Copenhagen: Munksgaard. pp.35–45.

Cotter, D. A., and Raper, K. B., 1966. Spore germination in *Dictyostelium discoideum. Proc. Natl. Acad. Sci. U.S.A.,* 56:880–87.

Cotter, D. A., and Raper, K. B., 1968a. Properties of germinating spores of *Dictyostelium discoideum. J. Bacteriol.,* 96:1680–89.

Cotter, D. A., and Raper, K. B., 1968b. Spore germination in strains of *Dictyostelium discoideum and other members of the Dictyosteliaceae. J. Bacteriol.,* 96:1690–95.

DeHaan, R. L., 1959. The effects of the chelating agent ethylenediamine tetra-acetic acid on cell adhesion in the slime mould *Dictyostelium discoideum. J. Embryol. Exptl. Morphol.,* 7:333–43.

Durston, A. J., 1973. *Dictyostelium discoideum* aggregation fields as excitable media. *J. Theoret. Biol.,* 42:483–504.

Durston, A. J., 1974. Pacemaker activity during aggregation in *Dictyostelium discoideum. Dev. Biol.,* 37:225–35.

Farnsworth, P., 1973. Morphogenesis in the cellular slime mould *Dictyostelium discoideum:* the formation and regulation of aggregate tips and the specification of developmental axes. *J. Embryol. Exptl. Morphol.,* 29:253–66.

Gerisch, G., 1968. Cell aggregation and differentiation in *Dictyostelium. Current Topics in Dev. Biol.,* 3:157–97.

Gingle, A. R., 1976. The critical density for signal propagation in *Dictyostelium discoideum. J. Cell Science,* 20:1–20.

Gingle, A. R., and Robertson, A., 1976. The development of the relaying competence in *Dictyostelium discoideum. J. Cell Science,* 20:21–27.

Goodwin, B. C., and Cohen, M. H., 1969. A phase shift model for the spatial and temporal organization of developing systems. *J. Theoret. Biol.,* 25:49–107.

Gregg, J. H., 1965. Regulation in the cellular slime molds. *Dev. Biol.*, 12(3):377–93.

Gregg, J. H., and Badman, W. S., 1970. Morphogenesis and ultrastructure in *Dictyostelium. Dev. Biol.*, 22(1):96–111.

Hohl, H. R., and Hamamoto, S. T., 1969. Ultrastructure of spore differentiation in *Dictyostelium:* the prespore vacuole. *J. Ultrastruc. Res.*, 26:442–53.

Konijn, T. M.; Barkley, D. S.; Chang, Y. Y.; and Bonner, J. T., 1968. Cyclic AMP: a naturally occurring acrasin in the cellular slime molds. *Am. Naturalist,* 102:225–33.

Konijn, T. M., and Raper, K. B., 1961. Cell aggregation in *Dictyostelium discoideum. Dev. Biol.*, 3:725–56.

Krinskii, V., 1968. Fibrillation in excitable media. *Probl. Kibern.*, 20:59–80.

Maeda, Y., and Takeuchi, U., 1969. Cell differentiation and fine structures in development of cellular slime molds. *Dev. Growth Diff.*, 11(3):232–45.

Nanjundiah, V. 1973. Chemotaxis, signal relaying and aggregation morphology. *J. Theoret. Biol.*, 42:63–106.

Newell, P. C., 1971. The development of the cellular slime mould *Dictyostelium discoideum:* a model system for the study of cellular differentiation. *Essays in Biochem.*, 1:87–126.

Newell, P. C.; Telser, A., and Sussman, M., 1969. Alternative developmental pathways determined by environmental conditons in the cell slime mold. *J. Bacteriol.*, 100:763–68.

Olive, L. S., 1975. *The Mycetozoans.* New York: Academic Press. 293pp.

Potts, G., 1902. Zür Physiologie des *Dictyostelium mucoroides. Flora,* 91:281–347.

Raman, R. K.; Hashimoto, Y.; Cohen, M. H.; and Robertson, A.; 1976. Differentiation for aggregation in the cellular slime moulds: the emergence of autonomously signalling cells in *Dictyostelium discoideum. J. Cell Science* 21:243–59.

Raper, K. B., 1935. *Dictyostelium discoideum,* a new species of slime mold from decaying forest leaves. *J. Agr. Res.*, 50:135–47.

Raper, K. B., 1940. Pseudoplasmodium formation and organization in *Dictyostelium discoideum. J. Elisha Mitchell Scient. Soc.*, 56:241–82.

Raper, K. B., and Fennell, D. I., 1952. Stalk formation in *Dictyostelium discoideum. Bull. Torrey Bot. Club,* 79:25–51.

Robertson, A., 1972. Quantative analysis of the development of cellular slime molds. In: *Some Mathematical Questions in Biology,* Lectures on Mathematics in the Life Sciences, vol. 4, J. D. Cowan, ed. Providence, R. I.: American Mathematical Society, pp.47–73.

Robertson, A., 1974. Information handling at the cellular level: intercellular communication in slime mould development. In: *The Biology of Brains.* London: Academic Press.

Robertson, A., and Cohen, M. H., 1972. Control of developing fields. *Ann. Rev. Biophys. Bioengnr.*, 1:409–64.

Robertson, A.; Cohen, M. H.; Drage, D. J.; Durston, A. J.; Rubin, J.; and Wonio, D.; 1972. Cellular interactions in slime-mould aggregation. In: *Cell Interactions: Proceedings of the Third Lepetit Colloquium.* L. G. Silvestri, ed. Amsterdam: North-Holland, p.299.

Robertson, A.; Drage, D. J.; and Cohen, M. H.; 1972. Control of aggregation in *Dictyostelium discoideum* by an external periodic pulse of cyclic adenosine monophosphate. *Science,* 175:333–35.

Ruch, T. C., and Patton, H. D., 1966. *Physiology and Biophysics.* Philadelphia: Saunders. 1244pp.

Rubin, J., and Robertson, A., 1975. The tip of the *Dictyostelium discoideum* pseudoplasmodium as an organizer. *J. Embryol. Exptl. Morphol.*, 33:227–41.

Shaffer, B. M., 1957. Properties of slime mould amoebae of significance for aggregation. *Quar. J. Micro. Sci.*, 98:377–92.

Shaffer, B. M., 1962. The Acrasina, part I. *Adv. Morphogen.*, 2:109–82.

Shaffer, B. M., 1965. Cell movement within aggregates of the slime mould *Dictyostelium discoideum* revealed by surface. *J. Embryol. Exptl. Morph.* 13(1):97–117.

Shante, V. K. S., and Kirkpatrick, S., 1971. An introduction to percolation theory. *Adv. in Phys.*, 20:325–57.

Sussman, M., and Ennis, H. L., 1959. The role of the initiator cell in slime mold aggregation. *Biol. Bull.*, 116:304–17.

Sussman, M., and Sussman, R., 1969. Patterns of RNA synthesis and of enzyme accumulation and disappearance during cellular slime mold cytodifferentiation. *Symp. Soc. Gen. Microbiol.*, 19:403–35.

Takeuchi, I. 1969. Establishment of polar organization during slime mold development. In: *Nucleic Acid Metabolism, Cell Differentiation and Cancer Growth,* E. V. Cowdrey and S. Seno eds., Oxford: Pergamon Press, p.297.

Wiener, N., and Rosenblueth, A., 1946. The mathematical formulation of the problem of conduction of impulses in a network of connected excitable elements, specifically in cardiac muscle. *Arch. Inst. Cardiologia de Mexico,* 16:105–58.

Winfree, A., 1972. Spiral waves of chemical activity. *Science,* 175:634–36.

Winfree, A., 1973. Scroll-shaped waves of chemical activity in three dimensions. *Science,* 181:937–39.

Wolpert, L., 1969. Positional information and the spatial pattern of cellular differentiation. *J. Theoret. Biol.,* 25:1–47.

Wolpert, L.; Hicklin, J.; and Hornbruch, A.; 1971. Positional information and pattern regulation in regeneration of Hydra. *Symp. Soc. Exptl. Biol.* 25:391–416.

Wright, B. E., 1973. *Critical Variables in Differentiation.* Englewood Cliffs, N.J.: Prentice Hall.

Zaikin, A. N., and Zhabotinsky, A. M., 1970. Concentration wave propagation in two-dimensional liquid-phase self-oscillating system. *Nature,* 225:535–37.

Chapter 4

THE EVOLUTION OF COMMUNICATION

Peter Marler

Some Primordia for Communication

It is not easy to draw a firm line between stimuli produced by animals that are truly communicative and others that are not. The difficulty arises in part because so many cases fall on the borderline, having some of the attributes that we require for communication while lacking others. One elementary requirement is illustrated by the observation that telling a man to jump off a bridge is an act of communication, while pushing him is not (Cherry, 1966). The point here is not just that the use of force is hardly communicative in the fullest sense, but also that communicative behaviors are specialized for their function.

There are many examples of stimuli passing from one organism to another that qualify as specialized signals. Illustrations are by no means restricted to interactions between members of the same species. For example, many animal secretions are highly specialized for the chemical repulsion of predators (Eisner, 1970). Some are generalized irritants that cause pain when a predator touches them. Others are "for all intents and purposes natural imitations of histological fixatives, as for instance the spray of a whip scorpion, which contains 84% acetic acid" (Eisner, 1970). Insects subject to predation may produce remarkably pure and concentrated secretions of such compounds as aliphatic acids, aldehydes, aromatics, quinones, and terpenes. The repellent effects on predators such as mice and toads are obvious, and sufficient to inculcate future avoidance of the prey. The specialization of secretions for such functions is clear, yet we may hesitate to label them as signals for communication in the fullest sense.

Clearly signal specialization can emerge in interactions between distantly related organisms. Plants have long been known to produce an enormous variety of chemicals, some shed into the environment, others retained within tissues of the plant. While some chemicals are involved in basic metabolic processes, many were long regarded as no more than excretory by-products. We now realize that many are specialized for functions in the realm of what we can begin to think of as "chemical ecology" (Whittaker, 1970).

Some plants secrete compounds that hinder the germination, growth, and survival of other plants. Leaves, fruits, and twigs of walnut trees, for example, produce hydroxyjuglone, which is carried from the leaves in raindrops to be released in the soil in the oxidized form juglone. This compound has the effect of inhibiting growth of seedlings of many other plant species, hence the sparseness of undergrowth under walnut trees. Another example of "allelopathic" compounds are the volatile terpenes produced by sagebrush and salvia in California chaparral. They too are said to discourage growth of other plants. Some chemical plant products, poisons, or unpleasant flavors are designed to discourage phytophagous predators. Such defenses may be aimed at vertebrate and invertebrate animals, and even at fungi and bacteria, the classical example being, of course, penicillin, evolved by a fungus as protection against bacteria.

Perhaps the most intricate case of chemical interchange between species was discovered by Brower in butterflies. Birds find monarch butterflies distasteful, and similar mimic butterflies gain vicarious protection as a result. Brower demonstrated that the protective agent, a cardiac glycocide, derives originally from milkweed, the food plant of the monarch larvae. Immune to the poison, they sequester it and pass it in turn to the imago in concentrations large enough to cause vomiting in a bluejay that eats the butterfly (Brower and Glazier, 1975).

Ironically, compounds probably evolved by plants originally for defense may eventually become key stimuli by which highly specialized phytophagous insects locate particular species as food for themselves and their larvae. In fact it seems to be the rule more than the exception that food plant location by insects depends on such idiosyncratic chemical characteristics of a host plant and not on a more direct perception of its nutritive characteristics (Dethier, 1970). Here we see specialization on the sensory side of the system, with insect chemical receptors especially tuned to compounds identifying their host plant.

Williams (1970) has discovered what is perhaps the ultimate in specialized chemical plant defense. Unsuccessful attempts to raise moth larvae to maturity in the laboratory were traced to a particular kind of hand towel paper. A compound eventually isolated from paper derived mainly from pulp of the balsam fir *(Abies balsamea)* proved to be a close replica of an insect growth hormone, withdrawal of which is necessary for metamorphosis from larva to adult—presumably evolved by the tree for its own chemical defense against plant-eating insects. "Present indications are that certain plants and more particularly the ferns and evergreen trees have gone in for an incredibly sophisticated self-defense against insect control that we are just beginning to comprehend" (Williams, 1970).

This brief review of chemical stimulus exchanges between plants and animals demonstrates the widespread evolution of highly specialized stimuli, exchanged even between distantly related organisms. One criterion for social communication is surely satisfied—that there be evidence of evolutionary specialization of stimulus production and design for evoking particular responses from some other organism—but others are not. In particular, communication in the fullest sense implies evolutionary specialization of a mutualistic, cooperative nature lacking in the cases cited.

Communication as a Mutualistic Phenomenon

The richest elaboration of systems of social communication should be expected in intraspecific relationships, especially where trends toward increasing interindividual cooperation converge with the emergence of social groupings consisting of close kin. However, a close genetic relationship is not a prerequisite for the mutualistic evolution of systems of communication. The point is made by further consideration of relationships between plants and animals, on a more cooperative plane than those just considered. The relationship between plants and those animals that pollinate their flowers or disperse their fruits, with gain to both, is a cooperative undertaking.

The dispersal of many fruits is aided by their bright coloration, serving as a signal that the dispersing animals can detect at a distance. The adaptation of flowers to animal pollinators is ramified endlessly. Flower design and coloration are often well matched to the visual physiology of particular pollinators. Hummingbird-pollinated flowers are often red, a color attractive to those birds (Grant and Grant, 1968) but less so to bees. Thus, blue and yellow are more common in bee-pollinated flowers. Many bee-pollinated flowers are ornamented with ultraviolet designs, invisible to us though it is the color to which the bee's eye is most sensitive. The patterns serve to lead bees to pollen and nectar, rewarding them with food while pollinating the flower. Many insect-pollinators are also attracted to flowers by their scent. Scent is less common in flowers pollinated by the relatively anosmic birds. However, bats have a keen sense of smell, and flowers specialized for pollination by them often have a strong musky odor (Baker, 1963).

While there is extensive convergence in the design of flowers pollinated by the same classes of pollinators, there are counter-selection pressures for species-specific flower characteristics. This results from competition between flowering plants for the favors of pollinators that are in demand. Diversity in flower design encourages specialization in particular pollinating species, sometimes proceeding so far that plant and insect assume complete mutual interdependence. Alternatively, specific flower appearance and odor may encourage individual pollinators to persist in visiting one flower type for a period of time, whether a season or an hour of the day. This trend toward diversity in the flowers of animal-pollinated plants is evident in the use of flower features in botanical taxonomy. The classification of plants pollinated by birds or bees leans more heavily on the flowers than does the taxonomy of wind- or water-pollinated species—the latter depending more on vegetative characteristics.

Plants may encourage specialization of a particular pollinator by evolving a floral structure that excludes other pollinators, insuring in return for pollination a guaranteed food supply for the particular symbiont. Adaptations to different pollinators occur even within closely related plants such as the phlox family (Grant and Grant, 1965).

All of these interspecific relationships are mutualistic in the sense that cooperative adaptations have been evolved in both partners, another step toward communication in the fullest sense. We should distinguish another kind of species relation in which one is essentially parasitic on the communication system of the other. The orchid family is rich in species, with intense competition for pollinators. Perhaps more than any other plant family, orchids have gone to excess in encouraging specializing by particular pollinators. Their flowers may differ not only in structure but also in odor, known to attract particular species of bees (Dodson, 1970). Some that are pollinated only by males of certain bee species mimic the female bee in flower structure, inducing pollination by the attempts of the insect to mate with the flower. Some of the flowers go so far as to mimic not only female appearance but also odor, varying with the species of the bee (Kullenberg, 1956).

Some of the insects that live parasitically within the colonies of social insects go to remarkable lengths in mimicking signaling behaviors of the host (Hölldobler, 1967, 1970). Perhaps the most remarkable example of parasitization, by one insect (in this case a predator) on the communication system of another has been discovered in fireflies, some of which prey on other fireflies (Lloyd, 1975). Females of one species, *Photuris versicolor,* attract males of other species to their death by mimicking the flash responses of the prey's own female. They are even able to vary their flash patterns according to the species of the prey. Commenting on the efficiency of the process, Lloyd notes that a female seldom answers more than ten males without catching at least one of them.

Genetics of Social Cooperation

Imagine a hypothetical "kinship" series of animal interactions ranging from distant relatives at one extreme to very close relatives at the other. At one end are relationships between distinct and distant species. Next comes interplay between related species, then conspecific but interracial interactions. Relationships between a single population of a species will follow, then those within family or kinship groupings, with the other extreme represented by within-group interactions of social insects such as honeybees, where most members of a society are sisters.

The more similar two animals are to each other, the more likely they are to depend on similar resources—to have similar hiding and nesting places, to obtain similar foods in the same ways at the same times. Since genetic similarity implies sharing of morphological, physiological, and behavioral characteristics, the hypothetical series indicated above manifests a trend toward increasing competition, though not necessarily toward increasing competitiveness, in the sense of active interference of one individual's resource exploitation by another. A close genetic resemblance also has other implications for social evolution.

The proper focus for theories of social evolution is not on individual animals but on genotypes. If an animal is living together with close genetic relatives, then natural selection will tend to minimize aggressive interference with them despite their dependence on virtually identical resources and their potentially potent competitive relationship. Here then is a countertrend to the effect of competition, exerting an opposite influence on the likelihood of cooperative relationships emerging. At some point along the kinship series the trends are likely to achieve a kind of balance, varying with circumstances, but nevertheless tending to hover around a level that will be characteristic of a species, and perhaps different from one species to another, depending upon the degree of intraspecific genetic diversity. In a genetically homogeneous species cooperative relationships are likely to prevail in all intraspecific relationships. In a species that is genetically more heterogeneous, the balance is more likely to be found with smaller social groupings, perhaps at the level of a social group or family.

Communicative behavior will be of paramount importance in achieving and modulating cooperative relationships. Thus, the genetic makeup of a typical social group is likely to bear on the degree of elaboration that the communication system of a species exhibits. The most advanced accomplishments should evolve in animals whose societies are so constructed that groups of very close genetic relatives live together in social contact. Where then should we look but among the social insects, where entire colonies may be composed of the children of a single mother, as in the honeybee (Wilson, 1971)? We should not be surprised in hindsight that one of the most remarkable examples of social communication was discovered by von Frisch (1967) in the dancing behavior of the honeybee. Remarkable for the precise correlations between circumstances of food found and various characteristics of the dance, conclusive proof of its communicative significance, independent of olfactory characteristics of the food source and its environs (Wenner, 1971), was obtained only recently. By ingenious experimentation Gould (1974) was able to modify the dance in such a way as to send foragers in new directions where none had been before, thus establishing that they can be guided by the dance alone.

It is hard to know where to look among vertebrates for comparable examples, for so little is known about the genetic structure of their populations in nature. Theorizing about the origins of vertebrate social behavior is seriously hindered as a result. There are suggestions that some vertebrates, such as house mice and some other rodents, live in quite inbred social groupings (Selander and Yang, 1969; Anderson, 1970). There is evidence that troops of wild baboons with adjacent home ranges are somewhat separated genetically (Buettner-Janusch, 1963, 1965). However, we need data on species differences in genetic composition before theorizing can advance significantly.

Given the importance of kin groupings in the evolution of social communication and other aspects of social cooperation, it would be advantageous for populations to evolve means of discriminating kin from others, of refraining from interaction with non-kin, or even of repelling them aggressively from the social group.

For species such as primates that are born into a close-knit group in which everyone knows everyone else and that grow up within it, discrimination of strangers is a step toward discrimination of kin from non-kin. It is surely more than coincidental that strange conspecific individuals provide the strongest stimuli for aggressive repulsion in a variety of species (e.g., Bernstein, 1964; Southwick, 1967, 1970; Rosenblum et al., 1968; Scruton and Herbert, 1972; Marler 1976).

A more complex vehicle for discriminating kin from others is provided by learned dialects in bird song (Marler and Mundinger, 1971). Several attempts have been made to explain their functional significance, such as the theory that population markers encourage birds to settle and mate with members of the birthplace population, perhaps perpetuating local physiological races (e.g., Marler and Tamura, 1962; Nottebohm, 1969, 1975; Nottebohm and Nottebohm, in press; Nottebohm and Selander, 1972). Learned bird song dialects may also favor kinship groupings as a step in social evolution. There is evidence that both male and female white-crowned sparrows are more sensitive to the dialect they learned in youth and that a dialect boundary does indeed block gene flow to some extent (Baker, 1974, 1975; Marler, 1970b; Milligan and Verner, 1971). Playback studies with recorded bird songs also reveal a tendency for strangers' songs to evoke more intense attack and repulsion than the familiar songs of immediate neighbors (e.g., Falls, 1969; Kroodsma, 1976).

Assuming that young birds are prone to settle near their birthplace, repulsion of birds with strange songs may also favor the integrity of local kinship groupings. One may hypothesize that the cooperative interactions between neighbors that are thus favored increase efficiency of the local population in competition with others, and in turn contributes to the increased fitness of its members. The same may be true of primate troops. It is conceivable that a significant amount of social behavior is directly or indirectly concerned with the genetic structuring of local populations. Only long-term studies of the genetics of natural vertebrate populations can test the validity of this proposition.

Choosing a Sensory Modality for Communication

Depending on the biology of the species, signals of one animal may help others to reproduce effectively, to avoid predation, to get more efficiently to food and water, or simply to find their way about in their environment. By controlling the pattern of individual interactions, and tendencies of animals to cluster or disperse depending on what they are doing, communication will also serve the particular societal structure of the species.

Spacing is a critical issue. Imagine the requirements for a signal that aids in spacing apart adjacent social groups. Efficient exchange requires signaling over a distance with a stimulus that a respondent can localize in space. The possibilities will depend in part on the particular sensory modality chosen—whether visual, auditory, chemical, or even electrical. The phylogenetic history of species will involve many such evolutionary choices, made within limits set by the basic sensory and motor endowment. The nature of the environment will play a role. For forest birds or monkeys that require signals to maintain intergroup spacing, auditory signals are appropriate. Auditory exchange is less hindered by obstacles than is vision, and sound frequencies and temporal coding of auditory signals offer many features on which distinctive categories may be based. Playback of recorded vocalizations of the African mangabey *Cercocebus atrogularis* has shown that one of them, the distinctively patterned "Whoop-gobble," does indeed play a role in mutual avoidance of adjacent groups (Waser, 1975). The "Whoop-gobble" provides for identification of both species and individual, as well as remarkably accurate localization of the caller, as Waser has been able to demonstrate under field conditions.

For communication within a monkey troop rather than between groups, the odds seem to shift in favor of vision as the best medium for diurnal signaling, though sounds serve as well. In a crowd, visual signals are easier to locate than sounds; it is hard for an animal with any but the most primitive visual receptors to avoid locating the source of a visual stimulus in the very process of perceiving it (Marler and Hamilton, 1966).

The elaborate spatial coding that is readily achieved with visual stimuli as a result of their inherent directionality and separability is harder to envision with sounds. Simultaneous signaling is possible with several independent varying expressions of face and body, and perhaps with independent elements within an area such as the face, and human ethologists and anthropologists are beginning to explore the nature and communicative role of such visual signals (Eibl-Eibesfeldt, 1970; Ekman, 1973).

Vision obviously has limited value in dim light or at night, except for special cases such as fireflies, which provide their own light source (Lloyd, 1966). In the many organisms that are active in the dark or that lack well-developed directional eyes, chemical signals are more important. The lack of dependence on ambient light is not the only advantage. Chemical signals have a durability that sounds lack. Durable visual signals are rare in animals, except for trails, rubbing posts, nests, and the like, although our own species makes extensive use of them.

The potential control of rates of fading, which adds an important dimension to chemical signal diversity, is another reason for the choice of a chemical signal when animals require durable signals. There is an abundant literature on animal marking behavior, by which chemical secretions placed on some other animal or on objects in the environment continue to be transmitted in the absence of the original marking individual (Ralls, 1971; Eisenberg and Kleiman, 1972). Chemical secretions on an animal's own body may have a profound impact on its social interactions with others. An aggression-eliciting pheromone is produced by male mice whose testes are actively producing androgens, radically changing their valence as stimuli for aggression in other mice (Lee and Brake, 1971; Lee and Griffo, 1973; Mugford and Nowell, 1971). The remarkable energetic efficiency of chemical communication is also relevant. It has been calculated that a receptor cell on the male silkworm moth's antenna requires only one molecule of the female sex substance, bondecal, for her presence to be detected (Schneider, 1965; Wilson, 1971).

When it comes to location of chemical signals, however, there are problems. It is one thing for a male to detect a female's presence and another to locate her. As long as the intervening medium is still, whether it is air or water, location of the source of a chemical signal is only possible by reference to the diffusion gradient. At more than a meter or so from the source the gradient is so gradual that orientation is difficult. With a moving medium it is much easier, and animals using olfactory signals over a distance make use of such movement. Release of a female moth's sexual attractant may be delayed until there is a breeze blowing of a certain speed, not strong enough to generate excessive turbulence, but strong enough to carry the scent downwind (Kettlewell, 1946). As the searching male moves upwind, backing when he loses her, he can get close enough to switch to other strategies.

Thus, each sensory modality has drawbacks and advantages. For certain fish there is still another choice, electrical signaling, originally discovered to be a system of object location and general orientation to the environment (Lissmann, 1958). More recently it has been shown to sustain a system of social communication (Hopkins, 1972, 1974) permitting electric fish to identify species, sex, and age of a partner and also providing some information about its motivational state.

When communicants are close to one another or in contact, other sensory modalities such as touch and taste become available for communication, with distinctive properties of their own. Perhaps for the reason that it requires contact, touch often seems to assume special social functions. Violent tactile contact may cause pain and social disruption. Gentle touching, on the other hand, especially as manifest in social grooming, recurs repeatedly among animals in situations where it can promote peaceful contact and at the same time reduce the likelihood of social violence (Sparks, 1964, 1967; Morris, 1971; Simpson, 1973).

Are Animal Signals Arbitrary?

The words in our language are arbitrary in two senses. Most words are semantically arbitrary, bearing no resemblance to their symbolic referent, other than in onomatopoeia. The word *food* does not resemble anything we eat. Thus, there is no obstacle other than inertia to the exchange of meaning of two words in the course of cultural evolution. Words exhibit another kind of arbitrariness in that there is nothing intrinsic to the process by which a word conveys information about food. Our speech does not require a word to have any particular acoustical properties, even those affecting locatability, for example.

Semantically, many animal signals are also arbitrary. The grunting sound by which chimpanzees announce discovery of a choice meal and readiness to share it has no morphological relationship to its referent—perhaps a cluster of ripe palm fruits.

African vervet monkeys have a repertoire of distinct alarm calls for different types of predators (Struhsaker, 1967). While some of the associations are rather general, others are specific enough to invite the speculation that they have wordlike properties. In particular, of the three that announce danger from (1) a venomous snake crossing the territory, (2) an eagle overhead, and (3) an approaching leopard, each evokes a different response functionally appropriate to the particular threat. Such calls have no iconic resemblance to the predators they symbolize. In this sense many though not all animal signals are semantically arbitrary (cf. Altmann, 1967).

Nevertheless, animal signals often possess physical characteristics that make them uniquely suited for the particular function they serve. Thus the exchange of meaning between words that we can easily imagine taking place in the evolution of speech patterns is often less likely with animal signals. In this sense the physical structure of many animal signals is far from arbitrary.

Some sounds require transmission over considerable distances, while others operate at close range, and their loudness varies accordingly. Among forest monkeys of East Africa, sounds males use to space troops apart are louder than those used within the group for coordinating foraging movements (Marler, 1973). The transmission distances of different sound frequencies vary with different habitat structure, and there is some evidence that male birds compose their songs of sounds that travel farthest in their habitual environments (Morton, 1975; Chappuis, 1971). Woodland bird songs, often consisting of medium-pitched pure tones, probably illustrate such adaptations.

Whales provide what is perhaps the most remarkable case of adjustment of vocal behavior to long-distance transmission. By using very low-pitched sounds, barely audible to our ears, and placing themselves at an intermediate depth in the ocean, in the so-called "deep sound channel," humpback whales are thought to be able to hear each other calling over distances measured in hundreds of miles. Because of the refraction of sounds by the thermocline near the ocean's surface and by compression layers deep in the ocean, sound waves are trapped within the channel, retaining much more of their energy than they would if they impinged on the surface or floor of the ocean (Payne and McVay, 1971; Payne and Webb, 1971).

The physical structure of the animal signals is related to their function in many ways. The importance of sound localization has already been indicated, and many bird signals are designed to provide an abundance of cues that permit localization by companions to be accomplished quickly and accurately. Other sounds are structured to minimize the clues available for localization, still serving to spread alarm to companions, while giving only a minimum of localization cues to a predator (Marler, 1955). By experimenting in the laboratory with owls trained to strike at sounds, Konishi (1973) could demonstrate that sounds like the hawk alarm calls of small birds are hard for an owl to locate.

The chemical signals of insects illustrate other kinds of adaptation to signal transmission. The molecular properties of pheromones often correspond nicely with functional requirements (Wilson, 1970). Using a ratio of the emission rate in molecules per unit time in relation to the behavioral threshold concentration as a frame of reference, a low ratio of the former to the latter indicates a slowly emitted signal with a very small "active space"—a three-dimensional volume enclosing the space where the threshold level for behavioral response is exceeded by the pheromone concentration. Such characteristics would fit well with a substance to be used in trail marking, by fire ants for example (Wilson, 1970). For a pheromone designed to transmit alarm different characteristics would be appropriate, with a larger active space and a more rapid diffusion rate, a prediction that can also be verified. The largest active space is found in the sex attractants of female moths, required to function over distances much greater than would be appropriate for communicating danger.

One could continue recounting examples of adaptations of signal structure to function. This review is not intended to be exhaustive, only to establish the point that in this rather special sense the structure of many animal signals is by no means arbitrary.

The Triggering of Signal Production

In spite of the logical difference in the symbolic or semantic significance of an alarm call and a sexual call, one with an external referent, the other without, the production of both must be associated with particular physiological states of the signaling animal, one externally triggered, the other generated endogenously. The process of symbolization would be easier to grasp if we knew more about the nature of these physiological states and the way they are engendered. If the soliciting female bird is signaling about physiological events that preface ovulation, such events are largely endogenous and cyclic in nature, arguing against the need to invoke an immediate trigger from the environment that one could then think of the solicitation call as symbolizing. The same might be said of other signals such as the begging call of a nestling bird. While this call may be triggered by approach of a parent, it often occurs without external provocation, as the nestling becomes hungry. The same duality is found in the singing of many birds, sometimes externally triggered but often starting without any precipitating event in the environment.

This distinction between signals that are externally and internally triggered, at first sight rather basic, may be less radical than it seems. Consider an alarm signal, usually triggered by an environmental event. One might be tempted to think that the role of internal physiology in its production is minimal. However, a bird that is alert or nervous is more likely to give alarm when danger threatens than one that is sleepy or relaxed. Thus, the physiological state at the time of confrontation with a predator cannot be ignored if we are to understand the process of alarm call production. Similarly the response evoked by an alarm signal in another bird will vary according to its state of arousal at different times or its previous learning experiences.

One may carry this line of thinking further and suggest that *no* reaction of an organism to an external stimulus can be understood without taking its current physiological state into account. Some stimuli evoke similar reactions across an array of physiological states so broad that one is tempted to ignore the significance of internal events. But these are cases where the particular reaction has high biological priority, so that the necessary physiological machinery must be in readiness much of the time. One may assert that the proper way to think of *all* stimulus-response relationships is as interactions between the organism and its environment (Marler and Hamilton, 1966). While the environment may provide a trigger responsible for orientation and timing of a reaction, the particular structure of the response, its coordination, its internal timing, and often even its orientation and intensity can be understood only by reference to the prior history of the subject (cf. Bullock, 1961). The responsiveness of an animal at the time of a particular stimulus event is the result of convergence of a multitude of past events reflecting the previous experience the individual, its genotype, and the particular pattern of environmental interaction experienced up to the instant of stimulus confrontation.

According to this view, physiological considerations are as important in the interpretation of a signaling action that has an external trigger as they are with internally triggered signals. One can also assert the converse, that signals endogenously triggered should not be thought of as completely independent of the external world. They are independent only in the sense that no immediate outside trigger is detectable. The precipitating endogenous event in the signaler's physiology reflects a history of innumerable prior environmental interactions, some closely related to the signal, others only remotely.

The sexual calling of a receptive female bird may lack an immediate environmental trigger, but it is nevertheless dependent on many prior environmental interactions, some with only a general bearing on the timing of ovulation, such as the requirements for normal growth, and others with a specific bearing such as increasing day length, social stimulation, and sensory feedback from nest-building activities.

The conclusion may be drawn that to speak of a signaling act as being "internally" or "endogenously" triggered implies no more than that its timing and structure are functions of prior events in the history of the individual, their influence being conveyed to the present by physiological means.

"Affective" Signaling: A Valid Category?

It is a commonly expressed view that the signaling behavior of animals is more susceptible to control by the kind of physiological states known in common parlance as "affective" or "emotional." The judgment that a state is emotional can be based on a number of different interpretations. Recurring themes are that the states are generalized, affecting many patterns of behavior; that autonomic arousal is often involved; that there is often some connotation of emergency in the tempo, intensity, and demeanor of the signaling animal; that there is often strong "momentum" to the behavior so that once begun it tends to continue for a period of time, resisting rapid change; that it is involuntary, or toward that end of a continuum with voluntary actions; that it is less susceptible to modification by learning or conscious effort; and that it can be placed somewhere on a dimension of pleasantness to unpleasantness.

Human signaling behavior, or at any rate, speech, is said to contrast with animal signaling in that it is voluntary, detached from any necessary linkage with pleasantness or unpleasantness, readily modified by learning or conscious decisions, not necessarily tied to autonomic arousal or other generalized physiological states, not necessarily associated with behavioral emergencies, and involved with physiological states that can change very rapidly.

The contrasts listed above, attempting a brief summary of a complex and difficult subject (e.g., Arnold, 1970), undoubtedly point to a significant difference between animal and human signaling behavior. In most of the circumstances in which animal signaling occurs, one detects urgent and demanding functions to be served, often involving emergencies for survival or procreation. To the extent that physiological states of emotion or affect are indeed distinguishable from other substrates for behavior, they have surely evolved to organize actions and responsiveness to stimulation in complex spatiotemporal terms, serving a variety of functions (e.g., Plutchik, 1970), not the least of which is to avoid inefficient vacillation in the face of conflicting environmental demands (cf. Pribram, 1970). Their pervasive influence on our own behavior is obvious, especially in social situations.

Firm resolution of this problem is difficult, but it is worthwhile to consider whether all animal signaling is as dominated by affect as is supposed. To take one example, we have already considered the vervet monkey's sound using different alarm calls for different types of predator (Struhsaker, 1967). The calls do not intergrade, and it is not obvious that they fall on a continuum of differing levels of arousal. To explain the vervet monkey's complex alarm-signaling behavior several different underlying physiological states must be postulated, seemingly more specific in nature than the emotional condition usually denoted as "fearful." However, these alarm calls do conform with another common attribute of emotional behavior, namely regular association with a complex of other, more directly functional behaviors. But it is usually the case that an animal signaler is itself engaged in the same functional behavior that is being signaled about. Insofar as monkeys are not known to engage in relaxed discourse about events in the distant past or future, there will be few if any occasions when a signal is likely to be dissociated from the other overt responses to a situation. Indeed, a monkey prone to such reflective signaling in the presence of a predator would probably not survive very long. Thus the issue of signaling about events remote in time, for which our speech seems more specialized than animal signaling, is much entwined with the issue of freedom from affective physiological control.

One circumstance in which such dissociation can be detected in animals is during play. Among the diagnostic criteria for play behavior are several that are reminiscent of distinctions between human speech and animal sounds (Marler and Hamilton, 1966). Separation of the behavior from its normal emotional substrate is sometimes mentioned, and there is correspondingly more freedom in switching from one pattern to another or reversing social roles than in the adult emotional version of the behavior. During play, signals are sometimes separated from the other ongoing behaviors that normally accompany them. Investigation of the distinctive physiological characteristics of play may well throw light on this distinction between affective and nonaffective signaling behavior.

One may also ask whether our speech is altogether free from emotional constraints. Autonomic arousal is a common accompaniment of speaking behavior, as is evident from the use of polygraphic lie detectors. Some psychoanalytic practices rest on similar assumptions. One does not need a galvanometer or an electrocardiograph to be convinced that speech uttered in social circumstances often has strong emotional components. Furthermore, some research on the physiology of human emotions seems to contradict the old notion that only a few simple physiological dimensions are involved. Two distinct components have been proposed, even for the most basic emotions (e.g., Schachter, 1970). One component incorporates many of the autonomic and hedonic functions imputed to emotion, and it is associated with the level of arousal; the other "cognitive" component seems to specify more precisely the particular emotion that will be subjectively experienced. While the cognitive is to some extent restricted to a particular emotion, the other may be shared by several emotions. To the extent that this model is indeed a correct one for human emotions, it may also be relevant to animals—at least the autonomic physiology of higher vertebrates seems similar to our own.

Thus, the distinction between generalized and specific physiological substrates for signaling behavior is not always clear. Whatever distinction we are groping for between speech and animal signaling (and it is hard to be sure that we are even asking the right questions), it is my own conviction that the underlying physiology of signaling will prove to be different in degree rather than in kind between animals and ourselves. If only to provoke some reappraisal, I would assert that no firm proof has yet been advanced of any fundamental differences between animals and man in this regard.

Signal Variation: Discrete and Graded Repertoires

That many patterns of animal behavior are stereotyped, especially those with communicative functions, is a deep-rooted notion in ethology, as shown by the concept of "fixed action patterns" (Lorenz, 1935, 1950; Schleidt, 1974). The stereotypy of some is indeed remarkable. Variability of the strutting display of the sage grouse is unusually low by some measures, as are the claw-waving displays of fiddler crabs, and some duck displays (Dane and Van der Kloot, 1964; Hazlett, 1972; Wiley, 1973). Some animal sounds such as certain bird songs and the loud calls of some adult male monkeys are also highly stereotyped (Marler, 1973).

Presumably this unusually narrow range of biological variation reflects some functional requirement. Accurate identification of a signal at a distance, without support by other cues and under noisy conditions, must be easier with stereotyped than with variable signals. However, quantitative description has revealed that extreme stereotypy is by no means a general rule. While some signals are almost fixed, others are exceedingly variable, as some monkey calls (e.g., Rowell, 1962, Rowell and Hinde, 1962; Green, 1975; Marler, 1965, 1970a; Marler and Tenaza, chapter 36, this volume).

What interpretation can be placed on this variation in degrees of stereotypy? Should it be viewed as a consequence of poor developmental control, or can some communicative significance be attributed to it? The problem that exists in interpreting variation in individual signals recurs at the level of entire signal repertoires, some of which are organized discretely, while others are highly graded. Within the repertoire of the African blue monkey, loud calls of the adult male, though superficially similar, proved to be categorically distinct (Marler, 1973). "Growls" and "pulsed grunts" grade into one another, however, by a series of intermediates.

The sound repertoires of certain animals contain few if any discrete sound types at all. The most extreme cases studied thus far are certain primates, notably the red colobus (Marler, 1970a), rhesus, and Japanese macaques (Rowell, 1962; Green, 1975), the talapoin monkey (Gautier, 1974), and the chimpanzee (Marler and Tenaza, chapter 36). In these species a major part, if not all, of the vocal repertoire consists of a single graded acoustical system. The grading may occur in several acoustical dimensions independently, such as frequency, tonal structure, and duration, each varying continuously, making it unrealistic to subject the sounds to a strict categorical classification.

Even in discrete signals, fine variation in morphology probably has communicative significance. The "chip" alarm call of female blue monkeys is discretely separate from other sounds in the repertoire. Nevertheless variations in intensity, frequency, morphology, and timing have the potential for conveying information to others, although presumably animals will be less sensitive to within-category variations than to variations between categories. The extent of within-category variation may differ from one signal to another in the same repertoire. The point is illustrated by vocalizations of the black and white colobus (Marler, 1969, 1972). The roaring of the male probably serves functions similar to the male loud calls of the blue monkey, namely the maintenance of territorial spacing and the rallying of group members. It is discretely separate from, for example, a system of squeak-screams used by adults and juveniles. Unlike roaring, where variability in structure and timing occurs but within limits, squeaks and screams vary along a number of dimensions. It is notable that they are used primarily for communication within the troop, whereas roaring includes an inter-troop function. Similarly the growl-pulsed grunt continuum of the blue monkey functions at close range within the troop, while that part of the repertoire especially concerned with distance communication, whether within or between troops, tends to fall into discontinuous, discrete categories.

If a species living in dense forest is socially organized in territorial groups, with a significant part of the vocal repertoire addressed to problems of intertroop communication, signaling must take place over appreciable distances in environments noisy from wind and sounds of insects, birds, and other primates. There must be strong selection pressures in such circumstances in order for a discrete type of signal organization to operate as the most efficient means of unequivocal conveyance of information to an adjacent troop. Any potential that graded signals might otherwise have for communication of more refined information over such long distances would surely be lost by signal degradation and masking. Pressures for specific distinctiveness would also favor discrete acoustical morphology and patterns of delivery.

Within the troop circumstances differ. Signaling is likely to occur over a shorter range. Even in a forest full of obstacles communicants in the same troop can often see as well as hear one another, and visual signals emitted in tandem with sounds may aid communicants in detecting and identifying the subtleties of graded signaling.

This line of argument can be brought to bear on those species that show excessive emphasis on signal grading in their vocal repertoire. In the absence of territoriality, greater group size, and more complex troop organization resulting from the presence of several adult males, we see a shift in emphasis toward intratroop communication, and a corresponding increase in the degree of grading of vocal repertoires. Far from representing disorderly erratic variation, as though from poor developmental control or relaxation of the relationship between vocal morphology on the one hand and ongoing behaviors and their physiological substrates on the other, this variation is in fact highly ordered, as demonstrated in both the Japanese macaque and the West African talapoin monkey (Gautier, 1974; Green, 1975). There can be no doubt that the graded repertoire of the Japanese macaque has the potential for conveying subtle and complex information about the circumstances of sound production. Further work is required to establish whether or not these variants are responded to differently.

It is intriguing that the constellation of behavioral and ecological traits that, with some exceptions, tends to characterize those primates with a graded repertoire—large, nonterritorial, multi-male groups with a tendency to move on the forest floor and to invade open country—is consistent with speculations about the probable ecology and social organization of early man (Washburn, 1961; Campbell, 1972). Notably the list of primates with a predominately graded vocal system includes the chimpanzee, the closest of all other surviving primates to human ancestry in its behavior, social organization, and temperament, as well as its tool preparation and use and its habit of hunting and eating mammalian prey (van Lawick-Goodall, 1971; Teleki, 1973). It is all the more intriguing to note that sound spectrographic descriptions of the structure of speech reveal that many adjacent speech sounds do in fact grade into one another in their acoustical morphology although they are heard as discretely distinct (e.g., Lisker and Abramson, 1964).

Specific Distinctiveness of Animal Signals

Behavior can be as revealing as external morphology in the diagnosis of difficult species, as every naturalist knows. However, the application of behavior to taxonomy is not easy. While their behavior may differ in some respects, species can be exceedingly similar in others. In trying to understand why some signals are so much more specifically distinct than others, it is important to remember that species do not live alone but in communities that include many other organisms. While it is an advantage for a species to possess "private" signals for functions most efficiently performed in interaction with conspecific animals, as with reproduction and often with aggression, the kind of "privacy" achieved by a high degree of signal species-specificity can also be a disadvantage. Alarm calls are often similar in groups of species living together. In both birds and monkeys interspecific communication has been found to occur frequently (Marler, 1957, 1973). Interspecies similarities in the calls used can only facilitate such interchange, thus serving a definite function. By contrast, the songs of male birds and the loud calls of male forest monkeys, serving reproductive isolation and spacing apart of conspecific troops, are specifically distinct from those of cohabiting relatives. The point serves as a reminder that the characteristics of the community in which a species lives may bear on its communication system. Thus, the resemblance in male songs and aggressive display calls of certain cohabiting bird species makes sense when one appreciates the possibility of special cases of interspecies competition as an influence on communication system design (Marler, 1960; Cody, 1973).

The Experimental Value of Modified or Synthetic Signals

Although few zoological studies of animal communication have gone farther than signal description, some investigations point the way to more analytical approaches. Once the functional role of a communication signal has been established, it is desirable to establish which components of the signal are necessary for a given response and which are redundant. Having answered this for one context, one must explore others because it is likely that a different subset of components assumes significance in other situations, with other recipients, which has proved to be the case with bird song (e.g., Falls, 1969; Emlen, 1972). The ideal approach to such problems is to synthesize signals artificially so that their structure can be changed systematically in one direction or another, along natural or unnatural lines of variation, to establish the limits of effectiveness in different situations. The approach has proved fruitful in studies of speech perception (e.g., Liberman et al., 1967) and may work as well in studies of animal communication.

An illustration is provided by Hopkins' (1972, 1974) studies of communication in electric fish in Guyana. In one of ten species living together, *Sternopygus,* he found the discharge pattern to be distinctly different from those of other fish present, the pulse repetition frequency being unique. Male and female frequencies were found to differ consistently, and, by working in the field during the rains when the fish breed, he found males embarking on courtship as the electric signals of an approaching female *Sternopygus* became detectable. Having hypothesized that the frequency difference was fundamental, he synthesized songs in which the only remaining natural property was the frequency, carried by a sine wave. Males courted the electrodes as long as the frequency fell within normal female range. If it was too low, approaching that of male *Sternopygus,* or too high and approaching the range of another species, *Eigenmannia,* the courtship ceased. Thus, although the electric discharges have other distinctive properties such as pulse shape, frequency seems to be the key property in this case (Hopkins, 1974).

The croaking of a bullfrog is another synthesizable animal signal. Capranica (1965, 1966) found that male frogs in a terrarium would readily croak in reply to recordings of their species, but not to sounds of thirty-three other frogs. Choosing to concentrate on the spectral structure and wave-form periodicity of the croaking, Capranica experimented with many synthetic calls. He demonstrated three key frequency characteristics that an optimal call must satisfy. Two spectral peaks are needed for maximum responsiveness, one around 200 Hz and another at 1,400 Hz. A sound with energy in only one of these regions is less than optimal in evoking a male bullfrog's response. In addition the call should have a minimal amount of energy in the mid-frequency region, around 500 to 600 Hz. The optimum periodicity in the temporal waveform should be around 100 per second. A mating call with all of these spectral and temporal features will evoke the greatest response from another male bullfrog.

Rigorous definition of the significant stimulus parameters prepares the way for exploring the physiological mechanisms underlying such specific responsiveness (Capranica and Ingle, in press). A fascinating correspondence is found with the pattern of sensitivity of the peripheral receptor systems in the bullfrog's auditory system. Detection of the mating call seems to involve excitation of both the amphibian and the basilar papilli of the inner ear. The low-frequency peak around 200 Hz seems to be the one that best excites the complex sensory units tuned to this part of the spectrum in the amphibian papilla. The high-frequency peak, around 1,400 Hz, excites the simple units of the basilar papilla. The further requirement that the optimal call should lack energy in the mid-frequency region, around 500–600 Hz, was identified with the inhibition of the low-frequency complex units of the amphibian papilla by sounds of this frequency. The maximum response of both simple and complex units to pulsed stimuli was obtained with a periodicity of 100 pulses per second, thus explaining the temporal wave-form of the optimal bullfrog call.

The bullfrog's detection of the species call is a direct reflection of the response characteristics of the amphibian and basilar papilli of his ear. Carrying this approach to study of sensory mechanisms in other frog species with different patterns of calling has revealed species differences in the sensitivity of these same sensory units. Thus, species differences in the peripheral stimulus filtering properties of the ear go far to explain the species-specificity of their calling behavior. As Capranica and Ingle indicate, such species differences in the sensitivity of hearing are less obvious in the peripheral auditory organization of birds and mammals, probably because their biology requires the detection of a greater variety of sounds from the environment than the biology of frogs does.

The general functions required of sense organs will have an inevitable influence on the type of physiological stimulus filtering evolved to meet the requirements for specific stimulus responsiveness. If such functions are simple and restricted, dominated by some particular function such as seeking out the opposite sex, then highly specific responsiveness can be imposed at the level of the receptors. Where more versatile sense organs are required, then more complex solutions to the physiological problem of stimulus filtering must be found (Marler, 1961).

Animal Signals as Predictors

The question of how animal signals came to have survival value for members of a species can be approached in different ways. One is to assume that a signal from animal A helps animal B to "anticipate" or "predict" future events. As a simple illustration, an alarm call given by a bird that sees a hawk permits other birds within earshot to behave as though they were anticipating or predicting a future approach of the predator, so they also rush for cover. If this reaction evoked by the signal has some regular and exclusive relationship to a particular environmental situation or referent—a hunting predator in this case—one may think of the signal as serving as a "sign" or "symbol" for it. The food call of chimpanzees, known as "rough grunting" (Marler and Tenaza, chapter 36, this volume), can be interpreted similarly. Other chimpanzees that hear it approach quickly, eager to partake of a preferred food, as though the call serves as a symbol for it.

Such a semantic interpretation by an animal seems appropriate when production of the signal is contingent upon a particular environmental situation or object that serves as an external referent. However, many animal signals are produced in circumstances where no external referents exist, such as those signals associated with agonistic behavior or copulation. Responses to such signals seem better interpreted on the basis of a prediction of how the signaling animal is likely to react on approach of the respondent. If the utterance of a sexual solicitation call by a female bird triggers full courtship in a responding male, it seems more appropriate to interpret his actions as based not on some external circumstance she is signaling about but rather on her anticipated or predicted response to his approach—assuming that the solicitation call is unique to the period of female sexual receptivity.

Selection of a Subset of Respondents by a Signal

When a respondent receives a given signal from a distance, its behavior may change in a specific, qualitative fashion. But often the first response that an observer can detect is no more than a change in its spatial relationship to the signaler. Thus, two signals that may eventually elicit very different responses—attack or copulation, for example—may first elicit an identical response, namely approach to the signaler. Sometimes a keen ethological eye detects other behaviors that permit a more specific prediction of the final response; an aggressive approach might be distinguishable from a sexual one for example. Such additional cues are often lacking, however, which makes the specific end point of a sequence initiated by signal-elicited approach difficult to predict. It may become predictable only later in the sequence, after further stimuli have been received. The culmination of an aggressive approach, for example, can vary widely, depending on further stimuli exchanged between sender and receiver.

Thus, while it is sometimes useful to speak of sexual signals, aggressive signals, alarm signals, and so on, it is often difficult to distinguish the responses that such categories of signals do in fact elicit. Both aggressive and alarm signals may elicit withdrawal of a respondent in some circumstances. In spite of the difficulties it is intriguing to consider approaching the analysis of signal function by a classification of the types of behavior that are evoked (Marler, 1967).

If a respondent withdraws in response to a signal, we usually classify this withdrawal as a form of escape behavior, our confidence in this judgment increasing if we see signs of autonomic arousal and excitement. Locomotion may be followed by tense immobility in a place of concealment. It remains as something of a paradox that the active and inactive phases of withdrawal are classified in the same behavioral category. Nevertheless, the number of respondent behaviors that succeed withdrawal from a signaler is relatively small.

By contrast, approach to a signaler can be followed by many possible types of respondent behavior, including genital contact; suckling, nursing, and other parent-young behaviors; food sharing, exchange, or stealing; competition for resting or breeding sites, or sharing of them; attack on the signaler or on another animal close by, such as a predator; or a variety of social activities such as standing close, sleeping together, grooming, and so on. We can exclude activities like foraging that may continue irrespective of changes in relative spacing of signaler and respondent. It is implicit in much of our thinking that the alternative response selected by the receiver of the signal is specified by that same signal that elicited the initial approach. In fact it seems likely that the specification is often partly or largely a function of further signals received during the approach or after it.

If the first response to many signals is approach with other signals responsible for further specification, one might question what advantage a species gains from having many different signal types for long-distance signaling. Would not one signal type suffice to elicit the approach? However, we must bear in mind that not all who hear a given signal will respond by approaching —this is obvious if one thinks of an infant signaling for its mother or a female soliciting for copulation. Diverse signals are still required to specify the appropriate class of respondent. The need for such diversity stems directly from the many different communicatory roles that individuals may play in a society, such roles being by no means interchangeable.

The thrust of this discussion is that some signals function not so much to impose a qualitative change on the behavior of respondents but rather to select a particular class of respondents that may be already predisposed to perform the response in question. A female monkey who has recently given birth will have a different set of response predispositions than an adult male engaged in consortship behavior. This fact greatly complicates the experimental analysis of communicative behavior, requiring exploration of the presence or absence of responses to a signal in all possible classes of recipients, some responding, others not, some inclined to respond in one mode, others differently. One may imagine the appropriate respondent being specified along several dimensions, including species, sex, age class, dominance status, individuality, and recent social history. The specification might also be made indirectly, addressing potential respondents that find themselves in a particular environmental context, as when vervet monkey alarm calls elicit different responses from animals out in the open and others already deep in cover. The specification might also be made according to a transitory physiological state—e.g., a food signal evoking a response from hungry animals but not from satiated ones.

Natural selection is likely to favor contrasting trends in the evolution of signals functioning to select different classes of respondent. Where restriction to members of the same species is favored there will be a strong tendency for emphasis on species-specificity. The converse will be true when the facilitation of interspecific communication conveys some advantage, the specification of appropriate respondents being broadened to cover several species. The similarity of alarm calls across species of birds and monkeys has been interpreted in this way (Marler, 1957, 1973). If specification of a particular class of respondents is facilitated by use of a signal that shares attributes with signals used by that class, then we can see how the specification of sex, individual, or age class of respondent may become reflected in the type of signal used for this function.

In the course of this brief discussion of respondent specification, we have concentrated on signals evoking approach from a distance. As the distance between signaler and respondent shrinks, the specification of alternate response patterns is likely to narrow. At closer range some of the difficulties of communication are eased, with less chance of error in identifying signals. The opportunity to receive compound signaling through several sensory modalities is increased. And if problems of species or individual specificity have been resolved earlier in the sequence, less emphasis is needed at closer range, and the release from stringent demands for species-specificity may permit further exploitation of other signal characteristics. I have argued in an earlier section that the increased exploitation of signals that are highly graded in structure rather than discrete is favored in such circumstances.

Signal Development: Genetic Control and Learning

Modifiability through learning, for which our speech has such rich potential, is sometimes thought to play no significant role in animal signaling behavior, and in some interpretations innateness becomes coupled with the presumption of an emotional basis. Genetic programs for neural outflow from the central nervous system to the signaling equipment have been reported in some organisms, the most striking example coming from the development of calling songs in crickets. In an elegant series of experiments Bentley and Hoy (1972) have shown that hybrid male crickets produce songs distinctly different from either parental song in a pattern that is directly attributable to genetic factors. Even more remarkable is the predisposition of hybrid females to be more responsive to the calling song of their hybrid sibling males than to the song of either parental species (Hoy, 1974), raising the question whether sensory mechanisms may not be involved in the motor development as well. However, there is no evidence that sensory control in insect song development goes so far as to permit modification through learning.

Birds, like humans, rely heavily on communication by sounds in maintaining the structure of their societies. Many songbirds learn their song (Marler and Mundinger, 1971) and at least some other vocalizations in their repertoires as a matter of course (Mundinger, 1970; Marler and Mundinger, 1975). Much has been learned in recent years about the nature and significance of vocal learning processes in birds.

One revealing approach has been study of the effects of deafening a bird surgically early in life upon its vocal development. A dove or chicken deafened soon after birth vocalizes at the normal time, and analysis of the sounds reveals a normal morphology (Konishi, 1963; Nottebohm and Nottebohm, 1971). Thus, a dove or chicken needs no access to an external model to develop normal vocalizations. Nor does such a bird need to hear its own voice in order to generate the normal vocal repertoire.

A contrast is struck with the song sparrow. Taken from the nest and reared as a group in acoustical isolation, young males of this species develop normal song, notwithstanding its greater complexity as compared with dove and chicken vocalizations. Like them, male song sparrows have the ability to generate the complex motor output of singing without the prerequisite of an external model, even though abnormalities are sometimes apparent (Kroodsma, in press). However, if a young male song sparrow is deafened early in youth, his subsequent singing, unlike that of doves and chickens, will be highly abnormal. All of the fine morphology is lost, and instead there is a burst of about two seconds of very noisy, erratic pulsed sounds with a rather insectlike quality. The song sparrow must hear its own voice if normal development is to occur (Mulligan, 1966).

Yet another condition is illustrated by the white-crowned sparrow. Here a young male taken from the nest and reared in isolation in a soundproof chamber will develop a highly abnormal song. Although this song is outside the set of normal patterns for the species, certain qualities of the species' typical song persist. Playback of a recording of normal song to a young male at a certain phase of life, between ten and fifty days of age in this species, results in the subsequent production of a close copy of the external model presented (Marler, 1970b). If a male white-crowned sparrow is deafened early in youth, the song he then develops is rather like that of a deafened song sparrow. It is much more abnormal than that of an intact male reared in social isolation. Almost all species-specific characteristics are lost, including those few that are still retained by an intact isolated male (Konishi, 1965). When deafened, both song sparrow and white-crowned sparrow males behave as though their songs were reduced to the lowest common denominator, perhaps the basic output from the passive syringeal apparatus with air flowing through it. This interpretation is reinforced by the similarity of songs of early deafened white-crowned sparrows, song sparrows, and another relative, the Oregon junco (Konishi, 1964), three species whose normal songs are highly divergent. The inference is drawn that hearing is significant in the divergent pathways normally taken by song development in these three species, not only for hearing external models but also for hearing their own voices.

The discovery that species differences in sparrow songs seem to originate with sensory mechanisms rather than with motor ones led to speculation about the existence of auditory templates. Visualized as lying in the neural pathways for auditory processing, embodying information about the structure of vocal sounds, and having the capacity to guide vocal development, they appear to have a more dominant influence on vocal development than structure of the sound-producing equipment or the characteristics of hearing in general, although they too can have an effect (Konishi, 1970). According to this view, the young male beginning to sing strikes a progressively closer match between his vocal output and the dictates of the auditory template. The transitions he goes through from subsong, to plastic song, and finally to full song are consistent with this interpretation. Species are thought to differ in the competence of the auditory template to guide song development in a fully normal fashion. In the song sparrow the template seems more or less adequate to guide normal development. However, in the white-crowned sparrow the template of a naive male is less adequate, although it may well still be sufficient to focus the male's attention on an appropriate class of external models, thus explaining the finding that a male will reject inappropriate models while he is learning. While the selectivity of learning might depend on a different mechanism, it seems economical to assume that the same one produces both effects. We presumably see in the song of an intact socially-isolated male white-crowned sparrow a picture of what the unimproved template of a naive bird embodies.

Given access to an appropriate external model during the critical period, the template becomes more highly specified, eventually embodying all of the instructions necessary for normal singing, including the characteristics of the particular dialect to which the male was exposed. Note that this learning precedes singing by a hundred days or more, permitting Konishi (1965) to deafen males both before and after learning, but before singing. The outcome was the same, the very elementary song of the trained and then deafened bird revealing no trace of the auditory learning that had already taken place. Thus, hearing is still required for the information incorporated in the improved template to be translated into motor activity. One may also postulate the existence of a similar sensory mechanism in the female white-crowned sparrow, who doesn't normally sing but who is responsive to the male song at the time of sexual pairing (Milligan and Verner, 1971). Konishi (1965) demonstrated that a female induced to sing by injection with androgens is in possession of the same information about song as the male. Not only will she sing, but, if exposed to normal song during early life, she will sing the particular dialect to which she was exposed.

While we can conceptualize song templates as single functional mechanisms, they may involve several physiological components that serve together as stimulus filters. Components that are modifiable through experience might be separate from components that underlie the selective perception of a naive, untrained male. The two sets might operate in series or in parallel, with control shifting from one to the other after training. There may be species differences in the nature, number, and mode of coupling of templates. As with other "feature detectors," one should be prepared for the likelihood that similar behavioral ends may be achieved by different physiological mechanisms.

Most intriguing of all is the possibility that a similar mechanism might underlie the learned development of speech. The studies of Eimas and his colleagues (1971, in press) have shown that some normal perceptual processing of speech sounds occurs in infants as young as a month of age, long before they have begun to speak or even to babble. This result suggests that human infants may possess auditory templates for certain speech sounds. Although they may have heard a lot of speech even by the time they are a month old, the early age raises the possibility that certain speech sounds can be processed without the need for prior exposure to them.

Auditory templates for certain speech sounds could serve a child well in more than one respect. They would focus an infant's attention on an appropriate class of external stimuli for social responsiveness, much as the auditory templates of the white-crowned sparrow restrict responsiveness to members of its own species when they are living in a community where many others are present. Auditory templates could also provide an orderly frame of reference for the infant's developing responsiveness to the speech of others, drawing attention to the particular subset of speech properties that retain valence into adulthood (Mattingly, 1972). The templates would become both modified and multiplied as a result of experience gained in the very process of aiding the infant's perception and analysis of the sounds of the language in which it participates. The number of parallels between song learning in birds and the acquisition of speech by a child is striking. We may press the parallel further and suggest that speech sound templates also function in the development of speaking. Improvements in a child's babbling, as with a bird's subsong, perhaps reveal growing skill in matching vocal output to auditory templates. Possibly most remarkable is the discovery by Nottebohm (1971, 1972) that the neural control of some bird songs is lateralized, with a tendency for one side of the brain to assume dominant control, an echo of the dominance of the left hemisphere in the control of our speech.

There may in fact be a basic set of rules for the organization of vocal learning to which any species might be expected to conform if the design of its societies depends on a series of complex, learned traits. Though full exploitation of the advantages of learning requires freedom, the provision of too much latitude in the morphology of signaling behavior would result in patterns so divergent that communication would be impeded and the structure of a society disrupted. Therein, perhaps, lies the survival value of genetic predispositions that a species brings to the task of vocal learning from its past history, interacting with environmental stimuli to extract and abstract from those models presented by experienced adults from which it must derive the norms for its own social behavior. Notwithstanding the fact that a child and a young bird put their capacities for vocal learning to entirely different functions, the processes underlying learning still have many attributes in common (Marler, 1970c).

Animals with Language?

Though the process of learning to speak is paralleled in many respects by avian vocal learning, it is obvious that birds lack language. A search for primordia for this attribute of our communication system, regarded by many as uniquely human, would surely require investigation of monkeys and apes, which have so much else in common with us. However, laboratory and field study seems to confirm that their patterns of vocal development are very different from our own. Whereas children and birds begin to show an almost irrepressible tendency toward vocal imitation at a certain age, no one has yet discovered a comparable tendency in any other mammal. In contrast with human children and young song birds, other primate young are not known to "babble."

It is true that when raised in a home like children are, and after much time and effort on the part of both subject and experimenter, chimpanzees learned to utter a few words (Hayes and

Hayes, 1952; Hayes and Nissen, 1971). But the process of acquisition, requiring laborious step-by-step assemblage of the necessary mouth movements with rewards at each stage, had little in common with the vocal imitation of bird and child. Remarkable though chimpanzee Vicki's breathy and unvoiced renditions of "cup," "papa," and "mama" were, they served as further confirmation of the gap between ape and man.

It was tempting to infer that the chimpanzee's inability to imitate speech reflects its lesser intelligence. Lieberman (1968) and Lieberman, Crelin, and Klatt (1972) analyzed the chimpanzee vocal tract and concluded that it would not be capable of producing the full array of human sounds; hence the failure to imitate. However, if an inappropriately structured vocal tract were the only obstacle, chimpanzees would attempt imitation, but the renditions would be imperfect, as occurs with the abnormal but still intelligible speech of persons suffering from laryngectomy or cleft palate. But chimpanzees make no attempt at all. For an explanation one must look rather to deficiencies in neural mechanisms that engender the predisposition for vocal learning.

The fact that no simple intelligence deficit was responsible emerged from several remarkable investigations setting out to teach chimpanzees language like communication systems that did not require the imitation of sounds. Aware of the extent to which chimpanzees use their hands in natural communication, Gardner and Gardner (1971, 1975) and Fouts (1973) have used the hand sign language of the deaf, American Sign Language. With various training techniques, including shaping, guidance, and observational learning, as well as imitation, they were able to teach the young female chimpanzee Washoe to perform eighty-five signs, each equivalent to a word, in a three-year period. Included were many nouns, such as *flower, dog,* and *toothbrush,* adjectives such as *red* and *white,* prepositions such as *up* and *down,* verbs such as *help, hug,* and *go.* Many words were used in appropriate combinations such as the invitation for a walk, *You me go out hurry,* or the request *Please gimme sweet drink.* The appropriateness of combinations of actions and objects indicates a grammar not very different from that of young children in early two-word sentences (Brown, 1970; Gardner and Gardner, 1974). Recently another young chimpanzee, Lana, has demonstrated similar prowess with a languagelike system based on keyboard signals to a computer, which talks back to her in a similar fashion (Rumbaugh et al., 1973).

The third chimpanzee, Sarah, accomplished in the use of a languagelike system, was trained by Premack (1971) to use colored plastic shapes instead of words, these shapes serving as symbols for objects and actions. A blue plastic triangle served as the symbol for *apple.* The one for *banana* was a red square, and so on. The relation between symbol and referent was noniconic, the shape lacking any physical resemblance to the object to which it referred. After Sarah was trained to present the appropriate shapes when she wanted a piece of fruit, other nouns and then verbs were introduced such as *give, wash,* and *insert,* each performed by the experimenter when Sarah presented the appropriate symbol.

Within her repertoire of about 130 words were not only many nouns, verbs, and adjectives, but also more complex constructions such as *same,* and *different,* questions, and the conditional *if–then.* A particular word order was required of Sarah in arranging the symbols on a board. Premack aimed more to test the conceptual abilities of Sarah than to see whether she could use language, reasoning that in our own species the one is closely mirrored in the other.

Can one infer that Sarah thinks in the language of these plastic shapes? Premack says yes. One test, he feels, is the ability "to generate the meaning of words in the absence of their internal representation." Premack asked Sarah to perform a feature analysis of an apple, using the plastic words to name its color and shape, the presence or absence of a stalk, and so on. Asked to perform a similar analysis on the plastic word for *apple,* the blue triangle, she answered by describing an apple once more and not the blue shape. This test bears on a further point, Sarah's ability to consider something that is not there at the moment—an illustration of the critical language requirement of displacement in time.

The importance of appreciating the natural motives of a subject in trying to understand its use of language is well illustrated by errors Sarah made in the use of shapes for different fruits. Required to present the appropriate shape for a fruit before she was allowed to eat it, she chose the wrong word surprisingly often. In a moment of inspiration Premack wondered whether Sarah was asking for what she preferred rather than for what was before her. An independent series of tests on her fruit preferences provided the explanation. The word for banana offered when confronted with an apple was not an error but an attempt to get the experimenter to give her something else, suggesting again that she truly understood the symbolic significance of the shapes.

The accomplishments of chimpanzees using languagelike systems of signaling to converse with an experimenter are surely the highest animal attainments demonstrated so far. Yet they also raise a curious dilemma. If a chimpanzee can indeed achieve some elementary competence with language when provided with an appropriate vehicle, why has this not been demonstrated in nature? It may well be that our knowledge of natural communication in animals is in such infancy that we can hardly judge whether such abilities are demonstrated in nature or not. However, it is also possible that the social organization and ecology of wild animals is so structured that they have little use for the special patterns of communication that our language makes possible.

From a biological viewpoint, symbolic communication is highly specialized, working most efficiently with particular kinds of problems. For many of the uses to which animals can put their signals—largely social in nature and taking place within groups in which members are familiar with one another over a long history of acquaintanceship—other kinds of signals can probably do the job better. Indeed it is conceivable that other types of communication other than language in the purest sense play a much more important role in our own biology than we are inclined to acknowledge. One of the benefits of a comparative approach to human communication may be a better appreciation of the rich potential of affective signals in performing the great variety of functions that sustain the organization of a complex society.

References

Altmann, S. A., 1967. The structure of primate social communication. In: *Social Communication among Primates,* S. A. Altmann, ed. Chicago: University of Chicago Press.

Anderson, P. K., 1970. Ecological structure and gene flow in small mammals. In: *Variation in Mammalian Populations,* R. J. Berry and H. N. Southern, eds. *Symp. Zool. Soc. Lond.,* 26:299–325.

Arnold, M. B., ed., 1970. *Feelings and Emotions.* New York: Academic Press.

Baker, H. G., 1963. Evolutionary mechanisms in pollination biology. *Science,* 139:877–83.

Baker, M. C., 1974. Genetic structure of two populations of white-crowned sparrows with different song dialects. *Condor,* 76:351–56.

Baker, M. C., 1975. Song dialects and genetic differences in white-crowned sparrows *(Zonotrichia leucophrys). Evolution,* 29:226–41.

Bentley, D. R., and Hoy, R. R., 1972. Genetic control of the neuronal network generating cricket *(Teleogryllus gryllus)* song patterns. *Anim. Behav.,* 20:478–92.

Bernstein, I. S., 1964. The integration of rhesus monkeys introduced to a group. *Folia Primat.,* 2:50–63.

Brower, L. P., and Glazier, S. C., 1975. Localization of heart poisons in the monarch butterfly. *Science,* 188:19–25.

Brown, R., 1970. The first sentences of child and chimpanzee. In: *Psycholinguistics,* R. Brown, ed. New York: The Free Press.

Buettner-Janusch, J., 1963. Hemoglobins and transferrins of baboons. *Folia Primat.,* 1:73–87.

Buettner-Janusch, J., 1965. Biochemical genetics of baboons in relation to population structure. In: *The Baboon in Medical Research,* H. Vagtborg, ed. Austin: University of Texas Press.

Bullock, T. H., 1961. The origins of patterned nervous discharge. *Behaviour,* 17:48–59.

Campbell, B., 1972. Man for all seasons. In: *Sexual Selection and the Descent of Man, 1871–1971,* B. Campbell, ed. Chicago: Aldine.

Capranica, R. R., 1965. *The Evoked Vocal Response of the Bullfrog: A Study of Communication by Sound.* Research Monograph 33. Cambridge: M.I.T. Press.

Capranica, R. R., 1966. Vocal response of the bullfrog to natural and synthetic mating calls. *J. Acoust. Soc. Am.,* 40:1131–39.

Capranica, R. R., and Ingle, D., in press. Sensory processing in the auditory and visual systems of anurans. In: *Physiology of the Amphibia,* vol. 3, B. Lofts, ed. New York: Academic Press.

Chappuis, C., 1971. Un exemple de l'influence du milieu sur les émissions vocales des oiseaux: l'évolution de chants au forêt equatoriale. *Terre et Vie,* 25:183–202.

Cherry, C., 1966. *On Human Communication,* 2d ed. Cambridge: M.I.T. Press.

Cody, M. L., 1973. Character convergence. *Ann. Rev. Ecol. Syst.,* 4:189–212.

Dane, B., and Van der Kloot, W. G., 1964. Analysis of the display of the golden eye duck. *Behaviour,* 22:283–324.

Dethier, V. G., 1970. Chemical interactions between plants and insects. In: *Chemical Ecology,* E. Sondheimer and J. B. Simeone, eds. New York: Academic Press.

Dodson, C. H., 1970. The role of chemical attractants in orchid pollination. In: *Biochemical Coevolution,* K. L. Chambers, ed. Corvallis: Oregon State University Press.

Eibl-Eibesfeldt, I., 1970. *Ethology: The Biology of Behavior.* New York: Holt, Rinehart and Winston.

Eimas, P. D.; Siqueland, E. R.; Jusczyk, P.; and Vigorito, J.; 1971. Speech perception in infants. *Science,* 171:303–306.

Eimas, P. D., 1975. Speech perception in early infancy. In: *Infant Perception,* L. B. Cohen and P. Salapetek, eds. New York: Academic Press.

Eisenberg, J. F., and Kleiman, D. G., 1972. Olfactory communication in mammals. *Ann. Rev. Ecol. Syst.,* 3:1–32.

Eisner, T., 1970. Chemical defense against predation in arthropods. In: *Chemical Ecology,* E. Sondheimer and J. B. Simeone, eds. New York: Academic Press.

Ekman, P., 1973. *Darwin and Facial Expression.* New York: Academic Press.

Emlen, S. T., 1972. An experimental analysis of the parameters of bird song eliciting species recognition. *Behaviour,* 41:130–71.

Falls, J. B., 1969. Functions of territorial song in the white-throated sparrow. In: *Bird Vocalizations,* R. A. Hinde, ed. Cambridge: Cambridge University Press.

Fouts, R. S., 1973. Acquisition and testing of gestural signs in four young chimpanzees. *Science,* 180:978–80.

Frisch, K. von, 1967. *The Dance Language and Orientation of Bees.* Cambridge: Harvard University Press.

Gardner, B. T., and Gardner, R. A., 1971. Two-way communication with an infant chimpanzee. In: *Behavior of Nonhuman Primates,* vol. 4, A. Schrier and F. Stollnitz, eds. New York: Academic Press.

Gardner, B. T., and Gardner, R. A., 1974. Comparing the early utterances of child and chimpanzee. In: *Minn. Symp. on Child Psychol.,* A. Pick, ed. 8:3–24. Minneapolis: University of Minnesota Press.

Gardner, R. A., and Gardner, B. T., 1975. Early signs of language in child and chimpanzee. *Science,* 187:752–53.

Gautier, J. P., 1974. Field and laboratory studies of the vocalizations of talapoin monkeys *(Miopithecus talapoin). Behaviour,* 60:209–273.

Gould, J. L., 1974. Honey bee communication. *Nature,* 252:300–301.

Grant, K. A., and Grant, V., 1968. *Hummingbirds and Their Flowers.* New York: Columbia University Press.

Grant, V., and Grant, K. A., 1965. *Flower Pollination in the Phlox Family.* New York: Columbia University Press.

Green, S., 1975. Variation of vocal pattern with social situation in the Japanese monkey *(Macaca fuscata):* A field study. In: *Primate Behavior: Developments in Field and Laboratory Research,* vol. 6, L. A. Rosenblum, ed. New York: Academic Press.

Hayes, K. J., and Hayes, C., 1952. Imitation in a home-raised chimpanzee. *J. Comp. Physiol. Psychol.,* 45:450–59.

Hayes, K. J., and Nissen, C. H., 1971. Higher mental functions of a home-raised chimpanzee. In: *Behavior of Nonhuman Primates,* A. M. Schrier and F. Stollnitz, eds. New York: Academic Press.

Hazlett, B. A., 1972. Ritualization in marine crustacea. In: *Behavior of Marine Animals,* vol. 1., H. E. Winn and B. L. Olla, eds. New York: Plenum Press.

Hölldobler, B., 1967. Zur Physiologie der Gast-Wirt-Beziehungen (Myrmecophilie) bei Ameisen. I. Das Gastverhältnis der *Atemeles-* und *Lomechusa*-Larven (Col. Staphylinidae) zu *Formica* (Hym. Formicidae). *Z. Vergl. Physiol.,* 56:1–21.

Hölldobler, B., 1970. Zur Physiologie der Gast-Wirt-Beziehungen (Myrmecophilie) bei Ameisen. II. Das Gastverhältnis des imaginalen *Atemeles pubicollis* Bris. (Col. Staphylinidae) zu *Myrmica* und *Formica* (Hym. Formicidae). *Z. Vergl. Physiol.,* 66:215–50.

Hopkins, C. D., 1972. Sex differences in electric signalling in an electric fish. *Science,* 176:1035–37.

Hopkins, C. D., 1974. Electric communication in fish. *Amer. Sci.,* 62:426–37.

Hoy, R. R., 1974. Genetic control of acoustic behavior in crickets. *Amer. Zool.,* 14:1067–80.

Kettlewell, H. B. D., 1946. Female assembling scents with reference to an important paper on the subject. *Entomologist,* 79:8–14.

Konishi, M., 1963. The role of auditory feedback in the vocal behavior of the domestic fowl. *Z. Tierpsychol.,* 20:349–67.

Konishi, M., 1964. Effects of deafening on song development in two species of juncos. *Condor,* 66:85–102.

Konishi, M., 1965. The role of auditory feedback in the control of vocalization in the white-crowned sparrow. *Z. Tierpsychol.,* 22:770–83.

Konishi, M., 1970. Comparative neurophysiological studies of hearing and vocalizations in songbirds. *Z. Vergl. Physiol.,* 66:257–72.

Konishi, M., 1973. Locatable and nonlocatable acoustic signals for barn owls. *Amer. Nat.,* 107:775–85.

Kroodsma, D., 1976. Effects of large song repertoires on neighbor "recognition" in male song sparrows. *Condor,* 78:97–99.

Kroodsma, D., in press. A reevaluation of song development in the Song Sparrow and the *Junco-Zonotrichia-Melospiza* complex. *Anim. Behav.*

Kullenberg, B., 1956. Field experiments with chemical sexual attractants on aculeate hymenoptera males. *Zool. Biol. Uppsala,* 31:253–354.

Lee, C. T., and Brake, S. C., 1971. Reactions of male fighters to male and female mice, untreated or deodorized. *Psychon. Sci.,* 24:209–211.

Lee, C. T., and Griffo, W., 1973. Early androgenization and aggression pheromone in inbred mice. *Hormones and Behavior,* 4:181–89.

Liberman, A. M.; Cooper, F. S.; Shankweiler, D. P.; and Studdert-Kennedy, M.; 1967. Perception of the speech code. *Psychol. Rev.,* 74:431–61.

Lieberman, P., 1968. Primate vocalizations and human linguistic ability. *J. Acoust. Soc. Amer.,* 44:1574–84.

Lieberman, P.; Crelin, E. S.; and Klatt, D. H.; 1972. Phonetic ability and related anatomy of the newborn and adult human, Neanderthal man, and the chimpanzee. *Amer. Anthropologist,* 74:287–307.

Lisker, L., and Abramson, A. S., 1964. A cross-language study of voicing in initial stops: acoustical measurements. *Word,* 20:384–422.

Lissmann, H. W., 1958. On the function and evolution of electric organs in fish. *J. Exp. Biol.,* 35:156–91.

Lloyd, J. E., 1966. Studies in the flash communication system in Photinus fireflies. *Univ. Mich. Mus. Zool. Misc. Publ.,* 130:1–95.

Lloyd, J. E., 1975. Aggressive mimicry in Photuris fireflies: signal repertoires by femme fatales. *Science,* 187:452–53.

Lorenz, K. Z., 1935. Der Kumpan in der Umwelt des Vogels. *J. Ornithol.,* 83:137–215, 289–413.

Lorenz, K. Z., 1950. The comparative method in studying innate behavior patterns. *Symp. Soc. Exp. Biol., IV: Physiological Mechanisms in Animal Behaviour.* Cambridge University Press, pp.221–68.

Marler, P., 1955. Characteristics of some animal calls. *Nature* (London), 176:6–7.

Marler, P., 1957. Specific distinctiveness in the communication signals of birds. *Behaviour,* 11:13–39.

Marler, P., 1960. Bird songs and mate selection. In: *Animal Sounds and Communication,* W. Lanyon and W. Tavolga, eds. Washington, D.C.: American Institute of Biological Sciences.

Marler, P., 1961. The filtering of external stimuli during instinctive behaviour. In: *Current Problems in Animal Behaviour,* W. H. Thorpe and O. L. Zangwill, eds. Cambridge: Cambridge University Press.

Marler, P., 1965. Communication in monkeys and apes. In: *Primate Behavior,* I. DeVore, ed. New York: Holt, Rinehart and Winston.

Marler, P., 1967. Animal communication signals. *Science,* 157:769–74.

Marler, P., 1969. *Colobus guereza:* Territoriality and a group composition. *Science,* 163:93–95.

Marler, P., 1970a. Vocalizations of East African monkeys I. Red colobus. *Folia Primat.,* 13:81–91.

Marler, P., 1970b. A comparative approach to vocal development: song learning in the white-crowned sparrow. *J. Comp. Physiol. Psychol.,* 71:1–25.

Marler, P., 1970c. Birdsong and speech development: could there be parallels? *Amer. Sci.,* 58:669–73.

Marler, P., 1972. Vocalizations of East African monkeys II. Black and white colobus. *Behaviour,* 42:175–97.

Marler, P., 1973. A comparison of vocalizations of red-tailed monkeys and blue monkeys: *Cercopithecus ascanius* and *C. mitis* in Uganda. *Z. Tierpsychol.,* 33:223–47.

Marler, P., 1976. On animal aggression: The roles of strangeness and familiarity. *American Psychologist,* 31:239–46.

Marler, P., and Hamilton, W. J. III, 1966. *Mechanisms of Animal Behavior.* New York: Wiley and Sons.

Marler, P., and Mundinger, P., 1971. Vocal learning in birds. In: *Ontogeny of Vertebrate Behavior,* H. Moltz, ed. New York: Academic Press.

Marler, P., and Mundinger, P., 1975. Vocalizations, social organization and breeding biology of the twite *Acanthus flavirostris. Ibis,* 117:1–17.

Marler, P., and Tamura, M., 1962. Song dialects in three populations of white-crowned sparrow. *Condor,* 64:368–77.

Marler, P., and Tenaza, R., 1977. Signaling behavior of wild apes with special reference to vocalization. Chapter 36, this volume.

Mattingly, I. G., 1972. Speech cues and sign stimuli. *Am. Sci.,* 60:327–37.

Milligan, M., and Verner, J., 1971. Interpopulation song dialect discrimination in the white-crowned sparrow. *Condor,* 73:208–213.

Morris, D., 1971. *Intimate Behavior.* New York: Random House.

Morton, E. S., 1975. Ecological sources of selection on avian sounds. *Amer. Nat.,* 109:17–34.

Mugford, R. A., and Nowell, N. W., 1971. Endocrine control over production and activity of anti-aggression pheromone from female mice. *J. Endocrinol.,* 49:225–32.

Mulligan, J. A., 1966. Singing behavior and its development in the song sparrow, *Melospiza melodia. Univ. Cal. Publ. in Zool.* (Berkeley), 81:1–76.

Mundinger, P., 1970. Vocal imitation and individual recognition of finch calls. *Science,* 168:480.

Nottebohm, F., 1969. The song of the chingolo, *Zonotrichia capensis,* in Argentina: description and evaluation of a system of dialects. *Condor,* 71:299–315.

Nottebohm, F., 1971. Neural lateralization of vocal control in a passerine bird. I. Song. *J. Exp. Zool.,* 177:229–62.

Nottebohm, F., 1972. Neural lateralization of vocal control in a passerine bird. II. Subsong, calls, and a theory of vocal learning. *J. Exp. Zool.,* 179:35–50.

Nottebohm, F., 1975. Continental patterns of song variability in *Zonotrichia capensis:* some possible ecological correlates. *Amer. Nat.,* 109:605–24.

Nottebohm, F., and Nottebohm, M., 1971. Vocalizations and breeding behavior of surgically deafened ring doves, *Streptopelia risoria. Anim. Behav.,* 19:313–27.

Nottebohm, F., and Nottebohm, M., in press. Ecological correlates of vocal variability in *Zonotrichia capensis hypoleuca. Condor.*

Nottebohm, F., and Selander, R. K., 1972. Vocal dialects and gene frequencies in the Chingolo sparrow *(Zonotrichia capensis). Condor* 74:137–43.

Payne, R. S., and McVay, S., 1971. Songs of humpback whales. *Science,* 173:585–97.

Payne, R. S., and Webb, D., 1971. Orientation by means of long range acoustic signalling in baleen whales. *Ann. N.Y. Acad. Sci.,* 188:110–42.

Plutchik, R., 1970. Emotions, evolution, and adaptive processes. In: *Feelings and Emotions.* M. B. Arnold, ed. New York: Academic Press.

Premack, D., 1971. On the assessment of language competence in a chimpanzee. In: *Behavior of Nonhuman Primates,* A. Schrier and F. Stollnitz, eds. New York: Academic Press.

Pribram, K. H., 1970. Feelings as monitors. *Feelings and Emotions,* M. B. Arnold, ed. New York: Academic Press.

Ralls, K., 1971. Mammalian scent marking. *Science,* 171:443–49.

Rosenblum, L. A.; Levy, E. J.; and Kaufman, I. C.; 1968. Social behavior of squirrel monkeys and the reaction to strangers. *Anim. Behav.,* 16:288–93.

Rowell, T. E., 1962. Agonistic noises of the rhesus monkey *(Macaca mulatta). Symp. Zool. Soc. Lond.,* 8:91–96.

Rowell, T. E., and Hinde, R. A., 1962. Vocal communication by the rhesus monkey *(Macaca mulatta). Proc. Zool. Soc. Lond.,* 138:279–94.

Rumbaugh, D. M.; Gill, T. V.; and von Glasersfeld, E.

C.; 1973. Reading and sentence completion by a chimpanzee (Pan). *Science,* 182:731–33.

Schachter, S. 1970. The assumption of identity and peripheralist-centralist controversies in motivation and emotion. In: *Feelings and Emotions,* M. B. Arnold, ed. New York: Academic Press.

Schleidt, W. M., 1974. How "fixed" is the fixed action pattern? *Z. Tierpsychol.,* 36:184–211.

Schneider, D., 1965. Chemical sense communication in insects. *Symp. Soc. Exp. Biol.,* 20:273–97.

Scruton, D. M., and Herbert, J., 1972. The reaction of groups of captive talapoin monkeys to the introduction of male and female strangers of the same species. *Anim. Behav.,* 20:463–73.

Selander, R. K., and Yang, S. Y., 1969. Protein polymorphism and genetic heterozygosity in a wild population of the house mouse *(Mus musculus). Genetics,* 63:653–67.

Simpson, M. J. A., 1973. The social grooming of male chimpanzees. In: *Comparative Ecology and Behavior of Primates,* R. P. Michael and J. H. Crook, eds. New York: Academic Press.

Southwick, C. H., 1967. An experimental study of intragroup agonistic behavior in rhesus monkeys *Macaca mulatta. Behaviour,* 28:182–209.

Southwick, C. H., 1970. Genetic and environmental variables in influencing animal aggression. In: *Animal Aggression: Selected Readings,* C. H. Southwick, ed. New York: Van Nostrand Reinhold.

Sparks, J. H., 1964. Flock structure of the red avadavat with particular reference to clumping and allopreening. *Anim. Behav.* 12:125–36.

Sparks, J. H., 1967. Allogrooming in primates: a review. In: *Primate Ethology,* D. Morris, ed. Chicago: Aldine.

Struhsaker, T., 1967. Auditory communication among vervet monkeys *(Cercopithecus aethiops).* In: *Social Communication among Primates,* S. A. Altmann, ed. Chicago: University of Chicago Press.

Teleki, G., 1973. *The Predatory Behavior of Wild Chimpanzees.* Lewisburg, Pa.: Bucknell University Press.

van Lawick-Goodall, J., 1971. *In the Shadow of Man.* London: Collins.

Waser, P. M., 1975. Experimental playbacks showing vocal mediation of avoidance in a forest monkey. *Nature,* 255:56–58.

Washburn, S. L., ed., 1961. *Social Life of Early Man.* New York: Wenner Gren Foundation.

Wenner, A. M., 1971. *The Bee Language Controversy.* Boulder, Col.: Educational Programs Improvement Corp.

Whittaker, R. H., 1970. The biochemical ecology of higher plants. In: *Chemical Ecology,* E. Sondheimer and J. B. Simeone, eds. New York: Academic Press.

Wiley, R. H., 1973. The strut display of male sage grouse: a "fixed" action pattern. *Behaviour,* 47:129–52.

Williams, C. M., 1970. Hormonal interactions between plants and insects. In: *Chemical Ecology,* E. Sondheimer and J. B. Simeone, eds. New York: Academic Press.

Wilson, E. O., 1970. Chemical communication within animal species. In: *Chemical Ecology,* E. Sondheimer and J. B. Simeone, eds. New York: Academic Press.

Wilson, E. O., 1971. *The Insect Societies.* Cambridge: Harvard University Press.

Chapter 5

ONTOGENY OF COMMUNICATION

Gordon M. Burghardt

Ontogeny overall! he noisily confided.
As paupers we are born and then must die.
The coalescing forces fight—
or might, or may, or it is conceivable—
And do we not converse?

A glimmering of tuned machines that signal
and receive,
while bits and pieces drift across the wind.
How reciprocal is the dance with death?
Wander far, if you do wonder
What the hell it's all about.

nodrog

How does an individual animal arrive at the condition in which he or she initiates and responds appropriately to communicatory signals? How does the animal seemingly "understand" what is meant and "know" what to do? These questions juxtapose ontogeny and communication, two central and controversial concerns in the modern study of behavior. Since a theoretical or empirical synthesis is not possible at the present time, I will limit myself to discussing some selected ideas and research on ontogeny and communication and to introducing some of the methods and concepts employed in their study, while attempting to provide a framework for organizing our knowledge and asking questions. The reader is encouraged to consult other statements that present overlapping treatments with different emphases (e.g., Beer, 1971; Candland, 1971; Hebb, 1972; Hess, 1973; Hinde, 1970, 1971b; Kuo, 1967; Lehrman and Rosenblatt, 1971; Schneirla, 1965).

The specific emphasis will be placed on ethology because it provides the most comprehensive and adaptable approach for a holistic treatment of communication, ontogeny, and other important questions in animal behavior (Tinbergen, 1963), while at the same time incorporating data from other fields, including psychology, physiology, and ecology (e.g., Hess, 1962; Hinde, 1970). Ethologists have devoted most of their efforts to species-characteristic behaviors and the signals involved in communication.

Ontogeny

According to the *Oxford English Dictionary,* ontogeny is "the origin and development of the individual being" and "the history or science of the development of the individual being," including, but going beyond, embryology. In other words, it is synonymous with "development," as applied to organisms. In using the word, it is clear that no presuppositions are made about the factors that influence the development of the organism and, consequently, its behavior. To study ontogeny means to investigate the complex antecedents and consequences of the interaction of genetic, structural, physiological, and environmental events from fertilization until death.

All social behavior involves communication and takes place over time; any temporal process can be viewed as involving some developmental aspect. It would simplify our problem immensely if we could limit ontogenetic processes to changes that occur over a relatively long (in the life of the individual) time span. Yet imprinting can occur in a few minutes; a single exposure to tainted food can lead to long-term aversions by rats, and a few seconds' exposure to light can synchronize metamorphosis in insects. Fleeting events can have long-lasting effects and be important parts of normal ontogeny. Potentially all processes (not just the obvious ones) that result in behavior change can be seen as relevant to ontogeny.

Most ethologists select birth or hatching as the starting point for ontogenetic studies. A case can be made that this is appropriate when dealing with communication since it is largely a postnatal phenomenon. However, for answering certain questions it is essential to study embryos, and stimulating prenatal studies dealing with communication are being performed (e.g., Gottlieb, 1971a; Hess, 1973; Impekoven and Gold, 1973). Serious technical problems have hindered rapid advances in this area and limited most prenatal studies on communication to auditory signals in birds.

Communication

Definitions of communication usually exclude phenomena that should be encompassed or include phenomena (such as most predator-prey interactions) that are best excluded. It is easy, then, to reach the position that no definition can really differentiate communication from noncommunication or that "We all know what communication means." If communicatory processes have evolved from noncommunicatory ones, then we should expect intergradations of all sorts and an absolute distinction will be unattainable. This position is unfortunate. Often examples at the borderline between accepted distinctions lead to clarifications enhancing understanding of unequivocal examples in either category (e.g., living and nonliving, plant and animal, classical and instrumental conditioning), and this seems true of communication also.

Communication is usually treated as synonymous with social behavior. If we accept "communication" as equivalent to "social behavior," then we do not need the term "communication." However, "communication" is used to refer not so much to the behavior itself as to its signal function, its information content, and its reception and interpretation by other organisms. Most examples of communication are, however, inferred from intraspecific social interactions and take wondrously varied forms (Otte, 1974).

Elsewhere I have reviewed definitions of communication and rejected most of them. A general, but useful, characterization of communication is that it occurs when one organism emits a stimulus that, when responded to by another organism, confers some advantage (or the statistical probability of it) to the signaler or its group (Burghardt, 1970b). While the emphasis of this functional definition is based on the signaler, it does not deny that "signalers" and "receivers" can have mutually evolved and interfaced communicational systems, nor does it preclude the possibility that disparate mechanisms may be involved.

This view of communication seems to isolate the relevant aspects of communication (e.g., eliminating predatory encounters yet keeping other interspecific ones) more clearly than some other recent formulations (Hinde, 1974; Wilson,

1975). Understanding the diversity of communication systems in animals entails a close analysis of signal characteristics, habitat, and adaptive value.

Questions about Ontogeny

> Assuming the scientist has developed adequate techniques of observation, his first task is to discover which phenomena or events are related to which other phenomena and to ascertain the nature or form of these relationships. This first stage may be described as involving the discovery of *the low order empirical laws holding among particular observed events.* Playing important roles in the accomplishment of this task, of course, are the procedures or operations of experimentation and measurement. [Spence, 1953:283]

What are the ontogenetic factors, processes, or antecedent conditions shaping and altering the diverse kinds of communication in animals differing in almost every aspect: life span, speed of development, sensory capacities, motor abilities, ecological requirements, and complexity of communication? Obviously the ontogenetic mechanisms will differ a great deal. Not so obvious is the fact that psychologists and biologists have too often opted for only one or a few mechanisms, usually mutually exclusive ones at that. This seems to be due to the hope that most differences are superficial, not "real." Perhaps the success of genetics and physiology in elucidating principles (e.g., in blood circulation) that apply to entire orders, classes, and even phyla keep fertilizing the hope of harvesting specific yet widely applicable explanations. In relation to behavior, this may be true someday, perhaps never—but definitely not now.

The antecedents to communicatory acts must be characterized and a major aim of this chapter is such a brief characterization. As a starting point, a distinction between the species' evolutionary heritage and the individual's environment and consequent experience is useful if it is carefully and thoughtfully applied. Many words have been used to summarize the various aspects of this supposed nature-nurture dichotomy: innate, genetic, instinctive, unlearned, endogenous, constitutional, phylogenetically adapted, maturational, reflexive, stereotyped, and hereditary on one side; learned, experience, stimulation, modification, environmental, variable, conditioned, acquired, plastic, and emergent on the other. Failure to recognize that the terms in either group are not synonymous with one another, and that these two sets do not refer to necessarily mutually exclusive events, led to loose and often needless arguments. Scientists are now agreed that the behavior of any organism is a combined outcome of its heredity, environment, and sometimes active participation. Where workers differ is on the relative emphasis given one or the other, the questions asked, and the methods and terminology employed (see Burghardt, 1973; Gottlieb, 1973; Hebb, 1972; Hinde, 1970; Lehrman, 1970; Lorenz, 1965; Thorpe, 1974; and Whalen, 1971 for some contrasting views). Here we will try to short-circuit some of these issues by providing a framework that does not contain, or even imply, an inherent bias for either nature or nurture. The lack of such a framework for posing questions has led to many impatient workers' verbally ostracizing the entire issue of the causal analysis of ontogeny from modern ethology. But this stance is unacceptable if we truly want to understand behavior and its origins, regardless of whether we accept Tinbergen's (1963) characterization of the four classes of problems in behavior (ontogeny plus evolution, function, and causation) or Schneirla's (1966:284) more extreme position that "Behavioral ontogenesis is the backbone of comparative psychology."

The living organism we see, smell, and hear is a phenotype that results from a given constitution (the genotype) developing in a given environment. Anatomical, physiological, and behavioral characteristics of organisms are all phenotypic characters. Now note that the first set of "nature" terms above ("innate" et al.) traditionally refer to properties derived from the genotype, while the second, or "nurture," set ("learning" et al.) refer to postfertilization environmental effects on the phenotype. It is apparent that the two sets of terms cannot be in opposition with a sort of inverse relationship to each other. Obviously all genotypes need an environment and the environment can only act on phenotypes containing a given genetic constitution. But we cannot stop, as some would, with an "everything affects everything" view. It may, but not equally, and this must be explained.

The next step, then, is to posit that any character (behavior or structure) can be differentially affected by given ranges of environmental (developmental) variation. That is, different phenotypes can result from the same genotype. Similarly, different genotypes can result in phenotypes that are not readily discriminable from one another. Many of the issues in ontogeny revolve around the nature of the limits set by different genotypes on phenotypes. Today a rather "tight" genotype-phenotype correlation would lead to labeling a behavior "genetically fixed," "environmentally stable," or "innate," and a "loose" correlation would lead to the label of "environmentally labile," "learned," or "acquired." But where to draw the line in a continuous gradation is a worry of many people (e.g., Marler, 1973). This issue could be handled if determining the nature of the genotypic limits was straightforward. But not only do methods of accomplishing this vary, but also researchers often take different levels or parameters of behavior as their units of analysis and then generalize to a broader functional unit. For example, Hess (1973) and many classical ethologists view imprinting in ducks and geese as a highly specialized form of instinctive behavior, bearing little resemblance to "learning" as studied by animal psychologists. Others stress the modifiability of imprinting and the operational similarity of some object acquisition aspects of imprinting to traditional learning paradigms (e.g., Hinde, 1970). We can see the origin of a labeling conflict in this example. If one is forced to label imprinting as "innate" or "acquired," it is obvious who will put it where, albeit grudgingly today.

Mayr (1974) attempts to solve the problem of the genotype-phenotype link by classifying behavior as due to either "closed genetic programs," in which "nothing can be inserted" through experience, and "open genetic programs," which allow for "additional input during the life span of the owner." While this terminology has some value, being derived from molecular biology and information theory, it is sometimes not useful at the level Mayr applies it. The open programs encompass too much, and too large a functional unit is used. Thus he calls imprinting an "open" system, a label that tends to make us lose sight of species differences and the initial visual and auditory preferences that put constraints on even some of the most "open" aspects of imprinting involving species recognition.

As another example, consider food recognition in newborn snakes (Burghardt, 1970a). Snakes of several colubrid genera (e.g., *Thamnophis, Storeria*) are born with the ability to recognize and attack chemical stimuli from specific types of animals, such as fish, earthworms, leeches, and slugs, which represent species-characteristic prey. Such responses to prey cues are stereotyped, show evolutionarily understandable differences and similarities across species, are difficult to habituate, are seemingly impervious to maternal prenatal experience (Burghardt, 1971), and remain intact over prolonged postnatal stimulus and response deprivation. These findings more than meet the criteria for a closed program advanced by Mayr. Yet within several feedings on actual prey, the young snakes increase their relative preference for it, while a single "bad taste" or illness associated with prey can dramatically decrease the food's palatability (Burghardt, et al., 1973). Even brief exposure to prey odors without ingestion can alter responses to prey! Is food recognition in snakes innate or learned? Genetically fixed or environmentally labile? Controlled by a closed or open genetic program? We seem to be in a quandary.

The resolution may be simple. Let us formalize what many workers are actually doing empirically, but discussing loosely. We need to inquire *as to which aspects of a behavior (e.g., topography, patterning, stimulus control, duration) are determined by which genetic or environmental influences. The relevant phenotype is a parameter of the behavior, not necessarily a functional or topographic unit, regardless of the level at which it is defined.* We must work to determine the evolutionary antecedents of the behavior repertoire of an individual and their relation to the historical development of phenotypic appearance. We can apply this approach to the question of whether differences in phenotypes are associated with differences in genotype or environment (Whalen, 1971). It is also compatible with asking whether the behavioral "information" contained in the organism (usually its nervous system) originated in the genome (genetic program or blueprint) or was acquired through experience (Lorenz, 1965; Mayr, 1974; Thorpe, 1974). Both populational and individual approaches (Burghardt, 1973) can thus be accommodated.

This "parameter of behavior" approach largely eliminates the need for arbitrary cutoff points on the continuum between "tight" and "loose" genotype-phenotype correlations. It also helps us realize that there is no such thing as "learned behavior," "innate behavior," and so forth (Verplanck, 1955). These are merely shorthand terms to refer to the origin of the aspect of the behavior in which we are interested. By this view, initial recognition of chemical cues from species-characteristic prey in snakes seems innate. Nothing is implied about the modifiability of such recognition after birth. On the other hand, we must still retain the distinction between inherited and environmentally based factors shaping a behavior, which careful writers on behavior have not been able to avoid, regardless of their verbal inventiveness and clever rhetoric. Since the time of Darwin we have known that phenotypic variation based on some inherited processes is necessary for adaptiveness and evolution.

The use of the term "parameter" here extends and makes more explicit the distinction Lorenz (1965) wanted to make between "character," on the one hand, and "organ" and "behavior pattern," on the other. However, he did not define his usage with sufficient clarity to avoid confusion. The "polythetic" approach advocated by Jensen (1970) and derived from numerical taxonomy is also compatible with this view of using operational approaches to confront, rather than avoid, critical empirical issues.

Species, Operations, and Outcomes

> The development and use of instruments and experimental designs that provide for the isolation, control, and systematic variation of the factors in the situation under observation are an obvious requirement for the discovery of laws, especially when the situation is one in which a large number of factors are operating. [Spence, 1953:283]

Our formulation does not necessitate that extremes in the tightness or looseness of genotype-phenotype correlations exist, nor does it stress their relative prevalence; but it does indicate what to look for and how to interpret our findings. Only if the evidence shows that virtually all parameters of a response are "tight" or "loose" can we generalize to the entire functional unit. To illustrate these two points let us look at some experimental examples in a most well-studied area, auditory communication.

Crickets (*Gryllus, Acheta*) produce sounds through use of the forewing. This process, termed "stridulation," appears only in males. One of the sounds they produce is a species-characteristic call to attract females. Numerous studies have established the following ontogenetically important findings (references in Alexander, 1968, unless otherwise noted; see also Otte, this volume):

1. While external stimuli can trigger singing bouts, they can also occur without any known external clue.
2. Different cricket species may have different but similar calls. Such differences seem to be based largely on the number of separate sound pulses per trill, the number of trills per phrase, and the intervals between them. Females do not get confused and readily recognize their own species call.
3. Hybrid males are generally intermediate between the parent species in call parameters. Yet through backcrossing Bentley (1971) was able to show that some aspects of the call could be correlated to the presence or absence of a specific sex-linked chromosome.
4. Exposing various stages of the developing cricket to foreign sounds, including calls from related species, does not affect the calls of the male or the discrimination and preference for conspecific calls by females.
5. Male crickets deafened either before hearing their calls or as experienced adults still stridulated normally.
6. Severing the wings or otherwise altering proprioceptive feedback had no effect on male calls.

Such results indicate that in crickets the singing of males and the preference of females are stable in spite of a wide variety of environmental manipulations. If there are environmental events that would alter the singing, they have not yet been found. We have here a case of a very tight genotype-phenotype correlation, a nearly perfect example of Mayr's (1974) closed genetic program. Since virtually every aspect studied seems resistant to environmental modification, it is valid to refer to auditory courtship communication in crickets as genetically fixed, environmentally stable, or innate, but only as a shorthand summary of the specific experimental data. One might, however, make reasonable inferences about results to be expected from experiments using related species or alternative manipulations.

Surprisingly, mammalian vocalizations have been less well studied. However, a few examples can be given (see Salzinger, 1973):

1. Squirrel monkey (*Saimiri sciureus*) infants raised by surgically muted mothers and isolated from all species-characteristic sounds evidenced normal calls. Further, an infant deafened five days after birth also showed the normal vocal repertoire (Winter et al., 1973).
2. Salzinger and others have shown that barking in dogs, meowing in cats, and calls in monkeys can be brought under operant control; that is, the rate of vocalization can be modified by reinforcement (reward) and conditioned to a stimulus (such as a light) that is present during reinforcement. The form of the responses has not been shown to vary, but both the rate and the stimulus control are influenced by events occurring throughout the life of the individual.
3. Sea lions (*Zalophus californianus*) were conditioned to produce a "click" sound, never given previously in captivity, in the presence of a circular target (Schusterman and Feinstein, 1965). The sound was even used as a means of determining auditory thresholds (Schusterman et al., 1972).

4. Great apes that have been hand-reared from infancy have been taught to speak a few words of human speech in the proper context.

Thus, in some mammals we find a tight genotype-phenotype correlation in the form and pattern of species-characteristic sounds, but less so than with crickets, especially in relation to frequency and stimulus control.

Let us now turn to birds. References for the following summary of findings on birds can be found in Marler and Mundinger (1971) and Marler (1973):

1. Bird vocalizations can be divided into the longer "songs," occurring in territorial and mating behavior, especially by males, and the shorter "calls," used for alarm and flight. The calls are minimally modified by social isolation or cross-fostering with other species with different repertoires. This is true of all or most calls of eastern meadowlarks (*Sternella magna*), whitethroats (*Sylvia communis*), song sparrows (*Melospiza melodia*), chaffinches (*Fringilla coelebs*), ring doves (*Streptopelia risoria*), and domestic chickens.

2. Rearing birds in complete auditory isolation led to some species' developing normally (chickens) and some abnormally (robins [*Turdus migratorius*] and grosbeaks [*Pheucticus melanocephalus*]: Konishi, 1965).

3. Doves and chickens deafened a day or two after birth still vocalized normally at maturity. This did not hold for robins and grosbeaks (Konishi, 1965).

4. Ducklings devocalized in the egg and hence unable to hear themselves still have normal preferences for maternal calls (Gottlieb, 1971a).

5. Arizona juncos (*Junco phaeonotus*) reared from four to five days of age in individual isolation develop a much simpler song than is normal. Hearing the adult conspecific song, however, was *not* necessary for developing the normal song, as young males reared with an agemate developed songs that were indistinguishable from those of wild adults. In the chaffinch such joint experience only enhanced song complexity, while still leaving it abnormal.

6. The chaffinch will not develop the species-characteristic song unless it hears the adult conspecific song. But if exposed to a call slightly different from the normal song only a small amount of imitation will occur. Further, if the bird is exposed to both the normal and an alien song, it will selectively learn just the conspecific song. When the chaffinch is old enough to sing itself, at 300 days of age, its song is no longer capable of being changed by such experience and is virtually unmodifiable.

7. In the white-crowned sparrow *(Zonotrichia leucophrys)* brief exposure to the species-characteristic song for a three-week period (the "sensitive period") between the tenth and fiftieth day of life was all that was needed for normal song at the later appropriate time. Later experience had little effect (Marler, 1973). Exposure to only an alien call led to songs resembling those of social isolates.

8. Immelmann (1969), who deafened estrildid finches after the sensitive period, found that male zebra finches *(Taeniopygia guttata castanotis)* foster-reared by society finches *(Lonchura striata)* would later sing the foster parent song, even if conspecific calls were given in the area during the sensitive period. Hence the parental bond can be of importance in focusing attention on what is to be learned.

9. Minor, but consistent, population differences in the songs of white-crowned sparrows are due to exposure to those variants during the sensitive period.

10. Some birds (parrots, mynahs) readily imitate human words, calls of other animals, and even whistles as adults—sounds they presumably never heard earlier. In the case of whistling, the

addition of food reward had no effect on the speed or quality of imitation (Foss, 1964).

11. The frequency of sounds normally part of the animal's repertoire can be increased or decreased through standard conditioning techniques. Chirping in domestic chicks is a good example (Lane, 1961).

In birds, in contrast to crickets and nonhuman mammals, a large variety of modifications can be made in some species, but generally the acquisition and modification of species-characteristic sounds are possible only under special circumstances or antecedent conditions. Bird vocalizations demonstrate the wide variety of experimental operations that may or may not affect the outcome. This shows us that a simple correlation of a behavioral phenotype with genotype or environment, while necessary, is not sufficient to understand ontogeny. It does not distinguish for us the myriad ways in which diverse genotypic and environmental variables interact, and it fails to give us more than guideposts to the elucidation of *how* behavior arises in the development of the individual.

A Model of Life History

The genetic constitution of an organism remains constant throughout life, sort of an indelible serial number affixed at the factory (somatic mutations excepted). But from the moment egg and sperm unite, a complex interaction begins between the zygote's genotype and the diverse aspects of its external and internal environment. A dynamic system has begun, which will not end until death. But the types of processes that occur and their rates differ across species and individuals. Observation of animals developing, either prenatally or after birth, leads to descriptions of characteristic modes and levels of anatomical, neurological, and behavioral development. Cross-sectional (comparison of animals at different ages) and longitudinal (long-term studies on individual animals) studies are necessary to establish adequate descriptions and correlations. More loosely, references are made to stages or plateaus of development such as are found in embryological studies (e.g., blastula, type I motility) and in behavior studies (infant, juvenile). While the criteria used for distinguishing stages of behavioral ontogeny are often arbitrary, inconsistent, and not at all sharp, scientists often find them useful (e.g., Hinde, 1971a). The remarkable incorporation of Piagetian thought into all areas of child development seems primarily due to Piaget's codifying a plethora of stages and substages (Etienne, 1973). These useful, but often crude, classifications must remain only descriptive until they are tied to specific relationships.[1] The occurrence of discrete events is also important; hatching, birth, molting, pupating, and metamorphosis. In addition, many behaviors appear suddenly and seemingly "full blown," as is true of many fixed (modal) action patterns and complexes of seasonal behaviors like courtship or producing and caring for offspring. Prenatal studies are more often neurological or physiological, but that does not eliminate their relevance to later communicatory phenomena. A comprehensive but concise treatment of the well-studied prenatal behavior of the chicken (Oppenheim, 1974) is most useful in demonstrating the unification of descriptive and experimental approaches.

A vivid and accurate portrayal of ontogeny can be elaborated from Waddington's (1956) epigenetic landscape, in which a marble rolling down a hill with ridges and valleys represents the zygote from the moment of fertilization (the top of the hill) to the end of life. The width, depth,

1. The proliferation of developmental stages parallels that of "instincts" as carried out by McDougall's more enthusiastic followers (see Bernard, 1924). Doubtless it will lead to a corresponding, unwarranted discrediting of Piaget's constructive aspects in years to come.

and steepness of the terrain represent the speed and plasticity of development, and the almost level areas represent stages. Further, the mass, size, and shape of the "marble" determine its interactions with the environmental topography. Using this model, we can appreciate that the narrower the valleys and the higher the ridges the more difficult it is for the marble to shift from one development pattern (valley) to another.[2] The ontogeny of animal vocalizations in different species, we have seen, incorporates many such widely varying "terrains."

Now let us consider the life cycle of an individual in relation to studying communication. From the descriptive information about the communicatory behavior of the adult of the species, we should be able to find the moment when a given aspect of behavior is first shown in its communicatory context or is synthesized out of already extant elements. We then either focus our attention on this first occurrence or move in two directions—earlier (which usually means studying different individuals of the same species) or later (which often involves longitudinal observations). If we look earlier, we aim at uncovering those antecedent conditions that led to the later functional behavior (conditions I shall call "prerequisites"). If we stay at the moment the behavior first occurred and the specific internal and external stimuli involved, we are studying the "corequisites." Much of the ethological sign stimulus and "Innate Releasing Mechanism" work is done at this point of first occurrence. If we look at what happens, or can happen, after the behavior is in the animal's repertoire, we are dealing with change or modification. Here we will deal not with the immediate causation or corequisite question but only with points earlier and later than the first occurrence. However, the corequisites can be viewed as being closely related to the prerequisites.

Students on the ontogeny of communication, then, look for the *variables, prerequisites, or antecedent conditions that lead to the expression of species-characteristic communication in the individual at the proper time and situation (hereafter referred to as Question O for origins) or are capable of modifying, altering, or maintaining existing patterns of communication (hereafter referred to as Question M for modification).* This distinction between origins (first occurrence) and modification is important because processes involved in the developmental shaping of behavior may have little in common with those subsequently altering such behavior, but are often confounded in discussions. Studying the conditions that lead to the chaffinch's singing its species-characteristic song for the first time relates to Question O. Attempts to condition the frequency of chaffinch singing relate to Question M.

More on the Question of Origins: Prolegomenon to Precursors

The fact that a behavior occurs at a certain age does not mean we should no longer be concerned with its topography or function. First, it may not yet be in adult form. Longitudinal studies show that in jungle fowl *(Gallus gallus)* fighting develops out of hopping, which gradually, from one week of age, becomes directed toward other chicks and subsequently incorporates pecking and kicking (Kruijt, 1964). Similar processes of the phenotypic appearance of ritualized behavior have been shown in the facial expressions of canids (Fox, 1969). Second, even if the behavior is in its adult form and context, various refinements may be seen over time. Questions O and M, then, are specific to the age and response selected. Most studies that fall un-

2. W. S. Verplanck suggested to me the analogy of a snowball rolling down a snow covered landscape, rapidly increasing in size and weight. Thus changes in the marble occur which influence subsequent interaction with the environment. Taking this even further, large boulders and trees could play the obvious role of hazards.

der Question M (modification) deal with changing the frequency, duration, or stimulus control of a response that is already in its adult form, as in the barking of dogs (Salzinger, 1973).

The answering of Question O (origins) has suffered most from confusion dealing with the nature of the antecedent conditions leading to the communicatory act. A modification and extension of Gottlieb's (1973) discussion of "precursors" in behavioral embryology is useful here in characterizing the problems of postnatal as well as prenatal origins (Burghardt, 1975). Recall the opening question of this chapter: "How does an individual animal arrive at the condition in which he or she initiates and responds appropriately to communicatory signals?"

There is a grouping of precursors that influence the pace of ontogeny temporally or quantitatively: that hasten, retard, or prevent the development in the individual of its species-characteristic patterns of behavior. Such retardation can take the extreme form of completely abolishing certain abilities ("atrophy"). For instance, raising an animal in the dark may irreversibly destroy pattern perception. If the species involved is one in which visual cues play an important role in mate selection, it is clear that the animal may show deficiencies or even be an abject failure in courtship and the communication involved therein. Exposure to patterned light is thus a "facilitative precursor" in this hypothetical example and perhaps a maintenance factor too. But such evidence does not tell us why the species comes to communicate in one way and not another. It is the "determinative precursors" that actually "force or channel neurobehavioral development in one direction rather than another" (Gottlieb, 1973:6). They are the source of the "adaptive information" contained in the behavior (using the terminology of Lorenz and Thorpe). With facilitative precursors all we can say is that relatively nonspecific experience or stimulation of a certain kind is necessary (prerequisite) for certain abilities to develop and manifest themselves at the appropriate time. They are often likely to affect several behavior patterns. Note that in the courtship example both patterned light perception and courtship behavior are involved, and that a given event may have a determinative, facilitative, or maintenance effect depending upon the aspect of behavior being considered. Bateson (1976) discusses "specific" and "non-specific" determinants, both inherited and environmental, which may affect any given behavior pattern.

Gottlieb (1971a) demonstrated a somewhat specific prenatal facilitative precursor. He established that newly-hatched peking ducklings *(Anas platyrhynchos)* selectively approach the species-characteristic maternal call when simultaneously presented with the maternal call of related species. Even prior to hatching the ducklings selectively respond to the maternal call, as ingeniously measured by bill clapping and heart rate. But embryos that could hear the sounds made by other embryos showed a faster development than sound-isolated embryos. Nonetheless, postnatal behavior was unchanged in spite of the different history of occurrence. Ducklings auditorily isolated and devocalized in the egg, to prevent an embryo from even hearing itself, showed decreased discrimination after hatching; but they could still discriminate between mallard and wood duck maternal calls. However, at 48 hours discrimination between the mallard maternal call and the more similar chicken maternal call was seemingly abolished, although some improvement was noted later. Thus, we see that Gottlieb's analysis is an elegant example of what seems beyond quarrel: "Exposure to normally occurring stimulation helps to regulate the time of appearance of the perfected response, as well as the latency of the perfected response" (1971a: 134).

The danger of considering behavior a unitary

phenotype is made evident by the concept of facilitative precursors.[3] For example, a child's temper tantrum is a complex sequence of behavior that is all too perfect the first time it is performed. Hebb claims that to call the behavior instinctive or unlearned is wrong because "it is not possible without the learning required for the development of purposive behavior" (1953:43). He also claims that the dramatic onset of fear of strangers in a chimpanzee baby "is not learned: but it is definitely a product of learning, in part, for it does not occur until the chimpanzee has learned to recognize his caretakers" (ibid.). Such confusion and subsequent retreat to the "everything is everything" view are due to the error of considering a behavior as a unitary whole and not treating as separate questions which aspects (topography, rate of occurrence, etc.) are due to what. In the two examples above the topography of movements is not determined by the hypothesized prior learning, while the elicitation of the behavior definitely seems to be, that is, the latter is facilitated by a certain type of experience.

Facilitative precursors may either be necessary (but insufficient) for the manifestation of a behavior or, by their absence, cause deterioration of an already present (but yet undemonstrated) capability. The latter process occurs in cortical neurons that respond to specific types of visual cues in cats (see Blakemore, 1973). The indirect and subtle effects of many kinds of experience may thus influence behavioral development. Gross sensory and motor abilities, as well as certain experiences, may obviously be prerequisite for the manifestation of the complex and specific behavior and perception involved in communication.

Facilitative prerequisites based on sensory stimulation in the embryo seem much easier to demonstrate prenatally than determinative ones based on stimulation (Gottlieb, 1973). Prenatally, genes and hormones have been shown to be determinative. Postnatally, a variety of other processes may be at work (conditioning, imprinting) but often it is not easily shown that they are determinative. This contrasts with the expressed hopes of theorists such as Schneirla (1965) and Moltz (1965) that the environment is "actively implicated in determining the very structure and organization of each response system" (Moltz, 1965:44). In fact, most studies that have been presented as disproving traditional conceptions of innate behavior or instinct unfortunately do no more than prove the necessity of certain facilitative or maintenance precursors for normal behavioral ontogeny. Rarely do they treat the essential question of determinative precursors. Studies on Question M (modification) are even less related to determinative factors. Classical ethologists (e.g., Lorenz, 1965; Thorpe, 1974) and evolutionary theorists (e.g., Mayr, 1974) clearly view the determinative factors underlying communication as being either ultimately genetic or shaped strongly by evolved mechanisms. Too many authors limit nongenetic determinative precursors to "reinforcement learning" or a similar unitary process (e.g., Dawkins, 1968). A summarizing diagram sets forth the relationships developed here (Fig. 1).

3. Facilitative precursors as defined here are similar to the so-called nonspecific factors, used by some writers to discredit "instinctivists" by showing that in theory and practice some manipulation of organism or environment can abolish or alter any and all "innate" responses. The problems in distinguishing specific from nonspecific factors are almost insurmountable, but may be of heuristic value (see also Bateson, 1976).

Some Antecedents and Modifiers of Communication

In the early stages of development of knowledge in any field the scientist does not know all of the variables entering into a particular set of phenomena, and he must, therefore, hypothesize or hazard guesses as to what these unknown factors might be. [Spence, 1953:284]

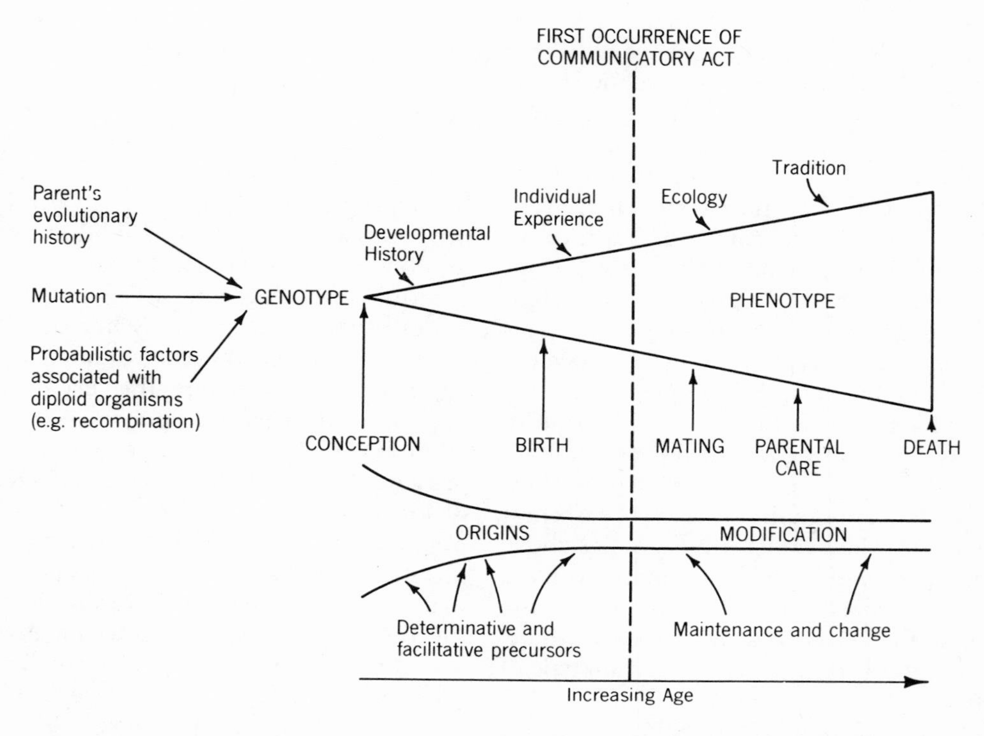

Fig. 1. An outline of the relations between genotype, phenotype, and the questions about antecedent conditions responsible for a given communicatory event. A typical but simplified ontogenetic sequence is incorporated into the diagram, which relates to the following generation. Mating feeds back to the genotype, and parental care to the phenotype.

It must be emphasized that the terms "factor," "precursor," or "variable" as applied to processes affecting communication are crude and oversimplified. Factors are often sets of heterogeneous mechanisms that are only superficially similar, while differences between the various factors may themselves be superficial. Here we will discuss some of these antecedent conditions. In actuality, any given act of communication can be affected by several of them. Most can conceivably act as facilitative precursors, determinative precursors, maintenance factors, or modifiers. Following is a catalogue of what seem to be the most debated and studied types of precursor and modifier.

GENOTYPE

Genetic differences are often associated with differences in communicatory behavior and hence survival and evolution, as in the cricket examples. Manning (1971) and Gould (1974) give excellent reviews of genetic mechanisms affecting behavior, with many examples from the social realm. But it must be remembered that any given aspect of adult behavior or structure is usually due to the action of several to many hundreds of different genes (alleles). This contrasts with the one- or two-gene models derived from simple Mendelian heredity, with clear-cut dominant and recessive factors. Similarly, a change in one or a few genes may affect several characters in the adult (pleiotropy). An especially pertinent case familiar to cat fanciers is the Siamese cat, in which a recessive gene yields individuals showing both a specific pigmentation and a distinctive vocalization with equal conspicuousness. Recent studies (Benzer, 1973) with *Drosophila* demonstrate that precise localization of behavior patterns to specific mutations can be attained.

Genotype is obviously prerequisite to the organism and hence to all communication, and in this sense Ginsburg (1958:404) is correct in stating that "all aspects of an organism may be thought of as 100 percent genetic but not 100 percent determined," genetically determined that is. But genotype is often a determinative factor in, for example, courtship communication, which can differ greatly between closely related species (e.g., crickets, ranid frogs). Such "isolating mechanisms" prevent similar species, with overlapping ranges, from mating successfully. However, most genetic studies necessarily focus on the often much less dramatic differences seen within or between populations of the same species. An exception is the discovery that different strains of mice refract identical experiences in opposite ways, becoming more or less aggressive, while still other strains are not affected at all (Ginsburg, 1966). This example also illustrates that genetic factors are prerequisite to and influence the modifiability of behavior ("learning")—a standard, if only newly rediscovered, point in many recent texts and symposia. But *how* genotype affects modifiability of any behavior, least of all communication, is too rarely studied.

DEPRIVATION OR AUGMENTATION

As an alternative to genetic experiments we may find that animals reared in complete social isolation (Kasper Hauser on deprivation experiments) show the normal repertoires of signaling behaviors and can respond properly to signals from conspecifics (Lorenz, 1965). These studies show that whatever events in the ontogeny of the individual may be necessary for the origin of these behaviors, they do not include commerce with other members of the species. Of course, abnormal rearing conditions are highly destructive where such normal experience is needed (facilitative or determinative) for critical environmental input. Rhesus monkeys are frequently cited as cases in which social deprivation can have an adverse effect on later social behavior (Harlow and Harlow, 1965), but one should remember the magnitude of the abnormal conditions necessary to produce such an effect. Even in rhesus monkeys, however, isolation-reared individuals recognize visual signals, such as threat (Sackett, 1966). When communication is unimpaired in spite of postnatal social deprivation, we can state that the determinative factor for information contained in the display is probably ultimately tied to the genotype and that its manifestation (phenotype) is impervious to certain kinds of experience. A distinction must be made between the performance of the display and the response to the display. These may be differentially affected by experience and deprivation. The commonly used term "expression" and the less-common companion term "impression" also get at the distinction (Leyhausen, 1973).

Prenatal facilitative factors may be involved but, as we have seen, evidence for experiential prenatal determinative factors is sparse indeed. As the bird examples show, the genetic variable can be manifested in various ways, ranging from a predisposition to learn the call exposed to in the nest (finch), through a predisposition to learn only the species-characteristic call (chaffinch), or to a mainly genotypic determinative factor that seems virtually impervious to all kinds of deprivation or augmented experience as well (chicken).

A final example will underline the importance of distinguishing facilitative and determinative precursors. In the oft-quoted controversy on nest building in naive virgin rats (see review in Burghardt, 1973), Riess (1954) argued, in effect, that prior manipulatory experience was the source of origin of the movements involved, as his rats reared without the opportunity to manipulate objects did not build nests. Eibl-Eibesfeldt's (1961) elegant demolishing of Reiss's position by taking into account the home environment only succeeded in bringing out another facilitative factor. Even if Eibl-Eibesfeldt's conditions had not led to nest building, we would still not know how the nest-building movements of rats originate ontogenetically.

EMBRYOLOGICAL FACTORS

Caspari (1971) cites the importance of differentiation. All the varied cells and structures of the adult organism begin from a fertilized egg with one combined set (from both parents) of genetic instructions. Yet while early cells seem to look the same, all the adult cells, regardless of their differences, still contain the same genes and therefore the same information. As Caspari (1971:3) puts it: "How do cells which contain the same genetic information assume a number of different biochemical and morphological states?"

In addition to differentiation, we also need to be concerned with "morphogenesis"—how and why differentiation occurs among groups of cells at different times and locations. Why does one sensory system develop before another (Gottlieb, 1971b)? How does a leg form differently from an arm? Although we will not dwell on such issues here, it is interesting to note that Caspari claims that the biological principles underlying differentiation and morphogenesis are sufficient to explain the development of behavior (see also McClearn and DeFries, 1973). Since most scientists consider the essence of communication to be behavior, the implication concerning the ontogeny of communication is obvious. But such an oversimplistic approach may lead to an ignoring of the special complexities and diversities found in the behavior of interacting organisms.

MATURATION

The question of origins often revolves around the issue of maturation. Animals and their behavior patterns change with age, and an older individual often evidences behavior not seen in the embryo or the neonate (newly born or hatched animal). But controversial issues involve the intervening steps between DNA and behavior: the influences, sources of information and variation, and reciprocal interactions that are responsible for an observed change in behavior. The phenotype changes, but the genotype does not. Is behavior that is seen in older organisms but not in the neonate, such as courtship, due only to maturational processes of the nervous system, physiology, and structure involving growth, proliferation, migration, and differentiation (Gottlieb, 1973)? If so, then the genotype-phenotype correlation is high given a minimal environment for survival. This phenomenon, called "maturation," is often posited as a factor on the basis of deprivation experiments. Calling in male crickets is as good an example as one could want. But while maturation is undoubtedly an important antecedent process in behavior, it is clearly not the only one even in crickets, and analyses then become snagged on conceptions of "experience" and "learning." The "closed genetic program" of Mayr (1974) closely resembles maturation. Because of the biological value of rapid and unambiguous species recognition, he concluded that communication behavior is much more closed than behavior patterns involving food or habitat, which involve ecological conditions that are much less stable and predictable for the species than are the appearance and behavior of conspecifics.

FUNCTION

While the function-structure issue can relate to broad evolutionary questions—such as whether a distinctive wing patch arose before or after the behavior of flashing that area of the wing (e.g., as shown by mockingbirds), its specific current referent is in ontogeny. During maturation, what facilitative or determinative role for the later mature system is played by the functioning of the immature system? Does structure lead function, or can function influence structure even while maturation is taking place? In our formulation, "mature system" refers to the communicatory act (Fig. 1), but most of the current lively debate on this issue is played in the embryological theater. There the issue deals with the effects of use or disuse of muscles or sensory stimulation on neurobehavioral development (see papers in Gottlieb, ed., 1973) and is the homology of the "nonspecific experience" issue in postnatal organisms.

The validity of the structure-function dichotomy seems meaningful to people on either side of the issue. But is it? As soon as we move out of the embryo, we have ample evidence that environmental events can facilitate or determine aspects of developing (maturing) systems, as in imprinting. Rather than allude to an almost mystical "function" (structure can similarly be unwisely invoked) we should instead isolate and investigate the role of the various possible precursors. One would certainly expect to find a range of genotype-phenotype correlations. While the experimental evidence may be in doubt in a given case, use of the framework outlined here precludes broad function-structure debates on inappropriate levels.

HORMONES

Secretions of hormones can determine the course of behavior development, most conspicuously in sex-related behavior, but elsewhere as well. Consequently, hormones may be determinative precursors. Many experiments have shown that castration and/or supplemental injections of either male or female hormones can lead to major changes in sexually dimorphic behavior in many and diverse vertebrates (Young, 1965). Normally, phenotypic sex is virtually always the same as genetic sex, but experimental work shows that hormone manipulation is one of the few known cases where a genetic determinative precursor can be qualitatively overridden (not just suppressed). The hormonal organizing effect may be limited to a certain period in the animal's life.

SENSITIVE PERIODS

In a number of communicatory phenomena, including classical imprinting, song learning, and hormonal organization, a limited time span is most important in the establishment or modification of certain modes of response or perception. These may be measured in hours, as in filial imprinting in ducks; minutes, as in maternal imprinting in goats; or weeks or months, as in sexual imprinting or song learning. Sensitive periods differ in their "criticality" and other attributes. Hess (1973) documents many examples and presents a convenient tripartite classification that helps to order the diversity. The examples of bird song learning clearly show the importance of sensitive periods.

It is useful to recognize two broad functions of sensitive periods. In one, the animal acquires new information necessary or useful for survival (e.g., species-characteristic song). In the other, experience during the sensitive period is necessary for abilities already prepotent or maturationally dependent to be expressed, perfected, or maintained (e.g., sensory effects). These functions can be combined, but often one or the other predominates. In other words, sensitive periods can be involved as both determinative and facilitative precursors.

ASSOCIATION

Association or contiguity refers to the pairing, or almost simultaneous occurrence, in time and place of two events. Obviously many of the experiential precursors to communication involve association, as in imprinting and song learning. Psychologists, unfortunately, have sought to elucidate the mechanism of association by elaboration of a highly specialized, abstract, and behaviorally impoverished procedure. If one of the two events (e.g., food) elicits a certain behavior (e.g., salivation) and the second (such as sound) normally does not, the second may begin to act as an effective stimulus even in the absence of the first. Such is the paradigm of classical or respondent conditioning (Type I or Type S).

While classical conditioning has assumed great theoretical importance in psychology, it has been considered much too narrowly, especially in relation to communication, and has mainly been applied to the modification of existing noncommunicatory processes. An exception is the history of attempts over more than forty years to condition the vocalizations of rats when shocked with electricity, which is nicely reviewed by David and Hubbard (1973). Successful conditioning is related to the animal's mobility and the use of a variable inter-trial interval. Thus, far too often, the "principles" emerging from psychological research on conditioning are artifacts of the particular events chosen for juxtaposition and of the convenient species whose members are studied. What else can we say about the history of attempts to demonstrate "conditioning" in embryos by the rigid application of an arbitrary paradigm? We know now that classical conditioning cannot override biological constraints on what stimuli can be associated with what responses, and that close temporal contiguity of UCS (unconditioned stimulus) and CS (conditioned stimulus) is not always necessary (see Hinde and Stevenson-Hinde, 1973; Seligman and Hager, 1972). Principles based on prenatal association have been invoked in ingenious (but empirically unsupported) ways by Schneirla (1965) in his approach-withdrawal theory of behavioral ontogeny.

REINFORCEMENT AND PUNISHMENT

Reinforcement (reward) refers to the presentation or withdrawal of events (e.g., food or shock, respectively) that lead to an increased frequency or probability of a response. Punishment refers to the presentation to the animal of an event (usually shock) that decreases responding. The animal must first respond, then a consequence follows, which increases or decreases the probability of its performing the behavior again under the same or similar circumstances. Skinner (1966) has drawn a parallel between reinforcement and natural selection, the former directly altering behavior patterns, and the latter the genotype. As with the link between association and classical conditioning, reinforcement is too often discussed only in the context of instrumental or operant conditioning, thus implying a certain ubiquitous procedure for studying it. Another danger is that all changes in behavior can be attributed to the action of some reinforcer somewhere. The recent views of Premack (1965) are helping us to realize that reinforcers are, in fact, responses, and he thus broadens the classical ethologists' view that the performance of fixed (modal) action patterns is reinforcing.

It is certainly true that rewards, such as food or water, and punishment, such as electric shock, can change the frequency and situation-specificity of behaviors, as the examples of chirping in chicks and barking in dogs suggest. Also, aggressive behavior in a male chicken can be eliminated quickly if he is shocked each time he attacks another chicken (Radlow, Hale, and Smith, 1958), while Thompson (1964) showed that the visual image of a rooster, in the form of a second animal or a mirror, would reinforce a key pecking response in an operant situation. Thompson (1966) also demonstrated parallels between ethological "releasers" and operant "reinforcers" in that the former can act as the latter in Siamese fighting fish *(Betta splendens)*. Melvin and Anson (1969) showed not only that male fighting fish would greatly increase their frequency of swimming through a narrow aperture when reinforced with a mirror to which they could display, but also that moderate electric shock punishment facilitated rather than suppressed the operant response. Such studies bring home the fact that the organism and the response need to be considered as important as the reinforcer.

But all such studies manipulate some aspects of the performance of the behavior and only deal with modification, usually with respect to frequency or stimulus control. They simply do not face the problem of the ontogenetic and phylogenetic origin or "roots" of the behavior (Question O). Some of the events that prove positively or negatively reinforcing to the individual prove to be both species-characteristic and response-specific.

There are two ways in which reinforcement may be involved as determinative precursors of communication. The traditional way that complex communication systems could be established, compatible with a strict application of reinforcement principles, is through "shaping." Here successive approximations to the desired final behavior are rewarded until what appears as qualitatively new responses and/or stimulus control are shown. While this view is elegant, its problems can easily be summarized: There is no direct evidence that this order of events occurs with respect to the normal ontogeny of communication in any species except, perhaps, in language in humans. A second and more recent approach to origins can be found in the concept of "autoshaping." Rats and pigeons exposed to stimuli and reinforcers noncontingently soon begin to respond to the stimulus in the usual "shaped" fashion (Hearst and Jenkins, 1974).

Self-reinforcement, however, may be an important factor in the ontogeny of communication. For instance, ducklings may quickly learn to depress a pedal to self-imprint in response to a flashing rotating light (Bateson and Reese, 1969), and song learning in some birds also involves the reward (or feedback) of hearing oneself sing.

Dimond (1970) has stayed close to traditional operant conditioning while emphasizing the action of reinforcement and other learning processes in animal social interactions and the answering of Question O. His premise is that "much of the social behavior must relate to the fact that one animal finds the presence of another animal rewarding. Within this system must surely lie the roots of gregariousness: aggregation, flock formation, and social grouping" (1970:118). He then suggests a framework involving reward systems of food, warmth, comfort, and shelter that interact with early experience to form the appropriate associations. Such processes may occur with mammals and birds, but it is as difficult to rule out "unlearned" social attractions as to prove them. In newborn snakes *(Thamnophis, Storeria, Natrix)* we now know that species-specific aggregations occur immediately after birth in the absence of variables involving food, warmth, or parental care (Burghardt, in press).

Dimond goes on to state, "Each animal learns to communicate with others. The suggestion arises that a sophisticated behavioral language is learnt from the earliest period of the neonate's existence . . . each animal acts to the other as an important communication source" (1970:118). Dimond accepts social contact as initially rewarding because it allows the animal to learn the communication system of its species; he implies that the social attractiveness is due to conditioning of primary biological needs (e.g., a secondary reinforcer). In fact, little ontogenetic work has actually been done on the role of such "social learning." A much-lauded study that concluded that conditioning was a determinative factor in regurgitation feeding of squabs by parent ring doves (Lehrman, 1955) surprisingly omitted a crucial ontogenetic variable—age of squab (Klinghammer and Hess, 1964). Thus, while the methods and precision of experimental psychology will be increasingly useful in the study of communication, liberation from narrow, theoretically oriented approaches seems essential if learning psychologists are to make substantive contributions.

HABITUATION

Habituation is considered the simplest and most basic form of learning: unpunished learning *not* to respond. Consider the gaping response of a young bird to cues associated with the arrival at the nest of a parent bringing food. If one presents those cues (such as branch shaking) often enough, the gaping response will wane if food is not forthcoming (Prechtl, 1953). Such habituation is very common and certainly seems relevant to communicatory signals, such as "shaping" through disuse. Some responses, once habituated, stay that way while others recover quickly. Sensory and motor factors need to be controlled. Further, in some short term cases, the opposite to habituation, sensitization, occurs, in which previously ineffective stimuli become effective through generalization. Habituation can also be viewed as a broad category, which includes extinction of conditioned behavior.

As Denny and Ratner (1970) point out, it seems easier to habituate preliminary appetitive behaviors (as in using a certain food-searching behavior) than consummatory ones. Communicatory gestures and responses to them are clearly subject to habituation. For instance, threat displays of male Siamese fighting fish habituate to live male fish as well as to mirror images (Baenninger, 1966; Baenninger, Bergman, and Baenninger, 1969), although exposure to a mirror can also act as a reinforcer in the acquisition of a new response (Thompson, 1966). Unfortunately, there is little information on habituation in actual social interactions, although it is undoubtedly involved in changes in intensity of courtship behavior or maternal care occurring over time. But there are too many other changes superimposed on any habituation effect to tell a clear story. In the ontogeny of communication it likewise seems clear that habituation should play a role and it is similarly unestablished.

OBSERVATIONAL LEARNING AND IMITATION

This process and the following two are controversial, but only in the sense that the mechanisms involved may or may not be basically different from the foregoing three, which psychologists have generally considered most basic. They have been largely ignored by ethologists and psychologists, but are included here to point out areas where some new thinking and methods may offer insights about communication (see also Griffin, this volume and 1976).

When a naive animal observes another engaging in some activity and later performs the same behavior in the absence of the model, some acquisition process is involved. Mainardi and

Pasquali (1968) have demonstrated this occurrence in mice. Traditionally it has been called imitation, with emphasis on problem solving: The subject observes a trained animal choosing the "right" stimulus or manipulating a lock correctly to get at food. Unfortunately, while this situation seems very common and obvious in people, the evidence in animals with arbitrary responses is controversial. One area of communication where it has been clearly established is in the vocal learning of songbirds discussed earlier. The fact that a white-crowned sparrow will show the song "dialect" it was exposed to during rearing leads us to conclude that certain "traditions" can be passed from one individual to another through social interaction (Wickler, 1965). This is especially true for responses to nonsocial objects such as food. Galef (1976) has shown that adult rats experiencing a sublethal dose of poison after eating highly palatable food can transmit an aversion for this food to offspring who have never experienced any illness because of eating it. Here communicatory processes between parents and young are important in modifying food preferences in early ontogeny, although classical imitation does not seem involved. Imitation of unnatural gestures used in sign language communication in chimpanzees has been demonstrated (see Fouts, volume 2), but the relevance to development of normal communication has yet to be demonstrated.

PERCEPTUAL LEARNING AND PRACTICE

Mere repeated exposure to an array of highly similar, but not identical, objects can lead to discrimination without the operation of traditional differential pairing or differential reinforcement. When students begin to watch a social group of animals, such as baboons, they want to mark them individually for easy recognition. But that type of aid soon becomes unnecessary as subtle differences in size, proportion, natural markings, scars, and behavior become clear. While reinforcement through information feedback is also involved, the case can be made that such reinforcement is not necessary (Hebb, 1972). Perceptual learning differs from the nonspecific facilitative precursors discussed earlier in that it does not refer to the experience necessary for the development of sensory abilities (e.g., acuity, pattern, odor, tones) but to the use of these abilities in making finer discriminations (cf. search image; Krebs, 1973). Practice is the motoric equivalent of perceptual learning.

Perceptual learning is of greater importance in socially stratified animals than is usually recognized and could be the basis for individual recognition. For instance, gull chicks quickly learn to recognize subtle differences in their own parents' calls. Indeed, exposure to the calls prenatally leads to discrimination (Impekoven and Gold, 1973). Clearly, an animal reared in a social group has ample opportunity for such perceptual learning. For example, as soon as a female northern elephant seal *(Mirounga angustirostris)* gives birth she emits a "warble" vocalization, which is answered by a distinctive sound by the pup. Such duets can occur throughout the nursing period (Le Boeuf et al., 1972). The phenomena of "selective attention" (Chance and Jolly, 1970) and "local enhancement" may facilitate perceptual learning and imitation by causing the animal to be attracted to important learning situations.

INSIGHT

A controversial legacy of the *Gestalt* psychologists, "insight" refers to the sudden apprehension of certain relationships that allows the animal to do something it previously struggled to accomplish rather unproductively. Chimps abruptly learn to stack up boxes to acquire food formerly out of reach (Köhler, 1927). Less dramatic are all the related instances of rapid learning after a period of poor responding. In fact, some psychologists argue that all learning takes place in discrete steps. Curiously, classical insight shares several attributes with instinct in that an integrated adaptive response is made without prior specific experience. But the behavior does seem purposeful because the animal uses it to gain an immediate goal or reward, while instinctive responses are performed for their own sake and only indirectly lead to adaptive results (e.g., ingestion of food, offspring).

The question is how much influence such innovative behavior has on communication and ontogeny. On one level it probably has very little effect. Even in mammals and birds it appears too variable, infrequent, and unstable to be a basis for the development of the ritualized communication involved in such fundamental and critical duties as courtship, maternal-offspring interaction, and defense. In developmental studies it is also difficult to separate insightful factors involved in the transition from one stage to another with that caused by "readiness" or "preparedness" due to maturation or prior experience, which may facilitate or even trigger such sudden progress.

PLAY

In recent years many investigators have turned to studying the rather poorly understood phenomena jointly termed play. Since it is generally found exclusively or primarily in infants and juveniles it is clearly ontogenetically relevant. We still do not know what play is, but we know it is important. But important for what? To develop communication gestures and skills, say some primate workers (Jolly, 1972; Mason, 1965), although several other functions of play have been mentioned in the literature, for example, to develop predatory skills. A stimulating symposium on play in mammals has recently been published (*Amer. Zool.* 1974), and some findings discussed there illustrate the problems in determining the relations between ontogeny and communication.

Animals reared in social isolation that, perforce, do not play, often show deficiencies in the patterning of later adult communicatory behaviors and in the "understanding" of the responses of others. Such evidence has been used to argue for the definitive importance of play in the later behavior of adults. Aside from the fact that social isolation deprives an animal of more than just play experience, it is not possible to state from the evidence whether play is a determinative or merely a facilitative factor in later communication. Play may exert primarily a refining influence in terms of response- and stimulus-specificity, patterning, and so forth. Evidence for the former would entail experiments such as this: Foster-rear an animal in a family of a related species with a different communicatory repertoire. If the foster animal later shows characteristics associated with the foster species, then to that extent play may be determinative. Note, however, the possible confusion with imitation or observation. To separate the two might entail prevention of physical interaction. Bekoff's retraction of earlier views on the necessity of play (facilitative or determinative) to an animal "in order to acquire species-typical social skills" (1974:337) is a recognition of the difficulties involved (see also Symons, 1974).

Detailed descriptive studies of the normal ontogeny of play are necessary prior to either speculation or gross experiments on the causal significance of play to later behavior. Some developmental studies of note involving monkeys, bears, canids, sea lions, ferrets, and meercats are presented in the above mentioned symposium.

The Interaction of Several Precursors

If . . . the system under study is a complex one involving a relatively large combination of interacting variables, the discovery of what the relevant factors are and the nature of the relations holding between them involves much more [theorizing]. . . . Theorizing at this stage, then, consists primarily in guesses as to what the unknown relevant factors might be and guesses as to the form the relations between them might take. [Spence, 1953:284]

The organism should not be viewed as an aggregation of precursors, but as a dynamic system with a history, a history that influences, to varying degrees, the individual's future commerce with the world. While environmental events in the ecological theater can have critical short-term individual as well as long-term evolutionary effects, internal control system properties involving feedback, maturation, self-reinforcement, and other self-correcting effects must not be ignored. There are three broad phenomena, touched on throughout this chapter, which explicitly involve several antecedents in a determinative fashion and bring out, to varying degrees, these self-adjusting mechanisms.

INSTINCT-CONDITIONING INTERCALATION

Early ethological theory held that much social communication of animals was based on releasers, fixed (modal) action patterns, and their reciprocal interactions. A morphological structure or behavior was the signal to a second individual, who responded with a behavior that either acted as a releaser or made prominent a structure (such as a colored feather), that was the releaser. A very useful way of viewing communication, this scheme made terminology clear and comparisons illuminating. The units of analysis were usually stereotyped, species-characteristic, and largely based on genetic differences and similarities. The signals, responses, and the code were prematurely viewed as innately determined and impervious to normal and even much abnormal experience (Lorenz, 1937).

The ethologists who formulated this theory in the 1930s and 1940s did not rule out learning as an important factor. The use of environmental factors, however, was clearly influenced by the conditioning approach then current in psychology. The ethologists knew that any sequence of behavior did not exist as a unit but as a series of apparently discrete parts. With the fixed (modal) action pattern (FAP) this is still controversial (Moltz, 1965; Barlow, this volume). But experience was often needed for optimal performance as indicated by the following evidence: (1) Deprived animals would often perform the behavioral units, but unrefined, or the arrangement would be out of sequence, as in nest building in virgin rats deprived of manipulatory experience (Eibl-Eibesfeldt, 1961). (2) Certain patterns would be entirely absent. This is especially true of orientation movements. For instance, social play is important for polecats (Eibl-Eibesfeldt, 1970) to direct the neck bite properly in mating and predation. (3) The releasers for social behavior could be directed normally to a small subset of potentially effective stimuli through specific early experience (e.g., imprinting, which will be discussed separately). (4) Some behavior patterns dropped out with repeated stimulation (habituation, discussed above).

Lorenz (1935) initially formulated the idea of instinct-conditioning intercalation to handle seemingly "blank" spaces in behavior sequences that could be filled in through conditioning or imprinting. Today, of course, we know that experience can have subtle effects and do more than "fill in the blanks." Indeed, fixed (modal) action patterns and releasers operate by diverse mechanisms and vary greatly in their parameters. Studies tracing the development of specific behavior patterns are needed, and too few of these exist (e.g., Kruijt, 1964; Williams, 1972). But we do know that even specific aspects of a behavior may be ontogenetically shaped through the involvement of several precursors.

IMPRINTING

Imprinting is the process whereby an animal attaches behavior to a subclass of potential stimuli in a manner seemingly different from the normal conditioning paradigm. As originally defined it was restricted to maternal and sexual objects (classical imprinting), but phenomena possessing some similar operational characteristics also seem involved in food choice, habitat and nest site selection, song learning and other behaviors (Burghardt, 1973). Nonetheless, classical imprinting is not a unitary concept, and even in birds the degree and mechanism of imprinting varies enormously. Recent reviews of imprinting are available and should be contacted for a thorough introduction to this complex area, which has gathered more experimental attention than any other problem in ethology (e.g., Hess, 1973; Sluckin, 1965, 1972; Bateson, 1971; Immelmann, 1972).

In classical imprinting the experiential determinative effect concentrates entirely on the releaser or perceptual side of behavior. When ducks imprint, it is to the object of filial or sexual behavior, not to the way the motor movements are manifested, which seem largely due to maturation.

Imprinting is important to communication because the animal can respond with filial or courtship behavior to signals emanating from distinctly unnatural objects. On the other hand, recent work indicates that imprinting is often strongest to objects that possess or accentuate some characteristic found in the natural parent or mate. Imprinting, therefore, is a phenomenon that combines several precursors, including genotype, association, self-reinforcement, and sensitive periods.

TEMPLATES

The classical notion of the Innate Releasing Mechanism (IRM) arose from the idea of an internal schema that, when matched, released the "proper" response. Today we know that aspects of such releasers, and hence releasing mechanisms or schemata, vary from tight to loose genotype-phenotype correlations, from innate to largely acquired, from rigid to easily modifiable, or from genetically closed to genetically open (e.g., Hailman, 1967; Schleidt, 1962).

Vocalizations, while often signals, are nonetheless motor responses of the signaler. The findings on bird songs, summarized earlier, allow us to see that several ontogenetic factors, including genotype, sensitive periods, auditory feedback, and conditioning, can be involved. These findings led to the template concept, a way of accounting for diverse findings by a hypothetical mechanism. A template is

> visualized as a mechanism or mechanisms residing at one or more locations in the auditory pathway which provide a model to which the bird can match feedback from its vocalizations. . . . In some species, such as the song sparrow, an individual raised in complete isolation from species members seems to possess a sufficiently well-specified template so that it can generate normal song, as long as the individual can hear itself. . . . However in other species, such as the white-crowned sparrow or the chaffinch, the initial specifications for the template are less complete. [Marler and Mundinger, 1971:428–29]

While the template model seems restricted to song learning, there is no reason why it cannot be much more widespread factor in behavioral development, as Marler (1973) himself postulates. However, it will be difficult to uncover such mechanisms in species that do not learn or perform such precise and relatively easily analyzed behaviors as songs. A major problem with the template idea is that it is the result of several experimental operations, each leading to somewhat different results depending on the species, age, sounds used, and so forth. The question is whether the template idea will actually allow us to make predictions or restrain us at the level of a summarizing concept that is tied closely to specific results.

Ontogeny as Town Meeting

This chapter has focused on the "how" of ontogeny and the asking of questions about developmental antecedents and modifiers of communication. Given the complexity of issues and data dealing with "how," only some of which could even be touched on here, it is not surprising that workers in development have rarely stepped back to ask the broader question of "why?" A few such questions would be: "Why do some animals develop quickly and some slowly?" "Why are there differences in the genotype-phenotype correlations?" "Why do different aspects of behavior have different ontogenetic histories in the same species?" "Why do some developmental processes show great variability and others little?" Such questions are most often posed by evolutionary theorists and ecologists and are answered by considering the life history of the organism in its ecological context. Many *post hoc,* but reasonable, explanations can be given, and have been given for years, with greater or lesser amounts of natural history data to support them (e.g., Wilson, 1975). They rarely deal with the exceptions or the detailed nature of the adjustments of ontogeny to ecological contingencies, and thus to natural selection for ontogenetic patterns in subsequent generations. This is not the place to examine in any depth the various generalizations that have been formulated in the past and keep recurring in almost equally uncritical guises. These include evolutionary "trends," such as those from simple to complex communication, rigid to variable communication, and tight genotype-phenotype correlations to loose correlations. What we need are detailed studies of ontogeny in situations where the adaptive significance of the various aspects of communication *and* their ontogenetic antecedents can be evaluated.

Genetic processes should not be viewed as immune to the experiences encountered by the animal. Behavior that appears genetically specified may have evolved from behavior originally shaped by specific experiences and learning. This is not a disavowal of natural selection but a key example of it, the theoretical implications of which do not seem to be commonly applied to behavior and rarely, if ever, to communication. Consider the phenomenon of "genetic assimilation" (Haldane, 1959; Waddington, 1953) or "threshold effect," as reinterpreted by Mayr (1963). The initial example involved a wing abnormality in fruit flies *(Drosophila)* but the process will be emphasized here. A given environmental input leads to the appearance of some atypical characteristic. Breeding selectively from such individuals (or relatives not showing the trait) leads to more individuals' responding to the environmental input and eventually showing the trait without the experience at all. This may almost seem like the inheritance of acquired characteristics but it is not. The animals were carrying the genotype that made the trait possible, but it was masked by a rather tight genotype-phenotype correlation. The unusual experience was strong enough to break the stabilizing effect in a rare individual and allow selection to operate on it, eventually bringing together enough of the relevant genes to enable the response to be more easily manifested. It is possible that such processes are involved in the evolution of complex behavior, showing again the intricate relationships of evolutionary and ecological factors in the ontogeny of communication. Indeed a conditioned stimulus in classical psychology may, by this process, become an unconditioned stimulus. The above example is one of many that could be given that suggest new directions in the study of antecedents to communication. But we can close on an optimistic note, as the work on ontogeny is constantly improving in quality, imagination, diversity of approaches, and scientific status. Indeed, of all the emphases of ethology, ontogeny is the most interdisciplinary and multifaceted.

Conclusions

We have now looked at a variety of factors involved in the ontogeny of communication.

While the list is not comprehensive and while more will undoubtedly be discovered in the near future, their range should be apparent. We can conclude, however, that processes that seem to be involved in the shaping and topography of communicatory behaviors and signaling systems may develop quite differently than those that can later modify such communication. Furthermore, a crucial distinction has been made between ontogenetic events that are often necessary but only as facilitative precursors and those that are responsible for the adaptive communication properties, the determinative precursors. Any classification has deficiencies; this one is no exception.

In communication, where we deal with social processes often very subtly and precisely adjusted with respect to the individual as well as to conspecifics, where the consequences of error and misunderstanding can be great, and where commonality of a "language" is both common and necessary, we need to be particularly careful that we do not let labels become explanations. This truism is one that many scientists would agree with when dealing with a dynamic system. What are needed now are longitudinal studies tracing the development of young animals (e.g., Kruijt, 1964; Williams, 1972). But we particularly need longitudinal studies with a point. Merely to describe the changes in behavior as an animal progresses from one age to another is an important, although tedious, job; and methods, problems of classification, and all the other difficulties of behavioral research are obstacles that lead to unfocused research making the wrong compromises. But even these are only first steps. Specific behavior patterns need to be not only followed but also experimentally manipulated in the field and laboratory. Only then will we find out the precise nature of the ontogenetic mechanisms for any given type of communication.

References

Alexander, R. D., 1968. Arthropods, In: *Animal Communication,* T. A. Sebeok, ed. Bloomington: Indiana University Press, pp.167–216.

Baenninger, L.; Bergman, M.; and Baenninger, R.; 1969. Aggressive motivation in *Betta splendons:* replication and extension. *Psychon. Sci.,* 16:260–61.

Baenninger, R., 1966. Waning of aggressive motivation in *Betta splendons. Psychon. Sci.,* 4:241–42.

Bateson, P. P. G., 1971. Imprinting. In: *The Ontogeny of Vertebrate Behavior,* H. Moltz, ed. New York: Academic Press, pp.369–87.

Bateson, P. P. G., 1976. Specificity and the origins of behaviour. *Advances in the Study of Behaviour,* 6:1–20.

Bateson, P. P. G., and Reese, E. P., 1969. The reinforcing properties of conspicuous stimuli in the imprinting situation. *Anim. Behav.,* 17:692–99.

Beer, C. G., 1971. Diversity in the study of the development of social behavior. In: *The Biopsychology of Development,* E. Tobach, L. R. Aronson, and E. Shaw, eds. New York: Academic Press, pp.433–55.

Bekoff, M., 1974. Social play and play-soliciting by infant canids. *Amer. Zool.,* 14:323–40.

Bentley, D. R., 1971. Genetic control of an insect neuronal network. *Science,* 174:1139–41.

Benzer, S., 1973. Genetic dissection of behavior. *Sci. Amer.,* 229(6):24–37.

Bernard, L. L., 1924. *Instinct: A Study in Social Psychology.* New York: Holt.

Blakemore, C., 1973. Environmental constraints on development in the visual system. In: *Constraints on Learning,* R. A. Hinde and J. Stevenson-Hinde, eds. New York: Academic Press, pp.51–73.

Burghardt, G. M., 1970a. Chemical perception in reptiles. In: *Communication by Chemical Signals,* J. W. Johnston, Jr., P. G. Moulton, and A. Turk, eds. New York: Appleton-Century-Crofts, pp.241–308.

Burghardt, G. M., 1970b. Defining "communication." In: *Communication by Chemical Signals,* J. W. Johnston, Jr., D. G. Moulton, and A. Turk, eds. New York: Appleton-Century-Crofts, pp.5–18.

Burghardt, G. M., 1971. Chemical cue preferences of newborn snakes: influence of prenatal maternal experience. *Science,* 171:921–23.

Burghardt, G. M., 1973. Instinct and innate behavior: toward an ethological psychology. In: *The Study of Behavior: Learning, Motivation, Emotion, and Instinct,* J. A. Nevin, ed. Glenview, Ill.: Scott, Foresman and Co., pp.322–400.

Burghardt, G. M.; Wilcoxon, H. C.; and Czaplicki, J. Z.; 1973. Conditioning in garter snakes: aversion to palatable prey indicated by delayed illness. *Animal Learning and Behavior,* 1:317–20.

Candland, D. K., 1971. The ontogeny of emotional behavior. In: *The Ontogeny of Vertebrate Behavior,* H. Moltz, ed. New York: Academic Press, pp.95–169.

Caspari, E., 1971. Differentiation and pattern formation in the development of behavior. In: *The Biopsychology of Development,* E. Tobach, L. R. Aronson, and E. Shaw, eds. New York: Academic Press, pp.3–15.

Chance, M., and Jolly, C., 1970. *Social Groups of Monkeys, Apes and Men.* New York: Dutton. 224pp.

Davis, H., and Hubbard, J., 1973. Conditioned vocalization in rats. *J. Comp. Physiol. Psychol.,* 82:152–58.

Dawkins, R., 1968. The ontogeny of a pecking preference in domestic chicks. *Z. Tierpsychol.,* 25:170–86.

Denny, M. R., and Ratner, S. C., 1970. *Comparative Psychology: Research in Animal Behavior,* rev. ed. Homewood, Ill: Dorsey. 869pp.

Dimond, S. J., 1970. *The Social Behavior of Animals.* New York: Harper. 256pp.

Eibl-Eibesfeldt, I., 1961. The interactions of unlearned behaviour patterns and learning in mammals. In: *Brain Mechanisms and Learning,* J. F. Delafresnaye, ed. Oxford: Blackwell, pp.53–73.

Eibl-Eibesfeldt, I., 1970. *Ethology: The Biology of Behavior.* New York: Holt, Rinehart and Winston. 530pp.

Etienne, A. S., 1973. Developmental stages and cognitive structures as determinants of what is learned. In: *Constraints on Learning,* R. A. Hinde and J. Stevenson-Hinde, eds. New York: Academic Press, pp.371–95.

Foss, B. M., 1964. Mimicry in mynas *(Gracula religiosa)* : a test of Mowrer's theory. *Brit. J. Psychol.,* 55:85–88.

Fox, M. W., 1969. The anatomy of aggression and its ritualization in canids. *Behaviour,* 35:242–58.

Galef, B. G., Jr., 1975. The social transmission of acquired behavior. *Biol. Psychiatry* (in press).

Ginsburg, B. E., 1958. Genetics as a tool in the study of behavior. *Pers. Biol. Med.,* 1958:397–424.

Ginsburg, Benson E., 1966. All mice are not created equal: recent finding on genes and behavior. *Soc. Ser. Rev.,* 40:121–34.

Gottlieb, G., 1971a. *Development of Species Identification in Birds: An Inquiry into the Prenatal Determinants of Perception.* Chicago: University of Chicago Press. 176pp.

Gottlieb, G., 1971b. Ontogenesis of sensory function in birds and mammals. In: *The Biopsychology of Development,* E. Tobach, L. R. Aronson, and E. Shaw, eds. New York: Academic Press, pp.67–128.

Gottlieb, G., 1973. Introduction to behavioral embryology. In Gottlieb, G., ed., 1973. pp.3–45.

Gottlieb, G., ed., 1973. *Behavioral Embryology,* vol. 1. New York: Academic Press. 369pp.

Hailman, J. P., 1967. The ontogeny of an instinct. *Behaviour Supplement,* no. 15.

Haldane, J. B. S., 1959. Natural selection. In: *Darwin's Biological Work,* P. R. Bell, ed. New York: John Wiley and Sons, pp.101–49.

Harlow, H. F., and Harlow, M. K., 1965. The affectional systems. In: *Behavior of Nonhuman Primates,* Vol. 2, A. M. Schrier, H. F. Harlow, and F. Stollnitz, eds. New York: Academic Press, pp.287–334.

Hearst, E., and Jenkins, H. M., 1974. *Sign-Tracking: The Stimulus-Reinforcer Relation and Directed Action.* Austin: Psychonomic Society. 49pp.

Hebb, D. O., 1953. Heredity and environment in mammalian behaviour. *Brit. J. Anim. Beh.,* 1:43–47.

Hebb, D. O., 1972. *Textbook of Psychology,* 3d ed. Philadelphia: W. B. Saunders. 326pp.

Hess, E. H., 1962. Ethology: an approach toward the complete analysis of behavior. In: *New Directions in Psychology,* R. Brown, E. Galanter, E. H. Hess, and G. Mandler. New York: Holt, Rinehart and Winston, pp.157–266.

Hess, E. H., 1973. *Imprinting.* New York: D. Van Nostrand. xvi+472pp.

Hinde, R. A., 1970. *Animal Behaviour,* 2d ed., New York: McGraw-Hill. 876pp.

Hinde, R. A., 1971a. Development of social behavior. In: *Behavior of Nonhuman Primates,* vol. 3: *Modern Research Trends,* A. M. Schrier and F. Stollnitz, eds. New York: Academic Press, pp.1–68.

Hinde, R. A., 1971b. Some problems in the study of the development of social behavior. In: *The Biopsychology of Development,* E. Tobach, L. R. Aronson, and E. Shaw, eds. New York: Academic Press, pp.411–32.

Hinde, R. A., 1974. *Biological Bases of Human Social Behaviour.* New York: McGraw-Hill. 462pp.

Hinde, R. A., and Stevenson-Hinde, J., 1973. *Constraints on Learning.* New York: Academic Press. 488pp.

Immelmann, K., 1969. Song development in the zebra finch and other Estrildid finches. In: *Bird Vocalizations,* R. A. Hinde, ed. London: Cambridge University Press, pp.61–74.

Immelmann, K., 1972. Sexual and other long-term aspects of imprinting in birds and other species. *Advances in the Study of Behavior,* 4:147–74.

Impekoven, M., and Gold, P. S., 1973. Prenatal origins of parent-young interactions in birds: a naturalistic approach. In: G. Gottlieb, ed., 1973, pp.325–56.

Jensen, D. D., 1970. Polythetic biopsychology: an alternative to behaviorism. In: *Current Issues in Animal Learning: a Colloquium.,* J. H. Reynierse, ed. Lincoln: University of Nebraska Press, pp.1–31.

Jolly, A., 1972. *The Evolution of Primate Behavior.* New York: Macmillan. xiii+397pp.

Klinghammer, E., and Hess, E. H., 1964. Parental feeding in ring doves *(Streptopelia roseogrisea)*: innate or learned. *Z. Tierpsychol.,* 21:338–47.

Köhler, W., 1927. *The Mentality of Apes,* 2nd ed. New York: Harcourt Brace. 336pp.

Konishi, M., 1965. Effects of deafening on song development in American robins and black-headed grosbeaks. *Z. Tierpsychol.,* 22:584–99.

Krebs, J. R., 1973. Behavioral aspects of predation. In: *Perspectives in Ethology,* vol. 1, P. P. G. Bateson and P. H. Klopfer, eds. New York: Plenum Press, pp.73–111.

Kruijt, J. P., 1964. Ontogeny of social behavior in Burmese red jungle-fowl *(Gallus gallus spadiceus)* Bonnaterre. *Behaviour, Supp.* 12. 201pp.

Kuo, Z. Y., 1967. *The Dynamics of Behavior Development: An Epigenetic View.* New York: Random House. 240pp.

Lane, H., 1961. Operant control of vocalizing in the chicken. *J. exp. Anal. Behav.,* 4:171–77.

Le Boeuf, B. J.; Whiting, R. J.; and Gantt, R. F.; 1972. Perinatal behavior of northern elephant seal females and their young. *Behaviour,* 43:121–56.

Lehrman, D. S., 1955. The physiological basis of feeding in ring doves *(Streptopelia risoria). Behaviour,* 7:741–86.

Lehrman, D. S., 1970. Semantic and conceptual issues in the nature-nurture problem. In: *Development and Evolution of Behavior,* L. R. Aronson, E. Tobach, D. S. Lehrman, and J. S. Rosenblatt, eds. San Francisco: W. H. Freeman, pp.17–52.

Lehrman, D. S., and Rosenblatt, J. S., 1971. The study of behavioral development. In: *The Ontogeny of Vertebrate Behavior,* H. Moltz, ed. New York: Academic Press, pp.1–27.

Leyhausen, P., 1973. The biology of expression and impression. In: *Motivation of Human and Animal Behavior: An Ethological View,* K. Lorenz and P. Leyhausen, eds. New York: D. Van Nostrand. 423pp.

Lorenz, K., 1935. Der Kumpan in der Umwelt des Vogels. *J. für Ornithologie,* 83:137–213. Translated in: K. Lorenz, *Studies in Animal and Human Behaviour.* Cambridge: Harvard University Press, 1:101–258, 1970.

Lorenz, K., 1937. Über das Bildung des Instinkbegriffes. *Die Naturwissenschaften,* 25:289–300, 324–31. Translated in: K. Lorenz, *Studies in Animal and Human Behaviour.* Cambridge: Harvard University Press, 1970, 1:259–315.

Lorenz, K., 1965. *Evolution and Modification of Behavior.* Chicago: University of Chicago Press, 121pp.

McClearn, G. E., and DeFries, J. C., 1973. *Introduction to Behavioral Genetics.* San Francisco: W. H. Freeman. 349pp.

Mainardi, D., and Pasquali, A., 1968. Cultural transmission in the house mouse. *Attidella Società Italina di Scienze Natural: e del Museo Civico di Storia Naturale di Milano,* 107:147–52.

Manning, A., 1971. Evolution of behavior. In: *Psychobiology: Behavior from a biological perspective,* J. L. McGaugh, ed. New York: Academic Press, pp.1–52.

Marler, P., 1973. Learning, genetics, and communication. *Soc. Res.* 40:293–310.

Marler, P. and Mundinger, P., 1971. The study of behavioral development. In: *The Ontogeny of Vertebrate Behavior,* H. Moltz, ed. New York: Academic Press, pp.389–450.

Mason, W. A., 1965. The social development of monkeys and apes. In: *Primate Behavior: Field Studies of Monkeys and Apes,* I. DeVore, ed. New York: Holt, Rinehart and Winston, pp.514–43.

Mayr, E., 1963. *Animal Species and Evolution.* Cambridge: Harvard University Press. 797pp.

Mayr, E., 1974. Behavioral programs and evolutionary strategies. *Amer. Sci.,* 62:650–59.

Melvin, K. B., and Anson, J. E., 1969. Facilitative effect of punishment on aggressive behavior in the Siamese fighting fish. *Psychon. Sci.,* 14:89–90.

Moltz, H., 1965. Contemporary instinct theory and the fixed action pattern. *Psychol. Rev.,* 72:27–47.

Oppenheim, R. W., 1974. The ontogeny of behavior in the chick embryo. *Advances in the Study of Behavior,* 5:133–72.

Otte, D., 1974. Effects and functions in the evolution of signaling systems. *Ann. Rev. Ecol. Syst.,* 5:385–417.

Prechtl, H. F. R., 1953. Zur physiologie der angeborenen auslösenden Mechanismen: I. Quantitative

Untersuchungen über die Sperrbewegung junger Singvögel. *Behaviour,* 5:32–50.

Premack, D., 1965. Reinforcement theory. *Neb. Symp. Motivation,* 13:123–80.

Radlow, R.; Hale, E. B.; and Smith, W. I.; 1958. Note on the role of conditioning in the modification of social dominance. *Psychol. Rep.,* 4:579–81.

Riess, B. F., 1954. Effects of altered environment and of age in the mother-young relationships among animals. *Ann. N. Y. Acad. Sci.,* 57:606–10.

Sackett, G. P., 1966. Monkeys reared in isolation with pictures as visual input: evidence for an innate releasing mechanism. *Science,* 154:1468–73.

Salzinger, K., 1973. Animal Communication. In: *Comparative Psychology: A Modern Survey,* D. A. Dewsbury and D. A. Rethlingshafer, eds. New York: McGraw-Hill, pp.161–93.

Schleidt, W., 1962. Die historische Entwichlung der Begriffe "Angeborener Auslösemechanismus" in der Ethologie. *Z. Tierpsychol.,* 19:697–722.

Schneirla, T. C., 1965. Aspects of stimulation and organization in approach/withdrawal processes underlying vertebrate behavioral development. *Advances in the Study of Behavior,* 1:1–74.

Schneirla, T. C., 1966. Behavioral development and comparative psychology. *Q. Rev. Biol.,* 41:283–302.

Schusterman, R. J., and Feinstein, S. H., 1965. Shaping and discriminative control of underwater click vocalizations in a California sea lion. *Science,* 150:1743–44.

Schusterman, R. J.; Balliet, R. F.; and Nixon, J.; 1972. Underwater audiogram of the California sea lion by the conditioned vocalization technique. *J. exp. Anal. Behav.,* 17:339–50.

Seligman, M. E. P., and Hager, J. L., eds., 1972. *Biological Boundaries of Learning.* New York: Appleton-Century-Crofts. 480pp.

Skinner, B. F., 1966. The phylogeny and ontogeny of behavior. *Science,* 153:1205–13.

Sluckin, W., 1965. *Imprinting and Early Learning.* Chicago: Aldine.

Sluckin, W., 1972. *Early Learning in Man and Animal.* Cambridge, Mass.: Schenkman.

Spence, K. W., 1953. Mathematical theories of learning. *J. Gen. Psychol.,* 49:283–91.

Sperry, R. W., 1971. How a developing brain gets itself properly wired for adaptive function. In: *The Biopsychology of Development,* E. Tobach, L. R. Aronson, and E. Shaw, eds. New York: Academic Press, pp.27–44.

Symons, D., 1974. Aggressive play and communication in Rhesus monkeys *(Macaca mulatta). Amer. Zool.,* 14:317–22.

Thompson, T. I., 1964. Visual reinforcement in fighting cocks. *J. exp. Anal. Behav.,* 7:45–49.

Thompson, T. I., 1966. Operant and classically-conditioned aggressive behavior in Siamese fighting fish. *Amer. Zool.,* 6:629–41.

Thorpe, W. H., 1974. *Animal Nature and Human Nature.* Garden City, N.Y.: Doubleday. xx+435pp.

Tinbergen, N., 1951. *The Study of Instinct.* Oxford: Clarendon Press. xii+228pp.

Tinbergen, N., 1963. On aims and methods of ethology. *Z. Tierpsychol.,* 20:410–33.

Verplanck, W. S., 1955. Since learned behavior is innate, and vice versa, what now? *Psychol. Rev.,* 62:139–44.

Waddington, C. H., 1953. Genetic assimilation of an acquired character. *Evolution,* 7:118–26.

Waddington, C. H., 1956. *Principles of Embryology.* London: Macmillan.

Wecker, S. C., 1963. The role of early experience in habitat selection by the prairie deer mouse, *Peromyscus maniculatus bairdi. Ecol. Monogr.,* 33:307–25.

Whalen, R. E., 1971. The concept of instinct. In: *Psychobiology: Behavior from a biological perspective.* J. L. McGaugh, ed. New York: Academic Press, pp.53–72.

Wickler, W., 1965. Über den taxonomischen wert homologen Verhaltensmerkmale. *Naturwiss.,* 52:441–44.

Williams, N. J., 1972. *On the Ontogeny of Behaviour of the Cichlid Fish Cichlasoma Nigrofasciatum (Günther).* Groningen: Rijksuniversiteit. 112pp.

Wilson, E. O., 1975. *Sociobiology.* Cambridge: Belknap Press (Harvard). 697pp.

Winter, P.; Handley, P.; Ploog, D.; and Schott, D.; 1973. Ontogeny of squirrel monkey calls under normal conditions and under acoustic isolation. *Behaviour,* 47:230–39.

Young, W. C., 1965. The organization of sexual behavior by hormonal action during the prenatal and larval periods in vertebrates. In: *Sex and Behavior,* F. Beach, ed. New York: Wiley, pp.89–107.

Chapter 6

MODAL ACTION PATTERNS

George W. Barlow

A cornerstone of ethological theory is the belief that behavior comes in discrete packets. It is thought that behavior can be segmented in time into actions that have a characteristic "morphology" such that each can be recognized and therefore named. They are the "natural" units of behavior. Lorenz (e.g., 1950) spelled out the central role they must play in any coherent theory of animal behavior.

Most quantitative investigations rely on such segments of behavior. They are particularly relevant to the analysis of communication. However, despite their obvious importance they have been little studied. While investigators may recognize some variation, they generally accept it as simply part of the "slop" in the system.

In order to discuss the role of these seemingly unitary events in communication it is first necessary to inquire into their fundamental nature. I will start with a consideration of the appearance of behavior that is recognizably patterned, followed by a discussion of stimulus control and the neurological basis of such behavior. Then I will treat some aspects of development as well as genetics before moving on to the role of patterned movements in communication.

Before doing so I have to take up the troublesome problem of terminology. The occurrence of relatively discrete motor patterns has long been recognized, for example, the bucking of a horse or the bowing of a pigeon. There is still no agreement, however, on a general term to categorize such behavior. Among the early terms that were used by Lorenz (reviewed by Baerends, 1957; Schleidt, 1974) are "Instinct," "Instinctive Action," "Instinctive Movement," and "Hereditary Coordination." At one point a group of ethologists held a conference on terminology and settled on "Fixed Action Pattern" (Thorpe, 1951). Tinbergen (1952) accepted that terminology but later (1964) employed the more open expression "Typical Form." Lorenz (1965) subsequently resorted to further variants such as "Fixed Motor Pattern" and later (1973) "Hereditary Motor Coordination" and "Innate Motor Pattern." Yet other terms emerge in the literature, such as "Species-Specific," "Species-Characteristic," and "Species-Typical Behavior." Then there are simply "Motor Pattern" or "Behavior Pattern." These are sometimes qualified, as in "Stereotyped Motor Pattern." It is small wonder, then, that individuals entering this area are sometimes uncertain as to which term to apply.

Labels such as "instinct," "innate," and "hereditary" should be avoided. As used here, they are not descriptive but rather are interpretive. "Fixed Action Pattern" (FAP) is a particularly unfortunate construction, although currently it is probably the most widely used term to refer explicitly to the kind of behavior under consideration here. "Fixed" denotes a degree of stereotypy that is seldom seen. The concept has also been tied historically to a number of presumed properties (see below) that may or may not apply.

In 1968 I proposed an alternative term, "Modal Action Pattern" (MAP), which gets around both difficulties. It was put forth as a practical substitute for Lorenz's unitary concept of behavior patterns, but without implicit presumptions about causation and control. It emphasizes the statistical aspect of patterning as opposed to the fixed. I give it the following postulational definition:

> (1) A Modal Action Pattern is a recognizable spatiotemporal pattern of movement that can therefore be named and characterized statistically. (2) It usually cannot be further subdivided into entirely independently occurring subunits, although some of its components may occur independently or in other MAPS. (3) It is widely distributed in similar form throughout an interbreeding population.

The definition stresses that the behavior is a normal part of the biology of interbreeding individuals and is not meant to include prevalent human gestures that are obviously learned, such as those expressed upon meeting or departing. It allows for possible behavioral polymorphism and for the influence of the environment. The acronym MAP also appeals to me more than FAP —MAP relates nicely to the notion of a spatiotemporal map of behavior.

Because of the process of recognizing and naming MAPS, to facilitate quantitative analysis, the trend has been to focus increasingly on their stereotypy to the exclusion of their other, more interesting properties proposed by Lorenz (see reviews in Baerends, 1957; Barlow, 1968; Lehrman, 1953; Moltz, 1965; Schleidt, 1974). Most but not all of these are condensed here: (1) The behavior patterns are stereotyped in appearance. (2) The pattern is produced by a functionally organized network in the central nervous system (CNS). (3) Once triggered, the behavior runs to completion without further environmental control, i.e., without exteroceptive feedback. (4) The behavior is spontaneous; its occurrence lowers the probability that it will soon recur, and its nonoccurrence increases its probability so much that the behavior may happen without apparent stimulation. (5) The directional (taxic) component accounts for much of the variation and is not a part of the core CNS coordination. (6) The behavior is heritable, its spatiotemporal patterning being little affected by experience.

It is a tribute to Lorenz that he formulated such clearly testable hypotheses about patterned behavioral output. Yet even though these hypotheses were put forth some time ago we still do not know, in a general sense, whether they are broadly applicable, whether they apply to some species but not to others, or, within a species, to some behavior and not to others. I would like to incite behavioral investigators to reexamine these hypotheses because they are important to fully understanding the mechanisms underlying communication. In the following therefore, I will summarize the usually inadequate information about these hypotheses. In doing so, I will play down findings on bird song; that information will appear elsewhere in this book, and it does not alter the conclusions reached here. First, however, a few words about the problem of delimiting which kinds of behavior are to be considered.

Recognizing MAPS

The problem of recognizing MAPS is that of deciding on the units. "The search for the units of behavior, their organization, and their empirical validation, . . . constitutes *the* central problem of behavioral analysis" (Condon and Ogston, 1967). Most observers rely on their own perceptual capability to select out those units (e.g., Altmann, 1965; Dawkins and Dawkins, 1974; Hailman, 1967; Lorenz, 1955, 1973; Tinbergen, 1960). It is a reasonable way to proceed and is at some point inevitable.

I shall not venture to prescribe just how one extracts units of behavior, for I think that there is no general answer and that specific answers are best dealt with by statistical procedures. I would like to mention, however briefly, that the difficulty lies in the complexity of motor output. Postures or movements may be graded, even in passerine birds (Blurton-Jones, 1968), although intermediates (Fig. 1) may be infrequent. The components of MAPs, often called "acts," may pop up in different MAPs (Stokes, 1960), requiring careful analysis and evaluation. Additionally, MAPs may overlap in various complex ways (Andrew, 1956; Barlow, 1962; Tinbergen, 1964).

An overriding consideration is that behavioral output is hierarchically arranged (DeLong, 1972; Hinde, 1953; Nelson, 1973; Tinbergen, 1950, 1951). An oversimplified view, but a helpful first approximation, is to think of neuromuscular effector groups organized into acts, which are further arranged into MAPs. The MAPs occur in stochastic sequences, which, if they become stereotyped, must be treated as units. At the other extreme, acts may appear alone and thus become simple MAPs, that is, units of behavior.

These difficulties, and more that I might mention, have been raised to bring out the imprudence of operationally defining MAPs. It may never be advisable to constrain the concept. The prudent approach is to seek the smallest distinguishable units of behavior, then to combine those that co-vary closely (Altmann, 1965; Cane, 1961; Dingle, 1972). In this regard, the statistical procedures developed by numerical taxonomists look promising (Sneath and Sokal, 1973). The

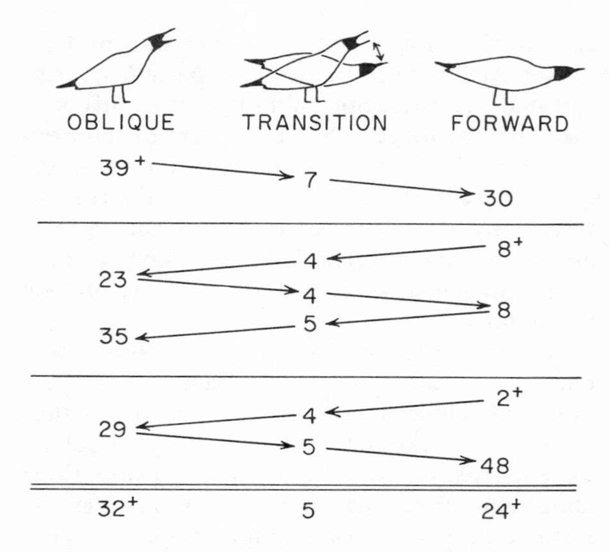

Fig. 1. Number of film frames (24 fps) in which the black-headed gull (*Larus ridibundus*) was in the oblique or the forward postures, respectively, and transitions between them. The action runs down the page, encompassing three film sequences. (From Tinbergen, 1960.)

techniques of ethologists for analyzing sequences may also be of use here (see review in Slater, 1973).

Analysis of MAPs

The realization that MAPs can be variable has been slowly growing; Cullen (1966) and Hinde (1966b) commented briefly on this as a general phenomenon. Examples in the earlier literature were provided in the study of wolves by Schenkel (1948), fruit flies by Bastock and Manning (1955) and Manning (1959), and primates by Hinde and Rowell (1962) and Rowell and Hinde (1962). A number of other papers have reported on the variability of displays and vocalizations in birds (Blurton-Jones, 1968; Hailman, 1967; Hinde, 1955–1956; Konishi, 1963; Stokes, 1960) and even displays in crustaceans (Crane, 1966).

Yet the general attitude remained that MAPs,

especially if used in communication (displays), are rigid in their performance. Thus when Dane et al. (1959) published the first truly analytical investigation of displays, based on films of ducks, their data were received as confirmation of the fixity of displays. But the displays only seemed invariable. The duration of each was so brief that the variance was of necessity small. By using the coefficient of variation (*CV*), which is relative variability, the displays could be shown to be not only variable but variable in their variability. In fact, their variability is of about the same order as that for other biological systems (Barlow, 1968).

Since then a number of quantitative studies have appeared, and the use of *CV* is a growing convention. In the analysis of MAPs most investigators have done the obvious and easy thing first; they have simply measured duration. Consequently we have a limited view of what is meant by the stereotypy.

Our thinking about MAPs is further limited by the choice of MAPs for analysis. Observers tend to select the more stereotyped cases for study (e.g., Bucholtz, 1970; Wiley, 1973) (see also Table 2). Further, most of our examples have come from birds (Schleidt, 1974).

It is too early to prescribe a standard statistical methodology to characterize stereotypy. We may not have the statistical tools to deal with the more complex multiparametric analyses of MAPs (Schleidt, 1974). Nonetheless, some suggestions are in order as to how to assess stereotypy in simpler cases. Two basic procedures will be mentioned here. One involves central-tendency statistics, the other information theory.

When the behavior can be measured on a continuous-interval scale, as in the case of time and amplitude, a central-tendency statistic is appropriate, such as the mean and its variance. The coefficient of variation *(CV)* is derived by dividing the standard deviation *(SD)* by the mean $(\bar{x})$. It is expressed as a percentage of the mean by multiplying by 100. As the stereotypy increases, *CV* decreases.

When analyzing MAPs it is convenient to speak of stereotypy rather than variability. *CV* can be converted to a statistic that I will call stereotypy, *ST*, which varies directly with the "true" stereotypy of the behavior:

$$ST = \frac{100}{\left(\frac{100\ SD}{\bar{x}}\right) + 1}.$$

The value 1 is added to the denominator to avoid the awkward situation of infinite *ST* when *CV* equals zero. The equation reduces to a more convenient form:

$$ST = \frac{\bar{x}}{SD + .01\bar{x}}.$$

A few landmark values are given in Table 1 to show the relationship between *CV* and *ST*.

Table 1

Equivalent reference values for the coefficient of variation (CV) and an index of stereotypy (ST).

ST	CV
100	0
50	1
25	3
10	9
5	19
2.5	39
1	99
0.99	100

The other approach is through the calculation of what has been called uncertainty, entropy, or simply, *H*. It is appropriate when the analysis produces discrete categories of countable elements. The use of *H* avoids the awkwardness of dealing with variance in a system that has few components. A short-cut tabular method of calculation has been given by Lloyd et al. (1968), together with suggestions for the particular formula to employ. For behavioral studies, Brillouin's formula seems the more appropriate:

$$H = \frac{c}{N} \log \left(\frac{N!}{n_1! n_2! \ldots n_s!} \right),$$

where the constant c is a scaling factor to convert from logarithm base 10 to a desired base such as 2 or e. N is the total number of occurrences, and n_1, n_2, etc., are the numbers in each category. (For applications to behavior studies, see Attneave, 1959; Dingle, 1972; Moles, 1963.) Limitations of the method have been spelled out by Quastler (1958).

H has at least two undesirable features. For one, it is a measure of uncertainty or disorder and is thus just another way to estimate variability. For another, the size of H depends on the number of categories. The latter shortcoming can be avoided by using another measure from information theory, redundancy *(R)*. In a general sense redundancy can be equated to structure (Moles, 1963). It is calculated from

$$R = 1 - \frac{H}{H_{max}},$$

where H_{max} is the maximum possible uncertainty. Redundancy has been employed as a measure of stereotypy by Altmann (1965), Dingle (1972), and Fentress (1972).

Another point needs making here about the analysis of stereotypy. The data should be sorted out according to individuals. The data from a number of individuals have sometimes been lumped because the mean value is of interest, as in comparative studies. Often the data are pooled because it is impossible to identify individuals (e.g., Marler, 1973). Yet Jenssen (1971) estimated that 98 percent of the variation in a population of iguanid lizards was between individuals, and therefore that only 2 percent was intraindividual. Such large differences between individuals are typical but that is not always the case. In the sage grouse *(Centrocercus urophasianus)* (Fig. 2), for instance, there is little difference between individuals in the temporal stereotypy of strut displays (Wiley, 1973).

There are other sources of error and of bias, but time does not permit their exploration. There are also other ways to explore the stereotypy of MAPs, as in the analysis of phase relationships between different components. Keeping this in mind, I will now review some of the literature that deals with a variety of ways in which patterned movement can be analyzed, such as through duration and amplitude, cyclicity, interrelationship of parts, completeness, and the separation of directional and core components.

DURATION AND AMPLITUDE

The literature reveals that MAPs may vary widely in their duration (Table 2). At one extreme are the exceedingly long static displays, as in the tail-spreading display of peacocks.

Fig. 2. A sage grouse at the "peak" of its strut display, with the inflated vocal sacs thrown forward. The face has sunk out of view, precluding visual input at that moment. (From Wiley, 1973.)

Schleidt (1974) suggested that the bulk of MAPs, nonetheless, last somewhere between 0.1 and 10 sec, with most lasting about 1.0 sec, which tends to be borne out in Table 2. The table also indicates that stereotypy is usually not great, ranging from about 2 to 8 (*CV* roughly 35 to 15 percent). Only rarely does it exceed 40 (*CV* less than about 1.5 percent). A note of caution, however: in most cases the data were pooled from a number of individuals, a procedure that produces higher variances.

The stereotypy of MAPs thus presents a fairly wide range across species. Even within a species, different MAPs can express noteworthy differences in stereotypy (Table 2). But then temporal stereotypy is not a necessary condition for a MAP, for static displays can be enormously variable in duration.

When MAPs are examined internally, the little evidence available suggests that different components of the same MAP may show varying degrees of temporal stereotypy. When chickens drink or eat, for instance, the down stroke is less stereotyped than the up stroke (Dawkins and Dawkins, 1973; Hutchinson and Taylor, 1962). The feeding movements of the gastropod *Urosalpinx cinerea follyensis* are comparably simple, and the different components are also differently sterotyped (Carricker et al., 1974).

Turning to displays, the masthead display of the goldeneye duck *(Bucephala clangula)* may be broken down into three main components that are individually more stereotyped than the total duration of the display (Table 2). Variation apparently arises in the coupling of the components or in the interaction of their variations. These data must be interpreted with some caution, however, because they were pooled from several males (Dane et al., 1959), and there is some evidence (Dane and Van der Kloot, 1965) for behavioral polymorphism. In the barbary dove *(Streptopelia roseogrisea),* however, the six components of the bowing display are less stereotyped than is the total time for its performance (Table 3). But like the goldeneye, the main source of variation lies in internal pauses (Davies, 1970).

Another aspect of internal variation was detected in the pushup display of the lizard *Anolis anaeus* (Fig. 3). The components of the display

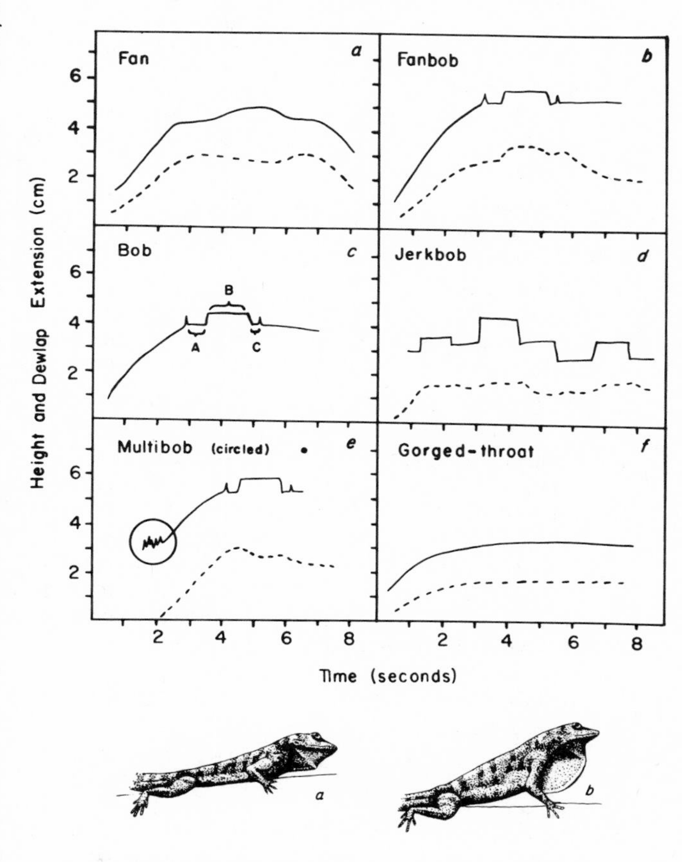

Fig. 3. The drawings of *Anolis anaeus* (bottom) depict two heights of pushing up and two types of dewlap extension; the display to the left is called gorged throat, that to the right, fan. The six panels above the lizards show the progressive changes in height (solid lines) and extension of the dewlap (dotted lines) in six different displays. In the panel for bob, the three elements of the signature bob are labeled A, B, and C. (After Stamps and Barlow, 1973.)

Table 2

Statistical parameters of the duration of some MAPs, and in a few cases of the intervals between them.

		Duration of MAP (sec)					Interval (sec)				
Species, MAP, and Location	**Film Speed (fps)**	**N**	**x**	**SD**	**CV**	**ST**	**x**	**SD**	**CV**	**ST**	**Reference**
Fiddler crab, waving											
Uca virens											
Texas	—	90	2.8	0.80	28.6	3.38	0.47	0.12	25.5	3.77	Salmon and Atsaides (1968)
Miss.		54	2.5	0.62	24.8	3.88	0.43	0.08	18.6	5.10	
U. longisignalis											
Miss.	—	80	5.6	1.50	26.8	3.60	4.41	2.80	63.5	1.55	
Florida		38	3.7	1.00	27.0	3.57	4.70	3.20	68.1	1.45	
U. pugnax											
Daytona, Fla.	—	42	2.3	0.31	13.5	6.91	1.40	0.70	50.0	1.96	
St. Augustine, Fla.		93	2.1	0.52	24.8	3.88	1.70	1.90	111.8	0.89	
Hermit crab											
Clibanarius vittatus											
Ambulatory rise	—	19	0.31	0.11	35.5	2.74					Hazlett (1972b)
Present cheliped		34	0.21	0.10	47.6	2.06					
Extend cheliped		14	0.27	0.11	40.7	2.40					
Lizards, pushup											
Anolis nebulosis+											
minimum*	18	—	3.6	—	—	—					Jenssen (1971)
"central"*			3.8	0.08	1.97	33.6					
maximum*			2.2	0.18	7.95	11.2					
Uta spp.											
Pyramid L., Cal.	38–70	20	0.1	0.03	30.0	3.22	0.6	0.2	33.3	2.91	Ferguson (1971)
Victorville, Cal.		22	0.3	0.07	23.3	4.11	0.8	0.2	25.0	3.85	
San Diego, Cal.		18	0.7	0.09	12.9	7.21	1.2	0.4	33.3	2.91	
Isla San Pedro Nolasco		25	0.8	0.10	12.5	7.41	1.3	0.3	23.1	4.15	
Isla Santa Catalina		22	1.0	0.90	90.0	1.10	1.3	0.3	23.1	4.15	
Isla San Lorenzo Norte		25	1.0	0.08	8.0	11.1	0.9	0.2	22.2	4.31	
Goldeneye duck											
Masthead	24	13	4.8	2.50	52.1	1.88					Dane, Walcott, and Drury (1959)
Up			0.21	0.03	14.3	6.54					
Pause			0.13	0.02	15.9	6.10					
Down			0.30	0.04	13.3	6.98					
Without pause			0.65	0.05	7.69	11.5					
Bowsprit		64	1.75	0.03	1.71	36.8					
Simple head throw		66	1.29	0.08	6.20	13.9					
Head flick		21	0.20	0.03	15.0	6.25					
Dip (♀)		39	1.79	0.75	41.9	2.33					
Display drinking		31	2.02	0.11	0.05	94.8					
Great cormorant, Wing waving											
England	16	13	0.77	—	54.	1.8					Van Tets (1965)
Netherlands		15	0.71	—	21.	4.5					

Table 2 (continued)

Species, MAP, and Location	Film Speed (fps)	Duration of MAP (sec) N	x	SD	CV	ST	Interval (sec) x	SD	CV	ST	Reference
Head wagging											
Northern gannet*	24	39	0.31	—	38.	2.6					Van Tets (1965)
Northern gannet*		21	0.37	—	26.	3.7					
Red-footed booby*	16	9	2.5	—	25.	3.8					
Wing bowing											
Northern gannet*	24	29	1.1	1.3	118.	0.84					Van Tets (1965)
Head throwing											
Red-footed booby*	16	21	1.4	0.4	40.	2.4					Van Tets (1965)
Sage grouse											
Strut display*											
Male D	18	26**	1.6+	—	3.3	23.2	154.9	29.6	19.1	4.98	Wiley (1973)
Male D	24	10**	1.7+	—	1.1	47.6	252.9	86.4	34.2	2.84	
Male A	18	16**	1.6+	—	1.7	37.0	170.0	25.6	15.1	6.21	
Male A	24	17**	1.7+	—	1.4	41.7	238.6	53.0	22.2	4.31	

NOTE: N = number of MAPs measured, x = mean, SD = standard deviation, CV = coefficient of variation, ST = stereotypy as defined in the text.

*Data from one individual.

**N slightly smaller for intervals.

+Estimated from bar diagram.

leading up to the terminal signature bob are relatively variable. But the signature bob is so stereotyped in a given individual that any variation it might have falls below the minimum resolvable difference, 0.02 sec, of film advancing at 24 frames per sec (Stamps and Barlow, 1973).

In comparison to the information on duration, that on amplitude or path movement is truly meager and has proved difficult and tedious to obtain. Rather than studying the path of movement, Hazlett (1972a) followed the change in angle relative to the body, made by the chelipeds and the first walking legs in a spider crab (*Microphrys bicornutus*) (Fig. 4). The chelipeds and legs were held at higher angles when displaying than when feeding, and the final angle was also more stereotyped when used in communication than in the noncommunicatory movements involved in feeding.

The study of temporal and spatial stereotypy is in its infancy and promises fertile ground for further research. There are large and unexplained differences in stereotypy between species, within species, and within the MAPs of individuals themselves. We now have the technology to move on.

CYCLICITY

It is important at least initially to remove cyclicity, or spontaneity, from the realm of causation and to treat it at first as a problem of description. This means working out the likelihood of occurrence of MAPs.

Most investigators would think first here of clearly repetitive, rhythmic MAPs. Yet many degrees of intermediacy exist between these and what seem to be arhythmic events. Turkey calls are given singly and do not seem to be periodic.

On a collapsed time scale, however, their calling is clearly rhythmic (Schleidt, 1964). Compressing time, as in fast-motion filming, reveals rhythmicity in many kinds of behavior that normally go undetected (see also Allison, 1971). In humans, this has been disclosed in religious ceremonies and in the pacing of a newspaper vendor (Eibl-Eibesfeldt, 1970).

There are two temporal aspects of repetitive behavior, inter-and intrabout intervals. The methodology for the analysis of one is applicable to the other since the one grades into the other. It is easy, then, to visualize how the compression of separate bits of behavior into bouts can lead to the evolution of new MAPs through rhythmic repetition of simpler elements, as in lizards (e.g., Ferguson, 1971; Gorman, 1968) and fiddler crabs (Crane, 1966, Salmon, 1967; Salmon and Atsaides, 1968).

Table 3

The temporal stereotypy of components of the bow display compared to that of the display as a whole in the barbary dove (Streptopelia roseogrisea).

	Number			
	Birds	**Displays**	**CV**	**ST**
Bow (total)	5	33	6.91	12.64
Components				
Up	5	33	15.78	5.96
Mid pause	3	24	35.02	2.78
Jump up	3	24	22.12	4.32
Top pause	5	33	47.94	2.04
Down	5	33	21.58	4.43
Bottom pause	5	33	18.05	5.25

NOTE: The coefficient of variation (CV) and stereotypy (ST) are unweighted means derived from data for individual birds.

Source: After Davies, 1970: Table 6.

The intervals between MAPs, not surprisingly, are generally more variable than the duration of the MAP itself (Table 2). In the sage grouse, for

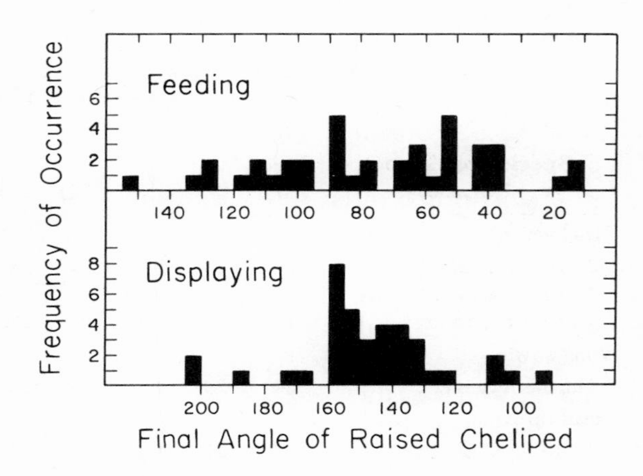

Fig. 4. The two histograms show the frequency of occurrence of the angle of the final position at which the cheliped is held by spider crabs. The upper histogram portrays data from feeding crabs. The lower, scattered data are from crabs in hostile interactions. (After Hazlett, 1972a.)

example, the strut display has a high mean *ST* of 32 as opposed to 5.9 for interstrut intervals. Intervals between drinks or food pecks are less stereotyped in chickens than are the durations of drinking or pecking (Dawkins and Dawkins, 1973, 1974; Hutchinson and Taylor, 1962). The same holds for the intervals between pushup displays in lizards (Ferguson, 1971).

INTERRELATION OF PARTS

The prevailing view has been that the components of a MAP are linked, or coupled, in a strict sequential relationship (Hinde, 1966a). The ordering and pattern of overlap are held to be centrally programmed and essentially invariant. Qualitative observations suggest that this is a good approximation for most MAPs. However, it is not a rule.

Quite early Hinde (1955–1956) noted that the internal ordering may vary in the display of one finch species while not in another (the components themselves were said to be variable in some species but, again, not in others). Hinde then took a clearly statistical view of those displays that were variable: "display patterns cannot always be considered as distinct entities—rather the postures . . . refer to combinations of components which often occur."

Whether the sequential model fits depends in part on the fineness of the elements chosen for analysis. For example, in drinking by chicks the downward movement of the head is usually taken as one component, which is then followed by drinking, which is then followed by raising the head. Dawkins and Dawkins (1973, 1974), however, discovered that both the downward movement and the upward recovery movement can be broken into yet smaller elements depending on whether the chicks pause in their paths. An analysis employing these finer elements produced variable sequencing.

The succession of events in the pushup display of lizards seems stereotyped to the eye of an untrained observer. The ordering of individual components, however, can be shown to vary (Stamps and Barlow, 1973). The linkage between movements of the dewlap, tail, tongue, and bobbing of the head is also subject to variation (Ferguson, 1971; Gorman, 1968; Jenssen, 1971; Stamps and Barlow, 1973).

Unfortunately, there has been little attempt to analyze quantitatively the internal sequencing of MAPS. The use of H or R from information theory would be a start, detailing the stereotypy of sequencing.

COMPLETENESS

Under different names completeness has long been recognized as a fundamental but accountable source of variation (Barlow, 1968). To illustrate completeness, consider a MAP consisting of three components occurring in the sequence *ABC*. In incomplete expressions the more terminal components drop out, as in *AB*, or simply *A*.

Schleidt (1974) has suggested a method for quantifying completeness. The number of acts in each performance is divided by the mean number of acts that occur in that MAP rather than by the maximum number ever observed. From repeated performances, the coefficient of variation is calculated, from which we may derive *ST*. The redundancy, R, of the different categories might prove to be more appropriate to this type of data, especially if the number of variants is small.

The concept of completeness need not be limited to MAPS with discrete components. It can also be applied to graded movements, as in Golani's (1973) analysis of the behavior of pairs of jackals *(Canis aureus)*. The graded movements were divided into a number of arbitrary steps. In the same way the graded "smile" of primates and humans can be segmented for analysis (Van Hooff, 1972).

TAXIC COMPONENT

In the traditional view the fundamental form of a MAP is independent of environmental influence (reviewed by Eibl-Eibesfeldt, 1970; Hinde, 1966a). This view derived from the assumption that there is a core coordination for MAPS provided by the CNS. MAPS are directed or steered, whether toward one's self, another animal, food, nest, light, or whatever. The relation to external objects is bound to vary, so the directing of the MAP will be to that degree variable. This taxic component has been regarded as a secondary or overlying feature of MAPS and hence separable from them.

The core coordination and the steering component can often be sorted out (see examples in Eibl-Eibesfeldt, 1970). The classic case is that of the greylag goose *(Anser anser)* retrieving an egg displaced from its nest. The goose places its bill on the far side of the egg and then pulls the egg back under its breast. The lateral swinging movement of the head is a taxic component compensating for the tendency of the eccentric egg to roll off the straight path (Lorenz and Tinbergen, 1939).

It is not always possible to make such a clear separation. Some MAPs can be recognized only by their orientation (Cullen, 1960). Many examples can be found in the ritualized paths of locomotion, as in the display flights of birds such as doves and hummingbirds, or in oceanic birds such as terns (Cullen, 1960; see also van Tets, 1965). In fishes, a good example is provided by display jumping (Fig. 5) in damselfishes (Holzberg, 1973). Stereotypy of interindividual orientation is illustrated in the aggressive interactions of the beautiful Asian fish *Badis badis* (Barlow, 1962) and the Arctic ground squirrel (*Spermophilus undulatus:* Watton and Keenleyside, 1974). Orientation is commonly ritualized in reproductive behavior, as in the spawning of the Florida flagfish (*Jordanella floridae:* Mertz and Barlow, 1966).

Fig. 5. The path of Signalsprung (display jumping) in a coral-reef damselfish (*Dascyllus marginatus*). Each figure is separated by 1/16 sec. (From Holzberg, 1973.)

In many cases, then, regularity of orientation itself may be the key element that evolution has exploited. Yet in most instances the basic coordination and the taxis of the behavior can be sorted out. Closer examination will doubtless reveal a continuum of examples between these poles.

Stimulus Control

Originally MAPs were thought to be unique in relation to the stimuli eliciting them. Once trigger they were believed to be no longer subject to external (exteroceptive) influence. That was not taken to mean that proprioceptive feedback was ruled out, although its role was seldom considered.

Two different conditions were confounded. The first is that once a MAP has started further input is not needed. Here the CNS is capable of producing patterned output, once stimulated, and without further guidance from the exterior. Note that external modulation is thus not ruled out if present and relevant.

The second condition is that once initiated further input is ignored. Even if the external situation changes significantly, the animal does not alter the performance of the MAP—it simply proceeds to completion.

Under either condition the effect of the stimulus is only to orient and release the response. Furthermore, the strength of stimulation is held to play little or no role in the form of the behavior itself. Thus any variation observed in the form of the MAP should be due to an internal change in motivational state.

Another aspect of stimulus control is that MAPs have been said to be set apart by virtue of their causation. Each MAP has common causal factors that differ from those of other MAPs.

IS FURTHER INPUT NEEDED?

Complex MAPs may be so independent of input that they can be performed in the absence of any apparent stimulation. This is the behavior

Lorenz (1937) called "vacuum activity" to stress the competence of the nervous system to execute MAPs without external guidance. While there are many examples of such behavior, it is difficult to say that no relevant stimuli were present because suboptimal stimuli may suffice (Bastock, Morris, and Moynihan, 1954). For the larger question, however, this makes little difference. The point is that external stimuli are normally overt and conspicuous. Yet they may be so reduced that an observer does not recognize them. They may even be absent. The animal may, nonetheless, still perform particular MAPs.

It has been shown in a few cases that once a MAP, or a series of MAPs, has been triggered, the stimulus can then be removed and the behavior will still go on to completion. The classic example is again that of the greylag goose (Lorenz and Tinbergen, 1939). If the egg is removed after the retrieval movement has been started, the goose will continue the movement as though the egg were still present. Other examples are rare. Tinbergen (unpublished) has found a comparable case in the courtship of a butterfly, the greyling (*Eumensis semele;* for details of behavior see Tinbergen et al., 1942). A parallel case occurs in the sexual behavior of the smooth newt (*Triturus vulgaris:* Halliday, 1975).

These examples and those of vacuum behavior indicate that many species have the ability to perform complex MAPs, even sequences of them, without guidance from the object triggering their performance. Nonetheless, this is also an area that has been little investigated. We need to know if MAPs are generally so independent. And the question remains as to whether changes in external stimulation can modify the course of the performance in spite of the obvious motor capability.

ARE MAPS ALTERABLE?

The question here is whether the spatiotemporal pattern of a MAP can change adaptively during its performance. Change, therefore, is limited to a degree of variation ordinarily observable in the MAP, not a fundamental restructuring or a long-term change. The scanty evidence suggests a spectrum of alterability.

For many MAPs, such as those used in noncommunicatory activites, it is clearly adaptive to respond to change. The form of feeding behavior employed by the fish *Badis badis* in extracting worms varies smoothly but radically depending on the resistance met (G. W. Barlow, 1961). But *Badis badis* feeds on a variety of prey and moves rather slowly. Thus there is both the need and the opportunity for exteroreceptive feedback.

In contrast, the feeding and drinking behavior of granivorous birds is simple and quick, so the upward and downward movements of the head have considerable momentum; the behavior is also more stereotyped (Hutchinson and Taylor, 1962). Dawkins and Dawkins (1974) tried to modify drinking in chicks by presenting a variety of aversive or disturbing stimuli. For the most part, the MAP was little affected if at all. But unusually shallow water caused the bill to be held in the water longer, a change the authors attributed to the lack of stimulation the fully immersed beak would ordinarily receive.

The dragonfly larva of *Aeschna cyanea* carries out high-speed strikes at its prey. Bucholtz (1970) reported that the strike could be stopped even after the fast phase had been entered, but she gave no substantiating data or details.

Alterability is a more contentious issue with MAPs that are clearly involved in communication. This is especially true for signals performed at high speed. It may be difficult for an animal to alter such MAPs because of inertia or momentum. Since the behavior is brief, it may be easiest for the animal to finish it, and at little peril, rather than to alter it. It may also be difficult to attend to the stimulus while performing, particularly if the head moves rapidly.

There is suggestive evidence, nonetheless,

that even brief rapid communicatory MAPs may be disruptable. The signature bob of the lizard *Anolis anaeus* is the most stereotyped component of its pushup display. If a male is performing a signature bob when another approaches, it completes the MAP before responding further (Stamps and Barlow, 1973). Now Stamps (pers. comm.) has done pilot experiments on tame males. One male is perched on one hand. When it starts to signature bob a rival male on her other hand is suddenly brought close to the performer. The displaying male usually completes the signature bob, but often it does break off the display.

The extent to which MAPs are alterable, within their normal scope of variation, is an open question and one that has not been much asked. And it is not just a matter of external stimulation, for motivation also plays a role (e.g., grooming in mice: Fentress, 1972). We need to know more about the susceptibility of ongoing MAPs to external stimulation, and the relation to their brevity, stereotypy, function, motivational state, and so forth.

STRENGTH OF STIMULATION

While there is not much information, it is becoming increasingly clear that the differences in the trigger, the releaser, may influence the MAP. The problem has been neglected in part because releasers have usually been considered only in their essential features; that is historically understandable. The releaser concept brought out important aspects of the appearance of signals and the nature of perception. Nonetheless, there has been a glossing over of relevant variation in the stimulus, such as rate of approach, proximity, size, and so forth.

A simple type of stimulus-dependent difference is one in which the performance will vary according to whether the appropriate stimulus is present. This is not as self-evident as it might seem when one considers species that advertise, as do male fiddler crabs awaiting females (Fig. 6). The tempo of the claw-waving display is faster when a female is present (Salmon, 1967), although each wave takes about the same amount of time (Table 4).

The relative size of the mate in a cichlid fish called the orange chromide *(Etroplus maculatus)* influences the coupling between head quivering and flickering of the pelvic fin (Barlow, 1968). Size of mate also affects the frequency of performance, i.e., the cyclicity, of a spectrum of MAPs in this and in other cichlid fishes (Barlow, 1970; Barlow and Green, 1970a, 1970b).

Proximity may alter the very form of the MAP. Hazlett (1972b) found that the height of the cheliped display in a hermit crab *(Pagurus longicarpus)* varies with the distance to its mirror image (Fig. 7). Similarly, Stamps and Barlow (1973) presented a territorial male lizard *(Anolis anaeus)* with a rival male at varying distances. Both the components of the pushup display and their interrelations varied as a function of distance (e.g., Fig. 8).

The strength of stimulation can also alter the sequence of MAPs employed in grooming behav-

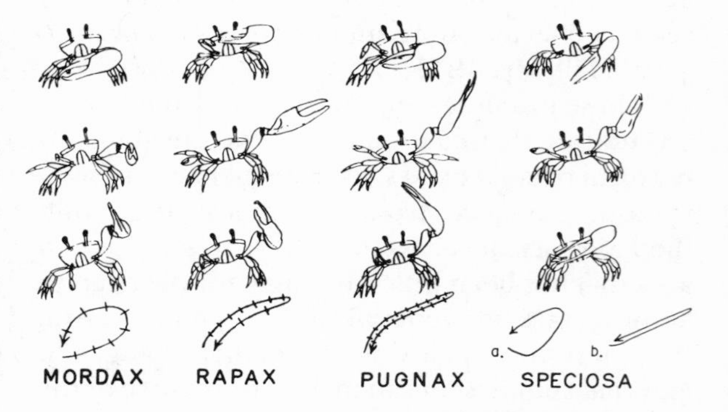

Fig. 6. Three consecutive stages, moving down the page, of claw waving in males of four species of fiddler crab (*Uca*). The bent arrow below each crab traces the position of the claw through one cycle; cross marks indicate a jerk. In *U. speciosa,* arrow (a) describes the claw's path in response to females, arrow (b) that toward intruding males. (From Salmon, 1967.)

Table 4

The effect of the presence of a female on the speed of a claw-waving display of males in two species of fiddler crab.

Species	Presence of ♀	N (♂♂)	Wave duration (sec) x	SD	CV	ST	Time (sec) to complete 5 waves in a series x	SD	CV	ST
Uca rapax	–	12	3.6	1.9	52.78	1.86	25.2	6.9	27.38	3.52
	+	12	2.5	1.2	48.00	2.04	15.8	5.7	36.08	2.70
Uca speciosa	–	12	0.5	0.01*	2.00	33.33	7.9	0.4	5.06	16.49
	+	12	0.3	0.01*	3.33	23.08	4.6	0.8	17.39	5.44

NOTE: x = mean, SD = standard deviation, CV = coefficient of variation, and ST = stereotypy as defined in the text.

*The variation was probably below that accurately resolvable by the rate of film advance, which was not given; presumably it was between 16 and 24 fps.

Source: After Salmon, 1967.

ior.[1] Typically a mouse starts with its face and finishes by grooming its back. A mild irritant on its back sets off an episode of grooming, yet the mouse starts with its face. But when a strong irritant is applied to the back the mouse then grooms the back first (Fentress, 1972). Thus if the stimulation is strong enough it overrides the typical program.

COMMON CAUSAL FACTORS

This issue was explored in detail by Russell et al. (1954; see also Hinde, 1966a; Hinde and Stevenson, 1969). It was proposed that all the components of a Fixed Action Pattern depend on the same causal factors. The idea is sound. If the response is unitary, its causation should be unitary. This may hold for many MAPS. But there is always the problem of the null hypothesis. One has to test all possible relevant stimuli to assure that there are not differential effects on the components (Barlow, 1968).

As seen, however, MAPS often are not as unitary as was formerly believed. One should expect differential effects, at least for the less-unitary MAPS. In one instance this has been demonstrated. The height of the pushup display and the extent to which the dewlap is extended are separately influenced in the fan bob display of the lizard *Anolis anaeus* (Stamps and Barlow, 1973). To a large extent, however, this display shares causal factors.

Generalizations are probably premature because there is so little to go on. The general impression that emerges, nonetheless, is that of a continuum. Some MAPS are cleanly triggered and relatively unitary, whereas others are to varying degrees under stimulus control. The original formulations were good first approximations. They brought out provocative and testable propositions by focusing on the more extreme cases. It is surprising that there has been so little attempt to test them critically.

1. It is difficult to apply a simple concept of MAPS to grooming because of its hierarchical organization. The distinction between acts, MAPS, and sequences is blurred. Each component is to some degree independent within the hierarchy. Yet there is so much patterning and close temporal linkage of components that they can be arranged into higher levels of organization that are reasonably stereotyped albeit stochastic. Furthermore, the choice of units of grooming influences one's conception of its higher-order organization (M. W. Woolridge, pers. comm.).

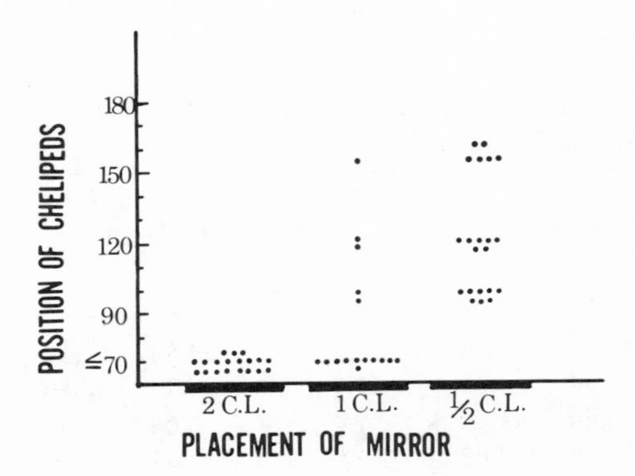

Fig. 7. The chelipeds of a hermit crab (*Pagurus longicarpus*) were raised to different heights (angles) depending on the closeness of a mirror in which it saw its image. C. L. stands for relative closeness in cephalothorax lengths of the subject. (From Hazlett, 1972b.)

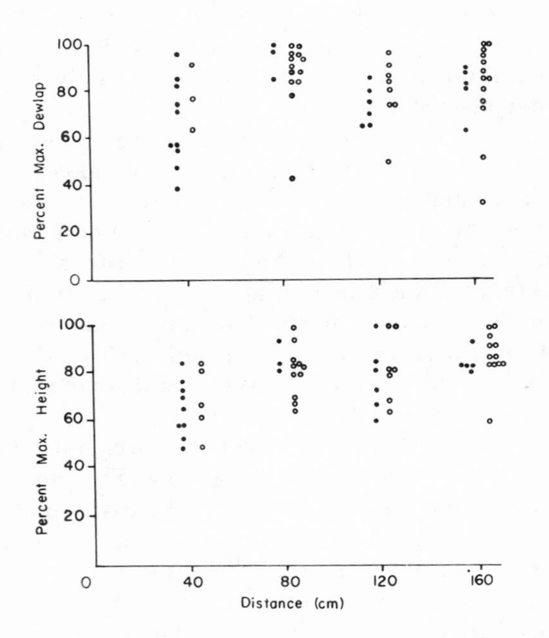

Fig. 8. Changes in the form of displays by male *Anolis anaeus* to "intruder" male lizards presented at various distances. Solid circles are for data from the display called fan, open circles for fanbob. (From Stamps and Barlow, 1973.)

Neurological Basis of MAPs

Out of necessity this section will be short and restricted primarily to a consideration of motor output. Obviously other aspects are important, such as sensory input and motivational states, plus learning, but they cannot be considered here. The proposition to be examined is that the spatiotemporal patterning of behavior, especially at the level of acts and MAPs, is a consequence of the functional organization of the nervous system (see Konishi, 1966).

Most of the relevant literature here deals with locomotion. But since many displays have been derived from locomotory behavior, and since some displays are little more than ritualized locomotion, the neurological basis of lomocotion is germane. The general conclusion is fairly clear. Locomotion is generated by central patterning, which is modified by both proprioceptive and exteroceptive feedback (Healey, 1957; Hinde, 1969; Hoyle, 1970; Wilson, 1966, 1967).

When one considers all motor output, a range of relationships can be seen (Hinde, 1969). At one extreme, swallowing (deglutition) and hind-leg scratching in mammals are machinelike in execution with almost no sensory modulation (references in Fentress, 1972). And Fentress (1972; see also Taub et al., 1973) concluded that "even complex movements in primates can be established and performed without any proprioceptive feedback"; he noted, however, that caution is called for in this interpretation (see Evarts, 1971).

There must be reasons why some MAPs are more dependent on peripheral modulation than others. Hinde (in Fentress, 1972) conjectured that those movements most involved in visual-motor coordination should be most severely handicapped by the loss of external information. Along similar lines, Konorski (1970) proposed that these more disruptable actions, requiring guidance, are more likely to encounter novel orientation. The obvious conclusion is that where it is adaptive to incorporate sensory information the system will have evolved to do so, and vice versa.

At one time it was thought that the organization of MAPs could be clarified through electrical stimulation of the brain. Unfortunately, the results are contradictory and difficult to interpret. Phillips and Youngren (1971), for example, concluded that "Complex 'natural' units of behavior are scare and fragments thereof are much more frequent." They were able, however, to produce sequences of aggressive behavior (see also Delius, 1973). The situation appears to be different in invertebrates such as insects (e.g., Loher and Huber, 1966) and gastropods (Willows, 1967). The CNS is more simply organized, and some MAPs are found to be controlled by command fibers (Kennedy, 1971), ganglia, or central "tapes" (Hoyle, 1970).

It may be, however, that the failure to evoke coherent MAPs in the CNS of vertebrates stems from a recent tendency to stimulate higher regions. Stephen Glickman (pers. comm.) suspects that the earlier workers had more success in this regard because they often stimulated lower regions of the CNS.

Even if diffusively structured, the organization of MAPs in the CNS can range from complex to simple. Brown (1969) made the attractive and, one hopes, testable point that the evolution of stereotyped behavior can be achieved, or is permitted, by simple neural nets. In contrast, variably adaptive behavior such as learning requires a large brain with specialized areas and complex circuitry. If so, this could account for the apparent progression from predominantly stereotyped behavior among the invertebrates to the progressively more graded and complex behavior of vertebrates. (This "apparent" progression could stand closer, more critical examination.)

The general situation seems to be that the CNS is functionally organized to produce patterned output. All motor output depends to some degree on central patterning. The organization is variously discrete to interwoven, and the output can be modified both by sensory information and by motivational state. Thus the mechanisms are present to produce a variety of motor outputs, ranging from stereotyped MAPs to smoothly graded and continuously modulated movements.

The Role of Experience

The neurological evidence indicates that MAPs reflect their functional organization in the CNS. That does not preclude a role for experience in that organization, however, particularly during ontogeny. Nor does it necessarily mean that once the organization is established it cannot be fundamentally changed through experience. By a fundamental reorganization I mean a statistically significant and persisting alteration of the spatiotemporal pattern of a MAP.

Now, no biological system develops independently of some type of experience, if experience be so broadly construed as to include one's own bodily tissues (Barlow, 1974b) and a variety of specific determinants (Bateson, 1974). It is therefore futile to propose a strict dichotomy between genetic and environmental factors (Lehrman, 1970). For analysis, however, it is convenient to disregard genetic variation and to ask rather if and how differences in experience might produce differences in MAPs.

The issue turns round the spatiotemporal pattern of MAPs themselves, not the development of associations between MAPs and stimuli. Historically there has been confusion on this point (e.g., Moltz, 1965), even though the distinction has been clearly spelled out (Eibl-Eibesfeldt and Wickler 1962; Marler and Hamilton, 1966; Tinbergen, 1960). The very basis of conditioning depends on the unconditioned response, often a MAP, forming some new association with a stimulus other than its appropriate (unconditioned) one.

The most open question in ethology is whether experience contributes to the organization of MAPs, and if so, how? There is little to go on at present. There are abundant fragments in behavioral studies. Unfortunately, most are highly qualitative although interesting and provocative. Lacking a guiding hypothesis, they have seldom dealt with the internal structure of a given MAP.

A reading of the literature on the ontogeny of behavior disclosed that experience plays a limited role, confined largely to facilitating the development of MAPs and to regulating their frequency of occurrence, their association with stimuli, and their integration into coherent sequences. In a few cases, animals have demonstrated the ability to mimic what are, for them, novel MAPs. Imitation, however, seems to be limited to primates (e.g., Gardner and Gardner, 1969; Hinde, 1970; van Lawick-Goodall, 1970), toothed whales (Tayler and Saayman, 1973), and to vocalizations in parrots and various other birds (Thorpe, 1961; for a general review see Davis, 1973).

Caged animals develop highly stereotyped, often seemingly idiosyncratic, bits of behavior (Holzapfel, 1939; Morris, 1964, 1966). But these movements may not be so idiosyncratic. The peculiar stereotyped activities of chimpanzees raised without mothers, for instance, have a degree of congruence, and they resemble stereotyped movements seen in some institutionalized humans. Such movements may be "a common primate reaction to a variety of disturbances in the environment" (Berkson and Mason, 1964). This suggests that even seemingly individualistic but patterned movements are based on functionally integrated components inherent in the CNS.

That the basic coordination of a MAP can be altered through experience remains scarcely demonstrated. A possible instance is the loss of one component of pecking in chicks of the laughing gull (*Larus atricilla:* Hailman, 1967). In most cases the basic coordination seems resistant to fluctuations in the environment. Nonetheless, environmental influences may be subtle or more indirect, or may come early, as in embryogenesis, and therefore be undetected. The problem is that little attempt has been made to disprove the hypothesis that the spatiotemporal pattern of MAPs is not affected by experience. If the attempt were made perhaps a different view would emerge.

Genetics

The growth of behavior genetics over the last ten to fifteen years has been extraordinary (Broadhurst et al., 1974, and Manning, 1975). It is now indisputable that genetic differences can produce phenotypic differences in behavior. Furthermore, in the broad usage of population genetics, behavior is heritable. As demonstrated through the technique of quantitative genetics, "heritability, which is a measure of the genetic variability accessible to selection, [is] relatively high for most [behavioral] characters and comparable to those calculated for morphological characters" (Ewing and Manning, 1967). As the field progresses, behavior is shown increasingly to be the product of the well-known and intricate complexities of genetic determination. These include pleiotropy, epistatic and additive effects, sex linkage, and the influence of the genetic environment on the expression of given alleles (Dobzhansky, 1972; Manning 1975).

Most of the analyses, however, have not dealt with the structure of MAPs. Investigators have centered on such things as rate of defecation or activity in an open field, frequency of courtship displays, susceptibility to audiogenic seizure, and many more (Manning, 1975). Yet MAPs seem especially well suited to the purposes of genetic analysis (Ewing and Manning, 1967).

MAPS appear to be polygenetically controlled (Manning, 1975). As a consequence, phenotypic variation between individuals is continuous. Differences of one or a few genes produce such small changes in the MAPS that they frequently go unnoticed (Franck, 1974). Thus qualitative differences within interbreeding populations are rare. (Of course, qualitative differences are merely large quantitative ones; it is a matter of degree.)

One of the more important small differences that can be produced by mutation is that of threshold of occurrence of the behavior. This is illustrated by Manning's (1959) observations on two closely related species of fruit fly (*Drosophila melanogaster* and *D. simulans*) that differ in only a few genes. They diverge in courtship behavior, but more in degree than in kind since they share the same displays. But the two species favor different displays. Manning contended that the divergence is due to differences in thresholds of the MAPS, perhaps as a consequence of different levels of arousal.

The threshold model is attractive. It could account for many differences that have arisen in the evolution of MAPS, as in the pushup display of an iguanid lizard (*Uta:* Ferguson, 1971, 1973). Franck (1974), however, warned that the threshold model should not be too generally applied "because it misleadingly implies a fundamentally uniform action of genes."

The most convincing demonstration of the genetic control of MAPS comes from the study of hybrids (see useful reviews by Franck, 1974; Marler and Hamilton, 1966). The most productive approach is to cross two closely related species that have distinguishable but homologous MAPS. Differences in the F_1 phenotypes, compared to the parents, can be attributed to genetic differences because they can be replicated. They can also be verified through classical genetic methods such as producing F_2 progeny and backcrosses. These studies confirm that MAPS are under polygenic control. In most of these studies, unfortunately, behavior is assessed by simply recording the presence or absence of the MAP, or the MAPS are described verbally with insufficient attention to essential detail.

Generally the behavior of the hybrid is intermediate, as is its morphology. However, the intermediacy is often less than perfect in that hybrids tend to favor one parent or the other (Davies, 1970; McGrath et al., 1972). Sometimes hybrids show intermediacy in one trait but apparent dominance in another, as in the display of a hybrid *Anolis* lizard (Gorman, 1969). If the species are sufficiently unrelated, as in a duck X goose cross (Poulson, 1950), the hybrid may fail to express many MAPS of its parent species. Of more general interest, if the parent species are quite but tolerably different, the behavior of the hybrids is often a mosaic of that of its parents rather than being intermediate (Buckley, 1969; Hinde, 1956; Leyhausen, 1950; Lind and Poulsen, 1963).

Uncoupling a MAP from its taxis has been seen in hybrid swordtail fish (Franck, 1970). Sequencing of MAPS, too, can become scrambled in closely related species with homologous displays (Lorenz, 1958; Ramsay, 1961; Sharpe and Johnsgard, 1966).

The evidence, then, reveals that behavioral traits are genetically determined in much the same way that morphological ones are. The genetics of MAPS can be studied with the same methodology and is subject to the same limitations of interpretation with regard to environmental interaction and analyses based on differences.

MAPS in Communication

So far I have seldom troubled to distinguish between communicatory and noncommunicatory MAPS. That has allowed us to consider patterned movement in general, to bring in all information available regardless of function. After all, one of the classic examples of stereotyped movement is that of the goose retrieving its egg, which is hardly a communicatory MAP. Additionally, it is often moot whether or when a MAP serves in communication.

Grooming and nest building, as two examples, might not seem communicatory. Yet preening behind the wing signals that copulation is imminent in pigeons (pers. obs.). Wing preening and other movements involved in bodily maintenance (Fig. 9) are an integral part of duck communication (Lorenz, 1941; McKinney, 1965). In penguins and other aquatic birds, a number of comfort movements may transmit information (Ainley, 1974). Allogrooming in primates has also evolved a communicatory function (Hinde, 1966b).

Communication is held to occur when a MAP performed by the sender changes the probability of subsequent behavior in a perceiver. Within this broad definition there are some important distinctions. If there is a communicatory system, then, in the evolutionary sense, communication should improve the fitness of both sender and receiver (Klopfer and Hatch, 1968). The evolution of communication therefore requires a parallel progression of the signal and the perception of it, whether within or between species.

This restriction, however, excludes informative interactions between predators and prey, where evolution may be acting in a comparable fashion on the spatiotemporal patterns of MAPs. Many sedentary species, for example, have evolved spectacular displays to put off their predators ("startle" displays of moths: Blest, 1957). There will be selection for those moths performing the most effective MAP, as well as for those predators that are not so easily startled. Likewise predators might evolve special movements or vocalizations that are more effective in driving or luring and capturing prey. The roaring of lions, if it confuses prey and facilitates their capture, could be considered a ritualized MAP. A number of fish predators (Shallenberger and Madden, 1973) and some turtles (Hediger, 1962) lure prey to them with stereotyped movements that connote food. Such cases present problems of signal detection, ambiguity, localization, and so on, that parallel those in mutualistic systems of communication.

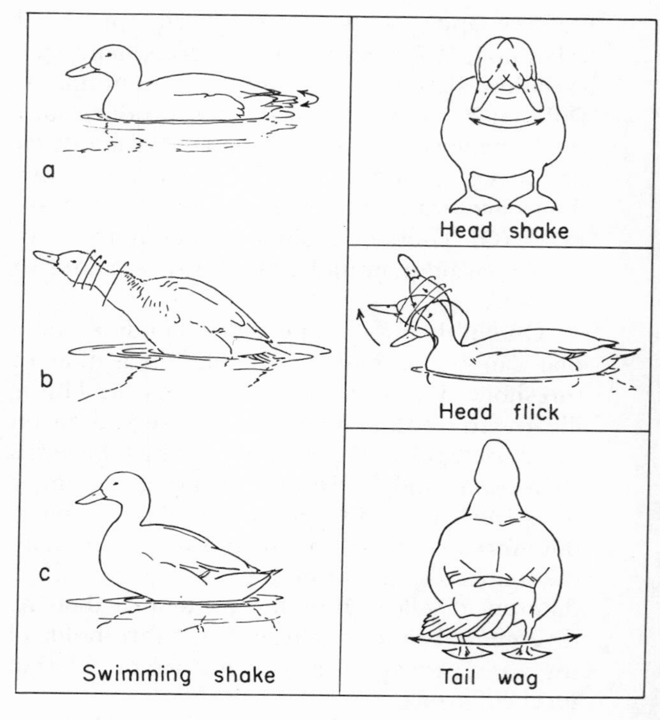

Fig. 9. Four different comfort movements in ducks that are simple, conspicuous, and may be involved in communication. (After McKinney, 1965.)

The discussion to follow is built around the evolution of MAPs, particularly as tools of communication. Evolving into a signal often means becoming stereotyped, and that will be the organizing theme. In some respects this is an unfortunate approach, though a reasonable and convenient one. It emphasizes stereotypy at the cost of other important aspects of MAPs in communication, and that is not my intent. In addition, it detracts from recognizing the gradation extending from the more variable to the more stereotyped signals.

I also sidestep a precise meaning of stereotypy. It is not easy to spell out exactly what is meant when a MAP, in its totality, is said to be stereotyped. Many MAPs are stereotyped along one dimension but variable along another. Or, using the same parameter, say, duration, one component may prove stereotyped and another not. There is so far no satisfactory measure of all aspects of stereotypy of a given MAP. Nor is the problem of level in the hierarchy of behavior easily resolvable, especially for MAPs such as those of complex grooming in rodents.

In the following I review first the problem of producing patterned output by the neuromuscular apparatus. Then I touch briefly on these properties of perception that feed back on that output. Finally, I run through the various ways stereotypy or variation might be favored, depending on signal function and environment. I avoid trying to explain the origin and derivation of displays and the many types of changes that are possibly involved. That area, the topic of ritualization, has been covered well and in detail elsewhere, following the pioneering paper of Daanje (1950; e.g., Blest, 1961; Cullen, 1966; Hinde, 1959; Lorenz, 1958; Marler, 1959; Morris, 1957, 1966; Tinbergen, 1964).

MOTOR OUTPUT

Just about any regularly performed behavior becomes stereotyped, whether it is clearly learned or not. Thus religious ceremonies of humans and the play of animals become simplified and hence stereotyped (Thorpe, 1966), as do acts of "superstitious" or interim behavior of animals in Skinner boxes (Millenson, 1967; Staddon, 1975). This progression toward economical output has been called the principle of least action (Wheeler, 1929), least effort (Tolman, 1932), or simply parsimony (Adams, 1931).

Spatiotemporal patterns of behavior characterizing populations or higher taxa doubtless become similarly parsimonious and therefore efficient. Since there is usually one most efficient method, the individuals of a population will converge on a particular method of performing the MAP. The capacity to develop such a stereotyped MAP will become genetically reinforced and common to the gene pool. The neural circuits underlying such stereotyped motor behavior are thought to be particularly susceptible to having their efficiency improved through natural selection (Brown, 1969).

If such behavior is especially frequent it tends to assume rhythmical repetition. Schleidt (1974) suggested that this is more likely if the behavior has been derived from originally rhythmic activity such as locomotion or respiration. It is probably also related to mechanical factors. Appendages of a certain mass and proportion, for example, will have an optimal period and amplitude for a given behavioral task. If the behavior needs to be repeated frequently it should assume an efficient rhythmic pattern. A good example is the fanning of eggs or larvae by a wide variety of teleost fishes. In *Badis badis* fanning is a stereotyped rhythmic behavior; its parameters such as tempo, bout length, and number of bouts are organized into rhythmic bursts in order to propel fresh water to the eggs (Barlow, 1964).

Mechanical factors are important in ways other than rhythmic repetition. Smooth, relatively quick motions are probably more efficient than slower, more erratic ones. Slowly performed movements, with many starts and stops, require more work to overcome inertia and momentum, respectively, and to hold the appendage or body in a particular position. Sheer size must also be considered. The smaller species or individuals are less restrained by mass and inertia. They can have quicker, more dynamic, and therefore more intricate movements (Cullen, 1966; Hutchinson and Taylor, 1962; Meyerriecks, 1960).

Irrespective of size, neuromuscular effectors tend to be conservative. Most communicatory MAPs have been derived from basic patterns of coordination, such as locomotion, respiration, feeding, elimination, and protective reflexes. The basic coordination, especially of locomotion, probably cannot endure the large changes that would be needed to produce radically different MAPs. The forms of MAPs are therefore limited to patterns similar to those of the more basic activities. Departures are obviously possible, but the nervous system has little scope for unique MAPs.

New MAPs are bound to share effector systems with more primary ones. They therefore have the potential to interfere with them. Sharing may occur quite far upstream. Consequently a basic coordination may service several behavioral tasks, so the same MAP may be employed to different ends. Context then becomes necessary in determining message and meaning (Smith, 1968, 1969).

Sharing of patterns of coordination in threat, attack, and predation is relatively common (Albrecht, 1966; Barlow, 1968; Blurton-Jones, 1968), although the MAPs may be to varying degrees distinguishable. A basic motor pattern may be used even more widely. Neck biting in felids is employed in the unrestrained biting of prey, by males when fighting with other males, by a male toward the female during copulation, and by a female when carrying kittens (Leyhausen, 1973). Some differences, such as the force exerted, are obvious, but differences in coordination are not fundamental.

In summary, the nervous system of animals favors stereotypy when faced with a familiar task. Regularly performed behavior is parsimonious and therefore stereotyped to some degree. The nervous system gives priority to its more essential patterns of coordination, which limits the extent to which secondary MAPs may differ from them. It also reduces, thereby, the number of displays that may be produced. Consequently context often must supply the missing information about the meaning of a MAP.

PERCEPTION

A consideration of the influence of perception automatically limits the discussion to MAPs used in communication. Right off, we encounter a chicken-and-egg argument. To become a signal a MAP should evolve into a form that is best perceived by the intended recipient. Conversely, the recipient should be adapted to perceive key MAPs. At the more general level, properties of the receptors are probably the more fundamental. For finer tuning, both receptor and signal must adapt to each other. But if there were one best signal for a receptor, in general, a number of difficulties would arise. For one, a species might have only one signal in that channel. For another, closely related sympatric species would converge on the same signal or would be forced to change their receptor systems. If that means an alteration of spectral sensitivity, for instance, it could leave the receptor less well adapted for other essential behavioral tasks, such as detecting items of food. One solution is to alter temporal patterning and make the adjustment in the receptor system farther upstream. Yet there must be some limit here. One predicts a compromise in the more complex species, with the MAPs being constrained by the motor system, although with a substantial amount of adjustment to maximize detection by the receptor system.

There has seldom been any attempt to examine how the spatiotemporal pattern of MAPs is related to perception. One exception is the work of Magnus (1958) on the fritillary butterfly *(Argynnis paphia)*. To reproduce, the male chases females, who are of a particular size and coloration and beat their wings at a characteristic rate. The faster the wings of a mechanical female beat the stronger the response of the male, until the rate exceeds the flicker-fusion frequency of the male's eyes. Then it stops responding.

Hailman (1967) followed the same logic when testing the pecking response of chicks of the laughing gull. In contrast, the perceptual system here is adapted to the MAP. The chicks responded best to an intermediate speed of the model of the parent's head.

An indirect confirmation of signals' being adapted to receptor systems lies in the relationship between threat and appeasement displays. Darwin (1872) pointed out that they are antithetical (see Marler, 1959; Barlow, 1962). What needs emphasizing here is the contrast between them in relation to perception. Threat displays involve conspicuous movements, increase in size (hair, fins, or feathers, etc., extended), and approach. Some species also change color, contrasting with the background and sharpening their profile. The receptor system is played to.

In the commonest type of appeasement display the opposite occurs. The animal becomes relatively if not absolutely motionless, small, and seems to melt into the background. It fits itself into the "blind spots" of the receptor system, so to speak, to minimize detection and thus the evoking of attack.

As one might expect, there are appreciable differences in visual filtering between such phyletically and ecologically divergent animals as frogs, rabbits, and cats. Nonetheless, a modestly coherent theme emerges (Barlow et al., 1972). Gone are arguments about central versus peripheral filtering of stimuli. There is a continual process of filtering and coding, starting at the retina and continuing on into the brain. Different "trigger features" are encoded (compressed) at different anatomical levels ascending the optic pathway. Salient attributes include dark-light contrast, bars, slits, or edges, corners, ends, and orientation. These are all aspects of form. Attributes of motion, such as movement away from or toward the fovea, expansion, and so forth, play a role but are less well understood.

Probably the most important general feature of visual perception is the reduction of redundancy that is inherent in the input (H. B. Barlow, 1961). The nervous system tends to respond to change, as in "on," "off," and "on-off" fibers. This makes possible lateral inhibition and the consequent significance of contours. It also means that the nervous system, after a while, ignores sustained input—it is neophilic.

Thus it is paradoxical that communicatory behavior of animals, the output, is so often redundant. Why should this be, if repetitious stimulation is likely to be ignored?

Animals have a limited number of signals, which they must therefore repeat often if communication is at all frequent. Further, multichannel redundancy is almost the rule. For example, a displaying fish may posture and move, change colors, and vocalize while giving off chemicals. It is as though the animal has little to say but insists on being heard, so it delivers basically the same message repeatedly and through different channels.

As Bateson (1968) put it, redundancy exists if missing items can be guessed with greater than random success. Redundancy in this sense is patterning sufficient to permit part-for-whole communication. This turns the issue around. If the output is redundant it does not necessarily follow that the intended receiver is observing all the output. Indeed, animals that repetitiously broadcast—for instance, lek species—usually change their behavior when they get feedback.

The receiver is faced with a continuous stream of behavior by others. It is also barraged by sensory stimulation from the environment as a whole (H. B. Barlow, 1961). It has to pick out the relevant signals (segments). Part-for-whole communication becomes significant. To deal with the flow of behavior the receiver must properly segment it to decode it. Here it gets help from the sender.

The sender breaks up its behavior into relatively unitary pieces, whether for communication or for other reasons that I have already discussed. It produces variously discrete MAPS.

If a MAP is to be attended to it must contrast with other stimuli reaching the receiver. Now, contrast can be a slippery concept (Andrew, 1964). Here I mean only that it must differ sharply from the commonplace. If it differs, it is to some degree novel and therefore detectable. To assure that the pattern is recognized and decoded, once detected it must be stereotyped (redundant) to conform to the code.

Communicatory MAPS should have sufficient redundancy, therefore, to be extracted from the continual stream of heterogeneous input. But if the signal is highly redundant and inescapable, it may be responded to only initially, then ignored. A method of "having one's cake and eating it" is to regularize some parameters of the output, facilitating detection and recognition, while making other parameters variable to maintain attention. This can also be accomplished by changing the order in which MAPS are presented, that is, through variable sequencing.

MAPS that have evolved into signals are commonly given in strings or sequences. Then they often overlap broadly and to varying degrees. I have no desire to explore here the different models of sequence analysis (see Nelson, 1973; Slater, 1973), but such models are germane to the question of how complex MAPS are organized and thus how they evolved.

Do animals first recognize redundant sequences of MAPS and learn to respond to them as units? Through experience, humans can organize words or numbers into "chunks" of increasing length (Simon, 1974). Possibly animals do something similar, as when shifting from heterogeneous summation to *Gestalt* perception during ontogeny. This ability has been reported by Hailman (1967) for laughing gull chicks (see also Bower, 1966, for human infants).

The maximum number of "pieces" per chunk evidently remains about the same, around five to seven in humans (Simon, 1974). This implies an upper limit to the number of units that can be organized into a new one. By coalescing smaller units into larger ones, however, greater complexity can be processed. Hence more information can be communicated in a given amount of time.

In animals, chunks would be equivalent to MAPS, or to groups of them, in a stream of behavior. The number of components making up a chunk in vertebrates must be relatively low. The few Markov chain analyses indicate, at best, second- or third-order dependence (Altmann, 1965; Fentress, 1972), and often less (Nelson, 1964). If animals evolved displays from sequences one would expect no more than two to four major components, and usually fewer, a seemingly reasonable first approximation. The displays of *Anolis* lizards are an excellent example (Gorman, 1968).

The anatid ducks, however, provide the best material here (Lorenz, 1941). Their displays are ritualized, rapidly performed sequences of up to three or four simpler acts. Differences between species consist in the main of re-ordered acts or the addition or loss of acts. "The elementary motor patterns are more widely spread among the species than their combinations" (Lorenz, 1955).

In fact, McKinney (1961) suggested that in some instances the sequences of MAPS, not just coupled acts, have become the displays. If so, the displays have moved up a level of organization in our conceptual hierarchy. McKinney speculated further that the highly stereotyped sequences of some species represent an evolutionary advance over the more "primitive" state, the relatively variable sequencing observed in other ducks.

This brief review of how the properties of the perceptual system might guide the evolution of MAPS as signals reveals how little is known in spite of progress in recent years. (A brief venture into the literature on pattern recognition by machines was of little help.) I feel the physiological information is there to bridge the gap, to say much more, but it has not been applied to this particular problem. It would thus be useful if physiologists were to pursue the question of how animals should structure signals to "exploit" the perceptual side of the system. What spatiotemporal patterns best lead to their detection, discrimination between them, and continued attention?

SIGNAL FUNCTION

This, the last major section, is a bit eclectic. It is meant to identify those factors favoring stereotypy or variability in MAPS serving as signals. It is often put that if a MAP is used in communication it must be stereotyped (Barlow, 1968; Hazlett, 1972b). But this should not be construed to mean simply that the more obviously a MAP is employed in communication the more it will be stereotyped. The inference is only that if there is a communication system involving a sender and a receiver, they must have an agreed-upon code. Therefore the signal must have a recognizable pattern, and so it must be to some degree stereotyped. But different needs will set, through natural selection, different degrees of stereotypy.

Morris (1957) coined the term "typical intensity" to characterize situations in which the need dictates stereotypy. If it is all-important that only one message be transmitted, then the MAP should remain about the same over a wide range of stimulation. In contrast is the system that utilizes an analog type of signal: The signal varies with the strength of stimulation. Here there is the potential to communicate more information. There is also more ambiguity because the signal is likely to be continuously graded or divisible into steps that differ little from the next one (see Marler, 1959). In either case the receiver must be able to distinguish small differences in the signal.

Working between these extremes, I will discuss the factors favoring stereotypy versus those favoring more graded, hence variable, MAPS. This argument is essentially an evolutionary one. For coherence, I will bring in other factors bearing on the adaptiveness and evolution of MAPS.

Basic to all considerations is ecology—the physical environment, interaction with other animals whether benign or predatory, and feeding behavior. While recognizing the ultimate role of bioenergetics, I will not attempt to develop the nexus between the animal's ecology and the consequent form of its displays. Suffice it to say, however, that in spite of earlier views to the contrary displays are seldom truly arbitrary in form (Andrew, 1956 ; Brown, 1963; Crook, 1964; Cullen, 1957; Gorman, 1968; Hinde, 1956; Tinbergen, 1964).

The course of an animal's development should also prove important. Alexander (1968) noted that nonsocial insects, for example, seem to fit all of Lorenz's original criteria for instinctive behavior, including highly stereotyped MAPS. Alexander interpreted this as an adaptation to the environment that prevails during development. First, no parents are present to exert their influence. Second, there is misleading environmental "noise" that must be overcome. Thus their MAPS should be resistant to irrelevant environmental stimulation and should not require parental molding.

Nonetheless, I suspect that the extreme stereotypy of arthropods is just as much attributable to the limitations of their simple central nervous systems. Note that many lower vertebrates, such as fishes, amphibians, and reptiles, experience a similar situation in ontogeny. While all these vertebrates show clearly stereotyped MAPS, they also perform graded, variable ones. Environment during ontogeny is probably a significant evolutionary factor in the development of stereotypy, but it cannot be considered of overriding importance.

A general factor influencing the evolution of stereotypy is noise. Interfering noise is usually proportional to the distance between sender and receiver. Different environments have different levels of noise so, for comparative purposes, both distance and environment need considering. The rule is that at greater distances and/or noise, signals are more redundant and individually stereotyped (Marler, 1965, 1968, 1973). This enables part-for-whole communication, so that the signal need not be received either completely or continuously.

Conversely those species that maintain close relationships tend to use signals that are more graded and flexible. This appears to hold in a variety of animals, such as primates (Marler, 1959, 1968, 1973), canids (Fox, 1970), and cichlid fishes (Barlow, 1968). There is a danger of oversimplification here, however.

Within species that associate closely, other factors may drive them toward stereotyped signals. Cullen (1966) noted that in colonial sea birds small territories necessarily heighten the stereotypy of signals involved in the reduction of attack. Another example is provided by hermit crabs. A relatively social species that lives in high population densities, and therefore has frequent contact with other members of its species, has highly ritualized aggressive displays. Species that live widely spaced are more aggressive and simply attack when they meet, rather than perform ritualized displays (Hazlett, 1972b).

The difference lies in the relationship between the animals in close contact. If they are members of a social group and are personally known to one another, the signals will tend to be graded. But if they do not live in social groups and are not likely to be known to one another, then stereotypy will be favored.

There are commonly differences within species in stereotypy of MAPs, even within the same signal category (Marler, 1968). These relationships are easier to visualize if categories of behavior are first compared. Crane (1966) stated that static threat displays of fiddler crabs are more stereotyped than those utilized to attract females. Hazlett (1972a) presented data to support just the opposite relationship in two other species of crustaceans: MAPs used in threat were found to vary more than those in mating behavior. This is more in keeping with Marler's (1959) observations on chaffinches and with my experience with cichlids and other teleost fishes. Such divergent conclusions suggest that the problem is one of secondary rather than primary correlations.

A universal factor is time. The briefer the encounter the more stereotyped the signal is likely to be, as in the mating of lek birds (Sibley, 1957). Wiley (1973) reemphasized the correlation after analyzing the strut display of the sage grouse, a lekking species.

The same conclusion derives from the stereotypy of sequences of displays: Camouflaged species of hermit crabs are secretive and expose themselves only briefly during interactions; their behavior is more stereotyped than that of related species that interact more frequently and openly (Hazlett and Bossert, 1965). Similarly, if there is great risk of being damaged, the encounter will tend to be brief; this may account for the stereotypy of behavioral sequences, and therefore reduction of ambiguity, in mantis shrimps (Dingle, 1972).

An important factor favoring brevity, hence simplicity and stereotypy, is exposure to predation (Moynihan, 1970). It accounts for the quickness of displays in the camouflaged hermit crabs. This conclusion, however, is largely inferential although it gets some support from observations on comfort behavior by Adélie penguins *(Pygoscelis adeliae)*. They are preyed upon by leopard seals *(Leo pardus)* mostly in the surf line. Young penguins bathe just beyond the surf and are easy prey. Experienced adults delay bathing until well beyond the surf. A most relevant factor here is that their bathing, unlike that of other waterfowl, is broken into short bouts, thus reducing the time they are continuously vulnerable (Ainley, 1974).

Frequency of display *per se* was mentioned earlier as a factor favoring stereotypy. But I have also maintained that MAPs less frequently performed must be more stereotyped. Both statements are true with more information, and for different reasons. Frequent behavior, whether or not employed in communication, will be stereotyped for reasons of neuromuscular economy. But if the MAPs are performed very frequently and in the service of close-up communication in a social group, there may even be a tendency for them to become less stereotyped (see below). Rarely occurring signals, on the other hand, must be stereotyped if the receiver is to detect and recognize them without fail. The evidence suggests that this is the case, as in seldom-emitted warning calls of songbirds (Marler, 1957, 1961b). Since such signals must often of necessity be decoded without benefit of much or any previous experience, it is also essential that they be relatively simple. Otherwise, it is difficult to encode the signal in the genome (H. B. Barlow, 1961). Additionally, simple signals are learned more quickly than complex ones.

The example of alarm calls brings into consideration intra- as opposed to extraspecific communication. Many passerine birds have evolved two types of vocalizations that are similar in their essential features. One is an alarm call. Its important feature is that it is difficult to locate, minimizing detection of the sender by the predator. In contrast, the other vocalization, the mobbing call, facilitates localization and brings other birds to mob the "enemy." The resemblance of these calls across species is due largely to convergence on the physical parameters of sounds that make them difficult or easy to locate, respectively (Marler, 1957).

In addition, communication between species favors selection for (1) a simple signal and (2) extraspecific stereotypy. The elaborate threat displays of surgeonfishes, for example, vary between the species and are largely exchanged between conspecific fish. Extraspecific aggression, in contrast, is usually simple and similar across species (Barlow, 1974a). Thus extraspecific communication encourages stereotypy.

Most of the foregoing has focused on factors increasing the stereotypy of MAPs, although the argument could have been developed the other way around. For example, more time permits signals that are more graded. In some instances it is easier to think about the advantages of lessened stereotypy. This applies to the more social animals that live in relatively stable groups.

Two factors operate here, and both are functions of time. First, prolonged association permits the opportunity to learn finer distinctions between signals. The receiver learns what follows what, and errors are corrected through reinforcement. There is more opportunity for signals to be graded or similar to one another. In addition, context can play a greater role.

The second factor is the inherent tendency of the nervous system to filter out redundant input (H. B. Barlow, 1961). This is particularly critical in the pair situation. If the pair has a prolonged courtship phase the probability is high that the partners will habituate to one another. If each partner is to stimulate the other's reproductive physiology to achieve sexual union, they must be able to hold one another's attention to effect some degree of arousal.

Novelty, or variation, is important in attention and responsiveness. The most celebrated example is the Coolidge effect: When a male rat is apparently no longer able to copulate with one female it is presented with a new female, and copulation again occurs (Dewsbury, 1973; Fisher, 1962).

It is possible to achieve, in motor output, both pattern and variation. The animal can vary the sequence of stereotyped MAPs. It can also produce variation by overlapping MAPs. Even the MAP can be varied. For example, the orange chromide can maintain a recognizable pattern of quivering while varying the duration of bursts, the coupling of head movements with the flickering of the pelvic fins, orientation relative to the mate, and the superposition of other MAPs (Barlow, 1968). This is an antimonotony strategy. It was first suggested by Hartshorne (1958, 1973; see also Nottebohm, 1972) for bird song and was later applied to visually perceived MAPs (Barlow, 1968).

I should like to digress here to return to the theme of rhythmic behavior. It is relevant to the problem of monotony. MAPs are so often derived from the coordination of respiration or locomotion that one expects them to be rhythmically performed (Schleidt, 1974). The fanning movements of parental fish, the wing waving of colonial sea birds, and the pushup displays of lizards are obvious examples of rhythmic behavior derived from locomotion.

Given this, rhythmic displays are actually not as frequent as one might expect. They should be encountered most where redundancy is needed, as in communication over a distance. This seems to be the case. Examples can be obtained from the claw-waving behavior of fiddler crabs, the almost dancelike behavior of lek birds, and advertisement songs of many passerine birds.

Otherwise, selection should favor patterned irregularity. A good case is found in the comparative study of pushup displays by iguanid lizards of the genus *Uta* (Ferguson, 1971). In the course of evolution the intervals between each pushup have assumed different lengths, as have the distances of the up-down excursions. The result is a syncopated progression rather than a rhythmic repetition.

Yet there are times when rhythmic behavior is highly evocative and apparently essential to the proper response. This is evident in the arousal and synchronization of sexual behavior. In cichlid fishes the variable courtship becomes repetitive and highly redundant shortly before spawning. The rhythmic aspects of copulation in other animals hardly needs mention. This reversal, the requiring of redundancy to evoke a response, is probably a "fail-safe" mechanism, insuring that both sexes will synchronize their ultimate reproductive act.

Returning to the main theme, care is needed when applying to all species the idea that a prolonged pair relationship will result in variable behavior. Some species that live in pairs actually have infrequent and brief contact. Their interactions are comparable to those of species that come together only to achieve fertilization. In many birds the male and female share a territory but they are only weakly bonded; personal recognition is minimal. If one is lost a new, strange bird of the same sex may enter and become the mate (Welty, 1963). One would predict rather stereotyped behavior in such situations, as compared to paired species that move about together and show evidence of personal recognition.

It has been noted by a number of authors (e.g., Thorpe, 1961) that the variability of displays is related to individual recognition. Most of the research here has been on bird song (Nottebohm, 1972). Thorpe listed two main functions of song: (1) It should be sufficiently stereotyped to characterize the species. (2) It should be variable enough for individual recognition. Marler (1961a, 1961b) suggested that different aspects of the song serve these two functions; Falls (1969) found some evidence for this (see also Hutchinson et al., 1968; Ingold, 1973).

It should be clear what is meant here. The variation serving recognition is between individ-

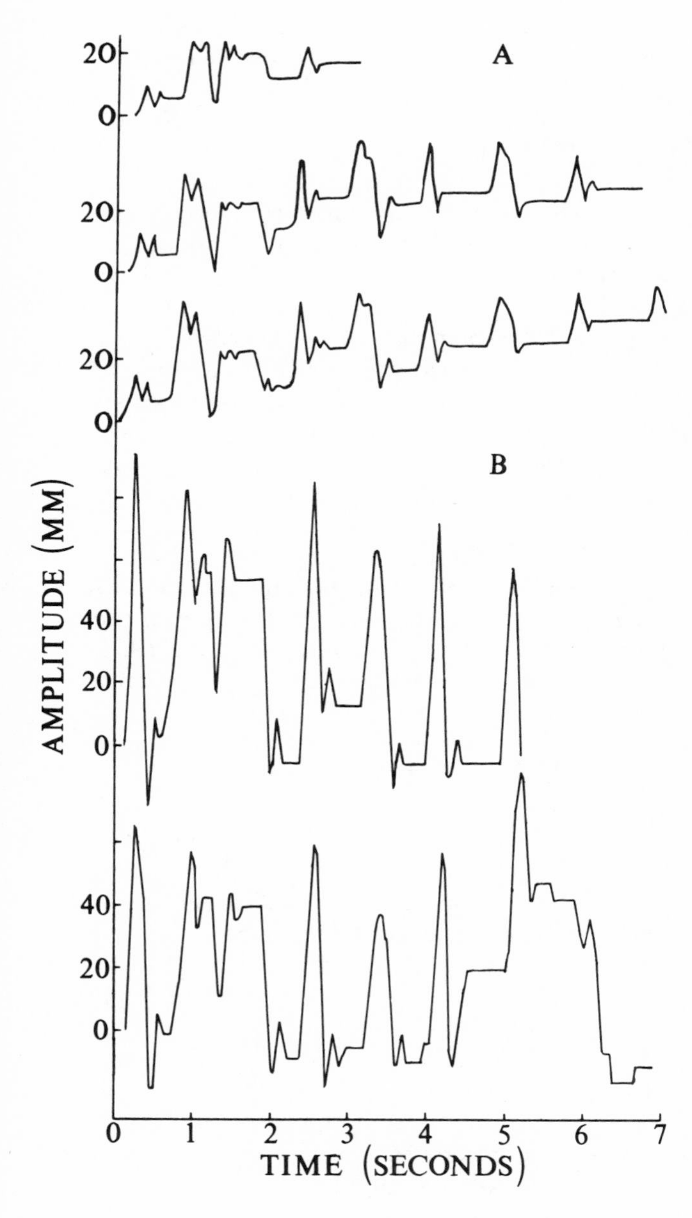

Fig. 10. The path of the head in the head bob display of the chuckwalla. The top three (A) are from one female, and the bottom two (B) are from one male. (From Berry, 1974.)

uals, not within them. The stereotypy of a MAP or of a component of it that is used in individual recognition should be higher than that shown for the group as a whole. Apparently it is. The variation of the means of temporal scores of displays of different individual *Anolis* is large, but it is small within individual lizards (Jenssen, 1971). In another lizard, the chuckwalla *(Sauromalus obesus obesus)*, the pattern of head bobbing is erratic and varies greatly between individuals (Fig. 10). However, "the first three to four seconds of the display are unique for each chuckwalla" and are essentially the same each time the display is given (Berry, 1974). Thus a portion of the behavior communicates individuality, as Marler (1961a, 1961b) suggested for bird song.

Marler (1961b, 1968) developed the view that when communicatory MAPs operate in the presence of a number of similar signals they tend to be more stereotyped, and there is evidence for this in bird and primate vocalization. This is a variation of the "noise" argument that was mentioned before. Marler had in mind closely related species occurring together. Ambiguity here must be avoided. "Key" signals should be executed with more precision, matching the "lock" in the perceptual apparatus of the receiver.

There is little evidence here for visual displays, and it does not clarify the issue. Isolated island populations of an iguanid lizard, *Uta,* are among the most- and the least-stereotyped populations (Table 2). In another case, the pushup displays of spiny lizards *(Sceloporus)* show character displacement where two species are sympatric (Ferguson, 1973). The expectation is that the displays should also be more stereotyped where the species overlap. That seems to be true for one species, but the opposite holds for the other. Thus there is no substantiation of greater precision, better fitting of the key to the lock, when visual signals are in competition. More studies are certainly needed.

This lock-and-key analogy is sometimes confused with a more general argument about signals without regard to competition between them. One should distinguish between two selective pressures. One is for detection of the signal from the background, which I have covered above. The other is for discrimination between comparably detectable signals. In some instances the two factors work in concert, but in many they do not. In any event, they are separable for purposes of analysis and exposition.

The discrimination of the correct signal from among others, as when closely related species coexist, leads to the evolution either of more complex or of more precise MAPS (Schleidt, 1974). Greater complexity permits less precision because it becomes increasingly improbable that competing signals will correspond as they become more complex. If the MAPS remain relatively simple there is a good chance of resemblance between them because of similar if not identical pathways of derivation. The small differences between them consequently should be more precisely maintained to be recognizably different.

Most species probably move toward increasing complexity when signal competition arises. There should be many instances, however, of isolated species with complex displays. They may have existed formerly in sympatry with a related species, or other considerations might be involved. One would be the need for subtler signaling in closed social groups.

Yet there must be an upper limit to complexity. At some point it becomes neurologically too expensive to produce. It taxes the perceptual capacity of the receiver as well (Moles, 1963). Excessively complex acts of communication would also become unduly long (Moynihan, 1970) and would therefore be costly in time and energy. They might also become so conspicuous and so preoccupying as to make the performer vulnerable to predation. Hence displays should generally be limited in their complexity. There is also reason to believe that the number of displays themselves is limited (Marler, 1959).

When species split off from one another and remain, or come in contact, the signals used in reproductive behavior should differ. Thus for each such MAP formerly in common there should now be two. This phenomenon attains central importance when it is realized that animals have a limited number of "major" displays (Moynihan, 1970). When new ones evolve old ones must be given up, whether through alteration or replacement. A methodological difficulty exists here, however. While Moynihan (1970) gave some criteria for deciding which MAPS are "major" displays, it is difficult to apply the criteria uniformly.

Radically different estimates of the number of signals can be obtained for the same species. The longer an investigator works with a species the more signals he recognizes (Altmann, 1968; Dane and Van der Kloot, 1965). And when a species is observed in captivity, as opposed to in the field, the number of "major" displays is apt to be underestimated (Moynihan, 1970).

Different observers can arrive at different numbers, especially if they have different views about what constitutes a unit of behavior. Altmann (1968) compared the classifications of vocal signals for the howler monkey *(Alouatta palliata)* contained in three different reports. Twenty different vocalizations were listed. Thirteen were reported by him (1959), ten by Carpenter (1934), and five by Collias and Southwick (1952). Only two signals were common to all three classifications, four appeared in two of them, and fourteen of the twenty were unique to one classification or another. The number of MAPS and vocalizations[2] reported for rhesus

2. In a strict sense vocalizations may be MAPS. In the present sense they are. I treat them as separate here only for convenience of exposition.

monkeys also varies drastically (Altmann, 1965, 1968; Hinde and Rowell, 1962; Rowell and Hinde, 1962).

Doubtless primates present the greatest difficulties here, and better agreement could be had in studies of lower vertebrates. While I have made no tallies, I generally find good agreement between the reports of different authors on the behavior of the same species of teleost fish, making allowance for differences in terminology.

Moynihan (1970) is doubtless correct in his thesis that animals have relatively few "major" displays, probably ranging from about 10 to 37, as he says. Nonetheless there is room for argument about the precision of the estimates, as noted. A quick histogram plot of his data discloses that most animals have about 13 to 21 "major" displays with a mode around 19 to 21. Using the same data, Wilson (1972) portrayed the number of "major" displays in animal groups, showing a progressive increase from fishes (mean = 17), to birds (21), to mammals (24).

Birdwhistell (1970) cataloged 50 to 60 "kines" for human beings. It is difficult to compare these directly to number of "major" displays in other animals because kines are recombined into a vast number of kinemorphs. If comparably fine distinctions were made, most animals would be found to have many more kines than displays. In fact, the communicatory acts of rhesus monkeys, as judged by Altmann (1965), are more comparable to kines or kinemorphs than to displays. This might account for the large discrepancy in number of signals reported by Hinde and Rowell (1962) and Rowell and Hinde (1962), whose approach was more clearly ethological in conception.

Closing Comments

I have made a number of points large and small in this article but the main ones are three. First, the highly patterned movements that I call MAPs are of central importance in animal behavior. They form the basis of almost all quantitative analyses, with ramifications extending into communication, development, neurophysiology, genetics, and the evolution of behavior. MAPs are a direct readout of the central nervous system. They have been valuable in understanding how the CNS works, and this should continue. But the physiological counterpart of overt behavior can never be observed directly in the nervous system. Only by careful and appropriate quantification of behavior can we expect to construct models that will illuminate the working of the CNS. The same applies to the analysis of communication. The very notion of quantification means ultimately that one must deal with units. That is why it is so important to understand the structure of MAPs. How stereotyped should they be to be considered unitary? If significantly variable, how can the variation best be quantified and incorporated into a model?

This brings us to the second point. In spite of the importance of MAPs we are poorly informed about their basic anatomy. Most of what has been written about MAPs has been based on presumptions because the MAPs are themselves so easily recognized. There is need for careful description of the infrastructure of MAPs. Most of this work will have to be done by means of film analysis. Electromyograms seem promising, though they are limited by the restraints of the methodology (e.g., Martin and Gans, 1972).

The third point is that we are almost as ignorant about the way MAPs act as stimuli. By the same token, little is known about the ways stimuli affect the performance of MAPs. Both points are important to understanding the role of MAPs in communication. The scientific method, of stating hypotheses and attempting to reject them, needs more exercise here.

This is not the place to enter into a discussion of experimental procedures, but it should be mentioned that the favored methodology of ethologists here has been the use of dummies, fake animals. These static models, no matter how useful in other situations (Leong, 1969), have proved a disappointment in the study of MAPS and their effects (Cullen, 1972; Gorman, 1968).

Three other approaches suggest themselves. One is to display motion pictures to the subjects. I have discussed film playback with many ethologists who have had only failures. A few investigators nonetheless, have had success (e.g., Jenssen, 1970; Turnbough and Lloyd, 1973). The difficulty may be that the individual pictures are perceived by the subject as disconnected images because the rate of advance lies below its critical fusion frequency (Jenssen, 1970).

Another technique is the use of animated models or simulations of essential moving features, as in the work of Magnus (1958), who produced a pattern resembling the beating wings of a female butterfly. The two chief deterrents are that only simple animals will respond to the motions of crude models, and models that convincingly mime MAPS are likely to be expensive technological achievements. Even then, one has no assurance they will work.

The third technique is electrical stimulation of the brain. Ideally, the wired-up animal can be forced to produce a faithful example of the desired MAP whenever a button is pressed. The ideal is not readily achieved. There are several reasons for this (Delius, 1973; Phillips and Youngren, 1971). While this technique is not an impossible one, its formidable difficulties account for its rare application to problems of communication (see also Cullen, 1972).

Whatever technique is employed, it will not be enough to deal with idealized MAPS. We have to push on, to explore the communicatory significance of variation in performance, intermediate MAPS, superposition, and more. Our thinking must not be dominated by digital models. Perhaps some MAPS, the more graded ones, act as analog senders. And of course the context will have to be brought in, for signals do not work in vacuums.

In closing I must return to a philosophical point. Is there any reality to MAPS or FAPS? I find it impossible to give an operational definition that is general enough to meet all contingencies. That which I have given is postulational. It relies on the animals showing a recognizable pattern of behavior, without a precise definition of what is meant by recognizable. Moreover, there may even be instances in which a population is polymorphic—a large portion showing one MAP, the other portion a different MAP.

The difficulty is that virtually all movements performed by animals are to some degree patterned. None are uniquely different from the highly patterned movements seen in displays. The concept of MAPS directs attention to the more highly patterned movements, movements that are often derived from more basic ones. The concept is therefore most easily applied to displays, MAPS that are adapted to serve in communication.

Obviously, however, there is a continuum. At one extreme is patterned behavior, like locomotion, which varies from moment to moment in harmony with the environment. At the other extreme are precisely patterned movements that are relatively independent of external modulation and may even fit the Lorenzian criteria of the Fixed Action Pattern. There is no boundary.

Yet the concept of MAP remains useful. Note the parallel in arbitrarily dividing the age of an organism into categories. When does a youth become an adult? Or when is an adult old? When is a motor pattern a MAP? When a movement has become ritualized into a display it is clearly a MAP. Where is the cutoff? I think it is too early and perhaps inadvisable to try to delimit rigidly the lower limit of MAPS. I would rather live with shades of grey and await future developments.

References

Adams, D. K., 1931. A restatement of the problem of learning. *Brit. J. Psychol.,* 22:150–78.

Ainley, D. G., 1974. The comfort behaviour of Adélie and other penguins. *Behaviour,* 50:16–51.

Albrecht, H., 1966. Zur Stammesgeschichte einiger Bewegungsweisen bei Fischen: untersucht am Verhalten von *Haplochromis* (Pisces, Cichlidae). *Z. Tierpsychol.,* 23:270–302.

Alexander, R. D., 1968. Arthropods. In: *Animal Communication,* T. A. Sebeok, ed. Bloomington: Indiana University Press, pp.167–216.

Allison, J., 1971. Microbehavioral features of nutritive and non-nutritive drinking in rats. *J. Comp. Physiol. Psychol.,* 76:408–17.

Altmann, S. A., 1959. Field observations on a howling monkey society. *J. Mammal.,* 40:317–30.

Altmann, S. A., 1965. Sociobiology of rhesus monkeys. II: Stochastics of social communication. *J. Theoret. Biol.,* 8:490–522.

Altmann, S. A., 1968. Primates. In: *Animal Communication,* T. A. Sebeok ed. Bloomington: Indiana University Press, pp.466–522

Andrew, R. J., 1956. Intention movements of flight in certain passerines, and their use in systematics. *Behaviour,* 10:179–204.

Andrew, R. J., 1964. Vocalization in chicks, and the concept of 'stimulus contrast.' *Anim. Behav.,* 12:64–76.

Attneave, F., 1959. *Applications of Information Theory to Psychology.* New York: Holt, Rinehart and Winston. 98pp.

Baerends, G. P., 1957. The ethological analysis of fish behavior. In: *The Physiology of Fishes,* M. E. Brown, ed. New York: Academic Press, 2:229–70.

Barlow, G. W., 1961. Ethology of the Asian teleost *Badis badis.* I. Locomotion, maintenance, aggregation and fright. *Trans. III. Acad. Sci.,* 54:175–88.

Barlow, G. W., 1962. Ethology of the Asian teleost *Badis badis.* III. Aggressive behavior. *Z. Tierpsychol.,* 19:29–55.

Barlow, G. W., 1964. Ethology of the Asian teleost *Badis badis.* V. Dynamics of fanning and other parental activities, with comments on the behavior of the larvae and postlarvae. *Z. Tierpsychol.,* 21:99–123.

Barlow, G. W., 1968. Ethological units of behavior. In: *The Central Nervous System and Fish Behavior,* D. Ingle, ed. Chicago: University of Chicago Press, pp.217–32.

Barlow, G. W., 1970. A test of appeasement and arousal hypotheses of courtship behavior in a cichlid fish, *Etroplus maculatus. Z. Tierpsychol.,* 27:779–806.

Barlow, G. W., 1974a. Contrasts in social behavior between Central American cichlid fishes and coral-reef surgeon fishes. *Amer. Zool.,* 14:9–34.

Barlow, G. W., 1974b. Fragen und Begriffe der Ethologie. In: *Grzimeks Tierleben, Erganzungsband, Verhaltensforschung,* K. Immelmann, ed. Munich: Kindler, pp.205–23.

Barlow, G. W., and Green, R. F., 1970a. The problems of appeasement and of sexual roles in the courtship behavior of the blackchin mouthbreeder, *Tilapia melanotheron* (Pisces: Cichlidae). *Behaviour,* 36:84–115.

Barlow, G. W., and Green, R. F., 1970b. Effect of relative size of mate on color patterns in a mouthbreeding cichlid fish, *Tilapia melanotheron. Comm. Behav. Biol.,* 4:71–78.

Barlow, H. B., 1961. The coding of sensory messages. In: *Current Problems in Animal Behaviour,* W. H. Thorpe and O. L. Zangwill, eds. Cambridge: Cambridge University Press, pp.331–60.

Barlow, H. B.; Narasimhan, R.; and Rosenfeld, A.; 1972. Visual pattern analysis in machines and animals. *Science,* 177:567–75.

Bastock, M., and Manning, A., 1955. The courtship of *Drosophila melanogaster. Behaviour,* 8:85–111.

Bastock, M.; Morris, D.; and Moynihan, M.; 1954. Some comments on conflict and thwarting in animals. *Behaviour,* 6:66–84.

Bateson, G., 1968. Redundancy and coding. In: *Animal Communication,* T. A. Sebeok, ed. Bloomington: Indiana University Press, pp.614–26.

Bateson, P. P. G., 1974. Specificity and the origins of behavior. In: *Advances in the Study of Animal Behavior,* Vol. 6, J. S. Rosenblatt, R. A. Hinde, C. Beer, and E. Shaw, eds. New York: Academic Press.

Berkson, G., and Mason, W. A. 1964. Stereotyped behaviors of chimpanzees. Relation to general arousal and alternative activities. *Percept. Motor Skills,* 19:635–52.

Berry, K. H., 1974. The ecology and social behavior of the chuckwalla, *Sauromalus obesus obesus* Baird. *Univ. Calif. Publ. Zool.,* 101:1–60.

Birdwhistell, R., 1970. *Kinesics and Context.* Philadelphia: University of Pennsylvania Press. 338pp.

Blair, W. F., 1955. Mating call and stage of speciation in the *Microhyla olivacea-M. carolinensis* complex. *Evolution,* 9:469–80.

Blair, W. F., 1956. The mating calls of hybrid toads. *Tex. J. Sci.*, 8:350–55.
Blest, A. D., 1957. The evolution of protective displays in the Saturnioidea and Sphingidae (Lepidoptera). *Behaviour*, 11:257–309.
Blest, A. D., 1961. The concept of ritualization. In: *Current Problems in Animal Behaviour*, W. H. Thorpe and O. L. Zangwill, eds. Cambridge: Cambridge University Press, pp.102–24.
Blurton Jones, N. G., 1968. Observations and experiments on causation of threat displays of the great tit *(Parus major)*. *Anim. Behav. Monogr.*, 1:75–158.
Bower, T. G. R., 1966. Heterogeneous summation in human infants. *Anim. Behav.*, 14:395–98.
Broadhurst, P. L.; Fulker, D. W.; and Wilcock, J.; 1974. Behavioral genetics. *Ann. Rev. Psychol.*, 25:389–415.
Brown, J. L., 1963. Ecogeographic variation and introgression in an avian visual signal: the crest of the Steller's jay, *Cyanocitta stelleri*. *Evolution*, 17:23–39.
Brown, J. L., 1969. Neuro-ethological approaches to the study of emotional behavior: stereotypy and variability. *Ann. N. Y. Acad. Sci.*, 159:1084–95.
Bucholtz, C., 1970. Der Beutefang der Larve von *Aeschna cyanea*. *Verhand. Deutsch. Zool. Gesell., 64. Tagung*, Frankfurt/Main: Fischer Verlag, pp.362–64.
Buckley, P. A., 1969. Disruption of species-typical behavior patterns of F_1 hybrid *Agapornis* parrots. *Z. Tierpsychol.*, 26:737–43.
Cane, V., 1961. Some ways of describing behaviour. In: *Current Problems in Animal Behaviour*, W. H. Thorpe, and O. L. Zangwill, eds. Cambridge: Cambridge University Press, pp.361–88.
Carpenter, C. R., 1934. A field study of the behavior and social relations of howling monkeys. *Comp. Psychol. Monogr.*, 10(48):1–168.
Carriker, M. R.; Schaadt, J. G.; and Peters, V.; 1974. Analysis by slow-motion picture photography and scanning electron microscopy of radular function in *Urosalpinx cinerea follyensis* (Muricidae, Gastropoda) during shell penetration. *Mar. Biol.*, 25:63–76.
Collias, N., and Southwick, C., 1952. A field study of population density and social organization in howling monkeys. *Proc. Amer. Phil. Soc.*, 96:143–56.
Condon, W. S., and Ogston, W. D., 1967. A segmentation of behavior. *J. Psychiatric Res.*, 5:221–35.
Crane, J., 1966. Combat, display and ritualization in fiddler crabs (Ocypodidae, genus *Uca*). *Phil. Trans. Roy. Soc. London, Ser. B*, 251:459–72.
Crook, J. H., 1964. The evolution of social organisation and visual communication in the weaver birds (Ploceinae). *Behaviour Suppl.*, 10:1–178.
Cullen, E., 1957. Adaptations of the kittiwake to cliff-nesting. *Ibis*, 99:275–302.
Cullen, J. M., 1960. The aerial display of the Arctic tern and other species. *Ardea*, 48:1–37.
Cullen, J. M., 1966. Reduction of ambiguity through ritualization. *Phil. Trans. Soc. London*, 251:363–74.
Cullen, J. M., 1972. Some principles of animal communication. In: *Non-Verbal Communication*, R. A. Hinde, ed. Cambridge: Cambridge University Press, pp.101–25.
Daanje, A., 1950. On locomotory movements in birds and the intention movements derived from them. *Behaviour*, 3:48–98.
Dane, B., and Van der Kloot, W. G., 1965. An analysis of the display of the goldeneye duck (*Bucephala clangula* (L.)). *Behaviour*, 22:282–328.
Dane, B.; Walcott, C.; and Drury, W. H.; 1959. The form and duration of the display actions of the goldeneye (*Bucephala clangula*). *Behaviour*, 14:265–81.
Darwin, C., 1872. *The Expression of Emotion in Man and Animals*. London: Murray. 397pp.
Davies, S. J. J. F., 1970. Patterns of inheritance in the bowing display and associated behaviour of some hybrid *Streptopelia* doves. *Behaviour*, 36:187–214.
Davis, J. M., 1973. Imitation: a review and a critique. In: *Perspectives in Ethology* P. P. G. Bateson and P. H. Klopfer, eds. New York: Plenum, pp.43–72.
Dawkins, M., and Dawkins, R., 1974. Some descriptive and explanatory stochastic models of decision-making. In: *Methods of Motivational Control Systems Analysis*, D. J. McFarland, ed. New York: Academic Press (in press).
Dawkins, R., and Dawkins, M., 1973. Decisions and the uncertainty of behaviour. *Behaviour*, 45:83–103.
Delius, J., 1973. Agonistic behaviour of juvenile gulls, a neuroethological study. *Anim. Behav.*, 21:236–46.
DeLong, A. J., 1972. The communication process: a generic model for man-environment relations. *Man-Environ. Systems*, 2:263–313.
Dewsbury, D. A., 1973. Copulatory behavior of montane voles (*Microtus montanus*). *Behaviour*, 44:186–202.
Dingle, H., 1972. Aggressive behavior in stomatopods and the use of information theory in the analysis of animal communication. In: *Behavior of Marine Animals*, H. E. Winn and B. L. Olla, eds. New York: Plenum, 1:126–56.
Dobzhansky, T., 1972. Genetics and the diversity of behavior. *Amer. Psychol.*, 27:523–30.
Eibl-Eibesfeldt, I., 1970. *Ethology. The Biology of Behavior.* New York: Holt, Rinehart and Winston. 530pp.

Eibl-Eibesfeldt, I., and Wickler, W., 1962. Ontogenese und Organisation von Verhaltensweisen. *Fortschr. Zool.,* 15:354–77.
Evarts, E. V. ed., 1971. Central control of movement. *Neurosci. Res. Prog. Bull.,* 9:1–170.
Ewing, A. W., and Manning, A., 1967. The evolution and genetics of insect behaviour. *Ann. Rev. Entom.,* 12:471–94.
Falls, J. B., 1969. Functions of the territorial song in the white-throated sparrow. In: *Bird Vocalizations,* R. A. Hinde, ed. Cambridge: Cambridge University Press, pp.207–32.
Fentress, J. C., 1972. Development and patterning of movement sequences in inbred mice. *Oregon St. Univ. Ann. Biol. Colloq.,* 32:83–132.
Ferguson, G. W., 1971. Variation and evolution of the push-up displays of the side-blotched lizard genus *Uta* (Iguanidae). *Syst. Zool.,* 20:79–101.
Ferguson, G. W., 1973. Character displacement of the push-up displays of two partially-sympatric species of spiny lizards, *Sceloporus* (Sauria, Iguanidae). *Herpetologica,* 29:281–84.
Fisher, A. E., 1962. Effects of stimulus variation on sexual satiation in the male rat. *J. Comp. Physiol. Psychol.,* 55:614–20.
Fox, M. W., 1970. A comparative study of facial expressions in canids: wolf, coyote and fox. *Behaviour,* 36:49–73.
Franck, D., 1970. Verhaltensgenetische Untersuchungen an Artbastarden der Gattung *Xiphophorus* (Pisces). *Z. Tierpsychol.,* 27:1–34.
Franck, D., 1974. The genetic basis of evolutionary changes in behaviour patterns. In: *The Genetics of Behaviour,* J. H. F. van Abeelen, ed. New York: American Elsevier, pp.119–40.
Gardner, R. A., and Gardner, B. T., 1969. Teaching sign language to a chimpanzee. *Science,* 165:664–72.
Golani, I., 1973. Non-metric analysis of behavioral interaction sequences in captive jackals (*Canis aureus* L.). *Behaviour,* 44:89–112.
Gorman, G. C., 1968. The relationships of *Anolis* of the *roquet* species group (Sauria: Iguanidae). III. Comparative study of display behavior. *Breviora Mus. Comp. Zool.,* 284:1–31.
Gorman, G. C., 1969. Intermediate territorial display of a hybrid *Anolis* lizard (Sauria: Iguanidae). *Z. Tierpsychol.,* 26:390–93.
Hailman, J. P., 1967. The ontogeny of an instinct. *Behaviour Suppl.,* 15:1–159.
Halliday, T. R., 1975. An observational and experimental study of sexual behaviour in the smooth newt, *Triturus vulgaris* (Amphibia: Salamandridae). *Anim. Behav.,* 23:291–322.
Hartshorne, C., 1958. Some biological principles applicable to song-behavior. *Wilson Bull.,* 70:41–56.
Hartshorne, C., 1973. *Born to Sing: An Interpretation and World Survey of Bird Song.* Bloomington: Indiana University Press. 304pp.
Hazlett, B. A., 1972a. Stereotypy of agonistic movements in the spider crab *Microphrys bicornutus. Behaviour,* 42:270–78.
Hazlett, B. A., 1972b. Ritualization in marine crustacea. In: *Behavior of Marine Animals,* H. E. Winn and B. L. Olla, eds. New York: Plenum, 1:97–125.
Hazlett, B. A., and Bossert, W. H., 1965. A statistical analysis of the aggressive communications systems of some hermit crabs. *Anim. Behav.,* 13:357–73.
Healey, H. G., 1957. The nervous system. In: *The Physiology of Fishes,* M. E. Brown, ed. New York: Academic Press, 2:1–119.
Hediger, H., 1962. Der Nahrungserwerb der nordamerikanischen Alligator-Schnappschildkröte (*Macrochelys temminki*). *Rev. Suisse Zool.,* 69:317–21.
Hinde, R. A., 1953. Appetitive behaviour, consummatory act, and the hierarchical organisation of behaviour, with special reference to the great tit (*Parus major*). *Behaviour,* 5:191–224.
Hinde, R. A., 1955–1956. A comparative study of the courtship of certain finches (Fringillidae). *Ibis,* 97:706–45; 98:1–23.
Hinde, R. A., 1956. The behaviour of certain cardueline F_1 inter-species hybrids. *Behaviour* 9:202–13.
Hinde, R. A., 1959. Behaviour and speciation in birds and lower vertebrates. *Biol. Rev.,* 34:85–128.
Hinde, R. A., 1966a. *Animal Behaviour.* New York: McGraw-Hill. 534pp.
Hinde, R. A., 1966b. Ritualization and social communication in rhesus monkeys. *Phil. Trans. Roy. Soc. London, Ser. B,* 251:285–94.
Hinde, R. A., 1969. Control of movement patterns in animals. *Quart. J. Exp. Psychol.,* 21:105–26.
Hinde, R. A., 1970. *Animal Behaviour,* 2d ed. New York: McGraw-Hill. 876pp.
Hinde, R. A., and Rowell, T. E., 1962. Communication by postures and facial expressions in the rhesus monkey (*Macaca mulatta*). *Proc. Zool. Soc. London,* 138:1–21.
Hinde, R. A., and Stevenson, J. G., 1969. Sequences of behavior. In: *Advances in the Study of Behavior,* D. S. Lehrman, R. A. Hinde, and E. Shaw, eds. New York: Academic Press. 2:267–96.

Holzapfel, M., 1939. Über Bewegungstereotypien bei gehaltenen Säugern. *Zool. Garten,* 10:184–93.
Holzberg, G., 1973. Beobachtungen zur Ökologie und zum Socialverhalten des Korallenbarsches *Dascyllus marginatus* Rüppell (Pisces; Pomacentridae). *Z. Tierpsychol.,* 33:492–513.
Hoyle, G., 1970. Cellular mechanisms underlying behavior—neuroethology. *Adv. Insect Physiol.,* 7:349–444.
Hutchinson, J. C. D., and Taylor, W. W., 1962. Mechanics of pecking grain. *12th Worlds Poult. Congr.,* pp.112–16.
Hutchinson, R. E.; Stevenson, J. G.; and Thorpe, W. H.; 1968. The basis for individual recognition by voice in the sandwich tern (*Sterna sandvicensis*). *Behaviour,* 32:150–57.
Ingold, P., 1973. Zür lautlichen Beziehung des Elters zu seinem Kueken bei Tordalken (*Alca torda*). *Behaviour,* 45:154–90.
Jenssen, T. A., 1970. Female response to filmed displays of *Anolis nebulosus* (Sauria, Iguanidae). *Anim. Behav.,* 18:640–47.
Jenssen, T. A., 1971. Display analysis of *Anolis nebulosus* (Sauria, Iguanidae). *Copeia,* 1971:197–209.
Kennedy, D., 1971. Nerve cells and behavior. *Amer. Sci.,* 59:36–42.
Klopfer, P. H., and Hatch, J. J., 1968. Experimental considerations. In: *Animal Communication,* T. A. Sebeok, ed. Bloomington: Indiana University Press, pp.31–43.
Konishi, M., 1963. The role of auditory feedback in the vocal behavior of the domestic fowl. *Z. Tierpsychol.,* 20:349–67.
Konishi, M., 1966. The attributes of instinct. *Behaviour,* 27:316–28.
Konorski, J., 1970. The problem of peripheral control of skilled movements. *Intern. J. Neurosci.,* 1:39–50.
Lehrman, D., 1953. A critique of Konrad Lorenz's theory of instinctive behavior. *Quart. Rev. Biol.,* 28:337–63.
Lehrman, D. S., 1970. Semantic and conceptual issues in the nature-nurture problem. In: *Development and Evolution of Behavior,* L. R. Aronson, E. Tobach, D. S. Lehrman, and J. S. Rosenblatt, eds. San Francisco: Freeman, pp.17–52.
Leong, C.-Y, 1969. The quantitative effect of releasers on the attack readiness of the fish *Haplochromis burtoni* (Cichlidae, Pisces). *Z. vergl. Physiol.,* 65:29–50.
Leyhausen, P., 1950. Beobachtungen an Löwen-Tiger-Bastarden mit einige Bemerkungen zur Systematik der Grosskatzen. *Z. Tierpsychol.,* 7:46–83.
Leyhausen, P., 1973. Verhaltensstudien an Katzen. *Z. Tierpsychol. Beiheft,* rev. ed., 2:1–232.
Lind, H., and Poulsen, H., 1963. On the morphology and behaviour of a hybrid between goosander and shelduck (*Mergus merganser* X *Tadorna tadorna*). *Z. Tierpsychol.,* 20:558–69.
Lloyd, M.; Zar, J. H.; and Karr, J. E.; 1968. On the calculation of information-theoretical measures of diversity. *American Midland Naturalist,* 79:257–72.
Loher, W., and Huber, F., 1966. Nervous and endocrine control of behaviour in a grasshopper (*Gomphocerus rufus* L., Acrididae). *Symp. Soc. Exper. Biol.,* 20:381–400.
Lorenz, K. Z., 1937. Über die Bildung des Instinktbegriffes. *Naturwissensch.,* 25:289–300, 307–18, 324–31.
Lorenz, K. Z, 1941. Vergleichende Bewegungsstudien bei Anatiden. *J. Ornithol.,* 89:194–294.
Lorenz, K. Z., 1950. The comparative method in studying innate behaviour patterns. *Symp. Soc. Exper. Biol.,* 4:221–68.
Lorenz, K. Z., 1955. Morphology and behavior patterns in closely allied species. In: *Group Processes,* B. Schaffner, ed. New York: Josiah Macy, Jr., Foundation, 1:168–220.
Lorenz, K. Z., 1958. The evolution of behavior. *Scient. Amer.,* 199(6):67–78.
Lorenz, K. Z., 1965. *Evolution and Modification of Behavior.* Chicago: University of Chicago Press. 121pp.
Lorenz, K. Z., 1973. The fashionable fallacy of dispensing with description. *Naturwissensch.,* 60:1–9.
Lorenz, K. Z., and Tinbergen, N., 1939. Taxis und Instinktbewegung in der Eirollbewegung der Graugans. *Z. Tierpsychol.,* 2:1–29.
Magnus, D., 1958. Experimentelle Untersuchungen zur Bionomie und Ethologie des Kaisermantels *Argynnis paphia* L. (Lep. Nymph.): Über optische Auslöser von Anfliegereaktionen und ihre Bedeutung für das Sichfinden der Geschlechter. *Z. Tierpsychol.,* 15:397–426.
Manning, A., 1959. The sexual behaviour of two sibling *Drosophila* species. *Behaviour,* 15:123–45.
Manning, A., 1975. Behaviour genetics and the study of behavioural evolution. In: *Essays on Function and Evolution in Behaviour: Festschrift for Professor Niko Tinbergen,* G. P. Baerends, C. R. Beer, and A. Manning, eds. Oxford: Clarendon.
Marler, P., 1957. Specific distinctiveness in the communication signals of birds. *Behaviour,* 11:13–39.
Marler, P., 1959. Developments in the study of animal

communication. In: *Darwin's Biological Work,* P. R. Bell, ed. London: Wiley, pp.150–206.

Marler, P., 1961a. The filtering of external stimuli during instinctive behaviour. In: *Current Problems in Animal Behaviour,* W. H. Thorpe and O. L. Zangwill, eds. Cambridge: Cambridge University Press, pp.150–66.

Marler, P., 1961b. The evolution of visual communication. In: *Vertebrate Speciation,* W. F. Blair, ed. Austin: University of Texas Press, pp.96–121.

Marler, P., 1965. Communication in monkeys and apes. In: *Primate Behavior,* I. DeVore, ed. New York: Holt, Rinehart and Winston, pp.420–38.

Marler, P., 1968. Visual systems. In: *Animal Communication,* T. A. Sebeok, ed. Bloomington: Indiana University Press, pp.103–26.

Marler, P., 1973. A comparison of vocalizations of red-tailed monkeys and blue monkeys, *Cercopithecus ascanius* and *C. mitis,* in Uganda. *Z. Tierpsychol.,* 33:223–47.

Marler, P., and Hamilton, W. J., 1966. *Mechanisms of Animal Behavior.* New York: Wiley. 771pp.

Martin, W. F., and Gans, C., 1972. Muscular control of the vocal tract during release signalling in the toad *Bufo valliceps. J. Morphol.,* 137:1–28.

McGrath, T. A.; Shalter, M. D.; Schleidt, W. M.; and Sarvella, P.; 1972. Analysis of distress calls of chicken X pheasant hybrids. *Nature* (London), 237:47–48.

McKinney, F., 1961. An analysis of the displays of the European eider *Somateria mollissima v. nigra* Bonaparte. *Behaviour Suppl.,* 7:1–124.

McKinney, F., 1965. The comfort movements of Anatidae. *Behaviour,* 25:121–217.

Mertz, J. C., and Barlow, G. W., 1966. On the reproductive behavior of *Jordanella floridae* (Pisces: Cyprinodontidae) with special reference to a quantitative analysis of parental fanning. *Z. Tierpsychol.,* 25:537–54.

Meyerriecks, A. J., 1960. Comparative breeding behavior of four species of North American herons. *Publ. Nuttal Ornithol. Club,* 2:1–158.

Millenson, J. R., 1967. *Principles of Behavioral Analysis.* New York: Macmillan. 489pp.

Moles, A., 1963. Animal language and information theory. In: *Acoustic Behaviour of Animals,* R. G. Busnell, ed. New York: Elsevier, pp.112–31.

Moltz, H., 1965. Contemporary instinct theory and the Fixed Action Pattern. *Psychol. Rev.,* 72:27–47.

Morris, D., 1957. "Typical intensity" and its relation to the problem of ritualization. *Behaviour,* 11:1–12.

Morris, D., 1964. The response of animals to a restricted environment. *Symp. Zool. Soc. London,* 13:99–118.

Morris, D., 1966. The rigidification of behaviour. *Phil. Trans. Roy. Soc. London, Ser. B,* 251:327–30.

Moynihan, M., 1970. Control, suppression, decay, disappearance and replacement of displays. *J. Theoret. Biol.,* 29:85–112.

Nelson, K., 1964. The temporal patterning of courtship behavior in the glandulocaudine fishes (Ostariophysi, Characidae). *Behaviour,* 24:90–146.

Nelson, K., 1973. Does the holistic study of behavior have a future? In: *Perspectives in Ethology,* P. P. G. Bateson and P. H. Klopfer, eds. New York: Plenum, pp.281–328.

Nottebohm, F., 1972. The origins of vocal learning. *Amer. Natural.,* 106:116–40.

Phillips, R. E., and Youngren, O. M., 1971. Brain stimulation and species-typical behaviour: Activities evoked by electrical brain stimulation of the brains of chickens (*Gallus gallus*). *Anim. Behav.,* 19:757–79.

Poulsen, H., 1950. Morphological and ethological notes on a hybrid between a domestic duck and a domestic goose. *Behaviour,* 3:99–103.

Quastler, H., 1958. The domain of information theory in biology. In: *Symposium on Information Theory in Biology,* H. P. Yockey, ed. New York: Pergamon, pp.187–96.

Ramsay, A. O., 1961. Behaviour of some hybrids in the mallard group. *Anim. Behav.,* 9:104–105.

Rowell, T. E., and Hinde, R. A., 1962. Vocal communication by the rhesus monkey (*Macaca mulatta*). *Proc. Zool. Soc. London,* 138:279–94.

Russell, W. M. S.; Mead, A. P.; and Hayes, J. S.; 1954. A basis for the quantitative study of the structure of behaviour. *Behaviour,* 6:153–205.

Salmon, M., 1967. Coastal distribution, display and sound production by Florida fiddler crabs (Genus *Uca*). *Anim. Behav.,* 15:449–59.

Salmon, M., and Atsaides, S. P., 1968. Behavioral, morphological and ecological evidence for two new species of fiddler crabs (genus *Uca*) from the Gulf Coast of the United States. *Proc. Biol. Soc. Wash.,* 81:275–90.

Schenkel, R., 1948. Ausdrucks-Studien an Wölfen. *Behaviour,* 1:81–129.

Schleidt, W. M., 1964. Über die Spontaneität von Erbkoordinationen. *Z. Tierpsychol.,* 21:235–56.

Schleidt, W. M., 1974. How "fixed" is the Fixed Action Pattern? *Z. Tierpsychol.,* 36:184–211.

Shallenberger, R. J., and Madden, W. D., 1973. Luring

behavior in the scorpionfish, *Iracundus signifer. Behaviour,* 47:33–47.

Sharpe, R. S., and Johnsgard, P. A., 1966. Inheritance of behavioral characters in F_2 mallard X pintail (*Anas platyrhynchos* X *Anas acuta*) hybrids. *Behaviour,* 27:259–72.

Sibley, C. G., 1957. The evolutionary and taxonomic significance of sexual dimorphism and hybridization in birds. *Condor,* 59:166–91.

Simon, H. A., 1974. How big is a chunk? *Science,* 183:482–88.

Slater, P. J. B., 1973. Describing sequences of behavior. In: *Perspectives in Ethology,* P. P. G. Bateson and P. H. Klopfer, eds. New York: Plenum, pp.131–53.

Smith, W. J., 1968. Message-meaning analysis. In: *Animal Communication,* T. A. Sebeok, ed. Bloomington: Indiana University Press, pp.44–60.

Smith, W. J., 1969. Messages of vertebrate communication. *Science,* 165:145–50.

Sneath, P. H. A., and Sokal, R. R., 1973. *Numerical Taxonomy: The Principle and Practice of Numerical Classification.* San Francisco: Freeman. 574pp.

Staddon, J. E. R., 1975. Schedule-induced behavior. In: *Handbook of Operant Behavior,* W. K. Honig and J. E. R. Staddon, eds. Englewood Cliffs, N.J.: Prentice-Hall, vol. 2 (in press).

Stamps, J. A., and Barlow, G. W., 1973. Variation and stereotypy in the displays of *Anolis aeneus* (Sauria: Iguanidae). *Behaviour,* 47:67–94.

Stokes, A. W., 1960. Agonistic behaviour among blue tits at a winter feeding station. *Behaviour,* 19:118–38.

Taub, E.; Perrella, P.; and Barro, G.; 1973. Behavioral development after forelimb deafferentiation on day of birth in monkeys with and without blinding. *Science,* 181:959–60.

Tayler, C. K., and Saayman, G. S., 1973. Imitative behaviour of Indian Ocean bottlenose dolphins (*Tursiops aduncus*) in captivity. *Behaviour,* 44:286–98.

Thorpe, W. H., 1951. The definition of terms used in animal behaviour. *Bull. Anim. Behav.,* 9:34–40.

Thorpe, W. H., 1961. Bird-song. The biology of vocal communication and expression in birds. *Cambridge Monog. Exper. Biol.,* 12:1–143.

Thorpe, W. H., 1966. Ritualization in ontogeny: I. Animal play. *Phil. Trans. Roy. Soc. London, Ser. B,* 251:311–19.

Tinbergen, N., 1950. The hierarchical organization of nervous mechanisms underlying instinctive behaviour. *Symp. Soc. Exper. Biol.,* 4:305–12.

Tinbergen, N., 1951. *The Study of Instinct.* Oxford: Clarendon. 228pp.

Tinbergen, N., 1952. "Derived" activities; their causation, biological significance, origin and emancipation during evolution. *Quart. Rev. Biol.,* 27:1–32.

Tinbergen, N., 1960. Comparative study of the behaviour of gulls (Laridae): a progress report. *Behaviour,* 15:1–70.

Tinbergen, N., 1964. The evolution of signalling devices. In: *Social Behavior and Organization among Vertebrates,* W. Etkin, ed. Chicago: University of Chicago Press, pp.206–30.

Tinbergen, N.; Meeuse, B. J. D.; Boerema, L. K.; and Varossieau, W.; 1942. Die Balz des Samtfalters, *Eumensis semele* (L.). *Z. Tierpsychol.,* 5:182–226.

Tolman, E. C., 1932. *Purposive Behavior in Animals and Men.* New York: Century. 463pp.

Turnbough, D. P., and Lloyd, K. E., 1973. Operant responding in Siamese fighting fish (*Betta splendens*) as a function of schedule of reinforcement and visual reinforcers. *J. Exper. Anal. Behav.,* 20:355–62.

van Hooff, J. A. R. A. M., 1972. A comparative approach to the phylogeny of smiling. In: *Non-Verbal Communication,* R. A. Hinde, ed. Cambridge: Cambridge University Press, pp.209–38.

van Lawick-Goodall, J., 1970. Tool-using in primates and other vertebrates. In: *Advances in the Study of Behavior,* D. S. Lehrman, R. A. Hinde, and E. Shaw, eds. New York: Academic Press, 3:195–249.

van Tets, G. P., 1965. A comparative study of some social communication patterns in the Pelicaniformes. *Amer. Ornithol. Union, Ornithol. Monogr.,* 2:1–88.

Watton, D. G., and Keenleyside, M. H. A., 1974. Social behaviour of the Arctic ground squirrel, *Spermophilus undulatus. Behaviour,* 50:77–99.

Welty, J. C., 1963. *The Life of Birds.* New York: Knopf. 546pp.

Wiley, R. H., 1973. The strut display of male sage grouse: a "fixed" action pattern. *Behaviour,* 47:129–52.

Willows, A. O. D., 1967. Behavioral acts elicited by stimulation of single, identifiable brain cells. *Science,* 157:570–74.

Wilson, D. M., 1966. Insect walking. *Ann. Rev. Entom.,* 11:103–23.

Wilson, D. M., 1967. An approach to the problem of the control of rhythmic behavior. In: *Invertebrate Nervous Systems,* C. A. G. Wiersma, ed. Chicago: University of Chicago Press, pp.219–29.

Wilson, E. O., 1972. Animal Communication. *Scient. Amer.,* 227(3):53–60.

Wheeler, R. H., 1929. *The Science of Psychology: An Introductory Study.* New York: Crowell. 556pp.

Part II

Some Mechanisms of Communication

Chapter 7

PHEROMONES

Harry H. Shorey

Modern man makes minimal use of chemical communication with others of his own species. Thus, he has little intuitive feel for the great reliance placed on this communication mode by much of the remainder of the animal kingdom. Correspondingly, until recent years, man's research into the chemical communication systems used by other animals has lagged behind research into visual and sonic means of communication. However, during the past few decades, advances in microanalytical chemistry, coupled with an increasing recognition that man may profit by controlling the chemical communication systems of certain animal species, has resulted in a great expansion in our knowledge in this previously neglected field.

The term "pheromone" was coined by Karlson and Lüscher (1959), and slight modifications in the definition were proposed by Kalmus (1965). A pheromone is a chemical or a mixture of chemicals which is released to the exterior by an organism and, which, upon reception by another individual of the same species, stimulates one or more specific reactions. Although this review is restricted to a consideration of animals, interorganism communication by pheromones also occurs in many species of plants.

Wilson and Bossert (1963) and Wilson (1963a) separated pheromones into two categories: Releaser pheromones are direct stimulators of behavior, often involving a classical stimulus-response sequence that is mediated entirely by the central nervous system; primer pheromones induce relatively long-lasting physiological changes, especially involving the endocrine and reproductive systems. Primer pheromones, for example, cause acceleration or inhibition of maturation of the reproductive systems of many vertebrate as well as invertebrate species. Because this chapter is concerned with communication systems in which a more immediate response to the chemical stimulus occurs, primer pheromones will not be discussed.

Sometimes when an animal is exposed to a pheromone, a stimulus-respone reflex is not seen. In much of the communication of vertebrates—as well as that of many invertebrates—the pheromone may cause subtler effects. It may, for example, sensitize the animal to other stimuli, such as the sight of a nearby potential mate. In other instances, the pheromone may be used to identify the other individual or to assess the physiological or social status of the other individual, with no overt behavioral reaction resulting at all.

The Pheromone-Communication System

Three components that make up any communication system are: an emitter of the message, a medium through which the message is transmitted, and a receiver of the message.

Pheromones are often emitted from glands that have become specialized through evolutionary processes for their communicative function. The nature of the glands and their locations on animals vary greatly among species. Frequently, specialized organs such as tufts of hair are associated with the organs to enhance transfer of the pheromone molecules into the medium. In some cases, specifically organized glands have not been located, and the pheromone may arise from cuticular secretions or as a by-product of digestive or excretory functions.

For animals that communicate over some distance, the medium for transmission of the pheromone is air or water. In a static environment, the chemical molecules spread from the release point by simple diffusion, creating a concentration gradient. However, the medium is rarely static. Air, especially, is almost continuously in motion. This motion may be caused by a general wind or by local convection currents attributable to unequally heated surfaces. In a flowing medium, the pheromone molecules are carried away from the source in a downwind or downstream direction, forming an elongate plume. The turbulence that is characteristic of moving air causes the molecular density in a cross section of the plume to be nonuniform. The plume is composed of filaments, with great discontinuities between areas of high and low density. Turbulence is a much more important factor than molecular diffusion in determining the dispersion or dilution of the pheromone molecules as the plume moves away from its source (Bossert and Wilson, 1963).

In some types of communication, the pheromones are not transmitted through the medium. They are retained on the surfaces of the emitting individuals and are perceived by direct contact by other individuals.

Pheromones may be perceived by either olfactory or gustatory receptor mechanisms. Gustatory receptors typically must come in contact with the chemical source, and they usually require relatively high chemical concentrations for stimulation. Gustatory receptors are usually located in the oral cavity of terrestrial vertebrates; they may also be located on various external portions of the bodies of fish and invertebrates. Olfactory receptors are specialized for detecting the chemical at some distance from its source and usually react at relatively low concentrations. Olfactory receptors of vertebrates are found in a nasal cavity, and those of invertebrates are typically located on external structures, such as antennae. Aquatic animals must be regarded as using olfaction, even though the transmitting medium is water. The olfactory receptors of fish, for example, occur in a distinct nasal cavity, through which a stream of water is circulated (Bardach and Todd, 1970).

Uses of Pheromones in Communication

Most reviewers have categorized pheromones according to their biological function (Butler, 1970; Shorey, 1973; Wilson and Bossert, 1963; Wilson, 1968). The same general scheme will be used here, with the following subsections dealing with the use of pheromones in (1) determining the identity of individual animals, of social groups of animals, or of the social or physiological status of animals of the same species; (2) stimulating aggregation; (3) stimulating dispersion; (4) stimulating sexual behavior; and (5) stimulating aggression. This type of categorization is not completely satisfactory. A number of behavioral activities involving pheromones are associated with more than one of the above categories. For instance, a pheromone that stimulates approach (aggregation) of sexual partners from a distance may then stimulate sexual behavior reactions at close range. Also, certain pheromones used in the marking of territories or home ranges or in the regulation of complex societies do not fit the above scheme directly. These complexities will be considered later.

IDENTIFICATION

Individual animals of many species of mammals and fish can determine the identity of other members of the same species by their odors (Bardach and Todd, 1970; Beauchamp, 1973; Müller-Schwarze, 1971; Mykytowycz, 1970). Whether any invertebrates are capable of this identification of individuals is doubtful; however, Linsenmair and Linsenmair (1971) noted that males and females of the desert wood louse, terrestrial crustaceans that pair for life, drive off from their burrows all intruders of either sex except their partners. The identification is apparently based on pheromones.

An ability to identify individual animals on the basis of pheromones alone infers that the pheromone of each individual consists of a complex blend of chemicals that differs in some way from that of other individuals. There is considerable evidence that such complex blends do occur. The chemicals secreted by the anal, chin, and inguinal glands of the rabbit *(Oryctolagus cuniculus)* vary both qualitatively and quantitatively from individual to individual (Goodrich and Mykytowycz, 1972). The number of different compounds in the pheromone blend may be very high; the castoreum gland of the beaver *(Castor canadensis)* contains approximately fifty different chemicals (Kingston, 1965).

Pheromones may also be used to identify the physiological or social status of an individual and the social group or colony to which that individual belongs. A semantic problem enters here; in some cases, the pheromone may not be used for actual "identification" of these characteristics, but may simply cause the release of appropriate behavior by the respondent, such as a copulatory reaction by a male when exposed to a pheromone released by a sexually receptive female. Especially in the vertebrates, there is probably a continuum between identification of the status of an individual and stimulation of appropriate behavioral acts by the "identifying" animal.

An extensive use of pheromones by mammals in determining the identity as well as the physiological or social status of their species mates is often inferred by the characteristic behavior displayed when two animals meet. A prominent part of this meeting behavior is the "sniffing" of certain portions of the body, especially the nasal, anal, genital, and various glandular areas (Schloeth, 1956).

Social insects, as well as many mammalian and fish species, differentiate between members of their own social group or colony and those of other social groups by odors (Mykytowycz, 1970; Wilson, 1971). Characteristic behavior may be displayed by the individual perceiving the odors; these responses will be discussed later with reference to behavior associated with territoriality, aggression, and marking of a home range. Mechanisms must exist for the establishment of a common odor on all members of the colony or social group. In certain mammalian species, the animals in a social group—especially the dominant males—mark other individuals within the group with a pheromone. This method of attaining a common group odor is seen in the gliding phalanger (*Petaurus breviceps papuanus:* Schultze-Westrum, 1965), the rabbit (Mykytowycz, 1965), and the tree shrew (*Tupaia belangeri:* Martin, 1968). Mutual scents of colony members in the social insects may be acquired in a number of ways, including: secretions produced by the queen and distributed throughout the colony; secretions produced by the colony members themselves, with each member having a common genetic complement provided by the queen-mother; the odor of the specific nest material in which the colony is housed; and a common odor due to the specific food on which the colony has fed (Hangartner et al., 1970; Morse, 1972; Wilson, 1963b).

AGGREGATION

An aggregation pheromone causes one or more responding individuals to become localized near the source of pheromone emission. General use of the term "attraction" to describe the process by which aggregation is accomplished should probably be avoided. According to the definition of Dethier et al. (1960), an attractant causes oriented locomotory responses toward the odor source. In addition to attraction, a number of other behavioral mechanisms, operating singly or in conjunction with each other, may bring about aggregation. For example, certain pheromones may inhibit locomotion and thus reduce dispersal of animals away from the vicinity of the odor source (Shorey, 1973).

Some pheromones appear to have no biological function other than promoting aggregation. In other cases, the same pheromone that stimulates the animals to aggregate then assumes an additional role in stimulating appropriate behavior, such as mating or aggression. This latter group of pheromones will be discussed here with reference to their aggregative properties as well as in later sections with reference to the stimulation of the additional behavioral activities.

Mechanisms of Aggregation

Most of the research on the behavioral mechanisms used by animals that aggregate in response to stimulatory odors has been conducted on insects. The movement toward the odor source may take place over tens, hundreds, or even thousands of meters. Many of the claims of great distances are probably exaggerated. For instance, if a male moth is marked and released at a great distance from a pheromone-emitting female, there is no way of knowing what proportion of its flight to the pheromone source might consist of a random, appetitive locomotory behavior, not stimulated by the pheromone.

The flow and turbulence characteristics of air probably destroy any recognizable chemical concentration gradient within a few centimeters of the source of the pheromone. For this reason, most investigators are agreed that the guidance mechanisms used by animals orienting toward an odor source from relatively great distances cannot be based simply on the following of an odor gradient (Bossert and Wilson, 1963; Shorey, 1973). The guidance is probably accomplished by a complex interaction of a number of behavioral mechanisms. One that has been demonstrated for certain flying insects is anemotaxis (wind steering), whereby the animal steers its body axis upwind when it is stimulated by an appropriate odor (Farkas and Shorey, 1974; Kennedy, 1939). An aquatic animal may also orient its body axis into the flowing current; in this case, the phenomenon is called "rheotaxis." Considerable evidence indicates that the upwind or upcurrent orientation is accomplished by the animal's steering its body axis so that the visual field of the substratum over which it is flying or swimming unreels caudally, from front to rear. If the animal were proceeding at an angle to the current, sideslippage would cause the visual field to appear to move at an angle relative to the body axis; presumably, the animal would then make corrective steering reactions that would orient its body axis upcurrent. If the animal loses the odorous current, it may make crosscurrent casts (Traynier, 1968), which probably maximize its likelihood of regaining the odor stimulation.

Another mechanism that might be used in distance orientation to a pheromone source is aerial odor-trail following (Farkas and Shorey, 1974). The cloud of molecules extending downcurrent from the odor source often can be visualized as forming a three-dimensional trail. Many species of flying insects exhibit lateral zigzag or sinusoidal oscillations across the breadth of the trail as they proceed toward the odor source. Some experimental evidence indicates that these oscillations may be an important factor in enabling the flying insect to maintain contact with the odorous trail. Perhaps appropriate turns, back toward the central axis of the trail, are stimulated when the insect senses that it is entering a lower pheromone density as it diverges to the right or left of the axis. This method for aerial trail following, then, would be analogous to a mechanism proposed for animals that follow terrestrial trails (discussed below). Possibly the phenomenon of aerial trail following, operating to cause the animal to maintain itself within the boundaries of the trail, interacts with anemotaxis, to orient each turn in an upcurrent direction.

The progressively increasing odor concentration encountered as an animal approaches a pheromone source may stimulate other behavioral reactions that result in the animal's localization near the source. The speed of forward progress may be reduced, and ultimately locomotion may be inhibited, as the animal encounters the higher concentrations (Bennett and Borden, 1971; Farkas and Shorey, 1974; Wood, 1970). Also, the animal may become sensitized by the high concentrations to respond to other environmental stimuli. For instance, a number of insect species, when flying in air containing a high concentration of an aggregation pheromone, tend to approach and land on vertical objects such as the trunks of trees (references in Farkas and Shorey, 1974). Presumably, this behavior is adaptive, because the pheromone emitter is likely to be located on such a vertical object.

When the animal is very close to the pheromone source and in a well-defined chemical concentration gradient, chemotactic mechanisms may steer it toward the highest pheromone concentration. Males of many moth species, when stimulated by the high pheromone concentrations found near receptive females, steer their bodies first to one side and then to the other, with wings vibrating. The air pulled by the vibrating wings, from front to rear and over the antennae, probably gives good cues as to the direction of the site of pheromone emission (Schneider, 1964).

Finally, visual orientation to the body of the pheromone emitter itself may be stimulated when a male insect is exposed to a high concentration of female pheromone (Daterman, 1972; Shorey and Gaston, 1970; Traynier, 1968).

Terrestrial odor trails differ from aerial odor trails in that the odor molecules do not emanate from a single point source. Rather, the pheromone emitter deposits the chemical as a series of discrete spots or as a continuous streak as it moves along the substratum. Such pheromone trails are widely used in aggregation behavior. Some examples, which will be discussed in later sections, include the pheromone trail deposited by a snake crawling along the ground, by a hoofed mammal through contact with the ground during locomotory behavior, and by workers of many ant species when they return to their nest after finding food. Mechanisms of terrestrial trail following have been best studied in ants. The following ants detect the small active space caused by the volatilization of the chemical into the air above the trail. Some evidence indicates that they maintain contact with the trail as they run along it by chemotactic mechanisms, which tend to steer them toward the central axis of the trail whenever they diverge to one side (Hangartner, 1967). Thus, the behavior of ants (and of dogs) following terrestrial odor trails is frequently characterized by a zigzag progression (Wilson, 1962; Wilson and Bossert, 1963).

Role of Aggregation Pheromones in Animal Behavior

As mentioned earlier, aggregation is typically a precursor of further behavioral activities such as aggression or mating. Those pheromones that stimulate both aggregation as well as the further behavioral activities will be considered in later sections. This section will be involved with pheromones whose roles are essentially restricted to causing aggregation, with further behavioral activities within the aggregations apparently being caused primarily by other stimuli.

Cellular slime molds live as single-celled amoebae, ingesting bacteria, during the feeding phase of their life cycle. After the food supply is exhausted, some of the amoebae of *Dictyostelium discoideum* (the most studied species) emit pulses of a pheromone identified as cyclic 3,'5' adenosine monophosphate (Konijn et al., 1968). The pheromone causes nearby amoebae to move toward the source and to emit pulses of the same chemical. The aggregating amoebae form a multicellular organism, which rises above the substratum and differentiates into a stalk and an apical fruiting body (Bonner, 1970).

Many animal species use pheromones to locate their home or nest site. Limpets (Mollusca) of a number of species have a specific area on a rock to which they return after feeding expeditions. Although not yet conclusive, considerable evidence indicates that the limpets find their "homes" by following a pheromone trail laid down by them on the substratum during previous excursions. Similarly, a number of social Hymenoptera (bees, ants, and wasps) recognize their own nest sites and are stimulated to enter the nest entrance by odors characteristic of their specific colony (Butler et al., 1969; Hangartner et al., 1970). The odor, which has sometimes been referred to as "hive atmosphere," may originate partly from pheromones produced by the colony mates and partly from the general odor of their food and nest material.

In some animal species that do not have a permanent home or resting site, pheromones may stimulate the formation of temporary aggregations in which the individuals spend certain times of the day or the year. Thus, pheromones promote aggregations of a number of arthropod species in sheltered places during their inactive time of day (Friedlander, 1965; Levinson and Bar Ilan, 1971). In certain coccinellid beetles (Hodek, 1960) and snakes (Dundee and Miller, 1968; Noble and Clausen, 1936), pheromones play a role in promoting aggregations prior to the time of hibernation. Noble and Clausen (1936) found that the common brown snake follows the trails laid by previous snakes moving toward hibernation sites in suitable cavities. Once in the cavities, a combination of visual and pheromonal stimuli maintains cohesiveness of the aggregations.

A number of marine animals that are sessile as adults produce free-swimming larvae. At the proper state for settling, the larvae often give a gregarious response to other members of their species. Thus, a pheromone released by barnacles that are successfully colonizing a surface causes larvae of their species to explore and settle on the same surface (Crisp and Meadows, 1963).

Many aquatic animals move together in aggregations called "schools." The function of the schools is not well understood; they might be of benefit in locating suitable food sources, for mutual protection, and in bringing together the sexes prior to mating. Vision is apparently the major sense involved in maintaining schools of fish. However, schooling fish have also been found to respond to the odor of conspecifics; this pheromone-induced aggregation may be a major mechanism for maintaining schooling behavior at night (Hemmings, 1966; Wrede, 1932).

Pheromones are characteristically used by the worker caste of social insects (bees, ants, and termites) to recruit colony mates to a supply of food (Moser and Blum, 1963; Stuart, 1970; Wilson, 1962, 1963b). There are many variations in the recruitment behavior; however, a typical sequence of events for many ant species is as follows: If an ant finds a source of food too large to transport to the nest in one piece, it returns to the nest while depositing a trail of pheromone on the ground. Other ants in the nest are stimulated to follow the trail. After arriving at the food source, they carry pieces to the colony while also depositing pheromone on the substratum. Thus, the terrestrial trails are fortified as long as food is available. If no food remains, the returning ants deposit no pheromone, and the trail fades out through volatilization. Some trails are very transitory; this is the case with the imported fire ant, which mainly forages for small items of food within a few meters of its nest (Wilson, 1962). Other ant species have highly persistent trails; an example is the Texas leaf-cutting ant, which may harvest leaves from plants hundreds of meters from its nest for several months (Moser and Blum, 1963).

Some ant species lay trails as they proceed outwards, in search of food. These have been called "exploratory trails" by Wilson (1963b) and are typical of army ants (genus *Eciton*). Army ants are true nomads, without a permanent nest. Their massive columns, either radiating from a central area at which the colony is temporarily bivouacked or during emigration of the entire colony, are guided by pheromone deposited by the individual workers as they proceed outwards.

Certain ants practice "slavery." They raid the nests of related species, kill or repel the workers, and capture the pupae, which they transport back to their nest. Slave-making ants of the *Formica sanguinea* group direct their raids to selected nests of other species by odor trails. Once in the raided nest, the ants discharge large amounts of pheromone consisting in part of decyl alcohol, dodecyl alcohol, and tetradecyl alcohol. The pheromone attracts further raiders and at the same time causes the defenders of the raided nest to disperse (Regnier and Wilson, 1971).

Workers of a number of stingless bee species, especially in the genera *Melipona* and *Trigona,* release from their mandibular glands large quantities of odor when they have located a suitable source of food. The pheromone attracts other workers, which collect the food and transport it to the nest (Lindauer and Kerr, 1960). Certain species make trails that are partly terrestrial and partly aerial. A scout bee of *Trigona postica,* after having collected nectar or pollen, flies from the flowers toward the nest, stopping at frequent intervals and leaving at each spot an odor mark from her mandibular glands. Recruited workers follow the resulting odor path to the food source. Another variation in stingless bee recruitment behavior is found in *Lestrimelitta limao,* a species that steals food from nests of other stingless bees. The first invading bees are almost invariably killed, resulting in the liberation of citral from their mandibular glands (Blum et al., 1970). This pheromone attracts large numbers of additional invaders, which raid the nest. Citral also causes a complete disruption in the social organization of the raided colony, with the resident bees often abandoning their nest.

A complex of pheromones secreted by the workers and the queen is used during swarming behavior of the honeybee (Butler and Simpson, 1967; Morse and Boch, 1971). Swarming bees usually form a cluster on a branch or some other substrate at least once during their migration from the parent hive to a new nesting site. Aggregation of the workers and the queen at a clustering site or at a new nest site is stimulated by a pheromone emitted by the scout bees that locate that site. The swarming bees are also attracted by a pheromone from the queen, 9-oxo-*trans*-2-decenoic acid, which enables them to relocate her if they should lose her. Another pheromone emitted by the queen, 9-hydroxy-*trans*-2-decenoic acid, acts as a behavioral stabilizer for the worker bees, causing them to alight and group into a quiet cluster when near her.

In many beetle species, either one sex or the other emits a pheromone that causes the aggregation of beetles of both sexes. In most cases, the biological function served by the aggregation has not been determined (Abdullah, 1965; Levinson and Bar Ilan, 1967). However, one might infer that two major functions often served are the enhancement of mating through the bringing of beetles of both sexes into close proximity, and the bringing of beetles to a suitable food source detected by the pheromone-emitting individual. Eisner and Kafatos (1962) assume that the large aggregations of *Lycus loripes* beetles that occur in response to a pheromone emitted by the males confer a survival benefit. These beetles are distinctively colored and possess a chemical defense mechanism against potential predators. A predator that has had a distasteful experience with one individual in the aggregation would have a reduced likelihood of attacking other individuals.

Aggregation of both sexes in response to a pheromone emitted by one sex is characteristic of the ambrosia beetles and bark beetles (family Scolytidae). These beetles are infrasocial, with large numbers colonizing a suitable host tree. The tree that is designated for colonization may be dead, dying, weakened, or healthy, depending on the species of beetle involved. Either males or females, depending on the species, initially invade a host tree. A pheromone, often consisting of a medley of chemicals, is secreted by these invaders, either when they first arrive at the tree or after they have constructed an entry tunnel through the bark and have commenced feeding on the phloem. This pheromone, often in association with volatile chemicals from the tree itself, causes the approach of others of both sexes.

Typically, the responding beetles of the same sex as the initial invaders also establish entry tunnels and release a pheromone. Beetles of the other sex, upon arriving at the tree, may also release a pheromone and enter the tunnels constructed by the initially invading sex. The beetles mate within the tunnels, and their offspring establish feeding galleries under the bark. Changes in the concentration and the blend of pheromone chemicals released as the tree is progressively attacked by more beetles may determine the course of colonization, being at first stimulatory and later inhibitory to aggregation. The complex of chemicals released early in the colonization process stimulates beetles to approach from a distance and often inhibits their tendency toward further locomotion after they land on the tree (Bennett and Borden, 1971; Vité and Pitman, 1969; Wood, 1970).

Later in the colonization process, chemicals may be released that cause beetles not to approach the tree or not to be arrested in their locomotion if they should land on the tree. This inhibition of aggregation is often seen when responding beetles of the opposite sex from that of the initial invaders enter the tunnels made by the latter (Rudinsky et al., 1973; Vité and Renwick, 1971). Inhibition of further colonization also appears to be due to the increasingly high concentration of previously attractive chemicals released by the aggregating beetles. Apparently, the concentration often rises to a sufficiently high level to stimulate beetles to orient visually to other, nearby trees (Gara and Coster, 1968). Thus, adjacent trees are often colonized.

Only a few additional examples of pheromone-mediated aggregation behavior will be given here so as to indicate the great diversity of biological functions that might be served by such behavior. Workers of certain termite species lay pheromone trails from a point of disturbance, such as a breach or an invader in the nest, to other places within the nest. Other termites follow the trail to the site of disturbance and then display appropriate behavior depending on the stimulus nature of the disturbance itself (Stuart, 1967). Young larvae of snails of the genus *Crepidula* aggregate in response to a pheromone released by mature females. They attach to a female and are induced, through the action of a primer pheromone produced by her, to transform into males, which later inseminate her

(Gould, 1952). Freshwater pulmonate snails *(Physa acuta)* follow trails made by conspecifics when they move to the surface of the water to replenish their air supplies (Wells and Buckley, 1972). A pheromone, 2-methoxy-5-ethylphenol, is given off from the feces of migratory locusts. The chemical serves both as an aggregation pheromone, promoting gregarious behavior of the immature locusts, and as a primer pheromone, inducing the physiological changes that are characteristic of the migratory phase of this species (Nolte et al., 1973).

DISPERSION

Pheromones that cause dispersion of conspecifics can be subdivided into several fairly distinct categories, dealing with maintenance of optimal interindividual spacing, inhibition of tendencies to aggregate, dispersion during times of danger, and maintenance of the integrity of territories.

Maintenance of Optimal Spacing

At particular times during the life cycle of some animal species, pheromones are apparently released by all the individuals and maintain an optimal distance between them. Amoebae of *Dictyostelium discoideum* were discussed earlier with reference to their behavior of aggregating after completion of feeding. Before the aggregation phase, the amoebae secrete a pheromone that repels others of the species. The pheromone thus causes the individual amoebae to remain dispersed and to optimally utilize the available food (Bonner, 1970). After the food has been depleted and the amoebae have come together into centers of aggregation, each center releases a pheromone that further maintains optimal spacing by inhibiting the formation of nearby centers. As a fruiting mass rises on a stalk from each center, above the substratum, the same spacing pheromone controls the direction of growth of the stalk so that it is maximally displaced from nearby stalks.

Mature larvae of a number of moth species that infest grain, when they meet, deposit on the substratum a pheromone from their mandibular glands (Mudd and Corbet, 1973). The pheromone causes the larvae to disperse, thereby increasing the likelihood that they will reach an area of low population density before they spin their cocoons and pupate.

Female flour beetles, when at high population densities, distribute themselves uniformly throughout their food medium. A pheromone secreted by the females themselves, as well as a pheromone secreted by their larvae, repels the females (Naylor, 1965). The resulting spacing is presumably adaptive, resulting in optimal colonization of the environment.

Inhibition of Tendency to Aggregate

Adults of many insect species aggregate in areas containing suitable food for their young. The females of certain of these species, after depositing their eggs in the food material, mark the external surface of the material with a pheromone that inhibits other females from depositing eggs in the same area. This behavior has been seen in certain hymenopterous parasites that lay their eggs in insects of other species (Rabb and Bradley, 1970) and in the apple maggot fly, which lays its eggs in various fruits (Prokopy, 1972).

When not receptive for mating, a female ground beetle *(Pterostichus lucublandus)* normally runs away from males. If pursued by a male, the female discharges a blast of liquid toward him from the tip of her abdomen. The male stops running, makes cleaning movements of his face and antennae, becomes uncoordinated, and may fall into a coma for several hours. Kirk and Dupraz (1972) feel that this defensive behavior may be used by the female while laying eggs. The male, which would eat the eggs of the species, is immobilized until the female completes oviposition and covers the entrance to the egg chamber.

Dispersion during Times of Danger

Pheromones released when an animal is threatened with danger or when injured are generally called "alarm pheromones." For many species, their primary role is the stimulation of dispersion of conspecifics. For other species, the alarm pheromone may induce aggregation plus aggressive behavior; these responses will be considered in a later section. And for still other species, the alarm pheromone may induce either dispersion or aggregation plus aggression, depending on the environmental context in which the animals perceive the pheromone.

Aquatic snails of nineteen species have been found to exhibit escape reactions when they perceive the juices of crushed conspecifics. The behavior usually consists of dropping from the substratum to the bottom of the water, followed by burrowing into the bottom material. However, certain air-breathing species react by crawling up out of the water (Snyder, 1967). Likewise, sea urchins *(Diadema antillarium)* move rapidly away from an area containing the juices of crushed conspecifics; they often mount on their ventral spines and "race" away for one or two meters (Snyder and Snyder, 1970). Alarm reactions are stimulated in a number of schooling fish species by a pheromone that is released when the skin of a conspecific is injured. The reactions are extremely variable, depending on the species, and range from avoidance of the area containing the pheromone and dispersal of the school to an increase in the cohesiveness of the school (Bardach and Todd, 1970; von Frisch, 1941). An increase in school cohesiveness might at first seem maladaptive; however, Pitcher (1973) indicates that certain predator fish are less successful in attacking schools of fish than individual prey.

Pheromones released by many animals when they are threatened or attacked by an enemy serve a dual function. They not only cause conspecifics to disperse but also may serve a defensive function, deterring or repelling the predator. Examples are seen in earthworms (Ressler et al., 1968), a number of aphid species (Bowers et al., 1972), the plant bug (*Dysdercus intermedius:* Calam and Youdeowei, 1968), the beetle *Blaps sulcata* (Kaufmann, 1966), and ants of a number of species.

Ant alarm pheromones are highly variable in their effect, depending on the species and the location and behavioral activities of the ants when they perceive the pheromone. At the nest, the pheromone typically triggers attraction and attack behavior (considered later). However, away from the nest, at the feeding place, ants perceiving the pheromone typically disperse (Maschwitz, 1966). Even at the nest, those species that construct small or diffusely distributed colonies often disperse (Wilson and Regnier, 1971).

Mammalian alarm pheromones may also be highly variable in their effect, depending to a large extent on the context in which they are perceived and on the previous experiences of the responders. However, repellency or dispersion is one of the possible effects, as demonstrated by the reactions of mice to the odor of stressed conspecifics (Rottman and Snowdon, 1972).

Territories

Individuals of many vertebrate species band together in social groups that occupy clearly demarked territories. Each territory is occupied solely by a given group and is defended against all entering foreign conspecifics. Occupied territories may be designated by the occupiers by means of visual, sonic, or chemical cues. Chemical cues are especially used by the mammals, which deposit odorous pheromone secretions on objects within the territory, especially near the borders of the territory. This scent marking seems to be especially advantageous, because ownership of the territory is advertised even when the residents are absent (Jones and Nowell, 1973). Male mice deposit a pheromone from the coagulating glands with the urine. Foreign males are deterred from investigating the marked areas; if an intruder enters a marked territory, the pheromone reduces his aggressive tendencies, thus tipping the scales in favor of a victory by the resident and resulting in the ultimate flight of the intruder (Jones and Nowell, 1973). Similar territorial behavior is seen in a number of other mammal species, including the rabbit (Mykytowycz, 1968), the tree shrew (Martin, 1968), and the black-tailed deer (Müller-Schwarze, (Müller-Schwarze, 1972).

SEXUAL BEHAVIOR

A group of diverse chemicals are collected under the term "sex pheromones." They are produced by either males or females and stimulate one or more behavioral reactions in the opposite sex. The reactions lead directly or indirectly to mating. The indirect reaction most frequently observed is aggregation of the opposite sex near the pheromone-emitting sex. After the two sexes have come together, a variety of other reactions involved with courtship or copulatory behavior may occur under the influence of sex pheromones. In some animal species, a pheromone might stimulate only the aggregation reactions; and other, nonpheromonal stimuli may be involved in stimulating close-range courtship and copulatory responses. In other species, pheromones might only be involved in courtship and copulatory behavior after the sexes have been brought together in response to other stimuli. In still other species, the same pheromone that causes the approach of a mating partner then stimulates appropriate close-range behavioral responses as well.

Aggregation Prior to Sexual Behavior

The role of pheromones released by females at or before the time of their receptivity for mating in causing male aggregation has been documented so frequently in the animal kingdom that only a few representative examples will be given here:

Phylum Protozoa
A ciliate (*Rhabdostyla vernalis:* Finley, 1952).
Phylum Aschelminthes
Rotifers (Gilbert, 1963), and a variety of free-living, plant-parasitic, and animal-parasitic nematode species (Greet et al., 1968; Salm and Fried, 1973).
Phylum Arthropoda
A variety of species of crabs (Kittredge et al., 1971; Ryan, 1966), amphipods (Dahl et al., 1970), copepods (Katona, 1973), spiders (Dondale and Hegdekar, 1973), ticks (Berger, 1972), mites (Beavers and Hampton, 1971; Cone et al., 1971), and a very large number of species of insects (see Jacobson, 1972, and Shorey, 1973, for partial reference lists).
Phylum Chordata
Various species of amphibians (Cedrini, 1971), snakes (Noble, 1937), and fish (Tavolga, 1956; Timms and Kleerekoper, 1972), and many species of mammals (Beauchamp, 1973; Davies and Bellamy, 1972; Lindsay, 1965; Schein and Hale, 1965).

Only a few examples of unusual pre-mating aggregation behavior will be mentioned here. Females of the spider *Pardosa lapidicina* secrete a pheromone into the silken strand that they leave on the substrate as they move about; males use the pheromone-treated strands in locating the females (Dondale and Hegdekar, 1973). The queen honeybee mates sequentially with a number of drones while flying several meters above the ground; therefore, the drones must orient to an aerial trail of pheromone that is emitted from a moving source (Butler and Fairey, 1964). In the solitary bee *(Adrena flavipes)*, an odor characteristic of the female's nest site, rather than one released by the female directly, causes males to search for females in that area (Butler, 1965). In a number of termite species, after reproductives of both sexes leave the nest in a dispersal flight, tandem pairs, each consisting of a male and a female, form on the ground. A male apparently recognizes a female by her pheromone and follows her closely as she runs off to a suitable area, where they excavate a cell in the soil or wood (Stuart, 1970). In some animals, including the crab-hole mosquito (Downes, 1966) and certain mites (Beavers and Hampton, 1971; Cone et al., 1971) and crabs (Kittredge et al., 1971; Ryan, 1966), the males are attracted to the vicinity of the female before she molts into her reproductive state; they remain in attendance for some time awaiting her emergence. In some crabs the pheromone that induces male aggregation is apparently identical to the molting hormone, crustecdysone (Kittredge et al., 1971). The male crab is stimulated to grasp the female with his chelae, hold her, and carry her beneath his body for some days before she molts, at which time copulation occurs.

In a number of animal species, the male releases a pheromone that attracts the female from a distance. Examples include:

Phylum Aschelminthes
 Several nematode species (Bonner and Etges, 1967; Salm and Fried, 1973).
Phylum Arthropoda
 A large number of insect species (see Jacobson, 1972, and Shorey, 1973, for partial reference list).
Phylum Chordata
 Mammals: mice (Davies and Bellamy, 1972) and rats (Carr et al., 1965).

Male bumble bees lay down scent trails on their mating flights. They fly around a circuit or loop, stopping at various points and depositing scent marks. A number of drones fly around the same circuit, continuously fortifying the scent marks. The queen bumble bees are apparently attracted to these scent paths and mate with the drones flying along them (Lindauer and Kerr, 1960).

In the ant *Camponotus herculeanus,* a male-produced pheromone has no aggregative properties *per se,* although it results in aggregation of the two sexes. The males take off from the nest in a swarming flight and then release from their mandibular glands a pheromone that stimulates the females to fly. Thus, the pheromone synchronizes the timing of flight of the two sexes (Hölldobler and Maschwitz, 1965).

Both males and females of some animal species emit pheromones that stimulate aggregation of the opposite sex. Examples include:

Phylum Aschelminthes
 Certain nematodes (Bonner and Etges, 1967; Salm and Fried, 1973).
Phylum Arthropoda
 Certain plant bugs (Mitchell and Mau, 1971) and beetles (August, 1971).
Phylum Chordata
 Mammals: mice (Davies and Bellamy, 1972) and rats (Carr et al., 1965).

Courtship and Copulation

Often, the same pheromone that stimulates the approach of a male to a female from a distance then induces his close-range courtship or copulatory behavior. This dual function of the pheromone is seen in such diverse groups as rotifers (Gilbert, 1963), insects (Shorey, 1973, and included references), fish (Losey, 1969; Tavolga, 1956), and mammals (Lindsay, 1965; Schein and Hale, 1965). Receptive females of many mammalian species deposit the pheromone with their urine (Beauchamp, 1973; Davies and Bellamy, 1972; Schein and Hale, 1965). The urine marks not only act as focal points for distance attraction of males but also at close range increase the intensity of courtship behavior patterns. In the gobiid fish *Bathygobius soporator,* a pheromone secreted by the female induces the approach of males, a change in the males' coloration to the "courtship phase," and courtship reactions composed of rapid fanning and gaping movements (Tavolga, 1956).

Among a number of insect species, the progression of behavioral steps involved in the approach of a male to the vicinity of the pheromone source and the following stimulation of his courtship and copulatory behavior are arranged in a hierarchy, with each successive step requiring a higher concentration for its release than the previous one (Bartell and Shorey, 1969; Traynier, 1968). For example, Bartell and Shorey (1969) found that the sequence of behavior exhibited by a group of males of the light brown apple moth that were stimulated by the odor of extracted female pheromone included antennal movements, whole-body movements, initiation of flight, orientation toward the pheromone source, and a complex series of activities near the source, including attempted copulation with another male moth. A greater than 100,000-fold increase in pheromone concentration was required to elicit the final step (copulatory attempts) as compared with the first step (antennal movements). The concentration required to elicit upwind orientation was intermediate.

In some cases, the female-produced pheromone apparently causes no aggregation behavior, and the only evident role for the pheromone is in the stimulation of courtship displays or direct copulatory reactions. The only reported response of males of the nematode *Nematospiroides dubius* following stimulation by the female sex pheromone is a flaring of the copulatory opening (Marchant, 1970). A male hermit crab, *Pagurus bernhardus,* perceives the pheromone upon contact with the exoskeleton of a receptive female. He then grasps the rim of the female's shell aperture with his minor cheliped and pulls her around for some hours or days, until she presumably arrives at the appropriate state for mating (Hazlett, 1970). In certain Diptera, including the sheep blowfly (Bartell et al., 1969) and *Drosophila melanogaster* (Shorey and Bartell, 1970), the odorous female pheromone apparently lowers the threshold for visually guided courtship behavior, which is directed toward a nearby fly. Additional pheromone stimulation, apparently obtained through direct contact, releases the copulatory behavior of the male flies. Male mosquitoes (*Aedes* spp.) are attracted by the sounds produced by the vibrating wings of flying females. The males are stimulated to copulate when they perceive a pheromone through direct contact with the females (Nijholt and Craig, 1971). Male tortoises (*Geochelone* spp.) also initiate courtship or copulatory attempts following pheromone stimulation (Auffenberg, 1965). As with much pheromone behavior, the stereotyped behavioral response is often displayed when the pheromone stimulus is presented in the wrong context; male tortoises have been observed to mount such inappropriate objects as a head of lettuce over which a female had recently clambered.

Although some male-produced pheromones stimulate the approach of females from a distance, the primary role for most such pheromones is in courtship. The details of the courtship behavior vary greatly from species to species. In certain millipede (Haacker, 1971) and cockroach (Barth and Lester, 1973) species, the male displays his pheromone glands when a female is nearby. The female is stimulated to feed on the glandular secretions, thus causing her to cease locomotion and to be positioned properly so that the male can copulate with her. In some tephritid fruit flies (Fletcher, 1969), as well as in swine (Hafez et al., 1962), a complex of chemical and sonic signals produced by the male causes the female to assume the mating position. The female fruit flies assume a stance with the reproduction segments extruded and directed toward the pheromone source, while pigs undergo an "immobilization reflex." The swine pheromone consists of a complex of 16-unsaturated C_{19} steroids, some of which are concentrated in the boar's sweat glands and salivary glands (Gower, 1972). A practical use for these chemicals has been found in animal breeding practices. Two of the chemicals, 5α-androst-16-en-3-one and 3α-hydroxy-5α-androst-16-ene, have been incorporated in an aerosol that is dispensed at the female pig prior to artificial insemination, causing her to assume the immobilization reflex and thus be positioned properly for artificial insemination (Melrose et al., 1971).

Unlike the well-known pheromones of female moths, which often attract males from considerable distances, the pheromones produced by males of many moth and butterfly species have the subtle (to humans) function of arresting locomotion of the female after the male has come in close proximity to her (Barth, 1958). The male-produced pheromones might also have a role in lowering the threshold of the female for accepting the male in copulation. Males of many butterfly species follow the females in a visually oriented aerial "dance." While airborne, they distribute their pheromone over the female's antennae from specialized glandular areas, usually located on the abdomen or wings. The female is thus stimulated to alight, whereupon the male may dispense more pheromone over her antennae before attempting to copulate with her (Myers and Brower, 1969; Pliske and Eisner, 1969). In many moth species, the males that have been attracted to the vicinity of the female evert brushlike scent-dispensing structures at the moment that they are about to attempt copulation (Aplin and Birch, 1968; Clearwater, 1972). The male moth pheromone, like that of butterflies, apparently functions mainly as a locomotory arrestant, inhibiting the normal flight response of the female.

Physiological Influences

Animals do not usually produce sex pheromones at a constant level, nor is their responsiveness to pheromones at a constant level, during their entire span of adult life. Rather, physiological mechanisms ensure that the pheromone communication system is operative at the appropriate stage of sexual maturity. For example, there is a direct link in rats and mice between the titre of estrogen in females and androgen in males (signifying reproductive maturity) and the attraction exerted by either sex for the opposite one. Likewise, the responsiveness of male and female rats or mice to the odor of sexually mature individuals of the opposite sex is linked to these same gonadal hormones (Davies and Bellamy, 1972; Caroom and Bronson, 1971).

Females of some insect species, exemplified by cockroaches, have repeated reproductive cycles analogous to those of mammals. Pheromone production in these insects has been found to be under the control of the juvenile hormone, produced in the corpora allata glands. On the other hand, hormonal control has not been implicated in pheromone production by some species of moths that are short-lived as adults and mate and lay their eggs within the first few days of adult life (Barth and Lester, 1973).

An additional physiological mechanism often controls the time of day at which pheromone communication between the sexes takes place. Most animals have their sexual activities compartmentalized into a time of day that is characteristic for a given species. In a number of insect species, the timing of both the release of pheromone by one sex and the maximal responsiveness to the pheromone by the opposite sex is controlled by a circadian rhythm that is entrained by the animal's previous exposure to the alternating cycles of light and dark during each twenty-four–hour day (Sower et al., 1970; Traynier, 1970).

Environmental Influences

Temperature, light intensity, wind velocity, and the nature of the surrounding vegetation play an important role in regulating pheromone communication between males and females of many animal species. Suboptimal levels of these factors may lead to greatly reduced communication efficiency or even to the abolishment of pheromone communication. A few examples, based on insects, will be given below.

In quantitative studies of moth species that mate at night, it has been found that a light intensity higher than that of full moonlight inhibits the tendency of females to release their pheromone (Sower et al., 1972) and of males to respond to the pheromone (Shorey and Gaston, 1964).

Wind velocity has multiple effects. Too low a velocity may create conditions in which a stable aerial trail of odor molecules cannot form, and too high a velocity may impede upwind orientation by responding flying insects. Thus, there are distinct upper and lower limits for many species within which successful communication can take place (Shorey, 1974). Females of certain moth species apparently can detect whether the air velocity is appropriate for pheromone communication. Females of the cabbage looper moth reduce their emission of pheromone when the velocity approaches the lower and upper limits of zero and four meters per second, respectively (Kaae and Shorey, 1972). The potential distance over which pheromone communication can occur is also profoundly affected by air velocity. The active space of the aerial pheromone trail, i.e., the volume within which the pheromone molecular density is above the behavioral threshold of the responder, is greatly reduced as the air velocity is increased (Bossert and Wilson, 1963). Using data concerning the release rate of pheromone by females, the average dilution of the pheromone molecules as they move downwind, and the threshold of pheromone molecular concentration needed to stimulate male responses, Sower et al. (1973) calculated that the theoretical maximum pheromone-communication distance for the cabbage looper moth is approximately 200 meters at a wind velocity of 0.3 meter per second and 50 meters at a velocity of 3 meters per second.

Some insect species are relatively monophagous, and it would appear adaptive for mating to take place on the same host plant species that is suitable for egg deposition and larval development. Thus, in certain moth species, either the females release their pheromone only when they sense the odor of the appropriate host plant (Riddiford, 1967) or male attraction to the pheromone is potentiated by the odor of the appropriate host plant (Brader-Breukel, 1969).

The spatial zone within which pheromone communication takes place may be influenced by the surrounding vegetation. For example, male moths are often best attracted to female pheromone sources that are located near the level of the top of the vegetative surface, whether the vegetation is cabbages or forest trees (Kaae and Shorey, 1973; Miller and McDougall, 1973); presumably, this is the elevation at which evolutionary selective processes have caused the males to search for female-produced odor trails.

Reproductive Isolation

Reproductive isolation among closely related animal species is often achieved by mechanisms that prevent individuals of one species from responding to the pre-mating communication signals emitted by the opposite sex of another species. The mechanisms may be related to the specificity of the signal itself, with the context of the message emitted causing appropriate responses by members of the opposite sex of the same species only. Or, alternatively, an identical communication signal might be used by more than one species, but the various species might be isolated geographically or according to different habitats or by different seasons or times of day during which pre-mating communication occurs. With regard to pheromones, these mechanisms for maintenance of reproductive isolation have been most studied in the moths, and the following discussion will be restricted to this group of animals.

The pheromones emitted by females of most moth species attract only males of the same species. There are, however, some exceptions to this statement (Barth, 1937; Götz, 1951), in which reproductive isolation mechanisms that operate at close range, after the male has been attracted to the vicinity of the wrong female, might be expected to operate. Until recently, most investigators believed that most female moths attract males of their own species only because of the utilization of a unique species-specific chemical in the pheromone communication of each species. To a certain extent, this molecular specificity hypothesis appears to be correct. Typically, any minor modification in the molecular structure of a sex pheromone results in a great loss of biological activity (Gaston et al., 1972; Schneider, 1967).

As more sex pheromones are identified, it becomes apparent that an identical chemical may often be produced by the females of a number of different species. Various mechanisms are apparently used by those species that "share" a common pheromone chemical to ensure the responsiveness of males to females of their own species only. One of these mechanisms is concentration specificity (Kaae et al., 1973a; Klun and Robinson, 1972). For instance, females of both the cabbage looper moth and the alfalfa looper moth apparently produce *cis*-7-dodecenyl acetate as their principal pheromone chemical. High release rates of this chemical, equivalent to those released by living cabbage looper females, attract primarily males of that species, while low release rates, apparently equivalent to those released by living alfalfa looper females, attract primarily males of that species.

For some closely related moth species, the qualitative, as well as the quantitative, blend of two or more chemicals constituting the pheromone of each species ensures species-specific responses. Females of more than one species may produce an identical chemical, attractive to males of all the species involved; but the females of each species may release additional chemicals in a pheromone blend. Some of the chemicals may potentiate the activity of the blend for males of the correct species, and others may cause an inhibition of responsiveness of males of the related species (Comeau and Roelofs, 1973; Klun et al., 1973). For a number of species, the same chemical both potentiates the activity for males of the correct species and inhibits responsiveness of males of the incorrect species. Many variations on this general theme are unfolding. In some cases, no single chemical in a pheromone blend will attract males when evaporated into the environment by itself. The multiple components of the blend are essential for biological activity, and specificity might be obtained by the related species' utilizing different, characteristic proportions of identical multiple components (Tamaki et al., 1973).

As mentioned earlier, some related species that are potentially cross-attractive may achieve reproductive isolation through spatial or temporal separation. For example, the circadian rhythms of timing of pheromone emission by females and of maximal pheromone-responsiveness by males are typically characteristic for a given species and often cause communication to be compartmentalized into a time of day that is different from that of a related species (Comeau and Roelofs, 1973; Kaae et al., 1973b).

STIMULATION OF AGGRESSION

Two general categories are recognized in which the pheromone released by one animal stimulates aggressive behavior by others of the species: the aggression may be directed either toward the pheromone-releasing animal itself or toward an animal of another species that is designated in some way by the pheromone emitter.

Aggression Directed toward a Conspecific

Among the vertebrates, dominant individuals in a social hierarchy frequently react to the pheromone released by subordinates by exhibiting aggressive behavior. Likewise, members of a social group that defend a particular territory may be stimulated to attack intruders when they detect a pheromone emitted by the strange individuals.

Pheromone-induced aggression against conspecifics has been best studied in mice, which establish territories as well as social hierarchies. The odor of a strange (to the social group) male increases the aggressive behavior of other males (Haug, 1971; Mackintosh and Grant, 1966). One pheromone is incorporated in the urine, and its production is under the control of androgen; thus, castrated males receive less aggression than do normal males or castrated males treated with androgen (Lee and Brake, 1972). Mugford and Nowell (1971) found that male mice also produce an aggression-promoting pheromone in the preputial glands; these glands apparently secrete the material directly to the exterior.

The simulation of male aggression by the odor of a strange male has been described for a number of other mammalian species, including guinea pigs (Beauchamp, 1973) and the rabbit (Mykytowycz, 1968).

Although a pheromone associated with the urine of female mice inhibits aggression by males (see below), the same or another urine-associated pheromone incites other females to respond aggressively to a strange female (Haug, 1972).

Pheromones are used in aggressive encounters among fish of certain species. The blind goby lives in burrows made by the shrimp *Callianassa affinis* (MacGinitie, 1939). A male and female remain paired for life. They recognize gobies invading their burrow by means of a pheromone. The resident male will fight to the death with an invading male, as will the resident female with an invading female. The yellow bullhead exhibits an elaborate social behavior, establishing both territories and hierarchies within its territories (Bardach and Todd, 1970). The pheromone from a strange fish invariably elicits an attack response from the occupier of a territory, and the pheromone from a subordinate fish elicits attack from a more dominant fish within a hierarchy.

Aggression against conspecifics is also stimulated by pheromones in certain social insect species. Honeybee colonies typically contain a single queen. If more than one queen is present, olfactory stimuli emanating from each of them stimulate aggression by the others (Riedel, 1972). Ants generally defend their nest and the area around it from conspecific ants of other colonies. In *Formica fusca,* the stimulus evoking aggression against foreign ants is believed to be an odor that is distinct for each colony (Wallis, 1963).

In some cases, the release of a pheromone may constitute aggression rather than incite it. Perception of a pheromone may in such cases lead to intimidation of the perceiving animal and possibly stimulate it to flee. Thus, Ralls (1971) observed that many mammals increase their frequency of marking the environment with a pheromone when they are intolerant of and dominant to other members of the same species and when they are likely to win if they attack.

Such a situation occurs in connection with territoriality, but it also occurs in other social situations. Especially frequent marking occurs when there is reason to infer that the animal is motivated to aggression. An example is the "stink fights" that occur between males of the ring-tailed lemur (Evans and Goy, 1968). Two males in a tree, engaged in aggressive interaction, direct pheromone odors toward each other from a variety of sources. A pheromone secreted by wrist glands is rubbed onto the tail. The males then wave their tails toward each other. In addition, a volatile pheromone is deposited on branches from the palms (palmar marking). First one male palmar marks and then the other, with pauses between. The more aggressive male moves forward and the other retreats. Also, the more aggressive male palmar marks branches that the other male has marked. It appears likely that the pheromone signals interact with other stimuli produced by the males in symbolizing their state of aggressiveness.

The use of pheromones as an aggressive stimulus and to advertise dominance is probably very common among mammals. The advertising is often done even in the absence of an antagonist, with the pheromone being deposited on objects in the environment from specialized skin glands or from glands associated with urine or feces. Circumstantial evidence for this role of pheromones is obtained from numerous morphological and behavioral observations. The largest glands and the highest rates of secretion of pheromones, as well as the highest frequencies of marking objects with a pheromone, are found in the dominant individuals in social hierarchies (Bronson and Marsden, 1973; Müller-Schwarze, 1972; Mykytowycz, 1968).

Finally, some additional comments are necessary concerning the inhibition of aggression among mammals. As mentioned earlier, the aggressive tendencies of animals that are foreign to a territory or subordinate within a social hierarchy are likely to be inhibited by the pheromone of a territory occupier or a more dominant animal, respectively. Also, animals that are familiar with each other and that coexist within a society are typically inhibited from aggression toward one another, with males being especially inhibited from exhibiting aggression toward females. This inhibition has been most studied in mice, where it has been found to be attributable, at least in part, to pheromones released by the various individuals (Dixon and Mackintosh, 1971; Haug, 1971; Mugford and Nowell, 1971).

Aggression Directed toward an Individual of Another Species

Pheromones that are used to elicit aggressive behavior of conspecifics toward an individual of another species are commonly found in the social Hymenoptera. Sometimes the biological function is the stimulation of colony mates to attack a prey designated by the pheromone or to plunder the colony of another species. This function was considered earlier. In other cases the pheromone, now referred to as an alarm pheromone, is involved with defense of the nest and its surroundings from attack by an enemy.

According to Maschwitz (1966), the glands in hymenopterous species that produce alarm pheromones are invariably associated with the organs of defense. Thus, the pheromones are secreted by various glands that occur in the vicinity of the mandibles or the sting. In fact, a number of ant and bee species have multiple glands that may secrete pheromones separately or simultaneously to incite defensive behavior by conspecifics.

The alarm pheromone often has a dual function, not only to attract colony mates and lower their threshold for attack behavior but also to act as a poisonous or repellent chemical that deters the enemy (Bergström and Lofqvist, 1973; Wilson and Bossert, 1963). The behavior stimulated by alarm pheromones may be arranged in a hierarchal sequence, with each behavioral step requiring a higher concentration. Thus, Moser et al. (1968) found that workers of the Texas leaf-cutting ant, when exposed to a low concentration of the pheromone 4-methyl-3-heptanone, raise their heads and antennae. At higher concentrations, they are stimulated to follow the molecular gradient to its source, and at still higher concentrations to become very active and to open their mandibles.

A honeybee worker that is disturbed at the hive releases an alarm pheromone, isopentyl acetate, from its sting chamber. Other bees are attracted by the pheromone and are stimulated to fly and to attack objects in the vicinity of the hive (Boch and Shearer, 1965). The attack behavior is guided by visual cues from the enemy (Maschwitz, 1966). Upon stinging the enemy, the recruited bees release isopentyl acetate from their sting glands and another pheromone, 2-heptanone, from their mandibular glands (Shearer and Boch, 1965). These pheromones stimulate and direct the attack of additional bees and are probably responsible for the phenomenon well known to beekeepers that more than one bee will often sting in the same spot.

FAMILIARIZATION WITH HOME RANGE

Mammals of many species deposit scent marks from specialized skin glands or from glands that are associated with urine or feces as they move about within their territories. This marking behavior is also seen in many species that do not defend well-defined territories but that have a home range within which they normally confine their activities. The precise role of the marking behavior has been little studied. The behavior may, in fact, serve a variety of interacting functions, which are listed in the following section. A number of investigators feel that a very important function served by scent marking in many species is the maintenance of the animal's familiarity with its home-range environment (Mykytowycz, 1970; Ralls, 1971). Thus, many mammals become especially active in marking objects with scent when they are displaced into a strange environment (Goddard, 1967; Martin, 1968).

As well as "reassuring" the animal that it is in a familiar environment, the pheromone scent marks may aid in orienting its activities in that environment (Goddard, 1967; Wynne-Edwards, 1962). Many mammals habitually follow the same paths as they move about within their home range (Wynne-Edwards, 1962); scent is deposited along the trails, sometimes by glands on the feet (as in ruminants: Bourlière, 1954) and sometimes by other methods. The black rhinoceros follows fecal scent trails (Goddard, 1967). The rhinos deposit feces in piles, which are located randomly over the home range. Any one pile may be used by a number of individuals. Before defecating, they sniff the pile and may sweep it with the anterior horn and shuffle through it with the feet. After defecating, they kick at the pile with their hind feet. The odor is apparently distributed by the feet as the rhino moves through its home range, creating a trail that can be followed by others of the species.

COMPLEX SOCIAL BEHAVIOR

Many of the behavioral activities of animals that live in organized societies appear to be regulated by pheromones. There have been few critical studies of this behavior. Mainly anecdotal evidence indicates that, within mammalian societies, pheromones often relate information concerning individual identity, group membership identity, age, social status, sex, and reproductive state. Also, pheromones often are part of a stimulus complex associated with "greeting" between animals, submission, dominance, attention seeking, gregariousness, signaling of danger or distress, trail following, territorial behavior, and identification of home range (Mykytowycz, 1970).

Likewise, many of the behavioral activities of social insects appear to be regulated by pheromones. Some of the more poorly understood behaviors are probably caused by the so-called surface pheromones, which may be absorbed on the body surfaces and be detected on contact or at extremely close range. These pheromones include the colony odors, the caste-recognition scents, and the releasers of grooming and food exchange (Wilson, 1971). The regulation of insect social organization is probably due in large part to pheromones released by the queen. Workers of a number of species of bees, ants, and wasps engage in "retinue" behavior. They encircle the queen, lick her body, and touch her with their antennae (Gary, 1970; Wilson, 1963b). The workers apparently are stimulated to engage in retinue behavior by pheromones secreted by the queen, and through this behavior they probably obtain from the queen the primer pheromones that inhibit development of their ovaries and control other aspects of their physiology and behavior.

Evolution of Pheromone Communication

Pheromone communication must have arisen very early during the evolution of primitive plants and animals. Indeed, the chemicals used for signaling between the cells of multicellular animals—the hormones and synaptic transmitter substances—probably evolved from the chemicals used for communication between individual free-living cells (Haldane, 1954). After all, a multicellular organism is an aggregation of cells designed to allow intercellular signaling, based mainly on chemicals. Subsequent evolutionary processes led to the elaboration of specialized groups of cells constituting glands for the synthesis and release of pheromones, as well as specialized sensory devices for the perception of pheromones; these developments enabled communication between multicellular animals. The course that this evolution has followed must vary considerably among different animal groups.

Similarities between the chemicals used as pheromones and as hormones by many animals are striking. Terpenoid or steroid compounds are used for these functions by a great variety of vertebrate and invertebrate species, with the same chemical sometimes having both functions in a given species. In various crab species, the process of evolution of hormonal systems from primitive pheromonal systems appears to have reversed (Kittredge et al., 1971). The female crabs release a sex pheromone into the water shortly before they molt. Apparently, the molting hormone, crustecdysone, functions also as the sex pheromone. Evolutionary processes probably led to relatively minor modifications of a preexisting hormone system to fulfil the pheromone communication function. These modifications included elaboration of a system for the release of the chemical into the external environment and of an external system on the antennules of the males for detection of the chemical.

Most pheromones have, perhaps, been derived by natural selection from metabolites that were initially produced for some other function. In mammals, the integumental glands that originally supplied wax or mucus to the skin have probably often been elaborated in this way to produce pheromones that are used for a variety of communication functions (Wynne-Edwards, 1962).

The aggregation pheromones released by a number of bark beetle species are remarkably similar to the terpenoid resins of the host trees that they attack (Hughes, 1973). The host terpenoids are in some cases attractive to the beetles, and attraction to the host may have been the only aggregative factor in primitive bark beetles. It seems reasonable that various beetle species later evolved mechanisms whereby metabolites of the host resins (thus, the resins modified chemically within the insects) stimulated aggregation of conspecifics. Similarly, Shorey et al. (1969) noted that males of two species of fly that congregate on certain food materials are stimulated by the odor of the food to court and attempt copulation with nearby flies. This response to a host odor may represent the primitive situation. Later stages of evolution may have resulted in metabolites of the food chemicals, produced by the insects themselves, fulfilling the sexual communication function. As pointed out by Moore (1967), the animals could now be independent of the original food or habitat odor. Further evolution of the scent-producing and receptor mechanisms would lead to the pheromones, often highly specific in terms of species as well as in their biological effect, that are characteristic of many animals today.

References

Abdullah, M., 1965. *Protomeloe crowson:* a new species of a new tribe (*Protomeloini*) of the blister beetles (*Coleoptera, Meloidae*) with remarks on a postulated new pheromone (cantharidin). *Entomol. Tidskr.,* 86:43–48.

Aplin, R. T., and Birch, M. C., 1968. Pheromones from the abdominal brushes of male noctuid Lepidoptera. *Nature,* 217:1167–68.

Auffenberg, W., 1965. Sex and species discrimination in two sympatric South American tortoises. *Copeia,* 1965:335–42.

August, C. J., 1971. The role of male and female pheromones in the mating behaviour of *Tenebrio molitor. Jour. Insect Physiol.,* 17:739–51.

Bardach, J. E., and Todd, J. A., 1970. Chemical communication in fish. In: *Communication by Chemical Signals. Advances in Chemoreception, Vol. I.,* J. W. Johnston, Jr., D. G. Moulton, and A. Turk, eds. New York: Appleton-Century-Crofts, pp.205–40.

Bartell, R. J., and Shorey, H. H., 1969. Pheromone concentration required to elicit successive steps in the mating sequence of males of the light-brown apple moth, *Epiphyas postvittana. Ann. Entomol. Soc. Amer.,* 62:1206–1207.

Bartell, R. J.; Shorey, H. H.; and Browne, L. B.; 1969. Pheromonal stimulation of the sexual activity of males of the sheep blowfly *Lucilia cuprina* (Calliphoridae) by the female. *Anim. Behav.,* 17:576–85.

Barth, R., 1937. Herkunft, Wirkung und Eigenschaften des weiblichen Sexualduftstoffes einiger Pyraliden. *Zool. Jb., Abt. Allg. Zool. u. Physiol.,* 58:297–329.

Barth, R., 1958. Estímulos químicos como meio de comunicacaõ entre os sexos em Lepidópteros. *An. Acad. Bras. Cienc., Rio de Janeiro,* 30:343–63.

Barth, R. H., and Lester, L. J., 1973. Neuro-hormonal control of sexual behavior in insects. *Ann. Rev. Entomol.,* 18:445–72.

Beauchamp, G. K., 1973. Attraction of male guinea pigs to conspecific urine. *Physiol. Behav.,* 10:589–94.

Beavers, J. B., and Hampton, R. B., 1971. Growth, development, and mating behavior of the citrus red mite (Acarina: Tetranychidae). *Ann. Entomol. Soc. Amer.,* 64:804–806.

Bennett, R. B., and Borden, J. H., 1971. Flight arrestment of tethered *Dendroctonus pseudotsugae* and *Trypodendron lineatum* (Coleoptera:Scolytidae) in response to olfactory stimuli. *Ann. Entomol. Soc. Amer.,* 64:1273–86.

Berger, R. S., 1972. 2,6-Dichlorophenol, sex pheromone of the lone star tick. *Science,* 177:704–705.

Bergström, G., and Lofqvist, J., 1973. Chemical congruence of the complex odoriferous secretions from Dufour's gland in three species of ants of the genus *Formica. Jour. Insect Physiol.,* 19:877–907.

Blum, M. S.; Crewe, R. M.; Kerr, W. E.; Keith, L. H.; Garrison, A. W.; and Walker, M. M.; 1970. Citral in stingless bees: isolation and functions in trail-laying and robbing. *Jour. Insect Physiol.,* 16:1637–48.

Boch, R., and Shearer, D. A., 1965. Alarm in the beehive. *Amer. Bee J.,* 105:206–207.

Bonner, J. T., 1970. The chemical ecology of cells in the soil. In: *Chemical Ecology,* E. Sondheimer and J. B. Simeone, eds. New York: Academic Press. 336pp.

Bonner, T. P., and Etges, F. J., 1967. Chemically mediated sexual attraction in *Trichinella spiralis. Exptl. Parasitol.,* 21:53–60.

Bossert, W. H., and Wilson, E. O., 1963. The analysis of olfactory communication among animals. *Jour. Theoret. Biol.,* 5:443–69.

Bourlière, F., 1954. *The Natural History of Mammals.* New York: Alfred A. Knopf, Inc. 363pp.

Bowers, W. S.; Nault, L. R.; Webb, R. E.; and Dutky, S. R.; 1972. Aphid alarm pheromone: Isolation, identification, synthesis. *Science,* 177:1121–22.

Brader-Breukel, L. M., 1969. Modalités de l'attraction sexuelle chez *Diparopsis watersi* (Roths.). *Coton et fibres tropicales* (Paris), 23, 24, 25.

Bronson, F. H., and Marsden, H. M., 1973. The preputial gland as an indicator of social dominance in male mice. *Behav. Biol.,* 9:625–28.

Butler, C. G., 1965. Sex attraction in *Andrena flavipes* Panzer (Hymenoptera:Apidae), with some observations on nest-site restriction. *Proc. Roy. Entomol. Soc. London* (A), 40:77–80.

Butler, C. G., 1970. Chemical communication in insects: behavioral and ecologic aspects. In: *Communication by Chemical Signals.* Advances in Chemoreception, Vol. I, J. W. Johnston, Jr., D. G. Moulton, and A. Turk, eds. New York: Appleton-Century-Crofts, pp.35–78.

Butler, C. G., and Fairey, E. M., 1964. Pheromones of the honeybee: biological studies of the mandibular gland secretion of the queen. *Jour. Apic. Res.,* 3:65–76.

Butler, C. G.; Fletcher, D. J. C.; and Watler, D.; 1969. Nest-entrance marking with pheromones by the honeybee—*Apis mellifera* L., and by a wasp, *Vespula vulgaris* L. *Anim. Behav.,* 17:142–47.

Butler, C. G., and Simpson, J., 1967. Pheromones of the queen honeybee (*Apis mellifera* L.) which enable her workers to follow her when swarming. *Proc. Roy. Entomol. Soc., London* (A), 42:149–54.

Calam, D. H., and Youdeowei, A., 1968. Identification and functions of secretion from the posterior scent gland of fifth instar larva of the bug *Dysdercus intermedius. Jour. Insect Physiol.,* 14:1147–58.

Caroom, D., and Bronson, F. H., 1971. Responsiveness of female mice to preputial attractant: effects of sexual experience and ovarian hormones. *Physiol. Behav.,* 7:659–62.

Carr, W. J.; Loeb, L. S.; and Dissinger, M. L.; 1965. Responses of rats to sex odors. *Jour. Comp. Physiol. Psychol.,* 59:370–77.

Cedrini, L., 1971. Olfactory attractants in sex recognition of the crested newt: an electrophysiological research. *Monit. Zool. Ital.,* 5:223–29.

Clearwater, J. R., 1972. Chemistry and function of a pheromone produced by the male of the southern armyworm *Pseudaletia separata. Jour. Insect Physiol.,* 18:781–89.

Comeau, A., and Roelofs, W. L., 1973. Sex attraction specificity in the Tortricidae. *Ent. Exp. Appl.,* 16:191–200.

Cone, W. W.; McDonough, L. M.; Maitlen, J. C.; and Burdajewicz, S., 1971. Pheromone studies of the twospotted spider mite. *Jour. Econ. Entomol.,* 64:355–58.

Crisp, D. J., and Meadows, P. S., 1963. Adsorbed layers: the stimulus to settlement in barnacles. *Proc. Roy. Soc. London* (B), 158:364–87.

Dahl, E.; Emanuelsson, H.; and von Mecklenburg, C., 1970. Pheromone transport and reception in an amphipod. *Science,* 170:739–40.

Daterman, G. E., 1972. Laboratory bioassay for sex pheromone of the European pine shoot moth, *Rhyacionia buolina. Ann. Entomol. Soc. Amer.,* 65:119–23.

Davies, V. J., and Bellamy, D., 1972. The olfactory response of mice to urine and effects of gonadectomy. *Jour. Endocrinol.,* 55:11–20.

Dethier, V. G.; Browne, L. B.; and Smith, C. N., 1960. The designation of chemicals in terms of the responses they elicit from insects. *Jour. Econ. Entomol.,* 53:134–36.

Dixon, A. K., and Mackintosh, J. H., 1971. Effects of female urine upon the social behaviour of adult male mice. *Anim. Behav.,* 19:138–40.

Dondale, C. D., and Hegdekar, B. M., 1973. The contact sex pheromone of *Pardosa lapidicina* (Araneida: Lycosidae). *Can. Jour. Zool.,* 51:400–401.

Downes, J. A., 1966. Observations on the mating behaviour of the crab hole mosquito *Deinocerites cancer* (Diptera: Culicidae). *Canadian Entomol.,* 98:1169–77.

Dundee, H. A., and Miller, M. C., III, 1968. Aggregative behavior and habitat conditioning by the ringneck snake, *Diadophis punctatus arnyi. Tulane Studies in Zool. and Bot.,* 15:41–58.

Eisner, T. E., and Kafatos, F. C., 1962. Defense mechanisms of Arthropods. X. A pheromone promoting aggregation in an aposematic distasteful insect. *Psyche,* 69:53–61.

Evans, C. S., and Goy, R. W., 1968. Social behavior and reproductive cycles in captive ring-tailed lemurs (*Lemur catta*). *Jour. Zool.,* 156:181–97.

Farkas, S. R., and Shorey, H. H., 1974. Mechanisms of orientation to a distant odor source. In: *Pheromones,* M. C. Birch, ed. Amsterdam: Elsevier Press: North-Holland Publishing Co., pp.81–95.

Finley, H. E., 1952. Sexual differentiation in peritrichous ciliates. *Jour. Morphol.,* 91:569–605.

Fletcher, B. S., 1969. The structure and function of the sex pheromone glands of the male Queensland fruit fly, *Dacus tryoni. Jour. Insect Physiol.,* 15:1309–22.

Friedlander, C. P., 1965. Aggregation in *Oniscus asellus* Linn. *Anim. Behav.,* 13:342–46.

Frisch, K. von, 1941. Über einen Schreckstoff der Fischaut und seine biologische Bedeutung. *Z. vergl. Physiol.,* 29:46–145.

Gara, R. I., and Coster, J. E., 1968. Studies on the attack behavior of the southern pine beetle. III. Sequence of tree infestation within stands. *Contrib. Boyce Thompson Instit. Plant Res.,* 24:77–86.

Gary, N. E., 1970. Pheromones of the honey bee. In: *Control of Insect Behavior by Natural Products,* D. L. Wood, R. M. Silverstein, and M. Nakajima, eds. New York: Academic Press, pp.29–53.

Gaston, L. K.; Payne, T. L.; Takahashi, S.; and Shorey, H. H., 1972. Correlation of chemical structure and sex pheromone activity in *Trichoplusia ni* (Noctuidae). In: *Olfaction and Taste IV, Proceedings of the Fourth International Symposium,* D. Schneider, ed. Stuttgart: Wissenschaftliche Verlagsgesellschaft, pp.167–73.

Gilbert, J. J., 1963. Contact chemoreception, mating behavior and sexual isolation in the rotifer genus *Brachionus. Jour. Expt. Biol.,* 40:625–41.

Goddard, J., 1967. Home range, behavior, and recruitment rates of two black rhinoceros populations. *East Afr. Wildl. Jour.,* 5:133.

Goodrich, B. S., and Mykytowycz, R., 1972. Individual and sex differences in the chemical composition of pheromonelike substances from the skin glands of the rabbit, *Oryctolagus cuniculus. Jour. Mammal.,* 53:540–48.

Götz, B., 1951. Die Sexualduftstoffe an Lepidopteren. *Experientia,* 7:406–18.

Gould, H. N., 1952. Studies on sex in hermaphroditic mollusc, *Crepidula plana.* IV. Internal and external factors influencing growth and sexual development. *Jour. Exp. Zool.,* 119:93–163.

Gower, D. P., 1972. 16-Unsaturated C_{19} steroids; a review of their chemistry, biochemistry and possible physiological role. *Jour. Steroid Biochem.,* 3:45–103.

Greet, D. N.; Green, C. D.; and Poulton, M. E.; 1968. Extraction, standardization and assessment of the volatility of the sex attractants of *Heterodera rostochiensis* Woll. and *H. schachtii* Schm. *Ann. Appl. Biol.,* 61:511–19.

Haacker, U., 1971. Die Funktion eines dorsalen Druesenkomplexes in Balzverhalten von *Chordeuma* (Diplopida). *Forma Functio,* 4:162–70.

Hafez, E.; Sumption, L.; and Jakway, J.; 1962. The behaviour of swine. In: *The Behaviour of Domestic Animals,* E. Hafez, ed. Baltimore: Williams and Wilkins, pp.334–69.

Haldane, J. B. S., 1954. La Signalisation animale. *Année Biol.,* 58:89–98.

Hangartner, W., 1967. Spezifitat und Inaktwierung des Spurpheromons von *Lasius fuliginosus* Latr. und Orientierung der Arbeiterinnen im Duftfeld. *Z. vergl. Physiol.,* 57:103–36.

Hangartner, W.; Reichson, J. M.; and Wilson, E. O.; 1970. Orientation to nest material by the ant, *Pogonomyrmex basius* (Latreille). *Anim. Behav.,* 18:331–34.

Haug, M., 1971. Rôle probable des vésicules séminales et des glandes coagulantes dans la production d'une phéromone inhibitrice du comportement aggressif chez la souris. *C. R. Hebd. Seances Acad. Sci.,* sér. D., sci. natur., 273:1509–10.

Haug, M., 1972. Effet de l'urine d'une femelle étrangère sur le comportement agressif d'un groupe de souris femelles. *C. R. Hebd. Seances Acad. Sci.,* sér. D., sci. natur., 275:995–98.

Hazlett, B. A., 1970. Tactile stimuli in the social behavior of *Pagurus bernhardus* (Decapoda:Paguridae). *Behaviour,* 36:20–48.

Hemmings, C. C., 1966. Olfaction and vision in fish schooling. *Jour. Exp. Biol.,* 45:449–64.

Hodek, J., 1960. Hibernation-bionomics in Coccinellidae. *Acta Soc. Entomol. Cech.,* 57:1–20.

Hölldobler, B., and Maschwitz, U., 1965. Der Hochzeitsschwarm der Rossameise *Camponotus herculeanus* L. (Hym. Formicidae). *Z. vergl. Physiol.,* 50:551–68.

Hughes, P. R., 1973. Effect of α-Pinene exposure on *trans*-verbenol synthesis in *Dendroctonus ponderosae* Hopk. *Naturwissenschaften,* 60:261–62.

Jacobson, M., 1972. *Insect Sex Pheromones.* New York: Academic Press. 382pp.

Jones, R. B., and Nowell, N. W., 1973. The coagulating glands as a source of aversive and aggression-inhibiting pheromone(s) in the male albino mouse. *Physiol. Behav.,* 11:455–62.

Kaae, R. S., and Shorey, H. H., 1972. Sex pheromones of noctuid moths. XXVII. Influence of wind velocity on sex pheromone releasing behavior of *Trichoplusia ni* females. *Ann. Entomol. Soc. Amer.,* 65:436–40.

Kaae, R. S., and Shorey, H. H., 1973. Sex pheromones of Lepidoptera. 44. Influence of environmental conditions on the location of pheromone communication and mating in *Pectinophora gossypiella. Environmental Entomol.,* 2:1081–84.

Kaae, R. S.; Shorey, H. H.; and Gaston, L. K.; 1973a. Pheromone concentration as a mechanism for reproductive isolation between two lepidopterous species. *Science,* 179:487–88.

Kaae, R. S.; Shorey, H. H.; McFarland, S. U.; and Gaston, L. K.; 1973b. Sex pheromones of Lepidoptera. XXXVII. Role of sex pheromones and other factors in reproductive isolation among ten species of Noctuidae. *Ann. Entomol. Soc. Amer.,* 66:444–48.

Kalmus, H., 1965. Possibilities and constraints of chemical telecommunication. *Proc. 2nd Int. Congr. Endocrinol.,* London 1964, pp.188–92.

Karlson, P., and Lüscher, M. 1959. "Pheromones": a new term for a class of biologically active substances. *Nature,* 183:55–56.

Katona, S. K., 1973. Evidence for sex pheromones in planktonic copepods. *Limnol. Oceanogr.,* 18:574–83.

Kaufmann, T., 1966. Observations on some factors which influence aggregation by *Blaps sulcata* (Coleoptera:Tenebrionidae) in Israel. *Ann. Entomol. Soc. Amer.,* 59:660–64.

Kennedy, J. S., 1939. The visual responses of flying mosquitoes. *Proc. Zool. Soc. London* (A), 109:221–42.

Kingston, B. H., 1965. The chemistry and olfactory properties of musk, civet and castoreum. *Proc. 2nd Int. Congr. Endocrinol.,* London 1964, pp.209–14.

Kirk, V. M., and Dupraz, B. J., 1972. Discharge by a female ground beetle, *Pterostichus lucublandus* (Coleoptera:Carabidae), used as a defense against males. *Ann. Entomol. Soc. Amer.,* 65:513.

Kittredge, J. S.; Terry, M.; and Takahashi, F. T.; 1971. Sex pheromone activity of the molting hormone, crustecdysone, on male crabs: (*Pachygrapsus crassipes, Cancer antennarius,* and *C. anthonyi*). *U.S. Fish Wild. Serv. Fish Bull.,* 69:337–43.

Klun, J. A.; Chapman, O. L.; Mattes, K. C.; Wojtkowski, P. W.; Beroza, M.; and Sonnet, P. E.; 1973. Insect sex pheromones: minor amount of opposite geometrical isomer critical to attraction. *Science,* 181:661–63.

Klun, J. A., and Robinson, J. F., 1972. Olfactory discrimination in the European corn borer and several pheromonally analogous moths. *Ann. Entomol. Soc. Amer.,* 65:1337–40.

Konijn, T. M.; Barkley, D. S.; Chang, Y. Y.; and Bonner, J. T.; 1968. Cyclic AMP: A naturally occurring acrasin in the cellular slime molds. *Amer. Naturalist,* 102:225–33.

Lee, C. T., and Brake, S. C.,1972. Reaction of male mouse fighters to male castrates treated with testosterone proprionate or oil. *Psychonomic Sci. Sect. Anim. Physiol. Psychol.,* 27:287–88.

Levinson, H. Z., and Bar Ilan, A. R., 1967. Function and properties of an assembling scent in the khapra beetle *Trogoderma granarium. Rivista di Parassitol.,* 28:27–42.

Levinson, H. Z., and Bar Ilan, A. R., 1971. Assembling and alerting scents produced by the bedbug *Cimex lectularius* L. *Experientia,* 27:102–103.

Lindauer, M., and Kerr, W. E., 1960. Communication between the workers of stingless bees. *Bee World* 41:29–41, 65–71.

Lindsay, D. R., 1965. The importance of olfactory stimuli in the mating behaviour of the ram. *Anim. Behav.,* 13:75–78.

Linsenmair, K. E., and Linsenmair, C., 1971. Paarbildung und Paarzusammenhalt bei der monogamen Wuesfenassel *Hemilepistus reaumuri* (Crustacea, Isopoda, Oniscoidea). *Z. Tierpsychol.,* 29:134–55.

Losey, G. S., Jr., 1969. Sexual pheromone in some fishes of the genus *Hypsoblennius* Gill. *Science,* 163:181–83.

MacGinitie, G. E., 1939. The natural history of the blind goby, *Typhlogobius californiensis* (Steindachner). *Amer. Mid. Naturalist,* 21:489–505.

Mackintosh, J. H., and Grant, E. C., 1966. The effect of olfactory stimuli on the agonistic behaviour of laboratory mice. *Z. Tierpsychol.,* 23:584–87.

Marchant, H. V., 1970. Bursal response in sexually stimulated *Nematospiroides dubius* (Nematoda). *Jour. Parasitol.,* 56:201–202.

Martin, R. D., 1968. Reproduction and ontogeny in tree-shrews (*Tupaia belangeri*), with reference to their general behaviour and taxonomic relationships. *Z. Tierpsychol.,* 25:404–95, 505–32.

Maschwitz, U. W., 1966. Alarm substances and alarm behavior in social insects. *Vitamins Hormones,* 24:267–90.

Melrose, D. R.; Reed, H. C. B.; and Patterson, R. L. S., 1971. Androgen steroids associated with boar odour as an aid to the detection of oestrus in the pig artificial insemination. *Br. Vet. Jour.,* 127:497–501.

Miller, C. A., and McDougall, G. A., 1973. Spruce budworm moth trapping using virgin females. *Can. Jour. Zool.,* 51:853–58.

Mitchell, W. C., and Mau, R. F. L., 1971. Response of the female southern green stink bug and its parasite, *Trichopoda pennipes,* to male stink bug pheromones. *Jour. Econ. Entomol.,* 64:856–59.

Moore, B. P., 1967. Chemical communication in insects. *Science Jour.,* 3:44–49.

Morse, R. A., 1972. Honey bee alarm pheromone: another function. *Ann. Entomol. Soc. Amer.,* 65:1430.

Morse, R. A., and Boch, R., 1971. Pheromone concert

in swarming honey bees (Hymenoptera:Apidae). *Ann. Entomol. Soc. Amer.*, 64:1414–17.

Moser, J. C., and Blum, M. S., 1963. Trail marking substance of the Texas leaf-cutting ant: source and potency. *Science,* 140:1228.

Moser, J. C.; Brownlee, R. C.; and Silverstein, R.; 1968. Alarm pheromones of the ant *Atta texana. Jour. Insect Physiol.,* 14:529–35.

Mudd, A., and Corbet, S. A., 1973. Mandibular gland secretion of larvae of the stored products pests *Anagasta kuehniella, Ephestia cautella, Plodia interpunctella* and *Ephestia elutella. Entomol. Exp. Appl.,* 16:291–93.

Mugford, R. A., and Nowell, N. W., 1971. The preputial glands as a source of aggression-promoting odors in mice. *Physiol. Behav.,* 6:247–49.

Müller-Schwarze, D., 1971. Pheromones in blacktailed deer (*Odocoileus hemionus columbianus*). *Anim. Behav.,* 19:141–52.

Müller-Schwarze, D., 1972. Social significance of forehead rubbing in blacktailed deer (*Odocoileus hemionus columbianus*). *Anim. Behav.,* 20:788–97.

Myers, J., and Brower, L. P., 1969. A behavioural analysis of the courtship pheromone receptors of the Queen butterfly, *Danaus gilippus berenice. Jour. Insect Physiol.,* 15:2117–30.

Mykytowycz, R., 1965. Further observations on the territorial function and histology of the submandibular cutaneous (chin) glands in the rabbit, *Oryctolagus cuniculus* (L.). *Anim. Behav.,* 13:400–12.

Mykytowycz, R., 1968. Territorial marking by rabbits. *Scientific Amer.,* 218:116–26.

Mykytowycz, R., 1970. The role of skin glands in mammalian communication. In: *Communication by Chemical Signals.* Advances in Chemoreception, Vol. I, J. W. Johnston, Jr., D. G. Moulton, and A. Turk, eds. New York: Appleton-Century-Crofts, pp.327–60.

Naylor, A. F., 1965. Dispersal responses of female flour beetles, *Tribolium confusum,* to presence of larvae. *Ecology,* 46:341–43.

Nijholt, H. F., and Craig, G. B., Jr., 1971. Reproductive isolation in *Stegomyia* mosquitoes: III. Evidence for a sexual pheromone. *Entomol. Exp. Appl.,* 14:399–412.

Noble, G. K., 1937. The sense organs involved in the courtship of *Storeria, Thamnophis,* and other snakes. *Bull. Amer. Mus. Nat. Hist.,* 73:673–725.

Noble, G. K., and Clausen, H. J., 1936. The aggregation behavior of *Storeria dekayi* and other snakes with especial reference to the sense organs involved. *Ecol. Monogr.,* 6:269–316.

Nolte, D. J.; Eggers, S. H.; and May, I. R.; 1973. A locust pheromone:locustol. *Jour. Insect Physiol.,* 19:1547–54.

Pitcher, T. J., 1973. The three-dimensional structure of schools in the minnow, *Phoxinus phoxinus* (L.). *Anim. Behav.,* 21:673–86.

Pliske, T. E., and Eisner, T., 1969. Sex pheromone of the queen butterfly: biology. *Science,* 164:1170–72.

Prokopy, R. J., 1972. Evidence for a marking pheromone deterring repeated oviposition in apple maggot flies. *Environmental Entomol.,* 1:326–32.

Rabb, R. L., and Bradley, J. R., 1970. Marking host eggs by *Telenomus sphingis. Ann. Entomol. Soc. Amer.,* 63:1053–56.

Ralls, K., 1971. Mammalian scent marking. *Science,* 171:443–49.

Regnier, F. E., and Wilson, E. O., 1971. Chemical communication and "propaganda" in slave-maker ants. *Science,* 172:267–69.

Ressler, R. H.; Cialdini, R. B.; Ghoca, M. L.; and Kleist, S. M.; 1968. Alarm pheromone in the earthworm *Lumbricus terrestris. Science,* 161:597–99.

Riddiford, L. M., 1967. *Trans*-2-hexenal: mating stimulant for polyphemus moths. *Science,* 158:139–41.

Riedel, S. M., 1972. Rapid adaptation by paired queens of the honey bee, *Apis mellifera. Ann. Entomol. Soc. Amer.* 65:825–29.

Rottman, S. J., and Snowdon, C. T., 1972. Demonstration and analysis of an alarm pheromone in mice. *Jour. Comp. Physiol. Psychol.,* 81:483–90.

Rudinsky, J. A.; Morgan, M.; Libbey, L. M.; and Michael, R. R.; 1973. Sound production in Scolytidae: 3-methyl-2-cyclohexen-1-one released by the female Douglas fir beetle in response to male sonic signal. *Environmental Entomol.,* 2:505–509.

Ryan, E. P., 1966. Pheromone:evidence in a decapod crustacean. *Science,* 151:340–41.

Salm, R. W., and Fried, B., 1973. Heterosexual chemical attraction in *Camallanus* sp. (Nematoda) in the absence of worm-mediated tactile behavior. *Jour. Parasitol.,* 59:434–36.

Schein, M. W., and Hale, E. B., 1965. Stimuli eliciting sexual behaviour. In: *Sex and Behavior,* F. A. Beach, ed. New York: John Wiley and Sons, pp.440–82.

Schloeth, R., 1956. Zur Psychologie der Begegnung zwischen Tieren. *Behaviour,* 10:1–79.

Schneider, D., 1964. Insect antennae. *Ann. Rev. Entomol.,* 9:103–22.

Schneider, D., 1967. Wie arbeitet der Geruchssinn bei Mensch und Tier? *Naturwiss. Rdsch.,* 20:319–26.

Schultze-Westrum, T., 1965. Innerartliche Verständigung durch Düfte beim Gleitbeutler, *Petaurus breviceps papuanus* Thomas (Marsupialia, Phalangeridae). *Z. vergl. Physiol.,* 50:151–220.

Shearer, D. A., and Boch, R., 1965. 2-Heptanone in the mandibular gland secretion of the honey bee. *Nature,* 206:530.

Shorey, H. H., 1973. Behavioral responses to insect pheromones. *Ann. Rev. Entomol.,* 18:349–80.

Shorey, H. H., 1974. Environmental and physiological control of sex pheromone behavior. In: *Pheromones,* M. C. Birch, ed. Amsterdam: Elsevier Press: North-Holland Publishing Co., pp.62–80.

Shorey, H. H., and Bartell, R. J., 1970. Role of a volatile female sex pheromone in stimulating male courtship behaviour in *Drosophila melanogaster. Anim. Behav.* 18:159–64.

Shorey, H. H.; Bartell, R. J.; and Browne, L. B.; 1969. Sexual stimulation of males of *Lucilia cuprina* (Calliphoridae) and *Drosophila melanogaster* (Drosophilidae) by the odors of aggregation sites. *Ann. Entomol. Soc. Amer.,* 62:1419–21.

Shorey, H. H., and Gaston, L. K., 1964. Sex pheromones of noctuid moths. III. Inhibition of male responses to the sex pheromone in *Trichoplusia ni* (Lepidoptera:Noctuidae). *Ann. Entomol. Soc. Amer.,* 57:775–79.

Shorey, H. H., and Gaston, L. K., 1970. Sex pheromones of noctuid moths. XX. Short-range visual orientation by pheromone-stimulated males of *Trichoplusia ni. Ann. Entomol. Soc. Amer.,* 63:829–32.

Snyder, N. F. R., 1967. An alarm reaction of aquatic gastropods to intraspecific extract. *Cornell Univ. Agr. Expt. Sta. Mem.,* 403:1–122.

Snyder, N., and Snyder, H.; 1970. Alarm response of *Diadema antillarum. Science,* 168:276–78.

Sower, L. L.; Kaae, R. S.; and Shorey, H. H.; 1973. Sex pheromones of Lepidoptera. XLI. Factors limiting potential distance of sex pheromone communication in *Trichoplusia ni. Ann. Entomol. Soc. Amer.,* 66:1121–22.

Sower, L. L., Shorey, H. H.; and Gaston, L. K.; 1970. Sex pheromones of noctuid moths. XXI. Light:dark cycle regulation and light inhibition of sex pheromone release by females of *Trichoplusia ni. Ann. Entomol. Soc. Amer.,* 63:1090–92.

Sower, L. L.; Shorey, H. H.; and Gaston, L. K.; 1972. Sex pheromones of Lepidoptera. XXVIII. Factors modifying the release rate and extractable quantity of pheromone from females of *Trichoplusia ni* (Noctuidae). *Ann. Entomol. Soc. Amer.,* 65:954–57.

Stuart, A. M., 1967. Alarm, defense, and construction behavior relationships in termites (Isoptera). *Science,* 156:1123–25.

Stuart, A. M., 1970. The role of chemicals in termite communication. In: *Communication by Chemical Signals.* Advances in Chemoreception, Vol. I, J. W. Johnston, Jr., D. G. Moulton, and A. Turk, eds. New York: Appleton-Century-Crofts, pp.79–106.

Tamaki, Y.; Noguchi, H.; and Yushima, T.; 1973. Sex pheromone of *Spodoptera litura* (F.) (Lepidoptera: Noctuidae):Isolation, identification, and synthesis. *Appl. Entomol. Zool.,* 8:200–203.

Tavolga, W. N., 1956. Visual, chemical and sound stimuli as cues in the sex discriminatory behavior of the gobiid fish *Bathygobius soporator. Zoologica,* 41:49–64.

Timms, A. M., and Kleerekoper, H., 1972. The locomotor responses of male *Ictalurus punctatus,* the channel catfish, to a pheromone released by the ripe female of the species. *Trans. Am. Fish Soc.,* 101:302–10.

Traynier, R. M. M., 1968. Sex attraction in the Mediterranean flour moth, *Anagasta kühniella: location of the female by the male. Canadian Entomol.,* 100:5–10.

Traynier, R. M. M., 1970. Sexual behaviour of the Mediterranean flour moth, *Anagasta kühniella:* Some influences of age, photoperiod, and light intensity. *Canadian Entomol.,* 102:534–40.

Vité, J. P., and Pitman, G. B., 1969. Aggregation behaviour of *Dendroctonus brevicomis* in response to synthetic pheromones. *Jour. Insect Physiol.,* 15:1617–22.

Vité, J. P., and Renwick, J. A. A., 1971. Inhibition of *Dendroctonus frontalis* response to frontalin by isomers of brevicomin. *Naturwissenschaften,* 58:418.

Wallis, D. I., 1963. A comparison of the response to aggressive behaviour in two species of ants, *Formica fusca* and *Formica sanguinea. Anim. Behav.,* 11:164–71.

Wells, M. J., and Buckley, S. K. L., 1972. Snails and trails. *Anim. Behav.,* 20:345–55.

Wilson, E. O., 1962. Chemical communication among workers of the fire ant *Solenopsis saevissima,* 1. The organization of mass-foraging. *Anim. Behav.,* 10:134–47.

Wilson, E. O., 1963a. Pheromones. *Scientific Amer.,* 208:100–14.

Wilson, E. O., 1963b. The social biology of ants. *Ann. Rev. Entomol.,* 8:345–68.

Wilson, E. O., 1968. Chemical systems. In: *Animal Communication,* T. A. Sebeok, ed. Bloomington: Indiana University Press, pp.75–102.

Wilson, E. O., 1971. *The Insect Societies.* Cambridge: Harvard University Press. 548pp.

Wilson, E. O., and Bossert, W. H., 1963. Chemical communication among animals. *Recent Progr. Horm. Res.,* 19:673–716.

Wilson, E. O., and Regnier, F. E., Jr., 1971. The evolution of the alarm-defense system in the formicine ants. *American Naturalist,* 105:279–89.

Wood, D. L., 1970. Pheromones of bark beetles. In: *Control of Insect Behavior by Natural Products,* D. L. Wood, R. M. Silverstein, and M. Nakajima, eds. New York: Academic Press, pp.301–16.

Wrede, W. L., 1932. Versuche über den Artduft der Elritzen. *Z. vergl. Physiol.,* 17:510–19.

Wynne-Edwards, V. C., 1962. *Animal Dispersion in Relation to Social Behaviour.* Edinburgh: Oliver & Boyd Ltd., 653pp.

Chapter 8

BIOLUMINESCENCE AND COMMUNICATION

James E. Lloyd

Nature never makes excellent things for mean or no uses.
John Locke

Light and Life—a catchy phrase of elegant simplicity once used for a symposium title—expresses a fundamental relationship of the natural world. Response to light, as well as ultimate economic dependence on it, is virtually a universal characteristic of life. Organisms capture light and make bigger molecules (photosynthesis), locomote or turn at rates dependent on its intensity (orthokinesis, klinokinesis), grow or locomote to and from it (phototropism, phototaxis), swim with their backs toward it (dorsal light reaction), go to sites without it (scototaxis), use it as a compass (menotaxis) with or without time compensation ultimately related to a twenty-four-hour light rhythm. Animals begin and end, or end and begin, daily activity by it; plants fold and open their leaves and blossoms in response to it; and both do these things in an experimentalist's darkness by means of a temperature-independent engram of the light rhythm previously experienced (circadian rhythm). Insects and birds begin developmental and reproductive cycles by it, using it as a token stimulus (photoperiod, diapause, migration). Life detects Light's presence, analyzes its spectral composition, responds to its polarization, filters it, and with simple and complex lenses (in trilobites even aspheric, aplanic lenses; Shawver, 1974) focuses and forms images on light-sensitive molecules and tissues of its own manufacture. Life generates Light and shines it in color and rhythm from a multitude of lantern types for obscure, yet probably simple, purposes. And the foregoing demonstrates that Light and Life have also been responsible for the generation of a specialized scientific lexicon.

Few organisms in the world are not somehow touched by light. Burrowing worms of abyssal-benthic ooze perhaps escape, but there are luminescent deep-sea animals in waters overhead. In the terrestrial environment subterranean animals, plants, and parts of plants live in darkness, but fruiting bodies of fungi appear at the surface and interact with organisms of the first world; and firefly larvae and pupae, fungal mycelia, and collembola shine light within the soil. Though cave animals usually live in complete darkness, so long as they stay in their caves, fungus gnats in New Zealand caverns luminesce. Darkness prevails in few places, and if organisms experience it at all, it is usually temporary. Even existence in the dark bole of a tree or in the gut or womb of a mammal is ephemeral.

The energy spectrum of biological chemiluminescence coincides generally with the action spectra of photoreceptors, but it never includes infrared or ultraviolet wavelengths. When emitted in a well-lit environment it will probably go undetected, though in special circumstances it perhaps can obliterate shadows and may be used for concealment (see Hastings, 1971). But an organism that emits light in an environment with low ambient lighting cannot very well remain unseen. Given darkness and living light, biological interaction of some sort is almost inevitable—virtually every organism is tuned in. What happens to an organism as a consequence of its own light depends on the relationship its coinhabitants have previously established with light. It seems unlikely that light emission can be adaptively neutral, even ignoring its expense and the relationship of energy budgets to differential survival and reproduction. The acceptance of adaptive neutrality for light emission in any organism precludes the conception of enticing new hypotheses and new knowledge. I disagree in particular and principle with the statement that "inasmuch as it is difficult to imagine any functional significance of bioluminescence in bacteria or fungi, we probably can assume that bioluminescence has arisen as a fortuitous correlate of the cellular oxidative mechanism, persisting in many animals, especially lower ones, despite no obvious survival value" (Brown, 1973).

The occurrence of bioluminescence in the living world is a tantalizing riddle in all its facets. It appears in bacteria, fungi, protozoans, balanoglossids, polychaetes, oligochaetes, nudibranchs, snails, squids, bivalves, ostracods, copepods, amphipods, shrimps, centipedes, collembolans, beetles, brittle stars, tunicates, fishes, and others (Harvey, 1952). Its presence or absence generally is of no value in phylogenetic classification whether one deliberates at the phylum or species level. E. N. Harvey (1952) said it best:

> It is apparent . . . that no clear development of luminosity along evolutionary lines is to be detected but rather a cropping up of luminescence here and there, as if a handful of damp sand has been cast over the names of various groups written on a blackboard, with luminous species appearing wherever a mass of sand struck. The Ctenophora have received the most sand. It is possible that all members of this phylum are luminous . . . . At the other extreme are very large groups in which only a few luminous animals are known, as in the gastropod and lamellibranch molluscs.

An explanation for the scattered occurrence of luminosity among contemporary organisms is perhaps found in an answer to the original enigma—the nature of the beginnings of photogenic chemistry in living systems—and combines modern chemistry and primordial ecology.

Originally it was speculated that the chemical pathways of the luminescent reaction evolved in the context of detoxification since it was known that oxygen is toxic to anaerobic organisms and life was thought to have developed under essentially anaerobic conditons. It was suggested that early in the history of life, oxygen, which in the light reaction is combined with substrates generically known as luciferin, was poisonous and the proto-organisms had to dispose of it. With the evolution of aerobic metabolism the oxygen-removing light reaction was not completely lost since it was intimately associated with the electron transport process (McElroy and Seliger, 1962). This hypothesis is appealing because it accounts for the phylogenetically widespread and presumably independent appearance of bioluminescence, as well as the chemical similarity. But it assumes that the partial pressure of oxygen was negligible in the ancient atmosphere, and it is now believed that oxygen levels were significant (Urey effect). Seliger (1975) has proposed a new and alternative hypothesis. He reasons that since a steady-state, low level of oxygen was present during early evolution it is unlikely that oxygen would at some point become toxic and require complete removal. He suggests that the enzymes of the light reaction (luciferases) were secondarily, and much later in evolution, derived from aerobic hydroxylases. The hydroxylases came into special significance in the "primitive soup" during a time of severe molecular competition for the readily oxidizable substrates, because they bestowed upon their possessors a trophic advantage—they permitted the breakdown of aromatic rings and long-chain alkanes, thus making it possible for the remaining open-chain carbon fragments to be handled with anaerobic enzyme systems. The critical chemical step was the splitting of the stabile C=C bond. The free energy derived from the splitting of the double bond was sufficient to leave the product in the excited state—this energy was lost to the organism, though ultimately it became the energy of bioluminescence. Hydroxylases have been retained, since the advent of more efficient oxidative pathways, for the metabolism of inert molecules. When, in more recent evolutionary history and after the evolution of photoreceptors, some ecological advantage resulted from the fortuitous occurrence of highly fluorescent product molecules (i.e., molecules whose released free energy from the decay of the excited state was in the visible spectrum), selection acted upon the light-emitting processes.

According to this alternative hypothesis, "bioluminescence, rather than being a vestigial process, is a ubiquitous phenomenon. It is the result of metabolic oxidation . . . yielding product molecules in excited electronic states . . . [which] . . . may fluoresce or, in the presence of a suitable energy acceptor, sensitize the fluorescence of the latter . . ." (Seliger, 1975).

It is important not only to explain the origin of bioluminescence but to account for its loss as well. The adaptation of stem organisms of antiquity explains the occurrence of comparable or identical photochemistry among evolutionarily divergent and distant taxa: recent adaptations explain the appearance of lightless species in otherwise luminescent taxonomic groups. A North American firefly *(Photinus indictus)* is virtually indistinguishable from several luminescent members of its species complex, except for its lack of a light organ. If, by reason of its lost lantern, this species were to be placed in a related genus of diurnal fireflies *(Pyropyga)* as it once was, the explanation for the detail of evolutionary convergence with *Photinus* spp. that would be required should be truly remarkable and deal with morphology of all its life stages and its chemical composition. The features by which this species is known to differ from its nearest *Photinus* relatives all concern mating behavior and communication and could derive from a single ecological factor—for example, cold nights. If its progenitor populations lived in bogs and marshes near the retreating Wisconsian glacier, members of these (chronological) populations that relied less and less on luminescent signals (and nocturnal activity) and more and more on pheromones (Lloyd, 1972) might have been more successful in reproducing. The chill of twilights under the influence of the great ice mass could have been genetically lethal for ectotherms dependent on flight for the functioning of their luminescent signals. The present ecological and geographical occurrence of these fireflies is commensurate with this scheme. (It can be conjectured that a firefly of northern Europe has responded to the same ecological factor in another way. In *Phosphaenus hemipterus* both males and females are flightless. If flight ability ceased to be of utility, yet luminescent signals were for some reason still operative, selection in other contexts could have broken up the gene complex required for wing development. *P. hemipterus* is exceedingly rare, perhaps near extinction.) The lack of adult luminescence in *Photinus cooki,* the only other species in *Photinus* (a genus of more than 240 species) known to be diurnal, appears to be a recent adaptation to signal-code competition among its close relatives (Lloyd, 1966:77).

To summarize, light is of significance in one context or another to most organisms. It is my belief that wherever bioluminescence occurs in the kingdoms of living things, it can be explained on the basis of adaptation and natural selection. Although the adaptive significance of the fundamental chemistry that evolved in Precambrian pools may not have centered on the release of photic energy, the maintenance and development of light-emitting behavior, when luminescence did finally appear in species of divergent lineages, depended on the new relationships that luminescent organisms had with other members of their biotic community. The explanation for the absence of luminescence in species whose close relatives are light emitters is to be found in geologically recent adaptations and may relate to a number of factors in their ecosystems, physical or biotic.

Origins of Bioluminescent Communication

Independently, in a thousand and more phyletic lineages, individuals became luminescent. (The alternative explanation that all luminescent organisms trace their luminescent geneology back in unbroken succession to a common luminescent ancestor and that all contemporary nonluminescent beings are descended from photic dropouts is unreasonable.) The alleles required for light production either were part of the gene complex that Seliger and McElroy suggested was put together long, long ago and maintained until recently, or recently were fortuitously added in some context other than light production. How many genes constitute a "luminescent package"? How many genes might be from the remote past, and how many were selected in a pleiotropic context? In any event it was probably a substitution at a single locus that finally turned each light on. Regardless of context or antiquity of origin, when the light came on, it gave its bearers a new ecological status. Of interest here are organisms whose Darwinian fitnesses were improved by light emission and the reasons for that.

ENHANCEMENT OF REFLECTED LIGHT SIGNALS BY MEANS OF LUMINESCENCE

Luminescence might simply have enhanced visual signals that were already important in behavioral interactions. If it appeared in or near appropriate anatomical sites, then the luminescence might have emphasized, amplified, or highlighted already-existing signals, such as postures, movements, gestures, areas of pale color, or reflective surfaces, that had hitherto depended entirely on reflected light for their efficacy. This effect would be significant in poorly lit habitats, and especially so in transitional ones such as diurnal and nautical twilight zones. In dark places there is no reflected light to enhance, and in well-lit ones bioluminescent amplification seldom can bestow selective advantage. Twilight zones combine essential elements—organisms with well-developed photoreceptors and the benefit from amplification of reflected light. Perhaps this partially accounts for the facts that two-thirds of the fish of the mesopelagic zone are luminescent and that there are no luminous freshwater forms.

The appeal of this model is that initially only instructions for luminescence itself were required of the genome. Already present, regardless of whether the signal recipient was of the same or different species, were the visual organs, data-processing neural circuits, appropriate scanning or search patterns, "attention to detail," and all the other essential ecological, physiological, and behavioral adaptations. It is obvious that the augmentation of visual signals by luminescence might have ultimately permitted an ecological shift into a nocturnal activity period or a truly dark habitat and, when viewed after the fact, can indeed be recognized as a transitional stage.

The light emitted by the fruiting bodies of several species of fungi is bright enough to be seen at distances of several meters. Since a number of insects feed and oviposit on fungi it is easily imagined that insects might use the luminescence as a beacon. Furthermore, natural selection might favor the maintenance of luminescence in the fungus because the fungus used the insects for some vital function, such as transporting spores to other accumulations of decomposing organic matter (Lloyd, 1974). Such a mutualistic relationship, *sans* luminescence, exists between stinkhorn fungi and greenbottle flies in European woods. The caps of fresh fruiting bodies of the fungus are covered with a green black, shiny layer of spore slime. The flies orient to the color and the sheen of the cap, land on it, and eat slime and spores. The slime is dissolved by the saliva and digested in the gut. The spores are later deposited in the feces (Wickler, 1968:155). Consider for a moment a nonluminescent fungus whose shiny white or greenish white cap was a signal to its insectan, spore-carrying symbiont. Would not the first appearance of luminescence in the fruiting cap enhance its signal and make it operable when reflected light was inadequate, such as after sunset or on the floor of a dark forest or cave entrance?

The light emitted by the larviform females of glowworm beetles (Phengodidae) is bright and can be seen for several meters, and it emanates from several sources over the surfaces of the insects. Females of the genus *Zarhipis,* from the western United States, are brightly luminescent. Workers there concluded that the luminescence is not associated with mating behavior and that males use only pheromones for locating the female. The antennae of male glowworm beetles are feathery and large, presumably greatly increasing the number of chemoreceptors and sensitivity, and perhaps permitting stereo-detection of molecules. Rivers (1886) found that *Zarhipis* males are attracted to females during daylight but that they fly only in temperate heat, from 9 A.M. to 4 P.M., so that in hot weather they do not appear "until the sun is declining." Tiemann (1967) noted that earlier workers had "observed that males will approach females during the day in the humid coastal area. Although males were not observed to approach females during the day in the low relative humidities of the desert environment, they did come to the female at dusk . . . males were attracted to the females within 10 minutes . . . the glow from the females was barely visible in the twilight." If males orient visually at close range to the pale-banded females and perhaps specifically to the pale-band pattern, luminescence would, at its first occurrence, in the low ambient light of dusk, enhance the female's visual signals (Lloyd, 1971).

There are many other potential examples of this origin for luminescent signals, but the articulation of complete models depends on detailed knowledge about the behavior of both luminescent and nonluminescent members of the taxon involved. It is inviting to speculate that the shimmers and reflections of the silvery and iridescent bodies and bright-colored markings of some fish, cephalopods, and crustaceans, all being organisms with well-developed eyes, predisposed members of these groups for a subsequent evolution of luminescent signals through the transitional stage of enhancement of reflected signals. The "brighter-than-life" reflections from the lateral lines of some fish, such as the neon tetra of home aquaria, shine as though reinforced by luminescence. Other fish do have lanterns distributed along their lateral lines. Body undulations or other communicatively significant body movements may be highlighted, or transmitted in their entirety in the case of luminescence, by these optical phenomena. Lanterns on other fish outline the body, mark the opercula and fins, or produce a (shimmery?) glow over the lateral or ventral surface. Perhaps circular light organs in the region of the anal fin ("egg mimics") stimulate and direct the activity of males during courtship and fertilization. Table 1 lists various displays employed during mating and aggressive encounters in nonluminescent fish, and suggests lantern positions that would amplify them.

About one-third of the known species of squid are luminescent. This may seem surprising at first, but not when one considers the nature of the skin of squids in general. In their mantles are pigment bodies (chromatophores) that can be expanded and contracted by means of muscle fibers. There are also reflectors or mirrors (iridophores, iridocytes), some of which lie above the chromatophores and some below. Color changes can be made rapidly and are under the control of the central nervous system. They sweep across the body with "rapidity and variety more like that of an electric sign than an animal" (Lane, 1960:94). "It pulsated slowly, while the colors came and went over its body in such a way that new adjectives will have to be coined adequately to describe it—reds, blacks, browns, yellows, rolling, surging, springing into vision as the pigment spots contracted or expanded, a living, liquid palette" (Beebe, 1926). This would seem to predispose squids for the incorporation of luminescent amplification when chemical and ecological opportunities occur.

Some peculiar patterns of photophore arrangement found on fish and squids may be fully explicable only when the natures of the light-analyzing mechanisms of the signal recipients are completely understood. Counterparts of the

Table 1

Some behavioral displays known from nonluminescent fish and the light-organ positions that could be significant if similar displays occur during mating, aggressive encounters, or other interactions of luminescent fish.

Displays of Nonluminescent Fishes	Luminescent Organ Position
Fish Viewed from Side:	
display side of body; zig-zag figure; side-by-side swimming; short, jerky motions; head-down position; raise and lower fins; body quivering; resplendent with iridescent colors and quivering with intense excitement; color contrasts; color changes; show off colors; hues intensified; swim around the female in circles; raise dorsal fin	at bases of paired fins; on back; on sides; in pectoral region; in front of eyes; on dorsal fin ray; along back from head to near tail; along lateral line system; at caudal fin; photophores usually lateral and ventral; tendency to form lines; over whole body; on fins; orbital region; upper side of peduncle (males); lower side of peduncle (females); on cheeks; on ventral fins; on anal fin;
Fish Viewed from Front:	
flare gill covers; open jaws; depress floor of mouth; mouth-to-mouth display; open mouth; quick breathing movements; raising opercula to look like eyes; jaw gaping; mouth-to-mouth throat displays	under eyes; spot on forehead; in front of eyes; in region of gill opening; on esca (lure) of angler fish and held near mouth; lines on jaws; on lip; on opercula; on pectoral fins; on tongue; at edge of eyes; on barbel extending from lower jaw; in post orbital region; at margin of tongue; on cheek; on lower jaw

NOTE: Similar displays which also depended on reflected light, may have been used by the ancestors of luminescent fishes when photophores and associated signaling behavior were first evolving. Luminescence could have amplified weak reflected light in the mesopelagic zone of perpetual twilight, where presently two-thirds of the piscine inhabitants have light.

Sources: Norman, 1947; Harvey, 1952; Marshall, 1966.

optical illusions of the human visual system are to be expected, and some lantern arrangements may be but abstractions of their phylogenetic precursors and their reflected-light analogues or homologues.

RITUALIZATION—THE ACQUISITION OF BIOLUMINESCENT SIGNALS FROM EMISSIONS OF OTHER CONTEXTS

A number of different situations in which emitters and receivers are involved must be distinguished. Of these (see next section for listing and discussion) I distinguish true communication or communication *sensu stricto* on the basis of the effects of natural selection: In communication *s.s.*, selection has brought about (enhanced and maintained) the emission and the mechanisms of production and reception in the specific context of information transfer that is being considered (Lloyd, 1971; Otte, 1974). An emission becomes ritualized—that is, adapted into a communicative context from another, stereotyped, and exaggerated—when, upon its detection, both the emitter and receiver benefit from the subsequent action the receiver makes on the basis of the information it derived from the emission. Or, when viewed from another perspective, when the emitter influences (manipulates) subsequent and mutually beneficial actions of the receiver. In the following examples it is speculated that communicative signals have been derived from emissions that were of original significance to the emitters in the context of illumination.

Fireflies of some species emit characteristic luminescence patterns when they land. In North America this is most commonly seen among fireflies of the genus *Photuris* (Lloyd, 1968), and it occurs in *Pteroptyx* and *Luciola* species in New Guinea and probably elsewhere (Lloyd, 1973a). Among *Photuris* spp. the sole function of the landing emissions seems to be illumination, and the use of such light is observed mainly in females. As a female approaches the ground her flash rate increases until finally the flashes fuse into a glow, which is discontinued only after landing. With practice an observer can quickly learn to recognize landing lights and will seldom confuse them with male advertisement patterns. Females land in areas where dozens of patrolling, advertising males are present, yet female landing flashes elicit no visible response from the males. The females are presumably not receptive to male sexual advances and probably have already mated. Landing flashes and glows could become incorporated into a signaling system if the females were sexually responsive: males that approached landing luminescence and emitted advertising flashes in the vicinity would have improved mating success.

Females of *Pteroptyx* fireflies, the synchronously flashing species of southeast Asia, emit light as they approach swarm trees and land on the foliage. These females are apparently unmated and upon entering the swarms find mates. In one species, the 3-flicker pteroptyx (no. 22, Lloyd, 1973a:1003), males sometimes pursue females in aerial chases over the foliage, bump or upset the aerodynamics of the females, force them to land, and land near them. A simple explanation for the origin of this behavior is that males that followed and landed near luminescing females increased their chances of mating with them.

Although the frequency of chases observed in the *Pteroptyx* species studied so far does not indicate that an aerial chase has become a ritualized and invariable part of their courtship, it has become so in the courtship of a species in a related genus. In another New Guinea firefly, the diamondback luciola *(Luciola obsoleta),* males and females perch in loose congregations, and each emits a variety of sexually distinct luminescent patterns. In late evening, about two hours after sunset and the onset of flashing activity, when a female takes flight she is pursued by flickering males. After traveling a few meters the female lands or is forced to land by the darting and bumping tactics of a pursuing male, which lands close by. Courtship then continues, additional luminescent signals are exchanged, and finally mounting and copulation occur.

Another possible example of ritualization of a noncommunicative emission concerns landing flickers of males of *Pteroptyx* fireflies. These flickers presumably function in illumination. Perched males of the 3-flicker pteroptyx begin emitting flickers when glowing females fly over them. The flicker that they emit at this time is different (in phrasing and flicker frequency) from the one used as an advertising pattern. A possible phylogeny for the evolution of the response flicker from a landing flicker is: (1) Males (of an ancestral population) emitted a flicker while landing. Males emitted the flicker also when landing near females that had answered their advertising flash pattern. (2) The courted females approached the landing males by orienting to their landing flickers. Males, approaching by foot or short hopping flights, continued to use the landing flicker for illumination. (3) Since females oriented to and/or approached the landing flickers of males, selection favored males that produced this flicker after each female answer, as well as when it was required for illumination during their locomotion, and/or during the time they were in visual, luminescent contact with females. (4) Perched males flickered, with the landing flicker, in response to the luminescence of approaching females with which they had no prior interaction, as now occurs in the 3-flicker pteroptyx.

BIOLUMINESCENT SIGNALS FROM BEHAVIORALLY SIGNIFICANT ENVIRONMENTAL PHOTIC STIMULI

Many organisms have evolved specific, yet poorly understood, positive responses to light. Hence, the success of luminescent, lochetic[1] fungus gnats (Fulton, 1939; Gatenby and Cotton, 1960), the popularity of light trapping among insect fanciers, certain poaching and hunting techniques, and the use of bright lights by marine biologists as well as anglers for attracting specimens and prey. A variety of behavior is undoubtedly involved. Some moths may use celestial light sources for bearings and thus maintain straight flight over some distance; if they take certain bearings on a streetlight they will fly a spiral route into the light. Artificial lights may activate neural circuits and behavior that evolved in the context of surface seeking in aquatic animals, entrance seeking in cave animals, or in relation to dawn and the beginning of flight activity in winged, diurnal organisms. If mates are brought together by mutual attraction to some form of natural illumination, then the advent of luminescence could lead to bioluminescent signaling. Luminescence in such cases might provide a concentrated light stimulus and focus the attraction of one of the pair. With the paucity of behavioral/ecological data on luminescent forms this phenomenon remains in the realm of speculation, and I am unable to find a suggestive example or to postulate one that is superior to explanations along other lines. The surface swarming of luminescent syllid worms such as the Bermuda fireworm might have originated in this manner, but the luminescent enhancement of a previously existing signal (say bright and dark bands of surface ripples as seen from below) seems more likely at present. Phototactic responses of shed gametes might predispose an organism to develop luminescent signals in this context.

1. From *lochan,* to ambush (Gr); Fulton, 1939.

Classification of Bioluminescent Emissions and Interactions

The term "communication" has been used loosely to include a number of different kinds of interactions (Sebeok, 1968; Otte, 1974). Still other relationships that organisms have with components of their environments, that no one would consider communication, involve similar or identical sensory, neural, and behavioral attributes and have importance in the evolution of communication. I believe that it is worthwhile to list these phenomena and try to distinguish precisely among them. A classification that focuses on adaptation and that is based on the action of natural selection is useful and relevant. Otte (1974) used this approach in a discussion of "communicative" interactions including mimicry, deceit, and intra- and interspecific signaling. For each interaction he considered the effects on *each participant's* survival or reproductive success, and scored it positive (+), negative (–), or no effect (0). Others (e.g., Odum, 1959:226) had previously used this sort of notation when considering various kinds of ecological interaction among organisms, but had focused, with an evolutionarily aloof perspective, on the consequences for population size. In the classification presented here primary attention is paid to the long-term effects of selection on the *emitting attributes (and emission) of the emitter, and the reception attributes of the receiver, in the specific context under consideration.* Attributes include sensory and effector mechanisms with their underlying neural circuits of emission analysis and production. The effects on individuals considered by Otte result in the population changes discussed by Odum and the long-term changes in the emissive-perceptive attributes of concern in this discussion; the mundane interactions of individual organisms determine differential genetic survival and long-term evolutionary changes.

I. SIGNALS—COMMUNICATION SENSU STRICTO (+/+)

The single category of interactions with which most researchers working on communication are concerned involves a transmitter and a receiver in separate individuals, and natural selection acts on the mechanisms of both to facilitate and enhance the transmission-reception in the given context (Table 2). This is the category for which I reserve the term "communication" (Lloyd, 1971), since it best agrees with previous expressions of selection-conscious workers and vernacular usage.

Bioluminescent examples of this category would include mating communication of fireflies (Lloyd, 1971), cuttlefish (*Sepia:* Girod, 1882), octopods (*Tremoctopus:* Lane, 1960), fish (Nicol, 1969:391), and fireworms (*Odontosyllis enolpa:* Harvey, 1952); shoaling in fish and crustaceans (Nicol, 1969:391); gamete orientation and attraction in sedentary or sessile marine organisms (hypothetical, occurrence unknown); emissions that are involved in interactions of mutual benefit between members of different species such as fungi and insects (discussed above); and warning lights in fireflies (undemonstrated) (Lloyd, 1971) and poisonous dinoflagellates (hypothetical).

II. SELF-SIGNALS—ILLUMINATION (+/+)

This category differs from the one above in that the transmitter and receiver mechanisms, in a given emission pathway, are in the same individual. Autocommunication can at least tentatively be distinguished from other intra-individual information-transferring phenomena, such as hormones and neural feedback, by the occurrence of the informationally significant alterations of the carrier energy during its passage between emitter and receiver mechanisms. Taxa for which illumination lights, the bioluminescent analogues of echolocation in bats and active electroreception in fish (Bullock, 1973), have been suggested are fireflies (Waller, 1685; Lloyd, 1968), squids (Lane, 1960:72, 113), and fish (Harvey, 1952:523).

III. FALSE SIGNALS—AGGRESSIVE MIMICRY (+/–)

The exploitation of a receiver in which an emission activates mechanisms that evolved and/or are maintained in the context of true communication is but one of several kinds of interactions in which the receiver is exploited. A false signal possesses those properties, and is presented in those circumstances that enable it to be sensed, neurally processed, and responded to in a manner appropriate to true signals. Selection acts upon the mechanisms of reception to promote the discrimination of false signals from true ones.

Female fireflies of the genus *Photuris* mimic the mating signals of females in the genera *Photinus, Pyractomena, Photuris,* and *Robopus,* attract the males of these species, seize them, and devour them (Lloyd, 1965; Farnworth 1973). The females of some species are able to mimic the flashed responses of more than one prey species, and individual females switch appropriately from one response to another, depending on the characteristics of flash pattern they are answering. Some males of the prey species respond to the false signals in the same manner that they would to true signals from their own females, and are caught (Lloyd, 1975).

If the luminescent lures of female angler fish are also used in courtship for signaling to their own tiny males (Harvey, 1952:529), if there is great competition among males for mates, and if species with similar signaling systems occur together, the occurrence of aggressive mimics among them would not be surprising. However, the attraction of other prey to the lure would not involve false signals, but false clues as discussed below.

IV. CUES—EMANATIONS OF INDIFFERENT EMITTERS (0/+)

Organisms perceive and process stimuli from sources that are themselves in no way affected by the outcome of the detection. Whether the emanation is detected is irrelevant to the emitter, and the emitter may even be inanimate (a cow pie that is used for food by scarab beetles), dead (a deceased cat that attracts staphylinid beetles that prey upon carrion-feeding larvae), or living (a leafy branch in a shaft of sunlight to which mating-swarms of insects are orienting). The biological significance of cues to the receiver may be that they are, after detection and neural processing, translated as "of no significance (now)." In other words, they can be disregarded. For example, a shore bird lands beside a log that emits stimuli that are translated as "a neutral object, a possible perch, of no negative value ('danger to me')." The detection by the receiver organism is of biological significance to the receiver in that context, and selection maintains and enhances the reception mechanisms. It would be unadaptive for a shore bird to respond to all such logs in the same manner as to a crocodile!

Pteroptyx fireflies of southeast Asia gather in trees, sometimes in great numbers. Males emit their flashes in synchrony with nearby males, and as a consequence, when there are enough males so that continuity is maintained over an entire tree, mass synchrony occurs. The flash rate is characteristic of a species, and emerging adults in the vicinity fly to trees that are pulsing with the appropriate characteristics; a pulsing tree is like a beacon. As flying fireflies approach a firefly-tree from a distance they will at first see only the entire tree. At some point during their approach their compound eyes will begin to resolve parts of the tree, then small clusters of flashing males, and finally individual males. Only when individual males, each with its own "halo" of neighbors with which it is interacting, are resolved by incoming females, are they finally in competition for them. Selection is not maintaining the beacon effect in the beacon context; selection produces group synchrony because it favors individuals that synchronize with their near neighbors. Mass synchrony is a consequence (effect) of natural selection and not a goal (in the sense of Williams, 1966:9; Lloyd, 1973a, 1973c). The emission of a treeful of fireflies is therefore a cue and not a signal. Each individual male,

Table 2

Emission-reception phenomena.

Designations			Definitions		
Emission Name	Type	Interaction Name (example)	Selection Sign Em	Selection Sign Rm	Clarification
Signal	I	Communication (firefly mating flashes)	+	+	Em and Rm in different individuals; same or different species.
Self-Signal	II	Auto-Communication (illumination flashes and glows)	+	+	Em and Rm in same individual.
False Signal	III	Signal Mimicry (*Photuris* firefly aggressive mimicry)	+	–	Emission activates Rm belonging to signal context.
Cue	IV	Cuing (pulsating *Pteroptyx* firefly tree in SE Asia attracting additional fireflies)	0	+	Emitter in inanimate or nonliving, or organism whose Em not affected by selection in this context.
False Cue	V	Cue Mimicry (hiding-luminescence of pony fish)	+	–	Emission activates Rm belonging to cue context.
Clue	VI	Cluing (firefly mating flashes when detected and used for attack orientation by predator)	–	+	Emission is from a living organism and is exploited by receiver organism.
False Clue	VII	Clue Mimicry (autotomy of luminous segment by polynoid annelid that diverts predator)	+	–	Emission activates Rm belonging to clue context.
Noise	VIII	(moonlight interfering with firefly receiving a luminescent signal)	0	–	Emitter is nonliving or an organism that is not affected by this interaction.
False Noise	IX	Jamming (luminous discharge of squid that hides the squid; see text)	+	–	Emission activates Rm, or at some level interferes with functioning, in a manner like that of noise.
Ambiguous Signal	X	(Cross-specific signaling, resulting in subsequent mating between fireflies of two sibling spp.)	–	–	Signal activates Rm belonging to another signal context, with mutual detriment resulting.
Indifferent Fazer	XI	(Lightning flash striking dark-adapted eyes of firefly)	0	–	Intrinsic nature of emission of abiotic or indifferent emitter causes receiver dysfunction.
Fazer	XII	(Shark spotlighting prey)	+	–	Intrinsic nature of emission causes receiver dysfunction, permits exploitation.
Scanner	XIII	(landing luminescence of firefly falling on leafy vegetation)	+	0	Emission of self-signal striking an indifferent receiver.
Locator	XIV	(light of bottom-dwelling fish striking shellfish prey)	+	–	Emission of self-signal striking exploitable receiver.
Quasi Signal	XV	(glow of single lantern of *Phengodes* female attracting tiny *Phausis* male)	0	–	Emission activates Rm belonging to signal context; emitter not affected by Rm.
Quasi Clue	XVI	("Froggs, . . . perhaps mistaking for fireflies . . . take in live coals," observation of colonial naturalist J. Banister; Ewan and Ewan 1970:296)	0	–	Emission of indifferent emitter activates Rm mechanisms belonging to clue context.

Table 2 (continued)

Designations			Definitions		
			Selection Sign		
Emission Name	**Type**	**Interaction Name (example)**	**Em**	**Rm**	**Clarification**
Quasi Cue	XVII	(streetlight attracting certain insects by activating their dawn responses)	0	–	Emission of indifferent emitter activates Rm mechanisms belonging to cue context.
Cipher	XVIII	(physiologist's "fortuitous correlate"—luminescence falling on a nonirritable, unresponsive, inedible, or solitary organism)	0	0	Emission from indifferent emitter striking Rm with no effect, influence, or significance.

NOTE: Emitters are living, dead, or inanimate; receivers are living organisms. Selection signs indicate positive (+), negative (–), or no effect (0) on the mechanisms in the specific context being considered (see text). "Em" indicates emission mechanisms and "Rm" receiver mechanisms, in the sense of Williams (1966:9) when applicable; otherwise they indicate emission and receiver mode.

within his own halo, is signaling. (A useful discussion of underlying genetic concepts and arguments may be found in Williams's [1966] extremely valuable little book.)

V. FALSE CUES—MIMICRY OF CUES (+/–)

The exploitation of a receiver in which an emission activates mechanisms that evolved and/or are maintained in the context of cuing (Table 1) has, like exploitation by false signals, the plus-minus polarity of selection signs. Selection on the emitter mechanisms favors the enhancement of the deception; selection on the receiver favors the discrimination of false cues from their true cue models. In sequel to the shore bird example, selection will favor those behavioral and morphological traits of crocodiles that promote their adjudication as logs by the reception mechanisms of the birds; and at the same time, in the race of measure-countermeasure, selection favors receiver mechanisms that do not make the mistake of computing crocs as a "something nothing."

Hastings (1971) presented suggestive evidence that ventral luminescence of the pony fish *(Leiognathus equulus)* simulates the shimmery water surface above and makes it difficult for predators below to detect; there are other explanations and considerations (Nicol, 1969: 392). In the special terminology of protective coloration literature (Robinson, 1969:229), this would be a form of eucrypsis, and more specifically homochromy.

Fireflies of some species pupate underground or in dead, rotting logs. They are luminescent and will, upon mechanical stimulation, turn on their lights. These lights may protect the fireflies by activating, in the potential predators that come upon them, the avoidance responses that normally keep the predators from moving into daylight (where desiccation or their own predators might overcome them; Lloyd, 1973d).

Beebe's (1926) description of prey capture by myctophid fish suggests a false-cue context, with the luminescence perhaps mimicking surface illumination: "Five separate times when I got fish quiet and wonted to a large aquarium, I saw good-sized copepods and other creatures come within range of the ventral light, then turn and swim close to the fish, whereupon the fish twisted around and seized several of the small beings."

VI. CLUES—EMISSIONS OF BETRAYAL (–/+)

Virtually every living thing is exploited by some other organism. Exploiters perceive and use emanations of their victims, at the least to make contact with them. It seems reasonable to use the term "clue" for such emissions since in detective fiction parlance a clue is evidence that leads to the undoing of the individual responsible for the clue's production. An insectivorous bird detects and computes visual properties of a caterpillar, then devours the insect. The cylindrical form itself is an important clue to the detection, as evidenced by the various ways it is disguised by protective coloration (Cott, 1957). On the other hand, emissions of predators, such as the circular eye form (Cott, 1957), that reveal their presence to quarry are also clues.

In the early evolution of flash communication in fireflies one selective agent that brought about the shortening of long glows, the simplest emission, into short pulses, could have been a visually orienting predator that approached perched, glowing individuals. Today wolf spiders prey upon fireflies and seem to orient to their lights (Lloyd, 1973b).

VII. FALSE CLUES—BAIT, BOGUS BONANZAS, ENTRAPMENT (+/–)

The exploitation of a receiver in which an emission activates mechanisms that evolved and/or are maintained in the context of cluing (Table 1) has, like exploitation by false signals and false cues, the plus-minus polarity of selection signs. False clues mimic clues. Selection on emitters favors enhancement of the deception, and on receivers favors its discrimination.

The luminescent lures of ceratoid and stomatoid fish may emit false clues by mimicking worms or other prey (Marshall, 1966:176, 303; Harvey, 1952:529). The autotomy of luminous segments by marine annelids (*Polynoe:* Harvey, 1952:208) seems to involve the use of false clues. Haswell's quote (Harvey, 1952:209) says it all:

> When certain species of *Polynoe* are irritated in the dark a flash of [bioluminescent] light runs along the scales, each being illuminated with a vividness which makes it shine out like a shield of light, a dark spot near the centre representing the surface attachment where the light-producing tissue would appear to be absent. The irritation communicates itself from segment to segment, and if the stimulus be sufficiently powerful, flashes of [bioluminescence] may run along the whole series of elytra, one or more of which then become detached, the animal meanwhile moving away rapidly and leaving behind it the scale or scales still glowing with [bioluminescent] light. The species in which the phenomenon of [bioluminescence] occurs are species characterized by the rapidity of their movements, and also by the readiness with which the scales are parted with; and it seems not at all unlikely that the [bioluminescence] may have a protective action, the illuminated scales which are thrown off distracting the attention of the assailant in the dark recesses which the Polynoidae usually frequent.

VIII. NOISE—HINDRANCES FROM INDIFFERENT EMITTERS (0/–)

Physical disturbances in the environment often interfere with the ability of receivers to receive and process the significant energy components that make up such phenomena as signals, cues, and clues. In the simplest meaning of "noise," the physical disturbance is in the same channel (sound, light, etc.) as the masked significant energy, and the noise jams the sensors by producing a background of some intensity.

Bright moonlight, as it shines on the vegetation where firefly females are perched, is a hindrance to males' perceiving luminescent responses of females. It probably also interferes with females' ability to detect flashes of males against the night sky. Selection has certainly resulted in adaptations such as filters, shades, or screens that reduce the reception of moonlight—that improve signal-to-noise ratios. Selection has brought about and maintains behavioral adaptations that result in the elimination of the noisy sun and moon—luminescent fireflies do their signaling after sunset and are far less active on moonlit nights.

IX. FALSE NOISE—EMITTERS WITH VESTED INTEREST (+/–)

False noise is emitted by living organisms that exploit or somehow place the receiver at a disadvantage and benefit themselves. It contrasts with noise in that the latter may be emitted by living, nonliving, inanimate, or dead emitters—in the case of noise, no advantage can or does accrue to the emitter. With false noise the emission interferes with the functioning of the receiver because it, like noise, produces a background energy level that disrupts the reception of important (to the receiver) energy phenomena.

Upon tactile stimulation the squid *Heteroteuthis dispar* discharges masses of mucus that become brilliantly luminescent. It has been suggested that this discharge may "baffle" pursuers and that it would be "disconcerting" to predators (Lane, 1960:107, 112). The luminous discharge may function in different modes of predator defense. If the secretion maintains cohesion and is mistaken for the body of the squid, while the squid itself escapes, we have a false clue. If the luminous cloud prevents the squid's detection because the resulting increased background light level exceeds the intensity of light clues (luminescent or reflected) that the predator would use for attack orientation, the luminous discharge is a false noise.

X. AMBIGUOUS SIGNALS—MISTAKEN IDENTITY WITH MUTUAL DETRIMENT (–/–)

In this interaction the receiver is negatively affected because the emission that is received mistakenly activates mechanisms that belong to another signaling context or species. In the subsequent interactions the emitter is also harmed. An important class of this category results in interspecific mating. Selection can be expected to improve mating-signal discrimination as well as to intensify existing but minor signal differences between species living in the same area and active at the same time. Such interactions commonly occur when sibling species[2] come into contact following (extrinsic) isolation that has led to their speciation and probably had an important role in the development of the timing differences found in mating signals of several species of fireflies (Fig. 1).

XI. INDIFFERENT FAZERS—DYSFUNCTION CAUSED BY INDIFFERENT EMITTERS (0/–)

The received energy of an emission, because of its intensity or chemical properties, may cause temporary or permanent impairment of function of receiver mechanisms. The response is not an evolved one—i.e., it is pathological. If the interaction is of no consequence to the emitter, selection will act only on the receiver. Bright rays of a flashlight will occasionally cause a flying firefly to become disoriented and spiral to the ground, temporarily out of competition for a mate. If an average male lives for but two evening flights, each thirty minutes long, and he loses ten minutes of flight time, he realizes an appreciable loss. Upon crashing to the ground, if he gets caught by a predator or trapped in a pool of water or in a spider web, his loss is even greater. Lightning flashes may have once brought about similar dysfunction, and selection could already have resulted in, and presently maintain, protective mechanisms for dealing with brief, high-intensity light bursts.

XII. FAZERS—EXPLOITATIVE DYSFUNCTION (+/–)

The effect of this emission on the receiver is the same as in the above category, but the emit-

2. Species that share an exclusive common ancestor and are difficult to distinguish on "conventional" grounds because of their homologous and extensive similarities.

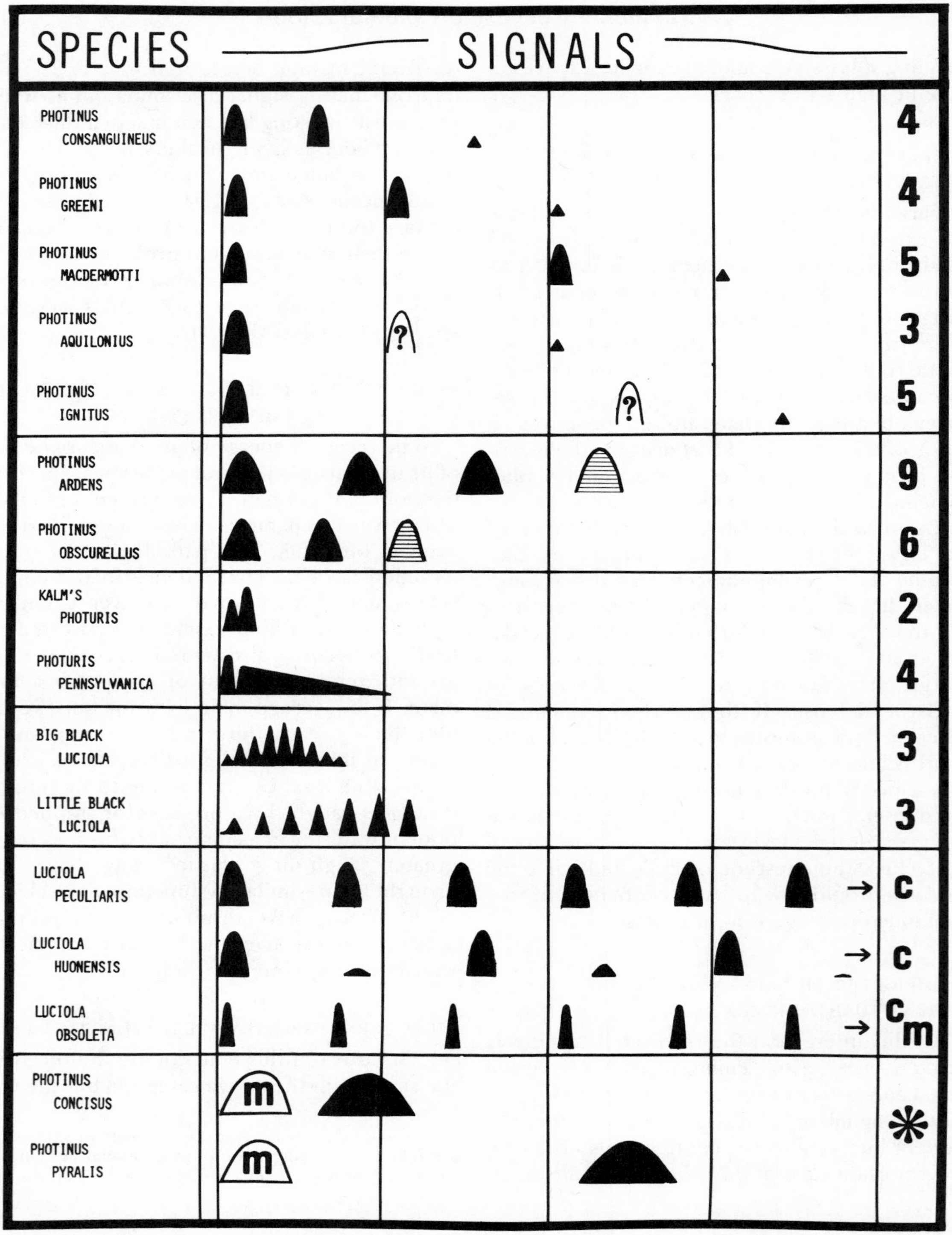

SPECIES	SIGNALS	
PHOTINUS CONSANGUINEUS		4
PHOTINUS GREENI		4
PHOTINUS MACDERMOTTI		5
PHOTINUS AQUILONIUS	?	3
PHOTINUS IGNITUS	?	5
PHOTINUS ARDENS		9
PHOTINUS OBSCURELLUS		6
KALM'S PHOTURIS		2
PHOTURIS PENNSYLVANICA		4
BIG BLACK LUCIOLA		3
LITTLE BLACK LUCIOLA		3
LUCIOLA PECULIARIS		→ c
LUCIOLA HUONENSIS		→ c
LUCIOLA OBSOLETA		→ cm
PHOTINUS CONCISUS	m	*
PHOTINUS PYRALIS	m	

ter is not indifferent to the effects upon the receiver—it causes them and capitalizes on them.

The ventral surface of the dogfish shark *(Spinax niger)* is covered with light organs. C. F. Hickling (1928, in Harvey, 1952:498) made the following observations and speculations, the latter of which place this predatory "tactic" here:

> When a luminescent specimen is held so that one's line of vision is perpendicular to the ventral surface of the fish, the luminescence is plainly visible. When the fish is then rotated slightly to left or right about its long axis, the light disappears. This observation seems to offer some explanation of the function of the luminescence of *Spinax.*
>
> The complex lantern-like structure of each individual organ seems designed to throw out a parallel beam of light, and to prevent scattering of the rays; the arrangement of the axes of all the organs parallel to the median vertical axis of the fish, seems to aim at precisely the effect described above, namely, that the luminescence will only shine upon objects immediately beneath the ventral surface.
>
> The mouth of *Spinax* is situated remarkably far behind the tip of the snout, so that *Spinax* can obviously only seize objects immediately beneath (in the relative sense) its mouth. But it is only when an object is immediately below the ventral surface of the fish that the light from the luminous organs flashes fully upon it. One may therefore suggest that the sudden flash of light, at the moment of attack, may cause the prey of *Spinax* to hesitate for just that fraction of a second in which the mouth can make a successful snatch.

Fig. 1. Firefly signals. Fireflies within each section are closely related. Their signals have diverged since speciation, probably in many cases because of inappropriate matings. Horizontal axis = time; vertical lines mark one-second intervals. Black symbols at left are male advertising flashes. Numbers indicate approximate length of intervals between consecutive male advertisements (in sec). Common names have been used for species not having latin binomens.

P. consanguineus group (eastern U.S.): The male pattern in three species is composed of two flashes, and the differences in their timing is critical and prevents interbreeding. Female answers occur 1 sec (black triangles) after the second flash of these patterns. Two other species in the group have single flashes and long female delays. The latter species may have recently lost the second flash. If one presumes the 1-sec female delay to have been derived from a common ancestor (it occurs in four of the six studied species of the group), the position of the "lost" male flash (hollow symbols with question mark) can be inferred on the basis of the timing of the female answers. The hypothetical ancestral signals are very much like signals of extant species. Unique codes among the species of this group may have been achieved by timing changes and flash losses.

P. ardens group (northeastern U.S.): Male advertisement patterns vary within individuals as indicated by solid and lined symbols. Note the mathematical relationship between the two species with respect to flash length, number, and period, and advertisement period. Since speciation have these parameters changed independently, or together as some basic component of their nervous systems was altered?

P. pennsylvanica group (northeastern U.S.): The timing of the two patterns is similar but flash duration and relative intensity differ.

Black luciola group (New Guinea): The flicker frequency of the little black luciola is about half that of its relative. The fireflies of the *L. peculiaris* group (New Guinea) emit continuous (c) trains of flashes. Males of the first two species give this pattern as they fly in search of mates, and they are attracted to penlight flashes that occur immediately after each of their flashes. The feeble nature of alternate flashes in *L. huonensis* suggests that these are being lost and that originally the timing was more like that of *L. peculiaris.* The flashes of *L. obsoleta* have a similar period, but they are even more interesting from another standpoint: *L. obsoleta's* signaling system is completely different, and these flashes do not appear in courtship until the male has reached the female and mounted her. The stages of courtship that precede mounting include aggregation of many individuals, sedentariness with flickering and flashing unlike the mounting (c_m) flashing pattern, aerial chases, and finally an intimate walking-luminescing interaction. If the c_m pattern is homologous with the flash trains of its relatives, how and whence have the complex preliminaries of *L. obsoleta* gotten into the act?

P. pyralis group (Texas): The structure of the flashes of males and females of both species are similar; it is the female response delay that prevents interbreeding. *Hollow symbols marked m indicate male advertisement flashes; solid symbols are the female answers.

XIII, XIV. SCANNERS AND LOCATORS—SELF-SIGNALS ONE WAY, CLUE PRECURSORS (+/0,+/–)

In the self-signal category (II above) the emitter and receiver mechanisms were in the same individual. The two categories considered here are identical with the self-signal during the outward leg of its propagation. In many situations the objects that are impinged upon are not really receivers—they have no receiver mechanisms and they merely intercept passing emissions (*scanners,* +/0). The *locator* (+/–) emission is a scanner that has been incorporated into a prey-tracking system and is of significance to both emitter and "receiver." It is important to consider these two categories, even though they do not involve true receivers, because of their evolutionary relationship to emissions and interactions that are of interest. A scanner becomes a locator—bats used their cries for orientation before they used them for tracking prey—and the selective pressure that a locator exerts on prey can lead to the development of receiver mechanisms in the prey. For example, a portion of the moth body that originally reflected the locator back to the bat ultimately became a detector (tympanum) in the moth. When the locator emission began to be detected and used by members of moth populations, it ceased to be solely a locator, but part of it also became a clue and balanced the selection signs—i.e., locator +/– versus clue –/+—of the bat-moth interaction.

It is common among some marine organisms such as squids, deep-sea fishes, and shrimps (Euphausiidae) to have photophores around, on, or in the eyes, and it has been suggested that these function in illumination. The eye lanterns of some squids and fishes shine into the eyes (Harvey, 1952:287; Lane, 1960:73). It would be interesting to know if the latter actually shine upon photosensitive tissue, or if they are instead reflected from a tapetum that somehow aims or focuses the light rays and permits accurate range or direction-finding of nearby prey.

XV–XVIII. QUASI EMISSIONS AND CIPHERS (0/–; 0/0)

Additional categories are theoretically plausible, if not always empirically identifiable. Quasi emissions are of no significance to emitters (0/–) in the contexts being focused upon. They inappropriately activate receiver mechanisms that belong to a diversity of contexts. Stray emissions could be mistaken for signals, clues, or cues (Table 1), and receivers could react to them as though they were the real thing.

For example, I once observed a male *Photuris congener* in the grasp of a large lycosid spider. While the spider devoured him, he continued to emit a flashing pattern that resembled that of *congener* females, and he attracted two more males. When the decoyed males arrived the spider seized them and subsequently ate them too. The consequences of the emission were irrelevant to the first captive and unfortunate for the respondents that mistakenly accepted quasi signals as signals. As a second example, consider the plight of the tiny male of the reticulate firefly *(Phausis reticulata)* that was attracted to the glow of a single lantern of the gigantic, relatively speaking, larviform female of the plumose glow-worm beetle (*Phengodes plumosa:* McDermott, 1958:15). The activated neural circuits of both individuals in this case were appropriate to a signal context. The male lost time and energy, but apparently the female, hardly more than tickled by the attentive male, merely continued to glow and advertise, unaffected—another example of a quasi signal (0/–).

COMBINATIONS

An emitter has simultaneous interactions of different kinds with other organisms. The flash of a female firefly given in response to the flash pattern of a male of her species is a signal, in that context; to a female perched below in dense vegetation that mistakenly flashes an answer it is a quasi signal; but if selection has favored this sort of response in females, because in a percentage of such circumstances they get an opportunity to get a mate that they otherwise would not have seen or begun courtship with, it is a clue. To an immigrating female firefly that, as a consequence of seeing the interaction, identified the locale as a potential oviposition site, it would be a cue.

Fig. 2. *Lycosa rabida* eating *Photuris* sp. Lycosid spiders are important predators of fireflies, and evidence suggests that they seek grounded luminescing individuals. Predators behaving in this manner could have brought about the evolution of the flash from the glow. The flashes of the firefly in the context in which the spider was the receiver were clues; the firefly's "intended" receiver may have been itself (self-signals), a mate (signals), or prey (false signals).

Fig. 3. Firefly melee. A grappling, struggling group of four fireflies, including two agressive mimic females and two prey males. Probably one female attracted both males (false signals) and the second female observed the interaction and attacked (clue?).

The wolf spider (*Lycosa rabida*) in Fig. 2 is grasping and sucking on a *Photuris* firefly that it probably located by cluing on the firefly's illumination self-signals, predatory false signals, or mating signals. The bunch of 4 fireflies in Fig. 3 was composed of two *Photuris* males of one species and a female of each of two other *Photuris* species. I found this group by investigating bright flares emitted by one of the males—the flares are characteristic emissions (function unknown) of these males when they are seized by predators. I believe that one of the females attracted both males by answering them with false signals, and that the other female observed the combination of emissions of a male and the female and launched her attack.

Male fireflies recognize courtship interactions taking place between other individuals (Lloyd, 1969, 1973c) and probably use the information for exploitation. In such cases there are two emitters, and both are essential for the receiver to obtain the information. In the case of sexual interloping, it is the female response to the other male that gives the interloper the essential information for identifying what is happening; the male pattern that had been observed was by itself not significant and was stored information at the time the female flash was emitted. With respect to the voyeur male, to what categories do the emissions of the courting pair belong? The courting male's pattern, when received by the voyeur, is a clue when the outcome of this is competition that diminishes his chances to copulate with the female. In some species selection has resulted in reducing this clue, and courting males greatly reduce the intensity of their mating flashes when they approach answering females. The category or identity of the flash of the courted female that reaches the voyeur male is unknown since its significance to the female is not understood. In attracting a second male either she might get the opportunity to select the "better" male (i.e., the one that would genetically bestow upon her sons the better attributes with respect to mate competition) or she might decrease her chances of getting either male because of her longer exposure to predators, getting knocked from her perch, or losing contact with both males, as a consequence of their fighting after both reached her. Until the statistical probabilities of these outcomes are known the interactions cannot be classified.

Similar uncertainties are involved in cases of the flare flashes of males captured by aggressive mimic females or spiders. If in a significant percentage of the encounters these flashes attract additional predators (and in a predator melee the firefly sometimes escapes), the selection signs for the prey–second-predator emissive-receptive interaction will be different from what it would be if the flare flash were simply a consequence of a poison that has a pathological effect on flash physiology.

Interacting organisms can have different interactions with each other simultaneously. The echo-locating cry of a bat is a locator when it strikes the body of the moth and the portion of the bat's emission that strikes the moth tympanum provides a clue for the moth.

EVOLUTIONARY RELATIONSHIPS

Quite obviously the interactions observed between organisms today have not always existed, but got that way via various evolutionary routes. Through mutual adaptation and unilateral exploitation and escape, interactions of one kind have passed, through time, into others. The fazer of the dogfish, before it became so intensely bright or well aimed, may have functioned in illumination of the sea-bottom (self-signal), or attracted prey (false cue?), or, like the light of the pony fish, have been used for hiding (false cue). It has perhaps also become involved in signaling systems. The false signals of *Photuris* females were probably derived from at least two sources, self-signals and their own sexual signals. An expected evolutionary source of false clues, such as those the self-fragmenting polynoid worm scatters in its wake when attacked, would be the clues their ancestors had emitted, to their detriment, in earlier times. Certain signals that fireflies use in courtship may have been adapted from illumination luminescence (self-signals), but the origin of firefly luminescent signaling, if the universality of larval luminescence among the lampyrids can be used as an indication, is obscured by our ignorance of the present function of larval lights. The most comprehensive statement that can be made about the origins and functions of bioluminescent emissions is that nothing is known of most of them, little is known of a few of them, and we have not learned of the best of them.

References

Beebe, W., 1926. *The Arcturus Adventure.* New York: Putnam. 290pp.

Brown, F. A., Jr., 1973. Bioluminescence. In: *Comparative Animal Physiology,* 3d ed., C. L. Prosser, ed. Philadelphia: W. B. Saunders, p.965.

Bullock, H. T., 1973. Seeing the world through a new sense: electroreception in fish. *Amer. Sci.,* 61:316–25.

Cott, H. B., 1957. *Adaptive Coloration in Animals.* London: Methuen. 508pp.

Ewan, J., and Ewan, N., 1970. *John Banister and His Natural History of Virginia 1678–1692.* Urbana: University of Illinois Press. 487pp.

Farnworth, E. G., 1973. Flashing behavior, ecology and systematics of Jamaican lampyrid fireflies. Ph.D. diss. University of Florida.

Fulton, B. B., 1939. Lochetic luminous dipterous larvae. *J. Elisha Mitchell Sci. Soc.,* 55:289–93.

Gatenby, J. B., and Cotton, S., 1960. Snare building and pupation in *Bolitophila luminosa. Trans. Roy. Soc. New Zealand,* 88:149–56.

Girod, P., 1882. Recherches sur la poche du noir des cephalopodes des cotes de France. *Arch. Zool. exp. gen.,* 10:1–100.

Harvey, E. N., 1952. *Bioluminescence.* New York: Academic Press. 649pp.

Hastings, J. W., 1971. Light to hide by: ventral luminescence to camouflage the silhouette. *Science,* 173:1015–17.

Lane, F. W., 1960. *Kingdom of the Octopus.* New York: Sheridan House. 299pp.

Lloyd, J. E., 1965. Aggressive mimicry in *Photuris:* firefly femmes fatales. *Science,* 149:653–54.

Lloyd, J. E., 1966. Studies on the flash communication system in *Photinus* fireflies. *Univ. Mich. Mus. Zool. Misc. Pub.,* no. 130. 95pp.

Lloyd, J. E., 1968. Illumination, another function of firefly flashes. *Entomol. News,* 79:265–68.

Lloyd, J. E., 1969. Flashes, behavior, and additional species of nearctic *Photinus* fireflies (Coleoptera: Lampyridae). *Coleop. Bull.,* 23:29–40.

Lloyd, J. E., 1971. Bioluminescent communication in insects. *Ann. Review Entomol.,* 16:97–122.

Lloyd, J. E., 1972. Chemical communication in fireflies. *Environ. Entomol.,* 1:265–66.

Lloyd, J. E., 1973a. Fireflies of Melanesia: bioluminescence, mating behavior and synchronous flashing. *Environ. Entomol.,* 2:991–1008.

Lloyd, J. E., 1973b. Firefly parasites and predators. *Coleop. Bull.,* 27:91–106.

Lloyd, J. E., 1973c. A model for the mating protocol of synchronously flashing fireflies. *Nature,* 245:268–70.

Lloyd, J. E., 1973d. Fireflies—commonplace beetles and larvae by day but tiny flashing lanterns on summer evenings. *Animals,* 15:220–25.

Lloyd, J. E., 1974. Bioluminescent communication between fungi and insects. *Florida Entomol.,* 57:90.

Lloyd, J. E., 1975. Aggressive mimicry in *Photuris* fireflies: signal repertoires by femmes fatales. *Science,* 187:452–53.

McDermott, F. A., 1958. The fireflies of Delaware. *Soc. Nat. Hist. Delaware.* 36pp.

McElroy, W. D., and Seliger, H. H., 1962. Biological luminescence. *Sci. Amer.,* Dec., 76–89.

Marshall, N. B., 1966. *The Life of Fishes.* New York: World Publishing Co. 402pp.

Nicol, J. A. C., 1969. Bioluminescence. In: *Fish Physiology,* W. S. Hoar, and D. J. Randall, eds. New York: Academic Press. pp.355–99.

Norman, J. R., 1947. *A History of Fishes.* London: Ernest Benn, Ltd. 463pp.

Odum, E. P., 1959. *Fundamentals of Ecology,* 2d ed. Philadelphia: W. B. Saunders Co. 546pp.

Otte, D., 1974. Effects and functions in the evolution of signaling systems. *Ann. Rev. Systematics and Ecology,* 5:385–417.

Rivers, J. J., 1886. Description of the form of the female lampyrid (*Zarhipis riversi* Horn). *Amer. Natur.,* 20:648–50.

Robinson, M. H., 1969. Defenses against visually hunting predators. In: *Evolutionary Biology, 3,* T. Dobzhansky, M. Hecht, and W. Steere, eds. New York: Appleton-Century-Crofts. 309pp.

Sebeok, T. A., ed., 1968. *Animal Communication.* Bloomington: Indiana University Press. 686pp.

Seliger, H. H., 1975. The origin of bioluminescence. *Photochemistry and Photobiology,* 21:355–61.

Shawver, L. J., 1974. Trilobite eyes: an impressive feat of early evolution. *Sci. News,* 105:72–73.

Tiemann, D. L., 1967. Observations on the natural history of the western banded glowworm *Zarhipis integripennis* (LeConte). *Proc. California Acad. Sci.,* 35:235–64.

Waller, R., 1685. Observations of the *Cicindela volans,* or flying glowworm. *Phil. Trans. Roy. Soc.,* 15:841–45.

Wickler, W., 1968. *Mimicry in Plants and Animals.* New York: McGraw-Hill. 155pp.

Williams, G. C., 1966. *Adaption and Natural Selection.* Princeton: Princeton University Press. 307pp.

Chapter 9

COMMUNICATION BY REFLECTED LIGHT

Jack P. Hailman

Optical communication is found in many species with well-developed vision and sociality, especially insects, crustaceans, cephalapods, fishes, amphibians, reptiles, mammals, and birds. Although a few animals generate their own light (see chapter 8 of this volume), most signal by modulating reflected sunlight. This short review concentrates on traditional concerns of ethology: the origins and structure of signals and the dynamics of the signaling process. A broader and more speculative discussion of optical signals is available as a monograph (Hailman, in press). To keep documentation short, examples are drawn chiefly from studies subsequent to Sebeok (1968), which may be used as an entry into older literature.

There is no unambiguous way for the casual observer to know that two animals are communicating. I believe that certainty requires comparison of two situations: one in which a reputed signal is given and another one in which it is not, other things being equal. Then if the behavior of the reputed receiver differs in the two situations, one can state operationally that communication occurred. In other words, communication is shown by the correlation between a difference in the behavior of the reputed sender and the reputed receiver. This chapter primarily considers conspecific animals sending signals to one another: reciprocal social communication.

It is difficult to specify what constitutes a visual signal. Virtually anything about an animal that can be perceived by another can be a signal: slight tensing of muscles, an action performed out of its usual sequence, or just normal, ongoing behavior. Even when a behavioral pattern obviously generates stimuli that affect another animal's behavior, one cannot be certain what aspect of the stimulus is effective. Many visual stimuli accompany the production of sound, marking by scent, or generation of tactile stimuli. We humans are so visually oriented that the tail slap of a fish may seem to be a visual signal until we realize that the displacement wave it causes is readily detected by another fish's lateral line system. I refer for convenience to behavioral patterns and morphological structures as signals, although it is actually the stimuli they generate that constitute the signal.

It is useful to discuss determinants of visual communication under the four classes of causes and origins that apply generally to behavior (Tinbergen, 1963; Hailman, 1967; Klopfer and Hailman, 1967). One may ask how a behavioral system works in the immediate sense of the relations among external inputs, internal mechanisms, and behavioral outputs—the determinants of *dynamic control*. The control system of a particular individual at a particular time of its life is structured by the past interactions of the organism with its environment, keyed by the organism's genetic endowment from its parents. These *ontogenetic origins* may involve particular experiences during development, or the ultimate control system may develop relatively similarly regardless of experience and environment. In turn, the genetic endowment of individuals in a population is structured by natural selection acting on phenotypic variation. Genes leading to beneficial phenotypes are preserved and those leading to other phenotypes are trimmed from the population: the *adaptive function* of a behavioral pattern depends on such natural selection. Finally, the evolutionary history of selection acting on the population also determines behavior, since the *phylogenetic origins* impart a directional impetus to evolutionary processes. One must ultimately understand all these behavioral determinants—control, ontogeny, function, and phylogeny—to understand communicative behavior.

Control of Visual Communication

A simple sort of reciprocal visual communication takes place as follows: Animal A sends a signal to animal B, and then animal B performs some act that constitutes a signal back to animal A. Animal A then responds with a second signal, and so on, with each receiving and replying in turn. It is the vogue to analyze animal communication as if that were the communicative interaction, but there seems to be no truly convincing case of animal communication's working in this way.

There are relatively few signals that yield a discrete and constant reply in the recipient. Indeed, the general lack of strict correlation between stimulus and "response" in animal behavior constantly challenges the ethologist. The lack implies either that the receiver is somehow different at different times of receipt or that the signal's effect depends on external factors apart from the signal itself. Both complications usually apply, the first being subsumed under "motivational" factors and the second under the "context" of communication (Smith, 1965, 1968). Therefore, recording the mere exchange of signals between two animals yields an incomplete picture of communication.

Another criticism of the simplistic model of communication was recently articulated by Schleidt (1973), who points out that many signals have long-lasting, not merely immediate, effects. One animal may incessantly signal to another to maintain some state of readiness in the recipient. There may be no obvious exchange of signals even though important communication transpires. When the recipient does deliver an identifiable signal it may be produced by the totality of ongoing external and internal processes —not merely in response to the other animal's immediately previous signal.

I skirt the difficulties in attempting to provide a complete framework for the communication control and concentrate instead on the visual signal itself: what kind of signals are utilized by animals and how are they to be descriptively classified?

EXTRINSIC AND INTRINSIC VISUAL SIGNALS

One may divide visual signals into those that have a physical existence apart from their creator, the sender (extrinsic signals), and those that are part of the animal itself (intrinsic signals). Bowerbirds decorate display structures called bowers with brightly colored objects such as flower petals (Marshall, 1954) (see Fig. 1). Many physical objects created or rearranged by one animal may serve as visual signals to another. Tracks in snow or mud, nests or burrows, browse

Fig. 1. Examples of extrinsic visual signals. Left: Bower of a regent bowerbird (*Sericulus chrysocephalus*) decorated internally with palm seeds; like other avenue builders, this species paints the inside walls with juice of plants and its saliva, but other species have far more elaborate display structures (after Marshall, 1954:plate 17). Right: A blackbuck (*Antilope cervicapra*) creating a visual mark of its territory by sweeping its horns through the grass. (After Schmied, 1973:165.)

marks or other evidence of feeding all could act as visual signals. Olfactory marking by dung or by rubbing on the bark of trees is common in mammals, but without experimentation one cannot know if the visual component of such signals is important. The blackbuck brushes its antlers through grass as a form of visual place-marking (Schmied, 1973) (Fig. 1), and red squirrels stack spruce cones as visual marks of their territorial boundaries (Kilham, 1954). Printed words such as these constitute a complex form of extrinsic visual signaling.

More extensively studied are the intrinsic or behavioral visual signals used by animals: postures, gestures, and other aspects of behavior that generate visual stimuli.

DIMENSIONS OF INTRINSIC VISUAL SIGNALS

Three primary dimensions describe an intrinsic visual signal: the orientation of the signaling animal or some part of it with respect to the intended receiver; the shape, or configuration, of the animal, which is the relative orientation of its parts; and the movement patterns of the animal or its parts. Some signals depend primarily on one dimension, whereas others utilize various combinations.

Mere orientation of one animal relative to its conspecifics constitutes an important visual signal in many animals (e.g., Scruton and Herbert, 1972; Figler, 1972; Stanley, 1971; Golani and Mendelssohn, 1971; Dunham, 1966), as illustrated in Fig. 2. Baylis (1974) notes that orientation of a cichlid fish with respect to the environment also carries information, an example of context, mentioned previously.

Fig. 2. Examples of orientational visual signals. Left: A male jackal (*Canis aureus*) stays behind and slightly to the side of a female during precopulatory behavior (after Golani and Mendelssohn, 1971:Fig. 35). Right: Rose-breasted grosbeak (*Pheucticus ludovicianus*) facing its opponent by turning the head to the side in a resting posture. (After Dunham, 1966:164.)

Specific shapes or postures also act as visual signals, even when there is no special orientation of the signaling animal toward its conspecifics. There are at least three mechanisms employed to create bodily shapes: motor adjustments of bodily parts such as appendages in different relations with one another (e.g., Tinbergen, 1959;

Fig. 3. Examples of signal shapes. Lett: A wildebeest (*Connochaetes taurinus*) assumes the "static-optic advertising display" by postural adjustments of body parts. (After Estes, 1969:313.) Right: A collared peccary (*Dicotyles tajacu*) creates the "intense curiosity" shape primarily through piloerection along its back. (After Schweinsburg and Sowls, 1972:142.) A third method of creating a specific shape is inflation of body parts with air.

Barash, 1973; Hall and Miller, 1968; R. R. Phillips, 1971; Spinage, 1969; Fox, 1969); pilomotor responses of fur and feathers (e.g., Fox, 1969; McBride et al., 1969; Rood, 1972; Krämer, 1969; Schweinsburg and Sowls, 1972; Ewer, 1971; Schmidt and van de Flierdt, 1973); and inflation of structures with air (e.g., Evans, 1961; Carpenter, 1963; Kahl, 1966); examples appear in Fig. 3. The wildebeest may hold the record static posture: it stands in the "static-optic advertising" shape for up to an hour (Estes, 1969).

Movements that occur without specific orientation or body shape may be difficult to recognize as visual signals. There are two classes: movements of the entire animal, such as incipient locomotion (Daanje, 1950; Andrew, 1956) and other movements (e.g., Walter and Hamilton, 1970; Figler, 1972; Saayman et al., 1973; Rood, 1972); and movements of part of an animal, such as its tail or an appendage (e.g., Smythe, 1970; Saayman et al., 1973; Spinage, 1969; Barash, 1973; Quanstrom, 1971; Cole and Ward, 1969; LaFollette, 1971). Some examples are shown in Fig. 4.

Fig. 4. Examples of signal movements. Left: "Rolling" by an Argentinian cavy (*Galea musteloides*) creates a visual signal by gross body movements. (After Rood, 1972). Right: "Tail flicking" of the Richardson's ground squirrel (*Spermophilus richardsonii*) creates a visual signal by movement of a body part, in which the tail is raised, lowered part way, moved in a circle, and then lowered to the ground. (After Quanstrom, 1971:647.)

Two of the three dimensions—orientation, shape, and movement—are often used together as elements of a unitary signal. Oriented static postures are common (e.g., Andrew, 1957; Estes, 1969; Steiner, 1971; Packard and Sanders, 1971; Wells and Wells, 1972; Frey and Miller, 1972; Dubost, 1971). Movement and shape are less-commonly combined (e.g., Walther, 1964, 1969; Estes, 1969). The combination of orientation and movement without a body shape different from normal is extremely common (e.g., Reese, 1962; Hazlett and Bossert, 1965; Hazlett and Estabrook, 1974a, 1974b; Hunsaker, 1962; Carpenter, 1963; Kühme, 1961; Otte, 1972; Markl, 1972; Rovner, 1968). Movement of part of the body does, of course, alter the animal's shape so that it becomes an empirical question as to whether shape is an important part of the signal. In other cases, the simple approach of one animal toward another involves no special shapes (e.g., Frey and Miller, 1972; Figler, 1972; Saayman et al., 1973; Sturm, 1973; Shank, 1972). Examples of combinational signals of two dimensions are given in Fig. 5.

The majority of intrinsic visual signals probably combine all three dimensions of orientation, shape, and movement. Many human facial signals are so composed, having orientation toward the intended receiver, changes in the shape of

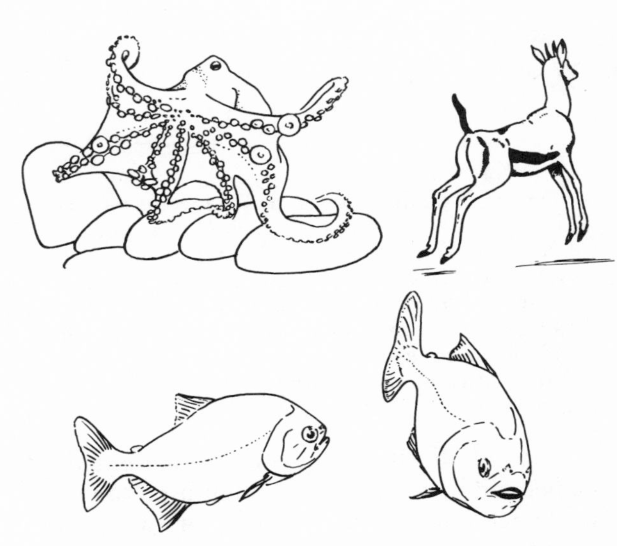

Fig. 5. Examples of combinational pairs of signal elements. Top left: A male Mediterranean octopus (*Octopus vulgaris*) assumes the "sucker display" involving a special posture and orientation toward the female, but no movement. (After Wells and Wells, 1972:300.) Top right: "Stotting" by a Thomson's gazelle (*Gazella thomsoni*) involves a special posture and movement, but no apparently specific orientation toward the intended receiver. (After Walther, 1964:872.) Bottom: "Transverse approach" by the piranha (*Serrasalmus nattereri*) involves oriented movement toward the opponent without a change from normal body shape. (After Markl, 1972:192.)

the mouth, eyes, or eyebrows, and dynamic movements such as eyelash fluttering or laughing. Examples of "three-dimensional" visual signals among animals are extremely common (e.g., van Lawick-Goodall, 1968; Andrew, 1963; Bovet, 1972; Stamps and Barlow, 1973; Zumpe and Michael, 1970; Lorenz, 1958); see Fig. 6.

In most studies of visual signals all the above types are described or implied. Examples come from all kinds of animals from cephalopods and arthropods to fish, birds, and mammals (e.g., van Rhijn, 1973; McBride et al., 1969; Krämer, 1969; Myrberg, 1972; Packard and Sanders, 1971; Ewer, 1971; Hall and Miller, 1968; Dingle and Caldwell, 1969; Gibson, 1968; Ewing and Evans, 1973; Spivak, 1971; Potts, 1973; R. E. Phillips, 1972; Wells and Wells, 1972; Kleiman and Eisenberg, 1973; Albrecht, 1969; Kahl, 1972; Winkler, 1972; Franck, 1968; Simpson, 1968; Tinbergen, 1953, 1959; Hinde, 1955/6; Andrew, 1957, 1963).

MORPHOLOGICAL ELEMENTS OF INTRINSIC VISUAL SIGNALS

Darwin (1871) observed that many animals possess elaborate morphological elements that are prominently displayed before conspecifics. Indeed, the ethological concept of a "display" stems directly from such phenomena. The elements may be structural shapes or specializations of the light-reflecting surface. These elements are so well known that only a few remarks are necessary.

Fig. 6. Examples of visual signals combining shape, orientation, and movement simultaneously. Left: Male chimpanzee (*Pan troglodytes*) "brandishing a stick prior to throwing it towards his mirror image." (After van Lawick-Goodall, 1968:240.) Right: The "fan" display of an anole lizard (*Anolis aeneus*), in which the legs are extended such that the head is raised, the dewlap is extended down, and the posture is displayed laterally to a conspecific. (After Stamps and Barlow, 1973:69.)

Structural elements, which alter the animal's shape, include those used for other functions as well as those that appear specialized for signaling. The former include horns and antlers of many ungulates, structures that intrigued Darwin and have recently been reviewed by Geist (1966). Used physically in fighting, especially among males during the breeding season, these weapons (Fig. 7) are also used as visual signals. Some elaborate plumages of birds, on the other hand, seem not to be used except for visual display. Wattles, crests, and other elements, although also possessing certain special surface structures for reflection, impart quite different shapes to various birds. Other such signals include extendible throat pouches in lizards (e.g., Crews, 1975), swordtails on fish (e.g., Hemens, 1966; Franck and Hendricks, 1973) and ear tufts on certain cats (e.g., Kleiman and Eisenberg, 1973); examples are shown in Fig. 8.

Fig. 7. Examples of structural elements secondarily elaborated for visual signaling. Top: Teeth and facial warts of a suid pig (*Phacechoerus*), showing structures greatly elaborated for visual display. (After Geist, 1966:194.) Bottom: Horns in the bovid sheep (*Ovis dalli*) on left and antelope (*Antelope*) on right, elaborated in different ways for visual display. (After Geist, 1966:203.)

Fig. 8. Examples of structural elements evolved primarily for visual signaling. Top: The swordlike tail of the swordtail (*Xiphophorus hellerii*), used in various displays. (After Hemens, 1966:293.) Bottom: Ear tufts of the caracal (*Felis caracal*), cited by Kleiman and Eisenberg (1973:646) as enhancing the "ear-flipping" signal. (Drawn from a photograph.)

Surface elements that reflect light in particular ways are divisible along two dimensions: the kind of stimulus they create and the relative permanence of the particular reflection. These elements may create specific brightness contrasts, colors, shapes, or orientations of shapes (Fig. 9).

Surface elements may also be classified according to their relative permanence along a continuum: permanent coloration, such as that of the zebra's stripes or the cardinal's red feathers (also see Noble, 1936); labile coloration of seasonal or relatively long duration, such as the starling's yellow beak during the breeding season (see also Marler, 1955; N. G. Smith, 1966); and modulated coloration that can be changed within the course of a single day, sometimes

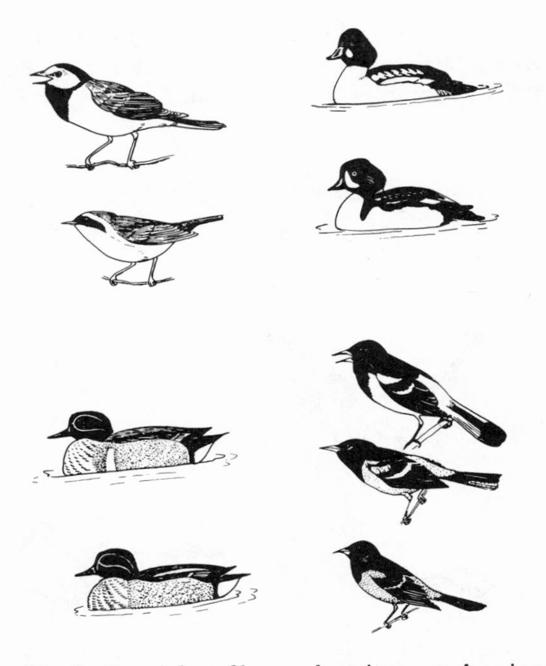

Fig. 9. Examples of how coloration encodes visual information. Top left: Contrast polarity is opposite in the male hooded warbler (*Wilsonia citrina*), which has a yellow mask on a black head, and the male yellowthroat (*Geothlypis trichas*), which is a warbler with a black mask on the yellow head. Top right: Differences in shape are evident in the white facial spot of the common (*Bucephala clangula*) and Barrow's goldeneye (*B. islandica*) male ducks. Bottom left: Differences in orientation of similar color markings are evident in the American green-winged teal (*Anas crecca carolinensis*), which has a vertical white stripe, and the European common teal (*A.c. crecca*), in which the white stripe is horizontal. Bottom right: Even when patterns of coloration are similar the color itself may differ, as in these three species of orioles (*Icterus*), from top: the yellow Scott's (*I. parisorum*), orange Baltimore (*I. g. galbula*), and russet orchard oriole (*I. spurius*). The communicative functions of coloration patterns in this figure have not been studied experimentally.

within seconds, such as blushing in humans and color changes in the octopus (Packard and Sanders, 1971). Cephalopods are particularly adept at rapid color changes (see also Wells and Wells, 1972; Warren et al., 1974), but bony fish may hold the record for both rapidity and diversity of modulated colors (e.g., Frey and Miller, 1972; Figler, 1972; Wickler, 1969; Myrberg, 1972; Markl, 1972; Sale, 1971; Gibson, 1968; Ewing and Evans, 1973; Machemer, 1970; Hamilton and Peterman, 1971; Keenleyside, 1972; Albrecht, 1969; Apfelbach and Leong, 1970; Noakes and Barlow, 1973). Some examples of colorations of varying permanence are illustrated in Fig. 10. Baylis (1974) provides an ex-

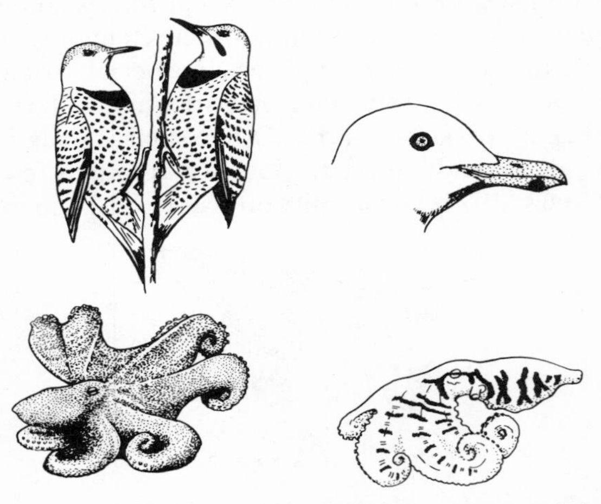

Fig. 10. Examples of relative permanence of signal coloration. Top left: The common flicker (*Colaptes auratus*) has many permanent plumage markings that may be visual signals. The female (left) lacks the male's moustache mark, which is black in some parts of the species' range and red in others. Noble (1936) showed the moustache mark to be a critical visual signal in sex recognition. Top right: The Kumlien's gull (*Larus glaucoides kumlieni*) possesses a red eye-ring and red beak-spot, both of which are dull during the nonbreeding season but intensify in color during breeding. N. G. Smith (1966) showed that eye-ring color is a visual signal in courtship and species recognition, and several studies have shown the red spot in related species to be a visual signal eliciting begging by the chicks. (After Smith, 1966:frontispiece.) Bottom: Two labile color patterns of the octopus (*Octopus vulgaris*) used in visual display. In the "fighting display" (left) the animal becomes entirely red, and in the "zebra crouch" (right) it assumes dark bars on a light background. (After Packard and Sanders, 1971:784.)

tensive discussion of the rapidity of color changes in a cichlid fish having elements of permanent coloration, a yellow ground color that requires several days to attain, coloration that is gained or lost in ten seconds, and an overall blanching that requires but two seconds.

OTHER ASPECTS OF INTRINSIC VISUAL SIGNALS

Structural and surface elements of visual signals are usually combined with behavioral elements of orientation, shape, and movement to produce unitary signals. The whippoorwill flashes its usually hidden white tail feathers (Bruce, 1973), and the orientation of attack and threat in canids is correlated with species-specific body markings (Fox, 1969). Morphological elements, such as a rack of antlers or the male cardinal's red plumage, are virtually always visible. It is not always evident whether the coloration is emphasizing behavior, or a movement is displaying a particular color. In some cases the former situation appears to hold (e.g., R. G. B. Brown et al., 1967; Gutherie, 1971a) and in other cases the latter (Otte, 1972; Kahl, 1966; Dingle and Caldwell, 1969; Dunham, 1966). Some examples are shown in Fig. 11.

Fig. 11. Examples of the use of color in visual display. Left: A mantis shrimp (*Gonodactylus bredini*) in the "meral spread display," in which the small, dark meral spots emphasize the posture. (After Dingle and Caldwell, 1969:120.) Right: Summary scheme of relation of body markings to social behavior in a stylized canid. Arrows point out color markings correlated with specific movements, postures, and orientations. (After Fox, 1969: plate XVII.)

The intensity of a signal usually refers to various levels or shapes along a single dimension, such as the angle to which a crest is raised in a jay (J. L. Brown, 1964). Such signals show "duality" of patterning (Hockett and Altmann, 1968) because they have qualitative (e.g., crest raised) and quantitative (angle to which raised) aspects. Morris (1957) noted that animals often show modal points of usage along such a continuum—a concept he calls the "typical intensities" of the signal. Varying two display elements along different continua creates a whole range of different visual signals, as in eye-color variation and dorsal-fin-raising in a damselfish (Rasa, 1969) (Fig. 12).

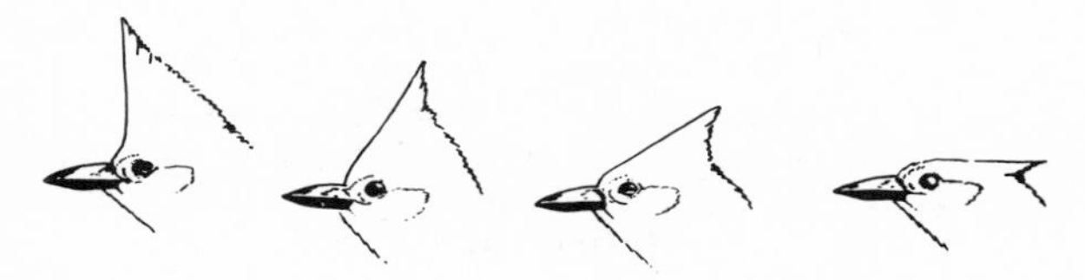

Fig. 12. Example of intensity of a visual signal. The crest of the Steller's jay (*Cyanocitta stelleri*) varies in angle from 90° (left) to 0° (right) along a continuum. (After J. L. Brown, 1964:296.)

Jenssen (1970) rearranged the sequential and temporal patterns of head movements and dewlap extensions in lizards by means of film loops and showed that atypical patterns reduced the effectiveness of the visual signal in inducing approach by females. The temporal and sequential aspects of signaling are therefore also important and deserve more attention.

PERCEPTION AND PHYSIOLOGICAL MECHANISMS

Unless the effectiveness of a reputed signal is investigated, the assertion that some orientation, posture, movement, structure, or coloration is actually a signal remains hypothetical. The "flehmen" posture of many male mammals is concerned with olfactory communication (Estes, 1972), but in the chamois it may actually be a visual signal as well (Krämer, 1969). On the other hand, many behavioral patterns that look like visual signals (Fig. 13) probably are not (examples in Bovet, 1972; Parker, 1972; Pilleri and Knuckey, 1969; Estes, 1969; Barth, 1970).

Fig. 13. Examples of action patterns that look like visual signals. Left: The "wing-raising display" of the male cockroach (*Periplaneta americana*) is believed to provide chemical-release and tactile signals to the female. (After Barth, 1970:725.) Right: Tail slapping by two fish provides tactile stimuli by means of displaced water waves. (After Tinbergen, 1951:25.)

G. K. Noble pioneered the use of models to test which elements of a reputed signal affect recipients (e.g., Noble, 1934a, 1934b, 1936; Noble and Vogt, 1935; Noble et al., 1938), and the tradition is still laudably active (e.g., Franck and Hendricks, 1973; Markl, 1972; Hailman, 1967, 1971; Ducker, 1970; Lill, 1968a, 1968b; Peeke, 1969; Stout and Brass, 1969; Crews, 1975; Payne and Swanson, 1972; D. G. Smith, 1972; Keenleyside, 1971; Fox, 1971; Potts, 1973; Youdeowei, 1969; Peeke et al., 1969; Peek, 1972; Deiker and Hoffeld, 1973; Jenssen, 1970; Grant et al., 1970); see Fig. 14.

The experimental analysis of reputed signals should lead to hypotheses about the underlying sensory mechanisms, but there is little progress to report (Hailman, 1970). Tinbergen (e.g., 1951) drew attention to "supernormal" experimental stimuli—those more effective than naturally occurring ones—and other workers continue to find new examples (e.g., Grant et al., 1970; Payne and Swanson, 1972). Such stimuli imply organizational principles about perception, but few studies have pursued the quest to actual mechanisms. Hazlett (1972) tried to relate hermit crab displays to the compound eye, and Fig. 15 summarizes some of my attempts to uncover visual mechanisms of gull chicks responding to parental signals of shape, orientation, and coloration (Hailman, 1967, 1970, 1971).

Fig. 14. Examples of models used to assess visual elements of a reputed signal. Top: Four of several models used to test for general shape and textural details of visual stimuli from glaucous-winged gulls (*Larus glaucescens*). Bottom: Four of several models used to test for effects of body postures. (After Stout and Brass, 1969:44–45.)

Ever since the classic studies of von Holst and Saint Paul (1963), attempts have been made to find brain areas that when stimulated elicit some signaling behavior. Åkerman's (1965a, 1965b) results may be the most convincing: he elicited normally appearing displays through stimulation of the preoptic nuclear complex and related brain areas of the pigeon, the behavior including bowing, nest demonstration, threat postures,

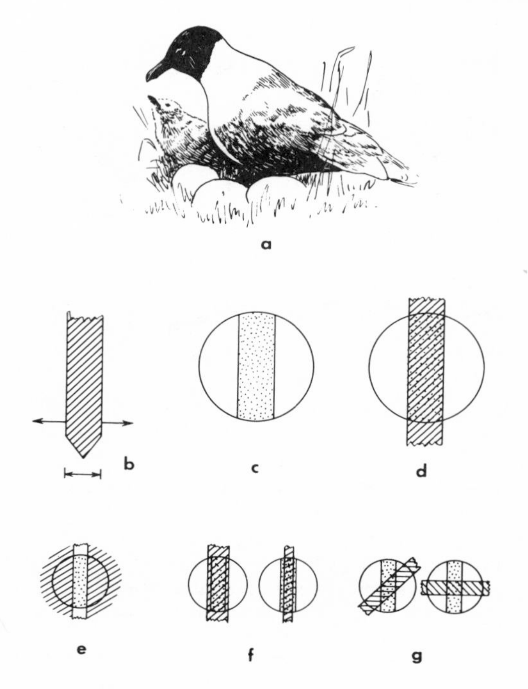

Fig. 15. Example relating perception of signals to vision physiology. The newly hatched chick of the laughing gull (*Larus atricilla*) begs by pecking at the red beak of its parent (a), which is pointed downward at feeding. The chick's perceptual ideal (b) is dark red or blue vertically elongated object on a light yellow or green background, of a certain width, moved horizontally and across its long axis. The receptive field of a cat's visual neuron (c) is optimally stimulated by dark bar on light background, and thus would encode chick's preference (d) with respect to contrast polarity (e), width (f), and vertical orientation (g). (Drawing a after Hailman, 1967: Plate I; c and d after Hailman, 1970:143; remaining parts after Hailman, 1971:330.)

Fig. 16. Examples of visual displays elicited by brain stimulation of the pigeon (*Columba livia*). Top row: Stimulation of the preoptic area elicits erection of head and body, ruffling of feathers, and movements of the crop, as well as walking, then walking in circles with bowing and lowering of fanned tail, and finally looking around. (After Åkerman, 1965a:326.) Second row: Nodding, chest lowering, and wing vibration of the "nest-demonstration" ritual elicited by stimulation of the preoptic area. (After Åkerman, 1965a:333.) Third row: Stimulation of the ventral diencephalis paraventricular gray elicits ruffling of feathers, depression of tail, crouching, and wing waving. (After Åkerman, 1965b:341.) Bottom row: Stimulation of the lateral hypothalamus elicits crouching, head turning, deep crouching ("cringing"), tail lowering, and even flight. (After Åkerman, 1965b:344.)

and various pilomotor responses of agonistic behavior (Fig. 16).

Although there is an active literature on hormonal bases of agonistic and reproductive behavior in general, there is little study of visual communication. Orcutt (1971) switched male pigeons from primarily bow-coo to the bow display more prevalently given by females through long-term injections of estrogen. Ducker (1970) injected estradiol into male birds that usually react to the red coloration of other males. The injected males behaved like females in not responding to red, but no treatment of females caused them to respond to red.

CONCLUSIONS CONCERNING CONTROL

In the widest sense there has been little study of the dynamic control of visual communication, either of behavior or underlying physiological mechanisms. Greatest emphasis has been accorded elements of behavior and morphology that act as individual visual signals. The diversity created by various extrinsic signals as well as by orientations, shapes, movements, structures, and colorations of intrinsic signals imparts to communication by reflected light a large informational capacity, which is further increased by "intensities" of signals and their sequential and temporal relations.

Fig. 17. Example of display of an interspecies hybrid. Left: The bowing of a barbary dove (*Streptopelia roseogrisea-risoria*) male. Right: The bowing of turtle dove (*S. turtur*). Center: The bow of an interspecies hybrid, which resembles the barbary dove in lack of neck plumage fluffing and the turtle dove in the posture at the beginning of the bow. However, the bottom of the bow goes below the horizontal, as in the barbary dove, so the hybrid shows elements of both parental displays. (After Davies, 1970.)

Ontogeny of Visual Communication

Ontogenetic determinants of behavior are often divided into genetic and experiential influences, although their interaction provides the most coherent understanding (e.g., Hailman, 1967). Since there is no overall understanding of dynamic control of visual communication, it is difficult to analyze how ontogenetic factors lead to developmental end points. This fact probably accounts for the relative paucity of studies on the ontogeny of optical communication.

GENETICS OF VISUAL COMMUNICATION

The fragmentary evidence about the genetics of specific displays comes primarily from studies of interspecies hybrids and cross-fostering. Gorman (1969) found that displays of a hybrid lizard filmed in the field resembled one parental species in total duration, the other parental species in number of head bobs, and both in aspects of the tail-flick components. Davies (1970) found in hybrid doves that bowing displays resembled one parent or the other, or were intermediate between the parental species, or showed a range of variation that exceeded that bounded by the two parental types (Fig. 17). Analyses of displays of hybrid ducks (Sharpe and Johnsgard, 1966; Kaltenhäuser, 1971) produced similar results. The results suggest that visual displays are polygenically controlled.

By rearing the young of one genotype with parents of another and then testing their behavioral choices in adulthood one can see whether genetic endowment or individual experience plays the major role in recognition of visual displays. Immelmann (1969) found that male estrildid finches courted females of their foster-parent species in preference to their genetic parents. Walter (1973) and Immelmann (1969) reared zebra finches with albino and normally pigmented parents and found that males reared by albinos chose albino mates. Walter further showed that males reared by normally pigmented parents preferred these, whereas those reared by mixed pairs showed no choice of mate color. However, females always preferred pigmented males. It appears in this case as if the male's preference is determined experientially and the female's genetically.

EXPERIENCE AND VISUAL COMMUNICATION

The foregoing results on cross-fostering demonstrate that early experience can affect mate preferences presumed to be primarily visually mediated. There is relatively little evidence, however, concerning experiential effects on the recognition of specific visual signals, on the production of signals, or how the signals are used in communicative behavior. We know from several studies (e.g., Peeke et al., 1969; Clayton and Hinde, 1968) that repeated presentation of a visual signal leads to decremental responses in the recipient, so that at least simple kinds of experience do affect responses to signals.

It is possible that an animal can respond at different ages to the same visual signal, but its perception of that signal changes as a function of experience. For instance, a gull chick will peck at red areas on the parent's beak. Perception in newly hatched chicks is simply coded (Fig. 15), but as chicks accumulate experience with the parents and are fed for responding to the stimulus, they develop a more highly structured, *Gestalt*-like preference (Fig. 18). At all ages chicks confine responses primarily to the same physical object (the parent's beak), but the ideal signal changes as a result of experience (Hailman, 1967)—a process I call "perceptual sharpening." Such perceptual sharpening will not be evident from mere field observation.

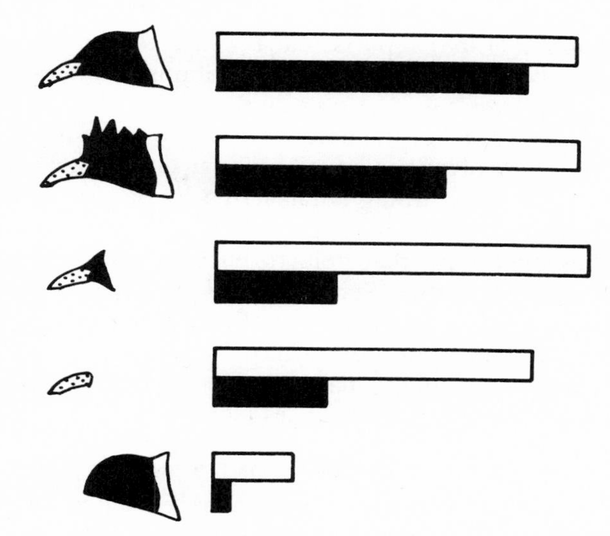

Fig. 18. Example of ontogenetic change in perception of visual signal. Pecking of laughing gull chicks (*Larus atricilla*) toward models of the adult parent shows little discrimination in newly hatched chicks (white bars), in which only the parent's beak is important in the signal (see Fig. 15). After a few days' experience in the nest, the presence and shape of the adult's head becomes very important (black bars). (After Hailman, 1967:89.)

The extent to which animals learn to produce visual signals is virtually unknown. Tayler and Saayman (1973) showed that a captive dolphin imitatively produced all kinds of behavioral patterns of another species, including visual displays. More complicated is the question of whether specific use of signals requires previous experience. Stephenson (1973) has given examples from Japanese macaques in which the same physical signal is used differently in different troops. Feekes (1972) reported experientially dependent development of ground pecking as a visual display in domestic fowl. Apparently a bird learns that feeding during tension-producing agonistic encounters lowers the psychological tension. Ground pecking then becomes incorporated into the ongoing behavior in this situation and thereby comes to act as a visual signal to the opponent. Furthermore, ground scratching also becomes incorporated by generalization from the related pecking behavior.

Many studies show that the social environment of rearing affects social relations. Most studies are quite general, but Fox (1971) found in canids that social experience facilitated inguinal presentations toward a visual model of a dog, and Fox and Clark (1971) separate general stages of development in which action patterns of display and other behavior become incorporated into increasingly complex sequences. Anthoney (1968) reports that lip smacking in baboons develops first from intention movements of sucking the mother's pink nipples. Young then respond to the mother's own lip smacking, which involves her visually similar pink tongue, so that the learning is facilitated by perceptual transfer. Furthermore, the young baboon's face has pink areas that help to elicit lip smacking from the mother. Social bonds built by the exchange of this signal later generalize to sexual visual communication, where the female's pink sexual skin and the male's pink penis further aid in the transfer of responses (see Fig. 19).

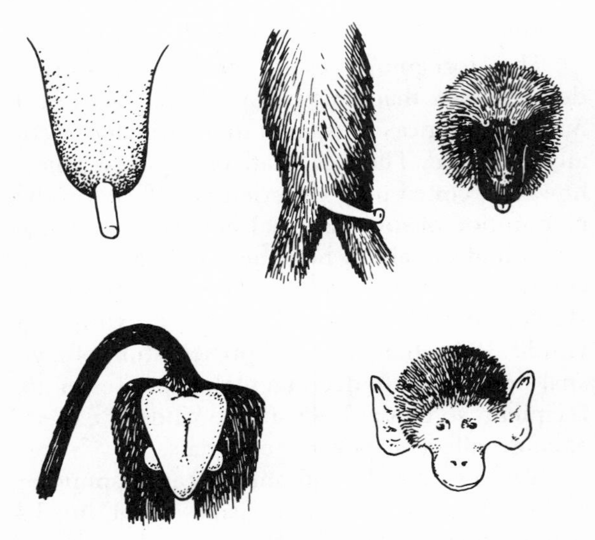

Fig. 19. Examples of similar visual features involved in ontogenetic transfer of communicative responses. The pink objects associated with elicitation of lip smacking in the baboon (*Papio cynocephalus*) include the nipple of the female (upper left), penis of the black infant male (upper middle), tongue of the lip-smacking adult female (upper right), sexual skin of the estrous female (lower left), and face of the black infant (lower right). (After Anthoney, 1968:363.)

CONCLUSIONS CONCERNING ONTOGENY

There exists only fragmentary evidence for both genetic and experiential effects on visual communication. Genetically determined propensities guide development by experience, perhaps by assuring certain initial behavior that changes through learning processes. Experience can help determine the form, use, or recognition of a visual signal. Probably both the taxonomic group and learning capacities of an animal, as well as the social structure and type of environment, influence the degree to which the development of various aspects of visual communication are experientially dependent.

Function of Visual Communication

The word "function" is used variously in biology (Hailman, 1975); I use it here as shorthand for "adaptive function" or "selective advantage." Were a detailed understanding of the dynamic control of visual communication possible, then one could ask after the pressures of natural selection that shape the total ontogenetic sequences leading to such control. Instead, goals must be limited to exploring the adaptiveness of two aspects of visual communication: situations that favor vision over other modalities, and selective pressures that act to structure the types of visual signals.

SELECTION PROMOTING VISUAL COMMUNICATION

At least three factors promote specific modalities: the environment in which signaling takes place, the kind of animals communicating, and the functional use of the communication. Communication via reflected light obviously requires light for reflection. An environmental factor given less attention is the transparency of the medium, since the turbidity of water or vegetation of a habitat may discourage communication by light. Wootton (1971) notes that visual signals are less well developed in a species of stickleback living in thick vegetation and tea-colored water than in its congeneric relatives in more transparent media. Similarly, Catchpole (1973) shows that the open-habitat sedge warbler has more visual displays than the congeneric reed warbler, which depends instead on vocal displays in its more vegetated habitat. Busnel (1968) points out that high ambient noise may discourage acoustic communication and thereby favor vision or other modalities.

Not all species in groups with well-developed vision use extensive visual communication; among insects, particularly, visual communication is relatively unusual. Perhaps more motile than sessile animals have developed visual signals because sessile animals cannot effectively modulate ambient light even if they have good photoreceptors. Even motile species that possess good vision may conduct important communication at night (e.g., orthopteran insects and anuran amphibians)—perhaps to escape predation and utilize humid conditions—so that their mating signals and other exchanges are largely acoustic. Some, diurnal insects do have visual signals (e.g., Waage, 1973; Otte, 1972), but others enigmatically lack them. Otte (1972) reports *Syrbula* grasshoppers have visual signals, whereas related genera living in the same habitat lack them. Similarly, Schremmer (1972) notes that male *Bombus confusus* bees perform looping flights before the female, whereas their congeners rely strictly on pheromonal courtship signals.

The visual mode may have certain advantages over communication by other means. Extrinsic visual signals, for instance, may have a persistence that is difficult to match in other modalities; olfactory signals may persist for hours or days but extrinsic visual signals may persist for years. The great diversity of visual signal elements impart a huge informational capacity to optical communication. Visual signals may be directed toward specific receivers, whereas many other signals (pheromones, sounds, electrical fields) tend to radiate indiscriminately. Visual signals also have high potentialities for indexical and representational qualities: that is, visual signals can point out a specific spatial locus or can mimic in form some physical object. Furthermore, since many visual signals are evolved from intention movements of nonsignal activities (see major section on phylogeny, below) such signals carry predictive information about the subsequent behavior of the sender, information that may be more difficult to encode in other modalities.

SELECTION FOR SIGNAL QUALITIES

It seems unlikely that visual signals are arbitrarily established in evolution. Natural selection is constrained to work on the available variation in a population, so that the physical attributes of a signal may be partly constrained by the phylogenetic origins of the signal (see major section on phylogeny, below). However, many qualities of a signal may be directly adaptive to signaling of certain kinds in certain environments.

The most emphasized signal attribute is species-specificity. It may be of advantage for each similar species of a monophyletic assemblage to have distinctive signals, either to insure monospecific flocking and aggregation or to prevent hybridization or gamete wastage in reproductive behavior. Tinbergen (1951) emphasized Lorenz's suggestion of species-specificity of speculum patterns in the wings of ducks (Fig. 20) as an example of the first case, and much literature emphasizes the species-specificity of courtship displays (e.g., Salmon, 1967; Purdue and Carpenter, 1972; Kroodsma, 1974); see Fig. 21. I doubt if selection for species-specificity is as important as is often believed. Convergence of signals among various species is known (e.g., Moynihan, 1968; Cody, 1969), and the attributes of the convergent signals are still to be explained. Furthermore, many closely related species have quite similar signals, suggesting that slight but consistent differences are sufficient for species recognition. A new search should be instituted for other factors that promote specific attributes of signals.

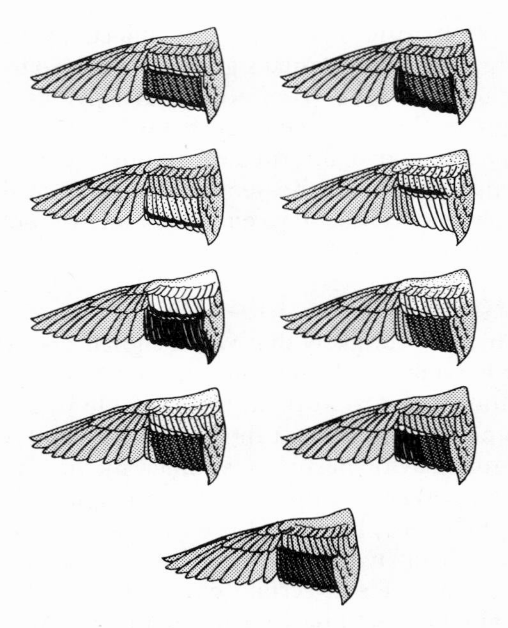

Fig. 20. Example of species-specificity in visual signals. Wing speculum patterns in some North American ducks of the genus *Anas.*

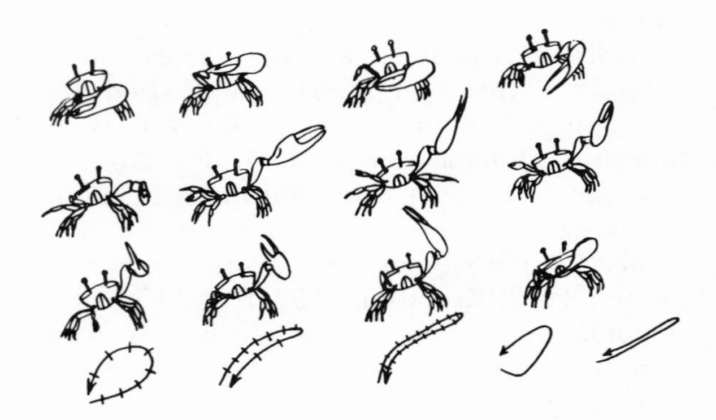

Fig. 21. Example of species-specificity in a visual display. The courtship claw-waving display of the fiddler crabs (genus *Uca*) involves different movement geometries and tempos in the four species pictured here; from left: *U. mordax, U. rapax, U. pugnax* and *U. speciosa.* Diagrams at bottom show spatial geometry of the movement with cross-marks indicating jerks (two smooth movement patterns in *U. speciosa* at right). (After Salmon, 1967:452.)

Experiments are necessary to see what aspect of a visual signal is effective. Crews (1975) found that movements and dewlap extension of the male Carolina anole were important in stimulating the female, but the color of the dewlap was not. The color might (for instance) be adaptive in promoting conspicuousness of the display in a particular kind of visual environment. Rand and Williams (1970) point out considerable redundancy among the visual signals of anole species on Hispaniola; each species differs from the others in many ways. They suggest that some signal features get the message through in one sort of environment, whereas other features are more effective in other habitats.

The effect of the medium on visual signaling in aquatic organisms has not been thoroughly explored. Luria and Kinney (1970) report that as turbidity increases, absorption of light becomes greater at short wavelengths (blue end of the spectrum). Baylis (1974) interprets the yellow (long wavelength) signal coloration of a cichlid fish as adaptive in combating its turbid medium.

Other aspects of signals appear explicable by factors having to do with neither species-specificity nor the signaling environment. The advantage of visual signals in having representational qualities is illustrated by Fig. 19. Gutherie and Petocz (1970) have generalized this notion of "automimicry," in which the visual signal mimics the appearance of some other feature of the animal. Wickler (1967) suggested that the face of the mandrill mimicks its genitals, although this suggestion is controversial (Anthoney, 1968; Dunbar and Dunbar, 1974). Gutherie and Petocz (1970) review various structures and color patterns that resemble canine teeth or antlers and horns in various mammals, and also note submissive signals of males that mimic postures of juveniles or estrous females (see Fig. 22). Gutherie (1971b) interprets rump-patch signals of mammals as a sort of elaborate automimicry of more

Fig. 22. Examples of automimicry in visual signals. Five artiodactyls in which ears, facial markings, or other signals mimic the horns, from left: *Sylvicapra grimmia, Acelaphus buselaphus, Artilocapra americana, Oreamnos americanus,* and *Oreotragus orcotragus.* (After Guthrie and Petocz, 1970:587.)

restrictive ano-genital display, which serves to remotivate the observer from aggression to sex.

Another possibility is that the form of the visual signal is directly related to the function served by the signal. The most frequently repeated example (e.g., Marler, 1968; Wilson, 1972) is that threat displays are selected to make the threatening animal appear larger (or perhaps closer) than it really is. Larger animals do more readily attack smaller individuals, but this fact does not bear on the structure of the signal. Hazlett (1970) provides the only direct experimental evidence: he increased the visual size of a hermit crab's shell and found that this change increased the probability of the crab's winning an encounter. Enigmatically, the same result obtained when the shell's weight (but not its size) was increased!

No discussion of signal qualities is complete without Darwin's (1873) principle of antithesis. Darwin believed that emotions were expressed outwardly as some inevitable result of neural activity, but did not quite realize that such expressions are often visual signals enhanced by natural selection for communication. He discovered several sets of opposite-appearing expressions and hypothesized that "opposite emotions" give rise to opposite expressions; he cited expressions of domestic animals, such as a dog with "hostile intentions" versus one "in a humble and affectionate frame of mind." Fig. 23 shows an example of two postures that appear quite different, if not opposite, and it is parsimonious to think in terms of distinctiveness of signals rather than oppositeness of postulated emotions. New examples of antithesis in visual signals continue to be reported (e.g., Krämer, 1969; Stanley, 1971; Ewer, 1971). Klopman (1968) reports that the Canada goose signals "high attack probabilities" with mouth open, neck coiled, and head aimed at the opponent, and "low attack probabilities" with mouth closed, neck extended horizontally, and head directed away from the opponent.

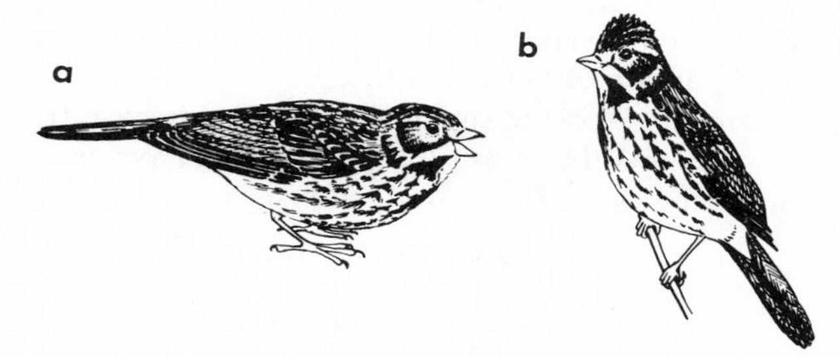

Fig. 23. Example of Darwin's principle of antithesis in visual displays. Left: The fox sparrow (*Passerella iliaca*) threatens a conspecific by crouching in a horizontal posture with dorsal feathers flattened and lateral feathers fluffed. Right: A "fearful" bird assumes a nearly opposite configuration, with vertical posture, dorsal raising, and lateral compression of feathers. (Drawn from field notes of the author.)

CONCLUSIONS CONCERNING FUNCTION

The selection pressures promoting visual communication over other types and the pressures that shape attributes of visual signals have not been studied extensively. Certain environmental variables encourage visual signaling, only certain animals are equipped to send and receive optically, and certain aspects of communication favor the use of vision. Species-specificity, although real, is probably overworked as an explanation for signal diversity, yet conditions of the environment that promote specific kinds of signals remain largely unexplored. Visual signals may sometimes have a direct relation with their apparent informational content, but this topic also presents mainly unsolved problems.

Phylogeny of Visual Communication

Natural selection shapes phenotypes of a population but is constrained by the starting materials. It may not be possible to understand the qualities of a visual signal unless one knows its phylogenetic origins.

The postures of gulls in Fig. 24 illustrate Darwin's antithesis principle discussed above. A comparison of Figs. 23 and 24, however, reveals a curious aspect of signals not explicable by antithesis: the threat posture of the sparrow is a head-forward, horizontal stance, whereas the threat posture of the gull is an upright, head-retracted stance. Similarly, the fear or anxiety posture of the sparrow is vertical, whereas that of the gull is horizontal. Although both sparrow and gull obey the antithesis principle, they do so with opposite polarities, so that further explanation is required. Since the sparrow attacks by pecking directly at the opponent and the gull attacks by pecking down on top of the opponent, it seems reasonable to assume that the threat postures of these two birds were evolved from their movements of attack. In other words, the probable phylogenetic origins of visual signals help to explain their qualities.

Fig. 24. Example of phyogenetic influence of visual displays. The upright threat posture (left) of the laughing gull (*Larus atricilla*) is antithetical to the submissive, hunched posture (right) of the young lava gull (*Larus fuliginosus*), so that gulls also illustrate Darwin's antithesis principle. However, comparison with Fig. 23 shows that the polarity of the antithesis is reversed in sparrows and gulls, a fact that is explained by considering phylogenetic origins of the displays. (Drawn from photographs by the author.)

Moynihan (1955) has suggested that the term "display" be restricted to those signals whose evolution has been influenced by their function as signals, even if they serve other functions as well. The evolutionary process by which noncommunicative behavior becomes a display has been called "ritualization" (Tinbergen, 1951, 1952). In this section I deal with noncommunicative behavioral patterns from which visual displays appear to have evolved: that is, with the phylogenetic origins of visual signals.

ORIGINS IN INTENTION MOVEMENTS

Animals may show incipient or incomplete performances of a motor act, often just prior to or just after the act itself. These incomplete acts are called "intention movements" (Daanje, 1950), without connotation of conscious intention. They often occur in behavioral situations that favor their elaboration and standardization as signals. A bird about to attack an opponent may open its beak in preparation for biting, and the movement, posture, and orientation of such a beak-open act can signal probable attack to an opponent; see Fig. 23.

There appear to be two major sources of visual displays among intention movements—agonistic and reproductive activities—although locomotion, feeding, and other activities have also been reported as origins. Many cases were reviewed by previous authors (e.g., Tinbergen, 1951), so I concentrate on illustrating the diversity of origins with recent examples.

"Agonistic" behavior includes all those activities associated with situations in which fighting, fleeing, or appeasement occur. Many visual displays that have an intimidating effect on the opponent ("threat" displays) bear a striking resemblance to the fighting methods of the species. Examples of sparrows (Fig. 23) and gulls (Fig. 24) were given above, and Fig. 25 illustrates other examples (see also Krämer, 1969; Tyler, 1972; Schweinsburg and Sowls, 1972; van Lawick-Goodall, 1968; Kahl, 1966; Allin and Banks, 1968). Protective responses or flight from the opponent can also lead to visual signals, as in the neck withdrawal of the Canada goose (Raveling, 1970) or leaning away from the opponent in the lemming (Allin and Banks, 1968). Eibl-Eibesfeldt (1970) asserts that ducking, combined with breaking eye contact, has become a submissive signal in human communication, its exact form (nodding, sweeping bows, etc.) being culturally determined.

Fig. 25. Examples of visual displays ritualized from agonistic behavior. Left: The "lateral display threat" of the chamois (*Rupicapra rupicapra*) ritualized from fighting movements. (After Krämer, 1969:918.) Right: The "erect gape" display of the Marabou stork (*Leptoptilos crumeniferus*) ritualized from escape behavior. (After Kahl, 1966: Plate V.)

Many examples of reproductive signals seem evolutionarily derived from noncommunicative reproductive activities (Fig. 26). Intention movements of copulation appear as signals in courtship of many animals, these having been reviewed for birds by Andrew (1957). Lordosis-like postures appear to be used not only as a signal of readiness by female mammals (e.g., Tyler, 1972), but also as general submissive signals by both sexes (e.g., van Lawick-Goodall, 1968). And penile erection is an evident display in many primates (e.g., van Lawick-Goodall, 1968). Intention movements of nest building have become visual signals in the courtship of many birds (e.g., Kahl, 1966; Kunkel, 1969; Güttinger, 1970; R. E. Phillips, 1972; Baltin, 1969). Güttinger (1970) reports courtship displays of finches evolved from behavior associated with parental care of the young, and submissive displays appear often to be derived from, or else mimic, the behavior of juvenile animals (e.g., Tinbergen, 1959; Güttinger, 1970; Anthoney, 1968).

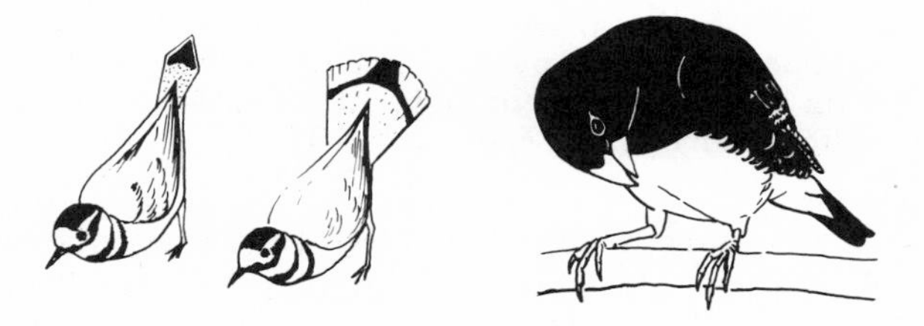

Fig. 26. Examples of visual displays ritualized from reproductive behavior. Left: The "scraping display" of the killdeer (*Charadrius vociferus*) ritualized from nest building. (After R. E. Phillips, 1972:3.) Right: Display element of the estrildid finch *Spermestes bicolor* ritualized from the movement of feeding begging young. (After Güttinger, 1970:1054.)

A rich source of signals is locomotory intention movements (Daanje, 1950). Since so many behavioral situations involve locomotion, this general activity is contextually placed so as to be easily ritualized (Fig. 27). Earlier works review many examples (e.g., Daanje, 1950; Andrew, 1956; Tinbergen, 1951, 1959) and new examples continue to appear (e.g., Myrberg, 1972; Fishelson, 1970; Sale, 1971; Ewer, 1971; Kunkel, 1967; Kahl, 1966).

Fig. 27. Examples of visual displays ritualized from locomotion. Above: Male bicolor damselfish (*Eupomacentrus partitus*) on left leads female by using swimming pattern with exaggerated tail movements. (After Myrberg, 1972:216.) Below: Display postures of the male house sparrow (*Passer domesticus*) (left) and European tree sparrow (*P. montanus*) (right) evolved from different phases of flight-intention movements. (After Daanje, 1950: Figs. 18 and 19.)

Foraging, feeding, and associated activities have also evolved into visual signals (Fig. 28): for instance, the pecking and ground scratching of fowl (Feekes, 1972) and the nursing movements that are the phylogenetic origins and also the ontogenetic origins of lip smacking in baboons (Anthoney, 1968). Smythe (1970) argues that white flash patterns of some mammals, used as warning signals to conspecifics, were originally evolved to entice a predator into betraying its hiding place or making a premature charge, so that antipredator behavior may also give rise to visual signals.

Responses involving orientation of the sense organs are situationally placed so as to be avail-

Fig. 28. Examples of visual displays ritualized from foraging and antipredator behavior. Left: A Burmese red junglefowl (*Gallus gallus spadiceus*), ancestral species to domestic fowl, pecks at nonedible particles as part of agonistic display interactions. (After Feekes, 1972:258.) Right: Tail flashing of the whitetail deer (*Odocoileus virginianus*), believed to be a social alarm signal; Smythe (1970) has proposed that many such signals may have evolved originally as antipredator behavior and then later as social signals.

able for evolution into signals themselves (Fig. 29). Looking at an opponent is a visual signal in primates, carnivores, and probably many other animals (e.g., van Lawick-Goodall, 1968; Kleiman and Eisenberg, 1973). The origin need not be visual itself, though, since the ears-up signal of ponies is an acoustical orientation that leads to a visual signal (Tyler, 1972).

Maintenance activities—those action patterns involved with preventive maintenance of the interior and exterior of the body, such as stretching, preening, scratching, grooming, shaking, yawning, etc.—appear in many communicational situations, although not always in an obviously ritualized form (Fig. 29). Sometimes actions of the entire animal are involved, as in rolling, which is elicited by wet fur but is also used as a signal to inhibit flight in conspecifics (Castell et al., 1969). Other times only a limb movement is

Fig. 29. Examples of displays ritualized from maintenance activities. Left: Squirrel monkey (*Saimiri sciureus*) rolls on back, displaying genitals with the same motor pattern as used when rolling out wet fur. (After Castell et al., 1969:490.) Right: Andean goose (*Chloëphaga melanoptera*) throws its head back in a display that closely resembles oiling movements. (After McKinney, 1965: Plate V.)

Fig. 30. Examples of visual signals evolved from autonomic responses. Above: Barbary dove (*Streptopelia risoria*), showing areas of body in which pilomotor responses were given in temperature and social experiments. (After McFarland and Baher, 1968:172.) Below: Female gelada baboon (*Theropithecus gelada*), showing beaded red chest that signals estrus. (After Dunbar and Dunbar, 1974: Plate VII.)

involved, as in scratching by chimpanzees (van Lawick-Goodall, 1968). The visual signal may in some cases be highly ritualized, as in the bowing and curtseying of waxbills, which evolved from bill wiping (Kunkel, 1967). McKinney (1965) has made an unusually complete analysis of comfort movements in waterfowl and their uses as unritualized and ritualized visual signals.

ORIGINS IN AUTONOMIC RESPONSES

Autonomic responses—those mediated by the sympathetic and parasympathetic nervous systems of vertebrates and usually involving smooth rather than striated muscle—often accompany more dramatic actions of animals and hence are in excellent behavioral situations to become used as signals.

Pilomotor actions—changing of the feather postures of birds or the fur of mammals—are parts of many visual displays (Fig. 30). Morris (1956) proposed that these display components originate in thermoregulatory responses: sleeking feathers decreases the thickness of the insulating layer, whereas fluffing increases it. In extreme cases, Morris (1956) says that feathers may be ruffled so that the air pockets are opened and the insulation made less effective. McFarland and Baher (1968) could not confirm the last point experimentally with ring doves, but did show temperature control of the other feather postures and also their use in presumptive communicative situations.

Tyler (1972) reports yawning (respiratory responses) in ponies as a visual signal, as does van Lawick-Goodall (1968) in chimpanzees. Vasoresponses—shunting the blood differentially to various parts of the body by control of restriction and dilation of blood vessels—has been known for a long time to carry visual consequences (e.g., Cannon, 1915) such as human blushing in embarrassment, flushing in anger, paling in fear, and so on. Longer-term vasoresponses are involved in sex skin coloration of some primates (e.g., Dunbar and Dunbar, 1974).

Stanley (1971) notes the importance of eye closure in hopping mouse behavior, and it is difficult to resist the speculation that the widespread colored eye-rings in birds accentuate the degree of eye closure. Pupillary responses are also used as visual signals, both in cats (Kleiman and Eisenberg, 1973) and in humans (Hess, 1965). Dr. R. Jaeger has pointed out to me that European women formerly used the drug from nightshade *(Atropa belladonna)* to dilate their pupils to make themselves more attractive to men.

OTHER ORIGINS OF VISUAL SIGNALS

A few phylogenetic origins of visual signals are not readily classified. Many animals possess the ability to match their background coloration (e.g., Gibson, 1968), even though the mechanisms for such color change may be quite different in different species. Perhaps this camouflage-related ability is the basis of many visual signals employing color changes.

There also exists "secondary ritualization," in which a signal evolved for communication in another sensory modality takes on visual properties and may be further changed to enhance the visual component. Several authors report visual signals that have evolved from scent-marking activities of mammals (Rood, 1972; Estes, 1969; Schmied, 1973), as noted in Fig. 31. The production of sound often involves assuming a specific posture, which may then become a visual signal, as in the head-tipping display of the chimpanzee ritualized from "soft-bark" calling (van Lawick-Goodall, 1968).

CONCLUSIONS CONCERNING PHYLOGENY

The evolutionary origin of a particular signal may never be identified with the same degree of certainty that other conclusions in ethology can be secured. Yet comparative study among species and the comparison of visual signals with noncommunicative behavior within a species offer convincing probabilities of many origins. Virtually any behavior that occurs in a situation of interaction between animals may be enhanced in some way to increase its value as a visual signal.

Fig. 31. Examples of visual signals secondarily evolved from signals in other modalities. Left: A wildebeest (*Connochaetes taurinus*) defecates during a social encounter, one form of scent marking that also serves as visual communication. (After Estes, 1969:322.) Right: Chimpanzee (*Pan troglodytes*) assumes particular facial expression while hooting. (After van Lawick-Goodall, 1968: Plate 9.)

Future Prospects

Clearly there remains much to be learned about the control, ontogeny, function, and phylogeny of optical communication. To judge from recent studies, the most immediately promising areas for new results may be motivational and contextual factors in communication, temporal and sequential patterns of signals, ontogenic and traditional determinants of all aspects, mechanisms of perception, and environmental structuring of signals. We are a very long way from a theoretical framework that will encompass the claw waving of a fiddler crab and the bower of the bowerbird under the same roof as the writing, painting, sculpture, and dance of man.

References

Åkerman, B., 1965a. Behavioural effects of electrical stimulation in the forebrain of the pigeon. I. Reproductive behaviour. *Behaviour,* 26:323–38.

Åkerman, B., 1965b. Behavioural effects of electrical stimulation in the forebrain of the pigeon. II. Protective behaviour. *Behaviour,* 26:339–50.

Albrecht, H., 1969. Behaviour of four species of Atlantic damselfishes from Colombia, South America (*Abundefduf saxatilis, A. taurus, Chromis multilineata, C. cyanea*; Pisces, Pomacentridae). *Z. Tierpsychol.,* 26:662–76.

Allin, J. T., and Banks, E. M. 1968. Behavioural biology of the collared lemming *Dicrostonyx groenlandicus* (Traill): I. Agonistic behaviour. *Anim. Behav.,* 16:245–62.

Andrew, R. J., 1956. Intention movements of flight in certain passerines,and their use in systematics. *Behaviour,* 10:179–204.

Andrew, R. J., 1957. The aggressive and courtship behaviour of certain emberizines. *Behaviour,* 10:255–308.

Andrew, R. J., 1963. The origin and evolution of the calls and facial expressions of the primates. *Behaviour,* 20:1–109.

Anthoney, T. R., 1968. The ontogeny of greeting, grooming, and sexual motor patterns in captive baboons (superspecies *Papio cynocephalus*). *Behaviour,* 31:358–72.

Apfelbach, R., and Leon, D., 1970. Zum Kampfverhalten in der Gattung *Tilapia* (Pisces, Cichlidae). *Z. Tierpsychol.,* 27:98–107.

Baltin, S., 1969. Zur Biologie und Ethologie des Talegalla-Huhns (*Alectura lathami* Gray) unter besonderer Bërucksichtigung des Verhaltens während der Brutperiode. *Z. Tierpsychol.,* 26:524–72.

Barash, D. P., 1973. The social biology of the Olympic marmot. *Anim. Behav. Monogr.,* 6:171–245.

Barth, R. H., 1970. The mating behavior of *Periplaneta americana* (Linnaeus) and *Blatta orientalis* Linnaeus (Blattaria, Blattinae), with notes on 3 additional species of *Periplaneta* and interspecific action of female sex pheromones. *Z. Tierpsychol.,* 27:722–48.

Baylis, J. R., 1974. The behaviour and ecology of *Herotilapia multispinosa* (Teleostei, Cichlidae). *Z. Tierpsychol.,* 34:115–46.

Bovet, J., 1972. On the social behavior in a stable group of long-tailed field mice (*Apodemus sylvaticus*). I. An interpretation of defensive postures. *Behaviour,* 41:43–54.

Brown, J. L., 1964. The integration of agonistic behaviour in the Steller's jay *Cyanocitta stelleri* (Gmelin). *Univ. Calif. Publ. Zool.,* 60:223–328.

Brown, R. G. B.; Blurton Jones, N. G.; and Russell, D. J. T.; 1967. The breeding behaviour of Sabine's gull, *Xema sabini. Behaviour,* 28:110–40.

Bruce, J. A., 1973. Tail flashing display in the Whip-poor-will. *Auk,* 90(3):682.

Busnel, R. G., 1968. Acoustic communication. In *Animal Communication,* T. A. Sebeok, ed. Bloomington: Indiana University Press, pp.127–53.

Cannon, W. B., 1915. *Bodily Changes in Pain, Hunger, Fear and Rage.* New York: W. W. Norton.

Carpenter, C. C., 1963. Patterns of behavior in three forms of the fringe-toed lizards (*Uma*-Iguanidae). *Copeia,* 1963:406–12.

Castell, R.; Krohn, H.; and Ploog, D.; 1969. Rückenwälzen bei Totenkopfaffen (*Saimiri sciureus*): Körperpflege und soziale Funktion. *Z. Tierpsychol.,* 26:488–97.

Catchpole, C. K., 1973. The functions of advertising song in the sedge warbler (*Acrocephalus schoenobaenus*) and the reed warbler (*A. scirpaceus*). *Behaviour,* 46:300–20.

Clayton, F. L., and Hinde, R. A., 1968. The habituation and recovery of aggressive display in *Betta splendens. Behaviour,* 30:96–106.

Cody, M. L., 1969. Convergent characteristics in sympatric populations: a possible relation to the interspecific territoriality. *Condor,* 71:222–39.

Cole, J. E., and Ward, J. A., 1969. The communicative function of pelvic fin-flickering in *Etroplus maculatus* (Pisces, Cichlidae). *Behaviour,* 35:179–99.

Crews, D., 1975. Effects of different components of male courtship behaviour on environmentally-induced ovarian recrudescence and mating preferences in the lizard, *Anolis carolinensis. Anim. Behav.,* 23:349–56.

Daanje, A., 1950. On locomotory movements in birds and the intention movements derived from them. *Behaviour,* 3:48–98.

Darwin, C., 1871. *The Descent of Man.* New York: D. Appleton.

Darwin, C., 1873. *Expression of the Emotions in Man and Animals.* New York: D. Appleton.

Davies, S. J. J. F., 1970. Patterns of inheritance in the bowing display and associated behaviour of some hybrid *Streptopelia* doves. *Behaviour,* 36:187–214.

Deiker, T. E., and Hoffeld, D. R., 1973. Interference with ritualized threat behaviour in *Cichlasoma nigrofasciatum. Anim. Behav.,* 21:607–12.

Dingle, H., and Caldwell, R. L., 1969. The aggressive and territorial behaviour of the mantis shrimp *Gonodactylus bredini* Manning (Crustacea:Stomatopoda). *Behaviour,* 33:115–36.

Dubost, G., 1971. Observations ethologiques sur le Muntjak (*Muntiacus muntjak*) Zimmermann 1780 et *M. reevesi* (Ogilby 1839) en captivite et semi-liberte. *Z. Tierpsychol.,* 28:387–427.

Ducker, G., 1970. Untersuchungen über die hormonale Beeinflubbarkeit der Farbbevorzugung von Feuerwebern (*Euplectes orix franciscanus*). *J. Ornithol.,* 111(1):19–29.

Dunbar, R. I. M., and Dunbar, P., 1974. The reproductive cycle of the gelada baboon. *Anim. Behav.,* 22:203–10.

Dunham, D. W., 1966. Agonistic behavior in captive rose-breasted grosbeaks, *Pheuticus ludovicianus* (L.). *Behaviour,* 27:160–73.

Eibl-Eibesfeldt, I., 1970. *Ethology: The Biology of Behavior.* Holt, Rinehart and Winston. 530pp.

Estes, R. D., 1969. Territorial behavior of the wildebeest (*Connochaetes taurinus* Burchell, 1823). *Z. Tierpsychol.,* 26:284–370.

Estes, R. D., 1972. The role of the vomeronasal organ in mammalian reproduction. *Mammalia,* 36:315–41.

Evans, L. T., 1961. Structure as related to behavior in the organization of populations in reptiles. In: *Vertebrate Speciation,* W. F. Blair, ed. Austin: University of Texas Press.

Ewer, R. F., 1971. The biology and behaviour of a free-living population of black rats (*Rattus rattus*). *Anim. Behav. Monogr.,* 4:127–74.

Ewing, A. W., and Evans, V., 1973. Studies on the behaviour of cyprinodent fish. I. The agonistic and sexual behaviour of *Aphyosemion bivittatum* (Lönnberg 1895). *Behaviour,* 46:264–78.

Feekes, F., 1972. "Irrelevant" ground pecking in agonistic situations in Burmese red junglefowl (*Gallus gallus spadiceus*). *Behaviour,* 43:186–326.

Figler, M. H., 1972. The relation between eliciting stimulus strength and habituation of the threat display in male Siamese fighting fish, *Betta splendens. Behaviour,* 42:63–96.

Fishelson, L., 1970. Behaviour and ecology of a population of *Abudefduf saxatilis* (Pomancentridae, Teleostei) at Eilat (Red Sea). *Anim. Behav.,* 18:225–37.

Fox, M. W., 1969. The anatomy of aggression and its ritualization in Canidae: a developmental and comparative study. *Behaviour,* 35:242–58.

Fox, M. W., 1971. Socio-infantile and socio-sexual signals in canids: a comparative and ontogenetic study. *Z. Tierpsychol.,* 28:185–210.

Fox, M. W., and Clark, A. L. 1971. The development and temporal sequencing of agonistic behavior in the coyote (*Canis latrans*). *Z. Tierpsychol.,* 28:262–78.

Franck, D., 1968. Weitere Untersuchungen zur vergleichenden Ethologie der Gattung *Xiphophorus* (Pisces). *Behaviour,* 30:76–95.

Franck, D., and Hendricks, R., 1973. Zur Frage der biologischen Bedeutung des Schwertfortsatzes von *Xiphophorus helleri. Behaviour,* 44:167–85.

Frey, D. F., and Miller, R. J., 1972. The establishment of dominance relationships in the blue gourami, *Trichogaster trichopterus* (Pallas). *Behaviour,* 42:8–62.

Geist, V., 1966. The evolution of horn-like organs. *Behaviour,* 27:175–214.

Gibson, R. N., 1968. The agonistic behaviour of juvenile *Blennius pholis* L. (Teleostei). *Behaviour,* 30:192–217.

Golani, I., and Mendelssohn, H., 1971. Sequences of precopulatory behavior of the jackal (*Canis aureus* L.). *Behaviour,* 38:169–92.

Gorman, G. C., 1969. Intermediate territorial display of a hybrid *Anolis* lizard (Sauria: Iguanidae). *Z. Tierpsychol.,* 26:390–93.

Grant, E. C.; Mackintosh, J. H.; and Lerwill, C. J.; 1970. The effect of a visual stimulus on the agonistic behaviour of the golden hamster. *Z. Tierpsychol.,* 27:73–77.

Gutherie, R. D., 1971a. The evolutionary significance of the cervid labial spot. *J. Mammal.,* 52:209–12.

Gutherie, R. D., 1971b. A new theory of mammalian rump patch evolution. *Behaviour,* 38:132–45.

Gutherie, R. D., and Petocz, R. G., 1970. Weapon automimicry among mammals. *Amer. Nat.,* 104:585–88.

Güttinger, H. R., 1970. Zur Evolution von Verhaltensweisen und Lautäusserungen bei Prachtfinken (Estrilidae). *Z. Tierpsychol.,* 27:1011–75.

Hailman, J. P., 1967. The ontogeny of an instinct. *Behaviour Suppl.,* 15:1–159.

Hailman, J. P., 1970. Comments on the coding of releasing stimuli. In: *Development and Evolution of Behavior,* L. R. Aronson, E. Tobach, D. S. Lehrman, and J. S. Rosenblatt, eds. San Francisco: W. H. Freeman, pp.138–57.

Hailman, J. P., 1971. The role of stimulus-orientation in eliciting the begging response from newly-hatched chicks of the laughing gull (*Larus atricilla*). *Anim. Behav.,* 19:328–35.

Hailman, J. P., 1976. Uses of the comparative study of behavior. In: *Evolution, Brain and Behavior: Persistent Problems,* R. B. Masterson, W. Hodos, and H. J. Jerison, eds. Hillsdale, N.J.: Erlbaum, pp.13–22.

Hailman, J. P. In press. *Optical Signals: Animal Communication and Light.* Bloomington: Indiana University Press.

Hall, D. D., and Miller, R. J., 1968. A qualitative study of courtship and reproductive behavior in the pearl gourami, *Trichogaster leeri* (Bleeker). *Behaviour,* 32:70–84.

Hamilton, W. J. III, and Peterman, R. M., 1971. Countershading in the colourful reef fish *Chaetodon lunula:* concealment, communication or both? *Anim. Behav.,* 19:357–64.

Hazlett, B. A., 1970. The effect of shell size and weight on the agonistic behavior of a hermit crab. *Z. Tierpsychol.,* 27:369–74.

Hazlett, B. A., 1972. Stimulus characteristics of an agonistic display of the hermit crab (*Calcinus tibicen*). *Anim. Behav.,* 20:101–107.

Hazlett, B. A., and Bossert, W. H., 1965. A statistical analysis of the aggressive communications systems of some hermit crabs. *Anim. Behav.,* 13:347–73.

Hazlett, B. A., and Estabrook, G. F., 1974a. Examination of agonistic behavior by character analysis. I. The spider crab *Microphrys bicornutus. Behaviour,* 48:131–43.

Hazlett, B. A., and Estabrook, G. F., 1974b. Examination of agonistic behavior by character analysis. II. Hermit crabs. *Behaviour.* 49:88–110.

Hemens, J., 1966. The ethological significance of the sword-tail in *Xiphophorus hellerii* (Haekel). *Behaviour,* 27:290–315.

Hess, E. H., 1965. Attitude and pupil size. *Sci. Amer.,* 212(4):46–54.

Hinde, R. A., 1955–56. A comparative study of the courtship of certain finches. *Ibis,* 97:706–45; 98:1–28.

Hockett, C. F., and Altmann, S. A., 1968. A note on design features. In: *Animal Communication,* T. A. Sebeok, ed. Bloomington: Indiana University Press, pp.61–72.

Holst, E. von, and St. Paul, U., 1963. On the functional organization of drives *Brit. J. Anim. Behav.,* 11:1–20.

Hunsaker, D., 1962. Ethological isolating mechanisms in the *Sceloporus torquatus* group of lizards. *Evolution,* 14:62–74.

Immelmann, K., 1969. Über den Einfluss frühkindlicher Erfahrungen auf die geschlechtliche Objektfixierung bei Estrildiden. *Z. Tierpsychol.,* 26:675–91.

Jenssen, T. A., 1970. Female response to filmed displays of *Anolis nebulosus* (Sauria, Iguanidae). *Anim. Behav.,* 18:640–47.

Kahl, M. P., 1966. Comparative ethology of the Ciconiidae. Part 1. The Marabou stork, *Leptoptilos crumeniferus* (Lesson). *Behaviour,* 27:76–106.

Kahl, M. P., 1972. Comparative ethology of the Ciconiidae. Part 4. The "typical" storks (Genera *Ciconia, Sphenorhynchus, Dissoura,* and *Euxenura*). *Z. Tierpsychol.,* 30:225–52.

Kaltenhäuser, D., 1971. Über Evolutionsvorgänge in der Schwimmentenbalz. *Z. Tierpsychol.,* 29:481–540.

Keenleyside, M. H. A., 1971. Aggressive behavior of male longear sunfish (*Lepomis megalotis*). *Z. Tierpsychol.,* 28:227–40.

Keenleyside, M. H., 1972. The behaviour of *Abudefduf zonatus* (Pisces, Pomacentridae) at Heron Island, Great Barrier Reef. *Anim. Behav.,* 20:763–74.

Kilham, L., 1954. Territorial behavior of the red squirrel. *J. Mammal.,* 35:252–53.

Kleiman, D. G., and Eisenberg, J. F., 1973. Comparisons of canid and felid social systems from an evolutionary perspective. *Anim. Behav.,* 21:637–59.

Klopfer, P. H., and Hailman, J. P., 1967. *An Introduction to Animal Behavior: Ethology's First Century.* Englewood Cliffs, N.J.: Prentice-Hall. 297pp.

Klopman, R. B., 1968. The agonistic behavior of the Canada goose (*Branta canadensis canadensis*). I. Attack behavior. *Behaviour,* 30:287–319.

Krämer, A., 1969. Soziale Organisation und Sozialverhalten einer Gemspopulation (*Rupicapra rupicapra* L.) der Alpen. *Z. Tierpsychol.,* 26:889–964.

Kroodsma, R. L., 1974. Species-recognition behavior of territorial male rose-breasted and black-headed grosbeaks (*Pheucticus*). *Auk,* 91(1):54–64.

Kühme, W., 1961. Beobachtungen am afrikanischen Elefanten (*Loxodonta africana* Blumenbach 1797) in Gefangenschaft. *Z. Tierpsychol.,* 18:285–96.

Kunkel, P., 1967. Displays facilitating sociability in waxbills of the genera *Estrilda* and *Lagonosticta* (Fam. Estrildidae). *Behaviour,* 29:231–61.

Kunkel, P., 1969. Zur Rückwirkung der Nestform auf Verhalten und Auslöserausbildung bei den Prachtfinken (Estrilididae). *Z. Tierpsychol.,* 26:277–83.

LaFollette, R. M., 1971. Agonistic behaviour and dominance in confined wallabies, *Wallabia rufogrisea frutica. Anim. Behav.,* 19:93–101.

Lill, A., 1968a. An analysis of sexual isolation in the domestic fowl: I. The basis of homogamy in males. *Behaviour,* 30:107–26.

Lill, A. 1968b. An analysis of sexual isolation in the domestic fowl: II. The basis of homogamy in females. *Behaviour,* 30:127–45.

Lorenz, K., 1958. The evolution of behavior. *Sci. Amer.,* 199(6):67–78.

Luria, S. M., and Kinney, J. A., 1970. Underwater Vision. *Science,* 167:1454–61.
McBride, G.; Parer, I. P.; and Foenander, F.; 1969. The social organization and behaviour of the feral domestic fowl. *Anim. Behav. Monog.,* 2:127–81.
McFarland, D. J., and Baher, E., 1968. Factors affecting feather posture in the Barbary dove. *Anim. Behav.,* 16:171–77.
Machemer, L., 1970. Qualitative und quantitative Verhaltensbeobachtungen an Paradiesfisch-♂♂, *Macropodus opercularis* L. (Anabantidae, Teleostei). *Z. Tierpsychol.,* 27:563–90.
McKinney, F., 1965. The comfort movements of Anatidae. *Behaviour,* 25:120–220.
Markl, H., 1972. Aggression und Beuteverhalten bei Piranhas (Serrasalminae, Characidae). *Z. Tierpsychol.,* 30:190–216.
Marler, P., 1955. Studies of fighting in chaffinches. 2. the effect on dominance relations of disguising females as males. *Brit. J. Anim. Behav.,* 3:137–46.
Marler, P., 1968. Visual systems. In: *Animal Communication,* T. A. Sebeok, ed. Bloomington: Indiana University Press, pp.103–26.
Marshall, A. J., 1954. *Bowerbirds: Their Displays and Breeding Cycles.* New York: Oxford University Press.
Morris, D., 1956. The feather postures of birds and the problem of the origin of social signals. *Behaviour,* 9:75–113.
Morris, D., 1957. "Typical intensity" and its relationship to the problem of ritualisation. *Behaviour,* 11:1–12.
Moynihan, M., 1955. Remarks on the original sources of displays. *Auk,* 72:240–46.
Moynihan, M., 1968. Social mimicry: character convergence versus character displacement. *Evolution,* 22:315–31.
Myrberg, A. A., Jr., 1972. Ethology of the bicolor damselfish, *Eupomacentrus partitus* (Pisces: Pomacentridae): a comparative analysis of laboratory and field behaviour. *Anim. Behav. Monog.,* 5:197–283.
Noakes, D. L. G., and Barlow, G. W., 1973. Cross-fostering and parent-offspring responses in *Cichlasoma citrinellum* (Pisces, Cichlidae). *Z. Tierpsychol.,* 33:147–52.
Noble, G. K., 1934a. Experimenting with the courtship of lizards. *Nat. Hist.,* 34:3–15.
Noble, G. K., 1934b. Sex recognition in the sunfish, *Eupomotis gibbosus* (Linne). *Copeia,* 1934:151–54.
Noble, G. K., 1936. Courtship and sexual selection of the flicker. *Auk,* 53:269–82.
Noble, G. K., and Vogt, W., 1935. An experimental study of sex recognition in birds. *Auk,* 52:278–86.
Noble, G. K.; Wurm, M.; and Schmidt, A.; 1938. Social behavior of the black-crowned night heron. *Auk,* 55:7–41.
Orcutt, F. S., Jr., 1971. Effects of estrogen on the differentiation of some reproductive behaviours in male pigeons (*Columba livia*). *Anim. Behav.,* 19:277–86.
Otte, D., 1972. Simple *versus* elaborate behavior in grasshoppers: an analysis of communication in the genus *Syrbula. Behaviour,* 42:291–322.
Packard, A., and Sanders, G. D., 1971. Body patterns of *Octopus vulgaris* and maturation of the response to disturbance. *Anim. Behav.,* 19:780–90.
Parker, G. A., 1972. Reproductive behaviour of *Sepsis cynipsea* (L.) (Diptera:Sepsidae). I. A preliminary analysis of the reproductive strategy and its associate behaviour patterns. *Behaviour,* 41:172–206.
Payne, A. P., and Swanson, H. H., 1972. The effect of a supra-normal threat stimulus on the growth rates and dominance relationships of pairs of male and female golden hamsters. *Behaviour,* 42:1–7.
Peek, F. W., 1972. An experimental study of the territorial function of vocal and visual display in the male red-winged blackbird (*Agelaius phoeniceus*). *Anim. Behav.,* 20:112–18.
Peeke, H. V. S., 1969. Habituation of conspecific aggression in the three-spined stickleback (*Gasterosteus aculeatus* L.). *Behaviour,* 35:137–56.
Peeke, H. V. S.; Wyers, E. J.; and Herz, M. J.; 1969. Waning of the aggressive response to male models in the three-spined stickleback (*Gasterosteus aculeatus* L.). *Anim. Behav.,* 17:224–28.
Phillips, R. E., 1972. Sexual and agonistic behaviour in the killdeer (*Charadrius vociferus*). *Anim. Behav.,* 20:1–9.
Phillips, R. R., 1971. The relationship between social behavior and the use of space in the benthic fish *Chasmodes bosquianus* Lacepede (Teleostei, Blenniidae). I. Ethogram. *Z. Tierpsychol.,* 29:11–27.
Pilleri, G., and Knuckey, J., 1969. Behaviour patterns of some Delphinidae observed in the Western Mediterranean. *Z. Tierpsychol.,* 26:48–72.
Potts, G. W., 1973. The ethology of *Labroides dimidiatus* (Cuv. & Val.) (Labridae, Pisces) on Aldabra. *Anim. Behav.,* 21:250–91.
Purdue, J. R., and Carpenter, C. C., 1972. A comparative study of the body movements of displaying males of the lizard genus *Sceloporus* (Iguanidae). *Behaviour,* 41:68–81.
Quanstrom, W. R., 1971. Behaviour of Richardson's

ground squirrel *Spermophilus richardsonii richardsonii. Anim. Behav.*, 19:646–52.
Rand, S. A., and Williams, E. E., 1970. An estimation of redundancy and information content of anole dewlaps. *Amer. Nat.*, 104:99–103.
Rasa, O. A. E., 1969. Territoriality and the establishment of dominance by means of visual cues in *Pomacentrus jenkinsi* (Pisces:Pomacentridae). *Z. Tierpsychol.*, 26:825–45.
Raveling, D. G., 1970. Dominance relationships and agonistic behavior of Canada geese in winter. *Behaviour*, 37:291–319.
Reese, E. S., 1962. Submissive posture as an adaptation to aggressive behavior in hermit crabs. *Z. Tierpsychol.*, 19:645–51.
Rhijn, J. G. van, 1973. Behavioural dimorphism in male ruffs, *Philomachus pugnax* (L.). *Behaviour*, 47:153–229.
Rood, J. P., 1972. Ecological and behavioural comparisons of three genera of Argentine cavies. *Anim. Behav. Monog.*, 5:1–83.
Rovner, J. S., 1968. An analysis of display in the lycosid spider *Lycosa rabida* Walckenaer. *Anim. Behav.*, 16:358–69.
Saayman, G. S.; Tayler, C. K.; and Brower, D.; 1973. Diurnal activity cycles in captive and free-ranging Indian Ocean bottlenose dolphins (*Tursiops aduncus* Ehrenburg). *Behaviour*, 44:212–23.
Sale, P. F., 1971. The reproductive behaviour of the Pomacentrid fish, *Chromis caeruleus. Z. Tierpsychol.*, 29(2):156–64.
Salmon, M., 1967. Coastal distribution, display and sound production by Florida fiddler crabs (Genus *Uca*). *Anim. Behav.*, 15:449–59.
Schleidt, W. M., 1973. Tonic communication: continual effects of discrete signs in animal communication systems. *J. Theor. Biol.*, 42:359–86.
Schmidt, U., and van de Flierdt, K., 1973. Innerartliche Aggression bei Vampirfledermäusen (*Desmodus rotundus*) am Futterplatz. *Z. Tierpsychol.*, 32:139–46.
Schmied, A., 1973. Beiträge zu einem Aktionssystem der Hirschziegenantilope (*Antilope cervicapra* Linne 1758). *Z. Tierpsychol.*, 32:153–98.
Schremmer, F., 1972. Beobachtungen zum Paarungsverhalten der Männchen von *Bombus confusus* Schenck. *Z. Tierpsychol.*, 31:503–12.
Schweinsburg, R. E., and Sowls, L. K., 1972. Aggressive behavior and related phenomena in the collared peccary. *Z. Tierpsychol.*, 30(2):132–45.
Scruton, D. M., and Herbert, J., 1972. The reaction of groups of captive talapoin monkeys to the introduction of male and female strangers of the same species. *Anim. Behav.*, 20:463–73.
Sebeok, T. A., ed., 1968. *Animal Communication.* Bloomington: Indiana University Press.
Shank, C. C., 1972. Some aspects of social behaviour in a population of feral goats (*Capra hircus* L.). *Z. Tierpsychol.*, 30:488–28.
Sharpe, R. S., and Johnsgard, P. A., 1966. Inheritance of behavioral characters in F_2 mallard X pintail (*Anas platyrhynchos* L. X *Anas acuta* L.) hybrids. *Behaviour*, 27:259–72.
Simpson, M. J. A., 1968. The display of the Siamese fighting fish, *Betta splendens. Anim. Behav. Monog.*, 1:1–73.
Smith, D. G., 1972. The role of the epaulets in the red-winged blackbird (*Agelaius phoeniceus*) social system. *Behaviour*, 41:251–68.
Smith, N. G., 1966. Evolution of some arctic gulls (*Larus*): an experimental study of isolating mechanisms. *Ornithol. Monogr.*, 4:1–99.
Smith, W. J., 1965. Message, meaning and context in ethology. *Amer. Nat.*, 99:405–409.
Smith, W. J., 1968. Message-meaning analyses. In: *Animal Communication*, T. A. Sebeok, ed. Bloomington: Indiana University Press, pp.44–60.
Smythe, N., 1970. On the existence of "pursuit invitation" signals in mammals. *Amer. Nat.*, 104:491–94.
Spinage, C. A., 1969. Naturalistic observations on the reproductive and maternal behaviour of the Uganda Defassa waterbuck *Kobus defassa ugandae* Neumann. *Z. Tierpsychol.*, 26:39–47.
Spivak, H., 1971. Ausdrucksformen und soziale Beziehungen in einer Dschelada-Gruppe (*Theropithecus gelada*) im Zoo. *Z. Tierpsychol.*, 28:279–96.
Stamps, J. A., and Barlow, G. W., 1973. Variation and stereotypy in the displays of *Anolis aeneus* (Sauria: Iguanidae). *Behaviour*, 47:67–94.
Stanley, M., 1971. An ethogram of the hopping mouse, *Notomys alexis. Z. Tierpsychol.*, 29:225–58.
Steiner, A. L., 1971. Play activity of Columbian ground squirrels. *Z. Tierpsychol.*, 28:247–61.
Stephenson, G. R., 1973. Testing for group specific communication patterns in Japanese macaques. In: *Symp. IVth Int. Congr. Primat., vol. 1: Precultural Primate Behavior*, W. Montagna and E. W. Menzel, Jr., eds. Basel: Karger, pp.51–75.
Stout, J. F., and Brass, M. E., 1969. Aggressive communication by *Larus glaucescens.* Part II. Visual communication. *Behaviour*, 34:42–54.
Sturm, H., 1973. Zur Ethologie von *Trithyreus sturmi*

Kraus (Arachnida, Pedipalpi, Schizopeltidia). *Z. Tierpsychol.,* 33:113–40.

Tayler, C. K., and Saayman, G. S., 1973. Imitative behaviour by Indian Ocean bottlenose dolphins (*Tursiops aduncus*) in captivity. *Behaviour,* 44:286–98.

Tinbergen, N., 1951. *The Study of Instinct.* London: Oxford Press. 228pp.

Tinbergen, N., 1952. "Derived" activities: their causation, biological significance, origin, and emancipation during evolution. *Quart. Rev. Biol.,* 27:1–32.

Tinbergen, N., 1953. *The Herring Gull's World.* London: Collins.

Tinbergen, N., 1959. Comparative studies of the behaviour of gulls (Laridae): a progress report. *Behaviour,* 15:1–70.

Tinbergen, N., 1963. On aims and methods of ethology. *Z. Tierpsychol.,* 20:410.

Tyler, S. J., 1972. The behaviour and social organization of the New Forest ponies. *Anim. Behav. Monog.,* 5:85–196.

van Lawick-Goodall, J., 1968. The behaviour of free-living chimpanzees in the Gombe Stream Reserve. *Anim. Behav. Monogr.,* 1(3):161–311.

Waage, J. K., 1973. Reproductive behavior and its relation to territoriality in *Calopteryx maculata* (Beauvois) (Odonata: Calopterygidae). *Behaviour,* 46: 240–56.

Walter, M. J., 1973. Effects of parental colouration on the mate preference of off-spring in the zebra finch, *Taeniopygia guttata castanotis* Gould. *Behaviour,* 46:154–73.

Walter, R. O., and Hamilton, J. B., 1970. Head-up movements as an indicator of sexual unreceptivity in female medaka, *Oryzias latipes. Anim. Behav.,* 18:125–27.

Walther, F., 1964. Einige Verhaltensbeobachtungen an Thomsongazellen (*Gazella thomsoni* Gunther, 1884) im Ngorongoro-Krater. *Z. Tierpsychol.,* 21:871–90.

Walther, F. R., 1969. Flight behaviour and avoidance of predators in Thomson's gazelle (*Gazella thomsoni* Guenther 1884). *Behaviour,* 34:184–221.

Warren, L. R.; Scheier, M. F.; and Riley, D. A.; 1974. Colour changes of *Octopus rubescens* during attacks on unconditioned and conditioned stimuli. *Anim. Behav.,* 22:211–19.

Wells, M. J., and Wells, J., 1972. Sexual displays and mating of *Octopus vulgaris* Cuvier and *O. cyanea* Gray and attempts to alter performance by manipulating the glandular condition of the animals. *Anim. Behav.,* 20:293–308.

Wickler, W., 1967. Socio-sexual signals and their intraspecific imitation among primates. In: *Primate Ethology,* D. Morris, ed. Chicago: Aldine, pp.69–147.

Wickler, W., 1969. Zur Soziologie des Brabantbuntbarsches, *Tropheus moorei* (Pisces, Cichlidae). *Z. Tierpsychol.,* 26:967–87.

Wilson, E. O., 1972. Animal communication. *Sci. Amer.,* 227(3):52–60.

Winkler, H., 1972. Beiträge zur Ethologie des Blutspechts (*Dendrocopos syriacus*). Das nicht-reproduktive Verhalten. *Z. Tierpsychol.,* 31:300–25.

Wooton, R. J., 1971. Measures of the aggression of parental male three-spined sticklebacks. *Behaviour,* 40:228–62.

Youdeowei, A., 1969. The behaviour of a cotton stainer, *Dysdercus intermedius* Distant (Heteroptera: Pyrrhocoridae) towards models and its significance for aggregation. *Anim. Behav.,* 17:232–37.

Zumpe, D., and Michael, R. P., 1970. Redirected aggression and gonadal hormones in captive rhesus monkeys (*Macaca mulatta*). *Anim. Behav.,* 18:11–19.

Chapter 10

TACTILE COMMUNICATION

Frank A. Geldard

PROLOGUE

Anyone immersed in the literature of contemporary communication theory and practice is well aware of the difficulty of traversing the bridge between human intercommunication and that of the subhuman species. When forms of behavior little used by either for elaborate intercourse, such as those employing somesthesis as a medium, become the focus of attention, perplexities abound, for the simple fact is that the phylogeny of tactile communication, far from having been rounded into a systematic division of knowledge, currently represents a substantial scientific void.

It is not that animals below man do not communicate tactually. Especially among the subhuman primates there is much in grooming and preening, and aggressive, copulatory, and suckling behavior that, at least by a loose definition of "communication," falls in this category. Although the vast majority of signals by which lower organisms impart information of importance to others of their own and related species tend to be auditory, visual, or olfactory, especially if released at a distance, there are clear instances of stroking, nudging, and other contact behavior short of the thigmotropic that might reasonably be regarded as communicative in nature. The difficulty is that records of such observations are scattered throughout a vast literature and are largely incidental to descriptions of more socially prominent (and, doubtless, significant) forms of conduct.

In what follows, therefore, the decision has been to bring together the current facts of human tactile communication, an area of almost continuous and intensive investigation for a half-century or more and one characterized by both persistent observation and ingenious experimentation. Moreover, it is one that bids fair to circumvent some old limitations of communication stemming from the handicaps of blindness and deafness.

Introduction

WHY TACTILE COMMUNICATION?

The vast majority of messages that flow through the human community are visual or auditory. They are generally designed to make an appeal to the eye if they involve many relational comparisons, if their content is complex or unfamiliar, if fine spatial discriminations are involved, or if they comprise large masses of reference data. Communication is bound to be visual if pictorial representation is demanded, for no other sense can rival vision in dealing with complex spatial patterns. Obviously, also, the medium of communication will typically be visual where there is auditory impairment or where custom or usage leads to the expectation of visual signaling.

Contrariwise, brief, simple, or transitory messages tend to be cast in auditory terms. Where one wishes to transmit, out of a larger context, only immediately relevant information or where rapid conveyance of the message is desired, the auditory channel is to be preferred. Especially where the recipient is preoccupied and we wish to "break in" on his attentional stream we do so by way of the ears. They are always open, so to speak. Moreover, audition is the channel of choice where some flexibility is required in framing a message; variations of emphasis and inflection carry shadings of meaning that can be entrusted to the eye only at the expense of time-consuming and inefficient circumlocution. Auditory messages tend to be employed where vision is overburdened or suffers outright impairment. The visual channel can be degraded through such deleterious influences as enforced mobility of the recipient or a variety of untoward environmental changes: ambient light variability, vibration, g-forces, hypoxia, or other defects arising from similar stressful alterations of the surroundings. And, of course, as with vision, there are situations leading normally, through the operation of usage or custom, to the expectation that auditory signals will be provided (Henneman, 1952).

With such complementariness between the realms of seeing and hearing, how can there be a role for the cutaneous senses in the world of communication? There is, to be sure, the obvious instance in which both major senses are unavailable; the deaf-blind constitute one population crying out for tactile aid. We shall consider some of the approaches to this problem, for it is an ancient one and one supporting an extensive literature. But, normally, as modern life brings on an overload of the ordinary channels of communication, it is also proper to inquire what the remaining sensory systems of the body have to offer by way of relief. Let us call the roll of possibilities.

The cutaneous channels, by physiological and psychological convention, are four in number: touch (pressure), pain, and the two temperature senses, warmth and cold. Additionally, there are the two chemical senses, smell and taste. What are the chances that any of these might substitute for hearing or seeing?

We may dispense with the last two rather quickly. Both gustation and olfaction normally come into operation through the triggering action of chemical stimuli: materials in solution in the case of taste, volatile substances for smell. In both instances there is the necessity of getting the stimuli from the source to the site of stimulation; typically there is a time-consuming transport problem. Odorants must be admitted to the nose by sniffing; they must then pass up to the sensitive epithelial patch high in the nostrils, chiefly by eddy currents swirling around the turbinate bones; and then they must be adsorbed on tiny fibrils immersed in mucus. All this takes time. Moreover, olfactory sensations, once aroused, have a slow subsidence rate. Again, it takes time for the chemical stimulus to be expended. The net result is a ponderous rise and fall of sensation, even though it seems probable that, at the site of transduction in the olfactory epithelium, the interchange of energy for a given smell molecule must be an exceedingly rapid affair. But each sniff involves millions, even trillions, of molecules, and not all follow the same time course through transport and adsorption.

The story for taste is much the same. Food particles and chemical substances, once led to the tongue and palate, must go into solution. It is not entirely clear yet whether the solvent is always simply water or whether enzymes, already present in the lingual tissues, must interact with them. In any case, there is the transport problem again—the process whereby dissolved taste stimuli are brought into juxtaposition with sensitive gustatory receptors—and the further question of how ions are able to release impulses in the cranial nerves subserving the tongue and palate. The whole chain of events is again time consuming and persistent. Neither of the chemical senses is really suitably constituted to relay promptly the messages available to them.

To return, then, to the skin, what of the thermal senses, warmth and cold? Somewhat similar considerations arise here. To be sure, low and high skin temperatures produce radically different effects, cold "flashing out," warmth "welling up." But the two responses have the common characteristic that, at least as related to the physical sources prompting them, they come out of relatively sluggish systems for information transmission. Again, the final word on the nerve-impulse generation mechanism is not yet in—we do not know whether impulses reporting on warmth and cold come from the same or different receptors, whether they are conducted over identical nerve fibers, or whether their central processing differs greatly one from the other—but we do know that the organism's integument is better designed to protect internal organs against surges in temperature than to provide faithful reports on thermal changes in the environment. To learn about heat interchanges in one's surroundings it is better to consult a thermometer than to heed the messages coming from one's own skin.

We are thus left with the prospect of utilizing the pressure and pain senses if cutaneous channels are to become effective in the transmission of messages, and since it will generally be conceded that pain, in the context of human intercommunication, is not a "consummation devoutly to be wish'd," we are, for all practical purposes, reduced to the use of the tactile sense in our search for an eye-ear substitute. At the same time, it should be hastily pointed out that the world of tactile sensation is a rich and versatile one. Of the six classes of physical stimuli that can arouse human senses—photic, acoustic, mechanical, electrical, chemical, and thermal—all but the first, light, can act upon the skin to produce somesthesis. Normally, to be sure, acoustic stimuli are not generated at power levels that will affect the skin appreciably, and the undesirable features of thermal and chemical stimulation for the purposes under discussion will be somewhat apparent from what has been said above about the senses to which these forms of energy are appropriate. Of the remaining two classes of stimuli, the mechanical and the electrical, we shall have much to say, for these provide the really live prospects for cutaneous communication.

The virtues (and some of the defects) of vision and audition as communication channels have been set forth. Does touch have comparable qualities to commend it? There are some features worthy of mention. To begin with, the skin is eminently available. Its area is somewhat over a thousand times that of the retina and is freely supplied with nerve endings, thus providing a nearly continuously sensitive surface. Moreover, it is an area that is relatively traffic-free. There is also a free orientation of the skin toward potential sources of information; bodily orientation toward the source is necessary for vision, less so for audition. Also, the skin is a flexible organ, rarely retaining a deformation for long. It thus has something of an advantage with respect to declining sensitivity from continuous stimulation, sensory adaptation. There are few accidents of stimulation where the skin is concerned; at least, there are rarely sustained accidental stimuli. On the contrary, unwanted lights and sounds often interfere with message reception in vision and audition. Finally, as with audition, there is little redundancy in cutaneous information; visual information, conversely, is commonly highly redundant. Thus, touch shares with hearing the advantage that information can be presented only when needed.

Tactile communication has been spoken of as if the only role it might play is that of a replacement for a missing visual or auditory system. Actually, except for the rare instances in which neither is available, there is no real expectation that somesthesis could furnish the richness of experience normally supplied by either sight or hearing. Broadly speaking, there are two great classes of perceptual discriminations made about things and events in the world; they involve distinctions of space, on the one hand, and of time, on the other. Vision is the great spatial discriminator; audition excels in the realm of time. Touch stands midway between the two major senses in the respect that it is endowed with some of each character.

Touch is better than vision at temporal discriminations and better than audition at spatial discriminations, but by the same token it is spatially inferior to vision and temporally inferior to audition. This means, however, that it can provide an avenue for the avoidance of mental bankruptcy for the deaf-blind, and while some spectacular steps have been taken—witness the Laura Bridgman–Helen Keller phenomenon—the vast potentialities of somesthetic substitution have not yet begun to be realized. As Cassirer has said:

> Vocal language has a very great technical advantage over tactile language; but the technical defects of the latter do not destroy its essential use. The free development of symbolic thought and symbolic expression is not obstructed by the use of tactile signs in the place of vocal ones. If the child has succeeded in grasping the meaning of human language, it does not matter in which particular material this meaning is accessible to it. [1944:36]

The truth of this assertion is amply evidenced by the clinical experience that at least a few of the deaf-blind can quite successfully employ the Tadoma, or "speech feeling," method to receive communications from their teachers. In this technique the "reader" places his thumbs on the instructor's lips, the index fingers to the sides of the nose, and the remaining fingers on the cheeks and upper throat. With the faint cues thus provided the total speech pattern of movement is remarkably faithfully interpreted (Kirman, 1973:59).

Currently, it is less as a surrogate than as an ancillary system that tactile communication should be viewed. Except for a few somewhat esoteric situations—to attract attention in emergencies, especially where the monotony of routine has dulled normal perception; to permit intercommunication where darkness and enforced quiet have supervened; to provide warnings of threatening events outside the visual and auditory fields; to preserve secrecy in clandestine operations—a tactile communication system's main prospect for service is as a cooperative supplement to other sense channels. Perhaps because we normally attend so little to what is going on in the skin, any startling stimulation of the integument immediately "cuts through" and comes to our attention. Panic buttons should be wired to tactile stimuli. For much the same reasons any visual and auditory overloading can best be relieved by appeal to the skin.

Cutaneous Channels of Communication

CLASSIFICATION OF TACTILE SYSTEMS

Attempts to deliver messages through the skin appear to have been made from very early times. However, there is no unbroken continuity of effort dating back to the pyramids of Egypt or the Golden Age of Greece, only isolated and sporadic attempts to deal with blindness and deafness at various periods of history. Thus, the Venerable Bede, in late seventh-century England, "master of all knowledge of his time," seems to have been greatly concerned about deaf-mutism and its attendant communication problems. George Dalgarno, Scottish protégé of Charles II, in his "deaf and dumb man's tutor" (1680), developed a whole touch alphabet for delivery by the fingers to another's hand. Jean-Jacques Rousseau incorporated a passage in his well-known educational treatise, *Émile* (1762), on the possibility of cutaneous communication and suggested that "if our touch were trained to note [natural vibratory] differences, no doubt in time we might become so sensitive as to hear a whole tune by means of our fingers . . . [and these tones] might be used as the elements of speech." To these scattered ideas there must be added those of string writing, a means of communication in which the "sender" tied a succession of distinctive knots, coded to the alphabet, to be felt by the recipient, who passed them through his fingers, and various systems of finger spelling.

In the late eighteenth century Valentin Haüy introduced raised letters embossed on paper (Farrell, 1950). Though extremely difficult, this system was the prototype for many similar tangible alphabets, most of which have gone out of use. So-called Moon type, an arrangement of embossed lines somewhat resembling the Roman alphabet, has persisted with some success, especially among those losing their sight late in life. The system of raised dots, now so widespread in use, was initially developed by Louis Braille from a previously existing military code (Bledsoe, 1972). The method of Braille, himself bereft of sight at an early age, has become the modern refuge of the blind, at least the more pertinacious of them. The coding of the braille language is an arbitrary one, developed along logical rather than psychological lines, and is very difficult to master. It exists in several forms, the variations stemming chiefly from different practices with respect to the formation of contractions of more elaborate language units. Withal, braille is by far the most widely used language for the blind and currently represents the most successful medium of their education.

It was earlier pointed out that both mechanical and electrical stimuli are capable of providing the signals from which cutaneous language symbols and other forms of communication can be constructed. A possible classification of tactile systems of communication follows almost immediately upon this distinction. One class, clearly the largest, would include all systems employing mechanical stimuli to move the skin, whether with single pulses, trains of them, steady vibration, or elaborately patterned energy fluctuations such as are found in the speech signal. A second class would involve all direct applications to the skin of electrical currents having temporal patterns corresponding to the mechanical ones, whether originating in direct, oscillating, or alternating potentials. Such a classification would have to make room for a third category, the electromechanical, for it is possible by taking advantage of the electrical capacitance properties of tissues to produce mechanical movements of the skin by this means. In such a system the skin becomes one plate of a condenser. With the application of suitable voltages it is possible to bring about substantial mechanical skin displacement.

Another way to classify tactile communication systems is in terms of their major mode of operation. Thus, some systems are designed for direct mediation of spoken speech, whether the final link with the body is mechanical, electrical, or electromechanical. Other systems involve pictorial displays. Many simply present a systematic array of discriminable signals that have either been coded to language elements or been given some other symbolic meaning. Still other devices are arranged for cutaneous monitoring or tracking of a continuously variable set of environmental events.

Actually, it is unnecessary to choose between these two modes of classification. The two dimensions of analysis practically force on one an Aristotelian cross-classification (see Table 1.) The suspicion is strong that, if we knew more about electromechanical possibilities, all the "pigeonholes" in the table would be filled. As we shall see, this area, which, for technological reasons that will become obvious, has been greatly neglected. Keeping the provisions of our table in mind as an orientation, let us take a closer look at the several operational problems represented in it.

DIRECT SPEECH MEDITATION: MECHANICAL METHODS

The modern period of research aimed at direct transmission of speech to the skin began with the work of David Katz (1930) and his coworkers in Germany, and of Robert Gault (1926) and his associates in the United States. Both lines of work initially made use of electromagnetic receivers (speaker units) held in contact with the

Table 1

Tactile communication systems and devices.

Mechanical	Electromechanical	Electrical
Direct speech mediation	**Skin capacitance effects**	**Direct speech mediation**
Speaking tube	Electrostatic "textured" matrix	Simple dermal electrodes
Electromagnetic receiver		Vocoder
Teletactor		
Vocoder		
von Békésy cochlear model		
Tadoma ("speech feeling") method		
Pictorial display		**Pictorial display**
Tactile TV		Tactile TV
Elektroftalm		Elektroftalm
Optacon		Optacon
Embossed legible print		
Coded language		**Coded language**
Vibratese		Katakana
Body braille		International Morse
Optohapt		Electrocutaneous vibratese
Polytap		
Air-blast symbols		
Visotactor		
Braille and derivatives		
Finger spelling		
String writing		
Tracking and monitoring		**Tracking and monitoring**
Vibrators and air jets on hands or forehead		Single and multiple active electrodes on fingers, arms, legs, or neck

skin while the amplified output of a microphone transmitted voice sounds or music to it. Thus, with fidelity limited only by the generally crude sound system then available and variable back action resulting from damping of the contactor by the skin, moment-to-moment variations of frequency and amplitude of the speech signal were imposed directly on the skin, typically that of the palm or fingertip. In Gault's work there was a precursor, a long, hollow tube held against the palm. The reinforced acoustic signal was thus impressed directly on the skin; with the tube's help (and with hearing masked) it was possible to distinguish among several of a group of simple words.

But the approach for which Gault's laboratory became best known centered on a new instrument, the teletactor, which was developed for the purpose by Bell Telephone Laboratories. This device consisted of a piezoelectric crystal of the type subsequently popular in sound systems (tweeters). It was small in size, little affected by skin damping, and would vibrate over a considerable range of frequencies. It could be held so as to affect the fingertip or it could readily be applied to other body areas. In one application of the teletactor (Gault and Crane, 1928) five of the instruments, dividing up the speech frequency band, were used simultaneously on the fingers of one hand; in effect this anticipated the tactual vocoder (see below).

Many patient experiments were conducted in an effort to transmit speech to the fingertip. Early results were promising. After 14 hours of well-distributed practice, the subjects learned to identify with about 75 percent accuracy which one of ten brief sentences had been spoken into the microphone, and with 30 hours of training behind them they could judge about half the time which one of more than fifty words, isolated from context, had been presented. But it was also found that a change in experimenters or in the rate of speaking by the same experimenter produced a collapse of this apparent ability to "hear through the skin." The subjects had presumably been relying on cues of emphasis and rhythm, in general the prosodic features of speech. The attack by way of the teletactor was eventually dropped, and it must be judged to have been a major disappointment.

Several lines of investigation have approached the problem of transmitting speech to the skin by way of variants of the vocoder, an acoustic analytic device dating from 1936 (Dudley). The prototype of this instrument in its tactual application is to be found in Project Felix of the Massachusetts Institute of Technology (Levine et al., 1949–1951). A development from it, a device designed at the Speech Transmission Laboratory, Stockholm (Rösler, 1957), broke up speech into ten channels having center frequencies ranging from 210 Hz to 7,700 Hz. The energy in each channel modulated a 300-Hz carrier that was common to all ten outputs, one for each finger. The net result of this arrangement was that patterns of vibration, analyzed with respect to the relative amplitudes of their components, could be presented in real time to different skin loci. Unfortunately, in a thorough study of performance with this vocoder (Pickett and Pickett, 1963) too few patterns could be discerned to yield anything but a hope that the device, through future refinements, might one day permit something approximating satisfactory recognition of tactual speech.

Subsequently there have been other attempts to utilize the vocoder principle. One device (Guelke and Huyssen, 1959), consisting of an array of twenty vibrating reeds contacting the several joints of the fingers at eight locations of one hand, delivers constantly changing patterns of vibration that follow those of the speech signal. The reeds perform a fairly discriminative frequency analysis that, by this arrangement, gets transformed to distinctive skin loci. Though some vowel sounds were easily distinguishable in early tests, as were characteristic temporal patterns of some diphthongs, great difficulty was experienced in discriminating among consonantal sounds. Also, as might have been predicted, there were difficulties in analyzing out separate loci when two or more were stimulated simultaneously.

Another variant of the vocoder is an instrument going under the name of Tactus (Kringlebotn, 1968). Somewhat like the early Gault instrument, it applies the speech signal to five bone conduction receivers, which contact the fingers of one hand. These are also energized simultaneously. Its unique feature is that, through the intermediation of a set of multivibrators, which fractionate the input in different ways, all the fingers receive the basic signal but with different portions stressed. No elaborate claims have been made for this device beyond the suggestion that it may help, in teaching the deaf child to speak, to reinforce the concept of rhythm and to provide some useful ancillary cues. Still other variants of the vocoder are known; the idea has been a persistent one. Kirman, reviewing the vocoder story (1973:59) was led to conclude: "The history of tactile vocoders indicates that simply providing the skin with such frequency-to-locus translators as have been tried does not enable it to comprehend speech."

It is not necessary to lead each frequency band to its own position on the skin. Up to some point of resolution determined by its "funneling" capacity, it is possible for a stretch of skin to distinguish among frequencies on the basis of the location each will seem to occupy when the area in question is broadly stimulated. This is the principle, so influential in auditory theory, on which the Békésy cochlear model operates (Békésy, 1955). An approach made by Keidel and his students at Erlangen, Germany, relies on the Békésy model. Their application has its source in the consideration that tactile response to vibration has much in common with that of the ear, except for three important features: (1) The two senses occupy overlapping, but quite different, frequency ranges. For the ear, the range important for speech extends from about 300 Hz to 3,000 Hz. The skin is most sensitive in the 25 Hz to 400 Hz range. (2) The skin's sensory response to a stimulus is relatively sluggish, requiring about 1.2 sec to reach a peak, while the ear does so in less than one-sixth that time (0.18 sec). (3) Differential frequency discrimination is sharp in the case of the ear, which in its best performance yields a Weber fraction of about 0.2 percent; the skin at its best does about 5 percent and, as frequency goes up, becomes 50 percent or more. Obviously, if spoken speech is to be apprehended by the skin with any fidelity at all, it is necessary to transpose frequencies downward into the region of the skin's optimal response. The Erlangen group went about it this way.

The basic experiment is that of Biber (1961). He first stored samples of speech—three classes of monosyllabic words having high-, low-, and middle-range characteristic frequencies, together with a selection of phonemes—on magnetic tape. The tape was then played back at reduced speed to a Békésy cochlear model on which the subject's forearm rested. Three different ratios of playback-to-recording speeds were tried: 1:8 (which should yield frequencies nearly ideal for the skin), 1:4, and 1:2. Endless tapes with known syllables or words permitted multiple training trials.

The 1:8 speed reduction proved unsatisfactory, mainly because the presentation time exceeded the subject's short-term memory; by the end of a word he would have forgotten how the word had started! (This perplexing problem has been encountered in other tactile performances and is far from solution.) The 1:4 time reduction was satisfactory, however, and with this amount of slowing in force Biber continued his training experiments. Indeed, he found that unknown words could be correctly recognized 83 percent of the time after only 18 hours of training, and performance became practically errorless after 32 hours.

The fault of Biber's system lies, of course, in the fact that the time necessary for the transmission of a given segment of speech has been multiplied by four. Since, in any normal operation of the senses or the response mechanisms triggered by them, there is likely to be a good deal of cooperation involving their acting together in time, artificial slowing of speech has to be put down as a serious disadvantage. Accordingly, the Erlangen laboratory (see Keidel, 1968, 1974) set about to remedy the defect.

The answer came in an ingenious application by Finkenzeller (1973) of still a different physical principle. Divers, working at great depths below the surface, can sustain normal breathing only in a helium atmosphere. But it has the disadvantage that the human voice, because of the high velocity of sound in helium, becomes so distorted as to be unintelligible. The way out of this dilemma has been found, quite recently, to be simple. Speech sounds are highly redundant; removing parts of them and providing continuity to the remainder leaves them quite intelligible. By a suitable computer program it is possible to extract every fourth wave in a speech signal, suppress the rest, and expand each remaining wave to fill up the time period originally occupied by all four. The only problem is that of joining the segments in such a way as to avoid repetitive clicks, and hence unwanted "periodicity pitch" from them. This, too, was accomplished by Finkenzeller by smoothing the connection between segments within a 400-microsecond period, thus obviating the click stimulus, which would otherwise constitute an interfering signal.

The technical difficulties of sound expansion having been overcome, there remains the important question of how well the skin is going to be able to utilize the transposed speech signal. To date one has only the report (Keidel, 1974:31): "The preliminary results are very promising, because (a) it now works in real time, (b) tactile memory is not overloaded, and (c) vibrotactile information can be combined with that from other sense modalities. . . ."

ELECTRICAL ANALOGS

Until now we have been considering only mechanical systems that deliver signals, chiefly those of speech, directly to the skin. There are electrical analogs, however, some that extend a considerable distance back in time. Early suggestions along these lines were made at the turn of the century by MacKendrick (see Lindner, 1936) and only a little later by DuPont (1907), who was impressed by the fact that electrical pulses corresponding to different musical tones from phonograph records yield some discriminably different "feels" and could especially transmit musical rhythms to the skin. Serious attempts to deliver speech signals to dermal electrodes have not been carried out at a frenzied pace, however, especially following the disappointment occasioned by the more or less complete failure of the electrocutaneous form of MIT's Felix (Levine et al., 1949–51). There have been some, however. Mention should be made of the work of Anderson and Munson (1951) at Bell Telephone Laboratories, which did much to establish limits for the avoidance of pain and to specify some conditions of successful discrimination of signals. Significant also are the experiments of Breiner (1968), in Germany. He devised a fourteen-element electrocutaneous vocoder, seven electrodes on each hand, which possessed the virtue that spatiotemporal patterns, marked by vivid perceived movement, could be created with word sounds, thus adding a possibly valuable element for use in cutaneous communication systems.

Perhaps the most persistent and thoroughgoing attempt to mediate speech sounds electrocutaneously was that of Lindner (1936). Indeed, after finding that direct feeding of speech signals into dermal electrodes gave little in the way of discrimination, he made a valiant effort to salvage what he could by devising an interesting, if primitive, vocoder that utilized both mechanical and direct electrical stimuli. The four fingers of the left hand divided up the task, the index and little fingers receiving electrical signals reproducing high and middle frequencies, while the middle and ring fingers got mechanical stimulation by low frequencies. There were no revolutionary findings. This somewhat analytic scheme overestimated the value of the vocoder principle, but it does represent a unique combining of mechanical and electrocutaneous approaches to tactile communication, perhaps the only such hybrid ever attempted.

The impression should not be left that efforts to communicate electrocutaneously have been entirely sterile. On the contrary, much has been learned about the basic psychophysical functions involved (lower and upper limits of intensity and frequency discrimination, etc.), relation to pain and discomfort, preferred electrode arrangements, and reaction time to such stimuli. Incomplete to scanty information has been acquired on such matters as the limits and possibilities of spatial discrimination, generation of movement patterns, masking, and temporal integrations in electrical stimulation. Much more needs to be learned about the conditions for the avoidance of pain in electrocutaneous communication. The reviews by Breiner (1968), Rollman (1974), and Sherrick (1975) point up the problems.

We shall be returning to the topic of electrical approaches, for there have been some interesting attempts to synthesize coded skin languages with this mode of stimulation.

PICTORIAL DISPLAYS IN TACTILE COMMUNICATION

It is not only the auditory environment that may be transmuted to somesthetic sensation; the visual world can be represented as well. Therein lies the possibility of alleviating some of the burden of blindness. Three lively systems, tactile television, the Optacon, and the Elektroftalm, have been devised to this end. Each will be dealt with briefly.

Tactile TV, a development carried out at the Smith-Kettlewell Institute of Visual Sciences, San Francisco, provides an elaborate array of 400 tiny electromagnetic vibrators that presses against a 10" X 10" area of a subject's back (Collins, 1970; White et al., 1970). The vibrator matrix reproduces with sufficiently fine grain the image picked up by a vidicon camera. Thus a vibrating pattern of a two-dimensional image in the camera's field of view is transferred to the subject's back. The subject can manipulate the vidicon, moving it vertically and horizontally to bring the image to any part of the matrix or, indeed, to take it "off screen." This is important, because it may well be that the transit of the image across the border of the field provides the salient cues that permit recognition of objects by way of their cutaneous patterns. That is strongly implied by the experiments of Craig (1974).

It is too early to evaluate tactile TV. The system has undergone several modifications, including departure from the mechanical mode of stimulation; e.g., an electrocutaneous matrix that contacts the abdomen has been substituted. Also, special arrangements with respect to electrode design and application have been introduced (Saunders, 1974). Whatever the outcome for a visual substitution system, it is already apparent that much remains to be learned about the spatial properties and inherent limitations of both mechanical and electrocutaneous displays.

The Optacon (*op*tical-to-*ta*ctual-*con*version), an instrument developed at the Stanford Research Institute, is also intended as a reading aid for the blind (Bliss et al., 1970). The device presents a set of tiny pins, 144 of them in a 24 X 6 array, to a single fingertip. Movement of the pins depends on vibration of piezoelectric Bimorph reeds to which they are rigidly attached, and these, in turn, are controlled by a bank of 24 phototransistors, each controlling a lineal array of 6 contactors. As the image of a printed letter, say, passes over the photosensitive surface, corresponding vibrators are set in motion, with the result that a vibratory "image" crosses the fingertip. The moving display, in principle, is exactly like the message that travels around the Allied Chemical Building in Times Square, New York; indeed, this kind of representation is familiarly known as a "New York Times display."

At least for some purposes, practical reading rates are possible with the Optacon, but as with tactile TV, full evaluation is not possible yet. The device has only relatively recently been made available commercially, but already scores of blind people have received training on it. It has an inherent limitation as a reading device in that it can comprehend only one letter at a time, but this is also true of many less "pictorial" systems that involve, additionally, the learning of a novel code. Moreover, written messages are not the only materials to which the Optacon may be adapted. Tracking experiments have been conducted with its aid, and there have been at least preliminary explorations of the possibility of using it as an environmental form detector (Bliss et al., 1970).

The third of this trio, the Elektroftalm, is a Polish invention (Starkiewicz and Kuliszewski, 1963). It was developed mainly to provide a mobility aid for the blind. Its "field of view," some 30° or more wide, is sensed by a camera, the "film" of which consists of 120 photocells mounted on top of the user's head. With suitable amplifying and switching elements in each circuit, changes in light intensity activate a mosaic made up of 120 small vibrators contacting the forehead. As with tactile TV (the Elektroftalm, first reported in use in 1962, is essentially its technological ancestor) sufficiently large and bright objects on a dark background may be detected. Assuming head-orientation cues to have been learned sufficiently well, the tactual "image" can aid in avoiding obstacles and in recognizing some familiar objects by their highlights. But, as its inventors say (Starkiewicz and Kuliszewski, 1965:32), "appreciation of a distinct plastic image is out of the question."

All three of the pictorial systems described above have embodiments involving direct electrical stimulation of the skin. We have seen that tactile TV has recently taken a turn in that direction with its electrocutaneous abdominal signaling system. In the course of its development, the Optacon in electrocutaneous form was given a brief trial, which proved generally unsatisfactory (Melen and Meindl, 1971), and the Elektroftalm was also converted to direct electrical stimulation of the forehead, but with results that seemed not to warrant continuation of the project.

It would be helpful if a set of crucial experiments could be devised and carried out to settle the long-standing question of the degree to which the human skin simulates the human eye in its capacity to make spatial distinctions. There is no difficulty in supplying a fine-grained mosaic to the skin, but to what extent is it a waste of effort? The skin is not innervated in the same way that the retina is; the suspicion is strong that tactual space is not just a vague replica of visual space. Presumably, pictorial systems of tactile communication stand or fall on the answer.

CODED CUTANEOUS LANGUAGES: MECHANICAL SYSTEMS

All the message systems we have considered thus far have involved imposition on the skin of language symbols developed throughout human history for either visual or auditory conveyance. The thought must have occurred that spoken or graphic stimuli may not be the most congenial for the skin to mediate. There is another way to go about solving the general problem of cutaneous communication. Instead of requiring the tactile mechanism to deal with signals unnatural to it, one may ask, "What discriminations are possible for the skin and its neural attachments?" If a substantial number of highly distinctive and manipulable signals can be discovered, it would then be possible to code them to symbols already in the receiver's possession.

There is one obvious existing code that presents itself as a viable candidate, International Morse. Anyone who has learned the Morse code for use in radio or telegraphy and is therefore able to receive it through audible dots and dashes or, as in shipboard signaling, by flashing lights, is equally capable of interpreting messages by way of mechanical taps or buzzes on the skin. Indeed, cutaneous Morse has benefited from a specialized miniature and portable instrument dubbed Taccon (Dalrymple, 1973), which facilitates coded communication in many situations. Interestingly, there have also been experiments aimed at ascertaining the feasibility of electrocutaneous Morse (Foulke and Brodbeck, 1968). The results of these were not unexpected. Subjects well versed in International Morse who could receive errorlessly at the rate of 20 five-letter words a minute when getting an audible signal were highly variable in performance on the dermal code, receiving at no better than half this speed when fully practiced. The main difficulty was in distinguishing the boundaries between dots and dashes; and adjustments of intensity, rise time, carrier frequency, etc., did not improve the situation. The very factors that bring about a deterioration of performance when electrocutaneous stimulation is substituted for mechanical in the various pictorial systems already considered appear to be responsible for a similarly poor outcome here.

Actually, despite its widespread use for a variety of purposes, chiefly because it is highly resistant to noise degradation, International Morse is really quite inefficient because it is very wasteful of time. Having only the two building blocks, dots and dashes, with which to work, strings of these symbols have to be separated by costly silences if the alphabet, numerals, punctuation marks, and conventional abbreviations are to be accommodated. Proficient Morse operators, those to whom important messages are entrusted, seldom exceed the rate of about 25 five-letter words per minute. Compared with ordinary reading or speaking rates, even with speeds achieved by practiced braille readers, Morse is really a very pedestrian language. Any scheme demanding fewer blocks of silence would be an improvement.

An approach to achieving a set of tactile coded languages was made by the author, his students, and colleagues over a span of years at the University of Virginia and, later, at Princeton University. Four systems of cutaneous communication were developed: vibratese, body braille, the optohapt, and polytap. Each has its own set of principles while sharing the common property that each depends on coding.

Vibratese

Vibratese was the outgrowth of a broad attack on the area of cutaneous communication that, reversing the earlier approach of trying to make the skin adapt to some previously existing communication hardware, asked what basic discriminations the tactile system was capable of making. In brief, it asked, "What is the tongue of the skin?" From the outset it was apparent that a dimensional analysis of the skin's discriminatory capacities was needed. Once the decision was made that sustained vibration was the stimulus of choice—superior to transient pressures, pokes, or jabs, which fail to take advantage of the excellent temporal distinctions of which the cutaneous system is capable—the dimensions involved, at least the first-order ones, became obvious. The discriminable dimensions of vibration, taken in relation to bodily stimulation, are: intensity (amplitude or energy), duration, frequency, and locus. There are several higher-order dimensions, such as wave form (simultaneously operating frequencies), movement (change of locus through time), "attack" (change of amplitude through time), and some others, but these are not at the heart of the matter and, for the most part, only provide shadings of perceptual patterns that, by extending coding possibilities, could enrich the vibratese language.

The fundamental experiments leading to vibratese are those of Spector (1954), who did a rough casting up of the roles of both intensity and duration; Howell (1956), who obtained the first important data on locus; and Goff (1967), whose measurements of frequency discrimination revealed why this vibratory dimension could not be readily accommodated. It remained for Howell to engineer a suitable code and to ascertain with what ease it could be learned (1956).

The initial attacks on intensity and duration were strictly psychophysical. Applying vibrators to the chest region—mainly because it provided a relatively traffic-free expanse of skin (we had possible vehicular use in mind)—there were first determined the number of discriminable steps ("just-noticeable differences") that could be felt between a point safely above threshold and one well below the discomfort level. On the average, fifteen steps could be appreciated. Similarly, there were charted perceived differences in vibratory duration between the briefest buzz identified as such and a relatively long-lasting one (2 sec). About twenty-five steps could be discriminated. But just-noticeable differences are not codable units; if a particular intensity or duration were to be presented in isolation it could not be identified without error (Miller, 1956). If code symbols are to be attached to vibratory intensities and durations it is not safe to go above four or five steps of each. Actually, not more than three steps were used in either dimension. More were not needed because, meanwhile, the decision had been reached to use five well-spaced loci on the chest and not to present more than one vibrator at a time. Since, where separate dimensions of the stimulus are combined, the internal relations are multiplicative, a total of forty-five signals (3 X 3 X 5) could be derived from the three steps of intensity (strong, moderate, weak), the three of duration (long, medium, short), and the five loci (separated by about 5–6 inches and arranged domino-fashion). Twenty-six specific combinations of intensity, duration, and locus were coded to letters of the alphabet, attention being paid to frequency of usage when assigning them; another ten were coded to digits; and a few were set aside for the most commonly recurring words in English prose (Geldard, 1962).

The vibratese language, having available a relatively large number of unique collocations of its three basic dimensions, proved to be far more efficient than simpler schemes, such as the telegraphic code. There are no wasteful silences in it. But could anyone learn it? The answer came quickly. Not only could it be mastered in a short series of training sessions, but two- and three-letter words could be introduced quite early in that learning. One subject acquired vibratese so rapidly that, within a matter of a few weeks, he was receiving almost without error at the rate of 38 five-letter words a minute, a speed that nearly doubles that of proficient Morse reception. When the experiment was discontinued it was not because the learning ceiling had been reached but because, in the precomputer period in which the first work with vibratese was done, signals could not be transmitted any faster!

Much later (Geldard and Sherrick, 1970), the presentation-rate difficulty was overcome by controlling the vibrators with eight-hole tapes punched for the Tally reader, locus being coded on five hole positions, intensity on two, and duration as either a one-column or a three-column pattern. This arrangement made possible speeds hitherto unobtainable. Also, by then, the more rugged Sherrick vibrators were available to replace the modified relays originally used in the Spector and Howell experiments, and easy attachment to the body could be effected with Velcro tapes (Sherrick, 1965). Accordingly, vibrator positions were changed from the chest region, where all but the weakest signals are readily conducted to the cochlea by way of the rib cage, to the fleshy surfaces of the upper and lower arms and to the pit of the stomach. The new placement, if anything, made acquisition of the vibratese alphabet easier than before, but it did not resolve the problem, alluded to previously, of the almost aphasia-like effect in which individual letters could be comprehended with ease but "blocking" intervened to prevent retention of letter order and hence word meaning. We shall encounter this phenomenon again.

Body Braille

Body braille—as it came to be called because it involved presentation of six stimulators widely spread out on the skin: two on the forearm at the wrist, two near the elbows, and two close to the shoulders—was devised less as a practical communication system than as an attempt to manipulate the variable of body locus in basic experiments (Virginia Cutaneous Project: 1948–1962:47). The confining of stimuli to the hand and especially the fingertips has become almost a fetish in tactile communication, with convenience for the experimenter as the chief factor recommending it. Since it had been found with vibratese that not more than seven contactors could be used and still preserve absolute identification of locus, even on the most expansive thorax, a natural conclusion was that intercontactor distance should never be allowed to limit discrimination in a tactile communication system.

In body braille uniformly brief, 60 Hz bursts of vibration were delivered to the six contactors on the arms. For the commonest letters (*E, T, A, O, I, N*) only a single vibrator was assigned; for all others double, triple, or quadruple contactor combinations were presented simultaneously. It was found that as many as four vibrators could be energized in a single pattern and still preserve perceptual uniqueness, provided only that three of them not be confined to one arm. The latter arrangement produces interesting inhibitory interactions in which wrist and shoulder loci combine to suppress the elbow vibration unless it, in turn, is raised in intensity. Then, indeed, if it is intensified sufficiently, it can inhibit the other two! The body braille alphabet was mastered in a few hours by its constructors (J. F. Hahn and the author), and, within a week of daily sessions, German proverbs were being successfully transmitted and deciphered, albeit a little ponderously.

The Optohapt

A third communication system characterized by coded signals is that furnished by the optohapt (Geldard, 1966), essentially a tactile reader that utilizes a large portion of the two square yards of skin constituting the human integument. It was originally devised to demonstrate the fundamental principle that spatial information, when cut off from its primary sensor, the eye, should be fed to the only other space receiver the body possesses, the skin. At the same time, the optohapt relies on temporal information as well, so that all signals generated by it are essentially spatiotemporal.

The reading element consists of a bank of nine photocells—it is, indeed, the reader unit of the Battelle optophone—which triggers nine vibrators distributed over the body (two on each arm and leg, one on the upper abdomen, care being taken to avoid corresponding body points). The language concocted for the optohapt was arrived at only after thorough investigation of all available symbols in the IBM library of type faces, for the material to be "read" was typed on paper carried by the platen of a long-carriage, accounting typewriter. If uniqueness of perceptual recognition is taken as the criterion, which it was, most of the letters of the English alphabet prove to be poor candidates. Only the letters, *I, J,* and *V* survived the weeding-out process, which involved a total of 180 characters and was conducted by the pair-comparison technique. The really discriminable signs are made up of punctuation marks, some mathematical and business symbols, and a few more esoteric forms, some found chiefly on branding irons!

Several subjects were trained up to a sufficiently high level of performance to demonstrate the adequacy of this system for general tactile communication (Geldard and Sherrick, 1968). Beginning with familiarization with isolated letters, then random alphabets at the pedestrian rate of 70 characters per minute, two-letter, then three-letter words were presented, until short sentences, containing only short words, could be received with ease. Concomitantly, as the "traffic would stand it," speed was moved up to 100 characters per minute, and for one subject, eventually, 125. This, of course, is not rapid communication (a maximum of about 25 words per minute), but performance left little doubt that more rapid rates would be possible if presentation speed could be increased. This was subsequently accomplished by punching into Tally tape a code having the greatest possible resemblance to that of the optohapt.

Polytap

Meanwhile, a fourth system, polytap, came under development, chiefly at the hands of Douglas Rohn, working in the author's laboratory. It has existed in several forms derived from the unexpected experimental result that radical changes could be made in vibrator number and position without seriously interfering with the subject's ability to retain and use the code. The question arose as to whether it might be possible to transfer optohapt patterns to the fingertips (for convenience) despite the generally poor discriminatory behavior shown by the fingers in earlier experiments with vibration (Gilson, 1968). Since the difficulty with prolonged vibratory patterns delivered to the fingertips seems to reside in the prodigious amount of stimulus spread within the bony hand, it was judged expedient to reduce the "buzz" to a "tap" and rely largely on locus (but somewhat on duration, supplied by short trains of well-spaced discrete taps) to give uniquely codable signals. The polytap instrument is thus equipped with twelve Bimorph benders, six for each hand, one for each finger, and one for the thenar eminence of the palm. This battery of contactors delivers either single (2-msec) taps or a train of three of them. Suitably coded—the commoner letters all represented by single taps—24 letters of the alphabet are accommodated. The two least-frequently encountered letters are signaled by a whole-hand "blast," *Z* on the right, *Q* on the left hand. The code is readily computerized, and was.

Experience with the polytap is very revealing (Geldard and Sherrick, 1972). Since the desire was to ascertain possibilities, not establish population norms for what was essentially an untried system, two subjects were given intensive training for a period of six months. After alphabet familiarization they were rapidly brought up to the handling of familiar four-letter words. When words, as contrasted with scrambled letters, were introduced, tactile reading speed doubled quite abruptly and both subjects felt strongly that, with intensive practice in that direction, speed could have been tripled. But, though reading rate got up to 24 four-letter words per minute, chains of words could not be sustained beyond about six or seven. Blocking occurred whenever there was a failure to assemble the elements of a word in visual imagery. Simple diphthongs could not be handled as units; even the word "the" had to be visualized letter by letter and the letters put together. We saw the same difficulty in Biber's experiment.

It is clear that polytap points up the necessity, in developing tactile communication systems, of placing a high priority on central somesthetic processing capabilities, not just the skin's capacity to discriminate signals. Toward the end of the six-month training period (and with something of a view to rallying flagging interest!) the polytap subjects were given less monotonous materials to read. The first part of *Madame Bovary* was edited to restrict word and sentence length but, of course, to retain inherent associative values and redundancy. Ceiling performance proved to be at the rate of 28 five-letter words per minute, but it simply could not be held at that speed for long; rate increases from daily practice and familiarity with context were nullified by failures of synthesis and consequent blocking.

Air-Jet Stimulator

It is not only contactor impacts and electrocutaneous substitutes for them that may be coded to form tactile languages. Another approach is to supply local pressures with the gentle gradients furnished by air puffs, as was done in the air-jet technique developed by Bliss and his colleagues (Bliss and Crane, 1969). This method proved to be a precursor to the Optacon, many features of the latter having been worked out with a matrix of 40 air jets (5 X 8). These, spaced on 1/4" centers (and subsequently packed together with only 1/16" spacing that would fit the fingertip) emitted brief puffs of air under 3-psi pressure. Air flow was interrupted 200 times a second and thus there was delivered an essentially vibratory stimulus. At first, full block letters were used in the matrix display, but subsequently better performance was realized by simplifying these to a set of characters more nearly like those of the optohapt.

A novel development with the air-jet stimulator was the introduction of translatory movement in the display. Two kinds were tried, a rotation in small circles of the jet tips and a "moving-belt" (Times Square) motion across the display. Both were effective in that they improved letter recognition and thus increased reading rate. Highly motivated blind subjects were the beneficiaries of the training supplied by these experiments; one of them is reported to have achieved a reading speed of forty words a minute.

The air-jet display also served as the medium for the investigation of another important aspect of tactile communication (Bliss et al., 1966). This concerns the amount of information obtained in brief tactile exposures as contrasted with the relatively meager amounts reported in immediate memory span experiments. The 24 interjoint "pads" of the fingers of the two hands (thumbs excluded) were stimulated by from 2 to 12 simultaneous air puffs having 2.5 msec overall durations. The observer was set to note certain specific loci and his time of report was controlled, as in comparable experiments in visual perception (Estes and Taylor, 1964). While showing the absolute inferiority of tactile to visual short-term memory, similar processing operations were judged to be in effect in that subjects had much more information in their possession at the time of reporting, providing time was not delayed more than about 0.8 sec beyond stimulus termination, than their relatively poor tactile immediate memory spans, measured conventionally, would seem to indicate. A model depicting tactile memory tasks was constructed. It calls for a sensory register having a duration of only a few seconds, which, however, has a storage perhaps 50 percent greater than the capacity of the short-term memory store and which decays exponentially with a time constant of 1.3 sec. Short-term capacity is held to differ widely in size for different people and to be limited chiefly by spatial, rather than temporal, resolution properties.

The Visotactor

Our inventory of coded tactile mechanical devices would be incomplete without mention of the Visotactor (Smith and Mauch, 1968), developed at Mauch Laboratories, Dayton, Ohio, and designed as a reading aid for the blind. This instrument is not unique in principle, but it involves an interesting adaptation to the reading task. The Visotactor consists of a molded, hand-held device that can be laid on a typed page and, guided by a so-called colineator, which insures that the instrument will be kept "on line," transformed by photocells typed or printed letters into patterns of tactual stimuli delivered through pins to the fingers; the resulting tactile complex is a lively one. We have encountered the principle, of course, in both the optohapt and the Optacon. It is the mobility and adaptability of the Visotactor that makes it different. It will accommodate type from 7 to 36 points in size and can be used by the blind to read unusual materials, such as envelopes, bank checks, and labels on jars, bottles, or cans, to mention a few. Other applications suggest themselves.

CODED CUTANEOUS LANGUAGES: ELECTRO-CUTANEOUS SYSTEMS

The coded systems based on electrocutaneous stimulation, of which there are several, should not detain us long. By and large, they have not met with much success. The difficulty seems always to be the same, the relatively unclear signal that fails in discreteness coupled with the excessive time required to process it, once received.

Continuous efforts, over a two-and-one-half year period, to develop an analog of vibratese led to such poor results as to have discouraged further exploration in that direction (Foulke, 1968). Loci at all ten fingertips, two intensities, and two durations were keyed to 26 letters, several punctuation marks, and some common short word endings. Though subjects attained speeds of 20 words per minute or more, it was at the cost of long reaction times to signals, well over a second on the average, and, in most instances, the necessity of analyzing the composition of the signal into its dimensional constituents. Whether the perceptual difficulties were inherent in the use of electrical stimuli or whether the dimension of locus was overloaded in this system is not clear. Such failures of synthesis were not present in vibratese once the learning process had advanced to a high enough level.

The same laboratory that attempted electrocutaneous vibratese also tried out a kind of body braille (Alluisi et al., 1965), with three electrodes on each side of the body, and also an electric braille confined to six fingers (Foulke, 1968). The coding principle of keying only to locus seems to have won out in this competition, and there was subsequently developed a novel system in which the characters to be coded, instead of being drawn from the English alphabet, were those belonging to the Katakana Syllabary (Foulke and Sticht, 1966). These language symbols, together with Hiragana, the script form, total 48 basic characters which, along with two diacritics that bring the number to 73, are well known to the average literate Japanese and are used in teaching reading and writing to children. Employing single-, double-, and triple-locus patterns with exclusively fingertip electrocutaneous stimulation, experiments were carried out with eight Japanese subjects familiar with kana. Early learning was rapid, and combinations of syllables forming words were ultimately mastered so that connected prose could be received successfully. But, as with so many other attempts to synthesize a facile cutaneous language, final word rates were disappointingly low and error rates were high; this promising effort has to be judged as having led to generally unsatisfactory performance.

ELECTROSTATIC POSSIBILITIES

The path through the forest of tactile language symbols has been lengthy and somewhat tortuous. It has, at every turn, offered choice points between two major approaches, mechanical stimulation of the skin or creation of sensory signals by direct application of electrical stimuli. There can be little doubt that, on the whole, the mechanical techniques have offered better solutions than have the electrocutaneous ones.

But, as was indicated at the beginning of our journey, there is a third possibility, that of creating mechanical motions of some complexity at the skin surface by appeal to the electrostatic properties of tissues. Experience here is not extensive, but some things are known, and there has even been a proposal for a working system that might mediate sophisticated communication.

Early interest in electrostatic stimulation of the skin was displayed in Piéron's Paris laboratory by Chocholle (1948). Some of the essential conditions for successful vibration of the skin were worked out by him, including the requirements that the skin be absolutely dry if the condenser property is to emerge. Moore (1968) showed how thresholds vary from finger to hand to arm and how electrode area was an important variable in such experiments. Recently, there has been a somewhat more elaborate application of the electrostatic principle by Strong and Troxel (1970), who devised a matrix of closely packed metal pins, a 10 × 18 array 1.0" wide and 1.8" long, which may be explored conveniently by the finger. Each of the 180 pins, the flat distal ends of which constitute a condenser "plate," separately completes a circuit by way of finger and hand to a common electrode at the heel of the hand. Bipolar square-wave pulses, repeated at the rate of 200 pps, elicit what is described as a "texture" sensation whenever the finger is used as a probe to explore the array of pins. The "feel" is dependent to some extent on the amplitude of the pulses and partly on the pulse repetition rate. No "texture" is felt unless the finger is moving.

Attempts to assess the possible utility of the texture patterns have involved, in addition to the obvious measures of absolute threshold and decrements in them with increase of electrode area, the size of the just-discriminable intensity increment (5–10 percent), and that of the frequency change (40 percent, with 200 pps as a base), experiments in which the internal spatial features of the matrix have been varied. It appears that two points or two lines within the array can be discerned as two if there are as few as three intervening "dead" pins. A point can be discriminated from a line if only two blanks separate them. The measurements were relatively crude by ordinary psychophysical standards, but the result that pattern differences of this degree of discriminability could be found at all holds out a hope that the electrostatic method may one day come into its own.

Tracking and Monitoring

Is it possible to track a moving target with the aid of purely tactile signals? The answer is certainly positive, but the qualifications that have to surround it depend on what, precisely, is meant by "track," "moving," and "target." Specification of one or more of these usually comes when the question is put in the form, "How does tactile tracking compare with visual or auditory tracking?"

Anyone with a predilection for a particular answer could find support in the literature for it. Cutaneous tracking has been shown to be as good as, better than, and worse than visual tracking; it has been demonstrated to be on a par with, better than, and worse than auditory tracking. It all depends on the kind of situation selected for test, how strictly analogous the tasks are in any two modalities being compared, and what aspects of the tracking response (time on target, errors, etc.) are selected as indicators of proficiency. In some situations, tactual tracking is as good as visual but only so long as target speed is low; let the task demand more rapid following and vision promptly exerts its superiority (Howell, 1960).

In one experiment (Hofmann, 1968) utilizing a simple task involving steering to avoid going off an erratic path and in which continuous error feedback was delivered either visually (two white lights—one to the left, one to the right), by auditory means (white noise in one ear or the other), or tactually (weak electrocutaneous pulse to one or the other side of the neck), an overall tracking efficiency score distinctly favored the auditory and tactile modes over the visual. In another experiment (Schori, 1970) in which fifteen subjects tracked by vision, another fifteen by hearing, and a third fifteen by feel, the three tasks being strictly analogous, there proved to be no reliable difference among the group scores. As a practical matter, one channel was as effective as any other. However, when a secondary task was set up, that of monitoring an extraneous signal that had to be responded to even as the tracking performance was continuing, there was an appreciable decrement in the primary task being carried out with tactile cues. Visual and auditory tracking performances were much less affected. The tactile channel apparently exacted much greater attentional effort than did the other two.

The role of cutaneous perception in tracking behavior is best approached from the standpoint of cooperation rather than substitution. Except for the predicament in which neither sight nor hearing is available to provide the needed cues, the tactile contribution presumably should be that of supplement rather than surrogate. As a matter of fact, few experiments on tracking behavior have come even close to replicating the complexities of real-life situations. There are a great many modern tasks, both public and private, that put unusual demands on flexibility of attention, rapidity of reaction, soundness of judgment, and coordination of muscular response. As a case in point, let us consider the task of the aircraft pilot. He "tracks" a great deal, considering that many of the data dictating an airplane's flight are nowadays processed automatically. The most important thing a pilot can do when flying an airplane is to insure that he will remain current with respect to air traffic conditions in both his near and remote environments. A survey of the circumstances surrounding mid-air collisions in military flying shows that four out of five of them occur in daylight and with normal visual ground contact in force. By and large, it is not material failure or poor visibility or even the fact that air speeds have been increased tremendously in recent years that accounts for collisions, as one might suspect.

While this is not the place to attempt a full-scale analysis of the pilot's task, it should be noted that the visual demands on him are prodigious, literally surrounded as he is with flight and engineering instruments, many of which have to be consulted with some frequency. If sensory information other than the visual could relieve some of the monitoring tasks, allowing more freedom for search of the air space, it would be a great boon. Or, quite apart from freeing vision, if other sensory data providing much the same information that only the eyes are privy to could be supplied in a supplementary way, just to provide confirmation and reassurance, there would undoubtedly be a net gain on the side of reduction of observational fatigue and increase of overall efficiency. Tactile communication could be of real help in cutting down the complexities and contributing to reduction of stress.

Some of the tactile tracking experiments have clearly shown how. Durr (1961) demonstrated that a five-vibrator signaling system, arranged domino-fashion on the chest (and, later, with an increase of accuracy, more widely spread out on the body), could furnish a suitable bidimensional display that could give much the same directions as are now supplied by the ILS (instrument landing system, a visual blind-flying guide). Triggs et al. (1974) have proven the efficacy of both mechanical and electrocutaneous displays to yield basic "pitch-and-roll" data. Hill (1970) has shown the relative superiority of a "ripple" type of display (rapidly successive pressures spread out over the skin, analogous to some automobile turn signals), superior at least in the promptness with which the receiver of the message responds to it. And there are a number of simpler things, e.g., provision of a tactual signal to replace or amplify a visual one in warning situations, such as "change gas tanks" and "too much pressure on the brakes" in taxiing operations.

It must be obvious that the guidance of aircraft does not constitute the sole area of application of cutaneous tracking in the service of vehicular and mechanical tasks. Cutaneous signaling can be employed successfully wherever amounts, directions, rates, or even forms of relational information, such as are revealed by reference to coordinates, are to be transmitted (Geldard, 1960). This means that the potentialities of tactile communication have not even begun to be realized. There is much on the shelf to be put to use, but precious few are casting glances at the shelf.

References

Alluisi, E. A., Morgan, B. B.; and Hawkes, G. R.; 1965. Masking of cutaneous sensations in multiple stimulus presentations. *Perceptual and Motor Skills,* 20:39–45.

Anderson, A. B., and Munson, W. A., 1951. Electrical excitation of nerves in the skin at audiofrequencies. *J. Acoust. Soc. Amer.,* 23:155–59.

Békésy, G. von, 1955. Human skin perception of traveling waves similar to those on the cochlea. *J. Acoust. Soc. Amer.,* 27:830–41.

Biber, K. W., 1961. Ein neues Verfahren zur Sprachkommunikation über die menschliche Haut. Thesis, Universität Erlangen, Germany.

Bledsoe, W., 1972. Braille: a success story. In: *Evaluation of Sensory Aids for the Visually Handicapped.* Washington, D.C.: National Academy of Sciences, pp.3–33.

Bliss, J. C., and Crane, H. D., 1969. Tactile perception. In: *Research Bulletin of the American Association for the Blind,* no. 19, L. L. Clark, ed., pp.205–30.

Bliss, J. C.; Crane, H. D.; Mansfield, P. K.; and Townsend, J. T.; 1966. Information available in brief tactile presentations. *Perception and Psychophysics,* 1:273–83.

Bliss, J. C.; Katcher, M. H.; Rogers, C. H.; and Shepard, R. P.; 1970. Optical-to-tactile image conversion for the blind. *IEEE Trans. Man-Machine Systems,* MMS-11:58–65.

Breiner, H. L., 1968. Versuch einer elektrokutanen Sprachvermittlung. *Z. exper. angewandte Psychol.,* 15:1–48.

Cassirer, E., 1944. *An Essay on Man.* New Haven: Yale University Press. ix+237pp.

Chocholle, R., 1948. Émission d'ondes acoustiques audibles par l'épiderme sous l'action d'un courant électrique alternatif. *C. r. Soc. Biol., Paris,* 142:469–71.

Collins, C. C., 1970. Tactile television—mechanical and electrical image projection. *IEEE Trans. Man-Machine Systems,* MMS-11:65–71.

Craig, J. C., 1974. Pictorial and abstract cutaneous displays. In: *Cutaneous Communication Systems and Devices,* F. A. Geldard, ed. Austin, Texas: The Psychonomic Society, pp.78–83.

Dalgarno, G., 1680. Didascalocophus, or the deaf and dumb man's tutor. In: *The Works of George Dalgarno.* Reprint ed., Edinburgh: T. Constable, 1834, pp.113–59.

Dalrymple, G. F., 1973. Development and demonstration of communication systems for the blind and the blind/deaf. In: *Final Report of Project 14-P-55016/1-03.* Cambridge: Sensory Aids Evaluation and Development Center, M.I.T., pp.17–29.

DuPont, M., 1907. Sur des courants alternatifs de périodes variées correspondant à des sons musicaux et dont les périodes des mêmes rapports que les sons; effets physiologiques de ces courants alternatifs musicaux rythmés. *C. r. Acad. Sci.,* 144:336–37.

Dudley, H. W., 1936. The Vocoder. *Bell Lab. Rec.,* 18:122–26.

Durr, L. B., 1961. The effect of error amplitude information on vibratory tracking. Master's thesis, University of Virginia.

Estes, W. K., and Taylor, H. A., 1964. A detection method and probabilistic models for assessing information processing from brief visual displays. *Proc. Nat. Acad. Sci.,* 52:446–54.

Farrell, G., 1950. Avenues of communication. In: *Blindness: Modern Approaches to the Unseen Environment,* P. A. Zahl, ed. Princeton: Princeton University Press, pp.313–45.

Finkenzeller, P., 1973. Hypothese zur Schallcodierung des Innenohres. Habilitationsschrift, Erlangen, Germany. 99pp.

Foulke, E., 1968. Communication by electrical stimulation of the skin. In: *Research Bulletin of the American Foundation for the Blind,* no. 17, L. L. Clark, ed., pp.131–40.

Foulke, E., and Brodbeck, A. A., Jr., 1968. Transmission of Morse code by electrocutaneous stimulation. *Psychol. Rec.,* 18:617–22.

Foulke, E., and Sticht, T. G., 1966. The transmission of the Katakana syllabary by electrical signals applied to the skin. *Psychologia,* 9:207–209.

Gault, R. H., 1926. Tactual interpretation of speech. *Sci. Mo.,* 22:126–31.

Gault, R. H., and Crane, G. W., 1928. Tactual patterns from certain vowel qualities instrumentally communicated from a speaker to a subject's fingers. *J. Gen. Psychol.,* 1:353–59.

Geldard, F. A., 1960. Some neglected possibilities of communication. *Science,* 131:1583–88.

Geldard, F. A., 1962. The language of the human skin. *Proc. XIV Internat. Congr. Appl. Psychol.,* 5:26–39.

Geldard, F. A., 1966. Cutaneous coding of optical signals: the optohapt. *Perception and Psychophysics,* 1:377–81.

Geldard, F. A., 1968. Pattern perception by the skin. In: *The Skin Senses,* D. R. Kenshalo, ed. Springfield, Ill.: Thomas, pp.304–21.

Geldard, F. A., 1970. Vision, audition, and beyond. In: *Contributions to Sensory Physiology,* vol. 4., W. D. Neff, ed. New York: Academic Press, pp.1–17.

Geldard, F. A., and Sherrick, C. E., 1968. Temporal and spatial patterning. *Princeton Cutaneous Research Project,* Report no. 12, pp.6–8.

Geldard, F. A., and Sherrick, C. E., 1970. Modified vibratese. *Princeton Cutaneous Research Project,* Report no. 16, pp.2–11.

Geldard, F. A., and Sherrick, C. E., 1972. Discriminability of finger patterns. *Princeton Cutaneous Research Project,* Report no. 20, pp.2–4.

Gilson, R. D., 1968. Some factors affecting the spatial discrimination of vibrotactile patterns. *Perception and Psychophysics,* 3:131–36.

Goff, G. D., 1967. Differential discrimination of frequency of cutaneous mechanical vibration. *J. exper. Psychol.,* 74:294–99.

Guelke, R. W., and Huyssen, R. M. J., 1959. Development of apparatus for the analysis of sound by the sense of touch. *J. Acoust. Soc. Amer.,* 31:799–809.

Henneman, R. H., 1952. Vision and audition as sensory channels for communication. *Quart. J. Speech,* 38:161–66.

Hill, J. W., 1970. A describing function analysis of tracking performance using two tactile displays. *IEEE Trans. Man-Machine Systems,* MMS-11:92–101.

Hofmann, M. A., 1968. A comparison of visual, auditory, and electrocutaneous displays in a compensatory tracking task. Ph.D. diss., University of South Dakota.

Howell, W. C., 1956. Training on a vibratory communication system. Master's thesis, University of Virginia.

Howell, W. C., 1960. On the potential of tactile displays: an interpretation of recent findings. In: *Symposium on Cutaneous Sensitivity,* USAMRL Report no. 424, G. R. Hawkes, ed., pp.103–13.

Katz, D., 1930. The vibratory sense and other lectures. *Maine Bulletin,* 32:90–104.

Keidel, W-D., 1968. Electrophysiology of vibratory perception. In: *Contributions to Sensory Physiology,* vol. 3, W. D. Neff, ed. New York: Academic Press, pp.1–79 (esp. pp.69–79).

Keidel, W-D., 1974. The cochlear model in skin stimulation. In: *Cutaneous Communication Systems and Devices,* F. A. Geldard, ed. Austin, Texas: The Psychonomic Society, pp.27–32.

Kirman, J. H., 1973. Tactile communication of speech. *Psychol. Bull.,* 80:54–74.

Kringlebotn, M., 1968. Experiments with some visual and vibrotactile aids for the deaf. *Amer. Ann. Deaf,* 113:311–17.

Levine, L., et al., 1949–1951. "Felix" (Sensory replacement). *Quart. Progress Reports, Research Lab. of Electronics, M.I.T.* Cambridge: Massachusetts Institute of Technology.

Lindner, R., 1936. Physiologische Grundlagen zum elektrischen Sprachtasten and ihre Anwendung auf den Taubstummenunterricht. *Z. Sinnesphysiol.,* 67:114–44.

Melen, R. D., and Meindl, J. D., 1971. Electrocutaneous stimulation in a reading aid for the blind. *IEEE Trans. Bio-Med. Engin.*, BME-18:1–3.

Miller, G. A., 1956. The magical number seven, plus or minus two: some limitations on our capacity for processing information. *Psychol. Rev.*, 63:81–99.

Moore, T. J., 1968. Vibratory stimulation of the skin by electrostatic field: effects of size of electrode and site of stimulation on thresholds. *Amer. J. Psychol.*, 81:235–40.

Pickett, J. M., and Pickett, B. H., 1963. Communication of speech sounds by a tactual vocoder. *J. Speech Hear. Res.*, 6:207–22.

Rollman, G. B., 1974. Electrocutaneous stimulation. In: *Cutaneous Communication Systems and Devices,* F. A. Geldard, ed. Austin, Texas: The Psychonomic Society, pp.38–51.

Rösler, G., 1957. Über die Vibrationsempfindung. Literaturdurchsicht und Untersuchungen in Tonfrequenzbereich. *Z. exper. angewandte Psychol.*, 4:549–602.

Rousseau, J-J., 1762. *Émile, ou de l'Éducation.* Oeuvres de Jean-Jacques Rousseau, tome VII. Amsterdam: Neaulme.

Saunders, F. A., 1974. Electrocutaneous displays. In: *Cutaneous Communication Systems and Devices,* F. A. Geldard, ed. Austin, Texas: The Psychonomic Society, pp.20–26.

Schori, T. R., 1970. A comparison of visual, auditory, and cutaneous tracking displays. Ph.D. diss., University of South Dakota.

Sherrick, C. E., 1965. Simple electromechanical vibration transducer. *Rev. Sci. Instr.*, 36:1893–94.

Sherrick, C. E., 1975. The art of tactile communication. *Amer. Psychologist,* 30:353–60.

Smith, G. C., and Mauch, H. A., 1968. The development of a reading machine for the blind. *Summary Report, VA Prosthetic and Sensory Aids Service.* iv+70pp.

Spector, P., 1954. Cutaneous communication systems utilizing mechanical vibration. Ph.D. diss., University of Virginia.

Starkiewicz, W., and Kuliszewski, T., 1963. Active energy radiating systems: the 80-channel Elektroftalm. In: *Proc. Internat. Congr. Technol. Blindness,* vol. I, L. L. Clark, ed. New York: American Foundation for the Blind, pp.157–66.

Starkiewicz, W., and Kuliszewski, T., 1965. Progress report on the elektroftalm mobility aid. In: *Proc. Rotterdam Mobility Research Conference,* L. L. Clark, ed. New York: American Foundation for the Blind, pp.27–38.

Strong, R. M., and Troxel, D. E., 1970. An electrotactile display. *IEEE Trans. Man-Machine Systems,* MMS-11:72–79.

Triggs, T. J., Levison, W. H.; and Sanneman, R.; 1974. Some experience with flight-related electrocutaneous and vibrotactile displays. In: *Cutaneous Communication Systems and Devices,* F. A. Geldard, ed. Austin, Texas: The Psychonomic Society, pp.57–64.

Virginia Cutaneous Project, 1948–1962, 1962. Ann Arbor, Mich.: University Microfilms (OP 16, 352).

White, B. W.; Saunders, F. A.; Scadden, L.; Bach-y-Rita, P.; and Collins, C. C.; 1970. Seeing with the skin. *Perception and Psychophysics,* 7:23–27.

Chapter 11

ACOUSTIC COMMUNICATION

René-Guy Busnel

In the total complex of animal behavior the isolation of a single system of communication is due to an arbitrary choice subject to a specific technology. Experimental results obtained by this method must be considered as incomplete, and they acquire their full value only when associated with results obtained from other means of information. They form a part of the puzzle of animal behavior, that field of synthesis which belongs to ethology. Animal acoustic communication should take its place in this framework of reference without preferential isolation, except in special cases. It can be considered as an entity only if its relative and hierarchical position in the polymorphic and polyvalent animal communication system is kept in mind.

During the last twenty years the study of acoustic communication has made a great leap forward, in part because of technical progress in recording, signal reproductions, and their graphic analysis. Thousands of publications and about fifteen synthetic books (see references) concerning different zoological groups form the present basic documentation and include almost all orders starting with invertebrates. From the first publications onward, animal acoustic communication has formed a relatively coherent ensemble drawing upon varied disciplines such as anatomy, physiology, neurophysiology, psychophysiology, and physical acoustics. The reader may wish to refer to some of the many exhaustive studies synthesizing this question and to other chapters in this book. In this chapter only those aspects of animal communication which seem to open new perspectives will be developed.

Reprinted from *Animal Communication: Techniques of Study and Results of Research,* Thomas A. Sebeok, ed. (Bloomington: Indiana University Press, 1968).

Hierarchical Position of Acoustic Signaling In Relation To Other Means of Communication

The hierarchic value of acoustic signaling differs, in relation to other means of communication, for each animal species employing it in various behavioral situations. In certain birds, it is of prime importance, as Brückner (1933) demonstrated in the hen and Schleidt (1960) in the turkey. In the latter species, for instance, the female, surgically deafened, laid and sat on her eggs normally, but after the young hatched, she did not seem to be able to differentiate them from predators and killed them as she would predators who approached the nest. The acoustic signal given by the young turkeys had a fundamental hierarchic value which induced recognition by the mother and suppressed her aggressiveness. Lacking this signal, the female turkey, in spite of visual information, was dominated by her aggressive behavior pattern. Similarly, the mother hen, unable to hear her chick, but seeing him isolated under a glass bell, abandoned him. In innate behavior in these species, acoustic signals thus have a privileged position. Lorenz (1950) and his students showed in the duckling and gosling that imprinting depends on acoustic and visual signaling.

In the sexual behavior of the insect Ephippiger (Busnel and Dumortier, 1954), the female is wholly oriented by the sexual signal, and goes toward a loudspeaker emitting signals of the male even when she is near a silent male. The sexual behavior of the sow is controlled by association of kinesthetic, optical, chemical, and acoustic signals; in the sequence of copulatory behavior, the experimentally isolated acoustic signal of the male, in 80 percent of the cases, is alone sufficient to ensure immobilization reflex behavior (Signoret et al., 1960). A porpoise, completely blinded with rubber cups, swims without error through a maze of diverse obstacles utilizing only his echolocation system (Norris, 1961, studying *Tursiops*), and he is able, under these same conditions, to detect metallic targets as small as 0.2 mm in diameter (Busnel et al., 1965a and b, studying *Phocaena*). The same is true for the bat (Griffin, 1944; Dijkgraaf, 1946). In some bat and dolphin species acoustic signaling by echolocation is hierarchically equal or superior to visual information. These examples may be multiplied. However, generalizations cannot be drawn, because in each species the situation depends upon sensory factors which predominate and induce one behavior or another at a given moment.

Finally, it must not be forgotten that some acoustic signals may be closely related to gesticulatory or kinesthetic signals such as in sea gulls (Tinbergen, 1959). In the case of honey bees, Wenner (1962, 1964) and Esch (1961, 1962, 1965) showed acoustic signals, as much as or more than chemical signals, to be inextricably linked to the signals of the dance. In these cases, communication is composed of a complex of factors, and it becomes increasingly difficult to find a nonrelative value in a given signal form.

Transmission Channel and Background Noise

Each animal species is surrounded by its own *Umwelt* (air, water, or solid) in which the transmission of vibrations follows specific physical rules. These vibratory phenomena support information during acoustic communication. The transmission channel, even when considered as homogeneous, is not inert, and it plays a role in signal structure or its perception that is independent of its total effect on the organism under consideration. In nature, the transmission channel always contains a background noise level which is statistically characteristic of the species' biomes and biotopes and is, of course, related to natural events such as rain, wind, and the breaking of waves. Moreover, when an acoustic signal is propagated in nature (which cannot rightly be considered as a free field), numerous acoustic wave reflections are produced. These, in their turn, constitute a special part of background noise and bring about both a lowering of signal intelligibility and a diminution in the signal carrying power. Returning signal waves are modified by obstacle impedance and volumetric form. This is perceived by certain species which employ this information for their autodirectivity (echolocation). Whatever the importance of the channel background noise, it can be said that in natural surroundings it almost always occurs and that, up to the present time, perception studies have too often neglected this signal-to-background-noise relationship from which the listener's peripheral analyzers have to extract whatever proper information is contained in the signal.

Another background noise whose importance has been underestimated is a specifically biological one which is superimposed on the natural background noise and from which message information must be extracted if the intelligibility coefficient is to remain acceptable or reach threshold perception. This biological noise, produced especially in a crowd or dense social grouping which is engaged in a more or less determined social activity (as in a reproduction area, a hive, a colony, a pack, or a fraternity), consists of a certain aggregate noise: the sum of all the emitted individual signals and the by-products of motor activities (wing or flight noise, noise of legs against the substratum, etc.). This latter carries no specific information. Such biological noises are emitted, for instance, in groupings of sea birds, bee colonies, porpoise schools, or seal harems. It covers exactly the same frequency band as that of the species' signal emission and would thus, theoretically, lead to an auditory saturation, making it impossible to transmit far-field any information other than specific information (such as the general noise characterizing the species). An animal must therefore have the capacity of perceiving a signal and extracting it from a random background noise. This signal individualizes communication between members of a species. The physical characteristics which allow this detection are varied, each animal using its own specific sensory organs to this purpose. In some cases these must even be able to detect an information-carrying signal whose intensity is below that of the background noise. Cherry (1957) has described this phenomenon in man as the "cocktail party effect." So far in animal communication it has received little attention from bioacousticians. Its origin can be related to a preferential motivation which triggers an acute selectivity of the nervous centers, the biophysics of which is still little known. However, it can be observed in quite a few higher animal groups, notably gregarious vertebrates, that members of a permanent or temporary couple are capable of personal recognition through individualized acoustic emissions, even though the number of such simultaneous signals sent produces a high background noise level [individual recognition experiments with the emperor penguin (Prevost, 1961) and with bats (Möhres and Kulzer, 1956; Griffin, 1958].

These problems of masking background noise in the transmitting canal should be more thoroughly studied. They would surely interest psychophysiologists as well as neurophysiologists, for whom this question should prove particularly interesting since it is not strictly auditory in the usual sense. That is to say, it does not concern just the eighth pair of nerves. The decoding of special information, selectively chosen from the background noise by means of a guiding motivation, is probably analogous to the mechanism used for specific vision. There it is the eye which, directed by a particular motivation, selects from the mass of data constituting the background noise such small informative details as a word on a page or a letter in a word which alone focus the attention. Possibly this is also the case in olfaction.

Communication Sound Source

Sound sources either have a mechanical origin by specialized or nonspecialized emission organs or result from vibrations imparted to the substrata by a part of the animal body acting as an amplifier. Percussion on the substratum as a sound source is illustrated by ground-tapping (usually with the hind feet as done by prairie hens, rabbits, mice, Neotema), by woodpecking (series of shocks produced by the beak of woodpeckers), etc.

The range of sound emission organs found in the animal kingdom is quite varied; they are usually bilateral in invertebrates and very often unpaired in vertebrates. They may be restricted to one sex, or they may present a considerable sexual dimorphism. They are found on all different parts of the body. For example, the following may be found functioning as sound emission organs in invertebrates: chitinous toothed files which, by friction, stimulate a vibrating body—wing, elytra, antenna, thorax, leg, abdomen (Orthoptera, Crustaceans); friction or vibration of nonspecialized organs such as the wings (mosquitoes and some moths); semirigid plates on a resonant cavity stimulated by neuromuscular contractions (Tymbal method—Cicada); reed-like organs which function by aspiration and expiration of air (death's-head hawk moth, *Sphinx atropos*). In Myriapoda two species have been found, *Scutigera* and *Rhysida,* which can automate legs. These species have no special stridulatory organ; however, when the legs are separated from the body, they emit sounds. When they are intact, they are silent. The hypothesis is that the noise emitted by the leg attracts predators, leaving the animal free to flee (Annandale et al., 1913; Cloudsley-Thompson, 1961).

In lower vertebrates, there is much polymorphism and a great variety of sound sources: nonspecialized organs may produce friction, as do vomerine teeth in certain fish; osseous, rattle-type apparatuses may be found, made up of moving, oscillating parts which knock each other when agitated, as in rattlesnakes; whistling or vibrating apparatuses which function by air expulsion through a more or less differentiated tube (larynx) ending in an aperture (glottis) with more or less functional lips. The expelled air is supplied by the lungs themselves or by being in contact with an air pocket reserve with or without diverticula (vocal sac of some amphibians); and finally, membranes may be stretched over resonating pockets (as is the swim bladder of fish). These apparatuses are activated either by external percussion (fin beating) or by contraction of muscles disposed in different ways around the cavity.

Sounds produced by nonspecialized organs are also found in higher vertebrates. These include breast-beating in the gorilla, organ-clapping such as wing-beating of the wood pigeon, drum-rolling in the hazel grouse and gold-collared manakins, and trembling of remiges (primaries) and rectrices (tail feathers) in the woodcock and snipe. Owls and storks use their beaks, and some bats (Möhres and Kulzer, 1956) and some insectivores, such as tenrecs (Gould, 1965), use their tongues. In many higher vertebrates specialized organs are found, usually working by propelled or aspirated air in a more or less differentiated tube equipped with modulating membranes or slit system. These organs are vocal cords, muscular glottal lips, the larynx of odontocetes, and the bird larynx and syrinx. These apparatuses often have additional organs which form air reservoirs or resonators (clavicular and cervical air sacs), as found in the bustard, ostrich, crane, and morse. In some monkeys these features are found in the thyroid cavity, as is the gibbon's vocal sac or the hyoid bone resonating chamber of the New World howler monkey. Curious peripheral sound organs are also found such as the fifteen-spined sound apparatus in tenrecs, the Madagascar *Hermicentetes, Centetes* (Gould, 1965), and the tail bell of the Bornean rattle porcupine, *Hystrix crassispinis.* Anatomists and morphologists have described the remarkable richness of the character of these structures in hundreds of animal species. However, much is still unknown. For example, it is not yet known what elements generate echolocation clicks in odontocete Cetaceans and in many bats. Phonation mechanics is a large field which still needs much exploring, and the microphysics of most phonatory apparatuses remains to be studied in many species.

Whether an animal that is provided with adequate sound emission organs will or will not activate them, depends upon physiological and psychophysiological factors particular to the species and its ecological conditions. Such factors may be abiotic—temperature [applies mostly to poecilothermal animals, but also to some vertebrates as, for instance, to the foot in the flycatcher bird (Curio, 1959)], humidity, or light —or they may be endogenous (hormonal state, age, psychological motivation, etc.). When a sound emission occurs, it follows a code in which different factors come into play: rhythm, pulse repetition rate, amplitude variation, intensity, frequency, etc. The periodicity of biological activity (circadian and seasonal rhythms under the influence of temperature and light) may also influence sound emission. The animal's age determines even the nature of the signals. Insect larva and nymphs do not generally have the sound apparatus which the imago has later. The signals of young mammals and birds do not have the same physical structures as do adult signals (subsong).

Other interesting aspects of studying sound emission apparatuses of animals may be looked at from a mechanical point of view. It can be considered that sound, when emitted by an organ having a more or less complex mechanism, is the product of this mechanism transmitted via the central nervous system. This system thus governs the physical structure, since it is a result of the structural dynamics of the apparatus under consideration. In this case, acoustic signals are the result of the activation of anatomical elements which are controlled by neuromuscular structures, in turn directed by neuromotor centers.

It is even more interesting to find that there is, in some species, specific programming of innate acoustic activity associated with other aspects of behavior and under control of the higher nervous structures. This is easier to demonstrate in invertebrates, where the acoustic activity is genetically fixed (Leroy, 1962) and where the nervous structures are more circumscribed and less diffuse than in higher vertebrates. It is not surprising to find cellular group localization in the brain which selectively commands the descending neuromotor channels, activating muscular fibers of the sound-producing organs. This coincides with data known about the role of more or less specialized centers of different levels of the central nervous system which command, by a given channel, a corresponding muscular reaction in voluntary or involuntary acts. The first example of this localization was experimentally obtained by Hüber (1952) on the field cricket. Interestingly enough, it was shown that the signal emission was regulated, not by one or more stimulating centers, but by the suppression of action from inhibiting centers. That is to say, as soon as these centers were destroyed, the insect transmitted its signal almost continuously. Depending upon the centers and the neuromotor coordination which these governed, corresponding activated muscles put the elytra into different stereotyped positions. These movements were in all respects comparable with those found in the different signaling pattern behavior of the species, such as sexual call, courtship, and rivalry signals.

The higher centers in the cricket controlling sound emission are localized in the pedunculate and central bodies. Fighting behavior was set off by the caulicles and the corpus callosum. The center of the pedunculate body was found to govern the courtship song. The central body directed sound apparatus vibrations, coordinated with the placing of the elytra in a "flat position." It is almost certain that the sounds emitted after these stimulations are not simple noises but a series of pulses organized into a signal and having all the features of natural and normal signals.

In amphibians, reflex croak of the "warning" type has been obtained in the leopard frog *Rana pipiens* by mechanically stimulating the anterior part of the spinal cord (Aronson and Noble, 1945). Working along the same lines, Schmidt (1965) has done excellent experimental research on central respiration mechanisms and their relation to acoustic signal transmission in different Anura, *Rana* and *Hyla.* In these species all vocalization control is localized in the trigemino-isthmic-tegmentum, which activates the phonatory motor mechanisms. In fact, the vocal cords and glottal system which regulate air flow are excited by the intermediary of the vagus; abdominal muscles making up the air pump system are controlled by the spinal nerve and the hyoid depressors by the hypoglossal nerve.

Mating calls can be obtained by nerve-impulse intensity or electrical stimulation when taking into account the animal's hormonal state, which may be modified with androgens. Schmidt also concludes that mechanisms concerned with phonoresponses, just like the sexual calls, appear to be localized in the preoptic zone. Recent research on birds indicates that stimulation of different nervous centers sets off signals which seem in all respects comparable to natural signals. In the male red-winged blackbird (*Agelaius pheniceus*) vocalizations were elicited from an area beneath the optic tectum including the torus semicircularis and the underlying gray matter. The vocalizations evoked by electric stimulation resembled the calls given at the approach of a potential predator or calls given by birds being caught in cages and held in the hand (distress call). Whistle-type calls were evoked from the hypothalamus (Brown, 1965). Different types of calls have been evoked by electrical stimulation of the brain in the cock or hen, but responsive areas were not delineated anatomically except in the hypothalamus, where androgen implants were used (von Holst and St. Paul, 1963). In ducks, quacking is governed by the archistriatum and the tegmental midbrain (Phillips, 1964). The mesencephalon directs the squawks of the southern lapwing (*Belonopterus chilensis*) (Silva et al., 1961) and the alarm call and "ruckcuh" threat coo of the feral pigeon (*Columba livia*) (Heinroth and Heinroth, 1949; Akermann et al., 1960; Rougeul and Assenmacher, 1961). Anatomically, there are similarities between the central gray matter of mammals and the torus of other vertebrate classes. In all cases, the sounds thus produced are innate signals; furthermore, they are associated with all the corresponding attitudes or postures attending these signals as for example, hair-bristling or feather-ruffling.

In mammals, knowledge acquired by electrical stimulation of the deep layers of the diencephalon in unanesthesized animals showed diffuse localization of the command of vocalization. These motor zones have never induced more than instinctive signals of the distress type, whether on cat, monkey, or porpoise. Among the first of many studies done in this field were those of Karplus and Kreidl (1909) and Hess (1928). Those of Hunsperger (1956) on the cat and of Lilly (1958a, 1958b and 1962) on porpoises can be more closely examined.

In the diencephalon and mesencephalon of the cat, Hunsperger described zones which set off defense or escape behavior accompanied or not by roaring, cries, and growling, constituting, however, a phonation rather than a message. Control of varying behavior connected with this defense is found in very different parts of the brain. The distress behavior scheme is localized in the mesencephalic central gray matter and in the perifornical region of the hypothalamus. This zone extends into the cortical region, especially in the periform lobe and the anteroventral parts of the temporal lobe. Thus, this structural entity extends from the mesencephalon to the anterior brain. The primary cat mewing which occurs in uneasy situations is governed by the mesencephalic level under the substratum and is a complementary of the defense reaction. The secondary mewing which appears after stimulation is set off from Papez' circuit, the hippocampus, and also from canals which connect these structures to the preoptic region. It is probable that these vocalizations are at least partially related to distress signals. As yet, the sounds thus elicited have never been really studied as communication.

According to Lilly, attaining the motor area in the monkey resulted in vocalization; however, as yet no true signal has been recorded, although distress cries were probably emitted during stimulation. Behind the parietal cortex in the dolphin, and above what corresponds to the orbital cortex in man, is the thalamic region at the caudal nucleus level which controls the distress signal and other sounds. These latter may be extremely varied owing to the different vocalization systems used by the animal (larynx, lips of the blowhole, etc.) (Lilly and Miller, 1962).

The higher we go in the animal kingdom, the more diffuse and heterogeneous become the motor zones, introducing a notion of degrees of freedom. The production of complex signals depends upon numerous centers which interfere with each other, and thus no longer permit the "all or none" responses of invertebrates or lower vertebrates. In mammals, zones corresponding to a specific signal are not found. Instead, generalized phonation zones can be described which are diversely activated by other centers concerned with different emotional behavior patterns.

The genetic transmission of anatomical structures of a species does not give rise to any particular problem as far as the phonating organs are concerned. It is logical to consider the nervous structures associated with these organs and, by the same token, the morphology of innate signals to be genetically controlled. This has been demonstrated to be true for a great number of species, whether invertebrates (Fulton, 1933–1952; Perdeck, 1958; Leroy, 1963–1966), lower vertebrates (Blair, 1956–1963) or certain birds (Thorpe, 1961; Thielcke, 1961–1962; Messmer, 1956; Sauer, 1954). However, if the activation of all phonation systems follows Mendelian laws, as does their morphology, and if innate sound emission follows the same rules, it is noted that besides the mechanical (and therefore cerebral) rigidity of the basic framework of physical signal production, a certain measure of choice eventually appears. This degree of freedom may be explained by a preestablished hiatus in the command center containing learned information memorized at certain stages of the individual's life.

An autocontrol, ensured by the ear through a feedback mechanism, must be added to the description of control of the physical characteristics of emitted sounds given above. Certain birds, for instance, deafened at birth, are no longer capable of developing the complete structure of the species' specific songs (Huchtker and Schwartzkopff, 1958; Konishi, 1964). Bred solitarily in soundproof rooms, they develop that portion of the complete species repertoire which is inbred, but none of the learned part.

Control of phonation and sound emission by the ear is also associated with learning mechanisms which occur in certain social conditions during the animal's development (imprinting in young birds, for instance) (Lorenz, 1950). This learning can eventually lead (especially in many birds) to great variations in the physical form of the signal, bringing about local dialects on the one hand and psittacism or mimetism on the other. In all instances studied, dialects had the value of a true signal no matter what their physical characteristics. Imitations, on the other hand, can probably be considered as a meaningful signal for most species except possibly for parasite birds, where they simply reflect learning and an empty phonatory activity.

In this learning process, genetic learning affects the structure of vocalization as well as the memorization processes. Here may lie one of the most fundamental differences between man and animals, as far as vocalization is concerned. In man the only aspects of vocalization under genetic control are those qualities of voice determined by the structure of the larynx. But language as such is usually learned through audition-phonation feedback.

The Message

The sound message, more than any other communication method, lends itself to comprehension and correct analysis by the experimenter, assuming of course, the electroacoustic equipment to be precise. Sonagram frequency spectra and oscillogram transcriptions have become classical in these studies, and they give us a rather precise picture of the signal's diverse physical parameters. There are hundreds of publications concerning this aspect of the acoustic communication problem.

Certain analyses may be done by series of elementary, orthogonal, translated functions instead of by harmonic analysis. These elementary functions are damped sinusoidals, synchronous with the pitch, and may be analyzed by autocorrelation. Taking human voice analysis as a basis, it is possible to determine the number of bits per second (or bauds) and obtain a partial quantification based on different metrologies. Putting sonagram-type signals on a digital machine is made possible by point analysis. A transcription into machine code may then be made, thus rendering the whole in a geometrical form easier to manipulate than analysis made directly from the sonagram. The original processes of form recognition are now rather well known, and we must think about applying them soon to animal acoustic signals. Probably, this field will be first approached through studies on echolocation signals of porpoises and bats. These signals are recorded by oscillographic traces which may be directly interpreted by computer.

Acoustic frequency bands vary widely according to species and zoological groups. Low-frequency signals are emitted by fish and certain amphibians and in flight noises. Very high frequency signals are emitted by many insects and some porpoises. The highest frequency recorded is 350 kHz for the echolocation clicks of *Steno bredanensis* (Norris and Evans, 1966).

For many of the lower species, the signal's morphology is a close physical expression of the mechanical structure of the emission apparatus. These signals have thus a sort of obligatory physical form rigidly determined by the elementary movement of the organs. Other signals, on the contrary, have a flexible physical structure due to the possibility of varied uses of the same organ (such as the bird syrinx, higher vertebrate vocal cords, and the delphinid larynx) and to a directing brain capable of making choices.

More than any other communication method, such as olfaction or vision, the acoustic signal's physical nature lends itself to reemission in rather satisfactory conditions. The experimenter can also intervene in the signal's structure by means of recording tapes. He may cut, fractionate, inverse, filter, and isolate certain factors. Thus, the signal's physical nature makes possible, and relatively easy, synthetic signal creation by means of electronic technology. Tectonics of the acoustic message may be composed of a simple, physically indivisible element called a phonatom (or pulse or note according to different authors) which forms a sort of basic molecule. This structure may be repeated a determined number of times in a given time period. A rhythmic element thus appears. There may be relative amplitude variations in some of these elements. The phonatom itself may be composed of a preferential frequency or of groups of frequencies. The signal may be continuous; however, it always has a defined length. The simplest (although not simple) structures are those of invertebrate and lower-vertebrate signals. Bird signals are much more complex, the same message having an extremely varied and heterogeneous physical configuration. Moreover, individual variations appear which have been brought to light by systematic studies done by many authors on certain birds such as the finch, in which Borror (1961) finds 13 themes and 187 variations. These individual variations may be personifications of the message, although it is not yet well known which physical structure is related to individualization. One exception may be *Laniarus erythrogaster* described by Thorpe (1963). In this species the temporal parameter between the messages exchanged by two individuals (i.e., the rhythmic pattern of silences) carries the information, and not the acoustic part of the signal itself.

Taking into account individual variations, whether minor or large, physical structure of the message is organized following a code particular to the species, and thus, from a purely zoological point of view, the structure has a specific character. Its physical variants may even characterize local populations (dialects). However, there may also be analogous structures in several species, thus inducing interspecific reaction as in the case of alarm signals in some Corvidae (Busnel et al., 1957; Frings et al., 1958) and sparrows (Marler, 1957; Marler and Tamura, 1964). Finally, certain animals are able, by memorization and imitation to copy the physical signal characteristic of other species. This has been shown to occur in birds (Armstrong, 1963) which imitate the songs of other species and even the voice of men and also in the porpoise (Lilly, 1962; Batteau, personal communication).

Graphic analysis of signals have not given bioacousticians as easily prehensible a result as they might have at first hoped. The spectrographic configuration, apart from artifacts which it sometimes induces, provides a general characterization of the phenomenon; however, in spite of near perfection in technique it is difficult to relate the image obtained after analysis to the semantic content of the different parts of the signal. Imperfect as it may be, it is certain that the tendency of the future will be to master all elements of the signal composition by using large computers (as is now done) in a rather satisfactory manner, with word synthesizers which are being developed following the series of vocoders. This technology, applied to human acoustic communication, is theoretically easy to conceive and use, as the experimenter directly controls the phenomenon by himself. The approach is quite different for animal acoustics, for there, the animal's responses are the only objective proof of a positive result. This adds an extra difficulty, but it is almost certain that in the near future research will be orientated in this direction.

The notion of syntax, used here in reference to human language, is acceptable to the biologist because it can be assimilated to that of gestalt, which takes into account the temporal form perception obligatorily included in every signal. The sound signal is a form, modeled after the space-time relationship, and syntax is related to recognition of the organizational value of this structure. This is the notion of "pattern recognition." If the signal, as such, has a recognizable temporal structure and thus a composite one, its structure is syntactic. This aspect of the problem has been little studied up to the present. Research has been done on the following species: insects (Busnel et al., 1956), amphibians (Capranica, 1964; Paillette, unpublished results), birds (Thielcke, 1962; Bremond, 1962; Falls, 1962), dogs (Busnel, unpublished results), and cats. Much new work is now under way in this field.

How can the notion of space configuration be approached? The answer is relatively simple in electroacoustics. The natural signal is taken as a starting point and is expanded, contracted, inversed, lengthened, shortened, etc. This type of transformation has many possibilities, as Bremond in our laboratory has shown, working with variable-speed tape records, heterodynes (for frequency transposition), ring modulators, frequency modulators, level modulators, and filters.

Message Content

An animal sound message contains a minimum of two types of information. The first intrinsic part indicates the presence of an individual of the species, his spatial position, and, in a number of species (birds and porpoises, among others), individualization—it is Peter or Paul. The hierarchical status is also present if a gregarious species is concerned. The second part of the message holds the semantic content which translates the transmitter's internal state in a given behavioral situation, that is to say, his motivation. It may also contain information relative to the milieu, such as localization of an individual, an object, a territory, or a predator. A thorough investigation of this subject has been done on birds in particular by Marler (1956) studying *Fringilla coeles.*

Signals are usually classed in function according to the behavior which they elicit in the receiver. In the animal kingdom signals are principally related to sexual life (calling the partner, courtship, and rivalry), family life (contact and reciprocal parent-young relationship signals), and social life (hierarchy, group activity, alarm, predator signaling, alimentary behavior, territorial behavior, etc.). Vocabularies in the animal kingdom vary widely in their complexity.

In many species the male sexual signal is different from that of the female. As a matter of fact, in many invertebrates and lower vertebrates, females are often mute. Experimentation has shown that sexual signals are recognized by partners of the same or of the opposite sex, or by both. Information indicating the transmitter's identity has been indisputably found in many, if not in all birds and in some mammals. When both sexes emit sounds, partner or parent recognition is accomplished by very small variations in the signal's physical constitution. As already noted above, a curious example of this is the bird *Laniarus,* which does not use the signal itself as a means of individual recognition, but instead uses the interval of time between calls and replies in a duo (Thorpe, 1963, antiphonal singing). The transmitter's social situation may be equally transmitted in the signal, although dominance in a group is a rather abstract notion. In reference to individual recognition, we may say there is parental, filial, and social information.

In social situations, we find contact, anxiety, and distress signals and also aggressive motivation signals such as territorial, rivalry, food, alarm, attention, and flying-away signals. There are different sounds for different types of alarm, graduated according to dominant motivation. Many birds and mammals are able to express the degree of alarm, imminence of potential danger, or excitement by varying the speed or intensity of the emitted signal (Lorenz's geese). Some bird territorial defense signals vary according to the intruder's distance away, and bee signals vary according to sugar concentration.

Interesting studies on sea gulls (Frings and Frings, 1956) and bees, *Apis* and *Melipone* (Wenner, 1962; Esch, 1962), have been made concerning transmitted signal information about food. The case of the honey-guide bird (Friedmann, 1955), holds special interest for its interspecificity. It conveys food-source information to any mammal showing an interest, such as the ratel or man, both of whom have been shown to integrate the message and follow the bird to its food emplacement. Signals have also been described for bees and birds which give information concerning the location of nests or nest-building materials.

Information concerning predators has been well described for birds and for some mammals such as the prairie dog (King, 1955). The signal may contain data indicating the spatial position of a danger, whether it be on the ground or in the air. These signals may be physically quite different, however, without being specific to a special danger. They may be understood by several species (Marler, 1956; Busnel and Giban, 1958). Wenner (1962, 1964) has shown in bees that acoustic signals can, through combination of simple signals, transmit several informative parameters. In some birds, and in particular in the robin, Busnel and Bremond (1962) have found signals transmitting a minimum of two sets of information. One of these relates to the transmitter's situation, the other designates by a message a particular individual. This important fact has been proved by Weeden and Falls (1959) studying the ovenbird (*Seiurus aurocapillus*) and by Bremond (1966) studying the robin, the wren (*Troglodytes troglodyte*), and the finch (*Fringilla coelebs*). The schamama (*Copsychus malabaricus*), studied by Gwinner and Kneutgen (1962), is particularly interesting in that it incorporates in its territorial defense signal an imitation of some of the motives of the song of the intruder, thus personalizing the threat. In the porpoise *Tursiops truncatus* Lang and Smith (1965) and Bastian (1967) also showed that acoustic signals could transmit complex data. Man excepted, these are so far the only known cases in the animal kingdom of a real combination of signals. This will pose anew the syntactic problem, as will the experiment with "crayfish counterpoint" (or cancrizan). The signal when read backwards shows the semantic contents to be related to the order of sounds perceived either phonetically or prosodically. This means that, in general, successive elements have a certain syntactic construction. In fact, if we consider the physical framework of the message and thus its organization, the semantic information is based on a structure which is sometimes simple, but more often complex. In many lower animals a very simple characteristic of the signal is reactogenic. The value of the signal's sharp wavefront has been shown to set off the attraction reaction in the female (Busnel and Dumortier, 1954). If a single artificial transient sound or sharp-wavefront sound, with a sufficient intensity is sent, it takes on the same value as the species' natural signal. This reaction to transients is interspecific.

On the contrary, pulse repetition rate is a very specific characteristic of a signal and may even serve in taxonomic determination (Dumortier, 1963a; Haskell, 1961). Variants of this character have been found in some amphibians (Busnel and Dumortier, 1955; Capranica, 1964). The entire signal is not always necessary as a support of information. In a fish, *Bathygobius soporator,* for instance, Tavolga (1965) found that the length of the signal could be reduced by 50 per cent and its rhythm and frequency modified in large measure without altering the information value, which seemed to remain unchanged if judged from the point of view of the receiver's behavioral response. Thus, information is not carried by the entire signal, and the richness of most acoustic structures should probably be considered redundant. This redundancy acts as a protection against background noise and increases the signal's intelligibility. Redundance is an important notion in biology, and its value has not yet been fully understood.

Falls (1962) and Busnel and Bremond (1962), in their studies on some birds, have found several reactogenetic signal characters in which there is a certain syntactic organization. This organization has not as yet been found to be a general rule, and each species must be considered as a special case and studied separately. The structure and semantics of animal signals are, nevertheless, quite rigidly fixed, and in the communication of most known species today there are only invariants. Thus, the signal repertoire seems limited, and in a way, stereotyped, since in the entire animal kingdom (bees, some birds, and possibly porpoises excepted) semantics is limited to objective situations evoking an instanteous or immediate-future type of event. Modifications in rhythm and dynamics indicate an evolution in the degree of motivation, but in most cases do not show a change in semantics. However there may be in a message, at least two, if not more, semantically significant units such as distance and designation of an intruder contained in the territorial defense signal or the kind of danger and distance from it contained in the danger signal. We do not yet know how to relate a given unit to a distinct message structure; but it seems certain that there is no true signal combination, and thus there can be no language, in the true sense of the word, since signal combination forms one of the bases of language (by signal combination is meant the possibility of associating basic acoustic structures, so as to form and transmit new information). There are probably degrees of complexity in the combinatory potentiality of an acoustic communication system which may be signs of complexity of the system itself. Animal vocabularies are rather poor and finite, the richest one having, as far as we know today, only thirty to fifty different signals.

Destination

The auditory system, which is the first physiological step in message reception, depends upon a series of functionally and anatomically complex organs. The ear does not present a unity of development. The external, middle, and inner ears differ as much by their embryological origins as by their chronological formation, and the three elementary germ layers of the first embryonic stages all play a role.

In vertebrates, the bilateral organs which perceive only excitations of a mechanical origin are always grouped in the head periphery, while, on the other hand, in invertebrates the system is composed of many receptors scattered over the body. These are mechanoreceptors, many of which are of the hair sensilla type. In vertebrates there is a pressure and time analyzer system which allows detection of vibrations as well as perception of spatial position and inertia. The reception organs, called stato-acoustic, can reach a great degree of sensitivity in higher orders. We have in the animal kingdom quite complete morphological data for a large number of receptor apparatuses, their peripheries, the nerve channels, and the higher centers of the diencephalon.

On the other hand, physiological theories of hearing are neither as definitive nor as satisfying. The phenomenon of nerve coding, by which information is transmitted to nerve centers, is still not clear. Besides, if hearing is taken as the basic physiological theme, message integration, as far as communication is concerned, is still only a part of behavioral studies. We mean here to summarize only some of the many aspects of hearing. The comparison that generally can be made between the frequency hearing range of the animal (whether it be obtained by electrophysiology or conditioning) and the frequency spectrum of the sound signal it emits shows that, in most cases, hearing ability extends much beyond the frequency field belonging to the species' acoustic communication. Actually, focusing our attention strictly on the frequency field leads to a narrow conception of auditory characteristics, since the hearing apparatus is a time gauge much more than it is a Fourier's series analyzer and especially measures the intensity-time relationship (Pimoniv, 1962).

Another essential point is the notion of feedback between the hearing system and the phonation system. Phonation control by hearing exists in practically all the animal kingdom starting from the evolutionary stage when the animal is capable of choice in the motives of his vocalization. There is absolute proof of this in studies done on some deafened birds (Konishi, 1964; Mulligan, 1964) and in *Kaspar-Hauser*-raised birds. In comparing message reception and spatial localization to emission systems, a second important point should be noted. "Binaurality" of the auditory system is currently found in the animal kingdom, but phonatory emission systems are often unpaired. Finally, it is to be remembered that message reception through the auditory canal is possible only in the presence of a central motivation which permits an objective selection of a given signal from the background noise. This is even possible during a partial loss of consciousness, as during sleep. There are many classical illustrations of this fact: after copulation many female animals are no longer attracted by the male signal which they nevertheless perceive; a satisfied animal will not react to a food call.

I believe that lack of understanding of this problem is one of the main causes for the intellectual gap which exists between hearing physiologists and behavior specialists. Central motivation is thus one of the main keys to a semantic integration of the message in communication. The starting point of the system is, of course, audition, but at a sensitivity level controlled by a number of physiological functions. Mere observation of animal behavior, although useful, is not sufficient for a correct interpretation of the message problem. It is necessary to be able to control the stimulus and thus follow eventual changes in the reactions, which can be the only criteria of the semantic value of the message. These reactions may be oriented movements, attraction, repulsion, flying, or running away and phonoresponse—all of which have been widely used by scientists experimenting on many animal species. One of the first experiments of this type which may be cited is that of Regen (1914) who, fifty years ago, attracted a female cricket to the telephone call of a male, and most recently the experiments of Lang and Smith (1965) and Bastian (1967), who, working along these lines, studied an eventual passage of information between two separated porpoises. It is in using such techniques that the semantic and syntactic value of a message or its structure may begin to be studied. This does not mean we should thus be able to understand all messages, because in any case the experimenter does not have access to that which we may term the innermost conscience of the animal. Thus, many real emotional factors are not perceived, since they are not manifested by coordinated motor reactions. Probably more specific information will be obtained in the future from chronic animals equipped with radio telemetric transmitters.

In some species reaction to a message is innate, automatic, and often of the reflex type. This is true particularly in invertebrates and in relation to alarm and distress messages in almost the entire animal kingdom. For this reason, these messages are often referred to as stimulus-sign, releaser, etc. In other kinds of messages, part of the information is learned. This is notably the case in individual recognition in some birds and mammals. This learning is associated with the receiver's general ability to learn the form of certain acoustic messages which he then integrates into his own communication vocabulary. The importance of signal-form learning in the acoustic signals of some species was first demonstrated by Koehler's German school (Sauer, Messmer, Thielcke), using the *Kaspar Hauser* technique of raising birds in acoustic isolation, and by Thorpe in England. According to Marler (in press), who has analytically reviewed this question, learning occurs at different ages or life periods, and is related in a certain way to Lorenz's imprinting. Learning seems to be important at different levels such as the group, species, family, couple. Although the signal may be locally modified in its physical characteristics, the semantic information usually remains unchanged. There may be an exception to this in signals newly learned at a certain stage in the individual's life which were not part of his prior vocabulary. Nicolaï (1959) describes such instances in the female bullfinch who learned certain signals from her first husband which were different from those she learned in her family surroundings. Another example is the robin's instantaneous imitation of an invader's signal. Thus, the invaded robin embodies its designation: "I am talking to you, invader of the moment" (Bremond, 1966).

Codification

The study of animal acoustic communication may be considered as an outgrowth of semiology. Therefore, Sebeok's term, "zoosemiotics" (1965), taken from de Saussure's (1916) "semiotics," seems quite adequate, since it is a study of a system based on signs whatever their origin may be. In most of the animal kingdom, a particular signal corresponds to a definite situation or to a given or innate experience according to the repertoire of the species. When there is a variant, it has not been proved whether it is only related to the physical aspect of the signal or whether it also concerns the semantic content. From a semiological point of view, if specific signals are total, complete messages, any mutilation destroys the meaning. It seems very probable, then, that according to the species the meaning or semantic content is rather strictly related to a series of phonic productions which would be distinctive, nonsignificant units. This makes most animal acoustic interaction truly semiotic, constituting a real sign communication system.

Language functioning is related to the fact that language may be decomposed into discontinuous, differential, or numerable discrete units. In the animal kingdom, can the total message be analyzed in smaller units, or is it itself the smallest communication unit? The message seems to lack first articulation units (since the smallest significant unit in each case is the indecomposable total message) and functions by second articulation units (since the least alteration of successive phonic productions changes, and probably destroys, the message in many species). Therefore, animal acoustic communication seems to be essentially a code composed of signals having as corollaries fixity of content, message invariability in relation to a single situation, irresolvability in the nature of what is said, and unilateral transmission. The significance of a statement is determined by the situation in which the speaker emits the statement, as well as by the auditor's behavioral response to that statement. Even in animals, relayed communications exist, at least at certain levels of communication such as those of birds and probably porpoises.

It must be remembered that the richness of a communication code is not composed exclusively of acoustic signals, and that is why a large number of animals have a more complete, complex code constituting a large number of other signal forms. It is in combining their usage that information corresponding to the species' social relationships can be transmitted. The more evolved an animal species, the more complex is its code of acoustic signals. However, some of these species—birds, bees, and porpoises—merit special attention in the light of recent studies. It is yet too early to judge the importance of results obtained from experiments on these animals, but it is possible that, in view of future experiments, the idea of communication systems, which is at present a little too mechanical, may have to be revised. Even if in the animal kingdom an ontogenesis and anatomical phylogeny of the phonatory and auditory organs can be found, thus increasing our knowledge, the enormous differences concerning acoustic communication systems between species must be sought essentially at the level of the functioning of the brain, which, after all, remains the principal organ directing the essential psychological activity of all animals.

References

Akermann, B., B. Anderson, E. Fabricius, and L. Svensson. Observations on the central regulation of body temperature and of food and water intake in the pigeon, *Columba livia*. *Acta Physiol. Scand.*, 1960, *50,* 328.

Annandale N., J. Coggin Brown, and F. H. Gravely. The Limestone Caves of Burma and the Malay Peninsula, *Myriapoda. J. Asiat. Soc. Bengal (Calcutta), (N.S.),* 1913, *9,* 415.

Armstrong, E. A. A. *Study of Bird Song.* Oxford Univ. Press, 1963, London.

Aronson, L. R., and G. K. Noble. The sexual behavior of Anura. II. Neural mechanisms controlling mating in the male leopard frog, *Rana pipiens. Bull. Amer. Nat. Hist.,* 1945, *86,* 87–139.

Bastian, J. Social communication content in the pulse outside the echolocation (in press).

Bastian, J. The transmission of arbitrary environmental information between Bottlenose Dolphins. In Les systèmes sonars animaux, R.-G. Busnel, ed., Frascati (Italy), Gap, France, 1967, pp. 803–873.

Blair, F. W. Call difference as an isolation mechanism in Southwestern toads (genus *Bufo*). *Texas J. Sci.,* 1956, *VII,* no. 1, 87–106.

Blair, F. W. Mating call and stage of speciation of two allopatric populations of spadefoots *(Scaphiopus). Texas J. Sci.,* 1958, *X,* no. 4, 484–488.

Blair, F. W. Isolating mechanisms and interspecies interactions in anurans. *Proc. XVI Int. Congr. Zool. (Washington, D.C.),* 1963, 315–319.

Borror, D. J. Intraspecific variation in passerine bird songs. *The Wilson Bull.,* 1961, *73,* no. 1, 57–78.

Borror, D. J. Songs of finches *(Fringillidae)* of Eastern North-America. *The Ohio J. Sci.,* 1961, *61,* no. 3, 161–174.

Bremond, J.-C. Paramètres physiques du chant de défense territoriale du rougegorge (*Erithacus rubecula* L.) *C. R. Acad. Sci.,* 1962, *254,* 2072–2074.

Bremond, J.-C. Thesis: Recherches sur la sémantique et les éléments physiques déclencheurs de comportements dans les signaux acoustiques du rouge-gorge (*Erithacus rubecula* L.), Fac. Sci. Paris, OA 1120, 1–112, Fig. 1–24, 1966.

Brown, J. L. Vocalization evoked from the optic lobe of a songbird. *Science,* 1965, *149,* no. 3687, 1002–1003.

Brown, J. L., and R. W. Hunsperger. Neurothology and motivation of agonistic behaviour. *Anim. Behav.,* 1963, *II,* no. 4, 439–448.

Brückner, J. H. Untersuchungen zur Tiersoziologie, insbesondere zur Auflösung der Familie. *Z. Tierpsychol.,* 1933, *128,* 1–105.

Bühler, K. Sprachtheorie. Fisher-Verlag, Iena, 1934.

Busnel, R.-G., ed. Colloque sur l'acoustique des Orthoptères. I.N.R.A., Jouy-en-Josas (5–8. avril 1954), I.N.R.A., ed., Paris, 1955.

Busnel, R.-G., ed. *Acoustic Behaviour of Animals.* Elsevier, 1963, Amsterdam.

Busnel, R.-G., and J.-C Bremond. Etude préliminaire du décodage des informations contenues dans le signal acoustique territorial du rouge-gorge (*Erithacus rubecula* L.). *C. R. Acad. Sci.,* 1961, *252,* 608–610.

Busnel, R.-G., and J.-C. Bremond. Recherche du support de l'information dans le signal acoustique de défense territoriale du rouge-gorge (*Erithacus rubecula* L.). *C. R. Acad. Sci.,* 1962, *254,* 2236–2238.

Busnel, M.-C., and R.-G. Busnel. La directivité acoustique des déplacements de la femelle d'*Occanthus pellucens* Scop. *C. R. Soc. Bio.,* 1954, *CXLVIII,* 830–833.

Busnel, R.-G., and B. Dumortier. Etude des caractères du signal du sifflet de Galton provoquant la phonotaxie de la femelle *d'Ephippiger bitterensis. C. R. Soc. Bio.,* 1954, *CXLVIII,* 1751–1754.

Busnel, R.-G., and B. Dumortier. Phonoréactions du mâle d'*Hyla arborea* à des signaux acoustiques artificiels. *Bull. Soc. Zool. France,* 1955, *LXXX,* no. 1, 66–69.

Busnel, R.-G., B. Dumortier, and M.-C. Busnel. Recherches sur le comportement acoustique des Ephippigères (Orthoptères, Tettigoniidae). *Bull. Bio. Fr. et Belg.,* 1956, *XC,* fasc. 3, 219–286.

Busnel, R.-G., A. Dziedzic, and S. Andersen. Rôle de l'impédance d'une cible dans le seuil de sa détection par le système sonar du Marsouin *Phocaena phocaena. C. R. Soc. Bio.,* 1965a, *159,* 69–74.

Busnel, R.-G., A. Dziedzic, and S. Andersen. Seuils de perception du système sonar du Marsouin *Phocaena*

phocaena L. en fonction du diamètre d'un obstacle filiforme. *C. R. Acad. Sci.*, 1965b, *260*, 295–297.

Busnel, R.-G. and J. Giban. La protection acoustique des cultures et autres moyens d'effarouchement des oiseaux. C. R. de l'I.N.R.A. (26.–27. nov. 1958) I.N.R.A., ed., Paris.

Busnel, R.-G. and J. Giban, ed. Le problème des oiseaux sur les aérodromes. C. R. de l'I.N.R.A., Nice (25.–27. nov. 1963), I.N.R.A., ed., Paris, 1965.

Busnel, R.-G., J. Giban, Ph. Gramet, H. Frings, M. Frings, and J. Jumber. Inter-spécificité des signaux acoustiques ayant une valeur sémantiqué pour des Corvidés européens et nord-américains. *C. R. Acad. Sci.*, 1957, *245*, 105–108.

Busnel, R.-G., and W. Loher. Sur l'étude du temps de la résponse du stimulus acoustique artificiel chez les *Chorthippus* et la rapidité de l'intégration du stimulus. *C. R. Soc. Bio.*, 1954. *CXLVIII*, 862–865.

Capranica, R. R. Evoked vocal response of the Bullfrog. J.A.S.A., 1964, *36*, no. 10, 2007.

Capranica, R. R., M. Sachs, and M. J. Murray. B. Auditory discrimination in the Bullfrog. *XVI. Comm. Biophys.*, 1964, no. 71, 245–249.

Cherry, C. *On Human Communication.* Wiley, New York, 1957.

Cloudsley-Thompson, J. L. A new sound-producing mechanism in Centipedes. *Entom. Month. Mag.*, 1961, *96*, 110–113.

Curio, E. Verhaltensstudien am Trauerschnäpper. *Z. Tierpsychol.*, 1959, *suppl.* 3, 1–118.

Delgado, J. M. R. Free behaviour and brain stimulation. *Int. Rev. Neurobiol*, 1964, *6*, 349–449.

Dijkgraaf, S. Over een merkwaardige functie van den gehoorzin bij vleermuizen. *Ned. Akad. v. Wetensch., Afd. Natuurk.*, 1943, *LII*, no. 9, 3–8.

Dijkgraaf, S. Die Sinneswelt der Fledermäuse. *Experientia*, 1946, *II*, no. 11, 1–31.

Dijkgraaf, S. Sinnesphysiologische Beobachtungen an Fledermäusen. *Acta Physiol. Pharmacol. Neerl.*, 1957, *6*, 675–684.

Dumortier, B. Etude expérimentale de la valeur interspécifique du signal acoustique chez les Ephippigères et rapport avec les problèmes d'isolement et de maintien de l'espèce (Orthopt., Ephippigeridae). *Ann. Epiphyties*, 1963a, *14*, no. 1, 5–23.

Dumortier, B. Morphology of sound emission apparatus in Arthropoda. *Acoustic Behaviour of Animals, Elsevier*, 1963b, Chapter 11, 277–345.

Dumortier, B. The physical characteristics of sound emissions in Arthropoda. *Acoustic Behavior of Animals, Elsevier*, 1963c, Chapter 12, 346–373.

Dumortier, B. Ethological and physiological study of sound emission in Arthropoda. *Acoustic Behaviour of Animals, Elsevier*, 1963d, Chapter 21, 583–654.

Esch, H. Uber die Schallerzeugung beim Werbetanz der Honigbiene. *Z. Vergl. Physiol.*, 1961, *45*, 1–11.

Esch, H. Uber die Auswirkung der Futterplatzqualität auf die Schallerzeugung im Werbetanz der Honigbiene. *Dtsch. Zool. Ges.*, 1962, 302–309.

Esch, H., and I. Esch. Sound: An element common to communication of stingless bees and to dances of the honey bee. *Science*, 1965, *149*, no. 3681, 320–321.

Falls, J. B. Properties of bird-song eliciting responses from territorial males. *Proc. XIII Int. Ornithol. Congr. Ithaca*, 1962, June 17–24, *1*, 259–271.

Friedmann, H. The honey guides. *U.S. Nat. Mus. Bull.*, 1955, *208*, 1–279.

Frings, H., and M. Frings. Auditory and visual mechanisms in food-finding behaviour of the Herring-gull. *Wilson Bull.*, 1956, *67*, 155–170.

Frings, H., M. Frings, J. Jumber, and R.-G. Busnel. Reactions of American and French species of *Corvus* and *Larus* to recorded communication signals tested reciprocally. *Ecology*, 1958, *39*, no. 1, 126–131.

Fulton, B. B. Inheritance of song in hybrids of two subspecies of *Nemobius fasciatus* (Orthoptera). *Ann. Ent. Soc. Amer.*, 1933, *XXVI*, no. 2, 368–376.

Fulton, B. B. Experimental crossing of subspecies in *Nemobius* (Orthoptera:Gryllidae). *Ann. Entomol. Soc. Amer.*, 1937, *XXX*, no. 2, 201–207.

Fulton, B. B. Speciation in the field cricket. *Evolution*, 1952, *VI*, no. 3, 283–295.

Gould, E. Evidence for echolocation in the Tenrecidae of Madagascar. *Proc. Am. Phil. Soc.*, 1965, *109*, no. 6, 352–360.

Griffin, D. R. Echolocation by blind men, bats and radar. *Science*, 1944, Dec. 29, *100*, no. 2609, 589–590.

Griffin, D. R. Bat sounds under natural conditions, with evidence for echolocation of insect prey. *J. Exp. Zool.*, 1953a, Aug., *123*, no. 3, 435–466.

Griffin, D. R. Acoustic orientation in the oil bird, *Steatornis. Nat. Acad. Sci.*, 1953b. Aug., *39*, no. 8, 884–893.

Griffin, D. R. *Listening in the Dark.* Yale Univ. Press, 1958.

Griffin, D. R., and A. D. Grinnel. Ability of bats to discriminate echoes from louder noise. *Science*, 1958, *128*, 145–147.

Griffin, D. R., J. H. Friend, and F. A. Webster. Target

discrimination by the echolocation of bats. *J. Exp. Zool.,* 1965, March, *158,* no. 2, 155–168.

Griffin, D. R., F. A. Webster, and Ch. R. Michael. The echolocation of flying insects by bats. *Anim. Behav.,* 1960, *VIII,* no. 3–4, 141–154.

Grimes, L. Antiphonal singing and call notes of *Laniarus barbarus. Ibis,* 1966, *108,* 122–126.

Gwinner, E., and J. Kneutgen. Uber die biologische Bedeutung der "zweckbedienlichen" Anwendung erlernter Laute bei Vögeln. *Z. Tierpsychol.,* 1962, *19,* 692–696.

Haskell, P. T. *Insect Sounds.* H. F. & C. Witherby, Ltd., 1961, London.

Heinroth, O., and K. Heinroth. Verhaltensweisen der Felsentaube (Haustaube). *Columbia livia livia* L. *Z. Tierpsychol.,* 1949, *6,* 153–201.

Hess, W. R. Stammganglien-Reizversuche. Verh. Dtsch. physiol. Ges. (Sept. 1927). *Ber. ges. Physiol.,* 1928, *42,* 554–555.

Hess, W. R. *Das Zwischenhirn.* 2nd Edition, Schwabe, 1954.

Holst, E. von and U. von Saint Paul. Vom Wirkungsgefüge der Triebe. *Naturwiss.,* 1960, *18,* 409–422.

Holst, E. von, and U. von Saint Paul. On the functional organisation of drives. *Anim. Behav.,* 1963, *11,* no. 283, 1–20.

Hüber, F. Verhaltensstudien am Männchen der Feldgrille (*Gryllus campestris* L.) nach Eingriffen am Zentralnervensystem. *Verhandl. Dtsch. Zool. Ges., Freiburg,* 1952, 138–49.

Hüber, F. Stiz und Bedeutung nervöser Zentren für Istinkthandlungen beim Männchen von *Gryllus campestris* L. *Z. Tierpsychol.,* 1955, *12,* no. 1, 12–48.

Hüber, F. Auslösung von Bewegungmustern durch elektrische Reizung des Oberschlundganglions bei Orthopteren (Saltatoria: Gryllidae, Acrididae). *Verhandl. Dtsch. Zool. Ges.,* 1959, 248–269.

Hüber, F. Experimentelle Untersuchungen zur nervösen Atmungsregulation der Orthopteren (Saltatoria:Gryllidae). *Z. Vergl. Phsyiol.,* 1960, *43,* 359–391.

Hüber, F. Untersuchungen über die Funktionen des Zentralnervensystems und insbesondere des Gehirnes bei Fortbewegung und der Lauterzeugung der Grillen. *Z. Vergl. Physiol.,* 1960, *44,* 60–132.

Hüber, F. Central nervous control of sound production in crickets and some speculations on its evolution. *Evolution,* 1962, *XVI,* no. 4, 429–442.

Hüber, F. The role of the central nervous system in Orthoptera during the coordination and control of stridulation. *Acoustic Behaviour of Animals,* Elsevier, 1963, Chapter 17, 440–484.

Huchtker, R., and J. Schwartzkopff. Soziale Verhaltensweisen bei hörenden und gehölosen Dompfaffen *(Pyrrhula pyrrhula). Experientia,* 1958, *XIV,* no. 3, 106–111.

Hunsperger, R. W. Affektreaktion auf elektrische Reizung im Hirnstamm der Katze. *Helv. Physiol. Pharmacol. Acta,* 1956, *14,* 70–92.

Kalmus, H. Analogies of language to life. *Language and Speech,* 1962, *5,* no. 1, 15–25.

Kanai, T., and S. C. Wang. Localization of a mechanism for vocalization in the hypothalamus and its descending pathways in the brain stem of the cat. *Feder. Proc.,* 1961, *20,* 331.

Kanai, T., E. A. Day, and C. B. Weld. A brain stem mechanism for vocalization in the cat. *Feder. Proc.,* 1964, *23,* no. 2, 1 p.

Karplus, J. P., and A. Kreidl. Gehirn und Sympathicus. *Pflügers Arch. ges. Physiol. Menschen u. Tiere,* 1909, *129,* p. 138, 1928, *219,* pp. 613–618.

King, J. A. Social behavior, social organization and population dynamics in a black-tailed prairiedog down in the black hills of South Dakota. *Contr. Lab. Vertebrate Bio.,* 1955, no. 67, 1–123. Univ. Michigan Press, Ann Arbor.

Konishi, M. Effects of deafening on song development in two species of juncos. *The Condor,* 1964, *66,* no. 22, 85–102.

Kulzer, E. Flughunde erzeugen Orientierungslaute durch Zungenschlag. *Naturwiss.,* 1956, *43,* 117–118.

Landois, H. *Thierstimmen.* Herdersche Verlagsbuchhandlung, Freiburg, 1874.

Lang, T. G., and H. A. P. Smith. Communication between dolphins in separate tanks by way of an electronic acoustic link. *Science,* 1965, *150,* no. 3705, 1839–44.

Lanyon, W. E., and W. N. Tavolga, ed. *Animal Sounds and Communication.* A.I.B.S., 1958, Washington, D.C.

Leroy, Y. Etude du chant de deux espèces de grillons et de leur hybride *(Gryllus commodus* Walker, *Gryllus oceanicus* Le Guillon, Orthoptères). *C. R. Acad. Sci.,* 1963, *256,* 268–270.

Leroy, Y. Transmission du paramètre Fréquence dans le signal acoustique des hybrides F_1 et P × F_1, de deux Grillons: *Teleogryllus commodus* Walker et *T. oceanicus* Le Guillon (Orthopt., Ensifères). *C. R. Acad. Sci.,* 1964, *259,* 892–895.

Leroy, Y. Essai de definition du comportement hierarchique chez les grillons (Insectes, Orthopteres). *C. R. Acad. Sci.*, 1966a, *263*, 1752–1754.

Leroy, Y. Signaux acoustiques, comportements et systématiques de quelques espèces de Gryllidés (Orthopt., Ensifères). *Fac. Sci.*, Paris, 1966b, Pierre Fanlac Perigneux.

Lilly, J. C. Correlations between neurophysiological activity in the cortex and short-term behavior in the monkey. *Bio. and Biochem. Bases of Behav.*, Harlow & Woolsey, Ed., Univ. Wisconsin Press, Madison, *1958a.*

Lilly, J. C. Some considerations regarding basic mechanisms of positive and negative types of motivations. *Am. J. Psychiatry*, 1958b, *115*, no. 6, 498–504.

Lilly, J. C. Vocal behavior of the bottlenose dolphin. *Proc. Am. Phil. Soc.*, 1962, *106*, no. 6, 520–529.

Lilly, J. C., and A. M. Miller. Sounds emitted by the bottlenose dolphin. *Science*, 1961a, *133*, no. 3465, 1689–1693.

Lilly, J. C., and A. M. Miller. Vocal exchanges between dolphins. *Science*, 1961b, *134*, no. 3493, 1873–1876.

Lilly, J. C., and A. M. Miller. Operant conditioning of the bottlenose Dolphin with electrical stimulation of the brain. *J. Comp. and Physiol. Psychol.*, 1962, *55*, no. 1, 73–79.

Lorenz, K. Beiträge zur Ethologie sozialer Corviden. *J. Ornithol.*, 1931, *LXXIX*, Heft 1, 67–127.

Lorenz, K. Der Kumpan in der Umwelt des Vogels: Der Artgenosse als auslösendes. Moment sozialer Verhaltensweisen. *J. Ornithol.*, 1935, *83*, 137–213; 289–413.

Lorenz, K. Vergleichende Bewegungsstudien an Anatiden. *J. Ornithol.*, 1941, *89*, no. 3, 194–294.

Lorenz, K. The comparative method in studying innate behavior patterns. *Symp. N.Y. Soc. exp. Bio.*, 1950, *4*, 221–268.

Marler, P. The voice of the chaffinch and its function as language. *Ibis*, 1956, *98*, 231–261.

Marler, P. Specific distinctiveness in the communication signals of birds. *Intern. J. Comp. Ethol.*, *XI*, Pt. I, 13–39, 1957.

Marler, P. Bird song and mate selection. *Animal Sounds & Comm., Am. Inst. Bio. Sci. (A.I.B.S.)*, 1960, no. *7*, 348–367.

Marler, P. Inheritance and learning in the development of animal vocalizations. In: *Communicative Behavior in Animals* (in press).

Marler, P., M. Kreith, and M. Tamura. Song development in hand-raised Oregon juncos. *The Auk*, 1962, *79*, 12–30.

Marler, P., and M. Tamura. Culturally transmitted patterns of vocal behavior in sparrows. *Science*, 1964, *146*, no. 3650, 1483–86.

Messmer, E., and I. Messmer. Die Entwicklung der Lautäusserungen und einiger Verhaltensweisen der Amsel (*Turdus merula merula* L.) unter natürlichen Bedingungen und nach Einzelaufzucht in schalldichten Räumen. *Z. Tierpsychologie*, 1956, *13*, no. 3, 341–441.

Möhres, F. P., and E. Kulzer, Über die Orientierung der Flughunde. *Z. Vergl. Physiol.*, 1956, *38*, 1–29.

Mulligan, J. A. Physical analysis of variation and development in the song of *Melospiza melodia*. University Microfilm, Inc., Ann Arbor, 1964.

Nicolai, J. Familientradition in der Gesangentwicklung des Gimpels *(Pyrrhula)*. *J. Ornithol.*, 1959, *100*, 39–46.

Nicolai, J. Der Brutparasitismus der *Viduinae* als ethologisches Problem. *Z. Tierpsychologie* 1964, *21*, no. 2, 129–204.

Norris, K. S., ed. *Whales, Dolphins and Porpoises.* U. of California Press, 1966, Berkeley.

Norris, K. S., and W. E. Evans. Variations in porpoise echolocation signals. *Symp. Marine Bio-Acoustics*, 1966, in press.

Norris, K. S., J. H. Prescott, P. V. Asadorian, et al. An experimental demonstration of echolocation behavior in the porpoise, *Tursiops truncatus* (Montagu). *Bio. Bull.*, 1961, *120*, no. 2, 163–176.

Perdeck, A. C. The isolating value of specific song patterns in two sibling species of grasshoppers (*Chorthippus brunneus Thunb* and *C. biguttulus* L.) *Behaviour*, Netherl., 1958, *12*, no. 1–2, 1–75.

Phillips, R. E. Wildness in the mallard duck: Effects of brain-lesion and stimulation on "escape behaviour" and reproduction. *J. Comp. Neurol.*, 1964, *122*, 139–155.

Pierce, J. R. *Symbols, Signals and Noise: The Nature and Process of Communication.* Harper & Bros., N.Y., 1961.

Pimoniv, L. *Vibrations en régime transitoire.* Dunod, 1962, Paris.

Prevost, J. Ecologie du manchot empéreur. *Actual. Sci. et Industr.*, Hermann Ed., Paris, 1961, no. 1291.

Randall, W. L. The behavior of cats *(Felis catus)* with lesions in the caudal mid-brain region. *Behaviour*, 1964, *23*, 107–139.

Regen, J. Über die Anlockung des Weibchens von *Gryllus campestris* L. durch telephonisch übertragene Stridulationslaute des Männchens. *Pflügers Arch. ges. Physiol. Menschen u. Tiere*, 1914, *155*, 193–200.

Rougeul et Assenmacher. Cited by Buser in *Brain*

Mechanisms and Learning. Blackwell Ed., Oxford, 1961.

Ruben, R. J., D. Warfield, and R. Glackin. Word discrimination in cats. *J.A.S.A.* 1965, 1204–1205.

Sauer, F. Le développement des signaux sonores de la fauvette *(Sylvia c. communis).* Comparaison entre des individus sauvages et des individus élevés isolés en chambre sourde depuis l'oeuf ou plus tardivement. *Z. Tierpsychol.,* 1954, *11,* 10–93.

Sauer, F. Behaviour of the young garden warblers. *J. Ornithol.,* 1956, *97,* 156–189.

Saussure, F. de. *Cours de linguistique générale.* Paris, 1916.

Schleidt, M. W., M. Schleidt, and M. Magg. Störung der Mutter-Kind-Beziehung bei Truthühnern durch Gehörverlust. *Behaviour,* 1960, *16,* no. 3–4, 254–260.

Schmidt, R. S. Central mechanisms of frog calling. *Behaviour,* 1965, *26,* 251–285.

Scott, J. P., and J. L. Fuller. *Genetics and the Social Behavior of the Dog.* U. of Chicago Press, 1965, Chicago.

Sebeok, Th. A. Coding in the evolution of signalling behavior. *Behavioral Sci.,* 1962, *7,* no. 4, 430–442.

Sebeok, Th. A. Zoosemiotics: Juncture of semiotics and the biological study of behavior. *Science,* 1965, *147,* no. 3657, 492–493.

Signoret, J. P., F. du, Mesnil du Buisson, and R.-G. Busnel. Rôle d'un signal acoustique de verrat dans le comportement réactionnel de la truie en oestrus. *C. R. Acad. Sci.,* 1960, *250,* 1355–1357.

Silva, C., J. Estable, and P. Segundo. Cited by Segundo in *Brain Mechanisms and Learning.* Blackwell Ed., Oxford, 1961.

Tavolga, W. N., ed. *Marine Bio-acoustics.* Proc. Symp., Lerner Marine Lab., Bimini, Bahamas. Pergamon Press, Ltd., 1964, Oxford.

Tavolga, W. N. Review of Marine Bio-acoustics (State of the Art: 1964), U.S. Nav. Traing. Device Center, 1965, *Navtradevcen 1212-1,* 1–100.

Tavolga, W. N., ed. *Marine Bio-acoustics.* Pergamon Press, Ltd., New York (in press).

Tembrock, G. Probleme der Bio-Akustik. *Wiss. Zeitschr.,* Humboldt-Univ., Berlin, 1958/59, *4–5,* 573–587.

Tembrock, G. Stridulation und Tagesperiodik bei *Cerambyx cerdo. Zool. Beitr., Neue Folge* (1960), *5,* 419–441.

Tembrock, G. *Tierstimmen.* Die neue Brehm-Büchere: Ziemen Verlag, Wittenberg-Lutherstadt, 1959.

Thielcke, G. Stammesgeschichte und geographische Variation des Gesanges unserer Baumläufer *(Certhia familiaris* und *c. brachydactyla) Z. Tierpsychol.* (1961), *18,* no. 2, 188–204.

Thielcke, G. Versuche mit Klangattrappen zur Klärung der Verwandtschaft der Baumläufer (*Certhia familiaris* L.), *(C. Brachydactyla Brehm),* und *(C. Americana Bonaparte). J. Ornithol.,* 1962, *103,* no. 2–3, 26–71.

Thorpe, W. H. The analysis of bird song. *Proc. Roy. Instn.,* 1954, *35,* no. 161, 1–13.

Thorpe, W. H. The process of song-learning in the chaffinch as studied by means of the sound spectrograph. *Nature,* 1954, *73,* no. 4402, 465–469.

Thorpe, W. H. The learning of song patterns by birds, with especial reference to the song of the chaffinch *(Fringilla coelebs Ibis),* 1958, *100,* 535–570.

Thorpe, W. H. Further studies on the process of song learning in the chaffinch *(Fringilla coelebs gengleri). Nature,* 1958, *182,* no. 4635, 554–557.

Thorpe, W. H. *Bird-song.* Cambridge Univ. Press, 1961, New York.

Thorpe, W. H. Antiphonal singing in birds as evidence for avian auditory reaction time. *Nature,* 1963, *197,* no. 4869, 774–776.

Thorpe, W. H. Aggression and fear in the normal sexual behavior of some animals. *Pathol. Treatm. Sex. Deviation,* 1964, 3–23.

Thorpe, W. H., and M. E. W. North. Origin and significance of the power of vocal imitation: with special reference to the antiphonal singing of birds. *Nature,* 1965, *208,* no. 5007, 219–222.

Tinbergen, N. Comparative studies of the behaviour of gulls *(Laridae):* a progress report. *Behaviour,* 1959, *XV,* no. 1–2, 1–70.

Tinbergen, N. The evolution of behaviour in gulls. *Scient. Am.,* 1960, no. *12,* 118–130.

Weeden, J. S., and J. B. Falls. Differential responses of male ovenbirds to recorded songs of neighboring and more distant individuals. *The Auk,* 1959, *76,* 343–351.

Wenner, A. M. Sound production during the waggle dance of the honey bee. *Anim. Behav.,* 1962, *10,* no. 1–2, 79–95.

Wenner, A. M. Sound communication in the honey bee. *Scient. Am.,* 1964, *181,* 1–7.

Chapter 12

ECHOLOCATION AND ITS RELEVANCE TO COMMUNICATION BEHAVIOR

Donald R. Griffin

Most bats and marine mammals orient themselves by emitting orientation sounds that are adapted for the purpose of locating objects at a distance by hearing their echoes (Griffin, 1958; Vincent, 1967; Airapet'yants and Konstantinov, 1970). They maintain normal orientation when vision is impossible, but become disoriented if deprived of hearing or if prevented from emitting orientation sounds. As far as we know echolocation is a sort of "solipsistic communication" between an animal and its environment. While the source of echoes may be the body of another animal, only passive physical reradiation of sound waves is involved rather than active reply by the second animal. Hence, echolocation does not properly fall within any reasonable definition of communication behavior, and its discussion in the present volume is justified only by its indirect relevance to the physiological and behavioral phenomena that may be important both in echolocation and in communication.

The echolocation of animals escaped notice until electronic technology permitted translation of the orientation sounds of bats from the ultrasonic range (20 to 150 kHz) into the frequency range of human hearing (Griffin and Galambos, 1941). The discovery of echolocation in whales and porpoises required conversion of underwater sounds into sounds conducted through air (Schevill and Lawrence, 1949, 1956; Kellogg et al., 1953; Kellogg, 1961; Norris, 1966; Airapet'yants and Konstantinov, 1970). While bats of the suborder Microchiroptera and marine mammals of the order Cetacea are the two groups in which echolocation is well developed and well studied, there are isolated cases in cave-dwelling birds (Griffin, 1958; Novick, 1959; Medway, 1967; Griffin and Suthers, 1970), and it occurs in a rudimentary form in terrestrial shrews (Gould et al., 1964; Buchler, 1972) and rodents (Rosenzweig and Riley, 1955). Even human beings are capable of a limited form of echolocation, although it is not known to be important except to the blind. For blind men, however, it is of the utmost importance (Supa, Cotzin, and Dallenbach, 1944; Griffin, 1958; Rice, 1967); I shall return below to further discussion of human echolocation.

Reliance on echolocation for rapid mobility under difficult conditions places severe demands not only on the auditory receptors but also, more importantly, on the analyzing capabilities of the animal's brain. Many bats fly in totally dark caves where irregular rocky obstacles are numerous and unpredictable. They also fly with equal skill in forested areas, avoiding branches, leaves, twigs, vines, and other small objects of considerable complexity. Some cetaceans swim in dark or highly turbid waters where vision is of little if any use. Often there are rocks or other underwater obstructions to be avoided, and some of the smaller porpoises live in muddy rivers and lakes where they must swim close to submerged vegetation, fallen trees, and roots. Two examples are especially impressive in the case of bats: landing on some suitable object and drinking water on the wing. The first maneuver involves flying close to the landing place, slowing, then rapidly turning the body upside down, and finally reaching for a toehold with the hind claws. Errors of a few millimeters produce either an unpleasantly hard collision or a fall and the need to repeat the whole procedure. Bats commonly fly low over the surface of water and dip the lower jaw or tongue just sufficiently to drink. An error of a few millimeters would result in either failure to reach the water or in a splashing submersion. Smooth floors often elicit this type of drinking behavior in captive bats, indicating that the specular reflection of orientation sounds from a horizontal surface is interpreted as a sign of water available for drinking.

The echolocation of stationary obstacles for many years appeared so incredible that no one even suggested that the same mechanism might also be used to locate small moving targets. Nevertheless convincing evidence has shown that at least under some conditions both bats and porpoises pursue insects and fish, respectively, largely by echolocation (Griffin, 1953, 1958; Griffin et al., 1960; Norris et al., 1961). Despite the sensitivity and acuity of echolocation, some insects are located by passive hearing of their flight sounds or sounds resulting from their movements on the ground or in vegetation (Kolb, 1961; Airapet'yants and Konstantinov, 1970). The accuracy and precision of echolocation implies that the auditory nervous system responds selectively to faint echoes from significant objects, despite a variety of other sounds competing for the animal's attention. Animals thus endowed might find the analytical requirements of an advanced communication system already at their disposal. While scarcely any solid evidence is available, it may be of interest to discuss the speculative possibility that both echolocation and communication share certain physiological and behavioral mechanisms.

Properties of Orientation Sounds

In all known cases the orientation sounds used for echolocation are quite brief—each one lasting only a small fraction of a second, in bats and cetaceans only one or a few milliseconds with the exceptions discussed below. It is convenient to refer to these sounds as pulses, since those of bats are not clicks with either simple or chaotic wave forms but orderly trains of about ten to several hundred sound waves. The ultrasonic orientation sounds of bats also have relatively faint audible components that can be heard under favorable conditions once one knows what to listen for (Galambos and Griffin, 1942; Dijkgraaf, 1943). Bats orient themselves in air with acoustic probes traveling at approximately 34 cm/msec, and porpoises use more rapidly traveling sound waves under water with velocities of about 155 cm/msec. Under some conditions echoes may return during the latter part of an emitted orientation sound, but many echolocating animals avoid this by adjusting the duration of emitted sounds in relation to the distance between themselves and the objects that are of immediate concern. This avoidance of overlap is almost universal in the best-studied bats of the family Vespertilionidae (Webster, 1966; Cahlander et al., 1964).

Two distantly related groups of bats, the family Rhinolophidae of the Old World and *Pteronotus parnellii* of the neotropical family Mormoopidae, have specialized orientation sounds containing relatively long constant-frequency portions in addition to frequency sweeps (Möhres, 1953; Griffin and Novick, 1955; Griffin, 1962). These constant frequencies are used to detect Doppler shifts due to relative motion of the bat and the target returning echoes (Schnitzler, 1968, 1970, 1973; Simmons, 1974; Simmons, Howell, and Suga, 1975). On the other hand the best-studied bats (the families Vespertilionidae, Molossidae, Natalidae, and Noctilionidae, which are all highly specialized for echolocation) employ a rapid downward sweep. This is also a very widespread, if not universal, component in the families Emballonuridae, Phyllostomatidae, and Desmodontidae (Simmons, Howell, and Suga, 1975).

The problems of accurate measurement of frequency patterns are complicated by the tendency of bats primarily concerned with large and motionless objects to use orientation sounds of relatively low intensity. Those species that are active predators, pursuing small moving insects or fish, use very considerably louder orientation sounds (Griffin and Novick, 1955; Griffin, 1962; Pye, 1967; Suthers, 1955). The advantage to the bat of the rapid frequency sweep is not entirely certain, but perhaps the most plausible explanation is that it enables a wide range of wavelengths to be reflected from small targets. With small objects approximating the wavelengths of the orientation sounds, the relative intensities of the various frequencies may well provide qualitative information about the nature of the reflecting object. This echo spectrum may be used to achieve discrimination between closely similar objects (Griffin et al., 1965; Griffin, 1967; Bradbury, 1970; Simmons, 1973). Another possible advantage is discussed in the next section.

Cetaceans, when concerned with difficult problems of orientation with respect to objects at short distances, also use very short-duration orientation sounds (Kellogg, 1961; Vincent, 1967; Norris et al., 1961; Norris, 1974). Individually, these are clicks of such short duration that their frequency spectrum is extremely broad. In the best-known cases both bats and porpoises shorten the duration of their orientation sounds to less than one millisecond when they are making close approaches to important objects such as morsels of food or, in the case of bats, landing places.

Orientation sounds also vary widely in other acoustic properties. The frequencies are generally high, or at least include wavelengths of a few millimeters or centimeters. This is presumably related to the greater magnitude of echoes from objects that are somewhat larger than the wavelengths of the sounds impinging on them. Another possible advantage is the rapid attenuation of high-frequency sound in air (Griffin, 1971; Evans and Bass, 1972). While restricting the range at which objects can be echolocated, this also limits the reverberations and "clutter" that would otherwise tend to interfere with the important echoes from objects at a close range.

Another important property of the orientation sounds used by both bats and cetaceans is the universal tendency for the rate at which they are repeated to increase sharply whenever the animal is faced with a difficult orientation problem—whether it be a bat landing on a crevice in the ceiling of a cave, a porpoise picking up a piece of fish floating close in front of the cement wall of a large tank, or either animal pursuing elusive moving prey. The pulse repetition rates rise from a very few isolated orientation sounds per second to brief bursts, in which they are repeated at rates up to 250 per second in bats and even higher in porpoises. When translated into audible clicks, this crescendo of orientation pulses becomes a buzz.

While the orientation sounds of most bats are above the frequency range of human hearing several species use the octave from 10 to 20 kHz and are audible even under natural conditions. In the tropics some insectivorous bats use frequencies that sweep as low as 4 or 5 kHz (Griffin, 1971). In dark caves two species of cave-dwelling birds (*Steatornis* and *Collocalia*) and the genus *Rousettus* among the Megachiroptera (Old World fruit bats with large eyes, which rely on vision under most circumstances) echolocate quite successfully with clearly audible clicks. These clicks contain frequencies from roughly 3 to 8 kHz (Griffin, 1958), but despite these relatively low frequencies *Rousettus* and some species of *Collocalia* can detect cylindrical obstacles with diameters as small as 1 to 3 mm (Griffin, Novick, and Kornfield, 1958; Griffin and Suthers, 1970; Fenton, 1975). Thus echolocation of small objects is biologically possible with frequencies audible to human ears.

Problem of Interference

In many situations the echo of an orientation sound, when it returns to the ears of an echolocating animal, will consist of sound waves reflected from different parts of a large object lying at sufficiently different distances so that interference between sound waves occurs. Except for very small objects or perfectly smooth surfaces such as calm water, such interference must occur even in the echoes of orientation sounds having very short durations. When single frequencies are employed, such destructive and constructive interference produces echoes that vary greatly in amplitude with small changes in the animal's relative distance from various parts of an echoing surface. These interference phenomena are similar in some ways to the familiar phenomenon of standing waves when a continuous tone is measured or listened to in a closed room. The principal difference is that interference patterns of echoes from different parts of a large object change rapidly with time. It seems likely that the variability of such echoes from short tones of constant frequency would make echolocation difficult, and this may help explain why so many echolocating animals use FM pulses as orientation sounds. This problem could theoretically be avoided by using other types of signals, such as bursts of random noise, but no animal highly specialized for echolocation has been found to use noise bursts. Some evidence suggests, however, that small terrestrial mammals use sounds of rather indefinite frequency in a marginal form of echolocation (Buchler, 1972). But these sounds are of such low intensity that it has been difficult to measure their acoustic properties accurately enough to throw much light on this question.

It is misleading to think of echolocation as merely the hearing of isolated echoes. In actual practice under natural conditions, echolocating animals must discriminate certain faint echoes from many other sounds of similar properties occurring at nearly the same time. These include sounds of outside origin and echoes of objects other than the target of immediate concern. Consider, for example, the problems faced by an insectivorous bat relying for its food supply entirely upon echoes from insects measuring only a fraction of the wavelengths of its orientation sounds. These are encountered at unpredictable times and places but, very often, close to other small objects such as leaves or twigs. Any one orientation pulse returns to the bat's ears echoes from a large number of objects at different distances and direction. Only one part of this complex of echoes is relevant. All other information must be ignored or used merely for the avoidance of stationary obstacles. Since most of the unimportant targets are larger than the critically important morsel of food, the interfering echoes will very often be more intense than those important to the animal. Yet, despite these difficulties, small insectivorous bats such as *Myotis lucifugus* capture insects measuring only a few millimeters in wingspread at rates of several per minute during their routine feeding behavior—often under conditions where there is extensive interference from echoes of other larger objects (Gould, 1955, 1959; Griffin et al., 1960; Webster, 1963).

The neurophysiological mechanisms employed by bats in such discriminative auditory orientation have been studied rather extensively in recent years by Grinnell (1963), Grinnell and Grinnell (1965), Suga (1964, 1965, 1973), Suga et al. (1975), Harrison (1965), and Henson (1965, 1970). There are some specializations for auditory sensitivity in the ultrasonic frequency range, apparently shared by other small mammals (Ralls, 1967; Sales and Pye, 1974). Auditory areas of bat brains are enlarged relative to those of other mammals, but this hypertrophy is present only posterior to the diencephalon. The medullary, midbrain, and, in particular, the collicular auditory areas are relatively enormous, whereas the auditory thalamus and cortex are only slightly enlarged in comparison to those of shrews and rodents.

Grinnell (1963) was the first to study the neurophysiological basis of discrimination against interference, and some of his experiments were directed toward explaining the demonstrated resistance of bats to jamming by broad-band interfering noise (Griffin et al., 1963). Evoked potentials from the posterior colliculus in response to short tone bursts similar to orientation sounds could be masked by simultaneous noise from a second high-frequency loudspeaker. In one typical experiment (Grinnell, 1963), the masking noise raised the threshold for a detectable evoked potential by 43 dB, provided the noise arrived from almost the same direction as the signal. When the noise-generating loudspeaker was moved 60°, however, the threshold fell by about 25 dB. Most of this effect was due to the directional differences in auditory sensitivity, but there was also evidence of neurophysiological interaction within the brain between the nerve impulses from the two ears, which served under some conditions to improve further the animal's ability to detect faint signals despite a masking noise. Implications of some recent neurophysiological experiments by Suga are discussed in the next section.

When flying bats were pressed to the limits of their abilities to detect fine wires in a severe jamming noise, they performed better than could be accounted for by signal detection theory if only a single communication channel was assumed to be operating between the wire obstacle and the bat's brain. The flight maneuvers of these bats showed a strong tendency to approach wires obliquely when the jamming noise became truly difficult, and this doubtless served to separate the faint echoes from the jamming noise by taking advantage of their differing angles of incidence (Griffin, McCue, and Grinnell, 1963). Neurophysiologically, the addition of information arriving via the second ear expanded the discriminative capabilities of the bat's information-processing system sufficiently to bring its performance well within the theoretical boundary conditions of signal-detection theory.

Simmons has recently developed a powerful new method for studying the capabilities of echolocation in bats. This method employs a modification of the Lashley jumping stand into a "Simmons flying stand," in which a bat is trained to fly from a starting platform to one of two other platforms. Blinded bats reinforced with food learned to choose the correct platform on the basis of echolocation. With this method Simmons (1973) has been able to demonstrate an ability to discriminate targets on the basis of their size, shape, angular position, and distance. These experiments and many of their ramifications are well reviewed by Simmons, Howell, and Suga (1975). Distance discrimination proved most illuminating. Differential distance thresholds prove to be only a few centimeters, considerably less than the lengths of the orientation sounds as they travel through the air. These results could be accounted for only by assuming that the auditory system of the bat made use of virtually all the information physically present in the echoes. Thus these experiments confirmed and greatly improved upon those of Griffin, McCue, and Grinnell with flying bats.

Simmons and his colleagues have gone on to experiment with the effects of interfering noise on distance discrimination, and the results, while too complex to discuss here, clearly support the conclusion that the auditory brains of bats approach very closely the theoretical limits set by signal-detection theory for an "ideal detector" (Simmons, Howell, and Suga, 1975). Another important conclusion that has resulted from Simmons's experiments has been the rigorous demonstration that bats determine distance by differences in the time required for an echo to return. This was established by substituting electronic delay lines for physical separation between the bat and its target. The two landing platforms were in fact equidistant, but when the echo from one was delayed by a small fraction of a millisecond the bat treated it as a more distant target.

Human Echolocation and Its Possible Relationship to Speech

This brief review of what is known about the capabilities for discriminative echolocation achieved by less than one gram of bat brain suggests that echolocating animals possess powers of auditory discrimination and information processing that are more than adequate for complex types of communication. But the extent to which bats actually utilize these capabilities for communication remains for future investigation to discover. Bats and cetaceans certainly do communicate by sound, as discussed elsewhere in this book. The communicative sounds of bats are mostly at frequencies lower than those of their orientation sounds (Gould, 1971; Bradbury, 1972); presumably the better carrying power of lower frequencies makes them superior for communication.

The search for other applications that might justify including a chapter on echolocation in this book has called to mind a puzzling aspect of human echolocation as it is practiced by the blind. A basic challenge is posed by the simple question: Why is it that blind men cannot echolocate as well as bats or dolphins? As more and more has been learned about the echolocating abilities of 7-gram bats, with brains weighing approximately 1 gram, the disparity has increased —to the embarrassing disadvantage of our 1,500-gram brains. Simple arithmetic of wavelengths suggests that a factor of five, or at the most ten, should separate the minimum sizes of objects discriminated because man is limited to lower frequencies of sound. But where bats catch fruit flies at rates of several per minute, blind men cannot safely drive automobiles, let alone fly airplanes to catch birds—which performance, of course, would be directly analogous.

This disparity of skill at echolocation can scarcely be due entirely to the hesitant recognition that "facial vision" or "obstacle sense" in the blind is largely, if not entirely, echolocation. Many direct investigations of human echolocation have been carried out, along with applied research attempting to improve the usefulness of echolocation to the blind (Zahl, 1950; Clark, 1963). Human echolocation has been studied surprisingly little in the thirty years since its existence and importance were conclusively demonstrated by Supa, Cotzin, and Dallenbach (1944), despite the obvious human importance of any improvements that might be achieved by understanding it better (Griffin, 1958). Rice, Schusterman, and Feinstein (1965) measured threshold diameters of disks that could be detected by blind human subjects at distances of 61 to 275 cm. Optimal conditions were provided; all targets were oriented so as to return maximum echoes, the room was quiet, and no other echo-generating surfaces were close to the test targets. The thresholds approximated a subtended angle of 4.6°, and size and distance discrimination were also quite good.

In later experiments, Rice (1971) found that especially proficient individual subjects could do considerably better. One blind man could detect 4 and 5 cm disks almost perfectly at 92 cm (subtended angle about 2.3°), and a young woman who had been blind from infancy could detect long cylindrical rods down to a diameter of about 6 mm (Rice, pers. comm.). In these experiments by Rice and his colleagues the subjects were allowed to make any sort of vocal sound they wished. All did make orientation sounds of some sort, but some used clicks, others longer-duration wide-band hisses, and still others repeated various words or syllables. Their performance did not differ significantly, indicating that the human auditory system can detect echoes almost equally well regardless of frequency pattern within the audible frequency range.

Except for Rice's careful, well-controlled experiments under optimal conditions, human echolocation has not been studied with appreciable success. Instead much effort has gone into the development and testing of guidance devices, small instruments for echolocation by means of high-frequency sound or light beams, which deliver signals to the user that are designed to warn him of approaching obstacles. In most devices these signals are delivered through earphones, and one of the best is binaural (Kay, 1966). In some instruments tactile presentation of the warning signals has been employed. Most of these devices work quite well in the laboratory, and in field tests under reasonably normal conditions they seem useful for blind persons. But I have been waiting in vain for many years to hear of a device so successful that the blind subjects were reluctant to give it back to the experimenter. A truly effective guidance device would so greatly improve a blind person's life that I can easily imagine him hiding it away to prevent its being taken from him. This would surely lead to urgent pleas for mass production, black markets, and so forth. Alas, nothing of the kind has been reported (Dufton, 1966).

One reason often suggested for the ineffectiveness of these devices is that the audible warning signals interfere with the normal use of ambient sounds that carry information about the environment. In theory the tactile presentation should avoid this difficulty, but here too the criteria of true success suggested above have yet to be reported. In this connection it should be borne in mind that echoes of sounds emitted by a blind person are only one class of ambient sounds that are doubtless useful for spatial orientation; sounds originating from other sources and their echoes from various objects all contribute to the audible sound fields through which we move. In one recent study newly blinded people learned to move about on city sidewalks and to cross streets more rapidly than usual when trained with tape recordings of the ambient sounds they were likely to hear (De l'Aune, Scheel, Needham, and Kevorkian, 1974). We almost certainly fail to pay attention to a variety of auditory information available to us because vision tells us what we need to know about most of our immediate surroundings.

It may well be that attempts to transfer responsibility for echolocation to the artificial guidance devices have deflected efforts away from the more tedious, but perhaps ultimately more rewarding, attempt to learn how a man could operate like a bat or a dolphin within the human range of hearing and with airborne sounds. But whatever deficiencies there may have been in research efforts on this front, thousands of intelligent and able-bodied blind people have been experimenting empirically along these lines for centuries. If it were easy some would have learned how to employ echolocation as skillfully as echolocating animals do. Yet efforts to improve acoustic orientation by the blind should certainly continue and should include both ambient sounds and echolocation.

Is there something qualitatively different about bat brains that allows discrimination of echoes imperceptible to much larger and more complex human brains? Extensive investigations by Grinnell and Suga (up to about 1970) failed to reveal any neurophysiological differences of a truly basic nature, although in bats and cetaceans a larger proportion of the auditory brain appears to be devoted to cells that recover sensitivity very rapidly after the end of a relatively loud sound. Bullock, Ridgway, and Suga (1971) and Bullock and Ridgway (1972) demonstrated this type of difference in marine mammals by comparing porpoises with sea lions, which have only a very limited capacity for echolocation, if any. In another type of neurophysiological investigation Henson (1965) has demonstrated that Hartridge (1945) was correct when he suggested that the relatively large middle-ear muscles of bats serve to reduce auditory sensitivity during the emission of each pulse of orientation sound. These muscles also relax rapidly enough that good sensitivity is regained in time to listen for echoes arriving after a few milliseconds. Human middle-ear muscles do not seem to operate with as short a latency, and probably they do not achieve as great a reduction insensitivity as Henson reports for bats (Wever and Lawrence, 1954). Our complete lack of any information about activities of the intra-aural muscles during human echolocation inhibits further speculation along these lines.

Suga and Shimozawa (1974) have recently demonstrated that in bats, along with the action of the middle-ear muscles, purely neural mechanisms attenuate the response to echoes in comparison to the response that would occur if the same sound were to arrive independently from an outside source rather than following a few milliseconds after the emission of an orientation sound. No comparable experiments have yet been carried out on other mammals, still less on men. Hence we cannot yet say whether this type of neural attenuation is an important factor that limits human echolocation in comparison to the superbly effective analysis of information contained in echoes that is achieved by the brains of bats, but the question clearly calls for further investigation.

Another approach to this question is stimulated by the subjective reduction in loudness of echoes of any sounds following closely after a much louder sound. This phenomenon can be demonstrated by making a tape recording of a loud sharp click that lasts only a few milliseconds. If the recording is made in any ordinary indoor situation, the click will be followed by a gradually decreasing series of echoes from the walls, floor, and other objects within a few feet of the recording microphone. When played back in normal fashion, these succeeding echoes will not be noticeable any more than they were with the original click, except for a slight dulling of its quality. But if the tape is played in reverse, so that the echoes precede the sharp click, their gradual buildup over several milliseconds is clearly audible as a hissing sound of growing loudness leading into the click itself. "Click" becomes "shhhick." Would a procedure as simple as time-inverted playback rescue the information so important to the blind?

It is not altogether clear how this reduction in subjective loudness of echoes is achieved in the human brain, or even at what neuroanatomical level it occurs. But the information of vital importance to echolocation is obviously contained in the same time span as that during which the suppression of echoes occurs. Perhaps this echo suppression is absent, or even reversed, in echolocating animals, as suggested by the recent experiments of Suga (1973). If so, this might explain their superior ability to react to echoes following within milliseconds after a loud orientation sound.

These compounded speculations can be useful only insofar as they stimulate new inquiry, preferably direct experimentation. But they lead to an inverse question: Why did human beings acquire the echo-suppression mechanism in the first place? Could it be helpful, if not essential, for the discrimination of speech, especially indoors or in situations where multiple echoes confuse the wave forms of impinging vocal signals that must be analyzed in order to extract meaningful information from messages of another member of the same species? Physicists have sometimes expressed surprise that we can understand speech at all when it arrives in such a jumble of interfering patterns that any direct oscillographic display makes sensible distinctions appear hopeless.

To be sure, the well-known emphasis of the auditory analyzing system upon Fourier analysis, and its almost total disregard of phase information, helps explain this discrepancy. Experiments by Batteau (1967, 1968), discussed by Freedman (1968), have shown that the human auditory system can respond differentially to clicks arriving at the external ear from different directions, and that this factor accounts in part for our ability to localize the direction of instantness of sounds even when only one ear is involved.

In Griffin (1968) I left to others to judge whether such conjectures could usefully be carried back into the evolutionary history of our remote ancestors at the stage when human speech first developed into a form that we would recognize as such. While I am not aware that my suggestion has yet led to any new experiments or new theoretical insights, I again suggest that the question remains significant for our understanding of human echolocation and perhaps also for theories about the evolution of human speech.

References

Airapet'yants, E. Sh., and Konstantinov, A. I., 1970. Echolocation in animals. Leningrad: Nauka. English translation, Jerusalem: Israel Program of Scientific Translations, 1973.

Batteau, D. W., 1967. The role of the pinna in human localization. *Proc. Roy. Soc. B.,* 168:158–80.

Batteau, D. W., 1968. The role of the pinna in localization: theoretical and physiological consequences. In: *Hearing Mechanisms in Vertebrates,* A. V. S. de Reuck, and J. Knight, eds. London: Churchill, pp. 234–39.

Bradbury, J. W., 1970. Target discrimination by the echolocating bat *Vampyrum spectrum. J. Exptl. Zool.,* 173:23–46.

Bradbury, J. W., 1972. The silent symphony: tuning in on the bat. In: *The Marvels of Animal Behavior.* Washington, D.C.: National Geographic Society.

Buchler, E., 1972. The use of echolocation by the wandering shrew *Sorex vagrans* Baird. Ph.D. thesis, University of Montana.

Bullock, T. H.; Ridgway, S. H.; and Suga, N.; 1971. Acoustically evoked potentials in midbrain auditory structures in sea lions (Pinnipedia). *Z. Vergl. Physiol.,* 74:372–87.

Bullock, T. H., and Ridgway, S. H., 1972. Evoked potentials in the central auditory system of alert porpoises to their own and artificial sounds. *J. Neurobiol.,* 3:79–99.

Cahlander, D. A.; McCue, J. J. G.; and Webster, F. A.; 1964. The determination of distance by echolocating bats. *Nature,* 201:544–546.

Clark, L., ed., 1963. *Proceedings of International Congress on Technology and Blindness.* New York: American Foundation for the Blind.

De l'Aune, W.; Scheel, P.; Needham, W.; and Kevorkian, G.; 1974. Evaluation of a methodology for training indoor acoustic environmental analysis in blinded veterans. In: *Proceedings of the 1974 Conference on Engineering Devices in Rehabilitation,* R. A. Foulds, and B. L. Lund, eds. Boston: Tufts University School of Medicine.

Dijkgraaf, S., 1943. Over een merkwaardige functie van den gehoorzin bij vleermuizen. *Verslagen Nederlandsche Akademie van Wetenschappen Afd. Naturkunde,* 52:622–27.

Dufton, R., 1966. *Proceedings of the International Conference on Sensory Devices for the Blind.* London: St. Dunston's.

Evans, L. B., and Bass, H. E., 1972. *Tables of Absorption and Velocity of Sound in Still Air at 68° F (20° C).* Huntsville, Ala.: Research Dept. WR 72-2, Wyle Laboratories.

Fenton, M. B., 1975. Acuity of echolocation in *Collocalia hirundinacea* (Aves:Apodidae), with comments on the distributions of echolocating swiftlets and molossid bats. *Biotropica,* 7:1–7.

Freedman, S. J., 1968. *The Neuropsychology of Spatially Oriented Behavior.* Homewood, Ill.: Dorsey Press.

Galambos, R., and Griffin, D. R., 1942. Obstacle avoidance by flying bats; the cries of bats. *J. Exp. Zool.,* 89:475–90.

Gould, E., 1955. The feeding efficiency of insectivorous bats. *J. Mammal.,* 36:399–407.

Gould, E., 1959. Further studies on the feeding efficiency of bats. *J. Mammal.,* 40:149–50.

Gould E., 1971. Studies of maternal-infant communication and development of vocalizations in the bats *Myotis* and *Eptesicus. Communications in Behavioral Biology,* 5:263–313.

Gould, E.; Negus, A.; and Novick, A.; 1964. Evidence for echolocation in shrews. *J. Exp. Zool.,* 154:19–38.

Griffin, D. R., 1950. Audible and ultrasonic sounds of bats. *Experientia,* 7:448–53.

Griffin, D. R., 1953. Bat sounds under natural conditions, with evidence for the echolocation of insect prey. *J. Exp. Zool.,* 123:435–66.

Griffin, D. R., 1958. *Listening in the Dark.* New Haven: Yale University Press. Reprint ed., New York: Dover Publications, 1974.

Griffin, D. R., 1962. Comparative studies of the orientation sounds of bats. *Symp. Zool. Soc.* (London), 7:61–72.

Griffin, D. R., 1967. Discriminative echolocation by bats. In: *Animal Sonar Systems,* R.-G. Busnel, ed. Jouy-en-Josas: Laboratoire de Physiologie Acoustique, pp.273–306.

Griffin, D. R., 1968. Echolocation and its relevance to communication behavior. In: *Animal Communication,* T. A. Sebeok, ed. Bloomington: Indiana University Press, pp.154–64.

Griffin, D. R., 1971. The importance of atmospheric attenuation for the echolocation of bats (Chiroptera). *Anim. Behav.,* 19:55–61.

Griffin, D. R.; Friend, J. H.; and Webster, F. A.; 1965. Target discrimination by the echolocation of bats. *J. Exp. Zool,* 158:155–68.

Griffin, D. R., and Galambos, R., 1941. The sensory basis of obstacle avoidance by flying bats. *J. Exp. Zool.,* 86:481–506.

Griffin, D. R.; McCue, J. J. G.; and Grinnell, A. D.; 1963. The resistance of bats to jamming. *J. Exp. Zool.,* 152:229–50.

Griffin, D. R., and Novick, A., 1955. Acoustic orientation of neotropical bats. *J. Exp. Zool.,* 130:251–300.

Griffin, D. R.; Novick, A.; and Kornfield, M.; 1958. The sensitivity of echolocation in the fruit bat *Rousettus. Biol. Bull.,* 115:107–13.

Griffin, D. R., and Suthers, R., 1970. Sensitivity of echolocation in the cave swiftlet. *Biol. Bull.,* 139:495–501.

Griffin, D. R.; Webster, F. A.; and Michael, C.; 1960. The echolocation of flying insects by bats. *Anim. Behav.,* 8:141–54.

Grinnell, A. D., 1963. The neurophysiology of audition in bats. *J. Physiol.,* 167:38–127.

Grinnell, A. D., 1973. Neural processing mechanisms in echolocating bats correlated with differences in emitted sounds. *J. Acoust. Soc. Amer.,* 54:147–56.

Grinnell, A. D., and Grinnell, V. S., 1965. Neural correlates of vertical localization by echo-locating bats. *J. Physiol.,* 181:830–51.

Harrison, J. B., 1965. Temperature effects on responses in the auditory system of the little brown bat *Myotis l. lucifugus. Physiol. Zool.,* 38:34–48.

Hartridge, H., 1945. Acoustic control of the flight of bats. *Nature,* 156:490–494.

Henson, O. W., Jr., 1965. The activity and function of the middle ear muscles in echolocating bats. *J. Physiol.,* 180:871–87.

Henson, O. W., Jr., 1970. The ear and audition. In: *Biology of Bats,* vol. II, W. A. Wimsatt, ed. New York: Academic Press, pp.181–263.

Kay, L. 1966. Ultrasonic spectacles for the blind. In: *Proceedings of the International Conference on Sensory Devices for the Blind,* R. Dufton, ed. London: St. Dunstan's, pp.275–92.

Kellogg, W. N., 1961. *Porpoises and Sonar.* Chicago: University of Chicago Press.

Kellogg, W. N.; Kohler, R.; and Morris, H. N.; 1953. Porpoise sounds as sonar signals. *Science,* 117:239–43.

Kolb, A., 1961. Sinnesleistungen einheimischer Fledermäuse bei der Nahrungssuche und Nahrungswahl auf dem Boden und in der Luft. *Z. Vergl. Physiol.,* 44:550–64.

Medway, Lord, 1967. The function of echonavigation among swiftlets. *Animal Behaviour,* 15:416–20.

Möhres, F. P., 1953. Über die Ultraschallorientierung der Hufeisennasen (Chiroptera-Rhinolophidae). *Z. Vergl. Physiol.,* 34:547–88.

Norris, K. S.; Prescott, J. H.; Asa-Dorian, P. V.; and Perkins, P.; 1961. An experimental demonstration of echo-location behavior in the porpoise, *Tursiops truncatus* (Montagu). *Biol. Bull.*, 120:163–76.

Norris, K. S., ed., 1966. *Whales, Dolphins, and Porpoises.* Berkeley: University of California Press.

Norris, K. S., 1974. *The Porpoise Watcher.* New York: W. W. Norton.

Novick, A., 1958. Orientation in palaeotropical bats. II. Megachiroptera. *J. Exp. Zool.*, 137:443–62.

Novick, A., 1959. Acoustic orientation in the cave swiftlet. *Biol. Bull.*, 117:497–503.

Pye, J. D., 1967. Synthesizing the waveforms of bat's pulses. In: *Animal Sonar Systems,* R.-G. Busnel, ed. Jouy-en-Josas: Laboratoire de Physiologie Acoustique, pp.43–67.

Ralls, K., 1967. Auditory sensitivity in mice: *Peromyscus* and *Mus musculus. Anim. Behav.*, 15:123–28.

Rice, C. E.; Schusterman, R. J.; and Feinstein, S. H.; 1965. Echodetection ability of the blind: size and distance factors. *J. Expl. Psychol.*, 70:246–51.

Rice, C. E., 1967. Human echo perception. *Science,* 155:656–64.

Rice, C. E., 1971. Early blindness, early experience and perceptual enhancement. *Research Bull., Amer. Foundation for the Blind,* 22:1–22.

Rice, C. E., 1973. Personal communication.

Rosenzweig, M. R., and Riley, D. A., 1955. Evidence for echolocation in the rat. *Science,* 121:600.

Sales, G., and Pye, J. D., 1974. *Ultrasonic Communication by Animals.* London: Chapman and Hall.

Schevill, W. E., and Lawrence, B., 1949. Underwater listening to the white porpoise *(Delphinapterus leucas). Science,* 109:143–44.

Schevill, W. E., and Lawrence, B., 1956. Food-finding by a captive porpoise *(Tursiops truncatus). Breviora, Mus. Comp. Zool., Harvard University,* 53:1–15.

Schnitzler, H.-U., 1968. Die Ultraschallortungslauts der Hufesienfledermäuse (Chiroptera-Rhinolophidae) in verschiedenen Orientierungssituationen. *Z. Vergl. Physiol.*, 57:376–408.

Schnitzler, H.-U., 1970. Echoortung bei der Fledermaus *Chilonycteris rubiginosa. Z. Vergl. Physiol.*, 68:25–38.

Schnitzler, H.-U., 1973. Control of Doppler shift compensation in the greater horseshoe bat, *Rhinolophus ferrum-equinum. J. Comp. Physiol.*, 82:79–92.

Schnitzler, H.-U., 1974. Response to frequency shifted artificial echoes in the bat *Rhinolophus ferrum-equinum. J. Comp. Physiol.*, 89:275–86.

Simmons, J. A., 1973. The resolution of target range by echolocating bats. *J. Acoust. Soc. Amer.*, 54:157–73.

Simmons, J. A., 1974. Response of the doppler echolocation system in the bat, *Rhinolophus ferrumequinum. J. Acoust. Soc. Am.*, 56:672–82.

Simmons, J. A.; Howell, D. J.; and Suga, N.; 1975. The information content of bat sonar echoes. *Amer. Scientist,* 63:204–15.

Suga, N., 1964. Single unit activity in cochlear nucleus and inferior colliculus of echo-locating bats. *J. Physiol.*, 172:449–74.

Suga, N., 1965. Functional properties of auditory neurones in the cortex of echolocating bats. *J. Physiol.*, 181:671–700.

Suga, N., 1973. Feature extraction in the auditory system of bats. In: *Basic Mechanisms in Hearing,* A. R. Moller, ed. New York: Academic Press.

Suga, N., and Jen, P. H.-S., 1975. Peripheral control of acoustic signals in the auditory system of echolocating bats. *J. Exp. Biol.*, 62:277–311.

Suga, N., and Schlegel, P., 1973. Coding and processing in the nervous system of FM signal producing bats. *J. Acoust. Soc. Amer.*, 54:174–90.

Suga, N., and Shimozawa, T., 1974. Site of neural attenuation of responses to self-vocalized sounds in echolating bats. *Science,* 183:1211–13.

Suga, N., Simmons, J. A., and Jen, P. H.-S., 1975. Peripheral specialization for fine analysis of Doppler-shifted echoes in the auditory system of the 'CF-FM' bat *Pteronotus parnellii. J. Exp. Biol.*, 63:161–192.

Supa, M.; Cotzin, M.; and Dallenbach, K. M.; 1944. "Facial vision." The perception of obstacles by the blind. *Amer. J. Psychol.*, 57:133–83.

Suthers, R. A., 1965. Acoustic orientation by fish-catching bats. *J. Exp. Zool.*, 158:319–48.

Vincent, F., 1967. Acoustic signals for auto-information or echolocation. In: *Acoustic Behaviour of Animals,* R.-G. Busnel, ed. Amsterdam: Elsevier, chap. 8.

Webster, F. A., 1966. Some acoustical differences between bats and men. In: *Proceedings of the International Conference on Sensory Devices for the Blind,* R. Dufton, ed. London: St. Dunstan's, 1966.

Wever, E. G., and Lawrence, M., 1954. *Physiological Acoustics.* Princeton: Princeton University Press.

Zahl, P. A., ed., 1950. *Blindness.* Princeton: Princeton University Press.

Chapter 13

ELECTRIC COMMUNICATION

Carl D. Hopkins

Communication among animals is a complex and highly refined process by which one individual's behavior, acting as a stimulus or signal, effects a change in the behavior of another. Mediated by the environment through which the stimulus must be transmitted, signals and the responses they elicit have evolved to the mutual benefit of both participants. But with each of the diverse energy modalities used for encoding stimuli, the problems associated with the creation, transmission, and final detection of communication signals are unique; it is logical to consider them separately. It is particularly interesting to explore the problems related to those modalities with which we have no personal experience because of our own sensory limitations. The electrical modality, used by relatively few aquatic species, is one such case. This chapter will discuss adaptation and refinements for communication using electric signals with the goal of developing an understanding of how signaling and receiving behavior have evolved to serve different functions and how the two types of behavior are designed to cope with peculiar properties of this modality.

While the ability to produce and to receive electric signals has evolved several times in different groups of fishes, its adaptive significance varies from species to species (Lissmann, 1958). Those with powerful Electric Organ Discharges (EOD), such as the electric eel, electric catfish, and Torpedo rays, are able to use their discharges for stunning prey or predators (Bauer, 1968; Belbenoit and Bauer, 1972). Those with both electric organs and electroreceptors can use their capabilities in the active detection of objects in the environment as distortions in a self-produced electric field (Lissmann and Machin, 1958). This ability, known as "electrolocation," is found even in species with discharges that are too weak to affect even very small prey. Some of these species also employ their capabilities for purposes of communication. For others, who are sensitive to electric fields but do not have electric organs themselves, passive detection of the electric fields surrounding most organisms in water is an effective prey-localization technique (Kalmijn, 1972; Kalmijn and Adelman, in prep.; Roth, 1972). And several authors have suggested that an electrical sense might also be useful in detecting the earth's magnetic field during navigation (Kalmijn, 1974; Rommell and McCleave, 1972).

A single species, in fact, frequently will demonstrate multiple uses for its electrical capabilities. Thus, it is important to keep in mind that the electrical emissions are not "specialized" for communication as are human vocal signals (Hockett, 1960; Hockett and Altmann, 1968) in that the energy of an electric signal may have a direct biological effect: either in the detection of an object in the environment or in affecting a prey item or a predator. Communication and electrolocation are two commonly shared but different functions. Electrolocation might be considered a solipsistic form of communication analogous to echolocation (Griffin, 1968); as such it differs markedly from exoteric communication, or communication with other organisms. Because natural selection has acted on both sensory and EOD capabilities of the electrical system for its multiple functions, we may expect to find some compromises in their shared design.

Electroreception is uncommon among aquatic organisms, so that communication using electric signals has the advantage of being relatively private, as are the visual and auditory channels whenever signals lie outside the usual spectral range (e.g., Eisner et al., 1969; Silberglied and Taylor, 1973). With a private channel, conspecifics can maintain contact with each other at a reduced risk of predation or can exchange cues regarding sex or species identity among themselves without revealing their identity to predators or competitors in cases where a mimicry complex is involved. Thus, it is important to appreciate the degree of privacy of the electric channel: which species are known to possess electroreceptors and which species can produce electric discharges?

Electric Signal Reception

An electric signal is perceived as current flows through the specialized low-resistance cutaneous sensory organs belonging to the lateral line system of certain fishes. All the known examples of electroreceptors may be classed as either ampullary or tuberous, based on their anatomical structure. In marine environments, electroreception is fairly widespread, but most of the species possess ampullary electroreceptors. The Ampullae of Lorenzini, found in nearly all sharks, skates, and rays, consists of a flask-shaped ampulla lying deep beneath the surface of the skin and connected to the exterior by a long neck or canal that may be as long as one-third of the fish's body length. The sensory cells lying embedded in the wall of the ampulla are responsive to low-frequency electrical stimuli. A marine catfish, *Plotosus,* is known to have similarly constructed ampullae that are presumably electroreceptive, and some migratory eels, such as *Anguilla rostrata,* found in salt water during part of their life cycle, show behavioral responses to weak electrical stimuli but are not yet known to have electroreceptors (Rommell and McCleave, 1972).

Two principal groups of freshwater fishes with well-developed electric capabilities are the gymnotid fish of South America and the mormyriform fishes of Africa. The Gymnotoidei, which are a characoid-related suborder of Cypriniformes (Ostariophysi), consist of four families of closely related fish estimated at between sixty and eighty species. The Mormyriformes, which belong to the suborder Osteoglossomorpha, are composed of two families with an estimated two hundred or more species. All known members of both these groups possess both ampullary and tuberous electroreceptors. Ampullary organs are basically the same flask-shaped structures but with short necks filled with a jelly-like substance extending to the surface; they are also known to be responsive to low-frequency electrical stimuli. Tuberous organs, consisting of a receptor cavity buried under layers of loosely packed epithelial cells with no canal to the outside, are responsive primarily to high-frequency stimuli (greater than 50 Hz). Sharing freshwater habitats with the mormyrids and the gymnotids is the very large group of electroreceptive catfish (Siluriformes, Ostariophysi), which are all thought to possess ampullary, or low-frequency, but not tuberous, electroreceptors. An occasional freshwater elasmobranch, such as the freshwater stingray *(Potamotrygon circularis),* also possesses ampullary electroreceptors (Szabo et al., 1972).

In all likelihood, additional species of electroreceptive fishes will be added to this list as researchers begin to look at different fishes' behavioral responses to weak electric fields. Although there is some confusion as to terminology, excellent reviews of the extensive work on the physiology and anatomy of electroreceptors may be found in Szabo (1965), Lissmann and Mullinger (1968), Bennett (1970, 1971b), Bullock (1973), Scheich and Bullock (1974), Kalmijn (1974), and Fessard (1974).

Signal Production

The production of electric currents is not as widespread as electroreception. In addition to the electrogenic gymnotids and mormyrids discussed above, there is one species of freshwater electric catfish (Malapteruridae). There are also several marine electric fishes, including the electric rays (Torpedinidae), electric skates (Rajidae), and stargazers (Uranoscopidae), as discussed in the reviews by Bennett (1971a), Lissmann (1958), Grundfest (1957, 1960), and Bullock (1973).

Electric currents are generated in specialized organs that are derived from either muscle or nerve, as shown from physiological, anatomical, and pharmacological studies of mature and developing tissue (see review in Bennett, 1971a). In muscle-derived organs, which are the most common, several long columns of multinucleated cells, called electrocytes (Bennett, 1971a), either run the length of the fish (e.g., in the gymnotids) or are localized in specific regions in the tail, head, skin, or pectoral fins (Grundfest 1957, 1960).

Fig. 1 illustrates the mode of action of a simple electric organ from the electric eel, *Electrophorus electricus* (family Electrophoridae, Gymnotoidei). In this species, placque-shaped cells, innervated on their posterior faces, lie within chambers of connective tissue to form a fairly accurately aligned column. When a nerve discharge excites the posterior face of the placque it generates a spike, which typically overshoots the zero potential by +50 mV. This can be demonstrated by recording the potential difference between two microelectrodes, one placed outside the posterior face of the cell and another placed inside (Fig. 1B). But the anterior face of this cell has a low resistance and is not excitable, even if stimulated electrically. When the microelectrode is advanced through the anterior face into the extracellular space, the discharge potential generated across the posterior face shows up across the entire cell as shown in the oscilloscope tracings in Fig. 1C. Because of the basic asymmetry of these electrocytes, the synchronous discharges from adjacent cells will summate to produce a relatively large voltage. Looking external to the electric eel, the discharge recorded in the water is simply a head-positive, monophasic spike, lasting several milliseconds, and attaining as much as several hundred volts.

The mechanism of activity of electrocytes in other species is usually more complicated. Biphasic discharges are produced by the gymnotid *Hypopomus artedi* (family Rhamphichthyidae), for example, because both the posterior and anterior faces of the electrocytes are electrically active, firing slightly out of phase with each other (Bennett, 1961). Other species exhibit other mechanisms that result in a variety of complex wave forms (see review in Bennett, 1971a).

All the electric organs known for members of the gymnotoid family Apteronotidae appear to be derived from neural tissue (de Oliveria Castro, 1955; Waxman et al., 1972). *Apteronotus albifrons,* for example, has an electric organ made up of enlarged loop-shaped spinal neurons, which are myelinated (Waxman et al., 1972). The apteronotids are interesting because they seem to

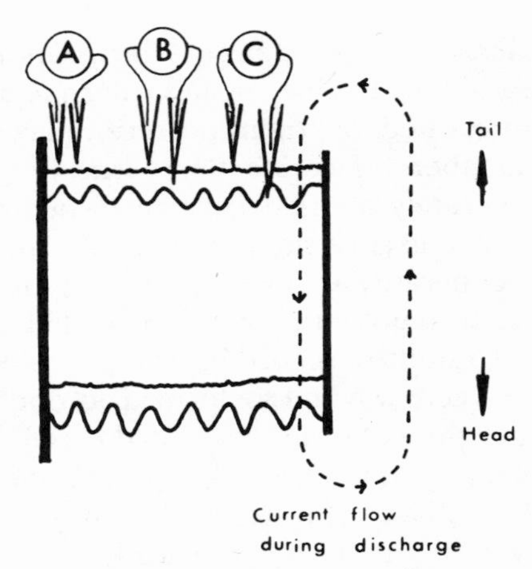

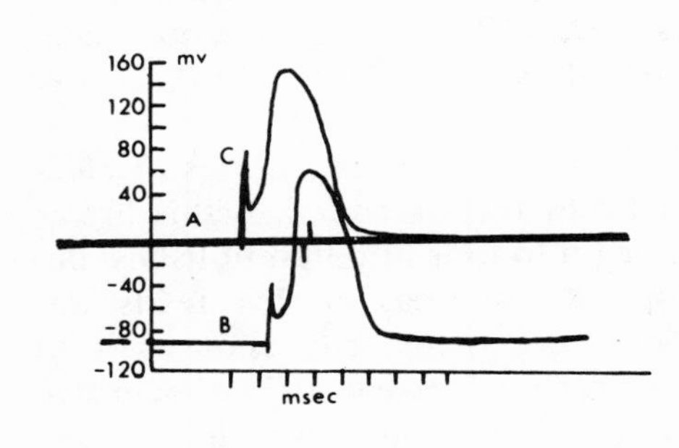

Fig. 1. The mechanism of additive discharge in the electrocytes of the electric eel, *Electrophorus electricus.* Left: Schematic diagram of two electrocytes in the column showing the orientation within the body and the direction of current flow. Right: Oscilloscope tracings of potentials recorded differentially between two electrodes. Positive voltages on the right-hand electrode go upwards.

A. When both electrodes are external to the posterior (innervated) face of the electrocyte, no potential can be seen, save a slight artifact.

B. When the right-hand electrode is advanced through the posterior membrane into the cell, a -90 mv resting potential can be seen. During the electric organ discharge (EOD), this potential changes to show a reversal potential of about +50 mv at the peak of the spike.

C. When the electrode advances through the anterior face (uninnervated), the resting potential disappears but the spike does not. Consequently, during the EOD there is an additive potential difference across each electrocyte. (After Keynes and Martins-Ferreira, 1953.)

be exploiting frequencies at the highest possible limit for electric signaling—1,800 pulses per second, as shown for at least one species (Bullock, 1969; Steinbach, 1970). Whereas sound signaling often occurs at much higher frequencies as small structures are set into vibration, electric signaling has an apparent upper frequency limitation imposed by neural activation and reactivation times of electric organs. There are no known examples of species that can de-couple part of their electric organs in order to achieve higher frequencies. This is probably because in so doing, the current-generating capability, which is directly related to the number of synchronized cells in the electric organ, would be significantly reduced, thereby severely affecting signal range.

In spite of a great deal of diversity in the structure and physiology of electric organs, all electric fish are capable of producing their own signal energy. In this regard, electric communication is distinguished from other communication modalities, which depend on available sources of energy such as sunlight. While this ability permits signaling at night when no external sources of energy are available, it also means that the evolution of signal amplitudes will be consistent with the presence of background noise in the environment.

Signal Transmission

Several peculiar properties of current flow in water are important to our understanding of the evolution of electrical communication. Some of these signal-transmission properties are considered here.

CONDUCTION VELOCITY

Electric signals travel so rapidly in water that conduction times may be considered instantaneous for most biological systems. In this respect the electric modality resembles the visual. Small time delays due to finite conduction times are biologically meaningful in other communication modalities. With sound detection, for example, differences in phase or time of arrival at the two ears provide cues about the location of the sound source (Steven and Newman, 1934; Marler, 1959; Konishi, 1974). Even monaural localization in vertebrates with a pinna depends on noninstantaneous conduction velocities (Batteau, 1967).

Electric fish probably cannot utilize time delays for signal localization, but they may use other mechanisms. Although no definitive work has been done to test the accuracy of spatial localization, Knudsen (1974) has been able to train gymnotid fish to make a choice between a sinusoidal electrical signal coming from a dipole on either the left or the right as they are free-swimming in a nylon mesh starting area in the center of an aquarium. Spatial localization presumably depends on a comparison of signal amplitudes at different parts of the body or on comparisons of amplitudes of different locations made sequentially while swimming. Signal amplitude differences at the skin are known to play an important role in the active detection of objects using electrolocation (Hagiwara and Morita, 1963; Hagiwara, Szabo, and Enger, 1965; Heiligenberg, 1973b, 1975).

The rapid signaling possible in the electric modality contrasts with that for the chemical modality—especially in water (Wilson, 1970).

SIGNAL RANGE

The range of electric signal transmission is limited; estimates vary between several cm and several meters. Granath et al. (1967, 1968) measured the strength of the electric field at various distances from the gymnotid *Apteronotus albifrons*, as well as its conditioned-response threshold for perception of an electric field. They then estimated that one electric fish should be able to sense another at approximately 3 m. Using a similar approach, Knudsen (pers. comm.) estimated the threshold for signal detection in another gymnotid, *Eigenmannia*, to occur between 25 cm and 200 cm. The situation is undoubtedly complex. Signal range probably depends on several factors: the size of the signaler and consequent amplitude of its EOD (Brown and Coates, 1952); the size of the receiver and consequent sensitivity of its electroreceptors (Bennett, 1971b); the angular orientation of the receiver's body with respect to the signaler's, and vice versa; the conductivity of the water; and the presence of nonconductors (the bottom, the surface, nonconducting objects) near the sender and receiver that might compress or distort the field. Other factors of importance would include the sensitivity of the receiver's receptors, the nature of the signal, the presence of noise, etc.

Moller and Bauer (1973) demonstrated that there are significant negative correlations between the discharge frequencies of two individual mormyrids (*Gnathonemus petersii*) when they are separated by distances smaller than 30 cm. When moved further apart, the EOD frequencies of the two fish were unrelated. Similar results were observed using a different technique by Russell et al. (1974). *Gnathonemus* produces an "echo" response to a conspecific's EOD after a characteristic delay. This response diminishes in intensity as the distance between the two fish increases, until at 30 cm it is no longer evident. The lack of responsiveness in these unconditioned behavioral tests may not be an accurate assessment of the maximum distance of communication; however, these estimates all imply a short-range system that is consistent with the severe rate of attenuation of an electric field surrounding a dipole source. The peak-to-peak electric potential surrounding an *Eigenmannia* falls off according to the inverse square of distance (Fig. 2A) when measurements are conducted in a large (3 m diameter, 1 m deep) tank

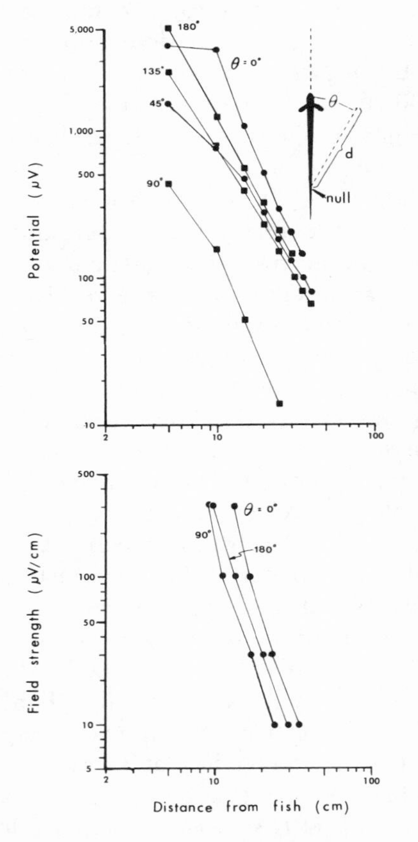

Fig. 2. The range of electric signaling is limited by the severe rate of signal attenuation with distance. Top: The peak-to-peak potential falls off according to the inverse square of distance from the null of the fish for various angles. Squares indicate negative values; circles, positive values. Bottom: The electric field strength falls off according to the inverse cube of distance for various angles. All measurements were made on an 18.6 cm-long gymnotid, *Eigenmannia virescens*, in a 3 m-diameter, 1 m-deep tank filled with water at a conductivity of 2.6 X 10^{-2} mho/m. (Data courtesy of Erik I. Knudsen, 1975; from Hopkins, 1974a.)

(Knudsen, 1975), and the electric field falls off according to the inverse third power of distance, as would be expected for a dipole source (Smyth, 1968). Measurements conducted in a small tank (1 m diameter, 16 cm deep) show the electric field falling off according to $r^{-2.3}$ instead of r^{-3} (Knudsen, pers. comm.). In confined areas the rate of signal decrement may be reduced by the formation of electrical images by the surface and bottom, and signals may therefore extend further horizontally.

DIRECTIONALITY

Electric currents lack the directionality characteristic of visual signals. At the source, an electric signal is broadcast in all directions in a typical dipole-shaped field (Hopkins, 1974). Even if the signaler bends its body in one way or another, there probably is not a significant narrowing of the beam. Electric signals are also capable of crooked-line transmission in that they flow around rocks or fallen trees in their path. Because of this, spatial localization is difficult and consequently signals involving spatial patterning are probably insignificant, except at extremely close range. Crooked-line transmission does allow signaling in dense vegetation, however; and suspended particulate matter—a common impediment to visual signaling in tropical fresh water—does not affect current flow.

FADE-OUT

An electric signal fades as soon as it is discontinued, thus necessitating that the recipient be present when the signal is emitted. In contrast to the lingering nature of an odor or visual mark, this property is something of a disadvantage; however, in combination with a rapid conduction velocity, it makes the electric modality ideal for transmitting information that is likely to fluctuate rapidly with time. Signals that allow predictions about an animal's motivation to attack or to flee, for example, need to be transmitted rapidly (Marler and Hamilton, 1966). Fast conduction and fade-out also permit the use of time-varying signals (see discussion in Wilson and Bossert, 1963), and, as shown in the next section, the information content of most electric signals depends on their temporal structure.

BACKGROUND NOISE

The ultimate determinant of the range of a communication signal is the nature and amplitude of background noise in the channel. The predominant source of electrical noise, aside from that from nearby electric fish, is the extremely low frequency (ELF) electromagnetic radiation of terrestrial origin, which appears to be caused principally by lightning (Soderberg, 1969; Watt, 1960; Liebermann, 1956a, 1956b). Because thunderstorms are extremely common in Africa and South America, where electric fish are found, and because electromagnetic waves from lightning travel long distances according to the inverse first power of distance (Watt, 1960), lightning discharges are frequently of sufficient amplitude to be detected by electric fish (Hopkins, 1973). Other inanimate sources of electrical noise may be magnetic storms, earthquakes, the movement of water through the earth's magnetic field (Kalmijn, 1974; Bullock, 1973). Biological sources, particularly other electric fish, add to this noise and create a substantial interference problem. Field observations show that electric fish are commonly found in groups, with many fish spaced only centimeters apart. Not infrequently these groups are composed of members of several species.

Functions of Electric Signals

Although our knowledge of the functions of electric signals is limited by comparison with other modalities, there appears to be a comparably rich repertoire of signals serving as designators and prescriptors (Marler, 1961): signals that dispose the interpreter to make responses appropriate for a particular species, sexual partner, individual, age class, or motivational state of signaler, or to some aspect of the environment.

SPECIES-SPECIFIC SIGNALS

If a signal evokes a response in members of only one species, then it suggests that it is species-specific and reduces the uncertainty about the species identity of the signaler. Such responses may play a crucial role in reproductive isolation or may aid in forming and maintaining social groups for protection or foraging. In searching for species-specific signals, we begin by demonstrating that the physical properties of the signal of interest are distinctive and characteristic in a given population. The resting discharges of the gymnotid fishes from Guyana show certain species-typical patterns, not only in wave form (Fig. 3) but also in frequency. Steinbach (1970) found similar differences among gymnotids in the Rio Negro, Brazil.

The regular and continuous electrical emissions of *Sternopygus macrurus* and *Eigenmannia virescens,* for example, are clearly distinguishable from all other sympatric species on this basis. Both produce "wave" or "tone" discharges in which the impulse is long compared to the interval between impulses, and when compared to other "wave" species, their frequencies are unique (Fig. 4). *Eigenmannia* in nonbreeding condition respond aggressively toward a Plexiglas fish model with electrodes playing tape recordings of their own species' discharges with head-butting attacks and electrical threat displays (see below), but respond less to recordings of other sympatric species, as shown in Fig. 5. In addition, sine waves of the characteristic frequency are as effective as tape recordings in eliciting agonistic behavior from *Eigenmannia,* whereas frequencies outside the species range are ineffec-

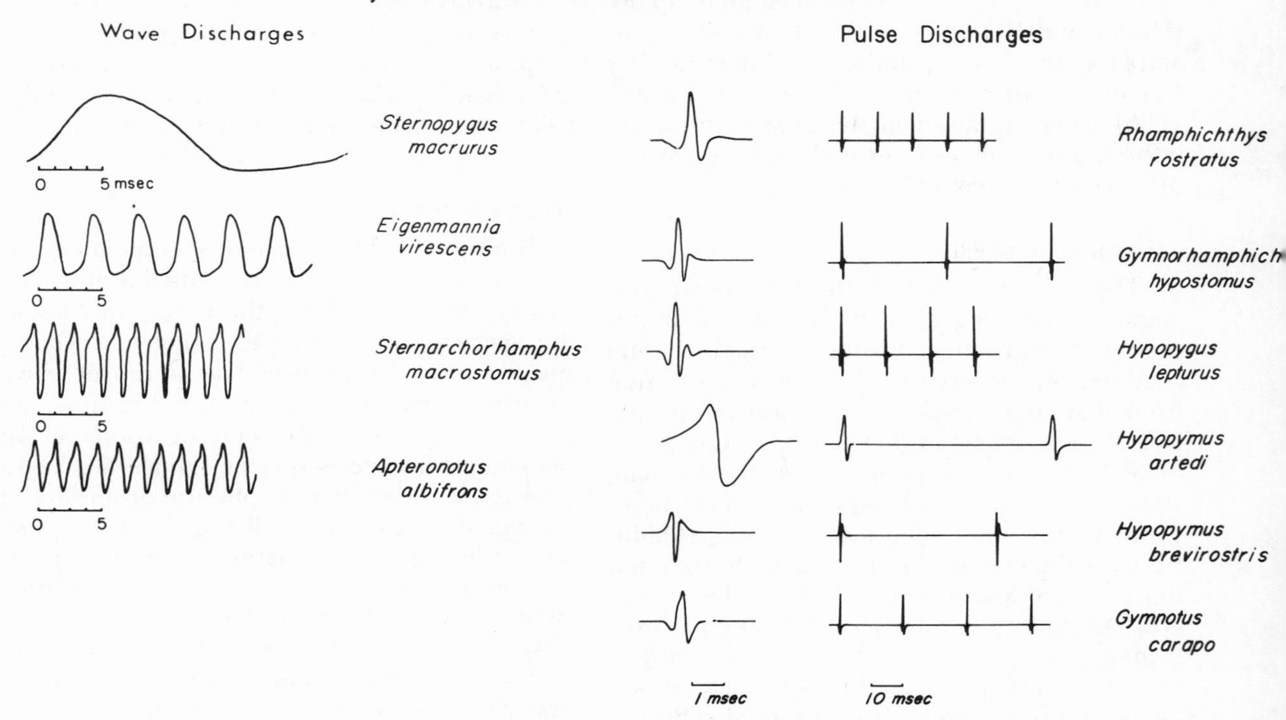

Fig. 3. Oscilloscope tracings of the wave forms of the ten most common gymnotids from Moco-moco Creek in Guyana. The records were obtained by placing one electrode near the head and one near the tail. Head positive signals are deflected upwards. (From Hopkins, 1974a.)

tive (Hopkins, 1974b). Similarly, *Sternopygus* males give courtship displays in response to sinusoidal electrical stimuli mimicking the discharge frequency of a female *Sternopygus,* but hardly respond at all to stimuli mimicking other sympatric species (Hopkins, 1972b, 1974c).

But not all wave species have characteristic frequencies. Those in the family Apteronotidae, in particular, show a great deal of overlap in pulse frequency, as shown for *Sternarchorhamphus macrostomus* and *Apteronotus albifrons* (Fig. 4), found sympatric in Guyana, frequently in the same habitat. These two species have discharges that are similar with respect to frequency, wave form, and polarity. It is not known by what mechanisms species recognition is accomplished in this case, but it does not appear to involve electrical characteristics of the undisturbed resting discharges. The resting discharges of many of the "pulse" species also do not appear to be species-specific. *Gymnotus carapo* produces a typical "pulse" discharge in which each impulse is separated by a relatively long interval. It responds aggressively to a wide variety of electrical stimuli but shows its lowest threshold for attack toward 1,000 Hz sinusoidal stimuli—the predominant component frequency of its own individual pulses (Black-Cleworth, 1970). *Gymnotus*

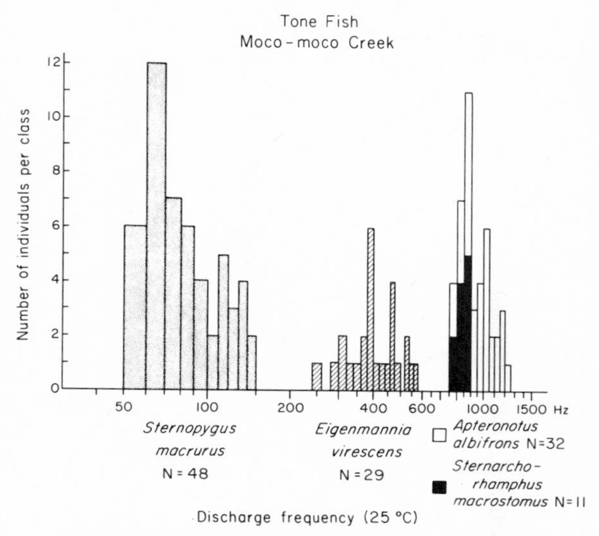

Fig. 4. Distribution of discharge frequencies, corrected to 25°C water, of the four species of wave-discharging gymnotids from Moco-moco Creek, Guyana. (From Hopkins, 1974c.)

is unspecific in its attacks directed toward other species of electric fish, showing only a moderate preference for attacking other *Gymnotus* or other pulse species with similar frequencies, e.g., *Steatogenes elegans,* as shown in a series of heterospecific aggression experiments. Further research is needed to determine if the characteristic low-frequency or long-duration pulses of *Hypopomus artedi* or the characteristic high-frequency pulses of *Gymnorhamphichthys hypostomus* effectively elicit species-specific responses.

The shape of the electric field, the use of multiple frequencies, time relationships between pulses of both sender and receiver, frequency modulations, and other time-varying signals might also serve as species-specific signals. The first two possibilities are characteristic of the resting discharge of several species; the others involve temporal modifications of the discharge. *Steatogenes elegans* and *Gymnorhamphichthys hypostomus* are known to possess specialized accessory electric organs located on the underside of the head in the region surrounding the urogenital papilla, which fire in synchrony with the main electric organ in the tail (Bennett, 1971a). These organs alter the shape of the electric field locally, and at a range of several centimeters they might serve to identify the signaler (Hopkins, 1974). *Adontosternarchus sachsii* is unique in South America because it produces two frequencies at once —one using its main electric organ in the tail and the other using its accessory organ in the chin (Bennett, 1971a). *Gnathonemus petersii*'s echo responses to artificial electrical pulses and to pulses from conspecifics, occurring after a delay of 12 to 14 msec (Russell et al., 1974) might also serve as a species-identification signal in a way analogous to the flash-answer system known for certain fireflies (Lloyd, 1966). This hypothesis needs to be tested by comparing the "echo" responses of sympatric species. As Bell et al. (1974) point out, echo responses could be adapted for preventing coincident pulses with neighbors.

The use of time-varying signals such as frequency modulations and discharge cessations—known to occur in most species of electric fish—is potentially a diverse way of encoding species-specificity. But it has not been shown conclusively that any species relies on these types of signals.

We might expect there to be selection pressure for optimization of certain characteristics of electrical pulses used for electrolocating that might tend to result in convergent evolution on a single type of resting discharge (e.g., Scheich and Bullock, 1974). This trend would counter the intrinsic signal value of species differences in resting discharges. But when two electric fish approach each other in a stream, their electrolocating pulses are likely to be the first indicators of each other's presence. If these pulses are species-specific, rapid and efficient species identification would be possible. Also, any physiological mechanism, such as stimulus filtering, that en-

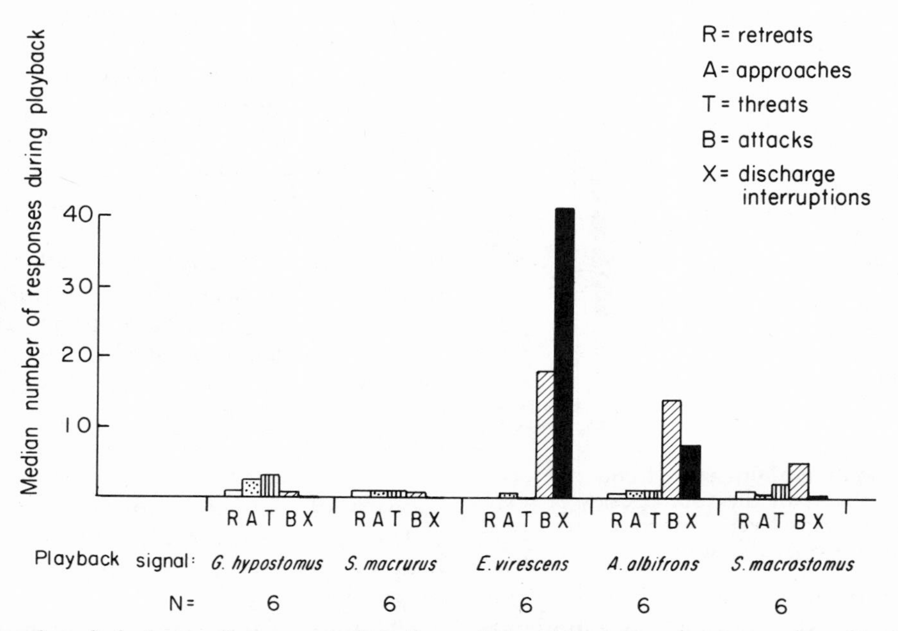

Fig. 5. Results of playback of tape-recorded signals to captive *Eigenmannia virescens.* Each bar indicates the median number of responses (per 2 minutes) recorded during playback for six fish. Responses include: R=retreats, A=approaches, T=threats, B=attacks, and X=discharge interruptions, directed at a Plexiglas model carrying the playback electrodes.

hances homospecific communication signals at the expense of extraneous noise produced by other species would be an obvious advantage; and if species-specificity and filtering depended on some property of the resting discharge, the signal-to-noise ratio of the electrolocating system would improve at the same time.

SEXUAL SIGNALS

As an organism comes into reproductive condition, it may respond to signals emitted by members of the opposite sex. While *Sternopygus macrurus* are reaching sexual maturity, for example, the discharge frequencies of males and females diverge, with males adopting low frequencies (50 to 90 Hz) and females adopting higher frequencies (100 to 150 Hz), with little overlap (Hopkins, 1972b). This frequency difference has communicative significance. As noted before, males respond to the resting electric discharge of females or to sine waves mimicking the female's frequencies by giving electrical displays thought to play a role in mate attraction or in courtship (Fig. 6). They do not respond to sine waves mimicking other males or to those mimicking other sympatric species with wave discharges.

In contrast to this example, electrical courtship displays consisting of brief interruptions in the otherwise continuous EOD can be experimentally elicited from a sexually mature male *Eigenmannia* during its breeding season by connecting

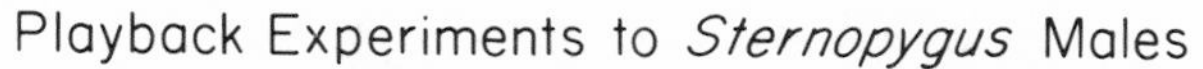

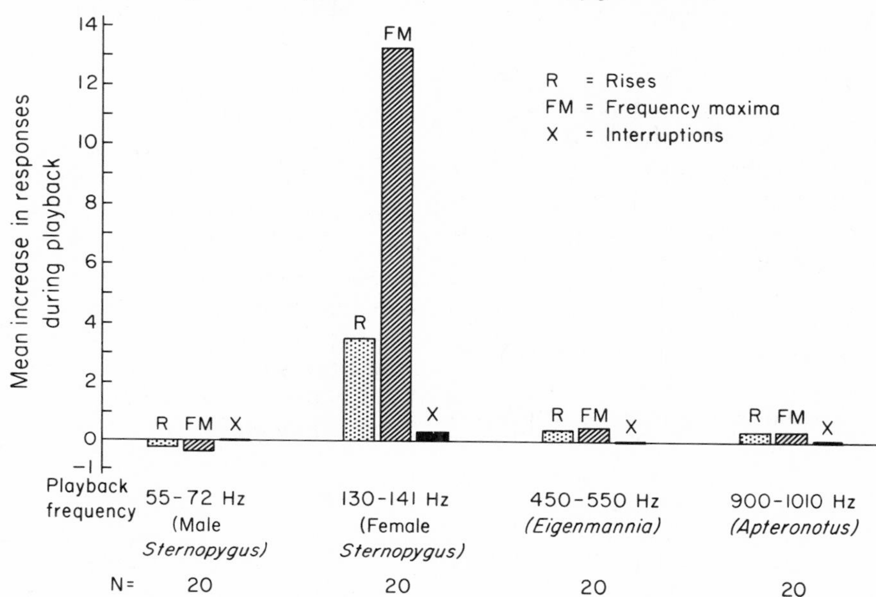

Fig. 6. Responses to sinusoidal stimuli by male *Sternopygus macrurus* in the field. The vertical scale represents the mean increase in the number of responses per one-minute stimulus period, minus the one-minute control period, for twenty trials at each stimulus, and for two males. The horizontal axis compares sine waves with frequencies (as shown) mimicking a male *Sternopygus*, a female *Sternopygus*, an *Eigenmannia virescens*, or an *Apteronotus albifrons*. Responses include the number of rises, or increases in frequency followed by decreases back to the resting frequency; frequency maxima, or points at which the EOD frequency goes through a maximum; and interruptions, or cessations in the discharge. (From Hopkins, 1972.)

wires to a tank containing another *Eigenmannia*. A male, a female, or sine waves are all equally effective in eliciting this response: there are no sex differences in the EOD frequency of *Eigenmannia*. However a male's courtship is reduced when he is presented with a tape recording of another courting male giving discharge interruptions, as it is for a sine wave interrupted artificially (Hopkins, 1974b). Whereas *Sternopygus* uses its resting discharge to evoke sexual responses, *Eigenmannia* depends on modifications in its discharge.

INDIVIDUAL SIGNALS

Responses evoked by signals emitted by a particular individual, a mate, a companion, or a rival, may play an important role in maintaining cohesive social groups of electric fish. Individual recognition requires signals for which there is great diversity within the species, in contrast to species recognition, which requires similarity among members of the species and lack of variation within the individuals. Because each individual fish of a wave-discharging species utilizes an extremely narrow frequency band in its normal resting EOD (as little as 0.5 percent during 10-minute sampling periods; Bullock, 1969) compared to the range for the species, frequency could easily encode information about individual identity, at least over short time spans. Evidence suggests that a male *Sternopygus* may be able to recognize his mate because of the unique frequency relationship between his mate and himself. During fieldwork in Guyana in 1971, two pairs of *Sternopygus* were discovered just prior to spawning, and in both cases the male's frequency was exactly one octave below that of the female (Hopkins, 1974c). When a male and a female *Sternopygus* come together for breeding, either one may change its frequency to be in an octave relation with the other, thereby imitating the partner and facilitating mate recognition.

Even in the mormyrids, which have extremely variable frequencies, individual recognition may be accomplished by an individual's use of different preferred discharge intervals or combinations of intervals. Malcolm (1972) has shown consistent individual differences in the interval histogram patterns of several isolated *Gnathonemus petersii,* but it remains to be shown whether conspecifics respond differentially to these differences.

AGE-CLASS SIGNALS

In several species of gymnotids, newly hatched fish discharge differently from adults. Of course, in all species there is a good correlation between the size of a fish and the amplitude of its discharge. In Trinidad, young *Gymnotus carapo* between 6 and 25 mm in length discharged at 15 to 35 pulses per sec; they then gradually adopted the adult frequency of 40 to 60 pulses per sec. *Hypopomus brevirostris* juveniles in Guyana discharged at higher frequencies (up to 90 per sec) than did the adults (30 to 40 per sec). Monophasic discharges produced by juvenile *Apteronotus albifrons* gradually change into the adult biphasic discharge by the time the fish are 50 mm long, but the frequency remains in the species-typical range throughout development (Hopkins, 1972a). Because there are no data showing age-specific responsiveness to any of these age-correlated signals, we cannot be sure that they convey age-specific information to the recipients. It is possible that they merely represent developmental changes in the electric organ or control centers.

MOTIVATIONAL SIGNALS

An interpreter may respond to certain signals in a way that is consistent with the most-probable following action of the signal emitter. Such signals appear to allow the receiver to make a prediction about the subsequent behavior of the signaler once the signal has been emitted, and they are therefore important in facilitating social interactions (Marler, 1961; Hazlett and Bossert, 1965; Nelson, 1964). They are exemplified in the electrical modality by signals for threat, submission, and courtship.

Threat

Black-Cleworth analyzed events during fighting behavior in *Gymnotus carapo* in seminatural aquaria and found several categories of electric signals that provided clues about the probability of forthcoming attacks from the signaler. One type, an SID display, consisting of a Sharp Increase in the EOD frequency followed by a Decrease back to the resting frequency (Fig. 6 in Black-Cleworth, 1970), usually accompanies and frequently precedes attacks or biting by dominant fish. This was also observed by Valone (1970). SIDs rarely accompany retreats and are consequently more common in the repertoire of dominant fish than of subordinates. Most likely, as a result of repeated temporal association between signal and action, the receiver comes to predict that an attack is highly likely following an SID display and responds appropriately. In an analysis of actions of recipients following SID displays—considered to be "responses," by definition—Black-Cleworth showed that recipients are unlikely to approach or attack but are likely to retreat. Furthermore, a resident *Gymnotus* in an aquarium is less likely to approach, and spends less time near, electrodes playing artificial SID displays than the same electrodes playing unmodulated pulses.

Analogous displays occur in other species; in fact, the SID format appears to be in widespread use by gymnotids and mormyrids with both wave and pulse discharges. During fighting behavior in *Gymnorhamphichthys hypostomus,* for example, aggressive individuals give frequent SID displays of varying amplitude and duration. *Hypopomus brevirostris* even has an exaggerated burst of high-frequency pulses called a "rasp" discharge, in which the normal pulse frequency of 50 Hz is suddenly elevated to several hundred Hz, in addition to its more typical SIDs, which function as aggressive threats (Hopkins, 1974a). While the wave discharge of *Apteronotus albifrons* is typically a very steady tone at about 1 kHz, an analogous "chirp" display, a sudden increase in frequency by as much as 30 percent, followed by a gradual return to the resting frequency, also serves as an aggressive threat (Black-Cleworth 1970; Hopkins, unpublished). Similarly, *Sternopygus macrurus* produces a brief SID when disturbed in its natural hiding places, a display that possibly functions as a heterospecific threat. *Eigenmannia virescens* occasionally produces brief SIDs—termed short rises—during attacks (Hopkins, 1974a).

Among the mormyrids, Möhres (1957), Szabo (in Lissmann, 1961), Bauer (1972), and Bell et al. (1974) found that smooth, rapid increases in frequency to a high level, sometimes followed by a cessation of the discharge, commonly accompanied attack, head butting, and a vigorous antiparallel fighting posture with several species, including *Gnathonemus petersii* (Fig. 7). The high-frequency bursts of pulses in *Gnathonemus* often cause a similar reciprocal discharge from a fish of equal aggressive motivation (Fig. 7A, B), but cause a cessation or reduction in frequency in a clearly subordinate fish (Bell et al., 1974). Thus, it appears that this display also serves as a threat by indicating that attack is imminent.

Several species of electric fish produce SID-like accelerations in discharge frequency while attacking prey. For electric rays, eels, and catfish, the long series of high-frequency pulses may actually stun the prey (Bauer, 1968, 1970; Belbenoit and Bauer, 1972), while in other species, such as the stargazer *(Astrocopus)* and *Gymnotus carapo,* its function is not known (Pickens and McFarland, 1964). Black-Cleworth has suggested that SIDs may have evolved in conjunction with their function in attack or biting prey, having evolved for display purposes through a process similar to the ritualization of intention movements (Daanje, 1950). If this is true, SIDs would be an example of what Darwin (1872) called "serviceable associated habits." Alternatively, SIDs could be a correlate of strong arousal and may have evolved a display function from this basis. While a sudden increase in frequency followed by a decrease may be an arbitrary representation of its designatum—which in this case is pending attack (see Hockett, 1960; Altmann, 1967)—the convergent evolution of the SID displays in gymnotids and mormyrids strongly supports the contention that the form of the display is not arbitrary but that there are selection pressures on its form that lead to some inevitable pathway of evolution. As yet, these selection pressures are not understood.

It is possible that tonic shifts in the EOD frequency could serve as a signal for threat, particularly in the mormyrids, where the discharge seems to alternate between a highly variable pulse frequency and a highly regular frequency (Moller, 1970). During dominance-determining antiparallel fights in *Gnathonemus petersii,* for example, the discharge sometimes alternates between one of several preferred intervals, 9 msec or 15 msec (Fig. 7), and in several instances, discharges appear to be delivered in pairs or triplets that result in discrete peaks in the interpulse

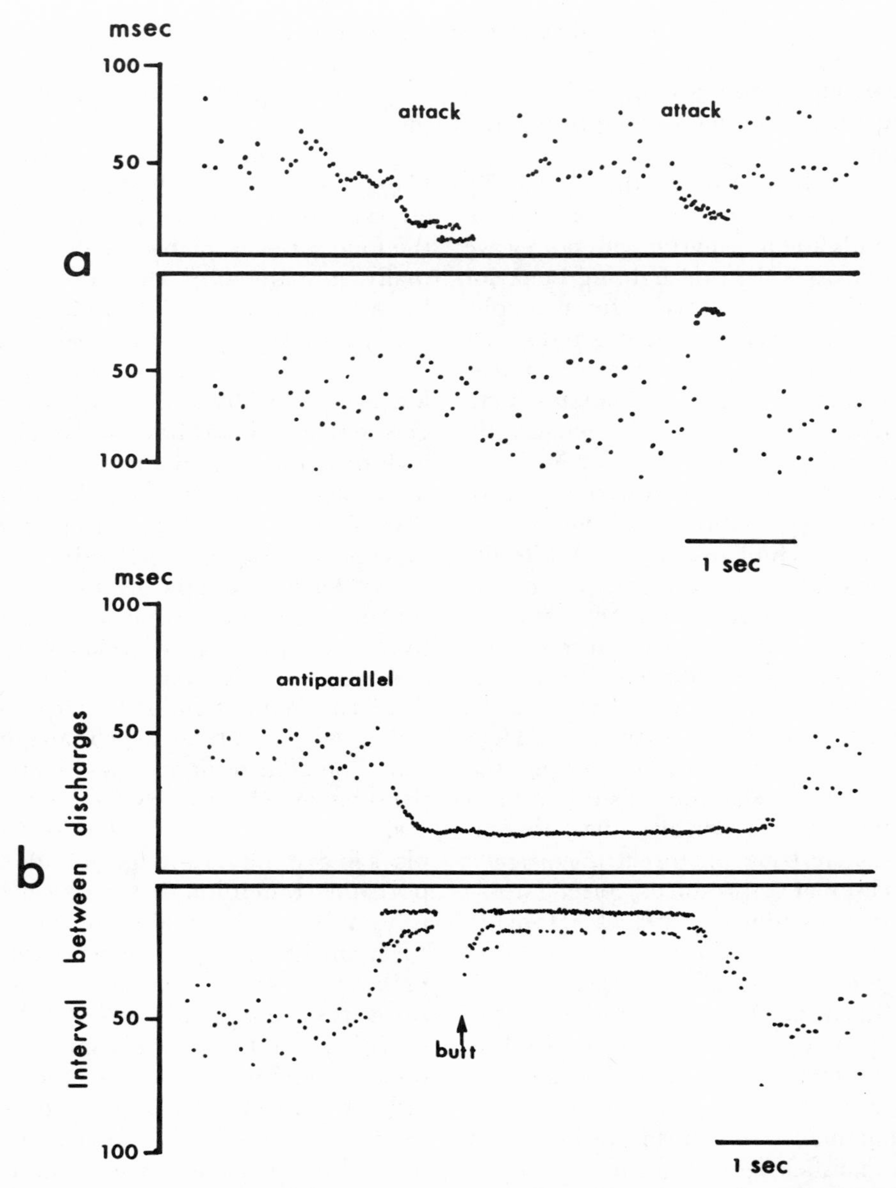

Fig. 7. Interval diagrams of electric discharges from two mormyrids, *Gnathonemus petersii*, during fighting. The interdischarge intervals of one fish are plotted as a function of time above the center lines; those of its opponent are plotted below. Very long intervals are off-scale and are not shown in this display.

a. The fish in the upper diagram gives two attacks accompanied by high-frequency (small-interval) discharges. The second attack evokes a brief acceleration in EOD rate in the second fish.

b. Both fish are involved in antiparallel fighting; the fish in the lower diagram gives a head butt that is accompanied by a brief slowing and then acceleration. This fish also shows an interesting alternation between 9-msec and 16-msec intervals.

interval histograms (Bell et al., 1974; Harder et al., 1964; Moller, 1970). Tonic shifts in frequency may also be related to dominance behavior in *Gymnotus carapo,* where Box and Westby (1970), Westby and Box (1970), and Westby (1975) found a correlation between the mean discharge frequency of an individual and the outcome of its aggressive encounters. Fishes with higher mean frequencies tended to be the more aggressive in their limited sample (Box and Westby, 1970). Tonic shifts in frequency are directly related to arousal among pulse fish, as has been shown in several studies of diurnal rhythmicity of the EOD frequency (Lissmann and Schwassmann, 1965; Schwassmann, 1971; Moller, 1970; Black-Cleworth, 1970).

Convergent evolution like that for SID displays is even more striking in another class of electrical threat signals consisting of brief interruptions in the otherwise steady discharge. Described in *Gymnotus* (Valone, 1970; Black-Cleworth, 1970; Box and Westby, 1970), discharge "breaks" or cessations lasting 1.5 sec or less frequently occur simultaneously with attack, rarely with retreat. They often precede attacks, but are more typical of individuals who eventually lose fights (Black-Cleworth, 1970). A respondent is more likely to retreat from a fish giving breaks than to approach—a response that is consistent with the attack motivation of the signaler. Discharge breaks are common in other pulse-emitting gymnotids such as *Hypopomus beebei* (Black-Cleworth, 1970), *Hypopomus artedi, Gymnorhamphichthys hypostomus,* and *Rhamphichthys rostratus* (Hopkins, unpublished) and appear to serve a similar function.

Discharge interruptions are also an important display in the wave species, *Eigenmannia virescens,* as shown in Fig. 8. They are frequently given at the same time as attacks, approaches, or darting-threat movements. Following an interruption, the recipient is likely to withdraw in retreat or else do nothing, but is unlikely to attack or to approach (Hopkins, 1974b). The African counterpart of *Eigenmannia, Gymnarchus niloticus,* also uses brief cessations (breaks) in its discharge for threats (Fig. 8). They are typically delivered at territorial boundaries in laboratory situations, while attacking, or while giving open-jawed threats. Breaks cause the interpreter to retreat or else defend its territorial boundary. And because the wave form, polarity, and frequency of *Eigenmannia* and *Gymnarchus* are so similar, the convergence in this electrical display used for threat seems even more remarkable. We are, once again, reminded of the possibility that this signal, which allows the receiver to predict motivation, may not be arbitrary. Yet, it is not clear that there is any iconic relationship between the signal and its designatum; rather it appears to be a case in which the signal is neither arbitrary nor iconic (Marler, 1961; Altmann, 1967) but somehow physically adapted to its function in the animal's social behavior. This adaptation is not understood.

Submission

Some signals evoke responses that are consistent with a reduced likelihood of attack or an increased likelihood of retreat, withdrawal, or quiet resting on the part of the signaler. These are submissive signals. Responses appropriate to this situation might vary, depending on the context. If an indicator of waning of aggressiveness were to be given during a serious fight, for example, the interpreter might renew fighting with increased vigor. With dominance clearly established, however, a submissive signal might result in a cessation of attack by the respondent. In this case, the submissive signal that reduces stimuli normally eliciting attack would be called an "appeasement display" (Moynihan, 1955; Tinbergen, 1959; Dunham et al., 1968).

One of the better-studied submissive signals is known from *Gymnotus carapo.* A discharge arrest, or a complete cessation in the EOD for up to

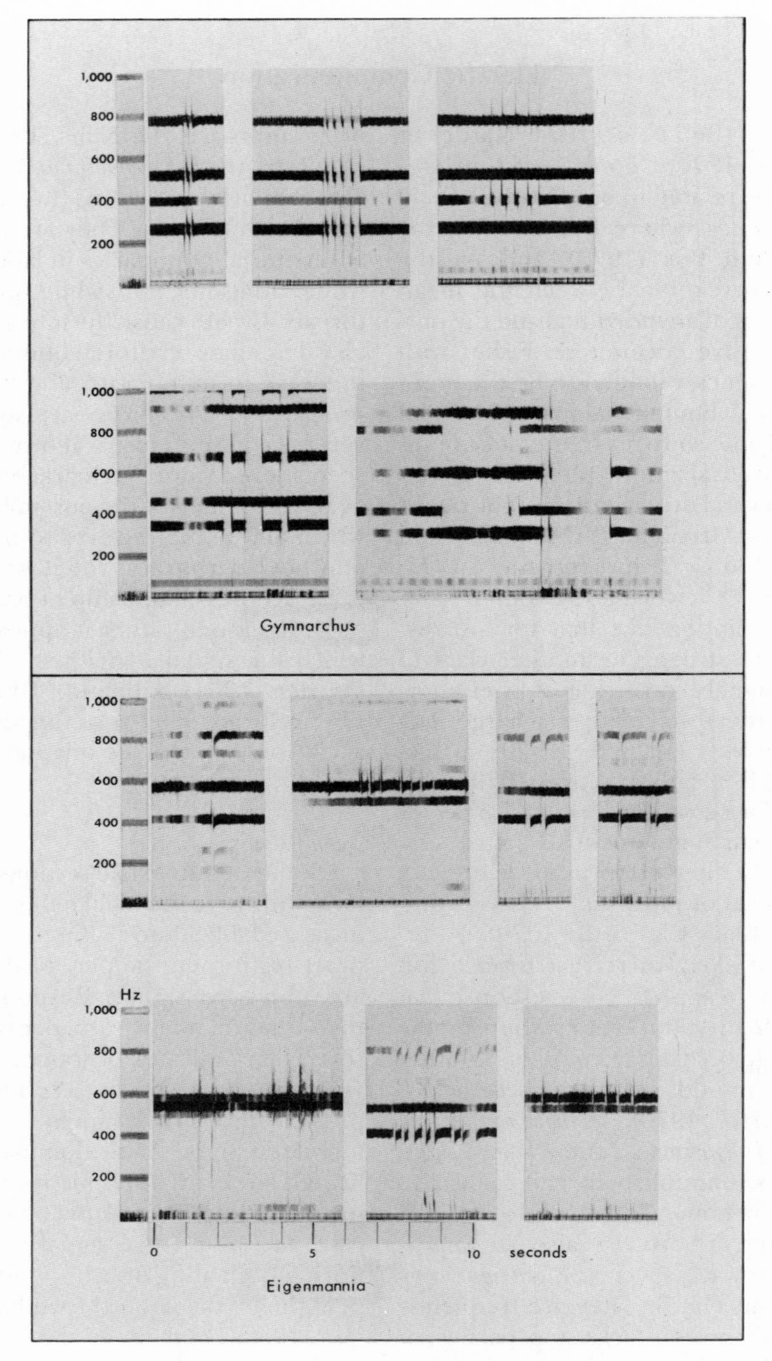

Fig. 8. Sound spectrograms of discharge interruptions as threat displays in the African species *Gymnarchus niloticus* and the South American species *Eigenmannia virescens*. Spectrograms were prepared using a Kay Electric 7030A Spectrum Analyzer, bandwidth = 37.5 Hz.

three minutes, fills the criterion for a display that reduces or is the antithesis of an attack-eliciting stimulus (Darwin, 1872; Tinbergen, 1959). Black-Cleworth considers any cessation longer than 1.5 sec to be an "arrest." She found that arrests were given exclusively by subordinate fish—those that had been defeated in aggressive encounters—and that they frequently were accompanied by retreat, but rarely by approach or attack. In her analysis of responses, arrests were likely to be followed by approach or by an absence of activity, but rarely by retreat; and biting attacks constituted a lower percentage of responses than expected. Because of this and because the normal EOD is typically an attacking-eliciting stimulus in *Gymnotus*, she concluded that arrests serve as appeasement signals.

Other pulse gymnotids such as *Hypopomus artedi, Gymnorhamphichthys hypostomus,* and *Hypopomus beebeii* produce discharge arrests, but the function is less well known for these species. Although they are given by subordinate fish during agonistic interactions, much like *Gymnotus,* arrests can be evoked by a wider variety of stimuli, including various frequencies of sine waves, single pulses, or even metal objects near the fish. Bennett (1968) has suggested that arrests might function in hiding; they might also provide a period of quiet listening to the environment.

A completely analogous discharge arrest occurs in the repertoire of subordinate *Gymnarchus niloticus.* As Szabo and Suckling (1964) and Harder and Uhlmann (1967) noted, *Gymnarchus* produces cessations lasting for periods up to 20 minutes at a time. Arrests in *Gymnarchus,* defined as cessations lasting longer than 1.5 sec, were given by subordinate individuals during agonistic interactions in which dominance was clearly established. Arrests accompanied retreats from the dominant, and once given, the dominant's attack level appeared to be reduced. Dominants usually ignored fish with an arrested discharge, only resuming their attacks when their opponent turned its discharge on again (Hopkins, unpublished).

Another class of submissive signals are encoded as frequency modulations in the resting EOD. *Eigenmannia* produces slight increases in its discharge frequency followed by a decrease back to the resting frequency (Fig. 9). The frequency change is typically on the order of 1 percent, and the duration can be as long as 40 sec. This display is given by subordinate fish—those who have given the least number of attacks in a standard watch and who have lost a competition for a hiding place during the daytime—and very rarely by dominants. These "long rises" are given at the same time as retreat from a dominant. Although this display reflects a clear lack of aggressiveness on the part of the signaler, the conditions under which it might serve as an appeasement display are unknown (Hopkins, 1974b). Remarkably enough, *Gymnarchus* also produces a submissive signal consisting of modulations in its EOD frequency. As can be seen in Fig. 9, these modulations usually consist of a decrease in the resting discharge followed by an increase back to the resting discharge. Often when a subordinate fish gives a discharge arrest, its discharge resumes and undergoes a long period of modulating frequency, as shown in the examples in Fig. 9. Frequency modulations are only given by subordinate fish, and they also appear to reduce the attack levels from the opponent.

Courtship

At certain times of the year, specialized signal exchanges between males and females appear to facilitate mating behavior in several ways: by reducing the distances between the male and the female, by overcoming aggressive tendencies of partners, by arousing sexual responsiveness, and by synchronizing spawning. Our knowledge of the reproductive behavior of this group of fishes

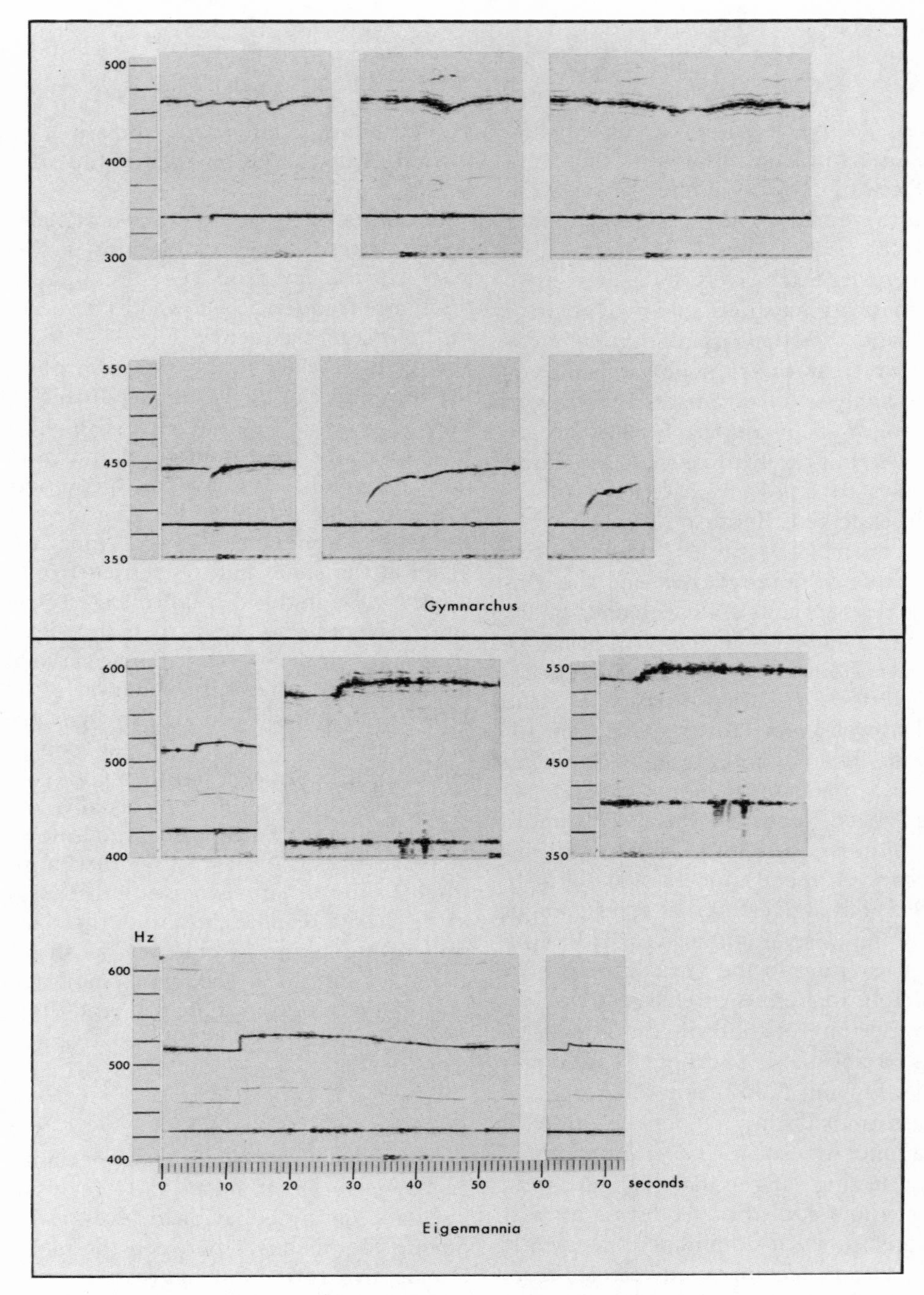

Fig. 9. Sound spectrograms of frequency modulations in the discharge of *Gymnarchus niloticus* and *Eigenmannia virescens*. Spectrograms were prepared using the Kay Electric 7030A Spectrum Analyzer, bandwidth = 1.1 Hz.

is unfortunately so limited that we cannot be sure of all the functions.

Sternopygus macrurus in the Rupununi District of Guyana come into breeding condition during the months of April and May—one month before the start of a three-month rainy season. At this time, when the sex of an adult can be identified by the frequency of its regular wave discharges (see above), large males take up residence in hiding places in the creeks that provide excellent cover and protection: under undercut banks, inside sunken logs or stumps, or under large rocks. Although they are not known to defend these hiding places, defense would be nearly impossible to observe under natural conditions. It is not uncommon for several such males to be hiding within meters of each other and, in some cases, for an occasional female or young individual to be present. These males produce a remarkable series of modifications in their discharge, consisting of both rises in frequency and interruptions in the discharge whenever a female passes their hiding place. Examples of this "song" from two males are shown in Fig. 10. There does not seem to be any strict temporal patterning to these discharge modifications, and one individual seems to be different from the next one in the patterns of rises and interruptions that they produce. Interruptions generally lasted between 0.3 and 1.7 sec, and rises sometimes reached 85 Hz above the resting frequency. The natural stimulus eliciting this response is a passing female, but sinusoidal stimuli with frequencies in the female range were just as effective. The female's response to these signals is uncertain, but because the modulated signals are only given by males, only during the breeding season, and only in the presence of a female, most likely the signal normally elicits approach by the female to the male's hiding place so that further courtship activities can take place. The signals may also have a sexually stimulating effect. To an observer, these signals serve to identify a male who is in reproductive condition who is in possession of an appropriate hiding place.

Female *Sternopygus* also produce electric signals at a later stage in the sexual behavior, when the male and female appear to be paired. In two cases in which an observed male and female were ready for spawning and in which the male's and female's discharges were one octave apart, the female produced slight modulations in her frequency, as shown in Fig. 11. These interesting signals illustrate a unique aspect of the electric modality not discussed earlier.

Because of the octave relationship between the male and the female, the second harmonic of the male's discharge adds to the fundamental of the female's to produce a resultant signal that is amplitude-modulated at the difference, or "beat," frequency of the two signals. The amplitude or depth of modulation is particularly pronounced when the two signals have similar amplitudes, that is, when the two fish are of similar size and are near each other. These beat-frequency amplitude modulations show up in the spectrograms in Fig. 11. When the female changes her frequency slightly, the beat frequency changes too, but the change in the beat frequency is proportionally much larger than the absolute change in the female's frequency. Thus, the two fish act as a "frequency amplifier."

Consider: A male discharge frequency of 65 Hz. His second harmonic at 130 adds to his mate's discharge at 132 Hz to produce an amplitude-modulated signal with a beat frequency of 2 Hz. Now, when the female increases her frequency by 4 Hz, up to 136 Hz, the beat frequency changes to a new value of 6 Hz (136–130 = 6 Hz) —three times its earlier value. Thus, comparatively small changes in a female's discharge produce relatively large changes in the beat frequency. Frequency amplification might be exploited to great advantage by *Sternopygus* for purposes of intra-pair communication. Scheich (1974) has demonstrated that cells in the Torus

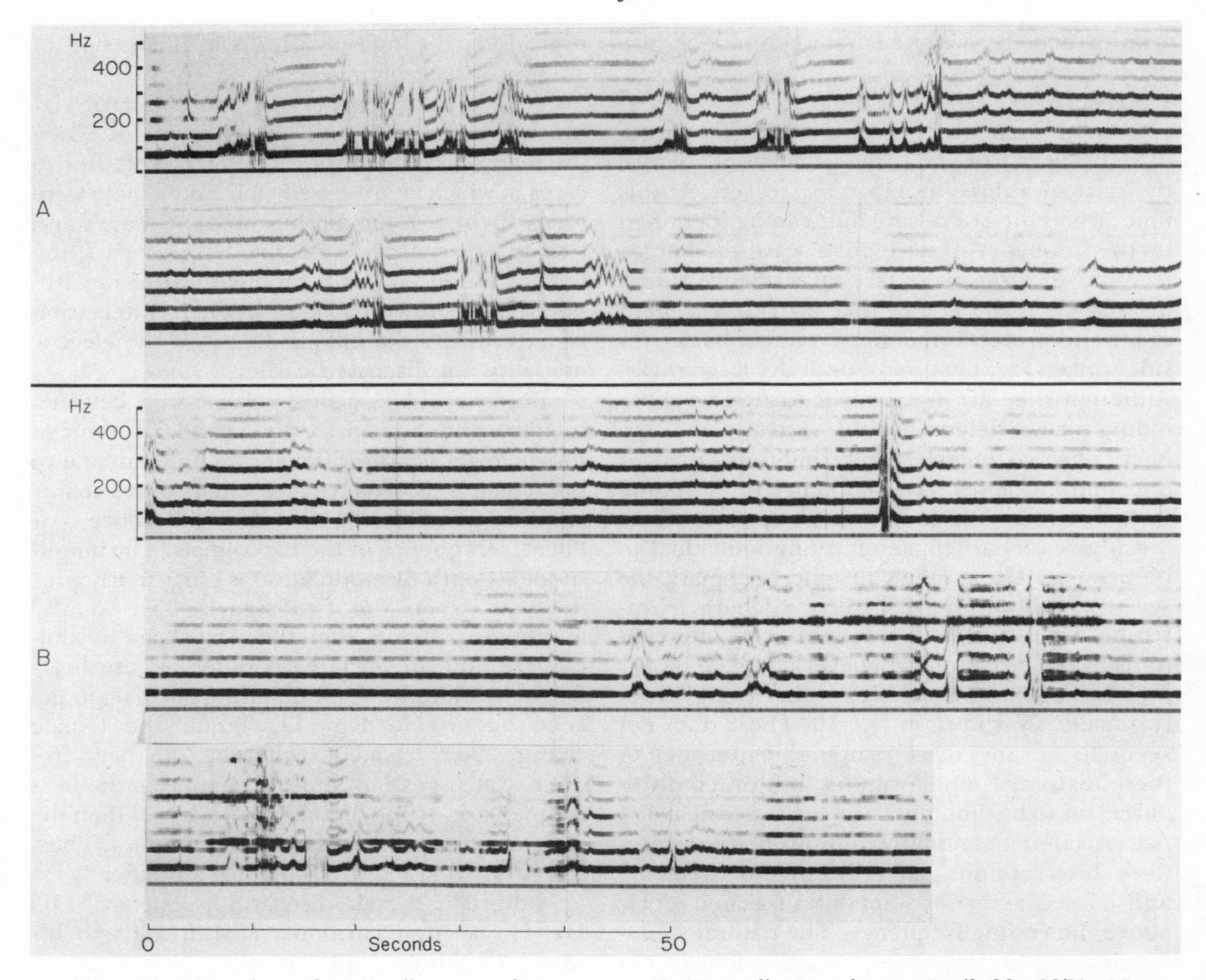

Fig. 10. Examples of naturally occurring sequences of rises and interruptions recorded in the field from two different male *Sternopygus macrurus*.

A. Recording made on May 1, 1970, Moco-moco Creek, Guyana.

B. Recording made on April 28, 1971, Moco-moco Creek, Guyana. Sonograms were prepared on a Kay Electric 7029A Sound Spectrum Analyzer, bandwidth = 19 Hz. (From Hopkins, 1974c.)

Semicircularis of the midbrain of the closely related *Eigenmannia* are sensitive to low-frequency beats produced by the addition of two signals of similar frequency. These cells respond differently to different wave form envelopes and to different beat frequencies—thus the neuronal mechanism for beat-frequency detection may also exist in *Sternopygus.*

Eigenmannia virescens also produce special signals during their breeding season. Unlike *Sternopygus, Eigenmannia* wait until the rainy season begins so that they may migrate into flooded

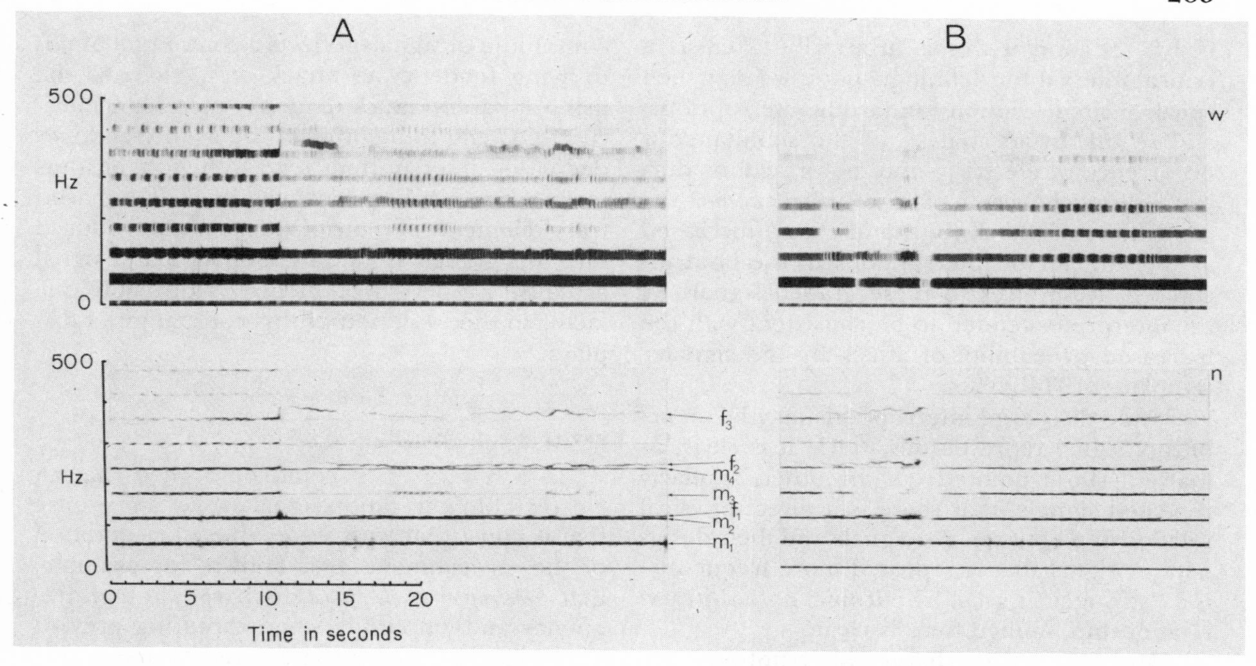

Fig. 11. A and B. Two examples of discharge variations produced by a female *Sternopygus macrurus* (f) held captive with her apparent mate (m). The fundamental frequency of the male's and the female's discharges are shown as m_1 and f_1, respectively. The second harmonics are labeled m_2 and f_2, and so on. Both wide bandwidth (W; bandwidth = 19 Hz) and narrow bandwidth spectrograms (N; bandwidth = 2.8 Hz) are shown for each sequence. (From Hopkins, 1974c.)

swamps and grasslands for breeding. No known sexual differences in frequency are known for this species, but sexually mature males respond to other *Eigenmannia* by giving many discharge interruptions—similar to those used for threat signals—but longer in duration (up to 0.5 sec) and at a much higher delivery rate—up to 50 per minute. Females also give discharge interruptions but do so at a reduced rate.

When a male and a female in breeding condition are placed together, the male continues to give long discharge interruptions, and then the female begins to produce modulations in her frequency, composed of long series of "rises" strung end to end. The functions of these signals are unknown. Nevertheless, the parallel between aggressive and submissive signals on the one hand, and male or female sexual signals on the other, is interesting and has been noted for many other species (Hinde, 1970).

Electric signals are not always rigidly stereotyped, but rather show considerable variation. The variations in motivational signals can sometimes be correlated with apparent changes in tendencies to perform certain actions. Thus, we can begin to think of a continuum of signals and even a continuum of responses.

The discharge interruptions produced by *Eigenmannia* vary in two ways: in repetition rate and in duration. During fighting behavior, *Eigenmannia* produces interruptions, sometimes singly, but more often in clusters or bouts. The median interval between discharge interruptions is 1.5 sec, and if this is arbitrarily taken as a critical interval for defining a bout, we may then speak of single, double, and triple interruptions, and so on. In an analysis of the simultaneous occurrence of electrical and motor actions during fighting between pairs of *Eigenmannia,* the probability of attack (butt and chase) increased as the number of interruptions in the bout increased. Responses to these graded signals by the interpreter tended to be consistent with the increased probability of attack by the signaler (Hopkins, 1974b).

While discharge interruptions may be an arbitrary signal representing attack, it is clear, as Marler (1961) pointed out for other, similarly repeated signals, that there is a direct physical relationship between the signals and their designata. A signal that is repeated more frequently, is a "stronger" signal. The iconic, not arbitrary, relationship implied here is clear.

Eigenmannia's discharge interruptions are also graded by duration. During the nonbreeding season, the median duration of interruptions produced by both males and females during fighting was 40 msec. Those produced by males during the breeding season while in the presence of a female had a longer median duration of 90 msec with an occasional male's interruptions lasting 500 msec. This observation suggests that as the tendency to attack or to court changes, so does the associated signaling.

Black-Cleworth (1970) found that SID displays were also graded. Small SIDs (small-frequency excursion and low-peak frequency) differed from large SIDs (large-frequency excursion and large-peak frequency) not only in the probability of being associated with attack but also in the responses that they elicited. Small SIDs were less effective in inhibiting approach than large SIDs.

Finally, the dichotomy between discharge arrests and discharge breaks in *Gymnarchus* and in *Gymnotus* can only be considered arbitrary. This continuum of signals reflects a continuum of underlying tendency to attack or to flee. As the fish's apparent attack tendency increases, interruptions tend to be briefer, and as the tendency decreases, interruptions tend to be longer, thus making the fish less conspicuous. The nonarbitrary element of conspicuousness introduced into this graded system by the fact that a signal actually is a period of silence, is probably one factor in the evolution of this continuum of displays.

ENVIRONMENTAL SIGNALS

In a few cases, it could be argued that an electric communication signal evokes a response that is consistent with some object or condition of the environment. It is known, for example, that *Electrophorus electricus* discharges at high frequency and amplitude when capturing prey or when disturbed. This characteristic burst of discharges causes a marked reaction in other electric eels nearby. They approach the disturbed or feeding eel. Bullock (1969) investigated the approach response and found that he could elicit it with a discharging eel in a net, with electrodes connected to a disturbed eel in an adjacent pond, or with electrodes connected to artificial electrical pulses. It is possible that the approach response represents an adaptive response to the presence of food.

Kastoun (1971) explored responses to environmental conditions in *Malapterurus electricus* by establishing electrical contact through wires between two fish in separate tanks. When one fish was disturbed by tapping on the side of the tank, it gave a short burst of 2 to 5 impulses and the second fish immediately fled to its hiding place. When the first fish was fed, its 14 to 562 strong discharges occurring during feeding caused the second fish to swim rapidly about the tank making circling motions. Finally, when the first fish was irritated with a small stick, its 21 to 113 impulses caused the second fish to approach and attack nearby objects.

NONSOCIAL SIGNALS

While communication involves adaptive responses to signals, some responses apparently have less social importance than others. As two fish approach each other, their electric signals may begin to interfere with each other's electrolocating systems. Heiligenberg (1973a, 1973b, 1974) has shown, in studies of the unconditioned following of laterally approaching and receding plastic objects, that the wave species *Eigenmannia is* jammed by signals having frequencies similar to its own. *Eigenmannia* discharging at about 500 Hz fail to respond to the approaching plastic objects when an external sine wave, ± 10 Hz from the fish's EOD frequency, is applied to the water. But, in a remarkable adaptive response to jamming signals such as these, discovered by Wantanabe and Takeda (1963) and explored by Bullock (1969) and Bullock et al. (1972a, 1972b), *Eigenmannia* shifts its own frequency either up or down by an amount sufficient to prevent interference: up to 20 Hz, depending on the amplitude of the stimulus. *Apteronotus albifrons* shifts up in frequency, but not down.

Pulse fish, like *Hypopygus lepturus,* are also jammed by extraneous pulses if they occur at the same time as the fish's EOD (Heiligenberg, 1974). Remarkably, several species of pulse fish produce a transient increase in EOD frequency when external stimuli overlap in time with their own pulses (MacDonald and Larimer, 1970; Heiligenberg, 1974). This response appears to help by lowering the probability of coincident pulses that are known to interfere with electrolocation.

While these electrical responses to the presence of jamming signals—conspecific or otherwise—appear to be adapted solely for preventing electrolocation interference, we cannot help noting the similarity between the Jamming Avoidance Response and Long Rises in *Eigenmannia,* and between phase-sensitive frequency increases and SIDs in pulse fish. It seems likely that these very basic responses could have played a role in the evolution of social signals. When two *Eigenmannia* with similar frequencies approach each other, for example, a frequency shift that allows continued approach *could* act as a signal indicating that aggregation will be permitted without conflict.

ELECTRIC COMMUNICATION IN NONELECTRIC FISH?

Lissmann (1958) suggested that electric organs and electroreceptors evolved because of advantages accrued by the ability to electrolocate in murky water. Wynne-Edwards (1962), on the other hand, argued that intraspecific communication could also have been the primary selective force. If the latter hypothesis were true, we might expect to find examples of primitive electrical communication systems in fish that are nonelectric. There are numerous examples of electrical emissions by nonelectric fish. Kalmijn (1972) has made preliminary measurements of electric field strengths surrounding many marine organisms. He has recorded DC potentials of up to 500 μV near the gills and mouth openings of several species of fish. Low-frequency AC fields attaining 500 μV were also strongest near the gills and head and tended to be synchronous with respiratory movements, while high-frequency AC fields were correlated with trunk or tail movements. Kalmijn (1971) also demonstrated that sharks are capable of homing in on fields like those produced by flatfish *(Pleuronectes platessa)* buried in sand. Other authors (Roth, 1972; Peters and Buwalda, 1972; Peters and Meek, 1973; Kalmijn and Adelman, in prep.) have evidence that similar phenomena occur in fresh water.

But Wynne-Edwards's hypothesis remains untested. While there is a limited knowledge about a variety of very weak electrical emissions that appear to be correlated with certain types of activity in nonelectric African catfish (Lissmann, 1963, and pers. comm.), there are no known cases of responsiveness to such signals.

Summary

Certain teleosts in both marine and freshwater environments are capable of producing and of receiving electric signals, thereby allowing electric communication—a new modality. Our most detailed knowledge concerns the gymnotid fishes of South America and the mormyriform fishes of Africa, two groups that are known to use their electrical capabilities for electrolocation as well. While electric signals are broadcast with little directionality within what appears to be a limited range, their rapid conduction velocity and fade-out make them ideal for encoding messages that fluctuate rapidly in time. Time-varying signals contribute the greatest diversity to the electric modality.

Electric communication serves many functions in the social behavior of these fish, with designators and prescriptors allowing species, sex, age-class, and individual recognition; in facilitating predictions about the motivation of a companion or a rival and about certain aspects of the environment. Motivational signals known for threat, submission, and courtship appear to be the most complex, with some evidence for grading.

References

Altmann, S., 1967. The structure of primate social communication. In: *Social Communication Among Primates,* S. Altmann, ed. Chicago: University of Chicago Press.

Batteau, D. W., 1967. The role of the pinna in human localizations. *Proc. Roy. Soc.,* B 168:158–80.

Bauer, R., 1968. Untersuchungen zur Entladungstätigkeit und zum Beutefangverhalten des Zitterwelses *Malapterurus electricus* Gemlin 1789 (Siluroidae, Malapteruridae Lacep. 1803). *Z. vergl. Physiol.,* 59:371–402.

Bauer, R., 1970. La décharge électrique pendant le comportement alimentaire de *Electrophorus electricus. J. Physiol.* (Paris), 62:341–2.

Bauer, R., 1972. High electrical discharge frequency during aggressive behavior in a Mormyrid fish, *Gnathonemus petersii. Experientia,* 28:669–70.

Belbenoit, P., and Bauer, R., 1972. Video recordings of prey capture behavior and associated electric organ discharge of *Torpedo marmorata* (Chondrichthyes). *Marine Biology,* 17:93–99.

Bell, C. C.; Myers, J. P.; and Russell, C. C.; 1974. Electric organ discharge patterns during dominance related behavioral displays in *Gnathonemus petersii. J. Comp. Physiol.,* 92:201–28.

Bennett, M. V. L., 1961. Modes of operation of electric organs. *Ann. N.Y. Acad. Sci.,* 94:458–509.

Bennett, M. V. L., 1968. Neural control of electric organs. In: *The Central Nervous System and Fish Behavior,* D. Ingel, ed. Chicago: University of Chicago Press, pp.149–69.

Bennett, M. V. L., 1970. Comparative physiology: electric organs. *Ann. Rev. Physiol.,* 32:471–528.

Bennett, M. V. L., 1971a. Electric organs. In: *Fish Physiology,* vol. 5, W. S. Hoar and D. J. Randall, eds. New York: Academic Press, pp.347–491.

Bennett, M. V. L., 1971b. Electroreception. In: *Fish Physiology,* vol. 5, W. S. Hoar and D. J. Randall, eds. New York: Academic Press, pp.493–574.

Black-Cleworth, P., 1970. The role of electric discharges in the non-reproductive social behaviour of *Gymnotus carapo. Anim. Behav. Monog.,* 3:1–77.

Box, H. O., and Westby, G. W. M., 1970. Behaviour of electric fish *(Gymnotus carapo)* in a group membership experiment. *Psychon. Sci.,* 21:27–28.

Brown, M. V., and Coates, C. W., 1952. Further comparisons of length and voltage in the electric eel *Electrophorus electricus. Zoologica,* 37:191–97.

Bullock, T. H., 1969. Species differences in effect of electroreceptor input on electric organ pacemakers and other aspects of behavior in Gymnotid fish. *Brain, Behav., Evol.,* 2:85–118.

Bullock, T. H., 1973. Seeing the world through a new sense: electroreception in fish. *Amer. Scientist,* 61:316–25.

Bullock, T. H.; Hamstra, R. H., Jr.; and Scheich, H.; 1972a. The jamming avoidance response of high frequency electric fish. I. General features. *J. Comp. Physiol.,* 77:1–22.

Bullock, T. H.; Hamstra, R. H., Jr.; and Scheich, H.; 1972b. The jamming avoidance response of high frequency electric fish. II. Quantitative aspects. *J. Comp. Physiol.,* 77:23–48.

Daanje, A., 1950. On the locomotory movements in birds and the intention movements derived from them. *Behaviour,* 3:48–98.

Darwin, Charles, 1872. *The Expression of the Emotions in Man and the Animals.* London: John Murray Publishers.

de Oliveria Castro, 1955. Differentiated nervous fibers that constitute the electric organ of *Sternarchus albifrons L. Ann. Acad. Basil. Cien.,* 27:557–60.

Dunham, D. W.; Kortmulder, K.; and Van Iersel, J. J. A.; 1968. Threat and appeasement in *Barbus stoliczkanus* (Cyprinidae). *Behaviour,* 30:15–26.

Eisner, T.; Silberglied, R. E.; Aneshansley, D.; Carrel, J. E.; and Howland, H. C.; 1969. Ultraviolet videoviewing: the television camera as an insect eye. *Science,* 166:1172–74.

Fessard, A., ed., 1974. *Handbook of Sensory Physiology.* Vol. III, pt. 3: *Electroreceptors and Other Specialized Receptors in Lower Vertebrates.* New York: Springer-Verlag.

Granath, L. P.; Erskine, F. T. III; Maccabee, B. S.; and Sachs, H. G.; 1968. Electric field measurements on a weakly electric fish. *Biophysik,* 4:370–72.

Granath, L. P.; Sachs, H. G.; and Erskine, F. T. III; 1967. Electrical sensitivity of a weakly electric fish. *Life Sci.,* 6:2373–77.

Griffin, D. R., 1968. Echolocation and its relevance to communication behavior. In: *Animal Communication,* T. A. Sebeok, ed. Bloomington: Indiana University Press, pp.154–64.

Grundfest, H., 1957. The mechanisms of discharge of electric organs in relation to general and comparative electrophysiology. *Prog. Biophys.,* 7:1–85.

Grundfest, H., 1960. Electric fishes. *Sci. Amer.,* October, pp.115–24.

Hagiwara, S., and Morita, H., 1963. Coding mechanisms of electroreceptor fibers in some electric fish. *J. Neurophysiol.,* 26:551–67.

Hagiwara, S., Szabo, T.; and Enger, P. S.; 1965. Electroreceptor mechanisms in a high frequency weakly electric fish, *Sternarchus albifrons. J. Neurophysiol.,* 28:784–99.

Harder, W.; Schief, A.; and Uhlemann, H.; 1964. Zur Funktion des electrischen Organs von *Gnathonemus petersii* (Gthr. 1862). *Z. für vergl. Physiol.,* 48:302–31.

Harder, W., and Uhlemann, H., 1967. Zum Frequenzverhalten von *Gymnarchus niloticus* (CUV.) (Mormyriformes, Teleostei). *Z. für vergl. Physiol.,* 54:85–88.

Hazlett, B. A., and Bossert, W. H., 1965. A statistical analysis of the aggressive communications system of some hermit crabs. *Anim. Behav.,* 13:357–73.

Heiligenberg, W. F., 1973a. "Electromotor" response in the electric fish *Eigenmannia* (Rhamphichthyidae, Gymnotoidei). *Nature,* 243:301–302.

Heiligenberg, W., 1973b. Electrolocation of objects in the electric fish, *Eigenmannia* (Rhamphichthyidae, Gymnotoidei). *J. Comp. Physiol.,* 87:137–64.

Heiligenberg, W. F., 1974. Electrolocation and jamming avoidance in a *Hypopygus* (Rhamphichthyidae, Gymnotoidei), and electric fish with pulse-type discharges. *J. Comp. Physiol.,* 91:223–40.

Heiligenberg, W., 1975. Theoretical and experimental approaches to spatial aspects of electrolocation. *J. Comp. Physiol.,* 103:247–72.

Hinde, R. A., 1970. *Animal Behaviour: A Synthesis of Ethology and Comparative Psychology,* 2d ed. New York: McGraw-Hill.

Hockett, C. F., 1960. Logical considerations in the study of animal communication. In: *Animal Sounds and Communication,* W. E. Lanyon and W. N. Tavolga, eds. Washington, D.C.: American Institute of Biological Sciences, pp.392–430.

Hockett, C. F., and Altmann, S. A., 1968. A note on design features. In: *Animal Communication,* T. A. Sebeok, ed. Bloomington: Indiana University Press, pp.61–72.

Hopkins, C. D., 1972a. Patterns of electrical communication among gymnotid fish. Ph.D. thesis, Rockefeller University.

Hopkins, C. D., 1972b. Sex differences in electric signaling in an electric fish. *Science,* 176:1035–37.

Hopkins, C., 1973. Lightning as background noise for communication among electric fish. *Nature,* 242:268–70.

Hopkins, C. D., 1974a. Electric communication in fish. *Am. Sci.,* 62(4):426–37.

Hopkins, C. D., 1974b. Electric communication: Functions in the social behavior of *Eigenmannia virescens. Behaviour,* 50:270–305.

Hopkins, C. D., 1974c. Electric communication in the reproductive behavior of *Sternopygus macrurus. Z. für Tierpsychol.,* 35:518–35.

Kalmijn, A. J., 1971. The electric sense of sharks and rays. *J. Exp. Biol.,* 55:371–83.

Kalmijn, A. J., 1972. Bioelectric fields in sea water and the function of the ampullae of Lorenzini in elasmobranch fishes. University of California, Scripps Inst. Oceanography, Ref. Series 72–83.

Kalmijn, A., 1974. The role of electroreceptors in the animal's life. I. The detection of electric fields from inanimate and animate sources other than electric organs. In: *Handbook of Sensory Physiology,* Vol. III, pt. 3, A. Fessard, ed. New York: Springer-Verlag, pp.145–200.

Kalmijn, A. J., and Adelman, R., in prep. I. The passive electric sense in siluroid, gymnarchid, and gymnotid fishes. II. The active electric sense in weakly electric fish.

Kastoun, E., 1971. Elektrische felder als Kommunikationsmittel beim Zitterwels. *Naturwissenschaften,* 58:459.

Keynes, R. D., and Martins-Ferreira, H., 1953. Membrane potentials in the electroplates of the electric eel. *J. Physiol.,* 119:315–51.

Knudsen, E. I., 1974. Behavioral thresholds to electric signals in high frequency electric fish. *J. Comp. Physiol.,* 91:333–53.

Knudsen, E. I., 1975. Spatial aspects of the electric field generated by weakly electric fish. *J. Comp. Physiol.,* 99:103–18.

Konishi, M., 1974. How the owl tracks its prey. *Am. Sci.,* 61:414–24.

Liebermann, L. N., 1956a. Extremely-low-frequency electromagnetic waves, I. Reception from lightning. *J. Applied Physics,* 27:1473–76.

Liebermann, L. N., 1956b. Extremely-low-frequency electromagnetic waves. II. Propagation properties. *J. Applied Physics,* 27:1477–83.

Lissmann, H. W., 1958. On the function and evolution of electric organs in fish. *J. Exp. Biol.,* 35:156–91.

Lissmann, H. W., 1961. Ecological studies on gymnotids. In: *Bioelectrogenesis,* C. Chayas, ed. New York: Elsevier.

Lissmann, H. W., 1963. Electric location by fishes. *Sci. Amer.,* 218:50–59.

Lissmann, H. W., and Machin, K. E., 1958. The mechanism of object location in *Gymnarchus niloticus* and similar fish. *J. Exp. Biol.,* 35:451–86.

Lissmann, H. W., and Mullinger, A. M., 1968. Organization of ampullary electric receptors in Gymnotidae (Pisces). *Proc. Roy. Soc.* B 169:345–58.

Lissmann, H. W., and Schwassmann, H. O., 1965. Activity rhythm of an electric fish, *Gymnorhamphichthys hypostomus,* Ellis. *Z. für Vergl. Physiol.,* 51:153–71.

Lloyd, J. E., 1966. Studies on the flash communication system in Photinus fireflies. *Misc. Publ. Mus. of Zool.,* Univ. Michigan, No. 130.

MacDonald, J. A., and Larimer, J. L., 1970. Phase-sensitivity of *Gymnotus carapo* to low-amplitude electrical stimuli. *Z. vergl. Physiol.* 70:322–34.

Malcolm, D. S., 1972. Differences in the resting pattern in weakly electric fish, *Gnathonemus petersii* (abstract only). *Bull. Ecol. Soc. Am.,* 53:(4)27.

Marler, P., 1959. Developments in the study of animal communication. In: *Darwin's Biological Work,* P. R. Bell, ed. London: Cambridge University Press.

Marler, P., 1961. The logical analysis of animal communication. *J. Theoret. Biol.,* 1:295–317.

Marler, P., and Hamilton, W. J., 1966. *Mechanisms of Animal Behavior.* New York: John Wiley.

Möhres, F. P., 1957. Elektrische entladungen im Dienste der Revierabgreuzung bei Fischen. *Naturwissenschaften,* 44:431–32.

Moller, P., 1970. "Communication" in weakly electric fish, *Gnathoneumus niger* (Mormyridae). I. Variation of electric organ discharge (EOD) frequency elicited by controlled electric stimuli. *Anim. Behav.,* 18:768–86.

Moller, P., and Bauer, R., 1973. "Communication" in weakly electric fish, *Gnathonemus petersii* (Mormyridae) II. Interaction of electric organ discharge activities of two fish. *Anim. Behav.,* 21:501–12.

Moynihan, M., 1955. Some aspects of reproductive behavior in the black-headed gull *(Larus ridibundus ridibundus* L.) and related species. *Behaviour Suppl.,* 4:1–201.

Nelson, K., 1964. The temporal patterning of courtship behavior in the glandulocaudine fishes (Ostariophysi, Characidae). *Behaviour,* 24:90–146.

Peters, R. C., and Buwalda, R. J. A., 1972. *J. Comp. Physiol.,* 79:29–38.

Peters, R. C., and Meek, J., 1973. Catfish and electric fields. *Experientia,* 29:299–300.

Pickens, P. E., and McFarland, W. N., 1964. Electric discharge and associated behaviour in the stargazer. *Anim. Behav.,* 12:362–67.

Rommell, S. A., and McCleave, J. D., 1972. Oceanic electric fields: perception by American eels? *Science,* 176:1233–35.

Roth, A., 1972. Wozu dienen die Elektrorezeptoren der Welse? *J. Comp. Physiol.,* 79:113–35.

Russell, C. J.; Myers, J. P.; and Bell, C. C.; 1974. The echo response in *Gnathonemus petersii* (Mormyridae). *J. Comp. Physiol.,* 92:181–200.

Scheich, J., 1974. Neuronal analysis of wave form in the time domain: midbrain units in electric fish during social behavior. *Science,* 185:365–67.

Scheich, J., and Bullock, T. H., 1974. The detection of

fields from electric organs. In: *Handbook of Sensory Physiology,* vol. III, pt. 3, A. Fessard, ed. New York: Springer-Verlag, pp.201–56.

Schwassmann, H. O., 1971. Circadian activity patterns in gymnotid electric fish. In: *Biochronometry.* Washington, D.C.: National Academy of Sciences, pp.186–99.

Silberglied, R. E., and Taylor, O. R., 1973. Ultraviolet differences between the sulphur butterflies, *Colias eurtheme* and *C. philodice,* and a possible isolating mechanism. *Nature,* 241:406–408.

Smyth, W. R., 1968. *Static and Dynamic Electricity.* New York: McGraw-Hill.

Soderberg, E. F., 1969. ELF noise in the sea at depths from 30 to 300 meters. *J. Geophys. Res., Space Physics,* 74:2376–87.

Steinbach, A. B., 1970. Diurnal movements and discharge characteristics of electric gymnotid fishes of the Rio Negro, Brazil. *Biol. Bull.,* 138:200–10.

Steven, S. S., and Newman, E. B., 1934. The localization of pure tones. *Proc. Nat. Acad. Sci.,* 20:593–96.

Szabo, T., 1965. Sense organs of the lateral line system in some electric fish of the Gymnotidae, Mormyridae and Gymnarchidae. *J. Morph.,* 117:229–50.

Szabo, T., Kalmijn, A. J.; Enger, P. S.; and Bullock, T. H.; 1972. Microampullary organs and a submandibular sense organ in the fresh water ray, *Potamotrygon. J. Comp. Physiol.,* 79:15–27.

Szabo, T., and Suckling, E. E., 1964. L'ârete occasionel de la décharge électrique continue du *Gymnarchus* est-il une réaction naturelle? *Naturwissenschaften,* 51:92.

Tinbergen, N., 1959. Comparative studies of the behavior of gulls (Laridae); a progress report. *Behaviour,* 15:1–70.

Valone, J. A., Jr., 1970. Electrical emissions in *Gymnotus carapo* and their relation to social behavior. *Behaviour,* 37:1–14.

Wantanabe, A., and Takeda, K., 1963. The change of discharge frequency by AC stimulation in a weak electric fish. *J. Exp. Biol.,* 40:57–66.

Watt, A. D., 1960. ELF electric field from thunderstorms. *J. Res. N. B. S.,* 64D:425–33.

Waxman, S. G.; Pappas, G. D.; and Bennett, M. V. L.; 1972. Morphological correlates of functional differentiation of nodes of ranvier along single fibers in the neurogenic electric organ of the knife fish *Sternarchus. J. Cell Biology,* 53:210–24.

Westby, G. W. M., 1975. Further analysis of the individual discharge characteristics predicting dominance in the electric fish, *Gymnotus carapo. Anim. Behav.,* 23:249–60.

Westby, G. W. M., and Box, H. O., 1970. Prediction of dominance in social groups of the electric fish, *Gymnotus carapo. Psychon. Sci.,* 21:181–83.

Wilson, E. O., 1970. Chemical communication within animal species. In: *Chemical Ecology,* E. Sondheimer and J. B. Simeone, eds. New York: Academic Press, pp. 133–56.

Wilson, E. O., and Bossert, W. H., 1963. Chemical communication among animals. *Recent Prog. Hormone Res.,* 19:673–716.

Wynne-Edwards, V. C., 1962. *Animal Dispersion in Relation to Social Behavior.* New York: Hafner.

Part III

Communication in Selected Groups

Chapter 14

COMMUNICATION, CRYPSIS, AND MIMICRY AMONG CEPHALOPODS

Martin H. Moynihan and Arcadio F. Rodaniche

Every individual organism, group, and species has its own characteristic distribution in space and time. Distributions are partly determined by behavior. Among the most important behavior patterns of animals are signals that transmit information (true or false) from one individual to one or more others (of the same or different species). Almost any act can be a signal while subserving other functions as well, but there are certain particular patterns that have become specialized, modified in form or frequency, expressly and only to facilitate the transmission of information. These are usually called "displays." They are related to, and may interact with, equally and similarly specialized patterns designed to prevent or confuse communication. These might be called "antidisplays." Some patterns can be displays in some situations, antidisplays in others. In either or both cases, the kinds of specialization involved are usually called "ritualization."

Cephalopods have many ritualized and unritualized patterns, arranged in adaptive sequences and repertoires.

The living cephalopods can be divided between two major systematic groups: One comprises a few species of *Nautilus*; the other includes the coleoids—many species of many genera of three or four orders—the squids, cuttlefishes, *Vampyroteuthis,* and octopi and argonauts. All are marine. Very little is known of the social behavior of *Nautilus* or of those coleoids that are confined to deep waters or open seas, but there have been fairly extensive studies of inshore and littoral forms, most notably species of *Sepia, Octopus, Sepioteuthis,* and *Loligo s.l.* (references in Moynihan, 1975). These animals not only share a common remote ancestry but also show similarities in ways of life (see Lane, 1957; Packard, 1972; Wells, 1962).

They are large (considering the animal kingdom as a whole), more or less active, comparatively intelligent, with large, complex brains (Young, 1964, 1971), predaceous upon smaller organisms and preyed upon by larger ones. They tend to have rapid rates of development. Most squids are conspicuously gregarious. Other species have social liens of less conspicuous but not necessarily simpler types. Many are extremely abundant. All seem to have elaborate "courtship" and other sexual or partly sexual reactions. Some are supposed to reproduce only once in a lifetime, in "big bangs" (Gadgil and Bossert, 1970). All make special provisions for their large eggs. They may "incubate" the eggs and/or lay them in selected sites and conditions. They seldom or never show strictly parental protection or training of hatched young, but adults and immatures may associate with one another according to regular rules. Many species live in rich and varied environments of great diversity and appreciable instability or unpredictability. Some have important interspecific relations in addition to or apart from ordinary run-of-the-mill predator-prey responses. All have superb eyesight, tactile sense organs, some chemical sense(s), and statocysts that could be, but probably are not, used for hearing.

These are the rather broad constraints within which their communication systems have had to operate.

Visual signals are predominant. They include movements and postures and changes in colors and color patterns. The great majority of cephalopods differ from the most nearly comparable vertebrates, the fishes, in having more independently moveable appendages—eight or ten arms and tentacles in addition to fins. Inshore and littoral species also have several kinds of chromatophores plus leucophores and iridocytes. These permit partial or complete color changes of remarkable speed and precision.

The only species that has been studied at length in both the field and laboratory is *Sepioteuthis sepioidea* (Boycott, 1965; Moynihan and Rodaniche, in prep.). It may be typical of many inshore cephalopods in some aspects of its communication behavior. Its unritualized signals include a host of minor movements and intention movements of a few or all of the arms, the fins, and/or the body as a whole. Its ritualized patterns include more exaggerated or stereotyped movements of the same parts, e.g., spreading, curling, raising, or lowering of the arms, and many color changes—including general lightening or darkening, flushes of yellow or lavender pink, and production of more or less broad transverse bars, longitudinal stripes, spots of differing sizes, iridescent ocelli, irregular blotches, and even semigeometric arrangements of a variety of tones and hues. Some of the more spectacular ritualized patterns of this and other species are shown in the accompanying illustrations.[1]

There are implications to be drawn from such behavior:

(1) The total numbers of "basic" or "major" ritualized patterns in the repertoires of many cephalopods seems to be of the same order of magnitude as in most bony fishes, birds, and mammals (approximately fifteen to thirty-five by rough definition). The reasons why kinds of display cannot proliferate endlessly without corresponding losses are discussed in Moynihan (1970). It is not surprising that cephalopods conform to the same general rules as other animals. Some or all of the inshore squids, cuttlefishes, and octopi are fortunate, however, in that the refinement of their chromatophores and associated organs permits unusual flexibility of combinations. A squid in the midst of a group, for instance, can transmit at least three or four different "messages" (in the sense of Smith, 1965), absolutely simultaneously, to completely different individuals and in different directions by assuming different color patterns on different parts of the body. It can also change any or all of the patterns and messages instantaneously whenever necessary or desirable. This may be the most flexible of ritualized visual systems. Perhaps only the most complex acoustic repertoires of certain birds and primates ("Songs," etc.) convey so much information so rapidly.

(2) Every species of cephalopod has some unique characteristics; but a few of the most elaborate and highly ritualized patterns (see legends of the figures) occur in essentially the same forms in all the coleoids so far studied in detail,

1. All photographs were taken in the field near the San Blas Islands along the Caribbean coast of Panama unless otherwise noted.

including members of orders that diverged in the Mesozoic Era (probably in the Triassic). These patterns would seem to have been extremely conservative during the course of evolution. Some of them may have been conservative because they were designed to influence a great variety of "receivers," individuals of many different species and/or different ages, social classes, and sexes of the same species. Signals adapted to many kinds of receiver would be expected to change less frequently or more slowly on the average than signals adapted to only one or a few kind(s) of receiver (Moynihan, 1975).

(3) Another distinctive and very widespread feature of the ritualized systems of inshore cephalopods is a subtle and problematical phenomenon, an apparent relation between overt behavior and the background against which it is performed. All ritualized patterns of all animals may be considered to be expressions of "drives." The term "drive" itself is loose, and open to criticism when used in analytical (physiological) studies of causation. It is convenient, however, as descriptive shorthand for preliminary or superficial accounts, if it is employed in the sense of Thorpe (1951) as "the complex of internal and external states and stimuli (usually or normally) leading to a given behaviour." The role of the external factors usually appears to be largely or completely indirect, the external situation merely stimulating or affecting internal factors that are the immediate "triggers" of the resulting performance. In most cases and for many purposes, it is sufficient for an observer to state that pattern *A* is produced by a certain strength or range of internal motivation *B,* or a combination of *B* with internal motivation *C.* Whenever the external situation is such as to stimulate *B* and *C* appropriately, pattern *A* is bound to appear. *Only A* appears, not *D* or *E* or anything else and irrespective of the minor details of the external situation. This sort of simplified schema may also be applied to the ritualized patterns of cephalopods, but it does not always "fit" particularly well. Most cephalopods appear to "choose" among available patterns more frequently than do most other animals. This probably means that they sometimes have to pay more conscious attention to a greater variety of aspects of their surroundings.

Examples from *Sepioteuthis sepioidea* may show what we mean.

Individuals of this species have many alarm patterns. Two of the most common are bold Transverse Bars and Longitudinal Streaks (Figs. 1–4 and 7). Both are produced when the escape tendency or drive is more or less strong but impeded by—in conflict with—a counteracting and incompatible tendency, e.g., attack, hunger (feeding drive), or a gregarious or sexual attraction. Rather surprisingly, the internal factors involved seem to be identical in some performances of either type. Some individuals in

Fig. 1. An adult *Sepioteuthis sepioidea*. The animal shows several special color patterns, including Transverse Bars, a Fin Stripe, and a trace of Longitudinal Streak on the back. It has also assumed a Downward Pointing posture.

Fig. 2. A "courting party" of three adult *Sepioteuthis sepioidea*. The animal on the upper right is a female in Pied coloration. This pattern is "discouraging." It is highly ambivalent, produced by both hostile and sexual tendencies. The female has also assumed a Downward Pointing posture, which in itself is hostile. The animal in the center is a male. He shows Longitudinal Streaks on the back. He must be somewhat alarmed, nervous, and intimidated by the female. The animal on the lower left may be another male. He is in a semi-Dark color pattern, possibly also slightly alarmed.

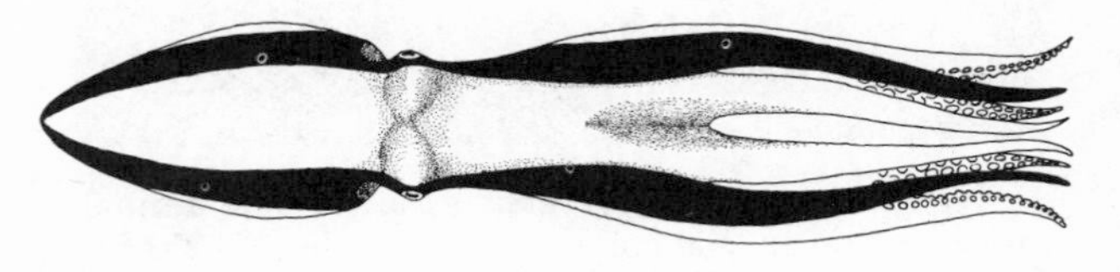

Fig. 3. An *Octopus vulgaris* with Longitudinal Streaks. (After Cowdry, 1911.)

Bar seem to be no more and no less likely to escape than are some individuals (of the same age and, presumably, sex) in Streak. The crucial causal difference between the two performances would appear, at times, to depend entirely on external circumstances. Bar patterns are assumed most frequently by individuals high up in open water. Streak patterns are relatively most common when individuals are low, near vegetation or coral or some other "broken" type of bottom. There are also intermediate situations: when a group of *sepioidea* is approached by some alarming stimulus in an intermediate habitat, some individuals may assume Bars and others Streaks before they all dash away together.

The partial interchangeability of Bar and Streak may be related to the fact that both patterns are conspicuous in some circumstances and cryptic or mimetic in others. They tend to be equally conspicuous whenever noticed, i.e., when functioning as conventional displays, in a wide range of environments. But their chances of being effectively cryptic or mimetic may be very

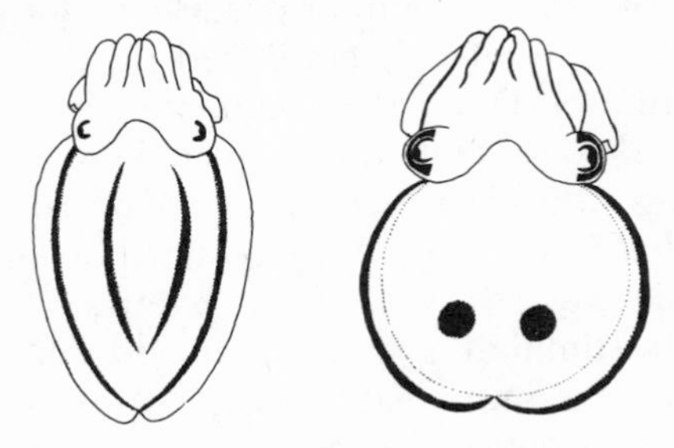

Fig. 4. Two color patterns of *Sepia officinalis.* (After Holmes, 1940.) Left: Longitudinal Streaks with Fin Stripes. Right: The so-called Dymantic display (a term applied to patterns that emphasize eyespots—in this and some other cases false eyespots). The Dymantic of *Sepia* is usually or often combined with dark borders to the fins.

Both Dymantic and Longitudinal Streaks are among the most obviously conservative of cephalopod displays. They occur in the repertoires of some species of at least three different orders: Teuthida, Sepiida, Octopida.

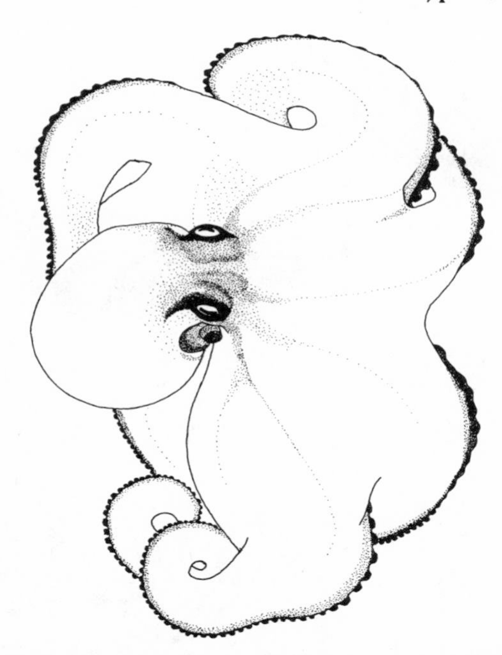

Fig. 5. The Dymantic of *Octopus vulgaris*. (After Wells, 1962.) Here the emphasis is on the real eyes. The dark borders to the mantle and arms probably are functionally equivalent and historically related to the corresponding features of the fins of *Sepia* and other forms. Homologies can be indirect and devious within the Cephalopoda.

different in different environments. The elaboration of mechanisms and strategies to permit crypsis and mimicry among cephalopods has been a subject for comment from the time of Aristotle (more recent accounts of interest include Lane, 1957; L. Tinbergen, 1939; Holmes, 1940; Packard and Sanders, 1971; Cousteau and Diolé, 1973). This elaboration may reflect the great vulnerability of cephalopods to predation (Clarke, 1966; Moynihan, 1973, 1975). It certainly has had many consequences for their display behavior.

Another pattern of *sepioidea,* the Dark (Figs. 2, 7, and 8), is remarkable in several ways. It is sometimes assumed by individuals resting or sleeping a few meters below the surface of the water during the middle of the day. In these circumstances it may, at least conceivably, have nothing to do with communication. It could be simply a means of maximizing heat absorption. (As far as we know, cephalopods are ectothermic. But recent discoveries of endothermic species among classes of vertebrates that were supposed to be "cold-blooded"—Chondrichthyes and Osteichthyes as well as Reptilia—may add a note of caution and indicate that methods of temperature regulation are varied even in the sea.) An apparently identical Dark pattern can also be assumed by disturbed or frightened individuals. It may seldom occur or never have as high an intensity as the most extreme Bars or

Fig. 6. A young adult or subadult *Sepioteuthis sepioidea*. The coloration of this individual is more or less "ordinary" (unritualized) over most of the body, but there are traces of a Transverse Bar across the front part of the back and two well-developed Dymantic spots on either side toward the rear.

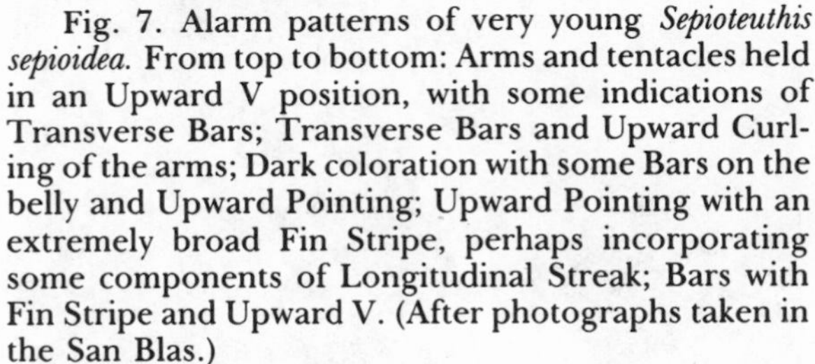

Fig. 7. Alarm patterns of very young *Sepioteuthis sepioidea.* From top to bottom: Arms and tentacles held in an Upward V position, with some indications of Transverse Bars; Transverse Bars and Upward Curling of the arms; Dark coloration with some Bars on the belly and Upward Pointing; Upward Pointing with an extremely broad Fin Stripe, perhaps incorporating some components of Longitudinal Streak; Bars with Fin Stripe and Upward V. (After photographs taken in the San Blas.)

This is a small sample of the possible combinations of signals.

Fig. 8. More alarm patterns, as shown by older *Sepioteuthis sepioidea* (again after photographs in the field). Left: Full Dark with Downward Curling. Right: Slight Dark with Upward Curling.

Fig. 9. An adult *Sepioteuthis sepioidea* with a "Rear Light" (a patch of pale color, not luminescent) at the "tail." This may be a low-intensity indication of the Pied pattern.

Streaks, but its motivation may overlap those of either one or both of the other two patterns at somewhat lower levels of alarm or anxiety.

If these suggestions are correct, then Darks can be produced by different factors at different times. The hypothesis is unattractive (too easy—explaining too much or too little), but it is plausible and certainly not logically impossible.

Darks tend to be rather conspicuous. They are cryptic only in situations that are very special

Fig. 10. A mixed school of squids, adult and subadult *Loligo* ("*Doryteuthis*") *plei* and *Sepioteuthis sepioidea.* From left to right: a *plei* with Center Light (paling) pattern in a Head-down posture; another *plei* with Center Light; a *plei* in the "ordinary" coloration of the species; a *sepioidea* in "ordinary"; a female *sepioidea* with a trace of Pied. The Pied and Center Light may be partly homologous.

Fig. 11. An adult male *Sepioteuthis sepioidea* in Lateral Silver color pattern. This pattern is sometimes assumed by courting males "defending" their females against intruders or possible rivals. It is the pale side of the body that is directed toward opponents. The performance is usually effectively repellent.

indeed. Disturbed squids often release blackish ink, which may hang in the water as a blob, somewhat contorted or "strung out" by the effects of currents. An individual that has inked usually shoots away from the scene in some disruptive or pale color pattern. Presumably predators often focus on the ink and fail to notice the escape of their quarry. Insofar as the ink serves as a decoy, something that looks like an animal or edible object, it is an image or mimic of the prey. On one occasion, one of us saw and blocked the retreat of two young *sepioidea* in very shallow water over pure white sand. They released ink and then, instead of going disruptive or pale, turned Dark and arranged their arms in "contorted" patterns. In these positions they looked like their own blobs of ink. Other predators might have been confused or distracted, not knowing which "blobs" were real or worth attacking. It would seem that some squids can mimic their own (self-produced) mimics!

Fig. 12. A Flamboyant display, with Upward V Curling, by a young *Octopus vulgaris*. Vs with Upward Curling or Pointing may be as widespread as Dymantic and Longitudinal Streak patterns. Compare with Fig. 7. (After Packard and Sanders, 1969, 1971.)

There probably are additional patterns of *sepioidea* and other cephalopods that are inter-

Fig. 13. A juvenile *Octopus chierchiae* eating a crab. This individual was captured on the Pacific coast of central Panama and maintained in the laboratory of the Smithsonian Tropical Research Institute on Naos Island. The drawing is based on a photograph taken in the laboratory.

Many cephalopods display while attacking and eating prey. Among the special patterns shown here are Stripes on the body, Darkening of the forward arms, and the beginning of a Longitudinal Streak along the side.

Fig. 14. Adult female *Octopus oculifer* (from the Pacific coast of central Panama, photographed in the laboratory at Naos). This color pattern is an extreme of conspicuousness, presumably homologous with the Zebra Stripes of other species. The raising of the forward arms may be an intention movement or low-intensity version of some sort of Upward V Curling. The performance as a whole is certainly hostile, probably aggressive.

Fig. 15. A high-intensity dispute between two adult male *Sepioteuthis sepioidea* in a courting group. Both animals have their arms in Spread positions. The lower individual shows Zebra Stripes. This type of coloration, with or without Spreads, is still another of the conservative patterns.

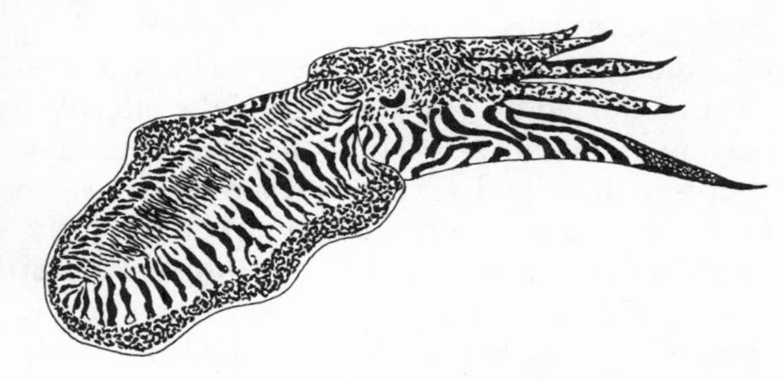

Fig. 16. Zebra Stripes of *Sepia officinalis.* (After Tinbergen, 1939.)

changeable in much the same way(s) and perhaps to the same extent as the Bar, Streak, and Dark. The quality or potential is not, however, characteristic of all cephalopod displays. Many are as largely determined by internal factors and as little interchangeable as are most ritualized patterns of vertebrates and arthropods. And, of course, even vertebrate displays, if not those of arthropods, show some variation in this respect. The primary difference between major systematic groups must be quantitative rather than qualitative. The peculiarity of cephalopods in this context can be summarized crudely. Some of the ritualized patterns of some cephalopods, perhaps most of the littoral and inshore species, seem to be more often or directly controlled by (adapted to) their physical environments than are most of the corresponding patterns of other animals. The performance of a ritualized pattern by a cephalopod may entail more calculation, the weighing of pros and cons, than is usually true of other animals.

This may be a reason why some or most coleoid cephalopods have relatively larger brains per body weight (Packard, 1972) than do the ecologically equivalent fishes.

The predominance of visual displays among coleoids is a problem in itself. As Tavolga (1968) and others have pointed out, aquatic environments are not ideal for visual communication. Light beams extinguish rapidly with distance in water and are scattered and obscured by suspended particles and plankton. It is obvious in the field that visual signals can be used only for communication over limited areas. The visual patterns of *S. sepioidea,* for instance, convey much information very rapidly, and sometimes broadly (in the sense of going off in different directions), but they do not transmit very far.

One may ask, therefore, why so many coleoids place such reliance on an imperfect channel, or why they do not make more use of other kinds of communication. There would appear to be two answers. First, visual signals are adequate. They work, and work well within their sphere. Second, they may have fewer drawbacks than the possible alternatives—for animals of coleoid habits and structures.

The deficiencies of visual communication are often minimized by two features. Many coleoids are so gregarious and cohesive that they do not need more than short- to medium-distance signals among themselves. Most of them also prefer the clearest waters available. The fact that cryptic or mimetic patterns cannot be seen at great distances is not a drawback.

Some of the possible alternatives, olfactory or other chemical signals, are slow and/or extremely diffuse. Tactile signals are very short range. Both kinds of signal may be too clumsy for animals as mobile as most coleoids, or be suitable for only a few types of social encounters. Coleoids do, in fact, use them for certain interactions, such as copulations, when a degree of close contact and a pause in other activities are perhaps inevitable.

Nautilus depends on touch, taste, and smell in a greater diversity of circumstances. It is also comparatively sluggish (Cousteau and Diolé, 1973).

The deep-water coleoids may have to use their chemical and tactile senses more frequently and extensively than do their littoral or surface relatives. It is suggestive, however, that many of them have evolved numerous or elaborate light organs. This would seem to indicate that they have retained at least part of the visual acuity and some of the preferences that are characteristic of the group as a whole.

Acoustic signals would have many theoretical advantages—they could be sent over long as well as short distances and refined to extreme precision—and it is not immediately evident why cephalopods do not use them. There may be an element of evolutionary "accident." Perhaps appropriate mutations did not occur among cephalopods at the right times. But there may also have been functional considerations. Much of the mobility of cephalopods is dependent on jet propulsion, a method of locomotion that seems to work best when the body is very flexible. Flexibility is made easier when hard parts are reduced. (This, in turn, may be a reason why most of the earlier shelled cephalopods were replaced by the largely unshelled coleoids.) In the absence of many hard parts, it may be difficult to evolve noise-making organs such as stridulating devices. Most coleoids lack structures like the "swim bladders" that are important in the sound reception of many fishes.

It is also possible that sounds are more likely than visual signals to attract the attention of potential predators. Some accounts of the acoustic behavior of fishes, e.g., Tavolga (1960), imply that many of the species that produce sounds most frequently are particularly well protected. The more vulnerable cephalopods may not have been able to take the same risks.

COMMENT

Two general aspects of the communication systems of cephalopods are of interest from a comparative point of view. (1) These systems are highly complex and sophisticated, probably as advanced as those of any other animals apart from man and (possibly) subhuman forms such as chimpanzees. Cephalopods are invertebrate, but they are not backward or primitive in their social behavior. (2) The repertoires of many coleoid cephalopods are distinguished by a maximum use and development of a single channel or medium of expression. They illustrate the intricacy and efficiency that limited systems can attain in conditions that are both favorable *per se* and narrowly bound in scope.

References

Boycott, B. B., 1965. A comparison of living *Sepioteuthis sepioidea* and *Doryteuthis plei* with other squids, and with *Sepia officinalis. J. Zool.,* 147:344–51.

Clarke, M. R., 1966. A review of the systematics and ecology of oceanic squids. In: *Advances in Marine Biology,* F. S. Russell, ed. New York: Academic Press, pp.91–300.

Cousteau, J.-Y., and Diolé, P., 1973. *Octopus and Squid.* London: Carsell.

Cowdry, E. V., 1911. The colour changes of *Octopus vulgaris. Univ. Toronto Stud., Biol. Ser.,* 10:1–53.

Gadgil, M., and Bossert, W. H., 1970. Life historical consequences of natural selection. *Amer. Natur.,* 104:1–24.

Holmes, W., 1940. The colour changes and colour patterns of *Sepia officinalis* L. *Proc. Zool. Soc. Lond.,* A 110:17–36.

Lane, F. W., 1957. *The Kingdom of the Octopus.* London: Jarrolds.

Moynihan, M., 1970. Control, suppression, decay, disappearance and replacement of displays. *J. Theor. Biol.,* 29:85–112.

Moynihan, M., 1973. The evolution of behavior and the role of behavior in evolution. *Brev. Mus. Comp. Zool.,* 415:1–27.

Moynihan, M., 1975. Conservatism of displays and comparable stereotyped patterns among cephalopods. In: *Essays on Function and Evolution in Behaviour: A Festschrift for Professor Niko Tinbergen,* G. Baerends and A. Manning, eds. Oxford, Clarendon Press.

Packard, A., 1972. Cephalopods and fish: the limits of convergence. *Biol. Rev.,* 47:241–307.

Packard, A., and Sanders, G. D., 1969. What the octopus shows to the world. *Endeavour,* 28:92–99.

Packard, A., and Sanders, G. D., 1971. Body patterns of *Octopus vulgaris* and maturation of the response to disturbance. *Anim. Behav.,* 19:780–90.

Smith, W. J., 1965. Message, meaning, and context in ethology. *Amer. Natur.,* 99:405–409.

Tavolga, W. N., 1960. Sound production and underwater communication in fishes. In: *Animal Sounds and Communication,* W. E. Lanyon and W. N. Tavolga, eds. Publ. 7, American Institute of Biological Sciences, Washington, D.C., pp.93–136.

Tavolga, W. N., 1968. Fishes. In: *Animal Communication,* T. A. Sebeok, ed. Bloomington: Indiana University Press, pp.271–88.

Thorpe, W. H., 1951. The definition of terms used in animal behaviour studies. *Bull. Anim. Behav.,* 9:34–40.

Tinbergen, L., 1939. Zur Fortpflanzungsethologie von *Sepia officinalis. Archs. néerl. Zool.,* 3:323–64.

Tinbergen, N., 1951. *The Study of Instinct.* London: Oxford University Press.

Wells, M. J., 1962. *Brain and Behaviour in Cephalopods.* London: Heinemann.

Young, J. Z., 1964. *A Model of the Brain.* London: Oxford University Press.

Young, J. Z., 1971. *The Anatomy of the Nervous System of* Octopus vulgaris. London: Oxford University Press.

Chapter 15

COMMUNICATION IN CRUSTACEANS AND ARACHNIDS

Peter Weygoldt

Crustaceans

The class Crustacea contains a very large number of species, which exhibit great diversity. Small forms, such as Branchiopoda, Ostracoda, and Copepoda, have simple, perhaps basic, patterns of behavior. Large forms, as exemplified by crayfish, crabs, and hermit crabs (Malacostraca), show complex social behavior and communication systems. Even truly social crustaceans have recently been discovered, forming closed societies, comparable to some of the more primitive societies of social insects like wasps. A number of crustaceans, belonging to different subclasses, have become parasites with reduced but highly specialized behavior patterns, and some of these have developed communicatory mechanisms not only affecting behavior but also development. It is thus difficult to draw a clear-cut picture of crustacean communication. However, since most studies have involved the Malacostraca, this chapter will mainly deal with this subclass.

METHODS OF COMMUNICATION

Communication signals may be transmitted by chemical, mechanical, acoustical, and optical means, or by combinations of these.

Chemical Communication

The use of chemical signals is probably the basic method of communication in crustaceans. Combined with tactile stimuli, it may govern the behavior in sexual recognition and pair formation in most of the Entomostraca and even in most Malacostraca. It has been substantiated, however, only in a few cases.

In barnacles, a chemical has been shown to induce settling and crowding in cypris larvae. This substance is contained in the shells of settled barnacles and acts quite specifically, for cyprids of *Balanus balanoides* react to the "settling factor" of conspecific adults to a greater extent than to that of other species. This factor is a protein probably similar to the arthropodin in the cuticle of other arthropods. Cypris larvae react only to surfaces coated with this substance, not to solutions of it (for references, see Frings and Frings, 1968). The function of this substance is to induce crowding, which may be necessary in sedentary animals to ensure cross fertilization.

Another substance is released by well-nourished barnacles, and this induces the newly hatched nauplii to become active and to leave the mantle cavity. It thus ensures that nauplii emerge when food is available.

Chemical stimuli are also responsible for aggregation in terrestrial isopods such as *Oniscus, Porcellio,* and *Armadillidium.* These animals react to the odor of conspecifics, especially under dessicated conditions (Kuenen and Nooteboom, 1963).

Pair formation and mating are in many crustaceans initiated by chemical signals. The substances by which males recognize receptive or precopulatory-molt females may be located on or in the surface of the female's cuticle or they may be released. The sense employed by the male is contact chemoreception in the former, but distance chemoreception in the latter. The factors involved have not been identified.

In the first case a male will recognize a receptive or potentially receptive female only after physical contact with her exoskeleton. This has been reported for the crabs *Callinectis sapidus* and *Carcinus maenas* and for the shrimps *Leander squilla*, *Palaemonetes vulgaris,* and *Pandalus borealis* (for references, see Frings and Frings, 1968), and proved for the hermit crab *Pagurus bernhardus* (Hazlett, 1970) and the wood louse *Porcellio dilatatus* (Legrand, 1958). In *Pagurus* the male immediately responds by grasping when his chelipeds or legs have touched a receptive female. This response is even elicited when the female has been wrapped in a piece of cloth, preventing adequate visual stimuli. In the wood louse *Porcellio* the male's ability to detect a receptive female is lost when the last segment of the antenna is removed. Contact chemoreception is also responsible for the recognition of family members in the social isopod *Hemilepistus* (see below).

Distance chemoreception in pair formation has been reported for the copepod *Labidocera aestiva,* some talitrids (Amphipoda), some mysid shrimps (Clutter 1969), the lobster *Homarus americanus,* the snapping shrimp *Synalpheus hemphilli,* and the crab *Hemigrapsus oregonensis* (for references, see Frings and Frings, 1968). It has been proved only by Ryan (1966) for the crab *Portunus sanguilentus.* In this crab, a chemical is present in the urine of receptive females. Males placed in water previously occupied by such a female exhibit searching and courting behavior.

Sex determination in a few crustaceans is phenotypic and strongly influenced by the presence or absence of conspecific animals. In some species of the parasitic isopod family Bopyridae the first larva to arrive at its host will develop into a female. Larvae that arrive later will settle on the female and transform into males (Reinhard, 1949; Reverberi and Pitotti, 1942). A complex sex determination has been found in the tanaid *Heterotanais oerstedi* (Bückle-Ramirez, 1965). Females may transform into males, and larvae kept together with an adult male or female will develop into females or males, respectively. The mechanisms involved in sex determination in these cases are unknown but it seems most probable that pheromones play an essential role.

Mechanical Communication

Communication by tactile stimuli is probably very important in many crustaceans; during courting, mating, or fighting, male and female tap or touch each other with their antennae, chelipeds, or legs. Mating and agonistic behavior involving such movements, in most cases combined with chemical or visual signals, have been described for numerous crustaceans of different orders. However, careful analysis of the communicative value of such movements are scarce. Usually it is difficult, if not impossible, to separate such stimuli from other communicative actions. Hazlett (1970) studied the responses of hermit crabs to tactile stimuli. These crabs can often be observed to move one or both cheliped mani out and back through an arc of about 30°. This "flicking" act is most common when a crab, which has withdrawn into its gastropod shell, is grasped by another crab. It thus probably represents a rather unspecific defense movement.

Hazlett was able to induce flicking in *Pagurus bernhardus* by touching various parts of the chelipeds or legs with a glass rod. He also showed that flicking is the response to tactile stimuli when the crab is withdrawn in its shell. It is inhibited by visual inputs and cannot be elicited when the crab is in the walking position with eyes uncovered. Flicking is further influenced by previous experience with an opponent. Crabs that have withdrawn in response to the aggressive display of another crab of equal size are much more likely to execute flicking than animals that have had an encounter with a much larger opponent. Another behavior elicited by tactile input is the "dislodging shaking." This is the movement performed by a hermit crab on whose shell a conspecific crab is crawling. It causes the offending crab to crawl off the shaking crab's shell or it dislodges the offender. Dislodging shaking can be elicited by placing a weight on the crab's shell, and this weight has to be larger in larger crabs or in crabs with larger shells.

The role of tactile stimuli in the communication system of fiddler crabs (genus *Uca*) was studied by Altevogt (1957), von Hagen (1970c), and Salmon (1965). Salmon, studying *U. pugilator,* showed that sexual discrimination depends on tactile stimuli during the night. A male gently touched with a grass leaf responds with courtship behavior, but more intense contact triggers aggressive display. Males of *U. vocator,* like those of other species, respond to dummies prepared of cardboard, but the response is greatly increased after the model has touched the male.

Communication by tactile stimuli, combined perhaps with chemical signals, is probably most important in many nocturnal and cavernicolous crayfish, isopods, and amphipods, but none of these have been investigated. The deep-sea galatheid *Munidopsis polymorpha* probably uses another, perhaps more common method of mechanical communication. When two animals approach each other they perform waving or trembling movements with their chelipeds that may cause water disturbances detectable over short distances. It seems that this behavior is a courtship display (Jakob Parzefall, pers. comm.).

Acoustical Communication

The use of acoustic signals is another method of mechanical communication. Many crustaceans are able to produce sounds. The methods by which these are emitted vary in different species, and one species may employ two or three different methods. For a review see Guinot-Dumortier and Dumortier (1960).

Hissing or rattling sounds may be produced by rubbing the second article of the antennae against the edge of the rostrum in spiny lobsters (*Panulirus argas, P. guttatus:* Moulton, 1957; Hazlett and Winn, 1962), by friction of the walking legs against the carapace in several species of freshwater crabs *(Potamon),* or by rubbing the maxillipeds against each other in the freshwater crab *Pseudothelphusa garmani* and in terrestrial hermit crabs *(Birgus, Coenobita).* These are stridulatory sounds. Similar sounds are produced in other crabs (several *Ocypode* species, *Sesarma angustipes, S. ricordi*) by forcing respiratory water through the exhalation opening beneath the epistome (burbling) (von Hagen, 1968). Spiny lobsters have been observed to respond to the stridulatory sounds of conspecific cage mates. In the crabs, however, these types of sounds are not used in intraspecific encounters, and their function may be to deter predators.

The most conspicuous sounds produced by crustaceans are those made by the snapping shrimps *Alpheus* and *Synalpheus.* In these shrimps, one of the chelipeds is enlarged and bears a huge chela. When widely opened a pair of smooth disks are held together by the cohesive forces of water. This allows the closer muscle of the claw to generate a large amount of tension before these forces are overcome. A rapid closing is thus facilitated (Ritzmann, 1973).

The claw with its specialized tubercle and depression on opposite parts (Volz, 1938) thereby produces a loud crack and a jet of water. The function of this is still uncertain, but Hazlett and Winn (1962) and Nolan and Salmon (1970) have shown that snapping plays an important role in agonistic encounters of homosexual individuals. Since the snapping frequency is increased at dawn and during the night, one of its functions may be to produce agonistic signals when the visual signal of the opened claw is not effective. A similar crack is emitted by the mantis shrimp *Gonodactylus oerstedii.* It is produced by striking the raptatorial appendage against something hard.

Whereas in the cases cited above, the communicatory role of sound production is still uncertain, recent studies by Altevogt (1966), von Hagen (1962, 1970), Salmon (1965, 1967, 1971), and others have indicated that acoustic communication is very important in terrestrial and semiterrestrial crabs of the families Ocypodidae and Grapsidae. Indeed, it has been shown that species of fiddler crabs and ghost crabs are very sensitive to substrate-borne vibrations, and *Ocypode quadrata* seems even able to detect airborne sounds of about 3 kHz (Horch and Salmon, 1969). The sense organs employed are the vibration receptors in the joints of the walking legs (Salmon and Horch, 1972, 1973). These crabs are able to use as sources of information a multitude of vibrations originating from conspecifics and perhaps from predators, too.

Thus, the substrate-borne vibrations produced by a rapidly fleeing *Uca tangeri* act as an alarm signal, inducing other crabs to retreat into their burrows or to stay there if already hidden (Altevogt, 1966). Another acoustic by-product is the sound produced by a male waving its large cheliped. At the end of each waving act the claw is moved downward and applies a vibration impulse to the substratum. This is perceived by hidden crabs and probably recognized by its rhythm, which, of course, is identical to the waving rhythm. It induces the crabs to leave their burrows and to start waving also. It may also be responsible for synchronous waving in fiddler crabs out of direct sight of each other (Gordon, 1958).

Besides these vibrations, which are epiphenomena of other (e.g., locomotory) activities, direct production of vibratory signals occurs in ghost crabs (genus *Ocypode*), fiddler crabs (genus *Uca*), and some other Ocypodidae (e.g., *Dotilla*), and in some Grapsidae. Though these are probably perceived as substrate-borne vibrations, they cause airborne sounds detectable by the human ear. The methods by which these sounds are emitted are different. A number of species drum on the substratum. "Honking" sounds are produced by several Uca species (e.g., *U. mordax* and *U. burgersi*) by convulsions of the large cheliped. Striking the walking legs against the ground also aids the sound production. Other species (e.g., *U. tangeri, U. pugilator*) push the base of the enlarged claw against the substrate or against one leg (percussion), thus emitting the well-known "rapping" sounds. *U. thayeri* is able to use both methods of sound production (von Hagen, 1973b). Drumming by striking the claws against the substratum has also been observed in *Dotilla blanfordi* and in *Ocypode quadrata. Ocypode ceratophthalmus* is able to produce three different types of vibrations (Hughes, 1966): rapping sounds by drumming on the ground, rasping sounds by stridulation, and hissing sounds by burbling (see p. 000). Stridulation is achieved by rubbing a ridge of tubercles on the inner side of one claw against a tubercle on the second article of the same appendage. The grapsids *Sesarma rectum* and *S. curacaoense* beat one claw against the other, which is placed on the ground (von Hagen, 1967).

The best-known cases of acoustic communication in crustaceans are those of *Uca tangeri* and some other *Uca* species. In *Uca tangeri* two different "rapping" signals are emitted by the males (Fig. 1). Short drumrolls, consisting of one to three pulses and repeated at intervals of about one second, represent the spontaneous activity, corresponding to the low intensity or spontaneous waving. Longer drumrolls, consisting of seven to twelve pulses, are produced by males sensing the substratum-borne vibrations of an approaching conspecific animal. Altevogt (1966) was able to simulate these vibration signals by drumming with his fingers on the substratum. This elicited typical behavior providing the vibrations were of the correct intensity and rhythm. When drumming near a male's burrow, the male answered, then appeared, took up an aggressive posture, but quickly disappeared at the sight of the author. When drumming near a female's burrow, the female appeared and, as long as her eyes were concealed, even allowed pulling on her leg and simulated carapace feeding, both components of the courtship behavior of this species.

Tropical mud flats and mangrove areas are usually inhabited by more than one species of *Uca* and *Ocypode,* all of which may produce sounds. These vibration signals are of special importance in species that are also active during the night and in those that inhabit the uppermost intertidal areas, with grassy vegetation reducing visibility. Salmon and Atsaides (1968) found that most *Uca* species, including *U. pugilator,* produce only one type of drumroll. The sounds of different species vary in the number of pulses, the time intervals between pulses, and the time intervals between consecutive drumrolls (Figs. 1, 2). All of these probably contribute to species recognition and may thus serve as isolating mechanisms. Even the frequencies of the sounds may be different. The rapping sounds of *Uca pugilator* contain maximal energies between 600 and 2,400 Hz, whereas those of *U. rapax* have maximal energies between 300 and 600 Hz. Whether these differences are detectable seems questionable. Salmon (1971) showed that *U. rapax* is more sensitive to vibrations between 480 and 1,000 Hz than is *U. pugilator.*

Visual Communication

Many crustaceans are able to produce light (Harvey, 1961). Indeed, the chemical basis of light production of the ostracod *Cypridina hilgendorfii* is one of the best-investigated examples of bioluminescence (Johnson et al., 1961; McElroy and Seeliger, 1962). In these and other species that emit luminescent clouds, bioluminescence may have no communicatory function but is used to blind predators temporarily and to conceal the animal. The most highly evolved light-emitting organs are found in the Euphausiacea. These organs resemble eyes in having, besides the luminescent layers, lenses, reflecting layers, and mechanisms analogous to eyelids to turn off the light. But even in these, the function of bioluminescence is unknown. Countershading is one of the possibilities suggested (Nicol, 1962; Clarke, 1963). Reports that shoals or swarms of euphausids often flash simultaneously only suggest that light production is seen by other members of the swarm and may, at least, be used to synchronize the animals. It may also serve to prevent dispersion of the animals and to facilitate species recognition.

Many crustaceans with well-developed eyes use conspicuous postures or movements, often combined with the display of brightly colored body parts for communication, and some even change the structure of their environment in the same context.

Interpretation of body postures or movements as displays with communicatory function must, however, be made with caution. Species that are active during the night also adopt certain postures during intraspecific encounters. Heckenlively (1970), for example, conducted a statistical study of body postures during agonistic

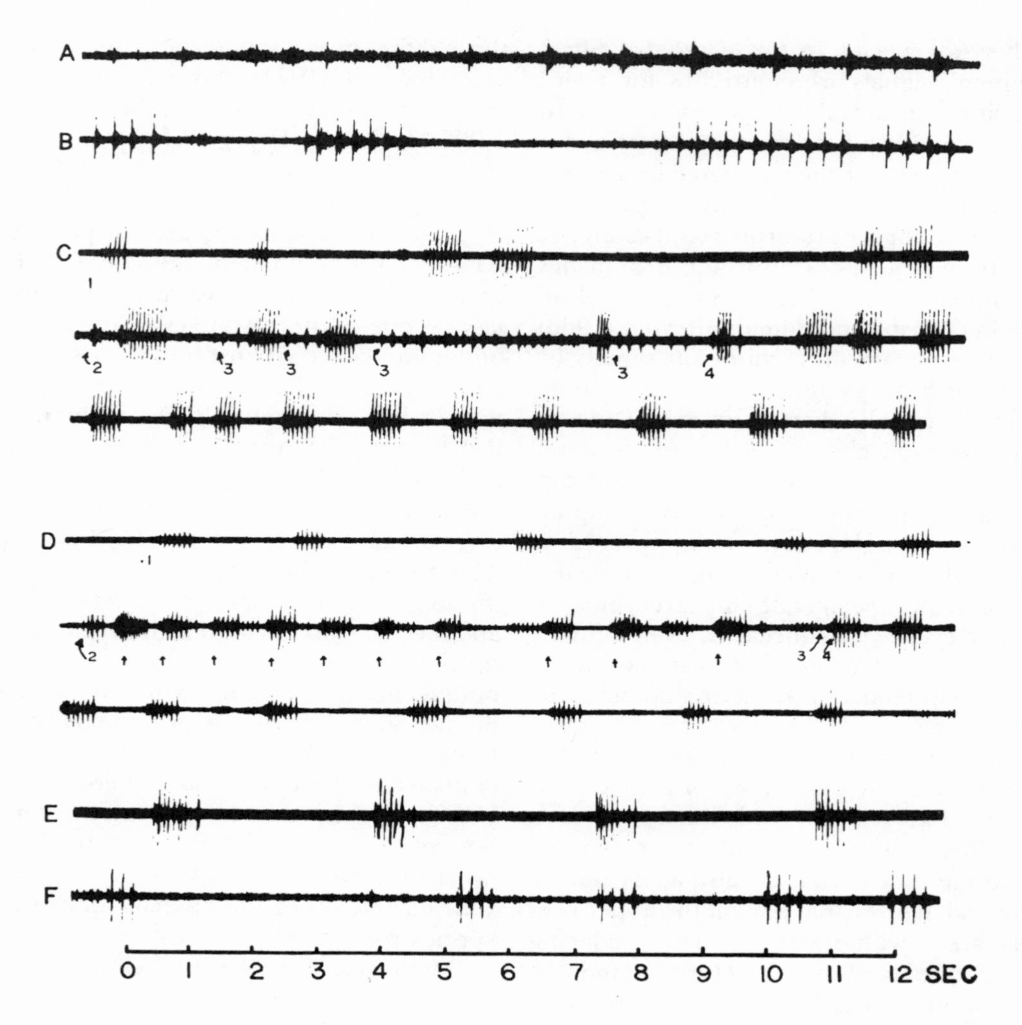

Fig. 1. Oscillograms of rapping sounds of different *Uca* species. A. *Uca tangeri,* short drumrolls, each consisting of one pulse. B. *Uca tangeri,* long drumrolls. C. *Uca pugilator*, response of a male to tactile stimulation by a female. 1 = before contact, 2 = female touches male, 3 = "stamping" sounds produced by walking, and exploratory "flicking" movements of the walking legs against the substratum, 4 = rapid series of sounds produced after female has left the area, gradually decreasing toward spontaneous rate. D. *Uca pugilator*, response of a male to playbacks of sounds from another stimulated male. 1 = spontaneous rate, 2 = first sound in playback series. Arrows between 2 and 3 indicate beginning of each sound produced by the test male. 4 = sounds produced after playback, with gradual return to spontaneous rate. E, F. Nocturnal sounds produced by undisturbed males of *Uca spinicarpa* and *Uca speciosa*, respectively. (From Salmon and Atsaides, 1968.)

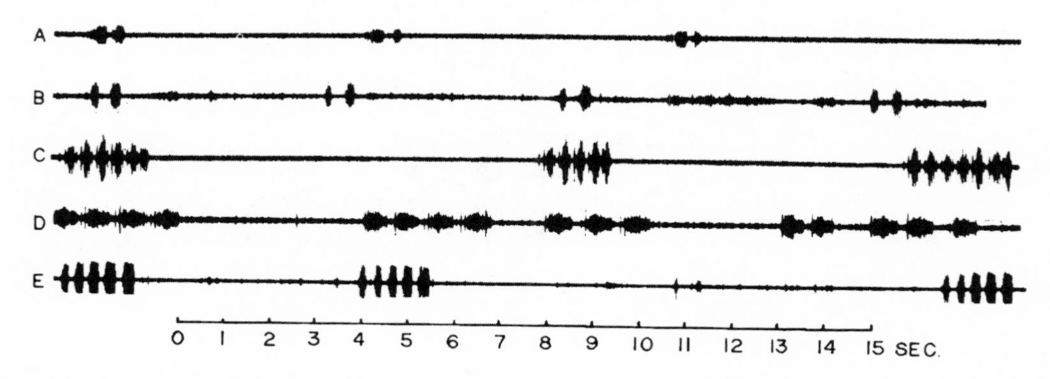

Fig. 2. Oscillograms of sounds produced by lone male fiddler crabs at night. A. *Uca pugnax*. B. *Uca virens*. C. *Uca longisignalis*. D. *Uca mordax*. E. *Uca rapax*. (From Salmon and Atsaides, 1968.)

interaction in the crayfish *Orconectes virilis* and came to the conclusion that "antennal position seems to be important in crayfish aggression, both as an indicator of the aggressive state of the individual and as a threat display to the opponent." It is not known, however, whether a crayfish is able to recognize the antennal position of its opponent and what features indicate its aggressive state. It was shown by Bovbjerg (1956) that blinded crayfish also engage in aggressive interactions.

To settle this question, dummies can be used. Von Hagen (1962, 1970c), studying *Uca tangeri* and *U. vocator,* conducted experiments with different models, crabs, and pieces of cork or cardboard of different forms and colors to study the features that elicit certain agonistic or courtship displays. Similarly, Hazlett (1969c) studied the communicatory role of cheliped presentation and leg postures in agonistic interactions of hermit crabs, using dummies prepared from dried specimens.

Visual displays have been most extensively studied in several hermit crabs (Hazlett, 1966a, 1966b, 1968a, 1968c, 1969a, 1969c, 1972a) and in fiddler crabs and a few other Ocypodidae (Altevogt, 1955, 1957, 1959, 1969; Crane, 1943, 1957, 1958, 1966, 1967; Griffin, 1968; von Hagen, 1962, 1970a, 1970b, 1972a, 1972b, 1973a, 1973b, 1973c; Salmon, 1965, 1967; Salmon and Atsaides, 1968; Warner, 1970; Yamaguchi, 1971).

In hermit crabs, visual displays are employed in agonistic encounters. Though these may lead to precopulatory behavior, this is triggered by contact chemoreception and consists mainly of tapping and stroking movements with the chelipeds or legs and by rocking or rotating the female's shell. The agonistic displays involve raising one or more ambulatory legs and presentation or extension of one or both chelipeds (Fig. 3).

The well-known conspicuous claw waving of fiddler crabs can lead to either aggression or mating, depending on whether a male or a female approaches the performing male. Here, sexual recognition depends on visual cues; the female has two small chelipeds, and in the male one of them is extremely enlarged and sometimes brightly colored.

Usually there are two, sometimes three, different forms of claw waving in one species, corresponding to differences in courting intensity. Spontaneous activity is performed whether other conspecifics are absent or present. It is changed into the courting activity at the sight of

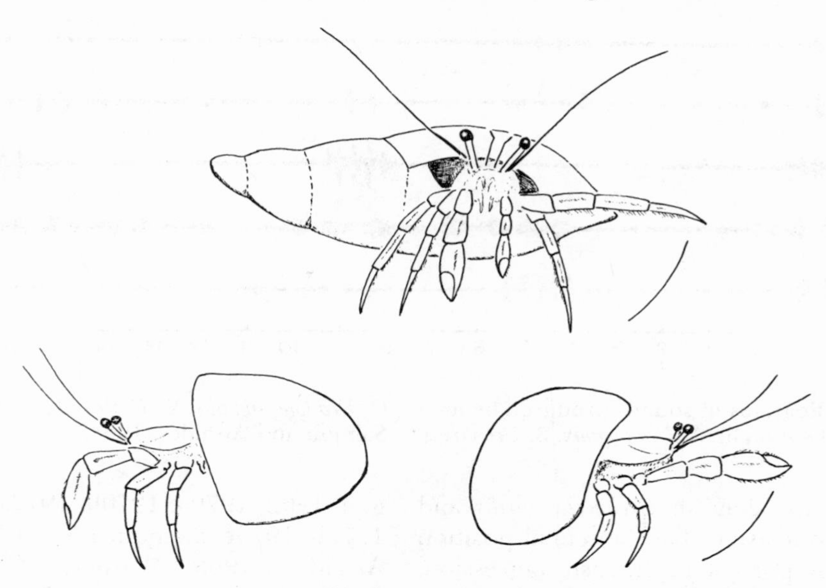

Fig. 3. Leg and cheliped presentations in hermit crabs. Top: A single ambulatory rise in *Clibanarius cubensis*. Left: Major cheliped presentation in *Pylopagurus operculatus*. Right: Major cheliped extension in *Pagurus pygmaeus*. (After Hazlett and Bossert, 1965.)

an approaching female. In this mating signal, the waving movements are performed faster and seem to be more exaggerated.

The waving movements vary in different species and thus they probably contribute to species recognition and isolation (Fig. 4). The waving act may be a smooth, continuous movement or it may be interrupted once or several times. It may be performed either while stationary, in front of the male's burrow, or during short, rhythmic bursts of locomotion. Coloration and color change contribute to the conspicuousness of the display. Many species are cryptically colored at the onset of each low-tide activity period, when the animals are feeding or constructing burrows. Later, when social activities become more frequent, the large claw and sometimes even the carapace and other appendages brighten up in some species.

These activities of different *Uca* species have been well described by the authors mentioned, and even excellent films on several species are now available (Altevogt, 1964a, 1964b; Altevogt and Altevogt, 1968a, 1968b, 1968c, 1968d, 1968e, 1968f; von Hagen, 1972a, 1972b, 1973a, 1973b, 1973c).

Claw waving has also been described for several other subsocial ocypodid crabs, e.g., *Dotilla blanfordi* (Altevogt, 1966) and *Heloecius cordifrons* and *Hemiplax latifrons* (Griffin, 1968). Simpler display patterns, usually cheliped presentation, occur in other Ocypodidae, Grapsidae, and among other crab families (Schöne, 1968; Warner, 1970; Wright, 1968).

Another method of courtship display and territory demarcation has evolved in *Ocypode saratan* (Linsenmair, 1967) and perhaps also in *O. quadrata* (Horch and Salmon, 1969). Like other ocypodid crabs, ghost crabs construct burrows. The sand, however, is not evenly distributed around the burrow by *O. saratan,* but used to build a pyramid. These pyramids attract both

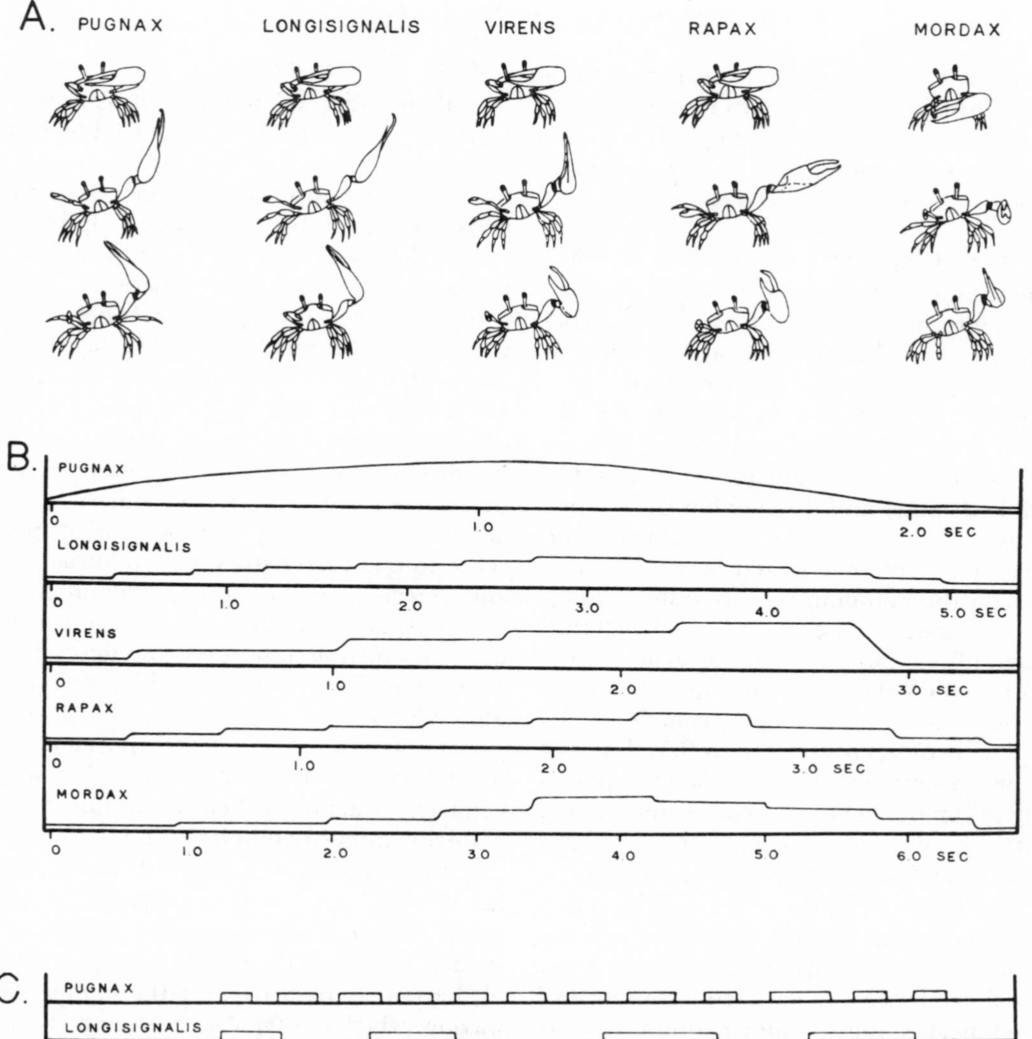

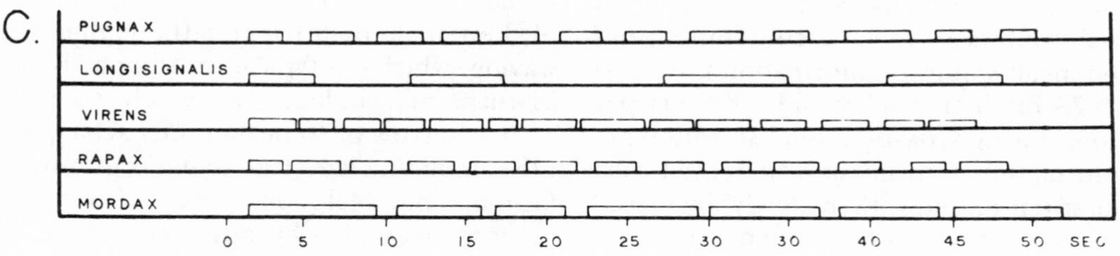

Fig. 4. Waving display shown by five species of fiddler crabs. A. Consecutive stages in the display of each *Uca* species. Top: beginning of the wave, middle: at the point of farthest lateral extension of the major chela, bottom: as the claw is flexed toward the body. B. Vertical position of the tip of the major chela (as it is raised and lowered) as a function of time for the completion of a single wave. Deflections, when present, indicate a "jerk." C. Recordings of a series of waves produced by individual males of each species. Each deflection indicates the beginning and end of a single wave. (From Salmon and Atsaides, 1968.)

females ready to mate and males searching for a territory. This was clearly shown by Linsenmair (1967), who was able, by building a few pyramids, to trigger the development of a new mating colony (see below).

INFORMATION TRANSFER BY DISPLAYING CRABS

Communication between two animals implies transfer of information. That communication occurs is often quite obvious because the animal to which the display is directed responds in a predictable way. The information content per display and the amount of information transferred per display is obviously a function of the number of possible displays and the number of possible responses. It is low if both are low. Hazlett and Bossert (1965, 1966) conducted statistical investigations on the communications system in aggressive encounters of hermit crabs. They found thirteen to fifteen possible behavioral acts, most of them visual displays. The average amount of information transferred ranged from 0.35 bits per display in *Clibanarius triclor* to 0.52 bits per display in *Pagurus marshi.* This latter species tends to be camouflaged by accumulated debris. The rate of information transmission ranged from 0.4 to 4.4 bits per sec in different species. Dingle (1969), studying the aggressive behavior of the mantis shrimp *Gonodactylus bredini,* arrived at similar values. He observed ten behavioral acts, visual displays as well as tactile stimuli, and calculated mean values of information transmission of 0.78 bits per display and 1.82 bits per interaction. During sixty-minute-observation periods, information transmission increased during the first ten minutes, then slowly decreased with the establishment of dominance-subordinate relationships. Similar values of 1.00 to 1.57 bits per interaction have also been found in the spider crab *Microphrys bicornutus* and in hermit crabs by character analysis (Hazlett and Estabrook, 1974a, 1974b). It was further shown that the uncertainty about a given act is reduced by the knowledge of the previous act.

THE MEANING OF DISPLAY

The meaning of spontaneous claw waving or sound production of ocypodid crabs has been discussed. These activities are sometimes performed in the absence of other conspecific animals. Some authors (Hediger, 1933, Peters, 1955; Vervey, 1930) have concluded that this behavior has a territorial function. Others (Altevogt, 1966; von Hagen, 1962) believe that its function is to attract females and that this behavior is therefore a low-intensity mating signal roughly comparable to the appetence behavior of other animals. They reject the term "territorial display" because it would characterize activities of negative social value. However, in many animals of different phyla the same display may serve to mark a territory and to attract females, and a clear-cut distinction and definition of both activities may be impossible. It may therefore be more interesting to observe and describe the responses to the displays by different animals of the community.

The assumption that the spontaneous claw waving and sound production in *Uca* species is primarily a mating signal, indicating the male's maturity and physiological state, is correct. The display is comparable to an advertisement, and the message may be circumscribed by the paraphase: "Here is a mature male." Females ready to mate will be attracted by this signal. The sight of the female then triggers the higher-intensity waving, which is a typical courting activity. This, in turn, may induce the female to enter the male's burrow or to permit the exchange of tactile stimuli that later lead to underground or surface mating. Males in search of a burrow or territory may also be attracted, and ritualized fighting may result. Further, crabs previously hidden may leave their burrows and, if they are males, start waving too.

The pyramid of *Ocypode saratan* is a similar advertisement. The male's mating territory consists of a spiral burrow, the pyramid, a path from the burrow entrance to the pyramid, and a small area surrounding these structures. The male usually waits inside the burrow, and the pyramid alone is sufficient to attract females. The pyramid also attracts males in search of a mating place. The males either try to take over the territory by stridulating in front of the burrow entrance until the owner appears and a ritualized fighting starts, or they may start building a new burrow and pyramid nearby. Thus, at the onset of the breeding season the first male that has a mating burrow and pyramid determines where a mating colony will develop. Each mating territory is occupied for four to eight days, during which time the male does not feed. Thereafter, he migrates to a feeding place, which may be a mile off. After some days of feeding the male searches for a mating place again. The population therefore always contains wandering males, and these may be attracted by the pyramids, throughout the breeding season. Clearly, the pyramid is at the same time a mating signal to receptive females, a territory signal to some males, and a social attractant to other males.

PHYLOGENETIC CONSIDERATIONS

The most conspicuous displays shown by many crabs are cheliped presentations. During agonistic encounters, some crabs and other decapods stretch their chelipeds toward the opponent. Other species maximally unfold their chelipeds (Fig. 5). At the end of the encounter the chelipeds may be folded again, with the claws pointing medially. It is most likely that unfolding and folding of the chelipeds was originally a defense movement, not a display. Defense and attack are most effective if executed with maximally spread chelipeds and the body held high on extended legs, whether this posture is seen by the opponent or not (see, for comparison, the similar defense postures in scorpions and whip scorpions, in which visual communication can be neglected). This behavior is also effective during defense against predators and will probably function in species with poor eyesight and in those that are active at night. In a number of crabs ritualized fighting has developed from this posture; the crabs push against each other with maximally spread chelipeds (for details see Schöne, 1961, 1968).

In many crabs with well-developed eyes this defense posture or the complete movement, or one or the other part of it, has become a signal, a threat display shown in agonistic interactions. In these, the chelae are not stretched toward the opponent but held in a posture that ensures maximal conspicuousness.

A complex courtship behavior is missing in many aquatic crabs; the male rapes the female, to which it has been attracted by contact or distance chemoreception. In some terrestrial crabs, especially fiddler crabs, claw waving is performed as a courtship signal. It is more likely that this movement has evolved from the greatly exaggerated and ritualized threat display, which is still performed in agonistic interactions (Schöne 1968; Wright 1968), than from an exaggerated locomotion (parade) (Altevogt, 1957). It has then triggered the development of the extreme cheliped asymmetry in fiddler crabs.

This type of cheliped presentation has been called "lateral merus" display by Wright (1968). Another type of display called "chela forward" has evolved in several unrelated ocypodid and grapsid crabs. Here, the merus of the cheliped is stretched forward and the chelae point downward (Fig. 5). In higher-intensity displays the chelae may move up and down or even rotate. In *Eriocheir* and *Pachygrapsus* the "chela forward" is used as a courtship display; in other grapsids and ocypodids it is an agonistic display. Wright assumes that this type of display has evolved independently in different groups, first as a mating signal that has subsequently changed its function to an agonistic display. This, however, does not explain the origin of the "chela forward" display. Perhaps the "chela forward" is a cheliped presentation, like the "lateral merus," which started from the feeding posture of those terrestrial

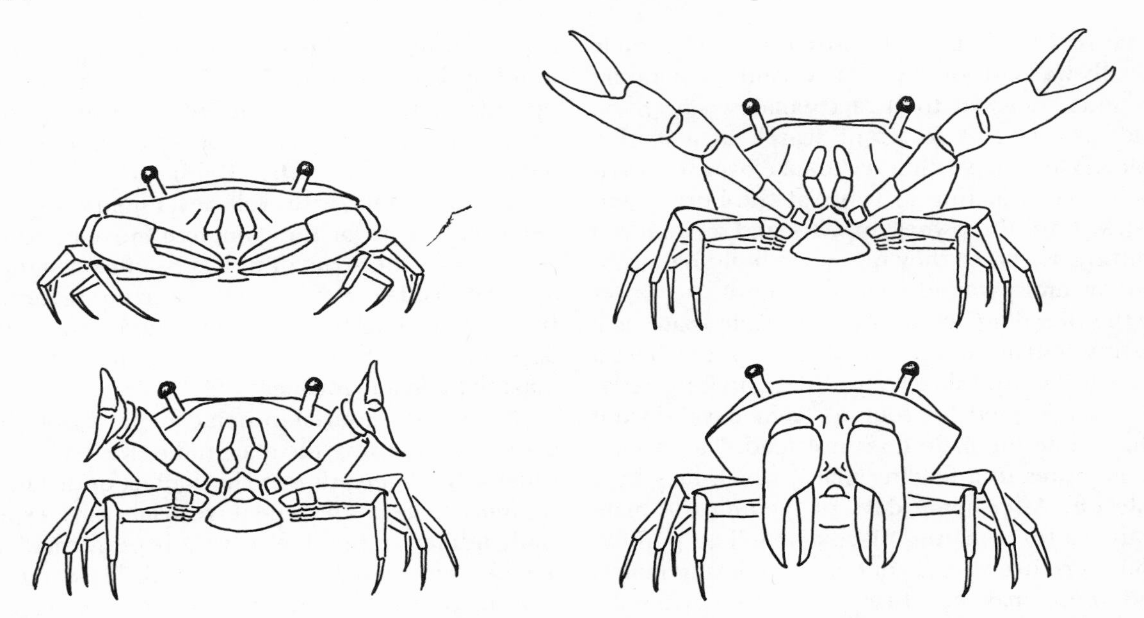

Fig. 5. Cheliped presentations in crabs. Top left: Resting position. Bottom left and top right: Two different forms of "lateral merus" display, with chelae stretched toward a possible opponent and maximally unfolded. Bottom right: "Chela forward" display.

crabs in which the chelae point downward in front of the body and are slowly moved alternately up and down to the mouth parts.

INDIVIDUAL RECOGNITION AND MONOGAMY

Individual recognition has been reported for a number of decapods: the crayfish *Orconectes* and *Cambarellus*, the cleaning shrimp *Stenopus hispidus*, and the hermit crab *Pagurus bernhardus*. It has been observed in the harlequin shrimp (*Hymenocera picta:* for references, see Wickler and Seibt, 1970; Seibt and Wickler, 1972) and in the social isopods *Hemilepistus reaumuri* (Linsenmair, 1971, 1972) and *H. aphghanicus* (Schneider, 1971).

Monogamy has also been reported for a number of decapods (Seibt and Wickler, 1972; Wickler and Seibt, 1970). Some of them live in sponges or burrows (*Alpheus, Synalpheus:* Alpheidae; *Spongicola:* Stenopidae), others in molluscs or tunicates (some Pontoniinae). *Periclimenes affinis* occurs in pairs on the anthozoan *Discosoma.* However, it is not known how long these pairings exist and whether they are facilitated by individual recognition or by the fact that both animals inhabit the same host or tube and each mate independently drives away other conspecifics of its own sex.

In the harlequin shrimp *(Hymenocera picta)* monogamy is due to individual recognition. The pair stays together for months, and if separated the male will select his former mate out of several other females. It is believed that chemical cues are most important in individual recognition.

In the social isopod *Hemilepistus reaumuri* monogamy lasts until the death of one member of the pair. These isopods inhabit some of the deserts and semideserts of northern Africa and Asia. They construct long, deep burrows in the hard desert soil. Each burrow is guarded by one member of the pair, and only the other member is allowed to enter. It is recognized by contact chemoreception and investigated with the antennae. These isopods form family groups that are closed societies, consisting of male, female, and their offspring. The young animals are first fed by food collected by the foraging parent. Later they forage also but freely enter their parents' burrow before noon and at night, when temperatures become unfavorable. The young animals also recognize their parents, but the parents do not recognize each individual young. Inside the burrow the offspring establish a family-specific chemical character that is recognized by the parents; young animals of other parents are devoured. The family odor is a mixture of the individual pheromones of the family. Under experimental conditions young animals of other parents can accept the family character. When kept together with the offspring of one particular pair for some days, during which time the pair has no access to the offspring, this pair will later tolerate the strangers also. However, if some young animals are removed from their parents for a period of weeks they lose the family odor and are subsequently treated as strangers. The source of the pheromone is not known. The individual and family-specific odors are learned, and the adults are able to learn characters throughout the breeding period. At the beginning of the next breeding season (spring), the families disperse and new societies are formed. The development of these behavioral and communications systems enable this isopod genus to live in arid areas that cannot be inhabited by any other crustacean.

ARACHNIDA

The courtship behavior of spiders has aroused interest since antiquity. Many spiders are especially ferocious predators, not hesitating to devour members of their own species. Males attempting to mate, therefore, have to use precise signals in order to be accepted as a mate instead of as a meal. The same problem seems to exist in some of the other arachnid orders, though to a much lesser extent. Mating has been observed in at least some members of all arachnid orders except the Palpigradi, the habits of which are literally unknown. Agonistic behavior, sometimes highly ritualized fighting, has been reported in some of the orders. However, spiders are the only arachnids in which attempts to analyze the communications systems have been made.

SCORPIONS

Scorpions perform complex mating dances. In *Euscorpius* the male grasps the female's chelae and steps back and forth, shows jerking or trembling movements, taps the female's genital area with his forelegs, and stings the female in the articular membrane at the base of her palpal chela (Angermann, 1957; Weygoldt, 1973a). Similar behavior patterns have been observed in other genera (Alexander, 1957, 1958a; Abushama, 1968; Garnier and Stockmann, 1972; Rosin and Shulov, 1963; Matthiesen, 1968). Finally, a spermatophore is deposited and the female is pulled over it. The mating dance seems to stimulate the female to accept the spermatophore. It is not known, however, which of the special movements of the dance (e.g., the sting or the tapping of the female gonopore) are essential signals. Nor is it known whether there is a mutual exchange of signals. To the observer it seems as if in many species the female does not respond to any of these movements and, at the end of the mating dance, is as reluctant as at the beginning. Mating usually takes hours, and the amount of information transmitted per action is probably very low.

How a female is recognized by a male is not known; the observations suggest that contact chemoreception is involved.

Some scorpions, when cornered, produce sounds. The methods of sound production vary in different species (Alexander, 1958b; Rosin and Shulov, 1961). It is not known whether any of these species use sound production as a method of communication.

PSEUDOSCORPIONS

Though not closely related, many pseudoscorpions perform mating dances similar to those of scorpions (Kew, 1912; Vachon, 1938; Weygoldt, 1969, 1970). But these have evolved within the order. Members of the more primitive families have no mating behavior at all. The males produce spermatophores in the absence of females (most families) or after short physical contact with a female, whether receptive or not (some Olpiidae). Females are most likely attracted to the spermatophores by chemical sex attractants. The male *Serianus carolinensis* (Olpiidae) surrounds its spermatophores with signal threads, which make it easier for the female to find the spermatophore and to approach it from the correct side. Even in some of these nonmating species, communication signals between individuals have been observed. A number of species are gregarious (Chthoniidae, Garypidae, Sternophoridae, Cheiridiidae). When one animal is closely approached by another, it may respond by vibrating or shaking movements of the pedipalps and rocking movements of the body. This behavior often spreads through the whole group and results in the usual spacing of the animals. A similar behavior has been observed in some Chernetidae. In *Lasiochernes pilosus* chemical communication is probably also involved. Excited animals produce a strong odor, which seems to stimulate other individuals.

The mating dances of most Cheliferoidea pose the same problems as those of the scorpions. They probably produce special sets of mechanical stimuli, combined perhaps with chemical signals. A courting male of *Withius subruber* (Withiidae), for example, rapidly vibrates his third pair of legs and thus perhaps stimulates the trichobothria on the female's palpal chelae. The Cheliferidae perform mating dances without physical contact of the mates. The males court by jumping or vibrating movements in front of the female, thus probably stimulating her trichobothria, and display their ram's-horn organs (Fig. 6). These carry apical glands (Heurtault, 1972) which most likely emit airborne chemical substances. A female thus courted approaches the male and keeps her palpal chelae close to the tips of his ram's-horn organs.

UROPYGI

Mating dances have also been observed in whip scorpions (Klingel, 1963; Sturm, 1958, 1973; Weygoldt, 1971, 1973b). The male grasps the antenniform legs of the female and rubs their tips with his chelicerae. The female is also stroked and tapped with the male's pedipalps and antenniform legs. Again, it is not known which of these actions are essential signals. The female is finally stimulated and embraces the male's opisthosoma (Thelyphonidae) or flagellum (Schizomidae). The spermatophore is deposited in this position. In *Mastigoproctus,*

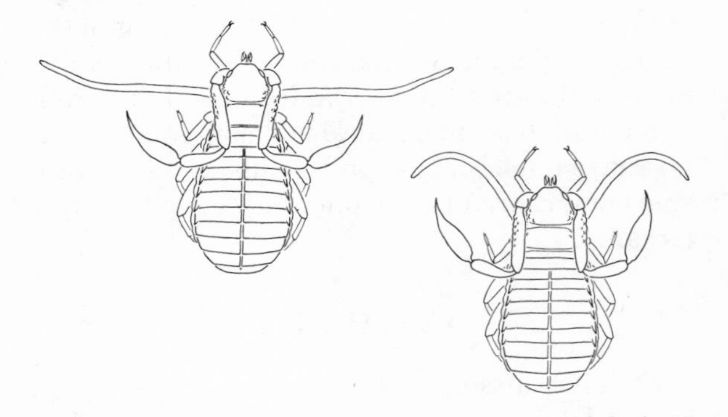

Fig. 6. Courting males of the pseudoscorpions *Hysterochelifer meridianus* (left) and *Hysterochelifer tuberculatus* (right), displaying their ram's-horn organs. (From Weygoldt, 1970.)

spermatophore formation takes two to three hours, during which time the female remains motionless.

Agonistic behavior between males has been observed in *Mastigoproctus.* With pedipalps widely opened the animals push against each other, at the same time tapping each other with the antenniform legs. Soon one animal gives up and performs rocking movements, thereby slowly retreating.

AMBLYPYGI

Whip spiders or tailless whip scorpions possess a complex arrangement of trichobothria on their walking legs, which provide a fine sense of distance mechanoreception, and extremely elongated antenniform first legs covered with mechanoreceptors and chemoreceptors (Foelix et al., 1975). During courting, a male sitting close to a female rapidly vibrates one or both antenniform legs *(Charinus, Tarantula)* or taps the female in a characteristic rhythm *(Heterophrynus = Admetus)* (Klingel, 1963; Weygoldt, 1972a, 1972b, 1972d, 1972f, 1974). Occasionally the male leaps or steps forward, his raptatorial pedipalps extended. The receptive female thereupon retreats but approaches again immediately afterward and stretches one antenniform leg toward the male. Thus, there is probably a mutual exchange of signals. After deposition of the spermatophore the male alters the movements of his antenniform legs or performs a regular dance, and the female approaches and steps over the spermatophore. However, the exact communicatory role of these movements is not known.

Male and sometimes even female whip spiders perform highly ritualized fighting when encountering another animal of their own sex and species (Weygoldt, 1972c, 1972e). In some species these are probably used to defend territories. After some irregular mutual tapping the combatants move apart until each animal can just reach its opponent with one antenniform leg (Fig. 7). Performing slow tapping or stroking movements, the animals coordinate with each other. Suddenly they stop and, with pedipalps widely opened, each steps forward. A short but vehement pushing-and-pulling struggle follows, after which the animals separate. Later a characteristic movement of one antenniform leg of the dominant animal is sufficient to induce retreat of the subordinate.

Fig. 7. Ritualized fighting in the whip spider *Charinus brasilianus*. Above: Prelude, each animal taps its opponent with one antenniform leg. Note characteristic posture with one pedipalp folded, the other unfolded. Below: The animals rush against each other with widely unfolded pedipalps. (From Weygoldt, 1972c.)

Some whip spiders are able to stridulate (Millot, 1949; Shear, 1970a), but the communicatory role of sound production is unknown.

ARANEAE

This is the largest order of the arachnids (except the mites) and the one with the highest degree of diversity. Numerous authors have studied the mating habits of different spiders (for a summary and a bibliography, see Platnick, 1971). Spiders have evolved a variety of communication systems in different families and species. Communication is achieved by chemical, mechanical, and visual signals.

Chemical Communication

The use of chemical signals is the most primitive means of communication in spiders. It has probably been retained in some form in all families. In many spiders courtship starts after the male has accidentally touched a female. This has been observed in theraphosids, in numerous haplogyne, and in some entelegyne spiders. Though conclusive evidence is scarce, it seems most likely that contact chemoreception is involved. For example, autotomized female appendages no longer elicit courtship responses from males when washed in ether and dried (Kaston, 1936). The tarsi of the pedipalps and the first pair of legs carry many chemoreceptive hairs (Foelix, 1970a, 1970b). Contact chemoreception also aids in the finding of a female by a male. Bristowe and Locket (1926) and many others have noticed that male lycosid spiders start courtship behavior when placed into a container previously occupied by mature females. In this case it is some factor of the drag line of the female that elicits the male behavior. The factor is species-specific. Males of the European species of *Trochosa* react only to the drag lines of conspecific mature females (Engelhardt, 1964). The substratum over which a female *Pardosa lapidicina* has walked does not trigger any courtship behavior if the female was prevented from producing silk (Dondale and Hegdekar, 1973). The presumed sex pheromone is quickly inactivated by water. This ensures that only fresh drag lines are attractive to males. The factor is not a necessary stimulus in the courtship behavior of lycosid spiders; it ensures that the sexes meet, but it can be bypassed by direct contact of a female by a male.

Similarly, the webs of mature females elicit courtship in conspecific males. In the theridiid spiders *Steatoda bipunctata* and *Teutana grossa,* a web washed in water and ether is no longer attractive to males (Gwinner-Hanke, 1970). In the agelenid *Tegenaria atrica* (K.-G. Collatz, pers. comm.) the factor is soluble in petroleum benzine and can thus be transferred to another, previously unattractive web under experimental conditions. Contact chemoreception is also the most important means of mutual recognition in social spiders (see below).

An airborne sex attractant is emitted by mature females of *Crytophora cicatrosa* (Blanke, 1973b). Its production starts a number of days after the female's final molt, reaches a peak at about the twentieth day, and thereafter slowly decreases. It also stops two days after copulation of the female. Crane (1949b) also found some evidence for the action of airborne pheromones in Salticidae.

Mechanical Communication

When a male spider has found the web of a mature female he starts emitting courtship signals. Agelenidae and Amaurobiidae drum on the web, and the female, if receptive, waits motionless or responds by shaking her web. Male orb weaver spiders (Araneidae, Tetragnathidae) at first remain at the periphery of the female's web and start plucking certain threads with their legs. In many species the males produce mating threads connected to the female's web and pluck or shake these in a certain rhythm, often combined with vibrating movements. These signals cause movements and vibrations of the web that can be perceived by the female and that are quite different from the more irregular struggling movements of a prey organism entangled in the web. The females respond by shaking or tapping on the web. There is clearly a mutual exchange of signals, which finally results in the female approaching and accepting the copulatory position. The male *Cyrtophora citricola,* for example, plucks the mating thread with his third legs, and the female, if receptive, grasps the mating thread with her third legs. Thus the male, when approaching, is guided directly toward the female's epigyne (Blanke, 1972). The plucking or vibrating movements applied to the mating threads or webs are probably species-specific. Males of other species may be chased away or devoured by the female. In fact, they seldom enter the web of another species because of different chemical sex attractants. A very interesting observation, however, is that of Czajka (1963) on *Ero furcata.* The female of this spider-eating spider is capable of imitating the courtship signals of the male of *Meta segmentata* and uses this to approach and capture the females of this species.

Similar plucking and shaking signals, sometimes combined with stridulatory vibrations, have been observed in Theridiidae, Linyphiidae, and Dictynidae. Many linyphiid males, instead of attaching mating threads, bite away threads or whole parts of the female's web (van Helsdingen, 1965). Males of *Linyphia triangularis* remain in the female's web for some days. During this time they are the dominant member of the pair. If a second male enters the web, threat display and fighting occurs. Several levels of aggressive behavior can be distinguished: approach with abdominal jerking, threat display with forelegs and chelicerae spread and abdominal whirring, and three phases of fighting (Rovner, 1968b).

Tactile stimuli are probably involved in the mating of all spiders. In many species the male, after approaching the female, taps or strokes her body, and in lycosid and salticid spiders the female responds by twisting her opisthosoma to facilitate insertion of the male copulatory organ. Tactile stimuli are also important in the brood care of wolf spiders. Newly hatched spiders settle on their mother's opisthosoma for a number of days. They fail to do so if the female abdomen has been shaved or covered with cloth. Hairy surfaces elicit attachment behavior, and the young spiders are more likely to settle on the back of a freshly killed spider abdomen of another species than on their own, shaved mother (Engelhard, 1964; Rovner et al., 1973).

Acoustical Communication

Many spiders are able to produce sounds, either by drumming on the substratum or by stridulation. Stridulatory structures have evolved on different parts of the body (for details, see Chrysanthus, 1953 and Legendre, 1963). In most cases the function of sound production is unknown. Some theraphosids, when cornered, produce a loud hissing sound resembling that of snakes, and it seems likely that this repels possible predators. Legendre, following Berland (1932), rejects this possibility and assumes that sound production is a by-product of other activities. His arguments, however, are not convincing.

Clearly, sound production was originally a by-product of movements of the palps, legs, or the opisthosoma. Such movements are performed by many spiders during courtship, and some of these use the vibrations caused by such movements as communicatory signals. This, in turn, probably triggered the evolution of special sound-producing structures and behavior mechanisms. Receptors may be the slit sense organs on the legs. In *Cupiennius salei* a single large slit sense organ on the tarsus of the legs is sensitive to surface vibrations and even to airborne sounds (Barth, 1967), and the lyriform organs of *Achaearanea tepidariorum* are sensitive to vibrations of the substratum (Walcott and van der Kloot, 1959). (For a discussion of sound perception in spiders see also Chryanthus, 1953; Legendre 1963; Frings and Frings, 1966; Liesenfeld, 1961; Walcott, 1963). The auditory function of the trichobothria, which are sensitive to minute air movements (Görner, 1965; Görner and Andrews 1969), is still a matter of speculation.

Acoustical communication has been studied in the theridiid spiders *Steatoda bipunctata* and *Teutana grossa* (Gwinner-Hanke, 1970) and in some Lycosidae (Buckle, 1972; Harrison, 1969; Rovner, 1967). It probably occurs in other species also. Although there is some evidence to suggest that the sounds may be perceived directly in the species mentioned above, it is more likely that the vibrations of the substratum or the webs are usually detected. Both theridiids, *Steatoda* and *Teutana,* possess stridulatory organs between prosoma and opisthosoma. The prosoma carries rows of parallel ridges, and the opisthosoma has a row of cuticular tubercles with hairs that can be moved over the ridges. Both species stridulate during courtship. In *Steatoda,* the female responds to the stridulation even when the male is not sitting in the female's web. Sound production is a necessary part of courtship behavior. Stridulation is also used in agonistic behavior between males. Males that are made unable to stridulate do not start courtship. In *Teutana* the communicatory role of sound production is more obscure. Females do not seem to respond to stridulation, and sometimes courtship may take place without any sound production. Fighting males never stridulate.

Sound production in wolf spiders has long been noticed by many naturalists (Rovner, 1967). Courtship in lycosids involves up-and-down movements of the pedipalps and first legs and rapid vibrations of the opisthosoma. From these movements different methods of sound production have evolved. The males of many species use their palps to drum on the substratum, usually on dry dead leaves in the natural habitat. This has been reported for *Cupiennius salei* (Melchers, 1963), *Lycosa rabida* (Rovner, 1967), *L. gulosa* (Lahee, 1904; Harrison, 1969), *Schizocosa avida* (Buckle, 1972), *S. crassipes* (Kaston, 1936), and several species of *Pardosa* (e.g., Hallander, 1967; Dumais et al., 1973). Other species, e.g., *Alopecosa aculeata,* scrape the surface of the ground with their pedipalps (Buckle, 1972). Vibrations of the opisthosoma may produce sounds when the abdomen hits the ground. This has been observed in *Alopecosa pulverulenta* (Bristowe and Locket, 1926), *Lycosa gulosa* (Harrison, 1969), and *Hygrolycosa rubrofasciata* (O. von Helversen, unpublished results) (Fig. 8). Males of this latter species possess an interesting sound-producing apparatus. The cuticle of the ventral side of the opisthosoma is thickened and hardened and covered with specialized knobbed hairs. In *Pardosa fulvipes,* on the other hand, a stridulatory apparatus is present, consisting of cuticular ridges on the surface of the book lung covers and specialized hairs on the coxae of the fourth legs (Kronestedt 1973). Whether this is used in courtship behavior is not known.

Oscillograms or sonagrams of the sounds of *Lycosa rabida, L. gulosa, Schizocosa avida,* and

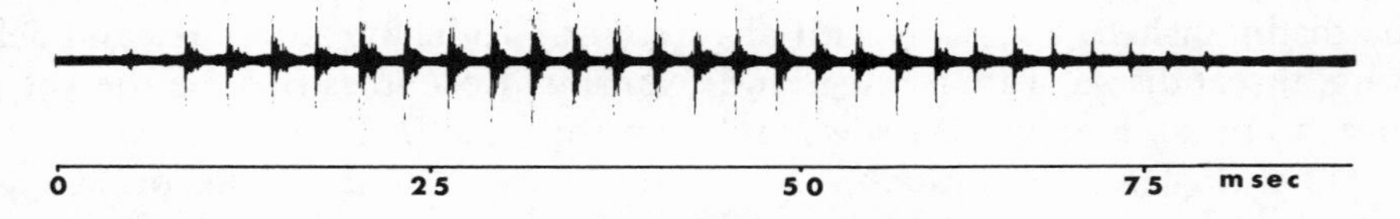

Fig. 8. Drumroll produced by a courting male of the wolf spider *Hygrolycosa rubrofasciata*. (Courtesy of O. von Helversen.)

Alopecosa aculeata have been published by Rovner (1967), Harrison (1969), and Buckle (1972). *Lycosa rabida* is the species that has been studied most intensively. Sounds are produced during courtship and agonistic behavior (threat display). Courtship signals consist of several brief bursts of pulses followed by a long, continuous train of pulses with a mean of 29 pulses per sec. Each pulse has a duration of 3 to 6 msec. The courtship sound ends abruptly and is followed by a period of silence. Females respond to the courtship sounds by leg-waving display and approach behavior, even when recorded sounds are played back through a loud speaker. *Lycosa rabida* is active not only under daylight conditions. When a male that has sensed the contact sex pheromone of a female starts drumming, it can be found by a receptive female during the night or in the dense undergrowth of the field. During threat display short bursts of pulses are emitted by the males.

Experiments with palpless males indicate that, under daylight conditions, the acoustic signals are not essential. Females readily respond by leg waving and will approach a male performing only the visual courtship displays.

Visual Communication

The use of visual signals in courtship and agonistic behavior has been studied in some lycosid spiders and in Salticidae, including Lyssomaninae, and in Oxyopidae.

Many wolf spiders are active during the day and, when sensing the contact pheromone of a female or when seeing a femalelike spider, will perform elaborate courtship movements involving waving of the first legs and pedipalps. In *Pardosa* and *Lycosa* species, especially, these appendages are conspicuously colored, often black. In *Lycosa rabida,* the courting male assumes a distinct posture with forelegs flexed. Then the palps, in alternation, are raised and waved in a circular path (palpal rotation: Kaston, 1936). After performing a number of palpal rotations, one of the forelegs is lifted and extended in a tapping movement, simultaneously with opisthosomal vibrations and palpal drumming. These courtship activities are alternated with periods of inactivity during which the female responds by waving her forelegs and approaching. Rovner (1968a) showed that palpal rotations are not essential signals, for receptive females also respond to performing palpless males. There is a mutual exchange of signals; in response to the leg waving of the female the frequency of the male's courtship activities increase.

Similar visual signals are employed in *Pardosa.* A male of *Pardosa amentata* stops at the sight of a female and raises one pedipalp. He then makes a step, simultaneously lowers both palps, stops again and raises the other palp. The palps, when lowered, vibrate, and the first legs and the abdomen vibrate at the same time. These are very distinctive movements, and the male walks around the female, step by step, approaches and mounts. In *Pardosa amentata* the approach behavior of the male is thus part of the courtship sequence (Vlijm and Dijkstra, 1966). Similar though different displays have been observed in *P. hortensis* and *P. nigriceps.* In *P. lugubris,* however, the male remains stationary and courts by raising the palps and the forelegs. Both are lowered stepwise; at the same time they vibrate and the opisthosoma also vibrates. After a number of such courtship sequences the male runs toward the female and mounts. There is often more than one species of *Pardosa* in the same environment, and the females are similar in appearance. The differences in male courtship behavior therefore probably contribute to species recognition and function as pre-mating isolating mechanisms.

Agonistic behavior among male wolf spiders can often be observed. Threat displays in *Lycosa rabida* (Rovner, 1968a) and *Schizocosa crassipes* (Aspey, 1974) involve a number of different movements of stepping, extension, and raising of the first pair of legs. It appears that these spiders have evolved a communication system similar to that described for some crabs but, until more evidence is available, this conclusion may be premature. Under laboratory conditions a dominance-subordinate relationship is soon established, and the subordinate male assumes a submission posture when attacked by a dominant male.

Males of the European species of *Trochosa* also perform courtship movements that include lifting and extending of their forelegs, vibrations of the palps, and quick up-and-down movements of the opisthosoma. However, these species are usually not active during the day, and copulation takes place even when the females are blinded. Engelhardt (1964), therefore, assumes that the courtship displays in these species do not have a communicatory function but are a by-product indicating the physiological state of the male or have the function of self-stimulation. One might conclude, then, that the courtship movements of the *Trochosa* species are vestigial behavioral patterns inherited from wolf spiders that were more active during the day. However, the displays are different in the four species, and it is not known whether they can be sensed by the female by means of her trichobothria or other mechanoreceptors.

Salticid spiders rely entirely on vision in their courtship and agonistic behavior. The males of most species are brightly colored and display the colored parts of the body in a species-specific manner (Bristowe, 1929; Drees, 1962; Legendre and Llinares, 1970; Plett, 1962). Usually different courtship and threat display movements are shown. Waving movements may be performed with the pedipalps, the first, or even the third legs (Figs. 9 and 10). The males may rock or jump sidewise in front of the females or perform other conspicuous movements. Receptive females usually wait nearly motionless. Crane (1948, 1949) has published a wonderful synthesis of salticid behavior, including the generalized scheme of salticid courtship display (1949b) shown in Table 1.

THE EVOLUTION OF DISPLAY

Mechanical, acoustical, and visual displays probably serve for species recognition and mutual stimulation (for discussions on this subject see Peckham and Peckham, 1889, 1890; Montgomery, 1910; Berland, 1912; Bristowe, 1929; and Crane 1949b). The display movements have probably evolved from activities with other functions. It is likely that the plucking or shaking signals of a male web spider arose from the orientation movements with which a spider investigates whether there is another spider or animal in the web. Similarly, visual displays most likely evolved from tactile, chemical, and tactochemical orientation movements and, perhaps also, from defense postures. Many spiders, especially theraphosids, when walking in a strange environment, perform waving movements with their first legs, and defense posture with the first or first and second legs raised is easily assumed. These very conspicuous postures, together with orientation leg waving, do not play a communicatory role since theryphosids are not visually oriented. However, with the acquisition of better vision these behavior patterns could have become communicatory signals that subsequently become more conspicuous through exaggeration of the whole movement or parts of it.

SOCIAL SPIDERS

Social phenomena exist in a number of spiders belonging to different families (Kullmann, 1968, 1969c, 1972; Shear, 1970b). The social phenomena involve mutual tolerance, interattraction (that is, the fact that individuals are attracted and remain close to conspecifics), and cooperation. Temporary societies are established in some spiders in which the offspring of

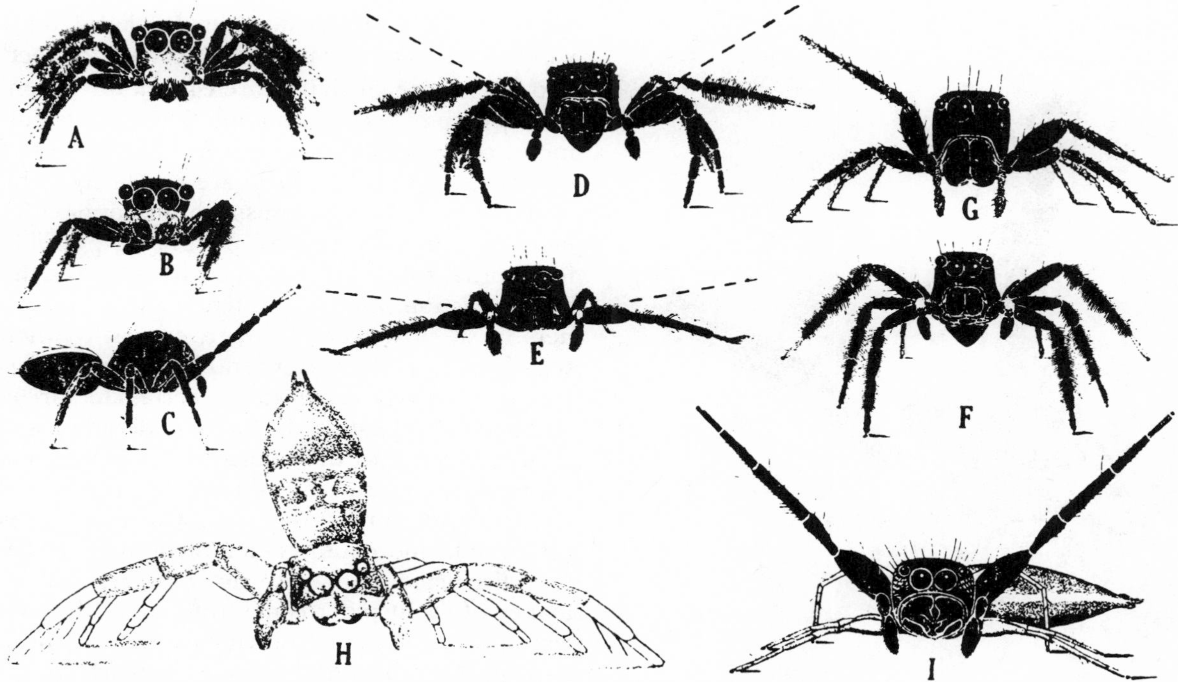

Fig. 9. Examples of display motions of salticid spiders. A. *Corythalia xanthopa*, threat; B, C. same, courtship; D. *Corythalia chalcea*, threat (dotted lines indicate peak position of legs during courtship); E. *Corythalia fulgipedia*, threat; F. same, courtship; G. *Mago dentichelis,* courtship; H. *Gertschia noxiosa*, courtship; I. *Ashtabula furcillata*, courtship. (From Crane, 1949b.)

a female remain with their mother and are nourished by her until they are able to take their own prey. This has been studied in the agelenid *Coelotes terrestris* (Tretzel, 1961a, 1961b), the theridiids *Theridion notatum* and *T. impressum,* and in a number of Eresidae (Kullmann, 1969a, 1969b; Kullmann et al., 1972). In *Coelotes terrestris* the young spiders remain in the female's web until they are half grown. They gather around their mother when she has prey. They are attracted by the movements of prey wrapping and feeding, and by stroking her first legs they "beg" for food. A warning signal is used by a female that has been disturbed. A few hard tapping movements on the web induce the young spiders to retreat into the web funnel.

In the subsocial Theridiidae and Eresidae young spiders are nourished by a fluid regurgitated from the female's gut (Kullmann and Kloft, 1969; Kullmann et al., 1972). No begging signals have been observed in these species. In the recognition of the young spiders by their mother, contact chemoreception is involved. The characteristic slow movements of the young spiders usually do not elicit aggressive behavior in the female, but when a young animal moves quickly it is touched and investigated with the forelegs. Conspecific young spiders of other females are accepted; other spiders are eaten or chased away. Tolerance and interattraction disappear later; the young spiders finally devour their dying mother, separate, and become mutually aggressive.

Fig. 10. The courting male of the salticid spider *Saites barbipes* displays its brightly colored third pair of legs. (Courtesy of C. Gack.)

Permanent spider societies have evolved in a number of species of different families (Kullmann, 1968, 1972; Shear, 1970b). The best-investigated species are *Agelena consociata* (Agelenidae: Krafft, 1966, 1969, 1970a, 1970b, 1971, 1972) and some *Stegodyphus* species (Eresidae: Kullmann, 1969a, 1970b; Wickler, 1973; Jacson and Joseph, 1973). Unlike insect and isopod societies, these spider societies are open. New conspecifics and even spiders of related species of the same genus are tolerated and sometimes integrated into the society. Experiments by Krafft (1970b, 1972) and Kullmann and Zimmermann (1972) indicate that mutual recognition is explained by contact pheromones and perhaps also by the surface structure of the spider's cuticle. Conspecific animals placed on a vibrator and thus behaving like prey are touched and investigated with the forelegs but not bitten. The source of the pheromone that inhibits killing of conspecifics is not known. Even pieces of cuticle turned upside down are less likely to be bitten than are prey organisms or pieces of elder marrow. This pheromone is not only responsible for mutual tolerance but probably also for the interattraction, that is, the tendency of the spiders to sit close to each other, especially when separated from the society and in a strange environment. In *Agelena consociata* a certain hierarchy has been demonstrated. Some spiders are less easily attracted to their conspecifics. One such animal alone carries a freshly killed prey animal into the interior of the web. Cooperation is achieved by more indirect interactions. The web is constructed cooperatively, but the performance of each spider is determined not by special signals of other individuals but by those parts of the web already present. Similarly, cooperation in prey capture is a function of the size, intensity, and duration of the struggling movements of the prey organism. The larger and more vigorous the prey the more spiders participate in its capture. When the prey is finally killed, however, only the dominant animal transports it to the young spiders.

OTHER ARACHNIDS

Little is known about communication in other arachnids. Mating preliminaries usually involve tapping or stroking movements or other tactile stimuli. In sopugids the female is raped by the male, embraced by his pedipalps and pinched with the chelicerae. If the assault is vigorous enough the female immediately assumes a certain posture and becomes completely immovable (Junqua, 1966; Muma, 1966). If not, the male is chased away or devoured. Pheromones, though probably generally present, have been demonstrated only in a few cases. Males of the harvest-

Table 1

Generalized course of display in Salticids.

Male	Female
Becomes aware of ♀, starts display, Stage I. (Minimal releaser: several sight factors; airborne chemical stimuli also usually involved.)	
	Retreats, or watches ♂, usually in braced, high position, often vibrating palps. Rarely attacks. (Minimal releaser and director: several sight factors.)
Approaches, in zigzags, or follows (if female retreats), continuing or resuming display. (Minimal releaser and director: above sight factors, plus type of female motion or lack of it.) Special signs, such as vibrating palps and light abdominal spots probably have directive value.	
	Becomes completely attentive; sometimes gives weak reciprocal display. (Minimal releaser: summative effect of display motions.)
Speeds up display tempo. (Releasers and directors: reduced motion of female, plus chemical stimuli. Self-stimulation is doubtless also a factor.)	
	Ceases motion and, usually, crouches low, legs drawn in.
Enters Stage II. (Releasers: primarily, proximity of female; also involved, usually, her lack of motion, low position, and doubtless reinforced chemical stimuli.) Copulation follows unless female withdraws. (Director: sometimes a pale abdominal crossbar.)	

man *Ischyropsalis hellwigi* and some other species of the same genus present cheliceral glands to the females during courtship, and the secretion probably stimulates chemoreceptors on the tips of the female chelicerae (Martens 1969a). In the tick *Argas persicus* an assembly pheromone has been demonstrated by Leahy et al. (1973). An interesting case of chemical communication has been observed in the mite *Myrmonyssus phalenodectes* by Treat (1958, 1969). This mite is a parasite of the noctuid moth *Pseudaletia* and destroys one of its host's tympanic organs. The first mite to arrive at its host invades one of the tympanic organs and marks a trail to this organ by walking back and forth and depositing a trail pheromone on the host's thorax. Mites that arrive later follow this trail and invade the same organ. It is thus ensured that the parasitized moth is not completely deafened, for hearing, in these moths, is used to escape bats.

References

Abushama, F. T., 1968. Observations on the mating behaviour and birth of *Leiurus quinquestriatus* (H. a. E.), a common scorpion species in the central Sudan. *Rev. Zool. Bot. Afr.,* 77:37–42.

Alexander, A. J., 1957. Courtship and mating in the scorpion, *Opisthophthalmus latimanus. Proc. Zool. Soc. London,* 128:529–44.

Alexander, A. J., 1958a. Courtship and mating in buthid scorpions. *Proc. Zool. Soc. London,* 133:145–69.

Alexander, A. J., 1958b. On the stridulation of scorpions. *Behaviour* (Leiden), 12:339–52.

Alexander, A. J., 1962a. Courtship and mating in amblypygids (Pedipalpi, Arachnida). *Proc. Zool. Soc. London,* 138:379–83.

Alexander, A. J., 1962b. Biology and behaviour of *Damon variegatus* Perty of South Africa and *Admetus barbadensis* Pocock of Trinidad, W. I. (Arachnida, Pedipalpi). *Zoologica* (New York), 47:25–37.

Altevogt, R., 1955. Beobachtungen und Untersuchungen an indischen Winkerkrabben. *Z. Morph. Ökol. Tiere,* 43:501–22.

Altevogt, R., 1957. Untersuchungen zur Biologie, Ökologie und Physiologie indischer Winkerkrabben. *Z. Morph. Ökol. Tiere,* 46:1–110.

Altevogt, R., 1959. Ökologische und ethologische Studien an Europas einziger Winkerkrabbe, *Uca tangeri* Eydoux. *Z. Morph. Ökol. Tiere,* 48:123–46.

Altevogt, R., 1964a. Ein antiphoner Klopfcode und eine neue Winkfunktion bei *Uca tangeri. Naturwissenschaften,* 51:644–45.

Altevogt, R., 1964b. *Uca tangeri* (Ocypodidae), Drohen und Kampf. Göttingen: Encycl. Cinem., Inst. Wiss. Film, Film E 692.

Altevogt, R., 1964c. *Uca tangeri* (Ocypodidae), Klopfen und Winken. Göttingen: Encycl. Cinem., Inst. Wiss. Film, Film E 693.

Altevogt, R., 1966. Vibration als semantisches Mittel bei Crustaceen. *Wiss. Z. Karl-Marx-Univ. Leipzig,* 15:471–76.

Altevogt, R., 1969. Ein sexualethologischer Isolationsmechanismus bei sympatrischen *Uca*-Arten (Ocypodidae) des Ostpazific. *Forma et functio* (Braunschweig), 1:238–49.

Altevogt, R., and Altevogt, R., 1968a. *Uca stylifera* (Ocypodidae), Balz. Göttingen: Encycl. Cinem., Inst. Wiss. Film, Film E 1268.

Altevogt, R., and Altevogt R., 1968b. *Uca princeps* (Ocypodidae), Kampf und Balz. Göttingen: Encycl. Cinem., Inst. Wiss. Film, Film E 1269.

Altevogt, R., and Altevogt, R., 1968c. *Uca insignis* (Ocypodidae), Balz. Göttingen: Encycl. Cinem., Inst. Wiss. Film, Film E 1288.

Altevogt, R., and Altevogt, R., 1968d. *Uca beebei* (Ocypodidae), Kampf und Balz. Göttingen: Encycl. Cinem., Inst. Wiss. Film, Film E 1289.

Altevogt, R., and Altevogt, R., 1968e. *Uca mertensi* (Ocypodidae), Balz. Göttingen: Encycl. Cinem., Inst. Wiss. Film, Film E 1290.

Altevogt, R., and Altevogt, R., 1968f. *Uca rapax* (Ocypodidae), Balz. Göttingen: Encycl. Cinem., Inst. Wiss. Film, Film E 1291.

Altevogt, R., and Altevogt, R. 1968g. *Uca batuenta* (Ocypodidae), Balz. Göttingen: Encycl. Cinem., Inst. Wiss. Film, Film E 1292.

Altevogt, R., and Altevogt, R., 1968h. *Uca terpsichores* (Ocypodidae), Balz. Göttingen: Encycl. Cinem., Inst. Wiss. Film, Film E 1293.

Angermann, H., 1957. Über Verhalten, Spermatophorenbildung und Sinnesphysiologie von *Euscorpius italicus* Hbst. und verwandten Arten (Scorpiones, Chactidae). *Z. Tierpsychol.,* 14:276–302.

Aspey, W. P., 1975. Agonistic behaviour and dominance hierarchy in the wolf spider *Schizocosa crassipes. Proceedings of the 6th Intern. Arachn. Congr.,* Amsterdam 1974, pp. 102–106.

Barth, F. G., 1967. Ein einzelnes Spaltsinnesorgan auf dem Spinnentarsus: Seine Erregung in Abhängigkeit von den Parametern des Luftschallreizes. *Z. vergl. Physiol.,* 55:407–49.

Berland, L. 1912. Observation sur l'accouplement des araignées. *Arch. Zool. exp. gen.,* 9:47–53.

Berland, L., 1932. *Les Arachnides (Scorpions, Araignées, etc).* Paris: Lechevalier et fils. 485pp.

Blanke, R., 1972. Untersuchungen zur Ökophysiologie und Ökethologie von *Cyrthophora citricola* Forskål (Araneae, Araneidae) in Andalusien. *Forma et functio* (Braunschweig) 5:125–206.

Blanke, R., 1973a. Neue Ergebnisse zum Sexualverhalten von *Araneus cucurbitinus* Cl. (Araneae, Araneidae). *Forma et functio* (Braunschweig) 6:279–90.

Blanke, R., 1973b. Nachweis von Pheromonen bei Netzspinnen. *Naturwissenschaften,* 60:481.

Bovbjerg, R. V., 1953. Dominance order in the crayfish *Orconectes virilis* (Hagen). *Physiol. Zool.,* 26:173–78.

Bovbjerg, R. V., 1956. Some factors affecting aggressive behavior in crayfish. *Physiol. Zool.,* 29:127–36.

Bovbjerg, R. V., 1960. Courtship behavior of the lined

shore crab, *Pachygrapsus crassipes* Randall. *Pacif. Sci.*, 14:421–22.

Bristowe, W. S., 1929. The mating habits of spiders, with special reference to the problems surrounding sex dimorphism. *Proc. Zool. Soc. London*, 1929:309–58.

Bristowe, W. S., and Locket, G. H., 1926. The courtship of British lycosid spiders, and its probable significance. *Proc. Zool. Soc. London*, 1926:317–47.

Buckle, D. J., 1972. Sound production in the courtship of two lycosid spiders—*Schizocosa avida* Walckenaer and *Tarantula aculeata* (Clerk). *The Blue Jay* (J. Nat. Hist. Soc. Saskatchewan), June 1972:110–13.

Bückle-Ramirez, L. F., 1965. Untersuchungen über die Biologie von *Heterotanais oerstedi*. *Z. Morph. Ökol. Tiere*, 55:714–82.

Caldwell, R. L. and Dingle, H. 1975. Ecology and evolution of agonistic behavior in Stomatopods. *Naturwissenschaften*, 62:214–22.

Chrysanthus, F., 1953. Hearing and stridulation in spiders. *Tijdschr. Entomol.*, 96:57–83.

Clarke, W. D., 1963. Function of bioluminescence in mesopelagic organisms. *Nature*, 198:1244–46.

Clutter, R. I., 1969. The microdistribution and social behavior of some pelagic mysid shrimps. *J. Exp. Mar. Biol. Ecol.* (Amsterdam), 3:125–55.

Crane, J., 1943. Display, breeding and relationships of fiddler crabs (Brachyura, genus *Uca*) in the northeastern United States. *Zoologica* (New York), 28:217–23.

Crane, J., 1948a. Comparative biology of salticid spiders at Rancho Grande, Venezuela. Part I. Systematics and life history in *Corythalia*. *Zoologica* (New York), 33:1–38.

Crane, J., 1948b. Comparative biology of salticid spiders at Rancho Grande, Venezuela. Part II. Methods of collection, culture, observation, and experiment. Zoologica (New York), 33:139–45.

Crane, J., 1949a. Comparative biology of salticid spiders at Rancho Grande, Venezuela. Part III. Systematics and behavior in representative new species. *Zoologica* (New York), 34:31–52.

Crane, J., 1949b. Comparative biology of salticid spiders at Rancho Grande, Venezuela. Part IV. An analysis of display. *Zoologica* (New York), 34:159–214.

Crane, J., 1957. Basic patterns of display in fiddler crabs (Ocypodidae, genus *Uca*). *Zoologica* (New York), 42:69–82.

Crane, J., 1958. Aspects of social behavior in fiddler crabs, with special reference to *Uca maracoani* (Latreille). *Zoologica* (New York), 43:113–30.

Crane, J., 1966. Combat, display and ritualization in fiddler crabs (Ocypodidae, genus *Uca*). *Phil. Trans. Roy. Soc. London*, B 251:459–72.

Crane, J., 1967. Combat and its ritualization in fiddler crabs (Ocypodidae) with special reference to *Uca rapax* (Smith). *Zoologica* (New York), 52:49–77.

Czajka, M., 1963. Unknown facts of the biology of the spider *Ero furcata* (Villers). *Polskie Pismo Entomol.*, 33:229–31.

Dingle, H., 1969. A statistical and information analysis of aggressive communication in the mantis shrimp *Gonodactylus bredini* Manning. *Anim. Behav.*, 17:561–75.

Dingle, H., and Caldwell, R. L., 1969. The aggressive and territorial behaviour of the mantis shrimp *Gonodactylus bredini* Manning (Crustacea: Stomatopoda). *Behaviour*, 33:115–36.

Dondale, C. D., and Hegdekar, B. M., 1973. The contact sex pheromone of *Pardosa lapidicina* Emerton (Araneida; Lycosidae). *Canad. J. Zool.*, 51:400–401.

Drees, O., 1952. Untersuchungen über die angeborenen Verhaltensweisen bei Springspinnen (Salticidae). *Z. Tierpsychol.*, 9:169–207.

Dumais, J.; Perron, J. M.; and Dondale, C. D.; 1973. Eléments du comportement sexuel chez *Pardosa xerampelina* (Keyserling) (Araneida: Lycosidae). *Canad. J. Zool.*, 51:265–71.

Engelhardt, W., 1964. Die mitteleuropäischen Arten der Gattung *Trochosa* C. L. Koch, 1848 (Araneae, Lycosidae). Morphologie, Chemotaxonomie, Biologie, Autökologie. *Z. Morph. Ökol. Tiere*, 54:219–392.

Foelix, R., 1970a. Structure and function of tarsal sensilla in the spider *Araneus diadematus*. *J. Exp. Zool.*, 175:99–124.

Foelix, R., 1970b. Chemosensitive hairs in spiders. *J. Morph.*, 132:313–34.

Foelix, R. F.; Chu-Wang, I-Wu; and Beck, L.; 1975. Fine structure of tarsal sensory organs in the whip spider *Admetus pumilio* (Amblypygi, Arachnida). *Tissue and Cell*, 7:331–46.

Frings, H., and Frings, M., 1966. Reactions of orb-Weaving spiders (Argiopidae) to airborne sounds. *Ecology*, 47:578–88.

Frings, H., and Frings, M., 1968. Other invertebrates. In: *Animal Communication*, T. A. Sebeok, ed. Bloomington: Indiana University Press, pp. 244–70.

Garnier, G., and Stockmann, R., 1972. Comportement de la reproduction chez le scorpion *Pandinus imperator*. *Fifth Intern. Congr. Arachn.*, Brno 1971, pp.13–21.

Gordon, H. R. S., 1958. Synchronous claw-waving of fiddler crabs. *Anim. Behav.*, 6:238–41.
Görner, P., 1965. A proposed transducing mechanism for a multiply-innervated mechanoreceptor (Trichobothrium) in spiders. *Cold Spring Harbor Symp. Quant. Biol.*, 30:69–73.
Görner, P., and Andrews, P., 1969. Trichobothrien, ein Ferntastsinnesorgan bei Webespinnen (Araneen). *Z. vergl. Physiol.*, 64:301–17.
Griffin, D. J. G., 1968. Social and maintenance behaviour in two Australian ocypodid crabs (Crustacea: Brachyura). *J. Zool.* (London), 156:291–305.
Guinot-Dumortier D., and Dumortier, B., 1960. La stridulation chez les crabes. *Crustaceana* (Leiden), 2:117–55.
Gwinner-Hanke, H., 1970. Zum Verhalten zweier stridulierender Spinnen, *Steatoda bipunctata* Linné und *Teutana grossa* Koch (Theridiidae, Araneae), unter besonderer Berücksichtigung des Fortpflanzungsverhaltens. *Z. Tierpsychol.*, 27:649–78.
von Hagen, H.-O., 1962. Freilandstudien zur Sexual- und Fortpflanzungsbiologie von *Uca tangeri* in Andalusien. *Z. Morph. Ökol. Tiere*, 51:611–725.
von Hagen, H.-O., 1967. Klopfsignale auch bei Grapsiden (Decapoda, Brachyura). *Naturwissenschaften*, 54:177–78.
von Hagen, H.-O., 1968. Zischende Drohgeräusche bei westindischen Krabben. *Naturwissenschaften*, 55:139–40.
von Hagen, H.-O., 1970a. Zur Deutung langstieliger und gehörnter Augen bei Ocypodiden (Decapoda, Brachyura). *Forma et functio* (Braunschweig), 2:13–57.
von Hagen, H.-O., 1970b. Anpassungen an das spezielle Gezeitenzonen-Niveau bei Ocypodiden (Decapoda, Brachyura). *Forma et functio* (Braunschweig), 2:361–413.
von Hagen, H.-O., 1970c. Die Balz von *Uca vocator* (Herbst) als ökologisches Problem. *Forma et functio* (Braunschweig), 2:238–53.
von Hagen, H.-O., 1970d. Verwandtschaftlichen Gruppierung und Verbreitung der karibischen Winkerkrabben (Ocypodidae, Gattung *Uca*). *Zool. Mededelingen* (Leiden), 44:217–35.
von Hagen, H.-O., 1972a. *Uca leptodactyla* (Ocypodidae), Balz. Göttingen: Encycl. Cinem., Inst. Wiss. Film, Film E 1421.
von Hagen, H.-O., 1972b. *Uca maracoani* (Ocypodidae), Balz. Göttingen: Encycl. Cinem., Inst. Wiss. Film, Film E 1423.
von Hagen, H.-O., 1973a. *Uca cumulanta* (Ocypodidae), Balz. Göttingen: Encycl. Cinem., Inst. Wiss. Film, Film E 1420.
von Hagen, H.-O., 1973b. *Uca thayeri* (Ocypodidae), Balz. Göttingen: Encycl. Cinem., Inst. Wiss. Film, Film E 1424.
von Hagen, H.-O., 1973c. *Uca major* (Ocypodidae), Balz. Göttingen: Encycl. Cinem., Inst. Wiss. Film, Film E 1422.
von Hagen, H.-O., 1975. Klassifikation und phylogenetische Einordnung der Lautäusserungen von Ocypodiden und Grapsiden (Crustacea, Brachyura). *Z. zool. Syst. Evolut.-forsch.*, 13:300–16.
Hallander, H., 1967. Courtship display and habitat selection in the wolf spider *Pardosa chelata* (O. F. Müller). *Oikos*, 18:145–50.
Harrison, J. B., 1969. Acoustic behavior of a wolf spider, *Lycosa gulosa. Anim. Behav.*, 17:14–16.
Harvey, E. N., 1961. Light production. In: *The Physiology of Crustacea*, vol. II, T. Waterman, ed. New York: Academic Press, pp.171–90.
Hazlett, B. A., 1966a. Factors affecting the aggressive behavior of the hermit crab *Calcinus tibicen. Z. Tierpsychol.*, 6:655–71.
Hazlett, B. A., 1966b. Observations on the social behavior of the land hermit crab, *Coenobita clypeatus* (Herbst). *Ecology*, 47:316–17.
Hazlett, B. A., 1968a. Size relationships and aggressive behavior in the hermit crab *Clibanarius vittatus. Z. Tierpsychol.*, 25:608–14.
Hazlett, B. A., 1968b. Effects of crowding on the agonistic behavior of the hermit crab *Pagurus bernhardus. Ecology*, 49:573–75.
Hazlett, B. A., 1968c. Communicatory effect of body position in *Pagurus bernhardus* (L) (Decapoda, Anomura). *Crustaceana* (Leiden), 14:210–14.
Hazlett, B. A., 1968d. Dislodging behavior in European Pagurids. *Publ. staz. zool. Napoli*, 36:138–39.
Hazlett, B. A., 1968e. Stimuli involved in the feeding behavior of the hermit crab *Clibanarius vittatus* (Decapoda, Paguridea). *Crustaceana* (Leiden), 15:305–11.
Hazlett, B. A., 1968f. The effect of shell size and weight on the agonistic behavior of a hermit crab. *Z. Tierpsychol.*, 27:369–74.
Hazlett, B. A. 1968g. The sexual behaviour of some European hermit crabs (Anomura, Paguridae). *Publ. staz. zool. Napoli*, 36:238–52.
Hazlett, B. A., 1968h. The phyletically irregular social behavior of *Diogenes pugilator* (Anomura, Paguridea). *Crustaceana* (Leiden), 15:31–34.
Hazlett, B. A., 1969a. "Individual" recognition and

agonistic behaviour in *Pagurus bernhardus*. *Nature,* 222:268–69.

Hazlett, B. A., 1969b. Stone fighting in the crab *Cancellus spongicola* (Decapoda, Anomura, Diogenidae). *Crustaceana* (Leiden), 16:219–20.

Hazlett, B. A., 1969c. Further investigations of the cheliped presentation display in *Pagurus bernhardus* (Decapoda, Anomura). *Crustaceana* (Leiden), 17:31–34.

Hazlett, B. A., 1970. Tactile stimuli in the social behavior of *Pagurus bernhardus* (Decapoda, Paguridae). *Behaviour* (Leiden), 36:20–48.

Hazlett, B. A., 1971. Interspecific fighting in three species of brachyuran crabs from Hawaii. *Crustaceana* (Leiden), 20:308–14.

Hazlett, B. A., 1972a. Stimulus characteristics of an agonistic display of the hermit crab *(Calcinus tibicen). Anim. Behav.,* 20:101–107.

Hazlett, B. A., 1972b. Stereotypy of agonistic movements in the spider crab *Microphrys bicornutus. Behaviour* (Leiden), 42:270–78.

Hazlett, B. A. 1975, Individual distance in the hermit crabs Clibanarius tricolor and Clibanarius antillensis. *Behaviour* 52:253–265.

Hazlett, B. A., 1975, Ethological analysis of reproductive behavior in marine Crustacea. *Publ. Staz. zool. Napoli* 39:Suppl. 677–695.

Hazlett, B. A., and Bossert, W. H., 1965. A statistical analysis of the aggressive communications systems of some hermit crabs. *Anim. Behav.,* 13:357–73.

Hazlett, B. A., and Bossert, W. H., 1966. Additional observations on the communications system of hermit crabs. *Anim. Behav.,* 14:546–49.

Hazlett, B. A., and Estabrook, G. F., 1974a. Examination of agonistic behaviour by character analysis I. The spider crab *Microphrys bicornutus. Behaviour* (Leiden), 48:131–43.

Hazlett, B. A., and Estabrook, G. F., 1974b. Examination of agonistic behaviour by character analysis II. Hermit crabs. *Behaviour* (Leiden), 49:88–110.

Hazlett, B. A., and Winn, H. E., 1962. Sound production and associated behavior of Bermuda crustaceans *(Panulirus, Gonodactylus, Alpheus,* and *Synalpheus). Crustaceana* (Leiden), 4:25–38.

Heckenlively, D. B., 1970. Intensity of aggression in the crayfish, *Orconectes virilis* (Hagen). *Nature,* 225:180–81.

Hediger, H., 1933. Notes sur la biologie d'un crabe de l'embouchure de l'Oued Boi Regreg, *Uca tangeri* (Eydoux). *Bull. Soc. Sci. Nat. Maroc.,* 13:254–59.

Hegdekar, B. M., and Dondale, C. D., 1969. A contact sex pheromone and some response parameters in lycosid spiders. *Canad. J. Zool,* 47:1–4.

Heurtault, J., 1972. Étude histologique de quelques caractères sexuels mâles des Cheliferidae (Pseudoscorpions). *Fifth Intern. Congr. Arachn.,* Brno 1971, pp.43–51.

Horch, K. W., and Salmon, M., 1969. Production, perception and reception of acoustic stimuli by semiterrestrial crabs (genus *Ocypode* and *Uca,* family Ocypodidae). *Forma et functio* (Braunschweig), 1:1–25.

Hughes, D. A., 1966. Behavioural and ecological investigations of the crab *Ocypode ceratophthalmus* (Crustacea: Ocypodidae). *J. Zool.* (London), 150:129–43.

Jacson, C. C., and Joseph, K. J., 1973. Life-history, bionomics and behaviour of the social spider *Stegodyphus sarasinorum* Karsch. *Insectes sociaux,* 20:189–204.

Johnson, F. H.; Sie, E. H.-C.; and Haneda, Y.; 1961. The luciferin-luciferase reaction. In: *Light and Life,* W. D. McElroy and B. Glass, eds. Baltimore: The John Hopkins Press, pp.206–18.

Junqua, C., 1966. Recherches biologiques et histophysiologiques sur un solifuge Saharien, *Othoes saharae* Panouse. Thés. Fac. Sci. Univ. Paris, series A, no. 4689 (5537), pp.1–124.

Kaston, B. J., 1936. The senses involved in the courtship of some vagabond spiders. *Entomol. Amer.,* 16:97–167.

Kew, H. W., 1912. On the pairing of pseudoscorpions. *Proc. Zool. Soc. London,* 25:376–90.

Klingel, H., 1963. Paarungsverhalten bei Pedipalpen (*Thelyphonus caudatus* L., Holopeltidia, Uropygi, und *Sarax sarawakensis* Simon, Charontinae, Amblypygi). *Verh. deutsch. zool. Ges.,* 1962:452–59.

Kuenen, D. J., and Nooteboom, H. P., 1963. Olfactory orientation in some land-isopods (Oniscoidea, Crustacea). *Entomol. Exp. Appl.* (Amsterdam), 6:133–42.

Krafft, B., 1966. Étude du comportement social de l'araignée *Agelena consociata* Denis. *Biol. Gabon.,* 2:235–50.

Krafft, B., 1969. Quelques remarques sur les phénomènes sociaux chez les araignées avec une étude particulière d'*Agelena consociata* Denis. *Bull. mus. hist. nat.* (Paris), 2d series, 41:Suppl. 1, 70–75.

Krafft, B., 1970a. Contribution à la biologie et à l'éthologie d'*Agelena consociata* Denis (Araignée sociale du Gabon). *Biol. Gabon,* 6:197–301.

Krafft, B., 1970b. Contribution à la biologie et à

l'éthologie d'*Agelena consociata* Denis (Araignée sociale du Gabon) II. Étude experimentale de certains phénomènes sociaux. *Biol. Gabon.*, 6:307–69.

Krafft, B., 1971. Contribution à la biologie et à l'éthologie d'*Agelena consociata* Denis (Araignée sociale du Gabon) III. Étude experimentale de certains phénomènes sociaux. *Biol. Gabon.*, 7:3–56.

Krafft, B., 1972. Les interactions entre les individus chez *Agelena consociata,* araignée sociale du Gabon. *Fifth Intern. Congr. Arachn.*, Brno 1971, pp.159–64.

Kronestedt, T., 1973. Study of a stridulatory apparatus in *Pardosa fulvipes* (Collet) (Araneae, Lycosidae) by scanning electron microscopy. *Zool. Scr.*, 2:43–47.

Kullmann, E., 1968. Soziale Phaenomene bei Spinnen. *Insects sociaux,* 15:289–98.

Kullmann, E., 1969a. Beobachtungen zum Sozialverhalten von *Stegodyphus sarasinorum* Karsch (Araneae, Eresidae). *Bull. mus. hist. nat.* (Paris), 2d series, 41: Suppl. 1, 76–81.

Kullmann, E., 1969b. Unterschiedliche Brutfürsorge bei den Haubennetzspinnen *Theridion impressum* (L. Koch) und *Theridion notatum* (Clerk) (Araneae, Theridiidae). *Verh. deutsch. zool. Ges.*, 1969:326–33.

Kullmann, E., 1969c. Soziales Verhalten bei Spinnen. *Phys. Med. Ges. Würzburg,* 77:1–12.

Kullmann, E., 1970a. Brufpflege mit Regurgitationsfütterungen bei Haubennetzspinnen (Araneae, Theridiidae). Erläuterungen zu einem wissenschaftlichen Film. *Verh. deutsch. zool. Ges.*, 1969:636–38.

Kullmann, E., 1970b. Beobachtungen zum Sozialverhalten von *Stegodyphus sarasinorum* Karsch. (Araneae, Eresidae). *Bull. mus. hist. nat.* (Paris), 2d series, 41:Suppl. 1, 76–81.

Kullmann, E., 1972. Evolution of social behavior in spiders (Araneae; Eresidae and Theridiidae). *Amer. Zool.*, 12:419–26.

Kullmann, E., 1974. *Theridion sisyphium* (Theridiidae) —Brutfürsorge und periodisch soziales Verhalten. Göttingen: Encycl. Cinem., Inst. Wiss. Film, Film E 1865.

Kullmann, E., and Kloft, W., 1969. Traceruntersuchungen zur Regurgitationsfütterung bei Spinnen (Araneae, Theridiidae). *Verh. deutsch. zool. Ges.*, 1968:487–97.

Kullmann, E.; Nawabi, S.; and Zimmermann, W.; 1972. Neue Ergebnisse zur Brutbiologie cribellater Spinnen aus Afghanistan und der Serengeti (Araneae, Eresidae). *Z. Kölner Zoo,* 14:87–108.

Kullmann, E.; Sittertz, H.; and Zimmermann, W.; 1971. Erster Nachweis von Regurgitationsfütterungen bei einer cribellaten Spinne (*Stegodyphus lineatus* Latreille, 1817, Eresidae). *Bonn. zool. Beitr.*, 22:175–88.

Kullmann, E., and Zimmermann, W., 1972. Versuche zur Toleranz bei der permanent sozialen Spinnenart *Stegodyphus sarasinorum* Karsch (Fam. Eresidae). *Fifth Intern. Congr. Arachn.*, Brno 1971, pp. 175–82.

Lahee, F. H., 1904. The calls of spiders. *Psyche,* 11:74.

Leahy, M. G.; Vandehey, R.; and Galun, R.; 1973. Assembly pheromone(s) in the soft tick *Argas persicus* (Oken). *Nature,* 246:515–16.

Legendre, R., 1963. L'audition et l'émission de sons chez les aranéides. *Ann. Biol.*, 2:371–90.

Legendre, R., and Llinares, D., 1970. L'accouplement de l'araignée salticide, *Cyrba algerina* (Lucas, 1846). *Ann. Soc. Hist. nat. Hérault,* 110:169–74.

Legrand, J.-J., 1958. Comportement sexuel et modalités de la fécondation chez l'Oniscoide *Porcellio dilatatus* Brandt. *C. R. Acad. Sci.* (Paris), 246:3120–22.

Liesenfeld, F. J., 1961. Über Leistung und Sitz des Erschütterungssinnes von Netzspinnen. *Biol. Zentralbl.*, 80:465–75.

Linsenmair, K. E., 1967. Konstruktion und Signalfunktion der Sandpyramide der Reiterkrabbe *Ocypode Saratan* Forsk. (Decapoda, Brachyura, Ocypodiae). *Z. Tierpsychol.*, 24:403–56.

Linsenmair, K. E., 1971. Paarbildung und Paarzusammenhalt bei der monogamen Wüstenassel *Hemilepistus reaumuri* (Crustacea, Isopoda, Oniscoidea). *Z. Tierpsychol.*, 29:134–55.

Linsenmair, K. E., 1972. Die Bedeutung familienspezifischer "Abzeichen" für den Familienzusammenhalt bei der sozialen Wüstenassel *Hemilepistus reaumuri* Audouin u. Savigny (Crustacea, Isopoda, Oniscoidea). *Z. Tierpsychol.*, 31:131–62.

Linsenmair, K. E., 1973. Die Wüstenassel: Sozialverhalten und Lebensraum. *Umschau,* 73:151–52.

Little, E. E., 1975. Chemical communication in maternal behaviour of crayfish. *Nature,* 255:400–401.

McElroy, W. D., and Seeliger, H. H., 1962. Biological luminescence. *Sci. Amer.*, 207:76–89.

Martens, J., 1969a. Die Sekretdarbietung während des Paarungsverhaltens von *Ischyropsalis* C. L. Koch (Opiliones). *Z. Tierpsychol.*, 26:513–23.

Matthiesen, F. A., 1968. On the sexual behaviour of some Brazilian scorpions. *Rev. Bras. Pesquisas Med. Biol.*, 1:93–96.

Melchers, M., 1963. Zur Biologie und zum Verhalten von *Cupiennius salei* (Keyserling), einer amerikanischen Ctenide. *Zool. Jb. Abt. Syst. Ökol. Geogr.*, 91:1–90.

Millot, J., 1949. *Ordre des Amblypyges.* In: *Traité de Zoologie,* vol. 6, P. P. Grassé, ed. Paris: Masson et Cie, pp. 563–88.

Montgomery, T. H., 1910. The significance of the courtship and secondary sexual characters of araneads. *Amer. Nat.,* 44:151–77.

Moulton, J. M., 1957. Sound production in the spiny lobster *Panulirus argas* (Latreille). *Biol. Bull.,* 113:286–95.

Muma, M. H., 1966. Mating behavior in the solpugid genus *Eremobates* Banks. *Anim. Behav.,* 14:346–50.

Nicol, J. A. C., 1962. Animal luminescence. *Adv. Comp. Physiol. Biochem.,* 1:217–73.

Nolan, B. A., Salmon, M., 1970. The behavior and ecology of snapping shrimp (Crustacea: *Alpheus heterochelis* and *Alpheus normanni*) *Forma et functio* (Braunschweig), 2:289–335.

Peckham, G. W., and Peckham, E. G., 1889. Observations on sexual selection in spiders of the family Attidae. *Oc. Pap. Nat. Hist. Soc. Wisconsin,* 1:1–60.

Peckham, G. W., and Peckham, E. G., 1890. Additional observations on sexual selection on spiders of the family Attidae, with some remarks on Mr. Wallace's theory of sexual ornamentation. *Oc. Pap. Nat. Hist. Soc. Wisconsin,* 1:117–51.

Peters, H. M., 1955. Die Winkgebärde von *Uca* und *Minuca* (Brachyura) in vergleichend-ethologischer, -ökologischer und -morphologisch-anatomischer Betrachtung. *Z. Morph. Ökol. Tiere,* 43:425–500.

Platnick, N., 1971. The evolution of courtship behaviour in spiders. *Bull. Brit. Arachn. Soc.,* 2:40–47.

Plett, A., 1962. Beobachtungen und Versuche zum Revier- und Sexualverhalten von *Epiblemum scenicum* Cl. und *Evarcha blancardi* Scop. (Salticidae). *Zool. Anz.,* 169:292–98.

Reinhard, E. G., 1949. Experiments on the determination and differentiation of sex in the bopyrid *Stegophryxus hyptius* Thompson. *Biol. Bull.,* 96:17–31.

Reverberi, G., and Pitotti, M., 1942. Il ciclo biologico et la determinazione fenotypica del sesso di *Jone thoracica* Montagu, bopiride parassita di *Callianassa laticaudata* Otto. *Publ. staz. zool. Napoli,* 19:111–84.

Ritzmann, R., 1973. Snapping behavior of the shrimp *Alpheus californicus. Science,* 181:459–60.

Rosin, R., and Shulov, A., 1961. Sound production in scorpions. *Science,* 133:1918–19.

Rosin, R., and Shulov, A., 1963. Studies on the scorpion *Nebo hierochonticus. Proc. Zool. Soc. London,* 140:547–75.

Rovner, J. S., 1967. Acoustic communication in a lycosid spider (*Lycosa rabida* Walckenaer). *Anim. Behav.,* 15:273–81.

Rovner, J. S., 1968a. An analysis of display in the lycosid spider *Lycosa rabida* Walckenaer. *Anim. Behav.,* 16:358–69.

Rovner, J. S., 1968b. Territoriality in the sheet-web spider *Linyphia triangularis* (Clerk) (Araneae, Linyphiidae). *Z. Tierpsychol.,* 25:232–42.

Rovner, J. S., 1971. Mechanisms controlling copulatory behavior in wolf spiders (Araneae, Lycosidae). *Psyche,* 78:150–65.

Rovner, J. S., 1972. Copulation in the lycosid spider (*Lycosa rabida* Walckenaer): a quantitative study. *Anim. Behav.,* 20:133–38.

Rovner, J. S., 1973. Copulatory pattern supports generic placement of *Schizocosa avida* (Walckenaer) (Araneae: Lycosidae). *Psyche,* 80:245–48.

Rovner, J. S.; Higashi, G. A.; and Foelix, R. F., 1973. Maternal behavior in wolf spiders: the role of abdominal hairs. *Science,* 182:1153–55.

Ryan, E. P., 1966. Pheromone: Evidence in a decapod crustacean. *Science,* 151:340–41.

Salmon, M., 1965. Waving display and sound production in the courtship behavior of *Uca pugilator,* with comparisons to *U. minax* and *U. pugnax. Zoologica* (New York), 50:123–50.

Salmon, M., 1967. Coastal distribution, display and sound production by Florida fiddler crabs (genus *Uca*). *Anim. Behav.,* 15:449–59.

Salmon, M., 1971. Signal characteristics and acoustic detection by the fiddler crabs, *Uca rapax* and *Uca pugilator. Physiol. Zool.,* 44:210–24.

Salmon, M., and Atsaides, S. P., 1968. Visual and acoustical signalling during courtship by fiddler crabs (genus *Uca*). *Amer. Zool.,* 8:623–39.

Salmon, M., and Atsaides, S. P., 1969. Sensitivity to substrate vibration in the fiddler crab, *Uca pugilator* Bosc. *Anim. Behav.,* 17:68–76.

Salmon, M., and Horch, K. W., 1972. Acoustic signalling and detection by semiterrestrial crabs of the family Ocypodidae. In: *Behavior of Marine Animals,* vol. 1, H. E. Winn and B. L. Olla, eds. New York: Plenum Press, pp.60–96.

Salmon, M., and Horch, K. W., 1973. Vibration reception by the fiddler crab, *Uca minax. Comp. Biochem. Physiol.,* 44A:527–41.

Salmon, M., and Stout, J. F., 1962. Sexual discrimination and sound production in *Uca pugilator* (Bosc.). *Zoologica* (New York), 47:15–21.

Schneider, P., 1971. Lebensweise und soziales Verhalten der Wüstenassel *Hemilepistus aphghanicus* Borutzky 1958. *Z. Tierpsychol.,* 29:121–33.

Schöne, H., 1961. Complex behavior. In: *The Physiology of Crustacea II,* T. Waterman, ed. New York: Academic Press, pp.465–520.

Schöne, H., 1965. Release and orientation of behaviour and the role of learning as demonstrated in Crustacea. *Anim. Behav. Suppl.,* 1:135–44.

Schöne, H., 1968. Agonistic and sexual display in aquatic and semi-terrestrial brachyuran crabs. *Amer. Zool.,* 8:641–54.

Schöne, H., and Schöne, H., 1963. Balz und andere Verhaltensweisen der Mangrovekrabbe *Goniopsis cruentata* Latr. und das Winkverhalten der eulitералen Brachyuren. *Z. Tierpsychol.,* 20:641–56.

Seibt, U., and Wickler, W., 1972. Individuen-Erkennung und Partnerbevorzugung bei der Garnele *Hymenocera picta* Dana. *Naturwissenschaften,* 59:40–41.

Shear, W. A., 1970a. Stridulation in *Acanthophrynus coronatus* (Butler) (Amblypygi, Tarantulidae). *Psyche,* 77:181–83.

Shear, W. A., 1970b. The evolution of social phenomena in spiders. *Bull. Brit. Arachn. Soc.,* 1:65–76.

Sturm, H., 1958. Indirekte Spermatophorenübertragung bei dem Geisselskorpion *Trithyreus sturmi* Kraus (Schizomidae, Pedipalpi). *Naturwissenschaften,* 45:142–43.

Sturm, H., 1973. Zur Ethologie von *Trithyreus sturmi* Kraus (Arachnida, Pedipalpi, Schizopeltidia). *Z. Tierpsychol.,* 33:113–40.

Treat, A. E., 1958. A case of peculiar parasitism. *Natur. History* (New York), 67:366–73.

Treat, A. E., 1969. Behavioural aspects of the association of mites with nocturid moths. In: *Proc. Second Intern. Congr. Acarology, 1967,* G. O. Evans, ed. Budapest: Akadémiai Kiadó, pp.275–86.

Tretzel, E., 1961a. Biologie, Ökologie und Brutpflege von *Coelotes terrestris* (Wider) (Araneae, Agelenidae). Teil I: Biologie und Ökologie. *Z. Morph. Ökol. Tiere,* 49:658–745.

Tretzel, E., 1961b. Biologie, Ökologie und Brutpflege von *Ceolotes terrestris* (Wider) (Araneae, Agelenidae) Teil II: Brutpflege. *Z. Morph. Ökol. Tiere,* 50:375–542.

Vachon, M., 1938. Recherches anatomiques et biologiques sur le réproduction et le développement de pseudoscorpions. Thés. Fac. Sci. Univ. Paris, series A, No. 1779 (2645), 1–207.

van Helsdingen, P. J., 1965. Sexual behaviour of *Lepthyphantes leprosus* (Ohlert, Araneida, Linyphiidae) with notes on the function of the genital organs. *Zool. Meded.,* 41:15–42.

Verwey, J., 1930. Einiges über die Biologie ostindischer Mangrovekrabben. *Treubia* (Bogor), 12:168–261.

Vlijm, L., and Borsje, W. J., 1969. Comparative research of the courtship behaviour in the genus *Pardosa* (Arach., Araneae) II. Some remarks about the courtship behaviour in *Pardosa pullata* (Clerk). *Bull. mus. hist. nat.* (Paris), 2d series, 41:Suppl. 1, 112–16.

Vlijm, L., and Dijkstra, H., 1966. Comparative research of the courtship behaviour in the genus *Pardosa* (Arach., Araneae). I. Some remarks about the courtship of *P. amentata, P. hortensis, P. nigriceps,* and *P. lugubris. Senckenbergiana Biol.,* 47:51–55.

Volz, P., 1938. Studien über das "Knallen" der Alpheiden. Nach Untersuchungen an *Alpheus dentipes* Guerin und *Synalpheus laevimanus* (Heller). *Z. Morph. Ökol. Tiere,* 34:272–316.

Walcott, C., 1963. The effect of the web on vibration sensitivity in the spider, *Achaearanea tepidariorum* (Koch). *J. Exp. Biol.,* 40:595–611.

Walcott, C., and van der Kloot, W. G., 1959. The physiology of the spider vibration receptor. *J. Exp. Zool.,* 144:191–244.

Warner, G. F., 1970. Behaviour of two species of grapsid crabs during intraspecific encounters. *Behaviour* (Leiden), 36:9–19.

Weygoldt, P., 1969. *The Biology of Pseudoscorpions.* Cambridge: Harvard University Press. 145pp.

Weygoldt, P., 1970. Vergleichende Untersuchungen zur Fortpflanzungsbiologie der Pseudoscorpione II. *Z. zool. Syst. Evolutionsforsch.,* 8:241–59.

Weygoldt, P., 1971. Notes on the life history and reproductive biology of the giant whip scorpion, *Mastigoproctus giganteus* (Uropygi, Thelyphonidae) from Florida. *J. Zool.* (London), 164:137–47.

Weygoldt, P., 1972a. Spermatophorenbau und Samenübertragung bei Uropygen (*Mastigoproctus brasilianus* C. L. Koch) und Amblypygen (*Charinus brasilianus* Weygoldt und *Admetus pumilio* C. L. Koch) (Chelicerata, Arachnida). *Z. Morph. Tiere,* 71:23–51.

Weygoldt, P., 1972b. *Charinus brasilianus* (Charontidae)—Paarungsverhalten. Göttingen: Encycl. Cinem., Inst. Wiss. Film, Film E. 1862.

Weygoldt, P., 1972c. *Charinus brasilianus* (Charontidae)—Kampfverhalten. Göttingen: Encycl. Cinem., Inst. Wiss. Film, Film E 1861.

Weygoldt, P., 1972d. *Admetus pumilio* (Tarantulidae)—Paarungsverhalten. Göttingen: Encycl. Cinem., Inst. Wiss. Film, Film E 1860.

Weygoldt, P., 1972e. *Admetus pumilio* (Tarantulidae)—

Kampfverhalten. Göttingen: Encycl. Cinem., Inst. Wiss. Film, Film E 1859.

Weygoldt, P., 1972f. *Tarantula marginemaculata* (Tarantulidae)—Paarungsverhalten. Göttingen: Encycl. Cinem., Inst. Wiss. Film, Film E 1863.

Weygoldt, P., 1973a. *Euscorpius italicus* (Chactidae)—Paarungsverhalten. Göttingen: Encycl. Cinem., Inst. Wiss. Film, Film E 1914.

Weygoldt, P., 1973b. *Mastigoproctus brasilianus* (Uropygi)—Balz und Spermaübertragung. Göttingen: Encycl. Cinem., Inst. Wiss. Film, Film E 1915.

Weygoldt, P., 1974. Kampf und Paarung bei der Geisselspinne *Charinus montanus* Weygoldt (Arachnida, Amblypygi, Charontidae). *Z. Tierpsychol.*, 34:217–23.

Wickler, W., 1970. Harlekin-Garnele und *Acanthaster*-Problem (Monogamie-Studien). *Naturw. Rundsch.*, 23:368–69.

Wickler, W., 1973. Über Koloniegründung und soziale Bindung von *Stegodyphus mimosarum* Paresi und anderen sozialen Spinnen. *Z. Tierpsychol.*, 32:522–31.

Wickler, W., and Seibt, U., 1970. Das Verhalten von *Hymenocera picta* Dana, einer Seesterne fressenden Garnele (Decapoda, Natantia, Gnathophyllidae). *Z. Tierpsychol.*, 27:352–68.

Wright, H. O., 1968. Visual displays in brachyuran crabs: field and laboratory studies. *Amer. Zool.*, 8:655–65.

Yamaguchi, T., 1971. Courtship behavior of a fiddler crab, *Uca lactea*. *Kumamoto J. Sci.*, Biol., 10:13–37.

Chapter 16

COMMUNICATION IN ORTHOPTERA

Daniel Otte

Introduction

The study of orthopteran signaling systems has become highly interdisciplinary, with research on communication being carried out from a number of points of view and by persons of varied backgrounds and training. Research on neuromuscular, developmental, and genetic aspects of signaling is being carried out in a number of laboratories around the world, sometimes concurrently at one place. Systematists continue to use signaling as a tool in distinguishing between species, to determine phylogenetic relationships among related species, and to provide basic information on the diversity of signaling systems. Orthoptera are also increasingly being used to study evolutionary and ecological principles, circadian rhythms, density dynamics, and population spacing patterns. Much information on this group is already available in several excellent reviews (Alexander, 1960a, 1962a, 1967, 1968; Walker, 1957, 1962). The present chapter reconsiders topics discussed previously, although from a slightly different viewpoint, as well as others not previously discussed. My own background in grasshopper and cricket biology unavoidably has resulted in a more comprehensive treatment of behavior in these groups. The review by Barth and Lester (1973) should be consulted for an introduction to the literature on blattid behavior.

Communication vs. Sensory Perception

On the basis of fitness changes that result from transfer of information between two organisms, three types of transmission can be recognized: (1) An act of information transfer that increases the fitness of the emitter and the receiver alike, causing selection to improve the mechanisms of emission and reception. (2) Emission as an incidental effect of other behavior or structures in which the fitness of the emitter remains unaltered. Under such conditions no evolution of emission occurs. (3) Emission incidental to other attributes, but which decreases the fitness of the emitter, causing selection to diminish the amount of information transmitted. Information emission by prey that informs predators of the prey's location is under negative selection, but the ability to receive information is under positive selection. A kind of chase between the emitters and the exploiting receivers ensues (Otte, 1975a, 1975b). Some highly interesting results in the evolution of cryptic coloration and posturing are possible

(Robinson, 1969a, 1969b). This chapter focuses principally on mutualistic interactions between emitters and receivers, but also considers cases of intraspecific signal exploitation.

Mechanisms of Signaling: Some Aspects of Evolution

Orthoptera have exploited most available modes of signaling potentially open to them. By most reasonable guesses, olfaction and touch probably preceded vision and hearing. As expected, nocturnally active groups tend to rely more on olfaction and hearing, while diurnal groups rely more on vision. Among nocturnal groups, cockroaches depend heavily on olfactory and tactile signals (Barth, 1965, 1970; Roth and Barth, 1964). Such a condition may be representative of the ancestral Orthopteroid stock. Gryllids and tettigoniids represent lineages that have entered the adaptive zone of acoustics. Certain nocturnal cricket genera (*Apterogryllus, Apteronemobius, Amusurgus*, and others) have secondarily lost their acoustic ability and may now rely mainly on olfactory cues (Alexander and Otte, in prep.). Among diurnal taxa, Acridoidea have evolved relatively acute vision and rely greatly on visual signaling, but subgroups of primarily nocturnal families have undergone secondary reversions from one signaling system to another. Certain diurnal crickets (for example, *Metioche, Homoeoxipha;* Fig. 1) have large eyes and have lost or partially lost acoustic signaling as a result of adopting a more diurnal habit (Alexander and Otte, in prep.). Implied here is the idea that the evolution of one system of signaling changes the relative importance of another—there occurs a tradeoff where one system increases in importance or acuity at the expense of another. No species appears to excel in all modes of signaling.

Fig. 2 is a partially hypothetical and partially factual representation of what the relative acuities of different groups may be like. The clustering of taxa is expected if a tradeoff between signaling modes is the rule. Orthoptera cannot

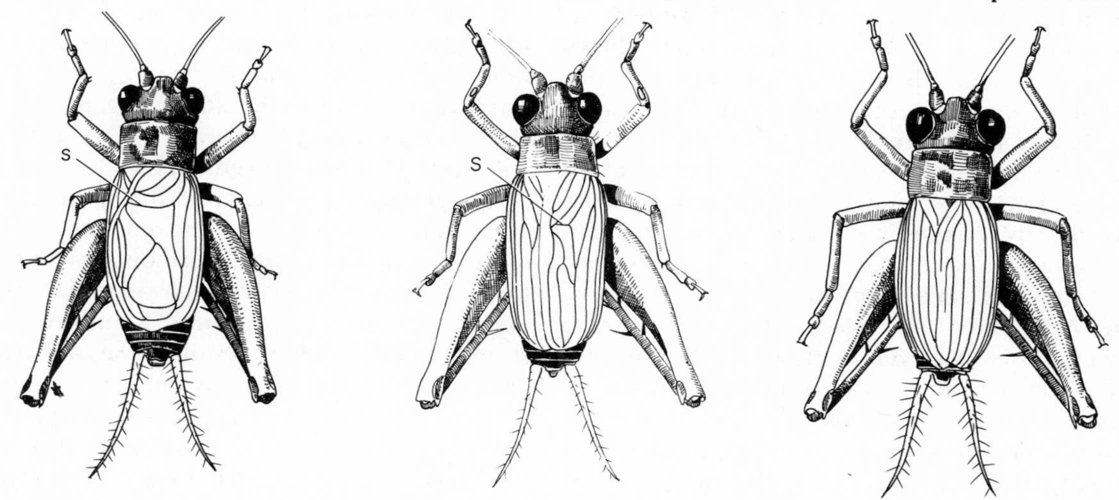

Fig. 1. Progressive loss of the stridulum (S) as shown in several related Australian crickets (Trigonidiinae). The first cricket evidently no longer produces sound even though the stridulum remains well developed. The stridulum of the second is a mere vestige of the normal condition in crickets. The third has lost all traces of a stridulatory file. (After Alexander and Otte, in prep.)

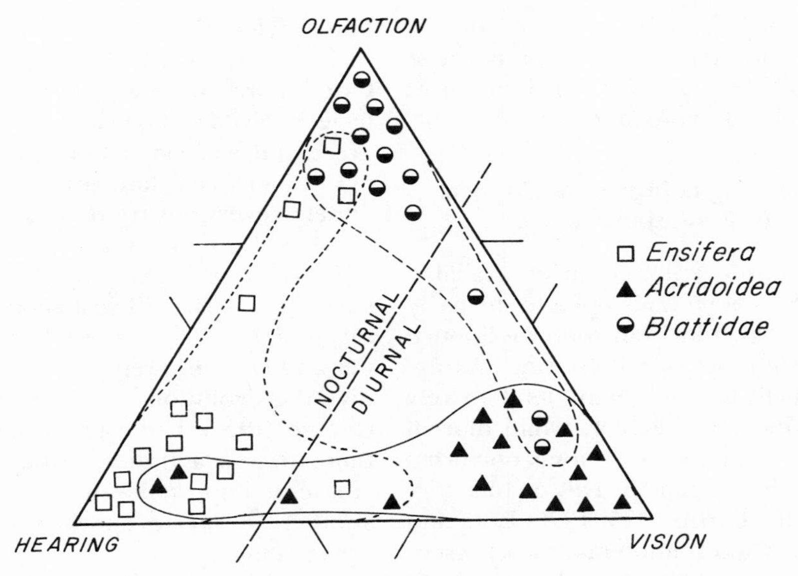

Fig. 2. Distribution of three major orthopteran groups on a triangle representing three major signaling modes. Each adaptive zone is dominated by one group.

(or at least do not) simultaneously have high visual, acoustic, and olfactory acuity. It also shows that each of the three adaptive zones is dominated by a major group. Diurnal and nocturnal zones can also be recognized, the latter having acoustic and olfactory subzones. Anatomical investigations into changes in neural networks that accompany changes in signaling would be of major interest.

From the standpoint of signaling methods, the Acridoidea are clearly the most diverse group. At least eleven methods of acoustic signaling are known in this group alone (Table 1). Crickets and tettigoniids, the other acoustic groups, possess only three known mechanisms of acoustic signaling: tegminal stridulation, in the vast majority of species; and femoro-abdominal stridulation (Richards, 1973) and antenno-frontal stridulation (Menon and Parshad, 1960) in one or two species each.

Some reasonable guesses can be made as to why grasshoppers should be so diverse. First, when they became diurnal and long-range vision was made possible, the stage was set for two classes of movements to evolve into visual signals: tactile signal movements such as jerking and repelling movements of the hind femora; and noncommunicative movements already closely associated with interindividual interactions, such as locomotary and orientation movements. Second, a specialized sound-receiving system (abdominal tympanum) was apparently acquired early and evidently only once by grasshoppers, perhaps in the context of predator avoidance; furthermore, it probably preceded most or all sound-producing mechanisms found today. Practically all members of large acridid subfamilies, even some silent subfamilies (Catantopinae), possess a tympanum. This indicates that the tympanum is ancient and that it could have evolved independently of specialized sound-producing mechanisms, since it can be maintained in their absence. Third, while the tympanum is very similar among widely divergent taxa, sound-producing mechanisms of the same taxa are very different, suggesting that they were not present in a common ancestor and must have been acquired independently. With a hearing mechanism already in existence, visual signals could repeatedly become acoustic with only slight modification of movements. The existence of a variety of sound-producing mechanisms in grasshoppers, some of them confined to a single species, suggests that visual signals and perhaps also various tactile movements became acoustic signals independently when the parts of the body involved came to rub against one another, probably accidentally at first. Subsequent specialization of these body surfaces into scrapers and files resulted in the mechanisms indicated in Table 1.

Table 1
Mechanisms of sound production in Acrididae.

A. Rubbing mechanisms
1. Femur III (F) — Forewing (S)
2. Femur III (S) — Forewing (F)
3. Femur III (F) — Abdomen (S)
4. Femur III (S) — Abdomen (F)
5. Tibia III (F) — Forewing (S)
6. Tibia III (S) — Forewing (F)
7. Tibia III (S) — Abdomen (F)
8. Forewing — Forewing
9. Hindwing — Hindwing

B. Striking mechanisms
1. Tibia III — Forewing
2. Tibia III — Substrate
3. Mandible snapping
4. Femur III — Forewing

C. Vibrating mechanisms
1. Femur III (rapid vibration)
2. Hindwing crepitation (vibration of membrane)

D. Expulsion of air from spiracles

NOTE: F denotes the file; S denotes the scraper. (After Kevan, 1954; Otte, 1970.)

The oedipodine *Encoptolophus sordidus* alone possesses four mechanisms (A2, B1, B2, and C2).

Once evolved, why did some crickets lose their songs? I suggested above that songs might have become superfluous when another mode of communication became more important. For example, olfaction might be a better system of signaling in subterranean forms, or vision may supercede hearing in diurnal forms, and so on. But Walker (1964) and Cade (1975) suggest that predation pressure might select against signaling by song. Walker demostrated that cats orient on and capture singing crickets using phonotaxis, and Cade showed that females of the tachinid fly *Euphasiopterx ochracea* locate male crickets (*Gryllus integer*) by their song and deposit larvae. The larvae burrow through the exoskeleton and feed internally.

There are various stages evident in the loss of calling songs. Cade showed that in *Gryllus integer* some males call while others do not. "For most of the night, cricket aggregations are composed of calling as well as noncalling males. Noncalling males, termed satellites, walk in the area occupied by calling males and attempt to intercept and copulate with females attracted by the calling males." He made the important discovery that noncalling males experience fewer parasite attacks. He also showed that if a calling male stops singing, one of the noncalling males begins to sing. In addition to being less susceptible to parasite attack, noncalling males are less likely to be attacked by the calling males themselves and may expend less energy in seeking mates by allowing calling males to attract the females (some of whom fly in to the calling male). The relative importance of these three factors (energy conservation, prevention of aggression, and avoidance of parasitic attack) in causing males to remain silent has not been assessed.

The next stage in song loss is evident in a Florida cricket *Gryllus ovisopis,* a relative of *G. integer* where the calling song is lost entirely; but males still perform courtship and aggressive songs. Why this species has lost its song while a sympatric relative retains its song is unknown. Walker and Mangold (in prep.) have attracted *E. ochracea* flies to mole cricket *(Scapteriscus acletus)* songs, and have raised adult flies from crickets infected artificially.

A third stage in song loss is evident in crickets that retain the acoustical apparatus, but do not sing (e.g. in some *Metioche*) (Alexander and Otte, in prep.). And a final stage is evident in some leaf-inhabiting and some burrowing species which have lost the sound-producing apparatus and, in some species, also the hearing organ (e.g., *Apterogryllus, Amusurgus*) (Alexander and Otte, in prep.).

What Walker's and Cade's studies suggest is that the predation costs attached to acoustical signaling may on occasion outweigh the female-attracting benefits. Whether predation or mate-theft ever selects for a genetically determined and frequency-dependent singing-male vs. silent-male dimorphism is not known, but it is conceivable.

Diversity of Signals

SIGNAL DIVERGENCE AMONG SPECIES

In a number of papers Alexander (1957, 1960a, 1962a), Walker (1957), Bigelow (1964), and others have emphasized the role of signal differences in preventing interspecific matings. Calling songs designate an individual's mating type and are, in effect, used by females to assess the genotype of potential mates. The precise stages by which signal differences arise remains problematical. Biologists generally agree that speciation is initiated through extrinsic separation of populations, usually geographically but occasionally seasonally (Alexander, 1968), and is completed upon evolution of appropriate identifying displays that prevent reproductive interactions between the diverged units. But the question of whether signal differences develop while populations are evolving independently of one another, as proposed by Mayr (1963), or whether the divergence takes place subsequently as a result of selection against interspecific interactions (Lack, 1947; Dobzhansky, 1970) requires further exploration.

Clear cases of song divergence in areas of sympatry have not been forthcoming in acoustic Orthoptera, a group ideally suited to detecting examples of it (Alexander and Walker, pers. comm.; Walker, 1974b). A careful analysis of Australian *Teleogryllus* species also failed to reveal differences between allopatric and sympatric populations (Hill, Loftus-Hills, and Gartside, 1972).

Walker (1974) discusses several possible instances of sympatric song divergence but finds the evidence too weak to conclude that divergence has occurred. He advances the following possible explanations for the rarity of this phenomenon:

(1) Sympatric divergence in song characteristics does not occur either because songs are not important, i.e., they are not under sufficient selection to produce divergence (he considers this unlikely) or because calling songs diverge sufficiently in allopatry that "when newly speciated populations become sympatric the songs are different enough so that no additional divergence occurs." He then cites examples of entirely allopatric species which have very similar songs and other pairs in which the songs are quite different.

(2) Sympatric divergence is difficult to detect because the critical song characters that diverge have not been examined, or because not enough songs have been analyzed to show a statistical difference between populations.

While the existing song patterns in relation to geography seem superficially to support the

Mayr model, i.e., that most species have already acquired song differences before they reestablish contact, Wallace (1970) presents a special case of the Dobzhansky model that might explain why few cases of character displacement are evident. According to this model one species arises from another as a small border population. Following contact with the parent population all individuals of the daughter population quickly come to possess song differences because they interact rather directly with members of the parental population. Subsequently the new species expands to become largely allopatric. Hill et al. (1972) advanced a similar hypothesis.

DIVERSIFICATION OF SIGNALS WITHIN SPECIES

In orthoptera the usual method by which a signal serving one function gives rise to several new signals, each with a different function, appears to be somewhat as follows: At an early stage a signal occurs in one context only; later the signal occurs in two contexts but remains structurally unchanged. Still later it becomes structurally distinct in the two functional contexts. The process may repeat itself as shown in Table 2. In crickets the calling song is probably ultimately derived from the courtship song. Some species have calling and courtship songs that remain structurally similar (stage 2), while others have signals that are very different (stage 3) (Alexander, 1962a). Alexander also postulates that aggressive songs originated as outgrowths of the calling song.

Table 2

Sequence of diversification of signal types in crickets, first through multiplication of function and then through structural specialization.

	Courtship	Calling	Agonistic
Stage 1	song A	—	—
Stage 2	song A	song A	—
Stage 3	song A	song B	—
Stage 4	song A	song B	song B
Stage 5	song A	song B	song C

Source: After Alexander, 1962b.

Rates of signal evolution and signal diversification seem to depend on signal function. Grasshopper courtship and calling signals have undergone extensive evolutionary changes, as though subject to particularly strong selection for change, while agonistic signals are more conservative (Table 3) (Otte, 1970, 1975). The table shows a net decrease in the number of signals in some taxa. Because reproductive signals are involved in species identification, unique codes are required for quick recognition of appropriate mating types. Cricket calling songs are more susceptible than courtship signals to change, evidently because selection operates more strongly on signals that promote early recognition (Alexander, 1967).

Many orthopteran species may possess several very different signals in certain contexts, while other species possess only one signal. For example, according to Spooner (1964), the katydid *Scudderia texensis* begins singing in the afternoon and stops late at night. Males sing different songs at different times of the day. Spooner shows some functional differences between signal types, but one wonders why some species manage to get along without them. Similar multiple signals are practically nonexistent in the Gryllidae but relatively common in Tettigoniidae and Acrididae.

Contexts of Signaling

PAIR FORMATION

Male and female Orthoptera form temporary pairs that last at best for several copulations. The precise sequence by which pairing is achieved varies from group to group, depending on the

Table 3

Courtship and agonistic signaling patterns in three groups of related oedipodine grasshoppers.

Species	Courtship	Agonistic
Trimerotropis pallidipennis	T — OS	T — S
T. maritima	T — OS — S	T — S
T. californica (Calif.)	T S	T — S
T. californica (Arizona)	S	T
Encoptolophus subgracilis	T — OS	T — S
E. costalis (Colorado)	T — VS — OS	T — S
E. costalis (Texas)	T — (TK) — VS — OS	T — S
E. sordidus	T — TK — VS — (OS) — VS'	T — S

NOTE: OS = ordinary stridulation, T = femur tipping, S = femur shaking, VS, VS' = vibratory stridulation, TK = ticking, ( ) = reduced expression. Each pattern is probably derived from the pattern preceding it. The top pattern is probably the most primitive since it is represented in many species. In some cases a single species, over its geographical range, provides several steps in a sequence.

method of signaling and on whether one or both partners signal during the process. Some of the principal pair formation sequences known are shown in Fig. 3. The figure emphasizes acoustic signaling because much is known about it. The main categories are: a. and b. A male signals and a female is attracted to him. c. Similar to b., but the female answers the male just prior to visual contact; the male then closes the gap. d.–g. Females answer signaling males, who then approach the female (in some phaneropterine katydids [Spooner, 1968] and in some pamphagid and pneumorid grasshoppers [Otte, 1970] males fly about in the field signaling and listening for female answers). h. Male and female wander about looking for one another in likely spots; they encounter one another by chance, and final recognition is achieved through visual and olfactory cues. Conditions clearly intermediate between these categories also exist.

COURTSHIP

Courtship behavior differs considerably among orthopteran families. Gryllids usually have distinct and elaborate courtship signals, especially in ground-dwelling groups, while tettigoniids do not (Alexander and Otte, 1967a; Spooner, 1968). Congeneric acridids have species-specific courtship behavior, sometimes elaborately developed (Otte, 1972a), but congeneric cricket species have quite similar patterns. The difference between crickets and grasshoppers in this regard may be due to the fact that accidental pairing in gryllids is less common. Since they are nocturnal for the most part, an accident in pair formation must involve actual contact. In grasshoppers accidental pairing is common because individuals can perceive moving individuals twenty five feet or more away and may be attracted to one another; but they are unable to distinguish between conspecific and heterospecific individuals at that distance. Thus, when accidents in pairing are common, courtship signals would be under stronger selection to become species-specific.

The considerable variation existing in courtship among subfamilies of crickets is described by Alexander (1962a), Alexander and Otte

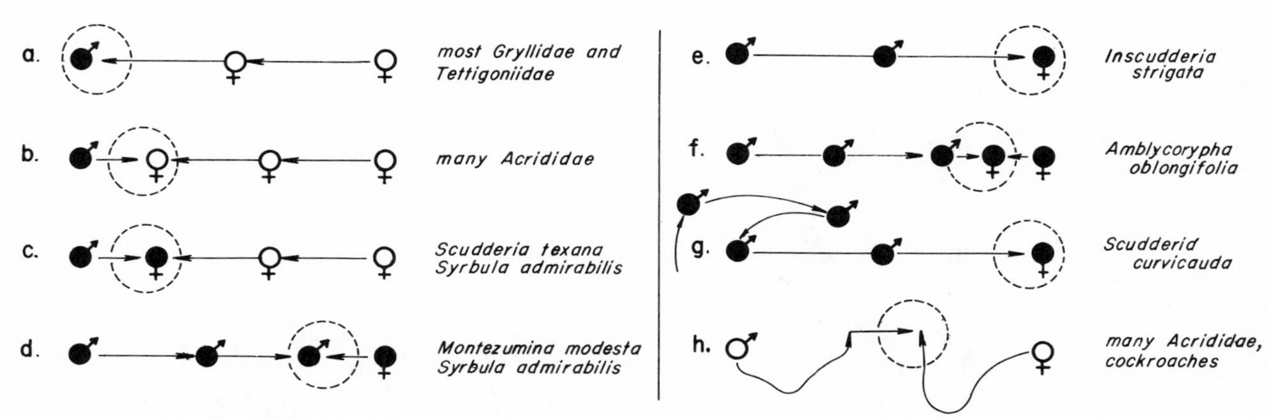

Fig. 3. Major categories of pair formation in Orthoptera. Black symbols denote acoustically active individuals; open symbols represent silent individuals. The broken circle indicates the zone of pair formation relative to initial positions.

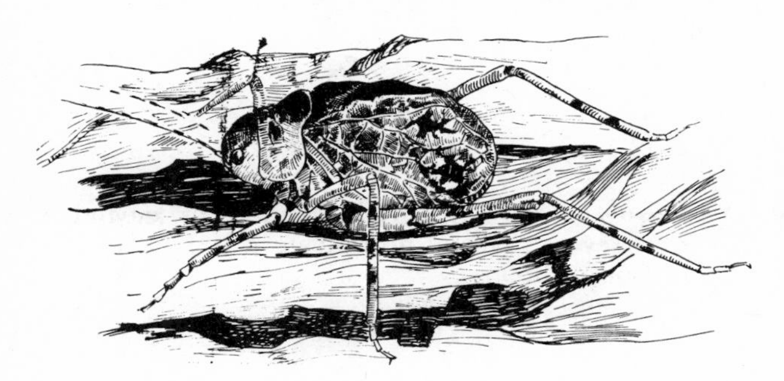

Fig. 4. Primary and secondary defense in an unidentified katydid collected on *Eucalyptus* bark in New South Wales. The raised-wing display reveals red, white, blue, and black coloration and was performed only after the insect was disturbed.

(1967a), and Otte (1970). In these papers the ecological significance of certain patterns is explored.

AGONISTIC AND AGGRESSIVE CONTEXTS

Most Orthoptera posess signals that promote spacing of individuals. Such signals are more elaborately developed in males than in females (for reasons outlined by Trivers, 1972). In some groups aggressive signals are mere elaborations of calling songs (e.g., gryllids: Alexander, 1962a; gomphocerine grasshoppers: Otte, 1970; Jacobs, 1953). In others, agonistic and pair-forming signals may be quite different (e.g., phaneropterine katydids: Spooner, 1968; oedipodine grasshoppers: Otte, 1970). In field crickets (Alexander, 1961), conocephaline katydids (Morris, 1971), and acridids (Otte and Joern, 1975) agonistic signals may act as threats that are backed up by physical attack, but in most Orthoptera the signals have at best a sex-identifying or spacing function.

In acridids, agonistic signals are evolutionarily more conservative than courtship or pair-forming signals, particularly in subfamilies where courtship and agonistic signals are quite

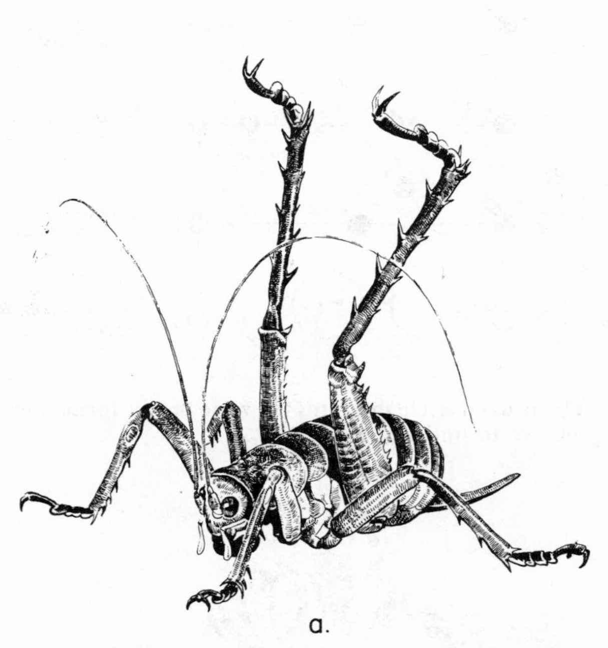

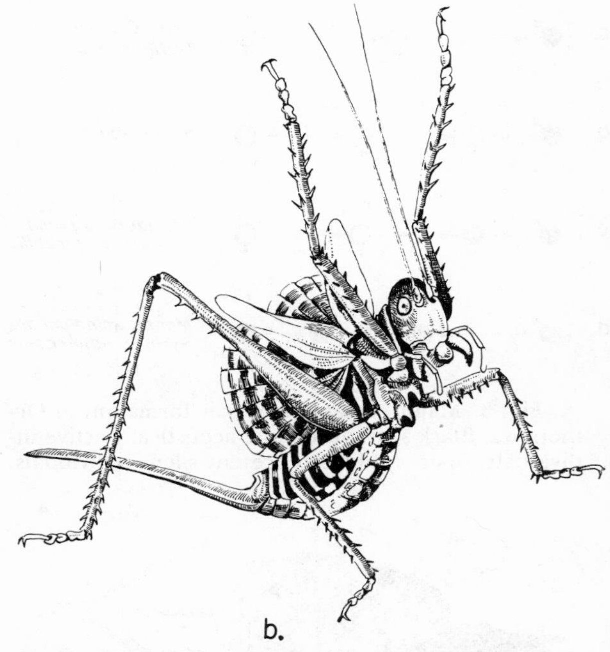

Fig. 5. a. Defensive posture in the giant weta (*Deinacrida*) of New Zealand. Individuals also stridulate by rubbing their hind femora against the abdomen. (After Richards, 1973.) b. Defensive posture in *Neobarrettia*, a North American predaceous katydid. Bites are severe. Males also stridulate while posturing. (After Cohn, 1965.)

different from one another and perhaps where they are not neurally linked.

DEFENSE

Orthopteran defense is of two kinds (Robinson, 1969a): primary defense, in which animals attempt to prevent attacks from being initiated, through hiding, cryptic behavior, and cryptic morphology (Fig. 4); and secondary defense, in which animals attempt to stop an attack that has been initiated by direct counteroffensive measures such as poisonous tissues, toxic sprays or liquids, or dangerous weapons such as spines or biting mouthparts. Secondarily, displays such as aposematic coloration and conspicuous and intimidating postures evolve, which warn predators of impending danger (Fig. 5).

Romaleine and pyrgomorphid grasshoppers are poisonous and discharge poisonous odors or froth, which deter attacks from predators. These groups are also characteristically brightly colored and tend to form tight aggregations, especially in the nymphal stages. In Argentina nymphs of the romaleine *Chromacris speciosa* are black with small red spots. When a tight cluster is disturbed the nymphs scatter, but reassemble within minutes. The members of a cluster are probably typically siblings, but clusters composed of two very different size classes are common. This species feeds on the poisonous solanaceous plant *Cestrum kunthi* in northern Argentina. Similar aggregations are known in the desert romaleine (*Taeniopoda eques:* Alcock, 1972).

In the African pyrogomorphid *Poekilocerus bufoni,* nymphs can eject poisonous secretions up to 60 cm by arching the back and spraying over their heads (von Euw et al., 1967). In adults the secretions run down the side of the body and over the spiracles. Air forced out of the spiracles is mixed with the gland secretion and forms a pungent-smelling froth. Toxic compounds are derived from their diet of Asclepiad plants, which are known to be rich in cardiac glycosides. These insects have several lines of defense: bright coloration, which deters predators from attacking; ejection of a jet or foam of defensive fluid containing cardenolides and histamine; possession of a penetrating and disagreeable odor perceived at several meters; and possession of cardenolides in their body tissues. Quite possibly, the second and last characteristics are functionally linked, and bright coloration evolved secondarily as a warning device.

The walking stick *Anisomorpha buprestoides* ejects a toxic spray toward birds in its vicinity even before being attacked. Once they have been sprayed, birds strongly avoid walking sticks (Eisner and Meinwald, 1966).

SOME GROUP-RELATED ACTIVITIES

Migratory locusts display various activities mediated by visual, tactile, and olfactory signals or cues that seem to promote formation and maintenance of large cohesive migratory swarms. Contact between developing locust nymphs promotes aggregation of individuals and leads to the development of conspicuous (black and yellow) coloration. The latter may further enhance aggregation through visual attraction (Ellis, 1963, 1964; Ellis and Hoyle, 1954). Black coloration is evidently not aposematic, but may be an energy color, which promotes faster development through greater absorbtion of solar energy (Hamilton, 1973). Locust odorants also have the effect of hastening or slowing development, resulting ultimately in nearly simultaneous maturation to the adult stage, an important requirement for the formation of mixed swarms in which mating takes place during or after migration. Nymphs exposed to the odors of adults grow more slowly in the presence of younger animals (Norris, 1954, 1964, 1970). Odors may affect gamete formation as well, causing changes in chiasma frequency and ultimately in the production of a more variable batch of offspring (Nolte, 1968). Ovipositing females of *Schistocerca gregaria* are also strongly attracted to one another during oviposition and lay their eggs in a group (Norris, 1970). Similar attraction of females can be achieved with immature and even dead animals. These studies indicate that it is important for *Schistocerca* females to lay eggs where other females lay them; this ensures that developing nymphs will group with numerous other individuals. Their chances of surviving and successfully migrating from unsuitable to suitable areas may be enhanced when they move as a group. Similar group effects have been demonstrated in house crickets (Chauvin, 1958).

Signal Interactions

Signaling individuals can interact in many ways (Alexander, 1975). Here I will concentrate on acoustic interactions among males attempting to maximize their female quota. The main categories are outlined in Table 4.

RANDOM ACOUSTIC ACTIVITY

The signals of males may be temporally and spatially relatively independent of one another. Yet, indirect interactions result when two males attempt to attract the same female.

UNSYNCHRONIZED CHORUSING

Female-attracting signals of many orthopteran species are temporally and spatially interdependent, resulting in the production of bursts of

Table 4

Categories of interaction between male Orthoptera involving acoustic signals.

I. Interaction between singing males
  A. Random acoustic activity; songs temporally independent; interaction indirect through a common receiver. (C)
  B. Nonrandom acoustic activity
    1. Unsynchronized chorusing; animals acoustically active at the same time and in bursts (Alexander, 1960a; Otte, 1970, 1972a). (C, AP, CP)
    2. Synchrony; individual song elements synchronized (Walker, 1969). (IR, C)
    3. Alternation (Shaw, 1968). (IR, C)
    4. Spacing (Alexander, 1961; Otte and Joern, 1975; Morris, 1971; Spooner, 1968). (C)

II. Interaction between singing and nonsinging males
  A. Mate theft (Spooner, 1968; Otte, 1972a; Cade, 1975). (C)
  B. Aggression and territoriality (Alexander, 1961; Morris, 1971; Otte and Joern, 1975; Cade, 1975). (C)

NOTE: The categories are not mutually exclusive. Symbols indicate possible relationships between interacting males. C = reproductive competition, AP = antipredator or defensive tactic, CP = cooperative, IR = interference reduction.

activity or in aggregations of males. Thus, in several species of oedipodine grasshoppers with loud, conspicuous flight displays, a number of males may become active all at once, and a hillside that has remained silent for many minutes suddenly becomes noisy with the loud buzzing sounds of dozens of flying males. In August 1973, a wheat field on the plains of Colorado was silent for more than thirty minutes after I arrived, then gave forth to flight displays by hundreds of males of the oedipodine grasshopper *Aerochoreutes carlinianus* over the entire field. A few minutes later all were silent again.

Both crepitating and stridulating species chorus. In the gomphocerine grasshopper *Syrbula admirabilis* males in the field and in the laboratory tend to sing at the same time so that silent periods alternate with bursts of activity (Otte, 1972a). The adaptive significance of such choruses is poorly understood. Experiments carried out on *S. admirabilis* suggest that such interactions are the result of intermale competition. Experiments show that two males singing at the same time and within hearing range of females divide the females between them, with the leading male having a slight advantage over the following one. It was also shown that two songs emanating simultaneously from the same loudspeaker are not any more or less effective in attracting females than the song of one male emitted by another speaker at the opposite end of the arena. Thus, males may attempt to interfere with other males' songs by making it more difficult for females to orient. The adaptive value of such simultaneous singing may lie in the ability of a male to cause females to remain available a little longer.

SYNCHRONY

Snowy tree crickets *(Oecanthus fultoni)* synchronize their chirps in such a way that a tree full of males can be perceived as a single rhythmically pulsating unit (Walker, 1969). The adaptive value of synchrony appears to be that a male by synchronizing reduces the interference of a neighbor with his chirp rhythm, the component of the song most significant to females searching for males. Whether or not synchrony also facilitates attraction of males from outside the chorus is problematical. If several males singing in unison increase their quota of females as against solitary or nonsynchronizing males, they can be viewed as cooperating with one another.

Synchrony between pairs of males depends on auditory stimuli, and the lead between males may change frequently. The mechanism of establishing synchrony involves either a temporary lengthening or shortening of both chirp intervals and chirp lengths by the individual attempting to synchronize with another male or with a chorus (Walker, 1969). The occurrence of lengthening or shortening depends on the phase relations between his song and that of the chorus (Fig. 6).

Alexander (1975) observed synchrony in two katydid species, but in both species it is evidently rare. On the Ohio State University campus he observed a very dense population of *Orchelimum vulgare* synchronizing in the same bed of tiger lilies day after day. Evidently the species synchronizes only under conditions of high density. In *Neoconocephalus ensiger* synchrony is also rare. In this species, males only synchronize at low temperatures when the chirp rate is greatly slowed down. Alexander (1960) reasons that:

> It scarcely seems likely that these males have been selected to synchronize when the conditions under which they can do so effectively are rarely encountered. Rather, their songs appear to become synchronizable at very low temperatures as an incidental effect of their structure at more usual singing temperatures.

What songs are synchronizable? According to Alexander (1960), for songs to be synchronizable they should contain "a precise or highly uniform chirp or phrase rate within the range of two to five per second. . . ."

There may be some interesting causal relations between synchrony (temporal clustering) and aggregation of individuals (spatial clustering) which need to be explored. Dense aggregations of males may be environmentally imposed (high overall population densities or the forced aggregation of males on resource patches), or males may voluntarily form aggregations. Synchrony is only possible when interindividual spacing is greatly reduced, but under those conditions it is expected to arise only when song elements are synchronizable and when there is some clear advantage to the participating members to retain a conspicuous rhythm. It is even conceivable that low synchronizability would sometimes impede the evolution of voluntary aggregation if the temporal rhythm were an important component of the signal. (Readers should consult Alexander, 1975, for a lenthy discussion of this and related points.)

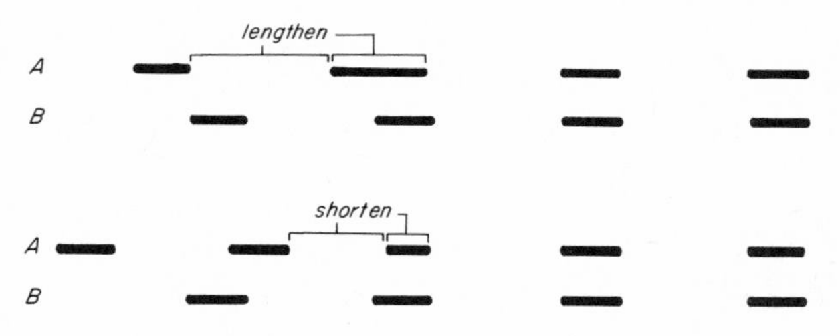

Fig. 6. Mechanism of synchronization in the cricket *Oecanthus fultoni*. Black marks denote entire chirps. The individual attempting to synchronize may either lengthen or shorten his chirp and chirp interval. Which he does depends on where his chirps fall in relation to the chirps of other males. (After Walker, 1969.)

ALTERNATION

In Goiania, Brazil, I listened to a species of tree cricket whose solitary song was superficially similar to that of *O. fultoni.* Adjacent males did not synchronize their chirps, but alternated instead. And, by roughly halving their chirp rate, they nearly maintained their original chirp rhythm. Shaw (1968) examined in detail a mechanism of chirp alternation in the katydid *Pterophylla camellifolia,* where males singing alone have a faster chirp rate than do males alternating with one another (Fig. 7). Most interactions consist of the entrainment of each katydid to a slower chirp rate because of inhibition by the other individual, plus intermittent escapes from entrainment. Alternation can be disrupted if the leader begins to solo before the termination of the follower's chirp.

In some grasshoppers, alternation occurs between aggressive songs of two males. When the pulse rate is very rapid (on the order of twenty or more per second), alternation occurs between successive songs, but when the gaps between pulses are great, alternation between pulses is possible (Otte, 1970).

SPACING

Signals that promote spacing of individuals are known in crickets (Alexander, 1961), conocephaline katydids (Morris, 1971), and Acrididae (Otte and Joern, 1975). Among gryllids, territorial defense and the signals that result in spacing are selected only in certain ecological situations (Alexander, 1961). The most strongly territorial species are those that live in burrows or on the ground. No arboreal species is known to defend territories, but some spacing may occur. Territorial defense is more prevalent among species that are sedentary and located in defensible regions, and it may be more likely to develop in predaceous species that are already equipped for attacking other insects. Morris (1971) describes aggressive interactions in conocephalines, where males do not restrict themselves to a given site, but attempt to clear their surroundings of other singing males.

The only territorial acridid known is a gomphocerine species, *Ligurotettix coquilletti,* which defends bushes of *Larrea divaricata* (creosote) in the Sonoran Desert (Otte and Joern, 1975) (Fig. 8). Proximate resources over which males fight are medium to large bushes. Ultimately such

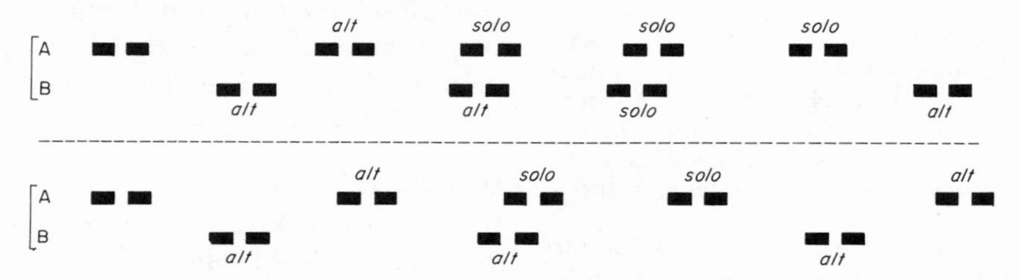

Fig. 7. Alternating and soloing in the true katydid (*Pterophylla camellifolia*). Each double unit denotes one chirp. (After Shaw, 1968.)

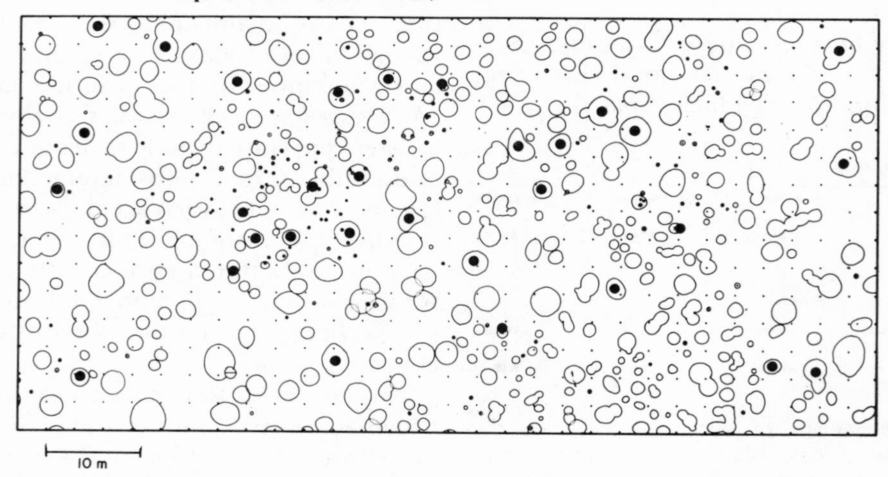

Fig. 8. Position of all resident males of the grasshopper *Ligurotettix coquilletti* on a 45 m X 90 m grid in the Sonoran Desert, on July 31, 1972. Individual males remain for as long as a month on one bush and defend it against other males. Black circles, males; large open symbols, creosote bushes. (After Otte and Joern, 1975.)

bushes are probably preferred because it is to these that females are most strongly attracted. Males of *Ligurotettix* begin singing in the morning and continue all day and into the night. Males that enter an occupied bush are sought out and attacked by the resident male, especially at low grasshopper densities, when the ratio of males to suitable bushes is low. By behaving territorially a male can, through defensive behavior, prevent mate theft and thereby increase his quota of females. Under high densities some abatement in territorial defense seems to set in, and it is possible to find bushes with several singing males. (See Otte and Joern, 1975, for a discussion of density-dependent aggressiveness).

Males of *Goniatron planum*, a close relative of *Ligurotettix,* are variably territorial in much the same way. In west Texas (near Marathon) host bushes *(Florensia cernua)* were small and males readily flew from the bushes when approached. In this region a single male sings in any one bush, but bushes frequently contain one to four silent males as well. We are uncertain of the tactics of these silent males, but it is clear they are in a good position to intercept females attracted to the bush. By remaining silent they might either reduce the chances of provoking attacks by the signaling male or they may be attempting to locate females without themselves having to expend energy in calling. Singing males failed to attack silent ones even though they seemed to be aware of them. The lower aggressiveness may have been due to the high density of males. In northern Mexico a sparse population of *Goniatron* was found in large bushes. Here males attacked one another when placed into the same bush, and no satellites were found.

MATE THEFT

Spooner (1968) describes attempted theft in several species of phaneropterine katydids. Males of this group are particularly susceptible to being robbed because females answer male songs during pair formation. *Scudderia texensis* males possess two female-stimulating songs, slow-pulse songs (sps) and fast-pulse songs (fps). Females respond to the low-intensity fps by approaching the male without answering, and to loud fps by stopping. When quite near the male, females respond to the sps by answering. The singing male then searches and approaches the female. When females answer males, other nonsinging males approach and attempt to steal copulations. However, singing males may have evolved a mechanism for circumventing theft. After the female has answered, but is still some distance from him, the singing male reduces the intensity of fps, thus causing the female to become silent and to approach still further, where he is more likely than the nonsinging males to find her.

In *Syrbula admirabilis* nonsinging males become highly excitable when they hear a female answering the song of another male. They rush about searching for her, occasionally reaching her before the calling male does. Theft in *S. admirabilis* and *S. fuscovitta* also occurs during courtship, which in these species is quite prolonged (Otte, 1972a). Courting males inadvertently attract other males, who assemble about the courting pair. When the female signals receptivity, noncourting males make a sudden rush, attempting to mount the female. Courting males sing very softly, perhaps to reduce the chances of theft to a minimum. Males also court much more vigorously when other males are about. In *S. fuscovittata,* wing flipping is usually absent when a male courts a female alone but is prevalent when other males are nearby. One can speculate that increased intensity advertises the fitness of the performing male and insures that the female perceives who the real performer is.

Cade (1975) has shown that singing cricket males *(Gryllus integer)* are frequently surrounded by silent (satellite) males, who may intercept females attracted to the calling males. He feels that such theft is at least partially accounted for by the fact that singing males are more prone to parasite attack (see p.337). Of course the selective effect of the parasitoids would depend on how soon they incapacitate a male. One might even predict that it would be advantageous for the first batch of larvae to be deposited on a male to silence him, thus ensuring the larvae of a greater share of the resource. Such a mechanism may operate in a cicada parasitized by the sarcophagid fly *Colcondamyia auditrix,* where the larvae silence singing males (Soper, Shewell, and Tyrell, in Cade, 1975).

Clustering in Space and Time

Dispersion of signaling animals in space varies from highly dispersed to strongly clustered. Likewise, signals themselves may be highly independent of one another or clustered in time. Describing an animal's position with respect to these two axes is of some interest (Fig. 9). I have placed no animals near the horizontal axis because some degree of mutual attraction should always be beneficial. An individual who finds that he is the only male displaying in an area may take this to mean that females are also absent, and consequently he may seek the location of other males. Lowering the density may have the effect of moving all species toward the origin. The models presented below are attempts to account for clustering in time (models 1, 2, and 3) and in space (models 4 and 5). Recall that clustering in time is of two kinds: unsynchronized chorusing and synchrony.

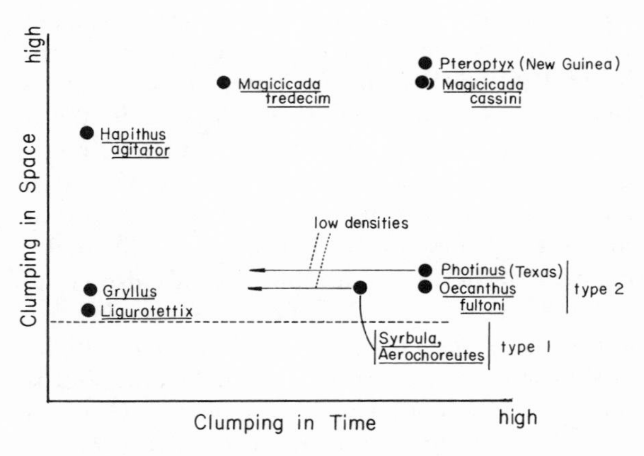

Fig. 9. A plot of the dispersion in space of singing animals and of their activities in time. Clumping imposed by extrinsic factors is omitted. Type 1, unsynchronized chorusing; type 2, synchrony. Arrows indicate changes expected to occur when densities are reduced.

THE INTERFERENCE REDUCTION MODEL

The precise synchrony in tree cricket songs may serve to reduce song interference. A male in effect synchronizes with another because in this fashion the other male interferes only minimally with the species-specific rhythm of his own song. Animals that synchronize for this reason would not be expected to form spatial aggregations, but if they are spatially aggregated for other reasons, they may be under greater pressure to synchronize.

THE INTERFERENCE OR INTERLOPING MODEL

Suppose the following conditions are met: (1) A lone male in the presence of receptive females attracts all receptive females capable of hearing him. (2) Each of two matched males singing independently of one another gets half the females (or has a 50 percent chance of attracting any given female). (3) Singing during the call of another male reduces the effectiveness of that particular call. (4) A male that occasionally sings alone, but interferes every time the other male sings, attracts more females than the other male. The following strategies by males might then obtain: (1) It is best to be alone among females (a tactic successfully employed by territorial species). (2) If other calling males are nearby, it would be advantageous for a male to call during silent periods, so that the location of his call remained clearly defined. (3) It would also be advantageous to call during the call of another male so as to interfere with that male's ability to attract females. Two males interacting and utilizing the same tactics will each attempt to sing alone and, whenever possible, to interfere with each other's singing. The inevitable result in a closely matched pair of males is an overlap of songs with perhaps some alternation in who sings first (Fig. 10).

THE ANTIPREDATION MODEL

This model simply says that a predator might have greater difficulty in locating one individual when many become active simultaneously than it has in locating an isolated individual. If predation were the primary force causing chorusing, males would not be expected to aggregate. The case of *Aerochoreutes carlinianus* discussed above fits this model.

THE LEK MODEL

Many animals display in groups (leks) rather than individually (Alexander, 1975). Such leks are aggregations of males mutually attracted to one another (in contrast to passive aggregations discussed below). Since degrees of aggregation vary widely, and since at some level all organisms are aggregated, one may have difficulty deciding in any particular case whether individuals have been attracted to one another. The following model might explain why males sometimes aggregate and sing simultaneously: Suppose that in an imaginary field receptive females search randomly for displaying males. Suppose also that once they are perceived by females several males acting together attract more than their share of females because of one or more of the following factors: (1) Their area of influence is larger and is also more likely to be encountered since it subtends here a larger angle (Fig. 11). (2) The surface of attraction is greater. (3) The larger area of influence is less likely to be overshadowed by a smaller area. (4) Two animals together constitute a supernormal stimulus and hence have a greater probability of attracting mates. Under these assumptions, one can construct the relations shown in Fig. 12. If a group of males attract more than their share of females, it becomes advantageous for males to associate with other males. When numerous males are at-

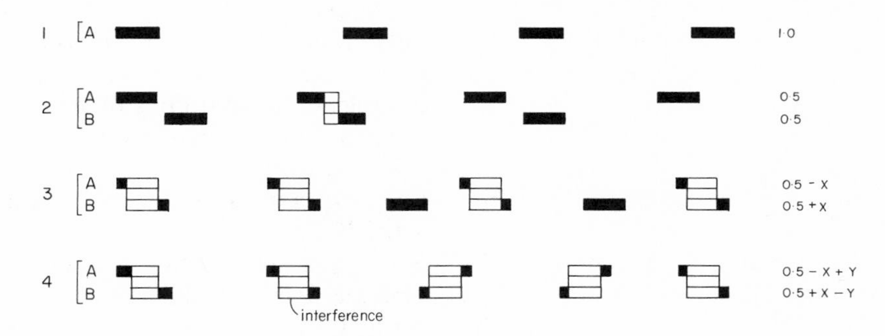

Fig. 10. Interference model of chorusing as may be employed by grasshoppers (*Syrbula*). If simultaneous singing constitutes interference, then interactions 2 and 3 are unstable, and in time each will tend to be replaced by the interaction beneath it. 1. Lone male receives all females. 2. Males A and B singing independently occasionally interfere with one another by chance. 3. Male B interferes with the songs of A and thereby gains a greater share of females. 4. A and B are equally matched; each attempts to interfere with the other. X=loss to A due to interference from B, Y=loss to B due to interference from A.

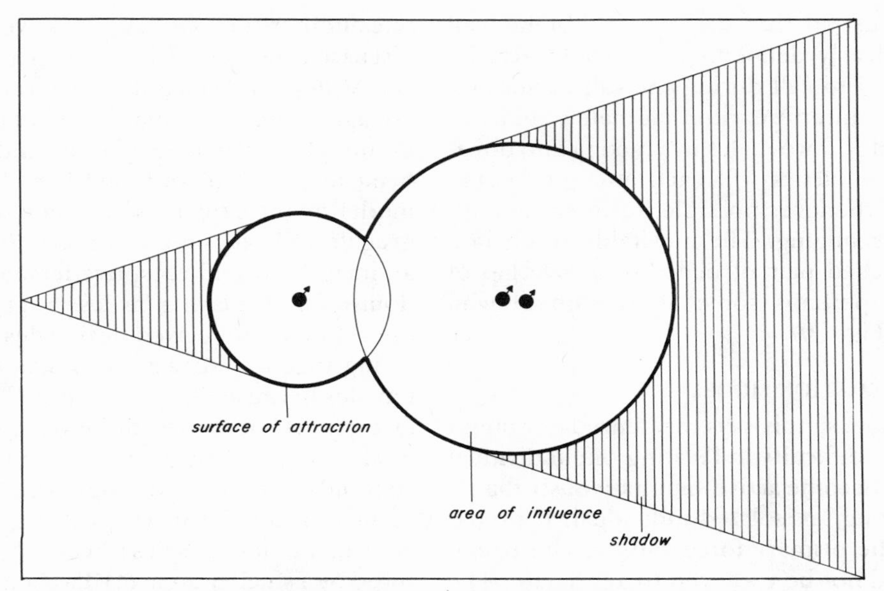

Fig. 11. Lek model of chorusing. Two males attract more than their share of females by having a larger area of influence, or a larger surface of attraction, or a greater chance of being encountered by randomly wandering females, or casting a larger shadow, which reduces the chances that a single male will be found.

tracted into the group, then the number of females attracted per male drops back to a low level, but it continues to be advantageous to aggregate if females refuse to approach lone males. If different females are restricted to different parts of the field several clusters of males could develop.

PASSIVE AGGREGATION MODEL

Males of some species are aggregated not because they are attracted to one another but because some aspect of the environment forces aggregation. Spooner (1968), for example, finds that groups of *Inscudderia strigata* males may be aggregated on their food plant, *Hypericum fasciculatum.* That the aggregation is passive is suggested by the fact that nymphs are also found aggregated on these plants. It is clear, however, that when forced together males behave differently than when they are alone.

Control of Signaling

NEUROMUSCULAR CONTROL OF EMISSION

Neuromuscular aspects of signaling have been analyzed rather extensively in crickets and grasshoppers. I can merely outline some of the findings of several lines of research.

Experimental work indicates that sound-producing movements are the expression of interaction between a series of thoracic muscles whose activities are coordinated centrally in ventral nerve ganglia. The motor activity is evidently generated by a small number of neurons that control the basic rhythm and coordinate the stridulatory muscles and that are little influenced by phasic feedback. In some species, the motor program is extraordinarily complex. Species in the European genera *Gomphocerippus* and *Myrmeleotettix* and the North American genus *Syrbula* utilize five to seven body parts and perform ten

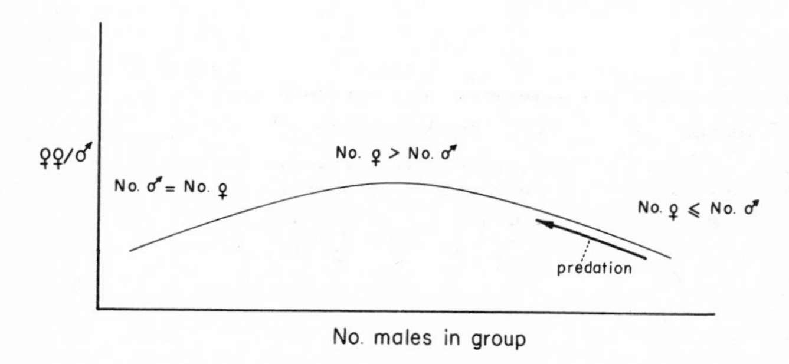

Fig. 12. If the chances of attracting females are greater the larger the aggregation of males, there will be an initial advantage to aggregating males, but when all the males are aggregated the advantage is lost. However, it remains advantageous to aggregate if isolated males attract no females. A return to solitary displays would be favored under severe predation.

to eighteen recognizably different movement patterns (Fig. 13). The functional aspects of complex courtship are discussed elsewhere (Otte, 1972a, 1975). In *G. rufus* hind femora, palpi, and antennae move synchronously, while the forepart of the body is raised and the head is moved from side to side periodically (Loher and Huber, 1966; Elsner, 1971). Courtship in *G. rufus* is normally released when a female is detected visually or acoustically. Blinded males may court after hearing a female's stridulation, but blind and deaf males may also court after tactile stimulation or "in vacuo." Thus different inputs can trigger the mechanisms that produce the motor output (Elsner, 1971). Several experimental techniques, including removal and immobilization of body parts, making central and peripheral lesions, and implantation of electrodes, have yielded what appear to be reasonably comprehensive pictures of neural control. It has been possible to implant as many as sixteen electrodes into freely moving grasshoppers without significantly affecting their behavior (Elsner, ms.).

In *G. rufus* coordination between the hind legs and the head changes in different subsequences. The motor systems of the head and hind legs are strongly coupled, and each chirp is accompanied by a burst of head muscle activity (Elsner, 1971). The motor pattern underlying courtship behavior in this species is programmed mainly in the CNS. Peripheral input has a minor influence on the quality of muscular activity. During the whole behavioral sequence, supra- and sub-esophageal ganglia and all three thoracic ganglia send coordinated motor commands to the muscles of the head, antennae, palpi, and legs. Participating interneurons are distributed over the cephalic and thoracic part of the CNS and synapse in all ganglia. These interneurons may determine not only the course patterns, i.e., the onset of different subsequences, but also the timing of chirps. The fine pattern, i.e., the pulse pattern within single chirps, is thought to be organized by local thoracic networks, which are driven by those interneurons. The hypothesis that individual command fibers time the start of the different parts of the motor patterns is appealing because gradual transitions between subunits cannot be observed.

The overall synchrony between body parts displayed by *Syrbula* and *Gomphocerippus* suggests how complex courtship patterns might have evolved. Increased complexity might have been produced by increasing the influence of single command fibers or various motor units. It has been postulated that the process may have occurred as follows: Initially only the hind legs were employed in signaling, but slight movements of other appendages were produced, perhaps because nervous commands loosely coupled with other motor units indirectly affected other motor patterns as they traveled from their origin to their destination in the third thoracic ganglion. Coupling between command fibers and various motor units that control palpi, antennae, and wings could have been under selection to improve if individuals displaying more movement were favored by females over individuals displaying less (Otte, 1972a).

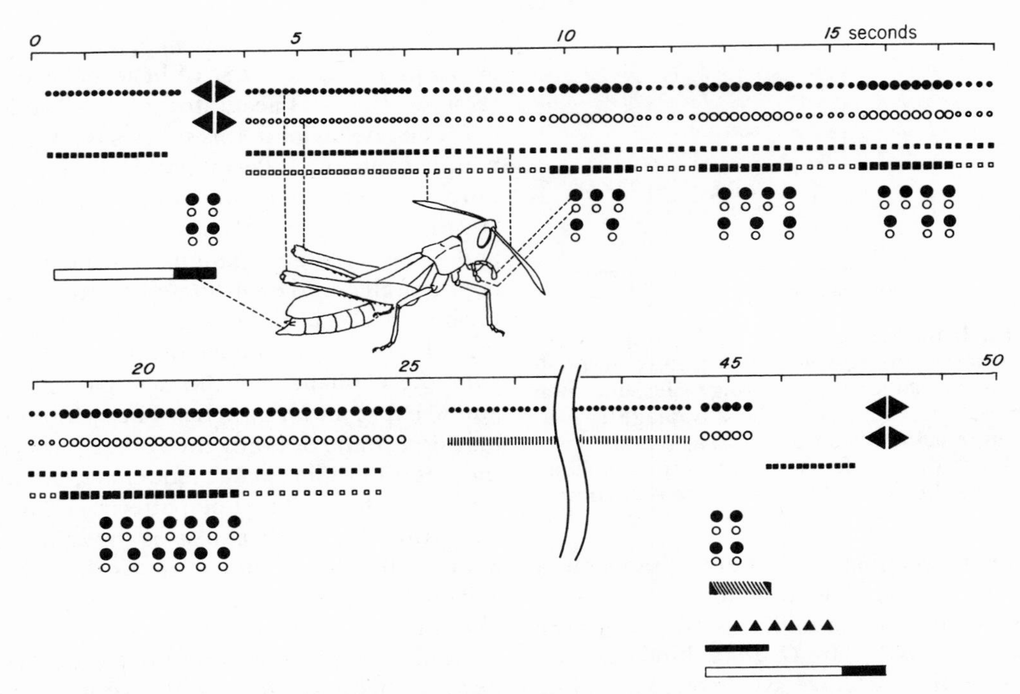

Fig. 13. A full cycle of courtship in the grasshopper *Syrbula admirabilis*, showing approximate temporal relations between various movements. Each cycle may be repeated twenty or more times in succession. Receptive females respond at the end of the sequence, allowing males to mount. (After Otte, 1972a.)

In crickets, brain commands may initiate or trigger song production, but the sequence and intensity of brain stimuli are variable. Commands affecting calling, aggression, and courtship are transmitted via separate fiber systems. Sound patterns themselves appear to be organized in the thoracic nervous system, since the calling song can still be generated when the head is removed (Otto, 1971). The calling song in *Gryllus campestris,* comprising a series of chirps, could result from the activities of a slow (3–4 Hz) thoracic oscillator that determined the chirp rate and a fast (30 Hz) oscillator that determined the pulse sequence. The structure of the oscillators is not known, but they appear to be located in the pro- and metha-thoracic ganglia. The rhythm of the slow oscillator coincides in *G. campestris* with the respiration cycle and with muscles involved in flight and walking; hence they are also believed to be influenced by these oscillators (Kutsch, 1969). But chirp and pulse rate vary widely among crickets, so the relationship between chirp and respiration cycles may be fortuitous or at least of no great consequence.

NEURAL CONTROL OF RECEPTION

Experiments by Stout and Huber (1972) indicated which components of male cricket chirps are transmitted to the female's brain. Recording from neurons in the cervical connectives (between subesophageal ganglion and brain) showed that several types of units are involved. Chirp coding units respond to entire chirps and therefore transmit information on chirp duration, while pulse coding units respond only to individual pulses (Fig. 14). In addition there ex-

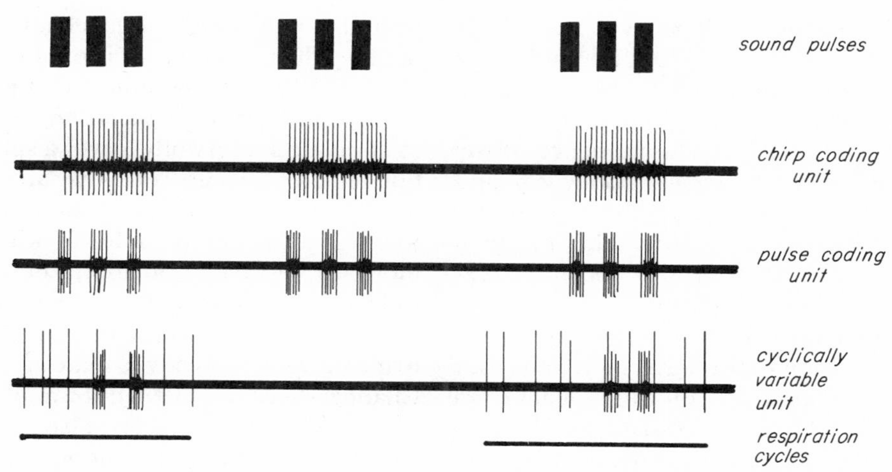

Fig. 14. Diagrammatic representation of recordings made in the neural connectives of the cervical region in *Gryllus campestris* exposed to the conspecific male call (see text for explanation). (After Stout and Huber, 1972.)

ist units that are variably responsive to chirps and fire only during respiratory cycles.

Cricket species with slightly different song parameters frequently coexist in the same habitat. While two species rarely if ever have the same song at the same temperature, there may be some overlap in songs over a range of temperatures (Walker, 1957). Thus, a female cricket may find the song of her own species on a cooler night to be the same as that of a related species on a warmer night. Since female responses are also temperature-dependent and a female is likely to be at the same temperature as the male she hears, this does not present a problem. But, given that the connectives to the brain transmit various song parameters, how is recognition of the song that is appropriate to a given temperature achieved? According to the Stout-Huber model, respiratory cycles, which are temperature-dependent, could act as timers against which song parameters are compared. Thus, the ratio of variable bursts per respiration burst could be the important cue. The model appears attractive in the case of *G. campestris* but suffers in at least two regards: in many cricket species songs are continuous trills with very fast pulse rates, and are therefore tenuously coupled to respiration cycles; and the coupling between respiration cycles and decision-making central neurons in females seems on intuitive grounds rather loose: a better mechanism might obtain if the receiving template itself were temperature-dependent.

NEUROENDOCRINE CONTROL OF SIGNAL EMISSION AND RECEPTION

The principal neuroendocrine elements controlling the onset of sexual activity and receptivity in Orthoptera are the neurosecretory cells (NSC) of the pars intercerebralis in the forebrain and certain cells lateral to the pars (Barth and Lester, 1973). Axons connect the NSC to the corpora cardiaca (CC), a pair of structures behind the brain. Attached to the posterior tip of the CC are the corpora allata (CA), which are also innervated by neurosecretory axons. The role of these various structures has been investigated in only a handful of species and appears variable among orthopteran taxa (Barth and Lester, 1973). In *Locusta migratoria* NSC comprise C and A+B cell types. C cells control sexual behavior directly by releasing hormones into the blood and indirectly through their effect on the CA, which in turn influences the intensity of mating activity (Pener, Girardie, and Joly, 1972). The influence of the CA on the behavior of *Syrbula fuscovittata* and *Gomphocerippus rufus* appears tighter than in *Locusta.* Allatectomy shortly after the imaginal moult prevents females from ever becoming sexually receptive (Loher, 1962, 1966).

Orthopteran neuroendocrine systems influence communication in two ways: indirectly, they influence the maturation of the gonads and accessory glands, which then cause individuals to engage in sexual activity; directly, they control the nature of behavior and chemical stimuli, which act as signals. Much variability exists between the few species examined in detail and even between the sexes of one species, making it difficult to set forth generalizations valid for large groups. In ovoviviparous cockroaches, female receptivity is correlated with oocyte maturation, but the CA appears to have far less influence on female receptivity than it does in grasshoppers. Allatectomy in the early stages in the grasshopper *G. rufus* results in females that never become receptive (it does not influence sexual behavior of males), but in some cockroaches allatectomy merely delays the onset of receptivity (Roth and Barth, 1964). Pheromone production in roaches is controlled by CA juvenile hormone, which stimulates female pheromone production.

Sperm of orthopteran males is generally transferred to females in a packet, the spermatophore. Insertion of the spermatophore into the genital tract may cause females to become unreceptive to male signals. In the grasshopper *G. rufus,* females become unreceptive to male signals immediately after mating. Cutting the nervous connection to the duct that receives the spermatophore causes females to copulate repeatedly, indicating that mechanical stimulation of the duct inhibits receptivity (Loher and Huber, 1966). In contrast, in the oedipodine grasshopper *Chimarocephala pacifica* the first male to copulate successfully leaves a spermatophore, but females remain receptive after copulation. Subsequent males are prevented from mating by the previous spermatophore, which acts as a block to further copulation. Between twenty-four and four hours before a female is to oviposit she becomes unreceptive, and an hour after ovipositing she is ready to mate again (Loher and Chandrashekaran, 1970). Stimuli that inhibit receptivity in roaches are also apparently mechanical in nature. Insertion of the spermatophore into the female bursa copulatrix inhibits further receptivity. Inhibition can also be produced artificially by inserting glass beads into the bursae of unmated females (Roth, 1962, 1964).

GENETICS AND DEVELOPMENT OF CRICKET SONG AND RESPONSE TO SONG

Bentley and Hoy (1970) have examined the appearance of song and flight motor patterns during development in the cricket *Teleogryllus commodus*. At hatching, chirp-eliciting neural circuits are not yet functional, but elements of the motor patterns gradually emerge in an ordered sequence over the course of the later nymphal stages. The circuit is completed before the molt to the adult stage. The last instar nymphs are able to generate nearly complete motor patterns for aggressive and courtship songs and portions of the calling song, but inhibition from the brain prevents the patterns from being elicited until after molting. Song patterns appear to be under genetic control and to be well isolated from environmental influences (Bentley and Hoy, 1972). In general, when parental song characteristics differ significantly, the hybrid characters are about intermediate between parental types. Each song parameter is evidently controlled by several genes. Also, song parameters may be sex-linked, since reciprocal hybrid songs (of ♂ A x ♀ B and ♂ B x ♀ A crosses) are quite different (Fig. 15). With each characteristic the song of the hybrid is more similar to the male of the maternal species than of the paternal species. Since some characteristics are sex-linked and others are not, genetic control of song is also multichromosomal.

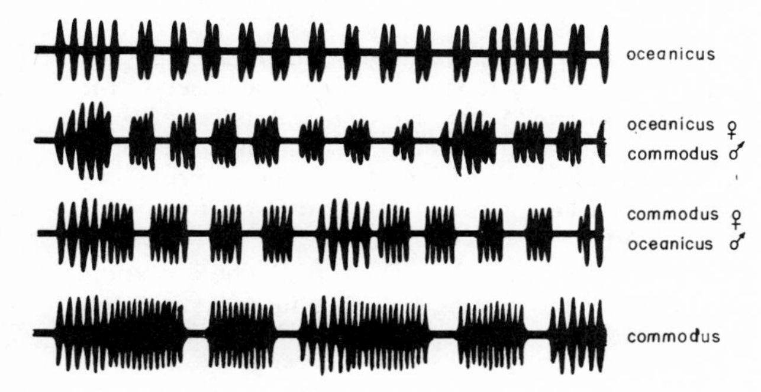

Fig. 15. Songs of *Teleogryllus commodus* and *T. oceanicus* and their hybrids. (After Bentley and Hoy, 1972.)

Hoy and Paul (1973) have also examined the genetic control of female responsiveness to male calls. The genetic differences that cause changes in male songs also appear to alter female responsiveness. Female responses were measured using a tethered female walking along a Y-maze globe held suspended beneath them. Recordings of the males of the parent species and of hybrid males were played through speakers to the right and left of the suspended females. The behavior of the females at choice points with respect to the sound source was measured by their turning tendency. Results indicated that female hybrids are more strongly attracted to the hybrid song than to the parental songs. Thus, it appears that the genetic coupling of the male's song generator and the female's sensory template is indeed close (see Alexander, 1962b).

A Partial Guide to Orthopteran Communication

Systematics. Alexander, 1957, 1960a, 1962a, 1967; Bigelow, 1960, 1964; Blondheim and Shulov, 1972; Jacobs, 1953; Leroy, 1966; Otte, 1970; Perdeck, 1958; Rentz, 1973; Shaw and Carlson, 1969; Walker, 1957, 1962.

Evolution. Alexander, 1960a, 1962a, 1975; Alexander and Otte, 1967a; Hill, Loftus-Hills, and Gartside, 1972; Jacobs, 1953; Otte, 1970, 1972a, 1975a; Perdeck, 1958; Spooner, 1968; Walker, 1957; Walker, 1962, 1974a; West and Alexander, 1963; Cade, 1975.

Comparative Ethology. Alexander, 1961, 1967, 1968, 1975; Alexander and Otte, 1967b; Barth, 1970; Barth and Lester, 1973; Busnel, 1954; Dumortier, 1963; Faber, 1953; Heiligenberg, 1966; von Hörman-Heck, 1957; Jacobs, 1953; Leroy, 1964; Loher and Chandrashekaran, 1970; Morris, 1971; Nielsen, 1971; Nielsen and Dreisig, 1970; Otte, 1970, 1972, 1975a; Roth, 1962, 1964; Shaw, 1968; Spooner, 1968; Willey and Willey, 1969, 1970, 1971; Young, 1971.

Defense. Alcock, 1972; Eisner and Meinwald, 1966; von Euw et al., 1967; Regen, 1913; Robinson, 1965, 1968a, 1968b, 1968c, 1968d, 1969a, 1969b; Rowell, 1967; Walker, 1964; Cade, 1975.

Environmental Control and Rhythms. Alexander and Meral, 1967; Cloudsley-Thompson, 1953; Cymborowski, 1973; Loher, 1957; Walker, 1962.

Neuromuscular Control. Bentley, 1969a, 1969b; Bentley and Kutsch, 1966; Busnel and Burkhardt, 1962; Dathe, 1972; Dietmar, 1971; Elder, 1971; Elsner, 1971; Huber, 1960, 1962, 1963; Kutsch and Huber, 1970; Kutsch, 1969; Leroy, 1964; Loher, 1966; Loher and Huber, 1966; Moss, 1971; Nocke, 1972; Otto, 1968, 1971; Shaw, 1968; Stout, 1970, 1971; Walker, 1969.

Physiology of Hearing. Adam, 1969; Busnel, Dumortier, and Pasquinelly, 1955; Haskell, 1956; Lewis, Pye, and House, 1971; McKay,

1969, 1970; Michelsen, 1966, 1968; Nocke, 1972; Popov, 1971; Regen, 1913; Rowell and McKay, 1969; Shaw, 1968; Stout and Huber, 1972; Suga, 1966; Suga and Katsuki, 1961; Zaretsky, 1971.

Hormonal Control. Barth, 1965, 1968, 1970; Barth and Lester, 1973; Blondheim and Broza, 1970; Highnam and Haskell, 1964; Loher, 1962, 1966; Loher and Huber, 1966; Pener, 1972; Pener, Girardie, and Joly, 1972; Pickford, Ewen, and Gillott, 1969; Roth, 1962; Roth and Barth, 1964.

Mechanisms of Signaling. Alexander, 1960b; Bailey and Broughton, 1970; Broughton, 1964; Kevan, 1954; Menon and Parshad, 1960; Morris and Pipher, 1967; Nocke, 1971; Richards, 1973; Otte and Cade, 1976.

Genetics. Bentley and Hoy, 1970, 1972; Fulton, 1933; Hoy and Paul, 1973; Leroy, 1965; Nolte, 1968.

Group Effects. Chauvin, 1958; Ellis, 1963, 1964; Ellis and Hoyle, 1954; Norris, 1954, 1964, 1970; Thomas, 1970.

Development. Bentley, 1969b; Bentley and Hoy, 1970; Loher 1957.

References

Adam, L. J., 1969. Neurophysiologie des Hörens und Bioakustik einer Feldheuschrecke. *Z. vergl. Physiol.,* 63:227–89.

Alcock, J., 1972. Observations on the behavior of the grasshopper *Taeniopoda eques* (Burmeister) (Orthoptera: Acrididae). *Anim. Behav.,* 20:237–42.

Alexander, R. D., 1957. The song relationships of four species of ground crickets (Orthoptera: Gryllidae: *Nemobius*). *Ohio J. Sci.,* 57:153–63.

Alexander, R. D., 1960a. Sound communication in Orthoptera and Cicadidae. In: *Animal Sounds and Communication,* W. E. Lanyon and W. N. Tavolga, eds. A.I.B.S. Publ. no. 7:38–92.

Alexander, R. D., 1960b. Communicative mandible-snapping in Acrididae (Orthoptera). *Science,* 132:152–53.

Alexander, R. D., 1961. Aggressiveness, territoriality, and sexual behavior in field crickets (Orthoptera: Gryllidae). *Behaviour,* 17:130–223.

Alexander, R. D., 1962a. The role of behavioral study in cricket classification. *Syst. Zool.,* 11:53–72.

Alexander, R. D., 1962b. Evolutionary change in cricket acoustical communication. *Evolution,* 16:443–67.

Alexander, R. D., 1967. Acoustical communication in arthropods. *Ann. Rev. Ent.,* 12:495–526.

Alexander, R. D., 1968. Arthropods. In: *Animal Communication,* T. A. Sebeok, ed. Bloomington, Indiana University Press.

Alexander, R. D., 1975. Natural selection and specialized chorusing behavior in acoustical insects. In: *Insects, Science and Society.* New York: Academic Press.

Alexander, R. D., and Meral, G. H., 1967. Seasonal and daily chirping cycles in the northern spring and fall field crickets, *Gryllus veletis* and *G. pennsylvanicus. Ohio J. Sci.,* 67:200–209.

Alexander, R. D., and Otte. D., 1967a. The evolution of genitalia and mating behavior in crickets (Gryllidae) and other Orthoptera. *Misc. Publ. Univ. Mich. Mus. Zool.,* 133.

Alexander, R. D., and Otte, D., 1967b. Cannibalism during copulation in the brown bush cricket, *Hapithus agitator* (Gryllidae). *Florida Entomol.,* 50:79–87.

Barth, R. H., Jr., 1965. Insect mating behavior: endocrine control of a chemical communication system. *Science,* 149:882–83.

Barth, R. H., Jr., 1968. The comparative physiology of reproductive processes in cockroaches. Part I. Mating behavior and its endocrine control. *Adv. Reprod. Physiol.,* 3:167–207.

Barth, R. H., Jr., 1970. Pheromone-endocrine interactions in insects. *Mem. Soc. Endocrinol.,* 18:373–404.

Barth, R. H., and Lester, L. J., 1973. Neuro-hormonal control of sexual behavior in insects. *Ann. Rev. Ent.,* 18:445–72.

Bailey, W. J., and Broughton, W. B., 1970. The mechanisms of stridulation in bush crickets (Tettigoniidae, Orthoptera). II. Conditions for resonance in the tegminal generator. *J. Exp. Biol.,* 52:505–17.

Bentley, D. R., 1969a. Intracellular activity in cricket neurons during generation of song patterns. *Z. Vergl. Physiol.,* 62:267–83.

Bentley, D. R., 1969b. Intracellular activity in cricket neurons during the generation of behavior patterns. *J. Insect Physiol.* 15:677–700.

Bentley, D. R., and Hoy, R. R., 1970. Postembryonic

development of adult motor patterns in crickets: a neural analysis. *Science,* 170:1409–11.

Bentley, D. R., and Hoy, R. R., 1972. Genetic control of the neuronal network generating cricket (*Teleogryllus, Gryllus*) song patterns. *Anim. Behav.,* 20:478–92.

Bentley, D. R., and Kutsch, W., 1966. The neuromuscular mechanism of stridulation in crickets (Orthoptera: Gryllidae). *J. Exp. Biol.,* 45:151–64.

Bigelow, R. S., 1960. Interspecific hybrids and speciation in the genus *Acheta* (Orthoptera: Gryllidae). *Can. J. Zool.,* 38:509–24.

Bigelow, R. S., 1964. Song differences in closely related cricket species and their significance *Aust. J. Sci.,* 27:99–102.

Blondheim, S. A., and Broza, M., 1970. Stridulation by *Dociostaurus curvicercus* (Orthoptera: Acrididae) in relation to hormonal termination of reproductive diapause. *Ann. Ent. Soc. Amer.,* 63:896–97.

Blondheim, S. A., and Shulov, A. S., 1972. Acoustic communication and differences in the biology of two sibling species of grasshoppers, *Acrotylus insubricus* and *A. patruelis. Ann. Ent. Soc. Am.,* 65:17–24.

Broughton, W. B., 1964. Function of the 'mirror' in Tettigonioid Orthoptera. *Nature* (London), 201:949–50.

Busnel, M. C., and Burkhardt, D., 1962. An electrophysiological study of the photokinetic reaction in *Locusta migratoria migratoriodes* (L). *Symp. Zool. Soc. Lond.,* 7:13–14.

Busnel, R. G., ed. 1954. Sur certain prapports entre le moyen d'information acoustique et le comportement acoustique des Orthopteres. *L'Acoustique des Orthopteres.* Paris: I.N.R.A., pp.281–306.

Busnel, R. G.; Dumortier, B.; and Pasquinelly, F.; 1955. Phonotaxie de ♀ d'Ephippiger (Orthoptere a des signaux acoustiques synthetiques). *C. R. Soc. Biol.* (Paris), 149:11–13.

Cade, W., 1975. Acoustically orienting parasitoids: fly phonotaxis to cricket song. *Science,* 190:1312–13.

Chauvin, R., 1958. L'action du groupement sur la croissance des grillons (*Gryllus domesticus*). *J. Ins. Physiol.,* 2:235–48.

Cloudsley-Thompson, J. L., 1953. Studies in diurnal rhythms—III. Photoperiodism in the cockroach *Periplaneta americana. Ann. Mag. Nat. Hist.,* 6:705–12.

Cohn, T. J., 1965. The arid-land katydids of the North American genus *Neobarrettia* (Orthoptera: Tettigoniidae): Their systematics and a reconstruction of their history. *Misc. Publ. Univ. Mich. Mus. Zool.,* 126:1–179.

Cymborowski, B., 1973. Control of the circadian rhythm of locomotor activity in the house cricket. *J. Insect Physiol.,* 19:1423–40.

Dathe, H. H., 1972. Akustische Beeinflussung von Bewegungs rhythmen by *Gryllus bimaculatus* (Insecta, Orthopteroidea). *Biol. Zbl.,* 91:579–96.

Dietmar, O., 1971. Untersuchungen zur zentralnervösen Kontrolle der Lauterzeugung von Gryllen. *Z. Vergl. Physiol.,* 74:227–71.

Dobzhansky, T., 1970. *Genetics of the Evolutionary Process.* New York: Columbia University Press.

Dumortier, B., 1963. Ethological and physiological study of sound emission in Arthropods. In: *Acoustic Behavior of Animals,* R. G. Busnel, ed. Amsterdam: Elsevier Publishing Co., pp.583–684.

Eisner, T., and Meinwald, J., 1966. Defensive secretions of arthropods. *Science,* 153:1341–50.

Elder, H. Y., 1971. High frequency muscles used in sound production by a katydid. II. Ultrastructure of the singing muscles. *Biol. Bull.,* 141:434–48.

Ellis, P. E., 1963. Changes in the social aggregation of locust hoppers with changes in rearing conditions. *Anim. Behav.,* 11:152–60.

Ellis, P. E., 1964. Marching and colour in locust hoppers in relation to social factors. *Behaviour,* 23:177–92.

Ellis, P. E., and Hoyle, G., 1954. A physiological interpretation of the marching of hoppers of the African Migratory Locust (*Locusta migratoria migratorioides* R and F). *J. Exp. Biol.,* 31:271–79.

Elsner, N., 1971. The central nervous control of courtship behavior in the grasshopper *Gomphocerippus rufus* L. (Orthoptera: Acrididae). *Symposium on Invertebrate Neurobiology,* Tihany, Hungary. Hungarian Academy of Sciences Publication.

Euw, J. von; Fishelson, L.; Parsons, J. A.; Reichstein, T.; and Rothschild, M.; 1967. Cardenolides (heart poisons) in a grasshopper feeding on milk weeds. *Nature* (London), 214:35–39.

Faber, A., 1953. *Laut- und Gebär densprache bei Insecten: Orthoptera.* Stuttgart: Staatliche. Museum für Naturkunde, pp. 1–198.

Fulton, B. B., 1933. Inheritence of song in hybrids of two subspecies of *Nemobius fasciatus* (Orthoptera). *Ann. Ent. Soc. Amer.,* 26:368–76.

Gillett, S., 1968. Airborne factor affecting grouping behavior in locusts. *Nature* 218:782–83.

Hamilton, W. J. III, 1973. *Life's Color Code.* New York: McGraw-Hill.

Haskell, P. T., 1956. Hearing in certain Orthoptera. *J. Exp. Biol.,* 33:756–76.

Haskell, P. T., 1957. Stridulation and associated behavior in certain Orthoptera. 1. Analysis of the stridulation of, and behavior between males. *Brit. J. Anim. Behav.*, 5:139–48.
Heiligenberg, W., 1966. The stimulation of territorial singing in house crickets (*Acheta domesticus*). *Z. Vergl. Physiol.*, 53:114–29.
Highnam, K. C., and Haskell, P. T., 1964. The endocrine systems of isolated and crowded *Locusta* and *Schistocerca* in relation to oocyte growth, and the effects of flying upon maturation. *J. Insect Physiol.*, 10:849–64.
Hill, K. G.; Loftus-Hills, J. J.; and Gartside, D. F.; 1972. Pre-mating isolation between the Australian field crickets *Teleogryllus commodus* and *T. oceanicus* (Orthoptera: Gryllidae). *Aust. J. Zool.*, 20:153–63.
Hörmann-Heck, S. von, 1957. Untersuchungen über den Erbgang einiger Verhaltensweisen bei Grillenbastarden. *Z. Tierpsychol.*, 14:137–83.
Hoy, R. R., and Paul, R. C., 1973. Genetic control of song specificity in crickets. *Science*, 180:82–83.
Huber, F., 1960. Untersuchungen uber die Function des Zentralnervensystems und insbesondere des Gehirns bei der Fortbewegung und der Lauterzeugung der Grillen. *Z. Verg. Physiol.*, 44:60–132.
Huber, F., 1962. Central nervous control of sound production in crickets and some speculation on its evolution. *Evolution*, 16:429–42.
Huber, F., 1963. The role of the central nervous system in Orthoptera during the coordination and control of stridulation. In: *Acoustic Behavior of Animals*, R. G. Busnel, ed. New York: Elsevier Publishing Co.
Jacobs, W., 1953. Verhaltensbiologische Studien and Feldheuschrecken. *Z. Tierpsychol. Suppl.* 1, pp.1–228.
Kevan, D. K. McE., 1954. Methodes inhabituelles de production de son chez les Orthopteres. In: *L'Acoustique des Orthopteres*, R. G. Busnel, ed. Paris: I.N.R.A. pp.103–41.
Kutsch, W., 1969. Neuromuskulare Aktivität bei verschiedenen Verhaltensweisung von drei Grillenarten. *Z. Vergl. Physiol.*, 63:335–78.
Kutsch, W., and Huber, F., 1970. Zentrale versus periphere Kontrolle des Gesänges von Grillen (*Gryllus campestris*). *Z. Vergl. Physiol.*, 67:140–59.
Lack, D., 1947. *Darwin's Finches.* New York: Harper.
Leroy, Y., 1964. Les caracteres asexuels et le comportement acoustique des mâles d'*Homoegryllus reticulatus* Fabricius (Orth., Ensiferes Phalangopsidae). *Bull. soc. entom. France*, 69:7–14.
Leroy, Y., 1965. Analyse de la transmission des divers parametres des signaux acoustiques chez les hybrides interspecifiques de Grillons (Orthopteres, Ensiferes). *Proc. Twelfth Int. Congr. Ent. London*, 1964.
Leroy, Y., 1966. Signaux acoustiques, comportement et systematique de quelque espèces des Gryllides (Orthopteres, Ensiferes). *Bull. biol. Fr. Belg.*, 100:63–134.
Lewis, D. B.; Pye, J. D.; and House, P. E.; 1971. Sound reception in the bush cricket (*Metrioptera brachyptera* L) (Orthoptera. Tettigonioidea). *J. Exp. Biol.*, 55:241–51.
Loher, W., 1957. Untersuchungen über den Aufbau und die Enstehung der Gesänge einiger Feldheuschreckenarten und den Einfluss von Lautzeichen auf das akustische Verhalten. *Z. Vergl. Physiol.*, 39:313–56.
Loher, W., 1958. An olfactory response of immature adults of the desert locust. *Nature*, 181:1280.
Loher, W., 1962. Die Kontrolle des Weibchensgesangs von *Gomphocerus rufus* L. durch die Corpus allata. *Naturwissenschaften*, 49:406.
Loher, W., 1966. Die Steurung sexueller Verhaltensweisen und der Oocytenentwicklung bei *Gomphocerus rufus* L. *Z. Vergl. Physiol.*, 53:277–316.
Loher, W., and Chandrashekaran, M. K., 1970. Acoustical and sexual behavior in the grasshopper *Chimarocephala pacifica pacifica* (Oedipodinae). *Ent. Exp. Appl.*, 13:71–84.
Loher, W., and Huber, F., 1966. Nervous and endocrine control of sexual behavior in a grasshopper (*Gomphocerus rufus* L.). *Symp. Soc. Exp. Biol.*, 20:381–400.
McKay, J. M., 1969. The auditory system of *Homorocoryphus* (Tettigoniidae, Orthoptera). *J. Exp. Biol.*, 51:787–902.
McKay, J. M., 1970. Central control of an insect auditory interneurone. *J. Exp. Biol.*, 53:137–46.
Mayr, E., 1963. *Animal Species and Evolution.* Cambridge: Belknap.
Menon, R., and Parshad, B., 1960. An interesting antenno-frontal stridulatory mechanism in a gryllid *Loxoblemmus equestris* Saussure (Orthoptera). *J. Anim. Morph. Physiol.* (Baroda), 7:167–69.
Michelsen, A., 1966. Pitch discrimination in the locust ear: observations on single sense cells. *J. Insect Physiol.*, 12:1119–31.
Michelsen, A., 1968. Frequency discrimination in the locust ear by means of four groups of receptor cells. *Nature*, 220:585–86.
Morris, G. K., 1970. Sound analysis of *Metrioptera sphagnorum* (Orthoptera: Tettigoniidae). *Canad. Entomol.*, 102:363–68.

Morris, G. K., 1971. Aggression in male conocephaline grasshoppers (Tettidoniidae). *Anim. Behav.,* 19:132–37.

Morris, G. K., 1972. Phonotaxis of male meadow grasshoppers (Orthoptera: Tettigoniidae). *J.N.Y. Entomol. Soc.,* 80:5–6.

Morris, G. K., and Pipher, R. E., 1967. Tegminal amplifiers and spectrum consistencies in *Conocephalus nigropleurum* (Bruner) (Tettigoniidae). *J. Insect Physiol.,* 13:1075–85.

Möss, D., 1971. Sinnesorgane im Bereich des Flügels der Feldgrille (*Gryllus campestris* L) und ihre Bedeutung fur die Kontrolle der Singbewegung und die Einstellung der Flügellage. *Z. Vergl. Physiol.,* 73:53–83.

Nielsen, E. T., 1971. Stridulatory activity of *Eugaster* (Orthoptera: Ensifera) Ent. Exp. et Appl. 14:234–44.

Nielsen, E. T., and Dreisig, H., 1970. The behavior of stridulation in Orthoptera Ensifera. *Behaviour,* 37:205–52.

Nocke, H., 1971. Biophysik der Schallerzeugung durch die Vorderflügel der Grillen. *Z. Verg. Physiol.,* 74:272–314.

Nocke, H., 1972. Physiological aspects of sound communication in crickets (*Gryllus campestris* L). *J. Comp. Physiol.,* 80:141–62.

Nolte, D. J., 1968. The chiasma-inducing pheromone of locusts. *Chromosoma* (Berlin), 23:346–58.

Norris, M. J., 1954. Sexual maturation in the desert locust (*Schistocerca gregaria* Forskal) with special reference to the effects of grouping. *Anti-Locust Bull.,* 18:1–44.

Norris, M. J., 1964. Accelerating and inhibiting effects of crowding on sexual maturation in two species of locusts. *Nature,* 203:784–85.

Norris, M. J., 1970. Aggregation response in ovipositing females of the desert locust, with special reference to the chemical factor. *J. Insect Physiol.,* 16:1493–1515.

Otte, D., 1970. A comparative study of communicative behavior in grasshoppers. *Misc. Publ. Univ. Mich. Mus. Zool.,* 141:1–168.

Otte, D., 1972a. Simple versus elaborate behavior in grasshoppers: an analysis of communication in the genus *Syrbula. Behaviour,* 42:291–322.

Otte, D., 1972b. Communicative aspects of reproductive behavior in Australian grasshoppers (Oedipodinae and Gomphocerinae). *Aust. J. Zool.,* 20:139–52.

Otte, D., 1975a. Effects and functions in the evolution of signaling systems. *Ann. Rev. Ecol. Syst.,* 5:385–417.

Otte, D., 1975b. On the role of intraspecific deception. *Amer. Nat.* 109:239–42.

Otte, D., and Cade, W., 1976. On the role of olfaction in sexual and interspecies recognition in crickets (*Acheta* and *Gryllus*). *Anim. Behav.,* 24:1–6.

Otte, D., and Joern, A., 1975. Insect territoriality and its evolution: population studies of desert grasshoppers on creosote bushes. *J. Anim. Ecol.,* 44:29–54.

Otto, D., 1968. Untersuchungen zur nervösen Kontrolle des Grillengesänges. *Zool. Anz. Suppl.* 31:585–92.

Otto, D., 1971. Untersuchungen zur zentralnervösen Kontrolle der Lauterzeugung von Grillen. *Z. Vergl. Physiol.,* 74:227–71.

Pener, M. P., 1972. The corpus allatum in adult acridids: the interrelation of its functions and possible correlations with life cycle. *Proc. Int. Study Conf. Current and Future Problems of Acridology,* London 1970, pp.135–47.

Pener, M. P.; Girardie, A.; and Joly, P.; 1972. Neurosecretory and corpus allatum controlled effects on mating behavior and color change in adult *Locusta migratoria migratorioides* males. *Gen. and Comp. Endoc.,* 19:494–508.

Perdeck, A. C., 1958. The isolating value of specific song patterns in two sibling species of grasshoppers (*Chorthippus brunnerus* Thunb. and *C. biguttulus* L.). *Behaviour,* 12:1–75.

Pickford, R.; Ewen, A. B.; and Gillott, C.; 1969. Male accessory gland substance: an egg-laying stimulant in *Melanoplus sanguinipes* (F) (Orthoptera: Acrididae). *Can. J. Zool.,* 47:1199–1203.

Popov, A. V., 1971. Synaptic transformation in the auditory system of insects. In: *Sensory Processes at the Neuronal and Behavioral Levels,* G. V. Gersuni, ed. New York: Academic Press, pp.301–20.

Regen, J., 1913. Über die Anlockungdes Weibchens von *Gryllus campestris* L. durch telefonish übertragene Stridulations laute des Männchens. Eine Beitrage zur Frage der Orientiering bei den Insecten. *Pflügers Arch. ges. Physiol.,* 155:3–10.

Rentz, D. C., 1973. The ecology, behavior and description of a new species of cricket from the Osa Peninsula of Costa Rica. *Ent. News,* 84:237–46.

Richards, A. M., 1973. A comparative study of the biology of the Giant wetas *Deinacrida heteracantha* and *D. fallai* (Orthoptera: Henicidae) from New Zealand. *J. Zool.* (London), 169:195–236.

Robinson, M. H., 1965. The Javanese stick insect, *Orx-*

ines macklotti De Haan (Phasmatodea, Phasmidae). *Ent. Mon. Mag.,* 100:253–59.

Robinson, M. H., 1968a. The defensive behavior of *Pterinoxylus spinulosus* Redtenbacher, a winged stick insect from Panama (Phasmatodea). *Psyche,* 75:195–207.

Robinson, M. H., 1968b. The defensive behavior of the Javanese stick insect, *Orxines macklotti* De Haan, with a note on the startle display of *Metriotes diocles* Westw. (Phasmatodea, Phasmidae). *Ent. Mon. Mag.,* 104:46–54.

Robinson, M. H., 1968c. The defensive behaviour of the Stick insect Oncotophasma martini (Griffini) (Orthoptera: Phasmatidae) *Proc. R. Ent. Soc. Lond.,* 43:183–87.

Robinson, M. H., 1968d. The startle display of *Balboa tibialis* (Bruner) (Orthoptera: Tettidonidae). *Ent. Mon. Mag.,* 104:88–90.

Robinson, M. H., 1969a. The defensive behavior of some orthopteroid insects from Panama. *Trans. R. Ent. Soc. Lond.,* 121:281–303.

Robinson, M. H., 1069b. Defenses against visually hunting predators. *Evol. Biol.,* 3:225–59.

Roth, L. M., 1962. Hypersexual activity induced in females of the cockroach *Nauphoeta cinerea. Science,* 138:1267–69.

Roth, L. M., 1964. Control of reproduction in female cockroaches with special reference to *Nauphoeta cineria.* I. First pre-oviposition period. *J. Insect Physiol.,* 10:915–45.

Roth, L. M., and Barth, R. H., Jr., 1964. The control of sexual receptivity in female cockroaches. *J. Insect Physiol.,* 10:965–75.

Roth, L. M., and Barth, R. H., Jr., 1967. The sense organs employed by cockroaches in mating behavior. *Behaviour,* 28:58–94.

Rothschild, M., and Parsons, J., 1962. Pharmacology of the poison gland of the locust *Poekilocerus bufoni* Klug. *Proc. R. Ent. Soc. Lond.* (C), 27:21–22.

Rowell, C. H. F., 1967. Experiments on aggregations of *Phymateus purpurescens* (Orthoptera, Acrididae, Pyrgomorphinae). *J. Zool.,* 152:179–93.

Rowell, C. H. F., and McKay, J. M., 1969. An acridid auditory interneurone. *J. Exp. Biol.,* 51:231–60.

Shaw, K. C., 1968. An analysis of the phonoresponse of males of the true katydid *Pterophylla camellifolia* (Fabricius) (Orthoptera: Tettigoniidae). *Behaviour,* 31:204–59.

Shaw, K. C., and Carlson, O. V., 1969. The true katydid, *Pterophylla camellifolia* (Fabricius) (Orthoptera: Tettigoniidae) in Iowa: two populations which differ in behavior and morphology. *Iowa State J. Sci.,* 44:193–200.

Spooner, J. D., 1964. The Texas bush katydid—its sounds and their significance. *Anim. Behav.,* 12:235–44.

Spooner, J. D., 1968. Pair-forming acoustic systems of Phaneropternie katydids (Orthoptera, Tettigoniidae). *Anim. Behav.,* 16:197–212.

Stout, J. F., 1970.Response of interneurons of female crickets (*Gryllus campestris*) to the male's calling song. *Am. Zool.,* 10:502.

Stout, J. F., 1971. A technique for recording the activity of single interneurons from free-moving crickets (*Gryllus campestris* L). *Z. Vergl. Physiol.,* 74:26–31.

Stout, J. F., and Huber, F., 1972. Responses of a central auditory neurons of female crickets (*Gryllus campestris* L) to the calling song of the male. *Z. Vergl. Physiol.,* 76:302–13.

Suga, N., 1966. Ultrasonic production and its reception in some neotropical Tettigoniidae. *J. Insect Physiol.,* 12:1039–50.

Suga, N., and Katsuki, Y., 1961. Central mechanisms of hearing in insects. *J. Exp. Biol.,* 38:545–58.

Thomas, J. G., 1970. Probable pheromone-secreting cells in the epidermis of mature males of *Schistocerca gregaria* Forskal (Orthoptera: Acrididae) *Proc. R. Ent. Soc. Lond.* (A), 45:125–35.

Trivers, R. L., 1972. Parental investment and sexual selection. In: *Sexual Selection and the Descent of Man, 1871–1971,* B. Campbell, ed. Chicago: Aldine, chap. 7.

Ulagaraj, S. M., and Walker, T. J., 1973. Phonotaxis of crickets in flight. Attraction of male and female crickets to male calling songs. *Science,* 182:1278–79.

Walker, T. J., 1957. Specificity in the response of female tree crickets to calling songs of the males. *Ann. Ent. Soc. Amer.,* 50:626–36.

Walker, T. J., 1962. Factors responsible for intraspecific variation in the calling songs of crickets. *Evolution,* 16:407–28.

Walker, T. J., 1964. Experimental demonstration of a cat locating orthopteran prey by the prey's calling song. *Fla. Entomol.,* 47:163–65.

Walker, T. J., 1969. Acoustic synchrony: two mechanisms in the snowy tree cricket. *Science,* 166:891–94.

Walker, T. J., 1974a. *Gryllus ovisopis* n.sp.: A taciturn cricket with a life cycle suggesting allochronic speciation. *Florida Entomol.,* 57:13–22.

Walker, T. J., 1974b. Character displacement and acoustical insects. *Amer. Zool.,* 14:1137–50.

Wallace, B., 1970. *Topics in Population Genetics.* New York: W. W. Norton.

West, M. J., and Alexander, R. D., 1963. Sub-social behavior in a burrowing cricket *Anurogryllus muticus* (De Geer) (Orthoptera: Gryllidae). *Ohio. J. Sci.,* 63:19–24.

Willey, R. B., and Willey, R. L., 1969. Visual and acoustical social displays by the grasshopper *Arphia conspersa* (Orthoptera: Acrididae) *Psyche,* 76:280–305.

Willey, R. B., and Willey, R. L., 1970. The behavioral ecology of desert grasshoppers. I. Presumed sex role reversal in flight displays of *Trimerotropis agrestis. Anim. Behav.,* 18:473–77.

Willey, R. B., and Willey, R. L., 1971. The behavioral ecology of desert grasshoppers, II. Communication in *Trimerotropis agrestis. Anim. Behav.,* 19:26–33.

Young, A. J., 1971. Studies on the acoustic behavior of certain orthoptera. *Anim. Behav.,* 19:727–43.

Zaretsky, M. D., 1971. Patterned response to song in a single central auditory neuron of a cricket. *Nature,* 229:195–96.

Chapter 17

COMMUNICATION IN THE LEPIDOPTERA

Robert E. Silberglied

Introduction

The Lepidoptera, or butterflies and moths,[1] combine aesthetic appeal with a diversity of problems of scientific interest that have kept their study at the forefront of evolutionary and behavioral biology. The varied tableaus of color and pattern on their wings provide rich material for the study of variation, polymorphism, and mimicry. In the nineteenth century they were widely used to develop, illustrate, and support the theories of evolution (Wallace, 1890, 1891) and sexual selection. They have been utilized in some of the earliest ethological studies (Tinbergen et al., 1942), and their pheromonal communication systems were among the first to be analyzed in detail. Interest in lepidopteran communication is greater today than ever before, stimulated in large part by the potential use of such knowledge in control programs for economically important species (Birch et al., 1974, and references therein).

1. An excellent general treatment of the order is that of Common (1970). The division of the Lepidoptera into butterflies (Papilionoidea and Hesperioidea) and moths (the remaining nine-tenths of the order) has no higher-level cladistic basis but is retained as an heuristic concept familiar to all (see also Bourgogne, 1951).

As holometabolous insects, lepidopterans develop in a series of distinct morphological stages. The adults are oviparous, and usually deposit their eggs on vegetation. The generally phytophagous larva, commonly referred to as a caterpillar, is a soft-bodied insect with a hydrostatic skeleton, a well-hardened head capsule bearing chewing mouthparts, and usually several pairs of abdominal ambulatory appendages called prolegs in addition to the usual three pair of thoracic legs. Upon completion of feeding, most larvae construct pupal enclosures, often using silk produced by labial salivary glands. Adults are characterized by a covering of scales over most of the body and wings, and in most cases an elongate proboscis used for nectar-feeding. The order shares a common ancestry with the caddisflies, and most probably arose and diversified concurrently with the evolution and diversification of angiosperms in the late Mesozoic and early Cenozoic eras (Common, 1975; MacKay, 1969, 1970; Skalski, 1973).

Even though lepidopterans constitute approximately one-tenth of all animal species, the preceding brief description indicates that at a gross level they exhibit surprisingly little morphological and ecological diversity. Behavior involving communication is similarly limited.

Communication between members of the different developmental stages of a species, dominance hierarchies and their associated behavior, and social behavior, even in its most rudimentary forms, are all rare or unknown. Most communication is limited to contexts of individual survival and reproduction, for which very few general patterns of communication have been adopted. For each of these communicative patterns, such as courtship behavior and mimicry, there exists an enormous wealth of detail at the specific level. It is neither possible nor desirable to attempt here a comprehensive survey of this information. Instead, this review will treat, somewhat superficially, the breadth of communicative interactions in which the Lepidoptera take part, with a limited number of examples chosen to illustrate them. The reader is referred to more detailed reviews of each individual topic, and bibliographic references have been kept to the minimum commensurate with access to the literature.

"Communication" is an elusive concept. The author has no desire to become mired in a discussion of the usage of this term, as he sees advantages to both narrow (Otte, 1974) and broad (Wilson, 1971, 1975) interpretations, depending on the emphasis intended by the writer. Readers who wish to pursue this matter will find ample material in the earlier sections of this book, and in the following references: Birch, 1974b; Brown et al., 1970; Burghardt, 1970; Marler, 1961; Morris, 1946; Otte, 1974; Sebeok, 1965; Whittaker and Feeny, 1971; and Wilson, 1971, 1975. In this article, Wilson's (1975) broad concept is employed: ". . . communication is an adaptive relation between the organism that signals and the one that receives, regardless of the complexity and length of the communication channel." However, due to space limitations, the author has arbitrarily excluded certain aspects of communication, such as most host (= prey) detection, most predator detection, and the use of simple physical defense or escape mechanisms. These are discussed only in those specific instances where it is necessary for the understanding of more elaborate communication systems.

The nature of the "complexity and length of the communication channel" may be illustrated by several examples. Communication between the male and female cecropia silkmoth (*Hyalophora cecropia,* Saturniidae) does not terminate with copulation. The sperm or some other substance produced by the male interacts with the bursa copulatrix of the female, which responds by releasing into the hemolymph a hormone that changes her oviposition rate (Riddiford and Ashenhurst, 1973). Male butterflies in several genera (*Parnassius, Acraea, Actinote, Amauris,* and others; see Scott, 1973) deposit a large structure (the sphragis) in the female copulatory opening. It is believed that the function of this "plug" is to prevent mating by other males, either by its physical presence or because it inhibits pheromone release by females (Eltringham, 1912; Labine, 1964). Gilbert (pers. comm. and cited in Scott, 1973) has discovered that in at least one species of heliconiine butterfly the male deposits on the female a pheromone that makes her unattractive to other males. Using radioactive tracers, Gilbert has also demonstrated that in *Heliconius* the male's spermatophore is partially metabolized and contributes nutritionally to the female and the eggs (pers. comm.). In all of these diverse instances a "signal" persists and functions long after the signaling individual has departed.

Any survey of communication must consider the sensory world of the animals concerned. As a result of the partial "overhaul" of the nervous system during "complete" metamorphosis, the successive developmental stages differ so markedly from one another in morphology, sensory physiology, and behavior that they must be considered as distinctly different organisms, each with its own *Umwelt,* defined to a first approximation by the physiology of its sense organs. For this reason the developmental stages are treated separately. The one feature that all the developmental stages of a species have in common, however, is the need for adequate defense. Therefore a consideration of defense is presented first.

Like most other arthropods, butterflies and moths have numerous and diverse defense adaptations. These include deciduous scales, urticating larval setae, regurgitation (by larvae), defecation (including the use of the meconium by the newly eclosed adult[2]), and compounds with noxious or toxic properties sequestered in the blood or released from specialized defensive glands (e.g., osmeteria of papilionid larvae) (Aplin and Rothschild, 1972; Bisset et al., 1960; Brower et al., 1968; Duffey, 1970; Edmunds, 1974; Ehrlich and Raven, 1965; Eisner, 1970; Eisner et al., 1971; Eisner and Meinwald, 1965; Eisner et al., 1970; von Euw et al., 1968; Frazier, 1965; Frazier and Rothschild, 1961; Jones et al., 1962; Pesce and Delgrado, 1972; Picarelli and Valle, 1972; Reichstein et al., 1968; Rotberg, 1972; Rothschild, 1971, 1972, 1973; Rothschild et al., 1970, 1972, 1973; and references therein). These features may be communicated to predators in an unambiguous manner by aposematic coloration or behavior. Space considerations prohibit more than the most superficial statement about this subject; the references cited immediately above and below are strongly recommended to readers interested in these aspects of lepidopteran communication.

The spectacular and diverse color patterns of moths and butterflies have long been used as evidence for the existence and mechanism of the evolutionary process (e.g., Brower, 1963; Creed, 1971; Ford, 1945, 1967, 1971; Kettlewell, 1973; Poulton, 1890; Rettenmeyer, 1970; Robinson, 1971; Rothschild, 1971, 1972, 1973; Wallace, 1890, 1891; Wickler, 1968; and references therein). Aposematic, startling, deflective, and mimetic patterns are clearly communicative, but what about crypsis? Camouflage and special protective resemblance (Robinson, 1969) clearly entail energy expenditures on the part of the insect. There is a biosynthetic and developmental cost to make cryptic features, as well as a behavioral one: the insect must be able to choose an appropriate background having little contrast with its own coloration, and it must posture there in an inconspicuous manner (Kettlewell, 1973; Sargent, 1973; and references therein). But is the cryptic individual signaling? Do predators perceive the prey but mistake it for something else, in the same way they discriminate against Batesian mimics by mistaking them for models known to be inedible? Or is no "signal" received at all, the predator being entirely unaware of the existence of the prey? Both probably occur in nature (depending to a large extent on the kind of predator); in either case the prey animal would survive. One may view crypsis as a form of communication in which the prey animal has been selected to decrease, rather than increase (as in aposematism), its signal-to-noise ratio. One would expect corresponding selection on the predator for better sensory and discriminatory abilities. The results of such escalating evolutionary exchanges are seen as the patterns on the wings and bodies of lepidopterans, and the forms they take depend on the strategies employed.

The various protective color patterns are traditionally grouped according to the manner in which they are usually presumed to function in communicative interactions with predators: crypsis (Robinson, 1969), disruptive coloration, disappearing or "flash colors" (Cott, 1940; Ford, 1967), deflective patterns (Blest, 1957; Poulton, 1890), startling or "novelty" coloration (Blest, 1957; Coppinger, 1970; Hinton, 1974), aposematism, Batesian and Müllerian mimicry, etc. Speculation as to the communicative signifi-

2. e.g., *Manduca sexta* (Sphingidae) (pers. obs.).

cance of color patterns of particular species is a common practice among lepidopterists, but only in a limited number of cases have these hypotheses been tested by experiment. The notable, pioneering studies of Kettlewell and others on color polymorphisms and camouflage, Blest on the function of "eyespot" patterns, the Browers and their colleagues on mimicry, and others are cited in Brower (1963), Brower et al. (1971), Edmunds, 1974; Ford (1971), Kettlewell (1973), Rettenmeyer (1970), Robinson, 1969, Rothschild (1971, 1972, 1973), Turner (1971a), and Wickler (1968). Since coloration plays other roles besides protection from predators, and since different predators may respond in different ways to the same pattern, generalization from one case to another should be done with caution and with the knowledge that it is only an heuristic exercise until the appropriate experiments have been performed.

The Egg

Lepidopteran eggs do not behave (in the conventional sense), yet in certain cases they may be communicative. Adult butterflies of many species lay eggs singly, often with considerable spacing and away from other, older eggs. This is adaptive because older eggs hatch first, and if food is limited, the second to hatch may be left hungry on a leafless twig. Larvae are often cannibalistic (Alexander, 1961a; Dethier, 1937; Turner, 1971a)—another disadvantage for the younger, hence smaller, larva. Some adult female *Heliconius* butterflies apparently scan host plants visually (*Passiflora* spp.) and do not oviposit near other eggs (Alexander, 1961a; Gilbert, 1975). To what extent are the often conspicuous colors of butterfly eggs, and color changes prior to hatching, of signal value to adult butterflies? While we do not know, it is clear that such signals would be adaptive to the sender as well, because they might prevent establishment of competing larvae on the same plant. Experimental testing of this possibility appears to have been done by certain *Passiflora* species that have stipular "egg-mimics," which may inhibit oviposition by female *Heliconius*—the only case known to date in which lepidopterans (and mimetic ones at that) may themselves be hoisted with the petard of mimicry (Gilbert, 1975)!

In contrast, some lepidopterans, particularly those with aposematic, aggregated larvae, lay their eggs in clutches. Ovipositing females of *Mechanitis isthmia* (Ithomiinae, Nymphalidae) can search for and relocate their egg clutches after being disturbed (Gilbert, 1969). Several females of *Heliconius sara* sometimes even lay their eggs together in mixed clusters (Turner, 1971a). Although the adaptive strategies of such species differ from the solitary egg layers, in both cases the female must be capable of recognizing eggs, and in both cases the subsequent behavior of the female is affected.

The Larva

Most larval behavior is related to growth, individual survival, and preparation for pupation. For the majority of species communication occurs only in the contexts of protective behavior and active defense. Sound production, known in lepidopterous larvae of several groups (references in Frings and Frings, 1970; Haskell, 1961), is generally believed to be defensive in function, but since many larvae respond to airborne sound (Hogue, 1972; Minnich, 1936), it may possibly be used for intraspecific communication as well. Larvae have poorly developed visual abilities but they can distinguish vertical from horizontal figures (Dethier, 1943; Hundertmark, 1937a; de Lépiney, 1928), possess the physiological basis for color vision (Ishikawa, 1969), and show color preferences (Götz, 1936; Hundertmark, 1937b). Their gluttonous appetites are subserved by well-developed senses of olfaction and taste (Schoonhoven, 1973, and references therein).

Most caterpillars live solitary lives, devoid of all but occasional interactions with other larvae and predators. It is among those larvae that have symbiotic relationships with other insects and among those that live together at high densities or in aggregations that we find sophisticated communication systems.

INTERSPECIFIC COMMUNICATION IN SOCIAL SYMBIOSES

Some lepidopterous larvae live among or in association with social insects: usually ants, and rarely with bees, wasps, or termites (Ford, 1945; Hinton, 1951; Wilson, 1971). Their habits include feeding on nest materials and detritus (some members of the families Tineidae, Pyralidae and Noctuidae) or on host brood (some Lycaenidae, Tineidae, Cosmopterygidae, Cyclotornidae, and Pyralidae), but in most cases (most Lycaenidae and some Pieridae) the larvae are simply phytophagous and are tended, guarded, and otherwise protected by the hosts. Some of the nest inhabitants are treated as invaders and suffer attacks; they survive because of their protective integument, silken webwork, or other defensive adaptations. But a few are closely attended, cleaned, and otherwise cared for; the host workers may even construct shelters for the phytophagous species, and some larvae are carried into the nest, where they may feed on ant brood or even solicit food from workers. Such habits are best developed among the "blues," "hairstreaks," "coppers," and "metalmarks" of the butterfly family Lycaenidae (in the broad sense of Ehrlich, 1958), of which most species are myrmecophilous (live in association with ants) in some sense, and a few are obligately so. The overwhelming diversity of these relationships at the species level has been reviewed by Balduf (1939), Hinton (1951), and Malicky (1969). (See also Clark and Dickson, 1971; Farquharson, 1921; Lamborn, 1913; Owen, 1971; Ross, 1966; and Wilson, 1971.)

Certain characteristics of these attended and "guest" larvae serve to distinguish them from others that would be attacked, killed, and eaten or discarded. These features appear to be chemical, tactile, and perhaps visual. Most such species have glandular setae, tubercles, or elaborate, often eversible glands that produce secretions the hosts find attractive and upon which they may feed. Consider, for example, the "Large Blue," *Maculinea* (=*Lycaena*) *arion,* the larvae of which undergo a change from ordinary phytophagy to carnivory at the fourth instar. After the third molt, the larvae leave the host plant and wander about. Ants find and stroke them with their antennae. A gland on the seventh abdominal tergite of the larva produces a secretion upon which the ant feeds. After a while the larva suddenly swells up its thoracic segments, markedly changing its form and perhaps providing other signals. The ant responds by picking up the larva with its mandibles and carrying it to the nest, where for the remainder of its larval life the caterpillar consumes ant larvae (references in Hinton, 1951).

Our knowledge of the complexity of these kinds of relationship is limited mostly to descriptions of observations made on a wide range of species, principally in the Lycaenidae. It is not clear to what extent, if any, the hosts benefit from the larval secretions, and for the majority of the larvae that are simply tended by ants on vegetation but which do not enter the nests, it is not clear if the larvae benefit by reduced aggression on the part of the ants (Lenz, 1917), protection from predators and parasitoids (Thomann, 1901), or both (Edmunds, 1974). The details of the communication between the larvae and their hosts are largely unknown. While the histology, ontogeny, and distribution of the glands that produce these "appeasement substances" have been described in considerable detail (Hinton,

1951; Malicky, 1969), none of the larval secretions have been chemically identified. That each species of lycaenid has but one or a few host species and that the larvae are attacked by "wrong" ants indicate a considerable level of complexity in the signals and responses. Malicky (1970a) has shown that even within one host genus (*Formica*), some species respond aggressively to the caterpillars in the vicinity of the nest entrance but not at a distance, others do not respond aggressively at all, and in one the "mood" of the ants was important but the distance from the entrance was not. In addition to the glands, certain other morphological and behavioral features of such larvae are evidently adaptations for living among or with ants (Malicky, 1970b). The behavior of some lycaenid larvae in soliciting food from ant workers may involve mimicry of the intraspecific food-solicitation signals of other workers or of larvae (Malicky, 1970b), as has been demonstrated for certain myrmecophilous beetles (Hölldobler, 1967, 1970, 1971). Mimicry of pheromones has also been suggested (Malicky, 1970b), but chemical evidence is wanting. The adults of nest-inhabiting lepidopterans lack those attributes that inhibit aggression by the hosts. However, the newly eclosed butterfly or moth bears a heavy coat of scales that readily come off in the jaws of attacking ants, facilitating escape from the nest.

Larvae of a few lycaenids have the remarkable habit of soliciting honeydew from various homopterans (references in Hinton, 1951). *Lachnocnema bibulus* does so by vibrating its elongated prothoracic legs over the dorsal surface of the membracid or jassid in a manner similar to that of an ant soliciting with its antennae.[3] It also solicits food from the ants that tend these homopterans. And larvae of another lycaenid, *Megalopalpus zymna,* use a similar tactic as a ruse to approach more closely membracids and jassids, which they suddenly attack and devour—a case of tactile aggressive mimicry.

3. The same trick is used by some *adult* lycaenids: *Allotinus horsfieldi* strokes aphids with its prothoracic legs to obtain honeydew, and *Miletus boisduvali* uses its proboscis in a similar manner on both aphids and scale insects (references in Hinton, 1951).

COMMUNICATION AMONG GREGARIOUS LARVAE

Larvae of a large number of lepidopterous species live in aggregations or at high densities. This habit is widely distributed among the various families. Usually all of the larvae in an aggregation have hatched from a single cluster of eggs.

Such larvae do many things together. They all begin and cease feeding at about the same time. When a predator threatens, they may all respond similarly and simultaneously, giving the impression of concerted defense. Many species produce highly ordered silk structures that, superficially at least, appear to involve coordinated activity employing communication. Unfortunately only a few species have been studied in detail.

One need not invoke complex communication systems to explain much of the seemingly coordinated behavior of such larvae. Their feeding times are usually regulated by extrinsic and intrinsic factors such as temperature, light level and photoperiod, and hunger. It is only to be expected that individuals of the same species, age, and usually parentage, would behave somewhat similarly in the same environment. Concerted defenses may in many cases simply be a simultaneous response of many or all larvae to some disturbing stimulus, and not to some alarm signal sent by the first larva that detects the predator. Larvae do of course respond to tactile stimuli of one another's movements, and this seems to be the means by which a disturbance may spread through some larval aggregations. Communicative synchronization of feeding and movement is known to occur in some species

(Alexander, 1961a; McManus and Smith, 1972; Symczak, 1950; Wellington, 1957, 1974). Coordinated responses, especially in defense, are certainly adaptive (Edmunds, 1974; Ford, 1945; Hogue, 1972; Poulton, 1898), but experimental study of this aspect of larval behavior has unfortunately been neglected.

Gregarious larvae in several unrelated groups have been reported to follow one another's paths on feeding excursions. This trail-following behavior has been attributed to odor (Symczak, 1950) and to the silk laid down by the larvae (Long, 1955; McManus and Smith, 1972; Wellington, 1974), but it is not known if the stimuli involved are chemical, tactile, or both. Individual larvae lay down silk lines for many purposes, including safety lines, secure footholds (e.g., Alexander, 1961a), tying together of food materials (Bell, 1920; Ford, 1945; Alexander, 1961a), and orientation to lead them back to resting places. It is hardly remarkable that groups of larvae show similar behavior, but it is of course more easily noticed, especially when their trails build up to form structures visible from a distance. Variation in the behavior of larvae that use silk trails for orientation has produced an interesting communication system in certain tent caterpillars (Lasiocampidae). Some larvae in each brood are more reluctant than others to explore new areas, and they follow the more adventurous individuals, which lay the first silk trails (Wellington, 1957, 1974). Selection probably operates strongly at the colony level against those broods that contain an imbalanced ratio of leaders to followers (Wellington, 1974). (Adult moths derived from these caterpillars also differ in behavior.)

Silk enclosures, trails, platforms, and other structures made by groups of larvae are often of elaborate construction. But so also are similar structures (molting and resting platforms, hibernacula, cocoons, etc.) built by single individuals. The behavior involved in making a complex cocoon is both complicated and relatively inflexible, but cocoons are rarely identical since their forms are affected to some extent by the physical limitations of the environment (Van der Kloot and Williams, 1953a, 1953b, 1954; Yokoyama, 1951). For larvae that live together, the environment includes silken structures already made by other larvae, and a succession of building activities by many larvae on the ever-enlarging construction may result in the spectacular enclosures of such species as tent caterpillars (*Malacosoma* spp., Lasiocampidae) and webworms (*Hyphantria* spp., Arctiidae). This process, in which the summation of relatively simple behavior patterns by individuals results in a complex construction, is similar in principle to the process of "stigmergy" hypothesized by Grassé (1959) to account for the complexity of nest construction by termites and other social insects (see also Wilson, 1971), a concept recently generalized by Wilson (1975) under the name "sematectonic communication." Such communication is characterized by individuals responding to the inanimate products of their labors, rather than directly with one another. (In this context it should be mentioned that an individual that interacts with its own constructions, such as a caterpillar building a cocoon, is in a similar sense communicating with itself.) Unlike social insects, caterpillars are not known to recruit other individuals to assist in building. But recruitment may be unnecessary for the production of community enclosures or multiple cocoons if, as is usually the case among gregarious lepidopterous larvae, their development and behavior is synchronized. Larvae that live communally are believed to gain a measure of protection from predators (Ford, 1945; Hogue, 1972; Tinbergen, 1958), and communal living is adaptive in other ways as well (Rathke and Poole, 1975). The tendency to build silken structures together with other larvae has been demonstrated as heritable (in the case of double-cocooned and polypupal-cocooned silk-moths: Yokoyama, 1959). Thus the prerequisites for the evolution of such behavior (selective pressure and heritability) are present, but no elaborate communication system need evolve. And in spite of the fact that tent caterpillars share a number of behavioral attributes with certain social insects—siblings living together, behavioral synchronization, polyethism (some larvae "adventurous"), (silk-) trail-following, the construction of elaborate dwellings by means of sematectonic communication, and probably strong selection at the colony level—they are in comparison with social insects merely "communal" (in the sense of Michener, 1969, as modified by Wilson, 1971).

Intraspecific competition is the context for interlarval communication of a rather different kind. Larval density can affect adult development, morphology, and physiology (Long, 1959; Long and Zaher, 1958). In some tortricid larvae spacing of individuals is achieved through defense of a feeding territory (Russ, 1969). Recent studies of *Ephestia* (=*Anagasta*) *kühniella* (Pyralidae) and some related species have revealed that antagonistic interactions between the competitive and agressive larvae, especially the release of a mandibular gland pheromone, affect larval spacing and survivorship (Corbet, 1971; Cotter, 1974; Mudd and Corbet, 1973). The pheromone therefore has been reported to have an "epideictic" effect, which is enhanced by the chemical's role as a host-detection kairomone (Brown et al., 1970) for parasitoid hymenopterans (Corbet, 1971). In contrast, among silkmoth larvae (*Bombyx,* Bombycidae), aggregation rather than spacing is mediated by pheromones (Okui, 1964).

The Pupa and Pharate Adult

Pupae, like eggs, seem to be relatively inert and devoid of communicative behavior. This is not universally true, however.

Pupae of some species have a number of active defensive adaptations (Cole, 1959; Hinton, 1955), certain of which can be regarded as communicative. (These adaptations are distinguished from passive defenses such as crypsis and warning colors of one sort or another, which may also have communicative functions.) Audible sounds are produced by stridulation or by knocking or scraping the body against the wall of the pupal cell (Hinton, 1948, 1955; Downey, 1966; Downey and Allyn, 1973; Hoegh-Guldberg, 1972). Similar sounds may also be produced by the pharate adult, using the appropriate structures on the overlying pupal cuticle (Hinton, 1948, 1955; Alexander, 1961b). The communicative value of these sounds has not been experimentally investigated; they are usually presumed to be defensive (Downey and Allyn, 1973). But as Gilbert (1975) has pointed out, the potential for intraspecific auditory communication between pupae and adults exists in at least one species (*Heliconius erato*), the pupa of which stridulates (Alexander, 1961b) and the adult of which can hear (Swihart, 1967a). It should be pointed out, however, that the lowest threshold for hearing in *Heliconius* adults as measured electrophysiologically by Swihart was about 60 db (at about 1.2 khz). The intensity of the pupal sounds has not been measured, but is probably below that level.

Heliconius pupae also emit species-specific odors, which have been interpreted as defensive (Alexander, 1961b) and pheromonal (Gilbert, 1975). Males of the *Heliconius erato* species group are attracted by pheromones to female pupae and await eclosion before attempting copulation (L. E. Gilbert, pers. comm.). But male *H. charitonia* "invade" the pupal integument with their genitalia and "rape the female pupa[4] as a routine

4. Actually the pharate adult. Relative to the behavior of most other Lepidoptera such behavior might be called "precocious promiscuity." In this situation the female cannot exercise male selection.

mating procedure" (Gilbert, 1975, and pers. comm.). These observations support the idea that such pupae (or pharate adults) have pheromonal (and perhaps sonic) means of indicating their presence and precise location to adult males. Sex-attractant pheromones are generally not released until after eclosion in most species (Jacobson, 1972).

Numerous instances of lycaenid pupae that are tended, protected, and sometimes even sheltered by ants have been documented and are reviewed by Hinton (1951, 1955) and Downey and Allyn (1973). Like the larvae (q.v.) of these and many other lycaenid species, such pupae are reported to secrete attractive substances; in addition, many lycaenid pupae produce audible sounds (Downey, 1966). Unfortunately, none of these relationships are understood in enough detail to be able to say more about the nature or significance of auditory or chemical communication by these fascinating insects.

The Adult

The behavioral repertoire of adult lepidopterans is far more complex than that of the immature stages. Reproduction and dispersal are added to defense and (in many cases) feeding as requisite activities of successful adults. Sex, flight, (usually) nectar location, and (for ovipositing females) larval host-plant identification all require a high degree of sensory capability and motor coordination. Before surveying adult communication we must first briefly examine the physiology of the senses involved.

SENSORY PHYSIOLOGY OF ADULTS

Vision

The visual *Umwelt* of lepidopterans differs significantly from our own. The visual spectrum of some butterflies appears to be the broadest in the animal kingdom, extending from the edge of terrestrial ultraviolet (around 300 nm) through the red (700 nm); it therefore includes our own spectral range plus 300 to 400 nm in the ultraviolet (Crane, 1955; Gilbert, 1975; Mazokhin-Porshniakov, 1969, and references therein; Obara, 1970; Petersen et al., 1952; Post and Goldsmith, 1969; Swihart, 1967b). Color vision has been demonstrated in both butterflies (Crane, 1955; Ilse, 1928, 1932a, 1932b, 1937, 1941; Ilse and Vaidya, 1956; Mazokhin-Porshniakov, 1969; Post and Goldsmith, 1969; Swihart, 1963, 1964, 1965, 1967b; C. Swihart, 1971; Swihart and Swihart, 1970) and moths (Knoll, 1922, 1925, 1927; Mazokhin-Porshniakov, 1964, 1969; Schremmer, 1941).

On the basis of anatomical, physiological, and behavioral studies, and by analogy with other terrestrial arthropods with well-developed compound eyes, lepidopterans are believed to be most behaviorally responsive to light of short wavelengths, able to adapt over a wide range of light intensities, and able to detect (if present) the plane of polarization. Significant morphological and physiological differences occur between species and are especially pronounced between nocturnal and diurnal forms (Autrum, 1965; Bernhard, 1966; Burkhardt, 1962, 1964; Dethier, 1963; Eltringham, 1919; Goldsmith, 1961; Goldsmith and Bernard, 1974; Mazokhin-Porshniakov, 1969; Miller et al., 1968; von Frisch, 1967; Wehner, 1972; Yagi and Koyama, 1963; and references therein).

Tapetal interference filters of unknown function (but believed to increase sensitivity to certain colors) have been reported in the eyes of some butterflies (Bernard and Miller, 1970; Bernhard et al., 1970; Miller and Bernard, 1968). (It has been suggested that some lepidopterans are sensitive to infrared light but not via the visual organs; see olfaction.) The role of vision in adult behavior is discussed under flower visitation and courtship.

Sound

Many lepidopterans are capable of hearing. Tympanic organs located on the metathorax or abdomen in diverse groups of moths are believed to have evolved independently at least ten times[5] (Kiriakoff, 1956, 1963; Treat, 1964, and pers. comm.). Saclike inflated structures located at the wing bases in some nymphalid butterflies (Swihart, 1967a) and organs associated with the mouthparts in some Sphingidae[6] (Roeder, 1971, 1972, 1974a; Roeder and Treat, 1970; Roeder et al., 1968, 1970) have also been identified as auditory in function. In addition to these organs, lepidopterans, like other insects, possess displacement-sensitive setae and subgenual and other scolopophorous organs that might act as receptors of air- or substrate-borne vibration. (See also Busnel, 1963; Frings and Frings, 1960.)

The "ears" of moths are most sensitive to ultrasound (Roeder, 1965, 1967a, 1971, 1972, 1974a, 1974b, 1975; Roeder and Treat, 1957; Sales and Pye, 1974; Schaller and Timm, 1950; Treat, 1964). It is now well known that the adaptive significance of the hearing of moths is that it enables them to detect echolocating insectivorous bats before they themselves are detected. Moths that hear bats perform a wide variety of defensive behavior, the nature of which depends on the distance at which the predator is detected and the species of moth concerned. Turning, looping, power diving or dropping to the ground, and other evasive tactics are used (Roeder, 1965, 1966, 1967a, 1967b, 1970, 1971; Roeder and Fenton, 1973; Treat, 1964).

Some arctiid and amatid moths, many of which are unpalatable or otherwise "protected" (Beebe and Kenedy, 1957; Blest, 1964; Eisner, 1970; Rothschild, 1965, 1973; Rothschild and Alpin, 1971) employ an additional strategy: they answer the ultrasonic cries of bats with aposematic clicking calls produced by a thoracic "microtymbal" (Blest et al., 1963; Fenton and Roeder, 1974; Dunning and Roeder, 1965). Bats confronted with such calls veer away from the prey (Dunning and Roeder, 1965). Some noisy palatable species are also avoided and are therefore Batesian mimics (Dunning, 1968). Probably some of the sounds produced by other moths (references in Frings and Frings, 1960; Haskell, 1961; see also Lloyd, 1974; Rothschild and Haskell, 1966) are defensive as well.

Auditory communication between lepidopterans has rarely been documented. Roeder and Treat (1957) suggested that ultrasonic components of wing sounds might be audible to other moths. Dahm et al. (1971) demonstrated that auditory signals produced by wing vibration are an important component of the mating system of the lesser waxmoth (*Achroia grisella,* Pyralidae). The males of this species (like the females of most moths) release a complex mixture of sex-attractant pheromones. Females are excited by the chemicals but do not orient to the source unless vibrations, such as the fluttering of a male's wings, are also present.

There are few other reports of even potential auditory communication between lepidopterans (Bourgogne, 1951). (Reference has already been made to the possibility of pupa-adult communication among certain heliconiine butterflies.) Perhaps the most widely cited case is that of the "cracker" butterflies of the genus *Hamadryas* (=*Ageronia* Nymphalidae), which produce (in an as yet undetermined manner) a loud, rapid series of clicking sounds during flight. Adult *Hamadryas* can hear (Swihart, 1967a). The behavioral significance of these sounds, often produced during close pursuit of other butterflies, is unknown; it has often been suggested

5. Geometridae, Pyralidae, Thyatiridae/Drepanidae, Epiplemidae/Uraniidae, Axiidae, Cossidae (all abdominal); Noctuidae/Agaristidae, Notodontidae, Amatidae, Arctiidae (all metathoracic).

6. K. D. Roeder (pers. comm.) has recently determined that some sphingids respond to sound even after destruction of the palp-pilifer region, indicating the existence of yet another lepidopteran "ear."

that they are involved with territoriality (q.v.) and/or courtship. A few other butterflies make audible sounds of unknown function during flight (e.g., some *Charaxes*) or while stationary (F. Scott, 1968). One of the most peculiar cases of sound production is that of the "Death's Head" sphinx moth (*Acherontia atropos,* Sphingidae), which has been reported to enter the hives of honeybees to obtain honey; when attacked by bees it emits a sound similar to that of "piping" by the queen (Bourgogne, 1951; see also Busnel, 1963).

Direct mechanoreception (not involving sound) is probably important for communication during contact between the sexes, but it remains uninvestigated (however, see Doane and Cardé, 1973).

Olfaction

Sensitivity to airborne chemical stimuli plays an important role in feeding and in the sexual lives of moths and butterflies. Floral odors are important orientation cues for flower visiting (q.v.), and pheromones are involved in the courtship of all intensively studied species.

The antennae are the primary olfactory organs. The frequent sexual dimorphism in these structures among moths (usually with greater surface area and receptor number in males) is generally believed to be related to their use as "odor filters" for the detection of (usually female) sex pheromones. Butterflies rely to a much greater extent on visual cues and exhibit little sexual dimorphism of antennal structure (Payne, 1974; Schneider, 1964).

The sensory physiology of insect olfaction has recently been reviewed by Kaissling (1971) and, with respect to pheromones, by Payne (1974). Sensilla basiconica and sensilla trichodea, located on the antennae, are the olfactory receptors (see Albert et al., 1974). Those involved in phermone reception are often highly specialized and respond only to a narrow range of chemical stimuli.[7] The response threshold for individual receptors is as low as a single molecule, and whole-organism behavioral responses are elicited with as few as 200 molecules (Kaissling and Priesner, 1970; Schneider, 1974). Comparative studies of response to pheromones and to various chemically related compounds (pheromone analogs, or "parapheromones," that differ in carbon chain length, location and orientation of unsaturated bonds, and attached functional groups) have revealed that even those that are stereochemically very similar to the natural pheromones are required in greater concentrations in order to elicit the same electrophysiological or behavioral responses, and that effectiveness decreases with increasing stereochemical discrepancy (e.g., Gaston et al., 1972; Payne et al., 1973; Roelofs and Comeau, 1971a; Schneider et al., 1967; and other references in Payne, 1974). The mechanism of transduction is not presently understood (Davies, 1971; Payne, 1974; and references therein). Theories currently in vogue differ in details but most suggest a chemical and/or physical interaction between pheromone molecules and matching acceptor (receptor) sites on the receptor cell membrane, which in some manner affects permeability to inorganic ions and thus initiates electrical events.

Early studies of sex pheromones concentrated on *the* pheromone of each species, since it was believed that species-specificity was conferred primarily or exclusively by chemical diversity. It is now clear that chemical specificity (hence reproductive isolation) is also conferred in many species by mixtures of two or more compounds. A compound that elicits behavioral and/ or electrophysiological responses may, when combined with others as a mixture, be more or

7. "Specialist" receptors of this type are contrasted with "generalist" receptors, sensitive to a broad spectrum of compounds. The latter are believed to be important as food odor detectors.

less effective (depending on the species and compounds concerned). Synergistic or inhibitory effects are believed to provide species-specificity in the communication system with a limited diversity of compounds (e.g., Comeau, 1971; Klun and Robinson, 1971; Minks et al., 1973; O'Connell, 1972; Roelofs and Cardé, 1974, and references therein; Roelofs et al., 1973; Roelofs and Comeau, 1968, 1971a, 1971b). Additional specificity is provided by concentration and by relative concentrations of components in mixtures (Bartell and Shorey, 1969a; Keae et al., 1973a; Klun and Robinson, 1972; Roelofs and Cardé, 1974, and references therein; Roelofs et al., 1971).

Behavioral responses to pheromones are also affected by previous exposure (Bartell and Lawrence, 1973; Bartell and Roelofs, 1973; Shorey, 1974, and references therein; Traynier, 1970), light intensity and photoperiod (Bartell and Shorey, 1969b; Shorey and Gaston, 1965), temperature (Batiste et al., 1973; Cardé and Roelofs, 1973; Collins and Potts, 1932; Klun, 1968; Shorey, 1966), and other factors (Jacobson, 1972; Shorey, 1974). These features of olfaction, together with others that surely remain to be discovered, are interrelated with one another and with similar factors affecting pheromone release by the opposite sex. The lack of diversity in a single variable (chemical structure) is compensated by tremendous complexity in the rest of the communication system.

The behavioral responses of insects to sex pheromones have been reviewed by Shorey (1973, 1974). With increasing concentrations of "attractant" pheromones a "hierarchy of responses" is elicited in males, which consists of antennal movements, increased activity, flight and orientation towards the source, followed by cessation of flight, localization of the source, release (in some species) of male-produced pheromones, and copulatory attempts (Bartell and Shorey, 1969a, 1969b; Daterman, 1972; Tranier, 1968). In some species additional chemical stimuli are needed at various points along the "hierarchy"; if they are not present the behavioral sequence is not completed (Cardé et al., 1975a). A "hierarchy" of responses has also been demonstrated among females in the "reversed-role" chemical communication system of the greater and lesser waxmoths, the males of which produce long-range chemical attractants (Dahm et al., 1971; Röller et al., 1968). Another "reversed-role" system has been reported to occur among certain ithomiine butterflies (Nymphalidae), the males of which produce pheromones that presumably function as intra- and interspecific attractants mediating aggregation (L. E. Gilbert, 1969 and pers. comm.; W. A. Haber, pers. comm.; but see also Pliske, 1975b). Orientation to a pheromone source during flight is probably mediated by anemotaxis combined with crosswind flights that are believed to enable the insect to remain within the active space; several other mechanisms of orientation have also been postulated (Farkas and Shorey, 1974, and references therein; Kennedy and Marsh, 1974).

In contrast, sex pheromones produced by males generally inhibit locomotion in females. The most extensively studied male pheromone system is that of the queen butterfly, *Danaus gilippus* (Danainae, Nymphalidae). The male queen butterfly overtakes the female in flight and disseminates (with everted and splayed brushlike "hair-pencils" extruded from his abdomen), a cuticular dust bearing a pheromone that induces her to land and become quiescent (Brower et al., 1965). Other danaine butterflies have similar structures, and in some cases the pheromones have been chemically identified (Brower and Jones, 1965; Edgar and Culvenor, 1974; Edgar et al., 1971, 1973; Meinwald and Meinwald, 1966; Meinwald et al., 1966, 1969a, 1969b, 1969c, 1971, 1974; Myers, 1972; Myers and Brower, 1969; Pliske and Eisner, 1969; Pliske and Salpeter, 1971; Schneider and Seibt, 1969; Seibt et al., 1972). Such "aphrodisiac" pheromones are believed to be of wide occurrence in the order (Birch, 1974c, and references therein). (See below.)

The idea that moths might orient to infrared radiation and specifically to the characteristic absorption and transmission energies of pheromones and other biologically relevant molecules (Callahan, 1965a, 1965b, 1965c, 1966, 1967, 1968, 1969a, 1969b, 1970, 1971; Callahan et al., 1968; Laithwaite, 1960; Wright, 1963) has not been supported by evidence gathered from controlled experiments (Griffith and Süsskind, 1970; Hsiao, 1972; Hsiao and Hackwell, 1970; Hsiao and Süsskind, 1970; Levengood and Limperis, 1967).

COMMUNICATIONS EQUIPMENT: SCALES AND PHEROMONES

Moths and butterflies are invested with a covering of flattened integumental outgrowth called scales. Each scale is produced during the pupal stage by a single epidermal cell, which usually dies before eclosion. Scales cover the entire body surface, except for the compound eyes (which may have a few scales or scalelike setae scattered between the ommatidia). (The structural diversity of wing scales is reviewed by Downey and Allyn, 1975.)

Scales serve many functions, including (1) aerodynamic: increasing lift during flight (Nachtigall, 1965, 1974); (2) sensory: acting as mechanoreceptors (trichogen cell derivatives of sensilla squamiformia: Dethier, 1963; Eltringham, 1933; Wigglesworth, 1972); (3) thermoregulatory: acting as insulation (Adams and Heath, 1964) or as solar-radiation-absorbing outgrowths (Kettlewell, 1973; Watt, 1968) that may also aid circulation by producing convection currents in the wing veins through uneven heat absorption (Bourgogne, 1951); (4) defensive: as the seat of most cryptic, startling, deflective, aposematic, mimetic, or other adaptive colors, patterns, and structures; as detachable and dispensable integumentary structures (Eisner, 1965; Eisner and Shepherd, 1965, 1966; Eisner et al., 1964; Hinton, 1951, and references therein); and (5) reproductive: as the seat of colors and patterns that play significant roles in courtship, and as a source of or disseminating organ for sex pheromones (references below). Thus, in considering the colors of lepidopterans and the structures of their scales, one must bear in mind that many, often conflicting, selective pressures have over the course of evolutionary time affected these features. The colors and structures of scales found on the bodies, wings, and legs of butterflies and moths thus represent compromises.

The communicative role of color in courtship is widely recognized as a major factor in the evolution of the diurnal Lepidoptera. But while conspicuous coloration is advantageous as a high-intensity sexual signal, it may be detrimental with respect to protection from predators. It is probably for this reason that the brilliant courtship colors of male butterflies are located on the upper surfaces of the wings, where they are exposed during flight but disappear when the insect comes to rest. (An adventitious benefit gained from such color distribution is that aerial predators may be left with a search image that "disappears" when the insect lands—so-called flash coloration). Shifting epigamic signals out of the sensory range of predators accomplishes the same function. The use of patterns that lie beyond the vertebrate-visible spectrum is one means of limiting sexual signals to "intended" receivers. Such ultraviolet signals are widely distributed among the diurnal Lepidoptera (Mazokhin-Porshniakov, 1957, 1969; Nekrutenko, 1964, 1968; Obara, 1970; Scott, 1973b; Silberglied, 1969, 1973; Silberglied and Taylor, 1973).

Color, produced by both pigmental and

structural means in wing scales, plays an important role in the courtship of diurnal species. Because of their finely divided morphology at the ultrastructural level, unpigmented scales are generally white due to surface scattering.[8] Their ridged, reticulated structure serves as a substrate for a wide range of pigments, including melanins in the case of very dark scales (Ford, 1945; Kolyer and Reimschuessel, 1970; Mason, 1926; Wigglesworth, 1972; Yagi, 1955). In addition, the integument constituting the scale ridges *(Morpho, Eurema, Colias)* or base *(Urania)* may have a regular laminated structure that acts as an optical interference filter that reflects either "visible" (to man) or ultraviolet light (Anderson and Richards, 1942; Eisner et al., 1969; Gentil, 1942; Ghiradella et al., 1972; Kinder and Süffert, 1943; Lippert and Gentil, 1959; Mason, 1927; Silberglied, 1969; Süffert, 1924).

The latter surfaces are called "iridescent" and occur widely as patterns of scales on the wings of butterflies and a few moths. The light reflected from such surfaces is generally of high intensity and spectral purity. The wavelengths, intensity, and polarization of the light reflected depend on the relative geometric positions of the light source, lamellar array, and observer. Crane (1954) pointed out that a chromatic modulation of the light reflected from the wing occurs with every wingbeat. In addition to the unusual physical properties of the reflected light, iridescence may have an advantage over pigment in that the color depends on the physical properties of the cuticle, over which the insect already has considerable control during development. The animal need not produce unusual pigments at high metabolic cost to achieve a brilliant color. Iridescent reflection may also be added to pigment-based color to produce combinations not readily achievable by pigment alone, as among those butterflies that combine iridescent ultraviolet reflection with "visible" color patterns of all kinds. But the behavioral significance of iridescent colors in general, and of the modulation of intensity, color, and polarization in particular, is poorly understood at present.

In addition to "ordinary" scales, trichogen cells form a diverse array of glandular cells specialized for sexual functions. These "androconia," "scent-scales," or "scent-hairs" [*sic*] occur on males (and on some female *Thyridia* spp., Ithomiinae: B. Drummond, pers. comm.) They may be scattered among the wing scales, or concentrated as special patches, tufts, "brands," "hair-pencils," etc., on eversible or inflatable sacs or tubes ("coremata") on various parts of the legs, wings, or body. Reference has already been made to the "hair-pencils" of male danaine butterflies; similar male organs and "sex scaling" occur in a wide array of butterflies and moths (Barth, 1960; Birch, 1972, 1974c; Jacobson, 1972; McColl, 1969; Percy and Weatherston, 1974; Varley, 1962). Observations on the use of these structures are lacking in the overwhelming majority of species (Varley, 1962), but in the few species studied the organs are exposed or everted during courtship and are considered to be the disseminating organs for "aphrodisiac" pheromones, which function (where known) by inhibiting locomotion of the female (Birch, 1974c; Brower et al., 1965; Pliske and Eisner, 1969; Tinbergen, 1968; Tinbergen et al., 1942).

Sex pheromones produced by females are also products of specialized epidermal cells. These cells are associated with intersegmental membranes that are everted (presumably by blood pressure), and the pheromones are released at the time of "calling." The morphology and histology of these glands have been reviewed by Percy and Weatherston (1974).

The chemistry of lepidopteran sex pheromones has received considerable attention in recent reviews (Beroza, 1970; Evans and Green,

8. Transparent scales occur in a few Castniidae and certain other "clearwing" forms (Poulton, 1898).

1973; Jacobson, 1972, 1974; Roelofs and Cardé, 1974; Roelofs and Comeau, 1971a) and will not be treated here in detail. The most interesting feature of these pheromones is the contrast between the low chemical diversity of female-produced sex pheromones and the high chemical diversity of the pheromones produced by males. Most of the former are C_{12}, C_{14} or slightly longer straight-chain, unsaturated (monoene or diene) alcohols, acetates and aldehydes (identified from members of the families Arctiidae, Bombycidae, Lymantriidae, Noctuidae, Pyralidae and Tortricidae). Differing only slightly from these are a hydrocarbon, an epoxide of a hydrocarbon, and a branched ester, of similar chain length (references in Evans and Green, 1973; Roelofs and Cardé, 1974). In contrast, some of the compounds isolated from male "androconia," scent-organs, etc., include small carboxylic acids, benzaldehyde, benzyl alcohol, and 2-phenethyl alcohol, various small terpenoids (all isolated from Noctuidae: Aplin and Birch, 1970; Birch, 1972, 1974c; Clearwater, 1972; Grant et al., 1972), citral (geranial and neral: Bergström and Lundgren, 1973), a bicyclic sesquiterpene alcohol (tentatively identified from a lycaenid butterfly: Lundgren and Bergström, 1975), and large heterocyclic ketones (from danaine butterflies:[9] Edgar and Culvenor, 1974; Edgar et al., 1974; Meinwald et al., 1966, 1969a, 1969b, 1969c, 1971). The larger molecular weight of the sex "attractant" pheromones (usually produced by females), which operate over long distances and persist in time, and the high volatility of many "aphrodisiac" pheromones (produced by males), which are used for a moment at close range, are well in accord with the theoretical constraints on molecular size in chemical communication systems (Bossert and Wilson, 1963; Wilson and Bossert, 1963).

ADULT BEHAVIOR

Courtship

With few exceptions courtship follows a single basic pattern throughout the order. Males are generally attracted to females by long-distance communication, either visual (as in most butterflies) or chemical (as in most moths). Close approach and persistent courting by males is mediated in many species by female pheromones. The male may then perform stereotyped behavior patterns, disseminating aphrodisiacs and/or presenting visual, auditory, or tactile signals, the response to which is inhibition of locomotion in receptive females. Females play an active role in acquiescing to males, and can usually reject inappropriate males (e.g., the wrong species)[10] or those that attempt to mate with them when they are not receptive. Rejection involves moving or flying away, or the assumption of a stereotyped "rejection posture." If the female acquiesces (by ceasing activity; sometimes lowering the abdomen and exposing the genitalia but often having no outward behavioral manifestation) copulation may occur. Various aspects of courtship and related activities have been reviewed by Birch

9. A terpenoid alcohol identified from the "hair-pencils" of the queen butterfly has been shown to affix cuticular dust particles bearing the ketone pheromone to the female's antennae; the alcohol is not active as a pheromone itself (Pliske and Eisner, 1969). It is probable that similar compounds found in several other danaines perform the same function. But in the monarch *(Danaus plexippus),* the males of which have small "hair-pencils" that lack detectable ketones (Meinwald et al., 1969a, 1969b), the function of the terpenoid compounds remains a mystery (Pliske, 1971a).

10. Teneral (freshly eclosed) females are occasionally raped or mated by the wrong species, but once the female's integument has hardened she can effectively reject males. Most interspecific matings probably happen during the teneral period, as was shown by Taylor (1972) in the oft-cited but frequently misinterpreted case of hybridization in the butterfly genus *Colias* (Pieridae). Copulation with teneral females is reported to be a normal occurrence in certain birdwing (*Ornithoptera,* Papilionidae: Borsch and Schmid, 1973) and heliconiine butterflies (*Heliconius,* Nymphalidae: Gilbert, 1975 and pers. comm.).

(1974), Farkas and Shorey (1974), Jacobson (1972, 1974), Miller and Clench (1968), Myers (1972), Roelofs and Cardé (1974), Scott (1973a, 1974), Shields and Emmel (1973), and Shorey (1973, 1974).

The courtship of butterflies (and some other diurnal forms) differs from that of most moths, primarily in its early stages. Male butterflies usually initiate courtship, but "solicitation" by receptive females has also been reported to occur (Crane, 1955; Scott, 1973a). Magnus (1963) distinguished between two strategies, "seeking" and "waiting,"[11] by means of which male butterflies locate potential mates. Approaches and subsequent courtship behavior by males are released by visual stimuli. Color (including ultraviolet components), motion, and size of the female have been shown to be important cues, while the details of pattern that enchant lepidopterists seem to play little or no role in courtship. Males usually distinguish conspecific females from other males on the basis of either color, odor, or both, but may be highly indiscriminate in their initial approaches (Brower et al., 1967; Crane, 1955; Johnson, 1974; Lederer, 1960; Magnus, 1950, 1958, 1963; Myers and Brower, 1969; Shapiro, 1972, 1973; Stride, 1956, 1957, 1958a, 1958b; Tinbergen 1968; Tinbergen et al., 1942). If the individual being courted turns out to be male, the sequence is usually terminated; "homocourtship" rarely goes so far as to end in "copulation."

Among sexually dimorphic species color plays an important role (Magnus, 1963; Stride, 1956, 1957, 1958a, 1958b), which must be somewhat diminished in species both sexes of which are similar or mimetic. As Poulton (1907) first pointed out, in species of the latter type, visual cues alone will not suffice to enable males to distinguish females, or even males, from members of other species in the mimicry complex. In sex-limited mimicry (where only females resemble other species), males cannot distinguish conspecific females on the basis of visual cues (except ultraviolet; see Remington, 1973), but females could still use them to discriminate among courting males. The reliability of visual cues may also be a problem among butterflies with seasonal forms and polymorphism (Burns, 1966; but see also Pliske, 1972). Brower (1963b) suggested that scent-dissemination organs and odors detectable to man, hence pheromonal means of communication, are more (?) common among Müllerian mimics (but see Vane-Wright, 1972). It is now apparent that the courtship of just about all lepidopterans studied involves pheromones at some stage. It is not surprising that Müllerian mimics are fragrant; that is indeed one of the means by which unpalatability is communicated to potential predators.

An alternative means by which males of mimetic species might differentiate conspecific females, and females recognize conspecific males, is by way of ultraviolet reflection patterns invisible to vertebrate predators. Survey of several mimicry complexes in the ultraviolet by Silberglied (1969, 1973, and unpublished) and C. L. Remington (1973, and pers. comm.), revealed differences between some species and between the sexes of some nondimorphic species, but in the absence of behavioral experiments these results are difficult to interpret. However, among the Pieridae, most species of which show strong dimorphism of ultraviolet reflection patterns not evident in visible light (Mazokhin-Porshniakov, 1957, 1969; Nekrutenko, 1964, 1968; Obara, 1970; Scott, 1973b; Silberglied, 1969, 1973; Silberglied and Taylor, 1973), these patterns are used both as sexual-recognition signals (Obara, 1970; Silberglied, 1973) and as partial isolating mechanisms (Silberglied, 1973).

11. "Seeking" males actively search the habitat in a stereotyped flight pattern, while "waiting" males simply sit and wait for females to pass by. Scott (1973a, 1974) adopted the terms "patrolling" and "perching," respectively, for these same activities.

The activities of moths during the early stages of courtship contrast strongly with those of butterflies in two ways: female moths initiate courtship and the first cues are generally olfactory rather than visual. Females "call" by releasing at the appropriate time sex pheromones[12] that elicit in males a "hierarchy of responses" (see above) that lead them to the source (Shorey, 1973, 1974). Orientation may be accomplished by means of anemotaxis and/or chemical cues (Farkas and Shorey, 1974), but sound (Dahm et al., 1971) and vision (Shorey and Gaston, 1970) are sometimes involved, especially for close-range orientation. Visual cues are certainly not necessary for some species as they are with butterflies; mating in complete darkness has been reported in *Catocala* (Noctuidae: Sargent, 1972). While a few moths are known to have color vision (q.v.), it is not known whether color plays any role in their courtship behavior. The role of color in the courtship of day-flying moths, many of which are bejeweled with iridescence (e.g., Uraniidae, Amatidae) and striking color patterns (e.g., Arctiidae), is an unexplored field.

Once in the immediate vicinity of a female, subsequent attentive behavior by the male depends in many cases on continued or additional olfactory cues. Magnus (1958) found that male *Argynnis paphia* (Nymphalidae) responding to moving female models by chasing would soon lose interest if the scent of a female were lacking. In the almond moth (*Cadra cautella:* Pyralidae), female-produced compounds different from the long-range sex-"attractant" are required for excitation of the male and a complete courtship sequence ending in copulation (Brady et al., 1971b; see also Cardé et al., 1975a). In many other moth species (e.g., *Ephestia kühniella,* Pyralidae: Traynier, 1968; *Porthetria dispar,* Lymantriidae: Brady et al., 1971a) a single pheromone serves both to attract and to excite males. However, close-range stimulant pheromones are not required in all species; in *Colias eurytheme* (Pieridae) males will court and attempt to mate with paper models in the absence of females (O. R. Taylor and R. E. Silberglied, unpublished).

In many butterflies the initial meeting of the two sexes is aerial, and the female must be induced to land before attempts at copulation can be made. Female moths generally "call" from a stationary position in an exposed place, but may take flight if disturbed. At this point "seduction" is in order. Males may release highly volatile, "aphrodisiac" pheromones (either airborne or on cuticular dust particles), inhibiting female locomotion. Reference has already been made to the "hair-pencilling" behavior of male danaine butterflies. Male *Eumenis* (Satyrinae, Nymphalidae) enfold the female's antennae between the forewings in an elaborate "bowing" display; there her antennae are exposed to a patch of "androconial" scales (Tinbergen, 1968; Tinbergen et al., 1942). Vane-Wright (1972) suggests that scales transferred from wings to abdominal scent-brushes of *Antirrhea* (Satyrinae, Nymphalidae) function in a manner similar to the cuticular "hair-pencil dust" of danaine butterflies. In *Eurema daira* (Pieridae) the male lowers one forewing and "buffs" the female's antennae with a patch of specialized scales (Silberglied, 1973, and unpublished). Male noctuid moths expose and splay their "brush-organs" immediately before attempting copulation, but no contact is ordinarily made with the female's antennae (Birch, 1970, 1974c; Grant, 1970, 1971); male pheromones are also required to elicit receptivity in female phycitid moths (Grant and Brady, 1975; Grant et al., 1975). Such behavior patterns are probably general throughout the order in most cases where special male scent organs or "sex scaling" is found, but are not

12. Gilbert (1969, and pers. comm.) and W. A. Haber (pers. comm.) suggest that male ithomiine butterflies (Nymphalidae) also "call" females by disseminating an "assembling" pheromone (but see Pliske, 1975b).

known to occur in well-studied species (e.g., in the Bombycidae and Saturniidae), which lack such organs.

Female moths have excellent control over their sex lives since they may "call" whenever receptive. Receptivity is governed by both internal (age, physiological condition, time since last mating) and external (time of day, light level) factors. If not receptive, they do not "call." Female butterflies, on the other hand, constantly exposed to the view of actively searching males, are often subject to close-range copulation attempts. In response to persistent males, unreceptive female butterflies either fly away, flap their wings, or assume stereotyped rejection postures. Scott (1973) has collated much of this information and should be consulted for details.

Aggregations

Some butterfly species (and more rarely certain moths) are occasionally or regularly found in dense aggregations. Individuals of aposematic species are believed to benefit from close proximity to one another, the general argument being that they provide a bigger (and perhaps more memorable) visual stimulus to predators. Aggregations of individuals also occur around food sources and mud puddles. *Heliconius* butterflies roost in groups at night, as do monarch butterflies (*Danaus plexippus,* Danainae, Nymphalidae) during migration (see Benson and Emmel, 1973; Turner, 1975). Bogong moths (*Argrotis infusa,* Noctuidae) aggregate by the thousands at their aestivation sites (Common, 1954), and dense clusters of inactive butterflies have also been reported (Muyshondt and Muyshondt, 1974).

Little is known about communication between individuals in such aggregations. In their choice of resting site they (i.e., the first to land) are certainly responding to various stimuli in their environment, but individuals also recognize others of their kind and orient to them. "Mudpuddling" by butterflies (and occasionally moths) is one of the more intriguing cases (Downes, 1973; Norris, 1936). The fact that such aggregations (mud-puddle "clubs") consist almost entirely of males has led to some interesting hypotheses; Wynne-Edwards (1962), for example, included them under "group nuptial displays." Recently, Arms et al. (1974) demonstrated that, in addition to visual cues, one proximate stimulus for mudpuddling by tiger swallowtail butterflies (*Papilio glaucus*) is sodium. Visual recognition is also involved in the attraction of numerous additional individuals to the site (Collenette and Talbot, 1928). However, it remains a mystery why males are so disproportionately overrepresented at mud puddles, carrion, urine, rotting carcasses, fruits, and other such sources. Another curious aggregation phenomenon not presently understood occurs only in those populations of the African butterfly *Acraea encedon* (Nymphalidae, Acraeinae) in which there is a highly imbalanced sex ratio with females predominating. Females aggregate and lay infertile eggs on non-host plants and on one another as well (Owen, 1971).

The communal roosting and "social chasing" of heliconiine butterflies is well known to lepidopterists who have worked in the Neotropics, and numerous other butterfly species have been reported to roost gregariously (e.g., Clench, 1970; Crane, 1955; Jones, 1930, 1931; Muyshondt and Muyshondt, 1974; Myers, 1930; Poulton, 1931a, 1931b; Poulton et al., 1933; Young, 1971). Benson (1971) and Turner (1971a, 1975) speculate that communal roosting, in addition to being protective ("it being usually believed that it helps a distasteful species to be gregarious"; Turner, 1971a), in *Heliconius* is part of a combination of behavioral features holding closely related individuals together in highly restricted home ranges and having some connection with "altruism" in the Hamiltonian sense (Hamilton, 1964). Gilbert (1974) combines these ideas with the consideration that the communal roosting habit in *Heliconius* may have evolved as an integral part of their coevolution with floral resources (see also below): young individuals perhaps learn roosting sites as well as "trap-lines" by following older, more experienced individuals. Crane (1955) performed extensive ethological experiments on *Heliconius erato* and discovered that while color plays an important role as a releaser of sexual behavior and "social chasing" in these butterflies, it is unimportant in roosting.

Flower-Visitation and Pollination

Butterflies and moths obtain their largely carbohydrate diet mostly from the nectaries of flowers, and in so doing often act as pollinators (Proctor and Yeo, 1973). The long evolutionary history of this relationship is reflected in the extreme morphological and behavioral adaptations of the organisms concerned. In the Lepidoptera, these include well-developed olfactory and visual senses used for locating flowers, the elongate, tubular proboscis (= haustellum or "tongue" [*sic*]) for feeding from nectaries deeply recessed within the flower, and in one instance an independent method of extra-oral pollen-feeding (Gilbert, 1972). Corresponding botanical developments include floral odors (Yeo, 1973), brilliantly colored perianths, nectaries (often concealed within tubular flowers or "spurred" organs, e.g., Emmel, 1971) that produce secretions rich in carbohydrates (and in some cases amino acids: Baker and Baker, 1973a, 1973b), and flowering phenologies corresponding to the activity periods of the insects concerned. Pollination relationships of great complexity (e.g., Gilbert, 1975) have evolved, occasionally to the point of obligate interdependence of both the plant and the lepidopteran, as in the oft-cited case of the yucca moths of the genus *Tegeticula* (= *Pronuba*) (McKelvey, 1947; Powell and Mackie, 1966; Riley, 1892).

Visual cues are the main signals used for flower-localization and identification, but in some species floral odor plays an important role as well (Ilse, 1928, 1941; Knoll, 1922, 1925, 1927; Lederer, 1951; Myers and Walter, 1970; Schremmer, 1941). For example, Schremmer (1941) found that the noctuid moth *Plusia gamma* uses odor cues to locate flowers for its first meal after eclosion, but thereafter it is imprinted with the visual pattern and will use it in addition to odor as an orientation cue. On the other hand, many species, including sphingid moths (Knoll, 1922, 1925, 1927) and some butterflies, rely entirely or almost exclusively on floral color and pattern. Crepuscular and nocturnal as well as diurnal species have been shown to use color vision in locating, identifying, and feeding from flowers. Most species have innate color preferences, but in some cases these may be modified by experience or training (Ilse, 1928; C. Swihart, 1971).

Detailed features of the flowers, such as the amount of "dissection" (the ratio of perimeter to area) of the corolla may be important identifying characteristics (Ilse, 1932a). Contrasting "guide-marks" (also known as "honey-guides," "nectar-guides," "Saftmale," and "Pollenmale"; see Proctor and Yeo, 1973) are used for location of the flower entrance (Knoll, 1922, 1925, 1927). Thus lepidopterans use both innate and imprinted "search images" (C. Swihart, 1971), which, together with olfaction and other behavioral adaptations, enable them to engage in a spectrum of floral-feeding relationships ranging from fortuitous, entirely facultative, situations to obligate mutualisms. Gilbert's (1972, 1975) fine study of coevolution of *Heliconius* (Heliconiinae, Nymphalidae) butterflies with their larval (Passifloraceae) and adult (certain Cucurbitaceae) food plants is illustrative of the complexity of such interrelationships. The great longevity of these butterflies (Gilbert, 1972; Turner, 1971b), made possible by their unpalatability to predators (Brower et al., 1963; Brower and Brower,

1964) and mutualism with cucurbit vines as nectar and pollen sources, appears to have evolved with behavioral sophistication unparalleled elsewhere in the Lepidoptera (Gilbert, 1975).

Territoriality and Antagonistic Behavior

Males of a large number of butterflies behave in a manner that has frequently been interpreted as territorial. Characteristically, a male chases other butterflies (or falling leaves, insects, birds, or lepidopterists) persistently until they leave the immediate area (e.g., Fleming, 1965; Hendricks, 1974; Pyle, 1972; Slansky, 1971). (Such species are often called "pugnacious." This aggressive behavior has been studied experimentally by Ross (1963), who found that marked *Hamadryas* butterflies (Nymphalidae) rarely remained in the same place for long periods. (See also Lederer, 1951, 1960; Swihart, 1967a). But recent studies of several species in this genus by D. Windsor (pers. comm.) indicate that while males are aggressive but not territorial at feeding sites, they will vigorously defend for weeks or even months perches from which females can be pursued. Baker (1972) has observed similar behavior in several other nymphalids (but see Scott, 1974), and L. E. Gilbert (unpublished, cited in Maynard Smith, 1976) and R. C. Lederhouse (pers. comm.) also have strong evidence for territoriality in certain swallowtail butterflies (*Papilio* spp., Papilionidae), the males of which defend hilltops to which females fly when receptive.

More experimental studies of territoriality are needed to determine the signals males use to recognize other males (see Stride, 1957). In many species it is not clear if and how aggressive behavior differs from "inspection" flights which characterize the earliest stage of courtship in many butterfly species (Swihart, 1967a). There are a number of situations that are often confused with territorial defense. Owen (1971) noted that some territories change "from day to day and even within a few minutes. The term 'individual distance' . . . is perhaps more appropriate than territory." (See also Scott, 1974.) Territoriality should also be distinguished from long-term residency within a small area, which is well known in a number of butterfly species. If this area is not defended it should be considered a home range (e.g., Ross, 1963; Turner, 1971a, 1971b). Males of some species actively search for females, flying ("seeking") often within a restricted area. More than one such male may be competing, not for the physical space but for females within it.

Lepidopterans seem at times to be more defensive of their food than of potential mates. Large butterflies such as *Anteos* sp. (Pieridae) and *Charaxes* spp. (Nymphalidae) often displace one another at food by shoving with the forewings, the front edges of which may be thickened and serrated (Owen, 1971). *Heliconius* butterflies sometimes defend the flowers at which they feed, but several male *H. charitonia* may wait on a female pupa with which one of them will eventually mate. As on flowers, males do attempt to prevent one another from landing, but once alighted little antagonistic behavior occurs (L. E. Gilbert, pers. comm.). Doane and Cardé (1973) have demonstrated that aggressive competition also occurs between male gypsy moths (*Porthetria dispar,* Lymantriidae) attempting to alight at a pheromone source. Two cases of a female butterfly coupled simultaneously with two males have recently been reported (Masters, 1974; Perkins, 1974). Intermale communication also occurs in instances where (1) one male rejects (David and Gardiner, 1961) or inhibits (Stride, 1956) "homocourtship" attempts by other males, (2) the male of a pair signals in a similar manner to other males that are attracted to his mate, possibly preventing "takeover" (Parker, 1970) by the intruders (O. T. Taylor and R. E. Silberglied, unpublished), and (3) the male transfers to a female a pheromone that renders her unattractive to other males (L. E. Gilbert, pers. comm. and cited in Scott, 1973).

Reproductive Isolation

Sexual reproduction is nearly universal in the Lepidoptera,[13] and most intraspecific communication between adults is sexual in nature. The highly stereotyped, species-specific courtship behavior patterns of adult lepidopterans serve not only to bring the sexes together; they function as one of the major reproductive isolating mechanisms between closely related species.

The breakdown of prezygotic isolating mechanisms results in hybridization when postzygotic isolating mechanisms are weak or absent (e.g., Ae, 1965; Cockayne, 1940; Taylor, 1972; Tutt, 1906). Since hybrids can often be readily obtained by hand pairing (Clarke and Sheppard, 1956; Lorković, 1954) but are rather exceptional occurrences in nature, prezygotic isolating mechanisms are believed to be of greater importance than genetic incompatibility or sterility as barriers to hybridization between closely related species.

Despite the extraordinary diversity of genitalic differences between males in most groups of Lepidoptera, physical or mechanical ("lock and key") isolating mechanisms do not appear to be of importance in the group (Jordan, 1905; Sengün, 1944).[14] Even intergeneric matings sometimes occur (e.g., Perkins, 1973). The major prezygotic isolating mechanisms are temporal (seasonal and circadian), ecological (e.g., habitat preferences), and ethological (courtship behavior). Often the effects of several different mechanisms, no one of which provides complete isolation, combine to isolate sympatric populations of different species (e.g., Petersen and Tenow, 1954).

13. Exceptions occur in some Psychidae (Seiler, 1923, 1961), Lymantriidae (Goldschmidt, 1917; Vandel, 1931), and a few other groups. See also Cockayne (1938) and Robinson (1971).

14. The incredible complexity of male genitalia is more likely a device that locks the pair together, preventing access by other males during copulation (Parker, 1970; Richards, 1927).

TEMPORAL AND ECOLOGICAL ISOLATION

Separation in time or space serves in many cases to isolate populations that might otherwise interbreed. Habitat preferences of adults may isolate closely related species (Petersen and Tenow, 1954; Shapiro and Cardé, 1970). Mating may be restricted to particular sites within the habitat, such as hilltops (Scott, 1970, 1974; Shields, 1968) or near food plants (Smith, 1953, 1954). Differences in altitudinal range as in *Pieris* (Petersen and Tenow, 1954) and among members of the *Papilio glaucus* group (Brower, 1959), not fundamentally distinguishable from grosser aspects of geographic isolation (often with physiological limits on range), may also be effective. While not obviously communicative in nature, the behavior that brings conspecific lepidopterans together for mating (or alternatively, restricts them from locations where the probability of locating conspecifics is low) is a necessary prerequisite before individuals can communicate directly with one another by means of shorter-range visual and chemical cues.

Similarly, temporal synchrony of reproductive behavior within a species is an important part of the mating system, and is sometimes of critical importance as an isolating mechanism. Seasonal isolation has been relatively little studied but is obviously important, especially in univoltine species. In the genus *Papilio,* closely related members of the *glaucus* group are to some extent seasonally isolated (Brower, 1959). Seasonal isolation may also result from reproductive diapause, which is especially common during the dry season in the tropics (O. R. Taylor, pers. comm.). Circadian rhythms of activity, pheromone release and receptivity to mating by females, and similar rhythms of activity and response to pheromones by males, are species-specific and have been shown to be controlled by photoperiod (George, 1965; Shorey and Gaston, 1965; Sower et al., 1970, 1971; Traynier, 1970) and temperature (Cardé and Roelofs, 1973; Cardé et al., 1975b). (Release of these behavior patterns is probably mediated by hormones; see Truman and Riddiford, 1974.) Such periodicity in reproductive behavior has been known since Fabre (1916); the subject has been surveyed by Jacobson (1972) and discussed recently by Roelofs and Cardé (1974) and Shorey (1974). Species-specific temporal patterning of butterfly courtship behavior is also known (e.g., Shields, 1968; Shields and Emmel, 1973; Miller and Clench, 1968), but in most cases does not differ significantly between closely related species within the same habitat.

Diel rhythms of reproductive behavior appear to be the main isolating mechanisms among sympatric *Callosamia* (Saturniidae: Ferguson, 1971–1972). Interspecific and in some cases intergeneric cross sensitivity of male antennae to female sex pheromones is common among saturniids (Priesner, 1968; Schneider, 1963); diel activity rhythms may therefore be expected to play a more important role in such groups than in those characterized by high pheromonal diversity. Unfortunately, little is known about the chemistry of saturniid sex pheromones, so the chemical basis of their cross sensitivity (e.g., a single component of species-specific mixtures?) remains unknown. It is indeed possible that chemical complexity yet to be discovered may provide additional specificity and effect, in part, reproductive isolation in this family.

One disadvantage to temporal partitioning among species using the same chemical communication system is that they are subject to competition for optimal times. Environmental conditions change during the day and night, and some times are better than others for communication. Reproductive activity periods (Watt, 1968), rhythms of pheromone release (Cardé, 1971; Comeau, 1971; Sanders and Lucuik, 1972; Sower et al., 1971), and male responsiveness (Batiste, 1970; Batiste et al., 1973; Cardé and Roelofs, 1973; Collins and Potts, 1932; Klun, 1968; Shorey, 1966) are all affected by temperature. Cardé et al. (1974) suggest that the mating period of the gypsy moth (*Porthetria dispar,* Lymantriidae) has become lengthened in North America (where it was introduced from Europe over a century ago) because of ecological release from competition for calling time. In Europe, where the gypsy moth coexists with the nun moth (*Lymantria monacha,* Lymantriidae), both use the same pheromone but are temporally isolated to some extent.

SPECIFICITY IN THE COMMUNICATION SYSTEM

The color patterns of diurnal lepidopterans undoubtedly play an important role in reproductive isolation. Only limited isolation results from the indiscriminate male responses to the patterns of females, especially among similarly colored sympatric congeners. But it is widely believed that mate choice (acquiescence or rejection) by females is determined in part by male color pattern. Unfortunately, as Brower (1963) points out, few experimental attempts have been made to test this hypothesis. In recent work with *Colias* butterflies (Pieridae), R. Silberglied and O. R. Taylor (data in Silberglied, 1973) "painted" male sulfur butterflies many different colors; they found no breakdown of reproductive isolation or discrimination by females against grossly miscolored males, except when the ultraviolet patterns were changed. Thus females may exert a selective force on male coloration, but this "coloration" may be outside the range of vertebrate vision. Coloration (especially iridescent colors) may vary in both time and space with respect to intensity, spectral quality, saturation, and polarization; future experimental work on visual signals should attempt to determine which of these parameters are most relevant to the communication system.

Chemical diversity of sex pheromones is only one of several possible means of conferring specificity within the pheromonal communication system. Some closely related moths use different geometric isomers (*cis* or *trans*) of the same long-distance sex attractants (e.g., two *Bryotropha* species, Gelechiidae: Roelofs and Comeau, 1969; two *Amathes* species, Noctuidae: Roelofs and Comeau, 1970). Concentration differences are a second possibility; for example the alfalfa looper (*Autographa californica,* Noctuidae) is attracted to lower concentrations of *cis*-7-dodecenyl acetate than the cabbage looper (*Trichoplusia ni,* Noctuidae) and is inhibited by high concentrations (Kaae et al., 1973). In areas of sympatry, males of the alfalfa looper might be expected to search for nonexistent ("phantom") conspecific females where the concentration of sex attractant (emitted at a high rate by female cabbage loopers) first becomes inhibitory. Since interspecific matings are presumed to be disadvantageous to both participants, one would expect strong selection for more specificity in the two communication systems. In the case of these noctuid moths, such specificity appears to be conferred by incomplete temporal partitioning (Kaae et al., 1973) and by the rejection behavior by females in response to the "wrong" male scent-organ pheromones disseminated later in courtship (Shorey et al., 1965).

Another means of adding specificity to the communication system involves the use of more complex pheromonal mixtures. The relative proportions of two or more compounds (an attractant plus "synergists" or "inhibitors") can be varied. Such systems effect isolation of several pairs of tortricid species (Minks et al., 1973; Roelofs and Comeau, 1971a). Other types of specificity were reviewed recently by Roelofs and Cardé (1974), to which readers are referred for details.

Reproductive isolation can also be based on a combination of visual signals in addition to pheromones. Our two commonest native North American sulfur butterflies, *Colias eruytheme* and *Colias philodice* (Pieridae), are so isolated (Taylor, 1970, 1972, 1973; Silberglied, 1973; Silberglied and Taylor, 1973).

THE ORIGIN OF PHEROMONES

Research on lepidopteran pheromones has largely been limited to chemical identification of active compounds and behavioral and physiological studies on their activity. Considerations of the sources of pheromones (whether biosynthesized or acquired from extraneous sources) and the evolutionary origin of pheromonal communication systems require different kinds of information and a broader data base, and consequently have until recently been neglected.

One likely possibility for the origin of certain pheromones and pheromone glands is that of defense. Several of the compounds identified as lepidopteran sex pheromones (e.g., citral) are chemically identical with defensive secretions, and there is little reason why, once evolved, such defensive compounds might not be used in intraspecific communication (Birch, 1970b, 1974c; T. Eisner, pers. comm.; M. Rothschild, pers. comm.). Birch (1974c) also argues that release of defensive compounds during mating, when the male and female are indisposed, would be protective of the pair.

It has long been known that certain male butterflies, especially many danaines and ithomiines (as well as certain moths), are attracted to certain kinds of vegetation. For example, Owen (1971) reported danaine butterflies gathering and feeding in large numbers on plant juices of *Heliotropium* (Boraginaceae) that had been damaged by the feeding activities of grasshoppers (see also Pliske, 1975b). Collectors in the tropics have for years used the technique of hanging heliotrope to dry in order to attract large numbers of butterflies and moths (e.g., Beebe, 1955; Gilbert and Ehrlich, 1970; Masters, 1968; Morrell, 1960; Pliske, 1975c). The similarity between this behavior and that of male orchid bees has often been pointed out; since it has been suggested that the bees use terpenes obtained from orchid flowers in some sexual context (Dodson et al., 1969; Vogel, 1963), a similar hypothesis has frequently been made about male ithomiines and danaines with whatever-it-is they get from borages. In a recent confirmation of ideas such as these (originally postulated by Miriam Rothschild, cited in Birch, 1970b), Edgar and Culvenor (1974) have shown that the hair-pencil-disseminated pheromones of both *Danaus* and *Euploea* (Danainae) are dihydropyrrolizines of plant origin. Their findings explain the deficiency of sex pheromone in male queen butterflies reared indoors by Pliske (Pliske and Eisner, 1969); it was due to a lack of contact with plants from which they might obtain the necessary compounds (see also Edgar et al., 1973, 1974; Pliske, 1975b; Schneider, et al., 1975). The monarch butterfly (*D. plexippus*) appears to have evolved independence from plant-derived pyrrolizine pheromones via a different courtship sequence involving only occasional "hair-pencilling" (Pliske, 1975a).

Hendry et al. (1975c and pers. comm.) have offered some speculations on the origin of sex pheromones produced by female moths. They reported that male oak leaf roller moths (*Archips semiferanus,* Tortricidae) became sexually active in the vicinity of oak leaves and "frequently attempted to copulate with host leaves that had been damaged by larval feeding." Chemical analysis of oak leaves revealed the presence of many compounds that had been identified as active components (?) of the complex sex pheromone mixture of female moths (Hendry et al., 1974a, 1974b, 1975a, 1975b). (These compounds were not detected in female moths reared on semisynthetic diets lacking oak leaves.) The compounds were also reported to be present in the immature stages.

Hendry et al. (1975c), interpreting these findings, suggest the possibilities that (1) these species might be deriving their pheromones directly from plants during their development, and that (2) the males become "imprinted" in some sense on the compounds they will later employ in courtship. Since different host plants have various combinations of such compounds, they suggest further that (3) populations that as larvae feed on different species might be reproductively isolated. These suppositions culminate with the suggestion that "diversification of insect species may be primarily due to the pheromone complexes available during evolution of host plants." (In a manner perhaps analogous, speciation via host-plant shifts has been reported in certain fruit flies [Bush, 1969a, 1969b] in which mating takes place at the oviposition sites.) It should be noted, however, that there is little evidence for most of Hendry's speculations at present. Pliske (1975c) feels that exogenous precursors should be unnecessary for the long-chain aliphatic compounds produced by female moths, and recent experiments by W. L. Roelofs, R. T. Carde, et al. (pers. comm.) shed doubt on some of his basic assumptions. For example, the fact that many apple-feeding tortricid moths have unique pheromonal systems is difficult to interpret with his model.[15]

Further studies of pheromone metabolism will be needed before such hypotheses can be evaluated, and a long time may elapse before a broad picture emerges of the origin of pheromones in lepidopterous species. Nevertheless, this remains one of the more exciting areas for

15. But see Hindenlang, D. M., and Wichmann, J. K., 1977. Reexamination of tetradecenyl acetates in oak leaf roller sex pheromone and in plants. *Science,* 193(4273):86–89.

research and should a plant origin for some lepidopterous pheromones turn out to be of general occurrence it would have important implications for the evolution of plant-insect interactions at the community level. It would certainly be ironic if Müller's (1883) suggestion, that floral odors attract lepidopterans because of a similarity to sexual attractants, were to be reversed with the implication that the odor of the opposite sex would excite a moth or butterfly through the remembrance of food fragrances past.

References

(Literature search completed in 1975.)

Adams, P. A., and Heath, J. E., 1964. Temperature regulation in the sphinx moth, *Celerio lineata. Nature* (London), 201(4914):20–22.

Ae, S. A., 1965. A study of classification by interspecific hybridization in *Papilio. Academia,* 45:221–37.

Albert, P. J.; Seabrook, W. D.; and Paim, U.; 1974. Isolation of a sex pheromone receptor in males of the spruce budworm, *Choristoneura fumiferana* (Clem.), (Lepidoptera: Tortricidae). *Journal of Comparative Physiology,* 91:79–89.

Alexander, A. J., 1961a. A study of the biology and behavior of the caterpillars, pupae and emerging butterflies of the subfamily Heliconiinae in Trinidad, West Indies. Part I. Some aspects of larval behavior. *Zoologica* (New York), 46(1):1–24 + 1 pl.

Alexander, A. J., 1961b. A study of the biology and behavior of the caterpillars, pupae and emerging butterflies of the subfamily Heliconiinae in Trinidad, West Indies. Part II. Molting, and the behavior of pupae and emerging adults. *Zoologica* (New York), 46(3):105–24 + 1 pl.

Anderson, T. F., and Richards, A. G., 1942. An electron microscope study of some structural colors of insects. *Journal of Applied Physics,* 13:748–58.

Aplin, R. T., and Birch, M. C., 1970. Identification of odorous compounds from male Lepidoptera. *Experientia,* 26(11):1192–93.

Aplin, R. T., and Rothschild, M., 1972. Poisonous alkaloids in the body tissues of the garden tiger moth (*Arctia caja* L.) and the cinnabar moth (*Tyria* (=*Callimorpha*) *jacobaeae* L.) (Lepidoptera). In: *Toxins of Animal and Plant Origin,* vol. 2, A. de Vries and E. Kochva, eds. London: Gordon and Breach, pp.579–95.

Arms, K.; Feeny, P.; and Lederhouse, R. C.; 1974. Sodium: stimulus for puddling behavior by tiger swallowtail butterflies, *Papilio glaucus. Science,* 185(4148):372–74.

Autrum, H., 1965. The physiological basis of colour vision in honeybees. In: *Colour Vision, Physiology and Experimental Psychology,* A. V. Dereuck and J. Knight, eds. Boston: Little, Brown and Co., pp.286–300.

Baker, H. G., and Baker, I., 1973a. Amino-acids in nectar and their evolutionary significance. *Nature* (London), 241:543–45.

Baker, H. G., and Baker, I., 1973b. Some anthecological aspects of nectar-producing flowers, particularly amino acid production in nectar. In: *Taxonomy and Ecology,* V. H. Heywood, ed., Systematics Association. New York: Academic Press, pp.243–64.

Baker, R. R., 1972. Territorial behaviour of the nymphalid butterflies, *Aglais urticae* (L.) and *Inachis io* (L.). *Journal of Animal Ecology,* 41:453–69.

Balduf, W. V., 1939. *Bionomics of entomophagous insects,* part 2. New York: Swift and Co. 384 pp. Reprint, 1974: England, Berks: Classey.

Bartell, R. J., and Lawrence, L. A., 1973. Reduction in responsiveness of males of *Epiphyas postvittana* (Lepidoptera) to sex pheromone following previous brief pheromonal exposure. *Journal of Insect Physiology,* 19:845–55.

Bartell, R. J., and Roelofs, W. L., 1973. Inhibition of sexual response in males of the moth *Argyrotaenia velutinana* by brief exposures to synthetic pheromone or its geometrical isomer. *Journal of Insect Physiology,* 19:655–61.

Bartell, R. J., and Shorey, H. H., 1969a. Pheromone concentrations required to elicit successive steps in the mating sequence of males of the light-brown apple moth, *Epiphyas postvittana. Annals of the Entomological Society of America,* 62:1206–1207.

Bartell, R. J., and Shorey, H. H., 1969b. A quantitative bioassay for the sex pheromone of *Epiphyas postvittana* (Lepidoptera) and factors limiting male responsiveness. *Journal of Insect Physiology,* 15:33–40.

Barth, R., 1960. *Orgãos odoríferos dos Lepidópteros.* Ministério da Agricultura Serviço Florestal, Parque Nacional do Itatiaia, Boletim Número 7.

Batiste, W. C., 1970. A timing sex-pheromone trap with special reference to codling moth collections. *Journal of Economic Entomology,* 63:915–18.

Batiste, W. C.; Olson, W. H.; and Berlowitz, A.; 1973.

Codling moth: the influence of temperature and daylight intensity on the periodicity of daily flight in the field. *Journal of Economic Entomology,* 66:883–92.

Beebe, W., 1955. Two little-known selective insect attractants. *Zoologica* (New York), 40:27–32.

Beebe, W., and Kenedy, R., 1957. Habits, palatability and mimicry in thirteen ctenuchid moth species from Trinidad, B. W. I. *Zoologica* (New York), 42:147–57.

Bell, T. R., 1920. The common butterflies of the plains of India. *Journal of the Bombay Natural History Society,* 27:29–32.

Benson, W. W., 1971. Evidence for evolution of unpalatability through kin selection in the Heliconiinae. *American Naturalist,* 105:213–26.

Benson W. W., and Emmel, T. C., 1973. Demography of gregariously roosting populations of the nymphaline butterfly *Marpesia berania* in Costa Rica. *Ecology,* 54:326–55.

Bergström, G., and Lundgren, L., 1973. Androconial secretion of three species of butterflies of the genus *Pieris* (Lep., Pieridae). *Zoon,* suppl. 1:67–75.

Bernard, G. D., and Miller, W. H., 1970. What does antenna engineering have to do with insect eyes? *I.E.E.E.* [Institute of Electrical and Electronics Engineers] *Student Journal,* Jan.–Feb., 1970:2–8.

Bernhard, C. G., ed., 1966. *The Functional Organization of the Compound Eye.* London: Pergamon Press.

Bernhard, C. G.; Boethius, J.; Gemne, G.; and Struwe, G.; 1970. Eye ultrastructure, colour reception, and behavior. *Nature* (London), 226:865–66.

Beroza, M., ed., 1970. *Chemicals Controlling Insect Behavior: Symposium of the American Chemical Society.* New York: Academic Press.

Birch, M. C., 1970a. Pre-courtship use of abdominal brushes by the nocturnal moth, *Phlogophora meticulosa* (L.) (Lepidoptera: Noctuidae). *Animal Behaviour,* 18:310–16.

Birch, M. C., 1970b. Persuasive scents in moth sex life. *Natural History,* 79(9):34–39, 72.

Birch, M. C., 1972. Male abdominal brush-organs in British noctuid moths and their value as a taxonomic character. *Entomologist,* 105(131):185–205, 233–44.

Birch, M. C., ed., 1974a. *Pheromones.* Amsterdam: North-Holland. xxi + 495pp.

Birch, M. C., 1974b. Introduction. In: *Pheromones,* M. C. Birch, ed. Amsterdam: North-Holland, pp. 1–7.

Birch, M. C., 1974c. Aphrodisiac pheromones in insects. In: *Pheromones,* M. C. Birch, ed. Amsterdam: North-Holland, pp.115–34.

Birch, M. C.; Trammel, K.; Shorey, H. H.; Gaston, L. K.; Hardee, D. D.; Cameron, E. A.; Sanders, C. J.; Bedard, W. D.; Wood, D. L.; Burkholder, W. E.; and Müller-Schwarze, D.; 1974. Programs using pheromones in survey or control. In: *Pheromones,* M. C. Birch, ed. Amsterdam: North-Holland, pp.411–61.

Bisset, G. W.; Fraser, J. F. D.; Rothschild, M.; and Schachter, M.; 1960. A pharmacologically active choline ester and other substances in the tiger moth, *Arctia caja* (L.). *Proceedings of the Royal Society. London* (B), 152:255–62.

Blest, A. D., 1957. The evolution of eyespot patterns in Lepidoptera. *Behaviour,* 11:209–56.

Blest, A. D., 1964. Protective display and sound production in some new world arctiid and ctenuchid moths. *Zoologica* (New York), 49:161–81.

Blest, A. D.; Collett, J. S.; and Pye, J. D.; 1963. The generation of ultrasonic signals by a new world arctiid moth. *Proceedings of the Royal Society* (B), 158:196–207.

Borsch, H. and Schmid, F., 1973. On *Ornithoptera priamus caelestis* Rothschild, *demophanes* Fruhstorfer and *boisduvali* Montrouzier (Papilionidae). *Journal of the Lepidopterists' Society,* 27(3):196–205.

Bossert, W. H., and Wilson, E. O., 1963. The analysis of olfactory communication among animals. *Journal of Theoretical Biology,* 5:443–69.

Bourgogne, J., 1951. Ordre des Lépidoptères. In: *Traité de Zoologie: Anatomie, Systématique, Biologie. vol. 10, Insectes supérieurs et hémipteroïdes.* P.-P. Grassé, ed. Paris: Masson et Cie, pp.174–448.

Brady, U. E.; Nordlund, D. A.; and Daley, R. C.; 1971a. The sex stimulant of the Mediterranean flour moth, *Anagasta kuehniella. Journal of the Georgia Entomological Society,* 6(4):215–17.

Brady, U. E.; Tumlinson, J. H.; Brownlee, R. G.; and Silverstein, R. M.; 1971b. Sex stimulant and attractant in the Indian meal moth and in the almond moth. *Science,* 171:802–804.

Brower, L. P., 1959. Speciation in butterflies of the *Papilio glaucus* group. II. Ecological relationships and hybridization. *Evolution,* 13:212–28.

Brower, L. P., 1963a. Mimicry; a symposium organized for the 16th International Congress of Zoology, Washington, D.C. *Proceedings of the 16th International Congress of Zoology,* 4:145–86.

Brower, L. P., 1963b. The evolution of sex-limited mimicry in butterflies. *Proceedings of the 16th International Congress of Zoology,* 4:173–79.

Brower, L. P.; Alcock, J.; and Brower, J. V. Z.; 1971. Avian feeding behaviour and the selective advantage of incipient mimicry. In: *Ecological Genetics and Evolution,* R. Creed, ed. Oxford, Blackwell Scientific Publications, pp. 261–74.

Brower, L. P., and Brower, J. V. Z., 1964. Birds, butterflies and plant poisons: a study in ecological chemistry. *Zoologica* (New York), 49:137–58.

Brower, L. P.; Brower, J. V. Z.; and Collins, C. T.; 1963. Experimental studies of mimicry. 7. Relative palatability and Müllerian mimicry among neotropical butterflies of the subfamily Heliconiinae. *Zoologica* (New York), 48:65–84.

Brower, L. P.; Brower, J. V. Z.; and Cranston, F. P.; 1965. Courtship behaviour of the queen butterfly, *Danaus gilippus berenice* (Cramer). *Zoologica* (New York), 50:1–39.

Brower, L. P., and Jones, M. A., 1965. Precourtship interaction of wing and abdominal sex glands in male *Danaus* butterflies. *Proceedings of the Royal Entomological Society of London,* 40:147–51.

Brower, L. P.; Ryerson, W. N.; Coppinger, L. L.; and Glazier, S. C.; 1968. Ecological chemistry and the palatability spectrum. *Science,* 161:1349–51.

Brown, W. L., Jr.; Eisner, T.; and Whittaker, R. H.; 1970. Allomones and kairomones: transspecific chemical messengers. *BioScience,* 20:21–22.

Burghardt, G. M., 1970. Defining "communication." In: *Advances in chemoreception,* vol. 1: *Communication by Chemical Signals,* J. W. Johnston, Jr., D. G. Moulton, and A. Turk, eds. New York: Appleton-Century-Crofts, pp. 5–18.

Burkhardt, D., 1962. Spectral sensitivity and other response characteristics of single visual cells in the arthropod eye. In: *Biological receptor mechanisms,* J. W. L. Beament, ed., Symposium of the Society for Experimental Biology no. 16. Cambridge: Cambridge University Press, pp.86–109.

Burkhardt, D., 1964. Colour discrimination in insects. *Advances in Insect Physiology,* 2:131–73.

Burns, J. M., 1966. Preferential mating versus mimicry: disruptive selection and sex-limited dimorphism in *Papilio glaucus. Science,* 153:551–53.

Bush, G. L., 1969a. Sympatric host race formation and speciation in frugivorous flies of the genus *Rhagoletis* (Diptera, Tephritidae). *Evolution,* 23(2):237–51.

Bush, G. L., 1969b. Mating behavior, host specificity and the ecological significance of sibling species in frugivorous flies of the genus *Rhagoletis. American Naturalist,* 103(934):669–72.

Busnel, R.-G., ed. 1963. *Acoustic behaviour of animals.* New York, Elsevier. xx + 933pp.

Callahan, P. S., 1965a. A photographic analysis of moth flight behavior with special reference to the theory for electromagnetic radiation as an attractive force between the sexes and to host plants. *Proceedings of the 12th International Congress of Entomology,* London, 1964, p.302.

Callahan, P. S., 1965b. Far infrared emission and detection by night-flying moths. *Nature* (London), 207:1172–73.

Callahan, P. S., 1965c. Intermediate and far infrared sensing of nocturnal insects. 1. Evidences for a far infrared (FIR) electromagnetic theory of communication and sensing in moths and its relationship to the limiting biosphere of the corn earworm. *Annals of the Entomological Society of America,* 58:727–45.

Callahan, P. S., 1966. Infrared stimulation of nocturnal moths. *Journal of the Georgia Entomological Society,* 4:6–14.

Callahan, P. S., 1967. Insect molecular bioelectronics: a theoretical and experimental study of insect sensillae as tubular waveguides, with particular emphasis on their dielectric and thermoelectret properties. *Miscellaneous Publications of the Entomological Society of America,* 5:315–47.

Callahan, P. S., 1968. Nondestructive temperature and radiance measurements on night flying moths. *Applied Optics,* 7:1811–17.

Callahan, P. S., 1969a. The radiation environment and its relationship to possible methods of environmental control of insects. *Proceedings of the Tall Timbers Conference on Ecological Animal Control by Habitat Management,* February 27–28, 1969, pp.85–108.

Callahan, P. S., 1969b. The exoskeleton of the corn earworm *Heliothis zea* Lepidoptera: Noctuidae with special reference to the sensilla as polytubular dielectric arrays. *University of Georgia Agricultural Experiment Station Research Bulletin,* 54:1–105.

Callahan, P. S., 1970. Insects and the radiation environment. *Proceedings of the Tall Timbers Conference on Ecological Animal Control by Habitat Management,* February 26–28, 1970, pp.247–58.

Callahan, P. S., 1971. Far infrared stimulation of insects with the Glagolewa-Arkadiewa "mass radiator." *Florida Entomologist,* 54(2):201–204.

Callahan, P. S.; Tachenberg, E. F.; and Carlysle, T.; 1968. The scape and pedicel dome sensors: a dielectric aerial waveguide on the antennae of night-flying moths. *Annals of the Entomological Society of America,* 61:934–37.

Cardé, R. T., 1971. Aspects of reproductive isolation in the *Holomelina aurantica* complex (Lepidoptera: Arctiidae). Ph.D. thesis, Cornell University.

Cardé, R. T.; Baker, T. C.; and Roelofs, W. L.; 1975a. Behavioural role of individual components of a multichemical attractant system in the Oriental fruit moth. *Nature* (London), *253*(5490):348–49.

Cardé, R. T.; Comeau, A.; Baker, T. C.; and Roelofs, W. L.; 1975b. Moth mating periodicity: temperature regulates the circadian gate. *Experientia,* in press.

Cardé, R. T.; Doane, C. C.; and Roelofs, W. L.; 1974. Diel periodicity of male sex pheromone response and female attractiveness in the gypsy moth (Lepidoptera: Lymantriidae). *Canadian Entomologist,* 106:479–84.

Cardé, R. T., and Roelofs, W. L., 1973. Temperature modification of male sex pheromone response and factors affecting female calling in *Holomelina immaculata* (Lepidoptera: Arctiidae). *Canadian Entomologist,* 105:1505–12.

Clark, G. C., and Dickson, C. G. C., 1971. *Life Histories of the South African Lycaenid Butterflies.* Cape Town: Purnell, xvi + 272pp.

Clarke, C. A., and Sheppard, P. M., 1956. Hand pairing of butterflies. *Lepidopterists' News,* 10:47–53.

Clearwater, J. R., 1972. Chemistry and function of a pheromone produced by the male of the southern armyworm *Pseudaletia separata. Journal of Insect Physiology,* 18:781–89.

Clench, H. K., 1970. Communal roosting in *Colias* and *Phoebis* (Pieridae). *Journal of the Lepidopterists' Society,* 24(2):117–20.

Cockayne, E. A., 1938. The genetics of sex in Lepidoptera. *Biological Reviews of the Cambridge Philosophical Society,* 13:107–32.

Cockayne, E. A., 1940. Hybrids. *Proceedings of the South London Entomological and Natural History Society,* 1939–1940:65–80.

Cole, L. R., 1959. Defenses of lepidopterous pupae against ichneumonids. *Journal of the Lepidopterists' Society,* 13(1):1–10.

Collenette, C. L., and Talbot, G., 1928. Observations on the bionomics of the Lepidoptera of Matto Grosso, Brazil. *Transactions of the Royal Entomological Society of London,* 76:391–414.

Collins, C. W., and Potts, S. F., 1932. Attractants for the flying gypsy moths as an aid in locating new infestations. *United States Department of Agriculture Technical Bulletin,* 336:1–43.

Comeau, A., 1971. Physiology of sex pheromone attraction in Tortricidae and other Lepidoptera (Heterocera). Ph.D. diss., Cornell University. (Not seen; cited in Roelofs and Cardé, 1974.)

Common, I. F. B., 1954. A study of the ecology of the adult bogong moth, *Agrotis infusa* (Boisd.) (Lepidoptera: Noctuidae) with special reference to its behaviour during migration and aestivation. *Australian Journal of Zoology,* 2:223–63.

Common, I. F. B., 1970. Lepidoptera (moths and butterflies). In: *The Insects of Australia, a Textbook for Students and Research Workers,* D. F. Waterhouse, ed. Carleton: Melbourne University Press, pp.765–866.

Common, I. F. B., 1975. Evolution and classification of the Lepidoptera. *Annual Review of Entomology,* 20:183–203.

Coppinger, R. P., 1970. The effect of experience and novelty on avian feeding behavior with reference to the evolution of warning coloration in butterflies. II. Reactions of naive birds to novel insects. *American Naturalist,* 104:323–35.

Corbet, S. A., 1971. Mandibular gland secretion of larvae of the flour moth, *Anagasta kuehniella,* contains an epideictic pheromone and elicits oviposition movements in a hymenopteran parasite. *Nature* (London), 232:481–84.

Cott, H. B., 1940. *Adaptive coloration in animals.* London: Methuen. 508pp.

Cotter, W. B., 1974. Social facilitation and development in *Ephestia kühniella* Z. *Science,* 183(4126):747–48.

Crane, J., 1954. Spectral reflectance characteristics of butterflies (Lepidoptera) from Trinidad, B. W. I. *Zoologica* (New York), 39(8):85–115.

Crane, J., 1955. Imaginal behavior of a Trinidad butterfly, *Heliconius erato hydara* Hewitson, with special reference to the social use of color. *Zoologica* (New York), 40:167–96.

Creed, R., ed., 1971. *Ecological Genetics and Evolution.* Oxford: Blackwell Scientific Publications. xxi + 391pp.

Dahm, K.; Meyer, D.; Finn, W.; Reinhold, V.; and Röller, H.; 1971. The olfactory and auditory mediated sex attraction in *Achroia grisella* (Fabr.). *Naturwissenschaften,* 58:265–66.

Daterman, G. E., 1972. Laboratory bioassay for sex pheromone of the European pine shoot moth, *Rhyacionia buoliana. Annals of the Entomological Society of America,* 65:119–23.

Davies, J. T., 1971. Olfactory theories. In: *Handbook of Sensory Physiology, Vol. 4: Chemical senses, 1. Olfaction,* L. Beidler, ed. New York: Springer-Verlag, pp.332–50.

Dethier, V. G., 1937. Cannibalism among lepidopterous larvae. *Psyche* (Cambridge), 44:110–15.

Dethier, V. G., 1943. The dioptric apparatus of lateral

ocelli. II. Visual capacities of the ocellus. *Journal of Cellular and Comparative Physiology,* 22:116–26.

Dethier, V. G., 1963. *The Physiology of Insect Senses.* London: Methuen. ix + 266pp.

Doane, C. C., and Cardé, R. T. 1973. Competition of gypsy moth males at a sex-pheromone source and a mechanism for terminating searching behavior. *Environmental Entomology,* 2(4):603–5.

Dodson, C. H.; Dressler, R. L.; Hills, H. G.; Adams, R. H.; and Williams, N. H.; 1969. Biologically active compounds in orchid fragrances. *Science,* 164:1243–49.

Downes, J. A., 1973. Lepidoptera feeding at puddle-margins, dung and carrion. *Journal of the Lepidopterists' Society,* 27(2):89–99.

Downey, J. C., 1966. Sound production in pupae of Lycaenidae. *Journal of the Lepidopterists' Society,* 20 (3):129–55.

Downey, J. C., and Allyn, A. C., 1973. Butterfly ultrastructure. 1. Sound production and associated abdominal structures in pupae of Lycaenidae and Riodinidae. *Bulletin of the Allyn Museum,* No. 14. 48pp.

Downey, J. C., and Allyn, A. D., 1975. Wing-scale morphology and nomenclature. *Bulletin of the Allyn Museum,* No. 31. 32pp.

Duffey, S. S., 1970. Cardiac glycosides and distastefulness: some observations on the palatability spectrum of butterflies. *Science,* 169:78–89.

Dunning, D. C., 1968. Warning sounds of moths. *Zeitschrift für Tierpsychologie,* 25:129–38.

Dunning, D. C., and Roeder, K. D., 1965. Moth sounds and the insect-catching behavior of bats. *Science,* 147:173–74.

Edgar, J. A., and Culvenor, C. C. J., 1974. Pyrrolizidine ester alkaloid in danaid butterflies. *Nature* (London), 248:614–16.

Edgar, J. A.; Culvenor, C. C. J.; and Pliske, T. E.; 1974. Co-evolution of danaid butterflies with their host plants. *Nature* (London), 250:646–48.

Edgar, J. A.; Culvenor, C. C. J.; and Robinson, G. S.; 1973. Hairpencil dihydropyrrolizines of Danainae from the New Hebrides. *Journal of the Australian Entomological Society,* 12:144–50.

Edgar, J. A.; Culvenor, C. C. J.; and Smith, L. W.; 1971. Dihydropyrrolizine derivatives in the "hairpencil" secretions of danaid butterflies. *Experientia,* 27:761–62.

Edmunds, M., 1974. *Defense in Animals.* Essex: Longman Group Ltd. xvii + 357pp.

Ehrlich, P. R., 1958. The comparative morphology, phylogeny and higher classification of the butterflies (Lepidoptera: Papilionoidea). *University of Kansas Science Bulletin,* 39(8):305–70.

Ehrlich, P. R., and Raven, P. H., 1965. Butterflies and plants: a study in coevolution. *Evolution,* 18:586–608.

Eisner, T., 1965. Insect's scales are asset in defense. *Natural History,* 74(6):26–31.

Eisner, T., 1970. Chemical defense against predation in arthropods. In: *Chemical Ecology,* E. Sondheimer and J. B. Simeone, eds. New York: Academic Press, pp.157–217.

Eisner, T.; Alsop, R.; and Ettershank, G.; 1964. Adhesiveness of spider silk. *Science,* 146(3647):1058–61.

Eisner, T.; Kluge, A. F.; Ikeda, M. I.; Meinwald, Y. C.; and Meinwald, J.; 1971. Sesquiterpenes in the osmeterial secretion of a papilionid butterfly, *Battus polydamas. Journal of Insect Physiology,* 17:245–50.

Eisner, T., and Meinwald, Y. C., 1965. Defensive secretion of a caterpillar. *Science,* 150:1733–35.

Eisner, T.; Pliske, T. E.; Ikeda, M.; Owen, D. F.; Vásquez, L.; Pérez, H.; Franclemont, J. G.; and Meinwald, J.; 1970. Defense mechanisms of arthropods. XXVII. Osmeterial secretions of papilionid caterpillars *(Baronia, Papilio, Eurytides). Annals of the Entomological Society of America,* 63(3):914–915.

Eisner, T., and Shepherd, J., 1965. Caterpillar feeding on a sundew plant. *Science,* 150(3703):1608–1609.

Eisner, T., and Shepherd, J., 1966. Defensive mechanisms of arthropods. XIX. Inability of sundew plants to capture insects with detachable integumental outgrowths. *Annals of the Entomological Society of America,* 59(4):868–70.

Eisner, T.; Silberglied, R. E.; Aneshansley, D.; Carrel, J. E.; and Howland, H. C.; 1969. Ultraviolet video-viewing: the television camera as an insect eye. *Science,* 166:1172–74.

Eltringham, H., 1912. A monograph of the African species of the genus *Acraea* Fab., with a supplement on those of the oriental region. *Transactions of the Royal Entomological Society of London* (1912), pp.1–374.

Eltringham, H., 1919. Butterfly vision. *Transactions of the Royal Entomological Society of London* (1919), pp.1–49..

Eltringham, H., 1933. *The Senses of Insects.* London: Methuen. ix + 126pp.

Emmel, T. C., 1971. Symbiotic relationship of an Ecuadorian skipper (Hesperiidae) and *Maxillaria* orchids. *Journal of the Lepidopterists' Society,* 25(1):20–22.

Euw, J. von; Reichstein, T; and Rothschild, M.; 1968.

Aristolochic acid-I in the swallowtail butterfly *Pachlioptera aristolochiae* (Fabr.) (Papilionidae). *Israel Journal of Chemistry,* 6:659–70.

Evans, D. A., and Green, C. L., 1973. Insect attractants of natural origin. *Chemical Society Reviews,* 2(1):75–98.

Fabre, J. H., 1916. *The Life of the Caterpillar.* New York: Dodd, Mead and Co.

Farkas, S. R., and Shorey, H. H., 1974. Mechanisms of orientation to a distant pheromone source. In: *Pheromones,* M. C. Birch, ed. Amsterdam: North-Holland, pp.81–95.

Farquharson, C. O., 1921. Five years' observations (1914–1918) on the bionomics of southern Nigerian insects, chiefly directed to the investigation of lycaenid life histories and to the relation of Lycaenidae, Diptera, and other insects to ants. *Transactions of the Royal Entomological Society of London* (1921), pp. 325–530.

Fenton, M. B., and Roeder, K. D., 1974. The microtymbals of some Arctiidae. *Journal of the Lepidopterists' Society,* 28(3):205–211.

Ferguson, D. C., 1971–1972. Fascicle 20.2. Bombycoidea (Saturniidae). In: *The Moths of America North of Mexico, Including Greenland,* D. C. Ferguson, J. G. Franclemont, R. W. Hodges, E. G. Munroe, R. B. Dominick, and C. R. Edwards, eds. London: E. W. Classey and R. B. D. Publications. xxi + 275pp. + 22 pl.

Fleming, R. C., 1965. An aggressive encounter between *Catocala cara* (Noctuidae) and *Polygonia interrogationis* (Nymphalidae). *Journal of the Lepidopterists' Society,* 19(1):52.

Ford, E. B., 1945. *Butterflies.* New Naturalist. London: Collins. xiv + 368pp.

Ford, E. B., 1967. *Moths,* 2d ed. New Naturalist. London: Collins. xix + 266pp.

Ford, E. B., 1971. *Ecological genetics,* 3d ed. London: Chapman and Hall. xx + 410pp.

Frazier, J. F. D., 1965. The cause of urtication produced by larval hairs of *Arctia caja* (L.) (Lepidoptera: Arctiidae). *Proceedings of the Royal Entomological Society of London* (A), 40:96–100.

Frazier, J. F. D., and Rothschild, M., 1961. Defense mechanisms in warningly-coloured moths and other insects. *Proceedings of the 11th International Congress of Entomology,* Vienna, 1960, 1:249–56.

Frings, M., and Frings, H., 1960. *Sound Production and Sound Perception by Insects: A Bibliography.* University Park: Pennsylvania State University Press. 108pp.

Frisch, K. von, 1967. *The Dance Language and Orientation of Bees.* Cambridge: Belknap Press of Harvard University Press. xiv + 566pp.

Gaston, L. K.; Payne, T. L.; Takahashi, S.; and Shorey, H. H.; 1972. Correlation of chemical structure and sex pheromone activity in *Trichoplusia ni* (Noctuidae). In: *Proceedings of the Fourth International Symposium on Olfaction and Taste, Seewiesen, Germany, 1971,* D. Schneider, ed. Stuttgart: Wissenschaftliche Verlagsgesellschaft MBH, pp.167–73.

Gentil, K., 1942. Elektronenmikroskopische Untersuchungen des Einbaues schillernder Leisten von *Morpho*-schuppen. *Zeitschrift für Morphologie und Ökologie de Tiere,* 38:344–55.

George, J. A., 1965. Sex pheromone of the oriental fruit moth *Grapholitha molesta* Busck (Lepidoptera, Tortricidae). *Canadian Entomologist,* 97:1002–1007.

Ghiradella, H.; Aneshansley, D.; Eisner, T.; Silberglied, R. E.; and Hinton, H. E.; 1972. Ultraviolet reflection of a male butterfly: interference color caused by thin-layer elaboration of wing scales. *Science,* 178:1214–17.

Gilbert, L. E., 1969. Some aspects of the ecology and community structure of ithomid butterflies in Costa Rica. Individual Research Project, Advanced Population Biology Course (July-August, 1969). Organization for Tropical Studies. Unpublished.

Gilbert, L. E., 1972. Pollen feeding and reproductive biology of *Heliconius* butterflies. *Proceedings of the National Academy of Sciences, U.S.A.,* 69:1403–1407.

Gilbert, L. E., 1975. Ecological consequences of a coevolved mutualism between butterflies and plants. In: *Coevolution of Animals and Plants,* L. E. Gilbert and P. H. Raven, eds. Austin: University of Texas Press, pp.210–40.

Gilbert, L. E., and Ehrlich, P. R., 1970. The affinities of the Ithomiinae and the Satyrinae (Nymphalidae). *Journal of the Lepidopterists' Society,* 24(4):297–300.

Goldschmidt, R. B., 1917. On a case of facultative parthenogenesis in the gypsy-moth *Lymantria dispar* L. with a discussion of the relation of parthenogenesis to sex. *Biological Bulletin* (Marine Biological Laboratory, Woods Hole, Mass.), 32:35–43.

Goldsmith, T. H., 1961. The color vision of insects. In: *Light and Life,* W. D. McElroy and B. Glass, eds. Baltimore: Johns Hopkins Press, pp.771–94.

Goldsmith, T. H., and Bernard, G. D., 1974. The visual system of insects. In: *The Physiology of Insecta,* vol. 2, 2d ed., M. Rockstein, ed. New York: Academic Press, pp. 165–272.

Götz, B., 1936. Beiträge zur Analyse des Verhaltens von Schmetterlingsraupen beim Aufsuchen des Futters und des Verpuppungsplatzes. *Zeitschrift für vergleichende Physiologie,* 22:429–503.

Grant, G. G., 1970. Evidence for a male sex pheromone in the noctuid *Trichoplusia ni. Nature* (London), 227:1345–46.

Grant, G. G., 1971. Scent apparatus of the male cabbage looper, *Trichoplusia ni. Annals of the Entomological Society of America,* 64:347–52.

Grant, G. G., and Brady, U. E., 1975. Courtship behavior of phycitid moths. I. Comparison of *Plodia interpunctella* and *Cadra cautella* and role of male scent glands. *Canadian Journal of Zoology,* 53:813–26.

Grant, G. G.; Brady, U. E.; and Brand, J. M.; 1972. Male armyworm scent brush secretion: identification and electroantennogram study of major components. *Annals of the Entomological Society of America,* 65(5):1224–27.

Grant, G. G.; Smithwick, E. B.; and Brady, U. E., 1975. Courtship behavior of phycitid moths. II. Behavioral and pheromonal isolation of *Plodia interpunctella* and *Cadra cautella* in the laboratory. *Canadian Journal of Zoology,* 53:827–32.

Grassé, P.-P., 1959. La reconstruction du nid et les coordinations interindividuelles chez *Bellicositermes natalensis* et *Cubitermes* sp. La théorie de la stigmergie: Essai d'interprétation du comportement des termites constructeurs. *Insectes Sociaux,* 6(1):41–83.

Griffith, P. H., and Süsskind, C., 1970. Electromagnetic communication versus oflaction in the corn earworm moth, *Heliothis zea. Annals of the Entomological Society of America,* 63:903–905.

Hamilton, W. D., 1964. The genetical evolution of social behaviour. *Journal of Theoretical Biology,* 7:1–52.

Haskell, P. T., 1961. *Insect Sounds.* Chicago: Quadrangle Books. xiii + 189pp.

Hendricks, D. P., 1974. "Attacks" by *Polygonia interrogationis* (Nymphalidae) on chimney swifts and insects. *Journal of the Lepidopterists' Society,* 28(3):236.

Hendry, L. B.; Anderson, M. E.; Jugovich, J.; Mumma, R. O.; Robacker, D.; and Kosarych, Z.; 1975a. Sex pheromone of the oak leaf roller—a complex chemical messenger system identified by mass fragmentography. *Science,* 187(4174):355–57.

Hendry, L. B.; Capello, L.; and Mumma, R. O.; 1974a. Sex attractant trapping techniques for the oak leaf roller, *Archips semiferanus* Walker. *Melsheimer Entomological Series,* 16:1–9.

Hendry, L. B.; Jugovich, J.; Mumma, R. O.; Robacker, D.; Weaver, K.; and Anderson, M. E.; 1975b. The oak leaf roller sex pheromone complex—field and laboratory evaluation of requisite behavior stimuli. Unpublished.

Hendry, L. B.; Jugovich, J.; Roman, L.; Anderson, M. E.; and Mumma, R. O.; 1974b. *Cis*-10-tetradecenyl acetate, an attractant component in the sex pheromone of the oak leaf roller moth (*Archips semiferanus* Walker). *Experientia,* 30:886–87.

Hendry, L. B.; Wichman, J. K.; Hindenlang, D. M.; Mumma, R. O.; and Anderson, M. E.; 1975c. Evidence for origin of insect sex pheromones—presence in food plants. *Science,* 188(4183):59–63.

Hinton, H. E., 1948. Sound production in lepidopterous pupae. *Entomologist,* 81:254–69.

Hinton, H. E., 1951. Myrmecophilous Lycaenidae and other Lepidoptera—a summary. *Proceedings of the South London Entomological and Natural History Society* (1949–1950), pp.111–75.

Hinton, H. E., 1955. Protective devices of endopterygote pupae. *Transactions of the Society for British Entomology,* 12(2):49–92.

Hinton, H. E., 1974. Lycaenid pupae that mimic anthropoid heads. *Journal of Entomology (A),* 49(1):65–69.

Hoegh-Guldberg, O., 1972(1971). Pupal sound production of some Lycaenidae. *Journal of Research on the Lepidoptera,* 10(2):127–47.

Hogue, C. L., 1972. Protective function of sound perception and gregariousness in *Hylesia* larvae (Saturniidae: Hemileucinae). *Journal of the Lepidopterists' Society,* 26(1):33–34.

Hölldobler, B., 1967. Zur Physiologie der Gast-Wirt-Beziehungen (Myrmecophilie) bei Ameisen. I. Das Gastverhältnis der *Atemeles*- und *Lomechusa*-Larven (Col. Staphylinidae) zu *Formica* (Hym. Formicidae). *Zeitschrift für vergleichende Physiologie,* 56:1–21.

Hölldobler, B., 1970. Zur Physiologie der Gast-Wirt-Beziehungen (Myrmecophilie) bei Ameisen. II. Das Gastverhältnis des imaginalen *Atemeles pubicollis* Bris. (Col. Staphylinidae) zu *Myrmica* und *Formica* (Hym. Formicidae). *Zeitschrift für vergleichende Physiologie,* 66:215–50.

Hölldobler, B., 1971. Communication between ants and their guests. *Scientific American,* 224(3):86–93.

Hsiao, H. S., 1972. The attraction of moths (*Trichoplusia ni*) to infrared radiation. *Journal of Insect Physiology,* 18:1705–14.

Hsiao, H. S., and Hackwell, G. A., 1970. The rôle of the antennae and compound eyes of *Heliothis zea* (Noctuidae) in its attraction to light. *Journal of Insect Physiology,* 16:1237–43.

Hsiao, H. S., and Süsskind, C., 1970. Infrared and

microwave communication by moths. *I.E.E.E.* [Institute of Electrical and Electronics Engineers] *Spectrum,* 7:69–76.

Hundertmark, A., 1937a. Das Formenunterscheidungsvermögen der Eiraupe der Nonne (*Lymantria monacha* L.). *Zeitschrift für vergleichende Physiologie,* 24:563–82.

Hundertmark, A., 1937b. Helligkeits- und Farbenunterscheidungsvermögen der Eiraupe der Nonne (*Lymantria monacha* L.). *Zeitschrift für vergleichende Physiologie,* 24:42–57.

Ilse, D., 1928. Über den Farbensinn der Tagfalter. *Zeitschrift für vergleichende Physiologie,* 8:658–92.

Ilse, D., 1932a. Zur "Formwahrnehmung" der Tagfalter. I. Spontane Bevorzugung von Formmerkmalen durch Vanessen. *Zeitschrift für vergleichende Physiologie,* 17:537–56.

Ilse, D., 1932b. Eine neue Methode zur Bestimmung der subjektiven Helligkeitswerte von Pigmenten. *Biologisches Zentralblatt,* 52:660–67.

Ilse, D., 1937. New observations on responses to colours in egg-laying butterflies. *Nature* (London), 140(3543):544–45.

Ilse, D., 1941. The colour vision of insects. *Proceedings of the Royal Philosophical Society of Glasgow,* 65:98–112.

Ilse, D., and Vaidya, V., 1956. Spontaneous feeding response to colors in *Papilio demoleus* L. *Proceedings of the Indian Academy of Science* (B), 43:23–31.

Ishikawa, S., 1969. The spectral sensitivity and the components of the visual system in the stemmata of silkworm larvae, *Bombyx mori* L. (Lepidoptera: Bombycidae). *Applied Entomology and Zoology,* 4:87–99.

Jacobson, M., 1972. *Insect Sex Pheromones.* New York: Academic Press. xii + 382pp.

Jacobson, M., 1974. Insect pheromones. In: *The Physiology of Insecta,* vol. 3, 2d ed., M. Rockstein, ed. New York: Academic Press, pp. 229–76.

Johnson, K., 1974. An attempted interfamilial mating (Lycaenidae-Nymphalidae). *Journal of the Lepidopterists' Society,* 28(3):291–92.

Jones, D. A.; Parsons, J.; and Rothschild, M.; 1962. Release of hydrocyanic acid from crushed tissues of all stages in the life-cycle of species of the Zygaeninae (Lepidoptera). *Nature* (London), 193: 52–53.

Jones, F. M., 1930. The sleeping Heliconias of Florida. *Natural History,* 30:635–44.

Jones, F. M., 1931. The gregarious sleeping habits of *Heliconius charithonia* L. *Proceedings of the Royal Entomological Society of London,* 6:4–10.

Jordan, K., 1905. Der Gegensatz zwischen geographischer und nichtgeographischer Variation. *Zeitschrift für wissenschaftliche Zoologie,* 83:151–210.

Kaae, R. S.; Shorey, H. H.; and Gaston, L. K.; 1973a. Pheromone concentration as a mechanism for reproductive isolation between two lepidopterous species. *Science,* 179:487–88.

Kaae, R. S.; Shorey, H. H.; McFarland, S. U.; and Gaston, L. K.; 1973b. Sex pheromones of Lepidoptera. XXXVII. Role of sex pheromones and other factors in reproductive isolation among ten species of Noctuidae. *Annals of the Entomological Society of America,* 66(2):444–48.

Kaissling, K.-E., 1971. Insect olfaction. In: *Handbook of sensory physiology,* vol 4: *Chemical senses, 1. Olfaction,* L. Beidler, ed. New York: Springer-Verlag, pp.351–431.

Kaissling, K.-E., and Priesner, E., 1970. Die Riechschwelle des Seidenspinners. *Naturwissenschaften,* 57:23–28.

Kennedy, J. S., and Marsh, D. 1974. Pheromone-regulated anemotaxis in flying moths. *Science,* 184:999–1001.

Kettlewell, B., 1973. *The Evolution of Melanism: The Study of a Recurring Necessity, with Special Reference to Industrial Melanism in the Lepidoptera.* Oxford: Clarendon Press. xxiv + 423pp.

Kinder, E., and Süffert, F., 1943. Über den Feinbau schillernder Schmetterlingsschuppen vom *Morpho*-Typ. *Biologisches Zentralblatt,* 63:268–88.

Kiriakoff, S. G., 1956. Sur l'origine et l'évolution des organes tympanaux phalénoïdes (Lépidoptères). *Bulletin et annales de la Société royale entomologique de Belgique,* 92(xi–xii):289–300.

Kiriakoff, S. G., 1963. The tympanic structures of the Lepidoptera and the taxonomy of the order. *Journal of the Lepidopterists' Society,* 17(1):1–6.

Klun, J. A., 1968. Isolation of a sex pheromone of the European corn borer. *Journal of Economic Entomology,* 61:484–87.

Klun, J. A., and Robinson, J. F., 1971. European corn borer moth: sex attractant and sex attraction inhibitors. *Annals of the Entomological Society of America,* 64:1083–86.

Klun, J. A., and Robinson, J. F., 1972. Olfactory discrimination in the European corn borer and several pheromonally analogous moths. *Annals of the Entomological Society of America,* 65:1337–40.

Knoll, F., 1922. Lichtsinn und Blumenbesuch des Falters von *Macroglossum stellatarum* (Insekten und Blumen III). *Abhandlungen der kaiserlich-königliche*

Zoologisch-botanischen Gesellschaft in Wien, 12:121–378.

Knoll, F., 1925. Lichtsinn und Blütenbesuch des Falters von *Deilephila livornica. Zeitschrift für vergleichende Physiologie,* 2:329–80.

Knoll, F., 1927. Über Abendschwärmer und Schwärmerblumen. *Berichte der Deutschen Botanischen Gesellschaft,* 45:510–18.

Kolyer, J. M., and Reimschuessel, A., 1970. Scanning electron microscopy on wing scales of *Colias eurytheme. Journal of Research on the Lepidoptera,* 8(1):1–15.

Labine, P. A., 1964. Population biology of the butterfly, *Euphydryas editha.* I. Barriers to multiple inseminations. *Evolution,* 18(2):335–36.

Laithwaite, E. R., 1960. A radiation theory of the assembling of moths. *Entomologist,* 93:113–17, 133–37.

Lamborn, W. A., 1913. On the relationship between certain west African insects, especially ants, Lycaenidae and Homoptera. *Transactions of the Royal Entomological Society of London* (1913), pp.436–520.

Lederer, G., 1951. Biologie der Nahrungsaufnahme der Imagines von *Apatura* und *Limenitis,* sowie Versuche zur Feststellung der Gustorezeption durch die Mittel- und Hintertarsen dieser Lepidoptera. *Zeitschrift für Tierpsychologie,* 18:41–61.

Lederer, G., 1960. Verhaltenswiesen der Imagines und der Entwicklungsstadien von *Limenitis camilla camilla* L. (Lepidoptera, Nymphalidae). *Zeitschrift für Tierpsychologie,* 17:521–46.

Lenz, F., 1917. Der Erhaltungsgrund der Myrmekophilie. *Zeitschrift für induktive Abstammungs- und Vererbungslehre,* 18:44–46.

Lépiney, J. de, 1928. Note préliminaire sur la rôle de la vision ocellaire dans le comportement des chenilles de *Lymantria dispar* L. *Bulletin de la Société zoologique de France,* 53:479–90.

Levengood, W. C., and Limperis, T., 1967. Infrared sensing of nocturnal moths. Final report, University of Michigan Report 742–32–F, University of Michigan. (Not seen; cited in Hsiao, 1972.)

Lippert, W., and Gentil, K., 1959. Über lamellare Feinstrukturen bei den Schillerschuppen der Schmetterlinge vom *Urania*- und *Morpho*-Typ. *Zeitschrift für Morphologie und Ökologie de Tiere,* 48:115–22.

Lloyd, J. E., 1974. Genital stridulation in *Psilogramma menephron* (Sphingidae). *Journal of the Lepidopterists' Society,* 28(4):349–51.

Long, D. B., 1955. Observations on sub-social behaviour in two species of lepidopterous larvae, *Pieris brassicae* L. and *Plusia gamma* L. *Transactions of the Royal Entomological Society of London,* 106(11):421–37.

Long, D. B., 1959. Observations on adult weight and wing area in *Plusia gamma* L. and *Pieris brassicae* L. in relation to larval population density. *Entomologia Experimentalis et Applicata,* 2(4):241–48.

Long, D. B., and Zaher, M. A., 1958. Effects of larval population density on the adult morphology of two species of Lepidoptera, *Plusia gamma* L. and *Pieris brassicae* L. *Entomologia Experimentalis et Applicata,* 1(3):161–73.

Lorković, Z., 1954. L'accouplement artificiel chez les Lépidoptères et son application dans les recherches sur la fonction de l'appareil genital des insects. *Revue Française de Lépidopterologie,* 14:138–39.

Lundgren, L., and Bergström, G., 1975. Sexual pheromones from the wings of *Lycaeides argyrognomon* Bgstr. (Lep., Lycaenidae). Manuscript.

McColl, H. P., 1969. The sexual scent organs of male Lepidoptera. M.Sc. thesis, University College Swansea, Wales. (Not seen; cited in Percy and Weatherston, 1974.)

MacKay, M. R., 1969. Microlepidopterous larvae in Baltic amber. *Canadian Entomologist,* 101:1173–80.

MacKay, M. R., 1970. Lepidoptera in Cretaceous amber. *Science,* 167(3917):379–80.

McKelvey, S. D., 1947. *Yuccas of the Southwestern United States,* part 2. Jamaica Plain, Mass: Arnold Arboretum. 192pp. + 45 pl.

McManus, M. L., and Smith, H. R., 1972. Importance of the silk trails in the diel behavior of late instars of the gypsy moth. *Environmental Entomology,* 1(6): 793–95.

Magnus, D. B. E., 1950. Beobachtungen zur Balz und Eiablage des Kaisermantels *Argynnis paphia* L. (Lep. Nymphalidae). *Zeitschrift für Tierpsychologie,* 7:435–49.

Magnus, D. B. E., 1958. Experimental analysis of some "overoptimal" sign-stimuli in the mating behaviour of the fritillary butterfly *Argynnis paphia* L. (Lepidoptera: Nymphalidae). *Proceedings of the Tenth International Congress of Entomology,* Montreal 1956, 2:405–18.

Magnus, D. B. E., 1963. Sex limited mimicry II—visual selection in the mate choice of butterflies. *Proceedings of the 16th International Congress of Zoology,* Washington, D.C., 1963, 4:179–83.

Malicky, H., 1969. Versuch einer Analyse der ökologischen Beziehungen zwischen Lycaeniden (Lepidoptera) und Formiciden (Hymenoptera). *Tijdschrift voor Entomologie,* 112:213–98.

Malicky, H., 1970a. Unterschiede im Angriffsverhalten von *Formica*-Arten (Hymenoptera, Formicidae) gegenüber Lycaenidenraupen (Lepidoptera). *Insectes Sociaux,* 17(2):121–23.

Malicky, H., 1970b. New aspects on the association between lycaenid larvae (Lycaenidae) and ants (Formicidae, Hymenoptera). *Journal of the Lepidopterists' Society,* 24(3):190–202.

Marler, P., 1961. The logical analysis of animal communication. *Journal of Theoretical Biology,* 1:295–317.

Mason, C. W., 1926. Structural colors in insects, I. *Journal of Physical Chemistry,* 30:383–95.

Mason, C. W., 1927. Structural colors in insects, II. *Journal of Physical Chemistry,* 31:321–54.

Masters, J. H., 1968. Collecting Ithomiidae with heliotrope. *Journal of the Lepidopterists' Society,* 22(2):108–110.

Masters, J. H., 1974. Unusual copulatory behavior in *Euphydryas chalcedonia* (Doubleday) (Nymphalidae). *Journal of the Lepidopterists' Society,* 28(4):291–94.

Maynard Smith, J., 1976. Evolution and the theory of games. *American Scientist,* 64:41–45.

Mazokhin-Porshniakov, G. A., 1957. Reflecting properties of butterfly wings and the role of ultra-violet rays in the vision of insects. *Biofizika,* 2:358–68 (in Russian). English translation: *Biophysics,* 2:352–62.

Mazokhin-Porshniakov, G. A., 1964. Methods for the study of insect color vision. *Entomologicheskoe Obozrenie,* 43:503–23.

Mazokhin-Porshniakov, G. A., 1969. *Insect Vision.* New York: Plenum. xiv + 306pp.

Meinwald, J.; Boriack, C. J.; Schneider, D.; Boppré, M.; Wood, W. F.; and Eisner, T.; 1974. Volatile ketones in the hairpencil secretion of danaid butterflies *(Amauris* and *Danaus). Experientia,* 30:721–22.

Meinwald, J.; Chalmers, A. M.; Pliske, T. E.; and Eisner, T.; 1969a. Identification and synthesis of *trans,trans*-3,7-dimethyl-2,6-decadien-1,10-dioic acid, a component of the pheromonal secretion of the male monarch butterfly. *Chemical Communications,* 1969:86–87.

Meinwald, J.; Chalmers, A. M.; Pliske, T. E.; and Eisner, T.; 1969b. Pheromones. III. Identification of *trans,trans*-10-hydroxy-3,7-dimethyl-2,6-decadienoic acid as a major component in "hairpencil" secretion of the male monarch butterfly. *Tetrahedron Letters,* 47:4893–96.

Meinwald, J., and Meinwald, Y. C., 1966. Structure and synthesis of the major components in the hairpencil secretion of a male butterfly, *Lycorea ceres ceres* (Cramer). *Journal of the American Chemical Society,* 88(6):1305–10.

Meinwald, J.; Meinwald, Y. C.; and Mazzochi, P. H.; 1969c. Sex pheromone of the queen butterfly: chemistry. *Science,* 164:1174–75.

Meinwald, J.; Meinwald, Y. C.; Wheeler, J. W.; Eisner, T.; and Brower, L. P.; 1966. Major components in the exocrine secretion of a male butterfly *(Lycorea). Science,* 151:583–85.

Meinwald, J.; Thompson, W. R.; Eisner, T.; and Owen, D. F.; 1971. Pheromones. VII. African monarch: major components of the hairpencil secretion. *Tetrahedron Letters,* 38:3485–88.

Michener, C. D., 1969. Comparative social behavior of bees. *Annual Review of Entomology,* 14:299–342.

Miller, L. D., and Clench, H. K., 1968. Some aspects of mating behavior in butterflies. *Journal of the Lepidopterists' Society,* 22(3):125–32.

Miller, W. H., and Bernard, G. D., 1968. Butterfly glow. *Journal of Ultrastructure Research,* 24:286–94.

Miller, W. H.; Bernard, G. D.; and Allen, J. L.; 1968. The optics of insect compound eyes. *Science,* 162:760–67.

Minks, A. K.; Roelofs, W. L.; Ritter, F. J.; and Persoons, C. J.; 1973. Reproductive isolation of two tortricid moth species by different ratios of a two-component sex attractant. *Science,* 180:1073–74.

Minnich, D. E., 1936. The responses of caterpillars to sounds. *Journal of Experimental Zoology,* 72:439–53.

Morrell, R., 1960. *Common Malayan Butterflies.* London: Longmans, Green and Co. xii + 64pp.

Morris, C. W., 1946. *Signs, Language and Behavior.* Englewood Cliffs, N.J.: Prentice-Hall.

Mudd, A., and Corbet, S. A., 1973. Mandibular gland secretion of larvae of the stored products pests *Anagasta kuehniella, Ephestia cautella, Plodia interpunctella* and *Ephestia elutella. Entomologia experimentalis et applicata,* 16(2):291–92.

Müller, H., 1883. *The Fertilisation of Flowers,* trans. by D'Arcy W. Thompson. London: Macmillan and Co. xii + 669pp.

Muyshondt, A., and Muyshondt, A., Jr., 1974. Gregarious seasonal roosting of *Smyrna karwinskii* adults in El Salvador (Nymphalidae). *Journal of the Lepidopterists' Society,* 28(3):224–29.

Myers, J. G., 1930. Epigamic behavior of a male butterfly *(Terias)* and the gregarious habit during rest of a heliconine butterfly, in Cuba. *Proceedings of the Royal Entomological Society of London,* 5:46–48.

Myers, J. H., and Walter, M., 1970. Olfaction in the

Florida queen butterfly: honey odour receptors. *Journal of Insect Physiology,* 16(4):573–78.

Myers, J. S., 1972. Pheromones and courtship behavior in butterflies. *American Zoologist,* 12:545–51.

Myers, J. S., and Brower, L. P., 1969. A behavioural analysis of the courtship pheromone receptors of the queen butterfly, *Danaus gilippus berenice. Journal of Insect Physiology,* 15:2117–30.

Nachtigall, W., 1965. Die aerodynamische Funktion der Schmetterlingsschuppen. *Naturwissenschaften,* 52:216–17.

Nachtigall, W., 1974. *Insects in Flight, a Glimpse Behind the Scenes in Biophysical Research.* London: George Allen and Unwin. 153pp.

Nekrutenko, Y. P., 1964. The hidden wing-pattern of some Palaearctic species of *Gonepteryx* and its taxonomic value. *Journal of Research on the Lepidoptera,* 3(2):65–68.

Nekrutenko, Y. P., 1968. *Phylogeny and geographical distribution of the genus* Gonepteryx *(Lepidoptera, Pieridae): an attemapt* [sic] *of study in historical zoogeography.* Kiev: Naukova Dumka. 128pp.

Norris, M. J., 1936. The feeding habits of the adult Lepidoptera Heteroneura. *Transactions of the Royal Entomological Society of London,* 85:61–90.

Obara, Y., 1970. Studies on the mating behavior of the white cabbage butterfly, *Pieris rapae crucivora* Boisduval. III. Near-ultra-violet reflection as the signal of intraspecific communication. *Zeitschrift für vergleichende Physiologie,* 69:99–116.

O'Connell, R. J., 1972. Responses of olfactory receptors to the sex attractant, its synergist and inhibitor in the red-banded leaf roller, *Argyrotaenia velutiana.* In: *Olfaction and Taste,* vol. 4, D. Schneider, ed. Stuttgart: Wissenschaftliche Verlagsgesellschaft MBH, pp.180–86.

Okui, K., 1964. Studies on aggregative behaviour of the silkworm, *Bombyx mori.* III. Behaviour patterns and interrelation between individuals. *Japanese Journal of Applied Entomology and Zoology,* 8:286–94,(in Japanese).

Otte, D., 1974. Effects and functions in the evolution of signalling systems. *Annual Review of Ecology and Systematics,* 5:385–417.

Owen, D. F., 1971. *Tropical Butterflies.* Oxford: Clarendon Press. xiv + 214pp.

Parker, G. A., 1970. Sperm competition and its evolutionary consequences in the insects. *Biological Reviews of the Cambridge Philosophical Society,* 45:525–68.

Payne, T. L., 1974. Pheromone perception. In: *Pheromones,* M. C. Birch, ed. Amsterdam: North-Holland, pp.35–61.

Payne, T. L.; Shorey, H. H.; and Gaston, L. K.; 1973. Sex pheromones of Lepidoptera. 38. Electroantennogram responses in *Autographa californica* to *cis*-7-dodecenyl acetate and related compounds. *Annals of the Entomological Society of America,* 66:703–704.

Percy, J. E., and Weatherston, J., 1974. Gland structure and pheromone production in insects. In: *Pheromones,* M. C. Birch, ed. Amsterdam: North-Holland, pp.11–34.

Perkins, E. M., Jr., 1973. Unusual copulatory behavior in the Nymphalidae and Satyridae. *Journal of the Lepidopterists' Society,* 27(4):291–94.

Pesce, H., and Delgrado, A., 1972. Poisoning from adult moths and caterpillars. In: *Venomous Animals and Their Venoms,* vol. 3, W. Bücherl and E. E. Buckley, eds. London: Academic Press, pp.119–56.

Petersen, B., and Tenow, O., 1954. Studien am Rapsweissling und Bergweissling (*Pieris napi* L. und *Pieris bryoniae* O.). Isolation und Paarungsbiologie. *Zoologiska Bidrag fran Uppsala,* 30:169–98.

Petersen, B.; Törnblom, O.; and Bodin, N.-O.; 1952. Verhaltenstudien am Rapsweissling und Bergweissling (*Pieris napi* L. und *Pieris bryoniae* Ochs.). *Behaviour,* 4:67–84.

Picarelli, Z. P., and Valle, J. R., 1972. Pharmacological studies on caterpillar venoms. In: *Venomous Animals and Their Venoms,* vol. 3, W. Bücherl and E. E. Buckley, eds. London: Academic Press, pp.103–18.

Pliske, T. E., 1972. Sexual selection and dimorphism in female tiger swallowtails, *Papilio glaucus* L. (Lepidoptera: Papilionidae): a reappraisal. *Annals of the Entomological Society of America,* 65(6):1267–70.

Pliske, T. E., 1975a. Courtship behavior of the monarch butterfly, *Danaus plexippus* L. *Annals of the Entomological Society of America,* 68(1):143–51.

Pliske, T. E., 1975b. Courtship behavior and use of chemical communication by males of certain species of ithomiine butterflies (Nymphalidae: Lepidoptera). *Annals of the Entomological Society of America,* 68(6):935–42.

Pliske, T. E., 1975c. Attraction of Lepidoptera to plants containing pyrrolizidine alkaloids. *Environmental Entomology,* 4(3):455–73.

Pliske, T. E., and Eisner, T., 1969. Sex pheromone of the queen butterfly: biology. *Science,* 164:1170–72.

Pliske, T. E., and Salpeter, M., 1971. The structure and development of the hairpencil glands in males of the queen butterfly, *Danaus gilippus berenice. Journal of Morphology,* 134:215–44.

Post, C. T., Jr., and Goldsmith, T. H., 1969. Physiological evidence for color receptors in the eye of a butterfly. *Annals of the Entomological Society of America,* 62(6):1497–98.

Poulton, E. B., 1890. *The Colours of Animals; Their Meaning and Use, Especially Considered in the Case of Insects.* London: Kegan Paul, Trench, Trübner and Co. xv + 360pp.

Poulton, E. B., 1898. Natural selection—the cause of mimetic resemblance and common warning colours. *Journal of the Linnean Society, Zoology,* 26:588–612 + pl. 40–44.

Poulton, E. B., 1907. The significance of some secondary sexual characters in butterflies. *Proceedings of the Royal Entomological Society of London* (1907), pp.xl–xliii.

Poulton, E. B., 1931a. The gregarious sleeping habits of *Heliconius charithonia* L. *Proceedings of the Royal Entomological Society of London,* 6:4–10.

Poulton, E. B., 1931b. The gregarious sleeping habits of a heliconian and ithomiine butterfly in Trinidad, observed by P. Lechmere Guppy. *Proceedings of the Royal Entomological Society of London,* 6:68–69.

Poulton, E. B. [et al.], 1933. The gregarious resting habits of danaine butterflies in Australia; also of heliconine and ithomiine butterflies in tropical America. *Proceedings of the Royal Entomological Society of London,* 7:64–67.

Powell, J. A., and Mackie, R. A., 1966. Biological interrelationships of moths and *Yucca whipplei* (Lepidoptera: Gelechiidae, Blastobasidae, Prodoxidae). *University of California Publications in Entomology,* 42:1–59.

Priesner, E., 1968. Die interspezifischen Wirkungen der Sexuallockstoffe der Saturniidae (Lepidoptera). *Zeitschrift für vergleichende Physiologie,* 61:263–97.

Proctor, M., and Yeo, P., 1973. *The Pollination of Flowers.* New Naturalist. London: Collins. 418pp.

Pyle, R. M., 1972. *Limenitis lorquini* (Nymphalidae) attacking a glaucous-winged gull. *Journal of the Lepidopterists' Society,* 26(4):261.

Rathke, B. J., and Poole, R. W., 1975. Coevolutionary race continues: butterfly larval adaptation to plant trichomes. *Science,* 187(4172):175–76.

Reichstein, T.; von Euw, J.; Parsons, J. A.; and Rothschild, M.; 1968. Heart poisons in the monarch butterfly. *Science,* 161:861–66.

Remington, C. L., 1973. Ultraviolet reflectance in mimicry and sexual signals in the Lepidoptera. *Journal of the New York Entomological Society,* 81(2):124.

Rettenmeyer, K., 1970. Insect mimicry. *Annual Review of Entomology,* 15:43–74.

Richards, O. W., 1927. Sexual selection and allied problems in the insects. *Biological Reviews of the Cambridge Philosophical Society,* 2:298–364.

Riddiford, L. M., and Ashenhurst, J. B., 1973. The switchover from virgin to mated behavior in female cecropia moths: the role of the bursa copulatrix. *Biological Bulletin* (Marine Biological Laboratory, Woods Hole, Mass.), 144(1):162–71.

Riley, C. V., 1892. The yucca moth and yucca pollination. *Report of the Missouri Botanical Garden,* 3:99–158.

Robinson, M. H., 1969. Defenses against visually hunting predators. In: *Evolutionary Biology, vol. 3,* T. Dobzhansky, M. K. Hecht, and W. C. Steere, eds. New York: Appleton-Century-Crofts, pp.225–59.

Robinson, R., 1971. *Lepidoptera Genetics.* Oxford: Pergamon Press. ix + 687pp.

Roeder, K. D., 1965. Moths and ultrasound. *Scientific American,* 212(4):94–102.

Roeder, K. D., 1966. Auditory system of noctuid moths. *Science,* 154(3756):1515-21.

Roeder, K. D., 1967a. *Nerve Cells and Insect Behavior,* 2d ed. Cambridge: Harvard University Press. xv + 238pp.

Roeder, K. D., 1967b. Prey and predator. *Bulletin of the Entomological Society of America,* 13(1):6–9.

Roeder, K. D., 1970. Episodes in insect brains. *American Scientist,* 58(4):378–89.

Roeder, K. D., 1971. Acoustic alerting mechanisms in insects. *Annals of the New York Academy of Sciences,* 188:63–79.

Roeder, K. D., 1972. Acoustic and mechanical sensitivity of the distal lobe of the pilifer in choerocampine hawkmoths. *Journal of Insect Physiology,* 18:1249–64.

Roeder, K. D., 1974a. Some neuronal mechanisms of simple behavior. *Advances in the Study of Behavior,* 5:1–46.

Roeder, K. D., 1974b. Responses of the less sensitive acoustic sense cells in the tympanic organs of some noctuid and geometrid moths. *Journal of Insect Physiology,* 20:55–56.

Roeder, K. D. Acoustic sensory responses and possible bat-evasion tactics of certain moths. *Canadian Journal of Zoology,* in press.

Roeder, K. D., and Fenton, M. B., 1973. Acoustic responsiveness of *Scoliopteryx libatrix* L. (Lepidoptera: Noctuidae), a moth that shares hibernacula with some insectivorous bats. *Canadian Journal of Zoology,* 51(7):681–85.

Roeder, K. D., and Treat, A. E., 1957. Ultrasonic reception by the tympanic organ of noctuid moths. *Journal of Experimental Zoology,* 134:127–58.

Roeder, K. D., and Treat, A. E., 1970. An acoustic sense in some hawkmoths (Choerocampinae). *Journal of Insect Physiology,* 16:1069–86.

Roeder, K. D.; Treat, A. E.; and Vande Berg, J. S.; 1968. Auditory sense in certain sphingid moths. *Science,* 159:331–33.

Roeder, K. D.; Treat, A. E.; and Vande Berg, J. S.; 1970. Distal lobe of the pilifer: an ultrasonic receptor in choerocampine hawkmoths. *Science,* 170: 1098–99.

Roelofs, W. L., and Cardé, R. T., 1974. Sex pheromones in the reproductive isolation of lepidopterous species. In: *Pheromones,* M. C. Birch, ed. Amsterdam: North-Holland, pp. 96–114.

Roelofs, W. L.; Cardé, R. T.; Benz, G.; and von Salis, G.; 1971. Sex attractant of the larch bud moth found by electroantennogram method. *Experientia,* 27: 1438–39.

Roelofs, W. L., Cardé, R. T.; and Tette, J.; 1973. Oriental fruit moth attractant synergists. *Environmental Entomology,* 2:252–54.

Roelofs, W. L., and Comeau, A., 1968. Sex pheromone perception. *Nature* (London), 220:600–601.

Roelofs, W. L., and Comeau, A., 1969. Sex pheromone specificity: taxonomic and evolutionary aspects in Lepidoptera. *Science,* 165(3891):398–400.

Roelofs, W. L., and Comeau, A., 1971a. Sex attractants in Lepidoptera. In: *Chemical releasers in insects,* vol. 3, A. Tahori, ed. New York: Gordon and Breach, pp.91–114.

Roelofs, W. L., and Comeau, A., 1971b. Sex pheromone perception: synergists and inhibitors for the redbanded leafroller attractant. *Journal of Insect Physiology,* 17:435–49.

Röller, H.; Biemann, K.; Bjerke, J. S.; Norgard, D. W.; and McShan, W. H.; 1968. Sex pheromones of pyralid moths. I. Isolation and identification of the sex attractant of *Galleria mellonella* (greater waxmoth). *Acta Entomol. Bohemoslov.,* 65:208–11.

Ross, G. N., 1963. Evidence for lack of territoriality in two species of *Hamadryas* (Nymphalidae). *Journal of Research on the Lepidoptera,* 2(4):241–46.

Ross, G. N., 1966. Life-history studies on Mexican butterflies. IV. The ecology and ethology of *Anatole rossi,* a myrmecophilous metalmark (Lepidoptera: Riodinidae). *Annals of the Entomological Society of America,* 59(5):985–1004.

Rotberg, A., 1972. Lepidopterism in Brazil. In: *Venomous Animals and Their Venom,* vol. 3, W. Bücherl and E. E. Buckley, eds. London: Academic Press, pp.157–68.

Rothschild, M., 1965. The stridulation of the garden tiger moth. *Proceedings of the Royal Entomological Society of London* (C), 30:3.

Rothschild, M., 1971. Speculations about mimicry with Henry Ford. In: *Ecological Genetics and Evolution,* R. Creed, ed. Oxford: Blackwell Scientific Publications, pp.202–23.

Rothschild, M., 1972. Some observations on the relationship between plants, toxic insects and birds. In: *Phytochemical Ecology,* J. B. Harborne, ed. London: Academic Press, pp.1–12.

Rothschild, M., 1973. Secondary plant substances and warning colouration in insects. In: *Insect/Plant Relationships,* H. F. van Emden, ed., Symposium of the Royal Entomological Society of London No. 6. London: Blackwell Scientific Publications, pp.59–83.

Rothschild, M., and Aplin, R. T., 1971. Toxins in tiger moths (Arctiidae: Lepidoptera). In: *Pesticide Chemistry,* A. S. Tahori, ed. London: Gordon and Breach, pp.177–82.

Rothschild, M., and Haskell, P. T., 1966. Stridulation of the garden tiger moth (*Arctia caja* (L.)) audible to the human ear. *Proceedings of the Royal Entomological Society of London* (A), 41:167–70.

Rothschild, M.; Reichstein, T.; Euw, J. von; Aplin, R. T.; and Harman, R. R. M.; 1970. Toxic Lepidoptera. *Toxicon,* 8:293–99.

Rothschild, M.; Euw, J. von; and Reichstein, T.; 1972. Aristolochic acids stored by *Zerynthia polyxena* (Lepidoptera). *Insect Biochemistry,* 2:334–43.

Rothschild, M.; Euw, J. von; and Reichstein, T.; 1973. Cardiac glycosides (heart poisons) in the polka-dot moth *Syntomeida epilais* Walk. (Ctenuchidae: Lep.) with some observations on the toxic qualities of *Amata* (= *Syntomis*) *phegea* (L.). *Proceedings of the Royal Society of London, (B),* 183:227–47.

Russ, K. 1969. Beiträge zum Territorialverhalten der Raupen des Springwurmwicklers, *Sparganothis pilleriana* Schiff. (Lepidoptera: Tortricidae). (Vorläufige Mitteilung). *Pflanzenschutzberichte,* 40:1–9.

Sales, G. D., and Pye, J. D., 1974. *Ultrasonic Communication by Animals.* London: Chapman and Hall, xi + 281pp.

Sanders, C. J., and Lucuik, G. S., 1972. Factors affecting calling by female eastern spruce budworm, *Choristoneura fumiferana* (Lepidoptera: Tortricidae). *Canadian Entomologist,* 104:1751–62.

Sargent, T. D., 1972. Studies on the *Catocala* of southern New England. III. Mating results with *C. relicta*

Walker. *Journal of the Lepidopterists' Society,* 26:94–104.

Sargent, T. D., 1973. Behavioral adaptations of cryptic moths. VI. Further experimental studies on bark-like species. *Journal of the Lepidopterists' Society,* 27(1): 8–12.

Schaller, F., and Timm, C., 1950. Das Hörvermögen der Nachtschmetterlinge. *Zeitschrift für vergleichende Physiologie,* 32:468–81.

Schneider, D., 1963. Function of insect olfactory sensilla. *Proceedings of the 16th International Congress of Zoology,* 3:84–85.

Schneider, D., 1974. The sex-attractant receptor of moths. *Scientific American,* 231(1):28–35.

Schneider, D.; Block, B. C.; Boeckh, T.; and Priesner, E.; 1967. Die Reaktion der Männlichen Seidenspinner auf Bombykol und seine Isomeren: Elektroantennogram und Verhalten. *Zeitschrift für vergleichende Physiologie,* 54:192–209.

Schneider, D.; Boppré, M.; Schneider, H.; Thompson, W. R.; Boriack, C. J.; Petty, R. L.; and Meinwald, J.; 1975. A pheromone precursor and its uptake in male *Danaus* butterflies. *Journal of Comparative Physiology,* 97:245–56.

Schneider, D., and Seibt, U., 1969. Sex pheromone of the queen butterfly: electroantennogram responses. *Science,* 164:1173–74.

Schoonhoven, L. M., 1973. Plant recognition by lepidopterous larvae. In: *Insect/Plant Relationships,* H. F. van Emden, ed., Symposium of the Royal Entomological Society of London no. 6. London: Blackwell Scientific Publications, pp.87–99.

Schremmer, F., 1941. Sinnesphysiologie und Blumenbesuch des Falters von *Plusia gamma* L. *Zoologische Jahrbücher* (Systematik), 74:375–434.

Scott, F. W., 1968. Sounds produced by *Neptis hylas* (Nymphalidae). *Journal of the Lepidopterists' Society,* 22(4):254.

Scott, J. A., 1968 (1970). Hilltopping as a mating mechanism to aid the survival of low density species. *Journal of Research on the Lepidoptera,* 7(4):191–204.

Scott, J. A., 1972 (1973a). Mating of butterflies. *Journal of Research on the Lepidoptera,* 11(2):99–127.

Scott, J. A., 1973b. Survey of ultraviolet reflectance of nearctic butterflies. *Journal of Research on the Lepidoptera,* 12(3):151–60.

Scott, J. A., 1974. Mate-locating behavior of butterflies. *American Midland Naturalist,* 91(1):103–17.

Sebeok, T. A., 1965. Animal communication. *Science,* 147:1006–14.

Seibt, U.; Schneider, D.; and Eisner, T.; 1972. Duftpinsel, Flügeltaschen und Balz des Tagfalters *Danaus chrysippus* (Lepidoptera: Danaidae). *Zeitschrift für Tierpsychologie,* 31:513–30.

Seiler, J., 1923. Geschlechtschromosomenuntersuchungen an Psychiden. IV. Die Parthenogenese der Psychiden Biologische und Zytologische Beobachtungen. *Zeitschrift für induktive Abstammungs- und Vererbungslehre,* 31:1–99.

Seiler, J., 1961. Untersuchungen über die Entstehung der Parthenogenese bei *Solenobia triquetrella* F. R. (Lepidoptera, Psychidae). III. Die geographische Verbreitung der drei Rassen von *Solenobia triquetrella* (bisexuell, diploid und tetraploid parthenogenetisch) in der Schweiz und in angrenzenden Ländern und die Beziehungen zur Eiszeit. Bemerkungen über die Entstehung der Parthenogenese. *Zeitschrift für Vererbungslehre,* 92:261–316.

Sengün, A., 1944. Experimente zur sexuell-mechanischen Isolation. *Rev. Fac. Sci. Istanbul* (B), 9:239–53.

Shapiro, A. M., 1972. An interfamilial courtship (Nymphalidae, Pieridae). *Journal of Research on the Lepidoptera,* 11:197–98.

Shapiro, A. M., 1973. An attempted interfamilial mating (Lycaenidae, Nymphalidae). *Journal of the Lepidopterists' Society,* 27(2):159.

Shapiro, A. M., and Cardé, R. T., 1970. Habitat selection and competition among sibling species of satyrid butterflies. *Evolution,* 24(1):48–54.

Shields, O., (1967) 1968. Hilltopping: an ecological study of summit congregation behavior of butterflies on a southern California hill. *Journal of Research on the Lepidoptera,* 6:69–178.

Shields, O., and Emmel, J. F., 1973. A review of carrying pair behavior and mating times in butterflies. *Journal of Research on the Lepidoptera,* 12(1):25–64.

Shorey, H. H., 1966. The biology of *Trichoplusia ni* (Lepidoptera: Noctuidae). IV. Environmental control of mating. *Annals of the Entomological Society of America,* 59:502–506.

Shorey, H. H., 1973. Behavioral responses to insect pheromones. *Annual Review of Entomology,* 18:349–80.

Shorey, H. H., 1974. Environmental and physiological control of insect sex pheromone behavior. In: *Pheromones,* M. C. Birch, ed. Amsterdam: North-Holland, pp.62–80.

Shorey, H. H., and Gaston, L. K., 1965. Sex pheromones of noctuid moths. V. Circadian rhythm of pheromone-responsiveness in males of *Autographa californica, Heliothis virescens, Spodoptera exigua,* and

Trichoplusia ni (Lepidoptera: Noctuidae). *Annals of the Entomological Society of America,* 58:597–600.

Shorey, H. H., and Gaston, L. K., 1970. Sex pheromones of noctuid moths. XX. Short-range visual orientation by pheromone-stimulated males of *Trichoplusia ni. Annals of the Entomological Society of America,* 63:829–32.

Shorey, H. H.; Gaston, L. K.; and Roberts, J. S.; 1965. Sex pheromones of noctuid moths. VI. Absence of behavioral specificity for the female sex pheromones of *Trichoplusia ni* versus *Autographa californica,* and *Heliothis zea* versus *H. virescens* (Lepidoptera: Noctuidae). *Annals of the Entomological Society of America,* 58(5):600–603.

Silberglied, R. E., 1969. Ultraviolet reflection of pierid butterflies. Phylogenetic implications and biological significance. M.S. thesis, Cornell University.

Silberglied, R. E., 1973. Ultraviolet reflection of butterflies and its behavioral role in the genus *Colias* (Lepidoptera-Pieridae). Ph.D. thesis, Harvard University.

Silberglied, R. E., and Taylor, O. R., 1973. Ultraviolet differences between the sulphur butterflies, *Colias eurytheme* and *C. philodice,* and a possible isolating mechanism. *Nature* (London), 241:406–408.

Skalski, A. W., 1973. Notes on the Lepidoptera from fossil resins. *Polskie Pismo Entomologiczne,* 43:647–54 (in Polish).

Slansky, F., Jr., 1971. *Danaus plexippus* (Nymphalidae) attacking red-winged blackbird. *Journal of the Lepidopterists' Society,* 25(4):294.

Smith, S. G., 1953. Reproductive isolation and the integrity of two sympatric species of *Choristoneura* (Lepidoptera: Tortricidae). *Canadian Entomologist,* 85:141–51.

Smith, S. G., 1954. A partial breakdown of temporal and ecological isolation between *Choristoneura* species (Lepidoptera: Tortricidae). *Evolution,* 8:206–224.

Sower, L. L.; Shorey, H. H.; and Gaston, L. K.; 1970. Sex pheromones of noctuid moths. XXI. Light:dark cycle regulation and light inhibition of sex pheromone release by females of *Trichoplusia ni. Annals of the Entomological Society of America,* 63:1090–92.

Sower, L. L.; Shorey, H. H.; and Gaston, L. K.; 1971. Sex pheromones of noctuid moths. XXV. Effects of temperature and photoperiod on circadian rhythms of sex pheromone release by females of *Trichoplusia ni. Annals of the Entomological Society of America,* 64:488–92.

Stride, G. O., 1956. On the courtship behaviour of *Hypolimnas misippus* L. (Lepidoptera, Nymphalidae), with notes on the mimetic association with *Danaus chrysippus* L. (Lepidoptera, Danaidae). *British Journal of Animal Behaviour,* 4:52–68.

Stride, G. O., 1957. Investigations into the courtship behaviour of the male of *Hypolimnas misippus* L. (Lepidoptera, Nymphalidae), with special reference to the role of visual stimuli. *British Journal of Animal Behaviour,* 5:153–67.

Stride, G. O., 1958a. On the courtship behaviour of a tropical mimetic butterfly, *Hypolimnas misippus* L. (Nymphalidae). *Proceedings of the 10th International Congress of Entomology,* Montreal, 1956, 2:419–24.

Stride, G. O., 1958b. Further studies on the courtship behaviour of African mimetic butterflies. *Animal Behaviour,* 6:224–30.

Süffert, F., 1924. Morphologie und Optik der Schmetterlingschuppen, inbesondere die Schillerfarben der Schmetterlinge. *Zeitschrift für Morphologie und Ökologie de Tiere,* 1:171–308.

Swihart, C. A., 1971. Colour discrimination by the butterfly *Heliconius charitonius* Linn. *Animal Behaviour,* 19:156–64.

Swihart, C. A., and Swihart, S. L., 1970. Colour selection and learned feeding preferences in the butterfly *Heliconius charitonius* Linn. *Animal Behaviour,* 18:60–64.

Swihart, S. L., 1963. The electroretinogram of *Heliconius erato* (Lepidoptera) and its possible relation to established behavior patterns. *Zoologica* (New York), 48:155–65.

Swihart, S. L., 1964. The nature of the electroretinogram of a tropical butterfly. *Journal of Insect Physiology,* 10:547–62.

Swihart, S. L., 1965. Evoked potentials in the visual pathway of *Heliconius erato* (Lepidoptera). *Zoologica* (New York), 50:55–62.

Swihart, S. L., 1967a. Hearing in butterflies (Nymphalidae: *Heliconius, Ageronia*). *Journal of Insect Physiology,* 13:469–76.

Swihart, S. L., 1967b. Neural adaptations in the visual pathway of certain heliconiine butterflies, and related forms, to variations in wing coloration. *Zoologica* (New York), 52(1):1–14.

Symczak, M., 1950. Orientation by memory and the social instinct in caterpillars of the large white butterfly (*Pieris brassicae* L.). *Bulletin international de l'Académie polonaise des Sciences et des Lettres* (B-II) (1949), pp.175–93.

Taylor, O. R., 1970. Reproductive isolation in *Colias eurytheme* and *Colias philodice* (Lepidoptera: Pieridae). Ph.D. thesis, University of Connecticut.

Taylor, O. R., 1972. Random vs. non-random mating in the sulfur butterflies, *Colias eurytheme* and *Colias philodice* (Lepidoptera: Pieridae). *Evolution,* 26(3): 344–56.

Taylor, O. R., 1973. Reproductive isolation in *Colias eurytheme* and *C. philodice* (Lepidoptera: Pieridae): use of olfaction in mate selection. *Annals of the Entomological Society of America,* 66(3):621–26.

Tinbergen, N., 1958. *Curious Naturalists.* London: Country Life Limited. 280pp.

Tinbergen, N.; Meeuse, B. J. D.; Boerema, L. K.; and Varossieau, W. W.; 1942. Die Balz des Samtfalters, *Eumenis* (=*Satyrus*) *semele* (L.). *Zeitschrift für Tierpsychologie,* 5:182–226. English translation: The Courtship of the grayling *Eumenis* (=*Satyrus*) *semele* (L.) (1942). In: *The Animal in Its World, Explorations of an Ethologist, 1932–1972,* vol. 1, *Field Studies.* Collected Papers. Cambridge: Harvard University Press, 1972, pp.197–249.

Thomann, H., 1901. Schmetterlinge und Ameisen. Beobachtungen über eine Symbiose zwischen *Lycaena argus* L. und *Formica cinerea* Mayr. *Jahresbericht der Naturforschenden Gesellschaft Graubündens,* 44:1–40.

Traynier, R. M. M., 1968. Sex attraction in the Mediterranean flour moth, *Anagasta kühniella:* location of the female by the male. *Canadian Entomologist,* 100:5–10.

Traynier, R. M. M., 1970. Sexual behaviour of the Mediterranean flour moth, *Anagasta kühniella:* some influences of age, photo-period, and light intensity. *Canadian Entomologist,* 102:534–40.

Treat, A., 1964. Sound reception in Lepidoptera. In: *Acoustic Behaviour of Animals,* R.-G. Busnel, ed. Amsterdam: Elsevier, pp.434–39, 800–801.

Truman, J. W., and Riddiford, L. M., 1974. Hormonal mechanisms underlying insect behaviour. *Advances in Insect Physiology,* 10:297–352.

Turner, J. R. G., 1971a. Studies of Müllerian mimicry and its evolution in burnet moths and heliconid butterflies. In: *Ecological Genetics and Evolution,* R. Creed, ed. Oxford: Blackwell Scientific Publications, pp.224–60.

Turner, J. R. G., 1971b. Experiments on the demography of tropical butterflies. II. Longevity and home-range behaviour in *Heliconius erato. Biotropica,* 3:21–31.

Turner, J. R. G., 1975. Communal roosting in relation to warning colour in two heliconiine butterflies (Nymphalidae). *Journal of the Lepidopterists' Society,* 29(4):221–26.

Tutt, J. W., 1906. *A Natural History of the British Lepidoptera,* vol. 5. London: Swan Sonnenschein and Co.

Vandel, A., 1931. La parthénogenèse. *Encyclopaedia Scientifique.* Paris.

Van der Kloot, W. G., and Williams, C. M., 1953a. Cocoon construction by the cecropia silkworm. I. The role of the external environment. *Behaviour,* 5:141–56.

Van der Kloot, W. G., and Williams, C. M., 1953b. Cocoon construction by the cecropia silkworm. II. The role of the internal environment. *Behaviour,* 5:157–74.

Van der Kloot, W. G., and Williams, C. M., 1954. Cocoon construction by the cecropia silkworm. III. The alteration of spinning behaviour by chemical and surgical techniques. *Behaviour,* 6:233–55.

Vane-Wright, R. I., 1972. Pre-courtship activity and a new scent organ in butterflies. *Nature* (London), 239(5371):338–40.

Varley, G. C., 1962. A plea for a new look at Lepidoptera with reference to the scent distributing organs of male moths. *Transactions of the Society for British Entomology,* 15:3–40.

Vogel, S., 1963. Das sexuelle Anlockungsprinzip der Catasetinen- und Stanhopeen-Blüten und die wahre Funktion ihres sogenannten Futtergewebes. *Österreichische botanische Zeitschrift,* 110:308–37.

Wallace, A. R., 1889 (1890). *Darwinism, an Exposition of the Theory of Natural Selection with Some of Its Applications,* 2d ed. London: Macmillan, xvi + 494pp.

Wallace, A. R., 1891. *Natural Selection and Tropical Nature, Essays on Descriptive and Theoretical Biology,* new ed. London: Macmillan. xii + 492pp.

Watt, W. B., 1968. Adaptive significance of pigment polymorphisms in *Colias* butterflies. I. Variation of melanin pigment in relation to thermoregulation. *Evolution,* 22(3):437–58.

Wehner, R., 1972. *Information Processing in the Visual Systems of Arthropods.* Berlin: Springer-Verlag. xi + 334pp.

Wellington, W. G., 1957. Individual differences as a factor in population dynamics: the development of a problem. *Canadian Journal of Zoology,* 35:296–323.

Wellington, W. G., 1974. Tents and tactics of caterpillars. *Natural History,* 83(1):64–72.

Whittaker, R. H., and Feeny, P. P., 1971. Allelochemics: chemical interactions between species. *Science,* 171:757–70.

Wickler, W., 1968. *Mimicry in Plants and Animals.* World University Library. London: Weidenfeld and Nicolson. 255pp.

Wigglesworth, V. B., 1972. *The Principles of Insect Physiology,* 7th ed. London: Chapman and Hall. viii + 827pp.

Wilson, E. O., 1971. *The Insect Societies.* Cambridge: Belknap Press of Harvard University Press. x + 548pp.

Wilson, E. O., 1975. *Sociobiology: The New Synthesis.* Cambridge: Belknap Press of Harvard University Press. ix + 697pp.

Wilson, E. O., and Bossert, W. H., 1963. Chemical communication among animals. *Recent Progress in Hormone Research,* 19:673–716.

Wright, R. H., 1963. Molecular vibration and insect sex attractants. *Nature* (London), 198:455–59.

Wynne-Edwards, V. C., 1962. *Animal Dispersion in Relation to Social Behaviour.* Edinburgh: Oliver and Boyd. xi + 653pp.

Yagi, N., 1955. Electron microscopical studies on the form of pterin pigments in the scale of Pieridae with reference to their bearing on systematics. *New Entomologist,* 4:1–20.

Yagi, N., and Koyama, N., 1963. *The Compound Eye of Lepidoptera, Approach from Organic Evolution.* Tokyo: Shinkyo-Press (Maruzen). 319pp.

Yeo, P. R., 1973. Floral allurements for pollinating insects. In: *Insect/Plant Relationships,* H. F. van Emden, ed. Symposium of the Royal Entomological Society of London no. 6. London: Blackwell Scientific Publications, pp.51–57.

Yokoyama, T., 1951. Studies on the cocoon-formation of the silkworm, *Bombyx mori* L. *Bulletin of the Sericultural Experimental Station* (Tokyo), 13(5):183–246 + pl. 2–8.

Yokoyama, T., 1959. *Silkworm Genetics Illustrated.* Tokyo: Japan Society for the Promotion of Science. 185pp.

Young, A. M., 1971. Notes on gregarious roosting in tropical butterflies of the genus *Morpho. Journal of the Lepidopterists' Society,* 25(4):223–34.

Chapter 18

COMMUNICATION IN DIPTERA

Arthur W. Ewing

Introduction

Except for investigations of fruit flies (genus *Drosophila*) and a few economically important species, rather little research has been done on behavior in Diptera, and investigation specifically on aspects of communication is limited to a few examples only. There are no social Diptera, and signals are used almost entirely in a sexual context. Thus a review of communication in Diptera entails a discussion of the stimuli involved in courtship and mating. Much of the information comes from qualitative descriptions of courtship behavior, and communication involving several sensory modalities is implicit in many of these. However, in the absence of experimentation, it is often difficult to identify the relevant signals and to assess their relative importance. Thus, for example, while the courtship behavior of various *Drosophila* species and in particular *D. melanogaster* has been described in detail starting with Sturtevant in 1915 and followed by Weidmann (1951), Spieth (1952), and Bastock and Manning (1955), only in 1962 did Shorey show that courting males produced songs, and it was in 1967 that the significance of these was demonstrated (Bennet-Clark and Ewing, 1967). The involvement of a pheromone in courtship was first described by Shorey and Bartell in 1970. In spite of intensive research carried out on the sexual behavior of *D. melanogaster,* a complete description of the signals involved in courtship is still not forthcoming.

As communication in *D. melanogaster* is probably better understood than that of any other dipteran species, except for those with very simple courtships, this example illustrates the fragmentary state of our knowledge. Even in those species with apparently simple behavior, the lack of complication may merely reflect our ignorance.

In this review I have used a certain amount of selectivity and have ignored some of the more anecdotal accounts that abound in the early literature. While some of these are of considerable potential interest, they are often difficult to interpret. Richards (1924) has compiled a bibliography of much of this work.

Mycetophilidae

There is one well-known example of light production in Diptera, which is in the mycetophilid midge *Arachnocampa luminosa* from New Zealand (Hudson, 1926), although complex bioluminescent behavior, as found in many Coleoptera where species-specific patterns of light production are used by both sexes, has not been described. These midges live in caves and other damp places, and their great numbers in, for example, the Glow-worm Grotto of Waitomo Cave constitute a tourist attraction. Larvae, pupae, and adults possess a light organ in the last abdominal segment. It is derived from the swollen distal ends of four malpighian tubules and, like analogous structures in other animals, has a reflector. Control of light production appears to be neural and is mediated by disinhibition from the brain (Gatenby, 1959). The carnivorous larvae construct mucous snares into which small phototactic insects are attracted. The larvae pupate suspended on a mucous thread, and female pupae luminesce particularly strongly when touched and also just prior to ecdysis. Males are attracted by the light to female pupae and have been observed waiting to mate as soon as the females emerge. Male pupae and adults also produce light, although not so readily as females, but the significance of this is unknown (Richards, 1960). Other luminescent Mycetophilids have been described from Australia and Tasmania, but their behavior has not been studied (Gatenby, 1960).

Culicidae, Chironomidae

One of the best-known examples of acoustic communication in insects is the flight tone in the mating of mosquitoes. Most mosquitoes, and indeed probably the majority of Diptera, form mating swarms. Roth (1948) demonstrated that the stimulus produced by the females of *Aedes aegypti* that acted as a sexual attractant for the male in these swarms was the flight sound. He further demonstrated that odor was not involved as an attractant. The mating swarms of mosquitoes and other Diptera do not appear primarily to be the result of any communication between individuals, but rather are due to species-specific preferences for particular swarm markers, which are often conspicuous objects in the environment such as the edge of a lake or an isolated tree (Downes, 1969).

Roth showed that the flight tone of males and females differed such that the frequency of the former was above the response threshold of males, and, under normal conditions, males were only attracted to the flight tone of the opposite sex. The flight tones of both sexes became higher with age, and those of newly emerged males were sufficiently low to attract other males. However, this occurred only in the abnormal laboratory situation, and immature males do not usually fly voluntarily. Further, in immature females the flight tone is below the threshold of male hearing, and the time at which it becomes audible coincides with the onset of female sexual receptivity. This synchronization of behavior and physiological state is found on the receptor side also. The flight sounds are perceived by Johnson's Organ, situated at the base of the antenna, and activated by vibration of the arista, which in males bears many fibrillae. Thus amputation or loading of the antennae renders males unresponsive to the female flight tone. In *Anopheles quadrimaculatus,* the fibrillae are retracted on eclosion and are only extended when males are 15-24 hours old. Males will not attempt to mate until this has occurred, and it coincides with rotation of the male genitalia, which is a prerequisite for successful copulation.

Flight tone rises markedly with temperature. Römer and Rosin (1969) have shown in *Chironomos plumosus* that the males' response curve follows the change in flight tone over the temperature range of 12° to 24°. The range of frequencies over which males will respond is quite large, presumably because of the changes in flight tone and in the males' response curve with age and because of the temperature effect which, in a very precise system, would reduce the probability of mating. Also, because of the breadth of frequency discrimination, the flight tone is probably not useful as a species-specific signal, and it seems probable that ecological factors are important in maintaining sexual isolation.

Not all mosquitoes and midges form mating swarms; there are a number of species in which males do not have plumose antennae, and in these, female flight tone does not appear to be an attractant. An example is *Culiseta inornata,* in which females produce a sex pheromone and which mate satisfactorily even if the wings of females or the antennae of males are removed (Kliewer et al., 1966). Similarly, in sabethine mosquitoes sex recognition is not auditory, and, as they are often brightly colored, visual components are probably involved (Haddow and Corbet, 1961). In the aberrant *Opifex fuscus* and *Deinocerties cancer,* males locate pupae on the water surface and attempt to copulate during emergence. Sex is discriminated only on attempted copulation. However, even in such an apparently simple mating system the published account suggests that during the brief "courtship," tactile, chemical, visual, and auditory stimuli may be involved (Provost and Haeger, 1967).

Empididae

One of the most widely known examples of visual signals employed in dipteran courtship is in the Empids. The males of many species present females with an insect or "balloon," which is constructed from foam or silk and which may contain an insect or inanimate object. Most Empids form mating swarms of one or both sexes, and those consisting of males carrying balloons are very conspicuous. Females approach males carrying the appropriate object and fly a few centimeters above; both then rise for a short distance, the female drops toward the male, and, as they both dive toward the ground, the male turns over and presents the female with the "gift" and immediately copulates with her. This description, which is fairly typical for many species, refers primarily to *Empis barbatoides* and *E. poplitea* (Alcock, 1973).

It is possible to construct a graded series from those that mate without the males' producing a courtship gift (e.g., *Tachydromia* spp.), through species in which the males provide an unadorned prey (e.g., *E. barbatoides, Ramphomya nigrita*), to those that wrap the prey in silk *(Hilara quadrivittata)* or wrap inanimate objects in silk (e.g., *H. maura*) or offer the females a silk or froth balloon alone (e.g., *H. sartor*: Hamm, 1928; Kessel, 1955; Alcock, 1973). Kessel (1955) recognizes eight discrete stages in this evolutionary process of ritualization of the behavior, concluding with the final emancipation of the signal from its original function. He suggests that the initial function of the behavior was to divert the predaceous intentions of the female, but there appears to be no evidence for this view. It seems probable, at least in some species, that only the males hunt, but do not themselves feed on the prey; while the females' only protein meal, which is necessary for maturation of the ovaries, comes from the prey presented during courtship (Downes, 1970). This ritual is similar in principle to the courtship feeding in some birds, which not only acts as an important stimulus in establishing and maintaining the pair bond but also provides necessary food for the female (Nisbet, 1973).

Recent descriptions of Empid courtship suggest that only a male carrying prey or prey surrogate is an adequate stimulus to the female although, as all aerial meetings do not result in copulation, some further degree of sexual selection must occur (Alcock, 1973). Also unclear is the basis of species recognition. Alcock reports that the two species *E. poplitea* and *E. barbatoides* fly in the same area but do not interact sexually; perhaps small differences in flight pattern are sufficient to discriminate between them. In some species males may recognize conspecific females because the females possess adornments such as fringed legs or eversible abdominal sacs *(H. flavinceris),* the latter suggesting production of a pheromone (Richards, 1924).

Some species of Empid, such as *R. ursinella,* mate on the ground without food transfer, and Downes (1970) suggests that this is an adaptation for life in arctic conditions. However, courtship of the aberrant *Xanthempis trigramma* is quite different. In this species females take up a station on the substrate and vibrate their wings. This signal is answered by males via a similar vibration when they approach to within 12-18 inches. The sexes then alternate wing vibration as the male approaches the female. The signals are presumably acoustic, as this behavior can occur with flies in visual isolation. This is followed by mutual tarsal contacting, and copulation occurs when the female signals her acceptance by turning (Hamm, 1933).

Trypetidae (= Tephritidae)

The mating behavior of members of several genera within this economically important family of fruit flies has been described. Olfactory and acoustic signals are probably important in the courtship of most species and, as many of these have patterned wings that may be displayed, visual stimuli are possibly also involved, although direct evidence for this is lacking.

The best-studied species is *Dacus tryoni,* the Queensland fruit fly, in which the sexually mature males produce a pheromone from sacs extruded from the posterior region of the rectum (Fletcher, 1969). This species also produces acoustic signals, which have been recorded by Monro (1953). They consist of 3 kHz tone bursts, each of approximately 8 cycles and with a repetition rate of 290 per sec. Monro suggests that the sound is produced by the edge of the wing, which is modified in males, being scraped across an abdominal comb. However, the wave form of the sound and the structure of the comb are not easily reconciled to this interpretation, and, as Keiser et al. (1973) state that the sounds produced by related species can still be heard after wing amputation, this hypothesis clearly requires further investigation.

As the acoustic signal is sufficiently loud to be heard by the unaided human ear and as the pheromone can also be detected by humans, it is probable that both stimuli could act as long-distance attractants for females. Males produce both stimuli at dusk, at which time the females are sexually receptive (Fletcher, 1969). Munro (1953) suggests that sound production is also involved in causing the males to aggregate.

Other species of *Dacus* produce pheromones via rectal glands, but it is not known to what extent these are species-specific. The possibility that acoustic signals are involved in sexual isolation has been investigated by Myers (1952), who states that the sounds produced by *D. cacuminatus* are higher in frequency than those of *D. tryoni* and that, in a choice situation, females of the former species selectively approach conspecific calling males. *D. tryoni* females, however, approached the nearest male and only rejected *D. cacuminatus* males after contact, presumably on the basis of contact chemoreception. Males attempt to copulate indiscriminately, and thus, as in most dipteran groups, it is the female that exercises choice of mate.

Pheromone production has been described in other *Dacus* species by Schultz and Boush (1971) and Economopoulos et al. (1971). The latter authors, working with *D. oleae* observed, as in *D. tryoni,* that the males produce the pheromone in a sexual context. They further note that a weak odor is produced by the females of this species. However, Haniotakis (1974), using bioassay techniques, has shown that it is the females of *D. oleae* that produce the chemical sex attractant and not the males (in contrast to the situation in other Trypetids so far investigated) and that the function of the pheromone produced by males is obscure. Female pheromone production commences on the third day after eclosion, when the ovaries are mature, and is switched off for seven days following mating.

The males of *D. oleae* also produce acoustic signals, which have been recorded by Féron and Andrieu (1962). These consist of an irregular song with tone bursts of up to four seconds' duration with a fundamental frequency of 320 Hz.

Another species shown to produce both a pheromone and an acoustic signal is the Mediterranean fruit fly, *Ceratitis capitata,* whose males have a rectal gland everted by haemolymph pressure (Lhoste and Roche, 1960; Féron, 1962). In this case the pheromone acts as a primary attractant for the female, and it is only after her approach that the male vibrates his wings. The sex pheromone of *C. capitata* has been identified and synthesized by Jacobson et al. (1973) and consists of two fractions, methyl (E)-6-nonenoate and (E)-6-nonen-l-ol. Bioassays showed that maximum attractiveness was found only if an acidic fraction were added.

In all the preceding species the wing vibrations of the males during courtship are probably primarily concerned with producing acoustic stimuli, although this requires formal proof in some of the species. However, Fletcher (1969) suggests that this may have been derived from wing movements whose initial function was to disperse a pheromone; this may occur in the Caribbean fruit fly *(Anastrepha suspensa).* This species releases a pheromone from the distended pleural region of the third, fourth, and fifth abdominal segments, and "cleaning" movements of the legs are interspersed with wing fanning. The leg movements possibly serve to spread the pheromone on the wings, which is then dispersed by fanning (Nation, 1972).

There is strong circumstantial evidence that wing movements in some species serve to provide visual stimuli. Thus in *A. suspensa,* following the approach of the female, the male may make wing-waving movements that are not the same as fanning. Similarly, in *Tephritis stigmatica* and *Euleia fratria,* during courtship one or both sexes make characteristic movements of the wings with the wings tilted so as to be visually oriented toward the partner. In the former species the wings are initially moved synchronously like windshield wipers, and this is followed by alternate wing extensions along the longitudinal plane of the body. At 45° of extension the wings are vibrated, suggesting an auditory component. In *E. fratria* the wings are alternately waved forwards and backwards, and, when the male is close to the female, he extends his wings 90° to the substrate and remains stationary for a period. Following that he runs toward the female, wings extended and vibrating (Tauber and Toschi, 1965a, 1965b).

In the island fruit fly *(Rioxa pornia)* also, following attraction of the female to the male by means of a pheromone, both sexes carry out movements of the patterned wings. The final stage in the courtship of this species is the production by the male of a mound of foam on which the female feeds while the male attempts to copulate (Pritchard, 1967). The production of similar gustatory stimuli has been recorded also in *Eutreta* species (Stolzfus and Foot, 1965) and in *Afrocneros mundus* (Oldroyd, 1964). It is unlikely that this can be considered a communicatory device; it probably functions to keep the female stationary while the male mounts.

Finally, visual stimuli may be important in both courtship and aggression within the genus *Rhagoletis* (Prokopy and Bush, 1973; Bush, 1966; Biggs, 1972). All the species examined have patterned wings, which may be waved during courtship by both sexes or by males during territorial fights. However, copulation often occurs in *R. pomonella* in the absence of the wing display and it is not therefore an essential component in courtship. It is of interest that members of the *pomonella* and *cingulata* species groups, which consist mainly of sibling species that are often sympatric, are very similar morphologically and sexual isolation is maintained by strict host-plant specificity. By contrast, members of the *suavis* species group infest only walnuts and the patterning of the wings is distinct in the different species, suggesting that visual stimuli may be involved in the maintainance of sexual isolation (Bush, 1969).

A representative pattern of normal courtship behavior within the Trypetidae might be as follows. Males take up station on the host plant, in particular, on the fruit; the host-plant specificity in itself is in many cases sufficient to ensure some degree of sexual isolation. The males produce a pheromone from abdominal glands, and pheromone dispersal may be facilitated by fanning movements of the wings. Alternatively or concurrently, a sound signal is produced, and these stimuli attract sexually receptive females. The signals are probably also species-specific and are thus factors in maintaining isolation. In almost all species the males will attempt to copulate without any further courtship and are apparently often successful. It is clear that contact chemoreception may play a part at this stage; the males, which are initially undiscriminating, will mate only with conspecifics, and the females will reject foreign males. In addition, further components of courtship containing visual or acoustic components may be interposed between the initial attraction and attempted copulation. Unfortunately, in the almost total absence of quantitative analyses of courtship behavior, it is impossible to evaluate the relative importance of the different stimuli or to speculate on what basis sexual selection may occur. Keiser et al. (1973) have shown that the mating success of *D. cucurbitae, D. dorsalis* and *C. capitata* is reduced in the dark and a further reduction occurs if the flies are also wingless. Removal of the wings alone depresses the success level of the last species only. These observations merely suggest that the relative importance of visual and acoustic stimuli may be different in the members of the two genera studied.

Drosophilidae

The courtship behavior of over three hundred species of Drosophilid have been described. While these qualitative descriptions, which are mainly by Spieth (1952, 1966, 1968, 1969), indicate an extraordinarily diverse and interesting repertoire of behavior, detailed studies are confined to a very few species. Indeed it is in only one species, *D. melanogaster,* that a comprehensive understanding of the various stimuli involved in courtship is emerging, and I will therefore concentrate on that species.

Most *Drosophila,* except for the Hawaiian species (Spieth, 1966), probably court and mate on the food source, where male courtship behavior is triggered by visual stimuli (Sturtevant, 1915; Milani, 1950; Spieth, 1974). Males orient to other flies or objects of approximately the right size. Before starting to court, males of most species tap the other fly with their fore-tarsi, presumably to receive chemotactile information concerning sex and conspecificity. This chemical specificity can take some time to develop after eclosion, and Manning (1959) has shown that *D. melanogaster* males will court newly emerged females of the sibling species *D. simulans,* but by the time females are four days old most males do not court following tapping. Males with their fore-tarsi removed, however, court a significantly greater number of mature *D. simulans* females. Tapping may also provide the male with information on the physiological state of the female in some species. Spieth (1969) states that *D. sulfurigaster* males turn away after tapping inseminated females who have not exhausted their sperm supply. It is probable that more than one surface pheromone is involved in these male responses.

Tapping, while seen in a majority of *Drosophila* species, is not the inevitable precursor of courtship even in *D. melanogaster,* and courting males will switch females without tapping again. This element must therefore be of minor importance in effecting sexual isolation, particularly as several species may be seen at the same time on a food source. Tapping may provide stimuli for females as well as males, since unreceptive females will repel males on being tapped. It is possible, however, that the discrimination by the female is made on some other basis and tapping merely triggers the repelling action. Airborne as well as contact pheromones are involved in the courtship of *D. melanogaster* and of other species. One of these is concerned in the "minority effect," first described by Petit (1958). She showed that when males of two genotypes were mixed, females preferentially mated with the rare genotype, and Ehrman (1969) has shown that male scents are responsible for the phenomenon.

Another pheromonal effect, similar to that found in *Musca domestica,* where male sexual activity is increased by a female odor, has been reported in *D. melanogaster* by Shorey and Bartell (1970). As it has also been found in *D. pseudoobscura* (Sloane and Spiess, 1971), it is likely to be a common phenomenon in the genus.

Subsequent to tapping, the male orients toward the female and attempts to follow her when she moves off. Orientation ensures contact without itself having any specific signal value. Visual stimuli are not important in the courtship of *D. melanogaster* and are probably unimportant in most *Drosophila* species. There are major visual components in the displays of some species, such as *D. subobscura* (Brown, 1965), *D. suzukii* (Manning, 1965), the Hawaiian species (Spieth, 1966), and in some members of the *nasuta* subgroup, which have silvery markings on the head that are probably conspicuous to the female. One of these species, *D. pulaua,* does not appear to produce auditory signals, although the related *D. albomicans* and *D. kepulauana* do. This suggests that *D. pulaua,* which will not mate in the dark, is highly dependent on visual stimuli (Wright, 1974). Mating in a large number of *Drosophila* species is light-dependent (see, e.g., Grossfield, 1966). However, the majority of experiments merely demonstrate that mating does not occur in the dark, a condition that may be due to a number of different causes. For example, I have observed that one light-dependent species, *D. auraria* (Spieth and Hsu, 1950), is totally inactive throughout the dark period.

Some form of wing display providing acoustic stimuli is seen in most *Drosophila* species. In *D. melanogaster* one wing is extended to 90° and vibrated horizontally through approximately 30° (Bennet-Clark and Ewing, 1968). The acoustic signal produced during vibration is the major one involved in sexually stimulating females. The sounds consist of a series of sinusoidal pulses of 3 ms duration and a pulse repetition rate of 30 per sec at 25°C (Shorey, 1962; Bennet-Clark and Ewing, 1967). The mating success of wingless males is increased by substituting artifically produced courtship songs, thus demonstrating that the pulse train is the effective stimulus. Further, we showed that songs with half and double the normal repetition rate were ineffective, suggesting that the songs may also be important as isolating mechanisms (Bennet-Clark and Ewing, 1969). The species-specific nature of the songs lends weight to this supposition (Ewing and Bennet-Clark, 1968; Ewing, 1970). Further, the songs of sibling or closely related species tend to be distinct (Waldron, 1964; Ewing and Bennet-Clark, 1968; Ewing, 1970; Patty et al., 1973). Thus, for example, the sibling pair *D. persimilis* and *D. pseudoobscura* have songs that differ with respect to intrapulse frequency as well as interval (Ewing, 1969), while within the *melanogaster* species group, the five sibling species *D. melanogaster, D. simulans, D. erecta, D.*

yakuba, and *D. tiessieri* produce pulsed songs with intrapulse intervals (at 25°C) of 34, 48, 42, 96, and 52 ms, respectively. Also, all the species, except possibly *D. yakuba,* produce "sine song" (see Fig. 1), and *D. erecta* has an additional pulsed song that is polycyclic and at a higher frequency (Ewing, unpublished).

The patterning of the bursts of song is also important, and intermittent song production is significantly more effective than continuous song. A pattern of two seconds of artifically produced song followed by three seconds of silence, which most closely mimics the normal situation, almost restores the courtship success of wingless males to that of normal flies (Bennet-Clark, Ewing, and Manning, unpublished).

Recently a clearer understanding of the acoustic properties of sound production and reception in small insects has led to more sensitive recordings of *Drosophila* courtship songs (Bennet-Clark, 1971, 1972). F. von Schilcher (pers. comm.) has recorded a second component in the song of *D. melanogaster,* which he calls "sine song"; it has also been found in *D. albomicans* and *D. kepulauana* (Wright, 1974). The significance of this component is unclear, and it may be a side effect of the mechanics of sound production and without any signal value, although this appears unlikely.

As in mosquitoes, the acoustic signals are perceived via the antennae. Immobilization of the arista of the antennae in females renders them sexually unreceptive (Manning, 1967).

The function of courtship songs is twofold: to stimulate sexually females that are initially unreceptive and to provide a specific signal that aids in species recognition. These two functions may be served by different components of the signals and, in those species with two distinct songs, by the different songs, but this is difficult to investigate. That a certain amount of song stimulation is necessary before females will copulate and that this stimulation is summated over a period of time have been demonstrated by the use of simulated songs (Bennet-Clark et al., 1973). Bennet-Clark and I suggested that one of the ways in which the courtship songs might act would be to inhibit walking by the female and thus facilitate copulation attempts by the male (Bennet-Clark and Ewing, 1967), and von Schilcher (pers. comm.) has demonstrated that simulated songs do indeed slow down the females. Most interestingly, the simulated songs also affect males but, in contrast to females, their

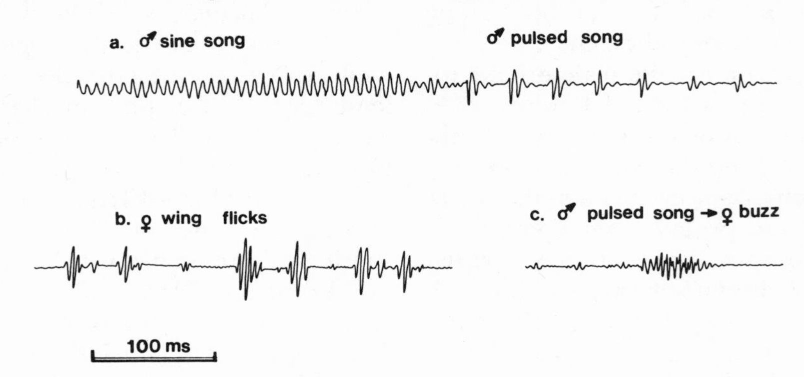

Fig. 1. Acoustic signals produced by *Drosophila melanogaster*. a. Male courtship song: sine song followed by pulsed song. b. Wing flicks made by unreceptive females that inhibit male courtship. Males courted by other males produce similar sounds. c. Part of male courtship song interrupted by a repelling buzz made by an immature female.

activity is greatly increased. In *D. melanogaster* and *D. simulans* the maximum effect occurs at the pulse repetition rate appropriate for that species. It is not clear if this has any natural significance, but is could have a social facilitatory effect that might be adaptive where flies are crowded together on a food source, or it could act as a positive feedback on the singing male and increase his sexual excitation.

Finally, prior to attempted copulation, males of many species including *D. melanogaster* extend the proboscis and lick the female's genitalia. It is not clear what information is conveyed by this movement. Certainly unreceptive fertilized females that extrude their ovipositors provide inhibiting stimuli for licking males. However it is probable that licking is not merely a test of the female's state of receptivity but provides positive stimuli, presumably chemical and tactile, for both sexes, as licking is repeated frequently throughout a courtship (Bastock and Manning, 1955). In many species licking is prolonged and is a major part of courtship (Spieth, 1952).

In *D. melanogaster* licking is often followed by attempted copulation. The females of this species do not have any acceptance posture that signals their readiness to mate. Successful attempts probably occur only if the female spreads her genital plates; however, this is unlikely to be perceived by the male. By contrast, females of some *Drosophila* species, such as members of the *D. nasuta* subgroup, do show an acceptance posture; they spread their wings to about 45°, depress their abdomens, and turn away from the male displaying in front of them (Spieth, 1969). However, males do not always attempt to copulate in response to this behavior, and it is not clear to what extent it is indeed a signal for the male.

Most of the signals involved in courtship are provided by males. Female signals are limited to a possible acceptance posture in some species, the production of one or more pheromones, and finally to the production of a "buzz" and a wing flick. The former sound is caused by a wing vibration and is made mainly by immature, unreceptive females when courted. Its effect is to inhibit male courtship, and, as it is very similar in several different species, it could function as an interspecific as well as an intraspecific signal (Ewing and Bennet-Clark, 1968). Males, on being courted by other males, also flick their wings, producing an irregular pulse train. The sound produced by female wing flicks is similar, and both have the effect of inhibiting courtship. Fig. 1 illustrates the different sounds produced by *D. melanogaster.*

The courtships of most Drosophila species show similarities to the patterns described above. One aberrant and very interesting group are the Hawaiian species described by Spieth (1966), which have evolved quite different patterns of courtship behavior. Most species do not court on the food source. Individual males take up small territories on vegetation, which they defend from other males. The flies are cryptic when feeding, and Spieth considers the behavior of the Hawaiian species to have evolved in response to the very intense predation to which they are subject. Males are often map-winged and may advertise their presence with visual signals. Some trail their abdomens along the substrate, depositing a pheromone (e.g., *D. grimshawi*). Others display a behavior similar to that described for *Anastrepha suspensa* (Nation, 1972), where the abdomen is raised, a drop of fluid is extruded, and the wings are vibrated, presumably for pheromone dispersal (e.g., *D. pilimana, Antopocerus tanythrix*).

The courtship behavior itself is extremely varied. Males possess epigamic features involving modifications of mouthparts, antennae, legs, and wings. Complex auditory, tactile, visual, and chemical signals are clearly involved in courtship; however, as the stimuli involved in courtship have not been investigated experimentally I shall not deal further with them.

Chloropidae

One of the clearest examples of a species-specific mechanical signal-response system is seen in the gall-forming flies of the genus *Lipara* (Mook and Bruggemann, 1968; Chvála et al., 1974). The larvae of these flies form galls on the reed *Phragmites communis,* and only one individual is found on a stem. Males fly from stem to stem and produce a substrate-transmitted vibration with a pattern of pulses characteristic of the species. If a virgin female is on the reed she will respond by producing a series of pulses that induces the male to search the stem. The flies then countersignal until the male finds the female (see Fig. 2).

Females respond to males up to 2 meters away. As the females are very static, this pattern of behavior is an efficient method of bringing the sexes together. Males of different species produce different patterns of pulses, and Chvála et al. (1974) have shown that females, whose signals are all similar, respond only to the signals of conspecific males. Mook and Bruggemann could find no stridulatory mechanisms. As the fundamental frequency within the pulses is low (about 300 Hz), the vibrations are probably caused by activation of thoracic flight mechanism, as in *Drosophila.* They are transmitted to the substrate via the legs and are almost certainly perceived by the sub-genual organs.

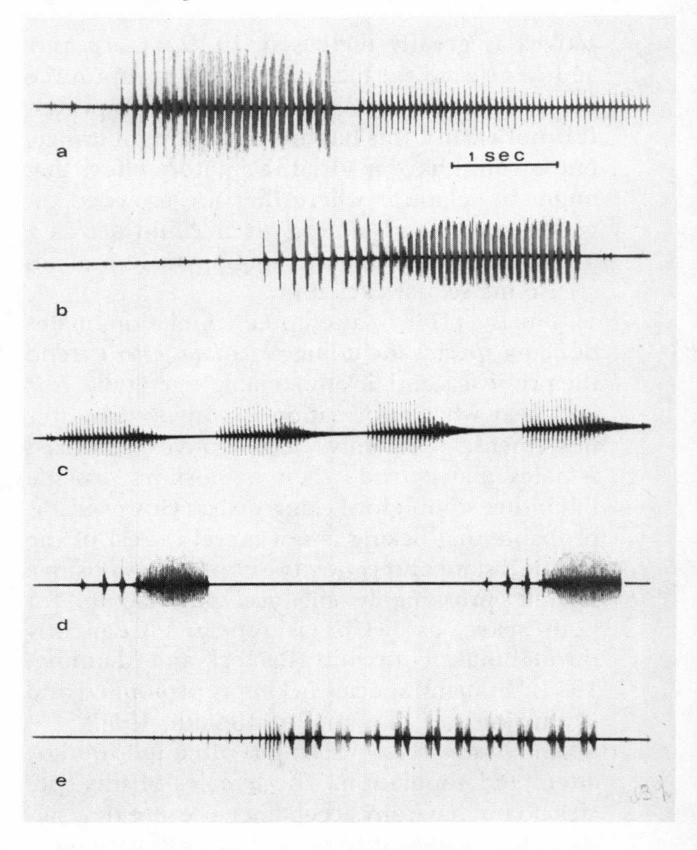

Fig. 2. Substrate-transmitted vibrations produced by gall-flies of the genus *Lipara*. The signals were recorded using a crystal gramophone pickup element in contact with the substrate. a. Signal of male *L. lucens* followed by part of the answering signal of a female. b-e: Signals of males of different species. b. *L. lucens*; c. *L. similis*; d. *L. pullitarsis*; e. *L. rufitarsis*. (From Chvála et al., 1974.)

It is of interest that while *Lipara* and *Drosophila* have similar methods of sound production, they utilize different methods of transmission and reception. The reed stem on which *Lipara* lives provides an excellent medium for the propagation of the sounds, but this is not so for the food sources on which *Drosophila* normally court. The latter therefore utilize airborne sounds, which are perceived by the antennae. However, one can still record the sounds produced by courting *Drosophila* whose wings have been totally removed if they court on the diaphragm of a crystal microphone. The use of substrate transmitted sounds is a possibility in at least some members of this genus also.

Calliphoridae, Muscidae

The courtship behavior of flies of these families provides a good example of apparent and misleading simplicity. In many species mounting by males is elicited by extremely generalized visual stimuli, such as any dark object of approximately the appropriate size (see, e.g. Vogel, 1957). Even the visual stimuli are not essential, as mating can occur in the dark. If the attempt is made on a conspecific female, copulation or rejection occurs within one or two seconds, and this does not appear to provide much time for the exchange of complex signals.

The females of *Musca domestica* and *Lucilia cuprina,* however, produce pheromones whose action is to stimulate sexual behavior in males (Rogoff et al., 1964; Bartell et al., 1969). Further, in the former species the same or another pheromone acts as a sex attractant. The pheromone can be extracted with benzene and is species-specific in its action, as extracts from *M. autumnalis* and *Stomoxys calcitrans* are ineffective (Rogoff et al., 1964). Recently Tobin and Stoffolano (1973a, 1973b) have filmed the mating behavior of *M. domestica* and *M. autumnalis.* They have shown that within the short period between mounting and copulation the male performs a series of complex actions that are probably concerned with tactile, chemical, and auditory stimuli and that the two species differed consistently in details of the behavior.

I have recorded the sounds produced by males of *M. domestica* during the brief courtship. These usually consist of a train of tone bursts of between 160 and 190 Hz and are produced by vibration of the partly folded wings. The first of these is both longer and more variable than the succeeding ones, with a mean duration of about 500 ms. Then follow up to six tone bursts of 240 ms, separated by 40 ms intervals. These sounds, while not as regular as many of those produced by *Drosophila* species, are patterned in such a way as to suggest that they have signal value. The flies also produce other sounds that may have an aggressive or warning function (Esch and Wilson, 1967).

Summary

This survey, although not exhaustive of the literature, demonstrates very diverse modes of communication within the Diptera, and yet nothing is known about the behavior of perhaps 99.5 percent of described species. This paucity of information is partly due to technical difficulties and to the enormity of the task. It is worthwhile both to consider these difficulties and to see, even with the limited information available, if any generalizations are possible.

It is immediately obvious that often more than one channel of communication is used by a single species, sometimes simultaneously. This makes analysis difficult, in comparison with stimulus-response chains. These are not common in Diptera, but one example is the courtship of *Tipula oleracea.* In this species the males approach and grab the forelegs of the female. The only relevant parameter that triggers the next stage of the male's courtship is the thickness of the female's legs. When the female raises her legs the male mounts; when the female's leg movements cease the male "kisses" the female's head, moves back, and in response to tactile stimuli from the tarsal contact of the female's abdomen, copulates. Stich (1963), in a series of simple but elegant experiments using models, has demonstrated that a specific stimulus is required at each step before the sequence can continue. Unfortunately, most behavior is not amenable to dissection in this manner.

Many species mate in swarms or require specific conditions not easily provided in the laboratory. In either case their behavior is difficult to observe, much less analyze experimentally. Many Diptera utilize sex pheromones to some extent, and of all sensory modalities, the chemosensory ones are probably the most difficult to work on. Without the use of chemical procedures to isolate and identify pheromones it is difficult to know whether the behavior effect under investigation is due to a single pheromone or to a medley: whether the pheromone has a unitary or multiple mode of action.

The use of pheromones is widespread in those species that do not form mating swarms, and even in the swarming species one cannot automatically discard the possibility that contact pheromones are being used. The pheromones can be classified on the basis of function; sex attractants (e.g., *Dacus* spp., *Lucilia cuprina*), sex stimulants or aphrodisiacs (e.g., *Drosophila* spp., *Musca domestica*), and repellents. The third class has been investigated less and possibly includes two types: those that repel members of other species and those that are produced by sexually unreceptive individuals, usually fertilized females. Repelling pheromones are probably produced by *Drosophila* species (Spieth, 1969; Cook, 1975) and by the gnat *Hippelates collusor,* whose females produce a sex attractant when receptive and switch to a repellent when their ovaries contain mature eggs (Adams and Mulla, 1968).

Acoustic signals are also common, but there are technical problems in recording and interpreting signals produced by small sound sources. Recording is difficult partly because acoustic power and distance follow an inverse sixth-power relationship where the wavelength of the sound is less than one-third the diameter of the source (Bennet-Clark, 1971). This is true of many Diptera, whose sound source, the wings, produce sounds of low frequency, in contrast to the majority of the better-known singing insects, such as crickets and cicadas, which produce high-frequency songs. A further complication is that both the type of microphone used and the recording mode of the tape recorder can affect the form of the signal.

The sounds produced by Diptera can be subdivided with regard to function in a manner similar to that used to classify pheromones. Sounds are probably used less as attractants than are pheromones because of the physical limitations mentioned above, but *Xanthempis trigramma* and some *Dacus* species produce sounds in this category. Sexual stimulation due to acoustic signals is also common, and both male and female *Drosophila* produce sounds that repel other flies.

All the communication that I have described occurs in a sexual context. As the females, at least, of many species mate only once or at long intervals, it is clearly adaptive for the sexual signals to be synchronized with the reproductive cycle. Thus the switching on and off of pheromone production at the appropriate time is a general feature of Diptera and of other insects, and the same is probably true of acoustic signals. The synchronization of ovarian development and various aspects of reproductive behavior, including pheromone production, has been shown to be under endocrine control in some insects, but the situation in Diptera awaits investigation (Barth, 1970).

The species-isolating function of communication is clearly seen in Diptera. The selective pressure acting to produce these isolating mechanisms is demonstrated by the extremely diverse song patterns recorded from different species of *Drosophila.* There are species that produce continuous songs, which may be of a single frequency, of two alternating frequencies, or frequency modulated. The majority of species, however, produce pulsed sounds of different pulse length, pulse frequency, and repetition rate, while some use more than one type of song (Ewing and Bennet-Clark, 1968; Ewing 1970).

An important criterion for signals used as isolating mechanisms is that they should be invariable within species. This is true of the pheromones that have been investigated and of acoustic signals. One possible exception is the scent that mediates the rare genotype advantage in *Drosophila.* Hay (1973) suggests that this may develop as a colony odor, similar to that found in some social Hymenoptera. However, this odor functions not as an isolating mechanism but as a promoter of genetic heterogeneity within a species.

References

Adams, T. S., and Mulla, M. S., 1968. Ovarian development, pheromone production, and mating in the eye gnat, *Hippelates collusor. J. Insect. Physiol.*, 14:627–35.

Alcock, J., 1973. The mating behaviour of *Empis barbatoides* Melander and *Empis poplitea* Loew (Diptera:Empididae). *J. Nat. Hist.*, 7:411–20.

Bartell, R. J.; Shorey, H. H.; and Barton Browne, L.; 1969. Pheromonal stimulation of the sexual activity of the males of the sheep blowfly *Lucilia cuprina* (Calliphoridae) by the female. *Anim. Behav.*, 17:576–85.

Barth, R. H., 1970. Pheromone-endocrine interactions in insects. Memoir Soc. Endocrin., 18. In: *Hormones and the Environment,* G. K. Benson, and J. G. Phillips, eds. Cambridge: Cambridge University Press, pp. 373–404.

Bastock, M., and Manning, A., 1955. The courtship of *Drosophila melanogaster. Behaviour,* 8:85–111.

Bennet-Clark, H. C., 1971. Acoustics of insect song. *Nature,* 234:255–59.

Bennet-Clark, H. C., 1972. Microphone and preamplifier for recording courtship songs of *Drosophila. Drosophila Inform. Ser.*, 49:127–28.

Bennet-Clark, H. C., and Ewing, A. W., 1967. Stimuli provided by courtship of male *Drosophila melanogaster. Nature,* 215:669–71.

Bennet-Clark, H. C., and Ewing, A. W., 1968. The wing mechanism involved in the courtship of *Drosophila. J. Exp. Biol.*, 49:117–28.

Bennet-Clark, H. C., and Ewing, A. W., 1969. Pulse interval as a critical parameter in the courtship song of *Drosophila melanogaster. Anim. Behav.*, 17:755–59.

Bennet-Clark, H. C.; Ewing, A. W.; and Manning, A.; 1973. The persistence of courtship stimulation in *Drosophila melanogaster. Behavioral Biology,* 8:763–69.

Biggs, J. D., 1972. Aggressive behavior in the adult apple maggot (Diptera: Tephritidae) *Can. Ent.*, 104:349–53.

Brown, R. G. B., 1965. Courtship behaviour in the *Drosophila obscura* group. Part II. Comparative studies. *Behaviour,* 25:281–323.

Bush, G. L., 1966. The taxonomy, cytology, and evolution of the genus *Rhagoletis* in North America (Diptera, Tephritidae). *Bull. Mus. Comp. Zool.*, 134(11):431–562.

Bush, G. L., 1969. Mating behaviour, host specificity, and the ecological significance of sibling species in frugivorous flies of the genus *Rhagoletis* (Diptera-Tephritidae). *Amer. Nat.*, 103:669–72.

Chvála, M.; Doskočil, J.; Mook, J. H.; and Pokorný, V.; 1974. The genus *Lipara* Meigen (Diptera, Chloropidae), systematics, morphology, behaviour, and ecology. *Tijdschrift voor Entomol.*, 117:1–25.

Cook, R., 1975. Courtship of *Drosophila melanogaster:* rejection without extrusion. *Behaviour,* 52:155–71.

Downes, J. A., 1969. The swarming and mating flight of Diptera. *Ann. Rev. Entomol.*, 14:271–98.

Downes, J. A., 1970. The feeding and mating behaviour of the specialized Empididae (Diptera); observations on four species of *Rhamphomyia* in the high arctic and a general discussion. *Can. Ent.*, 102:769–91.

Economopoulos, A. P.; Giannakakis, A.; Tzanakakis, M. E.; and Voyadjoglou, A. V.; 1971. Reproductive behavior and physiology of the Olive fruit fly. 1. Anatomy of the adult rectum and odours emitted by adults. *Ann. Ent. Soc. Am.*, 64:1112–16.

Ehrman, L., 1969. Simulation of the mating advantage of rare *Drosophila* mates. *Science,* 167:905–906.

Esch, H., and Wilson, D., 1967. The sounds produced by flies and bees. *Z. vergl. Physiol.*, 54:256–67.

Ewing, A. W., 1969. The genetic basis of sound production in *Drosophila pseudoobscura* and *D. persimilis. Anim. Behav.*, 17:555–60.

Ewing, A. W., 1970. The evolution of courtship songs in *Drosophila. Rev. Comp. Animal,* 4(4):3–8.

Ewing, A. W., and Bennet-Clark, H. C., 1968. The courtship songs of *Drosophila. Behaviour,* 31:288–301.

Féron, M., 1962. L'instinct de reproduction chez la mouche méditerranéene des fruits *Ceratitis capitata* Wied (Dipt. Trypetidae). Comportement sexuel. Comportement de ponte. *Rev. Path. veg. Ent. agric. France,* 41:1–129.

Féron, M., and Andrieu, A. J., 1963. Étude des signaux acoustiques du mâle dans le comportement sexuel de *Dacus oleae* Gmel (Dipt. Trypetidae). *Ann. Epiphyties,* 13:269–76.

Fletcher, B. S., 1969. The structure and function of the sex pheromone gland of the male Queensland fruit fly, *Dacus tryoni. J. Insect Physiol.*, 15:1309–22.

Gatenby, J. B., 1959. Notes on the New Zealand glowworm, *Bolitophila (Arachnocampa) luminosa. Trans. Roy. Soc. New Zealand,* 87:291–314.

Gatenby, J. B., 1960. The Australasian mycetophylid glowworms. *Trans. Roy. Soc. New Zealand,* 88:149–56.

Grossfield, J., 1966. The influence of light on the mating behavior of *Drosophila. Univ. Texas Publ.,* 6615:147–56.

Haddow, A. J., and Corbet, P. S., 1961. Entomological studies above a high tower in Mpanga Forest, Uganda. V. Swarming activity above the forest. *Trans. Roy. Entomol. Soc. London,* 113:284–300.

Hamm, A. H., 1928. On the epigamic behaviour of *Hilara maura,* Fab., and two allied species. *Proc. Roy. Soc.* B, 102:334–37.

Hamm, A. H., 1933. The epigamic behaviour and courtship of three species of Empididae. *Entomol. Monthly Mag.,* 69:113–17.

Haniotakis, G. E., 1974. Sexual attraction in the Olive fruit fly, *Dacus oleae* (Gmelin). *Environmental Entomol.,* 3 (1):82–86.

Hay, D. A., 1973. Recognition by *Drosophila* of individuals of other strains or cultures: support for the role of olfactory cues in selective mating? *Evolution,* 26:171–76.

Hudson, G. V., 1926. The New Zealand glow-worm. *Bolitophyla (Arachnocampa) luminosa:* summary of observations. *Ann. Mag. Natur. Hist.,* 17:228–35.

Jacobson, M.; Ohinata, K.; Chambers, D. L.; Jones, A. W.; and Fujimoto, M. S.; 1973. Insect sex attractants. 13. Isolation, identification, and synthesis of sex pheromones of the male mediterranean fruit fly. II. *Med. Chem.,* 16,3:248–51.

Keiser, I.; Kobayashi, R. M.; Chambers, D. L.; and Schneider, E. L.; 1973. Relation of sexual dimorphism in the wings, potential stridulation, and illumination to mating of Oriental Fruit Flies, Melon Flies, and Mediterranean Fruit Flies in Hawaii. *Ann. Entomol. Soc. Amer.,* 66:937–41.

Kessel, E. L., 1955. The mating activities of balloon flies. *System. Zool.,* 4:96–104.

Kliewer, J. W.; Miura, T.; Husbands, R. C.; and Hurst, C. H.; 1966. Sex pheromones and mating behavior of *Culiseta inornata. Ann. Entomol. Soc. Amer.,* 59:530–33.

Lhoste, J. and Roche, A., 1960. Organes odoriférants des mâles de *Ceratitis capitata* (Dipt. Trypetidae). *Bull. Soc. Entomol. Fr.,* 65:206–10.

Manning, A., 1959. The sexual isolation between *Drosophila melanogaster* and *Drosophila simulans. Anim. Behav.,* 7:60–65.

Manning, A., 1965. *Drosophila* and the evolution of behavior. In: *Viewpoints in Biology,* vol. 4, J. D. Carthy and C. L. Duddington, eds. London: Butterworths, pp. 125–69.

Manning, A., 1967. Antennae and sexual receptivity in *Drosophila melanogaster* females. *Science,* 158:136–37.

Milani, R., 1950. Release of courtship display in *subobscura* males stimulated with dummies. *Drosophila Inform. Ser.,* 24:88.

Mook, J. H., and Bruggemann, C. G., 1968. Acoustical communication by *Lipara lucens* (Diptera, Chloropidae) *Ent. Exp. and Appl.,* 11:397–402.

Monro, J., 1953. Stridulation in the Queensland fruit fly *Dacus (Strumeta) tryoni* Frogg. *Austr. J. Sci.,* 16:60–62.

Myers, K., 1952. Oviposition and mating behaviour of the Queensland fruit-fly *Dacus (Strumeta) tryoni* (Frogg) and the solanim fruit-fly (*Dacus (Strumeta) cacuminatus* (Hering)). *Aus. J. Sci. Res.* B, 5:264–81.

Nation, J. L., 1972. Courtship behaviour and evidence for a sex attractant in the male Carribean fruit fly, *Anastrepha suspensa. Ann. Ent. Soc. Am.,* 65:1364–67.

Nisbet, I. C. T., 1973. Courtship-feeding, egg-size and breeding success in common terns. *Nature,* 241:141–42.

Oldroyd, H., 1964. *The Natural History of Flies.* London: Weidenfeld and Nicolson.

Patty, R. A.; Goldstein, R. B.; and Miller, D. D.; 1973. Sonagrams prepared from *D. athabasca* male courtship sounds. *Drosophila Inform. Ser.,* 50:67–68.

Petit, C., 1958. Le déterminisme génetique et psychophysiologique de la compétition sexuelle chez *Drosophila melanogaster. Bull. Biol. Paris,* 92:248–329.

Pritchard, G., 1967. Laboratory observations on the mating behaviour of the island fruit fly *Rioxa pornia* (Diptera: Tephritidae). *J. Aust. Ent. Soc.,* 6:127–32.

Prokopy, R. J., and Bush, G. L., 1973. Mating behavior of *Rhagoletis pomonella* (Diptera: Tephritidae) IV. Courtship. *Can. Ent.,* 105:873–91.

Provost, M. W., and Haeger, J. S., 1967. Mating and pupal attendance in *Deinocerites cancer* and comparisons with *Opifex fuscus. Ann. Entomol. Soc. Am.,* 60:565–74.

Richards, A. M., 1960. Observations on the New Zealand glow-worm *Arachnocampa luminosa* (Skuse) 1890. *Trans. Roy. Soc. New Zealand,* 88:559–74.

Richards, O. W., 1924. Sexual selection and allied problems in the insects. *Biol. Rev.,* 2:298–360.

Rogoff, W. M.; Beltz, A. D.; Johnsen, J. O.; and Plapp, F. W.; 1964. A sex pheromone in the housefly, *Musca domestica* L. *J. Insect Physiol.,* 10:239–46.

Römer, F., and Rosin, S., 1969. Untersuchungen über die Bedeutung der Flugtöne beim Schwärmen von *Chironomus plumosus* L. *Revue Suisse Zool.,* 76:734–40.

Roth, L. M., 1948. A study of mosquito behavior. *Amer. Midland Naturalist,* 40(2):265–352.

Schultz. G. A., and Boush, G. M., 1971. Suspected sex pheromone glands in three economically important species of *Dacus. J. Econ. Entomol.,* 64:347–49.

Shorey, H. H., 1962. Nature of the sound produced by *Drosophila melanogaster* during courtship. *Science,* 137:677–78.

Shorey, H. H., and Bartell, R. J., 1970. Role of a volatile sex pheromone in stimulating male sexual behaviour in *Drosophila melanogaster. Anim. Behav.,* 18:159–64.

Sloane, C., and Spiess, E. B., 1971. Stimulation of male courtship behavior by female 'odor' in D. *pseudoobscura. Drosophila Inform. Ser.,* 46:53.

Spieth, H. T., 1952. Mating behavior within the genus *Drosophila* (Diptera). *Bull Amer. Mus. Nat. Hist.,* 99(7):395–474.

Spieth, H. T., 1966. Courtship behavior in endemic Hawaiian *Drosophila. Univ. Texas Publ.,* 6615:245–313.

Spieth, H. T., 1968. Evolutionary implications of mating behavior of the species of *Antopocerus* (Drosophilidae) in Hawaii. *Univ. Tex. Publ.,* 6818:319–33.

Spieth, H. T., 1969. Courtship and mating behavior of the *Drosophila nasuta* subgroup of species. *Univ. Texas Publ.,* 6918:255–70.

Spieth, H. T., 1974. Courtship behavior in *Drosophila. Ann. Rev. Entomol.,* 19:385–405.

Spieth, H. T. and Hsu, T. C., 1950. The influence of light on the mating behavior of seven species of the *Drosophila melanogaster* species group. *Evolution,* 4:316–25.

Stich, H. F., 1963. An experimental analysis of the courtship pattern of *Tipula oleracea* (Diptera) *Can. J. Zool.,* 41:99–109.

Stoltzfus, W. B., and Foote, B. A., 1965. The use of froth masses in courtship in *Eutreta* (Diptera: Tephritidae). *Proc. Ent. Soc. Wash.,* 67:263–64.

Sturtevant, A. H., 1915. Experiments on sex recognition and the problem of sexual selection in *Drosophila. J. Anim. Behav.* 5:351–66.

Tauber, M. J., and Toschi, C. A., 1965a. Bionomics of *Euleia fratria* (Loew) (Diptera: Tephritidae). I. Life history and mating behaviour. *Can. J. Zool.,* 43:369–79.

Tauber, M. J., and Toschi, C. A., 1965b. Life history and mating behavior of *Tephritis stigmatica* (Coquillett) (Diptera: Tephritidae). *Pan-Pacific Entomologist,* 41:73–79.

Tobin, E. N., and Stoffolano, J. G., 1973a. The courtship of *Musca* species found in North America. 1. The house fly, *Musca domestica. Ann. Entomol. Soc. Am.,* 66(6):1249:–57.

Tobin, E. N., and Stoffolano, J. G., 1973b. The courtship of *Musca* species found in North America. 2. The face fly, *Musca autumnalis,* and a comparison. *Ann. Entomol. Soc. Am.,* 66(6):1329–34.

Vogel, G., 1957. Verhaltensphysiolgische Untersuchungen über die den Weibchenbesprung des Stubenfliegenmännchens *(Musca domestica)* auslösenden optischen Faktoren. *Z. Tierpsychol.,* 14:309–23.

Waldron, I., 1964. Courtship sound production in two sympatric sibling *Drosophila* species. *Science,* 144:191–93.

Weidmann, U., 1951. Über den systematischen Wert von Balzhandlungen bei *Drosophila. Revue Suisse Zool.,* 54:502–11.

Wright, R. G., 1974. Some aspects of the courtship behaviour of three species from the *Drosophila nasuta* sub-group. B.Sc. Thesis, University of Edinburgh.

Chapter 19

COMMUNICATION IN SOCIAL HYMENOPTERA

Bert Hölldobler

Introduction

The truly social (eusocial) Hymenoptera include all ant species and the more highly organized bees and wasps. Wilson (1971), following Michener's definition, characterizes eusocial insects as follows:

> These insects can be distinguished as a group by their common possession of three traits: individuals of the same species cooperate in caring for the young; there is a reproductive division of labor, with more or less sterile individuals working on behalf of fecund individuals; and there is an overlap of at least two generations in life stages capable of contributing to colony labor, so that offspring assist parents during some period of their life.

The complex social life within the insect society depends on the efficiency of many different forms of communication, involving a diversity of visual, mechanical, and chemical cues. The basic social activities, such as gathering food, caring for offspring, defense against enemies, establishing dominance orders, searching for new nest sites, and territorial behavior, are regulated by the precise transmission of these signals in time and space.

Sex Communication

Males and females of social insects, no less than those of solitary insects, must communicate in order to find each other. We should therefore expect the most basic patterns of communication during courtship. Unfortunately, however, we have almost no information concerning which signals regulate sexual behavior in social wasps; and only recently have some of the signals involved during the nuptial flights and mating behavior in ants been analyzed.

In the carpenter ants *(Camponotus herculeanus)* it has been demonstrated that the nuptial flights of both sexes are synchronized by a strongly smelling secretion released from the mandibular glands of the males. The males release this synchronizing pheromone during the peak of the swarming activity, at which time the females are stimulated to take off too (Hölldobler and Maschwitz, 1965). Falke (1968) found six different compounds in the secretions of the male mandibular glands of *Camponotus herculeanus,* five of which he could identify: methyl-6-methyl-salicylate, 3,4-dihydro-8-hydroxy-2-methylisocoumarin, 2,4-dihydroxy-acetophenone, cholesterine, and 7-hydroxy-phthalide. None of these compounds released the swarming behavior in females. Perhaps the synchronization pheromone is identical with the sixth, not yet chemically identified substance. Recently, Brand et al. (1973) also found methyl-6-methylsalicylate and 3,4-dihydro-8-hydroxy-2-methylisocoumarin in the mandibular glands of males of *C. herculeanus* and *C. ligniperda.*

The mechanisms by which the males and females are attracted to one another after they have left the nest, as well as those controlling copulatory behavior, have remained unknown for most ant species. Haskins and Whelden (1965) described behavioral patterns of ergatoid *Rhytidoponera metallica* that suggest that these individuals attract males by chemical means. The wingless worker-females wait outside their nest for males from other nests for mating. They typically exhibit a "calling behavior," in which the abdomen is elevated to a slanting position, the sting is slightly extruded, and the last intersegmental membranes are dorsally extended (Fig. 1). Presumably the females discharge sex pheromones by which they attract males and stimulate copulatory behavior (Hölldobler and Haskins, unpublished). Buschinger (1968) described similar behavioral patterns of virgin *Harpagoxenus sublaevis,* and Kannowski and Johnson (1969) found circumstantial evidence for the existence of a female sex pheromone in *Formica montana* and *F. pergandei.*

Recently, we succeeded for the first time in locating the morphological origin of and in bioassaying a female sex pheromone in *Xenomyrmex floridanus,* the first such discovery for the ants as a whole (Hölldobler, 1971a). During the nuptial flight the males are strongly attracted to the females. When different glandular substances were tested, it was found that poison gland secretions of the females function as sex pheromones. The males gather on sticks contaminated with poison gland secretion and even try to copulate

Fig. 1. Ergatoid female of *Rhytidoponera metallica* in calling posture.

with the stick. If a female was contaminated with poison gland secretion she was highly attractive to the males, even before she had left the nest for the nuptial flight. Buschinger (1972a) demonstrated that in *Harpagoxenus sublaevis* the sex pheromone also originates from the poison gland.

Courtship in pharaoh's ant *(Monomorium pharaonis)* is regulated by several signals. First the males are chemically attracted and stimulated by a pheromone that originates from the Dufour's gland and the bursa pouches of the females. Males that are sexually stimulated by the pheromone attempt to copulate with any object of a suitable size (in particular, females, separated gasters of females, other males, and dummies out of filter paper: Hölldobler and Wüst, 1973). For a successful copulation, however, it is necessary that the female provides additional signals, such as touching the male with her antennae or presenting her gaster to him (Petersen and Buschinger, 1971; Hölldobler and Wüst, 1973).

These examples demonstrate that it is not unusual for myrmicine ant species to produce sex pheromones in one of their sting glands. Indeed, Buschinger (1971a, 1971b, 1972a, 1972b) has adduced circumstantial evidence that *Doromyrmex pacis* and *Leptothorax kutteri* also discharge a sex pheromone from the sting and that in those two species at least the signal is not species-specific.

It is further true that the females of the harvesting ant genus *Pogonomyrmex* produce a sex pheromone in the poison gland. Again, the pheromone is not species-specific. The sexual isolation of the species is accomplished instead by particular daily flight rhythms and by highly localized mating arenas. During the nuptial flight period the males from many different nests in the environment assemble at certain places (~50 m X 80 m in size) and stay there for up to six days. Every day during a short, specific period the females arrive at these arenas. By discharging the sex pheromone they stimulate mating behavior in the males. After mating they take off again, and only when they land a second time, often hundreds of meters from the mating arena, do they begin excavating soil to found a new colony (Hölldobler, unpublished).

Mating in bumblebees is regulated by visual and chemical cues. Schremmer (1972) reports that males of *Bombus confusus* select striking signposts on which they rest and from which they fly after any object that roughly resembles a female bumblebee. The mating behavior in other bumblebee species, such as *Bombus terrestris, B. pratorum,* and *B. lucorum,* is still more elaborate. Individual males of these species establish chemically marked flight routes by depositing spots of odorous secretions at intervals along the route (Frank, 1941; Haas, 1946). The height and location of these flight paths differ from species to species (Haas, 1949a, 1949b; Bringer, 1973); in addition the scents seem to be species-specific. According to Haas (1949a, 1949b, 1952) and Kullenberg (1956), males as well as females are attracted by these marking pheromones. When virgin females venture close enough, males recognize them by a specific female pheromone. This queen odor is also important for inducing copulatory behavior in the males (Free, 1971).

There is some uncertainty concerning the anatomical source of the marking pheromones. Haas (1952) found that the secretions originate from the mandibular glands, and this has been generally accepted by other investigators (Kullenberg, 1956; Stein, 1963; Bergström et al., 1968). Later, however, Kullenberg et al. (1973) reported that movie analyses of the marking behavior of the male bumblebees and refined dissecting methods have revealed that the pheromone is not produced in the mandibular glands but rather in the cephalic portion of the labial glands. If this is correct we probably have to assume that the major component of the marking secretions of *B. terrestris,* identified by Bergström et al. (1967) as 2,3-dihydrofarnesol, was not extracted, as reported, from the mandibular glands but from the labial glands. Calam (1969) has identified the main components in extracts of the heads of males of five other *Bombus* species and found them all different. Similar results were obtained by Kullenberg et al. (1970), who found that the marking pheromones differ among thirteen species of *Bombus* and six species of *Psithyrus.* Even within the species *B. lucorum* Bergström et al. (1973) discovered two forms in which the males clearly differ in their main cephalic volatile compounds. The "dark" form contains ethyl dodecanoate whereas the "blonde" form contains ethyl tetradec-cis-9-enoate as their main components. It is likely that these two varieties are in fact distinct sibling species. However, how far these chemical differences are effective as prezygotic isolating mechanisms remains to be tested by behavioral experiments.

In the honeybee *Apis mellifera* chemical communication plays a major role in regulating reproductive behavior. During the mating period drones usually assemble in large numbers in "congregation places." Every year the same localities are visited for this purpose (Ruttner and Ruttner, 1965, 1968, 1972). The specific cues by which these assembling areas are detected by the drones is still a mystery. No evidence exists that pheromones are involved, although Gerig (1972) recently reported that extracts from the heads of males attract flying males once they have arrived at the congregation places. When a virgin female appears, she is immediately pursued by a "swarm" of males. Multiple mating is commonplace in the honeybees.

The behavioral physiology of chemical communication between queens and drones was first studied by Gary (1962, 1963). Noting that drones do not respond to queen pheromones inside the hive, but do respond to virgin queens during the mating flights, Gary suspended virgin queens approximately 10–20 m high on helium-filled balloons and stationary towers. Such exposed queens were highly attractive to drones, which approached and even mated with them. A bioassay was then developed to determine the chemical cues of the attractiveness of the queens to drones. It turned out that the attraction pheromone originates from the mandibular glands of the queen. The primary active compound was identified as 9-oxo-2-decenoic acid, although the total mixture of the mandibular gland secretions was more attractive than pure 9-oxo-2-decenoic acid. These results have been confirmed by Pain and Ruttner (1963), who utilized Gary's bioassay with slight modifications. Butler and Fairey (1964) identified 9-hydroxydec-2-enoic acid as a second attractive compound in the mandibular gland secretions of honeybee queens, although Blum et al. (1971b) found that this substance did not release attraction in honeybee drones.

From these results it can be concluded that the mandibular gland substance of queens contains the sex pheromone and that its most effective component is 9-oxo-2-decenoic acid. Circumstantial evidence, however, indicates that there may be additional chemical signals involved during mating behavior in honeybees. Morse et al. (1962) extirpated the mandibular glands from virgin queens, but still a small group of these treated queens mated successfully. Butler (1971) suggests that at least in close range, additional pheromones from the abdominal tergites or the Koschnevnikov's gland in the sting chamber stimulate copulation behavior. It may well be that the main function of the mandibular gland pheromones is a long-distance attractant, while in close range additional signals become important. The sex pheromone of *Apis* is not species-specific. Receptor physiological investigations as well as behavioral observations have revealed that drones of *Apis mellifera* are attracted not only by queens of their own species but also by the mandibular gland secretions of *Apis cerana* and *Apis florea* (Butler et al., 1967; Ruttner and Kaissling, 1968). For further information the reader is referred to a recent review by Gary (1974).

Worker-Queen Communication

The division of labor in reproductive and nonreproductive castes is regulated by a variety of communicative signals. A honeybee society is constantly informed of the presence of their queen by chemical cues. Queen pheromones were found to originate from the queen's mandibular glands and were named by Butler (1954a) "queen substance." Its major component was identified independently by Callow and Johnston (1960) and by Barbier and Lederer (1960) as 9-oxodec-trans-2-enoic acid. As reported above, this substance functions outside the hive as the queen's sex pheromone. Although inside the hive the males are not responsive to the queen substance, this pheromone strongly affects the physiology and behavior of worker bees. More than a dozen other compounds of the mandibular gland secretions have been chemically identified, most of which remain unknown in function.

While stationary or slowly moving, the nest queen of a honeybee colony is usually surrounded by approximately eight to ten worker bees, the so-called court. Butler (1973) describes the forming of a court as follows:

> What seems to happen is that some of the household bees moving around the brood nest—the only place in an undisturbed colony where queen rearing occurs—happen to meet the queen, or to get within a few millimeters of her. The bees that do so, react in one of three ways: They ignore her, or they appear to be actually repelled by her and move rapidly away, or they join her "court" and stay with her for a period varying from a few seconds to half-an-hour or even more. Some of the bees that join a queen's "court" seem strongly stimulated by her and immediately begin examining her body with their antennae and often lick it too. If the queen moves and they lose contact with her, they often examine with their antennae the comb where she has been, apparently seeking some substance with which she contaminated it. Those that find her again usually examine and, perhaps, lick her. It seems probable that such bees are actively seeking queen substances.

In an attempt to analyze the communicative mechanism that leads to the court formation, Gary (1961a, 1961b) confined a queen in a cage with one wall made out of wire gauze. Many workers gathered at this wall. When a similar test was performed with a queen whose mandibular gland had been removed, only a few workers assembled at the cage. In contrast, however, Velthuis (1970a) reported that uncaged queens continued to attract workers even if their mandibular glands had been extirpated. Butler et al. (1973) conducted an additional series of experiments and confirmed that the scent of a mated laying queen as well as that of synthetic 9-oxo-2-decenoic acid cause an accumulation of workers on the cage, and that fewer workers gathered on the cage if the queen's mandibular gland had been removed. However, when the queen remained uncaged so that the workers could touch her, workers continued to assemble around a stationary queen, even when the mandibular gland had been extirpated. Since it could be shown that the heads of those queens still contained traces of 9-oxodecenoic acid, it is assumed that this major component of the queen substance releases the court formation. Yet it cannot be completely ruled out that additional unidentified substances, produced in other parts of the queen, such as the abdomen, may also be involved (Velthuis, 1970a). In any case it seems to be clear that in an undisturbed colony workers are attracted to their queen only over very short distances (Butler et al., 1973).

The effect of the queen pheromones can be tested by removing the queen from the colony. Shortly afterward the workers move around excitedly while showing increased fanning behavior. If the queen is not replaced within forty-eight hours, the workers begin construction of queen cells, in which new queens can be produced. This sequence suggests that the presence of a mated laying queen inhibits the rearing of new queens. And, indeed, Butler and Gibbons (1958) demonstrated that queen rearing can be inhibited even in the absence of a queen, merely by exposing the colony to queen substance extracted from the mandibular glands of mated, laying queens. Again there is some circumstantial evidence that additional pheromones, originating from the queen's abdomen, may contribute to this inhibitory process (Velthuis, 1970b).

The queen not only inhibits the production of new queens but also suppresses the ovary development of worker bees. When kept without a queen some workers undergo ovarian development. De Groot and Voogd (1954) and Voogd (1955) demonstrated that this growth can be prevented by exposing them to the queen substance. Queen rearing by workers and the development of their ovaries can be inhibited without the workers themselves having direct contact with a queen, provided these individuals have access to other workers that have recently been with a queen (Butler, 1954a; Pain, 1961). The precise mechanisms of transmission of the inhibitory signals is still very little understood. According to recent results of Velthuis (1972) it seems likely that a worker in contact with a queen becomes contaminated with traces of queen substance and probably transfers these pheromones when it contacts other worker bees. It is now suggested that the inhibitory effect is transmitted via the sensory channel and not via the alimentary channel, as previously assumed by Butler (1954a). This assumption is supported by the fact that some chemoreceptors on the antennae have been found to respond specifically to 9-oxo-2-decenoic acid (Beetsma and Schoonhoven, 1966; Kaisling and Renner, 1968).

By a variety of experiments it has been demonstrated that free movement of the queen over the brood combs is necessary to ensure an effective distribution of the inhibitory substances. This indicates that the queen herself actively takes part in distributing her queen substance and thereby suppresses the fertility of her daughters and the rearing of young queens.

Older queens tend to fail to inhibit queen rearing because their colonies have become too large or because the mobility of the old queen and the production of queen substance have decreased. This, in turn, leads either to a supersedure of the old queen by a young queen or the preparation by the colony for reproductive swarming, during which the old queen leaves with a group of workers to start a new colony. During swarming the major component of the queen substance, 9-oxo-decenoic acid, again plays an important role: it is the main signal by which the workers are kept close to their queen, and it also releases the clustering behavior around the queen after she has settled. According to Butler and Simpson (1967) a second component of the queen's mandibular gland secretions, the less-volatile 9-hydroxydecenoic acid, may function as an additional signal for tight clustering.

We have seen that 9-oxodecenoic acid serves many purposes: outside the hive it functions as a sex pheromone and as a powerful attractant during swarming behavior; inside the nest it is the most important signal for the social regulation of reproductive behavior in honeybees.

Honeybee societies are strictly monogynous. As just pointed out, the old queen leaves the nest with a swarm of workers before the young queens emerge. Since two or more queens do not tolerate one another it seems reasonable that eclosing and freshly hatched queens signal their presence. Indeed, Hansson (1945) found that young queens continuously exchange "quacking" and "piping" sounds. When he played the recorded sound back, he got a piping answer from a hatched queen and a quacking answer from queens that were still in their cells but close to eclosion. It appears plausible that this sound communication prevents premature emergence of a young queen before the older queen has left the hive. Hansson's experiments, however, have shown that the sound signals alone do not completely suppress the hatching of a young queen. Simpson and Cherryl (1969) report that piping sounds are also produced during swarming behavior. It was even possible to initiate swarming in honeybees by playing the piping sounds back to a colony. Of course, after the swarm has left the nest with the older queen, the absence of these piping sounds would indicate that the way is free for the eclosion of another queen.

Whereas it is apparent that only the substrate-borne vibrations of the sounds are perceived by the bees, there remains some confusion about the physical properties of the sounds. According to Hansson the frequency of the piping sounds averages 435–493 cycles per sec and the quacking sounds about 323 cycles per sec. Wenner (1962b) apparently was not aware of Hansson's work when he published pretty much the same biological findings. Contrary to Hansson, however, he characterized the piping sound with 1,300 cycles per sec and the quacking sound with 2,500 cycles per sec. Wenner also reported that only the piping sound, and not the quacking sound, released an answer in the young queens.

Besides these chemical and acoustical signals involved in communication between the castes, there exists at least one form of indirect communication between the workers and their queen. Workers are able to determine the sex of their queen's offspring by the size of the cells that they build. The queen lays unfertilized eggs in large hexagonal cells, out of which males develop; but the fertilized eggs, which she lays in small hexagonal cells, develop into workers. Koeniger (1970a, 1970b) was able to demonstrate experimentally that the queen measures the width of the cell with her front legs before she lays an egg. Thus the workers communicate to their queen via the cell size what kind of egg she should lay.

The mechanisms of communication between the castes has not been analyzed as well for other species of social Hymenoptera as it has been for honeybees. Some findings indicate that bumblebee queens produce a pheromone comparable to the honeybee's queen substance (Röseler, 1967, 1970). For more detailed information the reader is referred to Michener (1974). In ants only circumstantial evidence of the presence of special queen pheromones has been adduced (Stumper, 1956; Bier, 1958; Lange, 1958; Hölldobler, 1962; Watkins and Cole, 1966; Brian and Blum, 1969; Brian, 1970).

In some social wasps, such as *Vespa crabro* or *Vespa orientalis,* workers tend to form a court around their queen resembling that of honeybees. If the wasp queen is removed, a conspicuous unrest breaks out in the workers, but they immediately calm down again after the queen has been returned. Also, the development of the workers' ovaries is inhibited when a queen is present. All these observations strongly suggest that the wasp queens produce a queen substance. Indeed, Ikan et al. (1969) were able to isolate a substance from head extracts that showed in bioassays the effects of a queen substance. They identified it as δ-n-hexadecalactone.

In the more primitive social wasps, such as *Polistes,* dominance orders are established by certain overt behavioral patterns instead of by chemical signals. When the colony is founded by several females, only one of them becomes the queen, and the others become workers. As Pardi (1940, 1948) and Pardi and Cavalcanti (1951) first demonstrated, the egg-laying queen dominates the other females through her larger size and more aggressive behavior, which is mostly expressed through ritualized aggressive posturing. The dominant individual stands somewhat higher than the subordinate one, while the latter crouches and lowers its antennae. Pardi (1948) found that there is a correlation between the development of the ovaries and the position in the dominance order: females with the largest ovaries were also the most dominant ones. Similar dominance orders have been found in related genera such as *Mischocyttarus* (Jeanne, 1972) (Fig. 2). For more information on the establishment of dominance hierarchies and the evolution of queen control the reader is referred to Wilson (1971: 299–305), Evans and Eberhard (1970), and Spradbery (1973).

Alarm Communication

When a rapid exchange of information is crucial for the survival of a society, specialized, behaviorally very active signals are needed. The social insects are rich in such systems.

Like many solitary insects, social insects also

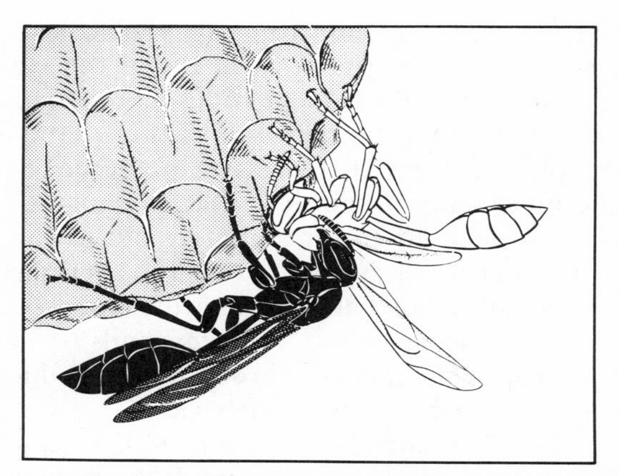

Fig. 2. Domination behavior in *Mischocyttarus drewseni.* The dominating wasp (black) is violently mouthing the thorax of the subordinate with her mandibles. The subordinate is responding with an extremely submissive posture: head down against the nest surface, abdomen raised, and wings spread. (From R. L. Jeanne, 1972.)

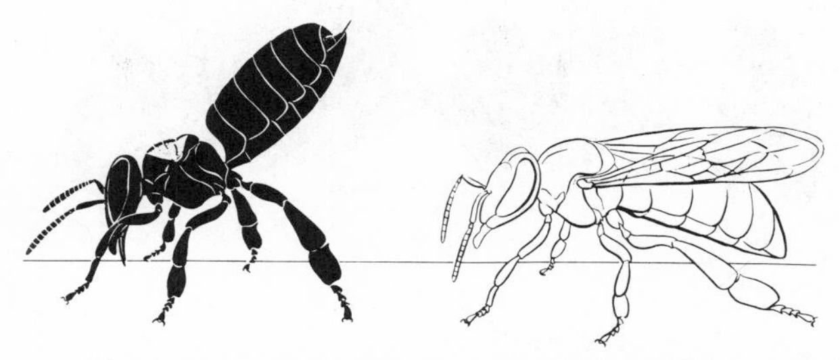

Fig. 3. The alarm-defense behavior (black) is contrasted with the normal posture (white). Top: *Formica polyctena;* middle: *Crematogaster ashmeadi;* bottom: *Apis mellifera.* (From Hölldobler, 1970b.)

use chemicals to repel predators. In social insects, however, defensive reactions are closely connected with alarm communication, and quite often both substances serve both functions. In many cases the discharge of alarm pheromones and defensive substances is accompanied by characteristic body movements and postures (Fig. 3). The species of *Formica* spray mixtures of formic acid and Dufour's gland secretions, both serving simultaneously as defensive substances and alarm pheromones (Maschwitz, 1964). During the emission the ants bend their gasters forward beneath their legs. Species of the myrmicine genus *Crematogaster* lift their abdomens to a characteristic vertical position or even forward over the head while releasing the defensive secretion through the sting and alarm pheromones from the mandibular glands (Blum et al., 1969). The same defensive behavior has been observed in Dolichoderinae (Goetsch, 1953), *Solenopsis fugax* and *Monomorium pharaonis* (Hölldobler, 1973b), and many other myrmicine species.

In addition to some early reports by Goetsch, Sudd's (1957b) observations on the pharaoh ant *(Monomorium pharaonis)* were among the first on chemical alarm communication. Workers of this species react with escape behavior when a nest mate is crushed nearby. Wilson (1958) and Butenandt, Linzen, and Lindauer (1959) carried out the first experimental investigations on alarm pheromones in ants. Butenandt et al. worked with the leaf cutter ant *Atta sexdens* while Wilson studied the harvester ant *Pogonomyrmex badius.* In both species workers discharge a strong-smelling substance from the mandibular glands (the morphological location of various pheromone glands is illustrated in Fig. 4) if they perceive some kind of threatening stimulus. McGurk et al. (1966) identified this alarm pheromone of *P. badius* as 4-methyl-3-heptanone.

Wilson and Bossert (1963) were able to study precisely the behavioral and physiological parameters of chemical alarm communication. By directly measuring the effects of the pheromone from whole crushed glands they found that workers respond to the threshold concentration averaging 10^{10} molecules/cc by moving toward the odor source. The total capacity of the gland reservoir is about 10^{15}–10^{16} molecules. As a consequence the entire content of the mandibular gland substance provides a brief signal. According to the experimental data acquired by Wilson and Bossert, the alarm pheromone of one ant expands in still air to its maximum radius of about 6 cm in 13 sec and fades out in about 35 sec. The lower concentration at the periphery releases attraction behavior; only the inner space of higher concentration, which expands to a radius of 3 cm and fades out in about 8 sec, induces real alarm and aggressive behavior.

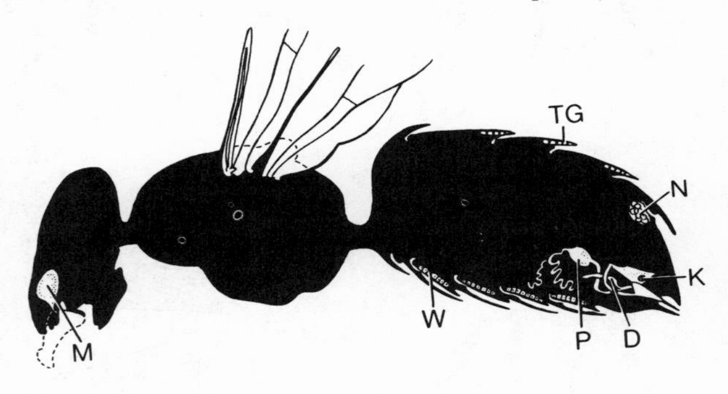

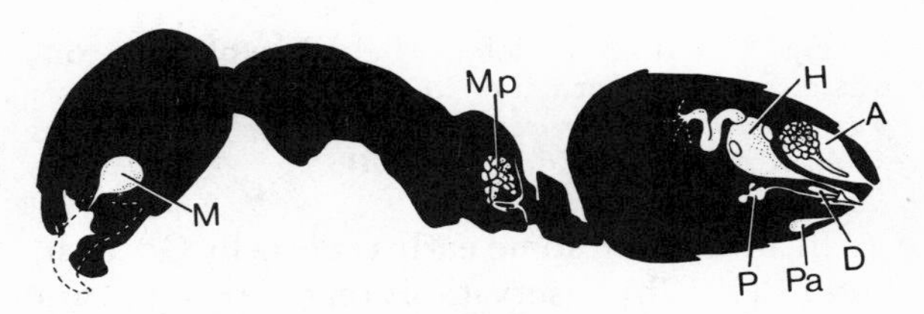

Fig. 4. Pheromone gland system of the honeybee *Apis mellifera* and the dolichoderine ant *Iridomyrmex humilis.* A = anal gland, D = Dufour's gland, H = hindgut, K = Koschnevnikov's gland, M = mandibular gland, Mp = metapleural gland, N = Nassanoff's gland, P = poison gland, Pa = Pavan's gland, TG = tergal glands (presumably scent glands), W = wax glands. (Based on Ribbands, 1953; Pavan and Ronchetti, 1955; Renner and Baumann, 1964; Wilson, 1971.)

These parameters seem very well designed for an economical alarm system. If the danger is local and only short-lived, the signal fades out quickly and only a small group of workers in the immediate vicinity are alerted. If, however, the danger is more persistent, the number of workers discharging the signal increases rapidly and the signal "travels" through the colony.

The alarm communication system of *Acanthomyops claviger* (Fig. 5) is another well-analyzed example. Regnier and Wilson (1968) found that undecane from the Dufour's gland and a number of terpenes produced in the mandibular glands release alarm response at concentrations of 10^{9}–10^{12} molecules/cc. The quantity of these substances altogether in one ant totals about 8 μg. Behavioral experiments have shown that the chemical alarm signal generated by all volatile substances of a single worker releases a response in nest mates up to a distance of about 10 cm. This defensive strategy is well adjusted to the structure of the large *Acanthomyops* colonies, which live widely expanded in subterranean nests. Also in this species, as in *P. badius,* the signal fades out rather quickly unless reinforced by other alarming ants.

Undecane, one of the alarm substances identified by Regnier and Wilson in the Dufour's gland of *Acanthomyops* has also been found in a number of other formicine species (Bernardi et al., 1967; Regnier and Wilson, 1969; Bergström

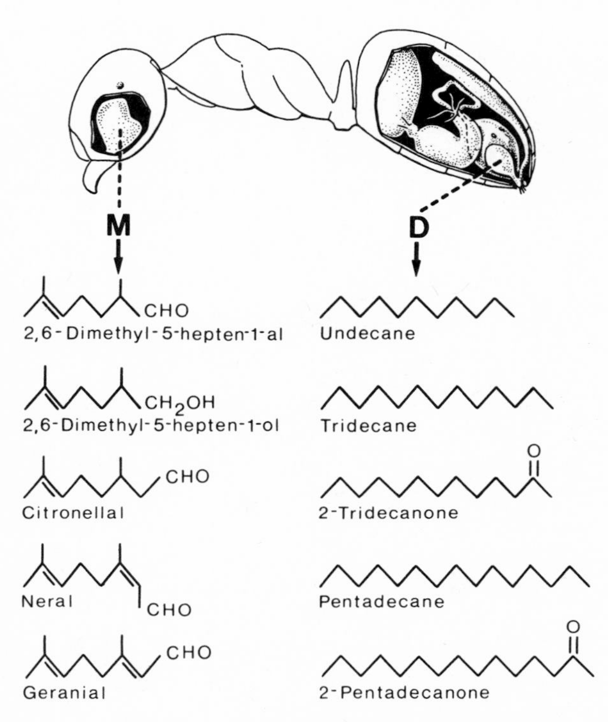

Fig. 5. Substances found in the mandibular gland and Dufour's gland of the ant *Acanthomyops claviger*. Undecane and the mandibular gland substances function both as defensive substances and as alarm substances. D = Dufour's gland, M = mandibular gland. (After Regnier and Wilson, 1968.)

and Löfquist, 1968, 1970, 1972, 1973). This indicates that alarm pheromones are not very species-specific. Indeed, using extracellular single-cell recordings, Dumpert (1972) found that the cells of the sensilla trichodea curvata on the antennae of *Lasius fuliginosus* react to twelve alarm substances produced by species of three different ant subfamilies. But Dumpert also found that some single cells of the sensilla trichodea curvata react most specifically to undecane, the alarm pheromone of *L. fuliginosus.* These results demonstrate that the relative specificity of alarm pheromones among different genera and subfamilies, revealed by behavioral experiments (Maschwitz, 1964), is probably achieved at the level of the central nervous system and not at the receptor level. It is therefore premature to speculate about the specificity of certain alarm pheromone receptors merely on the basis of behavioral specificity tests (Amoore et al., 1969; Blum et al., 1971). However, it should also be stressed that electrophysiological investigations alone are equally insufficient in proving the behavioral specificity of certain signals.

Nevertheless the efficiency of an alarm pheromone seems to depend on certain structural characteristics. Blum et al. (1966) tested a series of forty-nine ketones on *Iridomyrmex pruinosus* to find the relationship between chemical structure and alarm-inducing power. The natural alarm pheromone is 2-heptanone. By increasing the number of carbon atoms from three to thirteen a very low activity was elicited by the first (C_3–C_4) and the last (C_{11}–C_{13}) of the 2-alkanone series. An optimal reaction occurred between C_6 to C_9. Other structural variations, such as a displacement of the carbonyl group, the introduction of a second ketone group, or the presence of side-chain methyl groups, usually lowered the response-eliciting efficiency of the substance. Similar results were obtained by Regnier and Wilson (1968) for *Acanthomyops claviger.* They found that alkanes falling between C_{10} and C_{13} usually elicited good responses from the workers and showed excellent properties as alarm substances. As mentioned above, the main component of the natural alarm substances is undecane, a C_{11}-alkane.

These findings lead to the assumption that in most cases the size of a molecule is more important than a specific structure. Bossert and Wilson (1963) predicted that most alarm substances in social insects would have between five and ten carbon atoms and a molecular weight between 100 and 200. They speculated that this would be the ideal size of a molecule to meet the special requirements for an efficient chemical alarm communication. In fact, most of the alarm pheromones identified so far fall into these categories.

However, there are a few exceptions to this rule of a relative structural nonspecificity. Riley, Silverstein, and Moser (1974a, 1974b) found that workers of *Atta texana* and *A. cephalotes* produce only the (+) isomer of the alarm pheromone 4-methyl-3-heptanone. In behavioral tests it was apparently demonstrated that workers of *A. texana* distinguish the (+) isomer of this ketone from the (–) isomer.

Many alarm pheromones have been chemically identified (see reviews by Wilson, 1971; Gabba and Pavan, 1970; Pain, 1973; Blum, 1974). Most of them are ketones, aldehydes, acids, or hydrocarbons. They are produced in a variety of exocrine glands (Fig. 6). In summarizing the behavioral results we can say that most alarm pheromones in ants are not very specific. This is not surprising because there is little if any selective pressure to develop species-specificity of alarm communication. In fact, in many cases it seems even advantageous to be able to understand the alarm signals of a neighboring colony of another species. However, Regnier and Wilson (1971) demonstrated that this advantage can turn to a disadvantage under some circumstances. It is well known that certain ant species conduct "slave raids" on other ants. The raiders bring the pupae of the raided ant colonies into their own nest, and when the young workers eclose to adults, they function in the raiders' nest as brood tenders, nest builders, and foragers. The raider workers continue to conduct mainly slave raids. Often the raiders are obviously superior in fighting ability (Fig. 7). *Polyergus,* for example, has specially adapted saber-shaped mandibles. The slave-raiding species *Formica pergandei* and *F. subintegra* do not carry such armament but instead possess remarkably enlarged Dufour's glands.

Regnier and Wilson identified decyl acetate,

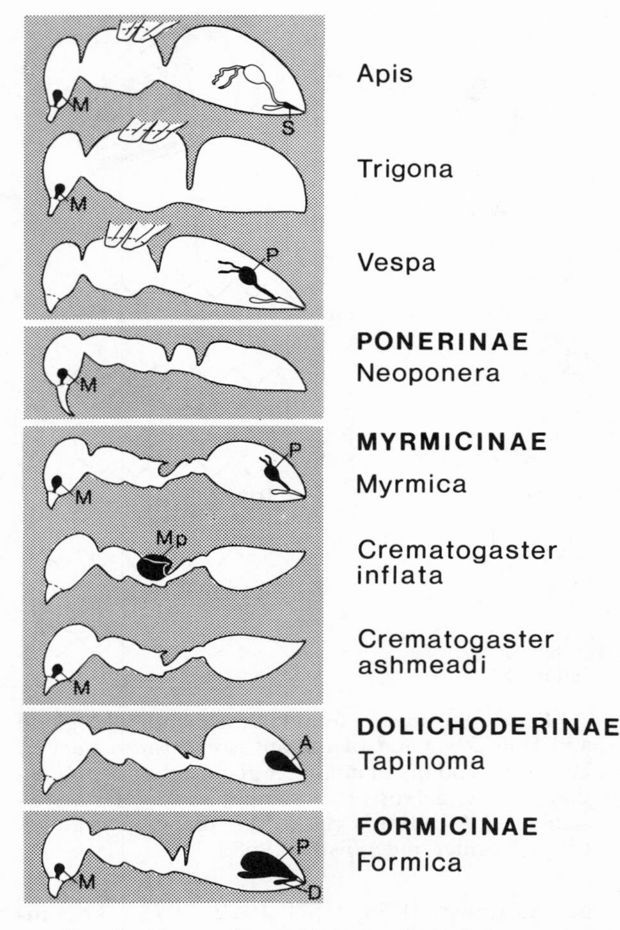

Fig. 6. Alarm pheromone glands in bees, wasps, and ants. A = anal gland, D = Dufour's gland, M = mandibular gland, Mp = metapleural gland, P = poison gland, S = sting chamber. (Based on Maschwitz, 1964, 1974; Duffield and Blum, 1973; Blum, 1966a.)

dodecyl acetate, and tetradecyl acetate as principal components of the glandular substances. One *F. subintegra* worker contains the relatively enormous amount of 700 μg of these substances. During the slave raids the raider ants discharge these substances upon encountering prey workers and apparently stimulate nest mates to join them in the fighting. In addition

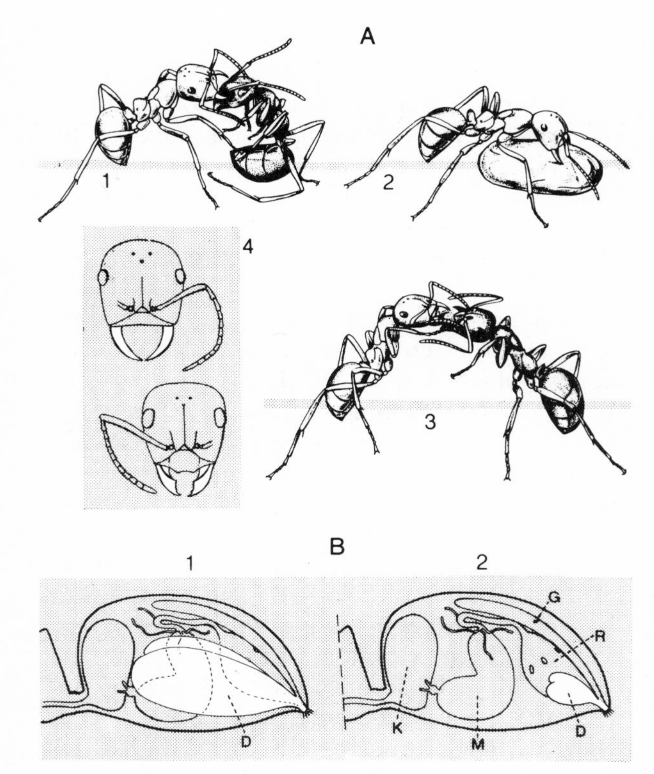

Fig. 7. A: 1. A worker of the slave raider ant *Polyergus rufescens* (left) attacks a slave ant *Formica fusca* (right). 2. The slave raider carries a pupa of *F. fusca* homeward. 3. A *Polyergus* worker is fed by a *F. fusca* slave ant, which has eclosed from a captured pupa. 4. The saber-shaped mandibles of *Polyergus* are contrasted with the "normal" mandibles of the slave ant species, *Formica fusca.* B: 1. Gaster of the slave raider ant *Formica subintegra,* showing the enormously developed Dufour's gland (D). 2. Gaster of the slave ant *Formica subsericea* with normal Dufour's gland. G = poison gland, R = hindgut, K = crop, M = midgut. (From Hölldobler, 1973a; based in part on Regnier and Wilson, 1971.)

they spray large amounts of the acetates on defending slave ants. It is interesting that these substances not only alarm and stimulate the raider species but also highly excite the slave ant species. The high concentration of the discharged acetate mixture, however, completely "confuses" the slave ants. They become disoriented, making it easy for the raiders to penetrate the slave ants' nest and remove the pupae. This grotesque exaggeration of a communication signal, resulting in misleading the society, is a fantastic analogy to the human propaganda technique. For this reason Regnier and Wilson called these substances "propaganda pheromones."

In addition to these pheromones, other modes of alarm communication have been discovered. Markl (1965, 1967, 1968, 1970) found that in leaf cutting ants (*Atta cephalotes* and *Acromyrmex octospinosa*) workers stridulate whenever they are prevented from moving freely, for instance, when they fight with workers of a neighboring colony or when they are trapped under sand after a cave-in of their nest. Nest mates are attracted by these stridulatory sounds from as far away as 8 cm. When the sound is emitted by a buried ant, the attracted workers begin to dig where the sound is loudest, and in a few seconds the trapped ant is rescued. The sounds are produced by special stridulatory organs. The posterior rim of the postpetiolar tergite acts as a scraper, while a field of parallel ridges at the anterior end of the first tergite of the gaster functions as the file. Markl was able to demonstrate that the ants respond only to vibrations conducted through the soil. Whereas the airborne stridulatory sounds extend far into the ultrasonic with a maximum between 20 and 60 kHz, the intensity spectrum of the soil-conducted vibrations does not contain frequencies above 6 to 8 kHz, with the intensity maximum concentrated around 3 kHz. The ants perceive the vibrations with receptors in the legs. Markl found that receptors of the forelegs are four to five times as sensitive as those of the middle and hind legs. It is also interesting to note that small workers are considerably more sensitive than the big soldier ants to the substrate-borne vibrations.

Since many ant species possess stridulatory organs, it can be expected that this kind of communication is more common than was previously assumed (Markl, 1973). But there are other kinds of vibrational warning and alarm signals in ants that are only little studied. In *Camponotus herculeanus,* for example, we observed that the sexual castes, especially the males that tend to leave the nest too early for the nuptial flight, are summoned back into the nest by workers by means of rapid oscillatory jerking movements (Hölldobler, 1965; Hölldobler and Maschwitz, 1965). A similar "warning" behavior has been observed in other species, such as *Lasius niger* and *L. alienus.*

In many arboreal ant species *(Camponotus, Polyrhachis, Hypoclinea, Dolichoderus)* vibrational jerking movements of workers can be readily observed when the nest is disturbed. Markl and Fuchs (1972) analyzed the signals produced by some of these movements in *Camponotus herculeanus* and *C. ligniperda.* They found that the ants actually rap on the substrate by hitting the ground alternately with the mandibles and the gaster. The hits follow in series of two to three, sometimes up to seven, with intervals of about 50 msec. These signals, which propagate in solid wood, have an intensity spectrum reaching from 100 Hz to 10 kHz and an energy maximum at 4–5 kHz. Circumstantial evidence indicates that one of the major biological functions of the sounds is to amplify or to modify the effect of other attack-releasing stimuli and alarm signals.

Finally, in the primitive Australian ponerine ant *Amblyopone australis,* we observed a remarkable vibrational alarm communication behavior that is apparently entirely transmitted by tactile contacts between nest mates; in this system substrate-borne vibrations could not be recorded (Hölldobler et al., unpublished).

In spite of these recent discoveries on mechanical alarm communication, it is still fair to say that chemical communication plays the major role in alarming and alerting behavior in social Hymenoptera. Maschwitz (1964, 1966a) demonstrated chemical alarm communication in twenty-three European species of social Hymenoptera, nearly all of the sample examined. To date, no alarm pheromones have been found in bumblebees or in wasps of the genus *Polistes.* As illustrated in Fig. 6, a variety of glands are involved in the secretion of these chemical alarm signals.

As in ants, honeybees display a typical behavioral pattern when discharging the alarm pheromones, which Maschwitz (1964) called "Giftsterzeln." The alarming bee brings its abdomen into a slanting position, opens its sting chamber, from which it releases the pheromone, and dispenses it into the air by rapid fanning with its wings (Fig. 3). The alarming effect can be impressively demonstrated by presenting a control odor followed by the crushed sting apparatus of a honeybee worker in front of the hive entrance. Whereas there is only a weak reaction to the control odor, the scent of the crushed sting immediately attracts workers from nearby, many of which assume alarm postures. In the next few minutes more than a hundred workers sometimes rush out of the hive. As Maschwitz (1964) demonstrated, the pheromone alone does not release aggression; additional cues characterizing an emeny are necessary to focus the defensive attack.

Although these experiments clearly indicate that the glandular source of the alarm pheromones is associated with the sting apparatus, the precise origin of the pheromones is not yet known. The main chemical component is isoamyl acetate (Boch et al., 1962). The first investigations were carried out with *Apis mellifera;* since then Morse et al. (1967) found the same alarm pheromone in *A. florea, A. cerana,* and *A. dorsata.* Maschwitz (1964) demonstrated that honeybee workers produce a second alarm pheromone in their mandibular glands. This substance has been identified as 2-heptanone (Shearer and

Boch, 1965). Boch et al. (1970) compared the efficiency of both alarm pheromones and found that 20–70 times more 2-heptanone is necessary to elicit an alarm effect comparable to that induced by isopentyl acetate. It is interesting to note that 2-heptanone also releases alarm responses in other *Apis* species, even though it is found only in *Apis mellifera* (Morse et al., 1967). This is another example of the general lack of species-level specificity in alarm pheromones.

It has been known for some time that the honeybee *Apis cerana* shows a social defense behavior that is accompanied by a peculiar hissing sound (Butler, 1954b; Sakagami, 1960). Only recently, however, Koeniger and Fuchs (1972, 1973) analyzed this behavior experimentally. They found that the short hissing sound (700 Hz) is emitted whenever the hive is mechanically stimulated, for instance, by shaking or knocking against the hive. The hissing is transmitted from one bee to the other with a transmission speed of 25 cm/sec, and it remarkably reduces aggressive behavior in the bees. Using Asiatic bears, the authors were able to demonstrate that the hissing sound of a bee colony functions as an effective acoustic repellent against large predators, and it is speculated that the bees might mimic the defensive hissing sounds of snakes, which are very common in the habitat of *Apis cerana.*

Multiple Functions of Alarm Signals

As just noted, alarm communication is closely meshed with defensive behavior. Not only are the behavioral patterns frequently identical but often the same substances function as both defensive secretions and alarm messengers. Maschwitz (1964) was able to show that in *Formica* species formic acid is used as a powerful defensive secretion, but its smell also effectively alarms nest mates, although this is not the case in *Acanthomyops claviger* (Regnier and Wilson, 1968). The myrmicine ants *Crematogaster* produce 2-hexenal in their mandibular glands (Bevan et al., 1961; Blum et al., 1969). This substance is the major component of the alarm pheromone but it also has a remarkable defensive power. As can be seen from Fig. 6, glands that produce alarm pheromones are associated with mandibles or the sting apparatus, morphological features that play a major role in aggressive and defensive behavior.

There seem to be a few exceptions to this rule: Sudd (1962) claims that the strong odor of the African stink ant *(Paltothyreus tarsatus)* originates from the metapleural gland and that this secretion functions as an alarm pheromone. The evidence, however, is not convincing, especially since Casnati et al. (1967) identified dimethyldisulfide and dimethyltrisulfide in the mandibular gland secretions of *P. tarsatus.* These secretions apparently serve as defensive substances and also as chemical alarm signals (Crewe and Fletcher, 1974). In *Crematogaster inflata* the metapleural gland is remarkably enlarged. Maschwitz (1974) found that the workers use the sticky substance as a defensive secretion and that in addition the fluid releases alarm behavior. In most of the other ant species, however, workers produce acidic secretions in the metapleural glands. For instance, the main component of the secretions of the leaf cutter ant *Atta sexdens* is phenylacetic acid. Since these secretions effectively suppress bacterial growth, it is believed that their main function is to suppress microorganisms in the interior of the nest (Maschwitz et al., 1970; Maschwitz, 1974).

The response behavior to the alarm signal varies in different groups and castes of the society, and it varies in time and space. For example, if the signal is discharged close to the nest, it releases aggressive behavior; but at a greater distance from the nest it elicits escape behavior (Maschwitz, 1964). Furthermore, young workers usually retreat into the nest when they smell the alarm signal, while older workers, especially those belonging to the soldier castes, move out and display aggressive behavior. Wilson (1958) showed that the alarm pheromone of the harvester ant *Pogonomyrmex badius* releases a variety of reactions. At low concentration it merely attracts nest mates, and in high concentration it releases aggressive behavior. If the high concentration persists the attracted workers start to dig where the concentration is highest. It was demonstrated that this signal elicits rescue behavior in *Pogonomyrmex,* for instance, if workers are buried under sand after a cave-in of their nest.

Circumstantial evidence shows that *Camponotus socius* uses low concentrations of formic acid, its major alarm-defensive secretion, to fortify its recruitment signals (Hölldobler, 1971c). Similarly the poison gland secretion of *Pogonomyrmex* is not only a defensive secretion but also a strong attractant. When an enemy is stung it is simultaneously marked with this attractant; thus more workers aim their attacks toward it. In other circumstances, however, the same substance functions as a very effective recruitment signal by which nest mates are attracted and guided to newly discovered food sources (Hölldobler and Wilson, 1971).

In stingless bees (Meliponinae), no less than in ants, alarm pheromones often have a double function. All alarm pheromones in this group appear to originate from the mandibular glands (Blum et al., 1970), and, as will be shown in the next section, often function also as trail pheromones.

Multiple functions of alarm pheromones are also known in honeybees. Morse (1972) has shown that the alarm pheromone (isopentyl acetate) released near queens in a swarm, causes a remarkable decrease of the discharge of the attractive Nassanoff gland pheromone. It has been speculated that bees use this inhibition as a mechanism to reject a foreign queen from a swarm. Simpson (1966) and Butler (1966) report that the mandibular gland secretion (2-heptanone) of honeybees, which can also serve as an alarm pheromone, has a strong repellent effect on foraging bees. Indeed, Nunez (1967) gave experimental evidence that honeybee foragers mark exhausted food sources with a repellent signal. It is most likely that this signal is identical with 2-heptanone.

Recruitment Communication

Although advanced social insects must alarm nest mates when danger threatens, it is of equal importance for them to transmit information about newly discovered food sources or better nesting sites. The rapid retrieval of food and the fast emigration to a better nest require an effective communication system.

The recruitment techniques employed by different groups of ant species vary considerably. The best-studied recruitment behavior is the chemical trail communication. Carthy (1950, 1951) was one of the first to conduct an experimental study on trail laying in *Lasius fuliginosus.* He found strong circumstantial evidence that in this species the trail pheromone originates from the hindgut. This suggestion was later confirmed by Hangartner and Bernstein (1964). Wilson (1959a), working with the fire ant *Solenopsis invicta* (=*S. saevissima*), provided the first bioassay methods to test trail-following behavior even in the absence of a trail-laying ant. He laid artificial trails of different glandular extracts away from the nest entrance and worker aggregations. By comparing the trail-following response of worker ants he was able to identify the Dufour's gland as the source of the trail pheromone of the fire ants. This technique was subsequently used by many investigators, which led to the discovery of a number of trail pheromone glands in different taxonomic groups of ants (Fig. 8).

Wilson's analyses (1962) also revealed for the first time the organization of chemical mass communication in fire ants. It was found that the

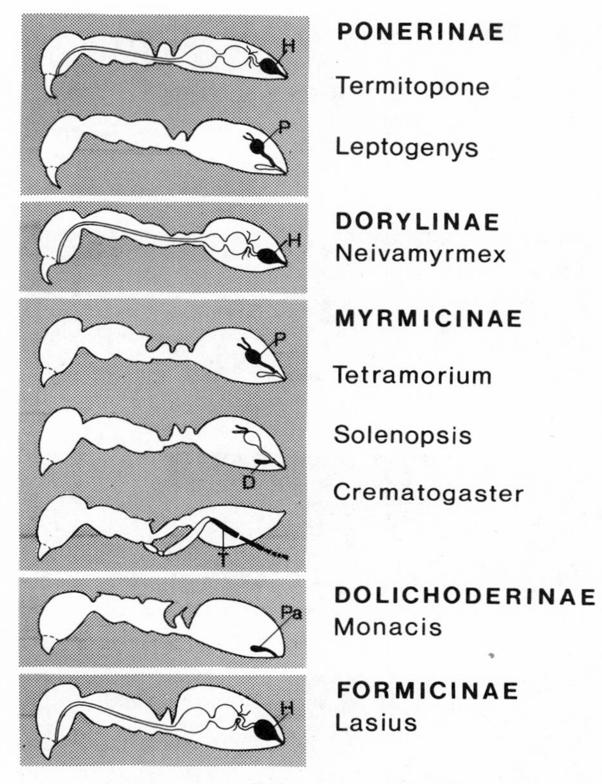

Fig. 8. Trail pheromone glands (black) in several species of five subfamilies of ants. H = hindgut, P = poison gland, D = Dufour's gland, T = tibial gland, Pa = Pavan's gland. (Based on Wilson, 1959a; Wilson and Pavan, 1959; Hangartner and Bernstein, 1964; Watkins, 1964; Blum and Ross, 1965; Blum, 1966b; Leuthold, 1968b; Fletcher, 1971.)

number of workers leaving the nest along the trail is controlled by the amount of trail substance discharged by workers already on the trail. Using the purified trail pheromone it was demonstrated that the number of ants drawn outside the nest is a linear function of the amount of the substance presented to the colony. This means that under natural conditions the number of workers being recruited can be accurately adjusted to the actual needs of recruits at the food source. In other words, the better the food source the more workers lay an odor trail when they return to the nest. This increases the amount of trail substance discharged and in turn draws more ants to the food source. As the food slowly diminishes fewer workers lay a trail, with the result that the concentration of the trail substance, which has a relatively high evaporation rate, decreases and, in turn, a smaller number of workers are stimulated to leave the nest. This phenomenon is called mass communication because it entails the transmission of information that is meaningful only with reference to larger groups and cannot be exchanged between mere pairs of individuals.

Subsequently Hangartner (1969a) demonstrated that even individual ants can contribute to the flexibility of this mass communication system. Individual workers of *Solenopsis* are apparently able to adjust the amounts of their own pheromone emissions to the specific food needs of their colony and to the quality of the food source. By inducing the homing foragers to lay their trail on a soot-coated glass plate, Hangartner found that the continuity of the sting trail increases with increasing starvation time of the colony, increasing quality of the food source, and decreasing distance between the food and the nest (Fig. 9).

This mass-communication system is certainly a highly advanced recruitment method. In an attempt to find out from which more-primitive forms of recruitment communication this system may have evolved, it is necessary to analyze and compare less-sophisticated modes of recruitment communication. The so-called tandem-running behavior is generally considered to be one of the most primitive recruitment methods. Only one nest mate is recruited at a time, and the follower has to keep close antennal contact with the leader ant. This behavior has been described in a phylogenetically scattered array of species including *Camponotus sericeus* (Hingston, 1929), *Ponera eduardi* (LeMasne, 1952), *Cardiocondyla*

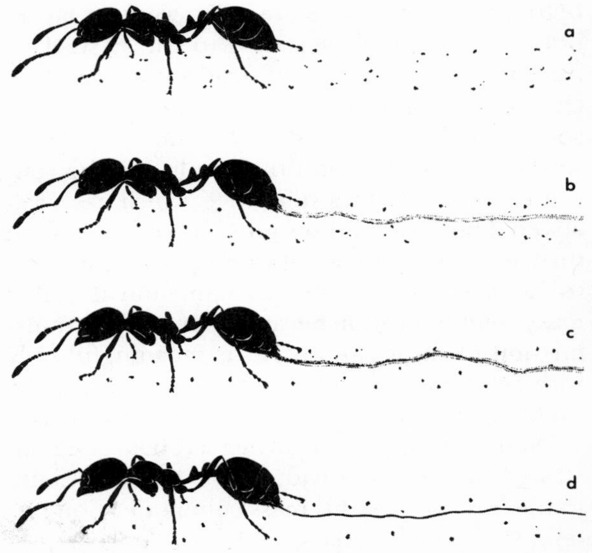

Fig. 9. A worker of the fire ant *Solenopsis geminata* running over a sooted glass plate and laying an odor trail from the extruded sting. a. If the food source is poor, the worker leaves only the tracks made by its feet on the glass plate. b.–d. The better the food source the more intense is the track made by the extruded sting. (From Hölldobler, 1970b, based on Hangartner, 1969c.)

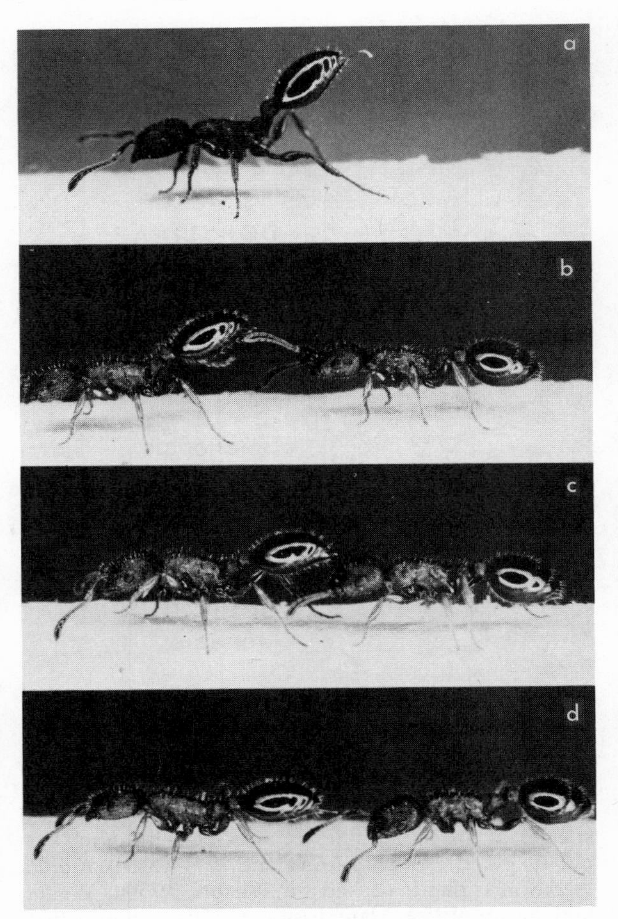

Fig. 10. Behavioral exchange of signals leading to tandem running in *Leptothorax acervorum.* a. A recruiting worker assumes the calling position. A nest mate arrives and touches the gaster b. and hind legs c. of the calling ant with its antennae. d. The calling ant lowers its gaster, and tandem running starts. The sting of the recruiting ant remains extruded, but is not dragged over the surface. (From Möglich et al., 1974b.)

venestula, C. emeryi (Wilson, 1959b), *Leptothorax acervorum* (Dobrzanski, 1966), and *Bothroponera tesserinoda* (Maschwitz et al., 1974b; Hölldobler et al., 1973). Until recently, however, nothing has been learned about the precise nature of the signals involved.

The analyses of the signals by which tandem running is organized in the myrmicine ant *Leptothorax acervorum* have now led to the discovery of a new kind of signal in ant communication, for which we proposed the term "tandem calling" (Möglich et al., 1974b). When a successful scouting forager of *Leptothorax acervorum* returns to the colony it first regurgitates food to several nest mates. Then it turns around and raises the gaster into a slanting position. Simultaneously, the sting is exposed and a droplet of a light liquid is extruded (Fig. 10). Nest mates are attracted by this calling behavior. When the first ant arrives at the calling ant, it touches it on the hind legs or gaster with its antennae, and tandem running starts. The recruiting ant leads the nest mate to the newly discovered food source. During tandem running the leader ant lowers its gaster, but its sting remains extruded. It is not dragged over the surface, however, as in those ant species that lay chemical trails from their stings. The follower keeps close antennal contact with the leader, continuously touching its hind legs and gaster. Whenever this contact is interrupted, for example, when the follower accidentally loses the leader or is removed experimentally, the leader immediately stops and resumes its calling posture. It may remain in this posture for several minutes, continuously discharging the calling pheromone. Under normal circumstances the lost follower rather quickly orients back to the calling leader ant, and tandem running continues. We have found the same tandem-calling behavior in *Leptothorax muscorum* and *L. nylanderi.*

The analyses of this interesting recruitment behavior has revealed two signal modalities by which tandem running is organized: (1) If a tandem pair has been separated the leader immediately stops and assumes the calling posture. However, when the ant is carefully touched with a hair on the hind legs or the gaster with a frequency of at least two contacts per second, the leader continues running to the target area. This experiment shows that the absence of the tactile signals normally provided by the follower ant is sufficient to release tandem calling by a leader ant. (2) The calling pheromone originates from the poison gland. In our studies workers were strongly attracted to dummies that had been contaminated with poison gland secretions but not to those bearing secretions of the Dufour's gland. Further experiments revealed that the poison gland substance not only functions as a calling pheromone but also plays an important role during tandem running itself by binding the follower ant to the leader. It was found that the leader could easily be replaced by a dummy contaminated with poison gland secretions. Gasters of freshly killed ants from which the sting with its glands had been removed could not replace a leader ant. However, when they were contaminated with secretions of the poison gland, they functioned effectively as leader dummies.

The discovery of a chemical tandem calling in *Leptothorax* throws considerable light on the evolution of chemical recruitment techniques in myrmicine ants. It now seems very plausible that the highly sophisticated chemical mass recruitment performed by *Solenopsis* and certain other myrmicine ants was derived from a more primitive chemical tandem-calling behavior of the *Leptothorax* mode. With the exception of *Crematogaster,* which produces a trail pheromone in the tibial glands of the hind legs (Leuthold, 1968b; Fletcher and Brand, 1968), all other myrmicine species generate the trail pheromone from one of the sting glands (Fig. 8). It is conceivable that a chemical calling behavior, during which an alerting and attracting pheromone is discharged through the sting into the air, was one of the first steps leading to chemical trail laying and mass communication in myrmicine ants.

In addition, the tandem-calling behavior is also relevant to the evolution of sex pheromones in myrmicine ants. As mentioned above (p. 419) it has recently been demonstrated that in several myrmicine species the pheromones originate from the sting glands (Hölldobler, 1971a; Hölldobler and Wüst, 1973; Buschinger, 1972a). It is interesting to note that in species in which wingless ergatoids attract males for mating, for example, *Harpagoxenus sublaevis* (Buschinger, 1971b), the females display sexual calling behavior apparently identical to the tandem-calling behavior of *Leptothorax.* This discovery supports the hypothesis that in at least some myrmicine ants sex attractants and recruitment pheromones had the same evolutionary origin. In fact, in some cases the same substances may function in specific situations as sex pheromones and in others as recruitment signals.

In formicine ants the trail pheromones originate from the hindgut (Blum and Wilson, 1964;

Hangartner and Bernstein, 1964; Hangartner, 1969a; Hölldobler, 1971c; Hölldobler et al., 1974). The analyses of the tandem-running technique in the formicine species *Camponotus sericeus* has similarly revealed some of the basic behavioral patterns out of which the more sophisticated methods of "group recruitment" and "mass recruitment" employed by other formicine species may have evolved (Hölldobler et al., 1974; Möglich et al., 1974a).

In *C. sericeus* the first scouting ant to discover the food source typically fills its crop and returns to the nest. As the worker heads home, it touches its abdominal tip to the ground for short intervals. Tracer experiments have shown that the ant is depositing chemical signposts with material from her hindgut. Inside the nest she performs short-termed fast runs, which are interrupted by food exchange and grooming. After several regurgitations, the recruiter ant now performs brief food offerings while facing nest mates head on. During one recruitment performance such "rituals" were observed to be repeated three to sixteen times. Apparently this behavior keeps nest mates in close contact with the successful scout ant. When the scout finally leaves the nest to return to the food source, those ants encountered by the recruiting ant usually try to follow the leader. But ordinarily only one ant, the one that keeps closest antennal contact with the leader, succeeds in following it. Most of the recruited ants, after feeding at the food source, turn straight back to the nest, where many of them start to recruit nest mates on their own. Experiments have shown that the hindgut trail, laid down by homing foragers, has no recruitment effect at all. Only experienced ants follow the trail and use it as an orientation cue. Similarly, during tandem running the presence or absence of the trail pheromone is insignificant. The leader ant and the follower are bound by a continuous exchange of tactile signals and by a very persistent surface pheromone.

We discovered that *Camponotus sericeus* also employs the tandem-running technique to recruit nest mates to new nesting sites. Since in this case a whole colony has to be recruited, the behavioral patterns initiating tandem running can be expected to be different from those used in recruitment to food sources. As Fig. 11 depicts, this is indeed the case. When facing the nest mate head on, the recruiter grasps it on the mandibles and pulls it forward heavily. Shortly afterward it loosens its grip, turns completely around, and presents its gaster to the nest mate. If the nest mate responds by touching the recruiting ant's gaster or hind legs, tandem running starts. This behavioral sequence is very stereotyped and is regularly employed when nest mates are invited to follow the signaler to a new nest. We have therefore called this behavior "invitation behavior."

It is interesting to note that some of the ants that fail to respond to the "invitation signals" are carried to the target area. The first behavioral sequences that initiate carrying behavior are almost identical with that of the invitation behavior. The main difference is that the recruiting ant keeps a firm grip when turning around. The nest mate is thereby slightly lifted, a movement that apparently causes it to fold its legs tightly to its body and roll its gaster underneath. In this posture it is carried to the target area (Fig. 12). For more details about social carrying behavior and the division of labor during nest movings in ants, see Möglich and Hölldobler (1974).

The analyses of the signals by which the tandem-running recruitment technique of *Camponotus sericeus* is organized have revealed that mechanical signals and motor patterns play an important role. Although chemical trails with hindgut contents are laid, they function only as orientation cues and do not release any recruitment effect. This brings us to the next higher organizational level of recruitment communication in formicine ants, "group recruitment."

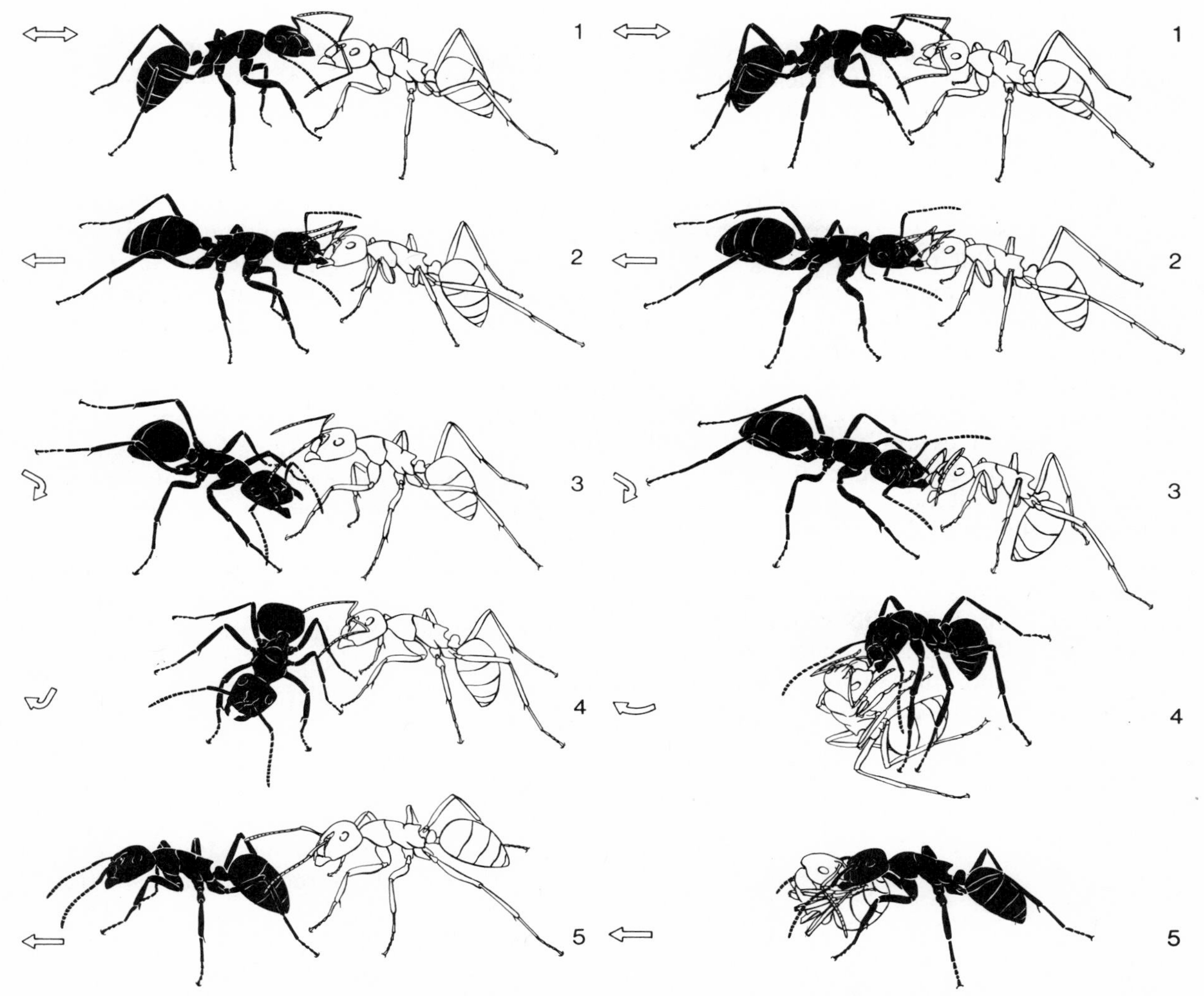

Fig. 11. Invitation behavior to tandem running in *Camponotus sericeus.* 1. The recruiter (black) approaches a nest mate (white) and displays a jerking behavior for about 2–3 sec. 2. The recruiting ant grasps the nest mate at the mandibles and pulls it at a distance of about 2–20 cm. 3. The recruiter loosens its grip and 4. turns around 180°. 5. The recruiter presents its gaster to the nest mate. The nest mate contacts the gaster and hind legs of the leader ant, then tandem running starts. The arrows indicate the direction of the movements. (From Hölldobler et al., 1974.)

Fig. 12. Behavioral sequences that initiate carrying behavior. 1. The recruiter ant (black) approaches a nest mate (white) and displays a jerking behavior for 2–3 sec. 2. The recruiter grasps the nest mate at the mandibles and pulls it a distance of about 2–20 cm. 3. When the recruiter turns it holds the nest mate with a firm grip; the nest mate is thereby slightly lifted. 4. The nest mate folds its legs and antennae tightly to its body and rolls its gaster inward. 5. In this posture it is carried to the target area. The arrows indicate the direction of the movements. (From Hölldobler et al., 1974.)

In this case one ant recruits about five to thirty nest mates at a time, and the recruited ants follow closely behind the leader ant to the target area. This behavior has been observed in *Camponotus campressus* (Hingston, 1929), *C. beebei* (Wilson, 1965), and *C. socius* (Hölldobler, 1971c). Because of its apparent intermediate stage between the tandem-running technique and the chemical mass communication, a detailed experimental analysis of this recruitment behavior was considered most desirable.

Working with *Camponotus socius,* I found that scouts set chemical signposts around newly discovered food sources and lay a trail with hindgut contents from the food source to the nest. The trail pheromone alone, however, does not release a recruitment effect. Inside the nest the recruiting ant faces its nest mates head on and performs a "waggle" display (Fig. 13). The vibrations with head and thorax last 0.5–1.5 sec with 6–12 strokes/sec. Nest mates are alerted by this behavior and subsequently follow the recruiting ant to the food source. The significance of the motor display inside the nest was demonstrated by closing the gland openings of recruiting ants with wax plugs. In this way it was possible to separate the waggle display from the chemical signals, and thus it could be shown that only ants stimulated by a recruiting ant would follow an artificial trail drawn with hindgut contents. For a complete recruitment performance, however, the presence of a leader ant was still essential. Freshly recruited ants without a leader would follow a hindgut trail for only about 100 cm. Essentially similar behavioral patterns are involved during recruitment to new nest sites. The main differences are that the motor display is frequently more a "jerking" movement, and, in contrast to recruitment to food sources, males respond to the signals and hence are recruited. In *Camponotus socius,* as in *C. sericeus,* the jerking movement appears to have been derived from an intention movement that precedes carrying behavior. Indeed, when nest mates do not respond to this signal, the jerking display initiates carrying behavior (Hölldobler, 1971c).

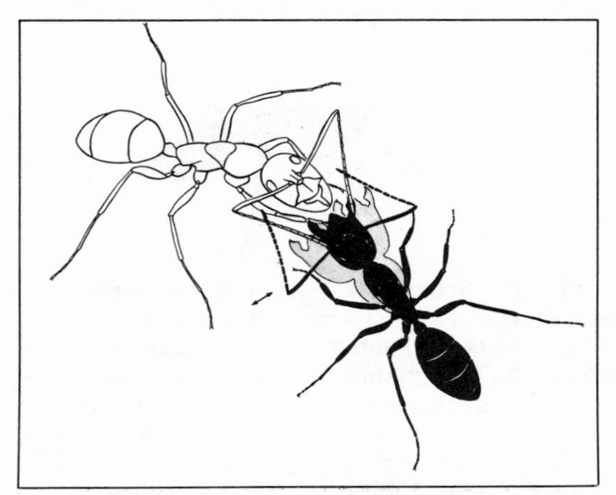

Fig. 13. Schematic illustration of the "waggle" movement of a recruiting ant (black) (*Camponotus socius*) upon encountering a nest mate. Arrow indicates the to-and-fro direction of the movement. (From Hölldobler, 1971c.)

The next organizational level within the formicine ants is represented by those species in which the trail pheromone alone does not also elicit a recruitment effect, but in which stimulated ants follow the trail to the food source even in the absence of the recruiting ant. We found this to be the case in *Formica fusca* (Möglich and Hölldobler, 1975). In this species successful scouts lay a hindgut trail from the food source to the nest. The trail pheromone has no primary stimulating effect. However, after the scout has performed a vigorous waggle display inside the nest, frequently interrupted by food exchanges, nest mates rush out and follow the trail to the food source without being guided by the recruiting ant. From here it is only a small step to chemical mass communication, where the trail pheromone alone releases a recruitment effect and where the outflow of foragers is controlled by the amount of pheromone discharged. This case is represented among the formicines by *Lasius fuliginosus* (Hangartner, 1967).

Cumulative studies have made clear that motor displays and mechanical signals play an important role in recruitment communication in many ant species (see also Sudd, 1957a; Szlep and Jacobi, 1967; Leuthold, 1968a; Szlep-Fessel, 1970). It appears, however, that during the evolutionary process of "designing" more efficient recruitment techniques, these signals became less important as the chemical recruitment system became more sophisticated.

There is another important result of these studies that provides a clue concerning the means by which hindgut material became involved in the recruitment process in formicine ants. Hindgut contents are necessarily frequently discharged by ants. A comparative study has revealed that in many species, ants do not defecate randomly but preferably visit specific locations. Besides certain sites inside the nest, the peripheral nest borders, garbage dumps, and trunk trails leading to permanent food sources or connecting two nest entrances are especially marked with hindgut material. Thus these disposal areas seem to be ideally suited to serve as chemical cues in home-range orientation, and, indeed, this has been documented in a number of species (Hölldobler, 1971c; Hölldobler et al., 1974; Hölldobler, unpublished). These results suggest that in formicine species the trail-recruitment communication behavior might have evolved by a gradual ritualization of the defecation process. We can speculate that in the first step hindgut material became an important cue in home-range orientation and was then transformed into a more specific orienting and stimulating signal used during recruitment behavior.

The specificity of trail pheromones in ants varies considerably. Most of our knowledge is based on behavioral investigations since almost nothing is known about the chemical nature of the trail pheromones. It was only recently that Tumlinson et al. (1971, 1972) chemically identified the first such pheromone. The trail substance of the leaf cutting ant *Atta texana* is evidently methyl 4-methylpyrrole-2-carboxylate; this substance has been isolated from the poison gland secretions and found to release a strong trail-following behavior in many attine species. Moser and Blum (1963) and Blum et al. (1964), working with poison gland extracts, had already shown that the trail pheromone of *Atta* releases trail following in many leaf cutting species. A much higher trail-pheromone specificity was discovered by Hangartner (1967) in *Lasius fuliginosus.* Although *L. fuliginosus* workers were able to "read" the trail pheromones of many formicine species (except for that from *L. flavus*), its own trail could be understood by none of the other species tested. Huwyler et al. (1973) identified as major components in the hindgut contents of *L. fuliginosus* hexanoic acid, heptanoic acid, octanoic acid, nonanoic acid, and decanoic acid. All these acids released trail-following behavior in *L. fuliginosus* workers.

Wilson (1962) compared the specificity of trail pheromones in fire ants *(Solenopsis)* that lay trails with secretions from the Dufour's gland. Artificial trails laid with the pheromone of *S. xyloni* released trail-following behavior in *S. invicta* (=*S. saevissima*) and *S. geminata,* but *S. geminata* trails had no effect on the other species. On the other hand, the secretions of *S. invicta* produced no response in *S. xyloni.*

A similar partial specificity of trail pheromones has been reported from other genera, such as *Eciton* (Torgerson and Akre, 1970) and *Monomorium* (Blum, 1966b). According to Blum's investigations the recruitment pheromone of the genus *Monomorium* originates from the poison gland. His specificity tests were therefore carried out with poison gland extracts. However, our own experiments demonstrated that *Monomorium pharaonis* discharges its recruitment pheromone from the Dufour's gland (Hölldobler, 1973b; Möglich, unpublished), whereas poison gland secretions release only a very weak trail-following response. These contradictive findings cannot be explained easily. In summarizing all these results we can say that although trail pheromones of ants are by no means strictly species-specific, they are generally more specific than alarm pheromones.

Among different ant species the persistency of chemical trails varies considerably. In those species that use less-permanent food sources (insect prey) the recruitment trails are usually short-lived, while in other species utilizing long-lasting food sources (especially honeydew plants) the trail pheromones are more persistent. Hangartner (1967) studied the physical nature of the relatively high persistency of the chemical trails in *Lasius fuliginosus* and found that in this species the persistency of a trail depends on the volume of substance discharged and on the porosity of the surface. In addition an inactivated trail can be reactivated after days by moistening it with water. Similar results were obtained for neotropical army ants *(Eciton),* the trails of which can persist for about one week when deposited during the dry season. During the rainy season the same trails are much less persistent (Torgerson and Akre, 1970). In some species, such as *Atta texana,* the trail pheromone contains a short-lived and a long-lived component (Moser and Silverstein, 1967).

This leads us to another important function of chemical trails in ants. As discussed above, some of the formicine trails composed of hindgut material contain relatively long-lasting trail substances that serve mainly as chemical cues in home-range orientation. These orientation trails, or trunk trails, as they are commonly called, can play a major role in regulating territorial behavior and in partitioning foraging grounds. This has recently been demonstrated for species of the myrmicine harvesting ant genus *Pogonomyrmex.*

Workers of *Pogonomyrmex* lay chemical trails with poison gland secretions to recruit nest mates to new rich seed falls (Hölldobler and Wilson, 1970). These recruitment pheromones are relatively short-lived. However, laboratory and field experiments revealed that in addition more enduring chemical signposts are concurrently deposited along the recruitment trails. The latter substances function as orientation cues, so that long after the recruitment signal has vanished, motivated foragers can still follow the same track (Hölldobler, 1971d). Circumstantial evidence indicates that these cues originate at least in part from the Dufour's gland. We have evidence of species-specificity in the mixture of compounds of the Dufour's glands of *Pogonomyrmex* (Regnier et al., 1973; Hölldobler and Regnier, unpublished). In addition Hangartner et al. (1970) showed that *Pogonomyrmex badius* workers are able to distinguish the odor of their own nest material from that of other nests. In our most recent laboratory experiments we found that even the trunk trails contain colony-specific chemical cues that enable the ants to choose the trails leading to their own nest as opposed to those leading to a neighboring colony.

In a recent analysis (Hölldobler, 1974) it was demonstrated that trunk trails used by *Pogonomyrmex barbatus* and *P. rugosus* during foraging and homing have the effect of avoiding aggressive confrontations between neighboring colonies of the same species. They channel the mass of foragers of hostile neighboring nests in divergent directions, after which each ant pursues its individual foraging exploration. This system subtly partitions the foraging grounds and allows a much denser nest-spacing pattern than does a foraging strategy without trunk trails, such as that employed by *P. maricopa.*

It is interesting to note that honeybees and wasps also use relatively persistent chemical orientation cues to locate the nest entrance (Butler et al., 1969, 1970). Apparently these species deposit the chemical signposts with footprints. It cannot be excluded that ants also employ the footprint technique for setting auxiliary chemical orientation cues. In fact, I frequently observed that foragers of *Pogonomyrmex* rub their legs over the abdominal tip before they leave the nest; an even more striking version of this behavior occurs in the slave-raiding ant *Polyergus* just before and during slave raids (Hölldobler, unpublished). Moreover, Torgerson and Akre (1970) have shown that workers of the army ants *Eciton hamatum,* which lay trails with hindgut contents, are able to set weak footprint trails after their gasters have been removed. Finally, Leuthold (1968b) provided circumstantial evidence that workers of *Crematogaster ashmeadi* can deposit chemical footprints without exhibiting the typical trail-laying behavior involving the hindlegs.

Chemical strategies during foraging also play an important role in interspecific competition among ants. I found that the subterranean species *Solenopsis fugax,* commonly called the thief ant, lays odorous trails in the tunnels leading into the brood chambers of neighboring ant species. The recruitment trail pheromone originates from the Dufour's gland. More important, however, is the fact that *Solenopsis,* when preying on the foreign brood, discharges a highly effective and long-lasting repellent substance from the poison gland. This material prevents brood-keeping ants from defending their own larvae against the predators. A very similar chemical offense is used by pharaoh ants *(Monomorium pharaonis).* In addition to the recruitment pheromone originating from the Dufour's gland, a repellent substance is discharged from the poison gland that enables the *Monomorium* to compete successfully with other ant species at the same food sources (Hölldobler, 1973b).

Very little is known about recruitment communication in social wasps. Hase's (1935) observations on *Polybia atra* indicate that this species alerts nest mates when a food source has been found. Similar observations were made by Lindauer (1961) on *Polybia scutellaris,* and Naumann (1970) got very good observational evidence that *Protopolybia pumila* conducts a sort of "departure dance," which stimulates nest mates to fly out for foraging. Although Kalmus (1954) claims that *Paravespula germanica* and *P. vulgaris* do not alert nest mates to new food sources, Maschwitz et al. (1974a) provided experimental evidence proving recruitment communication in these species. The main signal seems to be the scent of the food.

In sum, it appears on the basis of limited evidence that in at least some wasp species scouts have the ability to alert nest mates when new food sources are found, but these individuals do not transmit directional information about the food source.

A similar simple system of communication about food sources is possessed by bumblebees *(Bombus).* In these relatively primitive forms, social facilitation seems to occur after a successful forager has returned to the colony. Odors from the food source, clinging to the body of the scout bee, apparently provide some information about the food source (Free, 1970b).

Workers of some species of meliponine bees (stingless bees) employ a similar primitive recruitment communication. Kerr and Esch (1965) report that in *Trigona silvestris* an increasing number of bees fly out and search for food when a forager returns to the nest carrying a characteristic odor with the food. Lindauer and Kerr (1958, 1960) and Esch (1967b) studied the different organizational levels of recruitment communication in stingless bees and found a variety of recruitment techniques of increasing complexity.

In *Trigona droryana,* returning foragers alert nest mates by a buzzing sound and a characteristic zigzag run inside the hive; however, no information about the direction and distance of the food source is transmitted. A similar communication behavior has been found in *T. muelleri, T. jaty,* and *T. araujoi.* The next higher organizational level is represented by *Trigona (Scapto-*

trigona) postica. Recruiting bees are clearly able to transmit directional information about the food source in the following way: When a scout bee discovers a food source it usually makes several trips between the food and the nest. Then it begins to set chemical signposts at two-meter intervals on its way back to the nest, using a marking substance that originates from the mandibular glands. Having laid a chemical trail from the food source to the nest, the scout conducts zigzag runs and emits characteristic sounds, which apparently alert the nest mates and induce them to follow the scout along the trail to the food source. During one such guiding flight a bee can lead more than fifty recruited nest mates to the target area. Lindauer and Kerr proved that the trail pheromone alone is not enough to elicit trail-following behavior in newcomers. A guide bee is necessary to lead the recruits to the food source. Only after the newcomers have been led along the trail, do they follow the trail back and forth on their own.

There seems to exist some degree of specificity in these chemical orientation trails. *Trigona xanthotricha* follows trails of *T. postica,* but not vice versa. Neither *T. postica* nor *T. spinipes* follow each other's trails (Kerr et al., 1963). Blum et al. (1970) identified neral and geranial as major compounds of the mandibular gland secretions of *Trigona subterranea* and provided circumstantial evidence that these substances constitute the effective trail pheromone. Recently Kerr et al. (unpublished; cited in Blum, 1974) succeeded in inducing trail-following behavior in *T. spinipes* along an artificial trail of 2-heptanol, one of the major compounds in the mandibular gland secretions of this species. It is interesting to note here that a propaganda-pheromone technique has also been discovered in stingless bees comparable to that used by slave raiding ants (see p. 429). Stejskal (1962) observed that the robber bee *Lestrimelitta limao* lays chemical trails from its own nest to the host species' nests with mandibular gland secretions. Blum (1966a) identified citral as the major trail pheromone, and later (Blum et al., 1970) found that when citral is introduced into colonies of the hosts *Trigona* or *Melipona,* the behavior of these bees is completely disrupted; in particular the victims seem unable to launch a defense.

In general, meliponine recruits evidently have to be alerted before they follow a chemical trail. Lindauer and Kerr (1958) provided the first experimental proof that sounds, emitted by recruiting bees, constitute the important triggering signal. Lindauer (1961) described the decisive experimental procedure as follows:

> We divided the beehive into two compartments with a board. Through a sliding door at the entrance hole we could direct the marked collector bees into either one or the other compartment. After a fairly long feeding interval, we allowed a single scout to come to a known feeding place and could now establish that when we allowed her to return into compartment A she would also alert her colleagues in compartment B. In another experiment we combined two colonies in a single box, separated only by a wire screen. Now we could observe that the humming of a single collector bee of colony A would also alert novices of colony B. The result of the experiments was negative, however, when we padded the floor with foam rubber. The latter result indicates that the receptor mechanism for these humming sounds is not really hearing, but the vibrational sense. It thus seems to be proved that the humming of the collector bees has an alerting effect.

All species of *Trigona* studied produce such alerting sounds. A comparative investigation, however, revealed that the sounds of *Melipona* are more sophisticated. No odor trails are known in the genus *Melipona,* but the duration of the sounds produced by returning foragers varies directly with the distance to the food source (Kerr and Esch, 1965; Esch, 1967a). Thus the sounds not only alert nest mates but may also convey some information about the distance of the food source. Indeed, foragers of *Melipona quadrifasciata* that visited a nearby feeding station could be induced to fly out again to visit the station when a tape recording of their own sounds was played back to them (Esch, 1967a). However, it must be mentioned that this experiment worked only if the feeding station was not too far away. Furthermore, only experienced worker bees responded. In addition to this alerting and distance-indicating signal, recruiting bees of *Melipona* guide hive mates in a striking zigzag flight toward the food source. These guiding flights last only 30–50 meters, after which the followers usually lose contact with the leader bee. However, after twenty to thirty repetitions of such guiding flights, the recruited bees try to find the target area on their own. In some species, for example, *Melipona seminigra,* the guiding flights are even shorter, only 10–20 m in length.

Recruitment Communication in *Apis*

CHEMICAL SIGNALS

The celebrated "dance language" in honeybees is probably the most sophisticated and most thoroughly studied communication behavior in the animal kingdom. However, it should not be overlooked that chemical signals also play an important role in the recruitment communication of these insects. The chemical signals interact with the dance language, and when they are described separately from it, as will now be done, the reader should keep in mind that the separation is artificial.

Karl von Frisch, who was the first to understand the bees' dance language, also discovered that honeybees employ a variety of chemical cues in recruitment communication. As early as 1923 he demonstrated that environmental odors, such as floral scents, play an important role during the recruitment process. It seems that floral scents in particular are carried on the body surface of scout bees when they return to the hive, where, together with the taste of the nectar offered in regurgitation, it informs nest mates about the nature of the newly discovered food source. Bees that have already had a successful experience with a similar food source often become alerted by the familiar odor. Ribbands (1954, 1955a, 1955b) showed that some of these individuals are stimulated by the scents alone to fly out and search for food.

In addition to the environmental odors, pheromones are used during the recruitment of nest mates to certain target areas. As early as 1902 Sladen described "Sterzel"-behavior (chemical calling behavior) in honeybees. In many circumstances—for example, when bees attempt to attract lost foragers home to the hive or during swarming to a new nesting site—they bring their abdomen into a slanting position and by everting the Nassanoff's gland discharge an attraction pheromone, while simultaneously fanning their wings vigorously, thus accelerating the distribution of the pheromone (Renner, 1960) (Fig. 14). In 1923 von Frisch found that also during foraging worker bees discharge the Nassanoff gland secretions at feeding dishes containing highly concentrated sugar water. He found that more recruits arrive at a feeding station where scout bees were allowed to discharge the pheromone than at control stations where the

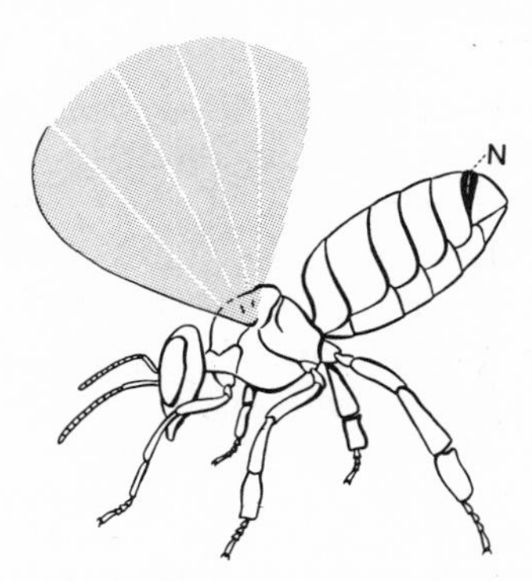

Fig. 14. A honeybee worker of *Apis mellifera* showing a recruitment calling posture with the Nassanoff gland exposed and fanning with the wings.

scout bees' scent organs had been sealed off. More recently Free and Williams (1970) pointed out that the role of the Nassanoff gland secretion as a recruitment signal is especially important during recruitment to new water sources. Since water is, in contrast to most of the natural food sources, all but scentless, it is more important to mark it so that it will be found. Indeed, bees commonly discharge the Nassanoff gland secretions after they have thoroughly inspected a new water source.

The Nassanoff gland opens dorsally between the last two abdominal tergites. Boch and Shearer (1962, 1964) identified geraniol, nerolic, and geranic acids as major compounds of the secretions, and their bioassays indicate that the mixture of these compounds is the attracting stimulus. In addition, Weaver et al. (1964) found citral to be a potent component, even though this substance constitutes only a minor fraction (Butler and Calam, 1969).

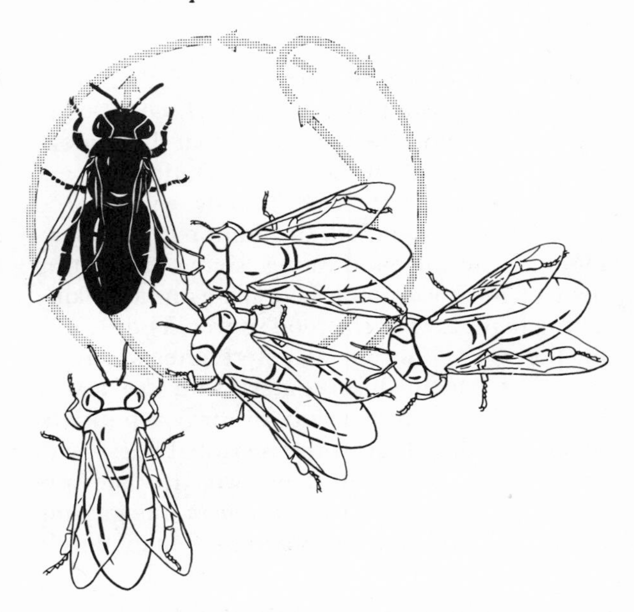

Fig. 15. The round dance of *Apis mellifera*. The recruiting worker (black) dances the figure indicated by the arrows. Stimulated nest mates follow closely. (Based on von Frisch, 1967a.)

DANCE COMMUNICATION

Von Frisch and his students described a variety of different dance forms conducted by honeybees in the hive (see von Frisch, 1967a). In the following brief account I will concentrate on the two most important dance patterns employed to alert and recruit nest mates to certain target areas.

When a scout bee of *Apis mellifera* discovers a new rich food source, say, about 15 m from the hive, she flies home, enters the hive, and regurgitates food to several nest mates. After several trips back to the food source she finally starts to conduct the "round dance" (Fig. 15), which von Frisch describes as follows:

> With swift tripping steps the forager bee runs in a circle, of such small diameter that for the most part only a single cell lies within it. She runs about over the six adjacent cells, suddenly reversing direction and then turning again to her original course, and so on. Between two reversals there are often one or two complete circles, but frequently only three-quarters or half of a circle. The dance may come to an end after one or two reversals, but 20 and more reversals may succeed one another; correspondingly, at times the dance lasts scarcely a second and at others often goes on for minutes. During dances of long duration the center of movement may shift gradually over the breadth of several cells. After the round dance has ended, food often is distributed again at this or some other place on the comb and the dance is then resumed; this performance may even be repeated thrice or (rarely) oftener. The dance ends unexpectedly as it began, and after a short period of cleaning and "refueling" the bee rushes hastily to the hive entrance and takes off on the next foraging flight.

The round dance contains no directional or distance information about the food source. It merely alerts and stimulates nest mates to fly out and search for the newly discovered food source. However, the alerted bees perceive the odor of the nearby source by antennating the dancer, and they also receive taste samples. These cues enable them to find the particular source. As von Frisch has shown, the better the food source the more vigorous, long lasting, and lively are the dances, and, in turn, the more bees are recruited to the food source. Although some of these parameters are difficult to measure, there exists good observational evidence that they increase with the quality of the food, for instance, the sugar concentration of the bait (Lindauer, 1948; Boch, 1956).

If the distance between feeding place and hive increases from 25 m to 100 m the round dance gradually changes into the "waggle dance" (Fig. 16), and at distances greater than 100 m the round dance is finally completely replaced by the waggle dance. Von Frisch (1967a) describes the waggle dance as follows:

> In the typical tail-wagging dance the bee runs straight ahead for a short distance, returns in a semicircle to the starting point, again runs through the straight stretch, describes a semicircle in the opposite direction, and so on in regular alternation. The straight part of the run is given particular emphasis by a vigorous wagging of the body. This results from rapid rhythmic sidewise deflections of the whole body that are greatest at the tip of the abdomen and least at the head. The axis about which the sidewise oscillation is to be envisaged lies close before the bee's head and perpendicular to the substrate. The movement to and fro is repeated 13–15 times in a second.

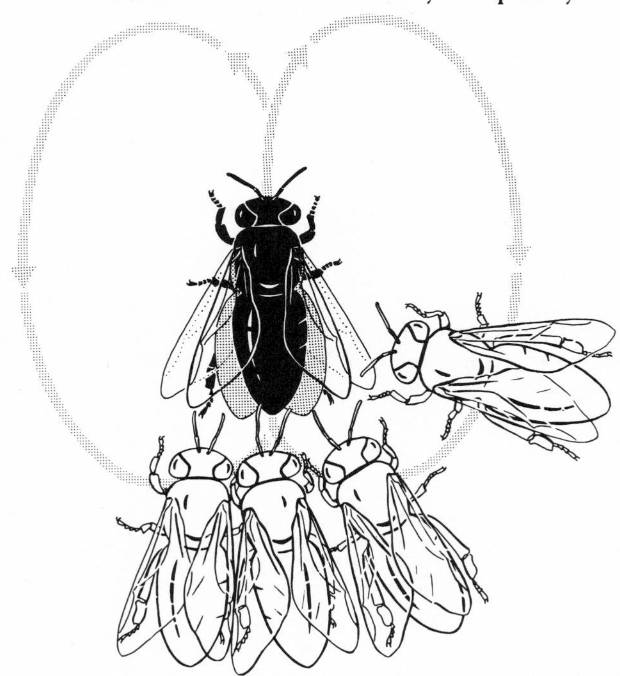

Fig. 16. The waggle dance of *Apis mellifera.* The recruiting worker (black) dances the figure indicated by the arrows. During the straight run it waggles vigorously with its body. The dancing bee is closely followed by stimulated nest mates. (Based on von Frisch, 1967a.)

Like the round dance the waggle dance announces the find of a new profitable food source, the kind of food (by odor and taste samples), and the productivity of the food (by vigor and liveliness of the dances). But, unlike the round dances, the waggle dances in addition transmit information about the distance and the compass direction of the target area. Alerted bees follow the dancing bee with close antennal contact, and thereby receive this information. The straight run seems to be the most important part of the waggle dance figure. Not only do the alerted bees pay closest attention to this part of the dance but also its features are most closely correlated with the specific distance and direction of the target area.

The greater the distance to the food source the longer the duration of the straight run. The straight run, however, is characterized not only by the vigorous wagging of the abdomen. Esch (1961, 1964) and Wenner (1962a) found that during the straight run, the dancer also emits a buzzing sound the duration of which is exactly the same as that of the straight run. Esch produced circumstantial evidence that the duration of the buzzing, which the follower bees perceive with their antennae and legs, is the most important distance-indicating cue. Occasionally, for example when the sugar concentration is low, recruiters conduct "silent" dances. Of 15,000 such dances observed by Esch, none succeeded in recruiting bees to the food source. The next question is: How does the dancer estimate the distance between the hive and the food source? Strong circumstantial evidence exists that the bees use the amount of energy consumed on the flight to the food source as a measure of the distance. For instance, if a bee has to fly to the food against a head wind, the distance she indicates in the dance is longer than the actual one. The same is true if the bee is somehow hindered by an artificial weight or some tinfoil mounted on her thorax to increase air resistance during her flight to the food. It is important to note that only one way is measured—only the energy required to fly to the food source is indicated in the dance.

As noted, the waggle dance contains not merely information about the distance of a food source but also the direction in which the recruits have to fly. If the weather is very warm and many bees are assembled outside the hive, one can frequently observe foragers regurgitating their crop contents even before they enter the hive; they also dance on a horizontal surface in front of the nest entrance under the open sky. In those cases the straight run of the dance always points in the direction of the food source. The dancing bee maintains the same angle relative to the sun as on the flight from the hive to the feeding place. In a series of ingenious experiments, von Frisch demonstrated that the bee orients just as well relative to the polarized light of the blue sky. When, however, the sky is completely clouded or the horizontal platform is placed in the dark, the dancers are disoriented and do not indicate a specific direction.

Inside the hive, where it is completely dark, the bees are therefore forced to use another cue to orient their dances. Now they dance on the vertical surface of the combs and translate the solar angle (azimuth) into the gravitational angle. The dancer changes the angle of the straight run with respect to the sun to an angle with respect to gravity. If the food source is located, say, 40° left of the sun, the dancing bee will orient its straight run at an angle of 40° to the left of the vertical. When the bee dances straight up, it is indicating a food source located on a straight line toward the sun. Similarly, if the scout dances straight down it means the goal is located in the opposite direction. In this way the scout is able to indicate any direction in the 360° around the nest (Fig. 17). However, she cannot signal "upward" or "downward" with reference to space outside the hive. Thus the hive mates are informed about the azimuthal angle and the distance but not about the elevation of the goal. A comparative analysis of successive dances has shown that there is a minor variation in the dance components. Therefore recruits following only a single dance would receive slightly different messages. However, Esch and Bastian (1970) found that recruits follow at least six dances before they fly out to the food source, while Mautz (1971) found them following 6.9–12.8 dances. Apparently the recruits integrate the information they receive from the different dances, an operation that accurately directs them to the goal. Lindauer (1955) discovered that honeybees employ waggle-dance communication to recruit nest mates not only to food and water sources but also to new nesting sites.

Wenner and his associates (see Wenner, 1967; Wenner et al., 1967, 1969; Johnson, 1967; Johnson and Wenner, 1970; Wells and Wenner, 1973) claimed new results that suggest that the bees do not understand the direction and distance information contained in the waggle dance. According to their hypothesis the recruited bees find their way to the food source entirely by means of other cues, such as odors. Their criticism of von Frisch's experiments in part reflects their incomplete interpretation of

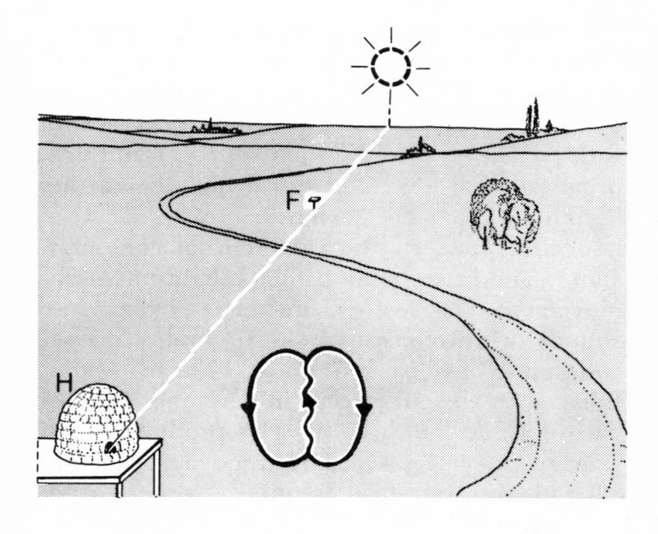

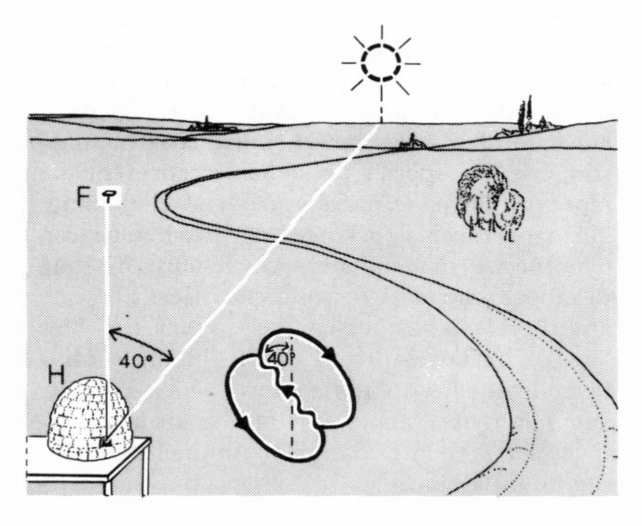

Fig. 17. Two examples illustrating the transfer of the angle between the sun, food source (F), and hive (H) to the gravitational dimension. Above: The food source is located in the direction of the sun; the straight waggle run of the dancing figure on the vertical comb points straight up. Below: The food source is located 40° to the left of the line drawn from the hive to the sun; therefore, the dancing bee orients its straight waggle run in an angle 40° to the left of the vertical line on the comb.

earlier, extensive literature. Although they criticized von Frisch's control experiments as being insufficient to prove direction and distance communication in honeybees, their own experimentations were far more seriously lacking in controls. For a critical evaluation of this controversy see von Frisch (1967b, 1968), Wilson (1971), Lindauer (1971), Michener (1974), Gould (1975), Gould et al. (1970), and Griffin (see this volume, p. 27). On the positive side, the skepticism of Wenner and his associates has stimulated new investigations by other research groups. New techniques and more-rigorous control experiments were applied, which finally led to the full confirmation of von Frisch's results.

The most important progress in the study of the honeybees' dance communication has very recently been achieved by Gould (1974, 1975). He succeeded in causing a "natural dummy" to dance and was thereby able to misdirect recruits into areas other than those where the feeding station was located. For the design of his experiments Gould exploited the following findings: (1) Von Frisch (1962) had reported that the honeybees tend to interpret a bright light in the hive as the sun and will orient their dances to it rather than to gravity. (2) Lindauer and Schricker (1963) demonstrated that bees with their ocelli painted over are far less sensitive to light. On this basis Gould argued:

> When a light of an appropriate brightness is used, foragers with painted ocelli will ignore the light and dance with respect to gravity, while untreated bees both dance and interpret dances as though the light were the sun. If the bees utilize the distance and direction information in the dance, it should be possible for ocelli-painted foragers to recruit bees to specific but incorrect locations.

Indeed, by this kind of experimentation, Gould demonstrated that a significant number of recruits used the false information about distance and direction transmitted by the dancing bee and arrived at the checkpoint but not at the feeding station.

EVOLUTIONARY ASPECTS OF DANCE COMMUNICATION

The search by von Frisch and his associates for more-elementary forms of communication in other bee species, including stingless bees, has resulted in the reconstruction of the possible evolution that led to the highly sophisticated dance language in the honeybee *Apis mellifera.* Boch (1957) demonstrated that even within the species *A. mellifera* there exist considerable variations in the dance language. Different geographical races communicate with different "dialects." For instance, in *A. mellifera carnica* the round dance changes to the waggle dance when the goal is about 85 m from the hive, whereas in *A. mellifera nigra* and *A. mellifera intermissa* this occurs at distances of about 65 m and in *A. mellifera fasciata* at only 12 m.

Lindauer (1956) studied dance communication in other *Apis* species. Probably the closest relative to *A. mellifera* is *A. indica.* It also nests in dark crevices, and like *A. mellifera* it translates the azimuthal angle into a gravitational angle on a vertical surface. However, the waggle dance is performed even when the goal is as close as 2 m from the nest. The giant honeybee *(Apis dorsata)* nests in the open under a rock or in a tree. Foragers dance on the vertical comb, but they need the sun or the blue sky to orient their waggle dance. The dwarf honeybee *(Apis florea)* also nests in the open. It communicates by the waggle dance but only performs on a horizontal platform. If it is forced to dance on a vertical surface, it either stops dancing or becomes disoriented. Obviously this species is not able to translate the azimuthal angle into the gravitational angle.

This brings us back to the stingless bees, discussed on p. 441–43. In the most advanced species, such as *Melipona quadrifasciata* and *M. merillae,* foragers not only show the direction to the goal by a short zigzag guidance flight but also indicate the distance by means of a sound code (Esch et al., 1965; Esch, 1967a). As in *Apis,* the duration of these particular sounds is correlated with the distance to the food source. From this organizational level it is not a big step to the "symbolic guidance behavior" of *Apis.* The honeybee runs in the direction of the food source and in so doing exhibits flight-intention movements. These are characterized by buzzing sounds, which consist of short vibrational episodes, one of which lasts about 15 msec. About thirty such episodes occur in a second. During each episode the flight muscles produce vibrations of 250 cycles/sec. This frequency is identical with the wing-beat frequency, although no wing stroke is actually executed (Esch, 1961). From these findings it is reasonable to conclude that the waggle-dance communication behavior in honeybees is a highly ritualized guiding flight to the target area. The simple motor displays, mechanical signals, and chemical cues of some stingless bee species presumably represent the more primitive mechanisms from which the waggle dance seems to have originated. One can hypothesize that in bees, which must fly long distances, chemical recruitment is less accurate and therefore has become less significant. This led to the development of the highly sophisticated, ritualized waggle dance, which not only stimulates nest mates but also transmits relatively accurate information about the location of the target area.

Communication during Trophallaxis

Wilson (1971) defines trophallaxis as the "exchange of alimentary liquids among colony members and guest organisms either mutually or unilaterally. In stomodeal trophallaxis the material originates from the mouth; in proctodeal trophallaxis it originates from the anus." Trophallaxis plays a central role in the social organization of most species of social insects. It is the major mechanism by which food is distributed in the society; but in addition it functions to transfer specific pheromones from one individual to another.

Usually trophallaxis is initiated by specific communication stimuli. Montagner (1966, 1967) studied the signals involved during food exchange between adults of the social wasps *Vespula (Paravespula) germanica* and *V. (P.) vulgaris.* He found that the soliciting wasp initiates regurgitation in a prospective donor by a series of tactile signals. When the begging wasp has approached the nest mate head-on, she lowers her body slightly, turns her head sideways, and strokes the mouthparts of the donor with her antennae and palpae (Fig. 18). This stimulation continues as long as the food exchange goes on. When regurgitation comes to an end, the donor pushes its antennae against the mandibles of the beggar, and the contact is interrupted.

Montagner describes a social hierarchy among the adults by which dominant individuals apparently receive more food than they give. The mother queen is on top of the hierarchical order, virgin queens are dominant over their sisters, and within the worker group there appears to exist a dominance relationship expressed by certain subtle behavioral patterns. For instance, a dominant worker does not lower its body when it solicits food from a subordinate. Occasionally, it even steps on the donor, stroking intensively with its antennae against the mouthparts of the subordinate and thereby "forcing" it to regurgitate its crop contents.

There also exists a reciprocal food exchange between wasp adults and larvae. Du Buyson (1903) and Janet (1903) were the first to describe the larval secretions of *Polistes* and *Vespula* as sweet substances. Roubaud (1911) suggested that larvae induce brood-tending behavior in adult wasps by offering them these secretions, and W. M. Wheeler (1918) finally proposed the

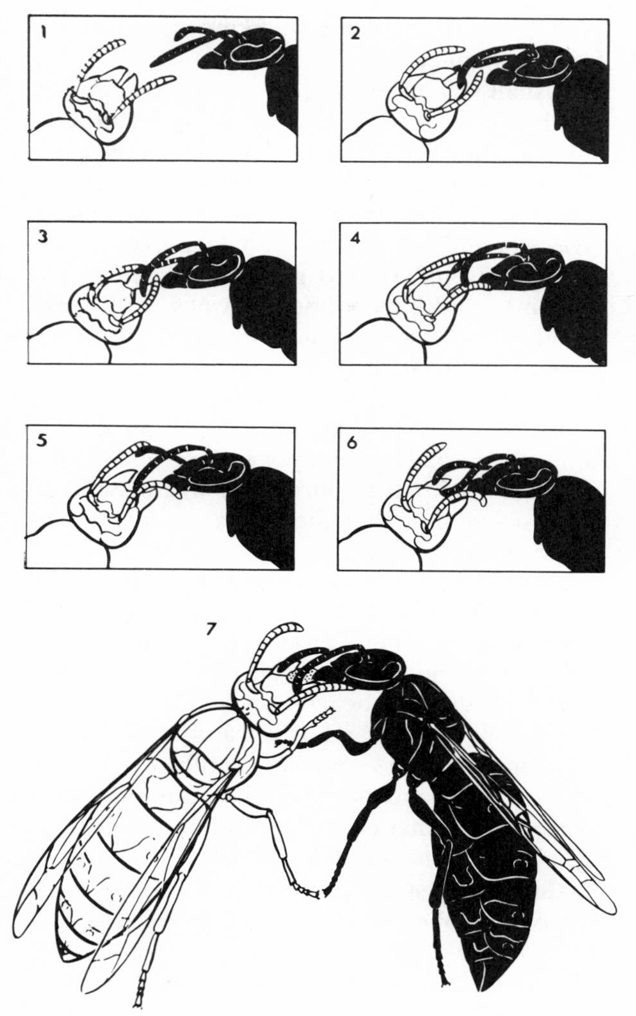

Fig. 18. The initiation of regurgitation between two workers of the wasp *Vespula (Paravespula) germanica.* 1.,2. The solicitor on the right approaches the donor and places the tips of her flexible antennae on the donor's lower mouthparts. 3. The donor responds by closing her antennae onto those of the solicitor, who then begins gently to stroke her antennae up and down over the lower mouthparts 4–7. If this interaction continues, the donor will begin to regurgitate, and the solicitor will be able to feed. (From Wilson, 1971, based on Montagner, 1966.)

term "trophallaxis" for this kind of reciprocal food exchange.

Usually wasp larvae give away these liquids, which are secreted from the saliva glands, as soon as they are stimulated by very unspecific tactile signals. Maschwitz (1966b) analyzed the larval salivary secretions of *Vespula (Paravespula) germanica* and found it to be an 8.9 percent sugar solution (trehalose and glucose). Amino acids, proteins, and proteolitic enzymes and small amounts of uric acid and ammonia are also present in the larval saliva secretions. Maschwitz provided convincing evidence that these larval secretions are used by workers as an energy source, especially when they are prevented from making foraging trips during bad weather and on other occasions. Thus the larvae can function as living storage containers.

According to Montagner's findings (1963, 1964) *Vespula* males obtain most of their food by milking larvae. Their "awkward" begging behavior toward adult workers solicits but little food; thus they depend heavily on larval secretions. Ikan et al. (1968) and Ishay and Ikan (1969), working with *Vespa orientalis,* confirmed Maschwitz's results and discovered in addition what Wilson (1971) has called "a biochemical division of labor." *Vespa orientalis* adults are apparently unable to digest proteins. Only larvae are capable of gluconeogenesis and convert proteins, which they are fed by workers, partly to carbohydrates, such as glucose, fructose, and sucrose, and return them to the adults during trophallaxis. This division of labor is not universal, however, because in other wasp species, such as *Vespula germanica,* proteolytic enzymes have been detected in the midguts of larvae and adults (Spradbery, 1973).

Trophallaxis is highly variable among species of bees and is most elaborately developed in the honeybees *Apis.* Michener (1974) describes a typical food exchange between forager and nest-worker in honeybees as follows:

> When a forager with a crop full of nectar arrives in the nest, she promptly approaches a receiver head-on, opens her mandibles, and regurgitates a drop of nectar onto the slightly projected base of her proboscis (upper surface of prementum and stipites), the rest of which remains folded back under the head. The receiver then extends her proboscis and, with the tip of it, takes nectar from the drop. Meanwhile the antennae of both bees are in continual stroking motion, an activity that probably keeps the two bees properly oriented for the transfer.

Montagner and Pain (1971), with the aid of high-speed motion picture analysis, recently described the movement of the antennae during the trophallactic act in more detail. Free (1956, 1959) studied the releasing stimuli of trophallactic behavior in honeybees with a series of ingenious dummy experiments. He was able to demonstrate that the head with antennae is clearly superior to any other part of the body in releasing the solicitation of food-offering behavior. If the antennae are removed, the dummy becomes less effective, but the effectiveness can be restored by inserting artificial antennae made of wire. In addition, odor seems to play a major role: heads belonging to nest mates were more effective as trophallactic releasers than those belonging to members of another colony. Free also found that a begging bee releases regurgitation in a nest mate by thrusting its proboscis between the mouthparts of a prospective donor.

The "eagerness" with which hive workers accept the crop load from a recruiting forager indicates to the forager the food need of the colony and may in fact determine whether the forager starts to recruit nest mates to the food source by dancing. A similar communication mechanism is applied during the "air conditioning" process in a beehive. Even if the outside temperature rises to more than 70°C, the internal temperature of the hive remains about 35°C. This is achieved by an increased evaporation of water.

> Water is carried into the hive and distributed on cells in tiny droplets. Droplets are deposited particu-

larly at the entrance of the open brood cells. At the same time a large number of bees can be seen hanging over the brood cells and continuously extending their proboscises back and forth. Each time they do this they press a drop of water from their mouths and spread it with the proboscis into a film, which has a large evaporating surface. When the water evaporates, the proboscis is retracted again and a new droplet spread out. [Lindauer, 1971]

This air-conditioning behavior is regulated by a highly organized division of labor. Only experienced bees function as water collectors; they give their water loads to the nest workers, which then spread the water around. However, the water collectors' activity has to be adjusted to the needs inside the hive, which in turn depends on the temperature. In other words, there has to be a sort of thermostat that turns the cooling system on or off. Lindauer (1954) discovered that this consists of a relatively simple "feedback" communication process.

Let us assume that water collecting is still in progress and the foragers are to be informed whether or not there is need for more water. To transmit this information the hive bees make use of the short moment when they have contact with the collectors; this is during water delivery at the entrance hole. As long as overheating exists, the home-coming foragers are relieved of their burden with great greed, three or four bees at once may rush up to a collector and suck from her the extruded water droplet. This stormy begging informs the collector bee that there is a pressing need for more water. When the overheating begins to subside, however, the hive bees show less interest in the water collectors. The latter now have to run around in the hive themselves, trying to find somewhere a bee that will relieve them of at least part of the water load. The delivery in such cases takes much more time, of course. This rejecting attitude contains the message "Water needs fulfilled," and the water collecting will thus stop, even though the collectors themselves have not been at the brood nests to experience the changed temperature situation. [Lindauer, 1971]

Lindauer could convincingly demonstrate that in honeybees "this delivery time is in fact an accurate gauge of water demand."

In ants the development of trophallactic food exchange is highly variable. In most of the species of the two more-primitive subfamilies Myrmeciinae and Ponerinae, trophallaxis is either completely lacking or only poorly developed. The rate of trophallactic food exchange can be measured quantitatively by labeling the food with radioactive tracers (Gösswald and Kloft, 1956, 1960; Wilson and Eisner, 1957). Many species of the subfamilies Aneuretinae, Dolichoderinae, Formicinae, and Myrmicinae show a relatively high food-exchange rate (Wilson and Eisner, 1957). In a few other species, however, trophallaxis apparently has become irrelevant with the development of very specialized feeding habits. Examples of these specialized groups include the harvesting ants and fungus-growing ants.

The rate of food exchange depends also on several environmental and physiological factors, such as temperature, humidity, and nutritional status, as well as on the size and social structure of the group (Kneitz, 1963; Lange, 1967). The radioactive tracer technique allows the direction of social food flow to be determined; it can further be used to identify which castes and age groups are participating. Lange (1967), for instance, discovered that in *Formica* species the flow of protein food is directed preferentially toward young workers and queens, whereas carbohydrates are more evenly distributed in the colony. In most species only female castes regurgitate food to other members of the society. In carpenter ants (*Camponotus herculeanus* and *C. ligniperda*) young males also take an active part in the social food distribution (Hölldobler, 1966). In these species the males live in the nest an unusually long time (more than nine months) before departing for the nuptial flight. During the first phase of their adult life especially, they receive large amounts of food from the workers. This intake apparently enables them to complete their spermatogenesis and to build up a rich fat body. It seems reasonable that during this period the males do not block the social food flow but rather participate actively in the social food distribution.

Foragers usually carry liquid food into the nest in their crops (foregut) and regurgitate part of the contents to individual nest mates (Fig. 19). There is strong circumstantial evidence that during trophallaxis secretions from the labial gland and postpharyngeal gland are also passed from one individual to the other. Recent results by Markin (1970) indicate that in the Argentine ant *Iridomyrmex humilis* secretions from the postpharyngeal gland are preferentially fed to queens and small larvae.

The social food flow in an ant society is organized by a variety of signals. Queen pheromones and specific cues by which young workers and brood are identified probably regulate the directed food flow. In addition certain behavioral patterns and tactile signals play a major role during trophallactic food exchange. Several attempts have been made to analyze these signals in ants (Kloft, 1959; Wallis, 1961; Hölldobler, 1966; Lenoir, 1972a, 1972b). Recently we applied high-speed motion picture analyses (200–450 f/sec), which, together with the facts already known, enabled us to synthesize a fairly complete picture of the food exchange behavior in *Formica* (Hölldobler, 1970a, 1973c).

Fig. 19. Schematic drawing illustrating the food flow from the crop (right) of the donor ant to the soliciting ant (left). K = crop, M = midgut, R = hindgut. (From Hölldobler, 1973c.)

The behavioral patterns of the donor and the solicitor are exceedingly different. Workers returning to the nest with a heavily filled crop approach nest mates head-on, with their mandibles wide open and their labia extended. If this results in a mouth-to-mouth contact and the labium of the food-offering ant is only slightly touched, regurgitation occurs instantly. If the food-carrying ant does not find a nest mate ready to accept the food, she will regurgitate a food droplet even without any tactile stimulation. After she has held it for a while between her mandibles, she will finally scrape it off on the ground or a wall of the nest chamber.

According to Wallis (1961), in about 90 percent of the cases the initiative during food-exchange behavior comes from the soliciting worker. The solicitor first antennates an approaching ant, and as soon as both ants stand head-on, the beggar conducts rapid strokes with its forelegs while simultaneously continuing to antennate the other ant's head (Fig. 20). The more intense the begging behavior the more precisely are the strokes of the forelegs aimed toward the mouthparts of the donor. Motion pictures taken from the underside clearly demonstrate that in these cases the strokes hit the labium of the donor. Indeed, these tactile signals seem to release regurgitation. The donor opens the mandibles, extrudes the labium, and regurgitates crop contents. Frequently, stroking with the forelegs ceases as soon as the food begins to flow. Nevertheless, the beggar continues to touch the head and to palpate the mouthparts of the donor with its maxillae. The donor, on the other hand, keeps its antennae folded backward, and only when the regurgitation comes to an end does it move them closer to the beggar's head. Often the beggar then provides another series of strokes with its forelegs, which may induce a sec-

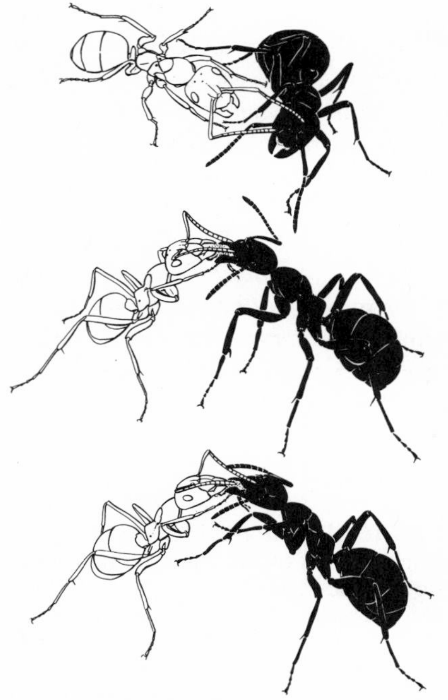

Fig. 20. Three behavioral steps that generally characterize trophallactic food exchange in the ant *Formica sanguinea.* Above: The begging ant (white) antennates a prospective donor ant. Middle: The solicitor stimulates the labium of the donor with its forelegs while continuing its antennation. This leads to regurgitation of crop contents by the donor. Below: As the crop of the donor gradually empties, the tendency to regurgitate crop contents weakens. The donor raises its forelegs and finally exhibits its own begging movements. This usually terminates the trophallactic food exchange. (From Hölldobler, 1973c.)

ond phase of regurgitation. But when the donor finally raises its forelegs and even conducts begging strokes, the trophallactic contact usually breaks up. These observations suggest that besides the other stimuli, the tactile stimulation of the mouthparts releases regurgitation in food-laden ants. This inference was confirmed experimentally. Ants with a full crop could be induced to regurgitate crop contents by artificially stimulating them at the labium.

Trophallactic relations exist not only between adults but also between larvae and adults. In species of the subfamilies Ponerinae, Myrmicinae, and Formicinae, especially, larvae stimulated by touch discharge small quantities of liquid, which are then readily licked up by the workers (Le Masne, 1953; Maschwitz, 1966b). Maschwitz found relatively high concentrations of amino acids but no carbohydrates in the stomodeal contents of larvae of *Tetramorium caespitum.* Wüst (1973) demonstrated that the stomodeal secretions originate from the labial gland. She found that the labial glands of *Monomorium pharaonis* contain amino acids and proteins, but no carbohydrates. Ant larvae also discharge a clear liquid from the anal region. Le Masne (1953) speculates that these substances originate from the Malpighian tubules and contain waste products. Wüst was able to show that the proctodeal substances are discharged from the rectal bladder and that they contain mainly amino acids. Wüst also provided experimental evidence that these larval secretions can play an important role in the "social food household" of an ant colony. In *Leptothorax curvispinosus,* these larval secretions seem to constitute the main food of the queens (Wilson, 1974).

Brood-Adult Communication

The preceding section presented a few examples of behavioral interactions between larvae and adults in the social Hymenoptera. I will now discuss some signals that have been found to regulate communication between brood and adults.

The fourth- and fifth-instar larvae of hornets (*Vespa crabro* and *V. orientalis*) produce sounds by extending and contracting their bodies rhythmically and thereby scraping their mandibles across the carton wall of the cells. Investigations by Schaudinischky and Ishay (1968) and Ishay and Landau (1972) indicate that these sounds may function as a "food-begging" signal. When the sounds were recorded and played back through a vibrator attached to an empty cell, workers oriented toward the cell and attempted to feed it. According to Ishay and Schwartz (1973), the sound frequencies produced by worker-, queen-, and male-larvae differ from one another. The same authors found species differences in the larval sounds of *V. orientalis* and *V. crabro.* In studying thermoregulation in the nests of hornets, Ishay and Ruttner (1971) discovered that female adults tend to warm older pigmented pupae, even when they are placed outside the nest. Ishay (1972) provided evidence that these pupal wasps emit a pheromone that not only attracts the adults but also evokes warming behavior.

Brooding behavior is also induced by chemical signals in young queens of the bumblebee species *Bombus vosnesenskii* and *B. edwardsii* by chemical signals. Heinrich (1974) demonstrated that in this case the pheromone apparently does not originate from the brood but is deposited onto the brood clump by the queen. The scent guides the queen and subsequently the workers to the site where the brood is located and induces them to provide warming behavior.

Although not much is known about brood-adult communication in honeybees, there is circumstantial evidence that honeybee workers can smell their larvae. Free (1967) has demonstrated that the odor of the brood alone causes honeybee workers to forage for pollen. Further, it has been shown that honeybee workers distinguish not only between worker larvae and male larvae (Haydak, 1958) but also between worker larvae and queen larvae (Woyke, 1971). The cues employed are still unknown.

Ants lick and tend their brood constantly. Numerous observations indicate that this intimate relationship between nurses and brood is based on chemical communication (Watkins and Cole, 1966; Hölldobler, 1965, 1967; Schneirla, 1971; Wilson, 1971). Glancey et al. (1970) claimed to have succeeded in extracting larval pheromones that release adoption behavior in adult ants from larvae of *Solenopsis invicta* (=*S. saevissima*). Walsh and Tschinkel (1974), unable to duplicate the rather unspecific bioassay, developed a more specific assay to test brood-recognition signals. They produced good evidence that in *S. invicta* a nonvolatile contact brood pheromone is distributed evenly over the whole cuticle of larvae, prepupae, and pupae. Although it was not possible to isolate the pheromone, the pupae could be deprived of their attractiveness by extraction. The authors concluded:

> There exists substantial evidence for a brood pheromone. The retrieval of skins and larval body contents on blotter, the persistence of the signal for such long periods after death (72 hours) despite disfigurement of the larval cuticle and the ability of organic solvents to destroy the signal without visibly altering the cuticle are compelling evidence for a pheromone.

I can fully confirm these results by my own independent investigations with *Camponotus ligniperda* (Hölldobler, 1965) and *Formica sanguinea* (Hölldobler, unpublished). In these species I found nonvolatile chemical components, attractive to adult ants, on the pupa's skin. I have circumstantial evidence that these pupal pheromones are at least in part contained in the exuvial liquid. During eclosion of the pupa the pheromones seem to stimulate nurse ants to aid the young in the eclosion process.

This result leads us to the formulation of another important problem. Ant larvae, pupae, and young callow workers can easily be transferred from one colony to another, often even from one species to another (K. Hölldobler, 1948; Plateaux, 1960; Hölldobler, 1967). After a certain age, however, adult workers are no longer accepted by foreign colonies. If, as is generally assumed, the colony odor is caused by the absorption of a specific mixture of environmental odorants into the cuticle, it is not clear why larvae, pupae, and callow workers should not carry the odor of the colony in which they are raised and therefore be just as subject to aggression from members of a foreign colony as their older nest mates are. To explain these contradictions we can hypothesize that in the brood stages the colony odor is masked by the brood-tending pheromones, which are not colony-specific. It is also conceivable that these pheromones have a high position in a hierarchical order of a pheromone system and "dominate" any other colony-specific odorous cues (Hölldobler, 1973a).

This absence of brood discrimination has actually been exploited by many social parasitic ants, which conduct so-called slave raids, during which they rob brood of closely related neighboring species. When these kidnapped pupae eclose in the slave raiders' nest, the young workers are "imprinted" with the odor of their captors' colony and in the future behave in a hostile manner toward their real sisters, who have remained behind in their mutual mother's nest. The assumption of a high position of the brood pheromone in a pheromone system would also imply that the Q/K ratio, i.e., the ratio of pheromone molecules released to the response-threshold concentration (Bossert and Wilson, 1963), should be very low. A high Q/K would saturate the nest with the dominant signal, and colony odors and other chemical signals would become almost ineffective. Indeed, the observations that the brood pheromones are nonvolatile and are effective only in very close ranges support this speculation.

Communication between Ants and Their Guests

We have seen that the complex life within the insect society depends on the efficiency of many different forms of communication. It is therefore notable that a large number of solitary arthropods have acquired the capacity to provide the correct signals to these social insects. They have "broken the code" and are thereby able to take advantage of the benefits of the societies. Ant colonies contain an especially large number of these solitary arthropods. The guests, which are commonly known as myrmecophiles, include many members of the order Coleoptera (beetles) but also many mites, collembolans, flies, wasps, and members of other insect groups. Different species of myrmecophiles occupy different sites within an ant colony. Some live along the trails of the ants, some at the garbage dumps outside the nest, others within the outermost nest chambers, while still others are found within the brood chambers (Hölldobler, 1971b, 1972, 1973a). In each case the requirements of interspecific communication are different.

Some of the most advanced myrmecophilic relationships are found in the staphylinid beetles *Lomechusa strumosa* and several species of the genus *Atemeles. L. strumosa* lives with the red slave-making ant *(Formica sanguinea)* in Europe. *Atemeles pubicollis,* also a European species, is normally found with the mound-making wood ant *(Formica polyctena)* during the summer. But in the winter it inhabits the nests of ants of the genus *Myrmica.* We know from Wasmann's observations, made sixty years ago, that these beetles are both fed and reared by their host ants. The behavioral patterns of the larvae of these beetles are similar for the various species; in particular the larvae prey to a certain extent on their host ants' larvae. It is therefore astonishing that the brood-keeping ants not only tolerate these predators but also feed them as they do their own brood.

Both chemical and mechanical interspecific communication is involved in these unusual relationships. The beetle larvae show a characteristic begging behavior toward their host ants. As soon as they are touched by an ant they rear up and try to make contact with the ant's head. If they succeed they tap the ant's labium with their own mouthparts (Fig. 21). This apparently releases regurgitation of food by the ant. The ant larvae beg for food in much the same way, but less intensely.

Fig. 21. Food-begging behavior of the larva of the myrmecophilous beetle *Atemeles pubicollis.* (From Hölldobler, 1967.)

By feeding ants on honey mixed with radioactive sodium phosphate it is possible to measure the social exchange of food in a colony. These experiments show that when myrmecophilous beetle larvae are present in the brood chamber they obtain a proportionately greater share of the food than the host-ant larvae receive. The presence of ant larvae does not affect the food flow to the beetle larvae, whereas ant larvae always receive less food when they compete with beetle larvae. This finding suggests that the releasing signals presented by the beetle larvae to the brood-keeping ants may be more effective than those presented by the ant larvae themselves.

The beetle larvae are also frequently and intensely groomed by the brood-keeping ants; thus it seemed probable that chemical signals are also involved in this interspecific relationship. The transfer of substances from the larvae to the brood-keeping ants could in fact be demonstrated by experiments with radioactive tracers. These substances are probably secreted by glandular cells, which occur dorsolaterally in the integument of each segment. The biological significance of the secretions was elucidated by the following experiments: Beetle larvae were completely covered with shellac to prevent the liberation of the secretion. They were then placed outside the nest entrance together with freshly killed but otherwise untreated control larvae. The ants quickly carried the control animals into the brood chamber. The shellac-covered larvae on the other hand were either ignored or carried to the garbage dump. It was found that for adoption to be successful at least one segment of the larva had to be shellac-free. Furthermore, it was possible to show that after all the secretions were extracted with acetone the larvae were no longer attractive. However, if the extracted larvae were contaminated with secretions from normal larvae they once again became attractive. Even filter paper dummies soaked in such secretions were carried into the brood chambers.

In sum, the experiments show that the adoption of the beetle larvae and their care within the ant colony depend on chemical signals. It may be that the beetle larvae imitate a pheromone that the ant larvae themselves use in releasing brood-keeping behavior in the adult ants. In obtaining food from the brood-keeping ants, however, the beetle larvae imitate and even exaggerate the food-begging behavior of the ant larvae (Hölldobler, 1967).

The question next arises of how the ant colony manages to survive the intense predation and food parasitism by the beetle larvae. Our observations have suggested a very simple answer. The beetle larvae are cannibalistic, and this factor alone is effective in limiting the number of beetle larvae in the brood chambers at any given time. *Lomechusa* larvae normally occur singly throughout the brood chambers, in contrast to the ant larvae, which are usually clustered together.

After a period of growth the beetle larvae pupate in the summer. At the beginning of autumn they eclose as adult beetles. The newly hatched *Lomechusa* beetles leave the ant nest and after a short period of migration seek adoption in another nest of the same host-ant species. *Atemeles* beetles, on the other hand, migrate from the *Formica* nest, where they have been raised, to the nests of the ant genus *Myrmica.* They winter inside the *Myrmica* brood chambers and in the spring return to a *Formica* nest to breed (Wasmann, 1910; Hölldobler, 1970a). The fact that the adult beetle is tolerated and fed in the nests of ants belonging to two different subfamilies suggests that it is able to communicate efficiently in two different "languages."

The *Atemeles* face a major problem in finding their way from one host species to another. *Formica polyctena* nests normally occur in woodland, while *Myrmica* nests are found in the grassland around the woods. Experiments have revealed that when *Atemeles* leave the *Formica* nest they show high locomotor and flight activity and orientate toward light. This may well explain how they manage to reach the relatively open *Myrmica* habitat. Once they reach the grassland the beetles must distinguish the *Myrmica* ants from the other species present and locate their nests. Laboratory experiments have revealed that they identify the *Myrmica* nests by specific odors. Wind-borne species-specific odors are equally important in the spring movement back to the *Formica* nests.

Having found the hosts, the beetles must secure their own adoption. The process involves the five sequential steps depicted in Fig. 22. First the beetle taps the ant lightly with its antennae and raises the tip of its abdomen toward the ant. The latter structure contains what I call the "appeasement glands." The secretions of these glands, which are immediately licked up by the ant, seem to suppress aggressive behavior. The ant is attracted next by a second series of glands along the lateral margins of the abdomen. The beetle now lowers its abdomen in order to permit the ant to approach. The glandular openings are surrounded by bristles, which are grasped by the ant and used to carry the beetle into the brood chamber. By experimentally occluding the openings of the glands, it could be shown that the secretion is essential for successful adoption. For this reason I have come to label them "adoption glands." Thus the adoption of the adult beetle, like that of the larva, depends on chemical communication. Again it is most probable that an imitation of a species-specific pheromone is involved (Hölldobler, 1970a).

Before leaving the *Formica* nest the *Atemeles* beetle must obtain enough food to enable it to survive the migration to the *Myrmica* nest. This it obtains by begging from the ants. The begging behavior is essentially the same toward both *Formica* and *Myrmica.* The beetle attracts the ant's attention by rapidly drumming on the ant with its antennae. Using its maxillae and forelegs it touches the mouthparts of the ant, thus inducing regurgitation (Fig. 23). As noted previously, the ants themselves employ a similar mechanical stimulation of the mouthparts to obtain food from one another. It is thus clear that *Atemeles* is able to obtain food by imitating these simple tactile food-begging signals.

Finally we can reflect on the significance of host changing, as seen in the beetle *Atemeles.* There are good reasons for believing *Atemeles* first evolved myrmecophilic relationships with *Formica.* We can hypothesize that the ancestral *Atemeles* beetles hatched in *Formica* nests in the autumn and then dispersed, returning to other *Formica* nests only to overwinter. This pattern is seen in *Lomechusa* today (Wasmann, 1915; Hölldobler, 1972). However, in the *Formica* nest, brood-keeping ceases during the winter, and consequently social food flow is reduced. In con-

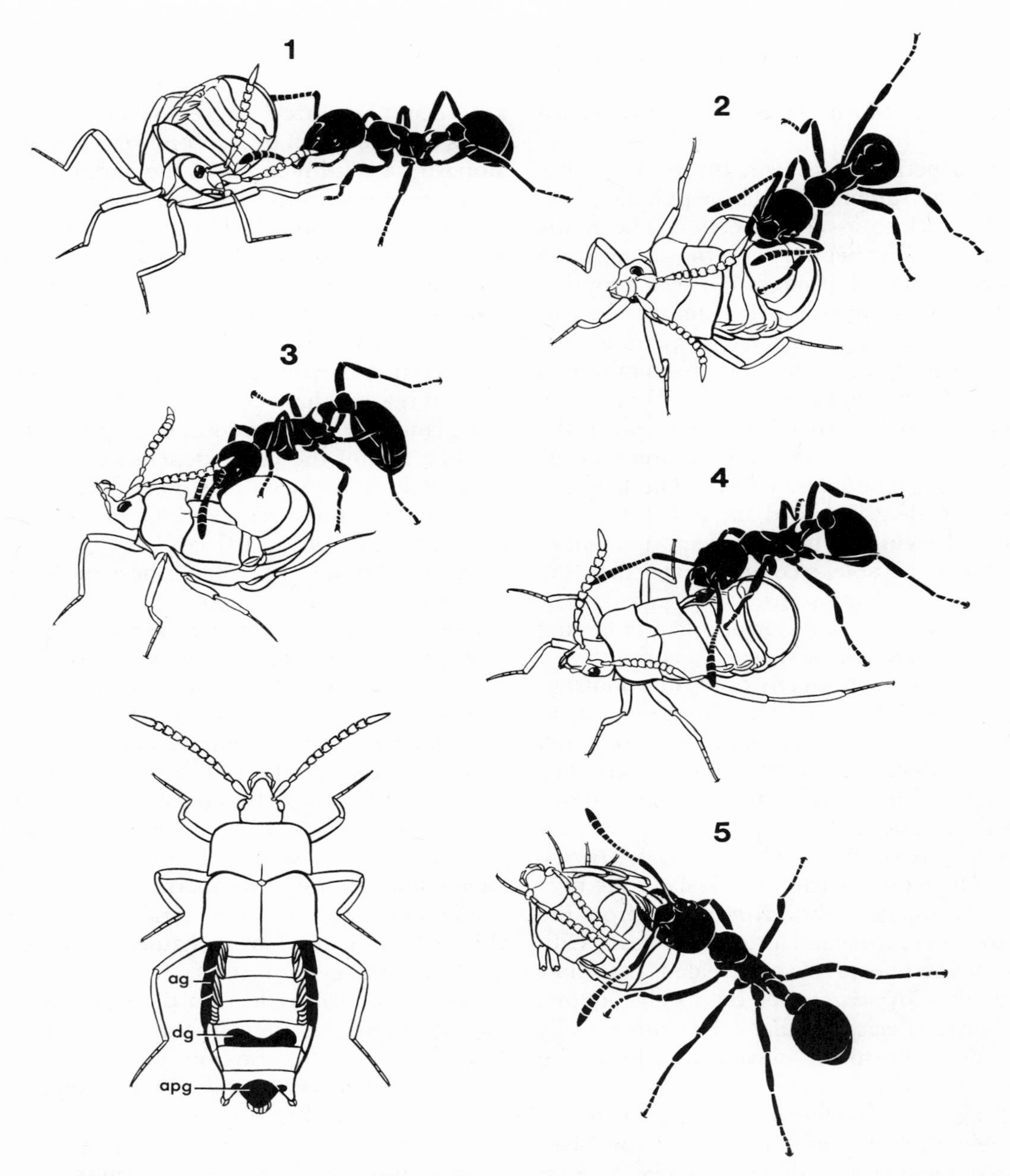

Fig. 22. Behavioral interactions between the beetle *Atemeles pubicollis* (white) and the ant *Myrmica laevinodis* (black) during the adoption process. 1.,2. The beetle antennates and presents its appeasement glands (apg) to the ant. 3. After licking, the ant moves around and licks the adoption glands (ag). 4. The beetle unrolls its abdomen, and the ant picks the beetle up by the bristles associated with the adoption glands. 5. The ant carries the beetle into the nest; the beetle assumes a typical transportation posture. (From Hölldobler, 1969.)

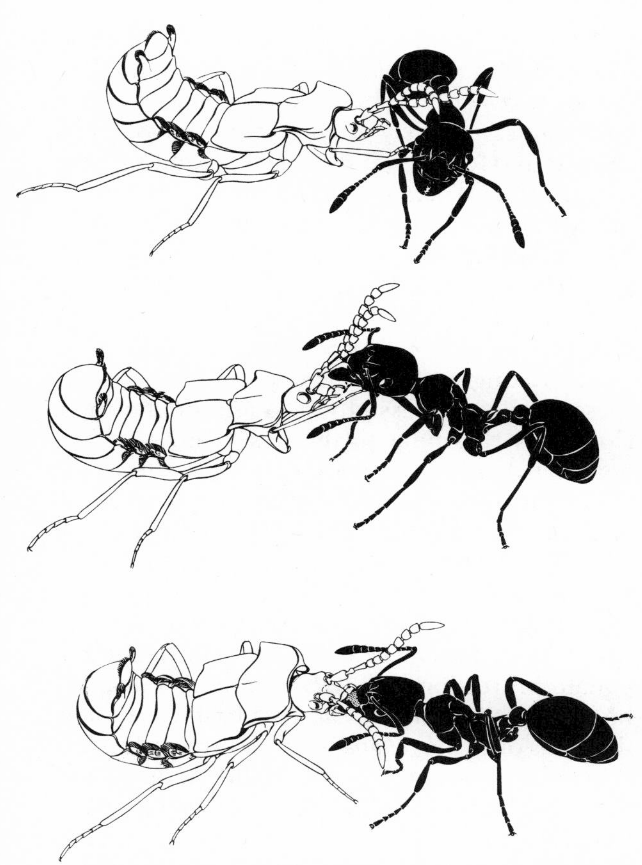

Fig. 23. The myrmecophilous beetle *Atemeles pubicollis* soliciting regurgitation in its host ant *Myrmica laevinodis.* Above: The beetle gains the attention of a worker ant by tapping it with its antennae and forelegs. Middle: The beetle then stimulates the labium of the ant, thereby releasing regurgitation (bottom). (From Hölldobler, 1970a.)

trast, the *Myrmica* colony maintains brood-keeping throughout the winter. Thus in *Myrmica* nests, larvae and nutrient from the social food flow are both available as high-grade food sources to the myrmecophiles. These observations coupled with the fact that the beetles are sexually immature when they hatch suggest why it is advantageous for the beetle to overwinter in *Myrmica* nests. In the *Myrmica* nest gametogenesis proceeds, and when spring comes the beetles are sexually mature. They then return to the *Formica* nest to mate and lay their eggs. At this time the *Formica* are just beginning to raise their own larvae and the social food flow is again optimal. The life cycle and behavior of *Atemeles* is thus synchronized with that of its host ants in such a manner as to take maximum advantage of the social life of each of the two species.

The North American staphylinid myrmecophile *Xenodusa* has a similar life history. The larvae are found in *Formica* nests, and the adults overwinter in the nests of the carpenter ants of the genus *Camponotus* (W. M. Wheeler, 1911). It is undoubtedly significant that *Camponotus,* like *Myrmica,* maintains larvae throughout the winter. It may well be that the host-changing behavior of *Xenodusa* has the same significance as that discussed in *Atemeles.*

All the myrmecophiles described so far possess the necessary repertoire to enable them to live in the brood chambers of the ants' nests. These chambers constitute the optimal niche in an ant colony for a social food-flow parasite. Other myrmecophiles, which lack the ability to communicate with their hosts to this degree, tend to occupy other parts of the colony. For example, staphylinid beetles of the European genus *Dinarda* are usually found in more-peripheral chambers of *Formica sanguinea,* where food exchange occurs between the foragers and the nest workers. It is here that *Dinarda* is able to participate in the social food flow. They obtain food in three ways. Occasionally they insert themselves between two workers exchanging food and literally snatch the food droplet from the donor's mouth (Fig. 24). They also use a simple begging behavior in order to obtain food from returning food-laden foragers. The beetle approaches an ant and touches its labium surreptitiously (Fig. 25). This usually causes the ant to regurgitate a small droplet of food. The ant,

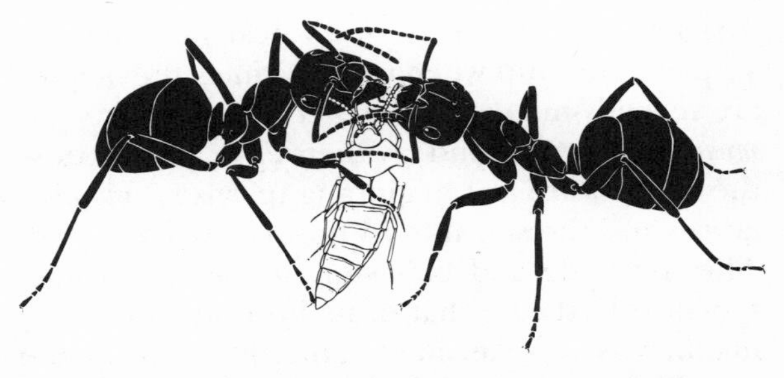

Fig. 24. The myrmecophilous beetle *Dinarda dentata* inserts itself between two ant workers exchanging food. (From Hölldobler, 1973a.)

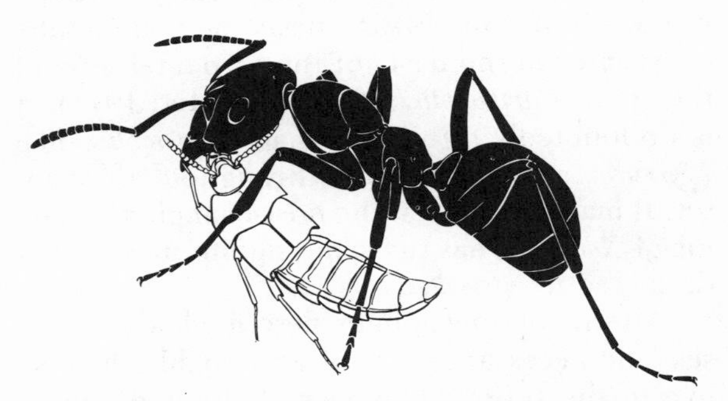

Fig. 25. The beetle *Dinarda dentata* approaches a food-laden ant and touches its labium surreptitiously. This usually causes the ant to regurgitate a small droplet of food. (From Hölldobler, 1973a.)

however, immediately recognizes the beetle as an alien and commences to attack it. At the first sign of hostility the beetle raises its abdomen and offers the ant the appeasement secretion, which is quickly licked up by the ant, and almost immediately the attack ceases. During this brief interval the beetle makes its escape. Other groups of staphylinid beetles, for example, those of the genus *Pella,* live outside the nest on the garbage dumps or along the trails of the ants. Such myrmecophiles have evidently not developed any of the interspecific communication signals that would permit them to live inside the nest chambers. They do, however, possess and use the abdominal appeasement glands when attacked by the ants.

Some of the myrmecophiles prey on ants. For example, *Pella laticollis* lives near the trail of *Lasius fuliginosus* and hunts ants. When attacked by the ants, it quickly provides the appeasement secretions. However, it uses the moment's pause to jump on the back of the ant and kill her by biting between the head and the thorax (Fig. 26). The beetle then drags the ant away from the trail and devours it (Hölldobler et al., unpublished).

Along the trails of *Lasius fuliginosus* the nitidulid beetle *Amphotis marginata* are also to be found. Acting as "highwaymen" in the ant world, these beetles successfully stop and obtain food from ants returning to the nest. Ants that are heavily laden with food are most easily deceived by the beetles' simple begging behavior. Soon after the beetle begins to feed, however, the ant realizes it has been tricked and attacks the beetle. The beetle then is able to defend itself simply by retracting its appendages and flattening itself on the ground. This mechanism gives the beetle adequate protection (Fig. 27). Laboratory experiments showed that *Amphotis* locates the nests and the trails of *Lasius fuliginosus* by recognizing host-specific odors and the trail pheromones laid down by the ants (Hölldobler, 1968; Hölldobler, unpublished).

Other myrmecophiles also utilize the chemically marked trails of their host species to locate

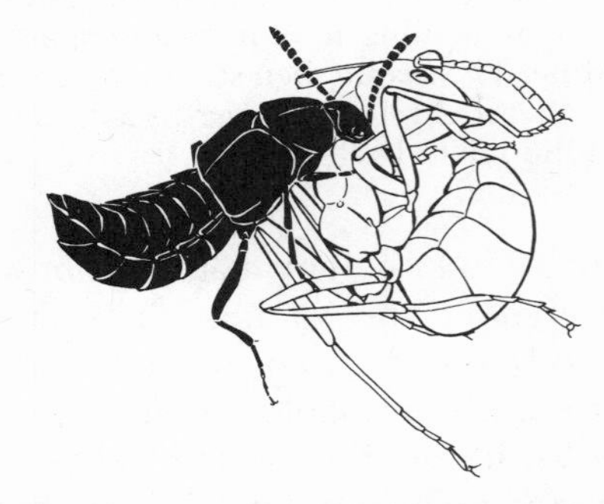

Fig. 26. The predatorial beetle *Pella laticollis* jumps on the back of the ant *Lasius fuliginosus* and kills her by a bite between the head and the thorax.

Fig. 27. The nitidulid beetle *Amphotis marginata* waits in ambush on the foraging trails of *Lasius fuliginosus* for food-laden workers. By stimulating the ant's mouthparts (top) the beetle causes it to regurgitate crop contents (middle). The robbed ant frequently reacts aggressively, but passive defense (bottom) enables the armored beetle to weather the attack. (From Hölldobler, 1971b.)

the host nests or to follow the colony during migrations. This is especially true for the myrmecophiles associated with army ants (Akre and Rettenmeyer, 1968). Moser (1964) reported that the myrmecophilic cockroach *Attaphila fungicola,* which lives in nests of the leaf cutter ant *Atta texana,* follows artificial trails laid down with the trail pheromone of the host ants.

In short, the success of the myrmecophiles depends largely on their ability to communicate with their hosts. Interspecific communication between a myrmecophile and its host might arise in evolution in two ways. First, we can think of the ant as a signal transmitter and the potential myrmecophile as a signal receiver. By the gradual evolutionary modification of its receptor system and behavior, the myrmecophile has succeeded in discriminating the transmitter's signals. In this way the myrmecophiles may have evolved the ability to recognize the odors of their specific hosts, the difference between host adults and larvae, and so forth. Second, the myrmecophile can be regarded as the signal transmitter and the potential host ant as the signal receiver. Beetle signals that induce social behavior in the ants have been favored in natural selection and very gradually improved. In both cases the ant's behavior serves as the model that the beetle mimics. The evolution of the myrmecophilous relationship therefore involves adaptive change in the potential myrmecophile only. By comparative analyses of the interspecific associations and communication mechanisms of closely related species it is possible to reconstruct a picture of the possible evolutionary pathways that led to the highly specialized social parasitic relationships in ant societies. The predatory behavior of *Pella laticollis* and the more primitive myrmecophilic behavior of *Dinarda dentata* may very well represent early evolutionary steps that have led in the end to the highly adapted myrmecophilic behavior of *Atemeles* and *Lomechusa.*

References

Akre, R. D., and Rettenmeyer, C. W., 1968. Trail-following by guests of army ants (Hymenoptera: Formicidae: Ecitonini). *J. Kansas Entomol. Soc.,* 41:165–74.

Amoore, J. E.; Palmieri, G.; Wanke, E.; and Blum, M. S.; 1969. Ant alarm pheromone activity: correlation with molecular shape by scanning computer. *Science,* 165:1266–69.

Barbier, J., and Lederer, E., 1960. Structure chimique de la substance royale de la reine d'abeille (*Apis mellifica* L.). *C. R. Acad. Sci.* (Paris), 250:4467–69.

Beetsma, J., and Schoonhoven, L. M., 1966. Some chemosensory aspects of the social relations between the queen and the worker in the honeybee. *Proc. Kon. ned. Acad. Wet.,* ser. C, 69:643–47.

Bergström, G.; Kullenberg, B.; and Ställberg-Stenhagen, S.; 1973. Studies on natural odoriferous compounds, VII: recognition of two forms of *Bombus lucorum* L. (Hymenoptera, Apidae) by analysis of the volatile marking secretions from individual males. *Chemica Scripta,* 3. Preprint, pp.1–9.

Bergström, G.; Kullenberg, B.; Ställberg-Stenhagen, S.: and Stenhagen, E.; 1968. Studies on natural odoriferous compounds, II: identification of a 2,3-dihydrofarnesol as the main component of the marking perfume of the male bumble-bees of the species *Bombus terrestris* L. *Arkiv Kemi,* 28:453–69.

Bergström, G., and Löfquist, J., 1968. Odour similarities between the slave-keeping ants *Formica sanguinea* and *Polyergus rufescens* and their slaves *Formica fusca* and *Formica rufibarbis. J. Insect Physiol.,* 14:995–1011.

Bergström, G., and Löfquist, J., 1970. Chemical basis for odour communication in four species of *Lasius* ants. *J. Insect Physiol.,* 16:2353–75.

Bergström, G., and Löfquist, J., 1972. Similarities between the Dufour's gland secretions of the ants *Camponotus ligniperda* (Latr.) and *Camponotus herculeanus* (L.). *Ent. scand.,* 3:225–38.

Bergström, G., and Löfquist, J., 1973. Chemical congruence of the complex odoriferous secretions from Dufour's gland in three species of ants of the genus *Formica. J. Insect Physiol.,* 19:877–907.

Bernardi, C.; Cardani, D.; Ghiringhelli, D.; Selva, A.; Baggini, A.; and Pavan, M.; 1967. On the components of secretion of mandibular glands of the ant *Lasius (Dendrolasius) fuliginosus. Tetrahedron Letters,* 40:3893–96.

Bevan, C. W. L.; Birch, A. J.; and Caswell, H.; 1961. An insect repellent from black cocktail ants. *J. Chem. Soc.,* part 1:488.

Bier, K. H., 1958. Die Regulation der Sexualität in den Insektenstaaten. *Ergebnisse der Biologie,* 20:97–126.

Blum, M. S., 1966a. Chemical releasers of social behavior, VIII: citral in the mandibular gland secretion of *Lestrimelitta limao. Ann. Entomol. Soc. Amer.,* 59:962–64.

Blum, M. S., 1966b. The source and specificity of trail pheromones in *Termitopone, Monomorium* and *Huberia* and their relation to those of some other ants. *Proc. Roy. Entomol. Soc.* (London), 41:155–60.

Blum, M. S., 1974. Pheromonal sociality in the hymenoptera. In: *Pheromones,* M. C. Birch, ed. Amsterdam: North-Holland Publishing Co., pp.223–49.

Blum, M. S.; Boch, R.; Doolittle, R. E.; Tribble, M. T.; and Traynham, J. G.; 1971b. Honeybee sex attractant: conformational analysis, structural specificity, and lack of masking activity of congeners. *J. Insect Physiol.,* 17:349–64.

Blum, M. S.; Crewe, R. M.; Kerr, W. E.; Keith, L. H.; Garrison, A. W.; and Walker, M. M.; 1970. Citral in stingless bees: isolation and function in trail laying and robbing. *J. Insect Physiol.,* 16:1637–48.

Blum, M. S.; Crewe, R. M.; Sudd, J. H.; and Garrison, A. W.; 1969. 2-Hexenal: isolation and function in a *Crematogaster (Atopogyne)* sp. *J. Georgia Entomol. Soc.,* 4:145–48.

Blum, M. S.; Doolittle, R. E.; and Beroza, M.; 1971a. Alarm pheromones: utilization in evaluation of olfactory theories. *J. Insect Physiol.,* 17:2357–61.

Blum, M. S.; Moser, J. C.; and Cordero, A. D.; 1964. Chemical releaser of social behavior II. Source and specificity of the odor trail substances in four attine genera (Hymenoptera: Formicidae). *Psyche* (Cambridge), 71:1–7.

Blum, M. S., and Ross, G. N., 1965. Chemical releasers of social behaviour, V: source, specificity and properties of the odour trail pheromone of *Tetramorium guineense* (F.) (Formicidae, Myrmicinae). *J. Insect Physiol.,* 11:857–68.

Blum, M. S.; Water, S. L.; and Traynham, J. G.; 1966. Chemical releasers of social behavior, VI: the relation of structure to activity of ketones as releasers of alarm for *Iridomyrmex pruinosus* (Roger). *J. Insect Physiol.,* 12:419–27.

Blum, M. S., and Wilson, E. O., 1964. The anatomical source of trail substances in formicine ants. *Psyche* (Cambridge), 71:28–31.

Boch, R., 1956. Die Tänze der Bienen bei nahen und fernen Trachtquellen. *Z. vergl. Physiol.,* 38:136–67.

Boch, R., 1957. Rassenmassige Unterscheide bei den Tänzen der Honigbiene *(Apis mellifica). Z. vergl. Physiol.,* 40:289–320.

Boch, R., and Shearer, D. A., 1962. Identification of geraniol as the active component in the Nasanoff pheromone of the honeybee. *Nature* (London), 194:704–706.

Boch, R., and Shearer, D. A., 1964. Identification of nerolic and geranic acids in the Nasanoff pheromone of the honeybee. *Nature* (London), 202:320–21.

Boch, R.; Shearer, D. A.; and Petrasovits, A.; 1970. Efficacies of two alarm substances of the honeybee. *J. Insect Physiol.,* 16:17–24.

Boch, R.; Shearer, D. A.; and Stone, B. C.; 1962. Identification of iso-amylacetate as an active compound in the sting pheromone of the honeybee. *Nature* (London), 195:1018–20.

Bossert, W. H., and Wilson, E. O., 1963. The analysis of olfactory communication among animals. *J. Theoret. Biology,* 5:443–69.

Brand, J. M.; Duffield, R. M.; McConnell, J. G.; and Fales, H. M.; 1973. Caste-specific compounds in male carpenter ants. *Science,* 179:388–89.

Brian, M. V., 1970. Communication between queens and larvae in the ant *Myrmica. Anim. Behav.,* 18:467–72.

Brian, M. V., and Blum, M. S., 1969. The influence of *Myrmica* queen head extracts on larval growth. *J. Insect Physiol.,* 15:2213–23.

Bringer, B., 1973. Territorial flight of bumble bee males in coniferous forest on the northernmost part of the island of Öland. *Zoon,* suppl. 1, pp.15–22.

Buschinger, A., 1968. "Locksterzeln" begattungsbereiter ergatoider Weibchen von *Harpagoxenus sublaevis* Nyl. (Hymenoptera, Formicidae). *Experientia,* 24:297.

Buschinger, A., 1971a. Weitere Untersuchungen zum Begattungsverhalten sozialparasitischer Ameisen (*Harpagoxenus sublaevis* Nyl. und *Doromyrmex pacis* Kutter, Hym. Formicidae). *Zool. Anz.,* 187:184–98.

Buschinger, A., 1971b. "Locksterzeln" und Kopula der sozialparasitischen Ameise *Leptothorax kutteri* Buschinger (Hym. Form.). *Zool. Anz.,* 186:242–48.

Buschinger, A., 1972a. Giftdrüsensekret als Sexualpheromon bei der Ameise *Harpagoxenus sublaevis. Naturwissenschaften,* 59:313–14.

Buschinger, A., 1972b. Kreuzung zweier sozialparasitischer Ameisenarten, *Doromyrmex pacis* Kutter und *Leptothorax kutteri* Buschinger (Hym. Formicidae). *Zool. Anz.,* 189:169–79.

Butenandt, A.; Linzen, B.; and Lindauer, M.; 1959. Über einen Duftstoff aus der Mandibeldrürse der Blattschneiderameise *Atta sexdens rubropilosa* Forel. *Arch. Anat. Microscop. Morphol. Exper.,* 48:13–19.

Butler, C. G., 1954a. The method and importance of the recognition by a colony of honeybees *(A. mellifera)* of the presence of its queen. *Trans. Roy. Entomol. Soc. London,* 105:11–29.

Butler, C. G., 1954b. *The World of the Honeybee.* London: Collins.

Butler, C. G., 1966. Mandibular gland pheromone of worker honeybees. *Nature* (London), 212:530.

Butler, C. G., 1971. The mating behavior of the honeybee *(Apis mellifera)* L. *J. Entomol.,* 46:1–11.

Butler, C. G., 1973. The queen and the "spirit of the hive." *Proc. Roy. Ent. Soc.* (London), 48:59–65.

Butler, C. G., and Calam, D. H., 1969. Pheromones of the honeybee: the secretion of the Nasanoff gland of the worker. *J. Insect Physiol.,* 15:237–44.

Butler, C. G.; Calam, D. H.; and Callow, R. K.; 1967. Attraction of *Apis mellifera* drones by the odours of the queens of two other species of honeybees. *Nature* (London), 213:423–24.

Butler, C. G.; Callow, R. K.; Koster, C. G.; and Simpson, J.; 1973. Perception of the queen by workers in the honeybee colony. *J. Apicult. Res.,* 12:159–66.

Butler, C. G., and Fairey, E. M., 1964. Pheromones of the honeybee: biological studies of the mandibular gland secretion of the queen. *J. Apicult. Res.,* 3:65–67.

Butler, C. G.; Fletcher, D. J. C.; and Watler, D.; 1969. Nest-entrance marking with pheromones by the honeybee *Apis mellifera* L., and by a wasp *Vespula vulgaris* L. *Anim. Behav.,* 17:142–47.

Butler, C. G.; Fletcher, D. J. C.; and Watler, D.; 1970. Hive entrance finding by honeybee *(Apis mellifera)* foragers. *Anim. Behav.,* 18:78–91.

Butler, C. G., and Gibbons, D. A., 1958. The inhibition of queen rearing by feeding queenless worker honeybees *(A. mellifera)* with an extract of "queen substance." *J. Insect Physiol.,* 2:61–64.

Butler, C. G., and Simpson, J., 1967. Pheromones of the queen honeybee *(Apis mellifera)* which enable her workers to follow her when swarming. *Proc. Roy. Ent. Soc.* (London), ser. A, 42:149–54.

Calam, D. H., 1969. Species and sex-specific compounds from the heads of male bumble bees (*Bombus* spp.). *Nature* (London), 221:856–57.

Callow, R. K., and Johnston, N. C., 1960. The chemical constitution and synthesis of queen substance of honeybees *(Apis mellifera). Bee World,* 41:152–53.

Carthy, J. D., 1950. Odour trails of *Acanthomyops fuliginosus. Nature* (London), 166:154.

Carthy, J. D., 1951. The orientation of two allied species of British ants, II: odour trail laying and following in *Acanthomyops (Lasius) fuliginosus. Behaviour,* 3:304–18.

Casnati, G.; Ricca, A.; and Pavan, M.; 1967. Sulla secrezione difensiva della glandole mandibolari di *Paltothyreus tarsatus* (Fabr.). *Chim. Ind.* (Milan), 49:57–61.

Crewe, R. M., and Fletcher, D. J. C., 1974. Ponerine ant secretions: the mandibular gland secretions of *Paltothyreus tarsatus* Fabr. *J. Entomol. Soc. Sth. Afr.,* 37:291–98.

Dobrzanski, J., 1966. Contribution to the ethology of *Leptothorax acervorum* (Hymenoptera: Formicidae). *Acta Biol. Exp.* (Warsaw), 26:71–78.

du Buysson, R., 1903. Monographie de Guêpe ou *Vespa. Ann. Soc. Entomol. France,* 72:260–88.

Duffield, R. M., and Blum, M. S., 1973. 4-Methyl-3-heptanone: identification and function in *Neoponera villosa* (Hymenoptera: Formicidae). *Ann. Entomol. Soc. Amer.,* 66:1357.

Dumpert, K., 1972. Alarmstoffrezeptoren auf der Antenne von *Lasius fuliginosus* (Latr.) (Hymenoptera, Formicidae). *Z. vergl. Physiol.,* 76:403–25.

Esch, H., 1961. Über die Schallerzeugung beim Werbetanz der Honigbiene. *Z. vergl. Physiol.,* 45:1–11.

Esch, H., 1964. Beiträge zum Problem der Entfernungsweisung in den Schwänzeltänzen der Honigbienen. *Z. vergl. Physiol.,* 48:534–46.

Esch, H., 1967a. Die Bedeutung der Lauterzeugung für die Verständigung der stachellosen Bienen. *Z. vergl. Physiol.,* 56:199–220.

Esch, H., 1967b. The evolution of bee language. *Scientific American,* 216:96–104.

Esch, H., and Bastian, J. A., 1970. How do newly recruited honeybees approach a food site? *Z. vergl. Physiol.,* 68:175–81.

Esch, H.; Esch, I.; and Kerr, W. E.; 1965. An element common to communication of stingless bees and to dances of honeybees. *Science,* 149:320–21.

Evans, H. E., and Eberhard, M. J. West, 1970. *The Wasps.* Ann Arbor: University of Michigan Press. 265pp.

Falke, J., 1968. Substanzen aus der Mandibeldrüse der Männchen von *Camponotus herculeanus.* Diss. University of Heidelberg.

Fletcher, D. J. C., 1971. The glandular source and social functions of trail pheromones in two species of ants *(Leptogenys). J. Entomol.,* ser. A, 46:27–37.

Fletcher, D. J. C., and Brand, J. M., 1968. Source of the trail pheromone and method of trail laying in the ant *Crematogaster peringueyi. J. Insect Physiol.,* 14:783–88.

Frank, A., 1941. Eigenartige Flugbahnen bei Hummelmännchen. *Z. vergl. Physiol.,* 28:467–84.

Free, J. B., 1956. A study of stimuli which release the food begging and offering response of worker honeybees. *Brit. J. Anim. Behav.,* 4:94–101.

Free, J. B., 1959. The transfer of food between the adult members of a honeybee community. *Bee World,* 40:193–201.

Free, J. B., 1967. Factors determining the collection of pollen by honeybee foragers. *Anim. Behav.,* 15:134–44.

Free, J. B., 1970b. The flower constancy of bumblebees. *J. Anim. Ecol.,* 39:395–402.

Free, J. B., 1971. Stimuli eliciting mating behavior of bumblebee *(Bombus bratorum)* males. *Behaviour,* 40:55–61.

Free, J. B., and Williams, J. H., 1970. Exposure of the Nasanov gland by honeybees *(Apis mellifera)* collecting water. *Behaviour,* 37:286–90.

Frisch, K. von, 1923. Über die "Sprache" der Bienen, eine tierpsychologische Untersuchung. *Zool. Jb.* (Physiol.), 40:1–186.

Frisch, K. von, 1962. Über die durch Licht bedingte "Missweisung" bei den Tänzen im Bienenstock. *Experientia,* 18:49–53.

Frisch, K. von, 1965. *Tanzsprache und Orientierung der Bienen.* Berlin: Springer-Verlag.

Frisch, K. von, 1967a. *The Dance Language and Orientation of Bees.* Cambridge: Belknap Press of Harvard University Press.

Frisch, K. von, 1967b. Honeybees: do they use direction and distance information provided by their dancers? *Science,* 158:1072–76.

Frisch, K. von, 1968. The role of dance in recruiting bees to familiar sites. *Anim. Behav.,* 16:531–33.

Gabba, A., and Pavan, M., 1970. Researches on trail and alarm substances in ants. In: *Communication by Chemical Signals,* J. W. Johnston, D. G. Moulton, and A. Turk, eds. New York: Appleton-Century-Crofts, pp.161–203.

Gary, N. E., 1961a. Mandibular gland extirpation in living queen and worker honey-bees *(Apis mellifera). Ann. ent. Soc. Amer.,* 54:529–31.

Gary, N. E., 1961b. Queen honeybee attractiveness as related to mandibular gland secretion. *Science,* 133:1479–80.

Gary, N. E., 1962. Chemical mating attractants in the queen honey bee. *Science,* 136:773–74.
Gary, N. E., 1963. Observations of mating behavior in the honeybee. *J. Apicult. Res.,* 2:3–13.
Gary, N. E., 1974. Pheromones that affect the behavior and physiology of honeybees. In: *Pheromones,* M. C. Birch, ed. Amsterdam: North-Holland Publishing Co., pp.200–21.
Gerig, L., 1972. Ein weiterer Duftstoff zur Anlockung der Drohnen von *Apis mellifica* (L.). *Z. angew. Entomol.,* 70:286–89.
Glancey, B. M.; Stringer, C. E.; Craig, C. H.; Bishop, P. M.; and Martin, B. B.; 1970. Pheromone may induce brood tending in the fire ant, *Solenopsis saevissima. Nature* (London), 226:863–64.
Goetsch, W., 1953. *Vergleichende Biologie der Insektenstaaten.* Leipzig: Geest u. Portig K. G. 482pp.
Gösswald, K., and Kloft, W., 1956. Untersuchungen über die Verteilung von radioaktiv markiertem Futter im Volk der kleinen Roten Waldameise *(Formica rufopratensis minor). Waldhygiene,* 1:200–202.
Gösswald, K., and Kloft, W., 1960. Neuere Untersuchungen über die sozialen Wechselbeziehungen im Ameisenvolk, durchgeführt mit Radio-Isotopen. *Zool. Beitr.,* 5:519–56.
Gould, J. L., 1974. Honey bee communication: misdirection of recruits by foragers with covered ocelli. *Nature* (London), 252:300–301.
Gould, J., 1975. Honey bee recruitment: the dance-language controversy. *Science,* 189:685–93.
Gould, J. L.; Henerey, M.; and MacLeod, M. C.; 1970. Communication of direction by the honeybee. *Science,* 169:544–54.
Groot, A. P. de, and Voogd, S., 1954. On the ovary development in queenless workerbees (*Apis mellifera* L.). *Experientia,* 10:384–85.
Haas, A., 1946. Neue Beobachtungen zum Problem der Flugbahnen bei Hummelmännchen. *Z. Naturforsch.,* 1:596–600.
Haas, A., 1949a. Artypische Flugbahnen von Hummelmännchen. *Z. vergl. Physiol.,* 31:281–307.
Haas, A., 1949b. Gesetzmässiges Flugverhalten der Männchen von *Psithyrus silvestris* Lep. und einiger solitärer Apiden. *Z. vergl. Physiol.,* 31:671–83.
Haas, A., 1952. Die Mandibeldrüse als Duftorgan bei einigen Hymenopteren. *Naturwissenschaften,* 39:484.
Hangartner, W., 1967. Spezifität und Inaktivierung des Spurpheromons von *Lasius fuliginosus* Latr. und Orientierung der Arbeiterinnen im Duftfeld. *Z. vergl. Physiol.,* 57:103–36.
Hangartner, W., 1969a. Trail laying in the subterranean ant *Acanthomyops interjectus. J. Insect Physiol.,* 15:1–4.
Hangartner, W., 1969c. Structure and variability of the individual odor trail in *Solenopsis* (Formicidae). *Z. vergl. Physiol.,* 62:111–20.
Hangartner, W., and Bernstein, S., 1964. Über die Geruchsspur von *Lasius fuliginosus* zwischen Nest und Futterquelle. *Experientia,* 20:392–93.
Hangartner, W.; Reichson, J.; and Wilson, E. O.; 1970. Orientation to nest material by the ant *Pogonomyrmex badius* (Latreille). *Anim. Behav.,* 18:331–34.
Hansson, A., 1945. Lauterzeugung und Lautauffassungsvermögen der Bienen. *Opuscula Entomol.,* suppl. 6, pp.1–124.
Hase, A., 1935. Über den "Verkehr" am Wespennest, nach Beobachtungen an einer tropischen Art. *Naturwissenschaften,* 23:780–83.
Haskins, C. P., and Whelden, R. M., 1965. "Queenless" worker sibship, and colony versus population structure in the formicid genus *Rhytidoponera. Psyche,* (Cambridge), 72:87–112.
Haydak, M. H., 1958. Do the nurse honeybees recognize the sex of the larvae? *Science,* 127:1113.
Heinrich, B., 1974. Pheromone induced brooding behavior in *Bombus vosnesenskii* and *B. edwardsii* (Hymenoptera: Bombidae). *J. Kansas Entomol. Soc.,* 47:396–404.
Hingston, R. W. G., 1929. *Instinct and Intelligence.* New York: Macmillan Co.
Hölldobler, B., 1962. Zur Frage der Oligogynie bei *Camponotus ligniperda* Latr. und *Camponotus herculeanus* L. (Hym. Formicidae). *Z. angew. Entomol.,* 49:337–52.
Hölldobler, B., 1965. Das soziale Verhalten der Ameisenmännchen und seine Bedeutung für die Organisation der Ameisenstaaten. Diss., University of Würzburg. 122pp.
Hölldobler, B., 1966. Futterverteilung durch Männchen im Ameisenstaat. *Z. vergl. Physiol.,* 52:430–55.
Hölldobler, B., 1967. Zur Physiologie der Gast-Wirt-Beziehungen (Myrmecophilie) bei Ameisen, I: Das Gastverhältnis der *Atemeles-* und *Lomechusa*-Larven (Col. Staphylinidae) zu *Formica* (Hym. Formicidae). *Z. vergl. Physiol.,* 56:1–21.
Hölldobler, B., 1968. Der Glanzkafer als "Wegelagerer" an Ameisenstrassen. *Naturwissenschaften,* 55:397.
Hölldobler, B., 1969. Orientierungsmechanismen des Ameisengastes *Atemeles* (Coleoptera, Staphylinidae) bei der Wirtssuche. *Zool. Anz.,* suppl. 33, pp.580–85.
Hölldobler, B., 1970a. Zur Physiologie der Gast-Wirt-

Beziehungen (Myrmecophilie) bei Ameisen, II: Das Gastverhältnis der imaginalen *Atemeles pubicollis* Bris (Col. Staphylinidae) zu *Myrmica* und *Formica* (Hym. Formicidae). *Z. vergl. Physiol.,* 66:215–50.
Hölldobler, B., 1970b. Chemische Verständigung im Insektenstaat. *Umschau,* 70:663–69.
Hölldobler, B., 1971a. Sex pheromone in the ant *Xenomyrmex floridanus. J. Insect Physiol.,* 17:1497–99.
Hölldobler, B., 1971b. Communication between ants and their guests. *Scientific American,* 224:86–93.
Hölldobler, B., 1971c. Recruitment behavior in *Camponotus socius* (Hym. Formicidae). *Z. vergl. Physiol.,* 75:123–42.
Hölldobler, B., 1971d. Homing in the harvester ant *Pogonomyrmex badius. Science,* 171:1149–51.
Hölldobler, B., 1972. Verhaltensphysiologische Adaptationen an ökologische Nischen in Ameisennestern. *Verh. Dtsch. Zool. Ges.,* 65:137–44.
Hölldobler, B., 1973a. Zur Ethologie der chemischen Verständigung bei Ameisen. *Nova Acta Leopoldina,* 37:259–92.
Hölldobler, B., 1973b. Chemische Strategie beim Nahrungserwerb der Diebsameise (*Solenopsis fugax* Latr.) und der Pharaoameise (*Monomorium pharaonis* L.). *Oecologia,* 11:371–80.
Hölldobler, B., 1973c. *Formica sanguinea* (Formicidae): Futterbetteln. *Encyclopaedia Cinematographica,* E 2.
Hölldobler, B., 1974. Home range orientation and territoriality in harvesting ants *(Pogonomyrmex). Proc. Nat. Acad. Sci.* (U.S.A.), 71:3274–77.
Hölldobler, B., and Maschwitz, U., 1965. Der Hochzeitsschwarm der Rossameise *Camponotus herculeanus* L. (Hym. Formicidae). *Z. vergl. Physiol.,* 50:551–68.
Hölldobler, B.; Möglich, M.; and Maschwitz, U.; 1973. *Bothroponera tesserinoda* (Formicidae): Tandemlauf beim Nestumzug. *Encyclopaedia Cinematographica* (E 2040/1973), 3–14.
Hölldobler, B.; Möglich, M.; and Maschwitz, U.; 1974. Communication by tandem running in the ant *Camponotus sericeus. J. Comp. Physiol.,* 90:105–27.
Hölldobler, B., and Wilson, E. O., 1970. Recruitment trails in the harvester ant *Pogonomyrmex badius. Psyche* (Cambridge), 77:385–99.
Hölldobler, B., and Wüst, M., 1973. Ein Sexualpheromon bei der Pharaoameise *Monomorium pharaonis* (L.). *Z. Tierpsychol.,* 32:1–9.
Hölldobler, K., 1948. Über ein parasitologische Problem: Die Gastpflege der Ameisen und die Symphilieinstinkte. *Z. Parasitenkunde,* 14:3–26.
Huwyler, S.; Grob, K.; and Viscontini, M.; 1973. Identifizierung von sechs Komponenten des Spurpheromons der Ameisenart *Lasius fuliginosus. Helvetica Chimica Acta,* 56:976–77.
Ikan, R.; Bergmann, E. D.; Ishay, J.; and Gitter, S.; 1968. Proteolytic enzyme activity in the various colony members of the oriental hornet, *Vespa orientalis* F. *Life Sciences,* 7:929–34.
Ikan, R.; Gottlieb, R.; Bergmann, E. D.; and Ishay, J.; 1969. The pheromone of the queen of the oriental hornet, *Vespa orientalis. J. Insect Physiol.,* 15:1709–12.
Ishay, J., 1972. Thermoregulatory pheromones in wasps. *Experientia,* 28:1185–87.
Ishay, J., and Ikan, R., 1969. Gluconeogenesis in the oriental hornet *Vespa orientalis* F. *Ecology,* 49:169–71.
Ishay, J., and Landau, E. M., 1972. *Vespa* larvae send out rhythmic hunger signals. *Nature* (London), 237:286–87.
Ishay, J., and Ruttner, F., 1971. Thermoregulation im Hornissennest. *Z. vergl. Physiol.,* 72:423–34.
Ishay, J., and Schwartz, A., 1973. Acoustical communication between the members of the oriental hornet *(Vespa orientalis)* colony. *J. Acoust. Soc. Amer.,* 63:640–49.
Janet, C., 1903. *Observations sur les guêpes.* Paris: C. Nand. 85pp.
Jeanne, R. L., 1972. Social biology of the neotropical wasp *Mischocyttarus drewseni. Bull. Mus. Comp. Zool. Harvard,* 144:63–150.
Johnson, D. L., 1967. Honeybees: Do they use direction information contained in their dance maneuver? *Science,* 155:844–47.
Johnson, D. L., and Wenner, A. M., 1970. Recruitment efficiency in honeybees: studies of the role of olfaction. *J. Apicult. Res.,* 9:13–18.
Kaissling, K. E., and Renner, M., 1968. Antennale Rezeptoren für Queen substance und Sterzelduft bei der Honigbiene. *Z. vergl. Physiol.,* 59:357–61.
Kalmus, H., 1954. Finding and exploitation of dishes of syrup by bees and wasps. *Brit. J. Anim. Behav.,* 2:136–39.
Kannowski, P. B., and Johnson, R. L., 1969. Male patrolling behavior and sex attraction in ants of the genus *Formica. Anim. Behav.,* 17:425–29.
Kerr, W. E., and Esch, H., 1965. Comunicasao entre as abelhas sociais brasileiras e sua contribuisao para o entendimento da sua evolvusao. *Ciencia e Cult.* (Sao Paulo), 17:529–38.
Kerr, W. E.; Ferreira, A.; and DeMattos, N. S.; 1963. Communication among stingless bees: additional data (Hymenoptera, Apidae). *J. N. Y. Entomol. Soc.,* 71:80–90.
Kloft, W., 1959. Versuch einer Analyse der trophobi-

otischen Beziehungen von Ameisen zu Aphiden. *Biol. Zentralbl.,* 78:863–70.

Kneitz, G., 1963. Tracerversuche zur Futterverteilung bei Waldameisen. *Symp. Gen. Biol. Ital.* (Pavia), 12:38–50.

Koeniger, N., 1970a. Über die Fähigkeit der Bienenkönigin *(Apis mellifica)* zwischen Arbeiterinnen und Drohnenzellen zu unterscheiden. *Apidologie,* 1:115–42.

Koeniger, N., 1970b. Factors determining the laying of drone and worker eggs by the queen honeybee. *Bee World,* 51:166–69.

Koeniger, N., and Fuchs, S., 1972. Kommunikativ Schallerzeugung von *Apis cerana* Fabr. im Bienenvolk. *Naturwissenschaften,* 59:169.

Koeniger, N., and Fuchs, S., 1973. Sound production as colony defense in *Apis cerana* Fabr. *Proc. Seventh Congr. IUSSI,* London, pp.199–204.

Kullenberg, B., 1956. Field experiments with chemical sexual attractants on aculeate Hymenoptera males. *I. Zool. Bidrag Upsala,* 31:253–54.

Kullenberg, B.; Bergström, G.; Bringer, B.; Carlberg, B.; and Cederberg, B.; 1973. Observations on scent marking by *Bombus* Latr. and *Psithyrus* Lep. males (Hym. Apidae) and localization of site of production of the secretion. *Zoon,* suppl. 1, pp.23–30.

Kullenberg, B.; Bergström, G.; and Ställberg-Stenhagen, S.; 1970. Volatile components of the cephalic marking secretion of male bumble bees. *Acta Chemica Scand.,* 24:1481–83.

Lange, R., 1958. Der Einfluss der Königin auf die Futterverteilung im Ameisenstaat. *Naturwissenschaften,* 45:196.

Lange, R., 1967. Die Nahrungsverteilung unter den Arbeiterinnen des Waldameisenstaates. *Z. Tierpsychol.,* 24:513–45.

Le Masne, G. M., 1952. Les échanges alimentaires entre adultes chez la fourmi *Ponera eduardi* Forel. *C. R. Acad. Sci.* (Paris), 235:1549–51.

Le Masne, G. M., 1953. Observations sur les relations entre le couvain et les adultes chez les fourmis. *Ann. Sci. Nat.,* 15:1–56.

Lenoir, M. A., 1972a. Note sur le comportement de sollicitation chez les ouvrières de *Myrmica scabrinodis* Nyl. (Hymenoptères, Formicidae). *C. R. Acad. Sci.* (Paris), 274:705–707.

Lenoir, M. A., 1972b. Sur la rôle de l'odorat dans le compartement de sollicitation chez les ouvrières de *Myrmica scabrinodis* Nyl. (Hymenoptères, Formicidae). *C. R. Acad. Sci.* (Paris), 274:906–908.

Leuthold, R. H., 1968a. Recruitment to food in the ant *Crematogaster ashmeadi. Psyche* (Cambridge), 75:334–50.

Leuthold, R. H., 1968b. A tibial gland scent-trail and trail-laying behavior in the ant *Crematogaster ashmeadi* Mayr. *Psyche* (Cambridge), 75:233–48.

Lindauer, M., 1948. Über die Einwirkung von Duft- und Geschmacksstoffen sowie anderer Faktoren auf die Tänze der Bienen. *Z. vergl. Physiol.,* 31:348–412.

Lindauer, M., 1954. Temperaturregulierung und Wasserhaushalt im Bienenstaat. *Z. vergl. Physiol.,* 36:391–432.

Lindauer, M., 1955. Schwarmbienen auf Wohnungssuche. *Z. vergl. Physiol.,* 37:263–324.

Lindauer, M., 1956. Über die Verständigung bei indischen Bienen. *Z. vergl. Physiol.,* 38:521–57.

Lindauer, M., 1961. *Communication among Social Bees.* Cambridge: Harvard University Press.

Lindauer, M., 1971. The functional significance of the honeybee waggle dance. *Amer. Nat.,* 105:89–96.

Lindauer, M., and Kerr, W. E., 1958. Die gegenseitige Verständigung bei den stachellosen Bienen. *Z. vergl. Physiol.,* 41:405–34.

Lindauer, M., and Kerr, W. E., 1960. Communication between the workers of stingless bees. *Bee World,* 41:29–41, 65–71.

Lindauer, M., and Schricker, B., 1963. Über die Funktion der Quellen bei den Dämmerungsflügen der Honigbienen. *Biol. Zbl.,* 82:721–25.

Markin, G. P., 1970. Food distribution within laboratory colonies of the Argentine ant, *Iridomyrmex humilis* (Mayr). *Ins. Soc.,* 17:127–57.

Markl, H., 1965. Stridulation in leaf-cutting ants. *Science,* 149:1392–93.

Markl, H., 1967. Die Verständigung durch Stridulationssignale bei Blattschneiderameisen, I: Die Biologische Bedeutung der Stridulation. *Z. vergl. Physiol.,* 57:299–330.

Markl, H., 1968. Die Verständigung durch Stridulationssignale bei Blattschneiderameisen, II: Erzeugung und Eigenschaften der Signale. *Z. vergl. Physiol.,* 60:103–50.

Markl, H., 1970. Die Verständigung durch Stridulationssignale bei Blattschneiderameisen, III: Die Empfindlichkeit für Substratvibrationen. *Z. vergl. Physiol.,* 69:6–37.

Markl, H., 1973. The evolution of stridulatory communication in ants. *Proc. Seventh Congr. IUSSI* (London), pp.258–65.

Markl, H., and Fuchs, S., 1972. Kopfsignale mit Alarmfunktion bei Rossameisen *(Camponotus)* (Formicidae, Hymenoptera). *Z. vergl. Physiol.,* 76:204–25.

Maschwitz, U., 1964. Gefahrenalarmstoffe und Gefahrenalarmierung bei sozialen Hymenopteren. *Z. vergl. Physiol.,* 47:596–655.

Maschwitz, U., 1966a. Alarmsubstances and alarm behavior in social insects. *Vitamins and Hormones,* 24:267–90.

Maschwitz, U., 1966b. Das Speichelsekret der Wespenlarven und seine biologische Bedeutung. *Z. vergl. Physiol.,* 53:228–52.

Maschwitz, U., 1974. Vergleichende Untersuchungen zur Funktion der Ameisenmetathorakaldrüse. *Oecologia,* 16:303–10.

Maschwitz, U.; Beier, W.; Dietrich, J.; and Keidel, W.; 1974a. Futterverständigung bei Wespen der Gattung *Paravespula. Naturwissenschaften, 61,* 506.

Maschwitz, U.; Hölldobler, B.; and Möglich, M.; 1974b. Tandemlaufen als Rekrutierungsverhalten bei *Bothroponera tesserinoda* Forel (Formicidae, Ponerinae). *Z. Tierpsychol.,* 35:113–23.

Maschwitz, U.; Koob, K.; and Schildknecht, H.; 1970. Ein Beitrag zur Funktion der Metathoracaldrüse der Ameisen. *J. Insect Physiol.,* 16:387–404.

Mautz, D., 1971. Der Kommunikationseffekt der Schwänzeltänze bei *Apis mellifica carmica* (Pollm.). *Z. vergl. Physiol.,* 72:197–220.

McGurk, D. J.; Frost, J.; Eisenbraun, E. J.; Vick, K.; Drew, W. A.; and Young, J.; 1966. Volatile compounds in ants: identification of 4-methyl-3-heptanone from *Pogonomyrmex* ants. *J. Insect Physiol.,* 12:1435–41.

Michener, C. D., 1974. *The Social Behavior of Bees.* Cambridge: The Belknap Press of Harvard University Press.

Möglich, M., and Hölldobler, B., 1974. Social carrying behavior and division of labor during nest moving. *Psyche* (Cambridge), 81:219–36.

Möglich, M., and Hölldobler, B., 1975. Communication and orientation during foraging and emigration in the ant *Formica fusca. J. Comp. Physiol.,* 101:275–88.

Möglich, M.; Hölldobler, B.; and Maschwitz, U.; 1974a. *Camponotus sericeus* (Formicidae): Tandemlauf beim Nestumzug. *Encyclopaedia Cinematographica* E 2039/1974, pp.3–18.

Möglich, M.; Maschwitz, U.; and Hölldobler, B.; 1974b. Tandem calling: a new kind of signal in ant communication. *Science,* 186:1046–47.

Montagner, H., 1963. Étude preliminaire des relations entre les adults et le couvain chez les guêpes sociales du genre Vespa, au moyen d'un radioisotope. *Insectes Soc.,* 10:153–66.

Montagner, H., 1964. Étude du compartement alimentaire et des relations trophallactique des mâles au sein de la societé des guêpes, au moyen d'un radioisotope. *Ins. Soc.,* 11:301–16.

Montagner, H., 1966. Le mécanisme et les consequences des compartements trophallactiques chez les guêpes du genre *Vespa.* Thesis, University of Nancy.

Montagner, H., 1967. *Comportements trophallactiques chez les guêpes sociales.* Paris: Service du Film de Recherche Scientifique, Film no.B 2053.

Montagner, H., and Pain, J., 1971. Étude préliminaire des communications entre ouvrières d'abeilles au cours de la trophallaxie. *Ins. Soc.,* 18:177–92.

Morse, R. A., 1972. Honeybee alarm pheromone: another function. *Ann. Entomol. Soc. Amer.,* 65:1430.

Morse, R. A.; Gary, N. E.; and Johanson, T. S.; 1962. Mating of virgin queen honey bees *(Apis mellifera)* following mandibular gland extirpation. *Nature* (London), 194:605.

Morse, R. A.; Shearer, D. A.; Boch, R.; and Benton, A. W.; 1967. Observations on alarm substances in the genus *Apis. J. Apicult. Res.,* 6:113–18.

Moser, J. C., 1964. Inquiline roach responds to trail-marking substance of leaf-cutting ants. *Science,* 143:1048–49.

Moser, J. C., and Blum, M. S., 1963. Trail marking substance of the Texas leaf-cutting ant: source and potency. *Science,* 140:1228.

Moser, J. C., and Silverstein, R. M., 1967. Volatility of trail marking substance of the town ant. *Nature* (London), 215:206–207.

Naumann, M. G., 1970. The nesting behavior of *Protopolybia punnila* in Panama (Hymenoptera, Vespidae). Ph.D. diss. University of Kansas, Lawrence.

Nunez, J. A., 1967. Sammelbienen markieren versiegte Futterquellen durch Duft. *Naturwissenschaften,* 54:322–23.

Pain, J., 1961. Sur la pheromone des reines d'abeilles et ses affects physiologique. *Ann. Abeille,* 4:73–152.

Pain, J., 1973. Pheromones and hymenoptera. *Bee World,* 54:11–24.

Pain, J., and Ruttner, F., 1963. Les extraits de glandes mandibulaires des reines d'abeilles attirent les mâles lors du vol nuptial. *C. R. Acad. Sci.* (Paris), 256:512–15.

Pardi, L., 1940. Ricerche sui Polistini, I: poliginia vera ed apparente in *Polistes gallicus* (L.). *Processi Verb. Soc. Tosc. Sci. Nat.* (Pisa), 49:3–9.

Pardi, L., 1948. Dominance order in *Polistes* wasps. *Physiol. Zool.,* 21:1–13.
Pardi, L., and Calvacanti, M., 1951. Esperienze su mechanismo della monoginia funzionale in *Polistes gallicus* (L.) (Hymenopt. Vesp.). *Boll. Zool.,* 18:247–52.
Pavan, M., and Ronchetti, G., 1955. Studi sulla morfologia esterna e anatomia interna dell'operaia di *Iridomyrmex humilis* Mayr e ricerche chimiche e biologiche sulla iridomirmecina. *Atti Soc. Ital. Sci. Nat.* (Milan), 94:379–477.
Petersen, M., and Buschinger, A., 1971. Das Begattungsverhalten der Pharaoameise *Monomorium pharaonis. Z. angew. Entomol.,* 68:168–75.
Plateaux, L., 1960. Adoptions expérimentales de larves entre des fourmis de genres différents: *Leptothorax nylanderi* Förster et *Solenopsis fugax* Latreille. *Ins. Soc.,* 7:163–70.
Regnier, F. E.; Nieh, M.; and Hölldobler, B.; 1973. The volatile Dufour's gland components of the harvester ants *Pogonomyrmex rugosus* and *P. barbatus. J. Insect Physiol.,* 19:981–92.
Regnier, F. E., and Wilson, E. O., 1968. The alarm defence system of the ant *Acanthomyops claviger. J. Insect Physiol.,* 14:955–70.
Regnier, F. E., and Wilson, E. O., 1969. The alarm defence system of the ant *Lasius alienus. J. Insect Physiol.,* 15:893–98.
Regnier, F. E., and Wilson, E. O., 1971. Chemical communication and "propaganda" in slave maker ants. *Science,* 172:267–69.
Renner, M., 1960. Das Duftorgan der Honigbiene und die physiologische Bedeutung ihres Lockstoffes. *Z. vergl. Physiol.,* 43:411–68.
Renner, M., and Baumann, M., 1964. Über Komplexe von subepidermalen Drüsenzellen (Duftdrusen ?) der Bienenkönigin. *Naturwissenschaften,* 51:68–69.
Ribbands, C. R., 1953. *The Behaviour and Social Life of Honeybees.* London: Bee Research Association, Ltd. 352pp.
Ribbands, C. R., 1954. Communication between honeybees, I: the response of crop-attached bees to the scent of their crop. *Proc. Roy. Entomol. Soc.* (London), ser. A, 29:10–12.
Ribbands, C. R., 1955a. Communication between honeybees, II: the recruitment of trained bees, and their response to improvement of the crop. *Proc. Roy. Entomol. Soc.* (London), ser. A, 30:1–3.
Ribbands, C. R., 1955b. The scent perception of the honeybee. *Proc. Roy. Soc.,* ser. B, 143:367–79.
Riley, R. G.; Silverstein, R. M.; and Moser, J. C.; 1974a. Biological responses of *Atta texana* to its alarm pheromone and the enantiomer of the pheromone. *Science,* 183:760–62.
Riley, R. G.; Silverstein, R. M.; and Moser, J. C.; 1974b. Isolation, identification, synthesis and biological activity of volatile compounds from heads of *Atta* ants. *J. Insect Physiol.,* 20:1629–37.
Röseler, P. F., 1967. Untersuchungen über das Auftreten der 3 Formen im Hummelstaat. *Zool. Jb.* (Physiol.), 74:178–97.
Röseler, P. F., 1970. Unterschiede in der Kastendetermination zwischen den Hummelarten *Bombus hypnorum* and *Bombus terrestris. Z. Naturforsch.,* 25:543–48.
Roubaud, E., 1911. The nature history of the solitary wasps of the genus *Synagris. Rept. Smith. Inst., 1910,* pp.507–25.
Ruttner, F., and Kaissling, K. E., 1968. Über die interspezifische Wirkung des Sexuallockstoffes von *Apis mellifica* and *Apis cerana. Z. vergl. Physiol.,* 59:362–70.
Ruttner, F., and Ruttner, H., 1965. Untersuchungen über die Flugaktivität und das Paarungsverhalten der Drohnen, 2: Beobachtungen an Drohnensammelplätzen. *Z. Bienenforsch.,* 8:1–18.
Ruttner, F., and Ruttner, H., 1968. Untersuchungen über die Flugaktivität und das Paarungsverhalten der Drohnen, 4: Zur Fernorientierung und Ortsstetigkeit der Drohnen auf ihren Paarungsflügen. *Z. Bienenforsch.,* 9:259–65.
Ruttner, H., and Ruttner, F., 1972. Untersuchungen über die Flugaktivität und das Paarungsverhalten der Drohnen, 5: Drohnensammelplätze und Paarungsdistanz. *Apidologie,* 3:203–32.
Sakagami, S. F., 1960. Preliminary report on the specific difference of behaviour and other ecological characters between European and Japanese honeybees. *Acta Hymenopterol.,* 1:171–98.
Schaudinischky, L., and Ishay, J., 1968. On the nature of the sounds produced within the nest of the oriental hornet *Vespa orientalis* F. *J. Acoust. Soc. Amer.,* 44:1290–1301.
Schneirla, T. C., 1971. *Army ants: A Study in Social Organization,* H. T. Topoff, ed. San Francisco: W. H. Freeman.
Schremmer, F., 1972. Beobachtungen zum Paarungsverhalten der Männchen von *Bombus confusus* Schenck. *Z. Morph. Tiere,* 72:263–94.
Shearer, D. A., and Boch, R., 1965. 2-Heptanone in the mandibular gland secretion of the honeybee. *Nature* (London), 206:530.
Simpson, J., 1966. Repellency of the mandibular

gland scent of worker honeybees. *Nature* (London), 209:531–32.

Simpson, J., and Cherryl, S. M., 1969. Queen confinement, queen piping and swarming in *Apis mellifera* colonies. *Anim. Behav.,* 17:271–78.

Spradbery, J. P., 1973. *Wasps.* Seattle: University of Washington Press. 408pp.

Stein, G., 1963. Über den Sexuallockstoff von Hummelmännchen. *Naturwissenschaften,* 50:305.

Stejskal, M., 1962. Duft als "Sprache" der tropischen Bienen. *Sudwestdeut. Imker,* 49:271.

Stumper, R., 1956. Sur les sécrétions des fourmis femelles. *C. R. Acad. Sci.* (Paris), 242:2487–89.

Sudd, J. H., 1957a. Communication and recruitment in Pharaoh's ant, *Monomorium pharaonis* (L.). *Anim. Behav.,* 5:104–109.

Sudd, J. H., 1957b. A response of worker ants to dead ants of their own species. *Nature* (London), 179:431–32.

Sudd, J. H., 1962. The source and possible function of the odour of the African stink-ant, *Paltothyreus tarsatus* F. (Hym. Formicidae). *Entomol. Mon. Mag.,* 98:62.

Szlep, R., and Jacobi, T., 1967. The mechanism of recruitment to mass foraging in colonies of *Monomorium venustum* Smith, *M. subopacum* ssp. *phoenicium* Em., *Tapinoma israelis* For. and *T. simothi v. phoenicium* Em. *Ins. Soc.,* 14:25–40.

Szlep-Fessel, R., 1970. The regulatory mechanism in mass foraging and recruitment of soldiers in *Pheidole. Ins. Soc.,* 17:233–44.

Torgerson, R. L., and Akre, R. D., 1970. The persistence of army ant chemical trails and their significance in the ecitonine-ecitophile association (Formicidae: Ecitonini). *Melanderia,* 5:1–28.

Tumlinson, J. H.; Moser, J. C.; Silverstein, R. M.; Brownlee, R. G.; and Ruth, J. M.; 1972. A volatile trail pheromone of the leaf-cutting ant, *Atta texana. J. Insect Physiol.,* 18:809–14.

Tumlinson, J. H.; Silverstein, R. M.; Moser, J. C.; Brownlee, R. G.; and Ruth, J. M.; 1971. Identification of the trail pheromone of a leaf-cutting ant, *Atta texana. Nature* (London), 234:348–49.

Velthuis, H. H. W., 1970a. Queen substances from the abdomen of the honeybee queen. *Z. vergl. Physiol.,* 70:210–22.

Velthuis, H. H. W., 1970b. Ovarian development in *Apis mellifera* worker bees. *Entomol. Exptl Appl.,* 13:377–94.

Velthuis, H. H. W., 1972. Observations on the transmission of queen substances in the honey bee colony by the attendants of the queen. *Behaviour,* 41:105–29.

Voogd, S., 1955. Inhibition of ovary development in worker bees by extraction fluid of the queen. *Experientia,* 11:181–82.

Wallis, D. J., 1961. Food-sharing behavior in the ants *Formica sanguinea* and *Formica fusca. Behaviour,* 17:17–47.

Walsh, J. P., and Tschinkel, W. R., 1974. Brood recognition by contact pheromone in the red imported fire ant, *Solenopsis invicta. Anim. Behav.,* 22:695–704.

Wasmann, E., 1910. Die Doppelwirtigkeit der *Atemeles.* Deut. eng. Nat., 1:1–11.

Wasmann, E., 1915. Neue Beitrage zur Biologie von *Lomechusa* und *Atemeles,* mit kritischen Bemerkungen über das echte Gastverhältnis. *Z. wiss. Zool.,* 114:233–402.

Watkins, J. F., 1964. Laboratory experiments on the trail-following of army ants of the genus *Neivamyrmex* (Formicidae: Dorylinae). *J. Kansas Entomol. Soc.,* 37:22–28.

Watkins, J. F., and Cole, T. W., 1966. The attraction of army ant workers to secretions of their queens. *Texas J. Sci.,* 18:254–65.

Weaver, N.; Weaver, C. C.; and Law, J. H.; 1964. The attractiveness of citral to foraging honeybees. *Prog. Rept. Tex. Agric. Exptl Stn.,* no. 2324, pp.1–7.

Wells, P. H., and Wenner, A. M., 1973. Do honeybees have a language? *Nature* (London), 241:171–75.

Wenner, A. M., 1962a. Sound production during the waggle dance of the honeybee. *Anim. Behav.,* 10:79–95.

Wenner, A. M., 1962b. Communication with queen honeybees by substrate sound. *Science,* 138:446–48.

Wenner, A. M., 1964. Sound communication in honeybees. *Sci. Amer.,* 210:117–23.

Wenner, A. M., 1967. Honeybees: do they use the distance information contained in their dance maneuver? *Science,* 155:847–49.

Wenner, A. M.; Wells, P. H.; and Johnson, D. L.; 1969. Honeybee recruitment to food sources: olfaction or language? *Science,* 164:84–86.

Wenner, A. M.; Wells, P. H.; and Rohlf, F. J.; 1967. An analysis of the waggle dance and recruitment in honeybees. *Physiol. Zool.,* 40:317–44.

Wheeler, W. M., 1911. Notes on the myrmecophilous beetles of the genus *Xenodusa,* with a description of the larva of *X. cava* LeConte. *N. Y. Ent. Soc.,* 19:164–69.

Wheeler, W. M., 1918. A study of some ant larvae with a consideration of the origin and meaning of social habits among insects. *Proc. Amer. Phil. Soc.,* 57:293–343.

Wilson, E. O., 1958. A chemical releaser of alarm and digging behavior in the ant *Pogonomyrmex badius* (Latreille). *Psyche* (Cambridge), 65:41–51.

Wilson, E. O., 1959a. Source and possible nature of the odor trail of fire ants. *Science,* 129:643–44.

Wilson, E. O., 1959b. Communication by tandem running in the ant genus *Cardiocondyla. Psyche* (Cambridge), 66:29–34.

Wilson, E. O., 1962. Chemical communication among workers of the fire ant *Solenopsis saevissima* (Fr. Smith): 1. The organization of mass-foraging; 2. An information analysis of the odor trail; 3. The experimental induction of social response. *Anim. Behav.,* 10:134–64.

Wilson, E. O., 1965. Trail sharing in ants. *Psyche* (Cambridge), 72:2–7.

Wilson, E. O., 1971. *The Insect Societies.* Cambridge: Belknap Press of Harvard University Press. 548pp.

Wilson, E. O., 1974. Aversive behavior and competition within colonies of the ant *Leptothorax curvispinosus. Ann. Entomol. Soc. Amer.,* 67:777–80.

Wilson, E. O., and Bossert, W. H., 1963. Chemical communication among animals. *Rec. Prog. Hor. Res.,* 19:673–716.

Wilson, E. O., and Eisner, T., 1957. Quantitative studies of liquid food transmission in ants. *Ins. Soc.,* 4:157–66.

Wilson, E. O., and Pavan, M., 1959. Source and specificity of chemical releasers of social behavior in the dolichoderine ants. *Psyche* (Cambridge), 66:70–76.

Woyke, J., 1971. Correlation between the age at which honeybee brood was grafted: characteristics of the resultant queens and results of inseminations. *J. Apicult. Res.,* 10:45–55.

Wüst, M., 1973. Stomodeale und proctodeale Sekrete von Ameisenlarven und ihre biologische Bedeutung. *Proc. Seventh Int. Congr. IUSSI,* London, pp.412–18.

Chapter 20

COMMUNICATION IN FISHES

Michael L. Fine, Howard E. Winn, and Bori L. Olla

One cannot separate social behavior and communication. Cherry (1957) stated that communication is "the establishment of a social unit from individuals by use of language or signs." In more detail Burghardt (1970) traced the various attempts at definition, spotted many shortcomings, and left us with his own attempt, namely that communication is the phenomenon of one organism producing a signal that, when responded to by another organism, confers some advantage (or the statistical probability of it) to the signaler or his group. The major factor in Burghardt's formulation is the "intent" of the signaler, where intent is viewed in the context of the sender's adaptive behavior based upon prior phylogenetic and ontogenetic events.

Marler (1961) suggested analyzing communication as follows: (1) determining whether the receiver of the signal is able to orient toward the signaler; (2) establishing the pattern of the response to the signal, e.g., sexual, aggressive, parental; (3) identifying various signals as stimulus situations that elicit recurring response patterns; (4) correlating variations in properties of the signal and variations in properties of the response; and (5) identifying the components of a signal (if it can be fragmented) to better understand how these units contribute to effective communication.

We have addressed ourselves to the above topics in this review, which has been divided into visual and chemical communication, both of which involve modalities with relatively long-lasting stimuli, and auditory and electrical communication, which are relatively instantaneous. We have attempted to analyze how information is coded within each of these modalities and then to discuss how the modality is used in actual communication. Redundancy is inherent in this approach, but it is, we hope, a way of gaining deeper understanding of the processes involved. Redundancy, although slowing rates of information transfer, can aid in detecting the signal from noise.

It is not surprising that the sections dealing with visual and acoustic communication are the largest two of this paper. The dearth of information on chemical communication probably reflects the fact that the species most likely to use this sensory modality in communication have not been studied much. In the case of electrical communication these systems appear to be highly specialized and restricted to a few groups, and at present they are not thought to be widespread.

Visual Communication

The importance of vision for moving, feeding, and communicating in the aqueous environment has resulted in well-developed eyes in many fish species (Walls, 1942; Yager, 1968; Ingle, 1971; Munz, 1971; Tomita, 1971). Color patterns and shapes may have evolved for environmental needs such as camouflage or locomotion rather than as signaling devices. For instance, Barlow (1974) generalized that cichlids living in clearer waters tended to be blue or green, while those in more turbid waters showed more yellow, orange, or red. Teleosts typically reflect the vertically oriented light regime of natural waters by being dark above and light below (countershading). Even brightly colored reef species are countershaded. At close range conspicuous markings render these fishes highly visible; but some colors, especially yellows, appear muted at greater distances. Thus, countershading may still be functional in forms such as butterfly fishes (Hamilton and Peterman, 1971). The cryptism of countershading can be forsaken for communicative needs; some cichlids display reverse countercoloring during the mating season, making them more conspicuous to potential partners (Albrecht, 1962; Barlow, 1974).

FUNCTION

Sexual Reproduction

Spawning is a major event in the life of a fish. Many species undergo extensive migrations to areas favorable to their young (Harden Jones, 1968). Unique visual signals have evolved in many cases to insure successful mating. Schooling species (e.g., many freshwater cyprinids) often swim to the bottom and stake out territories for reproduction. The details of spawning behavior in the 20,000 species of fish (Cohen, 1970) are largely unknown (Breder and Rosen, 1966); and the variety of social and sexual behaviors can be impressive even within single families, i.e., the pomacentrids (Reese, 1964; Albrecht, 1969; Fishelson, 1970; Swerdloff, 1970; Russell, 1971; Keenleyside, 1972a; Brown et al., 1973; Robertson, 1973).

Barlow (1970) has divided the functions of courtship into arousal and appeasement. According to the arousal hypothesis, courtship behavior stimulates the reproductive physiology of the animal to whom the behavior is directed. The appeasement role of courtship reduces the probability of attack by the mate. Both of these functions are minimized in fishes that spawn in the water column away from the bottom. Such fishes are nonterritorial and are often in groups that deny individuals the time and isolation necessary to develop a complex spawning ritual. Pelagic species are not typically known to form pairs of aggregations for extensive periods. In fact, many pelagic species have mass spawnings in which eggs and sperm are shed together. The cue for release of gametes may often be rapid swimming of the group. Here physical factors must control oogenesis and the female does not receive prolonged stimulation by a male. By nature, pelagic fishes are difficult to observe and many do not adapt readily to captivity. Our knowledge of their behavior is minimal, and generalizations are suspect because in at least certain instances transitory pairing and courtship behavior are known. Perhaps extended courtship is replaced with the self-stimulation of rapid swimming, and in some cases the continued attendance of several males stimulates the females (tactile). Magnuson and Prescott (1966) have described courtship behavior of captive Pacific bonito *(Sarda chiliensis).* The male appeared unable to identify the sex of a conspecific except by behavioral characteristics. A female would wobble, and one or more males would follow. If only one male was following, there would be a gradual transition from wobbling to circle swimming. The wobbles would become more and more pronounced, until the female continued one phase of the wobble into a circular path instead of turning back to the next phase of the wobble. Gametes were released in circular swimming. If two or more males followed a female, the males became involved in lateral threat displays inhibiting courtship. Brawn (1961a) also demonstrated pairing with courtship and agonistic behavior in the cod *(Gadus morhua)*, another schooling species.

Because of internal fertilization with claspers, it is necessary for sharks and rays to form pairs. Information on reproduction in sharks has been garnered from fisheries studies and indirect evidence, rather than from extensive observation (Springer, 1967). Sharks often assemble in unisexual groups of about the same size. Spawning migrations bring these groups together for reproduction. During courtship male sharks may not feed, but females do. Courtship may thus become a hazardous activity for the males of some of the larger species, for since they are inhibited from making strong attack, the females they are courting may sometimes respond by killing them. Among the large carcharhinid sharks, the female is aroused to mate by rough courtship of the male, who uses his teeth to slash the skin on the female's back.

Ethologists have devoted much more time to the study of freshwater and near-shore marine fish associated with the bottom than to typical pelagic species or deepwater forms. Spawning on the bottom typically involves claiming a territory, preparing or maintaining a nest, courting, laying and fertilizing eggs, and some guarding of the eggs. Cichlids and scattered species in other families form pair bonds for reproduction and care of eggs, but in many families the female either leaves the nest or is driven away from it by the male, who assumes parental duties. The amount of the male's courtship is highly variable. Three examples, a minnow, a nandid, and a stickleback, have been chosen to demonstrate the range in behavior.

Males of the fathead minnow *(Pimephales promelas)* are content merely to defend their territories, and it is the female who initiates spawning (McMillan and Smith, 1974). The male performs no displays outside of his territorial object. The female herself is weakly territorial and probably attracted by territorial objects. The bright banding of a male, along with his vigorous movements within the territory, are also stimuli attracting the female. The male is aggressive toward the female, butting, charging, and chasing. A female ready to spawn would try to remain close to the undersurface of the male's territorial object, despite his attacks. When the female succeeded in positioning herself laterally close to the male, butting would typically stop and spawning vibrations begin.

Badis badis provides an intermediate example (Barlow, 1962). Here, too, the female normally seeks out the male, but an unpaired male may look for a female and stimulate her to follow him back to his burrow. The male usually attacks the female when she first enters the burrow. The attack is diminished by two complementary actions of the female (appeasement). She remains nearly motionless and leans against the male with tautly spread median fins. The fishes start carouseling and butting before the complicated act of enfolding, in which the fishes wrap around each other and the female pulsates. Enfolding appears to determine the moment of ovulation in the female (arousal). The early enfoldings serve to marshal the eggs for subsequent enfoldings. Eggs fall from the pair as they disengage. While this complicated act is necessary to stimulate the female, it is not certain how important it is to the male. In general males are assumed to be able to spawn with less coordination than is required for females. For example, males of the colonially nesting sunfish *(Lepomis megalotis)* leave their nests for no more than two or three seconds to intrude on a neighbor's nest and try to fertilize freshly deposited eggs (Keenleyside, 1972b). It is interesting to contrast the complicated spawning behavior of *Badis* with poeceleiids. Liley (1966) felt that signals at the start of courtship activities show greater species specificity and are more divergent than the signals and responses which occur later in the sequence.

In the final example the male stickleback, *Gasterosteus aculeatus,* actively courts the female (Tinbergen, 1951). The female appears in the territory of a nesting male and approaches in reaction to the male's zigzag dance. The male leads to the nest, and the female follows. The male then shows the nest entrance, the female enters, the male quivers on the female's tail region, and the ritual terminates with spawning and fertilization. The male stickleback is exceptionally aggressive and undergoes an approach-avoidance conflict as the female approaches his nest. In a variant of the normal courtship the male uses a pattern known as dorsal pricking, a somewhat jerky pushing of the female with the dorsal side (Wilz, 1970). This behavior pushes the female away when she would normally be following the male to the nest. By inducing the female to cease following, the pricking display functions to facilitate the male's switch from high aggression to greater sexual motivation. According to Barlow's appeasement hypothesis, the male in this case actually affects a behavior that serves to appease him.

The primary function of courtship behavior is to bring males and females of the same species together and to avoid wasting gametes. Sunfish are often known to hybridize, and Keenleyside (1967) and Steele and Keenleyside (1971) have experimentally studied species recognition in *Lepomis megalotis* and *L. gibbosus,* the longear and the pumpkinseed. Longears nest in crowded colonies where they are constantly contending with intruders, leaving them little or no time for active discrimination between the two species. In an experimental chamber, male longears did not distinguish between the two species, though females did. Male pumpkinseeds nest individually, and consequently they could discriminate between approaching females before they reached the nest. Nesting males were shown to court conspecific females preferentially. Since both female longears and male pumpkinseeds chose conspecifics under experimental conditions, Steele and Keenleyside doubted that they hybridize. Rather, they assumed that hybridization occurs between female pumpkinseeds and male longears.

Territories and Dominance

Two primary functions of territory are to insure reproduction and survival of the young. Feeding and safety may provide proximal benefits, particularly for species in which the male assumes parental tasks. For instance, the male garibaldi *(Hypsypops rubicunda)* cares for and defends its nest site all year in order to assure it for himself during the breeding season (Clark, 1970). The male maintains his territory for many years, and the red algal nest he cultures requires two or more years to develop. To sustain his time and effort, he locates his nests in areas with abundant food and shelter. Clark assumed that the male's requirement for reproductive success must be important enough to offset the resulting disadvantages to females, which are forced to live in poorer areas and which probably have higher mortality rates.

Fishes hold territories for varying amounts of time (Reese, 1973), with a range in the pomacentrids of twenty to thirty minutes in *Chromis multilineata* (Myrberg et al., 1967) to years in *Hypsypops rubicunda* (Clark, 1970). Along with extremes in duration, there can be tremendous variation in social structure and use of territories even within a species. Typically, male *Chromis multilineata* leave their school or aggregation to defend spawning territories for short periods (Myrberg et al., 1967). Other males were chased from the territories, but females were led to the center if they resisted the males' initial chase by "holding ground." However, occasionally two members of the aggregation would actually pair in the water column and swim rapidly together onto a small undefended portion of the substrate to spawn. Another damselfish, *Dascyllus aruanus,* forms either pairs or harems of one male and two to five females (Fricke and Holzberg, 1974). The females form a linear size-dependent dominance hierarchy, and spawning takes place in order according to rank. The wrasses (Labridae) often undergo sex changes (Reinboth, 1972, 1973) that lead to interesting behavioral situations. Robertson (1972) studied harems of *Labroides dimidiatus* consisting of one territorial male, three to six mature females, and several immature individuals. Females are territorial and also have a size-related linear dominance hierarchy. When the male dies the α-female reverses her sex and starts showing the male aggressive display toward the other females within one and a half to two hours. The behavioral changeover can be completed within a few days, and sperm can be released fourteen to eighteen days after the start of the reversal. Robertson believes that dominance suppresses sex reversal and that release from aggressive signals is the hormonal trigger.

A fish is dominant in its territory, but it may find occasions for leaving it. Males of *Hypsypops rubicunda* court and pair in aggregations above their territories where aggressiveness is reduced. Spawning occurs on the male's territory (Clark, 1971). Various species of cichlids leave their territories without being vigorously attacked. They rise to the surface, turn pale, and depress their fins when swimming through neighboring territories (Baerends and Baerends-van Roon, 1950).

Not all aspects of dominance interactions are innate; experience may play a part. Habituation is involved in the termination of hostilities between territorial neighbors of the stickleback *Gasterosteus aculeatus* and of the convict cichlid *Cichlasoma nigrofasciatum* (van den Assem and van der Molen, 1969; Peeke et al., 1971) and most likely plays a similar role in all territorial species. The status of a green sunfish *(Lepomis cyanellus)* is determined, at least in part, by its experiences (McDonald et al., 1968). Similarly sized dominant-subordinate pairs of sunfish were broken up and manipulated so that dominant fishes were placed with larger conspecifics and subordinates with smaller ones. After five days of treatment the original pairs were reunited, and in fifteen of the twenty pairs the original dominance status, as indicated by coloration, was reversed.

Whether fish can recognize individual conspecifics is a question basic to the study of hierarchies. Although this question has not been answered thoroughly, there are apparently fish species whose members can recognize individual conspecifics and other species whose members lack this ability. Jenkins (1969) found evidence of nip-right relationships in salmonids, whereas Myrberg (1972c) found nip-dominance in pomacentrids. Although Jenkins's trout did form stabilized relationships among confined fish, the persistent occurrence of revolts, even in relatively peaceful groups with large size ranges, and the exceptional cases of rank change and social mobility suggested to him the presence of either a definite limit to the effectiveness of learning or an irreducible social instability in groups of stream resident trout. Another possibility, consistent with his suggestions, might be that individuals recognize each other imperfectly, if at all, so that the outcome of an interaction would depend on the prior successes (wins and defeats) of the fishes, coupled with their internal motivation at the moment. Also, working with damselfish in the wild, Myrberg (1972c) found that smaller fish challenged and even occasionally chased larger members of the colony. This again argues against the individual recognition one would assume in a nip-right hierarchy. However, Nelson (1964) seems to have established the occurrence of individual recognition in dominance hierarchies and pairs for some glandulocaudine fishes. Recently Fricke (1973) claims to have demonstrated individual recognition between members of breeding pairs of the anemone fish *(Amphiprion bicinctus)* and within groups of *Dascyllus aruanus* (Fricke and Holzberg, 1974). Catfish recognize individuals by chemical means (Bardach and Todd, 1970).

Fish territories, at least in some pomacentrids, appear to vary, depending on the species of intruding fish (Clark, 1970). *Hypsypops rubicunda* attacked bottom-grazing fishes frequently or almost every time. Predatory fishes were generally tolerated and plankton-eating forms were nearly ignored altogether. The number of interspecific attacks was higher when males were guarding eggs than at any other time. *Pomacentrus flavicauda* directed only one-fifteenth of its agonistic responses toward conspecifics (Low 1971); responses were typically directed toward competitors for algae rather than toward carnivores. Myrberg and Thresher (1974) attacked this problem experimentally by presenting various species held in jars to a territorial *Eupomacentrus planifrons* to see how close they could place the captives before eliciting attack. The maximum distance of attack was different for each species, even though individuals of various sizes were involved. Wrasses (male and female *Halichoeres garnoti*) were reacted to in a similar manner, even though their color patterns are highly divergent. The authors suggested that forms, rather than color pattern, allowed species discrimination, assuming of course that vision is important. They implied that the damselfish reacted to male and female wrasses as members of the same species. An alternate explanation is that the fish discriminated the two sexes as different entities but of equal threat potential, in which case color might still be important.

Cleaning Symbiosis

Cleaning has evolved in tropical and temperate freshwater and marine fishes from many families. In view of the polyphyletic origins and diverse situations involved, it is not surprising that early thoughts about cleaning (Limbaugh, 1961; Feder, 1966) have been questioned (Hobson, 1969). Cleaners are generally supposed to eat parasites, fungi, and necrotic tissue from their hosts; thus, cleaning is a case of mutualism with obvious benefits to both species. In an experimental demonstration Limbaugh (1961) removed all the known cleaning organisms from two small Bahamian reefs and found that within days the number of fishes on the reefs was drastically reduced. Many of the remaining fishes developed fuzzy white blotches, swellings, ulcerated sores, and frayed fins. Since *Hypsypops rubicunda* does not allow cleaners to enter its territory while it is guarding eggs, it provides a natural experiment. In southern California Hobson (1971) found an average of sixty-seven parasitic copepods *(Caligus hubsoni)* on egg-guarding males as opposed to four to eight on males outside the reproductive season. However, Youngbluth (1968) and Losey (1972) removed *Labroides phthirophagus,* an obligate cleaning wrasse, from patch reefs in Hawaii and found no dramatic effect on either the numbers of remaining fishes or on their health. Losey (1972) noted that *Labroides* eat not only ectoparasites but scales and associated dermal and epidermal tissues or mucus as well; thus, he suggested, the relationship could be considered parasitic or commensal. Perhaps these examples should not be thought of as conflicting but merely as reflecting various systems in which cleaning plays a role.

The process of bringing two species together for cleaning involves a highly developed form of communication. The fact that many different species allow themselves to be cleaned makes the process even more interesting. Losey (1971) listed the interactions involved in cleaning as follows:

Cleaner fish
(1) Inspect, swimming close to the host and apparently exploring its body surface;
(2) Clean, feeding on matter on the body surface of the host;
(3) Dance, a dorso-ventral oscillation of the body.

Host fish
(1) Pose, the position or orientation of the host's body and fins and the swimming movements frequently assumed during cleaning interactions;
(2) Body jerk, any short quick movement of the body or head;
(3) Attack, darting quickly toward the cleaner.

We will briefly highlight some of the factors involved in communication in *Labroides phthirophagus,* an obligate cleaner (Youngbluth, 1968; Losey, 1971), and *Oxyjulis californica,* a facultative cleaner (Hobson, 1971). *Labroides* maintains a cleaning station, the position of which is learned by the host fish. Either to pose or to inspect may be the initial action in a cleaning sequence. The wrasse recognizes individual species and has definite preferences. Large jacks and parrotfish may be pursued by the cleaner as they appear near his station, while smaller wrasses, butterfly fish, or damselfish may pose continually and be ignored by the cleaner. The señorita *(Oxyjulis californica)* maintains no station and only some individuals seem to clean. In addition, individuals appear to specialize, in that some clean the damselfish *Chromis punctipinnis* and others the topsmelt *(Atherinops affinis).* Since *Oxyjulis* obtains much of its food without cleaning, the average encounter with a host fish will not result in cleaning, and the hosts do not bother to pose until a cleaner initiates the activity.

The cichlid *Etroplus suratensis* has a more ritualized pose display than most other species (Wyman and Ward, 1972). While performing the head-up display, it rapidly flickers its dark pelvic fins while simultaneously quivering its entire body. This display may be necessary because it approaches the cleaner *(Etroplus maculatus)* in its territory; failure to emit the submissive display results in attack.

One of the most important behavioral questions about cleaning concerns mutual recognition. Fricke (1966) and Losey (1971) elicited posing by a presentation of cleaner models. Losey claimed that hosts show a graded response to increasingly real models of *Labroides phthirophagus,* but his results were not complete enough to explain the relative effectiveness of different elements of coloration and body shape. The conspicuous coloration of many cleaners prompted Eibl-Eibesfeldt (1955) to hypothesize a guild mark theory (occurrence of some similarities in color pattern). Recently, Ayling and Grace (1971) and Potts (1973) have described guild marks in cleaners. Hobson (1969) felt that evidence supporting a direct relationship between bright coloration and cleaning is still weak. Further, distinctive coloration will render a cleaner more conspicuous to predators, and Hobson (1971) doubted that cleaners are immune from predation during noncleaning situations. Darcy et al. (1974) experimentally demonstrated that gobies, which clean piscivores, will not usually be eaten by them, while a cleaning wrasse that does not normally service these fishes will be eaten.

Guild coloration or not, experience is probably a factor in the recognition of a cleaner by the many hosts involved. By presenting a moving model of a cleaner fish for positive reinforcement, Losey and Margules (1974) operantly conditioned a butterfly fish *Chaetodon auriga* to occlude a light beam. They believed that the reward is probably tactile stimulation of the model. In nature, hosts might learn to recognize a new cleaner as a source of tactile stimulation.

Schooling

The configurations or qualitative characteristics of schools of different species of fish are manifested in highly varying degrees of cohesion and polarization. Presumably for the initiation and maintenance of the school, each individual would be able to provide the appropriate stimuli (visual, olfactory, acoustic, etc.) as well as to perceive and respond to stimuli arising from conspecifics. Other factors determining the formation and the quality of a school are the life stages of the organism as well as its relationship with other environmental input, such as solar and lunar cycles. No definition of a school will be entirely satisfactory because of the varied forms that mutual attraction takes. However, for the purposes of this chapter a working definition of a school will be "an aggregation formed when one fish reacts to one or more other fish by staying near them" (Keenleyside, 1955), with further emphasis on the biosocial mutual attraction between individual fish (Shaw, 1970).

In spite of their movement and size schools often act as whole units, frequently exhibiting what appears to be instantaneous and synchronous reactions to a variety of stimuli. Within a school, specific stimuli or "messages" might convey information about attraction, changes in speed, cohesion, direction, presence of food, or approach of predators. To understand the mode of transmission and rapid integration of the information throughout an entire grouping of fish, it is necessary to identify these stimuli and determine how they are interpreted via the sensory components.

Vision appears to be one of the more important senses utilized by schooling species (for review and discussion see Morrow, 1948; Atz, 1953; Breder, 1959; Shaw, 1970; and Radakov, 1973). The dominant role played by vision was established in a study by Keenleyside (1955) in which purely visual cues were effective in stimulating the initial phases of intraspecific approach and attraction among stickleback, *Gasterosteus aculeatus,* roache, *Leuciscus rutilis,* rudds *Scardinius erythrophthalmus,* and the characid *Pristella riddlei.* For the characid, a conspicuous black patch on the dorsal fin was found to be an important visual signal used in the recognition and attraction of species mates. The responses of approach and attraction based solely on visual cues have also been reported for a sea catfish (*Plotosus anguillaris:* Sato, 1937), the atherinids (*Menidia menidia* and *M. beryllina:* Shaw, 1960), the mullets *Mugil cephalus* (Olla and Samet, 1974) and *M. chelo,* a roach (*Rutilus rutilus*), a pomacentrid (*Chromis chromis:* Hemmings, 1966), and a jack (*Caranx hippos:* Shaw, 1969). When the tuna *Euthynnus affinis* were separated by transparent barriers, and hence could rely only on visual communication, the long-term maintenance of attraction and the schooling tendency persisted for as long as nine days (Cahn, 1972). Topp (1970) has implicated a dramatic color change and an associated behavior in maintaining the integrity of schools of rudderfish (*Kyphosus elegans*).

Besides its function in the attraction response, vision appears to be important in the communication of changes in movements of the school. In studies on tuna (*E. affinis:* Cahn, 1972) and jacks (*C. hippos:* Shaw, 1969), fish separated by transparent barriers tended to turn simultaneously and react mutually to other positional changes. Hunter (1969) studied the communication of velocity changes among schooling jack mackerel (*Trachurus symmetricus*) by electrically stimulating one fish and observing the responses of others in the school. He suggested that the latency and the velocity of responses among the group of fish reacting to the stimulus fish depended upon their visual perception of movement although the roles of other sensory systems were not analyzed.

Despite the inaccessibility of many species that exhibit schooling, experimental work on selected groups has provided a basis for biologists to assess the value of schooling. Theoretically it would appear that the most beneficial and adaptive characteristic is the increased or more efficient performance of certain activities once the fish have formed into a schooling group. A recent study on social facilitation in feeding behavior (Olla and Samet, 1974) established that the initiation of feeding behavior of single mullet *(Mugil cephalus)* was greatly facilitated solely by visual cues arising from a feeding group. Additional studies (Welty, 1934; Uematsu, 1971) have also established the role of the group with respect to feeding facilitation. Other adaptive features of the school may be related to the facilitation of reproductive behaviors, lowered predation (Seghers, 1974), improvement in orientation during a migration, and energy conservation resulting from the hydrodynamic advantage of swimming behind other moving fish.

The existence of heterotypic aggregations or schools, such as Breder (1959), Collette and Talbot (1972), Ehrlich and Ehrlich (1973), and Hobson (1974) observed, leads to further questions about the degree to which communicative processes are used in social groupings of fish. If one is to rely on Shaw's (1970) analysis of schooling in which the critical factor is the biosocial mutual attraction among individuals, then these observed heterotypic groups may represent only the formation of temporary associations that may be highly adaptive in the natural environment, but in which mutual communication between species may also be less well developed or absent. For example, Collette and Talbot (1972) observed the transient appearance of several different species within the body of a resident school of bonnetmouths *(Inermia vittata)* and suggested "that a resident school can itself become a habitat for other schooling species." Breder (1959) suggested that heterotypic groupings of goatfishes and various gerrids may be viewed as feeding associations in which gerrids catch food particles missed by goatfishes as the latter feed along the bottom.

Other adaptive features of these groups may be that resident schools provide camouflages for certain species and hence serve as a predator defense mechanism. Since the formation of these groups may be based possibly on a mutual toleration between species, rather than on a mutual attraction, further investigations of the biological causal factors are needed to fully understand the formation and maintenance of these "schools."

Parent-Young Interactions

Except for the cichlids, information on parent-young interactions comes from a diffuse and incidental literature (Breder and Rosen, 1966), and even the cichlid literature focuses largely on behavior of adults (Noakes and Barlow, 1973). Many of the species that care for their young have evolved mouth breeding habits (Oppenheimer, 1970). Further work is needed on a wide variety of species from different families.

Parent cichlids appear to recognize their offspring by a combination of visual and chemical cues, although some species may accept young of other broods or even other species (Kühme, 1963; Myrberg, 1966; Noakes and Barlow, 1973; Barlow, 1974).

Parent cichlids of many species shepherd their young for a while and can alert them to danger (Noble and Curtis, 1939; Baerends and Baerends-van Roon, 1950; Künzer, 1962). Baerends and Baerends-van Roon's work on alerting in *Tilapia natatensis* will be described. The response of the young is released by disturbance of the water and not by visual stimuli; but once aroused, the young direct themselves toward the female by means of visual clues. Slow movement of the mother elicited following by the young, but violent movement accompanied by a color change to black caused them to scatter and head for the bottom. As danger became less imminent, the mother would retreat toward the young and assume a diagonal position, her longitudinal axis making an angle of about 10° or 20° below the horizontal. In this position she moved slowly backward and the young swarmed toward her head and into her mouth. The young were attracted to the trailing edge of the underside of disc models as they were moved away. Young would also be attracted to dark spots on a model, but would then wander along the surface of the disc looking for the mouth.

SIGNAL VARIATION AND CODING

Color Patterns

Coloration in fishes essentially represents a balance between those factors which maximize signal value for communication and those which function to make an animal less conspicuous, i.e., cryptic coloration for both predators and prey. The wide array of colors found in fishes suggests a use of visual signaling. Some fishes (notably cyprinids) are known to have good color vision (Beauchamp and Lovasik, 1973; Daw, 1973), and perhaps most teleosts have it to varying degrees; but most species have not been checked for the presence of cones. Although much work has been done on the possible uses of color in communication, few studies make the important distinction, either in the experimental design or in the interpretation of data, between color and brightness. It is always possible in some circumstances that varied color merely forms a pattern of contrasting brightness.

In their classic work on cichlids, Baerends and Baerends-van Roon (1950) designed experiments that included the consideration of discrimination based on color versus brightness. As an attempt to identify the visual components that were involved in the attraction of young *Aequidens latifrons* and *Cichlasoma bimaculatum* to adults, experiments were performed using discs of several different colors as well as grays of varying brightness. Both species demonstrated a preference for colors of shorter wave length with little discrimination of brightness. The authors concluded that color was an important releasing stimulus for following in these two species. However, in similar experiments on *Tilapia natalensis* attraction of young fish to the female was based perhaps on discrimination of brightness rather than of color.

Responses to brightness seem also to be involved in the experiments of Picciolo (1964), who found that the blue color pattern (contrasting brightness) displayed on the throat and breast of the male gourami *Colisa lalia* functions as a visual stimulus for sexual discrimination for both sexes. Model experiments demonstrated that the color pattern need not be blue but merely dark in order to attract and release aggressive responses in males.

Whether fish are responding to brightness or to color per se, the presence of the color enhances the communication of signals by vision. The examples cited below, although generally not considering the difference between brightness and color, show the communicative function of color even if it only enhances contrast.

The shallow inshore marine environment of the subtropical and tropical zones contains many colorful types. Lorenz (1966) hypothesized that bright, poster-colored fishes were more aggressive toward conspecifics than the more modestly colored species. More recent studies, however, demonstrate that drably colored fishes may defend territories and that territories are defended against other species (Rasa, 1969; Clark, 1970; Low, 1971; Myrberg and Thresher, 1974; Tavolga, 1974).

The degree of conspicuousness of coloration

in certain species has been shown to be important in communication. For example, Haskins et al. (in Liley, 1966) found that male guppies *(Poecilia reticulata)* that were conspicuously marked tended to have greater mating success when in competition with other males not so brightly marked. In another study Keenleyside (1971) found that the conspicuous opercula patch, black eye with light red iris patches, and black pelvic fins of *Lepomis megalotis* were the features that appeared to be most important in eliciting aggression from nest-guarding males.

An important aspect of the role that coloration plays in communication is its location on a fish. For example, the blue color pattern of the male *Colisa lalia* must be located on the anterior ventro-lateral surface in order to function as a cue for sex discrimination (Picciolo, 1964). When the color pattern was placed on the center of the lateral surface of a model, it did not particularly attract males or females. A model *Lepomis megalotis* with the eye and opercular patch placed near the tail was relatively ineffective in eliciting aggression although any aggression that was elicited was directed at the abnormally located eye and patch (Keenleyside, 1971). In a number of African mouth breeders of the genus *Haplochromis,* the males bear yellow or orange spots that look much like eggs (Fig. 1) on the anal fin (Wickler, 1962). The female lays her eggs in the male's pit, but takes them in her mouth before he can fertilize them. The male then spreads his anal fin and fertilizes the empty pit. The female grasps the egg dummies on his anal fin with her lips, engulfing the sperm ejected by the male. Another example of position of color is found in the blue-throated darters, a subgenus within the genus *Etheostoma.* These fish raise their heads when fighting, exposing their blue throats. In contrast, other members of the genus lack the blue throat and fight with their heads in the normal position (Winn, unpublished).

Fish coloration may be controlled by both the central nervous and the endocrine systems (Bagnara and Hadley, 1969; Fujii and Novales, 1969; Gentle, 1970a, 1970b). Many fishes take on breeding coloration during the spawning period. The classic example is the male stickleback *Gasterosteus aculeatus,* whose eye turns blue and whose belly becomes red (Tinbergen, 1951). Fishes may also exhibit transient, fairly rapid changes in color to communicate information about other behavior states. In *Pomacentrus jenkinsi* aggressive motivation is indicated by the darkening of the normally yellow eye to gray (Rasa, 1969). The amount of red color in the iris of *Lepomis megalotis* is related to dominance rank order (Hadley, quoted in Keenleyside, 1971). Nakamura and Magnuson (1965) observed the transient appearance of black spots ventral to the pectoral fins, faint vertical bars on the flanks, and a yellowish middorsal stripe in the tuna *Euthynnus affinis* during feeding. They interpreted this display as a possible social releaser signaling the presence of food to other members of the school. In further work with the related Pacific bonito *(Sarda chiliensis),* Magnuson and Prescott (1966) decided that this feeding display was agonistic. In the leaf fish *(Polycentrus schomburgkii)* a severely dominated male loses its dark coloration, becoming yellow-white, with the top of the head brown, a color pattern corresponding closely to that of a female ready to spawn (Barlow, 1967).

Leong (1969) performed a series of experiments on the cichlid *Haplochromis burtoni,* which exhibits rapid transient changes in coloration. In this study Leong was able to show the extreme specificity of visual signaling as well as indications of how the signal is being processed by the receiving fish. Only two components of the male's territorial coloration were found to affect attack readiness: the black vertical component of the head pattern and the orange patch above the pectoral fin (Fig. 1). The vertical component alone increased the attack rate by 2.79 bites/

min. The orange patch alone decreased the attack rate by 1.77 bites/min., while a dummy with both colorations present increased the attack rate by 1.08 bites/min., i.e., the sum of the effect of both components if presented separately (2.79 minus 1.77). Fig. 2 shows some of the models and their effects. Leong explained this cumulative effect by the rule of heterogeneous summation, which is represented by a unimodal peak. Had there been two peaks, each coinciding with one of the two stimuli, the fish might have been responding to either the vertical bar or the orange patch rather than a composite.

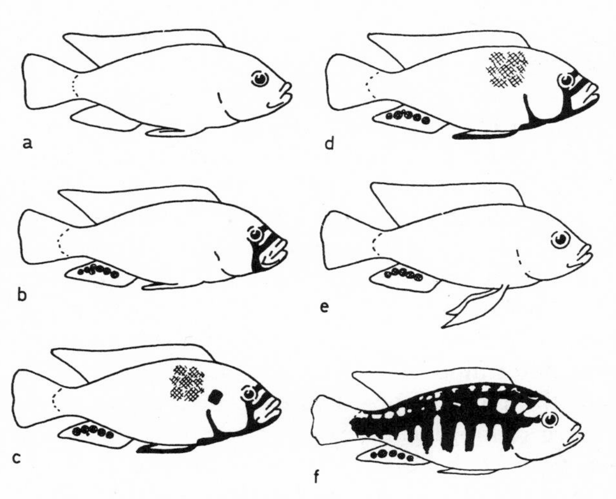

Fig. 1. Different color patterns of the male cichlid *Haplochromis burtoni:* a. Juvenile; b. adult ready to establish territory; c. territorial male; d. spawning male; e. and f. fleeing male. Crosshatch is orange and black is black. (From Leong, 1969.)

These two patterns were observed by Leong to appear and disappear within seconds; thus, they most likely reflect the internal state of the animal. He hypothesized that a combination of these patterns promotes a balanced level of aggression in established colonies, while an intruding male displaying only the vertical bar would rapidly be attacked.

Information about the role of bioluminescence in fish communication is meager and in most cases purely speculative (McAllister, 1967; Nicol, 1967; Tett and Kelly, 1973). Photophore patterns that are often species-specific and in some cases gender-specific suggest a recognition function. Further, it appears that the light emitted may be flashed on or off, increasing the possibilities of signal complexity and specificity as well as enhancing a signal.

The same balance that exists in coloration and shading in fish exists with bioluminescence, with a compromise necessary between those aspects that may be selected for on the basis of signal function and those that act to lessen the probability of attack from enemies. In this vein Clarke (1963) suggested that the ventrally located photophores may help to prevent detection inasmuch as the light emitted ventrally will match incident light, rendering the animal less conspicuous than if the light were emitted from the dorsal part of the body.

We know of only two cases in which bioluminescence has been observed in actual communication. Crane (1965) injected a gravid female *Porichthys notatus* with adrenaline, causing her to luminesce and turn pale. He placed the female in a tank with a nesting male, who courted her for about an hour, producing light intermittently in five- to ten-second displays, grunting, nudging, and grabbing her in his jaws. The nocturnal reef fish *Photoblepharon palpabratus* has a large bacterial organ under each eye (Morin, et al., 1975), which produces intense light. The fish is capable of covering the organ and flashing its light on and off. These fish spend the day in caves and come out on dark nights when they may use their lights for attraction, leading to group formation of three to twenty-five species mates. In addition, the defense of territory by male-female pairs from conspecific intruders is correlated with light emissions. When intruding *Photoblepharon* approached, the female swam back and forth

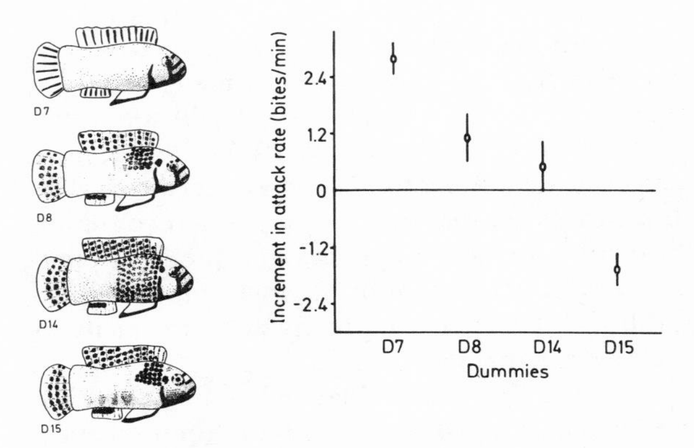

Fig. 2. Dummies and the effect of color patterns on the average increment in attack rate after their presentation. Vertical lines represent the variances of the mean. (From Leong, 1969.)

rapidly. She would then turn off her light, swim directly toward the intruder, and turn on the light when she was just next to the other fish. This was invariably effective in driving intruders away (Morin et al., 1975).

Movements

This section will be devoted to signal movements that are typically considered fixed action patterns. Barlow (1968) was dissatisfied with this term, feeling that it was applied uncritically to almost any behavior that has a degree of regularity sufficient to permit one to recognize it. He offered the term "modal action pattern" (MAP), feeling that it conveyed the essential features of the phenomenon without implying a degree of fixity that has seldom been tested.

Baerends and Baerends-van Roon (1950) divided the signal movements of various cichlid species into thirteen patterns: lateral displays, tail beating, tail fluttering, frontal display, mouth fighting, butting, swimming on the spot and tail wagging, attitude of inferiority, jerking and quivering, inviting, nipping of a substrate, skimming (pseudolaying and pseudofertilizing), calling the young, and jolting. Barlow (1974) recognized these categories and pointed out that the similarity between the different species is striking, although there are some statistical differences in the frequency of occurrence and sequencing of various MAPs and some patent but small differences in the MAPs used by the different species.

Frontal and lateral displays occur in agonistic encounters of many species. These displays are often bluffs, ritualized fighting, or the result of conflicting tendencies. In juvenile Atlantic salmon *(Salmo salar)* high attack tendency results in charging, nipping, and chasing, while fleeing is the result of high escape tendency (Keenleyside and Yamamoto, 1962). Frontal and lateral displays occur as a result of conflict between attack and escape with frontal display indicative of a relatively high level of attack tendency and lateral display indicative of escape tendency. The gray reef shark *(Carcharhinus menisoirah)* displayed toward divers with laterally exaggerated swimming and rolling along with spiral looping (Johnson and Nelson, 1973). The display occurred under approach-withdrawal conflict situations. It was elicited by rapid diver approach and was most intense when there was maximum escape-route restriction. More data are needed to determine objectively the underlying causes of the types of postures described above. Myrberg (1965) has interpreted some fin-spreading actions in cichlids as intention movements rather than as either fright- or attack-motivated behaviors.

Many fishes make themselves appear larger during agonistic displays by erecting fins and branchiostegals and by spreading the opercula. In most cichlids the intensity of a lateral display can be gauged by the extent of fin erection (Baerends and Baerends-van Roon, 1950), but even within the family there are exceptions. In the oscar *(Astronotus ocellatus)* full erection of the fins indicates a beaten or frightened fish. Rasa (1969) has separated fright and aggression in *Pomacentrus jenkinsi* on the statistical occurrence of external changes. Gray eyes and lowered dorsal fin are present in aggressive fish, while yellow eyes and raised dorsal fin are typical of frightened fish. Both of these conditions are components of agonistic interactions.

As with other aspects of visual communication, movements can be profoundly influenced by the environment. McKenzie and Keenleyside (1970) compared the reproductive behavior of the stickleback *Pungitius pungitius* from South Bay, Canada, with the behavior of the European form (Morris, 1958). The European form breeds in quiet, weedy streams, while the South Bay form breeds along rocky, barren, often turbulent lakeshores. South Bay males do not dance-jump when leading the female to the nest but, more like *Gasterosteus* males in open areas, swim directly and quickly to the nest after courtship jumping. Morris argues that jumping during the leading phase of courtship in the European form slows down the male's return to his nest, allowing the female to maintain visual contact as she follows him through the dense weeds.

Size

Generally, larger members of a species are dominant over smaller ones. For example, male *Tilapia mossambica* as little as 2 mm longer than conspecifics won aggressive encounters between them (Neil, 1964). However, size alone may not always be the only cue, but rather may act in concert with other signs to bring about a particular response pattern. Barlow (1970) demonstrated that visual perception of size was important in pair formation of the cichlid *Etroptus maculatus.* Males normally attack fish as large as or larger than themselves, making it difficult for a male to mate with a large female. Females generally attack fish that are smaller than themselves, making it difficult for a female to mate with a small male. Various chaetodontids (butterfly fish) show a great deal of intraspecific aggression toward equal-sized individuals, but little toward smaller or larger conspecifics (Zumpe, 1965). Hurley and Hartline (1974) showed that schools of *Chromis cyanea* escaped from larger geometric models at a greater distance than from smaller ones.

There have been cases in which factors other than size appear to be involved in dominance-hierarchical situations. DeBoer and Heuts (1973) contend that in *Hemichromis bimaculatus* stable dominance relationships need not be dependent on physical strength relationships between individuals but are probably determined by continuous stimulation from the environment and by mutual perception of each other's overt aggression and/or flight behavior. Keenleyside (1971) found the same level of agonistic response to small, medium, and large plywood models in the male longear sunfish *(Lepomis megalotis).* Myrberg and Thresher (1974) found that *Eupomacentrus planifrons* has a variable territory, the size of which depends on the particular species of intruder. But within the range tested, the size of the intruder does not affect the maximum distance of attack. *Pomacentrus jenkinsi* will attack any moving object, even as large as a human swimmer, that invades its territory (Rasa, 1969). As in other species, the possession of a territory by a small individual allows it to drive away larger nonresidents.

Shape

There is great diversity in body shapes of fishes, with the selection pressure for a particular configuration related to the environment. Fast-moving pelagic species have streamlined bodies selected mainly for hydrodynamic considerations, while the shapes of the more sedentary demersal species are selected for other adaptive reasons. Although it is clear that body shape evolved through selective pressures attuned to basic habitat and niche adaptation, we assume that shape is used in species recognition in many forms. Within a family of fish such as the sur-

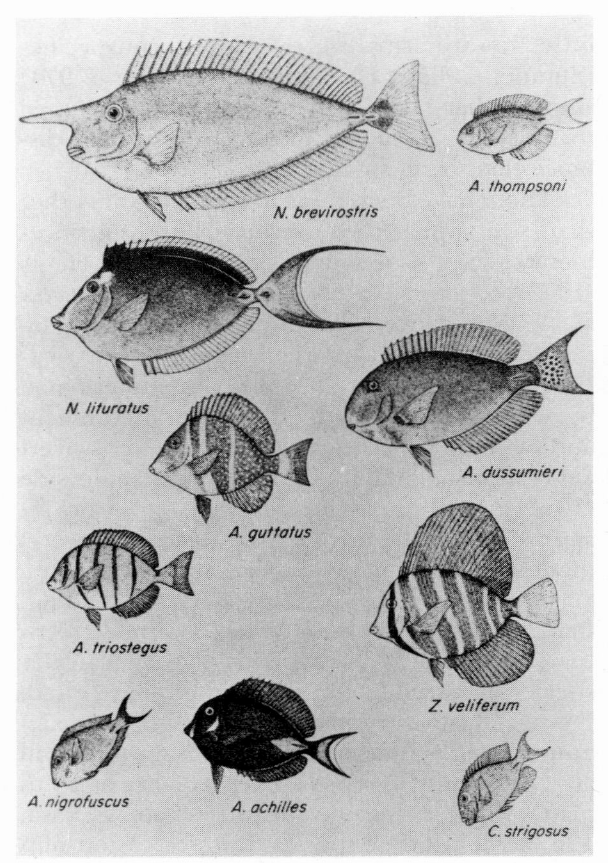

Fig. 3. Surgeonfishes, illustrating the diversity in body shapes. (From Barlow, 1974.)

geonfish (Fig. 3), there may be obvious differences in shapes between the different species (Barlow, 1974). However, it has not been demonstrated that fishes generally recognize members of their own and different species on the basis of form. For example, nest-guarding male longear sunfish *(Lepomis megalotis)* responded more aggressively to circular models than to models shaped like fishes (Keenleyside, 1971). Oval and triangular models were less effective, and the author suggested that "increasing the vertical dimension in relation to the horizontal enhances the stimulus value of the model." The results indicated, however, that there is a limit beyond which the vertical (or depth) feature of a model no longer serves as an effective stimulus in aggression.

The shape and size of many fishes are sexually dimorphic (Breder and Rosen, 1966). Experimental investigation of those secondary sex characters has been limited. Extreme attention to shape has been demonstrated in gouramis through the use of models (Picciolo, 1964). In *Trichogaster trichopterus* and *T. leeri* the male's dorsal fin is longer than the female's, and the fin appears to function as a visual cue for sex discrimination. A female model of *T. trichopterus* that displayed a swollen abdomen had a strong attraction value for males and was capable of evoking following behavior from them. The male sword-tail *(Xiphophorus hellerii)* develops a black-edged, tapering spike as the lower third of the caudal fin. Hemens (1966) investigated the behavioral significance of this "sword-tail" and found that it is an important visual stimulus in releasing aggression in other males but is not significant in male-female interactions.

Chemical Communication

Since living organisms produce biochemical products, it is not surprising that primitive plants and animals developed a chemosensory capability (Kittredge et al., 1974). The chemical senses of fishes have recently been reviewed by Kleerekoper (1969), Bardach and Todd (1970), Hara (1971), and Bardach and Villars (1974). This literature demonstrates the fundamental importance of the chemical senses for feeding, orientation, and communication of fishes. In some species the chemical sense is a primary modality mediating the way in which the organism perceives and reacts to its environment, while in others it may be auxiliary, acting in conjunction with other senses or perhaps as a "priming" mechanism before vision or other modalities become operative.

In the area of communication, studies equivalent to the playback of sounds or the presentation of visual models have succeeded in experimentally establishing that purely chemical cues, derived from skin washings or extracts, gonadal fluids, or urine, can elicit a variety of responses, including species recognition, intraspecific attraction (serving as a factor in schooling and aggregating behaviors), fright reactions, agonism, and various reproductive activities. Although the specific composition of these chemicals is not understood or documented for fish, these substances are, nevertheless, quite prevalent throughout the animal kingdom (Marler and Hamilton, 1966). These chemical stimuli were first designated as pheromones by Karlson and Luscher (1959) and were defined as "substances which are secreted to the outside by an individual and received by a second individual of the same species in which they release a specific reaction, for example, a definite behavior or a developmental process."

For fish whose behavior is not governed predominantly by visual cues, such as catfish of the genus *Ictalurus,* chemoreception is an important sensory modality for social behavior, particularly in the recognition of other individuals. Todd, Atema, and Bardach (1967) successfully conditioned blinded yellow bullheads *(Ictalurus natalis)* to discriminate between the odors of individuals. Test fish with nares cauterized were unable to discriminate between odors. When water containing only dermal mucus washings of donors was used as a test substance, normal test fish were able to distinguish the odors, but to a lesser degree than when tank water in which donor fish were maintained was used, suggesting that some factor in the excretory products is the major source of the chemical stimuli. Richards (1974) found that a chemical substance in the urine of blinded brown bullheads *(I. nebulosus)* is important in individual conspecific recognition. In this study another important chemical factor involved in the recognition process was found in extracts of the urophysis, a component in the caudal neurosecretory system. Göz (1941) demonstrated that *Phoxinus phoxinus* could chemically discriminate conspecifics.

In addition to their role in individual recognition, chemical secretions may also be utilized in the discrimination of another fish's hierarchal status (Bardach and Todd, 1970). Pairs of yellow bullheads were held in tanks and allowed to fight until one of each pair became dominant. Then each pair was separated. When water from a tank in which the dominant fish resided was added to the tank of the fish originally paired with it, the formerly subordinate individual responded by avoiding the area where water from the dominant's tank was introduced. In the converse experiment, a dominant fish swam toward the area of inflow of water from a subordinate's tank and sometimes exhibited aggressive behaviors at that point. In another series of experiments, when a previously dominant fish was returned to its tank after having unsuccessful encounters with more dominant fish, its formerly subordinate tank mate attacked it as if it were of inferior status. These results suggested that an alteration (decrease) in status may have been chemically communicated. In a final series of tests, when low-ranking bullheads with their nares cauterized were returned to their former tanks, they were unable to recognize their tank mates and immediately attacked them. Whereas normally low-ranking fish occupied particular areas of a community tank, the cauterized fish swam throughout the tank and attacked dominant fish in their shelters. These results indicated that, among bullheads, chemical stimuli appeared to mediate individual and status discrimination among species mates as well as to maintain and enhance normal social structuring of a community.

For fish in a reproductive state pheromones may serve as signals for recognizing appropriate mates as well as for identifying intruders in an established nest site. In the blind goby *(Typhlogobius californiensis)* male-female pairs, living as commensals in burrows of the shrimp *Callianassa affinis,* appear to retain their status for life, probably ten or fifteen years (MacGinitie, 1939). The gobies were found to be sensitive to intrusions of other individuals of the same sex through chemical cues. After an invasion, a fight between the stranger and the resident of the same sex ensued until one of the antagonists was killed or driven from the burrow. The opposite sexed resident fish became passive and accepted the victor (either the former resident or a stranger) as a mate. This suggested that pair formation may be dependent more on gender-specific rather than on mate-specific pheromones in this species. MacGinitie was also able to determine experimentally that odors of a strange fish introduced into the burrow of an established pair would elicit aggressive behavior by the inhabitant corresponding to the sex of the donor.

Pheromones may also stimulate nuptial behaviors. In the gobiid fish *Bathygobius soporator* gravid females produce a chemical secretion that stimulates courtship behavior in males even in the absence of visual cues (Tavolga, 1956). This substance was found to elicit courting movements by males within five to ten seconds after its introduction, and it continued to stimulate the male's response for approximately one hour. Tavolga tested extracts from various tissues and determined that the pheromone was produced in the ovaries and that it was sensed by olfaction.

Male glandulocaudine characids have a caudal gland that may produce a substance which, when directed toward the female during "dusting," increases the probability that she will pair (Nelson, 1964a, 1964b). Interestingly, the members of the genus *Glandulocauda* that do not appear to have an intact gland produce croaking sounds, which may have taken over much of the gland's function. Courting males of the genus *Hypsoblennius* produce a pheromone that attracts other ripe, nonparental males (Losey, 1969). Although not fully explained, these results suggest that possibly on population level the male pheromone may act to facilitate and enhance the sexual receptivity of other males during the breeding season. In a similar study by Leiner (1930) it was found that mucus secreted by a male stickleback *Gasterosteus aculeatus* elicited courtship behavior among other conspecific males. Visual cues are primarily responsible for pairing in the blue gourami (*Trichogaster trichopterus:* Cheal and Davis, 1974). Although chemical cues from a female of this species to some extent increases nest building in isolated males, the full expression of this response is mediated by a combination of chemical and visual stimuli. Similarly, in the angelfish *(Pterophyllum scalare)* chemical stimuli from males will increase spawning rates in females, but this response is increased further by the effects of visual and chemical cues (Chien, 1973). Additional examples of the role of chemical secretions in reproductive behavior are reviewed by Bardach and Todd (1970).

Many species of fish shed their gametes into the water, where fertilization takes place. But meeting of eggs and sperm is not totally haphazard. Spermatozoa generally move and may be attracted to eggs by a chemical message. Unlike that of most species, the sperm of the Pacific herring (*Clupea pallasii*) are almost motionless in the water (Yanagimachi, 1957). Once in the vicinity of the micropyle of a mature egg, however, the sperm move actively and instantly enter the micropyle. Yanagimachi found that the sperm attractant quality emanated from the egg membrane around the micropyle area—the interior of the egg is ineffective—and suggested that the essential groups of the sperm-activating factor either are proteins or are intimately associated with protein. Suzuki (1961) also implicated a messenger from the micropyle area for sperm attraction in the Japanese bitterling *(Acheilognathus lanceolata).*

Parental care and recognition of the young among several species of fish also appear to be mediated by chemical cues. Kühme (1963) studied adult jewel fish *(Hemichromis bimaculatus)* whose young had recently hatched. When water from the tank containing their young was introduced into a parent's tank, the adult oriented to that specific inflow locus and displayed fanning and other parental behavior. This response persisted for three weeks, which is the normal period of parental care. By exchanging these fry with younger or older offspring, Kühme was able to extend or shorten, respectively, the time period of parental care. He also established that parents could distinguish their young from those of other parents solely on the basis of chemical cues. Similar results of parental discrimination of and attraction to their own broods have been established in the dwarf cichlid (*Nannacara anomala:* Kunzer, 1964) and the Central American cichlid (*Cichlosoma nigrofasciatum:* Myrberg, 1966, 1975).

The utilization of chemical cues for the initiation and maintenance of schooling behavior in fish has been previously studied with varying interpretations of its degree of importance. Keenleyside (1955) studied blinded rudd *(Scardinius erythrophthalmus),* which perceived and were attracted to odors of their species mates. Following destruction of the olfactory epithelium, the fish ceased to respond to these stimuli. Keenleyside suggested that olfactory cues may keep these schools from scattering at night. The roach *Rutilus rutilus* exhibited comparably high levels of attraction to both visual and chemical stimuli of species mates (Hemmings, 1966). The author also hypothesized that mutual attraction and schooling maintenance may be mediated to a large degree by chemoreception during nocturnal conditions and by vision during the day, each being further integrated with stimuli received through the lateral line system. Kühme (1964) found that young jewel fish *(Hemichromis bimaculatus)* could orient positively to the odors of other comparably aged conspecifics. Although chemical cues appear to serve a function in the attraction phases of schooling behavior, as Shaw (1970) pointed out, other stimuli, particularly visual cues, must necessarily be integrated for the manifestation of all the finer spatial and positional adjustments of a school.

There is another group of chemical compounds among fish that have been found to stimulate alarm reactions or fright responses. The alarm substance, or *Schreckstoff* (von Frisch, 1938, 1941), is generally released from club cells in the epidermis of an injured fish (Pfeiffer, 1960; see Kleerekoper, 1969; Bardach and Todd, 1970; and Hara, 1971 for recent reviews).

Pfeiffer (1963) reported the presence of alarm substances and reactions in several species of North American Cyprinidae and Catastomidae and believed that only fish in the order Ostariophysi possessed these substances. Verheijen (1963) identified nine species of cyprinid fishes that also exhibited the flight reponse to intraspecific skin extracts. An alarm substance has been described for the top smelt (*Atherinops affinis:* Skinner et al., 1962), but Rosenblatt and Losey (1967) discounted it.

Reed (1969), following von Frisch's methods for measuring the alarm reaction, found that *Gambusia affinis* and *Fundulus olivaceus* (non-ostariophysans) and *Notropis venustus, N. texanus,* and *Hybopsis aestivalis* (Cyprinidae) exhibited fright responses to skin extracts from their own species. The fright reaction among *Fundulus* and *Gambusia* consisted of the fish becoming motionless. In some cases *Gambusia* also darted downward and began digging in the gravel as if attempting to hide. The Cyprinidae of this study, in general, became excited and formed tight schools that moved to the tank bottom. In the same study Reed also found that the same five prey species exhibited a fright response when exposed to the odors of three North American and two South American predatory species. The significance of this finding was that in several cases the prey and predator species are ecologically isolated and hence would normally not encounter each other in their natural environments.

The alarm substance has two beneficial effects for the species involved. The fright reaction should move fishes away from feeding predators as rapidly as possible. It may also be an anticannibalism device so that a fish preying on a young fish of its own species or a related species will be inhibited from further feeding upon release of the alarm substance.

Acoustic Communication

Unlike higher vertebrates, which typically share homologous sources of sound production, fishes early in evolution apparently did not possess specialized mechanisms to produce sound. Mechanisms of this sort appeared as later developments, evolving sporadically and independently in various fish taxa.

Fishes produce sounds in two basic ways: by stridulation of bony elements or by movement of the swimbladder (Barber and Mowbray, 1956; Burkenroad, 1931; Skoglund, 1961; Tavolga, 1962; Gainer and Klancher, 1965; Markl, 1971). Stridulation can be caused by grinding of the teeth, moving of the skull, pectoral girdle, or fins. The swimbladder can be set into motion by the rapid contraction of specialized intrinsic or extrinsic muscles. In addition the swimbladder can pick up and amplify vibrations produced by stridulation of other parts of the body.

Stridulatory sounds are usually of short duration and spread over a wider frequency range than swimbladder sounds. The croaking gourami *(Trichopsis vittatus),* for example, produces sound energy up to 12 kHz (Marshall, 1966), although it is doubtful that the fish can hear frequencies this high.

Sounds produced by swimbladder mechanisms generally contain energy in a frequency range from less than 100 Hz to several kHz with greatest amplitude in the lower frequencies. In many species the sound fundamental corresponds directly to the rate of muscle contraction (Packard, 1960; Skoglund, 1961; Winn and Marshall, 1963; Cohen and Winn, 1967; Markl, 1971).

Fishes also produce sounds incidental to swimming, known as hydrodynamic sounds (Moulton, 1960). The significance of these sounds in communication has not been well established but they may attract animals to sources of food in some instances. More extensive reviews on sound-producing mechanisms are available in Marshall (1962, 1967), Schneider (1967), Tavolga (1964, 1971a), Demski et al. (1973), and Fine (1975).

Sounds are usually named by either their function or their behavioral context (e.g., courtship sound or threat sound) or by how the observer may describe the sound (e.g., onomatopoeic description such as knock, thump, purr, staccato). That these representations of sound are, of course, subjective and will vary from observer to observer (Fish and Mowbray, 1970) makes it quite difficult to compare sounds from the literature where different investigators have named the same sound differently. Contributing to the problem is the failure of many investigators to publish the sonagrams or oscillograms of the fish signals or even to describe the signals completely.

With these limitations in mind, we have attempted to summarize in Table 1 the sounds and coincident behavior of various fishes. The fishes are listed taxonomically according to Greenwood et al. (1966). The examples used were of fishes that modulate their acoustic signals in some way, or conversely, those that fail to vary them in diverse situations. It is quite possible in many instances that signals are more variable or are graded in a finer manner than described in the original publications. Acoustic variations are presented in this table in an attempt to discern how fishes may code their sounds. Tavolga (1974) warned that parameters that are blatantly obvious to humans may be irrelevant to a fish. Meaningful parameters may be established experimentally by playing back whole sounds and their components while observing responses. (Playbacks will be reviewed later in this chapter.)

Hearing in fishes has been extensively reviewed (Moulton, 1963; Cahn, 1967; Enger, 1968; Flock, 1971; Lowenstein, 1971; Tavolga, 1971a; Erulkar, 1972; Hawkins, 1973; Popper and Fay, 1973). The problem is complicated by the dual nature of underwater sound; it is made of a vector velocity or displacement component and a scalar pressure component. Fishes receive sound vibrations through the lateral line and the labyrinth. Receptors respond to shearing forces that move the kinocilium on hair cells (Hilliman and Lewes, 1971). Lateral line neuromasts are deformed by near-field sounds and mechanical movements in the water, and are therefore tuned to have higher thresholds than units in the ear. Lateral line organs can act as directional sensors close to the emitting source (Harris and van Bergeijk, 1962). Recent evidence indicates that the labyrinth may also function in localization of sound (Schuijf and Siemelink, 1974).

FUNCTION OF SOUND

Fish and Mowbray's (1970) work on fishes of the western North Atlantic is indicative of the large number of fish species known to produce sounds. Impressive as this number is, it is important to realize that the behavioral significance of only a small fraction of these sounds is known. The sounds that observers have been able to identify as playing a role in specific behavioral acts are primarily related to either aggression or reproduction. In many cases the close relation between these two categories makes a complete dichotomy difficult.

Many species, such as the tiger loach (*Botia hymenophysa:* Klausewitz, 1958), use sound in defense of territory. In the toadfish *Opsanus tau* the boatwhistle, which can attract a female to a male's nest, may also indicate that the territory is occupied by a male, and hence other males are warned that they are unwelcome in the area (Gray and Winn, 1961; Winn 1964, 1967, 1972). Should another male approach the nest, the resident male will grunt at the intruder. The northern midshipman *(Porichthys notatus)* has three agonistic calls: a grunt spaced both regularly and irregularly and a buzz (Cohen and Winn, 1967). All three signals are elicited under identical conditions, making the individual message content for each sound unclear.

Perhaps uniquely, sea anemone fishes *(Amphiprion spp.)* make a sound that is specific for fighting and another that is used as a territorial threat sound (Schneider, 1964a). In addition these fishes have a submissive sound emitted by vanquished fish after a fight.

The female in a courting pair of the cichlid *Hemichromis bimaculatus* emits a "br-r-r" sound while aggressively holding ground after the male has bitten or rammed her or shows intentions of doing so (Myrberg et al., 1965). This behavior deters further attacks by the male, suggesting to Barlow (1970) that the action of the female induces fear in the male.

The squirrelfish *Holocentrus rufus* produces two calls of markedly different types (Winn et al., 1964). The grunt, although used when a territory is invaded by another species, is used chiefly against neighboring conspecifics. When the staccato sound is made, it acts primarily as a warning sound elicited when the territory is invaded by large fish, including predators, or by almost any fish that appears suddenly. Grunts appear to be associated primarily with aggression and experimentally were shown not to habituate readily, while the staccato is associated more with escape tendencies and did habituate readily.

A schooling, nonterritorial squirrelfish, *Myripristis berndti* produces both a grunt and a staccato sound (Salmon, 1967), just as the territorial species *Holocentrus rufus* does (Winn et al., 1964); but in *Myripristis* both sounds act as warning calls. In addition, *M. berndti* also produces two calls that could be termed agonistic against species members, but they serve the purpose of increasing distance between fish. This call might function in determining optimal spacing within the school rather than as territorial behavior. Salmon speculates from these observations that in various species of *Myripristis* schooling, rather than occupying a territory, encourages the evolution of a complicated vocabulary. Among different species of squirrelfish, at least, this conjecture is supported by the fact that the sounds emitted by *M. berndti* and *M. violaceus* (Salmon, 1967; Horch and Salmon, 1973) are more complicated than that of the territorial *Holocentrus rufus* (Winn et al., 1964). The problem may be more complex because *H. rufus* forms schools in certain habitats (Winn, pers. obs.). However, observations made of other schooling species appear not to support the idea as a generalization.

Cod *(Gadus morhua)* and haddock *(Melanogrammus aeglefinus),* which are schooling species, appear to have simple vocabularies (Brawn, 1961b; Hawkins and Chapman, 1966). The cod apparently uses the same grunt for agonism and courtship, as well as for the breakup of courtship (Brawn, 1961b), but possibly further study will show a more differentiated acoustic system. It is possible that this sound may function as an alerting or display-enhancing device and be devoid of a specific message content. In the sea catfish (*Galeichthys felis:* Tavolga, 1971b) and the croaking gouraim (*Trichopsis vittatus:* Marshall, 1966) the agonistic displays and vocalizations may have a hierarchical function. For example, once beaten, a gourami may remain near the dominant conspecific without eliciting further aggression.

Most soniferous fishes can be induced to vocalize when they are held or chased (Burkenroad, 1931; Fish, 1954; Fish and Mowbray, 1970; Horch and Salmon, 1973). We assume that the sounds may be produced in the wild when a fish is attacked by a predator and may act as warning calls. The squirrelfish *Myripristis violaceus* produces both a growl and a grunt (the latter were produced only when the fish were hand-held) when confronted by threatening situations. When these sounds were played back, the fish responded by showing attention to the sound source (a speaker) and producing growl sounds. On the basis of their observations, Horch and Salmon (1973) surmised that these sounds may be used in nature to alert species mates to danger.

Since Smith's work in 1905, sciaenid calls have been generally assumed to play a courtship role because the sounds occur only during the mating season. Dijkgraaf (1947) observed a group of four *Corvina nigra,* of which two appeared to be a pair. The smaller of the pair, probably the male, would swim closely behind and below the sluggish and larger female, with his head below her abdomen. At odd intervals and without apparent external motivation, the male would swim to one of the other fishes and chase it around the tank while emitting its knocking sound. This behavior occurred around twilight, when the apparently dominant male searched for the female and began his chasing.

Males are often more vocal than females during close courtship exchanges. In toadfish, as well as in many other species, there may be an endocrine as well as a physical basis for sexual differences in sound production. For example, courtship sounds are produced only by male toadfish *Opsanus tau.* However, both Demski and Gerald (1974) in *O. beta,* and Fine (unpublished) in *O. tau,* have elicited boatwhistle-like sounds in female toadfish by brain stimulation; though the female is not known to produce this sound in nature (Gray and Winn, 1961). Although the swim bladder intrinsic-muscle complex grows faster in the male than in the female, the bladder complex is equally developed in the two sexes, implying a hormonal rather than a morphological basis for differential sound production in the male (Fine, 1975).

Although not common, there are cases in which females have been observed to produce courtship calls. Delco (1960) described sounds by females during the courtship behavior of *Notropis lutrensis* and *N. venustus.* Stout (1963) was somewhat skeptical of this work since he found that sounds were largely, if not entirely, produced by males in the closely related species *N. analostanus.* Fish (1954), describing mating of a pair of sea horses *Hippocampus hudsonius* stated that "preliminary activity consisted of slow swimming, either together or apart, accompanied by occasional noisy snapping of the head. Clicks were often produced alternately by the two fishes, and during their actual embrace, these sounds were loud and almost continuous." Although no sounds were heard during nonaggressive courtship, spawning, or caring for the offspring, the "br-r-r" sound of the female *Hemichromis bimaculatus* seems to inhibit the male's aggression (Myrberg et al., 1965). In the croaking gourami *(Trichopsis vittatus)* both the male and female croak during lateral displays, aggressive interactions, and the early stages of courtship (Marshall, 1966). During courtship and spawning the female purrs in a headup posture. According to Marshall, this display operates as a distance-decreasing mechanism in contradistinction to the hypothesized function of the croaking sound. It is probable that in both the cichlid and the gourami courtship sounds have evolved from a system designed primarily for aggression.

There are few known examples of males producing courtship sounds during the reproductive season when no female is within visual distance. Toadfishes (*Opsanus tau:* Fish, 1954; Fish and Mowbray, 1959; Tavolga, 1958c, 1960; Gray and Winn, 1961; Winn, 1967, 1972; *O. beta:* Breder, 1941, 1968; Tavolga, 1958c, 1960; *O. phobetron:* Tavolga, 1968b) and *Porichthys notatus* (R. Ibara, pers. comm.) produce sounds of this type. Such a system of spontaneous calling is analogous to bird song and shares similar properties (Winn, 1964, 1972). Schleidt's (1973) concept of tonic communication should also be applicable to the toadfish. A background chorus of boatwhistles could enable females to enter final spawning readiness, at which time the call would become attractive and thereby serve as an orienting stimulus. Similarly, a priming function was shown by Marshall (1972), who played back male sounds to females of the cichlid *Tilapia mossambica,* causing them to lay eggs several days earlier than control females. There may be other examples where sounds act in communicating to distant fish. Although Gerald (1971) observed sound production associated only with active courtship behavior in various species of sunfishes, playbacks of the male's courtship grunts attracted both males and females. The sunfish call could be evolving for communication over longer distances or could simply have an incidental (to the emitter) attractive quality. Delco (1960) found that male and female *Notropis lutrensis* were attracted to a chamber from which female *N. lutrensis* sounds were produced. Only males of *N. venustus* responded to conspecific sound in a comparable experiment.

Schwarz (1974a) used information analysis, a fundamental tool for determining if communication occurs, for the first time on fish sounds. She showed that there were three lines of evidence that indicated a function for the sounds of *Cichlasoma centrarchus:* sound was associated with aggressive behavior; responses to silent behavior differed significantly from responses to behavior accompanied by sound; and finally, the amount of information transmitted when animals could hear as well as see one another differed from that transmitted when they could only see one another. A recipient conspecific responds to sound by ceasing aggressive acts and usually moving quickly away from the emitter. Aggressive displays not accompanied by sound emission have a much lower threat value. Breeding pairs do not use the sound for courtship, but rather as a threat. From introduction up to the time of spawning, the male produces most of the sounds, directing them toward the female. From this point on, the female makes most of the sounds, but directs them at fishes in adjacent tanks (Schwarz, 1974a, 1974b, in preparation).

SIGNAL VARIATION AND CODING

Most fish sounds are basically percussive, and the various messages sent between fishes are elaborations of pulses (Fig. 4). In an attempt to summarize how fishes code their signals, Winn (1964) categorized them in five basic ways (Fig. 5), which suggested to him that the temporal patterning of sounds was an important carrier of information. The first and second were variable-time-interval and fixed-time-interval signals, in which the time between units viewed as one sound on a spectrogram (grunt, knock, etc.) were variable or fixed. This scheme referred to whole sounds rather than to components of sounds related to individual muscle contractions. In the third and fourth types of signal, the duration of any signal is lengthened (unit-duration signal) or the amount of time during which units are produced is varied (time-length signal). The final category was harmonic-frequency signals of longer duration.

There are five basic ways in which a signal produced by a fish can be varied: amplitude, duration, repetition rate, number of pulses within a signal, and frequency. While in many cases, categories such as duration, repetition rate, number of pulses, and intervals are obviously related, at other times they can refer to quite disparate quantities.

Amplitude

By analogy with humans, we might expect fishes to reserve their louder sounds for higher emotional states or levels of arousal, and in fact such modulation does occur. The cod *(Gadus morhua,)* grunts more loudly when using sound as a determined threat and more faintly at the end of an aggressive encounter (Brawn, 1961b). Fish (1954) observed a sea robin, *Prionotus carolinus,* that would lie quietly on the bottom clucking softly while being gently stroked by the experimenters. However, if this stimulation was applied too long or too heavily, the fish would break away in apparent annoyance and emit a louder burst. The knocking signal of the sciaenid *Corvina nigra* becomes more energetic under higher stimulation (Dijkgraaf, 1947).

Amplitude is also used to differentiate sounds that have different meanings. The threatening sound of the sea anemone fish *(Amphiprion xanthurus)* is loud enough to be audible in air 10 m from the aquarium. The fighting sound of the fish is emitted much closer to a conspecific than the threatening sound and is of lower intensity (Schneider, 1964a). The courtship purr of the satinfin shiner *(Notropis analostanus)* is less intense than its agonistic vocalizations (Winn and Stout, 1960; Stout, 1963). The threatening sound in the tiger fish *Therapon jarbua* is louder and is elicited by a higher level of agonism than the drumming sound (Schneider, 1964b). The short grunt is quieter than the long, loud grunt in the sea catfish (*Galeichthys felis:* Tavolga, 1960). The squirrelfish *Myripristis berndti* produces grunts and staccatos as warning calls. The staccato, which is elicited under a more stressful situation than the grunt, is also louder (Salmon, 1967). Winn (1972) found that a toadfish boat-

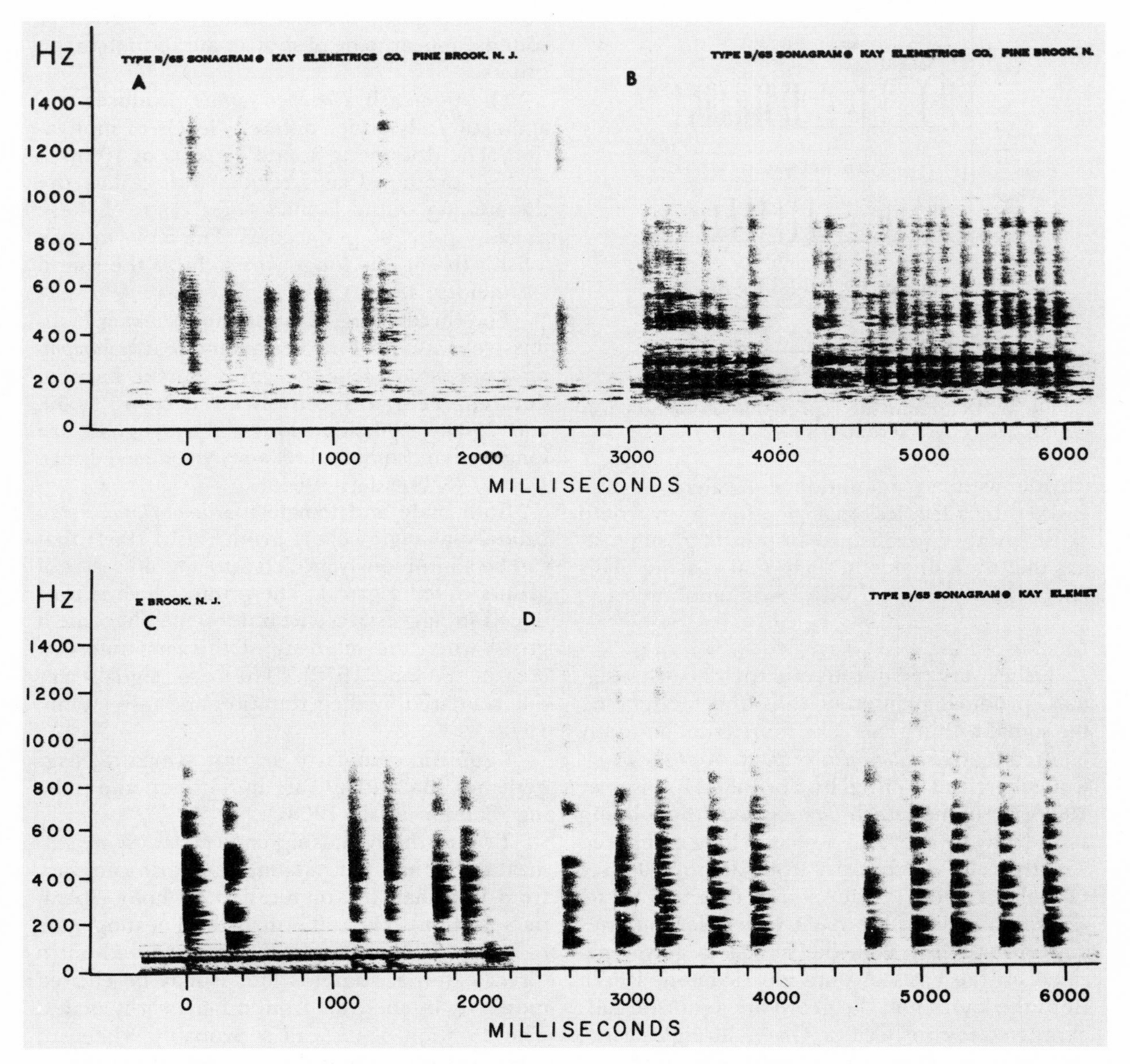

Fig. 4. Examples of fish sounds demonstrating their pulsed nature. a. rapid series of knocks of the satinfin shiner *Notropis analostanus;* b. staccato of the squirrelfish *Holocentrus rufus;* c. chorus of croakers *Micropogon undulatus* (note sounds occur in pairs); d. grunts of a toadfish *Opsanus tau.* Narrow-band filter.

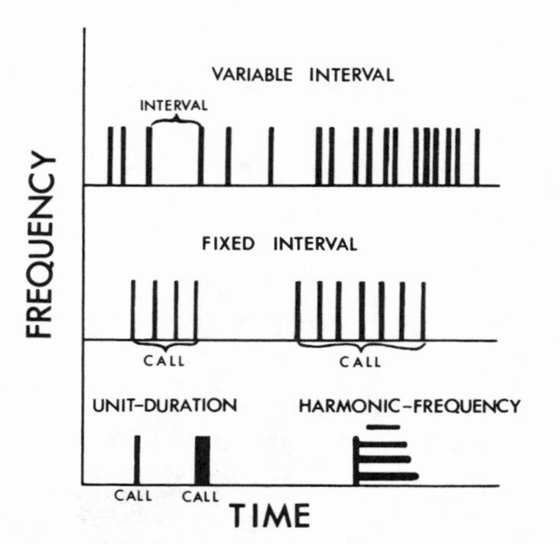

Fig. 5. Diagrammatic representation of how fish sounds are coded. (From Winn, 1964.)

whistle with an amplitude equivalent to that emitted by a fish less than one foot away would cause another toadfish to stop calling, suggesting that the call aids in territorial spacing. This effect is not obtained with lower amplitudes.

Duration

Fishes vary the duration of their acoustic signals, producing different calls and modulating the signal within a call. The short grunt naturally emitted by *Galeichthys felis* is 20 to 40 msec long, while the grunt emitted by a prodded fish is over 100 msec; in the catfish *Bagre marinus* the sobbing sound may be over half a second long compared with the yelp, which varies from 100 to 200 msec (Tavolga, 1960). The squirrelfish *Myripristis berndti* has one of the most varied vocabularies known for a fish, with the signals largely separated on the basis of duration (Salmon, 1967). Both the knock and the growl are agonistic calls expressed to conspecifics. Given during chasing, the knock is short, while the growl, produced only after physical contact, is many seconds long. This fish also produces grunts and staccatos in warning situations. While the staccato is considerably longer than the grunt, the individual components of the staccato, made of a series of sounds, appear to be of shorter duration than the grunts.

The tiger fish *Therapon jarbua* produces two agonistic calls under different levels of motivation. The drumming sound consists of 10 msec pulses, produced at irregular rates; and the threatening sound is made of a burst of these pulses, up to 200 per second. The more intense a fish's threat, the longer it produces the sound (Schneider, 1964b).

The threatening, fighting, and shaking (submissive) sounds of the sea anemone fish *(Amphiprion xanthurus)* separate easily on the basis of duration, being respectively 25 to 30, 45 to 60, and 250 to 400 msec long. These signals are longer than comparable ones produced by *A. polymus* (Schneider, 1964a).

Both male and female toadfish *Opsanus tau* produce a single coarse grunt (Fish, 1954) that can be continuously graded into a rapid series of grunts called a growl. The grunt, which is produced in aggressive encounters, may become a growl when the intensity of the encounter increases (Winn, 1972). The two signals are differentiated by their duration and rate (Winn, 1972).

Agonistic sounds of several sympatric triggerfishes (Balistidae) vary in duration and pulsing (Salmon et al., 1968).

The northern midshipman *(Porichthys notatus)* makes a continuous mating call that can vary from less than ten minutes to an hour (Ibara, pers. comm.). The call is made by a nesting male to attract a female and is produced without a direct external stimulus, but it may be elicited indirectly by the calls from other nearby males. The duration of the call is probably related to some aspect of the male's internal state.

Delco (1960) described a female courtship call for the minnows *Notropis leutrensis* and *N. venustus* having durations of 0.84 and 0.047 to 0.07 seconds, respectively. The duration of the components of the male satinfin shiner's *(Notropis analostanus)* courtship purr varies between 11 and 24 msec, with the purring sound that is directed at the female during courtship consisting of a more rapid series of lower intensity knocks (Winn and Stout, 1960; Stout, 1963).

The courtship grunt produced by the male damselfish *Eupomacentrus partitus* is variable in number of pulses and duration (Myrberg, 1972a, 1972b), with the variation apparently correlated with the amount of time the male spends swimming close to the female. Moulton (1958) observed a meeting of a pair of angelfish *Pomacanthus arcuatus* in which vocalizations changed from short grunts to longer moanlike sounds.

The freshwater drum *(Aplodinotus grunniens)* produces a drumming sound during a long session during the day (Schneider and Hasler, 1960). Early and late in the session it produces a main sound of three to five seconds with one to several short sounds preceding and following. During the peak hours of the session the sound series reaches a minute or longer before a break, and the short sounds decrease or disappear. Dijkgraaf (1947) described a similar phenomenon in another sciaenid, *Corvina nigra.* Under periods of increasing stimulation, the knocking signal became more frequent and more energetic, and the duration increased. However, the number of knocks in a signal varied from five to seven and did not increase as it did in *Aplodinotus.*

Sounds may function as an ethological isolating mechanism in fish, with duration being an important component of the system. Gerald (1971) has studied the male courtship grunts in six sympatric species of sunfish (Centrarchidae). He was able to separate the vocalizations on the basis of call duration and percentage of call pulsed (Fig. 6). Although he has some evidence that the sounds are discriminated by conspecific females, the system is not perfect and hybrids occasionally occur. Another interesting example

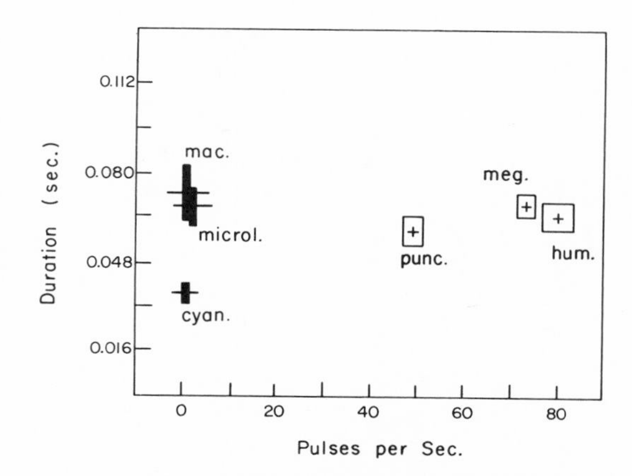

Fig. 6. Average grunt duration and pulses per second of six species of sunfish *(Lepomis).* The crosses represent the intersection of the means for each species, and the closed rectangles represent two standard errors on either side of the means. *L. macrochirus, L. microlophus,* and *L. cyanellus* had essentially no pulsation. (From Gerald, 1971.)

comes from three closely related Atlantic toadfishes, *Opsanus tau, O. beta,* and *O. phobetron* (Walters and Robins, 1961; Fish and Mowbray, 1959). The toadfish species share similar meristics and morphometrics and are separated primarily on size and color pattern. The boatwhistle calls of the three are distinct (Fig. 7). *O. beta* produces a double hoot, with the second sound shorter than the first, while *O. tau* and *O. phobetron* produce a single boatwhistle. While the call of an *O. tau* rarely exceeds half a second, *O. phobetron's* call may last up to a second (Tavolga, 1968b). Although the three species are currently allopatric, they may have been sympatric in the recent past and perhaps the divergence of their calls can be ascribed to character displacement. Duration is an important parameter in boatwhistle call recognition in *Opsanus tau* (see playback experiments).

Sounds of a nonsexual nature may also diverge among related forms. Such sounds could function for intraspecific species recognition, or they might be a reflection of a change in a fish's

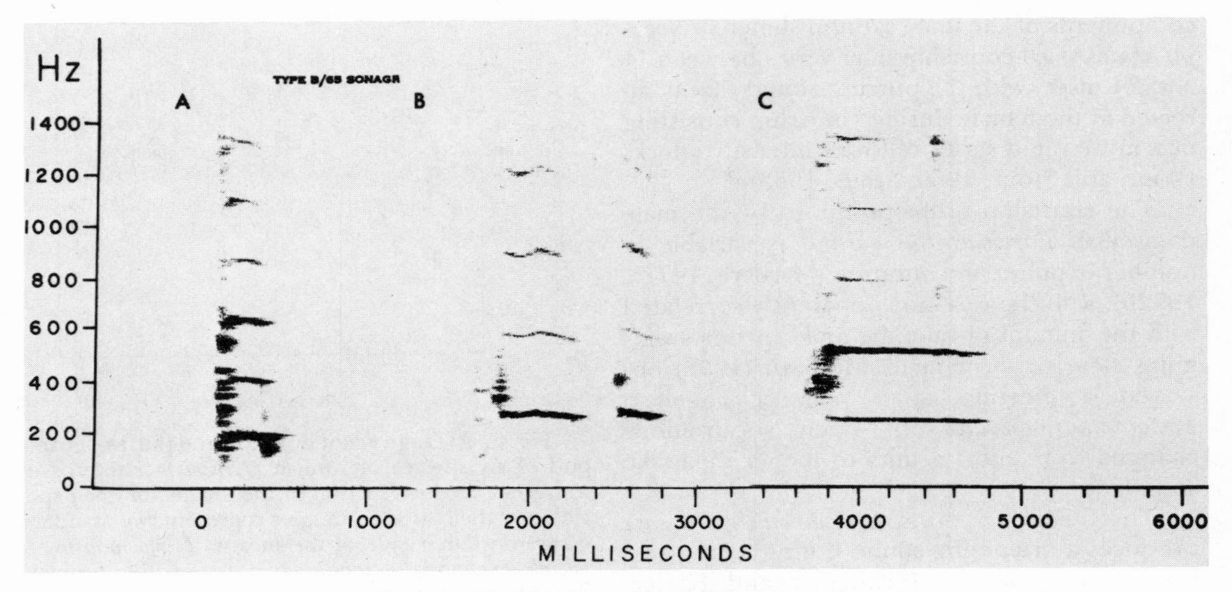

Fig. 7. Boatwhistle call of three species of toadfish. A. *Opsanus tau;* B. *O. beta;* C. *O. phobetron.* Narrow band filter. (B. and C. from tape accompanying Tavolga, 1968b.)

whole acoustic repertoire during speciation. For instance, the hand-held sounds of *Priacanthus meeki* range between 76 and 150 msec, while those of *P. cruentatus* go from 300 to 600 msec (Salmon and Winn, 1966).

Repetition Rate

The temporal patterning of fish sounds, a combination of duration and repetition rate (on and off times), is of paramount importance in their coding. Winn (1964, 1967, 1972) has focused on the idea embodied in repetition rate by dividing sounds into fixed and variable interval types. It is largely variable rates that make the drumming sounds of various sciaenids species-specific. *Corvina nigra* produces eight pulses a second (Dijkgraaf, 1947), *Cynoscion regalis* twenty-four (Tower, 1908), and *Aplodinotus grunniens* eighteen to twenty-seven (Schneider and Hasler, 1960). This sort of variation is shown by Fish and Mowbray (1970) in the sonagrams of a host of sciaenid species.

Individual species often produce different calls of varying repetition rates. The rapid series of knocks of *Notropis analostanus* is variably paced, while the purr has a constant but faster repetition rate (Winn and Stout, 1960; Stout, 1963), i.e., the fixed interval of Winn (1964, 1972). Likewise, squirrelfish of the genera *Holocentrus* and *Myripristis* produce variable-interval grunts and fast-paced equal-interval staccatos (Winn et al., 1964; Salmon, 1967). Toadfish grunts vary from individual pulses to rapidly pulsed growls (Winn, 1964). In fact, the boatwhistle of the toadfish probably represents an extreme within this spectrum, with the sonic muscles undergoing a sustained contraction at their maximal rate, which is considerably higher than that used in the grunt and growl. The thump produced during a fight between conspecific males in *Hemichro-*

mis bimaculatus has a pulse repetition rate of 12 Hz, while the "br-r-r" produced before attack is pulsed at thirty-five times a second (Myrberg et al., 1965).

Motivational states and their meaning can be graded by varying the rate of production of a sound. The courtship grunts of *Bathygobius soporator* increase in repetition rate during active courtship (Tavolga, 1958a). *Chasmodes bosquianus* produces its courtship grunt every second or two, but occasionally the male will emit a burst of three to four sounds in quick succession if a female approaches his shelter (Tavolga, 1958b). Likewise, male toadfish have been observed to increase their rate of boatwhistling when a gravid female approached (Gray and Winn, 1961; Fish, 1972). Experimental studies on rate are given in the playback section.

Recent work on the swimbladder's role in hearing supports our contention that the temporal nature of sound, rather than variations in its frequency content, is important for communication in fishes. The swimbladder presents an acoustic discontinuity to underwater sound, which causes it to resonate. This motion is reradiated and translated to the ears. Unlike an underwater bubble, which has a highly tuned resonant frequency determined by its size and depth, the swimbladder is highly damped within the body of the fish (Alexander, 1966; McCartney and Stubbs, 1970; Demski et al., 1973; Sand and Hawkins, 1973; Popper, 1974). Popper (1974) found that the response of the swimbladder of an insonified living goldfish was flat from 50 to 2,000 Hz, indicating that the bladder did not selectively favor any particular frequencies. A strongly tuned bladder would change a fish's hearing sensitivity with depth and size (age) and therefore be deleterious (Popper, 1971; Sand and Hawkins, 1973). The damped nature of the swimbladder makes it responsive to the time-based nature of a signal. A strongly resonant structure takes time to start responding and continues to respond after stimulation has terminated (Popper, 1974). Such a reverberating system would be poorly adapted to discriminate broad-band pulsed sounds.

Popper (1972) studied the auditory threshold in the goldfish as a function of signal duration and indicated that there were no differences between short pulses and continuous tones, and that thresholds were the same whether there is a long or short off-time between pulses. Popper concluded that temporal summation does not occur in goldfish and that the enhanced sensitivity for long sounds found in mammals is unnecessary for the short duration of fish sounds.

Number of Pulses or Units

Looking at the number of pulses or units in an acoustic display could be identical to focusing on the duration or repetition rate (or the on and off time). But, whereas the latter two categories are on an obvious continuum, the former may be looked at as a binary or incremental function. For instance, many fishes make sounds that contain one or multiple pulses (Table 1). If the sounds merely grade from uni- to multipulsed with some sort of proportional behavioral motivation, they do not belong in this category. This question cannot be answered in most cases. In *N. analostanus* single knocks are used in the initial stages of courtship and in chasing between males, while two different types of series of knocks are used for more intense fighting and courtship (Winn and Stout, 1960; Stout, 1963). The courtship purr of the female croaking gourami (*Trichopsis vittatus*) is separated from the male and female agonistic croaking sound by the number of pulses, averaging 2.3 pulses per croak and 4.5 pulses per croak respectively (Marshall, 1966). This type of coding reaches a peak in the damselfish *Eupomacentrus partitus,* where five different calls can be separated in this manner (Myrberg, 1972a, 1972b). The pop, chirp, long chirp, and burr have respectively 1, 3, 4–6, and

Table 1

Message and modulation of fish vocalization.

Species and Call	Function	Probable Coding
Gnathonemus petersii (Rigley and Marshall, 1973)		
Click sounds	Agonistic during lateral display or chasing of conspecific, model of conspecific, or a knife fish (*Gymnotus* sp.); also hand-held.	Single clicks or in series of 2–5 clicks with varying intervals. Average duration 25 msec.
Glandulocauda inequalis (Nelson, 1964b)		
Courtship croak	By male during hovering	Pulse train of varying duration, either grouped or relatively equally spaced. Frequency and amplitude variable. Some variation related to gulping of air bubbles and depth in the water.
Notropis analostanus (Winn and Stout, 1960; Stout, 1963)		
Knock	Male chasing male or by male during initial courtship contact with female	11–60 msec duration
Rapid series of knocks	Male fighting	Series of knocks, irregularly spaced.
Purring sound	Male courtship, during circling and "solo-spawning" behavior, occasionally during approaches to the female.	More rapid series, lower intensity and frequency range.
Galeichthys felis (Tavolga, 1960, 1971b)		
Type I	Possible echolocation	5–10 msec duration; 75–100 Hz rep. rate.
Type II (short grunt)	Dominance behavior	Soft, 10–40 msec duration; 200 Hz rep. rate.
Type III (long, loud grunt)	Prodding	Loud, 100–150 msec duration, 150 Hz rep. rate.
Percolator chorus	Possible function in the formation of schooling	Large chorus of animals producing short grunts at frequent intervals.
Bagre marinus (Tavolga, 1960)		
Sobbing sound	Associated with schooling or social behavior in some way, not understood	420–550 msec with decrease of fundamental, harmonics.
Yelp	As above	110–200 msec duration with decrease of fundamental, harmonics.
Grunt	Prodding	~ 100 msec, not explicitly stated.
Opsanus tau (Fish, 1954; Tavolga, 1958c; Gray and Winn, 1961; Winn, 1964, 1967, 1972)		
Grunt	Defense of nest by male or by male and female when threatened or prodded.	Single or multiple series of pulses, fundamental ~ 100 Hz.
Boatwhistle	Male courtship call, attracts females to nest, stimulates males to call when weak and inhibits them if too loud.	Sustained contraction with harmonics, duration, and frequency variable. 300 msec duration Solomons, Maryland; fundamental ~ 200 Hz.
Opsanus beta (Tavolga, 1958c, 1960, 1968b)		
Grunt	Similar to *tau.*	Similar to *tau.*
Boatwhistle	Similar to *tau.*	Higher fundamental 260–300 Hz. Durations up to 600 msec. More complicated call made of a double hoot preceded by a grunt.
Opsanus phobetron (Fish and Mowbray, 1959); Tavolga, 1968b)		
Grunt	Probably similar to *tau.*	Longer and higher frequency than in *tau* or *beta.*
Boatwhistle	Similar to *tau.*	Fundamental 260 Hz, long duration up to 1 sec and preceded by several pulses.

Table 1 (continued)

Species and Call	Function	Probable Coding
Porichthys notatus (Cohen and Winn, 1967; Ibara, pers. comm.)		
Irregular grunts	Prodded male, also recorded underwater when stimulus not seen.	Possible fundamental <85 Hz, 61–110 msec pulse duration.
Trains of regularly spaced grunts	As above.	As above, but in a series lasting between 41.5 and 120.8 sec. Intervals between grunts averaged 0.75 and 0.96 sec.
Buzz	As above; separate behavioral meanings for these three calls have not been found.	1.0–3.3 sec duration. Fundamental either ≤85 Hz or variably between 150 and 250 Hz.
Boatwhistle	Courtship call produced at night by nesting male.	105–110 Hz buzz ranging from <10 min to an hour.
Gadus morhua (Brawn, 1961b)		
Grunt	Courtship by male, aggression, breakup courtship between two males, or male and nonripe female.	Not specified; low frequency (≤50 Hz). Grunt louder when used as a determined threat and fainter at the end of the aggressive period.
Melanogrammus aeglefinus (Hawkins and Chapman, 1966; Hawkins, 1973)		
Grunts and rasplike noises	Aggression (both aggressor and victim), when chased or held by human, during approach by a large cod, competitive feeding.	Pulses produced singly in irregular intervals or in rapid sequences of 2–6 at a rate of 5/sec. Pulse duration 7.5–100 msec. Rasplike sound duration to 2.7 sec.
	Courtship.	Faster rate up to 100/sec.
Holocentrus rufus (Winn et al., 1964)		
Grunt	Intra- or interspecific invasion of territory; directed chiefly at neighboring conspecifics.	Single or in groups, long and variable time intervals between pulses.
Staccato	Warning call, approach of large predator.	Variable number of grunts repeated rapidly, uniform time interval between pulses.
Hand-held grunt	Prodding.	Grunt element longer than in first two calls, produced in series of varying intervals and rates.
Myripristis berndti (Salmon, 1967)		
Knock	Intraspecific chasing.	Short duration.
Growl	Agonistic, after physical contact sometimes resulting from chasing, usually between fishes of equal size.	Many seconds long.
Grunt	Warning call, approach of large predators.	Spaced irregularly.
Staccato	Warning call, higher motivation than grunt.	Rapid series of sounds, individuals of which appear to be shorter and louder than grunts.
Hand-held grunt	Prodding.	Not specified.
Myripristis violaceus (Horch and Salmon, 1973)		
Thump	Agonistic episodes of circling tail beating and chasing between pairs of conspecifics.	Low-pitched sounds in groups of 3–7, generally separated by intervals of less than 0.5 sec.
Knock	Behavior not yet separated from above.	Higher pitched and shorter duration than grunts. Series of up to 10 in succession at irregular intervals (0.3–2.0 sec).
Growl	Probably a warning call	Many knocklike pulses in rapid succession
Hand-held grunt	Prodding	Similar to thumps but louder and higher pitched.

Table 1 (continued)

Species and Call	Function	Probable Coding
Hippocampus hudsonius (Fish, 1953, 1954)		
Snaps	Courtship and when exploring new situations.	Not given.
Therapon jarbua (Schneider, 1964b)		
Trommellaute (drumming sound)	Agonistic.	10 msec pulse.
Drohlaute (threatening sound)	Agonistic but higher intensity, threatening or attacking.	Rapid series of drumming sounds up to 200 pulses/sec. The stronger a fish threatens, the longer the sound. Intensity higher than drumming sound.
Lagodon rhomboides (Caldwell and Caldwell, 1967)		
Click and double click	Chasing conspecifics during territorial defense.	In double click the second click has either a slightly higher or a slightly lower principal frequency than the first.
Corvina nigra (Dijkgraaf, 1947)		
Knocking sound	Male leaves female and chases other males during courtship.	2–7 knocks at a rate of 8/sec. Under more stimulation the knocking signal becomes more energetic and longer.
Aplodinotus grunniens (Schneider and Hasler, 1960)		
Growl sound	Prodding.	300–500 msec duration.
Drumming sound	Unknown.	3–5 sec to a minute or longer, pulse duration 12.4–28.9 msec at a rate between 18 and 27/sec.
Hemichromis bimaculatis (Myberg, Kramer, and Heinecke, 1965)		
"Br-r-r" sound (female)	Produced just before attacking an intruder; more sounds elicited by conspecifics than equal-sized fish of another species; max. production during egg incubation; also produced during courtship while aggressively holding ground after attack by male.	Middle frequency 400 Hz, pulse rep. rate 35/sec.
"Br-r-r" sound (male)	Produced in aggressive situations in early courtship and parental period.	Middle frequency 300 Hz, pulse rep. rate 35/sec, duration longer than in female.
Thump (male)	Produced in early stages of fighting between conspecifics during approach, lateral display, and tailbeating, often preceding the "br-r-r" sound.	3–5 pulses, pulse rep. rate 12 Hz.
Cichlasoma centrarchus (Schwarz, 1974)		
Low growling sound	Produced agonistically by male and female within two body lengths of opponent.	Frequency 100–1,300 Hz, pulses of uniform duration, sound duration from 0.5–4 sec; duration and pulse rep. rate increase with more highly aggressive actions.

Table 1 (continued)

Species and Call	Function	Probable Coding
Amphiprion xanthurus **(Schneider, 1964a)**		
Drohlaute (threatening sound)	In defense of anemone with tail and jerk movements before and between fights.	25–30 msec duration; loud, audible 10 m from aquarium in air.
Kampflaute (fighting sound)	In close proximity to opponent during fight.	45–60 msec, lower intensity.
Ruttellaute (shaking sound)	Submissive, combined with laterally directed movements of the head in the horizontal plane.	250–400 msec.
Amphiprion polymnus **(Schneider, 1964a)**		
Threat sound	Anemone defense.	12.5–20 msec.
Fighting sound	During fight.	22–30 msec.
Shaking sound	Submissive.	90–120 msec.
Eupomacentrus partitus **(Myrberg, 1972a, 1972b)**		
Chirp	Courtship during "dip."	3 pulses/sequence.
Long chirp	Courtship during "flutter."	4–6 pulses.
Grunt	Courtship during "close swim."	Variable number of pulses and duration.
Burr	Courtship, male moving rapidly to his nest when female nearby.	8–12 pulses, restricted frequency range.
Pop	Agonistic interactions involving head-on confrontations.	Single pulse.
Pomacanthus arcuatus **(Moulton, 1958)**		
Short grunt	Unknown.	Short, sharp sound.
Moanlike sound	Recognition signal, possibly of a pair.	Longer grunt.
Chasmodes bosquianus **(Tavolga, 1958b)**		
Grunting sound	Courtship by male.	155 msec average duration; usually produced at 1–2 sec intervals, occasionally in a burst of 3–4 sounds.
Gobius jozo **(Kinzer, 1961)**		
Snore	Courtship by male and as a threat by both sexes; can be made by an intruder in an established territory.	0.5 sec duration, but repetition variable.
Bathygobius soporator **(Tavolga, 1958a)**		
Grunt	Courtship by male.	Increase in frequency during active courtship.
Trichopsis vittatus **(Marshall, 1966)**		
Croaking sound (male and female)	During lateral displays of aggressive interactions and early stages of courtship.	19.6 msec average duration, average 4–5 single or double tone bursts/croak.
Purring sound (female only)	Courtship and spawning, during head-up posture.	22.4 msec average duration, average 2.3 pulses/purr.

8–12 pulses per sequence, while the grunt has a variable number of pulses. Each drumbeat sound of the black grouper *(Mycteroperca bonaci),* caused by opercular movements, consists of five sound pulses (Tavolga, 1960).

Repetition of entire sounds seems to be patterned in the clown fish (*Amphiprion xanthurus:* Schneider, 1964a). The submissive, shaking sound is given in groups of three to six, and the fish does not usually emit more than four attack sounds in one fight. The threatening sound varies with the situation: long series of ten or more sounds are emitted before a fight and shorter series of three to five are emitted after a lull in fighting, initiating a new phase of battle.

Frequency

The role of frequency in coding fish sounds is not well understood. Although some sounds (particularly stridulatory ones) have wide-frequency components, other sounds clearly show characteristic species differences in frequency (particularly fundamentals). The frequency spectrum is typically a property of the sound-producing mechanism and is sometimes a function of the sonic muscle contraction rate. Sustained muscle contractions often produce harmonics, but in some cases, as Tavolga (1962) has shown, harmonics may also be related to the acoustic properties of the environment. Playbacks to toadfish demonstrated that they respond positively to tones of 200–400 Hz without the presence of harmonics (Winn, 1972), and any harmonics above the second are beyond the toadfish's hearing range (Winn, 1972; Fish and Offutt, 1972). At this time there is no proof of relevance of harmonics to meaning in a fish signal.

In a system where cavity size and internal pressure remain relatively constant, frequency is not widely varied. Generally, certain frequencies or groups of frequencies are favored because of the physics of the system, but it is not typically known what portion of the acoustic energy fish focus on within their hearing range. In many instances the higher frequencies of a sound may be above the fish's hearing range.

Species-specific frequency differences occur because of morphometric and physiological changes in the sound-producing mechanism during evolution. There are differences in the fundamental frequency of the boatwhistle calls of the three species of toadfish that were discussed at length in the section on duration. In addition, there are indications of a possible clinal variation in the fundamental frequency of *Opsanus tau* boatwhistles up and down the Atlantic coast of North America, with frequency increases in the lower latitudes (Tavolga, 1968b, 1971a; Fish and Mowbray, 1970). However, the effects of temperature and fish size on the fundamental are unknown for this species. Schneider (1964b) has shown that increasing temperature increases the contraction rate of the sonic muscles in *Therapon jarbua.* In frogs, rising temperature generally causes an increase in the repetition rate and fundamental of a call, with a coincident drop in duration (Blair, 1958; Schneider, 1968, 1974; Lorcher, 1969). The rate of boatwhistling does increase with temperature in *Opsanus beta* (Breder, 1968), although the mechanism behind this correlation is unexplained.

The effect of animal size on frequency is an open question and may depend on the species. An inverse correlation between size and frequency was established in the croaking gourami (*Trichopsis vittatus:* Marshall, 1966), was assumed in the cichlids *Hemichromis bimaculatus* and *Cichlasoma nigrofasciatum* (Myrberg et al., 1965), and is probably true of many fishes. Myrberg et al. (1965) also found that the "br-r-r" sound of the female is higher pitched than the same sound of the larger male.

In species like the toadfish, where sonic muscle contraction rate governs the fundamental frequency (Skoglund, 1961), bladder size may not be an important consideration. In the Japanese gurnard *(Chelidonichthys kumu)* the dominant frequency of the "gu" call lies around 300 Hz in fishes of three different size groups, even though the sound energy ranges somewhat higher in smaller specimens (Bayoumi, 1970). The fundamental of low-pitched grunts from *Bagre* and *Galeichthys* is about 150 Hz, regardless of the size of the fish (Tavolga, 1962). The principal frequencies of the toadfish grunt do not appear to be related to size (Fish, 1954).

Frequency modulation within a call is common in birds and mammals (Armstrong, 1963; Poulter, 1968; Tembrock, 1968) but occurs rarely, if at all, in fishes. Caldwell and Caldwell (1967) found a double click in the pinfish *(Lagodon rhomboides),* where the second click has a slightly higher or lower principal frequency than the first. Several types of courtship sounds of the bicolor damselfish *(Eupomacentrus partitus)* have the same frequency range (Myrberg, 1972a, 1972b), except for the burr, which has a restricted and different peak frequency range. How the fish accomplishes this modulation is unknown. Tavolga (1960) shows several sonagrams of catfish and toadfish vocalizations in which the fundamental frequency decreases slightly through the course of a long sustained call. This decrease undoubtedly results from a diminution of muscle-contraction rate, perhaps due to fatigue, and is probably unrelated to communication. This change in contraction rate does provide a theoretical basis of frequency modulation.

Enger (1963) showed that some auditory neurons of *Cottus scorpius* responded to sounds up to 200 Hz and others up to 300–500 Hz. Within certain limits, frequency could be detected by the volley principle. Jacobs and Tavolga (1968) demonstrated good frequency discrimination in the goldfish. Given this and the fact that in some cases frequencies differ between species, frequency must be important in some cases (see playback section), although the fishes' system is primitive when compared with that of higher vertebrates. The ability to use frequency will vary from one group of fishes to the next, depending on the complexity of the auditory system.

THE PLAYBACK EXPERIMENT

Playing back their sounds to fish is an experimental method of unraveling messages, equivalent to presenting painted models in the visual modality. The experiment can be used for at least four distinct purposes: (1) to establish species recognition of a sound; (2) to separate different calls; (3) to observe the effect of the sound on behavior and discover or confirm how it is used; and (4) to establish how the call is coded or what parameters of the call are relevant to communication.

Moulton (1956) attempted one of the first playback experiments when he transmitted a staccato call and an electronic imitation to the sea robins *Prionotus carolinus* and *P. evolans.* Occasionally a fish would respond with a staccato call after the playback was turned off. Moulton was also able to suppress the staccato call by playing back 200–600 Hz signals for the approximate duration of the call. Recently Fish and Mowbray (1970) questioned Moulton's species determination, suggesting that on the basis of frequency range and pulse form his sounds came most likely from a sciaenid, probably *Cynoscion regalis.*

Another early work using the playback technique involved the playing of a tape loop of male courtship sounds of *Bathygobius soporator* to males and females (Tavolga, 1958a). Females responded within a minute by increasing their activity, respiration, and contacts between individuals (nipping, butting, and approaching), though they did not orient to the sound. When a male was confined in a flask during playback, the females oriented toward him, bumping the flask repeatedly. Male responses to playback were similar, except that they approached and remained near the sound if in a courtship set, i.e., isolated or recently exposed to females. Males in a combat set (exposed to other males) exhibited no response to the playback. Tavolga played back electronic pulses to the fish to explore the significant parameters of the call. Animals responded maximally to frequencies between 100 and 300 Hz, to pulse durations of 75–150 msec, and to variable repetition rates depending on duration. Pulses shorter than 75 msec were effective only at high repetition rates. Intensities higher than normal were also effective. Within these limits the sound was not specific and positive responses were elicited to the courtship sound of the blenny *Chasmodes* and to Tavolga saying "ugh-ugh."

Various aspects of the vocalization of the toadfish *Opsanus tau* have been investigated. Winn (1967) found that he could increase a fish's rate of boatwhistling by playing back boatwhistles at a rate of eighteen or more per minute. Grunts and slower paced boatwhistle playbacks (thirteen per minute and below) did not increase calling. In further work, Fish (1972) and Winn (1972) found that the maximum stimulatory threshold calling rate was one sound every 4.0 to 4.5 seconds. When sounds were played at this rate, the toadfish and the playback alternated. Faster playbacks and an experiment with a delayed playback of each sound the toadfish produced did not establish antiphony. Continuous tones suppressed calling, but when intermittent tones were played, toadfish placed their boatwhistles in the silent period. Winn (1972) played tone signals to toadfish to determine what part of the signals were relevant in communication. A frequency of 180 Hz was stimulating, but a sound with more energy at 100 Hz than at higher frequencies was not. Sounds too loud, equivalent to the call of a fish less than one foot away, caused a reduction in calling. Durations of 75 and 150 msec resulted in a loss of stimulatory value compared to a duration of 300 msec, the approximate average of natural calls from Solomons, Maryland. It was clear that amplitude, duration, repetition rate, and frequency are all important parameters in social facilitation of boatwhistling by toadfish.

Winn also showed that females were attracted to boatwhistles coming from cans in front of a speaker. Ripe females of *Porichthys notatus,* a member of the same family, were attracted to continuous tones of 105–120 Hz. The females became excited and swam around the tank and to the speaker (Ibara, pers. comm.). Nonripe females, a juvenile, and nest-guarding males were not attracted to the sound.

After a latency of thirty seconds to one minute, courtship sounds (chirps and grunts) of the bicolor damselfish *(Eupomacentrus partitus)* caused the male to take on color patterns of "white body and black mask" associated with courtship (Myrberg, 1972b). They also caused an increase in courtship behavior (tilt, dip, nudge, and lead) and vocalization. However, the fishes did not orient to the hydrophone. The agonistic pop of the bicolor and a squirrelfish staccato decreased courtship behavior compared to controls. Playbacks to bicolors of chirps from congeners *E. planifrons* and *E. leucostictus* were stimulatory to a lesser degree (Myrberg and Spires, 1972). In unpublished experiments (E. Spanier, pers. comm.) it was determined that the important parameter for species recognition of *Eupomacentrus* calls was the off-time or interval between sounds and not pulse duration and frequency.

Marshall (1966) performed a series of experiments with the croaking gourami *(Trichopsis vittatus).* Playback of croaking sounds to males produced no significant change in the rate of air-gulping or locomotion, and inconsistent increases in aggressive behavior. In some experiments behavior did change consistently between playbacks of croaks and background noise, but not between the croak and no sound control. Similarly, Winn (1967) inhibited toadfish boatwhistling through playback of background noise. It appears that such playbacks are suppressing behavior with an undefined stimulus. Marshall's (1966) reasons for the lack of positive results are instructive. The croaking sounds may have been only a minor part of the total complex of aggressive behavior and by themselves were not sufficient to elicit a consistent response. Playbacks are usually made from sound recorded on a loop and therefore are played without variation. This alone may make the sounds less realistic. The lack of correct temporal answering by the "playback" fish may be important. In several instances the dominant fish of a pair seemed aware that the sound was coming from the speaker and not his opponent. Habituation can quickly become a problem since there is no association of the sound with either a conspecific or specific postures of the conspecific. Finally, there is a problem with fishes maintaining variable levels of motivation. Some fish in Marshall's experiments increased and others decreased their aggressive activity during the playbacks. Preplayback and playback periods should be contiguous to minimize this problem.

Studies have been performed on several species of squirrelfish. Winn et al. (1964) caused *Holocentrus rufus* to enter their shelters by playing their warning staccato call, toadfish boatwhistles, and scad sounds. The fish stayed in their shelters for the first minute of sound emission, but after a while started investigating the speaker. When the sound was turned off, the fish became more active again. Lobster sounds and background noise had no effect. Playback of their grunts and staccatos to *Myripristis berndti* caused immediate orientation to the sound source (Salmon, 1967). The fish swam toward the speaker within five to ten seconds and ceased sound production. There was no response to playbacks of background noises or knocks. *Myripristis argyromus* will follow these same warning calls to their source. Thus, there does not appear to be discrimination between these calls by the two species (Popper et al., 1973). Playback of conspecific growls and hand-held grunts also caused *M. violaceus* to approach the speaker (Horch and Salmon, 1973). During playback of these sounds, knocks and thumps normally produced at irregular but frequent intervals ceased, and growl sounds not normally produced were elicited. Many fishes make sounds when held (Burkenroad, 1930; Fish and Mowbray, 1970), but Horch and Salmon have been the only ones to demonstrate that such a sound can have a communicatory function.

Delco (1960) demonstrated a positive approach to the call of a conspecific female in *Notropis lutrensis* and *N. venustus.* Stout (1963, 1966) played back the fighting and courtship calls of *Notropis analostanus.* With two males of unequal dominance together, the fighting sound caused an increase in number and duration of aggressive encounters. The purring sound also caused an increase in aggressive encounters and often caused males to exhibit solo spawning. With a male and two females together, the courtship purr increased the average duration and number of the male's courtships; the fighting knocks produced a decrease in courtship that was not statistically significant. Stout concluded that the differential response to the two sounds demonstrated that the fish discriminated them.

Gerald (1971) played courtship grunts from species pairs of sunfish out of underwater speakers adjacent to lift nets. A significant difference in the capture of *Lepomis megalotis* and *L. humilis* was found, with the calls attracting more conspecifics than heterospecifics. Note that the calls of these two species appear to be very much alike (Fig. 6). A somewhat similar test with *L. humilis* and *L. macrochirus* calls was not significant. However, samples were small, and further work might well establish the distinctiveness of the various sunfish calls.

Schwarz (1974b) played back the conspecific low growling sound to one individual of male-male and male-female pairs of the cichlid *Cichlasoma centrarchus,* which were acoustically but not visually isolated from each other. Playback of the growling sound to a male markedly lowered the number of highly aggressive encounters he directed at either his male or female partner, while playback of control noise or silence had no significant effect. She concluded that the sound functions to inhibit aggressive behavior in the recipient. In another cichlid, *Tilapia mossambica,* sounds appear to function as a priming stimulus; recordings of the male's low-pitched drum sounds caused females to lay eggs several days earlier than controls did (Marshall, 1972).

Electrical Communication

Various aspects of electric organs, receptors, and electrical communication have been recently reviewed (Black-Cleworth, 1970; Bennett, 1971a, 1971b; Bullock, 1973; Hopkins, 1974a, 1974b). Six groups of fishes have independently evolved a system of electric organs and specialized receptors (Rajidae, Torpedinidae, Mormyriformes, Gymnotoidei, Malapteruridae, and Uranoscopidae). Other fishes might be expected to respond to low electrical currents. Since Hopkins treats the subject in chapter 13 of this volume, only a short discussion follows.

The discharges are species-specific (Bullock, 1969) and are used in courtship and in heterospecific (Bauer and Kramer, 1974) and conspecific agonistic communication. According to Hopkins (1974a), the coding of electrical communication in various roles lies in the diversity of the signals, which can be classified according to certain parameters such as the shape of the electric field, the wave form of the electrical discharge, the discharge frequency, the timing patterns between signals from sender and receiver, the frequency modulations, and the cessations of the discharge.

Although electrical signals related to attack and threat behaviors appear to be species-specific, a common appeasement display has apparently evolved in several species. The cessation of discharge signaling by subordinates of *G. carapo* (Black-Cleworth, 1970), *Gymnarchus niloticus* (Hopkins, 1974a), and *Gnathonemus petersii* (Bell et al., 1974) reduced the rate of attack by the dominant. Such a display is effective in reducing aggression because the subordinate becomes electrically inconspicuous (Hopkins, 1974a).

Electrical discharges of *Sternopygus macrurus* have been shown to play a significant role in courtship of this species (Hopkins, 1972). During the breeding season the discharge frequencies of males and females are distinctly different. The male shows no response to the discharge of other males but responds to the female discharge with a distinct courtship signal.

Kastoun (1971) strung wire leads between two otherwise separated tanks, each containing an electric catfish *(Malapterurus electricus).* He stimulated one of the fish (stimulus fish) and monitored its signal output as well as the behavior of the fish in the adjacent tank (response fish). Knocking on the aquarium glass of one tank resulted in the emission of two to five impulses (defense or flight volleys) and the escape of the response fish to its living tube. By variously manipulating the stimulus fish, the researcher elicited from the response fish a search reaction and attacks on closely situated objects. The electrical volleys were discriminated through the number of impulses and the type of signal, i.e., impulse intensity, form, and grouping. The response fish was not observed to emit answering signals. In a similar experiment with electric eels *(Electrophorus electricus),* prodding or feeding the stimulus fish resulted in emission of impulses that attracted eels to electrodes in an adjacent tank (Bullock, 1969). Eels would cluster around and nuzzle a captive specimen lowered into their tank in a net, and they were attracted to electrodes emitting artificial pulses of variable parameters, including potentials up to 150 V.

Conclusion

Even using Burghardt's general concept of communication, as opposed to Tavolga's (1968a, 1970) more restrictive definition, it is obvious that fishes typically communicate for only short periods of their lives and about restricted subjects. Social behavior is much less complex than in higher vertebrates. There are certainly exceptions, such as schooling behavior, cleaning symbiosis, and defense of a feeding territory, all of which may last for long periods. Still, communication is a requisite for the continued existence of fishes, and it is used mostly during the mating season.

Fishes have undergone an extensive adaptive radiation unparalleled in other vertebrate classes. They are an old group with many phylogenetic lines. These two factors are reflected in the importance different modalities assume in the life styles and communicatory systems of various species. Even within a single medium, like sound, various groups appear to have evolved different strategies of detection.

The *Umwelt* of most fishes seems limited to a small surrounding pocket. Vision, though often limited by turbidity, is probably the primary sense of most species dwelling in surface-lit waters. Sound travels about five times as fast in water than in air and may be propagated for long distances. For these reasons it is often assumed that acoustic sensitivity is particularly important for aquatic vertebrates. This assumption is equivocal for most fishes, which cannot localize sounds more distant than a few meters. Banner (1972) felt lemon sharks could generally detect prey by sound at distances greater than their estimated visual range (still only about 5 m), but no one has demonstrated communication over long distances. The maximum distance a toadfish can hear the boatwhistle of another toadfish is 3 to 4 m. It is also noteworthy that sound signals have evolved in only a limited number of species. A superiority of light over sound was demonstrated in the pinfish *Lagodon rhomboides* by Jacobs and Popper (1968), who trained the fish in an avoidance task using either light or sound as the conditioned stimulus. Fish learned to avoid light in a median of four and a half training days, while it took twenty one days to avoid the sound. The authors felt these results were consistent with a primarily visual orientation of the fish. Squirrelfish and grunts learn a sound avoidance much more easily than *Lagodon,* though all three species are sound producers.

Olfaction is a highly developed sense for many species, but examples of its use in communication are not widespread. The number of examples should be readily extended by further studies. An electrical sense is developed in a few species. Fishes have been shown to react to an odor in water or to an electric current as if an individual emitting these stimuli were actually present.

Although we have discussed the subject on a topic by topic basis, communication typically involves several sensory channels. A courtship ritual might utilize visual, acoustic, and olfactory signals, but responses to signals of one modality indicate that redundancy, a safety factor, is built into the system.

References

Albrecht, H., 1962. Die Mitschattierung. *Experientia,* 18:284–86.

Albrecht, H., 1969. Behaviour of four species of Atlantic damselfishes from Colombia, South America

(*Abudefduf saxatiles, A. taurus, Chromis multilineata, C. cynaea,* Pisces, Pomacentridae). *Z. Tierpsychol.,* 26:662–76.

Alexander, R. McN., 1966. Physical aspects of swimbladder function. *Biol. Rev.,* 41:141–76.

Armstrong, E. A., 1963. A study of bird song. London: Oxford University Press.

Atz, J. W., 1953. Orientation in schooling fishes. In: *Proceedings of a Conference on Orientation in Animals,* T. S. Schneirla, ed. Washington, D.C.: Office of Naval Research, pp.115–30.

Ayling, A. M., and Grace, R. V., 1971. Cleaning symbiosis among New Zealand fishes. *New Zeal. J. Mar. Freshw. Res.,* 5:205–18.

Baerends, G. P., and Baerends-van Roon, J. M., 1950. An introduction to the study of cichlid fishes. *Behaviour,* Suppl. 1.

Bagnara, J. T., and Hadley, M. E., 1969. The control of bright colored pigment cells of fishes and amphibians. *Amer. Zool.,* 9:465–78.

Banner, A., 1972. Use of sound in predation by young lemon sharks, *Negaprion brevirostris* (Poey). *Bull. Mar. Sci.,* 22:251–83.

Barber, S. B., and Mowbray, W. H., 1956. Mechanism of sound production in the sculpin. *Science,* 124:219–20.

Bardach, J. E., and Todd, J. H., 1970. Chemical communication in fish. In: *Advances in Chemoreception,* vol. 1, J. W. Johnston, Jr., D. G. Moulton, and A. Turk, eds. New York: Appleton-Century-Crofts, pp.205–40.

Bardach, J. E., and Villars, T., 1974. The chemical senses of fishes. In: *Chemoreception in Marine Organisms,* P. T. Grant and A. M. Mackie, eds. New York: Academic Press, pp.49–104.

Barlow, G. W., 1962. Ethology of the Asian teleost, *Badis badis.* IV. Sexual behavior. *Copeia,* 1962:346–60.

Barlow, G. W., 1967. Social behavior of a South American leaf fish, *Polycentrus schomburgkii,* with an account of recurring pseudofemale behavior. *Amer. Midl. Nat.,* 78:215–34.

Barlow, G. W., 1968. Ethological units of behavior. In: *The Central Nervous System and Fish Behavior.* D. Ingle, ed. Chicago: University of Chicago Press, pp.217–32.

Barlow, G. W., 1970. A test of appeasement and arousal hypotheses of courtship behavior in a cichlid fish, *Etroplus maculatus. Z. Tierpsychol.,* 27:779–806.

Barlow, G. W., 1974. Contrasts in social behavior between Central American cichlid fishes and coral reef surgeon fishes. *Amer. Zool.,* 14:19–34.

Bauer, R., and Kramer, B., 1974. Agonistic behaviour in mormyrid fish: latency-relationship between the electric discharges of *Gnathonemus petersii* and *Mormyrus rume. Experientia,* 30:51–52.

Bayoumi, A. R., 1970. Underwater sounds of the Japanese gurnard, *Chelidonichthys kumu. Mar. Biol.,* 5:77–82.

Beauchamp, R. D., and Lovasik, J. V., 1973. Blue mechanism response of single goldfish optic fibers. *J. Neurophysiol.,* 36:925–39.

Bell, C. C.; Myers, J. P.; and Russel, J. C.; 1974. Electric organ discharge patterns during dominance related behavioral displays in *Gnathonemus petersii* (Mormyridae). *J. Comp. Physiol.,* 92:201–28.

Bennet, M. V. L., 1971a. Electric organs. In: *Fish Physiology,* vol. 5, W. S. Hoar and D. J. Randall, eds. New York: Academic Press, pp.347–491.

Bennet, M. V. L., 1971b. Electroreception. In: *Fish Physiology,* vol. 5, W. S. Hoar and D. J. Randall, eds. New York: Academic Press, pp.493–574.

Black-Cleworth, P., 1970. The role of electrical discharges in the nonreproductive social behaviour of *Gymnotus carapo* (Gymnotidae, Pisces). *Anim. Behav. Monogr.,* 3(1):1–77.

Blair, W. F., 1958. Mating call in the speciation of anuran amphibians. *Amer. Nat.,* 92:27–51.

Blair, W. F., 1974. Character displacement in frogs. *Amer. Zool.,* 14:1119–25.

Brawn, V. M., 1961a. Reproductive behaviour of the cod (*Gadus callarias* L.). *Behaviour,* 18:177–98.

Brawn, V. M., 1961b. Sound production by the cod (*Gadus callarias* L.). *Behaviour,* 18:239–55.

Breder, C. M., Jr., 1941. On the reproduction of *Opsanus beta* Goode and Bean. *Zoologica,* 29:229–32.

Breder, C. M., Jr., 1959. Studies on social groupings in fishes. *Bull. Amer. Mus. Nat. Hist.,* 117:397–481.

Breder, C. M., Jr., 1968. Seasonal and diurnal occurrences of fish sounds in a small Florida bay. *Bull. Amer. Mus. Nat. Hist.,* 138:327–78.

Breder, C. M., Jr., and Rosen, D. E., 1966. Modes of reproduction in fishes. New York: Natural History Press.

Brown, J. H.; Cantrell, M. A.; and Evans, S. M.; 1973. Observations on the behavior and coloration of some coral reef fish (family: Pomacentridae). *Mar. Behav. Physiol.,* 2:63–71.

Bullock, T. H., 1969. Species differences in effect of electroreceptor input on electrical organ pacemakers and other aspects of behavior in electric fish. *Brain Behav. Evol.,* 2:85–118.

Bullock, T. H., 1973. Seeing the world through a new sense: electroreception in fish. *Amer. Sci.,* 61:316–25.

Burghardt, G. M., 1970. Defining "communication." In: *Advances in Chemoreception,* vol. 1: *Communication by Chemical Signals,* J. W. Johnston, Jr., D. G. Moulton, and A. Turk, eds. New York: Appleton-Century-Crofts, pp.5–18.

Burkenroad, M. D., 1930. Sound production in the Haemulidae. *Copeia,* 1930:17–18.

Burkenroad, M. D., 1931. Notes on the sound-producing marine fishes of Louisiana. *Copeia,* 1931:20–28.

Cahn, P. H., 1967. *Lateral Line Detectors.* Bloomington: Indiana University Press.

Cahn, P. H., 1972. Sensory factors in the side-to-side spacing and positional orientation of the tuna, *Euthynnas affinis,* during schooling. *Fish. Bull.,* 70:197–203.

Caldwell, D. K., and Caldwell, M. C. 1967. Underwater sounds associated with aggressive behavior in defense of territory by the pinfish, *Lagodon rhomboides. Bull. S. Calif. Acad. Sci.,* 66:69–75.

Cheal, M., and Davis, R. E., 1974. Sexual behavior: social and ecological influences in the anabantoid fish, *Trichogaster trichopterus. Behav. Biol.,* 10:435–45.

Cherry, C., 1957. *On Human Communication: A Review, a Survey, and a Criticism.* New York: Wiley.

Chien, A. K., 1973. Reproductive behaviour of the angelfish *Pterophyllum scalara* (Pisces: Cichlidae). II. Influence of male stimuli upon the spawning rate of females. *Anim. Behav.,* 21:457–63.

Clark, T. A., 1970. Territorial behavior and population dynamics of a pomacentrid fish, the garibaldi, *Hypsypops rubicunda. Ecol. Monogr.,* 40:189–212.

Clark, T. A., 1971. Territory boundaries, courtship, and social behavior in the garibaldi, *Hypsypops rubicunda* (Pomacentridae). *Copeia,* 1971:295–99.

Clarke, W. D., 1963. Function of bioluminescence in mesopelagic organisms. *Nature,* 198:1244–46.

Cohen, D. M., 1970. How many recent fishes are there? *Proc. Calif. Acad. Sci.,* 38:341–46.

Cohen, M. J., and Winn, H. E., 1967. Electrophysiological observations on hearing and sound production in the fish, *Porichthys notatus. J. Exp. Zool.,* 165:355–70.

Collette, B. B., and Talbot, F. H., 1972. Activity patterns of coral reef fishes with emphasis on nocturnal-diurnal changeover. *Nat. Hist. Mus. Los Angeles Sci. Bull.,* 14:98–124.

Crane, J. M., Jr., 1965. Bioluminescent courtship display in the teleost *Porichthys notatus. Copeia,* 1965:239–41.

Darcy, G. H.; Maisel, E.; and Ogden, J. C.; 1974. Cleaning preferences of the gobie *Gobisoma evelynae* and *G. prochilos* and the juvenile wrasse *Thalassoma bifasciatum. Copeia,* 1974:375–79.

Daw, N. W., 1973. Neurophysiology of color vision. *Physiol. Rev.,* 53:571–611.

DeBoer, J. N., and Heuts, B. A., 1973. Prior exposure to visual cues affecting dominance in the jewel fish, *Hemichromis bimaculatus* Gill 1862 (Pisces, Cichlidae). *Behaviour,* 44:299–321.

Delco, E. A., Jr., 1960. Sound discrimination by males of two cyprinid fishes. *Tex. J. Sci.,* 12:48–54.

Demski, L. S., and Gerald, J. W. 1974. Sound production and other behavioral effects of midbrain stimulation in the free-swimming toadfish, *Opsanus beta. Brain Behav. Evol.,* 9:41–59.

Demski, L. S.; Gerald, J. W.; and Popper, A. N.; 1973. Central and peripheral mechanisms of teleost sound production. *Amer. Zool.,* 13:1141–67.

Dijkgraaf, S., 1947. Ein Tone erzeugender Fisch in Neapler Aquarium. *Experientia,* 3:493–94.

Eibl-Eibesfeldt, I., 1955. Über Symbiosen, Parasitismus, und andere besondere zwischenartliche Beziehungen tropischer Meeresfische. *Z. Tierpsychol.,* 12:203–19.

Enger, P. S., 1963. Single unit activity in the peripheral auditory system of a teleost fish. *Acta Physiol. Scand.,* 210 (Suppl.).

Enger, P. S., 1968. Hearing in fish. In: *Hearing Mechanisms in Vertebrates,* A. V. S. and J. Knight, eds. London: J. Churchill Ltd., pp.4–17.

Ehrlich, P. R., and Ehrlich, A. H., 1973. Coevolution: heterotypic schooling in Caribbean reef fishes. *Amer. Nat.,* 107:157–60.

Erulkar, S. D., 1972. Comparative aspects of spatial localization of sound. *Physiol. Rev.,* 52:237–360.

Feder, H. M., 1966. Cleaning symbiosis in the marine environment. In: *Symbiosis,* vol. 1, S. M. Henry, ed. New York: Academic Press, pp.327–80.

Fine, M. L., 1975. Sexual dimorphism of the growth rate of the swimbladder of the toadfish *Opsanus tau. Copeia,* 1975:483–90.

Fish, J. F., 1972. The effect of sound playback on the toadfish. In: *Behavior of Marine Animals: Current Perspectives in Research,* vol. 2: *Vertebrates,* H. E. Winn and B. L. Olla, eds. New York: Plenum Press, pp.386–432.

Fish, J. F., and Offutt, G. C., 1972. Hearing thresholds

from toadfish, *Opsanus tau,* measured in the laboratory and field. *J. Acoust. Soc. Amer.,* 51:1318–21.

Fish, M. P., 1953. The production of underwater sound by the northern seahorse, *Hippocampus hudsonius. Copeia,* 1953:98–99.

Fish, M. P., 1954. The character and significance of sound production among fishes of the Western North Atlantic. *Bull. Bingham Oceanogr. Coll.,* 14:1–109.

Fish, M. P., and Mowbray, W. H., 1959. The production of underwater sounds by *Opsanus sp.*, a new toadfish from Bimini, Bahamas. *Zoologica,* 44:71–76.

Fish, M. P., and Mowbray, W. H., 1970. *Sounds of Western North Atlantic Fishes.* Baltimore: Johns Hopkins Press.

Fishelson, L., 1970. Behaviour and ecology of a population of *Abudefduf saxatilis* (Pomacentridae, Teleostei) at Eilat (Red Sea). *Anim. Behav.,* 18:225–37.

Flock, A., 1971. The lateral line organ mechanoreceptors. In: *Fish Physiology,* vol. 5, W. S. Hoar and D. J. Randall, eds. New York: Academic Press, pp.241–63.

Fricke, H., 1966. Zum Verhalten des Putzerfisches, *Labroides dimidiatus. Z. Tierpsychol.,* 23:1–3.

Fricke, H. W., 1973. Behaviour as part of ecological adaptation. In situ studies in the coral reef. *Helgolander wiss. Merresunters.,* 24:120–44.

Fricke, H. W., and Holzberg, S., 1974. Social units and hermaphroditism in a pomacentrid fish. *Naturwissenschaften,* 61:367–68.

Frisch, K. von, 1938. Zur Physiologie des Fischschwarmes. *Naturwissenschaften,* 26:601–606.

Frisch, K. von, 1941. Über einem Schreckstoff der Fischaut und seine biologische Bedeutung. *Z. vergl. Physiol.,* 29:46–145.

Fujii, R., and Novales, R. R., 1969. Cellular aspects of the control of physiological color changes in fishes. *Amer. Zool.,* 9:453–63.

Gainer, H., and Klancher, J. E., 1965. Neuromuscular junctions in a fast-contracting fish muscle. *Comp. Biochem. Physiol.,* 15:156–65.

Gentle, M. J., 1970a. The central nervous control of color change in the minnow (*Phoxinus phoxinus* L.). 1. Blinding and the effects of tectal removal on normal and blind fish. *J. Exp. Biol.,* 54:83–91.

Gentle, M. J., 1970b. The central nervous control of color change in the minnow (*Phoxinus phoxinus* L.). 2. Tectal ablations in normal fish. *J. Exp. Biol.,* 54:93–102.

Gerald, J. W., 1971. Sound production during courtship in six species of sunfish (Centrarchidae). *Evolution,* 25:75–87.

Göz, H., 1941. Über den Art- und Individualgeruch bei Fischen. *Z. vergl. Physiol.,* 29:1–45.

Gray, G. A., and Winn, H. E., 1961. Reproductive ecology and sound production of the toadfish, *Opsanus tau. Ecology,* 42:274–82.

Greenwood, P. H.; Rosen, D. E.; Weitzman, S. H.; and Myers, G. S.; 1966. Phyletic studies of teleostean fishes, with a provisional classification of living forms. *Bull. Amer. Mus. Nat. Hist.,* 131:341–455.

Hamilton, W. J. III, and Peterman, R. M., 1971. Countershading in the colourful reef fish *Chaetodon lunula:* concealment, communication or both. *Anim. Behav.,* 19:357–64.

Hara, T. J., 1971. Chemoreception. In: *Fish Physiology,* vol. 5, W. S. Hoar and D. J. Randall, eds. New York: Academic Press, pp.79–120.

Harden Jones, F. R., 1968. *Fish Migration.* London: Edward Arnold Ltd. 325pp.

Harris, G. G., and van Bergeijk, W. A., 1962. Evidence that the lateral line organ responds to near-field displacements of sound sources in water. *J. Acoust. Soc. Amer.,* 12:1831–41.

Hawkins, A. D., 1973. The sensitivity of fish to sounds. *Oceanogr. Mar. Biol. Annu. Rev.,* 11:291–340.

Hawkins, A. D., and Chapman, C. J., 1966. Underwater sounds of the haddock, *Melanogrammus aeglefinus. J. Mar. Biol. Assoc. U.K.,* 46:241–47.

Hemens, J., 1966. The ethological significance of the sword-tail in *Xiphophorus hellerii* (Haekel). *Behaviour,* 27:290–315.

Hemmings, C. C., 1966. Olfaction and vision in fish schooling. *J. Exp. Biol.,* 45:449–64.

Hilliman, D. E., and Lewes, E. R., 1971. Morphological basis for a mechanical linkage in otolithic receptor transduction in the frog. *Science,* 174:416–19.

Hobson, E. S., 1968. Predatory behavior of some shore fishes in the Gulf of California. *U.S. Bur. Sport Fish. Wildl. Res. Rep.,* 73:1–92.

Hobson, E. S., 1969. Comments on certain recent generalizations regarding cleaning symbiosis in fishes. *Pac. Sci.,* 23:35–39.

Hobson, E. S., 1971. Cleaning symbiosis among California inshore fishes. *Fish. Bull.,* 69:491–523.

Hobson, E. S., 1974. Feeding relationships of teleostean fishes on coral reefs in Kona, Hawaii. *Fish. Bull.,* 72:915–1031.

Hopkins, C. D., 1972. Sex differences in electric signaling in an electric fish. *Science,* 176:1035–37.

Hopkins, C. D., 1974a. Electric communication in fish. *Amer. Sci.,* 62:426–37.

Hopkins, C. D., 1974b. Electric communication: functions in the social behavior of *Eigenmannia virescens. Behaviour,* 50:270–305.

Horch, K., and Salmon, M., 1973. Adaptations to the acoustic environment by the squirrelfishes *Myripristis violaceus* and *M. pralinius. Mar. Behav. Physiol.,* 2:121–39.

Hunter, J. R., 1969. Communication of velocity changes in jack mackeral *(Trachurus symmetricus)* schools. *Anim. Behav.,* 17:507–14.

Hurley, A. C., and Hartline, P. H., 1974. Escape response in the damselfish *Chromis cyanea* (Pisces: Pomacentridae). *Anim. Behav.,* 22:430–37.

Ingle, D., 1971. Vision: the experimental analysis of visual behavior. In: *Fish Physiology,* vol. 5, W. S. Hoar and D. J. Randall, eds. New York: Academic Press, pp.59–77.

Jacobs, D. W., and Popper, A. N., 1968. Stimulus effectiveness in avoidance behavior in fish. *Psychon. Sci.,* 12:109–10.

Jacobs, D. W., and Tavolga, W. N., 1968. Acoustic frequency discrimination in the goldfish. *Anim. Behav.,* 16:67–71.

Jenkins, T. M., Jr., 1969. Social structure, position choice and microdistribution of two trout species (*Salmo trutta* and *Salmo gairdneri*) resident in mountain streams. *Anim. Behav. Monogr.,* 2:57–123.

Johnson, R. H., and Nelson, D. R., 1973. Agonistic display in the gray reef shark, *Carcharhinus menisoirah,* and its relationship to attacks on man. *Copeia,* 1973:76–84.

Karlson, P., and Luscher, M., 1959. "Pheromones": a new term for a class of biologically active substances. *Nature,* 183:55–56.

Kastoun, E., 1971. Elektrische Felder als Kommunikationsmittel beim Zitterwels. *Naturwissenschaften,* 58:459.

Keenleyside, M. H. A., 1955. Some aspects of the schooling behaviour of fish. *Behaviour,* 8:133–247.

Keenleyside, M. H. A., 1967. Behavior of male sunfishes (genus *Lepomis*) towards females of three species. *Evolution,* 21:688–95.

Keenleyside, M. H. A., 1971. Aggressive behavior of male longear sunfish (*Lepomis megalotis*). *Z. Tierpsychol.,* 28:227–40.

Keenleyside, M. H. A., 1972a. Intraspecific intrusions into nests of spawning longear sunfish (Pisces: Centrarchidae). *Copeia,* 1972:272–78.

Keenleyside, M. H. A., 1972b. The behaviour of *Abudefduf zonatus* (Pisces, Pomacentridae) at Heron Island, Great Barrier Reef. *Anim. Behav.,* 20:763–74.

Keenleyside, M. H. A., and Yamamoto, F. T., 1962. Territorial behaviour of juvenile Atlantic salmon (*Salmo salar* L.). *Behaviour,* 19:139–69.

Kinzer, J., 1961. Über die Lautäusserungen der Schwarzgrundel *Gobius jozo. Aquar. Terras-Kunde,* 7:7–10.

Kittredge, J. S.; Takahashi, F. T.; Lindsey, J.; and Lasker, R.; 1974. Chemical signals in the sea: marine allelochemics and evolution. *Fish. Bull.,* 72:1–11.

Klausewitz, W., 1958. Lauterzeugung als abwehrwaffe bei der hinterindischen Tiger-schmerle (*Botia hymenophysa*). *Natur. Volk,* 88:343–49.

Kleerekoper, H., 1969. *Olfaction in Fishes.* Bloomington: Indiana University Press.

Kühme, W., 1963. Chemisch ausgelöste Brutpflege- und Schwarmreakionen bei *Hemichromis bimaculatus* (Pisces). *Z. Tierpsychol.,* 20:688–704.

Kühme, W., 1964. Eine chemisch ausgelöste Schwarmreaktion beim jungen Cichliden (Pisces). *Naturwissenschaften,* 51:120–21.

Künzer, P., 1962. Die Auslösung der Nachfolgreaktion durch Bewegungsreize bei Jungfishen von *Nannacara anomala* Regan (Cichlidae). *Naturwissenschaften,* 22:525–26.

Künzer, P., 1964. Weitere Versuche zur Auslösung der Nachfolgreaktion bei Jungfischen von *Nannacara anomala* (Cichlidae). *Naturwissenschaften,* 51:419–20.

Leiner, M., 1930. Fortsetzung der ökologischen Studien an *Gasterosteus aculeatus. Z. Morphol. Ökol. Tiere,* 16:499–540.

Leong, C. Y., 1969. Quantitative effects of releasers on the attack readiness of the fish *Haplochromis burtoni* (Cichlidae, Pisces). *Z. vergl. Physiol.,* 65:29–50.

Liley, N. R., 1966. Ethological isolating mechanisms in four sympatric species of poeciliid fishes. *Behaviour* (Suppl.), 13:1–197.

Limbaugh, C., 1961. Cleaning symbiosis. *Sci. Amer.,* 205:42–49.

Lorcher, K., 1969. Comparative bio-acoustic investigations in the fire-bellied toad, *Bombina bombina* (L.). *Oecologia,* 3:84–124.

Lorenz, K., 1966. *On Aggression.* New York: Harcourt, Brace, & World.

Losey, G. S., Jr., 1969. Sexual pheromone in some fishes of the genus *Hypsoblennius* Gill. *Science,* 163:181–83.

Losey, G. S., Jr., 1971. Communication between fishes

in cleaning symbiosis. In: *Aspects of the Biology of Symbiosis,* R. C. Chens, ed. Baltimore: University Park Press, pp.45–76.

Losey, G. S., Jr., 1972. The ecological importance of cleaning symbiosis. *Copeia,* 1972:820–33.

Losey, G. S., Jr., and Margules, L., 1974. Cleaning symbiosis provides a positive reinforcer for fish. *Science,* 184:179–80.

Low, R. M., 1971. Interspecific territoriality in a pomacentrid reef fish *Pomacentrus flavicauda* Whitley. *Ecology,* 52:648–54.

Lowenstein, O., 1971. The labyrinth. In: *Fish Physiology,* vol. 5, W. S. Hoar and D. J. Randall, eds. New York: Academic Press, pp.207–40.

McAllister, D. E., 1967. The significance of ventral bioluminescence in fishes. *J. Fish. Res. Bd. Can.,* 24:537–54.

McCartney, B. S., and Stubbs, A. R., 1970. Measurements of the target strength of fish in dorsal aspect, including swimbladder resonance. In: *Proceedings of an International Symposium on Biological Sound Scattering in the Ocean,* G. B. Farquhar, ed. Maury Center for Ocean Science MC Report 005. Washington, D.C.: Government Printing Office, pp.180–211.

McDonald, A. L.; Heimstra, N. W.; and Damkot, D. K.; 1968. Social modification of agonistic behavior in fish. *Anim. Behav.,* 16:437–41.

MacGinitie, 1939. The natural history of the blind goby, *Typhlogobius californiensis* Steindachner. *Amer. Midl. Nat.,* 21:489–505.

McKenzie, J. A., and Keenleyside, M. H. A., 1970. Reproductive behavior of ninespine sticklebacks (*Pungitius pungitius* (L.)) in South Bay, Manitoulin Island, Ontario. *Can. J. Zool.,* 48:55–61.

McMillan, V. E., and Smith, R. J. F., 1974. Agonistic and reproductive behaviour of the fathead minnow (*Pimephales promelas* Rafinesque). *Z. Tierpsychol.,* 34:25–58.

Magnuson, J. J., and Prescott, J. H., 1966. Courtship, locomotion, feeding and miscellaneous behaviour of Pacific bonito *(Sarda chiliensis). Anim. Behav.,* 14:54–67.

Markle, H., 1971. Schallerzeugung bei Piranhas (Serrasalminae, Characidae). *Z. vergl. Physiol.,* 74:39–56.

Marler, P., 1961. The logical analysis of animal communication. *J. Theoret. Biol.,* 1:295–317.

Marler, P., and Hamilton, W. J. III, 1966. *Mechanisms of Animal Behavior.* New York: John Wiley and Sons.

Marshall, J. A., 1966. The social behavior and role of sound production in *Trichopsis vittatus,* with a comparison to *Trichopsis pumilis.* Ph.D. diss., University of Maryland.

Marshall, J. A., 1972. Influence of male sound production on oviposition in female *Tilapia mossambica* (Pisces, Cichlidae). *Amer. Zool.,* 12:663–64.

Marshall, N. B., 1962. The biology of sound-producing fishes. *Symp. Zool. Soc. Lond.,* 7:45–60.

Marshall, N. B., 1967. Sound-producing mechanisms and the biology of deep-sea fishes. In: *Marine Bio-acoustics,* vol. 2, W. M. Tavolga, ed. Oxford: Pergamon Press, pp.123–33.

Morin, J. G.; Harrington, A.; Nealson, K.; Krieger, N.; Baldwin, T. O.; and Hastings, J. W.; 1975. Light for all reasons: versatility in the behavioral repertoire of the flashlight fish. *Science,* 190:74–76.

Morris, D., 1958. The reproductive behaviour of the ten-spined stickleback (*Pygosteus pungitius* L.). *Behaviour* (Suppl. 6).

Morrow, J. E., Jr., 1948. Schooling behaviour in fishes. *Quart. Rev. Biol.,* 23:27–38.

Moulton, J. M., 1956. Influencing the calling of sea robins (*Prionotus* spp.) with sound. *Biol. Bull.,* 111:393–98.

Moulton, J. M., 1958. The acoustical behavior of some fishes in the Bimini area. *Biol. Bull.,* 114:357–74.

Moulton, J. M., 1960. Swimming sounds and the schooling of fishes. *Biol. Bull.,* 119:210–23.

Moulton, J. M., 1963. Acoustic behavior of fishes. In: *Acoustic Behavior of Animals,* R. G. Busnel, ed. New York: Elsevier, pp.655–93.

Munz, F. W., 1971. Vision: visual pigments. In: *Fish Physiology,* vol. 5, W. S. Hoar and D. J. Randall; eds. New York: Academic Press, pp.1–32.

Myrberg, A. A., Jr., 1965. A descriptive analysis of the behavior of the African cichlid fish, *Pelmatochromis guentheri* (Sauvage). *Anim. Behav.,* 13:312–29.

Myrberg, A. A., Jr., 1966. Parental recognition of young in cichlid fishes. *Anim. Behav.,* 14:565–71.

Myrberg, A. A., Jr., 1972a. Ethology of the bicolor damselfish, *Eupomacentrus partitus* (Pisces: Pomacentridae): a comparative analysis of laboratory and field behaviour. *Anim. Behav. Monogr.,* 5:199–283.

Myrberg, A. A., Jr., 1972b. Using sound to influence the behaviour of free-ranging marine animals. In: *Behavior of Marine Animals,* vol. 2, H. E. Winn and B. L. Olla, eds. New York: Plenum Press, pp.435–68.

Myrberg, A. A., Jr., 1972c. Social dominance and territoriality in the bicolor damselfish, *Eupomacentrus partitus* (Poey) (Pisces: Pomacentridae). *Behaviour,* 41:207–31.

Myrberg, A. A., Jr., 1975. The role of chemical and

visual stimuli in the preferential discrimination of young in the cichlid fish, *Cichlasoma nigrofasciatum* (Gunther). *Z. Tierpsych.*, 37:274–97.

Myrberg, A. A., Jr.; Brahy, B. D.; and Emery, A. R.; 1967. Field observations on reproduction in the damselfish, *Chromis multilineata* (Pomacentridae), with additional notes on general behavior. *Copeia*, 1967:819–27.

Myrberg, A. A., Jr.; Kramer, E.; and Heinecke, P.; 1965. Sound production by cichlid fishes. *Science*, 149:555–58.

Myrberg, A. A., Jr., and Spires, J. Y., 1972. Sound discrimination by the bicolor damselfish, *Eupomacentrus partitus*. *J. Exp. Biol.*, 57:727–35.

Myrberg, A. A., Jr., and Thresher, R. E., 1974. Interspecific aggression and its relevance to the concept of territoriality in reef fishes. *Amer. Zool.*, 14:81–96.

Nakamura, E. L., and Magnuson, J. J. 1965. Coloration of the scombrid fish *Euthynnus affinis* (Cantor). *Copeia*, 1965:234–35.

Neil, E. H., 1964. An analysis of color changes and social behavior of *Tilapia mossambica*. *Univ. Calif. Publ. Zool.*, 75:1–58.

Nelson, K., 1964a. Behavior and morphology in the glandulocaudine fishes (Ostariophysi, Characidae). *Univ. Calif. Publ. Zool.*, 75:59–152.

Nelson, K., 1964b. The evolution of a pattern of sound production associated with courtship in the characid fish, *Glandulocauda inequalis*. *Evolution*, 18:526–40.

Nicol, J. A. C., 1967. The luminescence of fishes. *Symp. Zool. Soc. Lond.*, 19:25–55.

Noakes, D. L. G., and Barlow, G. W., 1973. Ontogeny of parent-contacting in young *Cichlasoma citrinellum* (Pisces, Cichlidae). *Behaviour*, 46:221–55.

Noble, G. K., and Curtis, B., 1939. The social behavior of the jewel fish, *Hemichromis bimaculatus* Gill. *Bull. Amer. Mus. Nat. Hist.*, 76:1–46.

Olla, B. L., and Samet, C., 1974. Fish-to-fish attraction and the facilitation of feeding behavior as mediated by visual stimuli in striped mullet, *Mugil cephalus*. *J. Fish. Res. Bd. Can.*, 31:1621–30.

Oppenheimer, J. R., 1970. Mouthbreeding in fishes. *Anim. Behav.*, 18:493–503.

Packard, A., 1960. Electrophysiological observations on a sound-producing fish. *Nature*, 187:63–64.

Peeke, H. V. S.; Herz, M. J.; and Gallagher, J. E.; 1971. Changes in aggressive interaction in adjacently territorial convict cichlids (*Cichlasoma nigrofasciatum*): a study of habituation. *Behaviour*, 40:43–54.

Pfeiffer, W., 1960. Über die Schreckreaktion bei Fischen und die Herkunft des Schreckstoffes. *Z. vergl. Physiol.*, 43:578–614.

Pfeiffer, W., 1963. Alarm substances. *Experientia*, 19:113–23.

Picciolo, A. R., 1964. Sexual and nest discrimination in anabantid fishes of the genera *Colisa* and *Trichogaster*. *Ecol. Monogr.*, 34:53–77.

Popper, A. N., 1971. The effects of size on auditory capacities of the goldfish. *J. Aud. Res.*, 11:239–47.

Popper, A. N., 1972. Auditory threshold in the goldfish (*Carassius auratus*) as a function of signal duration. *J. Acoust. Soc. Amer.*, 52:596–602.

Popper, A. N., 1974. The response of the swim bladder of the goldfish (*Carassius auratus*) to acoustic stimuli, *J. Exp. Biol.*, 60:295–304.

Popper, A. N., and Fay, R. R., 1973. Sound detection and processing by teleost fishes: a critical review. *J. Acoust. Soc. Amer.*, 53:1515–29.

Popper, A. N.; Salmon, M.; and Parvulescu, A.; 1973. Sound localization by the Hawaiian squirrelfishes, *Myripristis berndti* and *M. argyromus*. *Anim. Behav.*, 21:86–97.

Potts, G. W., 1973. Cleaning symbiosis among British fish with special reference to *Crenilabrus melops* (Labridae). *J. Mar. Biol. Assoc. U.K.*, 53:1–10.

Poulter, T. C., 1968. Marine mammals. In: *Animal Communication*, T. A. Sebeok, ed. Bloomington: Indiana University Press, pp.405–65.

Radakov, D. V., 1973. *Schooling in the Ecology of Fish*. New York: John Wiley and Sons.

Rasa, O. A. E., 1969. Territoriality and the establishment of dominance by means of visual cues in *Pomacentrus jenkinsi* (Pisces: Pomacentridae). *Z. Tierpsychol.*, 26:825–45.

Reed, J. R., 1969. Alarm substances and fright reaction in some fishes from the Southeastern United States. *Trans. Amer. Fish. Soc.*, 98:664–68.

Reinboth, R., 1972. Hormonal control of the teleost ovary. *Amer. Zool.*, 12:307–24.

Reinboth, R., 1973. Dualistic reproductive behavior in the protogynous wrasse *Thalassoma bifasciatum* and some observations on its day-night changeover. *Helgoländer wiss. Meersunters*, 24:174–91.

Reese, E. S., 1964. Ethology and marine zoology. *Oceanogr. Mar. Biol. Annu. Rev.*, 2:455–88.

Reese, E. S., 1973. Duration of residence by coral reef fishes on "home" reefs. *Copeia*, 1973:145–49.

Richard, I. S., 1974. Caudal neurosecretory system: possible role in pheromone production. *J. Exp. Zool.*, 187:405–8.

Rigley, L., and Marshall, J. A., 1973. Sound production by the elephant-nose fish, *Gnathonemus petersi* (Pisces, Mormyridae). *Copeia,* 1973:134–35.

Robertson, D. R., 1972. Social control of sex reversal in a coral-reef fish. *Science,* 177:1007–9.

Robertson, D. R., 1973. Field observations on the reproductive behavior of a pomacentrid fish *Acanthochromis polyacanthus. Z. Tierpsychol.,* 32:319–24.

Rosenblatt, R. H., and Losey, G. S., Jr., 1967. Alarm reaction of the top smelt, *Atherinops affinis:* reexamination. *Science,* 158:671–72.

Russell, B. C., 1971. Underwater observations on the reproductive activity of the damoiselle, *Chromis dispilus* (Pisces: Pomacentridae). *Mar. Biol.,* 10:22–29.

Salmon, M., 1967. Acoustical behavior of the menpachi, *Myripristis berndti,* in Hawaii. *Pac. Sci.,* 21:364–81.

Salmon, M., and Winn, H. E., 1966. Sound production by priacanthid fishes. *Copeia,* 1966:869–72.

Salmon, M.; Winn, H. E.; and Sorgente, N.; 1968. Sound production and associated behavior in triggerfish. *Pac. Sci.,* 22:11–20.

Sand, O., and Hawkins, A. D., 1973. Acoustic properties of the cod swimbladder. *J. Exp. Biol.,* 58:797–820.

Sato, M., 1937. On the barbels of a Japanese sea catfish, *Plotosus anguillaris* (Lacepede). Sci. Rep. Tôhuku Imp. Univ. Ser. 4, 11:323–32.

Schleidt, W. M., 1973. Tonic communication: continual effects of discrete signs in animal communication systems. *J. Theor. Biol.,* 42:359–86.

Schneider, H., 1964a. Bioakustiche Untersuchungen an Anemonenfishen der Gattung *Amphiprion* (Pisces). *Z. Morphol. Ökol. Tiere,* 53:453–74.

Schneider, H., 1964b. Physiologische und Morphologische Untersuchungen zur Bioakustik der Tigerfische (Pisces, Theraponidae). *Z. vergl. Physiol.,* 47:493–558.

Schneider, H., 1967. Morphology and physiology of sound-producing mechanisms in teleost fishes. In: *Marine Bio-acoustics,* vol. 2, W. N. Tavolga, ed. New York: Pergamon Press, pp.135–58.

Schneider, H., 1968. Bio-akustiche Untersuchungen am Mittelmeerlaubfrosch. *Z. vergl. Physiol.,* 61:369–85.

Schneider, H., 1974. Structure of the mating calls and relationships of the European tree frogs (Hylidae, Anura). *Oecologia,* 14:99–110.

Schneider, H., and Hasler, A. D., 1960. Laute und Lauterzeugung beim Süsswassertrommler *Aplodinotus grunniens* Rafinesque. *Z. vergl. Physiol.,* 43:499–517.

Schuijf, A., and Siemelink, M. E., 1974. The ability of cod *(Gadus morhua)* to orient towards a sound source. *Experientia,* 30:773–74.

Schwarz, A., 1974a. Sound production and associated behavior in a cichlid fish, *Cichlasoma centrarchus.* I. Male-male interactions. *Z. Tierpsychol.,* 35:147–56.

Schwarz, A., 1974b. The inhibition of aggressive behavior by sound in the cichlid fish, *Cichlasoma centrarchus. Z. Tierpsychol.,* 35:508–17.

Schwarz, A. Sound production and associated behavior in *Cichlasoma centrarchus.* II. Breeding pairs. In preparation.

Seghers, B. H., 1974. Schooling behavior in the guppy *(Poecilia reticulata):* an evolutionary response to predation. *Evolution,* 28:486–89.

Shaw, E., 1960. The development of schooling behavior in fishes. *Physiol. Zool.,* 33:79–86.

Shaw, E., 1969. The duration of schooling among fish separated and those not separated by barriers. *Amer. Mus. Novitates,* 2373. 13pp.

Shaw, E., 1970. Schooling in fishes: critique and review. In: *Development and Evolution of Behavior,* L. R. E. Tobach, D. S. Lehrman, and J. S. Rosenblatt, eds. San Francisco: W. H. Freeman, pp.452–80.

Skinner, W. A.; Mathews, R. D.; and Parkhurst, R. M.; 1962. Alarm reaction of the top smelt, *Atherinops affinis* (Ayres). *Science,* 138:681–82.

Skoglund, C. R., 1961. Functional analysis of swimbladder muscles engaged in sound production of the toadfish. *J. Biophys. Biochem. Cytol.* (Suppl.) 10:187–200.

Smith, H. M., 1905. The drumming of the drum-fishes (Sciaenidae). *Science,* 22:376–78.

Springer, S., 1967. Social organization of shark populations. In: *Sharks, Skates, and Rays,* P. W. Gilbert, R. F. Mathewson, and D. P. Dall, eds. Baltimore: The Johns Hopkins University Press, pp.149–74.

Steele, R. G., and Keenleyside, M. H. A., 1971. Mate selection in two species of sunfish *(Lepomis gibbosus* and *L. Megalotis peltastes). Can. J. Zool.,* 49:1541–48.

Stout, J. F., 1963. The significance of sound production during the reproductive behavior of *Notropis analostanus* (family Cyprinidae). *Anim. Behav.,* 11:83–92.

Stout, J. F., 1966. Sound communication in fishes with special reference to *Notropis analostanus. Proc. 3rd Annu. Conf. Biol. Sonar and Div. Mam.,* 3:159–77.

Suzuki, R., 1961. Sperm activation and aggregation during fertilization in some fishes. VI. The origin of the sperm-stimulating factor. *Annot. Zool. Japon.,* 34:24–29.

Swerdloff, S. N., 1970. Behavioral observations on Eniwetok damselfishes (Pomancentridae: *Chromis*) with special reference to the spawning of *Chromis caeruleus. Copeia,* 1970:371–74.

Tavolga, W. N., 1956. Visual, chemical and sound stimuli as cues in the sex discriminatory behavior of the gobiid fish *Bathygobius soporator. Zoologica,* 41:49–64.

Tavolga, W. N., 1958a. The significance of underwater sounds produced by males of the gobiid fish. *Bathygobius soporator. Physiol. Zool.,* 31:259–71.

Tavolga, W. N., 1958b. Underwater sounds produced by males of the blenniid fish, *Chasmodes bosquianus. Ecology,* 39:759–960.

Tavolga, W. N., 1958c. Underwater sounds produced by two species of toadfish, *Opsanus tau* and *Opsanus beta. Bull. Mar. Sci.,* 8:278–84.

Tavolga, W. N., 1960. Sound production and underwater communication in fishes. In: *Animal Sounds and Communication,* W. E. Lanyon and W. N. Tavolga, eds. Amer. Inst. Biol. Sci.: Publ. No. 7 (Washington, D.C.), pp.93–136.

Tavolga, W. N., 1962. Mechanisms of sound production in the ariid catfishes *Galeichthys* and *Bagre. Bull. Amer. Mus. Nat. Hist.,* 124:1–30.

Tavolga, W. N., 1964. Sonic characteristics and mechanisms in marine fishes. In: *Marine Bio-acoustics,* vol. 1, W. N. Tavolga, ed. New York: Pergamon Press, pp.195–211.

Tavolga, W. N., 1968a. Fishes. In: *Animal Communication,* T. A. Sebeok, ed. Bloomington: Indiana University Press. pp.271–88.

Tavolga, W. N., 1968b. Marine animal data atlas. Tech. Rep., Naval Training Device Center (Orlando, Fla.), 1212–2. 239pp.

Tavolga, W. N., 1970. Levels of interaction in animal communication. In: *Evolution and Development of Animal Behavior,* L. R. Aronson, E. Tobach, D. S. Lehrman, and J. S. Rosenblatt, eds. San Francisco: Freeman, pp.281–302.

Tavolga, W. N., 1971a. Sound production and detection. In: *Fish Physiology,* vol. 5, W. S. Hoar and D. J. Randall, eds. New York: Academic Press, pp.135–205.

Tavolga, W. N., 1971b. Acoustic orientation in the sea catfish. *Galeichthys felis. Ann. N.Y. Acad. Sci.,* 188:80–97.

Tavolga, W. N., 1974. Sensory parameters in communication among coral reef fishes. *Mount Sinai J. Med.,* 41:324–40.

Tembrock, G., 1968. Land mammals. In: *Animal Communication,* T. A. Sebeok, ed. Bloomington: Indiana University Press, pp.338–404.

Tett, P. B., and Kelly, M. G., 1973. Marine bioluminescence. *Oceanogr. Mar. Biol. Annu. Rev.,* 11:89–173.

Tinbergen, N., 1951. The study of instinct. Oxford: Clarendon Press.

Todd, J. H.; Atema, J.; and Bardach, J. E.; 1967. Chemical communication in social behavior of a fish, the yellow bullhead *(Ictalurus natalis). Science,* 158:672–73.

Tomita, T., 1971. Vision: Electrophysiology of the retina. In: *Fish Physiology,* vol. 5 W. S. Hoar and D. J. Randall, eds. New York: Academic Press, pp.33–57.

Topp, R. W., 1970. Behavior and color change of the rudderfish, *Kyphosus elegans,* in the Gulf of Panama. *Copeia,* 1970:763–65.

Tower, R. W., 1908. The production of sound in the drumfishes, the sea robin and the toadfish. *Ann. N.Y. Acad. Sci.,* 18:149–80.

Uematsu, T., 1971. Social facilitation and feeding behavior of the guppy. II. Experimental analysis of mechanisms. *Jap. J. Ecol.,* 21(1–2):54–67.

van den Assem, J. and van der Molen, J., 1969. Waning of the aggressive response in the three-spined stickleback. I. A preliminary analysis of the phenomenon. *Behaviour,* 34:268–324.

Verheijen, F. J., 1963. Alarm substance in some North American cyprinid fishes. *Copeia,* 1963:174–76.

Walls, G. L., 1942. The vertebrate eye and its adaptive radiation. Bloomfield Hills, Michigan: Cranbrook Inst. Sci.

Walters, V., and Robins, C. R., 1961. A new toadfish (Batrachoididae) considered to be a glacial relict in the West Indies. *Amer. Mus. Novit.,* 2047.

Welty, J. C., 1934. Experiments in group behavior of fishes. *Physiol. Zool.,* 7:85–128.

Wickler, W., 1962. "Egg-dummies" as natural releasers in mouth breeding cichlids. *Nature,* 194:1092–93.

Wilz, K. J., 1970. Causal and functional analysis of dorsal pricking and nest activity in the courtship of the three-spined stickleback *Gasterosteus aculeatus. Anim. Behav.,* 18:115–24.

Winn, H. E., 1964. The biological significance of fish sounds. In: *Marine Bio-acoustics.* vol. 1, W. N. Tavolga, ed. New York: Pergamon Press, pp.213–31.

Winn, H. E., 1967. Vocal facilitation and the biological significance of toadfish sounds. In: *Marine Bio-acous-*

tics, vol. 2, W. N. Tavolga, ed. New York: Pergamon Press, pp.283–304.

Winn, H. E., 1972. Acoustic discrimination by the toadfish with comments on signal systems. In: *Behavior of Marine Animals: Current Perspectives in Research,* vol. 2: *Vertebrates,* H. E. Winn and B. L. Olla, eds. New York: Plenum Press, pp.361–85.

Winn, H. E., and Marshall, J. A., 1963. Sound producing organ of the squirrelfish, *Holocentrus rufus. Physiol. Zool.,* 36:34–44.

Winn, H. E.; Marshall, J. A.; and Hazlett, B. A.; 1964. Behavior, diel activities, and stimuli that elicit sound production and reaction to sounds in the longspine squirrelfish. *Copeia,* 1964:413–25.

Winn, H. E., and Stout, J. F., 1960. Sound production by the satinfin shiner, *Notropis analostanus,* and related fishes. *Science,* 132:222–23.

Wyman, R. L., and Ward, J. A., 1972. A cleaning symbiosis between the cichlid fishes *Etroplus maculatus* and *Etroplus suratensis.* I. Description and possible evolution. *Copeia,* 1972:834–38.

Yager, D., 1968. Behavioral analysis of color sensitivities in the goldfish. In: *The Central Nervous System and Fish Behavior,* D. Ingle, ed. Chicago: University of Chicago Press, pp.25–33.

Yanagimachi, R., 1957. Some properties of the sperm-activating factor in the micropyle area of the herring egg. *Annot. Zool. Japon.,* 30:114–19.

Youngbluth, M. J., 1968. Aspects of the ecology and ethology of the cleaning fish *Labroides phthirophagus* Randall. *Z. Tierpsychol.,* 25:915–32.

Zumpe, D., 1965. Laboratory observations on the aggressive behaviour of some butterfly fishes (Chaetodontidae). *Z. Tierpsychol.,* 22:226–36.

Chapter 21

COMMUNICATION IN AMPHIBIANS AND REPTILES

A. Ross Kiester

One may view communication among members of a given species as being fundamentally about their ecology, that is, the environment in which they live and their relationship to it. Some species of amphibians and reptiles have communication systems that are more readily studied than others. Those whose ecology is similar to our own are more easily observed and in the end more easily understood, while those whose ecology is different are more difficult to study. We can find basic similarities between our own life histories and those of, say, iguanid lizards. Similarities in sensory capabilities such as a dependence on vision, goals of behavior, and important environmental variables make the lizard's life and hence the subjects about which lizards communicate more understandable to us than the life and communication of such species as caecilians and burrowing snakes, whose nature is almost completely foreign to us. Communication in such groups is little known, not only because they are difficult to study but also because their natural history is so poorly understood that the subjects and mechanisms of their communication often remain beyond our immediate comprehension. This is especially true of species that use chemical signals to a great degree. This bias in the study of communication in amphibians and reptiles must be kept in mind, especially when reviewing the evolution of communication.

I shall start from the premise that amphibians and reptiles communicate about their ecology, or, more precisely, about what they perceive their ecology to be. On that basis, I shall outline a model of the relationship of communication systems to ecology. This model is sketchy and primarily serves to define communication for our purposes. It considers behavior (in this case, communication behavior) as partly a series of mechanisms for statistically analyzing the spatial and temporal structure of the environment. The ecology of each species consists of a set of relevant environmental variables that vary as multiple four-dimensional time series. These variables are related to the individual environmental resources that the animal uses. Thus, an animal's perception of its ecology consists of the sum of its sampling of these time series by the various methods at its disposal.

Sampling may be done either directly, by repeated observation of the variables themselves, or indirectly, via the use of cues. Cues are other aspects of the environment that give information about the environmental variables. Formally, a cue may be considered a statistic on an environmental random variable. A cue provides information about the variation in some variable, averaged over some scale of space and time. Communication can then be considered to be statements by individual animals about these time series as they affect (potentially or actually) their fitness, and the perception of these statements by other individuals. In various ways and over various scales of time and space, these statements summarize the ongoing state of the environment and the ongoing relationship between the environment and fitness. From the point of view of the recipient, such statements are also cues that give information about the environment and the ways in which it may change in the future. The recipient uses this information in whatever way will enhance its fitness. From the point of view of the sender, communicatory statements may or may not be made in such a way as to affect fitness or potential fitness favorably. For both sender and receiver, particular communicatory statements may summarize different aspects of the environment and of the relationship between environment and fitness.

This view of communication is both narrower and broader than the conventional definition based on social interaction (Cherry, 1957; Marler, 1961); narrower in the sense that it applies only to those animals whose every activity can be thought of as an attempt to maximize fitness; broader in the sense that it can give a communicatory interpretation to many aspects of the lives of animals that are not usually included under the heading of communication. The advantage of this broader view is that the role of communication behavior in the life history strategy of a species may be more directly investigated.

More important, it allows the message to be analyzed; that is, by including the context of the communication in the communication itself we can, in some sense, describe its meaning. The subjects of animal communication can be studied from the point of view of evolution, that is, that the subjects (in general) about which animals communicate concern the features of the environment that affect their fitness. And from the ecology of particular species, one may infer the specific subjects of communication, namely, the particular resources that are important to fitness.

However, every communicatory act does not have to be interpreted in the light of a particular resource. Rather it is the whole pattern of the ecology and life-history strategy of a species and the different ways in which a given animal can perceive a particular communicatory statement as a cue that gives the ecological interpretation to communication behavior. In particular, I do not require that a given communicatory act elicit an immediate behavioral response; we are justified in assuming that there can be a long delay. Examples of this phenomenon will be seen later in the discussion of lizard communication.

Below, in reviewing the communication biology of each of the eight orders of amphibians and reptiles, I shall start with an explanation of the pattern of communication of a given species. For a large class of behavior related to communication, no one member of the class can be used to explain any other member of the class. Specifically I refer to the collection of behavior and senses that I call the "environmental sampling methods." This class consists primarily of the sensory abilities, methods of locomotion, and the pattern of movement throughout the environment on all scales of space and time (e.g., daily and annual movement patterns). Members of this class for any given species possess a certain correlation with each other. This correlation occurs because the determination of any one of the major elements of the class places constraints on the options for the other members of the class.

For example, the dominant sensory modality places constraints on the daily activity pattern. Thus, if vision is the dominant sensory modality, daily activity is limited to daylight hours. Or, the choice of a particular method of locomotion, such as saltation in frogs, may prevent the use of a particular sensory modality, such as odor-trail following. Conversely, one could argue that the dominant sensory modality is explained by the animal's diurnal activity pattern, and the fact that frogs do not use odor trails explains why saltation has evolved.

To repeat, no one member of the class in any way explains another; rather all members are interrelated and it is the entire class for each species that must be explained. Appreciation of this fact helps to avoid *ad hoc* explanations of various types of communication behavior. However, the description of the class of environmental sampling methods helps to explain the communication biology of a species because it is this class that constitutes the link between the environment and communication.

Amphibians

Most of the recent literature pertaining to communication and related biology in amphibians has been reviewed by Salthe and Mecham (1974). This review is quite thorough, and the reader is referred to it for a more complete survey of the literature. Yet, as Williams (1973) remarked, one must still turn to the classic work of Noble (1931) for the best overview of the biology of the amphibians.

CAECILIANS: APODA

Caecilians remain the least known of the eight orders of amphibians and reptiles, and are still resistant to further study. Taylor (1968) has provided a review of the species of the world. Caecilians possess eyes that are sometimes rudimentary and possibly without function. Presumably they are dependent on olfaction; they have a curious tentacle that may function as a chemoreceptive organ (Cochran, 1961). Taylor (1970) discusses the lateral line system of some caecilians, and Wake (1968) has given some evidence for seasonal cycles in breeding, but essentially nothing is known about communication in this group. This lack is not surprising in view of the habits of caecilians, which are primarily nocturnal burrowers. Even in captivity, if one is to keep them alive and well, they are almost impossible to observe.

FROGS AND TOADS: SALIENTIA

The vast majority of studies of communication behavior in amphibians have dealt with frogs and toads. Despite the fact that many species are nocturnal or cryptic or both, frogs are generally obvious creatures in the environment because of their striking vocalizations. With the advent of tape recorders and sound spectrographs, these vocalizations have become the subject of much study. Indeed, it is now commonplace to find sound spectrograms of mating calls as part of the conventional basic data of systematic studies including the description of new species (e.g., Pace, 1974). Much of the biology of calling in frogs has been reviewed by Bogert (1960). He provides a classification of frog calls that is still useful, despite the great amount of work in the field that has appeared since its publication. More recently Schiøtz (1973) has reviewed some of the ecological aspects of frog calls, Straughn (1973) has reviewed the broad-scale evolution of frog calls, and Martin (1972) has reviewed aspects of the evolution of calls in the toads *(Bufo).* In addition, Gans (1973) has reviewed the mechanisms of sound production in amphibians, and Schmidt (1973) has summarized his work on the nervous system control of calling.

Frogs are mostly visually oriented predators, and so one expects their vision to be rather acute. This is the case, but in a special way.

Maturana et al. (1960) found that the visual system of the leopard frog *(Rana pipiens)* responds to rather specific stimuli. Certain units in the retina respond only to small moving objects ("bug detectors") or only to large moving objects ("predator detectors"). Thus the frog's view of the world may be rather limited. It would be interesting to continue this kind of study on some of the small, highly territorial species, such as some of the dendrobatids. Here one might expect that detection of a moving object the same size as the frog (namely, another frog) would be important, and that the neurophysiology might have developed accordingly.

In a similar way, hearing in frogs is very specifically tuned. The classic work of Capranica (1965) on the territorial call of the bullfrog *(Rana catesbeiana)* showed that the auditory system possesses a frequency response narrowly tuned to the frequency of the call. This work has been extended to mating calls in other species, and even further to the geographic variation of calls within a single species. Capranica, Frishkopf, and Nevo (1973) found that geographic variation in the tuning of the frequency response of the auditory system of cricket frogs *(Acris crepitans)* exactly matches the variation in the frequency of the mating call. Recently Lombard and Straughn (1974) have reported on other species and have discussed aspects of the tuning of the ear in detail. Olfaction appears to play some role in orienting frogs toward breeding areas. The possible specificity of the responses has not been studied directly, although Savage (1961) argues that the smell of algae is the attracting factor.

Most frogs are true jumpers and therefore move through their environment in a more or less discontinuous fashion. Some, especially toads, might be better described as waddlers, and others such as the Surinam toad *(Pipa pipa)*, are aquatic. Frogs tend to be nocturnal, although there are many diurnal species as well. Many species aggregate seasonally to breed at particular localities, but again there are exceptions to this. Species such as the dendrobatid toads remain territorial and breed within their territories. The temporal pattern of aggregation can vary also. Salthe and Mecham (1974) review the various cases of the pattern of movement and aggregation for breeding, distinguishing cyclic and noncyclic patterns.

Some general patterns of frog life histories and environmental sampling methods seem to emerge. One pattern is shown by those frogs that are nocturnal, nonterritorial, aggregate breeders. Another pattern is apparent among diurnal, territorial, nonaggregate breeders. Examples of the first pattern are the classic frogs, such as the common European frog *(Rana temporaria)* or the American toad *(Bufo americanus)*. The second type would be represented by the dendrobatids, such as *Dendrobates pumilio.* However, many species, such as the green frog *(Rana clamitans)*, show elements of both. Male green frogs aggregate to breed but then set up territories along the edges of the breeding ponds and call frequently during the day.

Most species of frogs are not visually conspicuous. However, certain diurnal frogs, such as *Dendrobates auratus* and many of its relatives, are brightly colored. The brilliant coloration is often associated with extreme toxicity of the skin secretions (as in the poison-arrow frogs), and thus the coloration is believed to have an aposematic function (Daly and Myers, 1967). However, an aposematic function for a bright coloration does not prevent its use in species perception by other frogs. Here experimental analyses of the reactions of frogs and further neurophysiological studies would prove useful. At present there does not appear to be any evidence that frogs use chemical secretions or pheromones to affect species perceptibility, although this must be the case for at least some species.

It is, of course, the vocalizations of frogs that

are their most striking form of communication. As mentioned above, these calls have been the subject of much analysis. Such analyses have certain technical difficulties associated with them that were not always appreciated in earlier studies; foremost among the biological problems is the influence of temperature on call rate. All studies, especially if they are to be comparative, must correct for this factor (Zweifel, 1968). Other factors that affect the call are body size and whether the animal is immersed in water. On the purely technical side, there are the usual difficulties associated with the interpretation of sound spectrographs (see Martin, 1972).

That frog calls contain an important genetic component is demonstrated by the analysis of the calls of hybrid species (e.g., Gerhardt, 1974). However, studies on the ontogeny of calling are lacking.

Mates or potential mates appear to be the most important resource about which frogs communicate. Indeed, it is now generally recognized that the dominant vocalization in the repertoire of almost all species is the mating call. Despite that, the exact functions and mechanisms involved in their use are the subjects of much ongoing research: the scales of space and time over which mating calls function as identifiers of potential mates have not been completely worked out. Much of the problem in working out the details stems from the difficulty of determining the specific behavioral responses needed to demonstrate the function of the call at a particular point in the life history of the animal.

In most cases it is not known at what distance and how long before the actual act of mating frog calls can affect the behavior of frogs. Experimental evidence that in some species frog calls can attract males as well as females to the actual breeding site (in the case of aggregate breeders) has been provided by Bogert (1960) and others. On the other hand, as mentioned by Salthe and Mecham (1974) in their review of this subject, many species orient and move toward breeding sites by using other cues. In view of the importance of breeding-site location to the reproductive success of frogs, it is not surprising that frogs would use all available cues and not be dependent on any one cue. Some of the problems of experimental analysis of this question are due to the limited responsiveness of the animals outside of a particular stage of their reproductive cycle, as mentioned above.

Once a female frog has arrived in the general vicinity of a breeding site, she must identify and orient on a particular male frog as a potential mate. It is at this stage that the mating call plays its most conspicuous role. From an evolutionary point of view, this is when the mating call functions as an isolating mechanism, and much of the research has been concerned with analyzing the calls as such. Beginning with the work of A. P. Blair (1941, 1942) and W. F. Blair (1956, 1958, 1964), a whole school of study has arisen around this topic. Few workers today would dispute that frog mating calls function as isolating mechanisms. But it has taken a good deal of effort to unravel the exact mechanisms underlying this function, and especially to demonstrate that female frogs do in fact respond to the specific calls of the males of their own species. Through the use of tape-recorded playback experiments, female choice has now been demonstrated in a number of cases (see reviews in Martin, 1972; Schiøtz, 1973; Salthe and Mecham, 1974). As Schiøtz (1973) explains in his review, the female is responsive to the call of the male for only a short period. For example, Heusser (1968) found that female *Bufo bufo* were responsive for six to fourteen days. Other cases are even more striking. In a series of experiments on female choice in the frog *Physalaemus pustulosus,* Rand (pers. comm.) found that only females collected in amplexus the same night would respond consistently. Thus, negative experimental results require cautious interpretation.

At the closest range actual courtship takes over, and here the mating call does not seem to play a dominant role. Once a female is in the immediate vicinity of the male, he usually attempts amplexus directly. From the initial approach through ovulation and fertilization, there appears to be a series of tactile stimuli that organize the overall pattern of behavior. For example, Rabb and Rabb (1963) have detailed the courtship, amplexus, and ovipositional behavior of the pipid *Hymenochirus boettgeri.* They describe eleven kinds of behavior other than the calls used in the process of egg laying and fertilization in pipids. A complete review of this aspect of frog behavior may be found in Salthe and Mecham (1974).

Another important call related to courtship is the male release call, given by male frogs with which other male frogs have attempted amplexus. These calls are common, but not universal, in aggregate breeders (Salthe and Mecham, 1974). Brown and Littlejohn (1972), in a study of the male release call of the *Bufo americanus* species group, found that while variation between species did exist in certain components of the call, on the whole the release calls seemed more conservative in their evolution than the mating calls. They postulate that this conservatism may be related to the fact that the male release call serves a purpose in interspecific as well as intraspecific communication. Obviously, both parties benefit from the appreciation of a mistake, regardless of species.

Bogert (1960) lists in his classification of frog calls the female release call, and the ambisexual release vibration (which is not really a call) as two other kinds of behavior that serve a function similar to that of the male release call. The female release call is apparently used by females of some species that are not receptive to mating. The ambisexual release vibration is used by both sexes. In addition, Bogert (1960) lists the post-ovipositional call of *Phyllomedusa guttata* reported by Lutz (1947) as a separate call type associated with mating.

From the perspective of the evolutionist and evolutionary ecologist, frog mating calls are of great interest in ways other than their relation to female orientation and mate recognition. Some of these can only be sketched here. Since mating calls are important as isolating mechanisms, the patterns of variation in mating calls within a single species are of interest. Geographic variation in mating calls has been described for a number of species (Blair, 1974). Studies of geographic variation must be undertaken with care because of the effects of temperature on call rates and structure. Sometimes this variation is correlated with variation in other characters that are related to the quality of the call, such as body size. The function of whatever variation in mating calls that is not due strictly to other factors is related to the problems of species identification and character displacement. Another as yet uninvestigated possibility is that dialects, together with female choice, constitute a mechanism for controlling genetic variation within the population. This problem needs to be investigated both theoretically and empirically. Character displacement in mating calls has been reported for several species (Salthe and Mecham, 1974; Blair, 1974) although in some cases problems in interpretation have arisen because of variation in several characters at once.

The extension of the concept of character displacement between any two species to consideration of whole communities of frogs leads to the problem of species packing. This is a very real problem in some communities, where as many as fourteen species of frogs may be calling at one time (Bogert, 1960). In such situations the problems of acoustical interference and signal detection are considerable. Straughn (1973) argues that as more species are added to a community, the frequency band-width available to each species decreases. Thus in order to convey ample species-identification information, the individual species calls must then take on longer and more complex temporal patterns. Straughn bases his argument on a well-known theorem of communications engineering (Shannon, 1949). There are some subtleties in interpreting this theorem (Dym and McKean, 1972), but it does seem that this idea might be developed into a theory of frog species packing.

The necessity of adapting a call to the acoustical interference created by the calls of other species is only one of many constraints on the morphology of the call. Localizability and transmissability of calls in particular environments are questions that have as yet received little attention. Schiøtz (1973) discusses these constraints and mentions some possible patterns, among which is the apparent correlation of quiet calls with a generally quiet habitat in western Africa.

Many frogs are now known to be territorial, but the nature of the territoriality varies considerably. Perhaps the most strictly territorial species are some of the dendrobatids, such as *Dendrobates pumilio,* studied by Bunnell (1973). Males of this species appear more or less permanently territorial and breed within their territories. In addition, males care for the young tadpoles once they have hatched out. These males emit a distinct call, which serves to maintain the spacing pattern. The calling of individual males can be adjusted to the calls of other males by changing the rate and temporal pattern of the call. Thus, territories are maintained by a combination of call adjustment and attack and fighting. In addition, the calls apparently serve to attract females. So at least for males for periods of time, this species gives a more or less typical picture of a vertebrate territorial system.

However, the frogs that give the territorial call as listed in Bogert's (1960) review of call types show a rather different pattern. In some aggregating breeders of the genus *Rana,* males at the breeding sites show a form of territoriality and have a specific call associated with the advertisement and defense of the territory. Both green frogs *(Rana clamitans)* and bullfrogs *(Rana catesbeiana)* are good examples of this. The call of the bullfrog and the way in which it provokes the calling by another male has been the subject of a classic study by Capranica (1965). That the calls also elicit aggressive behavior has been demonstrated by tape-recorder playback experiments (e.g., Wiewandt, 1969; Emlen, 1968). Several other cases of frog territoriality have been recorded and are reviewed in Bunnell (1973) and Salthe and Mecham (1974). There appears to be a great variety of calls used by territorial males with some species having a territorial call in addition to a mating call and others not.

More complete studies on the life histories and vocalizations of territorial frogs are needed to understand the relationship between territoriality and call type diversity. It is important in such studies that the entire repertoire of call types be studied, otherwise the interpretation of a territorial call may be somewhat difficult.

Perhaps related to territorial calling is the phenomenon of chorus structure in frog calls. The existence of duetting and more complex chorus structures is now known in many frog families and has recently been reviewed by Wickler and Seibt (1974). Chorusing occurs when the calling of one individual affects the pattern of calling in another. Wickler and Seibt list some nineteen genera in which chorusing of some sort has been reported. As mentioned above, that chorusing exists in the territorial calls of *Dendrobates pumilio* (Bunnell, 1973). Wickler and Seibt (1974) found that in *Kassina senegalensis* (Rhacophoridae) males form groups of two or more in which one animal consistently sets the pace by calling once every two to eight seconds, thus eliciting replies from the other members of the chorus. On the other hand, these workers found that with *Bufo regularis* (Bufonidae) two males would alternate regularly. Duellman (1967) and

Wickler and Seibt (1974) review other patterns in other species. Wickler and Seibt conclude that the biological function of chorusing is unknown. Brattstrom (1962) indicates that the dominant (first-calling) male in a chorus of *Physalaemus pustulosus* enjoyed a comparative mating advantage, but this point needs more investigation.

When artificially penned together, some frogs do show feeding hierarchies (Boice and Witter, 1969; Boice and Williams, 1971), but whether hierarchies of this sort occur in nature is not known.

The last type of call listed in Bogert's (1960) classification are "rain calls." These are calls given by males away from the breeding site and often either before or after the breeding season. They have been recognized in some species of *Hyla* and possibly in other genera as well. There is no clear characterization of "rain calls," other than their time and place of occurrence. Bogert describes them as being "a chirping sound, a feeble rendition of sounds resembling the mating call, or a vigourous but recognizable modification of the mating call" (1960:198). Bogert recognizes and discusses the difficulty of adequately characterizing these calls and identifying them in any particular case. Often they may be simply described as premature renditions of the mating calls. Bogert does not feel that there is a clear biological function associated with these calls.

Finally, no account of communication in frogs would be complete without mention of the possibility of social behavior and communication in tadpoles. Wassersug (1973) presents some evidence that tadpoles do respond to each other when aggregating rather than just responding in some common fashion to a physical factor. He concludes that more investigation is needed to establish the existence of socially based schooling in tadpoles. As yet, nothing concrete is known about their potential mechanisms for communication.

SALAMANDERS: CAUDATA

Most salamanders possess the normal range of sensory modalities for a vertebrate. Vision generally appears well developed, although details of the functioning of the visual system are not known, especially with regard to form perception. Some cave-dwelling species have extremely reduced eyes, e.g., the blind salamanders *Typhlomolge* and *Hadieotriton,* and many aquatic forms have rather small eyes, e.g., hellbenders *(Cryptobranchus).* There is evidence that salamanders use extra-optic photoreception in orientation (Adler, 1970; Landreth and Ferguson, 1967). Twitty (1959) found that blinding *Taricha,* a terrestrial newt, had little effect on their ability to home. This leads one to wonder what the role of vision is in the overall life of salamanders. Studies on the auditory abilities of salamanders are rare. Ferhat-Akat (1939) presents some evidence that salamanders can hear, but the role of hearing seems quite small in the lives of most. On the other hand, olfaction seems to be very important. In plethodontid salamanders special nasolabial grooves appear to act as conduits by which samples of the substrate are carried into the nasal cavity (Brown, 1968). Good behavioral evidence for the importance of olfaction in homing by the newt *Taricha* is given by Grant, Anderson, and Twitty (1968). They found that anosmic animals were unable to home.

Salamanders are best described as crawlers. The major exceptions are some of the permanently aquatic forms, such as the congo eels *(Amphiuma)* and the sirens *(Siren).* Many other aquatic forms, such as mudpuppies *(Necturus)* and hellbenders *(Cryptobranchus),* however, crawl along the bottoms of the bodies of water in which they live. Crawling tends to keep salamanders in more or less continuous contact with their substrate, which is generally rather damp.

The vast majority of salamanders are nocturnal. The outstanding exceptions are the aposematically colored terrestrial newts, such as the red eft *(Notophthalmus)* and the various species of *Taricha.* Of course, many of the cave species are not nocturnal. Otherwise the only other generalization that can be made about salamander activity is that it is strongly related to humidity, with most species being more active on wet or rainy nights.

The most striking seasonal activity patterns are the aggregations formed for breeding by those terrestrial species that return to bodies of water. These include such well-known species as the spotted and tiger salamanders *(Ambystoma)* and the various species of newts. Some permanently terrestrial species may do a significant amount of moving about on a seasonal basis. Often this takes the form of moving away from streams, up the sides of small canyons in the wet season and back again as drier conditions prevail, e.g., *Ensatina* (Stebbins, 1954). Others seem to restrict seasonal movement to vertical migration within the soil, e.g., slender salamanders *(Batrachoseps),* or movement in and out of caves, e.g., the Shasta salamander *(Hydromantes shastae),* in accordance with seasonal rainfall patterns.

Many salamanders are always either cryptic or aposematic, but others develop special coloration at breeding time. Some newts, such as the European *Triturus,* develop crests during the breeding season. Males of the plethodontid *Aneides lugubris* develop large front teeth and masseter muscles as secondary sexual characters. Many salamanders can produce sound of some sort, and there are several reports in the literature of vocalizations by salamanders (Maslin, 1950). S. J. Arnold (pers. comm.) has reported that Pacific giant salamanders *(Dicamptodon ensatus)* regularly emit a characteristic "bark" when attacked by snakes. But vocalizations do not appear to be used for communication. The production of particular chemical signals appears to be common in salamanders, but the lack of precise behavioral responses to most stimuli makes their study difficult. Good circumstantial evidence comes from consideration of the many glands and skin secretions found on most salamanders. A complete review of the evidence for chemical signals can be found in Salthe and Mecham (1974). Cedrini and Fasolo (1971) demonstrated by electrophysiological studies that newts *(Triturus)* could detect odors given off into water by conspecifics.

Most communication in salamanders appears to be related to mating, although its role in locating potential mates in both aggregate and nonaggregate breeding species is unknown. This lack of knowledge is not surprising, for if communication of this type does in fact occur, it is almost certainly by chemical means.

Courtship, on the other hand, is much better known. A summary of reviews on this subject may be found in Salthe and Mecham (1974:365–79). As with frogs, salamander courtship involves a series of stimuli on the part of both males and females. But salamanders appear to use chemical signals as well as tactile and visual stimuli. This is especially true in aquatic breeding species, such as the newts of the genus *Triturus.*

Studies of salamander courtship have included work on the description of stimulus chains (Halliday, 1974) and phylogenetic analysis of the courtship patterns (Salthe, 1967). Arnold (1972) has examined some of the strategic aspects of courtship in certain salamanders. He distinguishes slow and fast courters and relates these two strategies to the competitive environment. Aggregate breeders, such as spotted salamanders *(Ambystoma maculatum),* court in groups, and competition among males for females is severe. In these species courtship is rapid (and in addition males have a series of behavioral tricks to induce other males to waste spermatheca). On the other hand, dispersed breeders, such as slimy salamanders *(Plethodon jordani),* usually court in single pairs and court more slowly and carefully.

Territoriality is known in some few species of salamanders, including the plethodontids *Hemidactylium scutatum* and *Eurycea bislineata* (Grant, 1955), several other plethodontids (Salthe and Mecham, 1974), and apparently the hellbender *Cryptobranchus:* Hillis and Bellis, 1971). Aggression has been observed in the field between males of some species, such as the plethodontid *Aneides lugubris,* in which the males possess enlarged teeth during the breeding season and may inflict injury on each other (Stebbins, 1951). However, the mechanisms of communication used in maintaining territoriality are little known. There is good reason to believe that many species possess no form of behavior similar to territoriality. Studies of artificially confined animals (which often show unexpected social behavior in frogs, turtles, and lizards) are few. Evans and Abramson (1958) report that a complex hierarchy is formed when individuals of *Notophthalmus (=Triturus)viridescens* are housed together. Such studies do not necessarily reflect any real behavior that may occur in the field, but they do indicate that further research may be warranted.

Reptiles

Brattstrom (1974) has reviewed social systems and much communication biology in reptiles. Mertens (1960) and Schmidt and Inger (1957) give good general accounts of the group. A more technical review of reptile biology may be found in Bellairs (1970). The multivolume series *The Biology of the Reptilia* will cover behavior and ecology in the near future.

TUATARAS: RHYNCHOCEPHALIA

Knowledge of the behavioral biology of this relict creature was reviewed by Wojtusiak (1973). Tuataras *(Sphenodon punctatus)* are primarily nocturnal, with good vision, are solitary, and live in burrows. Wojtusiak and Majlert (1973) report at least two distinct vocalizations given by a captive tuatara, but nothing appears known of the possible communicatory significance of these sounds. In general, the natural history of these animals in the wild, particularly their individual interactions, is little known.

LIZARDS: SAURIA

Lizards are usually credited with excellent vision. A complex radiation in the structure of the eye has occurred. Thus, the nocturnal geckos that have evolved from diurnal ancestors have developed highly specialized eyes, which allow them to remain largely dependent on vision even though they are normally active at low levels of illumination. Some other groups, such as the amphisbaenids and some species of fossorial skinks, have lost the emphasis on vision. Even within a group of species of the genus *Anolis,* Jenssen and Swenson (1974) found considerable variation in the function of the visual system. They studied the flicker-fusion frequency of seven species and found that the frequency varied from 26 to 42 cycles per sec. They were able to correlate the variation in frequency with the amount of light normally available in the habitats of the lizards. Those lizards that lived in more open, and hence more illuminated, habitats had higher frequencies. Thus, even at the generic level, a considerable amount of variation can occur. Benes (1969) gives behavioral evidence that whiptailed lizards *(Cnemidophorus)* can discriminate colors, and color vision probably occurs in other species as well. Extraoptic photoreception may occur in some lizards by way of the parietal eye (Stebbins and Eakin, 1958), but much remains to be learned about this system, and it is doubtful that it is ever used in communication.

Hearing also appears to be good in many species of lizard. Wever, Crowley, and Peterson (1963) have studied the auditory sensitivity of several species, and Campbell (1969) found definite temperature effects on auditory sensitivity. He found that the temperature of maximum sensitivity coincided with the preferred body temperature. This brings up the general question of temperature dependence of sensory functions in lizards. It is probable that most sensory systems function best at certain temperatures. Thus any interspecies comparison must be corrected for variation in preferred temperature. Olfaction and taste are not well known in lizards, but there is some anatomical and anecdotal behavioral evidence that it is important in many species (Bellairs, 1970). Kroll and Dixon (1972) have given evidence that the preanal patches of some geckos *(Phyllodactylus)* may function as a sense organ receptive to heat.

Daily activity patterns are extremely variable in lizards, ranging through all degrees of nocturnality and diurnality. The main requirement appears to be a certain amount of heat; consequently, few species are active in the late night or very early morning, while many are active during the day or early evening and night. Annual and seasonal cycles, especially in reproductive activity, may be very complex. Fitch (1970) reviews much of the literature on this question. As an example, Licht and Gorman (1970) and Gorman and Licht (1974) have found great variation and complexity in the reproductive cycles of several species of tropical and subtropical *Anolis.* Here different patterns of reproductive activity relative to weather may be found among rather closely related species, and within a single species great variation may be found over short distances.

Appearance in lizards is often extremely complicated. Complex color patterns and bizarre ornamental structures are common. Of course, many forms, especially those that are fossorial, are dull and cryptic. The well-known, brightly colored lizards are the diurnal iguanids and agamids, but certain species in many other groups are also brilliantly colored. For example, skinks are often thought to be predominantly dull and cryptically colored, but such species as *Lamprolepis smaragdinum* (Greer, 1970) and *Oelofsea laevis,* of which Steyn and Mitchell (1965) give a color photograph, are very bright. Ornamental structures include nuchal, dorsal, and caudal crests; cephalic spines, knobs, and horns; various enlarged scales; and gular dewlaps.

Mechanisms to modify appearance include the remarkable color change of many species, the most famous being the chameleons (e.g., Parcher, 1974; Burrage, 1973), but geckos and many iguanids, such as *Anolis,* have this ability also. Many species can alter the appearance of the whole body by flattening either laterally or dorso-ventrally, and by inflating themselves with air, as in the chuckwalla (*Sauromalus:* Berry, 1974). In many ways the most-developed case of modifiable appearance occurs in the dewlap of anoline lizards, a highly modified structure of skin and cartilage that hangs below the throat. It can be retracted until it is almost invisible or extended until it is larger than the head of the lizard (Fig. 1). These dewlaps are often brilliantly colored and in strong contrast to the colors of the body.

Fig. 1. An adult male *Anolis cybotes* with its dewlap extended.

In addition to varying the appearance of the body, lizards also often possess a series of postures and stereotyped movement patterns that are used in communication. They include the "nods," "bobs," and "pushups" of iguanids and agamids. These curious movements have been studied by herpetologists for many years. C. C. Carpenter and his students (for reviews see Carpenter, 1967; Brattstrom, 1974; Stamps, 1975) have brought the study of these motions into the realm of quantitative description. They film the lizards and then use frame-by-frame analysis to graph the vertical displacement of the head as a function of time. When working with *Anolis,* the extension of the dewlap is graphed in parallel as well. The resulting graphs, called DAP (=display action pattern) graphs, provide a description of several important aspects of the pattern. Using this technique, Carpenter (1962, 1966) and others (e.g., Clarke, 1963; Carpenter, Badham, and Kimble, 1970) have attempted to work out the phylogeny of several groups of lizards and of the displays themselves. For example, Gorman (1968) traced the evolution of the *roquet* species group of *Anolis,* using the male display as one important character. Other studies have examined the role of these display patterns in the organization of social systems in these lizards (see below).

These display patterns appear to have an important genetic component and are highly stereotyped. Gorman (1968) showed that the display of a natural hybrid of *Anolis aeneus* x *trinitatis* was intermediate between the two parental types. Although most of these displays are highly stereotyped, recent studies have begun to document individual variation within a species. Jenssen (1971), working with *Anolis nebulosus,* and Stamps and Barlow (1973), with *Anolis aeneus,* have analyzed variation in several components of the displays and have discussed the origin and function of the variation. According to Cooper (1971), hatchling *Anolis carolinensis* show components of the display almost immediately upon emergence from the egg. However, no detailed studies have been performed on the ontogeny of the displays. Such studies will probably be needed to unravel the source as well as the function of the variation in these display patterns.

Vocalizations among lizards occur most frequently among the geckos and are associated with their nocturnal habits (Evans, 1936). A well-known case is that of the barking gecko *(Ptenopus garrulus)* of the Kalahari Desert, whose din can keep travelers from sleeping at night. These lizards live in individual burrows and call from the entrances (Haacke, 1969). Another recently studied case is that of the genus *Ptyodactylus* in Israel (Frankenberg, 1974). Many species of several other families of lizards are listed in the literature as producing vocalizations or sounds of one sort or another, but most of these are very poorly known and appear to be little used.

Chemical communication of some sort almost certainly occurs in some species of lizards. A well-known problem in lizard biology concerns the role of the femoral pore secretions found in such iguanids as *Sceloporus.* The femoral pores of males of many species of iguanids become enlarged during the breeding season and exude a waxy substance that apparently gets deposited on the rocks and other surfaces where the lizards are active. Lizards are sometimes seen to tongue or lick these areas, e.g., the chuckwalla (*Sauromalus obesus:* Berry, 1974), but no one, to my knowledge, has been able to demonstrate a specific behavioral response to these secretions. The situation may be even more complex in species of lizards like the skink *Tribolonotus*, in which several other skin glands are prominent (Greer and Parker, 1968).

Lizards rarely aggregate for the specific purpose of mating. Thus males and females tend to find each other either by individual attraction or by similarities in habitat selection. Male-female association in nature does occur as in the rusty lizard (*Sceloporus olivaceous:* Blair, 1960). Laboratory studies by Pyburn (1955) showed that individuals of *S. olivaceous* tend to associate more frequently with other individuals of the same species than with individuals of the related *S. poinsetti.* Kiester (ms.) showed that individuals of *Anolis auratus* would initially move toward other members of the species, regardless of sex, in an experimental choice apparatus. Since conspecific association occurred without regard to sex, it may have an adaptive value in habitat selection, rather than just mate selection. But certainly the two resources of habitat and mates are correlated. Long-term pair bonds may occur in some lizards, but this has yet to be demonstrated.

Recent investigations have demonstrated several effects of behavior, including communication behavior, on the hormonal and reproductive cycles of lizards. Crews (1974) and Crews, Rosenblatt, and Lehrman (1974) have analyzed the effect of various social behaviors on ovarian recrudescence in female *Anolis carolinensis.* During winter months ovarian activity is normally shut down; however, ovarian recrudescence can be induced by the application of gonadotropin or by providing the appropriate unseasonal environment. This environmentally induced recrudescence can be influenced by the behavior of a group of conspecifics. Crews (1974) found that a stable dominance hierarchy of adult males facilitates recrudescence, while an unstable hierarchy inhibits it. Courtship and male-male aggression are specifically responsible for this effect. In addition, ovarian development was graded in accordance with the amount of male courtship to which the female was exposed.

Courtship in lizards varies from direct attempts at copulation by the male to elaborate display action patterns used only in the context of courtship. These patterns often bear some resemblance to the male aggressive patterns, but are usually quite distinguishable. Ferguson (1966, 1970) has given a detailed analysis of the courtship display of the iguanid *Uta stansburiana.* Crews (1975) showed that the ability of a male *Anolis carolinensis* to extend its dewlap was important in facilitating ovarian recrudescence in winter-dormant females, while the bobbing and strutting patterns were not.

Hunsaker (1962) showed that females of various species of the *Sceloporus torquatus* species group could recognize males of their own species partly by their display action pattern. He used mechanical models to simulate the various patterns of movement and then observed female response. Jenssen (1970), testing the response of female *Anolis nebulosus* to filmed displays of the males, found that the females would approach a film of a male displaying, and further that they could discriminate between the normal male display and one that was artificially altered. Thus female choice may be based on the display. On the other hand, Kiester (ms.) found that female *Anolis auratus* would choose live individuals of their own species over individuals of the similar-appearing *Anolis tropidogaster* in the absence of displays.

Another aspect of female choice is certain nonreceptivity signals given while a female is gravid. These appear after a female has copulated and seem to indicate nonreceptivity to courting males. In species such as *Crotaphytus collaris* (Cooper and Ferguson, 1971) and *Sceloporus virgatus* (Vinegar, 1972) females take on a distinct coloration during the time that they are gravid. On the other hand, Crews (1973) found that female *Anolis carolinensis* exhibited nonreceptivity by attempting to hide from males, and by not assuming a particular arched-neck stance that usually indicates receptivity to copulation. He was also able to demonstrate that this nonreceptivity was coition-induced. The causes of the variation among species in the mechanisms of demonstrating nonreceptivity are not clear, but appear worthy of future study.

In addition to the review by Brattstrom (1974), social systems in lizards have recently been reviewed by Stamps (1975), who gives a complete survey of the literature and proposes a model of the evolution of social systems in relation to phylogeny and sensory modalities. Many lizard species, especially iguanids and agamids, are normally territorial, while many others do not seem to show much social behavior at all. However, within a single normally territorial species a great deal of variation can often be seen. For example, Stamps (1973) found female *Anolis aeneus* in both territorial and hierarchical situations. She further found that particular display action patterns were associated with one or the other type of social system. Such variation between territorial and hierarchical systems appears common in iguanids (Kiester and Slatkin, 1974). Hunsaker and Burrage (1969) have argued that there is a continuum of social system types from territoriality to hierarchy in iguanids. However, evidence such as Stamps's (1973) seems to indicate that there are qualitative changes in behavior associated with the transition from territoriality to hierarchy. However this variation does occur, it appears to be a complicated phenomenon.

Colnaghi (1971) found that dominant males in an artificial hierarchy of *Anolis carolinensis* had greater access to food than subordinates. Kiester and Slatkin (1974) proposed a model of iguanid behavior in which all conspecific interactions are used as part of a strategy to estimate patterns of environmental variability and to structure daily movement patterns. Thus, interactions between individual lizards may have direct and indirect ecological effects as well as being acts of communication.

As with the mating calls of frogs, the display action patterns of lizards, especially the male territorial display, seem to function, in part, as isolating mechanisms. Thus the patterns of variation in these displays are of interest to evolutionists. Geographic variation in the display has been most thoroughly analyzed for *Uta stansburiana,* a wide-ranging iguanid of central and western United States (Ferguson, 1966, 1971; McKinney, 1971a, 1971b). Here a great deal of variation in the display has been demonstrated although its biological function is unknown. As with frogs, the possibility that the variation functions together with differential female choice to control variation and to permit local adaptation, needs investigation. Character displacement in displays has been demonstrated for *Sceloporus undulatus* and *S. graciosus* by Ferguson (1973). An interesting failure of species recognition has been reported by Gorman (1969) for *Anolis aeneus* and *A. trinitatis.* These two species are very closely related, and both have recently been introduced by man to the island of Trinidad. *A. trinitatis* originally comes from the island of St. Vincent, while *A. aeneus* comes from Grenada. Despite the fact that the lizards differ somewhat in color and in the form of the male territorial display, a large hybrid population has developed where the two species have been artificially brought into contact, although the hybrids are completely sterile. Thus, difference in the male display alone may not be sufficient to insure premating reproductive isolation.

The role of species-recognition mechanisms in the problem of species packing has been considered for a community of eight *Anolis* species by Rand and Williams (1970). They attempt an analysis of the information content and redundancy of coding of species identification in the dewlaps. They conclude that species indentification is encoded in many ways, that is, redundantly. They conjecture that in habitats of poor visibility, such as forests, redundancy will have to be high to permit the same level of species packing as in open habitats. They also conjecture that the problems of encoding species identification may be one of the factors that limit the number of coexisting species. These are interesting research projects, but the data will be difficult to analyze quantitatively because of the problem of assessing the amount of information in a pattern as perceived by a lizard.

A further analysis by Williams and Rand (ms.) of both large and small communities of *Anolis* has revealed that species recognition in species that live in communities with few or no congeners is not encoded in the dewlap. Rather, in these simple fauna recognition is achieved through characteristics such as size, shape, color, and pattern. They argue that it is thus only in larger communities (six or seven species) that the dewlap functions in species recognition. Therefore only when the presence of many species makes encoding species identification difficult is the dewlap incorporated as an additional signal for species recognition.

SNAKES: SERPENTES

In terms of number of species, snakes constitute the largest order of amphibians and reptiles. They have undergone an extensive ecological radiation, occurring in a remarkable variety of habitat from below leaf litter to the tops of trees and from sand deserts to the open ocean. Coincident with this radiation is a considerable variation in sensory abilities. However, despite their diversity, abundance, and human interest, snakes are poorly known. Most herpetologists probably have an interest in the biology of snakes, and yet very few work on them. They are simply too difficult. Thus any generalizations about sensory or species-perceptibility systems in snakes are bound to be suspect.

Vision often appears to be the dominant sensory modality in snakes. Nocturnal or crepuscular species, such as the cat-eyed snakes *(Leptodeira),* frequently have distinctively developed eyes with vertically elliptical pupils. Many of the predominantly subterranean forms, such as the blind snakes *(Leptotyphlops)*, have small or vestigial eyes. Snakes are capable of hearing, but only over certain frequency bands, and in general do not appear to depend on hearing (Wever and Vernon, 1960). They are, in addition, sensitive to low-frequency ground vibrations. On the basis of anatomical features the senses of taste and smell appear well developed and include such specialized organs as that of Jacobson (Bellairs, 1970). The constant use of the tongue in many species of snakes, of course, is well known. Bogert (1941) demonstrated experimentally that rattlesnakes *(Crotalus)* could recognize kingsnakes *(Lampropeltis),* which prey on them, by odors given off by the kingsnakes. Gehlbach, Watkins, and Kroll (1971) have shown that several species of snakes are capable of following pheromone trails. For example, blind snakes *(Leptotyphlops dulcis)* are able to follow the pheromone trails laid down by army ants *(Neivamyrmex nigrescens),* on which they prey. The pits of the pit vipers form a special sense organ used for the detection of infrared radiation and the tracking of warm-blooded prey. But there is no indication that this modality is used in communication.

The daily activity patterns of different species of snakes are as variable as those of lizards. Snakes appear to tolerate a greater range of environmental temperatures in their activity than do lizards. Annual activity patterns are generally poorly known. Some species of temperate zone snakes aggregate at "dens" for hibernation, and mating may take place near these dens (Evans, 1961). Information on the reproductive cycles of snakes has been summarized by Fitch (1970).

The visual appearance of snakes is as varied as their habits and habitats (see Schmidt and Inger, 1957; Mertens, 1960), ranging from cryptic to aposematic. Many species of snakes can inflate themselves or flatten the neck region. Sometimes this action causes the appearance of a particular pattern, as in the cobras and the African bird snake (*Theletornis:* Blair, 1968). In general, modifiable appearance mechanisms are not as well developed or common as in lizards. Vocalizations are restricted to hissing, which does not appear to serve a communicatory function. Other sounds produced by snakes, such as the rattle of the rattlesnake, also appear not to function in communication (Gans and Maderson, 1973).

Snakes produce a variety of smells and secretions, many of which are powerful and disagreeable to human beings and serve a defensive function (Mertens, 1946). The best-known case of chemical communication in snakes is that of the blind snake *Leptotyphlops dulcis,* studied by Watkins, Gehlbach, and Kroll (1969). These workers have demonstrated that blind snakes show a complex set of behavioral responses to their cloacal sac secretions. They are attracted to the secretion, which may cue both food sources and potential mates. On the other hand, several genera of snakes, which are sympatric with blind snakes *(Sonora, Tantilla, Virginia, Diadophis,* and *Lampropeltis)* and which include both potential competitors with and predators of blind snakes, are repelled by the secretion. In addition, army ants *Neivamyrmex nigrescens,* on which the blind snakes prey, as well as other species of ants are repelled by the secretion. The secretion itself is a mucuslike glycoprotein suspended in free fatty acids (Blum et al., 1971). It is quite likely that these studies represent only the tiniest tip of the iceberg of chemical communication in snakes.

The meager information available on social systems and intraspecific interactions and communication is summarized by Evans (1961) and Brattstrom (1974). One of the few obvious intraspecific interactions of snakes is the male combat dance, which usually takes the form of a ritualistic pushing match between males, although entwining may occur in some cases. Bogert and Roth (1966) list twenty six species in four families (Colubridae, Elapidae, Viperidae, and Crotalidae) for which male combat has been reported. They also give a detailed description of the combat of male gopher snakes *(Pituophis melanoleucus).* Several interpretations have been given to this behavior by various authors, including the idea that these combats represent attempts at homosexual matings. These interpretations have been reviewed by Brattstrom (1974), who concludes that more information is needed before the problem can be solved. Lowe and Norris (1950) review the known cases in which aggressive behavior in snakes is known to be associated with defense of a particular area, most notably in the cobras and their allies (Elapidae). They conclude that one of the functions of aggressive behavior in snakes may be the maintenance of territories. However, they too caution that insufficient information is available to determine the function or functions of aggressive behavior in snakes. Clearly, the relationships of aggressive behavior and movement and spacing patterns represent an outstanding problem in snake biology.

TURTLES: TESTUDINES

Virtually all turtles possess well-developed eyes and visual acuity. Hearing in turtles has been found to be fairly good in some species (Wever and Vernon, 1956). Many species of chelids possess large inner-ear structures, which may indicate the importance of hearing. As with snakes, turtles appear sensitive to low-frequency ground-transmitted vibrations. Olfaction also seems important in turtles. Eglis (1962) has described the motor patterns associated with sniffing behavior in several species of tortoise. This behavior may be quite stereotyped and is associated with the habit of sniffing at many objects in the environment. Some species, such as mud turtles *(Kinosternon)* and their relations and many side-necked turtles (Chelidae and Pelomedusidae), possess barbels—papillae on the chin or throat—which may serve a chemoreceptive function.

Most turtles' shells are basically cryptic in appearance. Even the brightly colored black and yellow shells of species like the star tortoise of India *(Geochelone elegans)* are cryptic in the habitats in which they live. On the other hand, the head and limbs of many species are quite distinctively and obviously colored. Thus turtles have the opportunity of controlling their appearance by withdrawing into or coming out of their shells. In addition, many possess stereotyped movement patterns consisting of head-nods and particular movements of the forelimbs, which are used in communicatory situations such as courtship and agonistic behavior. The iris of the eyes of male box turtles *(Terrapene carolina)* develops a bright red color during the breeding season. Evans (1952, 1953) found that this coloration is controlled by the hormone testosterone and functions as a releaser in courtship. Many turtles can produce vocalizations of some sort, and these cases have been reviewed by Gans and Maderson (1973). Social vocalizations have been found in groups of aggregated *Geochelone travancorica,* a tortoise of India (Campbell and Evans, 1972), in situations other than courtship. But the biological function of the sounds remains unknown. Several turtles produce distinctive odors and have special glands to do so. The best-known examples are the musk turtles or stinkpots (*Sternotherus:* Carr, 1952). But whether these odors are used other than in defense is unknown.

Most turtles are diurnal, and many water turtles commonly aggregate to bask. Annual cycles of movement are poorly known except in the sea turtles, such as the green turtle *(Chelonia mydas)* and the ridley *(Lepidochelys kempi)*, which migrate to particular beaches and form great aggregations for breeding (Carr, 1962, 1963). Many species of water turtles show seasonal movements overland (Gibbons, 1970), but the reasons for these movements are not well known.

Courtship behavior in turtles has been described for several species (Ernst and Barbour, 1972). More detailed studies of several species of tortoise *(Gopherus* and *Geochelone)* have been provided by Auffenberg (1964, 1965, 1966) and of members of the freshwater genus *Pseudemys* by Davis and Jackson (1970, 1973) and Jackson and Davis (1972). In most cases the male display consists of patterns of head movement ("nods") and attempts to bite or bump the female. Male water turtles also use the feet and claws to stimulate the female. Auffenberg (1966) suggests that glands on the chin of male *Gopherus* may produce a pheromone involved in courtship. Auffenberg (1965) found that individuals of two sympatric South American tortoises *(Geochelone carbonaria* and *G. denticulata)* could discriminate sex and species, in part, on the basis of the courtship display. Male tortoises, such as the Galapagos tortoise *(Geochelone elephantopus)* sometimes vocalize ("roar") while mating (Campbell and Evans, 1967; Gans and Maderson, 1973). Otherwise, few studies of species recognition for mating have been carried out on turtles. Some male sea turtles are notoriously nondiscriminatory during the breeding season. Male green turtles *(Chelonia mydas)* will often attempt to mate with almost any object of the appropriate size, including wooden decoys placed by fishermen (Carr, 1952).

Social organization and spacing mechanisms are not obviously well developed in turtles or tortoises. It is sometimes claimed that turtles possess no social spacing mechanisms, a characteristic Legler (1960) attributes to ornate box turtles *(Terrapene ornata)* on the basis of field observations. Yet in a laboratory situation under conditions of artificial crowding Harless and Lambiotte (1971) found evidence for a social hierarchy in this species. Hierarchical behavior has also been reported for captive eastern box turtles (*Terrapene carolina:* Boice, 1970), and Galapagos tortoises (*Geochelone elephantopus:* Evans and Quaranta, 1951). Combat between males in the field has been reported for the Texas tortoise

(*Gopherus berlandieri:* Weaver, 1970) and desert tortoise (*Gopherus agassizi:* Patterson, 1971). In addition, Patterson (1971) found evidence for genuine territoriality in the desert tortoise, which marks out territories with the use of urine and feces.

CROCODILES AND ALLIGATORS: CROCODILIA

The biology of crocodilians has been reviewed by Neill (1971) and Guggisberg (1972), but only little progress has been made in recent years in understanding communication among these animals. These rather generalized reptiles possess the usual range of vertebrate sensory modalities. They do not seem to have any ability to change their appearance. Males of many species, including the American alligator *(Alligator mississipiensis),* produce striking vocalizations ("bellows" and "roars"), which appear to be used in part as spacing mechanisms (Beach, 1944; Campbell, 1973). Lee (1968) advances the hypothesis that the noises made by unhatched alligators serve a communicatory purpose. In some preliminary experiments he found that individual eggs within an artificially composed clutch would hatch out synchronously, although the eggs came from different clutches laid at different times. He suggests that the noises and possibly the movements made by the unhatched alligators may be the mechanisms by which synchronization is achieved, and that synchronous hatching helps to avoid predators. This line of investigation bears following up. Crocodilians of both sexes also produce strong distinctive odors by means of special musk glands.

The existence of maternal behavior in alligators has been the subject of some debate (Neill, 1971). Kushlan (1973) reports a case of a female American alligator's retrieving young and suggests that maternal behavior definitely exists in this species. Female alligators are also reported to guard their nests and to uncover the eggs at the time of hatching. Although there is good evidence that some kind of maternal behavior does occur in this species, the mechanisms by which the mothers recognize the young and the role of communication between mother and young remain unknown.

Discussion

Two points seem to stand out. First, I would agree with and generalize from Brattstrom's (1974:45) remarks that "there is more to snake social behavior than has been assumed." It is probably the case that except for certain well-known lizards there is more to the social behavior and hence communication of amphibians and reptiles than has been assumed. Even in iguanid lizards, whose social and communication systems are probably better known than those of any other amphibians and reptiles, there are still questions about the existence of social organization in some genera. For example, Lynn (1965) reported that territoriality did not exist in horned lizards *(Phrynosoma)* and that displays were "weak" by iguanid standards. On the other hand, Whitford and Whitford (1973) report on actual combat in horned lizards. If combat is at all frequent, it is difficult to imagine that some sort of organization does not exist among populations of these lizards. In general, most amphibians and reptiles have been so poorly studied that the extent of social and communication behavior is in most species only dimly appreciated. Leyhausen (1965) has emphasized that many solitary species of mammals whose members encounter each other only infrequently have, nonetheless, rather complex social organizations. Thus communication between members of these species, no matter how fleeting or subtle, may have significant social and hence ecological consequences. I would expect that similar considerations may apply to many amphibians and reptiles.

Second, I would agree with and generalize from the remarks of Gans and Maderson (1973:1201), who conclude that the "sound-producing mechanisms [of reptiles] here described represent a random assemblage, with no central evolutionary tendency." In general, the communication mechanisms and responses of amphibians and reptiles also appear as a random assemblage. Stated more precisely, there is no clear phylogenetic model of communication behavior in these groups, partly because they do not constitute a natural phylogenetic assemblage (herpetology as a discipline thus sometimes seems to be a bit ill conceived). In addition, both groups have undergone extensive ecological and adaptive radiation.

The attempt to analyze communication behavior in amphibians and reptiles from an ecological point of view has not made great progress, because of the reasons discussed above. However, such an attempt does emphasize a class of questions and problems to which future research may profitably be devoted. I suggest that communication behavior should be studied in connection with the entire life history and life-history strategy of the species in question. Below are a number of areas of interest in the communication biology of amphibians and reptiles that are best studied in this fashion.

LONG-TERM AND STATISTICAL RESPONSES

A central problem in the communication behavior of amphibians and reptiles is that in many cases no immediate response is evidenced on the part of the supposed recipient, or there may be no obvious recipient at all. Frogs, geckos, and alligators all call, lizards give signature bobs, and so forth, without necessarily provoking any immediate response in a particular conspecific. However, there are examples of long-term responses to short-term communicatory acts: female *Anolis carolinensis* are induced to undergo ovarian recrudescence, in part, by viewing communicatory behavior by the male. I would expect that a similar phenomenon may occur as a result of calling by some species of frogs; that is, frog calls may affect hormonal and reproductive cycles in some species. If this hypothesis is true, it would help to explain the existence of some frog calls that are not given at the exact time or site of breeding (Bogert, 1960). Further, it would provide a mechanism for the synchronization of breeding cycles. Cunningham and Mullally (1956) have hypothesized that male Pacific treefrogs *(Hyla regilla)* are synchronized and that calls are, in part, responsible. These ideas need to be tested experimentally, and hypotheses of long-term response to communication need more attention generally.

Also, one can easily imagine that many territorial signals are given on the statistical expectation that a conspecific may be nearby and will perceive it. It may well be the case that much of the communication behavior of amphibians and reptiles may be of this statistical nature and not necessarily adapted only for direct, one-to-one encounters with individuals of their own species. If so, then the communication behavior can be understood only in the context of the statistical structure of the environment, in particular, in the context of the statistical expectation of the presence of conspecifics and their expected correlation with the resources in the environment.

DISPLAY CATALOGS

Unless communication behavior of amphibians and reptiles is studied in its ecological context, any attempt to compare displays between species or to compile a complete catalog of displays may lead to error. For example, there is some confusion in the literature in listing different types of calls emitted by frogs. Whether a certain species of frog possesses a call type previously described for another species or whether a given call is a mating call, a variation on the mating call, a territorial call, or a rain call can only be determined by studying the use of the call in the life history of the frog. Further, as Atz (1970) has emphasized, the homologies of behavior are difficult to determine, and comparisons of displays and their ecological contexts may help us understand the extent of this problem. Finally, only if the catalogs of the displays of any one species are compiled in reference to a complete knowledge of the life history of the species can the catalog be complete. It is only when complete catalogs are available that such information can be used to attack problems such as those posed by Moynihan's (1970) analysis of the evolution of display repertoires.

EFFECTS OF ENVIRONMENTAL VARIATION

Such a study of communication in an ecological context could also reveal the effects of environmental variation. Temperature variation is undoubtedly the best example of such effects. It is well known that temperature affects both sensory and species-perceptibility mechanisms. Even the frequency of the rattlesnake's rattle is temperature-dependent (Martin and Bagby, 1972). To understand these effects, both the temperature dependence of the physiological systems and the distribution of temperatures normally encountered in the environment must be determined. In addition, there is evidence that temperature variation affects more general behavior, such as learning ability (Krekorian, Vance, and Richardson, 1968). It is possible that other physical factors may affect communication systems as well.

INDIVIDUAL AND SPECIES RECOGNITION

The study of the mechanisms of species recognition has been a goal of many analyses of communication behavior in amphibians and reptiles. This has been true of studies of both frog calls and lizard display action patterns. However, it seems that species recognition is a complex of phenomena rather than single phenomenon for each species. The functions of species recognition are to find mates, select mates, synchronize breeding cycles, estimate patterns of environmental variability, and structure daily movement patterns (Kiester and Slatkin, 1974), select habitats (Kiester, ms.), and allow for maternal behavior.

Connected with the multiplicity of functions of species recognition is the fact, emphasized to me by Stanley Rand, that species recognition does not work with the same precision at all times. For instance, an individual may sometimes react to a member of another species as if it were a conspecific. Thus we may expect that the precision may vary from species to species and depend on both the ecological context and the strategic use to which the information gained by recognition is put. The ecological context includes the other species in the environment that may be confused with conspecifics or whose presence may sometimes impart the same information as a conspecific. The strategic considerations may include the degree to which conspecifics influence such activities as daily movement patterns and habitat selection.

Although no unequivocal evidence exists to show that individual recognition does occur in amphibians or reptiles, it is a possibility. Evans (1951) reported that subordinant individuals in a hierarchy of Mexican black iguanas *(Ctenosaura pectinata)* would react in a characteristic fashion to the approach of the dominant "tyrant" male. It is possible that they were responding simply to his size rather than to him *per se.* However, if individual recognition does occur, many of the complexities described for species recognition will have to be investigated.

References

Adler, K., 1970. The role of extraoptic photoreceptors in amphibian rhythms and orientation: a review. *J. Herpetology,* 4:99–112.

Arnold, S. J., 1972. *The Evolution of Courtship Behavior in Salamanders.* Ph.D. diss., University of Michigan. 570pp.

Atz, J. W., 1970. The application of the idea of homology to behavior. In: *Development and Evolution of Behavior,* L. R. Aronson et al., eds. New York: W. H. Freeman & Co., pp.53–74.

Auffenberg, W., 1964. Notes on the courtship of the land tortoise *Geochelone travancorica* (Boulenger). *T. Bombay Natur. Hist. Soc.,* 61:247–53.

Auffenberg, W., 1965. Sex and species discrimination in two sympatric South American tortoises. *Copeia,* 1965:335–42.

Auffenberg, W., 1966. On the courtship of *Gopherus polyphemus. Herpetologica,* 22:113–17.

Beach, F. A., 1944. Responses of captive alligators to auditory stimulation. *Amer. Natur.,* 78:481–505.

Bellairs, A., 1970. *The Life of Reptiles,* 2 vols. New York: Universe Books. xii+570pp.

Benes, E. S., 1969. Behavioral evidence for color discrimination by the whiptail lizard, *Cnemidophorus tigris. Copeia,* 1969:707–22.

Berry, K. H., 1974. The ecology and social behavior of the chuckwalla. *Univ. Calif. Publ. Zool.,* 101. vi+60pp.

Blair, A. P., 1941. Isolating mechanisms in tree frogs. *Proc. Nat. Acad. Sci.,* 27:14–17.

Blair, A. P., 1942. Isolating mechanisms in a complex of four species of toads. *Biol. Symposia,* 6:235–49.

Blair, W. F., 1956. Call difference as an isolation mechanism in southwestern toads (Genus *Bufo*). *Texas J. Sci.,* 8:87–106.

Blair, W. F., 1958. Mating call in the speciation of anuran amphibians. *Amer. Nat.,* 92:27–51.

Blair, W. F., 1960. *The Rusty Lizard, A Population Study.* Austin: University of Texas Press. 185pp.

Blair, W. F., 1964. Isolating mechanisms and interspecies interactions in anuran amphibians. *Quart. Rev. Biol.,* 39:334–44.

Blair, W. F., 1968. Amphibians and reptiles. In: *Animal Communication,* T. A. Sebeok, ed. Bloomington: Indiana University Press, pp.289–310.

Blair, W. F., 1974. Character displacement in frogs. *Amer. Zool.,* 14:1119–25.

Blum, M. S.; Byrd, J. B.; Travis, J. R.; Watkins, J. F. II; and Gehlbach, F. R.; 1971. Chemistry of the cloacal sac secretion of the blind snake *Leptotyphlops dulcis. Comp. Biochem. Physiol.,* 38B:103–107.

Bogert, C. M., 1941. Sensory cues used by rattlesnakes in their recognition of ophidian enemies. *Ann. N. Y. Acad. Sci.,* 41:329–44.

Bogert, C. M., 1960. The influence of sound on the behavior of amphibians and reptiles. In: *Animal Sounds and Communication,* W. E. Lanyon and W. N. Tavolga, eds. *Amer. Inst. Biol. Sci. Publ.* No. 7 (Washington, D. C.), pp.137–320.

Bogert, C. M., and Roth, V. D., 1966. Ritualistic combat of male gopher snakes, *Pituophis melanoleucus affinis* (Reptilia, Colubridae). *Amer. Mus. Novit.,* 2245:1–27.

Boice, R., 1970. Competitive feeding behaviors in captive *Terrapene c. carolina. Animal Behaviour,* 18:703–10.

Boice, R., and Williams, R. C., 1971. Competitive feeding behavior of *Rana pipiens* and *Rana clamitans. Animal Behaviour,* 19:544–47.

Boice, R., and Witter, D. W., 1969. Hierarchical feeding behavior in the leopard frog (*Rana pipiens*). *Animal Behaviour,* 17:474–79.

Brattstrom, B. H., 1962. Call order and social behavior in the foam-building frog, *Engystomops pustulosus. Amer. Zool.,* 2:394.

Brattstrom, B. H., 1974. The evolution of reptilian social behavior. *Amer. Zool.,* 14:35–49.

Brown, C. W., 1968. Additional observations on the function of the nasolabial grooves of Plethodontid salamanders. *Copeia,* 1968:728–31.

Brown, L. E., and Littlejohn, M. J., 1972. Male release call in the *Bufo americanus* group. In: *Evolution in the Genus Bufo,* W. F. Blair, ed. Austin: University of Texas Press, pp.310–23.

Bunnell, P., 1973. Vocalizations in the territorial behavior of the frog *Dendrobates pumilio. Copeia,* 1973:277–84.

Burrage, B. R., 1973. Comparative ecology and behaviour of *Chamaeleo pumilus pumilus* (Gmelin) and *C. namaquensis* A. Smith (Sauria: Chamaeleonidae). *Ann. So. African Mus.,* 61:1–158.

Campbell, H. W., 1969. The effects of temperature on the auditory sensitivity of lizards. *Physiol. Zool.,* 42:183–210.

Campbell, H. W., 1973. Observations on the acoustic behavior of crocodilians. *Zoologica* (New York), 58:1–11.

Campbell, H. W., and Evans, W. E., 1967. Sound production in two species of tortoises. *Herpetologica,* 23:204–209.

Campbell, H. W., and Evans, W. E., 1972. Observations on the vocal behavior of chelonians. *Herpetologica,* 28:277–80.

Capranica, R. R., 1965. *The Evoked Vocal Response of the Bullfrog.* Res. Monograph 33. Cambridge: M.I.T. Press. 110pp.

Capranica, R. R.; Frishkopf, L.; and Nevo, E.; 1973. Encoding of geographic dialects in the auditory system of the cricket frog. *Science,* 182:1272–75.

Carpenter, C. C., 1962. A comparison of patterns of display of *Ursosaurus, Uta* and *Streptosaurus. Herpetologica,* 18:145–52.

Carpenter, C. C., 1966. Comparative behavior of Galapagos lava lizards (*Tropidurus*). In: *Proc. Galapagos Int. Sci. Proj.,* R. I. Bowman, ed. Berkeley: University of California Press, pp.269–73.

Carpenter, C. C., 1967. Aggression and social structure in iguanid lizards. In: *Lizard Ecology: A Symposium,* W. H. Milstead, ed. Columbia: University of Missouri Press. pp.87–105.

Carpenter, C. C.; Badham, J. A.; and Kimble, B.; 1970. Behavior patterns of three species of *Amphibolurus* (*Agamidae*). *Copeia,* 1970:497–505.

Carr, A. F., 1952. *Handbook of Turtles.* Ithaca, N.Y.: Comstock. 542pp.

Carr, A. F., 1962. Orientation problems in the high seas travel and terrestrial movements of marine turtles. *Amer. Sci.,* 50:359–74.

Carr, A. F., 1963. Panspecific reproductive convergence in *Lepidochelys kempi. Ergebnisse der Biol.,* 26:298–303.

Cedrini, L., and Fasolo, A., 1971. Olfactory attractants in sex recognition of the crested newt. An electrophysiological approach. *Monitore Zool. Ital.* (n.s.), 5:223–29.

Cherry, C., 1957. *On Human Communication.* New York: Wiley. xiv+333pp.

Clarke, R. F., 1963. *An Ethological Study of the Iguanid Lizard Genera Callisaurus, Cophosaurus, and Holbrookia.* Ph.D. diss., University of Oklahoma. Ann Arbor, Mich.: University Microfilms 64–2610.

Cochran, D. M., 1961. *Living Amphibians of the World.* London: Hamish Hamilton. 199pp.

Colnaghi, G., 1971. Partitioning of a restricted food source in a territorial Iguanid (*Anolis carolinensis*). *Psychon. Sci.,* 23:59–60.

Cooper, W. E., Jr., 1971. Display behavior of hatchling *Anolis carolinensis. Herpetologica,* 27:498–500.

Cooper, W. E., Jr., and Ferguson, G. W., 1971. Steroids and color change during gravidity in the lizard, *Crotaphytus collaris. Gen. Comp. Endocrinol.,* 18:69–72.

Cooper, W. E., Jr., and Ferguson, G. W., 1972. Relative effectiveness of progesterone and testosterone as inductors of orange spotting in female collared lizards. *Herpetologica,* 28:64–65.

Crews, D., 1973. Coition-induced inhibition of sexual receptivity in female lizards (*Anolis carolinensis*). *Physiology and Behavior,* 11:463–68.

Crews, D., 1974. Effects of group stability, male-male aggression, and male courtship behaviour on experimentally-induced ovarian recrudescence in the lizard *Anolis carolinensis. J. Zool. Lond.,* 172:419–41.

Crews, D., 1975. Effects of different components of male courtship behaviour on environmentally-induced ovarian recrudescence and mating preferences in the lizard, *Anolis carolinensis. Animal Behaviour,* 23:349–56.

Crews, D.; Rosenblatt, J. S.; and Lehrman, D. S.; 1974. Effects of unseasonal environmental regime, group presence, group composition and males' physiological state on ovarian recrudescence in the lizard, *Anolis carolinensis. Endocrinology,* 94:541–47.

Cunningham, J. D., and Mullally, D. P., 1956. Thermal factors in the ecology of the Pacific treefrog. *Herpetologica,* 12:68–79.

Daly, J. W., and Myers, C. W., 1967. Toxicity of Panamanian poison frogs (*Dendrobates*): Some biological and chemical aspects. *Science,* 156:970–73.

Davis, J. D., and Jackson, C. G., Jr., 1970. Copulatory behavior in the red-eared turtle, *Pseudemys scripta elegans. Herpetologica,* 26:238–40.

Davis, J. D., and Jackson, C. G., Jr., 1973. Notes on the courtship of a captive male *Chrysemys scripta taylori. Herpetologica,* 29:62–64.

Duellman, W. E., 1967. Social organization in the mating calls of some neotropical anurans. *Amer. Midl. Nat.,* 77:156–63.

Dym, H., and McKean, H. P., 1972. *Fourier Series and Integrals.* New York: Academic Press. x+295pp.

Eglis, A., 1962. Tortoise behavior: a taxonomic adjunct. *Herpetologica,* 18:1–8.

Emlen, S. T., 1968. Territoriality in the bullfrog, *Rana catesbeiana. Copeia,* 1968:240–43.

Ernst, C. H., and Barbour, R. W., 1972. *Turtles of the United States.* Lexington: University Press of Kentucky. x+347pp.

Evans, L. T., 1936. The development of the cochlea in the gecko, with special reference to the cochlealagena ratio and its bearing on vocality and social behavior. *Anat. Rec.,* 64:187–201.

Evans, L. T., 1951. Field study of the social behavior of the black lizard, *Ctenosaura pectinata. Amer. Mus. Novit.,* No. 1493:1–26.

Evans, L. T., 1952. Endocrine relationships in turtles. III. Some effects of male hormone in turtles. *Herpetologica,* 8:11–14.

Evans, L. T., 1953. The courtship pattern of the box turtle, *Terrapene c. carolina. Herpetologica,* 9:189–92.

Evans, L. T., 1961. Structure as related to behavior in the organization of populations in reptiles. In: *Vertebrate Speciation,* W. F. Blair, ed. Austin: University of Texas Press, pp.148–78.

Evans, L. T., and Abramson, H. A., 1958. Lysergic acid diethylamide (LSD-25): XXV. Effect on social order of newts, *Triturus v. viridescens* (Raf.). *J. Psych.,* 45:153–69.

Evans, L. T., and Quaranta, J. V., 1951. A study of the social behavior of a captive herd of giant tortoises. *Zoologica,* 36:171–81.

Ferguson, G. W., 1966. Releasers of courtship and territorial behavior in the side-blotched lizard *Uta stansburiana. Animal Behavior,* 14:89–92.

Ferguson, G. W., 1969. Interracial discrimination in male side-blotched lizards, *Uta stansburiana. Copeia,* 1969:188–89.

Ferguson, G. W., 1970. Mating behavior of the side-blotched lizards of the genus *Uta* (Sauria: Iguanidae). *Animal Behavior,* 18:65–72.

Ferguson, G. W., 1971. Variation and evolution of pushup displays of the side-blotched lizard *Uta* (Iguanidae). *Syst. Zool.,* 20:79–101.

Ferguson, G. W., 1973. Character displacement of the push-up displays of two partially sympatric species of spiny lizards, *Sceloporus* (Sauria: Iguanidae). *Herpetologica* 29:281–84.

Ferhat-Akat, S., 1939. Untersuchungen über den Gehörsinn der Amphibien. *Z. Vergleich. Physiol.,* 26:253–81.

Fitch, H. S., 1970. Reproductive cycles in lizards and snakes. *Univ. Kansas Mus. Nat. Hist. Misc. Publ.,* 52:1–247.

Frankenberg, E., 1974. Vocalization of males of three geographical forms of *Ptyodactylus* from Israel (Reptilia: Sauria: Gekkonidae). *J. Herpetology,* 8:59–70.

Gans, C., 1973. Sound production in the Salientia: mechanism and evolution of the emitter. *Amer. Zool.,* 13:1179–94.

Gans, C., and Maderson, P. F. A., 1973. Sound producing mechanisms in recent reptiles: review and comment. *Amer. Zool.,* 13:1195–1203.

Gehlbach, F. R.; Watkins, J. F. II; and Kroll, J. C.; 1971. Pheromone trail-following studies of Typhlopid, Leptotyphlopid, and Colubrid snakes. *Behaviour,* 40:282–94.

Gerhardt, H. C., 1974. The vocalizations of some hybrid tree-frogs: acoustic and behavioral analyses. *Behaviour,* 49:130–51.

Gibbons, J. W., 1970. Terrestrial activity and the population dynamics of aquatic turtles. *Amer. Midl. Nat.,* 83:404–14.

Gorman, G. C., 1968. Relationships of *Anolis* of the roquet species group (Sauria: Iguanidae). III. Comparative study of display behavior. *Breviora,* No. 284, 1–31.

Gorman, G. C., 1969. Intermediate territorial display of a hybrid *Anolis* lizard (Sauria: Iguanidae). *Zeits. f. Tierpsychologie,* 26:390–93.

Gorman, G. C., and Licht, P., 1974. Seasonality in ovarian cycles among tropical *Anolis* lizards. *Ecology,* 55:360–69.

Grant, D.; Anderson, O.; and Twitty, V.; 1968. Homing orientation by olfaction in newts (*Taricha rivularis*). *Science,* 160:1354–56.

Grant, W. C., Jr., 1955. Territorialism in two species of salamanders. *Science,* 121:137.

Greer, A. E., 1970. The relationships of the skinks referred to the genus *Dasia. Breviora,* No. 348:1–30.

Greer, A. E., and Parker, F., 1968. A new species of *Tribolonotus* (Lacertilia: Scincidae) from Bougainville and Buka, Solomon Islands, with comments on the biology of the genus. *Breviora,* No. 291:1–23.

Guggisberg, C. A. W., 1972. *Crocodiles.* Harrisburg, Pa.: Stackpole Books. x+195pp.

Haacke, W., 1969. The call of the barking geckos (Gekkonidae: Reptilia). *Scient. Pap. Namib Desert Res. Stn.,* 46:83–93.

Halliday, T. R., 1974. Sexual behavior of the smooth newt, *Triturus vulgaris* (Urodela, Salamandridae). *J. Herpetology,* 8:277–92.

Harless, M. D., and Lambiotte, C. W., 1971. Behavior of captive ornate box turtles. *J. Biol. Psychol.,* 13:17–23.

Heusser, H., 1968. Die Lebensweise der Erdkote *Bufo bufo* (L.); Laichzeit: Umstimmung, Ovulation, Verhalten. *Vierteljahres. Naturforsch. Ges. Zurich,* 113:257–89.

Hillis, R. E., and Bellis, E. D., 1971. Some aspects of the ecology of the hellbender, *Cryptobranchus al-*

leganiensis alleganiensis, in a Pennsylvania stream. *J. Herpetology,* 5:121–26.
Hunsaker, D., 1962. Ethological isolating mechanisms in the *Sceloporus torquatus* group of lizards. *Evolution,* 16:62–74.
Hunsaker, D., and Burrage, B., 1969. The significance of interspecific social dominance in Iguanid lizards. *Amer. Midland Nat.,* 81:500–11.
Jackson, C. G., Jr., and Davis, J. D., 1972. A quantitative study of the courtship display of the red-eared turtle *Chrysemys scripta elegans* (Wied). *Herpetologica,* 28:58–64.
Jenssen, T. A., 1970. Female response to filmed displays of *Anolis nebulosus. Animal Behavior,* 18:640–47.
Jenssen, T. A., 1971. Display analysis of *Anolis nebulosus. Copeia,* 1971:197–209.
Jenssen, T. A., and Swenson, B., 1974. An ecological correlate of critical flicker-fusion frequencies of some *Anolis* lizards. *Vision Res.,* 14:965–70.
Kiester, A. R. (ms.). Conspecifics as cues: a mechanism for habitat selection in the Panamanian grass anole (*Anolis auratus*).
Kiester, A. R., and Slatkin, M., 1974. A strategy of movement and resource utilization. *Theoret. Pop. Biol.,* 6:1–20.
Krekorian, C. O'N.; Vance, V. J.; and Richardson, A. M.; 1968. Temperature-dependent maze learning in the desert iguana, *Dipsosaurus dorsalis. Animal Behavior,* 16:429–36.
Kroll, J. C., and Dixon, J. R., 1972. A new sense organ in the Gekkonid genus *Phyllodactylus* (Gerrhopugus group). *Herpetologica,* 28:113–21.
Kushlan, J. A., 1973. Observations on maternal behavior in the American alligator, *Alligator mississipiensis. Herpetologica,* 29:256–57.
Landreth, H. F., and Ferguson, D. E., 1967. Newts: sun-compass orientation. *Science,* 158:1459–61.
Lee, D. S., 1968. Possible communication between eggs of the American alligator. *Herpetologica,* 24:88.
Legler, J. M., 1960. Natural history of the ornate box turtle, *Terrapene ornata ornata* Agassiz. *Univ. Kansas Publ. Mus. Nat. Hist.,* 11:527–669.
Leyhausen, P., 1965. The communal organization of solitary mammals. *Symp. Zool. Soc. London,* 14:249–63.
Licht, P., and Gorman, G. C., 1970. Reproductive and fat cycles in Caribbean *Anolis* lizards. *Univ. Calif. Publ. Zool.,* 95:1–52.
Lombard, R. E., and Straughn, I. R., 1974. Functional aspects of anuran middle ear structures. *J. Exp. Biol.,* 61:71–93.
Lowe, C. H., and Norris, K. S., 1950. Aggressive behavior in male sidewinders, *Crotalus cerastes,* with a discussion of aggressive behavior and territoriality in snakes. *Nat. Hist. Misc.* (Chicago Acad. Sci.), No. 66:1–13.
Lutz, B., 1947. Trends toward aquatic and direct development in frogs. *Copeia,* 1947:242–52.
Lynn, R. T., 1965. A comparative study of display behavior in *Phyrnosoma* (Iguanidae). *Southwest. Nat.,* 10:25–30.
McKinney, C. O., 1971a. Individual and intrapopulational variation in the push-up display of *Uta stansburiana. Copeia,* 1971:159–60.
McKinney, C. O., 1971b. An analysis of zones of intergradation in the side-blotched lizard, *Uta stansburiana* (Sauria: Iguanidae). *Copeia,* 1971:596–613.
Marler, P., 1961. The logical analysis of animal communication. *J. Theor. Biol.,* 1:295–317.
Martin, J. H., and Bagby, R. M., 1972. Temperature-frequency relationship of the rattlesnake rattle. *Copeia,* 1972:482–85.
Martin, W. F., 1972. Evolution of vocalization in the genus *Bufo.* In: *Evolution in the Genus Bufo,* W. F. Blair, ed. Austin: Unviersity of Texas Press, pp.279–309.
Maslin, T. P., 1950. The production of sound in caudate amphibia. *Univ. Colorado Studies, Ser. Biol.,* 1:29–45.
Maturana, H. R.; Lettvin, J. Y.; McCulloch, W. S.; and Pitts, W. H.; 1960. Anatomy and physiology of vision in the frog (*Rana pipiens*). *J. Gen Physiol.,* 43(6 part 2):129–75.
Mertens, R., 1946. Die Warn- und Droh-reaktionen der Reptilien. *Abh. Senckenberg. Naturf. Ges.,* 471:1–108.
Mertens, R., 1960. *The World of Amphibians and Reptiles.* London: George G. Harrap & Co. 207pp.
Moynihan, M., 1970. Control, suppression, decay, disappearance, and replacement of displays. *J. Theor. Biol.,* 29:85–112.
Neill, W. T., 1971. *The Last of the Ruling Reptiles.* New York: Columbia University Press. xvii+486pp.
Noble, G. K., 1931. *Biology of the Amphibia.* New York: McGraw-Hill. xiii+577pp.
Pace, A. E., 1974. Systematic and biological studies of the leopard frogs (*Rana pipiens* Complex) of the United States. *Misc. Publ. Mus. Zool.* (Univ. Mich.), No. 148:1–140.
Parcher, S., 1974. Observations of the natural histories of six Malagasy Chamaeleontidae. *Z. Tierpsychol.* 34:500–23.

Patterson, R., 1971. Aggregation and dispersal behavior in captive *Gopherus agassizi. J. Herpetology,* 5:214–16.

Pyburn, W. F., 1955. Species discrimination in two sympatric lizards, *Sceloporus olivaceous* and *S. poinsetti. Texas J. Sci.,* 7:312–15.

Rabb, G. B., and Rabb, M. S., 1963. On the behavior and breeding biology of the African pipid frog *Hymenochirus boettgeri. Zeits. f. Tierpsychologie,* 20:215–41.

Rand, A. S., and Williams, E. E., 1970. An estimation of redundancy and information content of anole dewlaps. *Amer. Nat.,* 104:99–103.

Salthe, S. N., 1967. Courtship patterns and the phylogeny of the urodeles. *Copeia,* 1967:100–17.

Salthe, S. N., and Mecham, J. S., 1974. Reproductive and courtship patterns. In *Physiology of Amphibia,* Brian Lofts, ed. New York: vol. II, Academic Press, pp.309–521.

Savage, R. M., 1961. *The Ecology and Life History of the Common Frog (Rana temporaria temporaria).* London: Sir Isaac Pitman & Sons, Ltd. vii+221pp.

Schiøtz, A., 1973. Evolution of anuran mating calls. Ecological aspects. In: *Evolutionary Biology of the Anurans,* J. L. Vial, ed. Columbia: University of Missouri Press, pp.311–19.

Schmidt, K. P., and Inger, R. F., 1957. *Living Reptiles of the World.* Garden City, N. Y.: Doubleday & Co. 287pp.

Schmidt, R. S., 1973. Central mechanisms of frog calling. *Amer. Zool.,* 13:1169–77.

Shannon, C. E., 1949. Communication in the presence of noise. *Proc. IRE,* 37:10–21.

Stamps, J. A., 1973. Displays and social organization in female *Anolis aeneus. Copeia,* 1973:264–72.

Stamps, J. A., 1975. Social behavior and spacing patterns in lizards. In: *Biology of the Reptilia,* C. Gans, ed. New York: Academic Press, in press.

Stamps, J. A., and Barlow, G. W., 1973. Variation and stereotypy in the displays of *Anolis aeneus* (Sauria: Iguanidae). *Behaviour,* 47:67–94.

Stebbins, R. C., 1951. *Amphibians of Western North America.* Berkeley: University of California Press. ix+539pp.

Stebbins, R. C., 1954. Natural history of the plethodontid genus *Ensatina. Univ. Calif. Publ. Zool.,* 54:47–124.

Stebbins, R. C., and Eakin, R. M., 1958. The role of the "third eye" in reptilian behavior. *Amer. Mus. Novit.,* No. 1870:1–40.

Steyn, W., and Mitchell, A. J. L., 1965. A new Scincid genus: and a new record from South West Africa. *Cimbebasia,* No. 12:2–12.

Straughn, I. R., 1973. Evolution of anuran mating calls. Bioacoustical aspects. In: *Evolutionary Biology of the Anurans,* J. L. Vial, ed. Columbia: University of Missouri Press, pp.321–27.

Taylor, E. H., 1968. *The Caecilians of the World: A Taxonomic Review.* Lawrence: University of Kansas Press. 848pp.

Taylor, E. H., 1970. The lateral-line sensory system in the caecilian family Ichthyophidae (Amphibia: Gymnophiona). *Univ. Kansas Sci. Bull.,* 48:861–68.

Twitty, V. C., 1959. Migration and speciation in newts. *Science,* 130:1735–43.

Vinegar, M. B., 1972. The function of breeding coloration in the lizard, *Sceloporus virgatus. Copeia,* 1972:660–64.

Wake, M. H., 1968. Evolutionary morphology of the caecilian urogenital system. I. The gonads and the fat bodies. *J. Morphol.,* 126:291–333.

Wassersug, R. J., 1973. Aspects of social behavior in Anuran larvae. In: *Evolutionary Biology of the Anurans,* J. L. Vial, ed. Columbia: University of Missouri Press, pp.273–97.

Watkins, J. F. II; Gehlbach, F. R.; and Kroll, J. C.; 1969. Attractant-repellent secretions of blind snakes (*Leptotyphlops dulcis*) and their army ant prey (*Neivamyrmex nigrescens*). *Ecology,* 50:1098–1102.

Weaver, W. G., 1970. Courtship and combat behavior in *Gopherus berlandieri. Bull. Florida State Mus., Biol. Sci.,* 15:1–43.

Wever, E. G.; Crowley, D. E.; and Peterson, E. A.; 1963. Auditory sensitivity in four species of lizards. *J. Audit. Res.,* 3:151–57.

Wever, E. G., and Vernon, J. A., 1956. Sound transmission in the turtle's ear. *PNAS,* 42:292–99.

Wever, E. G., and Vernon, J. A., 1960. The problem of hearing in snakes. *J. Audit. Res.,* 1:77–83.

Whitford, W. B., and Whitford, W. G., 1973. Combat in the horned lizard, *Phrynosoma cornutum. Herpetologica,* 29:191–92.

Wickler, W., and Seibt, U., 1974. Rufen und Antworten bei *Kassina senegalensis, Bufo regularis* und anderen Anuren. *Z. Tierpsychol.,* 34:534–37.

Wiewandt, T. A., 1969. Vocalization, aggressive behavior, and territoriality in the bullfrog, *Rana catesbeiana. Copeia,* 1969:276–85.

Williams, E. E., 1973. The study of amphibia. *Science,* 182:376–77.

Williams, E. E., and Rand, A. S. (ms.). Species recognition, dewlap function and faunal size.

Wojtusiak, R. J., 1973. Some ethological and biological observations on the tuatara in laboratory conditions. *Tuatara,* 20:97–109.

Wojtusiak, R. J., and Majlert, Z., 1973. Bioacoustics of the voice of the tuatara, *Sphenodon punctatus punctatus. New Zealand J. Sci.,* 16:305–13.

Zweifel, R. G., 1968. Effects of temperature, body size, and hybridization on mating calls of toads, *Bufo a. americanus* and *Bufo woodhousei fowleri. Copeia,* 1968:269–85.

Chapter 22

COMMUNICATION IN BIRDS

W. John Smith

Birds have been very popular behavioral subjects with naturalists long before ethology began. Many birds are bold, busy, engaging creatures, so readily observed that they can scarcely be ignored. There are skulkers, of course, and birds that fly softly by night or sit apart, alone and uncommunicative. But so many species are conspicuous, active, and social that it was inevitable that the study of communication by ethologists should invest heavily in them. This it has done, and the development of most of its central concepts owes much to work with bird behavior.

The ways in which avian studies have contributed to these concepts are the main focus of this chapter. First, it considers the principal tools, behavioral and otherwise, with which birds communicate, and the apparent evolutionary origins of some of these tools. Second, it reviews the ways in which ethologists use birds to study the motivational and interactional mechanisms underlying the process of communicating. Finally, it reviews a particular kind of communicative behavior that has been the focus for intensive study: bird song. A descriptive summary of the now massive literature on bird communication is not attempted, and could not be done with a useful amount of detail in any brief account.

The Tools of Communicating

As birds interact they provide each other with information in many ways. Every feature of each individual's appearance, its every action, could in principle be informative for another individual. This is not to say that every such source of information is equally relevant at any given time, or even that all the incidental aspects of appearance and demeanor provide enough information to keep interactions from becoming chaotic. It merely means that birds bring abundant sources of information to their interactions. Among these sources are some that have evolved to be informative. That is, there are tools specialized for communicating.

Of these tools, the behavioral unit that has been central to ethological studies of communication is the "display"; displays and simultaneous or sequential compounds of displays are specialized acts that are performed by individuals. There are also behavioral units with repeatable, formal patterns that can be performed only through the cooperation of two or more participants; these "formalized interactions" have been studied less thoroughly than have displays. Other communicative specializations are objects produced by behavior, and yet others are not behavioral at all: plumage patterns and colors, for instance.

DISPLAYS

The concept of displays, or at least of behavior adapted to provide recipient individuals with information that is not otherwise readily apparent was used by Darwin (1872), who saw such acts as means of making available information about an individual's internal emotional state. The term "display" began to be used more or less consistently in reference to these units when Huxley's (1914) studies of great crested grebes *(Podiceps cristatus)* appeared. It was not until relatively recently that it was defined as an act specially adapted in "physical form or frequency to subserve social signal functions" (Moynihan, 1956, 1960, in the course of work with gulls and flocking passerines). Although the term connotes visible behavior, Moynihan recognized that the concept to which it applies is much more general, and that the behavior can be suited to reception by any sensory modality.

As a class of animals, birds are known to have a remarkable diversity of display behavior. This enormous diversity can be categorized, when this is useful, in terms of different sensory modalities. Visible displays, for instance, abound, and the conspicuous and bizarre posturings and movements of ducks, geese, grebes, and gulls figured importantly in the work of such pioneer ethologists as Heinroth, Lorenz, Huxley, and Tinbergen. Displaying birds may wave their wings, waggle their tails, stretch their necks, point with or deflect their bills, crouch, cower, leap, flutter, raise or lower crests or other tracts of feathers, manipulate token nest material or food, or do these and other things in complex combinations. The number of distinctive possibilities is limited, but large. The number of possible audible displays is extremely large, encompassing clear musical whistles, trills, harsh rasps, hoots, bell-like vocalizations, and many other products of the respiratory tract; as well as mechanical snaps, winnowings, and drumming sounds produced with wing or tail feathers; and sounds made by rapping the bill on a resonant object. Opportunities for tactile displays are less readily available to birds than to, say, mammals or social insects, and yet a diversity of special touching patterns has evolved, ranging from allopreening and various forms of bill-to-bill touching and even nipping, to bodily pushing and treading on the partner's back during mounting for copulation. Not surprisingly, the main sensory modality for which evolution has been much less productive of displays for birds than for other animals has been olfaction. Birds are not known to have displays involving the release of chemicals as scents.

The countless displays, like any other products of evolutionary processes, did not arise *de novo.* For visible displays, at least, the close observation of birds has provided very important clues about the kinds of acts that probably serve as precursors. Thus Daanje (1950) recognized in many of the stereotyped postural displays of birds positions that were parts of other acts; in particular, of acts with which a bird prepared to take flight, strike, turn away, etc. For instance, before taking flight a bird will first draw back its head and neck, lower its breast toward the ground, raise its tail, flex its legs, and begin to extend its carpals (wrists) outward. It may pause there, ready to spring, or begin to spring by extending its legs, head, and neck, depressing its tail, and spreading its wings. Flight does not always follow such preparations, even the second set, and so Daanje called them "intention movements" and showed how positions very similar to them occurred very commonly in postural displays. That is, the movements performed in preparing to fly are informative and are highly suited for evolutionary elaboration that enhances their effectiveness as signals. Intention movements appear to be a major evolutionary source of visible display patterns, not just in birds but in most other animals as well.

Other evolutionary sources of display behavior that have been postulated for birds also appear to be significant for other kinds of animals. Morris (1956), for instance, saw in the thermoregulatory lifting or depressing of body feathers, which are autonomic responses in tense situations, the sources of displays in which birds lift their crests, ruffle their flanks, or take on a sleeked appearance. Grooming and some other acts that occur during aggressive encounters with surprising regularity are seemingly irrelevant; the internal causes of these "displacement" activities, as they have been termed, remain controversial but the class is recognized as yet another evolutionary source of displays (Tinbergen, 1952, 1959). Bastock, Morris, and Moynihan (1953) saw in the Grass Pulling of disputing gulls a display that probably arose from yet another kind of source, the regular redirection onto substitute targets of striking and tearing acts in circumstances in which a bird was highly motivated to attack and yet afraid or for some other reason inhibited from directing the attack to its rival.

Ethologists have been less successful in proposing the evolutionary origins of vocal and other kinds of displays (except for most tactile acts, which appear to have obvious origins), but have made much use of research on birds in describing the ways in which displays come to diverge from their predecessor acts as evolution proceeds: increased conspicuousness, uncoupling and differential modification of component movements and their motivational thresholds, increasing the amount of use of an act beyond that which is directly functional, and so forth (see Hinde and Tinbergen, 1958; Tinbergen, 1959; Morris, 1966; and W. J. Smith, in press). And, finally, as comparisons of the displays of diverse species of birds have also contributed to our understanding of the kinds of selection pressures that mold the physical forms of displays, it can fairly be said that the study of bird communication has had a very pervasive influence on ethological concepts of display evolution.

While the concept of displays as formal units has been extremely productive in the study of communication, it is not without its problems (these are discussed relatively fully by W. J. Smith, in press). It can be very difficult, for instance, to demonstrate that an act has been specialized to facilitate the sharing of information. In practice, we tend to accept relatively stereotyped acts that appear to have no direct function as being displays if it seems likely that they will be of use in interactions. For vocal displays this criterion is probably sufficient, but for visible displays it is sometimes hard to be sure. For example, many birds flash their wings open while foraging, an act that looks very much like display behavior but that may be specialized to startle prey. Other acts—slightly rigid postures, small amplitude movements of the wings, or slight sleeking or fluffing of feathers, for instance—are so little specialized that it is very difficult to know whether they should be classified as displays. Such problems are not unexpected and for many analytic purposes may not be very important. Greater difficulties are posed by the fact that no display is fully constant; many are quite variable in form and may even intergrade with other displays. This can make it very difficult, at least by criteria of form, to define display units. Comparable definitional problems arise when specialized acts are used only in combinations: are the display units the combinations or the recombinable components? I shall suggest below an amplification of the traditional display concept that permits a practical resolution of such difficulties (see section on Interpretation of Bird Communication).

FORMALIZED INTERACTIONS

There remains the problem that acts specialized to be informative do not all fall within the capacities of single individuals. There are behavioral units that are essentially cooperative, that can only be performed by the combined actions of two or more participants. If we reserve the term "display" for formalized acts of individuals, we can refer to these cooperative performances as "formalized interactions." For instance, when two gulls (Laridae) greet each other on a male's pairing territory, they tend to go through a predictable behavioral sequence (reviewed by Tinbergen, 1959; Moynihan, 1962a). As the female arrives, the male emits Long Calls in the Oblique posture, facing toward her. They then align in parallel with each other, and both utter a vocal display known as the Mew Call. Remaining in parallel, they go through a somewhat variable succession of progressively less agonistic postural displays (displays less likely to be followed by attack or escape). The interaction is often terminated with mutual Head Flagging, in which they stand side by side, stiffly turning their heads away from each other. The exact sequence of displays is subject to omissions and some changes in order, but the greeting participants cooperate and each accommodates by tending to perform the same display as its partner most of the time, shifting displays when the other does. Similar greeting ceremonies, involving fixed mutual orientations and simultaneous performance of displays, are known in many other kinds of bird, although few are as complex as the gulls'.

Formalized interactions are not limited to greetings but may occur in appeasing and reassuring, courting, and other encounters, usually those that are rich in uncertainty. Their main characteristic is the recurrence of a pattern of cooperative behavior. Each participant has a definable, preestablished part to play in an interactional pattern of regular, classifiable moves and responses. Participants may play their parts somewhat differently from event to event, but even as each affects the others it is also constrained by the formalized pattern to accommodate to their acts.

OTHER FORMAL SOURCES

Behavior is a transient source of information, but birds can act to make information more persistently available. They may do so by making objects that will subsequently have direct functions, for example, the nests that males of savannah-dwelling weaverbirds (e.g., *Quelea* spp.: Crook, 1964) and diverse other species build and show to prospective mates as they attempt to form pair bonds; or their constructions may serve a similar communicative function but lack direct utility, as do the bowers and stages of bowerbirds (Ptilonorhynchidae: Gilliard, 1956, 1963) and the cleared lek areas of manakins (Pipridae: Chapman, 1938; Sick, 1959, 1967). Birds are not known to scent-mark sites, a common practice used by mammals and invertebrates for leaving a source of information to act in a communicator's absence. But at least one species, the Japanese quail *(Coturnix coturnix japonica),* does have a special chemical product with which it marks its droppings. It produces a foam that may be a visible rather than an olfactory marker (Schleidt and Shalter, 1972).

The principal specializations of birds that serve as persisting sources of information are neither behavior nor behavioral products, however. They are the colors and patterns of plumage, bare skin, epidermal outgrowths, or other "soft parts" that birds wear as badges or uniforms. The plumage of a male bird may be more distinctive and hence perhaps more important than his displays in identifying him to unmated females (Hinde, 1956, 1959). For instance, N. G. Smith (1966) found that four species of arctic gulls (*Larus* species) that are very similar in plumage and displays are differentiated primarily by the colors of their irides, the fleshy rings around their eyes, and the extent of the contrast these make with their white heads. By altering eye-ring colors Smith demonstrated that both males and females are prepared to respond to these characteristics of their partners in the period before egg laying; the badges are "prezygotic isolating mechanisms" preserving the integrity of each species' gene pool (see Mayr, 1963). That the identifying information provided by some badges may also be important for interactions other than pair bonding has been shown by experiments such as those done by Peek (1972) with red-winged blackbirds *(Agelaius phoeniceus).* He removed the bright red epaulets from seventeen otherwise normal males and found that although the birds displayed and appeared otherwise normal, they could not maintain their territories against other males.

Badges are not important solely because they persist. Indeed, some badges are kept hidden or inconspicuous much of the time, then revealed in conspicuous acts. Various species of tyrannid flycatchers, such as the kingbirds (genus *Tyrannus:* W. J. Smith, 1966) have a brilliant reddish or yellowish concealed crown patch that they rapidly uncover and may even erect when displaying while preparing to attack. Many birds have crests or other elongate feathers that are conspicuous primarily when erected. Gregarious species often have patches of white or bright colors on their wings, back, or tails that become visible as "flash patterns" when the birds take flight (Moynihan, 1960, 1962b). Canada geese *(Branta canadensis),* which have a flash patch at the base of the tail, have a white chin patch that is not concealed but that becomes very noticeable when the geese perform a Head-Tossing display before flying (Raveling, 1969).

Displays, their compounds, formalized interactions, constructions, and badges are all tools evolved by birds to facilitate information sharing. They provide formal means of contributing information to events. Communication, however, depends on further sources of information, sources that are not specialized to this end but are simply inherent in the structure of all events and entities. Some of these sources are public, accessible to all participants. Some are private, part of the genetic and experiential stores of information carried by each participant. Such sources can be thought of as acting contextually to the formal sources, and their role is considered below in discussing the evolution of messages.

The Interpretation of Bird Communication

Ethologists have had a number of goals in interpreting bird communication. A central one has always been to understand the motivational causes that underlie the performance of displays: the physiological states and their stimulus relationships that lead a bird to respond to a particular circumstance by displaying. The objective of such work is to understand the internal mechanisms that control the behavior of an individual bird, a goal that is not central to the study of communication *per se.* Communication, in the sense that the term is used in this book, is an interactional process, a procedure by which information is shared between or among individuals. The goal of defining what kinds of information are made available by what sources (i.e., by displays, formalized interactions, badges, etc., and by incidental nonformalized sources) is more pertinent, as is the study of responses to this information and the kinds of function that are generated.

MOTIVATIONAL CAUSATION

The earliest trends in the interpretation of bird communication were "causal" and functional, and they remain the predominate trends. The focus, not inappropriate for a branch of evolutionary biology arising from comparative physiology and comparative anatomy, has been on the mechanisms and adaptiveness of individual behavior. Tinbergen described this perspective succinctly when he stated that the "central" problem in analyzing the behavior of a communicator in events involving social cooperation is "what urges the actor to signal?" (1953:73).

The mechanisms of internal causation are not directly accessible through the study of behavior and have usually been represented by hypotheses about emotional states or motivation. The various behavioral criteria used to determine motivational causation include primarily the following (abstracted from accounts by Moynihan, 1955, and Tinbergen, 1959): (1) If a display posture or movement resembles other motor patterns it may share causal states with them. Thus if it resembles acts used in striking, it is assumed that it may be aggressively motivated. (2) If a display is performed simultaneously or in quick succession with the behavior of, say, approaching or turning away, it may share motivational causes with these acts. (3) If a display is performed when the displaying individual is somehow balanced by circumstances between two courses of action that are more direct, the display may be caused by a conflict of two motivational states. For example, if a bird that will attack intruders in its territory but will flee if attacked when it is an intruder in theirs does neither, but instead displays when facing a rival at their common territorial border, then those displays are probably caused by the conflicting motivational states of aggression and fear.

The concept of motivational "conflict," seen in the third situation above, has been one of the most useful contributions of this research. Most of the display behavior of birds appears to be performed when the communicators are in conflict states. Observational techniques for revealing the different absolute and relative strengths of conflicting motivations were developed by Moynihan (e.g., 1955, 1962a) in the course of work with gull displays. The existence of such differences in conflict states underlying different displays was subsequently verified experimentally by Blurton Jones (1968), working with threat displays of the great tit *(Parus major).*

Behavioral studies have provided better evidence for the existence of motivational conflict and for intensity differences among the conflicting motivations that cause the performance of different displays than they have for describing the motivational systems themselves. In practice, although not always in theory, there has been a predominating trend to lump the hypothetical causes of bird displays into three "unitary" motivational systems: aggression, fear, and sexual motivation (e.g., Tinbergen, 1959). Yet these gross, functionally oriented categories are each far from unitary. As Hinde has argued, variables such as the persistence, directiveness, and temporal clustering of the behavior patterns each category is said to control do not always change in concert (summarized in Hinde, 1970). Further, there are diverse behavior patterns assigned to each causal category, and at least in the case of "sex" these are remarkably heterogeneous (Moynihan, 1962a): at one extreme is copulation, at the other a tentative associating without contact.

Three motivational systems are in fact inadequate to explain the internal causation of bird displays. How many systems should be recognized is by no means clear, however, as the observation of behavior alone does not permit discrete physiological mechanisms to be identified (Hinde, 1970). Hinde has suggested that for some purposes it is appropriate to proceed by recognizing a class of "intervening variables," which he calls "tendencies." For instance, observations suggest that there is a tendency to drink a certain amount of water (measurable as a dependent variable) after an individual has been deprived for a certain period (measurable as an independent variable). The "tendency" to drink implies an internal mechanism that somehow relates the other two variables, but it does not postulate properties of this mechanism—e.g., it does not require it to be unitary. A tendency postulated as a cause of a display would be described in terms of the behavior correlated with the use of that display, be it attack, escape, or mounting, or some other activity not adequately described by the traditional motivations. For example, tendencies such as "staying," "flying," "approaching," and "being gregarious" have all been described for bird displays (see review by Hinde, 1970, and such studies as Moynihan, 1960; Andrew, 1961a; Stokes, 1962; Delius, 1963; Crook, 1963, 1964; Tinbergen, 1964; Fischer, 1965).

KIND OF INFORMATION MADE AVAILABLE

Correlations such as those used in research on causal tendencies are also useful as indications of the kinds of information that a display can make available about the communicator's behavior. The analytic perspective shifts very markedly, however. The superordinate behavioral unit is no longer an act of an individual bird, but an interaction based on the behavioral contributions of participating individuals. The important mechanisms at this level of integration are found less within individuals than among individuals in their moves, countermoves, and accommodations.

In most of their encounters, at least in natural circumstances, birds do not interact chaotically. The relevance of a display, a badge, or any comparable specialization is in the information it can contribute to these interactions, information that makes it easier for each participant to anticipate the actions of the others, to be more prepared to respond in ways that can further develop or stabilize their interaction.

Analysis of the behavioral information content of a display (its "messages" about behavior) requires that the range of circumstances in which the display naturally occurs be found and thorough observations be made of the behavior with which its use correlates throughout this range. Many displays correlate with different acts in different events, but unless a display varies consistently as these activities change it cannot provide detailed information about them. For example, the vocal display called song in many species of birds is uttered by an individual that is prepared to attack, greet, attempt to form a pair bond or copulate, or join another individual in close association without making contact. The most that can be learned from such a display is that the communicator will probably seek interaction if an opportunity arises, but just what kind of interaction a recipient should expect in any particular event cannot be known from the utterance of song alone—although it may be readily predicted from other sources of information that are available in the event.

Analyses of this sort are clearly at variance with traditional causal analyses, especially if the latter are phrased in terms of a small set of *a priori* motivational variables. Inasmuch as causal interpretations continue to adhere largely to this tradition, it may be useful to compare the results of the two approaches in some detail by attempting a provisional reinterpretation of some gull displays. Thanks primarily to the very careful work of Tinbergen and his students, gull displays are among the most intensively studied of any group of birds, and they have provided the main model for most current research. Further, Tinbergen's (1959) review of this work provides a detailed example of the usefulness and shortcomings of interpretation based on the motivational trio of aggression, fear, and sex.

Of the nine display patterns analyzed in that review, the most convincing interpretations involve the simpler conflicts of motivational states. Consider, for instance, the Aggressive Upright, a posture in which the neck is extended upward, the bill is angled downward, and the carpals are lifted away from the body. A communicator in this posture may approach and then attack an opponent; or it may hesitate, cease approaching, and then withdraw. Tinbergen indicates its motivational state as a conflict of aggression and fear.

An interpretation of the behavioral messages encoded would be closely comparable: one can expect attack and escape activities or indecisive acts while the gull chooses. However, in two closely related postures the carpals are not lifted and the bill angle differs, being horizontal in the Intimidated Upright and upward in the Anxiety Upright. Tinbergen's interpretation that these postures are a result of less aggression and more fear yields only a partial motivational description, which, moreover, does not tell us what limits to expect of the posturing gull's behavior. The existing descriptions imply that gulls adopt these postures when in a state of conflict between fear and motivations for virtually any behavior from foraging (e.g., when intruding on a shoreline territory) to joining and associating (e.g., when forming a pair bond) to attack (see Tinbergen, 1965, and the evidence he presents from his student Manley, 1960).

This large range of alternatives to escape is not adequately characterized by the traditional motivational terms, and what is needed is a concept that implies simply that the displays are performed when some unspecifiable motivation conflicts with fear. The motivation is unspecifiable because it can be very different at different times, and what is consistent is simply the conflict it engenders. A behavioral message interpretation of these two displays would recognize that the communicator will either escape or select some activity from a broad set of alternatives incompatible with escaping. That its motivational state varies widely simply means that the behavioral alternatives to escape are not specified precisely by the display. A lack of precision is by no means uncommon in behavioral messages, and in gulls this very broad message (termed the "general set" of incompatible alternatives by W. J. Smith, 1969a) (see Table 1) appears to recur in the songlike Oblique-cum-Long Call display, there indicating acts that are incompatible not with escape but with attack.

There are several display patterns of gulls for which the traditional trio of motivational causes encounters more obvious difficulties. Prominent among these are Choking and Mew Calling. Tinbergen interpreted the former as caused by a very strong conflict between defensive aggression and some motivation to remain with the nest site or territorial boundary, but he recognized that Choking can be performed in both "friendly" and "hostile" circumstances. Neither the behavioral messages nor the motivational causes of Choking are fully evident from existing descriptions, although the display apparently does correlate with the tendency of the communicator to remain at a special site. In the case of Mew Calling, however, it is evident that aggression, fear, and sexual motivation are not consistent causes of the display. The first two can be part of the motivational state at times; for instance, when a male herring gull turns from a hostile encounter at a territorial boundary and proceeds to walk in parallel with his mate, both of them Mew Calling. The parallel walking suggests a balanced conflict between approach and withdrawal, and, at least for the male, part of the motivation to approach could be aggressive (this is less obvious for the female). Yet neither attack nor escape is seen in many situations in which one individual approaches its mate or its offspring on the nest, e.g., a male bringing nest material to his mate, or an individual of either sex coming to relieve its partner from incubation or to feed the small chicks.

There are interesting behavioral correlations for Mew Calling, however, that involve interacting and also flying, walking, or, rarely, swimming. For instance, in the circumstances just described, Mew Calls usually just precede alighting in the territory and are then used while walking toward the mate or chicks. A parent herring gull returning to a territory when its chicks have left the nest and hidden in the vegetation usually stands and Mew Calls; its chicks then run to it

Table 1

Behavioral selection messages widely encoded in the displays of diverse species of birds.

A behavioral selection message is information that indicates what activity, from the whole behavioral repertoire available to it, a displaying animal is making or is possibly about to make. A much more detailed account of these messages is provided by W. J. Smith (in press).

Locomotion: walking, hopping, flying, swimming, etc., with no specification of particular functional classes such as flying to escape or to join.
General set: activities that are specified only insofar as they are diverse and usually or always incompatible with other behavioral selections that are encoded by a display.
Attack: attempts to inflict injury.
Escape: attempts to withdraw or avoid (can include "freezing" behavior).
Copulation: attempts to mount and copulate.
Indecisive: vacillating or hesitating between other selections.
Interaction: attempts to interact, or to avoid interaction, of any of several kinds including attack, greeting, association, copulation, etc., with no specification of the kind of interaction in any specific event. (Whether interaction is sought or avoided is indicated by the probability assigned to this selection as it is encoded by any particular display.)
Association: remaining in the company of another individual.
Remaining with site: restricting movements to a particular neighborhood.
Seeking: attempts to gain an opportunity to perform some other selection, such as interaction or escape.
Receptive: prepared to accept some other selection such as copulation or giving care (the latter may not itself be a widespread behavioral selection message).
Attentive: paying attention to a stimulus; monitoring.

and are fed. It seems characteristic of all cases in which herring gulls perform this display that the communicator is prepared to interact, either agonistically or not, and that its locomotory behavior is likely to be indecisive—slowing as it shifts from one form to another in alighting, approaching partway to another individual, stopping short, or deviating from an approach to walk in parallel. With hidden chicks the parent stops and Mew Calls when it cannot approach them further because it does not know where they are within the territory. Note that the traditional motives of aggression, fear, and sex are inadequate causal explanations for this display not because they never occur, but because they do not occur consistently and because many other motives may be causal to it in different cases. The information that the display makes available about behavior, however, includes at least that the communicator is prepared to interact and to locomote, although it will most likely do so in some indecisive fashion. Such information is probably useful to recipients of the display, especially since they will usually have a reasonable expectation of the most likely kind(s) of interaction. Such an expectation, however, is based on sources of information other than the display.

Bird displays very commonly encode the information that the communicator will behave indecisively, alternately starting different incompatible activities or pausing, and this undoubtedly relates to the finding that a conflict of motivational causes is often characteristic of the states of displaying individuals. Other displays, such as much of the singing referred to above, appear to carry information less about indecisive acts than about the behavior of seeking the opportunity to be decisive. In many species, singing customarily occurs when no other individuals are present (see review by Andrew, 1961b). Singing individuals of species as phylogenetically diverse as the chaffinch (*Fringilla coelebs:* Marler, 1956), European blackbird (*Turdis merula:* Snow, 1958), green-backed sparrow (*Arremonops conirostris:* Moynihan, 1963), Carolina chickadee (*Parus carolinensis:* S. T. Smith, 1972), and eastern phoebe (*Sayornis phoebe:* W. J. Smith, 1969b, 1970) behave as if seeking other birds: they take high, conspicuous perches, or in some species sing in special display flights, or may actively patrol. Song ceases immediately and some form of interaction is usually attempted if a suitable recipient becomes available; for instance, when a field sparrow *(Spizella pusilla)* hears a recording of its species' song played from a territorial border or beyond, it increases its singing in reply; but if the recording is played within the territory the owner does not sing—it flies directly toward the source of the sound (Goldman, 1973). (Birds do countersing across territorial boundaries that neither will cross to attack the other, and are then inaccessible to each other because of the boundary convention. The countersinging is apparently a less-intense interaction than they are seeking, but the boundary thwarts closer approach.) Still other displays provide information less about what a communicator will attempt to do or seek opportunity for than about the kinds of behavior it is prepared to accept. A female bird who is receptive to copulation, for instance, will solicit a male by adopting a posture that facilitates his mounting.

The information made available by a Mew Call, indicating that a bird will evidence indecision about locomotory behavior, is widely encoded by the displays of diverse other species. Many New World flycatchers (family Tyrannidae) have vocalizations that are employed as they alight, make flight-intention movements, or slow down or veer in flight in circumstances in which indecisive flying and perching may alternate: e.g., approaching an agitated mate or potential mate, including mutual greeting performances; begging by an offspring following parents whose tempers have worn thin; veering off from pursuit or attack of a predator; choosing between taking a station or continuing to patrol a territory, or between flying to a singing perch and foraging, or between remaining in a territory and following a foraging mate or a fleeing intruder out of it; staying with the mate or flying to a border in response to the calls of a neighbor in that region; or leaving the nest in the absence of its mate (e.g., W. J. Smith, 1966, 1969a). The flightless Adelie penguin *(Pygoscelis adeliae)* has vocal displays with remarkably comparable employment. One is used by a penguin as it approaches its territory after a long absence, on leaving its nest after incubating, when responding to fights in nearby territories, during stalemated aggressive encounters, after unsuccessfully threatening an intruding crowd (e.g., a creche of chicks), and in response to slowly approaching humans or predatory skuas (*Stercorarius skua:* Ainley, 1974). The other display is uttered while a penguin is walking between its nest and the beach, standing or walking by the water's edge (particularly as individuals begin to dive from a flock into the water), swimming at sea, and apparently when being chased.

The Carolina chickadee utters a fixed sequence of three distinct vocal displays as it alights in various circumstances in which it appears somewhat indecisive: a High Tee as it approaches the perch, a Chick on alighting, and a Dee immediately afterward. It also has a Lisping Tee, which is used primarily when in flight or when flight is very imminent (S. T. Smith, 1972). Many other flocking species, from red-legged partridges (Goodwin, 1953) to species of *Chlorospingus* tanagers (Moynihan, 1962c), also have displays used as flight becomes likely. In Canada geese, the Head-Tossing display mentioned above in discussing badges is performed by members of a family group as they get ready to fly. If the gander performs the display the whole family flies shortly afterward, but if other members begin it they may have to continue it for tens of minutes before the gander is ready to lead them in flight—if he gets ready at all (Raveling, 1969).

Such displays provide more information than just the likelihood of flight, of course. The Head Tossing of geese, for instance, is performed by individuals who would fly if that did not require them to leave their families, and the Lisping Tee of chickadees is often emitted in flights that sever association with their mates or flocks. Both displays provide information about the readiness of the communicators to associate with their companions, although the relative probabilities of flying away differ for the two species. Displays providing information about the behavior of associating are customarily used only if something makes that association difficult, anything from an incompatible behavioral alternative or the absence of companions to an obscuring environment. Green-backed sparrows have a Plaintive Note that mates utter when they become separated; they usually reestablish association immediately afterward (Moynihan, 1963). Ornithologists have described what they often call "contact calls" in the repertoires of many species that customarily forage actively and socially, the calls apparently conveying information about association behavior and helping the birds to maintain the coherence of their social groups.

While other messages about behavior are made available by the displays of some birds, most species appear to encode primarily the dozen messages listed in Table 1. The table may not contain all the messages that are widespread among the displays of diverse species, but it contains all those that appear at present to be widespread, and the eventual list may not be a great deal longer. (There are indications that the same list is also widespread among other classes of vertebrate animals; W. J. Smith, in press). Note that many of the messages are only broadly predictive of a communicator's behavior. The message of interaction, for instance, does not specify what kind of interaction to expect, that of locomotion does not specify the functional class of the locomotion, and the so-called general set message merely indicates that some of the acts that may be selected when a communicator displays are incompatible with other acts whose possibilities are also encoded by the display. In contrast, messages such as attack, escape, and copulation are much more narrowly predictive. It would appear that the evolution of the displays of most birds has responded to pressures favoring either messages that are serviceable in many different situations and very dependent on contextual sources of information to elicit appropriate responses, or messages that are less widely useful but that are more capable of eliciting precise responses immediately, with minimal dependence on other sources of information. It would also appear that only a few messages are either suitable to be very broadly useful or are needed in relatively narrowly defined circumstances.

A basic feature of bird communication that was probably crucial to the evolution of these limitations on the kinds of message that are widely used is the limited size of each species' display repertoire (W. J. Smith, 1969a). By traditional criteria for the recognition of display units, no bird appears to have more than about forty to forty-five displays, and many have fewer (Moynihan, 1970). Further, these tallies include both audible and visible displays, and the two classes usually overlap very considerably in the kinds of information they make available. With such small display repertoires, it is evident that a species that ties down too many of its displays with narrowly predictive messages will be unable to display at all in a great many circumstances. Evolution should favor precise messages in special circumstances, but it appears that in most species this sacrifice of displays for other than attack, escape, association, and copulation messages is not adaptive.

Because this argument turns on the fact that birds (and other animals) have only limited repertoires of display behavior, the problems discussed earlier that are inherent in the traditional display concept become a matter of concern. That some acts may be only slightly or partially specialized to be informative may not be crucial, as these are necessarily poorly distinguished from nondisplay acts and thus probably not as efficient in providing information as are more striking displays; nonetheless, it would be useful to have an estimate of the magnitude of this problem. The chief difficulty may lie in the fact that displays vary in form and even intergrade. Yet the concept of displays as vehicles for information suggests a means of resolving this issue. When a communicator is found to use markedly variable or intergrading displays, their different forms are usually found to vary in correlation either with differing probabilities of intensities of communicator acts (see below) or with different kinds of acts (different behavioral selections). If the specification of a particular set of behavioral messages is the fundamental task of a display, then we might accept a shift in the kinds of message encoded as marking the boundary between related displays and amplify the traditional definition of display behavior accordingly. Using this criterion we can determine the number of behavioral packages for different sets of information that are available to an individual of any species. Present indications are that the number of displays estimated by this criterion is usually very similar to the numbers estimated by somewhat more arbitrary criteria of form.

Displays provide more information about communicator activities than simply an indication of the selections that an individual may make from its behavioral repertoire. Supplemental information (see Table 2) is made available for each selection that gives at least the probability of its being performed. Although a display may occur simultaneously with a behavioral selection it is also likely to precede it or to occur when there is some possibility that the selection will follow. And since many, perhaps most, displays provide information about two or more incompatible selections it is obvious that only one can follow, or at least be the first to follow. For instance, the green-backed sparrow has at least three vocal and three visible displays that are performed in agonistic circumstances: some when attack is more likely than escape, some when attack and escape are about equally probable, and some when escape is more likely than attack; in most of these circumstances, indecisive behavior is more likely than either attack or escape (Moynihan, 1963).

It is easier to compare the probabilities of different behavioral selections from among these displays than it is to measure the probability for any one display, because the use of each display can change the circumstances of the interaction. After a severe threat, for instance, an opponent may flee and in fleeing obviate the need for attack by the communicator. Severe threats may indicate a high probability of attack if the situation remains unchanged, but an observer may rarely see that eventuality in a close interaction. Other kinds of circumstances may change much less abruptly, on the average, than do agonistic encounters; but in principle the problem remains, and even a continuously static situation may alter a communicator's readiness to act. The probability that each behavioral selection encoded by a display will be performed is in effect only at the instant of displaying and cannot, therefore, be measured accurately; the probability information made available by displays is to

Table 2

Widespread behavioral supplemental and nonbehavioral messages.

A behavioral supplemental message provides information about a behavioral selection that is effectively adverbial, the "how" of the behavior. Nonbehavioral messages specify "who" and in some cases "where" the communicator is. A much more detailed account is provided by W. J. Smith (in press).

Behavioral supplemental messages

Probability: the likelihood that a given behavioral selection will be made.
Intensity: the forcefulness, rapidity, etc., with which a given selection will be performed; not a unitary category, and will eventually require subdivision.
Relative stability: a measure of the expected persistence of a given behavioral selection.
Direction: the direction a given behavioral selection will take, e.g., toward or away from some other individual.

Nonbehavioral messages

Identifying messages

Population classes: species, subspecific or local populations, individual.
Physiological classes: maturity, breeding state, sex.
Bonding classes: pairs, families, troops, and the like.

Location message

A single category of information that enables the source of a display to be pinpointed.

some degree indeterminate for an observer. Nonetheless, it is of crucial importance to the displaying birds.

In addition to indicating the probability that a behavioral selection will be performed, many displays appear to provide some information about the expected intensity of the activity. Thus some displays providing information about escape behavior may correlate primarily with slight avoidance movements or local withdrawal, others with headlong fleeing. Different displays with association messages may correlate with different distances among the participants or perhaps different degrees of readiness to let association be interrupted. Very little study of such correlations has been undertaken, however, and this message, although it appears to exist independently of the probability message and need not vary in parallel with it, is still largely unknown.

In being repeated in a predictable pattern, displays can also provide information about the stability of a communicator's behavior or behavioral predispositions. As songbirds utter bouts of singing each morning and evening twilight, for instance, they reaffirm their continuing readiness to defend their territories. Schleidt (1973) has termed this procedure "tonic" communicating.

Finally, some displays also provide information about the direction to be taken by the communicator's attending or other behavior. That is, without specifically naming its intended recipients, the whereabouts of a predator, or other external entities such as a nest site, a bird can provide effectively comparable information by indicating the direction in which it will attack, from which it will flee, etc. Much of this is done by employing formally restricted angles of orientation, as in the mutual adoption of a parallel relationship between two greeting gulls. A male green-winged teal *(Anas carolinensis)* indicates that a female is the object of his attentions by orienting broadside to her when performing seven different displays and swimming directly away from her when performing an eighth (McKinney, 1965). When he is distracted by other males and is trying to avoid them to align with the female he uses yet another display in a more variable orientation that reflects his divided attention. A female mallard *(Anas platyrhynchos)* will perform Nod swimming to a group of males, but with the Inciting display takes a fixed bodily orientation with respect to one individual she has chosen as her mate (Weidmann and Darley, 1971). Both displays appear to indicate that she is ready to interact, but the latter indicates that she will take a greatly narrowed direction. Among many other possible examples of the importance of formally restricted orientations and the directional information they make available are two cases in the Triumph Ceremony of the gray-lag goose (*Anser anser:* Fischer, 1965). In the first phases of this display a gander displays and orients directly toward the gander of another family and approaches to attack him. In the second phase he returns to his own family using the same display postures and movements, but carefully maintaining an oblique orientation to each bird; they respond by similar displaying and adoption of mutually oblique orientations.

If the information provided by displays were entirely concerned with behavior then most birds would be awash in sources of information, relevant and irrelevant, and not know to which to attend. In fact, however, ethological studies of bird communication have demonstrated that a great deal of the information provided identifies communicators.

Most vocal displays identify the species of the bird that uses them, particularly the louder displays, which are effective over large distances. Visible displays may not need to do this if they reveal or are seen along with badges that accomplish the same task, and there are also some vocal exceptions. For instance, among species that flock together, the "alarm calls" that are uttered on sighting a predator may converge onto a common form, usually one with physical characteristics that make the source of the sound difficult to locate by a binaural recipient (Marler, 1955a). Identifying information is not wholly lost in such cases, but it specifies membership in an assemblage of interdependent species rather than in a single species.

Within a species some aspects of the form of displays may vary regionally and so identify local populations. Detailed mapping of the precise geographic distributions of such dialectal differences has very rarely been done with samples sufficiently large to show whether variation is continuous or stepped, however, and much remains to be learned about the limits of dialectal regions (Thielcke, 1969). The exact amount of variation that is truly geographical is also not known for most existing samples, as little attention has been paid in most cases to the behavior of the birds that were being recorded (see critique by S. T. Smith, 1972).

Many displays have pecularities of form that are specific not to populations of individuals but to single individuals. (For a detailed review of the numerous observations and experiments demonstrating this phenomenon in the vocal displays of birds see Beer, 1970.) Individual identifying information permits the formation of bonds between mates, parents and offspring, and even territorial neighbors. In some species, infants still in the egg learn to identify the pecularities in the voices of their incubating parents (Tschanz, 1968; Norton-Griffiths, 1969; Beer, 1970).

Displays may make available information that identifies not only individuals and populations but also the sex and, sometimes, the age classes to which communicators belong. For instance, in some birds song is used by only one sex, usually the male. Care has to be taken before concluding this for any species, however, as members of the opposite sex may sing on the relatively rare occasions that they take on the role of advertising.

And there are species for which the use of song identifies the communicator as a female, e.g., the red phalarope (*Phalaropus fulicarius:* Tinbergen, 1935).

Finally, in some species displays take forms indicating that the communicator is a member of a particular group of bonded individuals. In the American goldfinch *(Spinus tristis),* for instance, one vocalization is adjusted in the course of the year to identify the individual's bonds in a pair, a family, and a small flock, as each of these successive groups becomes important (Mundinger, 1970). In a Trinidadian hummingbird known as the little hermit *(Phaethornis longuemareus),* the males who gather near each other as a distinct group on a larger lek all share a single song form, different from that of other groups singing on the same lek (Wiley, 1971). In many species, members of mated pairs come to share combinations of vocalizations, which they use in duetting performances that are distinct from the performances of neighboring pairs (discussed below).

One other category of nonbehavioral information is provided by vocal displays: information that enables a binaural recipient to locate the source of the vocalization by comparing temporal changes in the phase and amplitude of the sound reaching its two ears (Marler, 1955a). Except in the relatively few cases in which these clues are reduced by natural selection, the complexity of bird vocalizations usually appears sufficient to provide this information in abundance. The complexity of form of a vocalization determines both its locatability and how many kinds of identifying information it can carry, how much of the hierarchy of classes can be represented. Within the limits of form, each display provides sufficient identifying information to permit it to be useful in many kinds of circumstances (Marler, 1959).

Another category of information that has often been said to be encoded in some bird displays is information about the "environment," for example, about hawks or habitats. However, there is still no convincing evidence that referents external to the communicator are encoded by formal behavior in any species except humans (see W. J. Smith, in press). Environmental information is often available in events when birds are communicating, but its sources would not appear to include their formal acts.

RESPONSES AND FUNCTIONS

Communication occurs only when information is shared between or among individuals. While it is essential to determine the informative potential of displays, badges, and other sources, it is also necessary to determine the kind of response to which this information contributes if the process of communicating is to be understood. Further, it is the responses that lead to the adaptive advantages of sharing information, the functions of the process, for both the suppliers and the recipients of the information. As a branch of evolutionary biology, ethology has been very interested in adaptive significance, and hence concerned with the functions of communicating. Unfortunately, research on functions and the responses that generate them is not easy. When a bird that is apparently a recipient of, say, a display does something, it is often very difficult to know to what the bird is responding, since many sources of information are available to it. Further, its response may be largely inaccessible to an observer, for instance, a slight alteration of the bird's state of responsiveness to subsequent stimuli but not an immediate and overt change in its behavior.

There has been some tendency to oversimplify the role of recipient individuals in events in which communication occurs. In part this is due to the ethological theory that displays act as "releasers" of responses (see Lorenz, 1950; Tinbergen, 1959). The theory is soundly based on the observation that animals do treat some stimuli as being especially significant for particular classes of responses, and on the knowledge that perceptual processes organize stimuli according to various preestablished criteria of salience. Carried to extremes, however, this perspective underestimates the dependence of recipients on the relationships among stimuli. Thus, while a particular badge—for instance, the red spot on the mandible of an adult herring gull *(Larus argentatus)*—may be very important in eliciting and directing pecks by its hungry chicks (Tinbergen and Perdeck, 1950), as the chicks mature and gain experience they learn additional characteristics of their parents' heads and come to demand much more detail of the visual stimuli at which they will peck (at least in laughing gulls *Larus atricilla:* Hailman, 1967, 1969). The releaser is not a key to unlock responses, but a stimulus specialized to be particularly noticeable to recipients, and to be accepted as that stimulus to which other stimuli have a contextual relationship in some particular frame of reference (W. J. Smith, in press).

Much experimental work has been done with birds in testing the responses elicited by badges and displays. Birds' appearances have been modified by removing or altering badges by clipping feathers or painting fleshy areas (see references to studies by Peek and by N. G. Smith under the initial discussion of badges, above) and by dying feathers. For instance, Marler (1955b) gave female chaffinches reddish breast feathers matching those of the males, and found that the altered birds usually became dominant over other females and under some conditions were even successful in agonistic encounters with males. Stuffed birds or portions of mounts have been used to elicit aggressive responses (e.g., by Lack, 1940, with the European robin, *Erithacus rubecula*) or sexual display and mounting (e.g., MacDonald, 1968, with the spruce grouse, *Canachites canadensis*). Models of birds have been used to test responses to particular postures; for example, Stout and Brass (1969) sought differential responses by placing two models in different postures within territories of breeding glaucous-winged gulls *(Larus glaucescens).* However, all these methods of providing specific visible stimuli share in some degree the problem of appearing unnatural. A female colored to look like a male will not always act like one, and a bird altered to look like another species may persist in trying to mate with members of its own. More serious, a stuffed mount or model is unresponsive. Even if it is built so that its poses can be changed, it cannot be manipulated to respond as naturally as its recipients would expect of a real bird. Static models elicit responses, but beyond the initial moves of their recipients the circumstances are difficult to interpret.

Yet another procedure is to record and play back vocal displays. This technique has been used frequently to test whether birds can discriminate their own song (or sometimes just an unstudied and variable sample of their vocal repertoire) from that of related species (e.g., Dilger, 1956; Thielcke, 1962; Lanyon, 1963; Gill and Lanyon, 1964; Thompson, 1969), or can recognize the song of their own subspecies (e.g., Thönen, 1962) or local dialect (e.g., Lemon, 1967). Tests of this sort have even shown individual recognition, as territorial males distinguish between the songs of their neighbors and of more distant males (e.g., Weeden and Falls, 1959; Falls, 1969; S. T. Emlen, 1971; Goldman, 1973). Although the responsiveness of test individuals to playback lessens with habituation and varies with such factors as time of day and phase of the nesting cycle (Verner and Milligan, 1971), and the strength of a response can be difficult to measure (see S. T. Emlen, 1971, 1972), the technique has revealed a good deal about the identifying information made available by vocalizations. Further, because sound recordings can readily be modified by filtering frequencies, rearranging sequences, and altering intervals it has been possible to ask what physical components of bird song carry the identifying information (e.g., Busnel and Brémond, 1961, 1962; Brémond, 1968a, 1968b; Falls, 1963, 1969; Schubert, 1971; S. T. Emlen, 1972; Helb, 1973). The pure tones of species such as the white-throated sparrow *(Zonotrichia albicollis)* can even be reproduced with audio oscillators and then modified in predetermined ways to study which characteristics provide species- and which individual-identifying information (Falls, 1969).

Playback has also been used to test the responses of females to song (e.g., Falls, 1969; Milligan and Verner, 1971; Payne, 1973b), of parents and their offspring to each other's vocalizations (see review by Beer, 1970), and of various species to their own and each other's alarm calls (e.g., Brown, 1962; Curio, 1971; Thielcke, 1971). Where responses to more than just the identifying information are sought, however, the technique encounters the same problem as does the use of models—the sound source is unresponsive to the actions of its recipients. Further, it is necessary to accompany the sound with a model or a mirror (e.g., Stout et al., 1969) to provide a visible participant, and this involves adding the information from a static posture or a careful mimic. This is not to argue that playback and the use of models and mounts are not necessary procedures, but that their employment introduces information that makes the experimental circumstances unnatural and hence difficult to interpret. The further development of techniques for presenting paired stimuli, done with careful attention to maximizing the naturalness of at least the initial impingement of these stimuli on the test animal, may contribute a good deal to our understanding of responses.

If we do not experiment we are left with the problem of deciding what responses accrue to displays and other formal information sources in natural events in which we lack control over many relevant variables, and we may find it difficult to quantify observations. Nonetheless, natural events do provide very useful clues, and under some circumstances particular encounters may recur very frequently. Tinbergen (1959) has described a very fruitful procedure, which he refers to as seeking "natural experiments." In one of the examples he gives, an act is repeated several times without eliciting overt responses from nearby individuals, and then a largely similar act is performed with a display and they do respond: A male black-headed gull *(Larus ridibundus)* gathering nest material on his territory repeatedly approached five individuals who were resting just outside. While he was preoccupied they continued to rest. Then he stopped, preened near his nest and again approached them, now in an Aggressive Upright posture. When he came within three meters they all adopted Anxiety Uprights and walked away. Since an observer sees the Aggressive Upright primarily or only in circumstances in which the communicator appears to be threatening, such an instance tends to confirm the suspected functions of intimidation and territorial defense and the expected response of withdrawal. Other responses are also seen when the recipients are different; e.g., counterthreat would be one response of a territorial neighbor.

Slight manipulations of natural situations can help an observer obtain repeated samples of a given kind of event. For instance, in order to increase the incidence of agonistic encounters that he could observe in a winter flock of blue tits *(Parus caeruleus),* Stokes (1962) provided a rich food source, and by stringing peanuts and chunks of coconut on a wire, he forced individuals to remain at it while eating. He counted the number of times communicators used particular displays in the presence of one other individual and tallied the subsequent behavior of each participant as attack, escape, or staying. The recipients behaved appropriately following displays that correlated with different communicator activities; e.g., the probability of the recipient's fleeing after three of the displays was about twice what it was when none of them was used, even when Stokes analyzed only events in which the communicator did not actually attack. This suggests that the displays and/or such other sources of similar information as intention movements and known status relationships were influencing recipient behavior.

Natural experiments and slight manipulations of natural circumstances will probably continue to provide the least-distorted insights about the responses made to displays and other formal sources of information. The advantages of more-controlled experiments will be necessary, however, and techniques need to be developed that do not grossly violate the expectations of the recipient birds. For instance, if models are placed in a territory they should mainly be in postures that a territorial intruder might adopt, and these are typically not those correlated with the highest probabilities of attack. By using more than one experimental technique in conjunction with natural observations the contributions of each procedure can be pooled and their disadvantages minimized.

The study of functions is also based on observation of natural events and involves consideration of the ways in which the responses made by the recipients of displays are adaptive for them and for the individuals who make the information available. Some functions are obvious. A bird that flees into cover when a companion utters an "alarm call" may avoid being caught by a hawk. Many other functions are less obvious. In the same example, for instance, the call may help the bird that utters it to escape if the sudden scattering of its companions confuses the hawk, but it may not help otherwise. Still, the call may function for that communicator in various ways; for example, by protecting the investment it has made in its offspring, if they are in the group, and by helping to keep alive members of the group who in subsequent events may give it warning when hawks approach.

The functions of a display in any given event may be quite different for a communicator and for a recipient, as they are in the above example. They may also differ among recipients, and differ from the same display in different kinds of situations—particularly when the displays encode broader behavioral selection messages. Further, the functions of a single communication event may have to be stated for more than one time span. Thus the immediate functions of, say, a male bird's song may be to attract an unmated female and to repel a neighboring male. In the longer view, the attraction of a female is a necessary step if the male is to pass on his genes to a future generation, and the repelling of the neighbor may help train that individual to remain away, thus reducing subsequent competition for the female. Thus, the functional assessment of displays is a complex task.

Many functions have been proposed for the displays of birds, although few have been thoroughly studied or experimentally tested. Cautioning that much more study is needed, Thielcke (1970) has reviewed the literature on bird vocalizations and offered the following descriptive list of functions: territorial defense, attraction of a mate, maintenance of a pair bond, mutual stimulation of mates, synchronization of pair activities, facilitation of the simultaneous hatching of the eggs in a clutch, familial recognition, group coherence, assembly of roosting groups, attraction to feeding sites, control of agonistic encounters, maintenance of pair relations and facilitation of the activities in which mates cooperate, raising of young, and alarming of young and other conspecific individuals.

Bird Song

Although much research has been done on display postures and movements, and on the vocalizations usually referred to as "calls" by ornithologists, a great deal of attention has centered on a kind of display that is complexly developed and frequently used by many species of birds: song. Other animals sing, of course, from cicadas and frogs to humpback whales (*Megaptera novaeangliae:* Payne and McVay, 1971) and titi monkeys (*Callicebus moloch:* Moynihan, 1966), but bird songs have attracted human attention for a very long time, even if initially primarily through their aesthetic appeal. They are interesting for reasons beyond aesthetics, however. Studies have now revealed that in their organizational complexity and in the ways in which they are learned, the songs of some birds show very interesting parallels with human speech behavior, parallels further extended by the use of song in formally patterned performances that require more than one participant.

Song is not a precisely defined behavioral category (see Thorpe, 1961; Armstrong, 1963). Although complexity of form is an important attribute of most of the songs that are intensively studied, many species have comparable performances in which the songs are very simple (e.g., the chipping sparrow, *Spizella passerina:* Borror, 1959; Marler and Isaac, 1960; and the chiffchaff, *Phylloscopus collybita:* Schubert, 1971). Further, some species have complex vocalizations that would not usually be called songs or that comprise a series spanning the distinction between songs and "call notes" (e.g., in the genera *Chlorospingus:* Moynihan, 1962c; *Tyrannus:* W. J. Smith, 1966; and *Parus:* S. T. Smith, 1972), and in many species the song vocalizations are partly or wholly specialized amalgams of call notes (Howard, 1920; Thorpe, 1961; Immelmann, 1969). The singing performance, in fact, is a better indicator than are the forms of the vocalizations. The most useful working definition of song seems to be a vocalization or a set of vocalizations that is repeated in more or less continuous, regular patterns, often in sustained bouts.

Recent research on bird song has concentrated on a number of issues. One is the identifying information songs carry, as is discussed above. Another is the temporal organization of the complexity, which makes many of them so distinctive among vocal signals. Because portions of this complexity may be typical only of local populations, the question has arisen as to how it is passed from generation to generation, leading to very detailed studies of song learning. Finally, the specialized behavior of duet singing has brought considerable attention to an interesting set of formalized interactions.

Before reviewing this research, a caution is in order. While the repetitive use of patterned vocalizations is an extremely interesting phenomenon, it should not be assumed (as it very often is) that the remainder of the vocal repertoire is necessarily of less interest. In fact, in most species the non-song vocalizations are used much more frequently than the songs, and are much more likely to be employed during active interactions. Birds communicate with whole repertoires of displays, not just with their most complex vocalizations.

COMPLEXITY AND TEMPORAL PATTERNING

The song units of many bird species are complex in that they are made up of more than one sort of component, arranged in an orderly fashion (see, e.g., the descriptions of chaffinch song and its three subsections in Thorpe, 1961). In some species each male has only one such complex song unit, but in others he characteristically has two or more, sometimes a great many. The different song forms or song types in a male's repertoire typically consist of rearrangements of a limited set of the basic components, and in several species at least some of these rearrangements have been shown to be rule-bound. For example, in the mistle thrush (*Turdus viscivorus:* Isaac and Marler, 1963) most of the components are not repeated while other components always are; in the olive-backed thrush (*Hylocichla ustulata:* Nelson, 1973) each component tends to be a variant or an elaboration of the preceding one; in the rose-breasted grosbeak (*Pheucticus ludovicianus:* Lemon and Chatfield, 1973) the components of songs occur in regular couplet and triplet combinations; in the yellow-throated vireo *(Vireo flavifrons)* approximately nineteen kinds of components can be divided into three categories on the basis of gross form, and the various positions members of these categories take within different song forms are limited in several ways (Smith, Pawlukiewicz, and Smith, ms.). Because an extremely large number of recorded samples must be obtained, the songs of none of these species have yet been described in sufficient detail to exhaust the limits of the variability and permit us to write rules that are characteristic of the entire species population. Nonetheless, the phenomenon is well established. It is interesting that it is comparable to one of the grammatical levels of human speech: the basic, distinctive, vocal units (songs and words or morphemes) are in both cases constructed of a limited number of components that are sequentially arranged according to systems of rules.

In human speech the grammatical level of which we are most aware is the organization of words into phrases, so the question arises whether the sequences of bird songs are also patterned and, if so, how complexly. Evidence for sequential patterning of songs is again available from diverse species. Ohio song sparrows, for instance, go through their entire song repertoires before they repeat runs of any one kind of song (Nice, 1943), and members of the species from central California, who have larger song repertoires, usually go through about ten different song forms before repeating a run of any one (Mulligan, 1966). Cardinals (Lemon and Chatfield, 1971) show a first-order Markovian relationship between successive song forms; i.e., each is largely determined by the one that went before, if they are not separated by a relatively long pause. In the eastern wood pewee (*Contopus virens:* Craig, 1943) the long daily bout of predawn singing comprises three song forms arranged in a number of orderly sequences, and singing during the day employs two of the songs according to yet other rules (W. J. Smith, 1968, and in prep.). Members of another tyrannid genus, the phoebes (*Sayornis,* three species), employ either two or three different units—one a very variable vocalization—in patterned bouts governed by rules similar to those apparent in the daytime singing of the pewee: one song form occurs primarily singly and terminates runs of the other unit(s). The runs vary in length in a nonrandom fashion (W. J. Smith, 1969b, 1970). The yellow-throated vireo, however, has much greater freedom: individual males use up to eight song forms in a number of couplet combinations, most of which have members that overlap with those of one or more other couplets and so can be used to produce recurrent runs of triplets; a few longer sequences are also found (Smith, Pawlukiewicz, and Smith, in press.). In populations of other vireo species, such as *V. solitarius* in the Chiricahua Mountains of Arizona, individual males may have as many as seventy-five different song forms (Smith and Smith, in prep.).

Since at least simple combinations and recombinations of song form are rule-bound in these species, birds can be said to have singing grammars that are in part analogous to the phrase-structure grammars of human speech. In one very important respect, however, they appear to fall short. There is no reason to suspect, on the basis of existing evidence, that the various song forms of any species are differentiated into lexical categories such as the nouns and verbs with which we are familiar in English. (This lack of categorical distinction greatly limits the possible informative diversity of singing.) Rather, it appears that each song, like each call, encodes information about the same referent classes as in every other avian vocalization: behavioral selections and supplemental behavioral classes, classes of membership that identify the singer, and the class of "location" information.

The considerable experimental evidence for the nonbehavioral classes of information that are made available by bird song is reviewed briefly in the section on responses and functions, above. Experimental testing of the behavioral message classes is much more difficult, however. Further, since birds very often sing when they are not interacting, it is difficult to formulate precise and appropriate descriptions of the behavioral messages from observations of naturally occurring events. Nonetheless, a number of studies have reported correlations between the use of different song forms and different activities of the singer (see Armstrong, 1963, for a review), and in some cases in sufficient detail to suggest how messages differ among the song forms or their formal combinations.

Carolina chickadees, for instance, have two different song forms, either of which can be used loudly or faintly (S. T. Smith, 1972). The more commonly used form is sung by territorial individuals seeking any of a considerable range of kinds of interactions with their mates or neighbors; with use of its fainter version the singer is less likely to attack. The other song form correlates almost entirely with the most active kinds of seeking behavior, e.g., patrolling and intruding into neighboring territories.

In Africa, males of the parasitic indigobirds (genus *Vidua*) have both nonmimetic songs, which are peculiar to the genus, and mimetic songs, with which each male mimics a host species that is used by indigobirds to raise their young. The nonmimetic songs correlate with agonistic behavior in which the singer chases other males from his singing station, and the mimetic songs correlate with visits by females, becoming especially frequent if a female flies from a male's station without having completed a mating sequence (Payne, 1973a).

Various studies have shown that the two or more songs of many species of New World warblers are used in different circumstances (e.g., Ficken and Ficken, 1965, 1966; Morse, 1966, 1967, 1970; Lein, 1972). The most detailed study is that of Lein (1973) on the chestnut-sided warbler *(Dendroica pensylvanica),* which he has shown to have at least five song forms, each of which can also be shortened or muted. One is used primarily in territorial encounters; two are used primarily at territorial borders, or at least away from the center of the territory; and the remaining two are used almost exclusively in the interior. The distance from a singing neighbor also influences the choice of song. As the probability of a territorial encounter increases or decreases while a male is singing (e.g., by the onset or cessation of a neighbor's singing, movement of either male toward a border), he switches song forms in a very predictable fashion.

The two song forms of the eastern phoebe (*Sayornis phoebe:* W. J. Smith, 1969b) may be alternated, or one may occur in runs terminated by single occurrences of the other, so that their proportional representation in singing bouts varies; the varying proportions correlate with different activities of the communicator. At one extreme the bird is very likely to interact if it gets the opportunity. At the other it is more likely to forage or preen, and if it can interact it will usually just associate, not attempting to make contact (e.g., to fight or to copulate). In the daytime singing of the eastern wood pewee, which also comprises differing proportions of two song forms, runs of different lengths of one of the forms correlate with different probabilities of the communicator's flying, apparently providing a measure of its activity level.

The story for the yellow-throated vireo is much more complicated. Each male has a number of patterns of singing and uses different selections of patterns when patrolling his territorial borders, attempting to confront a neighbor, foraging away from the nest but not in a boundary region, approaching the nest, or remaining on or in the vicinity of the nest. Patterns correlated with the last activity are the simplest, those used in patrolling the most complex. Unmated males patrol and tend to sing complex patterns that resemble those used by mated males when the latter are patrolling. Of the eight males studied, no two sang identical selections of song forms or identical patterns (Smith, Pawlukiewicz, and Smith, in press.). However, they had many similarities and did not contradict the uses of one another's patterns, and all differed in important details of their circumstances, such as the presence or absence of neighboring males, having mates, the attentiveness of their mates, and their successes in nesting. Although the analyses in that study were based on 9,419 songs recorded during over forty-one hours of observations, the variation among the males and the complexity of the correlations between behavior and singing patterns suggest that a much larger sample will be required to reveal more precisely the messages of the song forms and combinations and to clarify the species' grammatical rules.

SONG LEARNING

The basic forms of most bird displays develop even in individuals experimentally reared in isolation. Thus they do not involve learning based on hearing and seeing other members of the species perform. Remarkably, this is true even for quite complex songs, for example, those of the song sparrow (*Melospiza melodia:* Mulligan, 1966). However, there are species in which development of the full form of complex songs does require such learning and also subsequent learning through practice.

In these species young birds must usually hear appropriate songs, which will act as models, during a critical period early in their lives, well before they start to sing. By experimentally restricting what potential models they can hear, it has been shown that what they will accept is limited. In chaffinches and white-crowned sparrows *(Zonotrichia leucophrys),* for instance, it has to be the song of their own species or of a species with a similar song pattern. In nature, of course, they will almost always have the opportunity to hear their own species sing during the period in which they are fledglings and, in fact, will obtain as models songs that are typical of their own dialect group. In the bullfinch (*Pyrrhula pyrrhula:* Nicolai, 1959) and in various grass finches (Estrildidae: Immelmann, 1969) it is the song of the male that helped rear them: the father or, in experimental conditions, a foster father of another species. (In parasitic indigobirds what is learned appears to be the song of the host species raising the young; whether their species-specific songs require hearing a model has not yet been tested, but the appropriate model might not be available to a young bird; see Payne, 1973b, and above.) There tends to be a second learning period in which singing is practiced, and a variable, rambling "subsong" is perfected into a full adult song. This may come months after the auditory model is stored. The experience of countersinging during this practice enables members of some species (e.g., the chaffinch) to elaborate on their learned model. Inventive elaboration has also been found even in isolated cardinal males (*Richmondena cardinalis:* Dittus and Lemon, 1969), but it is not characteristic of other species such as the estrildids. The latter have an early end to the critical period in which they can learn by hearing other individuals, and this is correlated with a very early onset of sexual maturity.

There have been many studies of song learning in which young birds of various ages have been acoustically isolated. By controlling their exposure to sounds this research has determined many of the characteristics of the sounds that the birds will accept as models. By deafening and other procedures the roles of auditory and proprioceptive feedback have been studied, and by adopting procedures from psychology the reinforcing properties of song have been investigated. The following reviews provide a detailed coverage of these experiments: for the chaffinch, Thorpe (1958, 1961), and Hinde (1969); for the white-crowned sparrow, Marler (1970); for several species of sparrows, Marler (1967); for estrildid finches, Immelmann (1969); and for general coverage Hooker (1968), Konishi and Nottebohm (1969).

Among the most interesting things to emerge are a number of similarities between the learning of song by birds and of speech by humans. Marler (1970) reviews these, pointing out that both involve critical periods before sexual maturity in which young individuals select a set of sounds that they will learn from the many available to them in natural circumstances. They are predisposed to recognize the appropriate set and will work at reproducing it without extrinsic reinforcement, practicing through a stage known as "subsong" or "babbling," respectively, until they achieve an adult form that they recognize through a process that requires auditory feedback. (Female birds may produce little song, but it has been shown that they do learn it, for testosterone injections will elicit malelike singing behavior; presumably their learning serves primarily to help them identify males of their own population with whom to form pair bonds.) Marler has suggested that these parallels are not unexpected in any system of vocal learning in which what is learned cannot be left to chance and which requires the development and refinement of skill. Why some species learn their songs and others do not is a more difficult question, although part of the answer would appear to lie in the use of songs by some species as means of identifying local populations.

The rhythm and phrase structure of bird song are also part of what is copied by many species (see, e.g., Güttinger, 1973), just as human neonates copy with their body movements the rhythmic organization of the human speech they hear (Condon and Sander, 1974; these patterns of body movement continue to be a part of listening behavior and become part of speaking behavior as the infants mature). In humans the learning of patterns of organization in this fashion may facilitate the learning of hierarchical grammatical patterns, but it is too early even to speculate whether this is necessary in the ontogenetic development of hierarchically patterned singing by birds.

DUETTING

It is not wholly appropriate to consider the topic of duetting under bird song as in many species the performances are based on non-song vocalizations, or even on such nonvocal sounds as drumming (e.g., in the hairy woodpecker, *Dendrocopus villosus:* see Kilham, 1960). Nonetheless, the recent increase of research activity on the topic has centered largely on studies of song duets, especially those of African shrikes of the genus *Laniarius* (beginning with Thorpe, 1963, and Thorpe and North, 1965). Birds in this genus also provide a case of the learning of pair-specific song patterns, as pairs develop repertoires of duets in which some of the songs are peculiar to individuals and each mate knows and responds to its partner's special forms (most recently reviewed by Thorpe, 1972). Learning, like song, is not essential to the duetting performances of many other kinds of birds, however. The basis of duetting is simply a formally patterned mutual performance that, by Thorpe's definition, is characteristically used by paired individuals.

Because duets cannot be performed by single individuals they are formalized interactions in the sense described at the beginning of this chapter, and are distinctive within that category primarily by being audible (usually vocal) performances by pairs. In principle, visible duets would be largely comparable, but the term has not been applied to postures and movements. When mates duet one leads and the other comes in either with the same vocalization, creating a polyphonic performance, or a different vocalization, yielding antiphony. Other individuals, offspring for instance, may sometimes join in. The duet then becomes a trio, quartet, or "communal" performance (Thorpe, 1972). In many species a mate of either sex will initiate a duet, although the male does so much more often than the female. In *Laniarius* and some other genera, if one mate is missing the other sometimes sings both contributions (the vocalizations do not overlap and the performance is no longer a duet, but it may serve to recall the missing partner).

Duetting performances have not been distinguished by precise criteria from formalized vocal interactions used as greetings or as appeasing-reassuring ceremonies, and probably should not be, although Thorpe (following Armstrong, 1963) has ruled that patterned countersinging between neighbors of the same sex is not duetting. Because of the overlap between duetting and related performances, attempts to catalogue the taxonomic families of birds in which duetting occurs remain provisional. The present evidence, however, indicates that many families are involved. Some species that duet can be found in virtually every bird fauna throughout the world, although they may be proportionately more common in tropical than in temperate regions. Duetting species are typically birds that pair for life and are territorial for prolonged periods (although males and females may hold adjacent territories; see Kilham, 1960); often the sexes are very similar in appearance, and in many, but by no means all, cases the habitats are so densely vegetated that continuous visual monitoring of one mate by the other is hindered (Thorpe, 1972).

Thorpe's (1963) initial interest in duetting was spurred in considerable part by the precise timing with which some shrikes (genus *Laniarius*) and sylviid warblers (genus *Cisticola*) answer their mates, and by the brevity of the interval between the onset of the first bird's call and the onset of the answering call. He proposed that these intervals can be used to measure the auditory reaction time of birds, a physiological parameter that appears to be briefer than the corresponding reaction time of humans. It has subsequently been found, however, that duetting birds are frequently within sight of each other and use visible cues (Hooker and Hooker, 1969). Thus until a reasonably large sample has been recorded in which it is known that the answering bird is unable to see the initiator the value of this procedure for measuring auditory reaction time cannot be fully assessed. In addition, Payne (1970, 1971, 1973c) has argued that throughout a duetting sequence of vocalizations it is possible that each individual is following its own, autochthonous calling rhythm, a rhythm to which it can and does adhere even if its mate fails to call. He notes, for instance, that when duet sequences were recorded from a pair of *L. barbarus* over a period of more than an hour, the interval between a call by the first bird and the answer differed from sequence to sequence, which would not be expected if the interval were set largely by the auditory reaction time.

Birds do answer their mates' duetting calls much of the time, however, and there has been much speculation over the functions of a vocal ceremony that recurs as frequently as this one does. Thorpe has pointed out that for those species in which mates do duet when out of each other's sight in dense vegetation the vocalizations should help them remain in contact and aware of each other's whereabouts. The same applies to nocturnal duets of owls (e.g., J. T. Emlen, 1973) and perhaps to birds living in the dense mists and clouds of tropical mountaintops (Thorpe and Hall-Craggs, in Thorpe, 1972). Duetting must have other functions, however, as duetting vocalizations are much more complex than most "contact" calls, and because duetting birds are often in each other's sight: in some species they are usually perched beside each other, or within the same bush, or out in the open.

For species that form large flocks, one function of duetting with group-specific vocalizations might be to enable bonded individuals to remain together in the crowd. This has not been tested, although this possibility is suggested by the fact that pairs of such flocking species as orange-chinned parakeets (*Brotogeris jugularis:* Power, 1966) duet when agonistically aroused and that common crows *(Corvus brachyrhynchos)* duet when in large roosting or foraging flocks (Chamberlain and Cornwell, 1971).

Many duetting species that do not inhabit dense vegetation or join large flocks live in regions in which the seasonal changes that precede breeding are rapid and unpredictable in onset (Diamond and Terborgh, 1968), or where seasons are not sharply differentiated (Kunkel, 1966). Changes in their patterns of duetting might help mates to synchronize their physiological states and to change together rapidly enough to meet the demands of such circumstances. Yet the duetting continues all year, and so it must do more than that.

Duetting is just one of several kinds of mutual, formalized exchanges of display behavior that are characteristic of animals with persistent bonds, patterns that range from brief greetings to prolonged allopreening or allogrooming. Whatever else these exchanges accomplish, they probably serve the continuously necessary function of reaffirming the adherence of both individuals to their bonded relationships. Subsumed by this function is the mutual checking of each other's states or conditions—of the physiological "status quo," as Estes (1969) has put it. This may well be what has too often been described as the "identifying" function of duetting, as individual identification in any simple sense undoubtedly need not be repeated many times a day by formal behavior patterns between birds who know each other intimately. Not identification, then, but reaffirmation may be the most widespread function of duetting.

In addition to reaffirming their relationships, duets can serve more-immediate functions for a pair. They may enable the birds to coordinate their activities in such events as nest relief, in which the birds take turns incubating the eggs. Duets may often serve as greetings and as appeasing-reassuring performances in other circumstances, as seems to be the case in diverse species of tyrannid flycatchers (W. J. Smith, 1971, and research in progress). Further, in many cases duetting vocalizations must function in the wider social sphere beyond the pair. They are, for instance, audible to pairs on neighboring territories. Duetting by one pair can be answered by the duets of a neighboring pair, with the subsequent development of territorial countersinging. At least in the bell shrike *(Laniarius aethiopicus)* the timing of the duets of each pair degenerates within such encounters, and all the birds begin to countersing "more or less at random" (Thorpe, 1972). Yet in some species countersinging itself may be patterned in ways that are comparable to duetting patterns. For instance, neighboring pairs of the slate-colored bou-bou shrike *(Laniarius funebris)* tend to imitate each other's duets in countersinging (Wickler, 1972). Conspecific neighboring males of cardinals (Lemon, 1968a), pyrrhuloxias (*Pyrrhuloxia sinuata:* Lemon and Herzog, 1969), and black-crested titmice (*Parus atricristatus:* Lemon, 1968b) will match each other's selections of song forms when countersinging. Pairs of these species do not duet in such territorial disputes, but when mated cardinals do duet the female matches the male's song forms in replying to him.

No published studies of duetting include sufficient observations of the behavior of the duetters to permit analyses of the behavioral messages of the vocalizations, although some information about readiness to interact is being made available by the duetting vocalizations. Species in which individuals have large repertoires of duetting vocalizations may be able to provide different behavioral messages by selecting different vocalizations. Individuals of *Laniarius aethiopicus,* for instance, each have several flutelike notes as well as a "snarling" vocalization (Thorpe, 1972). The last seems to be used in duets by individuals who are "aggressive," although details about the likelihood of attack behavior have not been published. On the other hand, it is amply evident that duetting vocalizations provide nonbehavioral information about each communicator's location and about several features of its identity: its species and sometimes its individual identity, in some cases its sex (see Hooker and Hooker, 1969; Todt, 1970; Thorpe, 1972), and (by being used in duets with particular other individuals) its bonded relationships.

References

Ainley, D. G., 1974. Displays of Adelie penguins: a re-interpretation. In: *The Biology of Penguins,* B. Stonehouse, ed. London: MacMillan.

Andrew, R. J., 1961a. The motivational organisation controlling the mobbing calls of the Blackbird *(Turdus merula):* I, II, III, and IV. *Behaviour,* 17:224–46, 288–321; 18:25–43, 161–76.

Andrew, R. J., 1961b. The displays given by passerines in courtship and reproductive fighting: a review. *Ibis,* 103a:549–79.

Armstrong, E. A., 1963. *A Study of Bird Song.* New York: Oxford.

Bastock, M.; Morris, D.; and Moynihan, M.; 1953. Some comments on conflict and thwarting in animals. *Behaviour,* 6:66–84.

Beer, C. G., 1970. Individual recognition of voice in the social behavior of birds. *Advances Study Behavior,* 3:27–74.

Blurton-Jones, N. G., 1968. Observations and experiments on causation of threat displays of the Great Tit *(Parus major). Animal Behavior Monographs,* 1:75–158.

Borror, D. J., 1959. Songs of the Chipping Sparrow. *Ohio J. Science,* 59:347–56.

Brown, R. G. B., 1962. The reactions of gulls (Laridae) to distress calls. *Ann. Épiphyties,* 13:153–55.

Brémond, J.-C., 1968a. Valeur spécifique de la syntaxe dans le signal de défense territoriale du Troglodyte *(Troglodytes troglodytes). Behaviour,* 30:66–75.

Brémond, J.-C., 1968b. Recherches sur la sémantique et les éléments vecteurs d'information dans le signaux acoustiques du Rouge-gorge (*Erithacus rubecula* L.). *La Terre et la Vie,* 2:109–220.

Busnel, R.-G., and Brémond, J.-C., 1961. Étude préliminaire du décodage des informations contenues dans le signal acoustique territorial du Rouge-gorge (*Erithacus rubecula* L.). *Comptes rendus des Séances, Acad. Sci.,* 252:608–10.

Busnel, R.-G., and Brémond, J.-C., 1962. Recherche du support de l'information dans le signal acoustique défense territoriale du Rouge-gorge (*Erithacus rubecula* L.). *Comptes rendus des Séances, Acad. Sci.,* 254:2236–38.

Chamberlain, D. R., and Cornwell, G. W., 1971. Selected vocalizations of the Common Crow. *Auk,* 88:613–34.

Chapman, F. M., 1938. *Life in an Air Castle.* Nature Studies in the Tropics. New York: Appleton-Century.

Condon, W. S., and Sander, L. W., 1974. Neonate movement is synchronized with adult speech: interactional participation and language acquisition. *Science,* 183:99–101.

Craig, W., 1943. The song of the wood pewee *(Myochanes virens). N.Y. State Mus. Bull.,* 334:6–186.

Crook, J. H., 1963. Comparative studies on the reproductive behaviour of two closely related weaver bird species *(Ploceus cucullatus* and *Ploceus nigerrimus)* and their races. *Behaviour,* 21:177–232.

Crook, J. H., 1964. The evolution of social organisation and visual communication in the weaver birds (Ploceinae). *Behaviour,* Suppl. 10:1–178.

Curio, E., 1971. Die akustiche Wirkung von Feindalarmen auf einige Singvögel. *J. Ornithol.,* 112:365–72.

Daanje, A., 1950. On the locomotory movements in birds and the intention movements derived from them. *Behaviour,* 3:49–98.

Darwin, Charles, 1872. *The Expression of the Emotions in Man and Animals.* London: Appleton.

Delius, J. D., 1963. Das Verhalten das Feldlerche. *Zeits. Tierpsychol.,* 20:297–348.

Diamond, J. M., and Terborgh, J. W., 1968. Dual singing by New Guinea birds. *Auk,* 85:62–85.

Dilger, W. C., 1956. Hostile behavior and reproductive isolating mechanisms in the avian genera *Catharus* and *Hylocichla. Auk,* 73:313–53.

Dittus, W. P. J., and Lemon, R. E., 1969. Effects of song tutoring and acoustic isolation on the song repertoires of cardinals. *Anim. Behav.,* 17:523–33.

Emlen, J. T., 1973. Vocal stimulation in the Great Horned Owl. *Condor,* 75:126–27.

Emlen, S. T., 1971. The role of song in individual recognition in the Indigo Bunting. *Zeits. Tierpsychol.,* 28:241–46.

Emlen, S. T., 1972. An experimental analysis of the parameters of bird song eliciting species recognition. *Behaviour,* 41:130–71.

Estes, R. D., 1969. Territorial behavior of the wildebeest (*Connochaetes taurinus* Burchell, 1823). *Zeits. Tierpsychol.,* 26:284–370.

Falls, J. B., 1963. Properties of bird song eliciting responses from territorial males. *Proc. 13th Int. Orn. Congr.,* 259–71.

Falls, J. Bruce, 1969. Functions of territorial song in the White-throated Sparrow, In: *Bird Vocalizations,* R. A. Hinde, ed. Cambridge: Cambridge University Press, pp.207–32.

Ficken, M. S., and Ficken, R. W., 1965. Comparative ethology of the Chestnut-sided Warbler, Yellow Warbler, and American Redstart. *Wilson Bull.,* 77:363–75.

Ficken, M. S., and Ficken, R. W., 1966. Singing behavior of Blue-winged and Golden-winged Warblers and their hybrids. *Behaviour,* 28:149–81.

Fischer, H., 1965. Das Triumphgeschrei der Graugans *(Anser anser). Zeits. Tierpsychol.,* 22:247–304.

Gill, F. B., and Lanyon, W. E., 1964. Experiments on species discrimination in Blue-winged Warblers. *Auk,* 81:53–64.

Gilliard, E. T., 1956. Bower ornamentation versus plumage characters in bowerbirds. *Auk,* 73:450–51.

Gilliard, E. T., 1963. The evolution of bowerbirds. *Scientific American,* 209(2):38–46.

Goldman, P., 1973. Song recognition by Field Sparrows. *Auk,* 90:106–13.

Goodwin, D., 1953. Observations on voice and behaviour of the Red-legged Partridge *Alectoris rufa. Ibis,* 95:581–614.

Güttinger, H. R., 1973. Kopiervermögen von Rhythmus und Strophenaufbau in der Gesangsentwicklung einiger *Lonchura*-Arten (Estrildidae). *Zeits. Tierpsychol.,* 32:374–85.

Hailman, J. P., 1967. The ontogeny of an instinct. *Behaviour,* Suppl. 15.

Hailman, J. P., 1969. How an instinct is learned. *Scientific American,* 221(6):98–106.

Helb, H.-W., 1973. Analyse der artisolierenden Parameter im Gesang des Fitis *(Phylloscopus t. trochilus)* mit Untersuchungen zur Objektivierung der analytischen Methode. *J. Ornithol.,* 114:145–206.

Hinde, R. A., 1956. A comparative study of the courtship of certain finches (Fringillidae). *Ibis,* 98:1–23.

Hinde, R. A., 1959. Behaviour and speciation in birds and lower vertebrates. *Biological Reviews,* 34:85–128.

Hinde, R. A., 1969. *Bird Vocalizations: Their Relation to Current Problems in Biology and Psychology.* New York: Cambridge University Press.

Hinde, R. A., 1970. *Animal Behaviour: A Synthesis of Ethology and Comparative Psychology,* 2d ed. New York: McGraw-Hill.

Hinde, R. A., and Tinbergen, N., 1958. The comparative study of species-specific behavior. In: *Behavior and Evolution,* A. Roe and G. G. Simpson, eds. New Haven: Yale University Press, pp.251–68.

Hooker, B. I., 1968. Birds. In: *Animal Communication,* T. A. Sebeok, ed. Bloomington: Indiana University Press.

Hooker, T., and Hooker, B. I. (Lade), 1969. Duetting. In: *Bird Vocalizations,* R. A. Hinde, ed. New York: Cambridge University Press.

Howard, E., 1920. *Territory in Bird Life.* London: John Murray.

Huxley, J., 1914. The courtship habits of the Great Crested Grebe *(Podiceps cristatus);* with an addition to the theory of sexual selection *Proceedings Zoological Soc.,* 35.

Immelmann, K., 1969. Song development in the Zebra Finch and other estrildid finches. In: *Bird Vocalizations,* R. A. Hinde, ed. New York: Cambridge University Press.

Isaac, D., and Marler, P., 1963. Ordering of sequences of singing behaviour of Mistle Thrush in relationship to timing. *Animal Behav.* 11:179–88.

Kilham, L., 1960. Courtship and territorial behavior of Hairy Woodpeckers. *Auk,* 77:259–70.

Konishi, M., and Nottebohm, F., 1969. Experimental

studies in the ontogeny of avian vocalizations. In: *Bird Vocalizations,* R. A. Hinde, ed. New York: Cambridge University Press.

Kunkel, P., 1966. Quelques tendances adaptives du comportement de certains oiseaux tropicaux. *Chronique de l'Inst. Rech. Sci. Afrique Central,* 1(3):29–37.

Lack, D., 1940. The releaser concept of bird behaviour. *Nature,* 145:107–108.

Lanyon, Wesley, E., 1963. Experiments on species discrimination in *Myiarchus* flycatchers. *Amer. Mus. Novitates,* 2126:1–16.

Lein, M. R., 1972. Territorial and courtship songs of birds. *Nature,* 237:48–49.

Lein, M. R., 1973. The biological significance of some communication patterns of wood warblers (Parulidae). Ph.D. diss., Harvard University.

Lemon, R. E., 1968a. The relation between organization and function of song in Cardinals. *Behaviour,* 32:158–78.

Lemon, R. E., 1968b. Coordinated singing by black-creased titmice. *Canadian J. Zool,* 46:1163–67.

Lemon, R. E., 1967. The response of cardinals to songs of different dialects. *Anim. Behav.,* 15:538–45.

Lemon, R. E., and Herzog, A., 1969. The vocal behavior of Cardinals and Pyrrhuloxias in Texas. *Condor,* 71:1–15.

Lemon, R. E., and Chatfield, C., 1971. Organization of song in Cardinals. *Anim. Behav.,* 19:1–17.

Lemon, R. E., and Chatfield, C., 1973. Organization of song of rose-breasted grosbeaks. *Anim. Behav.,* 21:28–44.

Lorenz, K., 1950. The comparative method in studying innate behaviour patterns. *Symp. Soc. Experimental Biol.,* 4:221–68.

MacDonald, S. D., 1968. The courtship and territorial behavior of Franklin's race of the Spruce Grouse. *Living Bird,* 7:5–25.

McKinney, F., 1965. The displays of the American Green-winged Teal. *Wilson Bull.,* 77:112–21.

Manley, G. H., 1960. Agonistic and pair formation behaviour of the black-headed gull. D. Phil. diss., Oxford University.

Marler, P., 1955a. Characteristics of some animal calls. *Nature,* 176:6–8.

Marler, P., 1955b. Studies of fighting in chaffinches. (2) The effect on dominance relations of disguising females as males. *Brit. J. Animal Behaviour,* 3:137–46.

Marler, P., 1956. The voice of the chaffinch and its function as a language. *Ibis,* 98:231–61.

Marler, P., 1959. Developments in the study of animal communication. In: *Darwin's Biological Work,* P. R. Bell, ed. New York: Wiley.

Marler, P., 1967. Comparative study of song development in sparrows. *Proc. 14th Internat. Ornithol. Congr.* Oxford: Blackwell's.

Marler, P., 1970. A comparative approach to vocal learning: song development in White-crowned Sparrows. *Journal Comp. Physiol. Psychol.* 71, Monograph 2(2):1–25.

Marler, P., and Isaac, D., 1960. Physical analysis of a simple bird song as exemplified by the Chipping Sparrow. *Condor,* 62:124–35.

Mayr, Ernst, 1963. *Animal Species and Evolution.* Cambridge: Harvard University Press.

Milligan, M. M., and Verner, J., 1971. Inter-populational song dialect discrimination in the White-crowned Sparrow. *Condor,* 73:208–13.

Morris, D. J., 1956. The feather postures of birds, and the problem of the origin of social signals. *Behaviour,* 9:75–113.

Morris, D., 1966. Abnormal rituals in stress situations. *Phil. Trans. Roy. Soc.* B, 251:327–30.

Morse, D. H., 1966. The context of songs of the Yellow Warbler. *Wilson Bull.,* 78:444–55.

Morse, D. H., 1967. The contexts of songs in Black-throated Green and Blackburnian Warblers. *Wilson Bull.,* 79:64–74.

Morse, D. H., 1970. Territorial and courtship songs of birds. *Nature,* 226:659–61.

Moynihan, M., 1955. Some aspects of reproductive behavior in the Black-headed Gull (*Larus ridibundus* L.) and related species. *Behaviour,* Suppl. IV.

Moynihan, M. H., 1956. Notes on the behavior of some North American gulls. I. Aerial hostile behavior. *Behaviour,* 10:126–78.

Moynihan, M. H., 1960. Some adaptations which help to promote gregariousness. *Proc. 12th Internat. Ornithol. Congr.:* 523–41.

Moynihan, M., 1962a. Hostile and sexual behavior patterns of South American and Pacific Laridae. *Behaviour,* Suppl. VIII: 1–365.

Moynihan, M., 1962b. The organization and probable evolution of some mixed species flocks of neotropical birds. *Smithsonian Misc. Coll.,* 143(7):1–140.

Moynihan, M., 1962c. Display patterns of tropical American "nine-primaried" songbirds. I. *Chlorospingus. Auk,* 79:310–44.

Moynihan, M. H., 1963. Display patterns of tropical American "nine-primaried" songbirds. III. The Green-backed Sparrow. *Auk,* 80:116–44.

Moynihan, M., 1966. Communication in the Titi monkey, *Callicebus. J. Zool., Lond.,* 150:77–127.
Moynihan, M., 1970. The control, suppression, decay, disappearance and replacement of displays. *J. Theoretical Biology,* 29:85–112.
Mulligan, J. A., 1966. Singing behavior and its development in the song sparrow, *Melospiza melodia. Univ. Calif. Publ. Zool.,* 81:1–76.
Mundinger, Paul C., 1970. Vocal imitation and individual recognition of finch calls. *Science,* 168:480–82.
Nelson, K., 1973. Does the holistic study of behavior have a future? In: *Perspectives in Ethology,* P. P. G. Bateson and P. H. Klopfer, eds. New York: Plenum Press.
Nice, M. M., 1943. Studies in the life history of the song sparrow. *Trans. Linn. Soc. N.Y.,* 6:1–328.
Nicolai, J., 1959. Familientradition in der Gesangsentwicklung des Gimpels (*Pyrrhula pyrrhula* L.). *J. Ornithol.,* 100:39–46.
Norton-Griffiths, M., 1969. The organisation, control and development of parental feeding in the oystercatcher *(Haematopus ostralegus). Behaviour,* 34:55–114.
Payne, R. B., 1970. Temporal pattern of duetting in the Barbary Shrike *Laniarius barbarus. Ibis,* 112:106–108.
Payne, R. B., 1971. Duetting and chorus singing in .African birds. *Ostrich,* Suppl., 9:125–46.
Payne, R. B., 1973a. Behavior, mimetic songs and song dialects, and relationships of the parasitic indigobirds (*Vidua*) of Africa. *A.O.U. Ornithol. Monogr.* 11.
Payne, R. B., 1973b. Vocal mimicry of the paradise whydahs (*Vidua*) and response of female whydahs to the songs of their hosts (*Pytilia*) and their mimics. *Anim. Behav.,* 21:762–71.
Payne, R. B., 1973c. (Review of) Duetting and antiphonal singing in birds, its extent and significance. *Auk,* 90:451–53.
Payne, R. S., and McVay, S., 1971. Songs of humpback whales. *Science,* 173:585–97.
Peek, F. W., 1972. An experimental study of the territorial function of vocal and visual display in the male red-winged blackbird *(Agelaius phoeniceus). Animal Behav.,* 20:112–18.
Power, D. M., 1966. Antiphonal duetting and evidence for auditory reaction time in the Orange-chinned Parakeet. *Auk,* 83:314–19.
Raveling, D. G., 1969. Preflight and flight behavior of Canada Geese. *Auk,* 86:671–81.
Schleidt, W. M., 1973. Tonic communication: continual effects of discrete signs in animal communication systems. *J. Theor. Biol.,* 42:359–86.
Schleidt, W. M., and Shalter, M. D., 1972. Cloacal foam gland in the quail *Coturnix coturnix. Ibis,* 114:558.
Schubert, M., 1967. Probleme der Motivwahl und der Gesangaktivität bei *Phylloscopus trochilus* (L.). *J. Ornithol.,* 108:265–94.
Schubert, G., 1971. Experimentelle Untersuchungen über die Artkennzeichnenden Parameter in Gesang des Zilzalps, *Phylloscopus c. collybita* (Vieillot), *Behaviour,* 38:289–314.
Sick, H., 1959. Die Balz der Schmuckvögel (Pipridae). *J. Ornithologie,* 100:269–302.
Sick, H., 1967. Courtship behavior in the manakins (Pipridae): a review. *Living Bird,* 6:5–22.
Smith, N. G., 1966. Evolution of some arctic gulls (*Larus*): an experimental study of isolating mechanisms. *A.O.U. Ornithol. Monogr.,* 4:1–99.
Smith, S. T., 1972. Communication and other social behavior in *Parus carolinensis. Nuttall Ornithol. Club Publ.* (M.C.Z., Cambridge, Mass.), 11:1–125.
Smith, W. John, 1966. Communication and relationships in the genus *Tyrannus. Nuttall Ornithol. Club Publ.* (M.C.Z., Cambridge, Mass.), 61:1–250.
Smith, W. John, 1968. Message-meaning analyses. In: *Animal Communication,* T. A. Sebeok, ed. Bloomington: Indiana University Press.
Smith, W. John, 1969a. Messages of vertebrate communication. *Science,* 165:145–50.
Smith, W. John, 1969b. Displays of *Sayornis phoebe* (Aves, Tyrannidae). *Behaviour,* 33:283–322.
Smith, W. John, 1970. Song-like displays in the genus *Sayornis. Behaviour,* 37:64–84.
Smith, W. John, 1971. Behavioral characteristics of serpophaginine tyrannids. *Condor,* 73:259–86.
Smith, W. John, in press. *The Behavior of Communicating: An Ethological Approach.* Cambridge: Harvard University Press.
Smith, W. J.; Pawlukiewicz, J.; and Smith, S. T.; in press. Kinds of activities correlated with singing patterns of the yellow-throated vireo. *Anim. Behav.*
Snow, D. W., 1958. *A Study of Blackbirds.* London: Allen and Unwin.
Stokes, Allen W., 1962. Agonistic behaviour among Blue Tits at a winter feeding station. *Behaviour,* 19:118–38.
Stout, J. F., and Brass, M. E., 1969. Aggressive communication by *Larus glaucescens.* Part II. Visual communication. *Behaviour,* 34:42–54.

Stout, John F.; Wilcox, C. R.; and Creitz, L. E.; 1969. Aggressive communication in *Larus glaucescens.* Part I. Sound communication. *Behaviour,* 34:29–41.

Thielcke, G., 1962. Versuche mit Klanattrapen zur Klärung der Verwandtschaft der Baumläufer *Certhia familiaris* L., *C. brachydactyla* Brehm und *C. americana* Bonaparte. *J. Ornithol.,* 103:266–71.

Thielcke, G., 1969. Geographic variation in bird vocalizations. In: *Bird Vocalizations,* R. A. Hinde, ed. Cambridge: Cambridge University Press.

Thielcke, G., 1970. Die sozialen Funktionen der Vogelstimme. *Vogelwarte,* 25:204–29.

Thielcke, G., 1971. Versuche zur Kommunikation und Evolution der Angst-, Alarm- und Rivalenlaute des Waldbaumläufers *(Certhia familiaris). Zeits. Tierpsychol.,* 28:505–16.

Thompson, W. L., 1969. Song recognition by territorial male buntings *(Passerina). Anim. Behav.,* 17:658–63.

Thönen, W., 1962. Stimmgeographische, ökologische und verbreitungsgeschlichtliche Studien über die Mönchmeise. *Ornithologische Beob.,* 5/5:101–72.

Thorpe, W. H., 1958. The learning of song patterns by birds, with especial reference to the song of the chaffinch, *Fringilla coelebs. Ibis,* 100:535–70.

Thorpe, W. H., 1961. *Bird Song: The Biology of Vocal Communication and Expression in Birds.* Cambridge: Cambridge University Press.

Thorpe, W. H., 1963. Antiphonal singing in birds as evidence for avian auditory reaction time. *Nature,* 197:774–76.

Thorpe, W. H., 1972. Duetting and antiphonal singing in birds. Its extent and significance. *Behaviour,* Suppl. 18:1–197.

Thorpe, W. H., and North, M. E. W., 1965. Origin and significance of the power of vocal imitation with special reference to the antiphonal singing of birds. *Nature,* 1208:219–22.

Tinbergen, N., 1935. Field observations of east Greenland Birds. I. The behaviour of the Red-necked Phalarope (*Phalaropus lobatus* L.) in spring. *Ardea,* 24:1–42.

Tinbergen, N., 1952. 'Derived' activities; their causation, biological significance, origin and emancipation during evolution. *Quarterly Rev. Biol.,* 27:1–32.

Tinbergen, N., 1953. *Social Behaviour in Animals.* London: Methuen.

Tinbergen, N., 1959. Comparative studies of the behaviour of gulls (Laridae): a progress report. *Behaviour,* 15:1–70.

Tinbergen, N., 1964. Aggression and fear in the normal sexual behaviour of some animals. In: *The Pathology and Treatment of Sexual Deviation,* I. Rosen, ed. New York: Oxford University Press.

Tinbergen, N., 1965. Some recent studies of the evolution of sexual behavior. In: *Sex and Behavior,* F. A. Beach, ed. New York: Wiley.

Tinbergen, N., and Perdeck, A. C., 1950. On the stimulus situation releasing the begging response in the newly hatched Herring Gull (*Larus argentatus argentatus* Pont). *Behaviour,* 3:1–39.

Todt, D., 1970. Die antiphonen Paargesänge des osafrikanischen Grassängers *Cisticola hunteri prinoides* Neumann. *J. Ornithol.,* 111:332–56.

Tschanz, B., 1968. Trottellummen. *Zeits. Tierpsychol.,* Suppl. 4.

Verner, J., and Milligan, M. M., 1971. Responses of male White-crowned Sparrows to playback of recorded songs. *Condor,* 73:56–64.

Weeden, J. S., and Falls, J. B., 1959. Differential responses of male Ovenbirds to recorded songs of neighboring and more distant individuals. *Auk,* 76:343–51.

Weidmann, U., and Darley, J., 1971. The role of the female in the social display of mallards. *Animal Behaviour,* 19:287–98.

Wickler, W., 1972. Aufbau und Paarspezifität des Gesangduettes von Laniarius funebris (Aves, Passeriformes, Laniidae). *Zeits. Tierpsychol.,* 30:464–76.

Wiley, R. H., 1971. Song groups in a singing assembly of Little Hermits. *Condor,* 73:28–35.

Chapter 23

COMMUNICATION IN METATHERIA

John F. Eisenberg and Ilan Golani

Introduction

PHYLOGENY AND ADAPTIVE RADIATION

The infraclass Metatheria represents a unique mammalian radiation that can serve as a basis for comparison with the infraclass Eutheria, or placental mammals. Classically the Metatheria are considered as a single order, the Marsupialia (Simpson, 1945); however, Ride (1964) has argued that the metatherian families may be realistically grouped in four orders. Such a grouping reflects the antiquity of the early adaptive radiation of this taxon. Active investigation on the ethology of marsupials has begun only recently (Marlow, 1961; Grant, 1974; Sharman and Calaby, 1964; McManus, 1967, 1970; Ewer, 1968, 1969; Sorenson, 1970; Kaufman, 1974; Heinsohn, 1966; Russell, 1970, 1973; Stodart, 1966a, 1966b), although anatomical, ecological, and physiological studies have a longer tradition (see Tyndale-Biscoe, 1973, for an excellent review). Yet many of the more profound questions concerning the evolution of mammalian behavior will be answered only by continued comparisons between those eutherians and marsupials that have evolved convergent adaptations for similar ecological niches. The marsupials represent the only "control" group to test our hypotheses concerning the evolution of behavior within the eutherian mammals. The Metatheria deserve a rigorous treatment and we hope further efforts will be made to study the ethology of this group.

It is generally conceded by paleontologists that the Marsupialia and the Eutheria split off as two independent lines of mammalian evolution from the now extinct order Pantotheria. The marsupials as a recognizable group may have separated from this parental stock over 120 million years ago. Although fossil marsupials from the late Cretaceous are distributed in Europe and North America, it would appear that the major adaptive radiations of marsupials occurred in the geographically isolated land masses of Australia and South America. Marsupial evolution in South America involved the development of some rather large carnivorous forms: the Borhyaenidae, which occupied ecological niches similar to eutherian carnivores on the larger continental land masses. Other South American forms remained rather small (the Didelphidae and Caenolestidae), preying upon arthropods and smaller vertebrates.

In Australia the radiation of marsupials involved the production of herbivorous as well as carnivorous forms. The family Dasyuridae maintains a basic carnivore-insectivore adaptation. Some genera, such as *Planigale,* are rather small and resemble the Holarctic shrews. Other genera either are adapted for feeding on small invertebrates (e.g., *Sminthopsis*) or are adapted as more general predators (e.g., *Dasyurus, Sarcophilus,* and *Thylacinus*). The Australian bandicoots, family Peramelidae, are generalized omnivore-insectivores resembling the continental hedgehogs and armadillos in their feeding and foraging strategies.

Two major herbivorous taxa show extensive adaptive radiations in Australia, including the grazing, browsing kangaroos (family Macropodidae) and the arboreal folivores, comprising the gliders, koalas, and phalangers (superfamily Phalangeroidea). Thus, in the evolutionary history of the marsupials, adaptive radiation produced mammalian forms that replicate in their niche occupancy the major feeding strategies of eutherian mammals, which evolved on the contiguous land masses.

A MORPHOLOGICAL AND DEVELOPMENTAL COMPARISON BETWEEN METATHERIA AND EUTHERIA

These two major taxa of mammals differ in several fundamental respects. The brain structure of the Marsupialia shows some important differences, especially in the lack of a corpus callosum. In addition the relative brain size in many marsupials is considerably less than the brain size of eutherians having comparable body dimensions. The reasons for this discrepancy are difficult to pin down, but Andrew (1962) has hypothesized that brain size tends to be relatively smaller in taxa that have long been isolated from the major continental land masses (see also Jerison, 1973).

The single most profound difference between the Marsupialia and the Eutheria, however, involves structural differences in the reproductive tract (see Tyndale-Biscoe, 1973: 6–8). In addition to the differences in reproductive tract morphology, the marsupials have not evolved a complicated placenta. Most marsupial females develop a yolk sac placenta with a short functional life, although the bandicoots have evolved a placental structure similar to the eutherians'. One could say, as a generalization, that in eutherian mammals nutrition of the young within the reproductive tract of the female by means of a placenta is more prolonged; hence, the young of eutherians tend to be born in a somewhat more advanced condition than the young of the metatherians are.

Upon being born, the marsupial young transports itself into a teat area, which in most marsupial females is enclosed by a fold of skin, the so-called marsupium. This is not universal, however, since many species of marsupials in the families Dasyuridae and Didelphidae do not show a true pouch development. In all cases, however, the rather altricial marsupial young attaches to a teat and undergoes a great part of its early development *extra-utero,* reaching postpartum developmental stages over a longer period of time than do the young of comparably sized eutherians. These extended patterns of development in marsupials have some bearing upon the evolution of the signal systems between mother and young.

A Review of Marsupial Interaction Patterns

In order to acquaint the reader with the forms of interaction shown by marsupials, it is essential to offer a few selected examples of marsupial behavior. The maternal neonatal development cycle will be illustrated with *Didelphis marsupialis,* and the male-female courtship bouts will be illustrated with examples from *Didelphis, Phalanger,* and *Macropus.*

A GENERALIZED DEVELOPMENTAL CYCLE FOR *DIDELPHIS*

In the Virginia opossum *(Didelphis marsupialis)* the life history sequence appears to exhibit the following pattern (Reynolds, 1952). At the time of parturition the female opossum assumes a characteristic posture, squatting on the heels with the head lowered, licking at the pouch area. She remains with the long axis of her body in a vertical plane. As the young are born, they crawl unaided to her pouch. The female *Didelphis* does nothing to assist the movement of the young to the pouch other than to hold her position relatively constant. Similar postures during parturition have been noted for macropod marsupials (Sharman and Calaby, 1964). The tendency on the part of the female to keep her body axis in a constant alignment seems to be important in the initial orientation of the young as they move against gravity to the pouch, but olfactory and tactile cues seem to help the neonate locate the teat area.

During the neonatal and transitional phases of development for the young, they are associated intimately with the mother, being firmly attached to the teats during the neonatal phase. The mother licks the neonates and cleans the pouch throughout this and subsequent phases of development. Thus, at the transition period (Williams and Scott, 1953), when the sense organs become functional, the mother's body is the primary environment for the young. Although the female may have a nest into which she retires, the nest does not have a strong initial valance for the young. It is the female's body to which they are attached and to which they direct all their activity.

During the so-called socialization period of development (Williams and Scott, 1953), the young become capable of some locomotion and begin to move about and interact with the mother and with their litter mates. The young may now be transported on the mother's body. They can detach from the teat, crawl on her body, and return to the pouch unassisted. As this period proceeds, the young may be left in the nest while the mother forages independently, although they will still attempt to get into the pouch whenever possible. Eventually the young become so large that all of the litter cannot crowd into the pouch. They will, however, continue to nurse while partly outside the pouch; and at this time the mother still serves as the primary focus of social interaction. She not only comes to nurse them, she may in fact carry some of them in her pouch when she leaves the nest or permit them to ride on her body.

During the early parts of the socialization period, the young may occasionally become detached from the teat. When they are detached and uncomfortable, the young give a high-pitched, chirping cry. This cry causes the female to approach and stand near them until they climb on her body. In *Didelphis* there are no stereotyped retrieving movements by the female to pick up a young in the mouth and transport it to a nest because in a sense the mother is the nest. Rather, there is a very stereotyped reaction whereby the mother responds to the cry of the young, approaches while clicking, and stands near it, thus permitting it to climb on her or into the pouch. Although *Didelphis* mothers do not often touch the young with the forepaws, females of *Marmosa, Sminthopsis, Dasyuroides,* and many of Macropodidae typically draw the young under them with their forepaws. This movement may "direct" the young to the pouch area. When the youngsters are left alone in the nest, they may interact with one another and become familiar with smells, textures, and postures. Although the young may follow each other and the mother at the time of weaning, no prolonged following tendency appears to persist, and play is minimal.

The foregoing synopsis of development in *Didelphis* highlights the stages of development and the relationship between mother and young.

The nest phase for *Didelphis,* which follows the pouch phase of early development, may not be expressed in those marsupial species that build no nest or have no permanent shelter area (e.g., *Phascolarctos,* and the larger macropods). Although true play is seldom observed in the young opossum, many dasyurid marsupials show play behavior, including *Sarcophilus* (Hediger, 1958). Although following the mother may be shown only weakly by subadult opossums and an early dispersal of the young is the rule, many marsupial species show the capacity to form sustained family groupings (e.g., *Petaurus breviceps*). The young of many macropods may remain with the mother through the birth of a subsequent young. Interaction rates, following tendencies and allogrooming between mother and young, have been analyzed by Russell (1973) for *Macropus eugenii* and *Megaleia rufa.*

A REVIEW OF INTERACTIONS AMONG ADULTS

Didelphis marsupialis has been studied by Reynolds (1952) and McManus (1967, 1970). When moving into a new environment, adults mark by dragging the cloacal region on the ground. The male (and occasionally the female) will also mark by licking selected points in the living space and then rubbing his cheek on the same spot. Rubbing and licking can alternate for several minutes.

Upon approaching a female, the male generally attempts to mount after sniffing her cloacal orifice. During his approaches to the female he will emit a click sound (see McManus, 1970), which is very stereotyped in its temporal patterning. If the female is unreceptive, she will threaten or move away. Threat in the adult consists of several components, including opening the mouth wide (gape threat), hissing, growling, and biting. Similar displays play a role in antipredator behavior, including the emission of foul-smelling secretions from the anal glands and the so-called "death feigning" reaction (Fancq, 1969). Should the female threaten the male, he may remain in front of her, but turn his muzzle away so that his cheek is opposite her face (McManus, 1967). If the female is receptive and stands, the male will attempt to mount her. He may rub his cheeks on the female's body before and during the mount. The male grips the female's hind legs with his feet while assuming the mount. While attempting intromission, the pair falls to one side and completes copulation while lying on the substrate. The male uses a neck grip to restrain the female while mounting and intermittently during copulation. While mounted the male rubs his lower mandible on the neck of the female and frequently rubs the sides of his snout and cheek on her neck. The mount with intromission may last more than twenty minutes.

In the Metatheria the deposition of chemical traces with potential value in communication often involves urine and saliva, as well as specialized marking postures. In desert-adapted species, sandbathing may serve as a means of chemical marking. Sandbathing involves both rolling over and dragging the ventrum in the same locus. Sandbathing probably serves the dual function of dressing the pelage and leaving chemical traces on the substrate. Such patterns have been noted for *Sminthopsis* (Ewer, 1968), *Dasycercus, Dasyuroides,* and *Antechinomys.*

Head turning by the male or facing away from a threatening female have also been described for *Sminthopsis* (Ewer, 1968:349). This pattern will be analyzed in the section for *Sarcophilus.*

In some marsupial species allogrooming may be shown during an interaction. Mutual muzzle licking may terminate an encounter in *Sarcophilus,* while in *Dasycercus cristicaudata* the female is especially likely to groom the male around his cheek, while the male is likely to groom the female behind the ear, on the cheek, in the vicinity of the muzzle, and in the cloacal region.

In arboreal forms, such as *Phalanger gymnotis,* courtship is conducted with extreme deliberation, much of it often taking place on tree branches. After an encounter, the male may often sit to one side and exhibit a gape display, licking his lips afterward. This animal may also hang upside-down, clinging with the tail and/or the hind feet above a social partner. During social encounters both *Phalanger* and *Trichosurus vulpecula* produce short clicklike sounds. In the former it is a repetitive puff-puff while in the latter it is a sharper click apparently involving the teeth.

Before mounting, many marsupials show a typical pattern of male pursuit of a female moving rapidly ahead. In the macropods the male may grasp at the female's tail with his forepaws while in slow pursuit. This movement may involve almost a patting motion and has been noted for *Bettongia* (Stodart, 1966b), *Macropus,* and *Megaleia* (Sharman and Calaby, 1964).

During mounting, the males of many didelphid and dasyurid marsupials employ a neck grip, with the mouth seizing the nape of the female. *Sminthopsis crassicaudata* is an apparent exception (Ewer, 1968). In *Perameles nasuta* the male adopts an almost vertical posture during intromission, which precludes a neck grip (Stodart, 1966a), and the larger macropods have lost the male neck grip during mating. Most marsupial matings are characterized by prolonged single mounts with continuous intromission. Ewer records mounts in excess of eleven hours for *Sminthopsis,* while the larger macropods show shorter mounts of less than twenty minutes. *Perameles* shows extremely brief mounts with intromission, but several intromissions in succession precede ejaculation. For a discussion of copulatory behavior, see Eisenberg (in press).

The Analysis of Interaction Patterns

INTRODUCTION

Interactions between adults or between mother and young often involve the adoption of characteristic postures. Stereotyped postures are often considered "displays" and are thought to be components of a communication system permitting the transfer of information through the eye of the presumptive receiver. In the following section we shall suggest a method for describing the presumed displays of quadrupedal mammals. We shall use our studies of courtship and copulation in the Tasmanian devil *(Sarcophilus harrisii)* as an introduction to our method of analysis (see Golani, in press; Eisenberg, Collins, and Wemmer, 1975).

Some familiarity with the social behavior of *Sarcophilus* in the wild is a prerequisite to understanding the behavior patterns shown in captive encounters. The Tasmanian devil apparently does not form a cohesive social grouping beyond that of the female and her offspring. In nature, adults tend to move alone with no fixed denning site unless a female is in the process of rearing young. Although adults move alone, there is considerable overlap among the home ranges of neighbors (Guiler, 1970a, 1970b). In an encounter between a pair of devils, the male after contacting the female generally attempts to mount her. The development of all courtship interaction between adults is a function of (a) the degree of familiarity the partners have with each other, and (b) the stage of the female's estrous cycle.

INTERACTION VIEWED AS THE PROCESS OF SUCCESSIVE "JOINT" FORMATION

Motor interaction sequences may be studied by describing the consequences of movement as changes of contact points on the animals' own bodies (Golani, in press). For example, instead of noting that an animal faces its partner and shifts its body axis 90° clockwise, we may note that the animal has in effect shifted its partner from its front to its left side. The smallest distance between the two animals can define imaginary contact points on the two animals' own bodies, and in the above example we have described the shift of this contact point on one animal's own body. A "contact point" is defined as such whether or not actual tactile input occurs. The idea implicit in this form of description is that during interaction both partners manipulate each other around their own bodies, shifting each other around various parts of their head, torso, and pelvis. Any change in the relationship between the two animals is described in terms of its consequences on the animals' own bodies. A female may be shifted from one imaginary contact point to another on the male's body through any combination of (1) the male's movements around the female, (2) the male freezing and waiting for the female to move around, or (3) both animals moving simultaneously or successively. In this form of description the motor patterns by which specific contact points are achieved and maintained become secondary to the establishment and maintenance of the contact points themselves.

The analysis of Sequence I of *Sarcophilus* precopulatory behavior (Fig. 1) can serve as an example. From the point of view of the shift of the contact point on the male's own body, the sequence starts with the formation of a contact point on the male's snout. This contact point is maintained so long as the male is approaching the female (Fig. 1a, and Fig. 2♂, phase 1). When the distance between the snouts of the two animals is diminished to a few inches, the presumptive contact point shifts to the male's mouth (Fig. 1b, Fig. 2♂, phase 2), and immediately along the right lateral side of the face (Fig. 2♂, phase 3 and 4), to the posterior lateral side of the head, where it is steadily maintained (Fig. 1c, d, e, f; Fig. 2♂, phase 5). From that point, the contact point shifts along the ventral side of the head to the male's snout (Fig. 1g; Fig. 2♂, phases 6, 7, and 8), where it is again steadily maintained (Fig. 1h; Fig. 2♂, phase 9).

When the same motor interaction sequence is analyzed in terms of the shift of the contact point on the female's own body, it turns out that all the contact points are concentrated around her mouth. At first the contact point is steadily maintained on the female's mouth. This steady maintenance is achieved by her freezing while the male is approaching (Fig. 1a; Fig. 2♀, phase 1). Then the contact point shifts to her mouth, where it is steadily maintained until the end of the interaction (Fig. 1b, c, d, e, f, g, h; Fig. 2♀, phases 3, 4, 5, 6, and 7). Resulting from the series of movements and postures, the female's contact point with the male is maintained constant as if there were a joint between her snout and the male's head.

Once motor interaction sequences are described in these terms, it soon turns out that the behavior consists of continuous fast shifts of the contact points on the part of the two interacting animals from one steadily maintained position to another. When two Tasmanian devils encounter each other, they "slide" along each other's bodies, shifting each other toward particular positions on their own bodies where contact is kept steady for longer time periods, by either freezing or moving in such a way that the particular contact points on their own bodies are maintained constant. Subsequently, contact is released, and each animal slides again on the other's body to new contact points, which are again steadily maintained until the next shift.

The steadily maintained contact points can be conceived of as imaginary "joints" between the two interacting animals: as long as such a joint is maintained, the two animals seem to form a "superorganism," which moves as one unit,

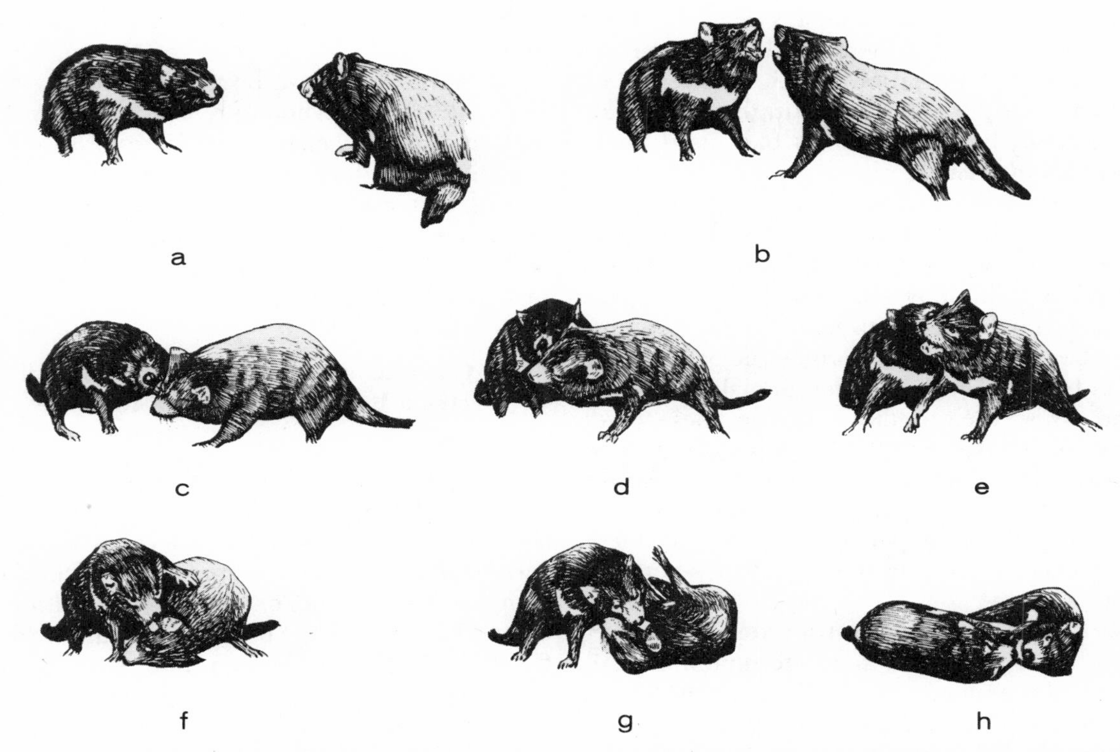

Fig. 1. Single frame drawings extracted from a film of one precopulatory motor interaction sequence for the Tasmanian devil, *Sarcophilus harrisii.* The male is on the right. The figures were extracted on an arbitrary basis in order to give some notion of the movements performed during the interaction. The film was taken at a speed of 24 frames per second. Letters were placed in the same absolute location on the floor of the cage to indicate lateral movements. The joints and the trajectories that were established in the sequence above are summarized in Fig. 2.

subject only to the constraints of the joint. Whereas the actual contact point is fixed, the angle between the two animals may vary. The bounds of each such joint can be rigorously specified, and once the bounds of such a joint are reached due to the movements of one or both animals, the joint either "breaks" and the two animals shift to a new joint, or, as often happens, by exerting further "force" on the joint, one animal may twist the other animal and make it roll on its back. This is analogous to the twisting of someone's arm, which forces him to bend to his knees or lie on the ground. When the contact

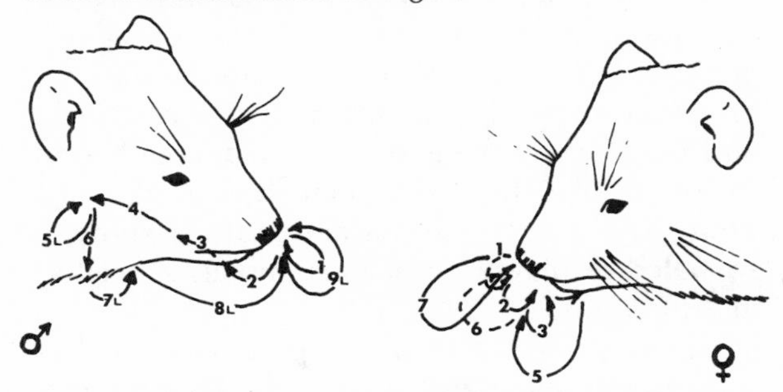

Fig. 2. Diagrams of the shift of the contact point with the partner on the animals' heads during the motor-interaction sequence presented in Fig. 1. Arrows that start and end at the same spot indicate steady maintenance of contact point (joint).

point is with the other animal's head, the head twists first, then the neck, and only then the torso. Once the rotation ability of these three body parts is pushed to its limit, the animal further rotates by rolling on its back as if being pinched and twisted by its snout. The exertion of force on the joint is not necessarily associated with actual contact or mechanical force and may be exerted from a distance. Once such a joint is established, the movements of the two animals become constrained between the contact points with the ground and the joint on the partner (see Figs. 1 and 3). All that the animals can do is either twist their bodies between these pivot points or break the joint with either the ground or the partner. Thus movements and postures are of secondary importance, in the sense that they ensue from the particular location of these interpartner joints and the trajectories between them.

It is the joints and the trajectories of movement to the next joint that determine the positions and the movements of the animals' heads. The rest of their bodies are carried along with their heads (except during mounting and copulation). The heads either are articulated to rigid joints or move along specified trajectories, and to a great extent they determine body configuration. The reader is referred to another paper (Golani, in press), where trajectories and joints are described in terms of the movements that establish and maintain them. The orderliness and simplicity of the patterning in a metatherian carnivore like the Tasmanian devil suggest its utility as a reference or baseline for the study of so-called displays in other marsupials as well as in eutherians.

It should be borne in mind that a joint can in principle be maintained through: (1) some mechanical connection such as a continuous grasp, bite, or sustained contact between two limb segments of two animals; and/or (2) the steady maintenance of sensory input, such as tactile input from vibrissae or even visual or olfactory input.

In the following sections we shall try to review the available literature concerning metatherian display by examining classes of so-called signals grouped according to the various sense organs that may be involved in the process of communication. Wherever possible we shall attempt to discuss each system in terms of the formation and maintenance of joints.

A Classification of Presumptive Signal Systems Based on Sensory Modalities

INTERACTION FORMS AND TACTILE INPUT

During various phases of interaction between adults and between adults and young animals, various parts of the body are sniffed or touched with the muzzle and/or vibrissae. In addition, certain key areas of the body are licked. The data for several species broken into sniffing and touching (Table 1) and licking (Table 2) are given for comparison. When maintained for relatively long time spans, these activities are involved in the formation of joints. It should be obvious that often tactile and chemical signals cannot be separated in this context.

When a series of species are compared during intraspecific encounters between adult males and females, sniffing and touching in the vicinity of the mouth or muzzle appear to be the most frequent occurrence, while sniffing in the cloacal area appears to be the second most frequent configuration. Thus, head to head joints are more frequent in dasyurid marsupials than head to cloaca joints. Licking or allogrooming appear to be less frequent in the Marsupialia and, when they do occur, these actions seem to involve mainly licking the muzzle or cheek.

In *Sarcophilus* mutual licking occurs on the muzzle below and up to the line connecting genal vibrissae and the mystacial vibrissae (see

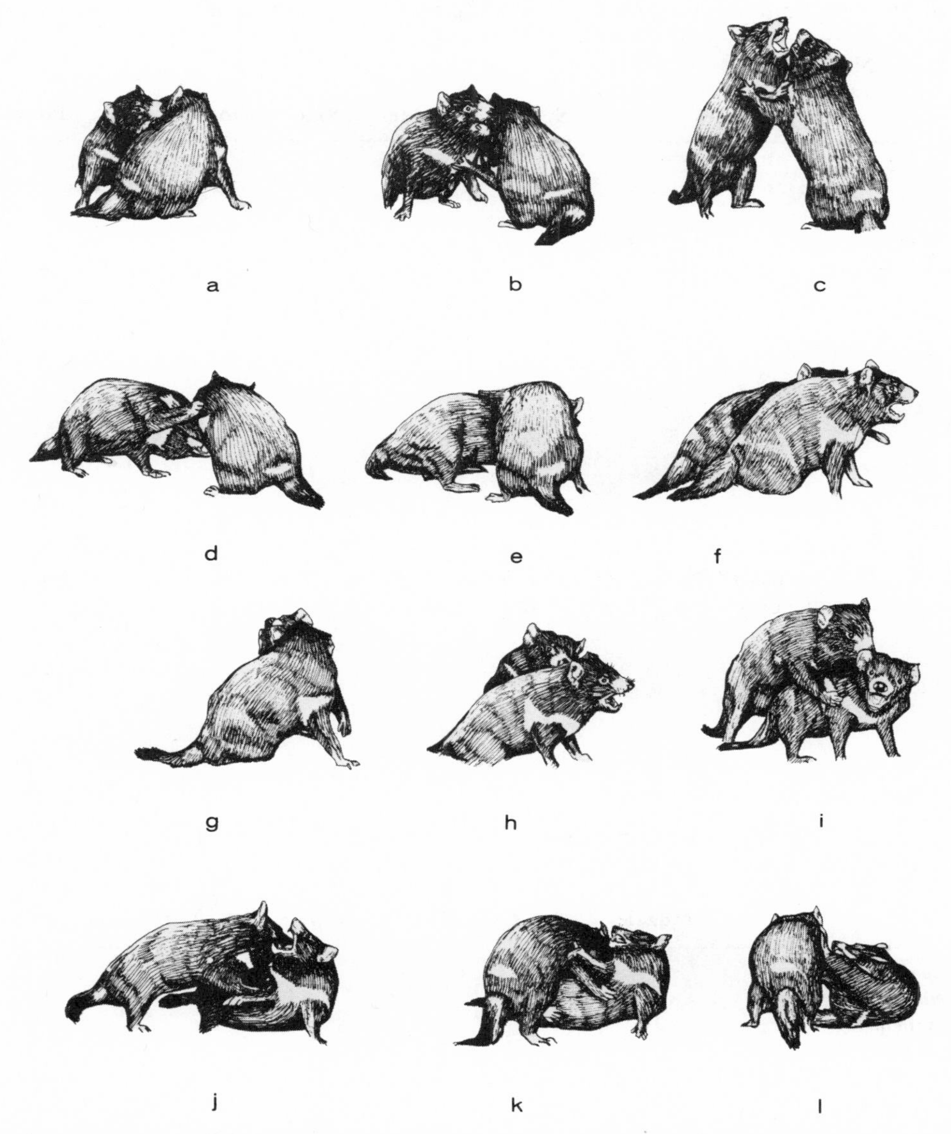

Fig. 3. A mouth to mouth joint (c) leads to a mutual upright. A (♀) mouth to (♂) cheek joint (d, e) results in neck and upper torso rotation by the male, which does not terminate in a roll onto back, as in Fig. 1 f, g. A (♂) mouth to (♀) nape joint (h) precedes and occurs together with mounting (i). This joint is broken, and the contact shifts to a (♀) mouth to (♂) cheek joint when the female rolls on her back (j, k, l). For further explanations, see Fig. 1.

Table 1

*Areas of the body sniffed or touched with the rhinarium during male-female encounters.**

	Muzzle and mouth	Cheek	Ear	Neck	Shoulder	Side	Cloaca	Tail	Pouch	Urine**
Didelphis	++						+			
Marmosa	++						+			
Antechinomys	++		+			+		+		
Sminthopsis	+++		+		+		+			+
Dasyuroides	++	+	+	+			+			
Dasycercus				+			++			
Antechinus	+					+	++			
Dasyurus	+++	+	+++	+			+		+	
Sarcophilus	++	+	+	+						++
Petaurus						+	+			+
Trichosurus								+		++
Phalanger	++		+		+	+	+	+		++
Macropus giganteus	+						+	+	+	++
Megaleia rufa	+	+	+				+	+	+	++

*In Tables 1, 2, and 3 (+) means observed; (++) means frequently observed; no sign means not observed to date.
**Refers to sniffing and/or licking urine deposited during the course of an encounter.

Table 2

Areas of body licked or allogroomed during male-female encounters.

	Muzzle	Cheek	Nape	Ear	Cloacal area
Didelphis					++
Marmosa					++
Antechinomys					
Sminthopsis			+		
Dasyuroides		+			+
Dasycercus	+	+	+	+	+
Antechinus	+	+			
Dasyurus	+				
Sarcophilus	++	+		+	
Petaurus			+		
Trichosurus					
Phalanger					+
Macropus giganteus					?
Megaleia rufa		+			?

Figs. 1 and 2). The partners alternate licking while lying on their sides. Such a bout of allogrooming may last from thirty to forty seconds or longer.

Allogrooming between a female and her offspring is very frequent in marsupials. During the pouch phase of development the young marsupial has a high stimulus valance for the female. In addition to licking and cleaning the pouch area, the female licks and cleans her young. Contact and grooming relationships may persist beyond the pouch phase and into later developmental phases for the young animal. This phenomenon has been reviewed in the research on the larger macropods *Macropus eugenii* and *Megaleia rufa,* by Russell (1973).

Tactile input is, of course, generated during all of the interaction and mating patterns displayed by the various marsupial species. The assumption of the T position by *Sarcophilus* pairs during courtship involves an actual prolonged mechanical contact between the male's cheek and the female's snout. Similar configurations are shown during the interactions of *Sminthopsis* (Ewer, 1968), *Antechinus,* and *Phalanger gymnotis,* to mention a few. A very pronounced hip-slamming interaction is shown in *Vombatus* (Wünschmann, 1970).

INTERACTION FORMS AND THE OLFACTORY CHANNEL

The importance of olfactory signals in the coordination of mammalian behavior cannot be overemphasized and the subject has been recently reviewed (Eisenberg and Kleiman, 1972). The most extensive investigation of olfactory communication in the Marsupialia was conducted by Schultze-Westrum (1965, 1969), who studied the communication system of the sugar glider *(Petaurus breviceps).* This particular species tends to form small, communal groups based on a mated pair and their descendent offspring. In his investigations Schultze-Westrum pointed out that marking with glandular secretions can often be dimorphic; that is, in the male the sternal gland and frontal glands are active and are rubbed on the partner to promote a "community odor." On the other hand, the female sugar glider has glands associated with her pouch area that are important in attracting the newborn young to the pouch on their journey from the cloacal orifice. In addition to secretions from glandular areas, urine, saliva, and feces are important components of the chemical communication system. Special glands, such as the paraproctal gland, may be involved in antipredator behavior and may be equally functional in both sexes.

Ewer (1968) points out that in *Sminthopsis crassicaudata* saliva may be important in promoting individual recognition as well as sexual identity and that sniffing at the muzzle or corner of the mouth is a primary interaction pattern exhibited by this species and takes precedent over sniffing at the cloacal opening.

Urine is obviously of primary importance in certain species, and feces appear to be of secondary importance as sources of chemical information. Often the fecal material itself is of less importance than the glandular deposits left on the feces (Schultze-Westrum, 1965).

Secretions from specialized glandular areas may be deposited with specialized marking movements, including dragging the cloacal region on the substrate, rubbing the chin or cheek, rubbing the sterum or ventrum, and, in those species which practice sandbathing, combining a ventrum rub with side rubs to incorporate at one and the same time a marking movement with a pelage dressing movement. Glands on the forehead are known only for *Petaurus.* The marking movements involving glandular areas are summarized in Table 3.

Saliva would appear to be of extreme importance, although the exact information potentially

Table 3

*Modes of chemical deposition in selected metatheria.**

	Urine dribble	Cloacal drag	Chin rub	Cheek rub	Sternal rub	Sandbathing: Ventral rub	Sandbathing: Side rub	Forehead rub	Distribution of saliva: Bite bark or twigs	Distribution of saliva: Mouth wipe or drool	Distribution of saliva: Face wash
Didelphis			++[1]	+						+	++
Marmosa	+	++	++	+					+		++
Antechinomys		+	++			+	+				++
Sminthopsis	+	+	+[2]			+	+		+		++
Dasyuroides	?	+	+	+		++	+				++
Dasycercus	?	++	+	+		+					++
Antechinus	?	+			+	+			+		++
Dasyurus	+			+							++
Sarcophilus	+	++									++
Petaurus		+			+			+			++
Trichosurus	+	+	+	+	+				+	+	++
Phalanger	+	+							+	+	++
Macropus giganteus	+				+[3]					+	modified
Megaleia rufa	+[3]				+[3]					++	modified

1. On female's neck.
2. On female's back.
3. Specialized, see text.

*In Tables 1-3 the data include the author's own data and the following sources: *Megaleia rufa* (Frith and Calaby, 1969); *Perameles* (Stodart, 1966b; *Bettongia* (Stodart, 1966a); *Didelphis* (Reynolds, 1952); *Sminthopsis* (Ewer, 1968).

available in saliva remains to be experimentally verified. Nipping or biting at bark or the surface of twigs is a common occurrence in marsupials such as *Marmosa robinsoni, Sminthopsis crassicaudata, Dasyurus viverinus, Trichosurus vulpecula,* and *Phalanger gymnotis.* A social partner coming upon such "marks" often pauses to sniff. It is suggested that salivary residues remain at these points although the possibility that actual wounds in the plant surface are responsible for odor production cannot be ruled out. Often, however, excessive salivation and drooling occur during encounters with conspecifics, e.g., *Didelphis, Macropus,* and *Megaleia.* Furthermore, spreading of saliva on the chest and then rubbing the chest area on the substrate is commonly used by male *Macropus giganteus* and *Megaleia rufa* during threatening encounters.

It may be argued that the so-called face wash of the marsupials, which is done by licking the forepaws and then sweeping simultaneously on both sides of the face with the forepaws, is not only a cleaning movement but also a form of self-marking. In fact, similar "washing" patterns in eutherians such as rodents may well subserve a dual function: impregnating the face with saliva and removing foreign matter in the vicinity of the eyes and vibrissae.

During social contacts the partners may sniff at various parts of each other's body and/or lick and nibble at the same areas. This is secondary evidence for the transfer of secretions and hence some form of chemical communication. Table 1 summarizes the data for those species we have studied. A simple inspection will indicate that mutual muzzle sniffing or sniffing at the corners of the mouth are of primary importance in a great many marsupial species. Special attention is often given to the vicinity of the ear, the cloaca, and occasionally the pouch. As pointed out earlier, prolonged examination of these areas often results in the formation of a joint. In some species sniffing the cloacal region appears to be very infrequent. Often, however, urine is actively investigated. This is especially true for *Sarcophilus harrisii,* which appears to spend more time sniffing urine, and then the muzzle or cheek than sniffing any other part of the body.

The postures assumed during interaction often reflect the presence of glandular areas, but the correlation is not perfect. Joint formation between a pair of *Sarcophilus* may involve mouth to mouth or mouth to cheek contact. Odor perception may be involved during the maintenance of these contact configurations; however, the presence of an odor-emitting structure does not necessarily imply joint formation. Cloacal dragging is an important form of marking behavior in *Sarcophilus* (Eisenberg et al., 1975), and yet we have never observed joint formation between the muzzle and the cloaca.

Extensive studies with eutherian mammals have indicated that olfactory signals can carry information concerning species identity, sexual identity, reproductive condition, individual identity, and prevailing motivational tendencies. It is further known that olfactory stimuli can both release sexual behavior and prime sexual behavior. Odors are important in eliciting maternal behavior and may be indirect indicators of dominance and state of arousal. All of the foregoing aspects of olfactory communication remain to be experimentally investigated in the Marsupialia, but they have been reviewed for the eutherians several times (Eisenberg and Kleiman, 1972).

INTERACTION FORMS AND THE AUDITORY CHANNEL

Recently the genesis of sounds by various marsupial species has been surveyed in Eisenberg, Collins, and Wemmer (1975). The function of several of these vocalizations may be inferred from their context. We have found it useful to classify the sounds produced by marsupials into four basic syllable types: (1) Tonal syllables have energy organized into narrow frequency bands. (2) A noisy syllable does not exhibit discrete energy bands but has energy widely distributed. (3) A mixed syllable is a sound that appears as a superimposition of noise on a harmonic series. (4) The "click" is a syllable exhibiting little harmonic structure and lasting less than .02 second.

The intermediate intensity clear or tonal calls tend to be involved in courtship and during mother-infant interactions. Infants that fall out of the pouch frequently emit a chirping call to which the mother responds by approaching the infant and allowing it to either climb back into the marsupium or onto her body. Clear calls of a loud intensity are frequently involved in a context where an animal is moderately aroused and is attempting to avoid a stimulus situation.

Hisses, screams, and growls, all noisy sounds, are widely used by marsupials in threat contexts. Clicks or clicklike sounds are often produced during the initial phases of an encounter or by females in response to displaced young. Clicks appear to give information about the exact position of the sender. They are often shown in the initial phases of courtship in a wide variety of marsupial species.

Table 4 portrays the number of syllable forms identified for a series of marsupials studied by the authors. Further details considering the physical properties of the vocalizations and contexts may be found by referring to the paper by Eisenberg, Collins, and Wemmer (1975).

The best-studied marsupial from the standpoint of vocalizations is the Tasmanian devil *(Sarcophilus harrisii).* Clicklike sounds are produced by *Sarcophilus* when the jaws are clapped together during threat; however, a click form of sound production can be generated by snorting or huffing. These sounds are often made before a physical encounter between two partners or upon separation after an encounter. *Sarcophilus* also produces a bark vocalization during nest defense and in other thwarting contexts. The hiss to growl to whine (or whine-growl) and terminal shriek appear to be a graded series and to indicate roughly the state of arousal on the part of the sender from mild irritation on the one hand to protest at the other extreme. Fig. 4 portrays the relative frequency of sound production during different configurations assumed during the social interaction between a male and a female *Sarcophilus.* It is a first approximation in the direction of correlating vocalization with motor interaction dynamics.

It should be noted that the graded series of vocalizations produced by *Sarcophilus* tends to parallel in intensity the form of the interaction. There appears to be a partial correlation between the amount of sound energy produced per unit time and the distance between two animals' heads and their angular relationship. For example, mutual uprights, which are associated with baring of teeth and mouth to mouth contact, are accompanied by the loudest sounds—shrieks or high-intensity whine-growls. The breaking of contact is accompanied by the production of soft, clicklike syllables which could function as position indicators. The conclusion may be tentatively advanced that certain forms of contact and maintenance involve the production of various classes of sounds that are not context-specific but are correlated with shifts in intensity of mood (see Fig. 5). Other interaction forms (e.g., mutual muzzle licking) are not accompanied by vocalization (see Eisenberg, Collins, and Wemmer, 1975).

INTERACTION FORMS AND THE VISUAL CHANNEL

Most of the species of Marsupialia are nocturnal. It would seem logical that visual display in the conduct of intraspecific encounters would be minimal. A variety of simple patterns may be shown interspecifically in the form of various antipredator mechanisms. In most of these cases, specific movements involved in threat are accompanied by vocalizations (hisses or growls).

Table 4

Some auditory signals for marsupials.

Species (Authority)	Context and call type							
	Agonistic							
	Threat or alarm at a distance	Low-intensity arousal	Intermediate arousal	High arousal	Male approaching or following female	Female responding to displaced young	Juvenile displaced from mother	Special context
Didelphis marsupialis (McManus, 1970)		Hiss	Growl	Screech	Click	Click	Chirp	
Marmosa robinsoni (Eisenberg et al., 1975)		Hiss			Click	?	Chirp	
Dasyuroides byrnei (Eisenberg et al., 1975)	Chit	Hiss	Chur	Chatter	?		Chirp	
Sminthopsis crassicaudata (Ewer, 1968)			Hiss	Churr, squeak (juv.)	"di-di-di"	?	Buzzy call	
Sarcophilus harrisii (Eisenberg et al., 1975)	Bark	Growl or hiss	Whine, whine-growl	Shriek	Huff or clap	?	?	Foot stamp
Trichosurus vulpecula (Winter, J. W., as cited in Tyndale-Biscoe, 1973)	Chatter, "buck-buck"	Hiss	Loud hiss	Screech	Click	?	"Zook-zook"	Post-coital chatter (♂)
Petaurus breviceps (Smith, 1973; Collins, unpublished)	"Wok-wok"		Hiss		?	?		Rattle call
Megaleia rufa (Frith and Calaby, 1969; Russell, 1970)	Snort, "ha"		Hiss	Growl	Cluck	Soft cluck	Squeak	Alarm call, "cough," foot thump

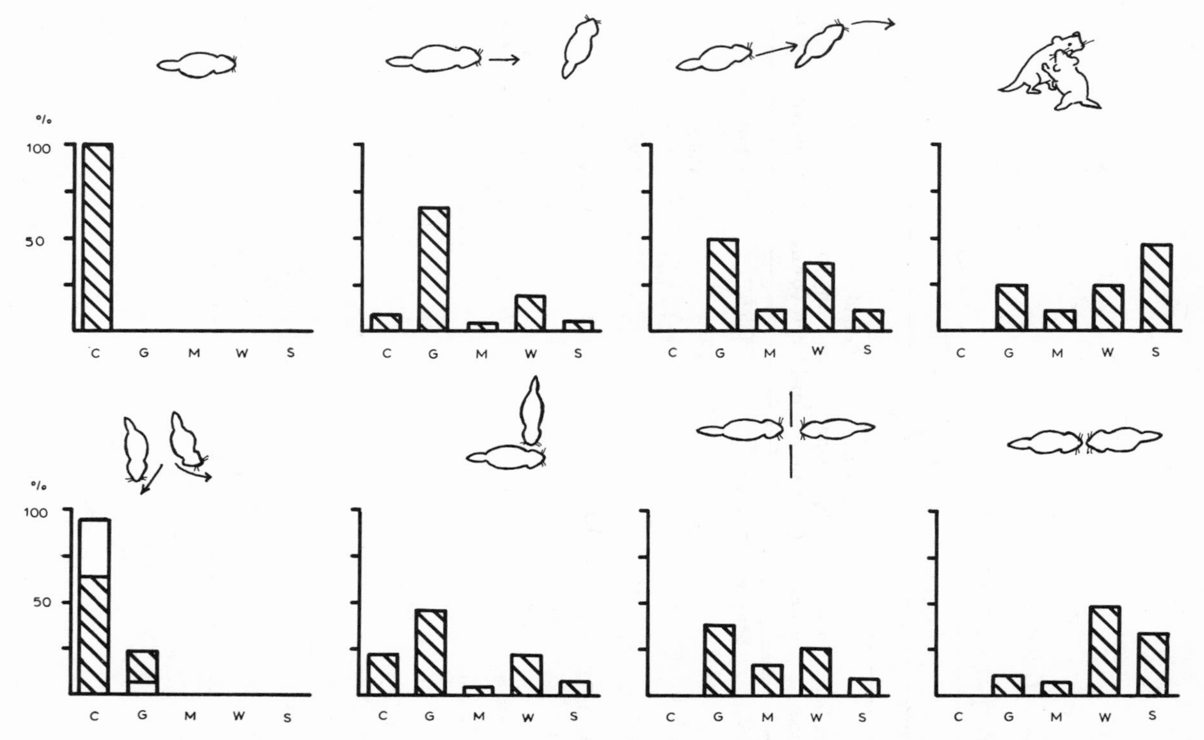

Fig. 4. Sound production and contexts for *Sarcophilus.*

Top row, left to right: Clicklike sounds (c) predominate during exploration; during approach growls (g) increase; during a chase more whining (w) and screaming (s) are added to growl sequences; during an upright whines and screams predominate.

Bottom row, left to right: After a mutual separation, clicklike sounds are prominent either within ten seconds (crosshatch) or during a two-minute period (white); in the T posture growls are prominent with whines; during nest box defense, growls, whines, and moans (m) are frequent; during face to face contact whines (w) and screams (s) are predominant.

Data abstracted from taped encounters; each continuous vocalization counted as one bout. Percentages calculated from total number of signals recorded in a defined context. With the exception of the upright context which accounted for only nine bouts of vocalization, all other contexts ranged from sixty-four to twenty-seven bouts. Vocalization bouts recorded totaled 367.

Given the preceding limitations, a wide variety of nocturnal marsupial species exhibit some form of upright posture, standing bipedally and often exposing a white ventrum to an opponent. Assumption of the bipedal posture may also involve the production of sounds. The gape display, where the mouth is opened widely exhibiting the teeth, is common in didelphine marsupials, the family Dasyuridae, and the family Peramelidae. In some species, such as *Dasyuroides byrnei,* the tail is ornamented, terminating in a black brush. This terminal brush of erectile hairs may be flashed in front of a social partner and may also coordinate pursuit during courtship chases. Aslin (1974) has analyzed this display.

When the nocturnal marsupials are observed under red or dim lighting, it is clear that auditory and olfactory cues are involved in the coordina-

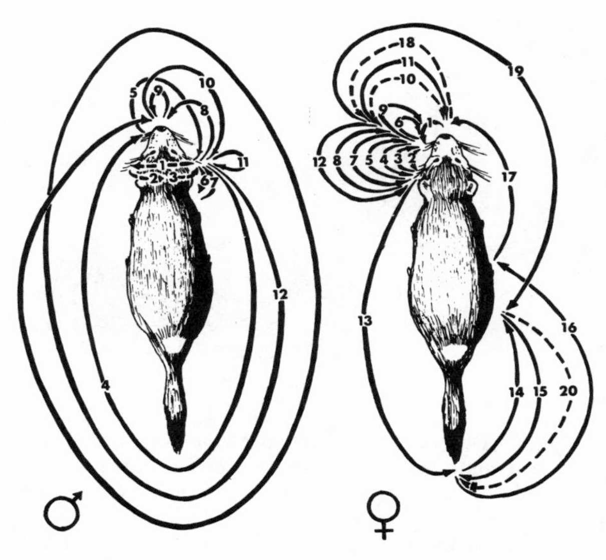

Fig. 5. Diagrams of the horizontal aspect of the shift of the contact point on the animals' own bodies during one precopulatory motor-interaction sequence. Arrows that start and end at the same spot indicate steady maintenance of contact (joints). Note that all joints are located on the head. Full lines describe an active shift. Broken lines indicate a shift resulting from the partner's movements. Broken lines on the female's head indicate a shift of the contact point on the male's throat resulting from rotation and rolling on back. Male symbol (♂) refers to the movement of the male's muzzle on the female's head. Female symbol (♀) refers to the movement of the female's muzzle on the male's head.

tion of interaction patterns. Tactile cues through actual body contact or vibrissae contact appear to assist in the integration of movements. Yet the presence of highly patterned motor behavior suggests that visual input plays some role in the coordination of interaction. It is a mistake, however, to assume that, given what appears to be a display posture, the "display" functions as a visual signal.

In a paper concerning communication in the insectivore genera *Suncus, Blarina,* and *Cryptotis,* Gould (1969) points out that sound production figures prominently in the orientation and interaction sequences of these small mammals. Although the eye reduction of shrews may be a specialization and may not reflect a primitive condition, in many respects the shrews' behavior patterns are conservative and perhaps reflect the repertoire of early nocturnal eutherian mammals (Eisenberg, 1975). Throughout the interaction of shrews, postures are assumed similar to the postures of eutherian mammals, which possess highly developed eyes. The suggestion was made by Gould that these postures are not displays in themselves; "there is no apparent sign that these postures serve any visual function in communication during aggression or courtship." No true exchange of information could take place between interacting shrews in the dark unless the vibrissae touched or the animals came into physical contact (if one disregards the production of vocalizations). Yet the animals stand on their hind legs. They exhibit a turning movement, orienting the side of the head toward an opponent. They lift the head when confronted with an opponent, as if exposing the throat (see Gould, 1969: Fig. 7b, p. 25), and they may roll on the back holding the limbs upright while urinating.

The similarity of many turning movements and postures shown by small nocturnal rodents or insectivores, whose interaction often takes place in the dark or under very dim illumination, has led us to the conclusion that the evolution of visual display patterns in many mammals is much less predominant than the evolution of such action patterns in diurnal reptiles and birds. Rather, in many mammals the emphasis has been to refine aspects of auditory and chemical communication.

Perhaps it is the formation of joints between two interacting nocturnal mammals that determines in a large measure what the form of the subsequent movements will be. Raising the head and neck while approaching an opponent, or rotating the head, or rolling on the back may all result from attempts to control and keep steady some sensory input other than visual, as we discussed for *Sarcophilus* in the previous section.

In diurnal species such as the larger macropods display may be more complicated (Veselovsky, 1969; Grant, 1974) and associated with visual communication in the sense described for reptiles and birds. For example, during the interaction between two males of the species *Megaleia rufa,* the males will assume a quadrupedal strut posture while emitting a low cluck-cluck-cluck vocalization. Assuming an upright posture, the male will then lick the forearms and chest, breaking off to scratch the earth with his forepaws. He may develop an erection and urinate on his ventrum and chest, continuing to wipe at his chest and lick while salivating profusely. With an erect penis the male may begin to advance bipedally upon the opponent.

During approach by a male toward a female red kangaroo, the female will often crouch almost prone on the ground while wiggling her ears. The exact contribution by these various movements to the communication system between two individuals is very difficult to assess. As can be seen from the brief description for the red kangaroo, potential visual components in a display are intermingled with vocalizations and chemical signals.

Fig. 6. Diagrams of joints and trajectories of contact on a male and a female *Sarcophilus* during two precopulatory motor-interaction sequences. Note that all trajectories and joints are located below an imaginary line from cheek to muzzle. Shift of a contact point to the female's nape occurs characteristically by moving behind her ear.

Discussion

THE FORMATION AND MAINTENANCE OF CONFIGURATIONS IN MARSUPIAL INTERACTION SEQUENCES

As pointed out in previous sections of this chapter, it is the location of joints that shapes a large component of marsupial motor interaction sequences. Such joints are formed as the result of an attempt on the part of one or both interacting animals first to establish and then to maintain a specific contact point on the partner's body. Figs. 5 and 6 describe the location of joints on Tasmanian devils. It is clear that once the partners establish contact with each other, they tend to shift each other along their bodies to the head, and once head to head contact is established, both animals attempt to maintain head to head contact below an imaginary line drawn below the eye connecting the cheek and the muzzle. The male can establish muzzle contact with the nape of the female as long as he moves his muzzle from a position below her eye to a position behind her ear and then ultimately to her nape. Only after establishing this muzzle to nape contact, generally through a neck bite, can the male establish cloaca to cloaca contact and ultimately intromission (Fig. 3h, i; Fig. 6♀).

Mutual upright, rolling on the back, hip-slamming, parallel alignment while lying on the ground, and other groupings of movements in *Sarcophilus* (Figs. 1 and 3) may be interpreted as resulting from an attempt on the part of the partners to either establish or maintain specific joints. The location of these joints and the trajectories between them is largely determined by the attempts to control the location of the mouth and the teeth of the partner at specific joints on the individual's own body.

A brief inspection of Table 5, which presents a listing of frequent contacts and interaction forms assumed during adult marsupial encounters, will indicate that the interaction forms in the Tasmanian devil involve the assumption of postures that control the position of the partners' mouths with respect to the rest of their bodies. The most frequent form of contact-promoting behavior involves mutual muzzle licking. Once more, the position of the mouth, armed with the large incisor teeth and surrounded by groups of vibrissae, is under mutual control. To an extent the didelphids and dasyurids have developed mechanisms for controlling the positions of the partner's mouth similar to those described for *Sarcophilus.* Indeed, turning the head away from the partner may not only reduce the potential intimidating effect of the mouth, thus acting as a form of cutoff (Chance, 1962), but it may also present a tactile receptor area, the genal vibrissae, to the social partner.

One can note from this rather simplified presentation (Table 5) that the morphology of the animals determines to a great extent how they will be constrained in the conduct of an intraspecific encounter. The family Macropodidae, with its specialization for browsing and grazing, has lost stabbing canine teeth and the mouth is not used in fighting to any appreciable extent (LaFollette, 1971; Grant, 1974). Conspicuous turning of the head during encounters is reduced and instead the interaction patterns are designed to position the animal for either delivering a forward kick or cuff with the forepaws to the partner or protecting itself from such activities by appropriate warding behavior with the limbs. The upright posture has taken on new significance in macropod interactions because the animals can kick one another in this posture. By the same token, they control their positions relative to one another by clasping with the forepaws in a mutual upright. In the case of the rat kangaroo *(Bettongia)* such kicking may take place even when the animals are lying on their sides clasping one another (Stodart, 1966a). The similarity of both the mutual upright and lying parallel on the ground to those postures illustrated for *Sarcophilus* is noteworthy.

Many marsupials are capable of considerable manual dexterity. This is especially true for phalangerids, dasyurids, and didelphids. It should not be surprising then that to reach out and grasp a partner in order to restrain it is not only physically possible but common. Even in the macropods, clutching at the tail of the female may be part of the courtship ritual, as the male follows an estrous female and prepares to mount her (Veselovsky, 1969).

Gripping the hind legs of the female by employing a prehensile grip with the hind foot is common in didelphine marsupials but lost in those forms that have evolved a hind foot adapted more to cursorial running than to grasping. Once more it is the morphology of the animals that shapes a large component of their interaction forms.

COMPARISONS OF THE METATHERIA WITH THE EUTHERIA

Since the analysis of marsupial interaction sequences and communication processes is still in its infancy, only a few limited conclusions can be drawn from a comparison of these two taxa. As previously stated, however, any future attempt to make generalizations concerning the evolution of eutherian behavior will have to be based on a comparison with metatherians—the only available "control" group. Any future comparisons will have to take into account the size, morphology, ecology, and behavioral capacity of the compared species. Otherwise, fallacious conclusions might be drawn. For instance, Ewer (1968) concluded that the behavioral repertoire of mar-

Table 5

Frequent contact and interaction forms during adult encounters.[1]

Species[2]	Primarily agonistic (defensive or submissive)									Contact promoting		
	Turn broadside	Turn head and expose cheek	Hip slam	Gape	Upright	Kick (back or forward)	Bite or nip	Lock mouths	Cuff	Sniff cloaca	Sniff other body area	Mutual licking (prolonged)
Didelphidae												
Didelphis marsupialis	+	♂		+	Rare	N.S.	+			♂		
Marmosa robinsoni		?		+	Rare	N.S.	+	+		♂ ♀		
Dasyuridae												
Sminthopsis crassicaudata		♂		+	+	?	+		Warding off	♂		
Antechinus flavipes		?		+	+		+			♂		
Sarcophilus harrisii[3]	+	♂	+	+	+	–	+	+		–	+	Muzzle lick
Peramelidae												
Perameles nasuta				+			+					
Phalangeridae												
Phalanger gymnotis				+	+	+	+		+			
Trichosurus vulpecula				+	+	?	+		+			
Vombatidae												
Vombatus ursinus	+		+	?	–	+	+	?				
Macropodidae												
Setonyx brachyura				–	+	+	–	–	+	?		
Bettongia leseuri				–	+	+	–	–		?		
Megaleia rufa	♂			–	+	+	–	–	+	♂	+	

Table 5 (continued)

Species[2]	Preparatory to mount								Mount leading to intromission			
	Follow	Step on tail	Grasp tail	Grasp body	Neck grip and drag	Rub cheek	Rub chin	Neck grip	Press torso and/or palpate with forelimbs	Grip ♀'s legs with hind feet	Rub cheek or chin on neck of ♀	Fall to one side
Didelphidae												
Didelphis marsupialis	♂			♂		♂	♂	♂	♂	♂	♂	+
Marmosa robinsoni	♂			♂				♂	♂	♂		+
Dasyuridae												
Sminthopsis crassicaudata	♂			♂			♂	?	♂	–	♂	–
Antechinus flavipes	♂				♂		♂	♂	♂	–	♂	–
Sarcophilus harrisii[3]	♂			♂ ♀	♂			♂	♂	–	N.S.	–
Peramelidae												
Perameles nasuta	♂	♂						–	–	–	–	–
Phalangeridae												
Phalanger gymnotis	♂			♂				?	?			
Trichosurus vulpecula	♂							?	?			
Vombatidae												
Vombatus ursinus	♂							?	?	–		
Macropodidae												
Setonyx brachyura	♂		♂					–	♂	–	–	–
Bettongia leseuri	♂		♂					–	♂	–	–	–
Megaleia rufa	♂		♂					–	♂	–	–	–

1. + = recorded; ♂ = noted for male; N.S. = not stereotyped; ♀ = noted for female; ? = not certain; – = not regularly observed; blank = not recorded to date.
2. Includes unpublished observations by the authors.
3. See The Analysis of Interaction Patterns.

References in addition to authors' observations include: *Marmosa* (Barnes and Barthold, 1969); *Antechinus* (Woolley, 1966); *Setonyx* (Packer, 1969); *Vombatus* (Wünschmann, 1970).

supials appeared to be simplified when compared with that of eutherians. She felt that marsupials responded very much to stimuli immediately impinging on them and seemed to be less controlled by centrally programmed patterns of coordination. She based these conclusions, however, on an examination of *Sminthopsis,* one of the smallest members of the family Dasyuridae, and then generalized by comparing its behavior with eutherians, which were not only larger but also in some respects more morphologically specialized. Had she confined her comparisons of *Sminthopsis* to small eutherians adapted to a similar ecological niche, she might have arrived at different conclusions.

Thus, the behavioral repertoire of *Sminthopsis,* based on a close comparison with the Tenrecidae (Eisenberg and Gould, 1970), seems no more simplified than that of this eutherian insectivore family. Indeed, one might suspect that extremely small nocturnal mammals, whether they are eutherian or metatherian, may give every appearance of being very much under the immediate control of current sensory input.

A comparison of metatherians and eutherians is thus very risky, since we are dealing with two highly complex groups and our observational methods are only now being refined to the point where we may be in a position to tease apart the significance of the various differences we observe. A comparison of motor interaction sequences of the Tasmanian devil and a carnivore of comparable size such as the jackal *(Canis aureus)* must be undertaken only after relevant anatomical differences have been elucidated.

In *Sarcophilus* the forelimbs are relatively longer than the hind limbs, unlike *Canis* (Moeller, 1968). Thus, during rapid locomotion the forelimbs of *Sarcophilus* appear to bear the major thrust, especially during a gallop. The hand of *Sarcophilus* shows a certain amount of dexterity. The Tasmanian devil can both manipulate objects (Eisenberg and Leyhausen, 1972; Ewer, 1969) and grasp the body of a partner, which is not the case for *Canis.* On the other hand, compared with *Canis, Sarcophilus* has a limited mobility of the head. The neck is relatively short and somewhat inflexible. As a result, the animal must shift direction by moving the whole forepart of the body, while *Canis* can shift by moving the head and neck alone. Also unlike *Canis, Sarcophilus* has a very limited ability to flex its torso in the horizontal plane. Thus, in encounters between two Tasmanian devils the use of the forepaw in grasping at a partner, a limited mobility of the head and torso, and certain differences in the gait—such as shifting front by shifting weight backward to the rigid tail and then using the hindquarters as an axis for the shift of direction—all significantly alter the form of interaction when compared with the highly articulated, cursorial *Canis.*

Another factor that has to be taken into account in a search for a comparison between motor interaction sequences of the two genera is the difference in social structure. Typically a male and female wild canine court each other for a prolonged period of time and in the process establish a pair relationship. In wild canids such as the jackal *(Canis aureus)* the pair relationship persists for several years at least. On the other hand, *Sarcophilus* apparently shows no enduring pair bond, nor is there necessarily any evidence that the male devil provisions the female during her lactation period.

Thus, in an interaction sequence between two canines much time may be spent in establishing a synchronized relationship—an entrainment of motor behavior. In an encounter between a pair of *Sarcophilus* the male after contacting the female generally attempts to mount her. The quality and duration of preliminary interaction between a male and female canid is a function of the degree of familiarity the partners have with each other and the stage of sexual cycling for the male and female. In an encounter between a male and female *Sarcophilus* the degree of familiarity is also significant, but in general the male consistently attempts to mount the female, while she, depending on her estrous cycle, may be making attempts to control the form of the male's activity. The female controls the form of the male's input by rolling on her back, assuming a mouth to mouth joint through mutual upright, assuming a mouth to cheek joint, or running away. Jackal females may assume many other postures and movements which are partly dependent on transitory motor habits that persist for several days and then give way to new motor habits (Golani, 1973). Yet, both *Sarcophilus* and *Canis* assume mutual upright, crouching, so-called T postures, head and neck rotations which might lead to torso rotations which might eventually lead to rolling on the back, hip slamming, head lifting while approaching each other frontally, and several other postures and movements that are, as every student of canine behavior knows, widespread with variations in many mammalian carnivores (see Eisenberg, Collins, and Wemmer, 1975; Golani and Mendelssohn, 1971).

As pointed out throughout this chapter, such postures and movements could be interpreted as means to establish and maintain specific joints during a motor interaction sequence. One notable difference between the two genera is that vocalizations are prominent in the initial phases of a *Sarcophilus* interaction (see Table 4) but are limited to growls and low whines in the interactions of *Canis*.

Thus, if one considers postures and interaction forms as static entities, there appears to be little difference in the complexity of the repertoire of *Canis* and *Sarcophilus*. Yet, by focusing on the fine-grain dynamics of motor interaction sequences of these two genera, we could discern three major differences in motor behavior. Whether these features reflect genuine differences of the levels of neural organization and control of the relatively "primitive" marsupial compared to the relatively advanced eutherian, or whether they simply reflect the poor visual capacity of *Sarcophilus* is still an open question. These differences are:

(1) Whereas in *Sarcophilus* joints are maintained through actual mechanical contact, such as grasping or neck gripping, or through mild tactile contact; in *Canis* joints may be maintained visually from a distance. The only instances in which *Sarcophilus* did maintain a joint from a distance was in a context in which the partner was stationary.

(2) While maintaining a joint with a stationary partner, *Sarcophilus* fluctuates around the joint by performing minimal shifts of front and weight (Golani, in press). *Canis* may maintain such joints without fluctuating around them. Since joints are maintained through a homeostatic motor activity, these differences could reflect a difference in the level of motor control shown by a marsupial, on the one hand, and an eutherian on the other.

(3) In attempting to perform a hip-slamming motion, *Sarcophilus,* if he fails to establish contact with the female, may continue to rotate around his center of gravity as many as five full circles while still shifting weight in the direction of the female. Such a phenomenon has never been described in the interaction sequences of *Canis.* This peculiarity in behavior might reflect a difference in the level of motor control shown by the two respective genera and an inability on the part of *Sarcophilus* to shift immediately to a more appropriate behavior.

In summary, it seems to us that the more interesting differences between eutherians and metatherians should be looked for in the fine detail of the dynamics of motor behavior rather than in the simple comparison of static forms of mutual postures and configurations. We believe that once the dynamics of marsupial behavior are studied in some detail, their behavior could then serve as a baseline for inferences concerning the evolution of mammalian display.

References

Andrew, R. J., 1962. Evolution of intelligence and vocal mimicking. *Science,* 137:585–89.

Aslin, H., 1974. The behaviour of *Dasyuroides byrnei* (Marsupialia) in captivity. *Z. Tierpsychol.,* 35:187–208.

Barnes, R. D., and Barthold, S. W., 1969. Reproduction and breeding in an experimental colony of *Marmosa mitis* Bangs (Didelphidae). *J. Reprod. Fert.,* 6:477–82.

Chance, M. R. A., 1962. An interpretation of some agonistic postures: The role of "cut-off" acts and postures. *Symp. Zool. Soc. Lond.,* 8:71–89.

Collins, L. R., 1973. Monotremes and Marsupials: A Reference for Zoological Institutions. *Smithsonian Institution Publication* no. 4888. Washington, D.C.: Smithsonian Press.

Eisenberg, J. F., 1975. Phylogeny, behavior and ecology in the Mammalia. In: *Phylogeny of the Primates: An Interdisciplinary Approach,* F. S. Szalay and W. P. Luckett, eds. New York: Plenum Press, pp.47–68.

Eisenberg, J. F., (in press). The evolution of the reproductive unit in the Class Mammalia. In: *Lehrman Memorial Symposium,* J. Rosenblatt and B. Komisaruk, eds. New Brunswick, N.J.: Rutgers University Press.

Eisenberg, J. F.; Collins, L. R.; and Wemmer, C.; 1975. Communication in the Tasmanian devil (*Sarcophilus harrisii*) and a survey of auditory communication in the Marsupialia. *Z. Tierpsychol.,* 37:379–99.

Eisenberg, J. F., and Gould, E., 1970. The tenrecs: a study in mammalian behavior and evolution. Smithsonian Contributions to Zoology, No. 27. 137pp.

Eisenberg, J. F., and Kleiman, D. G., 1972. Olfactory communication in mammals. *Ann. Rev. Ecol. & System.,* 3:1–32.

Eisenberg, J. F., and Leyhausen, P., 1972. The phylogenesis of predatory behaviour in mammals. *Z. Tierpsychol.,* 30:59–93.

Ewer, R. F., 1968. A preliminary survey of the behavior in captivity of the dasyurid marsupial, *Sminthopsis crassicaudata* (Gould). *Z. Tierpsychol.,* 25:219–65.

Ewer, R. F., 1969. Some observations on the killing and eating of prey by two dasyurid marsupials: the mulgara, *Dasycercus cristicaudata,* and the Tasmanian devil, *Sarcophilus harrisii. Z. Tierpsychol.,* 26:23–38.

Francq, E. N., 1969. Behavioral aspects of feigned death in the opossum, *Didelphis marsupialis. Amer. Midland Nat.,* 81:556–68.

Frith, H. J., and Calaby, J. H., 1969. *Kangaroos.* London: C. Hurst; New York: Humanities Press.

Golani, I., 1973. Non-metric analysis of behavioral interaction sequences in captive jackals (*Canis aureus* L.). *Behav.,* 44 (1–2):89–112.

Golani, I., (in press). Mechanisms of motor homeostasis in mammalian display. In: *Perspectives in Ethology,* II, P. Klopfer and P. P. G. Bateson, eds. New York: Plenum Press.

Golani, I., and Mendelssohn, H., 1971. Sequences of precopulatory behaviour of the jackal (*Canis aureus* L.). *Behav.,* 38:169–92.

Gould, E., 1969. Communication in three genera of shrews (Soricidae): *Suncus, Blarina,* and *Cryptotis. Communication in Behavioral Biology,* Part A, 3:11–31.

Grant, T. R., 1974. Observations of enclosed and free-ranging grey kangaroos, *Macropus giganteus. Z. Säugetierkunde,* 39:65–78.

Guiler, E. R., 1970a. Observations on the Tasmanian devil, *Sarcophilus harrisii* (Marsupialia: Dasyuridae). I. Numbers, home range, movements, and food in two populations. *Australian J. Zool.,* 18:49–62.

Guiler, E. R., 1970b. Observations on the Tasmanian devil, *Sarcophilus harrisii* (Marsupialia: Dasyuridae). II. Reproduction, breeding, and growth of pouch young. *Australian J. Zool.,* 18:63–70.

Hediger, H., 1958. Verhalten der Beuteltiere Marsupiala. *Handbuch der Zoologie,* 8:1–27.

Heinsohn, G. E., 1966. Ecology and reproduction of the Tasmanian bandicoots (*Perameles gunni* and *Isoodon obesulus*). *Univ. Calif. Pubs. Zool.,* 80:1–96.

Jerison, H. J., 1973. *Evolution of the Brain and Intelligence.* New York: Academic Press.

Kaufman, J. H., 1974. The ecology and evolution of social organization in the kangaroo family Macropodidae. *Amer. Zool.,* 14:50–62.

LaFollette, R. M., 1971. Agonistic behaviour and dominance in confined wallabies, *Wallabia rufogrisea frutica. Anim. Behav.,* 19:93–101.

McManus, J. J., 1967. Observations on sexual behavior of the opossum, *Didelphis marsupialis. J. Mammal.,* 48(3):486–87.

McManus, J. J., 1970. Behavior of captive opossums, *Didelphis marsupialis virginiana. Amer. Midland Nat.,* 84:144–69.

Marlow, B. J., 1961. Reproductive behaviour of the marsupial mouse, *Antechinus flavipes* (Waterhouse) (Marsupialia) and the development of pouch young. *Australian J. Zool.,* 9(2):203–18.

Moeller, H., 1968. Zur Frage der Parallelerscheinungen bei Metatheria und Eutheria. *Zeit. für Wissenschaft. Zool.,* 177:282–392.

Packer, W. C., 1969. Observations on the behavior of the marsupial *Setonix brachyurus* (Quoy and Gaimard) in an enclosure. *J. Mammal.,* 50:8–21.

Reynolds, H. C., 1952. Studies on reproduction in the opossum (*Didelphis virginiana virginiana*). *Univ. Calif. Pubs. Zool.,* 52(3):223–84.

Ride, W. D. L., 1964. A review of Australian fossil marsupials. *J. Royal Soc. Western Australia,* 47(4):97–131.

Russell, E., 1970. Observations on the behaviour of the red kangaroo (*Megaleia rufa*) in captivity. *Z. Tierpsychol.,* 37(4):385–404.

Russell, E., 1973. Mother-young relations and early behavioural development in the marsupials, *Macropus eugenii* and *Megaleia rufa. Z. Tierpsychol.,* 33:163–203.

Schultze-Westrum, T. G., 1965. Innerartiliche Verstandigung durch Düfte beim Gleitbeutler *Petaurus breviceps papuana* Thomas (Marsupialia: Phalangeridae). *Zeit. für vergleichende Physiologie,* 50:151–220.

Schultze-Westrum, T. G., 1969. Social communication by chemical signals in flying phalangers. In: *Olfaction and Taste,* C. Pfaffmann, ed. New York: Rockefeller University Press, pp.268–77.

Sharman, G. B., and Calaby, J. H., 1964. Reproductive behavior in the red kangaroo, *Megaleia rufa,* in captivity. *CSIRO Wildlife Research* 9(1):58–85.

Simpson, G. G., 1945. The principles of classification and a classification of mammals. *Bull. Amer. Mus. Nat. Hist.,* 8:1–350.

Smith, M., 1973. *Petaurus breviceps. Mammalian Species,* 30:1–4. Amer. Soc. Mammalogists.

Sorenson, M. W., 1970. Observations on the behavior of *Dasycercus cristicaudata* and *Dasyuroides byrnei* in captivity. *J. Mammal.,* 51:123–31.

Stodart, E., 1966a. Management and behaviour of breeding groups of the marsupial *Perameles nasuta* Geoffroy in captivity. *Australian J. Zool.,* 14:611–23.

Stodart, E., 1966b. Observations on the behaviour of the marsupial *Bettongia lesueuri* (Quoy and Gaimard) in an enclosure. *CSIRO Wildlife Research,* 11(1):91–101.

Tyndale-Biscoe, H., 1973. *Life of Marsupials.* New York: Elsevier.

Veselovsky, Z., 1969. Beitrag zur Kenntnis des Fortpflanzungsverhaltens der Känguruhs. *Der Zool. Garten,* 37:93–126.

Williams, E., and Scott, J. P., 1953. The development of social behavior patterns in the mouse in relation to natural periods. *Behaviour,* 6:35–64.

Woolley, P., 1966. Reproduction in *Antechinus* spp. and other dasyurid marsupials. In: *Comparative Biology of Reproduction in Mammals,* I. W. Rowlands, ed. London and New York: Academic Press, pp. 281–94.

Wünschmann, A., 1970. *Die Plumpbeutler (Vombatidae).* Wittenberg Lutherstadt: A. Ziemsen Verlag.

Chapter 24

INSECTIVORE COMMUNICATION

Walter Poduschka

The seven families of Insectivora recognized at the present time (Erinaceidae, Soricidae, Talpidae, Tenrecidae, Solenodontidae, Chrysochloridae, Potamogalidae) are a zoological order that remains a highly rewarding field of investigation not only in communication problems. Their behavior patterns and the physiology of the senses of many of them—for example, most of the African and Asiatic shrews, the southern and east African golden moles, the Echinosoricini, etc.—have not yet been examined at all. These research complexes rather have been neglected in favor of mere taxonomic or anatomical studies. At present, only the modest beginnings of a comparative insectivorology are available, presumably since our knowledge of many of the species of Insectivora can be supported only by fragmentary reports and papers that are scattered throughout a diffuse literature on the subject.

One of the reasons for the current gap in our knowledge is that the Insectivora, which are placed at the root of the mammals' genealogical tree, are by no means an homologous group. Their phylogenetic age leads us to suspect an especially large number of long-extinct families and offshoots, whose fossil remains are more difficult to find and interpret than are those of larger animals, simply because of their small size. This also explains the differences noticeable among present-day insectivores. Many intermediary forms that might offer an explanation for and insight into modern species have not been discovered or are as yet incompletely known. Single, sharply distinguishing features have evolved at different times among insectivores, though they are considered the least specialized order of mammals.

The overall picture is, therefore, that of a mosaic: progressively differentiated, archaic, and conservative characteristics can be stated and have developed independently at various times (Thenius and Hofer, 1960). This of course is valid not only for anatomical but also for ethological details. It leads us to propose two objectives: parallel to a phenomenological presentation of the single species we should also undertake a program of more extensive criticism and survey. Admittedly this is difficult with Insectivora, which are tiny, often very fast-moving, and for the most part nocturnal animals. They live a great deal of their lives under cover and can only be observed with extreme difficulty. Our attempts to explain and understand the extraordinary importance of acoustic communication; their capacity for perceiving ultrasonics, noticeable within the whole order; the presence of numerous actively working glands that produce their own kind of communication; the sense of smell, illustrated anatomically by an enormously well-developed bulbus olfactorius; and finally the fact that almost all insectivores are amazingly well equipped tactilely reveal just how difficult a study is. In addition, whereas we humans tend to rely heavily on visual signs, these stimuli are only very poorly developed and appear to have the least importance in the communication systems of insectivores.

For this reason any study should be based to a far greater extent on the second objective stated above; the ensuing multiplicity of problems and varying points of view must simply be accepted as unavoidable. Another factor that makes our task so difficult is that most insectivores are solitary animals, which means we cannot assume with certainty that interspecific communication exists except during certain isolated periods in the animal's lifetime: e.g., among mother-young units, among members of the same litter, in courtship behavior, in mating behavior, and in encounters between conspecifics. Active intraspecific communication confines itself to defense against or warding off of predators; the ability to recognize allomones, the communication signals of non-conspecifics—especially when they are transmitted by predators—is known with some insectivores and can be presumed with others.

Happily we possess sufficient facts about Erinaceidae and Tenrecidae. The knowledge we have of solenodons is relatively encouraging when one considers the rarity of the material available for examination; whereas almost no such examinations have been conducted on Echinosoricini, Potamogalidae, and Chrysochloridae. A few good but in no way exhaustive works exist on some Talpidae and even on the semi-aquatic desmans; but the greatest part by far of the shrews, the family with the largest number of species, remains virtually unknown apart from a few excellent detailed works.

Since the smell epithelium has the same basic structure in all mammals, it had been assumed that because of their relatively small size, insectivores are not able to smell well (Müller-Velten, 1966). This assumption has been disproved to a large degree by the discovery, in a few insectivores at least, of the vomeronasal organ (Jacobson's organ), which has been recognized as an actively functioning and extremely precise sensory organ and which apparently corresponds to or derives from the phylogenetical old water-tasting organ of fishes (Poduschka and Firbas, 1968). The ability to echolocate, which was proved to exist among insectivores in shrews and tenrecs and suspected in hedgehogs, Echinosoricini and solenodons, does not have any immediate relevance to communication in itself. We can, however, discuss it here since it has indirect relevance to the physiological and behavioral phenomena that may be important not only in echolocation but also in communication (Griffin, 1968). Echolocating animals must at any rate be capable of complex types of communication.

Because, apart from flying and gliding forms, the order Insectivora shows a radiation that is remarkable in its completeness, certain quantitative studies of the brains of the insectivores that have adapted to semi-aquatic life (*Limnogale, Neomys, Nectogale elegans, Potamogalidae, Chimmarogale,* and both Desmaninae) can also help us when we are studying higher forms. The comparison of results with those from more evolved species shows that this type of adaptation leads to the following modifications: regression of the olfactory centers, enlargement of the auditory centers and of those of the tactile trigeminal system, enlargement of the centers that are related to motricity and correlation of motricity, and enlargement of neocortical regions, especially of the centers of association; on the whole, there is an increase in brain weight (Bauchot and Stephan, 1968).

The present lack of any comparative insectivorology seems to justify an attempt at a survey of the hitherto known patterns of communicative behavior divided into the four communication mechanisms: chemical, acoustic, visual, tactile. A survey of communication functions taking place for territorial, sexual, social, submissive, or aggressive motives would necessitate, in order to form any kind of firm basis for the arguments, many more specialized studies, the essence of which can only be hinted at here. Because of the phylogenetic age of insectivores, and because some of their features, as a result of their age, have remained largely unaffected by time, a thorough investigation of communicative ability in insectivores and the means of communication they have at their disposal ought to provide us with valuable clues in the study of certain higher orders of mammals. Finally, because of the central phylogenetic position occupied by the insectivores, this study may even serve as a basis for research in primatology.

Hedgehogs

FAMILY: ERINACEIDAE; SUBFAMILY: ERINACEINAE (SPINY HEDGEHOGS)

5 genera. Europe, Asia, Africa.

All hedgehogs are either nocturnal or crepuscular animals. Except for those forms that are capable of running very fast *(Hemiechinus)* they are seemingly slow-moving, cryptophile, solitary creatures. Group tendencies are nonexistent, except in mothers with litters. Males and females remain together only during the copulation period, about ten to twelve days.

Optical Communication

The effectiveness of this form is questionable. When hedgehogs are slightly aroused, we notice a slow, continuous raising of the spines that normally lie flat when the animals are undisturbed. When aroused more intensely, the spines are raised with a jerk so sudden that a photograph taken with a shutter speed of 1/125 sec can only produce a blurred picture. It is possible that the erection of the spines may serve as a visual warning, analogous to the attempts of certain animals to make themselves appear as large as possible to impress adversaries—a very common phenomenon in animals. I have the notion, however, that this is not the case among hedgehogs. Rather, the erection of the spines is a means of enabling this attentive and cautious animal to adopt a defensive state. The mere increase in size of the silhouette never results in the retreat of another hedgehog; only taking a fighting position (after first pushing the erected spines on the head forward over the eyes and pressing the head against the substratum) and simultaneously emitting sharp snorts can cause a retreat. If the erection of the spines can be regarded as an optical threat at all, it is only in combination with acoustic signals.

Solitary females in estrus use the shoe of a human foster parent as a substitute for a mate when mounting; this seems to indicate that as far as mating behavior is concerned the appearance of the partner can be considered an optical signal, at least initially. There is, however, never more than the mere hint of mounting, which would mean that optical discrimination is very poor and that probably olfactory and/or auditory stimuli from the partner are necessary to trigger the ritually fixed behavior patterns of mounting proper. We should probably not consider the opalescent secretion from the eye glands as an optical signal; it is far more likely to be a chemical communication act which occurs in hedgehogs in only a rudimentary way; by contrast, tenrecs show the complete behavior pattern.

Threatened hedgehogs show threats of their own with a gaping mouth and thereby reveal their intention of biting, an optical signal that has been proved to exist in all insectivores studied so far.

Acoustic Communication

The subfamily spiny hedgehogs (Erinaceinae) has a vast repertoire of sounds at its disposal, ranging from quiet snorts to loud screams. They vary interspecifically, however, within the subfamily. A comparison reveals a concomitant series, not only of "dialects" with basic similarities that are comprehensible to all hedgehogs but also of completely different signals that are peculiar to and comprehensible to certain genera only, e.g., "schnalz" and squeaking noises in Algerian hedgehogs *(Aethechinus algirus).*

The contact or hunger signal of neonate hedgehogs is a quiet squeaking with high ultrasonic components, very low motorlike "tuckering," and smacking of the lips and/or tongue, the highest frequency components of which reach 37 kHz. Hungry young hedgehogs about three or four weeks old, having lost their own mother, run after strange females, squeaking in a similar way. Whether this is understood as a stimulus for adoption has not yet been tested fully.

The commonest acoustic signal in hedgehogs is a sharp, rapidly repeated snorting. When behavior turns aggressive, this normal quiet sound turns into a vigorous staccato and, when the animal is aroused further, into piercing cries very like those of hedgehogs in great pain or in the throes of death.

All acoustic signals in hedgehogs have strong ultrasonic components, which show easily visible harmonics on the sonagram (Poduschka, 1968, 1969). Frequencies lying within the human hearing range appear to be of little interest to hedgehogs, as filtering them out and playing them back to hedgehogs has proved. For technical reasons it has not yet been possible to discover exactly which harmonics present in the ultrasonic signals are actually relevant for the animals. Experiments indicate that the optimum hearing range for the Egyptian eared hedgehog *(Hemiechinus auritus aegypticus)* is about 40 kHz (Poduschka, unpubl.). Ravizza, Heffner, and Masterton (1969) have proved that this species is able to detect signals up to 45 kHz, presumably even up to 60 kHz.

This ability to perceive ultrasonics is of great advantage to hedgehogs, as indeed it is to all insectivores when searching for prey because they can hear and interpret the mechanical signals of insects, which form their main diet. Common European hedgehogs *(Erinaceus europaeus),* eared hedgehogs *(Hemiechinus auritus),* and long-eared hedgehogs *(Hemiechinus megalotis),* are also capable of roughly locating the source of ultrasonics emitted by insects.

Evidence has been found for echolocation in total darkness. An especially vigorous exhalation of breath occurred when the hedgehog was confronted closely with solid objects. It is still not clear whether the "rusty hinge creaking" emitted by young hedgehogs in dense cover, a sound used by many echolocating animals, which appears as a rattle on the sonagram, is also used by them as a means of echolocation.

Tactile Communication

Until they are about four weeks old, baby hedgehogs seek contact with the mother as often as possible, trying continually to crawl underneath her. When they touch her stomach hairs, which protrude from under the spiny coat of her sides, the mother reacts by raising her body slightly to allow the babies to reach her teats or her body's protecting warmth. If the baby hedgehogs are left alone, they crawl underneath one another; in the resulting pyramid formation, which is also common to many more highly evolved mammals, the most coveted position is at the bottom. This crawling underneath appears to be an intentional tactile stimulus, aimed at inducing the mother to raise her body; baby hedgehogs never react by lifting the body when touched in this way, but, on the other hand, they do not attempt to avoid their brothers and sisters when they do it.

In the courtship of *Hemiechinus auritus* the two partners nestle up against each other and touch each other's body with the side of the head by stretching it forward. This behavior pattern is largely identical with that of the tenrecs *Hemicentetes* and *Microgale* (Eisenberg and Gould, 1970) and that of *Setifer* and *Echinops* (Eisenberg and Gould, 1970; Poduschka, 1974a), whose numerous active gland areas also have communicative importance for the whole mating complex, but it is not like the courtship behavior of the more closely related common European hedgehog. It is not clear whether this can be explained by a mere retrogressive development of these glands in *Hemiechinus,* whereby only the tactile actions have remained. A female long-eared hedgehog in estrus rests her chin for a short period on the foot of the human keeper, which she accepts as a substitute, leading us to assume a tactile function of the intermandibular wart, from which sprout some sinus hairs. It is unknown whether there is an increase in the secretion of the glands at the roots of the intermandibular vibrissae which could possibly afford some kind of chemical communication in genuine mating behavior.

Chemical Communication

When the self-anointing process was understood, it was discovered that hedgehogs can "flehmen" and that they possess an active vomeronasal organ, which they use to define strongly irritant or new and unknown smells and tastes, particularly the individual stimuli encountered in their sexual lives (Poduschka and Firbas, 1968; Poduschka, 1970, 1973). The stimulating odors that enter the mouth while the hedgehog is flehming, as well as the tastes acquired by chewing or licking, are brought into contact with the spittle, which is increased in volume by chewing until it turns to foam. The odors and tastes travel via the ductus nasopalatinus to the sensitive epithelium of the vomeronasal organ, where they are registered and identified. Thus, both gaseous and solid stimuli induce involving a liquid medium. Olfaction in hedgehogs is therefore not merely a specialized form of chemical detection; the stimulus molecules do not necessarily have to reach the nasal cavity and the receptive epithelium through the air. In the vomeronasal organ the hedgehog has at its disposal a sensory organ of considerable communicative importance, which appears to function far more precisely than those sensory receptors used to identify lesser or more-usual smells and taste impressions. The externally visible process of self-anointing *per se* is not a form of communication but simply the last link in a chain of actions that has already reached its peak and fulfilled its purpose in the registration of the sense impression and its transfer into the central nervous system, and that now serves only to clean the vomeronasal organ and render it capable of further function (Fig. 1).

During courtship rutting males often leave scent marks and secretions on the substratum. In doing so, the hind legs are drawn close together and the spine is arched convexly upward, while the partially protruding penis exudes a sometimes whitish secretion, as the body is rocked gently to and fro and from side to side (Poduschka, 1969, 1976). This secretion probably is secerned by the accessory sexual glands, which are well developed in hedgehogs (Ottow, 1955), and seems to be diluted in urine. However, there is also the possibility that it may be some sort of innersecretory steroid diluted in the male urine. Without doubt this is a chemical form of communication that has a stimulating effect on the female. The smell of the secretion is also clearly discernible to the human nose and is completely different from the body and urine odors of a

Fig. 1. Self-anointing of a male subadult European hedgehog. Note the protruding tongue, which deposits the spittle on the spiny coat.

male hedgehog that is not in rut. The scent markings are always deposited behind the female, who usually allows herself to be driven around by the male in zigzag lines or in circles—the "hedgehog roundabout"—but always within a limited area of scarcely more than 100 square yards. The roundabout lasts several hours each night. The female crosses over the scent markings of the male again and again and by doing so comes into olfactory contact. Since before final intromission she allows herself to be mounted several times, from the side and head as well as in the usual manner, traces of the secretion on the underside of the male are deposited on the female's back. It can be safely assumed that this chemical message acts as the necessary release mechanism for her actual readiness to mate, i.e. as a true releaser pheromone. There are even hints that it may act additionally as a primer pheromone by modifying the estrous cycle (Poduschka, 1976).

Meaningful chemical signals derived from feces and urine of females not in estrus or showing mating appetite, as well as from customary defecation areas, have not been proved to exist among hedgehogs. They defend their normal hunting area against intruders, but do not set limits to their territory with scent markings. Juvenile hedgehogs, however, leave scent markings by occasionally pressing their feces against vertical objects at a height of about 2½ inches, even if they have been reared on their own. This is therefore an innate form of communicative behavior, which is not in any way triggered by the presence of a partner or a competitor. Presumably what occurs here is a combination of the chemical scents present in the feces and in the circumanal sebaceous glands, which, on the side toward the intestines, almost touch the especially large protodeal glands.

Individual scents in the nests of hedgehogs do not repel other hedgehogs. A strange nest will be occupied at any time, and the sex of the previous tenant is of no importance.

The ability, so noticeable in tenrecs, to produce a secretion from special eye glands is only weakly developed in hedgehogs and has only been observed so far in *Erinaceus europaeus* (Poduschka, 1969). On the other hand a rubbing of the oral angle (with its many glands) on the substratum by the Persian eared hedgehog *(Hemiechinus auritus persicus)* and the Syrian eared hedgehog *(Hemiechinus auritus syriacus)* has been observed, which can be assumed to be the placing of individual scent markings outside the mating season (cf. Quay, 1965).

SUBFAMILY: GALERICINAE; TRIBE: ECHINOSORICINI (HAIRY HEDGEHOGS)

5 genera. Indonesia, southeastern Asia, Philippines.

There is hardly anything known about the behavior of these animals. We can therefore only attempt to explain the sporadic observations made on living Echinosoricini—which are all from the largest genus, the moon rat *(Echinosorex gymnura)*—by drawing analogies with the behavior of their nearest relatives. *Echinosorex gymnura* is a strictly nocturnal animal, is not too rare but is locally spotty (Davis, 1962), and lives an ostensibly solitary existence (Lim Boo Liat, 1962). We can therefore only expect to find forms of communication between mother and young and during the mating season. Visual threatening consists of gaping with open mouth and producing a very low moaning sound (Davis, 1962). While doing this the animal adopts a crouching pose. Whether this can be called an optical signal is unknown. Acoustic emissions consist of snarls, growls, and groans when the animal is aroused or angry. Whether this "groaning" is identical with the well-known "rusty hinge creaking" referred to in connection with the hedgehogs is not known. Eisenberg (pers. comm.) stated that when *Echinosorex gymnura* explores a strange environment it emits ultrasonic clicks similar to those of the Haitian solenodon *(Solenodon paradoxus)*. When searching for moving prey the moon rat moves its ears individually; this possibly means that it is able to locate the source of sounds through the sound waves that reach each ear independently.

The well-developed sinus hairs on the snout and a strip of hair on the lower side of the naked tail lead us to believe that the animals are very dependent on tactile stimuli. The moon rat catches small fish, but it is not known how it is able to locate them in the water—optically, by tactile stimulus through the vibrissae, or by changes in water pressure that reveal the position of moving prey.

The moon rat deposits fecal matter within one specific area, perhaps as a form of chemical communication. A very pungent odor emanates from the animal, apparently from the anal glands (Eisenberg, pers. comm.). It has none of the musky quality that is associated with the scent of shrews (Davis, 1962).

Nothing at all is known about any behavior or communication patterns among the other four genera: the lesser gymnure *(Hylomys suillus)*, the shrew hedgehog *(Neotetracus sinensis)*, the Hainan gymnure *(Neohylomys hainanensis)*, and the Mindanao gymnure *(Podogymnura truei)*.

Moles

FAMILY: TALPIDAE (MOLES)

5 subfamilies, which include 12 genera and 19 species. Europe, Asia, North America.

Observation of the mostly fossorial or semiaquatic Talpidae is especially difficult and time-consuming. Existing reports probably differ so much because it is almost impossible to witness an entire behavior complex like reproduction behavior or ontogeny of the young. In communication we have to confine ourselves to forming hypotheses about the possibilities available for study, such as the moles' sensory powers or their vocal utterances, which have not yet been studied in detail.

Field observations confirm that the effective range of the mole's senses is very short (Godfrey and Crowcroft, 1960). On the other hand, the cutaneous senses are presumably more complexly developed in the moles than in any other animals, equipping them with an unusual tactile sense and possibly teletactile potential. In addition to the "special senses"—olfaction, taste, hearing, and vision (Quilliam, 1966)—the mole also has (for lack of a better term) the cutaneous senses at its disposal: touch, heat, cold, pain, and vibration. Quilliam emphasizes that the mole must also possess other sensory equipment, the location and structure of which remain a matter of conjecture; certain respiratory problems seem to indicate the presence of a well-developed baroreceptor system.

Most moles are nonsocial, mutually intolerant creatures. The common Eurasian mole *(Talpa europaea)* appears to know no exact limits to its territory. Encounters with other moles in the partly communally used tunnels occur over and over again. In most cases, if one of the animals does not immediately retreat, these encounters lead to serious fights that often end fatally. Talpidae apparently become aware of the presence of adversaries at the very last moment and blunder into them; this does not seem to say much for the presence or development of a communication apparatus.

If two moles in captivity are kept together in a space too confined or if they both fall into the same trap, as a rule one of them will be killed. Thus, a social life with communication can be expected to exist only between mother and young for a brief period, lasting from birth to a few weeks after weaning, and for the unknown but undoubtedly short period during which copulation occurs (Mellanby, 1971). At present we do not know if baby moles emit signals, or, if they do, what kind they emit in order to communicate with the mother. *Talpa europaea* breeds only once, in spring or early summer; the rest of the year it is in an asexual state (Matthews, 1935) in which interspecific communication appears to be even less distinct then during the mating season and rearing of the young.

An exception in puncto sociability is the American starnose mole *(Condylura cristata)*, which cannot be considered solitary: it can be found in small groups (Hamilton, 1931), and we can therefore assume that it possesses a more distinct but unfortunately so far undefined communication system. *Condylura cristata* seems to be far better equipped for communication than most of the other Talpidae since it can see and hear better than they can. Another exception is the shrew mole *(Neurotrichus gibbsi)*, the least fossorial of the five genera of American moles: it seems to be quite gregarious and apparently travels in flexible groups.

Acoustic Communication

If any sort of acoustic communication does take place, it presumably does so within the range of ultrasonics. Although several components of the mole's signals can also be heard by the human ear, it can be assumed that the ultrasonic components serve as an interspecific means of communication: those ultrasonic elements that occur in the shrill squeaking sounds, the very harsh guttural squeaks, the short snorting sounds, and the noises made by the harsh grinding of the teeth, as described by Eadie (1939) in specimens of excited or frightened hairytail moles *(Parascalops breweri)*. Godfrey and Crowcroft (1960) stated that their hearing is acute at close quarters: live earthworms were apparently detected by the sound of their movements. Just how difficult it is to determine acoustic communication in moles is shown by comparing Reed's discoveries (1944), which tell us that the shrew mole *(Neurotrichus gibbsi)* is for the most part mute; but it once emitted a faint, high-pitched, rather musical chattering, audible at short intervals for more than a minute. This is corroborated by Dalquest and Orcutt (1942), who were never able to observe a sound made by this species but suspected ultrasonic signals, since they were able to discern a reaction to noises having a frequency between 8,000 and 30,000 Hz. This is also confirmed by the results obtained by Quilliam (1966), who also noticed reactions in the common Eurasian mole within the normal hearing range and in ultrasonic frequencies. As a result of his anatomical investigations Quilliam suspects, however, that the ultrasonics are not registered via the cochlea, which in moles is the shortest in all of mammalia.

Chemical Communication

Since the bulbus olfactorius is very strongly developed in Talpidae (Godet, 1951; Stephan and Bauchot, 1968b) it can be assumed that moles are greatly influenced by chemical-olfactory stimuli—also having a communicative character—which is indicated by the frequent sniffing of moles with their snouts raised. Godfrey and Crowcroft (1960), however, suspected a range of detection by smell of only six to seven cm. The vomeronasal organ seems of questionable importance to me: Godet (1951) stated that it is clearly formed in the embryo, but degenerates later, and afterward it has no further function. This ought to be tested with modern means and methods of investigation; fresh studies of the vomeronasal organ in various animals during the past few years have resulted in a wealth of new facts and possibilities.

It has been assumed that the various scent glands play an especially large part in the communication of Talpidae. The eastern mole *(Scalopus aquaticus),* the hairytail mole, the Pacific mole *(Scapanus orarius),* the Townsend mole *(Scapanus townsendi),* the California mole *(Scapanus latimanus),* and the starnose mole *(Condylura cristata)* have well-developed skin glands, which produce secretions that leave visible stains on the animals' fur. Of course, this cannot be regarded as an optical signal in an animal with such poor vision, but it could easily be a kind of olfactory signal. Eadie (1939, 1947, 1948) writes about additional glands on the ventral body surface of the hairytail mole and of a large medial, perineal gland having both holocrine and merocrine secretions in the starnose mole.

Optical Communication

Although some moles are capable of using their eyes, sight does not seem to play a role in communication. Presumably *Talpa europaea* is not able to detect static objects (Quilliam, 1966), while the Mediterranean mole *(Talpa caeca)* is completely blind and cannot distinguish moving objects; the same is true of *Neurotrichus,* which shows not the slightest reaction to a sudden strong light (Dalquest and Orcutt, 1942).

Tactile Communication

Touch seems to be very highly developed. The vibrissae are arranged in rows and get longer the further they are from the tip of the proboscis; they are only slightly movable; each hair has its own innervation. Around the nostrils are numerous Eimer's organs for tactile and chemical use. Godfrey and Crowcroft (1960) assume that these organs are able to register mechanical pressure, temperature, humidity, and vibrations. The stiff hairs on the ears are also extremely sensitive, as are the bristles on the tail, which may be tactile but are less actively so than the vibrissae (Dalquest and Orcutt, 1942). Godfrey and Crowcroft (1960) assume that it is probably vibration that gives information about the general direction of a communication partner. The vibrissae on the face are so sensitive that they could probably be used as receptors for air pressure (Quilliam, 1966). Whether they also have a feasible function in a kind of echolocation process is still unknown, since the necessary experiments have not yet been conducted. To what extent the tactile hairs on the outer side of the digging paws of these nearly blind animals, described in more detail in Godet (1951), aid communication during an interspecific encounter is likewise still unknown.

SUBFAMILY: DESMANINAE (DESMANS)

2 genera.

The semi-aquatic Desmaninae occupy a special position in the family of moles. The modification of the sensory organs and the means of communication conditioned by life in another medium have resulted in a specialization that, at least from a cerebral-anatomical viewpoint, has been examined in a series of papers by Stephan and Bauchot (1959, 1968a, 1968b).

The genus *Desmana moschata* (western Russia) lives an apparently social life, which suggests some form of communication. In captivity the animals can be kept in pairs. Unfortunately I have not been able to locate any investigation of their ability to react to external stimuli of any kind.

The southwest European genus *Galemys pyrenaicus,* which in contrast to *Desmana* has never been seen in the wild with members of the opposite sex, appears to be very aggressive toward conspecifics. Only in recent years did Richard and Valette Viallard (1969) succeed in keeping specimens of the Pyrenean desman of both sexes together in captivity without mishap. This leads us to suspect the presence in these animals, too, of communication methods still unknown to us. Ritualized patterns of behavior that render fights harmless, at least between the sexes, and at the same time mechanisms that make possible successful courtship must exist. In the mother-young unit, too, there must be some kind of communication that we still know absolutely nothing about.

Acoustic Communication

The only acoustic signal known in *Desmana moschata* occurs, when the animal is excited, as a metallic squeak (Ognev, 1928). Whether this is an active warning signal or a passive signal of fear is unknown. We know rather more about *Galemys pyrenaicus:* Niethammer (1970) states that it is for the most part silent and possesses only a limited repertoire of sounds—chirping and cheeping. When frightened or suddenly confronted by a conspecific, it emits a high-pitched, loud scream, which unfortunately has not yet been analyzed by an ultrasonic receiver. On the other hand, it does not react to loud noises within the human hearing range like, for example, a loud clapping of the hands, but it does show a strong reaction to infrasonics and noises with ultrasonic components (chirping, clicking, rustling of leaves) and especially whenever large or small objects—a drop of water is enough—fall into the current of the small, fast-flowing, and thus relatively loud Pyrenean streams in which it lives (Richard, 1973). This gives evidence of extraordinarily acute hearing or rather discriminating ability to hear noises and signals that occur within the animal's micro-habitat. Richard also suspects the possibility of echolocation, produced by a "tambourinage" (drumming or loud paddling) with the forefeet on the surface of the water, especially in the vicinity of unexpected obstacles. This action can, of course, also produce a movement in the water, the reflection of which helps orientation.

Optical Communication

The only fact known at present is that the Pyrenean desman is able to register differences in light intensity.

Chemical Communication

Stephan and Bauchot (1968a) were able to discern an enlargement of certain brain structures in *Galemys* and *Desmana* and attribute it to the adaptation to semi-aquatic life. Unfortunately no one has yet investigated whether the ability of the Desmaninae to recognize the precise scent and taste of water (Richard and Valette Viallard, 1969) is connected with the activity of the vomeronasal organ, which is the mammalian equivalent of that special organ in fish that enables them to smell in water. Possibly we should look here for chemoreceptors for use in water, as described in the hypothesis of Bauchot, Buisseret, Leroy, and Richard (1973). As far as *Galemys* is concerned, the quality of the water is a matter of life and death, and precisely because of that the functioning of such an organ is vital to the preservation of its species. In addition, it could indicate, as a kind of communication organ, the presence of conspecifics, prey, or predators.

The anal and subcaudal glands are used as tools for chemical communication. Feces are deposited in specific areas, which are inspected olfactorily and tactilely with the vibrissae and the Eimer's organs of the proboscis. The same thing occurs after micturation. Whenever *Galemys* of different sexes are put together, each one inspects the feces of the other with great interest. This seems to indicate the presence of an intersexual, chemico-olfactory communication process. Chemical scent markings are left by the subcaudal glands, especially by males in spring, the time when reproduction takes place. These subcaudal glands are constantly in contact with the substratum. The smell of the individual scent markings is hardly detectable by the human nose. The nest and the immediate area around it are, however, so strongly impregnated with the gland secretion of the male that it is possible to detect a definite smell of "game." These gland secretions leave black, glistening streaks on the substratum (Richard, 1973).

Tactile Communication

Here, too, we can only draw conclusions from the development of the sensory tools and the functions that they presumably have. Those structures that serve the oral sense of touch (nervus trigeminus, sensitive trigeminal centers, and through them enlargement of the medulla oblongata) are more strongly developed in the Desmaninae than in *Talpa.* The vibrissae on the upper lip, which are unusually developed (Stephan and Bauchot, 1968a, 1968b), are innervated through the trigeminal system. The fact that *Galemys* possesses a proboscis extravagantly equipped with vibrissae and Eimer's organs, which underlines the importance of the tactile sense, was pointed out by Argaud (1944).

Shrews

FAMILY: SORICIDAE (SHREWS)

3 subfamilies with 20 genera and more than 265 species. Europe, Asia, Africa, North America, northern South America.

Our knowledge of the behavior and particularly the communication patterns among most shrews is very fragmentary, if not nonexistent. It is therefore only possible to try to grasp the complexity of communication among shrews, using the results obtained so far. Further work to fill the gaps in our knowledge would not only contribute to answering many unsolved questions but also reveal that shrews, as far as communication patterns are concerned, in no way form an homologous group, so that considerable interspecific variations in communicative ability and the necessary anatomical requirements for communication have to be reckoned with. It would also help to explain, or rather clarify, the numerous contradictions that exist at present. In his excellent paper on communication in three genera of shrews Gould (1969) makes comparative observations on the communication complexes of shrews, using modern research techniques and technical apparatus for the first time.

The hitherto universally accepted idea that Soricidae were solitary animals, does not coincide with the facts, either under good conditions, where adequate food and sufficient space play a major role, or in captivity. So many exceptions are now known that the assumption of a general aggressiveness among conspecifics cannot be defended any longer. There are a great many graduated variations here, which are reflected in the numerous forms and means of communication. They make an overall survey extremely difficult.

The bicolour white-toothed shrew *(Crocidura leucodon),* the lesser white-toothed shrew *(Crocidura suaveolens),* the common European white-toothed shrew *(Crocidura russula),* the musk shrew *(Suncus murinus),* and the least shrew *(Cryptotis parva)* can even be regarded as partially social (semisocial) creatures (Vogel, 1969; Gould, 1969). On the other hand, the genera *Soricinae* (red-toothed shrews) and *Neomys* (European water shrew) are, according to Crowcroft (1955), definitely nonsocial. If several specimens of the least shrew are well looked after in captivity, they are compatible (Conaway, 1958), and as many as twelve sleep and even eat together (Davis and Joeris, 1936). According to Crowcroft (1955), shrews do not generally defend a specific territory but merely the spot where they happen to be at the time and the space as far as they can see. This is not very far, and is, in effect, even more restricted by the fact that they live under dense cover most of the time.

Optical Communication

In *Sorex vulgaris* (= *Sorex araneus*) and *Corcidura coerulea* (= *Suncus caeruleus*?) vision seems to be poor. The optic regions of the brain are small and poorly developed (Clark, 1932). The least shrew (Hamilton, 1944) and the shorttail shrew (*Blarina brevicauda:* Rood, 1958) have weak eyesight, but Blossom (1932) reports good eyesight in the masked shrew *(Sorex cinereus).*

If a male and a female shorttail shrew are put together, the hair above the area of the side glands in the male parts to reveal an apparently bare patch (Eadie, 1938). Since this species is noted for its poor vision, this change cannot be taken as an optical signal but rather as hypertrophy of a secerning gland—similar to the appearance of the swollen naked rings around the eyes in tenrecs, which become especially noticeable during the increase in secretion in the rutting period. Theoretically, however, it is possible that in shrews these patches function as a form of optical communication intended for the female in close proximity and supplement any olfactory-chemical forms.

The common shrew *(Sorex araneus)* is also shortsighted, but the action of rearing up observed in this species is an optical threat recognized and understood by the adversary. This is also true of the action of throwing itself on its back and displaying the light-colored ventral surface, which occurs during an aggravation of the quarrel. Both movements are accompanied by an increase in the volume of screaming. We are able to observe here, therefore, an amalgamation of different communication systems, because screaming does not depend entirely on a visual stimulus: an excited common shrew will also scream in reply to a scream from another, unseen shrew in the immediate vicinity (Crowcroft, 1955). On the other hand, when Gould (1969) separated specimens of shorttail shrew and house shrew by placing a sheet of glass between them, he was not able to observe any signs of visual recognition. Only when they simultaneously placed their noses under the glass plate at the same point did each produce a high-intensity chirp. This also led Gould to the conclusion that "either shrews [of these species!] do not see each other under these circumstances or visual imput must be coupled with tactile and/or olfactory stimuli before recognition will occur."

Tactile Communication

Many shrews do not react to conspecifics until they make vibrissal contact. We know that tactile communication is of great importance to the common shrew, northern water shrew *(Sorex palustris),* European water shrew *(Neomys fodiens),* shorttail shrew, masked shrew, smoky shrew *(Sorex fumeus),* least shrew, and *Crocidura olivieri.*

One of the few ethological examinations of African shrews showed that the African bicolor white-toothed shrew *(Crocidura bicolor)* is "very sensitive to the movements of insects" (Ansell, 1964). This species is reported to rely more on touch and hearing than on any other sense. However, it has not been studied whether this sensitivity to the movements of insects produces results with the aid of hearing or tactile orientation—possibly the teletactile reception of air-pressure waves caused by the wing movements of the insects.

Acoustic Communication

As might be expected among these small insectivores, the relevant acoustic communication signals lie partly in the ultrasonic range. According to Gould's observations one can detect, in addition to ultrasonics up to 107 kHz, low-intensity sounds of 500–1,200 Hz. Clicks have a high localization valence and seem to be used as contact calls. High-intensity chirps and buzzes repel an approaching conspecific; aggressive encounters evoke mixed and graded sounds (Gould, 1969). Buchler (in press) observed, however, that the sound pressure of a bat's signals is 4,000 times greater than that of a vagrant shrew *(Sorex vagrans).* The Herero musk shrew *(Crocidura flavescens herero)* threatens with "a single sharp metallic squeak" (Marlow, 1954/55). The description of this acoustic threat signal indicates the prevalence of ultrasonic elements.

Gould (1969) heard and investigated seven sound types of different intensity in the musk shrew, shorttail shrew, and least shrew: chirps, clicks, twittering, "put," buzzes, chirp buzzes, and putter twitter. These terms, as used by Gould, are an onomatopoeic approximation of the actual signals. He also detected the source of ultrasonic clicks in infant *Blarina* by pulling open the lower jaw, and he was able to see the forward movement of the tongue as it was pressed against the upper palate at the moment of the click emission. Infant house shrews emit a whistle, which is considered a possible variant of the twittering mentioned above. Gould elaborated a clearly arranged synopsis of sound patterns together with typical contexts in which they might occur. One of the most fascinating results is that sometimes completely different signals are emitted, despite the fact that the causes remain the same. Obviously it is not possible to classify the various possibilities of acoustic communication in animals as strictly separate types of sound, closely linked to predictable stimuli or situations.

Neonate Soricidae possess a repertoire of sounds that increases in variety from day to day. They are extremely sensitive to sounds with ultrasonic components, e.g., humans chirping with the lips, as soon as the meatus is open (Dryden, 1968). The signals given by young common European white-toothed shrews and shorttail shrews (Gould, 1969) stimulate the mother to start searching for her young. This cheeping, which can be regarded as a signal for being lost, can cause the father, who has remained with the litter—a fact that proves that the common European white-toothed shrew is by no means a solitary animal—to carry the baby back into the nest from which it has crawled. Interestingly enough, this rescue instinct triggered by the acoustic communication of the baby has no effect on nonpregnant females but does have an effect on suckling females; they will carry a baby that has been deserted, even though it is not one of their own, into the nest and adopt it (Vogel, 1969).

The shorttail shrew possesses a large repertoire of acoustic communication: sharp, high-pitched squeaks when irritated; shrill, piercing squeaks when frightened followed by loud bird-like chatter; and rapid squeaks when contented. As a response to a challenge or as a warning it emits rapid unmusical clicks, like the chatter of teeth. When two aggressive males were confined together, one approached the other and made this peculiar chattering sound, whereupon the latter retreated. Males are more aggressive than females, juveniles less so than adults (Rood, 1958).

As a result of Blossom's investigations (1932) we know that the masked shrew is also capable of producing different signals that can be used as a means of communication. However, the situation that causes them and the purpose they serve are not known. While eating or when searching for food, the masked shrew is reported to utter a succession of faint twittering notes. They are very soft and produced so rapidly that they have a quality somewhere between a purr and a soft twittering. Very similar signals with similar causes are known to exist among other insectivores, e.g., the Haitian solenodon and many of the tenrecs, and are presumed to be phylogenetically very old. Possibly the twittering is similar to the echolocating rattle produced by many bats and is indeed used by shrews for echolocation purposes too. We know from the excellent work by Gould, Negus, and Novick (1964) and by Buchler (in press) that shrews are very adept at echolocation. Pearson (1944) describes the sounds made by male shorttail shrews when pursuing estrous females as "a stream of dry, unmusical clicks like a chitter, similar to the sounds of a twig brushing against a bicycle wheel," which could just as easily be equivalent to the series of impulsive clicks used by bats for echolocating purposes.

When common shrews fight, acoustic signals play an extraordinarily large and complex role. The occupier of a territory raises its muzzle, opens its jaws, and screams. Squeaking contests may follow between the adversaries. Staccato squeaks identify the male sex. The female's voice has less of a barking quality; the individual beats are closer together and at a higher pitch, forming a more continuous scream. As a weapon, the female voice, or rather the manner of its use, seems more effective than that of the male. To human ears the male scream is the more violent and aggressive, but that of the female is more piercing and persistent (Crowcroft 1955, 1957).

If two or more Savi's pigmy shrews *(Suncus etruscus)* are kept together in a container too small, an acoustically varied communication pattern is released, coupled with strong aggressive tendencies (Vogel, 1970). This behavior becomes more intense in close quarters than when the animals have sufficient space.

Chemical Communication

Shrews are equipped with scent-producing glands that vary numerically and potentially according to species. Contrary to earlier assumption, olfactory stimuli are definitely not without a certain significance in such fast-moving animals as shrews. At any rate the degree of olfactory refinement in shrews apparently varies from species to species: it seems to be bad in the least shrew, smoky shrew, gray shrew *(Notiosorex crawfordi),* masked shrew, and shorttail shrew, but quite good in *Crocidura olivieri* and in the common shrew. Current information, however, is self-contradictory or merely descriptive (Buchler, in press).

There is still a great deal of disagreement about the function of the side glands of shrews. Marlow (1954/55) oberved that male Herero musk shrews were frequently seen to rub their sides against the walls of the cage, leaving traces of the oily secretions from their musk glands at various points. As the words "musk glands" show, Marlow believed the characteristic shrew smell was attributable to the secretion from these glands. However, we have since learned, through the work of Dryden and Conaway (1967)—at least for the musk shrew—that the side glands are not responsible for the characteristic musk production. Apparently, concentrations of sweat glands in the throat and behind the ears produce this odor. If a male musk shrew is pursued or otherwise disturbed, he discharges scent while rubbing his sides along the cage wall and his throat and belly on the floor. The characteristic musky odor remains on such objects for several days. This rubbing of the belly seems to be a behavior pattern parallel to that of disturbed Tenrecidae (Eisenberg and Gould, 1970; Poduschka, 1974e) and may be considered phylogenetically a very old behavior pattern in mammals. Pearson (1946) found that the scent glands decrease in size during captivity: after a month they become thinner than in wild animals, making an examination of these glands difficult. Thus a continuous observation of these animals with a specific study of these glands in mind, which is possible only in captivity, is just not feasible.

In addition to the side glands, the shorttail shrew possesses yet another large medial ventral gland, oval in shape, 30 mm long and 9 mm wide. The secretion it produces has a very distinct odor, yet it has not been proved that it is able to ward off enemies, as has sometimes been maintained. This gland is more strongly developed in males during the rutting season than in females. But if a female becomes pregnant, her ventral gland decreases in size (Eadie, 1938). The decrease in, or complete disappearance of, scent emanation due to the reduction in size of the gland shortly before parturition possibly protects the female from discovery by rutting males or even by predators. Another gland, which presumably has some communicative function too, was discovered beneath the tail in the common European white-toothed shrew by Niethammer (1962). Histologically it is very similar to the side glands of this species.

A white secretion that issues from the eyelids in moments of excitement, similar to the way it does among tenrecs, is known at present to exist in the bicolor white-toothed shrew (from a conversation with E. von Lehmann, 1973), in the African forest shrew (*Crocidura giffardi* = *Praesorex goliath:* Vogel, pers. comm.), and in the lesser shrew *(Sorex minutus)* (from a conversation with R. Hutterer, 1974). Whether the secretion also produces olfactory stimuli is still unknown but very probable.

The significance of fecal deposits as a means of chemical communication is still a subject of disagreement; some authors consider them to be communication media, others doubt their importance.

Since a better and more exhaustive presentation of the mating behavior of shrews, which includes an illustrative explanation of the closely interwoven tactile, acoustic, and chemical communication patterns, is hardly imaginable, I should like to take the liberty of quoting Gould (1969) verbatim:

> During initial phases of courtship in both *Suncus* and *Blarina,* the male appears to play a passive role—approaching the female, rubbing the substrate, tolerating bites from the female without biting her—while rubbing and exuding odor in new areas and gradually increasing the receptivity of the female. The female repels the male with high intensity chirps and buzzes; the male is easily repelled by the female's loud vocalizations and bites. Some males emit frequent "put" while courting. Orientation of the female's head and body toward the male was particularly prominent when the male emitted frequent and loud "put." The male responds to bites and loud chirps by closing his eyes and ears and exposing his gland-covered neck. The male rubs his venter over the substrate and simultaneously over his body. His glandular odor, immediately detectable by the observer, is emitted after 2 or 3 minutes (Dryden and Conaway, 1967). The female's body is pervaded with the male's odor through the following means: rubbing of the substrate by the male followed by the female walking over the rubbed areas and toileting herself; occasional fights; the female rubbing her tail against the male's neck as he positions himself behind her. Female *Suncus* reduce biting after the male fur is covered by glandular secretions. (Distortions of lips and tongue-smacking after biting indicate that glandular secretions have a noxious taste to the female.) The male bites the female on the flanks, rump and tail and as she becomes more receptive he follows closely behind her, oriented in a manner similar to the caravan formation prevalent during infancy. Continual advances and increases in click rates by the male (*Suncus* only) are followed by a receptive chirp or twitter by the female *Suncus* and a series of clicks by the female *Blarina.* Copulation follows.

This report is valuable since it shows not only similarities but also differences among members of different genera.

Tenrecs

FAMILY: TENRECIDAE

2 subfamilies with 9 genera comprising 29 species. Madagascar and adjacent islands.

Most tenrecs do not display any deep-rooted grouping tendencies. They live a solitary existence, well dispersed within their environment. Single species, especially the lesser hedgehog tenrec *(Echinops telfairi)*, can sometimes be found in twos and threes, but it is not certain whether these are merely the remaining members of mother-young units. The streaked tenrec *(Hemicentetes)* is the only tenrec genus in which we find colony formation. *Hemicentetes,* whose equipment for acoustic communication is the most specialized and best developed among tenrecs, has a special stridulation organ in the mid-dorsal region. Since a similar organ has not been developed elsewhere other than among juvenile specimens of the tailless tenrec *(Centetes ecaudatus)*, the genus that produces the largest number of young among recent mammals (up to 32 in one litter), it is obvious that the acoustic interspecific method of communication is linked with the more gregarious way of life led by *Hemicentetes* and *Centetes.*

The perineal drag used for leaving chemical scent markings seems to be prevalent in all species of tenrec. The same is true for gaping with the mouth as an optical form of communication. Furthermore, the varied use of ultrasonics in communication is remarkable, as is the complex function of the white secretion produced from specially developed glands situated in the eyelids. This has been the subject of closer study in *Setifer, Echinops, Microgale dobsoni,* and *Microgale talazaci* and is adjudged to be, among other things, a form of chemical communication (Poduschka, 1972b, 1974b).

Optical Communication

Whether the erection of the prickles, bristles, or quills, peculiar to all Tenrecinae when aroused, is an optical signal is not clear. It could be a sign of defense posture, similar to that among hedgehogs, especially since a conspecific only reacts to it when it is accompanied by acoustic signals. As a result, we can say with some degree of certainty that it is used as an interspecific optical threat only within a whole communication complex containing acoustic, possibly optical, and almost certainly chemical forms. Intraspecifically it can be the equivalent of a threatening gesture designed through the erection of the prickles or bristles to make the animal appear larger and, therefore, more capable of successful defense.

When a female greater hedgehog tenrec *(Setifer setosus)* is willing to mate she lifts the perineum from the ground while simultaneously curving her spine concavely downward (lordosis), thereby presenting an unmistakable optical mating signal, which could possibly be strengthened by the emanation of chemical stimuli from the perineal glands, but which is understood visibly by the male (Poduschka, 1972a, 1974a).

The eye-gland secretion, described in more detail in several specific papers, could in certain circumstances act as an antipredator mechanism: the dazzling white patches that suddenly appear when the animal is excited or afraid, in place of the tiny and inconspicuous eyes, could possibly aid a nocturnal or crepuscular creature to scare off a predatory enemy.

Tactile Communication

As nocturnal or crepuscular animals, Tenrecidae possess long and numerous vibrissae (Fig. 2). They appear, however, to be used only for orientation and not in tactile communication. In an encounter other forms of communication are used. In the marsh tenrec *(Limnogale mergulus)*, the only semi-aquatic form of Tenrecidae, we find—presumably as a kind of practical adaptation for life in water—the same numerous vibrissae on the snout as among the equally semi-aquatic *Talpidae Galemys pyrenaicus* and *Desmana moschata;* these vibrissae increase uniformly in length the further they sprout from the tip of the snout, and they are innervated by unusually strong nerve cords. There is no doubt of their primary use as sensory tools (Bauchot and Stephan, 1968). Whether they have a communication value we cannot say, since nothing is

Fig. 2. Adult male *Setifer setosus.* The facial vibrissae extend far beyond the body width.

known about the ethological details of this species, presumably the rarest of tenrecs.

During the courtship behavior of all the Tenrecidae studied so far (tailless tenrec, streaked tenrec, greater hedgehog tenrec, lesser hedgehog tenrec, microgale dobsoni tenrec, microgale talazaci tenrec), there occurs a ritual nestling up to each other with those parts of the body that have concentrations of glands. Here the nose-to-nose, nose and eye area–side, nose–anal/genital contact is merely the beginning of the ritual, or rather the most stereotyped form. Following it, the male and female tenrec rub their sides against each other or crawl over and under each other. All this indicates a combination of tactile communication with stimuli from gland areas in the partner; in addition there is an individual olfactory-chemical stimulus caused by the animal's own secretions, which could be considered a feedback system.

The mounting that follows is soon accompanied by the male's scratching the female's sides with his hind legs, aimed at stimulating her, which acts as a release mechanism for the presentation of her perineum. The male avoids a premature release from the mating "tie" (lasting in *Setifer* for more than two hours, presumably because of the strong accessory erectile tissues in the penis or by an intravaginal retention of the vine-shoot-like glans penis: Poduschka, 1974a) by allowing his hands to rest on the female's sides during the whole tie period. If he gets tired and slides off, he rests his hands on her back. Whenever the female tries to break away from the tie position, the mere hint of a grasp from the male, which is never so firm that the female cannot escape, is enough to indicate that the tie between the genitalia is not yet broken.

During courtship the male lesser hedgehog tenrec repeatedly bites the spines on the female's side. By doing so he seems to stimulate her readiness to mate. This pattern is even more clearly and efficiently developed among the greater hedgehog tenrec: every time the female tries to escape—which is part of her mating ritual—she is held back by the male's energetic biting into the sagittal lower region of her back and drawn back to him again (Poduschka, 1972, 1974a). This bite is not identical with the copulation bite, which, among *Setifer* at least, is not meant to keep the female still so that mating can take place but is only delivered when intromission has already taken place, in order to improve the lordosis of the female. This bite is therefore a tactile signal, to which the female responds by trying to evade the copulation bite. As she slips away, she raises the perineum, thus improving presentation and the tie of the genitalia (Fig. 3).

Acoustic Communication

All tenrecs possess a large and varied repertoire of signals, which seems all the more remarkable when tested by an ultrasonic receiver. Throughout its whole lifetime the streaked tenrec *(Hemicentetes)* even has a special, mid-dorsal stridulation organ consisting of several rows of

Fig. 3. Copulating *Setiferes.* Note the male's stimulant scratching with the hind foot and the stimulant bite on the female's back.

specially formed quills, which are moved by special muscles in such a way that signals are produced by the resulting friction. These stridulation sounds show little harmonic structure. They form a broad band from about 2–200 kHz in adult animals. Young *Hemicentetes* produce stridulation sounds of lower intensity when they are between eleven and seventeen days old. When they are about seventeen days old the intensity of stridulation is very near adult level (Eisenberg and Gould, 1970). This is a phenomenon parallel to that observed in young *Setiferes,* where the signals of the young become stronger and higher in the ultrasonic range the older the babies get (Poduschka, 1974c); and certainly very dissimilar to that in young rodents, whose ultrasonic signals are noticeable just after birth, but gradually become softer or even disappear completely after a few days as soon as the meatus acusticus and the eyes are open, but, on the other hand, will continue for several more days if the animals are handled (Noirot, 1968; Noirot and Pye, 1969; Sales, 1972).

Hemicentetes stridulates almost continuously when active and on the move. Stridulation occurs while feeding, during social contact, during courtship and mounting, during exploration, when escaping pursuers, or when merely moving away. Generally speaking, low stridulation occurs when the animal is normally active. When the animal is aroused intensely, crest erection occurs (presumably more as a sign of readiness for defense than as an active optical signal) together with an increase in stridulation, which does not remain at a peak but decreases in waves after the excitement is over. It is difficult to decide what can be considered communication with a partner, with young, or with threatening predators, and what can be considered a reaction to the stimulus that caused the excitement.

When foraging, young *Hemicentetes* move about nine to ten feet away from the mother, who keeps them close to her by stridulation. Thus, stridulation serves in the mother-young unit not only to identify the female's whereabouts (Eisenberg and Gould, 1970), but also to indicate to the mother where the young are, since they too are able to stridulate.

Young tailless tenrecs also possess a stridulation organ in the mid-dorsal region. It is not such a specialized one, however, and it disappears during individual ontogenesis. It consists of two rows of white spines, which by means of a special dermal musculature vibrate together and produce sounds. Some of these spines may remain in the subadult animal and still produce signals but they are gradually lost and do not get replaced (Gould, 1965). Gould was also the first to find out that stridulation among young tailless tenrecs, a pulsating sound varying in intensity between 12 and 15 kHz, is associated with high levels of excitement, and keeps mother and young together.

It is probable that stridulation was originally a warning signal to a predator but later became an interspecific signal indicating position. Stridulation in *Centetes* is different from that among *Hemicentetes.* Here stridulation occurs in conjunction with crest erection and erection of the center quills and a simultaneous hiss from a half-open mouth. Stridulation seems to occur when there is the inclination to attack coupled with an equally strong inclination to withhold. It was also found that stridulation serves as a warning signal to other members of the group, resulting in arousal and attentiveness. It may also indicate the identity and position of a juvenile that has been startled. It could also help in the location of the young by the mother and/or the location of young by other young. The exact function is unknown (Eisenberg and Gould, 1970).

Echolocation in tenrecs has been proved to exist among *Hemicentetes, Echinops,* and *Microgale dobsoni.* Contrary to the obvious assumption, stridulation is not essential for echolocation; the species studied by Gould (1965) echolocate by means of clicks produced with the tongue.

The vocal emissions of Tenrecidae take place mostly within the range of ultrasonics—at least so far as they are relevant for the animals. When they were filtered out, it was noticed that sounds within the human hearing range were of little interest to the animals.

During mating *Echinops* females utter snapping signals with a frequency of up to 51 kHz and twittering noises up to 37 kHz. These are used not only as a defensive signal in response to the continual insistence of the male but also as an aural threat: Sometimes an unwilling female leaves the nest where she has been urged to mate by the stimulating scratches and mounting attempts of the male; she turns toward him energetically with gaping mouth, and emits these same twittering noises. We can assume, therefore, that these twittering noises serve a number of purposes. *Echinops* also possess other belligerent sounds, including an unmistakable hiss. *Echinops* also reacts to the high-frequency signals emitted by its prey, which are thus recognized as allomones. Mealworms crawling over each other produce a sound with a highest frequency of 42 kHz, which is recognized by *Echinops* as a signal from a well-known prey and which results in a direct search for it.

During mating the greater hedgehog tenrec produces snapping signals up to 83 kHz and squeaks whose strongest sound pressure is between 50 and 60 kHz.

In Gould's (1965) opinion the clicks are produced among *Echinops, Hemicentetes,* and *Microgale* by the lips or the tongue; according to my studies of *Setifer* and *Echinops,* however, they seem to be produced more by the root of the tongue or the soft palate or in the larynx.

The small, soft-furred Tenrecidae *(Oryzorictinae)* emit acoustic signals far less frequently. When defending themselves, they are mostly silent and threaten by gaping with the mouth. Many emit squeals or a long squealing trill, a scream, a wail, or a buzz. When an encounter with a strange conspecific occurs, they utter a soft squeaking sound, which in Eisenberg and Gould's opinion (1970) is meant to prevent any aggressive behavior in the possible adversary.

Chemical Communication

Chemical communication in tenrecs is a subject that has hardly been studied exhaustively. Until now only one aspect has been examined in any detail: the white secretion from the lid glands that was described so far among *Setifer, Echinops,* and the two Oryzorictinae, *Microgale dobsoni* and *Microgale talazaci.* This phenomenon can without doubt be observed in most, if not all, Tenrecidae, possibly in various forms or stages of development; its existence is even more probable in the light of our knowledge of at least the anatomical features necessary for communication that are present in several more species than those mentioned above (Cei, 1946). The chemical examination of this secretion, exuded from special eyelid glands, which exist in addition to the lacrymal glands, is extremely difficult since only a very small quantity can be obtained, but some preliminary results are available.

Eisenberg and Gould (1970) have described additional gland areas in *Hemicentetes, Setifer, Echinops,* and *Microgale* that emit olfactory stimuli. Their exact nature as far as communication is concerned has still to be investigated in detail. The glands in question are in the axillary, inguinal, head, ear, and caudal areas. A further study of the sternal and possible ventral gland areas is already under way (Poduschka, 1974e, 1974f).

The secretion from the lid glands also has communicative character in certain circumstances, since the exuded secretion gives off a strong and long-lasting smell. The odor is understood by conspecifics, which respond by getting very agitated and eventually rubbing off the secretion now being exuded from their own eyelids and nostrils: quite often they also rub the area around their own eyes on vertical objects but never on the substrate. These scent markings are not to be regarded in the same way as territorial markings used in the defense of the area in which an animal lives, since tenrecs, as far as we know, do not have any territorial possessions. They can serve the purpose of self-assertion on the spot where the animal happens to be at the time, a phenomenon already observed among many mammals and well documented by Eisenberg and Kleiman (1972) and Kleiman (1966).

Even if a male *Echinops* in an enclosure that is strange to him has neither smeared eye-gland secretion on the wall nor attempted to place sternal markings (see below), has neither defecated nor micturated, another male put into the enclosure with him will become very excited by scent deposits made by the first male—a phenomenon which has not been explained. The second male will attack as soon as he sees the first male, but will not attack a female in the enclosure or one placed there simultaneously with him. This proves the existence of sexually differentiated scent emanations. Besides the eye-gland, sternal, or ventral secretions, others that could help determine the presence of a male conspecific are pheromones in the spittle, the sweat glands, and other outlets in the skin; the breath; or digestion gases. It must be emphasized that it is not just dominant males that mark in this way, a behavior that Ralls (1971) assumes to be the general norm in mammals.

As a form of chemical communication the eye-gland secretion has various purposes and/or functions:

1. Active: marking behavior, warding off adversaries, suppression of own unease in strange surroundings, possible stimulation of the female during courtship through a smell that acts as an olfactory signal. Whether the eye secretion acts as a primer pheromone and induces ovulation by altering the physiology of the reproduction system is still unknown but not unlikely. Induced ovulation is presumed to exist in tenrecs; fertilization takes place within the ovary (Strauss, 1939, 1942). An optical significance of the white secretion among conspecifics cannot be said to exist, since the female does not look at the male during the preliminary and actual mating behavior; it is just as possible to assume that the male is unable to see the female during the actual mating activity since his eyes might be completely covered by the white secretion (Fig. 4).

2. Passive: excitement during mating due to chemical stimuli from the secretion of the partner, also accompanied by tactile and acoustic stimuli. Several other stimuli, e.g., strong pungent odors or the occurrence of secretion during the very last minutes of a tenrec's life are passive reactions, which have no real bearing on communication.

I have been able to observe and film a special kind of communication or marking among the greater hedgehog tenrec. In strange surroundings or when the animal is unsure of itself be-

Fig. 4. *Echinops* during courtship. The male's eyes are nearly completely covered by his eye-gland secretions.

cause of the presence of several strange conspecifics, the upper part of the body is pressed down flat on the substratum and the forelegs are stretched out passively sidewards so that only the hind legs push the body forward (Poduschka, 1974e). To judge from appearance, it is the sternal glands that are being used here (a behavior pattern similar to those of some more-evolved mammals) or possibly the ventral glands, the presence of which has also been discovered in Soricidae and suspected in solenodons. This apparent method of marking can also take place on a three-dimensional object, for example, on a piece of wood in the enclosure: The animal slides over the top so that its ventral surface is pressed firmly on the object. If the lower side of the object is not lying on the ground (in the case of a large branch or bough of a tree) the animal crawls underneath it, and, lying on its back, the animal presses its ventral area against the object to mark it. The suspected secretion must exude from the gland areas in the breast, possibly from those in the throat or the ventral region. The communicative function of this act is not yet clear. Presumably it is similar to that produced by the rubbing off of the eye-gland secretion, but I have not been able to detect any olfactory stimulus. According to Eisenberg and Gould (1970), the streaked tenrec also leaves scent signals not only by means of perineal drag but also through rubbing the venter by extending and flexing the body on the substratum and by twisting the body when lying on its side. All this has definite potential significance in chemical communication.

Chemical communication also takes place among the two Oryzorictinae, *Microgale dobsoni* and *Microgale talazaci,* as it does in *Setifer* and *Echinops* among the Tenrecinae through markings that exude from the cloacal region. Markings are made by pressing the perineal area on the substratum: while moving forward, the cloaca is repeatedly pressed down on the substratum. In captivity lactating female *Setiferes* deposit feces and urine in one place in the enclosure. This action is then copied by the young, who also use this spot for defecation. The chemico-olfactory stimuli released by the mother's excrement can thus be considered a kind of communication leading to a closely related imitation (Poduschka, 1974c).

The salivating of the lesser hedgehog tenrec, which Eibl-Eibesfeldt (1965) has interpreted as a form of marking behavior, should not be considered as such, but rather as the last link in a chain of actions that reaches its peak in the registration and identification of smells and tastes in the vomeronasal organ. It is thus an equivalent to the self-anointing of the hedgehogs, which, like *Echinops* and *Setifer,* possess an actively functioning vomeronasal organ. Similar to that of hedgehogs, this behavior pattern takes place only when a strange smell or taste has been detected by the animal, and therefore it has no relevance to any active communication process of its own. This salivation also occurs among *Setiferes,* and as far as communication is concerned, it is only a reaction to an extraordinarily strong and stimulative allomone (Poduschka, 1974c).

Otter Shrews

FAMILY: TENRECIDAE; SUBFAMILY: POTAMOGALIDAE (OTTER SHREWS)

3 genera. West to East Africa.

There is hardly anything known about the communication behavior of these animals. The lesser otter shrew *(Micropotamogale lamottei)* in captivity sometimes emits a high-pitched, sharp, loud scream at intervals of about 2 sec (Kuhn, 1964). These acoustic signals have unfortunately never been measured or recorded. Judging from the results obtained from the study of other insectivores, especially the closely related tenrecs, we can assume that these sounds contain ultrasonic components that are important for interspecific communication.

The eyes are remarkably small and presumably have little communicative significance. On the other hand, the vibrissae are very numerous and extraordinarily well developed. Even in the newborn we can clearly see the sinus hair warts from which the strong vibrissae protrude. When the animal is resting, these vibrissae point backward, but they can be spread out sideways and forward when the animal is attentive, even when it is just a few days old (Vogel, pers. photos). Among adults rhythmical movements of the vibrissae backward and forward can be detected (Kuhn, 1964).

Among the Ruwenzori otter shrew *(Mesopotamogale ruwenzorii)* the use of specific defecation areas has been reported (Rahm, 1961), which could possibly have communicative importance.

According to Cei (1946) the big otter shrew *(Potamogale velox)* possesses special glands in the lids, which are presumably equivalent to those studied in Tenrecidae (Poduschka, 1974b). It is questionable, when one considers the otter shrew's semi-aquatic way of life, whether they have similar functions. Among Tenrecidae their communicative functions are limited to the emanation of individual- or at least sexually specific smells and to possible visual signals aimed at warding off inter- and intraspecifics.

Solenodons

FAMILY: SOLENODONTIDAE; GENUS: *SOLENODON*

2 species. Hispaniola: Haitian Solenodon *(Solenodon paradoxus);* Cuba: Cuban Solenodon *(Solenodon cubanus [Atopogale cubana,* sensu Cabrera syn.])

At present we are still not in a position to distinguish between the behavior patterns of the two species. The Cuban form was considered extinct several times during this century, but according to the latest reports, this is untrue.

Thanks to the work of the late Erna Mohr, who was lucky enough to be able to keep and observe more living specimens of solenodons than anyone else (fifteen in all), we know relatively much about the behavior of these extremely rare animals. Of course, modern demands for more detailed information extend beyond the scope of the reports she produced at the time. They were restricted for the most part to a phenomenological inventory of behavior patterns. This inventory has in the meantime been complemented by a few ethological works, which I have specified here.

Solenodons are not solitary animals. They have been found in groups of up to eight animals sleeping in the same hole.

Acoustic Communication

Mohr (1936a) has already described the varied repertoire of sounds that the Haitian solenodon *(Solenodon paradoxus)* possesses. It consists of a "mournful sounding tone like that of kittens prior to the opening of the eyes"; gurgles; shrill screams lasting up to five seconds (?); and a melodious "Strophe," which also lasts a good five seconds, resembling that of a robin. It is often repeated using the same pattern of notes. Mohr assumed at first that this was produced only by young, unweaned animals but later revised her opinion (Mohr, 1936b) when she again heard this Strophe, this time during courtship and mating of sexually mature Haitian solenodons.

Using modern methods of detection, Eisenberg and Gould (1966) registered some chewing and digging sounds also. These were emitted while the animals were walking or running, and were used in particular by the young animals as a source of sound or as a signal to approach. These authors also carried out the first successful ultrasonic tests. My own studies in this field have so far produced varying results, insofar as I have been able, with the aid of a Holgate Ultrasonic Receiver, to detect both clicks with frequencies up to 74.8 kHz and noises produced by the exhalation of breath up to 73 kHz. Of course, this does not prove that these particular signals or their highest components have any relevance for the animals, but it does give some indication: By producing strange noises of a high frequency, which cause solenodons to get very frightened, it was possible to show that this animal is very sensitive to ultrasonic signals between 65 and 75 kHz. The highest components of the emitted signals do not remain constant.

It is also possible to record similar signals of up to 40 or 50 kHz, leading us to suspect varied communicative content. Solenodons are just as sensitive to shrill human voices and loud laughter (Mohr, 1936a) as to mechanical noises with high ultrasonic components (Poduschka, 1974d). Solenodons are able to detect low-pitched ultrasonics well and move toward them, indicating that the ultrasonic signals emitted by insects or small rodents, for example, have interspecific communication value for them. The highest frequencies recorded in the "tuckering" (motorlike) "schnalz" sounds, also mentioned by Mohr, were 28 to 40 kHz. I believe that these sounds are produced mechanically in the larynx or the oral cavity.

The clicks mentioned above can be heard especially when the Haitian solenodon is confronted with a strange conspecific or during the exploration of strange territory. It is not possible to say with certainty whether they are used in echolocation, or as a signal (Eisenberg and Gould, 1966), or could in strange territory be an acoustic parallel, aimed at self-assertion, to the olfactory-chemical behavior pattern of marking already known. These vocalizations of Haitian solenodons are similar to the echolocation pulses of shrews (Gould, Negus, and Novick, 1964).

The Haitian solenodon shows a well-developed appetence for nooks and crannies and employs a special kind of breath exhalation when exploring impenetrable cracks and crevices, which could very well be used as a form of echolocation similar to that detected in hedgehogs.

Chemical Communication

According to Mohr (1936b), the Haitian solenodon reveals a pattern of marking during mating or courtship that includes the use of ventral glands. Unfortunately these glands have not yet been studied. The solenodon slides on his ventral area around and alongside the female, pushing himself along with his forefeet and dragging his hind legs behind him. This reminds us strongly of a special kind of marking behavior observed in the tenrec *Setifer setosus* (Poduschka, 1974e, 1974f), but the solenodon does not press the pectoral area on the substratum; he presses down the venter.

The function of the secernent side glands, which are said to produce odorous substances, has been subject to interpretations that are partially contradictory. So rarely have a male and a female been kept together in captivity that it has not yet been possible to prove that these secretions have any communicative value. According to Mohr (1936b) the side glands only begin to secrete when the animals are six to eight months old; in her opinion the secretion indicates sexual maturity. On the other hand, I have not yet been able to detect any secretion from these glands in a male that was put in with a female after spending five and a half years as a solitary. It could be that the secretion from these glands only indicates sexual activity, which was no longer the case in this particular male—perhaps because of his long period of abstinence or perhaps because solenodon males are sexually active only during the first few years of their lives. The male and female lived together for almost three and a half years, yet never mated. This may indicate that the secretion from the side glands is of essential communicative importance, necessary for the release of a complete pattern of mating behavior in this species.

According to Mohr (1936b) the axillary and ventral areas of adult Haitian solenodons are continually moist. Therefore, glands that could have communicative character must be present. Ignoring for the moment the more abundant material available to Mohr for observation, I must report that I have never been able to detect such an unmistakable secretion in the male and female solenodons I have studied in the past three years. Variations must therefore exist among individuals, conditioned by age or the environment in which the animals are kept.

Fecal deposits are made at random while the animal is on the move. Even in the wild no specially reserved defecation areas have been reported. Eisenberg and Gould (1966) noticed perineal drag immediately after defecation. I have been able to observe and film this only after micturation.

Optical Communication

The only optical signal so far observed among these animals, which live together peacefully with members of their own species and whose tiny eyes and nocturnal habits preclude vision as an important means of communication, is a threatening gesture with gaping mouth. The solenodons were very frightened by an electronic flash. A reduction in the intensity of the flash and the less fearful reaction from the animal that resulted could mean that the startle reaction was indeed the result of an optical stimulus and not of an acoustic one caused by the noise of the camera shutter. I suspect, however, a combination of both sensory impressions (Poduschka, 1974d).

Tactile Communication

Since the solenodon leads a nocturnal life, the vibrissae presumably act as tactile organs. They are found in abundance on the head (Fig. 5), where they sprout in especially large numbers

Fig. 5. Adult male *Solenodon paradoxus*. Note the vibrissae on the head and throat.

on the sides and on the mandible pointing downward, and are also found on the ventral surface, where they stretch from extended embryonic *Milchleiste* up into the axillary regions. There are also carpal vibrissae, like those found in ground squirrels (Poduschka, 1971). The facial vibrissae can be moved in the skin by muscular movements, which are also referred to by Gundlach (quoted by Barbour, 1944) in the Cuban solenodon.

While observing the communal existence of the male and female solenodon, I have never been able to describe with any certainty a particular behavior pattern as an interspecific act of tactile communication. The animals are quite uninhibited in touching each other with various parts of the body, sometimes vigorously sometimes gently, and seem not to be influenced by the partner while absorbed in their remarkably energetic, continual activities, except that they understand the actions of the partner as an optical signal inviting them to join in or to imitate. They do, however, snuggle up to each other, a behavior pattern common in Tenrecidae and many Erinaceidae; and sometimes they push the nose and cheek area along the partner's body, an action that could be considered the beginning of a tactile communication pattern that changes into a chemical one influenced by the gland concentrations of the partner. It is, of course, possible that this is an action combining both kinds of communication, an assumption made even more feasible by the fact that when solenodons meet they nudge each other with the point of the nose. One pushes its nose into the ears or the axillary areas of the other; that is, into parts of the body that give off strong olfactory stimuli (Eisenberg and Gould, 1966). I have experienced this particular behavior on my own person by a specimen that was especially familiar with me. It seems to me, therefore, to be an integral part of their ethogram.

Golden Moles

FAMILY: CHRYSOCHLORIDAE

5 genera. Southern and eastern Africa.

I have not been able to locate any papers on items of the behavior and/or communication of golden moles. It seems, therefore, that nothing at all is known about these very interesting animals, which, because of their solitary and cryptophile way of life, are very difficult to study.

So far only anatomical or taxonomic studies have been produced. The bulbus olfactorius is very large (Stephan and Bauchot, 1960), indicating the importance of the sense of smell. It is questionable, however, whether one can expect to find chemical forms of communication as a result, since olfactory stimuli soon disappear in the dry air of the extremely arid habitat of these animals and cannot therefore be detected easily. Such forms are at least fairly probable, however, within the mother-young unit and indeed are very necessary for these completely blind animals as an indication of readiness to mate.

Results

In the following attempt at a comparative survey of the presently known communication systems of insectivores, I have listed two subfamilies under headings of their own. I want to emphasize that I am quite aware of contravening the normal practice of systematic zoological observation in doing so. This survey, however, is concerned with only one section of ethology: communication, along with the anatomical features necessary for it to take place. It must therefore deal separately with these two subfamilies, since the one differs so completely in its behavior patterns, depending on conditions in its Umwelt, and the other's behavior is virtually unknown.

The first subfamily is the semi-aquatic Desmaninae. Because of adaptation to another medium, the Desmaninae brain differs greatly from that of other Talpidae: modifications include regression of the olfactory centers and enlargement of the auditory, trigeminal, and motor centers. There is also enlargement of neocortical regions, especially of the centers of association. On the whole, there is an increase in brain weight and of the index of encephalization (Bauchot and Stephan, 1968). These differences make it impracticable to compare the communicative behavior and relevant sensory powers of the Desmaninae to those of the other Talpidae.

The second subfamily taken out of its normal systematic context is made up of the five genera of Gymnures (Echinosoricini). So little is known about them that practically nothing can be said about their ability to communicate. Because of the dearth of reports on their behavior, we can merely draw analogies from the study of other insectivores. Nevertheless, we know more about Echinosoricini than about Chrysochloridae. The latter is a zoological family in its own right and ought therefore to be given a rubric of its own, even if it remains unstudied.

When studying the four systems of communication—chemico-olfactory, visual-optical, acoustic-auditory, and tactile—the means by which they are produced, the manner in which they are used, and the anatomical features necessary for their existence, we come across several forms that have parallel, convergent, or analogous functions. In mentioning them we find that a division into completely separate systems is a limitation that cannot be maintained. The discovery, for example, of an actively functioning organ that reacts to two completely different stimuli (the vomeronasal organ in Erinaceidae and Tenrecidae) or of the hitherto unforeseeable complex significance in a few insectivore families of supplementary eye-gland secretions and their functional importance as a means of communication indicates that a combination of various sensory functions can often occur even within the field of communication. This would corroborate the well-represented view that, as a rule, whole complexes of methods are employed in order to achieve communication inter- and intraspecifically. The diversity of these methods and the number of possibilities in combination with others cannot be foreseen. Moreover, we cannot exclude the future discovery of other functioning organs or abilities in insectivores that have long since been discarded by higher orders of mammals or are now present only in rudimentary form and thus extremely difficult to interpret.

In the following tables are listed some of the individual abilities or behavioral patterns that have been found to exist in at least a few genera of the families concerned. The question then arises whether signs of these abilities can be found among other families. Since the data available to us at the moment are still very incomplete, the result is merely an approximate survey of those communication methods common to all insectivores, using the few details that our present knowledge affords us. Because of lack of space some of the details listed in the following tables have not been described in the preceding survey, but they are to be found in various other works on the subject, which in most cases I have listed here.

All the insectivores studied so far emit and react to ultrasonic signals. Five of the nine families (and/or subfamilies) listed in Table 1 emit ultrasonic clicks; four have not been examined. Four use other ultrasonic signals or signals with strong ultrasonic components; in three of them the use of ultrasonics is suspected; two have not been examined. A positive answer to the question of reactions within the human hearing range can be given with certainty only among Potamogalidae, Soricidae, and solenodons; among the others these frequencies seem to be of lesser significance. Echolocation has been observed in Tenrecidae and Soricidae and suspected in Erinaceidae, Desmaninae, and

Table 1

Aspects of acoustic communication.

	Hedgehogs	Gymnures	Moles	Desmans	Shrews	Tenrecs	Otter shrews	Solenodons	Golden moles
Emission of signals	yes	yes	yes	yes	yes	yes	yes	yes	unknown
Emission of ultrasonic clicks	yes	yes	unknown	unknown	yes	yes	unknown	yes	unknown
Emission of other ultrasonic signals	yes	yes	suspected	suspected	yes	yes	very probable	yes	unknown
Reaction to signals within human hearing range	hardly any	unknown	suspected	hardly any	yes	questionable	yes	yes	unknown
Reaction to ultrasonics	yes	unknown	yes	yes	yes	yes	unknown	yes	unknown
Echolocation	suspected	unknown	suspected	suspected	yes	yes	unknown	suspected	unknown

Solenodontidae. To sum up we can say that insectivores are a group of animals that have adapted an ability to detect and emit ultrasonic signals and thus possess the necessary anatomical requirements for echolocation. The question of receptors for the reflected signals has not yet been clarified.

We have not yet exhausted the whole repertoire of signals. Their vocalizations seem—generally speaking—to fall into three groups: some combined with inhalation or exhalation through the nose; others derived from clicks with the tongue (Gould, 1969); and others probably originating in the larynx. These three types of sounds are known at present to exist in Tenrecidae, solenodons, and hedgehogs. The continuously emitted sounds, combined with a changing state of agitating and locomotive activity, are rather similar in *Hemicentetes* (Gould and Eisenberg, 1966; Eisenberg and Gould, 1970), in *Suncus* and *Blarina* (Gould, 1969), as well as in *Setifer,* which emit a series of "put" signals. *Tenrec ecaudatus,* too, emits a variety of respiratory sounds that are comparable to "puts" (Eisenberg and Gould, 1970). "Put" sounds can also be heard in *Echinops* and may have the same significance as the low sniffing sounds and low chuckling of a hedgehog that is only slightly agitated.

The broad perspective of chemical communication (Table 2) is too large a field to be treated exhaustively here, since present knowledge has to confine itself, for the most part, to single aspects and single observations of a relatively small number of species.

The deposition of feces as a communication signal is not common to all families and/or subfamilies. On the other hand, the emission of chemico-olfactory stimuli from various glands occurs in all families studied so far. Especially significant seems to be the eye-gland secretion, which apart from its other functions also has communicative character. As far as we now know, it could be regarded as a fully developed behavioral complex; at least the anatomical and histological requirements for such a complex do exist. The exact function of the evolutionary archaic vomeronasal organ has still to be tested in most insectivores. The salivating of *Echinops,* which Eibl-Eibesfeldt (1965) has described as marking behavior, is not to be considered as such, but as the last link in a chain of actions that reaches its peak in the registration and identification of a sensory impression of taste or smell in the vomeronasal organ. Just as in hedgehogs, it only takes place after the animal has detected a strange smell or taste and, therefore, it is in no way an active communication signal in its own right. It also occurs among *Setifer,* where it can be regarded only as a reaction to an extraordinarily strong stimulant or completely new allomone (Poduschka, 1974c).

As predominantly nocturnal or crepuscular creatures with, for the most part, very poor eyesight, insectivores depend a great deal on their tactile abilities (Table 3), which—apart from the vibrissae that are well developed and numerous in all families—have led to the development in Talpidae of especially effective, and apparently extremely versatile, tactile organs on the proboscis (Eimer's organs). Roughly speaking, we can say that the importance of tactile communication is indirectly proportional to the optical ability of insectivores. Investigations have shown that tactile communication is common to all insectivores during courtship. The Tenrecidae occupy a somewhat special position with their unique use of stimulant scratching. Stimulant bites during courtship and mating have been observed in Tenrecidae, Erinaceidae, and Soricidae. The male Soricid offers the female some of his own gland areas for her to bite and thus succeeds in getting her to detect the secretion or at least in transferring some of it onto her body.

Visual gestures or changes in appearance among insectivores (Table 4) are to be regarded for the most part less as active communication

Table 2

Aspects of chemical communication.

	Hedgehogs	Gymnures	Moles	Desmans	Shrews	Tenrecs	Otter shrews	Solenodons	Golden moles
Fecal deposits	no	yes	unknown	yes	data contradictory	yes	unknown	no	unknown
Rubbing of side glands	no	unknown	yes	unknown	yes	yes	unknown	suspected	unknown
Olfactory signals from other glands (ventral, sternal, perineal, etc.)	yes	unknown	yes	yes	yes	yes	unknown	yes	unknown
Eye-gland secretions	yes	unknown	unknown	unknown	yes	yes	unknown	unknown	unknown
Anatomical adaptions for eye-gland secretion	yes	unknown	questionable	unknown	unknown	yes	yes	unknown	unknown
Rubbing of head on partner or on objects	yes	unknown	unknown	unknown	yes	yes	unknown	yes	unknown
Perineal drag	yes	unknown	unknown	unknown	yes	yes	unknown	yes	unknown
Actively functioning vomeronasal organ	yes	unknown	unknown	unknown	unknown	yes	unknown	unknown	unknown
Investigation of feces by conspecific	no	unknown	unknown	yes	probable	no	unknown	no	unknown
Fecal deposits pressed on vertical objects	yes	unknown	unknown	unknown	yes	yes	unknown	unknown	unknown

Table 3

Aspects of tactile communication.

	Hedgehogs	Gymnures	Moles	Desmans	Shrews	Tenrecs	Otter shrews	Solenodons	Golden moles
Contact with facial vibrissae	yes	probably	unknown	very probable	yes	yes	very probable	yes	unknown
Rubbing body on conspecific	yes	unknown	unknown	unknown	yes	yes	unknown	yes	unknown
Stimulant scratching	no	unknown	unknown	unknown	no	yes	unknown	unknown	unknown
Stimulant or mating bite	yes	unknown	unknown	unknown	yes	yes	unknown	unknown	unknown
Other taste organs	side hairs	unknown	yes	yes	unknown	no	unknown	unknown	unknown
Sniffing or licking partner's genitalia	yes	unknown	unknown	unknown	yes	yes	unknown	unknown	unknown

Table 4

Aspects of visual communication.

	Hedgehogs	Gymnures	Moles	Desmans	Shrews	Tenrecs	Otter shrews	Solenodons	Golden moles
Alterations in appearance	yes	unknown	unknown	unknown	yes	yes	unknown	no	unknown
Gaping	yes	yes	unknown	unknown	yes	yes	unknown	yes	unknown
Eye-gland secretions	yes	unknown	unknown	unknown	unknown	yes	yes	unknown	unknown
Lordosis	yes	unknown	unknown	unknown	yes	yes	unknown	unknown	unknown
Acceptance of substitute object as sexual partner	yes	unknown	improbable	unknown	unknown	no	unknown	unknown	unknown
Recognition of gland secretions in conspecific	improbable	unknown	improbable	unknown	unknown	questionable	unknown	unknown	unknown
Recognition of light intensity	yes	unknown	unknown	yes	partial	no	unknown	unknown	unknown
Optical recognition of glands in conspecific	no	unknown	unknown	unknown	questionable	questionable	unknown	unknown	unknown
Recognition of change in appearance in conspecific	probable	unknown	improbable	unknown	yes	probable	unknown	unknown	unknown

systems than as passive reactions to the active stimuli of other systems, since eyesight among insectivores is generally poor. Gaping with the mouth seems to be an action peculiar to all insectivores. In Soricidae, however, we know that it is accompanied by ultrasonic emissions. Whether the remarkable eye-gland secretion of the tenrecs is in fact understood by conspecifics as a visual signal has not been proved either way, but it is certainly of less importance than the chemical message thus conveyed. Lordosis, as a signal inviting intromission of the penis, is common to all insectivores, but presumably even this action is accompanied by chemico-olfactory exudation. Since these animals have such poor eyesight, the odor is much more effective than a visible change in appearance or body position.

References

Ansell, W. F. H., 1964. Captivity behaviour and post-natal development of the shrew *Crocidura bicolor. Proc. Zool. Soc. London,* 142:123–27.

Argaud, R., 1944. Signification anatomique de la trompe du Desman des Pyrenées. *Mammalia,* 8:1–6.

Barbour, T., 1944. The Solenodons of Cuba. *Proc. New Engl. Zoölog. Club,* 28:1–8.

Bauchot, R.; Buisseret, C.; Leroy, Y.; and Richard, P. B.; 1973. L'équipement sensoriel de la trompe du Desman des Pyrenées (*Galemys pyrenaicus,* Insectivora, Talpidae). *Mammalia,* 37:17–24.

Bauchot, R., and Stephan, H., 1968. Étude des modifications encéphaliques observées chez les insectivores adaptés à la recherche de nourriture en milieu aquatique. *Mammalia,* 32:228–75.

Blossom, P. M., 1932. A pair of long-tailed shrews *(Sorex cinereus cinereus)* in captivity. *J. Mammal.,* 13:136–43.

Buchler, E. R., in press. Experimental demonstration of echolocation by the wandering shrew *(Sorex vagrans).*

Buchler, E. R., in press. Echolocation by the wandering shrew *(Sorex vagrans)*: Discrimination and potential adaptiveness.

Cei, G., 1946. Morfologia degli organi della vista negli insettivori. *Archivio ital. Anat. Embriol.,* 52:1–42.

Clark, W. E. Le Gros, 1932. The brain of the insectivora. *Proc. Zool. Soc. London,* 1932:975–1013.

Conaway, C. H., 1958. Maintenance, reproduction and growth of the least shrew in captivity. *J. Mammal.,* 39:507–12.

Crowcroft, P., 1955. Notes on the behaviour of shrews. *Behaviour,* 8:63–81.

Crowcroft, P., 1957. *The Life of the Shrew.* London: Max Reinhardt.

Dalquest, W. W., and Orcutt, D. R., 1942. The biology of the least shrew-mole. *Neurotrichus gibbsi minor. Amer. Midl. Nat.,* 27:387–401.

Davis, D. D., 1962. Mammals of the lowland rain-forest of North Borneo. *Bull. Nat. Mus. Singapore,* 31:1–130.

Davis, W. B., and Joeris, L., 1936. Notes on the life-history of the little short-tailed shrew. *J. Mammal.,* 26:136–38.

Dryden, G. L., 1968. Growth and development of *Suncus murinus* in captivity on Guam. *J. Mammal.,* 49:51–62.

Dryden, G. L., and Conaway, C. H., 1967. The origin and hormonal control of scent production in *Suncus murinus. J. Mammal.,* 48:420–28.

Eadie, W. R., 1938. The dermal glands of shrews. *J. Mammal.,* 19:171–74.

Eadie, W. R., 1939. A contribution to the biology of *Parascalops breweri. J. Mammal.,* 20:150–73.

Eadie, W. R., 1947. The accessory reproductive glands of *Parascalops* with notes on homologies. *Anat. Rec.,* 97:239–52.

Eadie, W. R., 1948. The male accessory reproductive glands of *Condylura* with notes on a unique prostatic secretion. *Anat. Rec.,* 101:57–79.

Eibl-Eibesfeldt, I., 1965. Das Duftmarkieren des Igeltanrek (*Echinops telfairi* Martin). *Z. Tierpsychol.,* 22:810–12.

Eisenberg, J. F., and Gould, E., 1966. The behavior of *Solenodon paradoxus* in captivity with comments on the behavior of other Insectivora. *Zoologica,* 51:49–58.

Eisenberg, J. F., and Gould, E., 1970. *The Tenrecs: A Study in Mammalian Behavior and Evolution.* Washington, D.C.: Smithsonian Institution Press.

Eisenberg, J. F., and Kleiman, D. G., 1972. Olfactory communications in mammals. *Ann. Rev. of Ecology and Systematics,* 31:1–32.

Firbas, W., and Poduschka, W., 1971. Beitrag zur Kenntnis der Zitzen des Igels (*Erinaceus europaeus* Linné). *Säugetierkundl. Mitt.,* 19:39–44.

Godet, R., 1951. Contribution à l'éthologie de la

Taupe (*Talpa europaea* L.). *Bull. Soc. Zool. de France,* 76:107–28.

Godfrey, G., and Crowcroft, P., 1960. *The Life of the Mole.* London: Museum Press.

Gould, E., 1965. Evidence for echolocation in the Tenrecidae of Madagascar. *Proc. Amer. Phil. Soc.,* 109:352–60.

Gould, E., 1969. Communication in three genera of shrews (Soricidae): Suncus, Blarina, and Cryptotis. *Communications in Behavioral Biology,* part A, vol. 3:263–313.

Gould, E., and Eisenberg, J. F., 1966. Notes on the biology of the Tenrecidae. *J. Mammal.,* 47:660–86.

Gould, E.; Negus, N. C.; and Novick, A.; 1964. Evidence for echolocation in shrews. *J. Exper. Zool.,* 156:19–37.

Griffin, D. R., 1968. Echolocation and its relevance to communication behavior. In: *Animal Communication,* T. A. Sebeok, ed. Bloomington: Indiana University Press, pp.154–64.

Hamilton, W. J., 1931. Habits of the star-nosed mole, *Condylura cristata. J. Mammal.,* 12:345–55.

Hamilton, W. J., Jr., 1944. The biology of the little short-tailed shrew *Cryptotis parva. J. Mammal.,* 25:1–7.

Kleiman, D. G., 1966. Scent marking in the Canidae. *Symp. Zool. Soc. London,* 18:167–77.

Kuhn, H.-J., 1964. Zur Kenntnis von Micropotamogale lamottei Heim de Balsac, 1954. *Z. Säugetierk.,* 29:152–73.

Liat, Lim Boo, 1967. Note on the food habits of *Ptilocercus lowii* Gray (pentail tree-shrew) and *Echinosorex gymnurus* (raffles) (moonrat) in Malaya with remarks on "ecological labelling" by parasite patterns. *J. Zool. London,* 152:375–79.

Marlow, B. J. G., 1954/55. Observations on the Herero musk shrew *Crocidura flavescens herero,* St. Leger, in captivity. *Proc. Zool. Soc. London,* 124:803–08.

Matthews, L. H., 1935. The oestrus cycle and intersexuality in the female mole (*Talpa europaea* Linné). *Proc. Zool. Soc. London,* 1935:347–83.

Mellanby, K., 1971. *The Mole.* London: Collins.

Mohr, E., 1936a. Biologische Beobachtungen an *Solenodon paradoxus* Brandt in Gefangenschaft. I. *Zool. Anz.,* 115:177–88.

Mohr, E., 1936b. Biologische Beobachtungen an *Solenodon paradoxus* Brandt. II. *Zool. Anz.,* 116:65–76.

Müller-Velten, H., 1966. Über den Angstgeruch bei der Hausmaus (*Mus musculus* L.). *Z. vergl. Physiol.,* 52:401–29.

Niethammer, G., 1962. Die (bisher unbekannte) Schwanzdrüse der Hausspitzmaus *Crocidura russula* (Hermann, 1780). *Z. Säugetierk.,* 27:288–34.

Niethammer, G., 1970. Beobachtungen am Pyrenäen-Desman, *Galemys pyrenaicus. Bonner zool. Beitr.,* 18:157–82.

Noirot, E., 1968. Ultrasounds in young rodents. II. Change with age in albino rats. *Anim. Behav.,* 16:129–34.

Noirot, E., and Pye, D., 1969. Sound analysis of ultrasonic distress calls of mouse pups as a function of their age. *Anim. Behav.,* 17:340–49.

Ognev, S. I., 1928. Desmaninae. In: *Mammals of Eastern Europe and Northern Asia.* Vol. I, *Insectivora and Chiroptera.* Moscow-Leningrad: Glavnauka, 1928; Jerusalem: Israel Program for Scientific Translations Ltd., 1962.

Ottow, B., 1955. *Biologische Anatomie der Genitalorgane und der Fortpflanzung der Säugetiere.* Jena: G. Fischer.

Pearson, O. P., 1944. Reproduction in the shrew (*Blarina brevicauda* Say). *Amer. J. Anatomy,* 75:39–93.

Pearson, O. P., 1946. Scent glands of the short-tailed shrew. *Anat. Rec.,* 94:615–29.

Poduschka, W., 1968. Über die Wahrnehmung von Ultraschall beim Igel, *Erinaceus e. roumanicus. Z. vergl. Physiol.,* 61:420–26.

Poduschka, W., 1969. Ergänzungen zum Wissen über *Erinaceus europaeus roumanicus* und kritische Überlegungen zur bisherigen Literatur über europäische Igel. *Z. Tierpsychol.,* 26:761–804.

Poduschka, W., 1970. Das Selbstbespeicheln der Igel. Film CT 1320 der Bundesstaatl. Hptst. f. wiss. Kinemat., Vienna.

Poduschka, W., 1971. Zur Kenntnis des nordafrikanischen Erdhörnchens, *Atlantoxerus getulus* (F. Major). *Zool. Garten* (Leipzig), N.F., 40:211–26.

Poduschka, 1972a. Das Paarungsverhalten des Grossen Igel-Tenrek, *Setifer setosus* (Froriep, 1806). Film CTF 1451 der Bundesst. Hptst. f. wiss. Kinemat., Vienna.

Poduschka, W., 1972b. Augendrüsensekretionen der Tenreciden *Setifer setosus, Echinops telfairi, Microgale dobsoni* und *Microgale talazaci.* Film CTF 1520 der Bundesstaatl. Hptst. f. wiss. Kinemat., Vienna.

Poduschka, W., 1973. *Erinaceus europaeus, Hemiechinus auritus* (Erinaceidae): Selbstbespeicheln. *Encycl. Cinemat.* E 1962/1973, Göttingen.

Poduschka, W., 1974a. Das Paarungsverhalten des Großen Igel-Tenrek (*Setifer setosus,* Froriep, 1806) und die Frage des phylogenetischen Alters einiger Paarungseinzelheiten. *Z. Tierpsychol.,* 34:345–58.

Poduschka, W., 1974b. Augendrüsensekretionen bei den Tenreciden *Setifer setosus* (Froriep, 1806), *Echinops telfairi* (Martin, 1838), *Microgale dobsoni* (Thomas, 1918) und *Microgale talazaci* (Thomas, 1918). *Z. Tierpsychol,* 35:303–19.

Poduschka, W., 1974c. Fortpflanzungseigenheiten und Jungenaufzucht des Großen Igel-Tenrek, *Setifer setosus* (Froriep, 1806). *Zool. Anz.,* Jena 193 (314): 145–80.

Poduschka, W., 1974d. Sinnes- und verhaltensphysiologische Beobachtungen an *Solenodon paradoxus* Brandt. Manuscript.

Poduschka, W., 1974e. Die verschiedenen Markierungsarten des Großen Igel-Tenrek, *Setifer setosus* (Froriep, 1806). Manuscript.

Poduschka, W., 1974f. Markierungsverhalten des Grossen Igel-Tenrek, *Setifer setosus* (Froriep, 1806). Film CTF 1562 der Bundesst. Hptst. f. wiss. Kinemat., Vienna.

Poduschka, W., 1976. Das Paarungsvorspiel des Osteuropäischen Igels (Erinaceus europaeus roumanicus) und theoretische Überlegungen zum Problem männlicher Sexualpheromone. (Manuscript.)

Poduschka, W., and Firbas, W., 1968. Das Selbstbespeicheln des Igels, *Erinaceus europaeus* Linné, 1758, steht in Verbindung zur Funktion des jacobsonschen Organes. *Z. Säugetierk.,* 33:160–72.

Quay, W. B., 1965. Comparative survey of the sebaceous and sudiferous glands of the oral lips and angle in rodents. *J. Mammal.,* 46:23–36.

Quilliam, T. A., 1966. The mole's sensory apparatus. *J. Zool. London,* 149:76–88.

Rahm, U., 1961. Beobachtungen an dem ersten in Gefangenschaft gehaltenen *Mesopotamogale ruwenzorii* (Mammalia—Insectivora). *Rev. Suisse Zool.,* 68:73–90.

Ralls, K., 1971. Mammalian scent marking. *Science,* 171:443–49.

Ravizza, R. J.; Heffner, H. E.; and Masterton, B.; 1969. Hearing in primitive mammals. II: Hedgehog *(Hemiechinus auritus). J. Audit. Res.,* 9:8–11.

Reed, C. A., 1944. Behaviour of a shrew-mole in captivity. *J. Mammal.,* 25:196–98.

Richard, P. B., 1973. Le Desman des Pyrenées *(Galemys pyrenaicus).* Mode de vie. Univers sensoriel. *Mammalia,* 37:1–16.

Richard, P. B., and Valette Viallard, A., 1969. Le Desman des Pyrenées *(Galemys pyrenaicus):* Premières notes sur sa biologie. *La Terre et la Vie,* 3:225–45.

Rood, J. P., 1958. Habits of the short-tailed shrew in captivity. *J. Mammal.,* 39:499–507.

Sales, G. D., 1972. Ultrasound and aggressive behavior in rats and other small mammals. *Anim. Behav.,* 17:542–46.

Stephan, H., and Bauchot, R., 1959. Le cerveau de *Galemys pyrenaicus* Geoffroy, 1811 (Insectivora Talpidae) et ses modifications dans l'adaption à la vie aquatique. *Mammalia,* 23:1–18.

Stephan, H., and Bauchot, R., 1960. Les Cerveaux de *Chlorotalpa stuhlmani* (Matschie) 1894 et de *Chrysochloris asiatica* (Linné) 1758 (Insectivora, Chrysochloridae). *Mammalia,* 24:495–510.

Stephan, H., and Bauchot, R., 1968a. Gehirn und Endocranialausguß von *Desmana moschata* (Insectivora Talpidae). *Morphol. Jb.,* 112:213–25.

Stephan, H., and Bauchot, R., 1968b. Vergleichende Volumenuntersuchungen an Gehirnen europäischer Maulwürfe (Talpidae). *J. Hirnforschung,* 10:247–58.

Strauss, F., 1939. Die Befruchtung und der Vorgang der Ovulation bei Ericulus aus der Familie der Centetiden. *Biomorphosis,* 1:281–312.

Strauss, F., 1942. Vergleichende Beurteilung der Placentation bei den Insectivoren. *Rev. Suisse Zool.,* 49:269–82.

Thenius, E., and Hofer, H., 1960. *Stammesgeschichte der Säugetiere.* Berlin, Göttingen, Heidelberg: Springer.

Vogel, P., 1969. Beobachtungen zum intraspezifischen Verhalten der Hausspitzmaus (*Crocidura russula* Hermann, 1870). *Rev. Suisse Zool.,* 76:1079–86.

Vogel, P., 1970. Biologische Beobachtungen an Etruskerspitzmäusen (*Suncus etruscus* Savi, 1832). *Z. Säugetierk.,* 35:173–85.

Chapter 25

COMMUNICATION IN LAGOMORPHS AND RODENTS

John F. Eisenberg and Devra G. Kleiman

The Lagomorpha

INTRODUCTION

The order Lagomorpha is divided into two families, the Ochotonidae (pikas) and Leporidae (hares and rabbits). The lagomorphs are characterized by having ever-growing incisors with enamel on both the posterior and anterior surfaces. They differ from the rodents in that there are two pairs of upper incisors, with the second pair located directly behind the first. It is a small order of only nine genera but with an extremely wide distribution. The Leporidae were found over the entire world except for Australia and southern South America, where they have been introduced by man. The pikas have a more limited distribution, being confined to montane areas in eastern Asia and western North America. It is generally conceded by paleontologists that the lagomorphs have been phylogenetically distinct from the rodents for a considerable period of time, but the two orders share certain ancient affinities, which led Simpson (1945) to place both rodents and lagomorphs in the same cohort—the Glires.

The order is characterized reproductively by induced ovulation (Asdell, 1964) and a trend toward the production of precocial young. Although the young of *Ochotona* are born sparsely haired with the eyes closed, as are the young of the European rabbit *(Oryctolagus)*, the young of most species of North American rabbits *(Sylvilagus)* are born well furred but with the eyes closed. The young of hares *(Lepus)* are born with the eyes open and are fully furred.

Female pikas typically bear their young in nests within the burrows built in the rock slides they inhabit. Young European rabbits are born in nests generally situated in a rather deep burrow and constructed of fur plucked from the mother's chest. On the other hand, young of the genus *Sylvilagus* are born in nests that are merely shallow depressions in the ground, although they are generally lined with fur plucked from the mother's chest as well as with a loose covering of vegetation gathered by the female after parturition. Lying in a shallow depression which may be partly covered with grasses, the young of hares are protected even less than the young of *Sylvilagus*.

Characteristically, female hares and rabbits suckle their young at long intervals, usually once a day (Southern, 1948; Denenberg et al., 1969; Sorenson et al., 1972). The female is thus not in continuous attendance upon the young, although, if the young are disturbed and emit a sharp squeal, the female may return. Moreover, antipredator behavior may be exhibited if the nest has been disturbed by a snake, weasel, or other predator (Marsden and Holler, 1964). There is no recorded retrieving on the part of the females of the genera *Sylvilagus, Oryctolagus,* and *Lepus* (Sorenson et al., 1972; Denenberg et al., 1969). The absence of a retrieval response may be attributed in part to the fact that in *Lepus* the young are precocial enough to flee on their own, whereas in *Sylvilagus* the whole strategy of antipredator behavior is based on concealment, which is reminiscent of the pattern of hiding young in many species of ungulates. The absence of the retrieval response in *Oryctolagus* is a little more difficult to account for. Retrieving has not been tested for in *Ochotona.*

THE OCHOTONIDAE

The pikas in both the New World and Old World are found in alpine habitats and, in particular, choose rock slides or tallus slopes. They differ from other genera of the Lagomorpha in that they gather a variety of grasses and forbs during the summer months and lay them on rocks in piles to dry in the sun. This dried herbaceous material is then cached in rock crevices for use as fodder during the winter. The natural history of pikas has been well documented by Severaid (1956) in North America and by Kawamichi (1968, 1970, 1971) and Haga (1960) for the Asiatic species. Pikas appear to be diurnal and, although colonial, they are spaced with individuals occupying their own home ranges. In the Japanese pika *(Ochotona hyperborea)* a pair can occupy the same home range through a breeding season, but the North American pika *(Ochotona princeps)* usually shows separate centers of activity for the male and female throughout the annual cycle (Kilham, 1958). Males enter a female's home range only to court and mate. In all pikas males and females are extremely intolerant of members of their own sex.

Communication by means of chemical signals undoubtedly takes place in colonies of pikas. The animals typically defecate at one place in their home range and this could serve as a source of chemical information for neighbors or strangers intruding on the home range; however, we lack experimental data. Pikas have a conspicuous gland on the cheek (Harvey and Rosenberg, 1960), apocrine in structure, which shows shifts in activity through the reproductive cycle. The animals have been noted to rub the cheeks at various points within their territories, and sniffing of the glandular areas is frequent during male-female interaction prior to mating.

In contradistinction to the Leporidae, the pikas are rather vocal. So conspicuous are the vocalizations that most studies of communication in the Ochotonidae have concentrated on auditory communication. Somers (1973) defines two loud calls used in distance communication by members of a given family group or neighbors in a colony. The short call is used when the territory is invaded by a conspecific or when an aerial or terrestrial predator is sighted. Repetitive and quite harmonic in its structure, the call appears to alert colony members. It is reminiscent of similar warning calls given by diurnal montane rodents, such as marmots *(Marmota)* and the Andean viscacha *(Lagidium).* In addition to the short call with its warning function, pikas typically produce a "song," which is a series of short notes varying in duration as the song sequence progresses. Each song sequence can last for twenty to thirty-five seconds. In *O. princeps* it appears to be given predominantly by males, although the Asiatic species appear to show a similar song form that is not so pronounced in its dimorphism (Kawamichi, 1968, 1970). Somers suggests that the song may have both a territorial and a reproductive significance. It is similar to calls given by males of arboreal and semiarboreal rodents, including *Erethizon* and *Dinomys* (Eisenberg, 1974). Dialect variations between separate populations of pikas have been described for the short calls (Somers, 1973).

THE LEPORIDAE

The rabbits and hares do not produce the striking song and antipredator vocalizations found in the Ochotonidae. Most calls of rabbits and hares are used for short-range communication. Antipredator behavior in hares and rabbits often involves visual display, including the conspicuous white patches on the underside of the tail, which have evolved as a form of colony warning and individual distraction display during antipredator behavior. As a result of the less-conspicuous vocal repertoire in the leporids, most researchers have confined themselves to studies of nonvocal communication and, in particular, olfactory communication (Coujard, 1947; Mykytowycz, 1965, 1966a, 1966b, 1966c, 1967).

The behavior of the European rabbit *(Oryctolagus cuniculus)* is perhaps the best studied of all lagomorphs (Southern, 1948; Myers and Mykytowycz, 1958; Myers and Poole, 1958; Mykytowycz, 1958, 1959, 1960, 1961). In brief, the European rabbit lives in organized social units called warrens. Activity centers around a series of burrows used for several generations. A warren generally contains several breeding adult females who can and do form a dominance order that restricts the breeding behavior of younger females by restricting their access to high-quality nesting sites. Females construct burrows for rearing the young and do so unaided by males.

Males form a distinct dominance hierarchy with a dominant adult male ranging over the entire warren and having access to several breeding females. A dominant male has active anal glands that impart an odor to the hard fecal pellets deposited in specific dung piles. Such fecal pellets are to be distinguished from those feces formed during an initial passage through the gut which are reingested and then passed in the hard form. The size of a male's anal glands correlates with his dominance status (Mykytowycz, 1966a). It has been suggested that the dung piles serve as indicators of adult male occupancy to any strange males wandering into the area.

The inguinal glands of adult male rabbits may play a role in sexual attraction (Mykytowycz, 1966b). Mykytowycz (1966c) has also noted that Harder's gland is dimorphic in the European rabbit, being larger in the male, and, furthermore, that this gland is dependent on androgen levels. Thus, in castrated males the Harder's gland diminishes in size.

Dominant adult male rabbits also possess an active submandibular gland. Secretions from this gland are deposited by "chinning" behavior, where the male rubs the chin on conspicuous objects in the environment or on does (Heath, 1972). Chinning is frequently exhibited by dominant males within a rabbit warren. It would appear that marking behavior in the wild rabbit insures that strange males are excluded from the warren and females are covered with the dominant male's scent, thus maintaining group integrity (Myers and Mykytowycz, 1958; Heath, 1972).

The European hare *(Lepus europaeus)* does not show the strong dimorphism or seasonal change in the size of anal glands that the European rabbit shows. However, conspicuous dimorphism and seasonal activity in Harder's glands and inguinal glands have been noted (Mykytowycz, 1966a, 1966b, 1966c). Hares typically do not form warrens, and their activities appear to be much more individualistic. Adults are generally well spaced, although temporary associations can be formed when does come into heat (see Lechleitner, 1958; O'Farrell, 1965).

Studies on North American rabbits have been carried out by Marsden and Holler (1964). The two species *Sylvilagus floridanus* and *S. aquaticus* were studied in confined populations. The swamp rabbit *(S. aquaticus)* displayed territoriality, while the cottontail *(S. floridanus)* did not. Both species demonstrated structured dominance hierarchies among conspecific males. *S. aquaticus* males actively marked their territories with the submandibular gland.

During courtship in rabbits and hares, a complex series of events occur whereby males attempt to approach females coming into estrus and a female generally responds by boxing or lunging at the male. The male may then turn and dash past her or leap over her, often urinating on her. Enurination behavior by males during approaches has been described for *Oryctolagus* (Southern, 1948; Heath, 1972), *Sylvilagus* (Marsden and Holler, 1964), and *Lepus* (Forcum, 1966; Lechleitner, 1958). Enurination while leaping or combined with flashing of the white underside of the tail (Southern, 1948) clearly involves visual as well as chemical communication. It has been described also for numerous caviomorph rodents (Kleiman, 1971, 1974).

Although auditory communication is not so pronounced in leporids as with pikas, rabbits and hares produce sounds in a variety of contexts. A sharp thump with the hind foot produced when the animal is startled can serve as a warning signal to colony members in *Oryctolagus* and *Sylvilagus.* A throaty growling sound may be produced by males and females when disturbed in their burrows. Females of the genus *Sylvilagus* produce a similar sound when they are disturbed on the nest. A graded series of squeaking sounds are produced by rabbits during courtship and copulation. Males will produce a squeak sound when approaching a female and females may produce a similar sound when approached by a male. A modified squeak, labeled the chirp by Marsden and Holler (1964), may be produced by a female while a male is driving her preparatory to mounting. A high loud squeal is given during copulation by male rabbits of the genus *Oryctolagus* (Myers and Poole, 1961). Marsden and Holler report a similar sound for *Sylvilagus* which they believe is produced by the female. A two-syllable call may be given by a startled cottontail rabbit who has made a run and then turns to observe the source of the disturbance. It can serve to alert colony members. All lagomorphs apparently produce a distress cry, which is a high-pitched screaming note generally given by an animal that has been captured by a predator.

The Rodentia

INTRODUCTION

The rodents form a distinct order of mammals having an ancient lineage. In fact sciurids are clearly recognized from the fossil records of the Oligocene. The group is characterized by having a single pair of rootless, ever-growing incisors in both the upper and lower jaws. The incisors have enamel only on the anterior surface. Since the posterior surface of the incisors wears more rapidly than the anterior, a chisel-like cutting tool results, which provides rodents with their key adaptation, the ability to gnaw into hard surfaces and make effective use of a variety of plant parts for foodstuffs. There are forty-three living families of rodents, grouped into fifteen superfamilies, including over 1,680 species; it is the most diverse order of living mammals.

We will first briefly review rodent communication mechanisms and then discuss some key evolutionary trends resulting in convergence in communication patterns of different species. The subordinal classification of rodents into Sciuromorpha, Myomorpha, and Hystricomorpha is somewhat artificial and such taxonomic grouping has been criticized, but no alternative subordinal classification has been universally accepted (Wood, 1965; Simpson, 1959). We will therefore refer mainly to superfamilies when discussing adaptive trends within this diverse order.

The basis for much of our knowledge of mammalian physiology, psychology, and psychophysics has resulted from intensive studies on a few selected species of muroid rodents. Behavioral investigations on *Rattus norvegicus* have been carried out for seventy years and much of the work was summarized by Munn (1950). Similar to investigations on the laboratory rat are those on the laboratory mouse *(Mus musculus).* In recent years the Syrian golden hamster *(Mesocricetus auratus)* and the Central Asian jird *(Meriones unguiculatus)* have become popular rodents for behavioral studies. The only two genera of wild muroid rodents for which intensive studies have given us a reasonable picture of the communication systems are *Microtus* and *Peromyscus.* Recent research on *Peromyscus* has been summarized in the volume edited by King (1968).

Comparisons of rodent behavior repertoires were summarized by Eibl-Eibesfeldt (1958), Eisenberg (1967), and Grant and Mackintosh (1963). The behavior of hystricomorph rodents has been reviewed by Kleiman (1974), and the vocal repertoires of South American hystricomorphs (= caviomorphs) are the subject of a review by Eisenberg (1974).

SIGNALS AND COMMUNICATION—A BRIEF REVIEW

The sensory capacities of *Rattus norvegicus* and the role of the brain in sensory integration were painstakingly studied by Lashley (1950). Although his studies did not intentionally analyze communication by rats in their normal social environment, Lashley's theories were pervasive. Lashley concluded that many complex activities of rats did not depend either on a specific area of the rat's sensory cortex or on a specific form of sensory input. Strongly influenced by Lashley's conclusion, Beach studied the sexual response of the male rat and the retrieval response of the lactating female. After a series of ablation experiments, Beach and Jaynes (1956) concluded "It appears that the female's retrieving behavior, like the sexual behavior of the male rat or like the maze-learning performance of both sexes, normally involves multi-sensory controls." There is no doubt that several sensory inputs may be involved in the integration of behavior between two interacting rodents, but it is equally evident that certain behavior patterns may be released by very specific signals during interactions. The discovery of infant ultrasonic sounds as releasers of retrieval behavior in rodents (Zippelius and Schleidt, 1956) and a renewed interest in olfaction as a communication channel have dominated recent research on rodent communication.

Utilizing the golden hamster *(Mesocricetus auratus)* as an experimental species, Murphy and Schneider (1970) demonstrated that male copulatory behavior will not be exhibited unless the male can perceive the odor of the estrous female. The vaginal secretions of the estrous female have now been implicated as the source of the chemical signal (Murphy, 1973). Thus, the presence of a stimulus from a single source is necessary for the release of sexual behavior in the male hamster. Obviously care must be exercised in the application of general theories to both new social contexts in well-studied species and analysis of interaction forms in species that are newly studied. Some forms of social interaction in rodents may indeed depend on a multiplicity of stimulus inputs, whereas other signal systems may in fact conform to the classifical concepts of releaser and innate release mechanisms (Lorenz, 1950).

Although considerable attention has been devoted to the analysis of those behavior patterns thought to have communicatory significance, the analysis of the reception and processing of stimulus inputs has proceeded more slowly, although the patterning of social interactions is very dependent on the perception of conspecific signals.

This differential emphasis is noticeable in the field of hormone-behavior research. Although there are numerous reports indicating that physiological state affects behavior output (e.g., see Levine, 1972), the converse area of study has been almost ignored. The most direct proof that hormonal state affects stimulus reception has been provided by Komisaruk et al. (1973) for the tactile sensory system. They have shown that the sensory field of the pudendal nerve (the perineum) is significantly increased in female rats given estrogen treatment. Thus, there are changes at the level of the peripheral receptors due to gonadal hormones. Since the lordosis posture is stimulated by both the mounting and the pelvic thrusting of the male, it would appear that increased sensitivity of the perineum might facilitate the female's response during copulatory attempts by the male and thus improve the chances of successful intromission.

Within the olfactory system there is increasing evidence that odor detection depends, to some degree, on physiological state. For example, Sakellaris (1972) has shown a decreased threshold to the odor of pyridine in adrenalectomized rats when compared with normal rats. Corticosterone administration raised the threshold to normal levels. Recently Pietras and Moulton (1974) have reported increased performance in odor detection (using mainly cyclopentanone) in rat females during natural estrus when compared with other stages of the estrous cycle or pseudopregnancy. There is every reason to believe that the sensitivity to conspecific odors would also be affected.

Using enclosed conspecifics as the odor source, Carr and Caul (1962) could not detect any differences in the abilities of normal and gonadectomized male and female rats to discriminate the odors of sexually active or inactive members of the opposite sex. However, recent studies have indicated that hormonal state as well as experience affect odor preferences. For example, Carr et al. (1966) have shown that sexually active intact male rats prefer the odors of receptive females to those of nonreceptive females, while both sexually inactive and castrated male rats show no preference, as measured by the time spent investigating the different odors.

These studies clearly suggest that hormones are affecting the sensory systems, but where such changes occur, i.e., at the peripheral receptors or at other levels in the central nervous system, should be examined more closely.

Visual Input

Numerous studies of rodent behavior have resulted in the description of characteristic postures, movements, and configurations, many of which are common to a wide range of rodent species (Grant and Mackintosh, 1963; Eibl-Eibesfeldt, 1958; Eisenberg, 1962, 1963a, 1967; Kleiman, 1974). (See Fig. 1.) The description of postures and movements has allowed certain forms of comparison within or between species when the postures and movements are quantified under controlled test situations and expressed as frequencies or ratios. Some movements or postures are associated with the genesis of auditory, olfactory, or tactile input; other postures could serve solely as visual signals. (See Fig. 2.) Yet proof of stereotyped movements or postures serving as visual signals is lacking. Many of the commonly studied rodent species are nocturnal or crepuscular. It is supposed that the role of visual communication in these species (e.g., *Rattus, Mus, Peromyscus, Mesocricetus*) is minimal, yet diurnal rodents have a rich repertoire of postures and movements, some of which may be true visual displays (Steiner, 1970; Horwich, 1972; Kleiman, 1971, 1972, 1974).

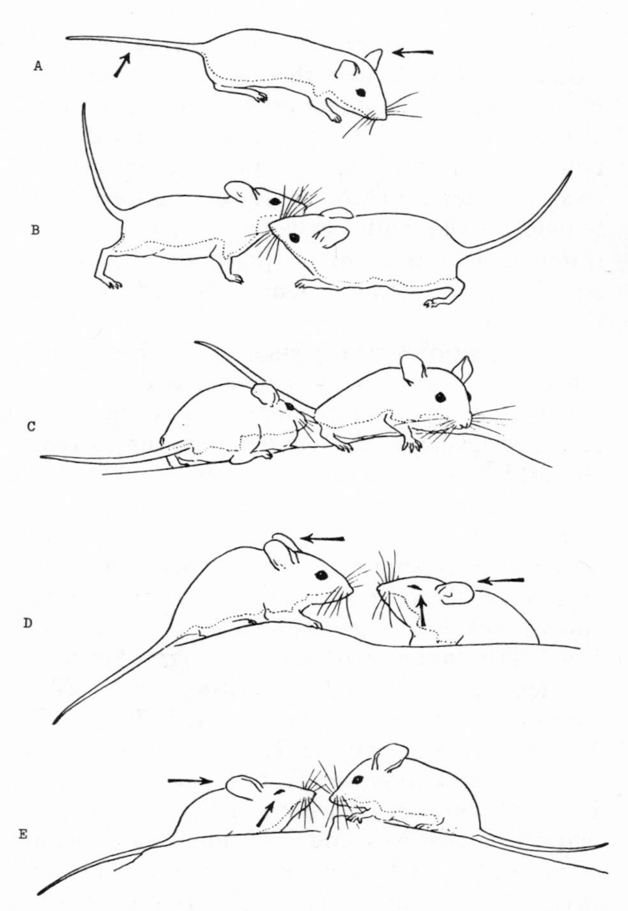

Fig. 1. Contact postures during encounters with *Peromyscus maniculatus.* The arrows indicate important features and the dotted lines demarcate the white ventrum from the brown upperparts. A. Elongate posture during exploration or prior to contact—note the erect ears and tense, extended tail. B. Naso-nasal contact—note the juxtaposition of the mouths and depression of ears. C. Naso-anal contact. D. Acceptance of contact by the animal on the right—note the closed eyes and flattened ears. E. Submission by the animal on the left —compare with D. (From Eisenberg, 1962.)

Fig. 2. Threat in *Peromyscus maniculatus.* Animal on the left shows expanded pinna and open mouth as he emits a threat squeak. The animal on the right is in an elongate posture with the ears slightly folded. (From Eisenberg, 1968.)

Even in nocturnal rodents such as *Rattus* and *Peromyscus* visual acuity may be quite well developed (Munn, 1950; Lashley, 1932, 1938; Vestal, 1973). Grant, Mackintosh, and Lerwill (1970) offer evidence that the black chest patch in *Mesocricetus* displayed during the assumption of an upright posture may intimidate a conspecific. In follow-up experiments, Johnston (1973) has not been able to determine why or how the chest patch might function during agonistic encounters; he has questioned the interpretation of Grant et al., since postures involving the display of the patch during fights are normally shown by subordinate males. The white venter of many nocturnal rodents could also serve as a signal during the assumption of upright postures in a thwarting context. No doubt many of the markings or coat color patterns of various rodent species are also the result of predator selection and thus function either as cryptic patterns or during antipredator display and as such may be examples of interspecific communication (see Benson, 1933; Kaufman, 1974).

Surely in nocturnal and fossorial rodents or in rodents with reduced eyes, olfaction and audition must be the primary input channels for distance communication. Tactile and olfactory input become extremely important during encounters at close range, and the interaction patterns may well be considered as the establishment and maintenance of "joints" in the manner outlined in the chapter by Eisenberg and Golani in this volume. The establishment of joints between partners can result in extremely stereotyped configurations and postures with a minimum of visual input.

Auditory Input and Perception

The study of the genesis of sounds produced by rodents is still in its early stages, but an excellent analysis for *Cavia* was recently completed (Arvola, 1974). Comparisons of audiograms for selected rodent species is much more advanced. It is clear that many rodent species show more than one sensitivity peak when cochlear microphonics are measured (Ralls, 1967; Brown, 1973), suggesting that certain frequencies are more important for survival than others. Whether optimum sensitivities are related to predator detection, prey detection, or the perception of conspecific signals remains to be investigated.

One consistent pattern that has emerged is that many muroid rodents *(Peromyscus, Apodemus, Rattus, Mus)* can perceive sounds well above the range of human hearing (> 18 Khz–<70 Khz; see Zippelius and Schleidt, 1956; Ralls, 1967; Price, 1970; Brown, 1973). On the other hand, many rodent species seem to have sensitivity ranges not too different from the human range (e.g., *Cavia:* Strother, 1967).

In addition, those rodent species with an enlarged mastoid bulla typically show a peak of maximum auditory sensitivity for rather low-frequency sounds (Webster, 1962), and the larger the relative bullar inflation, the greater the sensitivity to selected lower frequencies (Lay, 1972). Some species having slight bullar inflation retain a sensitivity peak for frequencies greater than 20 Khz (Brown, 1973), but those species exhibiting extreme bullar inflation appear to have maximum sensitivities for frequencies less than 20 Khz. This sensitivity shift is not correlated with a particular rodent taxon since it includes the caviomorph genus *Chinchilla* (Rothenberg and Davis, 1967); the muroid subfamily Gerbillinae (Lay, 1972), and the heteromyid genus *Dipodomys* (Strother, 1967; Webster, 1961, 1962).

For those species of rodents that show auditory sensitivity above 25 or 50 Khz, a number of important signal forms have been identified which are inaudible to humans and may be either inaudible to their mammalian predators or difficult to localize. The production of ultrasonic signals in muroid rodents was recently reviewed in Sewell (1970) and Sales (1972a, 1972b). Production of ultrasonic cries by young rodents displaced from the nest and/or handled has received the most attention to date (*Mus, Rattus:* Noirot, 1966, 1968; *Mus, Rattus, Mesocricetus, Clethrionomys* and *Apodemus:* Okon, 1970a, 1972; *Peromyscus:* Smith, 1972). Yet adult sounds in the ultrasonic range are receiving increasing attention (Barfield and Geyer, 1972; Sales, 1972a, 1972b; Whitney et al., 1973).

A wide variety of rodent sounds are produced with frequencies below 20 Khz. These include sounds produced mechanically by foot stamping, tooth chattering, or quill rattling. Unvoiced exhalations include hissing, while true vocalizations exhibit astonishing variety and forms of modulation. The form and function of vocalizations have been analyzed for *Cavia* (Arvola, 1974; Coulon, 1973); selected caviomorph rodents (Eisenberg, 1974); the lemming genera *Dicrostonyx* and *Lemmus* (Brooks and Banks, 1973; Arvola et al., 1962); *Sciurus carolinensis* (Horwich, 1972); *Cynomys* (Waring, 1970); *Citellus ornatus* (Balph and Balph, 1966); and *Dipodomys, Perognathus,* and *Liomys* (Eisenberg, 1963a).

Rodent species from diverse taxa produce sounds that are physically similar. These sounds are often associated with contexts similar enough to suggest behavioral homologies. For example, tooth chattering is widespread in the Rodentia and generally accompanies aggressive

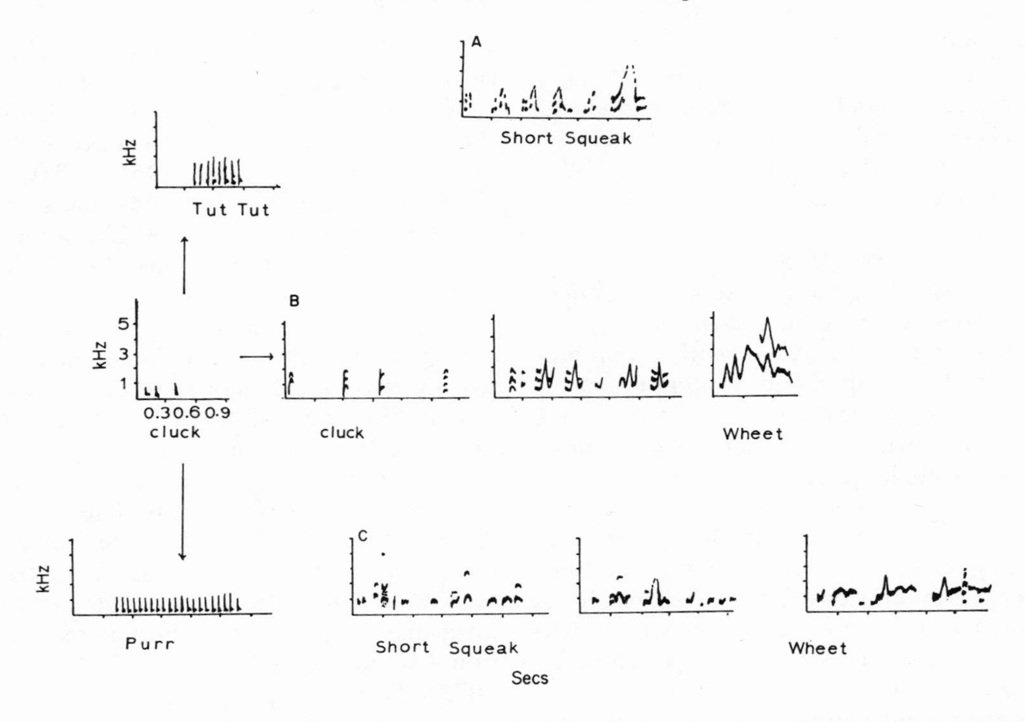

Fig. 3. Gradation of some syllable types in *Cavia porcellus.* The basic cluck (Type II syllable) can be uttered with a specific temporal patterning to yield the tut-tut or purr. Longer clucks (Type I syllable) may transform to short squeaks or loud wheets. Series C was produced by scratching a male on the neck and then on the groin. Series B was produced by a two-week-old juvenile exploring an open area but separated from the mother and siblings. Series A is an inflected squeak series produced by "grooming" the young around the neck. The ordinates are in 1 kHz increments; the abscissae are in 0–3 sec increments. (From Eisenberg, 1974.)

arousal and threat behavior. Low-intensity, repetitive calls may occur during courtship (Eisenberg, 1974). Calls may be produced by the male following ejaculation. In *Rattus* these calls are ultrasonic (Barfield and Geyer, 1972), while in *Octodon, Octodontomys,* and *Myoprocta,* they have audible components (see Kleiman, 1974; Eisenberg, 1974).

Many rodent species produce a graded series of calls which involve, on the one hand, emphasis on single frequencies to, on the other hand, emphasis on a wide range of frequencies that may approximate noise. Such a series of calls may reflect subtle changes in motivation from low-intensity arousal to extreme arousal with a high tendency to avoid a conspecific (for a discussion of *Cavia,* see Arvola, 1974; for *Dicrostonyx,* see Brooks and Banks, 1973). (See Fig. 3.) Graded series of vocalization forms that parallel motivational shifts in thwarting contexts have been described by Dunford (1970) for *Tamias striatus* and Coulon (1973) for *Cavia*.

The young of rodents born altricially generally have a stereotyped call given when they are displaced from the nest. In muroid rodents the call is generally in part or wholly ultrasonic (Sewell, 1970). On the other hand, precocially born rodents may produce short, audible contact notes that allow them to remain near or locate the mother when they are moving together (Eisenberg, 1974).

Sounds produced as part of a species' antipredator strategy often vary greatly in form and pitch (*Marmota:* Barash, 1973; Waring, 1966; *Spermophilus:* Melchior, 1971; *Dasyprocta:* Eisenberg, 1974; *Lagidium:* Eisenberg, 1974). These pitch differences may reflect different selective pressures, which have acted to produce calls audible at different distances in habitats differing in their efficiency of sound propagation (Eisenberg, 1974).

Finally, some rodent species appear to produce calls that are related to the establishment and maintenance of spacing and/or the attraction of sexual partners. These calls include the territorial calls of the North American red squirrels (*Tamiasciurus:* Smith, 1968) and the prairie dogs (*Cynomys:* Waring, 1970), and the "song" of *Dinomys* (Eisenberg, 1974).

Chemical Signals

In the early studies of rodent interaction patterns too little attention was paid to the role of olfaction in the coordination of social behavior. In an exhaustive analysis of filmed encounters of *Mus musculus,* Banks (1962) was unable to define any unique set of postures or movements which could reliably indicate that a bite with subsequent fighting would be delivered. Yet it is now known that an attack by one male mouse upon another is profoundly influenced by odor (Mackintosh and Grant, 1966). Olfactory communication in mammals has been the subject of several recent reviews (Ralls, 1971; Eisenberg and Kleiman, 1972; Johnson, 1973), and only a brief summary for rodents will be included in this section.

Chemical signals in rodents are derived from many sources. Vaginal secretions of *Mesocricetus* release sexual behavior in the male (Murphy, 1973). Specific glandular areas may be found in most rodent species, e.g., flank glands in *Mesocricetus* and *Cricetus* (Lipkow, 1954; Eibl-Eibesfeldt, 1953; Dieterlen, 1959), ventral glands in *Meriones* and *Peromyscus* (Thiessen et al., 1968; Doty and Kart, 1972), and supracaudal glands in *Cavia* (Martan, 1962). Urine, also a source of information, has been extensively studied in *Mus, Peromyscus,* and *Rattus* (see Mackintosh and Grant, 1966; Doty, 1973; Carr et al., 1965; Lydell and Doty, 1972).

Movements associated with the deposition of chemical traces may be more or less elaborate, depending upon the odor source. Flank marking and ventral gland marking are usually visually conspicuous acts while urine deposition on the substrate or even on a conspecific (which occurs while crawling over in *Rattus*) may go unnoticed by a human observer. Urine marking on a conspecific in most caviomorphs, however, is associated with elaborate postures (Kleiman, 1974). (See Fig. 4.) An unusual source of odor with associated marking movements has been described by Collins and Eisenberg (1972) in *Dinomys.* Eye gland secretions drain from the external nares onto the rhinarium, which is then rubbed on various points in the living space.

For many rodents both gland size and marking frequency are sexually dimorphic, the male exhibiting a larger gland and higher levels of marking (e.g., *Mesocricetus:* Vandenbergh, 1973; *Meriones:* Thiessen et al., 1968; *Cavia:* Martan, 1962).

Chemical signals function in a variety of ways. Godfrey (1958) demonstrated that male bank voles *(Clethrionomys)* could distinguish between the odor of their own subspecies and that of a closely related subspecies. The ability to select the appropriate species for mating may be critical in areas of sympatry. Moore (1965) demonstrated that *Peromyscus maniculatus* males discriminated between odors of *maniculatus* females and other species. Sympatric populations of *P.*

eremicus and *P. californicus* also showed an ability to discriminate on the basis of odor (Smith, 1965). Recently Doty (1972) has shown that male urine alone can serve as an attractant to estrous females of *P. maniculatus* and that these females can discriminate between the odors of male *P. maniculatus* and *P. leucopus,* which occur in sympatry over a wide range of habitats. (For a full discussion, see Doty, 1972, 1973.) Thus sexual isolation in sympatric populations may be mediated by olfactory cues.

The sex of an individual can be discriminated on the basis of odor alone in *Mus, Meriones,* and *Rattus* (Bowers and Alexander, 1967; Dagg and Windsor, 1971; LeMagnen, 1952). Furthermore, the relative age of an individual may be assessed on the basis of odor since immature animals do not yet exhibit the hormone-dependent changes in glandular size and marking.

Fig. 4. Enurination (*Harnspritzen*) postures in some hystricomorph rodents. a. *Myoprocta* male, bipedal. b. *Octodon* male, tripedal. c. *Cavia* male, tripedal. d. *Dolichotis* male, bipedal. e. *Cuniculus* male, bipedal. (From Kleiman, 1974.)

Reproductive behavior is profoundly influenced by chemical signals. Some chemicals act almost as "releasers" in that *normal* sexual behavior will not proceed if they are absent or if the olfactory nerve function is impaired (Doty and Anisko, 1973; Heimer and Larsson, 1967). Other chemical substances act as "primers" since their effects are delayed. Chemical signals or pheromones that act as primers may in *Mus* inhibit pregnancy (Parkes and Bruce, 1962); induce estrus (Whitten, 1966); or advance puberty (Vandenbergh, 1967). The nature of these priming and releasing effects has only begun to be investigated, and considerable variation with respect to the nature and degree of these effects will no doubt be shown when the phenomena are compared over a range of species drawn from several families.

Several recent studies have shown that odors can indicate the mood of the animal, e.g., fear in *Mus* (Müller-Velten, 1966).

Tactile Input

Close-range behavior involves touching, grasping, opposing or locking incisors, allogrooming, and a host of other interaction forms. Olfaction is strongly involved in many forms of interaction involving tactile stimulation, and extremely stereotyped configurations can result during initial encounters between two conspecifics (see Eibl-Eibesfeldt, 1958, for a summary of his work; Eisenberg, 1962, 1968, for *Peromyscus;* Eisenberg, 1963a, 1963b, for the Heteromyidae; Kleiman, 1971, 1972, for *Myoprocta;* Steiner, 1970, 1971, for *Spermophilus colombianus;* Stanley, 1971, for *Notomys;* Horwich, 1972, for *Sciurus;* Koenig, 1960, for *Glis glis;* Ewer, 1971, for *Rattus rattus;* Wilsson, 1968, for *Castor;* Dieterlen, 1959, for *Mesocricetus*).

The extent to which tactile input can promote the assumption of specific postures has been experimentally analyzed only within the context of mating behavior. Because many species of rodents show a specific mating posture (lordosis) assumed by the female, the analysis of the stimuli necessary to induce lordosis has been the subject of some study. Females of *Rattus* and *Cavia* assume lordosis in response to the mounting of the male. The study of the stimuli necessary to induce lordosis in *Mesocricetus* indicates that olfactory and perhaps visual stimuli from the male increase the ease with which lordosis is elicited by tactile stimuli, but ultimately tactile stimuli are both sufficient and necessary to elicit lordosis in the female of *Mesocricetus auratus* (Murphy, 1974).

The ubiquity of allogrooming in the sexual and maternal behavior of rodents attests to its important role in promoting exchange of tactile stimuli. It is entirely possible that both gustatory and olfactory stimuli are perceived by the groomer; however, this complex of possible stimulus exchanges has not been analyzed to date.

EVOLUTIONARY TRENDS IN RODENT COMMUNICATION SYSTEMS

Auditory Communication and Predator Detection in Nocturnal Desert Rodents

An anatomical peculiarity shared by many arid-adapted, nocturnal rodents is the presence of an inflated middle ear cavity (Howell, 1932). Externally this anatomical peculiarity is reflected in the expansion of the mastoid bulla. Extreme inflation of the mastoid bulla is occasionally accompanied by an enlargement of the external ear (Ognev, 1959), but often in those species having the largest bullae, the pinna is small (e.g., *Dipodomys* and *Microdipodops*). Some species of nocturnal desert rodents show little bullar expansion but extremely hypertrophied pinnae (e.g., *Alactaga* and *Euchoreutes*). For these species it is assumed that the large pinna increases sensitivity to low-amplitude sounds by focusing sound energy at the meatus. The propagation of sound in desert air involves considerable energy loss, especially at frequencies > 10 Khz (Knudsen, 1931, 1935). The resulting signal attenuation may have necessitated the evolution of anatomical adaptations to increase sensitivity to low-amplitude sounds. Pinna enlargement is one way, bullar inflation another.

The expanded middle ear cavity has been correlated with structural modifications in the cochlea; and the most extreme cochlear modifications are found in those species exhibiting the greatest bullar inflation (Pye, 1965; Lay, 1972). Generally speaking, those species from the most arid habitat with the least vegetational cover show the greatest bullar expansion (Petter, 1961). Indeed, a whole range of bullar expansions can be demonstrated within a single genus such as *Meriones* (Lay, 1972), and environmental correlates with respect to aridity and ground cover can be made.

One outstanding acoustical feature of the rodents possessing an expanded bulla is the enhanced sensitivity for low-frequency sounds (less than 2 Khz). Wisner et al. (1954) and Legouix et al. (1954) hypothesized that the sensitivity of hearing was maximized toward values near the resonant frequencies of the ossicles. Lay (1972) demonstrated for a series of gerbilline rodents that a shift in auditory sensitivity for lower frequencies parallels increased bullar expansion. Webster (1960, 1962) demonstrated for *Dipodomys* that not only does the inflated tympanic bulla maintain sensitivity for low-frequency sounds, but, furthermore such sounds are often generated by the predators themselves, e.g., snake movement or the wing beats of owls. Webster went on to show that the ability to avoid the strike of a snake on the part of *Dipodomys* was dependent on intact mastoid bullae (Webster, 1962).

Petter (1961) felt that the inflated mastoid bullae of desert rodents could enhance their sensitivity for the low-frequency sounds produced during various aspects of intraspecific communication. In particular he noted the ubiquity of foot stamping or drumming as a mode of signaling in desert rodents. He further noted that the largest tympanic bullae occurred in those species most strongly adapted to extremely arid habitats and thus living at very low densities where the carrying capacity is understandably low.

Sparse cover, low carrying capacity, and enforced foraging for seeds in open areas all intercorrelate. Webster would argue that predator selection is primarily responsible for the inflated mastoid bullae and the resulting shift in auditory sensitivity. Lay (1972) agrees in general with Webster's hypothesis. Both workers cite reports indicating that paradoxically many vocal sounds of gerbillines and *Dipodomys* are high pitched.

While it is agreed that predator selection may be decisive in the evolution of the expanded middle ear cavities of nocturnal, arid-adapted rodents, selection may have also acted on the sound-producing mechanisms of these species to produce call forms that must function over a considerable distance and that furthermore emphasize frequencies to which this auditory apparatus is maximally sensitive. Close-range sounds during fighting, sexual behavior, or distress might not be under such selective constraints, but the cry of a displaced young could be so modified. Thus, any similarities in syllable structure or pitch of the young animal's calls could be the result of evolutionary convergence.

In a detailed study of the family Heteromyidae (Eisenberg, 1963b) it was noted that young kangaroo rats when displaced from the nest do not emit ultrasonic pulses but instead emit a repetitive buzzy sound that emphasizes low frequencies. The threat growl of adult *Dipodomys nitratoides* also emphasizes frequencies < 2 Khz. Recently Owings and Irvine (1974) have replicated these observations with *Dipodomys merriami.* In a recent review (Eisenberg, 1975) the comparison of "abandoned cries" for a series of young nocturnal desert rodents suggested that the overall pitch may conform to the optimum sensitivity of the adult cochlea. Thus, a gerbilline such as *Tatera indica* exhibiting little bullar inflation has neonate young that give calls with energy at 4 to 6 Khz, while *Meriones hurrianae* with a larger tympanic bulla has an abandoned cry pitched from 2 to 3 Khz. At the other extreme the young of *D. nitratoides* and *D. merriami* call with energy less than 2.5 Khz.

In conclusion, then, it would appear that the restrictions of open, arid environments favor selection for enhanced sensitivity to low-frequency sounds on the part of nocturnal rodents. That predators have been the primary selective force is undoubtedly true; however, it would seem reasonable to assume that certain classes of auditory signals (e.g., cry of the displaced young) have undergone selection to exhibit a pitch conforming to the optimal sensitivity of the adult cochlea.

Sandbathing as a Form of Chemical Marking

The development of increased secretory activity in the sebaceous glands associated with the hair follicles of arid-adapted rodent species is a widespread phenomenon (Sokolov, 1962). In addition, many species have evolved specialized gland fields on the ventrum (*Gerbillus* and *Meriones*) or mid-dorsal region *(Dipodomys)* in addition to the classical glandular areas of the anogenital region (see Quay, 1953; Fiedler, 1974). The sebaceous glands associated with the hair follicles act as epidermal lubricants to reduce drying of the skin. The pelage will generally become quite oily if excess depositions of sebum are not removed through dust bathing. Obviously chemical substances in the sebum as well as depositions from other skin glands and even urine at sandbathing loci could serve in chemical communication (Eisenberg, 1963a and 1963b). Rodents from diverse families adapted to arid habitats show sandbathing behavior, including the Gerbillinae (Fiedler, 1974); the Heteromyidae (Eisenberg, 1963a), the Dipodidae (Eisenberg, 1967), and several genera of caviomorph species, including *Chinchilla, Octodon, Octodontomys,* and *Pediolagus* (Wilson and Kleiman, 1974). (See Fig. 5.)

Although the movement patterns in sandbathing may vary widely from one genus to another in a species-specific manner (Eisenberg, 1967), all species tend to sandbathe at specific loci and these same spots are utilized by conspecifics ranging within the same living space. The potential for communicating by chemical signals by means of such sandbathing spots is strongly implicated (Eisenberg, 1967). Such spots may be used by species living in family groups (e.g., *Pediolagus*) as a means by which group odor can be maintained through successive use by all colony members (Kleiman, 1974). Indeed, a class of play movements called "locomotor-rotational movements" are released by conspecific odors. Such odors may be exchanged either while sniffing a partner or while sniffing a sandbathing spot (Wilson and Kleiman, 1974); the arid-adapted forms (e.g., *Pediolagus* and *Octodontomys*) exhibit locomotor-rotational movements more often in response to sniffing sandbathing loci.

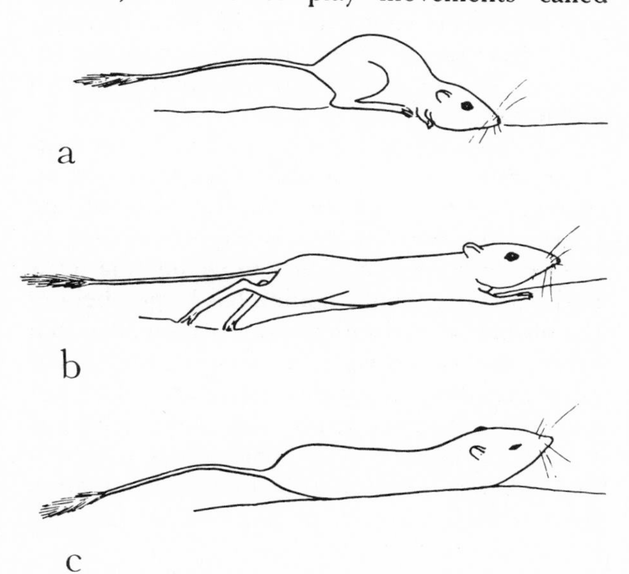

Fig. 5. *Dipodomys* sandbathing. a. Flexion of the body followed by b. extension in performing the ventral-rubbing component. c. Side rubbing by extending with the side pressed into the substrate. (From Eisenberg, 1963a.)

Differences in the patterns of rubbing, extension, and flexion as well as differences in the specific body areas rubbed in sand are to be found when a series of species are compared (Eisenberg, 1967). Chemical marking and pelage dressing appear to have been combined into a single movement complex in many desert-adapted species, although pure marking movements may be retained without a necessary pelage dressing function. Comparative studies strongly suggest that pelage dressing movements in the form of sandbathing have evolved independently within several lines of rodent evolutionary descent and that such patterns should be considered examples of convergent behavioral evolution.

Visual Signals and Diurnality

In the evolution of diurnality in the Rodentia, several types of movement patterns have evolved that imply visual communication. The tail movements of the diurnal, arboreal sciurids, often accompanied by vocalizations, could serve to accentuate the position of the sender as well as to communicate varying degrees of arousal to a potential receiver (Bakken, 1959; Horwich, 1972). In the aggressive acouchi, piloerection of the rump hair serves a similar function (Fig. 6).

In some diurnal caviomorphs, e.g., the acouchi *(Myoprocta pratti),* tail wagging combined with body trembling and alternate stepping movements of the forefeet are important components of courtship which may indicate the approach-withdrawal tendencies of the courting male (Kleiman, 1971, 1974). Such tail and body move-

Fig. 6. Piloerection in *Myoprocta pratti.* Erection of the rump hair is a ritualized aspect of aggressive behavior in the acouchi. Note that the apparent size of the animal is exaggerated. (From Kleiman, 1972.)

ments are reminiscent of similar movements by diurnal ungulates, e.g., the Uganda kob (*Adenota kob:* Buechner and Schloeth, 1965) during courtship, and some diurnal macropods, e.g., the whiptail wallaby (*Macropus parryi:* Kaufmann, 1974), and have probably evolved from intention locomotor movements (Andrew, 1972). The stereotypy of such displays in diurnal caviomorphs, macropods, and ungulates is probably the result of evolutionary convergence resulting from a similarity in selective pressures within the three groups. Other convergences in morphology and behavior can be discerned when the forest-adapted cursorial caviomorphs (*Dasyprocta, Myoprocta,* and *Cuniculus*) are compared with their ungulate counterparts in the Old World tropical forests (Dubost, 1968; Eisenberg and McKay, 1974).

During the course of the evolution of steppe-adapted cursorial caviomorphs, such as *Dolichotis,* a striking form of antipredator behavior has evolved, which bears a strong resemblance to similar patterns shown by the smaller antelope genera of East Africa. The pattern involves a peculiar gait, "stotting," which serves to display the white rump markings prominently. Such a signal pattern could serve to induce a predator to launch a futile attack because the display is always given when the sender is well outside the normal attack range of the predator. At the same time the signal appears to alert mates or young of the predator's presence in the living space (for a full discussion, see Smythe, 1971). Similar movements are also employed during the play of certain caviomorphs (Kleiman, 1974; Wilson and Kleiman, 1974).

Conspicuous movement patterns are involved in the challenge and territorial defense of the colonial prairie dog (*Cynomys ludovicianus:* King, 1955; Smith et al., 1973). The territorial call is given during the course of a complicated movement sequence when the caller "throws its body upwards and rises on its hind legs with nose pointed straight up and with forefeet thrust out from the body, and then returns to its normal quadrupedal position" (King, 1955). In this classical description, a vocalization and a stereotyped movement sequence are combined in a single display.

The preceding examples of movement patterns shown by diurnal rodents suggest that information is transferred via the eye of the presumptive receiver. Yet the experimental analysis of the exact role of such movement patterns in the transfer of information has lagged behind the original descriptions. Auditory signals often accompany stereotyped movements, which suggests that the presumptive signal can often not be reduced to a single physical component. For the nocturnal rodents visual signals probably play a much reduced role in the information transfer system.

The Antipredator Calls of Colonial Rodents

One of the more conspicuous examples of convergent behavioral evolution in the Rodentia is the development of specific vocalizations that are emitted by colonial rodents either when a potential predator is perceived or when the subjects are disturbed by a novel stimulus input. This behavioral phenomenon is typical of rodents that live in rather open habitats where the predator can be kept in view by the calling animal; however, similar warning calls or sounds have been evolved by species in forested habitats. During responses to a mobile predator, the calls of the latter are often not repetitive but rather are given before or during a directed flight from the predator (Eisenberg and McKay, 1974). These forms of antipredator strategy are not confined to rodents but have evolved convergently in ungulates, lagomorphs, primates, and some small carnivores. The specific form and pitch of the call apparently involve a complex of factors including the size of the species, the vocal apparatus, and those physical features of the habitat that affect sound propagation (Eisenberg, 1974).

The evolutionary adaptations leading to coloniality and ultimately to the formation of communal groups defending a group territory in diurnal rodents have been reviewed by Barash (1973, 1974) for the genus *Marmota.* The use of warning calls as an antipredator strategy has been described for *Marmota olympus, M. caligata,* and *M. flaviventris* (Barash, 1973; Waring, 1966; Armitage, 1962). Convergent trends have been noted for *Cynomys* (King, 1955; Waring, 1970) and *Spermophilus* (Balph and Balph, 1966). Within the colonial Caviomorpha, similar calls have been noted for *Lagidium* (Pearson, 1948), *Spalacopus* (Reig, 1970), *Ctenomys* (Pearson, 1959), and *Lagostomus* (Hudson, 1872).

Specific distinctiveness exists when calls are compared from one species to the next, but the ecological sources of such variations have not been explored. These kinds of adaptive aspects in the varying forms of "warning cries" in colonial rodents will surely prove to be rewarding areas of study for future students of rodent communication.

CONCLUSIONS

It is evident from the preceding discussion that the rodents display significant diversity in their communication systems. The basis for the diversity can be found in the variety of habitats and niches to which members of this order have become adapted. We did not attempt initially to outline the natural history of selected species since such an approach would have necessitated another chapter to deal with an order containing so many species. However, it must be emphasized that the analysis of communication can yield biologically significant results only if a study is conceived and executed with regard to a species' natural history. That this has not always been done in the past is evidenced by the fact that the first complete natural history study of the Norway rat appeared in 1962 (Calhoun, 1962), and Ewer (1971) has only recently published on *Rattus rattus.* Guinea pig research has flourished since the early part of this century, but the ecology and social behavior of *Cavia* and related genera were ignored until Rood published in 1972. There is, as yet, no detailed study of the natural history of *Meriones unguiculatus,* although the gerbil is being used increasingly in research on communication. Experimental analysis of a species' communication system must proceed in step with field research dealing with the adaptive nature of the behavioral repertoire. It is heartening to note that this unified approach is now under way.

References

Andrew, R. J., 1972. The information potentially available in mammalian displays. In: *Non-verbal Communication,* R. A. Hinde, ed. New York: Cambridge University Press, pp.179–204.

Armitage, K. B., 1962. Social behavior of a colony of the yellow-bellied marmot (*Marmota flaviventris*). *Anim. Behav.,* 10:319–31.

Arvola, A., 1974. Vocalization in the guinea-pig, *Cavia porcellus* L. *Ann. Zool. Fennici,* 11:1–96.

Arvola, A.; Ilmen, M.; and Koponen, T.; 1962. On the aggressive behaviour of the Norwegian lemming (*Lemmus lemmus*), with special reference to the sounds produced. *Arch. Soc. "Vanamo"* 17(2):80–101.

Asdell, S. A., 1964. *Patterns of Mammalian Reproduction.* Ithaca: Cornell University Press.

Bakken, A., 1959. Behavior of gray squirrels. In: *Symposium on the Gray Squirrel,* V. Flyger, ed. Contribution 162, Maryland Department of Research and Education, pp.393–407.

Balph, D. M., and Balph, D. F., 1966. Sound communication of Uinta ground squirrels. *J. Mamm.,* 47:440–50.

Banks, E. M., 1962. A time and motion study of prefighting behavior in mice. *J. Genetic Psych.,* 101:165–83.

Barash, D. P., 1973. The social biology of the Olympic marmot. *Anim. Behav. Monographs,* 6(3):171–245.

Barash, D. P., 1974. The evolution of marmot societies; A general theory. *Science,* 185:415–20.

Barfield, R. J., and Geyer, L. A. 1972. Sexual behavior; Ultrasonic post-ejaculatory song of the male rat. *Science,* 176:1349–50.

Beach, F., and Jaynes, J., 1956. Studies of maternal retrieving in rats. III. Sensory cues involved in the lactating female's response to her young. *Behaviour,* 10:104–25.

Benson, S. B., 1933. Concealing coloration among some desert rodents of the southwestern United States. *Univ. Calif. Publ. Zool.,* 40:1–70.

Bowers, J. M., and Alexander, B. K., 1967. Mice: Individual recognition by olfactory cues. *Science,* 158:1208–10.

Brooks, R. J., and Banks, E. M., 1973. Behavioural biology of the collared lemming (*Dicrostonyx groenlandicus,* Traill): An analysis of acoustic communication. *Anim. Behav. Monographs,* 6(1):1–83.

Brown, A. M., 1973. High levels of responsiveness from the inferior colliculus of rodents at ultrasonic frequencies. *J. Comp. Physiol.,* 83:393–406.

Buechner, H. K., and Schloeth, R., 1965. Ceremonial mating behavior in Uganda kob (*Adenota kob thomasi* Neumann). *Z. Tierpsychol.,* 22:209–25.

Calhoun, J. B., 1962. The ecology and sociology of the Norway rat. U.S. Publ. Hlth. Serv., Publ. No. 1008, pp.1–288.

Carr, W. J., and Caul, W. F., 1962. The effect of castration in the rat upon the discrimination of sex odours. *Anim. Behav.,* 10:20–27.

Carr, W. J.; Loeb, L. S.; and Dissinger, M. L.; 1965. Responses of rats to sex odors. *J. Comp. Physiol. Psychol.,* 59:370–77.

Carr, W. J.; Loeb, L. S.; and Wylie, N. R.; 1966. Responses to feminine odors in normal and castrated male rats. *J. Comp. Physiol. Psychol.,* 62:336–38.

Collins, L. R., and Eisenberg, J. F., 1972. The behavior and breeding of pacaranas (*Dinomys branickii*) in captivity. *Internat. Zoo Yb.,* 12:108–14.

Coujard, R., 1947. Study of the odoriferous glands of the rabbit and the influence of sex hormones on them. *Rev. Canad. Biol.,* 6:3–14.

Coulon, J., 1973. Le répertoire sonore du cobaye domestique et sa signification comportementale. *Rev. Comp. Animal,* 7(2):121–32.

Dagg, A. L., and Windsor, D. E., 1971. Olfactory discrimination limits in gerbils. *Can. J. Zool.,* 49:283–85.

Denenberg, V. H.; Zarrow, M. X.; and Ross, S.; 1969. The behaviour of rabbits. In: *The Behaviour of Domestic Animals,* E. S. E. Hafez, ed. Baltimore: William and Wilkins, pp.417–37.

Dieterlen, F., 1959. Das Verhalten des syrischen Goldhamsters (*Mesocricetus auratus* Waterhouse). Untersuchungen zur Frage seiner Entwicklung und seiner angeborenen Anteile durch geruchsisolierte Aufzuchten. *Z. Tierpsychol.,* 16(1):47–103.

Doty, R. L., 1972. Odor preferences of female *Peromyscus maniculatus bairdi* for male mouse odors of *P. m. bairdi* and *P. leucopus noveboracensis* as a function of estrous state. *J. Comp. Physiol. Psychol.,* 81:191–97.

Doty, R. L., 1973. Reactions of deer mice (*Peromyscus maniculatus*) and white-footed mice (*Peromyscus leucopus*) to homospecific and heterospecific urine odors. *J. Comp. Physiol. Psychol.,* 84:296–303.

Doty, R. L., and Anisko, J. J., 1973. Procaine hydrochloride olfactory block eliminates mounting in the male golden hamster. *Physiol. and Behav.,* 10:395–97.

Doty, R. L., and Kart, R., 1972. A comparative and developmental analysis of the midventral sebaceous glands in 18 taxa of *Peromyscus,* with an examination of gonadal steroid influences in *Peromyscus maniculatus bairdii. J. Mamm.,* 53:83–99.

Dubost, G., 1968. Les niches écologiques des forêts tropicales Sud-Américaines et Africaines, sources de convergences remarquables entre rongeurs et artiodactyles. *La Terre et la Vie,* 1:3–28.

Dunford, C., 1970. Behavioral aspects of spatial organization in the chipmunk, *Tamias striatus. Behav.,* 35:215–32.

Eibl-Eibesfeldt, I., 1958. Das Verhalten der Nagetiere.

Handbuch der Zoologie, 8 Band, 12 Lieferung; 10(13): 1–88.

Eisenberg, J. F., 1962. Studies on the behavior of *Peromyscus maniculatus gambelii* and *P. californicus parasiticus. Behav.,* 19(3):177–207.

Eisenberg, J. F., 1963a. A comparative study of sandbathing behavior in heteromyid rodents. *Behav.,* 22:16–23.

Eisenberg, J. F., 1963b. The behavior of heteromyid rodents. *Univ. Calif. Publ. Zool.,* 69:1–100.

Eisenberg, J. F., 1967. Comparative studies on the behavior of rodents with special emphasis on the evolution of social behavior, Part I. *Proc. U.S. Nat. Mus.,* 122(3597):1–55.

Eisenberg, J. F., 1968. Behavior patterns. In: *Biology of Peromyscus (Rodentia),* J. A. King, ed. Amer. Soc. Mammalogists, Special Publication 2.

Eisenberg, J. F., 1974. The function and motivational basis of hystricomorph vocalizations. In: *The Biology of Hystricomorph Rodents,* I. W. Rowlands and B. Weir, eds. Symp. Zool. Soc. London, No. 34:211–44.

Eisenberg, J. F., 1975. The behavior patterns of desert rodents. In: *Rodents in Desert Environments,* I. Prakash and P. K. Ghosh, eds. Monographae Biologicae. The Hague: W. Junk, pp.189–221.

Eisenberg, J. F., and Kleiman, D. G., 1972. Olfactory communication in mammals. *Ann. Rev. Ecol. & Systematics,* 3:1–32.

Eisenberg, J. F., and McKay, G. M., 1974. Comparison of ungulate adaptations in the New World and Old World tropical forests with special reference to Ceylon and the rainforests of Central America. In: *The Behaviour of Ungulates and Its Relation to Management,* V. Geist and F. Walther, eds. Morges: IUCN, pp.585–601.

Ewer, R. F., 1971. The biology and behaviour of a free-living population of black rats (*Rattus rattus*). *Anim. Behav. Monographs,* 4(3):127–74.

Fiedler, U., 1974. Beobachtungen zur Biologie einiger Gerbillinen, insbesondere *Gerbillus (Dipodillus) dasyurus* (Myomorpha, Rodentia) in Gefangenschaft. II. Ökologie. *Z. Säugetierk.,* Bd. 39, 1:24–41.

Forcum, D. L., 1966. Postpartum behavior and vocalizations of snowshoe hares. *J. Mamm.,* 47:543.

Godfrey, J., 1958. The origin of sexual isolation between bank voles. *Proc. Roy. Phys. Soc. Edinburgh,* 27:47–55.

Grant, E. C., and Mackintosh, J. H., 1963. A comparison of social postures of some common laboratory rodents. *Behav.,* 21:246–59.

Grant, E. C.; Mackintosh, J. H.; and Lerwill, C. J.; 1970. The effect of a visual stimulus on the agonistic behaviour of the golden hamster. *Z. Tierpsychol.,* 27:73–77.

Haga, R., 1960. Observations on the ecology of the Japanese pika. *J. Mamm.,* 41(2):200–13.

Harvey, E. B., and Rosenberg, L. E., 1960. An apocrine gland complex in the pika. *J. Mamm.,* 41(2): 213–20.

Heath, E., 1972. Sexual and related territorial behavior in the laboratory rabbit (*Oryctolagus cuniculus*). *Lab. Anim. Sci.,* 22:684–91.

Heimer, L., and Larsson, K., 1967. Mating behavior of male rats after olfactory bulb lesions. *Physiol. Behav.,* 2:207–209.

Horwich, R. H., 1972. The ontogeny of social behavior in the gray squirrel (*Sciurus carolinensis*). *Z. Tierpsychol.,* Beiheft 8:1–103.

Howell, A. B., 1932. The saltatorial rodent, *Dipodomys:* The functional and comparative anatomy of its muscular and osseus systems. *Proc. Amer. Acad. Arts. Sci.,* 67:377–536.

Hudson, W. H., 1872. On the habits of the vizcacha (*Lagostomus trichodactylus*). *Proc. Zool. Soc. London,* 1872:822–33.

Johnson, R. P., 1973. Scent marking in mammals. *Anim. Behav.,* 21(3):521–36.

Johnston, R. E., 1973. Determinants of dominance in hamsters. I. Do chest markings have a threat function? *Amer. Zoologist,* 13:1264.

Kaufman, D. W., 1974. Adaptive coloration in *Peromyscus polionotus:* Experimental selection by owls. *J. Mamm.,* 55(2):271–84.

Kaufmann, J. H., 1974. Social ethology of the whiptail wallaby, *Macropus parryi,* in northeastern New South Wales. *Anim. Behav.,* 22:281–370.

Kawamichi, T., 1968. Winter behaviour of the Himalayan pika, *Ochotona roylei. J. Fac. Sci.,* Series VI, Zoology, 16(4):582–94.

Kawamichi, T., 1970. Social pattern of the Japanese pika, *Ochotona hyperborea hesoensis:* Preliminary report. *J. Fac. Sci.,* Series VI, Zoology, 17(3):462–73.

Kawamichi, T., 1971. Daily activities and social patterns of two Himalayan pikas, *Ochotona macrotis* and *O. roylei,* observed at Mt. Everest. *J. Fac. Sci.,* Series VI, Zoology, 17(4):587–609.

Kilham, L., 1958. Territorial behavior in pikas. *J. Mamm.,* 39:307.

King, J. A., 1955. Social behavior, social organization, and population dynamics in a black-tailed prairiedog town in the Black Hills of South Dakota. *Contr. Lab. Vert. Biol.*, no. 67, Univ. of Michigan.

King, J. A., ed., 1968. Biology of *Peromyscus* (Rodentia). Special Publication No. 2. Amer. Soc. of Mammalogists.

Kleiman, D. G., 1971. The courtship and copulatory behaviour of the green acouchi, *Myoprocta pratti. Z. Tierpsychol.*, 29:259–78.

Kleiman, D. G., 1972. Maternal behaviour of the green acouchi (*Myopracta pratti* Pocock), a South American caviomorph rodent. *Behav.*, 43(1–4):13–48.

Kleiman, D. G., 1974. Patterns of behavior in hystricomorph rodents. In: *The Biology of Hystricomorph Rodents,* I. W. Rowlands and B. Weir, eds. Symp. Zool. Soc. London, No. 34:171–209.

Knudsen, V. O., 1931. The effect of humidity upon the absorption of sound in a room, and a determination of the coefficients of absorption of sound in air. *J. Acoust. Soc. Amer.*, 3:126–38.

Knudsen, V. O., 1935. Atmospheric acoustics and the weather. *Sci. Monthly,* 40:485–86.

Koenig, L., 1960. Das Aktionsystem des Siebenschläfers (*Glis glis* L.). *Z. Tierpsychol.*, 17:427–505.

Komisaruk, B. T.; Adler, N. T.; and Hutchinson, J.; 1973. Genital sensory field: Enlargement by estrogen treatment in female rats. *Science,* 178:1295–98.

Lashley, K. S., 1932. The mechanism of vision. V. The structure and image-forming power of the rat's eye. *J. Comp. Psychol.*, 13:173–200.

Lashley, K. S., 1938. The mechanism of vision. XV. Preliminary studies of the rat's capacity for detail vision. *J. Genet. Psychol.*, 18:123–93.

Lashley, K. S., 1950. In search of the engram. In: *Physiological Mechanisms in Animal Behaviour.* Cambridge: S.E.B. Symposia No. 4.

Lay, D. M., 1972. The anatomy, physiology, functional significance and evolution of specialized hearing organs of gerbilline rodents. *J. Morphol.*, 138(1):41–93.

Lechleitner, R. R., 1958. Certain aspects of behavior of the black-tailed jack rabbit. *Am. Midl. Nat.*, 60:145–55.

Legouix, J. P.; Petter, F.; and Wisner, A.; 1954. Étude de L'Auditien chez des Mammiferes a Bulles Tympaniques Hypertrophis. *Mammalia,* 18(3):262–71.

LeMagnen, J., 1952. Les phenomenes olfacto-sexuels chez la rat blanc. *Arch. Sci. Physiol.*, 6:295–331.

Levine, S., ed., 1972. *Hormones and Behavior.* New York: Academic Press.

Lipkow, J., 1954. Über das seitenorgan des goldhamsters (*Mesocricetus auratus auratus* Waterh.). *Z. Morph. Oekol. Tiere,* 42:333–72.

Lorenz, K. Z., 1950. The comparative method in studying innate behaviour patterns. In: *Physiological Mechanisms in Animal Behaviour.* Cambridge: S.E.B. Symposia No. 4, pp.221–68.

Lydell, K., and Doty, R. L., 1972. Male rat odor preferences for female urine as a function of sexual experience, urine age, and urine source. *Hormones and Behavior,* 3:205–12.

Mackintosh, J. H., and Grant, E. C., 1966. The effect of olfactory stimuli on the agonistic behaviour of laboratory mice. *Z. Tierpsychol.*, 23(5):584–87.

Marsden, H. M., and Holler, N. R., 1964. Social behavior in confined populations of the cottontail and swamp rabbit. Wildlife Monograph No. 13. Wildlife Society.

Martan, J., 1962. The effect of castration and androgen replacement on the supracaudal gland of the male guinea-pig. *J. Morph.*, 110:285–93.

Melchior, H. R., 1971. Characteristics of Arctic ground squirrel alarm calls. *Oecologia,* 7:184–90.

Moore, R. E., 1965. Olfactory discrimination as an isolating mechanism between *Peromyscus maniculatus* and *Peromyscus polionotus. Am. Midl. Nat.*, 73:85–100.

Müller-Velten, H., 1966. Über den Angstgeruch bei der Hausmaus. *Z. Vergl. Physiol.*, 52:401–29.

Munn, N. L., 1950. *Handbook of Psychological Research on the Rat.* Boston: Houghton Mifflin.

Murphy, M. R., 1973. Effects of female hamster vaginal discharge on the behavior of male hamsters. *Behav. Biol.*, 9(3):367–75.

Murphy, M. R., 1974. Relative importance of tactual and nontactual stimuli in eliciting lordosis in the female golden hamster. *Behav. Biol.*, 11:115–19.

Murphy, M. R., and Schneider, G. E., 1970. Olfactory bulb removal eliminates mating behavior in the male golden hamster. *Science,* 167:302–4.

Myers, K., and Mykytowycz, R., 1958. Social behavior in the wild rabbit. *Nature,* 181:1515–16.

Myers, K., and Poole, W. E., 1958. Sexual behaviour cycles in the wild rabbit, *Oryctolagus cuniculus* (L.). *CSIRO Wildlife Res.*, 3(2):144–45.

Myers, K., and Poole, W. E., 1961. A study of the biology of the wild rabbit, *Oryctolagus cuniculus* (L.), in confined populations. II. The effects of season and population increase on behavior. *CSIRO Wildlife Research,* 6(1):1–41.

Mykytowycz, R., 1958. Social behaviour of an experimental colony of wild rabbits. I. Establishing of the colony. *CSIRO Wildlife Res.*, 3:7–25.

Mykytowycz, R., 1959. Social behaviour of an experimental colony of wild rabbits. II. First breeding season. *CSIRO Wildlife Res.,* 4:1–13.

Mykytowycz, R., 1960. Social behaviour of an experimental colony of wild rabbits. III. The second breeding season. *CSIRO Wildlife Res.,* 5:1–20.

Mykytowycz, R., 1961. Social behaviour of an experimental colony of wild rabbits, *Oryctolagus cuniculus* (L.). IV. Conclusion: Outbreak of myxomatosis, third breeding season, and starvation. *CSIRO Wildlife Res.,* 6:142–55.

Mykytowycz, R., 1965. Further observations on the territorial function and histology of the sub-mandibular cutaneous (chin) glands in the rabbit, *Oryctolagus cuniculus* (L.). *Anim. Behav.,* 13:400–412.

Mykytowycz, R., 1966a. Observations on odoriferous and other glands in the Australian wild rabbit, *Oryctolagus cuniculus* (L.), and the hare, *Lepus europaeus* (P.). I. The anal gland. *CSIRO Wildlife Res.,* 11:11–29.

Mykytowycz, R. 1966b. Observations on odoriferous and other glands in the Australian wild rabbit, *Oryctolagus cuniculus* (L.), and the hare, *Lepus europaeus* (P.). II. The inguinal glands. *CSIRO Wildlife Res.,* 11:49–64.

Mykytowycz, R., 1966c. Observations on odoriferous and other glands in the Australian wild rabbit, *Oryctolagus cuniculus* (L.), and the hare, *Lepus europaeus* (P.). III. Harder's, lachrymal, and submandibular glands. *CSIRO Wildlife Res.,* 11:65–90.

Mykytowycz, R., 1967. Communication by smell in the wild rabbit. *Proc. Ecol. Soc. Aust.,* 2:125–31.

Noirot, E., 1966. Ultra-sounds in young rodents. I. Changes with age in albino mice. *Anim. Behav.,* 14:459–62.

Noirot, E., 1968. Ultra-sounds in young rodents. II. Changes with age in albino rats. *Anim. Behav.,* 16:129–34.

O'Farrell, T. P., 1965. Home range and ecology of snowshoe hares in interior Alaska. *J. Mamm.,* 46:406–18.

Ognev, S. I., 1959. *Saugetiere und Ihre Welt.* Berlin: Akademie-Verlag.

Okon, E. E., 1970. The effect of environmental temperature on the production of ultrasounds by isolated non-handled albino mouse pups. *J. Zool. Lond.,* 162:71–83.

Okon, E. E., 1972. Factors affecting ultrasound production in infant rodents. *J. Zool. Lond.,* 168:139–48.

Owings, D. H., and Irvine, J., 1974. Vocalization in Merriam's kangaroo rat. *J. Mamm.,* 55:465–66.

Parkes, A. S., and Bruce, H. M., 1962. Pregnancy-block in female mice placed in boxes soiled by males. *J. Reprod. Fert.,* 4:303–308.

Pearson, O. P., 1948. Life history of mountain viscachas in Peru. *J. Mamm.,* 29:345–74.

Pearson, O. P., 1959. Biology of the subterranean rodents, *Ctenomys,* in Peru. *Mems. Mus. Hist. Nat. "Javier Prado,"* 9:1–56.

Petter, F., 1961. Répartition géographique et écologie des rongeurs désertique. *Mammalia,* 24:1–219.

Pietras, R. J., and Moulton, D. G., 1974. Hormonal influences on odor detection in rats: Changes associated with the estrous cycle, pseudopregnancy, ovariectomy, and administration of testosterone propionate. *Physiol. Behav.,* 12:475–91.

Price, G. R., 1970. Sensitivity of the rat ear re-examined through the cochlear microphonic. *J. Audit. Res.,* 10:340–48.

Pye, A., 1965. The auditory apparatus of the Heteromyidae (Rodentia, Sciuromorpha). *J. Anat. Lond.,* 99:161–74.

Quay, W. B., 1953. Seasonal and sexual differences in the dorsal skin gland of the kangaroo rat, *Dipodomys. J. Mamm.,* 34:1–14.

Ralls, K., 1967. Auditory sensitivity in mice, *Peromyscus* and *Mus musculus. Anim. Behav.,* 15:123–28.

Ralls, K., 1971. Mammalian scent marking. *Science,* 171:443–49.

Reig, O. A., 1970. Ecological notes on the fossorial octodont *Spalacopus cyanus* (Molina). *J. Mamm.,* 51:592–601.

Rood, J. P., 1972. Ecological and behavioural comparisons of three genera of Argentine cavies. *Anim. Behav. Monographs,* 5:1–83.

Rothenberg, S., and Davis, H., 1967. Auditory evoked response in chinchilla: Application to animal audiometry. *Perception and Psychophysics,* 2(9):443–47.

Sakellaris, P. C., 1972. Olfactory thresholds in normal and adrenalectomized rats. *Physiol. Behav.,* 9:495–500.

Sales, G. D. [Sewell], 1972a. Ultrasound and aggressive behaviour in rats and other small mammals. *Anim. Behav.,* 20(1):88–101.

Sales, G. D. [Sewell], 1972b. Ultrasound and mating behaviour in rodents with some observations on other behavioural situations. *J. Zool. Lond.,* 168:149–64.

Severaid, J. H., 1956. The natural history of the pikas. Ph.D. diss., University of California, Berkeley.

Sewell, G. D., 1970. Ultrasonic signals from rodents. *Ultrasonics,* January:26–30.

Simpson, G. G., 1945. The principles of classification and a classification of the mammals. *Bull. Am. Mus. Nat. Hist.*, vol. 85.

Simpson, G. G., 1959. The nature and origin of supraspecific taxa. *Cold Spring Harbour Symposium on Quantitative Biology*, 24:255–72.

Smith, C. C., 1968. The adaptive nature of social organization in the genus of tree squirrels *Tamiasciurus. Ecological Monographs*, 38:31–63.

Smith, J. C., 1972. Sound production by infant *Peromyscus maniculatus* (Rodentia: Myomorpha). *J. Zool. Lond.*, 168:369–79.

Smith, M. H., 1965. Behavioral discrimination shown by allopatric and sympatric males of *Peromyscus eremicus* and *Peromyscus californicus* between females of the same two species. *Evolution*, 19(3):430–35.

Smith, W. J.; Smith, S. L.; Oppenheimer, E. C.; deVilla, J. G.; and Ulmer, F. A.; 1973. Behavior of a captive population of black-tailed prairie dogs. Annual cycle of social behavior. *Behav.*, 46:189–220.

Smythe, N., 1971. On the existence of "pursuit invitation" signals in mammals. *Amer. Nat.*, 104:491–94.

Sokolov, W., 1962. Skin adaptations of some rodents to life in the desert. *Nature*, 193:823–25.

Somers, P., 1973. Dialects in southern Rocky Mountain pikas, *Ochotona princeps* (Lagomorpha). *Anim. Behav.*, 21(1):124–38.

Sorenson, M. F.; Rogers, J. P.; and Baskett, T. S.; 1972. Parental behavior in swamp rabbits. *J. Mamm.*, 53(4):840–50.

Southern, H. N., 1948. Sexual and aggressive behaviour in the wild rabbit. *Behav.*, 1:173–94.

Stanley, M., 1971. An ethogram of the hopping mouse, *Notomys alexis. Z. Tierpsychol.*, 29:225–58.

Steiner, A. L., 1970. Étude descriptive de quelques activities et comportements de base de *Spermophilus columbianus columbianus* (Ord.). *Rev. Comp. Animal.*, 4:23–42.

Steiner, A. L., 1971. Play activity of Columbian ground squirrels. *Z. Tierpsychol.*, 28:247–61.

Strother, W. F., 1967. Hearing in the chinchilla (*Chinchilla lanigera*): I. Cochlear potentials. *J. Audit. Res.*, 7:145–55.

Thiessen, D. D.; Friend, H. C.; and Lindzey, G.; 1968. Androgen control of territorial marking in the Mongolian gerbil. *Science*, 160:432–34.

Vandenbergh, J. G., 1967. Effect of the presence of a male on the sexual maturation of female mice. *Endocrinology*, 81:345–49.

Vandenbergh, J. G., 1973. Effects of gonadal hormones on the flank gland of the hamster. *Hormone Res.*, 4:28–33.

Vestal, B. M., 1973. Ontogeny of visual acuity in two species of deermice *(Peromyscus). Anim. Behav.*, 21(4):711–20.

Waring, G. H., 1966. Sounds and communication of the yellow-bellied marmot *(Marmota flaviventris). Anim. Behav.*, 14:177–83.

Waring, G. H., 1970. Sound communications of black-tailed, white-tailed, and Gunnison's prairie dogs. *Amer. Midl. Nat.*, 83(1):167–85.

Webster, D. B., 1960. Auditory significance of the hypertrophied mastoid bullae in *Dipodomys. Anat. Rec.*, 136:299.

Webster, D. B., 1961. The ear apparatus of the kangaroo rat, *Dipodomys. Am. J. Anat.*, 108:123–48.

Webster, D. B., 1962. A function of the enlarged middle ear cavities of the kangaroo rat, *Dipodomys. Physiol. Zool.*, 35:248–55.

Whitney, G.; Coble, J. R.; Stockton, M. D.; and Tilson, E. F.; 1973. Ultrasonic emissions: Do they facilitate courtship of mice? *J. Comp. Physiol. Psychol.*, 84:445–52.

Whitten, W. K., 1966. Pheromones and mammalian reproduction. *Adv. Repro. Physiol.*, 1:155–77.

Wilson, S. C., and Kleiman, D. G., 1974. Eliciting play: A comparative study (Octodon, Octodontomys, Pediolagus, Phoca, Choeropsis, Ailuropoda). *Amer. Zool.*, 14:341–70.

Wilsson, L., 1968. *My Beaver Colony.* New York: Doubleday.

Wisner, A.; Legouix, J. P.; and Petter, F.; 1954. Étude Histologique de l'Orielle d'un Rongeur a Bulles Tympaniques Hypertrophies, *Meriones crassus. Mammalia*, 18:371–74.

Wood, A. E., 1965. Grades and clades among rodents. *Evolution*, 19(1):115–30.

Zippelius, H. M., and Schleidt, W., 1956. Ultrashall-Laute bei jungen Mäusen. *Naturwissenschaften*, 43:502.

Chapter 26

ARTIODACTYLA

Fritz R. Walther

Introduction

The order Artiodactyla comprises three suborders: Nonruminantia, Tylopoda, and Ruminantia (Haltenorth, 1963). The Nonruminantia comprise three families: Suidae (five genera and eight species), Tayassuidae (one genus and two species), and Hippopotamidae (two genera and two species). The Tylopoda consist of only one family: Camelidae (two genera and four species). The Ruminantia comprise five families: Tragulidae (two genera and four species), Cervidae (eleven genera and thirty-two species), the Giraffidae (two genera and two species), the Antilocapridae (one genus and one species), and the Bovidae (forty-two genera and ninety-nine species). Numbers of subfamilies, genera, and species vary somewhat with different classification systems; the figures given above represent the minima.

At present, the discussion of communication in Artiodactyla suffers from certain difficulties. The Nonruminantia and the Ruminantia have rather different physical structures and means of communication. Also, there are considerable differences within Ruminantia with respect to size, physical structure, habitat, life habits, and social organization.

Furthermore, our present knowledge of behavior of Artiodactyla is limited. Some information is available on behavior of Hippopotamidae, Tayassuidae, and a few Suidae species. Virtually nothing is known about behavior of the Tragulidae. In only about seven cervid species has behavior been studied intensively enough to allow description and discussion of the phenomena and problems of communication. Good information is available on the behavior of Tylopoda, Giraffidae, and Antilocapridae; however, they comprise relatively few species. More investigations of communicative behavior have been carried out on bovid species than on other groups of Artiodactyla, but the approximately thirty species investigated make up less than one-third of all bovid species.

Information from studies on behavior of artiodactyl species is usually rather good on visual displays (postures and gestures), considerably less so on acoustical and olfactory behavior, and poor on tactile communication. (This is provided that one takes the term "communication" seriously and does not consider any form of physical contact to be a communication.)

Further difficulties arise from the general problems of expressive behavior and intraspecific communication. Expression can be a phenomenon or an epiphenomenon (Eibl-Eibesfeldt, 1957). As a phenomenon it is a special and well-defined display (movement, posture, or vocalization), like the threatening presentation of horns toward an opponent in many bovid species. In the case of an epiphenomenon, a basically nonexpressive behavior is performed in a special manner. For example, in a stiff-legged walk, the special manner (stiff-legged), not the behavior pattern *per se* (walking), adds an expressive character to the performance. Such expressive epiphenomena can, at least occasionally, be attributed to almost any behavior. A discussion of them could easily lead to a discussion of behavior in general. For this reason, it appears advisable to focus this presentation on expressive phenomena. On the other hand, we cannot completely exclude expressive epiphenomena since some of them are important in communication and/or may contribute to a better understanding of comparative and evolutionary aspects of certain (special and well-defined) displays.

Furthermore, the realm of expressive behavior is not confined to social communication (Leyhausen, 1967). In other words, there is also expressive behavior without function in intraspecific communication. For example, "flehmen" (Fig. 1) is a very common expressive behavior (especially in males) of many artiodactyl species (Schneider, 1930, 1931, 1934), but, except for occasional contagion, it has no effect on conspecifics. An expressive behavior with a clear function in social communication is addressed (but not necessarily directed) to a definite conspecific recipient (occasionally also to an animal of another species when the latter is treated more or less as a conspecific partner by the sender) and it releases a definite response by the recipient (provided the latter had become aware of the sender's action, or does not deliberately ignore it). It would follow from these statements that expressive behavior without function in social communication is not addressed to and does not release a definite response from a partner. Generally, this is correct; however, there are behavior patterns which are important in communication but which are not addressed to definite partners but to potential recipients, "to whom it may concern." Other displays are clearly addressed to definite partners, but do not release marked responses by the latter. In certain situations, of course, no response is a response. This seems to be especially true in mating rituals where many artiodactyl females show no special reactions toward certain courtship displays of the males. Sometimes it may be difficult to distinguish such displays from expressive behavior having no function in social communication.

Fig. 1. The male Uganda kob (*Adenota kob*) shows the lip curl (flehmen) after having smelled the female's urine. (After photo by H. Buechner.)

Another difficulty arises from contagion. This means that an animal that is (presumably) in the right mood for a given behavior performs it when this behavior is exhibited by another animal close by. This contagion has to be distinguished from a response by the same behavior (for example, a threatened recipient returning the threat using the same display as the sender). Mere contagion, however, can hardly be said to be a response. One animal simply does the same thing as the other (yawning, eating, lying down, grooming, urinating), and neither addresses this behavior to the other; nor does the behavior itself call for a response. Occasionally, almost any behavior can have contagious effects. Thus, as with expression as an epiphenomenon, a presentation of contagious effects would inflate this paper to a general discussion of behavior. On the other hand, contagion can contribute quite remarkably to intraspecific communication in certain cases, as in the coordination of group activities. Therefore, it cannot be completely excluded from this discussion.

The statement that there is expression without function in communication can also be reversed, as communication does not necessarily depend on expressive behavior. For example, tracks, excrement, and other spoor that indicate that an animal of a given species, sex, and age is or has been present in a given area may have communicative functions, but they are not expressive behaviors. Moreover, at least in the broadest sense, the term "transmission of communication" could even be extended to cases where the behavior of the partner is influenced by merely mechanical means, as in a fight. When one thinks of the transitions in fighting behavior from all-out fights to ritualized fights, to playful sparring, and to gentle, but slightly aggressive pushing of the partner (tactile communication!), one may easily understand that it is hard to make a clear-cut distinction, even when one thinks of "communication" as primarily implying the imparting of information by signs and signals. In effect, this would mean the inclusion of fighting behavior, copulatory behavior, nursing, cleaning the young, social grooming, etc., in our discussion. This again would lead to an unwieldy inflation of this presentation. However, one has to be aware that there are such transitions and that the boundary between communicative and noncommunicative behavior is sometimes vague.

In short, the following presentation will focus on the elaborate and well-pronounced displays (expressions as phenomena) that release definite and clear responses in (conspecific) recipients. Expressions without a clear function in social communication, expressive epiphenomena, and behavior patterns that have only contagious effects on conspecifics will be discussed only as far as it is necessary for a better understanding of certain general problems of communication. Also, the discussion of communication without expressive behavior will be restricted to a minimum.

A last difficulty arises with respect to classification of the behavior patterns under discussion. An approach frequently used in the literature is the presentation and classification according to sense impressions (visual, acoustical, olfactory, and tactile behavior). This approach is clear and simple, but is also superficial and unsatisfactory, somewhat resembling a "classification" of plants by the colors of their flowers. Moreover, this approach cuts natural and well-integrated units of communication into pieces (although an occasional separation of behavior patterns that belong together can hardly be avoided by any kind of classification). Another approach is the classification of displays according to functional circles (*Funktionskreise:* von Uexküll, 1921). Thus, one may speak about aggressive displays, sexual displays, alarm signals, etc. This approach appears to agree better with biological situations than one using the sense impressions; however, it also has its problems. In the artiodactyles expressive displays often have one basic meaning and message, two (or a few more) effects on the recipient, and multiple social functions. For example, threat displays of certain species express the readiness of the sender to become aggressive. Depending on how equal the addressee and the sender are, such threats may have either challenging or intimidating effects on the recipient. In differing situations the same threat display may be used to establish or maintain a territory, establish or maintain a position in a social hierarchy, coordinate group activities, reject sexual approaches (as when used by a female toward immature males), or prevent strange young from suckling and soliciting milk. Thus, a classification of displays according to functional circles can also result in considerable and poorly founded separations between similar or even identical behavior patterns. Much the same is true of a classification based on involved partners (♂:♂, ♂:♀, ♀:♀), as was suggested by Carpenter (1942).

Another approach would be a classification based on phenomenological characteristics of behavior patterns such as head-up postures, head-down postures, broadside positions, etc. This approach is certainly a good one, but difficulties arise from the position of a given behavior pattern within the entire behavioral inventory of a species. Since behavioral inventories vary with the species, the meaning and message of phenotypically very similar behavior patterns can be different and even opposite in different species. For example, in certain bovid species, the opponents routinely drop to their "knees" (carpal joints) during a fight. If an animal drops to its knees in an agonistic encounter without establishing horn contact with its opponent, this action can be an intention movement for fighting and can thus be a threat. Other bovid species do not drop to their knees in fighting; when one of them exhibits this behavior during an agonistic encounter it is usually an intention movement for lying down, i.e., a submissive behavior. Thus, the meaning and message of the same behavior pattern (dropping to the knees) in the same situation (agonistic encounter) can be very different depending on its "bedding" in the species-specific behavior inventories. (By the way, this is also the root for misunderstandings in encounters between animals of different species.) Occasionally, similar problems may arise with respect to certain behavior patterns even within the same species. In short, a classification under merely phenotypical aspects can easily unite behavior patterns that are different in meaning, origin, and function, especially when the discussion includes many different species.

Possibly a classification according to phylogenetic origin of expressive displays would be helpful. However, our present knowledge of this subject is limited and often still in the stage of speculation.

The best approach may be to try a classification of displays with communicative functions according to their meanings and messages by using some aspects of functional circles for the outlines and, as much as possible, incorporating phenomenological and phylogenetical aspects (Walther, 1974) and limiting the use of categories derived from sense impressions. I propose that communication of artiodactyles be broken into the following categories: (1) advertising presence, position, state, and status, (2) excitement activities, (3) alarm and flight signals, (4) advertising readiness for social contact and group cohesiveness, (5) mother-offspring signals, (6) orientation relative to the partner and signals of direction, (7) threat displays, (8) space-claim displays, (9) dominance displays, (10) courtship displays, (11) submissive and appeasement behavior.

A few remarks on organs and parts of the body that play a role in the expressive behavior of artiodactyles may complete this introductory discussion. Artiodactyles are primarily pantomimers. In particular, the position of the neck relative to the body (stretched forward, erected, lowered) and the position of the head relative to the neck (head held in one line with the neck, chin tucked in toward the throat) often have very definite meanings and can signal a multitude of information to conspecific partners (Schloeth, 1961). In some species the torso can also show special postures (e.g., lordosis or kyphosis of the back).

Two aspects of the physical structure of animals of the artiodactyl type are noteworthy with respect to orientation of the entire animal, or at least its head, toward the addressee. First, the eyes of these animals are located much more laterally on the head than, say, in humans, monkeys, and certain carnivores. For this reason, a broadside position or a sideward turn of the head in Artiodactyla does not necessarily result in the loss or avoidance of eye contact with the partner. It often is one-eyed fixation of the other. Even when such an animal turns its body or head so far that it almost faces away from its partner, this movement is functionally comparable to a slight sideward inclination of the human head.

Second, these animals present their full breadth when standing in lateral position to an addressee, whereas they offer a relatively small silhouette when standing in frontal orientation. In humans the situation is just the opposite. Linked with the broadside position is the presentation by some species of striking color patterns (such as black or white stripes and bands) and/or additional structures such as beards and manes (which sometimes are extended over the entire back and can be erected).

Movements of the legs come after the postures of the head, the neck, and the torso with respect to importance in communication. Bending or stretching the legs can contribute to the appearance of body postures; and a slow-motion walk—sometimes combined with an exaggerated lifting of the forelegs—and various forms of symbolical kicking with the forelegs and stamping and scratching the ground are used as means of expression.

Facial expression is not lacking in artiodactyles, but apparently it does not play as great a role as it does, for example, in primates. Movements of the mouth and the mouth organs are quite common. For example, wide-open mouths in hippos and symbolic biting in Suidae and Tayassuidae ("squabbling": Schweinsburg and Sowls, 1972) are threats. Tongue flicking occurs in courting males of quite a number of species (especially in the cervids, but also in certain bovids). Folding the skin of the nose or inflating the nose region is found in certain species, such as gazelles (Walther, 1958, 1966a, 1968a). Eye movements—including a straight look, a look from the corner of the eye, and a pop-eyed look—do occur in artiodactyles. However, they have not been thoroughly investigated, and thus virtually nothing is known about their effects on recipients.

Ear movements occur rather frequently in connection with expressive displays in artiodactyles (Freye and Geissler, 1966). Examples of common ear movements are: the laying back of the ears in threat and courtship behavior; the "ear drop" (similar to the permanent ear attitude of Indian cattle) in courting males of Indian blackbuck *(Antilope cervicapra),* as well as of the gnus and hartebeests; the "pointing" with one ear toward the opponent in certain threat and dominance displays of Antilopinae and Hippotraginae species; and holding the ears sideways during fights in many bovid species. However, it is uncertain that ear movements are signals to conspecifics and release responses in them since they usually are combined with other, more striking postures or gestures of the head, neck, torso, or legs. Thus, the probability is great that the recipients may react primarily to these other behavior patterns.

The same may be said for most tail movements. Striking movements of the tail are combined with flight behavior in many artiodactyl species. However, they occur either when the animal is already fleeing or immediately preceding flight. Thus, it is hard to say whether conspecifics become alarmed by these tail movements or by the first animal's running away, its alarm posture, or its alarm calls. The communicative role of tail movements is even more dubious in courtship and threat displays. Here, the displaying animal is frequently frontally oriented to the recipient, and the latter cannot see the sender's tail.

According to Tembrock (1959, 1963, 1964, 1965), who has done pioneer work in studies of vocalization in mammals, the acoustical transmission of information in Artiodactyla has developed to different degrees. Species living in dense vegetation possess more differentiated sound patterns than those living in open landscapes, where optical information prevails. In the Suidae (and possibly other groups of Artiodactyla) three trends in differentiating vocalization seem to be significant in the development of information transmission: transmission of short sounds to long sounds, adding rhythm to the short sounds, and transformation of the frequency range. Besides a relatively few loud, striking sounds such as whistling, roaring, and barking, vocalizations are often very soft in the artiodactyles and can only be heard at very close range.

Kiley (1972) has emphasized that vocalizations in horses, pigs, and cattle are generally not discrete displays conveying specific messages but rather convey information on the general motivational state of the animal (Table 1). Certainly, there is much truth in these statements (which in part refer to certain general problems of expressive behavior discussed above). However, it would be too sweeping a generalization to deny the display character of all vocalizations and their conveyance of specific messages.

Vocalization can be brought about by the combined activities of the larynx and the mouth organs. Some sounds appear to be closely related to belching. Sometimes, special postures of the neck and/or head appear to be necessary to produce certain sounds (for example, the rutting call of red deer). In a number of artiodactyles (such as gazelles and some other bovids), sounds uttered through the nose are quite common. In this case, certain cartilaginous structures of the nose (vibrating organs), skin folds and skin bags in the nose region that can be enlarged, and the opening or closing of the nostrils produce these vocalizations. Some noises can also be made with the teeth.

Artiodactyla are considered to be macrosmatic. Besides urine and feces (and possibly

Table 1

a) Syllable	Frequency	Amplitude	Tonality
'm'	50 - 125 cps.	l - m	h
'en'	125 - 300 cps.	m - h	m
'en'	500 - 800 cps.	h	m - h
'h'	-	l - m	l
Inspiration (Open mouth)	100 -1250 cps.	m - h	l - m

b) List of calls

mm ; men ; menh ; (m)enh ; menenh.

See-saw Type A: menenh - (m) enenh -

See-saw Type B: menenh - insp - enenh - insp -

A

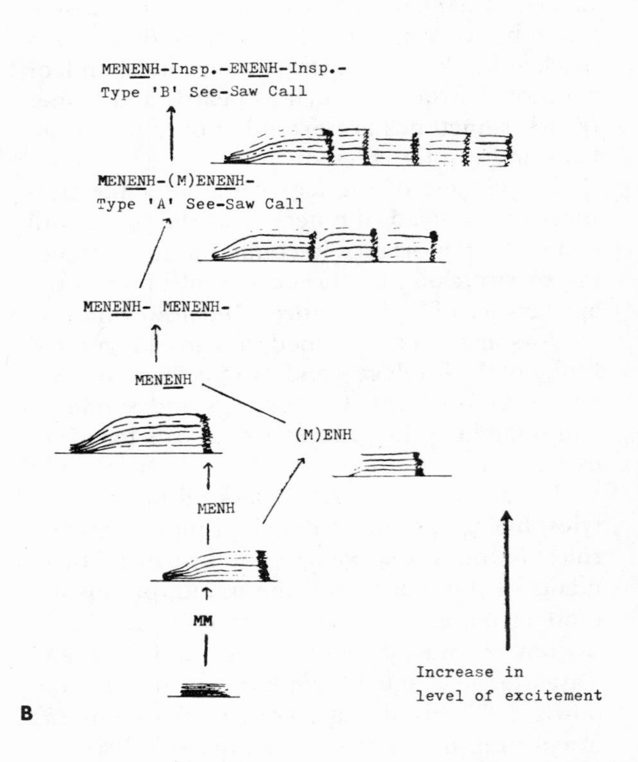

A. Syllables of cattle calls and the way they combine to form the calls (Kiley, 1972). 1 = low, m = medium, h = high. B. Interrelationships of the calls in cattle (Kiley, 1972). Note that amplitude, frequency, length, and/or repetition increase with excitement.

saliva), secretions (pheromones) of skin glands are supposed to be important in intraspecific communication. Skin glands are frequently found in artiodactyl species. However, it has to be emphasized that in the majority of cases, the importance of these glands and their secretion in intraspecific communication is at present postulated on the fact that artiodactyles have such glands and a keen sense of smell. Except for the occasional sniffing at glands or secretions, and adding of urine, feces, or secretion to an already existing marking site, convincing observations or responses to these scents are very rare. Assumptions on repellent effects of secretion marks to potential territorial competitors should be treated with particular caution. The studies of Müller-Schwarze (1967, 1969, 1971) have recently brought a clearer picture of the social functions and effects of certain pheromones in mule deer (i.e., in one species of Artiodactyla). As far as I can see no statements or only very limited ones can presently be made on the communicative functions of the following glands: mental, crural, circumcaudal, infracaudal, circumanal, proctodeal, prevulval, preputial, inguinal, tibial, parungular, interdigital, and occipital. However, all of them may possibly play a role in communication of certain artiodactyl species.

In tactile encounters such as nosing, licking, or rubbing, the epiphenomenal mode (gentle or violent, brushing or knocking, with increasing or decreasing pressure, etc.) appears to be of greater importance with respect to transmission of communication than are the behavior patterns themselves, but no studies on this subject are available at present. For technical reasons discussed above, it appears inopportune to include in the transmission of communication other tactile stimuli, such as horn or antler contact, biting, shoving, or throwing the body on the partner.

I will now discuss the displays according to the categories suggested.

Advertising Presence, Position, State, and Status

It is a basic presupposition for any communication that both partners are aware of each other's presence and position, and often also their social status (such as territorial or nonterritorial, high ranking or low ranking in a social hierarchy) or of their general motivational state (such as alarmed, relaxed, in migratory mood). Apparently, the mere presence of an animal is not necessarily sufficient for these purposes, but additional advertising devices are required. Thus, the basic message of such advertisement behavior is, "Here is an animal of this given species," often combined with the modification "And it is up and doing" and/or "It is in this particular mood, social status, sex, or age."

Emphasizing presence always means self-exposure of the animal, either relative to other conspecifics (social self-exposure) or to the environment or a combination of the two. Visual self-exposure relative to conspecifics is only found in gregarious or semigregarious artiodactyles, mainly in adult males. The self-exposing animal separates itself from the group most of the time. In species that form temporary and/or seasonal harem groups, such as Grant's gazelle *(Gazella granti)*, pronghorn *(Antilocapra americana)*, topi *(Damaliscus lunatus)*, and impala *(Aepyceros melampus)*, it is commonly the male who, linked with his "shepherd" role, keeps himself separated from the females. This means that he usually stands, moves, and rests at the periphery of the group, often ten to thirty meters or more from the females. Thus, his presence can easily be recognized by other males. Visual self-exposure is sometimes combined with courtship or dominance displays. It appears likely that the self-exposure forms the base for certain dominance displays (p.690).

In the case of nonterritorial species, such as red deer *(Cervus elaphus)*, or in species where territorial behavior is combined with harem behavior and males have very large territories, such as Grant's gazelle, the male remains separated from the females but moves with them. In species such as wildebeest (*Connochaetes taurinus:* Estes, 1969), Thomson's gazelle (*Gazella thomsoni:* Walther, 1964a; Estes, 1967), and Uganda kob (*Adenota kob:* Buechner, 1961; Leuthold, 1966), territories are comparatively small and female herds visit the territorial males for only a few hours per day (pseudo-harems). In this case, territorial males do not participate in daily movements of female herds (or, of course, in those of bachelor groups), but they remain behind as solitary individuals. This separation is characteristic of the territorial status of these males (but it is neither the only nor an unmistakable indication of territoriality).

Visual self-exposure relative to the environment means that an animal does not make use of cover (otherwise a very common strategy in animals), but instead stands on the rim of a slope or on top of a rock, hill, or termite heap. This standing freely on elevated ground can serve several functions. It may sometimes be used for better observation of the surrounding area. In hot climates, this position may allow the wind to cool the legs. On the other hand, this position also emphasizes the animal's presence, allowing some communication—be it repellent to rivals and competitors or attractive to potential sexual partners.

Social and environmental exposure can also be combined. For example, in chamois *(Rupicapra rupicapra)* and in pronghorn the male often separates himself somewhat from a female group during rutting season, but he stands or rests above them on the slope of a mountain or hill.

Acoustical advertising of presence and position is sometimes combined with visual displays or advertising, as when territorial bulls of brindled wildebeest *(Connochaetes taurinus)* combine their groaning-croaking calls ("ugh") with a head-up posture, and black wildebeest *(Connochaetes gnou)* males combine their shrill, blaring call ("he-it") with a throwing upward of the head. In some species, acoustical advertising may possibly be used because visual displays are not effective enough in the habitat (water, high grass, thicket, forest) in which these animals live. Examples of this kind might be the trumpeting grunts of hippo *(Hippopotamus amphibius),* the roaring rutting call of red deer *(Cervus elaphus hippelaphus),* the bugling of elk *(Cervus elaphus nelsoni),* and the long roaring of bulls in certain bovines (Schloeth, 1961).

In a few species, such sounds are apparently uttered primarily by territorial males and may advertise their territorial status. Possible examples are the whistling of Uganda kob (Buechner and Schloeth, 1965), the burring and rhythmically repeated "pferrr" (usually five to a stanza) in Grant's gazelle, or the loud, strophic panting in pronghorn.

Perhaps most acoustical signals function as vocal contacts that work on the principle of feedback. In this case, the advertisement of presence and position ("I am here!") calls for the partner's answer ("Where are you?"). In other words, the advertisement of presence and position extends into the realm of advertising the readiness for social contact.

Olfactory advertisement of presence has two principal aspects. First, many artiodactyles can emit scent from skin glands, some of which, the inguinal glands, for example, are open and produce secretion more or less constantly. Thus, this scent is always with the animal. Other glands secrete only during certain periods of the animal's life, as when the belly gland of musk deer *(Moschus moschiferus),* ejects a strong-smelling secretion during rut. Again, other glands are opened automatically when the animal performs certain movements; for example, the interdigital glands are more or less closed while the two hooves of one foot are close together, but they open when the hooves are spread, i.e., when the feet are abruptly forced against the underlying surface during galloping, jumping, stamping, or pawing the ground (commonly with one foreleg in artiodactyles). They are also spread when the animal emphatically stretches a foreleg, as in the foreleg kick (p.705). Certain glands are sometimes opened in combination with visual and acoustical displays, as when a roaring red deer automatically opens its preorbital glands when opening its mouth and stretching its head and neck forward and upward in the typical rutting-call posture. On the other hand, males of certain species such as Thomson's gazelle and Indian blackbuck can open their preorbital glands at will. In this case, the (presumable) emission of scent can be timed and restricted to special occasions, primarily agonistic and sexual encounters.

Another possibility of olfactorily emphasizing presence is self-impregnation with urine (sometimes also with sperm). This is used by the males of some artiodactyl species mainly during rut. Males of Gray's waterbuck *(Onotragus megaceros)* and *Capra* spp. splash urine into the hair (beard) of the throat region. *Odocoileus* spp. (Fig. 2a) as well as moose (*Alces alces;* Geist, 1966b) and reindeer (*Rangifer tarandus;* Espmark, 1964), and possibly all cervids within the Telemetacarpalia group urinate on their hind feet (besides "normal" urination). The urine runs over the tarsal glands, and the animal rubs its hind feet together, apparently mixing the urine scent with that of the secretion from these glands (Müller-Schwarze, 1971). This seems to be especially common in dominant individuals. Thus, another animal can learn something about the social status of the bearer when sniffing at its tarsal glands. The European bison *(Bison bonasus)* will wallow in its own urine (Hediger, 1949). During rutting season, male moose may lie on the spot where they have previously urinated (Kakies, 1936), and territorial males of blesbok *(Damaliscus dorcas phillipsi)* often rest in the middle of their dung piles (Walther, 1969a).

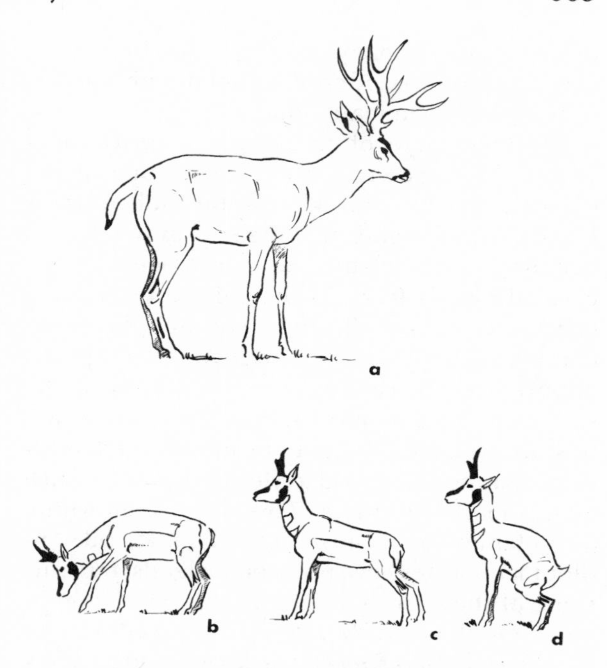

Fig. 2. a. Mule deer (*Odocoileus hemionus*) in rub urination. b–d. Typical sequence of (b) pawing the ground, (c) urination, and (d) defecation in a male pronghorn (*Antilocapra americana*). The hoof scratches on the ground, the urine, and the feces are all deposited at the same spot.

The second method for olfactory advertisement of presence and position is by object marking. In this case, the animal deposits odoriforous substances on definite spots of the environment. Since these marks remain for some time after the animal has left, their message is not only "I am here!" but also "I was here!" Again, the secretion of skin glands as well as urine and feces are used by certain artiodactyl species for this purpose. The remains of other activities such as pawing the ground and rubbing or goring of ground and vegetation with horns or antlers may also serve a similar purpose. Pawing the ground is often combined with urination and/or defecation in males of many bovid species. In cervids, gland marking is often combined with aggression against inanimate objects.

Besides some of the glands already mentioned in another context (e.g., interdigital glands), of the skin glands the subauricular glands (in pronghorn), dorsal glands (e.g., in peccary), frontal glands (e.g., in roe deer), postcornual glands (e.g., in chamois and mountain goat), and, above all, the preorbital glands (in many bovid and cervid species) are of special importance in marking, and deposition of the secretion often requires special movements and postures (Fig. 3a). It has to be emphasized, however, that not all artiodactyles that possess such glands mark with them. This varies even within a subfamily or genus. For example, all Antilopinae species have preorbital glands, but only some of them, such as Thomson's gazelle, goitered gazelle *(Gazella subgutturosa),* red-fronted gazelle *(Gazella rufifrons),* blackbuck, and gerenuk *(Litocranius walleri),* mark objects with them, whereas other Antilopinae species, such as Grant's gazelle, Soemmering's gazelle *(Gazella soemmeringi),* and dorcas gazelle *(Gazella dorcas),* do not. In mountain gazelle *(Gazella gazella)* there are even variations within the same species (provided that the classification system presently used is correct): *Gazella gazella benetti* marks, *Gazella gazella gazella* does not. In all Antilopinae that do mark, only the males do so, and the same is true for a great number of other bovids and cervids. However, in klipspringer *(Oreotragus oreotragus)* and blesbok the females also mark occasionally.

Territorial animals mark their territories, but gland marking does not necessarily imply that the individual is territorial. For example, many cervid species mark rather intently with their preorbital glands. Axis deer *(Axis axis)* do it in a very striking fashion (often combined with object aggression) by rising on their hind feet and depositing the mark on a branch, as high above the ground as possible. However, the majority of cervid species are not territorial. A number of them have—obviously prematurely—been assumed to be territorial, but territoriality is a proven fact at present only in roe deer (Hennig, 1962). Even within the same species, object marking is not restricted to territorial individuals. For instance, the territorial males of Thomson's gazelle mark their territories with preorbital gland secretion, but nonterritorial

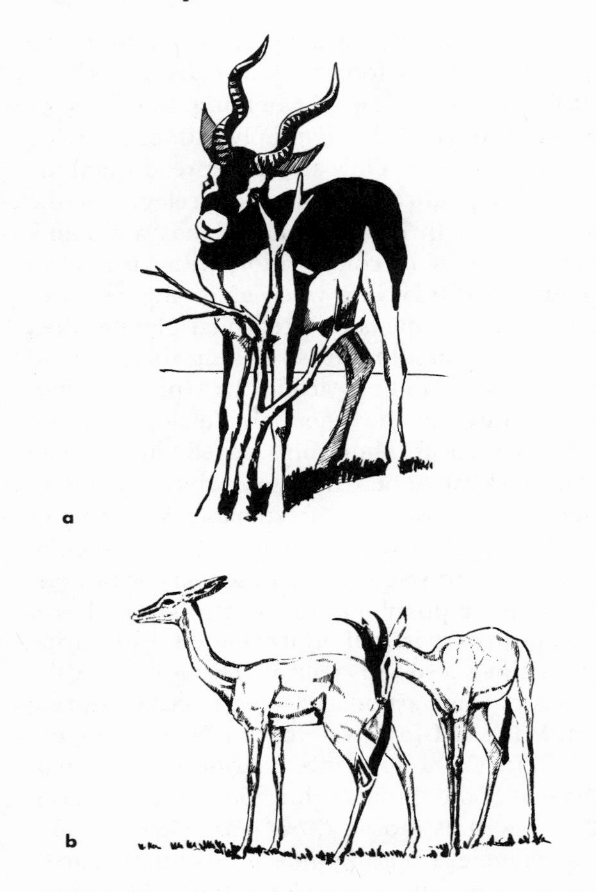

Fig. 3. Marking with preorbital glands. a. Indian blackbuck (*Antilope cervicapra*) marking the top of a branch. b. Male dibatag (*Ammodorcas clarkei*) marking the female.

males in bachelor groups and mixed migratory herds show the same—if less frequent—behavior (Walther, 1964a). Thus, the mark *per se* only indicates the presence or past occurrence of a male but does not necessarily indicate his territorial status. It is the concentration of marks by the same individual within a limited area and the specific marking system, such as a belt of marks along the territorial boundary, that make marking indicative of territoriality (Walther, 1964a).

It is tempting to assume that these secretion marks would have an intimidating or repellent effect on potential competitors. Occasionally one can observe such reactions; however, on the whole they are rare. In Thomson's gazelle, nonterritorial males as a matter of course enter well-marked territories without paying any attention to the marks. In territorial marking, the marks are possibly more important for the orientation of the owner of the territory than for other animals. The most commonly observed reaction by an animal to another's secretion mark is to sniff the marked object; sometimes the newcomer will mark it, too, or mark another one close by. Thus, the first animal leaves an indication of his presence, and the newcomer adds his own "visiting card."

Almost all that has been said about effects and functions of secretion marks is also true of marking with urine and/or feces. Urination and defecating are normal physiological processes in all mammals. Therefore, one can speak about a special function of excrements only when other conspecifics clearly react to them, and/or when an animal deposits urine and feces on definite places (resulting in dung piles in the case of defecation), and/or when urinating and/or defecating behavior shows features exceeding the mere need to excrete digestive waste. For example, the tylopods of the New World have a certain "defecation ritual" (Pilters, 1954): an animal will sniff at a fixed dung pile, stamp and paw it, then add its own droppings to it. Sometimes several animals do so simultaneously. Male hippo and pigmy hippo *(Choeropsis liberiensis)* splash their feces around with swirling movements of their tails. In some artiodactyles (such as *Gazella* and *Antilocapra*), urination and defecation postures of males are exaggerated (Fig. 2c); scratching the ground with a foreleg often precedes the process, and defecation follows immediately after urination. Thus, urine, feces, and tracks of scratching (and possibly, some secretion of interdigital glands) are deposited at the same spot. In other genera, such as *Oryx, Addax,* and *Damaliscus,* urination and defecation are separated from one another. Only defecation is preceded by pawing the ground and, in adult male oryx and addax, it is performed in a deeply crouched posture. Again, in other species and/or in other situations only the urine conveys information on state of the animal. This is especially true for urinating by females sexually driven by males. By sniffing or licking the female's urine, followed by flehmen in most of the Tylopoda and Ruminantia, the males can obviously find out whether the female is in or close to estrus.

To establish relatively large dung piles, an animal must be in the same area for some time. This is easily achieved in territorial species. In many, as in vicuna *(Lama vicugna),* Kirk's dikdik *(Rhynchotragus kirki),* topi *(Damaliscus lunatus topi),* and the *Gazella* species, dung piles are found either in the approximate center of the territory (apparently close to the owner's preferred resting place) or along the boundary (linked with agonistic encounters). Sometimes several individuals may use the same dung pile, making it larger than one animal alone could do. This may happen with the owners of neighboring territories, but may also occur with apparently nonterritorial species such as nilgai *(Boselaphus tragocamelus).* Generally, there are several tendencies that may contribute to the establishment of dung piles in artiodactyles: Animals often defecate after a long rest. Thus, when a species has fixed resting places in a territory or a home range, there is often an accumulation of dung near each. Furthermore, animals often defecate in excitement. Thus, animals often defecate at the same spot in an area where agonistic encounters frequently take place (such as along the boundaries of a territory) or where a frequently used trail leaves a relatively safe area to enter more dangerous terrain. In quite a number of artiodactyl species there is also a tendency to place droppings where another conspecific has previously defecated. Finally, these animals often show a preference to establish dung piles on bare ground, such as on their trails and even on human trails, where they are very visible and exposed. All these points indicate that dung piles play a certain role in the social communication of these animals. However, it seems to be a rather "anonymous" kind of information, which is not commonly addressed to definite recipients but to potential partners, "to whom it may concern" (except for marking in connection with agonistic encounters).

Excitement Activities

The term "excitement activities" refers to movements and vocalizations that indicate that an animal is in a state of agitation. They all occur in several heterogenous situations, and some occur in many such situations—when an animal watches an enemy at a distance, when it is forced to cross unfavorable terrain, when it is separated from a familiar group or mate, when it is involved in an agonistic encounter, or when it is expecting food or is prevented from getting food in captivity—but do not contribute to the solution of these situations.

Excitement activities in the described sense, are often termed "displacement activities" (Tinbergen, 1940). I hesitate to use this term for a number of reasons: It does not cover the whole range under discussion; the theoretical conception underlying this term implies that the animal is in an inner conflict that in some cases, at least, cannot be observed and thus remains an open question; and finally, when using this term one has to take the conception of functional circles (or major instincts) so literally and stick with it so tightly that it does not appear to be justified in the light of certain facts, such as the multiple functions of expressive behavior.

Excitement activities frequently seen in Artiodactyla are self-grooming; scratching and shaking (not as a reaction to itching or insects); the volte (stepping around in a narrow circle); stationary vertical jumps; and, in some species, also stamping (with a foreleg). Kiley (1972) lists a considerable number of vocalizations (grunts, squeals, snorts, "mm" calls) in various artiodactyl species that generally fall in the category of excitement activities. She also uses the term "excitement" and states that these vocalizations reflect its level (Table 1). Thus, the excitement activities that mainly advertise the general motivational state of the animal have a certain importance for communication insofar as they can be contagious and can bring a conspecific animal into the same mood as the sender. In general, however, they are not addressed to a definite partner, and even in situations where this could be possible, they do not elicit special and definite responses (above the level of mere contagion).

Alarm and Flight Signals

Alarm signals can be considered as special kinds of excitement activities, as they correspond to very high levels of excitement. This excitement, however, is often not just of a general and unspecific nature. I think it is necessary to look at these behavior patterns in an evolutionary context. Apparently, they are "on their way" from mere expressions of high excitement to becoming special displays; however, none have yet completely reached this final state in the artiodactyles. It follows, then, that all these behavior patterns occur at least occasionally in situations having nothing to do with alarm, and merely express a high level of excitement. For example, this is probably true in the cases of so-called displacement alarm (Estes, 1969; David, 1973) when behavior patterns commonly seen in alarm situations also show up in agonistic encounters of certain species. The degree of perfection in this change from unspecific excitement activities to specific alarm signals varies within different species as well as with respect to single behavior patterns. Thus, these behavior patterns are probably not alarm signals by nature and origin, nor do they exclusively convey and release alarm; however, the latter is one of their striking and common functions. In this aspect, it is not out of place to use the term "alarm signals" (provided one is aware of the relativity of this terminology).

At least in certain situations, every fast-running or leaping animal—not necessarily even a conspecific—may attract the attention of other animals and release alarm or flight reactions in them. This "running away" can be made more conspicuous by striking locomotor patterns (such as stotting), special movements or postures of the tail, ruffling hairs of the (white) rump patch, emitting scents from certain skin glands, and distress cries. The last should be distinguished from alarm calls (p.668). Distress cries are uttered only when an animal is captured or is very close to being captured. In most artiodactyl species, distress cries sound like a roaring or bleating "aaaaa" (*a* as in h*a*re) or "uuuuu" (*u* as in m*u*rder). They are long, loud sounds made with an open mouth, are absolutely situation-specific, and usually have a strong alarming effect on other animals. However, release of alarm in conspecifics is clearly a secondary effect of a distress cry since it is also uttered when a pursued or captured animal is completely alone. Although distress cries are sometimes apparently caused by pain, they generally appear more closely related to fear. For instance, some of my captive dorcas gazelle regularly gave loud and persistent cries when being captured for veterinary treatment but very seldom cried out during the treatment itself, although it was often likely to be painful. Although distress cries may be uttered by adults in mortal fear, they are more commonly heard from young animals. Distress cries of the young release a mother's defense against a predator, provided it is not too big and dangerous. In the latter case, the mother may perform distracting maneuvers, such as crossing several times at full gallop between her fleeing young and the pursuing predator. In certain species mothers often cease their defense when the young is killed and is no longer crying (Walther, 1969b).

In mule deer *(Odocoileus hemionus),* metatarsal scent is discharged in fear-inducing situations (Müller-Schwarze, 1971); and in springbuck *(Antidorcas marsupialis),* an emission of scent from the dorsal gland is likely during stotting (or "pronking," as the modified form of stotting of the springbuck is often termed: Bigalke, 1972). Other olfactory alarm signals appear to be possible in artiodactyles; however, at present, this is more or less subject to speculation.

Among alarm-releasing locomotor patterns, the so-called stotting (Fig. 4c) deserves mention. This special, striking kind of jumping (usually not used for clearing obstacles), often results in a chain of leaps, during which the animal bounces up and down with all four legs rather stiffly stretched. Stotting is common in Antilopinae species but is also found in pronghorn and in certain cervid species such as fallow deer *(Dama dama)* and mule deer. It apparently corresponds to a high—but not the highest—level of running and flight excitation. Thus, it occurs when excitation is rising or falling (Walther, 1964a, 1969b). It is predominantly seen at the beginning of flight (provided the pursuer is not

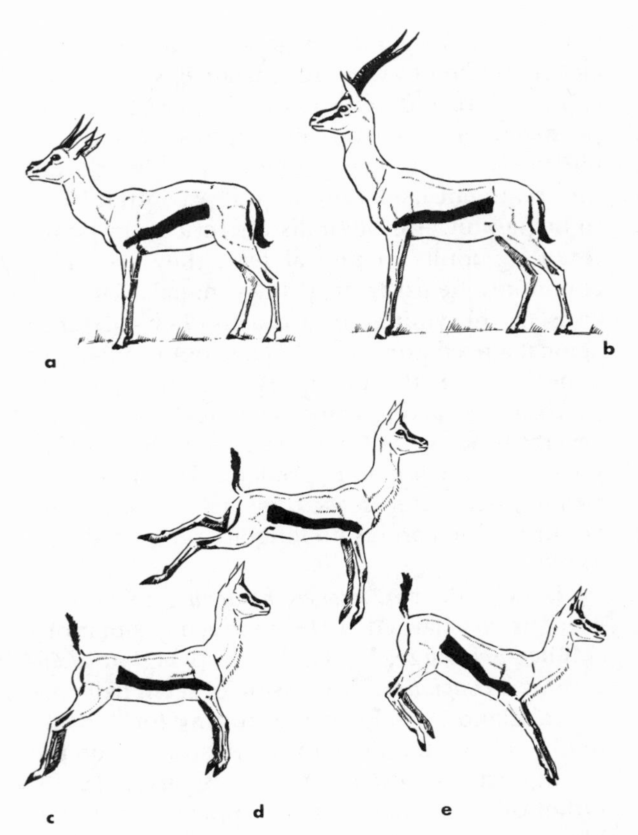

Fig. 4. Alarm behavior in Thomson's gazelle (*Gazella thomsoni*). a. Looking around, relaxed (adult female). b. Long-neck posture of alertness (adult male). c.–e. Stotting (adolescent female): c. Normal stotting gait. d. Paddling with the hind legs in extremely high stotting. e. Landing from high stotting.

too close) and at the end (when the enemy has ceased pursuit). It is more frequent in young animals and in females than in adult males.

Ruffling of rump patch hairs (springbuck also ruffle the white hair in the pouch of the croup, which has a dorsal gland inside) is combined with flight (and also stotting) or may precede flight in most species having a rump patch. The same is true for certain tail movements of some species. A relatively common movement is the vertical erection of the tail, found in various artiodactyl species such as warthog *(Phacochoerus aethiopicus),* mountain gazelle, dibatag *(Ammodorcas clarkei),* and white-tailed deer *(Odocoileus virginianus).* During flight *Tragelaphus* species curl the tail up so that the tip almost touches the root. An important aspect of the signal character of such tail movements in some species (e.g., in *Tragelaphus* and *Odocoileus*) is the exposure of the white underside of the tail when it is erected or curled.

Of course, there are also other tail and body movements linked with flight. For example, giraffe *(Giraffa camelopardalis)* presses its curved tail laterally to its hindquarters at the beginning of flight, and certain other species, such as Thomson's gazelle, Grant's gazelle, and Kirk's dikdik, often shake their flanks before flight.

However, as stated above—with the exception of distress cries—all these behavior patterns can also occur in situations other than flight, where they do not release alarm in conspecifics and sometimes do not even attract attention.

When watching a potential danger, all artiodactyles tense their muscles and stand as motionless as a statue (posture of alertness). They are oriented frontally toward the dangerous object, with ears turned forward. In many species, the animal stiffly erects its neck to its maximum height ("long-neck" posture, Fig. 4b). It sometimes also stamps with a foreleg and frequently utters special sounds (alarm calls). In some artiodactyl species, alarm calls are produced in the mouth, as in the loud, doglike barking of many cervids and of the genus *Tragelaphus.* Others are produced in the nose (which, by the way is also true for quite a number of other sounds, especially in bovids). In the latter case, alarm calls can also be very loud, like for example, the whistling of ibex *(Capra ibex),* chamois, and reedbuck *(Redunca redunca);* or they can be of medium volume, like the snorting of wildebeest and topi; or they can even be rather soft, like the "quaking" alarm calls of some gazelle species. Especially in

Thomson's gazelle, the alarm call, "quiff" (*i* as in h*i*ll), is so soft that a human cannot hear it beyond a distance of 30 m. Nose calls are apparently vibration sounds the timbre of which may vary considerably with distance.

The significance for communication of an animal standing in the posture of alertness, uttering alarm calls, and conspecifics reacting by becoming alert, is naturally much clearer than for many other behavior patterns that may also occur in situations of alarm and flight. On the other hand, the posture of alertness and, at least in certain species, the vocalizations under discussion may also occur in certain other (exciting) situations, and even when the animal is alone. Thus, in most cases, they do not appear to be addressed to a definite partner.

Advertising Readiness or Need for Social Contact and Group Cohesiveness

Signals advertising readiness or need for social contact or group cohesiveness can be considered as modifications of signals advertising presence, position, and general mood. In common with the latter group of signals as well as with those indicating alarm and flight, they are usually not addressed to definite partners.

Perhaps the closest relation between signals of presence and those facilitating social contact is found in the visual field. This is especially true in open plains areas, where the figure of an animal (of the size of an ungulate) is often the most striking sight in its vast and uniform surroundings. Here, body markings (such as black stripes or bands, white rump patches) and sometimes tail movements (like the almost perpetual tail wagging of certain gazelles) may play a role in making an animal more recognizable to conspecifics. Like landmarks, the striking figure of an animal on the open plain attracts the attention of other animals, which often turn toward it and approach it (Walther, 1972). In this way, conspecifics join to form groups and, eventually, the large herds so typical of many artiodactyl species (bison, wildebeest, springbuck, gazelles, reindeer) on open plains.

Because of their annual and circadian rhythms, all the animals in such herds are in approximately the same mood, and many behavior patterns can become contagious, contributing to the coordination of activity within the group. This synchronization of group activities is of special importance during moves and migrations. Obviously, a moving conspecific easily causes others to follow. One may think here of a modification of the infantile following reaction. In this regard, it is certainly important that in most artiodactyles, herds march in file, with one animal behind the other, at least during moves of some length. This means that the animal behind always has the preceding animal's rump in front of him. Presumably, rump patches and tail movements play an additional role in releasing this following reaction. This appears valid even in some solitary-living species, where social attraction is generally restricted to sex partners or offspring. An example is the blue duiker *(Cephalophus monticola),* which flips its little tail up and down, almost like a reflector when moving. Since the underside of the tail is white, it may act like an intermittent light in the dim forests in which these little creatures live.

Sounds used in vocal contact are also related to signals advertising an animal's presence and position. In some cases, they are more or less identical with the latter. An acoustical signal for social contact easily releases and, in a sense, calls for a partner's vocal response, which is often given in the same sound as that uttered by the sender. These sounds are frequently repeated, sometimes in a rhythmical manner. Relatively soft grunting sounds in pigs (Hainard, 1949; Snethlage, 1957), cattle (Schloeth, 1961; Kiley, 1972), red deer (Darling, 1937; Burckhardt, 1958a; Kiley, 1972), fallow deer (Gilbert, 1968), and axis deer (Schaller, 1967); roaring and growling in camels; and bleating in llamas (Pilters 1954, 1956) may be mentioned as examples. In large herds of certain gregarious artiodactyl species, these vocal contact sounds are not addressed to definite partners, but are expressions of particular moods (general motivational states) of an animal. Herd members may generate and sustain a particular mood throughout the group by reciprocal uttering and repeating of these sounds. They may also be important in species recognition and cohesiveness of conspecifics. Perhaps the most impressive example of this is provided by migratory herds of wildebeest, which are almost constantly vocalizing relatively loud croaking calls, so that the herds are enveloped in "clouds" of (familiar) noises.

In some species, the members of a group may olfactorily impregnate each other. For example, in peccary *(Tayassu tajacu),* members of the same group pair up in reverse-parallel position and rub their heads on each other's dorsal glands. Since it is likely that all members of a group eventually exchange and mix their individual scents in this way, such behavior may result in creating a mutual "group scent." Male gerenuk and dibatag (Fig. 3b) mark females with their preorbital glands, and male Gray's waterbuck urinate on the long hair of their throats and rub this wet region on the backs of females. In these last cases, the communication tends to become more specifically addressed than the other, rather "anonymous" actions discussed above.

Signals in Mother-Offspring Relations

The communication between mother and offspring adds few new aspects to the discussion of advertising presence and position, readiness or need for social contact, and alarm signals. The means used in mother-offspring relations are only special cases and are often identical with the signals mentioned above. Thus, a difference is not so much in the signals *per se* but in their being clearly addressed to a definite partner and releasing rather pronounced and special responses of that partner. This is closely linked with the bond between and the individual recognition of mother and young. Although these topics exceed the realm of a discussion on communication, they are so important to an understanding of communication between mother and offspring that one must at least describe them briefly.

In all artiodactyl species, the young follow the mother (Fig. 5b), but there are species-specific differences in form and intensity of this following behavior. Neonate artiodactyles follow moving objects close to them and larger than themselves. Under natural conditions, it is highly probable (although not absolutely certain) that the object will be the mother. Later the young show a strong preference for following their mothers. The mother usually has only to move away to get her young to follow. If, for some reason, this does not work, she may call her young, or walk back to it and touch it with her nose, or circle around and pass it in a fast gait from behind. The last method appears to be most effective in releasing the following reaction (Walther, 1969a).

Apparently, the individual bond of a young artiodactyle to its mother is due to imprinting-like processes. It is uncertain whether there is visual imprinting in artiodactyles, but acoustic and olfactory imprinting to an individual maternal partner is more certain. The young apparently learn to recognize their own mother's voice, and, at least in certain situations, they react to it by approaching (Walther, 1959). Sounds used by mothers in calling their young—like that for nursing—do not seem to differ from vocalizations commonly used in group contact by such species as blackbuck and dorcas gazelle. In other species, special maternal calls have been described: growling and bleating in camel (Pilters, 1954, 1956), bleating in fallow deer (Tem-

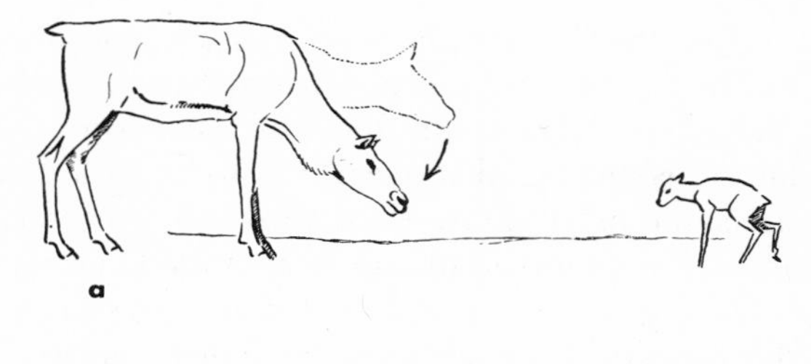

Fig. 5. a. Head bobbing of a female caribou (*Rangifer tarandus*) to a neonate calf. (After W. D. Berry in Pruitt, 1960.) b. Chamois (*Rupicapra rupicapra*) fawn following the mother.

brock, 1968), guttural grunting in cattle (Schloeth, 1958, 1961).

Visual displays may serve the same or similar purposes as these maternal calls. For example, in reindeer, Pruitt (1960) describes how a mother can lure her neonate calf by head bobbing (Fig. 5a). Also, I could make my tame blesbok calf approach me by (silent) bowing movements, similar to those that adult blesboks frequently perform with the head and neck (Walther, 1969a). Apparently, in these cases, the effects of vocal and visual "calling" of the young act cumulatively when displayed simultaneously or alternately.

The young may also contact its mother vocally, and there is often a "question-answer" vocalization between mother and offspring. Sounds made by the young are usually higher in pitch than the calls of the adult animals, but they may often be the infantile forms of adult vocal contact sounds. The mother reacts to the call of her young by calling back and/or by approaching it. When moving, she may stop and wait for it. Reactions of mothers to distress calls of their offspring have already been mentioned.

A blesbok calf I raised (Walther, 1966a, 1969a) was strongly imprinted to me (individually) in the olfactory realm. One may assume that in natural conditions, this is also true with the mother, and it probably holds true in a number of other artiodactyl species.

In certain duikers, mothers have been observed marking their young with their preorbital glands (Frädrich, 1964). This may facilitate recognizing their own offspring and distinguishing them from strangers. The licking of the young by the mother (i.e., wetting the young with the mother's saliva and, thus, possibly impregnating them with her scent), which is widespread in cervids and bovids, may also have a similar result (as a secondary effect—the primary functions of licking the young are different). Generally, licking may contribute to the establishment and maintenance of the bond between mother and young in many Ruminantia. It may also convey certain messages. However, little is known as to the identity of these messages.

Orientation Relative to the Partner and Direction Signaling

All expressive behaviors indicate momentary psychosomatic states of the sender. Whether and how the recipient responds depends, at least to some extent, on its central-nervous evaluating mechanisms and psychosomatic state. With respect to communication, an additional problem arises in addressing the partner. Apparently, there are three possibilities in artiodactyl communication:

(1) The behavior is not addressed to definite partners, but only to potential ones. It is left

almost completely to the potential partners whether they relate the performer's behavior to themselves and whether to respond to it. This "to-whom-it-may-concern" type of communication is found in most of the signals advertising presence, position, state, and status; excitement activities; alarm and flight signals; and in many signals advertising readiness or need for social contact and group cohesiveness.

(2) An individual bond is established between the partners (for example, by imprinting). Based on this special attachment, each partner automatically relates the signals to itself and responds to them. This is true for some signals advertising readiness or need for social contact and, above all, for signals used in mother-offspring relationships. (In this latter case, however, orientation components are sometimes involved; see below.)

(3) The sender clearly addresses a definite partner. The "I-mean-you" component is largely brought about by a change in the sender's orientation relative to the addressee (usually) at close range. Frequently (but by no means necessarily), the sender approaches the recipient before or after the new orientation is achieved. In certain cases, the addressee's position relative to the sender also plays a role. Having moved to a new position, the sender remains in it for a while. Thus, the previous movement contributes to addressing the partner, but apparently the sender's orientation *per se* also has a definite meaning and message.

The three basic orientations of the sender to the recipient are: frontal, reverse, and lateral. In each case, the sender may orient its whole body toward the recipient. Sometimes, however, the animal only brings its head into the corresponding position, in which case, the head obviously substitutes for and represents the body *(pars pro toto).*

Especially when the sender is moving, orientation of the hindquarters toward the recipient may easily release the following reaction of the latter, as discussed above. It has to be added that the same orientation when shown by an inferior animal in agonistic situations is suitable either to releasing the superior combatant's pursuit or to diminishing its aggressiveness. There can also be a sexual component in it, since it is the usual orientation of the female toward the male in mounting and copulation, and females of many artiodactyl species walk in front of the male in the course of the mating ritual. Generally, in artiodactyles, the orientation of the hindquarters toward the partner frequently expresses peaceful intentions or even inferiority.

Frontal orientation is often indicative of hostile intentions. (The most remarkable exception to this rather rough and general rule is, in many species, the frontal orientation of the mother toward her young when calling to nurse.) This is very understandable since most organs used for fighting (teeth, tusks, horns, antlers, neck, forelegs) are located on the anterior part of the body in the Artiodactyla. Thus, when turning, approaching, or standing in frontal orientation toward a conspecific, an artiodactyl has directed practically all its potential weapons toward the partner.

Expressis verbis or implicitly, the broadside position has sometimes been considered as resulting from a conflict between aggression (frontal orientation) and escape tendencies (hindquarters position) (e.g., Fraser, 1957; Ewer, 1968). However, at least in Artiodactyla, this is rather unlikely. There are a few artiodactyl species, such as mountain goat (*Oreamnos americanus:* Geist, 1965), giraffe (Backhaus, 1961), and Barbary sheep (*Ammotragus lervia:* Haas, 1959), that fight either regularly or occasionally in parallel or reverse-parallel position. In these species, the broadside position is clearly related to aggressive behavior only. Geist (1966a) has expressed the opinion that fighting in parallel or reverse-parallel position is a phylogenetically old fighting technique in artiodactyles. Under this assumption, it is also possible to consider broadside orientation as a phylogenetic relic of aggressive intentions in those recent species that no longer fight in this position.

Above all, however, an animal of the physical constitution of Artiodactyla can block a partner's or opponent's path by assuming the broadside position in front of him. This blocking of the path is not a theoretical assumption or postulation, for it can actually be observed in quite a number of the species. Apparently in a relatively few artiodactyles, such as cattle (Schloeth, 1958), greater kudu *(Tragelaphus strepsiceros),* and sitatunga (*Tragelaphus spekei:* Walther, 1964b), the young may block its mother's path so that it may suckle. More numerous, by far, are the cases where a superior animal blocks an inferior's path by broadside position. This has been observed in lesser *(Tragelaphus imberbis)* and greater kudu (Walther, 1958, 1960a), oryx (*Oryx gazella beisa:* Walther, 1958), Grant's gazelle (Walther, 1972, 1974), Thomson's gazelle (Walther, 1974), mountain gazelle, hartebeest (*Alcelaphus buselaphus:* Gosling, 1974), tsessebe (*Damaliscus lunatus lunatus:* Joubert, 1972), warthog (Frädrich, 1965), and others. By taking this lateral position in front of a recipient, the performer forces the other to stop and often also to withdraw or at least deviate from its original course.

All these orientations can occur together with special displays. Typically, frontal orientation is frequently combined with threat and courtship displays, broadside position with dominance and, to a lesser extent, with courtship displays, and hindquarter orientation with behavior of inferiority. This will be discussed in detail later. However, it must be emphasized that these positions *per se* (i.e., without additional displays) can, in certain situations, release responses in conspecifics. Apparently, their major effect is to inform the recipient of the direction it is expected to take or not take. Presuppositions, facilitating such direction signaling, are that the sender is superior or at least not clearly, inferior to the recipient, and that the latter is somewhat ready to "obey" the sender's intentions. To date, these problems have been studies in detail for only one species, Grant's gazelle. However, there is reason to assume that the results obtained are valid for quite a number of other artiodactyl species.

In Grant's gazelle, direction signaling is especially obvious when a relatively stable group of females (and their offspring) remains for several months in the (large) territory of a male (Walther, 1972). In such a harem group, only the male is territorial and only he is aware of territorial boundaries, which do not exist for the females. Thus, the females will transgress the boundaries without hesitating if the male does not prevent them from doing so. When females are at the point of leaving the territory, the male's efforts to block their path and to herd them back are very striking (Fig. 6).

This situation, however, is an extreme case. More commonly, the male tries to direct their course long before they near the boundary, simply by using his position relative to the females and placing himself between the harem and the boundary. He shows no particular display, but stands, moves, or, most commonly, continues grazing. He often directs and relates his position to only one of the females, the one that has moved farthest in a given direction. Her position relative to him can also modify the meaning of his position. The females definitely react to the male's behavior (as long as they do not try to leave him and the area deliberately). Such permanent direction signaling works so inconspicuously and effectively that it took me a shamefully long time to become aware of it. Although it may

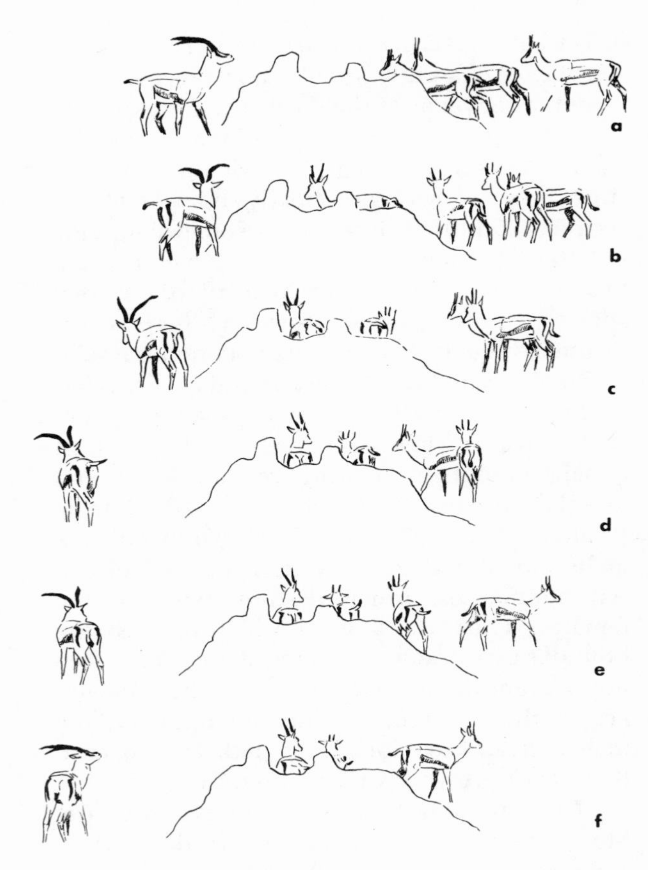

Fig. 6. Blocking the way in Grant's gazelle (*Gazella granti*). (After 16 mm film at 24 frames/sec.) a. Frame 1: Several females and their offspring approach the boundary of a territory (marked by a termite heap), coming from the center. The territorial male stands at the opposite side of the termite heap, at one side of the females' path. b. Frame 32: The male turns and starts walking to a position in front of the females. c. Frame 72: The male moves into broadside position. The females stop or turn 90°. d. Frame 152: The male is in full broadside position, blocking the females' path. e. Frame 164: The females walk back or move parallel to the boundary. The male is still in broadside position. f. Frame 182: Still in broadside position, the male assumes an erect posture and turns his head in the direction of the females (head flag). Most of them have turned and walked back to the center of the territory.

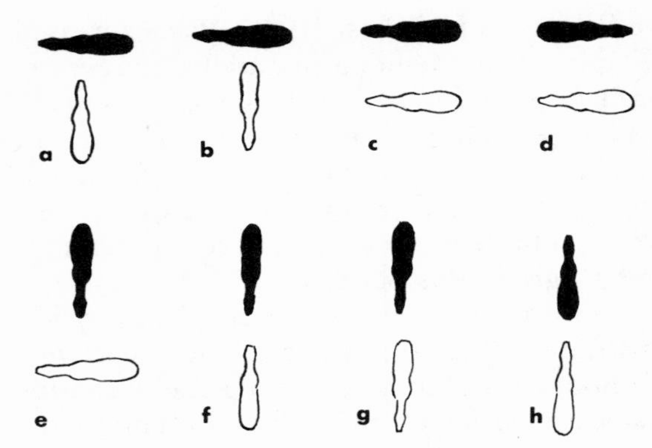

Fig. 7. Direction signaling by the position of a (territorial) male Grant's gazelle (black) relative to that of a female (white) of his harem—"translated" into human language. a. "Stop! Do not advance in this direction!" b. "Do not turn around!" c. "Stop or go ahead, but do not turn in my direction!" d. "Continue in your direction!" e. "Continue in your direction and speed up!" f. "Turn around and withdraw!" g. "Go ahead!" h. "Follow me!"

appear to be somewhat childish and anthropomorphical, the simplest and perhaps even the only way to characterize meanings and messages of these positions is to "translate" them into words of the human language, as is done in Fig. 7. This figure, of course, gives only the major positions that may occur in such situations. Transitions between these positions (e.g., between male frontally oriented toward the female and male in broadside position) are possible.

This example of ("silent") herding behavior in Grant's gazelle may demonstrate the importance of mere orientation between partners in social communication. Such orientations are also involved in all the displays discussed below. In some of these displays, the role of the orientation component is very significant; in others it appears to be of minor importance, but it is never lacking in any that are aimed and addressed to definite recipients. The reader is asked to keep this in mind since I will describe these displays without coming back to details of the previously discussed orientation components.

Threat Displays

Threat displays indicate readiness for fighting ("I am going to fight you!"). This distinguishes them from dominance displays (p.690). However, there are some transitional cases that may be termed "threat-dominance displays" (*Droh-Imponieren* in German literature), in which features of both dominance and threat displays are combined. One may distinguish between "symbolic" actions, in which an animal uses the same behavior patterns as in fighting but does not touch its opponent, and more or less ritualized intention movements, where the performance is restricted to the very initial movements of beginning a fighting action. Most ritualized are those threat displays where intention movements are "frozen" into postures. Since there are offensive and defensive fighting techniques, there are corresponding offensive and defensive threat displays, beside others that can be used both ways. In an agonistic encounter, when only one opponent shows an offensive threat and the recipient responds with a defensive threat, the latter is a sign of inferiority (Fig. 8).

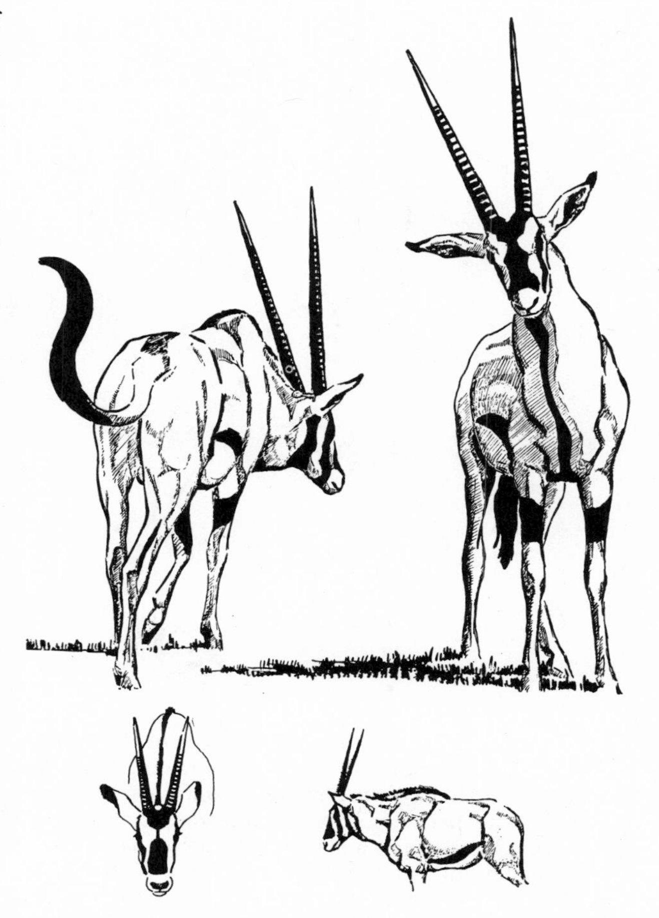

Fig. 8. Dominant bull (right) of oryx antelope (*Oryx gazella beisa*) in broadside position, displays with (erected) head-up posture, angling the horns and "pointing" the ear toward a subordinate bull (left), who responds by lowering the head and withdrawing. Lower left: Frontal view of an oryx in (pronounced) head-low posture. Lower right: Oryx antelope in relaxed "normal" posture (for comparison).

Whether threat, dominance, and space-claim displays (p.687) challenge or intimidate the recipients depends on whether they are equal or inferior to the senders and also on the situation. For example, during migration even recipients equal to the sender will often simply "obey" the latter's threat without any counterdisplay. In a very considerable number of encounters the threat remains one-sided (Fig. 9). This means that only one of the animals involved shows a threat display. In a relatively few cases the other animal may immediately attack the sender or simply ignore its threat. Usually, however, a recipient will withdraw, or show submissive behavior, or even flee in a one-sided threat encounter. Since threat displays occur under the same conditions that can lead to overt aggression, they can substitute for and save fighting in such cases. Especially in encounters between peers, the sender's threat releases the addressee's counterdis-

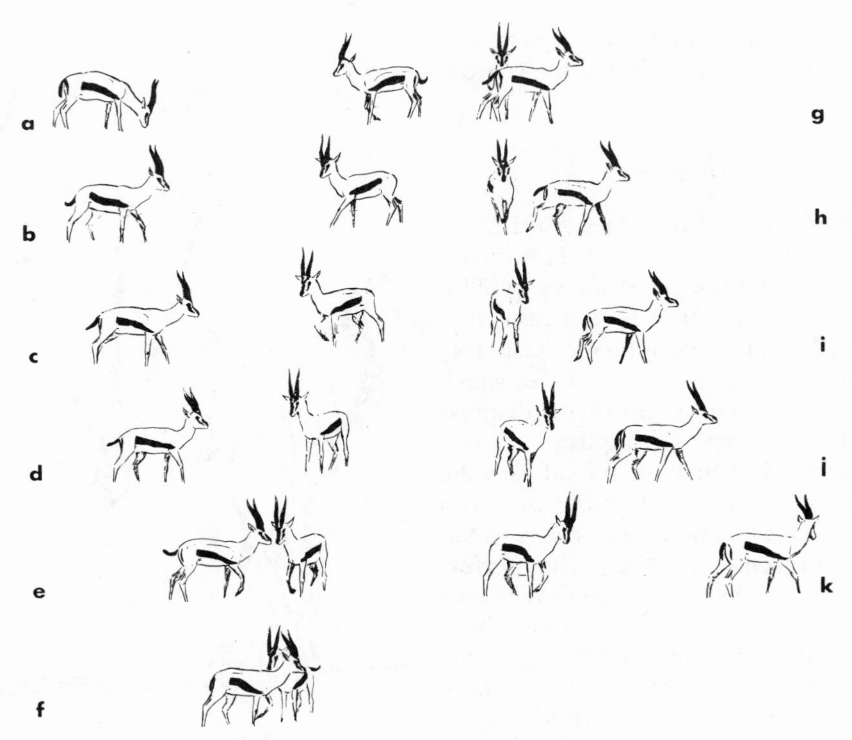

Fig. 9. Coordination of group activity by horn threat and pursuit march in adult male Thomson's gazelles. (After 16 mm film at 24 frames/sec.) a. Frame 1: Within a bachelor group (which was changing from grazing to moving), a male (right) approaches one of the males still grazing (left). b–c. Frames 16 and 24: The challenger (right) threatens with a high presentation of horns. The recipient (left) ceases grazing and starts moving, passing in front of the threatening male at an angle of 90°. d–g. Frames 32, 48, 56, and 72: Without interrupting the high presentation of horns, the challenger stands and lets the addressee pass in front of him. h–k. Frames 80, 88, 96, and 120: Still threatening continuously, the challenger places himself behind the withdrawing recipient and follows him in a pursuit march.

play. Also, in some of these reciprocal threat encounters, one opponent may eventually give in and withdraw. However, the probability that reciprocal threat encounters will end in a fight is high in the artiodactyles. The aggressiveness of both opponents obviously is heightened by the reciprocal displays, finally culminating in overt fighting. Thus, fights are not prevented in such cases. However, each opponent has become aware of the other's hostile intentions by the previous displays, and both are prepared to fight. Surprise attack, the most dangerous form of aggression, is effectively avoided by these reciprocal threats.

The statements above, especially the thesis on the intimidating and challenging effects of threat displays, may be substantiated by the example of a quantitative analysis on the outcomes of one-sided and reciprocal threat encounters in Thomson's gazelle (Table 2). The data were collected during a two-year study (1965–66) in Serengeti National Park. In the table, the term "horn threats" refers predominantly to medial and high presentation of the horns and to sym-

Table 2

Intimidating and challenging effects of threat displays in agonistic encounters of Thomson's gazelle.

	One-sided horn threats	Reciprocal horn threats
Number of observed cases	1,680	738
Percentage ending		
with fight	1.5	70.7
with other forms of aggression	4.0	11.3
with withdrawal, flight, or submission of one of the opponents	84.0	13.3
in other ways	10.5	4.7

bolic butting, but it also includes (rarer) cases of symbolic downward and sideward blows and of head-low postures. "Fight" means any form of horn contact. "Other forms of aggression" mainly include air-cushion fights and grazing rituals (see below), but also the rarer cases of (one-sided or reciprocal) aggressions toward inanimate objects and the very exceptional cases of body attacks. "Withdrawal, flight, or submission of one of the opponents" are listed according to the relative frequency in which they occur after a threat. In one-sided threats, it is, of course, always the (nonthreatening) addressee who withdraws or flees or shows submissive behavior. "Ending in other ways" refers to those cases in which the recipients did not show any reaction to the threats and/or the sender(s) eventually ceased threatening, and in which both animals involved continued with clearly nonagonistic activities such as herding females, running plays, or relaxed standing.

The one-sided threat encounters ended with withdrawal, flight, or submissive behavior of the addressee in 84.0 percent of the 1,680 observed cases, clearly demonstrating the intimidating effect of the threats. On the other hand, the reciprocal threat encounters (both opponents displaying, usually with the same form of threat) led to fights in 70.7 percent of the 738 observed cases. (This proportion is even greater in encounters between completely equal opponents.) These statistics clearly demonstrate the challenging effect of these displays. While the intimidating effect of threat displays has been acknowledged frequently and readily in ethological literature, the challenging effect has rarely been pointed out *expressis verbis*. However, it definitely exists, it is by no means rare, and it should be distinguished from the intimidating effect since it does not make sense to speak about intimidating when the threats lead to fighting and obviously none of the opponents have been intimidated.

Threats, as well as fighting, may serve a multitude of social functions in artiodactyles. The most important, many of which occur in a single species, are: territorial establishment and ratification, defense against territorial invasion, maintaining or enlarging individual distance (especially in grazing), coordination of group activities (especially when a group changes from one activity to another, e.g., from resting to moving), "voting" to determine marching direction and order, pushing during movement (i.e., keeping the migration going), establishment and maintenance of social hierarchies, herding (male:female), defense against sexual approaches (female:male), and soliciting milk (young:mother) and defense against it (female: young). Also in the mating rituals, threat and dominance displays of the courting male play an important role in many species; while in others, male courtship displays are apparently related to agonistic behavior. In short, there is hardly any realm of the social life of artiodactyles in which aggressive displays are not involved. With little exaggeration, one may say that the entire social organization and communication of these animals is based on aggression.

Among "symbolic" actions (= full performance of the aggressive action without touching the opponent) we may first mention two forms of so-called redirected aggression (Moynihan, 1955): aggression against an inanimate object and aggression against an animal other than the one that released this aggression (= *"Radfahrer"-Reaktion:* Grzimek, 1949). Neither type necessarily consists of addressed threat displays. For example, the object aggression may sometimes be a play with inanimate objects. Moreover, it appears that there is a connection between object aggression and marking behavior and, thus, as in all marking activities, object aggression frequently occurs when no potential addressee is present. Even when an addressee is present, the second animal often does not react to the performer's action but either ignores it or simply watches it. Thus, the effect of object aggression on a recipient is sometimes dubious. On the other hand, there are cases in which object aggression is addressed to a definite recipient and clearly releases reactions in that animal.

Common objects to be attacked are trees, branches, bushes, rocks, grass, the ground, and, in captivity, fences and feeders. Object aggression is often due to rather complex situations. For example, when a territorial male or a "master" of a harem group sees a potential rival at a distance, it would be very much against his "interests" to leave his territory or his females in order to threaten or attack his rival. But since his aggressiveness is stimulated by his adversary's presence, he remains where he is and begins fighting an inanimate object close by. In the same situation, a male may also become aggressive toward a group member, for example, an immature male, whose presence was tolerated or ignored before. These actions are usually rather striking, so that the potential rival releasing them can notice them and, subsequently, will also notice the presence of a potential opponent even at a considerable distance. The rival may then stay away.

Redirected aggression against conspecifics is present in all artiodactyl species. Object aggression, however, has yet to be reported from Hippopotamidae, Suidae, and Tayassuidae. There are only a few reports of object aggression in captive Camelidae and Giraffidae (Pilters, 1954, 1956; Backhaus, 1961). In these groups it apparently has no function in social communication. Cervidae and Bovidae frequently perform object aggression with their antlers and horns. Object aggression may possibly be of greater importance to social communication in cervids than in bovids. In some cervid species such as mule deer, the sound caused by beating trees or bushes with the antlers may release flight in immature or subordinate males during rut (Geist, pers. comm.). I do not know of any corresponding events in bovids. In some bovid species, however, object aggression obviously has become ritualized. Beating and goring the grass and the ground, alternately to the right and left, has led to a rhythmical and persistent "weaving" (Fig. 10a) in certain species such as Grant's gazelle (Walther, 1965). On the whole, it seems that redirected aggression represents an intermediate stage between addressed and unaddressed behavior as well as between threat displays and dynamic visual marking (Hediger, 1954), on the

Fig. 10. Some "symbolic" threat behaviors. a. Object aggression ("weaving") in Grant's gazelle. b. Dropping down on the carpal joints ("knees") in wildebeest (*Connochaetes taurinus*). (After photo by R. D. Estes.) c. Rising on the hind legs in ibex (*Capra ibex*). d. Undirected neck winding and snapping of a female situtunga (*Tragelaphus spekei*) in response to the male's sexual approach.

one hand, and advertising presence, position, state, and status, on the other.

The following behavior patterns are clearly addressed to definite partners in the overwhelming majority of cases. Chasing occurs in its severest form after one of two combatants has been completely defeated in a fight, but it can hardly be considered a means of communication. However, there is also a kind of "symbolic" chasing in certain artiodactyl species. With no previous fighting, one animal will run after another as if the latter had previously been defeated. In a sense, this symbolic chasing anticipates victory. Typically, it is most frequently used by animals of a high social status, such as territorial males, toward hopelessly inferior partners, especially females or nonterritorial, immature males. In some species the males show no intention of actually attacking the addressee during these symbolic chases. They may utter certain sounds, however, such as roaring in impala (Schenkel, 1966) or, in Thomson's gazelle, a very typical "chasing call," a strophic "pshorr-pshorr-pshorr," uttered through the nose, which almost always causes the inferior addressee to flee as fast as it can.

Related to the symbolic chase is the feint attack, in which one animal approaches the other in a rush. This action is usually combined with other species-specific intention movements for attacking (e.g., open mouth, lowered horns or antlers), but the animal stops just before touching its opponent (if it has not already fled). Feint attacks occur at least occasionally in practically all artiodactyl species.

When both opponents perform feint attacks and continue with such offensive and corresponding defensive maneuvers as would occur in a true fight but without touching each other, one may speak about an "air-cushion fight," in which it appears as if there is an invisible cushion between the opponents. These air-cushion fights are very frequent in certain artiodactyles, such as gazelles and topi. They can occur as intermezzos between true fighting, but may also substitute completely for an overt fight. Air-cushion fights as well as feint attacks and other symbolic actions can be combined with all the sounds commonly heard during the fighting of certain species: loud roaring in hippo; loud growling in warthog (Frädrich, 1967) and peccary (Schweinsburg and

Sowls, 1972); growling, gargling, and roaring in tylopods (Pilters, 1954, 1956); and rather soft growling sounds in certain bovid and cervid species.

Biting as a threat behavior in the Artiodactyla commonly takes the form of symbolic snapping (i.e., snapping in the direction of an opponent without touching it). It is very pronounced in tylopods, especially in camels (Pilters, 1956). In hippo (Verheyen, 1954) and peccary (Frädrich, 1967), the opening of the mouth as a threat display may lead to a yawning-like performance. Squabbling, tooth clicking, and tooth chattering, which in severe threat encounters are intensified to a staccato snapping of the jaws, have been described in peccary (Schweinsburg and Sowls, 1972), and similar phenomena have been observed in wild boar (*Sus scrofa:* Frädrich, 1967). Symbolic snapping is also quite common in certain cervid species, but apparently not in all of them. I have seen symbolic snapping quite frequently in red deer, axis, and fallow deer, but never in white-tailed deer. Symbolic (as well as actual) snapping is found in only some of the Bovidae species. Here, it is apparently negatively correlated with the presence, development, and use of horns (Walther, 1960a). Symbolic snapping occurs mainly in bovids with small horns, such as the Cephalophinae, and in hornless females of certain species (Fig. 10d), such as *Tragelaphus* and *Kobus.* In sitatunga it was also observed in young males as long as they had very small horns or no horns at all. The only bovid species I know of in which females with large horns occasionally show symbolic snapping is the eland antelope *(Taurotragus oryx)*—interestingly enough, this species is very closely related to *Tragelaphus* (a genus in which the females have no horns).

The grinding of teeth brought about by exaggerated sideward movements of the lower jaw in several cervid species (Schneider, 1930) is probably also a ritualized form of biting behavior.

Symbolic (and actual) pushing with the mouth shut is apparently closely related to snapping. In the Suidae (Fig. 11a), symbolic pushing is a more or less pronounced, relatively slow upward movement of the head in the direction of an opponent, usually causing him to withdraw (Frädrich, 1967). In the Tylopoda, Cervidae, Giraffidae, and a number of Bovidae species, the movement is a short, but relatively violent, horizontal push forward with the snout and head in the direction of an opponent. As with symbolic snapping, this pushing with mouth shut is especially frequent in female bovids without horns. In the genus *Tragelaphus* it is the most common defense of the female against an approaching or driving bull at the beginning of the mating ritual (Walther, 1958, 1964b). Occasionally, pushing with the mouth shut (as well as snapping) may be combined with stretching the whole neck forward in the direction of the opponent. This is possibly the origin of the head-and-neck-stretched-forward posture, which is a common

Fig. 11. a. Symbolic pushing with mouth shut (right) in wild boar (*Sus scrofa*). (After Frädrich, 1967.) b. Threatening with wide-open mouth in hippo (*Hippopotamus amphibius*).

dominance or courtship display in many artiodactyles (see below).

In connection with mouth and head movements, we may mention spitting of stomach contents, which is exclusively found in the New World tylopods (Pilters, 1954, 1956). They spit only when an adversary is close enough, and then they aim for the opponent's head. Having received the "full load" in its face, the recipient may show a grimace of "loathing" (Pilters, 1954, 1956).

Species that practice neck fighting (pressing down or lifting up of the opponent with the neck) may show corresponding symbolic neck movements—forward stretching alternating with lowering and steeply erecting and winding of the neck (Fig. 10d). They are especially evident in females of sitatunga (Walther, 1964b) and, to a lesser extent, in greater kudu (Walther, 1958, 1964b). Postures derived from neck fighting (p. 683), however, are obviously commoner than symbolic movements.

In fighting, certain artiodactyl species rise on their hind feet (Fig. 12a). Then the animal either throws its body on its opponent (Suidae and Tylopoda; also female nilgai: Frädrich, 1967; Pilters, 1954, 1956; Walther, 1966a), or beats the opponent with its forelegs (especially the Cervidae: Müller-Using and Schloeth, 1967), or "dives down" into a horn clash (many Caprinae species: Fig. 10c) (Walther, 1960b, 1966a; Geist, 1966a). The symbolic form, occurring at a distance from which the animal cannot reach its adversary, is a rising on the hind legs or a more or less pronounced jump with the anterior part of the body (*Drohsprung:* Walther, 1960a) in the direction of an adversary (the hind feet remain on the ground). This behavior is shown by the species mentioned above and by others, including Soemmering's gazelle, Grant's gazelle, chamois, and the *Tragelaphus* species, which have none of the aforementioned fighting techniques in their recent behavior inventories.

Fig. 12. a. An aggressive jump with biting intentions (right) is countered by an extreme nose-up posture (left) in guanaco (*Lama guanicoë*). (After Pilters, 1954.) b. Rising on the hind legs and (symbolic) beating with the forelegs in red deer (*Cerbus elaphus*).

Especially in cervids, where striking out with the forelegs (with or without rising on the hind feet) is common in agonistic encounters, symbolic beating or kicking with the forelegs (Fig. 12b) can be used as a threat. Kicking with the hind feet is, on the whole, rare in the fights of the Artiodactyla (it plays a certain role in the fights of the tylopods), and the corresponding symbolic performance is even rarer.

Some species (especially wildebeest, hartebeest, topi, nilgai, oryx, addax, and roan and sable antelope) tend to drop down on the "knees" (carpal joints) of their forelegs during fighting. This can also occur symbolically when the animal is still some distance from its rival (Fig. 10b). In this case, it is often, although not necessarily, combined with goring the ground (object fighting), or it may pass into grazing.

In Cervidae and Bovidae (i.e., species in which at least the males have antlers or horns), symbolic butting (Fig. 14a), i.e., a pronounced nodding movement of the head in the direction of the addressee, is widespread. It is more frequent in females and juveniles (animals with no horns or smaller horns) than in adult males. Head throwing, i.e., nodding head movements (like the exaggerated affirmation of humans), apparently is a rhythmically repeated form of the butt. Head shaking (like that of humans in negation) can be considered the symbolic form of twisting the head and horns (Fig. 14b) left and right (horns interlocked in fighting animals). All these threat movements occur in rather similar forms in almost all bovid and cervid species; however, there are considerable differences as to the frequency of the movement and how pronounced it is. For example, in gazelles head shaking is rare and head throwing even more so, but these movements are frequent in wildebeest, hartebeest, and topi.

Other behavior patterns that are used as fighting techniques but may also occasionally occur as symbolic threat movements are the downward and sideward blow (Fig. 14c, d). In the symbolic downward blow, the animal brings its head and horns or antlers down from an upright position in a violent movement. The forehead may touch the ground, and the horns (or antlers) then point forward (toward the opponent). Occasionally, the head and horns may be kept at the lowest point of this movement for several seconds (low presentation of horns). In the sideward blow, the head and horns are rapidly moved sideward from either an upright or a lowered position. These two movements occur with some frequency only in relatively few species—for example, downward blows in blackbuck (Walther, 1959; Schaller, 1967) and sideward blows in oryx *(Oryx gazella)*. The horn sweep is a combination of the symbolic downward and sideward blow. The whole performance resembles weaving, but is shorter and more violent. I have seen horn sweeps frequently only in adult male Grant's gazelle, where they occurred predominantly in very severe threat encounters between peers.

The threat displays discussed above are movements. Other threat displays are postures in which an animal—in contrast to symbolic actions—does not show the entire aggressive action but just the first intention of it, although often in an exaggerated form. Moreover, these intention movements are "frozen" for several seconds, in extreme cases for one or even several minutes. Vocalizations are not commonly heard in connection with these postures, but when there are any, they are usually rather soft growling sounds. When animals have skin glands that can be opened at will—the preorbital glands of (males of) certain bovid and cervid species are especially important in this regard—they are widely opened. Apparently, there are few threat postures in Nonruminantia—one could think of the opening of the mouth in hippo (Fig. 11b) or of a nose-up posture with turning of the cheek toward the opponent in peccary, which Schweinsburg and Sowls (1972) interpret as an intention movement for biting with the side of the mouth. In Tylopoda and Ruminantia, however, threat postures are well pronounced, frequently used, and obviously the most important means of threat.

In the head-low posture (= *Kopf-tief-Halten* or *Kopf-tief-Drohen:* Walther, 1958, 1964b, 1966a), the neck and head are stretched downward and forward (Fig. 14h). The nose is close to the ground in an attitude similar to that of grazing, and sometimes the animal may switch to grazing. Apparently, meaning and origin vary with the species, and the head-low posture is of a somewhat ambivalent nature even within the same species. For example, male guanaco *(Lama guanicoe)* may approach a rival with a head-low posture (Pilters, 1956), which appears to be a rather offensive threat in this species. It has possibly evolved from biting an adversary's forelegs (a common fighting technique in tylopods) and/or from a special form of neck fighting (getting under the opponent's body). In moose (Geist, 1963) the head-low posture combined with raising the hair on the neck, withers, and rump and holding the ears down (inside toward the opponent), and, sometimes, with a very loud roar, appears to be a defensive threat, frequently used in intra- and interspecific encounters. Also, in many bovid species, the head-low posture is apparently a defensive threat. The horns, which are directed backward and upward and more or less parallel to the neck, are in an ideal position to parry an opponent's butt or downward blow. At least in some bovids, such as *Gazella* and *Oryx,* the head-low posture may turn into a submissive posture (Fig. 14i) that is similar or almost identical with it. On the whole, here, the head-low posture often expresses some kind of inferiority, especially when used as a response to an offensive threat display of a challenger.

The head-and-neck-stretched-forward posture (Fig. 15a) is common in courting artiodactyl males (p.702), but is much rarer in agonistic encounters. It occurs sporadically over a range of some rather different species. It also has different origins and meanings. For example, in white-tailed deer, this posture has been termed the "hard look" (Thomas, Robinson, and Marburger, 1965). However, the more we learn about this behavior in white-tailed deer, the more it appears to be a display of inferior animals toward superior opponents. (Geist, pers. comm.), possibly related to the head-low posture of other artiodactyl species. Thus, it may not be such a "hard look" as it was initially assumed to be. In species that practice neck fighting, such as giraffe (Backhaus, 1961) and nilgai (Walther, 1958), the head-and-neck-stretched-forward posture can be considered an intention movement for a stroke with the neck (in giraffes) or for neck fighting since these animals can and do place the neck over the opponent's from this posture and press the rival down (Fig. 20a). Here, this posture is a display between peers, but it is also frequently used by superior animals toward inferiors.

Closely related to neck fighting are the head-and-neck-forward/upward posture and the nose-up posture. The latter posture (Fig. 13b) is a mirror image (upward) of the head-low posture. In a pronounced nose-up posture, the neck and head are stretched upward stiffly, the nose pointing skyward (Fig. 12a). In llama, the ears lie back, and both postures show a relation to pushing with mouth shut (Pilters, 1954, 1956). Pilters even considers pushing with mouth shut to be the origin of at least the head-and-neck-forward/upward posture. The major point, however, seems to be that the head-and-neck-upward/forward posture is the perfect initial posture and the swing-out movement *(Ausholbewegung)* for placing the neck over an opponent's neck or body. This becomes clear from descriptions and pictures of behavior of tylopods (Pilters, 1954, 1956) and giraffe (Backhaus, 1961), and, evidently, is true of okapi (*Okapia johnstoni:* Walther, 1962). Thus, in these animals the head-and-neck-upward/forward posture is an offensive threat, usually used by a superior opponent or in encounters between peers.

In tylopods and in (hornless) nilgai females (Walther, 1960a, 1966a), the nose-up posture was observed as a fighting technique. Here it is used to parry an aggressor's jumping attack (throwing the anterior part of the body on the opponent) and immediately push it back with the chest and long side of the neck, in a special kind of neck fighting. Thus, it is a basically defensive maneuver, but one that allows the defender to counterattack immediately. These features also determine the character of the corresponding symbolic performance (threat display), which Pilters (1954, 1956) very adequately interprets as the expression of strong resistance in *Lama.* Much the same is true in okapi (Walther, 1960c, 1962); whereas in giraffe (Backhaus, 1961), the nose-up posture (also shown in lateral position to a rival) may possibly be more offensive in nature relative to the particular fighting technique of this species (sideward strokes of the neck and head against an adversary's neck, shoulder, body, or hindquarters: Backhaus, 1961).

Phenotypically, the erect posture (Fig.13a) is intermediate between head-and-neck-forward/upward and the nose-up posture. In the erect posture, the animal stretches its neck straight upward as in the nose-up posture, but the head and nose point forward or forward/upward, in similar or even identical fashion to the head-and-neck-forward/upward posture. Pilters (1954, 1956) interprets it in llama as the "utmost readiness for defence." However, I have frequently seen it preceding the jump attack in llama, and, thus, I consider it more offensive in nature. Of course, "utmost readiness for defence" can be said to be the point where defense verges on offensive action. Being an intermediate stage between an (offensive) head-and-neck-forward/upward posture and a (defensive) nose-up posture, it is very possible that the erect posture is used both offensively and defensively. In addition, very slight changes in this posture (neck somewhat more forward or backward) may shift its meaning more to the offensive or the defensive side, respectively. In Marco Polo sheep (*Ovis ammon poli:* Walther, 1960b), as well as in a number of cervid species (Müller-Using and Schloeth, 1967; Geist, 1966b), the erect posture precedes an (aggressive) rising on the hind feet, and I am inclined to interpret the erect posture generally and primarily as an intention movement for rising on the hind legs (Walther, 1960a). This derivation, however, excludes neither a neck-fight component nor a connection to pushing with mouth shut nor beating with the forelegs in certain species, since all these behavior patterns may occur in combination with rising on the hind

Fig. 13. a. Erect posture as intention movement for rising on the hind legs in guanaco. (After Pilters, 1956.) b. Head-and-neck-forward/upward posture (right) and turning away (left) in erect posture in llama (*Lama guanicoë glama*). (After Pilters, 1956.)

legs. The similarity to the head-and-neck-forward/upward and the nose-up posture has occasionally led to mistaking one for the other or, in the earlier literature, to not clearly distinguishing between them. (I readily admit being among the "sinners.") Despite the doubtlessly close relationship and transitional stages between these postures, one should, as far as possible, try to distinguish them, in the interest of a better analysis.

In species with horns or antlers, i.e., all the Bovidae and Cervidae, presentation of horns (*Hörnerpräsentieren:* Walther, 1958) and antlers toward an opponent is a frequent and very important form of threat. It can occur in four forms: low, medial, and high presentation and sideward-angling of horns or antlers. In a pronounced low presentation, the horns are held parallel to and on the ground, with the tips pointing toward the opponent. As a posture (i.e., not as a momentary phase in a movement such as a downward blow or horn sweep), this pronounced form is rare. However, when it occurs, it is a very severe threat. In a very common but less-pronounced form of the low presentation, the head is not held so low, but is still clearly below body level, and the chin is tucked in toward the throat so that the horns point forward/upward. This posture is sometimes similar to the head-low posture, and the two may change into one another. Interestingly enough, in pronghorn, a posture frequently seen in herding males corresponds to this less-pronounced, low presentation of bovids and cervids with respect to the height at which the head is held. Otherwise, however, it is a head-low posture, i.e., the chin is not tucked in toward the throat, apparently because of the position of the horns relative to the skull axis. In pronghorn, the horns "lean" forward more than those of bovids. Thus, the horn tips point forward when a male pronghorn holds his head downward/forward, whereas a bovid or cervid (in which the horns or antlers stick straight upward or, in some species, even "tilt" backward) has to tuck in his chin for the same effect.

In the medial presentation (Fig. 14e), the neck is held forward at body level, the chin is

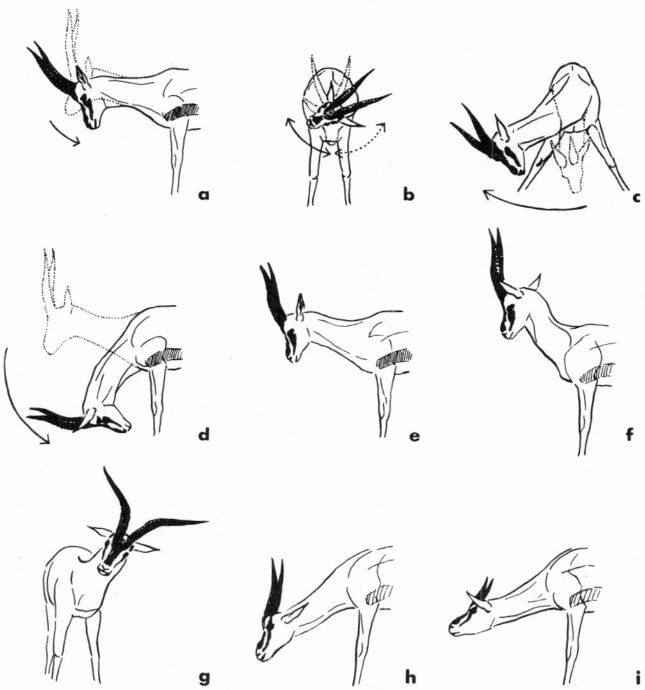

Fig. 14. Some common threat displays in Bovidae and Cervidae, demonstrated by the examples of Thomson's gazelle (a–f.) and Grant's gazelle (g–i.). a. (Symbolic) butting. When repeated and exaggerated, this results in head throwing. b. (Symbolic) head twisting. When repeated (to the right and the left) this results in head shaking. c. (Symbolic) sideward blow. d. (Symbolic) downward blow. The end phase corresponds to a low presentation of horns. e. Medial presentation of horns. f. High presentation of horns. g. Sideward angling of horns. h. Head-low posture (subadult male). i. Submissive posture (adolescent male).

Imagine that the addressee is standing in front of the threatening animal (in b. and c. the reader is in the position of the adversary), except in g., where the recipient is to the side of the displaying animal (in the position of the animal in h.).

tucked in toward the throat so that the nose points approximately vertically to the ground, and the horns or antlers subsequently point upward or somewhat forward. This kind of presentation is the commonest and is used by almost all bovid and cervid species.

In the high presentation of horns (Fig. 14f) the neck is held erect so that the head and horns are carried distinctly above body level and the chin is tucked strongly in toward the throat. The horns, which tower above the head, point upward or upward/forward. The high presentation of horns or antlers occasionally occurs in many bovid and cervid species. As a frequent and pronounced display, however, it is found in relatively few, the most important of which are the gazelles (Walther, 1958, 1964a, 1968a), but it is also described in oryx (Walther, 1958) and in sable (*Hippotragus niger:* Huth, 1970) and roan antelope (*Hippotragus equinus:* Joubert, 1970). In the *Gazella* species it is a very common and rather serious, offensive threat, used predominantly by adult and, in particular, territorial males. It is an open question whether the high presentation of horns is to be considered a pure threat (swing-out movement for a downward blow) or a combination of a threat (horn presentation) and a dominance display (erect posture, p.697).

Sideward angling of horns (Fig. 14g) in broadside position as well as the broadside position itself are threat displays in those artiodactyl species (p.673) that fight from a lateral (parallel or reverse-parallel) position. However, the combination of broadside position and sideward angling of horns also appears in a number of species that do not do so, but in which the rivals fight in frontal orientation to each other. In these cases, obviously, a dominance display (broadside position, p.693) and a threat display (angling of horns toward an opponent) are combined. Such a sideward angling of horns or antlers is occasionally found in almost all bovid and cervid species. However, I have only observed it to be a frequent and elaborate display (always combined with an erect or head-up posture) in a few species, including Grant's gazelle (Walther, 1965), oryx (Walther, 1958, 1966a), and waterbuck *(Kobus defassa).* Joubert (1970) recorded it in roan antelope, and Estes (1969) in wildebeest. In medial and high presentation, as well as in sideward angling of horns, one ear is often emphatically turned ("pointing") in the addressee's direction. In presentation of horns, it is usually directed forward; in sideward angling of horns it points sideward toward the opponent.

To the human observer, all these forms of presentation of horns and antlers appear to refer strongly to the presence and potential use of these weapons. Observations by Hediger (1946) of certain cervids and by Geist (1971) of bighorn sheep *(Ovis canadensis)* support the view that this is also true for a conspecific recipient. In other species, however, there is evidence to the contrary. In a captive, dehorned oryx bull, the postures of high presentation and sideward angling of horns had strongly intimidating effects on conspecifics (Walther, 1958), and I have had occasion to observe precisely the same thing in a dehorned male dorcas gazelle. Thus, the role of horns in these displays should be considered with caution. At least in certain species, the recipients may react more, and perhaps even primarily, to the corresponding postures of the head and neck and pay little attention to the challenger's horns in threat encounters.

With the exception of sideward angling of horns or antlers, the sender is usually frontally oriented toward an addressee in all the threats described above. Sometimes the animal will not turn its whole body but only its head and neck in the opponent's direction. The addressee also often stands frontally to the sender. In principle, however, the addressee's position does not matter. In other words, its flank or hindquarters can also face the sender. This last orientation is regularly found in pursuit marches, where the inferior animal walks away in a "normal" or a submissive attitude while the superior one follows with a threat or a dominance display. Occasionally, all the threats described above can be combined with broadside displays (p.693). For example, the sender may block the opponent's path in lateral position and may show a high presentation of horns (without turning its head and horns in the opponent's direction). Such cases must be distinguished from the combination of threats and inferior behavior patterns, e.g., when the inferior animal stands frontally toward its challenger and responds to the latter's threat with a counter display, at the same time turning its head away from the challenger and avoiding a clearly directed threat; or when a withdrawing opponent continues to threaten forward while the pursuing dominant marches behind it. We must emphasize that these cases are mosaiclike combinations of threat displays with tendencies for withdrawal or turning away. The latter are expressed by the orientation component. However, the threat displays themselves cannot be explained as a result of an inner conflict between aggression and escape tendencies as certain authors (Tinbergen, 1952; Hinde, 1966, Ewer, 1968) have assumed. They are expressions of pure aggression (Walther, 1974).

Displays of Space-Claim

Besides threat and dominance displays (p. 690), some additional behavior patterns, which are certainly not forms of aggression, may show up in agonistic encounters of artiodactyles. On the other hand, they do fit meaningfully into an agonistic situation. They are connected to behavior patterns used in relation to space, the environment, inanimate objects, or the ground. Thus, they are not restricted to hostile encounters but they may also occur (even primarily) in other situations. When used in hostile encounters, the animal may claim occupancy of a place —literally, in the case of a territorial animal; symbolically and only very temporarily, in the case of a nonterritorial animal. The intimidating or challenging effects on recipients are incomparably less pronounced than in threat and dominance displays. One can imagine that these behavior patterns may have a somewhat self-insuring effect on the performer and may perhaps signal to the opponent: "This is my place (you'd better stay away)." This view is supported by the absence of these displays in situations where other threat and dominance displays may be used but where a space-claim would not agree with the requirements of the situation (as in courtship, mother-infant relationships, coordinating group activities). Such behavior patterns are predominantly object aggression, marking with skin glands, urination and/or defecation, pawing the ground, and possibly also grazing in agonistic encounters.

Object aggression is so closely related to aggressive behavior that it was discussed in connection with threat displays. However, its occurrence in nonagonistic situations, its relative ineffectiveness as a threat in some species, the possibility that in a few species the secretion of certain glands may be deposited in connection with this behavior, and the possibility that it may leave visually recognizable marks on vegetation or the ground make probable a connection with marking behavior.

Gland marking is quite common in connection with agonistic encounters (when such glands are present). The way in which certain species, such as chamois (Hediger, 1949) or mountain goat (Geist, 1965), deposit the secretion of their postcornual glands (rubbing the object with the forehead or the region between the horns) makes even an original connection with object aggression rather likely. However, in many species that mark with their preorbital glands, deposition of secretion shows no resemblances to aggressive behavior. On the other hand, the frequency of preorbital gland marking often increases strikingly just before, after, and sometimes during agonistic encounters. Furthermore, the males of some species, such as Thomson's gazelle, goitered gazelle, and blackbuck, open their preorbital glands wide during fights and threat and dominance displays (and during courtship). Finally, these and other species, such as pronghorn, blesbok, and axis deer, frequently rub their horns or antlers more or less intensively on the same object before or after marking it with their glands. Thus, it does not appear too farfetched to assume a connection with agonistic behavior in these animals as well. In the case of territorial males, this marking in agonistic encounters may serve to make the position of the boundary more recognizable to an opponent and to the owner himself. In a hostile encounter, this means the owner can, if necessary, retreat to the marked area, as this is his territory, into which no opponent will easily follow. If an animal is not territorial, it may create a fixed starting point for the combat by marking —comparable to the "corner" in a boxing ring.

Practically the same aspects are true for urination and/or defecation in connection with agonistic encounters. In species in which urine and/or feces have a role in marking territory, these behavior patterns occur more frequently in encounters of territorial than nonterritorial individuals. On the other hand, urination and/or defecation in agonistic encounters is not restricted to territorial individuals or species, and, since it is not dependent on the presence of certain glands, it is more widespread than gland marking in Artiodactyla. However, it is not used as frequently by the same individual in the same encounter, probably because the "material" is not as easily at hand as gland secretions.

In some species, pawing the ground with a foreleg in agonistic encounters shows a relation to pawing before urination and/or defecation. In species that do this, pawing is neither obligatory (an animal may also urinate and/or defecate without pawing) nor restricted to hostile encounters. However, it is apparently somewhat more frequent in agonistic situations than in connection with "normal" urination/defecation. There is an interesting trend in this behavior, which may be demonstrated by its occurrence in Thomson's gazelle, Grant's gazelle, and oryx antelope. In agonistic encounters of territorial male Thomson's gazelle, pawing the ground precedes urination/defecation. Occasionally, a male may paw the ground without urination/defecation following. In all the observed cases, however, there was strong circumstantial evidence that the buck had urination/defecation "in mind" but did not actually perform it because of the pressure of the situation. The same is true for Grant's gazelle. In some relatively rare cases, however, this explanation is unlikely; for example, in a very severe fight between adult Grant's males, one or both combatants may paw the ground while their horns are interlocked. In these and a few other cases, pawing the ground does not seem to be linked to urination/defecation, but appears to be a display of its own. In oryx antelope, pawing the ground during threat encounters without defecation following (urination is not combined with defecation in the oryx) is even rather frequent. Thus, there is obviously an increasing separation (or, depending on the direction in which this trend is considered to go, an increasing combination) of pawing the ground and urination and/or defecation. With this separation, pawing the ground is more likely to become a special means of threat.

A last behavior possibly belonging to the category of space-claims is grazing in agonistic encounters, in which a mechanism similar to or identical with that of a transitional action (Lind, 1959) seems to be involved. The head of an ungulate is close to the ground in a number of aggressive behavior patterns such as the head-low posture, low presentation of horns, downward blow, and actual fighting. Thus, the head is generally in a position proper for grazing, and the animal may then sometimes switch to grazing, which in certain situations and/or species appears to be a behavior of an inferior animal; in others there are no indications of inferiority. Unfortunately, its relation and significance to agonistic interactions has not been realized until recently, apparently because observers always assumed that grazing would not belong to an agonistic interaction and that the encounter was over when one or both combatants started grazing. So, at present, although this behavior is certainly widespread in artiodactyles, only the very thorough studies of Estes (1969) on agonistic grazing in wildebeest, David's (1973) very similar results in bontebok *(Damaliscus dorcas dorcas),* and my own observations in Thomson's and Grant's gazelle are available for a somewhat detailed discussion.

In Grant's gazelle, grazing in connection with agonistic encounters occurs relatively often on the part of inferior opponents. They frequently respond by grazing to a superior challenger's approach and/or threat and dominance displays before they withdraw. Occasionally, however, a superior combatant or both opponents may also graze. When both rivals graze at the end of an encounter, they may move away from each other while doing so. Thus, the meaning of agonistic grazing appears to be somewhat ambivalent in Grant's gazelle. On the whole, however, one may say that it frequently, although not necessarily, indicates inferiority in this species.

The situation is clearly different in Thomson's gazelle. Here, agonistic grazing is primarily a behavior of adult males, especially of territorial males, and both opponents generally perform it simultaneously. What makes me use the term "grazing ritual" in the case of this species (Walther, 1968a) is a very pronounced and predictable change of position between rivals (territorial neighbors). Grazing almost uninterruptedly, they go through a frontal position, a parallel or reverse-parallel position, and a reverse position (hindquarters to hindquarters). These three changes in position are the minimum. Sometimes the opponents may graze side by side along the entire boundary of a territory, or one or both may turn again to a frontal position after having grazed for a while in parallel or reverse-parallel position. In short, there can be many variations and repetitions in such a grazing ritual, and it may last from a few minutes up to half an hour or longer. In Thomson's gazelle, a grazing ritual rarely precedes a fight. Incomplete attempts are frequent during pauses in fights. Especially when grazing in frontal position, the rivals can easily change from grazing back to fighting. The complete, pronounced ritual, however, is seen only following or in place of a fight.

When grazing in frontal position after a clash, male Thomson's gazelle immediately enlarge the distance between one another by stepping backward. In this species, as in many artiodactyl species, individual distance (Hediger, 1954) depends on age, sex, and activity. It reaches its largest extent in adult males during grazing. Thus, by simultaneous frontal grazing immediately after a horn clash, the opponents signal to one another for a return to the (large) grazing distance. If necessary, the grazing posture allows them to begin fighting again immediately. The parallel or reverse-parallel position probably has the same origin and basic meaning as other broadside displays—blocking the path—as it is rather obvious in grazing along the territorial boundary. Finally, in the hindquarters-to-hindquarters position, agonistic grazing may allow an exit without "losing face," the battle, or territorial status for both combatants. Thus, agonistic grazing is closely related to fighting in Thomson's gazelle, is most commonly shown by territorial peers (i.e., nothing speaks for a behavior of inferiority), and appears to belong to that category of expressive behavior that is connected with space relations (taking place right on the boundary or, at least, in the boundary zone of neighboring territories).

In wildebeest (Estes, 1969), one territorial male may approach another and may intrude into his territory while permanently grazing. (This was never observed in Grant's or Thomson's gazelle.) When they approach each other, the owner of the territory often threatens the intruder by sideward angling of his horns. The intruder keeps his present grazing attitude, i.e., behaves like a subordinate male Grant's gazelle. On the other hand, the owner of the territory may also start grazing, and both animals may go through changes in position similar to those of territorial "Tommy" males (except that this occurs inside the territory belonging to one of them and not at the territorial boundary). They may even circle around each other in grazing, something gazelle males do not do. Furthermore, in wildebeest, grazing apparently does not turn as easily and as readily into fighting (or vice versa) as in Thomson's gazelle. Thus, agonistic grazing of wildebeest offers an even more complicated picture than that of the two gazelle species; however, it also seems to be restricted to territorial males in wildebeest (Estes, 1969) and in bontebok (David, 1973). Thus, it could have something to do with space relations in these two species as well.

Possibly, the key to a general interpretation may be seen in the close relationship of agonistic grazing to horn fighting, as it was found in Thomson's gazelle. Biting and snapping as threat and fighting behavior are found in a number of artiodactyl species as discussed above. One may assume that snapping generally was a fighting technique of the ancestors of recent artiodactyles. It even (still?) occurs in hornless or antlerless females of certain bovid and cervid species. Interestingly enough, symbolic snapping (directed toward the ground) was observed in combination with boxing the opponent with the forehead in greater kudu and sitatunga females (Walther, 1964b). Thus, one could imagine that in some of the recent bovid species also (which primarily fight with their horns), the biting could become activated in agonistic encounters and (since the head is lowered according to the head-low posture or the low or medial presentation of horns in agonistic encounters) that it may turn into grazing (= biting the grass). At first sight, this may appear to be a rather bold speculation, but it would not only explain the occurrence of grazing in agonistic encounters but also make understandable the numerous species-specific variations of agonistic grazing: It is likely that such a common ancestral behavior has become differentiated in a variety of ways in the single species during phylogentic evolution. It is also understandable that such a relatively mild form of aggressiveness can sometimes occur in a dominant animal, sometimes in an inferior one, depending on the situation. Of course, more data are required to substantiate these assumptions.

Dominance Displays

In contrast to threat displays, dominance displays (= *Imponieren* in German literature; display threat: Lent, 1965; bravado display: Geist, 1966b; present threat: Geist, 1971) do not indicate immediate readiness to fight. Commonly an animal demonstrates its height and/or breadth or shows other striking postures or movements, none of which are related to recent fighting techniques of the species. Sometimes, weapons may also be presented, but not in a position suitable for fighting. Thus, these displays indicate the claim of superiority over the addressee ("I am the boss!") without showing fighting intentions. On the other hand, dominance displays have the same effects—intimidation or challenge—on the recipient as threat displays, and they occur in the same situations. Typically, however, they are infrequent in females and juveniles (animals that often play a subordinate role in artiodactyl societies) and in agonistic situations predominantly or exclusively found in these age or sex classes (soliciting milk and defense against it, defense against sexual approaches, playful encounters). In encounters between unequal opponents, these displays are typical of superior, stronger, older, or higher-ranking (also territorial) animals, but are rarely used by inferiors, whereas the latter may quite frequently show (defensive) threat displays. When the encounter cannot be settled by dominance displays (a situation especially frequent in reciprocal encounters, as when both rivals show the same displays), the opponents may change from dominance displays to threat displays and in some cases from these to fighting. The rivals rarely move immediately from reciprocal dominance displays to fighting, whereas that direct change is very common in reciprocal threat displays. Thus, in such cases, the reciprocal dominance displays contribute to the prolongation of the opening phase before a fight. When the display remains one-sided, the (inferior) recipient usually withdraws upon receipt of a dominance display by the superior sender.

These statements may be substantiated by some quantitative records on certain reciprocal threat and dominance displays in Grant's gazelle, a species that has threat displays (the following presentation refers predominantly to medial and high presentation of horns); at least two dominance displays (head-sideward inclination in frontal approach or in parallel walk, and the head flag in erect posture and in broadside position—for detailed discussion of these displays see below); and an intermediate form, a "threat-dominance display," in which features of dominance displays (broadside display and circling) are combined with a threat component (sideward angling of horns toward the laterally standing opponent).

Table 3 shows which behavior followed immediately after these displays. Cessation of aggression by both opponents (including cases where both are grazing) is relatively rare after reciprocal displays. It is rarest after threats (1.1%). Withdrawal or (the even rarer) submissive behavior of one of the opponents is relatively infrequent (in *reciprocal* encounters) after the head-sideward inclination (10.8%). Here, however, one has to take into account the fact that the approach with head-sideward inclination is especially frequent in the opening phase of an agonistic encounter in this species, and, thus, the encounter is likely to be continued after this display. With this exception, encounters are most frequently settled and decided (i.e., one of the opponents gives in) after the head flags, i.e., dominance displays (28.8%), and rarest after horn threats (10.1%). Sideward angling of horns as an intermediate behavior is right in the middle (19.5%). As mentioned above, encounters are frequently continued with other displays—such as head flag, sideward angling or presentation of horns, horn sweep, and object aggression—after an approach or parallel walk with head-sideward inclination (75.4%). The same is true, although to a lesser extent, for *(reciprocal)* head flags (60.8%) and sideward angling of the horns (53.7%). It is considerably rarer after horn threats (11.2%). The most conclusive figures are provided by the frequency of fights immediately following the displays under discussion: In the two dominance displays none occurred after head-sideward inclination and fights followed only 2.6% of the reciprocal head flags. However, fights followed "threat-dominance displays" (sideward angling of horns) in 24.2% of the cases, but they followed reciprocal pure threat displays (horn threats) in frontal position in 77.5% of the cases. These figures clearly demonstrate in Grant's gazelle that threat displays are significantly closer than dominance displays to

Table 3

Ending or continuation of (reciprocal) agonistic encounters after certain threat and dominance displays in Grant's gazelle.

	Reciprocal head-sideward inclination in frontal or broad-side position	**Reciprocal head flag in erect posture and broadside position**	**Reciprocal sideward angling of horns in broadside position**	**Reciprocal presentation of horns in frontal position**
Number of observed cases	65	153	149	89
Percent followed by				
other threat or dominance displays	75.4	60.8	53.7	11.2
fight	0.0	2.6	24.2	77.5
withdrawal, flight, or submission of one of the opponents	10.8	28.8	19.5	10.1
cessation of aggression by both opponents	13.8	7.8	2.7	1.1

overt fighting, and the same appears to be true in other Artiodactyla.

It is certainly noteworthy that an animal exposes itself to a possible attack ("daring") by the rival much more during dominance displays than during threat displays. In this context, there is a direct connection between the dominance displays and advertising and emphasizing presence. On the other hand, the commonest and most important dominance displays of artiodactyles are not without connection to threats and fighting behavior. Sometimes features of threat and dominance displays occur in combination (*Droh-Imponieren* = threat-dominance displays), as mentioned above. Moreover, the alternative "threat or dominance display" does not depend on the posture or movement *per se,* but on its relation to the recent fighting behavior of a given species. It follows from this that the same posture can be a threat display in one species and a dominance display in another. For instance, in species like mountain goat, which fight from a broadside position, assuming a lateral position is an intention movement for fighting and, thus, a threat display. However, when a broadside display appears in the hostile encounters of a species like Grant's gazelle, which fights only from a frontal position, the broadside position is a dominance display.

In Artiodactyla, the relationship between threat and dominance displays and the evolution of the latter are rather obvious. Paleontologists (e.g., Thenius and Hofer, 1960) generally agree that the recent artiodactyles have evolved from forms that had not developed special organs for fighting. There are some recent species, such as the Tylopoda, that still lack special armament and use their legs, teeth, and neck in fighting. Others, like the Suidae and hippos, have specialized already existing organs (the teeth) for use as weapons. Still others, mainly Cervidae and Bovidae, have developed special organs (antlers and horns) for intraspecific aggression. However, these species have also evolved from hornless/antlerless ancestors (*Archeomeryx optatus* or related prehistorical ungulates).

It is certainly correct to presume that these "unarmed" ancestors had their fights and threat displays, but they could not use horns or antlers since they had none. It is not out of place to assume that their fighting techniques and threat displays were similar, if not identical, to those shown by recent "unarmed" artiodactyles (e.g., tylopods). With the development of horns and antlers as special means of intraspecific aggression and with the corresponding development of phylogenetically "new" fighting methods and threat displays, ancestral fighting techniques have been replaced in the bovids and cervids. Of course, this has varied among the species according to differences in advancement (size, mass, shape, permanent or temporary usefulness as weapons) of horns and antlers (Walther, 1960a, 1966a). In some species, apparently, although the ancestral fighting techniques have more or less completely disappeared, the corresponding threat displays have remained. Hence, even though these ritualized intention movements originated in fighting techniques, they now have no connection with the recent fighting behavior of these species and are now dominance displays in the defined sense. Because of their separation from recent fighting techniques, they are milder forms of challenge or intimidation compared to the "new" forms of threat, which refer directly to actual fighting behavior. When, in a species, both "modern" threats and displays derived from ancestral aggressive behavior coexist, a more subtle gradation in the forms of challenge and intimidation becomes possible, which is certainly advantageous (allows more shades) to social communication and makes the "survival" of these phylogentically old displays understandable.

We will come back to this hypothesis in the discussion on courtship displays. At the moment, it is sufficient that the reader is aware of the possible relationships between threat and dominance displays, and can understand why certain postures or movements are considered threat displays in some species and dominance displays in others, and why dominance displays generally are more frequently found in the more highly advanced artiodactyles than in the more primitive forms.

The broadside position was mentioned previously as a means of blocking another's path and as an initial position for fighting in lateral (parallel or reverse-parallel) position—two aspects that are not mutually exclusive. Apparently both gave rise to the broadside attitude as a dominance display. One can speak of a special broadside display only when additional behavioral features are added to the basic broadside orientation, making it more striking to conspecifics (Fig. 15c). This can be achieved in many different ways: In warthog, bushbuck *(Tragelaphus scriptus),* and nyala *(Tragelaphus angasi),* the animal ruffles up the manes on its neck and back. In gaur *(Bibos gaurus),* cattle *(Bos primigenius),* and greater kudu, the displaying animal may hump its back by extending its hind feet further under its belly than usual. And/or it may stretch its head and neck forward (e.g., nilgai) or forward/downward (e.g., greater kudu and gaur), or in some species (e.g., blackbuck, Grant's gazelle, and lesser kudu) the broadside position may be combined with an erect posture or even with a nose-up posture.

An exaggeration of the broadside position by a sideward inclination of the head (see below) away from the opponent is very frequently seen, e.g., in nilgai, mountain gazelle, Grant's gazelle, red deer, and elk. This movement is often related to sideward angling the horns, e.g., in wildebeest, oryx, and roan antelope.

When both rivals show a broadside display, they are automatically brought into a parallel or (more often) a reverse-parallel position (Fig. 17). In many species, they then begin to circle, always keeping their flanks toward each other. As the encounter becomes more severe, they may change from circling in reverse-parallel position

Fig. 15. Some of the dominance (or threat-dominance) displays that are used as fighting techniques or threat displays in other, behaviorally more primitive, artiodactyl species. a. Neck stretch of a female caribou directed frontally toward the addressee, corresponds to a bite or a (very intensive) push with mouth shut and possibly to neck fighting. (After Berry, in Pruitt, 1960.) b. Manchurian sika (*Cervus nippon dybowskii*), with nose-up posture and grinding of teeth, directed frontally toward the addressee, corresponds to hooking with (elongated) upper canines and/or to neck fighting. (After Schneider, 1930.) c. Broadside display (i.e., the reader is in the position of the addressee) of a greater kudu bull (*Tragelaphus strepsiceros*), combined with head-low posture, humped back, and bent hind legs positioned under the belly. This corresponds to fighting in lateral position (jostling with the rump or shoulders, sideward blow with the head and neck) and possibly also to a swing-out movement to an aggressive jump (throwing the body on the opponent while simultaneously turning toward the adversary).

to frontal orientation and may eventually attack one another frontally (provided they do not belong to a species that fights from a lateral position anyhow). In some species, such as Grant's gazelle, circling can also be seen in one-sided encounters. This means the displaying animal moves around the (nondisplaying) recipient. Apparently, this "encircling" has a strongly intimidating effect.

Some Suidae species, such as wild boar, fight in a lateral position (Frädrich, 1965). With respect to a derivation of broadside displays from ancestral fighting behavior, it is noteworthy that pressing and jostling with the shoulder or the whole side of the body against the opponent's flank plays a large role in these fights. In this species, however, assuming the broadside position is a threat rather than a dominance display. On the other hand, one could definitely think of the broadside position as a dominance display in other Suidae, such as bush pig *(Potamochoerus porcus)* and warthog, which commonly fight in frontal position but which show the broadside attitude as an expressive behavior in agonistic encounters (Frädrich, 1965, 1967). The broadside position is also present in at least some tylopods, such as vicuna (Pilters, 1954, 1956), and in the Giraffidae. However, in these species it is more a threat than a dominance display because of its close relationship to the fighting behavior of these animals, as previously discussed. A few Bovidae fight in parallel or reverse-parallel position (Fig. 17a), but since the majority of bovids and cervids do not when broadside displays are found in a considerable number of bovids and cervids (as listed above), they are dominance displays in the outlined sense (Fig. 17c).

In the head-sideward inclination (Fig. 16b), the head is turned sideward at an angle of about 45°, and in the head-sideward turn (Fig. 16a) at an angle of about 90° or even slightly more. Apparently, these two behavior patterns have not been distinguished from each other in previous

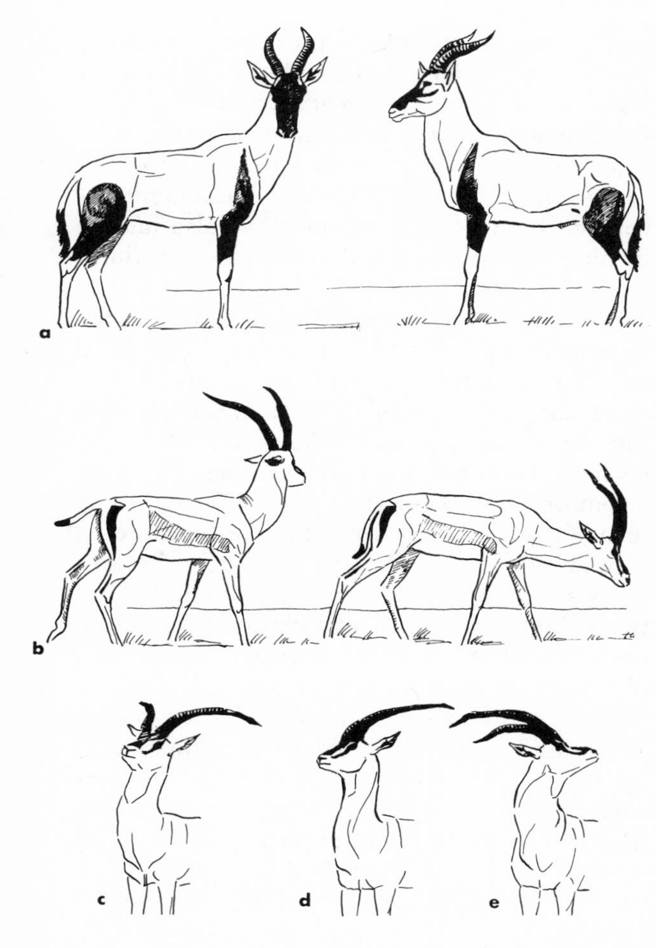

Fig. 16. Several forms of dominance displays. a. (Erected) head-up posture (right) and sideward turn of the head (left) in an encounter between two territorial topi bulls (*Damaliscus lunatus topi*). b. Sideward inclination of the head (left) in a dominant male Grant's gazelle during a pursuit march behind the inferior opponent (right) in submissive posture. c.–e.: Sequence of behavior patterns in the head-flag display of Grant's gazelle. (The position of the addressee is marked by X at the right of the picture.) c. The buck assumes an erect posture in broadside position. d. Head-sideward inclination as a swing-out movement away from the recipient. e. Head flag (*hohes Kopf-Zuwenden*) toward the recipient, after which the animal turns its head back to the initial position (c).

literature. However, it appears that one is considerably more frequent in a given species than the other; for example, the head-sideward inclination is frequent and the head-sideward turn rare in Grant's gazelle and mountain gazelle, while the opposite is true in Uganda kob and impala. This fact and the possibility that there may be some differences in messages and meanings make a distinction desirable, although the two behavior patterns are closely related and transitional cases are possible. Both can be combined with other species-specific dominance displays, such as the erect posture in mountain gazelle, Grant's gazelle, and topi, the head-and-neck-stretched-forward posture in nilgai (Fig. 17b), and, occasionally, with threat displays such as the high presentation of horns. Obviously, the head-sideward inclination has a tendency to turn into the sideward angling of horns. This tendency is less pronounced in the head-sideward turn.

Generally, when an animal makes a turn, the head precedes the body and the movement or the posture of the head alone is often sufficient for and of special importance to social communication. In this respect, the sideward inclination and sideward turn of the head can be considered as intention movements for turning to a broadside position (when the animal stands with its body frontally oriented toward its rival before turning) or as an emphasis or exaggeration of the latter position (when the animal is already oriented with its flank toward the addressee). In combination with the broadside position, the head is always turned away from the opponent. As it became clear from many observations of different species, however, the animal always keeps its opponent in sight, watching the rival carefully from the corner of its eye, and immediately reacts to the slightest movement of its opponent (one-eyed fixation and lurking watch, respectively).

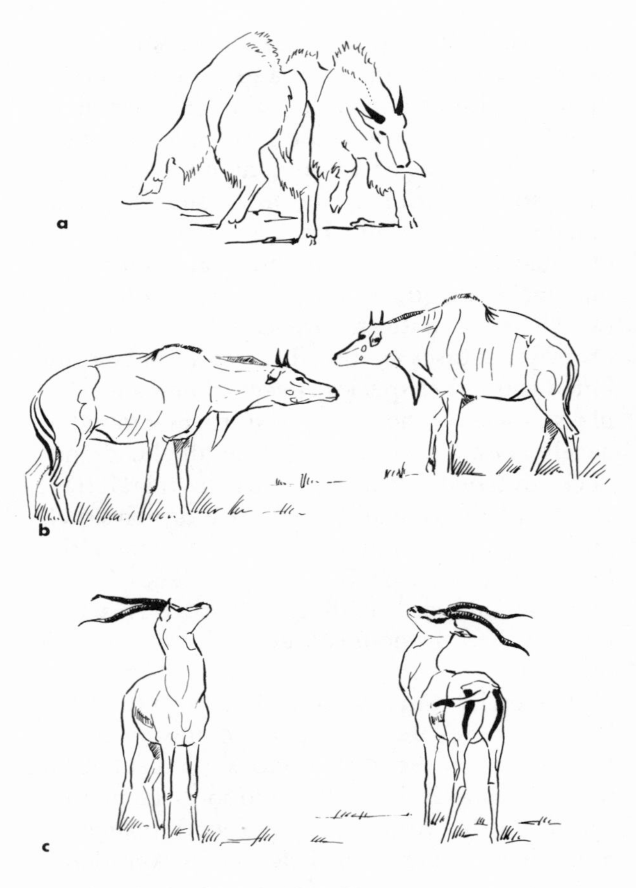

Fig. 17. Some examples of reciprocal broadside positions in agonistic encounters. a. Fighting mountain goat (*Oreamnos americanus*). Note the ruffling of the hair on the back, especially on the withers and the croup, and the swing-out movement of the lowered head for a sideward blow against the opponent's flank. (After Geist, 1966a.) b. Nilgai bulls (*Boselaphus tragocamelus*) circling each other. The neck stretch can be considered an intention movement for the neck fight, which actually occurs in this species. Thus, the neck stretch in combination with the broadside position is probably a threat-dominance display in nilgai. c. Reciprocal head flag in two (standing) bucks of Grant's gazelle. Since neither broadside position and erect posture, nor the head flag, itself, belong to the (recent) fighting techniques of Grant's gazelle, this is an example of a pure dominance display.

As with the broadside position, certain authors seem inclined to consider every form of sideward turning of the head as indicative of involvement of escape tendencies and thus consider that such displays are brought about by inner conflict between aggression and flight. Of course, these head movements can also initiate turning around for withdrawal, and sometimes this is definitely the case in artiodactyles. However, this is only one possibility. It is as possible —and this happens fairly frequently—that the animal turns into a full broadside display from a head-sideward inclination or a head-sideward turn in, at first, frontal orientation. One may even see a very effective swing-out movement for a reinforced turn toward the rival in the initial turn of the head away from him. Furthermore, as with all dominance displays, both behavior patterns are primarily displayed in encounters between high-ranking or territorial peers, and, in encounters between unequal opponents, are displayed by the clearly superior one. Last, but not least, in certain species, such as Grant's gazelle and mountain gazelle, a sideward inclination of the head is rather persistently shown by the pursuer (!) when following a withdrawing inferior. This truly does not speak in favor of the involvement of escape tendencies in this display. On the whole, I am inclined to consider the head-sideward inclination as well as the head-sideward turn abbreviations of the broadside display. The head-sideward inclination is possibly a somewhat more offensive form and more closely related to threat behavior in the strict sense than the sideward turn of the head.

One common principle in dominance displays is to make the animal appear as long and broad as possible. This is verified by broadside displays and related behavior patterns. Another widespread principle is to make an animal appear as tall as possible. This is mainly achieved by the erect posture (Fig. 16a, c), the nose-up posture, and the head-and-neck-forward/upward posture. Because these three are closely related and their differences were pointed out in the discussion on threat displays, I will lump them here and refer to them as "erect displays." Erect displays can be combined with broadside displays or either may occur independently.

Since most artiodactyl species rise up on their hind feet in sexual mounting, one could consider the erect postures to be intention movements for the latter. However, one can also think of a development parallel to that of the broadside displays, inferring that these erect postures are phylogenetic relics of ancestral fighting behaviors. A point that speaks in favor of a connection with sexual behavior is the fact that a combatant sometimes mounts its rival from an erect posture in an agonistic encounter. On the other hand, *Tragelaphus* species show erect postures as dominance displays, but do not mount in erect postures. The male puts his head and neck on the female's back during sexual mounting (Walther, 1958, 1964b). Hence, their erect displays are obviously not closely related to sexual mounting. Or, in certain highly advanced cervid species, such as red deer, Manchurian sika (*Sika nippon dybowskii:* Schneider, 1930), and barasingha (*Cervus duvauceli:* Schaller, 1967), an erect posture with almost vertically raised nose (Fig. 15b) and gnashing teeth apparently corresponds to the attitude adopted by related primitive species *(Moschus, Hydropotes, Muntiac)* when threatening or attacking an opponent with their prolonged upper canines (Antonius, 1939).

Finally, it is possible that sexual mounting—of the artiodactyl type—itself is related to and has originated from aggressive jumping and throwing the body on the partner. In this connection, it may be mentioned that all Tylopoda copulate lying down. Thus, their copulatory posture has nothing to do with rising on the hind legs. A male tylopod, however, frequently jumps at and throws his body on the female to force her to the ground—a behavior that is similar to or even identical with that shown in fights between rivals. It is this aggressive jumping at the female, not the copulatory posture of tylopods, that corresponds to the mounting behavior of other artiodactyles. It is possible that tylopods, which appear to be behaviorally primitive artiodactyles in more than one regard, may have kept this general feature of ancestral artiodactyl behavior. In short, it is impossible at present to decide conclusively whether the erect dominance displays of certain artiodactyles have evolved from sexual mounting or from ancestral fighting techniques. However, the odds appear to be in favor of the latter.

Since a connection between erect displays and recent fighting techniques is very likely in Tylopoda (jumping at the other, neck fighting), Giraffidae (pronounced neck fighting in several forms), and Cervidae (rising on the hind legs and beating with the forelegs), erect postures as true dominance displays occur mainly in certain Bovidae. A possible exception appears in the genus *Ovis,* where erect postures are apparently related to rising on the hind feet as a fighting technique (Walther, 1960b), bringing them more on the side of threat displays. In discussing certain other bovids, some difficulties arise from the major role of the neck in erect postures. In a species with a relatively long, movable neck, its erection is an essential component in these displays and makes their recognition very easy. In a species with a relatively short, massive neck, however, it cannot be erected to any great extent. It is then doubtful whether one can speak about an erect posture and, if so, whether it can be directly compared to the displays of long-necked species.

In the wild oxen, for example, the anatomical structure of the neck largely inhibits the performance of pronounced erect displays. Raising the head in these animals usually results in a nose-forward posture, and it is doubtful whether it can be directly compared to erect postures of other species. Erect postures as clear dominance displays, often combined with a more or less pronounced lifting of the nose, have been observed in species such as impala (Schenkel, 1966), greater kudu, lesser kudu, topi, blackbuck, and Grant's gazelle. I do not hesitate to add to this list the "head-up" displays of species with less movable necks, such as wildebeest (Talbot and Talbot, 1963; Estes, 1969), oryx antelope (Walther, 1958), and tsessebe (*Damaliscus lunatus lunatus:* Joubert, 1972).

In all species under discussion, the erect displays occur in frontal orientation toward an addressee as well as in combination with the broadside position. Tsessebe bulls displaying broadside in erect posture with lifted nose in front of an opponent sometimes jump up into the air with their forelegs (Joubert, 1972). This example is of special interest, since one might doubt that erect postures are intention movements for rising to the hind feet, arguing that this view makes no sense when the erect posture is displayed in a lateral orientation toward the recipient. Regardless of whether it is logical in human terms, however, jumping up with the forefeet from an erect posture while in broadside position is a fact.

Some additional behavior patterns can be considered dominance displays (for example, the protruding of a "goulla bag" from the mouth of camels), but I will discuss only one more in detail—the unique modification of the erect posture in Grant's gazelle. Standing in broadside position with vertically erected neck, the displaying animal turns its lifted head and nose sideward toward the addressee (*hohes Kopf-Zuwenden:* Walther, 1965) and then forward again. The tail is often horizontally stretched during this display and may swing to the right or the left. This head flag (Figs. 16c–e, 17) is commonly initiated by a head-sideward inclination or sometimes even a sideward turn of the head away from the rival as a swing-out movement. The tension of neck muscles (Estes, 1967) and/or the white throat patch may have an additional effect in this display. However, I think that the term "neck-intimidation display," suggested by Estes (1967), conceals the fact that the head turn toward the rival is its most important component. The head flag is always delivered while standing and may be repeated several times. The releasing situation apparently comes when the sender is at the point of passing the receiver or, vice versa, when the receiver is passing or has just passed the sender. The sender can stand broadside in front of or parallel to the recipient or, most frequently, in reverse-parallel position (Fig. 16c). In reciprocal encounters, the reverse-parallel position of the opponents is almost obligatory. It provides a good chance for one of the opponents to cease displaying after a while and walk forward in a normal, common way. Thus, the display provides a "golden bridge" for an inferior to retreat "without losing face." One may speculate that this "golden bridge" is a principle in all highly ritualized dominance displays (Walther, 1965).

Interestingly enough, a perfect analogy occurs in human behavior when we signal—in a rather rude and arrogant way—to a (usually inferior) person to go away by a sideward swing of the head. In Grant's gazelle, the head flag can be understood as a (phylogenetically) "new" orientation ("pointing") movement from an erect posture, adapted and specialized to a situation in which the recipient is standing in reverse-parallel position to the sender. Thus, the sender anticipates and demands the recipient's withdrawal in the direction opposite to its own. Because of the highly stereotyped nature of the head flag, "failures" occasionally occur, i.e., when the sender's position relative to the recipient is inadequate for the orientation of the head flag (comparable to "failures" in orientation of the inciting behavior of female mallards: Lorenz, 1963). Possibly, such a "failure" gave rise to the nonobligatory but relatively frequent "undirected" head flag of female Grant's gazelle during the mating ritual (when she is standing before the male and facing in the same direction as he), which apparently provokes the male's mounting (Walther, 1965).

In barren ground caribou (*Rangifer tarandus arcticus*) a head-and-neck-stretched-forward posture (Fig. 15a) in agonistic encounters (Pruitt, 1960) could be considered a dominance display (provided that reindeer do not bite, an action that I have not seen mentioned in the literature). It fits the picture that male reindeer show the same posture in courtship (since dominance displays can be used in the courtship of certain species). On the other hand, some of the situations that are described for this behavior (females' warding off strange fawns, females' warding off other adults from their fawns) are atypical for dominance displays and agree better with threat behaviors. Possibly, this is a case of a threat behavior "on its way" to becoming a dominance display. Also, in the *Ovis* species, the head-and-neck-stretched-forward posture and the kick with the foreleg can occur in agonistic encounters (Walther, 1960b; Geist, 1968, 1971). Geist (1971), who did a very intensive study on these behaviors, considers them primarily courtship displays in *Ovis*. Therefore, they will be discussed more in detail in this context. It may be mentioned, however, that certain species rotate the head about its long axis in this posture so that one cheek almost points toward the ground. In bighorn sheep, this behavior was termed the "twist" by Geist (1971), and he considers it a horn display (the horns expand laterally in sheep rams). However, the fact that practically the same behavior can also be seen in the hornless females of greater kudu (Walther, 1964b) contradicts this interpretation. At least, it cannot be generally applied to all artiodactyl species that show this behavior. It may also be that this "twist" (as a phylogenetic relic) is more closely related to biting behavior since in many mammalian species turning the head about its long axis is frequently combined with biting.

Since dominance displays and certain other agonistic behavior patterns (such as offensive threats, symbolic chases, pursuit marches, and mounting) are frequently shown by superior animals, and since similar (even identical in some species) behavior patterns occur in males of certain species in sexual encounters, the question may be raised whether these behavior patterns are basically sexual in nature and whether a superior male may treat an inferior as if it were an (estrous) female. In a study on the behavior of bighorn sheep, Geist (1971) has strongly argued in favor of this view. In studies on the behavior of oryx (Walther, 1958), several *Tragelaphus* (Walther, 1964b), and several Antilopinae species (Walther, 1968a), however, I came to an almost opposite conclusion. I fully agree with Geist that in a number of artiodactyl species an adult male treats inferior conspecifics more or less alike, regardless of sex (as far as they are of interest to him—in many species, adult males do not pay attention to the young, and in some species they are not interested in females either if they are not in estrus). Since females are smaller and lighter and usually have smaller horns (in some species, no horns at all), they are naturally inferior partners; thus, the adult males treat them as inferiors in sexual encounters, just as they treat younger and weaker males as inferiors in agonistic encounters. Although the addressee frequently accepts the inferior role when challenged by a superior partner, there are cases in which he (sometimes also she) does not do so, but reacts with (defensive) counterdisplays or even fights back. This behavior is easily understood as a reaction by a recipient to a basically agonistic behavior of a challenger (i.e., treating the addressee—also the estrous female in sexual encounters—as an inferior opponent). What is very difficult to understand is that an addressee

—male or female—would react by submission or defense to a basically sexual behavior, i.e., being treated like a female by a superior companion (it is unlikely that these animals share the "male bias" of certain humans).

Courtship Displays

Like "alarm signals," the term "courtship displays" refers to only one function, but one that is particularly important. These displays are used almost exclusively by males and addressed almost exclusively to females. Obviously, there are great differences within the Artiodactyla. In the tylopods, for example, the behavior of males toward females strongly resembles male behavior in encounters with other males and may even be more or less identical with it. In species such as blackbuck, males use the same (or almost the same) dominance displays toward females in courtship as they use in agonistic encounters with (male) rivals, but they do not use the fighting techniques and the threats. In still other species, for example, Thomson's gazelle, the behavior inventory of a male encountering other males is (almost) entirely different from that occurring in encounters with females. I speak about "courtship displays" here to distinguish them from threat and dominance displays, which either occur in encounters between partners of the same sex or are used toward partners of both sexes.

Interestingly enough, there are situations in which these courtship displays also work like threat or dominance displays. For example, when Thomson's gazelle are migrating (in large, mixed herds, and usually in file), if the animal in front of an adult male stops or slows down it is often "pushed" from behind by the adult male in an attempt to speed it up and keep the migration going. If the animal in front is another male, the "pusher" will threaten him with high or medial presentation of horns. Since in Thomson's gazelle these displays are not normally used by a male toward a female, if the animal in front is a female, the male will not threaten her but will show "courtship" displays (head-and-neck-stretched-forward posture, nose-up; see below). A "Tommy" male uses these behavior patterns only very rarely toward another male. In this situation (and in a few others), the courtship displays affect females as threats do males.

It follows that (1) the so-called courtship displays can serve functions other than mating (they can be especially important in herding, soliciting urine, coordinating group activities, and soliciting milk); and (2) although courtship displays can be phenotypically different from (recent) threat and dominance displays in certain species, there are unmistakable connections between these courtship displays and agonistic behavior. One may say that the most elaborate courtship displays in artiodactyles are special kinds of dominance displays that probably originated in ancestral fighting behavior. As discussed in the section on dominance displays, the hypothesis states that, especially in the bovids, ancestral fighting techniques were largely replaced by the "modern" horn fight during evolution, whereas expressive displays (intention movements) related to ancestral fighting techniques have remained. Such displays owe their aggressive nature to their origin, but they are milder forms of aggression than threat displays, which refer to the recent fighting techniques of these species. Expressive behavior patterns that mildly challenge or intimidate the partner—just enough to diminish somewhat its aggressiveness (intimidation) or avoidance tendencies (challenge) released by the other's close approach, but not enough to release serious aggression or flight—are tailor-made for mating rituals of the artiodactyl type, in which the male has to approach the female and establish some kind of dominance over her. Male dominance is apparently a prerequisite for successful mating in these animals.

If, for some reason (e.g., the male is too young) the female turns out to be superior, that usually means the end of the mating activity. In the more highly-evolved artiodactyl species, male dominance is achieved by the male's courtship displays and in some species by dominance displays as well; in more primitive artiodactyles, true threats or fighting may serve the same purpose.

The reactions of the females are in full agreement with the basically aggressive nature of the male courtship displays. Most commonly, the females respond by withdrawal; in some species or in certain situations also by flight at a gallop, submissive behavior, defensive threats, sometimes even fighting, and, in rare cases, offensive threats or dominance displays. All these are clearly responses to aggressive displays on part of the males. It is noteworthy that there are no genuine female courtship displays or sexual behavior patterns in the artiodactyles, except for some directly related to copulation, such as lifting the tail, standing for the male's mount, and leaning into the male's mount. What artiodactyl females show during courtship rituals are either male behavior patterns (e.g., the female's mounting the male, which, by the way, is not very frequent in nondomesticated artiodactyles), or agonistic displays, or behavior of inferiority as discussed above.

A few examples may substantiate these statements. In the tylopods, the males use the same displays toward females in courtship as they show in agonistic encounters with male rivals (Pilters, 1954, 1956). Also neck fighting, jumping at the female and throwing the body on her to force her down to the ground, and biting her forelegs are more or less obligatory. Sometimes a tylopod female may defend herself by the same behavior patterns; then, the mating ritual, at best, differs from a true fight only in intensity. In wild boar, the male may put his snout under the female's belly and lift her up (Frädrich, 1956), a behavior that is also known as a fighting technique in this species. Driving giraffe males push the female's shoulder, flank, and hindquarters with their "horns" (Backhaus, 1961). Even in certain bovid species, for example, mountain gazelle, the driving male uses his horns in a ritualized manner. Chasing the female at a trot or gallop occasionally occurs in the majority of artiodactyl species. In some, such as roe deer *(Capreolus capreolus),* it is obligatory, and in others, such as pronghorn, it is very frequent. More commonly, the male drives the female ahead at a walk (Fig. 19), as he would an inferior male opponent after a hostile encounter. This mating march, i.e., the ritualized withdrawal of the female from the pursuing male (which can often turn into her true withdrawal or even flight, especially when the female has not yet reached the peak of the heat), is a basic component of the mating rituals in many artiodactyl species. Obviously, the partners synchronize their readiness for mating in this way.

In oryx, addax, sable and roan antelope, the mating march has been largely replaced by a more stationary performance, the mating whirl-around (*Paarungskreisen:* Walther, 1958), in which the male and female step around each other in reverse-parallel orientation (Fig. 18b) (a behavior that also occurs in agonistic encounters between male opponents in these species, see Fig. 8)—the female showing a head-low posture all the time (like an inferior male opponent in an agonistic encounter). Especially in oxen, the mating ritual has become even more stationary. Here, male and female stand in reverse-parallel position throughout most of the ritual ("guarding" =*Hüten:* Schloeth, 1961). Generally, broadside displays are frequently used in courtship rituals of Artiodactyla—be it in reverse-parallel position (e.g., in guarding or in the mating whirl-around), or with the male blocking the female's path by standing in broadside position in front of her (observed, for example, in bison, bush pig,

Fig. 18. Some of the agonistic elements in the courtship rituals of certain Bovidae. a. Erect posture of the male (right) and intention movement for pushing with mouth shut of the female (left) in an early phase of the mating ritual of lesser kudu (*Tragelaphus imberbis*). b. Mating whirl-around in oryx antelope. Note the head-up posture of the male and the head-low posture of the female (compare Fig. 8).

Fig. 19. Frequent male courtship displays in Bovidae, demonstrated by the example of Thomson's gazelle. a. Neck stretch. b. Head-and-neck-forward/upward posture. c. Nose-up movement with foreleg kick.

and greater kudu) or walking in lateral escort with the female.

Also, in those species that show special courtship displays in the defined sense, there are hardly any that do not occur as a threat or dominance display in other artiodactyl species. The difference is that in certain species, these displays are almost exclusively used by males toward females. As is true of dominance displays, this is most pronounced in the Bovidae.

The head-and-neck-stretched-forward posture (Fig. 19a, 20c, 21a,b) or—as an abbreviation of this precise, but very long term—the neck stretch (=*Kopf-Hals-Vorstrecken:* Walther, 1964a, 1968a; *Überstrecken:* Walther, 1958; low stretch; Geist, 1971) is one of the commonest attitudes of courting artiodactyl males. In wild boar and warthog, the male tries to keep his nose in contact with the female's genitals. This is accompanied by rhythmical sounds resembling the starting of an outboard motor (Frädrich, 1967) and by rhythmical tongue flipping from the slightly opened mouth. Finally, the male warthog may put his chin on the female's hindquarters (precopulatory posture: Simpson, 1964). In Cervidae, the males approach the females from behind with the neck-stretch posture and tongue flicking in order to touch the females' genitals (naso-genital testing: Müller-Using and Schloeth, 1967; Geist, 1963, 1966b).

In the Bovidae, the neck-stretch is also very common in courtship, but with certain species-specific differences. Probably in all Bovidae species, the male may occasionally show such a posture when touching, licking, or sniffing a female's vulva. In oryx, roan, sable antelope, and Grant's gazelle, this movement is only used to reach the female's genitals. It is neither an obligatory component of the courtship ritual nor a special display. In sheep and goats, nilgai, hartebeest, impala, and mountain goat, the neck-stretch is definitely a display, frequently combined with tongue flicking or with various species-specific additions: in nilgai with vertical erection of the tail; in ibex with flapping the tail over the back (Fig. 21a); in sheep ("low stretch": Geist, 1971) with uttering roaring sounds and rotating the head around its long axis (Fig. 21b); in *Capra* and *Ovis* species and in dorcas gazelle with foreleg kicks (see below).

In *Tragelaphus* species, the neck stretch (*"Überstrecken"*: Walther, 1958), combined with very soft sounds (like "imm—imm—imm") and occasional tongue flicking, is an intention movement for neck fighting (Fig. 20). This is very clear in greater kudu, where a driving male accompanies the female in lateral escort, or may frontally approach and place his neck over her neck from the neck-stretch posture (Walther, 1964b). This ritualized neck fight during courtship is less frequent and pronounced in other *Tragelaphus* species; however, the males frequently rub the sides of their stretched necks with winding movements on the females' hindquarters, and also show a pronounced neck stretch during the lateral escort. This connects this posture to neck fighting,

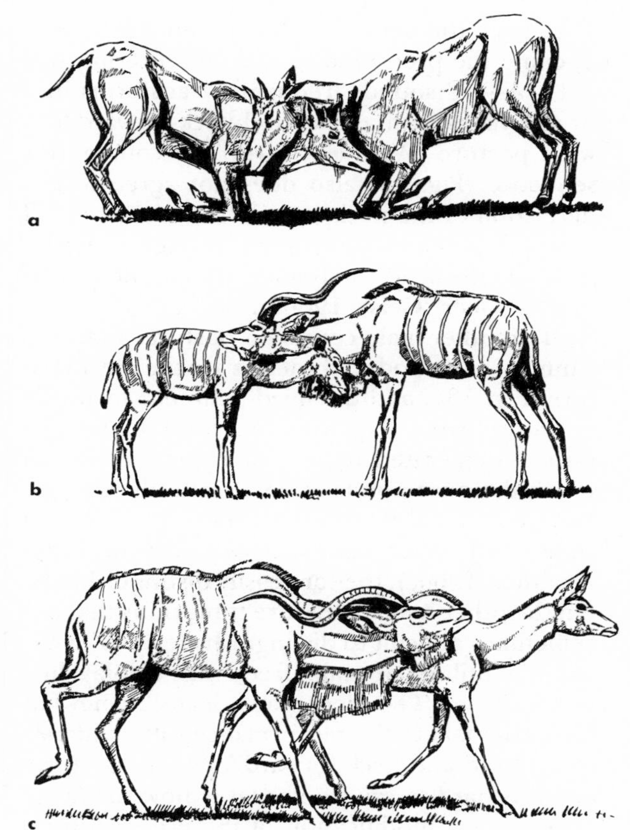

Fig. 20. Ritualization of the neck fight (*Halskampf*) as a courtship display. a. Neck fight in an agonistic encounter between a subadult (left) and an adult (right) bull of nilgai antelope. b. Neck fight (strongly diminished in intensity) between male and female (here, in frontal position) in an early phase of the mating ritual of greater kudu. c. Neck stretch as an intention movement for neck fighting in a greater kudu bull (here, in lateral escort to the female).

and it makes a connection to naso-genital testing unlikely. The relationship between neck stretch and neck fighting is not impossible in other bovids, of course, although it is not as clear as in *Tragelaphus* and *Taurotragus* (Walther, 1960a, 1966a). In *Gazella* species, tongue flicking during the neck stretch is rare. In Thomson's gazelle

(Fig. 19a), the neck stretch in combination with opening the preorbital glands and uttering soft "bl—bl—bl" sounds frequently transforms into or alternates with a head-and-neck-forward/upward posture (Walther, 1964a). Of course, this sequence (Fig. 19) also does not agree with a derivation of the neck stretch from naso-genital testing. A somewhat similar change between lowering the head and bobbing it up is described in mule deer (Geist, 1966b).

Two other male courtship postures found in a number of artiodactyl species are the erect posture (Fig. 18a) and the head-and-neck-forward/upward posture (Fig. 19b). Both can be temporarily exaggerated by a nose-up movement in which the nose points almost vertically upward. In principle, all three are displays in their own right, and in certain species, such as okapi (Walther, 1960c), they are easily distinguishable. On the other hand, there are transitions that are sometimes difficult to distinguish clearly. For example, Antilopinae males frequently display during walking. The neck then leans somewhat forward, making the performance rather similar to a head-and-neck-forward/upward posture, and it is hard to say whether it is now the latter posture or a modification of the erect posture. Perhaps an even more difficult distinction to make is the one between the head-and-neck-forward/upward posture and the nose up; often the latter appears to be nothing but an exaggerated form of the former (Fig. 19c).

The erect posture may occasionally be seen in mating rituals of many artiodactyl species since, being an intention movement for rising onto the hindlegs, it is related to mounting. As a truly elaborate display (one that not only precedes mounting but also is retained during large parts of the mating ritual) it is not so widespread.

Erect posture, head-and-neck-forward/upward postures, and/or nose-up movements are typical in courting males of guanaco (combined with laying back the ears and grunting: Pilters, 1954, 1956), giraffe (Backhaus, 1961), okapi (Walther, 1960c), and Antilopinae species (Walther, 1968a). In erect posture, the males of chamois utter strange grunting sounds (the *"Blädern"* in the terminology of German hunters). The erect posture also appears in certain cervids, such as red deer; however, here, strictly speaking, it probably belongs more to herding than to courtship.

Antilopinae are especially interesting in this respect. For example, in blackbuck, an erect display—apparently closer to the head-and-neck-forward/upward posture than to the erect posture in the strict sense—(with ear dropped and tail flapped over the back) is more or less the same in courtship and in agonistic encounters between male rivals. In Grant's gazelle, the erect posture occurs in courtship and in agonistic encounters; however, there are some differences (Walther, 1965, 1968a). For example, in courtship, the erect posture frequently transforms into a head-and-neck-forward/upward posture, and both are displayed almost continuously (i.e., easily for quarter of an hour to one hour and longer); whereas in agonistic encounters, the erect posture does not transform into a head-and-neck-forward/upward posture, and it usually precedes or follows head flagging (i.e., it is held only for seconds). In Thomson's gazelle, it is difficult to speak about an erect posture in the strict sense; however, the head-and-neck-forward/upward display is very common (it is very briefly held in this species and is more a movement than a posture) and is restricted to encounters with females. Thus, there is no similar display in the agonistic encounters between males.

Frequently the nose-up movement is combined with foreleg kicks (see below) in the Antilopinae (Fig. 19c). In some other bovid species, the erect displays are combined with exaggerated walking. The most pronounced case of this kind is perhaps the prancing with high lifting of angled forelegs in topi (Walther, 1968b).

Head-sideward inclinations and/or head-sideward turns (frequently in combination with erect postures) of the driving male when standing behind a female may occasionally occur in many artiodactyl species. In some such as eland, they are frequently and even regularly seen during courtship. I am aware of the sideward turn of the head as a truly striking and pronounced courtship display only in pronghorn (Fig. 22b), where males show this behavior when approaching a female and when following her. Possibly this behavior is linked to the black patch on the cheek and/or the presence of a cheek patch gland in this species.

Another typical male courtship display is kicking (Fig. 21) with the forelegs (*Laufschlag* or *Laufeinschlag:* Walther, 1958), which is found in okapi and a large number of Bovidae species (Walther, 1960c). With respect to a possible origin of courtship displays in ancestral fighting techniques in combination with and in correlation to the latter's replacement by "modern" fighting techniques, it is certainly of interest that the foreleg kick as a courtship display is not found in those groups of recent artiodactyles that often fight with their forelegs, i.e., the Tylopoda and, above all, the Cervidae. In some cervids, such as fallow deer, white-tailed deer, and axis deer, kicking with the forelegs occurs (apparently without function) in the young during suckling—not for soliciting milk when the young is following behind its mother (which happens in certain bovids) but while the fawn is actually nursing. It is presently unknown whether

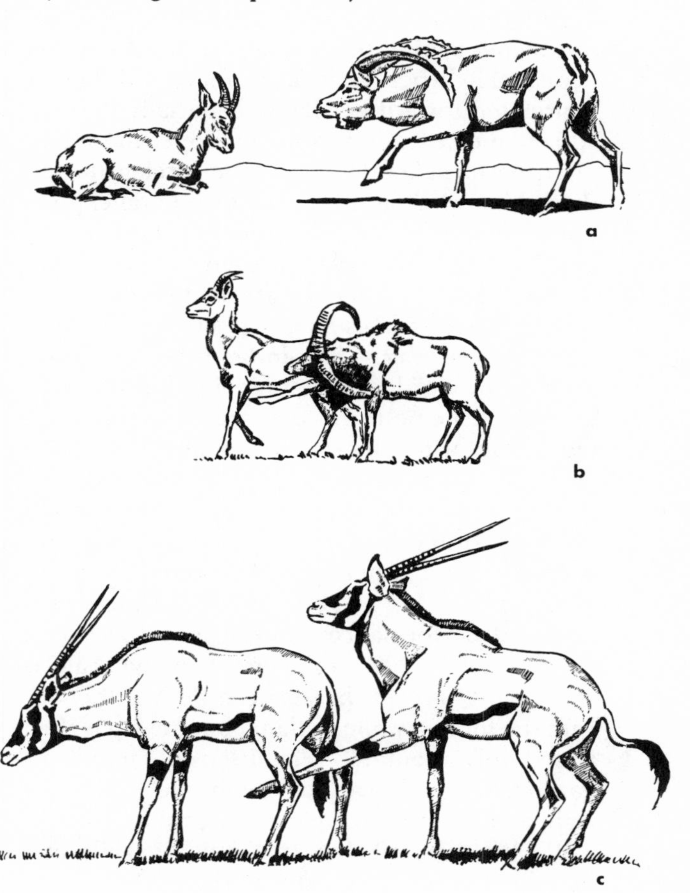

Fig. 21. Several forms of the foreleg kick (*Laufschlag*) and its combination with some other displays. The postures in combination with the foreleg kick are typical for these species (except for the mounting intention in oryx, which is not necessarily combined with the foreleg kick); however, the kick with the foreleg can also be performed in "normal" posture in all of the species. The positions of the females relative to the males and the distances between the sex partners shown in these pictures do not imply species-specific differences.

a. (Knock-kneed) foreleg kick with neck stretch, tongue flipping, and tail flapped over the back in ibex. The (resting) female responds with symbolic butting in this case. b. Foreleg kick with neck stretch and "twist" of the head in Punjab urial (*Ovis orientalis cycloceros*). The male is in lateral escort to the female in this case. c. Foreleg kick (between the female's hindlegs) with head-up (erect) posture and mounting intentions (bending the hind legs, characteristic tail posture) in oryx antelope.

this behavior of young cervids might be related to the foreleg kick in the courtship of certain bovids, and, if so, how to interpret this relationship.

Bovidae groups in which the kick with the foreleg is not found are the bovines, *Tragelaphus* species, *Taurotragus, Boselaphus, Connochaetes, Aepyceros,* and *Ammotragus.* It is also lacking (or only infrequent and weakly pronounced) in *Alcelaphus, Rupicapra,* and some of the *Damaliscus* species. However, it frequently occurs in the majority of Caprinae, Antilopinae, Neotraginae, Hippotraginae, Reduncinae, Cephalophinae, and in *Oreamnos* and *Damaliscus lunatus topi.* It is most frequently seen when the male is standing or walking behind the female, but it also occurs occasionally when he is face to face with her or is oriented toward her flank. Especially in oryx, addax, and roan and sable antelope, the males often perform the kick with the foreleg during the mating whirl-around in reverse-parallel position. Sometimes the male touches the female's hind legs and, occasionally, even her belly, with his foreleg. More often the female is not touched —the male moves his hind leg between her hind legs (*Laufeinschlag* = kick-in-between) in certain species, or he does not raise his foreleg high enough, or he performs the foreleg kick at a distance from which he cannot possibly touch her.

Sloppy "knock-kneed" performances (Fig. 21a) occasionally happen in any species. In scimitar-horned oryx *(Oryx gazella tao),* they are even the rule; otherwise the foreleg is rather stiffly stretched in a "good" kick. The most pronounced performances (raising one foreleg to approximately 90°) are seen when the male is standing. However, he also can deliver the kick with the foreleg while walking, but the leg is then raised only about 45° (Fig. 19c). In Thomson's gazelle, the male can kick alternately with the right and the left foreleg during walking ("drumroll": Walther, 1964a). In Grant's gazelle and blackbuck, the foreleg kick is reduced to a big, stiff-legged step (but the full foreleg kick was recently observed in juvenile males of blackbuck: Benz, 1973). In Soemmering's gazelle only an occasional tripping of the courting male resembles the foreleg kick of other Antilopinae species.

In all bovid species under discussion, the foreleg kick can be delivered when the male is standing or walking in a "normal" or an erected attitude. In the *Hippotragus* and *Kobus* species, this is the rule. In some species, such as topi and oryx antelope (Fig. 21c), it can be combined with mounting intentions (bending the hind legs). In Caprinae (Fig. 21a, b) and in dorcas gazelle, it is frequently combined with the neck stretch. In many Antilopinae species, there is a strong tendency to combine it with a nose-up movement. In particular, the foreleg kick is used to make the female continue after she has stopped walking during the mating march or the mating whirl-around. Linked to this, it can also be used as a "last inquiry" before mounting: when the female does not react (by walking ahead) to the foreleg kick, she is ready to accept and tolerate the male's mounting.

As mentioned above, the kick with the foreleg is often combined with the neck stretch in sheep. The two behavior patterns are also frequently used in this combination as well as independently between males in sheep (Geist, 1971). This is the only presently known case in which behavior patterns primarily or exclusively serving as courtship displays in most bovid species play an important role in agonistic encounters. Here, they obviously serve the same function as dominance displays since they typically are shown by dominant males in encounters with inferior ones.

Visual courtship displays appear to be important in artiodactyles, but other types of stimuli play significant roles too. In Suidae, acoustical displays are perhaps as important or even more important than the visual ones; playing back the tape-recorded mating grunts of a male can release "immobilized" standing of a female in heat (Signoret and du Mesnil, 1960). The tylopods are also rather noisy during courtship. In most Ruminantia, however, either courtship is silent or the calls of the sex partners are soft and apparently occur only in combination with and in addition to pronounced visual displays. Scents are probably emitted by glands opened during courtship (preorbital glands, possibly also subauricular glands, and interdigital glands in connection with the foreleg kick), again in addition to the visual displays. Tactile stimulation of the female by licking and touching her, especially her genitals and hindquarters, with the tongue, mouth, chin, or neck is widespread. For example, in Suidae it apparently plays a major role; and in Cervidae, at least, the licking of the female is very pronounced. In Bovidae, there is a whole range —from species where courting males frequently and intensively touch females (for example, *Tragelaphus* with its neck fight and related behavior) to species that hardly touch the female at all during the mating ritual (for example, Grant's and Thomson's gazelle), except, of course, in mounting, but even this is restricted more or less to contact of the genitals in Antilopinae and Neotraginae. In short, the importance of tactile stimulation during courtship varies widely in artiodactyl species, whereas visual displays are found in almost all of them.

In view of the relevance of visual courtship displays, it is surprising that the list of the most important ones is so short: neck stretch, head-and-neck-forward/upward posture, erect posture, nose-up movement, head-sideward turn, and foreleg kick. Obviously, the species-specific character of courtship rituals in artiodactyles is not demonstrated by a multitude of different displays but by differences in frequency of single displays (ranging down to the absence of certain displays in certain species), by differences in the elaboration, and by combinations within these relatively few displays. The degree of specialization achieved in this simple way is astonishing. Of course, the species-specific differences become more pronounced when one takes into account additional features, such as ear and tail movements and tongue flicking. However, it is possible to characterize the courtship behavior of many species simply by using the postures and movements listed above. For example, in eight Antilopinae species investigated (Table 4), there are no two with courtship behavior that is completely alike, although the basic components are only neck stretch, erect posture, head-and-neck-forward/upward posture (and/or nose up), and foreleg kick. In Table 4, horn threat and sideward turn of the head are added for completion, although the courtship behavior of a single species can be distinguished without them.

Furthermore, as can be seen from Table 4, two or three (sometimes even four or five) of these displays usually occur in the courtship of a single species. This is valid for many but not all artiodactyles, for example, courting bontebok males show only one of them, the neck stretch (David, 1973).

Besides these male displays that apparently (mildly) intimidate (or sometimes also challenge) the female, appeasing behavior plays a certain role in courtship of some artiodactyl species. Appeasement is mainly achieved by licking the partner's head, neck, or shoulders (Fig. 22a), and the males are at least as active, if not more active, in this regard as the females. Licking the female's genitals or her croup and touching the female's croup with the chin by the male may contribute to the female's sexual arousal. The majority of displays in courtship of artiodactyles, however, are either clearly aggressive behavior or have possibly/probably originated from such behavior. In any case, the small proportion of genuine sexual displays in the mating rituals of these animals is surprising. On the whole, here, the relationship of sexual drive to courtship dis-

Table 4

Distribution of (male) courtship displays in the Antilopinae species investigated

	a	b	c	d	e	f	g	h
Gazella gazella	+1)	?		++		++		
Gazella dorcas	?	++	++	++		+	?	
Gazella thomsoni	–	++	–	++	++	++	?	
Gazella subgutterosa		++		+		+	++	
Gazella granti	–	?	–	–	++	++	++	(+)2)
Gazella soemmeringi		+	–	?3)	–	?	++	
Litocranius walleri	–	(+)4)	–	++	++	++	+	?
Antilope cervicapra	(+)5)		–	–	+	++	?	

a = posture or movement similar to medial presentation of horns or butting, b = neck stretch, c = neck stretch + foreleg kick, d = foreleg kick in "normal" or moderately erect posture, e = nose up + foreleg kick (including drumroll and big step), f = nose-up movement and/or head-and-neck-forward/upward posture, g = erect posture (in the strict sense), h = sideward turn of the head, ++ = pronounced and frequent, + = pronounced and moderately frequent, (+) = aberrant performance or situation, ? = not well pronounced and/or questionable whether a special display or accidental, blank = not observed, but possibly exceptionally and/or accidentally occurring, – = not observed, probably lacking, 1) = gentle touching of the female with the horns, 2) = head flag (belongs more to herding than to courtship in the strict sense, occasionally also during courtship), 3) = inconspicuous tripping steps, 4) = occasionally when marking the female with preorbital glands, 5) = threatening and beating the female with the horns in herding (not during courtship in the strict sense).

plays is analogous to that between the French nation and the Foreign Legion: The legionaries served and fought for France, but most of them were not French. Correspondingly, the courtship displays occur in the service of sexual drive, but most of them are not sexual behavior.

Submissive and Appeasement Behavior

Most submissive displays in artiodactyles are in every way the antithesis of dominance and offensive threat displays. Sometimes, there are connections with defensive threats. Submissive displays indicate the acceptance of an inferior role. In a sense, they anticipate defeat and lack features that could possibly challenge an opponent and release its aggression. The effect on a (superior) recipient may range from a diminution to complete cessation of aggression. Submissive displays and appeasement behavior enable an inferior animal to remain with a group and/or in a familiar terrain despite the presence of superior and aggressive conspecifics. This is very important for females in the courtship rituals; and for juvenile animals, it may often be essential to their survival.

According to Frädrich (1967), submissive gestures are unknown or dubious in Suidae and

Fig. 22. a. Licking to appease the female (left) by the courting male (right) in okapi (*Okapia johnstoni*). b. Head-sideward turn (alternately to the right and the left) in the courting male pronghorn.

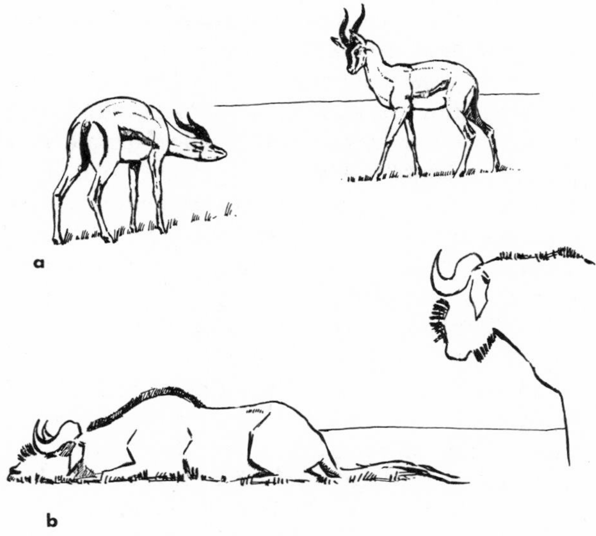

Fig. 23. Submissive behavior. a. Head-foreward-downward posture in a male dorcas gazelle (*Gazella dorcas*) in response to the threat of a superior opponent (right). b. Lying down with the head and neck on the ground and with the hindquarters toward the superior in black wildebeest (*Connochaetes gnou*).

Hippopotamidae. They appear in the other groups of artiodactyles; however, their intensity, frequency, and importance vary considerably with the species. The commonest forms that can occur singly or in combination are lowering the head (Fig. 23a), turning 180°, and lying down (Fig. 23b). Following one after the other, they may express increasing degrees of submission.

In agonistic encounters, turning 180° (with the hindquarters toward an opponent) is usually an intention movement for withdrawal or flight. One can consider it a special display only when it is combined with other features of submission and/or when the animal does not withdraw or flee but remains in the superior opponent's vicinity. Since females are oriented with their hindquarters toward the males in sexual encounters, one might also think that an inferior male might mimic a female through this behavior in an agonistic encounter. However, the same submissive behavior can also be seen in encounters between females. It is unlikely for a (subordinate) female to "mimic" a female—which she is, after all. With respect to the 180° turn and other submissive behavior patterns, the opposite is more probable, i.e., that they are behavior patterns of inferiority by origin and nature and that they may also appear on the part of a female in sexual encounters because she is inferior to (adult) males in most of these species, as discussed above.

Lying down with the head and neck stretched forward on the ground, often with the hindquarters toward the opponent, apparently occurs considerably more frequently in captivity than in the wild. In free-ranging artiodactyles one may see it most frequently when females presumably not yet in heat are sexually approached by males. In some species, such as black wildebeest, lying down in submission is sometimes accompanied by uttering sounds frequently heard from calves (Walther, 1966b). This and the resemblance to the infantile lying-out (lying in seclusion) behavior of certain artiodactyl species make it tempting to assume that an animal lying down in submission may be mimicking a baby and that this infantile behavior may stop the superior's aggression (Burckhardt, 1958b).

However, there are several objections to this hypothesis. It is obvious that most artiodactyles can distinguish an adult animal from an infant (by visual appearance and scent) regardless of its posture and position. Thus, it is unlikely an adult animal can assume a baby's identity simply by lying down. It is also doubtful that certain vocalizations in artiodactyles are so typical of juveniles that they definitely identify an animal as an infant. The sounds uttered in connection with submissive behavior are often distress cries. These, of course, are relatively frequently heard from juveniles, but they are neither genuine infantile vocalizations nor restricted to juveniles. Moreover, submissive lying down as well as other forms of submission are by no means necessarily combined with sounds. On the contrary, silent submission is very common in Artiodactyla. Finally, submissive lying down is a behavioral response to a threat by very superior conspecifics, who, very frequently, are adult males. In most artiodactyles, however, adult males have hardly any contact with the young and pay little attention to them. Thus, it remains obscure why such a male should react to infantile behavior. Finally submissive lying down is also shown by species, such as black wildebeest, whose young do not lie out. In short, it is more probable that lying down with the head and neck stretched forward functions as a submissive behavior simply because it is the perfect antithesis to the self-exposure and emphasizing of an animal's presence in dominance and offensive threat displays, and because the inferior animal blends into the ground and, in a sense, leaves little for its opponent to attack.

Lowering the head ranges from performances identical to the head-low posture described as a defensive threat to postures in which the animal stretches its head and neck more forward-downward (e.g., dorcas and mountain gazelle), or turns its horns somewhat away from the challenger (e.g., oryx antelope), and/or holds its neck in a rather strikingly curved fashion (e.g., guanaco: Pilters, 1956).

The back often appears slightly humped. As in the head-low posture, submissive lowering of the head may sometimes change into grazing. It is often shown during withdrawal and sometimes during flight. The most pronounced and even exaggerated performances, however, can be seen when an animal does not withdraw but remains close to its superior partner. In this case, the lowering of the head can be combined with a 180° turn, but it may also be displayed in any other orientation relative to the superior, including frontal orientation.

Besides these submissive displays, there is also appeasing by licking in a number of artiodactyl species. Typically, the inferior animal grooms the superior one, especially its head, sometimes also the neck and shoulders, as Schloeth (1961) has pointed out in an excellent study on the behavior of Camargue cattle. Of course, this behavior only occurs when the hostilities are not too severe.

The possibility, or even probability, of a relationship between submissive displays and threat behavior, as well as the occasional switch by an inferior animal from a submissive behavior to a threat display and vice versa, have been mentioned several times in the literature (Lorenz, 1935; Tinbergen, 1959). The comparative study of artiodactyl behavior may possibly provide further insight into this mechanism. As mentioned above, the head-low posture occurs as a defensive threat display in certain bovids and cervids. When an animal challenged by another's offensive threat or dominance display (high presentation of horns, erect posture, etc.) shows a defensive head-low threat, this means that it does not ("dare") respond to the challenge in an equivalent way and is not ready to attack its opponent, but is only ready to defend itself if attacked. Hence, a defensive threat when used in response to a challenger's offensive threat or dominance display, comes very close to a behavior denoting inferiority (Fig. 8). In the particular case of the head-low posture, furthermore, the animal has only to stretch its lowered head and neck somewhat more forward and/or turn the horns away from the opponent to assume a more pronounced submissive attitude. Thus, this difference is only marginal. It may be mentioned here that in the behavioral inventory of a species there are also other relationships between threat and submissive displays, which apparently depend on the position of certain displays. For example, in hartebeest and topi, instead of a head-low posture, an attitude very similar to medial presentation of horns serves as an expression of submission; and in fallow deer, submission is expressed by fast, repeated snapping movements of the mouth in combination with lowering the head or lying down. Finally, there are cases of mosaiclike combinations of threat and submissive displays—as in humans, when an inferior individual may withdraw and/or bow when charged by a feared superior, but may curse and clench the fists at the same time. For example, in okapi (Walther, 1962), submissive lying down may occasionally be combined with vertical erecting of the head and neck, which is a (defensive) threat behavior (related to neck fighting) in this species.

In conclusion, I wish once more to emphasize the prevailing role of agonistic displays and of aggressively tinted behavior patterns in the communication and social life of Artiodactyla. If it were possible to eliminate the intraspecific aggression from the behavioral repertoires of these animals, their communication as well as their social organization would break down.

References

Antonius, O., 1939. Über Symbolhandlungen und Verwandtes bei Säugetieren. *Z. Tierpsychol.,* 3:263–78.

Backhaus, D., 1961. *Beobachtungen an Giraffen in Zoologischen Gärten und in freier Wildbahn.* Brussels: Inst. Parc. Nat. Congo et Ruanda Urundi.

Benz, M., 1973. Zum Sozialverhalten der Sasin (Hirschziegenantilope, *Antilope cervicapra* L. 1758). *Zoolog. Beitr.,* 19:403–66.

Bigalke, R. C., 1972. Observations on the behaviour and feeding habits of the springbok *(Antidorcas marsupialis). Zoolog. Afric.,* 7:333–59.

Buechner, H. K., 1961. Territorial behavior in Uganda kob. *Science,* 133:698–99.

Buechner, H. K., and Schloeth, R., 1965. Ceremonial mating behavior in Uganda kob (*Adenota kob thomasi* Neumann). *Z. Tierpsychol.,* 22:209–25.

Burckhardt, D., 1958a. Observations sur la vie sociale du cerf *(Cervus elaphus)* au Parc National Suisse. *Mammalia,* 22:226–44.

Burckhardt, D., 1958b. Kindliches Verhalten als Ausdrucksbewegung im Fortpflanzungszeremoniell einiger Wiederkäuer. *Rev. Suisse Zoolog.,* 65:311–16.

Carpenter, C. R., 1942. Societies of monkeys and apes. *Biol. Symposia,* 8:177–204.

Darling, F. F., 1937. *A Herd of Red Deer.* London: Oxford University Press.

David, J. H. M., 1973. The behavior of the bontebok, *Damaliscus dorcas dorcas* (Pallas 1766), with special reference to territorial behaviour. *Z. Tierpsychol.,* 33:38–107.

Eibl-Eibesfeldt, I., 1957. Die Ausdrucksformen der Säugetiere. *Handb. Zoolog.,* 8, 10(13):1–88.

Eibl-Eibesfeldt, I., 1967. *Grundriss der vergleichenden Verhaltensforschung.* Munich: R. Piper & Co.

Espmark, Y., 1964. Rutting behaviour in reindeer *(Rangifer tarandus). Anim. Behav.,* 12:420–26.

Estes, R. D., 1967. The comparative behavior of Grant's and Thomson's gazelles. *J. Mammal.,* 48:189–209.

Estes, R. D., 1969. Territorial behavior of the wilde-

beest (*Connochaetes taurinus* Burchell, 1823). *Z. Tierpsychol.*, 26:284–370.
Ewer, R. F., 1968. *Ethology of Mammals.* London: Logos Press.
Frädrich, H., 1964. Beobachtungen zur Kreuzung zwischen Schwarzrückenducker, *Cephalophus dorsalis* Gray, 1846, und Zebraducker, *Cephalophus zebra* Gray, 1838. *Z. Säugetierk.*, 29:46–51.
Frädrich, H., 1965. Zur Biologie und Ethologie des Warzenschweines (*Phacochoerus aethiopicus* Pallas), unter Berücksichtigung des Verhaltens anderer Suidae. *Z. Tierpsychol.*, 22:328–93.
Frädrich, H., 1967. Das Verhalten der Schweine (Suidae, Tayassuidae) und Flusspferde (Hippopotamidae). *Handb. Zoolog.*, 8, 10(26):1–44.
Fraser, A. F., 1957. The state of fight or flight in the bull. *Brit. J. Anim. Behav.*, 5:48–49.
Freye, H. A., and Geissler, H., 1966. Das Ohrenspiel der Ungulaten als Ausdrucksform. *Wiss. Z. Univ. Halle,* 5:893–915.
Geist, V., 1963. On the behavior of the North American moose (*Alces alces andersoni* Peterson 1950) in British Columbia. *Behav.*, 20:377–416.
Geist, V., 1965. On the rutting behavior of the mountain goat. *J. Mammal.*, 45:551–68.
Geist, V., 1966a. The evolution of hornlike organs. *Behav.*, 27:175–214.
Geist, V., 1966b. Ethological observations on some North American cervids. *Zoolog. Beitr.*, 12:219–50.
Geist, V., 1968. On the interrelation of external appearance, social behavior and social structure of mountain sheep. *Z. Tierpsychol.*, 25:199–215.
Geist, V., 1971. *Mountain Sheep.* Chicago: University of Chicago Press.
Gilbert, B. K., 1968. Development of social behavior in the fallow deer *(Dama dama). Z. Tierpsychol.*, 25:867–76.
Gosling, L. M., 1974. The social organization of Coke's hartebeest *(Alcelaphus buselaphus cokei).* In: *The Behaviour of Ungulates and its Relation to Management.* V. Geist and F. R. Walther, eds. Morges, Switzerland: IUCN publications, no. 24, vol. 1:488–511.
Grzimek, B., 1949. Die "Radfahrer-Reaktion." *Z. Tierpsychol.*, 6:41–44.
Haas, G., 1959. Untersuchungen über angeborene Verhaltensweisen beim Mähnenspringer (*Ammodorcas lervia* Pallas). *Z. Tierpsychol.*, 16:219–42.
Hainard, R., 1949. *Les mammifères sauvages d'Europe* Paris: Délachaux & Niestlé.
Haltenorth, T., 1963. Klassifikation der Säugetiere: Artiodactyla. *Handb. Zoolog.*, 8, 10(26):1–167.
Hediger, H., 1946. Zur psychologischen Bedeutung des Hirschgeweihs. *Verh. Schweiz. Naturf. Ges. Zürich,* 1946:162–63.
Hediger, H., 1949. Säugetierterritorien und ihre Markierung. *Bijdr. Dierk.*, 28:172–84.
Hediger, H., 1954. *Skizzen zu einer Tierpsychologie im Zoo und im Zirkus.* Stuttgart: Europa.
Hennig, R., 1962. Über das Revierverhalten der Rehböcke. *Z. Jagdwiss.*, 8:61–81.
Hinde, R. A., 1966. *Animal Behaviour.* London: McGraw-Hill.
Huth, H. H., 1970. Zum Verhalten der Rappenantilope (*Hippotragus niger* Harris 1838). *D. Zoolog. Gart.*, 38:147–70.
Joubert, S. C. J., 1970. A study of the social behaviour of the roan antelope (*Hippotragus equinus equinus* Desmarest, 1804) in the Kruger National Park. Masters thesis, University of Pretoria.
Joubert, S. C. J., 1972. Territorial behaviour of the tsessebe (*Damaliscus lunatus lunatus* Burchell) in the Kruger National Park. *Zoolog. Afric.*, 7:141–56.
Kakies, M., 1936. *Elche zwischen Meer und Memel.* Giessen: Brühl.
Kiley, M., 1972. The vocalizations of ungulates, their causation and function. *Z. Tierpsychol.*, 31:171–222.
Lent, P. C., 1965. Rutting behavior in a barren-ground caribou population. *Anim. Behav.*, 13:259–64.
Leuthold, W., 1966. Variations in the territorial behaviour of Uganda kob (*Adenota kob thomasi,* Neumann 1896). *Behav.*, 27:214–57.
Leyhausen, P., 1967. Biologie von Ausdruck und Eindruck. *Psychol. Forsch.*, 31:113–227.
Lind, H., 1959. The activation of an instinct caused by a "transitional action." *Behav.*, 14:123–35.
Lorenz, K., 1935. Der Kumpan in der Umwelt des Vogels. *J. Ornith.*, 83:137–413.
Lorenz, K., 1963. *Das sogenannte Böse.* Vienna: Borotha-Schoeler.
Moynihan, M., 1955. Some aspects of reproductive behaviour in the black-headed gull (*Larus ridibundus* L.) and related species. *Behav.*, Suppl. 4:1–201.
Müller-Schwarze, D., 1967. Social odours in young mule deer. *Am. Zoolog.*, 7:807.
Müller-Schwarze, D., 1969. Complexity and relative specificity in a mammalian pheromone. *Nature* (London), 223:525–26.
Müller-Schwarze, D., 1971. Pheromones in black-tailed deer *(Odocoileus hemionus columbianus). Anim. Behav.*, 19:141–52.
Müller-Using, D., and Schloeth, R., 1967. Das Verhalten der Hirsche. *Handb. Zoolog.*, 8, 10(28):1–60.

Pilters, H., 1954. Untersuchungen über angeborene Verhaltensweisen bei Tylopoden, unter besonderer Berücksichtigung der neuweltlichen Formen. *Z. Tierpsychol.,* 11:213–303.

Pilters, H., 1956. Das Verhalten der Tylopoden. *Handb. Zoolog.,* 8, 10(27):1–24.

Pruitt, W. O., 1960. Behavior of the barren-ground caribou. *Biol. Pap. Univ. Alaska,* 3:1–43.

Schaller, G. B., 1967. *The Deer and the Tiger.* Chicago: University of Chicago Press.

Schenkel, R., 1966. On sociology and behaviour in impala (*Aepyceros melampus suara* Matschie). *Z. Säugetierk.,* 31:177–205.

Schloeth, R., 1958. Über die Mutter-Kind-Beziehungen des halbwilden Camargue-Rindes. *Säugetierk. Mitt.,* 6:145–50.

Schloeth, R., 1961. Das Sozialleben des Camargue-Rindes. *Z. Tierpsychol.,* 18:574–627.

Schneider, K. M., 1930. Das Flehmen (I. Teil). *D. Zoolog. Gart.,* 3:183–98.

Schneider, K. M., 1931. Das Flehmen (II. Teil). *D. Zoolog. Gart.,* 4:349–64.

Schneider, K. M., 1934. Das Flehmen (V. Teil). *D. Zoolog. Gart.,* 7:182–201.

Schweinsburg, R. E., and Sowls, L. K., 1972. Aggressive behavior and related phenomena in the collard peccary. *Z. Tierpsychol.,* 30:132–45.

Signoret, J. P., and du Mesnil, F., 1960. Rôle d'un signal acoustique de verrant dans le comportement reactionnel de la truie en oestrus. *C. rendues séances Acad. sci.,* 250:1355–57.

Simpson, C. D., 1964. Observations on courtship behaviour in warthog (*Phacochoerus aethiopicus* Pallas). *Arnoldia,* 1:1–4.

Snethlage, K., 1957. *Das Schwarzwild.* Hamburg: Paul Parey, p.3.

Talbot, L. M., and Talbot, M. H., 1963. The wildebeest in western Massailand, Tanganyika. *Wildl. Monogr.* 12.

Tembrock, G., 1959. *Tierstimmen.* Wittenberg: A. Ziemsen.

Tembrock, G., 1963. *Grundlagen der Tierpsychologie.* Berlin: Akademie.

Tembrock, G., 1964. *Verhaltensforschung.* Jena: Gustav Fischer, p.2.

Tembrock, G., 1965. Untersuchungen zur intraspezifischen Variabilität von Lautäusserungen bei Säugetieren. *Z. Säugetierk.,* 30:257–73.

Tembrock, G., 1968. Artiodactyla. In: *Animal Communication,* T. A. Sebeok, ed. Bloomington: Indiana University Press, pp.383–404.

Thenius, E., and Hofer, H., 1960. *Stammesgeschichte der Säugetiere.* Berlin: Springer.

Thomas, J. W.; Robinson, R. M.; and Marburger, R. G.; 1965. Social behavior in a white-tailed deer herd containing hypogonadal males. *J. Mammal.,* 43:462–69.

Tinbergen, N., 1940. Die Übersprungbewegung. *Z. Tierpsychol.,* 4:1–40.

Tinbergen, N., 1951. *The Study of Instinct.* London: Oxford University Press.

Tinbergen, N., 1952. "Derived" activities, their causation, biological significance and emancipation during evolution. *Rev. Biol.,* 27:1–32.

Tinbergen, N., 1959. Einige Gedanken über "Beschwichtigungsgebärden." *Z. Tierpsychol.,* 16:651–65.

Uexküll, J. von, 1921. *Umwelt und Innenwelt der Tiere.* Berlin.

Verheyen, R., 1954. *Monographie éthologique de l' Hippopotame (Hippopotamus amphibius L.).* Brussels: Inst. Parc. Nat. Congo et Ruanda Urundi.

Walther, F. R., 1958. Zum Kampf- und Paarungsverhalten einiger Antilopen. *Z. Tierpsychol.,* 15:340–80.

Walther, F. R., 1959. Beobachtungen zum Sozialverhalten der Sasin (Hirschziegenantilope, *Antilope cervicapra* L.). *Jahrb. G. v. Opel-Freig.,* 2:64–78.

Walther, F. R., 1960a. Entwicklungszüge im Kampf- und Paarungsverhalten der Horntiere. *Jahrb. G. v. Opel-Freig.,* 3:90–115.

Walther, F. R., 1960b. Einige Verhaltensbeobachtungen am Bergwild des Georg von Opel-Freigeheges. *Jahrb. G. v. Opel-Freig.,* 3:53–89.

Walther, F. R., 1960c. "Antilopenhafte" Verhaltensweisen im Paarungszeremoniell des Okapi (*Okapia johnstoni* Sclater, 1901). *Z. Tierpsychol.,* 17:188–210.

Walther, F. R., 1962. Über ein Spiel bei *Okapia johnstoni. Z. Säugetierk.,* 27:245–51.

Walther, F. R., 1964a. Einige Verhaltensbeobachtungen an Thomsongazellen (*Gazella thomsoni* Günther, 1884) im Ngorongoro-Krater. *Z. Tierpsychol.,* 21:871–90.

Walther, F. R., 1964b. Verhaltensstudien an der Gattung *Tragelaphus* De Blainville, 1816, in Gefangenschaft, unter besonderer Berücksichtigung des Sozialverhaltens. *Z. Tierpsychol.,* 21:393–467.

Walther, F. R., 1965. Verhaltensstudien an der Grantgazelle (*Gazella granti* Brooke, 1872) im Ngorongoro Krater. *Z. Tierpsychol.,* 22:167–208.

Walther, F. R., 1966a. *Mit Horn und Huf.* Berlin: Paul Parey.

Walther, F. R., 1966b. Zum Liegeverhalten des

Weissschwanzgnus (*Connochaetes gnou* Zimmermann, 1780). *Z. Säugetierk.* 31:1–16.

Walther, F. R., 1968a. *Verhalten der Gazellen.* Wittenberg: A. Ziemsen.

Walther, F. R., 1968b. Kuhantilopen, Pferdeböcke und Wasserböcke. In: *Grzimeks Tierleben,* B. Grzimek, ed. Zurich: Kindler, 13:437–71.

Walther, F. R., 1969a. Ethologische Beobachtungen bei der künstlichen Aufzucht eines Blessbockkalbes (*Damaliscus dorcas philippsi* Harper, 1939). *D. Zoolog. Gart.,* 36:191–215.

Walther, F. R., 1969b. Flight behaviour and avoidance of predators in Thomson's gazelle (*Gazella thomsoni* Günther 1884). *Behav.,* 34:184–221.

Walther, F. R., 1972. Social grouping in Grant's gazelle (*Gazella granti* Brooke 1872) in the Serengeti National Park. *Z. Tierpsychol.,* 31:348–403.

Walther, F. R., 1974. Some reflections on expressive behaviour in combats and courtship of certain horned ungulates. In: *The Behaviour of Ungulates and its Relation to Management:* V. Geist, and F. R. Walther, eds. Morges, Switzerland: IUCN publications, no. 24, vol. 1:56–106.

Chapter 27

COMMUNICATION IN PERISSODACTYLA

Hans Klingel

The order consists of three families with fifteen species in six genera: Equidae, one genus, six spp.; Rhinocerotidae, four genera, five spp.; Tapiridae, one genus, four spp. The equids are inhabitants mainly of open grasslands, and they live socially in large or small groups. The tapirs are solitary animals that live in dense forests. The different species of rhinoceroses occur in habitats ranging from rain forest to grasslands; they are solitary or live in small groups.

There is evidence that the degree of complexity of communication systems is correlated with the respective biological requirements, i.e., life habits and habitat. However, information now available is not sufficient to allow final statements.

The various modes of communication are dealt with by families. Tactile communication is disregarded in this context because of the difficulties in assessing the communicative significance of bodily contacts.

Equidae

Two distinct forms of social organization have evolved in the equids. Plains zebra (*Equus quagga:* Klingel, 1967), mountain zebra (*E. zebra:* Klingel, 1968, 1969a; Joubert, 1972), and horse (*E. przewalskii:* Bruemmer, 1967; Feist, 1971; Tyler, 1972) live in family groups consisting of one stallion, one or several mares, and their young, and in stallion groups. Both types of group are nonterritorial; they are characterized by their stability and coherence based on mutual bonds between the group members. This requires the capability of individual recognition of group members.

Grevy's zebra (*E. grevyi:* Klingel, 1969b, 1974), African wild ass (*E. africanus:* Klingel, in prep.), feral donkeys (Moehlman, 1974), and probably Asiatic wild ass (*E. hemionus:* Klingel, in prep.) live singly or in unstable anonymous groups of either one or both sexes. Some of the stallions are territorial and defend their huge territories only under certain conditions and only against their territorial neighbors, but are otherwise tolerant toward other male conspecifics. Only mares and foals know and recognize each other individually, and they are the only members of a population that establish mutual bonds that are maintained for some time.

SPECIES AND INDIVIDUAL RECOGNITION

In overlap areas, e.g., South West Africa *(E. quagga* and *E. zebra)* and in northern Kenya *(E. quagga* and *E. grevyi)* the members of the respective species tend to stay together in monospecific associations. Although they frequently use the same water holes, graze close together and even flee together, they do not contact individuals of the other species (Klingel, 1968, 1974). This indicates that species-specific differences are recognized. No information is available on how this is performed; it can, however, be assumed that knowledge of their own species' appearance is learned by the young through an imprinting process. On the other hand, in captivity it is possible to interbreed all the equine species. So obviously there are genus-specific characteristics which are recognized and which can act as adequate stimuli in abnormal situations. Individual recognition is established between a mare and her foal during the first few days after birth. According to Tyler (1972) the horse mare learns to know her foal, obviously by imprinting, when she licks it for twenty to thirty minutes after birth.

The foals have a following reaction as soon as they are able to walk. The imprinting of the foals on their mothers takes place after one to several days. It is enhanced by agonistic behavior of the mare toward all the other conspecifics during the first few days after birth (*E. quagga:* Walther, 1962; Klingel and Klingel, 1966a; Klingel, 1969c; *E. grevyi:* Klingel, 1972, 1974; *E. przewalskii f. caballus:* Tyler, 1972). In the nonterritorial equids the group members know each other individually by sight, voice, and smell (Klingel, 1967, 1968, 1969a, 1972; Tyler, 1972).

OPTICAL COMMUNICATION

The visual expressions of the equids are numerous. They consist of postures; movements of legs, tail, and head; and facial expressions effected by opening the mouth and by various positions and movements of the ears and lips.

Facial Expressions

The estrous mare, when approached and contacted by the stallion, shows the estrous face: she makes exaggerated chewing movements but without closing her mouth altogether; at the same time her incisors are partly bared, the corners of her mouth are pulled back, and her ears are folded back (Antonius, 1940; Trumler, 1958, 1959a; Klingel, 1972). This expression is regularly observed in all the species with the exception of the horse, where it occurs only rarely and only in young mares (Antonius, 1951) (Fig. 1).

A similar expression is the snapping (Zeeb, 1959) of young horses when they are threatened (Tyler, 1972). In snapping, the jaw movements are faster than in the estrous expression, and the teeth meet each time. Plains zebra, Grevy's zebra foals, and, rarely, adult plains zebra stallions make similar faces during greeting ceremonies with adult stallions; the position of the ears, however, is more variable. The significance of these three facial expressions is difficult to assess. They can all be considered submissive expressions, as they are always performed by the weaker partners. Tyler (1972), however, observed that young horses were frequently attacked and bitten or kicked in spite of having made the snapping response. Zeeb (1959) interprets the snapping as ritualized grooming movements, and Tyler (1972) observed that occasionally mutual grooming ensued from the snapping response (Fig. 2).

Fig. 1. Estrous face (donkey). Note position of ears.

Fig. 2. Snapping response: horse foal; plains zebra two-year-old stallion, with adult stallion.

The greeting expression (Antonius, 1940; Trumler, 1959a; Klingel, 1967, 1968, 1972; Dobroruka, 1961; Tyler, 1972) can be observed, when two adult partners, e.g., two stallions or a stallion and a mare, meet and establish naso-nasal contacts. The representatives of all species that have been investigated extend their heads, usually have their ears directed forward, and draw the corners of their mouths up in a jerking movement, except for the horse, where this last movement rarely occurs (Tyler, 1972). Plains zebra often open their mouths and make chewing movements with bared teeth; mountain zebra and donkey chew with their lips closed; Grevy's zebra and horse do not move their jaws (Klingel, 1972). This expression obviously denotes friendliness (Fig. 3).

Fig. 3. Greeting expression (plains zebra). Note position of ears.

The threat expression (Antonius, 1937; Trumler, 1959a; Klingel, 1967, 1968, 1972; Tyler, 1972; Moehlman, 1974) is similar to the estrus expression, but without the chewing movements. There are several degrees: when threatening lightly, the ears are only slightly laid back; at higher intensities the ears are straight back, the head is lowered and swayed from side to side, and the mouth is partly opened. This expression occurs regularly before and during fights and in stallions when driving mares or other conspecifics. Its communicative significance is much more obvious than that of the previous expressions: it demonstrates the readiness to fight, especially to bite, and conspecifics accordingly move out of the way, flee, or get ready to meet the attack (Fig. 4).

The flehmen expression consists of the animal's lifting its head up high, curling back the upper lip to expose the teeth, and lowering the lower lip; the jaws remain closed (Schneider, 1930, 1934; Klingel, 1972; Trumler, 1959a; Zeeb, 1959; Tyler, 1972). It is usually a reaction to some smell, usually urine, dung, or female genitalia, and facilitates the entrance of odorants into the vomeronasal (Jacobson's) organ (Estes,

Fig. 4. Threat expression: Grevy's zebra ♂; horse ♀ (note position of ears and tail).

1972). In a few instances it could be observed that flehmen can act as a signal and elicit responses by conspecifics (Dobroruka, 1961; Klingel, 1974; Moehlman, 1974) (Fig. 5).

Other facial expressions with doubtful or no communicative significance are lip clapping in plains zebra (Klingel, 1972), yawning in all the species, and extension of the upper lip in feral ass males when approaching females (Moehlman, 1974).

Leg Movements

Equids kick during fights with both front and hind legs. Ritualized kicking serves as a visual threat and is carried out in various degrees. It ranges from only slightly lifting one hind foot from the ground at low intensity to actually kicking with one or both hind feet in the direction of a pursuer at high intensity. Stallions display ritualized kicking with their forelegs during breaks between fighting bouts (Klingel, 1967).

Tail Postures and Movements

The communicative significance of tail postures and/or movements has hardly been investigated. During flight young foals hold their tails straight up, and this releases flight behavior in other foals (*E. przewalskii f. caballus:* Tyler, 1972). Grevy's zebra stallions, when driving mares or other stallions, raise their tails (Fig. 4a); during copulation they swing them from side to side as do horse stallions (Klingel, 1972; Antonius, 1937).

Body Postures and Movements

During estrus mares of all the species stand with their hind legs apart and their tails raised at an angle of about 45°, especially when being courted by a stallion. This posture is most conspicuously displayed by young plains zebra (Klingel, 1967) and by horse mares (Tyler, 1972) even when there is no stallion in attendance. Stallions are attracted by this posture even from a distance, and in plains zebra this results in the abduction of the young mare from her maternal group. It could be proved that the posture and not the smell is the decisive stimulus: drugged plains zebras, under the influence of Etorphin, Acetylpromazine, and Hyoscine displayed the same posture, and they were courted and even mounted by stallions (Klingel, 1967; Klingel and Klingel, 1968). Horse stallions are also attracted by urinating mares, as their posture is similar to the estrous posture (Tyler, 1972).

In the territorial equids the territorial stallions are distinguishable from other males by a number of behavior patterns, which can also be considered an advertisement of their status to conspecifics: their posture is generally upright; they keep some distance from groups of conspecifics in the territory, which makes them very visible; they drive stallions and mares; they mount mares without erection, prior to copulation.

During the last two displays territorial stallions roar. This underlines the possible function of these two displays as signals, although they primarily serve to demonstrate the stallions' dominance to the driven animals or to test the receptiveness of the mares, respectively (Klingel, 1972, 1974, and in prep.).

When migrating, adult plains zebra often move their heads up and down conspicuously. This gives the impression of exaggerated walking and could serve as a following stimulus for group members or other conspecifics (Klingel, 1972).

Fig. 5. Flehmen (horse ♂).

ACOUSTICAL COMMUNICATION

Two vocalizations seem to be more or less identical and to have the same significance in all species. Both are snorts, but they differ in their duration. The short snort is obviously a warning sound, emitted when the animals are disturbed. The long snort is an expression of well-being (e.g., when the animals are eating). Its communicative significance is doubtful.

Contact calls serve for maintaining or reestablishing contact between two partners: mare and foal in all species, group members in the nonterritorial species. In the territorial species the same type of call is used by the stallions to mark their territories as well as in other situations (Klingel, 1972, 1974, in prep.).

In *E. quagga,* the contact call consists of one- to three-syllable barking sounds, which are produced during exhalation and which can be transcribed by "ha, haha, hahaha." The frequency is ten to eighteen sounds in five seconds. During inhalation the animals often produce an "i" sound. These calls are individually different and are assumed to effect individual recognition (Klingel, 1967).

In *E. zebra* the contact call is probably the long squeal, which is repeated several times (Blaine, 1922; Antonius, 1930, 1951; Klingel, 1968).

The contact call of *E. przewalskii f. caballus* is the whinny or neigh, which is individually different (Tyler, 1972; Ödberg, 1969; Kiley, 1972; Tembrock, 1965; Waring, 1971) (Fig. 6a).

In *E. grevyi* the contact call is an intermittent roar, emitted when exhaling; it is succeeded by an "i" sound during inhalation (Antonius, 1951; Klingel, 1969b, 1972, 1974).

The contact call of *E. africanus* is the bray, a most varied and complex vocalization composed of thirty frequency bands. It consists of two syllables and is highly variable in temporal patterning, frequency bands, intensity, and other characteristics (*E. africanus f. asinus:* Moehlman, 1974) (Fig. 7).

The call of *E. hemionus* is described as a monosyllabic "kiang" emitted in fast succession; it is somewhat similar to the whinny of the horse and the bray of the ass (Antonius, 1937).

The contact calls of the various species, which are so different to the human ear, seem to

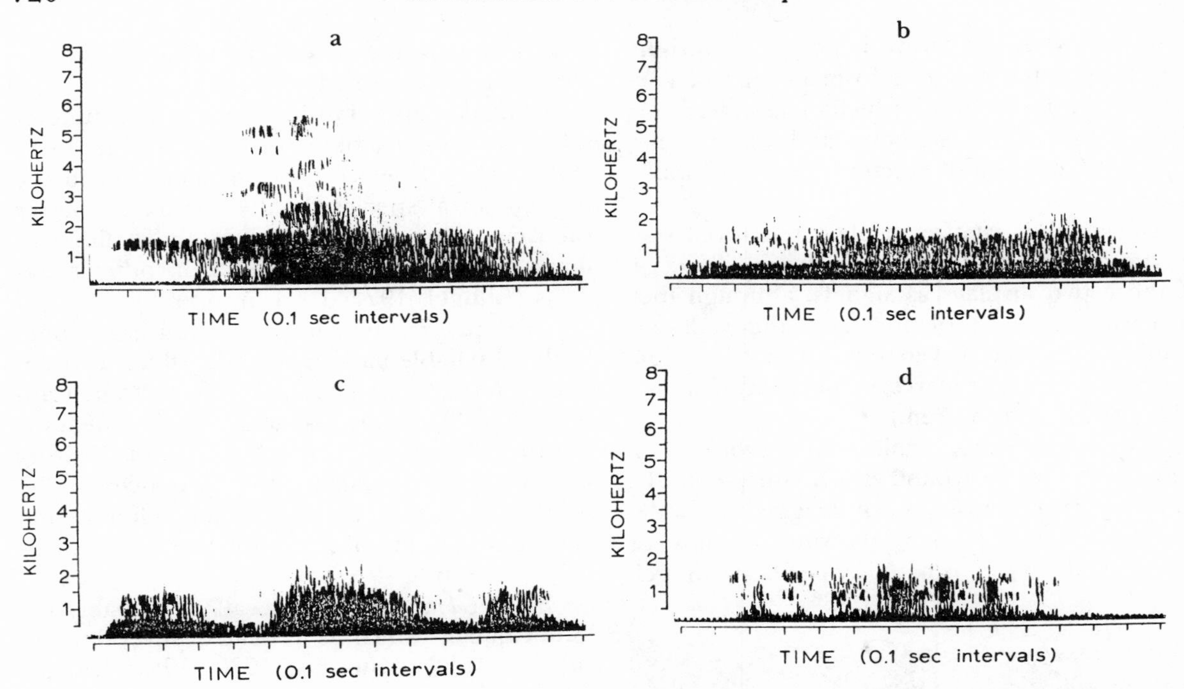

Fig. 6. Vocalizations of the horse: a. whinny; b. nicker emitted prior to feeding; c. nicker emitted by stallion during sexual behavior; d. nicker emitted by mare when concerned for the foal's well-being. All these sounds are variable. (Photo courtesy G. H. Waring.)

have some typical, genus-specific properties that are recognized and responded to by congenerics, who, however, do not react to calls of nonequine species (Antonius, 1951). Ödberg (1969) changed the whinny of horses by replaying tapes 50 to 75 percent faster (and therefore higher) and 50 percent slower (and lower). In all cases his horses still recognized the sound as a whinny and reacted accordingly. The calls of mules and hinnies are intermediate between those of the horse and the ass (Antonius, 1934, 1950).

The remaining vocalizations of the equids are listed by species. Their communicative significance is not obvious in all the cases:

E. quagga: (1) dissyllabic "i-ha" (warning); (2) high-pitched squeal (pain, distress); (3) long squeaking (pain, distress in foals). (Klingel, 1967.)

E. zebra: (1) short bark, usually repeated once (warning); (2) two-syllable "i-ho" (warning?); (3) high-pitched squeal (pain, but also emitted during greeting displays and play fights). (Klingel, 1968.) Joubert (1972) describes a total of four vocalizations, but does not compare them with previous descriptions.

E. przewalskii: (1) short squeal, emitted by a mare when approached by a stallion and when fighting (distress, pain?); (2) nicker, a low-pitched guttural sound (contact call, greeting call, alarm call: Fig. 6b, c, d), (3) roar, recorded from stallions when meeting a nonreceptive estrous mare; (4) grunt (sign of content?). (Kiley, 1972; Tyler, 1972; Ödberg, 1969; Hafez et al., 1962; Waring,

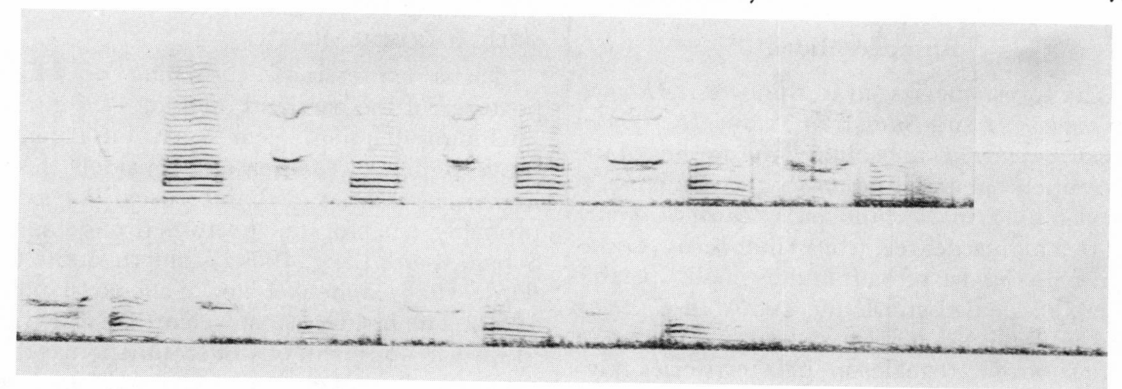

Fig. 7. Donkey (♂) bray sequence, consisting of eleven exhalations. Note the large number of frequency bands of the exhalations. (Photo courtesy P. Moehlman.)

1971.) Ödberg (1969) distinguished three basic types of vocalization from his sonograms: whinny, grunt, and squeal with variations and intermediates.

E. grevyi: (1) grunt (excitement). (Klingel, 1974.)

E. africanus: (1) grunt; (2) growl, atonal vocalizations of different duration (agonistic context); (3) whuffle, atonal and of low intensity (approach, searching). (*E. africanus f. asinus:* Moehlman, 1974.)

E. hemionus: (1) squeal (defense). (Antonius, 1937.)

OLFACTORY COMMUNICATION

In equids communication by odors works only over short distances, when the animals engage in naso-nasal and naso-genital or other close contacts and when smelling urine and dung. Olfactory contact with conspecifics serves to effect individual recognition (in the territorial species probably only of mare and foal, in the nonterritorial species also of group members) and, with stallions, to test the reproductive state of the mares. The roles of circumanal, circumoral, and perineal glands are unknown (Schaffer, 1940). Equine dung and urine transmit information by scents to conspecifics. This is most pronounced in the reproductive context: the stallions smell the dung piles and urination spots, particularly those of estrous mares, and thereby receive the relevant messages. (They subsequently display the flehmen, which enhances olfactory perception, defecate or urinate on the dung, and urinate on the urination spots. The biological significance of this "marking" behavior remains obscure; Klingel, 1972.) Horse stallions and mares seem to recognize the dung of the group members by smelling, and use this capability when searching for their groups (Tyler, 1972).

In the territorial species, the territorial stallions mark their territories with dung piles, which are used for years and thereby reach considerable size. They do not serve to advertise the territories for other conspecifics, but obviously function as marks for the orientation of the territorial animal itself (*E. grevyi:* Klingel, 1969b, 1974; *E. africanus f. asinus:* Moehlman, 1974; *E. africanus somaliensis:* Klingel, in prep.; *E. hemionus:* Klingel, in prep.).

Rhinocerotidae

The forest species Javan rhinoceros *(Rhinoceros sondaicus)* and Sumatran rhinoceros *(Didermoceros sumatrensis)* are solitary; the species of the more open habitats, Indian rhinoceros *(Rhinoceros unicornis),* black rhinoceros *(Diceros bicornis),* and to a higher degree, white rhinoceros *(Ceratotherium simum)* are often found in small, unstable groups (Schenkel and Lang, 1969). In *C. simum* the adult bulls are territorial, but tolerate one or more subordinate males in their territories (Owen-Smith, 1971). In *D. bicornis* the bulls live in well-defined, undefended, overlapping home ranges (Klingel and Klingel, 1966b; Schenkel, 1966; Goddard, 1966; Schenkel and Schenkel-Hulliger, 1969). A similar system seems to exist in *R. unicornis* (Schenkel and Lang, 1969), but Ullrich (1964) considers it to be territorial. The two remaining species are insufficiently known in this respect (Schenkel and Lang, 1969). In all the rhinoceros species the only permanent associations are those between a female and her young. In *D. bicornis* rather long-lasting associations between several adults and young have been recorded in the comparatively resident population of the Ngorongoro Crater (Klingel and Klingel, 1966b).

SPECIES AND INDIVIDUAL RECOGNITION

Rhinoceroses have poor eyesight. *D. bicornis,* for example, occasionally mistake a motorcar for a conspecific. The voices seem to play a certain role in intraspecific recognition, but as far as is known they are not sufficiently varied to be used for individual recognition (Schenkel and Lang, 1969). In an overlap area of *C. simum* and *D. bicornis* mixed grazing communities and interspecific play were observed (Steele, 1960), but such incidents occur only very rarely. They may be based on a "mistake," i.e., the members of the other species are taken for conspecifics. Dung piles are used by both species (Player and Feely, 1960).

OPTICAL COMMUNICATION

Facial expressions, consisting of different postures of the ears and opening of the mouth (including flehmen) seem to have no communicative significance, when one considers the limited visual abilities of rhinoceroses. The same is probably true for tail posture (for details see Schenkel and Lang, 1969; Schloeth, 1956; Goddard, 1966; Schenkel and Schenkel-Hulliger, 1969). The head-up posture and the fast charge toward an opponent obviously function as threat displays (Schenkel and Lang, 1969; Owen-Smith, 1971) (Fig. 8).

ACOUSTICAL COMMUNICATION

Several vocalizations have been described for rhinoceroses but no detailed investigations have yet been published. It is therefore only possible to compile the various quotations:

C. simum: (1) snorting in intraspecific combats; (2) grunting of males when courting; (3) trumpeting of males when being rejected; (4) long, rumbling bellow as a threat; (5) squealing when fleeing and of young when in distress; (6) bass bellow; (7) shrieks when fighting. (Player and

Fig. 8. Ritualized fight in the white rhinoceros, consisting of fast charges. Note position of ears.

Feely, 1960; Schenkel and Lang, 1969; Owen-Smith, 1971.)

D. bicornis: (1) puffing snort when alarmed, also when charging; (2) low-pitched squeal of females during copulation; (3) growling, and (4) grunting when fighting; (5) shrill scream (whistle); (6) crying during courting activities, possibly a contact call of the young (when this vocalization is imitated by the experimenter it can be used to attract female rhinoceroses). (Ritchie, 1963; Goddard, 1966; Schenkel and Schenkel-Hulliger, 1969; Klingel, pers. obs.)

R. unicornis: (1) bleating of the young; (2) bleating of the females to attract their young; (3) snorting as threat; (4) short grunting as a warning sound when fleeing; (5) high-pitched squealing (whistling: Schenkel and Lang, 1969) of females when being courted; (6) extended grunting of males when driving males or females; (7) extended squealing of driven males; (8) purring as a contact call. (Ullrich, 1964; Lang, 1961.)

R. sondaicus: (1) harsh blowing sound (snorting?) when disturbed; (2) grunting when attacking, also when wounded; (3) squealing (whistling), produced when inhaling and exhaling as a contact call; (4) blowing of young. (Schenkel and Lang, 1969; Sody, 1959.)

D. sumatrensis: (1) snorting when alarmed; (2) quacking when fleeing; (3) squeaking when feeding undisturbed; (4) "low and rather plaintive" noise when wallowing; (5) squealing when wounded. (Hubback, 1939.)

OLFACTORY COMMUNICATION

Dung and urine are used for communication in all the species. Preputial scent glands have been described in *C. simum* (Cave, 1966), but their function is unknown. No particular scent glands and no behavior related to such glands have been described. Cows in estrus attract bulls by scent; bulls are capable of following a scent spoor made by a cow.

In *C. simum* the territorial bulls mark their territories by exhibiting specialized techniques of defecation and urination. Before and after defecation they perform backwardly directed kicking movements with the hind legs, which result in breaking up and scattering the dung. Dung is deposited at particular places only, and thus twenty to thirty dung heaps per territory are maintained. Urine is distributed in the form of a fine spray in three to five spasmodic bursts over vegetation or the ground, impregnating the territory with the characteristic odor. Cows, nonterritorial bulls, and calves use the permanent dung piles, but in a nonritualized manner (Owen-Smith, 1971) (Fig. 9).

Fig. 9. Ritualized defecation in the white rhinoceros: above, defecation on permanent dung pile; below, spreading the dung with hind legs.

In *D. bicornis* ritualized defecation and urination are carried out by both sexes, but ritualized urination is carried out by cows only during estrus. This display serves to inform all the passing members of the population of the presence and physiological status of the marking individual and will accordingly result in spacing or attraction (Schenkel and Schenkel-Hulliger, 1969).

In *R. unicornis,* dung is placed on permanent dung piles, but only rarely is it scattered by kicking movements of the defecating animal. Only dominant bulls urinate in a ritualized manner. During pre-mating activities both partners repeatedly display ritualized urination and thereby mark the mating area (Schenkel and Lang, 1969).

R. sondaicus defecate in an unritualized manner, i.e., without kicking movements, either on dung piles or in wallows. In the latter case the smell will be attached to the skin of the animal and from there to the tracks and vegetation. Urine is squirted onto the vegetation while the animals are walking. They also urinate when wallowing, thus impregnating themselves with the scent as well (Schenkel and Schenkel-Hulliger, 1969).

D. sumatrensis defecate into water (wallows) as well as on land, but normally seem to use permanent dung piles. However, in denser populations than now exist dung piles have been observed, and there may therefore be no qualitative difference between this species and the others. Bulls urinate in a ritualized manner, and the same has been observed in captive cows (Hubback, 1939; Strickland, 1967; Schenkel and Lang, 1969). Bulls rub their horns against trees, thereby wearing off the bark. This is frequently accompanied by pawing up the earth and sprinkling the surrounding vegetation with urine (Hubback, 1939).

Tapiridae

Very little is known of the life habits in the wild of this group; and only two species, the lowland tapir *(Tapirus terrestris)* and the Malayan tapir *(T. indicus),* have been studied in some detail in captivity (von Richter, 1966; Schneider, 1936). The two remaining species are the Central American tapir *(T. bairdi)* and the mountain tapir *(T. pinchaque).*

In the tapirs the only stable association is that of a female and her young. Even when several individuals are kept together in captivity they do not seem to pay any attention to each other, except when a female is in heat. The social organization and behavior of free-ranging tapirs are still unknown. They can be assumed to be either territorial or to space in large, but undefended, home ranges.

SPECIES AND INDIVIDUAL RECOGNITION

Little information is available on this subject. It can be assumed that a female and her young recognize each other mainly by smell. Hunsaker and Hahn (1965) consider the clicking noise as an aid to species identification (see below).

OPTICAL COMMUNICATION

In all the tapir species the tips or edges of the ears are conspicuously white and may enhance the perceptibility of ear positions and movements (von Richter, 1966). Only a few facial expressions can be distinguished, and they can only rarely be observed. The threat expression is similar to that of the Equidae: the ears are held back, and the lips are opened so that the teeth become visible (canine in the lower jaw, caniniform incisor in the upper jaw, both sexes). Sometimes a fight ensues (von Richter, 1966). Flehmen and yawning do not seem to have a communicative significance.

ACOUSTICAL COMMUNICATION

In a detailed investigation of vocalization in *T. terrestris* four sounds were recorded (Hunsaker and Hahn, 1965). (The four sounds described by von Richter, 1966, in *T. terrestris* and *T. indicus* are obviously the same ones.) (1) Shrill, fluctuating squeal as a response to fear and pain, but also when fleeing and in appeasement behavior; present in both sexes and all ages. It could serve as a warning call. (2) A sliding squeal is utilized in exploratory activities and is assumed to advertise the presence of the animals. This call is mainly uttered by the dominant individuals. It probably functions as a contact call and can be used by humans to attract tapirs in the wild. (3) A clicking noise observed in young and adult animals during exploratory behavior and when two conspecifics approach each other. In the majority of cases two clicks are produced at an interval of 0.125 sec. (4) A snort with a duration of 0.25 sec was recorded when the animals were threatening or charging.

OLFACTORY COMMUNICATION

T. indicus and *T. terrestris* in captivity were always found to defecate on the same spot. Defecation is often ritualized, as the animals scrape the ground with their hind legs, but usually without touching their dung. Whenever possible, the tapirs defecate when wallowing or bathing (von Richter, 1966; Krieg, 1948; Kuelhorn, 1955). Urination takes place in the water, and, in a ritualized manner, on the land. Males squirt urine onto certain marking spots; both males and females do so during pre-mating activities and in other situations of excitement. Urine and, to some extent, dung seem to be used by tapirs to mark their home ranges and, in females, to inform conspecifics of their physiological state.

Naso-genital contacts occur during precopulatory display; they most certainly have a communicative significance (von Richter, 1966).

References

Antonius, O., 1930. Beobachtungen an Einhufern in Schönbrunn, V. Bergzebras, Grevyzebras und Zebroide. *Der Zoolog. Garten* (n.F.) 2:261–74.

Antonius, O., 1934. Beobachtungen an Einhufern in Schönbrunn, X. Zebroid und Maulesel, XI. Über den zweiten Schönbrunner Maulesel. *Der Zoolog. Garten* (n.F.) 7:165–79.

Antonius, O., 1937. Über Herdenbildung und Paarungseigentümlichkeiten der Einhufer. *Z. Tierpsychol.,* 1:259–89.

Antonius, O., 1940. Über Symbolhandlungen und Verwandtes bei Säugetieren. *Z. Tierpsychol.,* 3:263–78.

Antonius, O., 1950. Fruchtbare Maultiernachzucht. *Zoolog. Anzeiger,* Erg. Bd., 145:28–33.

Antonius, O., 1951. *Die Tigerpferde.* Frankfurt/Main: Schöps.

Blaine, G., 1922. Notes on the zebras and some antelopes of Angola. *Proc. Zool. Soc. London,* 22:317–39.

Bruemmer, F., 1967. The wild horses of Sable Island. *Animals,* 10:14–17.

Cave, A. J. E., 1966. The preputial glands of Ceratotherium. *Mammalia,* 30:153–59.

Dobroruka, L. J., 1961. Eine Verhaltensstudie des Przewalski-Urwildpferdes (*Equus przewalskii Poliakow 1881*) in dem Zoologischen Garten Prag. *Equus,* 1:89–104.

Estes, R. D., 1972. The role of the vomeronasal organ in mammalian reproduction. *Mammalia,* 36:315–41.

Feist, J. D., 1971. Behaviour of feral horses in the Pryor Mountain Wild Horse Range. M.Sc. thesis, University of Michigan.

Goddard, J., 1966. Mating and courtship of the black rhinoceros (*Diceros bicornis* L.). *East African Wildlife J.,* 4:69–75.

Hafez, E. S. E.; Williams, M.; and Wierzbowski, S; 1962. The behaviour of horses. In: *The Behaviour of Domestic Animals,* E. S. E. Hafez, ed. London: Bailliere, Tindall & Cassell, pp.370–96.

Hubback, T., 1939. The Asiatic two-horned rhinoceros. *J. of Mammology,* 20 (1):1–20.

Hunsaker, D. II, and Hahn, T. C., 1965. Vocalization of the South American tapir, *Tapirus terrestris. Animal Behaviour,* 13:69–74.

Joubert, E., 1972. Tooth development and age determination in the Hartmann zebra, *Equus zebra hartmannae. Madoqua,* ser. 1, no. 6:5–16.

Kiley, M., 1972. The vocalizations of ungulates, their causation and function. *Z. Tierpsychol.,* 31:171–222.

Klingel, H. 1967. Soziale Organisation und Verhalten freilebender Steppenzebras, *Equus quagga. Z. Tierpsychol.,* 24:580–624.

Klingel, H., 1968. Soziale Organisation und Verhaltensweisen von Hartmann- und Bergzebras (*Equus zebra hartmannae und E. z. zebra*). *Z. Tierpsychol.,* 25:76–88.

Klingel, H., 1969a. Dauerhafte Sozialverbände beim Bergzebra. *Z. Tierpsychol.*, 26:965–66.

Klingel, H., 1969b. Zur Soziologie der Grevyzebras. *Verh. d. dtsch. Zool. Ges., Würzburg, Zool. Anzeiger*, Suppl. Bd., 33:311–16.

Klingel, H. 1969c. Reproduction in the plains zebra, *Equus burchelli boehmi:* behaviour and ecological factors. *J. Reproduction and Fertility*, Suppl. 6:339–45.

Klingel, H., 1972. Das Verhalten der Pferde *(Equidae). Handbuch der Zoologie*, 8, 10 (24):1–68.

Klingel, H., 1974. Soziale Organisation und Verhalten des Grevy-Zebras *(Equus grevyi). Z. Tierpsychol.*, 36:37–70.

Klingel, H., Social organisation, behavior, and ecology of African and Asiatic Wild Asses (in prep.).

Klingel, H., and Klingel, U., 1966a. Die Geburt eines Zebras *(Equus quagga boehmi Matschie). Z. Tierpsychol.*, 23:72–76.

Klingel, H., and Klingel, U., 1966b. The rhinoceroses of Ngorongoro Crater. *Oryx*, 8, 5:302–306.

Klingel, H., and Klingel, U., 1968. *Equus quagga:* Paarungsverhalten. *Encyclopaedia cinematographica*, E 1044 *Publ. Wiss. Film*, Göttingen, A II:461–66.

Krieg, H., 1948. *Zwischen Anden und Atlantik.* Munich: Hanser. 492pp.

Kuelhorn, F., 1955. Säugetierkundliche Mitteilungen aus Süd-Mattograsso. 3. Teil. *Säugetierkundl. Mitteilungen*, 3:77–82.

Lang, E. M., 1961. Beobachtungen am indischen Panzernashorn (*Rhinoceros unicornis). Zoolog. Garten* (n.F.), 25:369–409.

Moehlman, P., 1974. Behaviour and ecology of feral asses *(Equus asinus).* Ph.D. diss., University of Wisconsin.

Ödberg, F. O., 1969. Bijdrage tot de Studie der Gedragingen van het Paard (*Equus caballus L.).* Lic. thesis, Ghent.

Owen-Smith, N., 1971. Territoriality in the white rhinoceros (*Ceratotherium simum) Burchell. Nature*, 231:294–96.

Player, I. C., and Feely, J. M., 1960. A preliminary report on the square-lipped rhinoceros *(Ceratotherium simum simum). The Lammergeyer*, 1:3–24.

Ritchie, A. T. A., 1963. The black rhinoceros *(Diceros bicornis). East Afr. Wildl. J.*, 1:54–62.

Richter, W. von, 1966. Untersuchungen über angeborene Verhaltensweisen des Schabrackentapirs *(Tapirus indicus)* und des Flachlandtapirs *(Tapirus terrestris). Zoologische Beiträge*, 12:67–159.

Schaffer, J., 1940. *Die Hautdrüsenorgane der Säugetiere.* Berlin-Wien: Urban und Schwarzenberg.

Schenkel, R., 1966. Zum Problem der Territorialität und des Markierens bei Säugern - am Beispiel des Schwarzen Nashorns und des Löwen. *Z. Tierpsychol.*, 23:593–626.

Schenkel, R., and Schenkel-Hulliger, L., 1969. Ecology and behaviour of the black rhinoceros. In: *Mammalia depicta*, Wolf Herre, ed. Hamburg, Berlin: Parey.

Schenkel, R., and Lang, E. M., 1969. Das Verhalten der Nashörner. *Handbuch der Zoologie*, 10 (25):1–56.

Schloeth, R., 1956. Zur Psychologie der Begegnung zwischen Tieren. *Behaviour*, 10:1–80.

Schneider, K. M., 1930. Das Flehmen, I. *Der Zoolog. Garten* (n.F.), 3:183–98.

Schneider, K. M., 1934. Das Flehmen, V. *Der Zoolog. Garten* (n.F.), 7:182–201.

Schneider, K. M., 1936. Zur Fortpflanzung, Aufzucht und Jugendentwicklung des Schabrackentapirs. *Der Zoolog. Garten* (n.F.), 8:83–99.

Sody, H. J. V., 1959. Das Javanische Nashorn *(Rhinoceros sondaicus)* historisch und biologisch. *Z. Säugetierkde.*, 24:109–240.

Steele, N. A., 1960. Meeting of rhinos of two species. *The Lammergeyer*, 1:40–41.

Strickland, D. L., 1967. Ecology of the rhinoceros in Malaya. *Malay Nat. J.*, 20:1–17.

Tembrock, G., 1965. Untersuchungen zur intraspezifischen Variabilität von Lautäusserungen bei Säugetieren. *Z. Säugetierkunde*, 30:257–73.

Tembrock, G., 1968. Perissodactyla. In: *Animal Communication*, T. A. Sebeok, ed. Bloomington: Indiana University Press, pp.377–82.

Trumler, E., 1958. Beobachtungen an den Böhmzebras des "George-v.-Opel-Freigeheges für Tierforschung e.V." Kronberg im Taunus, 1. Das Paarungsverhalten. *Säugetierkundliche Mitteilungen*, 6 (Sonderheft):1–48.

Trumler, E., 1959a. Das "Rossigkeitsgesicht" und ähnliches Ausdrucksverhalten bei Einhufern. *Z. Tierpsychol.* 16:478–88.

Trumler, E., 1959b. Beobachtungen an den Böhmzebras des "George-v.-Opel-Freigeheges für Tierforschung e.V." Kronberg im Taunus, 2. Die Hautpflege. *Säugetierkundliche Mitteilungen*, 7 (Sonderheft):104–25.

Tyler, S. J., 1972. The behaviour and social organisation of the new forest ponies. *Animal Behaviour Monographs*, 5, 2:87–196.

Ullrich, W., 1964. Zur Biologie der Panzernashörner *(Rhinoceros unicornis)* in Assam. *Zoolog. Garten (n.F.)*, 28:225–50.

Walther, F., 1962. Beobachtungen über die Mutter-Kind-Beziehungen und die Rolle des Hengstes bei der Aufzucht eines Fohlens von *Equus quagga boehmi. JB. George-v.-Opel-Freigehege Tierforsch.,* 3:44–51.

Waring, G. H., 1971. Sounds of the horse (*E. caballus*). A.I.B.S. Sept. 1971. Paper read at meeting of Ecol. Soc. of America. Colorado State University, Fort Collins, Colorado.

Zeeb, K., 1959. Die "Unterlegenheitsgebärde" des noch nicht ausgewachsenen Pferdes (*Equus caballus*). *Z. Tierpsychol.,* 16:489–96.

Chapter 28

CANID COMMUNICATION

Michael W. Fox and James A. Cohen

Introduction

Studies on the development of communication and its role in social organization in canids have been conducted primarily on captive animals. Schenkel (1947, 1967) and Zimen (1974) provide the most detailed observations on communication and social organization in the wolf *(Canis lupus),* and Fox (1971a) reviews many aspects of the ontogeny and phylogeny of communication in several canid species. These studies focus principally on visual displays, and with the exception of Tembrock's research on the vocalizations in *Vulpes vulpes* and overview of other species (Tembrock, 1968), there is a dearth of literature dealing with vocal, tactile, and olfactory communication in this family. Kleiman (1966) gives some observations of scent-marking behavior in a few captive species, but no systematic studies have been reported on this important aspect of communication. Recent interest in mammalian pheromones and in the possible application of antipredator chemicals to protect livestock from coyotes may rectify this gap in our knowledge. Canids do have well-developed scent glands and vomeronasal, or Jacobson's, organ, and it is surprising that to date no studies have been conducted on this well-defined and accessible organ.

In summary, most work has been done on the visual modality of communication (i.e., displays) in the *Canidae.* An exhaustive survey by the authors of field studies of various species (many of which are reviewed by various authors in Fox, 1975a) unearthed little new material to add to our knowledge of the comparative ethology and communication in canids. It is unfortunate that few field biologists are trained to perceive and record significant communication phenomena and that most controlled studies are on captive specimens. One exception is L. D. Mech (pers. comm.), who is now investigating in detail both vocal and chemical communication in the wolf in northern Minnesota.

Before considering various modalities of communication, it should be emphasized that often more than one channel of communication may be utilized at the same time, e.g., vocalization plus visual display, such as a growl and threat gape. Also the same signal(s) may be given in different contexts (growl as a threat, or as a warning of danger) and signals of one modality may be simultaneously combined (superimposition), such as in the growl/scream and defensive gape, or they may be successively combined as in growl-whine and approach-withdrawal movements in play soliciting. Important variables to consider are age, sex, season, prior experience, and expectations and relations between interactees. Also, the terrain may influence the modality of communication utilized and the intensity or amplitude and frequency of the signal, as well as the proximity of conspecifics.

SOCIO-ECOLOGY

The *Canidae* include a wide range of species that differ in degrees of sociability, which is related to their socioecology and communication patterns. Species such as the wolf *(Canis lupus)*, jungle whistling dog, or dhole *(Cuon alpinus)*, and Cape hunting dog *(Lycaon pictus)* are highly gregarious and hunt in packs. These have been designated Type III canids and have a complex repertoire of subtly graded signals, which complement a more complex social organization than in other canids (Fox, 1975b). Type II canids have permanent pair bonds but do not usually form packs, the young dispersing at around ten months of age, as in the coyote *(C. latrans)* and golden jackal *(C. aureus)*. The least gregarious, the Type I canids, as exemplified by the red fox *(Vulpes vulpes)*, in which there is usually only a temporary pair bond and the young disperse around five to six months of age, have a more stereotyped communication repertoire than the more gregarious types (for further details see Fox, 1975b), especially of close-proximity visual displays. With this perspective in mind, the various modalities of communication in canids will now be reviewed and compared.

Visual Communication

Included in this category are body postures often combined with different tail positions and movements, and facial expressions, including various positions of the ear and orientations of the eyes. Body markings are also important in enhancing certain displays.

Various gross body postures associated with changes in emotional/motivational state are schematized in Fig. 1. Darwin's principle of antithesis is exemplified by the stiff, upright threat posture and the crouched posture of submission. Changes in the direction of the lean of the body, distribution of weight in fore and hind legs, extension or flexion and turning of head and neck and movements of the tail can be identified as component units of these displays. Golani (1973) has developed a method for analysis of such movements and for interaction sequences between subjects.

Forepaw raising may be a submissive signal, expressing the intention to roll over (see Fig. 2), or a defensive warding-off reaction. Pawing directed at a conspecific is also associated with play soliciting.

Back arching as a threat display is well developed in red and gray foxes (i.e., in vulpine or

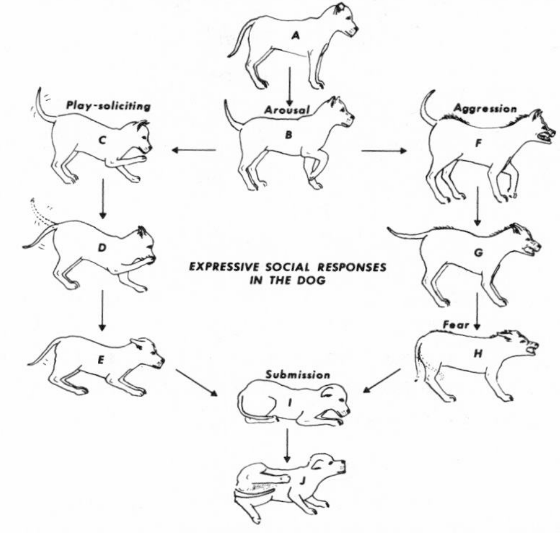

Fig. 1. Schema of gross body postures in the dog related to changes in motivational state. This schema illustrates how consecutive changes in social displays may occur. (From Fox, 1971a.)

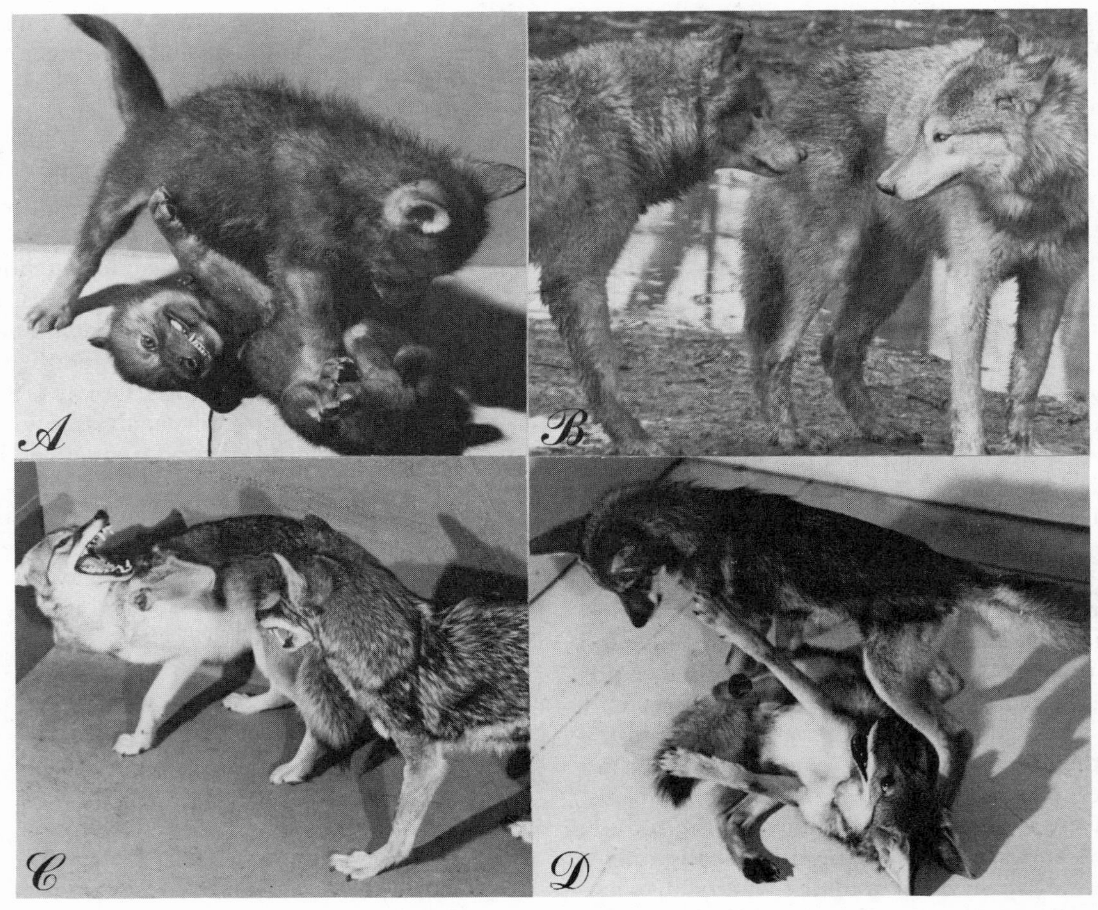

Fig. 2. Agonistic behaviors: A. Standing over by dominant wolf cub. Note submissive grin in passive recumbent subordinate. B. Direct stare by dominant wolf (on left) evokes submissive expression in lower ranking wolf, which remained immobile and eventually broke eye contact. C. Aggressive offensive expression (comparable to #3 in Fig. 5) in dominant coyote and defensive gape (comparable to #9 in Fig. 5) in subordinate, coupled with forepaw raising, an incomplete or intention movement of rolling over. White chest and belly enhance this display. D. Homologous behavior in adult coyotes, but tail position is horizontal in dominant coyote and vertical in wolf.

foxlike canids), where the body orientation may be side-on to the rival. Possibly correlated with more frontal face-to-face threat orientation in wolves and dogs (and other more doglike canids), back arching is less frequently seen and is usually of low amplitude. It is more common in the coyote.

It is important to note not only the kind of signal that is being given but also the orientation of interactees where a "T" posture may be assumed in threat or courtship and an "L" posture in active submission. A subordinate wolf usually approaches the head of a higher-ranking wolf side-on, while a coyote threatening another will

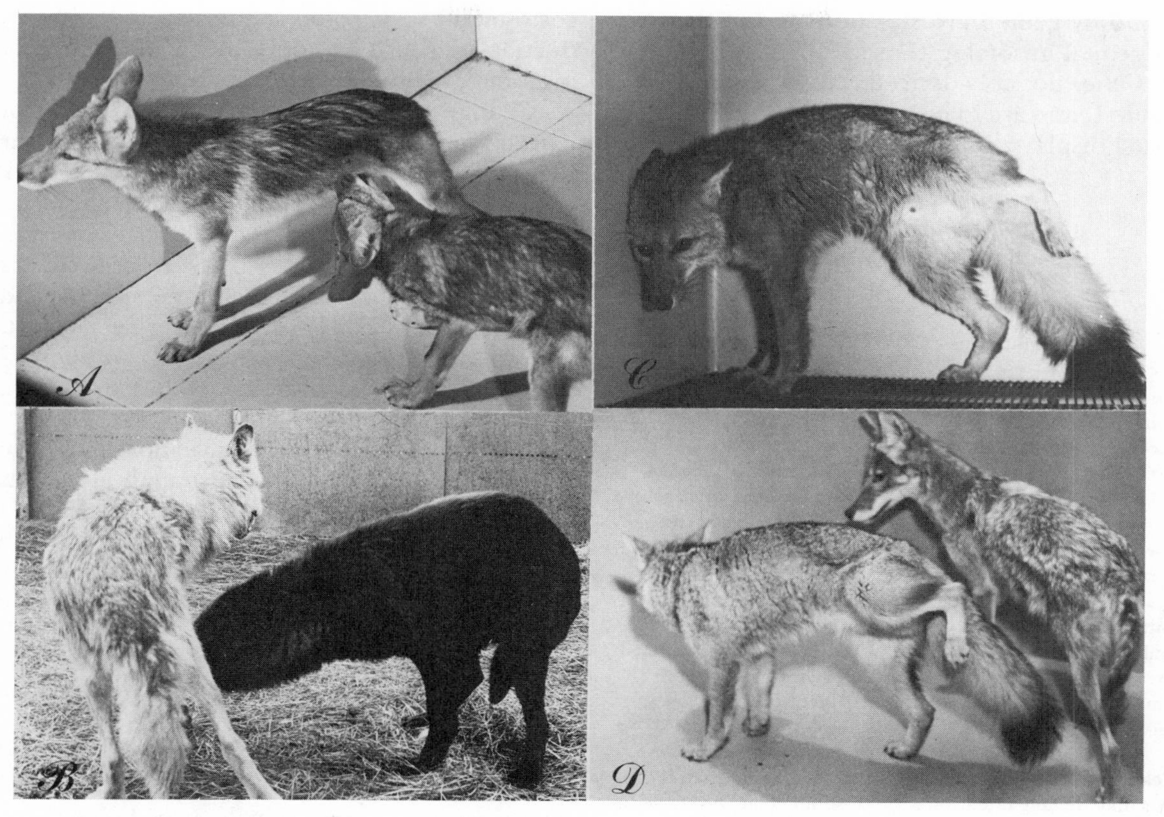

Fig. 3. A. Passivity and inguinal presentation during social investigation in coyotes. B. Inguinal presentation by alpha wolf (on left) for genital grooming by subordinate. C. Inguinal display as a submissive gesture toward handler in Culpeo (*Dusicyon culpeolus*) and in the same animal (D) toward a coyote. Passive submissive lateral recumbency was not seen in this animal.

approach its side and may place its chin or forefeet on the other's shoulders.[1]

1. Standing-over is an agonistic posture in the coyote (Fox and Clark, 1971) and is also seen in the golden jackal, wolf, and domestic dog. It is to be differentiated from clasping, where one seizes the other around the waist or chest between its forelimbs. This frequently occurs in aggressive contexts but is most often identified in sexual contexts. Reciprocal clasping or "hugging" is seen during contact-play or play-fighting and courtship in "canine" types, while rearing up together and pushing with both forelimbs is common to both vulpine (foxlike) and canine (doglike) canids in agonistic interactions.

A common action in greeting is to twist the trunk so that the groin region is presented to a conspecific, the "C" posture, and in some species such as the gray fox *(Urocyon cinereoargenteus)*, coyote, and culpeo *(Dusicyon culpaolus)* the hind leg nearest the partner may be raised (see Fig. 3). This is the inguinal response, a derived socio-infantile display (Fox, 1971b). In the wolf, dog, and coyote it may be followed by rolling over into lateral recumbency (passive submission) and submissive urination. The interactee may respond by genital investigation, and in the wolf, by genital grooming (Fox, 1971b).

Other possibly derived activities are listed in Table 1, and are to be added to the repertoire of visual displays. It is to be noted that all derived socio-infantile activities are associated with reduction of social distance and maintenance of proximity in adults.

Table 1

Neonatal- and infantile-derived activities in adult canids.

Neonatal and infantile patterns	Adult behavior
Contactual circling	Circling, leaning
Chin-resting	"T" posture
Successive paw raising (nursing)	Playful approach and play soliciting? (or merely exaggerated approach)
Side-to-side lateral head swinging–rooting to nurse	Social greeting, active submission
Vertical, upward head movements, "butting," nose stabbing–nursing	Social greeting, play soliciting
Unilateral paw raising, to reach up to teat, to paw mother's face	Play soliciting, as directed pawing or pawing intention
Face-mouth oriented licking to solicit food from mother	Social greeting, active submission
Licking intention prior to feeding: licking intention while approaching to lick face of mother	Social greeting, active submission and slower in passive submission
Passivity during anogenital stimulation by mother, and passivity with inguinal contact during such stimulation	Passivity during social investigation. Submission, social greeting with inguinal presentation ("C" posture), submissive urination
Visually guided approach to inguinal area of mother to feed	Inguinal orientation and contact during social investigation
Distress vocalizations (screams and yelps) when in pain	Passive submissive and defensive vocalizations (may "cut-off" aggressor)
Distress vocalizations (whines) when cold or hungry (care-soliciting)	Care-soliciting vocalizations, active submission, social greeting (may "remotivate" aggressor)

Source: From Fox, 1971a.

Changes in body position and movements of extremities are enhanced by differences in fur length and color, e.g., longer hairs on back ("hackles" in agonistic piloerection); pale hairs in external auditory meatus and black or pale tail tip enhance movements; pale-colored cheeks may orient ritualized attacks in some species; and the pale ventral body surface along neck, chest, and abdomen may enhance signals of submission, such as looking away and rolling over (Fox, 1969).

The more foxlike canids have more felinelike displays than other canids. Although all canids wag or flag the tail in greeting, the red fox has vertical and "J"-shaped positions associated with intense arousal, threat, and play-chasing very similar to the cat (Fox, 1974). Also foxes have not only the play-soliciting bow but also a rolling-over display in play analogous to that of the domestic cat. Coyotes have a variation of this—a combined play-bow–diving roll as a play-soliciting signal (see Fig. 4A).

Head movements are associated with forward bite intention threat signals (see below) and turning away and breaking eye contact, as in submission, or in a higher-ranking wolf's "daring" a subordinate to attack (Schenkel, 1967), or simply ignoring it (Fox, 1971a).

Some body postures may be interpreted as compromise postures, as in the play-bow where approach and avoidance may be simultaneously combined and "frozen" into a distinct attitude. Another interpretation of the play-bow is that it is a highly ritualized form of stretching, which communicates a "relaxed" state. Such actions as yawning, grooming, preening, blinking, and eye closure in other species may be associated with relaxation and may serve secondary social functions in ritualized forms.

Stretching and yawning may be relaxed ac-

Fig. 4. Play-bow in domestic dog (B), Arctic fox (C), Culpeo (D), and rolling following play-bow in coyote (A) and rolling without play-bow in kit fox (*Vulpes macrotis*) as a play-soliciting signal (E).

tions preceding greeting and may contribute independently to the ontogeny of the play-bow. This display may be coupled with incomplete forward and backward leaps or intention movements, i.e., approach/withdrawal ambivalence. The metacommunicative "mood" that "this is play"[2] is enhanced by the open-mouth play face (see below) and vocalizations (play panting or barking in *C. familiaris*).

Stretching and looking away may be signals of ambivalence, the former being a displacement like the ground sniffing or brief grooming of a cat facing an adversary (Fox, 1974). Licking (i.e., repeated extrusion of the tongue) may be directed toward a conspecific in greeting in the "canine" species but is a sign of ambivalence if

2. See Bekoff (1972) for discussion of metacommunication and play in canids and other animals.

the tongue is curled back and touches the nose or lips.

Certain expressive movements of canids may have an habitual quality and give the animal in question a certain style or character, i.e., the "hang dog" expression of the obsequious pet dog or the "perpetual puppy" behavior of an indulged lap dog. This capacity to assume a particular demeanor in certain contexts or social situations is a quality of the more social canids and is best exemplified in the wolf pack, where the omega or lowest-ranking wolf can be easily identified by its carriage and behavior. Interestingly, when there is a change in rank, there is a change in the individual's behavior and character or body demeanor. Thus in a relatively stable social situation, certain individuals may take on a characterological set of expressive movements, a point which has considerable diagnostic significance in the somatic analysis and treatment of psychiatric disorders in man (Lowen, 1970; Fox, 1976).

FACIAL EXPRESSIONS[3]

Some facial expressions may vary in intensity, as in threat, where the aggressive pucker or mouth-closed, lips-forward expression gives way to a more open-mouth expression with the lips retracted vertically to display the teeth. This latter action is absent in "vulpine" canids. Besides successive shifts in intensity, facial expressions associated with different contexts or emotional states may also occur simultaneously, as in the superimposition of the submissive grin (horizontal retraction of lips) with an agonistic gape and snarl (see Fig. 5, 9).

In the vulpines, the facial expressions lack the subtle gradation of intensity of the more doglike canids, being more stereotyped and with a lower incidence of simultaneously combined expressions. Tables 2 and 3 list the components of vari-

3. For details of the ontogeny of facial expressions in canids, see Fox (1970, 1971a).

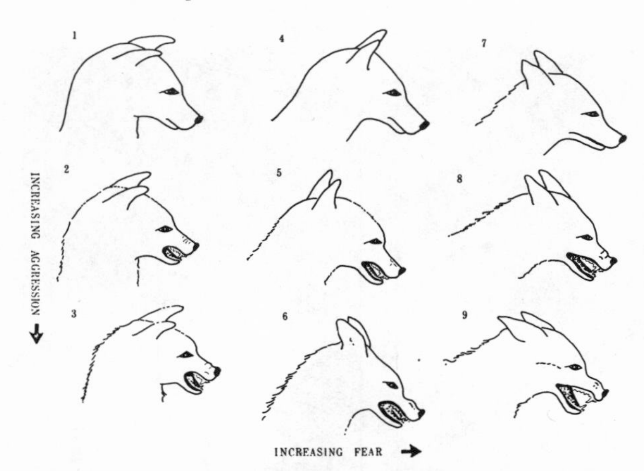

Fig. 5. Successive shifts in intensity (amplitude) of fear/submission (1, 4, 7) and aggression (1, 2, 3) in coyote. Greater range of expressions is afforded by simultaneous combination or superimposition, e.g., of 3 + 7 in 9, an expression associated with fear-biting and defensive aggression. (From Fox, 1971a.)

Table 2

Components of expressions associated with increase of social distance and aggression.

	R	A	G	C	W	D
Head high and neck arched	+	±	+	+	+	+
Gape*	+	+	+	±	−	−
Growl	±	+	+	+	+	+
Bark	±	±	±	±	±	+
Agonistic "pucker" (horizontal contraction of lips)	+	+	+	+	+	+
Agonistic baring of teeth (vertical contraction of lips)	−	−	−	+	+	+
Snapping of teeth	−	−	−	±	+	+
Ears erect and forward	+	+	+	+	+	+
Direct stare	+	+	+	+	+	+
Eyes large	+	+	+	+	+	+
Ears flattened and turned back	+	+	+	+	+	+

R = red fox C = coyote
A = Arctic fox W = wolf
G = grey fox D = dog
*Occasional spitting in red fox.
+denotes frequent expression
±denotes infrequent expression
−denotes expression not observed
Source: From Fox, 1971a.

Table 3

Components of expressions associated with decrease of social distance and submission.

	R	A	G	C	W	D
Head lowered and neck extended horizontally (crouch)	+	+	+	+	+	+
Ears flattened and turned down to sides	+	+	+	+	+	+
Submissive "grin" (horizontal retraction of lips)	+	+	±	+	+	+
"Play face"	+	+	±	+	+	+
Licking (cut off)	–	–	–	+	+	+
Licking (social greeting)	±	±	±	+	+	+
Licking (intention)	–	–	–	+	+	+
Nibbling	–	–	±	±	+	+
Jaw wrestling (play)	–	±	±	+	+	±
Whining and whimpering	–	–	±	+	+	+
Looking away	±	±	±	+	+	+

R = red fox C = coyote
A = Arctic fox W = wolf
G = grey fox D = dog
+denotes frequent expression
±denotes infrequent expression
–denotes expression not observed
Source: From Fox, 1971a.

ous expressions observed in canids together with their occurrence in different species.

The variety of facial expressions that canids can display is considerable; at least ten distinct categories have been identified—almost as many as described by van Hooff (1967) in primates (see Table 4).

As in most displays, these expressions serve to increase, decrease, or maintain a certain distance from one or more conspecifics (see Figs. 6 and 7). Some may also serve as alarm signals in certain contexts. In general they serve to express the intentions of the animal. The complete lack of any overt signal may also have significance. Fox et al. (1974) use the term "passive" indifference for the lack of any overt reaction by a dominant (alpha) wolf when threatened by the alpha wolf of another pack in the latter's territory. Other wolves displayed submission or defensive aggression.

Vocal Communication

On the basis of spectrographic evidence (Cohen and Fox, 1976), eight basic sound types of canids have been identified: whines (including shorter "yips" and yelps, and longer, softer whimpers), screams, barks, growls, coos, howls, mews, and grunts. Variation within these sound categories may occur at both the individual and the interspecies levels, although the degree of such variation must be largely dependent on the structural limitations of the vocal apparatus. The sounds may vary in duration, frequency, intensity, cyclicity, and context.

Not all these basic sound types are included in the vocal repertoire of every canid species, and the same sound may be used by different species in very different contexts (see Table 5). Foxes, for example, are the only canids known to emit a pure scream while greeting conspecifics, and domestic dogs will bark in many situations (e.g., threat, play soliciting, contact seeking), while foxes bark only in threat.

The following brief descriptions of the eight major canid sounds will point out some of the similarities and differences between certain physical properties of the sounds:

Whines. Whines are commonly heard in wolves, coyotes, foxes, and domestic dogs. They are wide-banded, cyclic sounds of short duration and moderate frequency variations. The clear horizontal stratification typical of the whine is evident in Figure 8A.

Screams. The scream, like the whine, is common among canids. Although it too is a wide-banded sound, it differs in that its frequency variations are much greater, it may be of longer duration, it is noncyclic, and it is delivered with much greater intensity. It also exhibits strong horizontal stratification (see Fig. 8B).

Table 4

Comparison of facial expression in primates (after van Hooff 1967) and canids.

Primate	Canid*	Situation/Motivation
Tense-mouth face	1. Agonistic "pucker." Vertical retraction of lips in W, C and D	Tendency to attack
Open-mouth face	2. Threat gape (marked in foxes)	Tendency to attack. Signals bite intention
Staring bared-teeth scream face	3. W.C.D. Threat gape with vertical and horizontal retraction of lips	Signals bite intention with some flight tendency
Frowning bared-teeth scream face	4. W.C.D. Above with wider gape and greater vertical lip retraction	Defensive threat when escape is blocked
Silent bared-teeth face	5. Submissive grin (horizontal retraction of lips)	Low tendency to flee: Ritualized appeasement
Bared-teeth gecker face	–	In infants when disturbed. Low tendency to approach and flee in adults
Teeth-chattering face	6. W.C.D. Agonistic tooth snapping	Tendency to flee
Lip-smacking face. Tongue-smacking face	7. W.C.D. Licking intention	Strong approach tendency and weaker flight tendency
Chewing-smacking face. Snarl-smacking face	8. W.C.D. Nibbling intention	Strong approach tendency and weaker flight tendency
Protruded-lip face	–	Approach (often to mate)
Pout face	9. W.C.D. Submissive rooting approach and forepaw raising	Approach (mother-infant) intention movement to take nipple in mouth, and between adults
Relaxed open-mouth face	10. Play face	During play or leads to play

*Except where indicated, expressions are seen in all canids.
NOTE: W = only in wolf. D = only in domestic dog. C = only in coyote.

Vocalizations accompanying above expressions: 1. Low growl. 2. Louder growl or explosive "Tch Tch" in fox. 3., 4. Silent or successive growl and whine. 5. Silent, or whining; winnowing call in foxes. 6. Silent or successive whining and growling. 7. Silent or whining. 8. Silent or whining. 9. Silent or whining. 10. Silent; barking only in domesticated dog.

Source: From Fox, 1971a.

Table 5

Canid sound types (not mixed†)

	Mew	Grunt	Whine** (whimpers)	Scream	Yip	Yowl-howl	Coo	Growl	Bark	Click	Tooth-snap	Pant
Greeting	F	WD††	WCD	F	C	WD	F	WD	D	—	—	F
Play solicit	—	—	D	—	—	—	—	—	D	—	WD	FD
Submission	F	—	WCD	WCD	—	—	—	—	—	—	—	—
Defense	—	—	WCD	CFW	—	W	—	WCDF	WD	F	WCD	—
Threat*	—	—	—	—	—	—	—	WCDF	WCDF	F	WCD	—
Care- or contact-seeking	NF	DW	N DWC	F	C	—	F	—	D	—	—	—
Distress (Pain)	N	—	N WCDF	N WCDF	—	—	—	D	D	—	—	—
Contact-seeking (lone calls)	—	—	N WCD	—	—	WCD	F	—	D	—	—	—
Group vocalization	—	—	WCD	—	C	WCD	—	WCD	WCD	—	—	—

NOTE: D = dog; W = wolf; C = coyote; F = red fox; N = neonates of above species.
*May serve as warning to others.
**Includes "yelps" and "yips."
†Some of above sounds may be "mixed" simultaneously, or successively, e.g., whine-growl, bark-howl.
††Also "contentment grunts" of contact in neonate W, C, and D.

Barks. Barking may also be heard in most canids, although the context may vary greatly. The bark is a short wide-banded sound with few frequency variations. It is often cyclic in wolves, coyotes, and especially dogs, but is noncyclic in foxes (see Fig. 8C).

Growls. The growls of most canids are wide-banded, with the lower frequencies often carrying more energy than the higher ones. The duration and intensity of the growl may vary greatly with motivation. It is noncyclic and exhibits strong vertical stratification (see Fig. 8D).

Coos. Foxes are the only canids whose vocal repertoire is known to include the "coo."[4] This sound is typically characterized as a short call with a wide frequency range and short (vertical) frequency variations of moderate degree. Although the sound may be repeated, it is not of a cyclic nature and often occurs when the animal is socially isolated (see Fig. 8E).

Howls. The howl is commonly heard in wolves, coyotes, and some breeds of domestic dog. It is not heard in foxes. Lower frequencies are more prominent than higher frequencies. It is characterized as a long-to-extended sound (i.e., 1.0+ sec) with relatively few frequency variations. It may be repeated but is of a noncyclic nature (see Fig. 8F).

Mews. The mew is heard in neonates of wolves, dogs, coyotes, and foxes, generally when the animal is in some way distressed (i.e., cold, hungry, etc.). It is a short, wide-banded, cyclic sound with moderate-to-great frequency variations. Foxes are the only common canids known to produce this sound in the adult stage (see Fig. 8G).

Grunts. Grunts are often heard in wolves, coyotes, and dogs. The frequency range may

4. The Asiatic wild dog *(Cuon alpinus)* emits a whistle call for contact or assembly, but without spectrographic analysis we cannot determine if this sound is unique or a higher-frequency form of the "coo."

Fig. 6. Facial expressions associated with reduction of social distance in canids: A. Open mouth play face in wolf during courtship hug by malemute dog showing more submissive "grin." Open mouth play face in red fox and gray fox (B), Arctic foxes (C), F_2 coyote x beagles (D), and maned wolf (*Chrysocyon brachyurus*) (E). Greeting grin of domestic dog (F) specifically directed to human beings and possibly a mimic of the human grin.

vary from narrow to wide. The sound is very short (usually less than 0.5 sec), cyclic, and with few frequency variations (see Fig. 8H).

Further description and discussion of these eight sound types may be found in Tembrock (1960, 1968), Theberge and Falls (1967), and Cohen and Fox (1976).

In addition to these vocal sound types, all species communicate to some extent by means of more "mechanical" sounds such as the guttural "clicking" heard in foxes, or the tooth-snapping heard in wolves, coyotes, and dogs. Because these "mechanical" sounds are of relatively low volume and do not carry very far, they are re-

Fig. 7. Facial expressions associated with increasing social distance in canids: A. Agonistic display of coyote while wolf is showing active submissive greeting. B. Reciprocal threat gape in Arctic foxes. C. Threat gape of gray fox. D. Defensive-submissive gape of jungle whistling dog (*Cuon alpinus*). E., F. Increasing intensity of threat display of golden jackal. G. Offensive and defensive (on right) agonistic facial expressions in wolves.

served for close-contact situations and are often heard during agonistic encounters.

The above sound types may be mixed in one of two ways: either by superimposition of two vocal sounds or by successive emission of two or more types (vocal and/or mechanical) (see Fig. 9A-F). A combination of these two processes is also possible; for example, a mixed sound might follow a pure sound. The phenomenon of sound mixing correlates well with the superimposition of "body language" postures and facial expressions described earlier. A wolf faced with a situation simultaneously eliciting fear and aggression, for example, may concurrently display facial and body characteristics expressive of both motivations (Schenkel, 1947). The accompanying vo-

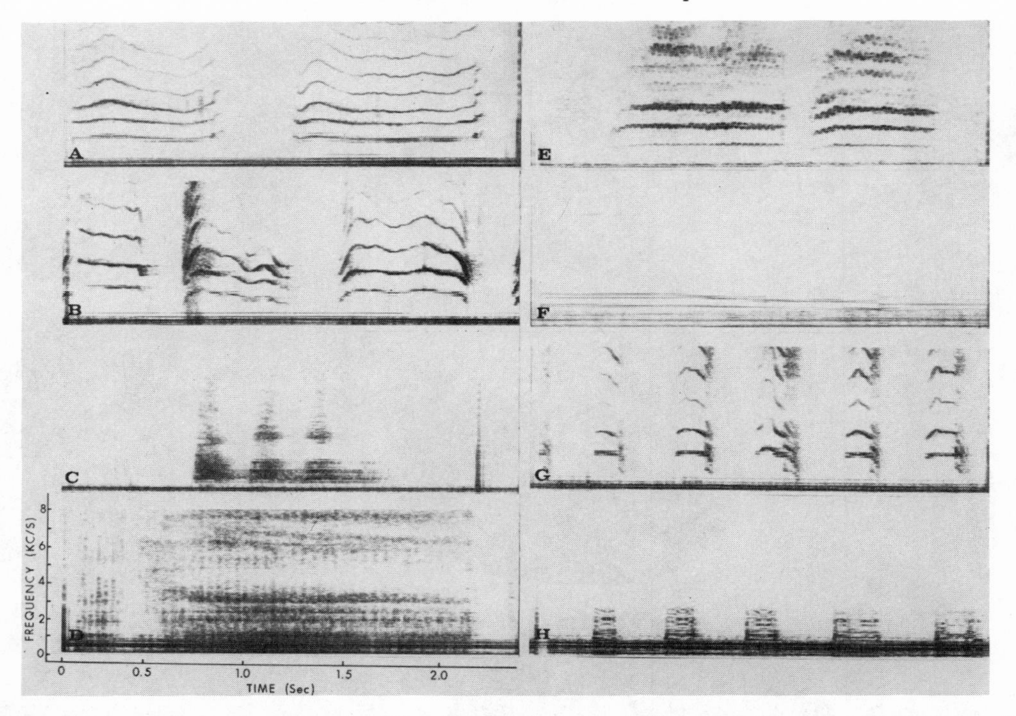

Fig. 8. A. Whines of a 1-day-old male chihuahua. B. Screams of a 6-day old male chihuahua. C. Barks of an adult female dingo. D. Growls of an adult male wolf. E. Coos of 33-day-old female red fox. F. Excerpt from howl of an adult female wolf. G. Mews of 5-week-old male chihuahua. H. Grunts of an 8-day-old male Irish setter and Doberman pinscher hybrid.

calizations are likely to include components that may be heard separately in pure form in fearful and aggressive contexts, respectively.

Tactile Communication

The ontogeny of tactile communication involving contact-seeking behavior for warmth and care (grooming, etc.) has not been systematically studied in canids. Contactual circling (Fox, 1971b) and body leaning have been identified as a derived socio-infantile action in wolves and dogs; adult wolves may run together, for example, frequently in contact or bumping each other. The inguinal response, a socio-infantile derived activity, may evoke or solicit genital grooming by a conspecific in the wolf. Wolves of a particular allegiance may engage in this activity (Fox, 1972), or it may occur in the absence of any identifiable sexual or filial bond. The groomer may assume the role of "parent" (or care giver) in this ritual, irrespective of rank since the alpha wolf may solicit genital grooming from a subordinate.

Although we have not yet quantified the frequencies of bouts of face and body grooming in captive canids, we have the impression that the vulpines engage in more social grooming, especially during the breeding season, than do the canine types.[5]

5. A red fox caged temporarily with a golden jackal groomed the latter's eyelids until they were raw and hairless. Whereas in the red fox the groomee stops the groomer when it has had enough, the opposite is true in the jackal, who remains motionless until the groomer has finished.

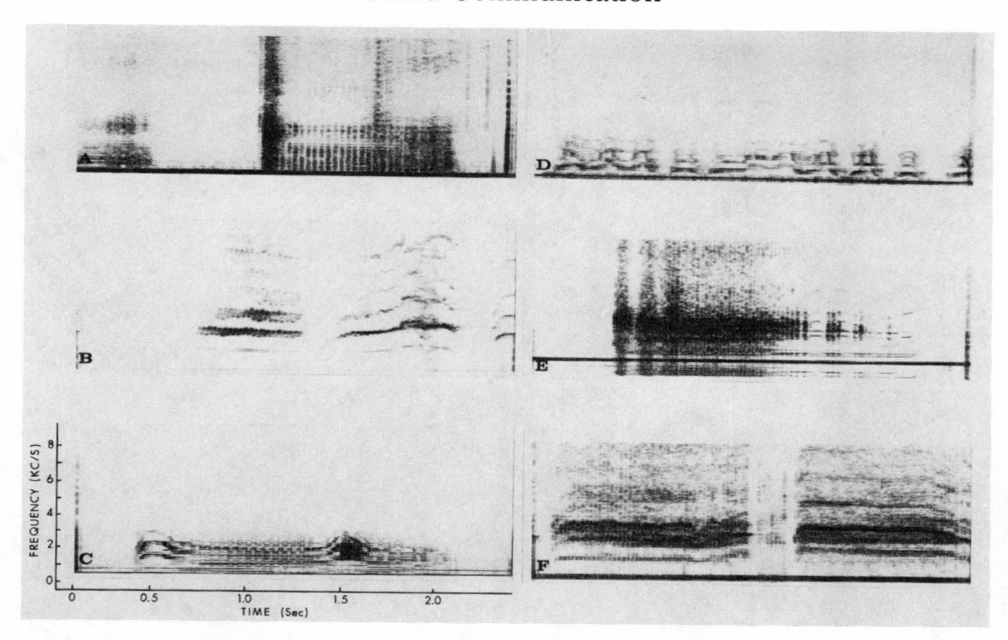

Fig. 9. Mixed sounds. Successive: A. Bark → growl of 4-week-old male chihuaha. B. Pure coo and coo → scream of 33-day-old female red fox. C. Yelp → growl → bark → growl of 10-day-old male Irish setter and Doberman pinscher hybrid. Simultaneous: D. bark/ howls of 1-year-old female wolf. E. Growl/ scream of Culpeo. F. Growl/scream of 5-month-old female Arctic fox.

Forelimb pawing and upward pushing with the muzzle or nose are friendly gestures that are visual displays at a distance but at closer proximity lead to contact. The body region of the conspecific that is touched may have some significance. Touching the shoulder region may be intimidating or a threat gesture and is used by males to "test" the readiness of the estrus female to stand for copulation. The sides of the mouth and groin region are associated with submissive and friendly contact in canids (associated with derived infantile food soliciting [greeting] and genital grooming).

Touching during social investigation is usually confined to certain discrete body regions (see Fig. 10). Some of these regions may be rich in scent or apocrine glands, notably the perioral, anal, genital, and aural regions, and dorsum of the tail (supracaudal gland). In some dogs (malemute, husky) and some wolves, a submandibular gland has been detected, and this gland, in the region of the cheek, may evoke cheek-oriented investigation by conspecifics. Little is known about the social significance of these glands and the area is open for further studies.

Chemo-Olfactory Communication

Probably one of the least-understood phenomena in the study of mammalian behavior is the role of the olfactory sense. Tembrock (1968), Ewer (1968), Gleason and Reynierse (1969), and Eisenberg and Kleiman (1972) have reviewed various aspects of scent marking and olfactory communication in vertebrates. Intraspecific odors may give information about individual identity, sex, tribe or pack, age, and physiological status; and in marking behavior one sees self-marking, partner-marking, and marking of territory. To this list of potential information

Fig.10. A. Several wolves rolling in scent of beaver "castors" while the gray fox (B) rubs the chin and chest only and does not roll over. Social investigation in foxes includes orientation toward genital-groin region (C), supracaudal tail gland (D), and perioral region (E), as well as to the ears and anal regions.

transmissions we might add that social status or rank as well as the emotional state of the animal might be identifiable, for many mammalian species deposit various odors when they are alarmed. They are not usually deposited at a specific marking spot, although the actual composition of the secretions deposited on a traditional or specific marking spot may be altered by such emotional reactions.

Ewer (1968) proposes that the animal's own smell restricts it to a familiar home range and scent marking tends to space animals out: a stranger's odor may be threatening and the animal's own odor reassuring. One of the few works dealing with the subject in *Canidae* is by Kleiman (1966), who presents the following basic theories regarding the action of scent marking in canids:

(1) Scent marking originated as a device for familiarizing and reassuring the animal when it entered a strange environment.

(2) Secondary, associated functions have arisen, such as the bringing together of the sexes and the maintenance of territory, which are important to the survival of the species.

(3) There exist species-selective patterns of response to visual and odor-bearing objects; i.e., some species may use one marking method to the exclusion of others. Species also differ in their marking postures.

(4) Scent marking may serve to inform animals of the population density of conspecifics and the number of rivals, estrus females, etc.

(5) Scent-marking postures in conjunction with other behavioral and anatomical character-

istics can be used to establish taxonomic relationships. Sprague and Anisko (1973) have studied elimination behavior in laboratory beagles and present useful observations on various patterns of elimination, but add little to our understanding of the role of marking behavior *per se.*

The reactions of captive wolves and red, gray, and Arctic foxes to propionic acid (a major component of anal gland secretion in red fox; Albone and Fox, 1971), skatole, and urine and feces of familiar and strange conspecifics, other canid species, rat, and impala were studied in our laboratory. Findings are summarized in Table 6.

THE WOLF

The response of defecation during the tests with conspecific feces and urine exclusively (Table 6) can be interpreted in several ways. Defecation may be a means of intraspecific communication, a scent deposited on the strange conspecific odor to warn an intruder away from the den site. Another interpretation has it that such behavior is instrumental in the making of new acquaintances (i.e., leaving a "calling card"). The action could also simply mask a "threatening" strange conspecific odor by covering it with the animal's own odor; a reassurance function might then be proposed, as was done by Ewer (1968). All these interpretations are warranted when the wolf is compared with the other canids in which defecation occurred in response to a wider range of stimuli; a social function may therefore be subserved by fecal marking in the wolf and to a lesser extent in the less-social canids such as the red fox (Table 6). In the latter species this form of marking is more generalized and includes the marking of novel objects with feces.

THE GRAY FOX

The gray foxes were noted for their use of rubbing almost exclusively as a scent-marking device (see Table 6). No urination on the test plate was ever observed, and the sole occurrence of defecation was in response to their own urine,

Table 6

Frequency of occurrence of various responses to different odors in canids.

Species	Control*	Own urine	Own feces	Strange conspecific urine	Strange conspecific feces	Rat feces	Impala feces	Propionic acid	Skatole
Wolf ♂	–	–	–	U	–	R	R	R	–
♀	–	–	–	D	D	U	U	U	U
Arctic fox ♂	–	–	–	U	U	U	D	U	–
♀	–	–	–	U	U	U	–	D	–
Gray fox ♂	–	D	–	R	–	–	R	R	–
♀	–	–	–	R	–	–	R	R	–
Red fox ♂									
♀ (small)	D	–			U	–	R D	R	R
♀	–		–	C	D	–	–	–	–

*Clean glass only.
Test materials presented on glass plate in home cage for 15 min.
NOTE: R = rubs or rolls, D = defecates, U = urinates, C = covers or buries, – = no response.

perhaps just a reflex response to a novel stimulus, the plate. Rubbing was used extensively in the tests with strange conspecific odors. In this species the scent-marking behavior of rubbing may have selectively evolved as a specialized pattern. As Kleiman (1966) points out, some canid species tend to use one marking method to the exclusion of others. Our contention that the gray fox rubbing behavior is socially based rather than being a simple reduction of strangeness is supported by two observations. One is that in most instances, rubbing was followed by mutual sniffing and licking of the neck, face, and anal regions. Secondly, there is the absence of behavior such as urination and defecation and burying, which can be more easily interpreted as actions to hide or mask the odor or as simple responses to the novelty of the stimulus. Further support of a specialized social function comes from the observations that gray foxes, unlike other fox species studied, engage in much social grooming and appear to maintain the pair bond outside the breeding season. Behavior such as mutual licking and grooming (as occurs after a strange scent is added by rubbing) is well known among many species of mammals to have the function of renewing and strengthening social bonds.

THE RED FOX

In the test of the red foxes with strange conspecific feces, the smaller female urinated on the plate immediately after sniffing the feces of the larger female, who had defecated on the plate. It is our interpretation that this urination is specifically a scent-marking behavior elicited in response to the conspecific's action of defecating rather than being just a response to the odor on the plate. This is the sole occurrence of urination in all our tests with the red foxes. The larger female also defecated on the smear sample of impala feces. It is perhaps relevant here that Burrows (1968) in his studies of the red fox in England found that novel objects such as a trap or a dead dog would be covered with fox feces; this can only be interpreted as a marking response evoked by a novel stimulus in an otherwise familiar territory. Arousal that triggers marking may be a generalized phenomenon in some mammalian species. It might be argued that as arousal increases toward the edge of the territory or in unfamiliar territory or in the presence of novel objects, the probability of marking with urine, feces, or some specific glandular secretion is increased.

THE ARCTIC FOX

The Arctic foxes *(Alopex lagopus)* differed from the wolves and the gray and red foxes in their overall use of urination and, to a lesser extent, defecation as a marking method. Why the Arctics should prefer this one pattern and the reds and grays should not is a matter for further study.

Defecation was only used as a response in two tests, one with impala feces and one with the artificial scents. Rubbing was not observed in this species in any of the tests, but in a later study, rubbing was elicited in Arctic foxes by gray fox feces. The low frequency of rubbing in this species might be construed as evidence that there is little direct social function in the Arctic's rubbing behavior. We have never observed these foxes engaged in social grooming, and social investigation is of much lower frequency than in the wolf and gray fox.

Discussion

The most striking observation that can be drawn from our tests with the wolves is the exclusive occurrence of defecation in response to strange conspecific odors. This sharply contrasts the use of urination and rubbing for interspecific, prey, and artificial odor tests. We think that this distinction may be related to the highly developed sociability of the wolf. Rubbing may enhance social integration and communication in the pack. One could hypothesize that there could be less tendency to add to the body the scent of a conspecific, foreign to the pack.

In contrast, the distinction in the red and gray foxes was less clear-cut. Defecation occurred in response to the foxes' own odors, the control plate, conspecific tests, and prey scents, but there was no pattern or regularity as seen in the wolf. We tend to ascribe less of a social function to defecation by the foxes, and interpret the actions as a response to a novel stimulus but which may also serve to mask a strange odor.

Kleiman (1966) has defined scent marking in canids as follows: "Scent marking has been defined as urination, defecation, or rubbing of certain areas of the body which is (1) oriented to specific objects, (2) elicited by familiar conspicuous landmarks and novel objects or odors, and (3) repeated frequently on the same object." Using the data from our study and the framework of the above definition, we would define scent marking in terms of our experiment as (1) urination, defecation, and/or rubbing of the body oriented toward the test plate, (2) response to a novel, unfamiliar object or scent (note that while not denying that scent marking is elicited by familiar, conspicuous objects, this part of Kleiman's definition has little bearing on our study). The frequently repeated response to familiar objects was observed during the testing situations in our subjects but had no relation to the test stimulus. The male gray fox, for example, was observed to urinate continually on the middle, back steel part of his cage, resulting in an accumulation of crusted matter, probably a combination of urine and sawdust.

Kleiman (1966) also observes that

> Although dogs and foxes could use feces as material for marking, defecation is not used by the members of most species as a means of depositing odor on a specific object or landmark. The fennec fox and Arctic fox were occasionally observed defecating in response to novel stimuli: however, the act was never repeated more than once. . . . In the case of the Arctic foxes and fennec fox, feces were considered to be a marking substance because they were directed onto novel objects.

We would like to point out that in contrast with Kleiman's statement, our wolves and red and gray foxes did use defecation as a scent-marking response; in addition, in most cases we ascribe more significance to this action than to a simple response to a novel stimulus. The response of defecation may well have some value in intraspecific communication, especially in the wolf. This hypothesis requires further investigation.

To explain the rubbing and rolling of canids (see Fig. 10), we offer four possible hypotheses: (1) The animal, wolf or otherwise, rolls in the novel odor to familiarize itself with it; by rubbing, the animal adds the strange odor to its own body odor in order to make it less strange. This hypothesis does not account for the fact that the animal is attracted to the odor; also, an underlying motivation to become familiarized with the novel odor by "wearing it" is less tenable than the second hypothesis. (2) Rolling and rubbing may have important social consequences. In the wolves, adding a new scent to the body might influence within-pack social interactions. D. Guthrie (pers. comm.) proposes that "wearing" an odor (either a strange odor or the animal's own urine, as in ungulates) may be related to status-identity and self-assertive ("this is me") display. The odor might increase social investigative behavior and attention, and reduce aggression or assertion of dominance by socially superior individuals. (3) The animal imparts some of its own body odor onto the material in which it rolls, and, as happens when a conspicuous object (a rock or a tree) is marked, other individuals will be attracted to the material and learn of the first animal's presence. This hypothesis is dubious in that a more efficient way of marking in canids is to use urine or feces. (4) This behavior is simply a response to a pleasurable stimulus: wolves will often roll and rub against a deer carcass before they eat.

One interesting ontogenetic phenomenon is the control of urination. During the first three weeks in canids it is under maternal control, for the cub does not urinate until it is licked by the mother, who ingests urine and feces. This must certainly help in keeping the den clean and dry while the mother is away. When the neonate's bladder becomes extremely distended it will be spontaneously evacuated, however. With the subsequent development of voluntary control of urination and defecation, the young canid begins to evacuate independently at some distance from the den, often at a particular spot. Possibly as a consequence of emancipation and ritualization (Fox, 1971b), urination can be evoked by a conspecific; usually the latter is older and socially dominant. Submissive urination by a subordinate is seen during greeting or active submission (Schenkel, 1967) and is often associated with passive submission. This has been observed by the authors while being greeted by domestic dogs and hand-raised coyotes, golden jackals, and wolves. An intriguing, possibly partially emancipated submissive urination is seen in hand-raised red foxes. While greeting each other or their handler, they will briefly squat and urinate. This action is very brief, lacking the clear lateral recumbency or sideways twist with genital presentation seen in the aforementioned more social canids, and appears to be less of a social display than a nonritualized arousal or excitement reaction. The latter interpretation is supported by the fact that one female red fox not only urinates but also invariably defecates in a greeting-play soliciting display to one particular handler.

After urinating, many canid species will scrape[6] the ground with the feet. The probability of occurrence of this behavior is increased by the presence of a rival conspecific. Scraping may add a visual signal to the mark (analogous to tree clawing in the leopard), or an additional olfactory cue may be deposited on the ground by interdigital scent glands, which are present in red foxes (Tembrock, 1968) and possibly in other canids.[7]

Summary and Conclusions

Details of visual, auditory, tactile, and chemo-olfactory communication in various canid species are reviewed from an ethological standpoint. Comparisons between species add to our understanding of socio-ecology, where the behavior patterns of communication are related to the type of social organization that has evolved as an adaptation to a particular niche or life style (Fox, 1975b).

Future research could be advantageously directed toward more detailed studies of chemo-olfactory communication, an area where the least amount of work has been done in canids and in mammals in general. Rather than describe differences in behavior in various contexts (sexual, maternal, agonistic, etc.) this review focuses on comparisons between species of various modalities of communication since the same signals may occur in different contexts. In studying canid communication, it is considered important for the observer to know the nature of the relationships between interactees as well as their prior experiences, and imperative that the context be clearly defined.

Since many canid species will hybridize, cross-breeding studies may throw further light

6. Scraping may also occur in the absence of urination in the dog, wolf, and coyote during agonistic (threat) interactions.

7. Canids may also mark over, that is, they urinate where a mate or companion has just marked (Fox et al., 1975).

on the inheritance of certain actions and communication signals; the role of experience and of genetic factors in the encoding and decoding of visual, auditory, and chemical signals await future research. At this stage it may be concluded that while the expression of most signals (displays and vocalizations) are experience-independent (i.e., relatively environment-resistant), their decoding may be more experience-dependent.

In other words, how a canid communicates may indeed be innate, but how it responds and to whom and when may be significantly modified by early experience. This variable of early experience must be considered in studying captive and/or hand-raised animals, and, whenever possible, field studies should complement those of the laboratory. A wild canid also knows many of the signals of other species sharing its habitat and is exposed to a much more complex and variable environment in infancy than one raised in captivity. In the more social species, protocultural influences must be considered, where individual differences reflect differences in rank and social role, as differences between species reflect social adaptations to different sets of ecological factors.

References

Albone, E., and Fox, M. W., 1971. Anal gland secretion in the red fox. *Nature,* 233:569–70.

Bekoff, M., 1972. The development of social interaction, play, and metacommunication in mammals: an ethological perspective. *The Quarterly Review of Biol.,* 47:412–34.

Burrows, R., 1968. *Wild Fox.* Newton Abbot, England: David and Charles.

Cohen, J. A., and Fox, M. W., 1976. Vocalizations of wild canids and possible effects of domestication. *Behavioural Processes* 1:77–92.

Eisenberg, J. F., and Kleiman, D. G., 1972. Olfactory communication in mammals. *Annual Rev. Ecology and Systematics,* 3:1–32.

Ewer, R. F., 1968. *Ethology of Mammals.* New York: Plenum Press.

Fox, M. W., 1969. The anatomy of aggression and its ritualization in canids. *Behaviour,* 35:242–58.

Fox, M. W., 1970. A comparative study of the development of facial expressions in canids, wolf, coyote and foxes. *Behaviour,* 36:49–73.

Fox, M. W., 1971a. *Behavior of Wolves, Dogs, and Related Canids.* New York: Harper & Row.

Fox, M. W., 1971b. Socio-infantile and socio-sexual signals in canids: a comparative and ontogenetic study. *Z. Tierpsychol.,* 28:185–210.

Fox, M. W., 1972. The social significance of genital licking in the wolf, *Canis lupus. J. Mammal.,* 53:637–39.

Fox, M. W., 1974a. *Understanding Your Cat.* New York: Coward McCann.

Fox, M. W., 1974b. The behaviour of cats. In: *The Behaviour of Domestic Animals,* 3d ed., E. S. E. Hafez, ed. London: Bailliere.

Fox, M. W., ed., 1975a. *The Wild Canids.* New York: Van Nostrand Reinhold.

Fox, M. W., 1975b. Evolution of social behavior in canids. In: *The Wild Canids,* M. W. Fox, ed. New York: Van Nostrand Reinhold.

Fox, M. W., 1976. *Between Animal and Man.* New York: Coward McCann.

Fox, M. W.; Beck, A. M.; and Blackman, E.; 1975. Behavior and ecology of a small group of urban dogs. *Applied Animal Ethol.,* 1:119–37.

Fox, M. W., and Clark, A. L., 1971. The development and temporal sequencing of agonistic behavior in the coyote *(Canis latrans). Z. Tierpsychol.,* 28:262–78.

Fox, M. W.; Lockwood, R.; and Shideler, R.; 1974. Introduction studies in captive wolf packs. *Z. Tierpsychol.,* 35:39–48.

Gleason, K. K., and Reynierse, J. H., 1969. The behavioral significance of phereomones in vertebrates. *Psychological Bull.,* 71:58–73.

Golani, I., 1973. Non-metric analysis of behavioral interaction sequences in captive jackals (*Canis aureus* L.). *Behaviour,* 44:89–112.

Kleiman, D., 1966. Scent marking in the Canidae. *Symp. Zool. Soc. London,* 18:167–77.

Lowen, A., 1970. *Depression and the Body.* New York: Coward, McCann.

Schenkel, R., 1947. Expression studies of wolves. *Behaviour,* 1:81–129.

Schenkel, R., 1967. Submission: its features and functions in the wolf and dog. *Amer. Zool.,* 7:319–30.

Sprague, R. H., and Anisko, J. J., 1973. Elimination patterns in the laboratory beagle. *Behaviour,* 47:257–67.

Tembrock, G., 1960. Spezifische Lautformen beim rotfuchs *(Vulpes vulpes)* und ihre beziehungen zum verhalten. *Säugertierkunde. Mitt.,* 8:150–54.

Tembrock, G. 1968. Land mammals. In: *Animal Communication,* T. A. Sebeok, ed. Bloomington: Indiana University Press, pp.338–404.

Theberge, J. B., and Falls, J. B., 1967. Howling as a means of communication in wolves. *American Zoologist,* 7:331–38.

van Hooff, J. A. R. A. M. 1967. The facial displays of catarrhine monkeys and apes. In: *Primate Ethology,* D. Morris, ed. Chicago: Aldine, pp.7–68.

Zimen, E., 1974. Social dynamics of the wolf pack. In: *The Wild Canids,* M. W. Fox, ed. New York: Van Nostrand Reinhold.

Chapter 29

COMMUNICATION IN THE FELIDAE WITH EMPHASIS ON SCENT MARKING AND CONTACT PATTERNS

Christen Wemmer and Kate Scow

Introduction

It is ironic that students of social behavior and communication have not shown as much interest in cats as the general public has. Fewer than a third of the thirty-six living species of cats have been studied from an ethological standpoint. Because of the scarcity of information, most studies have been of a general nature, and few have been addressed to particular aspects of communicative behavior. Critical, detailed investigation of auditory, visual, and tactile modes of interaction is to our knowledge greatly underrepresented. At the time of writing there are only a handful of studies published on olfactory aspects of communication (Fiedler, 1957; Palen and Goddard, 1966; Verberne, 1970). For reviews of cat behavior the reader is referred to Leyhausen's pioneering studies (1956, 1960) and the more recent works of Schaller (1967, 1972). This paper is not intended as such a review, but reports on certain aspects of communication involving contact directed to objects or companions. It is hoped that our statements concerning the ecological determinants and design features of felid communication will stimulate more complete analysis than we can now offer.

Methods

Our studies have taken place at the Chicago Zoological Park (Brookfield Zoo). Unsystematic observations have been made during the past two years on the following species: African lion *(Panthera leo),* Indian tiger *(P. tigris),* leopard *(P. pardus),* jaguar *(P. onca),* snow leopard *(P. uncia),* clouded leopard *(Neofelis nebulosa),* cheetah *(Acinonyx jubatus),* puma *(Puma concolor),* lynx *(Lynx lynx canadensis),* sand cat *(Felix margarita),* wildcat *(F. silvestris),* Pallas' cat *(F. manul),* golden cat *(Profelis temmincki),* leopard cat *(Prionailurus bengalensis),* fishing cat *(P. viverinus),* margay *(Pardofelis wiedii),* Jaguarundi *(Herpetailurus yagouaroundi).* A focal animal method of observation (Altmann, 1974) has been employed to monitor social interaction in two litters of Pallas cat containing four (one male, three female) and five (one male, four female) young, respectively. Daily observations of varying length (ten to eighty minutes) were made on two litters of leopard cat (two male, two female, respectively) and one litter of sandcats (four male). In these cases we continuously sampled patterns of contact in all animals.

A brief review of felid natural history is necessary to understand the ecological setting to which the felid mode of communication is adapted.

The Felid Habitus

The felid body plan is progressive, but its uniformity between species is striking when compared with most other families of Carnivores. The small size and forest habitats of most living cats are probably primitive adaptations for utilizing the relatively diverse small vertebrate fauna inhabiting such regions (Kleiman and Eisenberg, 1973). The most economical scheme for exploiting such prey is a system of solitary land tenure. The cat occupies a more or less exclusive hunting ground and probably encounters other community members infrequently. The mother family, the most complex social unit, is but a brief, usually seasonal association. High-intensity vocalization and locus-specific marking with scent are the two predominant methods by which various species space themselves and avoid confrontations (Muckenhirn and Eisenberg, 1973; Schaller, 1967, 1972). In more open habitats movement may be regulated by vision (Leyhausen, 1965a). It is also likely that neighbors recognize one another, and that the "brotherhood" is characterized by a loose social hierarchy in which the territory insures even the lowest-ranking cat of priorities in resources and space (Leyhausen 1965a).

Since nearly all cats can kill prey as large as themselves, hostilities between conspecifics are potentially lethal (Leyhausen, 1960, 1965b; Schenkel, 1968; Schaller, 1972). This fact has been instrumental in evolving a more stereotyped repertory of distance signals on the one hand, and a highly graded repertory of proximal signals on the other. Furthermore, there is a parallel between the dependence on vision and audition in the localization and capture of prey, and the linkage of visual (facial expression) and auditory signals in proximal agonistic interactions.

Secondly, the feline estrous cycle and induced ovulation place several constraints on courtship. During proestrus the female becomes hyperactive and announces her emerging receptivity by rubbing and calling (Rabb, 1959; Michael, 1961). This attracts a number of males, and severe rivalry, which is manifested by a seemingly disproportionate amount of display relative to combat, gives rise to a dominant animal (Ewer, 1973). The female's repulsion of the male diminishes after further repeated intense and often strikingly dramatic transactions. Induced ovulation requires that the female copulate repeatedly, often over a period of days before ova are produced and conception can occur (Ewer, 1973; Schaller, 1972). Often as the sexual motivation of the female waxes, that of the male wanes (Ewer, 1974). Sexual exhaustion of the first male may result in his retirement from further involvement, and another male may then step in. The observation that males are more tolerant of one another than are females may in part be explained by the adaptiveness of nonfatal inter-male competition during these circumstances (Leyhausen, 1965a; Berrie, 1973; Provost et al., 1973). In summary, the usual solitary existence of the mature but sexually inactive cat breaks down when females come into estrus and become highly attractive to males. The proximity of rival males provokes intensive display and fighting. Further agonistic behavior develops between the dominant male and female. In these contexts and in occasional territorial disputes motivational differences between animals are resolved through intense and highly modulated interactions.

Some General Features of Felid Communication

In view of the paucity of information for a variety of species, it is difficult to make generalizations. However, the available information suggests several parallels and contrasts with other carnivore groups.

There are six calls that are common to the repertories of most small cats that have been studied and these can be grouped into categories of discrete and graded call types. Some physical properties of these calls are listed in Table 1. All of these vocalizations, with the exception of spitting, vary in intensity, duration, and emission rate. Spitting and hissing share features of broad nontonal energy distribution. Sonographed examples of the two calls can be arranged on a continuum from long, moderate intensity hisses to brief, loud, and explosive sounding spits. However, most examples fall at the extremes of the continuum, and therefore the labeling of two basic calls seems justified. Whether the variants are lumped or split, the call(s) cannot be considered graded because the variation between extremes does not involve qualitative differences. Intermediates exhibit quantitative variation within the same physical parameters.

Graded calls among the small cats display several characteristics. Call transitions often occur without an interruption in the air column. This produces a usually short intermediate segment of sound that shares certain characteristics of the preceding and following calls. This kind of noninterrupted, inter-call gradation has been described for a number of small carnivores (Wemmer, in press) and primates, particularly open-habitat terrestrial forms (Marler, 1965, 1967). The features of this kind of acoustical grading are to be distinguished from the temporally discrete (interrupted) but graded calls of other mammals (e.g., red colobus monkey: Marler, 1970).

A second feature of these graded vocalizations is that gradation often seems to be one-directional. For example, howling (an harmonic call) may arise from growling (a pulsed call) and terminate with a scream (a noisy, high-frequency call) followed by a rapid volley of spitting. Rising excitation and sound intensity seem to be common concurrent features of such sequences. When growling is resumed it is usually after an interruption in sound production. Extensive sonographic analysis is needed, however, to confirm this observation, for each call type may be considerably modulated in frequency, intensity, and noise level. As Schaller (1972) noted, the repertory of cats is smaller than it seems because discrete signals are uncommon.

These calls share certain features with the graded vocalizations of terrestrial primates discussed by Marler (1965, 1967). They are addressed to conspecifics at close range and are accompanied by highly varying facial expressions. The potential of such visual-vocal signal systems for communicating fine-grain motivational changes has been discussed by Marler. In cats these signals occur mainly during proximal agonistic interactions, territorial skirmishes, and preambles to copulation. The sounds may broadcast and attract neighboring cats, but this seems to be a secondary or inadvertent side effect. The discrete, high-intensity calls that serve to attract or space neighbors seem to be lacking, or are at most only poorly developed, among the small cats that have been studied. Toms are certainly attracted by the discrete and repeated miau of the estrous domestic cat, but this call cannot be compared on relative grounds with the roaring and sawing calls of the great cats.

This leads us to consider the sensory assortment of the feline signal repertory. Based on Leyhausen's (1960) studies of the domestic cat, Eisenberg (1973) tabulated twenty-five visual patterns (facial expressions, tail and body postures), sixteen of which occur in combination. There were eight vocal patterns and three olfactory patterns, and to this list can be added about seven contact patterns (body rubbing, clasp, mount, bite, lick, pat, hind leg pump). A similar profile emerges from Schaller's (1972) lion study; there are at least seventeen visual patterns, thirteen vocalizations (including graded series), seven contact patterns, and five olfactory patterns. On the basis of this cursory examination, vision and audition seem to be the two top-ranking signal modalities, a pattern that agrees with the general impression that cats are essentially sight- and sound-oriented animals.

Scent Marking

Urine, feces, and glandular exudates of the skin are potential carriers of chemical information in nearly all terrestrial mammals. In the Felidae feces seem to lack the widespread communicative significance that they have in the canids, but urine is undoubtedly an important information carrier. The five patterns of object-oriented contact that can be distinguished vary in expression between species and possibly between sexes; interspecific postural differences and variation in sequencing also exist, but detailed studies on these aspects are lacking.

Urination occurs in a squatting or standing position. The latter is nearly a universal male felid trait, but it is also seen in the females of some species (Table 2). It is assumed that retromingent urination against upright objects (*Harn-Spritzen* of Leyhausen, 1956; spraying or urine spraying of Schaller, 1967, 1972) has evolved specifically as a scent-marking pattern. The urine is spread over a larger area than if deposited on the ground; it can be sniffed at head level; and according to the diffusion model of Bossert and Wilson (1963), a point source of scent above ground can produce an active space as much as twice as large as a scent source at ground level.

Ewer and Wemmer (1974) have also pointed out that the height of a scent mark can be used to judge the sex of the owner; however, this information might be redundant to sex identifiers in the scent itself. The upheld tail at times touches part of the object to be sprayed, suggesting that it assists in orienting the direction of the spray (Fig. 1a). More often than this, however, the terminal part of the tail undergoes a marked erratic twitching lasting several seconds. The pattern occurs in snow leopard, lynx, leopard cat, and domestic cat, but is apparently absent in the other Pantherinae. While Schaller states that spraying is purely an olfactory pattern in lions, the conspicuous character of this movement in the above-mentioned cats implies it may also have a visual signal function. Alternately it may simply be an autonomic manifestation of urine emission. Schaller (1972) states that in lions one to twenty jets of urine are emitted during the assumption of the characteristic stance and that it travels three to four meters.

Defecation is similar between sexes and species, but its association with other patterns of behavior differs between species. In many small felids (particularly the genus *Felis*) feces are deposited in areas where they can be covered by repeated scratching motions of a forefoot. Larger cats *(Panthera)* make no attempt to cover feces; defecation does not occur in conspicuous areas and is not locus-specific, as it is in many other carnivores. Lindemann (1955) reported that both the European wildcat and the lynx bury their feces at specific localities within their territories, but leave them uncovered on stones and tree stumps in the spaces between territories *(Niemandslandstreifen)*. In the lynx these "rendezvous" sites are eagerly sought out during the mating period by all adult conspecifics (Lindemann, 1955).

While anal scent glands occur in most small cats, the burying of feces suggests that anal scent secretion does not take place during defecation, as it does in certain rodents. Adamson (in Schaller, 1967) reported that anal scent was voided during urine spraying in her female lion Elsa, and Schaller (1967) observed that tiger urine deposited by spraying has a strong musky odor compared with urine deposited in a squat-

Fig. 1. Scent-marking patterns in a male snow leopard (*Panthera uncia*): a. Urine spraying (notice position of tail). b. Scuffing with the hind feet. c. Sniffing of leg prior to head and neck rubbing. d. Head and neck rubbing. The animal must rise up on its hind legs to touch the elevated branch.

Table 1

Some physical properties of vocalizations common to members of the genera Felis, Prionailurus, and Lynx.

Vocalization	Harmonic structure	Duration	Relative intensity	Frequency modulation
Hiss	–	Variable but usually brief	Moderate	–
Spit	–	"Fixed"	Moderately loud	–
Purr	–	Variable, but usually long and repeated	Soft	–
Growl	–	Variable, but usually long and repeated	Moderate	+
Miau	+	Variable	Moderately loud	+
Scream	+	Variable	Loud	+

Table 2

Distribution of scent-marking patterns in various Felidae.

Species	Urine spraying M	Urine spraying F	Feces	Scuffing (scraping)
African lion (*Panthera leo*)	+	–	Scattered haphazardly, not buried	+
Tiger (*P. tigris*)	+	+	Scattered and covered	+
Leopard (*P. pardus*)	+	–	Occasionally deposited on a scrape made by scuffing	+
Jaguar (*P. onca*)	+	–		
Puma (*Puma concolor*)	–	–	Deposited on scrapes	+
Snow leopard (*Panthera uncia*)	+	–	Scattered, not buried	+
Clouded leopard (*Neofelis nebulosa*)	+			–
Cheetah (*Acinonyx jubatus*)	+	–	Occasionally deposited on scrapes	+
Canada lynx (*Lynx lynx*)	+	+	Localized and covered and uncovered	?
Bobcat (*Lynx rufus*)	+	–	Localized deposits	?
Fishing cat (*Prionailurus viverinna*)	+	?	In water ?	–
Leopard cat (*P. bengalensis*)	+	–	Scattered, buried	–
Margay (*Pardofelis wiedii*)	+	–	Not buried	+
Golden cat (*Profelis temmincki*)	+	?	?	?
Jaguarundi (*F. yagouaroundi*)	+	–	?	+
Pallas cat (*F. manul*)	+	–	Feces covered	–
Sand cat (*F. margarita*)	+	–	Feces covered	–
Domestic cat (*F. catus*)	+	–	Covered	
Wildcat (*F. silvestris*)	+	–	Locus specific	–

ted posture. He attributed this to a granular white precipitate, but even in captivity where spraying can be viewed at close range it is not possible to witness the emission of the anal scent, and it is possible that the white precipitate seen was actually the glycerides excreted in the urine by the cats.

Scuffing (scraping, treading) with the hind feet is commonly associated with urination (Fig. 1b). The animal squats on its hindquarters with the entire length of its hind feet touching the ground. Then each foot is alternately thrust backward while the claws are usually extended. Upon completion of this scuffing motion the foot is almost always lifted as it is brought forward. Occasionally one foot may repeat the scuffing motion up to six times before the other foot comes into action (snow leopard). The emphasis of the contact along the length of the foot probably varies between species, as a number of varia-

Table 2 (continued)

Head rubbing	Claw raking	Recumbent head rubbing	Reference
+	+	+	Fiedler (1957), Schaller (1972)
+	+	+	Fiedler (1957), Schaller (1967)
+	+	+	Eisenberg (1970), Fiedler (1957), Muckenhirn and Eisenberg (1973), Schaller (1972)
+	+		Fiedler (1957), Wemmer and Scow (pers. obs.)
+	+	+	Fiedler (1957), Hornocker (1969), Seidensticker et al. (1973)
+	+	+	Hemmer (1968, 1972)
+	+	?	Hemmer (1968)
+	+	+	Eaton (1970)
+	+	+	Wemmer and Scow (pers. obs.), Lindemann (1955)
—	—	?	Provost et al. (1973)
+	+	+	pers. obs.
+	+	+	pers. obs.
+	+	+	M. Peterson (pers. comm.)
+	+		pers. obs.
+	+	+	Ewer (pers. comm.)
+	+	+	pers. obs.
+	+	+	Hemmer (1974), Wemmer and Scow (pers. obs.)
+	+	+	Lindemann (1955)

tions have been described in the Viverridae (Wemmer, in press). The tempo of the movement is moderately fast, but varies somewhat with the size of the species.

It can be seen in Fig. 2 that the variation within an individual is slight and manifests a typical intensity in rate of delivery, but not in bout length. The result of the movement is a characteristic scrape in which loosened soil is heaped at the posterior end. The movement may also scrape bark loose or scratch the surface of horizontal or diagonal logs (jaguarundi). Urination occurs for brief periods during or near the termination of the scuffing bout or after the movement has stopped. Urination during scuffing may be associated with a slight deceleration or even a brief pause in leg motion. The churning effect of the feet mixes urine into the soil, and distributes it along the length of the foot. The fur on the sides of the feet of captive snow leopards may acquire a yellowish tinge from the activity, but this may result from both unusually high levels of scuffing and its repeated occurrence at one site in captivity. In the African lion the urine often wets the hind legs (Schaller, 1972). Thus it may act as a solvent transferring pedal scent to the substrate.

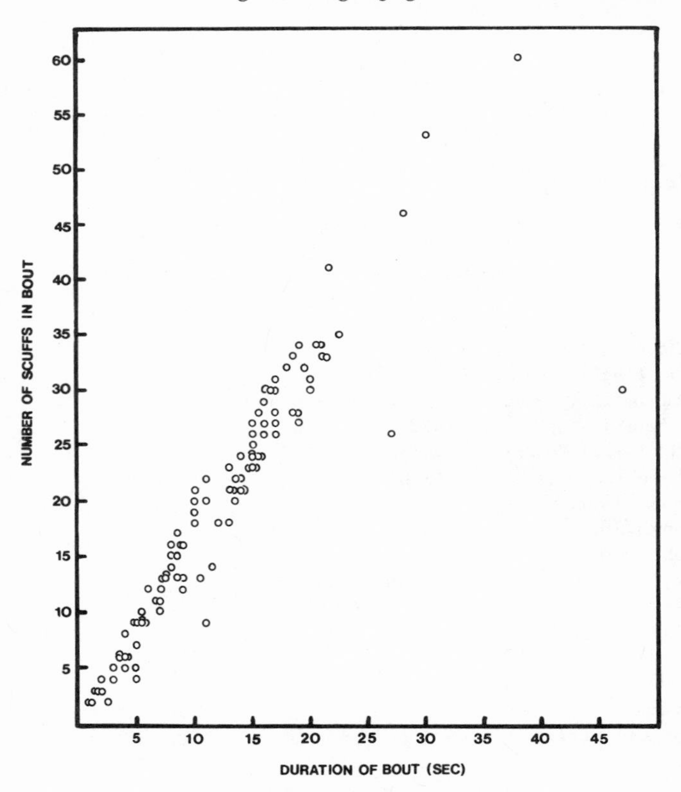

Fig. 2. The relationship between the number of scuffs composing a scuffing bout and the duration of the bout in a male snow leopard (*Panthera uncia*). Bouts varied from 1 to 47 sec and contained 2 to 60 individual leg movements (scuffs).

Seidensticker et. al. (1973), who studied the mountain lion in the Idaho primitive area, found that nearly all the 86 scrapes examined were made by resident adult males, and that the frequency of scraping was highest in the overlap zone between the home ranges of two radio-tracked males. From one to six scrapes were found within a small area at each site; they were usually placed near but not on animal trails, and only 11 of the 86 sites were revisited. Feces or urine was detected at 17 of the sites, usually on the soil and needles heaped up by the scraping action. Whether scrapes are made with the fore- or hind feet is not known. The authors in this study concluded that scrapes demark the home ranges of adult males by indicating that the area is occupied.

Schaffer (1940) described glands in the feet of the domestic cat, but the extent and type of glandular development in the feet of other species is not known. Scrapes made by tigers are also defecated upon if only in small amounts; these scrapes are more evident and perhaps more commonly performed during the monsoon, while the feces persist longer than the scrape during the dry season (Schaller, 1967).

Head rubbing is associated with several postural variants and is almost always preceded by sniffing (Fig. 1c, d) or licking and biting of the focal object, and by flehmen. Most commonly the face is rubbed as the animal stands beside a branch or rock at head level. The contact phase is often one-directional; it occurs as the body leans toward the object with the cheek or neck serving as the point of contact. Forward movement of the body is achieved by a combination of neck extension, walking, and leaning forward. Upon breaking contact the cat may assume a different position in relation to the object or may repeat the movement from the same starting point. The head, cheeks, and neck may also be turned and rotated as the animal stands still, sometimes securing the branch with the claws of a forefoot. A similar style is employed by a sitting cat. If the branch is elevated the cat will grasp it with one or both forefeet while standing upright on the hind legs. Rubbing and sniffing very often lead shortly to salivation. A clear watery saliva appears on the closed lips and is wiped onto the object and the cheeks and neck. A vigorous bout of rubbing virtually soaks these areas.

The cheeks and neck are also rubbed against novel or odoriferous substances on the ground. Here the cat reclines with the object between the forefeet. After the object is sniffed and licked, the sides of the head and neck are pressed against it and extended. The movement often alternates with sniffing, or licking and flehmen, and there is a tendency for the same side of the head to be used several times before switching to the opposite side. For example, the sequence of actions in an adult male snow leopard in response to an unknown scent in the soil was: sniff, right, right, left, sniff, left, pause (22 sec) sniff, right, sniff and lick, right, right, left, lick, left, stand. This recumbent rubbing, which is also characterized by salivation, may lead to rolling and writhing on the sides and back. There are two basic differences between recumbent head rubbing and the other versions. The former is clearly evoked by strong novel odors such as carrion, vomit, the feces of strange animals, and catmint. The latter pattern, which is part of the daily routine, characteristically occurs against objects that are also sprayed with urine.

Depending on the position of the tree trunk, claw raking is performed in an upright or a horizontal position. The cat generally grips the trunk with extended forelegs and depressed body, and the claws are then drawn backward simultaneously or alternately in strokes of variable length and speed. The motion often has a jerky quality that results from the intermittent snagging of the claws and raking. The action serves to remove loose claw sheaths but also leaves a visual and possibly an olfactory trace having social significance to other cats.

The occurrence of these patterns in different felid species is presented in Table 2, but there have been few studies on the temporal organization of scent-marking behavior. Urine spraying occurs by itself and in association with other patterns. Eisenberg (1970) reported that in the Ceylon leopard scraping (scuffing) and urine spraying occurred near the sloping trunks of trees in which clawing with fore- and hind feet (scuffing?) and cheek rubbing took place. In the snow leopard, cheek rubbing, scuffing, and spraying are often associated acts (Fig. 3); sniffing preceded cheek rubbing, scuffing, and spraying in a decreasing proportion of cases, while urine spraying and cheek rubbing were the most common and second most common terminal acts. In Pallas cats clawing and urine spraying are coupled together, and Schaller (1972) reports that head rubbing may precede spraying by male Serengeti lions, particularly at spraying sites used by other males.

Table 3 compares some attributes of the six different scent-marking patterns. Urine spraying is distinctive in being a one-directional marking pattern, which by itself does nothing to modify the sender's body odor. Clawing involves forelimb contact, but it is doubtful that much of the

Table 3

A comparison between characteristics of scent-marking and anointing patterns common to the Felidae.

Pattern	Material deposited	Substrate disturbance	Anointing material	Habitual site usage	Possible information potential
Urine spray	Urine	None	None	Yes	Individual identity, sexual identity and condition.
Head and neck rubbing	Saliva, skin exudates	None	Saliva, glandular exudates	Yes	Individual identity.
Scuffing	Urine, feces, skin exudates	Considerable	Urine	Yes	Individual identity, sexual identity and condition.
Clawing	Skin exudates	Considerable	None	Yes	Individual identity.
Recumbent head and neck rubbing	Saliva, skin exudates	Minimal	Carrion, feces, vomit, catmint	No	Individual identity, sexual identity and condition.
Rolling	Skin exudates	Minimal	Carrion, feces, vomit, catmint	No	?

cat's personal odor is imparted to the substrate. Like scuffing, however, the behavior may considerably modify the substrate, and it is likely that the odors emanating from the lacerated bark and scraped soil modify and present a "disturbance" to the olfactory landscape. This substrate modification probably creates a detectable secondary cue that intensifies perceptibility of the signal's location.

Cheek rubbing and scuffing, associated with urination, however, transfer body secretions (saliva and urine) to other body regions, namely, the head, cheeks, neck, shoulders, chest, and hind feet. These parts of the body are also targets of companion-oriented contact, a problem to be considered in the following pages. Recumbent rubbing and rolling, on the other hand, clearly differ from the above patterns in being directed to decomposing animal matter. They differ from each other in the extent to which the body is covered with the foreign scent, but it is important to remember that these behaviors may occur in tandem. In any case, an overriding result of either act is that the cat's body odor is radically changed, and any trace of the animal's scent left at the site must be slight compared with the intensity of the foreign scent source.

In passing, mention should be made of motivation and function. The difficulty in understanding these aspects of felid scent marking is that marking is often not characteristically linked to other motivationally distinctive (aggressive, sexual, or fearful) behavior. Recumbent head rubbing and rolling is an exception. Todd (in Palen and Goddard, 1966) studied this pattern in domestic cats as a response to catmint; he showed that as a specific response to this plant its expression is controlled by a single allele, and he regarded it as a sexual display.

Palen and Goddard revealed that as a reaction to refined catmint (trans-cis-nepetalactone) rubbing and rolling were independent of sex and gonadal state. The occurrence of the pattern in the estrus female and as a reaction to this chemical led them to conclude that both conditions increase skin sensitivity on the head. Thus, the occurrence of catmint-evoked rubbing and rolling is noncyclical in both sexes, while in the

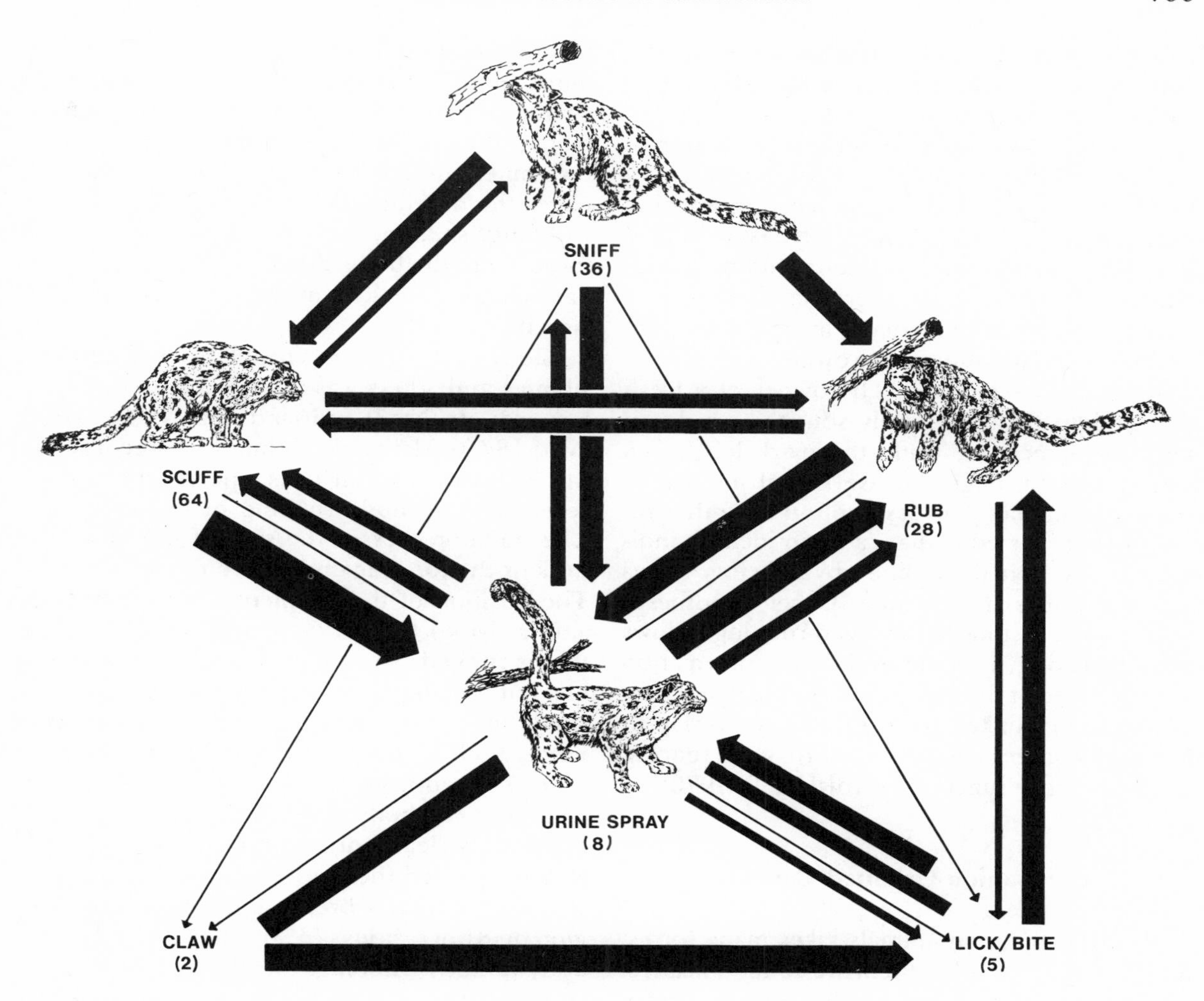

Fig. 3. Sequencing of scent-marking patterns in a male snow leopard (*Panthera uncia*). The thickness of the arrows is proportional to the percentage of transitions from one pattern to another. For example the scuff-to-urine-spray arrow represents 66% of all transitions originating with scuff.

female spontaneous rubbing and rolling cycles with the estrous period. If the pattern evolved as an olfactory signal the male would be expected to sniff sites where the female has rubbed.

Palen and Goddard did not investigate the pattern from a zoosemiotic standpoint; however, Michael (1961) reported that the male watches the proestrous female during this activity and attempts to mount her. The female's orgiastic postcoital rolling frequently provokes forelimb sparring with the male. Therefore the display appears to be visual rather than olfactory; when sniffing and licking occur in the sexual context they are directed to the female's head and neck, and the vulva. Possibly the orientation to the anterior body areas is reinforced by heightened rubbing during proestrus.

Less is known concerning the other patterns. Leyhausen (1965a) remarked that while territory marking is often interpreted as having a warning function in solitary mammals, there is no indication that a cat is intimidated upon sniffing a companion's scent mark. The receiver's response is neither overt nor immediate, and probably depends on the identity of the owner and its relationship to the receiver. The owner of a fresh mark might be aggressively sought by a dominant male receiver, while the mark left by an estrous female might arouse sexual interest in the same animal. The assumption here that remains to be tested is that cats can identify individuals and sexual condition by olfactory traces in urine and skin secretion. Other hypotheses also require testing. When head rubbing follows urine spraying, is some additional information incorporated at the site? It is possible that different sequence-linked patterns simply overlap in message content and serve as cross-referencing or redundancy function (Birdwhistell, 1970)?

Companion-Oriented Contact

Contact between animals takes many forms, and probably not all of them are communicative. At one extreme there are the fleeting, infrequent, and seemingly inadvertent contacts that result from proximity due to a common concern or activity. An example is the crowding and bumping of bodies that accompany the flight of a family group into a burrow. Sender and receiver are difficult to delineate, and the points of contact often vary without a predominating pattern. Most contact, however, is clearly intentional, of variable duration, and specifically oriented to a part of the companion's body. It may be a single brief, one-directional act, but if several points of contact exist between two or more animals, the bodies are usually aligned in characteristic configurations. The configurations restrict the range of targets selected for contact (Wemmer and Fleming, 1974). In the following pages the distributions of four contact patterns (rubbing, sniffing, biting, and patting with the forepaw) to the body targets of siblings are compared among the leopard cat, sand cat, and Pallas cat.

Body rubbing between siblings and between siblings and adults was infrequently seen in the leopard cats (N=7) and sand cats (N=14). However, 187 incidents of rubbing were recorded in the Pallas cats. In all three species the pattern resembles the body rubbing of domestic cats. The initiator presses the side of the head and neck or the torso against the companion's body. The position of the recipient's body (standing, sitting, lying) to an extent determines the area that is rubbed. Several juxtapositions are possible, but for simplicity we can consider four basic situations ranked in decreasing order of occurrence: body to body (79%), body to head-neck (8.5%), head-neck to head-neck (6.4%), and head-neck to body (5.3%). The body-to-body category is clearly divisible into two groupings. Kittens pressed their sides against the sitting or standing mother's breast and lower throat region on 81 occasions (43%), while the remaining body-to-body rubbing for the most part occurred between siblings and consisted of one cat's rubbing its side against the other's.

Of the remaining three patterns, sniffing was exhibited least often (Fig. 4). Though it is difficult to assess the relative importance of companion-oriented sniffing in different species, it is our impression that it is not as prevalent in these small cats as it is in other carnivores. For example, in meerkats observed under similar conditions the overall sniffing rate was 7.9/hr (data from Wemmer and Fleming, 1974). In the leopard cats the rate was 2.3/hr and in the Pallas cats

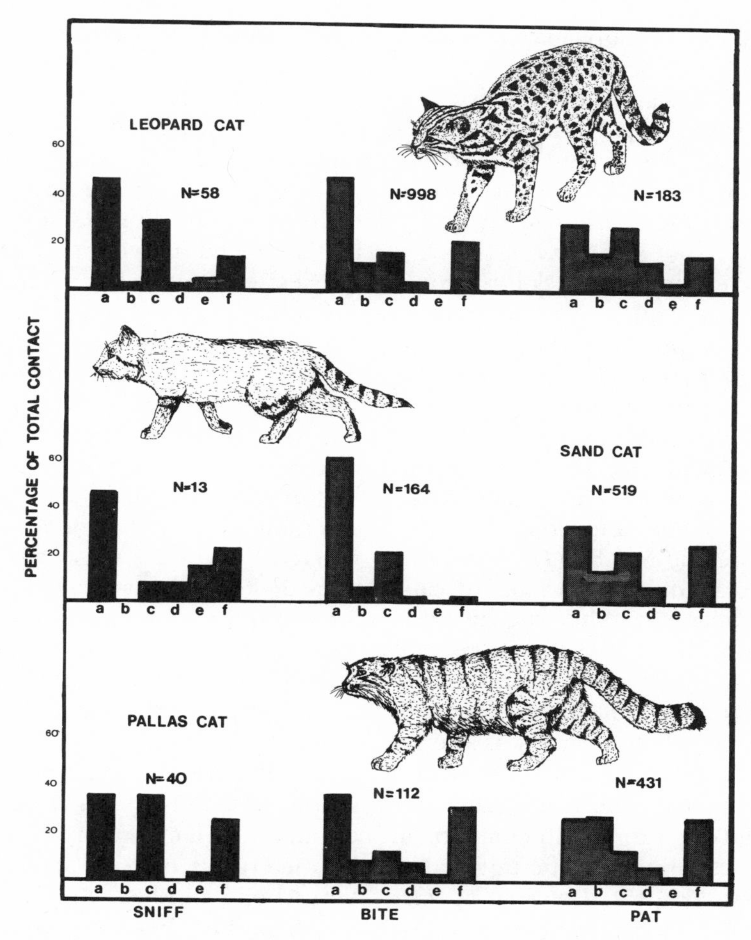

Fig. 4. The percentage distribution of sniffing, biting, and patting to general body areas of siblings in the leopard cat, sand cat, and Pallas cat. a = head-neck, b = forelimbs (including feet), c = torso, d = hind limbs (including feet), e = ano-genital region, f = tail.

2.1/hr (data for sand cats are not in comparable form).

In all three species the head-neck and tail rank high among the general body areas sniffed (Fig. 4); however, there are differences among species. In the leopard cats and sand cats the head-neck was the area most often sniffed, while this area and the torso received equal attention in the Pallas cats. The torso was highly sniffed in the leopard cats and least sniffed in the sand cats, but the sample size for the latter species was small. In all species the ano-genital region was infrequently checked, which is in marked contrast to other species of carnivores, particularly canids and viverrids (Ewer, 1973; Wemmer, in press; Wemmer and Fleming, 1974). Michael (1961) has observed in fact that the ano-genital region of anestrous domestic cats has a repelling effect when sniffed by sexually mature males. When the histograms for these three species are compared statistically a significant positive correlation is found for the leopard cats and Pallas cats ($r_s = .940$, $P < .05$, Spearman rank correlation coefficient). Unfortunately, the small number of sand cat observations are inadequate for statistical treatment.

In all three species the head-neck is the most frequent biting target, with the torso ranking second and third in importance (Fig. 4). Pallas and leopard cats differ from the sand cat in that the tail is the second most common biting target. All other areas (limbs and ano-genital region) receive a relatively small portion of bites compared with these three areas. The ranked distribution of biting in the Pallas and leopard cats are identical, and a correlation of .829 ($P = .05$) was found between the sand cats and leopard cats, and sand cats and Pallas cats. When specific areas of the torso are considered there is no general agreement in biting (Fig. 5). Only in the sand cat did one body area receive more than 27% of all bites. In the leopard cat the rump was the favored target, and in the Pallas cat the side and belly. In all three species chest and haunches were always low-ranking targets. In comparison with the general picture no specific target predominates the torso (Fig. 5). Between species tests of specific head-neck and body targets (Fig. 5) produce relatively high correlations, but only Pallas cat and sand cat exhibit a statistically significant correlation at $P = .01$.

Fig. 4 shows that patting in the three species is directed mainly to the forebody, infrequently to the hind legs and ano-genital region, but more often to the tail. Ear, cheek, and neck stand out among the specific head-neck targets, while back and side predominate in the torso (Fig. 5). There are no significant correlations among head regions of the three species; however, values for leopard cats and Pallas cats, and leopard cats and sand cats were significantly correlated ($P > .01$, $> .05$, respectively).

Discussion

Scent transfer between cat and environment is mediated through elimination patterns and body contact with inanimate objects. Not all parts of the body are used in scent marking through contact. For example, the head and neck are employed in two distinct patterns by all the species reviewed, while scuffing seems to be restricted at least to the Pantherinae and cheetah, but is probably absent in the genus *Felis* (Table 2). The zone of contact between body and object has properties similar in some ways to a mechanical joint or an articular facet (I. Golani, pers. comm.). The relationship between the object and other, more distal body parts can vary considerably, but movement at the joint is more restricted. Contact between animals can be regarded in a similar way. The body has certain focal points for contact, and indeed it can be visualized as a field of valences; the shape or "relief" of the field may differ for different types of contact (sniffing, biting, or licking). The distribution of valences (or the form of the relief) is probably determined on the one hand by topographical features such as the location of sense organs and glandular areas, and on the other by characteristics of body movement and orientation to one another.

Recumbent head and neck rubbing in the felids has analogs of varying similarity in certain members of all but one family of the Carnivora, including three species of civet (Viverridae: Wemmer, in press; Ewer and Wemmer, 1974), the spotted hyena (Kruuk, 1972), various dogs (Fox, 1971), the polar bear, and the tayra (a member of the Mustelidae, pers. obs.). The specific use of the head and neck in these patterns suggests the hypothesis that the additional scent may enhance the attractiveness of those regions for certain types of contact received from companions. In all three cat species the head-neck region was the predominant sniffing target, though the tail also received considerable atten-

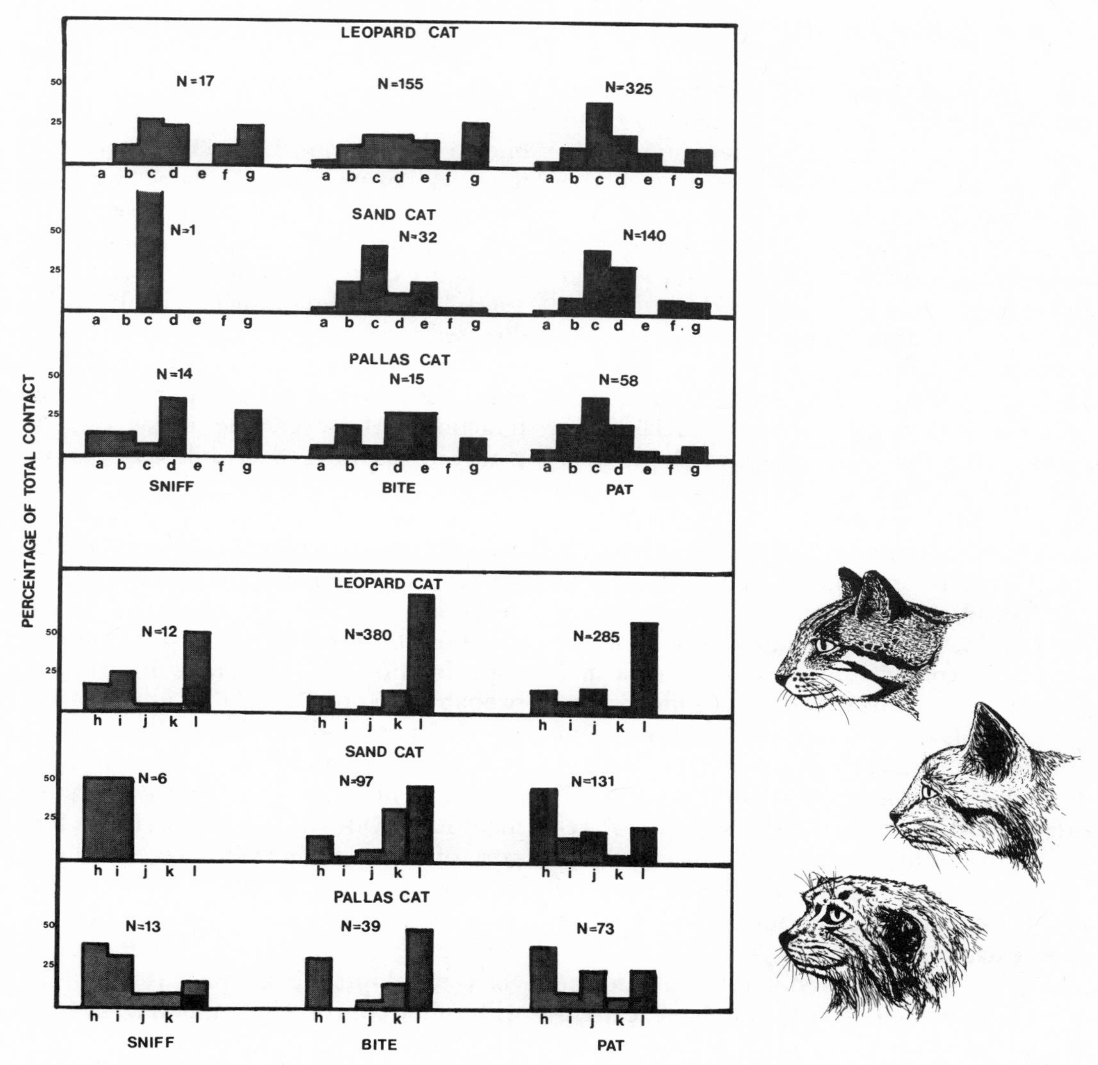

Fig. 5. The percentage distribution of sniffing, biting, and patting to specific regions of the torso (upper three lines) and to specific head regions (lower three lines). a = chest, b = shoulder, c = back, d = side, e = belly, f = haunch, g = rump, h = ear, i = nose, j = cheek, k = throat, l = neck.

tion. The head-neck zone also received most of the bites, as well as a substantial proportion of pats from the forelimbs. The data do not make a compelling defense for the hypothesis. Furthermore, anointing is only occasionally seen even in captive situations, and rubbing, which was frequently observed only in the Pallas cats, involved the head-neck zone but minimally (about 20%). In other words, support the hypothesis gains from the sniffing data is weakened by the biting and patting data.

Other species show similar trends whether

head and neck anointing is in their repertory or not. The African civet frequently and vigorously anoints the cheeks and neck with a variety of substances, including its food; up to 80% of all snaps and bites are directed to this contrastingly marked region, which is also presented to the biting animal. However, the same area receives less than 40% of all sniffing (Ewer and Wemmer, 1974; Wemmer, in press). In the genet, a viverrid carnivore in which anointing is weakly developed and neck presentation is absent, these areas receive about 70% of the bites and about 25% of the sniffing (Wemmer, in press). Head and neck rubbing is absent in the meerkat (a social-living mongoose), but these areas receive about 40% and 35% of all bites and sniffs (Wemmer and Fleming, 1974). In most of the examples the head is the most common sniffing target and the head and neck the prevailing biting target. The anterior location of the sense organs undoubtedly determines the predominance of these regions during contact, but skin secretions and possibly strange scents may provide olfactory information. This is supported by the observation that the ear is the most commonly sniffed part of the head in the Pallas cats and sand cats (Fig. 5). Schaffer (in Kleiman and Eisenberg, 1973) reports that the auditory meatus in cats is glandular, but his observations are based on the domestic cat, and interspecific differences are likely to exist.

The anal glands of the domestic cat and the sand cat are voided during traumatic experiences, and the scent has a pungent, unpleasant odor. There is no indication that the glands are used in scent marking, either alone or in combination with urine spraying. The anal region is glandular in many mammals, and in some species it is highly attractive to conspecifics (Fox, 1971). There are two situations in which cats are attentive to the ano-genital region. Mothers sniff and lick this region in their infants and ingest the milk feces, and males respond in a similar way to the vulva of proestrous and estrous females. On the whole, however, cats are not particularly oriented to this part of the body.

Small felids dispatch their prey with a fatal bite to the nape of the neck; the orientation to the neck constriction is innate, but the precise orientation to the nape of the neck and the necessary pressure for a lethal bite must be learned through experience with live prey (Leyhausen, 1960, 1965b). The young cats of all three species directed their bites to the neck more often than to other parts, a situation that implies that this innate orientation is also operant in sibling interactions. However, the Pallas cats directed most bites to the throat, while the other two species oriented more strongly to the nape (Fig. 5). Differences in neck structure undoubtedly contribute to this behavioral contrast. The Pallas cat is a relatively short-necked species, and the condition is particularly pronounced in the kittens. The throat is probably more vulnerable than the nape, but postural differences during interaction probably also play a role.

There are other similarities and contrasts among species and patterns for which explanations are not apparent. The contact mapping method provides a refined description for a category of behavior that often receives little attention, but the general picture of contact we have presented is the sum of occurrences from various contexts and postural configurations during early development. Contact is also integrated with other concurrent communicative activities. No doubt a better understanding of target selection will be gained by relating mode and target of contact to these variables.

References

Altmann, J., 1974. Observational study of behavior sampling methods. *Behaviour,* 197:227–67.

Berrie, P. M., 1973. Ecology and status of lynx *(Lynx canadensis)* in interior Alaska. In: *The World's Cats,* vol. 1, R. Eaton, ed. Laguna Hills, Cal.: Lion Country Safari, pp.4–41.

Birdwhistell, R. L., 1970. *Kinesics and Context.* New York: Ballantine Books.

Bossert, W. H., and Wilson, E. O., 1963. The analysis of olfactory communication among animals. *J. Theoret. Biol.,* 5:443–69.

Eaton, R. L., 1970. Group interactions, spacing and territoriality in cheetahs. *Z. Tierpsychol.,* 27(4):481–91.

Eisenberg, J. F., 1970. A splendid predator does its own thing untroubled by man. *Smithson.* 1:48–53.

Eisenberg, J. F., 1973. Mammalian social systems: are primate social systems unique? *Symp. Fourth Int. Congr. Primat.,* 1:232–49.

Ewer, R. F., 1973. *The Carnivores.* Ithaca: Cornell University Press.

Ewer, R. F., 1974. Panel discussion. In: *The World's Cats,* vol. 2, R. Eaton, ed. Seattle: Woodland Park Zoo.

Ewer, R. F., and Wemmer, C. 1974. The behaviour in captivity of the African civet, *Civettictis civetta* (Schreber). *Z. Tierpsychol.,* 34:359–94.

Fiedler, E., 1957. Beobachtungen zum Markierungsverhalten einiger Säugetiere. *Z. Säugetierk,* 22:57–76.

Fox, M. W., 1971. *Behaviour of Dogs, Wolves and Related Canids.* London: Jonathan Cape.

Hemmer, H., 1968. Untersuchungen zur Stammesgeschichte der Pantherkatzen (Pantherinae). Teil II, Studien zur Ethologie des Nebelparders *Neofelis nebulosa* (Griffith 1821) und des Irbis *Uncia uncia* (Schreber 1775). *Veröffentlichungen der Zoologischen Staatssammlang München,* 12:155–242.

Hemmer, H., 1972. *Uncia uncia. Mammalian Species,* No. 20:1–5.

Hemmer, H., 1974. Studien zur Systematik und Biologie der Sandkatze (*Felis margarita* Loche, 1858). *Z. des Kölner Zoo.,* 17(1):11–20.

Hornocker, M. G., 1969. Winter territoriality in mountain lions. *J. Wildl. Mgt.,* 33:457–64.

Kleiman, D. G., and Eisenberg, J. F., 1973. Comparisons of canid and felid social systems from an evolutionary perspective. *Anim. Behav.,* 21(4):637–59.

Kruuk, H., 1972. *The Spotted Hyena.* Chicago: University of Chicago Press.

Leyhausen, P., 1956. Das Verhalten der Katzen. *Handbuch der Zoologie,* 8:1–34; *Z. Tierpsychol.,* 16:666–70.

Leyhausen, P., 1960. Verhaltensstudien an Katzen. *Z. Tierpsychol.* Beiheft 2:1–120.

Leyhausen, P., 1965a. The communal organization of solitary mammals. *Symp. Zool. Soc. London,* 14:249–63.

Leyhausen, P., 1965b. On the function of the relative hierarchy of moods (as exemplified by the phylogenetic and ontogenetic development of prey-catching in carnivores). In: *Motivation of Human and Animal Behavior: An Ethological View,* K. Lorenz and P. Leyhausen, trans. B. A. Tonkin. New York: Van Nostrand Reinhold, 1973, pp.120–43.

Lindemann, W., 1955. Über die Jugendentwicklung beim Luchs (*Lynx L. lynx.* Kerr.) und bei der Wildkatze (*Felis s. silvestris* Schreb.). *Behaviour,* 8:1–45.

Marler, P., 1965. Communication in monkeys and apes. In: *Primate Behavior: Field Studies of Monkeys and Apes,* I. DeVore, ed. New York: Holt, Rinehart, and Winston, pp.544–84.

Marler, P., 1967. Animal communication signals. *Science,* 157:769–74.

Marler, P., 1970. Vocalizations of East African monkeys. I. Red. colobus. *Folia Primat.,* 13:81–91.

Michael, R. P., 1961. Observations upon the sexual behaviour of the domestic cat (*Felis catus* L.) under laboratory conditions. *Behaviour,* 18:1–24.

Muckenhirn, N. A., and Eisenberg, J. F., 1973. Home ranges and predation of the Ceylon leopard. In: *The World's Cats,* vol. 1, R. Eaton, ed. Laguna Hills, Cal.: Lion Country Safari, pp.142–75.

Palen, G. F., and Goddard, G. F., 1966. Catnip and estrous behavior in the cat. *Anim. Behav.,* 14:372–77.

Provost, E. E.; Nelson, C. A.; and Marshall, A. D.; 1973. Population dynamics and behavior in the bobcat. In: *The World's Cats,* vol. 1, R. Eaton, ed. Laguna Hills, Cal.: Lion Country Safari, pp.42–67.

Rabb, G. B., 1959. Reproductive and vocal behavior in captive pumas. *J. Mammal.,* 40:616–17.

Schaffer, J., 1940. Die Hautdrüsenorgane der Säugetiere. Berlin: Urban and Schwarzenberg.

Schaller, G. B., 1967. *The Deer and the Tiger.* Chicago: University of Chicago Press.

Schaller, G. B., 1972. *The Serengeti Lion.* Chicago: University of Chicago Press.

Schenkel, R., 1968. Toten Löwen ihre Artgenossen? *Umschau in Wissenschaft und Technik,* 6:172–74.

Seidensticker, J. C.; Hornocker, M. G.; Wiles, W. V.; Messick, J. P.; 1973. Mountain lion social organization in the Idaho primitive area. Wildlife Monographs No. 35:1–60.

Verberne, G., 1970. Beobachtungen und Versuche über das Flehmen katzenartiger Raubtiere. *Z. Tierpsychol.,* 27:802–27.

Wemmer, C., in press. Comparative ethology of the large spotted genet, *Genetta tigrina,* and related viverrid genera. Smithsonian Contributions to Zoology.

Wemmer, C., and Fleming, M. J., 1974. Ontogeny of playful contact in a social mongoose, the meerkat, *Suricata suricatta. Amer. Zool.,* 14:415–26.

Since the writing of this article, three important articles have appeared on the topic of felid olfactory communication that should be consulted by interested readers.

Verberne, Gerda, 1976. Chemocommunication among domestic cats, mediated by the olfactory and vomeronasal senses. II. The relation between the function of Jacobson's organ (vomeronasal organ) and flehmen behavior. *Z. Tierpsychol.,* 42:113–28.

Verberne, Gerda, and Leyhausen, P., 1976. Marking behaviour of some Viverridae and Felidae time-interval analysis of the marking pattern. *Behaviour,* 58(3–4):192–253.

Verberne, Gerda, and de Boer, Jaap, 1976. Chemocommunication among domestic cats, mediated by the olfactory and vomeronasal senses. I. Chemocommunication. *Z. Tierpsychol.,* 42:86–109.

Chapter 30

COMMUNICATION IN TERRESTRIAL CARNIVORES: MUSTELIDAE, PROCYONIDAE, AND URSIDAE

Cheryl H. Pruitt and Gordon M. Burghardt

The order Carnivora is usually divided into two large superfamilies: the Feloidea, composed of the catlike species (chap. 29), and the Canoidea, composed of the doglike species. In contrast to the recent widespread interest in the social behavior and communication of various wild and domestic canids (chap. 28), there has been relatively little study of their relatives in the Canoidea. The focus in this chapter will be on communication between conspecifics in those families typified by weasels (Mustelidae), raccoons (Procyonidae), and bears (Ursidae). The three families are taxonomically discrete and noncontroversial, except for the two panda species, which are usually placed as a separate subfamily (Ailurinae) within the Procyonidae (Ewer, 1973; Morris, 1965). To place the giant panda *(Ailuropoda melanoleuca)*, with its large size and general resemblance to the bears, here is debatable. Recently scientists have argued that giant pandas are aberrant bears (Davis, 1964; Morris and Morris, 1966; Sung, Chang-kun, and Sarich, 1973). For convenience we will discuss both panda species with the procyonids, although it is a moot point because virtually nothing is known about their communication.

All three groups of animals are primarily terrestrial with several species also arboreal or burrowing. The three families vary from the generally small mustelids through the medium-sized procyonids to the much larger bears. In general all the animals discussed here have well-developed visual, auditory, and olfactory senses. Vision has not generally been considered as important in these carnivores as in those mammals with some convergent habits, like the primates. As in most mammals, coloration does not vary widely between the sexes or age classes, or with the seasons, except for some mustelids. Neither does body shape vary, although size may be a dimorphic characteristic in several species. Because of this sameness in coloration and size, we may miss the potential significance of visual cues.

Each family will be discussed separately within each of the major communication categories, and in all of these discussions we will bring together information by species. Since there are more mustelids than the other two families combined, discussion of them will be broader than of the other groups. In numerous instances we will refer to anecdotal information; it is hoped that this will help point the way to systematic studies.

Mustelidae

The mustelids are primarily small- to medium-sized forest-dwelling animals. Many have short legs, long bodies, and powerful jaws. They are among the most carnivorous of the carnivores. As expert predators with stealthy, "slinky" appearances and clever ways, they earn emnity from farmers. Their fur, on the other hand, is prized. Neither human passion aids their survival. Fur industries in countries such as the United States and the Soviet Union are now raising many mustelids, particularly of the weasel subfamily (Mustelinae: marten, sable, mink, etc.), in captivity and much information concerning husbandry and breeding is being obtained. Field observation of these secretive animals has been mostly unsuccessful. In addition to the afore-mentioned weasel group, there are four other subfamilies including the badgers (Melinae), skunks (Mephitinae), otters (Lutrinae), and honey badgers, or ratels (Mellivorinae). Though the mustelids are found in all continents except Australia, they seem more common in temperate zones. The aquatic otters will not be covered here.

Within the mustelids are found numerous examples of coloration unusual for mammals. Pocock (1908) noted that, contrary to the color patterning of most other mammals, the underside of the body in many of the Mustelidae is darker than the upper, which enhances its conspicuousness. A predatory mammal strives for concealment as it stalks its prey; to do otherwise spoils the success of the hunt. Pocock also discussed the atypical coloration of the ratel *(Mellivora capensis)*, grison *(Grisen vittatus),* skunk *(Mephitis mephitis),* and African weasel *(Poecilogale albinucha)* and its possible adaptive significance in these mustelids:

> animals which are coloured so as to be conspicuous in their natural surroundings are very often protected from their enemies by distastefulness arising from a nauseating flavour or odor, . . . which make them dangerous to meddle with. They also as a very general rule have no need of procryptic coloration to enable them to capture wary or keen-sensed prey. Their movements are usually slow and deliberate, and instead of avoiding they seem rather to court observation, some indeed attracting attention by the emission of characteristic sounds. [Pocock, 1908: 945–46]

Pocock countered the notions that the coloring of these species has no warning significance but rather aids them in the capture of small ground prey. "Animals which require sharp hearing either to escape enemies or capture prey usually at all events have large ears; and the fact that the animals forming the subject matter of this paper have small external ears is in keeping with the theory that they have no enemies to fear" (1908:951). The theory Pocock proposes is an interesting one, lending a novel slant to mustelid interspecific behavior. Specifically, Pocock theorizes that the role of mustelid coloration is more one of warning than of aid in prey capture. It reveals the possibility of highly distinctive communicative behavior that sets the mustelids apart from most other carnivores, especially the procyonids and ursids, which are not nearly so vivid in their coloration.

In his discussion of various subspecies of spotted skunks *(Spilogale putorius)* Mead proposes that behavior and coloration may play another communicative role. "Behavioral differences may also act as isolating mechanisms and indeed lack of, or reduction in amount of, white on the tip of the tail of *interrupta* may play an important role in 'species' recognition, thereby acting as a barrier to gene flow" (1968:388).

Hall (1974) found that seasonal change to a white winter coat in several north temperate weasels is related to protective coloration. There is an east-to-west band in North America above which all *Mustela frenata* molt into white in winter and below which all molt brown winter coats.

The use of the anal glands as warning devices is quite characteristic of mustelids. Reports of musky odors can be found for most members of the family. A dearth of behavioral data on the visual, auditory, chemical, and tactile sensitivity of mustelids creates a problem in any attempt to discuss possible communication channels in these animals; unfortunately, this is true also for the ursids and procyonids. Training experiments have shown that both polecats (*Mustela putorius:* Neumann and Schmidt, 1959) and mink (*Mustela vison:* Sinclair et al., 1974) have reasonable visual acuity with a resolvable angle of 16.2 and 15.1 min, respectively. This compares with a value of 5.5 for the domestic cat and 20 for the rat (see table in Sinclair et al.). But unlike the polecat, the mink is somewhat aquatic, often swimming deep in water to chase and capture fish. In comparison with the more completely aquatic otters, its acuity in water is about double that in air, suggesting that some mustelids are intermediate or transition species between terrestrial and aquatic environments. Olfaction has been shown to be instrumental in the prey selection and predatory behavior of young polecats (Apfelbach, 1973). A sensitive period for olfactory stimulation apparently occurs between two and three months of age and determines the species of prey to be sought as the animal matures. As for tactile cues, some information is available on young ermine *(Mustela erminea).* The litter keeps huddled together by fastening their tails to each other ("coupling reflex") until one month of age (Ternovskiy, 1974). This may serve to keep the pups from crawling away when the mother is absent and also to keep them warm, as isolated pups are not able to effectively regulate their body temperature. A role of this behavior in the ontogeny of other types of communication would fit speculative theories such as Dimond's (1970).

MOVEMENTS AND EXPRESSIONS

In most mustelid research reports examples of reproductive, maternal, aggressive, and play behavior are documented, but the precise movements constituting these behaviors have not been delineated with any consistency. While each species may act in certain ways, the exact body movements that signal the intentions are usually unknown. We will report here the types of social behaviors and their expressive components which have been reported for each species.

Polecat (Mustela putorius)

The expressive behavior of the polecat is perhaps the best documented of the mustelids. Ferrets *(Mustela furo)* are the generally albino domesticated counterpart. Poole's several studies (1966, 1967, 1972, 1973, 1974) have contributed substantially, particularly in describing aggression and aggressive play behavior. Additional information on this species has been gathered by Eibl-Eibesfeldt (1955) and Lazar and Beckhorn (1974).

Play. Conspecific play develops in young polecats after the age of six weeks and the concurrent opening of the eyes (Poole, 1966). Interestingly, the characteristic neck biting, seen especially in aggressive play and predation, does not develop out of generalized biting; instead, both types of biting occur similtaneously at the age of four weeks. Lazar and Beckhorn (1974) confirmed Poole's observations on the emergence and topography of play in young polecats although they "feel that neck bites differentiate out of general kit biting" (p. 411). Inhibited biting, observed in differences in play with humans and conspecifics, was displayed by these animals. Initiation of aggressive play involved one or more of three techniques: (1) mounting and neck biting until the partner reciprocated; (2) dancing up to and jumping on the partner; and (3) chasing and biting the partner's hindquarters. At high intensities the initiator shook its head as it kept a neck bite on the partner, or it dragged the partner through the area by the neck. Poole described additional factors that inhibited play. If fear-eliciting stimuli or prey objects were present, the polecats adopted defensive threat postures and hid or chased the prey, respectively.

Agonistic behavior. Aggression appeared to take different forms, depending on whether or not it occurred during the breeding season and which sexes were involved. Poole (1967, 1973) regarded nonbreeding season agonistic behavior as "ritual aggression," which took place whenever two polecats unfamiliar to each other were placed together. Two particular movements, "inhibited chin biting" and "dancing," were characteristic of ritual aggression. The former was composed of mutual gentle pawing and muzzle biting. If a polecat "danced," it faced its opponent and leapt off the ground, shaking its head while snapping its jaws at the partner.

When males fought during the breeding season, the interaction began with mutual anal sniffing. An attack then commenced without other threats or warnings (Poole, 1967). During the average ten- to fifteen-minute fights, "sustained biting" and "flank shielding" were observed, the latter when losers left the area. Biting composed 41 percent of the fighting interactions, and bites delivered to the neck were longest in duration (Poole, 1974). Once a fight was terminated the loser displayed an "aggressive threat" (Fig. 1b) to further discourage the winner from continuing the fight. Both animals assumed a prone, outstretched position at the termination of the encounter. The winner usually "belly-crawled," moving with the forefeet propelling him forward (Fig. 1c).

Use of defensive threats and sideways attacks characterized male agonistic interactions; however, they were never observed in juvenile aggressive play (Poole, 1967). In a defensive threat (Fig. 1a) the polecat refused to fight by arching his back, orienting his head horizontally toward the opponent, baring his teeth, hissing, and holding his neck vertically. A sideways attack was directed toward a defensive or frightened polecat during a high-intensity encounter. This movement serves to "initiate or prolong a fight with an unwilling partner. This further terrorizes the opponent who adopts an attitude which has been called defensive threat..." (Poole, 1966:30). The sideways attack, which occurred only after the outcome of the fight was decided, was always displayed by the winner, which probably remained in a highly aroused aggressive state. All threats, whether defensive or aggressive, were observed only after the aggressive interaction had begun (Poole, 1967).

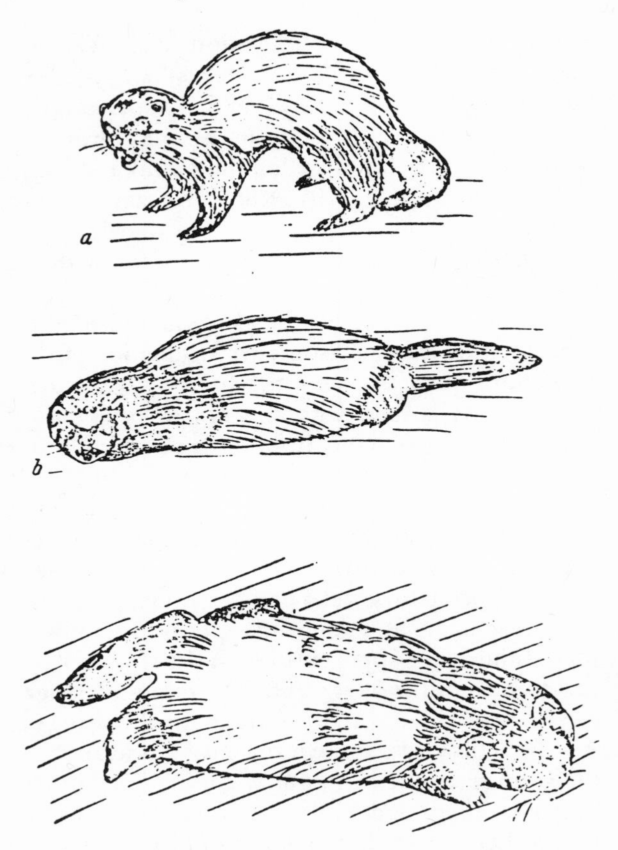

Fig. 1. Defensive threat (a) and aggressive threat (b) in the polecat. (c) Prostration in the polecat. The hind legs are splayed outward and the feet are directed laterally. (From Poole, 1967.)

Estrous females were not so aggressive as

males, although their agonistic behavior level did increase from the non-estrous state. The usual female-female interaction was characterized by ritual aggression and low-intensity fights in which one of the polecats ignored the partner's inhibited chin bites (Poole, 1967). If one of the females was pregnant, the nonpregnant female clucked and danced around the pregnant polecat, which reciprocated with a "lunge-hiss." A nonpregnant female displayed an arched back trot in reaction to the pregnant female's defense of an area. This motion consisted of short, quick steps around the pregnant female, possibly followed by the nonpregnant female's biting of her own tail.

During aggressive activities by either sex the angle of the neck was an important signaling device (Poole, 1967). Through film analysis Poole discovered that the neck was never attacked if, from the aggressor's vantage point, it appeared convex. When the neck assumed a concave or linear appearance, an attack was likely to occur. Poole further described two displacement activities that were linked with aggressive behavior. Chasing one's own tail occurred if a polecat's attack attempts were rebuffed. A polecat "scrabbled," or performed digging movements on the floor, if it had been in an attack-avoidance conflict situation.

Other behaviors. Aspects of aggressive behavior also were seen in polecat mating attempts (Poole, 1967). Anal sniffing and neck nuzzling were included in preliminary mating movements if the female was in estrous. If she was unreceptive, she delivered inhibited chin bites and neck-nips to discourage the male's mounting attempts. If those movements failed, she employed aggressive threats and lunge-hisses.

Finally, Goethe (1940) reported a captive female's attempt to "teach" her young to approach food objects by dragging food into the nest and then pulling it out of the nest, while still in full view of the young.

Weasels

While play has been reported for juvenile captive common weasels *(Mustela nivalis)* and the closely related (if not conspecific, Ewer, 1973) least weasels *(Mustela rixosa),* descriptions of play have not been made. East and Lockie (1964) noted that infant common weasels played together only in the nest; after fourteen weeks they were no longer tolerant of one another. Hartman (1964) observed interspecific aggression when a male common weasel shook a male ferret *(Mustela furo)* by its nose. Threat behavior and associated vocalizations were observed in young common weasels when they were nine to ten weeks old (East and Lockie, 1964). As the weasels matured, the threats were applied in varying contexts, including aggressive play. The female weasel initiated mating behavior by leaping around the male, which then mounted her (Ewer, 1968). The pair then alternated between bouts of play-fighting and sexual behavior for several hours. Ambiguous observations were recorded for the stoat *(Mustela erminea).* Some female stoats carried their young by the mane at the base of the neck, while others carried their young by the middle of the back (Ewer, 1973).

Instances of conspecific tail biting were observed in American mink *(Mustela vison)* when sexually excited males tried to bite unreceptive females at the base of their tails (Poole, 1967). Enders further described movements typical of courtship in the mink:

> Copulation in the mink is often furious. It may be preceded by a "courtship" of varying length which resembles a rough and tumble fight . . . After a longer or shorter struggle the male secures a firm hold on the skin of the neck immediately back of the head. If after being caught the female is reluctant to breed, she arches her back. [1952:721]

The ontogenetic development of stages of defensive behavior has been described by Ternovskaya (1974).

Sable (Martes zibellina)

Individual sable differ in their aggressiveness toward humans, and in females there is a correlation of this hereditary component with fertility (Belyaev and Ternovskaya, 1973). Stages in the ontogeny of aggressiveness have been described by Ternovskaya (1974).

Marten (Martes americana)

Nonserious fighting during the breeding season was initiated by the female, and the male responded with inhibited biting (Ewer, 1968). Ewer speculated that agonistic behavior sexually stimulated the male. Teeth baring, chasing, and growling were the primary components of conspecific aggression, as noted by Herman and Fuller (1974). Much of the aggression they observed occurred near food.

Tayra (Tayra barbara)

Kaufmann and Kaufmann (1965) noted instances of unsolicited female-male grooming, in which the pair was situated head to head as she licked his ears and scratched his head with her forepaws. The male did not reciprocate nor did either tayra ever invite grooming from humans.

Grison (Grison vittatus)

In conspecific play captive grisons shook their partners by the scruff of the neck (Dalquest and Roberts, 1951). One grison exhibited similar behavior with old socks (Kaufmann and Kaufmann, 1965). The Kaufmanns also reported, however, that another pair of their captive grisons never interacted aggressively. On the other hand, Dalquest and Roberts described a posturing that occurred during grison fights in which the submissive animal flattened its entire body flush to the ground. When scolding their grisons for attacking humans, the Kaufmanns also observed this submissive posture. Dalquest and Roberts further noticed that during male-female "arguments," the female arched the posterior portion of her body and raised the tail.

African Weasel (Poecilogale albinucha)

Three types of play were observed in Alexander and Ewer's (1959) African weasels. The animals dug and burrowed in tanks of dirt for periods of up to thirty minutes. Additionally, they "romped" with one another and played with humans. During the breeding season the male bit the female at the nape of the neck and then tried to mount her. At his mounting attempts she rolled over.

Ansell (1960) observed that the African weasel's alarm reaction was to fluff the tail hair. If the weasel discovered an "interesting scent," the tail was carried vertically.

Ratel (Mellivora capensis)

A captive ratel's behavior fell into two categories, a "relaxed" state and a "fury mood" (Sikes, 1964). When relaxed, the ratel was affectionate, playful, and responsive to human voices. As the ratel played with objects, it shook them from side to side with its jaws. In the fury mood "all recognition seems to be forgotten, as the state of blind aggressiveness and ferocity mounts" (1964:32).

Badger (Meles meles)

A captive badger extended play invitations to humans by approaching them with a stiff-legged walk, "barking" all the while. He stopped short in front of a human, shook his head from side to side, and ran away (Eibl-Eibesfeldt, 1950). This badger also was reported to somersault occasionally.

Skunks

The spotted skunk *(Spilogale putorius)* gave a threat-bluff by charging its opponent, throwing the body forward, and moving the tail off to the side to allow spraying from the anal sacs (Johnson, 1921; Gander, 1965). The skunk did not actually spray in this position; rather, as it sprayed all four feet were on the ground, the back was arched, and the tail curved over the body (Walker, 1930) (Fig. 2).

The striped skunk *(Mephitis mephitis)* similarly threatened although it did not perform the handstand (Ewer, 1973; Verts, 1967). As in the spotted skunk, the back was arched and the tail was raised. Additionally, the striped skunk may shuffle backward and stamp the forefeet on the ground (Verts, 1967). Ewer (1973) added that on occasion the striped skunk rocked back and forth on the fore and hind legs.

Ritualized fighting and biting were observed prior to actual mating in the striped skunk (Verts, 1967). During mating the male scratched the female's genital area with his feet. Verts suggested that this highly unusual behavior may function to stimulate the female to become more receptive and to assume a more typical lordosis posture.

VOCALIZATIONS

Vocalizations have been determined for many of the mustelids, but the number of known calls for each species is relatively limited. Spectographic and fine-grained contextual analyses of calls have been all too rare, although call development has been studied in weasels especially. Exact descriptions of vocalizations are difficult to draw from the various studies, because of the relatively subjective names assigned to the calls themselves.

Fig. 2. Handstand threat posture of the spotted skunk. (From Eibl-Eibesfeldt, 1957, after Bourlière.)

Weasels

Hartman (1964) made the primary contribution to the knowledge of weasel vocalizations by delineating six separate sounds made by common weasels. He heard a high-pitched squeak, which was especially "long-drawn and plaintive" in the female. East and Lockie (1964) noted that infant weasels squeaked when any movement occurred near them. The frequency of this squeaking decreased after eighteen to nineteen days of age. Neonate least weasels (Heidt et al., 1968) and ermine (Ternovskiy, 1974) also squeak.

When they were twenty-four days old, common weasels responded to squeaks with an open-mouthed hiss (East and Lockie, 1964). Hartman explained the hiss as a response made partly out of fear and partly as a threat. Huff and Price (1968) concur. Ternovskiy (1974) did not note the typical hissing and "clicking" of annoyed ermine until fifty days of age. Threatening with "short sharp sounds" by adults is present in common weasels (Hartman, 1964) and this may be homologous with the clicking of ermine and the chirping of least weasels (Huff and Price, 1968; Heidt et al., 1968). Huff and Price theorized that the chirp was made when the least weasel was disturbed. Heidt's group, however, also noted chirps when juveniles initially approached live mice, when they first attempted to interact with the mice as prey objects, and when the mother threatened the youngsters, if they bothered her.

Both sexes of common weasels trilled when meeting in a friendly fashion. Both male and female least weasels also trilled prior to mating activities and females trilled when calling to the young (Heidt et al., 1968). Huff and Price (1968) suggested that the one- to two-second duration trill occurred in friendly and maternal interactions. Goethe (1950) reported another agonistic sound, an "r-r-r" vocalization, but he did not specify the context in which it was heard. When East and Lockie (1965) bred common weasels in captivity, they noticed that unreceptive females screamed when approached by the male but, if they were in estrus, they chattered to the male. At fifty-six to sixty days ermine begin to utter the "rumbling" of adults during sexual games (Ternovskiy, 1974).

"Twittering" was noted in infant common weasels (Hartman, 1964) and a perhaps homologous melodious chattering in young ermine thirty-three to forty days old, about the time the eyes first opened (Ternovskiy, 1974).

Progulske (1969) briefly mentioned two vocalizations that he heard in a captive black-footed ferret *(Mustela nigripes)*. It chattered (in bursts of six or seven chirps) when humans approached its quarters. It also threatened humans by hissing.

If aggressive play is too rough, a polecat may squeak, cry, or yelp (Poole, 1966, 1972). While generally there is no vocalization during high-intensity agonistic encounters (Poole, 1972), the animal may scream when severely frightened during a fight (Poole, 1966). A clucking sound was also heard, but never by an "intimidated" animal (Poole, 1966). Conspecific threats were accompanied by hissing and teeth baring during aggressive play (Poole, 1966).

Marten

Vocalizations known to be given by American martens occurred during the breeding season; vocalizations at other times have not been reported (Markley and Bassett, 1942; Ewer, 1973) except by Herman and Fuller (1974) in their report of growling during non-breeding season aggression. Female American martens clucked during estrus, "apparently to attract the male" (Markley and Bassett, 1942:609). Ewer (1973) further theorized that the cluck served to signal a "readiness for social contact" and physiologically to activate the male to copulate. During mating the female squealed when the male was too rough, and both sexes tended to growl or purr during copulation itself (Markley and Bassett, 1942).

Tayra

Kaufmann and Kaufmann (1965) described four distinct vocal signals and their variations in captive tayras. A high-pitched "yowl" (similar to that of male domestic cats) was heard only during preliminary mating movements. Juveniles produced low-pitched, open-mouthed "b-a-a-a" distress sounds. As the tayra matured, the frequency of the distress call decreased. If alarmed, the tayra "snorted" and jumped backward, away from the source of the alarm. Finally, three gradations of "clicking" responses were heard during aggression. Mild aggression evoked clicks lasting about three seconds, while more intense encounters produced lower-pitched, slower clicks. At the highest intensity of aggression, the tayra emitted a low-pitched, closed-mouth, teeth-baring snarl. This was the last signal preceding an actual attack.

Grison

A nasal "anh-anh" sound, seemingly a distress or separation call, was heard in captive juvenile grisons when they were separated from humans (Kaufmann and Kaufmann, 1965). The same study reported a snort of alarm in the grisons similar to that in the tayras. In agonistic interactions of low intensity, the grison produced a sound that resembled a "low motor" (Kaufmann and Kaufmann, 1965). As the intensity of the encounter increased, the vocalizations became slower, higher-pitched, louder, and barklike. If the level of aggression further increased, the grison made high-pitched barks and concurrently held the tail in an S-shaped curve. The grison screamed loudly with the mouth open and teeth bared if the fight progressed to its highest intensity. Screaming occurred immediately preceding and during attack.

African Weasel

Although Ansell (1960), Alexander and Ewer (1959), and Rowe-Rowe (1969) have noted vocalizations in the African weasel, the only call which all researchers reported in common was a characteristic half-scream, half-growl. Ansell heard this sound in response to the presentation of a freshly killed bird to the weasel, but Alexander and Ewer speculated that it might be a threat vocalization, and Rowe-Rowe interpreted it as a sound of surprise or alarm. In their 1959 study Alexander and Ewer observed that if a female emitted the scream-growl, an approaching male left her alone. Rowe-Rowe delineated three other calls in this species: a quick, high-pitched growl; a soft, self-directed grunting as the animal explored; and a low growl if disturbed. Alexander and Ewer also reported hearing a "rumbling" vocalization during male sexual behavior that may be similar to Rowe-Rowe's reported low growl of disturbance.

Ratel

A captive ratel (Sikes, 1964) was reported to give two vocalizations: a repeated "h-r-r-r—h-r-r-r" call when disturbed or when engaging in running play with humans; and a snarl, heard if the animal was teased into aggressive play with its human caretakers.

Striped Skunk

Verts's (1967) study of the striped skunk included a list of vocalizations he had observed. Perhaps because the thrust of his study was not behaviorally oriented, he did not discuss possible conspecific connotations of those sounds. However, he did show that the striped skunk "churrs," "growls," "screeches," "twitters," "coos," and "hisses." He also found that the young were more vocal than the adults. Additionally, a pregnant female hissed when disturbed, and she screeched at males that were placed in an enclosure with her. Wight (1931) observed mating calls in the striped skunk. During conspecific agonistic behavior, the striped skunk was reported to growl and spit (Cuyler, 1924).

PHEROMONES

Mustelids are distinctive among mammals in their defensive use of anal scent glands in interspecific situations. Located just within the anus, the glands are contracted by muscular action, propelling the contents toward the opponent. Although the same glands are readily available for conspecific communication purposes, they are rarely put to that use (Ewer, 1973). In addition to the anal gland, most of the mustelids also possess abdominal glands, which are dragged over surfaces or objects within a territory. The function of the abdominal gland is probably the marking of areas within a territory or home range; therefore, we may attach greater importance for conspecific communication to it. Although the anal gland may be involved in territorial pursuits, it is more likely that the abdominal gland is dominant in that role. Hall (1926) could not locate abdominal glands on the fisher, mink, or striped and spotted skunks. Developmental studies of gland morphology and secretions are rare. Specific glands and their communicative uses are discussed separately for each species.

Weasels

Ewer (1973) reported that when alarmed, various weasels (stoat, common weasel, and least weasel) emitted the contents of their anal glands. It does not appear that the abdominal glands are utilized in alarm or active defense situations; rather, as the ventral surface is dragged, the substrate is marked.

> Any defense of an area by females by setting scent or otherwise cannot be a signal between adjacent territory-holding females, although scent could well act as a sign to wandering females. I suggest that female weasels and stoats defend their small territory mainly against the male owner of the territory in which they live. [Lockie, 1966:157]

After antagonistic encounters among polecats during the breeding season, the victor explored the area in which the fight took place and marked objects by sliding, rolling, and rubbing its back on the ground or the object (Poole, 1967). The loser similarly marked and rubbed the area after the victor had completed its marking. Lockie (1966) noted that polecats typically deposited urine and feces at specific spots in their territories.

Vaginal discharges of female American mink are claimed to be important to male mink in determining the female's sexual state. Volatile amines have been isolated that vary with the female's estrous cycle and hence could be the signal substances used by the male (Sokolov et al., 1974). Azbukina (1970) found several skin glands in the American mink: anal, plantar, nape, gluteal, and caudal. The anal glands are delayed in their development. This and the temporary appearance of supplemental apocrine glands (nape, gluteal, caudal) are related to the licking and massaging of pups by parents and mutual parent-pup smelling. Azbukina (1972) also described in detail the complex morphology of the anal sacs in American mink. They consist of two types of glands (apocrine and holocrine), a cavity for storing the musk, and a central excretory duct.

Marten

In captivity female American martens urinated on feces, stones, and objects within their enclosure (Markley and Bassett, 1942; Herman and Fuller, 1974). Although Seton (1929) was not dealing with captive animals, he believed that the urine marking was associated with the anal glands. Use of the abdominal glands was observed in both male and female captives, particularly as they rubbed their abdomens over branches (Markley and Bassett, 1942). The exact size of male and female abdominal glands was delineated earlier by Hall (1926), who found the female gland to be 1.0 X 6.4 cm and the male gland .8 X 4.0 cm. Both abdominal gland and urine marking increased at the onset of the breeding season (Ewer, 1973), which coincided with the increase in marking observed after the family groups had disbanded (Goethe, 1964; Herter and Ohm-Kettner, 1954).

Wolverine (Gulo gulo)

When alarmed, the wolverine normally released a particularly foul-odored anal sac secretion. Typical of most mustelids, this species marked by rubbing surfaces with the abdominal gland (Krott, 1959).

Tayra and Grison

The only reports for these species (Dalquest and Roberts, 1951; Kaufmann and Kaufmann, 1965) indicated that both used the anal gland to mark objects. This behavior first appeared in the tayra as it approached maturity (Kaufmann and Kaufmann, 1965). The yellow-green grison musk secretion was deposited after the grison raised its tail and then brushed it against the object to be marked (Kaufmann and Kaufmann, 1965).

African Weasel

In mustelid fashion the African weasel released its sweet and pungent musk from the anal glands (Alexander and Ewer, 1959; Rowe-Rowe, 1969; Ansell, 1960). Ansell argued that since the odor was not particularly foul, the species-characteristic coloration served as a warning device more than the anal secretion did. Before releasing the musk, the weasel raised the tail to a vertical position and fluffed the tail fur. The actual range of the musk spray was from 20 to 100 cm (Alexander and Ewer, 1959). Although the species never released its musk in a conspecific interaction, on one occasion a male was attracted to a female that had just sprayed in an interspecific situation (Alexander and Ewer, 1959). Urination and defecation habits of the African weasel showed that it eliminated in specific areas of its home environment. Alexander and Ewer believed that this habit "serves to show whether the sleeping quarters are or are not occupied" (1959:316).

Ratel

The anal secretions of the ratel, copiously applied to trees and stones, may serve to stimulate courtship and to establish the animal's territory (Sikes, 1964). African hunters believed that the breeding of ratels was accomplished only if the pair hunted for and attacked a beehive together. Sikes elaborates:

> Perhaps a mutual-fury reaction is the necessary primary stimulus required to trigger off the hormone sequence necessary to the attainment of successful copulation and embryo implantation. Possibly there may be other essential factors to success such as the need to allow the scent to lie in the cage undisturbed for a suitable period. . . . [1964:36]

Badger (Meles meles, Taxidea taxus)

Ewer's (1968, 1973) descriptions of pocketed abdominal glands were the only sources of information available. The invertible, pouched gland presumably played a role similar to that in other mustelids, i.e., territorial marking through "anal drag" movements.

Skunks

Both the striped and the spotted skunks' anal glands were specially modified to store the foul, musky fluid (Ewer, 1973; Verts, 1967). Blackman (1911) drew the analogy of a syringe and bulb to the anal sac mechanism in these species. As the skunks sprayed, separate streams from each of two anal sacs fused and were directed toward the opponent (Cuyler, 1924). The tail was held over the skunk's back, perhaps to avoid fouling its own body. As might be expected, the release of anal fluids was limited to defensive, interspecific situations. However, it has been noted that a spotted skunk in estrus was capable of releasing a characteristic nest scent which presumably encouraged nursing behavior in the young (Tembrock, 1968).

The most recent study on the chemical composition of the male odorous component of striped skunk musk is that it contains *trans*-2-butene-1-thiol, 3-methyl-1-butanethiol, and *trans*-2-butenyl methyl disulfide, but not the commonly cited 1-butanethiol (Andersen and Bernstein, 1975).

Procyonidae

The procyonids are typically small- to medium-sized omnivores of the forest. Many are arboreal. The main subfamily (not including the pandas) is found only in the New World, particularly Central and South America. This subfamily includes such species as the raccoons, coatis, and kinkajous. They have well-developed forepaws with separated digits which allow them to manipulate objects in a superficially primate-like fashion. The pandas, in contrast, are purely vegetarian forest animals of eastern Asia. Their forepaws are more bear- or doglike. Davis (1964) and Sung, Chang-kun, and Sarich (1973) argue that the giant and lesser pandas are not closely related.

When procyonids are viewed as a family, only

a very sketchy picture of their communication channels can be drawn. For example, Ewer (1973) delineated specific auditory ranges and limits only for the ringtail (*Bassaricus astutus:* 45 kHz), the ringtail coati (*Nasua nasua:* 45 kHz), and the raccoon (*Procyon lotor:* 35 kHz). The auditory capabilities of the other species have yet to be determined.

Ferron (1973) noted that the olfactory apparatus of the raccoon was highly developed, particularly when compared to the domestic cat and the mink. Ferron's results seemed to refute Cole's earlier observation that olfaction was not well developed for locating food items (Cole, 1912); but it does support Stuewer (1943), who noted that the raccoon was capable of locating food items buried in two inches of sand.

The tactile abilities of the procyonids have been reported only for the raccoon and the coati (Welker and Seidenstein, 1959; Welker, Johnson, and Pubols, 1964; Cole, 1912; Zollman and Winkelmann, 1962; and Ewer, 1973). The paws and nose of the raccoon have the greatest amount of somatic tissue represented in the cortex (Welker et al., 1964). The coati *(Nasua narica),* not so adept as the raccoon in the use of the paws, utilized the snout (and its olfactory abilities) to a greater degree (Kaufmann, 1962).

The only record of procyonid visual abilities was found in the raccoon (Dücker, 1965; Michels, Fisher, and Johnson, 1960; and Cole, 1912). While Cole believed that the raccoon's vision was quite good, he made no mention of color vision in that species. Using experimental techniques, both Michels et al. and Dücker could find no evidence that the raccoon did discriminate colors.

MOVEMENTS AND EXPRESSIONS

Coati (Nasua narica)

Kaufmann (1962), who performed a fairly complete study of the coati, listed six primary visual signals used in its social organization. A nose-up posture was prevalent in adult male encounters (Fig. 3a). If the head was oriented downward, the situation was clearly agonistic (Fig. 3b). Head jerking and inhibited biting were seen in juveniles as they approached adult females and solicited grooming from them. Mutual grooming (between adult males and females) probably maintained the band structure by providing frequent, socially positive interactions. Tail switching occurred when two coatis faced each other. The "tail-to" position, which was performed by both sexes as the breeding season approached, seemed to serve as an appeasement gesture in reducing overt hostility.

Although fights involving physical contact were rare, they did occur, and the expressive components included nose-up postures, squeals, charging, and jockeying for position. Gilbert (1973) described a "defiant" posture used in "warning-threat-challenge" encounters. The coati spreads and bows its forelegs, raises the tail over the back, lowers and turns the head to one side, opens the mouth, and exposes teeth. The lower the head the more intense the signal. Aggression with contact, which occurred during the breeding season, could mean harsh wounds to the animals. Inhibited fighting between the sexes also occurred during the mating season. The animals would direct their paw swats toward the head, open the mouth, and jockey the head back and forth. These inhibited fights were usually terminated mutually by nose-up postures and squeals. This kind of fighting "seems to play a significant part in their relationship" (Kaufmann, 1962:133) and "helps preserve band structure by serving as a positive outlet for latent or developing hostility" (1962:161). That band structure is strong is supported by the observation of a joint attack by coatis on a boa constrictor when one of their number was attacked (Janzen, 1970).

Finally, the coati used an upright, bipedal

posture when exploring visually and olfactorily. Kaufmann mentioned in passing that this bipedal investigatory stance was also observed in raccoons, kinkajous, and olingos.

Raccoon (Procyon lotor)

A primary source of information concerning expressive behavior in the raccoon is Cole's study of captive animals (1912). Aggression, which first appeared at the age of twelve weeks, involved the laying back of the ears, a lowered head, and "humped" hindquarters. The teeth were bared as the animal growled. When introducing wild male raccoons to each other, Barash (1974) noted initial hissing and tail lashing followed by a posture involving flattened ears, an elevated tail, and raised shoulder hackles. A dominance-subordination relationship was indicated by "the subordinate lowering his chin, neck, ventral body surface, and tail to the floor, and then retreating to a far corner of the cage, giving the dominant animal free access to the enclosure" (p. 795). If no such relationship developed, an arched back, retracted lips, bared teeth, and growling were added to the aggressive display.

Cole described play of long duration in his captive raccoons (up to seventy-five minutes), self-play, and conspecific types of play. In each type of play the raccoon manipulated objects or the other animal with the forepaws. When playing with each other, animals tended to grasp with their teeth and to engage in rough-and-tumble rolling play. Cole further stated that "these animals pay almost no attention to one another. Though one raccoon is retreating from you, growling and snapping, the others are in nowise disturbed. Their indifference to each other's behavior could hardly be more marked, and this fact must be taken account of in considering the question of imitation" (1912:308).

While adult raccoons, like many carnivores, are generally solitary outside of breeding and litter rearing activities, a social organization based on individual recognition nonetheless exists, as neighboring male raccoons have initial dominance-subordination relationships not found in paired raccoons from distant locales (Barash, 1974).

Red Panda (Ailurus fulgens)

The red panda's behavior has not been well documented. Morris and Morris (1966) reported that during a captive panda-human aggressive interaction, the panda charged the human in a bipedal stance, "raised like a bear." As the panda charged, it hissed.

Giant Panda (Ailuropoda melanoleuca)

Morris and Morris (1966) documented the movements of the giant panda. It used a rolling, head-swaying, diagonal walk and trotted at higher speeds. Gallops and bipedal walks, however, have not been observed in this species. During interspecific agonistic episodes the panda swatted at the enemy with a front paw. In higher-intensity encounters the panda charged, pulled the opponent to itself with the front feet, and then bit (Morris and Morris, 1966). Only a preliminary note on play is available (Kleiman and Collins, 1972).

VOCALIZATIONS

As might be expected, vocalization among the relatively solitary procyonids is predominant during infancy and adolescence and again during breeding seasons. The sounds given in the first months of life are probably most vital in maintaining mother-infant contact and in signaling alarm or hunger states. Additionally, several authors designated certain sounds emitted by infant coatis and raccoons as nonspecific distress calls (Kaufmann, 1962; Welker and Seidenstein, 1959; and Cole, 1912). Specific vocalizations in each species were enumerated as follows.

Coati

Kaufmann (1962) listed eight distinct vocalizations given by coatis, probably the most social procyonid. While grunting was heard most often in females with their young, low-intensity grunts were made by all members of a band as they fed and traveled. In certain situations grunting served both as an alarm call and as maintenance for band contact (Kaufmann and Kaufmann, 1963; Smythe, 1970). Very young coatis chittered as they approached females, soliciting grooming. The chittering continued as they were groomed. Kaufmann hypothesized that the chitter is a generalized distress call. He based his theory on the return of band members to trapped and chittering juveniles and on the occurrence of this sound during play fights and mounting by subadults. Additionally, band members responded by grunting if a coati was lost and chittered. Juveniles were heard to whine after being attacked or while hiding in vines. Squealing occurred during agonistic encounters, accompanied by a nose-up posture unless physical contact had been made. During very serious aggression in the absence of physical contact, growls and snarls could be heard in adult males. Barks were linked to alarm or startle responses. A chop-chop vocalization was made by adult males only during the breeding season. Kaufmann believed this sound was related to a ritualization of biting, inasmuch as the chuckle occurred in the presence of other males concurrently with urine rubbing and assuming a head-down, "appeasement" posture. Adult males and females emitted chuckling calls, particularly when approaching each other after brief separations. Additionally, Sunquist and Montgomery (1973) observed two coatis mating 20 m above ground and emitting short high-pitched vocalizations. In a more anecdotal vein Gilbert (1973:101–15) discusses coati vocalizations.

Raccoon

Cole (1912) and Welker and Seidenstein (1959) reported a generalized distress whine by captive young. In addition Cole noted a growl while feeding, a whimper if hungry, and barks and growls during agonistic encounters. Tevis (1947) reported a "low purr" by the mother to her young, ostensibly for maintenance of contact; a "soft snore and throaty growl" when animals recognized each other; and a "snort-growl" as an invitation to aggression. Gander (1965) noticed increased vocalization during the breeding season. Juvenile screeches and female and infant twitters were observed by Stuewer (1943), and Schneider (1973) claimed that four-week-old cubs "cry very loudly, emit twittery noises, and even growl" (p. 71).

Ringtailed Cat or Cacomistle (Bassaricus astutus)

Grinnell, Dixon, and Linsdale (1937) described only two vocalizations for this species, a "snarl" made when it was disturbed and a "normal" foxlike vocalization. Unfortunately the conditions under which the foxlike calls were made were not given.

Olingo (Bassaricyon gabbii)

Both sexes of this species have a "characteristic mating call" (Poglayen-Neuwall and Poglayen-Neuwall, 1965). No other mention of vocalizations can be found in the sparse literature.

Kinkajou (Potos flavus)

Grinnell et al. (1937) reported a short whistle vocalization when young animals were separated from one another. They also noted that when the kinkajou became excited, the pitch and volume of vocalizations changed.

Red Panda

B. H. Hodgson (in Morris and Morris, 1966) made the major contribution to the literature on this species in the mid-1800s. He noted that the red, or lesser, panda hissed and spitted if provoked and displayed a "short deep grunt like that of a young bear" (p. 14).

Giant Panda

In captivity the panda "shrieks" if threatened or surprised by humans or unidentified noises (Morris and Morris, 1966). During estrus the female "bleats" and "calls" with greater than normal intensity as she scent marks her enclosure. Chinese zoo officials (Morris and Morris, 1966) reported that the male emits barks during the breeding season as he runs through the enclosure. This vocalization has not been noticed by European and American zoo personnel, however.

PHEROMONES

Communication by chemical means in the procyonids usually takes the form of scent marking. Glandular secretions may serve several complex functions, including the more obvious marking of a route or territory boundary as well as merely making the area familiar odoriferously. Ewer (1968) underscored the value that a scent may have for the animal doing the marking. In establishing an area as familiar, the animal may be enhancing its own degree of confidence while at that location. As Ewer emphasized, an animal will less readily leave an area that contains its scent. In several cases, as a procyonid left an anal secretion, it also urinated, doubly insuring that it had left a distinct mark.

Olingo

A single report indicated that the olingo rubbed its anal sacs over tree stumps and branches. As the olingo crouched over the area to be marked, rubbing and urination were accomplished (Poglayen-Neuwall and Poglayen-Neuwall, 1965).

Kinkajou

The kinkajou compensates for a lack of anal sacs by possessing paired manidular glands and enlarged skin glands on the chest and abdomen (Poglayen-Neuwall, 1962). The same report noted the male's habit of biting these glands on the female's jaw and throat during preliminary mating procedures. The biting of these areas presumably further excited the male and he consequently mounted the female. No mention was made of the use of the abdominal and chest glands in this species.

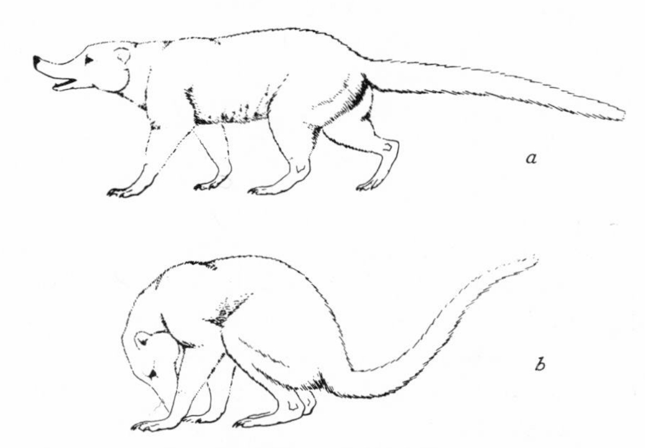

Fig. 3. (a) Typical high-intensity nose-up posture in the coati. (b) Typical high-intensity head-down posture. (From Kaufmann, 1962.)

Coati

The coati also lacks the typical procyonid anal sacs. Glands located along the dorsal edge of the anus appear to replace them (Ewer, 1973). During the breeding season, abdomen and hindquarter rubbing in the coati occurred, accompanied by urination (Kaufmann, 1962). Typically, the male engaged in urine rubbing, performing this act on trees 5 to 8 cm in diameter or larger. If two adult males meet during the breeding season, they urine rub and emit squeals or chop-chop vocalizations as they assume nose-up aggressive postures. Kaufmann believed that this ceremony implied an interaction of a high-intensity attack drive (noses up), a slightly lower-intensity escape drive (chop-chop vocalization), and a sex drive (urine rub). Kaufmann did not notice any evidence that urine rubbing kept other males out of an area. He hypothesized that current urine odors kept male agonistic encounters low by warning subordinates of a dominant male's presence in the area. A second means of chemical communication noted in the coatis was the perineal sniff (Kaufmann, 1962). During the mating season the male was able to determine which females were in heat by sniffing near the dorsal glands surrounding the anus.

Giant Panda

In the giant panda anal glands are located on a very large naked anogenital skin area. Captive female pandas may back up to preferred areas within their enclosures and rub their anogenital areas against the object with their backs arched and their tails raised (Morris and Morris, 1966; Kleiman and Collins, 1972). Although rubbing has been noted throughout the entire year, it was especially prevalent during breeding seasons. As the panda marked the area, she also shook and tossed her head and exhibited an open-mouthed grin face (Morris and Morris, 1966).

Ursidae

There are seven species of bears, among them the largest land carnivores of the world, ranging from about 30 to 800 kilograms at maturity. They are heavily built plantigrade animals and, with the exception of the polar bear, more vegetarian than carnivorous. While they have adapted to climates from arctic to tropical, they are basically north temperate forms; only the little-studied spectacled bear of the Andes mountains gets into the southern hemisphere. Bears have never lived below the Sahara desert in Africa and that, along with their bipedal and arboreal habits, omnivorous feeding, intelligence, and strength, has led to speculation that the center of hominid and ape evolution in Africa was not unrelated to the absence of bears (Kortlandt, 1972). The ecological convergence of species such as the Mountain gorilla and the eastern North American black bear, except for social organization, is quite dramatic.

Our knowledge of methods of communication in the ursids is rather limited at present. Anecdotal information (Meyer-Holzapfel, 1947; Seton, 1929; Couturier, 1953) stressed the auditory and olfactory abilities of the bears but placed much less emphasis on vision and expressive movements. Experimental evidence for sensory capabilities has been restricted to two major research efforts. Kuckuk (1937) noted that all sensory systems of the brown bears *(Ursus arctos)* appeared to be well developed. He further elaborated on that species' auditory and visual abilities. Visual recognition of the experimenter was performed at a distance of 15 m, and auditory signals were responded to at a distance of 150 m. Training of simple circus tasks was accomplished for these brown bears through the use of hidden food items (Kuckuk, 1937). Olfactory investigation appeared to be the major factor in locating the hidden foods.

Bacon (1973), Burghardt (1975), and Bacon and Burghardt (in press, a, b) further explored the olfactory and visual aspects of black bear behavior, and they concluded that these two perceptual systems are closely coordinated during foraging, feeding, and rare predatory behaviors. It appears that ursid vision is far more developed than anecdotal sources would have us believe; training experiments demonstrated the ability of captive subjects to discriminate hues and patterns.

Distinct vocalizations have been reported for the bears (Meyer-Holzapfel, 1957; Garrison, 1937; Stonorov, 1972; Jonkel, 1970; Pruitt, 1974; and Wemmer, Von Ebbers, and Scow, in press). The range and use of vocalizations do not appear to differ widely from what is reported for the mustelids and procyonids.

Expressive body movements appear regularly during social interactions (Burghardt, 1975; Henry and Herrero, 1974; Egbert and Stokes, in press; Egbert and Luque, 1975; Pruitt, 1974; Jordan, in press; and Burghardt and Burghardt, 1972). Unlike the mustelids, during interspecific aggression the bears do not couple bodily movements with musky anal secretions. Use of pheromones in ursids may be limited to urine marking and body rubbing (Tschanz et al., 1970).

Although spectacular coloration patterns are not so numerous in the ursids as in the mustelids and procyonids, the sloth, sun, and spectacled bears have prominent light fur patterning. The Asiatic black bear, sloth bear, and occasionally the American black bear have white V's on their chests. The protective, social, or warning function of these lighter patches of hair can hardly be surmised at this time.

MOVEMENTS AND EXPRESSIONS

Although the expressive capabilities of the ursid family have been discussed and disputed (Krott, 1961; Krott and Krott, 1963; Ewer, 1968), the literature now is reflecting the evidence that bears do possess stereotyped behavior patterns that convey intraspecific social messages. In those species investigated, it appears that bears use body postures, facial expressions, and movements to signal social intentions. Unlike their many relatives, bears have tails that are too short to be important in communication. Meyer-Holzapfel noted that "the threatening motion of baring the teeth is common to the bear, as to other animals of prey, although it lacks the 'crescendo of hissing' characteristic to all species of cats" (1957:23, trans.), and further, that "motions and attitudes are more capable of carrying expression [than the face]" (1957:23, trans). Behavioral components of fear evidenced in bears (Meyer-Holzapfel, 1957) included a "cowering attitude," an outstretched head, "wild and restless" eyes, the hissing and gnashing of teeth, and repeated withdrawal movements. Fortunately, research focusing on the presence and types of bear expressions has increased in the past few years, further expanding the very brief behavioral outlines provided by Meyer-Holzapfel and her predecessors.

Brown Bear (Ursus arctos)

In his investigation of grizzly bear *(Ursus arctos horribilis)* behavior, Hornocker (1962) determined that adult males bluffed other bears during aggressive episodes. The bears vocalized prior to the threatening bluff. Hornocker further noted a distinct dominance hierarchy within the group of free-roaming grizzlies he observed. Although the attainment of rank was not described, Hornocker did delineate the effect of that rank on conspecifics. Adult males occupying the top ranks of the hierarchy were more aggressive than their subordinates. Cautious and defensive males avoided the dominant and aggressive males, behaving agonistically only if attacked or surprised. Females' ranks were temporary and relative to the ranks held by the males.

Meyer-Holzapfel (1968) noted that dominant brown bears became aggressive if they were forced to defend a tree from other bears. In their study of free-roaming Alaskan brown bears at McNeil river falls, Egbert and Stokes (in press) found that episodes of aggression and play were linked closely to the salmon level. If the salmon were plentiful, aggression was low and play increased. Subadults, who did not occupy high ranks in the group, generally were tolerant of other bears and reserved aggression for defensive situations. Other supporting evidence for these patterns of aggression has been reported (Egbert and Luque, 1975; Stonorov, 1972; Stonorov and Stokes, 1972).

During interspecific aggression Kuckuk's

(1937) captive brown bears lowered their heads and slanted their snouts and muzzles toward the human. Kuckuk also utilized plywood models and stuffed brown bears to elicit aggression and exploratory behavior. He found that the stuffed specimens elicited aggressive behavior with physical contact in the male (olfactory investigation and paw swats to the head), but the female only threatened, assumed a bipedal stance, and then ran from the dummy. Both bears sniffed the plywood model and then ignored it. Heran (1966) described a defensive posture, seen when a female spotted a male in another section of a new enclosure. That posture combined "tense" ears with an arched back and bipedal stance. The head was stretched forward and the forepaws extended from the body. Heran also observed play behavior in the male-female pair after initial exploratory behaviors were forgotten. Finally, Eibl-Eibesfeldt (1957) reported that captive brown bears exhibited at least three separate movements as they begged for food from visitors (see also Hediger, 1950).

Black Bear (Ursus americanus)

Aspects of play, threat, and aggressive behavior have been described variously by several authors, each of whom dealt with different aspects of play or aggression. Several conclusions can be drawn.

Play. The most definitive aspect of play behavior was the absence of any vocalization as opposed to the very vocal aggressive behavior (Burghardt and Burghardt, 1972; Henry and Herrero, 1974; Pruitt, 1974 and in press). Additionally, two ear postures, crescent and partially flattened, were particularly characteristic of play, whether solitary, with conspecifics, or with humans (Henry and Herrero, 1974; Pruitt, 1974). Locomotor initiatory movements of social play included circling, chasing, sideways walks, rolling over (rare), and face-to-face approaches, all of which were related to the initial intensity of the consequent play (Pruitt, 1974; Henry and Herrero, 1974; Leyhausen, 1948).

Agonistic behavior. During aggression black bears threatened conspecifics by lunging or charging, stopping short, and swatting the ground or the other bear (Jonkel and Cowan, 1971; Jordan, in press; Pruitt, 1974). Threats to humans were quite similar, involving both the sudden lunge and the swat to the ground or an enclosure fence (Jordan, in press). In an experimentally manipulated situation and in isolated naturally occurring episodes, the animal threatened from the rear (Pruitt, 1974) (Fig. 4). Redirection and displacement of aggression were also noted in these situations (Pruitt, 1974). An extension of the upper lip was exhibited during threats and aggressive play (Jonkel and Cowan, 1971; Henry and Herrero, 1974; Burghardt and Burghardt, 1972; Pruitt, 1974). In agonistic encounters the threatening bear typically assumed a lowered and outstretched head and neck posture, coupled with the eyes oriented toward the opponent (Jonkel and Cowan, 1971; Frame, 1974; Jordan, in press; Pruitt, 1974). A bipedal stance was observed in captive bears during aggression and aggressive play by Pruitt (1974), Jordan (in press), and Leyhausen (1948); how-

Fig 4. Threat and accompanying lip extension in the black bear. (From Pruitt, 1974.)

ever, it was not observed by Herrero (unpubl.) in free-roaming bears (Fig. 5). Urination has reportedly occurred in conjunction with intensely fearful or agonistic encounters (Frame, 1974; Jordan, in press; Pruitt, 1974; Herrero, unpubl.). Both Herrero and Pruitt attributed the unique urination and fleeing pattern to animals placed in subordinate positions, both conspecifically and interspecifically. Jordan (in press) further described a stiff-legged, sliding walk during indirect threats, although both Pruitt and Bacon (pers. comm.) observed this behavior pattern to occur when captive animals were waking or descending a tree, even in the absence of threatening stimuli.

Mating. Ludlow provided an excellent description of courtship and mating behaviors in the black bear, finding that they showed "little courtship behavior prior to copulation" (1974: 29). The copulation behavior that she reported was quite similar to that observed in the brown bear by Meyer-Holzapfel (1957). The predominant behavior in mating was the male's neck biting and mounting after the receptive female lifted her tail (Ludlow, 1974, and in press).

On the whole, isolated ear, head, and mouth positions were not viewed as accurate indicators of impending behavior (Egbert, pers. comm.; Pruitt, in press). Instead, various combinations of facial and body expressions served as valid signals for behavior (Henry and Herrero, 1974; Pruitt, 1974). For example, if a bear approached another bear that was displaying flattened or frontally positioned ears, the initiator discontinued its approach (Henry and Herrero). Flattened ears appearing in the midst of an agonistic encounter did not necessarily dissuade an aggressor, however (Pruitt, 1974).

Fig. 5. Bipedal stance and lip extensions during black bear play. (From Pruitt, 1974.)

Polar Bear (Thalarctos maritimus)

Expressive systems in the polar bear have not been investigated to any great extent. We do know however, that captive polar bears lie curled around their young and hold the youngsters between their arms, keeping them off the den floor (Meyer-Holzapfel, 1957). Reports of social and solitary play in polar bears have been made (Meyer-Holzapfel, 1957); Vlasak, 1950), but precise movements involved in that play have not been determined. Observations of stereotyped movements in captive polar bears are abundant (Eipper, 1928; Schloeth, 1954; Holzapfel, 1939), perhaps because of the popularity of the species in zoos. The typical stereotyped movements were figure 8s, pacing, and head swinging.

Malayan Sun Bear (Helarctos malayanus)

In his observations of a captive sow and her cub, Dathe (1970) observed that the sow maintained normal contact and carried the cub in her mouth for the first seven weeks. At seven weeks, however, she came into estrus and rejected the cub completely.

VOCALIZATIONS

Meyer-Holzapfel's 1957 monograph on the Ursidae was the earliest attempt to provide a thorough understanding of bear behavior. Unfortunately she often did not delineate the species to which a behavior applied, perhaps assuming that most bears engaged in that activity. This was also true of her report on bear vocalizations. The six vocalizations that Meyer-Holzapfel (1957) reported were growling, dull purring, howling, roaring, blowing, and, in cubs, the emitting of a humming sound as they nursed. Garrison (1937) also did not specify the species involved when he recorded the clicking vocalization by a sow to tree her cubs and a "muffled" sound to get them down. Instances in which the species was identified are as follows.

Brown and Grizzly Bears

Stonorov (1972) and Stonorov and Stokes (1972) reported the presence of at least three sounds in Alaskan brown bears during agonistic encounters. They heard "chomping," a rapid opening and closing of the mouth, in high-intensity aggression; "bawling," thought to act as a threat to subordinates; and "roaring," heard during aggressive threats. Couturier (1954) observed that female brown bears uttered threatening growls toward males at the conclusion of the mating season. Vocalizations were never heard during episodes of conspecific social play in free-roaming grizzly bears (Henry and Herrero, 1974).

Black Bear

Black bear cubs characteristically "purr" (Jonkel and Cowan, 1971), "hum" (Meyer-Holzapfel, 1957), or "churckle" (Bacon, 1973; Pruitt, 1974) as they nurse from the sow. This sound also has been heard when captive cubs attempted to suckle the feet, arms, and hands of their human caretakers, or parts of siblings' bodies (Burghardt and Burghardt, 1972; Leyhausen, 1948; Meyer-Holzapfel, 1957; Pruitt, 1974; Seitz, 1952; and Vlasak: 1950). A short, open-mouthed, low moan (Pruitt, 1974), characteristic of juvenile discomfort or unrest, has been termed a "bleat" (Leyhausen, 1948) and a "grunt" (Jordan, in press). Huffing (Jordan, in press; Jonkel and Cowan, 1971) was a rapid exhalation of a single breath of air and may be closely akin to the anecdotal "woof" described by trappers and woodsmen (e.g., Seton, 1929). The huffing and its in-out variation occur during aggressive threat. A "squall" (Jonkel and Cowan, 1971), heard when cubs were frightened, may be similar to the long, low, closed-mouth moan observed by Pruitt (1974). A hoarse, pulsing bellow, described by Jordan (in press) and Leyhausen (1948), occurred during threat. Both Leyhausen (1948) and Pruitt (1974) described growls emitted by their captive cubs, but this sound was heard by each author only once and only in an agonistic context. During the jaw pop (Jordan, in press; Pruitt, 1974), the snapping shut of the lips caused a popping sound. The jaw pop was heard near the end of an aggressive encounter, perhaps as a defensive threat. Just prior to aggressive physical contact black bears occasionally gurgled (Herrero, ms.; Pruitt, 1974, in press). This low rumbling may serve as a final warning of serious aggressive intentions.

Polar Bear

Chuffing vocalizations and their meanings in various contexts were discussed by Wemmer, Von Ebbers, and Scow (in press). Most frequently the chuff was made by sows to their very young cubs. As the cubs matured, the frequency of sow chuffing decreased. When a sow and her cubs were separated at the age of fifteen months, all the bears chuffed and roared. A male and two females chuffed while they were being reintroduced after an eighteen-month separation. Two thirds of the observed chuffing instances were associated with transitions between one behavioral sequence and another in the female or her cubs. The cubs reacted to maternal chuffing by establishing contact with her or by ceasing their own moaning vocalizations. The chuff call was rarely given by adults. When observed in adults, it was likely to occur in a stressful situation.

In addition to mother-cub chuffing and the cubs' moaning, polar bears also blow when threatened (Wemmer, Von Ebbers, and Scow, pers. comm.) and growl and huff when captured in traps (Jonkel, 1970).

Spectacled Bear (Tremarctos ornatus)

Wemmer, Von Ebbers, and Scow (in press) reported that the spectacled bear emitted a chuffing vocalization, particularly between mother and cubs.

Sloth Bears (Melurus ursinus)

An open-mouthed, in-out exhalation termed chuffing was also reported for this species (Wemmer, Von Ebbers, and Scow, in press).

PHEROMONES

Whether bears use pheromones is highly speculative, since the size of the typical carnivore anal sacs is greatly reduced in the ursids (Ewer, 1973) and the presence of neck or shoulder glands has been disputed (Schumacher and von Marienfried, 1930; Hediger, 1949).

Brown Bear

Trees on which bears repeatedly rub have been considered as conspecific information points (Mills, 1919; Seton, 1929), as a declaration of property "ownership" (von Uexküll, 1934), as points of sexual advertisement during the breeding season (von Jacobi, 1957), and as a grooming method and pastime (Krott, 1962). Tschanz et al. (1970) investigated this tree marking-rubbing behavior in captive brown bears and noted six forms of rubbing against trees or inanimate objects. Tree-rubbing behavior occurred more frequently in adult males than females and peaked just before the onset of the breeding season. Subadults assumed defensive, bipedal stances when they approached within one meter of a tree rubbed by an adult male. The Tschanz group manipulated the location and introduction of new trees into the enclosure and then noted the bears' behavioral reactions. They observed that

> the attempt of the bears to carry over the location of the rubbing places on the trunk of the old tree to that of the new can have been induced neither by its appearance nor by the scent attached with it. Rather the spatial relationship to the rest of the characteristics of the pit must have been the determining factor. According to this, the rubbing places do not only serve as a basis for rubbing, but also as points of orientation in the inhabited area. [1970:61–62, trans.]

Additionally, the males were particularly attentive to those areas previously rubbed by females. It was also noted that males sniffed the urine and feces left by females more frequently than the eliminations of other males. Tschanz et al. determined that the primary functions of rubbing were to differentiate areas used by each sex and to offer general information.

Meyer-Holzapfel (1957) observed anal sniffing of the female just before copulatory activities. Meyer-Holzapfel (1968) later described rubbing and biting of trees by brown bears and noted their particular attraction to strong, resin-odored objects.

Black Bear

Both free-roaming and captive male black bears have been observed to sniff the female's anal-genital region while she was in estrus (Herrero, ms.; Ludlow, 1974). Published reports of black bear reactions to conspecific urine and feces traces differed from the report by Tschanz et al. (1970) on brown bears. Herrero (ms.), Leyhausen (1948), and Bacon (pers. comm.) reported that free-roaming and captive black bears did not pay special attention to the eliminations of other bears. However, captive bears did seem to have minimally overlapping areas for defecation (Bacon, pers. comm.).

Tree rubbing has been documented for black bears (Pruitt, 1974), but evidence for the function of "bear trees" is lacking for this species. Establishment of tree preferences was observed in two captive juveniles, but these preferences were not mutually exclusive (Pruitt, 1974).

Spectacled Bear

Tschanz et al. (1970) noted that captive spectacled bears tore the bark from trees in much the same manner as did brown bears.

Conclusions

The relatively solitary procyonids, ursids, and mustelids do not associate in nuclear family groups throughout the year. Instead, during the breeding season the males associate with solitary females or bands of females and juveniles. Because there is little group hunting, feeding, or living, extensive and complex social communication patterns have not evolved in these families to the same extent that they have in the more social carnivores. Yet even this conclusion should not be accepted too quickly. Vocal repertoires may, when well studied, be found to generally contain several discrete or graded calls. The coatis have the most social life style of the three families (Kaufmann, 1962; Gilbert, 1973). Tribes of females, subadults, and juveniles travel, sleep, and forage together. The males are solitary for much of the year but join the matriarchal groups for breeding and stay several months thereafter. The two cited references contrast the behavior of this fascinating procyonid in two differing environments: tropical rain forest and temperate arid mountains.

Scent marking has been a predominant means of social signalling in the mustelids and in several of the procyonids. The scent mark effectively reduces group contact by establishing home ranges or territories outside the breeding season. During estrus, scents may advertise the presence of a receptive female. The abdominal gland and urine-feces marking are the primary means for marking nests, familiar objects, and home ranges in conspecific communication. The extensive use of musky anal secretions in defensive warning and alarm situations is the most distinctive aspect of mustelid communication. The anal spray, however, serves as a warning message or even a weapon, as in skunks, that is easily interpreted and probably evolved from social usage. The procyonids have not relied so heavily upon musk emission during interspecific threat as have the mustelids. Although anal sprays and musks are not emitted by the ursids, urine and feces are available for use in intraspecific communication. Additionally, rubbing behaviors may communicate sexual and territorial information.

Coloration of a species may also function in mediating warnings or in soliciting attention. Light on dark color patterns are characteristic of the mustelids. The skunk's white stripes or spots, the African weasel's stripe, and the grison's lighter dorsal surface are all highly visible, particularly during the evening hours when the animals are most active. In the procyonids the lighter areas encircling the raccoon's eyes and the distinctive white of the giant panda may bring attention to expressive areas of these species. Light areas of fur accentuate the eyes, muzzle, and snout of several of the Ursidae, and rust-colored patches appear inside the relatively large ears of black bear cubs (Burghardt and Burghardt, 1972). These coloration patterns may further accent the facial expressions in intraspecific communication for the ursids and procyonids.

When species members do interact, the established expressive signals are probably interpreted in a situational context. For example, the procyonids do not rely on elaborate tail and facial signals. This is understandable, considering that many of the family members are arboreal, and any such signals would be often obscured from view. Data that enumerate methods of expressive communication are noticeably lacking in the mustelids. This is not to suggest that the family uses visual displays less than its carnivore relatives, but that, with few exceptions, there is little evidence of the active use of such a communication channel. Bear studies, however, have given support to the possibility of an active, if limited, use of expressive communication. Facial and body movements, as well as vocal signals, are characteristic of intraspecific communication in the brown, black, and polar bears. Behavioral data is extremely limited for all three of these families, however, and conclusive statements cannot presently be made.

Quite often the social behavior of the mustelids, procyonids, and ursids is compared to the social behavior of canids and felids without considering the vast differences in the ecologies and life histories of the various carnivore families. Because we observe a greater number of expressive signals in the canids and felids, we may conclude falsely that the mustelids, procyonids, and ursids are not capable of performing those expressive communicative acts. It must be remembered that the social organization of the group-living canids and felids probably requires a more refined communication system. The mustelids, procyonids, and ursids, however, have less need for extraneous communicative behaviors. Instead, their need is for a system that imparts precise meanings to discrete and simplified signals. Flexibility in social organization is found in these groups, however, as shown by Schneider's (1973) finding that raccoons at the northern limit of their range reverse the fall family disbanding process that occurs in more southern populations as winter approaches. The author concludes:

> It may be that cubs need adult guidance to survive their first winter under the severe conditions found this far north in the species' range. In the south raccoons do not den for the winter and cubs become independent at different times within their southern range. This suggests that family behavior is as flexible and adaptive to changing conditions as their diet and habitat requirements. [1973:71]

Similar studies with other members of these three families would be most useful.

Species differences may also be approached through comparative studies, and they show that environmental effects on behavioral organization may be more pervasive than previously realized. Herrero (1972) emphasized differences in degrees of aggressive behavior in black and grizzly bears, which he attributed to ecological and evolutionary influences. The smaller, more agile black bear is able to utilize trees as protective areas, thereby decreasing the necessity for any overt defense. The grizzly bear, on the other hand, does not climb trees so easily and may need to resort to the infliction of injury when threats are not enough to defend its young. In all likelihood these behavioral traits and communication signals are, through selection, transmitted to succeeding generations (Herrero, ms.). Herrero's theory may be applied to the procyonids and mustelids as well. Environmental requirements vary widely among the species within the procyonid, mustelid, and ursid families. To be most effective a discussion of carnivore communication must include an analysis of the species' environment and its interaction with that environment. The limited accessibility and observability of many of the species in these three families has discouraged observations of their social behaviors and ecologies. Until we have more behavioral data, our understanding of these families' communication patterns will remain limited.

References

Alexander, A. J., and Ewer, R. F., 1959. Observations on the biology and behaviour of the smaller African polecat. *African Wildlife,* 13:313–20.

Andersen, K. K., and Bernstein, D. T., 1975. Some chemical constituents of the scent of the striped skunk *(Mephitis mephitis). J. Chem. Ecol.,* 1:493–99.

Ansell, W. F. H., 1960. The African striped weasel, *Poecilogale albinucha* (Gray). *Proc. Zool. Soc. Lond.,* 134:59–64.

Apfelbach, R., 1973. Olfactory sign stimulus for prey selection in polecats (*Putorius putorius* L.). *Z. Tierpsychol.,* 33:270–73.

Azbukina, M. D., 1970. Specific skin glands of American mink puppies *(Mustela vison). Collection of Scientific Technical Information (Hunting-Fur Trapping-Wildlife),* Kirov, No. 30:60–67 (in Russian).

Azbukina, M. D., 1972. Morphology of the musk glands of the American mink. *Collection of Scientific Technical Information (Hunting-Fur Trapping-Wildlife),* Kirov, Nos. 37–39:127–35 (in Russian).

Bacon, E. S., 1973. Investigation on perception and behavior of the American black bear *(Ursus americanus).* Ph.D. diss., University of Tennessee.

Bacon, E. S., and Burghardt, G. M., in press a. Ingestive behaviors of the American black bear. In: *Proc. of the Third Internat'l. Conference on Bear Research and Management,* M. R. Pelton, J. W. Lenfer, and G. E. Folk, eds. Morges, Switzerland: IUCN.

Bacon, E. S., and Burghardt, G. M., in press b. Learning and color discrimination in the American black bear. In: *Proc. of the Third Internat'l. Conference on Bear Research and Management,* M. R. Pelton, J. W. Lenfer, and G. E. Folk, eds. Morges, Switzerland: IUCN.

Barash, D. P., 1974. Neighbor recognition in two "solitary" carnivores: the raccoon (*Procyon lotor*) and the red fox *(Vulpes fulva). Science,* 185:794–96.

Belyaev, D. K., and Ternovskaya, Yu. G., 1973. Behaviour and reproductive functions of animals. VI. Correlation between defensive behaviour and reproductive functions in sable. *Genetika* (Acad. Sci. USSR), 9:53–62 (in Russian, English summary).

Blackman, M. W., 1911. The anal glands of *Mephitis mephitica. Anat. Rec.,* 5:491–515.

Burghardt, G. M., 1975. Behavioral research on common animals in small zoos. In: *Research in Zoos and Aquariums,* Washington, D.C.: National Academy of Sciences, pp.103–133.

Burghardt, G. M., and Burghardt, L. S., 1972. Notes on the behavioral development of two black bear cubs: The first eight months. In: *Bears—Their Biology and Management,* S. M. Herrero, ed. Morges, Switzerland: IUCN, pp. 207–20.

Cole, L. W., 1912. Observations of the senses and instincts of the raccoon. *J. Anim. Beh.,* 2:299–309.

Couturier, M. A. J., 1954. *L'Ours Brun.* Grenoble: L'imprimerie Allier. 955pp.

Crandall, L. S., 1964. *The Management of Wild Mammals in Captivity.* Chicago: University of Chicago Press. xv+769pp.

Cuyler, W. K., 1924. Observations on the habits of the striped skunk *(Mephitis mesomelas varians). J. Mammal.,* 5:180–89.

Dalquest, W. W., and Roberts, J. H., 1951. Behavior of young grisons in captivity. *Amer. Midl. Nat.,* 46:359–66.

Dathe, H., 1970. A second generation birth of captive sun bears, *Helarctos malayanus,* at East Berlin Zoo. *Int. Zoo Yb.,* 7:131.

Davis, D. D., 1964. The giant panda. A morphological study of evolutionary mechanisms. *Fieldiana Zool. Mem.,* 3:1–339.

Dimond, S. J., 1970. *The Social Behavior of Animals.* New York: Harper Colophon Books. 256pp.

Dücker, G., 1965. Colour vision in mammals. *J. Bombay Nat. Hist. Soc.,* 61:572–86.

East, K., and Lockie, J. D., 1964. Observations on a family of weasels *(Mustela nivalis)* bred in captivity. *Proc. Zool. Soc. Lond.,* 143:359–63.

East, K., and Lockie, J. D., 1965. Further observations on weasels *(Mustela nivalis)* and stoats *(Mustela erminea)* born in captivity. *J. Zool.,* 147:234–38.

Egbert, A. L., and Luque, M., 1975. Among Alsaka's brown bears. *Nat. Geogr. Mag.,* 148(3):428–42.

Egbert, A. L., and Stokes, A. W., in press. The social behavior of brown bears *(Ursus arctos)* on an Alaskan salmon stream. In: *Proc. of the Third Internat'l Conference on Bear Research and Management,* M. R. Pelton, J. W. Lenfer, and G. E. Folk, eds. Morges, Switzerland: IUCN.

Eibl-Eibesfeldt, I., 1950. Über die Jugendentwicklung des Verhaltens eines männlichen Dachses (*Meles meles* L.) unter besonderer Berücksichtigung des Spieles. *Z. Tierpsychol.,* 7:327–55.

Eibl-Eibesfeldt, I., 1955. Biologie Iltis. Wiss. Film C 697, Göttingen.

Eibl-Eibesfeldt, I., 1957. Ausdrucksformen der Säugetiere. *Handb. Zool.,* 8(10)2:1–26.

Eibl-Eibesfeldt, I., 1970. *Ethology—The Biology of Behavior.* New York: Holt, Rinehart and Winston.

Eipper, P., 1928. Tiere sehen dich an. Berlin (not seen, cited in Meyer-Holzapfel, 1957).

Enders, R. K., 1952. Reproduction in the mink *(Mustela vison). Proc. Amer. Philos. Soc.,* 96:691–755.

Ewer, R. F., 1968. *Ethology of Mammals.* New York: Plenum Press. 418pp.

Ewer, R. F., 1973. *The Carnivores.* Ithaca, N.Y.: Cornell University Press. 494pp.

Ferron, J., 1973. Comparative morphology of the olfactory organ in some carnivorous mammals. *Nat. Can.* (Que.), 100:525–41.

Frame, G. W., 1974. Black bear predation on salmon at Olsen Creek, Alaska. *Z. Tierpsychol.,* 35:23–38.

Gander, F. F., 1965. Spotted skunks make interesting neighbors. *Animal Kingd.,* 68:104–108.

Garrison, L., 1937. Bears vocabulary. *Yosemite Nature Notes,* 16:78–79.

Gilbert, B., 1973. *Chulo.* New York: Knopf. xviii+290pp.

Goethe, F., 1940. Beiträge zur Biologie des Iltis. *Z. Säugetierk.,* 15:180–223 (not seen; cited in Ewer, 1968).

Goethe, F., 1950. Vom Leben des Mauswiesels (*Mustela n. nivalis* L.). *Zool. Garten,* 17:193–204 (not seen; cited in Ewer, 1973).

Goethe, F., 1964. Das Verhalten der Musteliden. *Kükenthal: Handb. Zool., VIII,* 37:1–80 (not seen; cited in Ewer, 1973).

Grinnell, J.; Dixon, J. S.; and Linsdale, J. M.; 1937. *Fur-Bearing Mammals of California. Their Natural History, Systematic Status and Relations to Man,* vol. I. Berkeley: University of California Press.

Hall, E. R., 1926. The abdominal skin gland of *Martes. J. Mammal.,* 7:227–29.

Hall, E. R., 1974. The graceful and rapacious weasel. *Nat. Hist.,* 83(9):44–49.

Hartman, L., 1964. The behaviour and breeding of captive weasels *(Mustela nivalis* L.). *New Zealand J. Sci.,* 7:147–56.

Hediger, H., 1949. Säugetierterritorien und ihre Markierung. *Bijdr. tot de Dierkde,* 28:172–84 (not seen; cited in Eibl-Eibesfeldt, 1970).

Hediger, H., 1950. *Wild Animals in Captivity.* London: Butterworth.

Heidt, G. A.; Peterson, M. K.; and Kirkland, G. L.; 1968. Mating behavior and development of least weasels *(Mustela nivalis)* in captivity. *J. Mammal.,* 49:413–19.

Henry, J. D., and Herrero, S. M., 1974. Social play in the American black bear: Its similarity to canid social play and an examination of its identifying characteristics. *Amer. Zool.,* 14:371–90.

Heran, I., 1966. Zum Verhalten in Gefangenschaft aufgewachsener Braunbären. *Zool. Garten,* 33:132–37.

Herman, T., and Fuller, K., 1974. Observations of the marten, *Martes americana,* in the Mackenzie district, Northwest Territories. *Can. Field Nat.,* 88(4):501–503.

Herrero, S., 1972. Aspects of evolution and adaptation in American black bears (*Ursus americanus* Pallus) and brown and grizzly bears (*Ursus arctos* Linné) of North America. In: *Bears—Their Biology and Management,* S. Herrero, ed. Morges, Switzerland: IUCN, pp.221–31.

Herrero, S. M., unpubl. ms. Black bear behavior. University of Calgary.

Herter, K., and Ohm-Kettner, I. D., 1954. Über die Aufzucht und das Verhalten zweier Baummarder (*Martes martes* L.). *Z. Tierpsychol.,* 11:113–37.

Holzapfel, M., 1939. Weben bei zwei Lippenbären. *Z. Tierpsychol.,* 3:151.

Hornocker, M. G., 1962. Population characteristics and social and reproductive behavior of the grizzly bear in Yellowstone National Park. Masters thesis, Montana State University.

Huff, J. N., and Price, E. O., 1968. Vocalisations of the least weasel, *Mustela nivalis. J. Mammal.,* 49:548–50.

Jacobi, R. von, 1957. Bären und Bärenbäume. *Monatsschrift über Natur und Gesellschaft,* 20:397–99.

Janzen, D. H., 1970. Altruism by coatis in the face of predation by boa constrictor. *J. Mammal.,* 51:387–89.

Johnson, C. E., 1921. The 'hand stand' habit of the spotted skunk. *J. Mammal.,* 2:87–89.

Jonkel, C., 1970. The behavior of captured North American bears. *BioScience,* 20:1145–47.

Jonkel, C. J., and Cowan, I. McT., 1971. The black bear in the spruce-fir forest. *Wildl. Monogr.,* 27:1–57.

Jordan, R. H., in press. Threat behavior of the black bear *(Ursus americanus).* In: *Proc. of the Third Internat'l. Conference on Bear Research and Management,* M. R. Pelton, J. W. Lenfer, and G. E. Folk, eds. Morges, Switzerland: IUCN.

Kaufmann, J. H., 1962. Ecology and social behavior of the coati, *Nasua narica,* on Barro Colorado Island, Panama. *Univ. Calif. Publ. Zool.,* 60:95–222.

Kaufmann, J. H., and Kaufmann, A. 1963. Some comments on the relationship between field and laboratory studies of behaviour, with special reference to coatis. *Anim. Behav.,* 11:464–69.

Kaufmann, J. H., and Kaufmann, A., 1965. Observations on the behavior of tayras and grisons. *Z. Säugetierkunde,* 30:146–55.

Kleiman, D. G., and Collins, L. R., 1972. Preliminary observations on scent marking, social behavior and play in the juvenile giant panda, *Ailuropoda melanoleuca*. *Amer. Zool.*, 12:644.

Kortlandt, A., 1972. *New Perspectives on Ape and Human Evolution.* Amsterdam: Stiching voor Psychobiologie.

Krott, P., 1959. Der Vielfrass (*Gulo gulo* L. 1958). *Monogr. Wildsäuget.*, 13:1–159 (note seen; cited in Ewer, 1973).

Krott, P., 1961. Der gefährliche Braunbär (*Ursus arctos* L. 1758). *Z. Tierpsychol.*, 18:245–56.

Krott, P., 1962. A key to ferocity in bears. *Nat. Hist.*, 71:64–71.

Krott, P., and Krott, G., 1963. Zum Verhalten des Braunbären (*Ursus arctos* L. 1758) in den alpen. *Z. Tierpsychol.*, 20:160–206.

Kuckuk, E., 1937. Tierpsychologie beobachtungen an zwei jungen braunbären. *Z. Vergl. Physiol.*, 24:14–41.

Lazar, J. W., and Beckhorn, G. D., 1974. Social play or the development of social behavior in ferrets *(Mustela putorius)? Amer. Zool.*, 14:405–14.

Leyhausen, P., 1948. Beobachtungen an einen jungen Schwarzbären. *Z. Tierpsychol.*, 6:433–44.

Lockie, J. D., 1966. Territory in small carnivores. *Symp. Zool. Soc. Lond.*, 18:143–65.

Ludlow, J. C., 1974. A preliminary analysis of the activities of captive black bears *(Ursus americanus)*—locomotion and breeding. Masters thesis, University of Tennessee.

Ludlow, J. C., in press. Observations on the breeding of captive black bears, *Ursus americanus.* In: *Proc. of the Third Internatl. Conference on Bear Research and Management,* M. R. Pelton, J. W. Lenfer, and G. E. Folk, eds. Morges, Switzerland: IUCN.

Markley, M. H., and Bassett, C. F., 1942. Habits of captive marten. *Amer. Midl. Nat.*, 28:604–16.

Mead, R. A., 1968. Reproduction in western forms of the spotted skunk (genus *Spilogale). J. Mammal.*, 49:373–90.

Meyer-Holzapfel, M., 1957. Das Verhalten der Bären *(Ursidae). Handb. Zool. Berl.*, 8(10):1–28.

Meyer-Holzapfel, M., 1968. Zur Bedeutung verschiedener Holz- und Laubarten für den Braunbären. *Zool. Garten,* 36:12–33.

Michels, K. M.; Fischer, B. E.; and Johnson, J. I.; 1960. Raccoon performance on color discrimination problems. *J. Comp. Physiol. Psychol.*, 53:379–80.

Mills, E. A., 1919. *The Grizzly.* Boston and New York (not seen; cited in Tschanz et al., 1970).

Morris, D., 1965. *The Mammals.* New York: Harper and Row. 448pp.

Morris, R., and Morris, D. 1966. *Men and Pandas.* London: Hutchinson.

Neumann, F., and Schmidt, H. D., 1959. Optische differenzierungsleistungen von Musteliden versuche an Frettchen und Iltisfrettchen. *Z. Vergl. Physiol.*, 42:199–205.

Pocock, R. I., 1908. Warning colouration in the musteline Carnivora. *Proc. Zool. Soc. Lond.*, 1908:944–59.

Poglayen-Neuwall, I., 1962. Beiträge zu einem Ethogram des Wickelbären (*Potos flavus* Schreber). *Z. Säugetierk.*, 27:1–44.

Poglayen-Neuwall, I., and Poglayen-Neuwall, I., 1965. Gefangenschaftsbeobachtungen and Makibären (*Bassaricyon* Allen 1876). *Z. Säugetierk.*, 30:321–66.

Poole, T. B., 1966. Aggressive play in polecats. *Symp. Zool. Soc. Lond.*, 18:23–44.

Poole, T. B., 1967. Aspects of aggressive behaviour in polecats. *Z. Tierpsychol.*, 24:351–69.

Poole, T. B., 1972. Diadic interactions between pairs of male polecats (*Mustela furo* and *Mustela furo* X *M. putorius* hybrids) under standardised environmental conditions during the breeding season. *Z. Tierpsychol.*, 30:45–58.

Poole, T. B., 1973. The aggressive behaviour of individual male polecats (*Mustela putorius, M. furo* and hybrids) toward familiar and unfamiliar opponents. *J. Zool.*, 170:395–414.

Poole, T. B., 1974. Detailed analyses of fighting in polecats (Mustelidae) using ciné film. *J. Zool.*, 173(3):369–93.

Progulske, D. R., 1969. Observations of a penned, wild-captured black-footed ferret. *J. Mammal.*, 50:619–20.

Pruitt, C. H., 1974. Social behavior in young captive black bears. Ph.D. diss., University of Tennessee.

Pruitt, C. H., in press. Play and agonistic behavior in young captive black bears. In: *Proc. of the Third Internatl. Conference on Bear Research and Management,* M. R. Pelton, J. W. Lenfer, and G. E. Folk, eds. Morges, Switzerland: IUCN.

Rowe-Rowe, D. T., 1969. Some observations on a captive African weasel, *Poecilogale albinucha. Lammergeyer,* 10:93–96.

Schneider, D., 1973. The adaptable raccoon. *Nat. Hist.*, 82(7):64–71.

Schloeth, R., 1954. Biespiele von Stereotypien bei den Bären in Zoo Basel. *Leben und Umwelt,* 10:100–104 (not seen; cited in Meyer-Holzapfel, 1957).

Schumacher, E., and von Marienfried, S., 1930. Jagd und Biologie. *Verständl. Wiss.* (Berlin), 44 (not seen; cited in Tschanz et al., 1970).

Seitz, A., 1952. Eisbärenzucht in Nürnberger Tiergarten. *Zool. Garten,* 19:180–89.

Seton, E. T., 1929. *Lives of Game Animals,* vol. II (1). Garden City, N.Y.: Doubleday, Doran. 949pp.

Sikes, S. K., 1964. The ratel or honey badger. *Afr. Wild Life,* 18:29–37.

Sinclair, W., Dunstone, N., and Poole, T. B., 1974. Aerial and underwater visual acuity in the mink *Mustela vison* Schreber. *Anim. Beh.,* 22:965–74.

Smythe, N., 1970. The adaptive value of the social organisation of the coati *(Nasua narica). J. Mammal.,* 51:818–20.

Sokolov, V. Ye.; Khorlina, I. M.; Golovnya, R. V.; and Zhuravleva, I. L.; 1974. Changes in the amine composition of volatile substances of vaginal discharges in American mink *(Mustela vison)* in relation to the sex cycle. *Doklady Acad. Sci. USSR,* 216:220–22 (in Russian).

Stonorov, D., 1972. Protocol at the annual brown bear fish feast; McNeil River Falls, Alaska. *Nat. Hist.,* 81:66–73.

Stonorov, D., and Stokes, A. W., 1972. Social behavior of the Alaska brown bear. In: *Bears—Their Biology and Management,* S. M. Herrero, ed. Morges, Switzerland: IUCN, pp.232–42.

Stuewer, F. W., 1943. Raccoons: Their habits and management in Michigan. *Ecol. Monogr.,* 13:203–57.

Sung, W.; Chang-kun, L.; and Sarich, V.; 1973. Giant pandas in the wild and in a biological laboratory. *Nat. Hist.,* 82(10):70–73.

Sunquist, M. E., and Montgomery, G. G., 1973. Arboreal copulation by coatimundi *(Nasua narica). Mammalia,* 37(3):517–18.

Tembrock, G., 1968. Land Mammals. In: *Animal Communication,* T. A. Sebeok, ed. Bloomington: Indiana University Press, pp.338–405.

Ternovskaya, Yu. G., 1974. Ontogenesis of defensive behavior of martin-like animals under experimental conditions. In: *Ecological and Evolutionary Aspects of Animal Behavior,* A. V. Severtsov, ed. Moscow: Nauka, pp.109–19 (in Russian).

Ternovskiy, D. V., 1974. Postnatal development of the ermine (*Mustela erminea* L.). *Teriologiya* (Acad. Sci. USSR, Siberian Branch) 2:305–12 (in Russian).

Tevis, L., 1947. Summer activities of California raccoons. *J. Mammal.,* 28:323.

Tschanz, B.; Meyer-Holzapfel, M.; and Bachmann, S.; 1970. Das Informations-system bei Braunbären, *Z. Tierpsychol.,* 27:47–72.

Uexküll, J. von, 1934. *Streifzüge durch die Umwelten von Tieren und Menschen.* Berlin: Springer, translated in C. H. Schiller, ed., *Instinctive Behavior.* New York: International Universities Press, 1957, pp.5–80.

Verts, B. J., 1967. *The Biology of the Striped Skunk.* Urbana: University of Illinois Press.

Vlasak, J., 1950. Über Kunstliche aufzucht eines Eisbären. *Zool. Garten,* 16:149–79.

Walker, A., 1930. The 'hand-stand' and some other habits of the Oregon spotted skunk. *J. Mammal.,* 11:227–29.

Welker, W. I.; Johnson, J. I.; and Pubols, B. H.; 1964. Some morphological and physiological characteristics of the somatic sensory system in raccoons. *Amer. Zool.,* 4:75–94.

Welker, W. I., and Seidenstein, S., 1959. Somatic sensory representation in the cerebral cortex of the raccoon *(Procyon lotor). J. comp. Neurol.,* 111:469–501.

Wemmer, C.; Von Ebbers, M.; and Scow, K.; in press. An analysis of the chuffing call and other vocalizations in the polar bear (*Thalarctos maritimus* Phipps). *Z. Tierpsychol.*

Wight, H. M., 1931. Reproduction in the eastern skunk *(Mephitis mephitis nigra). J. Mammal.,* 12:42–47.

Zollman, P. E., and Winkelmann, R. K., 1962. The sensory innervation of the common North American raccoon *(Procyon lotor). J. Comp. Neurol.,* 119:149–57.

Chapter 31

CETACEANS

David K. Caldwell and Melba C. Caldwell

Cetacean communication is a field that has received considerable attention in the popular press but has had comparatively little attention from experimental researchers. The smaller odontocete (toothed) cetaceans include, among others, the dolphins, the porpoises, and some of the larger dolphins that are called "whales": the killer whale, false killer whale, and pilot whale. These species do lend themselves, with varying degrees of success, to experimental studies because they are small enough to be maintained in captivity, they can be fed in a practical manner, and they in fact seem to adapt well to captivity. On the other hand, the mysticete (baleen) whales are usually too large to keep in captivity, their preferred foods are impractical to handle, and with few exceptions they are not considered suitable for captive studies in general. Behavioral studies on any wild cetacean are difficult because of the problems inherent in working at sea, and particularly in finding ways of studying highly mobile animals without disturbing their normal behavior. Therefore, our discussions of cetacean communication must be based primarily on studies of the captive, smaller odontocetes and on the understandably less-structured studies of wild groups of odontocetes and mysticetes.

It is impossible to discuss communication systems without also taking note of the senses that make the systems work. We will touch on the senses appropriate in each section.

Evans and Bastian (1969), D. K. Caldwell and M. C. Caldwell (1972a), and, most recently, Kinne (1975) provided illustrated summaries and literature compilations on communication in marine mammals, including cetaceans, pinnipeds, and others. This chapter is restricted to the cetaceans. We feel that we can best contribute to a discussion of the cetaceans by offering generalizations that indicate the manner and degree to which cetaceans are believed to communicate. We will also update recent summaries by pointing out further advances in the field.

We are most familiar with communication in the Atlantic bottlenosed dolphin (see, for example, M. C. Caldwell and D. K. Caldwell, 1967, and D. K. Caldwell and M. C. Caldwell, 1972b, and 1972c), and our discussions will be based heavily on personal experiences with this species and on the literature about it. Because this dolphin is the easiest to maintain in captivity, it is the species most often studied in the captive environment. In the areas of sensory and communicative processes, the amount of published material equals that for all other species combined. There have also been numerous observations of this species in the wild. Therefore, we suggest that the Atlantic bottlenosed dolphin serves as the model of choice for a study of the communication system of the odontocete cetaceans, although other species obviously differ in the fine points.

Surprisingly, some of the large mysticete whales have been studied in the wild even more than the smaller odontocetes have, probably because their geographical locales are more predictable and they move slower. Logically, then, they are often subjects more practical for wild study than the fast-moving and often wide-ranging small odontocetes are. When it is appropriate and possible, we will make comparisons both between odontocetes and mysticetes and between species in either or both groups, but our main emphasis will remain on the Atlantic bottlenosed dolphin. It is interesting to note that recent studies on the Indian Ocean bottlenosed dolphin have given good comparative results (Tayler and Saayman, 1972; Saayman, Bower, and Tayler, 1972; Saayman, Tayler, and Bower, 1973).

Acoustic Communication

Although communication in cetaceans is by no means limited to the acoustic mode, it is the one most often mentioned, and is probably the primary sensory system in this group of animals. In fact (see Wood, 1973, for a good discussion of the subject), the large, well-developed brain of cetaceans is almost surely related more to the acoustic system (probably echolocation in particular) than to the high degree of "intelligence" often ascribed to these mammals. Also, acoustic data are gathered on a quantitative basis more easily than data concerning other sensory and communication systems. Thus, more is known about acoustic systems of cetaceans than about their other sensory systems such as touch and vision.

From the outset we want to make it clear that we do not believe that dolphins or any other cetaceans "talk." They certainly communicate via several sensory modalities, including sound, but despite a rather extensive popular literature to the contrary, they do not talk in the sense that humans talk (i.e., passing abstract information and ideas by the use of a language consisting of words that have specific meanings). There is good evidence that at least some cetacean sounds provide general information to members of a group, and they probably convey general states of emotion as well. There is also reason to believe that individuals recognize other individuals through the acoustic mode, but there is no evidence yet to support the concept of a true language used by dolphins or any other cetacean. As Norris (1974:196–97) noted, this human "method of acoustic communication is almost grotesquely clumsy and difficult"—probably arising of necessity out of our use of tools.

Useful reviews of cetaceans known to produce sounds have been provided by Evans (1967) and Poulter (1968), and other shorter reports have added a number of species and even representatives of larger systematic groups to their lists. Sounds of the mysticetes include chirps, cries, moans, and clicks, and in addition to these the minke whale *(Balaenoptera acutorostrata)* may produce a whistle. These all have been ably reviewed to species recently (Winn and Perkins, 1976). Cummings (1971) has prepared a useful tape recording of undersea sounds that includes many emitted by both mysticetes and odontocetes. Species or representatives of larger groups that still have not been recorded undoubtedly will be found to produce some degree of sound as opportunities are found to make sound recordings. It is unlikely that any species of cetacean is mute. When cetaceans have been recorded, they have all been found to vocalize to some degree, and it is likely that at least some of these vocalizations are emitted in social situations and therefore have some communicative function. We know, for example, that the Atlantic bottlenosed dolphin is a very social species, and many of its sounds are clearly emitted in a social context—but other sounds are not.

Cetacean vocal emissions may be broadly divided into two categories: so-called pure-tone sounds, which are usually termed whistles or squeals (see Tavolga, 1968, for examples), and pulsed sounds. The latter in turn may be divided generally into (a) the trains of regular clicks emitted in exploratory or environmental search situations (echolocation; see Norris, 1969, for examples) and (b) those sounds emitted in an emotional context (usually termed burst-pulse sounds that are variously described as barks, yelps, squeaks, squawks, grunts, moans, and so on). (See M. C. Caldwell and D. K. Caldwell, 1967, for examples.) As Schevill (1964) pointed out, all odontocetes probably produce pulsed sounds and there is good evidence that mysticetes produce only this kind of sound. Whistles are known to be produced by dolphins and "whales" of the families Delphinidae, Monodontidae, Ziphiidae, and Steniidae (if one recognizes the family Steniidae as valid), and possibly by the minke whale (see above). The Phocoenidae and Platanistidae are not known to whistle although, like the others, they are known to produce pulsed sounds, which in many cases are complex and surely must have value for communication. The Physeteridae certainly produce pulsed sounds, but whether or not they whistle is still a matter of some controversy.

There is a considerable literature (see D. K. Caldwell and M. C. Caldwell, 1972a) in which attempts have been made to contrive some sort of complicated language from dolphin whistle contours. We have found instead that normally each individual dolphin has its own distinctive whistle contour, which we have termed its "signature" whistle (Caldwell and Caldwell, 1965). We have demonstrated the existence of this signature whistle primarily in the Atlantic bottlenosed dolphin (Caldwell, Caldwell, and Turner, 1970; Caldwell and Caldwell, unpubl.). There is evidence for it also in the saddleback dolphin (Caldwell and Caldwell, 1968), in the Atlantic spotted dolphin (Caldwell, Caldwell, and Miller, 1973), and to a lesser extent in the eastern Pacific white-sided dolphin (Caldwell and Caldwell, 1971), as well as other dolphin species. Although the rapidity, loudness, and/or duration of each whistle emission may vary, the basic contour for most animals remains much the same in a variety of situations in the life of the animal (after the whistle has developed; see Caldwell and Caldwell, in press). It is probably the variations in the signature whistle of an individual that carry a good share of the communication load, rather than the use of many different contours by the same individual. The individual contour, on the other hand, tells the rest of the community which animal is whistling. This identifying function may be even more important socially than a whistle language. We have found that on occasion a dolphin may emit only a portion of its signature whistle or may repeat it several times without stopping. Also, underwater sound recordings by hydrophone do not indicate which of several animals is vocalizing, and these tapes can indeed sound complex.

Excited dolphins tend to produce shorter whistles or an almost continuous whistling. The short ones usually occur when the animal hears other animals and stops to listen, or when it is closely attending to its environment. For instance, delayed playback (0.175 sec) of their own sounds to killer whales in the wild were found to inhibit rather than stimulate ongoing vocalizations (Spong, Bradford, and White, 1971). This pausing makes good biological sense because it is doubtful that dolphins can vocalize and listen attentively simultaneously. Our own experiments have shown, however, that a dolphin can distinguish between individuals' signature whistles by hearing only a small portion of one. One-half second is enough to identify one whistling dolphin to another.

Several investigators have examined the possibility that dolphins may transfer abstract information by electronic acoustic links or sound playback systems. Wood (1973:108ff.) reviewed the published literature in this area and concluded that the experiments did not prove that dolphins have any language ability but rather only that dolphins tend to respond to another's whistle and that they can be conditioned to respond to sound cues that are sometimes subtle.

We know from behavioral observations and experimental work that a dolphin can determine with great precision the direction from which it hears most pure tones. (Caldwell et al., 1971; Renaud and Popper, 1975). Presumably it can do at least as well for pulsed sounds. Thus, in our experience with whistles in particular and with pulsed sounds from presumption, we can say, for example, that a dolphin whistling with frequent, loud, short-clipped emissions apparently conveys the following information to another dolphin: "Attention! I am a dolphin with which you are (or are not) familiar. I am located to your left, and I am to some degree excited." Contrary to the beliefs of other students of dolphin sounds, we believe that distress is indicated in this manner rather than by a special "distress whistle." Dolphins that we have recorded under all kinds of stress (distress), including capture, intensive medical examinations, and grounding, have produced no whistles other than excited variants such as vocal "quaver" or "breaks" in the signature whistle. When we have examined data presented by other researchers that is intended to demonstrate a distress whistle, we have found evidence only of the signature whistle of an individual that happened to be recorded in a very stressful situation.

It is apparent that the whistle sounds emitted by one dolphin have a noticeable effect on others within hearing range. Inasmuch as the whistle is more easily quantified than pulsed sounds, the effect of this type of sound emission has been studied more closely than the effects of pulsed sounds. We do believe, however, that both pulsed and whistle sounds are communicative in nature and, further, that they have such value from the day the dolphin is born.

We do not know if a cetacean calf is born with the ability to echolocate effectively, but we do now have hard data for several species that show that the animal is able to produce what appear to be good trains of echolocation clicks on the first day of its life. We have found that several species of cetaceans are capable of producing whistles on the first day of life as well, and although the whistles are subject to modification to the signature whistle for that individual (Caldwell and Caldwell, in press), they do appear to be recognized by the mother and of use to her in identifying and locating her calf if she should become separated from it.

After the sound production for a given animal is established in its own repertoire, there is still a question of the degree of communication provided, by the pulsed sounds in particular. We have already noted that we believe the whistle is used primarily to identify an individual, assist in locating it, and provide some indication of its emotional state. We do not know if there is some signature value also in the pulsed sounds produced by individuals. Pulsed sounds are so complex that it is hard to say at what mathematical point some of them become useless for obtaining environmental information (i.e., as echolocation signals) and must therefore be emotional. Nor do we yet know the limits of the acoustic analyzer of the animal in distinguishing the variety of pulsed sounds that are produced. We have recorded some twenty general kinds of pulsed sounds made by Atlantic bottlenosed dolphins that seem to be correlated with general kinds of behavior, but only one of these sounds (a so-called male sexual yelp) has shown an indication of being behavior specific (Puente and Dewsbury, 1976), and even then it certainly is not always emitted at times when an observer would expect it. Most pulsed sounds seem to intergrade when large numbers of them are considered, and it is difficult to make clear-cut statements about them. For example, the "squawk" produced during a play chase is difficult to distinguish from the squawk made during an aggressive fight chase. Likewise, it is difficult to say exactly when a play chase develops into a fight chase. Additional study might resolve a few of the problems regarding the specificity of pulsed sounds as they relate to behaviors, but we can add little to what we wrote about the matter some years ago, summarized here. Captive Atlantic bottlenosed dolphins can voluntarily shape both whistles and pulse-type sounds (M. C. Caldwell and D. K. Caldwell, 1967, 1972). This behavior can be conditioned or it can result from an extended isolation from other dolphins. We have found no evidence for it in either undisturbed captive communities or wild populations.

In recent years considerable field work has been conducted on wild populations of the humpback whale (see, for example, Payne and McVay, 1971; Winn, Perkins, and Poulter, 1971). When they are in the southern parts of their range (in Bermuda and the West Indies), these whales produce long series (lasting many minutes) of different pulsed sounds that have come to be known as "songs." The series of sounds may be repeated on a predictable basis over and over again, and apparently each individual has its own song, which resembles the signature whistle in dolphins. There may be geographical dialects as well (Winn, pers. comm., 1973), so that a listener may be able to distinguish not only an individual but also the shallow West Indian bank from which that whale comes. We have not been able to demonstrate such geographical dialects in dolphins, but they are possible.

Echolocation is primarily a sound system used for environmental exploration (Norris, 1968, 1969) that is utilized by a wide variety of cetaceans (Evans, 1973). We have also noted that it is not always clear just when echolocation grades into pulsed sounds that are strictly emotional. However, even when these sounds are being used for environmental exploration, they may still have some communicative value to other members of a cetacean herd. Extensive echolocation, for example, may indicate to another animal that there is a strange object or animal in the immediate vicinity, or that food may be present and is being actively hunted. Hunts are often accompanied by considerable echolocation even when the prey is visible.

There is still controversy regarding the site of vocalization in cetaceans, but a growing literature (see, for example, Norris et al., 1972; Hollien et al., 1976) demonstrates that both clicks (pulses) and whistles are produced in a series of air sacs that lie above the bony cranium in the soft-tissue region around the blowhole on the top of the head. There is also evidence to suggest that the sounds may be directed forward and out of the head through a "tunnel" formed by lipids of varying densities that make up the soft fatty melon (see, for example, Norris, 1968; Litchfield and Greenberg, 1974; Litchfield, Karol, and Greenberg, 1973; Norris and Harvey, 1974; Litchfield et al., 1975). This feature of the melon has sometimes been referred to as an "acoustical transducer" and a "sonic lens" (see, for example, Wood, 1961, and Norris, 1968). The top of the bony cranium may also serve as a sound reflector, and there seems to be a very generalized correlation between the shape of this bony surface and the ecological need of the species to echolocate. For example, the cranium is frequently more cupped in contour when a more precise echolocation beam would seem to be required in a murky-water habitat where vision would be less useful (Wood, 1964).

In addition to the whistles produced by some cetaceans and the pulsed sounds produced by probably all of them, most cetaceans produce a variety of other sounds that have communicative value. They can slap their tails, flippers, heads, or entire bodies vigorously on the surface of the water. This results in a loud sonic report that can be heard both in and out of water. The context appears to be one in which the animal is disturbed or "angered" by some outside stimulus, or in the case of some captive animals by a trainer's tardiness in bringing the food bucket. Dolphins often snap their jaws together with such force that it makes a noticeable sound, and they may do this either out of the water or under water. This behavior, too, appears to indicate "displeasure."

Atlantic bottlenosed dolphins in captivity might squirt or splash water on a trainer and even strike out at him with the side or tip of the snout. These actions are often accompanied by the loud, raucous pulsed sound, made with the blowhole out of the water, that is sometimes called a "Bronx cheer."

In Atlantic bottlenosed dolphins, and probably other cetaceans as well, excitement may result in a loud, explosive exhalation of air through the blowhole out of water. Like the slapping of the water's surface, this sound is made when the animal is disturbed. Normal exhalation during regular breathing (blowing) is fairly quiet, and explosive exhalations appear to be communicative. A loud blow by one animal is frequently followed by the same type of exhalation by other animals nearby. In captivity blowhole responses may be elicited from animals in adjacent tanks which are neither in visual contact nor under the same stressful stimuli. Dolphins may also emit a large bubble of air from the blowhole under water without vocalizing. This makes an audible sound that behaviorally seems to suggest inquisitiveness or, in some cases, surprise. Amundin (1974) suggested that this release of air by the harbor porpoise may be displacement behavior.

Like those of land mammals, the digestive systems of cetaceans frequently rumble. We record these sounds most often when the animals have just been fed, and it is likely that other cetaceans recognize the rumbles as indicators of recent feeding and, therefore, the presence of food.

The very act of swimming may convey a message to other individuals in a herd, particularly if the swimming is rapid and accompanied by "porpoising" movements at or near the surface of the water. The sounds of rapid swimming might signal escape from danger, a chase for food, sexual activity, or merely play. Less-excited movements perhaps indicate that all is well and normal.

We have "recorded" many times yet another factor that surely has communication value to a group of cetaceans. When a strange animal or object enters the environment, an abrupt silence usually results, just as the forest goes silent when a hunter first enters. Vocalizations cease, breathing can hardly be heard, and swimming is almost noiseless. This sudden absence of sound can no doubt be an alert for danger as meaningful as a positive signal. Vocalization in the dolphin community resumes only gradually, usually with short echolocation bursts (apparently used in an exploratory manner) or with whistles so brief that they are only "chirps."

If we are to prove that sound plays a part in the communication process in cetaceans, we must clearly show that the animals are able to perceive the sounds at the levels at which they are being produced and to which behavioral responses (i.e., communication) should therefore be expected.

The potential methods by which cetaceans receive sound have been studied rather extensively from an anatomical point of view (Tavolga, 1965), but only recently has there been much experimental work on the subject (Norris, 1968, 1974; McCormick et al., 1970). The results of these experiments (often corroborated by simple behavioral observations) now suggest that the lower jaw of dolphins may function as the primary acoustic receiver for at least the sounds in the higher frequencies (50 kHz or more). Most of this work has been done by recording responses to sound stimuli directly from the auditory centers in the brain (Bullock et al., 1968; Bullock and Ridgway, 1972). The midbrain has loci responsive to echolocation-type click sounds; totally separated, the whistle-sensitive loci are primarily cortical.

Using behavioral techniques with free-swimming Atlantic bottlenosed dolphins, Johnson (1967) has shown that these animals are able to hear effectively frequencies from about 200 Hz up to about 150 kHz, or in the range where they produce the most sounds. Although most of the experimental work has been done with dolphins of this species, similar studies have used other odontocetes with similar (within variable species limits) general results. Findings indicate that cetaceans of a given species should be expected to hear the sounds that others of their kind produce and that many of the sounds might therefore be communicative.

It is important to remember that cetaceans are intensely aware of any sound they can perceive, and they are easily conditioned to it. Because they perceive and discriminate a wide range of sound (see, e.g., Herman and Arbeit, 1972), there is a strong probability that they use sound in the communication of a variety of emotions in addition to relaying such information as the presence of food or danger. Again, however, there is no evidence that with their repertoire of sounds cetaceans of any kind communicate in an abstract language, nor, as we are even sometimes asked, by mental telepathy. Norris (1974:197) introduced an idea worthy of consideration. He suggested that dolphins may offer considerable passive acoustic information to other dolphins. While an individual may be making no sounds of its own, another dolphin may be able to scan acoustically this individual. Thus, by using its highly refined echolocation system, the scanning dolphin may extract information from the body and nasal cavities of other dolphins that reveals their emotional state.

Visual Communication

It has long been suspected from behavioral observations, especially when they are made in captivity, where good visual acuity is required for the execution of many conditioned behaviors, that the Delphinidae have good vision. Only recently have experimental, detailed anatomical studies supported this theory (see, for example, Hall et al., 1972; Dawson, Birndorf, and Perez, 1972; Perez, Dawson, and Landau, 1972; Schusterman, 1973; Dawson and Perez, 1973; Dral, 1972, 1974). There is now good evidence that dolphins see well both in and out of water (Herman et al., 1975), and studies of their anatomy further suggest that they may have color vision as well as black and white. In our early research (D. K. Caldwell and M. C. Caldwell, 1972a, 1972b) we referred to "stereoscopic" vision in the Atlantic bottlenosed dolphin. Dawson, Birndorf, and Perez (1972) have used the more correct term "binocular" vision for this species (Fig. 1).

From both behavioral observations and anatomical data, then, there is good evidence that, except for the blind Ganges River dolphin, cetaceans use vision to some degree in communication. The visual stimuli to which they respond may be divided into two general groups: active and passive.

ACTIVE

We classify as active those visual stimuli that are produced under muscular control by one or

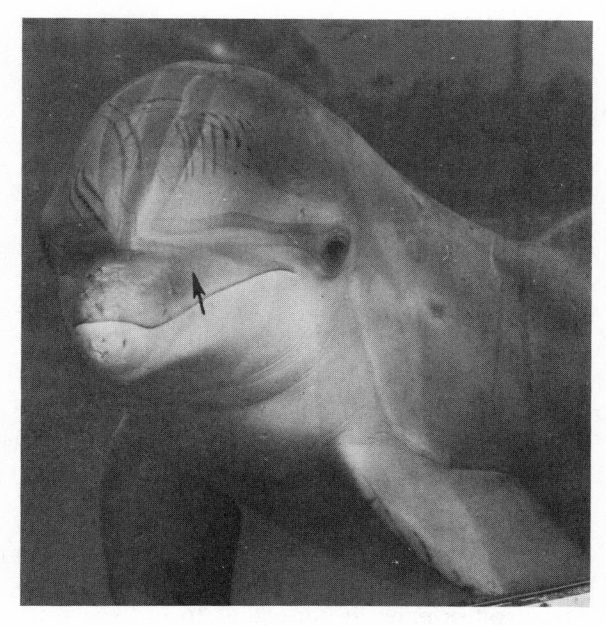

Fig. 1. Juvenile male Atlantic bottlenosed dolphin (MLF 232). The dark parallel marks on the forehead (melon) were inflicted by another juvenile male when the two were first placed together about one and a half years before this picture was made. At this writing, some five years later, the scars remain. The light gray markings on the head and between the eye and flipper are typical pigmentation for this species, but the degree and intensity may vary between individuals and thus provide cues for visual individual recognition. The light line on the upper jaw at the tip of the arrow marks the row of hair pits mentioned in the text that possibly have a sensory function. Even from this angle, the binocular vision in the species is evident. (Photograph courtesy Marineland of Florida.)

more animals and that are in turn reacted to by one or more other animals in the social group.

Dolphins do not have the mechanics for facial expressions found in many other mammals like primates, dogs, and cats. Nor do they have available to them the variety of moveable body extremities, such as hands, ears, and tails, utilized by some mammal groups in communication. Visual signals are nonetheless evident among cetaceans.

These signals may take the form of open-mouth threats or body postures, which probably are most useful in the maintenance of dominance hierarchies, although it has been suggested that some body postures are related to sexual signaling (see below). An Atlantic bottlenosed dolphin may indicate threat by facing another dolphin and opening its mouth, thereby exposing its teeth, or by arching its back slightly and holding its head downward. Submission for dolphins, as for many mammals, may take the form of a display that is the opposite of the threat; in this case the mouth is kept closed and the side of the body is turned to the threatening dolphin. It is usually the dominant animal, most often the largest and oldest of either sex, that displays such threats in a dolphin community. We should note, however, that established dominants can control a group using only the slightest gesture, to which other members seem to be very sensitive.

Sexual solicitation, which may be displayed by an animal of any age, including infants, often involves the soliciting dolphin swimming ahead of the other animal, looking back, and rolling onto its side or back to display the genital region. Tavolga and Essapian (1957:12–13, Fig. 1) have suggested that a soliciting male may also assume a particular S-shaped posture. We have not observed this to any great extent, nor did Puente and Dewsbury (1976) find a correlation between the posture and copulation. Vision may also aid in special positioning between individuals in a group—again, behavior related to social hierarchies. And finally, although it is not true vision, some dolphins (and probably most, if not all, cetaceans) certainly "see" by means of their ability to echolocate, and this too may be used in spatial positioning, at least in murky waters. It is even possible that the configuration of the body, such as an open or a closed mouth, might be "seen" in this manner.

Speed and direction of motion are almost certainly perceived visually. Even a low-intensity movement of the flukes by a lead animal serves as a perfectly adequate intention movement for maintaining synchronous swimming (Caldwell and Caldwell, 1964).

PASSIVE

We classify as passive those visual stimuli that are not under the muscular control of the animal producing them, although they may well be received by another animal to the same degree that active signals are.

As noted by Norris (1967), sexual dimorphism in relative size occurs in a few cetacean species, and the development of the teeth varies conspicuously in some (with the teeth of the male much more developed and prominent). Although they are probably not primarily visual stimuli, these dimorphisms are surely visually perceived by other cetaceans. For most cetacean species, however, the only visible difference in sexes is in the genital region. When a new animal is introduced to a group, this area receives detailed examination by the others (see illustration in D. K. Caldwell and M. C. Caldwell, 1972b:50). The recipient of the scrutiny passively accepts examination even though it is surrounded by investigating animals. Although other sensory systems may be activated during this procedure (echolocation, tactile, and possibly even gustatory), visual information probably provides additional if not primary sensory input.

The young of some cetaceans are marked in a manner quite different from the adults (compare, for example, young and adult pigmentation in the Atlantic spotted dolphin as illustrated by Caldwell and Caldwell, 1966; the adult is heavily spotted while the young is not spotted at all). Such differences certainly help visually distinguish a young animal from an adult, which in turn induces a different behavioral response. A juvenile spotted dolphin, for example, may be just getting its spotted pigmentation even though it is nearly as large as some well-spotted adults. Norris (1967) noted that sexual and pattern dimorphisms occur in odontocetes. In our own experience there appears to be a tendency for Atlantic bottlenosed dolphin males to develop a very pink belly during certain times of the year, and this change in coloration may be related to breeding activity (Caldwell, 1960). If this is true and if Atlantic bottlenosed dolphins have the capability for color vision (see Perez, Dawson, and Landau, 1972), it is likely that other members of the school would perceive that the animal is in breeding condition rather than in the ever-ready sexual behavior pattern typical of most male dolphins.

Evans and Bastian (1969) also pointed out that many delphinids have areas of white or very light pigmentation which are vivid and visible at great distances under water, and that these visible displays may be important to the social interaction of these dolphins. Although it has never been tested, we concur with this hypothesis in general. There must also be some evolutionary basis for some of the bold markings on cetaceans like the killer whale and the Dall porpoise, and it seems quite likely that their markings are related to visual communication. Conspecific identification in these animals seems likely; individual recognition is potentially possible when, for example, animals have the individual markings noted in the Dall porpoise by Norris and Prescott (1961) and in the Atlantic bottlenosed dolphin by D. K. Caldwell and M. C. Caldwell (1972b). There are many other reports of unusually marked individuals of both odontocete and mysticete cetaceans, and all of them must provide visual identification (and communication) signals to other members of the herd.

Atlantic bottlenosed dolphins and sperm whales, for example, are known to have complex social organizations, and recent field studies have shown that many baleen whales do too. We can therefore presume from even limited studies that cetaceans also have complex social signals, many of which appear to be visual. Visual signals are much more difficult to study and analyze than acoustic, and scientists may never recognize more than a few of the more obvious and pronounced ones. From our own human experience we know that many of our visual signals are obscure and ambiguous. One individual must know another intimately before subtle nuances are recognized, and similar signals from another person may not mean the same thing at all, especially when the cultures of the two humans are different. We believe, therefore, that there must be many more visual signals in cetacean communication than we recognize. We suggest that while vision is probably of less importance than the use of sound, it is perhaps of more importance than recent literature suggests.

Tactile Communication

Behavioral observations quickly lead one to the conclusion that touch is one of the most important means by which cetaceans communicate. First observed in captive dolphins of many species, touch has also been demonstrated in some of the larger baleen whales, as shown by underwater studies in recent years.

All cetaceans that we have observed in captivity seek and are receptive to gentle body contact (for photographs see Evans and Bastian, 1969, and D. K. Caldwell and M. C. Caldwell, 1972a, 1972b). Students of wild behavior have observed and illustrated body contact in both odontocetes (see, for example, Evans and Bastian, 1969) and mysticetes (see, for example, Cousteau and Diole, 1972, for illustrations of this behavior in the humpback whale; and Saayman and Tayler, 1973, for the southern right whale).

Some of the body contact is obviously related to sexual activity. Copulation is most frequently preceded by gentle mouthing and nipping. Particularly in mature animals the precopulatory play may become progressively more violent, with the participants even leaping from the water and diving forcefully at each other, sometimes only glancing bodies but at other times ramming melons so hard that it causes loud reverberations in a tank (D. K. Caldwell and M. C. Caldwell, 1972b:52). One would think that such violence would fracture their skulls, but instead it seems to function as a sexual stimulus because it almost always ends in copulation.

Most tactile stimulation is of a more delicate nature, involving as little as simply touching flippers as the animals slowly swim together for long periods. At the experimental level Pepper and Beach (1972) found that mild tactile stimulation from humans served as a reward to a captive Atlantic bottlenosed dolphin, a finding suggesting that such stimulation by another dolphin might also be rewarding.

On the obverse side of the coin, tactile communication is not always used to promote closer interpersonal relationships. Dolphins follow the normal mammalian pattern in displaying aggression (see photographs of aggressive behavior by the Atlantic bottlenosed dolphin in D. K. Caldwell and M. C. Caldwell, 1967, 1972a, 1972b; M. C. Caldwell and D. K. Caldwell, 1967; Evans and Bastian, 1969). Fights may occur over objects, space, proximity to other individuals, food, or for no apparent reason at all. Fighting dolphins slash and bite with the teeth, slash and ram with the jaws, and strike with the flukes. The encounters sometimes result in injuries with permanent scarring (Fig. 1). Mothers may punish their young by holding them down (Fig. 2), biting them, or even holding them out of water.

Although almost any animal may be provoked into an attack, large males are the most aggressive. Immature males are the most frequent recipients of attack. Norris (1967) pointed out the abundance of scars in the urogenital area

Fig. 2. Adult female Atlantic bottlenosed dolphin punishing her infant male calf by holding him down on the bottom. The calf usually vocalized loudly when such punishment took place. (Photograph by David K. Caldwell at Marineland of the Pacific.)

of immature males, an indication of the age-old struggle by maturing groups for sexual rights.

Sexual aggression by large males has been discussed as a husbandry problem (Caldwell, Caldwell, and Townsend, 1968). Homosexuality in dolphins has also been seen and discussed at some length by several writers. We have a single example of homosexuality used solely in the context of communicating and establishing dominance. The incident occurred between two large male Atlantic bottlenosed dolphins when they were first put together. Although the encounter began with the usual open-mouth threats, it ended with sexual pursuit and two successful intromissions by the victor, and a somewhat reluctant submission by the loser. The entire episode lasted one hour, terminated, and was not seen again. The loser, who had been the dominant animal in the community, never again established himself, nor was he seen to try. This behavioral pattern of rape as an expression of aggressive dominance behavior is one of the unfortunate consequences of human penal institutions; whether it occurs in wild dolphins, as it does in other human contexts, we cannot say. In this instance, however, it was a clear example of tactile aggression.

On the anatomical level, the skin of several odontocetes had been examined histologically. Most recently studied by Simpson and Gardner (1972), dolphin skin appears richly innervated, particularly in the regions of the jaw, flukes, vulva, and perineum, which suggests a greater sensitivity in these areas.

A few rudimentary or vestigial hairs are present in cetaceans. Located on the upper jaw, they are variously termed bristles in the mysticetes; bristles, hairs, or even vibrissae in the Amazon River dolphin; or pits (Fig. 1) in the Atlantic bottlenosed dolphin (which bear hairs in the newborn). The terminology depends on the level of evolutionary regression of the species, but they most likely retain a tactile function in all species (see Simpson and Gardner, 1972, and D. K. Caldwell and M. C. Caldwell, 1972a, for illustrations and interpretations).

Chemical Communication

In an earlier summary (D. K. Caldwell and M. C. Caldwell, 1972a) we included what was then known about the olfactory and gustatory senses in cetaceans. Behaviorally, it appears that dolphins may be able to taste, although there is no evidence from a behavioral or anatomical point of view that they can smell. There is some anatomical evidence that mysticetes may be able to do the latter, although it has not been demonstrated behaviorally. While there is behavioral evidence for taste in dolphins (they will eat one kind of fish and not another, for instance, even if the fish appears to be of the same texture and is cut into small pieces), the anatomical evidence is confusing (Jansen and Jansen, 1969; Morgane and Jacobs, 1972). In Atlantic bottlenosed dolphins we were unable to demonstrate taste receptors (D. K. Caldwell and M. C. Caldwell, 1972a:479), although Suchowskaja (1972) reported their presence in *Tursiops truncatus* and *Delphinus delphis.* Kruger (1959:177) suggested that the excellent development of the nucleus ventralis medialis in the Atlantic bottlenosed dolphin might indicate the presence of the ability to taste.

The work of Sokolov and Kuznetsov (1971) indicates behavioral conditioning by Black Sea dolphins to chemical stimuli. The levels of success achieved by the dolphin on the initial tests (74 percent and 78 percent) suggest that they can discriminate. The behavioral breakdown that followed leaves room for doubt. As the writers pointed out, the test design itself was one that animals have difficulty solving. Our own experimental dolphin has shown difficulty solving similarly designed problems.

Further experiments in this area would be desirable. It would seem that cetaceans would be sorely disadvantaged by the loss of all chemoreception. The loss of smell was probably a necessary concomitant to the movement of the blowhole to the top of the head during the transition from land to water. On the other hand, we can think of no comparable anatomical reason for evolutionary pressure to have forced these animals to discard something so basic as the sense of taste.

If indeed cetaceans do have chemoreceptors, they are capable of chemical signaling by waste products and glandular secretions. Male Atlantic bottlenosed dolphins (and other dolphin species as well, we have observed) do have two small openings located just anterior to the anus (Fig. 3). These openings lead to glandular tissue via a large duct. The base of the gland extends into small tubules that have a single row of secretory epithelium. Although it is possible that the structures are nothing more than undeveloped mammary glands, biochemical or behavioral work on their secretions might prove useful.

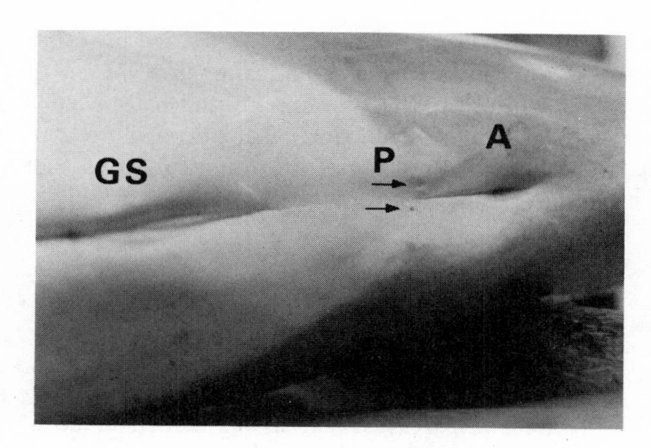

Fig. 3. Posterior ventral surface of an adult male Atlantic bottlenosed dolphin (MLF 214) showing paired anal pores (P). The pores are posterior to the male genital slit (GS) and just anterior to the anus (A). (Photograph by David K. Caldwell.)

Sensory Coordination

As behaviorists, we rely heavily on simple sensory clues that appear to elicit a particular kind of behavior. On the other hand, as cetologists, we have learned that cetaceans simply do not react to seemingly all-or-none signals as birds and insects, for example, do. A particular stimulus, whether it is color, shape, or sound, does not automatically elicit the initiation of a behavior such as courtship or feeding. As we have said before (D. K. Caldwell and M. C. Caldwell 1972a:485), in cetaceans "a single cue may increase the probability of a behavior's occurring, but somewhere in the brain all simultaneously incoming stimuli are being processed and balanced one against the other. It is the summation of all internal and incoming stimuli, plus individual experience, that determines the final behavior. . . ." We have not changed in our thinking on this point. After our considerable studies of dolphins in recent years, we find that one dolphin may react to another in a variety of ways although the same communicatory cues were presented each time. But perhaps we cannot perceive the subtle differences in cues that the attending dolphin can.

References

Amundin, M., 1974. Some evidence for a displacement behaviour in the harbour porpoise, *Phocoena phocoena* (L.); a casual analysis of a sudden underwater expiration through the blow hole. *Rev. Comp. Animal,* 8(1):39–45.

Bullock, T. H.; Grinnell, A. D.; Ikezono, E.; Kameda, K.; Katsuki, Y.; Nomoto, M.; Sato, O.; Suga, N.; and Yanagisawa, K.; 1968. Electrophysiological studies of central auditory mechanisms in cetaceans. *Zeitschrift für vergleichende Physiologie,* 59:117–56.

Bullock, T. H., and Ridgway, S. H., 1972. Evoked potentials in the central auditory system of alert porpoises to their own and artificial sounds. *J. Neurobiology,* 3(1):79–99.

Caldwell, D. K., 1960. Notes on the spotted dolphin in the Gulf of Mexico. *J. Mammal.,* 41(1):134–36.

Caldwell, D. K., and Caldwell, M. C., 1966. Observations on the distribution, coloration, behavior and audible sound production of the spotted dolphin, *Stenella plagiodon* (Cope). *Los Angeles County Mus., Cont. in Sci.,* 104:1–28.

Caldwell, D. K., and Caldwell, M. C., 1967. Dolphins, porpoises and behavior. *Underwater Nat.,* 4(2):14–19.

Caldwell, D. K., and Caldwell, M. C., 1972a. Senses and communication. In: *Mammals of the Sea; Biology and Medicine,* S. H. Ridgway, ed. Springfield, Ill.: Charles C. Thomas, pp.466–502.

Caldwell, D. K., and Caldwell, M. C., 1972b. *The World of the Bottlenosed Dolphin.* Philadelphia: J. B. Lippincott.

Caldwell, D. K., and Caldwell, M. C., 1972c. Dolphins communicate—but they don't talk. *Naval Res. Rev.,* 25(6–7):23–27, cover illus.

Caldwell, M. C., and Caldwell, D. K., 1964. Experimental studies on factors involved in care-giving behavior in three species of the cetacean family Delphinidae. *Bull. So. Calif. Acad. Sci.,* 63(1):1–20.

Caldwell, M. C., and Caldwell, D. K., 1965. Individualized whistle contours in bottlenosed dolphins *(Tursiops truncatus). Nature,* 207:434–35.

Caldwell, M. C., and Caldwell, D. K., 1967. Intraspecific transfer of information via the pulsed sound in captive odontocete cetaceans. In: *Les Systemes Sonars Animaux, Biologie et Bionique,* R.-G. Busnel, ed. Jouy-en-Josas, France: Laboratoire de Physiologie Acoustique, 2:879–936.

Caldwell, M. C., and Caldwell, D. K., 1968. Vocalization of naive captive dolphins in small groups. *Science,* 159:1121–23.

Caldwell, M. C., and Caldwell, D. K., 1971. Statistical evidence for individual signature whistles in Pacific whitesided dolphins, *Lagenorhynchus obliquidens. Cetology,* 3:1–9.

Caldwell, M. C., and Caldwell, D. K., 1972. Vocal mimicry in the whistle mode by an Atlantic bottlenosed dolphin. *Cetology,* 9:1–8.

Caldwell, M. C., and Caldwell, D. K., in press. The whistle of the Atlantic bottlenosed dolphin (*Tursiops truncatus*). 2. Ontogeny. In: *Natural History of Whales,* H. E. Winn and B. L. Olla, eds. New York: Plenum Publishing.

Caldwell, M. C., and Caldwell, D. K., unpubl. The whistle of the Atlantic bottlenosed dolphin (*Tursiops truncatus*). 1. General description for the species and statistical analysis of stereotypy in individuals.

Caldwell, M. C.; Caldwell, D. K.; and Miller, J. F.; 1973. Statistical evidence for individual signature whistles in the spotted dolphin, *Stenella plagiodon. Cetology,* 16:1–21.

Caldwell, M. C.; Caldwell, D. K.; and Townsend, B. C., Jr.; 1968. Social behavior as a husbandry factor in captive odontocete cetaceans. In: *Proceedings of the Second Symposium on Diseases and Husbandry of Aquatic Mammals.* St. Augustine, Fla.: Marineland Res. Lab., pp.1–9.

Caldwell, M. C.; Caldwell, D. K.; and Turner, R. H.; 1970. Statistical analysis of the signature whistle of an Atlantic bottlenosed dolphin with correlations between vocal changes and level of arousal. *Los Angeles County Mus. Nat. Hist. Found., Tech. Rept. 8, LACMNHF/MRL ONR contract N00014-67-C-0358,* 40 pp. (processed).

Caldwell, M. C.; Hall, N. R.; Caldwell, D. K.; and Hall, H. I.; 1971. A preliminary investigation of the ability of an Atlantic bottlenosed dolphin to localize underwater sound sources. *Marineland Research Laboratory (Florida), Tech. Rept. 6, ONR contract N00014-70-C-0178.* 6pp. (processed).

Cousteau, J.-Y., and Diole, P., 1972. *The Whale: Mighty Monarch of the Sea.* Garden City, N.Y.: Doubleday.

Cummings, W. C., 1971. Sonar training tape: non target sounds. *Naval Acoustic Sensor Training Aids Dept., Fleet ASW School, San Diego, and Applied Bioacoustics Branch, Naval Undersea Center, San Diego, Calif. Biological Tape no. 12-3522. 5TR-P015.*

Dawson, W. W.; Birndorf, L. A.; and Perez, J. M.; 1972. Gross anatomy and optics of the dolphin eye *(Tursiops truncatus). Cetology,* 10:1–12.

Dawson, W. W., and Perez, J. M., 1973. Unusual retinal cells in the dolphin eye. *Science,* 181:747–49.

Dral, A. D. G., 1972. Aquatic and aerial vision in the bottle-nosed dolphin. *Netherlands J. Sea. Res.,* 5(4): 510–13.

Dral, A. D. G., 1974. Problems in image-focusing and astigmatism in Cetacea—a state of affairs. *Aquatic Mammals,* 2(1):22–28.

Evans, W. E., 1967. Vocalization among marine mammals. In: *Marine Bio-acoustics,* W. N. Tavolga, ed. New York: Pergamon Press, vol. 2, pp.159–86.

Evans, W. E., 1973. Echolocation by marine delphinids and one species of fresh-water dolphin. *J. Acoustical Soc. Amer.,* 54(1):191–99.

Evans, W. E., and Bastian, J., 1969. Marine mammal communication: social and ecological factors. In: *The Biology of Marine Mammals,* H. T. Andersen, ed. New York: Academic Press, pp.425–75.

Hall, N. R.; Hall, H. I.; Caldwell, M. C.; and Caldwell, D. K.; 1972. Visual acuity under conditions of near darkness and discrimination learning in the Atlantic bottlenosed dolphin *(Tursiops truncatus). Cetology,* 12:1–7.

Herman, L. M., and Arbeit, W. R., 1972. Frequency difference limens in the bottlenose dolphin: 1-70 KC/S. *J. Auditory Res.,* 2:109–20.

Herman, L. M.; Peacock, M. F.; Yunker, M. P.; and Madsen, C. J.; 1975. Bottlenosed dolphin: double-slit pupil yields equivalent aerial and underwater diurnal acuity. *Science,* 189:650–52.

Hollien, H.; Hollien, P.; Caldwell, D. K.; and Caldwell, M. C.; 1976. Sound production by the Atlantic bottlenosed dolphin *Tursiops truncatus. Cetology,* 26:1–8.

Jansen, J., and Jansen, J. K. S., 1969. The nervous system of Cetacea. In: *The Biology of Marine Mammals,* H. T. Andersen, ed. New York: Academic Press, pp.175–252.

Johnson, C. S., 1967. Sound detection thresholds in marine mammals. In: *Marine Bio-acoustics,* W. N. Tavolga, ed. New York: Pergamon Press, vol. 2, pp.247–60.

Kinne, O., 1975. Orientation in space: animals: mammals. In: *Marine Ecology,* O. Kinne, ed. New York: Wiley, vol. 2, Physiological Mechanisms, part 2, pp.702–852.

Kruger, L., 1959. The thalamus of the dolphin *(Tursiops truncatus)* and comparison with other mammals. *J. Comp. Neurology,* 111(1):133–94.

Litchfield, C., and Greenberg, A. J., 1974. Comparative lipid patterns in the melon fats of dolphins, porpoises and toothed whales. *Comp. Biochemistry and Physiology,* 47B:401–7.

Litchfield, C.; Greenberg, A. J.; Caldwell, D. K.; Caldwell, M. C.; Sipos, J. C.; and Ackman, R. G.; 1975. Comparative lipid patterns in acoustical and nonacoustical fatty tissues of dolphins, porpoises and toothed whales. *Comp. Biochemistry and Physiology,* 50B:591–97.

Litchfield, C.; Karol, R.; and Greenberg, A. J.; 1973. Compositional topography of melon lipids in the Atlantic bottlenosed dolphin *Tursiops truncatus:* implications for echo-location. *Marine Biology,* 23:165–69.

McCormick, J. G.; Weaver, E. G.; Palin, J.; and Ridgway, S. H.; 1970. Sound conduction in the dolphin ear. *J. Acoustical Soc. Amer.,* 48(6–2):1418–28.

Morgane, P. J., and Jacobs, M. S., 1972. Comparative anatomy of the cetacean nervous system. In: *Functional Anatomy of Marine Mammals,* R. J. Harrison, ed. London: Academic Press, pp.117–244.

Norris, K. S., 1967. Aggressive behavior in Cetacea. In: *Aggression and Defense: Neural Mechanisms and Social Patterns,* C. D. Clemente and D. B. Lindsley, eds. Berkeley: University of California Press, UCLA Forum in Medical Sciences no. 7, pp.225–41.

Norris, K. S., 1968. The evolution of acoustic mechanisms in odontocete cetaceans. In: *Evolution and Environment,* E. T. Drake, ed. New Haven, Conn.: Yale University Press, pp.297–324.

Norris, K. S., 1969. The echolocation of marine mammals. In: *The Biology of Marine Mammals,* H. T. Andersen, ed. New York: Academic Press, pp.391–423.

Norris, K. S., 1974. *The Porpoise Watcher.* New York: W. W. Norton.

Norris, K. S.; Dormer, K. J.; Pegg, J.; and Liese, G. J.; 1972. The mechanism of sound production and air recycling in porpoises: a preliminary report. In: *Proceedings of the Eighth Annual Conference on Biological Sonar and Diving Mammals,* 1971. Fremont, Calif.: Biological Sonar Laboratory, pp.113–29.

Norris, K. S., and Harvey, G. W., 1974. Sound transmission in the porpoise head. *J. Acoustical Soc. Amer.*, 56(2):659–64.

Norris, K. S., and Prescott, J. H., 1961. Observations on Pacific cetaceans of Californian and Mexican waters. *Univ. Calif. Publ. in Zool.*, 63(4):291–402, plates 27–41.

Payne, R. S., and McVay, S., 1971. Songs of humpback whales. *Science,* 173:585–97.

Pepper, R. L., and Beach, F. A., III, 1972. Preliminary investigations of tactile reinforcement in the dolphin. *Cetology,* 7:1–8.

Perez, J. M.; Dawson, W. W.; and Landau, D.; 1972. Retinal anatomy of the bottlenosed dolphin *(Tursiops truncatus). Cetology,* 11:1–11.

Poulter, T. C., 1968. Marine mammals. In: *Animal Communication,* T. A. Sebeok, ed. Bloomington: Indiana University Press, pp.405–65.

Puente, A. E., and Dewsbury, D. A., 1976. Courtship and copulatory behavior of bottlenosed dolphins *(Tursiops truncatus). Cetology,* 21:1–9.

Renaud, D. L., and Popper, A. N., 1975. Sound localization by the bottlenose porpoise, *Tursiops truncatus. J. Exper. Biol.,* 63:569–85.

Saayman, G. S.; Bower, D.; and Tayler, C. K.; 1972. Observations on inshore and pelagic dolphins on the south-eastern cape coast of South Africa. *Koedoe,* 15:1–24.

Saayman, G. S., and Tayler, C. K., 1973. Some behaviour patterns of the southern right whale *Eubalaena australis. Zeitschrift für Saugetierkunde,* 38(3):172–83.

Saayman, G. S.; Tayler, C. K.; and Bower, D.; 1973. Diurnal activity cycles in captive and free-ranging Indian Ocean bottlenose dolphins (*Tursiops aduncus* Ehrenburg). *Behaviour,* 44(3–4):212–33, pl. 6.

Schevill, W. E., 1964. Underwater sounds of cetaceans. In: *Marine Bio-acoustics,* W. N. Tavolga, ed. New York: Pergamon Press, vol. 1, pp.307–16.

Schusterman, R. J., 1973. A note comparing the visual acuity of dolphins with that of sea lions. *Cetology,* 15:1–2.

Simpson, J. G., and Gardner, M. B., 1972. Comparative microscopic anatomy of selected marine mammals. In: *Mammals of the Sea; Biology and Medicine,* S. H. Ridgway, ed. Springfield, Ill.: Charles C. Thomas, pp.298–418.

Sokolov, V. E., and Kuznetsov, V. B., 1971. Chemoreception in the Black Sea dolphin (*Tursiops truncatus* Mont.). *Doklady Akad. Nauk SSR, ser. biol.,* 201:998–1000.

Spong, P.; Bradford, J.; and White, D.; 1971. Field studies of the behaviour of the killer whale *(Orcinus orca).* In: *Proceedings of the Seventh Annual Conference on Biological Sonar and Diving Mammals,* 1970. Menlo Park, Calif.: Stanford Research Institute, pp.169–74.

Suchowskaja, L. I., 1972. The morphology of the taste organs in dolphins. *Investigations on Cetacea,* 4:201–4, 2 plates.

Tavolga, M. C., and Essapian, F. S., 1957. The behavior of the bottle-nosed dolphin *(Tursiops truncatus):* mating, pregnancy, parturition and mother-infant behavior. *Zoologica,* 42(1):11–31, plates 1–3.

Tavolga, W. N., 1965. Review of marine bio-acoustics; state of the art: 1964. Port Washington, N.Y.: U.S. Naval Training Device Center, Tech. Rept.: NAVTRADEVCEN 1212-1: i-v, 1–100.

Tavolga, W. N., 1968. Marine animal data atlas. Orlando, Fla.: U.S. Naval Training Device Center, Tech. Rept.: NAVTRADEVCEN 1212–2: i-x, 1–239.

Tayler, C. K., and Saayman, G. S., 1972. The social organisation and behaviour of dolphins *(Tursiops aduncus)* and baboons *(Papio ursinus):* some comparisons and assessments. *Ann. Cape Provincial Mus.* (*Nat. Hist.*), 9(2):11–49.

Winn, H. E., and Perkins, P. J., 1976. Distribution and sounds of the minke whale, with a review of mysticete sounds. *Cetology,* 19:1–12.

Winn, H. E.; Perkins, P. J.; and Poulter, T. C.; 1971. Sounds of the humpback whale. In: *Proceedings of the Seventh Annual Conference on Biological Sonar and Diving Mammals,* 1970. Menlo Park, Calif.: Stanford Research Institute, pp.39–52.

Wood, F. G., 1961. (No title.) In: *Man and Dolphin,* J. C. Lilly, Garden City, N.Y.: Doubleday, p.236.

Wood, F. G., 1964. (No title, in General Discussion section.) In: *Marine Bio-acoustics,* W. N. Tavolga, ed. New York: Pergamon Press, vol. 1, p.395.

Wood, F. G., 1973. *Marine Mammals and Man.* Washington, D. C.: Robert B. Luce.

Chapter 32

COMMUNICATION IN SIRENIENS, SEA OTTERS, AND PINNIPEDS

Howard E. Winn and Jack Schneider

Introduction

The three groups of marine mammals discussed in this chapter are the Sirenia, the Pinnipedia, and one mustelid,[1] the sea otter *(Enhydra lutris).* The Sirenia are coastal marine, estuarine, and river animals, whereas the sea otter is coastal. In general, the groups have retained their own characteristic modes of communication with little convergence. Their communication is limited and fashioned by the physical and biological characteristics of the environment. These families communicate in water and in air, and the effects of the media on communication must be known to understand signal adaptation.

Visual and acoustic communication are important in both media, depending on the amount of ambient interference. The greatest noises encountered by these marine mammals are surf noise and particulate turbidity (both biological and physical), which are found near the land-water boundary. This interference does not affect all the families equally since in general their habitat selection is varied; however, the otarids and the phocids inhabit similar areas. Northern species contend with less turbidity and, during much of the year, with less biological noise (i.e., snapping shrimp). Pagophilic species live with optimal visual and acoustic conditions during much of the year, except during plankton blooms and the semiannual absence of sunlight. Except for the extinct family Hydrodamalidae, the species of Sirenia live in tropical estuaries and rivers with constant high particulate densities and rather low noise levels. The Hydrodamalidae lived in kelp beds near the margins of subarctic islands, a habitat where good vision and hearing could be adaptive. The sea otter lives in an environment similar to that reported for Hydrodamalidae, and also along the coast from the Kuril and Aleutian islands to southern California (Kenyon, 1969), with similar environmental demands of good vision and hearing.

The use of olfaction in water may be ruled out because of the slow rate of molecular diffusion in the medium plus the necessity of closing the nares under water. However, both pinnipeds and (only briefly) the sea otter haul out on land, where olfaction could be used, though phylogeny and behavioral need would also be determining factors. Gustatory and tactile signals depend on contact transmission and could be of use in either medium.

The general behavior of sea cows and sea otters in relation to communication will be de-

1. A second marine otter *(Lontra felina)* lives in the coastal waters of Peru and Chile.

scribed separately, followed by a more in-depth discussion of signaling by pinnipeds. As there have been few natural experimental studies, the function of signals and the relative strength of the communicative channels must be assumed from correlations of signal and behavior.

Sirenia and Sea Otter

SIRENIA

Sirenia are unaggressive grazers that inhabit slow-moving, silty, eutrophic, fresh and marine coastal waters of tropical and subtropical areas (Bertram, 1964; Moore, 1956). They are normally dispersed individually or in small groups, except during mating and cold spells (Moore, 1956). They are totally aquatic, and during normal activity only the dorsally located nostrils (during breathing) and the arch of the back break the surface (Bertram, 1964). Under these conditions, we would expect vision to be less useful than in other environments. Bertram (1964) and True (1884) suggested that vision is poor, though Bertram mentioned that manatees sometimes lift the head quite high out of the water near the bank, perhaps to see.

The importance of the acoustic and tactile senses is stressed in general descriptions of sirenien behavior. Gohar (1957) described the hairs of *Dugong dugong* as being most numerous around the mouthparts, the chin, the muzzle, and the dorsal tail ridge. The bristles around the mouthparts are oriented proximally and may serve to direct food into the mouth. He did not speculate on a function for the trunk hairs. Murie (True, 1884) described manatee feeding: "occasionally it would sniff or examine [various vegetables] by snout and lips without chewing or swallowing. . . ." Descriptions of greeting, play, and mating include accounts of tactile communication, presumably involving the vibrissae.

Mother-young interactions include the mother's supporting the young on her back as she rests and, occasionally, as she moves. This behavior may be important as a survival mechanism and may also be important in socialization, for it occurs during adult play behavior as well, when nuzzling and body contact result from an animal's rising in the water column, colliding with and raising a second animal (Moore, 1956).

The most complete descriptions of presumed manatee courtship are described by Moore (1956) and confirmed by Caldwell and Caldwell (1972a) as occurring during social aggregations. These authors reported that courtship is a stereotyped progression of tactile interaction.

> An animal nuzzled the other animal's side, put its flipper on the other's back, then rolled until its venter was towards the object of its intentions. For once no avoidance move was made by the animal being approached. After a pause the aggressor rolled gently venter down and let his flipper slide off into the water. Paddling softly with his flippers, he explored with his muzzle along the other manatee's side until at about its midlength, during which time his own body had moved away from the other's to a right angle so that the other's long axis crossed his own like the top of a "T." The male then rolled over on its own long axis until venter up (thus revealing his sex) and in this position carried his nuzzling down the other's side to under its belly and explored with his muzzle along towards its genital area. [Moore, 1956]

Copulation was not observed. The use of the foreflipper in maintaining body contact is similar to behavior observed in mating pinnipeds.

Moore (1956) termed muzzle-to-muzzle contact between animals as "greeting." This stereotyped behavior is accompanied by the animals' rising in the water column and maintaining contact with their muzzles above the water surface. Moore speculated that if airborne scent is important in this ritual, it is extremely interesting to find this in-air gesture retained in this former terrestrial animal. He hypothesized that this ceremony facilitates individual recognition. It seems analogous to pinniped behavior, and would seem to be useful to these normally nonaggressive inhabitants of cloudy water. Manatee breath is said to be sulfurous (Parker, 1922).

Results of behavior studies and acoustic monitoring of manatees suggest the importance of acoustic communication and environmental sensing, although very little is known. Bertram (1964) stated that tame manatees are attracted through conditioning by the human whistle more readily than by other signals; wild manatees are alarmed by the human voice and startled by unusual noises (Barrett, 1935). In their normally quiet, opaque, natural habitat, hearing would be a selectively advantageous channel of communication. Knowing the habits of the manatee, we would expect it to communicate vocally. In a list of the sounds of various mammals, Tembrock (1963) stated that *Trichechus manatus* produces a 4,304 Hz call. This result is similar to those of the more comprehensive study by Schevill and Watkins (1965), who found that the fundamental tones of *T. manatus* vocalizations are at 2.5 to 5 kHz, but may be as low as 600 Hz. They described the calls as squeaky and rather ragged, and stated that they lasted 0.15 to 0.5 sec and were 10–12 db above background at distances of 3 to 4 m. Evans and Herald (1970) reported similar results from their studies of *T. inunguis.* The major differences between the calls of the two species appear to be the fundamental frequency, which is 6 to 8 kHz for *T. inunguis* as compared to 2.5 to 5 kHz for the Florida species, and the occurrence of pulses associated with some calls produced by the Amazon species. These observations were made in captivity, and no evidence was given for the function of the calls.

Hartman (1969) reported that the underwater sounds of the manatee are highly variable and include chirp-squeaks, squeals, and screams, all produced in a variety of unrelated circumstances. They seem to be associated with emotional states, especially alarm, and are not used in echolocation. One predictable vocal reaction is the alarm duet between a mother and her calf as she calls it to her side before fleeing (Hartman, 1969). It should be noted that when startled, the manatee will plunge into the water using the force of its tail and create great turbulence and noise, which may have communicative function (True, 1884). Krumholz (1943) described the behavior of a startled group in which the male reared out of the water and headed with two females for deeper water, while another female and a pup fled in the opposite direction. Arthur Myrberg (pers. comm.), in a recent study of sounds made by a male, a mother, and a baby, heard sounds from each individual, and the vocabulary seemed limited. Manatee sounds are much like gull shrieks, and are produced in social interactions. The baby made quite a few sounds when isolated.

SEA OTTER

In Alaska, the mature sea otter is normally solitary or found in small groups (Kenyon, 1969, 1972). In California, Fisher (1939) observed groups of sixty to eighty. The otter is rarely found on land, except when nursing, pupping, or resting (Kenyon, 1969; Barabash-Nikiforov, 1947). Sandegren et al. (1973) never observed the otter on land in California, though Vandevere (1971) did.

The sea otter spends the majority of its life in the water, swimming, grooming, feeding, and mating (Kenyon, 1969). The animals may habitually frequent the same areas to feed (Limbaugh, 1961; Fisher, 1940). There is evidence that male sea otters breeding in Alaska do not hold aquatic or terrestrial territories (Kenyon, 1969), although aquatic territoriality has been observed in the sea otter in California (Vandevere, 1970), and Fisher's (1939) observations imply a dominance hierarchy in the California sea otter. Recently, Calkins and Lent (1975) have observed territoriality in some Alaskan sea otters; its expression may be related to topographic factors. These authors describe the male patrolling his territory and chasing intruders, and an occasional fight. Autogrooming by the defending male typically followed a chase. The male patrolled on his back, vigorously kicking and splashing, thus providing a highly audible and visible display. The absence of rigid aquatic territoriality may account for the lack of anal scent glands, which are found in other mustelids (Kenyon, 1969).

Smell is important for environmental sensing and for intraspecific communication in air, but it may not be used under water. In an early evaluation of sea otter senses, Elliot (1887) wrote:

> The quick hearing and the acute smell possessed by the sea otter are not surpassed by any other creatures known to sea or land. They will take alarm and leave from the effects of a small fire as far as 4 or 5 miles to the windward of them, and the footsteps of a man must be washed by many an ebb and flood before its traces upon the beach cease to alarm this animal and drive it from landing there, should it happen to approach for that purpose.

Barabash-Nikiforov (1935) observed that the sea otter bed is located in the shelter of some spur or projecting rock, and the direction of the exit tracks seems to indicate that the animal lies with its head pointing up wind. By Kenyon's (1969) account, smell may function in individual recognition or to indicate estrus. He observed a precopulatory male searching for the female by sniffing. During courtship the male will monitor the air at the water's surface and will change direction as much as 130° when arriving downwind from a feeding female (Vandevere, 1970). This is accompanied by anogenital inspection of animals by mature males (Fisher, 1939; Kenyon, 1969; Vandevere, 1970), which suggests pheromone secretion. In this behavior, the male cruises among the raft of animals, inspecting each one until his advances evoke a response that leads to mating.

During courtship, the male seems to propel the female by pushing with his nose against her anogenital region (Vandevere, 1970); he possibly receives an olfactory cue from this action, and he may be stimulated by vision and taste. Kenyon (1969) suggested that sea otters can distinguish food items by taste, and it would seem likely that estrus would be accompanied by a combination of cues including taste, odor, and vision, as in other mammals.

Kenyon (1969) and Fisher (1939) reported that mating behavior includes the male's positioning himself on top and biting the female's head region; in pinnipeds, this causes the female to become rigid (Kenyon, 1969) or limp (Vandevere, 1970), depending on the population of animals. Intromission follows. Should the male lose his grip, it must be reestablished for female acceptance. It would seem probable that this action arises ontogenetically from infancy, for the female carries her pup on land by biting its head (Kenyon, 1969).

Sea otter communication is dominated by hearing, vision, and touch. Tactile communication involves licking; contact with general body surface, forepaws, and vibrissa; and copulation.

Licking has been observed in maternal grooming of the young's pelage and anogenital region, where it probably stimulates defecation and prevents soiling of the fur of the mother and the young (Sandegren et al., 1973). This behavior may have an important function in communication, not only in signaling estrus but also in maintaining the mother-young bond and in providing comfort to the distressed young. Licking the head in a stereotyped manner seems to bear no relation to cleanliness or hunger and appears to comfort the pup in a stressful situation (Sandegren et al., 1973). Body contact and contact maintained by the forepaws may also serve the same ends (Fisher, 1940; Kenyon, 1969; Sandegren et al., 1973), as pups crying in distress cease calling when contact with the mother is made.

The adults' ability to find and identify food depends on tactile identification of food objects at the depths where food is found (Kenyon, 1969; Sandegren et al., 1973). Food is also identified by vibrissa contact; Kenyon (1969) stated that the vibrissae are abraded off in wild, foraging adults but not in captive animals. In Kenyon's (1969) photograph of a sea otter being offered food, the vibrissae are extended forward and are apparently touching the food object, in a manner identical to the vibrissa action in feeding seals. The vibrissae are voluntarily controlled, and when extended forward they serve as a sensory aid when the otter is walking among rocks or examining a strange object (Kenyon, 1969). The use of vibrissae in individual recognition or in sexual-agonistic signaling is postulated.

In some circumstances vision seems less important to the sea otter than hearing or olfaction (Kenyon, 1969), although obviously these animals have good vision, which is continually used in communication. Certain postures convey alarm and warning. When alarmed, the sea otter will rise halfway out of the water (Scammon, 1874; Kenyon, 1969; Vandevere, 1970; Fisher, 1939). Whether this maneuver merely affords the animal a visual vantage or whether it contains message value for other sea otters is not known, but it is an oft-observed and stereotyped action.

Mating animals have been observed to swim in coordination, the male following the dives and surfacing pattern of the female (Kenyon, 1969). The importance of this behavior in communication and reproduction is not mentioned, though its stereotyped character would suggest its use in communication.

Though the sea otter lacks piloerector muscles (Kenyon, 1969), it displays a defensive posture similar to that found in other mammalian species. According to Kenyon's (1969) photograph and comment, the threatened animal hunches up, and may hiss and attempt to bite and push the intruder with its paws. The male's mating posture, with feet held high out of the water (Fisher, 1939), and his red penis are highly visible signals.

Vision is important in maintaining the mother-young bond. The mother watches the pup constantly (Kenyon, 1969), and when the mother and pup are out of sight of each other, both of them will cry until they are in contact (Kenyon, 1969; Sandegren et al., 1973; Fisher, 1939). The pup cry can be heard by humans at such a distance that twelve-power binoculars were not sufficient to locate the pup (Fisher, 1940).

A variety of sounds are made by sea otters (Kenyon, 1969; Fisher, 1940; Sandegren et al., 1973; Limbaugh, 1961; Vandevere, 1970), including the baby cry, the adult scream under stress, the adult female scream when separated from her young, the whistle or whine under conditions of frustration or mild distress, the contentment cooing of females during premating and postmating behavior, aggressive snarls or growls when trying to escape, hissing, grunts during feeding, and aggressive or frustration barks.

In summary, the sea otter depends primarily on hearing, olfaction, vision, and touch for communication. Because of the lack of experimental data, we have relied on field observation of social behavior, inasmuch as there is no social behavior without communication.

Pinnipedia

Pinnipeds divide into three families: Otariidae (sea lions and fur seals), Odobenidae (walrus), and Phocidae (northern true seals, antarctic seals, monk seals, hooded seals, and elephant seals). All the otarids and the walrus are polygynous. The phocids have a few polygamous social groups (elephant seals and some populations of grey seals), but most are assumed to be monogamous.

We have used common names for seals as follows: Steller or northern sea lion *(Eumatopias jubatus),* California sea lion *(Zalophus californianus),* southern sea lion *(Otaria byronia),* Australian sea lion *(Neophoca cinerea),* Hookers sea lion *(Phocarctos hookeri),* northern fur seal *(Callorhinus ursinus),* South American fur seal *(Arctocephalus australis),* South African fur seal *(A. pusillus),* Kerguelen fur seal *(A. tropicalis),* Guadalupe fur seal *(A. philippii),* Australian fur seal *(A. doriferus),* Tasmanian fur seal *(A. tasmanicus),* New Zealand fur seal *(A. forsteri),* walrus *(Odobenus rosmarus),* bearded seal *(Erignathus barbatus),* grey seal *(Halichoerus grypus),* harbor seal *(Phoca vitulina),* ringed seal *(P. hispida),* Caspian seal *(P. caspia),* Baikal seal *(P. sibirica),* harp seal *(P. groenlandicus),* ribbon seal *(P. fasciata),* Weddell seal *(Leptonychotes weddelli),* crabeater seal *(Lobodon carcinophages),* leopard seal *(Hydrurga leptonyx),* Ross seal *(Ommatophoca rossi),* Mediterranean monk seal *(Monachus monachus),* West Indian monk seal *(M. tropicalis),* Hawaiian monk seal *(M. schauinslandi),* hooded seal *(Cystophora cristata),* southern elephant seal *(Mirounga leonina),* and northern elephant seal *(Mirounga angustirostris).*

The nomenclature is from King (1964), except for uniting the genera *Pusa, Pagophilus,* and *Histriophoca* into the genus *Phoca* (Burns and Fay, 1970).

PINNIPED RECEPTORS

Vision

The eye of the seal is well developed (Walls, 1963; Lavigne and Ronald, 1972; Jamieson and Fisher, 1970, 1971; Piggins, 1970; Nagy and Ronald, 1970; Hobson, 1966; Johnson, 1893; Landau and Dawson, 1970; King, 1964; Wilson, 1970). The large spherical lens is adapted to aquatic vision, the refractive index of the cornea being similar to that of water. On land, under lighted conditions, pinniped vision is good. The vertical slitlike pupil adapts the eye to aerial vision by reducing astigmatism, sharpening the focus, and reducing the light. Under low-light conditions in air, the pupil is dilated and the animals have poor form vision (Schusterman, 1968, 1972; Schusterman and Balliet, 1971). Because of the general absence of land predators, vision need not be too acute at night on land. The predominantly rod-dominated retina and *tapetum lucidum* adapt the eye to the low light levels found in the water. In the harp seal, color vision is present, with the greatest sensitivity in the green and blue-green region. The placement of the eyes forward and high on the head suggests an adaptation to looking upward and forward, with considerable binocular ability for judging distance. This may aid in locating breathing holes in ice (Kooyman, 1968), in hauling out on rocks, and in capturing prey (Hobson, 1966).

That vision may not be an essential sense in some species is suggested by statements that blind grey seals successfully feed and pup. Furthermore, the fact that Weddell seals feed in the almost lightless winter of the Antarctic suggests that sound may have increased importance. It is possible that the presence of a high level of phosphorescence allows Weddell seals to communicate and locate prey visually (Kooyman, 1968). Ponting (in Kooyman, 1968), while watching Weddell seals during the winter night at McMurdo Sound, stated, "a seal emerged, its beautiful head all blazing with phosphorescence." It seems reasonable that any available light is utilized by the Weddell seal for orientation and hunting (Kooyman, 1968), especially in clear water. Kooyman (1975) has shown that Weddell seals dive deeper, more frequently, and longer during the day than at night, again suggesting the importance of vision. Kooyman (1968) suggested that seals can see as deep as 1,000 m using both phosphorescence and ambient light. Lythgoe and Dartnell (1970) suggested that the rhodopsins of the elephant seal eye are adapted to the bioluminescence of prey squid (see also Jamieson and Fisher, 1972). Also,

Schusterman (1967) characterized the California sea lion as primarily a visual animal.

A variety of studies have demonstrated good vision in pinnipeds (Schusterman, 1965, 1967, 1968; Schusterman and Balliet, 1970, 1971; Schusterman and Feinstein, 1965; Schusterman and Thomas, 1966). California sea lions, Steller sea lions, and harbor seals are capable of discriminating a size-difference ratio as small as 1.06:1 under water, an ability similar to that of several species of monkey. Under dim illumination the seal's visual acuity is superior to man's, and visual shape and spacing discrimination in seals is highly efficient. It has been hypothesized that seals produce clicks when visual cues are scarce or unavailable, but that they are normally dependent on the visual sense under water.

Schusterman and Thomas (1966) suggested that the visual perceptual organization of seals may be quite different from that of terrestrial animals because seals perform many tasks while upside down or on their sides. A variety of acuity tests suggest that the eye has excellent resolution both in air and in water under lighted conditions, but is much better in water under low-light conditions. Some apparent species differences in reaction to objects (e.g., man) on land may be due to variable selective attention factors rather than to visual acuity *per se* (Schusterman and Thomas, 1966).

Audition

Hearing and sound production both under water and in air are well developed. Acoustic signals are varied in both media, but are still poorly known under water.

Audiograms have been made in both air and water for several species and are summarized in Fig. 1. A recent in-air audiogram of the California sea lion (Schusterman, 1974) does not differ fundamentally from the audiogram made from an evoked potential study (Bullock et al., 1971). Not included in Fig. 1 is the underwater hearing curve of the ringed seal (Terhune and Ronald, 1975). Seals have good hearing under water and raised thresholds in air, although the grey seal seems to be an exception at 2 to 4 kHz. Where there are lowered thresholds they are variable from 2 to 4 kHz and 15 to 30 kHz. The lowest thresholds in the high frequencies are near or centered at 20 to 35 kHz for the grey seal, near 15 kHz for the harp seal, and between 12 and 40 kHz for the harbor seal. The ear seems best adapted to hearing in water, with some accommodation for in-air hearing. Interestingly enough, much of the prime energy in many vocalizations is from 0.5 to 4 kHz (in the range of best aerial hearing), except for the clicklike sounds discussed below, which have their main energy in the second low-threshold area of the

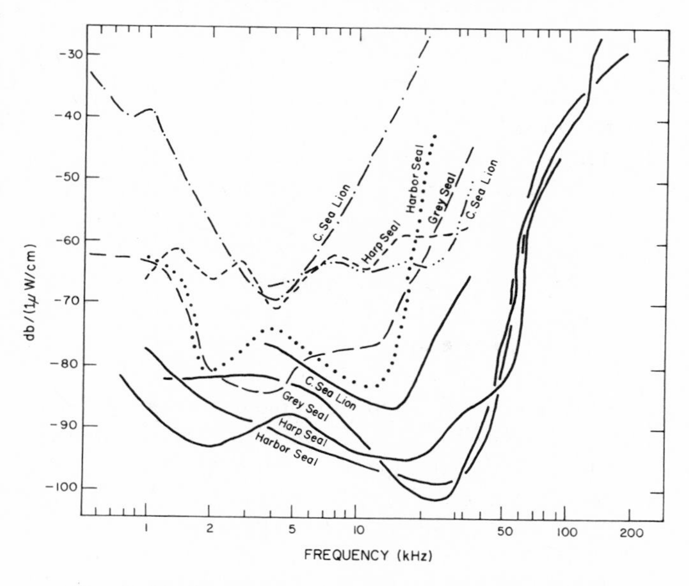

Fig. 1. Underwater (solid lines) and in-air (other lines) audiograms of some seals: harp seal in air (from Terhune and Ronald, 1971), under water (from Terhune and Ronald, 1972); harbor seal in air and under water (from Møhl, 1968a, 1968b); California sea lion in air (from Bullock et al., 1971; Gentry, 1966), under water (from Schusterman et al., 1972); grey seal in air and under water (from Ridgway, 1973).

audiograms at the higher frequencies. Other detailed comparisons of audiograms may be found in Schusterman (1974) and Terhune and Ronald (1975).

One must ask why there is a second area of decreased thresholds at high frequencies. Although it is not our intention to review the problem of echolocation in seals, a few observations should be made. Poulter (1963) reported that seals echolocate using clicks. Others (Evans and Haugen, 1963; Schusterman, 1974) could not find any evidence for echolocation. Certainly, on the basis of limited evidence, seals do not appear to produce clicks as frequently as do odontocetes. To clarify one problem, we need a definition of a click. Winn and Perkins (1976) defined it as a sound of less than 5 msec in duration, although it may be useful to include some pulsed sounds lasting up to 20 msec. Although it is not absolutely documented, there seems to be a break in the durations of pulsed sounds up to a next class, with durations of 50 to 200 msec. These longer sounds have frequencies only up to 5 to 7 kHz normally, whereas short "clicks" have frequencies up to 30 to over 100 kHz. Therefore, it is clear that the sounds that Schusterman (1967) discussed are not the ones expected to be involved in a refined echolocation system. They are too long and of too low a frequency (up to 4 kHz).

Clicks as defined here have been described for the California sea lion, the harp seal, the hooded seal, the grey seal, and the harbor seal (Mohl and Ronald, 1970; Poulter, 1963; Schevill et al., 1963; Schneider, 1974; Schusterman et al., 1970; Terhune and Ronald, 1973). Schevill et al. (1963) pointed out that seal clicks were much less intense than odontocete clicks, suggesting that their usefulness at a distance must be limited. Much of the equipment used to record clicks has been inadequate for resolving amplitudes at various high frequencies; in fact, there is much distortion (Diercks et al., 1973), and the great variability in the results of physical analyses of the sounds is yet to be adequately explained. Watkins (1973) discussed this problem in detail.

Seals can localize sounds under water as well as or better than can the harbor porpoise (Table 1) (Gentry, 1967; Møhl, 1964, 1967, 1968a, 1968b.) Localization is improved at higher frequencies, but recently Moore (1975) showed that this is not true for the California sea lion. The harbor seal could detect an average Weber fraction of 0.013 for frequencies of 1 to 60 kHz, but above that, frequency discrimination was lost (Møhl, 1967, 1968b). The minimum audible angle perception for click trains was $9° \pm 4°$ under water and $3° \pm 4°$ in air for the harbor seal (Terhune, 1974).

Touch

The tactile sense is well developed in seals, largely because of the presence of vibrissae around the mouth. These vibrissae are highly developed and have complicated innervations (Cajal, 1952; Ling, 1966; Scheffer, 1962; Stephens et al., 1971). Poulter (1972) suggested that vibrissae are an acoustic sensor and that they respond to low-amplitude high-frequency signals. However, knowledge of the functions of vibrissae of terrestrial animals argues against this hypothesis. Dykes (1972) stated that vibrissae are adapted to discriminate texture and shape of objects within the immediate proximity of the animal's face. Fibers could respond to frequencies up to 500 Hz; a few fibers could respond above that and once even above 1 kHz, where the amplitude had to be increased beyond the usual biological reality. At 20 Hz vibrissae are not sensitive enough to respond to the loudest biological sound ever recorded, namely that of the blue whale (Dykes, 1972).

Chemoreception

Little is known about the senses of taste and smell in seals (King, 1964). Taste buds, although present, are scarce, and a small olfactory bulb exists in the brain. Taste, at least, may be utilized

Table 1

An overview of the reproductive social organization and related events of living seals.

Family and species	Reproductive* social organization	Reproductive habitat	Harem herding	Territoriality	Copulation habitat and time	Where young born	Nursing times and weaning time
Otariidae Sea lions: 5 species	Polygynous.	Intertidal zone and each side of it in water and on land. Sandy and smooth rocky beaches.	No. Only estrous females gregarious.	Male topographic territories. More rigid on rocky than on sandy terrain. C.S.L.	On land and in water. S.S.L.: 6–16 days post partum; S.Am. S.L.: few days post partum.	On beaches, 2 hours to 2 days after arrival.	S.S.L.: go to sea, every 3 days spend one day at sea, weaning about a year or more. C.S.L.: wean at 3–4 months. S.Am.S.L.: wean at 5–6 months. A.S.L.: wean at 3–4 months.
Fur seal: 8 species	Polygynous.	On land; rocky rugged habitats.	Yes. Strong herding but not absolute. Females not as gregarious as above. Herding weaker in S.A.F.S.	Strong male topographic territories.	On land. N.F.S.: 4-7 days post partum. S.A.F.S.: 6 days post partum. S.Am. F.S.: several days post partum.	On beaches shortly after arrival.	N.F.S.: leave 5–10 days post partum for 7–10 day periods, return for 2 days of nursing. Wean at 3–4 months. N.Z.F.S.: wean at 6 months.
Odobenidae Walrus	Polygynous.	On ice and in water.	No.	Female nursing herds. Separate male and young herds.	In water. Perhaps after weaning.		Wean at 16 or more months.
Phocidae Northern elephant seal	Polygynous.	On land.	Yes. Very weak herding.	Male territories.	On land. Several days before weaning.		On land. Wean at 4–6 weeks.
Southern elephant seal	Polygynous.	On land.	Yes. Very weak herding.	Male territories.	On land. Suggested first copulation in water. 2 weeks post partum.		On land. Wean at about 3 weeks.
Grey seal	Polygynous and monogamous depending on population.		Harems but no herding.	Male territories.	On land, ice, occasionally in water.	On land.	On land, less common in water. Wean at about 3 weeks.

Family and species	Reproductive* social organization	Reproductive habitat	Harem herding	Territoriality	Copulation habitat and time	Where young born	Nursing times and weaning time
Hooded seal	Family units of larger male, female, and pup.	On ice and in water.	No.	Family territories.	In water. 12–14 days post partum.	On ice.	On ice. Wean at 10–12 days.
Bearded seal	Monogamous, essentially nongregarious.		No.		In water; perhaps a year after birth of pup.	On ice floes.	On ice and probably later in water. Wean at several weeks.
Harbor seal	Monogamous, essentially nongregarious.	In water, intertidal rocks, on ice; population differences.	No.		In water.	Edge of shore, on ice in some populations.	Enter water after birth. Wean at 3 weeks; one population weans at 3 weeks on ice.
Ringed seal Related Caspian and Bailsal seals			No.		In water.	In ice or snow lairs; landfast ice only.	In lair, female goes into water periodically. Ringed weans at 2 months. Caspian weans at 4–5 weeks.
Harp seal	Monogamous.	On ice and in water.	No.	Perhaps no strict underwater territories.	Mostly in water; at weaning.	In large female aggregations on ice.	On ice. Wean at 2–4 weeks.
Banded seal	Presumed monogamous.	In water.	No.		In water, 1–2 months post weaning.	On ice.	On ice. Wean at 6 weeks.

Table 1 (continued)

An overview of the reproductive social organization and related events of living seals.

Family and species	Reproductive* social organization	Reproductive habitat	Harem herding	Territoriality	Copulation habitat and time	Where young born	Nursing times and weaning time
Weddell seal	Polygynous	On ice and in water.	No. Groups of females up on ice.	Male underwater territories.	In water; end of weaning.	On ice.	On ice; wean at 6 weeks.
Crabeater seal	Perhaps nongregarious (gregarious nonbreeding).	On ice.	No.		Presumed in water; end of weaning.	Presumed on ice.	Presumed on ice. Wean at 2–5 weeks.
Leopard seal	Apparently solitary.	On ice.	No.		In water; end of weaning.		Wean possibly at 2 months.
Ross seal	Apparently solitary.	On ice.	No.		Presumed in water; end of weaning.		
Monk seals, 2 living species	Small groups; not polygynous.	On beaches, and now on sand spits.			Presumed in water; end of weaning.	On land.	On beaches. Wean at about 5–6 weeks.

NOTE: Some details of variation due to habitat, subspecies, ages, etc. are not shown. See text for scientific names. C.S.L. = California sea lion; S.S.L. = Steller sea lion; S.A.F.S. = South African fur seal; N.F.S. = northern fur seal; S.Am.F.S. = South American fur seal; N.Z.F.S. = New Zealand fur seal; S.Am. S.L. = South American sea lion; A.S.L. = Australian sea lion.

*The documentation for presumed monogamy is poor for some species of phocids.

in the final selection of food and, perhaps, minimally in communication. Final identification of pups by a mother may involve odors; and northern fur seal, California sea lion, ringed seal males, and probably other species have strong odors that could communicate information about sexual status and even be used in individual recognition. No exocrine glands have been described for seals. Apparently the scent of man is detected by seals on land, and it has been suggested that a seal can smell the presence of a trainer not in view (Evans and Bastian, 1969). It is presumed that the nares are closed under water, so that olfaction cannot be used. The involvement of chemical cues in mother-pup recognition is discussed in a later section.

SOCIAL ORGANIZATION

Seals generally have short, well-defined breeding seasons on traditional grounds and are dispersed to varying extents at other times of the year. Insofar as social systems are concerned, of the 32 species of seals, 14 are social (usually haremlike) breeding otarids (sea lion, fur seals, and walrus). Walrus, although polygamous, do not have harems, but nursery herds, separate from males and nonparturient females, are formed (Burns, 1970). Mating in the polygynous otarids occurs shortly after parturition. Of the 18 phocid seals, two elephant seal species and some grey seal populations are also social haremlike breeders, while most others maintain loose aggregations or are more solitary and only come together briefly to mate.

Male Weddell seals maintain an underwater territory below a breathing hole and presumably an above-water territory near perennial cracks in the ice in the vicinity of an aggregation of territorial females with pups (Cline et al., 1971; Mansfield, 1958; Isenmann, 1970). They are polygynous, with mating taking place in the water. Most of the other phocids are less social and usually monogamous; they mate under water but haul out frequently to form small or large aggregations. Mating occurs at weaning. For instance, the ringed seal female and pup have snow or ice lairs on landfast ice. Harp seal females haul out on ice in large aggregations to give birth; the males are in the water in groups but start to come up on the ice three to four days after parturition, before mating, probably monogamously, in the water at weaning time (Sergeant, pers. comm., and 1963; Terhune, pers. comm.). The bearded seal pups on ice floes. Hooded seals mate on land and organize into family units of a large male, a female, and a pup (King, 1964) and are much more spatially separated than harp seals. Leopard seals appear more solitary, as is probably the case with the Ross seal (King, 1964); whereas crabeater seals seem partially gregarious, although little is known about pupping. What little is known of Hawaiian monk seals indicates that loose aggregations of females with pups are found and that sometimes a male is with them, thus suggesting some sort of pairing (Rice, 1964; Wirtz, 1968; Kenyon and Rice, 1959). Harbor seals mate monogamously in the water and are essentially nongregarious in the mating season. The general social organization and related behavior are summarized in Table 1.

In the polygynous otarids, the fur seals, except for the African fur seal, exhibit the strongest herding behavior. The northern fur seal is the most active herder; the remaining fur seals are not quite as active. The more sexually dimorphic, larger sea lions and the similar African fur seal (Rand, 1967) show no, or very weak, herding behavior (Miller, 1974). In phocids the highly sexually dimorphic elephant seals weakly herd females, and the grey seals show almost no herding behavior. Miller (1974) has related weak herding behavior with exaggerated sexual dimorphism (large male size), need for less locomotion, and habitat differences. It is easier to herd on rocky rugged terrain. Perhaps herding is relatively ineffective for reproductive success but conserves energy when landmarks are abundant.

The polygynous, harem-forming seals exhibit numerous adaptive variations in their social organization. While northern elephant seal females stay with their pups during a thirty-day nursing period, fur seal mothers leave their pups for six-to-seven-day periods of feeding at sea. Thus, fur seal pups can fast for long periods; while in the elephant seal the pups must feed almost every day, but the cows can fast for long periods. Communication for reunification after long and distant separation must be more evolved in fur seals. Steller sea lion females leave their pups and go to sea for a few hours to a few days, then return to nurse for two days (average cycle three days) (Gentry, 1970).

Other social systems exist in addition to the terrestrial harem organization of the elephant seals and fur seals. In Weddell seals, different phases of breeding are consummated in two media: the females aggregate and the pups are born along perennial cracks in fast ice, while polygynous males defend breathing holes against subordinate males and mate in the water. Less-dominant males and nonreproductive females are distant from the rookery (Lindsey, 1937). This situation, in addition to freeing space for parturient females that would normally be occupied by males, minimizes pup disturbances caused by intermale conflict, so characteristic of restricted breeding grounds. Furthermore, acoustic and postural displays are required by Weddell seal males when approached by other males under water.

In Steller sea lions, on the other hand, first the males occupy terrestrial, aquatic, and semiaquatic territories. Then the females arrive and establish favorite spots in aggregations, particularly in the semiaquatic areas (Gentry, 1970; Sandegren, 1970), so that several females are within one male's territory. The males' territorial boundaries are well defined by ritualized threats, fighting, and geographic demarcations (Gentry, 1970). The females tend to move between two or three favored resting sites rather than randomly across the reproductive area, but they do cross some male territories. In addition, herding bulls sometimes try to prevent cows from leaving territories (Orr and Poulter, 1967), but the response is weak when compared to that of strongly herding species. Herding is directed exclusively toward cows with young pups or estrous cows, and may be initiated by the cow. The females form dominance hierarchies, in which the more aggressive secure and defend the more favorable pupping sites, have greater nursing vigor, and, through activation behavior, aid in the development of the pup.

Peterson (1965) stated that "fur seal harems are aggregations of otherwise solitary individuals." In this situation, successful breeding can result if aggression due to crowding is reduced. In the northern fur seal, this is achieved through strict territoriality, harem maintenance, and agonistic signaling. The timing of social events is briefly: Northern fur seal bulls come ashore early in the season and establish territories near the water's edge. As the season progresses, areas more inland become territorialized, and the size of those territories already established is forcibly reduced by crowding. The harem bulls usually maintain positions throughout the season and repulse nonterritorial bulls. The majority of the pregnant females return after the males establish territories. The females, like the males, normally return to the same location every year. During the next eight days, the females pup, come into estrus, and mate. The females then return to the sea for food, returning to the rookery every seven to nine days to nurse. The breakup of the breeding assembly is caused by the females' desertion and by the waning of territoriality in the male. Except for this breeding period from May to August, Northern fur seals are at sea.

In the elephant seals, several factors affect communication: exaggerated male size, leading to poor locomotion; increased female gregariousness; increased female aggression during the breeding season, due to the males' defense of their proximal position near the females; and maternal protection of pups. Because of these factors, pressures develop to restrict violence, to allow effective territorial behavior, and to limit the need for terrestrial locomotion.

Territorial behavior is almost ubiquitous in seals, although in some phocids where fixed geographic cues are lacking, only individual distances are maintained. In the polygamous forms, a territory may be as small as 10 X 10 m; whereas in the ringed seals (Olds, 1950), family units space out at least 50 m or more. In phocids, most females with pups defend their territories, as the males of most species probably do during the mating season. The male territories are frequently in the water, but it should be emphasized that little is known, and individual spacing without reference to an area may be the rule. A few species are solitary to a certain extent, although haul-out areas are reported, and normally they are spaced out individually (leopard seal: Marlow, 1967; Gwynn, 1953; probably Ross seal: Laws, 1964). Related to this is the fact that males are larger than the females. Males of many species have secondary sexual devices that, in some cases, may amplify sounds. They consist of nasal sacs in elephant seals and in hooded seals, large neck "shields" in fur seals, and an elongated nose in grey seals. Pharyngeal pouches occur in the walrus, and similar, less well developed devices have been suggested for ribbon seal, northern sea lion, bearded seal, ringed seal, and harbor seal (Schevill et al., 1966). Some males, such as the harp seal, may be nonterritorial (Sergeant, pers. comm.).

The development of communication systems that enhance adaptation of individuals result in social organizations in seals that are critical for at least three functional needs: the maintenance of close proximity between mother and young during the suckling stage and their reunification after the mother has gone to sea to feed (particularly for otarids); the maintenance of the harem (or social pair in some species) by male-to-male agonistic signals and by the herding of females; and courtship and copulatory signaling. After the seals finish reproductive activities and leave for sea to feed for the remainder of the year, they are generally more solitary, and complex signaling systems are not known, although in some seals pairs or small groups are maintained.

The interaction of the above communication needs is intimately tied to pup survival. This is true for the harem- and social-breeding species, which include all otarids and a few phocids. Much of the pup mortality in some species is socially induced (Le Boeuf, 1972). Thus, although communication reduces mortality, it is not efficient enough to prevent significant mortality. For instance, Le Boeuf (1972) gave four prime causes of death of northern elephant seal pups, each of which involves a breakdown in communication (social disorganization): mother-pup separation without reunion, failure of some females to nurse alien pups, female aggression toward alien pups, and bulls trampling young pups. These same problems occur with other social pinnipeds, e.g., grey seal (Coulson and Hickling, 1964), monk seal (Wirtz, 1968), northern fur seal (Anonymous, 1971). Less is known about ice-breeding species.

In the polygynous species, the communicative system must deal with several factors, including increased male size and aggression, female gregariousness, female defense of pups, and exclusion of nonreproductive individuals. Communication must therefore function to restrict violence, maintain spatial relationships, and limit the need for terrestrial locomotion so that the individual can function efficiently.

COURTSHIP AND MATING SIGNALS

The structure of the mating system develops out of a need for competing males to have access to females. It is represented by two situations:

that in which a male has access to many females (polygyny), and that in which only one male has access to one female (monogamy). In many species mating occurs exclusively in water, in others only on land, and in some species in both environments.

As a consequence of a structured social organization, and perhaps because they are accessible for study, seals breeding in environments restricted by breathing holes (on ice) or by topography (on land) are reported to have complex signaling and breeding systems. A variety of signals, including odor, color, posture, taste, and sound, are involved in courtship behavior.

Courtship and mating occur in the water for such species as the Weddell seal (Cline et al., 1971), harbor seal (Venables and Venables, 1957), harp seal (Silvertsen, 1941), and leopard seal (Marlow, 1967). Although complete behavioral analyses are lacking, we do know that mating is preceded by various displays. Male Weddell seals call actively beneath the ice. This may induce nursing females above the ice to peer down the breathing hole (Kaufman, in press), and the females' behavior may have communicative function as well.

Courtship is poorly known in the harbor seal since it occurs offshore between individuals that are not sexually dimorphic. Courtship signals are apparently produced while the seals are swimming, blowing bubbles, and vocalizing (Venables and Venables, 1957). In the harp seal and the Weddell seal, the male coaxes the female into the water with locomotory and acoustic displays. Female grey seals may solicit copulation by nuzzling the male (James, pers. comm.). Information on phocid courtship and copulatory behavior is lacking because they occur in water, where few observations have been made.

Males assess estrus by olfactory investigation. Territorial male New Zealand fur seals sniff rocks used by females as well as the females themselves (Miller, 1974). Miller stated that assessment of the female reproductive state by chemical cues may be absent in northern fur seals, but it is the rule in otarids. Scent glands are unknown in seals. What role the musky odor of male California sea lions and northern fur seals plays is unknown (Peterson and Bartholomew, 1967). Male ringed seals (Kenyon, 1962) and harp seals exude strong odors. Perhaps this odor is used in territorial marking.

Otarids, excluding the walrus, normally mate on land (Sandegren, 1970; Gentry, 1970), as do two socially similar phocids, the elephant seal and the grey seal, which occasionally also mate at sea.

Otarid males investigate the anogenital area of the female before copulation. Their interest may be increased by the swollen red vulva, which sometimes induces them to lick the region (Bartholomew, 1953; Bartholomew and Hoel, 1953; Gentry, 1970).

The female otarid often solicits copulation in a much more active and elaborate manner than do female phocids. Most prominent are the exaggerated walk of the estrous display in female northern fur seals (Peterson, 1968) and the solicitous precopulatory display of female Steller sea lions, which may be essential to induce the male to mount (Gentry, 1970; Sandegren, 1970). Signaling in the fur seal (Peterson, 1965) during "estrous" displays consists of nose and head rubbing and nipping the bull's neck and mandible. The females become less and less aggressive. The vibrissae are usually erect. The male produces the extended low roar, while the female sometimes hisses. When not in estrus the female gives the "evasive" display typical of all nonestrous male-female encounters. It is a combination of avoidance and threat, with the vibrissae erect and the neck arched. The bull attempts to keep the female in his territory while threatening vocally. A female may face the bull and grip his mandible or the underside of his neck in her teeth. This seems to limit his nipping at her back since she is firmly locked to him and moves with him during attempted lunges. Sometimes the female does not establish a firm grip and simply counters the bull's lunges at her with aggressive, open-mouthed threats of her own.

If the female proves to be sexually unreceptive, the "whicker" and the extended low roar are used in alternation by the male. If aggressive threatening and physical blocking are not sufficient to stop an escaping female, a bull can forcibly bend her over until he can grasp the skin of her back in his teeth. Then he may lift her bodily and throw her into his territory.

When ready to accept mounting, a female permits the bull to nuzzle her perineum and rub his whiskers along her back, which is arched upward, while the hind flippers are spread. Sometimes a female seems to be suddenly intolerant of a bull and breaks into evasive display. A female's estrous display always induces a bull to mount.

Northern elephant seal bulls sometimes mate immediately after a successful agonistic behavioral interaction (Bartholomew, 1952; James, 1970) and may be only secondarily stimulated by the presence of females. James (1970) hypothesized that this is a result of general arousal or, more probably, the disinhibitory effect of stimulation provided by an intruder on a beach master habituated to female presence.

Precopulatory displays by the New Zealand fur seal are much less frequent than those of sea lions (Miller, 1974). Miller states that the tactile components are much less developed than for sea lions. Token female resistance and passivity carry stimulus values to male seals.

Copulatory behavior itself is quite stereotyped. Basically the male positions himself behind and on top of the female, thus controlling her with his weight and/or foreflippers. The posture in some species changes so that the male is on the female's side (grey seal: Hewer, 1957). Rubbing of partners just before and during copulation occurs more often in the thigmotactic sea lions and South African fur seal than in other fur seals (Miller, 1974).

During aquatic mating (Weddell seal: Cline et al., 1971; harbor seal: Venables and Venables, 1957; Bishop, 1967; leopard seal: Marlow, 1967), the male grasps the axillary area of the female with both flippers. In terrestrial mating of other species one foreflipper is positioned over the female dorsal, and the male's weight is both on the female and on the substrate, so that the male is partially on his side, though during the early stages, the male may be on top briefly (Carrick et al., 1962; Schneider, pers. obs.). Neck grasping by the male induces the female to be more passive. She may bite his neck, possibly encouraging him to dismount at the end of mating.

Terhune (pers. comm.) has seen harp seal cows with cuts and scratches on various anterior parts of the body, some of which were thin lines one to six inches long running along the shoulder anterior to posterior. They could have been made by the nails of the foreflippers and by biting the female's neck if the male had mounted from behind under water.

Herding of females by males to maintain harems is not generally very effective in pinnipeds. Sea lions, the South African fur seal, and the northern elephant seal are less-successful herders than the smaller fur seals, perhaps because their large size limits their locomotion. In all cases, the tendency of the females to stay in one place is as important as efficient herding to maintaining a harem. As a result, Miller (1974) suggested that the most important function of herding behavior is to communicate criteria for normal males in a group of animals where mate choice is apparently lacking. In a sense, it may be functional, like courtship activities.

MOTHER-PUP SIGNALS

Just prior to parturition, the females establish a territory. In otarids, the northern elephant seal, and the Weddell seal (Kaufman, 1975), the females may aggressively clear areas. Though the duration of nursing varies among the different species, there are several advantages to maternal maintenance of territory and establishment of a consistent place for suckling. First, adoption by violence is discouraged in stable nursing situations (Le Boeuf et al., 1972), although exceptions to this rule exist (Burns et al., 1972; Marlow, 1972). Second, the female's territory provides a defending area in which disturbance of the pup by other females, and of the mother by males and females, is minimized. Third, female territoriality lays the groundwork for precision in geographic reunification of nursing otarid pairs.

In otarids, the mothers leave the young to go to sea, presumably to feed, and then return periodically to nurse their pups. Northern fur seals, for example, leave five to ten days post partum for periods of seven to ten days and return for two days of nursing (Peterson, 1965). Pups generally remain near the birth site, and the mother returns to shore near the site; thus, geography and spatial memory play an initial role in reunification. Successful reunification also depends on the development of signals—acoustic, visual, olfactory, and tactile—for mutual recognition between the mother and the pup.

Acoustic signaling in some species may start when the female directs calls at the anogenital region just before the birth of the pup. This behavior may be the beginning of the establishment of the mother-pup bond. In otarids, mutual vocalization starts immediately after birth (for example, the northern fur seal: Peterson, 1965). Calls made by nursing pairs have been termed mother attraction and pup attraction (Table 2). Though both types are reported for most pinnipeds, the grey seal and the harbor seal use only the mother-attraction call. This call is vocalized by young pups when distressed, when being abused physically (in elephant seals), or when apparently hungry. Pup-attraction calls, emitted by the mother, seem to contain more specific meanings and are used only during reunification or activation (Sandegren, 1970). The northern elephant seal mother starts calling as soon as she comes ashore. This sound is easily heard above all the other sounds in the colony (Bartholomew and Collias, 1962). Maternal calls are individually identifiable, but pup recognition of the call may take varying amounts of time to learn, from three days, in the northern sea lion (Sandegren, 1970), to two months, in Hooker's sea lion (Stirling, 1972). Rand (1967) noted learning in the South African fur seal, in which pups initially respond to pup-attraction calls of all females, only later answering to the call of a specific cow. In the northern fur seal, pup-attraction calls cease when the seals are reunified, whether or not suckling follows. Several authors (Wilson, 1973a; Fogden, 1971; Evans and Bastian, 1969) concluded that harbor seals, which nurse primarily in the water, attract pups through smacking the water, circling and nuzzling the pup, or whole-body contact. The pup does not recognize the mother, and pups initiate suckling by calling and nuzzling on approach. Terhune (pers. comm.) stated that in the harp seal only the pups call. As many vocalizing pups are rejected as are accepted after being sniffed—another instance that suggests that odor is the final cue for identification.

Olfactory and tactile signals may well be used in final identification when the phocid mother approaches the pup (Fig. 2). With one exception, there have been no experimental studies on mother-pup recognition signals. In a series of preliminary experiments, Kaufman (in press) determined that odor was important in Weddell seal reunification. In one experiment, the skin of a dead pup was attached to a live pup. The live pup was accepted by the mother of the pup that had died, from which the skin had been taken. Kenyon and Rice (1959) related an incident in which a Hawaiian monk seal mother, separated from her pup, heard the cry of an isolated, nearby pup and rapidly crawled to it. She sniffed

Table 2

Use and description of pinniped vocalizations.

Vocalization	Use	Description
	Northern fur seal (Peterson, 1965)	
Male trumpeted roar	Territorial	Loud, prolonged call; variable in pitch, rate, and volume; 5-sec burst with rising pitch; rapid repetition when intensity is high; volume varies with intensity.
Low roar (male)	Immediate	Short, loud, not pulsed; intergrades with trumpeted roar; duration 1 or 2 sec.
Whicker (male)	Mild threat	Resembles clicking; little variability in pitch or volume; constant repetition rate, but slower toward end of series.
Boundary puffing (male)	Territorial boundary display	Harsh, panting; a sharp exhalation repeated several times in rapid succession; diminishing volume.
Whine (male)	Submission	Loud, high-pitched squeal; duration 3–5 sec.
Open mouth threat (female)	Threat	Hiss.
	New Zealand fur seal (Miller, 1971; Stirling and Warneke, 1971)	
Trumpeted roaring (male full threat call)	Threat; boundary display	Pulsed, growllike sound, followed by screamlike portion; mean duration 1.4 sec.
Male low-intensity threat	Threat	Pulsed growllike sound; major energy below 1.7 kHz.
Male moan	Not clear; probably agonistic	Cowlike moo; duration 2 sec.; major energy <1,000 Hz; side bands present.
Male gutteral challenge	Threat; territorial	Low-pitched growl followed by single pugg.
Male barking	Sexual interest; territorial	Repeated high-pitched call.
Jaw clapping	Agonistic	"Fwapp."
Snort	Weak Threat	Snort.
Submissive call (male and female)	Submission	Modulated high-pitched call; 2–4 kHz; duration ≃ 1½ sec.
Pup attraction (female)	To attract pup	(1) High-pitched whine, similar to horse whinny; 1.5 kHz. (2) Lower-pitched monotonic moan; 0.15 kHz; pulsed.
Female attraction (by pup)	To attract female	High-pitched call; first section has low fundamental frequency (as low as 2.5 kHz); second section has higher frequency (1.6 kHz); duration to 15 sec; first section always 0.5 sec.
Chung	Threat	
Open-mouth threat (female)	Threat	
Bleat and bawl (pup)	Female attraction	
Chirp	Inquiry	Downward-swept frequency varying between 30 and 1 kHz, often repeated in series.
Low pulses	Threat	Less than 200 Hz, duration 0.5–1.0 sec.

Table 2 (continued)

Vocalization	Use	Description
	Hawaiian monk seal (Kenyon and Rice, 1959)	
Soft bubbling sound	Alarm	Sounds like "bgg-bgg-bgg-bgg"; originates deep in throat.
Grunting bawl (female)	Threat	Expelled air forms snort, snort bellow, or "mrraugh" sound; similar to that made by *Eumetopias jubata.*
Bleat (pup)	Female attraction	Similar to adult bellow; sounds like "mwaa-mwaa-mwaa."
Grunting bawl (pup)	Threat	Diminutive version of adult grunting bawl; "aaah" or "gaah."
Growl or moan (female)	Pup attraction	Hoarse, throaty.
	Harbor seal (Scheffer and Slipp, 1944; Evans, 1967)	
Various sounds	Not known	Snort, squall, bawl, throaty grunt, doglike bark.
Flipper slapping	Alarm; pup attraction	Flipper slaps water surface during diving.
Pup calls	Mother attraction; distress	Sounds like "maa" or "kroo-roo-uh"; about 500 Hz.
	Bearded seal (Ray, Watkins, and Burns, 1969)	
Male song	Territorial	First section: modulated warble; long, oscillating frequency; may be 1 min in duration. Second section: moan; short, unmodulated frequency. Call starts at about 2,000 cps and ends as low as 200 cps.
	Ringed seal (Stirling, 1973)	
Bark	Direct threat	Low-pitched.
Yelp	Probably submissive	High-pitched, sometimes modulated.
Growls	Threat	Low- and high-pitched.
Chirp		Short, descending frequency.
	California sea lion (Peterson and Bartholomew, 1969)	
Bark (male)	Aggression; territorial	Duration .2-.3 msec; 3 barks/sec; fundamental 200 Hz; produced almost continuously at height of breeding season.
Female threat	Threat	Bark, squeal, and growl; used with open-mouth threat.
Mother response	Response to pup attraction	Individually varied; similar to bawling.
Pup bark	Play; mild threat; when disturbed	Higher frequency than adult bark; 3–8 barks in succession, pulsed.
Pup attraction (female)	To attract pup	Individually varied; duration 1–2 sec; bawling sound; may be pulsed.
	Walrus (Schevill, Watkins, and Ray, 1966)	
Bell	Sexual behavior	1.0–1.5 sec; fundamental 400–1,200 Hz.
Rasps and clicks		

Table 2 (continued)

Use and description of pinniped vocalizations.

Vocalization	Use	Description
	Northern elephant seal (Bartholomew and Collias, 1962)	
Clap threat	Threat	Loud, resonant clapping sound with metallic quality, like exhaust noise of a locomotive; frequency $<$2,500 cps; extremes recorded of bursts of 3–7 pulses at a rate of 15–25/sec with 0.1 sec between bursts, and single claps spaced at 1-sec intervals.
Snort.	Threat	Snort.
Whimper.	Submission	Whimper.
Female threat	Threat	Harsh, deep, belching roar; frequency $<$700 Hz; pulse rate changes from 100/sec to 50/sec; duration $>$3 sec.
Pup attraction (female)	To attract pup	High-pitched bark; frequency 500–1,000 Hz; duration 0.2 sec; repetition rate and intervals between barks not constant.
Female attraction (by pup)	To attract female	(1) Long, puppylike yelp; duration 0.3–0.5 sec; repeated several times; ascends to 1,000 Hz; pulsed 80–90 /sec. (2) More fluctuations in pitch than type 1; ascends from 100 to 1,000 Hz.
	Weddell seal (Kooyman, 1968; Schevill and Watkins, 1965; Watkins and Schevill, 1968)	
Chi-chi-chi	Threat	Series of short-duration pulses; series as long as 42 sec; frequency and repetition rate start high but drop gradually during series.
Trill	Threat; dominance	Descends continuously from 6 kHz to 0.5 kHz in 1.5 sec.
Eeeyo	Threat	Frequency stays constant at 3,500 Hz for 2 sec, then changes to 250 Hz.
Teeth clattering	Threat	Jaws open and close rapidly; maximum gape 5 cm.

it several times, left, and continued to search. Odor recognition may also be important in the grey seal (Burton et al., 1975). The subject of mother-pup signals is obviously wide open to experimentation in various species.

The pinniped nursing period is prolonged, particularly in the socially breeding otarids. Steller sea lion may nurse for a year, northern fur seal for three months, and northern elephant seal (a phocid) for about a month. The California sea lion nurses on land for about twenty days and then continues for up to six months in the water. Phocids generally nurse for only three or four weeks.

Sandegren (1970) and Gentry (1970) described interesting female Steller sea lion behavior termed activation, which was used to induce, by aversive conditioning, suckling and other behavior by the pup. The mother lifts and drops the pup. Its response (activation) is intense irritation, shown by fast movements, head shaking, and bleating, which is transformed into a long, loud scream. Soon, merely a movement of the mother evokes the same reaction. Other activating behavior by the mother included nipping, slapping, pressing, stroking with foreflipper, and nose pushing. These actions apparently stimulate activity, locomotion, and movement, resulting in nursing and thus encouraging rapid growth (Sandegren, 1970).

AGONISTIC BEHAVIOR

The social land-breeding seals, including otarids and some phocids, have evolved highly stereotyped and ritualized signal systems. These signals may help separate the individuals, as in male territories, or they may act to keep the animals close to one another, as in harems. In some species, the fighting and rushing of males across areas occupied by females and pups cause considerable pup mortality. Thus, the signaling systems must be efficient enough to reduce mortality to a level where the population is maintained; and successful mating must be ensured. Although little is known about the underwater ritualized behavior of many phocids, we do know that some species, at least, perform highly ritualized aquatic displays.

Many of the threat and attraction displays include visual and acoustic elements; at times, the tactile and olfactory senses are involved as well. Acoustic-visual complexes change with motivation. In ritualized fighting behavior, body postures and sounds are produced. The vibrissae are erect, and contact is made. Several authors have noted that "vibrissae contact" in greeting is an agonistic signal (Bonner, 1968; Bonner and Laws, 1964; Orr and Poulter, 1967).

Fig. 2. Final pup identification by mother grey seal, presumably by vibrissal contact information and odor. (Photo by Jack Schneider.)

Miller (1971) systematized the communication of threat by territorial male New Zealand fur seals (Fig. 3). First, a territorial male perceives another male approaching at a distance. If his vocal and visual displays do not cause retreat, the territorial male then approaches the other closely. If a retreat still does not occur, the territorial male repeats and elaborates upon the signaling. If no submissive behavior or retreat is demonstrated by the alien male, a fight may result. Although fighting is discouraged, the fact that threat displays occasionally terminate with a fight ensures that the more ritualized visual and acoustic threat signals are reinforced. Descriptions of the territorial behavior of many species of social land-breeding otarids and of a few phocids generally agree with Miller's system based on the New Zealand fur seal (northern elephant seal: Bartholomew, 1952; northern sea lion: Gentry, 1970; grey seal: Schneider, unpublished ms.).

In addition to the ubiquitous open-mouth threat (Fig. 4), males of species that maintain territories have other intense and elaborate inter-male warning calls, sometimes more than one per species (Fig. 5). The warbling song of the bearded seal may be the most elaborate. It is emitted by reproductively active males presumably maintaining below-ice territories, and is unlike the "belch roar" of the northern elephant seal. Physically, the frequency-modulated bursts of the bearded seal's call may be analogous to the pulsed nature of the belch roar. The territorial calls of otarids are of low frequency, pulsed, and normally associated with an elevated posture. Since the first threat calls are produced when the interacting animals are relatively far apart, coupling with a posture must occur within visual range, as in the land-breeding otarids, in

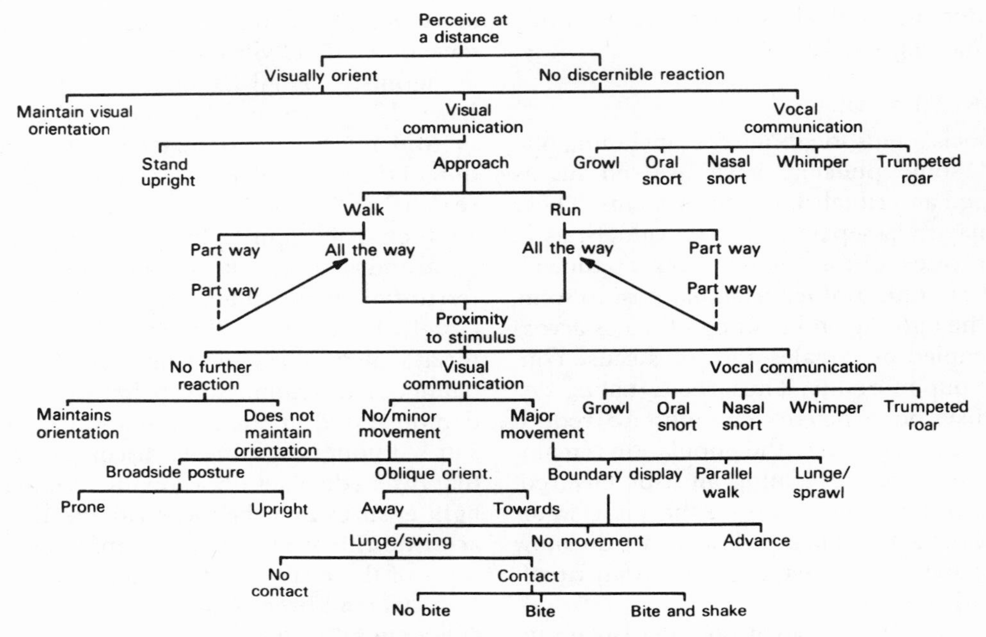

Fig. 3. Communication of threat by territorial males. (From Miller, 1971.)

the northern elephant seal, and under water by the Weddell seal (Ray and DeCamp, 1969; Kaufman, in press). Calling without visual contact stimulates territorial calling in Weddell seals (Watkins and Schevill, 1968) and elephant seals (Bartholomew, 1952). In elephant seals, call amplification conveys dominance; thus large resonating proboscises are positively selected (Bartholomew, 1952). Recent ideas suggest that the call merely identifies the caller and that the proboscis is selected for as a visual stimulus (Le Boeuf and Schusterman, pers. comm.). The barking of dominant male California sea lions restricts the movement and barking by other, smaller males (Schusterman and Dawson, 1968).

Should social relationships of males not be settled by display, fighting occurs. Fighting, perhaps because of seal anatomy, is directed only at certain areas. Biting and height (or getting on top) are fight strategies. During terrestrial combat, biting is directed at the foreflippers, face, neck, and head, though a retreating animal must protect its hind flippers and its posterior dorsal portion from attack. Wounds and scars found on Weddell seals (Kaufman, in press), harbor seals (Naito, 1973), and leopard seals (Marlow, 1967) suggest that the generalized points of aggression are similar in underwater disputes, though the genital region is also a target. In fur seals fighting is ritualized to a point where the animals cease activity if a bite holds (Peterson, 1965). Le Boeuf and Peterson (1969) showed that northern elephant seals actually have a social hierarchy maintained by stereotyped threat displays. Copulation is most frequent by males of the highest status (Le Boeuf, 1974).

Submissive behavior has been described in some otarids (Peterson, 1965; Miller, 1971) and in the elephant seal. The southern elephant seal signals submission by "deflating his proboscis

Fig. 4. Open-mouth threat of a female grey seal toward a mature bull. (Photo by Jack Schneider.)

and backing away, uttering short high-pitched cries" (Carrick et al., 1962). High-pitched whimpers are typical of submissive calls.

In the northern elephant seal, female aggressive behavior, expressed by brief conflicts and vocalizations, is evoked by intrusions of females, alien pups, yearlings, and any male. If a pup is attacked, it cries and the mother chases and attacks. If her pup is pinned down inadvertently by a male, the female threatens vocally (Bartholomew and Collias, 1962). Much of this signaling serves only to protect the pup. Christenson (1974) has shown that pup survival is related to the aggressiveness of the mother: strong signaling ensures survival, while weakly aggressive mothers frequently lose their pups. With some variations, the social otarids exhibit similar behavior.

Other types of behavior have various signal values. A snort may have universal significance in low-intensity warning. Flipper waving (Schusterman, 1968) is a low-intensity visual warning display that precedes open-mouth threat in grey seals (Schneider, pers. ob.). Flipper waving also occurs in feral harp seals and in captive harp and ringed seals (Terhune, pers. comm.).

Alert behavior is common. Many of the first signals produced when one animal becomes aware of another serve this function. Underwater clicks may express a mood of alertness or fear (Schusterman, 1967).

Acoustic signals of the social breeders fall into two general groups (northern elephant seal: Bartholomew and Collias, 1962). Threat sounds are loud, harsh, and segmented, whereas attractive calls are less harsh, unsegmented, of variable pitch, and of higher frequency (Fig. 5 and Table 2). Some sounds, such as the male territorial calls, are related to motivational state. For instance, the snort of the northern elephant seal is the lowest-intensity threat sound likely to be the first sound produced in any aggressive situation (Bartholomew and Collias, 1962). It is produced in any position and can cause subordinate males to leave. The clap threat is a signal of incipient attack. It is of higher intensity and is always produced in a stereotyped posture, with the forequarters elevated to the maximum and the inflated proboscis extended into the fully opened mouth. If the intruder does not leave, a fight ensues, usually only when the dominance relation of the bulls has not been established. Subdominant adult males also use these sounds. They are illustrated in Fig. 5, and a summary of the sounds and functions is given in Table 2.

Yearlings produce a hiss and a roar in aggressive situations. The roar is similar to one of the two vocalizations of the female: the belch-roar threat or the high-pitched attraction call, the bark used for maintenance of or finding contact with the pup. The belch-roar threat of the female causes varying responses, depending on the status of the individual being threatened: a yearling immediately retreats; a subordinate female may retreat or reply with similar threat calls; or, if the animals are of equal status, a formalized postural fighting bout may ensue. Bulls are seemingly indifferent to female vocal threats. Bartholomew and Collias (1962) concluded that

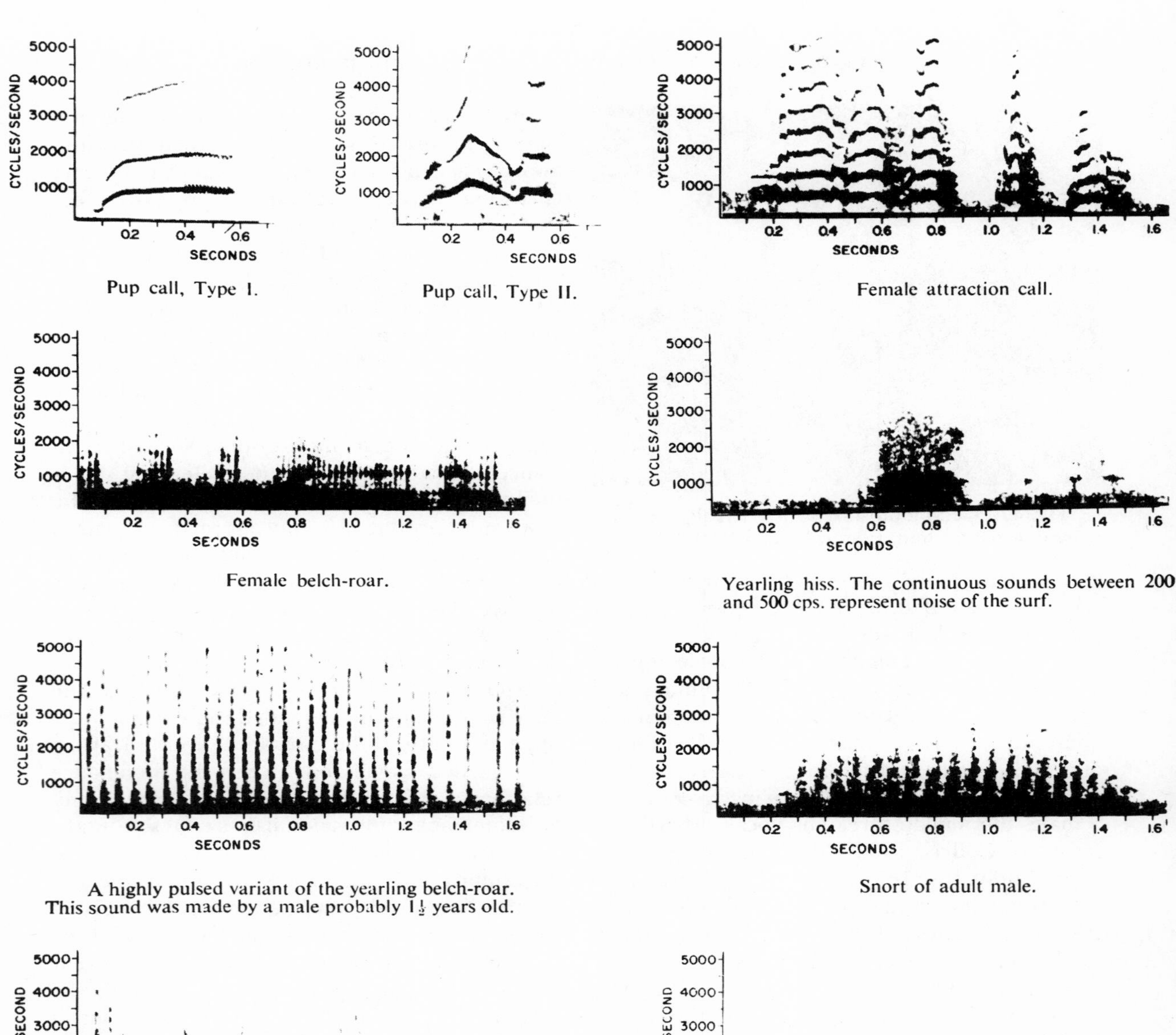

Pup call, Type I.

Pup call, Type II.

Female attraction call.

Female belch-roar.

Yearling hiss. The continuous sounds between 200 and 500 cps. represent noise of the surf.

A highly pulsed variant of the yearling belch-roar. This sound was made by a male probably $1\frac{1}{2}$ years old.

Snort of adult male.

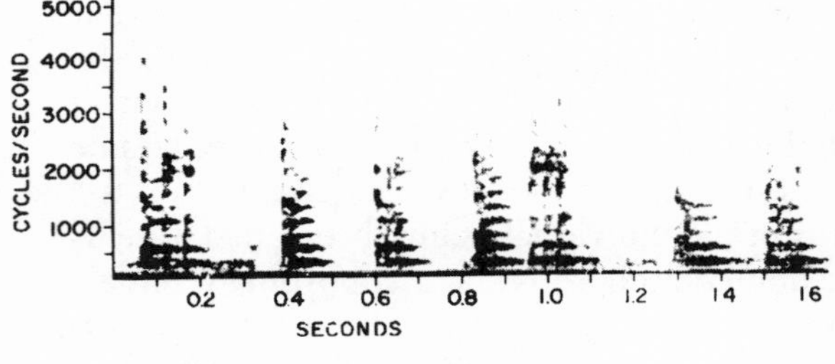

Clap-threat of adult male.

[illegible] of adult male [illegible] into a [illegible]

Fig. 5. Various sounds of the northern elephant seal. It can be seen that aggressive calls are harsher and pulsed, whereas attractive calls are less harsh and harmonic. (From Bartholomew and Collias, 1962.)

vocal communication is unusually important in the social behavior of the northern elephant seal. They further suggested an ontogeny of the pup protest to the yearling threat, and finally to either the female belch-roar or the mature male clap threat. Other species of social pinnipeds have similar call organization and development.

The threat sounds of the southern elephant seal were found to be quite different from those of the northern elephant seal, two species separated since the Pleistocene (Le Boeuf and Petrinovich, 1974a).

Tactile and olfactory signals are poorly understood. We do know that the erect vibrissae are used in greetings. Bonner (1968) calmed down an aggressive Kergulen fur seal by stroking its vibrissae with a long stick. There is much evidence that males have strong odors, and it has been suggested that pheromones may act in territorial behavior (northern fur seal: Peterson, 1965; ringed seal: Kenyon, 1962; California sea lion: Peterson and Bartholomew, 1967; New Zealand fur seal: Miller, 1971).

Visual signals are many. They consist of postures, low for submissive behavior, high for aggressive behavior. Open mouths in various threat situations (Fig. 4), with a strikingly colored pink mouth, and the various developments of proboscis and vocal sacs in some adult males are all visual signals.

The existence of dialects in male threat vocalizations of the northern elephant seal has been studied in some detail by Le Boeuf and Petrinovich (1974b). Essentially, they have shown that the call pulse rates of animals from an island that has a few animals is increasing and is correlated with the immigration of animals from another island where animals have a higher pulse rate. The pulse rate of calls from individuals does not vary systematically from year to year. The pulse rates of animals from a rookery with few immigrants has remained relatively constant over a four-year period. Although the functional significance of pulse rates is unknown, the authors felt that the differences arose as the result of isolation. The differences may well represent the raw material for future speciation.

PLAY

Play may be with objects, self, or other animals, and has been described for the northern elephant seal (Schusterman, 1968; Rasa, 1971), the Steller sea lion (Schusterman, 1968; Farentinos, 1971), the California sea lion (Schusterman, 1968; Peterson and Bartholomew, 1967), the South African fur seal (Rand, 1967), the grey seal (Hewer, 1957; Wilson, 1973b), the Weddell seal (Kaufman, in press), and the harbor seal (Schusterman, 1968; Wilson, 1973a, 1973b). Patterns of play are sex-specific in southern elephant seal and Steller sea lion pups (Rasa, 1971; Gentry, 1974) and frequently include sham fighting, pseudo-copulation, herding (in northern fur seal: Peterson, 1965), and swimming. In the northern fur seal (Bartholomew, 1959), the most frequent type of pup play changes from pseudo-copulation, during the reproductive season, to sham fighting and swimming in autumn. Bartholomew related the frequency of reproductive play to the visual presence of reproducing animals. Peterson (1962) found that sham fighting became more adultlike by fall. Farentinos (1971) made the interesting observation that normally aggressive nursing female Steller sea lions, after leaving their pups, gather and engage in play, which includes mock fighting, wrestling, and synchronous swimming. It appears that play is important to socialization, and thus the ontogeny of communicative behavior (see also Gentry, 1974).

Wilson (1973b), in summarizing play in harbor seals, said that similar dyadic play occurs between mother and pup, small juveniles, adolescents, and adults, commonly as a prelude to coitus, and occasionally outside the mating season. Group and dyadic play occur in the harbor seal, but only dyadic play was found in the grey seal. Although the motor patterns found in adult agonistic behavior generally compose play, this was not the case in the harbor seal. The grey seal exhibits a species-specific signal to invite and maintain play not found in the common seal. Wilson concludes that the differences found in the common seal may relate to mating without competition (monogamous).

INTERSPECIFIC COMMUNICATION

Birds often communicate danger to and induce fleeing responses by seals. Bartholomew (1952) describes how gulls first react to a human by mobbing and calling; this in turn alerts the cormorants, which upon the closer approach of man take to the air; and their flight causes many of the elephant seals to take to the water without even determining what the danger is. Gentry (1970) suggested that northern sea lions fled to the water when the alarm cry of the western gull was produced, but the response was not elicited by the cries of other bird species. These relations have not been studied in any detail. Interspecies communication of various seals is known to occur, but will not be covered here.

Summary

Sea cows, sea otters, and pinnipeds each have solved their communicatory needs in different ways. Sea cows are aquatic, and little is known about their communication; whereas sea otters relate more to other otters than to sea cows and pinnipeds. Seals must go onto ice or land to give birth and may be diphyletic. The otarids are polygamous social breeders and usually mate on land. Phocids have some polygamous forms, but are generally monogamous and usually mate under water.

Recent reviews of various aspects of reproduction, social organization, and communication of pinnipeds can be consulted for other references and details not covered here (Peterson, 1968; Caldwell and Caldwell, 1972a, 1972b; Poulter, 1968; Evans and Bastian, 1969; Schusterman, 1968; Ronald and Mansfield, 1976). The more social polygamous species and those that are economically important have been studied the most. Many of the phocids carry out activities in the water and in very rigorous environments, where their accessibility to study is limited.

The pinnipeds exhibit a wide variety of reproductive and social organizations, ranging from solitary to highly gregarious forms. All seals must come up on ice and land to give birth, apparently for thermoregulatory reasons. This fact determines, in some cases, the kind of communication system that is needed. During the remainder of the year seals are generally less gregarious and, in some cases, even solitary.

It is clear that visual, auditory, and to a lesser extent tactile senses are dominant in seals. Vision and hearing are adapted for use under water, where seals spend most of their time, but are adequate on land. This fact correlates with feeding and the presence of predators in water. Colonial seals on land appear to have few predators, but the senses are still adequate in air. Sea lions appear to rely heavily on vision, although sound is useful over much greater distances. Circumstances exist in which the emphasis must shift to the acoustic mode, such as during the antarctic night or in murky waters. Because blind seals frequently appear healthy and are able to pup, we can conclude that the acoustic channel is developed sufficiently to solve their needs in the absence of vision.

Sound and visual posturing are highly developed during mating seasons, and both channels appear useful in maintaining territories, courtship, mating, and mother-young interactions. Broad underwater areas of the arctic and probably the antarctic are completely saturated with mating calls. These "songs" are probably functionally similar to bird song or humpback whale song.

The visual and acoustic senses of seals in water are as well developed as those of primates. Vision seems more important in seals than in the acoustically dominated odontocetes. The placement of the eyes on the side and far back of the head in porpoises leads one to believe that the primary uses of vision may be to maintain coordination of swimming groups and to see to the side where sound is reduced; whereas the forward and upward placement of the eyes in seals adapts them to their particular needs, such as feeding. It is interesting to note that Schevill and Watkins (1971) suggested that the Weddell seal's sounds are directed downward and forward, thus covering an area not within the visual cone; but the projected sounds were not typical echolocation clicks. Although one thinks of sounds as more important in the land-breeding species than in the aquatic mating phocids, Mohl et al. (1975) have tentatively identified sixteen types of sounds produced by the phocid harp seal.

Our knowledge of the chemical senses of pinnipeds is rudimentary at best; but sufficient observations exist to suggest that olfaction is used on land in mother-pup recognition, recognition of musky-smelling males, and recognition of estrous females.

In general, the strategy taken by many pinnipeds, especially the otarids and a few phocids, is to aggregate on barren beaches and to establish social groups. These frequently involve a harem organization. Ritualized acoustic and visual displays have been developed to help maintain the groups, to reduce aggression, to reduce harm to pups, and to ensure continued contact or reunification of mother and pup.

Some species of phocids do not form breeding aggregations; thus there are brief periods of underwater courtship that have only been fleetingly observed in a few species. Mother-young communication is brief, and little is known about the relations between mother and young once pupping areas on ice or land are deserted. In some species, such as the Weddell and harp seals, a profusion of underwater sound signals have been heard. In fact, courtship may be more highly evolved in the underwater mating phocids. Some phocids also form female aggregations or even family units on ice, and mating more often takes place in water (Table 1).

Territorial signaling is ubiquitous to varying degrees in the social breeders and for those species that have topographic cues under water (e.g., Weddell seal). Under water, many phocid males probably shift to individual spacing, again using acoustic and visual displays. However, the grey seal bull uses sexual activity rather than territorial fighting and boundary displays as a strategy to ensure that he mates with those cows in his area of influence (Anderson et al., 1975).

In general, topographic (returning to and staying in a specific area), acoustic, and olfactory signals of mother and pup are designed to assure reunion in those species where the mother leaves for the sea periodically. Courtship in the polygamous forms is limited, but in some species, such as the northern fur seal, females have estrous displays. Courtship in the supposedly monogamous phocids may be more elaborate under water, but our information is still incomplete. Again, in polygynous species male signaling during reproduction is designed to protect a space, fixed or not, to maintain a harem. Female aggression is primarily designed to maintain spacing before estrus and to protect pups.

Play is sometimes species-, age-, and sex-specific and appears to be important to the proper development of adult signaling systems.

It appears that with the development of polygamy in the otarids and a few phocids, a positive selection for sexual dimorphism exists. The large size of males (three to six times as large as the females in some cases) allows for fasting while holding territories; but more important, sexual dimorphism (vocal sacs, teeth, size) strengthens communication signals. Limited terrestrial locomotion probably requires emphasis on acoustic displays associated partly with stationary visual displays. The more effective the signals the more females are fertilized; thus, sexual dimorphism is enhanced through positive feedback (see Bartholomew, 1970, for discussion of polygamy). This development in turn requires that signals become highly ritualized and effective, in order to reduce pup mortality from crushing by males. Although some of the more monogamous phocids have limited sexual dimorphism, which is probably useful for territorial maintenance, it does not have such a strong positive selective value.

More experimental research is needed on the roles of sensory inputs in reproductive behavior. The recent expansion of studies on seals should greatly increase our understanding of communication in these animals.

References

Anonymous, 1971. *Fur Seal Investigations, 1969.* Special Sci. Rept., Fisheries, no. 628. Seattle, Washington. 90pp.

Anderson, S. S.; Burton, R. W.; and Summers, C. F.; 1975. Behaviour of grey seals *(Halichoerus grypus)* during a breeding season at North Rona. *J. Zool.* (London), 177:179–95.

Barabash-Nikiforov, I. I., 1935. The sea otters of the Commander Islands. *J. Mammal.,* 16:255–61.

Barabash-Nikiforov, I. I., 1947. The sea otter *(Enhydra lutris* L.): biology and economic problems of breeding. In: *The Sea Otter,* P. Cohen, ed. Jerusalem: Program for Scientific Translations.

Barrett, D. W., 1935. Notes concerning manatees and dugongs. *J. Mammal.,* 16:216–20.

Bartholomew, G. A., Jr., 1952. Reproductive and social behavior of the northern elephant seal. *Univ. Calif. Publs. Zool.,* 47:369–472.

Bartholomew, G. A., Jr., 1953. Behavioral factors affecting social structure of the Alaska fur seal. *Trans. 18th N. Am. Wildl. Conf.,* pp.481–502.

Bartholomew, G. A., Jr., 1959. Mother-young relations and the maturation of pup behavior in the Alaska fur seal. *Anim. Behav.,* 7:163–71.

Bartholomew, G. A., Jr., and Collias, N. E., 1962. The role of vocalization in the social behavior of the northern elephant seal. *Anim. Behav.,* 10:7–14.

Bartholomew, G. A., Jr., and Hoel, P. G., 1953. Reproductive behavior of the Alaska fur seal, *Callorhinus ursinus. J. Mammal.,* 34:417–36.

Bertram, G. C. L., 1964. *In Search of Mermaids: The Manatees of Guiana.* New York: T. Y. Crowell. 183pp.

Bishop, R. H., 1967. Reproduction, age determination, and behavior of the harbor seal, *Phoca vitulina* L., in the Gulf of Alaska. Masters thesis, University of Alaska.

Bonner, W. N., 1968. The fur seal of South Georgia. *Brit. Antarctic Surv. Sci. Rept.,* 56:1–81.

Bonner, W. N., and Laws, R. M., 1964. Seals and sealing. In: *Antarctic Research,* R. Priestly, R. J. Adie, and G. de Q. Robins, eds. London: Butterworths, pp.163–90.

Bullock, T. H.; Ridgway, S. H.; and Suga, N.; 1971. Acoustically evoked potentials in midbrain auditory structures in sea lions (Pinnipedia). *Z. vergl. Physiol.,* 74:372–87.

Burns, J. J., 1970. Remarks on the distribution and natural history of pagophilic pinnipeds in the Bering and Chukchi Seas. *J. Mammal.,* 51:445–54.

Burns, J. J., and Fay, F. H., 1970. Comparative morphology of the skull of the ribbon seal, *Histriophoca fasciata,* with remarks on systematics of Phocidae. *J. Zool.,* 161:363–94.

Burns, J. J.; Ray, G. C.; Fay, F. H.; and Shaughnessy, P. D.; 1972. Adoption of a strange pup by the ice-inhabiting harbor seal, *Phoca vitulina largha. J. Mammal.,* 53:594–98.

Burton, R. W.; Anderson, S. S.; and Summers, C. F.; 1975. Perinatal activities in the grey seal *(Halichoerus grypus). J. Zool.* (London), 177:197–201.

Cajal, R., 1952. *Histologie du système nerveux de l'homme et des vertébrés.* Madrid, pp.469–73.

Caldwell, M. C., and Caldwell, D. K., 1972a. Behavior of marine animals. In: *Mammals of the Sea,* S. H. Ridgway, ed. Springfield, Ill.: Charles C. Thomas, pp.419–65.

Caldwell, D. K., and Caldwell, M. C., 1972b. Senses and communication. In: *Mammals of the Sea,* S. H. Ridgway, ed. Springfield, Ill.: Charles C. Thomas, pp.466–502.

Calkins, D., and Lent, P. C., 1975. Territoriality and mating behavior in Prince William Sound sea otters. *J. Mammal.,* 56:528–29.

Carrick, R.; Csordas, S. E.; and Ingham, S. E.; 1962. Studies of the southern elephant seal, *Mirounga leonina* (L.). IV. Breeding and development. *C.S.I.R.O. Wildl. Res.,* 7:161–97.

Christenson, T. E., 1974. Aggressive behavior of the female Northern elephant seal, *Mirounga angustiros-*

tris. Abstr. Proc. Anim. Behav. Soc., Urbana–Champaign, 1974.

Cline, D. R.; Siniff, D. B.; and Erickson, A. W.; 1971. Underwater copulation of the Weddell seal. *J. Mammal.,* 52:216–18.

Coulson, J. C., and Hickling, G., 1964. The breeding biology of the grey seal *H. grypus,* on the Farne Islands, Northumberland. *J. Anim. Ecol.,* 33:485–512.

Diercks, K. J.; Trochta, R. T.; Greenlaw, C. F.; and Evans, W. E.; 1973. Delphinid sonar: measurement and analysis. *J. Acous. Soc. Am.,* 54:200–204.

Dykes, R. W., 1972. What the seal's vibrissae tell the seal's brain. *Proc. 9th Ann. Conf. on Biological Sonar and Diving Mammals,* pp.123–36.

Elliot, H. W., 1887. The sea otter fishery. In: *The Fisheries and Fishery Industries of the United States,* sect. V, vol. 2, G. B. Goode, ed. Washington, D.C., Government Printing Office, pp.483–91.

Evans, W. E., 1967. Vocalization among marine mammals. In: *Marine Bio-Acoustics,* vol. 2, W. N. Tavolga, ed. New York: Pergamon Press, pp.159–86.

Evans, W. E., and Bastian, J., 1969. Marine mammal communication: social and ecological factors. In: *The Biology of Marine Mammals,* H. T. Andersen, ed. New York: Academic Press, pp.425–75.

Evans, W. E., and Haugen, R. M., 1963. An experimental study of the echolocation ability of a California sea lion, *Zalophus californianus* (Lesson). *Bull. South. Calif. Acad. Sci.,* 62:165–75.

Evans, W. E., and Herald, E. S., 1970. Underwater calls of a captive Amazon manatee, *Trichechus inunguis. J. Mammal.,* 51:820–23.

Farentinos, R. C., 1971. Some observations on the play behavior of the Steller sea lion *(Eumetopias jubata). Z. Tierpsychol.,* 28:428–38.

Fisher, E. M., 1939. Habits of the southern sea otter. *J. Mammal.,* 20:21–36.

Fisher, E. M., 1940. Early life of a sea otter pup. *J. Mammal.,* 21:132–38.

Fogden, S. C. L., 1971. Mother-young behavior at grey seal breeding beaches. *J. Zool.* 164:61–92.

Gentry, R. L., 1966. Some aspects of underwater hearing by a California sea lion. Masters thesis, San Francisco State College.

Gentry, R. L., 1967. Underwater auditory localization in the California sea lion *(Zalophus californianus). J. Auditory Res.,* 7:187–93.

Gentry, R. L., 1970. Social behavior of the Steller sea lion. Ph.D. diss., University of California.

Gentry, R. L., 1974. The development of social behavior through play in the Steller sea lion. *Amer. Zool.,* 14:391–403.

Gohar, H. A. F., 1957. The Red Sea dugong. *Publs. Mar. Biol. Stn. Ghardaga,* 9:3–50.

Gwynn, A. M., 1953. The status of the leopard seal at Heard Island and Macquarie Island 1948–1950. *Interm Rep. Aust. Natur. Antarct. Res. Exped.,* 3:1–33.

Hartman, D. S., 1969. Florida's manatees, mermaids in peril. *Nat. Geogr.,* 136:342–53.

Hewer, H. R., 1957. A Hebridean breeding colony of grey seals, *Halichoerus grypus* (Fab.), with comparative notes on the grey seals of Ramsey Island, Pembrokeshire. *Proc. Zool. Soc. Lond.,* 128:23–66.

Hobson, E., 1966. Visual orientation and feeding behavior in *Phoca vitulina* and *Zalophus californianus. Nature* (London), 210:326–27.

Isenmann, P., 1970. Contribution a l'étude de la zone de velage du phoque de Weddell *(Leptonychotes weddelli)* à Pointe Géologie, en Terre Adélie. *Mammalia,* 34:573–84.

James, H., 1970. Sexual arousal in the northern elephant seal *(Mirounga angustirostris). Proc. 7th Ann. Conf. on Biological Sonar and Diving Mammals,* pp. 115–22.

Jamieson, G. S., and Fisher, H. D., 1970. Visual discrimination in the harbor seal, *Phoca vitulina,* above and below water. *Vision Res.,* 10:1175–80.

Jamieson, G. S., and Fisher, H. D., 1971. The retina of the harbor seal, *Phoca vitulina. Can. J. Zool.,* 49:19–23.

Jamieson, G. S., and Fisher, H. D., 1972. The pinniped eye: a review. In: *Functional Anatomy of Marine Mammals,* R. J. Harrison, ed. New York: Academic Press, pp.245–361.

Johnson, G. L., 1893. Observations on the refraction and vision of the seal's eye. *Proc. Zool. Soc. Lond.,* 719–23.

Kaufman, G., in press. Colony behavior of Weddell seals, *Leptonychotes weddelli,* at Hutton Cliffs, Antarctica. In: K. Ronald and A. W. Mansfield, eds. *Symposium on the Biology of the Seal,* Conseil International pour l'Exploration de la Mer.

Kenyon, K. W., 1962. Notes on phocid seals at Little Diomede Island, Alaska. *J. Wildl. Mgmt.,* 26:380–89.

Kenyon, K. W., 1969. The sea otter in the eastern Pacific Ocean. *N. Am. Fauna,* 68:1–352.

Kenyon, K. W., 1972. The sea otter. In: *Mammals of the Sea,* S. H. Ridgway, ed. Springfield, Ill.: Charles C. Thomas, pp.205–14.

Kenyon, K. W., and Rice, D. W., 1959. Life history of the Hawaiian monk seal. *Pacif. Sci.,* 12:215–52.

King, J. E., 1964. *Seals of the World.* London: Brit. Mus. Nat. Hist. 154pp.

Kooyman, G. L., 1968. An analysis of some behavioral

and physiological characteristics related to diving in the Weddell seal. *Antarctic Res. Ser.,* 11:227–61.

Kooyman, G. L., 1975. A comparison between day and night diving in the Weddell seal. *J. Mammal.,* 56:563–74.

Krumholz, L. A., 1943. Notes on manatees in Florida waters. *J. Mammal.,* 24:272–73.

Landau, D., and Dawson, W. W., 1970. The histology of retinas from the Pinnipedia. *Vision Res.,* 10:691–702.

Lavigne, D. M., and Ronald, K., 1972. The harp seal, *Pagophilus groenlandidus* (Erxleben 1777) XXIII. Spectral sensitivity. *Can. J. Zool.,* 50:1197–1206.

Laws, R. M., 1964. Comparative biology of Antarctic seals. In: *Biologie Antarctique.* Paris: Hermann.

Le Boeuf, B. J., 1972. Sexual behavior in the northern elephant seal *Mirounga angustirostris. Behaviour,* 41:1–26.

Le Boeuf, B. J., 1974. Male-male competition and reproductive success in elephant seals. *Amer. Zool.,* 14:163–76.

Le Boeuf, B. J., and Peterson, R. S., 1969. Social status and mating activity in elephant seals. *Science,* 163:91–93.

Le Boeuf, B. J., and Peterson, R. S., 1974. Elephant seals: interspecific comparisons of vocal and reproductive behavior. *Mammalia,* 38(1):16–32.

Le Boeuf, B. J., and Petrinovich, L. F., 1974a. Elephant seals: interspecific comparisons of vocal and reproductive behavior. *Mammalia,* 38(1):16–32.

Le Boeuf, B. J., and Petrinovich, L. F., 1974b. Dialects of northern elephant seals, *Mirounga angustirostrus:* origin and reliability. *Anim. Behav.,* 22:656–63.

Le Boeuf, B. J.; Whiting, R. J.; and Gantt, R. F.; 1972. Perinatal behavior of northern elephant seal females and their young. *Behaviour,* 43:121–56.

Limbaugh, C., 1961. Observations on the California sea otter. *J. Mammal.,* 42:271–73.

Lindsey, A. A., 1937. The Weddell seal in the Bay of Whales. *J. Mammal.,* 18:127–44.

Ling, J. K., 1966. The skin and hair of the southern elephant seal, *Mirounga leonina* (Linn). I. The facial vibrissae. *Aust. J. Zool.,* 14:855–66.

Lythgoe, J. N., and Dartnell, H. J. A., 1970. A "deep sea rhodopsin" in a mammal. *Nature* (London), 227:955–56.

Mansfield, A. W., 1958. The breeding behavior and reproductive cycle of the Weddell seal (*Leptonychotes weddelli* Lesson). *Sci. Rep. Falkld. Isl. Depend. Surv.,* 18:1–41.

Marlow, B. J., 1967. Mating behavior in the leopard seal, *Hydrurga leptonyx* (Mammalia: Phocidae), in captivity. *Aust. J. Zool.,* 15:1–5.

Marlow, B. J., 1972. Pup abduction in the Australian sea-lion, *Neophoca cinerea. Mammalia,* 36:161–65.

Miller, E. H., 1971. Social and thermo-regulatory behaviour of the New Zealand fur seal, *Arctocephalus forsteri* (Lesson, 1828). Masters thesis, University of Canterbury, Christchurch, New Zealand.

Miller, E. H., 1974. Social behaviour between adult male and female New Zealand fur seals *Arctocephalus forsteri* (Lesson) during the breeding season. *Aust. J. Zool.,* 22:155–73.

Møhl, B., 1964. Preliminary studies on hearing in seals. *Vidensk. Meddr. dansk naturh. Foren.,* 127:283–94.

Møhl, B., 1967. Frequency discrimination in the common seal. In: *Underwater Acoustics,* vol. 2, V. M. Albers, ed. New York: Plenum Press, pp.43–54.

Møhl, B., 1968a. Hearing in seals. In: *The Behavior and Physiology of Pinnipeds,* R. J. Harrison, R. C. Hubbard, R. S. Peterson, C. E. Rice, and R. J. Schusterman, eds. New York: Appleton-Century-Crofts, pp.172–95.

Møhl, B., 1968b. Auditory sensitivity of the common seal in air and water. *J. Auditory Res.,* 8:27–38.

Møhl, B. and Ronald, K., 1970. The harp seal, *Pagophilus groenlandicus* (Erxleben, 1777). IV. Underwater phonations with special references to the click. *Int. Comm. Explor. Sea,* Copenhagen, 5:1–6.

Møhl, B.; Terhune, J. M.; and Ronald, K.; 1975. Underwater calls of the harp seal, *Pagophilus groenlandicus.* In: *Biology of the Seal,* K. Ronald and A. W. Mansfield, eds. Inter. Comm. for the Exploration of the Sea—Verbaux Réunion, vol. 169, pp.533–43.

Moore, J. C., 1956. Observations of manatees in aggregations. *Am. Mus. Novit.,* 1811:1–24.

Moore, P. W. B., 1975. Underwater localization of clicks and pulsed pure tone signals by the California sea lion *(Zalophus californianus). J. Acoust. Soc. Am.,* 57:406–10.

Nagy, A. R., and Ronald, K., 1970. The harp seal, *Pagophilus groenlandicus* (Erxleben, 1777). VI. Structure of retina. *Can. J. Zool.,* 48:367–70.

Naito, Y., 1973. Comparison in color pattern of two species of harbour seal in adjacent waters of Hokkaido. *Sci. Rep. Whales Res. Inst., Tokyo,* 25:301–10.

Olds, J. M., 1950. Notes on the hood seal *(Cystophora cristata). J. Mammal.,* 31:450–52.

Orr, R. I., and Poulter, T. C., 1967. Some observations on reproduction, growth and social behavior in the Steller sea lion. *Proc. Calif. Acad. Sci.,* 35:193–226.

Parker, G. H., 1922. The breathing of the Florida manatee. *J. Mammal.,* 3:127–35.

Peterson, R. S., 1962. *Behavior of Fur Seal Pups during Autumn.* Seattle: Marine Mammal Biological Laboratory. 59pp.

Peterson, R. S., 1965. Behavior of the northern fur seal. Ph.D. diss., Johns Hopkins University. 214pp.

Peterson, R. S., 1968. Social behavior in pinnipeds with particular reference to the northern fur seal. In: *The Behavior and Physiology of Pinnipeds,* R. J. Harrison, R. C. Hubbard, R. S. Peterson, C. E. Rice, and R. J. Schusterman, eds. New York: Appleton-Century-Crofts, pp.3–53.

Peterson, R. S., and Bartholomew, G. A., 1967. The Natural History and Behavior of the California Sea Lion. *Amer. Soc. Mammal. Spec. Publ.,* no. 1.

Peterson, R. S., and Bartholomew, G. A., 1969. Airborne vocal communication in the California sea lion, *Zalophus californianus. An. Behav.,* 17:17–24.

Piggins, P. J., 1970. Refraction of the harp seal, *Pagophilus groenlandicus* (Erxleben, 1777). *Nature* (London), 227:78–79.

Poulter, T. C., 1963. Sonar signals of the sea lion. *Science,* 139:753–55.

Poulter, T. C., 1968. Marine mammals. In: *Animal Communication,* T. A. Sebeok, ed. Bloomington: Indiana University Press, pp.405–65.

Poulter, T. C., 1972. Sea lion vibrissae—an acoustic sensor. *Proc. 9th Ann. Conf. on Biological Sonar and Diving Mammals,* pp.95–105.

Rand, R. W., 1967. The Cape fur seal *(Arctocephalus pusillus).* 3. General behaviour on land and at sea. *South Afr. Div. Sea Fish. Invest. Rep.,* no. 61.

Rasa, A. O., 1971. Social interaction and object manipulation in weaned pups of the northern elephant seal. *Z. Tierpsychol.,* 29:82–102.

Ray, C., and DeCamp, M. E., 1969. Watching seals at Turtle Rock. *Nat. Hist.,* 78:26–35.

Ray, C.; Watkins, W. A.; and Burns, J.; 1969. The underwater song of *Erignathus* (bearded seal). *Zoologica,* 54:79–83.

Rice, D. W., 1964. The Hawaiian monk seal. *Nat. Hist.,* 73:48–55.

Ridgway, S. H., 1973. Control mechanisms in diving dolphins and seals. Ph.D. diss., Cambridge University.

Ronald, K., and Mansfield, A. W., eds., 1976. *Biology of the Seal.* Inter. Comm. for the Exploration of the Sea, Rapports et Procés—Verbaux Réunion, vol. 169. 560pp.

Sandegren, F. E., 1970. Breeding and maternal behavior of the Steller sea lion *(Eumatopias jubatus)* in Alaska. Masters thesis, University of Alaska.

Sandegren, F. E., Cher, E. W., and Vandevere, J. E., 1973. Maternal behavior in the California sea otter. *J. Mammal.,* 54:668–79.

Scammon, C. M., 1874. *The Marine Mammals of the North-Western Coast of North America, Described and Illustrated: Together with an Account of the American Whale Fishery.* San Francisco: John H. Carmany.

Scheffer, V. B., 1962. Pelage and surface topography of the northern fur seal. *N. Am. Fauna,* 64. 206pp.

Scheffer, V. B., and Slipp, J. W., 1944. The harbor seal in Washington state. *Amer. Biol. Nat.,* 32:50–53.

Schevill, W. E., and Watkins, W. A., 1965. Underwater calls of *Trichechus* (Manatee). *Nature* (London), 205:373–74.

Schevill, W. E., and Watkins, W. A., 1971. Directionality of the sound beam in *Leptonychotes weddelli* (Mammalia: Pinnipedia). In: *Antarctic Pinnipedia,* W. H. Burt, ed. Washington, D.C.: American Geophysical Union.

Schevill, W. E.; Watkins, W. A.; and Ray, C.; 1963. Underwater sounds of pinnipeds. *Science,* 141:50–53.

Schevill, W. E.; Watkins, W. A.; and Ray, C.; 1966. Analysis of underwater *Odobenus* calls with remarks on the development and function of the pharyngeal pouches. *Zoologica,* 51:103–106.

Schneider, J., 1974. Description and behavioral significance of grey seal *(Halichoerus grypus)* vocalizations. Masters thesis, University of Rhode Island. 38pp.

Schusterman, R. J., 1965. Orienting responses and underwater visual discrimination in the California sea lion. *Proc. 73rd Ann. Conv. Am. Psychol. Assoc.,* 1:139–40.

Schusterman, R. J., 1967. Perception and determinants of underwater vocalization in the California sea lion. In: *Les Systèmes sonars animaux,* R. G. Busnel, ed. Jouy-en-Josas, France: Laboratoire d'Acoustique Animale, pp.535–617.

Schusterman, R. J., 1968. Experimental laboratory studies of pinniped behavior. In: *The Behavior and Physiology of Pinnipeds,* R. J. Harrison, R. C. Hubbard, R. S. Peterson, C. E. Rice, and R. J. Schusterman, eds. New York: Appleton-Century-Crofts, pp.87–171.

Schusterman, R. J., 1972. Visual acuity in pinnipeds. In: *Behavior of Marine Mammals,* vol. 2, H. E. Winn and B. L. Olla, eds. New York: Plenum Press, pp. 469–92.

Schusterman, R. J., 1974. Auditory sensitivity of a California sea lion to airborne sound. *J. Acoust. Soc. Am.,* 56:1248–51.

Schusterman, R. J., and Balliet, R. F., 1970. Visual acuity of the harbor seal and the Steller sea lion under water. *Nature* (London), 226:563–64.

Schusterman, R. J., and Balliet, R. F., 1971. Aerial and underwater visual acuity in the California sea lion *(Zalophus californianus)* as a function of luminance. *Ann. N.Y. Acad. Sci.,* 188:37–46.

Schusterman, R. J.; Balliet, R. F.; and Nixon, J.; 1972. Underwater audiogram of the California sea lion by the conditioned vocalization technique. *J. Exp. Analysis of Behav.,* 17:339–50.

Schusterman, R. J.; Balliet, R. F.; and St. John, S.; 1970. Vocal displays under water by the gray seal, the harbor seal, and the Steller sea lion. *Psychon. Sci.,* 18:303.

Schusterman, R. J., and Dawson, R. G., 1968. Barking, dominance, and territoriality in male sea lions. *Science,* 160:434–36.

Schusterman, R. J., and Feinstein, S. H., 1965. Shaping and discriminative control of underwater click vocalizations in a California sea lion. *Science,* 150:1743–44.

Schusterman, R. J., and Thomas, T., 1966. Shape discrimination and transfer in the California sea lion. *Psychon. Sci.,* 5:21–22.

Sergeant, D., 1963. Harp seals and the sealing industry. *Can. Audubon,* 25: 29–35.

Silvertsen, E., 1941. On the biology of the harp seal. *Hvalrad Skr.,* 26.

Stephens, R. J.; Beebe, I. J.; and Poulter, T. C.; 1971. Innervation of the vibrissae of the California sea lion, *Zalophus californianus. Proc. 8th Ann. Conf. on Biological Sonar and Diving Mammals,* p.111.

Stirling, I., 1972. Observations on the Australian sea lion, *Neophoca cinerea* (Peron). *Aust. J. Zool.,* 20:271–79.

Stirling, I., 1973. Vocalization in the ringed seal *(Phoca hispida). J. Fish. Res. Bd. Canada,* 30:1592–94.

Stirling, I., and Warneke, R. M., 1971. Implications of a comparison of the airborne vocalizations and some aspects of the behavior of the two Australian fur seals (*Arctocephalus* spp.) on the evolution and present taxonomy of the genus. *Aust. J. Zool.,* 19:227–41.

Tembrock, G., 1963. Acoustic behavior of mammals. In: *Acoustic Behavior of Animals,* R. G. Busnel, ed. New York: Elsevier, pp.751–86.

Terhune, J. M., 1974. Directional hearing of a harbour seal in air and water. *J. Acous. Soc. Am.,* 56:1862–65.

Terhune, J. M., and Ronald, K., 1971. The harp seal, *Pagophilus groenlandicus* (Erxleben, 1777). X. The air audiogram. *Canadian J. Zool.,* 49:385–90.

Terhune, J. M., and Ronald, K., 1972. The harp seal, *Pagophilus groenlandicus* (Erxleben, 1777). III. The underwater audiogram. *Canadian J. Zool.,* 50:565–69.

Terhune, J. M., and Ronald, K., 1973. Some hooded seal *(Cystophora cristata)* sounds in March. *Can. J. Zool.,* 51:319–21.

Terhune, J. M., and Ronald, K., 1975. Underwater hearing of two ringed seals *(Pusa hispida). Can. J. Zool.,* 53:227–31.

True, F. W., 1884. The sirenians or sea cows. In: *The Fisheries and Fishery Industries of the United States,* sect. I, G. B. Goode, ed. Washington, D.C.: Government Printing Office, pp.114–36.

Vandevere, J. E., 1970. Reproduction in the southern sea otter. *Proc. 7th Ann. Conf. on Biological Sonar and Diving Mammals,* pp.221–27.

Vandevere, J. E., 1971. Fecal analysis of the southern sea otter. *Proc. 8th Ann. Conf. on Biological Sonar and Diving Mammals,* pp.97–103.

Venables, U. M., and Venables, L. S. V., 1957. Mating behavior of the seal *Phoca vitulina* in Shetland. *Proc. Zool. Soc. Lond.,* 128:387–96.

Walls, G. L., 1963. *The Vertebrate Eye and Its Adaptive Significance.* New York: Hafner.

Watkins, W. A., 1973. Bandwidth limitations and analysis of cetacean sounds, with comments on "Delphinid sonar: measurement and analysis." *J. Acous. Soc. Am.,* 55:849–53.

Watkins, W. A., and Schevill, W. E., 1968. Underwater playback of their own sounds to *Leptonychotes* (Weddell seals). *J. Mammal.,* 49:287–96.

Wilson, G. S., 1970. Vision of the Weddell seal *(Leptonychotes weddelli).* In: *Antarctic Ecology,* M. W. Holdgate, ed. New York: Academic Press, pp.490–94.

Wilson, S., 1973a. Mother-young interactions in the common seal, *Phoca vitulina vitulina. Behaviour,* 48:23–36.

Wilson, S., 1973b. Juvenile play of the common seal *Phoca vitulina vitulina,* with comparative notes on the grey seal *Halichoerus grypus. Behaviour,* 48:37–60.

Winn, H. E., and Perkins, P. J., 1976. Distribution and sounds of the minke whale with a review of mysticete sounds. *Cetology.*

Wirtz, W. O., 1968. Reproduction, growth, development, and juvenile mortality in the Hawaiian monk seal. *J. Mammal.,* 49:229–38.

Chapter 33

COMMUNICATION IN PROSIMIANS

Peter H. Klopfer

Communication?

Do prosimian primates communicate? A cursory glance at the relevant literature might well raise doubts. "Olfactory information may be expected to play a more important role . . . ," "patterns of scent deposition are perhaps . . . ," "a gesture which may originally . . ." are common statements. A more general, if equally cautious, contention is that

> Communication is, in fact, such a problematical phenomenon that its exact meaning with respect to animals needs very careful statement. A correct formulation of the terms in which it is to be conceived is likely to hold implicitly the concepts that will automatically rearrange our empirical and theoretical views of mental evolution. The numerous inconclusive attempts to link human and non-human mentality in a continuum, which fill many serious books today, are perhaps all obstructed by the same weakness, the lack of adequate ideas of the animalian forms of perception and action from which our own development has taken off in its great expansion. A coherent view of animal mentality, without immediate reference to our own, might provide a foundation for surer insights into the fateful evolutionary shift that has taken place in the hominid stock. [Langer, 1973:108]

Is it really so difficult to formulate the terms? The simplest procedure is to consider all instances of information transfer as communication (Sebeok, 1968). However, simple though this is, it eclipses the very distinctions we would make. "Interactions" include phenomena as diverse as the control of protein synthesis by chromosomal DNA and echo location by bats. It is more useful and no more difficult to note finer differences and to consider the various classes of interactions separately. Mere perception by one organ or organism of signals emanating from another is thus best not termed communication. It is more useful to restrict the term to interactions dependent on a shared code, with the further provision that this sharing be of mutual benefit (note Klopfer and Hatch, 1968). This excludes prey-predator interactions. When a predator "communicates" his presence to his prey, only the latter benefits. We should also distinguish between a command, "Jump in the lake," and a push off the pier. This can be done on the basis of the energy required to achieve the identical results, for the command requires very much less than the push.

Other distinctions could, of course, be made (see Hinde, 1972), but these seem as useful and adequate as any. Of course, we do not thereby distinguish between intentional and nonintentional communication. Fortunately, that distinction becomes important only with human linguistic behavior. The implication of this last point—that mechanisms can be ignored when we consider function—needs to be stressed. If the fitness (in the Darwinian sense, i.e., the proportion of one's genes carried by future generations) of an organism is enhanced by an integration or synchronization of its behavior with that of others, synchronizing or integrating mechanisms are likely to arise. Where the organisms concerned are consanguinous, i.e., have some of their genes in common, the selective pressure for such mechanisms will be even greater. Thus, while my fitness is partially measured by the proportion of my offspring in future generations, it may also depend on the proportion of my brother's children in that generation. (My brother and I hold approximately 50 percent of our genes in common. Hence, his offspring will also contribute to my fitness, albeit it will take four of them to equal only one of mine.) This notion of "inclusive fitness" (Hamilton, 1964) provides a basis for altruistic behavior (since the survival of three of my siblings, even at the cost of my own life, will still increase my fitness to 3 X 50 percent) and, coincidentally, for the evolution of codes that are mutually useful and thus serve social ends.

The design features of the resulting communication systems may even include feedback controls that correct signals that are misunderstood, or adjust for inadequate signal/noise ratios, or switch signals or even codes as contexts change. Such controls may produce results that appear similar to the cognitively controlled, intentional acts involved in a conversation between two of us. No matter, the issue of intentionality may still be ignored so long as a communicative function is served. Natural selection could produce automatons whose behavior is functionally identical to that of willful men even while underlying mechanisms differ. If behavior is classified in terms of its function, the fact of a difference in ontogeny or mechanism is irrelevant. A leaf-eating insect may be cryptically colored and thus equally well protected whether it is born with green pigment or with scales that differentially refract different wavelengths, or whether it wraps a green leaf around itself.

It must be added that the issue of similarity in behavioral mechanisms may be important to those concerned with the reconstruction of the history or evolutionary past of an organism. Phylogenetic studies have traditionally leaned heavily on the identification of homologies, though some biologists as well as philosophers consider this a risky technique (Klopfer, 1973; Langer, 1973). Fortunately, we can sidestep this issue here, while admitting to its existence, and focus on the question of whether extant primates, particularly the prosimian primates, do communicate, how they do so, and the sorts of information they can convey. The issue of how they came to do so and what their relatives made of their talents we shall ignore.

Indications of Communication

The fact that many prosimian primates (Table 1) do show integrated patterns of social behavior may be considered a *prima facie* ground for suspecting the occurrence of some kind of communication. The existence of seemingly stereotyped (species-characteristic) movements, odor-producing glands, and distinctive sounds further suggest that these are the vehicles for a communicative code. A detailed list of the sounds, gestures, and odor source of prosimians, and the context in which they occur, is provided by Doyle (1974). It is as long and varied as that for any other primate. Note Table 2. The more insistent questioner, however, will demand to know how we can be certain that the behavior we assume to be communicative truly meets the criteria for communication, i.e., that (1) it entails a code (a stereotyped convention); (2) the "mean-

Table 1

Suborder Prosimii.

The following classification of the prosimians is generally accepted, though there are disagreements as to the status of particular groups—whether *L. fulvus* is a distinct species or a subspecies of *L. macaco*. These issues are not important here, but they need to be kept in mind in using this table. Species marked with an asterisk are mentioned in the text.

Family:	Lemuridae	
Genus:	*Lemur*	
	*catta**	ringtailed lemur
	*variegatus**	variegated lemur
	macaco (includes *fulvus**)	brown or black lemur
	mongoz	mongoz lemur
	rubriventer	
	Hapalemur	gentle lemur or hapalemur
	griseus	
	simus	
	Lepilemur	sportive lemur or lepilemur
	mustelinus	
Family:	Cheirogaleinae	small nocturnal lemurs
Genus:	*Cheirogaleus*	dwarf lemurs
	major	greater dwarf lemur
	medius	fattailed dwarf lemur
	trichotis	hairyeared dwarf lemur
	Microcebus	mouse lemur
	*murinus**	
	coquereli	Coquerel's mouse lemur
	Phaner	forked lemur
	furcifer	
Family:	Indriidae	
Genus:	*Indri*	
	indri	indri
	Avahi	woolly lemur, avahi
	laniger	
	Propithecus	sifakas
	diadema	diademed sifaka
	verreauxi	white sifaka
Family:	Daubentoniidae	
	Daubentonia madagascariensis	aye-aye
Family:	Lorisidae	
Genus:	*Loris*	
	tardigradus	slender loris
	Nycticebus	
	coucang	slow loris
	Arctocebus	
	calabarensis	golden potto
	Perodicticus	
	potto	potto
	Galago (Galago)	galagos, bushbabies
	senegalensis	Senegal or lesser bushbaby
	crassicaudatus	thicktailed or greater bushbaby
	alleni	Allen's bushbaby

Table 1 (continued)

Family:	Lorisidae (cont.)	
Genus:	*Galago (Euoticus)*	needleclawed bushbaby
	elegantulus	
	inustus	
	Galago (Galagoides)	Demidoff's or dwarf bushbaby
	demidovii	
Family:	Tarsidae	
Genus:	*Tarsius*	tarsiers
	spectrum	spectral tarsier
	bancanus	Horsfield's tarsier
	syrichta	Philippine tarsier

Table 2

Some common communicative signals of Lemur catta.

Activity	Accompanying vocalization	Context or apparent significance
Stare	—	Mild threat given to unusual objects or to predators.
Eyelids lowered	—	Response when stared at or when sleepy.
Lips pursed forward	–, meow, howl	Occurs with bouts of chewing and yawning, always with meow.
Open mouth, teeth covered	bark	Given when mobbing ground predators.
Grin	–, spat, meow	Occurs with play bite, with all spat calls, and in first syllable of meow.
Flared lip	–, purr, squeal	Occurs while marking tail and with tail waving.
Chewing	—	Occurs while scent marking.
Yawning	—	Occurs during stink fight.
Ears flattened	–, spat	Performed with stare, in all aggressive encounters, and while tail marking or tail waving.
Swagger	—	Response of superior male toward inferior.
Tail pendulum swing	–, clicks	Occurs with stare at ground predator.
Touch noses	—	Greeting.
Nose poke	—	Occurs when female dislodges juvenile or when preventing an animal from grooming an infant.
Cuff	–, spat	Occurs in threat, spats, jump fights.
Nose-genitalia	—	Done by males to females in breeding season.
Genital mark	—	Done by males or females at any time.
Rub brachial gland on axillary gland	—	Done by males before tail or palmar marking.
Palmar mark	—	Done by males any time, in stink fights.
Tail mark	–, purr, squeal	Done by males, any time after rubbing glands and before tail waving.
Tail wave	–, spat	Done by males, any time especially in breeding season and in stink fights.

Source: After Jolly, 1966.

ing" is at least specific in particular contexts (the effects on conspecifics are predictable); (3) the exchange mutually enhances fitness (i.e., this must not be a zero-sum game); and (by way of further distinguishing those instances of communication that appear to involve cognitive or purposive elements) (4) it entails a feedback control such that the sender adjusts his signal as he perceives the recipient is unable to receive it or is responding inappropriately.

Other important questions concern the communicative patterns that serve particular functions—can one generalize about signals that enhance reproduction or group movements and distinguish them, as a class, from those involved in communicating the presence of predators or of food? For instance, Marler (1955) has proposed that the calls produced by songbirds in response to aerial predators share characteristics that hinder localization of the source of the sound, while calls given during casual feeding share different features, ones that aid in localization. Are there evolutionary trends or patterns with respect to the degree to which "codes" are linked to morphologic structures or perceived atomistically (the specific "releaser" of the ethologist) rather than holistically (note Nelson, 1973)?

The evidence for communication among nonhuman primates is inferential, though compelling. An excellent example of its character is to be found in van Hooff's (1972) account of the phylogeny of laughter and smiling. The movements or displays in question are seen to be stereotyped, with slight variations from species to species; there is a close correlation between the appearance of the display and certain other behavior patterns (i.e., the context of the act is predictable); there is a further correlation between the display and the response evoked in other animals. The specificity of this result and its dependence on the display itself are further supported by the experiments of Miller et al. (1964), who demonstrated that a televised facial expression was sufficient (at least in macaques) to convey affect.

Among the prosimian primates, unfortunately, correlational studies as exemplified by van Hooff (1972) are as yet unknown. As previously mentioned, Doyle (1974) has listed in considerable detail all the acts, movements, sound production, and scent deposition committed by prosimians that might be involved in communication. He summarizes as well the situations in which these acts most frequently occur. But, to date, there have been few rigorous experimental studies of prosimian communication. One exception is Harrington's (1971) analysis of olfactory communication in *Lemur fulvus*, while others have noted responses to particular calls (Andrew, 1964; Jolly, 1966; Doyle, 1974). All the criteria listed above have not been shown to apply.

Evidence for Olfactory Communication

The olfactory modality has been assumed to be of particular importance to prosimians because of (1) the nocturnal habits of many species (which presumably reduce the effectiveness of visually perceived signals in favor of acoustic and olfactory cues); (2) the abundance of skin glands, some with strong-smelling (to us!) secretions; (3) the presence of a substantial olfactory lobe as part of the brain; and (4) behavioral responses to objects marked or areas traversed by other individuals. It is the existence of the behavioral responses that has permitted tests to demonstrate the existence of a communicative system based on scents.

The anatomy of the olfactory organs of the Prosimii is described by Hill (1953); that of the specialized skin glands by a variety of authors, but particularly Montagna and his coworkers (Montagna, 1962). Deposition of scents by prosimians is described by Andrew (1964),

Doyle et al. (1967), Ilse (1955), Jolly (1966), and Petter (1962). A review is provided by Johnson (1973). The often elaborate rituals associated with marking behavior inevitably suggest that there is some significance attached thereto, such as a role in territorial demarcation (Jolly, 1966), synchronization of breeding (Doyle et al., 1967), spatial orientation (Seitz, 1969), and individual identification (Clark, in prep.).

Specific tests of the ability of individual *Lemur fulvus* to discriminate among individuals of the same and different subspecies and species, and between sexes, as well as of the territorial function of scents, were conducted by Harrington (1971). His subjects were, for the most part, captive-reared *Lemur fulvus.* Scents were collected from the animals by rubbing sterile gauze over their glands or by allowing spontaneous marking of gauze or lucite rods left within the animals' quarters. Scents were presented to the animal being tested by successively placing scent-impregnated gauzes into its cage for 30 seconds at a time, until the subject ceased to respond or its responses (specifically, sniffing or marking of the gauze) waned. Once this criterion for habituation had been attained, the test animal

Fig. 2. *Lemur fulvus.* (Photo by R. Haeckel.)

was either presented with a new series of marked gauzes (from a different donor) or, in the case of the "control" subjects, another series from the original donor. If discrimination of scents was occurring, responses would be expected to reappear in the first instance, but not in the second. Apparently, individuals, sexes, and taxa can be identified by scent, though which scents are most important (assuming the products of different glands differ), or whether sexual condition or season alters the result could not be stated. *Lemur catta* respond differently to the secretions of the antebrachial (forearm) gland of different males (A. Rosenkoetter, pers. comm.).

Fig. 1. Anal marking by *Lemur catta.*

Harrington also substituted clean and scent-marked rods on sections of rod paths on which the animals traveled within their large enclosures. These substitutions had no apparent effect on the manner in which the animals used their space, raising doubt that scent trails are of particular importance. Of course, in a natural setting, scent trails may be of consequence. This is currently under investigation by Rosenkoetter (pers. comm.), who is seeking to correlate marking sites, territorial boundaries, and paths of movement of *Lemur catta* (living free in Madagascar!). No data are available as of this writing, however.

In sum, olfactory signals could be used for communication, but the evidence in hand supports no more than this possibility. Perhaps olfactory and visual information must be presented simultaneously in order for the former to be employed. Marking behavior, as noted before, is associated with conspicuous, often stereotyped, movements. Perhaps, too, the role of olfaction is evident only in particular contexts or particular locales. In galagos, A. Clark (pers. comm.) reports that urine marking is performed on conspicuous branches, while chest rubbing occurs against vertical trunks. The context-dependence of so much behavior should alert us to the importance of spatial and temporal context (Shettleworth, 1972).

Other considerations also lend credence to the view that olfaction is relatively important to the communicative systems of prosimians. Five of these have been summarized by Harrington (1971):

(1) The relatively small home range of prosimians as opposed to other primates enhances the effectiveness of olfactory signals. An *L. fulvus,* for instance, has a range of circa 7 ha, compared with 18-78 ha for *Cercopithecus aethiops* (which, in turn, has a much smaller range than most other monkeys or apes).

(2) The predominantly nocturnal habit of prosimians, compared to other primates, also would favor olfaction in signaling.

(3) The more highly seasonal breeding behavior of prosimians would necessitate synchronization of behavior within troops, which could be assisted by olfactory cues (note Michael et al., 1971).

(4) The relatively simple forms of agonistic behavior shown by prosimians, which are largely confined to the four weeks of the breeding season, are less dependent on a complex (multimodal) signaling mechanism.

(5) Finally, the habit of mutual oral grooming, so marked in prosimians, whose teeth and tongue appear to have been especially adapted therefore, provides a ready mechanism for the transmission of scents.

If these correlatives of olfactory communication do stand in a causal relation to one another, one would expect those ceboidea that most resemble lemurs in having a highly developed olfactory communication system, e.g., the Hapalidae, to resemble lemurs in these other features, too (Harrington, 1971). This point has yet to be investigated.

Evidence for Acoustic Communication

Several descriptions of the vocalizations of prosimians are available (note Jolly, 1966), but systematic studies of their role in communication are far more rare. It was shown that *Lemur catta* could be made to vocalize in an operant-conditioning paradigm (Wilson, ms.), supporting the view that acoustic signaling is not altogether alien to these animals. Though this has been generally accepted, there is some further significance to Wilson's study. Studies of vocalization in experimenter-controlled situations have many advantages over *post hoc* analyses of spontaneous utterance, particularly for the study of the behavioral and neural organization of vocalization. Hence, Wilson's demonstration that discriminative vocal conditioning is possible in lemurs is of substantial methodological significance. To date, however, this approach has yet to be exploited.

One attempt to correlate laboratory and field findings, though not utilizing operant methods, is being made with the mouse lemur, *Microcebus murinus* (McGeorge, 1973, and work in progress). McGeorge first described the full range of sounds produced by captive animals and correlated them with the social situation and behavior that they accompanied. She then reasoned that there must exist a maximum distance between animals beyond which a sound signal would not be perceived. This distance would be expected to vary with seasonal (and other) changes in ambient noise and foliage density. Hence, if mouse lemur sounds are communicative, entailing both an emitter and a receiver/responder, then there should be systematic variations in the frequency and amplitude (i.e., carrying capacity) of the sounds that are related to the sound-propagation (or, conversely, sound-absorption) characteristics of the animal's environment. The specific tests involve measuring the acoustic properties of the environment, the animal's vocalizations, and the distances between (and reactions of) the individual animals. This work is in progress at this time.

In *Lemur variegatus* there is a form of ritualized, synchronized "duetting," which also implies communication, or at least a mutual responsiveness to calls. In pairs or small troops of animals (wild or captive), a seemingly spontaneous, raucous, and very loud call erupts a dozen or so times in a day. While several voices repeat a basso cadence for ten to thirty seconds, a single voice provides a tenor counterpoint. The timing and rhythm of the voices appears to be fixed, though this has yet to be confirmed by audiospectographic analysis. Curiously, in one group of two animals and one group of six, it was generally (solely?) a particular male who chimed in with the tenor line. For instance, in the larger group, of 168 recorded occasions of duets, "Mars" took the solo line 164 times; on the remaining four occasions the solo singer could not be identified with certainty but might have been "Mars." Of the pair, "Mercury" took the tenor line 99 times out of 100; on the hundredth occasion, the identity of the singer was unconfirmed. During the song, the members of the chorus are often lying prone on branches, muzzles pointed upwards, while the solo singer points his downward. No particular function or relation of the song to other activities is known.

Fig. 3. *Lemur variegatus.*

The Need for Holistic Analysis

In most experimental studies of communication, including those described here, the "signals" are treated atomistically: a particular scent or sound produces a certain response. Occasionally, one finds that the signal is compounded of two or more sensory modalities, or has meaning only in a given context. Thus, it would not be surprising to discover that the scent from a *Lemur catta's* brachial gland has one effect when exuded while the tail is elevated and shaken behind its head, and another when the tail is tucked against the back (Jolly, 1966; the general effect of "context" on meaning has been dealt with extensively by Smith, 1968). The issue I wish to raise here is more complex. It is the problem of devising research strategies for dealing with complex systems that must be treated holistically.

To begin, it is necessary to document the claim that the communicative system of prosimians does indeed require treatment as a whole, rather than analytically. Apart from prejudices favoring a *Gestalt Weltanschauung,* are there empirical grounds for a holistic approach? The outstanding successes of analytic and synthetic approaches in the study of the "language" of birds (e.g., Thorpe, 1961; von Frisch, 1965) certainly would support a negative response. However, some studies of mammalian communication lead to a different conclusion, in particular a study of the precopulatory display of jackals (Golani, 1973). A jackal's behavior patterns can be described as a sequence of configurations, each composed of a group of discrete and simultaneously occurring events (position of ear or tail, body orientation, etc.). Any particular configuration recurs infrequently, though within brief periods of time there is a higher degree of regularity. Golani explored the degree to which the "events" of a particular configuration are seen as discrete phenomena by the jackals themselves. He found the significance of specific events to lie in their relation to other simultaneously occurring events and in their temporal relations. However, the shifts in the composition of configurations over longer stretches of time imply a change in the significance of specific events. In Golani's words, "This indicates that it is necessary to trace the nature of the change from one significance to another, rather than to look for stable, unchanging significance" (1973:111).

The sophisticated computer techniques (Guttman-Lingoes Multidimensional Scalogram Analysis) employed by Golani have not been applied to other mammals. The assertion that the peculiar "stable instability" of the jackal communication system is true of mammals in general, or even of another mammal, let alone prosimians, is perhaps premature. Yet I would not gainsay the likelihood of this assertion's being validated, not the least because of the great plasticity displayed by primates faced with novel communicative tasks (Premack, 1970).

Holistic systems are those "whose behavior is constrained by important nonlinearities," wrote Nelson, in a provocative essay on the future of holistic studies of behavior. "We should try to understand its components, of course, but if we are to understand the system we must consider it as a totality or not at all" (1973:310).

> This means that an enormous amount of care is going to be needed, both to arrive at the specific hypothetical mode of functioning and to determine just how it is to be sought in the physiology of the neuromuscular system. One of the implications, I believe, is that the "controlled experiment" will be of limited usefulness in elucidating neural function, and it is going to play a much-reduced role in the behavioral study of the near future. This need not be entirely a bad thing. Controlled experiments are costly, usually, in time, money, animals, and effort which might at this stage of our knowledge be applied more productively in finding out just what it is that is worth doing experiments on. I hesitate to use the examples of astronomy and geophysics, but they may prove to provide better research paradigms for us than physics has. The difficulty inherent in subjecting the earth and stars to controlled experiments has perhaps made astronomy a different sort of science, but not necessarily one lacking in rigor or success.
>
> We may, if we like, subject animals to controlled experiments, but there is no law that they must divulge their secrets to us thereby. My personal opinion is that the greater priority is on the development of new modes of behavioral description, based on rigorous notions (where such are possible) of harmony and conflict, part and whole, simple and complex, containment and change. . . .
>
> I expect considerable dissent from some quarters over these points. It seems not to be generally realized that the subjects of most controlled behavioral experiments are analog models of "real animals in real situations" just as a collection of neuromimes is. As a minimal assumption, the experimenter counts on Nature's having no intentions of resisting his invasion of her privacy, and we really have no means of determin-

ing how often he mistakes her intentions. Kavanau (1967) has catalogued as a cautionary tale for us some instances of the contrariness of well-bred mice. [1973:311]

It is a tale all of us might ponder.

References

Andrew, R. J., 1964. The displays of the primates. In: *Evolutionary and Genetic Biology of Primates,* vol. II, J. Buettner-Janusch, ed. New York: Academic Press, pp.227–309.

Doyle, G. A., 1974. Behavior of prosimians. In: *Behavior of Nonhuman Primates,* vol. 5, A. M. Schrier and F. Stollnitz, eds. New York: Academic Press.

Doyle, G. A.; Pelletier, A.; and Bekker, T.; 1967. Courtship, mating, and parturition in the lesser bushbaby under semi-natural conditions. *Folia Primatologica,* 7:169–97.

Frisch, K. von, 1965. *Tanzsprache und Orientierung der Bienen.* Berlin: Springer-Verlag.

Golani, I., 1973. Non-metric analysis of behavioral interaction sequences in captive jackals. *Behaviour,* XLIV:89–112.

Hamilton, W. D., 1964. The genetical evolution of social behavior, I and II. *J. Theoretical Biology,* 7:16, 17–52.

Harrington, J., 1971. Olfactory communication in *Lemur fulvus.* Ph.D. diss., Duke University.

Hill, W. C. O., 1953–60. *Primates: Comparative Anatomy and Taxonomy,* vols. I–IV. Edinburgh: University of Edinburgh Press.

Hinde, R. A., 1972. *Nonverbal Communication.* Cambridge: Cambridge University Press.

Ilse, D. R., 1955. Olfactory marking of territory in two young male lorises kept in captivity in Poona. *British J. of Anim. Behav.,* 3:118–20.

Johnson, P. R., 1973. Scent marking in mammals. *Animal Behav.,* 21:521–35.

Jolly, A., 1966. *Lemur Behavior.* Chicago: University of Chicago Press.

Kavanau, J. L., 1967. Behavior of captive white-footed mice. *Science,* 155:1623–39.

Klopfer, P. H., 1973. Does behavior evolve? *Ann. N.Y. Acad. Sci.,* 223:113–19.

Klopfer, P. H., and Hatch, J., 1968. Experimental considerations. In: *Animal Communication,* T. A. Sebeok, ed. Bloomington: Indiana University Press, pp.31–43.

Langer, S., 1973. *An Essay on Human Feeling,* vol. 2. Baltimore: Johns Hopkins University Press.

McGeorge, L., 1973. A study of the vocalizations of the lesser mouse lemur, *Microcebus murinus.* Manuscript, Department of Zoology, Duke University.

Machida, H., Giacometti, L., 1967. The anatomic and histochemical properties of the skin of the external genitalia of the primates. *Folia Primatologica,* 6:48–69.

Marler, P., 1955. Characteristics of some animal calls. *Nature,* 176:6–8.

Michael, R. P.; Keverne, E. B.; and Bonsall, R. W.; 1971. Pheromones: isolation of male sex attractants from a female primate. *Science,* 172:964–66.

Miller, R. E.; Banks, J., Jr.; and Ojawa, N.; 1964. Role of facial expression in "cooperative avoidance conditioning" in monkeys. *J. Abnormal and Social Psychol.,* 67:24–30.

Montagna, W., 1962. The skin of lemurs. *Ann. N.Y. Acad. Sci.,* 102:190–209.

Nelson, K., 1973. Does the holistic study of behavior have a future? In: *Perspectives in Ethology,* P. Bateson and P. Klopfer, eds. New York: Plenum Press, pp.281–328.

Petter, J. J., 1962. Recherches sur l'écologie et l'éthologie des lemuriens malagaches, *Memoires du Museum National d'Histoire Naturelle,* Series A (Zool.) 27:1–115.

Premack, D., 1970. The education of Sarah: a chimp learns the language. *Psychology Today,* 4:55–58.

Sebeok, T. A., ed. 1968. *Animal Communication.* Bloomington: Indiana University Press.

Seitz, E., 1969. Die Bedeutung geruchlicher Orientierung beim Plumploris. *Zeitschrift f. Tierpsychologie,* 26:73–103.

Shettleworth, S., 1972. Constraints on learning. In: *Advances in the Study of Behavior,* D. S. Lehrman, R. A. Hinde, and E. Shaw, eds. New York: Academic Press, pp.1–68.

Smith, W. J., 1968. Message-meaning analysis. In: *Animal Communication,* T. A. Sebeok, ed. Bloomington: Indiana University Press, pp.44–60.

Thorpe, W. H., 1961. *Bird Song.* Cambridge: Cambridge University Press.

van Hooff, J., 1972. A comparative approach to the phylogeny of laughter and smiling. In: *Nonverbal Communication,* R. A. Hinde, ed. Cambridge: Cambridge University Press, pp.209–41.

Wilson, W., submitted for publication. Discriminative conditioning of vocalizations in *Lemur catta.* Manuscript, Department of Zoology, Duke University.

Chapter 34

COMMUNICATION IN NEW WORLD MONKEYS

John R. Oppenheimer

Introduction

Primate communication has been the subject of a number of reviews and books. Marler (1965, 1968) has discussed the various principles and functions of communication in primates, and Altmann (1967) has described the theoretical structure of primate communication in terms parallel to those used in linguistics in his book on communication among primates. Altmann (1968) has also written an extensive review of social communication in all primates, which covers the literature up to 1966. More recently Ploog and Melnechuk (1969) have written a review that concentrated on visual and vocal communication in squirrel monkeys *(Saimiri)*, including their work and that of their colleagues on brain stimulation. Jolly (1972) has compared communication systems in a number of New and Old World primate species. Peters and Ploog (1973) have written a general review of this area, which covers the literature up to 1972. See also the chapters in this volume on the Old World monkeys by Gautier and Gautier and on the apes by Marler.

The New World species have been looked at specifically in three papers. Moynihan (1967) covered specific behavior patterns in detail and introduced a number of hypotheses on the evolution and function of certain behavior patterns and displays. Data to test these hypotheses are only now beginning to appear in the literature. Snyder (1972) has reviewed briefly the literature for the marmosets and tamarins, and Epple (1975) has covered this area in much greater depth.

In this chapter I will primarily cover the literature from the mid-1960s to 1974 for the platyrrhine species that have been studied in some detail. In Table 1 are listed the platyrrhine genera and species so the reader can easily determine which species still need attention. In Tables 2 through 5 the behavior patterns used by the New World primates are arranged according to their most prominent channel of communication: tactile (information conveyed while in physical contact), olfactory and visual (information conveyed up to intermediate distances), and acoustic (information conveyed even when out of visual contact). The species on which extensive information was available are included in the tables, and other species about which less is known are mentioned in the appropriate section in the text. The information presented has been gleaned from the literature, which varies tremendously in its amount of detail. For instance, a statement may be made that an individual of one species makes "threats," whereas another paper on a different species may go into great detail on each type of threat display, its channel, its orientation, and the circumstances that elicit it, and may even include information on threshold levels and associated behavior patterns. Thus in some places I may have unintentionally read more than what was intended and in other cases I have purposely eliminated a great deal of additional information, particularly in relation to associated behavior patterns. Nonetheless, the reader should be able to obtain a representative picture of our knowledge on communication in the platyrrhines as it exists in the present literature.

Our knowledge about communication in the primates, as in other animals, depends on our ability to observe and then later on our success in correctly transmitting this information. Others have pointed out (Lancaster, 1968; Marler, 1965; Ploog and Melnechuk, 1969) that language is a specialized ability limited to man, which allows information to be communicated about the environment, as I am doing in this chapter. It is separate and different from the vocal communication system possessed by all primate species, including man, which communicates emotion. Tool use has been suggested to be a possible reason for the development of language in man (Lancaster, 1968), but tool making and use occur in other species (Alcock, 1972; Eisenberg, 1973), for instance, chimpanzees (van Lawick-Goodall, 1968). I suspect that language evolved in man because division of labor prevented some individuals from obtaining personal knowledge of the environment, which they would need at some later time. My point in making this digression is to stress that nonhuman primates need and obtain the same information about their environments, but each individual gains this information first hand, in many cases with the help of other troop members or members of its family. Obtaining information about the environment by observational learning is entirely ignored here, even though the same communication channels are used by the receiver. However, the store of this knowledge and of knowledge about other individuals and neighboring troops permits a primate, or other animal, to make appropriate responses to the subtlest of cues or stimuli, ones that a human observer might be entirely unaware of. This is particularly important in captive studies, where normal behavior may be elicited, for a number of reasons, by inappropriate stimuli, and for that matter abnormal or inappropriate behavior patterns may be elicited by normal stimuli.

I have included in Table 1 a common name for each of the species discussed below for the reader's convenience; however, in the text I use primarily the scientific name of the species. The reason is that there are frequently several common names for each species and in some cases the common name may be applicable to more than one species. Readers who are interested in a particular species but are only acquainted with a common name other than that used here can consult Napier and Napier (1967:355–70).

Group Size and Social Structure

The size and social structure of monkey groups affect aspects of the communication system employed, and in turn the communication system helps to maintain the optimum group size and structure. The size of the group determines how many individuals need to be interacted with, and the social structure, including dominance hierarchies, determines how each individual to be communicated with should be addressed. The communication system helps to attract or repulse individuals, often of specific age or sex, to and from the group.

The marmosets and tamarins live in family groups including the adult pair and their young.

Twins are usually born each year, and the male takes part in parental care. The adults dominate the young of their own sex, and attack and drive other adults away from the group (Epple, 1967, 1975). In captivity the dominant female may inhibit reproduction in subordinate females, even though these females have been copulated with. Though there is strong pair bonding between the adults, the male and female of neighboring pairs may copulate with one another. The pair bond is maintained by aggressive competition for the attention of the mate (Epple, 1975). Evidently, the young of successive years may stay with the parents so that family sizes of up to nine have been reported for *Saguinus geoffroyi* (Moynihan, 1970; Muckenhirn, 1967), six for *S. midas* (Thorington, 1968b), and eight for *Leontopithecus rosalia* (Coimbra-Filho and Mittermeier, 1973). These larger families may break up into smaller subgroups during the day (Muckenhirn, 1967; Thorington, 1968b), but several families may come together while feeding in large fruit trees (Coimbra-Filho and Mittermeier, 1973; Muckenhirn, 1967; Thorington, 1968b).

Callimico goeldii females, like the cebid females, give birth to one infant (Heltne et al., 1973), and thus may have slightly smaller family groups than do the other callitrichids.

Aotus trivirgatus, which also lives in a simple, parental family group, is the only New World species that is active at night, or more specifically just after sunset and before dawn, and sleeps in tree holes during the day (Moynihan, 1964; Perachio, 1971). Possibly because this species is most active in the dark or in poor light it has fewer visual signals than do other platyrrhine species. In captivity adults of the same sex "fight savagely" when placed together, and this aggressiveness probably helps to maintain the family unit by keeping it separate from other families (Moynihan, 1964).

Callicebus also lives in family groups, including the adult pair and one or two young. Neighboring families usually meet at specific sites along boundaries of their one-acre territories in the early morning and perform elaborate vocal and visual displays. Aggressive contact is rare. A female in estrus may briefly slip away from her mate to visit a neighboring male (Mason, 1966, 1968; Moynihan, 1966).

Squirrel monkeys *(Saimiri)* live in troops of up to three hundred individuals in undisturbed forests, but in the small patches of forest, which are common today, they live in troops of ten to thirty-five. The adult males live at the periphery of the troop, which consists of adult females and their young (Baldwin, 1968). During the mating season the males become "fatted" and sexually active, both physiologically and behaviorally (DuMond, 1968). The "fatted" males travel together and interact with one another agonistically. Their social structure is based on a linear dominance hierarchy, where dominance is most frequently expressed in penile displays. During the nonbreeding season the males travel together less often, but still give penile displays to one another. The adult males try to approach females by making rapid dashes into the troop, but they are chased off by the females and/or young, including infants (Baldwin, 1968). During the day pregnant females may form a subgroup separate from the main body of females with young (Thorington, 1968a). Play interactions may be rare or absent in some troops (Baldwin and Baldwin, 1973b). Allogrooming is rare (Moynihan, 1967), even between females in the main body of the troop (DuMond, 1968). Moynihan (1967) has suggested that allogrooming is rare in this species because penile displays, rather than grooming, are used in precopulatory behavior and because the individuals stay close to one another (see below); however, I suspect that cohesiveness in the main body of the troop is brought about by the pressures of the surrounding males, and that it is this pressure rather

Table 1

Scientific classification, common names, weight, and data on social structure.

Family and generic names (1)	*Number of species in genus, major species studied (1)*	*Common name (1)*	*Weight of adults in kg (1)*	*Home range in km^2*	*Average troop size*	*Largest troop seen*	*Size of feeding aggregations or subgroups*	*Reference (see below)*
Callitrichidae								
Cebuella	(1 species)		0.1?		3.5?		Same as family?	
	pygmaea	Pygmy marmoset						
Callithrix	(8 species)		0.4		3–8	8	Same as family?	(1)
	jacchus	Common marmoset						
	argentata	Silvery marmoset						
	geoffroyi	Geoffroy's marmoset						
Saguinus	(22 species)		0.5					
	fuscicollis	Brown-headed tamarin						
	midas	Golden-handed tamarin		0.10+	3.4	6	2 to 7	(2)
	tamarin	Black tamarin		0.10+	3.4	6	2 to 7	(2)
	geoffroyi	Rufous-naped tamarin			3.4	9	3 to 13	(3)
	oedipus	Cotton-top tamarin			3.5?		Same as family?	
Leontopithecus	(3 species)		0.6					
	rosalia	Golden lion tamarin			3.5	8	up to 16	(4)
Callimico	(1 species)							
	goeldii	Goeldi's marmoset or monkey	0.4		3.0?		Same as family	
Cebidae								
Aotus	(1 species)							
	trivirgatus	Night or owl monkey	1.0		3.0?		2 or more families	(5)
Callicebus	(3 species)		0.7					
	moloch	Dusky titi		0.004	3.4	4	Same as family	(6)

Pithecia	(2 species)		1.6		3.4?	5	Same as family	(1)
	monachus	Monk saki						
Chiropotes	(2 species)	Bearded sakis	3.1					
Cacajao	(3 species)							
	rubicundus	Red uakari						
Saimiri	(2 species)		1.1					
	sciureus	Common squirrel monkey		0.15	18–22	300	5 to 8	(7)
	oerstedii	Red-backed squirrel monkey		0.18	23	27		(8)
Cebus	(4 species)		3.3					
	capucinus	White faced capuchin		0.86	15–20	30+	Same as troop	(9)
	nigrivittatus	Weeper capuchin				33	Same as troop	(10)
	albifrons	White-fronted capuchin						
	apella	Brown or tufted capuchin	3.5		8–18			(11)
Alouatta	(5 species)		7.4					
	villosa	Mantled howler		0.24+	16	40	Same as troop	(12, 13)
	seniculus	Red howler		0.04	6.8	16	Same as troop	(13)
Ateles	(4 species)		6.9	3.00+	18.5			(14)
	geoffroyi	Black-handed spider monkey		0.60	12–20	33	3 to 17	(15)
	fusciceps	Brown-headed spider monkey					Smaller than troop?	(16, 17)
	belzebuth	Long-haired spider monkey					Smaller than troop?	
Brachyteles	(1 species)	Woolly spider monkey	10.0					
Lagothrix	(2 species)		10.0		15–25+			(1)
	lagothricha	Brown woolly monkey						

(1) Napier and Napier, 1967; (2) Thorington, 1968b; (3) Muckenhirn, 1967; (4) Coimbra-Filho and Mittermeier, 1973; (5) Moynihan, 1964; (6) Mason, 1966, 1968; (7) Baldwin, 1968; Thorington, 1968a; (8) Baldwin and Baldwin, 1972; (9) Oppenheimer, 1969a; (10) Oppenheimer and Oppenheimer, 1973; (11) Kühlhorn, 1939; (12) Chivers, 1969; (13) Neville, 1972a; (14) Durham, 1971; (15) Eisenberg et al., 1972; (16) Carpenter, 1935; (17) Klein and Klein, 1971.

than a high level of gregariousness (Moynihan, 1967) that keeps the troop together.

In the capuchins *(Cebus),* troop size averages about twenty (Table 1), but varies depending on the stage in the population growth cycle (Oppenheimer, unpublished data). Adult males are part of the troop structure and receive grooming from all other members of the troop; females also groom each other and their young (Oppenheimer, 1969b). Adult females determine much of what the troop does and may even initiate troop splitting and formation (Oppenheimer, 1969a). The adult male or males interact with their counterparts of other troops during intertroop encounters. Agonistic behavior within a troop is rare, but may occur as an outgrowth of play or during the weaning period. Subgroup formation, which eventually leads to troop division, is probably a result of decreased affiliative interactions, rather than an increase in agonistic ones (Oppenheimer, 1968, 1969a, and unpublished data).

Howler monkeys *(Alouatta)* live in troops of ten to twenty individuals (Table 1), with twice as many adult females as adult males (Carpenter, 1965; Chivers, 1969; Neville, 1972a). They are the only New World primates that are solely vegetarian and are adapted to eating leaves. They rest much of the day, probably in order to digest the vegetable matter, and usually move only short distances (Richard, 1970). They have, particularly in the males, an enlarged saclike hyoid apparatus that acts as a resonating chamber (Schön, 1971), which allows production of vocalizations used to achieve intertroop spacing (Altmann, 1967; Chivers, 1969). In *A. villosa* the amount of time devoted to social interactions within the troop is low, primarily because so much time is spent resting (Richard, 1970). More social interactions, including allogrooming, occur in the red howler, *A. seniculus* (Neville, 1972b).

The spider monkeys *(Ateles)* live in troops of up to thirty-three individuals (Carpenter, 1935). The troops break up into subgroups of three, four, or more individuals during the day; the subgroups vary in composition from all male, to adult females and young, and/or a combination of both (Carpenter, 1935; Klein and Klein, 1971; Eisenberg and Kuehn, 1966). Eisenberg and Kuehn (1966) observed that the adult male of an *A. geoffroyi* group released on Barro Colorado Island spent only seven of twenty-six days with the adult females. This adult male, and some of the young males born after the release, spent many daylight hours traveling as a subgroup with a capuchin troop. Sometimes just before dusk he would give long calls that might bring the females to him (Oppenheimer, pers. obs.). The young of *Ateles* stay close to their mothers until two years of age and may continue to be nursed and carried by their mothers during this period. It may be that both the long association between mother and young and the tendency of males to attack females (Klein and Klein, 1971) influence the formation of separate subgroups.

Tactile Communication

The importance of tactile communication in primates, as well as in other animals, has been indicated by the intensive research of the Harlows and their colleagues on rhesus monkeys (Harlow, Harlow, and Suomi, 1971). The opportunity to receive positive feedback through tactile communication with the mother allows the young to develop normal patterns of social behavior and the requisite physiological responses that underlie them. The behavioral units of tactile communication are usually characterized in terms of their visual components, rather than in relation to the message conveyed. This is probably because a number of tactile behavioral units may occur under similar circumstances, and the difference in message content between them, if any, is not immediately obvious to the observer.

See, for instance, the paper by Maurus and Pruscha (1973), where cluster analysis is used to determine which behavioral units are similar.

The smaller species—the marmosets, tamarins, and the night monkey *(Aotus)*—probably all sleep in tree holes or crotches of trees at night (Table 2). In such small spaces, probably all members of the family sleep in contact with one another. Moynihan (1967) has suggested that this passive contact may fulfill, at least in part, the social function of allogrooming (see below). He suggested this because his initial observations of *Aotus* and *Saguinus geoffroyi* indicated that they did very little allogrooming; however, more recently (Moynihan, 1970) he has observed that individuals of *Saguinus geoffroyi* do engage in social allogrooming. Thus the social function of passive contact while sleeping is now questionable and would be best studied under laboratory conditions. Probably all species, large or small, seek contact with specific members of their family or troop while resting or sleeping. Such contact, which is actively allowed, probably does play a role in developing cohesiveness in the group.

Huddling is related and may be of shorter duration than the above behavior, but the relative positions of the individuals and the extent or type of contact have not been adequately described in the literature (Table 2).

Hugging or embracing is a much more specific type of contact, which has different forms and functions among the different species. Probably in all species it is unilaterally performed by a mother to her infant; it has been described for *Alouatta villosa* as cuddling (Baldwin and Baldwin, 1973a). Mutual cuddling, or hugging, has been observed in three species (Table 2). It serves as a greeting or contact-promoting behavior that tends, at least in *Ateles,* to reduce agonistic tendencies (Klein and Klein, 1971). It is an adult behavior pattern in *Callimico* (Lorenz, 1972) and *Ateles* (Eisenberg and Kuehn, 1966) and an infant behavior pattern in *Cebus capucinus* (Oppenheimer, 1968, 1973). In *Ateles,* and possibly in *Cebus,* one of the individuals sniffs the sternal gland of the other. Thus this behavior pattern has both tactile and olfactory components and will be discussed further in the next section.

Carrying is primarily a parental behavior pattern. In the monogamous species—the marmosets, tamarins, and the owl and titi monkeys (Table 1)—the male parent does most of the carrying (Eisenberg and Kuehn, 1966; Epple, 1975; Heltne et al., 1973; Snyder, 1972), but juveniles may also eagerly carry young (Kleiman, pers. comm.). Although this behavior on the part of the male may have been selected for because marmosets and tamarins usually give birth to twins (Epple, 1975), the behavior also occurs in *Aotus, Callicebus,* and *Callimico,* where the female usually gives birth to only one infant at a time. It is more likely that the monogamous social structure of these species has selected for behavior patterns that strengthen the cohesiveness of the family unit, i.e., paternal carrying. Capuchin *(Cebus)* infants are often carried by juveniles during play, and on rare occasions adult male capuchins and adult male spider monkeys will carry an infant for short periods (Oppenheimer, pers. obs.; Eisenberg, 1976). After a howler monkey *(Alouatta)* female has been shot, an adult male will retrieve and carry the infant (Carpenter, 1934). While two individuals are in such intimate contact, other signals may well occur. Captive capuchin infants have been reported to squeeze or knead the skin of their carriers when they were disturbed (Oppenheimer, 1973). This has also been observed in free-living *C. capucinus* (Curt Freese, pers. comm.).

While riding on the back of another individual, an infant usually curls its tail around the base of the carrier's tail, as in *Cebus* (Oppenheimer, 1968), or the body, as in *Saimiri* (Baldwin, 1969), in order to secure its position. When giving assis-

Table 2

Tactile communication

	(1) *Cebuella pygmaea*	(2) *Callithrix jacchus*	(3) *Callithrix geoffroyi*	(4) *Saguinus fuscicollis*	(5) *Saguinus geoffroyi*	(6) *Leontopithecus rosalia*	(7) *Callimico goeldii*	(8) *Aotus trivirgatus*
Sleep in contact	In tree hole				In tree hole?			In tree hole
Huddle				Rest in contact				
Hug							M and F S, PC	
Carry	Infant on M	Infant on M	Infant on M	Infant on M	Infant on M	Infant on M	Infant on M	Infant on M
Tail contact					With body, PB			M and F tail twine
Allogroom	X?	Use tongue and teeth: S, C		C	PB, C, S (M to F)	All; S, C	S	PC, side by side
Touch mate or partner with hand								
Nuzzle								
Lick, manipulate, touch genitalia							F to M S	
Female mount and thrust					F to F		On human, S	
Male mount and thrust					M to F; S	M to F; S	M to F; S	S, 2 min
Wrestle (grapple, push, pull)		?A	Play	?A	A	Play	Play	M to M, A; F to F, A
Tag								
Hit				?A	A		M to F, D; F to M, A	A
Kick with feet					A			
Hip thrust								
Chest to back contact								
Mouth; warning, sham bite								
Bite		?A		?A	A		M to F; D	A

Note: A = Agonistic, C = Contact promoting, D = Dominance, F = Female or Mother, G = Greet or appease, M = Male, PB = Strengthen pair bond, PC = Precopulatory, S = Sexual, T = Territorial.

(1) Moynihan, 1967; Napier and Napier, 1967; (2) Epple, 1967; Moynihan, 1970; (3) Snyder, 1972; (4) Epple, 1972b, 1975; (5) Epple, 1967; Moynihan, 1970; (6) Snyder, 1972; (7) Lorenz, 1972; Snyder, 1972; Heltne et al., 1973; (8) Moynihan, 1964,

	(9)	(10)	(11)	(12)	(13)	(14)	(15)	(16)	(17)	(18)
	Callicebus moloch	*Saimiri sciureus*	*Saimiri oerstedii*	*Cebus capucinus and nigrivittatus*	*Cebus albifrons*	*Cebus apella*	*Alouatta villosa*	*Alouatta seniculus*	*Ateles geoffroyi*	*Lagothrix lagotricha*
		Infant on F		Infant on F					Young on F	Young on F
		Infant play	Yes							
				Infants			F to Infant, G		G, sniff sternal	G, sniff sternal
	Infant on M	Infant on F		Infant on F			Infant on F		Infant on F	Infant on F
	M and F tail twine	Infant on F		Infant on F			Infant on F		On body or tail	
	M and F; C	F to Infant		All; C	Yes	Yes	Rare	F does most, C	Adults, F to Infant	Adults, F to Infant
	During T dispute M and F	Play, threat	C	Assist during A	C; assist during A		Play, C	F to Infant on F	C	
							F to Infant	M and F perineum		
	M to F	F to M	M to F				M to F, F to M	M to F	M to F, S; F to M, C	
		Play	Play	Play and S					Modified "mount"	
	S	Play and S	S; 20–70 sec	Play and S			S, 20–45 sec		S, 5–10 min	
	Push, T dispute	Play; pull, A	Push, threat	Play, A			Play; pull, A	Play	Play, S, PC	Play
				Mild A			Play, mild A	Play, A	A, Play	
	T dispute			Mild A			Admonish, mild A	Play	Mild A threat A	Mild A
		Medium threat								
		High threat		Assist during A	Assist during A				F sit in M lap, S	
		Play	Admonish	Play admonish			M to F, G; play, mild A	Play, A	Play and mild A	Play and mild A
	T dispute	A		A		A	A		A	A

1966; (9) Mason, 1966; Moynihan, 1966; (10) Baldwin, 1968, 1969; Latta et al., 1967; (11) Baldwin, 1968; Baldwin and Baldwin, 1972; (12) Oppenheimer 1973; Oppenheimer and Oppenheimer, 1973; (13) Bernstein, 1965; (14) Nolte, 1958; (15) Altmann, 1959; Baldwin and Baldwin, 1973a, 1976; Bernstein, 1964; Carpenter, 1934; (16) Neville, 1972b; (17) Eisenberg and Kuehn, 1966; Klein, 1971; Klein and Klein, 1971; (18) Williams, 1967.

tance during an agonistic encounter in *Ateles,* the assister will wrap its tail around the assistee's body or tail (Klein and Klein, 1971). In *Saguinus geoffroyi,* one adult may loop its tail over its partner's body while they are sitting together; this additional type of contact may help to strengthen the pair bond (Moynihan, 1970). Ritualized tail twining occurs between paired adults in *Aotus* and *Callicebus* (Moynihan, 1966) and would seem to function as a positive tactile communication.

In monogamous species, allogrooming is interpreted as being, in part, sexually motivated, because it occurs primarily between the two adults, who are of opposite sex, and may be associated with sexual activities between these two individuals (Table 2). In *Aotus* allogrooming specifically occurs just prior to and after copulatory interactions, when the mates are sitting side by side in contact; however, in addition to its proposed function to stimulate sexual motivation, it still plays its general role of reducing hostility between the mates (Moynihan, 1964). In those species where there are two or more adult females in the social unit, allogrooming tends to be primarily a contact-promoting behavior, which helps to set up and maintain the social structure of the troop. Adult males may receive a lot of grooming attention from the other troop members, but most grooming is done by the adult females and is directed to specific other adults, or to young, particularly newborn infants (Oppenheimer, 1969b). Allogrooming in *Ateles* appears to be a means of maintaining a relationship once it has been established (Eisenberg and Kuehn, 1966).

In addition to the species mentioned in Table 2, allogrooming has been observed in both species of *Pithecia* and in *Cacajao rubicundus;* but it has not been observed in *Cebuella pygmaea,* where captive conditions may well inhibit such activity (Moynihan, 1967). Allogrooming tends to be rare in *Saimiri,* where it is primarily a maternal activity, and in *Alouatta,* both of which genera have evolved specialized precopulatory displays (Moynihan, 1967). It is interesting to note that social allogrooming is more frequent in *A. seniculus,* which has a less highly evolved precopulatory tongue display, than does *A. villosa* (Neville, 1972b). It may also be that in *Alouatta* the vocal interactions between troops act as an outside force that contribute to intratroop cohesiveness, and this in turn lessens the need for social allogrooming. Put another way, selection in *Alouatta* may have led to a "social" strategy, where energy is expended in repulsing individuals outside the troop, rather than expended by attracting individuals within the troop, but both strategies would result in intratroop cohesiveness. My personal observations of *A. seniculus* in Venezuela suggest that its intratroop vocal interactions are less frequent than those of *A. villosa* on Barro Colorado Island in Panama, a condition that could explain the different frequencies of social allogrooming in the two species. More detailed discussions of allogrooming can be found in Moynihan (1967) and Sparks (1967).

Touching another individual with the hand seems to reassure one or both individuals (Table 2). In *Callicebus* and *Cebus* it is used to reassure the partner while both threaten a third party. In *Cebus* the hand touch usually includes draping the arm around the partner's shoulders and may shift into chest-to-back contact (Fig. 1; Oppenheimer, 1968, 1973). Touching with the hand may also occur during play, or when an immature *C. albifrons* approaches an adult male and touches the male's face (Bernstein, 1965). In *Saimiri sciureus* touching may occur during play, but a heavier hand touch or possibly a grasp, described as "putting hand on," occurs when a male threatens another with a genital display (Baldwin, 1968). In this case, the hand is probably used to support the displayer as well as to hold the second party in an appropriate position (see genital present below).

Placing the face against another and moving

Fig. 1. *Cebus capucinus* assistance position showing adult female with chest in contact with adult male's back. Usually the dominant, i.e., the male, is on top. Female making medium-intensity (only lower worn canines exposed) open-mouth-bared-teeth face threat toward observer. Male in display posture, frowning and staring at observer. Male's face square and forehead almost naked with some very short white hairs. Female's face vertically oval with long white and dark hairs on forehead. Photo taken of members of troop A_1 (Oppenheimer, 1968, 1969) on Barro Colorado Island, Panama Canal Zone.

it back and forth, nuzzling, has been reported for three species (Table 2), but probably is more widespread. It tends to be either a maternal or an adult sexual act.

Licking, touching, and manipulating the genitalia of the opposite sex (Table 2) usually follow sniffing, which will be discussed later (Table 3). Whereas sniffing has been reported for most New World species, contact between the tongue or hand and the genitalia has only been reported for the Cebidae and for *Callimico,* which is an intermediate form that is sometimes placed in this family (see Napier and Napier, 1967:371). If this difference does exist, it may indicate an underlying hormonal-pheromonal and/or neurological difference between the two New World families. In some species only one sex, either the male or the female, performs the behavior, whereas in other species both sexes are reported to do it. It is characterized as a sexual or contact-promoting behavior and probably causes an increase in sexual motivation when the female is in estrus.

Female mounting and thrusting tends to be primarily a play or homosexual activity (Table 2); however, a *Cebus capucinus* female has been observed in nature to alternate her mounting and thrusting with that of a male (Oppenheimer, 1968, 1973).

Adult male mounting and thrusting obviously occurs in all species, even though observations are lacking for some. Among juveniles it occurs during play, but among adults it is sexually motivated. Bouts of thrusting between males and estrous females may last from twenty seconds to ten minutes (Table 2). Subadult male capuchins *(C. capucinus)* have been observed to mount and thrust upon each other, and on adult female *Ateles geoffroyi* in the wild on Barro Colorado Island (Oppenheimer, unpublished data; Richard, 1970). The use of visual sexual displays in dominance interactions does occur in some species (see below, Table 4), and thus may preclude the use of a tactile sexual behavior for this purpose. Possibly one could view the combined "put hand on" and genital display of a male *Saimiri* threatening another (Baldwin, 1968) as a ritualized, redirected, intention movement to mount, but it seems more likely that this combined display has evolved for the purpose of transmitting an olfactory signal.

Wrestling, which includes grappling, pushing, pulling, and mouthing, occurs primarily during play, but also during agonistic encounters, where mouthing turns to biting. The information obtained during play from performing these behavioral acts allows an individual to learn the extent of his or her abilities, whereas the information received from immediate feedback, in terms of the partner's response, places general, as well as specific, limits to the expression of these acts and helps to establish the social structure of the group (Carpenter, 1934). See Loizos (1967) for a detailed discussion of play behavior, and Baldwin and Baldwin (1973b) for a discussion of the role of play in squirrel monkeys. In many species these play activities are most obvious in the young, particularly the males, but they disappear as the play gets more intense (Baldwin, 1969; Carpenter, 1934). Vocal and visual displays then become the main channels of low- to medium-intensity communication, at least among the males, whereas tactile communication in the form of allogrooming may become more prominent among the females. Agonistic wrestling has been observed in a number of the species, particularly in adults (Table 2); however, in adult spider monkeys *(Ateles)* these behavior patterns retain their play aspects during greeting interactions (Klein, 1971). Pushing and pulling in some species have become semiritualized forms of threat (Table 2).

Tagging, a repeated behavior that is intermediate between touching and hitting, is reported to occur in howlers *(Alouatta)* during play (Bernstein, 1964) and during agonistic encounters (Neville, 1972b). It has been observed in *C. capucinus,* particularly when a juvenile is harassing an adult female (Oppenheimer, 1973).

Hitting is a frequently observed agonistic behavior pattern among the species (Table 2), but it has also been reported to occur during play in *Alouatta seniculus* (Neville, 1972b). Kicking, another agonistic act, has only been reported in two species, and hip thrusting in one (Table 2).

Chest-to-back contact, in contexts other than mounting and carrying, has only been reported for four species of cebids. It was observed incidentally in *Ateles* when a captive adult female solicited mounting from a male by sitting in his lap (Eisenberg and Kuehn, 1966). It functions as an extreme threat to the individual on the bottom in *Saimiri sciureus* when it follows "put hand on" during a genital display and hip thrusting (Baldwin, 1968). In *Cebus* (Fig. 1) it occurs as an assistance behavior when one individual reinforces the threat of another toward a third party (Bernstein, 1965; Oppenheimer, 1973; Oppenheimer and Oppenheimer, 1973). In other species assistance behavior may only involve touching with the hand or tail (Table 2). In *Saimiri sciureus* contact involves threat so that assistance takes the form of merely standing side by side while performing threatening genital displays toward the third party (Baldwin, 1968). Noncontact assistance is the mildest form of assistance in other species, such as *C. capucinus* (Oppenheimer, unpublished data) and *C. nigrivittatus* (Oppenheimer and Oppenheimer, 1973).

Although biting, the ultimate agonistic act, has been reported for most species, mouthing, warning, and sham bites have only been reported for the Cebidae (Table 2). Most of the mouthing occurs during play, whereas the warning or sham bites tend to be mild agonistic forms.

Olfactory Communication

Two recent reviews of scent marking in mammals (Eisenberg and Kleiman, 1972; Ralls, 1971) covered a number of points that are relevant here. Feces, urine, and secretions from specialized skin glands do play a role in social communication. The resulting odors or pheromones are used to mark an area, to facilitate individual or group recognition, to indicate the sexual status of females, and to inhibit the sexual development of other members of the group; a single odor can have one or more functions. Releasing of the scent, or scent-marking behavior, can be brought about by the approach of a subordinate individual or one that is not accepted, such as a stranger; and the scent can function to increase the confidence of the scenter by surrounding the scenter with an odor familiar to it, as well as to threaten or make less confident nearby individuals by disrupting the odor field around them (Eisenberg and Kleiman, 1972). More specifically, for the platyrrhines the olfactory system can be used to communicate information such as the "sex, age, social rank, territory, and more detailed biological facts" (Epple and Lorenz, 1967).

In some cases special movements have been evolved for distributing the scent (Ralls, 1971), and it is these visual components of the behavior or the movements of the recipient of the odor that are most obvious to the human observer and which are primarily indicated in Table 3. The absence of visual components of marking behavior is not proof that olfactory communication is lacking in the species. In my review of the literature only two species appear to lack the specialized behavioral components (Table 3), *Pithecia monachus* and *Cacajao rubicundus* (Moynihan, 1967), but the behavior of neither species has been studied intensively.

Sternal glands and/or the related gular and epigastric glandular areas have been found in almost all the New World species, including *P. monachus* and *C. rubicundus* (Epple and Lorenz, 1967), and use of these glands has been documented in some. Epple and Lorenz (1967) have indicated the morphological differences between the species and have found that the glandular area of the Callitrichidae (excluding *Leontopithecus*) and *Aotus* is situated "at the articulation of the sternum and claviculae," a brush of hairs just caudad to it, and caudad of the brush "a ribbon-like field of nearly naked glandular skin. . . ." The sternal glands seem to reach their fullest development in the dominant males. For other details see Epple and Lorenz (1967). Some species also have suprapubic and circumgenital scent glands (Epple, 1971, 1972a). Epple (1972a) has reviewed the olfactory behavior in marmosets and tamarins and her paper should be consulted.

Although most of the information to be presented below will deal with the visual components of marking behavior, two studies of the ability of platyrrhines to discriminate between odors need to be mentioned. Epple (1971) presented adult *Saguinus fuscicollis* of both sexes with wooden perches that had been marked with urine and glandular secretions of a separate set of adults. The nine test animals spent the same amount of time sniffing the male and female marked perches, but marked with their own urine and glandular secretions those perches having the odor of strange males more frequently than the perches having the odor of strange females. Thus they were able to discriminate between the odors of the two sexes. In addition, the most aggressive animals showed the most interest in the perches, whereas the less aggressive tended to avoid marked perches. Another study with infant *Saimiri sciureus* four to twelve weeks of age showed that they preferred surrogate mothers having their own odor on them to clean surrogate mothers, and were able to distinguish their own odor from that of another. Thus odors can be important to squirrel monkey infants and may help them to distinguish their mothers from other individuals (Kaplan and Russell, 1974).

Release of feces and urine are basically autonomic responses elicited by fear (Eisenberg and Kleiman, 1972). They have been reported to occur in four species of cebids. In addition to urinating and defecating when frightened, *Aotus* has been observed to vomit (in captivity) under the same conditions, but none of these behaviors were considered a signal (Moynihan, 1964). Re-

Table 3

Visual components of olfactory communication

	(1)	(2)	(3)	(4)	(5)	(6)	(7)	(8)	(9)
	Cebuella pygmaea	***Callithrix jacchus***	***Callithrix argentata***	***Saguinus fuscicollis***	***Saguinus geoffroyi and oedipus***	***Leontopithecus rosalia***	***Callimico goeldii***	***Aotus trivirgatus***	***Callicebus moloch***
Defecate								Fright, not signal	
Urinate								Fright, not signal	
Urine rub or wash								Yes	
Anal-genital rub and urinate on tail					F, (M), S		F, S		
Female sex attractant		Urine or scent		Urine or scent	Urine or scent		Urine or scent		
Anal-genital display: urine and/or scent	M and F threat	M and F threat	M and F threat						
Anal-genital rub on substrate	M and F	M and F; S, D, T	M and F	M and F; S, D, A	Sit-rub; S, D, Fr	M and F	M and F	M and F	A, not signal
Suprapubic rub				Pair-bonding	Pull-rub D	Yes			
Rub or squeeze chest gland	X	OS	OS	OS	?	Mostly M, T			OS, T
Rub neck gland on substrate									
Sniff sitting spot and/or urine spot		M and F; S	M and F; S	M and F; S	M and F; S	M, S			
Sniff or lick genitalia		M and F; S		M and F; S	M and F; S		M to F; S	M and F; precopulation	Mutual
Sniff genitalia during display	Sub. to Dom.	Sub. to Dom.	Sub. to Dom.		Sniff stranger				
Sniff sternal gland									
Sniff female's body				M; S					
Sniff armpit								Yes	
Sniff face		Stranger's face			Stranger's face			M to F precopulation; stranger's face	

NOTE: A = Autogroom, D = Dominance, Dom = Dominant, F = Female, Fr = Frustration, M = Male, OS = On substrate, S = Sexual, Sub = Subordinate, T = Territorial, X = Stated not to occur.

(1) Epple, 1972a, 1975; Moynihan, 1967; (2) Epple, 1967, 1972a, 1975; (3) Epple, 1972a; (4) Epple, 1971, 1972a, 1975; (5) Epple, 1967, 1972a, 1975; Moynihan, 1970; Muckenhirn, 1967; (6) Epple, 1972a, 1975; Epple and Lorenz, 1967; Snyder, 1972; (7) Epple, 1975; Lorenz, 1972; (8) Hill, 1960; Moynihan, 1964; (9) Mason, 1966; Moynihan, 1966, 1967;

(10)	(11)	(12)	(13)	(14)	(15)	(16)	(17)	(18)	(19)	(20)
Pithecia monachus	*Cacajao rubicundus*	*Saimiri sciureus*	*Saimiri oerstedii*	*Cebus capucinus* and *nigrivittatus*	*Cebus albifrons*	*Cebus apella*	*Alouatta villosa*	*Alouatta seniculus*	*Ateles geoffroyi* and *fusciceps*	*Lagothrix lagothricha*
				X		Strong excitement	Excitement		Alarmed	Alarmed
				X		OS, identity	Excitement F to M, S	Marking ?	Alarmed	Alarmed
X	X	M and F	F	Assert identity	Assert identity		Excitement, S		X	
							M sniff F urine		M drink F urine	
		M urinates; T, D, S	M							
X	X	Rare					F, S, mark?	A		
X	X	M to F, urine rub		?	Yes	OS, with threat	OS, comfort, mark?		Assert identity	Assert identity
						Urinate OS, and rub OS		Prior to approach		
			M lick urine	Yes (pers. obs.)			M to F		M sniff F urine	
		M to F	M to F				M to F; F to M	M to F; S	M to F	M to F
									Juv. M to adult M	
				During hug?					During embrace	During embrace
									During embrace	During embrace
							F lick M precopulation			

(10) Moynihan, 1967; (11) Moynihan, 1967; (12) Baldwin, 1968; Latta et al., 1967; Moynihan, 1967; (13) Baldwin, 1968; Baldwin and Baldwin, 1972; (14) Oppenheimer, 1973; Oppenheimer and Oppenheimer, 1973; (15) Bernstein, 1965; (16) Dobroruka, 1972; Moynihan, 1967; Nolte, 1958; (17) Altmann, 1959; Carpenter, 1934; Collias and Southwick, 1952; (18) Neville, 1972b; (19) Carpenter, 1935; Eisenberg and Kuehn, 1966; Klein and Klein, 1971; Moynihan, 1967; (20) Epple and Lorenz, 1967; Williams, 1967.

lease of urine and feces in the other three species is triggered by strong excitement or alarm (Table 3). It certainly could be debated whether urinating and defecating are ritualized in these circumstances, but just as certainly their odoriferousness makes it reasonable to assume that their occurrence is noticed by nearby individuals and that their predictability would allow a distress or presence-of-disturbance signal to be conveyed.

Ritualized urine washing or rubbing, where the urine is applied to the palms and rubbed on the soles, and in *Cebus capucinus* (Oppenheimer, 1973) to the tip of the tail, is also reported to occur only in the Cebidae (Table 3). In *Saimiri sciureus* adult males and females urine rub after doing genital displays during a sexual sequence (Latta et al., 1967; see also Castell and Maurus, 1967). In *Cebus capucinus* and *C. nigrivittatus* it is most frequently the adult males that urine rub, usually when they have been disturbed by a conspecific or a human observer, and it seems to have both self-reassuring and identifier functions (Oppenheimer, 1968, 1973; Oppenheimer and Oppenheimer, 1973).

A somewhat similar behavior of rubbing substances into the fur has also been noted for free and captive individuals of *C. capucinus* (Oppenheimer, 1968) and captive *C. apella* (Nolte, 1958). In captivity a wide range of strong-smelling substances, like vinegar, onions, and tobacco or tobacco smoke, can elicit rubbing and are rubbed into the fur. In the free-ranging *C. capucinus* such behavior was rare, but on two occasions strong-smelling plants were vigorously rubbed into the fur on all parts of the body. In one situation several individuals did it together, and in the other it was performed by a female that was alone (except for the observer) at the time (Oppenheimer, 1968). In terms of the Kleiman hypothesis (Eisenberg and Kleiman, 1972:26) these individuals rubbed to reestablish their optimum odor field, or possibly to create a supraoptimum one, which gave them a sense of security.

Estrous females of two species, *Saguinus geoffroyi* (Epple, 1967) and *Callimico goeldii* (Lorenz, 1972), urinate on their tails, which they curl up between their legs. Moynihan (1970) did not observe the urine wetting, but did observe tail coiling prior to and in between mounting and thrusting by a male (see visual displays below). This display could be used to help disperse pheromones that indicate that the female is in estrus and receptive. The urine of estrous females is reported to be a sex attractant in a number of species and probably is for all (Table 3).

Genital displays, including release of urine or a glandular secretion, are reported for members of both New World families. The marmosets *Cebuella pygmaea, Callithrix jacchus, C. argentata,* and *C. geoffroyi,* all lift their tails and expose the anal-genital area when threatening a subordinate who approaches and sniffs; adults of both sexes perform this display (Epple, 1972a). Adult male *Saimiri* present their genitalia frontally by lifting one leg away, and release a small amount of urine. This may occur as a threat or after licking a female or after a female has urinated or urine rubbed (Baldwin and Baldwin, 1972). When used as a threat the dominant male will approach, display, and place his hand on the subordinate's head, which apparently forces the subordinate to sniff the dominant's genitalia (Baldwin, 1968).

Rubbing of the anal-genital area on the substrate is primarily a behavior pattern of the marmosets and tamarins, though it has been observed in the cebids as well (Table 3). Such rubbing is usually done while in a sitting position, which would place the circumgenital and circumanal glands in contact with the substrate or a conspecific (Epple, 1972a). This type of behavior is elicited in sexual, dominance, threatening, and frustrating circumstances. It is rare in *Saimiri* (Moynihan, 1967) and is thought to be a type of autogrooming or cleaning behavior in *Alouatta seniculus* (Neville, 1972b) and *Callicebus moloch,* though on one occasion an adult *Callicebus* rubbed while carrying an infant (Moynihan, 1967).

Pull rubbing (Moynihan, 1970) or rubbing the suprapubic glands on a substrate occurs in *Saguinus* and *Leontopithecus* and appears to help strengthen the pair bond or to assert dominance. It may occur along with anal-genital rub and with chest rubbing (Epple, 1972a).

Chest rubbing or rubbing of the sternal and/or epigastric gland(s) occurs in many of the platyrrhines (Table 3). It is usually done by the dominant male and is interpreted to be a threat in some species and an assertion of the individual's identity in others (see the Kleiman hypothesis above). The marmosets and tamarins seem to rub the gland directly on the substrate; the cebids may do this, but they usually rub the gland with a hand or foot. *Callicebus* may squeeze the gland with a hand, as well as rub it (Moynihan, 1966), and *Saimiri* may touch it with a foot (Baldwin, 1968). In addition, three cebid species mix the sternal gland secretions with either urine or saliva. The fatted males (in sexual condition) of *Saimiri sciureus* urine rub and then lift the wet foot to the sternal gland. This behavior usually occurred when a male was about to approach a group of females who were likely to chase him away (Baldwin, 1968). *Lagothrix* (Epple and Lorenz, 1967) and *Ateles* (Klein and Klein, 1971) both rub their chests with their hands and then may raise the hand to the mouth for saliva, after which they continue to rub the chest. In some cases *Ateles* drools saliva onto the chest or onto the object being rubbed on. This behavior was also elicited in *Ateles* by mildly or strongly stressful situations or exposure to celery or green onions (Klein and Klein, 1971), which suggests a similarity to urine rubbing and rubbing of plants into the fur, as occurs in other species, such as *Cebus capucinus* (see above).

There are conflicting reports as to whether certain species chest rub. Epple (1972a) has observed it in *Saguinus geoffroyi,* but Moynihan (1970) has not. Epple and Lorenz (1967) have observed it in three captive *Cebus capucinus,* but I have never observed it in the wild; the only indication of functioning sternal glands that I observed was when an infant dropped its head to the chest of another infant during a hug, and even this behavior was rare. Similar sniffing occurs frequently in *Ateles* during embraces (Eisenberg and Kuehn, 1966). Moynihan (1967) had not observed chest rubbing in *Ateles geoffroyi,* but others have (Eisenberg and Kuehn, 1966; Klein and Klein, 1971; pers. obs.).

Alouatta seniculus, which has a gular scent gland (Epple and Lorenz, 1967), has been observed to rub this gland on the substrate prior to approaching a conspecific (Neville, 1972b). *Cebus apella,* which lacks a gular gland, has been observed to urinate on an object and then to rub its neck on the object (Dobroruka, 1972).

All these glands, as well as urine and urine-marked spots, are sniffed by conspecifics. In some cases after sniffing the conspecific may mark (Epple, 1971). Sniffs are also directed at the body in general, at the armpit and at the face (Table 3). Moynihan (1964) notes that *Aotus* has large apocrine and sebaceous glands on either side of the nose. Since adult *Aotus* sniff each other's faces just prior to copulating, it seems likely that these glands also function in olfactory communication.

The genitalia of the platyrrhine species are quite varied in external morphology. One of the most striking, and initially deceiving, forms is the long pendulous clitoris of the *Ateles* female. Klein and Klein (1971) have suggested that this elongated clitoris, which tends to retain a small amount of urine, has been evolved to mark the environment wherever it touches, and that this in turn makes it easier for the male to locate the female. Also, when males make contact with the clitoris with their hands they are more likely to obtain a urine sample on their fingers, which they can then smell and lick.

In addition to the information conveyed over the short term by olfactory communication, there are hints of long-term effects or responses. Baldwin (1968) suggests that hormonal activity in submissive male squirrel monkeys *(Saimiri)* might be inhibited by pheromones released in the dominant male's urine during a penile display, which the submissive male is forced to inhale. The presence of a dominant male *Callimico goeldii* retards sexual development in other males (Lorenz, 1972). Similarly the presence of a dominant female marmoset may prevent other females from producing young, even though copulation takes place. Once removed from the presence of the dominant female, these females may produce young (Epple, 1975). See Eisenberg and Kleiman (1972) for the pertinent details on olfactory inhibition of reproduction in rodents.

In summary, it appears that most platyrrhines have at least two, and some have as many as four, types of olfactory display. These displays have sexual, alarm, threat, and self-identification functions. They may also have physiological effects on the hormonal systems of other group members. It is hoped that the isolation of specific substances and the elucidation of their activity will occur soon. But even if it does we still need more observations and quantitative analyses of olfactory communication.

Visual Communication

Visual communication can be divided into three categories: (1) postures, and movements of part or all of the body; (2) piloerection displays; and (3) facial expressions (Moynihan, 1967). In some situations all three types can occur as part of a complex display, such as when directing a threat display toward a predator or a human. Moynihan (1967, 1970) has reviewed the occurrence of gestural and piloerection displays in a number of New World species, and Andrew (1963) and van Hooff (1967) studied the facial expressions of some platyrrhines in their studies of communication in primates.

GESTURES

All species have "look at" and most likely "look away" behavior patterns, though it is unritualized in its general form. Prolonged looking at or staring is usually interpreted as a threat, but it should be regarded as an expression of interest that may have a number of different motivations, including aggression. These motivations are expressed by additional facial expressions, body gestures, or acoustic signals. Looking away can indicate alarm or a desire to avoid contact. *Aotus* and *Callicebus* have a specialized form, which involves lowering the head (Moynihan, 1964, 1966). Repeated looking at or glancing and looking away (at another monkey or a human) in *Cebus capucinus* usually induces approach and assistance by a third, usually more dominant, individual during an agonistic interaction (Oppenheimer, 1973); a dominant male will look away when stared at by a human observer (Oppenheimer, pers. obs.).

Submissive crouches or huddles occur during dominance interactions in *Saimiri* and prior to copulation in *Alouatta villosa* females (Table 4).

Freezing is a related behavior, which occurs when an individual is alarmed, or in the case of *Callimico goeldii* females, it occurs just prior to copulation. It is not clear from the descriptions in the literature whether freezing also includes a lowering of the body. This may vary from species to species or with the type of motivation. Crouching or freezing as an alarm or submissive response occurs in species of both New World families.

"Lie on back" may be another common behavior pattern, but it has only been noted in the literature for three species. It is an appeasement gesture that inhibits attack, or in juvenile *Cebus capucinus* during play it encourages contact (Oppenheimer, pers. obs.). This behavior has apparently been ritualized in squirrel monkeys and has the added component of side-to-side rolling. This back rolling occurs during play and is performed by the dominant animal in order to induce others to approach and make contact (Castell, 1969). Back rolling is also performed by female howler monkeys after copulation (Carpenter, 1934), but the subsequent behavior has not been noted; and it was performed by a captive female *Saguinus geoffroyi* after a male failed to respond to her invitations to copulate (Moynihan, 1970:43).

Soliciting allogrooming consists of an individual's positioning itself in front of another and sitting face to face with eyes averted or lying prone and presenting the back. *Saguinus geoffroyi* solicits by erecting the hair on the nape (Moynihan, 1970).

Soliciting for play probably takes many forms. During sexual play female *Saimiri sciureus* entice the male partner by assuming odd postures, such as looking between the legs or hanging by the feet. The females may allow the male to make contact, but then dart away (Latta et al., 1967). Play among juveniles and subadult white-faced capuchins *(Cebus)* is solicited by similar patterns, as well as by lying on the back (Oppenheimer, pers. obs.).

Soliciting sexual mounting by the female takes a wide variety of forms in the different species (Table 4), including crouching with the tail aside, backing into the male, mounting the male *(C. capucinus),* curling the tail up between the legs, and presenting the tongue (see below). More than one of these patterns may occur in a single species: a female *Alouatta villosa* may tongue flick, present her rump, urinate, and take a submissive crouch posture (Carpenter, 1934). Tongue protrusion or flicking is a common sexual behavior that is performed by both sexes (see below).

Head shaking can occur from side to side, back and forth, or up and down and may include shaking of the body, depending on the species (Table 4). In most cases the individual giving the display is approaching a dominant individual, either a conspecific or a human. *Saguinus geoffroyi* may in addition head shake at a predator (this may actually be swaying, see below) or at its female mate during copulation (Moynihan, 1970). This behavior can be described as appeasement during an ambivalent approach, one which may include components of escape. In *Cebus capucinus* the head shake is accompanied by grinning, frowning, and a vocalization called guttural chatter (Oppenheimer, 1973). Andrew (1963) describes such a display complex, plus tongue protrusion and eye closure, as being or having evolved from a protective response to a noxious odor or taste, and states that these responses can be elicited when an individual approaches another who may attack. In *C. capucinus* the dominant individual ignores the head shaker and thus allows the approach to continue (Oppenheimer, 1968).

Tail-rump shake involves side-to-side movement of the tail. It is done by *Aotus* and *Saguinus geoffroyi* only during sit rubbing, though sit rubbing can occur without tail movement. In *S. geoffroyi* sit rubbing occurs when an individual is frustrated or engaged in a sexual or dominance interaction; in *Aotus* it occurs under similar circumstances (Moynihan, 1964:52, 1970:20). In the other species tail-rump shaking occurs without anal-genital rubbing and is associated with strong excitement, tantrums, and sexual and agonistic behavior. It is also a signal or part of a movement by a mother to get an infant off her back (Table 4).

Bringing the tail forward under the body is associated with alarm in *Callithrix jacchus* (Le-

Table 4

Visual communication

Gestures:	(1) *Cebuella pygmaea*	(2) *Callithrix jacchus*	(3) *Callithrix argentata*	(4) *Saguinus geoffroyi*	(5) *Saguinus midas* and *tamarin*	(6) *Leontopithecus rosalia*	(7) *Callimico goeldii*	(8) *Aotus trivirgatus*
Look at	Yes	Mobbing					A	
Look or turn away	Yes	Alarm		Head down, alarm			Head down, alarm	
Submissive crouch or huddle								
Freeze				Fright			F to M; S	Alarm
Lie on back				Stops attack, S				
Solicit allogroom				Raise hair on nape		Lie down		X
Solicit play								
Solicit mount				Curl tail up			Freeze crouch	X
Head shake				Alarm, Ap		Jerk back at human		
Tail-rump shake				With sit-rub			F, S	With sit-rub
Tail forward		Alarm		Alarm at human				
Upward tail curl				Alarm; F, Sexual			Wet with urine, S	
Genital display	AT	M & F AT	M & F AT	M & F AT		X		X
Display posture		Arch P.; DT	Arch P.; DT	Stand-up threat		Yes	Yes	Threat
Stiff walk in arch posture		Sex play, D	D	X			D	
Bounce (Bo), Intent. to Lunge (Lu), Sway (Sw)	Sw at human	Sw at predator, DT		Sw at human		Sw at human		Sw at predator
Lunge at				A		Threat		A
Chase		A	A	A	A	Plav	Play, A	
Bang objects								X
Branch shake (Sh) and break (B)								
General piloerection	Several kinds	DT, mobbing		Threat			Yes	

(9)	(10)	(11)	(12)	(13)	(14)	(15)	(16)	(17)	(18)
Callicebus moloch	*Saimiri sciureus*	*Saimiri oerstedi*	*Cebus capucinus and nigrivittatus*	*Cebus albifrons*	*Cebus apella*	*Alouatta villosa*	*Alouatta seniculus*	*Ateles geoffroyi*	*Lagothrix lagothricha*
	Yes		Observe or threat		Threat	AT, S, observe		Yes	Yes
From human, head down	From partner		From human			Avoid contact		Avoid contact	Avoid
	Yes	Yes				S, crouch		Yes	Yes
							Fright in young	Play	Play
	F, PC, roll		(play, Ap.)			Roll: S, comfort			
	Infant to mother		Sit, lie, avert eyes					Present back	Lie, present back
	Look betw. legs, S								
	Tail aside, look, crouch		F back into M			Present rump, & tongue	Tongue protr., F	Sit in M lap	
Captive at human	Bobbing, M at M		At dominant		Greeting	Head, play; body, A		Ambivalent approach	Yes; context ambivalent
Tail-lash, threat			Tantrum, thwarted		Strong excitement		F, remove infant		
	M and F	M, A	X		Bipedal AT	Scrotal contractions?	M & F, S	? M Pelvic thrust at F	
Arch P., threat			Threat		X	M at M			
	M at F; S, "Pacing"		M assert D			M assert	?	X	X
	Sw at predator		Bo and Lu threat	Lu, threat	Lu and Bo threat	Lu	?	Lu	Lu
A, intertroop	A	Mild threat	A			A		Yes	
A, intertroop	Play? A	A	Play, A			Play, A	Play in young	Play, A	Play, A
	M, excited		M; AT	M					
	Sh, M at F thwarted		Sh and B, AT		Sh; A, D	Sh, M at M		Sh and B, at human	Sh at foreign stimulus
Threat	X	X	AT		Excitement			Alarm	Alarm

Table 4 (continued)

Gestures:	(1) *Cebuella pygmaea*	(2) *Callithrix jacchus*	(3) *Callithrix argentata*	(4) *Saguinus geoffroyi*	(5) *Saguinus midas* and *tamarin*	(6) *Leontopithecus rosalia*	(7) *Callimico goeldii*	(8) *Aotus trivirgatus*
Erect hair on tail		Alarm		Alarm				
Facial expressions:								
Grin		DT	DT	DT		?	DT	X
Open mouth								
Open mouth bared teeth		Threat				At human		?
Lip smack		M, Sex play					M, S	
Protruded lips				With frown threat				X
Tongue protrusion (sin=single; R=repeated)		R; Sex play	R; DT	R, in copulation		R; S	R; S, Ap	
Eyebrows lowered (L), raised (R)		L, threat	L, threat	L, threat		R, at human	L, threat	X
Eyes closed (P=partial; C=complete)		Submissive ?		Yes				C,
Flatten ears		DT	DT	DT?			DT	Alarm

NOTE: X = Stated not to occur, AT = Aggressive, Ap = Appeasement, DT = Defensive threat, PC = Promote contact, A = Agonistic, M = Male, F = Female, D = Dominance, S = Sexual.
(1) Epple, 1975; Moynihan, 1966, 1970; (2) Epple, 1967, 1968; Epple-Hösbacher, 1967; LeRoux, 1967; Moynihan, 1970; (3) Epple, 1967, 1968; (4) Epple, 1967, 1975; Moynihan, 1970; (5) Thorington, 1968b; (6) Andrew, 1963; Epple, 1967, 1968; Snyder, 1972; (7) Epple, 1967; Lorenz, 1972; (8) Andrew, 1963; Moynihan, 1964; (9) Mason, 1971; Moynihan, 1966,

Roux, 1967) and *Saguinus geoffroyi* (Moynihan, 1970). With the tail in this position, the tip may be coiled (tail tip coiling), a response that indicates a higher level of alarm and may function as "a warning and/or appeasement . . . display" in captive *S. geoffroyi* housed in crowded conditions (Moynihan, 1970).

Upward tail coiling, where the complete tail, rather than just the tip, is coiled and is held between the legs, is performed most frequently by sexually motivated *S. geoffroyi* females and less frequently by sexually thwarted males in captivity. At least in females the behavior is thought to indicate sexual receptivity (Moynihan, 1970). Upward tail coiling plus wetting the tail with urine has been observed in captive female *S. geoffroyi* (Epple, 1967) and *Callimico goeldii* (Lorenz, 1972) during sexual interactions. Moynihan (1970:48–52) discusses this behavior pattern in some detail and gives suggestions as to how it might have evolved into a sexual soliciting act. I have also observed this behavior, without urine wetting, in captive *S. geoffroyi* and thought it indicated alarm on the part of both male and female performers during sexual encounters. Since the lower-level tail tip coiling is done in response to alarm, and urine release may also be stimulated by alarm or invasion of one's individual distance, further studies would be of interest.

(9)	(10)	(11)	(12)	(13)	(14)	(15)	(16)	(17)	(18)
Callicebus moloch	*Saimiri sciureus*	*Saimiri oerstedi*	*Cebus capucinus* and *nigrivittatus*	*Cebus albifrons*	*Cebus apella*	*Alouatta villosa*	*Alouatta seniculus*	*Ateles geoffroyi*	*Lagothrix lagothricha*
			Ap, PC	Fright PC	Submissive, PC, Ap		?		Yes PC
Yes	? Threat		Play AT	Play	Threat (several)	Play AT, and rough play	Rough play, A	Yes Grimace, A	
				Approach M, PC	PC		F, S, and tongue p.	X	F to M, PC, S
To con-specific			M at F, S					PC, Ap	PC, Ap
			Sin, PC		While lip smacking	R, M and F; S	sin, f, S, & lip smack	X	X
			L, search, protect		R, threat				L
P					P, sub-missive			P, Ap	P
					Threat				

1967; (10) Baldwin, 1968, 1969; Castell, 1969; Latta et al., 1967; Moynihan, 1967; Winter et al., 1966; (11) Baldwin and Baldwin, 1972; (12) Oppenheimer, 1973; Oppenheimer and Oppenheimer, 1973; Oppenheimer, pers. obs.; (13) Andrew, 1963; Bernstein, 1965; (14) Dobroruka, 1972; Nolte, 1958; Weigel, 1974; (15) Altmann, 1959; Baldwin and Baldwin, 1973a, 1976; Carpenter, 1934; Collias and Southwick, 1952; (16) Neville, 1972b; (17) Carpenter, 1935; Eisenberg and Kuehn, 1966; von Wagner, 1956; (18) Andrew, 1963; Williams, 1967.

There are two major types of genital displays. In the family Callitrichidae species that display their genitals (scrotal sac and labia) do so by lifting their tails and orienting their rump toward the intended receiver, while keeping hands and feet in contact with the ground (see Epple, 1967: Figs. 8 and 9). After the tail is raised pheromones may be released (Epple, 1972a). This display is done by both sexes and is thought to be an aggressive threat (Epple, 1967, and 1975). In the family Cebidae the genital present is done frontally, with the chest raised, legs spread, and penis or clitoris erect (Castell, 1969; Eisenberg and Kuehn, 1966; Winter, 1968). In *Saimiri* the genital display is done at a distance, from 10 cm to 4 m, and the displayer is erect; it is also done in contact, with the displayer bending over his partner and seeming to thrust his penis in the partner's face (Castell, 1969) while pressing the partner's head down (Winter, 1968). The distant or open position is used by young and by adult females and appears to indicate frustration or self-defense. The contact or closed position is used by males during dominance and courtship interactions (Ploog, 1967). A captive *Ateles* male was observed to perform pelvic thrusts with an erect penis in front of a female before copulation (Eisenberg and Kuehn, 1966); however, in the wild such behavior was not associated with sexual behavior, but rather with resting in a group or alone, while being groomed, while playing, or after homosexual hugging (Klein, 1971). In *Alouatta seniculus* both the male and female may have erections during a sexual interaction (Neville, 1972b).

The display posture has been observed during an agonistic interaction (Table 4) in most species, and involves orienting toward and looking at another individual with the back slightly arched. While in the display posture other aggressive displays may be performed, such as bouncing or the open-mouth–bared-teeth face. Though the display posture is usually done while on all fours, *Aotus* and *Callicebus* may sometimes stand up to transmit their threat (Moynihan, 1964, 1966, 1967). *S. geoffroyi* always stands up during display posture so that its intimidating white venter is fully exposed toward the second party (Moynihan, 1970).

Although related because of the arched back, stiff walking usually occurred at different times, at least in *Cebus capucinus;* it occurred most often when free-living subadult or adult males came to the caged capuchins and strutted across the tops of their cages or along a limb in a nearby tree. The display seemed to indicate unchalleneged dominance (Oppenheimer, 1973). Such a display has been observed in captive marmosets, where it indicates dominance, and in *Callithrix jacchus* it may also be performed during sex play (Epple, 1967). A possibly related display, pacing, is done by *Saimiri sciureus* males in front of a receptive female prior to copulation (Baldwin, 1968); however, no mention is made as to whether the male's back is arched, and the stationary display posture is apparently lacking from the *Saimiri* behavioral repertoire (Table 4).

Bouncing and intention movements to lunge are performed by *Cebus capucinus* while in the display posture (Oppenheimer, 1973), and swaying is performed by marmosets while sitting (Moynihan, 1970). In all cases these movements are done toward a human or a predator and may function as a defensive threat or alarm. Swaying is a side-to-side motion, though in *Callicebus* there may also be a vertical component (Moynihan, 1966). Young capuchins *(Cebus)* may make short forward and backward movements as if they were going to lunge, whereas in adults the movement, bouncing, is in the vertical plane (Oppenheimer, 1974). Such behavior, which may be accompanied by vocalizations, attracts the attention of the predator to the performer and may also alert other troop members to the presence and location of danger and allow them time to escape. Moynihan (1966) suggests that swaying may inform the predator that it has been seen and thus for the moment terminate the hunt.

Lunging at is an agonistic behavior pattern reported for a number of species and may function as a threat (Table 4). Lunging at can be considered a preliminary movement to grappling or chasing.

Chasing is a play activity, at least in the young of some species, but in adults it is primarily aggressive (Table 4). Running away is assumed to occur in all species, particularly those where chasing is reported.

Banging of objects, as a signal, has only been reported for three species of cebids (Table 4). It appears to be a male aggressive behavior. Free-living *Cebus capucinus* males on Barro Colorado Island did this when they visited the cages where capuchins were kept captive; they lifted and dropped sheet metal on the cage roof and jumped up and down on it. This behavior, in part, would seem to have the same motivation as branch shaking, which occurs in the trees (Oppenheimer, unpublished data).

Branch shaking has only been reported for cebids, and it functions as a threat display (Table 4). *Saimiri sciureus* males branch shake at sexually nonreceptive females (Baldwin, 1968). In *Cebus capucinus* branch shaking is generally stimulated by the presence of a human observer, but less frequently it may also be used as a general threat within the troop; branch breaking, without branch shaking, was primarily observed to occur at the start of intertroop encounters (Oppenheimer, 1973). Moynihan (1964) has suggested that branch shaking and breaking may not occur in *Aotus* because they are too small; however, they have approximately the same weight as *Saimiri* (Table 1). This display appears to be absent in all species that live in monogamous pairs. It may be that in such species the family members tend to disperse less and thus displays that attract a predator's attention might endanger the whole group.

Fig. 2. *Cebus capucinus* subadult male staring at and making protruded lips face toward adult female approaching on the ground. Female made no obvious response. Photo taken at Curt Freese's study site in dry forest at Santa Rosa National Park, Costa Rica.

PILOERECTION

Piloerection is a common display that accompanies other types of threat behavior or may be triggered by alarm (Table 4). Moynihan (1967, 1970) discusses this behavior in detail. Piloerection on the tail is a specialized alarm display that has been observed in *Callithrix jacchus* and *Saguinus geoffroyi* (Moynihan, 1970).

FACIAL EXPRESSIONS

All species do have a "relaxed face" (van Hooff, 1967), but there is usually little mention of it in the literature.

Grinning in the Callitrichidae in combination with flattening of the ears appears to be a defensive threat (Epple, 1967), and in the Cebidae it is an appeasement or submissive display that may be used by the dominant or the subordinate animal (Table 4). In *Cebus capucinus,* infants may combine the grin with a frown and a lateral head shake when approaching a dominant animal, usually the troop's adult male (Oppenheimer, 1968, 1973). In *C. apella* the grin is accompanied by raising of the eyebrows (Weigel, 1974). The grin involves pulling back the corners of the mouth while keeping the jaws shut. There may be some confusion with low-intensity open-mouth–bared-teeth face. Moynihan (1966) has described a "baring the teeth" face in *Callicebus,* which is similar to the *Cebus* grin, except that the corners of the mouth are not retracted.

The open-mouth face usually occurs during play in some cebids (Table 4). The mouth is open, the corners are partially retracted, but the teeth are covered by the lips. It is similar to the "relaxed open-mouth face" that has been described by van Hooff (1967) to occur during play.

The open-mouth–bared-teeth face occurs in many of the New World species as a threat display (Table 4). It has been observed in *Aotus* by Andrew (1963), but not by Moynihan (1964). In *Cebus capucinus* (Fig. 1) the mouth is open and the lips are pulled back so that the canines are exposed (Oppenheimer, 1968, 1973). This fits in part the "staring open-mouth face" and "staring bared-teeth scream face" of van Hooff (1967). Weigel (1974) has studied the facial expressions of *Cebus apella* and found that they have an open-mouth–bared-teeth face that is silent, one that is accompanied by a "staccato beep," and a bared-teeth scream face that may be similar to that described by van Hooff (1967) mentioned above.

Lip smacking is performed by marmosets, tamarins, and cebids during sexual or friendly encounters (Table 4). In *Alouatta seniculus* it may be performed with tongue protrusions by a female (Neville, 1972b).

Protruded-lips face has been observed in a number of species. In combination with a frown it functions as an aggressive threat in *Saguinus geoffroyi* (Epple, 1967). In the Cebidae (Fig. 2) it is a friendly gesture that promotes contact and may be used by an adult male when approaching or being approached by an adult female (Oppenheimer, 1968, 1973). Van Hooff (1967) has also observed this face in an Old World species *(Macaca nemistrina)* when a male approached a female in heat.

Tongue protrusion is used most often during sexual or friendly encounters to promote contact, though in other species it is reported to be a defensive threat (Table 4). It may accompany lip smacking, and the tongue may either be extended and held out as a single protrusion or be rhythmically flicked, i.e., repeated protrusion. It occurs in marmosets, tamarins, and cebids. In *Alouatta villosa* the male and female display repeated tongue protrusions toward each other (Fig. 3), which may eventually lead the female to lick the male on the face or body, and finally end in copulation (Carpenter, 1934). In *Alouatta seniculus* only the female has been observed to protrude the tongue as a single, not repeated, movement (Neville, 1972b). In *Saguinus geoffroyi* the tongue may be protruded and held in one position, or moved up and down, and with lower motivation it may be moved in and out. It may be accompanied by wrinkling of the nose, partial closure of the eyes, and head flicks. It tends to be primarily a male behavior pattern, used with

Fig. 3. Howler monkey (*Alouatta villosa*) adult male, partially suspended by tail, making rhythmic tongue protrusions at female. Photo taken on Barro Colorado Island.

sniffing in an encounter with a stranger, and during copulation, at which time it may be performed by the female as well (Moynihan, 1970). The behavior suggests that in some species the tongue performs some olfactory or taste function under these circumstances, or at least that the visual display has been ritualized from tasting behavior.

Raising and lowering of the eyebrows has been noted in a number of species (Table 4). Usually the eyebrows are lowered in threat (Fig. 1), but in *Cebus apella* the eyebrows and forehead are raised during open-mouth–bared-teeth face, which is a threat, and the forehead may be raised briefly with the mouth closed as a submissive gesture (Weigel, 1974). Lowering of the eyebrows is also done while searching for an object or as a protective response (Andrew, 1963).

Partial or complete closure of the eyes is also part of the protective response (Andrew, 1963), and it may indicate submissiveness when an individual is alarmed by some general danger or a threat (Table 4).

Another aspect of the protective response in most mammals is flattening or pulling back of the ears or ear tufts (Andrew, 1963), and it is interpreted as being part of the defensive threat display in some of the New World primates (Table 4).

PELAGE MARKINGS

A fourth type of visual communication has been reported for *C. capucinus* (Oppenheimer, 1969a). It involves pelage markings or differences on the forehead that indicate the age and sex of the individual (Fig. 1). These signals may be similar to dimorphic differences in body size and shape, size of canines, or shape of the genitalia, all of which remain the same for long periods and change gradually during the lifetime of the individual, but allow rapid recognition of status or potential status even between strangers.

Acoustic Communication

Vocal communication plus some nonvocal sounds like tooth grinding are included in Table 5. Other sounds that accompany visual displays, such as branch shaking, are ignored here. Since each researcher has used his or her own set of names for vocalizations of the species they studied, I have arranged the vocalizations according to the general situation in which they were elicited. The motivation for calls given in one situation by different species may also be different. Thus, I have used the name of the vocalization for each species as used in the literature by the individual investigators. Use of the same name by different investigators, such as "twitter," does not mean that the calls are necessarily the same in physical structure, nor that they are homologous, though they may be. I have tried to come up with a set of generalized circumstances that would fit most calls; however, in doing so some information has been lost and I may have placed some vocalizations in inappropriate categories (a question mark has been inserted where entries were most likely to be wrong). Thus, Table 5 only indicates in what situations New World primates emit vocalizations and which calls of the different species may have similar functions.

Sonograms, such as in Fig. 4, have been made of the vocalizations of a number of species (Andrew, 1963; Baldwin and Baldwin, 1976; Eisenberg, 1976; Eisenberg and Kuehn, 1966; Epple, 1968; Moynihan, 1964, 1966, 1970; Oppenheimer, 1968, 1973; Ploog, 1967; Winter, 1972; Winter et al., 1966), but the vocalizations of other species are just now being analyzed or are yet to be done. Thus, although information is available in part, I will not attempt to discuss the physical structures of the calls in a thorough way. For discussion of the physical structure of primate calls see Marler (1965); however, it should be mentioned that ultrasonic calls do occur, particularly in the marmosets (Epple, 1975).

One needs to be cautious in comparing species studied by different investigators. I have included in Table 5 two studies of the vocal system of *Saguinus geoffroyi,* one by Epple (1968) and one by Moynihan (1970). Differences in terminology and in vocal situations occurred, and Moynihan has discussed the areas of agreement and disagreement with both Epple and Andrew (1963). Other differences tend to be subjective, for instance "Muckenhirn (1967) . . . reports that wild *Saguinus o. geoffroyi* are highly vocal while Moynihan (1970) points out that they are actually quite silent" (Epple, 1975).

The vocal systems of the New World primates differ in a number of more specific ways. The simplest is the number of vocalizations: *Aotus trivirgatus* has 10 (Moynihan, 1964); *Ateles geoffroyi* has 16, of which 10 are heard frequently (Eisenberg and Kuehn, 1966); *Alouatta villosa* has 20

(Altmann, 1959); and *Saimiri sciureus* has 26 (Winter et al., 1966). Possibly with additional observations more call types will be found for species that now appear to have a low number. Another difference is that some species have discrete vocal systems, where each call is distinct and is elicited by specific stimuli, as in *Aotus trivirgatus* (Moynihan, 1967). In other species the vocalizations may be variable in form, as in *Callimico goeldii* (Epple, 1968), or may grade into one another by way of frequent vocalizations of intermediate type, as in *Callicebus moloch* (Moynihan, 1966, 1967). Such differences may also occur between populations of the same species, as in *Saimiri sciureus,* where the Gothic race has a more discrete vocal system than does the Roman (Winter, 1969b). Even within one species one part of the vocal system may be discrete and another graded, as in *S. sciureus* (Winter, 1969a) and *C. capucinus* (Oppenheimer, 1968); in *C. jacchus* the vocalizations of the infant are more variable than those of the adult, and the infant system seems to be graded (Epple, 1968, 1975). Also some species may combine two vocalizations to make up a third, as in *Cebus* and *Saimiri* (Oppenheimer, 1968, 1973; Winter et al., 1966), or may combine two or more vocalizations into a sequence, as in the bark-roar-oodle of *A. villosa* (Altmann, 1959), the dawn song and gobbling phrases of *C. moloch* (Moynihan, 1966), and the long call ending in a "coda" of *A. fusciceps* or the roaring whoop of *A. geoffroyi* (Eisenberg, 1976).

Calling at dawn or at the initiation of a rain or thunderstorm has been reported for a number of species (Table 5). In *A. geoffroyi* the call is given with the onset of a storm and acts as a group cohesion call (Eisenberg, 1976). In the other three species mentioned in Table 5, plus *Saguinus midas* (Thorington, 1968b), the calls are given at dawn, but *A. villosa* also calls at the initiation of storms or at airplanes flying overhead. In all four species the calls act as an intertroop spacing mechanism. Chivers (1969) has made a detailed study of the dawn calls of *A. villosa* on Barro Colorado Island, and his data suggest that the daily movements of a troop are set in relation to the location of neighboring troops established at the time of the dawn chorus.

Six species, including the four above that give a dawn chorus, use their long loud call when interacting with another troop nearby or in sight (Table 5). In this capacity the vocalization also functions as an intertroop spacing mechanism. one that uses considerably less energy than chasing and/or fighting.

Most of the New World species have a vocalization that is used when one or more individuals are separated from the rest of their troop. The same call may be used when the troop responds to the call of the lost individual. As in *S. geoffroyi,* some species may use the same vocalization for a lost call as they use for the intertroop spacing call (Table 5). These vocalizations are usually given in bouts of one to six calls: *S. geoffroyi,* one to three (Epple, 1968); *A. trivirgatus,* two to three (Moynihan, 1964); *C. capucinus:* on Barro Colorado, one to five with an average of three (Oppenheimer, 1968, 1973), and at Santa Rosa National Park in Costa Rica, an average of two (Curtis Freese, pers. comm.); *C. nigrivittatus,* one to six with an average of 3.2 (Oppenheimer and Oppenheimer, 1973); *A. villosa,* one to five calls (Baldwin and Baldwin, 1973a); *A. geoffroyi,* three times; and *A. fusciceps,* four to five times per bout (Eisenberg, 1976). The use of a number of calls per bout as well as repeated bouts probably help the receiver determine the location of the caller and possibly the caller's identity. For instance, a young capuchin that calls just a few trees away from the troop does not elicit any response, nor do the calls of a nontroop member (Oppenheimer, 1968, and unpublished data). Thus, the long loud calls of each species have the potential to serve as both intratroop cohesion calls and intertroop spacing calls, and in some species they do (Table 5).

The intertroop spacing calls or dawn calls and the lost calls of the platyrrhine species may all have arisen from a similar call of a common ancestor, as they are similar in physical structure, or possibly this similarity was brought about by convergence. The top row in Fig. 4 shows a single "arrawh" call (for comparison with the *Ateles* call below) and an "arrawh" triplet bout for *C. capucinus,* as well as the calls given by a captive *C. apella* in response to the "arrawh." In the second row only the second call of an *Ateles* triplet is shown, and these calls were given by the male in order to call the other troop members to him. The *Ateles* call is twice as long as that of *C. capucinus,* and it has two narrow-frequency bands instead of the one in the *Cebus* arrawh; however, it did sound similar to the arrawh. The *Alouatta* call was recorded on Barro Colorado Island from a captive male who was responding to the roars of a nearby troop. Most of the first and third, and all of the second call of the howler triplet bout are shown, and they are about the same length as those of *C. capucinus.* Most of the energy is in two frequency bands as in *Ateles.* The *Saguinus* contact call was also recorded from a captive individual. The frequency range is much higher and the energy is in five frequency bands, rather than one or two. The higher pitch correlates with the smaller body size of the species (Tembrock, 1963), but Moynihan (1970) also suggests that in smaller species, where danger is greatest from predators, there may have been selection for the higher frequencies, which do not travel as far as the low ones. The sonogram at the left shows parts of three calls from a triplet bout, and it can be seen that the calls are only slightly longer in duration than those of *C. capucinus.* The sonogram at the bottom right shows a complete call. The *C. apella* sonogram shows two complete calls and parts of two others. They are shorter than in the other species and show more modulation. They show a greater resemblance to the *C. capucinus* "huh" call, which is a lower-intensity "arrawh." If this sonogram is representative of the *C. apella* lost calls, it may be that this atypical call has been selected for because of sympatry with at least one other species of the genus in much of the northern part of its geographic range. Except for the *C. apella* call, these intratroop cohesion calls and intertroop spacing calls are similar in structure to the long-distance contact call of the timber wolf (*Canis lupus:* Theberge and Falls, 1967). This suggests that the similarity of function of these calls, to communicate over long distances, is responsible for the similarity in physical structure (Oppenheimer, 1968). For other factors that may influence the physical structure of calls see Marler (1965).

A number of species have a medium-intensity contact call that is elicited as the other troop members move out of sight or when the movement of the caller is blocked by a gap that cannot easily be jumped. In young *C. capucinus* the "huh" call may grade into "arrawh" calls if visual contact is lost and/or assistance is slow in coming (Oppenheimer, 1968, 1973). It looks as though this may also be true for *S. geoffroyi* and *A. villosa* (Table 5).

Most of the species seem to have a contact call that is emitted while in visual contact with other troop members. It probably functions to maintain contact as well as a minimal spacing between members of the troop. This may also be true of the troop feeding call. This call occurs in captives at feeding time or when the human who does the feeding approaches, and in free-ranging capuchins it usually occurs at fruit trees where several individuals feed at once. The call may also draw nearby troop members to the food source (Oppenheimer, 1968, 1973).

A number of vocalizations are given in response to the return of or calls from a separated troop member. In part this may have to do with the particular individual involved, his distance from the troop, the length of the separation period, or the mood of the other troop members.

Table 5

Acoustic communication

	(1)	(2)	(3)	(4)	(5)	(6)	(7)	(8)	(9)
					Saguinus geoffroyi				
	Cebuella pygmaea	*Callithrix jacchus*	*Callithrix argentata*	*Saguinus fuscicollis*	*Study 1*	*Study 2*	*Leontopithecus rosalia*	*Callimico goeldii*	*Aotus trivirgatus*
Dawn, thunder, rain, airplane		Phee							
Another troop		Phee				Long whistle			Resonant grunt, roar
Individual out of visual contact		Loud phee, twitter	Loud phee, twitter		Long te	Long whistle	Pü Pü Pü + Whee, twitter	Repeated, high pitch	Hoot, gulp?
Individual loosing visual contact (or unable to proceed)					Te				
Individual in visual contact		Phee	Phee	Short whine	Te	Twit or short whine	Pe	Te, tschog	Sneeze, grunt? gulp?
Troop feeding		Faint phee, twitter	Faint phee, twitter						
Response to returning troop member					Te	Twitter, long whistle			Moan
Infant moving near mother									
In physical contact									
Approach for nurse, groom, ride, food						Twitter, squeak, infantile rasp	Twitter	Whistle	
Approach dominant slowly or avoid	Squeal	Squeal	Squeal		Squeal (rasping screech)			Whistle	Low trill, moan
Approach subordinate slowly							Twitter		Low trill, moan
Approach peer slowly									
Easy play									
Rough play		Birdlike chirps	Birdlike chirps						
Thwarted approach								(See text)	
Male approaching estrous female									Squeak (M and F)
Receive possible threat								Te	
Receive mild threat		Chatter	Chatter	Rattle	Chatter	Twitter	Whee	Tschog	

(10)	(11)	(12)	(13)	(14)	(15)	(16)	(17)	(18)	(19)
Callicebus moloch	*Saimiri sciureus*	*Saimiri oerstedii*	*Cebus capucinus & nigrivittatus*	*Cebus albifrons*	*Cebus apella*	*Alouatta villosa*	*Alouatta seniculus*	*Ateles geoffroyi & fusciceps*	*Lagothrix lagothricha*
Song, moan						Roar (A1-3,B), oodle		Long call, whinny	EE ee oolk
(a)						Roar (A1-3,B), oodle, bark			
(b)	Isolation peep	Isolation peep	Arrawh	Caw	Fuh	Wrah-ha, caw (infant)		Long call	
			Huh	Caw	Mik	Bah-caw, yelp, whimper		Trill awk, eeah, squeal	Eeeoolk-awk
	Peep, chirp					Cluck		Whinny	
	Twitter, trill		Huh, yip		Mik, whine			Whinny	eeolk
Moan	Isol. peep, twitter		Arrawh, twitter		Mik	Whimper		Tee tee	
	Tuck		Yip		Rolling peep	Eh			Click?
Purr	Chuck, churr, purr		Purr, chirp			Purr		Gutteral whinny	Cluck
	Purr, oink grumble	Peep, milk purr	Twitter, chirp			Explosive bark, eh			Sob
			Guttural chatter	Click	Chirp, grunt		Meow-whine	(d)	Cluck, squeak
	Purr, churr, girren	Purr girren						Whinny	
	Twitter				Chatter	Whimper			
	Play peep					Chirping, squeal		Ook ook, twitter, trill	Ooh-ooh
	Shriek					Grunting, squeak		Ak ak, growl	Hugh-hugh
Chirrup ?	Frust. peep, chomp, yow		Twitter, trill			Whimper, caw, explosive bark		Chitter, squeak	Screech
Moan	Squeal, purr		Warble		Flutelike call			Ook ook ?	
Squeak	Crackle, arr	Peep	Trill		Igk	Bark, eh		Squeak	
Whistle	Scream, yow, squawk	Shriek	Whistle			Whimper, heh	Squeak		

Table 5 (continued)

	(1)	(2)	(3)	(4)	(5)	(6)	(7)	(8)	(9)
	Cebuella pygmaea	*Callithrix jacchus*	*Callithrix argentata*	*Saguinus fuscicollis*	*Saguinus geoffroyi* Study 1	Study 2	*Leontopithecus rosalia*	*Callimico goeldii*	*Aotus trivirgatus*
Receive intense threat		Squeal	Squeal	Loud sharp notes	Trill	Long sharp notes	Trill, squeal	Chatter	Scream
Chased or contacted	Rasping screech	Scream	Scream	Long rasps, whistles	Squeal	Long and broken rasps	Scream		Scream
Seek assist from 3rd party				Long rasp		Long rasps	Squeal		
Give mild threat									
Give medium threat		Tsee, crackle, cough	Tsee, crackle, cough				Trill		Gruff grunt, sneeze?
Give intense threat or chase		Chatter							
General alarm or threat				Loud sharp notes		Tsit, loud sharp notes, t chuck			
Mob other species		Crackle, cough, tsik	Crackle, cough, tsik, noise		Tsik, trill		Tsik		
Threaten other species and alert troop		Faint whistle	Trill	Trill	Te, trill	Twitter, trill	Trill, whistle	Te, trill	
Alert troop to predatory bird		Faint whistle		Trill	Whistle	Trill	Whistle		
(a) Chuck, moan, resonant note, gobbling phrases.									
(b) Whistle, trill, chuck, resonant note.									
(c) Oodle, grunt, roar, bark (C1 and 2, D1 and 2).									
(d) Twitter, trill, whinny, squeak.									

(1) Andrew, 1963; Epple, 1975; (2) Epple, 1968, 1975; (3) Epple, 1968, 1975; (4) Moynihan, 1970; (5) Epple, 1968, 1975; Andrew, 1963; (6) Moynihan, 1970; (7) Andrew, 1963; Epple, 1968; (8) Epple, 1968; Lorenz, 1972; (9) Moynihan, 1964;

(10) *Callicebus moloch*	(11) *Saimiri sciureus*	(12) *Saimiri oerstedii*	(13) *Cebus capucinus & nigrivittatus*	(14) *Cebus albifrons*	(15) *Cebus apella*	(16) *Alouatta villosa*	(17) *Alouatta seniculus*	(18) *Ateles geoffroyi & fusciceps*	(19) *Lagothrix lagothricha*
Trill	Kecker		Chatter scream		Kecker ?	Metallic clack, heh, yelp			
Scream	Shriek	Chitter, chutter	Scream	Scream	Scream			Squeals, screams	Scream
Trill, whistle			Scream			Infant wails			Scream
	Arr, spit	Spit					Growl	Growl	
Gnashing	Err		Tooth grind		Tooth grind		Grunt	Champing, roar	
Chuck, moan	Churr				Kecker ?		Screech ?	Hiss	Harff
			Chortle		Iku	Cluck		Cough	Harff
	Yap	Chaun	Grrah			Bark, roar		Bark	Myonk, myonk
Chuck, sneeze, scream, grunt?	Alarm peep, yap	Alarm peep	Grrah	Bark	Ika	(c)	(Bark, roar)	Bark	
	Alarm peep	Alarm peep, chitter	Grrah						

(10) Mason, 1966; Moynihan, 1966; (11) Baldwin, 1968; Latta et al., 1967; Winter, 1968, 1972; Winter and Ploog, 1967; Winter et al., 1966; (12) Baldwin and Baldwin, 1972; (13) Oppenheimer, 1973; Oppenheimer and Oppenheimer, 1973; (14) Andrew, 1963; Bernstein, 1965; (15) Dobroruka, 1972; Kühlhorn, 1939; Nolte, 1958; Weigel, 1974; (16) Altmann, 1959; Baldwin and Baldwin, 1973a, 1976; Carpenter, 1934, 1965; Collias and Southwick, 1952; (17) Neville, 1972b; Oppenheimer, pers. obs.; (18) Carpenter, 1935; Eisenberg, 1976; Eisenberg and Kuehn, 1966; Klein and Klein, 1971; (19) Andrew, 1963; Williams, 1967.

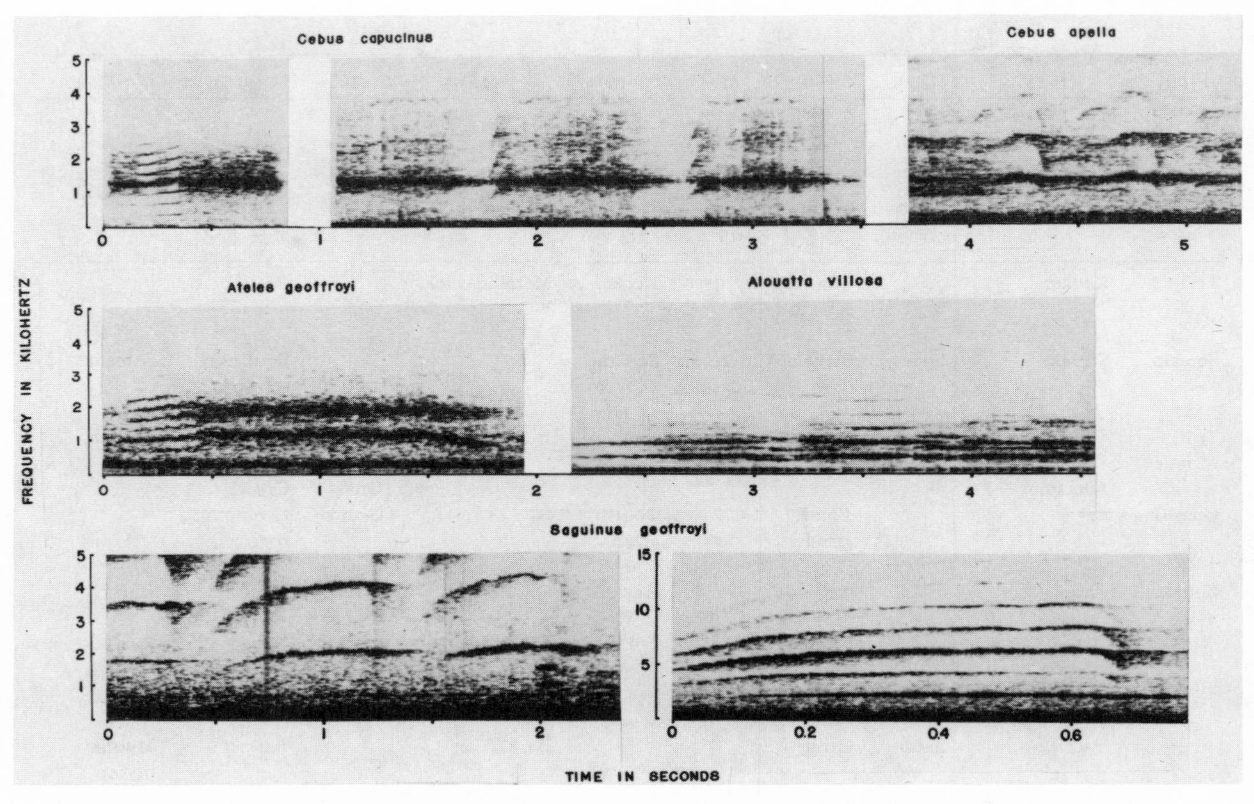

Fig. 4. Sonograms of intratroop cohesion or "lost" calls and/or intertroop spacing calls of five New World species. All recordings made on Barro Colorado Island: *C. capucinus* and *A. geoffroyi* calls from free-living individuals, the remainder from captives.

Only in species that have more than one adult female in the troop are the infants reported to emit contact calls when moving on or near their mothers (Table 5). If this is a true difference it suggests that the call functions to identify the infant to the mother when she may not be paying specific attention, as there should also be visual and olfactory cues. In the monogamous species there would usually be a maximum of two infants and they would belong to the same mother.

A contact call given while in physical contact has only been reported from cebid species. These calls are frequently of very low volume, and thus they may have been overlooked in the Callitrichidae, which are all small species (Table 1). Although one might assume that all these calls would be friendly in nature and would tend to prolong contact, this may not be true for *S. sciureus.* In this species the infant gives milk purrs to its mother while nursing, and adults may give purrs during a genital display, while huddling, or during interruptions in play. Winter et al. (1966) have concluded that it indicates an aggressive motivation. Additional data to support this conclusion would be most welcome, but apparently the social structure of *Saimiri* troops places a great deal of emphasis on aggressive interactions.

Infants of most species emit calls when they approach their mother to nurse or to be groomed, or when they approach any individual who might give them a ride or some food. These calls most likely promote contact and elicit the appropriate response from the individual approached. Although again in *Saimiri* the call is supposed to have a slightly aggressive motivation (see above).

Subordinates in most species are reported to emit calls when approaching a dominant animal. In *C. capucinus* these calls are given by an infant or juvenile who is approaching an adult male. The call is accompanied by head shaking, grins, and a frown. The male makes no response (Oppenheimer, 1968, 1973). The "meow-whine" of *A. seniculus* is given by a female in estrus when she approaches an adult male (Neville, 1972b). These calls identify the caller as being duly submissive and should appease any aggressive tendencies in the dominant animal.

In fewer species the dominant animal emits a call when approaching a subordinate. Such vocalizations are reported for five species (Table 5). The vocalizations used are contact or greeting calls, except for the two species of *Saimiri,* which have in addition a call specifically for this purpose. It is interesting to note that the *Saimiri* species seem not to have a call given by a subordinate to a dominant animal (possibly this is due to an error on my part in interpreting the literature).

Only two species are noted to use a call when individuals approach a peer (Table 5). This is probably because the category is too general and such encounters occur under specific conditions, such as play and grooming. Nonetheless, it may be that there is less need for peers to communicate with one another in the vocal channel.

Play, particularly easy play, seems to be silent in most species. In the three species that are reported to vocalize during easy play, primarily contact calls are used. In rough play the vocalizations tend to be alarm or fear calls that may elicit assistance from other troop members and/or terminate the play interaction, at least for the moment.

Thwarted approaches occur when the approacher has a strong motivation to approach and is prevented from doing so by the approachee. In capuchins this occurs during the weaning period when the infant tries to nurse or to ride on its mother. The rejected infant gives contact twitters, distress trills, and tail-rump waggles (Oppenheimer, 1968, 1973). In *C. goeldi* when an estrous female rejects a male the male makes "a sharp smacking noise with his teeth (or tongue) . . ." (Lorenz, 1972).

Though in all species males must approach estrous females, vocalizations are reported to occur during these encounters only in the Cebidae. These calls on the part of the male seem to have an appeasement function and should encourage contact. The "ook ook" call of *A. geoffroyi* is given by males in captivity when approaching an estrous female (Eisenberg and Kuehn, 1966), but Klein and Klein (1971) say that in the wild this call is part of play behavior.

In all species, individuals respond to threats with vocalizations, and most species have more than one type of vocalization. The different vocalizations are usually keyed to the intensity of the threat, though the threshold levels are yet to be established for some species. The response to the most intense threat is in most species a scream, screech, or squeal.

It is usually the scream or other vocalization used to respond to intense threat that is used to elicit assistance from a third party. This call may be given even though the caller has received only a mild threat or none at all, as when an adult female capuchin approaches a human observer long after visual contact has been made (Oppenheimer and Oppenheimer, 1973). Thus there may be a certain amount of flexibility in the amount of stimulus needed to elicit the call. An individual easily excited may respond with a scream even though the stimulus is of a low level, and/or an individual may scream in order to obtain positive feedback from the assisting individual.

In many species the intraspecific threats, which may be primarily visual (also olfactory and tactile), are accompanied by vocalizations or other sounds like tooth grinding and gnashing. In the Callitrichidae the intraspecific threat vocalizations may be the same as those used to mob another species or to alert the troop to the presence of a predator. In *C. capucinus* the only threatening sound is that of tooth grinding, and it is used only in intertroop or interspecific (at human) encounters (Oppenheimer, 1973). In *Saimiri* and *Ateles,* the two most aggressive species, there are clearly three levels of vocal threats (Table 5).

General alarm calls are reported for a number of species. These calls may be the same as a vocalization given in response to a mild threat or a contact call, or they may be specialized calls.

Mobbing calls are elicited by snakes or other terrestrial animals and are given by several troop members. In the Callitrichidae the mobbing calls are different from the alarm calls (Muckenhirn, 1966), whereas in the Cebidae the mobbing and alarm calls may be the same (Table 5).

The alarm vocalizations elicited by terrestrial animals and birds flying overhead are usually the same; however, there may be temporal differences. For instance in *C. capucinus* and *C. nigrivittatus* the "grrah" is repeated by an individual many times when the stimulus is a terrestrial animal, but it is given only once to a bird overhead (Oppenheimer, 1968, 1973; Oppenheimer and Oppenheimer, 1973). Although the vocal repertoires for many species are incomplete, it is interesting to note that the nocturnally active *Aotus* and the diurnally active, and large, *Alouatta* and *Ateles* have not been reported to give alarm calls to birds overhead. To complete a full circle, the alarm calls of some species *(S. geoffroyi, C. moloch, Saimiri, A. villosa)* are similar to or the same as the vocalizations used by these species as intertroop spacing calls, lost calls, or general contact calls. Epple (1975) and Moynihan (1970) have both heard the lost call during intertroop encounters. Epple states that the vocalization draws the other troop members to the caller and thus serves as a "call for assistance."

Comments

The studies of communication in the New World primates range from cursory observations, with consequently little descriptive information, to detailed studies of a specific behavior pattern. Representative of the latter type of study is the work of Symmes and Newman (1974), where variants of the isolation peep are played back to squirrel monkeys *(Saimiri)* in order to determine what physical part of the call is functional.

Between these two extremes are the studies that present a complete or partial list or narration of the behavioral patterns of a species. In many cases when motivations or functions are attributed to the behavioral patterns, these determinations have been arrived at on purely subjective grounds. Listing and describing behavior patterns are certainly basic and necessary steps toward understanding how the New World primates communicate, but they are only the beginning. What we need to do now is to determine quantitatively the relationships between the various behavior patterns of a species, as has been started for *Saimiri* (Winter, 1968), *Cebus* (Oppenheimer and Oppenheimer, 1973), and *Ateles* (Eisenberg, 1976). Such an analysis provides a more objective method for determining motivation and function. Once these studies are completed the structure of primate communication systems (or ethograms) can be compared among related species, as well as among the different genera. Such comparisons will yield a much greater understanding of communicatory behavior and of the evolution of such behavior than do comparisons of individual behavior units of a number of species, as have been presented here.

References

Alcock, J., 1972. The evolution of the use of tools by feeding animals. *Evolution,* 26:464–73.

Altmann, S. A., 1959. Field observations on a howling monkey society. *J. Mammal.,* 40:317–30.

Altmann, S. A., 1967. The structure of primate social communication. In: *Social Communication among Primates,* S. A. Altmann, ed. Chicago: University of Chicago Press, pp.325–62.

Altmann, S. A., 1968. Primates. In: *Animal Communication,* T. A. Sebeok, ed. Bloomington: Indiana University Press, pp.466–522.

Andrew, R. J., 1963. The origin and evolution of the calls and facial expressions of the primates. *Behaviour,* 20:1–109.

Baldwin, J. D., 1968. The social behavior of adult male squirrel monkeys *(Saimiri sciureus)* in a seminatural environment. *Folia primat.,* 9:281–314.

Baldwin, J. D., 1969. The ontogeny of social behaviour of squirrel monkeys *(Saimiri sciureus)* in a seminatural environment. *Folia primat.,* 11:35–79.

Baldwin, J. D., and Baldwin, J. I., 1972. The ecology and behavior of squirrel monkeys *(Saimiri oerstedi)* in a natural forest in Western Panama. *Folia primat.,* 18:161–84.

Baldwin, J. D., and Baldwin, J. I., 1973a. Interactions between adult female and infant howling monkeys *(Alouatta palliata). Folia primat.* 20:27–71.

Baldwin, J. D., and Baldwin, J. I., 1973b. The role of play in social organization: comparative observations on squirrel monkeys *(Saimiri). Primates,* 14:369–81.

Baldwin, J. D., and Baldwin, J. I., 1976. The vocalizations of howler monkeys *(Alouatta palliata)* in southwestern Panama. *Folia primat.* 26:81–108.

Bernstein, I. S., 1964. A field study of the activities of howler monkeys. *Anim. Behav.,* 12:92–97.

Bernstein, I. S., 1965. Activity patterns in a cebus monkey group. *Folia primat.,* 3:211–24.

Carpenter, C. R., 1934. A field study of the behavior and social relations of howling monkeys. *Comp. Psychol. Monogr.,* 10:1–168.

Carpenter, C. R., 1935. Behavior of red spider monkeys in Panama. *J. Mammal.,* 16:171–80.

Carpenter, C. R., 1965. The howlers of Barro Colorado Island. In: *Primate Behavior-Field Studies of Monkeys and Apes,* I. DeVore, ed. New York: Holt, Rinehart and Winston, pp.250–91.

Castell, R., 1969. Communication during initial contact: a comparison of squirrel and rhesus monkeys. *Folia primat.,* 11:206–14.

Castell, R., and Maurus, M., 1967. Das sogenannte Urinmarkieren von Totenkopfaffen *(Saimiri sciureus)* in Abhangigkeit von umweltbedingten und emotionalen Faktoren. *Folia primat.,* 6:170–76.

Chivers, D. J., 1969. On the daily behaviour and spacing of howling monkey groups. *Folia primat.,* 10:48–102.

Coimbra-Filho, A. F., and Mittermeier, R. A., 1973. Distribution and ecology of the genus *Leontopithecus* Lesson, 1840 in Brazil. *Primates,* 14:47–66.

Collias, N., and Southwick, C., 1952. A field study of population density and social organization in howling monkeys. *Proc. Amer. Phil. Soc.,* 96:143–56.

Dobroruka, L. J., 1972. Social communication in the brown capuchin, *Cebus apella. Int. Zoo Yearb.,* 12:43–45.

DuMond, F., 1968. The squirrel monkey in a seminatural environment. In: *The Squirrel Monkey,* L. A. Rosenblum and R. W. Cooper, eds. New York: Academic Press, pp.88–146.

Durham, N. M., 1971. Effects of altitude differences on group organization of wild black spider monkeys *(Ateles paniscus). Proc. 3rd Int. Congr. Primat.,* Zurich 1970, vol. 3. Basel: Karger, pp.32–40.

Eisenberg, J. F., 1973. Mammalian social systems: Are primate social systems unique? *Symp. 4th Int. Congr. Primat.,* vol. 1. Basel: Karger, pp.232–49.

Eisenberg, J. F., 1976. Communication mechanisms and social integration in the black spider monkey, *Ateles fusciceps robustus,* and related species. *Smithson. Contrib. Zool.,* no. 213:1–108.

Eisenberg, J. F., and Kleiman, D. G., 1972. Olfactory communication in mammals. *Ann. Rev. Ecol. Syst.,* 3:1–32.

Eisenberg, J. F., and Kuehn, R. E., 1966. The behavior of *Ateles geoffroyi* and related species. *Smithson. Misc. Coll.,* 151(8):1–63.

Eisenberg, J. F.; Muckenhirn, N. A.; and Rudran, R.; 1972. The relations between ecology and social structure in primates. *Science,* 176:863–74.

Epple, G., 1967. Vergleichende Untersuchungen über Sexual- und Sozialverhalten der Krallenaffen (Hapalidae). *Folia primat.*, 7:37–65.

Epple, G., 1968. Comparative studies on vocalization in marmoset monkeys (Hapalidae). *Folia primat.,* 8:1–40.

Epple, G., 1971. Discrimination of the odor of males and females by the marmoset *Saguinus fuscicollis* spp. *Proc. 3rd Int. Congr. Primat.,* Zurich 1970, vol. 3. Basel: Karger, pp.166–71.

Epple, G., 1972a. Social communication by olfactory signals in marmosets. *Int. Zoo Yearb.,* 12:36–42.

Epple, G., 1972b. Social behavior of laboratory groups of *Saguinus fuscicollis.* In: *Saving the Lion Marmoset,* D. D. Bridgwater, ed. Wheeling, W. Va.: Wild Animal Propagation Trust, pp.50–58.

Epple, G., 1975. The behavior of marmoset monkeys (Callitrichidae). In: *Primate Behavior: Developments in Field and Laboratory Research,* vol. 4, L. A. Rosenblum, ed. New York: Academic Press. pp.195–239.

Epple, G., and Lorenz, R., 1967. Vorkommen, Morphoiogie und Funktion der Sternaldrüse bei den Platyrrhini. *Folia primat.*, 7:98–126.

Epple-Hösbacher, G., 1967. Soziale Kommunikation bei *Callithrix jacchus* Erxleben, 1777. In: *Neue Ergebnisse der Primatologie: 1st Congr. Int. Primat. Soc.*, Frankfurt 1966, D. Starck, R. Schneider, and H.-J. Kuhn, eds. Stuttgart: Gustav Fischer Verlag, pp.247–54.

Harlow, H. F.; Harlow, M. K.; and Suomi, S. J.; 1971. From thought to therapy: lessons from a primate laboratory. *Amer. Sci.*, 59:538–49.

Heltne, P. G.; Turner, D. C.; and Wolhandler, J.; 1973. Maternal and paternal periods in the development of infant *Callimico goeldii. Amer. J. Phys. Anthropol.*, 38:555–60.

Hill, W. C. Osman, 1960. *Primates. IV. Cebidae,* Part A. Edinburgh: Edinburgh University Press, pp.1–523.

Jolly, A., 1972. *The Evolution of Primate Behavior.* New York: Macmillan. 397pp.

Kaplan, J., and Russell, M., 1974. Olfactory recognition in the infant squirrel monkey. *Dev. Psychobiol.*, 7:15–19.

Klein, L. L., 1971. Observations on copulation and seasonal reproduction of two species of spider monkeys, *Ateles belzebuth* and *A. geoffroyi. Folia primat.*, 15:233–48.

Klein, L., and Klein, D., 1971. Aspects of social behaviour in a colony of spider monkeys, *Ateles geoffroyi,* at San Francisco Zoo. *Int. Zoo Yearb.*, 11:175–81.

Kühlhorn, F., 1939. Beobachtungen über das Verhalten von Kapuzineraffen in freier Wildbahn. *Z. Tierpsychol.*, 3:147–51.

Lancaster, J. B., 1968. Primate communication systems and the emergence of human language. In: *Primates: Studies in Adaptation and Variability,* P. C. Jay, ed. New York: Holt, Rinehart and Winston, pp.439–57.

Latta, J.; Hopf, S.; and Ploog, D.; 1967. Observation on mating behavior and sexual play in the squirrel monkey *(Saimiri sciureus). Primates,* 8:229–46.

LeRoux, G., 1967. Contribution a l'étude des moyens d'intercommunication chez le Ouistiti à Pinceaux *(Hapale jacchus).* Thesis, Faculté des Sciences, Rennes.

Loizos, C., 1967. Play behaviour in higher primates: a review. In: *Primate Ethology,* D. Morris, ed. Chicago: Aldine, pp.176–218.

Lorenz, R., 1972. Management and reproduction of the Goeldi's monkey *Callimico goeldii* (Thomas, 1904) Callimiconidae, Primates. In: *Saving the Lion Marmoset,* D. D. Bridgwater, ed. Wheeling, W. Va.: Wild Animal Propagation Trust, pp.92–109.

Marler, P., 1965. Communication in monkeys and apes. In: *Primate Behavior: Field Studies of Monkeys and Apes,* I. DeVore, ed. New York: Holt, Rinehart and Winston, pp.544–84.

Marler, P., 1968. Aggregation and dispersal: two functions in primate communication. In: *Primates: Studies in Adaptation and Variability,* P. C. Jay, ed. New York: Holt, Rinehart and Winston, pp.420–38.

Mason, W. A., 1966. Social organization of the South American monkey, *Callicebus moloch:* a preliminary report. *Tul. Studies Zool.*, 13:23–28.

Mason, W. A., 1968. Use of space by *Callicebus* groups. In: *Primates: Studies in Adaptation and Variability,* P. C. Jay, ed. New York: Holt, Rinehart and Winston, pp.200–16.

Mason, W. A., 1971. Field and laboratory studies of social organization in *Saimiri* and *Callicebus.* In: *Primate Behavior,* vol. 2, L. A. Rosenblum, ed. New York: Academic Press, pp.107–37.

Maurus, M., and Pruscha, H., 1973. Classification of social signals in squirrel monkeys by means of cluster analysis. *Behaviour,* 47:106–28.

Moynihan, M., 1964. Some behavior patterns of platyrrhine monkeys. I. The night monkey *(Aotus trivirgatus). Smithsonian Misc. Coll.*, 146(5):1–84.

Moynihan, M., 1966. Communication in the Titi monkey, *Callicebus. J. Zool., Lond.*, 150:77–127.

Moynihan, M., 1967. Comparative aspects of communication in New World primates. In: *Primate Ethology,* D. Morris, ed. pp.236–66.

Moynihan, M., 1970. Some behavior patterns of plattyrhine monkeys. II. *Saguinus geoffroyi* and some other tamarins. *Smithson. Contrib. Zool.*, no. 28:1–77.

Muckenhirn, N. A., 1966. Vocalizations of the marmosets, *Saguinus o. oedipus* and *S. o. geoffroyi. Am. Zool.*, 6:550.

Muckenhirn, N. A., 1967. The behavior and vocal repertoire of *Saguinus oedipus* (Hershkovitz, 1966) (Callitrichidae, Primates). Master's Thesis, University of Maryland.

Napier, J. R., and Napier, P. H., 1967. *A Handbook of Living Primates: Morphology, Ecology and Behaviour of Nonhuman Primates.* New York: Academic Press, pp.1–456.

Neville, M. K., 1972a. The population structure of red howler monkeys *(Alouatta seniculus)* in Trinidad and Venezuela. *Folia primat.*, 17:56–86.

Neville, M. K., 1972b. Social relations within troops of red howler monkeys. *Folia primat.*, 18:47–77.

Nolte, A., 1958. Beobachtungen uber das Instinktverhalten von Kapuzineraffen (*Cebus apella* L.) in der Gefangenschaft. *Behaviour,* 12:183–207.

Oppenheimer, J. R., 1968. Behavior and ecology of the white-faced monkey, *Cebus capucinus,* on Barro Colorado Island, C.Z. Ph.D. diss., University of Illinois. 179pp. Ann Arbor, Mich.: University Microfilms (Diss. Abstr. 30, No. 1).

Oppenheimer, J. R., 1969a. Changes in forehead patterns and group composition of the white-faced monkey *(Cebus capucinus). Proc. 2nd Int. Congr. Primat.,* Atlanta, Ga., 1968, vol. 1. Basel: Karger, pp.36–42.

Oppenheimer, J. R., 1969b. *Cebus capucinus:* play and allogrooming in a monkey group. *Amer. Zool.,* 9:1070 (Abstr.).

Oppenheimer, J. R., 1973. Social and communicatory behavior in the *Cebus* monkey. In: *Behavioral Regulators of Behavior in Primates,* C. R. Carpenter, ed. Lewisburg, Pa.: Bucknell University Press, pp.251–71.

Oppenheimer, J. R., 1974. *Cebus* monkeys of Barro Colorado Island: Ecology and behavior, 16mm film, color. University Park, Pa.: Psychological Cinema Register.

Oppenheimer, J. R., and Oppenheimer, E. C., 1973. Preliminary observations of *Cebus nigrivittatus* (Primates: Cebidae) on the Venezuelan llanos. *Folia primat.,* 19:409–36.

Perachio, A. A., 1971. Sleep in the nocturnal primate, *Aotus trivirgatus. Proc. 3rd Int. Congr. Primat.,* Zurich 1970, vol. 2. Basel: Karger, pp.54–60.

Peters, M., and Ploog, D., 1973. Communication among primates. *Ann. Rev. Physiol.,* 35:221–42.

Ploog, D. W., 1967. The behavior of squirrel monkeys *(Saimiri sciureus)* as revealed by sociometry, bioacoustics and brain stimulation. In: *Social Communication Among Primates,* S. A. Altmann, ed. Chicago: University of Chicago Press, pp.149–84.

Ploog, D., and Melnechuk, T., 1969. Primate communication. *Neurosci. Res. Prog. Bull.,* 7:419–510.

Ralls, K., 1971. Mammalian scent marking. *Science,* 171:443–49.

Richard, A., 1970. A comparative study of the activity patterns and behavior of *Alouatta villosa* and *Ateles geoffroyi. Folia primat.,* 12:241–63.

Schön, M. A., 1971. The anatomy of the resonating mechanism in howling monkeys. *Folia primat.,* 15:117–32.

Snyder, P. A., 1972. Behavior of *Leontopithecus rosalia* (the golden lion marmoset) and related species: a review. In: *Saving the Lion Marmoset,* D. D. Bridgwater, ed. Wheeling, W. Va.: Wild Animal Propagation Trust, pp.23–49.

Sparks, J., 1967. Allogrooming in primates: a review. In: *Primate Ethology,* D. Morris, ed. Chicago: Aldine, pp.148–75.

Symmes, D., and Newman, J. D., 1974. Discrimination of isolation peep variants by squirrel monkeys. *Exp. Brain Res.,* 19:365–76.

Tembrock, G., 1963. Acoustic behaviour of mammals. In: *Acoustic Behaviour of Animals,* R. G. Busnel, ed. New York: Elsevier, pp.751–86.

Theberge, J. B., and Falls, J. B., 1967. Howling as a means of communication in timber wolves. *Amer. Zool.,* 7:331–38.

Thorington, R. W., Jr., 1968a. Observations of squirrel monkeys in a Columbian forest. In: *The Squirrel Monkey,* L. A. Rosenblum and R. W. Cooper, eds. New York: Academic Press, pp.69–85.

Thorington, R. W., Jr., 1968b. Observations of the tamarin *Saguinus midas. Folia primat.,* 9:95–98.

van Hooff, J. A. R. A. M., 1967. The facial displays of the Catarrhine monkeys and apes. In: *Primate Ethology,* D. Morris, ed. Chicago: Aldine, pp.7–68.

van Lawick-Goodall, J., 1968. The behaviour of free-living chimpanzees in the Gombe Stream Reserve. *Anim. Behav. Monogr.,* 1:161–311.

Wagner, H. O. von, 1956. Freilandbeobachtungen an Klammeraffen. *Z. Tierpsychol.,* 13:302–13.

Weigel, R. M., 1974. Facial expressions in the brown capuchin *(Cebus apella).* M.S. thesis, University of Illinois, pp.1–66.

Williams, L., 1967. *Man and Monkey.* London: Deutsch.

Winter, P., 1968. Social communication in the squirrel monkey. In: *The Squirrel Monkey,* L. A. Rosenblum and R. W. Cooper, eds. New York: Academic Press, pp.235–53.

Winter, P., 1969a. The variability of peep and twit calls in captive squirrel monkeys *(Saimiri sciureus). Folia primat.,* 10:204–15.

Winter, P., 1969b. Dialects in squirrel monkeys: vocalization of the Roman Arch type. *Folia primat.,* 10:216–29.

Winter, P., 1972. Observations on the vocal behaviour of free-ranging squirrel monkeys *(Saimiri sciureus). Z. Tierpsychol.,* 31:1–7.

Winter, P., and Ploog, D., 1967. Social organization and communication of squirrel monkeys in captivity. In: *Neue Ergehnisse der Primatologie,* D. Starck, R. Schneider, and H.-J. Kuhn, eds. Stuttgart: Gustav Fischer Verlag, pp.263–71.

Winter, P., Ploog, D., and Latta, J., 1966. Vocal repertoire of the squirrel monkey *(Saimiri sciureus),* its analysis and significance. *Exp. Brain Res.,* 1:359–84.

Chapter 35

COMMUNICATION IN OLD WORLD MONKEYS

J-P. Gautier and A. Gautier

From a theoretical point of view, various authors have emphasized the difficulties in selecting the features to be considered as communicative acts. The selection can be restricted to include only those signals emitted by an individual that elicit an overt response in a congener, insuring that the message was received. Nevertheless, as Smith (1965) notes, "the response to a signal need not be overt," but includes, according to Cherry, "total change of state, mental and physical." The ethologist, moreover, may broaden his selection to the point where everything has signal value, including the differential aspect of morphology of the members of a social group, which, as Marler (1965) points out, has a direct influence on the spatial distribution of individuals within the group. Thus, signaling displays in any way constituting active communication are accompanied by a sort of communication that could be called passive, and which, moreover, alone acts uninterruptedly throughout the group. We shall see in the course of this survey that a straightforward distinction between these two types of communication is not always possible.

Smith (1965) has shown how a message included in a signal emitted by an individual is, for the receiver, charged with supplementary information provided by the context of emission, including both the immediate and the historical contexts. The historical context assumes a greater importance when one considers species with extended ontogenetic development and long life-span. Thus, understanding communicative exchanges that coordinate the social life of a group of monkeys requires very long-term longitudinal studies. Such studies have been carried out in only a very few species.

Various surveys of communication in the primates and a number of specific studies conducted in the field and the laboratory have been completed in the last few years. Altmann (1968) has compiled an exhaustive bibliography of them. The difficulties encountered in developing a comparative synthesis of these works has often been pointed out: many studies are incomplete, accentuating visual repertoire in one instance and vocal exchanges in another. The situation is further complicated by the diversity of research methods. Certain authors adopt a strictly structural approach, others a situational point of view; some have set up elaborate functional categories. Though the structural approach describes the signals objectively, it cannot by itself satisfy the biologist. On the other hand, a classification of the functional type necessitates a schematization, as each type of signal occurs in a variety of contexts. Thus Green (1975) has recently shown that classifying the calls of the Japanese macaque according to a functional method provided a poorer correlation than did one based on the "emotional" approach, which takes into account the demeanor and the degree of excitation of the vocalizers.

We are really dealing here with different levels of study. The structural point of view allows for the establishment of a "morphological" classification of signals, which could be completed only after an ontogenetic study of their evolution. This first stage is not simple as far as the discrimination of elementary "units" (see Altmann, 1965) is concerned. The emotional and situational point of view should show us, with some elementary patterns, how signals are differentiated by the variations of the internal state of the vocalizers and the context of emission. Although this level of study is particularly interesting for understanding the nature of the transmitted message and the rules of a communicative system, it does not imply that one must drop the functional point of view. Whatever the nature and specificity of the mechanisms on which the nature and structure of these signals depend, it is important to understand how these signals play a more or less long-term role in social regulation.

The nature and frequency of different kinds of communicative exchanges evidently depend on the sensory capacities of the species in question. In the superfamily Cercopithecoidea, which encompasses all Old World monkeys, vision is a particularly well-developed mode of perception. It is commonly held that it predominates over the auditory and olfactory senses (Schultz, 1969). These monkeys possess binocular vision, "normal" color vision, and remarkable visual acuity (see Devalois and Jacobs, 1971). The absolute spectrum of auditory perception and sensation of monkeys is also close to ours, although certain species surpass us in the ultrasonic range (Stebbins, 1971). Similarly, their phonation organs are among the most highly evolved (Kelemen, 1963) and a number of species possess extremely developed laryngeal annexes (Starck and Schneider, 1960; Gautier, 1971). Their tactile sensory capacities are also highly developed, and Schultz (1969) speaks of the "admirable mechanism for tactile discrimination, which has become best perfected in the higher species of primates as a vital addition to the usefulness of their hands." The olfactory sense, on the contrary, has undergone a progressive atrophy among the simians.

Despite the abundance of fieldwork, the preliminary analysis of all kinds of signals has been accomplished in only a few species. In the last few years, however, studies on forest monkeys, many of which are still in progress, have become increasingly important. Thus, in this survey of communication in Old World monkeys, it will be possible for us not only to speak of baboons and macaques, species from which there has been a tendency to draw generalizations for application to the Cercopithecoidea, but also to dwell on various forest species, principally African guenons, with which we are particularly familiar.

This study is divided into three broad categories that take into account both the different levels of organization in a monkey population and the spatial distribution of its members. The first deals with the signals arising from exchanges between groups and tending to individualize groups within the population. The second brings to light the reactions of the group as a whole in relation to stimuli that are not conspecific and are outside the social environment. The third subsumes signals exchanged in interindividual relationships within a group. Although artificial, this classification permits one to approach a level of increasing complexity, paired with an ever-greater "personalization" of exchanges.

We shall try to emphasize the comparative

approach, but we must always keep in mind that although on the basis of anatomical and morphological criteria the family of Cercopithecidae constitutes a particularly homogeneous group among the primates (Schultz, 1970), it nevertheless occupies quite disparate habitats, ranging from the equatorial rain forests to the temperate zones. It also shows great diversity in social organization, from simple groups of four or five members, as in de Brazza's monkey (*Cercopithecus neglectus:* pers. obs.), to complex social organizations with several hundred individuals, as in the hamadryas baboon (*Papio hamadryas:* Kummer, 1968).

If, as Kummer (1971b) notes, the direct cause of social structures is the behavior of adult troop members, one should try to grasp the reciprocal relations that exist between the nature and frequency of interindividual exchanges within groups of a given species and its mode of social organization. In the following pages we shall see whether the habitat of a species, its patterns of interaction, and the nature of the signals exchanged can be meaningfully correlated.

Signals That Individualize The Social Units within the Population

Distribution of monkey populations into social units is the rule among Old World species, even though the concept of the group as a definite entity is not always easy to grasp. The existence of relatively stable groups in a population, in contrast to a nomadic life with labile groups, as in chimpanzees (Goodall, 1965; Sugiyama, 1968a), does not mean that these social units have no exchanges among themselves: most studies show, on the contrary, that a fairly important part of specific signals occurs in intergroup relationships. As Rowell (1972) notes, monkeys of neighboring groups "are intensely interested in whatever they can see and hear of them; in a communicative sense, each monkey belongs to the whole local population of his species as well as to his immediate group."

This spatial distribution presupposes a certain number of mechanisms by which groups are individualized. According to Kummer (1971b), three imperatives are required: "to keep members of each group together; to separate the groups; and to tie each group to a piece of land." They are assumed by communication mechanisms.

The cohesion of group members can rest on two different modalities of signal action: (1) All individuals exchange signals permitting them to maintain contact; this type functions at close range (see intragroup communication). (2) Wide-range signals, uttered by a few individuals, gather the members of the group around the vocalizers. This pattern of signal generally has a sufficiently wide range to be perceived by neighboring bands; thus it simultaneously assumes the intergroup communication function of maintaining distance (Marler, 1968). Increase in distance can be accomplished either by wide-range signals, if the groups do not come into visual contact, or by proximity exchanges, if the nature of the relationships tends to draw the groups together.

Generally speaking, these signals are characteristic of adult males. Their nature depends on the distance at which the exchange takes place and on the structure of the habitat in which the species is evolving. It is possible to classify monkeys into two categories: (1) Species in which there is no visual contact between groups, either because of the size and/or shape of their home ranges or because of habitat density. In the former there are few or no exchanges of species-specific signals (patas [*Erythrocebus patas*]: Hall, 1968b; talapoin [*Miopithecus talapoin*]: Gautier-Hion, 1971a); in the latter the exchanges are essentially of an acoustic nature (as in most forest guenons). (2) Species for which the wanderings of individuals from different groups lead actively or passively to a closed proximity; such encounters, favored by open habitats, permit all sorts of exchanges, particularly visual ones. The existence of species-specific signals depends on the individualization of groups and their mutual tolerance, and it becomes difficult to be precise about the concept of social unit when the exchange and mingling of individuals happens without any specific manifestation (the case in the geladas [*Theropithecus gelada*]: Crook, 1966; and in the hamadryas: Kummer, 1968). In other cases, individualization depends on mutual avoidance (olive and chacma baboons [*P. anubis* and *ursinus*]: DeVore and Hall, 1965) or on an intergroup hierarchy of dominance (rhesus monkey [*Macaca mulatta*]: Southwick et al., 1965): the signals exchanged are those of the usual aggressive repertoire.

The olfactory capacities seem little used in intergroup relations (see below). Olfactory markings appear to be a form of vestigial behavior and operate with other kinds of signals.

VOCAL SIGNALS

Carpenter (1934) was the first to point out the vocal mechanisms used for group spacing, along with the roaring of male howlers *(Alouatta palliata).* Subsequently, many authors have emphasized the existence of powerful vocalizations acting in intergroup communication in Old World monkeys. Such signals exist in langurs (hanuman langur and ceylon gray langur [*Presbytis entellus*]: Jay, 1965a; Sugiyama et al., 1965; Ripley, 1967; Yoshiba, 1968; Vogel, 1973; nilgiri langur [*P. johnii*]: Tanaka, 1965; Poirier, 1968, 1970a, 1970b; lutong [*P. cristatus*]: Bernstein, 1968), in the African *Colobus* spp. (black and white colobus [*Co. guereza*[1]]: Ullrich, 1961; Schenkel and Schenkel-Hulliger, 1967; Marler, 1969, 1972; black colobus [*Co. satanas*]: Sabater Pi, 1970; angolan colobus [*Co. angolensis*]: Groves, 1973), in the macaques (Japanese macaque [*M. fuscata*]: Itani, 1963; lion-tailed macaque [*M. silenus*]: Sugiyama, 1968b; crab-eating monkey [*M. fascicularis* or *M. irus*]: Shirek-Ellefson, in Marler, 1968; Kurland, 1973; pig-tailed macaque [*M. nemestrina*]: Chivers, 1973), in the baboons (*P. ursinus:* Hall, 1968a), in the patas (Hall, 1965, 1968b), in the mangabeys (grey-cheeked mangabey [*Ce. albigena*]: Chalmers, 1968; Waser, 1975; agile mangabey [*Ce. galeritus*]: Quris, 1973), and finally in a large number of other species of *Cercopithecus* (spot-nosed guenon [*C. nictitans*], moustached monkey [*C. cephus*], crowned guenon [*C. pogonias*]: Gautier 1969; Struhsaker, 1969; mona monkey [*C. mona*], campbell's monkey [*C. campbelli*], red-eared guenon [*C. erythrotis*], l'hoest's monkey [*C. l'hoesti*], lesser spot-nosed guenon [*C. petaurista*]: Struhsaker, 1970; *C. campbelli:* Bourlière et al., 1970; red-tailed monkey [*C. ascanius*], blue monkey [*C. mitis*]: Aldrich-Blake, 1970; Marler, 1973; de Brazza's monkey [*C. neglectus*]: Gautier, 1971).

The most obvious general characteristic of these calls is their exceptional intensity and range: the roaring of the guereza can be heard for more than a mile, and the "Pm1" of the spot-nosed guenon carries as much as a kilometer. Despite their diversity, these vocalizations are all of a pure, low-pitched structure and seem particularly adapted for penetrating a woodland habitat.

They are uttered primarily by adult males. Ontogenetic development studies show that they do not appear in males before sexual maturity, between five and seven years of age (*C. nictitans, pogonias, neglectus, petaurista, Ce. albigena, galeritus:* Gautier, 1971, 1973, and pers.obs.). Fig. 1 illustrates these facts in several species raised in captivity.

Depending on the species, there can be from one to four vocalizations of high intensity; in

1. To avoid confusion, we will indicate the genus *Cercopithecus* by *C.*, *Cercocebus* by *Ce.*, and *Colobus* by *Co.*

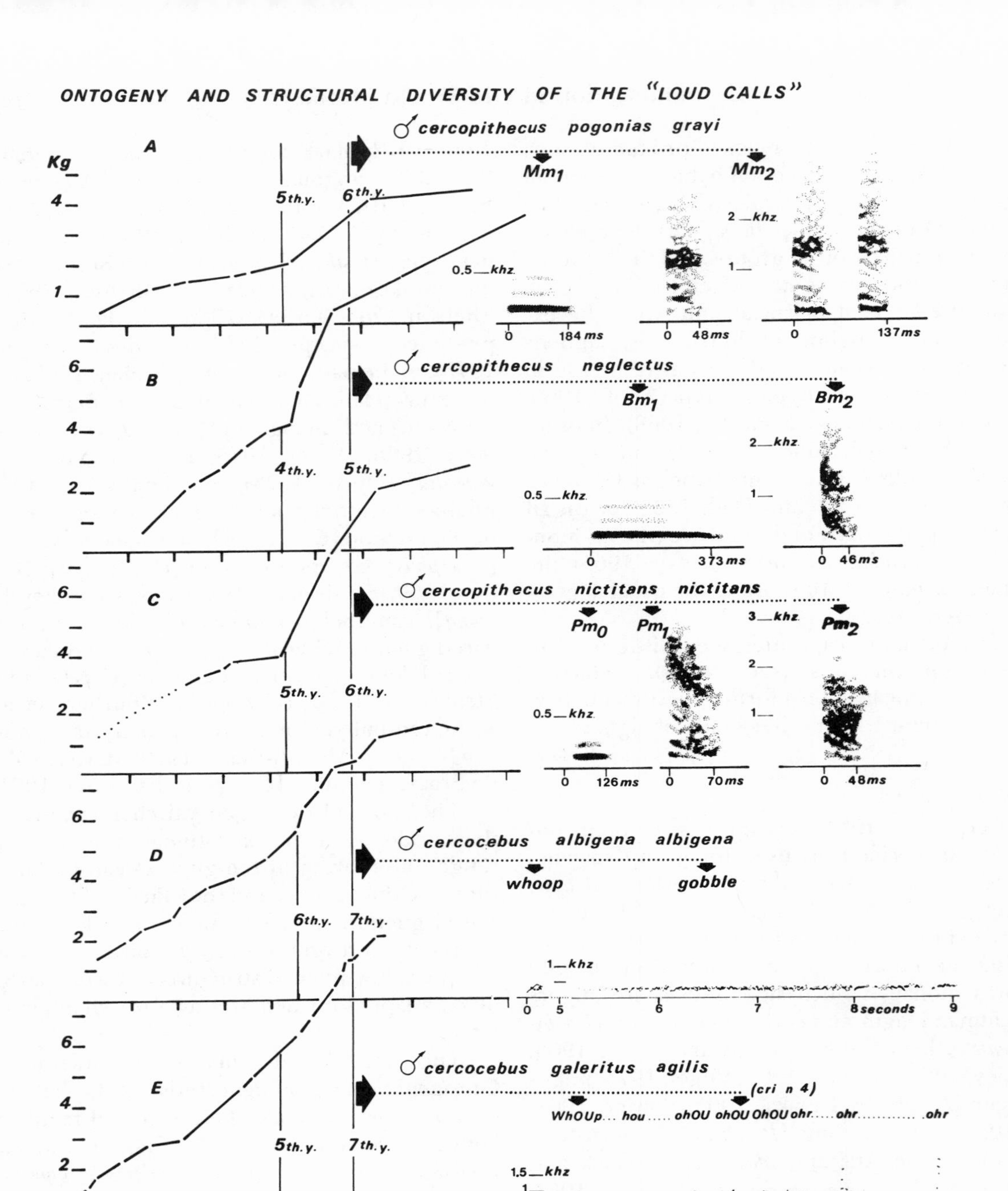

Fig. 1. Ontogeny and spectrographic analyses of the "loud calls" of five *Cercopithecus* species. The period of sexual maturation in males is indicated by a break in the weight curve. Loud calls are emitted for the first time at the end of this period, be it at 5, 6, or 7 years of age. A: *C. pogonias:* Mml, boom-type call, type 1 loud call; Mm2, unitary and binary barks, type 2 loud call. B: *C. neglectus:* Bml, boom-type call, type 1 loud call; Bm2, unitary barks, type 2 loud call. C: *C. nictitans:* PmO, boom-type call, seldom manifested; Pm1 and Pm2, barks, types 1 and 2 loud calls. D, E: *Ce. albigena* and *Ce. galeritus:* whoop gobble, type 1 loud call.

intergroup relations, however, only one or two types are used.

Type 1 Loud Calls

In addition to their intensity, these calls are notable for their structural originality, in contrast to other vocalizations of specific repertoires, and for their slight variability in structure, which is stereotyped and discrete (Marler, 1972), as much at the species as at the individual level. Thus the "booms" of a male *C. pogonias* or the barks of a *C. nictitans* maintain great stability in their fundamental structure, in spite of aging of the animals, the only change being an increase in intensity and a modification of timbre. In some species, on the other hand, the calls of different individuals can be distinguished by ear (*Cercopithecus* spp.: pers.obs.). This interindividual variability, which supposedly enables congeners to identify the vocalizers, rests on slight transformations of fundamental or accessory structures. Waser (1975) has shown through playback experiments that in the grey-cheeked mangabey males can distinguish "whoop gobbles" uttered by stranger group males from those emitted by members of their own band. Other monkeys, like the guereza, have such stereotyped vocalizations that individual recognition of emitters seems improbable (Marler, 1972).

The diversity of structures at the interspecific level is great, but schematically four categories can be recognized: (1) calls of the "boom" type, given principally by representatives of the superspecies *mona* (Mm1, Fig. 1 A); (2) "barks" of various guenons and some macaques (Pm1, Fig. 1 B); (3) "whoops," sometimes followed by "hiccups" (langurs) or by "gobbles" (mangabeys; Fig. 1 D, E); (4) "roaring" of the colobus. The category of barks is by far the most heterogenous.

Proceeding from these categories one notes: (1) Calls of similar structure occur simultaneously in phylogenetically close species (same genus) and relatively remote species (different genera), e.g., the booms of the superspecies *mona* are observed in de Brazza's monkey and to a lesser degree in *C. mitis* and *C. nictitans* (Gautier, 1971, 1973; Marler, 1973) (Mm1, Pm1; Fig. 1 A, B, C); and the bark of *C. nictitans* is practically indistinguishable from the "pyow" of *C. mitis* (Gautier, 1969; Aldrich-Blake, 1970; Marler, 1973) and seems close to the D1 calls of the Japanese macaque (Itani, 1963) if one refers to Itani's phonetic transcription. Furthermore the complex call structure of the mangabeys described by Chalmers (1968) for *Ce. albigena* and by Quris (1973) for *Ce. galeritus,* and which probably exists in *Ce. torquatus* (Struhsaker, 1970), includes a whoop of pure structure (pitch: 200–300 Hz) that seems to correspond to the whoop of the langur, is close in pitch (300–400 Hz: Vogel, 1973, for *P. entellus*), and is perhaps comparable to that of *M. silenus* (Sugiyama, 1968b). Furthermore, the hiccup that follows the whoop in *P. johnii* (Poirier, 1970a, 1970b) may be related to the gobble of the mangabeys.

(2) Conversely, phylogenetically close species can possess very different loud calls. This is the case in the two monkey pairs *C. nictitans–C. cephus* and *C. mitis–C. ascanius,* the usual vocalizations of which are very similar (Gautier, 1969; Marler, 1973). Generally speaking, species possess increasingly differentiated loud calls in direct proportion to their degree of sympatry, and *a fortiori* when actually living in polyspecific association (Gautier and Gautier-Hion, 1969; Gautier, 1973). These calls would therefore have an important role in speciation.

In addition, species differentiation is brought about through temporal patterns and emission sequences. In various species, loud calls are in effect uttered in series whose temporal organization is often elaborate (Marler, 1968). These sequences can include from one to three different vocalizations, sometimes associated with nonvocal sounds. Fig. 2 illustrates these sequences in

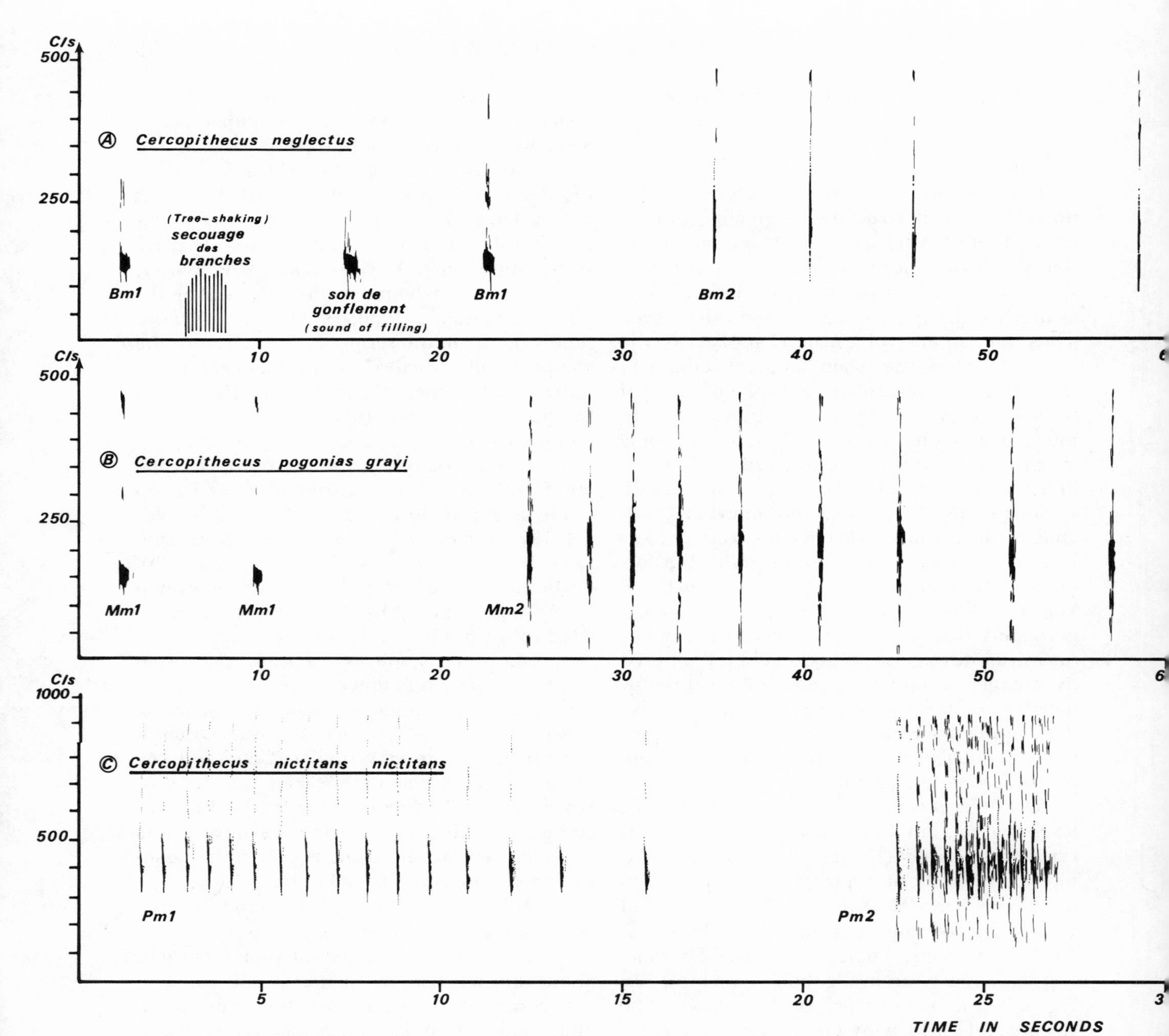

Fig. 2. Spectrographic analyses of the emission sequences of loud calls in *Cercopithecus neglectus, C. pogonias,* and *C. nictitans.* A, B: *C. neglectus* inserts tree shaking and a sound produced by inflating the vocal sac between two booms (see also Fig. 3), whereas *C. pogonias* can emit only one or several booms. The barks (type 2 loud calls, Bm2 and Mm2) also have different structures and temporal emission patterns, which accentuate sequence-specificity. C: In *C. nictitans,* type 1 loud calls (Pm1) and type 2 loud calls (Pm2) each have different emision rhythms during stereotyped sequences.

three sympatric species, *C. nictitans, neglectus,* and *pogonias.* In the last two, the calls are similar, species differentiation being based on emission sequences.

Another frequently emphasized characteristic is the diurnal periodicity in the emission of these calls. The rhythm may be monophasic, with calls being uttered predominantly in the morning: *P. entellus* ("morning whoops," Jay, 1965a), *Ce. albigena* (Waser, pers. comm.), *Ce. galeritus* (Quris, 1973); or in the evening: *C. mona* (Struhsaker, 1970). It can be biphasic, the calls being more frequently given morning and evening, as in *C. nictitans* (Gautier [Belinga region], 1969) and *Co. guereza* (Marler, 1972); or polyphasic, as in *C. nictitans* and *C. pogonias* living in a mixed troop (Gautier-Hion and Gautier [M'passa region] 1974). Thus the same species can, according to the region, present different emission rhythms or none at all (cf. Struhsaker [Cameroun], 1970 for *C. nictitans).* The increase in the frequency of the emission of loud calls in the morning and evening shows that they are principally uttered close to sleeping sites (Hall, 1968b; Aldrich-Blake, 1970; Gautier-Hion and Gautier, 1974).

From the work of Stark and Schneider (1960) and Kelemen (1963) it is known that the phonation organs of the primates are among the most highly developed. Hill and Booth (1957), working with *Colobus* spp., and Gautier (1971), studying various African guenons, have offered evidence suggesting that males of species that emit strong vocalizations possess either a well-developed larynx (*Colobus* spp. except the red colobus, *Co. badius,* which does not utter any loud calls: Marler, 1970), or hypertrophied vocal annexes (other species).

Itani (1963) has noted a distension in the submaxillary region in Japanese monkey males during the emission of D1 calls. Our observations in *C. neglectus, pogonias,* and *nictitans* show that vocal sacs in effect play an active role in the emission of some loud calls. Fig. 3 analyzes the emission behavior of boom-type calls in a *C. neglectus* male, in which the inflation of the vocal sac precedes the emission. If a hole is punched in this sac, the call will be of low intensity for the sac acts as a resonator and amplifier (Gautier, 1971). The participation of these annexes is less evident in other species. In most cases, however, the loud vocalization, preceded by a lowering of the head, is uttered when the head is thrown back violently (Fig. 4).

Such complex vocal behavior necessitates a certain amount of "concentration" from the vocalizer and indicates, on the behavioral level, the relative quietude and weak motility of the animal. These characteristics differentiate loud calls of the first type from those of the second.

Each monkey group has one or several male emitters, e.g., there are several in *Ce. albigena* (Waser, 1975), various langurs, and *M. fuscata,* but only one in groups of *Colobus* or *Cercopithecus* spp. There is no doubt that these individuals have the status of social leaders. Hunkeler et al. (1972) estimate that in *C. campbelli,* the emitter of loud calls must be the exclusive genitor. In fact, when dominance relations are established between two adult males in species where a single vocalizer is observed *in natura,* the loud calls of the subordinate male can be totally or partially inhibited (Gautier, 1971). This throws a new light on the social structure of harems, which would only in some cases correspond to one-male "status" group.

The ritual character of loud calls, resulting from their stereotyped nature and their daily rhythm of occurrence, is reinforced by the fact that it is often difficult, especially for the morning and evening emissions, to find evidence of any stimuli provoking them (Marler, 1968). For this reason, numerous authors employ the term "spontaneous," an adjective used to describe other calls of mammals (e.g., the bell of the deer: Kiley, 1972).

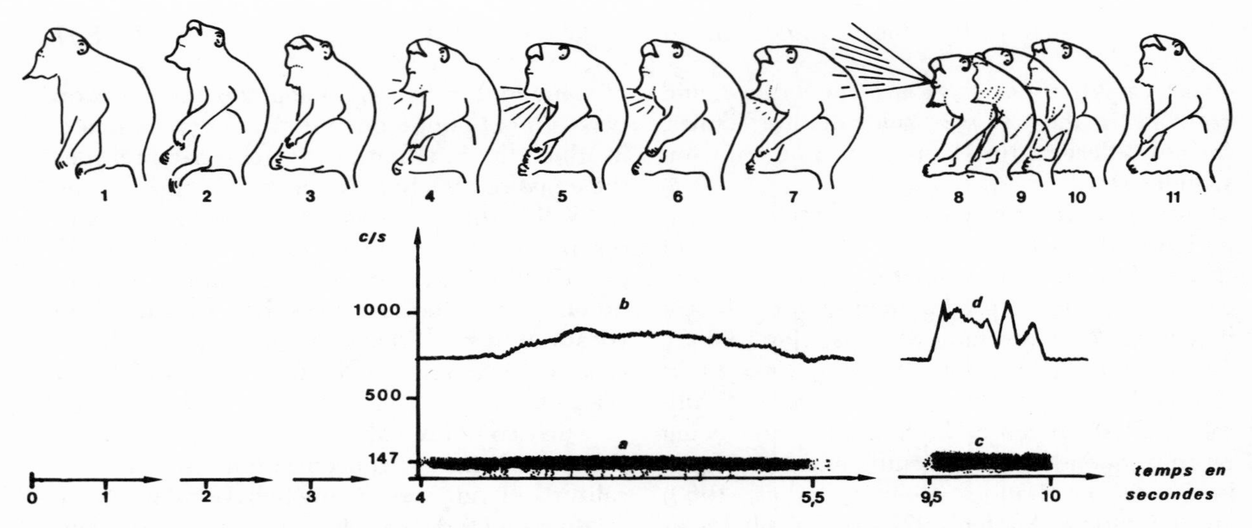

Fig. 3. Behavioral sequence accompanying the emission of booms (type 1 loud call) in *Cercopithecus neglectus.* 1–3: preparatory phases; 4–7: inflation of vocal sac accompanied by sound (spectra a, b); 8–10: emission of boom with deflation of the vocal sac (spectra c, d). (From Gautier, 1971, in *Biologia gabonica.*)

Fig. 4. Emission behavior of type 1 loud calls in an adult male *Cercopithecus nictitans.* (Paimpont; from photographs.)

For most species, what most often provokes the emission of type 1 loud calls is the production of the same calls by males from neighboring groups. Nevertheless, the existence of phonoresponses varies. They have never been observed in *C. cephus* or *C. ascanius* (Gautier, 1969; Marler, 1973) and are also rare in *C. pogonias.* These variations are a result of both population density and differences in call function (e.g., absence of territoriality in *C. cephus:* Struhsaker, 1969).

Type 1 loud calls are also heard at the time of close-range intergroup exchanges, in which they are produced predominently at the end of an encounter, when calm has been restored. Thus the roaring of guereza, like the Pm1 of *C. nictitans,* appears after interactions between groups, when the vocalizer has left the area of conflict (Marler, 1972; Gautier, 1969). Similarly, the D1 of the Japanese macaque (Itani, 1963) is uttered by the male only after the other members of the group have taken cover.

Furthermore, these vocalizations are given following numerous stimuli that are not specific; the lists of these stimulations, given by various authors, are astonishingly similar. Some are social, such as those produced by alarms from congeners; while others result from various happenings—breaking branches, falling trees, thunderclaps, rifle reports, birds of prey, approach of humans—prompting Struhsaker (1970) to state that these calls are evoked by situations "all of which seem to have in common, the potential to disrupt the coherence of the social group."

Here again, these loud calls appear only as deferred responses to those stimuli that are immediately followed by type 2 loud calls or various aggressive alarms (see below). Type 1 calls are therefore uttered by animals that have returned to a lower state of excitation.

Experiments aimed at provoking these vocalizations in captivity have shown that any disturbance threatening the integrity of the group will release them. This always happens if a female is removed from the group. Type 2 loud calls are uttered during her capture, and upon her reintroduction, the male leader immediately tries to copulate with her and emits a long series of type 1 loud calls.

According to the literature, loud calls seem to fulfill two types of function (see Marler, 1968): (1) An intragroup function of cohesion and rallying. Not only do such calls permit localization of the vocalizers but also, by provoking a new outbreak of contact calls from congeners, they facilitate interindividual localization, especially after a disturbance. Furthermore, the calls given in the morning and evening clearly have the function of orienting movement—they are followed by the group's either getting under way or stopping its activity (Itani, 1963; Jay, 1965a; Shirek-Ellefson, in Marler, 1968; Gautier, 1969; Gautier-Hion and Gautier, 1974). (2) A function of maintaining intergroup distance, assumed principally by the phonoresponses between males of adjacent groups. Thus in areas of high density, as is the case for groups of *C. nictitans* in northeastern Gabon, it is not rare for the males of three or four adjacent groups to exchange their loud calls upon awakening. In this way the groups are informed of the position of their neighbors before beginning any activity. Aldrich-Blake (1970) notes than in *C. mitis* the response of a male varies according to whether he is at the center of his home range. In the langurs, on the other hand, there is an active search for close contacts between groups, the males going several hundred meters away from their own groups in order "to challenge each other" vocally and visually (Ripley, 1967; Poirier, 1970b).

The fact that loud calls depend on hormonal mechanisms (see above, ontogeny) implies a more or less direct correlation with the phenomenon of reproduction. Hill and Booth (1957) thought that the loud calls of the colobus were an "assertion of status." Going back to the ideas of

Darling, Poirier (1970b) suggests that intergroup spacing displays are simultaneously the social stimuli necessary for success in reproduction. This seems confirmed by the observations of Hunkeler et al. (1972, see below) with a group of monas. The vocal "outbidding" observed in males of various species in mixed troops (Gautier-Hion and Gautier, 1974) should also be understood as an assertion of status. With these males, it is not a question of distancing and spacing, since they live in permanent association, but rather a mutual yet somewhat competitive affirmation of their identity (in terms of species, age, sex, and status) (see also Marler, 1973).

The strict species specificity of the loud call structures and/or sequences for sympatric species and especially for those living in association, leads one to think that these vocalizations are behavior patterns adapted to assure individualization of social groups and species isolation in reproduction. The loud calls of male monkeys who are social leaders of their groups would thus play a role close to that of the songs of birds, in which females select males on the basis of vocalizations (Hinde, 1970).

Type 2 Loud Calls

In the well-known calls of the howlers, Altmann (1959) makes a distinction between roaring and male barks. The first is a "proclamation of an occupied area," the second a sort of alarm call uttered during disturbances. These two categories exist in the Japanese macaque (Itani, 1963) with the D1 calls (see above) and C5 barks, the latter constituting an aggressive vocalization in males.

These two examples give the basic features that characterize the type 2 loud calls, which are found in most forest guenons, in some macaques (*M. fuscata, M. mulatta*: Altmann, 1962; Southwick, 1962; Lindburgh, 1971 ["soft bark"]; and perhaps *M. fascicularis:* Kurland, 1973 [*"kra vocalizations"*]; and finally in the patas monkey (Hall, 1965) and the vervet monkey (Struhsaker, 1967b).

These calls can occur in the same sequence as those of type 1; they are of comparable intensity but they do not have such original structures and are most often related to the aggressive alarm vocalizations of the given species' repertoire. This relationship varies according to the species; in some, the only originality lies in the particularly low voice of the male vocalizers (e.g., *M. fuscata*); in others, such as *C. neglectus* and *pogonias,* the gradation is total. In *C. pogonias,* loud calls and the usual aggressive alarm calls coexist in the adult male and can evolve toward each other (Fig. 5) (pers.obs.).

Because of such partially graded structures, these calls are less stereotyped and more variable than those of the first type. Marler (1973) remarks that the "pyow" of *C. mitis* and the "hack" of *C. ascanius* (type 1) diverge much more than do the "ka calls" (type 2) of the two species. The same thing is found in *C. nictitans, cephus, pogonias,* and *neglectus* (cf. Fig. 1: Mm2, Bm2, Pm2).

In addition, species distinction is gradually reduced as the excitement level of the vocalizers rises. In an excited male *C. nictitans,* for example, the Pm2 calls uttered in series can give rise to phrases including inspirated and expirated units (similar to the ka calls of *C. mitis* and *ascanius*), which also appear in *C. neglectus.* Thus two species (*nictitans* and *neglectus*), whose usual repertoires are of fundamentally different structures, can find common vocal structures for type 2 loud calls when the arousal level of the vocalizer is raised.

Species distinction can, however, be brought about on the basis of sequencing. Two cases are to be considered: (1) Type 2 loud calls occur in the same sequence as those of type 1. The sequence is then stereotyped, type 2 calls following the others according to a rather stable temporal pattern. This is the case for the spontaneous

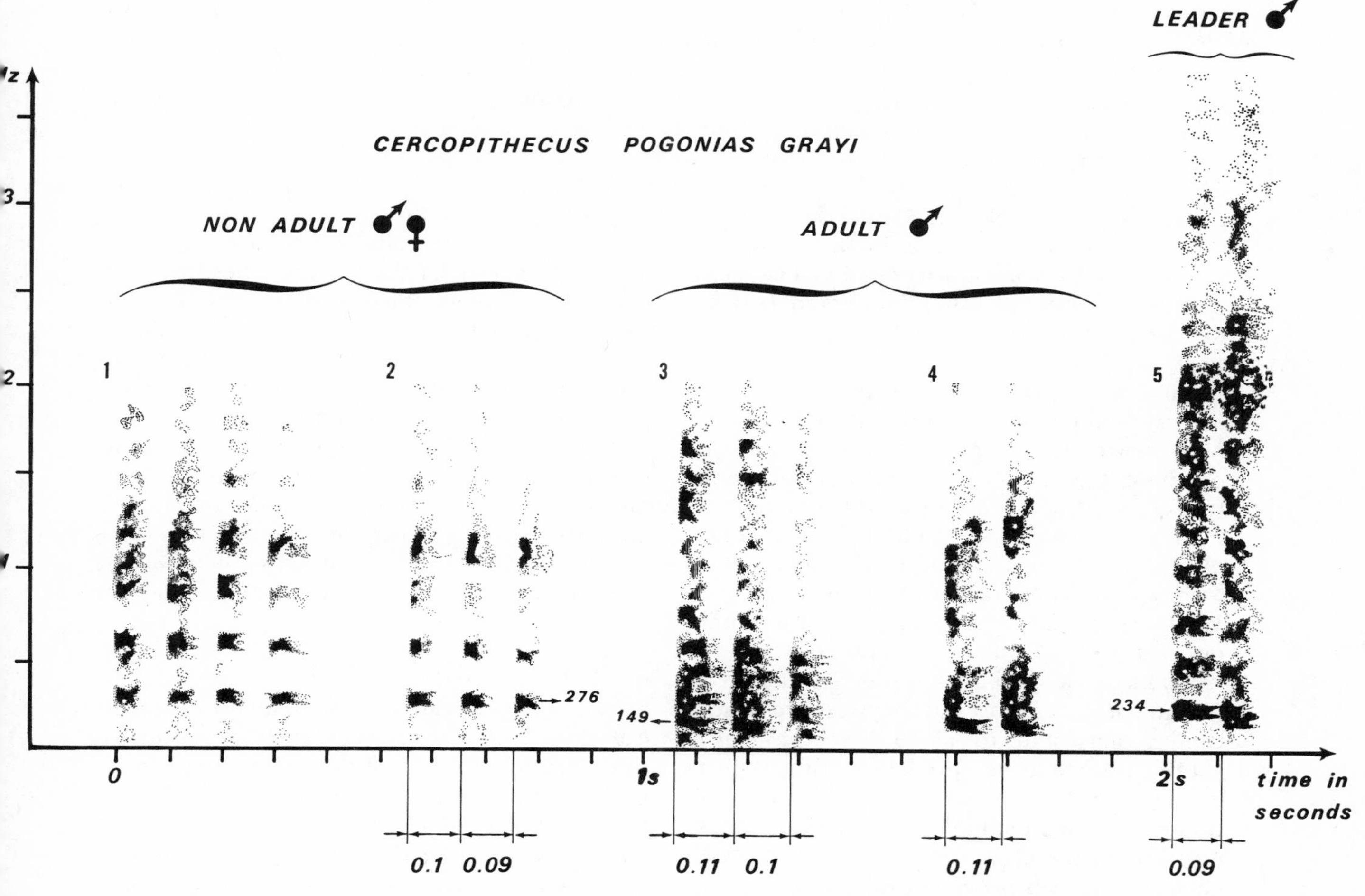

Fig. 5. Aggressive alarms of *C. pogonias* and their development into type 2 loud calls. 1, 2: Quaternary and ternary low-pitched rhythmic calls of non-adult individuals. 3, 4: Ternary and binary low-pitched rhythmic calls of an adult male (note the drop in the fundamental sound from 276 to 149 Hz). 5: Binary loud call (Mm2) of the adult male, which can develop into calls 3 and 4.

morning and evening emissions of the forest guenons. (2) Type 2 loud calls are emitted alone in response to various stimuli (*Cercopithecus* spp., macaques). The emission patterns become less stable and the variability in calls increases with the excitation level of the vocalizers.

Type 2 calls, which do not appear to set the vocal sacs actively into play (Gautier, 1971), are associated with more or less rapid movements of the uttering monkeys, which jump from branch to branch and can interrupt their sequences with tree shaking. Such behavior indicates a high excitation level.

The emitters are adult males, but the exclusiveness of loud calls, which in type 1 are reserved for a group's male social leader, is not as imperative for type 2. Thus in *C. pogonias,* a single male per group utters booms but he is often seconded by another male during emissions of type 2 calls (pers.obs.).

In the course of individual maturation, type 2 loud calls appear in males before those of type 1. They occur just at the moment of sexual maturity even though the individual may not have acquired any particular status in the group that would confer social maturity on him. The observations of loud-call inhibition noted between male competitors do not apply to type 2 vocalizations.

In stereotyped sequences type 2 calls can have a certain spontaneous character. In the guenons, however, during intergroup vocal battles, one notes that the relative percentage of the two types of loud calls is determined by the distance between uttering males. Table 1 shows that in *C. nictitans* the number of type 2 calls varies directly with closeness of adjacent groups (Gautier, 1969). When the two groups are in proximity only this type of emission is uttered.

In macaques and guenons, type 2 calls occur as immediate responses to the presence of adjacent groups and to all non–species-specific stimuli noted above (predators, rifle reports, etc.) that indicate potentially predatory situations. In the mountainous regions of northeastern Gabon, the flight of a crowned-hawk eagle effectively reveals groups of monkeys, since the male of each group releases his loud calls as the eagle flies over. On the other hand, the bark of the male patas is rarely heard and occurs only during exchanges between groups (Hall, 1968b).

These vocalizations are followed by alarm calls, aggressive alarms, or by defensive or aggressive calls from group members, according to the nature and perception of the danger. If it becomes imminent, all the monkeys may flee except for the one or more males who remain to face the danger (e.g., Japanese macaque: Itani, 1963; *M. fascicularis,* Kurland, 1973). In this case, male calls play a role in alarm diffusion and congener defense. Gabonese hunters are acquainted with this behavior and imitate type 2 calls of *C. nictitans,* thereby attracting the male and killing it while its congeners flee.

Between groups, loud calls provoke phonoresponses associated with various aggressive displays—e.g., visual threats, tree shaking, jump display, aggressive calls—which are then followed by spacing of the groups. These aggressive vocal displays thus function effectively in increasing intergroup distances. Nevertheless, their functional potentiality varies according to species and population density and can act either at a certain distance, without visual cues (as in many forest monkeys) or with the reinforcement of other aggressive displays of various natures (as in colobus and macaques). These exchanges are thus less stereotyped and can differ according to the reciprocal familiarity of adjacent groups and their mutual tolerance.

NONVOCAL ACOUSTIC SIGNALS

These signals can occur separately or in association with loud calls. The most important is tree shaking or branch shaking. It is widespread in the following species: *Macaca radiata* (Nolte, 1955), *M. mulatta* (Southwick, 1962; Altmann, 1962; Lindburgh, 1971), *M. fuscata* (Imanishi, 1957; Itani, 1963), *M. nemestrina* (Bernstein, 1967), *M. irus* (Shirek-Ellefson, in Marler, 1968), stumptail macaque (*M. speciosa:* Bertrand, 1969), barbary macaque (*M. sylvana:* Deag, 1973), *M. fascicularis* (Kurland, 1973), mangabeys (*Ce. albigena, Ce. torquatus, Ce. galeritus:* pers.obs.), colobus (*Co. guereza:* Marler, 1972), and guenons (*C. nictitans, cephus, pogonias, neglectus:* pers.obs.; *C. aethiops:* Struhsaker, 1967b; *C. campbelli:* Bourlière and al., 1970). In Lowe's mona, tree shaking is associated with "jumping around." Only this form exists in the langurs ("jump display," *P. entellus:* Ripley, 1967; Jay, 1965a; *P. johnii:*

Table 1

Type 1 and type 2 loud calls exchanged between two male C. nictitans leaders, according to the distance separating their two groups.

Inter-♂♂ distance	No. of loud calls of the initiator ♂		No. of loud calls of the respondent ♂		% of loud calls exchanged between ♂♂	
	Type 1	Type 2	Type 1	Type 2	Type 1	Type 2
200 m	15	21	12	15	43%	57%
400 m	10	3	7	1	85%	17%

Poirier, 1970a, 1970b), and the signals emitted are simultaneously acoustic and visual.

In *C. neglectus,* the sequences of type 1 loud calls necessarily include violent branch shaking intercalated between booms (Fig. 2). Identical phenomena are observed in the langurs, where the jump display is intermingled with the whoops and hiccups of the males.

Association of shaking with type 2 loud calls is even more frequent (*C. campbelli, nictitans,* or *pogonias*). Tree shaking is also associated with the C5 call of the Japanese macaque and with the "ho-ho-ho" of the rhesus (Itani, 1963; Altmann, 1962; Southwick, 1962). In the absence of strong calls, it can occur alone, as in the guereza.

The manifestation of such displays in intergroup relations indicates an aggressive tendency in the animals giving them and seems to be, as Southwick (1962) thought, "threatening gesture to members of adjacent groups." They reinforce the action of other types of interspacing behavior through their emphatic manifestations.

VISUAL SIGNALS

The most common intergroup visual signal is the jump display of the langurs or the jumping around of the guenons, which has an obvious acoustic component (Ripley, 1967) and constitutes an intimidation pattern (Poirier, 1970a). Other rare postures are also observed, such as the tail-erected behavior of male vervets (Struhsaker, 1967a) or the penile display of the guereza (Marler, 1972), vervet, and proboscis monkey (*Nasalis larvatus:* Wickler, 1967). According to Wickler, these postures function "as optical markers of the presence of the group or of its territorial boundaries, largely as a warning to conspecifics." In *P. johnii,* Poirier (1970a) also points out stopping postures in which the male turns his tail down over his head.

In some species, the conjoint approach of two conspecific groups provokes vigilance behavior between the males of the different units. This is the "observational tonus" of *P. entellus* (Ripley, 1967) sometimes associated with aggressive displays (Fig. 6) or the "vigilance behavior" of *P. johnii* males (Poirier, 1970a). Simonds notes similar behavior in *M. radiata,* but the most precise description has been offered by Deag (1973) in the barbary macaque under the name "monitoring behavior."

When groups of langurs approach each other, conflicts arise in which the exchanged signals are those from the aggressive repertoire. Ripley (1967) notes that in the langur of Ceylon (*P. senex*) the occurrence of agonistic behavior is almost exclusively limited to these encounters. In the rhesus of India (Southwick et al., 1965) and the provisionized Japanese macaques (Kawanaka, 1973) aggressive behaviors are particularly numerous and intergroup hierarchies may be established (Fig. 7). This kind of ex-

Fig. 6. Seated "on the lookout," apart from his own group, a male langur (*P. entellus*) turns toward a neighboring group, passing by on the ground at 50 m, and manifests an aggressive display. (Polonnaruwa, Ceylon; photo by C. M. Hladick.)

Fig. 7. Aggressive exchanges between two groups of Japanese macaques. Young adult males and adult females of group B (left) face the male leader of group A (right). (Takasakiyama; photo by K. Kawanaka, 1973, in *Primates.*)

change seems to be the direct consequence of overpopulation.

OLFACTORY SIGNALS

The importance of olfactory signals in territorial markings, well known in the prosimians and some New World monkeys, seems reduced in those of the Old World. Only Gartlan and Brain (1968) have observed sequences of stereotyped markings that appear to act in territorial spacing in the vervet of Lolui Island. In these animals, whose territory is well defined, markings made by adult males and females are observed principally near territorial limits: one kind is effected by rubbing the corner of the jaw and cheek against a prop; another kind, more intense, is done with the chest. Each rubbing is followed by long and careful sniffing.

Curiously, identical markings are observed in de Brazza's monkey in captivity (pers.obs.) (Fig. 8) when they are introduced into a new cage, or when a new branch is put into a familiar enclosure. Adult males generally mark first and are followed by the females, who either simply sniff or sniff and mark the same spot.

Fig. 8. Adult male *C. neglectus* marking a branch newly introduced into its cage. (Paimpont; photo by A. R. Devez, CNRS.)

The role of these olfactory markings in intergroup relations *in natura* remains to be demonstrated. In captivity a branch marked by a male of one group and introduced into an adjacent group cage does not seem to provoke marking (observation in progress). These experiments seem to make conspicuous a "confidence-giving effect" and perhaps a "repelling effect" (see Mykytowycz, 1972).

CONCLUSION

Various detailed and/or comparative surveys dealt with the question of modes of intergroup relations occurring in Old World monkeys (Southwick, 1962; Ripley, 1967; Marler, 1968; Washburn and Hamburg, 1968; Gautier, 1969; Bates, 1970; Kummer, 1971b; Wilson, 1972; Kawanaka, 1973; Deag, 1973). The different species may possess several types of species-specific signals or none at all.

The mode of action and the nature of the habitat influence signal structure. At close range, they are multimodal; consequently their structure is variable and graded. On the other hand, the greater the distance the exchanges must carry and the more optically dense the habitat the more the signals tend to be unimodal and stereotyped. This duality is reinforced by another imperative: species specificity. At close range animals are in visual contact and there is no problem of species identification but at greater distances signals must convey a species-specific message.

The ritualization of exchanges is particularly obvious in the forest *Cercopithecus,* the signals of which are essentially vocal, are of a stereotyped nature, and occur in vocal battles at wide range, the occurrence frequency of which varies with population density. In the African and Asian colobus, vocal exchanges are coupled with intergroup contact searching, in which a variety of nonvocal sounds and visual exchanges are displayed.

Some macaques living in at least partially woodland habitats seem to have retained the stereotyped vocal signals found in *Cercopithecus* (e.g., *M. fuscata:* Itani, 1963). Others, whose environment is more open, offer an extremely rich variety of exchanges (*M. sylvana:* Deag, 1973).

In provisionized bands or in those living in high density in habitats that are not completely natural (e.g., some Japanese macaques, the rhesus of India), exchanges are of an essentially aggressive nature and give rise to an obvious hierarchization between groups.

Lastly, in an open environment, as for the baboons, either mutual avoidance based on visual cues (*P. anubis* and *ursinus*) or mingling of groups (*P. hamadryas* and *Th. gelada*) seems to be the rule, and few or no species-specific signals are exchanged.

Multimodality of signals, predominance of visual patterns, and lack of stereotyping in intergroup exchanges clearly seem to be correlated with openness of the milieu. The stereotyped character of the wide-ranging vocal sounds of the *Cercopithecus* living in dense zones, on the contrary, is dictated by the needs of species specificity.

Signals That Assume the Protection of the Social Unit as a Whole

This section considers the totality of signals evoked in the group by any modification of the nonconspecific environment, be it the manifestation of an unusual noise, a sudden movement in the brush, or the approach of some animal either innocuous or potentially predatory. A variety of responses have been catalogued; they are principally vocal and depend on the intensity, duration, and suddeness of the perturbation.

One type corresponds to vocalizations derived, from a structural point of view, from calls that occur during peaceful group progression. Fig. 9A, B, C shows this evolution in the

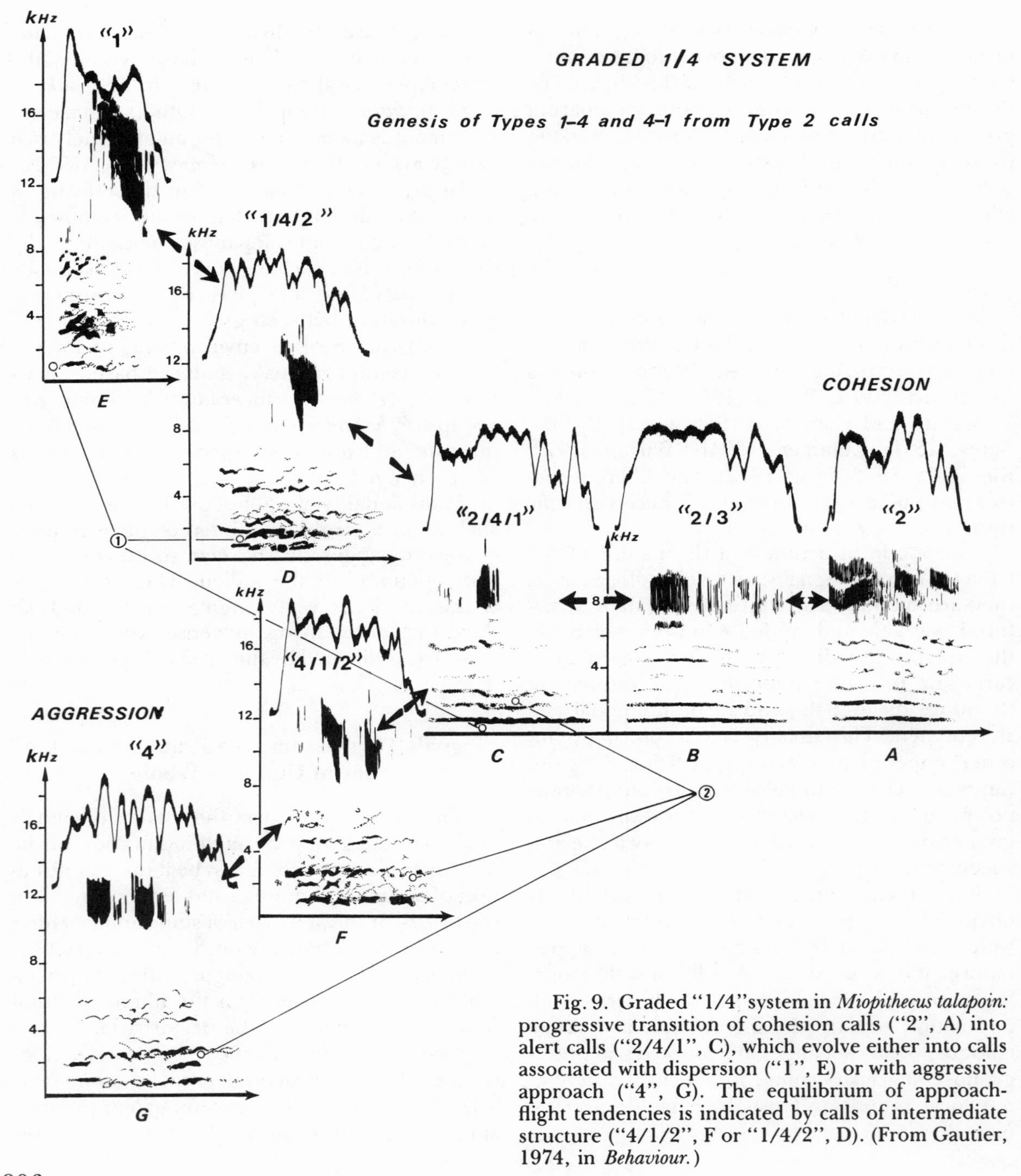

Fig. 9. Graded "1/4" system in *Miopithecus talapoin:* progressive transition of cohesion calls ("2", A) into alert calls ("2/4/1", C), which evolve either into calls associated with dispersion ("1", E) or with aggressive approach ("4", G). The equilibrium of approach-flight tendencies is indicated by calls of intermediate structure ("4/1/2", F or "1/4/2", D). (From Gautier, 1974, in *Behaviour.*)

talapoin (Gautier, 1974). In *Macaca fuscata,* Itani (1963) notes that the A-4 call heard in slightly disturbed situations is similar to those heard when the group is on the move, does not provoke a vocal response in congeners but attracts the attention of adjacent monkeys. In *Ce. galeritus,* call 2 (Quris, 1973) appeared under the same conditions as call 1, linked to phases of the group's activity, but more precisely following slight disturbances. Call 2 derives gradually from call 1 through an elevation of the frequency band used, a concentration of energy in the higher frequencies, and an increase in intensity (Fig. 10a, b, c, d; from Quris, 1973).

All these calls indicate a slight increase in the attention of the animals, which is translated on the structural level by a progressive concentration of call energy frequencies. They are "alert calls," uttered by one or several members or by the entire group. Any slight modification of the environment will provoke them: in the forest, a sudden clouding over the sky or a gust of wind rustling in the brush is sufficient. As a result they attract the attention of adjacent monkeys, raising their arousal level; secondarily they act as a source of cohesion for the group, in the same way as the calls linked with progression.

The second kind of signal occurring in disturbing or uncertain situations consists of high-frequency calls, often with a broad frequency range and a short length, which can be uttered in rapid and sometimes long series. The frequency of emission, the duration of the sequence, and the call intensity increase with the duration and intensity of the stimulus. These vocalizations may derive from agonistic calls (Rowell, 1972) or from those associated with progression, through the intermediate stage of alert calls described above, as shown on Fig. 9 for *M. talapoin* and in Fig. 10(a), (d), (f) for *Ce. galeritus.* They often evoke identical phonoresponses in congeners and are thus rapidly propagated throughout the entire group. They are very widespread, e.g., the "shrill barks" of the baboons (Hall and DeVore, 1965) and of *M. mulatta* (Rowell and Hinde, 1962; Lindburgh, 1971), the "high-pitched alarm calls" of *M. radiata* (Sugiyama, 1971), the "snick alarms" of *P. cristatus* (Bernstein, 1967), the alarm calls of *P. johnii* (Poirier, 1970a; 1970b), the "chist calls" of *Co. badius* (Marler, 1970), the "karaou" or "chuckles" of *Ce. albigena* (Malbrant and Maclatchy, 1949; Chalmers, 1968); the "kakou" of *Ce. galeritus* (Malbrant and Maclatchy, 1949; Quris, 1973), the "high-

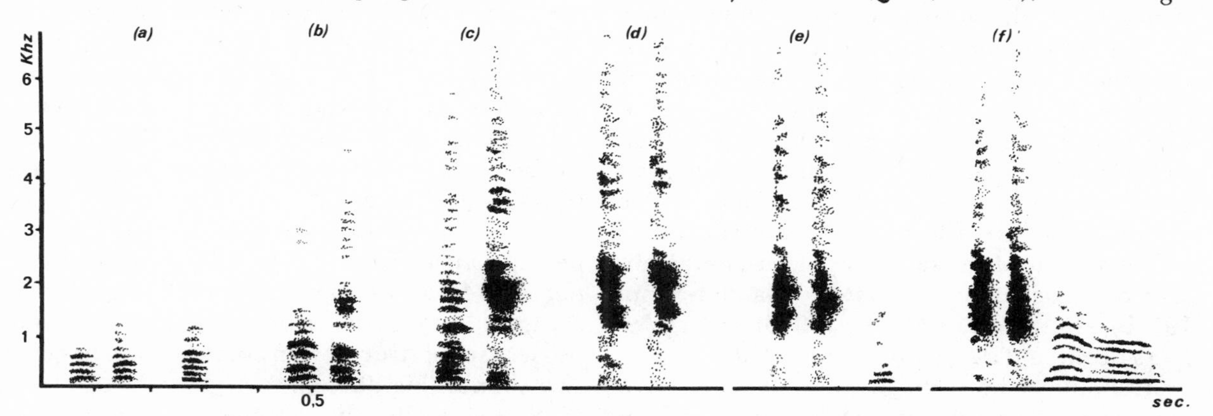

Fig. 10. Gradual structural evolutions between calls 1(a), 2(d), and 3(f) in *Cercocebus galeritus.* Transition calls (b) and (c) show a progressive increase of the frequency band used. Starting from the intermediate call (e), a tonal unit is added, which is broadly developed in call 3. (Paimpont; from Quris, 1973, in *La Terre et la Vie*.)

pitched chittering" of the patas (Hall, 1968b), and finally the "chirps" of the *Cercopithecus* species (Aldrich-Blake, 1970; Struhsaker, 1967b, 1969; Gautier, pers.obs.) and the "ists" of the talapoin (Gautier, 1967, 1974).

These vocalizations are generally uttered by all members of the group except the infants, but principally by the females. In many species the adult males are never, or only rarely, emitters (*Papio* spp., *Cercopithecus* spp.: *C. aethiops, C. nictitans, C. mitis).*

The calls are produced following any sudden, relatively intense disturbance, such as a violent noise, an abrupt movement in the brush, the sighting of a strange animal or predator, or any uncertain situation. They consist of rough reactions by individuals and contain no information about the nature of the stimulus. Thanks to the phonoresponses, all the congeners, including those who know nothing of the original situation, reach a comparable vigilance level. These emissions are generally termed "alarm calls" or "social alarms."

The most drastic reaction that they evoke is dispersion or flight, but there may be only an increase in vigilance followed by a return to previous activities. If the stimulus persists without posing a real threat, the animals maintain their state of excitement by uttering calls in series. In some cases, they converge around the source of the stimulus, vocalizing without interruption. The fundamental structure of the calls is then altered by the rapidity of the rhythm of emission, or they are partially transformed and may evolve toward aggressive calls. This harassment, which is reminiscent of the mobbing of birds, was observed especially in *M. talapoin* (Gautier-Hion, 1971b), *C. ascanius* (Marler, 1973), *C. cephus* (pers.obs.), and *M. fuscata* (Green, 1975).

It should be noted that many authors discussing alarm calls state that certain emissions provoke no reaction whatsoever in the group (e.g., *C. mitis:* Aldrich-Blake, 1970; *M. mulata:* Lindburgh, 1971). In captivity, the karaou uttered by a member of a *Ce. albigena* group is generally followed with upward flight by its congeners; nevertheless, if the call is uttered by an individual who is eating or sitting peacefully, no reaction is observed (pers. obs.). Poirier (1970b) also notes that a group of langurs responds "in a more positive manner" to the alarm calls emitted by males than to those uttered by females, as if their reactions were taken more seriously. Some authors think that such vocalizations, which do not necessarily provoke phonoresponses, may function as cohesion calls (Struhsaker, 1967b; Marler, 1970; Lindburgh, 1971; Quris, 1973).

In fact, alarm vocalizations that sometimes appear to be uttered spontaneously often occur in situations of frustration. Thus a *Ce. albigena* female who has been rejected by a male to whom she has presented herself may emit a karaou close to that used as an alarm. Similarly, a young talapoin whose mother has refused it her breast utters vocalizations that in the course of maturation will evolve toward alarm calls (Gautier, 1974). Green (1975) also calls attention to an "A trill" uttered by an infant *M. fuscata* whose mother has refused it contact. The structure of this trill is close to that of the chirps.

Various species possess calls characteristics of certain more specific stimuli. The guenons like the Japanese macaque, emit a particular vocalization upon sighting a flying predator (Itani, 1963; Struhsaker, 1967b; Gautier, 1967, 1974). In fact, in *Miopithecus,* as in *Cercopithecus,* the stimulus that releases this call can be any object in flight of which the animal has a sudden visual perception: a dead leaf, a bird, an airplane. Gautier (1974) terms these emissions "aggressive alarms": they derive from cohesion calls and possess some structures in common with aggressive sound. (Fig. 9A, F).

Subsequently, the response becomes more specific. If the bird approaches and is perceived as a true predator, the calls evolve toward typical social alarms and are followed by dispersion (Fig. 9E). If, however, the bird alights, one can observe mobbing behavior, as in the talapoin, in which the calls resume their double alarm and aggression structure (Fig. 9F, G). In the Japanese macaque, the C4 calls uttered in response to a flying predator are also, according to Itani (1963), aggressive sounds.

In the vervet, Struhsaker (1967b) differentiates several other vocalizations, linked with either a minor mammalian predator, a sudden movement by it, or the presence of a snake or a human being. For him, these calls are different, and they convey specific information about the origin of the danger. It is possible, however, that instead of specific reactions, one is dealing with varied intensity responses that depend on differences in the excitation levels produced by the stimuli (see also Bertrand, 1971).

In most species there is, in addition, a third type of call limited either to adult males (sometimes subadults) or to adult males and females, and which occurs in cases of severe disturbance. In guenons and various macaques, type 2 loud calls occur in these situations (see above). In the mangabeys, baboons, and langurs, one also finds loud calls of the bark type, which have many characteristics similar to those of *Cercopithecus* although they do not arise in intergroup relations. This is the case with the "two-phase barks" of the *Papio* spp. (Hall and DeVore, 1965; Rowell, 1966), the "gruff barks" of *P. johnii* (Poirier, 1970a), the "alarm barks" of *P. entellus* (Jay, 1965a), the "threat alarm barks" of *C. aethiops* (Struhsaker, 1967b), the "kra-ing" of *M. fascicularis* (Kurland, 1973), the "barking" of *M. sylvana* (Deag, 1973), and the 3bis call of *Ce. galeritus* (Quris, 1973), a call very similar to that of *Ce. albigena* (pers. obs.). A disyllabic emission has also been described in a group of captive gelada (von Spivack, 1971).

In addition to having similar intensities, all these vocalizations have structures close to those of the calls linked with intraspecific aggression, from which they more or less directly derive (see Fig. 5). Even in species in which they do not occur in intergroup relationships, their intensity enables them to attract the attention of adjacent bands: in *P. johnii,* the gruff barks provoked by extragroup dangers sometimes evoke hiccups in males of adjacent groups. They can also be interrupted by the whoops of the males, another characteristic that tends to relate them to type 2 loud calls of guenons and macaques.

These vocalizations are uttered either in direct response to a stimulus (real or potential predator, dog, human, falling tree, etc.) or following and in association with alarm calls from group members. Thus, in response to the same disturbance, calls uttered by females and juveniles are high-pitched social alarms, whereas males bark loudly. The barks only can be a response to the increasing level of excitation in members of the troop, which is indicated by alarm calls. Generally, one or more males continue to vocalize while approaching the danger, while the congeners take cover or flee.

While social alarm calls divert the predator's attention, male barks focus it and thus have secondary protective value for other members of the group. They are often reinforced by other noise-making activity such as tree shaking, bouncing, or jumping. The alpha male of the group generally shows a greater facility for response. Females can also be vocalizers if the excitation conditions are sufficiently strong (langurs, *Cercopithecus*).

Thus, signals manifested by the members of the group taken as a whole during any sort of perturbation are essentially individual emotional responses that secondarily play a role in group defense. The quality of the calls uttered, their intensity, their number, and the duration of each sequence vary according to the intensity of the stimulation. The calls convey no specific message but indicate the excitation level, the more so as these calls evoke similar emissions in congeners. In that way, the degree of vigilance rapidly reaches equilibrium throughout the group, making possible a coordinated response by all the members. The reactivity threshold of the animals can be modified in many ways: by nonbiotic factors, like the variations in light intensity for forest monkeys, or biotic ones, such as the vocalizers' sex, age, experience, or status in the group. This does not mean that social alarms are wrongly named, for although they constitute individual reactions to a given situation in a given context, they nevertheless contain a social component.

Thus in the forest a solitary *Cercopithecus* does not respond to the presence of an observer with alarm vocalizations but rather with silent flight. Likewise, the guenon group decimated by hunting in certain regions of Gabon rarely emit alarm calls in disturbed situations. Furthermore, in de Brazza's monkey, which naturally lives in small groups, the social alarm call does not exist in the species repertoire. In an alarming situation the monkeys respond with either silent flight close to the ground or freezing behavior in which the animal lowers its head and curls up against a branch, concealing its beard and white markings (Gautier-Hion, 1973). This freezing has also been described in *Procolobus verus* (Booth, 1957), for which the author does not indicate any alarm calls. It can also occur in *C. cephus* living in reduced groups, although members living in larger bands (as in mixed troops) respond to disturbed situations with numerous alarm calls. The manifestation of such types of signals seems therefore to depend on the social environment. As Itani (1963) has said, a solitary monkey is a mute monkey; consequently calls are social behavior.

The powerful calls that are more or less aggressive manifestations of adult males appear to have the double role of alarm transmission and group defense by focusing the attention of the predator. They are reinforced by impressive displays, which are visual or visual and acoustic (tree shaking or jumping). Among the strictly visual displays, one finds "yawning," stopping postures, and penile displays. Thus a male *Ce. albigena* may jump violently from branch to branch, stop while rapidly erecting his tail, start up again, shake branches, sit down again, exhibit his penis, and yawn (pers. obs.). A similar sequence occurs in *M. nemestrina* (Bernstein, 1967). Penile display has also been noted in *C. aethiops* (Wickler, 1967), *Co. guereza* (Marler, 1972), and *Ce. galeritus* (Quris, 1973). Yawning, often called "tension yawning," is also a manifestation provoked principally by stimulations exterior to the group, as in the patas (Hall, 1968b) or *Ce. albigena* (pers. obs.). Stopping postures, which are reminiscent of presentation postures, doubtlessly have a role in species recognition. The male of *Presbytis johnii* strikes one between his whoop sequences, pulling his tail over his head, just as the crested mangabey does when facing an observer. The white-collared mangabey, on the other hand, points its tail vertically, touching only the tip to its back. All these seem to be more or less aggressive displays that indicate the species, sex, and status of the displaying monkey (Figs. 11, 12, 13), and may have the same role in individualization and maintenance of group integrity as type 1 loud calls.

Some species also exhibit "prevention" behavior for potential disturbances which may serve to transmit a certain state of vigilance. An erect individual "standing bipedally" has been described in baboons (Hall, 1965), the rhesus (Lindburgh, 1971), the patas (Hall, 1965), the vervets (Struhsaker, 1967a), and the talapoin (Gautier-Hion, 1971b). Vigilance behavior also exists in males of various species (e.g., "watchful behavior" of the male patas: Hall, 1965, 1967), indicating the particular roles played by males in group protection.

Fig. 11. Stopping posture taken facing the observer in *Cercocebus galeritus*. (Gabon; photo by R. Quris.)

Fig. 12. Penile display, exhibited facing the observer in *Cercocebus galeritus*. (Gabon; photo by R. Quris.)

Fig. 13. "Yawning" in response to the presence of an observer in *Cercocebus galeritus*. (Gabon; photo by R. Quris.)

CONCLUSION

The social alarm signals have common characteristics in most species: (1) At the structural level, their species specificity is sometimes so unclear that proper identification is often difficult; for instance, the chirps of *C. nictitans, C. cephus,* and *C. erythrotis,* like those of *C. mitis* and *C. ascanius,* are practically indistinguishable (Gautier, 1969; Struhsaker, 1970; Marler, 1973). This phenomenon is very general. Fig. 14 shows the astonishing similarity between the alarm call of a forest bird *(Trichophorus calurus)* sympatric with the *Cercopithecus* spp. of Gabon and that of a *C. cephus.* Interspecific and even intergeneric alarm reactions are consequently highly developed. This is particularly obvious in mixed troops (Gautier and Gautier-Hion, 1969), and Booth (1957) points out that the green colobus, which does not possess alarm calls, uses those of *Cercopithecus* with which it associates.

(2) From a causal and functional point of view, homogeneity is also apparent. Few calls

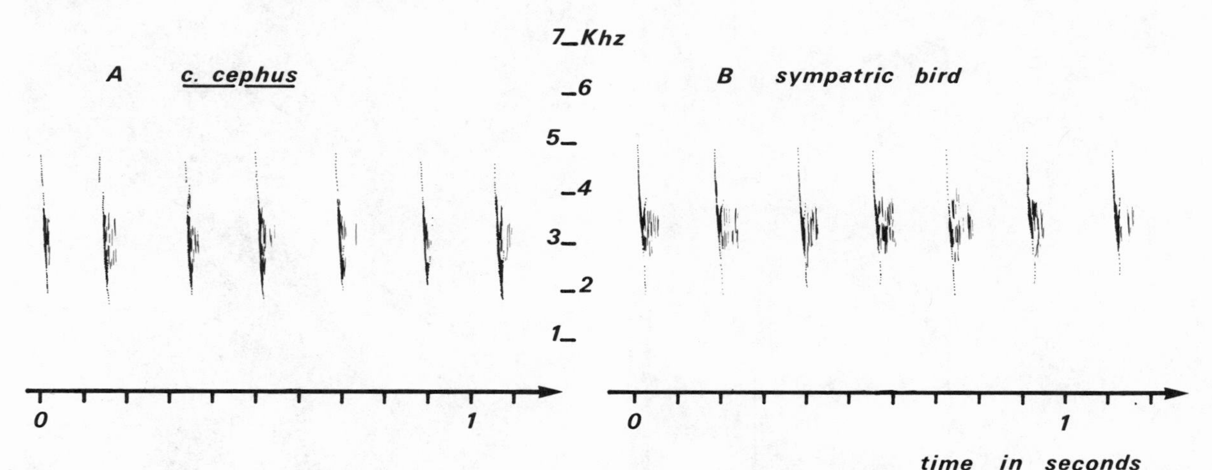

Fig. 14. Spectrographic analyses of the "bird-like alarm calls" of *Cercopithecus cephus* (A), compared with those of the vocalization of a sympatric bird (*Tricophorus calurus*). The similarity of the two calls is observed both at the structural level and in the emission rhythm.

convey specific messages but they all indicate a certain excitation level in the vocalizers correlated to the degree of perturbation. Except in the vervet, a single clear distinction appears in forest or semi-woodland monkeys between visually perceived "flying" stimuli and auditorily or visually perceived situations of uncertainty. In addition, calls close to those used in alarm situations frequently occur in a frustrated monkey prevented from performing some activity.

The existence of signals charactersitic of adult males also seems very widespread and indicates the particular role played by these monkeys in group protection. All the species for which we have information show a mixed sonographic structure for these calls, in which vocalization features of the aggressive repertoire occur.

Signals that Coordinate the Routine Activity of the Social Unit and the Interrelations of Its Members

Signals acting within the group are those that regulate daily life—those that involve cohesion and spatial distribution of members, their dominance relationships, sexual behavior, and raising of the young, as well as those expressing interindividual affinities.

Two essential modes of signal action can be recognized: In one, the exchange does not imply any particular receiver and can take place at medium range ("one-to-many" communication: Itani, 1963). Such communication involves principally acoustic signals and some visual displays. In the other, the emitter is oriented toward a particular receiver. Visual exchanges with subtle changes in expression and tactile and olfactory signals then play a more important role ("one-to-one" communication).

VOCAL SIGNALS

Some vocalizations linked with the general routine of the group act without the support of visual information. In interindividual close-range exchanges, on the other hand, the overlapping of vocal, visual, and tactile signals is great.

Signals Associated with the Routine Activity of the Social Unit

A group of monkeys is most often silent during rest. Vocal activity resumes slightly before individuals begin to stir and continues during their movements. The calls emitted are often nasal, of short range, and uttered without modification of facial expression. They do not appear to be addressed to specific receivers, but nevertheless evoke identical vocal responses in adjacent monkeys. Occurring as individual reactions, they are diffused throughout the group thanks to phonoresponses, which permit reciprocal individual localization and the general cohesion of the social unit. This kind of signal has been reported in most species, except the African colobus, which are particularly silent (Booth, 1970, 1972). They are the "grunts" of the baboons (Hall, 1962; Andrew, 1963b; Hall and DeVore, 1965; Rowell, 1966; Aldrich-Blake et al., 1971), the drills and mandrills (Struhsaker, 1969; Gartlan, 1970), the langurs (Jay, 1965a; Poirier, 1970a; Vogel, 1973), the "deep muffled gruff" vocalizations of *M. mulatta* (Altmann, 1962), "the calling sounds" of *M. fuscata* (Itani, 1963), the "basic grunts" and "A-calls" of *M. speciosa* (Bertrand, 1969), and the "grunts" of the mangabeys (Andrew, 1963b; Chalmers, 1968; Quris, 1973; Deputte, 1973), the guenons (Haddow, 1952; Struhsaker, 1967b, 1970; Aldrich-Blake, 1970; Marler, 1973) and the talapoin (Gautier, 1967, 1974; Gautier-Hion, 1971b; Wolfheim and Rowell, 1972).

In the forest monkeys, one notes that the frequency of call occurrence follows the activity rhythm of the animals, increasing with the potential risk of losing contact between congeners, e.g., with decrease in luminosity, increase in speed of movement or density of the environment, or state of insecurity after a disturbance (Itani, 1963; Gautier-Hion, 1970). Other situations, such as the group's coming into the presence of a fruit tree or the sight of a congener, also provoke calls. In most species these calls are given principally by adult females and juveniles. If the vocalizers are male, their vocalizations can easily be differentiated either through structural differences or because of their maturational transformation.

We have seen in the guenons, macaques, langurs, and colobus that type 1 loud calls uttered by males precede, especially in the morning, group movement and punctuate locomotor activity, playing a rallying and coordinating role. A similar system seems to exist in some baboons. Aldrich-Blake et al. (1971) note, for example, that in *P. anubis* the "two-phase wa hoo bark" in males precedes and coordinates group passage in difficult situations such as the crossing of a road (see also Gautier, 1969). Gartlan (1970) also points out a "mobilizing two-phase grunt" in the drill, similar to that of mandrill (P. Jouventin, pers. comm.). On the other hand, males can retain the same basic call as females and juveniles, but its structure is profoundly modified. Fig. 15 shows the evolution of calls linked with cohesion in a male *C. pogonias,* the growth of which was followed for seven consecutive years.

Originally, the call was composed of a stable low-pitched unit, followed by a shrill component, modulated in frequency (Fig. 15, C1). Common in young animals, this call becomes rare in adult males; during ontogeny, the shrill component has a tendency to disappear, being replaced by a low-pitched unit of similar duration (Fig. 15, compare C1 and C4), while the average pitch of the low structure can go as low as 170 Hz. Note that the curve depicting the decrease in pitch (Fig. 15A) is symmetrical to that showing the increase in weight (Fig. 15B). Both show a marked variation at the beginning of puberty. During the prepuberty period (which lasts five years) the animal gains 2,250 g (an average of 43 g per month) whereas its call decreases by 76 Hz (average of 1.45 Hz per month). On the other hand, during the puberty phase (about

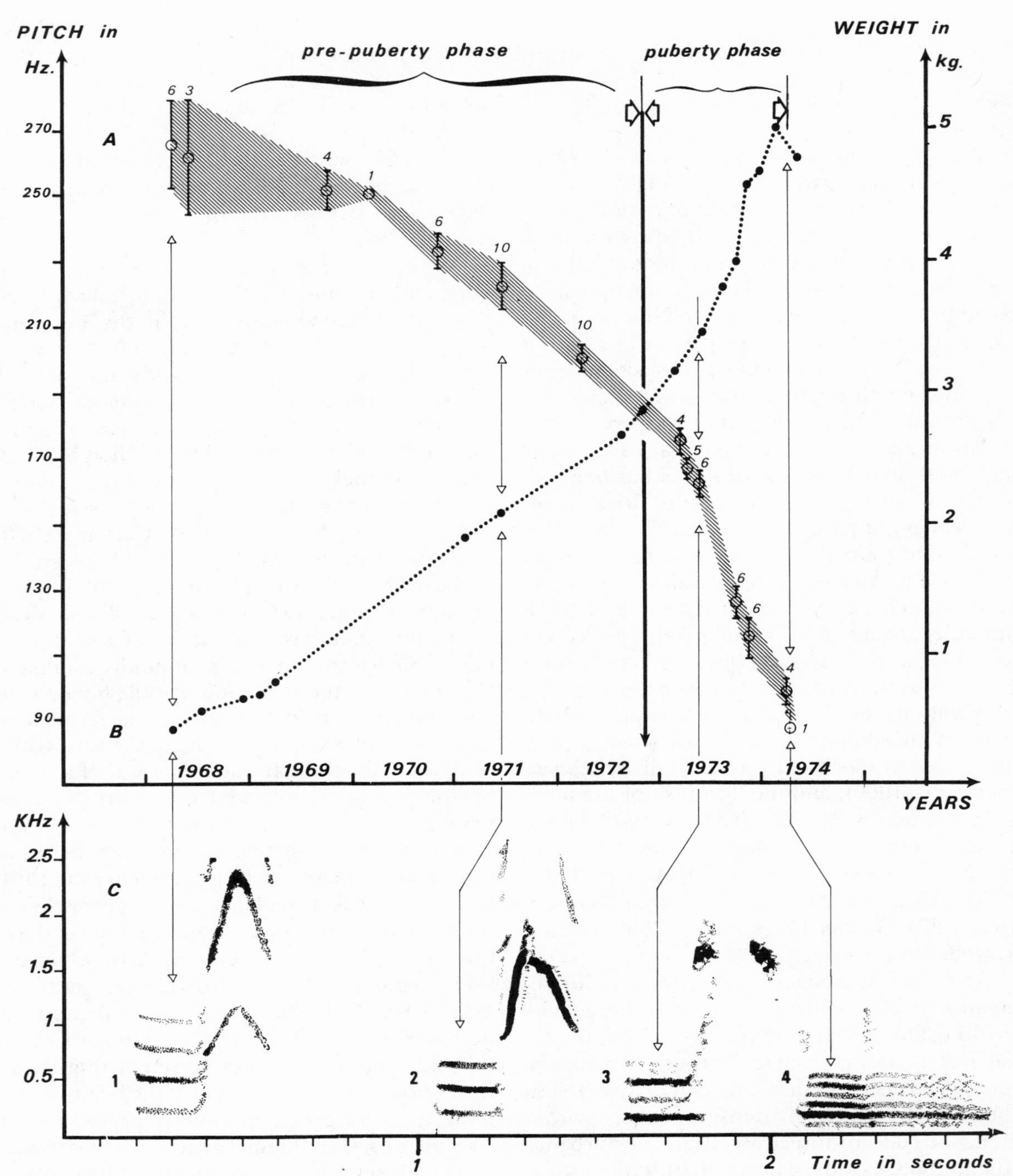

Fig. 15. Maturational transformations of the structure of cohesion calls of a male *C. pogonias.* A: Average pitch of the call's low-pitched component as a function of age (the hatched area corresponds to the standard deviation). B: The animal's weight during the same period. Note the symmetric inflections of the two curves starting at puberty. C: Spectrographic analyses of cohesion calls at different ages. Note the disappearance, between ages 6 (3) and 7 (4), of the high-pitched component, which is progressively replaced by a low-pitched component.

Table 2

Quantitative data for Fig. 15.

	(1) Prepuberty phase	(2) Puberty phase	(2) ÷ (1)
Duration	52 months	19 months	0.36
Increase in weight	2,250 g (420 g to 2,670 g)	2,200 g (2,670 g to 4,870 g)	1
Gain per month	43 g	112 g	2.61
Decrease in pitch	76 Hz (266 Hz to 190 Hz)	92Hz (190 Hz to 98 Hz)	1.2
Drop per month	1.45 Hz	4.7 Hz	3.24

nineteen months), corresponding to descent of the testicles and the growth of adult canines, the monkey gains 2,200 g (average of 112 g per month) and the pitch of its call drops by 92 Hz (average of 4.7 Hz per month). Therefore during this period the animal's weight increases as much as in the previous five years and its voice falls 1.2 times as much as it did in those five years (see Table 2). Note also that variability in the pitch of the call is greater when the monkey is young (cf. standard deviation: Fig. 15A). In the female, the weight curve does not have the same variation at the beginning of puberty; likewise, the decrease in the pitch of the call is regular and much smaller than in the male.

The break in the slope of the male weight curve of various guenons (*C. nictitans, pogonias, cephus, petaurista, neglectus; Ce. galeritus* and *albigena:* Gautier and Gautier-Hion, in prep.), which seems absent in a Platyrrhine such as the squirrel monkey (Rosenblum and Cooper, 1968) resembles that of humans. It is accompanied by an abrupt transformation in the voice, related to the breaking in the voice in man (Gautier, 1974).

These features are important at the communication level. If, all conditions being equal, each age and sex class possesses different structural characteristics and emission frequencies for the same fundamental vocal type, one can assume that an individual wandering in a dense environment, periodically receiving cohesion calls, could immediately recognize the age and sex of its closest congeners.

In forest guenons cohesion calls also vary according to a more or less graded system as a function of distance between vocalizer and congeners and of the general excitation level. This system, described for the talapoin (Gautier, 1974) is found in a variety of species (Fig. 16) and schematically includes four types of vocalization. The first corresponds to the calls uttered during the usual progression of the group. The second is heard particularly if interindividual distance increases or if the excitation level of the monkeys rises subsequent to a perturbation. The third is typically given by an infant that has lost its mother or a monkey that has lost contact with the group. The fourth variation is uttered by an infant that has reestablished contact with its mother.

The call gradation of this cohesion-distancing-isolation system is total in the talapoin and mona (Fig. 16) but only partial in others such as *C. cephus,* in which the number of intermediate calls is smaller and the structural diversity greater. The same kind of system seems present in most forest monkeys. In *C. aethiops* (Struhsaker, 1967b) the "eh eh calls," "progression

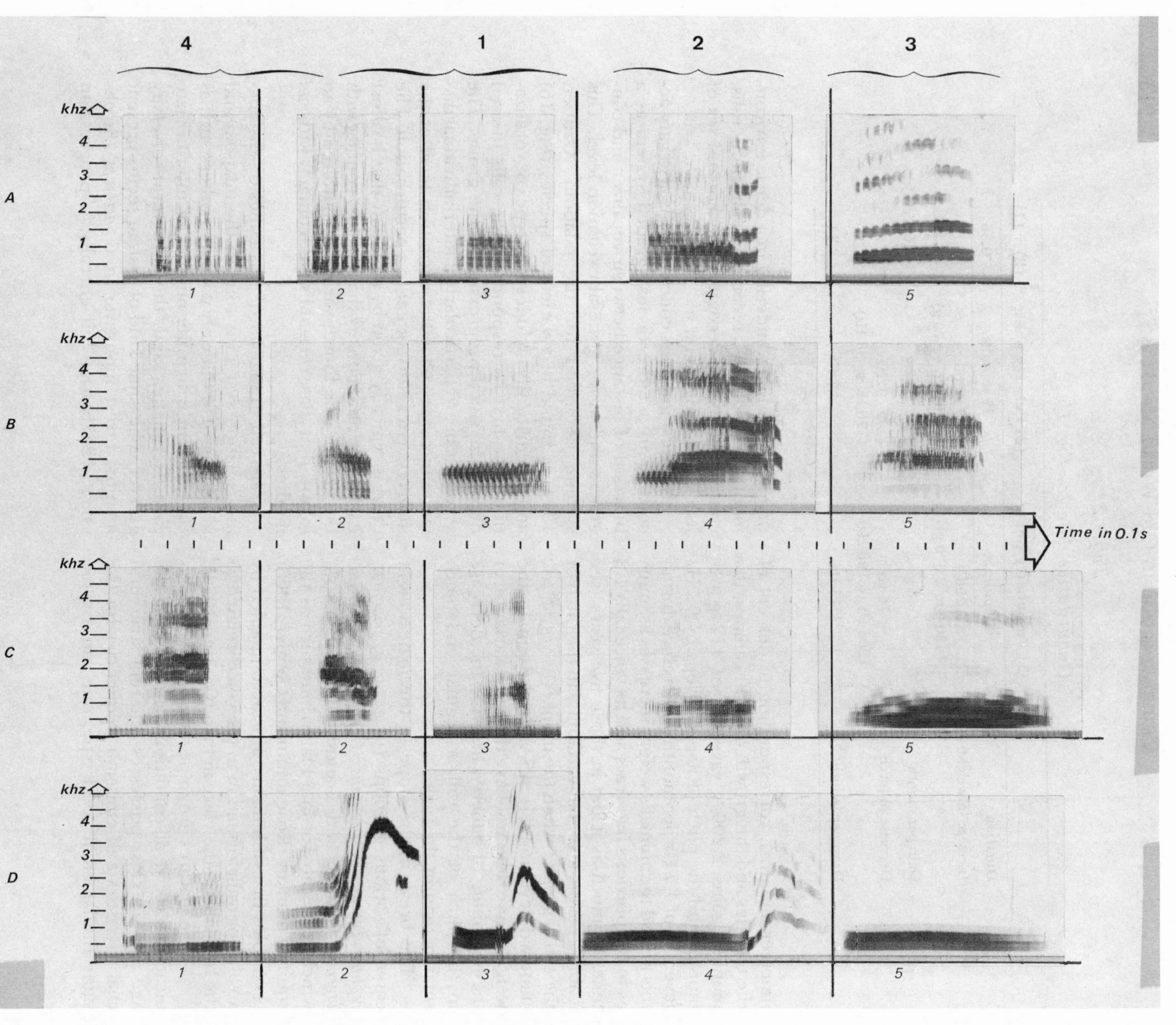

Time in 0.1s

Fig. 16. Comparison of the vocal systems of "cohesion" in *C. cephus* (A), *C. nictitans* (B), *C. neglectus* (C), and *C. pogonias* (D). Each column contains calls of comparable emission context and function.

Column 1: Calls emitted during routine progression. For A, B, C these calls have only a low-pitched unit; for D the low- and high-pitched components are associated (D3; see also Fig. 15C and 42a, b). Column 2: Calls emitted after a slight disturbance, exhibiting a discontinuous transition (A4, B4) or a gradual one (C4, D4) toward calls in Column 3. Column 3: "Isolation" calls, with structures new and pure in *C. cephus* (A5) or noisy in *C. nictitans* (B5), whereas they derive from preceding calls in *C. neglectus* (C5) and *C. Pogonias* (D5). Note the disappearance of the high-pitched component. Column 4: "Huddling calls" of low intensity, characterized by a vibrating sonority, with structures derived from calls in Column 1 (A2, B2, C2, D2).

grunts," and "most rrr calls" are probably comparable.

Throughout this graded system, the conditions under which one vocal type passes into another, if they generally imply variations in interindividual distance, are sometimes provoked by variations in "psychological" distance, which depend on the excitation level of the animals. Thus, an increase in habitat density, which favors loss of contact, is translated by an evolution of the first type of call into the second. Similarly, a young monkey huddled in its mother's arms might utter an isolation call if its mother refuses to nurse it. Thus we heard isolation calls uttered by most of the members of a mona group an hour after nightfall after a crowned eagle had flown over.

Under such conditions of excitement, mutual vocal enlistment resulting in a chorus can be heard in various species. This is the case in baboons (Hall and DeVore, 1965; Rowell, 1966), mangabeys (Chalmers, 1968; Gautier and Deputte, 1975), the macaques (Rowell and Hinde, 1962; Itani, 1963), and the talapoin (Gautier, 1974).

More or less varied cohesion calls have been noted in macaques, baboons, and mangabeys. In the Japanese macaque, "A-calling sounds" are all used for group cohesion and control of progression. Itani (1963) differentiates them according to their structure, the animals uttering them and the activity in progress: A2 are characteristic of certain females, A9 are emitted when the troop is calm or in slow movement, and A10 are used for departing.

Variations of the same type may exist in other species. Hall and DeVore (1965) mention a great variety of grunts uttered by baboons during routine social behavior and a "doglike bark" emitted by a monkey temporarily separated from the group. The same is true of the mangabeys. Nevertheless, if it seems that all species possess calls used for social cohesion, the mangabeys as savannah monkeys appear freer from environmental constraints, and their cohesion vocalizations have evolved mainly on the basis of more specialized interindividual exchanges in which vocal signals are associated with other sensory modalities.

Signals Associated with Peaceful Interindividual Relations

In A-group sounds of Japanese macaques, Itani (1963) distinguishes "calling sounds" from "muttering." The two have similar structures but the latter are addressed to a particular congener. This type of signal includes all the vocalizations leading to peaceful interindividual close exchanges. In all species for which they have been described, these signals derive, from a structural point of view, from calls linked with cohesion of the group as a whole. Unfortunately the literature offers few data that could permit interspecific comparisons.

In the vervet, Struhsaker (1967b) points out several vocalizations derived from the progression grunt, which are correlated with close interindividual exchanges. Generally speaking, "aar-raugh calls" occur when dominant monkeys and subordinates are at close range: at the approach of a dominant male, a subordinate one gives a "woof-woof," "wa," or "woof-wa" call. A female approached by a dominant female will utter a "wa-waa" call and a juvenile a "rraugh" call.

In the baboons, Hall and DeVore (1965) point out the "high-frequency grunting" characteristic of adult males being approached by an infant. In *Ce. albigena* and *galeritus,* adults also emit a particular grunting, derived from the progression grunt, when they come close to an infant. Furthermore, the adult male may utter a slow-rhythm grunting (Fig. 17) (pers. obs.), which does not exist in the female. In the mangabeys in general, a great number of variations based on grunting occur (e.g., in a juvenile when a male approaches, in females preceding mutual sniffing). In the stumptail macaque, Bertrand (1969) also notes a number of variations in these types of calls: "greeting grunts" directed toward infants, "B calls" characteristic of females whose children are out of sight, and "coos" uttered at the sight of a congener or after a separation.

At present the greatest diversity has been noted in the Japanese macaque, apparently due to the quality of the studies rather than to any species-specific peculiarity. Itani (1963) shows evidence for A5 calls uttered between adult males, A6 given by a subordinate male toward a dominant one, and D4 characteristic of receptive subordinate females. Green (1975), in his very detailed study of the coo sounds of this same species, throws new light on the extreme variability of these vocalizations given at close range. He distinguishes seven categories of coos of very similar sonographic structure. One promotes the general progression of the group, and the six others include close-range and "personalized" signals. Thus the "double coo" is characteristic of solitary or "depressed" monkeys (cf. isolation calls of the *Cercopithecus*) and especially of females who have lost their young, whereas the "dip early high coo," the "dip late high coo," and the "smooth late high coo" are uttered by subordinate individuals in response to dominant ones. These same sounds are also often given by juveniles communicating with their mothers.

In *Cercopithecus,* infants also give calls with structures close to those used for cohesion but less variable than the coo sounds of the macaques. In some species, however, the young utter shrill calls of a specific structure, which may be trilled (e.g., in *C. nictitans, mitis, cephus, ascanius*) or not trilled (*C. pogonias* and *neglectus*), and which can be associated with cohesion calls in either a discrete or a graded manner (e.g., *C. pogonias,* Fig. 16).

Their structure is highly variable, especially in intensity and frequency modulation, giving them tones that are sometimes plaintive, sometimes interrogatory and "peremptory," and sometimes violent, according to the vocalizer's excitation level and the context of emission. Four situations may be distinguished: (1) Calls are exchanged between an infant and its mother, rather spontaneously and in the absence of any apparent disturbance. The mother responds to these mother-young contact calls with the same sounds or, more often, with the usual cohesion calls. (2) The calls become more intense when the infant solicits closer contact (carrying or grooming) or suckling. (3) If the mother accepts contact, the call becomes modulated and plaintive, its intensity diminishes, and only a rhythmic and atonal breathing may be uttered. (4) If the mother refuses contact (especially during weaning) the calls become intense, their frequency modulation attenuates, and finally the quavering disappears. These sharp whistlings can become noisy and evolve toward screams (Fig. 18). As in the young of many species, these vocalizations are associated with spasmodic head and tail movements.

The frequency of emission of these calls decreases rapidly in the course of ontogeny, especially in males. The adult female gives them

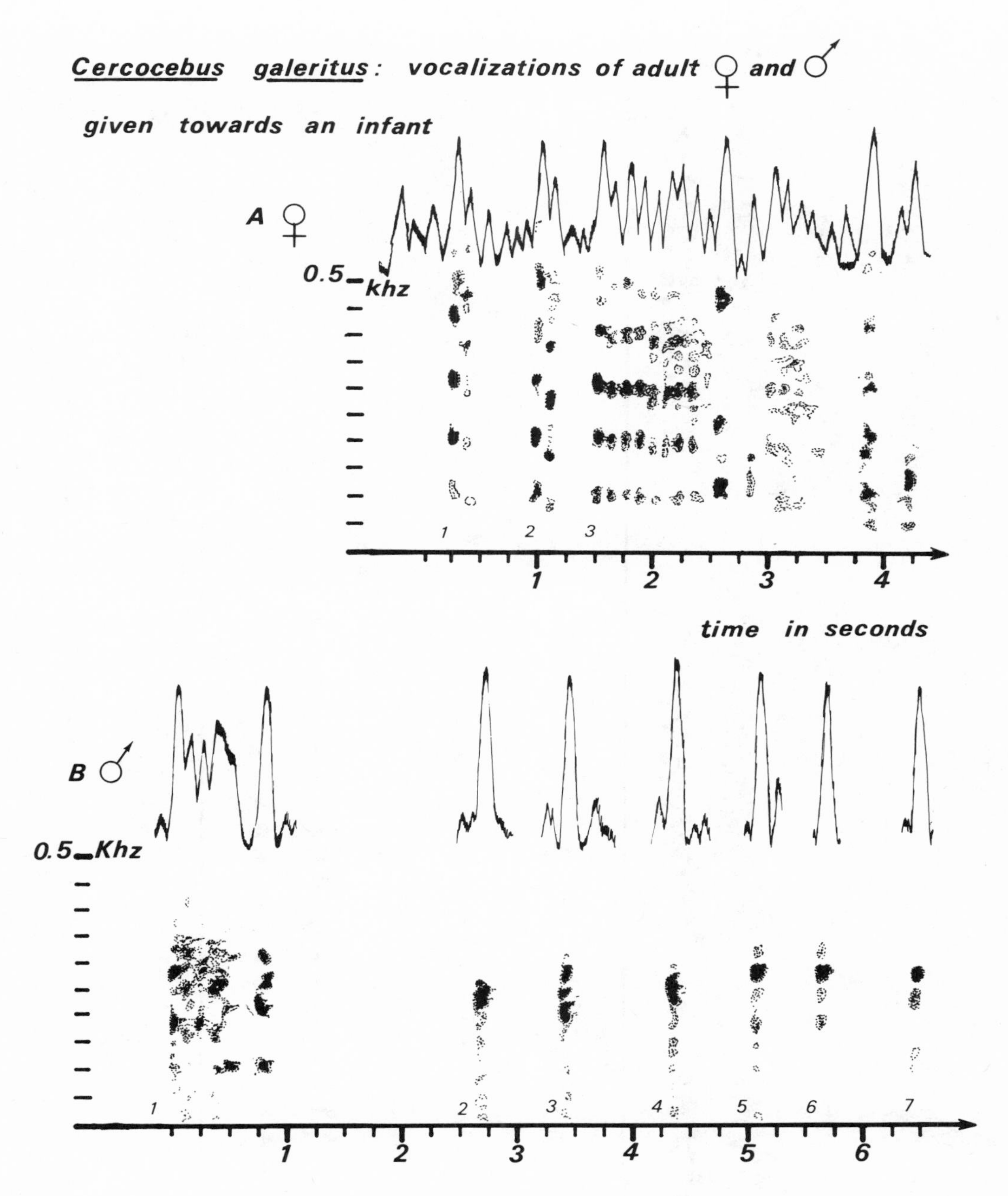

Fig. 17. Gruntings of the female (A) and male (B) *Cercocebus galeritus,* emitted while approaching an infant. A1, 2: binary forms; A3, B1: unitary forms of call 1 (cf. Fig. 10); B2-B7: grunting in slow rhythm (1.32 units/sec) characteristic of the adult male.

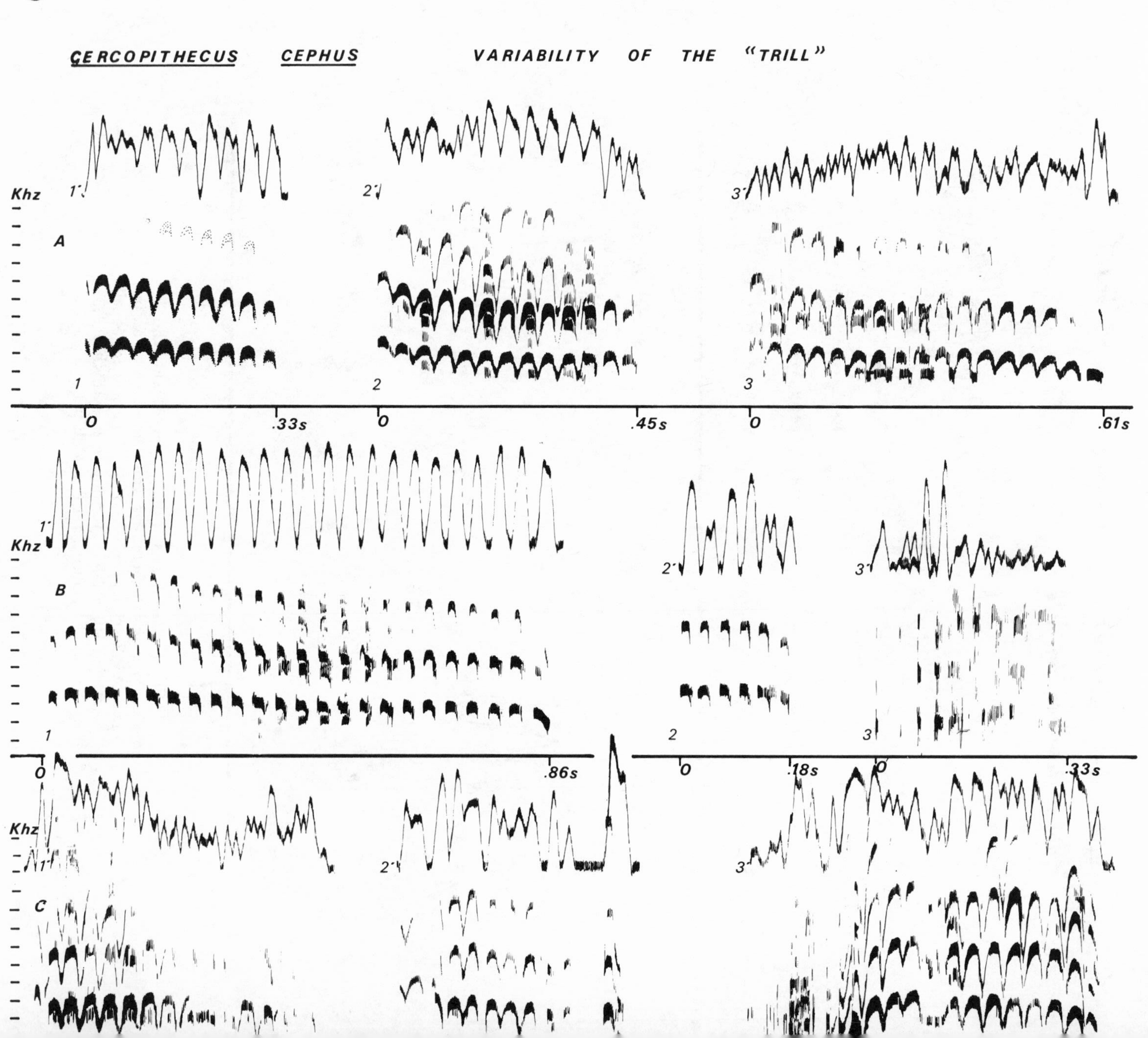
CERCOPITHECUS CEPHUS VARIABILITY OF THE "TRILL"
Khz
A
1
1'
2
2'
3
3'
0
.33s
0
.45s
0
.61s
B
.86s
.18s
.33s
C

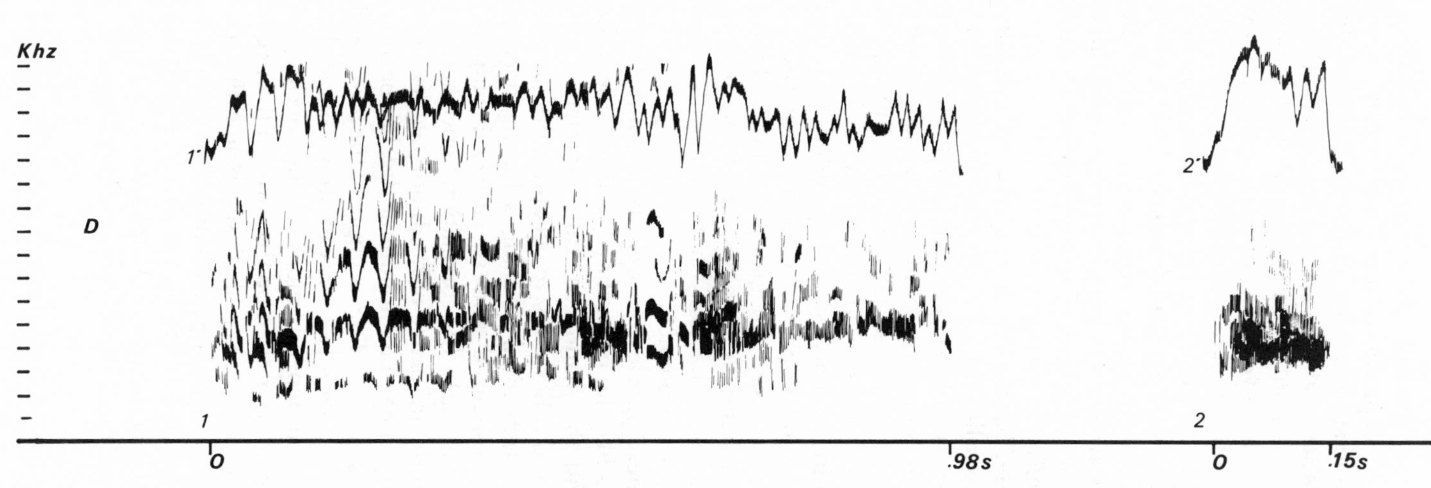

Fig. 18. Trill variations in a young female *C. cephus.*

A1: "normal" contact calls; A2, 3: calls emitted by a female rejoining a congener after separation; B1, 2, 3: contact solicitation calls after a slight frustration. Note the separation of the constituent units of the trills (1, 2), which are quite visible on the amplitude modulation spectra (1', 2'), as well as the loss of tonality in call 3. The latter can be emitted by a juvenile reestablishing contact with its mother. C: Calls given after an important frustration. C1: "Quasi-normal" trill with maximum energy concentrated in the second harmonic; C2: "trill-chirp," the last pulse of which is transformed into an alarm call; C3: "scream-trill," opening with a noisy component. Note the reinforcement of harmonics 2, 3, 4 in proportion to the fundamental sound. D: Distress calls. D1: "Trill-scream" with disappearance of quavering in favor of noise; D2: short "scream" without quavering (see 1' and 2').

again during contact with infants. Similarly, these vocalizations reappear in juveniles and subadults at the birth of a sibling. This regressive phenomenon occurs frequently in captivity (see Gautier-Hion, 1971b; Gautier, 1974).

All these signals can be classified under a general rubric: lack of, search for, solicitation of, or maintenance of contact. Thus, they incontestably play a role in general social cohesion, but they also have a more short-term, immediate significance concerning the emotional state of the uttering monkey. In effect, Green (1975) remarks that the low-pitched coos of the macaques are used in low-level apprehension and agitation situations in vocalizing animals, whereas the high-pitched coos are given much more when the monkey is agitated and probably apprehensive. This same remark, correlating call structure with the vocalizer's emotional state, was made by Kiley (1972) for the ungulates and Gautier (1974) for the talapoin. Thus, as Green points out, signal morphology being a direct reflection of the uttering monkey's internal state, is sufficient to inform the receiver about it.

Green shows that, starting from the same basic sound this mechanism permits a great variety of combinations; the precision of the message received is increased by close-range visual communication and knowledge of the sex, age, and sexual state of the vocalizer. Our observations on the variations in call structures due to ontogeny show that vision is not always necessary for the identification of the vocalizer's age or sex since the signal morphology alone can be sufficient.

The observer must be conscious of this maturational approach to vocalizations, seldom emphasized at the present time in studies on Old World monkeys, before claiming that the calls given by a juvenile and by an adult male under the same stimulation are of different natures. It is often the same call, structural characteristics of which have evolved with age.

Whether the calls are different by nature or because of maturational modifications, the result is the same at the operational level as far as the identification of the message is concerned. However, distinction is fundamental for understanding the evolution of a communication system. Ontogenetic modifications in calls (which differ according to sex) that allow the identification of age and sex classes and modifications caused by the emotional state of the monkeys are two complementary processes that permit the achievement of an extremely varied communication system.

On the basis of these remarks, we can conclude the following: All species (except perhaps the African colobus) have developed calls whose more or less immediate function is maintenance of cohesion of the social unit. This function is particularly obvious in forest guenons, where maintaining contact is a real problem (e.g., for the cryptic talapoin, which lives in a particular dense environment in groups of more than a hundred members). It "blurs," however, in monkeys in an open habitat, where numerous visual exchanges are possible.

In these species, as well as in some forest monkeys like the mangabeys (and doubtless the drills and the mandrills), calls associated with contact and cohesion have become more diversified, interindividual exchanges having become more personalized and sociability having increased. Similarly, the correlation between these calls and the influence of nonbiotic factors, such as density of habitat and luminosity, has become much less evident. The example of the forest and arboreal *Ce. albigena* seems to show that such a vocal system depends more on the species' degree of evolution than on the direct influence of the environment. We shall see below that exchanges of close-range calls are often associated with visual signals like lip-smacking or presentation, and that they precede or accompany a great number of tactile and olfactory exchanges.

Signals Associated with Agonistic and Sexual Interindividual Relations

An increase in the frequency of interindividual contacts increases the probability of conflicts in which the slight distance separating protagonists permits exchanges of all kinds. Although visual signals play a predominant role in agonistic relations (Marler, 1965), they are frequently accompanied by a great number of vocal and nonvocal sounds. This multimodality consequently increases the structural variability of the signals exchanged. This, taken with the fact that protagonist roles can be reversed rapidly, makes it difficult to separate signals linked with attack from those correlated with flight. Rowell and Hinde (1962) were the first to show that aggression-flight signals taken as a whole constitute an almost perfectly graded system. This statement probably applies to most, if not all, species. Vocal signals that accompany threat behaviors are most generally low-pitched; those linked with submission and flight are shrill. Vocalizations emitted by a monkey conflicting with these two tendencies have a composite structure, blending low-pitched and shrill components (Fig. 9, D, F).

One category of sounds accompanying aggressive behavior consists of brief, low-pitched calls with rolling sonority, which derive from the various grunts acting in cohesion or friendly interindividual relations throughout the group. Such calls are observed in baboons (Hall and DeVore, 1965), macaques (e.g., *M. mulatta:* Rowell and Hinde, 1962; *M. fuscata:* Itani, 1963; *M. nemestrina:* Grimm, 1967; *M. speciosa:* Bertrand, 1969; *M. sylvana:* Deag, 1974), langurs (*P. entellus:* Jay, 1965a; Vogel, 1973; *P. johnii:* Poirier, 1970a), mangabeys (Quris, 1973, and pers. obs.), and the patas ("whoo-wherr growl" of the male: Hall, 1968b).

In various *Cercopithecus* spp. with a cohesion grunt of the vervet type, an obvious gradation also exists toward the aggressive calls. In certain species (e.g., *C. cephus* and to a small extent *C. nictitans,* pers. obs.), these two signals are very close and cannot be distinguished easily without the postural behaviors that acompany them. *C. neglectus,* curiously enough, which uses no grunt-type call during progression, has an aggressive grunt close to that of *C. cephus* and *nictitans* (pers. obs.) (Fig. 19).

When the threat intensifies, the call structure becomes diversified. The calls remain low-pitched but their temporal parameters are modified, probably in proportion to the movements associated with the behavior and with physiological components such as respiratory rhythm. Roars and barks can be distinguished, as well as calls of intermediate structure emitted rhythmically, the last being the commonest in many species. The roars derive from grunts, and Hall and DeVore (1965) term them "loud two-phase grunting" in the baboons. Grunts increase in duration and intensity in guenons also, and such roarings are present in most macaques.

The barks are also generally widespread in the macaques (*M. irus:* Goustard, 1963; *M. mulatta; M. fuscata; M. nemestrina; M. speciosa*), the baboons ("two-phase barks"), the mangabeys, the langurs ("barking" and "coughing"), the patas, and the guenons. We have seen that different types of loud barking can occur in intergroup relations (cf. type 2 loud calls, Fig. 1) and in alarm situations. Roars and barks are emitted more frequently than grunts by adult males.

Still commoner are the low-pitched rhythmic calls associated with threat, such as the "pant threats" of *M. mulatta* and *M. speciosa,* the "chutters" of *M. fuscata,* the "intention notes" or "staccato growls" of *M. nemestrina,* and the discontinuous barking of *M. irus.* The repetition of the "huh-huh sounds" in the patas and the "two-phase uh-uh" in the baboons is another example of the rhythmic emission form. This is also the case in the mangabeys, the langurs, the talapoin, and the guenons ("chutters" of the vervet: Struhsaker, 1967b). Some nonvocal sounds ac-

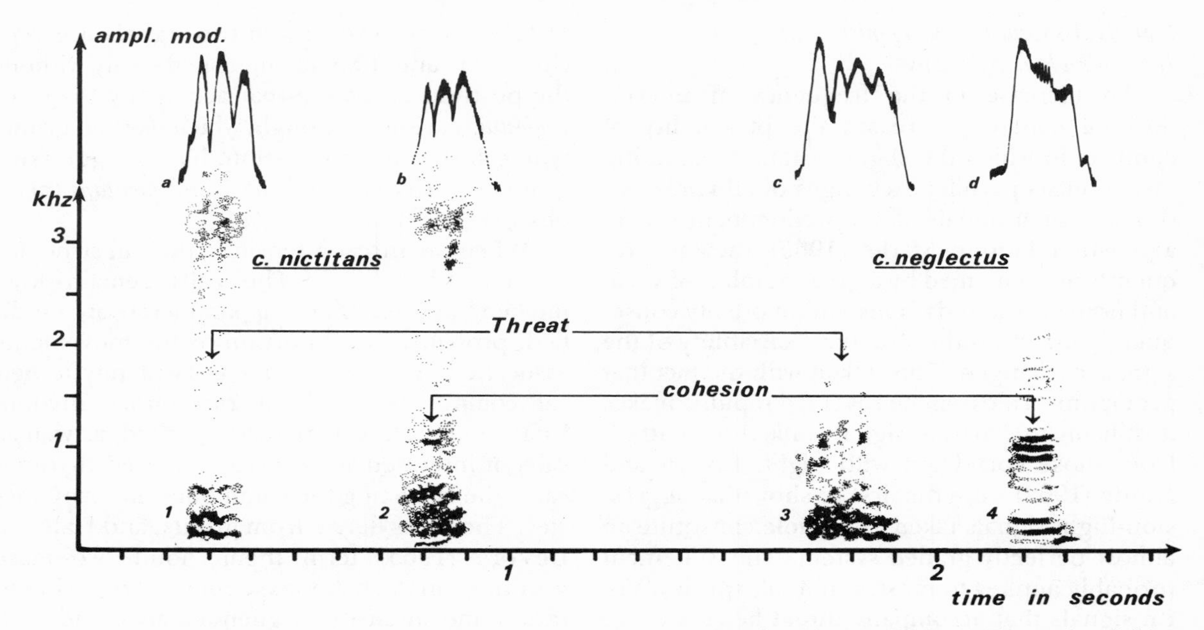

Fig. 19. Spectrographic analyses of threat and cohesion calls in *Cercopithecus nictitans* and *C. neglectus*. Threat gruntings of the two species (1, 3) and the cohesion grunting of *C. nictitans* (2) have close structures, characterized by a rolled sonority, quite visible on the amplitude modulation spectra (a, b, c). The cohesion call of *C. neglectus* (4,d) does not have this characteristic.

company threat behavior: tree shaking, canine grinding (langurs, mangabeys), gnashing (rhesus), or tongue clicking *(Colobus guereza)*, and can for the most part occur in intergroup conflicts as well as in other situations (see below).

Rhythmic calls constitute a link between aggression and flight vocalizations. In the latter there is a distinction between those occurring preferentially in response to alarm situations, emitted singly or in chorus (see above), and those occurring more specifically in interindividual agonistic relations. At the start of agonistic encounters, adult females and immature subordinates often give noisy, more or less low-pitched, rhythmic vocalizations. This type of call is emitted by all the *Cercopithecus* and all the *Cercocebus* spp. we have studied (*C. nictitans, cephus, pogonias, neglectus, ascanius, mona,* and *Ce. albigena, galeritus,* and *torquatus*). In most of them these vocalizations derive maturationally from the "gecker" of infants and juveniles; like the "yakking" of the *Papio* spp. (Hall and DeVore, 1965) derives from "chirplike clicking sounds." These calls are present in infants of almost all species; for example, the "gecker" of *M. mulatta* and *M. speciosa* (Rowell and Hinde, 1962; Bertrand, 1969), the "long cry" of *M. nemestrina* (Grimm, 1967), or the type 5 call of the talapoin (Gautier, 1974). Their occurrence in agonistic relations in older animals has been noted as the "yak" ("harsh staccato barking": Hall et al., 1965) in the patas, as the B1 call in *M. fuscata* (Itani, 1963), and in the langurs ("subordinate segmented sound" of *P. johnii:* Poirier, 1970a).

At a more advanced stage of agonistic relations, these rhythmic vocalizations become shriller and develop into calls of higher frequency. The rhythmic component may either

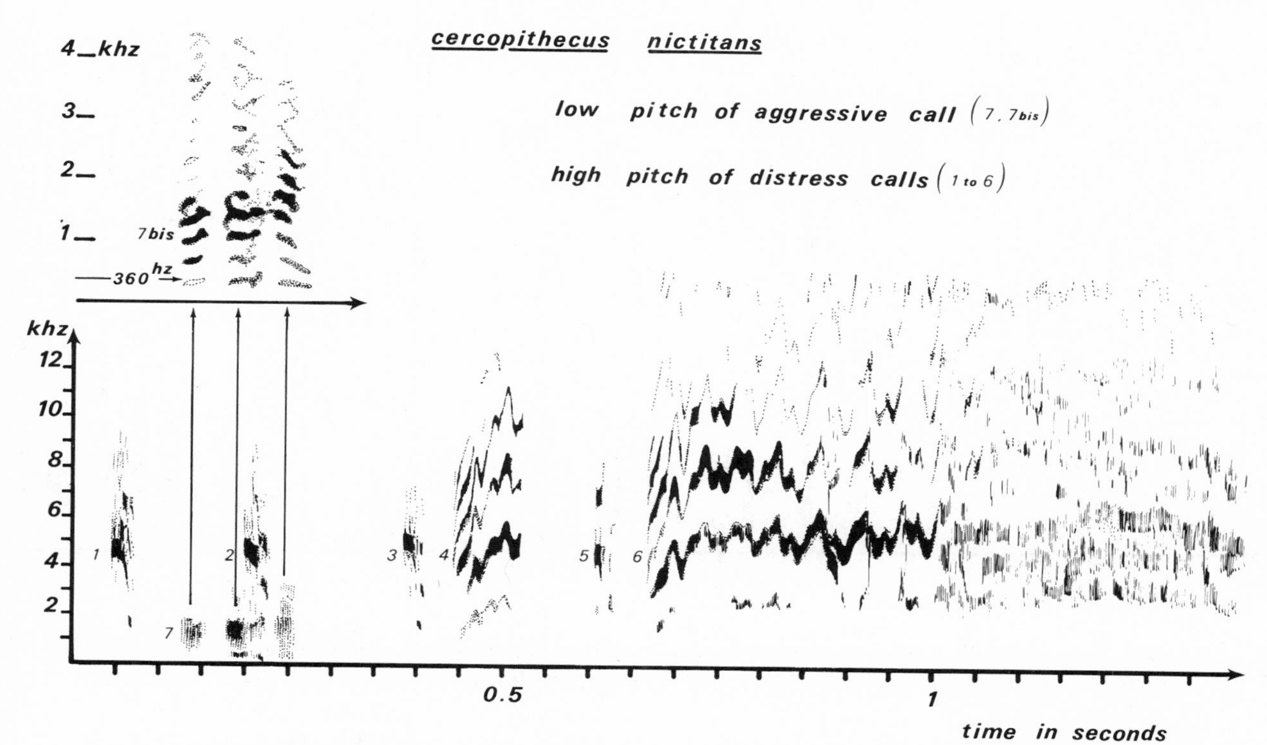

Fig. 20. Spectrographic analyses of a series of calls exchanged between two young *Cercopithecus nictitans* females during a conflict. Calls 1–6: High-pitched vocalizations of a threatened animal. Calls 1, 2, 3, 5: high-pitched rhythmic calls; call 4: pure quavered whistling; call 6: pure, then noisy quavered whistling. Calls 7 and 7 bis: Low-pitched rhythmic vocalization of the menacing animal. Call 7: included in the sequence; call 7 bis: detailed analysis.

disappear or persist, as in the "chutter squeal" of the vervet (Struhsaker, 1967b). The structure of these vocalizations is either pure ("whistle," "squeal") or somewhat rough ("screech," "scream") (Fig. 20). It would be useless to enumerate the species in which they are present, as they have been noted in practically all. Most authors agree that they are manifested preferentially by females and by the young. Furthermore, it seems difficult to relate their structural diversity as revealed by sonographic analysis to qualitatively different functions. It is generally believed that these vocalizations indicate a high arousal level and function as defensive signals, either in interindividual conflicts or in extragroup situations (e.g., the presence of a predator). They possess, even more so than do aggressive signals, a great power to evoke reactions in congeners. Depending on the context, congeners may converge on the emission place and threaten the protagonists or potential predators. These phenomena play a very important role in the defense of infants. In the young, the set of defensive vocal reactions constitute responses to increasingly frustrating situations. Thus, for the infant talapoin (Gautier, 1974), these reactions begin with emissions of the gecker type when mother/young contact is

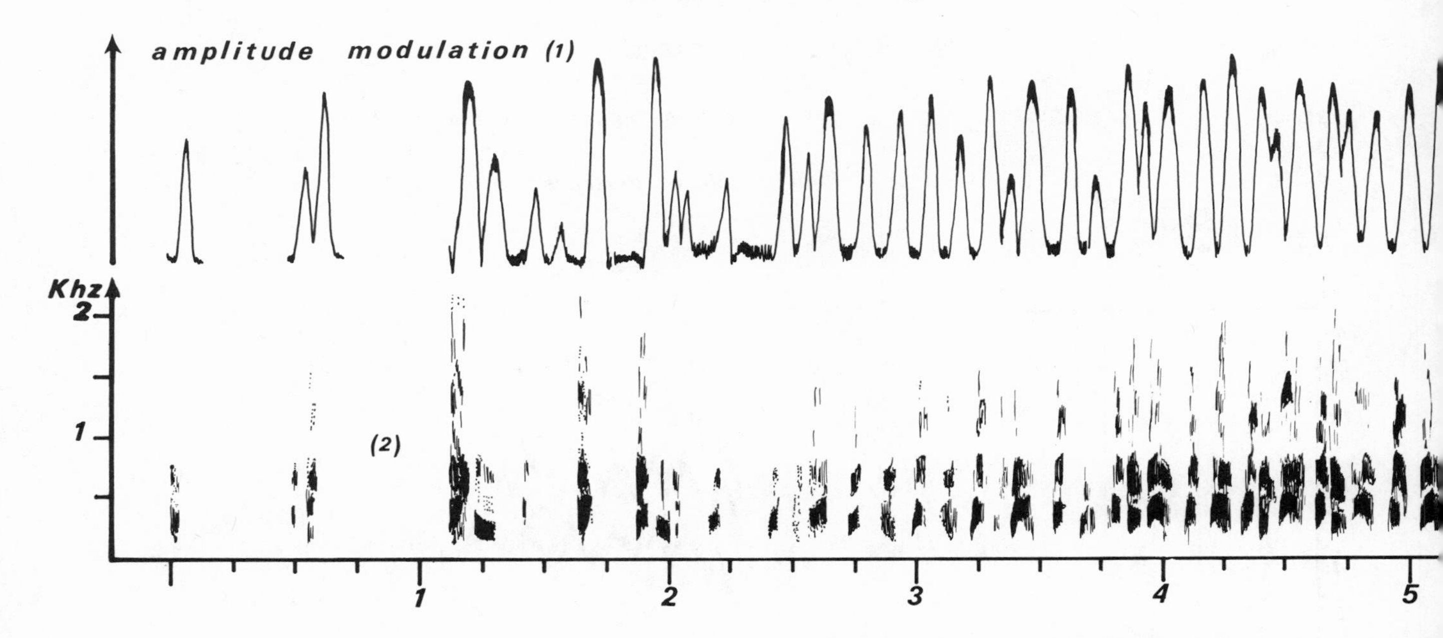

Fig. 21. Rhythmic call of low pitch emitted by a female *Cercocebus torquatus* during copulation. The call is a succession of inhaled (atonal) and exhaled (tonal) units associated with cheek movements.

slightly disturbed. If the break in contact is more substantial, the gecker yields to pure, type 6A whistlings (associated or not with type 3 lost calls). If, in the end, the situation becomes serious or if the young animal is threatened, the whistlings are replaced by type 7 screams. As far as species-specific variations are concerned, this chain of infant vocal reactions is common to many species of Old World monkeys.

Certain nonvocal sounds express a state of ambivalence and high excitation. Their limited range of action and their association with particular facial expressions cause them to function also as visual signs. Thus the "teeth chattering face" of *Macaca speciosa* (Bertrand, 1969) may be silent or accompanied by "teeth clicking sounds." This kind of signal, like the widespread lip smacking, will be discussed under visual communication (see below). The fact that they indicate a high level of excitation is revealed by their association with homosexual and heterosexual mounting behavior and sexual behavior in general. They may or may not occur simultaneously with vocalizations related to aggressive calls.

In *M. radiata* (Sugiyama, 1971), a dominant male soliciting a mount will "grin" and "tongue click." This same sound, associated with "clonic jaw movements," is given by males soliciting copulation with females (Kaufman and Rosenblum, 1966). Males of *M. fuscata* (Green, 1975) engage in teeth chattering or lip smacking (Tokuda, 1961–62) with females soliciting "consortship." Their homosexual mounting is accompanied by lip smacking, teeth chattering, and vocalizations. In *M. sylvana* (Deag, 1974), teeth chattering accompanies mounting and precedes copulation in a male looking at a female in estrus. Teeth gnashing is manifested by *M. mulatta* and *M. speciosa* males during copulation. In the talapoin (Gautier, 1974), mounting between males is also accompanied by lip smacking and a

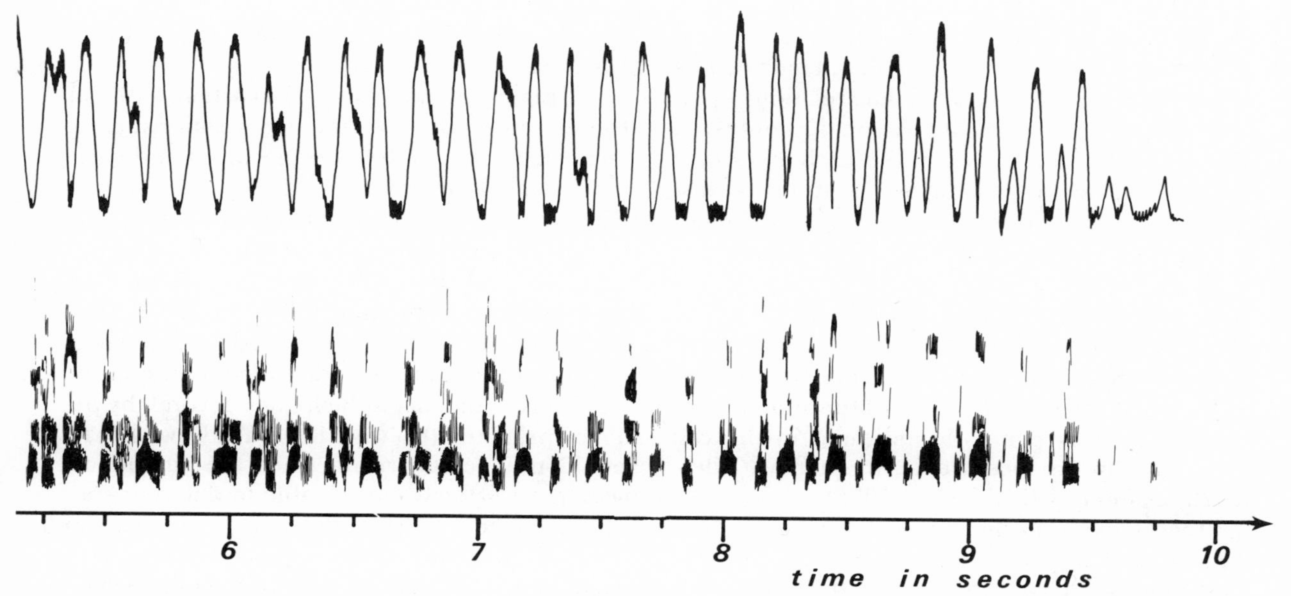

complex vocalization (type 10) close to that emitted by two partners copulating.

In some species, it is the males that vocalize during copulation. In *M. mulatta* (Southwick et al., 1965), the male utters a "high pitched staccato note"; in *M. speciosa,* a "vibrato scream" (Bertrand, 1969) or a "rhythmic expiration vocalization" (Chevalier-Skolnikoff, 1974). However, it is more generally the females that vocalize, whether before copulation, during the entire duration of thrusting, or at the end of the mount.

Among calls manifested by receptive females, one must distinguish between signals commonly used in interindividual encounters, the manifestation and structure of which indicate above all the status of the emitting individual (e.g., the gruntings or "soft squealing" of female langurs: Jay, 1965a), and those signals specific to encounters between the two sexes. Thus in the patas (Hall, 1965), *M. fuscata* ("estrus call": Tokuda, 1961–62), *M. nemestrina* or *radiata* (Kaufman and Rosenblum, 1966), *M. sylvana* (Deag, 1974), and *Ce. galeritus, torquatus,* and *C. neglectus* (pers. obs.), receptive females may emit particular calls which are indicative of their receptivity and which potentially play a role in the solicitation of copulation. The same type of call can occur in the course of copulation in all the species mentioned, as well as in other macaques (e.g., *M. irus:* Goustard, 1963) and in baboons (Hall and DeVore, 1965; Saayman, 1971). In *M. radiata* and *nemestrina,* as in the talapoin, both males and females can vocalize. Generally speaking, these calls are strictly linked with respiratory rhythm, and their length and intensity seem to depend on the females' state of excitation (Fig. 21).

If vocalizations preceding copulation play a role in sexual encounter, those appearing during mounting have a less obvious function. However, they draw congener attention to the coupling partners and can provoke harassment from other males or juveniles. In this sense they can contribute to the social regulation of reproduction in certain species.

Among the vocal repertoires of various species, vocalizations associated with agonistic behavior seem to be the least specific and the most graded. This is the case in woodland monkeys as well as those of open habitat. The relationship of call structures seems to depend on the context in which the calls occur: when the exchanges take place at close range and are continually sustained or relayed by visual signals, the needs of species specificity are reduced. The variability in calls and the possibility that threat calls may gradually change into those of subordination or flight are also the result of close-range exchanges and rapid reversals of the protagonist roles. Furthermore, elevation of the animal's excitation level is generally indicated in the call structure by an increase in frequency (Kiley, 1972; Gautier, 1974; Green, 1975).

Ontogenetic studies show that various shrill calls appear in frustration situations among infants and occur very early (from the first day of life in the talapoin and mangabeys, pers. obs.). On the other hand, low-pitched threat calls are not uttered until much later (generally at about one year).

VISUAL SIGNALS

Since human vision, like that of the monkeys, is excellent, it is not surprising that visual signals have been described many times and are of particular importance in general works dealing with communication in monkeys (see Andrew, 1963a, 1963b, 1964; Marler, 1965, 1968; Lancaster, 1968; Vine, 1970; Bertrand, 1971).

Despite the difficulty of comparing different studies, it is obvious that the numerous catalogues of visual signals compiled for various species reveal an astonishing similarity. Compare, for example, Altmann (1962) and Hinde and Rowell (1962) for *M. mulatta,* and Struhsaker (1967a) for *C. aethiops*. It is quite difficult on this basis to bring out specific differences. Thus any primatologist beginning work on a new species is almost sure to find the quasi-totality of facial displays described in the classic works of Van Hooff (1962; 1967).

In his synthetic essay, Bernstein (1970a) stresses the great fundamental relationship of visual signals in the Cercopithecoidea and tries, with justified precautions, to separate signals that would be "species-specific." Quite often one discovers that the designation "species-specific" is the result of anatomical peculiarities modifying or enriching the appearance of a facial display, or of the rarity of the signal's occurrence, or simply of the lack of studies of certain species.

This resemblance, which goes beyond the order of primates (Bolwig, 1962; Andrew, 1972), should not be surprising since the anatomical bases are relatively comparable in different species and since a considerable part of mammalian facial display and posturing appears to have originated from defense and protection behaviors (Andrew, 1963b). At the functional level, furthermore, this non–species-specificity is easily explained. In visual communication, by definition, animals see each other and are closely spaced. Therefore there is little or no problem of species recognition. This implies a second characteristic: at close range, visual signals can be modulated and extremely refined without the risk that the received message will be incomplete or poorly interpreted (Marler, 1965; Lancaster, 1968).

Starting from a common basis, differentiation is nevertheless carried out through anatomical specializations, e.g., presence of a more or less long and supple tail; more or less striking colorations in facial masking or the anogenital regions; presence or absence of cyclical intumescence of the sexual skin; differences in facial hair covering, development of manes, cranial toupets, or side-whiskers; accentuation of facial wrinkles. Forest monkeys generally possess either brillantly colored faces (*Cercopithecus,* mandrills), or highly contrasting colors *(Co.*

guereza). In *Cercopithecus,* these colorations are coupled with well-developed facial hair, which is reduced only around the eyes and the mouth. The rigidity and contrast of these "clownlike" masks do not permit subtle modifications of expression, but do assure immediate species recognition in closed habitats and doubtlessly play a role in reproductive isolation, especially in the numerous species living in association (Gautier-Hion and Gautier, 1974). Among woodland monkeys there are some notable exceptions in coloration: the talapoin, the crested mangabey, and the olive colobus, all of which live near water, are of a rather uniform greenish gray. The mangabeys, macaques, and baboons have the least facial hair. Whereas the often sombre faces of the mangabeys and colobus tend to "obscure" expressions, the dull and often pale faces of monkeys from open habitats permit subtle variations in expression.

Whatever the similarity in basic facial display and posturing of the Old World monkeys, it is clear that the visual communication within a group of forest guenons is fundamentally different from that manifested by a group of baboons. The difference lies partly in the variability and subtlety of the signals displayed (looks, facial expressions, gestures) and partly in the intensity and frequency of exchanges. Even though it may be possible to describe the subtlety of an almost imperceptible eyelid movement, no one seems to have tried to analyze the elements of the look (form, surface, and brilliance of the eye, or dilation of the pupil), which Chance (1967) in Vine (1970) justifiably thinks may constitute "the major stabilizing influence in primate groups." Although *a priori* it may seem easier to measure the occurrence frequency of behavior patterns, some methodological problems arise. In order to compare baboon communication with that of de Brazza's monkey, should one choose a group in each species of identical size and composition or rely on the fact that the one lives in groups of several tens of individuals while the other lives in social units of from three to six? This choice presupposes two different procedures as far as the causality of the modes of social organization is concerned: in the one, it is hypothesized that the means of communication of a species are directly responsible for its "sociability"; in the other, it is supposed that life in large groups develops and reinforces the probability and nature of exchanges.

In the group, a lot of visual signals act at medium range. This is the case with those that operate passively and are linked to individual morphology and anatomy. In the guenons, the bearing and color of the tail, like the facial coloring, are thus the first indices of species recognition in a woodland habitat. Similarly, all visible signals associated with age and sex can be communicated at a distance. Thus, the ease of recognizing an adult male varies directly with the difference in size between the sexes.

The passive aspect of signals linked to species-specific and individual conformity is enriched by the animals' status or self-confidence. Several types of gait have been described. The confident animal moves in a relaxed manner, with straight back, legs, and arms (e.g., "confident walk" of the vervet: Struhsaker, 1967a; "strutting gait" of the langur: Poirier, 1970a), whereas the subordinate has a hunched back, slightly bent legs and arms, and a low or curved tail. A jumping gait is characteristic of animals at play. Upon stopping, a confident individual will sit down with outstretched legs; a male will spread his legs and exhibit his penis. An individual that is often threatened will hunch its back, hold its head to the side, and gaze all around.

In most species, the one or more dominant males give particular emphasis to their posturing behavior. They sit down with ostentation, and when they stop in a quadrupedal position their legs and arms are stretched out. In woodland monkeys the tail is pulled up over the back in a conspicuous manner. If, as in certain species, the tail is needed for balance, it assumes a signal value in various specific stopping postures: the tail of *Ce. galeritus* touches its head (Fig. 11); that of *Ce. albigena* is brought up vertically, the tip curved toward the back; and that of *C. cephus* forms a backwards question mark. The tail-on-head stopping posture of *Ce. galeritus* is so frequent that Quris (pers. comm.) thinks that it replaces the function of color markings of *Cercopithecus* in species recognition.

In monkeys of open habitats, the tail also plays a role in social signaling, even in species with reduced tails, like the stumptail, for which Bertrand (1969) recognizes several positions. "Tail up" is an assertion of dominance, "tail curled up" characterizes a high degree of excitation, whereas "tail curled down" indicates a startled animal. Bertrand remarks that the erected tail is generally manifested by excited and confident animals, whereas the curved tail (like the lowered tail in the dog) is characteristic of threatened individuals. However, in baboons (Hall and DeVore, 1965) and in *M. fascicularis* (Angst, 1974), the "tail erect" characterizes subordinate animals. Some attitudes or gaits may have in addition more differentiated roles; e.g., the "swinging gait" of the male hamadryas preceding troop departure (Kummer, 1968) or the "drunken gait" of young *Cercopithecus* and the talapoin, which serves as an invitation to play (Gautier-Hion, 1971b).

Visual signals acting at short range between individuals, bringing them together or separating them, are quite varied (see Marler, 1968). Some are neatly classified in one or the other category, indicating either a friendly or an agonistic tendency; others have more ambivalent significations. The change from the former to the latter depends on the individuals' self-confidence and the variations in excitation level in relation to approach-flight behaviors.

Signals Associated with Peaceful Interindividual Relations

These signals often precede or occur during tactile and olfactory close-range exchanges (mounting, smelling, grooming; see below). The most frequent are lip smacking and teeth chattering, postures of presentation and of invitation to grooming and play, and the facial displays and gestures preceding sexual encounters. These signals are frequently combined; lip smacking and grimacing may be associated with presentation and various tactile exchanges, and they can occur together in peaceful, agonistic, or sexual contexts.

Lip smacking is known in the various baboons, the gelada, the macaques, the African and Asian colobus, the mangabeys, the patas, the guenons, and the talapoin and had been described by van Hooff (1962, 1967) in most of these species before it was observed in natural groups. He describes the different forms that it can assume and distinguishes functional lip smacking, associated in grooming with the ingestion of skin flakes and originating from suckling movements (Anthoney, 1968), from "vacuum lip smacking," in which this lip-tongue movement is not followed by any ingestion. Lip movement and tongue protrusion may vary according to the species, the latter being particularly evident in the langurs (Jay, 1965a) and the gelada (von Spivak, 1971).

Lip smacking is sometimes associated with staring and raising of the eyelids and eyebrows, as in threat. A relationship with threat behavior is particularly obvious when the behavior is intense, as in *Ce. albigena,* for example, where the upper lip may reveal the teeth and the animal shakes its head vigorously from side to side.

Lip smacking appears during or as a prelude to grooming. Acting as an invitation, it is used especially by individuals whose relations are "strained." It also precedes other types of approach and may be used by a dominated animal as well as by a dominant one. Used by a receptive female going toward the male, it may be followed by copulation; it may precede grooming or sitting in proximity and, in agonistic behavior, it is often associated with presentation in the threatened animal. Dominant males may also lip smack or teeth chatter while approaching subordinate females who are soliciting copulation (e.g., Chacma baboon: Bolwig, 1959; the hamadryas: von Spivak, 1971; the white collared mangabey: pers. obs.; and the Japanese macaque: Green, 1975). Similarly, a dominant female approaching a mother lip smacks before touching the infant. The Barbary macaque also engages in lip smacking and teeth chattering before sniffing an infant and taking it from its mother (Lahiri and Southwick, 1966). In the gelada, the female induces a juvenile to hang from her back by touching it and addressing it with lip smacking.

This behavioral pattern seems therefore to be a positive social signal (see van Hooff, 1962, 1967; Hinde and Rowell, 1962) that may either appease the aggressive behavior of a dominant animal or diminish the flight reaction of a subordinate, thus permitting the fulfillment of subsequent behavior. Its meaning is nevertheless sometimes ambiguous, notably in *Cercocebus* or *Cercopithecus,* in which it is often associated with threats. This aspect is especially frequent in agonistic exchanges between a pair of animals of adjacent or poorly established status. The two individuals address each other with head bobbing and lip smacking. In *Ce. albigena,* intense lip smacking ("snarl-smacking face" of van Hooff, 1967), associated with threats and head shaking, appears to indicate a state of uncertainty in the animal. Hall et al. (1965) also attributes elements of threat and fear to the "gnashing of the teeth" of the patas.

In some species lip smacking appears only in social grooming (as in the mandrills: van Hooff, 1967), but it plays an important social role in interindividual encounter in baboons, macaques, and mangabeys, notably between animals who are poorly acquainted or who have unstable dominance relations. Although present in *Cercopithecus* (as well as in the patas) during grooming and agonistic manifestations, lip smacking has rarely been observed as a preliminary to encounters between two individuals or as a friendly solicitation.

Most species have several postures of invitation to grooming, indicating a desire either to groom or, more often, to be groomed. Individuals will present preferentially the chest, the back, or the genitals. The ritualization of these postures generally becomes more obvious in direct proportion to the animals' lack of mutual familiarity and the degree to which dominance relations are strained. By contrast, this ritualization does not appear between mother and infant (Gautier-Hion, 1971b).

Genital presentation, which is often associated with lip smacking, occurs in both sexual and various other social contacts. From the qualitative point of view, "normal" presentation, with the animal erect on its paws, is distinguished from crouched presentation, with the individual on bent legs and with a curved back (Fig. 22). The latter is generally found in threatened or strongly dominated animals. It seems to be the rule in female patas and hamadryas (Hall, 1965; Kummer and Kurt, 1965).

"Copulatory" presentation preceding genital sniffing, mounting, or copulation by the male is frequent and obvious in the baboons, macaques, and mangabeys, and also exists in the *Colobus* spp. It is less distinct in the guenons, in which it does not receive the emphasis that it gets in the baboons. In forest monkeys, presentation is often made with the body arched and the tail curled on the side *(C. neglectus)* or in the form of a question mark *(C. cephus, C. ascanius);* the ani-

Fig. 22. "Crouched" presentation in *Colobus badius:* a female carrying a newborn presents herself to an adult male (Gombe National Park; photo by T. H. Clutton-Brok.)

Fig. 23. A dominant female of a *Cercocebus albigena* group presents her genitals to a young infant held by its mother before entering in contact with the former. (Paimpont; photo by B. Deputte.)

mal looks at its partner with a pout, which may be accompanied by a call.

Nonsexual presentation, although indeed manifested by males, is essentially a female behavior (Hall, 1962). It can precede grooming, passing by or sitting down in the proximity of a congener, or handling of an infant held by its mother (Fig. 23). When displayed by a dominant animal, it indicates friendly intentions and seems to inhibit the flight reaction of the one being approached. Mutual presentations between males can thus be observed. Presentation is also frequent during agonistic behavior in which the pursued animal stops and exhibits its genitals. Most authors consider it as having a role in the cessation of aggressive manifestations, in progress or potential (Carpenter, 1942; Hall, 1962; Altmann, 1962; Poirier, 1970a). Chalmers (1968) has shown with a band of *Ce. albigena* in the wild that interindividual approaches accompanied by presentation significantly reduced the probability of aggressive behavior.

As Carpenter (1942) noted, presentation is a "greeting gesture." In subordinate monkeys it is a gesture acknowledging the status of a congener, and in dominant animals it is a manifestation of momentary "friendly intentions." This seems obvious, the more so since the presenting animal is quite vulnerable. Hall (1962) stresses that this pattern has a role in status recognition among males in baboon groups, in which the dominant individuals are those to which the most presentations are addressed. In forest guenons presentation rarely occurs as an invitation to interindividual encounters, except occasionally during play or as a prelude to mounting in young males. It is also observed in young males toward an adult male (*C. nictitans, C. cephus, C. pogonias,* pers. obs.) and can be associated with pouting (which is found in receptive females) and/or trilled calls (Fig. 24).

The grimace can also precede certain peaceful encounters and assume a meaning rather close to lip smacking, with which it can be asso-

Fig. 24. "Arched" presentation, associated with pouting in a young male *Cercopithecus pogonias*. Note the position of the left rear paw. (Paimpont.)

Fig. 25. An adult male *Theropithecus gelada* curls his upper lip during a peaceful contact. (Photo by H. Spivak.)

ciated in the same sequence. It appears sometimes in threatened animals (see below) and sometimes in dominant individuals; thus a male gelada going toward a female exhibits a grimace with the upper lip raised, to which he may add lip smacking (von Spivak, 1971) (Fig. 25). Altmann (1962) points out a similar behavior in rhesus males.

Various facial expressions, which may or may not be specific variations of the grimace, have been described. A notable case is the mandrill's "eight-smile face," which is associated, like the snarl-smacking face of *Ce. albigena,* with head shaking. Sometimes considered to be a friendly mode of approach (Bernstein, 1970a), this smile may be an ambivalent approach-flight manifestation, similar to the behavior of *Ce. albigena.* Note also the "high grin face" of *Cynopithecus niger* (Bernstein, 1970a), which may correspond to a species-specific exacerbation of the grimace.

Head wagging is also present in forest *Cercopithecus* like *C. nictitans* or *C. cephus* and appears in ambivalent situations. In *C. neglectus* it is associated with closed-mouth chewing movements.

Lip smacking, presentation, various grimaces, and head wagging are all displays indicating social tendencies in which elements of threat are rare or absent. They are followed by an absence of partner reaction or by a reduction of interindividual distance and an increase in the probability of short-range exchanges. Displayed at medium or close range, they are often a prelude to tactile and olfactory exchanges in which the animals, being mutually vulnerable, need previous assurances of the peaceful intentions of the partner. Like the majority of close-range signals, whether vocal, visual, or tactile, these behavioral patterns are increasingly ritualized in direct proportion to the degree to which partner relations are unstable or strained. This may be one of the reasons why lip smacking and grimaces appear to have ambivalent meanings in which approach-flight tendencies are closely associated.

It is notable that visual exchanges that indicate the mood of the partner are essentially present in guenons in their original function (lip smacking and grooming; presentation and copulation), but rarely occur outside of these functional contexts, as though they had not acquired any ritualized social function in these species. The infrequency of these signals parallels the underdevelopment of "personalized"-contact vocal signals. Rowell (1971) justly thinks that this "lack of appeasement and policing are probably the fundamental differences between the behavior of guenons and the baboon-mangabey-macaque genera."

In various species, sexual encounter is accompanied by differentiated visual exchanges. The grimace may appear during copulation or mounting between males. It can be associated with lip smacking or teeth chattering (*P. ursinus, P. anubis:* Hall, 1962; Hall and Devore, 1965; Saayman, 1970; *M. mulatta:* Southwick et al., 1965; Lindburgh, 1971; *M. fuscata:* Itani, 1963; *M. fascicularis:* Angst, 1974; *M. sinica:* C. M. Hladik, pers. comm.; *M. talapoin, Ce. albigena, Ce. galeritus:* Gautier-Hion, 1971b and pers. obs.). These facial displays, which may be associated with calls (see above), indicate the individual excitation level: the visual signal does not seem very communicative since the partners do not usually perceive it (Fig. 26).

The facial displays and gestures that precede copulation are of a more obvious function, seeming to facilitate sexual encounter. Pouting with protruding lips appears quite frequently in receptive females. This pouting with or without calls exists in *M. fuscata* (Itani, 1963), *P. ursinus* (Hall, 1962; Saayman, 1970), *Ce. torquatus, Ce. galeritus, Ce. albigena, C. nictitans, C. cephus, C. pogonias* (pers. obs.). In the patas, Hall (1965) describes a similar facial display with protruding jaws and swollen cheeks, accompanied by wheezing and chortling sounds. An apparently identical behavior has been observed in *C. neglectus* (pers. obs.). Finally, the well-known "jaw thrust" (Kaufmann and Rosenblum, 1966) of *M. nemestrina* and *M. radiata* seems to be an exacerbation of this same pattern. In *M. nemestrina,* this display is frequent, especially in males, and appears in varied contexts. In this species, as in the bonnet macaque, the male accompanies it with a lateral head shaking reminiscent of a male *Ce. albigena* inviting a receptive female to approach.

In *Cercopithecus,* the receptive female combines this pout with an "arched" presentation, so that the face and posterior are simultaneously directed toward the partner. The same pout, with or without calls, occurs in most species during copulation (Fig. 27).

Finally, there is the play face ("relaxed open mouth face" of van Hooff, 1967), which appears either as an invitation from a distance or during the acting out of the behavior. Andrew (1963a) has commented on its relationship to aggressive faces. As in the latter, the teeth are uncovered but there is an obvious difference in the eyes, which are often "slitted" or partially closed (Fig. 28).

Signals Associated with Agonistic Interindividual Relations

A particularly large number of visual exchanges occur in agonistic behaviors, for which they constitute the essential basis of signalization, enriched by a great number of graded vocalizations (see above). The descriptions of these offensive or defensive signals have been attempted (see van Hooff, 1962, 1967; Hinde and Rowell, 1962), but remain unsatisfactory. The problem is that it is necessary to classify the signals into separate patterns, which do not take account of their extreme variability or the behavioral dynamics. In this respect, the methods sug-

Fig. 26. Grimace during copulation in *Macaca sinica.* (Ceylon; photo by C. M. Hladick.)

Fig. 28. "Play face" in a young female *Cercopithecus nictitans.* (Paimpont.)

Fig. 27. Pout associated with inhaled-exhaled call during copulation in an adult female *Cercocebus albigena.* (Paimpont; photo by A. R. Devez, CNRS.)

gested by Altmann (1965) should permit great progress in understanding behavioral sequence chains.

Many parts of the body are activated either simultaneously or separately in signals linked with aggression-flight: eyes, eyelids, eyebrows, scalp, ears, lips, tail, legs and arms. Because of this fact, the same open-mouth face, with teeth exhibited, may have different meanings depending on whether the animal's ears are erect or lying down, whether the animal is leaning forward or apparently recoiling, whether its posture is straight or crouched, or whether its tail points up or is bent down.

An animal engaged in an agonistic action is rarely completely confident. If it is, there is no real agonistic exchange but rather a defined and stable dominance relation in which a simple glance is sufficient to "repress" a partner. In a real agonistic exchange, on the contrary, each animal must at every instant take into account the slightest behavioral variations of the other. Therefore, no matter what species is under consideration, it appears that signals associated with aggression-flight generally include the most graded ones of the species repertoires, dealing as much with vocalizations (see above) as with visual exchanges.

Several general traits can nevertheless be

shown: The look can be fixed and intense. The emitting animal stares directly at its partner, its eyes only slightly widened, with the eyelids barely raised and the upper lip perhaps pouting slightly (e.g., *M. mulatta:* Hinde and Rowell, 1962; and *C. nictitans,* Fig. 29). This type of look is characteristic of a threatening monkey. An animal never stares at a partner in peaceful group exchanges, which are made with simple glances or vague looks (except lip smacking). On the other hand, an animal that has been stared at or threatened turns away and deliberately looks in the opposite direction from the partner. This looking away can indicate both submission or an attempt at nonparticipation. It can be observed in a dominant animal in response to a fearful glance from a dominated monkey, in which case it indicates nonaggression.

Fig. 29. "Stare" associated with a slight pout in a male sub-adult *Cercopithecus nictitans.* (Paimpont; photo by A. R. Devez, CNRS.)

Between these two types, numerous other looks are manifested. Altmann (1962) and Bertrand (1969) mention the fearful look characteristic of a potentially threatened animal. In *Ce. albigena* a female in estrus assumes this fearful look while glancing at the dominant male if another male follows her with his eyes (pers. obs.). Bertrand also notes a look of feigned indifference when a dominated individual pretends not to see the threat or stare of the male, and a look of feigned interest when the threatened monkey intensely busies itself with tail grooming or a random object. These subtle looks can evidently be expressed by all monkeys and appear to be capable of reducing congener aggression.

The stare is intensified by a number of gradual facial movements: raising the eyelids, which is particularly obvious and repeatable in fairly rapid succession in the macaques, baboons, and mangabeys; raising the eyebrows, which reveals the supra-oculary region with perhaps colored eyelids, often of a striking white (*Th. gelada:* von Spivak, 1971; *M. fascicularis:* Angst, 1974; and Fig. 30: *Ce. torquatus*); retraction of the scalp, which reinforces the disclosure of the eyelids; and pulling the ears back, which accentuates the stretching of the face.

All these elements are particularly clear in monkeys of an open habitat, as well as in the mangabeys and the talapoin. In the *Cercopithecus* spp., however, movement of the ears is hidden by the abundant hair surrounding them, and so are raising of the eyelids and retracting of the scalp. These actions are perceptible in *C. cephus* and *C. pogonias* and to a slight degree in *C. nictitans,* but are almost nonexistent in de Brazza's monkeys. On the other hand, in various mangabeys and baboons, scalp movements are accentuated by the presence of toupets or side-

Fig. 30. "Stare," exhibition of white eyelids, "open mouth threat face," associated with grunting, and hair erection in a young male *Macaca fascicularis* during a chase. (Photo by W. Angst, 1974, in *Fortschritte der Verhaltensforschung*.)

whiskers that can be turned down or flaunted.

Various mouth expressions have been described by van Hooff (1962, 1967), who compares their appearance in the different species. The staring open mouth face, or open mouth threat face, appears very generally. It is associated with the stare; the stretching of the eyelids, scalp, and ears toward the back; and a more or less open mouth, frequently in an "O" shape, is particularly obvious in the macaques, but it is just as obvious in a *Cercopithecus* like *C. cephus,* for example. In *C. diana* (van Hooff, 1967), the patas (Hall et al., 1965), and *C. neglectus* (pers. obs.), the open mouth face appears repeatedly, the mouth opening widely without the teeth being exposed. Poirier (1970a) also notes "air biting" in the langurs, which seems comparable. Sudden openings and closings of the mouth have also been pointed out in the gelada (von Spivak, 1971).

The open mouth threat face appears in most species as an expression of threat, although Struhsaker (1967a), who terms it "gaping," attributes to it a sense of defensive threat, and Deag (pers. comm.) points out that it is rarely followed by attack in the barbary macaque. This face is frequently associated with a lowering of the head (baboons, macaques) or "head bob" (*P. hamadryas,* macaques, langurs, mangabeys, guenons). In the forest guenons, head bob and forearm jerking, which are both associated with the stare, are so rigid and jerky that the animals look like jacks in the box. Thus the name "hocheur," given in French to *C. nictitans,* could as well be applied to *C. cephus.*

During a threat, animals can jump in place (baboons, guenons), slap the ground, strike in the direction of a partner (baboons, macaques, langurs, mangabeys, patas, guenons), make sudden forward movements (macaques, patas, guenons). These threats are sometimes followed by real pursuits, with grasping of the scalp and biting, or false pursuits, where the partner is never caught (see Kummer, 1957; Kummer and Kurt, 1965; Bertrand, 1969).

Hair erection appears during agonistic behavior in several baboons (notably in hamadryas, gelada), in *M. speciosa, M. fascicularis* (Fig. 30), *Ce. galeritus,* and *C. neglectus.* It indicates an "emotional" activation of the autonomous system and does not seem to be connected with approach any more than with flight. In the stumptail, Bertrand (1969) considers it to be a "mixture of aggression and fear."

In many species a "bared-teeth threat face" (van Hooff, 1962) associated with the stare prolongs the "O"-shaped mouth (Fig. 31). It can be considered offensive or defensive. In the talapoin, it is the usual threat face. When associated with scalp and ear retraction, it is often followed by attack. It becomes defensive when the scalp and ears come forward (Gautier-Hion, 1971b).

One or more facial displays derived from the

Fig. 31. "Bared teeth threat face" in an adult male *Cercopithecus neglectus.* (Paimpont: photo by A. R. Devez, CNRS.)

Fig. 32. Facial display associated with a "scream" in an infant *Cercopithecus nictitans.* (Paimpont.)

grimace are generally manifested by threatened animals. The lips are more or less turned up, revealing the clenched teeth (macaques, baboons, langurs, mangabeys, guenons, patas). These grimaces are either silent or associated with many geckers, screams, and screeches, which, as we have seen, are common to a number of different species. The lifting of the lip is particularly obvious in the gelada (von Spivak, 1971; Fedigan, 1972), where the lip can be rolled completely back. This rolling up is also observed in a threat. It is not very apparent in guenons, and especially rare in infants, whose facial displays associated with geckers or screams are mostly characterized by enlargement of the eyes and lifting of the eyelids (Fig. 32).

A threatened animal may also respond by directing a threat face to a third animal. These redirected threats, particularly well described in the hamadryas (Kummer and Kurt, 1965) but also present in many species (e.g., gelada: von Spivak, 1971; mangabeys: pers. obs.; guenons: Gautier-Hion, 1971b; Rowell, 1971), seem particularly frequent in captivity. Under these conditions, chain reactions can involve the entire group of animals (see Rowell, 1971).

Special mention should be made of yawning, which so many authors have noted in so many species: the baboons (Bolwig, 1959; Hall and DeVore, 1965; Hall, 1968a; Kummer, 1968; Stoltz et al., 1970), gelada (von Spivak, 1971; Fedigan, 1972), macaques (Hinde and Rowell, 1962; Simonds, 1965; Kaufman and Rosenblum, 1966; Bertrand, 1969; Angst, 1973, 1974), langurs (Poirier, 1970a), *Cercocebus* spp. (Chalmers, 1968; Bernstein, 1970a; Quris, 1973), *Cercopithecus* spp. (Gautier, 1971 and pers. obs.), talapoin (Gautier-Hion, 1971b; Wolfheim and Rowell, 1972). In *Papio anubis* and *ursinus,* as well as in *Ce. albigena* (Chalmers, 1968), yawning is considered a threat display. Most authors, however, observing that yawning is often undirected and can even be done with closed eyes, think that it has no more than very low communicative value and constitutes rather an individual expression of an animal, hence its name: "tension yawning" (Fig. 33).

Fig. 33. Nondirected "tension yawning" in a male *Macaca fascicularis.* (Photo by W. Angst.)

Studies in progress in the cause and function of this behavior on two captive groups of *Ce. albigena* (Deputte, pers. comm.) show that yawning is quite generally evoked in males by situations exterior to group life, coming either from exchanges between two groups (see also Angst, 1973) or disturbed situations (potential or real predator). They also show that yawning by males elicits no response within the group, even though it may evoke a kind of outbidding between males of the same group when associated with shaking in ritualized sequences, or more rarely between males of neighboring groups. Rather than a threat, so-called tension yawning appears to be a display that expresses status among males.

These generalities about visual signals occurring in agonistic contexts show the close relationship between basic signals in different species. There are, however, profound differences in sequence chains. A *Cercopithecus* making a threat, stares, opens its mouth in a figure "O," violently wags its head, and jumps up and down on its hind legs. These different elements are repeated in turn, intensely and rigidly, without any gradual blending from one to the other. Under the same conditions and beginning with the same basic elements, a *Ce. galeritus* will modulate and refine these various patterns. It does not sustain the stare, but raises its eyelids, relaxes its eyes, threatens with open mouth, then makes a defensive face, approaches its partner while lip smacking, and then stops with tail erect before making a false pursuit.

Schematically, one could say that a *Cercopithecus* attacks or is attacked; a baboon, macaque, or mangabey has more subtle individual relations, which take into account the slightest changes in partner attitude, associated with both threat and appeasement behavior, and they sometimes exaggerate various reactions to the extent of a bluff. A female *Cercopithecus* pursued by a male flees, lies down, and grimaces and screams when the male reaches her. A female mangabey multiplies defensive expressions, which are mixed with lip smacking and presentations and combined with excessively violent screams, the intensity of which appears to bear little relation to the potential threat. This emphasis is absent in guenons, where aggressive manifestations are rare and of short duration anyway.

In baboons, macaques, and mangabeys, regulation of intragroup relations by visual mechanisms is ceaseless and subtle, linked with frequent looks that yield continuous information on the activity of neighbors. In woodland guenons, the lack of facial expressiveness and the passivity of exchanges causes reactions to be less predictable and subtle, and less susceptible to modulation. These features incontestably indicate a less-developed socialization, correlating with a lesser degree of exchange ritualization. In species with more open habitats, many signals have developed increasing ritualization, liberating them from their original functional context in order to give them an increasingly social significance.

OLFACTORY AND TACTILE SIGNALS

Olfactory communication in Old World monkeys has received relatively little attention, in part because of the relative absence of specialized skin glands in these species and in part because olfactory signals do not appear to be an indispensable means of communication in animals with vision as highly developed as that of the monkeys. This excellent vision has warped human observation; thus all primatologists dealing with genital presentation posture treat it as an essentially visual signal because that is what the observer effectively perceives (see Bertrand, 1971). Nevertheless, Michael and Keverne (1970) have shown in the rhesus, that although the sexual skin of a female is sufficient to attract the attention of a male, it is still ineffective in activating his sexual behavior.

As Moynihan (1967) points out for New World monkeys, the importance of olfactory communication is doubtless greater than the studies would suggest. In these monkeys, as in those of the Old World, any object or individual that is new or has returned after an absence is first sniffed or touched and then intensively sniffed by its congeners. Visual recognition of the new arrival seems to be only a quick means of orienting exploration, which is subsequently made precise and refined by olfaction.

Sniffings are directed to various parts of the body: genitals, abdomen, snout, armpit, or fur. They occur in relatively stereotyped postures, which often imply tactile exchanges probably resulting from the olfactory capacities of the individuals, which do not operate at a distance, as is the case for other mammals. Most observers have therefore retained only the tactile signal. Thus Marler (1965), going back to the list of behavioral features that appear when a stranger congener is introduced into a baboon group (Hall, 1962), emphasizes the wealth of tactile stimuli provoked by the introduction: genital-stomach nuzzling, rump nuzzling, back-fur nuzzling, and mouth-head kissing are considered principally contact exchanges.

At present, no experimental proof permits us to assign a particular role to olfaction in these activities, but the fact that they are all accomplished with the nose and directed preferentially to specific parts of the body suggests that the olfactory element plays some role. Thus, we will deal with tactile and olfactory signals simultaneously when behavioral patterns appear to imply both types of signal.

Olfactory or Olfactory and Tactile Signals

Sniffing of genitals is without doubt the most widespread of these signals. It occurs primarily in sexual encounters, provoked either by female presentation (Fig. 34) or by the initiative of a male forcing a female to stand up (*M. radiata:* Nadler and Rosenblum, 1973; *Cercopithecus* spp.: Booth, 1962; *M. talapoin:* Scruton and Herbert,

Fig. 34. A young male *Macaca fascicularis* inspects the genitals of a female in presentation. (Photo by W. Angst, 1974, in *Fortschritte der Verhaltensforschung*.)

1970; Gautier-Hion, 1971b; *Ce. albigena:* pers. obs.). It can also occur temporarily as a check on the sexual state of the female. Thus in *M. radiata* (Simonds, 1965) or *M. sinica* (Jay, 1965b) during the course of the day, in a veritable patrol, the males sniff the genitals of all the females in the group.

Sniffing is carried out directly or by means of sniffing a finger that has touched the vaginal orifice. These sniffings are rather infrequent in the guenons, but occur very often in mangabeys and certain macaques (Chalmers, 1973) and are also present in the baboons (Hall and DeVore, 1965) and gelada (von Spivak, 1971). They are rare in langurs, and Poirier (1970a) only observed four in *P. johnii* during 1,250 hours of observation.

Michael and Keverne (1968) have demonstrated the existence of a pheromone of vaginal origin ("short-chain aliphatic acids": Michael et al., 1971) that acts on the sexual activity of male rhesus through the olfactory pathway. It is probable that similar mechanisms are present in other species and that the function of genital sniffing is to collect information about the state of the females' cycles.

In several species genital sniffing is mutual and assumes a stereotyped form. In *C. mitis* (Rowell, 1971), *C. nictitans, C. pogonias, C. cephus* (pers. obs.), a male and a female face in opposite directions and circle while sniffing each other's genitals. In the mangabeys, a number of postures imply mutual sniffing. Bernstein (1970a) describes the side-to-side posture in *Ce. atys,* where the two congeners mutually sniff their ano-genital parts, with arms and legs passed over the back of the congener. A similar posture is present in *Ce. albigena* (Fig. 35,c,f), especially among females. Reciprocal ano-genital contacts predominantly between males have also been described in *M. radiata* (Kaufmann and Rosenblum, 1966).

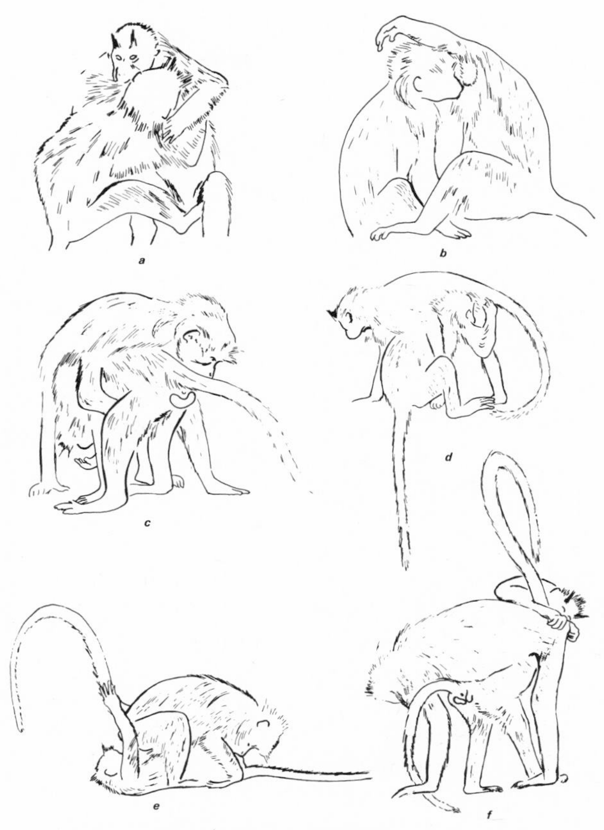

Fig.35. Various sniffing postures in *Cercocebus albigena* (see text). (Paimpont.)

Further variations of these sniffings exist in *Ce. albigena* (pers. obs., Fig. 35,d,e); one individual may sit astride another or lie on it in reversed posture, putting its genitals on the nose of the other while it sniffs the congener's genitals. Fig. 36 shows a female carrying an infant while engaged in such behavior with another female.

Sniffings with or without manipulation of the ano-genital parts is also very widespread with respect to newborns, especially during their first months (baboons: DeVore, 1963; Hall and DeVore, 1965; langurs: Jay, 1963, 1965b; *M. radiata:* Rosenblum and Kaufman, 1967; *M. speciosa:* Bertrand, 1969; *Cynopithecus niger:* Poirier,

1970b; *M. talapoin:* Gautier-Hion, 1971b; *M. sylvana:* Deag and Crook, 1971; *C. aethiops:* Struhsaker, 1967a; *Ce. albigena, Ce. torquatus, C. ascanius:* pers. obs.).

In *Ce. albigena, M. talapoin,* and *M. speciosa,* the mother and the other females frequently examine an infant's genitals. They lift the juvenile's tail and visually, tactilely, and/or olfactorily inspect its genitals. The role of olfaction is very clear when a mother turns her infant upside down, putting its genitals next to her nose (Fig. 37).

In *M. radiata,* these examinations are reserved to male infants. In *C. aethiops,* only females participate in them, whereas in the stumptail, the mangabeys, and the talapoin, sniffings attract the attention of the entire group.

Muzzle contact—an individual touching its nose to the muzzle of a congener—has been described in most species (e.g., "mouth-kissing" in *P. ursinus:* Hall, 1962; "mouth to mouth" in *P. anubis:* Hall and DeVore, 1965; "maul beriechen" in *Th. gelada:* von Spivak, 1971; "sniff at" in *M. mulatta:* Altmann, 1962; "sniffing at face" in *M. radiata:* Simonds, 1965; "sniffs at" in *P. hamadryas:* Kummer, 1957, 1968; and *Co. guereza:* Horwich and Lafrance, 1972; "muzzle-muzzle" in *E. patas:* Hall, 1965; and *C. aethiops:* Struhsaker, 1967a; "museau-museau" in talapoin: Gautier-Hion, 1971b; "nose to mouth" in *Cercopithecus albogularis:* Rowell, 1971).

In the baboons, as in the mangabeys, this behavior is relatively rare and is often associated with alimentary behavior. In the *Cercopithecus* spp., on the other hand, muzzle-muzzle examination is frequent (Fig. 38), and Struhsaker (1967a) notes that in the vervet it can be associated with grooming, play, and penile displays, and can precede a sniffing of the perineum or a heterosexual mounting as well as follow intense agonistic encounters.

Several other modes of sniffing occur. Sniffing of the ventral part of the abdomen exists

Fig. 36. Female *Cercocebus albigena* carrying an infant and engaged in reciprocal ano-genital sniffings with another female. (Paimpont; photo by A. R. Devez, CNRS.)

Fig. 37. Sniffing of a young male infant's genital organs by its mother, *Cercocebus albigena.* (Paimpont; photo by A. R. Devez, CNRS.)

Fig. 38. A juvenile male *Cercopithecus nictitans* comes to sniff the muzzle of an adult male. (Paimpont; photo by A. R. Devez, CNRS.)

in the baboons ("genital nuzzling" of *P. ursinus:* Hall, 1962), the vervet ("face in inguinal region": Struhsaker, 1967a), the gelada (where the partner examines the "abdominal beads": Bernstein, 1970a), and *C. nictitans* (pers. obs.).

Other sniffing, mouthing, or chewing, in which a partner's muzzle "digs" in the fur of a congener, occur either in the dorsal fur ("back-fur nuzzling" of the chacma baboon: Hall, 1962; sniffing of the back in *Colobus guereza:* Horwich and Lafrance, 1972) or behind the head or on the side of the neck (*M. talapoin:* Gautier-Hion, 1971b). Subsequently, they are included in mutual embracing behavior along with mouthing and chewing of the neck skin (cf. Fig. 35a).

Kissing is an equally widespread behavior that seems to be more tactile than olfactory. In *Miopithecus* and *Cercopithecus* spp., however, kissing often implies sniffing; in these species it appears to derive from the neck-sniffing behavior of a mother toward an infant huddling in her arms (Gautier-Hion, 1971b).

Kissing can be ventro-dorsal (e.g., baboons: Hall and DeVore, 1965; langurs: Jay, 1965b), but ventro-ventral kissing is more generally observed with the partners face to face, holding each other (various macaques, baboons, mangabeys, *Cercopithecus* spp., *Erythrocebus, Miopithecus*). It is especially frequent between adult males and can be accompanied by grinning, lip smacking, tongue-clicking (*M. radiata, M. nemestrina,* talapoin, etc.), or vocalizations (*P. johnii,* talapoin). In the gelada (von Spivak, 1971), it is accompanied by lifting of the upper lip. In various species, kissing is followed by mouthing, sniffing or chewing of the neck or sometimes the shoulder skin, but it never goes as far as real biting (langurs, macaques, talapoins, baboons).

It must be noted that the guenons possess a characteristic odor in the fold of the neck behind the head. Perhaps it is simply caused by an accumulation of sweat; it seems partially species-specific. In *Ce. albigena,* a particularly distinct odor of flour is released from the armpit, which is sniffed by females, one lifting her arm at the approach of another (Fig.35b).

Thus, although Old World monkeys are considered to make little use of olfaction, they have developed an important series of signals concerning regions that release particular odors, signals that set in play relatively stereotyped postures that are common throughout several species. It is clear that the distinction between the tactile and olfactory parts of a signal is more or less arbitrary for most of the behavior patterns we have described, and it is quite difficult to know which takes the lead at the level of message transmission.

The function of these olfactory and tactile exchanges is not always obvious. Nevertheless, most authors consider them to be ritualized greeting behavior, of which certain ones are clearly linked to dominance: in *M. radiata* (Simonds, 1965; Sugiyama, 1971) sniffing of the muzzle can occur between animals of equal rank, but is more often performed by a dominant animal on one of lower rank. In *C. albogularis* (Rowell, 1971), in 61 percent of all cases this sniffing is directed toward superior individuals in the hierarchy and principally toward adult males. It is the same in *C. nictitans,* where juvenile males come to sniff the muzzle of an adult male after it has emitted loud calls (pers. obs.). Similarly, Kummer (1968) considers it to be a gesture of "cautious approach" in *P. hamadryas.*

This sniffing of the muzzle may have originally been a way of exploring what a partner was eating, or more generally whatever it may have been doing (vocal emissions, for example). Hall (1962) rightly thinks that it serves in the acquisition of alimentary habits in young individuals. Depending on the species, it has acquired a more or less important ritualization. It is curious that in forest guenons it is the most frequently used pattern for interindividual encounters, whereas in the baboons, which have developed numerous signals for this purpose, the muzzle-muzzle seems to have maintained its "primitive" function.

Examination of the conditions under which these various olfactory and tactile exchanges occur yields complementary information about their role. These signals are generally activated by the presence of a new individual in the group (newborn or stranger), after any strong disturbance and especially after agonistic behaviors, or during encounters between adult males during a particular excitation and especially in the presence of females in estrus.

Thus, the birth of an infant provokes sniffing from part or all of the group for several weeks or months. This signal, considered by various authors to be a greeting, seems originally to have been a way to recognize and integrate a newcomer. Such signals yield indications about its sex and eventually about its personal odors. Individuals, but principally the mother, can with these sniffings, make an "olfactory identification card for the infant," the odor perhaps playing a role in reinforcing maternal ties. In the talapoins and mangabeys these sniffings are reactivated by any disturbance and are frequent after the child has been carried by another female. Horwich and Lafrance (1972) make the same remark for the back sniffings of the infant *Colobus.*

The introduction of a new individual into the group provokes a recrudescence of tactile and olfactory exchanges in the same way; this has also been clearly shown by Hall (1962) in the chacma baboon, where the stranger is sniffed; touched on the genitals, the back, and the muzzle; mounted; groomed; and kissed. Identical results were obtained upon introduction of new individuals into a group of talapoins (pers. obs.). Furthermore, an individual that has been temporarily separated from its group is treated in the same way, as though visual recognition were insufficient for its reintegration.

In a stable group, sniffing or kissing occurs during situations of high excitation. Thus male kissing takes place in the presence of females in estrus, after conflict, or when males unexpectedly run into each other. This behavior reduces social tension, making a sort of truce between two congeners. For Kaufman and Rosenblum (1966), kissing reflects a "temporary state of dominance equivalence." In the lutong, kissing occurs in periods of apparent distress. It sometimes becomes a collective behavior, with all the group members precipitating toward each other, kissing and giving piercing calls (Bernstein, 1968).

These exchanges therefore constitute ritualized gestures in a stable group. Originally, and under changing conditions, they probably retain a function in species and individual recognition among group members. Odor may play a role here in the creation and reinforcement of social ties.

The use of odors for individual recognition and integration of group members is a universal trait of insect societies (Wilson, 1968) and is also well known in a number of mammals (Mykytowycz, 1972). Perhaps primates retain primitive traits to varying degrees. It remains to be seen through experimentation at what point the ritualization of exchanges accompanies a decrease in the role played by olfaction.

The recent experiments of Kaplan and Russel (1974) with infant squirrel monkeys show that olfaction plays an important role in the mother-infant social attachment and seems to be more active than the visual information furnished by a surrogate. This seems to be an unexplored avenue in the understanding of primate societies.

Tactile Signals

Tactile exchanges are particularly numerous in a group of monkeys in active (e.g., grooming, kissing) or passive situations ("huddling together"). None of these exchanges should be underestimated for the understanding of social regulation because the fact that one individual accepts contact with a congener whereas another refuses it constitutes the key to the social organization of a group and the spatial distribution of its members. Unfortunately, not being able to understand the message received by an individual in contact with a congener, we are reduced to describing the differentiated behavioral patterns (Marler, 1965). In several tactile exchanges, these different patterns do not exist: two individuals simply sit down side by side or back to back. The best measure we have is to note the frequency and duration of these postures.

Marler (1965) distinguishes two broad categories of tactile stimuli on the basis of the behavior they induce: those that produce aggregative tendencies and those that provoke dispersive tendencies. Nevertheless, certain patterns can assume different significations according to the identity of the displaying monkey, and their classification is not simple. Thus the same push with the hand by an adult *M. speciosa* manifested in order to take the place of a congener and bringing on a distancing between individuals, is used by juveniles as an invitation to play, inducing a reduction in interindividual distance (Bertrand, 1969).

"Negative" tactile exchanges include all noxious signals that occur in agonistic behaviors. They range from hair grasping to slapping and biting and are often associated with visual threats or calls. These various elements are common to all species, but some of them have developed stereotyped behavior in which the gesture is preferentially directed toward a particular area of the body. Thus the male hamadryas "symbolically" bites his females at the base of the neck during his "herding" behavior (Kummer, 1959), whereas in aggressive behavior, female and juvenile vervets preferentially bite their partners' tails (Struhsaker, 1967a). Symbolic biting of the tail is also common in *Ce. galeritus* and *Ce. albigena* males. The latter, following olfactory control, may "bite" the sexual skin of a female (Fig. 39).

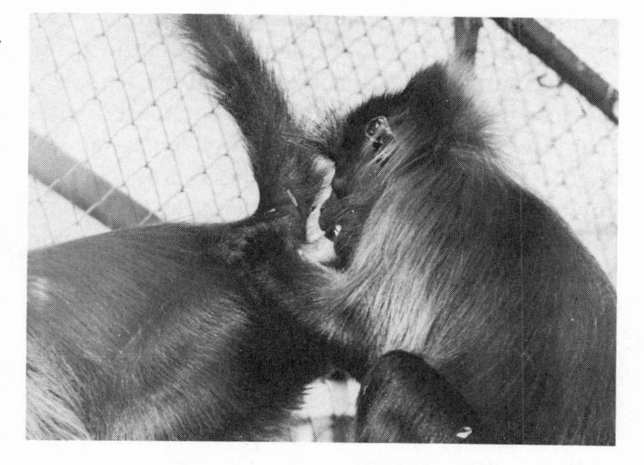

Fig. 39. An adult male *Cercocebus albigena* "bites" the sexual skin, after sniffing it, of a female in presentation. (Paimpont; photo by B. Deputte.)

"Neck-biting" or "chewing" is frequent in the baboons (Hall and DeVore, 1965), the bonnet macaque (Simonds, 1965; Sugiyama, 1971), the vervets (Struhsaker, 1967a), and the talapoins (Gautier-Hion, 1971b). It occurs mostly during kissing in the last three species.

Mounting behavior constitutes complex exchanges during which tactile signals occur, frequently associated with characteristic facial displays and vocal or nonvocal sounds. Thus, in various species, male mounting is accompanied by a "grin face" and the emission of complex phrases (e.g., *P. ursinus:* Hall, 1962; *M. fuscata:* Itani, 1963; *M. talapoin:* Gautier-Hion, 1971b; Gautier, 1974). Male mounting is observed especially in moments of tension, for example, in the presence of females in estrus (Hall and DeVore, 1965; Gautier-Hion, 1971b). Heterosexual mounting is the most frequent type (see Bernstein, 1970b); mounting between adult females is rare, but can be observed when one or both females are in heat (e.g., in talapoins and mangabeys: pers. obs.), and a mother will mount a female who presents herself (baboons: Hall and DeVore, 1965).

Mounting frequently occurs before or after grooming, in play, and during agonistic exchanges. It is generally considered an indication of dominance (see Bernstein, 1970b). In some species, its frequency is clearly correlated with group hierarchy, based on aggressive exchanges (e.g., *P. anubis*). In others the correlation is less obvious, and Simonds (1965) remarks that in *M. radiata,* inferior individuals can mount dominant ones and sometimes the latter force the former to do so.

Contacts bringing the hand into play are numerous (see negative contacts). "Touching" is described in *P. johnii* as a "mildly assertive gesture," but it also acts as a gesture of reassurance after a fight. Similarly, in *M. radiata,* "hand stretching" and "gentle touching" of one individual by another seems to appease the aggression of the receiver. According to Rowell (1971), such appeasement gestures do not exist in arboreal guenons but are frequent in baboons and macaques.

Touching and manipulating the genitals, a behavior described in numerous baboons and macaques, is addressed most generally to males (*Papio ursinus:* Hall, 1962; *Papio anubis:* Hall and DeVore, 1965; *M. radiata:* Rosenblum and Kaufman, 1967; Sugiyama, 1971; *M. Mulatta:* Altmann, 1962; *M. fuscata:* Kawai, 1960). This behavior is also present in the patas (Hall et al., 1965) and *C. aethiops* (Struhsaker, 1967a), and we have observed it in *Ce. galeritus* and *C. nictitans.* In *C. nictitans,* juvenile males come to touch and sniff the genitals of an adult male notably after hearing loud calls from the latter. Most authors think that this gesture corresponds to a recognition of status, as Kawai (1960) has said.

Grasping hair, pushing, biting the neck or tail, and pulling or holding the tail of a partner are all tactile signals appearing in the course of play, where the list of behavioral patterns involving contacts is endless. A number of these signals serve as invitations to play; characteristically, they are not stereotyped and bear no relation to any sort of dominance status. These contact plays seem to be of capital importance for the normal social development of juveniles (Harlow and Harlow, 1965; Mason, 1965). In the same way, numerous tactile exchanges, either passive ("passive support") or active ("cradling," "starting push") occur between mother and infant—all contacts that act to protect and reassure the young or coordinate its activities (see, e.g., Bobbit et al., 1964). Thus light hand touching exists in most species and serves to invite the infant to hang on its mother.

Social grooming is probably the most important kind of tactile exchange among Old World monkeys, and many authors have discussed its function and distribution throughout age classes (see review by Sparks, 1967). Motor patterns are generally the same for all species, all parts of the body being submitted to grooming. Grooming is broadly distributed through age and sex classes; nevertheless, adult females groom more frequently and longer than do males. In the macaques and baboons, however, adult males are active in social grooming (notably *P. hamadryas:* Kummer, 1968). In the mangabeys and guenons adult males receive most of the grooming but they groom each other very little.

The correlation of grooming with social rank has given rise to controversies: the results vary according to species and observation conditions (see Bernstein, 1970b). According to Rowell (1971), grooming is particularly directed to the top of the hierarchy in the *Cercopithecus* spp. (like *C. albogularis* and *aethiops*), and to the bottom of the hierarchy in the macaques and baboons. Taking into account species-specific variations, one can say with Sugiyama (1971) that grooming is generally not a "one-way social behavior."

Grooming of an infant by a mother is an important activity in baboons (DeVore, 1963), langurs (Jay, 1963), and macaques (Bertrand, 1969). In the *Cercopithecus* spp., *M. talapoin,* and *Ce. albigena,* on the contrary, grooming of young infants is rare and not intense during the first months of life but develops at a later time (pers. obs.). Frequency of grooming among adult animals may vary according to female cycles. Michael and Herbert (1963) have shown that the quantity of grooming given by rhesus females reaches a minimum at the middle of their cycle whereas that given by males reaches a maximum at that time. In *P. ursinus,* Saayman (1971) shows that it is females in the deflating phase of the sexual skin (follicular phase of the menstrual cycle) that most frequently groom males. Hall and DeVore (1965) also note the increase in grooming by males of females in estrus. On the contrary, in *C. aethiops,* Rowell (1971) finds that grooming is most frequent with respect to pregnant females.

All the authors recognize within grooming social activities that go beyond the function of cleaning. It plays a role in the establishment or maintenance of interindividual contact. Thus an individual rejected by another has access to the latter through grooming (*M. speciosa:* Bertrand, 1969). Rhesus females of inferior status groom males when they are in estrus in order to establish contact (Lindburgh, 1971). The frequent increase in grooming by males during periods of female receptivity also seems to facilitate encounters, the avoidance response of females being reduced. Similarly, rhesus females groom other mothers in order to be able to touch their offspring (Rowell, Hinde, and Spencer-Booth, 1964). Grooming also plays a role in appeasement and tension reduction. Poirier (1970b) notes that in *P. johnii,* 45 percent of all grooming sequences occur after agonistic exchanges and thus help to reduce interindividual tension. Similarly, in a group of talapoins, a female of high rank can interrupt an agonistic sequence by grooming the dominant male when the latter is threatening another individual of the group (Gautier-Hion, 1971b). The appeasement role of grooming also appears clearly during the weaning of an infant. A macaque or talapoin female, harassed by an infant seeking to be nursed, grooms it attentively, thus seeming to calm it and distract its attention.

As Poirier (1970b) notes, the tactile stimuli produced during grooming seem to be "a very pleasurable experience" for the animals. Nevertheless, Bernstein (1970b) citing Falk (1958) remarks that the act of grooming is just as much a social reward as that of being groomed. At any rate, grooming relations are the most likely to be reciprocal. Grooming serves to create, express, and reinforce interindividual social ties through these positive rewards.

Atypical forms of grooming are observed in several species. Kummer (1957) describes the formal grooming with a finger in the hamadryas.

In *C. mitis* and *C. neglectus* (Booth, 1962), and *C. nictitans, Ce. albigena* and *M. talapoin* (Gautier-Hion, 1971b), an unusual form of grooming is observed especially in the male; after having been groomed, he will "pay back" the grooming by rapidly and sometimes brutally hitting his partner's coat without paying any attention to it. Poirier describes a similar pattern in *P. johnii,* and Rowell (1972) speaks of "aggressive" grooming in the talapoin. No flight reaction is observed in the partner, and it seems that these various kinds of grooming deserve to be termed formal: they are clearly devoid of any function in the cleaning of the coat.

Several more or less active postures are observed in which the animals are in close contact: sitting together, huddling, and tail twining. The "chin on nape" posture described by Bertrand (1969) in various macaques, baboons, and the gelada, also exists in the hamadryas (Kummer, 1957) and in various *Cercopithecus* spp. and mangabeys (Gautier-Hion, 1971b and Fig. 40). They are used at night or among subgroups during their daytime rests; they clearly indicate particular affinities between individuals. Partners in these subgroups remain the same in a stable social unit, even if through maturation of individuals the evolution of the society transforms other types of relations, such as aggression and sex. A sort of competition may rise in which individuals of low rank try to gain access to these subgroups (pers. obs.). In *C. albogularis,* Rowell (1971) observes that it is mainly dominant individuals that tend to sit together (82 percent of all cases), whereas others frequently remain alone.

In *M. radiata,* "huddling" is particularly well developed (Kaufman and Rosenblum, 1966), and Sugiyama (1971) observes groups of up to ten individuals. Baboons and certain macaques like *M. nemestrina,* however, are not "huddling species" but have developed formal hand contacts and kissing (Rowell, 1972) instead of general and passive body contacts. In the hamadryas, though, members of each "one-male unit" may huddle against each other during the course of the day, the different units being broadly spaced (Kummer, 1968: Fig. 16, p. 34).

Fig. 40. Sleeping group in *Cercopithecus pogonias;* note the "chin on nape" and "tail twining" postures. (Paimpont.)

In species with long tails, sitting together and huddling are frequently accompanied by tail twining (*C. campbelli:* Bourlière et al., 1970; *M. talapoin, C. nictitans, C. cephus, C. pogonias, C. neglectus, C. petaurista, C. ascanius, Ce. albigena:* Gautier-Hion, 1971b and pers. obs.; *C. aethiops, C. neglectus:* Chalmers, 1973). Although it probably plays a role in the equilibration of these arboreal animals, tail twining in most cases is an active search for contact (Fig. 40). In infants, tail twining seems to depend on the coiling reflexes of the tail. Even in species in which this behavior has become rare among adults (like *Ce. albigena*), it is a general behavioral component of infants huddled in their mothers' arms.

This survey has tried to bring out the great similarity of basic patterns occurring in various species during close exchanges. Species differentiation should be made afresh on the basis of frequency of appearance, which seems to be correlated with the increasing social function of exchange.

Thus the side-to-side sniffing posture is present in the *Cercopithecus* spp., in the mangabeys, and at least in *M. radiata.* But while it is rarely observed in guenons and then only during sexual encounters, it is a very frequent form of social approach among female mangabeys.

Muzzle contact, on the other hand, for which there is little evidence in the baboons and macaques (where it is linked with feeding), is the essential mode of interindividual encounter in the *Cercopithecus* spp. If one admits that olfaction plays a role in this behavior, its frequency indicates a primitive trait, this pattern being frequent in carnivores especially.

Kissing and social presentation seem especially to characterize baboons, macaques, and mangabeys, in which they occur with numerous other acoustic and visual appeasement signals. In these species, hierarchization within the group is more obvious than in the guenons, and signals that serve to assure status, such as mounting, occur frequently.

Discussion

This chapter is a survey of the most usual signals manifested by Old World monkeys, but it is obvious that the social regulation of populations does not depend solely on these exchanges of "specialized" communicative acts. For Altmann (1967) "social communication is a process by which the behavior of an individual affects the behavior of others." This mode of action is virtually unlimited and does not necessitate individualized signals. As Bertrand (1971) notes, to a congener the sight of a monkey bringing something out of his cheek pouches is a sign that the individual is eating. This behavior can induce the partner to approach and to perform tactile and olfactory exchanges (e.g., muzzle-muzzle) or even some kind of dominance behavior in order to gain access to the food.

This passive communication is not limited to visual signals. An active monkey makes all sorts of noises characteristic of its activities. In forest monkeys, vocal responses are evoked by the noises of movement among the branches in the same way as by cohesion calls. Similarly, the noises of urination and defecation upon awakening are the first indications of reciprocal localization of group members and are followed by vocal contact calls. Some of these nonvocal sounds have become ritualized, e.g., tree shaking or jumping around; on the other hand, some calls must ultimately be considered as passive acts of communication, such as the trills exchanged "spontaneously" between infant and mother without any apparent stimulation and without modification of subsequent behavior.

The production of certain odors undoubtedly also plays a role in passive communication. This is probably the case with the sexual pheromones diffused during the female's period of receptivity, whether she is presenting or eating.

There exists therefore a gradual passage from passive to active communication, correlated with a growing ritualization of signals. Both act simultaneously or in relay, and it is always a little arbitrary to establish a classification separating acts that are communicative, or signals, from those that are not.

It is just as artificial to treat signals of vocal, visual, tactile, or olfactory nature separately, since in most situations these various sensory modalities occur simultaneously (Marler, 1965), especially in close exchanges. Thus in kissing, tactile signals blend with olfactory cues (neck sniffing), sounds (repeated gruntings), and appeasement faces (grimace or lip smacking). At medium distance, visual and auditory signals take turns or work together. This is the case in approach-flight behavior, where sound reinforces gesture by focalizing the partner's attention, as well as in peaceful exchanges where the calls precede contacts.

Visual signals occur alone principally in stereotyped displays like penis exhibition and stopping postures. At a greater distance or when the exchange does not require that the congener be informed of the emitter's identity, sounds act alone, especially in alarm situations or exchanges regulating the spatial distribution of the population.

CHARACTERISTICS OF THE SIGNALS

Throughout this survey, we have noted the great similarity among the visual, olfactory, and tactile signals encountered in all species considered. The more a given species is the object of long-term observation, the more this relationship is confirmed. Thus, after several years of contact with a captive group of *C. nictitans,* we have just observed for the first time the very characteristic posture of mutual sniffing in the inverted position, which is common in the mangabeys (see Fig. 36). Ultimately it seems conceivable that no truly unique facial display or posture exists (except perhaps stopping postures): those that are species-specific result from morphological structures characteristic of a given species which emphasize the appearance of the displays (see Kummer, 1970); others stem from our own ignorance due to the small number of social exchanges manifested by certain species under observation.

Thus, although Kummer (1970) underlines the difficulties of using frequency as a taxonomic criterion, it seems obvious that the frequency of occurrence of social patterns is one of the first observational features differentiating the baboons, macaques, and mangabeys from the guenons. The structural relationships are less obvious when dealing with vocal signals, for which the parallels that one can establish from species to species are complex. Forest guenons, which include a great variety of species, offer a particularly interesting field of study for this subject, and Struhsaker (1970) has stressed the importance of their vocalizations as a phylogenetic indicator. But this problem is complicated, because on common phylogenetic bases, sympatric species seem to diverge in proportion to the needs of species specificity. We have previously noted that *C. neglectus,* which is not included in the mona group, nevertheless has boom-type loud calls that are not easily distinguished from those of *C. pogonias* except on the basis of sequential emission. These two species live in western Africa and have numerous similar calls, but their ecological niches and their different ways of life preclude frequent encounters between them.

Conversely, *C. nictitans* and *C. cephus* of western Africa frequently live in association (80 percent of all cases) and have very similar general repertoires, but adult males possess completely different loud calls. The situation is the same for *C. mitis* and *C. ascanius* in eastern Africa. Furthermore, as we have already noted, the loud calls of male *C. nictitans* and male *mitis* are extremely similar, but curiously enough, the latter has a boom-type call (Rowell, 1971; Marler, 1973). Yet *C. mitis* does not cohabit with boom emitters. On the contrary, in Gabon, where *C. nictitans* frequently live with *pogonias,* booms are reserved to the latter. Recent observations have shown, however, that *C. nictitans* may occasionally emit them, particularly in captivity (Gautier, 1973); the call appears to be rather like a relic from the ancestral stock.

Thus, starting from a general "pool," close species seem to be able to retain part or all of the vocalizations held in common, or if species specificity necessitates it, to differentiate certain types of calls that then assume highly stereotyped forms. This is the case of loud calls among males of the *Cercopithecus* spp., which appear to have a role in reproductive isolation. For alarm signals, on the contrary, natural selection has favored the development of very similar sounds for many species, the only imperative seeming to be that the structure of the calls permit rapid maximal diffusion without indicating the position or identity of the uttering monkey. This characteristic is even more obvious for calls linked with approach-flight behavior; infant defensive calls (geckers, screams, screeches, whistles) seem to be especially widespread in Old World monkeys.

The multimodality of exchange is correlated with an increasing variability of signals. In unimodal exchanges, signals are maximally stereotyped, can be perceived clearly despite the distance between emitter and receiver, and imply a species-specific message. At close range, on the other hand, constant shifts through different sensory modalities permit subtle variations. These variations are the result of several phenomena: lack of stereotyped character (with the same basic signal, an animal can modulate individual variations, which are probably due to anatomic peculiarities or to tiny alterations in its excitation level; e.g., trill variations in a juvenile *Cercopithecus,* Fig. 18); the presence of gradation (structural variations permit passage from one type of signal to another; e.g., graded aggression-flight system in the talapoin, Fig. 9).

The phenomenon of gradation demonstrated by Rowell and Hinde (1962) for the acoustic signals of the rhesus ("graded sounds") is also evident in visual signals (e.g., for approach-flight signals: see also Vine, 1970). This notion is the opposite of that of discrete signals, for which there is no gradual passage from one type to the other (e.g., type 1 loud calls or penile display). In vocal repertoires, the gradation may be total, with all the elements of the system capable of gradually developing into one another: this is probably the case with the baboons, the red Colobus, and (except for loud calls) the mangabeys. The gradation is only partial when there is a structural discontinuity between groups of graded calls in the same repertoire: this is the case for certain macaques and for the talapoin, in which several graded systems coexist (an aggression-flight system, a cohesion system, and a high-pitched system: Gautier, 1974). Finally, in the forest *Cercopithecus* spp., only a few types of calls can develop imperceptibly from one kind to another, notably defensive signals.

The gradation of a communicative system is therefore a question of degree (Marler, 1973). In order for it to be real and functional, the evolution of one structure into another must be rapid and reversible, implying situational and motivational changes. In the opposite case, the analysis of certain structures may show that they are derived from one another, yielding the general characteristic of what one might call the "voice" of a species, without implying a real functional signification in communicative acts (Gautier, 1974).

DETERMINISM AND GENESIS OF THE SIGNALS

A number of authors have emphasized that communicative acts constitute stable behavioral elements within a species (see e.g., Bernstein, 1970a). The works of Kummer and Kurt (1965), for example, have shown that captive hamadryas behave very similarly to wild ones. The manifestation of some particular patterns in each of these environments shows that the nonoccurrence of a behavior does not mean that it is absent from the species repertoire, but rather that it depends on environmental influence, social or nonsocial (cf. the boom of *C. nictitans*).

Genetic determinism of fundamental call structure seems obvious, although few works make direct reference to it. The recent crossing of two *Cercopithecus* spp. offers a confirmation. The mother, *C. ascanius,* can emit two calls during peaceful exchanges—one low-pitched and discontinuous, the other shrill and quavered—both of which retain their own structures when associated in one sequence (Fig. 41, 2 and 2 bis). In *C. pogonias* (the father), however, the frequent association of the two calls is made by transitional structures (Fig. 41, 1 and 1 bis), and the quaver does not exist. At the age of one month, the young hybrid, raised with both parents, emits several kinds of calls: Trilled, shrill calls identical to those of its mother (Fig. 41, compare 3 and 8) and others with a less clear quaver; low-pitched trills in which the structural discontinuity is less apparent than in the mother's (Fig. 41, compare 2 and 2 bis to 4 and 4 bis); and calls resulting from the more or less total association of the two preceding types: the call sometimes possesses just a slight transformation of modulation (Fig. 41, 7); at a subsequent stage, low-pitched and shrill cries are linked by a transitional structure similar to that of the father when it was immature (Fig. 41, compare 1 and 1 bis with 5 and 5 bis). Both maternal and paternal influences appear quite clearly in these signals: passing from a substantial quavering (mother) to a reduced one (father); passing from two distinct calls (mother) to a single pattern through the presence of transitional structures (father).

In the infant, most signals occur as immediate reactions to anything that upsets its contact with its mother, especially with her breast. This is particularly clear for vocalizations. If a talapoin about ten minutes old is removed from its mother and denied any possibility of contact, it gives almost all the basic calls of its species' repertoire, especially the distancing-cohesion-isolation system calls, the flight system calls, and the shrill calls that subsequently develop into alarm vocalizations (Gautier, 1974). Similarly, Winter et al. (1973) have shown that almost all the vocal signals of the squirrel monkey are present at birth.

Threat calls, however, generally occur later than defensive sounds. In the Japanese macaque, they have been observed in the twenty-fourth week (Kawabe, 1965, in Nishimura, 1973) in the nineteenth week (Takeda, 1965, 1966), and in the sixth month (Nishimura, 1973); in the talapoin at more than one year of age (pers. obs.), and in the grey-cheeked mangabey at ten to twelve months (pers. obs.). It is the same for alarm calls or aggressive alarms. These tardy manifestations seem to be due to the particular status of the infants. They are protected by their mothers and turn toward them at the slightest disturbance, and they only begin to individualize their reactions as they develop their independence. One particular case in the development of signals concerns male loud calls. Their manifestation clearly seems to depend on sex hormones, that of type 1 loud calls being strictly correlated with the social status of the vocalizer.

In the course of ontogenesis, calls are submitted to three essential modifications:

1. Their frequency of occurrence increases rapidly during the first months of life (up to six months in the Japanese macaque: Nishimura, 1973), and subsequently diminishes in the course of individual maturation. This decrease in frequency, however, is less obvious in females (pers. obs.).
2. The thresholds of appearance of different types of calls are subject to modifications (Gautier, 1974); this phenomenon is a corollary of variations in emission frequency. In juveniles, very slight alterations in excitation level provoke the emission of a call. In adults, the animal must be submitted to a much more substantial stimulation for the same signal to be uttered.
3. The structure of the calls is subject to transformations. Different parameters are involved (intensity, duration, rhythm, noise), but the most obvious and measurable is drop in pitch. This last phenomenon is accentuated in males of the guenons and mangabeys (and no doubt in some other species) because of the break in voice at sexual maturity.

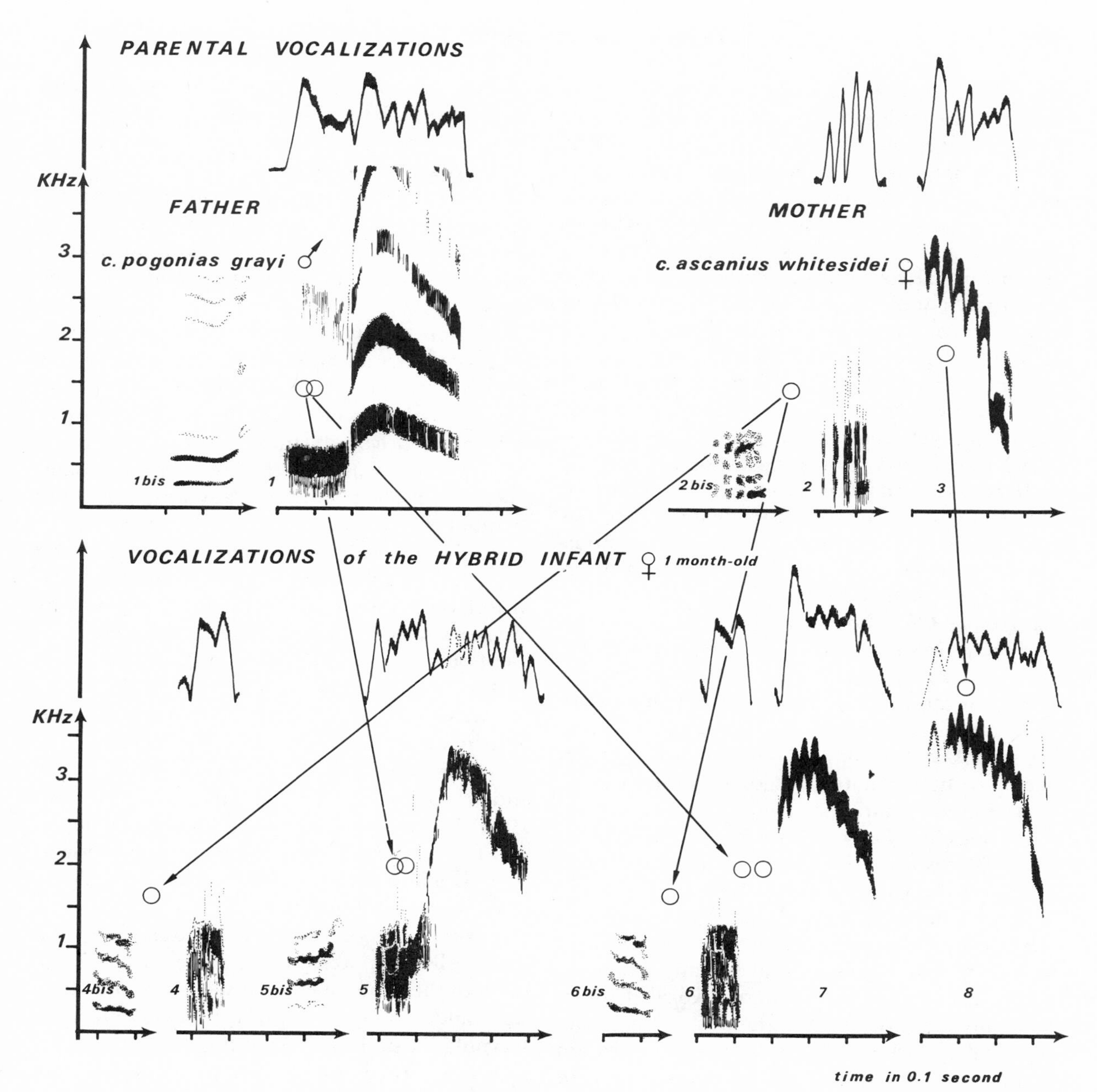

Fig. 41. Comparison of the calls of a young hybrid with those of its parents.

Parental vocalizations. 1, 1 bis: Father's calls (*C. pogonias*) before sexual maturity. 2, 2 bis, 3: Mother's calls (*C. ascanius*) as an adult.

Infant vocalizations. 4, 4 bis: Low-pitched call, 8: high-pitched call, both similar to its mother's. 6, 7: Association. 5: Juxtaposition of low- and high-pitched calls as in the father. ("Bis" indicates analyses done with the help of narrow 45 Hz filters; the other analyses were done with broad 300 Hz filters.)

These developments are not, however, irreversible. An adult monkey, even if it is male, will revert to juvenile calls under conditions in which the stimulations are sufficient to increase its excitation level in an unaccustomed way. Thus, although each age, sex, and status class possesses differentiated vocalizations for a given situation under normal conditions, it appears that under disturbed conditions any adult individual can emit any type of call (Gautier, 1974) except those that are hormonally determined.

The appearance of facial displays and gestures happens more progressively. The first to occur are generally defense or retreat faces, which may or may not be associated with calls. Others observed very early are the play face (fifteen to thirty days for the mangabeys and talapoins, pers. obs.), yawning (about one month in the same species), as well as a slow form of lip smacking, which is observed from the very first days of life. Only the play face is manifested at an early stage in social exchanges. Lip smacking and presentation are integrated into a social context later (one month for lip smacking and more than four months for presentation in *Ce. albigena:* pers. obs.). The appearance of presentation seems to be facilitated by genital examinations by the mother and is actively adapted to social exchanges by congeners' influence: a male mangabey forces a passing juvenile to present itself by lifting the young animal's tail.

Facial displays, gestures, and postures generally seem to depend more on learning for their normal integration into social exchanges than do acoustic signals. Thus, although vocal displays suddenly occur in males at the onset of maturity (see Fig. 1), visual displays are present in rough form in infants and require a long period of learning. In *Ce. albigena,* juveniles engage in yawning from the age of four weeks and do a rough job of tree shaking at about two and a half or three months. But the association of these two kinds of behavior does not occur until eighteen months, and penile display is not added until about three years. The complete sequence of tree shaking, stopping posture, penile display, and yawning is not linked together until the juvenile male reaches sexual maturity, at about five years (pers. obs.). At this point the sub-adult intensely surveys the activities of adult males and performs its displays by imitating them and using the same props.

Winter et al. (1973) find in their ontogenetic study of the vocalizations of the squirrel monkey that there is no difference between the manifestations of calls produced by animals acoustically isolated from birth and those of normally raised monkeys. Before affirming that the manifestation of calls is subject to any sort of learning in monkeys, however, it must be seen experimentally how well the signals given by isolated juveniles are adapted to suitable contexts. Empirical observation shows that calls uttered by monkeys raised in impoverished social environments are not as modulated and variable as those of monkeys raised in social groups, and that the majority of their calls are given without clear discrimination in all frustrating situations.

On the contrary, the experiments of Harlow and his coworkers (see, e.g., Harlow and Harlow, 1965) have shown that social learning is necessary for the normal and adapted manifestation of a great many kinds of postural behavior. Similarly, Mason (1960, 1961a, 1961b), after numerous experiments of social deprivation conducted on the rhesus, notes that "the effective development of the elementary forms of social coordination and communication is dependent upon learning." Conducted according to an extremely rigorous procedure in which monkeys were placed in the presence of photographs or films of congeners, the studies of Miller and his coworkers (see Miller, 1971) have also demonstrated that socially isolated monkeys do not adequately respond to the nonverbal expressions of other monkeys and that these animals seem to be "defective transmitters."

CAUSALITY: TRANSMITTED MESSAGE

According to general opinion, the signals emitted by monkeys essentially indicate the emotional state of the displayers; they rarely if ever convey specific information about the environment. Thus the calls given by a group of monkeys finding a fruit tree are the same as those uttered during their normal progression or manifested by an individual upon the arrival of a congener; only their quantity increases. Similarly, the nature, intensity, and frequency of emission of alarm calls is modified more by the intensity of the stimulation than by the nature of the danger. In a parallel way, displays like shaking, yawning, or penile display appear in situations of high excitation, whether the stimulation is provided by an encounter between two groups or by a non-conspecific potential danger.

Visual signals also convey indications about the emotional state of the emitter. Using methods of interanimal conditioning, Miller and his coworkers (1967, 1971) have shown that the behavior of a "receiver" monkey is modified by very subtle alterations in the face of the displaying monkey. In general, visual, acoustic, and olfactory signals all appear in diverse situations, and they seem to convey no specific indication of the nature of the stimulus, whether it is the presence of a congener, familiar or strange, or an element in the exterior environment.

Green (1975) has classified ten types of fundamental calls of the Japanese macaque's repertoire according to "attributes indicating demeanor and internal state," and has shown that there is a correlation between the excitation level of the animals and the structure of the calls uttered. For Green, arousal is one of the essential components of internal state that determine the form of the signal. Above all, the message is emotional and predisposes the partner to respond from a given behavioral range, the response being modulated according to the context of emission. Green's work, which systematizes what other authors had previously suggested or mentioned (e.g., Rowell and Hinde, 1962; Gautier, 1974), clearly shows that a given type of call does not always occur in a precise context and therefore cannot offer any information peculiar to the situation.

It does not appear, however, that this work provides the "consistent framework" wished for, which would permit comparative studies. The classification of situations according to the demeanor and arousal of emitting individuals remains somewhat subjective insofar as it does not use physiological techniques that would allow, for example, the measurement of variations in arousal before, during, and after vocal emissions. Furthermore, if excitation level partially determines the nature and structure of the emitted signal, one must be conscious of the fact that variations in the arousal do not seem to occur on a comparable scale for all age classes. The thresholds for the appearance of various vocalizations are particularly low in a juvenile, for example; thus, the same stimulation that evokes a scream in an infant will provoke only an alert call in the adult. Group members are implicitly aware of this phenomenon since they take the alarm calls of juveniles "less seriously" than those of adults.

Moreover, in response to a given stimulation, juveniles and adult females will emit social alarm calls, whereas males give barks that can be followed by rallying loud calls uttered by just one individual. We have also seen that the social environment influences the type of reaction evoked in an individual by any sort of stimulation: an isolated animal will react to a disturbing situation by hiding, but one in a group will emit calls. For a given situation, we can conceive on the one hand that alterations of individual internal states will differ according to age, sex, experience, status, and the environment taken as a whole; and on the other hand that other, more specific influences, notably social ones, ultimately determine the form of the vocal response. In fact, it is difficult to establish correlations between signal type and arousal level without dealing separately with the individual classes. In this view, the ontogenetic transformations of the call structures, which give the perceptual bases for identification of the age classes, certainly play a major role in the interpretation of the signals by the congeners.

Mason (1965) has also shown that in juvenile chimpanzees the appearance of some types of social behavior, like play, can be correlated with arousal levels. Such a concept permits an explanation of changes in activity pattern without resorting to independent motivations. Mason points out, however, that some kinds of behavior may overlap in their correlation with arousal level, and that they are differentiated by other factors, especially the surrounding context.

It is generally admitted that the manifestation of signals is involuntary and uncontrolled, and Bertrand (1971), quoting Andrew (1964), points out that monkeys are incapable not only of emitting sounds voluntarily but also of witholding them. Yamaguchi and Myers (1972) also conclude that there is a lack of voluntary control over vocalizations in the rhesus. For Sutton et al. (1973), who have conditioned rhesus to give vocalizations in order to obtain a reward, to select low-pitched sounds, to increase their duration, and to reduce the frequency of calls uttered, emotion is not the only variable capable of determining everything in vocal emissions. On the contrary, these authors think that their experiments, indicating a learning controlled by the discrimination of unfamiliar stimuli, offer an example of voluntary vocal behavior.

The important role played by learning in the adapted manifestation of facial and postural signals can be correlated with a certain control of their expression. Some observations would lead one to believe that, and anyone who has seen a female monkey trying to touch an infant in the arms of its mother has surely been astonished by the subtlety of the behavior. The female grooms the mother with great attention, then surreptitiously slips her hand over the mother's abdomen in order to touch the infant's fur. All the while she pretends to continue grooming with one hand, remaining prepared to return all her attention to the latter activity if the mother intercepts her gesture. All primatologists have observations of this kind in their notebooks, but possessing only a very anthropomorphic vocabulary, they would rather remain silent about them.

Thus, neither the establishment of lists of species-specific signals, nor their correlation with simple variations in arousal, nor even attempts to comprehend their adaptive functions permit us at present to take into account the subtlety of the interindividual exchanges that regulate the life of monkey groups.

CORRELATIONS WITH HABITAT AND MODE OF SOCIAL ORGANIZATION

It is customary to point out that in open habitats visual exchanges predominate over vocal (e.g., Andrew, 1963b, 1964; Marler, 1965; Moynihan, 1967). This is well illustrated by the differences in the nature of intergroup exchanges in woodland and savannah environments. Correlated with the increasing frequency of visual exchanges, an increase in their variability and gradation is to be expected. In a closed habitat, on the contrary, passive visual communication is replaced by vocal emissions punctuating group activity, and active exchanges are more stereotyped and less frequent. Nevertheless, the frequency and the subtlety of visual exchanges also depend on the degree of evolution of the species. A good proof of this is the exam-

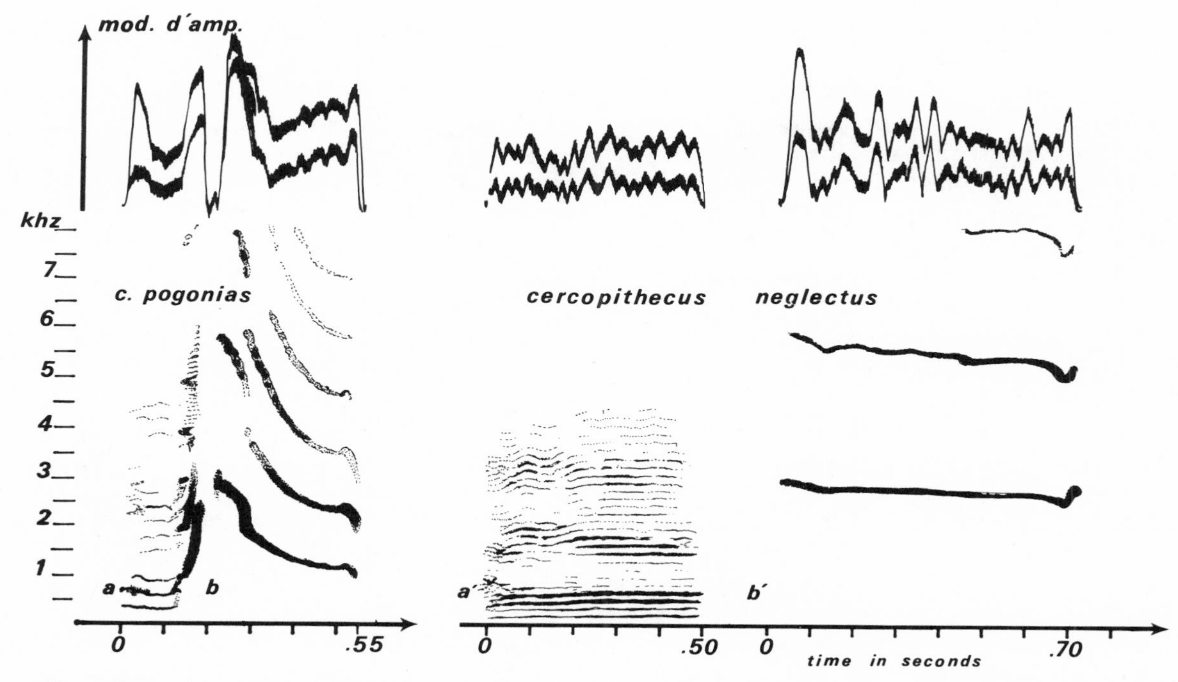

Fig. 42. Spectrographic analyses showing the difference in variability for the same type of fundamental call in two *Cercopithecus.* In *C. pogonias,* the low- and high-pitched structures are associated through transitional elements (a, b); in *C. neglectus,* on the other hand, they are separate (a', b'). Furthermore, in *C. pogonias* the high-pitched structure possesses a significant frequency modulation, which is very reduced in *C. neglectus* (compare b and b').

ple of the forest mangabeys, the complexity of whose visual signals approaches that of the baboons, and who also have developed numerous olfactory and tactile exchanges.

The gradation of vocal systems has also often been linked to the openness of the environment (e.g., Marler, 1965; Moynihan, 1964, 1966). It was thought that discrete signals were necessary in the forest for the transmission of unambiguous messages (see Altmann, 1967). Studying the Colobinae, Marler has subsequently offered evidence that this correlation is not obvious. Of two species of Colobus, both of which live in woodland habitats, one (*C. badius:* Marler, 1970) possesses a graded vocal repertoire, the other (*C. guereza:* Marler, 1972) various discrete signals. In the same way, although it lives in very dense forests, the talapoin has an almost fully graded repertoire (Gautier, 1974). Finally, the arboreal mangabeys (*Ce. albigena*) as much as the semiterrestrial ones (*Ce. galeritus*), and the drills and mandrills, all have vocal repertoires in which the gradation is probably close to that of baboon calls.

The gradation of a communicative system is more closely related to frequency of contact than to openness of environment and seems to be correlated, as Marler (1970) emphasizes, with a more complex social organization. The latter occurs especially in relatively large groups, since an increase in the number of individuals, which necessarily augments the general activity level (see

Gautier-Hion and Gautier, 1974), can favor the development of exchanges. Thus *Co. badius* forms groups of thirty to fifty individuals, whereas *C. guereza* groups seldom go beyond about ten animals. Similarly, the talapoin lives in bands that sometimes reach more than a hundred head, whereas those of the *Cercopithecus* spp. are under twenty.

A supplementary example is offered by two forest species, between which we have already pointed out a relationship in vocal repertoires: *C. neglectus,* which lives in groups of three to six animals, and *C. pogonias,* in which groups reach fifteen individuals. In captivity, as in nature, de Brazza's monkey is a discrete, passive animal that can remain immobile for hours. On the other hand, mona groups have a high level of general activity. Although *C. neglectus* and *C. pogonias* have a common fundamental call structure, the former emits calls that are not very variable, while the latter's are more modulated (Fig. 42). This example comes close to that of the fox and the wolf (Kortland, 1965, in Vine, 1970). The wolf's repertoire, although basically similar to the fox's, is much larger. Kortland attributes this to differences in social organization, the wolf's methods of hunting in cooperation implying more highly developed means of signalization.

Thus, variability, gradation, and frequency in social exchanges characterize species whose social organization is more complex. These species include essentially monkeys of open habitat: macaques and baboons, which have particularly numerous peaceful or agonistic short-range exchanges. Nevertheless, much more familiarity is required with the forest mangabeys, and even more with the mandrills and drills, before it is possible to judge the subtlety of the messages conveyed by their communicative acts.

References

Aldrich-Blake, F. P. G., 1970. The ecology and behaviour of the blue monkey, *Cercopithecus mitis stuhlmanni.* Ph.D. thesis, University of Bristol.

Aldrich-Blake, F. P. G.; Bunn, T. K.; Dunbar, R. I. M.; and Headley, P. M.; 1971. Observations on baboons, *Papio anubis,* in an arid region in Ethiopia. *Folia primat.,* 15:1–35.

Altmann, S. A., 1959. Field observations on a howling monkey society. *J. Mammalogy,* 40:317–30.

Altmann, S. A., 1962. A field study of the sociobiology of rhesus monkeys, *Macaca mulatta. Ann. N.Y. Acad. Sci.,* 102:338–435.

Altmann, S. A., 1965. Sociobiology of rhesus monkeys. II—Stochastics of social communication. *J. Theoret. Biol.,* 8:490–522.

Altmann, S. A., 1967. The structure of primate social communication. In: *Social Communication among Primates,* S. A. Altmann, ed. Chicago: University of Chicago Press, pp.325–69.

Altmann, S. A., 1968. Primates. In: *Animal Communication,* T. A. Sebeok, ed. Bloomington: Indiana University Press, pp.466–522.

Andrew, R. J., 1963a. Evolution of facial expression. *Science,* 142:1034–41.

Andrew, R. J., 1963b. The origin and evolution of the calls and facial expressions of the Primates. *Behaviour,* 20:1–109.

Andrew, R. J., 1963c. Trends apparent in the evolution of vocalizations in the old world monkeys and apes. *Symp. Zool. Soc. Lond.,* 10:89–101.

Andrew, R. J., 1964. The displays of the primates. In: *Evolutionary and Genetic Biology of the Primates,* vol. II, J. Buettner-Janusch, ed. New York: Academic Press, pp. 227-309.

Andrew, R. J., 1972. The information potentially available in Mammal displays. In: *Non-verbal Communication,* R. A. Hinde, ed. Cambridge: Cambridge University Press, pp.179–203.

Angst, W., 1973. Pilot experiments to test group tolerance to a stranger in wild *Macaca fascicularis. Amer. J. Phys. Anthrop.,* 38:625–28.

Angst, W., 1974. Das Ausdrucksverhalten des Javaneraffen *Macaca fascicularis* Raffles, 1821. *Fortsch. der Verhalt.,* 15. P. Varey, ed. (Berlin and Hamburg). 90pp.

Anthoney, T. R., 1968. The ontogeny of greeting, grooming and sexual motor patterns in captive baboons (superspecies *Papio cynocephalus). Behaviour,* 31:358–72.

Bates, B. C., 1970. Territorial behavior in primates: a review of recent field studies. *Primates,* 11:271–84.

Bernstein, I. S., 1967. A field study of the pigtail monkey. *Primates,* 8:217–28.

Bernstein, I. S., 1968. The lutong of Kuala Selangor. *Behaviour,* 14:136–63.

Bernstein, I. S., 1970a. Some behavioral elements of Cercopithecoidea. In: *Old World Monkeys: Evolution, Systematics and Behaviour,* J. R. Napier and P. H. Napier, eds. New York: Academic Press, pp.263–95.

Bernstein, I. S., 1970b. Primate status hierarchies. In: *Primate Behavior: Developments in Field and Laboratory Research,* vol. I, L. A. Rosemblum, ed. New York: Academic Press, pp.71–104.

Bertrand, M., 1969. The behavioral repertoire of the Stumptail Macaque. *Bib. Primat.,* 11. Basel: Karger. 273pp.

Bertrand, M., 1971. La communication chez les Primates supérieurs. *J. de Psychol.,* 3:451–73.

Bobbit, R. A.,; Jensen, G. D.; and Gordon, B. N.; 1964. Behavioral elements (taxonomy) for observing mother-infant peer interactions in *M. nemestrina. Primates,* 5:72–79.

Bolwig, N., 1959. A study of the behavior of the Chacma baboon, *Papio ursinus. Behaviour,* 14:136–63.

Bolwig, N., 1962. Facial expressions in Primates with remarks on a parallel development in certain Carnivores. *Behaviour,* 22:167–93.

Booth, A. H., 1957. Observations of the natural history of the Olive colobus monkey, *Procolobus verus* (van Beneden). *Proc. Zool. Soc. Lond.,* 129:421–31.

Booth, C., 1962. Some observations of the behaviour of *Cercopithecus* monkeys. *Ann. N.Y. Acad. Sci.,* 102:477–87.

Bourlière, F.; Hunkeler, C.; and Bertrand, M.; 1970. Ecology and behavior of Lowe's guenon (*Cercopithecus campbelli lowei*) in Ivory Coast. In: *Old world Monkeys: Evolution, Systematics and Behaviour,* J. R. Napier and P. H. Napier, eds. New York: Academic Press, pp.297–350.

Carpenter, C. R., 1934. A field study of the behaviour and social relations of howling monkeys. *Comp. Psychol. Monogr.,* 10:1–148.

Carpenter, C. R., 1942. Sexual behaviour of free-ranging rhesus monkeys, *Macaca mulatta. J. Comp. Psychol.,* 33:113–42.

Chalmers, N. R., 1968. The visual and vocal communication of free-living mangabeys in Uganda. *Folia primat.,* 9:258–80.

Chalmers, N. R., 1973. Differences in behaviour between some arboreal and terrestrial species of African monkeys. In: *Comparative Ecology and Behaviour of Primates,* R. P. Michael and J. H. Crook, eds. London: Academic Press, pp.69–100.

Chance, M. R. A., 1967. Attention structure as the basis of primate rank orders. *Man,* 2:503–18.

Chevalier-Skolnikoff, S., 1974. Male-female, female-female and male-male sexual behavior in the Stumptail Monkey, with special attention to the female orgasm. *Arch. Sex. Behav.,* 3:95–115.

Chivers, D., 1973. An introduction to the socio-ecology of Malayan forest Primates. In: *Comparative Ecology and Behavior of Primates,* R. P. Michael and J. H. Crook, eds. London: Academic Press, pp.101–46.

Crook, J. H., 1966. Gelada baboon herd structure and movement: a comparative report. *Symp. Zool. Soc. Lond.,* 18:237–58.

Deag, J. M., 1973. Intergroup encounters in the wild Barbary macaque, *Macaca sylvanus L.* In: *Comparative Ecology and Behaviour of Primates* R. P. Michael and J. H. Crook, eds. London: Academic Press, pp.316–73.

Deag, J. M., 1974. A study of the social behaviour and ecology of the wild Barbary macaque, *Macaca sylvanus L.* Ph.D. Thesis, University of Bristol.

Deag, J. M., and Crook, J. H., 1971. Social behavior and agonistic buffering in the wild Barbary macaque, *Macaca sylvanus. Folia primat.,* 15:183–200.

Deputte, B., 1973. Etude d'un comportement vocal chez un groupe captif de Mangabeys *(Cercocebus albigena)*: méthode télémétrique d'enregistrement individuel des vocalisations. Rennes: DEA, Eco-Ethologie.

Deputte, B., and Gautier, J-P., in prep. Mise au point d'une méthode télémétrique d'enregistrement individuel des vocalisations dans un groupe social de singes.

Devalois, R. L., and Jacobs, G. H., 1971. Vision. In: *Behavior of Nonhuman Primates,* A. M. Schrier and F. Stollnitz, eds. New York: Academic Press, pp.107–57.

DeVore, I., 1963. Mother-infant relations in free-ranging baboons. In: *Maternal Behavior in Mammals,* H. L. Rheingold, ed. New York: Wiley and Sons, pp.305–35.

DeVore, I., and Hall, K. R. L., 1965. Baboon ecology. In: *Primate Behavior: Field Studies of Monkeys and Apes,* I. DeVore, ed. New York: Holt, Rinehart and Winston, pp.53–110.

Fedigan, L. M., 1972. Roles and activities of male geladas *(Theropithecus gelada). Behaviour,* 26:82–90.

Gartlan, J. S., 1970. Preliminary notes on the ecology and the behaviour of the Drill, *Mandrillus leucophaeus* Ritgen, 1824. In: *Old World Monkeys: Evolution, Systematics and Behaviour,* J. R. Napier and P. H. Napier, eds. New York: Academic Press, pp.445–80.

Gartlan, J. S., and Brain, C. K., 1968. Ecology and social variability in *Cercopithecus aethiops* and *C. mitis.*

In: *Primates: Field Studies in Adaptation and Variability,* P. Jay, ed. New York: Holt, Rinehart and Winston, pp.253–92.

Gautier, J-P., 1967. Emissions sonores liées à la cohésion du groupe et aux manifestations d'alarme dans les bandes de Talapoins *(Miopithecus talapoin). Biol. gabon.,* 2:17–30.

Gautier, J-P., 1969. Emissions sonores d'espacement et de ralliement par deux Cercopithèques arboricoles. *Biol. gabon.,* 5:117–45.

Gautier, J-P., 1971. Etude morphologique et fonctionnelle des annexes extra-laryngées des Cercopithecinae; liaison avec les cris d'espacement. *Biol. Gabon.,* 7:229–67.

Gautier, J-P., 1973. Influence éventuelle de la vie en associations polyspécifiques sur l'apparition d'un type inhabituel d'émission sonore chez les mâles adultes de *Cercopithecus nictitans. Mammalia,* 37:371–78.

Gautier, J-P., 1974. Field and laboratory studies of the vocalizations of talapoin monkeys *(Miopithecus talapoin)*; structure, function, ontogenesis. *Behaviour,* 49:1–64.

Gautier, J-P., and Deputte, B., 1975. Mise au point d'une méthode télémétrique d'enregistrement individuel des vocalisations: application à *Cercocebus albigena. La Terre et la Vie,* 29, pp.298–306.

Gautier, J-P., and Gautier-Hion, A., 1969. Les associations polyspécifiques chez les Cercopithecidae du Gabon. *La Terre et la Vie,* 2:164–201.

Gautier-Hion, A., 1970. L'organisation sociale d'une bande de Talapoins dans le N-E du Gabon. *Folia primat.,* 12:116–41.

Gautier-Hion, A., 1971a. L'écologie du Talapoin du Gabon. *La Terre et la Vie,* 25:427–90.

Gautier-Hion, A., 1971b. Répertoire comportemental du Talapoin du Gabon, *Miopithecus talapoin. Biol. gabon.,* 7:295–391.

Gautier-Hion, A., 1973. Social and ecological features of Talapoin monkey; comparisons with other Cercopithecines. In: *Comparative Ecology and Behaviour of Primates.* R. P. Michael and J. H. Crook, eds. London: Academic Press, pp.148–70.

Gautier-Hion, A. and Gautier, J-P., 1974. Les associations polyspécifiques de Cercopithèques du plateau de M'passa, Gabon. *Folia. primat.,* 22:134–77.

Goodall, J., 1965. Chimpanzees of the Gombe Stream Reserve. In: *Primate Behavior: Field Studies of Monkeys and Apes.* I. DeVore, ed. New York: Holt, Rinehart and Winston, pp.425–73.

Goustard, M., 1963. Introduction à l'étude de la communication vocale chez *Macaca irus. Ann. Sci. Nat. Zool.,* 12:707–47.

Green, S., 1975. Variation of vocal pattern with social situation in the Japanese monkey *(Macaca fuscata):* a field study. In: *Primate Behavior: Development in Field and Laboratory Research,* vol. IV, L. A. Rosenblum, ed. New York: Academic Press.

Grimm, R. J., 1967. Catalogue of sounds of the pig-tailed macaque *(Macaca nemestrina). J. Zool. Lond.,* 152:361–73.

Groves, C. P., 1973. Notes on the ecology and behaviour of the Angolan colobus (*Colobus angolensis* P. L. Sclater, 1860) in N-E Tanzania. *Folia primat.,* 20:12–26.

Haddow, A. J., 1952. Field and laboratory studies on an African monkey, *Cercopithecus ascanius schmidtii Matschie. Proc. Zool. Soc. Lond.,* 122:297–394.

Hall, K. R. L., 1960. Social vigilance behaviour in the Chacma baboon, *Papio ursinus. Behaviour,* 16:261–94.

Hall, K. R. L., 1962. The sexual, agonistic and derived social behaviour patterns of the wild Chacma baboon, *Papio ursinus. Proc. Zool. Soc. Lond.,* 139:283–327.

Hall, K. R. L., 1965. Behaviour and ecology of the wild Patas monkeys, *Erythrocebus patas,* in Uganda. *J. Zool.* (London), 148:15–87.

Hall, K. R. L., 1967. Social interactions of the adult male and adult females of a Patas monkey group. In: *Social Communication among Primates,* S. A. Altmann, ed. Chicago: University of Chicago Press, pp.261–80.

Hall, K. R. L., 1968a. Social organization of the old world monkeys and apes. In: *Primates: Studies in Adaptation and Variability,* P. Jay, ed. New York: Holt, Rinehart and Winston, pp.7–31.

Hall, K. R. L., 1968b. Behaviour and ecology of the wild Patas monkey. In *Primates: Studies in Adaptation and Variability,* P. Jay, ed. New York: Holt, Rinehart and Winston, pp.32–119.

Hall, K. R. L.; Boelkins, R. C.; and Goswell, M. J.; 1965. Behaviour of Patas monkeys, *Erythrocebus patas* in captivity, with notes on the natural habitat. *Folia primat.,* 3:22–49.

Hall, K. R. L., and DeVore, I., 1965. Baboon social behaviour. In: *Primate Behaviour: Field Studies of Monkeys and Apes,* I. DeVore, ed. New York: Holt, Rinehart and Winston, pp.53–110.

Harlow, H. F., and Harlow, M. K., 1965. The affectional systems. In: *Behavior of Nonhuman Primates,* vol. II, A. M. Schrier, H. F. Harlow, and F. Stollnitz, eds. New York: Academic Press, pp.287–334.

Hill, O. W. C., and Booth, A. H., 1957. Voice and larynx in African and Asiatic Colobinae. *J. Bomb. Nat. Soc.,* 54:309–21.

Hinde, R. A., 1970. *Animal Behaviour: A Synthesis of Ethology and Comparative Psychology,* 2d ed. New York: McGraw-Hill. 871pp.

Hinde, R. A., and Rowell, T. E., 1962. Communication by postures and facial expressions in the rhesus monkey *(Macaca mulatta). Proc. Zool. Soc. Lond.,* 138:1–21.

Horwich, R., and Lafrance, L., 1972. The mountain Guereza. *Field Mus. Nat. Hist. Bull.,* 43:2–5.

Hunkeler, C.; Bertrand, M.; and Bourlière, F.; 1972. Le comportement social de la Mone de Lowe *(Cercopithecus campbelli lowei). Folia primat.,* 17:218–36.

Imanishi, K., 1957. Social behavior in Japanese monkeys, *Macaca fuscata. Psychologia,* 1:47–54.

Itani, J., 1963. Vocal communication of the wild Japanese monkey. *Primates,* 4:11–66.

Jay, P., 1963. Mother-infant relations in langurs. In: *Maternal Behavior in Mammals,* H. L. Rheingold, ed. New York: Wiley and Sons, pp.305–35.

Jay, P., 1965a. The common langur of North India. In: *Primate Behaviour: Field Studies of Monkeys and Apes,* I. DeVore, ed. New York: Holt, Rinehart and Winston, pp.197–249.

Jay, P., 1965b. Field studies. In: *Behavior of Nonhuman Primates,* vol. II, A. M. Schrier, H. F. Harlow, and F. Stollnitz, eds. New York: Academic Press, pp.525–91.

Kaplan, J., and Russel, M., 1974. Olfactory recognition in the infant squirrel monkey. *Develop. Psychobiol.,* 7:15–19.

Kaufman, I. C., and Rosenblum, L. A., 1966. A behavioral taxonomy for *Macaca nemestrina* and *M. radiata,* based on a longitudinal observation of family groups in the laboratory. *Primates,* 7:205–58.

Kawai, M., 1960. A field experiment on the process of group formation in the Japanese monkey *(Macaca fuscata)* and the releasing of the group at Ohirayama. *Primates,* 2:181–253.

Kawanaka, K., 1973. Intertroop relationships among Japanese monkeys. *Primates,* 14:113–59.

Kelemen, G., 1963. Comparative anatomy and performance of the vocal organ in Vertebrates. In: *Acoustic Behaviour of Animals.* R. G. Busnel, ed. New York: Elsevier Publishing Co., pp.489–518.

Kiley, M., 1972. The vocalizations of Ungulates; their causation and function. *Z. Tierpsychol.,* 31:171–222.

Kummer, H., 1957. Soziales Verhalten einer Mantelpavian-Gruppe. *Schweiz. Z. Psychol. Beiheft.,* 33:1–91.

Kummer, H., 1968. Social organization of hamadryas baboons; a field study. *Bib. primat.,* no. 6. Basel: Karger. 189pp.

Kummer, H., 1970. Behavioral characters in primate taxonomy. In: *Old World Monkeys: Evolution, Systematics and Behaviour,* J. R. Napier and P. H. Napier, eds. New York: Academic Press, pp.25–36.

Kummer, H., 1971a. Spacing mechanisms in social behavior. *Soc. Sci. Inform.,* 9:109–22.

Kummer, H., 1971b. Immediate causes of primate social structures. *Proc. Third Congr. Primat., Zurich, 1970,* vol. III:1–11.

Kummer, H., and Kurt, F., 1965. A comparison of social behavior in captive and wild hamadryas baboons. In: *The Baboon in Medical Research,* H. Vagtborg, ed. Austin: University of Texas Press, pp.65–80.

Kurland, J. A., 1973. A natural history of the Kra macaques (*Macaca fascicularis* Raffles, 1821) at the Kutai Reserve, Kalimantan Timur, Indonesia. *Primates,* 14:245–62.

Lahiri, R. K., and Southwick, C. H., 1966. Parental care in *Macaca sylvana. Folia primat.,* 4:257–64.

Lancaster, J., 1968. Primate communication systems and the emergence of human language. In: *Primates: Studies in Adaptation and Variability* P. Jay, ed. New York: Holt, Rinehart and Winston, pp.439–57.

Lindburgh, D. G., 1971. The rhesus monkey of North India: an ecological and behavioral study. In: *Primate Behavior: Developments in Field and Laboratory Research,* vol. 2, L. A. Rosenblum, ed. New York: Academic Press, pp.1–106.

Malbrant, R., and Maclatchy, A., 1949. *Faune de l'Equateur africain français, vol. II: Mammifères.* Paris: P. Lechevallier. 323pp.

Marler, P., 1965. Communication in monkeys and apes. In: *Primate Behavior: Field Studies of Monkeys and Apes,* I. DeVore, ed. New York: Holt, Rinehart and Winston, pp.236–65.

Marler, P., 1968. Aggregation and dispersal: two functions in primate communication. In: *Primates: Studies in Adaptation and Variability,* P. Jay, ed. New York: Holt, Rinehart, and Winston, pp.420–38.

Marler, P., 1969. *Colobus guereza:* territoriality and a group composition. *Science,* 163:93–95.

Marler, P., 1970. Vocalizations of East African monkeys: I—Red colobus. *Folia primat.,* 13:81–91.

Marler, P., 1972. Vocalizations of East African monkeys: II—Black and White colobus. *Behaviour,* 42:175–97.

Marler, P., 1973. A comparison of vocalizations of Red-tailed monkeys and Blue monkeys, *Cercopithecus ascanius* and *C.mitis* in Uganda. *Z. Tierpsychol.,* 33:223–47.

Marler, P., and Hamilton, W. J., 1967. Mechanisms of animal behavior. New York: Wiley and Sons. 771pp.

Mason, W. A., 1960. The effects of social restriction on the behavior of rhesus monkeys: I—Free social behavior. *J. Comp. Physiol. Psychol.,* 53:582–89.

Mason, W. A., 1961a. The effects of social restriction on the behavior of rhesus monkeys: II—Tests of gregariousness. *J. Comp. Physiol. Psychol.,* 54:287–90.

Mason, W. A., 1961b. The effects of social restriction on the behavior of rhesus monkeys: III—Dominance tests. *J. Comp. Physiol. Psychol.,* 54:694–99.

Mason, W. A., 1965. The social development of monkeys and apes. In: *Primate Behavior: Field Studies of Monkeys and Apes,* I. DeVore, ed. New York: Holt, Rinehart and Winston, pp.514–43.

Michael, R. P., and Herbert, J., 1963. Menstrual cycle influences grooming behaviour and sexual activity in the rhesus monkey. *Science,* 140:500–501.

Michael, R. P., and Keverne, E. B., 1968. Pheromones in the communication of sexual status in primates. *Nature,* 318:746–49.

Michael, R. P., and Keverne, E. B., 1970. Primate sex pheromones of vaginal origin. *Nature,* 225:84–85.

Michael, R. P.; Keverne, E. B.; and Bonsall, R. W.; 1971. Pheromones: isolation of male sex attractants from a female primate. *Science,* 172:964–66.

Miller, R. E., 1967. Experimental approaches to the autonomic and behavioral aspects of affective communication in rhesus monkeys. In: *Social Communication among Primates,* S. A. Altmann, ed. Chicago: University of Chicago Press, pp.125–34.

Miller, R. E., 1971. Experimental studies of communication in the monkey. In: *Primate Behavior: Developments in Field and Laboratory Research,* L. A. Rosenblum, ed. New York: Academic Press, pp.139–75.

Moynihan, M., 1964. Some behavior patterns of Platyrrhine monkeys: I—The night monkey *(Aotus trivirgatus). Smithson. Misc. Coll.,* 146:1–84.

Moynihan, M., 1966. Communication in the titi monkey, *Callicebus. J. Zool.* (London), 150:77–127.

Moynihan, M., 1967. Comparative aspects of communication in New World primates. In: *Primate Ethology,* D. Morris, ed. London: Weidenfeld and Nicolson, pp.236–65.

Mykytowycz, R., 1972. The behavioral role of the mammalian skin glands. *Naturwissenschaften,* 59: 133–39.

Nadler, R. D., and Rosenblum, L. A., 1973. Sexual behavior during successive ejaculations in bonnet and pigtail macaques. *Amer. J. Phys. Anthrop.,* 38:217–20.

Nishimura, A., 1973. Age changes of the vocalization in free-ranging Japanese monkeys. *Symp. Fourth Int. Congr. Primat.,* vol. I: *Precultural Primate Behavior.* Basel: Karger, pp.76–87.

Nolte, A., 1955. Field observations on the daily routine and social behavior of common Indian monkeys, with special reference to the bonnet monkey (*Macaca radiata* Geoffroy). *J. Bomb. Nat. Hist. Soc.,* 53:177–84.

Poirier, F. E., 1968. Nilgiri langur (*Presbytis johnii*) territorial behavior. *Primates,* 9:351–64.

Poirier, F. E., 1970a. The communication matrix of the Nilgiri langur (*Presbytis johnii*) of South India. *Folia primat.,* 13:92–136.

Poirier, F. E., 1970b. The Nilgiri langur *(Presbytis johnii)* of South India. In: *Primate Behavior: Developments in Field and Laboratory Research,* vol. I, L. A. Rosenblum, ed. New York: Academic Press, pp.251–383.

Quris, R., 1973. Emissions sonores servant au maintien du groupe social chez *Cercocebus galeritus agilis. La Terre et la Vie,* 27:232–67.

Ripley, S., 1967. Intertroop encounters among Ceylon gray langurs. In: *Social Communication among Primates,* S. A. Altmann, ed. Chicago: University of Chicago Press, pp.237–53.

Rosenblum, L. A., and Cooper, R. W., 1968. *The Squirrel Monkey.* London: Academic Press. 451 pp.

Rosenblum, L. A., and Kaufman, I. C., 1967. Laboratory observations of early mother-infant relations in pigtail and bonnet macaques. In: *Social Communication among Primates,* S. A. Altmann, ed. Chicago: University of Chicago Press, pp.33–41.

Rowell, T. E., 1966. Forest living baboons in Uganda. *J. Zool.* (London), 149:344–64.

Rowell, T. E., 1971. Organization of caged groups of *Cercopithecus* monkeys. *Anim. Behav.,* 19:625–45.

Rowell, T. E., 1972. *Social Behaviour of Monkeys,* B. M. Foss, ed. London: Penguin Science of Behaviour. 202pp.

Rowell, T. E., and Hinde, R. A., 1962. Vocal communication by the rhesus monkeys *(Macaca mulatta). Proc. Zool. Soc. Lond.,* 138:279–94.

Rowell, T. E.; Hinde, R. A.; and Spencer-Booth, Y.,

1964. Aunt-infant interaction in captive rhesus monkeys. *Anim. Behav.*, 12:219–26.

Saayman, G. S., 1970. The menstrual cycle and sexual behaviour in a troop of free-ranging Chacma baboons *(Papio ursinus). Folia primat.*, 12:81–110.

Saayman, G. S., 1971. Behaviour of the adult males in a troop of free-ranging Chacma baboons *(Papio ursinus). Folia primat.*, 15:36–57.

Sabater Pi, J., 1970. Aportacion a la ecologia de los *Colobus polykomos satanas*, Waterhouse 1838, de Rio Muni (Republica de Guinea Ecuatorial). *P. Inst. Biol. Apl.*, 48:17–32.

Schenkel, R., and Schenkel-Hulliger, L., 1966. On the sociology of free-ranging Colobus (*Colobus quereza caudatus*, Thomas 1885). *First Congr. Int. Primat. Soc.*, 185:215–32.

Schultz, A. H., 1969. *The Life of Primates.* London: Weidenfeld and Nicolson. 281pp.

Schultz, A. H., 1970. The comparative uniformity of the Cercopithecoidea. In: *Old World Monkeys: Evolutions, Systematics and Behavior,* J. R. Napier and P. H. Napier, eds. New York: Academic Press, pp.39–52.

Scruton, D. M., and Herbert, J., 1970. The menstrual cycle and its effect on behaviour in the talapoin monkey (*Miopithecus talapoin*). *J. Zool.* (London), 162:419–36.

Simonds, P. E., 1965. The Bonnet macaque of South India. In: *Primate Behavior: Field Studies of Monkeys and Apes,* I. DeVore, ed. New York: Holt, Rinehart and Winston, pp.175–86.

Smith, W. J., 1970. Message, meaning and context in ethology. *Amer. Naturalist,* 99:405–409.

Southwick, C. H., 1962. Patterns of inter-group social behavior in primates, with special reference to Rhesus and Howling monkeys. Ann. N.Y. Acad. Sci., 102:436–54.

Southwick, C. H.; Beg, M. A.; and Siddiqui, M. R.; 1965. Rhesus monkeys in North India. In: *Primate Behavior: Field Studies of Monkeys and Apes,* I. DeVore, ed. New York: Holt, Rinehart and Winston, pp.111–59.

Sparks, J., 1967. Allogrooming in primates: a review. In: *Primate Ethology,* D. Morris, ed. London: Weidenfeld and Nicolson, pp.148–75.

Starck, D., and Schneider, R., 1960. Respirationsorgane. A-Larynx. *Primatologia,* 3:423–590.

Stoltz, L. P., and Saayman, G. S., 1970. Ecology and behaviour of baboons in the Northern Transvaal. *Ann. of Transv. Mus.*, 26:100–43.

Stebbins, W. C., 1971. Hearing. In: *Behavior of Nonhuman Primates.* A. M. Schrier and F. Stollnitz. New York: Academic Press, pp.159–92.

Struhsaker, T. T., 1967a. Behavior of Vervet monkeys (*Cercopithecus aethiops*). *Univ. Calif. Publ. Zool.*, 82:1–74.

Struhsaker, T. T., 1967b. Auditory communication among Vervet monkeys *(Cercopithecus aethiops).* In: *Social Communication among Primates.* S. A. Altmann, ed. Chicago: University of Chicago Press, pp.238–324.

Struhsaker, T. T., 1969. Correlates of ecology and social organization among African Cercopithecines. *Folia primat.*, 11:80–118.

Struhsaker, T. T., 1970. Phylogenetic implications of some vocalizations of *Cercopithecus* monkeys. In: *Old World Monkeys: Evolution, Systematics and Behavior.* J. R. Napier and P. H. Napier, eds. New York: Academic Press, pp.365–444.

Sugiyama, Y., 1968a. Social organization of chimpanzees in the Budongo forest, Uganda. *Primates,* 9:225–58.

Sugiyama, Y., 1968b. Ecology of the lion-tailed macaque *(Macaca silenus Linnaeus);* a pilot study. *J. Bomb. Nat. Hist. Soc.*, 65:283–93.

Sugiyama, Y., 1971. Characteristics of the social life of Bonnet macaque (*Macaca radiata*). *Primates,* 12:247–66.

Sugiyama, Y.; Yoshiba, K.; and Parthasarathy, M. D., 1965. Home range, mating season, male group and inter-troop relationship in Hanuman langurs *(P. entellus). Primates,* 6:73–106.

Sutton, D.; Larson, C.; Taylor, E. M.; and Lindeman, R. C.; 1973. Vocalization in rhesus monkeys: conditionability. *Brain Res.*, 52:225–31.

Takeda, R., 1965. Development of vocal communication in man-raised Japanese monkeys. I—From birth until the sixth week. *Primates,* 6:337–80.

Takeda, R., 1966. Development of vocal communication in man-raised Japanese monkeys. II—From the seventh to the thirtieth week. *Primates,* 7:73–116.

Tanaka, J., 1965. Social structure of Nilgiri langur. *Primates,* 6:107–22.

Tokuda, K., 1961–62. A study on the sexual behavior in the Japanese monkeys troop. *Primates,* 3:1–40.

Ullrich, W., 1961. Zur Biologie und Soziologie der Colobusaffen (*Colobus guereza caudatus,* Thomas 1885). *Zool. Garten,* 25:305–68.

van Hooff, J. A. R. A. M., 1962. Facial expressions in higher primates. *Symp. Zool. Soc. Lond.*, 8:97–125.

van Hooff, J. A. R. A. M., 1967. The facial displays of the Catarrhine monkeys and apes. In: *Primate*

Ethology, D. Morris, ed. London: Weidenfeld and Nicolson, pp.7–68.

Vine, I., 1970. Communication by facial visual signals. In: *Social Behaviour of Birds and Mammals.* J. H. Crook, ed. New York: Academic Press, pp.279–354.

Vogel, C., 1973. Acoustical communication among free-ranging common Indian langurs *(Presbytis entellus)* in two different habitats of North India. *Amer. J. Phys. Anthrop.,* 38:469–80.

von Spivak, H., 1971. Ausdrucksformen und soziale Beziehungen in einer Dschelada -Gruppe *(Theropithecus gelada)* in Zoo. *Z. Tierpsychol.,* 28:279–96.

Waser, P. M., 1975. Experimental playbacks show vocal mediation of intergroup avoidance in a forest monkey. *Nature,* 255:56–58.

Washburn, S. L., and Hamburg, D. A., 1968. Aggressive behavior in old world monkeys and apes. In: *Primates: Studies in Adaptation and Variability.* P. Jay, ed. New York: Holt, Rinehart and Winston, pp.458–78.

Wickler, W., 1967. Socio-sexual signals and their intraspecific imitation among primates. In: *Primate Ethology,* D. Morris, ed. London: Weidenfeld and Nicolson, pp.69–147.

Wilson, E. O., 1968. Chemical systems. In: *Animal Communication,* T. A. Sebeok, ed. Bloomington: Indiana University Press, pp.75–102.

Wilson, C., 1972. Spatial factors and the behavior of non-human primates. *Folia primat.,* 18:256–75.

Winter, P.; Handley, P.; and Schott, D.; 1973. Ontogeny of squirrel monkeys calls under normal conditions and under acoustic isolation. *Behaviour,* 47:230–39.

Wolfheim, J. H., and Rowell, T. E., 1972. Communication among captive talapoin monkeys. *Folia primat.,* 18:224–56.

Yamaguchi, S. I., and Myers, R. E., 1972. Failure of discriminative vocal conditioning in rhesus monkeys. *Brain Res.,* 37:109–14.

Yoshiba, K., 1968. Local and intertroop variability in ecology and social behavior of common Indian langurs. In: *Primates: Studies in Adaptation and Variability.* P. Jay, ed. New York: Holt, Rinehart and Winston, pp.217–42.

Chapter 36

SIGNALING BEHAVIOR OF APES WITH SPECIAL REFERENCE TO VOCALIZATION

Peter Marler and Richard Tenaza

The Social Group and Its Living Space

To the extent that the social organization of a species is a consequence of interaction between its members, the communication system is a basic component in social design. Studies of animal communication should eventually help to explain how the diversity of animal social systems is maintained. Conversely, the social system of a species provides the background against which one may hope to interpret the function that its communication signals serve. Information that field studies have provided on the behavior of the great apes is now sufficient for us to begin appraising the differences in organization that they exhibit and speculating about their evolvement.

The first problem is to define the basic social units, the spatial distribution of individuals, their patterns of contact and separation, and the ways in which they interact with one another—either competitively, thus limiting one another's access to resources, or cooperatively, thus aiding one another in reproduction, resource exploitation, and the avoidance of threats to life and health. Defining social groupings and distinguishing competitive from cooperative social units is less easy in practice than it might first appear. The difficulties are well illustrated by the great apes.

In one cluster of species, the gibbons and siamang, the social units are readily discernible. So far as is known, all species exhibit the same reproductive groupings. Monogamous pairs with long-lasting pair bonds live with their offspring in tropical rain forest, occupying territories from which adjacent families and mature offspring are excluded.

Of the other types of social organization represented in the great apes, the gorilla's pattern is most like the gibbons'. While there are some solitary adult males, most animals live in a social group.

These gorilla groups are perhaps five times larger than those of gibbons, averaging fifteen members in the nine troops studied by Fossey (1972), as compared with an average of 3.4 members in Kloss's gibbon (Tenaza, 1975), with one to three fully mature silverbacked males per group. The balance is made up of blackbacked males, adult females, and their offspring. The overlap between the extensive home ranges of gorilla groups is considerable, but each has exclusive use of a part, apparently as a consequence of an admixture of mutual avoidance and aggressive repulsion of neighboring groups (Schaller, 1963; Fossey, 1972, 1974).

The patterns of social organization in the chimpanzee and the orang-utan are harder to discern. Rather than forming durable groupings that move synchronously as coherent units, individuals and smaller groups tend to separate and coalesce in combinations that vary in composition from day to day. The periods of individual separation seem to be much longer for the orang-utan than for the chimpanzee. Although little is known about social groupings of the orang-utan, the most solitary of the ape species, studies by Mackinnon (1974, 1975) suggest a pattern of social organization in at least one of its populations (in Sumatra) that is common in many other mammals (Eisenberg, 1966; Fisler, 1969) but is most unusual for higher primates. The majority of sightings are either of single animals or of adult females with one or two offspring. Single adult males occupy a large home range that encompasses the ranges of several adult females and their offspring. If the inhabitants of this shared living space are regarded as a group, it is not very different in size and composition from groups of the mountain gorilla. One well-counted group of seventeen animals included two adult males, two subadult males, six adult females, two adolescents, two juveniles, and three infants (Mackinnon, 1974). The home range of one group overlaps relatively little with ranges of adjacent groups. There is evidence of active exclusion and avoidance by aggression and vocal signaling. Although all members may be seen occasionally in close proximity, they never seem to move around as a coherent durable group, as the gorilla does. Instead the members are dispersed most of the time, having only transient contacts with one another, except for relationships of mothers and offspring and consortships between an adult male and an estrous female. In other orang-utan populations group members were more dispersed, and the definition of a grouping depends on a distinction between the brief, peaceful encounters occurring between members as compared with the aggressive encounters with members of other groupings. As Mackinnon (1974) summarizes a complex situation, "It would seem that links of familiarity and general tolerance exist between animals that range over the same area."

Something similar may be true of the more social chimpanzee. This species was most commonly seen by van Lawick-Goodall (1968b, 1971) in groups of two to six animals, but more recent data suggest even smaller subgroupings. Wrangham (pers. comm.) found mean group sizes of 1.1 in one season and 2.0 in another, sampling those in which at least one member was an adult male. The subgroupings of chimpanzees in the Gombe Stream population are constantly changing in composition, apart from the durable subgroups that consist of mothers and their immature offspring. In striking contrast with the orang-utan and all of the other apes, peaceful subgroupings of chimpanzees often include more than one adult male at a time, and more than one adult female. If it is indeed possible to define a larger group on the basis of the predominantly peaceful and cooperative interactions of subgroupings when they meet (there is in fact a good deal of aggression when reunions occur), this grouping is probably larger than in the orang-utan and may amount to as many as fifty animals. Early evidence suggested that such a "community" at the Gombe National Park has north and south neighboring "communities" with which there is both peaceful intermingling, from which high-ranking adult males may perhaps exclude themselves (van Lawick-Goodall, 1971), and also a degree of mutual avoidance (Nishida and Kawanaka, 1972). More recent data have confirmed that adult male members of the community defend a territory against adjacent group males. This territory encompasses many overlapping adult female home ranges, and contacts between communities may involve mainly females with home ranges straddling the boundary (Wrangham, 1975).

Thus, the apes offer an unusually broad spectrum of organizational patterns with durable heterosexual pairs and dependent young, represented by the gibbons at one extreme, and at the other the chimpanzee, with a more labile social organization and several adult males living peaceably together. Having one-male groupings (i.e., one fully mature male) with several adult females and their offspring, the orang-utan and the gorilla fall somewhere between.

In all species the evidence reported suggests some degree of exclusive use of living space by resident groups. It is not always easy to discern whether adjacent communities simply avoid invading parts of one another's home ranges, perhaps aided by long-range vocal and nonvocal sounds, or whether there is active aggressive exclusion. At present there is evidence for territorial behavior in male and female gibbons, and for male group territoriality in the chimpanzee, with suggestions of territoriality in adult male orangutans and, perhaps, of a less well defined form of group territoriality in the gorilla.

Communication Systems and Social Organization

Basic to the complexities of the frequency and quality of different kinds of social interaction that may be mediated by communication signals is the elementary question of signaling distance. The dimensions of the space and the nature of the terrain over which signaling must occur to function normally are important in understanding the structure of the signal system that is used. The social systems of the great apes are diverse enough that we may be prepared for a variety of special requirements for distance signaling among species. In the gibbons for example the requirements of territoriality loom large in both sexes. Having signals that can be received over a distance approximating the normal spacing of neighboring troops permits the maintenance of spacing patterns with a significant economy of effort, as compared with a signal system that only permits location and identification of neighboring groups at close range (Marler, 1968). Vocal signals lend themselves well to such a function, and they are convenient for distance signaling in animals that dwell in dense rain forest. Thus, in the gibbon repertoire we should be prepared for loud vocalizations that are designed for the maintenance of intergroup spacing. Given the small size and relatively coherent nature of the social group in gibbons and given a social network that is presumably simple, at least with regard to number of participants, the requirements for complex intragroup signaling may not be great (cf. Chivers 1974). Thus, the dictates of intergroup signaling might be expected to dominate the gibbon vocal repertoire.

The emphasis on distance signaling may be even greater in the orang-utan, where the requirements for territoriality are added to those of a highly dispersed group organization, though there will still be demands for close-range signaling within subgroups and consortships. By contrast, coherent social groups of the gorilla, which are larger than the gibbons' and have more participants in the network of social interactions, should lead to emphasis on close-range signaling, though in adult males there are still requirements for the maintenance of spacing between groups. The strongest emphasis of all on complex, relatively close-range, within-group signaling might be expected in the labile social system of the chimpanzee, though here too there are requirements for long-range communication between separated group members and between different social groups.

The apes are well provided with the classical sensory modalities for the detection of stimuli from sources at a distance from the body. Following sections will review the use of sounds in ape communication and, in less detail, visual signals (see reviews in Andrew, 1963; van Hooff, 1962, 1967, 1970, 1971; van Lawick-Goodall, 1968a; Wickler, 1967).

There is also evidence that specialized chemical signals are emitted in social situations. The most obvious circumstance, well studied in recent years in macaques by Michael and his colleagues, is found in sexual relationships between estrous females and adult males (Michael, 1969). Sexual pheromones may be as important in apes as in macaques, though they are yet to be studied. Certainly there is every sign of keen olfactory interest among males when a female is in estrus. A finger is often touched to the genital area of the female chimpanzee and then brought to the nose and sniffed. In chimpanzees the olfactory investigation of the female genital area is most frequent at the first sign of sexual swelling and during her detumescence, and is less common when the female has a large sexual swelling (van Lawick-Goodall, 1968b). Similar behavior by captive female gorillas toward infants has been recorded (Hess, 1973). Apart from genital secretions, there seems to be no evidence of the discrete glands specialized to produce chemical signals that are commonly found in prosimians and are also present in both platyrrhine and catarrhine monkeys.

Production and Perception of Sounds

Apes have ample means for producing a variety of well-structured sounds. All of them make some use of nonvocal sounds, notably the drumming of male chimpanzees on the ground and on buttresses of large trees, and the chest beating by the gorilla, augmented in silverbacked and large blackbacked males by the resonant properties of inflated air sacs on the chest (Schaller, 1963). Vocalizations nevertheless provide the main medium for auditory communication. The basic structure of the larynx of the great apes is similar to that of man, and it probably operates on the same fundamental principles (Negus, 1949). However, the post-laryngeal tract tends to lack a pharyngeal region. Together with differences in the range of tongue movements that are possible, this lack would restrict the capacity of apes to produce certain of the sounds of human speech (Lieberman, 1968; Lieberman, Crelin, and Klatt, 1972).

Of greater potential biological significance is the presence in all of the apes, other than some gibbons, of laryngeal air sacs, especially marked in adult males. In the adult male chimpanzee, for example, small paired air sacs communicate directly with the laryngeal ventricle. After the lungs are filled, the sacs can be inflated by closing the mouth and nose or the ventricular bands of the larynx. In the gorilla these sacs are larger still, even in adult females, and the sacs of the adult male orang-utan are huge, with a capacity as great as seven liters (Huber, 1931) and extend into the armpit as far as the shoulder blades. Although Negus (1949) attributes a respiratory function to these sacs, which he thinks would permit the recirculation of breath without the inhalation of new air, a function in sound production seems more likely. The sacs are inflated before and during the production of certain loud vocalizations. Possibly the sacs serve as resonators. Another possibility is that flow of air between the sacs and the pharynx over the ventricular ligaments sets them into vibration, thus providing a secondary sound source. Such sounds have been recorded (Brandes, in Huber, 1931), but it is likely that in the apes they are no more than by-products of the process of inflation and deflation of the sacs. Gautier (1971) has analyzed the role of the laryngeal sacs of Cercopithecus monkeys in sound production, and since the morphology is similar to that in the great apes, we may take his analysis as a provisional model.

His studies of the "double boom" of the adult male *Cercopithecus neglectus* show that air expelled from the sac does not excite the laryngeal cords, and probably not the false cords either. Inflation of the sac immediately before the first vocalization is silent, though if the sac is punctured, then a sound is generated at the time when inflation would normally begin. Before the second boom there is an audible sound during inflation of the sac. Although air is forcefully expelled from the sac during sound production, the fundamental frequency of the sound produced is unchanged from that resulting without sac involvement. By puncturing and resealing the sac of an adult male, Gautier demonstrated that it serves as a resonator and amplifier, selectively emphasizing the fundamental frequency at the expense of higher harmonics, which become more evident when the sac is experimentally opened.

Probably the laryngeal sacs possessed by apes serve a similar function of selectively emphasizing certain frequencies that are produced in the larynx by the vocal cords in the normal way. Thus, the relatively narrow range of frequencies characterizing the loud pant-hooting of chimpanzees, as compared with their other vocalizations, may be attributable to involvement of the laryngeal sacs, which are visibly inflated during this vocalization. To judge by the sound spectograms that Mackinnon (1974) presents of the long calls of adult male orang-utans, again with a narrower range of frequencies than some of the other vocalizations, the same may apply.

Data on hearing are available only for the chimpanzee (Stebbins, 1971). On the basis of what is known about this species, we can assume provisionally that they hear their vocalizations in much the same way we do. The auditory threshold curve for the chimpanzee resembles our own, with the greatest sensitivity between about 1 kHz and 8 kHz, though the range does extend into the ultrasonic, to about 30 kHz (see Stebbins, 1971, for details). Their capacity to discriminate between different frequencies is also much like our own, so that our ears are probably a good basis for at least a preliminary analysis of chimpanzee sounds and probably of the sounds of the other species of great ape as well.

Auditory Signals

NONVOCAL SIGNALS

The apes commonly use nonvocal sounds in their displays, the most elementary being branch shaking, breaking, and dropping. The noise of gibbon brachiation displays is audible for 100 m or more. The hand may be used to thump on the ground or on the bole of a tree. This action has become specialized in the chimpanzee as stereotyped drumming behavior in which a chimpanzee will

> leap up at a tree and drum on the trunk or buttresses with his feet—usually the two feet pound down one after the other in quick succession making a double beat; there is then a slight pause before the next double beat. From one to three double beats are normal. . . . Frequently, when a number of chimpanzees travelling together come upon some favored 'drumming tree' along the track (a tree with wide buttresses) each male in turn drums in this manner. This results in a whole series of one to three double beats with irregular intervals between each. [van Lawick-Goodall, 1968b:273].

The gorilla creates a nonvocal sound with rather similar structural and temporal patterning by a quite different means, namely chest beating, documented in detail by Schaller (1963). Chest beating occurs as one component in a sequence that typically includes eight other distinct acts.

> The hands are almost invariably held flat while beating the chest; that is, with fingers extended and the palm often slightly cupped. The animal tends to hold its hand within six inches or less of the chest and the alternate beats are rapid and direct. The fingers are

> often spread on females and young animals, but those of adult males tend to touch each other. The sound produced when a silverbacked or large blackbacked male beats his chest may be described as a hollow "pok-pok-pok" somewhat resembling the noise produced by rapping an empty gourd with sticks. Under favorable conditions the sound carries for as much as a mile. In silverbacked males the prominent air sacs presumably act as resonators, for their sudden inflation on each side of the throat is sometimes readily apparent before the chest beats. Small blackbacked males, females and youngsters produce a mere slapping sound when beating the chest. [Schaller, 1963:225]

Although as many as twenty slaps may be given in gorilla chest beating, the modal value is between three and four, not too different from the number of chimpanzee drumming beats that is typical (from one to three double beats are normal; van Lawick-Goodall, 1968b:273). Both are effective long-distance signals, audible in the forest at considerable distances.

Orang-utans and gibbons spend most of their time in the forest canopy, where sound attenuation is less than on the ground. They have no nonvocal signals equivalent in carrying power to drumming and chest beating. However, the noisy branch breaking by disturbed orang-utans (Schaller, 1963; Mackinnon, 1974) may be communicative. Female Kloss's gibbons brachiate among branches of tall trees while singing (Tenaza, 1976), but the noises of rustling vegetation thus produced are audible for much shorter distances than the songs occurring with them. This form of branch shaking occurs in all gibbons, and is probably more a visual than an auditory signal.

VOCALIZATIONS OF THE CHIMPANZEE

Although the vocal behavior of the chimpanzee has been studied more than that of any other ape (Reynolds and Reynolds, 1965; van Hooff, 1962, 1967; van Lawick-Goodall, 1968a, 1968b), including some sound spectrographic analyses (van Hooff, 1971; Marler, 1965, 1969), a comprehensive acoustical description has yet to be published. Based on long and intimate study, the verbal descriptions of vocalizations of chimpanzees in the Gombe National Park in Tanzania by van Lawick-Goodall provide the basis for the present account, augmented with data from a ten-week study of behavior of the Gombe population (Marler, 1969, 1976; Marler and Hobbet, 1975).

The catalog of vocalizations prepared by van Lawick-Goodall (1968a) included twenty-four classes of vocal sounds. For purposes of comparison with other ape species the list was reduced to thirteen. Chimpanzee vocalizations are highly graded and the list of twenty-four includes several obvious variations on a common theme (e.g., pant-hoots and pant-shrieks; barks and shrieks), which were merged into single categories. Sounds distinguished primarily by context were merged wherever acoustical analysis failed to reveal reliable characters for distinguishing them (e.g., hoo and hoo-whimper and whimper, scream calls and screaming, panting and copulatory panting). Sounds distinguished by the age and sex of the vocalizer were also grouped, even though acoustical differences were detectable (e.g., screaming and infant screaming). Finally, two sounds occurred so rarely that their acoustical identity was not established (soft grunt, groan). Since each of the two sounds is similar to another category (grunting, rough grunting), they are not treated separately here. In addition to these changes some new names have been coined either where acoustical analysis suggests a more appropriate term or where sounds had been defined by their accompanying posture or facial expression (e.g., bobbing pants: changed to pant-grunting; cf. also mixed names used by van Hooff, 1971).

The thirteen remaining basic vocal categories are listed in Table 1 together with four nonvocal

Table 1

A catalog of chimpanzee vocalizations.

	Vocalization	Circumstances
1.	Pant-hoot [pant-shriek and roar]	Hearing distant group; rejoining group; meat eating; in nest at night; general arousal.
2.	Pant-grunt (bobbing pants) [pant-shriek]	Subordinate approaching or being approached by dominant.
3.	Laughter	Playing, especially being tickled.
4.	Squeak	Being threatened; submission; being close to dominant.
5.	Scream [infant scream, screaming]	Fleeing attack; submission; when lost; while attacking dominant; copulating female.
6.	Whimper [hoo-whimper, quiet huu]	Begging; infant-parent separation; strange sound or object.
7.	Bark (soft bark)	Vigorous threat.
8.	Waa Bark	Threat to other, often dominant, at distance.
9.	Rough Grunt (grunts) [barks, shrieks, groans]	Approaching and eating preferred food.
10.	Pant [copulatory pants]	Copulating male; grooming; meeting another as prelude to kissing, etc.
11.	Grunt [soft grunt]	Feeding; mild general arousal; social excitement.
12.	Cough	Mild, confident threat to subordinate chimpanzee, baboon.
13.	Wraaa	Detection of human or other predator, also dead chimpanzee; may be threat component.
Non-vocal	Lip Smack and Teeth Clack	Allo-grooming.
Non-vocal	Drumming	Hearing distant group; rejoining group.
Non-vocal	Ground Thumping	Threat to predator.
Non-vocal	Branch Shaking	Aggressive display to chimpanzee or predator.

NOTE: ( ) = synonyms with van Lawick-Goodall. [ ] = calls designated separately by van Lawick-Goodall, here lumped with the call above either because they seem acoustically indistinguishable, because they intergrade without clear separation, or because they are different renditions of the same basic call by male or female or by particular age classes. (Condensed and modified from van Lawick-Goodall, 1968b.)

sounds ranging in loudness from lip smacking and teeth clacking, which are audible only up to a distance of two or three meters, branch shaking, which is somewhat louder, ground thumping, which we hear at many meters distance, and drumming, audible several hundred meters away. The frequency with which each type of vocalization and one nonvocal sound—lip smacking—were used in one study (Marler, 1976) by each of ten classes of individuals identified by sex and age is shown in Table 2 (center figure in each cell). Also included are the per-

Table 2

Frequency of use of chimpanzee vocalizations. I.

	Number of individuals of given age, sex, and status in population											
	(9)	(3)	(4)	(2)	(4)	(7)	(8)	(3)	(1)	(3)		
	♀M	♀O	♀A	♀J	♀I	♂1	♂2	♂A	♂J	♂I	Total	Percentage
Pant-hoot	17 [109] 23	8 [51] 33	5 [30] 13	0.6 [4] 8	0.1 [1] 5	34 [219] 51	27 [175] 44	7 [44] 25	0.6 [4] 11	2 [11] 3	648	28
Pant-grunt	42 [120] 25	9 [26] 17	3 [8] 3	[0]	[0]	16 [46] 11	22 [62] 16	7 [21] 12	[0]	0.7 [2] 0.6	285	12
Laughter	1 [4] 0.8	[0]	0.7 [2] 0.8	9 [25] 47	4 [11] 52	5 [14] 3	9 [24] 6	0.7 [2] 1	2 [5] 14	68 [184] 57	271	12
Squeak	21 [56] 12	14 [36] 24	25 [65] 27	4 [10] 19	1 [3] 14	3 [7] 2	11 [29] 7	12 [33] 18	2 [4] 11	8 [21] 7	264	11
Scream	27 [59] 12	7 [15] 10	18 [39] 16	3 [6] 11	1 [3] 14	7 [16] 4	13 [29] 7	11 [23] 13	4 [9] 24	8 [17] 5	216	9
Whimper	21 [37] 8	3 [4] 2	10 [18] 8	3 [5] 8	1 [2] 10	6 [11] 3	7 [13] 3	7 [13] 6	5 [8] 22	37 [64] 20	175	8
Bark	16 [15] 3	6 [6] 4	44 [42] 18	2 [2] 4	1 [1] 5	13 [12] 3	9 [9] 2	3 [3] 2	[0]	5 [5] 2	95	4

Waa bark	26 [22] 5	5 [4] 3	17 [14] 6	1 [1] 2	[0]	10 [8] 2	14 [12] 3	18 [15] 8	2 [2] 5	7 [6] 2	84	4
Rough grunt	13 [11] 2	1 [1] 0.7	2 [2] 0.8	[0]	[0]	52 [43] 10	13 [11] 3	8 [7] 4	[0]	10 [8] 2	83	4
Pant	39 [32] 7	5 [4] 3	[0]	[0]	[0]	26 [21] 5	17 [14] 4	4 [3] 2	5 [4] 11	4 [3] 0.9	81	4
Grunt	12 [5] 1	9 [4] 3	14 [6] 3	[0]	[0]	21 [9] 2	30 [13] 3	12 [5] 3	2 [1] 3	[0]	43	2
Cough	21 [7] 1	[0]	24 [8] 3	[0]	[0]	45 [15] 3	3 [1] 0.2	3 [1] 0.5	[0]	3 [1] 0.3	33	1
Wraaa	17 [3] 0.6	11 [2] 1	[0]	[0]	[0]	56 [10] 2	11 [2] 0.5	6 [1] 0.5	[0]	[0]	18	0.8
Lip smack	[0]	[0]	18 [3] 1	[0]	[0]	12 [2] 0.5	24 [4] 1	47 [8] 4	[0]	[0]	17	0.7
Total	480	153	237	53	21	433	398	179	37	322	2313	
Percentage	21	7	10	2	0.9	19	17	8	2	14		

NOTE: Usage by chimpanzees at Gombe National Park of thirteen vocal and one nonvocal sound (lip smacking) by male and female adults, juveniles, and infant chimpanzees. In each cell the middle figure is the number of vocalizations; bottom is the percentage of vocalizations of that class of animals; top is the percentage of renditions of that type of vocalization. ♀M–adult female with infant. ♀O–adult female without infant. ♀A–adolescent female. ♀J–juvenile female. ♀I–infant female. ♂1–older adult male. ♂2–younger adult male. ♂A–adolescent male. ♂J–juvenile male. ♂I–infant male.

centage that each makes up of the total recorded for that class of vocalization (top figure in each cell) and the percentage of all vocalizations recorded for that class of individuals (bottom figure in each cell). At the top of each column the number in parentheses indicates the number of individuals representing each class in the study population at Gombe at the time. These numbers range from one (juvenile male) to nine (mother with infant), some individuals making a disproportionally large contribution to the total number of vocalizations recorded. To correct for this Table 3 was prepared by dividing the frequencies with which each vocalization was recorded by the number of individuals in each class, providing a clearer picture of differences in frequency of usage between sex and age classes.

The following brief account of thirteen basic chimpanzee sounds illustrates the kind of communicative sounds that an ape may use. Vocal sounds are considered in order of the frequency of use by the Gombe study population as a whole (Marler, 1976).

(1) *Pant-hoot* is by far the most frequent vocalization heard at Gombe Stream. Every entry in Tables 1 and 2 denotes a pant-hooting sequence lasting from seven to eleven seconds, each with many parts. Pant-hooting is one of three vocalizations that are typically voiced on both exhalation and inhalation, the others being laughter and pant-grunting. Fig. 1 shows a portion of a typical pant-hooting sequence by an adult female. Aligned with it are single frames from synchronized movie film, illustrating the configurations of the mouth during pant-hooting. One can see that the frequency spectrum is broader in exhalation sounds than in sounds produced on inhalation and that this in turn relates to the degree of mouth opening. One exhalation sound in Fig. 1 is given with the mouth opened less widely than usual and the frequency spectrum is correspondingly narrowed.

Although both sexes and all ages were recorded pant-hooting, the vocalization is given most often by males (males, 453; females, 195). It is by far the most common sound of an adult male and constitutes 50 percent of all sounds that an older male typically made in the Gombe study (Table 3). In both males and females it is used more frequently with increasing age and makes up an increasing proportion of the sounds uttered as a chimpanzee matures. There are consistent differences in the structure of male and female pant-hooting, readily audible to the unaided human ear. There are also individual differences that permit an experienced human listener to identify a distant pant-hooting individual (van Lawick-Goodall, 1968a; Marler and Hobbet, 1975). Fig. 2 illustrates the kind of differences by which the pant-hooting of three adult males can be distinguished.

Pant-hooting is given in many situations. It frequently occurs in response to pant-hooting of distant individuals, often after the chimpanzee has listened with obvious close attention. On many occasions it seems to occur spontaneously as animals are feeding, or during the night, a feature it shares with other primate calls that serve a long-distance spacing function (Marler, 1968). It may also accompany special occasions such as the eating of prey after capture (Teleki, 1973b) or the rejoining of groups after separation (van Lawick-Goodall, 1968a). It is a loud sound that carries far, and its functions, perhaps multiple, include the long-range announcement of an individual's presence and sex, which permits the continued separation of some individuals and groups and the reunion of others.

(2) *Pant-grunt* is another vocalization with sound on both inhalation and exhalation. It is quieter and faster than pant-hooting, with breathing cycles averaging about five per second, as compared with one per second in pant-hooting. It is a low-pitched sound, with little energy above 1.5 kHz. As can be seen in Fig. 3, exhalation sounds are lower-pitched than inhala-

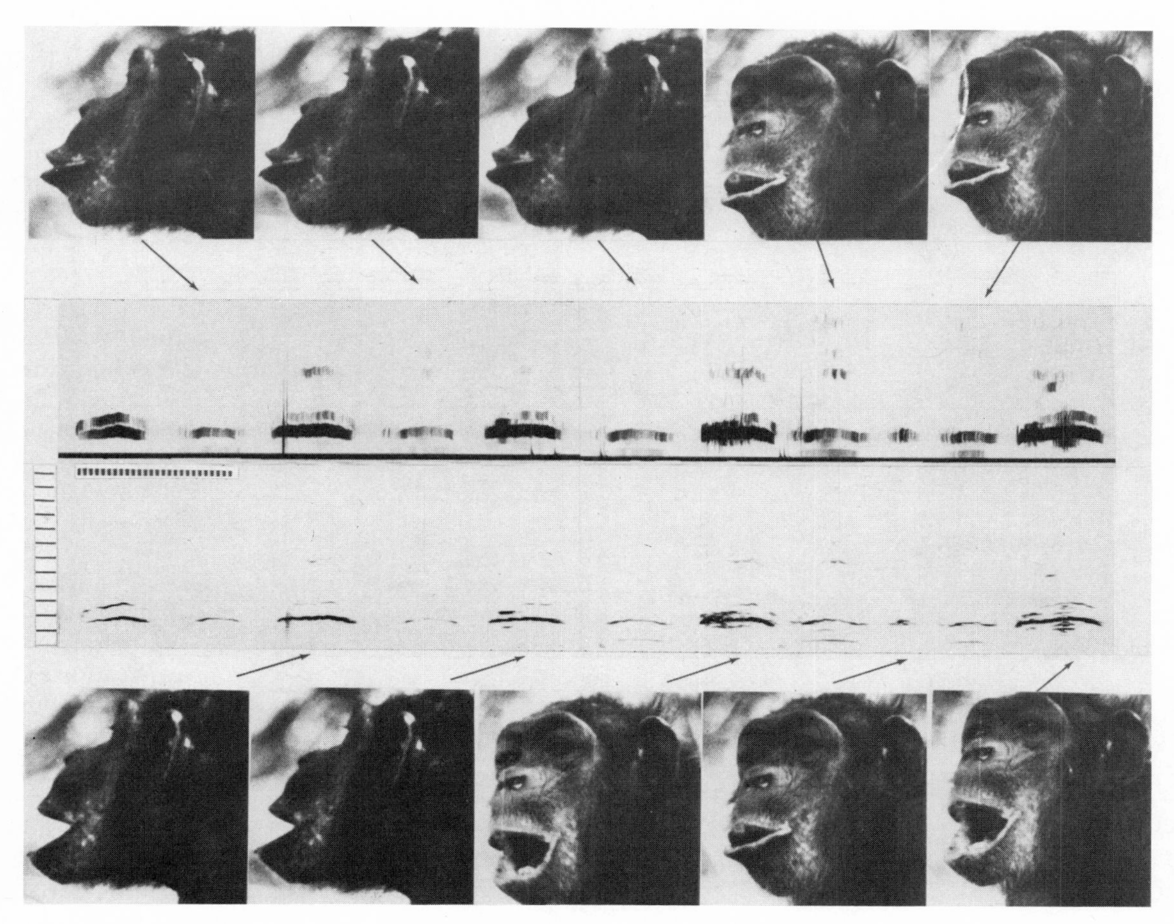

Fig. 1. Movie stills of an adult female chimpanzee, Flo, pant-hooting with a synchronized sound recording. The sound spectrograms show in wide-band (300 Hz) and narrow-band (40 Hz) displays the structure of pant-hooting with arrows indicating the points represented by each photograph. The pictures in the upper row illustrate the inhalation phase, and those in the bottom the exhalation phase, in which the mouth is usually opened more widely. The frequency scale indicates 500 Hz intervals. The time marker is 0.5 seconds.

tion sounds, the former usually below 500 Hz. The latter often have breathy overtones. Rare or absent at younger ages, pant-grunting first becomes evident in adolescence and its frequency increases with maturity. It was recorded in roughly equal numbers in males and females (males, 131; females, 154).

Pant-grunting is typically given by a subordinate approaching or being approached by a dominant animal. As proximity is reduced, the pant-grunting individual may crouch in the bobbing movement described by van Lawick-Goodall (1968a), or it may lose its nerve and flee, and the vocalization may grade into pant-screaming. Though not always successful, it seems to function as a submissive signal, at least

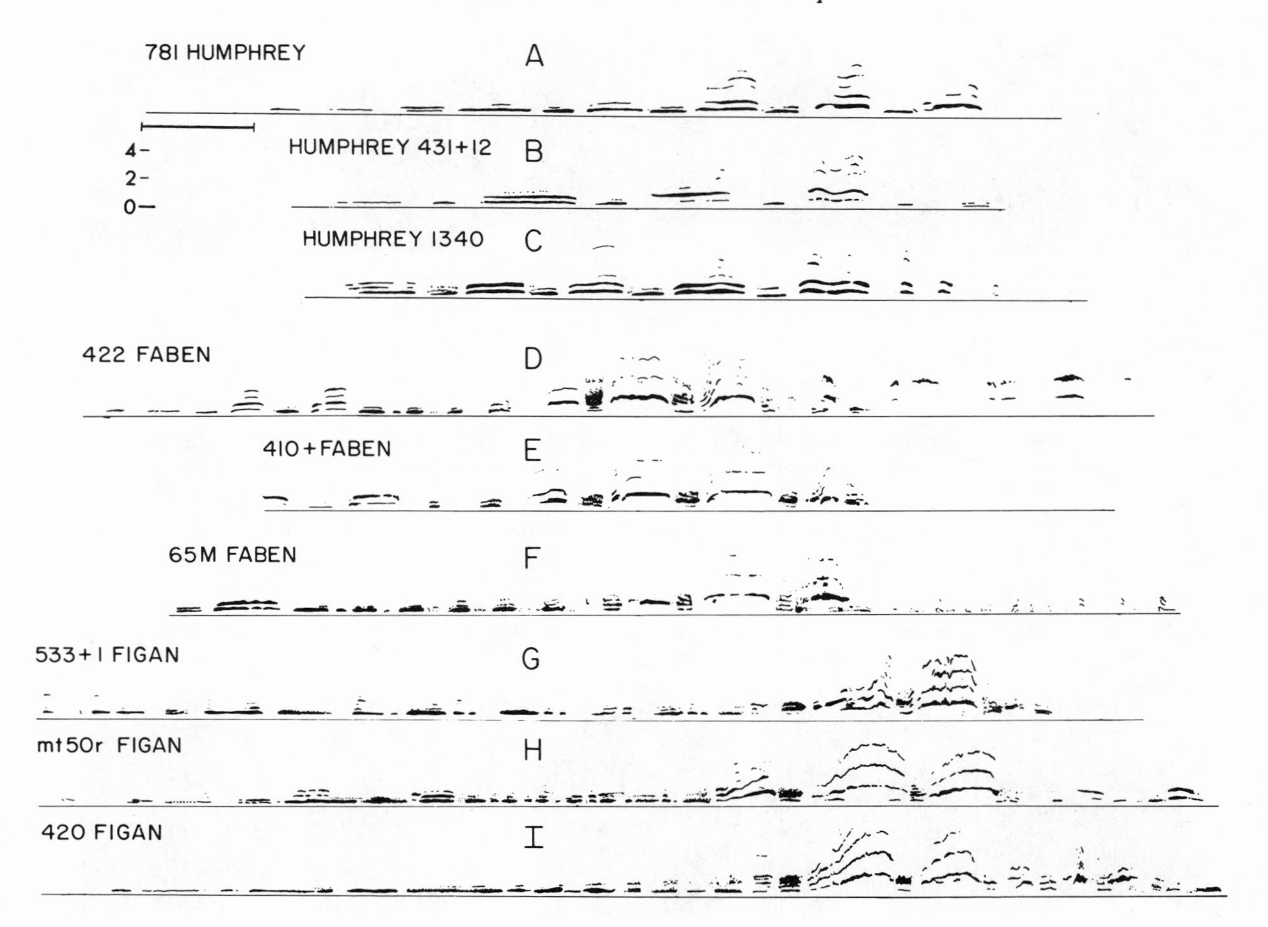

Fig. 2. Examples of the pant-hooting of three adult male chimpanzees, with three samples from each, illustrating the consistency of patterning in one animal and the ways in which the pant-hooting of individuals may differ. The frequency scale indicates kHz. The time marker is 0.5 seconds in duration.

on occasion facilitating the establishment of peaceful proximity with the dominant partner.

(3) *Laughter* was so named for its resemblance to human laughter, though it is by no means identical (Fig. 4). Van Hooff (1971) refers to it as ah-grunting. It is a highly variable sound with at least three distinct forms: steady exhaled laughter (A), pulsed exhaled laughter which has the quality of a chuckle (B), and wheezing laughter with phonation on both exhalation and inhalation (C). Type A elements may be given singly in the early stages of a laughter sequence or in rhythmical succession, each element lasting a quarter of a second or so. As a play bout proceeds and its vigor increases, type A may grade into type B, as a single exhalation becomes broken into a series of bursts followed by a pause for silent inhalation and then another burst of pulses or chuckles. This in turn may grade into the C form as tickling and biting reach their high-

Table 3

Frequency of use of chimpanzee vocalizations. II.

	♀M	♀O	♀A	♀J	♀I	♂1	♂2	♂A	♂J	♂I	Total
Pant-hoot	10.6 [12.1] 22.7	14.8 [17.0] 34.0	6.6 [7.5] 12.6	1.7 [2.0] 7.6	0.3 [0.3] 4.2	28.0 [31.3] 50.6	19.1 [21.9] 44.1	12.8 [14.7] 24.6	3.5 [4.0] 10.8	3.2 [3.7] 3.4	114.5
Pant-grunt	29.0 [13.3] 24.9	19.0 [8.7] 17.1	4.4 [2.0] 3.4	[0]	[0]	14.4 [6.6] 10.7	16.6 [7.6] 15.3	15.2 [7.0] 11.7	[0]	1.5 [0.7] 0.7	45.9
Laughter	0.5 [0.4] 0.7	[0]	0.6 [0.5] 0.8	14.2 [12.5] 47.7	3.2 [2.8] 38.9	2.3 [2.0] 3.2	3.4 [3.0] 6.0	0.8 [0.7] 1.2	5.7 [5.0] 13.5	69.5 [61.3] 57.0	88.2
Squeak	9.3 [6.2] 11.6	17.9 [12.0] 23.6	24.4 [16.3] 27.4	7.5 [5.0] 19.1	1.2 [0.8] 11.1	1.5 [1.0] 1.6	5.4 [3.6] 7.2	16.4 [11.0] 18.4	6.0 [4.0] 10.8	10.5 [7.0] 6.5	66.9
Scream	12.0 [6.6] 12.4	9.1 [5.0] 9.8	17.8 [9.8] 16.5	5.4 [3.0] 11.4	4.5 [2.5] 34.7	4.2 [2.3] 3.7	6.5 [3.6] 7.2	13.9 [7.7] 12.9	16.3 [9.0] 24.3	10.3 [5.7] 5.3	55.2
Whimper	8.4 [4.2] 7.9	2.6 [1.3] 2.6	9.2 [4.6] 7.7	5.0 [2.5] 9.5	1.0 [0.5] 6.9	3.2 [1.6] 2.6	3.4 [1.7] 3.4	8.8 [4.4] 7.4	15.9 [8.0] 21.6	42.6 [21.4] 19.9	50.2
Bark	8.1 [1.7] 3.2	9.5 [2.0] 3.9	50.0 [10.5] 17.6	4.8 [1.0] 3.8	1.4 [0.3] 4.2	8.1 [1.7] 2.7	5.2 [1.1] 2.1	4.8 [1.0] 1.7	[0]	8.1 [1.7] 1.6	21.0
Waa bark	12.6 [2.4] 4.5	6.8 [1.3] 2.6	18.4 [3.5] 5.9	1.1 [0.2] 0.8	[0]	5.8 [1.1] 1.8	7.9 [1.5] 3.0	26.3 [5.0] 8.4	10.5 [2.0] 5.4	10.5 [2.0] 1.9	19.0
Rough grunt	8.3 [1.2] 2.2	2.1 [0.3] 0.6	3.4 [0.5] 0.8	[0]	[0]	42.1 [6.1] 9.9	9.7 [1.4] 2.8	15.9 [2.3] 3.8	[0]	18.7 [2.7] 2.5	14.5
Pant	22.9 [3.6] 6.7	8.3 [1.3] 2.6	[0]	[0]	[0]	19.1 [3.0] 4.9	11.5 [1.8] 3.6	6.4 [1.0] 1.7	25.5 [4.0] 10.8	6.4 [1.0] 0.9	15.7
Grunt	6.7 [0.6] 1.1	14.4 [1.3] 2.6	16.7 [1.5] 2.5	[0]	[0]	14.4 [1.3] 2.1	17.8 [1.6] 3.2	18.9 [1.7] 2.8	11.1 [1.0] 2.7	[0]	9.0
Cough	14.3 [0.8] 1.5	[0]	35.7 [2.0] 3.4	[0]	[0]	37.5 [2.1] 3.4	1.8 [0.1] 0.2	5.4 [0.3] 0.5	[0]	5.4 [0.3] 0.3	5.6
Wraaa	10.0 [0.3] 0.6	23.3 [0.7] 1.4	[0]	[0]	[0]	46.7 [1.4] 2.3	10.0 [0.3] 0.6	10.0 [0.3] 0.5	[0]	[0]	3.0

Table 3 (continued)

	♀M	♀O	♀A	♀J	♀I	♂1	♂2	♂A	♂J	♂I	Total
Lip smack	[0]	[0]	18.6 [0.8] 1.3	[0]	[0]	7.0 [0.3] 0.5	11.6 [0.5] 1.0	62.8 [2.7] 4.5	[0]	[0]	4.3
Total	53.4	50.9	59.5	26.2	7.2	61.8	49.7	59.8	37.0	107.5	

NOTE: Usage by chimpanzees at Gombe National Park of thirteen vocal and one nonvocal sound per individual member of each age, sex, and status class. Of the three numbers in each cell the center one is the product of the number of vocalizations of that type divided by the number of individuals in that class in the study population. The other two are percentages, either of the total for that type of sound (top) or of the number of vocalizations by that class of animal (bottom).

est intensity. At this stage phonation on inhalation begins and the laughter assumes a labored wheezing quality, and it may be accompanied by efforts to ward off the tickler. When given rhythmically, the rate varies from about two to five breath cycles per second. The frequency characteristics vary widely. The maximum emphasized frequency is higher in young animals than in adults though a high-frequency, breathy component is often present irrespective of age.

Though recorded from all age classes other than nulliparous adult females, laughter is much more common in young animals than in adults. Males laughed more than females in the Gombe sample, and male infants may well be more outgoing and prone to engage in the vigorous play that generates laughter. However, this sample is probably biased by overrepresentation of laughter by one male infant (Flint) that spent much time in the study area with three siblings and the mother. All were prone to spend much time in play and many sound recordings were made. Not only did young animals make the biggest contribution to laughter, but it was also the most frequent vocalization recorded from infants of both sexes and from juvenile females, though it made a lesser contribution to the vocal output of juvenile males. The context for laughter was always some variety of play, usually with physical contact in the forms of tickling by the hand and biting with the teeth. On a significant number of occasions it was recorded in the absence of physical contact, such as when two animals chased each other around a bush in play. The closing scene in the film "Vocalizations of Wild Chimpanzees" (Marler and van Lawick-Goodall, 1971) illustrates such a chase in play between two infant males competing for possession of a palm flower. There is a close correlation between play and laughter, and laughter terminates without transitional forms when play is interrupted.

(4) *Squeak* is a short shrill call, one to two tenths of a second in duration, with a fundamental that is frequency-modulated, peaking in midcall at 0.5 to 1 kHz and then falling (Fig. 5). There is a rich array of four to eight or more harmonics often with significant energy up to 6 to 8 kHz. Squeaks are usually given in series at two to five per second, often grading into other sounds in the course of a sequence. Although squeaking was recorded from all age and sex classes, it was more frequent in females than in males in the main sample (female, 170; male, 94). It is used especially often by adolescent animals of both sexes, and in the Gombe study made up a high proportion of the average output of both (Table 3). There was a general tendency for frequent use by animals of intermediate age, which perhaps indicates a correlation with lability of dominance relations. Among males it is notably less common in older adults than in any

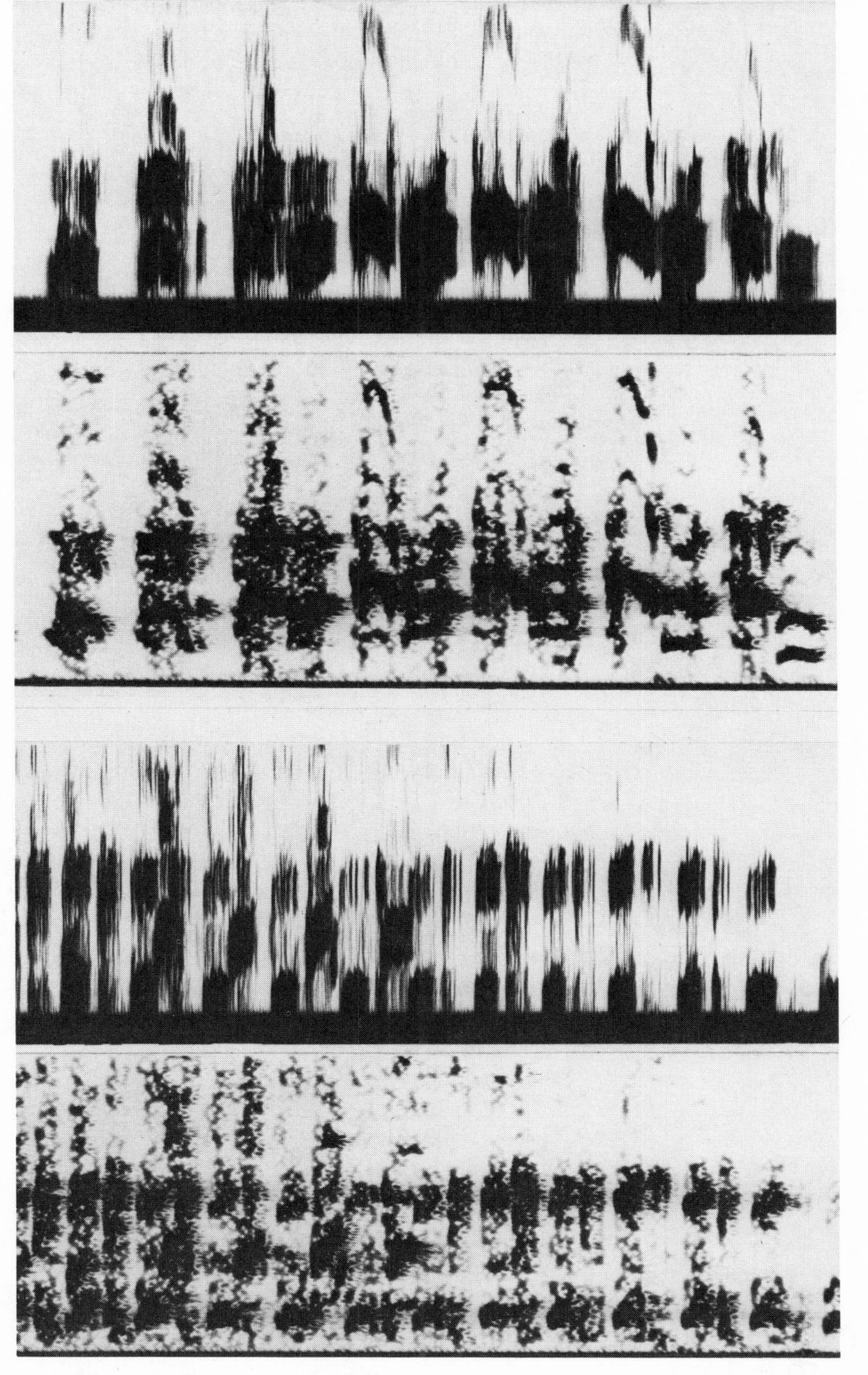

Fig. 3. Pant-grunting by a male and a female chimpanzee. At the top in both wide- and narrow-band displays is an example of pant-grunting by an adult male, Hugo. Below is an example from the adult female Flo. In each case the rapid alternation of inhalation and exhalation phases can be seen. The frequency markers indicate 500 Hz intervals. In both cases a segment about 2 seconds in duration is shown.

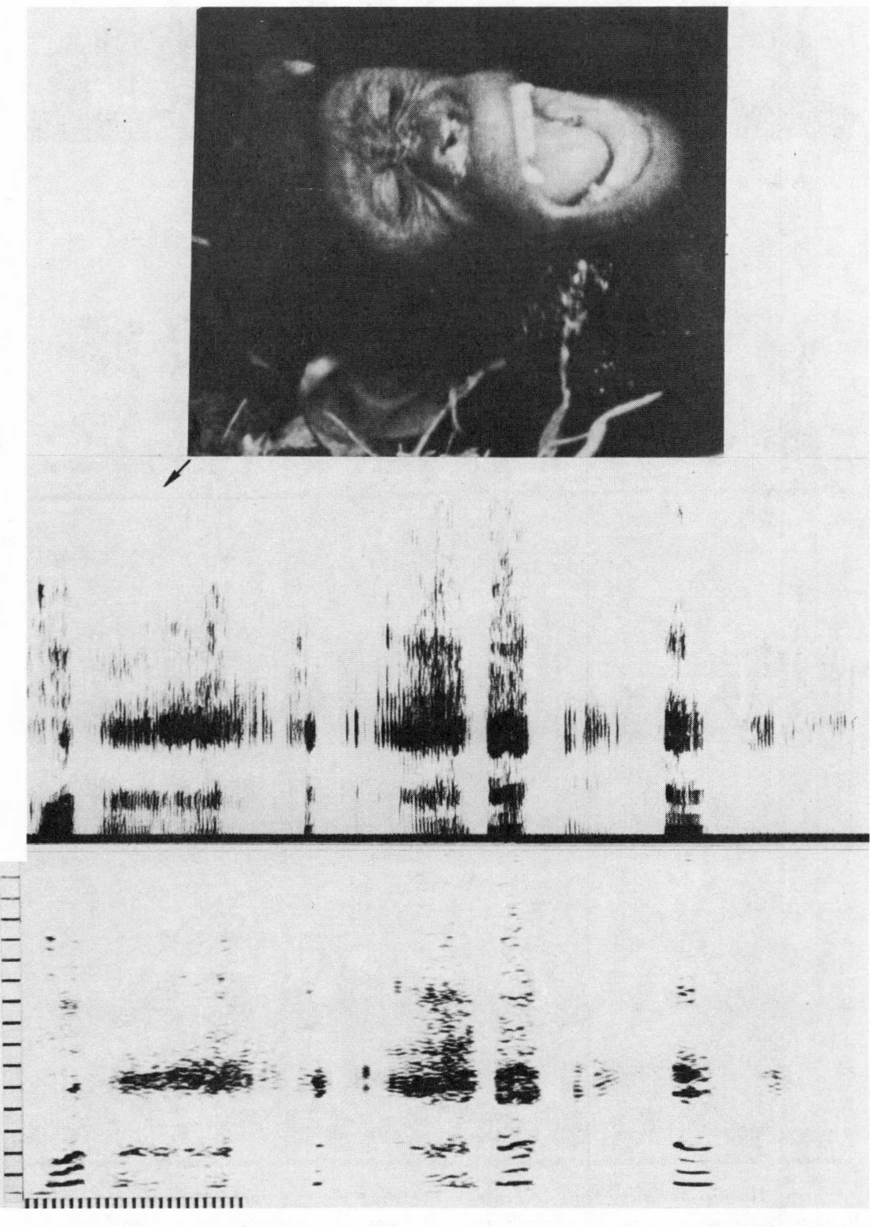

Fig. 4. An example of laughter from an infant male, Flint. The time of the movie still is indicated by the arrow. Both wide- (top) and narrow-band displays are shown. Scales as in Fig. 1.

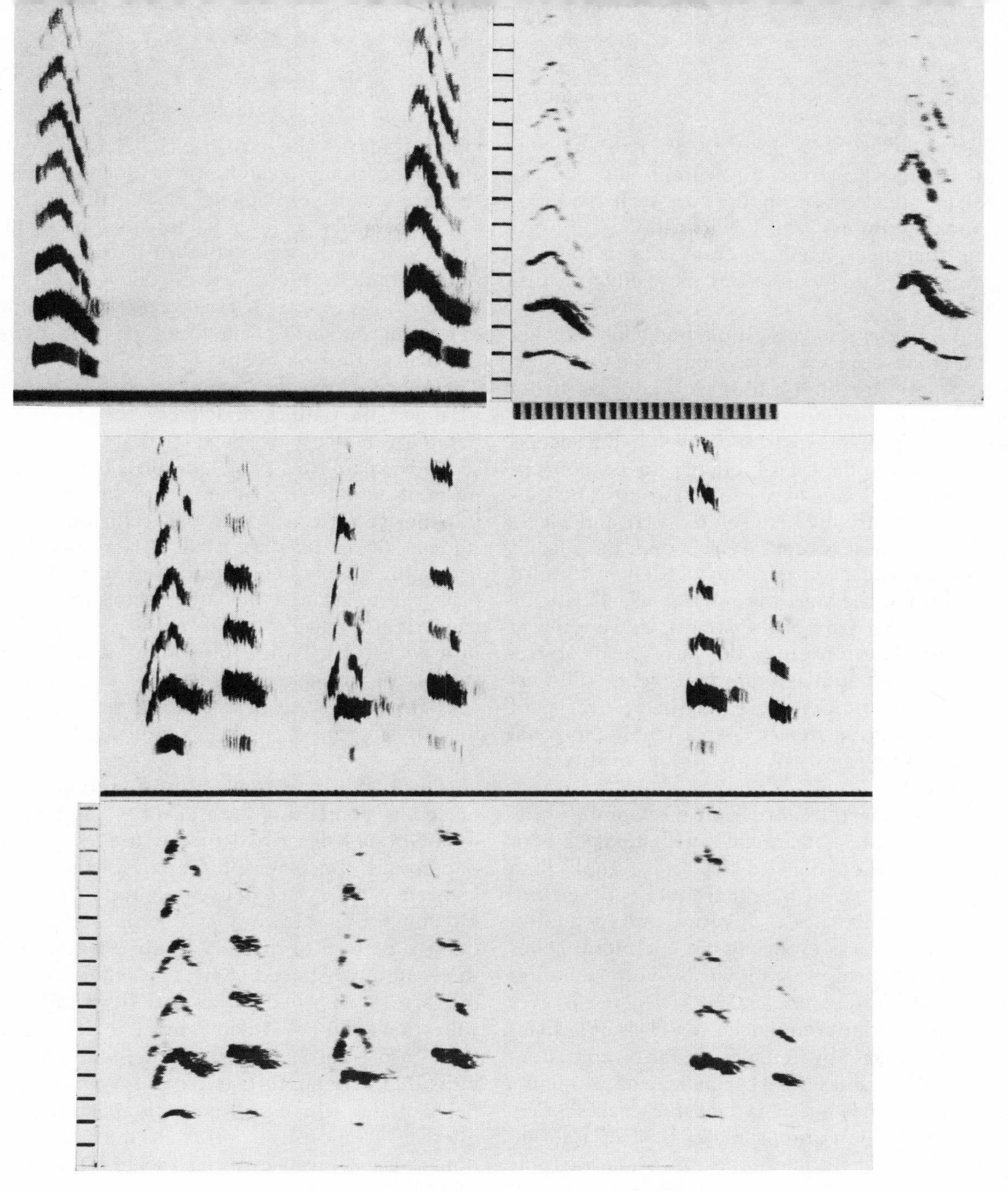

Fig. 5. Examples of squeaks from two adult female chimpanzees, Fifi (top) and Melissa (bottom). Scales as in Fig. 1. The narrow-band displays are on the right and at the bottom.

other age class. It is given by animals showing other signs of distress, for example after an aggressive attack or during the approach of a higher-ranking animal. Van Lawick-Goodall (1968a) notes that the touching or embracing of the subordinate evoked by squeaking may reduce its distress.

(5) *Scream* is a high-pitched, far-carrying sound that represents 9 percent of the total sample. There are intergrades with squeaks, from which it can be distinguished by the longer duration—more than a quarter or even half a second—and the tendency for energy to be concentrated in the lower harmonics (Fig. 6). The fundamental is usually between 0.5 kHz and 1 kHz or above, but it is often weaker than the second harmonic and even the third. After an upward inflection at the start the pitch is usually steady for the major part of the sound, descending at the end. The tone may be fairly pure, or frequency modulated at a rapid rate (e.g., 100 warbles per second), or there may be an overlay of noise of varying density. As a result the tonal quality of screams can vary widely in the same individual.

Screams were recorded in the Gombe study from all classes of animals, with females contributing rather more to the total sample than males (122:94). The overall frequency of screaming by adult females with young and by adolescent females was especially high. Among males the overall peak was shown by younger adult males and adolescent males. When the figures are converted to frequencies per class of individual, the usage by females is fairly consistent across age classes, apart from a peak in infants that is probably an artifact of the small sample. The distribution among males is more uneven, with the oldest adults screaming considerably less than the adolescent and juvenile males. Like squeaking, screaming is given by victims of social strife associated with dominance relations—hence the infrequent use by high-ranking adult males. It may evoke reassurance behavior in others or, according to van Lawick-Goodall (1968a), recruitment of help against an aggressor.

(6) *Whimper* is a soft, low-pitched sound about one tenth of a second in duration with a fundamental frequency usually between 200 and 350 Hz (Figs. 7 and 8). This is either a pure tone or a harmonic series, sometimes with only two or three overtones, sometimes a dozen or more. The spectral structure seems to correlate somewhat with the pattern of delivery, the purer form being given singly or spaced at least one or two seconds apart, the more complex forms being more often delivered in series at a rate of two to four per second. It was recorded for both sexes (female, 66; male, 109). All age classes used it at a roughly similar level with the exception of a higher rate in juvenile and infant males, for whom it also makes up a larger proportion of the utterances of a typical individual. The circumstances of whimpering are varied, and van Lawick-Goodall (1968a) lists three separate calls for this category occurring in rather different situations.

One call is given on hearing a strange sound or seeing something strange (Fig. 7). Van Lawick-Goodall described this as the "hoo" call, but sound spectrographic analysis reveals no consistent difference between it and other types of whimpering. However, we confirmed the distinctive pattern of delivery in this context of either single or spaced calls. The other forms of call are given when an animal begs for food, when a frightened infant clings to its mother (Fig. 8) or searches for a nipple, and when an individual is being threatened or is having a temper tantrum, which produces the most rapid sequences. Responses by others to the call include donating food, retrieving or suckling the infant, and offering reassurance gestures; or the call may be ignored.

(7) *Bark* is a loud sharp call that varies considerably in pitch (Fig. 9). The duration ranges

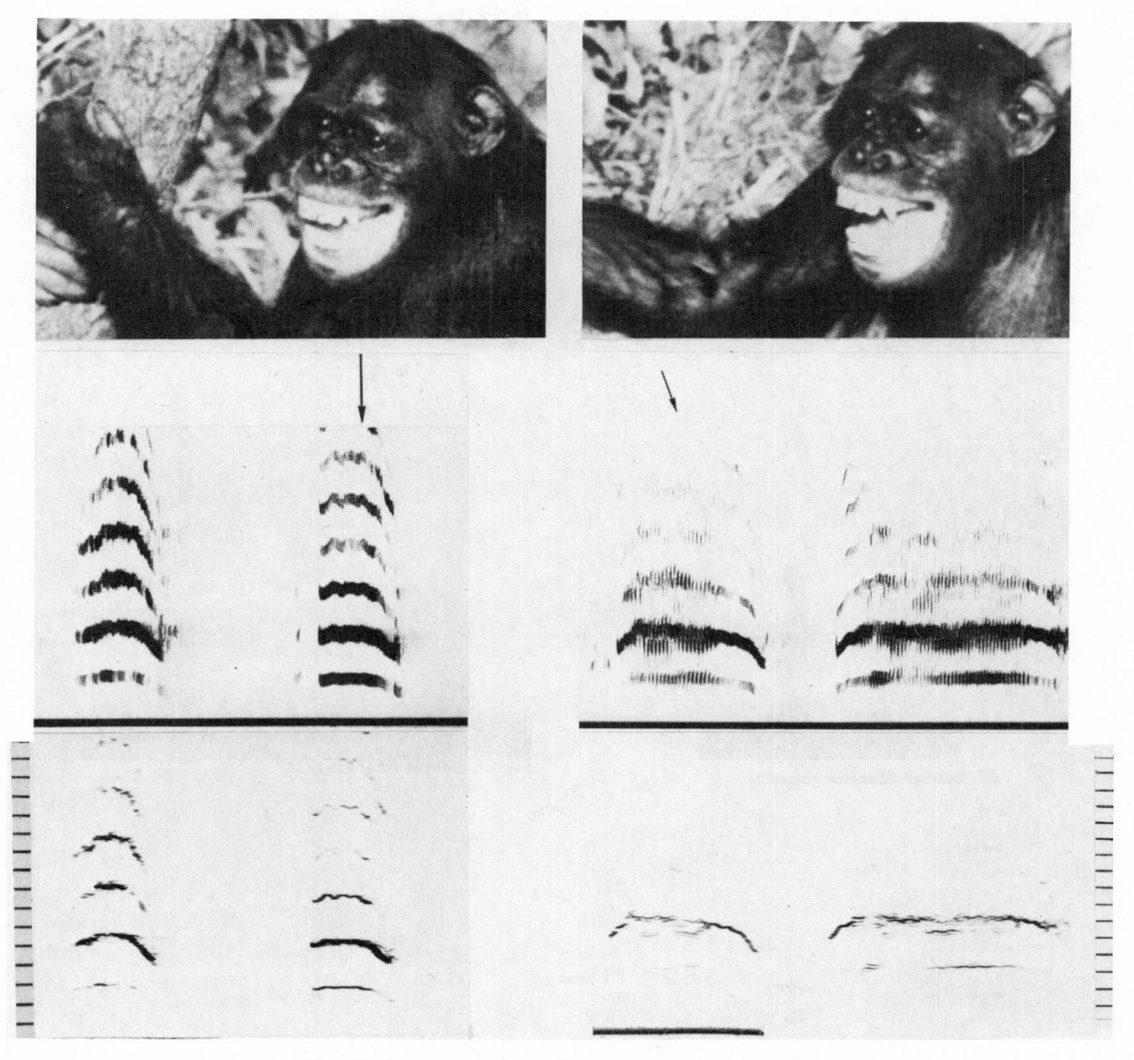

Fig. 6. Two samples of screaming by the same adult female chimpanzee, Pooch. Times of the movie stills are indicated by the arrows. Wide-band displays are above. Scales as in Fig. 1.

from .1 to .25 seconds with a fundamental between 300 Hz and 1 kHz. So far the description is very like that of a squeak. Barks differ from squeaks in their abrupt onset and their tendency to be noisy. This may take the form of a noise overlay or sudden breaks in pitch with a step change of .1 kHz to .5 kHz. The modulation of frequencies in the bark is often steeper than in

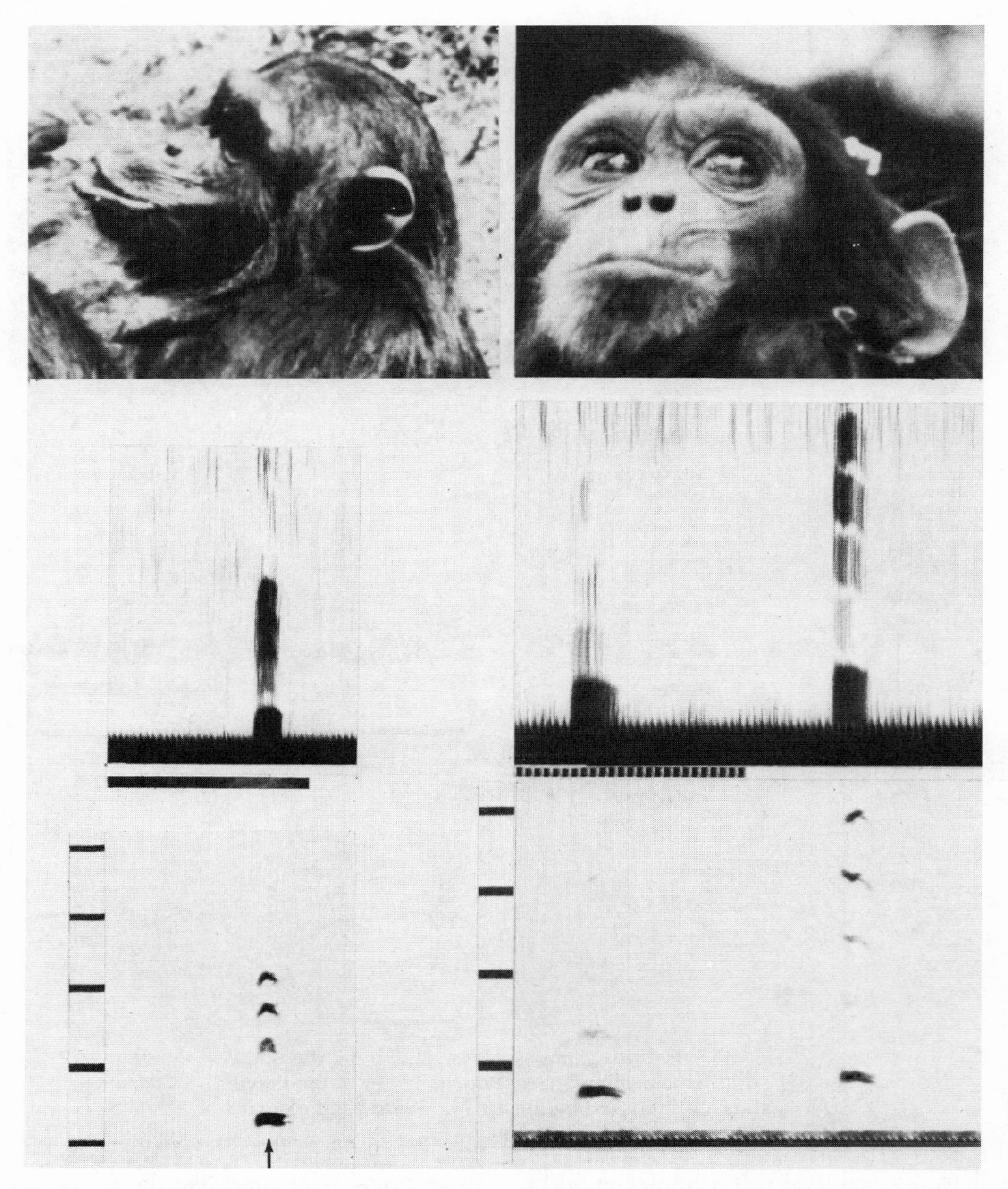

Fig. 7 Examples of whimpering by an adolescent female, Fifi, on the left, and by an infant male, Goblin, on the right. Wide-band displays are above. Timing of the one movie still is indicated by an arrow in the left example, taken during begging for food. The infant was apprehensive of a microphone brought too close. Scales as in Fig. 1.

Fig. 8. Two samples of whimpering by an adult male, Faben, given after glimpsing his reflection in a mirror. Precise timing of the single frame photographs is not indicated. Wide-band displays above, narrow-band below. Scales as in Fig. 1.

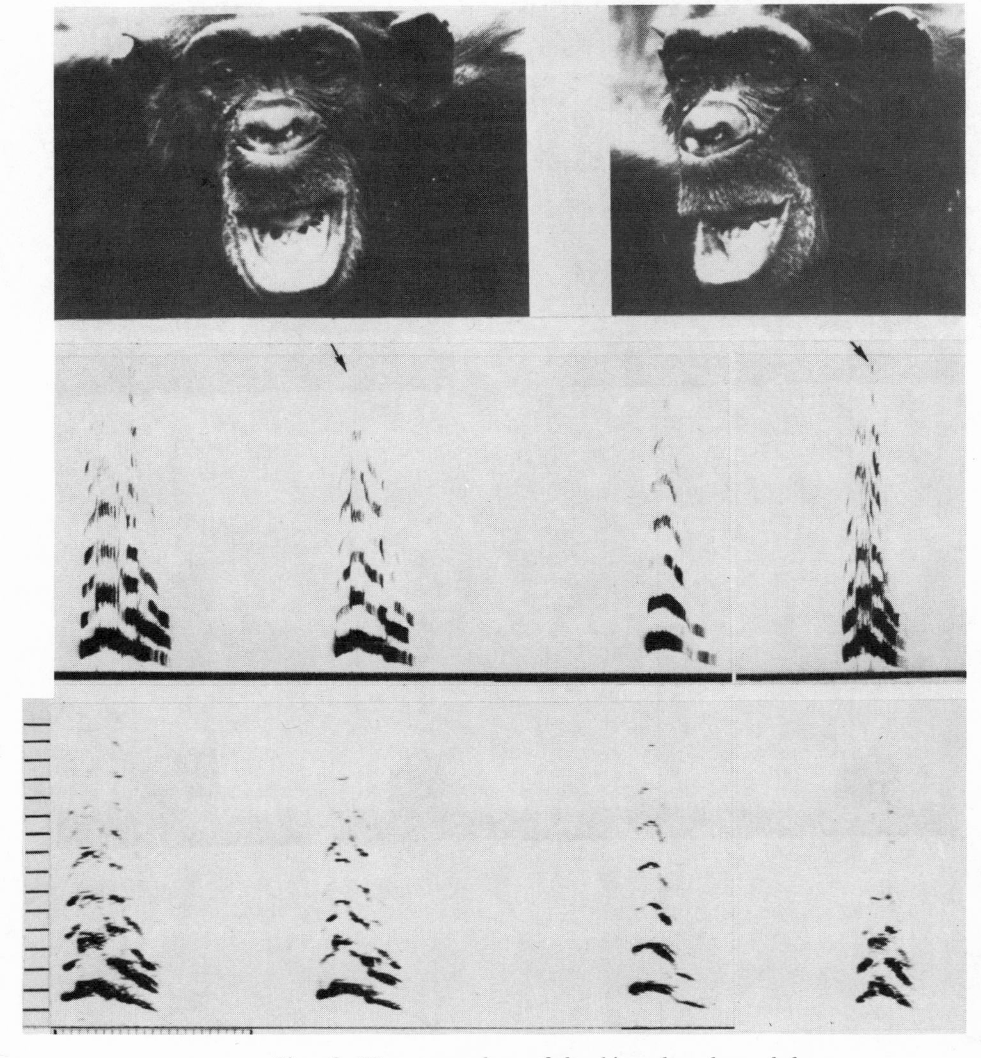

Fig. 9. Two samples of barking by the adult female Flo. Times of the movie stills are indicated by arrows. Wide-band displays above, narrow-band below. Scales as in Fig. 1.

the squeak. Finally barks differ from squeaks in that, although there may be a broad array of up to eight or more harmonics, the main energy is usually concentrated in the first one or two so that the call tends to be rather low-pitched. Barks grade into squeaks at the higher-pitched end and into grunts at the lower end.

While barks were not recorded for juvenile males, the numbers for other classes are so small that this lack is hardly significant. Most notable is the high frequency of barking by adolescent females. In the Gombe study barking makes up nearly 20 percent of the vocal output of a typical adolescent female. Although some of this excess results from the indulgence in frequent barking by a single animal (Gigi), adolescent females do seem to bark more than other classes. Barking is associated with annoyance or mild aggressiveness toward another individual, though it may also occur in generalized excitement or while threatening others at a distance. The response of others is not clear although, as van Lawick-Goodall (1968a) notes for the waa bark, others may be prone to look toward the caller and to bark themselves.

(8) *Waa bark.* Van Lawick-Goodall (1968a) does not distinguish between barks and waa barks, lumping both under the latter heading. However, with practice a distinction can be made both by sound spectrograph and by ear between the shorter bark and the waa-barks, which are more drawn out, often lasting half a second or longer. The mouth tends to be open more widely and conspicuous frequency breaks are consistently present (Fig. 10). There is an overall tendency for the pitch to rise and fall audibly during the course of the sound, but often with a heavy overlay of noise, usually restricted to the early or middle section of the call. Given singly or in series, the call often grades into another call such as a bark or a scream.

The waa bark was recorded in similar numbers from males and females (43:41) with no striking preponderance in any age class other than a rather frequent use by adolescents of both sexes (Tables 2 and 3). Some complementarity between the use of barks and waa barks by a typical individual adolescent male and female (Table 3) suggests that adolescent males are prone to use waa barks in situations in which adolescent females use barks instead. The use of the waa bark when animals are both aggressive and apprehensive, as when a subordinate threatens a dominant animal at a distance, is perhaps consistent with frequent use by an age class whose dominance relations are not yet fully crystallized. Van Hooff (1971) also interprets barking (shrill bark = bark + waa bark) as associated with balanced aggressive and submissive motivation.

(9) *Rough grunt.* A wide array of sounds ranging from squeaks to barks, with a pulsed grunt as the most typical form, is given by animals eating a favored food. By comparison with other grunt-like sounds of chimpanzees, the tempo of rough grunting is distinctively slow: a typical rate of delivery would be two to four calls per second. However, in a group, with the attendant social excitement commonly generated, the rate may be higher and the structure of the call may also vary. In adult animals that are relaxed and feeding, each call consists of up to about ten glottal pulses with a wide frequency range, given in rapid succession, producing a sound something like a groan (Fig. 11). In younger animals the higher frequency emphasis and pulse rate result in the "tonal grunt" of van Hooff (1971) and an intermediate form that is somewhat barklike (the barking sounds of van Lawick-Goodall, 1968a). With increased excitement the pulse rate may increase still further and the result is a squeak or a "shriek" (ibid.). When food is present in a complex social situation, many different forms are often intermingled. However, in the spaced calls of a relaxed individual the pulsatile structure of a call is usually evident; hence the name for the category as a whole.

Rough grunting was heard more from males than from females in the Gombe study (69:14). Within each sex it was used more frequently by adults, and in males the more frequent use of rough grunting by older adults was striking (Table 3). This call also made up a significant portion of the vocal output of a typical older adult male—almost 10 percent. The close association with preferred food suggests that it functions as

Fig. 10. Three examples of the waa bark given by the adult female Gigi. Timing of the movie still is indicated by an arrow. Wide-band displays above, narrow-band below. Scales as in Fig. 1.

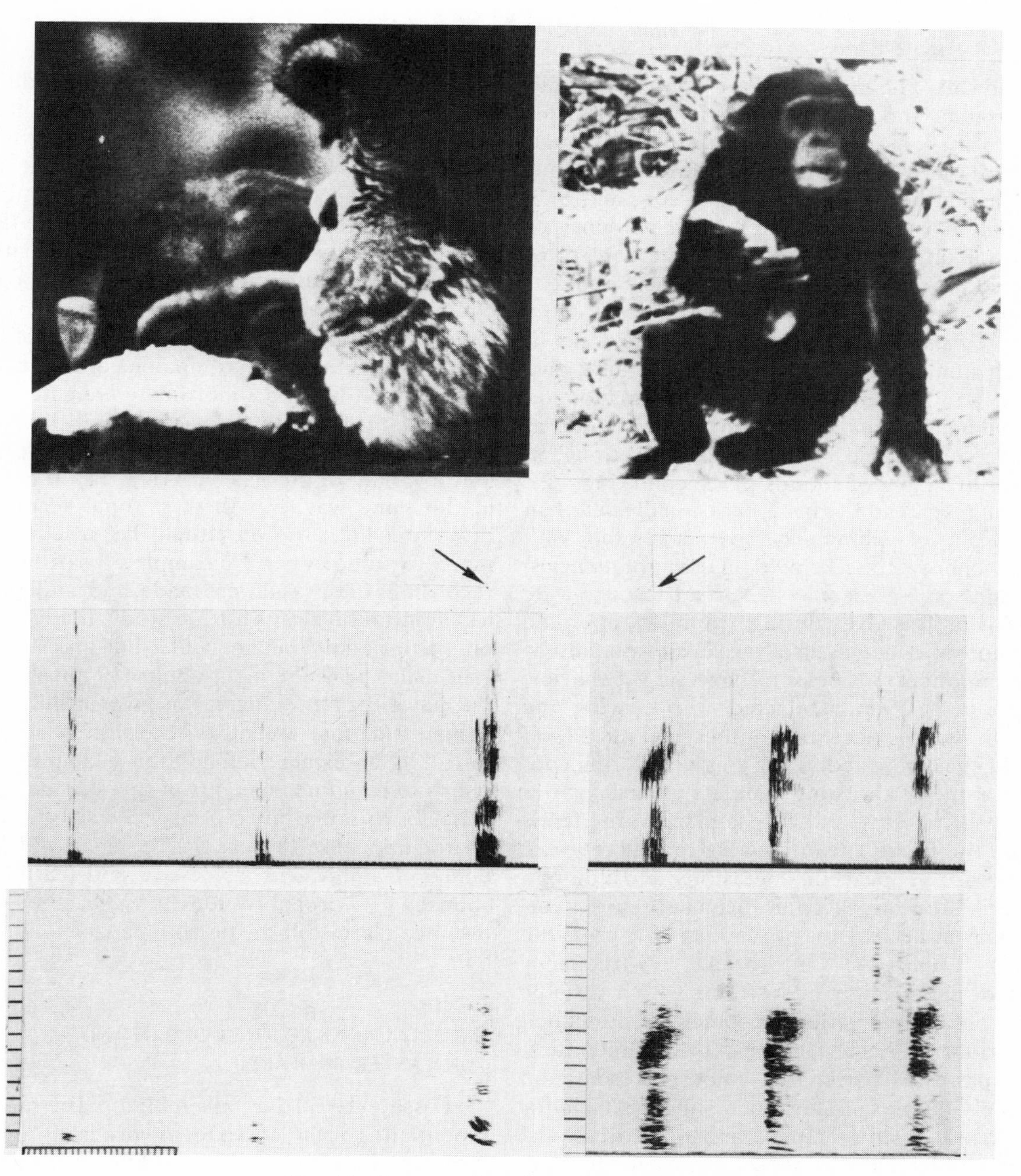

Fig. 11. Two samples of rough grunting given by an old adult male, David (left), and by an infant male, Flint (right). Arrows indicate timing of the stills. Wide-band displays above, narrow-band below. Scales as in Fig. 1.

a food call. The transfer of attention to the calling animal and the often hurried approach of others are consistent with the notion that rough grunting does in fact convey information about the presence of food. The call is soft, and only animals close by can hear it. The tendency of high-ranking adult males to assume control of limited food sources may explain their frequent use of this call.

(10) *Pant.* There are several situations in which animals of either sex, typically adults, give rapid panting sounds. There is phonation on both inhalation and exhalation, cycling rapidly at a rate comparable to that of pant-grunting, at about five pairs of sounds per second (Fig. 12). Voicing is rare, though it is occasionally heard on the expiratory phase. The sounds are soft, with most energy concentrated in low frequencies, though some noise may be spread across a wide spectrum. It is given during grooming, upon the reunion of two animals after separation, and as a component in peaceful greeting behavior, when it is often associated with placing the mouth on the face of another, as van Hooff (1971) has indicated. It is also given during copulation by the thrusting male. Its intensity varies in these contexts, and copulatory panting tends to be the loudest (van Lawick-Goodall, 1968a). These are all "peaceful" occasions, and although its communicative significance is obscure, panting may indicate a low probability of aggression.

(11) *Grunt.* A soft, brief, low-frequency sound, given singly or in trains with a variable rate, occurs in a variety of somewhat ill-defined situations (Fig. 12). The difficulty of identification may well have led to its underrepresentation in the Gombe sample, which suggests rarity or absence in young animals and only occasional use by older ones. It occurs during feeding behavior and during occasions of social excitement, but the contexts for this call are not yet well defined. It often grades into other calls as the situation changes, with transitions into rough grunting, barking, pant-grunting, and other calls. It is an acoustically simple sound produced on a single exhalation through a mouth that is closed or only slightly open.

(12) *Cough* is similar to grunting but unvoiced, produced by a rush of air out through a more open mouth, giving it a breathy tone (Fig. 13). It is usually given only once, in association with a brief threatening hand gesture to a companion. Adult males and adult and adolescent females account for almost all utterances of this call, which generally accompanies a mild confident threat toward a subordinate animal.

(13) *Wraaa.* Most rarely heard of all chimpanzee sounds in our study is this unusual call, probably a variant of the waa bark (Fig. 14). It begins in the same way, but the last tonal section is drawn out into a howl that may last a second or more, to judge by ear. No completely satisfactory recordings of this call were made, and a full physical description awaits further study. It was noted only from adult females and adult and adolescent males, with the older adults responsible for the majority of utterances. It is given in response to man and other predators or to strange animals (once to a distant buffalo, for example) and seems to combine elements of threat and alarm. It has been noted in response to discovery of a dead companion (Teleki, 1973), again with evidence of ambivalent aggressive and fearful responses. It probably functions as the only distance alarm call of the chimpanzee.

VOCALIZATIONS OF THE GORILLA AND CHIMPANZEE COMPARED

Fossey (1972) has assembled a list for the mountain gorilla of sixteen vocalizations, indicating where she concurs or differs with Schaller's (1963) classification. A summary of her final list is presented in Table 4. Brackets around three groups of calls suggest where categories might be further lumped, because the calls "could be grouped together on the bases of simi-

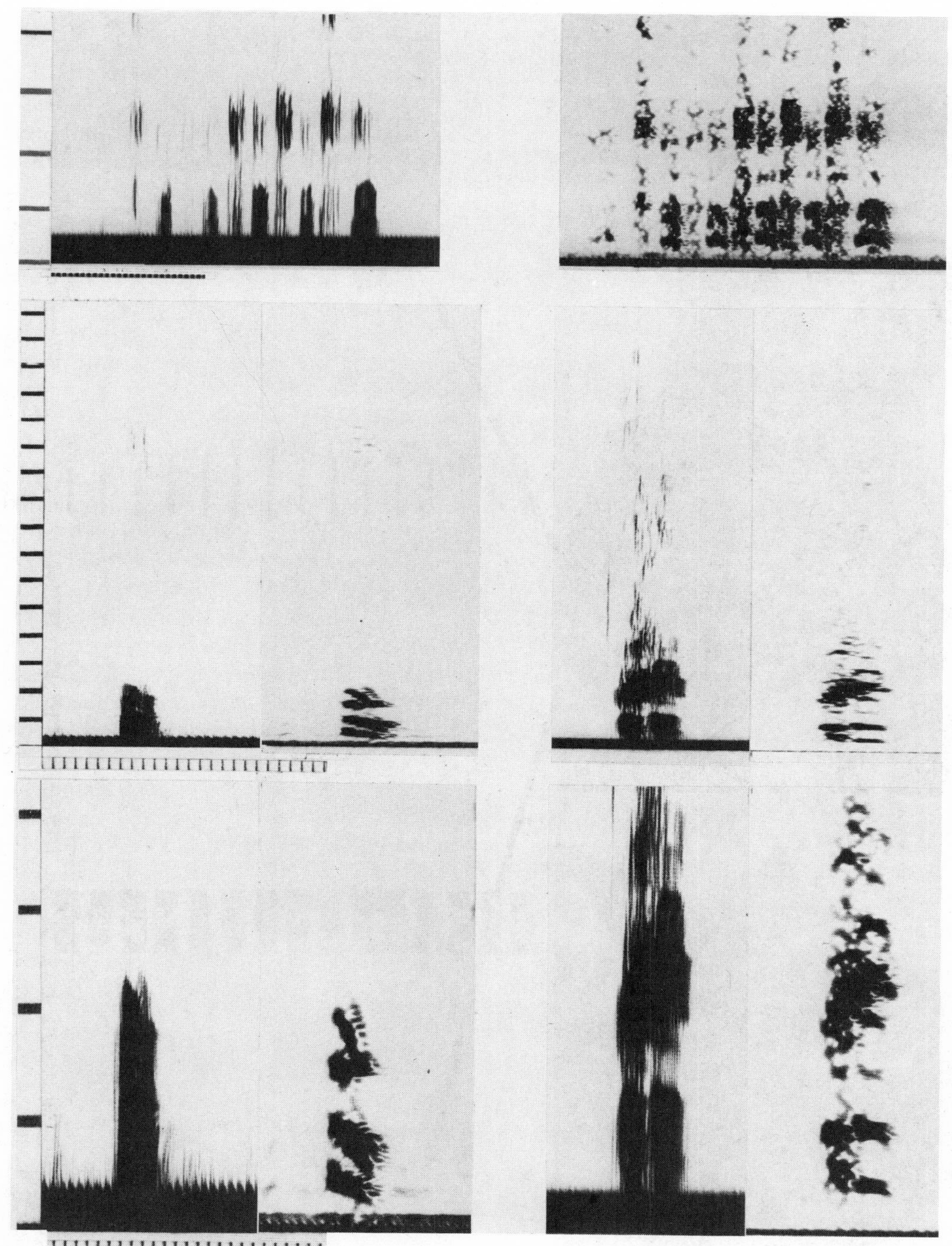

Fig. 12. An example of panting by an adult male, Hugo, during copulation (top). Below are examples of grunts given by an adult female, Pooch, on the left, and by an adult male, Worzle, on the right. Each is given with a compressed (middle) and an expanded frequency scale (bottom), with wide-band displays on the left and narrow-band on the right. Scales as in Fig. 1.

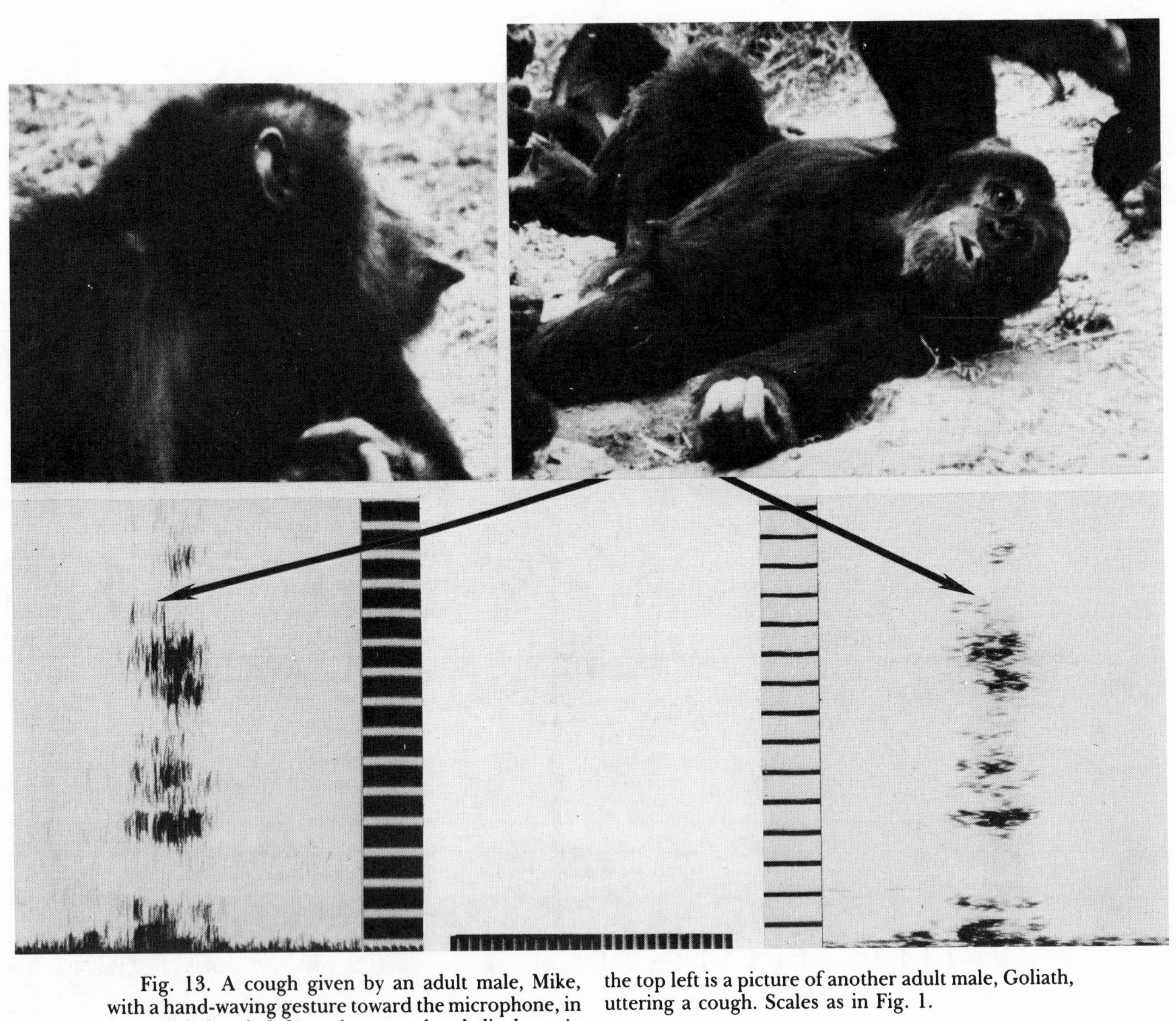

Fig. 13. A cough given by an adult male, Mike, with a hand-waving gesture toward the microphone, in both wide-band (left) and narrow-band displays. At the top left is a picture of another adult male, Goliath, uttering a cough. Scales as in Fig. 1.

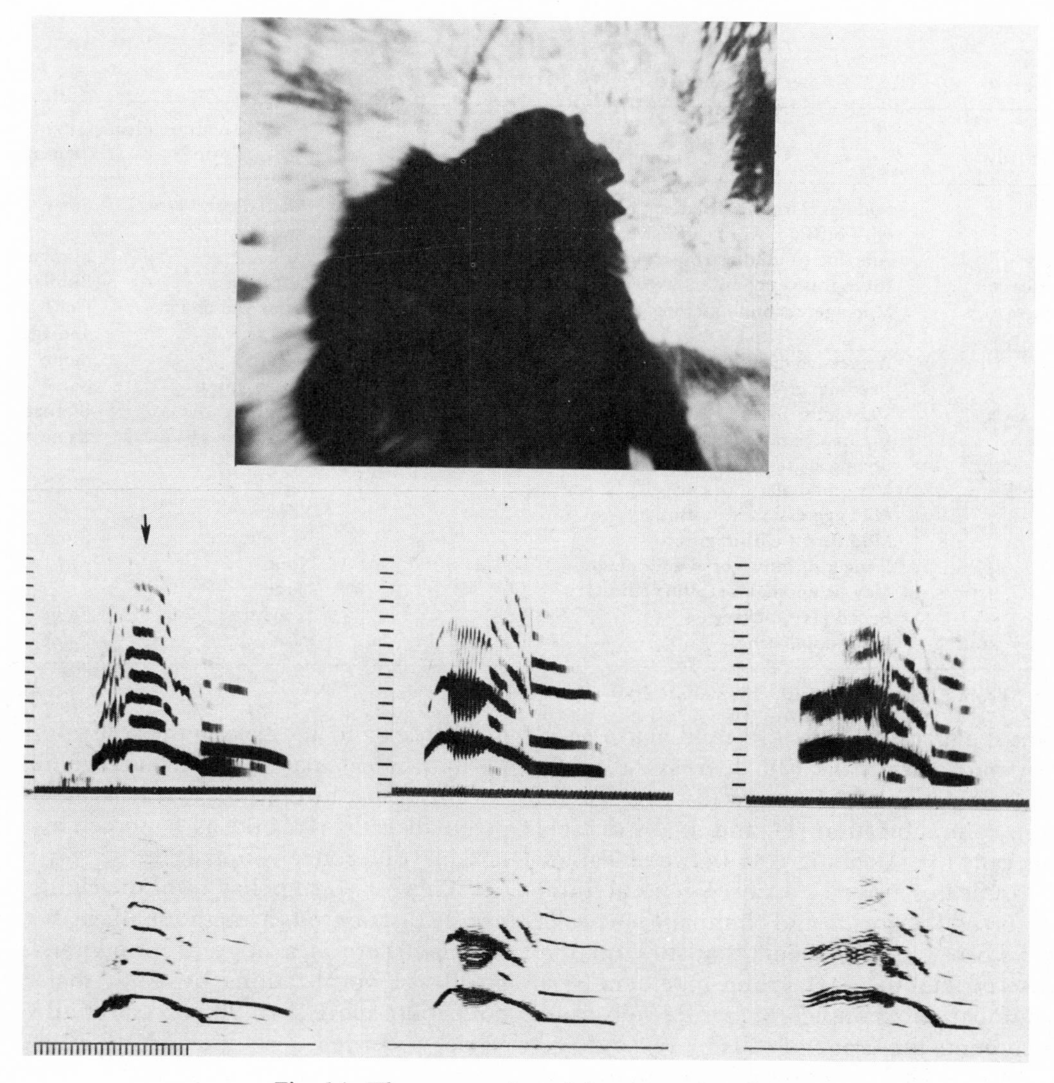

Fig. 14. Three examples of the wraaa given by an adult male, Hugo. The arrow indicates the timing of the movie still. Wide-band displays above, narrow-band below. Scales as in Fig. 1.

larities in their physical structure, a subjective impression of the sounds, the context in which they occurred and the responses they elicited." However, the sixteen types present a compromise between lumping and splitting similar to that used in chimpanzee classification, making it most useful for comparative purposes. Fossey divides the vocalizations into seven functional

Table 4

A catalog of gorilla vocalizations.

Vocalization	Circumstances	Equivalent chimpanzee call Type	Equivalent chimpanzee call Circumstances
Wraagh	Sudden alarming situation; loud noise; unexpected contact with buffalo, with aggressive elements.	Waa bark and Wraaa	Same
Hoot bark	Alerting to mild alarm; group movement initiation.	Bark ?	Similar
Hoot series	Intergroup encounters with aggressive component.	Pant-hoot	Similar
Pig grunt	Mild aggression in moving group.	Grunt and cough	General arousal and aggression
Scream	Aggressive disputes within group; copulating female.	Scream	Same
Belch	Feeding; group contentment.	Rough grunt	Same
Question bark	Very mild alarm or curiosity.	Bark ?	Similar
Cries	Infant separated, in difficulty.	Whimper and squeak	Same
Roar	Strong aggression of silverback ♂ to predator or other group.	None	
Hiccup bark	Very mild alarm or curiosity.	Bark ?	Similar
Growl	Mild aggression in stationary group.	None	
Pant series	Mild threat within group.	Pant-grunt?	Similar
Whine	Danger of injury or abandonment (?).	None	
Whinny	May be anomalous; ailing animal.	None	
Chuckles	Social play, tickling.	Laughter	Same
Copulatory panting	Male, copulating.	Pant	Same

Source: From Fossey, 1972; with chimpanzee equivalents.

groupings: aggressive calls (3), mild alarm calls (2), fear and alarm calls (2), distress calls (2), group coordination vocalizations (3), calls for intergroup communication (1), and finally miscellaneous calls (3). A comparison between Tables 1 and 4 indicates a similar array of general functions inferred for gorilla and chimpanzee vocalizations. Fossey also presents statistics on the response evoked in other group members by a large number of examples of gorilla calls, and she documents the usage of calls by sex and age class as well. The latter permits a number of illuminating comparisons with the chimpanzee.

Table 5 presents a digest of the numbers of vocalizations recorded by Fossey from four classes of individuals (Fossey, 1972, Tables 2 and 3, excluding data from unidentified individuals). The data on chimpanzee vocalizations have been rearranged to fit the same categories. The table also shows the number of individuals in each sex and age class in the study population, permitting calculation of the average number of vocalizations uttered by each class member, given directly (D) and as a percentage (E). A striking difference is apparent.

Whereas vocal behavior of the chimpanzee is evenly distributed throughout all classes of individuals, there is a huge preponderance in the gorilla of vocalizations by adult males, which contribute more than 90 percent of all vocal behavior recorded. This is true both as overall figures and as data reduced to the output of an average class member. Fossey's data also show that there is a further asymmetry in the vocal behavior of adult males, silverbacked males making a much greater contribution than blackbacked males (1490:73).

Using the physical descriptions of gorilla calls presented by Fossey (1972), we have attempted to compare them with chimpanzee vocalizations.

Table 5

A comparison of the frequency of gorilla and chimpanzee vocal behavior by sex and age classes.

	Gorilla*					Chimpanzee				
	Ad♂	**Ad♀**	**Juv.**	**Inf.**	**Total**	**Ad♂**	**Ad♀**	**Juv.**	**Inf.**	**Total**
Number of individuals in study population	31	36	20	19	106	15	12	10	7	44
Number of vocalizations of known individuals	1,583	77	7	33	1,700	831	633	506	343	2,313
Percentage of total vocal output	93	5	0.5	2		36	27	22	15	
Average vocal output per class member (B/A)	50.1	2.1	0.35	1.7	54.25	55.4	52.7	50.6	49	207.7
Percentage each class member makes of total	92	4	0.6	3		27	25	24	24	

*After Fossey, 1972; tables 2 and 3.

Table 4 indicates that all thirteen chimpanzee calls have plausible acoustical parallels in the gorilla. The circumstances in which they are given also seem to be similar, indicating a surprising degree of correspondence in vocal behavior of the two species.

Retabulation of Fossey's data in a format corresponding to that used for the chimpanzee data permits further comparisons (Tables 6 and 7). Having arranged the calls in order of the frequency of use in the general sample, we can see that the hoot series, which corresponds to the most frequently used chimpanzee vocalization, pant-hooting, ranks only third in the gorilla. However, the circumstances for the two species were not strictly comparable. Whereas the chimpanzee study population at Gombe National Park is fully habituated to human observers, Fossey's study population was less well habituated, especially in the early phases of study. She noted that while "alarm calls" (wraaghs) were the most frequent vocalization, it was probable that if more data had been collected from the same groups later, the alarm calls would have been relatively less frequent and the belch or other group coordination vocalizations would be the most frequent for nearly all age and sex classes (Fossey, 1972:40).

Working with the same study population as Fossey, but now better habituated, Harcourt (pers. comm.) suggests that the rank order of usage is now the following: (1) belch, (2) chuckles, (3) pig grunt, (4) hoot bark or hiccup bark (not distinguished), (5) hoot series, (6) whine and cries, (7) question barks, (8) screams, (9) pant series, (10) copulatory pants (not quantified in Fossey's sample), (11) growl, (12) wraagh, (13) roar, and (14) whinny. Harcourt expressed less confidence in the relative rankings of 4–14 than in that of the first three, and notes that chuckles might rank first on a duration measure though third on an onset measure.

The change in estimated rank of the gorilla wraagh from 1 to 12 is as Fossey had predicted. The frequent calling in the early study was presumably triggered by the observer. If Harcourt's data are confirmed quantitatively, then this becomes another case of close resemblance be-

Table 6

Frequency of use of gorilla vocalizations. I.

	Number of individuals of given age and sex in population						
	14	17	36	20	19		
Call	**♂ Silverback**	**♂ Blackback**	**♀**	**Juvenile**	**Infant**	**Total**	**Percentage**
Wraagh	92 [503] 34	7 [39] 42	0.7 [4] 5	0.1 [1] 14	[0]	547	32
Hoot bark	93 [408] 27	3 [15] 16	3 [13] 17	0.4 [2] 29	[0]	438	26
Hoot series	98 [274] 18	1 [3] 3	0.4 [1] 1	0.4 [1] 14	[0]	279	16
Pig grunt	89 [108] 7	6 [7] 8	5 [6] 8	[0]	[0]	121	7
Scream	56 [58] 4	8 [9] 10	31 [33] 43	2 [2] 29	4 [4] 12	106	6
Belch vocalization	79 [44] 3	18 [10] 11	4 [2] 3	[0]	[0]	56	3
Question bark	84 [41] 3	4 [2] 2	12 [6] 8	[0]	[0]	49	3
Cries	[0]	[0]	[0]	[0]	100 [24] 73	24	1
Roar	94 [17] 1	6 [1] 1	[0]	[0]	[0]	18	1
Hiccup bark	88 [15] 1	[0]	12 [2] 3	[0]	[0]	17	1
Growl	87 [13] 0.9	7 [1] 1	7 [1] 1	[0]	[0]	15	0.8
Pant series	23 [3] 0.2	[0]	69 [9] 12	8 [1] 14	[0]	13	0.7
Whine	11 [1] 0.06	44 [4] 4	[0]	[0]	44 [4] 12	9	0.5

Table 6 (continued)

	Number of individuals of given age and sex in population						
	14	17	36	20	19		
Call	**♂ Silverback**	**♂ Blackback**	**♀**	**Juvenile**	**Infant**	**Total**	**Percentage**
Whinny	71 [5] 0.3	29 [2] 2	[0]	[0]	[0]	7	0.4
Chuckles	[0]	[0]	[0]	[0]	100 [1] 3	1	0.06
Total	1,490	93	77	7	33	1,700	
Percentage	88	5	5	0.4	2		

NOTE: Of the three numbers in each cell, the center one is the number of vocalizations; the top one is the percentage of that type of vocalization; the bottom one is the percentage of vocalizations of that class of animals. (From Fossey, 1972: Table 3, excluding data from unidentified individuals.)

tween chimpanzee and gorilla. The estimated decline in rank of gorilla cries from 8 to 6 in frequency of use is perhaps another manifestation of the habituation process, as is the decline in ranking of screams from 5 to 8. The increased ranking of chuckles as estimated by Harcourt to 2 from 15 perhaps reflects the other side of the habituation coin, namely that certain activities would have been inhibited during observation of nervous animals, notably play behavior. This presumably became relatively more common as habituation proceeded.

Perhaps the most intriguing of these changes in relative frequency of use of gorilla calls is the estimated rank increase of the belch from 6 to 1, bringing the gorilla data into even more marked contrast with that for the chimpanzee, where the equivalent call, "rough grunting," ranked 9 in the overall Gombe sample. The situations that Fossey (1972) describes for gorilla belching include not only feeding, but also sunning, grooming, and play, with calling of one individual often evoking belching in several others—a broader array of contexts than those in which chimpanzee rough grunting occurs. It seems likely that this is a significant difference between the two species.

One consequence of the provisioning of the chimpanzee population during gathering of the Gombe sample is the presence of satiated animals, possibly favoring the occurrence of play (Wrangham, pers. comm.). Gathering of animals in the provisioning area may have made agemates more readily available for activities such as play. Furthermore, the presence in the camp area of the Flo family—an unusually large and coherent family group—may also have favored the occurrence of play around the camp. Thus chimpanzee laughter, a subject of special study, is surely overrepresented in the Gombe sample. It now seems conceivable that it may actually be less frequent than in the gorilla in normally dispersed chimpanzee populations.

There was probably an increase in aggressive activities in the Gombe Stream study population as a result of provisioning (Wrangham, 1974), and it is possible that the high ranking in fre-

Table 7

Frequency of use of gorilla vocalizations. II.

	Number of individuals of given age and sex in population					
	14	**17**	**36**	**20**	**19**	
Call	**♂ Silverback**	**♂ Blackback**	**♀**	**Juvenile**	**Infant**	**Total**
Wraagh	93.5 [35.9] 33.8	6.0 [2.3] 42.6	0.3 [0.1] 4.5	0.1 [0.05] 14.3	[0]	38.4
Hoot bark	95.4 [29.1] 27.4	3.0 [0.9] 16.7	1.3 [0.4] 18.2	0.3 [0.1] 28.6	[0]	30.5
Hoot series	98.5 [19.6] 18.4	1.0 [0.2] 3.7	0.1 [0.02] 0.9	0.3 [0.05] 14.3	[0]	19.9
Pig grunt	92.8 [7.7] 7.2	4.8 [0.4] 7.4	2.4 [0.2] 9.0	[0]	[0]	8.3
Scream	70.7 [4.1] 3.9	8.6 [0.5] 9.3	15.5 [0.9] 40.9	1.7 [0.1] 28.6	3.4 [0.2] 11.4	5.8
Belch vocalization	82.7 [3.1] 2.9	16.0 [0.6] 11.1	1.3 [0.05] 2.3	[0]	[0]	3.75
Question bark	90.6 [2.9] 2.7	3.1 [0.1] 1.8	6.3 [0.2] 9.0	[0]	[0]	3.2
Cries	[0]	[0]	[0]	[0]	100.0 [1.3] 74.3	1.3
Roar	96.0 [1.2] 1.1	4.0 [0.05] 0.9	[0]	[0]	[0]	1.25
Hiccup bark	95.7 [1.1] 1.0	[0]	4.3 [0.05] 2.3	[0]	[0]	1.15
Growl	92.8 [0.9] 0.8	5.2 [0.05] 0.9	2.1 [0.02] 0.9	[0]	[0]	0.97
Pant series	36.4 [0.2] 0.2	[0]	54.5 [0.3] 13.6	9.1 [0.05] 14.3	[0]	0.55
Whine	14.9 [0.07] 0.07	42.6 [0.2] 3.7	[0]	[0]	42.6 [0.2] 11.4	0.47

Table 7 (continued)

Call	♂ Silverback	♂ Blackback	♀	Juvenile	Infant	Total
	Number of individuals of given age and sex in population					
	14	17	36	20	19	
Whinny	80.0 [0.4] 0.4	20.0 [0.1] 1.9	[0]	[0]	[0]	0.50
Chuckles	[0]	[0]	[0]	[0]	100.0 [0.05] 2.9	0.05
Total	106.3	5.4	2.2	0.35	1.75	

NOTE: Data recalculated from Fossey, 1972, by dividing the number of vocalizations of each type from identified individuals by the number of individuals of that class in the study population. Of the three numbers in each cell, the center one is the number of vocalizations; the top one is the percentage of that type of vocalization; the bottom one is the percentage of vocalizations of that class of animal. (Fossey, 1972; tables 2 and 3.)

quency of use of pant-hooting is to some extent attributable to the high level of general arousal maintained in the camp area.

Although there is a clear and intriguing contrast in the ranking of the chimpanzee rough grunt and the gorilla belch in frequency of use, perhaps the most striking conclusion to be drawn from the data is again the surprising degree of correspondence between the two species in the rank order of use of corresponding calls.

The whimpers and squeaks of the chimpanzee probably correspond with the "cries" of the gorilla. Whereas cries are confined to infant gorillas, the corresponding chimpanzee calls are used by all ages and both sexes, even though infants still whimper more. This may be taken as a sign of the greater emotionality and expressiveness of the chimpanzee at all ages as compared with the gorilla. However, it may be noted that screams, which rank fifth in frequency of general usage for both species, are used by gorillas of all ages and both sexes, most frequently by silverbacked males. There is a contrast in the chimpanzee in that although all classes of individuals engage in screaming, it is less frequent in males than females and is notably infrequent in the older, higher-ranking males. By contrast silverbacked male gorillas scream the most frequently.

Thus, comparisons of temperament in the two species, which tend to emphasize the phlegmatic nature of the gorilla, should not overlook the remarkable vocal range of silverbacked males. If vocal output is indeed to be used as an index of temperament, then silverbacked males can hardly be viewed as any less expressive and emotional than chimpanzees, however phlegmatic other classes of individuals may appear. One wonders whether their restraint might not be correlated with the relative exuberance and assertiveness of the silverbacked male. Certainly the fact that silverbacks assume prime responsibility for intergroup spacing correlates with their high vocal output and with the existence of at least one vocalization that is unique to adult males (in addition to the roar, Fossey also mentions the whinny as restricted to adult males but notes that this may be an anomalous call associated with sickness). A territorial organization places heavy responsibilities on the class of individuals that must maintain it. More complex within-group organization in the chimpanzee perhaps helps to explain the lack of any vocalizations that are unique to adult males (see below, p.1027).

VOCALIZATIONS OF THE ORANG-UTAN

The social life of the orang-utan, long considered the most mysterious of the apes, is becoming known from the research of Harrisson (1960), Horr (1972), Rodman (1973), Mackinnon (1974, 1975), and Galdikas-Brindamour (1975). In a preliminary account of vocalizations Mackinnon lists sixteen different types and illustrates some of them with sound spectrograms. While the data permit no detailed comparison with the chimpanzee and gorilla, some comments can be made. It can be seen from Table 8, which lists the vocalizations and a summary of the circumstances in which they are given, that vocalization occurs in much the same general situations as it does in the other two apes. According to Mackinnon, some sounds are given on exhalation, others on inhalation, the latter perhaps occurring more frequently than in other species. One interesting sound is the kiss squeak, associated with a sharp intake of air through pursed trumpet-lips. In one population in Sumatra "animals frequently held the knuckles or back of the hand to the lips while making this noise and this had the affect of deepening the tone." Most distinctive of all is the "long call" of adult, aggressively dominant males. This call consists of a long train of low-pitched calls which the inflated laryngeal pouch imbues with a rich deep tone, the series lasting from one to three minutes. There is a bubbly introduction building up to a climax of roars, then trailing off into a series of sighs. The number of call units varies from twenty-five to fifty, with differences between Sumatran and Bornean populations in length of the units and duration of the series.

Table 8

A catalog of orang-utan vocalizations.

Vocalization	Circumstances
1. Kiss squeak (and wrist kiss)	Both sexes, being chased or responding to observer.
2. Grumph	As (1).
3. Gorkum	Moderate intimidation display call.
4. Lork	With intense threat, usually adult female.
5. Raspberry	Both sexes, being chased or observer response.
6. Ahoor	Threat, usually adult male.
7. Bark	Threat or warning, adult male.
8. Complex calls	With vigorous intimidation display.
9. Chomp	Associated with (1) and (2).
10. Play grunts	In intense play.
11. Hoots and whimpers	Frightened infant.
12. Fear squeak	Frightened young.
13. Crying and screaming	Frightened or pained young and also adult females.
14. Frustration screams	Captive juvenile when food withheld.
15. Mating cries	Copulating female.
16. Long calls	Dominant adult male advertisement or disturbance, with aggressive components.

Source: From Mackinnon, 1974.

Mackinnon says that adult males give long calls irregularly, sometimes several in one day, sometimes none over several days. Evidently this call is given less frequently than chimpanzee pant-hooting or gorilla pant series. In Sumatra the calling peaks sharply in the early morning, and Mackinnon points out an intriguing complementarity with the loud calling of other sympatric species, each with species-specific timing.

According to Mackinnon, adult male long calls are a major vehicle for long-range interaction between orang-utans, sometimes occurring spontaneously, other times triggered by sudden noises such as a tree falling, a branch breaking, or most often by the calling of other orang-utans. Behavior accompanying calling such as hair erection, laryngeal pouch inflation, rocking, and branch shaking—also associated with aggressive display—suggest a corresponding motivation for calling. Males may react similarly to calling by rushing and branch shaking or calling in return, though at times males seem to become silent or to withdraw from another's calling. Mackinnon concludes that long calls are important for spacing out adult males; as a consequence face to face encounters between them are rare.

Although males are highly competitive, the resource at issue is not clear—food, space, or females. There are suggestions that the long call attracts sexually receptive females (Mackinnon, 1974; Horr, 1972), though females may withdraw and hide in response to calling and may sometimes ignore it In any case, these observations suggest that the male long call functions to coordinate both intragroup and intergroup interactions. The initiative taken by adult male orang-utans in this regard is reflected in the extreme sexual dimorphism of this species, both in size and external morphology, and in the vocal repertoire, the long call being unique to the adult male. In this respect the orang-utan is closer to the gorilla than to the chimpanzee or the gibbons.

VOCALIZATION OF GIBBONS

Despite the contrasts in social organization between gibbons and other apes, gibbon vocal repertoires are similar in size to those of the other apes. The most complete descriptions of vocalizations by wild gibbons are for the Kloss's gibbon, and white-handed gibbon (Tables 9 and 10), and the siamang (Chivers, in press). Ellefson's list of twelve vocalizations compiled during sixteen months of detailed observation of four white-handed gibbon families may be more complete than our list of ten Kloss's gibbon vocalizations, which is based on a fourteen-week study of thirteen family groups. Chivers (in press) describes eight siamang vocalizations after seventeen months' study of two groups. Boutan (1913) describes fifteen vocalizations used by a white-cheeked gibbon that he kept as a pet for more than five years, and he notes that he heard no vocalizations by other wild or captive members of this species that were not used by his captive specimen. We can conclude for the present that the vocal repertoires of gibbons include something in the vicinity of eight to fifteen vocalizations, possibly with minor differences among species. The most frequently occurring vocalizations in the repertoire of Kloss's gibbon are illustrated in Figs. 15, 16, and 17.

All classes of gibbon vocalizations are used by both sexes and, as might be predicted from the intrasexual nature of gibbon territorial aggression, sexual dimorphism occurs only in the songs.

Similarly, most vocalizations are not confined to particular age classes, though in the siamang (Chivers, in press), Kloss's gibbon, and the white-handed gibbon one vocalization is given only by juveniles (Tables 9 and 10). Although primarily the prerogative of adults, spacing calls are used even by young gibbons still with their parent (Chivers, 1972; Ellefson, 1974). However, such singing by young gibbons might in

Table 9

A catalog of Kloss's gibbon vocalizations.

Vocalization	Circumstances
1. Hoo	Langur or intergroup gibbon encounters; meeting a (human) predator.
2. Howl	Same.
3. Whup	By females before or after a song or both; preceding a scream (below); occasionally by males before a postdawn bout of singing; occasionally upon initial detection of a (human) predator.
4. Sirening	Sirening and alarm trills occur only when a (human) predator is encountered. Sirening normally precedes alarm trills.
5. Alarm trill	Same.
6. Whistles (one- and two-syllables)	Precede bouts of male predawn singing; uttered by both sexes when an undisturbed group is traveling leisurely or foraging; occasionally upon first detection of a (human) predator; by males during intervals between successive songs of their mates. In the last circumstance whistles usually or always are one-syllabled but in other circumstances they may be either one- or two-syllabled.
7. Songs	Produced by both sexes but with pronounced sexual dimorphism. Song begins (1) spontaneously, (2) upon hearing another of the same sex singing, or (3) in the case of females, when meeting a female neighbor upon a shared territorial boundary.
8. Whoo	Produced only by females, normally just prior to beginning a bout of singing, but occasionally whoo's occur without singing.
9. Quivering squeal	Observed only one occasion, when two males were on the ground fighting. Presumed due to its plaintive quality to have been uttered by the submissive male.
10. Whistle-howl	Heard once from adult females of two groups while the adult males of the groups were fighting.
11. Loud, prolonged squeal	Uttered by juveniles, normally when separated from their mothers but occasionally when carried by mothers.

NOTE: This list is thought to be less complete than that for the white-handed gibbon. All categories are used by both sexes, other than the whoo, a feature of female singing. (From Tenaza, unpubl.)

some cases be inhibited in the presence of an adult of the same sex (Tenaza, 1976).

If long-distance vocalizations are to assist in maintaining of spatial organizations based on established social relationships among individuals and groups, as apparently they do, they will be more efficient if they identify vocalizing individuals to others. Countersinging interactions of male Kloss's gibbons occur between males separated by distances of 150–500 m (Tenaza, 1976) and songs of neighboring males predictably show a high degree of individuality (Fig. 18).

Evidence from field studies shows that gibbon songs serve primarily as intergroup rather than intragroup signals, inducing close neighbors of the same sex to sing (Fig. 19). Furthermore, about half of all singing bouts for both sexes of Kloss's gibbons are by individuals engaged in dyadic countersinging with adjacent neighbors (Tenaza, 1976). White-handed gibbons (Carpenter, 1964; Ellefson, 1974) and siamangs (Chivers, in press) behave in a basically similar fashion, though in these species, unlike *H. klossii,* bisexual choruses also occur.

Table 10
A catalog of white-handed gibbon vocalizations.

Vocalization	Circumstances	Comparable Kloss's gibbon vocalization	
1. Hoo	Short distance intragroup separation in dense vegetation; mild upset due to presence of a human observer as the gibbons hide.	Hoo	No attempt was made to define the various hoo's of Kloss's gibbons.
2. Glug-hoo	Initial sight of preferred food. Similar to hoo but "throatier, more mezzo."	Hoo (?)	
3. Hoo-sigh	Emitted by an animal separated from the group (usually an immature) or by all group members when disturbed by an observer.	Hoo (?)	
4. Grunt-squeal	Emitted by young animals during contact play ("middle pitch range").	? (play not observed)	
5. Squeak-grimace	"Given by subordinate decreasing distance toward a dominant" ("high-pitched").	? (comparable situation not observed)	
6. Hoo-ooo	"Animal about to move aggressively toward a co-dominant or subordinate animal at food. It is a threat. . . ."	Howl (?)	
7. Grunt-whine	Emitted a few seconds prior to the end of a copulation.	? (copulation not observed)	
8. Morning calls	Emitted spontaneously, upon hearing morning calls of gibbons in other groups, or upon intergroup meeting upon a territorial boundary; sometimes upon detecting a potential predator (e.g., a tiger or a human). Given by both sexes but sexually dimorphic.	Songs	
9. Conflict hoo	Adult males in intergroup encounters upon shared territorial boundaries. Similar to male morning calls (songs) but "temporal patterning and varying intensity of conflict hoo's is unique."	Howl (?) The howl is softer than the conflict hoo but occurs (with hoo's) when males meet upon territorial boundaries.	
10. Screech	"an animal in danger of being caught during an agonistic chase. . . the loud, shrill end point of a squealing continuum. . . ."	Quivering squeal (?)	
11. Bark-hoot	Detecting a potential predator (human).	Whup	

Source: From Ellefson, 1967.

When an individual or a pair of gibbons breaks silence with its song or duet it usually stimulates others to begin singing, and a chorus of temporally overlapping bouts of singing by two or more individuals or pairs of gibbons develops. In Kloss's gibbon 96–97 percent of song bouts by both sexes occur in such choruses, while only the remaining 3–4 percent are produced by gibbons singing alone (Tenaza, 1976).

With information presently available from wild populations of gibbons we can distinguish three kinds of choruses, all based upon sex of participants: Choruses consisting entirely of males singing; choruses consisting entirely of females singing; and choruses consisting of duets sung by mated pairs of gibbons. Interspecific differences in the occurrence and co-occurrence of these chorus types are summarized in Table 11.

All-female choruses occur only in Kloss's gibbon and are confined to the four hours following sunrise after the gibbons have left their sleeping trees for the day. Choruses consisting entirely of males singing occur in the white-handed gibbon (Carpenter, 1964; Ellefson, 1974; Chivers, 1972), the Sumatran dark-handed gibbon (W. Wilson and C. Wilson, pers. comm.), the Bornean gibbon (N. Fittinghoff, P. Rodman, and D. Lindburg, pers. comm.), and Kloss's gibbon (Tenaza, 1976). In all but the white-handed gibbon, all-male choruses typically begin before dawn while the gibbons still are in their sleeping trees. In the white-handed gibbon they typically begin later in the morning, after a period of foraging (Carpenter, 1964; Chivers, 1972; Ellefson, 1974), though males occasionally sing from their sleeping trees just before dawn.

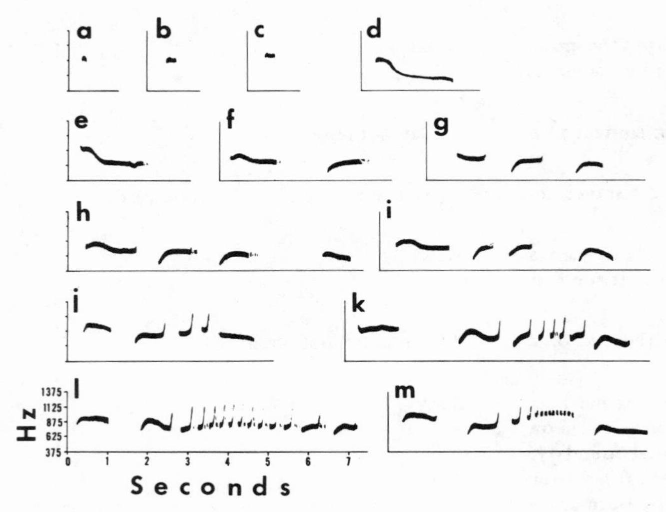

Fig. 15. Vocalizations of Kloss's gibbon. I. Male songs and whistles. a-c are examples of whistles that precede male bouts of singing. d-m are songs. Males normally begin their song bouts with one- or two-syllable songs (d-f), which gradually, over a period of about twenty-five minutes on the average, are progressively elaborated (g-k) until fully developed songs begin occurring (l-m). The songs presented here were selected from the output of several different males.

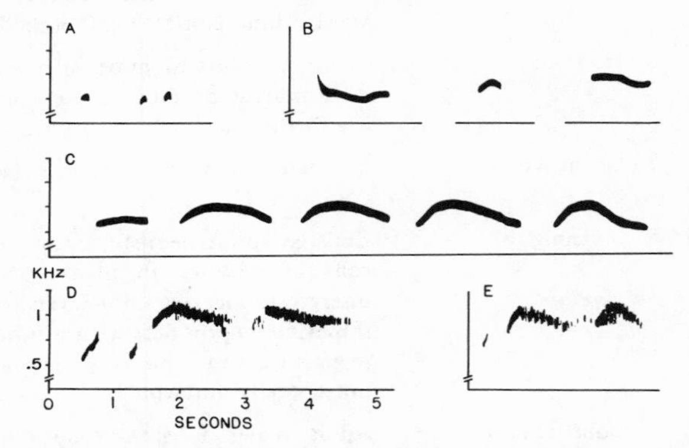

Fig. 17. Vocalizations of Kloss's gibbon. III. Alarm calls. A. Three consecutive whoo's by one gibbon. B. Three examples of howls illustrating the variable structure of this call. C. A five-syllable example of sirening, which may include from three to eight syllables. D. Two whup's followed by an alarm trill by the same gibbon and a second alarm trill by another gibbon. E. A similar sequence.

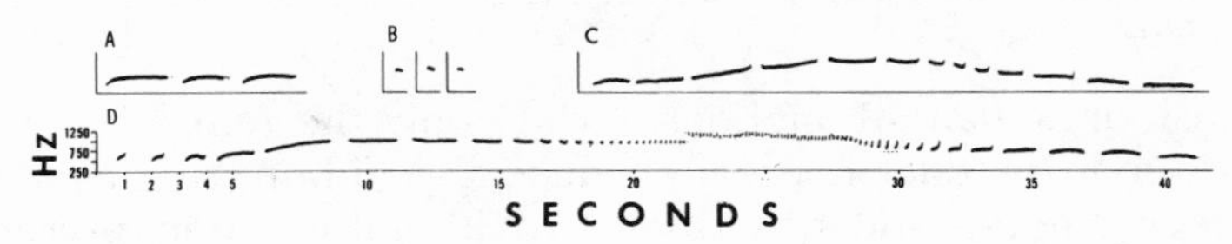

Fig. 16. Vocalizations of Kloss's gibbon. II. Female song and related vocalizations. A. Three consecutive whoo's by the same female. B. Three examples of whistles by males given during intervals between female songs. C. A female song without a trill. D. A typical female song with a prolonged trill.

Predawn chorusing by male Kloss's gibbons occurred on 90 percent of all observation days on Siberut Island (Tenaza, 1976), with choruses beginning as early as five hours before sunrise. The adaptive advantage of predawn chorusing in Kloss's gibbon may be related to the demonstrably short supply of safe sleeping trees in Siberut, and to consequent pressures upon males controlling desirable sleeping trees to warn others away by singing in them (Tenaza, 1975; Tenaza and Tilson, in prep.). Whether this hypothesis might be extended to other gibbons with predawn male choruses cannot be determined until the nature and abundance of their sleeping sites have been evaluated.

Nonvocal interactions between adult gibbons of neighboring groups are, with the exception of courtship activities, concerned totally with excluding one another from their respective territories (Carpenter, 1964; Ellefson, 1974; Tenaza, 1976). Hence it seems safe to assume that song serves a similar function. The fact that gibbons without adjacent neighbors sing less often than those holding territories contiguous with other territories (Chivers, 1971; Berkson et al., 1971) provides further evidence for the spacing function of gibbon songs.

The vocal repertoire of adult Kloss's gibbons includes two calls, sirening and alarm trills (Table 9 and Fig. 17), that occur only when a predator is detected. These alarm calls are loud enough to be heard by a man on the ground 800 m away, whereas members of a Kloss's gibbon family rarely are more than 10 m from one another (Tenaza, 1976). It therefore seems likely that these calls function to warn gibbons in adjacent territories of the presence and location of predators, as well as to warn immediate group members, for which the softer hoo and howl alarm calls (Tables 9 and 10 and Fig. 17) that are audible to about 100 m also suffice.

The frequency-modulated syllables of sirening and the rapidly repeated sound elements in the alarm trill (Figs. 17 and 20) make both of these calls easy to locate by binaural comparison of time, phase and intensity cues (cf. Marler, 1955). Hence they accurately reveal the location of a calling animal to predators as well as to neighboring gibbons. Evolution of such behavior is understandable in theory if animals benefit-

Table 11

Distribution among species of three kinds of gibbon choruses.

	Kinds of choruses present (+) or absent (0)		
Species of gibbon	**All-male**	**All-female**	**Male-female duets**
White-handed	+	0	+
Dark-handed	+	0	+
Bornean	+	0	+
Kloss's	+	+	0
Hoolock	0	0	+
Siamang	0	0	+
Pileated	?	?	+
White-cheeked	?	?	+
Javan	?	?	?

NOTE: Species for which information is incomplete are placed below the line.

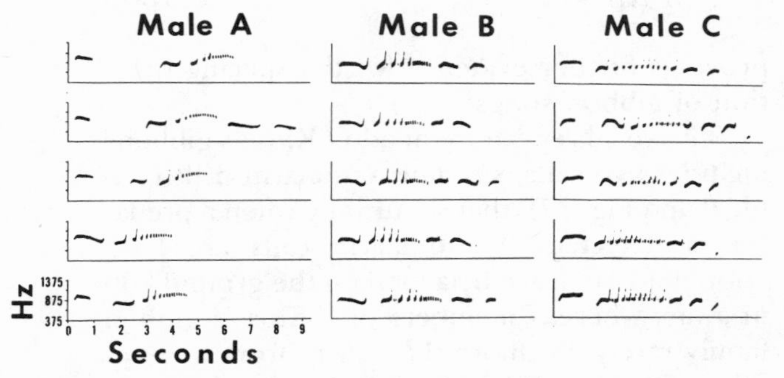

Fig. 18. Individuality in Kloss's gibbon songs, as illustrated by five consecutive songs by each of three males occupying territories within 500 m of one another on Siberut Island, Indonesia. Variation observed in any particular male is small in comparison to variation between males, allowing each individual to be recognized by his songs.

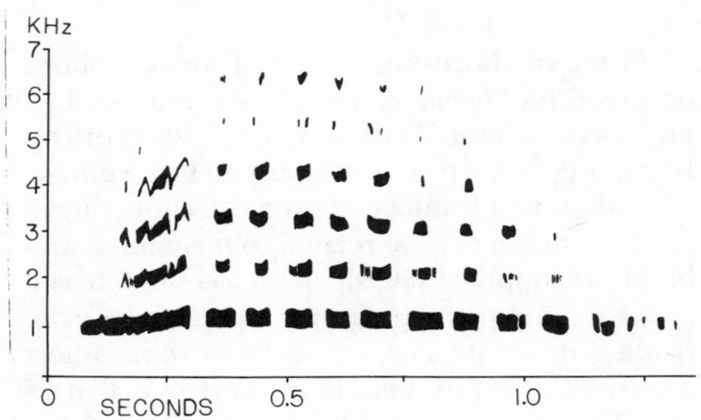

Fig. 20. An alarm trill of a Kloss's gibbon, illustrating the rapidly repeated sound elements, covering a wide frequency range. These are features making this loud call easy to locate at distances up to 800 m.

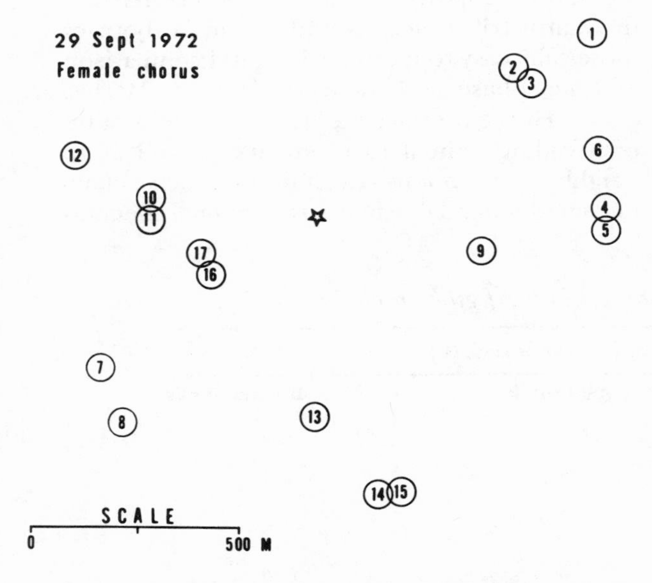

Fig. 19. The spatio-temporal relationships among gibbons joining a chorus. The map illustrates the tendency for chorusing to spread gradually from neighbor to neighbor. Circles indicate the location of gibbons engaged in the chorus, while numbers indicate the order in which each began singing.

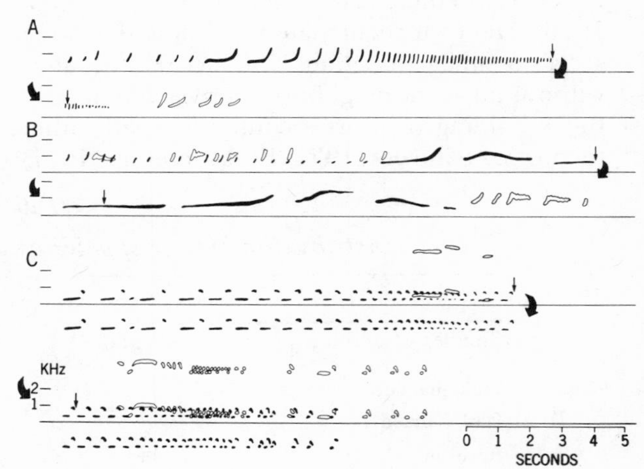

Fig. 21. This figure and the next indicate species differences in the form of male-female dueting in mated gibbon pairs. The four basic types of duet described in the text are shown. In each case male contributions to the duet are in outline, and female contributions are in black. A. Bornean gibbon. B. White-handed gibbon (dark-handed and pileated gibbon duets are similar. C. Siamang.

ing from one another's alarm calls, i.e., those occupying adjacent territories, have a high probability of being closely related (cf. Andrew, 1963; Smith, 1965).

There are suggestions that adult gibbons occupying adjacent territories might often be related as parents and offspring or perhaps as siblings. Aldrich-Blake and Chivers (1973) observed a male siamang establish a home range contiguous to that of his parents. Tenaza and Tilson (in prep.) witnessed four instances of settling by young adult Kloss's gibbons. In each case the gibbon (two females, two males) left its parents at maturity and settled in a territory contiguous to that of its parents. Three of the four definitely mated in those territories subsequently. Thus proximity of close kin can plausibly be invoked as a factor in the evolution of loud alarm calls.

Unique among vocalizations of higher primates are the duets in which paired males and females of several gibbon species engage. Songs and duets of captive siamangs (Fig. 21C) have been analyzed in detail by Lamprecht (1970). In all species of gibbons for which information is available, other than Kloss's gibbon, males and females sing duets with their mates (Table 11). Male and female songs overlap during duetting in the pileated gibbon (Marshall et al., 1972), as they do in the siamang and hoolock (Figs. 21 and 22). In other gibbons male and female alternate songs with one another.

In the white-handed gibbon the female may emit a few short notes but for the most part is silent while her mate is singing. She signals her readiness to give a full song by uttering a series of short, monotonous notes. When she begins her full song, the male normally stops singing until she has completed it, whereupon he adds a short coda (Fig. 21B), then pauses for several seconds before starting to sing again. Occasionally a male does not cease singing when the female begins her song; in these rare instances

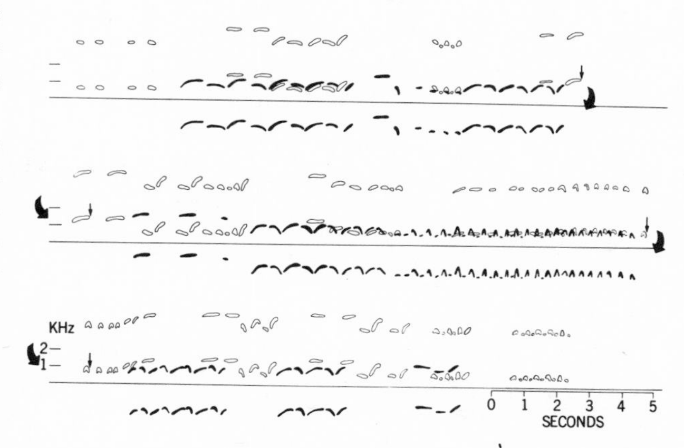

Fig. 22. Another example of male-female dueting, in the hoolock. Kloss's gibbon does not duet, and available tape recordings of the Javan gibbon suggest that it does not either (J. T. Marshall, pers. comm.).

the female does not complete her song but stops and makes another series of presong notes before starting another song (Tenaza, unpubl.). That the female presong notes serve to signal her mate is supported by our observations of a mated pair of white-handed gibbons at the San Francisco Zoo, made from September through November of 1970. About halfway through the observation period the male of this pair died. Although the female continued to sing normal songs after her mate's death, she stopped producing the presong notes. Similarly, an adult female pileated gibbon caged with a mangabey monkey at the San Francisco Zoo did not precede her songs with series of short notes (Fig. 23E), although mated pileated gibbon females in the wild do produce presong notes (Marshall, et al., 1972). In Kloss's gibbon, which does not duet, mated females do not produce long series of short notes before each song but make only three or four short whups before uttering the first long syllable of song (Fig. 23F).

Duetting by wild dark-handed gibbons, recorded in Malaya by D. J. Chivers and in Suma-

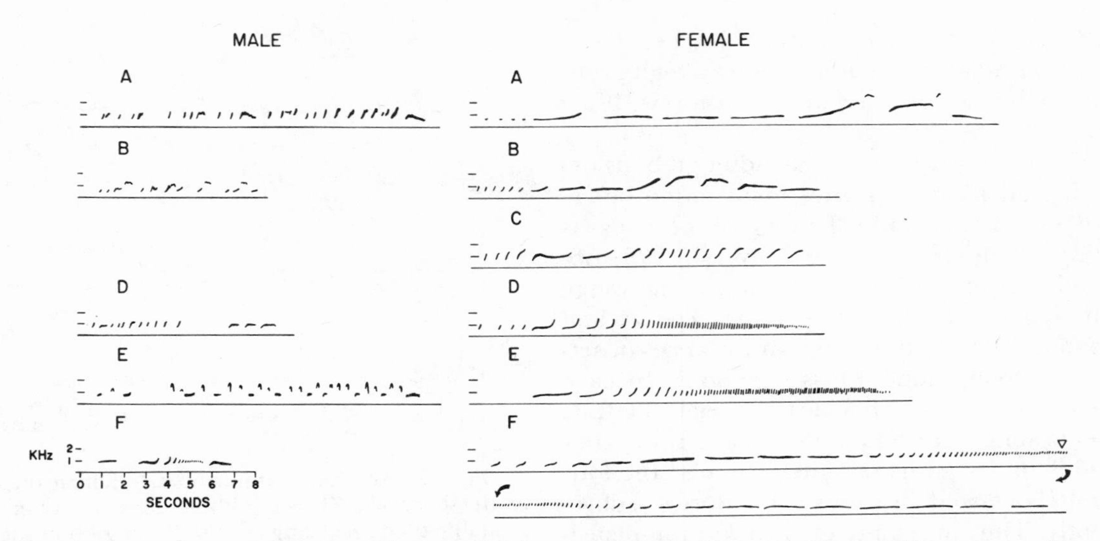

Fig. 23. Species and sex differences in songs of six species of gibbon. A. White-handed gibbon. B. Dark-handed gibbon. C. Javan gibbon. D. Bornean gibbon. E. Pileated gibbon. F. Kloss's gibbon. Two other species are shown in Figs. 21 and 22.

tra by W. Wilson and C. Wilson, is similar to the duetting of white-handed gibbons described above. Duetting by Bornean gibbons has been recorded in eastern Kalimantan by P. Rodman and N. Fittinghoff. It differs from duetting in other typical gibbons in that the male does not sing between female songs but simply adds a brief coda to the end of her song (Fig. 23A). In a continuous ten-minute recording made by J. T. Marshall of singing by a wild female Javan gibbon, (Fig. 23C), duetting did not occur.

In duetting by siamangs (Fig. 21C) the male sings stereotyped phrases that overlap particular, predictable portions of the female song (Lamprecht, 1970). Thus the sequence of events in a siamang duet is nearly as predictable as it is in the typical gibbons described above.

Singing by hoolocks contrasts with that of the siamang and with that of the typical gibbons in two respects. First, there is no striking sexual dimorphism in the structure of sound elements contributed to a duet. Although in the pair we tape recorded at the Vancouver Zoo (British Columbia) the male sounds tended to be lower in pitch than those of the female, both sexes produced basically the same sounds. Second, the course of events in a duet is not readily predictable. Thus the hoolock duet selected for illustration (Fig. 22) is not necessarily typical for the species or even for the pair that produced it. Duets of this pair varied in duration, in relative contributions of male and female, in the ways basic sounds were combined, and in the order of singing by male and female. Recordings of wild hoolocks in Assam by R. Tilson and in Burma by J. T. Marshall support this general picture.

Captivity seems to have no effect upon the song structure or the nature of duetting in gibbons. Six white-handed gibbon pairs that we recorded in various zoos all have the same basic song and duet patterns, with slight variations within and among individuals and pairs; they do not differ from those we recorded on the Malay Peninsula and heard in Thailand. Similarly, siamang pairs we have recorded in zoos and in the wild are basically alike and their patterns do not differ from those described by Lamprecht (1970).

Several species of gibbons have hybridized in captivity (see records in *The International Zoo Yearbook*). Duetting between mates of different species and the inheritance of songs by their hybrid progeny remain unstudied. In 1970 we tape recorded duets between a white-handed gibbon male and a Bornean gibbon female that had been caged together for nine years and had produced two viable offspring, a male and a female, at the Micke Grove Zoo (Lodi, California). Both individuals sang their sex- and species-typical songs. The male responded to his mate's songs in the manner typical of his species, although his mate was of another species. He would stop singing when the female began a song and would add the typical coda to the end of her songs. After death of the parents the hybrid offspring were kept together and began duetting late in 1975. They behaved like a mated pair. Songs of the hybrid female are structurally intermediate between songs of the parental species, demonstrating clearly the inheritance of these vocal patterns. Both song structure and duetting behavior of the male hybrid resemble those of the male white-handed gibbon.

The pelage of three species of gibbons—the hoolock, the pileated, and the white-cheeked—is sexually dimorphic: adult males are black and adult females are generally yellowish (cf. Fooden, 1969). Brockelman and his colleagues (1974) have raised the question of whether color dimorphism might substitute for song dimorphism to facilitate sexual recognition in these species. Sonagraphic analyses demonstrate a high degree of sexual dimorphism in pileated gibbon songs but little in hoolock songs (Figs. 22 and 23). Adult white-cheeked gibbons show a degree of sexual song dimorphism comparable to that in pileated gibbons (Marshall et al., 1972). Hence among the sexually dichromic gibbons we find two species with pronounced sexual dimorphism in songs and one with apparently slight dimorphism. Thus sexual dimorphism in color does not replace sexual dimorphism in song except perhaps in the hoolock. Instead it may complement song by facilitating instant visual recognition of an adult's sex in the same way that song allows rapid auditory recognition of sex in all species but the hoolock. None of the sexually dichromic gibbon species has yet been studied more than casually in the field.

Production and Perception of Visual Signals

The visual system of the great apes is very similar to our own, with extensive overlap of the visual fields, a mixed foveal retina with color vision, and high visual acuity. The correspondence between apes and ourselves in visual signaling equipment is also notable. The lack of a tail is compensated by extensive use of the hands in visual signaling. The facial musculature is elaborate and is undoubtedly specialized for the production of a variety of facial expressions (Huber, 1931; Andrew, 1963) made more visible by the lack of hair on the face. In gibbons conspicuousness of the dark, expressive area of the face is enhanced by a band of white hairs above the face *(H. hoolock),* beside it *(H. concolor),* or surrounding it *(H. lar, H. agilis, H. pileatus,* and *H. muelleri).* Piloerection provides another way of changing the contour of the trunk and limbs, and slower physiological changes are conspicuous in the visible genital swelling of female chimpanzees during estrus and in some cases during pregnancy (van Lawick-Goodall, 1968b, pers. comm.).

There is thus ample morphological equipment for generating a wide array of visual as well as auditory signals, with a somewhat lesser emphasis on olfactory signaling. There are parallel trends on the sensory side. Thus, we should prepare for the likelihood that vision and audition will be the most important sensory modalities for social communication, though, as indicated earlier, olfaction is probably important in sexual behavior. Finally, mention should be made of the tactile sense. Although sensory and motor specializations are hard to detect, there is evidence that certain parts of the body are well provided with tactile nerve endings, especially the face, hands, and genitalia. There is ample evidence that tactile signals are especially important in the kinds of social communication that emphasize contact with those parts of the body that are richly innervated (e.g., Simpson, 1973).

VISUAL SIGNALS

Of all parts of the ape body none is more concerned with the production of visual signals than the face. Some facial configurations occur only with particular vocalizations, which they structure by altering the size and shape of the mouth aperture and resonating cavities. Whatever information these vocalization-bound facial modifications might communicate to recipients about the performer's motivational state therefore appears to depend mainly on their accompanying vocalizations. Here we shall disregard these expressions and focus on others which, while they can be accompanied by vocalizations, apparently are independent enough in some if not all species of apes to convey information about what the performer is likely to do next even when given silently. Andrew (1963) considers most such facial expressions to have been freed or "facilitated" during their evolution from a former association with vocalizations. Ignoring unfacilitated facial configurations and lumping together those distinguished only by minor differences of context, intensity, or accompanying vocalizations, the repertoire of chimpanzee facial expressions can be reduced to six (Table 12).

Because chimpanzee facial expressions have been intensively studied under nearly ideal observation conditions in the field (van Lawick-Goodall, 1968a, 1968b, 1971) and in a captive colony (van Hooff, 1971), they provide the best baseline for comparison with other apes. Thus, we find that every general category of chimpanzee facial expression has been described in at least two of the other three apes (Table 13). Furthermore, the facial expressions of gibbon, gorilla, and orang-utan function within the same range of social circumstances in which comparable chimpanzee expressions occur (Table 12) (Baldwin and Teleki, 1976; Schaller, 1963; Mackinnon, 1974). Similar facial expressions apparently serve comparable social functions among the apes.

Ultimately the similarity of facial expressions among apes and between apes and monkeys (van Hooff, 1967; Chevalier-Skolnikoff, 1974) can be traced to similarities in facial anatomy. The simplest facial musculature among apes is that of the gibbon, its structural complexity somewhere between the simpler cercopithecoid monkeys and the more complex great apes. This primitive "ground plan" of ape facial musculature seen in the gibbon has increased in complexity along divergent lines leading to the orang-utan on one hand and to the chimpanzee, gorilla, and human on the other (Huber, 1931). Chimpanzee facial expressions are shown in relation to their effector muscles in Fig. 24.

Modifications of orang-utan facial musculature are most conspicuous in the platysma muscles over the cranium, and in the labio-buccal musculature. They are related to support and control of the greatly enlarged laryngeal air sacs, to support of the male cheek pads (Huber, 1931), and perhaps, as Chevalier-Skolnikoff (1974) has suggested for the chimpanzee, to elaboration of the rather prehensile lips for plucking and manipulating food items.

In the chimpanzee/gorilla/human line of facial development homologs of gibbon facial muscles, particularly in the midfacial region, are

Table 12
A classification of facial expressions of the chimpanzee.

Facial expression	Synonyms	Circumstances
A. Teeth bared by retraction of lips		
1. Grin		
(a) With mouth closed or only slightly open	Grin (van Hooff, 1962; Andrew, 1963); silent bared-teeth face, horizontal bared-teeth face, bared-teeth yelp face (van Hooff, 1967, 1971); silent grin, closed grins (van Lawick-Goodall, 1968a, 1971).	During submissive behavior and when an individual is frightened, e.g., following attack from a superior.
(b) With mouth wide open	Silent bared-teeth face, open-mouth bared-teeth face, bared-teeth scream face (van Hooff, 1967, 1971); scream call face, full open grin (van Lawick-Goodall, 1968a, 1971); threat face, high grin (Andrew, 1963).	During nonaggressive physical contact with other individuals; when threatening superior or animal of another species of which the chimpanzee is afraid.
B. Teeth mostly or entirely covered by lips		
2. Open-mouth threat face (mouth open to varying degrees)	Waa-bark face (van Lawick-Goodall, 1968a); Waow-bark face, staring open-mouth face (van Hooff, 1967, 1971).	When threatening subordinate or distant dominant or animal of another species of which the chimpanzee is not unduly afraid.
3. Tense-mouth face (mouth closed, lips pressed tightly together)	Tense-mouth face (van Hooff, 1967, 1971); bulging-lips face (van Hooff, 1971, van Lawick-Goodall, 1968a); glare, compressed-lips face (van Lawick-Goodall, 1968a, 1971).	Prior to or during chase or attack upon a subordinate; prior to copulation.
4. Pout face	Protruded-lips face, stretched-pout whimper, silent pout face, pout-moan face (van Hooff, 1967, 1971); whimper face, horizontal pout (van Lawick-Goodall, 1968a, 1971).	Upon detecting a strange object or sound; infant while searching for mother or her nipple; begging; after being rejected for grooming; after threat or attack; during juvenile "temper tantrums"; in other situations of "anxiety" or "frustration."
5. Play face	Play face (van Lawick-Goodall, 1968a); relaxed open-mouth face (van Hooff, 1971).	During playful physical contact with other individuals.
6. Lip-smacking face	Lip smacking (van Hooff, 1967, 1971).	While grooming another individual.

structurally differentiated into increasingly more and finer subunits proceeding from gibbon to chimpanzee to gorilla to human. The advantage of this increasing complexity might, as Huber (1931) suggested for humans, be related to an increased ability to produce subtle degrees of expression with nuances of meaning, but there is little evidence on the matter.

Lip smacking is presumed to have originated from jaw, tongue, and lip movements performed while eating ectoparasites and other foreign particles removed from body surfaces during auto- and allo-grooming (van Hooff, 1967). In the white-handed gibbon, orang-utan, and chimpanzee these movements, which have not been observed in gorillas, still occur in their original context of ingesting particles removed with the lips, teeth, and fingers from another's body during allo-grooming (Baldwin and Teleki, 1976; Carpenter, 1964; van Lawick-Goodall, 1968b;

Table 13

Probable homologs of chimpanzee facial expressions in other apes.

Chimpanzee facial expression	Probable homologs in other apes: Gibbon	Gorilla	Orang-utan
1a. Grin with mouth closed or only slightly open	Grimace	None described	Grimace, fear face
1b. Grin with mouth wide open	High intensity "grin"	Fear face and anger face	Wide-mouth grin†
2. Open-mouth threat face	Mouth-champing	Annoyance face staring	Bare-teeth threat face
*3. Tense-mouth face	None described	Tense-mouth face	Tense-mouth face
4. Pout face	Lip-pout	Light-distress face	Pout face
5. Play face	Low intensity "grin"	Pleasure face	Play face
6. Lip-smacking face	Opening and closing mouth and protruding tongue while grooming another gibbon (Tenaza, pers. obs.)	None described	Lip and mouth movements performed while eating particles removed from body of another individual during allo-grooming.

NOTE: Descriptions and terms for gibbon facial expressions (except lip-smacking) are from Baldwin and Teleki (1976); those for the gorilla from Schaller (1963); and those for the orang-utan from Mackinnon (1974).

*Although it has not yet been described in the literature, the tense-mouth face with bulging lips appears clearly on the male Stefi just prior to copulation with the female Kati in a film on sexual behavior of captive gorillas in the Basel Zoo by J. P. Hess. For the orang-utan tense-mouth face, see B. Galdikas-Brindamour (1975).

†See Mackinnon, 1975.

Mackinnon, 1974). In the orang-utan they have been described only in this functional context, whereas white-handed gibbons (Tenaza, pers. obs.) and chimpanzees (van Hooff, 1967, 1971; van Lawick-Goodall, 1968b) often perform lip smacking during grooming without taking any foreign particle into the mouth. Noningestive lip smacking in the chimpanzee "does not resemble the original functional smacking in that the protrusion of the tongue is barely noticeable," (van Hooff, 1967:41). In adult white-handed gibbons full downward protrusion of the tongue through parted lips occurs during non-ingestive lip-smacking, but the tongue is only slightly protruded when the behavior is ingestive (Tenaza, pers. obs.).

The Design and Function of Communication Signals in Apes

SPECIES-SPECIFICITY

Selective pressure for species-specificity in communication signals, such a powerful force in the evolution of avian vocalizations, has less effect on the signals of primates, though in certain groups such as the *Cercopithecus* monkeys (Marler, 1973) and gibbons (Figs. 21–23), its influence can be discerned. Although the chimpanzee and the gorilla are sympatric in the gross sense, they seem to live within earshot of one another only rarely (Gartlan, pers. comm.). They are presumably separated by differences in habitat selection. It is worth considering the possibility that there is active repulsion between them. Comparison of vocalizations of the two species has revealed a surprising degree of correspondence that includes some of those signals thought to be used for maintaining the spacing of conspecific groups. Chimpanzee pant-hooting and gorilla hoot series are somewhat similar acoustically, and one can hardly overlook the convergence by gorillas and chimpanzees on two quite different means of producing far-carrying drumming sounds, chest beating and tree drumming. Both species also use ground thumps and branch breaking. Areas where both species live in close proximity should be studied for possible interspecific reactivity.

Orang-utans and gibbons do live in close contact. While more complete descriptions of their vocal behavior are needed before we can speculate about possible evolutionary interactions, Mackinnon (1974) has an intriguing observation on the timing of adult male orang-utan long-calling in Sumatra in relation to that of loud calls of the white-handed gibbon, the siamang, and the leaf monkey *(Presbytis aygula)* (Fig. 25). The notion that the nocturnal peak of orang-utan calling is timed to avoid temporal overlap with the calling of other species is reinforced by the absence of such a peak in calling of Bornean orang-utans. There primate densities are lower, the siamang is absent, and the only other primate heard calling regularly was the Bornean gibbon *(Hylobates muelleri).* Irrespective of any detailed correspondence in the structure of vocalizations, the mere presence of loud calling by one species may add significantly to the background of noise against which another must make itself heard, hence presumably the adaptive value of avoiding temporal overlap.

On Siberut Island, Indonesia, both male Mentawai langurs *(Presbytis potenziani)* and male Kloss's gibbons produce loud predawn spacing calls. They overlap, with gibbon calling (songs) concentrated in the hours 0300–0600 and langur calling at 0400–0600. However, calling by the langurs does not occur in prolonged choruses like the singing of the gibbons but in bouts ranging from eight seconds to eight minutes duration, during which males from different groups take turns producing calls of three to four seconds duration spaced from one to sixty seconds apart. Even during the hour before dawn, when langur calling reaches its peak, these rounds of calling (including silent intervals between calls) occupy on the average only five and a half minutes, or less than 10 percent of the hour. Hence they interfere very little with the gibbons' choruses. Since among primates on Siberut only langurs and gibbons habitually make loud spacing calls, interspecific competition among primates for the auditory environment is negligible.

Dawn bird choruses on Siberut are, by contrast, continuous and loud, involving many individuals of several different species. Tenaza noted in the field that background noise generated by the dawn bird chorus not only reduces the audible range of gibbon songs but also makes it difficult and often impossible to determine by ear the location of a singing gibbon. It therefore is inefficient for gibbons to sing during the bird choruses, and it seems likely that the temporal separation of Kloss's gibbon choruses from bird choruses (Fig. 26) is an evolutionary consequence of interspecific competition for the auditory environment.

Species-specificity is a matter of special interest in gibbon songs. In a study of the structure of female songs, we found strong resemblances in six of the eight species analyzed, with extremes in the series connected by intermediates (Fig. 23). Female songs in this group are preceded by short, monotonously repeated notes, and begin with a prolonged syllable of rising inflection. In female white-handed and dark-handed gibbons this first syllable is followed by

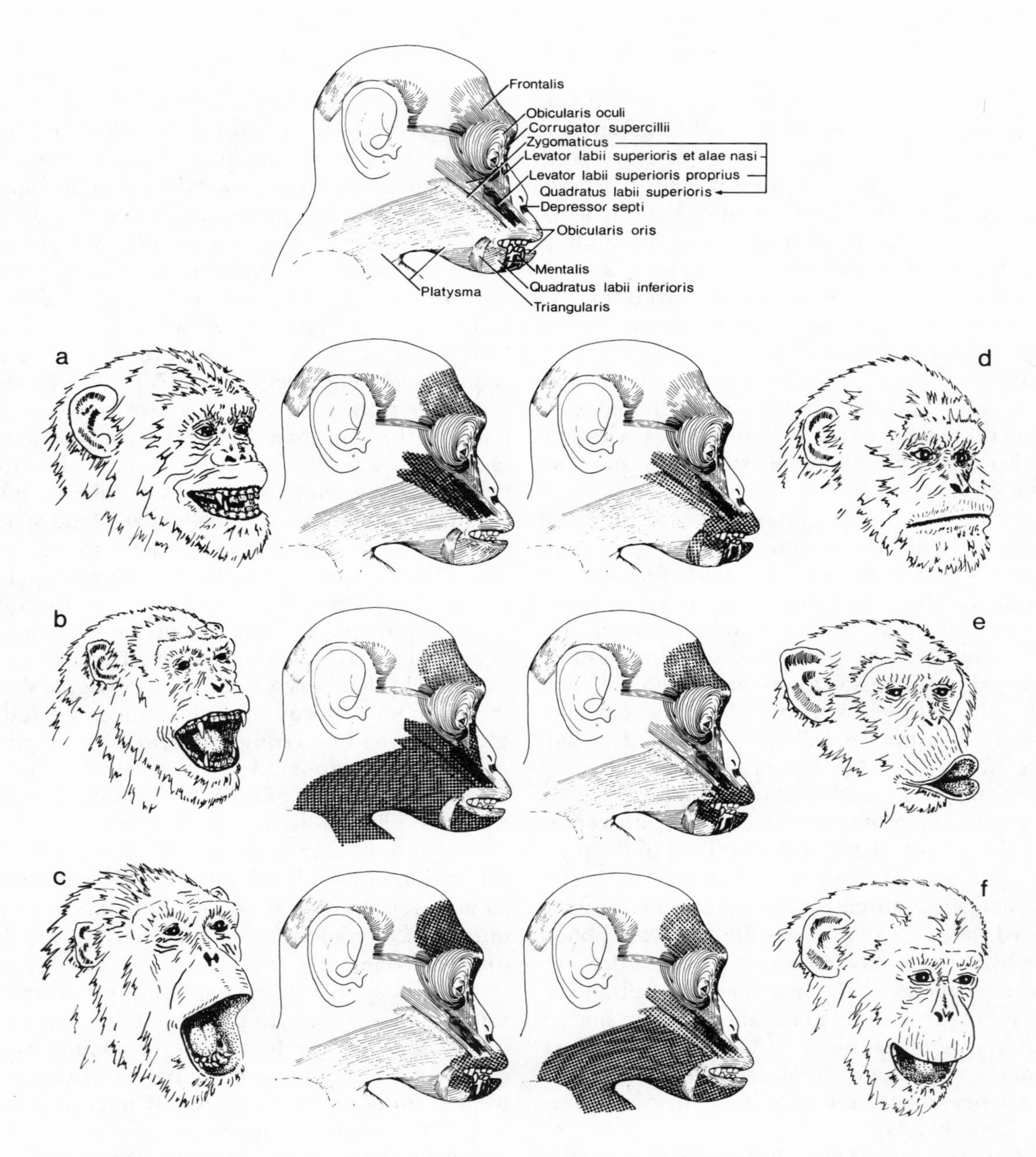

Fig. 24. Muscles of facial expression in the chimpanzee. Adjacent to each facial expression, the muscles responsible for the most characteristic features of the expression are darkly shaded and those of occasional or otherwise minor involvement are lightly shaded. Note that in the labeled diagram and for those expressions involving the mentalis muscle, a section of superficial musculature of the lower jaw is cut away to expose the mentalis. a. Grin with mouth slightly open. Corners of the mouth are retracted by zygomaticus

similarly prolonged notes, which rise in pitch near or following the middle of the song, then fall in pitch at the end. In other species of this group the prolonged notes initiating female song become progressively shorter, grading into a slow trill in the Javan gibbon and into a rapid trill in the Bornean, pileated, and Kloss's gibbons. The Kloss's gibbon female trill grades back into prolonged notes that terminate the song, but in the others song ends in the trill.

Despite basic similarity among female songs in this group of gibbons, each species has a species-typical female song structure. The greatest similarities are between songs of dark-handed and white-handed gibbon females on the one hand and between those of Bornean and pileated gibbon females on the other, but these two song types are bridged rather nicely by the slowly trilled song of the female Javan gibbon. The long, typical song of female Kloss's gibbons might seem widely separated from dark- and white-handed gibbon songs, but the gap is narrowed by its trill-less variant (Fig. 16C), which resembles them rather closely in both song duration and signal structure.

Among males in these species there is less interspecific similarity in songs than we find among females. While the songs of Bornean and pileated gibbon females are nearly identical, male songs of these species differ considerably; the same is true of the dark-handed and white-handed gibbons. This perhaps is related to present and past contact between populations and selection against hybridizing individuals, leading to interpopulation divergence in male songs. Available evidence suggests that male gibbon song functions in mate attraction, as well as in territorial interaction (Aldrich-Blake and Chivers, 1973; Chivers, 1974; Tenaza, 1976). White-handed and dark-handed gibbons presently have contiguously allopatric distributions in Malaya and Sumatra (Fooden, 1969). Paul Gittins recently discovered a zone of sympatry and three hybrid pairs, two with young, near the Mudah River in Malaya (D. J. Chivers, pers. comm.). Although Bornean gibbons are restricted to Borneo and pileated gibbons to southeastern Thailand and Cambodia (west of the Mekong River), these areas have been connected by extensive land bridges during the Pleistocene and also are connected by the major faunal migration route proposed by de Terra (1943). Hence it is possible that the similarity in female songs of the two species reflects close

contraction and brows may be elevated by frontalis contraction. b. Grin with mouth wide open. The entire quadratus labii superioris group contracts to retract the upper lip and mouth corners while the platysma and quadratus labii inferioris pull the lower lip down and back. Brows may be raised by frontalis. c. Open-mouth threat. Lips are tensed by obicularis oris contraction and lower lip is raised by mentalis contraction. Furrowing of the brow by corrugator supercillii contraction is counteracted somewhat by simultaneous contraction of frontalis, which acts to raise the brows. d. Tense-mouth face (bulging lips form). Lips are tensed by obicularis oris contraction and the lower lip is pushed up against the upper lip by mentalis contraction. Corners of the mouth appear to be further tensed by zygomaticus and triangularis contraction. Eyebrows are drawn medially downward into a frown by corrugator supercillii contraction. e. Pout face. Lips are tensed and protruded by action of obicularis oris and mentalis. Slight contraction of corrugator supercillii and frontalis may wrinkle the brow and forehead. f. Play face. Relaxed lowering of the mandible is the most characteristic feature of this expression, with some retraction of the mouth corners and depression of the lower lip resulting from contraction of the platysma. The outer corners of the eyes may be elevated by zygomaticus contraction, and the brows may be elevated and slightly wrinkled by contraction of the frontalis and corrugator supercillii. (Facial expressions are adapted from drawings and photographs in van Lawick-Goodall, 1968a, 1968b, 1971; van Hooff, 1971; Chevalier-Skolnikoff, 1974. Interpretations of muscle involvement are based on examination of photographs and on Andrew, 1963, and van Hooff, 1967. Facial musculature is adapted from Chevalier-Skolnikoff, 1974 [after Huber, 1931], and from Huber, 1931. Muscle terminology follows Huber, 1931.)

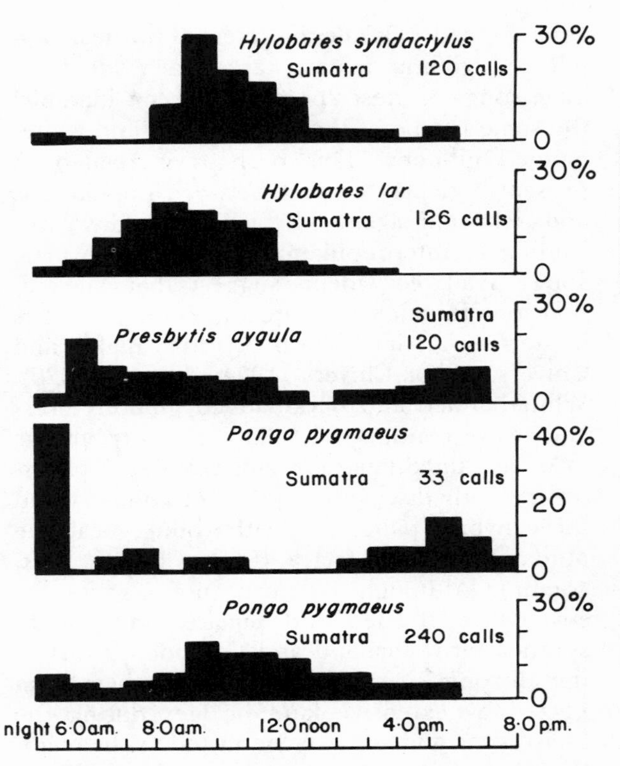

Fig. 25. Histograms of the relative frequencies of calling of several primate species with time of day. (From Mackinnon, 1974.)

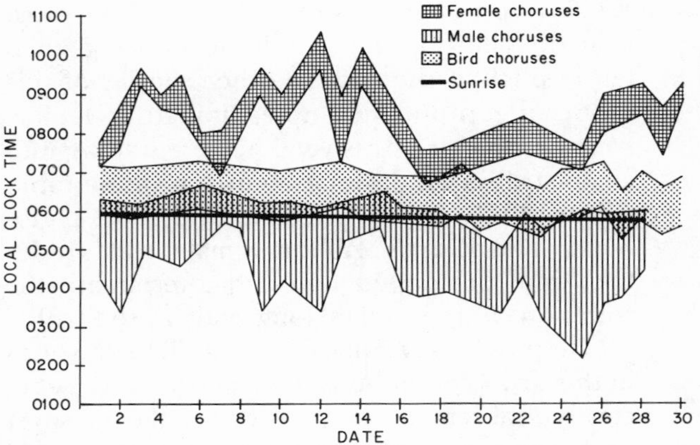

Fig. 26. Temporal separation of Kloss's gibbon choruses from dawn bird choruses on Siberut Island in September 1972. The graph for each chorus type is constructed by connecting chorus starting times below and chorus stopping times above and shading the area between. The beginning of bird choruses is set at the time the first diurnal bird was heard singing. Their termination is arbitrarily set at one hour after the start, by which time participation normally had waned to such an extent that the birds no longer seriously masked gibbon songs.

phylogenetic relatedness, rather than convergence, and that selection against hybridization has led to the divergence in male songs. Pileated and white-handed gibbons are for the most part allopatric but have a narrow zone of sympatry in southeastern Thailand (Marshall et al., 1972). W. Y. Brockelman and J. T. Marshall have discovered interspecific pairs and hybrid individuals in this area (Brockelman, pers. comm.).

Whatever the details of their recent evolutionary divergence, the basic similarity of their songs suggests that the gibbons illustrated in Fig. 23 are more closely related to one another than any of them is to the hoolock or to the siamang. It also supports the conclusion that Kloss's gibbon should be grouped with these "typical" gibbons (Chasen and Kloss, 1927; Schultz, 1932) rather than with the siamang, where originally it had been placed (Miller, 1903). As indicated earlier, the songs of the siamang and the hoolock differ considerably from one another and from the other species considered here.

AGE CLASS AND SEX-SPECIFICITY

Any specificity in vocal morphology that relates to sex results in the encoding of additional information of potential communicative value. Some vocal categories may be completely absent from one sex or the other. This is true of the gorilla and the orang-utan, with certain adult male vocalizations absent in the female repertoire. A predominantly male role in territorial defense perhaps explains this sexual asymmetry in vocalizations, a suggestion that seems at least plausible when we consider the sharing of territorial defense by male and female gibbons with similar male and female repertoires. However, the same applies to chimpanzees, in which males are territorial and females are not.

Sexual differences in vocal behavior may also occur in the frequency of use of similar categories of vocalization. Quantitative data on chimpanzee and gorilla calling reveal many such asymmetries. Most striking is the overwhelming domination of gorilla vocalization by adult males, which applies to all but one of the fourteen vocalizations used by adult gorillas (Fossey, 1972; and Table 7). Thus, only one vocalization is used more by adult females (69 percent). Even here adult males contribute 23 percent to the total, and the overall numbers are small, this being one of the least frequently used vocalizations. There is no record of a gorilla vocalization unique to adult females.

Although there are interesting differences in the details of vocal use between male and female chimpanzees, perhaps most striking is the extent of sharing of all vocal categories between the sexes. There is no call type unique to one sex, and males and females contribute roughly equal proportions to the overall frequency of vocal behavior.

A difference in adult sexual roles in the chimpanzee and the gibbons is implied by consistent differences in male and female renditions of certain calls. Although both male and female chimpanzees engage in pant-hooting, there are differences striking enough to permit a human observer to determine the sex of the vocalizer (Marler and Hobbet, 1975). The same is true of gibbon songs (Fig. 23). The contrast has been interpreted differently in the two genera. Both male and female gibbons defend their territory against intruders of their own sex, hence the selection pressure for sexual differences in calling. Within the highly dispersed social groupings of chimpanzees, the sexual difference in pant-hooting is presumably as much an issue of communication within groups as between them.

One way to measure the extent of sharing of a vocalization by different sex and age classes is with an index of diversity derived from information theory (Baylis, in press). Based on the formula indicated in the legend of Table 14, H_1max is the value if all contributed equally to a given vocalization. The extent of departure of the realized values of H_1 from the maximum reflects the degree of inequality of sex and age class contributions to that call type.

As can be seen from Table 13, the values for many chimpanzee vocalizations are close to the maximum and H_1 is less than half the maximum for only two of fourteen sounds, laughter and lip smacking. The situation in the gorilla is very different, with H_1 less than half of H_1max in twelve of fifteen sounds, an obvious reflection of the lower diversity of usage of most gorilla sounds. The usage diversity of the great majority of chimpanzee sounds exceeds that of even the most diversely used gorilla vocalization.

There is ample evidence that the social roles of primates vary with age, and data on chimpanzee and gorilla vocalization reveal striking differences in this regard. In the gorilla the domination of vocal behavior by adult males is due almost entirely to silverbacks, blackbacked males differing much less from adult females than silverbacks do in the frequency of use of different call types. Thus, there are many behavioral markers signifying the age class that silverbacked males represent. The gorilla data do not permit further characterization of intrasexual age classes, but they reveal some calls unique to infants (2) or used frequently by them (1).

Similarly in the Gombe sample of chimpanzee vocalizations, laughter is dominated by infants and juveniles of both sexes, though especially by males. Other differences in age

Table 14

Diversity of usage of chimpanzee and gorilla sounds by age and sex classes.

Chimpanzee vocalization type H_i		Gorilla vocalization type H_i		Age and sex classes	
(H_i max = 2.32)		(H_i max = 2.32)			
Scream	2.22	Whine	1.46	Chimpanzees	H_i
Waa Bark	2.22	Scream	1.34	♂ 1	2.64
Bark	2.20	Pant series	1.32	♂ 2	2.78
Pant	2.13	Belch vocalization	0.73	♀	2.89
Squeak	2.03	Whinny	0.72	Juvenile	3.28
Grunt	1.97	Question bark	0.53	Infant	2.10
Pant-hoot	1.95	Pig grunt	0.44	(H_i max = 3.81)	
Whimper	1.92	Growl	0.44	Gorillas	H_i
Pant-grunt	1.92	Wraagh	0.37	♂ Blackbacked	2.49
Rough grunt	1.77	Hoot bark	0.32	♂ Silverbacked	2.59
Cough	1.62	Hiccup bark	0.26	♀	2.55
Wraaa	1.45	Roar	0.24	Juvenile	2.24
Lip smack	1.45	Hoot series	0.12	Infant	1.18
Laughter	1.17	Cries	0.00	(H_i max = 3.91)	
		Chuckles	0.00		

NOTE: $-\Sigma_i P_i \log_2 P_i$ and H_1 max $= \log_2 a$, where "a" is the number of event categories in P_1 is the calculated relative frequency with which either the ith sex/age class uttered that vocalization or the ith vocalization was uttered by that age/sex class. The ratio of H_1/H_1 max indicates the degree to which the diversity of usage approaches the maximum possible. Items are listed in order of increasing diversity. (Measured by the diversity index of Baylis, in press.)

class usage of vocal types can be seen in Figs. 27 and 28. Thus, adolescent females tend to dominate the bark vocalization, older adult males the wraaa and rough grunting, infant males the whimpering, and older adult females the pant-grunting. However, the greater use of a call by a particular age may be only statistical in nature, and thus unreliable as a marker of that particular age class to other individuals. This is true of distinctions between older and younger adult male chimpanzees, none of which is as marked as the differences between blackbacked and silver-backed gorillas. The pattern of vocal usage by older and younger adult male chimpanzees does differ, however (Fig. 28), in ways that may again correlate with more subtle changes in social role with age. Thus, in the Gombe sample, older males use the aggressive "cough" more often than younger ones, as well as the wraaa call, which has both alarm and aggressive connotations. The same applies to the rough grunting given at food, which, since older males outrank younger males, they tend to control. These preliminary data suggest that quantitative records of vocal behavior may provide a sensitive and versatile tool for investigating the details of social role in ape societies.

INDIVIDUAL SPECIFICITY

An observer must be familiar with the vocal behavior of a particular population of a species before individuality becomes detectable unless

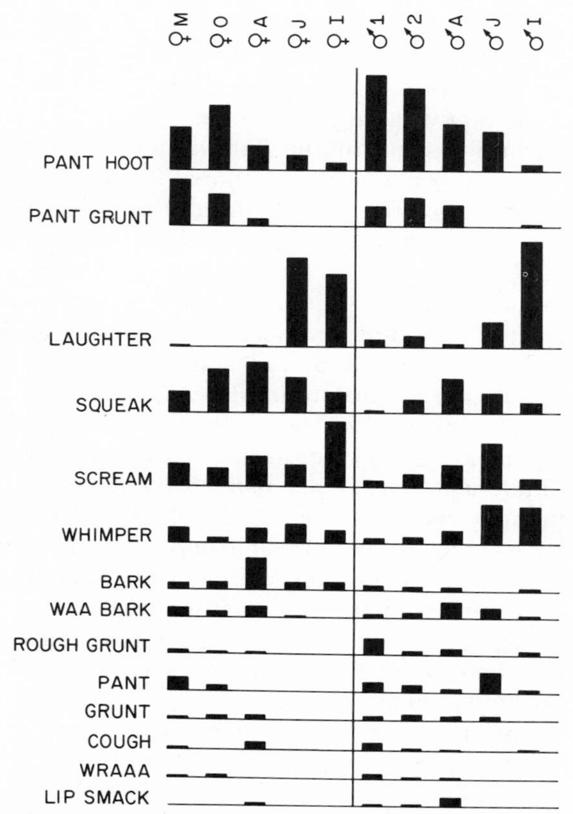

Fig. 27. Histograms of the average output by members of different age and sex classes of thirteen vocalizations and one nonvocal sound, lip smacking, represented as percentages of the output of an average member of that class. ♀ M – a maternal female with infant. ♀ O – an adult female without infant. ♀ A – an adolescent female. ♀ J – a juvenile female. ♀ I – an infant female. ♂ 1 – older adult males. ♂ 2 – younger adult males. ♂ A – an adolescent male. ♂ J – a juvenile male. ♂ I – an infant male. The figures derive from Table 3.

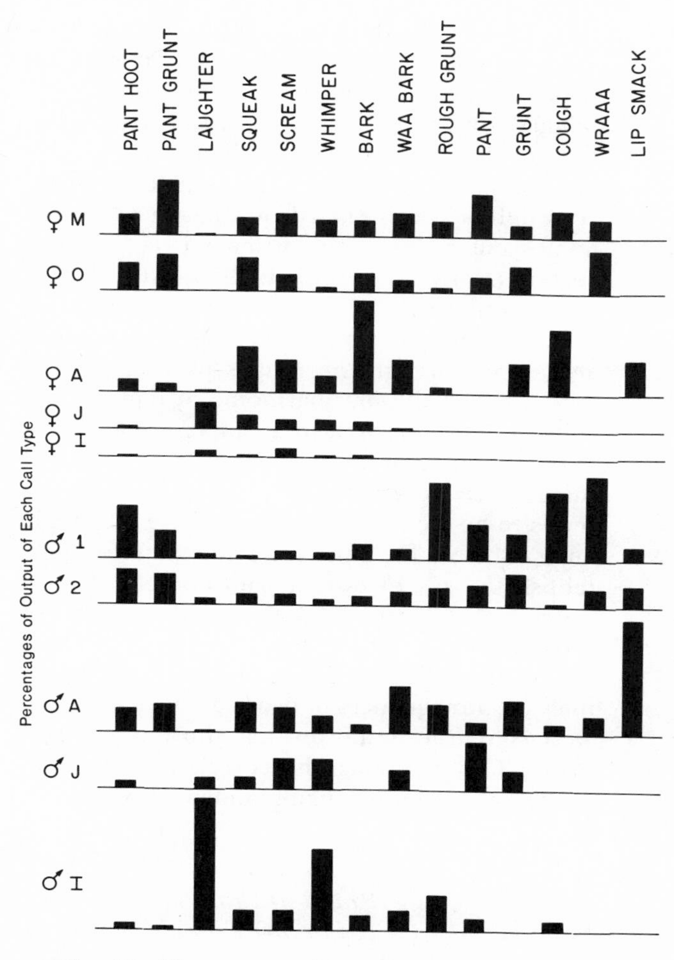

Fig. 28. Histograms of the frequency of use of chimpanzee sounds by different age and sex classes indicated as a percentage of the output of each call type. The data derive from Table 3.

individual differences are very prominent, as in many bird songs. Thus far individual specificity has been demonstrated in only two ape vocalizations, the pant-hooting of chimpanzees (Marler and Hobbet, 1975) and the "songs" of male Kloss's gibbon (Tenaza, 1976), though it undoubtedly occurs in vocalizations of other species of apes as well (e.g., Chivers, 1974:238). In both cases the vocalization is uttered by males and females, with individuality imposed on sex-specific variations.

In Kloss's gibbon males in adjacent territories countersing in the dark and through dense foliage over distances of 150–500 m. Females, on the other hand, meet at territorial boundaries in view of one another and engage in vigorous, mutual, visual displays while countersinging (Tenaza, 1976). Demands for individuality in female song might be less than in male song, though this has yet to be studied.

The pant-hooting of adult female chimpanzees differs consistently from that of adult and adolescent males in the absence of a "climax" section (Marler and Hobbet, 1975). Also, the average duration of female pant-hooting sequences studied was greater than that of male sequences, though the ranges overlap greatly. The pitch of the first harmonic of female loud calls—their equivalent to a climax—tends to be deeper than that of the climax calls of most males; their duration is shorter; and their shape tends to be more arched. Thus, there are ample cues available for a listening chimpanzee to establish the sex of distant pant-hooters, even if individual differences cannot be discerned. An intriguing relationship has been suggested between the duration of pant-hooting and age in males. The sequences of two older males studied were significantly shorter than those of younger males. This difference also correlates with dominance rank; higher-ranking animals had shorter pant-hooting sequences. The long sequences of the two females studied, lower than males in rank, conform to this relationship. More study is needed to establish its generality.

Individuality in the pant-hooting of seven chimpanzees at Gombe National Park was studied by calculating the significance of differences between all possible pairs (Marler and Hobbet, 1975). The four measures compared were (a) duration of pant-hooting sequences, (b) peak frequency of climax exhalation calls, (c) duration of climax exhalation calls and (d) interval between climax exhalation calls. All pairs differed significantly ($p < .05$, two tailed t-test) in at least one measure, and they averaged 2.3 significantly different measures per pair analyzed, out of a possible maximum of 4.0 (Table 15). On this basis there are cues available in pant-hooting for individual recognition, although the likelihood that some properties of pant-hooting are more conspicuous than others to listening chimpanzees limits the value of a simple statistical analysis of individual differences.

In both gibbons and chimpanzees field observers have been able to identify individuals by their calling. Whether species members do the

Table 15

A tally of those measures in which the pant-hooting of pairs of individual chimpanzees showed significant differences ($p < .05$, two-tailed t-test).

	Mike	Humphrey	Charlie	Faben	Figan	Nova	Flo
Mike		BCD	B D	B	C	ABCD	BCD
Humphrey			ABCD	ABCD	AB D	A CD	A C
Charlie				D	CD	BC	BCD
Faben					C	BCD	B
Figan						BCD	BC
Nova							D

NOTE: Female "loud" calls are equivalent to the male pant-hooting "climax." (A) is duration of complete sequences. (B) is peak frequency of climax of loud exhalation calls. (C) is duration of climax or loud exhalation calls. (D) is interval between climax or loud exhalation calls. (From Marler and Hobbet, 1975.)

same remains to be demonstrated, though it seems probable. The patterned sounds of pant-hooting and singing provide sufficient parameters to maintain individuality in addition to higher orders of distinctiveness at sexual and specific levels. Further study may reveal individuality in other simpler vocalizations.

GROUP SPECIFICITY

The abundant demonstrations of group specificity manifest in bird song as local dialects (Marler and Mundinger, 1971) have yet to be paralleled in any of the apes or other primates, apart from humans. The only suggestion of something equivalent comes from a demonstration of troop differences in one class of calls of the Japanese macaque (Green, 1975b). With this exception the absence of local dialects in ape vocalizations is consistent with the failure thus far to demonstrate that individuals learn from others in vocal development.

REPERTORIAL SPECIFICITY

No student of ape behavior has yet undertaken an analysis of what we might label repertorial specificity—the degrees to which different items in a repertoire differ distinctively from one another. The problem is complex since an observer's selection of measures on which an acoustical comparison of recorded vocalizations is based might not coincide with those parameters most salient to a conspecific listener. Human judgments as to which sound features are most conspicuous, used in studies of *Cercopithecus* vocalizations (Marler, 1973), are probably better than a random selection, but perceptual studies with conspecific subjects are urgently needed, as well as thorough descriptive analyses.

The first question is whether vocalizations are discretely separate from one another, and, if so, what the acoustical distance is between them. The situation is complicated by the abundance of intermediate forms between the categories of many ape calls, constituting vocal systems that are graded rather than discrete (Fossey, 1972; Marler, 1965, 1969, 1976; van Hooff, 1971). Methods for interpreting such graded vocal systems remain to be developed (but see Green, 1975a). From a functional viewpoint one might expect some intraspecific vocal distinctions to be more critical than others. The difference between one call designating predators and another designating food might be highly significant, whereas the potential confusability of calls associated with varying levels of intensity within a single system such as aggressive signaling would be less critical. Presumably discrimination will be most accurate and rapid between calls that are discretely different from one another and that have considerable acoustical space separating them. At present we lack the data to test this prediction.

DISCRETE AND GRADED SIGNAL SYSTEMS

Variation from one rendition of an item from an animal's repertoire to another may take several forms. A call type that is discretely separate from other types may exhibit within-category variation. Such variation might be accidental and random, or it might be orderly and even highly organized. Among chimpanzee vocalizations laughter is perhaps the most discrete, yet it varies considerably in ways that are likely to have communicative significance.

Categories may vary so much that different types become connectable by intermediate forms. Variation of this type raises a number of questions. In the first place it renders uncertain the meaning of the original categories. Even though field observers confronted with such a graded vocal system feel reasonably confident in discerning categories, their judgments may be based on frequent usage of modal forms, intermediates often being rarer (Table 16).

Table 16

Numbers of intermediate chimpanzee calls classified by the categories they fall between.

Call category	A (Intermediates)	B (Typical)	B/A
1. Waa bark (most variable)	92	84	0.9
2. Scream	173	216	1.25
3. Bark	54	95	1.76
4. Grunt	25	43	1.79
5. Squeak	146	264	1.81
6. Wraaa	7	18	2.57
7. Whimper	54	175	3.24
8. Pant-grunt	64	285	4.45
9. Pant	13	81	6.23
10. Rough grunt	7	83	11.86
11. Pant-hoot	52	648	12.0
12. Cough	0	33	33
13. Laughter (least variable)	0	271	271
Total	343 (×2) = 686	2,313	6.7

NOTE: (A) is two entries for each intermediate call, as compared with the number of typical examples of that category recorded (B). The ratio of B to A is one measure of the variability of each category. They are listed in order of decreasing variability. (From Marler, in press.)

The functional significance of vocal grading is still unclear. We can distinguish at least two ways in which it may emerge during a descriptive analysis of vocal behavior. There may be adjacent graded forms or separate ones. In the former case an individual produces a string of continuously changing variants so that calling changes quickly from type A to type B in a series of small steps rather than a single jump. In such a series the differences between adjacent pairs may be slight, and one may wonder whether they would be perceptible to another animal. But if they are given in a string with brief intervals separating them, then each provides a frame of reference for the next, and directions and rates of change of vocal morphology are thus more readily detectable. It may be that this is how some gradations come to assume communicative significance. If this argument were valid, one might predict a correlation between degrees of vocal grading, and the tendency for them to be uttered in rapid strings. There are hints that this may be true of vocalizations of the chimpanzee.

Table 16 classifies 343 chimpanzee calls judged to be intermediates and classified according to the two categories they fall between. Thus, there are two entries for each intermediate call. They make up 12 percent of the 2,656 calls analyzed. The ratio of typical renditions of a call type to those judged to be intermediate in form between it and another category, shown in the third column of Table 16, provides a rough measure of the degree of variability of that call type. At one extreme is the waa bark, for which there were actually more records of intermediates than of typical forms. At the other extreme are the cough and laughter, with no intermediates at all. Intermediates were recorded for eleven of the thirteen vocal categories altogether. Thus, the great majority of chimpanzee call types grade into one another through intermediates. Two seem exempt from such grading, and two have intermediates only rarely. The four that are either discrete or close to it are rough grunting, pant-hooting, coughing, and laughter.

The first nine calls in Table 16 are especially prone to gradations with other categories. While quantitative data have yet to be analyzed, these calls are often given in rapid trains in which there are changes to and fro between several call types. The transitions are commonly bridged by a series of continuously graded intermediates, so that changes are gradual rather than abrupt. This manner of delivery might prove to be a clue to the communicative significance of graded forms in the chimpanzee vocal repertoire.

In addition to grading of "adjacent" utterances, similar variation also occurs in vocalizations rendered separately. Thus, an observer recording vocalizations given singly at different times may, upon analysis, find them to form a graded series. Given the difficulty in identifying a single signal on a continuum, one finds it hard to imagine how graded sounds given in this fashion would be used in communication, unless additional cues were provided by the signaling animal. It may be that when two animals are communicating at close range, visual signals provided simultaneously or just before or after a graded vocal signal give the additional information needed for the accurate forecast of the signaler's future behavior that a respondent would require.

VISUAL AND AUDITORY SIGNALING

A relationship has been postulated between the distance over which a communication signal normally functions and its place on a continuum from gradedness to discreteness (Marler, 1973). The longer the distance, the more likely the signal is to be discrete. One would thus expect a relationship between the loudness of a given category of ape sounds and the degree of variability, but reliable measurements of neither have yet been made. One might also be prepared for the possibility that within a graded complex, modal forms might be louder than intermediates.

We have interpreted the discreteness of long-distance signals to be a compromise between information coding and the problems of accurate identification of a signal by another, at a distance, under noisy conditions. At closer range accurate identification of intermediate forms is easier to accomplish. Perhaps even more important is the supplementation of a graded sound signal at close range by visual information, and we have seen that the apes are provided with a rich array of facial expressions, gestures, and postural and morphological signals. Appropriate coordination between auditory and visual signaling should greatly enhance the accuracy with which subtle gradations in signal form, and presumably in corresponding encoded information, might be received and used by another animal.

Such speculations cannot be tested until we can determine how an animal such as a chimpanzee processes the stimuli that conspecific vocalizations provide. In recent years a contrast has been demonstrated in our own species between the continuous processing of non-speech sounds and the categorical processing of the sounds of speech, using series of synthetic speech sounds in which all parameters are under experimental control (Studdert-Kennedy, 1975). A similar approach to analysis of the perceptual processing of vocal sounds by apes is a necessary prerequisite to understanding the communicative significance of the unusual emphasis on graded vocal sounds among higher primates.

SIGNAL REPERTOIRE SIZES

No attempt has been made to estimate the sizes of the repertoire of visual signals that apes use, but we have estimated the numbers of distinct vocalizations. In all cases initial classifications were developed in the field by ear, with some subsequent refinement as a result of sound spectrographic analysis of recordings, especially in the chimpanzee. We have noted that the estimates of repertoire size, reached by several different observers, are rather similar for the ape species studied thus far (cf. Moynihan, 1970). They range between eight and sixteen. The former is Chivers's (in press) estimate of the call repertoire of the siamang. In Kloss's gibbon only ten vocalizations have been described thus far, but we estimate that a final total of about fifteen vocal categories is likely. More studies are needed of orang-utan and siamang vocalizations before we can be sure that there are no significant differences in overall vocal repertoire size among the apes. Also, it may turn out that these estimates derived from descriptive studies alone are deceptive. There may be overemphasis on louder sounds and on modal forms along graded acoustical continua.

It seems conceivable that there are limits to the number of long-distance signals a species is likely to evolve, perhaps because of the need for a discrete organization with sufficient acoustical space between categories for unequivocal identification under difficult conditions. The number of effectively different signals that a graded vocal repertoire provides for close-range communication may prove to be much greater. Analyses of the highly graded vocal repertoire of the Japanese macaque show that very small steps along some vocal continua are associated with different circumstances of vocalization, so encoding information about different situations (Green, 1975a). Thus the complete repertoire of sound signals, each having different behavioral significance to a listening Japanese macaque, may be much larger than an estimate based only on modal forms would lead us to expect.

SYMBOLISM AND SYNTAX

What is it that apes are signaling about, and can we infer that signals serve as signs or symbols? Two obvious external referents for vocalizations are food and danger. While the circumstances in which gorilla belching is given include not only food discovery but also generally pleasurable and relaxing situations, the rough grunting of the chimpanzee seems more closely tied to food alone (Fossey, 1972; van Lawick-Goodall, 1968a). Thus, in the sense of Smith (1969) one might consider that rough grunting incorporates a message about the signaler's discovery of a favored food. Moreover, subsequent events show that rough grunting also specifies readiness of the signaler to share the food, not a necessary concomitant of all food discovery. One might think of the alarm calls of chimpanzees (see Table 1, whimper, wraaa) as encoding information about the presence of danger in varying degrees, though in none of the apes is this information as specific as in the vervet monkey, where different classes of predators seem to be specified (Struhsaker, 1967; see p.51, this volume).

To establish that such signals do in fact function as signs we must examine a complete communicative interaction to determine whether another animal's behavior is changed by receipt of the signal in an appropriate fashion. Appropriateness here may be taken to imply correspondence between signal-evoked changes in the respondent's behavior and changes directly evoked by the referent in question—food for example. The hurried approach of other chimpanzees to a rough-grunting individual, often calling similarly as they come, implies a signlike function for the vocalization.

Such behavioral exchanges can often be interpreted in other ways, however. The first detectable behavioral response to many signals is a rather generalized change in behavior. Whether it is a movement oriented to the signal source—approach or withdrawal—or scanning in that direction, or something even less specific such as a change in vigilance, this initial response will in turn change the pattern of stimulation that the respondent experiences. It is often very difficult in practice to separate the role of such stimulation changes from that of initial signal reception in determining the recipient's eventual behavioral change.

Utterance of a similar call as an early response to rough grunting might seem to provide a basis for inferring that it is recognized as a food sign, since chimpanzees do not seem to give this call in any other circumstances. However, further reflection suggests that the capacity to perceive a signal and respond in kind demonstrates only an ability to identify the signal type. More information is needed to infer that it functions as a sign, just as a child must be required to do more than simply voice a written word to reveal that its symbolic meaning is understood.

Laboratory experimentation with chimpanzees taught to use languagelike systems of communication in human interchanges can overcome the uncertainties of field observation (Gardner and Gardner, 1971; Premack, 1971; Rumbaugh et al., 1974; Gill and Rumbaugh, in press). The results reveal a capacity to use artificial signals—whether hand signs, plastic word-tokens or computer keyboard symbols—as both signs and symbols. The experiments show not just a triggering of an appropriate behavioral event but also some demonstration that the underlying conception is understood (Langer, 1942; von Glasersfeld, in press; Premack, 1971). However, as Premack (1975) has indicated, "the whole topic of representational process or symbolization is in a highly unsatisfactory state and seems likely to remain so until some relevant operational criteria are provided." Research on the capacities of chimpanzees to use languagelike systems promises not only insights into the use of representational processes by animals, but also a more objective view than we have had yet of the nature and significance of such processes in our own species.

Preoccupation with animal symbolism should not lead us to neglect the social value of nonsymbolic communication to animals. In a discussion of communication by such means, Premack gives the following human analogy. He asks us to consider two ways in which we might benefit from his knowledge of present conditions at some place he has just visited and we have not.

> I could return and tell you, "The apples next door are ripe." Alternatively, I could come back from next door chipper and smiling. On still another occasion I could return and tell you, "A tiger is next door." Alternatively, I could return mute with fright, disclosing an ashen face and quaking limbs. The same dichotomy could be arranged on numerous occasions. I could say, "The peaches next door are ripe," or say nothing and manifest an intermediate amount of positive affect since I am only moderately fond of peaches. Likewise, I might report, "A snake is next door," or show an intermediate amount of negative affect since I am less shaken by snakes than by tigers. [Premack, 1974:1]

With foreknowledge of his fears and appetites we can learn much of value without any symbolic communication if we know where and when the situation mirrored in his affective signaling behavior was experienced. While the distinction between affective and symbolic communication may not be a sharp one (Marler, this volume, p.54), similar arguments help us understand the communicative value of signals with nonspecific referents (a possible paraphrase of "affective signals") as contrasted with signals that symbolize a specific class of reference.

Menzel (1971, 1973, 1974) has demonstrated the efficiency with which captive chimpanzees can choose which companion to follow to hidden food, the companion having been shown the site by the experimenter beforehand. Respondents even developed skill in overtaking the leader at the last minute and reaching the food first. However, it is not always clear that the term "communication" is appropriate for such interchanges. To judge from Menzel's descriptions, respondents may derive their cues not only from observing the leader's signaling behavior but also by directly perceiving its eager efforts to regain the hidden food itself.

The relationship between the two animals Rock and Bell is illuminating in this regard (Menzel, 1974:134–35). Rock was dominant over Bell and if she led him to food he would attack her and take it as soon as she uncovered it. As tests continued, Bell became more and more devious about approaching the food when she was the experimental "leader" and Rock was present. As Menzel describes later trials,

> Bell accordingly stopped uncovering the food if Rock was close. She sat on it until Rock left. Rock, however, soon learned this, and when she sat in one place for more than a few seconds, he came over shoved her aside, searched her sitting place, and got the food. Bell next stopped going all the way. Rock, however, countered by steadily expanding the area of his search through the grass near where Bell had sat. Eventually Bell sat farther and farther away, waiting until Rock looked in the opposite direction before she moved towards the food at all—and Rock in turn seemed to look away until Bell started to move somewhere. On some occasions Rock started to wander off, only to wheel around suddenly precisely as Bell was about to uncover the food. Often Rock found even carefully hidden food that was thirty feet or more from Bell, and he oriented repeatedly at Bell and adjusted his place of search appropriately if she showed any signs of moving or orienting in a given direction. If Rock got very close to the food, Bell invariably gave the game away by a "nervous" increase in movement. However on a few trials she actually started off a trial by leading the group in the opposite direction from food, and then, while Rock was engaged in his search, she doubled back rapidly and got some food. In other trials, when we hid an extra piece of food about 10 feet away from the large pile, Bell led Rock to the single piece, and while he took it she raced for the pile. When Rock started to ignore the single piece of food to keep his watch on Bell, Bell had temper "tantrums."

The actions of Bell to which Rock is responding seem in no way specialized for a communicative function, failing to satisfy what many regard as a critical criterion (Hockett, 1961). It is as appropriate to question whether Bell is "communicating" with Rock as to ask whether a mouse racing for its burrow "communicates" to an owl the appropriate trajectory for a successful attack upon it. Nevertheless, such interactions are fascinating for the perspicacity they reveal in one animal's ability to predict what another is going to do, surely a key element in the evolution of communication.

SIGNALS AND SPACING

Communicative interchange often has repercussions on the relative spacing of the participants. They may come together for a period of time, engage in some activity that requires proximity, then separate again afterward. Alternatively, the consequence of communication may be repulsion, or the maintenance of spacing that might otherwise vary. By its impact on spacing, social communication has immediate and far-reaching ecological consequences, becoming in turn the focus of strong selective influences. Yet we are still ignorant of the ways in which signaling behavior controls the spatial organization of ape populations. Approaches that show promise, such as Kummer's analyses of spacing mechanisms in baboons and Waser's demonstration by field playback techniques of the role of the grey-cheeked mangabey whoop-gobble in intertroop spacing, have yet to be systematically applied to studies of ape behavior (Kummer, 1971; Waser, 1975). However, gibbons have been induced to approach and withdraw by tape recorded calls played back in the field (Brockelman and Kobayashi, 1971; Mackinnon and Chivers, in prep.). The inferences to be drawn from the descriptive studies reviewed here are speculative and limited in scope.

Although the apes display a broad spectrum of patterns of social organization, the most striking conclusion to emerge from this survey of signaling behavior is the degree of correspondence between species. Their vocal repertoire sizes are similar. We have seen that the chimpanzee and the gorilla, whose social systems differ strikingly, share many sound signals so intimately that one can discern homologies between a significant portion of their repertoires. As other ethologists have concluded in comparative studies of communicative behavior, so in the apes one must look to changes in the pattern of usage of signals to explain the divergent communicative organization of species that share common origins.

If we compare the patterns of social organization in chimpanzee and gorilla in terms that might relate to their systems of vocal communication, they seem to have more in common in patterns of between-group than within-group relationships. There is territoriality in both species, and adult males take a prime role in territorial defense in both the chimpanzee and the gorilla. They differ in that the chimpanzee territory is maintained by a group of males while in the gorilla the onus tends to fall on one silver-backed male. The two species also differ in the closeness of identification of adult females with a particular group and its home range and territorial commitments. However, in neither species is there evidence as yet of a female role in territorial maintenance. Thus while intergroup relationships in chimpanzee and gorilla do differ in a number of significant respects they also have major features in common.

With regard to within-group relationships, the contrast between the two species is more striking. Groups of gorillas are relatively compact and coherent. Individuals may get out of visual contact with companions feeding in dense vegetation, and vocalizations, especially belching, are used to maintain contact in this circumstance (Harcourt, pers. comm.). The belching vocalization is a dominant component in the vocal behavior of habituated gorillas. The distances involved are small, and mechanisms for reestablishing contact seem relatively direct and uncomplicated.

Chimpanzee group members are often much more dispersed. Some adult males and females spend as much as 80 to 95 percent of their time alone (Wrangham, pers. comm.). To communicate with other group members, they must often signal vocally over great distances, with companions out of sight to them. During long periods of isolation any adult may encounter predators. In rejoining group members all must be ready to engage in the signal exchanges required for reestablishment of relationships with fellow group members after long periods of visual if not vocal separation. These are occasions of high arousal and involve a variety of extended vocal and visual signaling.

Thus patterns of chimpanzee and gorilla social organization differ most strikingly in the organization of within-group relationships. The chimpanzee has a large, dispersed social group, containing several adult males, with members recombining from day to day in different subgroupings of adult males, females, and young. They also spend much time alone. Group members are often separated by long distances. The gorilla has smaller, more coherent social groups, usually with only one fully adult male. Within-group vocal signaling is over much shorter ranges. The rate of within-group reunions, and the level of social arousal associated with them, is much lower than in the chimpanzee. The compactness of the group is such that they confront such exigencies as predator detection, dissemination of alarm, and defense as a group rather than on an individual basis.

Given these differences in social organization in the two species, can they be correlated with any of the differences in vocal behavior? If so, can a case be made for the relationship being a causal one? Given the compact gorilla social group, with relatively stable social composition, it is possible for one sex- and age-class to assume responsibility for many of the communicative decisions required for the maintenance of within-group social organization, defense against predators, as well as the maintenance of relations with other social groups and solitary individuals. An argument such as this makes the domination of gorilla social behavior by silverbacked males to some extent intelligible. However, although gorilla group composition and coherence *permits* the silverbacked male to assume the main prerogative of vocal signaling, it is by no means clear that this is a *required* consequence.

A significant and perhaps major proportion of chimpanzee vocal behavior is occasioned by renewed contacts between within-group social sub-units. The highly dispersed, fluid nature of chimpanzee society must favor competence to cope with major environmental and social contingencies on an individual basis in all sex and age groups other than dependent infants. The obvious exception here is territorial defense, which is a solely male responsibility, as in the gorilla. The major vocalization involved in territorial exchanges, pant-hooting, also serves other functions incorporating long-distance vocal signaling, hence its presence in the female repertoire as well. Moreover we have noted that the female form of this vocalization is discriminable from that of adult males. The other acoustical signal involved in between-community interactions, drumming, is restricted to adult males.

There is ample evidence that chimpanzee subgroupings signal vocally to one another over distances of hundreds of meters, something that probably never occurs within a gorilla group (Harcourt, pers. comm.). An obvious case is meat-eating, where loud sustained vocalizations attract distant group members to the event (van Lawick-Goodall, 1968a, 1971; Teleki, 1973b). Coordination of movements and other behaviors within the community is presumably modulated by these exchanges, in which all animals other than infants may participate on occasion, presumably to the benefit of the group as well as to themselves. The remarkable spread of the use of all vocalizations throughout the chimpanzee community membership is perhaps interpretable in these terms, in striking contrast to the domination of gorilla vocal behavior by one sex- and age-class, the silverbacked males.

We can also speculate about the adaptive significance of similarities that exist in the vocalizations of the chimpanzee and the gorilla. In both, the onus of territorial maintenance falls primarily on mature adult males. In both species, the vocalizations most commonly enjoyed in intergroup encounters, pant-series and pant-hooting, are given most frequently by the age-class responsible for territorial maintenance, mature males, even though the asymmetry in usage is much less striking in the chimpanzee than in the gorilla. The similarity of the non-vocal sounds used in such encounters, chest-beating and drumming, can only result from convergence. It remains to be determined whether such mechanical sounds have a unique advantage in this context, such as better carrying-power than vocalizations, or whether interactions between the two species at some stage of their history might have favored interspecific territoriality, and thus convergence on similar signals for intergroup spacing. If details of vocal behavior are shaped by adaptive social function with any degree of precision, and we should remember that this is by no means a proven assumption, then we might infer from the remarkable degree of correspondence in the morphology of vocalizations of the two species that the major environmental and social contingencies met in within-group behavior of the two species are similar in nature, although differing in the distribution of responsibilities through the group. Some of the quantitative differences in ranking of use of equivalent vocalizations may also be adaptive. The need to maintain greater group coherence in the gorilla perhaps explains the more frequent use and broader array of contexts of gorilla belching, as compared with the less frequently used and more restricted chimpanzee rough grunting. Belching often seems to serve foraging gorillas as a "keeping in contact" signal (Harcourt, pers. comm.). Chimpanzees have no call that functions as a general, close-range, contact signal. In spite of the differences in their social organization it is interesting to note that several equivalent vocal pairs in the two species rank rather similarly in frequency of usage.

These then are the kinds of correlations between social organization and vocal signaling that throw some light on the behavioral similarities and differences in the apes. While the data are still imperfect, we can begin to see more clearly what additional observations are desirable. Data on the intensity and range of reception of vocalizations with different functions are needed. Above all, new approaches should be sought to characterize the *functions* of different vocalizations, so that more subtle interspecies comparisons of the proportions of a signal repertoire devoted to different kinds of adaptive tasks may be possible. For instance we have noted that apart from the gorilla roar, the only other gorilla call that definitely seems to lack a counterpart in the chimpanzee repertoire is the growl. If an aggressive function can indeed be assigned to this call, can we infer that there is a need for a close-range within-group aggressive vocalization in the gorilla which is absent in the chimpanzee, or met in a different way? Within-group aggression is often a noisier and more highly aroused interaction in the chimpanzee, including elaborate and highly ritualized aggressive displays. Playback of recorded vocalizations applied recently to primates in the field by Waser (1974, 1975) in studies of vocal communication in mangabeys is one approach to such functional questions.

Another issue on which information is urgently needed is the distribution of kin relationships in ape societies. The extent of personal acquaintanceships is another issue. One could imagine that the chimpanzee's social system, more so than the gorilla's, requires familiar acquaintanceship to be spread farther through the population. Chimpanzee vocal behavior could be one means of establishing and maintaining this larger number of individual acquaintances. At least one call, which also happens to be the most frequently used, does bear information about individual identity. By contrast one might speculate that in gibbons, the sphere of acquaintanceship would be more limited than in the chimpanzee. However, observations of young Kloss's gibbons settling adjacent to their natal territories suggest that frequent use of loud vocalizations might serve to maintain kinship ties as well as mediating territorial competition. Thus relationships between neighboring gibbon groups may be more complex than those of the gorilla, with the social network in which an animal is involved with others on an individual basis extending beyond the family group into neighboring ones. We have noted in Kloss's gibbon the use of alarm calls so loud that they seem designed to reach farther than the limits of a typical home range. They effectively link otherwise competitive social units in a cooperative association for avoidance of predators. Knowledge of how kin are distributed in other ape populations will have a significant bearing on our understanding of how this and other signaling behavior affects social organization.

The list of important but unanswered questions about social communication in apes is a long one. Although enormous strides have been made in our knowledge about the behavior of apes, we are still far from achieving any kind of general theory about how variations in communicative behavior mediate variations in the social organization. Even our descriptions of signaling are incomplete, and we are largely ignorant of its effects on the behavior of others. The further analysis of ape communication presents challenges for which the next generation of field observers and experimenters will have to develop new methods of study and analysis if the many difficulties are to be overcome.

The difficulties in understanding communication in free-living primates are not only practical but logical. It is hard to understand a communicative system without participating fully in it oneself. In the process of disentangling the relationships between social organization and signaling behavior there is a prospect of learning more about the general principles underlying animal communication, undoubtedly more complex and subtle than is often supposed.

References

Aldrich-Blake, F. P. G., and Chivers, D. J., 1973. On the genesis of a group of siamang. *Am. J. Phys. Anthrop.*, 38:631–36.

Andrew, R. J., 1963. The origin and evolution of the calls and facial expressions of the primates. *Behaviour,* 20:1–109.

Baldwin, L. A., and Teleki, G., 1976. Patterns of gibbon behavior on Hall's Island, Bermuda. *Gibbon and Siamang,* 4:21–105.

Baylis, J., in press. A quantitative study of long-term courtship: II. The dynamics of courtship in two new world cichlids. *Behaviour.*

Berkson, G.; Ross, B. A.; and Jatinandana, S.; 1971. The social behavior of gibbons in relation to a conservation program. *Primate Behavior,* 2:225–55.

Boutan, L., 1913. Le pseudo-language, observations effectuées sur un anthropoide: le Gibbon (*Hylobates leucogenys* Ogilby). *Act. Soc. Linn.* Bordeaux, 47:5–81.

Brockelman, W. Y., and Kobayashi, N. K., 1971. Live capture of free-ranging primates with a blowgun. *J. Wildl. Mgmt.*, 35:852–55.

Brockelman, W. Y.; Ross, B. A.; and Pantuwatana, S.; 1974. Social interactions of adult gibbons *(Hylobates lar)* in an experimental colony. *Gibbon and Siamang,* 3:137–56.

Carpenter, C. R., 1964. *Naturalistic Behavior of Nonhuman Primates.* University Park, Pa.: Pennsylvania State University Press.

Chasen, F. N., and Kloss, C. B., 1927. Spolia Mentawiensia, mammals. *Proc. Zool. Soc. Lond.*, pp.797–840.

Chevalier-Skolnikoff, S., 1974. Facial expression of emotion in nonhuman primates. In: *Darwin and Facial Expression,* P. Ekman, ed. New York: Academic Press.

Chivers, D. J., 1971. Spatial relations within the siamang group. *Proc. 3rd Int. Cong. Primat.*, Zurich, 3:14–21.

Chivers, D. J., 1972. The siamang and the gibbon in the Malay Peninsula. *Gibbon and Siamang,* 1:103–35.

Chivers, D. J., 1974. *The Siamang in Malaya: A Field Study of a Primate in Tropical Rain Forest,* H. Hofer and A. H. Schultz, eds. Contributions to Primatology, vol. 4. Basel: Karger.

Chivers, D. J., in press. Communication within and between siamang family groups. *Behaviour.*

Eisenberg, J. F., 1966. The social organization of mammals. *Handbuch der Zoologie,* Band 8, Leiferung 39, 10(7):1–92.

Ellefson, J. O., 1967. A natural history of gibbons in the Malay Peninsula. Ph.D. diss., University of California, Berkeley.

Ellefson, J. O., 1974. A natural history of white-handed gibbons in the Malayan Peninsula. *Gibbon and Siamang,* 3:1–136.

Fisler, G., 1969. Mammalian organizational systems. *Contrib. Sci. Los Angeles Co. Mus. Nat. Hist.*, no. 167.

Fooden, J., 1969. Color-phase in gibbons. *Evolution,* 23:627–44.

Fossey, D., 1972. Vocalizations of the mountain gorilla *(Gorilla gorilla beringei). Anim. Behav.*, 20:36–53.

Fossey, D., 1974. Observations on the home range of one group of mountain gorillas *(Gorilla gorilla beringei). Anim. Behav.*, 22:568–81.

Galdikas-Brindamour, B., 1975. People of the forest. *Nat. Geogr. Mag.*, 148:444–73.

Gardner, B. T., and Gardner, R. A., 1971. Two-way communication with an infant chimpanzee. In: *Behavior of Nonhuman Primates,* A. M. Schrier and F. Stollnitz, eds. New York: Academic Press.

Gautier, J. P., 1971. Etude morphologique et fonctionnelle des annexes extra-laryngées des Cercopithecinae; liaison avec les cris d'éspacement. *Biol. gabon.,* 7:229–67.

Gill, T. V., and Rumbaugh, D. M., in press. Mastery of naming skills by a chimpanzee. *J. of Human Evol.*

Glasersfeld, E. von, in press. Signs, communication and language. *J. of Human Evol.*

Green, S., 1975a. Variation of vocal pattern with social situation in the Japanese monkey *(Macaca fuscata):* a field study. In: *Primate Behavior: Developments in Field and Laboratory Research,* vol. 4, L. A. Rosenblum, ed. New York: Academic Press.

Green, S., 1975b. Dialects in Japanese monkeys: vocal learning and cultural transmission of locale-specific vocal behavior. *Z. Tierpsychol.,* 38:304–14.

Harrisson, B., 1960. A study of orang-utan behaviour in semi-wild state, 1956–1960. *Sarawak Mus. J.,* 9:422–47.

Hess, J. P., 1973. Some observations on the sexual behaviour of captive lowland gorillas, *Gorilla g. gorilla* (Savage and Wyman). In: *Comparative Ecology and Behaviour of Primates,* R. P. Michael and J. H. Crook, eds. New York: Academic Press.

Hockett, C., 1961. Logical considerations in the study of animal communication. In: *Animal Sounds and Communication,* W. E. Lanyon and W. N. Tavolga, eds. Washington, D.C.: Am. Inst. Biol. Sci. Publ. no. 7.

Horr, D. A., 1972. The Borneo orang-utan. Population structure and dynamics in relationship to ecology and reproductive strategy. Unpubl. paper read to Am. Assoc. Phys. Anthrop.

Huber, E., 1931. *Evolution of Facial Musculature and Facial Expression.* Baltimore: Johns Hopkins Press.

Kummer, H., 1971. Spacing mechanisms in social behavior. In: *Man and Beast: Comparative Social Behavior,* J. F. Eisenberg and W. S. Dillon, eds. Washington, D.C.: Smithsonian Institution Press.

Lamprecht, J., 1970. Duettgesang beim Siamang, *Symphalangus syndactylus* (Hominoidea, Hylobatinae). *Z. Tierpsychol.,* 27:186–204.

Langer, S., 1942. *Philosophy in a New Key.* New York: Mentor Books.

Lieberman, P., 1968. Primate vocalizations and human linguistic ability. *J. Acoust. Soc. Amer.,* 44:1574–84.

Lieberman, P.; Crelin, E. S.; and Klatt, D. H.; 1972. Phonetic ability and related anatomy of the newborn and adult human, Neandertal man, and the chimpanzee. *Amer. Anthropologist,* 74:287–307.

Mackinnon, J., 1974. The behaviour and ecology of wild orang-utans (*Pongo pygmaeus*). *Anim. Behav.,* 22:3–74.

Mackinnon, J., 1975. *In Search of the Red Ape.* New York: Ballantine Books.

Marler, P., 1955. Characteristics of some animal calls. *Nature* (London), 176:6–7.

Marler, P., 1965. Communication in monkeys and apes. In: *Primate Behavior,* I. DeVore, ed. New York: Holt, Rinehart & Winston.

Marler, P., 1968. Aggregation and dispersal: two functions in primate communication. In: *Primates. Studies in Adaptation and Variability,* P. C. Jay, ed. New York: Holt, Rinehart & Winston.

Marler, P., 1969. Vocalizations of wild chimpanzees, an introduction. In: *Proc. 2nd Int Cong. Primat.,* Atlanta, 1968, 1:94–100. Basel: Karger.

Marler, P., 1973. A comparison of vocalizations of red-tailed monkeys and blue monkeys: *Cercopithecus ascanius* and *C. mitis* in Uganda. *Z. Tierpsychol.,* 33:223–47.

Marler, P., 1976. Social organization, communication and graded signals: vocal behavior of the chimpanzee and the gorilla. In: *Growing Points in Ethology,* P. Bateson and R. A. Hinde, eds. Cambridge: Cambridge University Press.

Marler, P., and Hobbet, L., 1975. Individuality in a long-range vocalization of wild chimpanzees. *Z. Tierpsychol.,* 38:97–109.

Marler, P., and van Lawick-Goodall, J., 1971. *Vocalizations of Wild Chimpanzees* (sound film). New York: Rockefeller University Film Service.

Marler, P., and Mundinger, P., 1971. Vocal learning in birds. In: *Ontogeny of Vertebrate Behavior,* H. Moltz, ed. New York: Academic Press.

Marshall, J. T., Jr.; Ross, B. A.; Chantharojvong, S.; 1972. The species of gibbons in Thailand. *J. Mammal.,* 53:479–86.

Menzel, E. W., 1971. Communication about the environment in a group of young chimpanzees. *Folia Primat.,* 15:220–32.

Menzel, E. W., 1973. Leadership and communication in young chimpanzees. In: *Precultural Primate Behavior,* E. W. Menzel, ed. *Symp. 4th Int. Cong. Primat.* 1:192–225. Basel: Karger.

Menzel, E. W., 1974. A group of young chimpanzees in a one-acre field. In: *Behavior of Nonhuman Primates,* A. M. Schrier and F. Stollnitz, eds. New York: Academic Press.

Michael, R. P., 1969. The role of pheromones in the communication of primate behaviour. In: *Proc. 2nd*

Int. Cong. Primat., Atlanta, 1968. 1:101–107. Basel: Karger.

Miller, G. S., 1903. Seventy new Malayan mammals. *Smithsonian Misc. Collns.,* 45:1–73.

Moynihan, M., 1970. Control, suppression, decay, disappearance and displacement of displays. *J. Theoret. Biol.,* 29:85–112.

Negus, V. E., 1949. *The Comparative Anatomy and Physiology of the Larynx.* New York: Grune and Stratton.

Nishida, T., and Kawanaka, K., 1972. Inter-unit-group relationships among wild chimpanzees of the Mahali mountains. *Kyoto Univ. Afr. Stud.,* 8:131–69.

Premack, D., 1971. On the assessment of language competence in the chimpanzee. In: *Behavior of Nonhuman Primates,* A. M. Schrier and F. Stollnitz, eds. New York: Academic Press.

Premack, D., 1974. Concordant preferences as a precondition for affective but not for symbolic communication (or how to do experimental anthropology). In: *Experimental Behaviour: A Basis for the Study of Mental Disturbance,* J. H. Cullen, ed. New York: John Wiley and Sons.

Premack, D., 1975. Symbols inside and outside of language. In: *The Role of Speech in Language,* J. F. Kavanagh and J. E. Cutting, eds. Cambridge: MIT Press.

Reynolds, V., and Reynolds, F., 1965. Chimpanzees in the Budongo Forest. In: *Primate Behavior: Field Studies of Monkeys and Apes,* I. DeVore, ed. New York: Holt, Rinehart & Winston.

Rodman, P. S., 1973. Population composition and adaptive organisation among orang-utans of the Kutai Reserve. In: *Comparative Ecology and Behaviour of Primates,* R. P. Michael and J. H. Crook, eds. New York: Academic Press.

Rumbaugh, D. M.; von Glasersfeld, E.; Warner, H.; Pisani, P.; and Gill, T. V.; 1974. Lana (chimpanzee) learning language: a progress report. *Brain and Language,* 1:205–212.

Schaller, G. B., 1963. *The Mountain Gorilla.* Chicago: University of Chicago Press.

Schultz, A. H., 1932. The generic position of *Symphalangus klossii. J. Mammal.* 13:368–69.

Simpson, M. J. A., 1973. The social grooming of male chimpanzees. In: *Comparative Ecology and Behaviour of Primates,* R. P. Michael and J. H. Crook, eds. New York: Academic Press.

Smith, J. M., 1965. The evolution of alarm calls. *Amer. Natural.,* 99:59–63.

Smith, W. J., 1969. Messages of vertebrate communication. *Science,* 165:145–50.

Stebbins, W. C., 1971. Hearing. In: *Behavior of Nonhuman Primates,* A. M. Schrier and F. Stollnitz, eds. New York: Academic Press.

Struhsaker, T., 1967. Auditory communication among vervet monkeys *(Cercopithecus aethiops).* In: *Social Communication Among Primates,* S. A. Altmann, ed. Chicago: University of Chicago Press.

Studdert-Kennedy, M., 1975. Speech perception. In: *Contemporary Issues in Experimental Phonetics,* N. J. Lass. ed. Springfield, Illinois: Thomas.

Teleki, G., 1973a. Group response to the accidental death of a chimpanzee in Gombe National Park, Tanzania. *Folia Primat.,* 20:81–94.

Teleki, G., 1973b. The predatory behavior of wild chimpanzees. Lewisburg, Pa., Bucknell University Press.

Tenaza, R. R., 1975. Territory and monogamy among Kloss' gibbons *(Hylobates klossii)* in Siberut Island, Indonesia. *Folia Primat.,* 24:60–80.

Tenaza, R. R., 1976. Songs and related behavior of Kloss' gibbons *(Hylobates klossii)* in Siberut Island, Indonesia. *Z. Tierpsychol.* 40:37–52.

Terra, H. de, 1943. Pleistocene geology and early man in Java. *Trans. Am. Phil. Soc.,* n.s., 32:437–64.

van Hooff, J. A. R. A. M., 1962. Facial expressions in higher primates. *Symp. Zool. Soc. Lond.,* 8:97–125.

van Hooff, J. A. R. A. M., 1967. The facial displays of the Catarrhine monkeys and apes. In: *Primate Ethology,* D. Morris, ed. London: Weidenfeld and Nicolson.

van Hooff, J. A. R. A. M., 1970. A component analysis of the structure of the social behaviour of a semi-captive chimpanzee group. *Experientia,* 26:549–50.

van Hooff, J. A. R. A. M., 1971. A structural analysis of the social behaviour of a semi-captive group of chimpanzees. In: Aspecten van het sociale Gedrag en de Communicatie bij Humane en Hogere Niet-humane Primaten. Diss., University of Utrecht, pp.11–127.

van Lawick-Goodall, J., 1968a. A preliminary report on expressive movements and communication in the Gombe Stream chimpanzees. In: *Primates. Studies in Adaptation and Variability,* P. C. Jay, ed. New York: Holt, Rinehart & Winston.

van Lawick-Goodall, J., 1968b. The behaviour of free-living chimpanzees in the Gombe Stream Reserve. *Anim. Behav. Monog.,* 1:161–311.

van Lawick-Goodall, J., 1971. *In the Shadow of Man.* London: Collins.

von Glasersfeld, E., in press. Signs, communication and language. *J. of Human Evol.*

Waser, P., 1974. Intergroup interactions in a forest monkey: the mangabey *Cercocebus albigena*. Ph.D. diss., The Rockefeller University.

Waser, P. M., 1975. Experimental playbacks show vocal mediation of intergroup avoidance in a forest monkey. *Nature,* 255:56–58.

Wickler, W., 1967. Socio-sexual signals and their intraspecific imitation among primates. In: *Primate Ethology,* D. Morris, ed. London: Weidenfeld and Nicholson.

Wrangham, R. W. (1974). Artificial feeding of chimpanzees and baboons in their natural habitat. *Animal Behaviour,* 22:83–93.

Wrangham, R. W. (1975). Behavioral ecology of chimpanzees in Gombe National Park, Tanzania. Ph.D. thesis. University of Cambridge.

Chapter 37

MAN–CHIMPANZEE COMMUNICATION

Roger S. Fouts and Randall L. Rigby

This chapter traces the scientific inquiries into two-way communication with chimpanzees from the early attempts to establish vocal communication to ongoing research in gestural and symbolic languages. We shall begin by selectively reviewing the historical speculation concerning the possibility of teaching chimpanzees to speak. This will be followed by a review of early experiments in raising chimpanzees in a human home environment, and by a discussion of the more recent experiments concerning the use of gestural languages and symbols in establishing two-way communication.

The chimpanzee is a nonhuman primate that is very similar to, and at the same time very different from, a human being. From either of these aspects we can obtain a wealth of information on the mental and behavioral capacities of the chimpanzee and, in addition, comparative data to assist in the understanding of human behavior. Since we are interested in two-way communication between man and chimpanzee, we will emphasize the similarities of the communication capabilities of the two species. It is obvious that a physical similarity exists, but it is important to stress that there are also some basic differences. Although man and chimpanzee had a common ancestor, many thousands of years of separate evolution and adaptation have endowed each species with unique physical and behavioral characteristics. However, some remarkable physiological similarities have been found in blood protein and type, chromosomal characteristics, structure, and behavior. The last two were observed by early researchers in man-animal communication.

Historical Developments and Speculation Concerning Communication

The Great Apes

Probably more than any single factor, the physical similarity between man and the great apes aroused the curiosity of those interested in teaching apes to behave in ways similar to man. The famous *Diary* of Samuel Pepys reflects this interest in an entry made in August 1661:

> By and by we are called to Sir N. Battens to see the strange creature that Captain Jones hath brought with him from Guiny; it is a great baboon, but so much like a man in most things, that (though they say there is a species of them) yet I cannot believe but that it is a monster got of a man and a she-baboon. I do believe it already understands much English; and I am of the mind it might be taught to speak or make signs.

A similar reaction was recorded by Julien Offray de La Mettrie (1709–1751), who, in *L'Homme machine* (1748), pondered the varying capacity of animals to learn.[1] La Mettrie, obviously attracted to the striking similarities between man and the apes, proposed teaching sign language to apes in a school for the deaf. His idea was to choose an ape with the most "intelligent face" and send him to school under the teacher Amman (an early writer of books on the education of the deaf). La Mettrie failed to distinguish between monkeys, apes, and orangs (he referred to them interchangeably), but his basic idea was clearly two centuries ahead of its time. It is apparent that he recognized the intellectual capacity of the ape when he wrote:

> Why should the education of monkeys be impossible? Why might not the monkey, by dint of great pains, at last imitate after the manner of deaf mutes, the motions necessary for pronunciation? . . . it would surprise me if speech were absolutely impossible in the ape.

Another distinction made by La Mettrie, which was later to become a cornerstone of controversy concerning language acquisition in chimpanzees, was that speech and/or communication with lower primates included the use of gestures. He was obviously influenced by Amman's works, including *Surdus loquens* (1692) and *Dissertatio de loquela* (1700), which contained plans for teaching signs, finger spelling, and lip reading. Training apes to use such communication methods, or at least gestural communication, was to La Mettrie a logical proposal, and he felt that they would be able to master it easily. Unfortunately, La Mettrie was unable to follow up on his idea, and the proposal that apes could learn to communicate with man lay dormant for years.

R. M. Yerkes (1925:53), who devoted his entire life to observing and writing about the great apes and other nonhuman primates, echoed La Mettrie's proposal almost two centuries later: "If the imitative tendency of the parrot could be coupled with the quality of intelligence of the chimpanzee, the latter undoubtedly could speak." He predicted future scientific trends when he stated,

> I am inclined to conclude from the various evidences that the great apes have plenty to talk about, but no gift for the use of sounds to represent individual, as contrasted to racial, feelings or ideas. Perhaps they can be taught to use their fingers, somewhat as does the deaf and dumb person, and helped to acquire a simple, nonvocal sign language. [Yerkes, 1925]

THE INFLUENCE OF THE HOME-RAISING EXPERIMENTS

With the beginning of the twentieth century and the application of scientific principles to psychological developments and discoveries, an innovative type of animal-training experiment developed. Using an evolutionarily close relative of man, the chimpanzee, the experimenters attempted to duplicate the environment of a human household, the conditions of which are most favorable to language acquisition in man. Because the requirements for human language were obviously met in such a setting, i.e., adequate social atmosphere, sufficient periods in which to practice babbling, and an appropriate model for vocal imitation, it was assumed that the chimpanzee would acquire vocal language. Lightner Witmer (1909) summarized the rationale for this type of experiment:

> While my tests of Peter give no positive assurance that he can acquire language, on the other hand they yield no proof that he cannot. If Peter had a human face and were brought to me as a backward child and this child responded to my tests as credibly as Peter did, I should unhesitatingly say that I could teach him to speak, to write and to read within a year's time.

1. The authors are grateful to Gordon Hewes, University of Colorado, for bringing to our attention this work of La Mettrie.

Peter has not a human form, and what limitations his ape's brain may disclose after a persistent effort to educate him, it is impossible to foretell. His behavior, however, is sufficiently intelligent to make this educational experiment well worth the expenditure of time and effort.

In general, the rationale for the home-raising experiments that followed were based on the above assumptions. If the necessary language-eliciting environment were provided, perhaps the puzzle of nonhuman primate language could be solved.

As pointed out by Kellogg (1968), keeping nonhuman primates as pets in the home is certainly not a novel idea and can be traced back several centuries. Such practices frequently occur today, but no instances of intellectual language use by such pets have been reported (Yerkes, 1925; Kellogg, 1968). A major problem is that the great majority of those who have nonhuman primates in the home are not familiar with language training and are ill equipped to observe and record the animals' reactions. However, as Kellogg (1968) observes, "It is quite another story for trained and qualified psychobiologists to observe and measure the reactions of a home-raised pongid amid controlled experimental home surroundings." We will describe several home-raising experiments, all of which were designed to determine the extent to which a nonhuman primate could acquire a vocal language capability.

Peter

The first chimpanzee to be mentioned was named Peter and was observed by L. Witmer (1909). Although this study was not a home-raising experiment, the rationale and desire to undertake such a task with an ape was inspired by this and similar reports found in the literature during this period.

Peter was a 4- to 6-year-old chimpanzee owned by a man named McArdles, who had trained him for 2½ years. Employed by Keiths Theatre in Philadelphia as an example of "a monkey who made a man of himself," the entertaining, humanlike chimpanzee aroused the curosity of Witmer, who made arrangements to test Peter to determine the extent of his intelligence. The tests were conducted in the fall of 1909 at the Psychological Clinic in Philadelphia. Most of them involved motor coordination and simple reasoning tasks—opening a box to obtain articles inside, unlocking locks with keys, and driving nails into a board with a hammer—all of which Peter was able to do with relative ease. It was observed that Peter showed only imitative writing movements and possessed no special writing ability. He was able to articulate the word "mama," however; it is noted that he did so with considerable effort and with apparent unwillingness. The articulation of the sound "m" was said to be perfect, but the second "ma" in "mama" sounded more like "ah" and was inaudible more often than not. Witmer noted that the sound was a hoarse whisper rather than an articulated word and that Peter always tried to speak with the inspired rather than the expired breath. Peter was trained to articulate the sound "p" in only a few minutes, leading Witmer to conclude:

This experiment was enough to convince me that Peter can be taught to articulate a number of consonantal sounds and probably to voice correctly some of the vowels. . . . If a child without language were brought to me and on the first trial had learned to articulate the sound "p" as readily as Peter did, I should express the opinion that he could be taught most of the elements of articulate language within six months' time.

Witmer also observed that although Peter was unable to speak, he was nonetheless able to understand spoken words. Using the analogy of Helen Keller, who first comprehended the use of symbols in the place of objects, Witmer proposed that in a similar manner Peter could be made to comprehend symbols as representing objects; and with further training to articulate these symbols, he would be able to communicate. Recognizing that early language training would be crucial in the event that vocal language could be taught to the chimpanzee, Witmer predicted that "within a few years, chimpanzees will be taken early in life and subjected for purposes of scientific investigation to a course of procedure more closely resembling that which is accorded the human child." Clearly a precursor for things to come, Witmer's prediction has been attempted on several occasions.

Joni

Joni (Kohts, 1935) was a male chimpanzee raised and observed by N. Kohts and her family from 1½ to 4 years of age. The period of observation was from 1913 to 1916, and the daily events pertaining to Joni's behavior were recorded and later compared with those of Kohts's own human child, Roody, during the period 1925 to 1929. Kohts's manuscript was prepared in the early 1930s, fifteen years after the chimpanzee had been observed and tested. The comparison of the developmental sequences of the home-raised chimpanzee and of the child probably reflects the influence of Yerkes. Only a small part of the report is devoted to language capacity in the chimpanzee, and this is in direct comparison to that of the human child. No special attempts were made to train Joni to use articulate language, the author's purpose being to record the language capability emitted in the home environment without any special training. Kohts reported that Joni was able to produce at least twenty-five sounds elicited by various stimuli within his environment, but they were clearly his own natural sounds to express emotions and desires. Our own observations of chimpanzees are to a very great extent in agreement with Kohts's, in that vocalizations in chimpanzees seem to be elicited by the environment. (Fouts, 1973).

During the period of observation the chimpanzee never attempted to imitate the human voice or to express himself by other than his own characteristic utterances. The lack of vocal language communication was interpreted by Kohts to mean that the mental capability of the chimpanzee was qualitatively different from that of the human child. This was possibly in reaction to Yerkes, who had published his book *Almost Human* in 1927. Kohts responded by saying: "Not only is it impossible to say that he is 'almost human'; we must go even further and state quite definitely that he is 'by no means human.' " This conclusion was based mostly on language capability, for Kohts was able to observe many instances of human-chimpanzee commonality in play behavior, emotional expressions, conditioned reflexes, and "a few" intellectual processes, including curiosity, recognition, identification, and "sounds of an undifferentiated nature." It was concluded that the more biologically important the function the greater the capability of the ape to approach or surpass that of man. Conversely, the higher the intellectual process involved (Kohts considered language the highest) the more dominant man became over the apes.

Kohts's observation that there is a qualitative difference between humans and chimpanzees stems from her emphasis on vocal language capability rather than language capability regardless of mode. To view intellectual processes solely from the position of vocal language ability, in our opinion, limits the possible avenues for two-way communication with nonhuman primates. We conclude from reading Kohts's account that the study was undertaken with the expectation of establishing two-way vocal communication with the chimpanzee within a matter of months. When this expectation was not met but was later achieved with Roody, Kohts's interpretation was that the chimpanzee was qualitatively different from humans in mental capabilities. A valuable lesson learned here is that physical similarity does not necessarily mean total similarity. The chimpanzee may look and act like man but its mode of communication is not necessarily the same. Kohts's conclusion that a qualitative difference in intelligence exists between humans and chimpanzees because of differing communication modes is a common, prejudicial misjudgment.

Gua

The experiment involving the infant chimpanzee Gua evolved as a direct result of the influence of Witmer (1909), and Yerkes (1925), and Yerkes and Yerkes (1929), who had expressed opinions concerning the feasibility of raising an infant chimpanzee in the home. Gua was 7½ months old when she was obtained by W. N. Kellogg and L. A. Kellogg (1967) from the Yerkes Experimental Station in Orange Park, Florida. Gua lived in the Kelloggs' home for nine months and was afforded the same surroundings and treatments as their similarly aged son, Donald. During this time the Kelloggs were able to distinguish four naturally occurring sounds made by Gua. All of them seemed either to be elicited by the external environment or to be the result of the emotional state of the chimpanzee. The sounds—barking, food bark, screech or scream, and the "oo-oo" cry—are similar to those reported by other observers of chimpanzees (Yerkes, 1925; Yerkes and Yerkes, 1929; Kellogg, 1968). In attempting to teach the two syllable word "pa-pa," the Kelloggs noted that Gua showed considerable curiosity in the facial movements although she never tried to imitate the sounds. Upon being encouraged to do so by manipulation of the lips, Gua made occasional lip reactions but did not attempt to produce the sound. The Kelloggs proposed that if the chimpanzee ever progressed to actual articulation of human sounds it would be under training circumstances similar to those experienced by Gua. At the same time the Kelloggs were attempting to teach their son to articulate spoken words, also without success. Although Donald was clearly making gurgling and babbling noises, Gua was not able to make such sounds.

Viki

One of the most successful attempts to train a chimpanzee to speak was conducted at the Yerkes Laboratories of Primate Biology in Florida by Keith Hayes and Catherine Hayes (1952). They obtained a female chimpanzee only a few days after birth and tried to provide a background of experience resembling that of a human infant as closely as possible. The chimpanzee, Viki, lived in the Hayeses' home for six years and learned to produce four words, recognizable as "mama," "papa," "cup," and "up" (Hayes and Hayes, 1952). In agreement with an earlier report (Witmer, 1909), the Hayeses found the chimpanzee's vocal expression hoarse and seemingly difficult for the animal to produce. Viki's language training, which extended over several years, consisted of manipulating her mouth and lips and subsequently rewarding approximations to desired sounds. As training progressed, fewer manipulations were required to obtain the desired syllables. In this manner Viki was able to produce the words "mama" and "papa" on demand, although her difficulty in speaking was apparent. The words "cup" and "up" were more readily added to her vocabulary as they closely resemble sounds naturally produced by a chimpanzee. In his review of home-rearing experiments, Kellogg (1968) indicates that this project represents the acme of chimpanzee vocal achievement in human sounds. But it must be noted that even after these words were learned they were often inaudible and used incorrectly to identify objects.

Other Experiments

It is generally recognized that of all the animals, the chimpanzee most closely resembles man in physiology, intelligence, and imitative ability (Yerkes and Yerkes, 1929). Attempts at home-raising experimental animals other than chimpanzees have been reported, but none have produced any notable differences in vocal language capability from that of the chimpanzee. Furness (1916) attempted to train a young orangutan in a home-type experiment but after extended training was able to report only limited success. The orangutan was able to say "papa" and "cup," but these words were hoarse and were produced with considerable difficulty. These findings are consistent with later ones concerning vocalizations in chimpanzees (Hayes and Hayes, 1952). For the orangutan to produce the sound "papa," Furness found it necessary to manipulate the animal's lips with his fingers. "Cup" was pronounced with relatively greater ease, probably because it is a more naturally occurring sound for such primates, as Hayes and Hayes (1952) noted.

Toto, a gorilla, was raised in the home of A. M. Hoyt (1941) in Havana, Cuba. Notable comparisons between Toto and other home-raised pongids were recorded, but if the gorilla produced any humanlike sounds they were not reported by Hoyt.

Conclusions about Home-Raising Experiments

The relative lack of success in home-raising experiments designed to teach vocal speech to nonhuman primates has led a number of people to conclude that there exist qualitative differences between man and apes that constitute a definition of man. According to this line of reasoning, Noam Chomsky (1968) and other linguists imply that man is unique because he uses language, which provides a linguistic reservoir from which he can structure thought. By rearranging the linguistic symbols he can alter his thought; he can use the symbols in relation to tense; he can abstract thought; and, ultimately, by modifying his concepts he can produce novel combinations of symbols, which he can relate to other humans by the use of language. According to Chomsky this capability makes man unique, the only animal capable of creative thinking.

Chomsky's argument is to a great extent based on a traditional point of view. His assumption is that if an event cannot be observed it does not exist. We consider this to be reasoning based on "negative evidence." In the present instance it is erroneous to assume that language *per se* does not exist because vocal language is not exhibited to any great extent by chimpanzees. If all the researchers had assumed that language ability did not exist in chimpanzees, scientific attempts to investigate communication with nonhuman primates might well have ended with the home-raising experiments. Instead, the rationale for such experiments was changed because it was felt that attempts to communicate vocally were leading nowhere. Kellogg and Kellogg (1967) summed up, "We feel safe in predicting . . . it is unlikely any anthropoid ape will ever be taught to say more than a half-dozen words, if indeed it should accomplish this remarkable feat."

The results of the home-raising experiments indicated two possibilities. The first was that vocal communication with lower primates on a significant level is impossible because it is beyond the capability of animals other than humans. The Hayes and Hayes (1951) study, which involved the most highly controlled and systematic attempt to teach vocal communication to a chimpanzee, has often been cited by those who take this position. The negative-evidence rule applies here: Those who hold this position see no communication; therefore the capacity for communication does not exist.

The second possibility (based on anatomical evidence recently confirmed by Lieberman, 1968) is that speech as commonly used by humans is not a suitable medium of communication for chimpanzees. Similarly, Gardner and Gardner (1971) have reported that the reason chimpanzees do not learn to speak is behavioral as well as anatomical. Some specific portions of an animal's behavioral repertoire are highly, and perhaps completely, resistant to modification; attempts to teach vocal language to chimpanzees have apparently failed because vocalization is resistant to modification (Gardner and Gardner, 1971).

Although attempts at vocal training were largely unsuccessful, chimpanzees use their limbs, particularly their hands, in a highly efficient and well-coordinated way (Witmer, 1909; Kohts, 1935; Hayes and Hayes, 1952; Riesen and Kinder, 1952). Having noted that a species-specific characteristic of using hand gestures to communicate either greeting or threat had been observed in both wild (van Lawick-Goodall, 1968) and captive chimpanzees (van Hooff, 1971), Gardner and Gardner (1971) proposed a method of communication using a form of gestural language. They selected the American Sign Language (Ameslan), which does not totally satisfy the linguistic criterion of language capability (Gardner and Gardner, 1971; Chomsky, 1968), but nonetheless provided an excellent means of determining linguistic abilities of chimpanzees. Ameslan was chosen because it is actually used as a language by a group of human beings (Gardner and Gardner, 1971). Widely employed by the deaf, it is composed of gestures, mainly executed by the fingers, hands, and arms, and it has specific movements and places where the signs begin and end in relation to the signer's body. The signs are analogous to words in a spoken language (Stokoe, Casterline, and Croneberg, 1965).

The results obtained by Gardner and Gardner with their chimpanzee, Washoe, indicated that their choice of language medium was a good one. By using the chimpanzee's natural ability to use gestures and her imitative and intellectual capacities, they were successful in teaching Washoe to use Ameslan. The Gardners provided evidence that a true two-way, nonvocal communication channel between man and chimpanzee can be established. Recently, using Ameslan, experimenters have begun to explore the intellectual capacities of chimpanzees (Mellgren, Fouts, and Lemmon, 1973; Fouts, Chown, and Goodin, 1973; Fouts, Mellgren, and Lemmon, 1973). This method of communication has begun to reveal conceptual processes in chimpanzees that were heretofore impossible to determine, and what has been accomplished shows that the nature of animal intelligence may at last be studied to an extent never before attempted.

Project Washoe

Project Washoe was the first successful attempt to teach a nonhuman primate in human language. The project was begun at the University of Nevada in Reno by R. A. Gardner and B. T. Gardner (1969, 1971) in June 1966 and was terminated in October 1970, by which time Washoe had acquired a vocabulary of over 130 signs. Although the number of signs is relatively small, Washoe was able to use her vocabulary very well. She could readily produce spontaneous combinations of signs that demonstrated her syntactic ability, and her combinations were contextually correct. She was also able to transfer her signs and combinations to novel situations with ease and with a high degree of reliability.

Washoe, an infant female chimpanzee estimated to be 8–14 months of age in June 1966, was obtained by the Gardners from a trader in the United States. It is assumed that she was born in the wild and was raised for several months by her natural mother before being captured. Washoe was raised in the Gardners' backyard, an area of 5,000 square feet. She lived in a completely self-contained house trailer (8x24 ft), which provided for her toilet, kitchen, and sleeping needs. Throughout the project her researchers used only Ameslan to communicate with Washoe, and, when in her presence, they used only Ameslan to communicate with one another.

McCarthy (1954) has indicated that the barren social and physical environments once common to institutions for the mentally retarded are not favorable conditions for the development of language in humans. Gardner and Gardner (1971) extended this hypothesis to the chimpanzee and claimed that the same could be said for the raising of chimpanzees in cages. Since one would expect less linguistic capability from chimpanzees raised under these conditions, Washoe was provided with an environment that was kept as interesting as possible. Her teaching program was made part of her environment and was maintained throughout her daily routine.

A member of the research team was with Washoe during all her waking hours. At the change of shift two researchers overlapped for an hour. She was often afforded additional companionship when visitors were present and during frequent outings in the nearby community. She played and climbed trees and playground equipment in the Gardners' backyard. The function of the researchers on the team was to keep Washoe totally immersed in Ameslan. Since their principal job was to be a conversing companion and a model for Washoe, they used Ameslan in normal daily routines, games, and general activities. They chatted with her in Ameslan while cooking meals, cleaning, brushing her teeth, playing with her in her sandbox, and correcting her lapses in toilet training.

METHODS OF ACQUISITION

Manual Babbling

A number of methods of acquiring the signs were examined. Manual babbling, which is considered analogous to vocal babbling in human infants, was infrequently observed early in the experiment, but as the project progressed, the occurrence of babbling increased. The increase continued until the end of the second year, after which time manual babbling was rarely observed. This decline is attributed to Washoe's progress in acquiring a vocabulary during this period, and it was concluded that the acceleration of signing may have replaced the babbling, just as babbling in humans decreases as vocal speech develops. Gardner and Gardner (1971) reported that the only sign attributable to this method of acquisition was the *funny* sign, which is produced by touching the nose with the index finger.

Shaping

Shaping procedures similar to those used in operant conditioning were used in teaching Washoe new signs. She was rewarded whenever she made an approximation to a sign in order to encourage her repeating it. Successively closer approximations were then rewarded, and in this manner several signs were acquired. For example, originally Washoe would bang on doors with her fists when she wanted the door to be opened. This natural response was shaped using rewards for successive approximations to the correct sign. The sign for *open* consists of placing the two open palms against the object to be opened and then moving them up and apart. Washoe quickly acquired the *open* sign and soon generalized it spontaneously from doors to other objects such as books, briefcases, boxes, and drawers. Although shaping procedure was used a great deal during the first year to introduce new signs, it soon became apparent to the Gardners that this technique was not as efficient as other methods.

Guidance

The Gardners considered guidance the most effective method of teaching a new sign to Washoe. It consists of physically molding the hands and arms in the appropriate position for the sign, usually in the presence of an object or action that represents the sign. An example is given by Gardner and Gardner (1971):

> The sixth sign that Washoe acquired was the sign *tickle.* It is made by holding one hand open with fingers together, palm down, and drawing the extended index finger of the other hand across the back of the first hand. We introduced this sign by holding Washoe's hand in ours, forming her hands with ours, putting her hands through the required movement, and then tickling her.

Molding

A more recent investigation into the optimal training method (Fouts, 1972) utilized a design intended to test three procedures: molding, which involved physically guiding Washoe's hands into the correct position and movement for the sign; imitation, which consisted of the experimenter's making the sign and Washoe's being required to imitate his example; and free style, which was a combination of molding and imitation. Under the conditions of the experiment it was found that the molding procedure produced the most rapid acquisition of signs, followed by free style (which is not significantly different from molding), and lastly by imitation. Although imitation showed the poorest performance in the experiment, it nonetheless resulted in the acquisition of signs.

Observational Learning

Gardner and Gardner have reported that during the project Washoe apparently learned to comprehend Ameslan as practiced by her researchers although no overt efforts were made to teach her the signs. Since it was the researchers' practice to converse in Ameslan in Washoe's presence, the Gardners called this process of learning "observational learning." It is analogous to Fouts's (1972) use of imitation, although in the Fouts experiment a deliberate effort was made to train the chimpanzee to produce a particular sign. For example, Washoe learned the sign for *sweet* (which is produced by touching the lower lip or the tongue with the extended index and second fingers of one hand while the remaining fingers are pressed into the palm) merely by observing her trainer. *Flower,* which Washoe at first signed incorrectly, was later corrected by observing how the researchers made the sign. Gardner and Gardner (1971) indicate that because Washoe was totally immersed in an environment of sign language, she often acquired signs after several months' exposure to them without any concerted effort on the part of the researchers to teach them to her. It would appear that she spontaneously acquired the signs.

RECORDING WASHOE'S SIGNS

The various signs given by Washoe were recorded each day. A major portion of the recording procedure was concerned with whether the signs were spontaneous or prompted. Spontaneous signs were noted when Washoe made a correct signing response in answer to a question or made one entirely on her own, such as *open.* Prompted signs were those for which a correct response required assistance from one of the members of the research team. Additionally, information concerning the correctness of the form of the response was recorded. The criterion for reliability of responding was whether Washoe could produce the response spontaneously and correctly on fifteen consecutive days. In addition to keeping a signing record, the researchers kept a daily diary of the uses Washoe made of various signs and combinations and the contexts in which she used them. Later in the project a daily tape recording was added to the recording procedures. One of the researchers observed the team in a training session and verbally recorded the training procedures and Washoe's responses using a whisper microphone.

The first thirty-six months of the project yielded notable results in Washoe's ability to acquire signs. By the end of this period she was using 85 signs reliably. By the time of this writing (June 1974), she was using over 160 signs reliably. A cumbersome recording process is required in a project such as this; therefore it should be stressed that the reported size of Washoe's vocabulary is limited more by the researchers' ability to handle the recording and testing of the vocabulary than by Washoe's ability to acquire signs.

Testing Procedures

Because the approach in this project was at variance with any previous attempts, new ways of testing had to be designed in order to quantify the accumulated data accurately. Several methods were tried before a test was found that satisfied the requirements of both the researchers and the chimpanzee. A double-blind procedure (a procedure in which the observer evaluating the behavior of the subject does not know what treatment the subject has received until after the evaluation is completed) was used to control for the possibility that the experimenter might cue the subject as to the correct answer.

The first test used flash cards. Washoe was shown pictures on large cards and then questioned about the cards. This test had some drawbacks, the main one being that the test was experimenter-paced and so required an excessive amount of discipline. The box test was then devised. Washoe was to identify three-dimensional objects placed in a box by the researcher. Although this test was a great improvement over the flash-card technique, it was logistically difficult to conduct. It was replaced by the slide test, which was similar to the box test except that 35 mm color transparencies were used as exemplars. The slide test was efficient and easy to administer, and it differed from the other two tests in that it could be paced by Washoe.

The Gardners report that in the initial slide test Washoe correctly identified 53 items out of a possible 99. This performance was considerably above the chance level, which was 3 correct responses out of a possible 99. Although the correct responses were obviously encouraging to the researchers, Washoe's errors produced equally interesting data in that the errors fit into meaningful conceptual categories. Conceptual categories are such things as animals, foods, or grooming articles. Instances of responding to conceptual categories occurred when Washoe signed *dog* in response to a picture of a cat, *brush* for a picture of a comb, and *food* for a picture of meat. Also it was reported that when pictures of three-dimensional replicas of objects (e.g., toy cats, toy dogs) were used in the box test, the *baby* sign occurred frequently among Washoe's errors. When the slides were of real items, the *baby* error did not occur; however, the *baby* error continued to occur when a picture of a replica was used. For example, when a slide of a doll resembling a cat was shown, Washoe made four errors out of ten trials, and all four were the *baby* sign. However, when a photograph of a real cat was shown, she correctly identified and signed seven out of eight trials, and the single error was not the *baby* sign.

One important criterion of language use is that words must be used with others to form phrases, or more appropriately, combinations. Washoe first signed the phrase *gimme sweet* and *come open* during the tenth month of the project. At this time she was approximately 18 to 24 months old (Gardner and Gardner, 1971). It is interesting to note that similar instances of language use appear in humans at approximately the same age. Fouts (1973) has recently found that the chimpanzee is capable of forming combinations at a much earlier age. The discrepancy between the new findings and those reported by Gardner and Gardner (1971) is most likely due to the comparatively late start in training Washoe to use Ameslan (8–14 months), whereas in Fouts's research the teaching of Ameslan was begun at a much earlier age. In a more recent project by Gardner and Gardner (pers. comm.), two young chimpanzees acquired their first signs at age 3 months. Schlesinger and Meadows (1972) have found a correspondingly early acquisition of signs in humans, in which deaf children exhibited the use of signs at 5 months of age. The two-month difference is probably not due to qualitative differences in intelligence but to comparatively faster motor-coordination development in chimpanzees (Kellogg and Kellogg, 1967).

The segmentation of combinations by Washoe was done in much the same manner as executed by human signers. Washoe would keep her hands raised in the signing space until she had completed the combination. She would terminate the combination by making contact with some object or surface, comparable to the human signers' hands in repose.

The contextual relevance in which Washoe used her signs was found to be very good. She would correctly sign intended destinations with phrases such as *go in, go out,* or *in down bed.* When playing with members of the research team she did not generalize but referred to each one by name, signing *Roger you tickle, you Greg peekaboo* (Gardner and Gardner, 1971). Washoe signed at a locked door on thirteen separate occasions, and each time her contextual relevance was appropriate and correct: *gimme key, more key, gimme key more, open key, key open, open more, more open, key in, open key please, open gimme key, in open help, help key in,* and *open key help hurry* (1971:167).

The Gardners analyzed Washoe's two-sign combinations according to a method proposed by Brown (1970) for use with children. It was found that her earlier combinations were comparable to the earliest two-word combinations of children in terms of expressed meanings and semantic classes. Longer combinations were often formed by Washoe by adding appeal signs, such as *please* and *come,* to shorter combinations. Between April 1967 and June 1969, 245 different combinations of three or more signs were recorded in the researchers' diary. About half were formed by the addition of an appeal sign. The remaining ones were introduced by new information and relationships among signs, such as pronouns or proper names.

In analyzing Washoe's combinations the Gardners found evidence that possibly indicates that Washoe had specific preferences for word order. The combination *you me* was preferred in over 90 percent of the samples taken, with *me you* used in the remaining 10 percent. Earlier instances of this combination showed a preference for *you-me*-action, but later this changed to a preference for *you*-action-*me* order. The Gardners were reluctant to accept this order as an indication of syntax in Washoe's manual language. They point out that it may merely be her imitation of the preferred order of the members of the research team. From a behavioral viewpoint, however, there appears to be little difference between Washoe's preferences in word order and language behavior in human children in learning syntax.

Fouts (pers. obs.) has also noted the use of sign order to express meaning in signing chimpanzees. One of his chimpanzees, Lucy, has a definite preference for the sign order *tickle Lucy* when asking someone to tickle her; and when the order is reversed to *Lucy tickle,* she correspondingly tickles her companion.

Washoe and the Oklahoma Chimpanzees

Project Washoe was terminated in October 1970, when several members of the research team were receiving their degrees and leaving the project. Washoe was then brought to the Institute of Primate Studies in Norman, Oklahoma, directed by W. B. Lemmon. The Institute has many chimpanzees in its main colony, and since Washoe's arrival, it has been directing research with chimpanzees reared in private homes around the area. In both the colony proper and the private homes, the animals have been taught to use Ameslan.

One of the first experiments using Ameslan conducted in Oklahoma (Fouts, 1973) was designed to determine the relative ease or difficulty of acquiring signs by four young chimpanzees, two males and two females. An interesting finding was that just as in humans, chimpanzees have different rates of acquiring signs. By means of a molding procedure (Fouts, 1972), a total of ten signs were taught to the chimpanzees in daily thirty-minute training sessions. The acquisition rate for each sign was compared on the basis of the number of minutes of training necessary to reach five consecutive unprompted responses. After the chimpanzees had acquired the ten signs, they were tested on nine of them (all nouns), with the sign for *more* (an adjective) excluded from the test. Testing was conducted using the double-blind box test procedure, similar to the test described by Gardner and Gardner (1971).

The results indicated that some of the signs were consistently easy or difficult for the chimpanzees to acquire. The mean times for acquiring the signs ranged from 9.75 minutes to 316 minutes. The mean times to reach criteria for each chimpanzee across signs were: 54.3, 79.7, 136.4, and 159.1 minutes. These differences may be partially due to the individual chimpanzees' behavior in the training sessions.

All the chimpanzees performed above the chance level during testing. The correct responses in the double-blind box were: 26.4%, 58.3%, 57.7% and 90.3% correct. The low score of 26.4% obtained by one of the chimpanzees may have been a result of the difference between acquisition and testing. This particular chimpanzee seemed to require much praise and positive feedback from the experimenter for her correct responses when acquiring the signs. However, in the double-blind testing situation the observers who recorded her scores were unable to give her any positive feedback, and, as a result, her performance would begin to deteriorate noticeably after the initial trials in the test were completed. Another important finding was that Washoe was not the only chimpanzee that had the capacity to use Ameslan.

Mellgren, Fouts, and Lemmon (1973) examined the conceptual ability of a chimpanzee in regard to the category of items by studying the relationship between generic and specific signs in Ameslan. The subject in this experiment was Lucy, a seven-year-old chimpanzee that had been raised in species-isolation (by humans without ever seeing another chimpanzee) in a human home since she was two days of age. She had been taught Ameslan for two years and had a vocabulary of seventy-five signs. Our objective in the experiment was to determine if a new sign would become generic or specific relative to a category of items. Lucy had previously learned five food-related signs: *food, fruit,* and *drink,* which she used in a generic manner; and *candy* and *banana,* which she used specifically. The sign we chose to teach her was *berry,* and the category of items consisted of twenty-four different fruits and vegetables, ranging from a quarter of a watermelon and a grapefruit to small berries and berrylike items such as blueberries, cherry tomatoes, and radishes. The exemplar for the *berry* sign was a cherry.

The twenty-four different fruits and vegetables were presented to Lucy in a vocabulary drill, and her responses were recorded. As each item was presented she was asked *what that?* She was allowed to handle the various foods and eat them if she wished. The order of presentation was varied from day to day. The food-related items were presented along with at least two other items that were not in the fruit and vegetable category but were items for which she had a sign in her vocabulary. For example, she would be asked to identify a shoe, a string, and then a fruit or vegetable. Following her response to the food items she was questioned about such things as a book or a doll. For the first four days of training data were collected to obtain a systematic baseline of her responding to the twenty-four items to determine her usual response to them. On the fifth day she was taught the *berry* sign, using the cherry as the exemplar, and for the second four days the *berry* sign remained entirely specific to cherries. On the ninth day she was taught the *berry* sign again, but this time blueberries were used as the exemplar. On the ninth and tenth days she called the blueberries *berry,* but on the eleventh and twelfth days she switched back to what she had previously called them; but throughout these four days she persisted in calling cherries the sign she had originally been taught for them, *berry.* By her responses it was apparent that she preferred to use the *berry* sign in a specific sense.

Lucy's conceptualizations of fruits and vegetables were also examined. She showed a preference for labeling the fruit items with the *fruit* sign in 85% of the trials and for using the *food* sign in 15% of the trials. A dichotomy of responding to the two categories was indicated because she preferred to refer to vegetables as *food* 65% of the time and as *fruit* 35% of the time.

In a very revealing finding, Lucy created novel combinations to describe her perception of the stimulus item. She preferred to call a watermelon *candy drink* or *drink fruit,* whereas the experimenters referred to it with entirely different signs, which Lucy did not have in her vocabulary (*water* and *melon*). Another, more striking example occurred with radishes. For the first three days of the experiment Lucy labeled them *fruit food* or *drink.* On the fourth day she bit into a radish, spat it out, and called it *cry hurt food.* She continued to use *cry* and *hurt* to describe the radish for the next eight days. Sixty-five percent of the *smell* signs were used to describe the four citrus fruits by labeling them *smell fruit,* probably referring to the odor released when one bites into the skin of citrus fruits. The spontaneous generation of novel combinations not only demonstrated Lucy's ability to form new combinations but also indicated her ability to use her existing vocabulary of signs to map various concepts she had about the categories of the fruits and vegetables she was presented with.

A good understanding of vocal English appears to exist in a number of the home-reared chimpanzees near the Institute. Some of them were exposed to vocal language before being taught Ameslan. To determine the relationship between their English vocabularies and their Ameslan vocabularies, a study was undertaken by Fouts, Chown, and Goodwin (1973) using Ally, a young male chimpanzee that was being home-reared. A pretraining test was administered to determine his understanding of ten vocal English words. He was given such vocal commands as "Give me the spoon," "Pick up the spoon," "Find the spoon." Ally had to obey the command by choosing the correct item from a group containing several other objects. Only after correctly obeying the command five consecutive times was he considered to have an understanding of the vocal English word. Training was begun by dividing the ten signs into two lists of five signs each. One experimenter attempted to teach Ally a sign using only the vocal English word as the exemplar. Following this a second experimenter, who did not know which words had been taught or if any had been acquired, would test Ally on all five, using the objects corresponding to the vocal English words. For the second list of five words the experimenters exchanged roles. Ally was able to transfer the sign he was taught to use for the vocal English word to the object representing that word. This finding is very similar to the acquisition of a second language in humans. A second possible implication is that the learning occurred via cross-modal means.

The initial research in gestural language ability in chimpanzees indicates that chimpanzees can produce novel combinations of signs in their existing vocabularies. Humans have this capacity also but in addition they are able to understand novel combinations produced by someone else. A recently completed experiment (Chown, Fouts, and Goodin, 1974) seems to indicate that chimpanzees are able to understand novel combinations when they are used as a command. In the first phase of the experiment Ally was taught to pick out one of five objects in a box and place it in one of three places. For example, a command might be *Put baby in purse.* When Ally was sufficiently adept at this, the second phase of the experiment began. New items that had not been used in training were placed in the box and a new place to put the object was added. A screen between the experimenter and Ally prevented the experimenter from giving helping cues. With five items to choose from and three places to put them, chance would produce one correct response in fifteen trials. But Ally's performance was far above the chance level. During one test session he responded correctly to 22 of 36 commands.

CHIMPANZEE-TO-CHIMPANZEE COMMUNICATION USING AMESLAN

Because a major criterion of language is that it be used by members of the same species, we have been interested in intraspecific communication using Ameslan in chimpanzees. Fouts, Mellgren, and Lemmon (1973) explored various conditions under which such communication might be observed. The experimental subjects were Booee and Bruno, two young male chimpanzees who had already acquired a vocabulary of 36 signs each. The experimenters intended to keep the chimpanzees' vocabularies small so that they could examine the animals' acquisition of

Fig. 1. Bruno signing *hat.*

Fig. 2. Bruno signing *look.*

Fig. 3. Bruno demonstrates his ability to sign *drink* with either hand.

Fig. 4. Booee signing *baby.*

new signs when Washoe was introduced into the signing dyad, but their vocabulary has already increased to over 40 signs. Activities such as tickling, play, mutual comforting, and mutual sharing are most conducive to communication in Ameslan between the chimpanzees. Booee and Bruno, however, seem to prefer their own natural communication over Ameslan, perhaps because of their relatively greater exposure to other chimpanzees. Washoe and Ally are quite different, and prefer to use Ameslan when communicating with humans or other chimpanzees. When Booee and Bruno reach a preset criterion of reliability in chimpanzee-to-chimpanzee communication, Washoe will be introduced into the dyad. In April 1974 Ally was introduced to Bruno, and a good deal of communication from Ally to Bruno has been observed and recorded, mostly signs referring to food or play.

Manny, a young chimpanzee in the colony, has acquired from Washoe the *come hug* sign, which is used correctly when the chimpanzees greet one another or are engaged in mutual comforting. Kiko, a three-year-old chimpanzee acquired the *food* and *drink* signs from Booee and Bruno and displayed them correctly, but he was stricken with pneumonia and died. Another mature chimpanzee, housed in close proximity to Washoe, has often shown the *food, drink,* and *fruit* signs, but when offered a drink or a piece of fruit he has failed to show the signs again. Apparently he has failed to make the important connecting link between the sign and the object it represents.

One thing that makes research with chimpanzees so interesting is their ability to use Ameslan in situations other than experiments and at times when it is not expected. Both Lucy and Washoe have spontaneously invented signs for objects that were not in their vocabulary. Washoe invented a sign for *bib* by making the outline of a bib on her chest (Gardner and Gardner, 1971), and Lucy invented a sign for *leash* by making a hook with her index finger on her neck. Washoe was observed making a new combination, *gimme rock berry.* When the experimenter approached to question her about the seemingly incorrect sign, he found that Washoe was pointing to a box of brazil nuts on the other side of the room. On another occasion Washoe referred to a rhesus monkey (with whom she had earlier had a fight) as a *dirty monkey.* Until this time she had used the sign *dirty* to refer to feces or soiled items, usually as a noun. She now uses it regularly as an adjective to describe people who refuse her requests.

Sarah and Communication via Plastic Objects

Premack (1970, 1971a, 1971b) and Premack and Premack (1972) were the first to devise an artificial system for two-way communication between two species by using variously colored and shaped pieces of plastic to represent words. Sarah, the subject used in their research, is a wild-born female chimpanzee estimated to be six years of age when the project was begun. Since the Premacks' artificial language is visual and written they decided to use the chimpanzee because of the similarities between its visual system and man's.

Pieces of plastic that vary in size, shape, texture, and color are used to represent words. The plastic pieces are backed with metal and can be arranged in lines on a magnetized board. In this manner, the pieces can be displaced in space but not in time on the board, thus avoiding the necessity of relying on memory (which is an integral part of human languages, gestural or vocal). The use of pieces of plastic allows the human experimenter to determine which pieces of plastic will be available to the subject at any time. Also, the design of the experiment controls for the individual difficulty of any problem, but at the expense of the spontaneity of usage typically found in human language. It should also be pointed out that when an experiment is so overcontrolled, it limits the possible findings to those preconceived by the human experimenter rather than explores the full mental capacities of the subject.

The Premacks' approach emphasizes the functional aspects of language by breaking into behavioral constituents and providing environmental contingencies for each constituent they selected for training Sarah. They briefly summarize their work as follows: "We have been teaching Sarah to read and write with various shaped and colored pieces of plastic, each representing a word; Sarah has a vocabulary of about 130 terms that she uses with a reliability of between 75 and 80 percent" (Premack and Premack, 1972:92).

The Premacks appear to work from the premise that the relational and logical functions of language are derived from operant procedures, and therefore they use training procedures based on operant methodology. The procedure is of the same type as that used by psychologists to train pigeons or rats to peck keys or press bars. They reduce the constituents to very simple steps and then use standard operant techniques to train the subjects. For example, training may be started by placing some fruit on a board and allowing Sarah to eat it. Next, Sarah is required to place a piece of plastic representing the same fruit on a magnetized board before being allowed to eat the fruit.

As training continues in this manner, new pieces of plastic are added simultaneously with new aspects of the situation, e.g., a different kind of fruit or a new person. For example, a new piece of plastic may be added when Sarah's reward is changed from bananas to apples; or she may be required to place a piece of plastic representing the new trainer's name (e.g., Mary or Jim) on the board. In the next step a piece of plastic representing the word *give* is introduced, and Sarah is required to place it in between the trainer's name and the name of the fruit. Finally, a piece of plastic representing Sarah's name is introduced, and it has to be placed at the end of the sequence. Sarah appears to have little difficulty making these conditional discriminations. She may also be induced to respond to a sequence like "Sarah give apple Mary" by offering her a piece of chocolate if she gives up her apple. By using a more preferred item (chocolate) they are able to induce Sarah to place the pieces representing "Mary give apple Jim."

Premack and Premack report that Sarah is able to use and understand the negative article, the interrogative "wh" (who, what, why), the concept of name, dimensional classes, prepositions, hierarchically organized sentences, and the conditional.

> Sarah has managed to learn a code, a simple language that nevertheless included some of the characteristic features of natural language. Each step of the training program was made as simple as possible. The objective was to reduce complex notions to a series of simple and highly learnable steps . . . compared with a two-year-old child Sarah holds her own in language ability. [Premack and Premack, 1972:99]

Lana and the Computer

Rumbaugh, Gill, and von Glasersfeld (1973) devised a computer-controlled training situation that objectively examined some of the language capacities in a chimpanzee. They used Lana, a 2½-year-old female chimpanzee, as their subject. After six months of training Lana was able to read projected word characters, complete an incomplete sentence based on its meaning and serial order, and reject an incomplete sentence that was grammatically incorrect.

Rumbaugh et al. are using a PDP-8 computer with two consoles containing twenty-five keys each. On each key is a lexigram in "Yerkish," an artificial language developed by the experimenters. The symbols are white geometric figures, created from nine stimulus elements, used singly or in combination. The keys, on which the symbols are displayed, have colored backgrounds made up of three colors used singly or in combination. When the key is available for use by Lana it is softly backlit. When Lana presses a key, it becomes brightly lit. When a key is not available for use, it has no backlighting. When Lana depresses a key, a facsimile of the lexigram appears in serial order on one of seven projectors above the console. The computer also dispenses appropriate incentives to Lana when she depresses the keys in the correct serial order in accordance with the grammar of Yerkish. She may ask for such things as food, liquids, music, movies, toys, to have a window opened, to have a trainer come in, and so on, when they are available (that is, when the appropriate keys are softly backlit). There is also a console available only to the human experimenters so that the computer can mediate conversations between the experimenters and Lana.

Lana's training was begun by requiring her to press a single key in order to receive an incentive. Next, she was required to begin each request with a *please* and end it with a *period.* The depression of the *period* key instructed the computer to evaluate the phrase for correctness of serial order. If it was correct a tone sounded and Lana would receive what she had asked for; if not the computer would erase the projector display and reset the keys on the console. Later, Lana was required to depress holophrases (e.g., *machine gives M & M*) in between a *please* and a *period.* Following this, she was taught to depress each key represented in the original holophrase (e.g., *Please/Machine/give/M & M/period*). Then the keys were randomized on the console and she had to select and press them in the correct serial order.

One very interesting finding was that Lana soon learned to attend to the lexigrams on the projectors without training. She would erase sentences in which she had made an error, by pressing the *period* key, rather than finish them. On the basis of Lana's spontaneously learning to attend to the projected lexigrams and their order, Rumbaugh et al. were able to examine her ability to read sentence beginnings, to discriminate between valid and invalid beginnings, and to complete sentences. In their first experiment Lana was presented with one valid sentence beginning *(Please machine give)* and six invalid beginnings. She could either erase them or complete them. If she completed them she had to choose from correct lexigrams (e.g., *juice, M & M, or piece of banana*) and incorrect lexigrams *(make, machine, music, Tim, movie, Lana). Music* and *movies* were incorrect since the computer was programmed to accept these with *make* rather than with *give.* Lana's performance on the various aspects of this test ranged from 88% correct (*please X give,* an invalid beginning) to 100% correct (*please machine give,* the valid beginning). The second experiment was the same except that *make* was substituted for *give* in the sentences with the valid beginning. Lana's performance was 86% or more correct in this experiment. The third experiment used only valid beginnings with varying numbers of words in them: e.g.: (1) *please,* (2) *please machine,* (3) *please machine give,* (4) *please machine give piece,* (5) *please machine give piece of,* and so on. Lana ranged from 70% to 100% correct on the various beginnings.

Rumbaugh et al. concluded that Lana accurately read and perceived the serial order in Yerkish and was able to discriminate between valid and invalid beginnings of incomplete sentences in order to receive an incentive.

Comparison of the Methods

Using artificial languages or Ameslan avoids the problem of vocal communication. In one project using an artificial language (Rumbaugh et al., 1973) the human element has been removed by using a computer as an intermediary, and because a computer is used the experimenters are able to keep an exact record of the chimpanzees' communication. However, this refinement is also expensive in that the chimpanzee is forced into a strict and rigid paradigm that allows only that behavior to appear that will fit into the experimental situation. For example, the computer is not programmed to accept novel or innovative uses of the language. Although this method has managed to find and confirm conclusively such behavior as responsiveness to word order (syntax), in exploring the mental capacities of the chimpanzee, the artificial language approaches are limited to examining only behavior conceived of by the experimenters, and not responses created by the chimpanzees.

It is our contention that when conducting initial experimental research, such as we are doing with chimpanzee language acquisition, the situation must be structured only to the extent that control of the experiment remains in the hands of the experimenter. This point was made by Köhler (1921):

> Lack of ambiguity in the experimental setup in the sense of an either-or has, to be sure, unfavorable as well as favorable consequences. The decisive explanations for the understanding of apes frequently arise from quite unforeseen kinds of behavior, for example, use of tools by the animals in ways very different from human beings. If we arrange all conditions in such a way that, so far as possible, the ape can only show the kinds of behavior in which we are interested in advance, or else nothing essential at all, then it will

become less likely that the animal does the unexpected and thus teaches the observer something.

We feel the gestural language approach more closely fits the idea expressed by Köhler. Using this method we can simultaneously conduct highly controlled experiments and allow the chimpanzee to show, spontaneously, many of its capabilities. This is possible because *the chimpanzee, not the experimenter, has the control of its language use.* The chimpanzee can from within itself make statements or do what it wishes without having to rely on an arbitrary symbol to do so. It is not bound to a computer program or by the limits of the experimenter in making vocabulary available.

For example, in the study done with Lucy on the twenty-four different fruit and vegetables (Mellgren et al., 1973), had we limited her possible responses to the five food-related signs in her vocabulary, we would not have discovered her conceptualizations of the items. Nor would Washoe be able to insult people by calling them *dirty* if she had not been allowed to change a noun into an adjective. Similarly, Washoe would not be able to refer to brazil nuts as *rock berry.*

Each approach has its advantages, and it is up to the scientists to decide whether they wish to examine only those things they are capable of conceiving of, or if they are willing to accept some help from the chimpanzee in examining the animal's mental capacities.

Conclusions

The language skills of the chimpanzee are similar to those displayed by humans, although many definitions of language have attempted to exclude the chimpanzee from the realm of language as used by humans. An often-quoted, popular paper, Bronowski and Bellugi (1970) lists five characteristics of language: delay between stimulus and utterance, separation of affect from content, prolongation of reference, internalization, and reconstitution. Bronowski and Bellugi contended that Washoe probably met the first four, but failed to demonstrate the last. They define the structure activity of reconstitution as consisting "of two linked procedures—namely, a procedure of analysis, by which messages are not treated as inviolate wholes but are broken down into smaller parts, and a procedure of synthesis by which the parts are rearranged to form other messages" (1970:670).

Bronowski and Bellugi conclude: "What the example of Washoe shows in a profound way is that it is the process of total reconstitution which is the evolutionary hallmark of the human mind, and for which so far we have no evidence in the mind of the nonhuman primate, even when he is given the vocabulary ready made" (1970:673). We disagree with their contention of "no evidence." Most certainly the empirical evidence presented earlier in this chapter demonstrates that chimpanzees have the capacity for reconstitution, particularly Lucy's reference to a radish as *cry hurt food* and Washoe's calling a brazil nut *rock berry.* We contend that these gestural utterances more than meet the most restricting definitions of reconstitution and represent a remarkable intellectual and linguistic accomplishment, given the limited vocabulary we have allowed the chimpanzees to learn. And if Bronowski and Bellugi's contention that reconstitution is the "evolutionary hallmark of the human mind" is correct, then we must assume that the capacity for language was in the repertoire of the species before the great apes split off from hominoid evolution. Another alternative may be that the basis of language is not unique to language *per se,* but may actually be the basis of other behavior of which language is just one product.

References

Amman, J. C., 1692. *Surdus Loquens.* London: Sampson Low, Marsten Low and Seale.

Amman, J. C., 1700. *Dissertatio de loquela.* Reprint. Amsterdam: North-Holland Publishing Company, 1965.

Bronowski, M., and Bellugi, U., 1970. Language, name, and concept. *Science,* 168:669–73.

Brown, R. W., 1970. The first words of child and chimpanzee. In: *Psycholinguistics: Selected Papers,* R. W. Brown, ed. New York: The Free Press, pp.208–31.

Chomsky, N., 1968. *Language and Mind.* New York: Harcourt, Brace, and World.

Chown, W. B.; Fouts, R. S.; Goodin, L. T.; 1974. Productive competence in a chimpanzee's comprehension of commands. Masters thesis, University of Oklahoma.

Fouts, R., 1972. The use of guidance in teaching sign language to a chimpanzee. *Journal of Comparative and Physiological Psychology,* 80:515–22.

Fouts, R., 1973. Acquisition and testing of gestural signs in four young chimpanzees. *Science,* 180:978–80.

Fouts, R.; Chown, W.; and Goodin, L.; 1973. The use of vocal English to teach American Sign Language (ASL) to a chimpanzee: Translation from English to ASL. Paper presented at the Southwestern Psychological Association Meeting, Dallas, Texas.

Fouts, R.; Mellgren, R.; and Lemmon, W.; 1973. American Sign Language in the chimpanzee: chimpanzee-to-chimpanzee communication. Paper presented at the Midwestern Psychological Association Meeting, Chicago, Illinois.

Furness, W. H., 1916. Observations on the mentality of chimpanzees and orangutans. *Proc. Amer. Phil. Soc.,* 55:281.

Gardner, B. T., and Gardner, R. A., 1971. Two-way communication with an infant chimpanzee. In: *Behavior of Non-Human Primates,* A. M. Schrier and F. Stollnitz, eds. New York: Academic Press.

Gardner, R. A., and Gardner, B. T., 1969. Teaching sign language to a chimpanzee. *Science,* 165:644–72.

Hayes, K., and Hayes, C., 1951. The intellectual development of a home-raised chimpanzee. *Proc. Amer. Phil. Soc.,* 95:105–109.

Hayes, K., and Hayes, C., 1952. Imitation in a home-raised chimpanzee. *Journal of Comparative and Physiological Psychology,* 45:450–59.

Hoyt, A., 1941. *Toto and I: A Gorilla in the Family.* New York: Lippincott.

Kellogg, W. N., 1968. Communication and language in the home-raised chimpanzee. *Science,* 162:423–27.

Kellogg, W. N., and Kellogg, L. A., 1967. *The Ape and the Child: A Study of Environmental Influence on Early Behavior.* New York: Hafner.

Köhler, W., 1921. Methods of psychological research with apes. In: *The Selected Papers of Wolfgang Köhler,* Mary Henle, ed. New York: Liveright, 1971, pp.197–223.

Kohts, N., 1935. Infant ape and human child (instincts, emotions, play and habits). *Scientific Memoirs of the Museum Darwinian* (Moscow).

La Mettrie, Julien Offray de, 1948. *L'Homme machine.* Reprint. Chicago: Opencourt, 1912.

Lenneberg, E. H., 1968. The natural history of language. In: *The Genesis of Language, a Psycholinguistic Approach,* F. Smith and A. Miller, eds. Cambridge: MIT Press, pp.215–52.

Lieberman, P., 1968. Primate vocalizations and human linguistic ability. *Journal of the Acoustical Society of America,* 44:1574–84.

McCarthy, D., 1954. Language development in children. In: *Manual of Child Psychology,* L. Carmichael, ed. New York: Wiley.

Mellgren, R.; Fouts, R.; and Lemmon, W.; 1973. American Sign Language in the chimpanzee: semantic and conceptual functions of signs. Paper presented at the Midwestern Psychological Association Meeting, Chicago, Illinois.

Pepys, S. *The Diary of Samuel Pepys,* vol. 3, R. Latham and M. Williams, eds., 1912. Reprint. Berkeley: University of California Press, 1970.

Premack, A. J., and Premack, D., 1972. Teaching language to an ape. *Scientific American,* 227:92–99.

Premack, D., 1970. A functional analysis of language. *Journal of the Experimental Analysis of Behavior,* 14:107–25.

Premack, D., 1971a. Language in chimpanzee? *Science,* 172:808–22.

Premack, D., 1971b. On the assessment of language competence and the chimpanzee. In: *Behavior of Nonhuman Primates,* vol. 4, A. M. Schrier and F. Stollnitz, eds. New York: Academic Press, chap. 4.

Riesen, A. H., and Kinder, E. F., 1952. *Postural Development in Infant Chimpanzees.* New Haven: Yale University Press.

Rumbaugh, D.; Gill, T. V.; and Glasersfeld, E. von.; 1973. Reading and sentence completion by a chimpanzee (Pan). *Science,* 182:731–33.

Schlesinger, H. S., and Meadows, K. P., 1972. *Sound and Sign, Childhood Deafness and Mental Health.* Berkeley: University of California Press.

Stokoe, W. C.; Casterline, D. C.; and Croneberg, C. G.; 1965. *A Dictionary of American Sign Language on*

Linguistic Principles. Washington, D.C.: Gallaudet College Press.

van Hooff, J. A. R. A. M., 1971. *Aspects of the Social Behavior and Communication in Human and Higher Non-Human Primates.* Rotterdam: Bronder-Offset H.V.

van Lawick-Goodall, J., 1968. The behaviour of free-living chimpanzees in the Gombe Stream Reserve. *Animal Behavior Monographs,* 1:163–311.

Witmer, L., 1909. A monkey with a mind. In: *The Psychological Clinic,* vol. 3., no. 7. pp.189–205.

Yerkes, R. M., 1925. *Chimpanzee Intelligence and Its Vocal Expression.* Baltimore: Williams and Wilkins.

Yerkes, R. M., 1927. *Almost Human.* New York: Century.

Yerkes, R. M., and Yerkes, A. W., 1929. *The Great Apes.* New Haven: Yale University Press.

Chapter 38

ZOOSEMIOTIC COMPONENTS OF HUMAN COMMUNICATION

Thomas A. Sebeok

Note: O. F. Kugelmass has written a brilliant paper about certain tribes in Borneo that do not have a word for "no" in their language and consequently turn down requests by nodding their heads and saying, "I'll get back to you." This corroborates his earlier theories that the urge to be liked at any cost is not socially adaptive but genetic, much the same as the ability to sit through operetta.
—Woody Allen, "By Destiny Denied," *The New Yorker,* February 23, 1976. By permission.

1. "Zoosemiotics": Notes on Its History, Sense, and Scope

The term *zoosemiotics* was launched in 1963 and initially proposed as a name "for the discipline, within which the science of signs intersects with ethology, devoted to the scientific study of signalling behavior in and across animal species" (Sebeok, 1972:61). It obviously satisfied a felt need, for—despite some initial resistance, as to any neologism, especially one with overtones of academic jargon—it rapidly diffused in two criss-crossing directions: multidisciplinary and multilingual. It has since been adopted by scholars in a variety of fields, notably biology; and it has penetrated many of the languages of Europe, East and West, and beyond, including Hebrew and Japanese. Outside of scientific writings, the word has cropped up in well-known newspapers, like *Le Monde,* and magazines, like *Il Mondo.* It was featured in at least one novel by a famous English author, as well as in a balloon emanating from the muzzle of that most distinguished of beagles, Snoopy. Discharging a professional obligation to lexicography, I endeavored until recently to keep track of these migrations, and, on occasion, published at least highlights from the progressive record (Sebeok, 1972: chap. 9; 1976a:57, 86ff.).

What was originally intended by this term and what it seems to have come to mean to many others is quite another story, and still a bit perplexing. In most instances *zoosemiotics* has been used, roughly, as a one-word equivalent for "the study of animal communication," particularly in explicit or at least implicit contrast with "the study of human communication." This restricted usage is, however, far from what the original

This chapter incorporates observations delivered in Milan at the concluding Plenary Session, on June 6, 1974, of the First Congress of the International Association for Semiotic Studies, in the course of an invited presentation on the state of the art of "Nonverbal Communication." Responses to the ensuing discussion from the floor by Geoffrey Broadbent, David Efron, Tomás Maldonado, Christian Metz, and Leo Pap have all been blended into the text. The argument, of course, has been greatly expanded, brought up to date, and refocused to fit the overall purposes of this book.

definition actually implied. In 1970, in a typology of semiotic systems in general, it was clearly specified that "Human semiotic systems are of two kinds: anthroposemiotic, that is, species-specific systems of man; and zoosemiotic, that is, those component sub-systems of human communication that are found elsewhere in the animal kingdom as well" (Sebeok, 1972:163). Because of a colossal accretion in semiotic theory and praxis in recent years, accompanied by a well-nigh unmanageable proliferation of literature in animal communication studies, a conceptual cleansing is called for. Is *zoosemiotics* a useful term? What does it cover? How does all this fit into the vaster framework of general semiotics? Some of the difficulties of terminology and classification, which confront everyone who enters the field of study covered in the preceding chapters, have worried the most thoughtful of its practitioners and observers, but the only thing that is absolutely clear is that we are far from having reached consensus in this area (Hinde, 1972:86–98, 395).

The subject matter of semiotics is, quite simply, messages—any messages whatsoever. Since every message is composed of signs according to some ordered selection, semiotics has been variously identified as the doctrine (Locke, Peirce), or the science (Saussure), or the theory (Morris, Carnap, Eco) of signs. Correspondingly, the study designated semiotics comprises the set of general principles that underlie the structure of all signs, constituting a code, which was defined by Cherry (1966:305) as "an agreed transformation, or set of unambiguous rules, whereby messages are converted from one representation to another." Further, semiotics aims to uncover the ways in which such principles are or may be manifested in diverse messages, and to identify the specifics of particular sign systems, with comparative (including cross-taxonomic) as well as typological, synchronic (both structural and functional) as well as diachronic (both phylogenetic and ontogenetic; cf. section 7, below) ends in view. Semiotics is concerned, successively, with the generation and encoding of messages, their propagation in any sensorially appropriate form of physical energy, their decoding and interpretation. The methods employed by some investigators are more empirical, those by others more analytical. Some prefer to study communication, others signification (Prieto, 1975, for instance, makes much of this distinction). Plainly, however, these tendencies are complementary, each implying the other. (Naturalists, as one would expect, by their inclination and training have leaned toward an empirical approach to animal communication, but solid foundations for an analytical approach to animal signification have also been laid in the classic literature of ethology, notably in von Uexküll's marvelous 1940 monograph, "Bedeutungslehre" [cf. Sebeok, 1977a, forthcoming].)

If the subject matter of semiotics encompasses any messages whatsoever, the subject matter of linguistics is confined to verbal messages only. The fundamental competence underlying verbal messages is generally assumed to be (1) species-specific, and (2) species-consistent. Species-specificity of the linguistic propensity means that the formal principles we deem sufficient to characterize natural languages (spoken or not) differ radically from those found sufficient to characterize any known system of animal communication, including especially man's so-called nonverbal communication systems. This does not necessarily imply, however, that the neural substrates and/or psychological processes involved need be substantially dissimilar—these are surely secondary and tertiary laminations that are each of a distinct order (cf. Dingwall, 1975). Moreover, this conception of species-specificity does not exclude the possibility of quite sophisticated, though always only partial, code sharing, and hence communication, between man and animal (Hediger, 1967, 1974 generally; Fouts and Rigby, in respect to the man–chimpanzee dyad, chap. 37, this book). Nor is species-consistency necessarily universal, for severely handicapped children may lack the capacity to master language in more than rudimentary fashion (Malson, 1964; Curtiss et al., 1975).

The situation of the verbal code in a semiotic frame has been considered by almost everybody who has written on the subject since Locke (1975 [1690]:721). Late in the seventeenth century, he asserted that articulate sounds are the signs "which Men have found the most convenient, and therefore generally make use of. . . ." Linguists—building upon Locke without attribution—generally flatter themselves by at least acquiescing in dicta like Bloomfield's that "Linguistics is the chief contributor to semiotic," or persisting in Weinreich's sentiment that verbal messages constitute "the semiotic phenomenon par excellence" (Sebeok, 1976a:11–12)—no doubt a conscious rephrasing of Sapir's (1931) "language is the communicative process par excellence in every known society"—just recently reechoed by Greimas (1976:9) in his remark that "la linguistique . . . est la plus élaborée des sémiotiques. . . ."

It was apparently Saussure who promoted linguistics to the status of a pilot science, or "le patron général de toute sémiologie" (Sebeok, 1976a:12), a programmatic statement which, when pursued blindly, can lead into many a cul-de-sac (cf. Marcus, in Sebeok, 1974:2871ff.; see also Polhemus, in Benthall and Polhemus, 1975:20ff.). I have referred to the principle that is usually invoked in this connection as one of "intersemiotic transmutability," which may have been first, or was, at any rate, most insistently enunciated by Hjelmslev (1953:70): "in practice, a language is a semiotic into which all other semiotics may be translated—both all other languages, and all other conceivable semiotic structures." Elsewhere, I have questioned whether this *ex cathedra* declaration has actual support or remains, as I think, although still much cherished by linguists, hardly more than unsubstantiated dogma. In particular, I tried to show that animal sounds are often incapable of being paraphrased: "one gropes in vain for a set of linguistic signs to substitute instead of the significative unit employed by the speechless creature both to refer to his scarcely understood species-specific code and to the context of delivery, or *Umwelt,* through which the message fragment is aligned within the observed sequence of signs emitted" (Sebeok, 1976b). Even the transmutation of certain categories of human nonverbal messages into linguistic expression is, at best, likely to introduce gross falsification, or, like most music, altogether defy comprehensible verbal definition. Sapir (1931) put his finger on a "more special class of communicative symbolism," such as the use of railroad lights, bugle calls in the army, or smoke signals, in which "one cannot make a word-to-word translation, as it were, back to speech but can only paraphrase in speech the intent of the communication."

Semiotic systems that are species-specific in man are, then, for convenience, categorized as *anthroposemiotic* (Sebeok, 1972:163ff., 1976a:3). Language clearly belongs here, not only in its global spoken form but also as a visible means of communication used by a small minority population among a minority of mankind with partial or total hearing impairment and by those associated with such persons (Stokoe, 1972: esp. chap. 1). Here are counted also a wide array of speech surrogates (Sebeok and Umiker-Sebeok, 1976), mute communication systems preserved in certain monasteries (Barakat, 1975), aboriginal sign languages used among native peoples of the Americas and Australia (Sebeok and Umiker-Sebeok, 1977b), complex (viz., non-isomorphic) transductions into parasitic or restricted formations, like script or other optical displays of the chain of speech signs (the Morse code, or any of the several acoustic alphabets designed to aid the blind, or sound spectrograms), optionally imposed upon chronologically prior acoustic patterns (Kavanagh and Mattingly, 1972), and more or less context-free artificial constructs developed for various scientific or technical purposes (see, e.g., the respective articles by Golopenţia-Eretescu, Gross, and Freudenthal in Sebeok, 1974).

Over and above such transfers (Sapir, 1931), transforms, derivatives, and substitutes, there are those macrostructures that are based, in the final analysis, on a natural language, the "primary system" on which culture is superimposed, "regarded as a hierarchy of semiotic systems correlated in pairs, realized through correlation with the system of natural language" (Sebeok, 1975:76–77, 1976a:23n38). Particularly, this is implied by the concept of "secondary modeling systems," propagated chiefly by the Moscow-Tartu School of semioticians (Eimermacher, 1974; Sebeok, 1975:57–83; Ivanov, 1976; Winner and Winner, 1976). All secondary modeling systems are, therefore, anthroposemiotic by definition.

In a third category that might be reckoned anthroposemiotic are sets of signs affirmed to be uniquely used by man independently of any linguistic infrastructure (although, of course, unavoidably intertwined with verbal effects), but one must exercise great caution with respect to this division. In 1968, I blithely declared that music is "a species-specific, but not species-consistent form of behavior" (Sebeok, 1972:164–65). There is ample cause for wonder now if the first part of this allegation is true, and in what way? The relation between human and avian music was thoughtfully reviewed by Joan Hall-Craggs (Hinde, 1969: chap. 16), with special regard to the nature of the esthetic content of bird song. She concluded that "the form of music remains the privilege of birds and men" (ibid.:380), but suggested that the resemblances between the two varieties of semiosis can best be understood in terms of analogous functional requirements, such as the need to signal to distant listeners. The philosopher Hartshorne (the same, incidentally, who had served as the senior editor of C. S. Peirce's selected papers) has since reexamined the material in even more detail (1973: chap. 3). He characterized bird song as "the best of the subhuman music of nature" (ibid.:39) and declared, "considering the enormous gap between the anatomies and lives of man and bird, it remains astonishing how much musical intelligibility the utterances of the latter have for the former" (ibid.:46).

Investigations in this area have even crystallized into a subdiscipline called "ornithomusicology" by Szöke (1963), who maintains that since birds evolved elaborate musical utterances before we appeared on the scene, it is reasonable to suppose that the development of primitive music was actually stimulated by hearing and mimicking bird vocalizations (cf. remarks by Hewes, Livingstone, and Lomax in Wescott, 1974). Some species of Mysticetes, notably *Megaptera novaeangliae,* also "produce a series of beautiful and varied sounds," likewise called songs, the function of which is still a matter for much speculation but is usually assumed to serve communicative ends, possibly over great distances (Payne and McVay, 1971:597); these prolonged vocalizations are frequently compared to bird songs, the chief difference being that the latter normally last only a few seconds, whereas those of the humpback whales have a cycling time of up to thirty minutes, their patterns being repeated by individuals with considerable accuracy. Whatever the ultimate merits may be of such cross-taxa comparisons and contrasts between distantly related species occupying only vaguely similar ecological niches (as considered, e.g., in terms of quite abstract geometric patterns by Nelson, 1973:299–300, 324ff.), the facile grouping of music among anthroposemiotic systems appears, in retrospect, to have been premature.

The same can be said, *mutatis mutandis,* about other nonverbal art forms, for instance, abstract picture-making, a behavior that has been induced in apes (Morris, 1962; Bourne, 1971: chap. 9), and even in capuchin monkeys, with some success. According to the ethologist Andrew Whiten, the taste exhibited by apes—their choice of color, brightness, composition—"provides a unique background against which we may try to understand the origins and fundamental nature of visual art in our species . . ." (in Brothwell, 1976:40). Nicholas Humphrey's experiments show that apes prefer blues and greens over yellows and reds, leading to speculations that they favor the safety of green trees as opposed to the perils of exposure against red or yellow earth. Whiten "explains" their liking for bright light by assuming that it helps them perceive potential danger and surmises that their predilection for regular pattern might have something to do with an aptitude for handling intricate spatial relationships required to move safely through forests (ibid.:32ff.). Apes do seem to enjoy what they are doing, but forms of life that are not our direct phylogenetic ancestors, like the bowerbirds, also exhibit significant traces of a visual esthetic sense (von Frisch, 1974:244ff.; Waddington in Brothwell, 1976:8; Griffin, 1976:76ff.); thus male black woodpeckers chisel out nests that no less a scientist than von Frisch has depicted as architectural "works of art" (1974:189). Other birds build elaborate nests that they continue to improve upon with practice, in the sense of imparting a heavier semiotic charge: their constructions become, at least in our eyes, tidier and more elegant, but not recognizably more useful by strictly biological criteria.

Because I now consider it increasingly doubtful that any sign system that is not manifestly language-related belongs with man's repertoire of anthroposemiotic devices, I provisionally conclude, as an heuristic tactic, that all other systems used by man are to be construed as zoosemiotic until demonstrated to be otherwise. This view represents a radical shift in my position over the last ten years, one that still preserves the established dichotomy but enlarges the biological base as against the cultural superstructure, encouraging the search for true antecedents (homologies), not just the sharing of traits. It also counsels caution about a saltatory "discontinuity theory" in the terms argued for by Eric H. Lenneberg (in Sebeok, 1968: chap. 21) and supported to a degree by some notable ethologists (e.g., Klopfer in Hahn and Simmel, 1976:7–21). The strategic anthroposemiotics/zoosemiotics dichotomy will stand just as long as the riddle of the origin of language remains unsolved (Hinde, 1972:75ff., 94ff; Wescott, 1974; Lieberman, 1975, and chap. 1, this book). Recent concerted efforts at experimentation with various Great Apes notwithstanding (Fouts, forthcoming), no breakthrough is in sight; indeed, Thorpe (in Hinde, 1972:174) fears that the solution is likely to elude us forever. It may well be the case, as Julian Huxley (1966:258) once remarked, and as I would very much like to believe, that language "can properly be regarded as ritualized (adaptively formalized) behaviour," but, unfortunately, he did not go on to spell out just how one could apply the essentially comparative methods of ethology to a phenomenon that stubbornly remains a singularity in our known universe. In brief, what zoosemiotics has hitherto failed to provide is a comparative perspective for language (Hinde in Benthall and Polhemus, 1975:107–40), particularly with appropriately correlated operational procedures. The importance of a comparative semiotics (called for in Sebeok, 1976a: chap. 3, 1977a) cannot be overestimated, so it is encouraging to know that at least a few animal behaviorists of the first rank are not only commencing to share this long-felt conviction but have actually concluded that "the road now seems open" to realize its goals (Griffin, 1976:95–106).

The relation between the mutually opposite categories in man is hierarchical, and can therefore productively be viewed in terms of a notion standard in linguistics, *markedness.* Anthroposemiotic systems are always *marked,* in contradistinction to the zoosemiotic systems that comprehend them. This means that a specific anthroposemiotic sign implies the presence of a certain property *X,* whereas a generic zoosemiotic sign implies nothing about the presence of *X* (it may, but need not, indicate the absence of *X*). The marked sign is always the negative of the unmarked sign: "statement of *X*" vs. "no statement of *X.*" Some major controversial issues of long standing can be clarified in this light, such as the much-debated question whether a particular facial expression signifies the same emotion for all peoples or whether its meaning depends on the culture of the expressor and the "expressee." Ekman's carefully wrought theory postulates "culture differences in facial expressions as well as universals" (1972:279). The pancultural expressions are plainly zoosemiotic—they reflect biological bias in human behavior; hence, in the technical sense of the term, they are unmarked. The consequences of social learning, which varies both from culture to culture and according to smaller groupings within a culture, include the acquisition of markedness for every possible transition state in terms of the gain or loss of whatever the feature under consideration.

"The relationships between verbal and nonverbal communication are rather tenuous," Hinde (1974:146) ruefully conceded in his latest excellent survey of human zoosemiotic techniques. Oddly, however, he has overlooked a pivotal article by Gregory Bateson (in Sebeok, 1968: chap. 22), which cogently and forcefully set forth the reasons why this must be so. There is a popular belief, Bateson said, "that in the evolution of man, language replaced the cruder systems of the other animals," but he believed this to be totally wrong, because, if "verbal language were in any sense an evolutionary replacement of communication by means of kinesics and paralanguage, we would expect the old . . . systems to have undergone conspicuous decay." Such is manifestly not the case: rather, "the kinesics of men have become richer and more complex, and paralanguage has blossomed side by side with the evolution of verbal language . . . [both of which] have been elaborated into complex forms of art, music, ballet, poetry, and the like, and, even in everyday life, the intricacies of human kinesic communication, facial expression and vocal intonation far exceed anything that any other animal is known to produce." In brief, the two kinds of sign systems, though they are often in performance subtly interwoven, serve ends largely different from one another; indeed, zoosemiotic devices perform functions that anthroposemiotic devices are unsuited for, and vice versa. An exquisite illustration of the "reconciliation of the human necessity of speaking with the spiritual need for silence . . . within a single behavioral frame in which both components, otherwise contradictory, were indispensable" (Bauman, 1974:159–60) is related from the life of Quakers, whose style of preaching, mixing a "bundle of words and heap of Non-sense," evoked astonished comment even in 1653 (ibid.:150).

2. Inner/Outer

Another coined term (Sebeok, 1974:213, 1976a:3), albeit proposed no more than half seriously, was *endosemiotics,* "which studies cybernetic systems within the body. . . ." Clearly, man's semiotic systems are characterized by a definite bipolarity between the molecular code at the lower end of the scale and the verbal code at the upper. Amid these two uniquely powerful mechanisms (Marcus, 1974; Sebeok, 1972:62, 1977a) there exists a whole array of others, ranging from those located in the interior of organisms (von Uexküll's *Innenwelt*) to those linking them to the external "physical world" (his *Umwelt*), which of course includes biologically and/or sociologically "interesting" other organisms, like preys and predators. Semiotic networks are thus established between individuals belonging to the same as well as to different species. Jacob, who has most succinctly stated that the "genetic code is like a language," goes further: if they are to specialize, he points out, "cells must . . . communicate with each other," and, at the macroscopic level, "evolution depends on setting up new systems of communication, just as much within the organism as between the organism and its surroundings" (Jacob, 1974:306, 308, 312). After the new integrations have occurred, such that the coordination of elements has progressed from molecular interaction to the exchange of verbal messages, a still more novel hierarchy of integrons is set up: "From family organization to modern state, from ethnic group to coalition of nations" (ibid.:320), a variety of elevated ("secondary") codes come into play—cultural, moral, social, political, economic, military, religious, ideological, etc. The genetic conception of integron—called "shred out" in general systems theory, in reference to evolution "from slow, inefficient, chemical transmission by diffusion at the cell level up to increasingly rapid and cost-effective symbolic linguistic transmissions over complicated networks at the higher levels of living systems" (Miller, 1976:227)—is equivalent to the semiotic notion of "radius of communication," the progressive widening of which mirrors the history of civilization (Sapir, 1931) as much as it marks stages in the maturation of every individual.

There is no absolute boundary where zoosemiotics abruptly turns into anthroposemiotics. Least of all is this a correlate of "the appearance of a new property: the ability to do without objects and interpose a kind of filter between the organism and its environment: the ability to symbolize," which Jacob (1974:319) ascribes to mammals in general. So does Washburn (1973:181), who refers to "the mammalian brain as a symbolic machine." In fact, the groundwork for the mosaic of changes that enable organisms to utilize symbols was prefigured much earlier, as Tomkins (1975) convincingly delineated, and was sketchily reviewed in the framework of Peirce's doctrine of signs in Sebeok (1977a). On the invertebrate side, insects, such as the balloon flies, have evolved a symbolizing capacity in one of their species, *Hilara sartor* (ibid.; for symbolic communication in bees, cf. Griffin 1976:19–25). Also, John Z. Young has recently shown that the octopus deals with the world in a manner that can only be described as "symbolic." In a lecture given at the American Museum of Natural History in 1976, he said: "The essence of learning is the attaching of symbolic value to signs from the outside world. Images on the retina are not eatable or dangerous. What the eye of a higher animal provides is a tool by which, aided by a memory, the animal can learn the symbolic significance of events." Cephalopod brains may not be able to elaborate complex programs—i.e., strings of signs, or what Young calls "mnemons"—such as guide our future feelings, thoughts, and actions, but they can symbolize at least simple operations crucial for their survival, such as appropriate increase or decrease in distance between them and environmental stimulus sources ("Withdraw" or "Approach": Schneirla, 1965). The use of symbols on the part of the alloprimates is, of course, a current commonplace, but it has been apparent to unbiased scientists at least since Wolfe's (1936) experiments with a group of young chimpanzees nearly a half century ago (Wolfe was an excessively timid reader of Charles Morris; ibid.: 70). As Katz (1937:237) then noted in a needless display of the double negative, "It appears that chimpanzees are not completely incapable of using non-linguistic symbols." A recent remark of Lévi-Strauss sums the matter up far more cogently: "Les animaux sont privés de langage, au sens que nous l'entendons chez l'homme, mais ils communiquent tout de même au moyen . . . d'un système symbolique" (Malson, 1973:20).

The genetic code and the metabolic code—which intimately couples the endocrine and nervous systems (Tomkins, 1975:763)—are obviously at once endosemiotic and zoosemiotic, but other intracorporeal sign processes, notably the phenomenon of "inner speech" (Egger, 1904; Vygotsky, 1962; Vološinov, 1973), may be at least partially anthroposemiotic. Thus memory experiments have convincingly shown that thinking has two richly interconnected components in man: one verbal, the other nonverbal, each with characteristic properties. The imaginal effects in this dual coding system are zoosemiotic. Neurological studies display in extreme form a functional separation between the verbal and nonverbal spatial systems (Bower, 1970:509). Further, at least two scholars have independently pointed to the evolutionary intermediacy of man's dreaming, focusing their arguments chiefly on one particular kind of semiotic entity, the icon (Bateson, in Sebeok, 1968:623; Thom 1975a:72–73; cf. Sebeok, 1976a). Moreover, tests conducted on patients with commissurectomies (in the so-called Bogen [1969] series) have also yielded rather conspicuous clues that the right half of the brain may be primarily responsible for imaging processes in dreaming. If confirmed by current sleep research experiments, these results will be highly interesting in view of the association of the right hemisphere with the normal imaging mechanisms implicated with the handling of visual-spatial tasks criterial of (nonvocal) nonverbal communication (cf. Ornstein, 1972:64–65, 235n17; see also section 6, below).

The field of transducer physiology studies the conversion of "outer" signs to their initial "inner" input and considers the relative or absolute contrasts between the pathways of information outside the body and the pathways deep inside it. Although one can but concur with Shands (1976:303ff.) that it is essential to grapple with "the human problem of the greatest moment . . . of so relating the outer to the inner that the minimal information derivable from inner sources comes to be a reliable index of the external situation," and that this bifurcation must eventually be dealt with in semiotic terms, this science, powerfully foreseen by Leibniz, is as yet barely developed. Its modern theoretical foundations were laid in Bentley's spellbinding paper (1941) on the human skin as philosophy's last line of defense, the argument of which rested on the semiotic of Peirce (ibid.:18). Beck, looking toward "a truly human sociobiology" (1976:157), reviewed recent work with specific regard to nonverbal communication in man.

It is, in fact, hard to ascertain where "inner" ends and "outer" begins. The human skin itself is a rich arena of momentous semiotic events throughout the life of each individual, not only within our species (Moles, 1964; Kauffman, 1971; Montagu, 1971) but, more fascinatingly and almost wholly out of awareness, also in intricate communicative interaction with the teeming faunal and floral inhabitants of that veritable microscopic dermal ecosystem (Marples, 1965). Beyond the skin toward the outside, as Hediger (e.g., 1968:83) has incontrovertibly been demonstrating since 1941, every individual, according to its species, moves in the interior of an invisible but nonetheless sharply defined insulating space circumscribed by that animal's "individual distance" (the minimum remove within which it may approach another) and its "social distance" (the maximum separation between the members of any group). These concepts are crucial in the management of animals in zoos and circuses, and in their handling in laboratory experiments, under conditions of domestication, or as pets. In a test in which the density of children was modified in a playroom, a similar process was observed in operation (Hutt and Vaizey, 1966). Evidence bearing on the structuring of space and time in animals, or having to do with territoriality, overcrowding, and other sorts of distance regulation, were later extrapolated to man's perception of space and cultural modifications of this basic biological structuration. The branch of anthroposemiotics that studies such behavior is sometimes called proxemics (Hall, 1959; Watson in Sebeok, 1974:311–44). Its subject matter falls between bodily contact (the most intimate involvement of the *ego* with the *alter*) and patterns of physical appearance such as facial postures and bodily position, eye movements, and the nonverbal aspects of vocal acts. All of these come into play, in the main, beyond the Hediger "bubble," a variably shaped sphere of personal space that admits no trespass by strangers and is defended when penetrated without permission (Fig. 1).

The distinction between anthroposemiotic and zoosemiotic events is thus not at all demarcated at the integumentary threshold. Both processes have important extensions past the skin, in either direction. These "boundary" communicative phenomena, to which Peirce drew our attention repeatedly (when discussing Secondness) as the shock of reaction between *ego* and *non-ego,* may prove particularly interesting for future semiotic and related researches.

3. Vocal/Nonvocal

Sound emission and sound reception are so much a part of human life that it comes as something of a surprise to realize how uncommon this prominence of the role of sound is in the wider scheme of biological existence. In point of fact, according to Huxley and Koch (1964 [1938]:26–27), "the great majority of animals are both deaf and dumb." Of the dozen or so phyla, "only two contain creatures that can hear or produce functional sound," namely, the Arthropods and the Chordates. Their respective situations are, how-

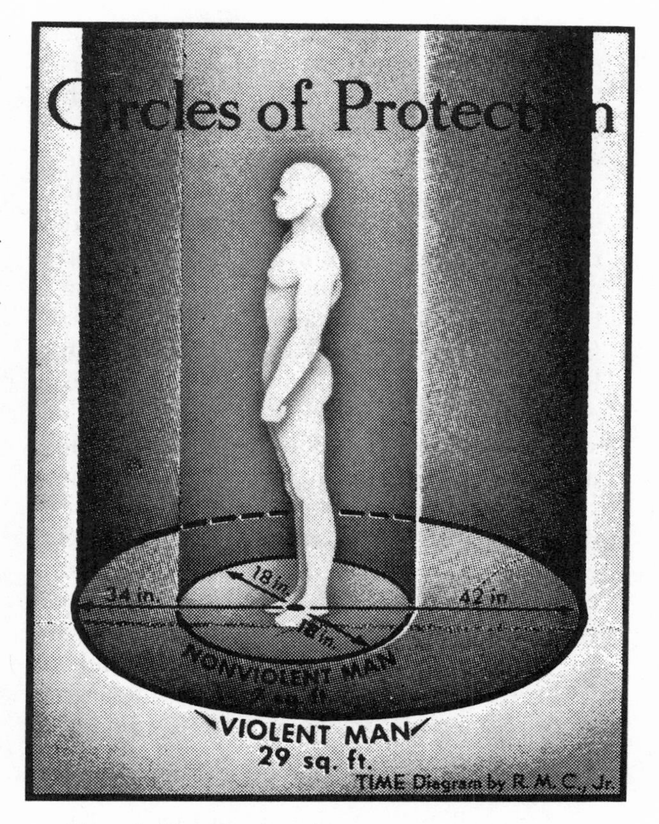

Fig. 1. Psychiatrist Augustus F. Kinzel has differentiated violent from nonviolent human subjects in terms of their communicative radius: on the average, the former stopped him at a distance of three feet, the latter at only half that distance. The two areas of insulating space differed, as well, in shape, from nearly cylindrical in nonviolent subjects to those bulging to the rear in violent ones—an avenue of approach interpreted as particularly menacing. Reprinted by permission from TIME, The Weekly Newsmagazine; Copyright Time Inc.

ever, quite different: while practically no members of the lower classes of Chordates are capable of sound production, the "highest three-and-a-half classes of the vertebrates are . . . unique in having all their members capable of sound-production, as well (save for the snakes) as of hearing." The methods of sound production, of course, vary enormously from group to group. Not only does our own method appear to be unusual, but Huxley and Koch (ibid.:32) confirm that it evolved only once in the stream of life. The vocal mechanism that works by means of a current of air passing over the cords and setting them into vibration seems to be confined to ourselves and, with distinctions, to our nearest relatives—the other mammals, the birds (since they possess a syrinx), the reptiles, and the amphibians (although some fish use wind instruments as well, they do so without the reed constituted by our vocal cords). So far as we know, no true vocal performances are found outside the land vertebrates or their marine descendants. Among many, notably ourselves, unarticulated vocalizations are used for status displays or to convey information about age, sex (Guthrie, 1976:33), and a host of specific characteristics about the state of the emitter-in-context (Lotz, 1956:212); usually, they are employed in the manner of icons (Sebeok, 1976c).

Humans communicate via many channels, only one of which is acoustic. Acoustic communication in man may be somatic (e.g., humming) or artifactual (e.g., drumming: Sebeok and Umiker-Sebeok, 1976). Acoustic somatic communication may be vocal (e.g., shouting for a waiter) or nonvocal (snapping one's fingers to summon him). Finally, acoustic somatic vocal communication may be verbal (speech) or nonverbal (Pike, 1943:32–41, 149–51, remains by far the best survey of such mechanisms), with the latter being either linked to or independent of speech (Argyle, 1975: chap. 18). On the other hand, by no means all verbal systems are manifested in the acoustic medium: Classical Chinese occurs only in written form; the American Sign Language (ASL) is encoded and decoded visually; and the mode in which man communes with himself, his thinking—which, as Peirce taught, "always proceeds in the form of a dialogue . . . between different phases of the *ego*. . ." (Sebeok, 1976a:28n45), and which constitutes one of man's unique uses of language (Bronowski in Sebeok, 1974:2539–40)—requires audible articulation but facultatively.

These observations, which underline obvious distinctions, are necessary because much conceptual confusion is engendered by the terminological disarray that bedevils the field of human zoosemiotics (Sebeok, 1976a:156–64). For example, at least two major books (Hall, 1959; Critchley, 1975) bear the title *Silent Language.* The attentive reader soon discovers that neither work deals with language, except in a misleadingly metaphoric sense, or even, strictly, with silence: thus Critchley declares in his very first paragraph, "Gesture may sometimes be audible though still unvoiced."

The foregoing is summarily depicted in Fig. 2, which substantially develops a single node from a tabular representation introduced in an earlier classification of zoosemiotic devices to il-

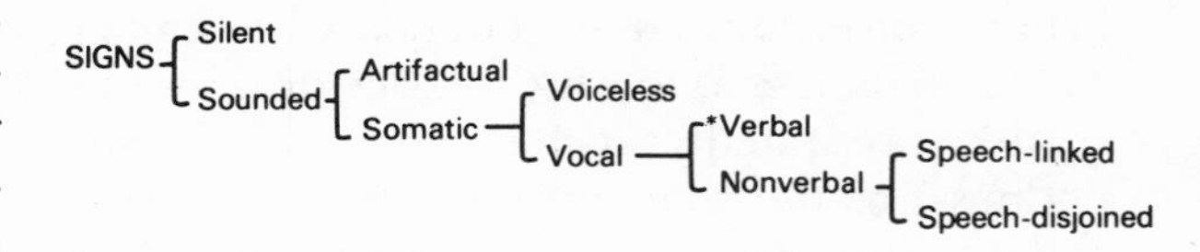

Fig. 2. An asterisk indicates the category assumed to be purely anthroposemiotic; the status of speech-linked phenomena, such as singing (i.e., a tune plus lyrics), is best reckoned as hybrid, or transitional, if not downright fuzzy (cf. Crystal, 1969:179–94). Terms to the right are progressively marked. Note that less than one percent of the information conveyed by speech is used for linguistic purposes as such (cf. Lotz, 1956:212).

lustrate the human production of signs according to the different communicative techniques involved (Sebeok, 1976a:30, Table 4; also in Eco, 1976:175, Table 34), but which also constitutes a considerable amplification of a clarifying figure with similar intent by Argyle (1975:346, Fig. 18.1).

The superficially similar classification by Wescott (1969:152), possibly the most tireless nomenclator in this field, is both incomplete and at least partially mistaken. He distinguishes, in the acoustic channel, among three communicative systems, which he labels, respectively, *language, phasis,* and *strepitus.* "Language and phasis are both vocal," he continues inaccurately (cf. Dingwall, 1975:32), the former being articulated, the latter consisting "solely of grunts and other vocalizations insusceptible to combination." Strepitus is then said to differ from both "in being nonvocal," e.g., hand clapping or foot stamping. Incidentally, there have been many efforts to design a system that describes all human movement in terms of the place and type of articulation of the segments of an idealized lay figure, that is, to devise a notation for muscular movement. The model, as construed in a volume of three-dimensional space so that the behavioral sequences can then be delineated as a syntagmatic concatenation of volumes, was envisaged by Bouissac (1973) and, in a measure, actually attempted by Schutz (1976; cf. Kelley, 1971, which, oddly enough, is not cited).

The taxonomic fragment sketched in Fig. 2 could be enriched in various intersecting ways. For instance, one may inquire to what degree the signs emitted by the source are "wanted," i.e., constitute its message for the destination, or have become increasingly "unwanted" (i.e., noisy) in the course of transmission. Or, one may focus on whether the emission and/or the reception of a given string is conscious or out of awareness (a distinction to be pursued further in section 5, below).

4. Verbal/Nonverbal

The subject matter of linguistics, as we all "know," is communication consisting of verbal messages and the undergirding verbal code enabling them. By contrast, the concept of nonverbal communication is one of the most ill defined in all of semiotics. No wonder the notion is often negatively formulated: as early as 1888, Kleinpaul (1972) paradoxically designated the topic of his classic manual as *Sprache ohne Worte,* or "wordless language." This concept recurs over and over in recent book titles, particularly since the appearance of the comprehensive, handsome treatise by Ruesch and Kees (1956), in the shape of "nonverbal" or "non-verbal." Sample listings, merely from this decade, include Bosmajian (1971), Davis (1971), Eisenberg and Smith (1971), Harrison (1974), Hinde (1972), Knapp (1972), Krames et al. (1974), Mehrabian (1972), Pliner et al. (1975), Poyatos (1976), Scherer (1970), and Weitz (1974). Countless monographs, special journal issues, and brief articles insist on viewing nonverbal communication as "communication minus language"; many of them are listed in Part 8 of the Eschbach-Rader bibliography of semiotics (1976:75–87), which is itself headed "Non-verbal communication" in the restricted sense. Such works tend to deepen the gulf segregating the "nonverbal" from the verbal; to paraphrase an amusing observation of Voegelin and Harris (1947:588), one might infer from some of them "that houses are built in sullen silence."

Two further interconnected problems immediately arise. On the one hand, the contents of these works so labeled encompass an astounding congeries of topics. On the other hand, the partial synonyms that have been devised to cope with this massive confusion have all proved unsatisfactory for one reason or another, as has already been adumbrated in Sebeok (1976a:158–62). Such competing though only partially overlapping terms and expressions include bodily communication, body language or body talk, coenetics, gesticulation, gesture, gSigns, kinesics, kinesiology, motor signs, pantomime, proxemics, silent language, tacesics, etc., and, of course, zoosemiotics with extensions thereof. This is not the place to enter the soggy quicksand of usages appropriate to this particular trade, but what about the nature of the traffic itself? What do we know about the commodity purveyed, beyond the presumed barring of what belongs strictly to linguistics?

Mehrabian's definition of what pertains to nonverbal behavior and what its functions are is fairly typical in that it starts out with an equivocation. He distinguishes (1972:chap. 1) between two senses, one narrow "and more accurate," the other broader but, while traditional, allegedly "a misnomer." The former embraces "facial expressions, hand and arm gestures, postures and positions, and various movements of the body or the legs and feet." The latter is equivalent to what has frequently been included under the subcategory of "paralinguistic or vocal phenomena" (also in Argyle, 1975:chap. 18). No mention is made of the obvious: that all animals communicate nonverbally, and that, in point of fact, other books, with identical titles, are devoted to just this, to the exclusion of virtually the entire human domain (e.g., Krames et al., 1974; Pliner et al., 1975; and Hinde, 1972, which is evenly balanced between the behavior of man and that of the creatures devoid of language). As to the concept of "paralanguage" (Crystal in McCormack and Wurm, 1976:13–27; Laver, 1976:347–54)—which ought, more accurately, to be termed "paraphonation," but which the innocent inquirer may reasonably assume to bear some relation to language—it further confounds the already frustrating jigsaw puzzle. Paraphonetic features may easily be homologized with aspects of animal communication, such as the conveyance of information about sex, age, and individual identity, much as song is assumed to do in many birds. However, in preparation for a state-of-the-art paper on the topic, David Crystal wrote a number of colleagues asking them what they understood to fall under this heading; he then reported his findings: "animal vocalization (or some aspect of it), memory restrictions on language, recallability for language, utterance length, literary analysis, environmental restrictions on language use . . . glossolalia, and emotional expression in general language disturbance—in effect, a fair proportion of sociolinguistics and psycholinguistics" (in Sebeok, 1974:269).

It is indeed very difficult, in practice, always to assign unambiguously what segment of a vocal encounter (conversation, state of talk) between people concerns linguists and what segment concerns "nonverbal" interactional analysts: "please remember that the integral role of gesture in speech is quite as important for our understanding of an utterance as the one or two significant movements or indications which actually replace an uttered word," Malinowski (1965[1935]:26) warned more than forty years ago. The borderline becomes more blurred the closer the focus of analysis gets. Some of the awesome entanglements of verbal responses with other kinds of acts were first heroically wrestled with by Malinowski, but he failed to achieve the integration he had preached because he lacked one indispensable analytic tool, an understanding of Kiriwinian verbal structure. Goffman dealt with them concretely in his masterful working paper (1975) on minimal dialogic units. They amply justify the research strategy increasingly being applied to the organization of conversations by workers like Duncan (1975), Poyatos (1976: see Fig. 4, p.66), and others under such labels as speaker synchrony, interactional synchrony, interactional equilibrium, and conveyance of indexical information (Laver, 1976:354–58).

In Fig. 3, the two oppositions Vocal/Nonvocal and Verbal/Nonverbal, as they are realized in a few of the aforementioned phenomena, are condensed into a sample distinctive feature chart. This matrix is merely meant to be illustrative: many other oppositions, as well as many other kinds of sign systems, could be adduced at will. (The assigned values are adopted from standard linguistic usage: + means that the feature is present, – that it is not, ± that both are co-present, and 0 that the distinction is inapplicable.)

But the conceptual chaos does not end there, because "nonverbal" of course subsumes a considerably vaster radius than the sphere of bodily communication as such. (Incidentally, one may well ask, the so-called organs of speech also being parts of the human body, why are the several manifestations of man's linguistic endowment, notably speech itself, generally not comprehended under this or similar rubrics?) Surely, music (Nattiez, 1975), the culinary arts (Barthes, 1967:27–28), a circus act (Bouissac, 1976), gardening (Malinowski, 1965[1935]), a floral arrangement (Cortambert, 1833), the application of perfumes (Sebeok, 1972:100–101), the choice and combination of garments (Bogatyrev, 1971; Guthrie, 1976:chap. 18) are only some random means among man's multiform options for communicating nonverbally. Accordingly, it would hardly be an exaggeration to claim that the range of the "nonverbal," thus conceived, becomes coincidental with the entire range of culture exclusive of language yet further encompassing much that belongs to ethology. But this way of looking at "nonverbal" seems to me about as helpful as the Kugelmass theories reported by Woody Allen in the epigraph to this chapter.

	VOCAL	VERBAL
Language	0	+
Speech	+	+
ASL	–	+
Babbling; paralanguage	+	–
'Fig'	–	–
Song (tune with or without lyrics)	+	±
Whistle	+	0

Fig. 3. The "fig" (thumb thrust between middle and ring fingers of fist) is an invitation to sexual intercourse in some cultures (Bäuml and Bäuml, 1975:72), a gesture randomly selected for this chart as an example of a soundless movement of the sort Efron (1972 [1941]:96), one supposes, might have called an intrinsically coded kinetograph. Whistle is "Verbal O" because it could represent merely a tune or be used as a speech surrogate (Sebeok and Umiker-Sebeok, 1976).

5. Witting/Unwitting

People are capable of encoding messages either deliberately or unwittingly, and to decode messages either with the knowledge that they are so engaged or without conscious awareness of what they are about. In a dyadic interaction between organisms, therefore, four possibilities exist in respect to this distinction.

The first possibility is that neither the emitter nor the receiver is able to identify the message, let alone restate it verbally. The pupil response furnishes a nice illustration of this: "While it is evident that men are attracted to women with large pupils their responses are generally at a nonverbal level. It seems that what is appealing about large pupils in a woman is that they imply a strong and sexually toned interest in the man she is with, as well as making her look younger. . . . The enlarged pupils, in effect, act as a 'signal' transmitted to the other person. Several observations made by others have indicated that this is what really can occur in the interpersonal relationship between a man and a woman, and apparently without conscious awareness" (Hess, 1975:95).

This state of affairs can, however, be altered either at the source or at the destination, or, of course, at both the input and the output end. For instance, a woman can deliberately dilate her pupils with one of several pharmaceuticals to enhance her appearance: such, indeed, was the custom in Central Europe in the interwar period (the drug used was a crystalline alkaloid derived from belladonna, which means "beautiful woman"). At the other end of the transmission chain, as Hediger (1976) has noted, the best circus animal trainers "haben schon längst erkannt, dass die Pupillenbewegungen z.B. ihrer Tiger wichtige Schlüsse auf deren Stimmung zulassen. . . ."

Signs that are normally unwitting, like the pupil response in man and animals, are regarded by most specialists (e.g., Peirce, Bühler, Jakobson, and American clinical semioticians generally) as constituting a subcategory of indices that, since Hippocrates, has comprehended symptoms and syndromes (Sebeok, 1976a:124–28, 181–82). These are of prime concern no less to semiotics than to medicine (Sebeok, 1977a). Instead of dwelling on these wide implications here, let us briefly reconsider the story of Clever Hans (Pfungst, 1965). One is compelled to concur with Katz (1937:5) that at the turn of the century this case was (and is now more than ever: Hediger, 1976) "a problem of first-rate importance, not merely of interest to special sciences, like zoology and psychology, but having some bearing on the deepest philosophical questions. . . ." Let us add that it is also of interest to the integrated science of communication, semiotics most particularly (cf. Rosenthal, in Pfungst, 1965[1911]:xxxiii, xxxix n11). This eponymous horse gave its name to one of the classic errors in the history of psychology after it was finally realized "how fatally the unintentional effect on the animal of the observer can influence the results" (Katz, 1937:7), and paradigmatically illustrated "the power of the self-fulfilling prophecy. Hans' questioners, even skeptical ones, expected Hans to give the correct answers to their queries. Their expectation was reflected in their unwitting signal to Hans that the time had come for him to stop his tapping. The signal cued Hans to stop, and the questioner's expectation became the reason for Hans' being, once again, correct" (Rosenthal, 1966:138).

It must be emphasized that one is concerned here with what Stumpf (in his original [1907] Preface to Pfungst, 1965) described as *minimale unabsichtliche Bewegungen,* or unintentional minimal movements of the horse's questioner, which occasioned this colossal case of self-deception. Máday soon refocused the whole phenomenon in its properly explicit frame, designating the proposition as a *Zeichenhypothese* (1914:12) and enumerating the proofs therefor (ibid:13–18). He also described the *unwillkürliche Zeichen,* or unwitting signs, in abundant detail (see esp. ibid.:chap. 6 and pp.247–59), dispelling whatever lingering doubts may have existed. To be sure, Pfungst tells of other clever animals, and there have been records of many since, not just "talking" equines but learned canines, reading pigs, and at least one "goat of knowledge." The most fascinating "talking" horse was Berto, which was blind yet gave excellent results when the attendant "thought that the questions had been written on its skin or uttered aloud" (Katz, 1937:17–18). All of them were assiduously coached performers intentionally cued by their trainers, who were entertainingly exposed by the prominent American illusionist and historian of conjuration, Christopher (1970:chap. 3; see also Máday, 1914:chap.15). In sum, the Clever Hans phenomenon lies at the very heart of zoosemiotics; the investigators were misled because they sought in the pupil, the horse—or the great spotted woodpecker (Griffin, 1976:25–26), or the porpoise (Sebeok, 1972:59ff.), or the chimpanzee—for what they should have looked for in the teacher, man: covert and unwitting message transmissions from man to man (Rosenthal, 1966:chap. 9) as well as from man to animal (ibid:chap. 10).

Much of the mystique that enveloped the Delphinidae in the 1960s is traceable to the Clever Hans delusion (Sebeok, 1972:59–60; Wood, 1973:chap. 1; Caldwell and Caldwell, chap. 33, this book). In the late 1970s, a more lively issue, which has hardly been faced up to yet (but see the inconclusive discussion by D. Premack, 1971:820–21; and cf. Brown, 1973:50–51), is the pervasive, insidious penetration of Clever Hans into all the attempts so far designed to erase the seemingly ineradicable linguistic barrier between man and the Great Apes. (Eccles, 1974:106, doubts "if these clever learned responses can be regarded as a language even remotely resembling human language.") The possibly insoluble dilemma that experimenters in this area must confront is that man can scarcely be eliminated from any conceivable variant of the requisite training procedures, because, to put it quite starkly, every such animal is critically dependent for emotional sustenance upon its trainer—whether informed or naive (A. J. Premack, 1976:101ff.), even when the system used is computer-enhanced (Rumbaugh and Gill, 1976). Deprived of social contact with a human partner, the home- or laboratory-raised ape perishes, but, when given the human contact, the experimenter's expectancy effects must be fully reckoned with (Timaeus, 1973).

Of Clever Hans, James R. Angell noted in 1911: "No more remarkable tale of credulity founded on unconscious deceit was ever told, and were it offered as fiction, it would take high rank as a work of imagination" (Pfungst, 1965). The lesson Hans tapped out as his legacy for science has not, however, been mastered even today. All efforts, without exception, that aimed to shape linguistic apes, whether the research design was quasinaturalistic or followed an essentially Skinnerian paradigm, not only offered "rich opportunities for drawing the wrong conclusions," in Brown's (1973:34) characteristically tactful parlance, but had their focus misplaced to begin with. The really interesting issue has to do with the nature of the communicative coupling between subject and object, hence that is precisely a semiotic problem; it was formulated by Rosenthal (in Pfungst 1965:xxxiii) in the following two sentences (and was expanded by this immensely insightful investigator in many of his other publications, where he drew on social psychological knowledge at large for fruitful hypotheses): "If we knew precisely by what means we unintentionally communicate our expectancies to our animal and human subjects, we could institute more effective controls against the effect of our expectancies. More generally, if we knew more about the modalities by which we subtly and unintentionally influence one another, we would then have learned a great deal that is new about human social behavior." The same point is made in Hediger's (1976:45–46) wide-ranging review of Clever Hans and its consequences, with even more explicit reference to zoosemiotics. The relevant question, he correctly emphasizes, is not how to eliminate human signs from the dyad, but how to—at long last—program a thorough investigation of all channels through which such signs actually are and might be transmitted, and thus to determine what really is happening in man–animal and man–man interaction: "Das ist eine Frage der Semiotik," Hediger insists, and, of course, this has become increasingly clear over the last decade; but the nitty-gritty of the search will also demand an exceptionally broad spectrum of cooperating disciplines.

6. Left/Right

There has been a great deal of agitation lately in the field of brain research about the question of lateralization of at least two modes of mentation in man (Dimond, 1972; Ornstein, 1972:chap. 3). The yeast that fuels this ferment derives from studies termed, as long ago as the 1870s (Sebeok, 1976a:57–58), *asemasia,* i.e., the impairment of nonverbal communicative functions (Zangwill, 1975:95–106), and, more immediately, from recent split brain experiments (Krashen, 1976:157–91). Some implications of this endeavor—essentially in its infancy—with the aim of moving toward a "true synthesis of biology and culture in the operation of human minds," were recently reviewed by Paredes and Hepburn (1976), giving rise to impassioned debate in ensuing issues of *Current Anthropology* and elsewhere. In general, however, the emerging picture seems to indicate that the left side of the brain processes verbal tasks better, while the right side deals more skillfully with visual-spatial tasks; this is underscored by a demonstration that the two hemispheres are not ontogenetically equal until the end of the first decade of life (Dennis and Whitaker, 1976). Another, broader way in which scientists, such as A. R. Luria (e.g., in Sebeok, 1974:2561–94; cf. Ornstein, 1972:67), prefer to describe the division of labor is to speak of two opposed but complementary ways of thinking, such that the left brain is more likely to deal with tasks seriatim ("logically"), while the right brain manages problems as a whole, perceiving their simultaneous relationships ("holistically").

This line of research has led to the following provisional conclusion, succinctly stated (though immediately, and properly so, hedged in with all sorts of qualifications) by Bogen (1969): "the left hemisphere is better than the right for language and for what has sometimes been called 'verbal activity' or 'linguistic thought'; in contrast we could say that the right hemisphere excels in 'non-language' or 'non-verbal' function." In the current secondary literature, especially on the part of anthropologists and members of the artificial intelligentsia, one finds that this duality of information handling is labeled more starkly as "verbal to nonverbal" (Tunnell, 1973:27), and such uncompromising assertions as "the appreciation of gestures . . . is the province of the right hemisphere" (Weizenbaum, 1976:220). Psychological tests tend to support the view that the right hemisphere appears somewhat more skillful than the left "in nonverbal reasoning and spatial abilities" (Bower, 1970:509), although both hemispheres are indubitably equipped for language representation in some ways and to some extent, the cerebral dominance seemingly involving, in Kinsbourne's dramatistic phrase, "active competition" between the two, such that "the left hemisphere is genetically destined to win" (1975:114).

"Dominance" refers to the processing of information by one hemisphere and its ability to control responding. This variable is likely to be independent of "capacity," or the performance of some task when required by the contextual contingencies of a hemisphere. Now, according to Levy (1973:158), the left hemisphere "simply does not bother to handle information which can be handled by the right," an observation that is in good conformity with the semiotic model espoused here, especially in respect to the hierarchical notion of "markedness" mentioned above (section 1). The pithy formulation of Eccles (1974:92) expresses this best of all by asseverating that the minor hemisphere resembles a very superior animal brain, which is to say that it provides the primary locus for the coding processes we term zoosemiotic but lacks the ability to report mental functions utilizing the verbal (or anthroposemiotic) code, at any rate, vocally (i.e., it is mute). Evolutionary continuities in semiosis from animals to man as well as sudden discontinuities are thus both accounted for, but in grossly different locations in the brain. The corpus callosum serves as the principal channel of intercommunication between the two hemispheres, insuring exact synchrony (unless, of course, the commissures are surgically or otherwise severed). The messages that flow back and forth are presumably all coded neurochemically, but whether the left-to-right commerce is verbal (or digital, as was suggested in 1960; cf. Sebeok, 1972:chap. 1; see also Bogen, 1969), while the right-to-left traffic is nonverbal (or analog), remains one of many intriguing problems winking at the edge of experimental palpability. It also remains to be seen whether the two opposing schools of belief in regard to the ritualization of man's overall semiotic competence alluded to by the English zoologist Pumphrey (Sebeok, 1976a:67, 142) will be content with some such heuristic model; and, still further down the road, whether a proper characterization of the left hemisphere will require a judicious application of catastrophe theory (Thom, 1975b; cf. Sebeok, 1976b, 1977a)—since that is the most sophisticated qualitative method designed so far to handle discontinuous phenomena—whereas the right hemisphere's smooth, continuously changing Gestalt configurations will stay amenable to traditional quantitative analysis.

7. Formation/Dissolution: Diachronic Glimpses

Diachronic considerations of two very different sorts are pertinent to zoosemiotic inquiry: one focuses on the evolution of signs and systems of signs in phylogeny (Hahn and Simmel, 1976; Marler, chap. 4, this book), the other considers their development in ontogeny (Sebeok, 1976a:98–99; Burghardt, chap. 5, this book) as well as impairment, leading to their ultimate dissolution in the life-span of individuals (Zangwill, 1975). The former constitutes the principal axis of synthesis in the entire field of the biological study of behavior (Sebeok, 1972:135, 1976a:85), or ethology, which has been insistently characterized as "hardly more than a special case of diachronic semiotics" (Sebeok, 1976a:156). This identification should disquiet no one who is cognizant of the common historical roots of the comparative method indispensably utilized in branches of both (Lorenz, 1966:275–76): they stem from Baron Cuvier, the founder of comparative anatomy (which, in his conception, studied the static interrelationships of immutable species, created and re-created several times over), now transformed into modern phylogenetic behavioral systematics, no less than, albeit indirectly through Friedrich von Schlegel's applications, into comparative grammar (cf. Sebeok, 1977a).

Any sensitive and observant caretaker is well aware that a "normal" infant is born with elaborate equipment for interacting with its human surroundings by means of a wide array of vocal and nonvocal signs. Indeed, its success in encoding and decoding vital messages is a most important measure of its very normalcy. Its semiotic growth and differentiation are undoubtedly best conceived of as a series of catastrophes (Thom, 1975b). Thom's topological model, following ideas originating in biology and pursued by creative thinkers since the likes of D'Arcy Thompson and C. H. Waddington, could account for successive stages of bifurcation where, much as the development of any cell in an embryo diverges from that of its immediate neighbors, sign functions become ever more specialized and cluster to form particular constellations in a dynamic semiotic system occurring and explicable at any given time. Verbal signs suddenly emerge, superimposed upon babbling—which itself usually functions as an insufficiently explored vocal but, of course, preverbal link between the baby and its caretakers (Bullowa, 1970, and, with others, in McCormack and Wurm, 1976:67–95). Language then continues to unfold (Brown, 1973) until the child acquires full mastery over the language of its native speech community along with the culturally appropriate nonverbal systems of signs.

Study of the latter—which also embody the rules of when and how to use the language in accordance with personal needs and social norms—on a scale comparable with the former has barely begun, notwithstanding the pioneering instigation of this sort of research launched by Darwin in a famed segment of the diary he started in 1840 (1877). Surprisingly enough, there exists no definitive treatise of this hardly negligible area comprehending a configuration of attributes in any individual, which Chance calls "primary-group relations," a type of communication that is concerned with associations of an addresser with addressees, a process that Chance further characterizes as the wholly "nonverbal" infrastructure of social cohesion and control (1975:100). There are only a few authoritative survey articles about the state of this art, even the best of which are now getting a bit dated, such as Brannigan and Humphries (1972) or N. G. Blurton Jones (in Hinde, 1972:271–96). These should be supplemented by special studies, e.g., of facial expressions in infants and children (Charlesworth and Kreutzer, 1973), including tongue showing (Smith et al., 1974:222–27), and the like. The paucity of really robust achievements in this domain of nonverbal infant competences—concerning the sights, sounds, smells, and overall body management appropriate to the survival of the baby in all cultures—is the most startling fact about it. One's sense of wonderment remains far from fulfilled, although the tempo of research has become much livelier of late.

If relatively little is known about the formation of sign systems in the course of a human life, the destructive effects of injury or disease remain hardly understood at all. What Steinthal, in 1881, dubbed *Asemie,* and Jackson, following Hamilton's independent coinage of 1878, then propagated under the accurate label of *asemasia,* comprehending "the loss of gesticulating power," or of pantomime (Critchley, 1975: chap. 3; Sebeok, 1976a:57–58), are, alas, too frequently experienced by patients and observed by attending physicians. An example of the extent to which pantomimic movement may be unattainable by victims of severe brain damage was recorded by Luria (1972:45) from his celebrated patient Zasetsky, who was wounded in war: "I was lying in bed and needed the nurse. How was I to get her to come over? All of a sudden I remembered you can beckon to someone and so I tried to beckon to the nurse—that is move my left hand lightly back and forth. But she walked right on by and paid no attention to my gesturing. I realized then that I'd completely forgotten how to beckon to someone. It appeared I'd even forgotten how to gesture with my hands so that someone could understand what I meant." At one time, Jackson hazarded a tripartite clinical classification of the aphasias, and the third of his categories was the most global, namely, the loss of language when pantomime and gesture as well as speech are annihilated, a tragic condition so devastating as to be tantamount to social extinction.

The communicational problems that beset the aging and the aged typically fall between the two stools of social gerontology and psychosemiotics; in consequence, they have been largely misinterpreted or altogether neglected. Philip B. Stafford (pers. comm.), for example, has studied closely one dominant sign of senescence in our culture, "repetitiousness," and showed that, contrary to the usual assumption that this tedious habit is simply a symptom of physiological deterioration in old folks, it is rather a semiotic manifestation of an adaptive strategy useful to the elderly in capturing an audience. Nor must one take it for granted that senescence is accompanied by a mere decrease in semiotic potency: on the contrary, a restocked ambry of nonverbal skills is often required and acquired in course of the aging process to cope with the usually, often dramatically, altered social environment. Just how this is accomplished has hardly been studied so far, but I am convinced that the semiotics of old age is one of the most promising research areas for the immediate future and that it will have great import for both applied gerontology and clinical geriatrics.

Since the application of the principle of ritualization to language was not proving feasible (section 1, above), Koehler (1956:85ff.) proposed "to seek for roots, initia, precursors" to language and thought he found eleven, but then wisely concluded that "No animal has got all those initia of our language together, they are distributed among very different species, this one having one capacity, that species another. We alone possess all of them . . ." (ibid.:87). Linguists like Hockett (with Altmann, in Sebeok 1968: chap. 5) and, to a lesser extent, Lyons (in Hinde, 1972: chap. 3) later tried similarly to disassemble the verbal code into a quasi-logical roster of components of varying numbers and to examine each function separately from a comparative standpoint, a mechanistic and desperate procedure that turned out to be a largely empty exercise, partly for the reason foreseen by Koehler, partly for others such as are given by Hewes (in Wescott, 1969:4ff.). Unless and until one or more semiotic systems utilizing coding methods comparable with that of language are discovered, this sort of quest seems futile to me. Moreover, it appears increasingly unlikely that a terrestrial language-like animal communication system will ever be located under natural conditions. The sole alternative lies, then, in the continued scanning—a long, arduous, and costly endeavor, the outcome of which is uncertain—for communicating intelligences on other planets (Arbib, 1974).

Zoosemiotic systems in man, on the other hand, are eminently amenable to comparative study (Pitcairn and Eibl-Eibesfeldt, in Hahn and Simmel, 1976: chap. 5) and will, no doubt, continue to produce worthwhile findings. An example that shows just how fascinating this line of inquiry can be is Ferguson's (1976:138) suggestive research proposal that certain verbal routines, such as greetings and thanks, are "related phyletically to the bowings and touchings and well-described display phenomena of other species." The most fertile ground for the application of the methods of ritualization is surely in the domain of interpersonal rituals for which politeness formulas furnish one attractive target.

References

Arbib, Michael, 1974. The likelihood of the evolution of communicating intelligences on other planets. In: *Interstellar Communication: Scientific Perspectives.* Boston: Houghton Mifflin, pp.59–78.

Argyle, Michael, 1975. *Bodily Communication.* New York: International Universities Press.

Barakat, Robert A., 1975. *The Cistercian Sign Language: A Study in Non-Verbal Communication.* Kalamazoo, Mich.: Cistercian Publications.

Barthes, Roland, 1967. *Elements of Semiology.* New York: Hill and Wang.

Bauman, Richard, 1974. Speaking in the light: the role of the Quaker minister. In: *Explorations in the Ethnography of Speaking,* Richard Bauman and Joel Sherzer, eds. Cambridge: Cambridge University Press, pp.144–60.

Bäuml, Betty J., and Bäuml, Franz H., 1975. *A Dictionary of Gestures.* Metuchen: N.J.: The Scarecrow Press.

Beck, Henry, 1976. Neuropsychological servosystems, consciousness, and the problem of embodiment. *Behavioral Science,* 21:139–60.

Benthall, Jonathan, and Polhemus, Ted, eds., 1975. *The Body as a Medium of Expression.* London: Allen Lane.

Bentley, Arthur F., 1941. The human skin: philosophy's last line of defense. *Philosophy of Science,* 8:1–19.

Bogatyrev, Petr., 1971. *The Functions of Folk Costume in Moravian Slovakia.* The Hague: Mouton.

Bogen, Joseph E., 1969. The other side of the brain. II: An appositional mind. *Bulletin of the Los Angeles Neurological Societies,* 34:135–62.

Bosmajian, Haig A., 1971. *The Rhetoric of Nonverbal Communication: Readings.* Glenview, Ill.: Scott, Foresman.

Bouissac, Paul, 1973. *La Mesure des gestes: Prolegomènes à la sémiotique gestuelle.* The Hague: Mouton.
Bouissac, Paul, 1976. *Circus and Culture: A Semiotic Approach.* Bloomington: Indiana University Press.
Bourne, Geoffrey H., 1971. *The Ape People.* New York: G. P. Putnam's Sons.
Bower, Gordon H., 1970. Analysis of a mnemonic device. *American Scientist,* 58:496–510.
Brannigan, Christopher R., and Humphries, David A., 1972. Human non-verbal behaviour, a means of communication. In: *Ethological Studies of Child Behaviour,* N. Blurton Jones, ed. Cambridge: Cambridge University Press, pp.37–64.
Brothwell, Don R., ed., 1976. *Beyond Aesthetics: Investigations into the Nature of Visual Art.* London: Thames and Hudson.
Brown, Roger, 1973. *A First Language: The Early Stages.* Cambridge: Harvard University Press.
Bullowa, Margaret, 1970. The Start of the Language Process. *Actes du Xe Congrès International des Linguistes.* Bucharest: Academy of the Romanian Socialist Republic, 3:191–98.
Chance, Michael R. A., 1975. Social cohesion and the structure of attention. In: *Biosocial Anthropology.* New York: John Wiley & Sons.
Charlesworth, William R., and Kreutzer, Mary Anne, 1973. Facial expressions of infants and children. In: *Darwin and Facial Expression: A Century of Research in Review,* Paul Ekman, ed. New York: Academic Press, pp.91–168.
Cherry, Colin, 1966. *On Human Communication: A Review, a Survey, and a Criticism.* 2d ed. Cambridge: MIT Press.
Christopher, Milbourne, 1970. *ESP, Seers & Psychics.* New York: Thomas Y. Crowell.
Cortambert, Louise, 1833. *Le Langage des fleurs,* 4th ed. Paris: Audot.
Critchley, Macdonald, 1975. *Silent Language.* London: Butterworths.
Crystal, David, 1969. *Prosodic Systems and Intonations in English.* Cambridge: Cambridge University Press.
Curtiss, Susan, et al., 1975. An update on the linguistic development of Genie. In: *Georgetown University Round Table on Languages and Linguistics,* Daniel P. Dato, ed. Washington, D.C.: Georgetown University Press, pp.145–57.
Darwin, Charles, 1877. A biographical sketch of an infant. *Mind,* 2:286–94.
Davis, Flora, 1971. *Inside Intuition: What We Know About Nonverbal Communication.* New York: McGraw-Hill.
Dennis, Maureen, and Whitaker, Harry A., 1976. Language acquisition following hemidecortication: linguistic superiority of the left over the right hemisphere. *Brain and Language,* 3:404–33.
Dimond, Stuart, 1972. *The Double Brain.* Edinburgh: Churchill Livingstone.
Dingwall, William Orr, 1975. The species-specificity of speech. In: *Georgetown University Round Table on Languages and Linguistics,* Daniel P. Dato, ed. Washington, D.C.: Georgetown University Press, pp.17–62.
Duncan, Starkey D., Jr., 1975. Language, paralanguage, and body motion in the structure of conversations. In: *Socialization and Communication in Primary Groups,* Thomas R. Williams, ed. The Hague: Mouton, pp.283–311.
Eccles, John C., 1974. Cerebral activity and consciousness. In: *Studies in the Philosophy of Biology, Reduction and Related Problems,* Francisco J. Ayala and Theodore Dobzhansky, eds. Berkeley: University of California Press.
Eco, Umberto, 1976. *A Theory of Semiotics.* Bloomington: Indiana University Press.
Efron, David, 1972. [1941]. *Gesture, Race and Culture.* The Hague: Mouton.
Egger, Victor, 1904. *La parole intérieure.* Paris: Alcan.
Eimermacher, Karl, 1974. *Arbeiten sowjetischer Semiotiker der Moskauer und Tartuer Schule.* Kronberg: Scriptor.
Eisenberg, Abne N., and Smith, Ralph R., Jr., 1971. *Non-Verbal Communication.* Indianapolis: Bobbs-Merrill.
Ekman, Paul, 1972. Universals and cultural differences in facial expressions of emotion. In: *Nebraska Symposium on Motivation 1971,* James K. Cole, ed. Lincoln: University of Nebraska Press, pp.207–83.
Eschbach, Achim, and Rader, Wendelin, 1976. *Semiotik-Bibliographie I.* Frankfurt a/M: Syndikat.
Ferguson, Charles A., 1976. The structure and use of politeness formulas. *Language in Society,* 5:137–51.
Fouts, Roger S., forthcoming. Capacities for language in Great Apes. In: *Sociology and Psychology of Primates,* R. H. Tuttle, ed. The Hague: Mouton, pp.371–90.
Frisch, Karl von, 1974. *Animal Architecture.* New York: Harcourt Brace Jovanovich.
Goffman, Erving, 1975. *Replies and Responses.* Urbino: Centro Internazionale di Semiotica e di Linguistica.
Greimas, Algirdas Julien, 1976. *Sémiotique et sciences sociales.* Paris: Seuil.
Griffin, Donald R., 1976. *The Question of Animal Awareness: Evolutionary Continuity of Mental Experience.* New York: Rockefeller University Press.
Guthrie, R. Dale, 1976. *Body Hot Spots: The Anatomy of*

Human Social Organs and Behavior. New York: Van Nostrand Reinhold.
Hahn, Martin E., and Simmel, Edward C., eds., 1976. *Communicative Behavior and Evolution.* New York: Academic Press.
Hall, Edward T., 1959. *The Silent Language.* Garden City, N.Y.: Doubleday.
Harrison, Randall P., 1974. *Beyond Words: An Introduction to Nonverbal Communication.* Englewood Cliffs, N.J.: Prentice-Hall.
Hartshorne, Charles, 1973. *Born to Sing: An Interpretation and World Survey of Bird Song.* Bloomington: Indiana University Press.
Hediger, Heini, 1967. Verstehens- und Verständigungsmöglichkeiten zwischen Mensch und Tier. *Schweizerische Zeitschrift für Psychologie und ihre Anwendungen,* 26:234–55.
Hediger, Heini, 1968. *The Psychology and Behaviour of Animals in Zoos and Circuses.* New York: Dover.
Hediger, Heini, 1974. Communication between man and animal. *Image,* 62:27–40.
Hediger, Heini, 1976. Der Kluge Hans: Möglichkeiten und Grenzen der Kommunikation zwischen Mensch und Tier. *Neue Zürcher Zeitung,* 156 (July 7):45–46.
Hess, Eckhard H., 1975. *The Tell-Tale Eye: How Your Eyes Reveal Hidden Thoughts and Emotions.* New York: Van Nostrand Reinhold.
Hinde, Robert A., 1974. *Biological Bases of Human Social Behaviour.* New York: McGraw-Hill.
Hinde, Robert A., ed., 1969. *Bird Vocalizations.* Cambridge: Cambridge University Press.
Hinde, Robert A., ed., 1972. *Non-Verbal Communication.* Cambridge: Cambridge University Press.
Hjelmslev, Louis, 1953. *Prolegomena to a Theory of Language.* Baltimore: Waverly Press.
Hutt, Corinne, and Vaizey, M. Jane, 1966. Differential effects of group density on social behaviour. *Nature,* 209:1371–72.
Huxley, Julian, 1966. A discussion of ritualization of behaviour in animals and men. *Philosophical Transactions of the Royal Society of London,* 251:249–71.
Huxley, Julian, and Koch, Ludwig, 1964 [1938]. *Animal Language.* New York: Grosset & Dunlap.
Ivanov, Vjačeslav V., 1976. *Očerki po istorii semiotiki v SSSR.* Moscow: Nauka.
Jacob, François, 1974. *The Logic of Living Systems: A History of Heredity.* London: Allen Lane.
Katz, David, 1937. *Animals and Men: Studies in Comparative Psychology.* London: Longmans, Green and Co.
Kauffman, Lynn E., 1971. Tacesics, the study of touch: a model for proxemic analysis. *Semiotica,* 4:149–61.
Kavanagh, James F., and Mattingly, Ignatius G., eds., 1972. *Language by Ear and by Eye.* Cambridge: MIT Press.
Kelley, David L., 1971. *Kinesiology: Fundamentals of Motion Description.* Englewood Cliffs, N.J.: Prentice-Hall.
Kinsbourne, Marcel, 1975. Minor hemisphere language and cerebral maturation. In: *Foundations of Language Development: A Multidisciplinary Approach,* Erik H. Lenneberg and Elizabeth Lenneberg, eds. New York: Academic Press, 2: chap. 7.
Kleinpaul, Rudolf, 1972 [1888]. *Sprache ohne Worte: Idee einer allgemeinen Wissenschaft der Sprache,* 2d ed. The Hague: Mouton.
Knapp, Mark L., 1972. *Nonverbal Communication in Human Interaction.* New York: Holt, Rinehart and Winston.
Koehler, Otto, 1956. Thinking without words. In: *Proceedings of the XIV International Congress of Zoology.* Copenhagen: Danish Science Press, pp.75–88.
Krames, Lester; Pliner, Patricia; and Alloway, Thomas; eds., 1974. *Nonverbal Communication.* New York: Plenum Press.
Krashen, Stephen D., 1976. Cerebral asymmetry. In: *Studies in Neurolinguistics,* Haiganoosh Whitaker and Harry A. Whitaker, eds. New York: Academic Press, 2: chap. 5.
Laver, John, 1976. Language and nonverbal communication. In: *Handbook of Perception 7: Language and Speech.* New York: Academic Press, chap. 10.
Levy, Jerre, 1973. Psychobiological implications of bilateral asymmetry. In: *Hemisphere Functions in the Human Brain,* Stuart J. Dimond and J. Graham Beaumont, eds. London: Elek, pp.121–83.
Lieberman, Philip, 1975. *On the Origins of Language: An Introduction to the Evolution of Human Speech.* New York: Macmillan.
Locke, John, 1975 [1690]. *An Essay Concerning Human Understanding,* Peter H. Nidditch, ed. Oxford: Clarendon.
Lorenz, Konrad, 1966. Evolution of ritualization in the biological and cultural spheres. *Philosophical Transactions of the Royal Society of London,* 251:273–84.
Lotz, John, 1956. Symbols make man. In: *Frontiers of Knowledge in the Study of Man,* Lynn White, Jr., ed. New York: Harper & Brothers, pp.207–31.
Luria, Aleksandr R., 1972. *The Man with a Shattered World: The History of a Brain Wound.* New York: Basic Books.
McCormack, William, and Wurm, Stephen A., eds.,

1976. *Language and Man: Anthropological Issues.* The Hague: Mouton.

Máday, Stefan von, 1914. *Gibt es denkende Tiere?* Leipzig: Wilhelm Engelmann.

Malinowski, Bronislaw, 1965 [1935]. Coral Gardens and Their Magic, II: *The Language of Magic and Gardening.* Bloomington: Indiana University Press.

Malson, Lucien, 1964. *Les Enfants sauvages: mythe et réalité.* Paris: Union Générale d'Éditions.

Malson, Lucien, 1973. Un entretien avec Claude Lévi-Strauss. *Le Monde,* Dec. 8.

Marcus, Solomon, 1974. Linguistic structures and generative devices in molecular genetics. *Cahiers de Linguistique Théorique et Appliquée,* 11:74–104.

Marples, Mary J., 1965. *The Ecology of the Human Skin.* Springfield, Ill.: Charles C. Thomas.

Mehrabian, Albert, 1972. *Nonverbal Communication.* Chicago: Aldine, Atherton.

Miller, James G., 1976. Second annual Ludwig von Bertalanffy memorial lecture. *Behavioral Science,* 21:219–27.

Moles, Abraham A., 1964. Les voies cutanées, compléments informationnels de la sensibilité de l'organisme. *Studium Generale,* 17:589–95.

Montagu, Ashley, 1971. *Touching: The Human Significance of the Skin.* New York: Columbia University Press.

Morris, Desmond, 1962. *The Biology of Art.* London: Methuen.

Nattiez, Jean-Jacques, 1975. *Fondements d'une sémiologie de la musique.* Paris: Union Générale d'Éditions.

Nelson, Keith, 1973. Does the holistic study of behavior have a future? In: *Perspectives in Ethology,* P. P. G. Bateson and Peter Klopfer, eds. New York: Plenum, chap. 8.

Ornstein, Robert E., 1972. *The Psychology of Consciousness.* New York: The Viking Press.

Paredes, J. Anthony, and Hepburn, Marcus J., 1976. The split brain and the culture-and-cognition paradox. *Current Anthropology,* 17:121–27. Discussion: ibid.: 318–26, 503–11.

Payne, Roger S., and McVay, Scott, 1971. Songs of the humpback whales. *Science,* 173:587–97.

Pfungst, Oskar, 1965 [1911]. *Clever Hans (The Horse of Mr. von Osten),* Robert Rosenthal, ed. New York: Holt, Rinehart and Winston.

Pike, Kenneth L., 1943. *Phonetics: A Critical Analysis of Phonetic Theory and Technic for the Practical Description of Sound.* Ann Arbor: University of Michigan Press.

Pliner, Patricia; Krames, Lester; and Alloway, Thomas; eds., 1975. *Nonverbal Communication of Aggression.* New York: Plenum Press.

Poyatos, Fernando, 1976. *Man Beyond Words: Theory and Methodology of Nonverbal Communication.* Oswego: The New York State English Council.

Premack, Ann J., 1976. *Why Chimps Can Read.* New York: Harper & Row.

Premack, David, 1971. Language in chimpanzee? *Science,* 172:808–22.

Prieto, Luis J., 1975. Sémiologie de la communication et sémiologie de le signification. In: *Études de linguistique et de sémiologie générales.* Geneva: Librairie Droz, pp.125–41.

Rosenthal, Robert, 1966. *Experimenter Effects in Behavioral Research.* New York: Appleton-Century-Crofts.

Ruesch, Jurgen, and Kees, Weldon, 1956. *Nonverbal Communication: Notes on the Visual Perception of Human Relations.* Berkeley and Los Angeles: University of California Press.

Rumbaugh, Duane M., and Gill, Timothy V., 1976. Language and the acquisition of language-type skills by a chimpanzee (*Pan*). *Annals of the New York Academy of Sciences,* 270:90–123.

Sapir, Edward, 1931. Communication. In: *Encyclopaedia of the Social Sciences.* New York: Macmillan, 4:78–81.

Scherer, Klaus R., 1970. *Non-verbale Kommunikation: Ansätze zur Beobachtung und Analyse der aussersprachlichen Aspekte von Interaktionsverhalten.* Hamburg: Helmut Buske.

Schneirla, Theodore C., 1965. Aspects of stimulation and organization in approach/withdrawal processes underlying vertebrate behavioral development. In: *Advances in the Study of Behavior.* New York: Academic Press, 1:1–74.

Schutz, Noel W., Jr., 1976. *Kinesiology: The Articulation of Movement.* Lisse: Peter de Ridder Press.

Sebeok, Thomas A., 1972. *Perspectives in Zoosemiotics.* The Hague: Mouton.

Sebeok, Thomas A., 1976a. *Contributions to the Doctrine of Signs.* Bloomington: Research Center for Language and Semiotic Studies.

Sebeok, Thomas A., 1976b. Marginalia to Greenberg's conception of semiotics and zoosemiotics. In: *Linguistic Studies Offered to Joseph Greenberg on the Occasion of His 60th Birthday,* Alphonse Juilland, ed. Saratoga, Calif.: Anma Libri.

Sebeok, Thomas A., 1976c. Iconicity. *Modern Language Notes,* 91 (6).

Sebeok, Thomas A., 1977a. Semiosis in nature and culture. In: *Proceedings of the International Symposium on*

Semiotics and Theories of Symbolic Behavior in Eastern Europe and the West, Thomas G. Winner, ed. Lisse: Peter de Ridder Press.

Sebeok, Thomas A., forthcoming. Neglected figures in the history of semiotic inquiry II. Jakob von Uexküll. Address prepared for delivery at the III. Wiener Symposium über Semiotik, August 1977.

Sebeok, Thomas A., ed., 1968. *Animal Communication: Techniques of Study and Results of Research.* Bloomington: Indiana University Press.

Sebeok, Thomas A., ed., 1974. *Current Trends in Linguistics 12: Linguistics and Adjacent Arts and Sciences.* The Hague: Mouton.

Sebeok, Thomas A., ed., 1975. *The Tell-Tale Sign: A Survey of Semiotics.* Lisse: Peter de Ridder Press.

Sebeok, Thomas A., and Umiker-Sebeok, Donna Jean, eds., 1976. *Speech Surrogates: Drum and Whistle Systems.* The Hague: Mouton.

Sebeok, Thomas A., and Umiker-Sebeok, Donna Jean, eds., 1977b. *Aboriginal Sign Languages: Gestural Systems Among Native Peoples of the Americas and Australia.* New York: Plenum.

Shands, Harley C., 1976. Malinowski's mirror: Emily Dickinson as Narcissus. *Comtemporary Psychoanalysis,* 12:300–34.

Smith, W. John, et al., 1974. Tongue showing: a facial display of humans and other primate species. *Semiotica,* 11:201–46.

Stokoe, William C., Jr., 1972. *Semiotics and Human Sign Languages.* The Hague: Mouton.

Szöke, Péter, 1963. Ornitomuzikológia. *Magyar Tudomány,* 9:592–607.

Thom, René, 1975a. Les Mathématiques et l'intelligible. *Dialectica,* 29:71–80.

Thom, René, 1975b. *Structural Stability and Morphogenesis.* Reading, Mass.: W. A. Benjamin.

Timaeus, Ernst, 1973. Some non-verbal and paralinguistic cues as mediators of experimenter expectancy effects. In: *Social Communication and Movement: Studies of Interaction and Expression in Man and Chimpanzee,* Mario von Cranach and Ian Vine, eds. New York: Academic Press, chap. 11.

Tomkins, Gordon M., 1975. The metabolic code. *Science,* 189:760–63.

Tunnell, Gary G., 1973. *Culture and Biology: Becoming Human.* Minneapolis: Burges Publishing Co.

Uexküll, Jakob von, 1940. Bedeutungslehre. *Bios,* 10.

Voegelin, Charles F., and Harris, Zellig S., 1947. The scope of linguistics. *American Anthropologist,* 49:588–600.

Vološinov, Valentin N., 1973. *Marxism and the Philosophy of Language.* New York: Seminar Press.

Vygotsky, Lev Semenovich, 1962. *Thought and Language,* E. Hanfmann and G. Vakar, eds. and trans. Cambridge: MIT Press.

Washburn, Sherwood L., 1973. The promise of primatology. *American Journal of Physical Anthropology,* 38:177–82.

Weitz, Shirley, ed., 1974. *Nonverbal Communication: Readings with Commentary.* New York: Oxford University Press.

Weizenbaum, Joseph, 1976. *Computer Power and Human Reason.* San Francisco: W. H. Freeman.

Wescott, Roger W., 1969. *The Divine Animal: An Exploration of Human Potentiality.* New York: Funk & Wagnalls.

Wescott, Roger W., ed. 1974. *Language Origins.* Silver Spring, Md.: Linstok Press.

Winner, Thomas G., and Winner, Irene P., 1976. The semiotics of cultural texts. *Semiotica,* 18 (in press).

Wolfe, John B., 1936. Effectiveness of token-rewards for chimpanzees. *Comparative Psychology Monographs,* 12 (5).

Wood, Forrest G., 1973. *Marine Mammals and Man: The Navy's Porpoises and Sea Lions.* Washington D.C.: Robert B. Luce.

Zangwill, Oliver L., 1975. The relation of nonverbal cognitive functions to aphasia. In: *Foundations of Language Development: A Multidisciplinary Approach,* Eric H. Lenneberg and Elizabeth Lenneberg, eds. New York: Academic Press, 2: chap. 6.

Index of Names

INDEX OF ANIMALS

Animals for which some behavioral information is given in the text are listed in taxonomic order. Common names have been supplied where possible.